BIOLOGY

The Unity and Diversity of Life

TENTH EDITION

CECIE STARR / RALPH TAGGART

LISA STARR
Biology Illustrator

THOMSON
———★———
BROOKS/COLE

Australia • Canada • Mexico • Singapore • Spain
United Kingdom • United States

BIOLOGY PUBLISHER: Jack C. Carey

EDITOR-IN-CHIEF: Michelle Julet

DEVELOPMENTAL EDITOR: Mary Arbogast, Peggy Williams

ASSISTANT EDITOR: Suzannah Alexander

EDITORIAL ASSISTANTS: Karoliina Tuovinen, Jana Davis

MEDIA PROJECT MANAGER: Pat Waldo

TECHNOLOGY PROJECT MANAGERS: Donna Kelley, Keli Amann

MARKETING MANAGER: Ann Caven

MARKETING ASSISTANT: Sandra Perin

ADVERTISING PROJECT MANAGER: Linda Yip

SENIOR PROJECT MANAGER, EDITORIAL/PRODUCTION: Teri Hyde

PRINT/MEDIA BUYER: Karen Hunt

PERMISSIONS EDITOR: Joohee Lee

PRODUCTION SERVICE: Lachina Publishing Services, Inc.;
Grace Davidson

TEXT AND COVER DESIGN: Gary Head, Gary Head Design

ART EDITOR AND PHOTO RESEARCHER: Myrna Engler

ILLUSTRATORS: Lisa Starr, Gary Head

OFFICE SUPPORT: Brad Griffin, Verbal Clark

COVER IMAGE: *From Central America, one of the tropical rain forests that may disappear in your lifetime.* Kevin Schafer/Getty Images

COVER PRINTER: Phoenix Color Corp (MD)

COMPOSITOR: Preface, Inc.; Angela Harris, John Becker

FILM HOUSE: H&S Graphics; Tom Anderson

PRINTER: Quebecor/World, Versailles

For more information about our products, contact us at:
Thomson Learning Academic Resource Center
1-800-423-0563
For permission to use material from this text, contact us by:
Phone: 1-800-730-2214   **Fax:** 1-800-730-2215
Web: http://www.thomsonrights.com

Library of Congress Control Number: 2002111898
Student Edition with InfoTrac College Edition: ISBN 0-534-38800-0
Student Edition without InfoTrac College Edition: 0-534-38801-9
Annotated Instructor's Edition: ISBN 0-534-38802-7

International Student Edition: ISBN 0-534-27413-7
(Not for sale in the United States)

BOOKS IN THE BROOKS/COLE BIOLOGY SERIES

Brooks/Cole—Thomson Learning
10 Davis Drive
Belmont, CA 94002
USA

Asia
Thomson Learning
60 Albert Street, #15-01
Albert Complex
Singapore 189969

Australia
Nelson Thomson Learning
102 Dodds Street
South Melbourne, Victoria 3205
Australia

Canada
Nelson Thomson Learning
1120 Birchmount Road
Toronto, Ontario M1K 5G4
Canada

Europe/Middle East/Afric
Thomson Learning
Berkshire House
168-173 High Holborn
London WC1 V7AA
United Kingdom

CONTENTS IN BRIEF

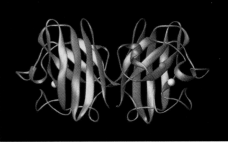

DETAILED CONTENTS

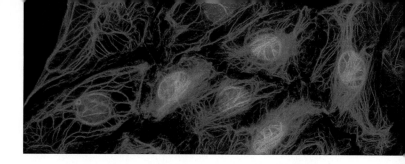

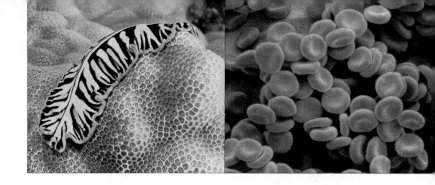

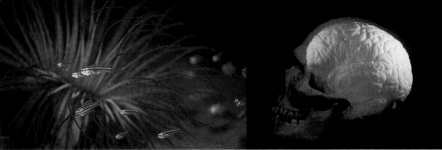

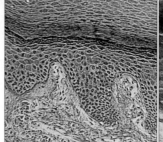
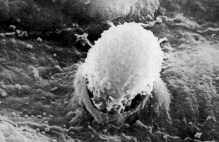

PREFACE

Successive revisions of this book span nearly thirty years and reflect feedback from many instructors and students. This new edition retains the concept spreads and other pedagogical features that are the hallmarks of the book.

CONCEPT SPREADS Reading and absorbing textbook assignments for multiple courses in the same timeframe can overwhelm students. We make it easier for them to read about and understand biology by focusing on one concept at a time. We list key concepts on the first page of each chapter. Then we organize text, art, and evidence in support of each concept on two facing pages, at most. As shown below, each *concept spread* starts with a numbered tab and ends with a boldfaced summary of the key points. Students can preview the on-page summary before they read the concept spread. They can read it again to check whether they understand the key points before turning to the next concept.

Concept spreads also offer teachers flexibility in assigning topics to fit their course requirements. For example, those who spend less time on photosynthesis may bypass the spreads on properties of light and the chemiosmotic theory of ATP formation. They may or may not assign the Focus essay on the global impact of photosynthesis. All spreads and essays are part of a chapter story, but some offer more depth.

We incorporate headings within concept spreads to help students keep track of the hierarchy of information. Transitions between spreads help them follow the story. So does setting aside some details in optional illustrations for motivated students.

With concept spreads, students find assigned topics fast, and they can focus on manageable amounts of information. This makes them more confident in their capacity to absorb the material. Our approach has a tangible outcome—improved test scores.

VISUALIZING CONCEPTS We continue to develop text and art together, as an inseparable whole. Our *"read-me-first diagrams"* are a prime example of this approach. They allow visual learners to build a mental image of a concept before reading the text details about it. Figure 34.5 in the sample pages at right shows how simple descriptions walk students step by step through these preview diagrams. Many of our millions of student readers have written in to tell

us that our approach helps them far more than a reliance on "wordless" diagrams.

Many *anatomical drawings* are integrated overviews of structure and function. Students need not jump back and forth from text, to tables, then to art, and back again to visualize how an organ system is put together and what its component parts do. We also use *zoom sequences*, from macroscopic to microscopic views, to move students visually into a system or process. For example, Figures 37.19 and 37.20 start with a ballerina's biceps and move on down through levels of skeletal muscle contraction.

Icons remind students of where art fits into the story line. For instance, icons of a cell remind students where

---

Gold numbered tabs, such as this one, identify the start of each new concept in a chapter. Other tabs (brown) in the chapter identify Focus essays. Many essays enrich the basic text by addressing medical, environmental, and bioethical issues. Others offer detailed examples of experiments to demonstrate the power of critical thinking.

34.2

HOW ARE ACTION POTENTIALS TRIGGERED AND PROPAGATED?

Action potential propagation isn't hard to follow if you already know something about the gradients across the neural membrane. And so we now build on Section 34.1.

Approaching Threshold

When you weakly stimulate a neuron at its input zone, you disturb the ion balance across its membrane, but not much. Imagine putting a bit of pressure on the skin of a snoozing cat by gently tapping a toe on it. Tissues beneath the skin surface have receptor endings—input zones of sensory neurons. Patches of plasma membrane at these endings deform under pressure and let some ions flow across. The flow slightly changes the voltage difference across the membrane. In this case, pressure has produced a graded, local signal.

influx of ions, the cytoplasmic side of the membrane becomes less negative. This causes more gates to open and more sodium to enter. The ever increasing, inward flow of sodium is a case of **positive feedback**, whereby an event intensifies as a result of its own occurrence:

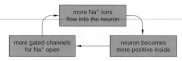

At threshold, opening of sodium gates no longer depends on the strength of the stimulus. The positive-feedback cycle is under way, and the inward-rushing sodium itself is enough to open the gated channels.

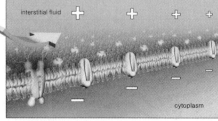

a Membrane at rest (inside negative with respect to the outside). An electrical disturbance (*yellow arrow*) spreads from an input zone to an adjacent trigger region of the membrane, which has a great number of gated sodium channels.

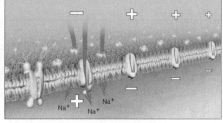

b A strong disturbance initiates an action potential. Sodium gates open. The sodium inflow decreases the negativity inside the neuron. The change causes more gates to open, and so on until threshold is reached and the voltage difference across the membrane reverses.

Figure 34.5 Propagation of an action potential along the axon of a motor neuron.

Graded means that signals arising at an input zone vary in magnitude. They are small to large, depending on the stimulus intensity or duration. *Local* means these signals do not spread far from the site of stimulation. Why? It takes certain kinds of ion channels to propagate a signal, and input zones simply don't have them.

When a stimulus is intense or long-lasting, graded signals spread from the input zone into an adjoining trigger zone. This patch of membrane is richly endowed with voltage-sensitive gated channels for sodium ions. *And this is where a certain amount of change in the voltage difference across the plasma membrane triggers an action potential.* The amount is the neuron's threshold level.

When these gates open, positively charged sodium ions flow into the neuron, as in Figure 34.5. With the

An All-or-Nothing Spike

Figure 34.6 shows a recording of the voltage difference across the plasma membrane before, during, and after an action potential. Notice how the membrane potential peaks once threshold is reached. All action potentials in a neuron spike to the same level above threshold as an *all-or-nothing* event. Once a positive-feedback cycle starts, nothing stops full spiking. Unless threshold is reached, the membrane disturbance subsides when the stimulation ends, and an action potential won't occur.

Each spike lasts for only a millisecond or so. Why? At the patch of membrane where the charge reversed, gated sodium channels close and shut off the sodium inflow. And about halfway into the reversal, potassium

Example of a cell icon

its organelles are located, as in Chapter 4. Other icons remind students of how reaction stages interconnect in a metabolic pathway, as in Chapter 7 (photosynthesis) and Chapter 8 (aerobic respiration). Others remind them of evolutionary relationships, as in Chapters 25 and 26. A multimedia icon directs students to art in the CD-ROM packaged at the back of every book. Others direct them to supplemental material on the Web and to InfoTrac® College Edition, an online database of full-length articles from 4,000 academic journals and popular sources.

Finally, we added hundreds of new *micrographs and photographs*. These are not window dressing, tacked on after the fact. They sharpen the meaning of "biodiversity" and hint at why it is worth preserving.

BALANCING CONCEPTS WITH APPLICATIONS   We draw students into each chapter with a lively or sobering application. That application gives way to a list of key concepts, an advance organizer. We attempt to maintain

their interest with focus essays, which provide depth on medical, environmental, and social issues without interrupting the conceptual flow. Where we sense that the core material needs to be livened up a bit, we weave briefer applications into the text proper. On the last pages of the book, a separate Applications Index affords fast reference to our many hundreds of applications.

FOUNDATIONS FOR CRITICAL THINKING   Like all textbooks at this level, ours helps students sharpen their capacity to think critically about nature. We walk them through experiments that yielded clear evidence in favor of or against hypotheses. The main index at the back of the book lists the experiments we selected (see the entries *Experiment, examples*, and *Test, observational*).

We also selectively use chapter introductions as well as entire chapters to show productive outcomes of critical thinking. The chapter introductions to Mendelian genetics (11), DNA structure and function (13), speciation (18), immunology (39), and behavior (46) are examples. The end of each chapter has a set of *Critical Thinking* questions. Numerous *Genetics Problems* at the end of Chapters 11 and 12 help students grasp principles of inheritance.

SUPPLEMENTS   The Instructors' Examination copy for this edition lists a comprehensive package of print and multimedia supplements, including online resources that are available to qualified adopters. Please ask your local sales representative for details.

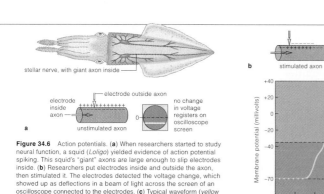

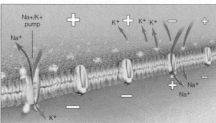

Figure 34.6   Action potentials. (**a**) When researchers started to study neural function, a squid (*Loligo*) yielded evidence of action potential spiking. This squid's "giant" axons are large enough to slip electrodes inside. (**b**) Researchers put electrodes inside and outside the axon, then stimulated it. The electrodes detected the voltage change, which showed up as deflections in a beam of light across the screen of an oscilloscope connected to the electrodes. (**c**) Typical waveform (*yellow* line) for an action potential on an oscilloscope screen.

This icon signifies that we explore the concept further on our interactive CD-ROM.

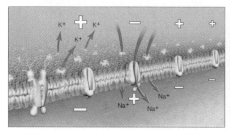

c With the reversal, sodium gates shut and potassium gates open (*red* arrows). Potassium follows its gradient out of the neuron. Voltage is restored. The disturbance triggers an action potential at the adjacent site, and so on, away from the point of stimulation.

d Following each action potential, the inside of the plasma membrane becomes negative once again. However, the sodium and potassium concentration gradients are not yet fully restored. Active transport at sodium–potassium pumps restores them.

This is an example of our "read-me-first" diagrams for visual learners. They are illustrated previews of the text material, complete with simple "a b c" descriptions that guide the student each step of the way.

channels opened, so potassium flows out. This restores the voltage difference at the patch but not the original gradients. Sodium–potassium pumps actively transport sodium back outside and potassium inside. Once this is done, most potassium gates close and sodium gates are in their original position—ready to be opened with the arrival of a suitable signal at the membrane patch.

The Direction of Propagation

During an action potential, the inward rush of sodium ions affects the charge distribution across the adjacent membrane patch, where an equivalent number of gated channels open. Gated channels open in the *next* patch, and the next, and so on. This positive feedback event is self-propagating and does not diminish in magnitude. You might be wondering: Do action potentials spread

back to the trigger zone? No. For a brief period after the inward rushing of sodium ions, the voltage-gated channels remain insensitive to stimulation, so sodium ions cannot move through them. This is one reason why action potentials do not spread back to the patch of membrane where they were initiated. It is why they propagate themselves away from it.

Ions cross the neural membrane through transport proteins that serve as gated or open channels. At a suitably disturbed trigger zone, sodium gates open in an all-or-nothing way, and the inward-rushing sodium causes an action potential. Sodium–potassium pumps restore the original ion gradients.

Sodium gates across the membrane are briefly inactivated after an action potential, which is one reason why an action potential is self-propagating away from a trigger zone.

Boldfaced key concepts end each tabbed section. Students can read them to preview the section's main points, then read them again to reinforce what they learn from the section.

All together, these highlighted statements are a running summary of the take-home lessons.

This icon reminds students to check out the website, which expands on the section's topic.

A MAJOR CONCEPTUAL SHIFT In the past, with each revision, we have updated details to reflect new work in biology's rapidly changing fields. This time we did more. We decided to undertake a major conceptual shift after concluding that one is long overdue in biology textbooks at this level.

The Emerging Big Picture

Biology's overriding paradigm is one of interpreting life's spectacular diversity as having evolved from simple molecular beginnings. For decades now, researchers have been chipping away at the structural secrets of biological molecules. They have discovered how different kinds are put together, how they function, and what happens when they mutate. They are casting light on how life originated, what happened during the past 3.8 billion years, and what the future may hold for humans and other organisms, individually and collectively.

This is profound stuff. *Yet introductory textbooks—including previous editions of this one—have not given students enough information to understand the remarkable connection between molecular change, evolution, and their own lives.* We rebuilt much of this tenth edition to show the connection more clearly, starting with the new models for biological molecules. The models can give students a deeper appreciation of how *structural* diversity translates into *functional* diversity.

Powerful Conceptual Tools for Students

Molecular models appear in books, newspapers, and popular magazines, but few students have a clue to what they represent. Why even ask them to memorize the fact that polypeptide chains form coils, sheets, and loops without giving clear examples of the outcomes? Why not show how different configurations form, say, membrane-spanning barrels, anchors, or jaws that grip enemy agents in the body? Why not complete the point and say that mutation may alter a configuration just enough to block or enhance a transport or anchoring or defensive function?

And why not connect that point with the paradigm? *Small changes in molecular sequences and functional domains give rise to variation in traits—the raw material of evolution.*

Once the connection becomes clear, students can apply what they learn about changes in molecular structure to many of their questions and concerns about life. Genetic disorders? Look to modifications in molecular structure that have altered how cells, organs, and the organism itself function. Events that long ago put chimpanzees and humans on their separate evolutionary paths? Look to transposons and other disruptions in shared ancestral DNA. Differences in the number of legs on a fly or centipede or human, or the number of petals on a flower? Look to small mutations in master genes that control development of the basic body plan.

And what about similarities in how, say, tulips and Tina Turner or any other organism function? Look to commonalities in molecular responses to environmental challenges.

Chapters That Develop the Conceptual Tools

Several chapters build on one another to help students develop knowledge of the molecular basis of life. The first chapter is a simple preview of the molecular basis of life's unity and diversity. Chapters 3, 5, and 6 have sections on the structure and function of enzymes and other molecules that prepare students for the important chapters on cell structure and metabolism, genetics, evolution, anatomy, and physiology.

For instance, with this background, students can sense the power of comparative molecular studies in clarifying evolutionary relationships. Among the outcomes of such studies are refined evolutionary tree diagrams and the three-domain classification system.

Chapter 28 is a new prelude to the units on anatomy and physiology. It points out that we sometimes forget how much plants and animals have in common at the molecular level, because their body plans are so different. For example, one of the essays in this chapter focuses on recurring challenges to survival of *all* individual cells in the multicelled body—specifically, the requirements for gas exchange and internal transport, for homeostasis in the internal environment, and for integration and control. This chapter also starts students thinking about how cells communicate with one another, using simple examples of signal reception, transduction, and response from plants and animals.

New Connections Essays That Reinforce the Big Picture

New to this edition are Connections essays that can help students "connect the dots" between text details and big-picture concepts. The facing page lists these essays.

Setting the Stage

Connections essay 1.4 answers a central question: How can life display both unity *and* diversity? It suggests that the theory of evolution by natural selection connects the two. Obviously, we cannot get into actual mechanisms of evolution until after the book's units on cell biology and genetics. But a simple outline of the theory can help students connect its basic premises with the topics of those units as well.

Essay 3.1 is a user-friendly introduction to the models that represent biological molecules. Most textbooks show molecules as sticks, balls, ribbons, and bubbled blobs. How many students stare blankly at them? We let them in on an unfortunately well-kept secret. Different models convey different information, such as mass, structural organization, reactivity, or function. That is why we use different kinds in different contexts. For example, the models for glycogen, hemoglobin, enzymes, and DNA tell us different things about how cells and organisms are put together and how they function.

Without this knowledge, can students extrapolate from molecular models to generic icons—to the circles and squares used to represent enzymes, ATP, and all the other biological molecules? Can they really extrapolate from generic icons alone to the concept of structure, function, and evolution? We think not.

Essay 5.2, for example, showcases some important membrane proteins. These stunning models make it easier for students to recognize that structural diversity translates into the functional diversity necessary for metabolism, gene function, integration, immunity, the formation of tissues and organs, and other tasks. They make it possible to comprehend how, say, cystic fibrosis arises from a mutation that changes the function of just one kind of membrane transporter. As another example, Essay 3.8 moves students from the molecular structure of HbA to HbS, and on to sickle-cell anemia.

Making Sense of Evolution and Biodiversity

Unless we explain continuity and change at the molecular level convincingly enough, many students will never become open to the idea of continuity and change in life itself—to the possibility that evolution is more than "just a theory."

Continuity and change—Essay 8.7 picks up on this thought. It ties together concepts from the cell biology unit into a view of how all of life connects as a result of evolution at the molecular level.

Essay 19.9 reinforces a major point—that interpreting the past scientifically requires a huge intellectual shift, from direct observation to inferences based upon molecular studies, the fossil record, and morphological comparisons. Essay 28.2 picks up on the same thought. The essay invites caution in interpreting the evolutionary history of life, because "adaptations" are not always what they seem. Its comparison of llama and camel hemoglobin underscores the point.

With tentative acceptance of evolutionary theory, students can think about applying it to their own lives. Essays 26.2, 26.11, 37.4, and 38.1 interconnect in this respect. They hint at where our amazing brains and hands came from, why we have such intricate highways for blood circulation through our body, and why we are prone to lower back pain. Essay 34.12 connects the brain's evolutionary history with its development. It shows how that connection helps explain why teenagers tend to fall asleep in class and engage in notably impulsive behavior.

Acceptance also gives broad insight into the source of biodiversity, into mass extinctions and slow recoveries. Essays 21.9, 22.1, 23.1 reinforce this point.

Connecting Our Lives to the Big Picture

Deeper understanding of how each of us connects with the sweeping story of life—that is what we hope students will take away from their introduction to biology. With this in mind, we use some essays to show how biology impacts our own lives and the choices that we make.

Essay 16.10, for example, reflects on how the ability to study and alter genomes rapidly, as with the use of DNA microarrays, is outpacing our attempts to assess its bioethical implications. What are the ramifications of human gene therapy? Cloning genetically engineered mammals? Manipulating genomes of crop plants? These are issues facing students today, and the citizens and leaders of tomorrow.

At the end of Unit V, which shows what it takes to be a plant, Essay 32.6 considers one of the costs of growing

CONNECTIONS ESSAYS

Vertical red bands down the edge of a text page flag these essays.

1.4	AN EVOLUTIONARY VIEW OF DIVERSITY
3.1	THE MOLECULES OF LIFE—FROM STRUCTURE TO FUNCTION
3.8	WHY IS PROTEIN STRUCTURE SO IMPORTANT?
5.2	A GALLERY OF MEMBRANE PROTEINS
6.9	LIGHT UP THE NIGHT—AND THE LAB
7.8	AUTOTROPHS, HUMANS, AND THE BIOSPHERE
8.7	PERSPECTIVE ON THE MOLECULAR UNITY OF LIFE
16.10	BIOTECHNOLOGY IN A BRAVE NEW WORLD
19.9	INTERPRETING AND MISINTERPRETING THE PAST
20.4	WHERE DID ORGANELLES COME FROM?
21.9	EVOLUTION AND INFECTIOUS DISEASES
22.1	AN EMERGING EVOLUTIONARY ROAD MAP
23.1	TRENDS IN PLANT EVOLUTION
26.2	TRENDS IN VERTEBRATE EVOLUTION
26.11	TRENDS IN PRIMATE EVOLUTION
27.1	ON MASS EXTINCTIONS AND SLOW RECOVERIES
28.2	THE NATURE OF ADAPTATION
28.6	RECURRING CHALLENGES TO SURVIVAL
32.6	GROWING CROPS AND A CHEMICAL ARMS RACE
34.12	REFLECTIONS ON THE NOT-QUITE COMPLETE TEEN BRAIN
37.4	EVOLUTION OF THE VERTEBRATE SKELETON
38.1	EVOLUTION OF CIRCULATORY SYSTEMS
43.8	DEATH IN THE OPEN
45.6	NATURAL SELECTION AND THE GUPPIES OF TRINIDAD
46.3	THE ADAPTIVE VALUE OF BEHAVIOR
47.5	AN EVOLUTIONARY ARMS RACE
49.14	RITA IN THE TIME OF CHOLERA
50.10	BIOLOGICAL PRINCIPLES AND THE HUMAN IMPERATIVE

enough plants to feed the human population. It invites students to think about how we have become locked into using herbicides, fungicides, and pesticides on a massive scale. It invites them consider the effect of these complex molecules on nontargeted organisms, including us.

As two more examples, Essay 49.14 recounts how Rita Colwell made a sweeping connection between copepods, a bacterial life cycle, sea surface temperature changes during El Niño episodes, and horrible cholera outbreaks in Bangladesh.

The essay in our concluding chapter 50 compares life's evolution over the past 3.8 billion years with the impact of relative latecomers—humans. It asks students to think about the bioethics of mitigating the disproportionate effect of we latecomers on the world of life.

OTHER MAJOR CHANGES   Now that systematists have reached consensus, we have subsumed the kingdoms of organisms into the three-domain classification system. Section 1.3 introduces the system; Sections 19.7 and 19.8 fill in details. We took a stronger cladistic approach and did major work on the chapters of the biodiversity unit. Section 22.1 gives the rationale for doing so. The outcome is most evident in how we present the protistans (Chapter 22) and vertebrates (Chapter 26).

Again, we moved away from generic boxes and circles for molecules and present more realistic models for membranes, protein transporters, enzymes, RNAs, and other cell components. You can check out some examples of how we put them to use in Chapters 3, 5, 14, and 34.

To give an idea of what went into the new graphics, Lisa Starr rendered our molecular models from primary structural data. She even tracked down researchers who are still constructing a model of the human cardiac gap junction and worked directly with them to convert their most recent data into graphic form for this book.

We rewrote Section 13.5 (mammalian cloning) and Chapter 16 (recombinant DNA and genetic engineering) to keep up with the rapid advances in biotechnology. Section 16.10 addresses some of the bioethical issues. So do some of the detailed *Critical Thinking* questions at the end of Chapter 16. These are serious issues that should provoke discussion among students. The Chapter 15 introduction and focus essay on cancer are updated.

Given the wealth of new information to be covered in the evolution unit, we decided to condense the history of evolutionary thought into a few introductory sections for Chapter 17 (microevolution). Chapter 19 (macroevolution) is heavily rewritten and reorganized, starting with a new introduction on issues associated with measuring geologic time. There is stronger treatment of the evidence from comparative biochemistry and systematics. This chapter concludes with a reflection on how we interpret the past.

I urge our adopters to review Chapter 28. Concepts that will help students get through the two units on plant and animal anatomy/physiology are presented here. We have made so many refinements and updates—for example, on human nutrition, embryonic development, and predator–prey interactions, but we are running out of room to highlight them all. Suffice it to say that we made an honest, solid attempt to keep up with biology and to do it justice all across the board.

Writing this preview of all the changes reminded me again of how fortunate I am to be under the Wadsworth umbrella. No author can execute a huge revision every single time without a dedicated production team and enlightened management. Gary Head, Lisa Starr, Diana Starr, Suzannah Alexander, Jana Davis, Teri Hyde, Karen Hunt, Grace Davidson, Myrna Engler, Angela Harris—this is my core team, the best of the best. Pat Waldo, Donna Kelley, Keli Amann, Chris Evers, Steve Bolinger—they create our stunning multimedia package. Susan Badger, Sean Wakely, Michelle Julet, Jack Carey, Kathie Head— these are anomalies in higher education, publishers who nurture authors with intelligence, strength, and grace.

Cecie Starr, October 2002
E-mail starr@brookscole.com

ACKNOWLEDGMENTS

This book is the current version of an educational effort that started nearly three decades ago. We thank all of the students and instructors who used and commented on previous editions. We thank the individuals listed here for their significant influence on the book's development. We give special thanks to this edition's overall advisor, E. William Wischusen, who helped formulate connections to remind students of where they have been and where they are going in the book. As he pointed out, if we expect them to understand the chapters on genetics, evolution, anatomy, and physiology, we must provide them with the intellectual tools and graphics to do so. We must clearly connect chapters back to the tools that guide students through biomolecules and membranes. And we must include a new chapter on the commonalities between plant and animal systems to help students see the big picture while learning the details.

We also thank Walter Judd, who guided our overhaul of the macroevolution and biodiversity chapters, with close attention to the protistans and vertebrates. The new evolutionary tree diagram (Sections 19.8 and 22.1) and classification system (Appendix I) are his contributions.

If teachers appreciate the new clarity of thought and currency of this new edition, they might give a special nod of recognition to Bill and Walt for urging us gently to give it our best shot—and then to keep on giving more.

Major Advisor for the Tenth Edition

E. WILLIAM WISCHUSEN   *Louisiana State University*

General Advisors/Contributors

JOHN ALCOCK   *Arizona State University*
GEORGE COX   *San Diego State University*
MELANIE DEVORE   *Georgia College and State University*
DANIEL J. FAIRBANKS   *Brigham Young University*
TOM GARRISON   *Orange Coast College*
DAVID GOODIN   *The Scripps Research Institute*
PAUL E. HERTZ   *Barnard College*
JOHN D. JACKSON   *North Hennipin Community College*
WALTER JUDD   *University of Florida*
EUGENE N. KOZLOFF   *University of Washington*
KAREN E. MESSLEY   *Rock Valley College*
ELIZABETH LANDECKER-MOORE   *Rowan University*
JON REISKIND   *University of Florida*
THOMAS L. ROST   *University of California, Davis*
LAURALEE SHERWOOD   *West Virginia University*
STEPHEN L. WOLFE   *University of California, Davis*

Contributors of Influential Reviews

ALDRIDGE, DAVID   *North Carolina Agricultural/Technical State University*
ANDERSON, ROBERT C.   *Idaho State University*
ARMSTRONG, PETER   *University of California, Davis*
BAJER, ANDREW   *University of Oregon*
BAKKEN, AIMEE   *University of Washington*
BARBOUR, MICHAEL   *University of California, Davis*
BARHAM, LINDA   *Meridian Community College*

BARKWORTH, MARY   *Utah State University*
BELL, ROBERT A.   *University of Wisconsin, Stevens Point*
BENDER, KRISTEN   *California State University, Long Beach*
BENIVENGA, STEPHEN   *University of Wisconsin, Oshkosh*
BINKLEY, DAN   *Colorado State University*
BORGESON, CHARLOTTE   *University of Nevada*
BRENGELMANN, GEORGE   *University of Washington*
BRINSON, MARK   *East Carolina University*
BROWN, ARTHUR   *University of Arkansas*
BUTTON, JERRY   *Portland Community College*
CARTWRIGHT, PAULYN   *University of Kansas*
CASE, CHRISTINE   *Skyline College*
CASE, TED   *University of California, San Diego*
CHRISTIANSEN, A. KENT   *University of Michigan*
CLARK, DEBORAH C.   *Middle Tennessee University*
COLAVITO, MARY   *Santa Monica College*
CONKEY, JIM   *Truckee Meadows Community College*
CROWCROFT, PETER   *University of Texas at Austin*
DABNEY, MICHAEL W.   *Hawaii Pacific University*
DAVIS, JERRY   *University of Wisconsin, La Crosse*
DeGROOTE, DAVID K.   *St. Cloud University*
DELCOMYN, FRED   *University of Illinois, Urbana*
DEMMANS, DANA   *Finger Lakes Community College*
DEMPSEY, JEROME   *University of Wisconsin*
DENGLER, NANCY   *University of California, Davis*
DENETTE, PHIL   *Delgado Community College*
DENNISTON, KATHERINE   *Towson State University*
DeSAIX, JEAN   *University of North Carolina*
DETHIER, MEGAN   *University of Washington*
DeWALT, R. EDWARD   *Louisiana State University*
DiBARTOLOMEIS, SUSAN   *Millersville University of Pennsylvania*
DIEHL, FRED   *University of Virginia*
DONALD-WHITNEY, CATHY   *Collin County Community College*
DOYLE, PATRICK   *Middle Tennessee State University*
DUKE, STANLEY H.   *University of Wisconsin, Madison*
DYER, BETSEY   *Wheaton College*
EDLIN, GORDON   *University of Hawaii, Manoa*
EDWARDS, JOAN   *Williams College*
ELMORE, HAROLD W.   *Marshall University*
ENDLER, JOHN   *University of California, Santa Barbara*
ERWIN, CINDY   *City College of San Francisco*
EWALD, PAUL   *Amherst College*
FALK, RICHARD   *University of California, Davis*
FISHER, DAVID   *University of Hawaii, Manoa*
FISHER, DONALD   *Washington State University*
FLESSA, KARL   *University of Arizona*
FONDACARO, JOSEPH   *Hoechst Marion Roussel, Inc.*
FROEHLICH, JEFFREY   *University of New Mexico*
FULCHER, THERESA   *Pellissippi State Technical Community College*
GAGLIARDI, GRACE S.   *Bucks County Community College*
GENUTH, SAUL M.   *Mt. Sinai Medical Center*
GHOLZ, HENRY   *University of Florida*
GOODMAN, H. MAURICE   *University of Massachusetts Medical School*
GOSZ, JAMES   *University of New Mexico*
GREGG, KATHERINE   *West Virginia Wesleyan College*
GUTSCHICK, VINCENT   *New Mexico State University*
HASSAN, ASLAM   *Univeristy of Illinois College of Veterinary Medicine*
HELGESON, JEAN   *Collin County Community College*
HUFFMAN, DAVID   *Southwest Texas State University*
INEICHER, GEORGIA   *Hinds Community College*
INGRAHAM, JOHN L.   *University of California, Davis*
JENSEN, STEVEN   *Southwest Missouri State University*
JOHNSON, LEONARD R.   *University of Tennessee College of Medicine*
JOHNSTON, TIMOTHY   *Murray State University*
KAREIVA, PETER   *University of Washington*
KAUFMAN, JUDY   *Monroe Community College*
KAYE, GORDON I.   *Albany Medical College*
KAYNE, MARLENE   *Trenton State College*
KENDRICK, BRYCE   *University of Waterloo*
KILBURN, KERRY S.   *Old Dominion University*
KILLIAN, JOELLA C.   *Mary Washington College*
KIRKPATRICK, LEE A.   *Glendale Community College*
KLANDORF, HILLAR   *West Virginia University*
KREBS, CHARLES   *University of British Columbia*

KREBS, JULIA E.   *Francis Marion University*
KUTCHAI, HOWARD   *University of Virginia Medical School*
LANZA, JANET   *University of Arkansas, Little Rock*
LASSITER, WILLIAM   *University of North Carolina*
LEVY, MATTHEW   *Mt. Sinai Medical Center*
LEWIS, LARRY   *Salem State College*
LITTLE, ROBERT   *Medical College of Georgia*
LOHMEIER, LYNNE   *Mississippi Gulf Coast Community College*
LUMSDEN, ANN   *Florida State University*
LYNG, R. DOUGLAS   *Indiana University, Purdue University*
MANN, ALAN   *University of Pennsylvania*
MARTIN, JAMES   *Reynolds Community College*
MARTIN, TERRY   *Kishwaukee College*
MASON, ROY B.   *Mt. San Jacinto College*
McKEAN, HEATHER   *Eastern Washington State University*
McKEE, DOROTHY   *Auburn University, Montgomery*
McNABB, ANN   *Virginia Polytechnic Institute and State University*
MENDELSON, JOSEPH R.   *Utah State University*
MICKLE, JAMES   *North Carolina State University*
MILLER, G. TYLER   *Wilmington, North Carolina*
MILLER, GARY   *University of Mississippi*
MINORKSY, PETER V.   *Western Connecticut State University*
MOISES, HYLAN C.   *University of Michigan Medical School*
MORENO, JORGE A.   *University of Colorado, Boulder*
MORRISON-SHETLER, ALLISON   *Georgia State University*
MORTON, DAVID   *Frostburg State University*
MURPHY, RICHARD   *University of Virginia Medical School*
MYRES, BRIAN   *Cypress College*
NAPLES, VIRGINIA   *Northern Illinois University*
NELSON, RILEY   *University of Texas at Austin*
NORRIS, DAVID   *Brigham Young University*
PECHENIK, JAN   *Tufts University*
PERRY, JAMES   *University of Wisconsin, Center-Fox Valley*
PETERSON, GARY   *South Dakota State University*
POLCYN, DAVID M.   *California State University, San Bernardino*
REESE, R. NEIL   *South Dakota State University*
REID, BRUCE   *Kean College of New Jersey*
RENFROE, MICHAEL   *James Madison University*
RICKETT, JOHN   *University of Arkansas, Little Rock*
ROSE, GREIG   *West Valley College*
ROST, THOMAS   *University of California, Davis*
SALISBURY, FRANK   *Utah State University*
SCHAPIRO, HARRIET   *San Diego State University*
SCHLESINGER, WILLIAM   *Duke University*
SCHNEIDEWENT, JUDY   *Milwaukee Area Technical College*
SCHNERMANN, JURGEN   *University of Michigan School of Medicine*
SCHREIBER, FRED   *California State University, Fresno*
SHONTZ, NANCY   *Grand Valley State University*
SELLERS, LARRY   *Louisiana Tech University*
SLOBODA, ROGER   *Dartmouth College*
SMITH, JERRY   *St. Petersburg Junior College, Clearwater Campus*
SMITH, MICHAEL E.   *Valdosta State College*
SMITH, ROBERT L.   *West Virginia University*
STEARNS, DONALD   *Rutgers University*
STEELE, KELLY P.   *Appalachian State University*
STEINERT, KATHLEEN   *Bellevue Community College*
SUMMERS, GERALD   *University of Missouri*
SUNDBERG, MARSHALL D.   *Empire State University*
SWEET, SAMUEL   *University of California, Santa Barbara*
TAYLOR, JANE   *Northern Virginia Community College*
TIZARD, IAN   *Texas A&M University*
TROUT, RICHARD E.   *Oklahoma City Community College*
TYSER, ROBIN   *University of Wisconsin, LaCrosse*
WAALAND, ROBERT   *University of Washington*
WAHLERT, JOHN   *City University of New York, Baruch College*
WALSH, BRUCE   *University of Arizona*
WARING, RICHARD   *Oregon State University*
WARNER, MARGARET R.   *Purdue University*
WEBB, JACQUELINE F.   *Villanova University*
WEIGL, ANN   *Winston-Salem State University*
WEISS, MARK   *Wayne State University*
WELKIE, GEORGE W.   *Utah State University*
WHITE, EVELYN   *Alabama State University*
WINICUR, SANDRA   *Indiana University, South Bend*

Introduction

Current configurations of the Earth's oceans and land masses—the geologic stage upon which life's drama continues to unfold. This composite satellite image reveals global energy use at night by the human population. Just as biological science does, it invites you to think more deeply about the world of life—and about our impact upon it.

CONCEPTS AND METHODS IN BIOLOGY

Why Biology?

Leaf through a newspaper on any given Sunday and you might get an uneasy feeling that the natural world is spinning out of control. Northern forests are burning fiercely (Figure 1.1). The Arctic ice cap is melting and Greenland's glaciers are turning into lakes. Heat waves and droughts are breaking records everywhere, and maybe the whole atmosphere is warming up. The last individual of an ancient primate lineage bit the dust just as the human population neared the 6.1 billion mark. Geneticists decoded yet another garbled message in human DNA that causes another horrible disease. People are still arguing about when, exactly, the mass of embryonic cells busily dividing inside a woman's

womb is "alive." In a bizarre application of genetic engineering, somebody decided to make an "artistic statement" by making a glow-in-the-dark bunny. The entire live bunny turns fluorescent green in black light.

It's enough to make you throw down the paper and go sit in a park. It's enough to make you wish you lived in the good old days when things were so much simpler.

Of course, read up on the good old days and you'll discover they weren't so good. For example, in 1918, maybe when your grandparents were alive, the Spanish flu swept around the world. Those infected often died within hours of the first symptoms. Before the pandemic subsided, *30 to 40 million* people had died. Unlike most

Figure 1.1 Biology starts with the premise that any aspect of nature—including this out-of-control forest fire in Montana—has one or more underlying causes. By giving us a way of thinking critically about life, biology helps us understand how to deal with nature, and with our place in it.

THEORY - The notion that Earth orbits around the Sun rather than vice versa, offered by Copernicus in 1543, is a theory. Continental drift is a theory. The existence, structure, & dynamics of atoms is atomic theory. Even electricity is a theoretical construct, involving electrons, which are tiny units of charged mass that no one has ever seen. Each of these theories is an explanation that

flu viruses, which mainly endanger infants, the elderly, and the very ill, the Spanish flu virus struck healthy young adults, the ones who support most of the social infrastructure. Then, as today, many thought the world was coming apart at the seams. They felt helpless and hopeless before a force of nature on the rampage.

What it boils down to is this: For a couple of million years at least, humans and their immediate ancestors have been trying to make sense of the natural world. We observe it, we come up with ideas, we test the ideas. However, the more pieces of the puzzle we fit together, the bigger the puzzle gets. We are now smart enough to know that it's almost overwhelmingly big.

You might choose to walk away from the challenge and simply let others tell you what to think. Or you might choose to develop your own understanding of the puzzle. Maybe you're interested in the pieces that affect your health, your home, the food you eat, and your offspring, should you choose to reproduce. Maybe you just find the connections among organisms and their environment fascinating. Regardless of the focus, **biology**—the scientific study of life—can help deepen your perspective on the world around you.

Start with a question that seems simple enough: *What is life?* Offhandedly, you might say you know it when you see it. Yet the question opens up a story that began at least 3.8 billion years ago.

From a biological perspective, "life" is an outcome of ancient events by which lifeless matter—atoms and molecules—became organized into the first living cells. "Life" is a way of capturing and using energy and raw materials. "Life" is a way of sensing and responding to the environment. "Life" is a capacity to reproduce. And "life" evolves, which simply means that the traits characterizing individuals of a population can change from one generation to the next.

Throughout this book, you will come across many examples of how organisms are constructed, how they function, where they live, what they do. The examples support concepts which, when taken together, convey what "life" is. *This chapter is an overview of the basic concepts.* It sets the stage for forthcoming descriptions of scientific observations, experiments, and tests that help show how you can develop, modify, and refine your views of life.

Key Concepts

1. Unity underlies the world of life, for all organisms are alike in key respects. They all consist of one or more cells made of the same kinds of substances, put together in the same basic ways. Their activities require inputs of energy, which they must get from their surroundings. All organisms sense and respond to changing conditions in their environment. They have a capacity to grow and reproduce, based on instructions contained in DNA.

2. The world of life shows immense diversity. Many millions of different kinds of organisms, or species, now inhabit the Earth. Many millions more lived in the past. Each species is unique in some of its traits— that is, in some aspects of its body plan, functioning, and behavior.

3. Theories of evolution, especially the theory of evolution by natural selection as formulated by Charles Darwin, help explain life's diversity. The theories unite all fields of biological inquiry into a single, coherent whole.

4. Biology, like other branches of science, is based on systematic observations, hypotheses, predictions, and observational and experimental tests. The external world, not internal conviction, is the testing ground for scientific theories.

It has been confirmed to such a degree, by observation and experiment, that experts accept it as fact, i.e. a theory is an explanatory statement that fits the evidence. They embrace such an explanation confidently but provisionally - taking it as their best available view of reality, at least until some severely conflicting data or some better explanation might come along. - See Structure Of Scientific Revolutions.

DNA, ENERGY, AND LIFE

Nothing Lives Without DNA

THE MOLECULES OF LIFE   Let's start this overview of life by picturing a frog busily croaking on a rock. Even without thinking about it, you know the frog is alive and the rock is not. Could you say why? After all, both consist only of protons, electrons, and neutrons, the building blocks for atoms. But atoms are building blocks for larger and more diverse bits of matter called molecules. It is at the molecular level that differences between living and nonliving things start to emerge.

You will never find a rock made of nucleic acids, proteins, carbohydrates, and lipids. In nature, only *cells* build these molecules, which you will read about later. And the signature molecule of a cell—the smallest unit that has the capacity for life—is a nucleic acid known as **DNA**. No chunk of granite or quartz has it.

DNA holds instructions for assembling a variety of proteins from smaller molecules, the amino acids. By analogy, if you follow suitable instructions and invest energy in the task, you might organize a pile of a few kinds of ceramic tiles (representing amino acids) into diverse patterns (representing proteins), as in Figure 1.2.

Enzymes are among the proteins. When these worker molecules get an energy boost, they swiftly build, split, and rearrange the molecules of life. Some work with the nucleic acids called RNAs to turn DNA instructions into proteins. Think of this as a flow of information, *from DNA to RNA to protein*. As you will see later, this molecular trinity is central to our understanding of life.

DNA AND INHERITANCE   We humans tend to think we enter the world abruptly and leave it the same way. But we are more than this. We and all other organisms

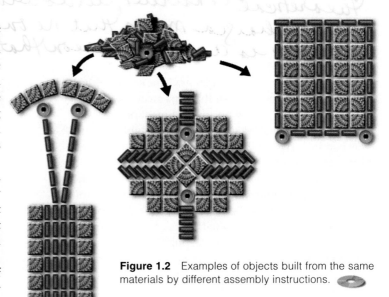

Figure 1.2   Examples of objects built from the same materials by different assembly instructions.

are part of an immense journey that began about 3.8 billion years ago with the chemical origin of the first living cells.

Under present-day conditions in nature, new cells and multicelled organisms inherit their defining traits from parents. **Inheritance** is simply the acquisition of traits through the transmission of DNA from parents to offspring. We reserve the term **reproduction** for the actual mechanisms of transmitting DNA to offspring. Why do baby storks look like storks and not pelicans? Because they inherit stork DNA, which isn't exactly the same as pelican DNA in its molecular details.

For frogs, humans, trees, and other large organisms, DNA also guides **development**. This term refers to the

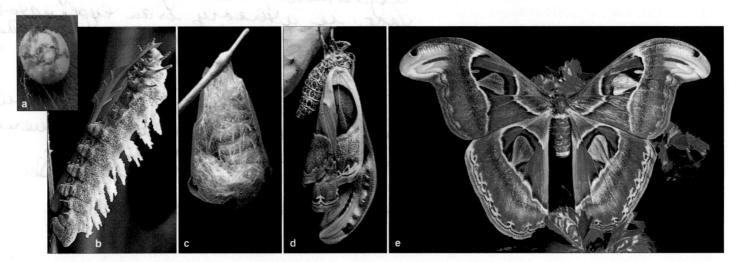

Figure 1.3   "The insect"—a series of stages of development guided largely by instructions in DNA. Different adaptive properties emerge at each stage. Shown here, a silkworm moth, from the egg (**a**), to a larval stage called a caterpillar (**b**), to a pupal stage (**c**), to the winged form of the adult (**d,e**).

Figure 1.4 Response to signals from pain receptors, activated by a lion cub flirting with disaster.

transformation of a new individual into a multicelled adult, typically with tissues and organs specialized for certain tasks. For example, the moth shown in Figure 1.3 is only the adult stage of a winged insect. It started out as a single cell, a fertilized egg that developed into a caterpillar. That immature larval stage ate soft leaves and grew rapidly until an internal alarm clock went off. At that time its tissues started to be remodeled into a different stage, a pupa. In time the adult, adapted to reproduce, emerged. It has special parts to make sperm or eggs; its wings have the color, pattern, and fluttering frequency that are adaptations for attracting a mate.

Like other animals, "the insect" is a series of stages, each of which must develop properly before the next begins. And instructions for forming every stage were written into moth DNA long before each time of moth reproduction—and so an ancient story of life continues.

Nothing Lives Without Energy

ENERGY DEFINED Being alive requires more than DNA. Moths and everything else in the universe also require **energy**, a capacity to do work. In their cells, work gets done as atoms give up, share, and accept electrons. It gets done as molecules are assembled, rearranged, and split apart. Energy drives these molecular events.

METABOLISM DEFINED Put it this way: Each living cell has the capacity to (1) obtain and convert energy from its surroundings and (2) use energy to maintain itself, grow, and produce more cells. We call this capacity **metabolism**. Think of one cell in a leaf that makes food by a process called photosynthesis. That cell intercepts sunlight energy and converts it to chemical energy, in the form of ATP molecules. ATP helps drive hundreds of metabolic events by transferring energy to reaction sites, where enzymes put together sugar molecules. In most cells, ATP also can form by aerobic respiration. This process can release energy that cells stored earlier, in the form of starch and other kinds of molecules.

SENSING AND RESPONDING TO ENERGY It's often said that only living things respond to the environment. Yet even a rock shows responsiveness, as when it yields to the force of gravity and tumbles down a hill or changes its shape slowly under the repeated battering of wind, rain, or tides. The difference is this: *Organisms can sense changes in their surroundings and then make compensatory, controlled responses to the changes.* How? Each organism has **receptors**, which are molecules and structures that detect stimuli. A **stimulus** is a specific form of energy that a receptor can detect. Examples are sunlight energy,

heat energy, a hormone molecule's binding energy, and the mechanical energy of a bite (Figure 1.4).

Cells adjust metabolic activities in response to signals from receptors. Each cell (and organism) can withstand only so much heat or cold. It must rid itself of harmful substances. It requires certain foods, in certain amounts. Yet temperatures do shift, harmful substances might be encountered, and food is sometimes plentiful or scarce.

Finish a snack, and sugars leave your gut and enter your blood. Blood, together with the tissue fluid that bathes your cells, makes up your internal environment. Unchecked, too much or too little sugar in the blood can cause diabetes and other problems. When the level rises, the pancreas, a glandular organ, secretes more insulin. Most of your body cells have receptors for this hormone, which stimulates cells to take up sugar. When enough cells do so, the blood sugar level returns to normal.

Organisms respond so exquisitely to energy changes that their internal operating conditions usually remain within tolerable limits. This state, called **homeostasis**, is one of the key defining features of life.

All organisms consist of one or more cells, the smallest units of life. Under present-day conditions in nature, new cells form only when existing cells reproduce.

DNA, the molecule of inheritance, encodes protein-building instructions, which RNAs help carry out. Many proteins are enzymes that speed up cellular work, which includes building all the complex molecules characteristic of life.

Cells live only for as long as they engage in metabolism. They acquire and transfer energy that is used to assemble, break down, stockpile, and dispose of materials in ways that promote survival and reproduction.

Single-celled and multicelled organisms sense and respond to environmental conditions in ways that help maintain their internal operating conditions.

ENERGY AND LIFE'S ORGANIZATION

Levels of Biological Organization

Through their individual and interactive uses of energy and materials, organisms contribute to a great pattern of biological organization. As Figure 1.5 shows, life's properties emerge when DNA and other molecules are organized as cells. A **cell** is the smallest organizational unit having a capacity to survive and reproduce on its own, given DNA instructions, building blocks, energy inputs, and suitable conditions. An obvious unit is an amoeba or another free-living cell. The definition also fits **multicelled organisms** consisting of specialized, interdependent cells, typically organized in tissues and organs. Do you question this? After all, your own cells could never live alone in nature, for body fluids must constantly bathe them, deliver substances to them, and remove their metabolic wastes. However, even isolated human cells remain alive under controlled conditions in laboratories around the world. Researchers routinely maintain isolated human cells in cultures for important experiments, as in cancer studies.

Cells and multicelled organisms usually are part of a **population**: a group of organisms of the same kind. A zebra herd is an example. The next level of organization is the **community**: all populations of all species living in the same area, such as the African savanna's bacteria, grasses, trees, zebras, lions, and so forth. The next level is the **ecosystem**: a community *together with* its physical and chemical environment. The **biosphere**, the highest level, includes all parts of the Earth's crust, waters, and atmosphere in which organisms live. Astoundingly, *this globe-spanning organization begins with the convergence of energy, certain materials, and DNA in tiny, individual cells.*

BIOSPHERE
All regions of the Earth's crust, waters, and atmosphere that sustain life

ECOSYSTEM
Community and its physical environment

COMMUNITY
Populations of all species occupying the same area

POPULATION
Group of individuals of the same kind (that is, the same species) occupying the same area

MULTICELLED ORGANISM
Individual consisting of interdependent cells typically organized in tissues, organs, and organ systems

ORGAN SYSTEM
Two or more organs interacting chemically, physically, or both in ways that contribute to organism's survival

ORGAN
Structural unit in which tissues, combined in specific amounts and patterns, perform a common task

TISSUE
Organized aggregation of cells and substances functioning together in a specialized activity

CELL
Smallest unit with the capacity to live and reproduce, independently or as part of multicelled organism

ORGANELLE
Membrane-bound internal compartment for specialized reactions (most prokaryotic cells have none)

MOLECULE
Unit of two or more bonded-together atoms of the same element or different elements

ATOM
Smallest unit of an element (a fundamental substance) that still retains the properties of that element

SUBATOMIC PARTICLE
Electron, proton, neutron, or some other fundamental unit of matter

populations of shrubs and trees

community (all of the populations living in the same area)

populations of grasses

population of zebras

multicelled organism

once-live multicelled organism that is about to revert to unorganized molecules and atoms

ecosystem (community together with its physical environment)

Figure 1.5 Levels of organization in nature.

beetle larva

Producers capture, convert, and use or store some energy from the sun.

PRODUCERS

NUTRIENT CYCLING

CONSUMERS, DECOMPOSERS

ONE-WAY FLOW OF ENERGY

Energy gets transferred from one organism to another; in time, all flows back to the environment.

Figure 1.6 Example of the one-way flow of energy and the cycling of materials through the biosphere.

(**a**) Plants of a warm, dry grassland called the African savanna capture energy from the sun and use it to build plant parts. Some of the energy ends up inside plant-eating organisms, such as this adult male elephant. He eats huge quantities of plants to maintain his eight-ton self. He produces piles of solid wastes—dung—that still hold some unused nutrients and energy. Although most organisms might not recognize it as such, elephant dung is an exploitable food source.

(**b**) And so we next have little dung beetles scrambling to the scene almost simultaneously with the uplifting of an elephant tail. Working fast, they carve fragments of moist dung into round balls, which they roll off and bury in burrows. In those balls the beetles lay eggs—a reproductive behavior that will help assure their forthcoming offspring (**c**) of a compact food supply.

Thanks to beetles, dung does not pile up and dry out into rock-hard mounds in the intense heat of the day. Instead, the surface of the land is tidied up, the beetle offspring get fed, and the leftover dung accumulates in beetle burrows—there to enrich the soil that nourishes the plants that feed (among others) the elephants.

Interdependencies Among Organisms

A great flow of energy from the sun into the world of life starts with **producers**—plants and other organisms that make their own food. Animals are **consumers**. Directly or indirectly, they use food energy that is stored in tissues of producers. For example, some energy gets transferred to zebras after they browse on the plants. It gets transferred again when a lion devours a zebra, as in Figure 1.5. And it is transferred again when fungal and bacterial decomposers extract energy from tissues and remains of lions, elephants, or any other organism. **Decomposers** break down sugars and other biological molecules to simpler materials, some of which is cycled back to the producers. In time, all of the energy that plants originally captured from the sun's rays returns to the environment, but that's another story.

For now, keep in mind that organisms connect with one another by a one-way flow of energy *through* them and a cycling of materials *among* them, as in Figure 1.6. Their interconnectedness affects the structure, size, and composition of populations and communities. It affects ecosystems, even the biosphere. Understand the extent of their interactions and you will gain insight into food shortages, cholera epidemics, acid rain, global warming, biodiversity losses, and other modern-day problems.

Nature shows levels of organization. The characteristics of life emerge at the level of single cells and extend through populations, communities, ecosystems, and the biosphere.

A one-way flow of energy through organisms and a cycling of materials among them organizes life in the biosphere. In nearly all cases, energy flow starts with sunlight energy.

IF SO MUCH UNITY, WHY SO MANY SPECIES?

So far, this overview of life has focused on its unity, on characteristics that all living things have in common. Think of it! They consist of the same lifeless materials. They stay alive by way of metabolism, ongoing energy transfers at the cellular level. They interact when using energy and raw materials. They sense and respond to the environment. They have the capacity to reproduce based on instructions in their DNA, which they inherit from individuals of a preceding generation.

Superimposed on this common heritage is immense diversity. You share the planet with many millions of kinds of organisms, or **species**. Many millions more preceded you over the past 3.8 billion years. However, their lineages vanished; they are extinct. For centuries, scholars have tried to make sense of the confounding diversity. One of them, Carolus Linneaus, promoted a classification scheme that assigns a two-part name to each newly identified species. The first part designates the **genus** (plural, genera). A genus encompasses all of the species that seem closely related because of their form, functioning, and ancestry. The second part of the name designates a particular species within a genus.

For instance, *Quercus alba* is the white oak's name. *Q. rubra* is the red oak's name. (A genus name can be abbreviated this way once it has been spelled out in a document.) Such names are passports to great worlds of information; try them out for yourself on the web.

Biologists group species that are related by descent from a common ancestor. They are still working out the groupings. For instance, scholars once recognized only two groups—animals and plants. Most biologists now favor a three-domain system of classification. The domains are called bacteria, archaea, and eukarya:

EUBACTERIA (BACTERIA)	ARCHAEBACTERIA (ARCHAEA)	EUKARYOTES (EUKARYA)

Within these domains are at least six groups that are still known as kingdoms—**archaebacteria**, **eubacteria**, **protistans**, **fungi**, **plants**, and **animals** (Figure 1.7).

Archaebacteria and eubacteria are all single cells of a type called *prokaryotic*. The word means they do not contain a nucleus. (In other cell types, this membrane-bound sac keeps the DNA separated from the rest of the cell's interior.) Different kinds of prokaryotic cells are producers, consumers, and decomposers. Of all the groups, theirs show the most metabolic diversity.

Archaebacteria live in hot springs, salt lakes, and other habitats as harsh as the ones that prevailed when cells originated. The eubacteria are more common and

Clockwise from top, a forest of giant kelp (*Macrocystis*), the shell of a coccolithophore, the migrating slug stage of a slime mold (*Dictyostelium discoideum*), a flagellate (*Trypanosoma brucei*) that causes African sleeping sickness, and a ciliated protozoan (*Paramecium*). Even this tiny sampling of diversity conveys why many biologists now believe that the "protistans" are not a single group but rather are many evolutionary separate lineages.

From the muck of an oxygen-free habitat, a microscopically small colony of archaebacterial cells of the genus *Methanosarcina*. Archaebacteria are evolutionarily closer to us than to true bacteria.

Figure 1.7 A few representatives of life's diversity.

widespread. Their name means true bacteria, so we will just refer to them as bacteria (singular, bacterium).

Like plants, fungi, and animals, the protistans are *eukaryotic*, meaning a nucleus houses their DNA. Most are larger and show far more internal complexity than prokaryotic cells. The grouping includes microscopic single-celled producers, consumers, and decomposers, and also giant multicelled "seaweeds." The differences among protistans are great, so this one group probably will be split into more informative ones soon.

Most fungi, such as the mushrooms sold in grocery stores, are multicelled decomposers or parasites that

Stinkhorn (*Dictyophora indusiata*) and trumpet chanterelles (*Craterellus*). Although certain fungi are parasites and cause diseases, these two species—and the vast majority of others—are decomposers. Without decomposers, communities would become buried in their own wastes.

Left: Coast redwoods (*Sequoia sempervirens*) growing in California. Like nearly all plants, redwoods are photosynthetic. *Right:* Flower of a hybrid *Gazania*. The patterns and colors of this plant's flowers guide bees to the nectar. Bees get food; the plant gets help reproducing. Like many other organisms, these two species interact in a mutually beneficial way.

Like the ancient fishes that were its ancestors, *Basiliscus basiliscus*, a basilisk lizard, has four limbs. Unlike fishes, it is adapted to land, although it runs fast across water by stomping its feet on the surface.

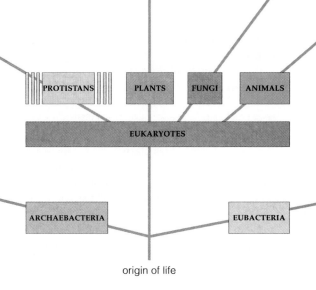

```
┌──────────┐  ┌────────┐  ┌────────┐  ┌──────────┐
│PROTISTANS│  │ PLANTS │  │ FUNGI  │  │ ANIMALS  │
└──────────┘  └────────┘  └────────┘  └──────────┘
        ┌────────────────────────────────┐
        │          EUKARYOTES            │
        └────────────────────────────────┘

┌────────────────┐              ┌──────────────┐
│ ARCHAEBACTERIA │              │ EUBACTERIA   │
└────────────────┘              └──────────────┘

              origin of life
```

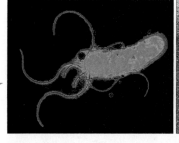
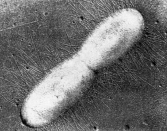

Left: *Helicobacter pylori*, a bacterium that can cause gastric upsets and ulcers. *Right:* *Escherichia coli*, a bacterial resident of the mammalian gut. It made a copy of its DNA and is now reproducing by dividing in two.

feed in a distinctive way. They secrete enzymes that digest food outside of the fungal body, then individual cells absorb the digested bits.

Plants are multicelled, photosynthetic producers. Photosynthesis means they make their own food by using sunlight energy and simple raw materials. They have cellular pipelines for water and substances that typically extend through roots, stems, and leaves.

Animals are multicelled consumers that ingest the tissues of other organisms. They include the herbivores that graze upon photosynthesizers, carnivores (meat eaters), parasites, and scavengers. All develop through

a series of embryonic stages, and they actively move about during at least part of their life.

Pulling this together, can you sense what it means when someone says life has unity *and* diversity?

Although unity pervades the world of life, great diversity threads through it, for organisms differ enormously in their form, in how body parts function, and in behavior.

We group organisms into three domains—archaebacteria, eubacteria, and eukaryotes. Protistans, fungi, plants, and animals are major groups of eukaryotes.

AN EVOLUTIONARY VIEW OF DIVERSITY

How can organisms be so much alike yet also show such staggering diversity? One explanation is known as evolution by means of natural selection. All through this book, you will be looking at many examples of its predictive power. For now, simply become familiar with its basic premises. They build on the simple observation that individuals of a population vary in the details of their shared traits.

Mutation—Original Source of Variation

DNA has two striking qualities. Its instructions work to ensure that offspring will resemble their parents, yet they also permit variations in the details of most traits. Example: Having five fingers on each hand is a human trait. Yet some humans are born with six fingers on each hand instead of five. This is an outcome of a **mutation**, a heritable change in DNA. *Mutations are the original source of variations in heritable traits.*

Many mutations are harmful, for they sabotage the body's growth, development, or functioning. Yet some are harmless or beneficial. One case is a mutation that results in dark-colored instead of light-colored moths.

Moths fly at night and rest in the day, when birds that eat them are active. Birds tend to miss the light-colored moths on light-colored tree trunks. But what if people build coal-burning factories and soot-laden smoke darkens all the tree trunks? Dark moths on dark trunks aren't as conspicuous to bird predators (Figure 1.8), so they are more likely to live and then reproduce. Where soot accumulates, then, the variant (dark) form of this trait is more adaptive than the light form. An **adaptive trait** is any form of a trait that helps an individual survive and reproduce under prevailing environmental conditions.

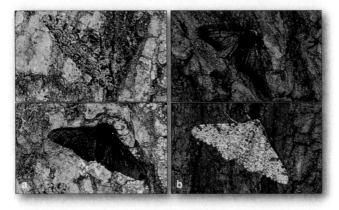

Figure 1.8 An outcome of natural selection involving two forms of the same trait—body surface coloration. (**a**) Light-colored moths (*Biston betularia*) on a nonsooty tree trunk are hidden from predators. Dark ones stand out. (**b**) The dark color is more adaptive in places where soot darkens tree trunks.

Figure 1.9 *Facing page:* One outcome of artificial selection practices: just a few of the 300+ varieties of domesticated pigeons. Breeders started with variant forms of traits in captive populations of wild rock doves.

WILD
ROCK
DOVE

Evolution Defined

Now think about variant moths in terms of the whole population. At some point, a DNA mutation arose and resulted in a moth of a darker color. When the mutated individual reproduced, some of its offspring inherited its mutant DNA and the dark color. Birds saw and ate many light-colored moths, but most dark ones escaped detection and lived long enough to reproduce. So did their offspring. So did *their* offspring, and so on. The frequency of one form of the trait increased relative to the frequency of the other in the population. If the dark form becomes the most common one, people may end up referring to "the population of dark-colored moths."

Like moths, individuals of most populations have different forms of many shared traits. The relative frequencies of those different forms typically change through successive generations. When that happens, **evolution** is under way. In biology, the word means heritable change in a line of descent over time.

Natural Selection Defined

Long ago, the naturalist Charles Darwin used pigeons to explain a conceptual connection between evolution and variation in traits. Domesticated pigeons differ in size, feather color, and other traits (Figure 1.9). Pigeon breeders, Darwin knew, select certain forms of traits. If they are after black, curly-edged tail feathers, they let only individual pigeons with the most black and most curly tail feathers mate and reproduce. So "black" and "curly" become the most common tail feathers in the captive pigeon population. Lighter and less curly feathers become less common or are eliminated.

Pigeon breeding is a case of **artificial selection**, for the selection among different forms of a trait is taking place in an "artificial" environment—under contrived, manipulated conditions. Yet Darwin saw the practice as a simple model for *natural* selection, a favoring of some forms of a given trait over others in nature.

Whereas breeders are "selective agents" promoting the reproduction of some individuals but not others in captive populations, different selective agents operate across the range of variation among pigeons in the wild. Pigeon-eating peregrine falcons are such agents. Swifter or more effectively camouflaged individuals of a pigeon

population have a far better chance of escaping from falcons and living long enough to reproduce, compared to the not-so-swift or too-conspicuous individuals among them. What Darwin identified as **natural selection** is simply a result of differences in survival and reproduction among individuals of a population that vary in one or more of their shared, heritable traits.

Unless you happen to breed pigeons, the variation among pigeons may not be a burning issue in your life. So think about something closer to home. **Antibiotics** are potent secretions of certain soil-dwelling bacteria and fungi; they kill bacterial and fungal competitors for nutrients. In the 1940s, we learned to use antibiotics to kill disease-causing bacteria. Doctors prescribed these "wonder drugs" even for mild infections. In some cases antibiotics were added to toothpaste and chewing gum.

As it turned out, antibiotics are powerful agents of natural selection. Consider how one of these, called streptomycin, shuts down certain essential proteins in bacteria that are its targets. Some bacterial strains have mutations that slightly changed the molecular form of the proteins—and streptomycin no longer can bind to them. Unlike nonmutated cells, the mutants survive and reproduce. Said another way, the antibiotic acts against the bacteria susceptible to it but *favors* variant strains that resist it!

Antibiotic-resistant strains are making it difficult to treat bacterial diseases, including tuberculosis, typhoid, gonorrhea, and staph infections. As resistance evolves by selection processes, strategies for using antibiotics must also evolve to overcome new defenses. That is why drug companies try to modify parts of antibiotic molecules. Such molecular changes in the laboratory might yield more effective antibiotics. These may work only until new generations of more resistant superbugs enter the evolutionary competition for nutrients.

Later in the book, we will consider mechanisms by which populations of moths, pigeons, bacteria, and all other organisms evolve. Meanwhile, reflect on the key points about natural selection. They are the bedrock of biological inquiry into the nature of life's diversity.

1. Individuals of a population vary in form, function, and behavior. Much of the variation is heritable; it can be transmitted from parents to offspring.

2. Some forms of heritable traits are more adaptive to prevailing conditions. They improve an individual's chance of surviving and reproducing, as by helping it secure food, a mate, hiding places, and so on.

3. Natural selection is the outcome of differences in survival and reproduction among variant individuals in a given generation.

4. Natural selection leads to a better fit with prevailing environments. The adaptive forms of traits tend to become more common than other forms. Thus the population's characteristics change; it evolves.

In the evolutionary view, then, *life's diversity is the sum total of variations in traits that have accumulated in different lines of descent generation after generation, as by natural selection and other processes of change.*

Mutations in DNA introduce variations in heritable traits.

Although many mutations are harmful, some give rise to variations in form, function, or behavior that are adaptive under prevailing environmental conditions.

Natural selection is a result of differences in survival and reproduction among individuals of a population that vary in one or more heritable traits. The process helps explain evolution—change in individual lines of descent which, over time, has given rise to life's rich diversity.

THE NATURE OF BIOLOGICAL INQUIRY

The preceding sections sketched out some key concepts. Now consider approaching this or any other collection of "facts" with a critical attitude. *"Why should I accept that they have merit?"* The answer requires insight into how biologists make inferences about observations and how they test the predictive power of their inferences against actual experience in nature or the laboratory.

Observations, Hypotheses, and Tests

To get a sense of "how to do science," start by following some practices that pervade scientific research:

1. Observe some aspect of nature, carefully check what others have found out about it, then frame a question or identify a problem related to your observation.

2. Develop **hypotheses**, or educated guesses, about possible answers to questions or solutions to problems.

3. Using hypotheses as a guide, make a **prediction**—that is, a statement of what you should observe in the natural world if you were to go looking for it. This is often called the "if–then" process. (*If* gravity does not pull objects toward Earth, *then* it should be possible to observe apples falling up, not down, from a tree.)

4. Devise ways to **test** the accuracy of your predictions, as by making systematic observations, building models, and conducting experiments. **Models** are theoretical, detailed descriptions or analogies that help us visualize something that has not yet been directly observed.

5. If tests do not confirm the prediction, check to see what might have gone wrong. For example, maybe you overlooked some factor that influenced the test results. Or maybe the hypothesis is not a good one.

6. Repeat the tests or devise new ones—the more the better, for hypotheses that withstand many tests are likely to have a higher probability of being useful.

7. Objectively analyze and report the test results and the conclusions drawn from them.

You might hear someone refer to these practices as "the scientific method," as if all scientists march to the drumbeat of an absolute, fixed procedure. They do not. Many observe, describe, and report on some subject, then leave the hypothesizing to others. Some are lucky; they stumble onto information they aren't even looking for. (Of course, chance does favor a mind that's already prepared to know what the information means.) It is not one single method scientists have in common. It's a critical attitude about being shown rather than told, and taking a logical approach to problem solving.

Logic encompasses thought patterns by which an individual draws a conclusion that does not contradict evidence used to support it. Lick a cut lemon and you notice it's mouth-puckeringly sour. Lick ten more. Each

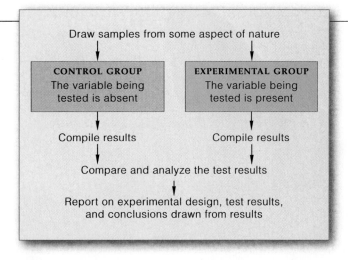

Figure 1.10 Generalized sequence of steps involved in an experimental test of a prediction based on a hypothesis.

time you notice the same thing, and you conclude that all lemons are mouth-puckeringly sour. You correlated one specific (lemon) with another (sour). By this pattern of thinking, called **inductive logic**, an individual derives a general statement from specific observations.

Express the generalization in "if–then" terms, and you have a hypothesis: "If you lick any lemon, then you will experience a sour taste." By this pattern of thinking, **deductive logic**, an individual makes inferences about specific consequences or specific predictions that must follow from a hypothesis.

You decide to test the hypothesis by sampling lemons. One variety, the Meyer, is relatively mellow. You also recall that some people can't taste anything. So you modify the hypothesis: "If most people lick any lemon *except* the Meyer lemon, they will experience a sour taste." Suppose you sample all known lemon varieties in the world and conclude the modified hypothesis is a good one. You'll never *prove* it; there may be lemon trees in places we don't even know about. You *can* say the hypothesis has a high probability of not being wrong.

Comprehensive observations are a logical means to test the predictions that flow from hypotheses. So are **experiments**. These tests simplify observation in nature or the laboratory by manipulating and controlling the conditions under which observations are made. When suitably designed, observational and experimental tests allow you to predict that something will happen if a hypothesis isn't wrong (or won't happen if it *is* wrong). Figure 1.10 gives a general idea of the steps involved.

AN ASSUMPTION OF CAUSE AND EFFECT Experiments start from this premise: *Any aspect of nature has one or more underlying causes.* With this premise, science is very different from faith in the supernatural ("beyond nature"). Scientific experiments deal with potentially falsifiable hypotheses, which means they are tested in the natural world in ways that might disprove them.

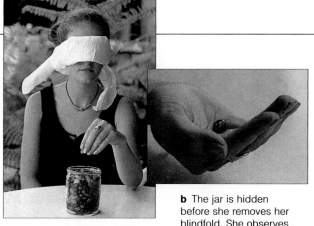
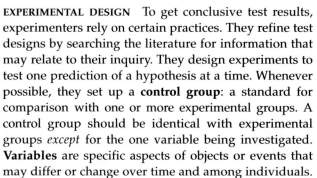
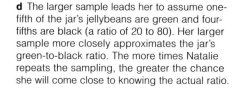

a Natalie, blindfolded, randomly plucks a jellybean from a jar of 120 green and 280 black jellybeans. That's a ratio of 30 to 70 percent.

b The jar is hidden before she removes her blindfold. She observes a single green jellybean in her hand and assumes the jar holds only green jellybeans.

c Still blindfolded, Natalie randomly picks 50 jellybeans from the jar and ends up with 10 green and 40 black ones.

d The larger sample leads her to assume one-fifth of the jar's jellybeans are green and four-fifths are black (a ratio of 20 to 80). Her larger sample more closely approximates the jar's green-to-black ratio. The more times Natalie repeats the sampling, the greater the chance she will come close to knowing the actual ratio.

Figure 1.11 A simple demonstration of sampling error.

EXPERIMENTAL DESIGN To get conclusive test results, experimenters rely on certain practices. They refine test designs by searching the literature for information that may relate to their inquiry. They design experiments to test one prediction of a hypothesis at a time. Whenever possible, they set up a **control group**: a standard for comparison with one or more experimental groups. A control group should be identical with experimental groups *except* for the one variable being investigated. **Variables** are specific aspects of objects or events that may differ or change over time and among individuals.

Section 1.6 shows the design of one experiment. As you will see, the experimenters directly manipulated a single variable in an attempt to support or disprove a prediction. They also tried to hold constant any other variables that could influence the results.

SAMPLING ERROR Rarely can experimenters observe *all* the individuals of a group. Instead, they must use subsets or samples of populations, events, and other aspects of nature. They must avoid **sampling error**, or risking tests with subsets that are not representative of the whole. Generally, distortion in test results is less likely when the samplings are large and when they are repeated (Figure 1.11).

About the Word "Theory" p^3

Suppose no one has disproved a hypothesis after years of rigorous tests. Suppose scientists use it to interpret more data or observations, which could involve more hypotheses. When a hypothesis meets these criteria, it may become accepted as a **scientific theory**.

You may hear someone apply the word "theory" to a speculative idea, as in the expression "It's only a theory." But a scientific theory differs from speculation for this reason: *Researchers have tested its predictive power many times and in many ways in the natural world and have yet to find evidence that disproves it*. This is why the theory of natural selection is respected. We use it successfully to explain diverse issues, such as how life originated, how plant toxins relate to plant-eating animals, what the sexual advantage of feather color might be, why certain cancers run in families, and why antibiotics are losing effectiveness. Together with the record of Earth history, the theory has even helped explain how life evolved.

Maybe an extensively tested theory is as close to the truth as scientists can get with available evidence. For instance, after more than a century of many thousands of different tests, Darwin's theory stands, with only minor modification. We can't prove it holds under all possible conditions; that would take an infinite number of tests. As for any theory, we can only say *it has a high probability of being a good one*. Even so, biologists keep their eyes open for any new information and new tests that might disprove its premises.

This last point gets us back to the value of thinking critically. Scientists must keep asking themselves: *Will observations or experiments show that a hypothesis is false?* They expect one another to put aside pride or bias by testing ideas even in ways that may prove them wrong. When someone doesn't or won't do this, others will— for science is a competitive yet cooperative community. Ideally, individuals share ideas, knowing it's as useful to expose errors as to applaud insights. They can and often do change their mind when shown contradictory evidence. This is a strength of science, not a weakness.

A scientific approach to studying nature is based on asking questions, formulating hypotheses, making predictions, devising tests, and objectively reporting the results.

A scientific theory is a longstanding hypothesis, supported by scientific tests, that explains the cause or causes of a broad range of related phenomena. It remains open to tests, revision, and tentative acceptance or rejection.

The Power of Experimental Tests

BIOLOGICAL THERAPY EXPERIMENTS   Worry about the new strains of pathogenic (disease-causing) bacteria that resist antibiotics, and you won't be alone. Bruce Levin of Emory University and Jim Bull of the University of Texas have been looking for alternatives to antibiotic therapy. They studied literature on a *biological therapy* that enlists bacteriophages—"bacteria eaters"—to fight infections. Bacteriophages, a class of viruses, attack a narrow range of bacterial strains. When they contact a target, they inject genetic material into it. What happens next is a hostile takeover: The cell's metabolic machinery makes new viral particles. It dies after viral enzymes rupture its outer membrane. Virus particles escape and infect new cells.

The biologists wondered, as others had decades ago, whether injections of bacteriophages could help people fight bacterial infections. The discovery of antibiotics had diverted attention from that idea, at least in the West. Now the bacteria eaters were looking good again. Levin

and Bull focused on duplicating earlier experiments by researchers H. Williams Smith and Michael Huggins.

They chose 018:K1:H7, a harmful strain of *Escherichia coli*. Bacteriophages utilized in laboratory work usually ignore 018:K1:H7. However, like a harmless strain that lives in the intestines of humans and other mammals, it exits the body in feces. So Levin and Bull looked for *E. coli* killers in samples from a sewage treatment plant in Atlanta. They isolated two kinds and named them *H* (Hero, an effective killer) and *W* (the less effective Wimp).

With graduate student Terry DeRouin and technician Nina Moore Walker, they grew a population of 018:K1:H7 in a culture flask. They used a strain of laboratory mice, all females of the same age. In one experimental group, fifteen mice were injected with 018:K1:H7 in one thigh and over 10^7 particles of *H* bacteriophage in the other. All survived. Fifteen mice of a control group received 018:K1:H7 injections only. None survived. Figure 1.12*a*

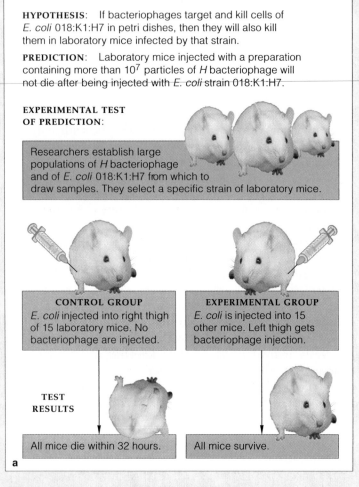

HYPOTHESIS: If bacteriophages target and kill cells of *E. coli* 018:K1:H7 in petri dishes, then they will also kill them in laboratory mice infected by that strain.

PREDICTION: Laboratory mice injected with a preparation containing more than 10^7 particles of *H* bacteriophage will not die after being injected with *E. coli* strain 018:K1:H7.

EXPERIMENTAL TEST OF PREDICTION:

Researchers establish large populations of *H* bacteriophage and of *E. coli* 018:K1:H7 from which to draw samples. They select a specific strain of laboratory mice.

CONTROL GROUP
E. coli injected into right thigh of 15 laboratory mice. No bacteriophage are injected.

EXPERIMENTAL GROUP
E. coli is injected into 15 other mice. Left thigh gets bacteriophage injection.

TEST RESULTS

All mice die within 32 hours.

All mice survive.

a

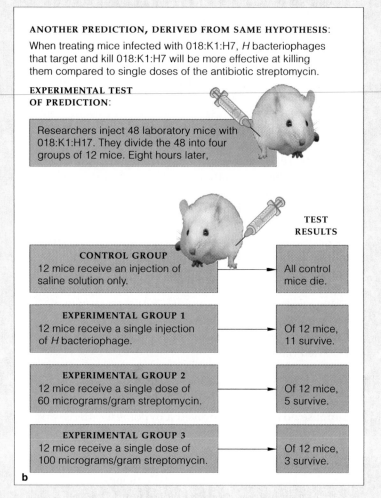

ANOTHER PREDICTION, DERIVED FROM SAME HYPOTHESIS:

When treating mice infected with 018:K1:H7, *H* bacteriophages that target and kill 018:K1:H7 will be more effective at killing them compared to single doses of the antibiotic streptomycin.

EXPERIMENTAL TEST OF PREDICTION:

Researchers inject 48 laboratory mice with 018:K1:H17. They divide the 48 into four groups of 12 mice. Eight hours later,

TEST RESULTS

CONTROL GROUP
12 mice receive an injection of saline solution only.

All control mice die.

EXPERIMENTAL GROUP 1
12 mice receive a single injection of *H* bacteriophage.

Of 12 mice, 11 survive.

EXPERIMENTAL GROUP 2
12 mice receive a single dose of 60 micrograms/gram streptomycin.

Of 12 mice, 5 survive.

EXPERIMENTAL GROUP 3
12 mice receive a single dose of 100 micrograms/gram streptomycin.

Of 12 mice, 3 survive.

b

Figure 1.12   Two experiments to compare the effectiveness of bacteriophage injections as opposed to antibiotics for treating a bacterial infection.

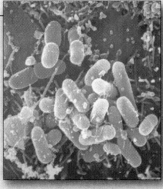

Figure 1.13 Some cells of a pathogenic strain of *Escherichia coli* (*pink*) adhering to a human intestinal cell in tissue culture.

outlines the experimental test. Figure 1.13 shows an example of the target pathogens.

Figure 1.12*b* outlines a second experimental test of another prediction derived from the same hypothesis. The results reinforced Smith and Huggins's conclusion that certain bacteriophages are as good as or better than antibiotics at stopping certain bacterial infections.

Each type of bacteriophage targets one or at most a few strains of bacterial species. It now takes too long to discover which bacterium is causing a severe infection, so the right bacteriophage might not be enlisted until it's too late. New procedures may speed up identification.

IDENTIFYING IMPORTANT VARIABLES   In nature, many factors can change the course of an infection. To give examples, genetic differences among individuals can lead to differences in immune responses to infection. Other factors are the individual's health, nutritional habits, and age. Some pathogens are deadlier than others, and so on.

That's why Levin and Bull simplified and controlled the variables. Variables, remember, are specific aspects of objects or events that may differ among individuals and over time. All cells in their experiments descended from the same parent cell. They received the same nutrients at the same temperature in a flask. They could expect them to respond the same way to an attack. They were the same age and sex, and they were raised under identical laboratory conditions. Each mouse in each experimental group was injected with the same amount of antibiotic or bacteriophages. Each mouse in a control group got the same injection of saline solution. The focus was on *one variable*—a specific bacteriophage versus a specific antibiotic—in a simple, controlled, artificial situation.

BIAS IN REPORTING RESULTS   Whether intentionally or not, experimenters run a risk of interpreting their data in terms of what they want to prove or dismiss. A few have even been known to fake measurements or nudge findings to reinforce their bias. That is why science emphasizes reporting test results in *quantitative* terms—with actual counts or some other precise form. Doing so will allow other experimenters to check or test the results readily and systematically, as Levin and Bull did.

THE LIMITS OF SCIENCE

The call for objective testing strengthens the theories that emerge from scientific studies. It also puts limits on the kinds of studies that can be carried out. Beyond the realm of science, some events remain unexplained. Why do we exist, for what purpose? Why does any one of us have to die at a particular moment? Such questions lead to *subjective* answers. These come from within, as an outcome of all the experiences and mental connections that shape human consciousness. Because people differ vastly in this regard, subjective answers do not readily lend themselves to scientific analysis and experiments.

This is not to say subjective answers are without value. No human society can function for long unless its members share a commitment to certain standards for making judgments, even subjective ones. Moral, aesthetic, philosophical, and economic standards vary from one society to the next. But they all guide people in deciding what is important and good, and what is not. All attempt to give meaning to what we do.

Every so often, scientists stir up controversy when they explain something that was thought to be beyond natural explanation—as belonging to the supernatural. This is often the case when a society's moral codes are interwoven with religious narratives. Exploring a long-standing view of the natural world from the scientific point of view might be misinterpreted as questioning morality, even though the two are not the same thing.

As one example, centuries ago in Europe, Nicolaus Copernicus studied the planets and concluded the Earth circles the sun. Today this seems obvious enough. Back then, it was heresy. The prevailing belief was that the Creator made the Earth—and, by extension, humans—the immovable center of the universe. Later a respected scholar, Galileo Galilei, studied the Copernican model of the solar system, thought it was a good one, and said so. He was forced to retract his statement publicly, on his knees, and put the Earth back as the fixed center of things. (Word has it that when he stood up he muttered, "Even so, it *does* move.") Later still, Darwin's theory of evolution ran up against the same prevailing belief.

Today, as then, society has sets of standards. Those standards might be questioned when some new, natural explanation runs counter to supernatural beliefs. This doesn't mean that scientists who raise the questions are less moral, less lawful, less sensitive, or less caring than anyone else. It simply means one more standard guides their work: *The external world, not internal conviction, must be the testing ground for scientific beliefs.*

Systematic observations, hypotheses, predictions, tests. In all these ways, science differs from systems of belief that are based on faith, force, or simple consensus.

SUMMARY

Gold indicates text section

1. There is unity in the living world, for all organisms share these characteristics: They consist of one or more cells. They are assembled from the same kinds of atoms and molecules according to the same laws of energy. They engage in metabolism, and they also sense and respond to specific aspects of the environment. Their DNA gives them a capacity to survive and reproduce. Table 1.1 lists these characteristics. *1.1*

2. The hierarchy of biological organization includes cells, multicelled organisms, populations, communities, ecosystems, and the biosphere. *1.2*

3. Many millions of species (kinds of organisms) exist. Many millions more existed in the past but are extinct. Classification schemes are ever more inclusive groupings of species, starting with genus and proceeding through family, order, class, and phylum, to kingdom. *1.3*

4. Life's diversity arises through mutations: heritable changes in the structure of DNA molecules. The changes are the basis for variation in heritable traits. These are traits that parents bestow on their offspring, including most details of body form and function. *1.4*

5. Darwin's theory of evolution by natural selection is central to biological inquiry. Its key premises are: *1.4*

   a. Individuals of a population differ in the details of their shared heritable traits. Variant forms of traits may affect the ability to survive and reproduce.

   b. Natural selection is the outcome of differences in survival and reproduction among individuals that differ in one or more traits. Adaptive forms of a trait tend to become more common; less adaptive ones become less common or vanish. Thus the defining traits may change over successive generations; the population may evolve.

6. The methods of scientific inquiry are diverse. These terms are important aspects of the methods: *1.5*

   a. Theory: Explanation of a broad range of related phenomena supported by numerous tests. An example is Darwin's theory of evolution by natural selection.

   b. Hypothesis: A proposed explanation of a specific phenomenon. Sometimes called an educated guess.

   c. Prediction: A claim about what can be expected in nature, based on premises of a theory or hypothesis.

   d. Test: An attempt to produce actual observations that match predicted or expected observations.

   e. Conclusion: A statement about whether to accept, modify, or reject a theory or hypothesis, based on tests of predictions that were derived from it.

7. Logic is a pattern of thought by which an individual draws a conclusion that does not contradict evidence used to support the conclusion. Inductive logic means individuals derive a general statement from specific observations. Deductive logic means individuals make inferences about particular consequences or predictions that must follow from a hypothesis. Such a pattern of thinking is often expressed in "if–then" terms. *1.5*

8. Predictions that flow from hypotheses can be tested by comprehensive observations or by experiments. *1.5*

9. Experimental tests simplify observations in nature or the laboratory because observations are made under precisely manipulated and controlled conditions. The tests are based on this premise: Any aspect of nature has one or more underlying causes, whether obvious or not. Hypotheses are scientific only when they may be tested in ways that might disprove them. *1.5, 1.6*

10. A control group is a standard against which one or more experimental groups (test groups) are compared. Ideally, it is the same as each experimental group in all variables except the one being investigated. *1.5, 1.6*

11. A variable is a specific aspect of an object or event that might differ over time and between individuals. Experimenters directly manipulate the variable they are studying to support or disprove a prediction. *1.5, 1.6*

12. Test results might be distorted by sampling error: chance differences between a population, event, or some other aspect of nature and the samples chosen to represent it. Distortion is less likely when samplings are large and when they are repeated. *1.5, 1.6*

13. Scientific theories arise from systematic observations, hypotheses, predictions, and experimental testing. The external world is the testing ground for theories. *1.7*

Table 1.1 *Summary of Life's Key Characteristics*

SHARED CHARACTERISTICS THAT REFLECT LIFE'S UNITY

1. Organisms consist of one or more cells.

2. Organisms are constructed of the same kinds of atoms and molecules according to the same laws of energy.

3. Organisms engage in metabolism; they acquire and use energy and materials to survive and reproduce.

4. Organisms sense and make controlled responses to their internal and external environments.

5. Heritable instructions encoded in DNA give organisms their capacity to grow and reproduce. DNA instructions also guide the development of complex multicelled organisms.

6. Characteristics that define a population of organisms can change through the generations; the population can evolve.

FOUNDATIONS FOR LIFE'S DIVERSITY

1. Mutations (heritable changes in the structure of DNA) give rise to variation in heritable traits, including most details of body form, functioning, and behavior.

2. Diversity is the sum total of variations that accumulated in different lines of descent over the past 3.8 billion years, as by natural selection and other processes of evolution.

For this chapter and subsequent chapters, *italics* after a review question identify the section where you can find answers. They include section numbers and *CI* (for *C*hapter *I*ntroduction).

1. Why is it difficult to formulate a simple definition of life? *CI*

2. Name the molecule of inheritance in cells. *1.1*

3. Write out simple definitions of the following terms: *1.1*
   a. cell   b. metabolism   c. energy   d. ATP

4. How do organisms sense changes in their surroundings? *1.1*

5. Study Figure 1.5. Then, on your own, arrange and define the levels of biological organization. *1.2*

6. Study Figure 1.6. Then, on your own, make a sketch of the one-way flow of energy and the cycling of materials through the biosphere. To the side of the sketch, write out definitions of producers, consumers, and decomposers. *1.2*

7. List the shared characteristics of life. *CI, 1.3*

8. What are the two parts of the scientific name for each kind of organism? *1.3*

9. List the six kingdoms of species as outlined in this chapter. Also list some of their general characteristics. *1.3*

10. Define mutation and adaptive trait. Explain the connection between mutation and the diversity of life. *1.4*

11. Write brief definitions of evolution, artificial selection, and natural selection. *1.4*

12. Define and distinguish between: *1.5*
    a. hypothesis and prediction
    b. observational test and experimental test
    c. inductive and deductive logic
    d. speculation and scientific theory

13. With respect to experimental tests, define variable, control group, and experimental group. *1.5*

14. What does sampling error mean? *1.5*

1. _____ is the capacity of cells to extract energy from sources in their environment, and to transform and use energy to grow, maintain themselves, and reproduce.

2. _____ is a state in which the internal environment is being maintained within tolerable limits.

3. The _____ is the smallest unit of life.

4. If a form of a trait improves chances for surviving and reproducing in a given environment, it is a(n) _____ trait.

5. The capacity to evolve is based on variations in heritable traits, which originally arise through _____ .

6. You have some number of traits that also were present in your great-great-great-great-grandmothers and great-great-great-great-grandfathers. This is an example of _____ .
   a. metabolism   c. a control group
   b. homeostasis   d. inheritance

7. DNA molecules _____ .
   a. contain instructions for traits
   b. undergo mutation
   c. are transmitted from parents to offspring
   d. all of the above

8. For many years in a row, a dairy farmer allowed his best milk-producing cows but not the poor producers to mate. Over many generations, milk production increased. This outcome is an example of _____ .
   a. natural selection   c. evolution
   b. artificial selection   d. both b and c

9. A control group is _____ .
   a. a standard against which experimental groups are compared
   b. identical to experimental groups except for one variable
   c. a standard with several variables against which an experimental group is compared
   d. both a and b

10. A specific aspect of an object or event that may change over time or change among individuals is a _____ .
    a. control group   c. variable
    b. experimental group   d. sampling error

11. The fewer the number of individuals from a population that are chosen at random for an experimental group, _____ .
    a. the greater the chance of sampling error
    b. the smaller the chance of sampling error
    c. the less likely that differences among them will distort the test results

12. Match the terms with the most suitable descriptions.
    _____ adaptive   a. statement of what you should find in
            trait   nature if you were to go looking for it
    _____ natural   b. proposed explanation; educated guess
            selection   c. improves chances to survive and
    _____ scientific   reproduce in prevailing environment
            theory   d. related set of hypotheses that form a
    _____ hypothesis   broadly useful, testable explanation
    _____ prediction e. outcome of differences in survival and
                           reproduction among individuals that
                           differ in details of one or more traits

Critical Thinking

1. Some spiders (*Dolomedes*) that feed on insects around ponds occasionally capture tadpoles and small fishes, as in Figure 1.14, and isn't that fun to think about? While they are immature, the female spiders confine themselves to a small patch of vegetation next to the pond. When sexually mature, they mate and store sperm that fertilize the eggs. Only then do they move out and occupy larger areas around the pond. Develop hypotheses to explain what might cause the spiders to live in different places at different times. Design an experiment to test each hypothesis.

Figure 1.14   A spider (*Dolomedes*) and its prey: a tiny minnow. The spider delivered paralyzing venom as well as digestive enzymes into its captive and is now sucking predigested juices from it. Such spiders can move about below the water's surface. Hairs on their body trap oxygen (for aerobic respiration) during hunting expeditions.

2. A scientific theory about some aspect of nature rests upon inductive logic. The assumption is that, because an outcome of some event has been observed to happen with great regularity, it will happen again. However, we can't know this for certain, because there is no way to account for all possible variables that may affect the outcome. To illustrate this point, Garvin McCain and Erwin Segal offer a parable:

Once there was a highly intelligent turkey. The turkey lived in a pen, attended by a kind, thoughtful master, and it had nothing to do but reflect upon the world's wonders and regularities. It observed some major regularities. Morning always began with the sky getting light, followed by the clop, clop, clop of its master's friendly footsteps, which was followed by the appearance of delicious food. Other things varied—sometimes the morning was warm and sometimes cold—but food always followed footsteps. The sequence of events was so predictable that it eventually became the basis of the turkey's theory about the goodness of the world.

One morning, after more than a hundred confirmations of the goodness theory, the turkey listened for the clop, clop, clop, heard it, and had its head chopped off.

The turkey learned the hard way that explanations about the world only have a high or low probability of not being wrong. Today, some people take this uncertainty to mean that "facts are irrelevant—facts change." If that is so, should we just stop doing scientific research? Why or why not?

3. Witnesses in a court of law are asked to "swear to tell the truth, the whole truth, and nothing but the truth." What are some of the problems inherent in the question? Can you think of a better alternative?

4. Many popular magazines publish an astounding number of articles on diet, exercise, and other health-related topics. Some authors recommend a specific diet or dietary supplement. What kinds of evidence do you think the articles should include so that you can decide whether to accept the recommendations?

5. Although scientific information is often used when making a decision, it can't tell an individual what's "right" or "wrong." Give an example of this from your own experience.

6. As the old saying goes, Everybody complains about the weather, but nobody does anything about it. Maybe you can start thinking about how we humans might change at least one aspect of local weather conditions. Two researchers at Arizona State University found that, at least for their investigation of the northeast coast of North America, it really does rain more on weekends, when we'd rather be out having fun!

As R. Cerveny and R. Balling, Jr. reasoned, if the amount of rain falling each day of the week is a random event, then rules of probability should apply. (*Probability* means the chance that each possible outcome of an event will occur is proportional to the number of ways it can be reached.) Thus, each day of the week should get one-seventh (14.3 percent) of the total rainfall for the week. But it turns out Monday is driest (13.1 percent). Days of the week are wetter, and Saturday is the wettest of all (with 16 percent of the total). Sunday is a bit above average.

What causes this effect? As the researchers hypothesized, if *air pollution* is greater on weekdays than weekends, then human activities might influence regional patterns of rainfall. Monday through Friday, coal-burning factories and gas-powered vehicles release quantities of tiny particles into the air. The particles can promote updrafts and act as "platforms" for the formation of water droplets. Air pollution must build up during the week and extend into Saturday. Over the weekend, fewer factories operate and fewer commuters are on the road. Therefore, by Monday, the air must be cleaner—and drier.

What kind of evidence do you suppose Cerveny and Balling gathered to test the hypothesis? Jot down a few ideas and check them against the researchers' article in the 6 August 1998 issue of *Nature*. Whether or not you continue in biology, this exercise will be good practice in searching through scientific literature for actual data that you can evaluate on topics of interest to you.

7. Scientists devised experiments to shed light on whether different fish species of the same genus compete in their natural habitat. They constructed twelve ponds, identical in chemical composition and physical characteristics. Then they released the following individuals in each pond:

Ponds 1, 2, 3:	Species A	300 individuals each pond
Ponds 4, 5, 6:	Species B	300 individuals each pond
Ponds 7, 8, 9:	Species C	300 individuals each pond
Ponds 10, 11, 12:	Species A, B, C	300 individuals of each species in each pond

Does this experimental design take into consideration all factors that can affect the outcome? If not, how would you modify it?

Selected Key Terms

For this chapter and subsequent chapters, these are the **boldface** terms that appear in the text, in the sections indicated here by *italic* numbers (or *CI*, short for Chapter Introduction). As a study aid, make a list of the terms, write a definition for each, then check it against the one in the text. You will use these terms later on. Becoming familiar with each one will help give you a foundation for understanding the material in later chapters.

adaptive trait *1.4*	ecosystem *1.2*	mutation *1.4*
animal *1.3*	energy *1.1*	natural selection *1.4*
antibiotic *1.4*	eubacteria *1.3*	plant *1.3*
archaebacterium *1.3*	evolution *1.4*	population *1.2*
artificial selection *1.4*	experiment *1.5*	prediction *1.5*
biology *CI*	fungus *1.3*	producer *1.2*
biosphere *1.2*	genus *1.3*	protistan *1.3*
cell *1.2*	homeostasis *1.1*	receptor *1.1*
community *1.2*	hypothesis *1.5*	reproduction *1.1*
consumer *1.2*	inductive logic *1.5*	sampling error *1.5*
control group *1.5*	inheritance *1.1*	species *1.3*
decomposer *1.2*	metabolism *1.1*	stimulus *1.1*
deductive logic *1.5*	model *1.5*	test, scientific *1.5*
development *1.1*	multicelled	theory, scientific *1.5*
DNA *1.1*	organism *1.2*	variable *1.5*

Readings

Carey, S. 1994. *A Beginner's Guide to the Scientific Method.* Belmont, California: Wadsworth. Paperback.

McCain, G., and E. Segal. 1988. *The Game of Science.* Fifth edition. Pacific Grove, California: Brooks/Cole. Paperback.

Moore, J. 1993. *Science as a Way of Knowing.* Cambridge, Massachusetts: Harvard University Press.

On-Line readings at Student Guide for InfoTrac:
www.brookscole.com/biology

I Principles of Cellular Life

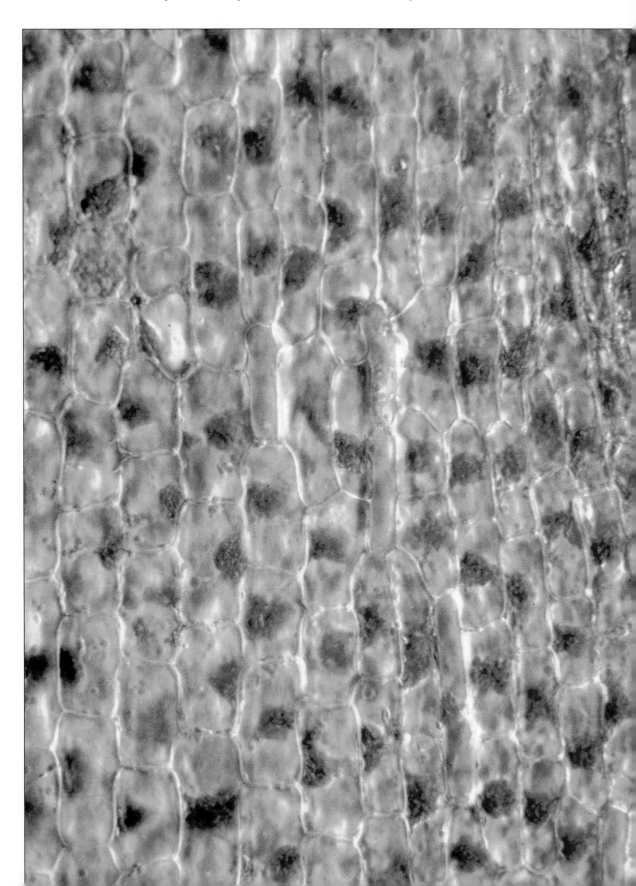

Living cells of one kind of plant (Elodea), observed with the aid of a microscope. Each of these rectangular cells contains efficient chemical factories called chloroplasts (the round, green structures inside).

2

CHEMICAL FOUNDATIONS FOR CELLS

How Much Are You Worth?

Hollywood thinks actress Julia Roberts is worth $20 million per picture, the Texas Rangers think shortstop Alex Rodriguez is worth $252 million per decade, and the United States thinks the average teacher isn't worth much more than $41,820 per year. From a biochemical perspective, though, how much is a human body really worth?

Think of it! Your body is a collection of elements. An **element** is a fundamental form of matter that has mass and takes up space, and at least here on Earth it can't be broken down into something else (except in physics laboratories). Break a chunk of copper into smaller and smaller bits and the smallest bit you end up with, even a lone atom, is still copper. Each solid, liquid, or gaseous substance consists of characteristic proportions of one or more such elements.

Ninety-two elements occur naturally on Earth. Four kinds—oxygen, hydrogen, carbon, and nitrogen —are the most abundant ones in your body, as they are in all other living organisms. Others in your body include phosphorus, potassium, sulfur, calcium, and sodium. Also present are *trace* elements, meaning that each represents less than 0.01 percent of body weight.

Figure 2.1a lists each element in your body. It also lists the cost per kilogram for that element. It all adds up to a paltry $118.63 for someone who weighs seventy kilograms, or 154.3 pounds.

Figure 2.1 **(a)** How much is a human being worth, chemically speaking? Shown here, the mass of each element in a seventy-kilogram human body and its commercial cost. Prices are based on the forms of elements most likely to be found in nature (for example, the hydrogen and oxygen atoms in molecules of water rather than in purified form). Note that the market price for gold changes monthly. The cost for uranium, no longer available through chemical suppliers, was obtained from E-Bay.

(b) Proportions of some elements in a human body and in the fruit of a pumpkin compared to proportions of elements making up the Earth's crust. How are the proportions similar? How do they differ?

Mass of Elements in a 70-Kilogram Human Body		Cost (Retail)
Oxygen	43.00 kilograms (kg)	$0.021739
Carbon	16.00 kg	6.400000
Hydrogen	7.00 kg	0.028315
Nitrogen	1.80 kg	9.706929
Calcium	1.00 kg	15.500000
Phosphorus	780.00 grams (g)	68.198594
Potassium	140.00 g	4.098737
Sulfur	140.00 g	0.011623
Sodium	100.00 g	2.287748
Chlorine	95.00 g	1.409496
Magnesium	19.00 g	0.444909
Iron	4.20 g	0.054600
Fluorine	2.60 g	7.917263
Zinc	2.30 g	0.088090
Silicon	1.00 g	0.370000
Rubidium	0.68 g	1.087153
Strontium	0.32 g	0.177237
Bromine	0.26 g	0.012858
Lead	0.12 g	0.003960
Copper	72.00 milligrams (mg)	0.012961
Aluminum	60.00 mg	0.246804
Cadmium	50.00 mg	0.010136
Cerium	40.00 mg	0.043120
Barium	22.00 mg	0.028776
Iodine	20.00 mg	0.094184
Tin	20.00 mg	0.005387
Titanium	20.00 mg	0.010920
Boron	18.00 mg	0.002172
Nickel	15.00 mg	0.031320
Selenium	15.00 mg	0.037949
Chromium	14.00 mg	0.003402
Manganese	12.00 mg	0.001526
Arsenic	7.00 mg	0.023576
Lithium	7.00 mg	0.024233
Cesium	6.00 mg	0.000016
Mercury	6.00 mg	0.004718
Germanium	5.00 mg	0.130435
Molybdenum	5.00 mg	0.001260
Cobalt	3.00 mg	0.001509
Antimony	2.00 mg	0.000243
Silver	2.00 mg	0.013600
Niobium	1.50 mg	0.000624
Zirconium	1.00 mg	0.000830
Lanthanum	0.80 mg	0.000566
Gallium	0.70 mg	0.003367
Tellurium	0.70 mg	0.000722
Yttrium	0.60 mg	0.005232
Bismuth	0.50 mg	0.000119
Thallium	0.50 mg	0.000894
Indium	0.40 mg	0.000600
Gold	0.20 mg	0.001975
Scandium	0.20 mg	0.058160
Tantalum	0.20 mg	0.001631
Vanadium	0.11 mg	0.000322
Thorium	0.10 mg	0.004948
Uranium	0.10 mg	0.000103
Samarium	50.00 micrograms (µg)	0.000118
Beryllium	36.00 µg	0.000218
Tungsten	20.00 µg	0.000007
Grand Total:		$118.63

a

All you have to do is briefly observe Julia or Alex or any teacher and you know that a human body is far more than a list of its ingredients. Even so, that list is a starting point for the body's distinct characteristics. For example, compare its elemental proportions with those of, say, the Earth's crust or a pumpkin (Figure 2.1b). Although it has many of the same elements as dirt and pumpkins, the proportions differ, and the kinds and amounts are integrated as a unique, highly organized, dynamic form.

Another point: That listing of ingredients is like a snapshot, frozen in time. Your 43 kilograms of oxygen atoms are bound to other atoms in body water or in molecules that cells use as building blocks or energy —but this isn't your lifetime supply. At any moment, uncounted numbers of oxygen atoms are leaving your body, as in the carbon dioxide being exhaled from your lungs, and new ones from air, food, and water are replacing them. Like all other organisms, you must continually take up sufficient amounts of oxygen and the other elements listed to build and maintain cells— and, ultimately, yourself. If you were to inhale merely a few breaths of oxygen-free gas, you would fall into a coma within five seconds. Deprived of oxygen for two to four minutes more, you would be dead.

Remember this example if someone tries to tell you "chemistry" isn't important. *You* are chemical, and so is every living and nonliving thing in the universe. Hamburgers and pasta, refrigerators and jet fuel, sickle-cell anemia and other genetic disorders, health and disease, pesticides and agriculture, acid rain and old-growth forests, global warming, nerve gas in the hands of terrorists—you name it, and "chemistry" has something to say about it. Study it well.

Key Concepts

1. All substances consist of one or more elements, most notably oxygen, carbon, hydrogen, and nitrogen. Each element consists of atoms, which are the smallest units of matter that still display the element's properties. The atoms making up each element consist of protons, electrons and, except for the hydrogen atom, neutrons.

2. The atoms that make up each kind of element have the same number of protons and electrons. They may differ slightly from one another in their number of neutrons. Variant forms of an element's atoms are called isotopes.

3. Atoms have no overall electric charge unless they become ionized—that is, unless they lose electrons or acquire more of them. An ion is an atom or molecule that has lost or gained one or more electrons and now carries an electric charge.

4. Whether a given atom will interact with other atoms depends on how many electrons it has and how they are arranged. When energetic interactions unite two or more atoms, this produces a chemical bond.

5. The molecular organization and activities of living things arise largely from ionic, covalent, and hydrogen bonds between atoms.

6. Life probably originated in water and is exquisitely adapted to its properties. Foremost among its properties are temperature-stabilizing effects, cohesiveness, and a capacity to dissolve or repel a variety of substances.

7. All organisms depend on the controlled formation, use, and disposal of hydrogen ions (H^+). The pH scale is a measure of the concentration of hydrogen ions in solutions.

Earth's Crust		Human		Pumpkin	
Oxygen	46.6%	Oxygen	65%	Oxygen	85%
Silicon	27.7	Carbon	18	Hydrogen	10.7
Aluminum	8.1	Hydrogen	10	Carbon	3.3
Iron	5.0	Nitrogen	3	Potassium	0.34
Calcium	3.6	Calcium	2	Nitrogen	0.16
Sodium	2.8	Phosphorus	1.1	Phosphorus	0.05
Potassium	2.1	Potassium	0.35	Calcium	0.02
Magnesium	1.5	Sulfur	0.25	Magnesium	0.01
		Sodium	0.15	Iron	0.008
		Chlorine	0.15	Sodium	0.001
		Magnesium	0.05	Zinc	0.0002
		Iron	0.004	Copper	0.0001
		Iodine	0.004		

b

2.1

REGARDING THE ATOMS

Structure of Atoms

What are the smallest particles that retain the properties of an element? **Atoms**. A line of about a million of them would fit in the period ending this sentence. Physicists have split atoms into more than 100 kinds of smaller parts. But the only subatomic particles we will need to consider are called **protons**, **electrons**, and **neutrons**.

All atoms have one or more protons, which carry a positive electric charge (p^+). Except for hydrogen, atoms also have one or more neutrons, which are uncharged. Protons and neutrons make up an atom's core region, or atomic nucleus (Figure 2.2). Zipping about the nucleus and occupying most of the volume of an atom are one or more electrons, which carry a negative charge (e^-). Each atom normally has as many electrons as protons. In this configuration, the atom carries no net charge.

Each element has a unique *atomic* number, based on the number of protons in its atoms. The hydrogen atom has one proton, so the number is 1. For the carbon atom, with six protons, it is 6 (Table 2.1). Each element has a *mass* number, or the combined number of protons and neutrons in the atomic nucleus. The carbon atom, with six protons and six neutrons, has a mass number of 12.

Why bother with atomic and mass numbers? They give you a sense of whether and how substances will interact. *This information can help you predict how a given substance might behave in individual cells, in multicelled organisms, and in the environment, under many conditions.*

Isotopes—Variant Forms of Atoms

Although all atoms of an element have the same number of protons, they do not all have the same number of

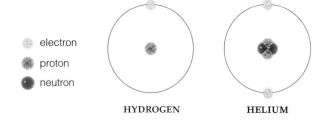

Figure 2.2 The hydrogen atom and helium atom according to one model for atomic structure. The model is highly simplified; the nucleus of these two representative atoms would actually be an invisible speck at the scale employed here.

neutrons. For each element, atoms that vary in neutron number are called **isotopes**. Most naturally occurring elements have two or more common isotopes. Carbon has three, nitrogen has two, and so on. A superscript number to the left of a symbol for an element signifies the isotope. Carbon's three isotopes are ^{12}C or carbon 12 (the most common form, with six protons and six neutrons), ^{13}C (six protons, seven neutrons), and ^{14}C (six protons, eight neutrons).

All isotopes of an element interact chemically with other atoms in the same way. This means that cells can use any isotope of an element for metabolic activities.

What about radioactive isotopes (radioisotopes)? A physicist, Henri Becquerel, discovered them in 1896 after he put a heavily wrapped rock on top of an unexposed photographic plate inside a desk drawer. The rock held energy-emitting isotopes of uranium. A few days after the plate was exposed to the emissions, a faint image of the rock appeared on it. Marie Curie, a coworker, gave the name "radioactivity" to this chemical behavior.

As we now know, a **radioisotope** is an isotope that has an unstable nucleus and becomes more stabilized by spontaneously emitting energy and particles. The process of radioactive decay transforms a radioisotope into an atom of a different element at a known rate. For instance, we know that over a predictable time span, carbon 14 becomes nitrogen 14 (Section 19.2).

Table 2.1 *Atomic Number and Most Common Mass Number of Selected Elements*			
Element	Symbol	Atomic Number	Mass Number
Hydrogen	H	1	1
Carbon	C	6	12
Nitrogen	N	7	14
Oxygen	O	8	16
Sodium	Na	11	23
Magnesium	Mg	12	24
Phosphorus	P	15	31
Sulfur	S	16	32
Chlorine	Cl	17	35
Potassium	K	19	39
Calcium	Ca	20	40
Iron	Fe	26	56
Iodine	I	53	127

Elements are forms of matter that occupy space, have mass, and cannot be degraded to something else by ordinary means.

Atoms, the smallest particles unique to each element, have one or more positively charged protons, negatively charged electrons, and (except for hydrogen) neutrons.

Most elements have two or more isotopes, which are atoms that differ in the number of neutrons. A radioisotope has an unstable nucleus. Radioactive decay transforms it into another element at a predictable and characteristic rate.

Using Radioisotopes to Track Chemicals and Save Lives

Radioisotopes are often used as **tracers**, or substances with an isotope attached to them. Researchers insert them into a cell, a multicelled body, an ecosystem, or any other system and use them like shipping labels. Devices detect the emissions from radioactive tracers and follow them through a pathway or pinpoint its destination.

For example, Melvin Calvin and other botanists used radioactive tracers to track the steps of photosynthesis. As they knew, all isotopes of an element have the same number of electrons, so all interact with other atoms the same way. Plants, they hypothesized, use any carbon isotope when they build carbohydrates. By putting plant cells in a medium enriched with a tracer (^{14}C rather than ^{12}C), they tracked the uptake of carbon through each reaction step in the formation of sugars and starches.

Botanists use a phosphorus 32 tracer to identify how plants take up and use soil nutrients and fertilizers. They apply such findings to work on improving crop yields.

Radioisotopes are tools for medical diagnosis as well as for research. When clinicians analyze, say, the human thyroid, they use a type that swiftly decays into a harmless element. The thyroid is the only gland of ours that uses iodine to build thyroid hormones, which greatly influence body growth and functions. If a patient's symptoms point to abnormal thyroid output, clinicians may inject a trace amount of iodine 123 into the patient's blood. They would then use a photographic imaging device to scan the gland. Figure 2.3 shows examples of the type of images they get.

PET (for *Positron-Emission Tomography*) also uses radioisotopes to form images of body tissues. Clinicians attach a tracer to glucose or another molecule. They inject the labeled glucose into the patient, who is moved into a PET scanner (Figure 2.4*a*). All body cells use glucose, so they take up the labeled kind, too. Uptake is greater in some tissues than others, depending on their metabolic activity at the time of examination. Laboratory devices detect the radioactive emissions and use them to form an image. Such images may reveal variations or abnormalities in metabolic activity, as in Figure 2.4*d*.

Radioisotopes energetic enough to kill cells still can be useful. In *radiation therapy*, radioisotopes stop or impair the activity of abnormal body cells. To give an example, emissions from a source of radium 226 or cobalt 60 may destroy small, localized cancers. As another example, the emissions from plutonium 238 drive artificial pacemakers that smooth out irregular heartbeats. The radioisotope is sealed in a case so its emissions won't damage body cells.

normal thyroid

enlarged

cancerous

Figure 2.3   Location of the human thyroid relative to the trachea (windpipe), and scans of the thyroid gland from three patients.

detector ring inside PET scanner   only body section inside ring

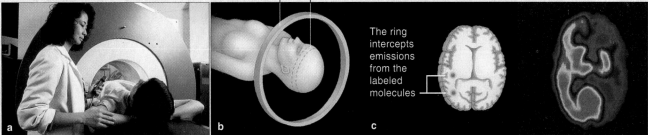

The ring intercepts emissions from the labeled molecules

d PET scan (*red* indicates areas of greatest activity; *blue*, of least activity)

Figure 2.4   (**a**) Patient moving into a PET scanner. (**b,c**) Inside, a ring of detectors intercepts radioactive emissions from labeled molecules that were injected into the patient. Computers analyze and color-code the number of emissions from each location in the scanned body region.

(**d**) Different colors in a brain scan signify differences in metabolic activity. Cells of this brain's left half absorbed and used the labeled molecules at expected rates. However, cells in the right half showed little activity. The patient was diagnosed as having a neurological disorder.

WHAT HAPPENS WHEN ATOM BONDS WITH ATOM?

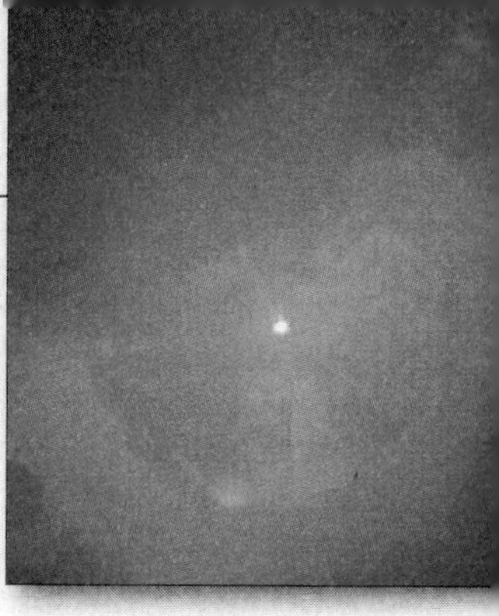

Electrons and Energy Levels

Cells stay alive because energy inherent in all electrons makes things happen (Figure 2.5). Countless atoms in cells acquire, share, and donate electrons. Atoms of some elements do this easily; others don't. What determines whether an atom will interact with others? *The outcome depends on the number and arrangement of its electrons.*

Tinker with magnets and you can get a sense of the attractive force between unlike charges (+ –) as well as the repulsive force between like charges (++ or – –). Electrons carry a negative charge. In atoms, the positive charge of protons attracts electrons, but they repel each other. They respond to the pulls and pushes by moving in different orbitals. An orbital is a volume of space around the atomic nucleus in which electrons are likely to be at any instant (Figure 2.6). Visualize two children circling a cookie jar, not yet expert in the art of sharing. Each is drawn to the cookies but dreads being hit on the head by the other. They bob about on opposite sides of the jar to avoid each other. They never occupy the same space at the same time. In this analogy, the orbital corresponds to the perimeter in which the children are moving around the cookie jar.

Each orbital can house one or at most two electrons. Atoms differ in how many occupied orbitals they have because they differ in their number of electrons.

Hydrogen is the simplest atom. It has one electron in a spherical orbital closest to the nucleus. The orbital corresponds to the *lowest available energy level.* In other atoms, two electrons fill the first orbital, and two more electrons occupy a second spherical orbital around the first. The larger atoms have more electrons in orbitals farther from the nucleus, at *higher energy levels.*

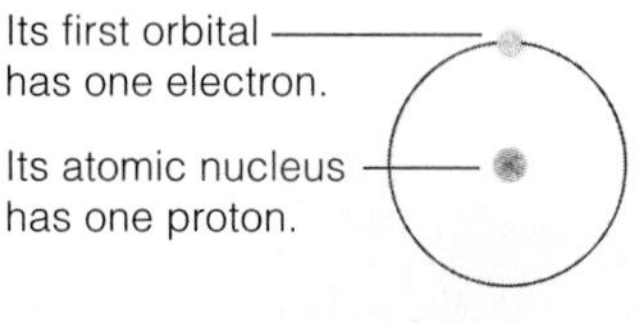

An electron at any energy level has a specific amount of energy. On its own, it cannot move to a different level. But

Its first orbital has one electron.

Its atomic nucleus has one proton.

a Shell model of an H atom

Figure 2.5 How a lone electron can make things happen. Electrons were confined in a bubble of liquid helium, then fluctuating sound waves popped the bubble. An electron escaped and made this white-centered red flash.

what if an atom absorbs extra energy from the sun? If the energy input is just the right amount, the electron will become excited enough to reach a higher energy level. Left to itself, it will quickly return to the lower energy level by emitting the extra energy. Most often, its emission is in the form of light. You will read about this event in Sections 7.3 and 7.4, on photosynthesis.

Think "Shells"

The **shell model** is a simple way to show how electrons are distributed in an atom. By this model, a "shell" encloses all orbitals available to electrons at the same energy level (Figure 2.7). The first, spherical orbital is in the first shell. Enclosing the first shell is a second shell at a higher energy level. The second shell has four more available orbitals. More orbitals fit in a third shell, a fourth shell, and so on through large, complex atoms of the heavier elements listed in Appendix VI.

From Atoms to Molecules

How can you predict whether two atoms will interact? Check for electron vacancies in their outermost shell. Vacancies mean an atom might give up, gain, or share electrons under suitable conditions, which of course will change the distribution or number of its electrons. You can use the shell model to picture what goes on here.

Figure 2.7 shows the electron distribution for some atoms. Electrons are assigned to an energy level (a shell). Count the vacancies—one or more unfilled orbitals in

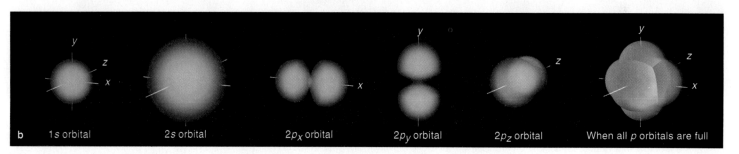

b | 1s orbital | 2s orbital | 2p$_x$ orbital | 2p$_y$ orbital | 2p$_z$ orbital | When all p orbitals are full

Figure 2.6 Electron arrangements in atoms with respect to three axes. Each axis (x, y, or z) is perpendicular to the other two. (**a**) A hydrogen atom's electron occupies a spherical 1s orbital, at the lowest energy level. (**b**) Every atom has a 1s orbital occupied by one or two electrons. Other orbitals include the 2s and p orbitals at the second energy level. Each can hold two electrons.

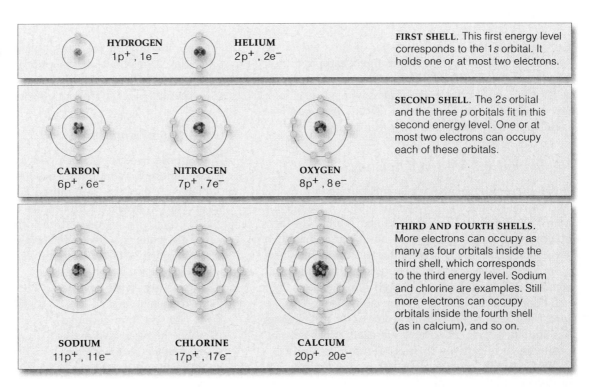

Figure 2.7 Examples of the shell model for the distribution of electrons in atoms. Successive shells correspond to higher energy levels.

Hydrogen, carbon, and other atoms with electron vacancies (unfilled orbitals) inside their outermost shell tend to give up, accept, or share electrons. Helium and other atoms with no electron vacancies in their outermost shell are inert; they show little if any tendency to interact with other atoms.

HYDROGEN
$1p^+$, $1e^-$

HELIUM
$2p^+$, $2e^-$

FIRST SHELL. This first energy level corresponds to the $1s$ orbital. It holds one or at most two electrons.

CARBON
$6p^+$, $6e^-$

NITROGEN
$7p^+$, $7e^-$

OXYGEN
$8p^+$, $8e^-$

SECOND SHELL. The $2s$ orbital and the three p orbitals fit in this second energy level. One or at most two electrons can occupy each of these orbitals.

SODIUM
$11p^+$, $11e^-$

CHLORINE
$17p^+$, $17e^-$

CALCIUM
$20p^+$ $20e^-$

THIRD AND FOURTH SHELLS. More electrons can occupy as many as four orbitals inside the third shell, which corresponds to the third energy level. Sodium and chlorine are examples. Still more electrons can occupy orbitals inside the fourth shell (as in calcium), and so on.

Figure 2.8 Chemical bookkeeping. We use symbols for elements when writing *formulas*, which identify the composition of compounds. For instance, water has the formula H_2O. The subscript number tells you two hydrogen (H) atoms are present for each oxygen (O) atom. Use such symbols and formulas when writing *chemical equations:* representations of the reactions among atoms and molecules. Substances entering a reaction (reactants) are to the left of the reaction arrow, and products are to the right, as shown by this chemical equation for photosynthesis.

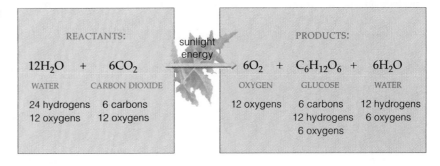

REACTANTS:

$12H_2O$ + $6CO_2$

WATER   CARBON DIOXIDE

24 hydrogens   6 carbons
12 oxygens   12 oxygens

sunlight energy →

PRODUCTS:

$6O_2$ + $C_6H_{12}O_6$ + $6H_2O$

OXYGEN   GLUCOSE   WATER

12 oxygens   6 carbons   12 hydrogens
      12 hydrogens   6 oxygens
      6 oxygens

each atom's outermost shell. As you can see, helium is one of the atoms with no vacancies. It is an *inert* atom; it shows little tendency to enter chemical reactions.

Now look again at Figure 2.1, which lists the most abundant elements that make up a typical organism. These elements include hydrogen, oxygen, carbon, and nitrogen. The atoms of all four elements have electron vacancies. *Such atoms tend to eliminate their vacancies by forming bonds with other atoms.*

Each **chemical bond** is a union between the electron structures of atoms. Take a moment to study Figure 2.8. It summarizes a few conventions we use when talking about metabolic reactions. A **molecule** forms when two or more atoms join up. Some molecules consist of only one element. Molecular nitrogen (N_2), with two nitrogen atoms, is like this. So is molecular oxygen (O_2).

The molecules of **compounds** consist of two or more different elements in proportions that never vary. Water is an example. Each water molecule has one oxygen atom and two hydrogen atoms. The water molecules in

rainclouds, the ocean, a Siberian lake, your bathtub, the petals of a leaf or flower, or anywhere else always have twice as many hydrogen as oxygen atoms.

In a **mixture**, two or more elements or compounds intermingle in proportions that can and usually do vary. Swirling water with the sugar sucrose (a compound of carbon, hydrogen, and oxygen) gives you a mixture.

Electrons occupy orbitals, or volumes of space around an atom's nucleus. By a simplified model, orbitals are arranged as a series of shells enclosing the nucleus. Successive shells correspond to levels of energy, which become higher with distance from the nucleus.

One or at most two electrons occupy any orbital. Atoms with unfilled orbitals in their outermost shell tend to interact with other atoms; those with no vacancies do not.

In molecules of an element, all atoms are the same kind. In molecules of a compound, atoms of two or more elements are bonded together, in unvarying proportions.

IMPORTANT BONDS IN BIOLOGICAL MOLECULES

Eat your peas! Drink your milk! For as long as you've been alive, you've been eating complex carbohydrates, proteins, and other "biological molecules." Only living cells make and use these molecules, which consist of a few kinds of atoms joined by a few kinds of bonds—mainly ionic, covalent, and hydrogen bonds.

Ion Formation and Ionic Bonding

An atom, recall, has just as many electrons as protons, so it carries no net charge. That balance can change for atoms having a vacancy—an unfilled orbital—in their outermost shell. For example, a chlorine atom has such a vacancy and can acquire another electron. A sodium atom has a lone electron in its outermost shell, and that electron can be pulled or knocked out. An atom that has either lost or gained one or more electrons is an **ion**. The balance between its protons and its electrons has shifted, so now the atom is ionized; it has become positively or negatively charged (Figure 2.9*a*).

In living cells, neighboring atoms commonly accept or donate electrons among themselves. When one loses an electron and another grabs it, both become ionized. Depending on cellular conditions, the two ions might not separate; they might stay together as a result of the mutual attraction of opposite charges. An association of two ions that have opposing charges is known as an **ionic bond**. You see one outcome of ionic bonding in Figure 2.9*b*, which shows a portion of a crystal of table salt, or NaCl. In such crystals, sodium ions (Na^+) and chloride ions (Cl^-) interact through ionic bonds.

Covalent Bonding

Suppose two atoms, each with an unpaired electron in its outermost shell, meet up. Each exerts an attractive force on the other's unpaired electron but not enough to yank it away. Each becomes more stable by *sharing* its unpaired electron with the other. A sharing of a pair of electrons is a **covalent bond**. Example: A hydrogen atom partly fills the electron vacancy in its outermost shell when it has bonded covalently with another hydrogen atom.

Two hydrogen atoms, each with a single proton, share two electrons

In structural formulas, a line between two atoms is a *single* covalent bond. It means an electron pair, as in molecular hydrogen (H_2, or H—H). In a *double* covalent bond, two atoms share two electron pairs, as in molecular oxygen (O=O). In a *triple* covalent bond, two atoms share three pairs, as in molecular nitrogen (N≡N). These examples are gaseous molecules. Each time you breathe in, a stupendous number of H_2, O_2, and N_2 molecules flow into your lungs.

Covalent bonds can be nonpolar or polar. Atoms exert the same pull on electrons and share them equally in a *nonpolar* covalent bond. "Nonpolar" means there is no difference in charge between the two ends of the bond—that is, at its two poles. Molecular hydrogen has a nonpolar covalent bond. Its two H atoms, each with one proton, attract the shared electrons equally.

electron transfer

SODIUM
ATOM

11 p$^+$,
11 e$^-$

CHLORINE
ATOM

17 p$^+$,
17 e$^-$

SODIUM
ION

11 p$^+$,
10 e$^-$

CHLORIDE
ION

17 p$^+$,
18 e$^-$

+   −

a Formation of a sodium ion and a chloride ion

Figure 2.9 **(a)** Ionization by way of an electron transfer. In this case, a sodium atom donates the single electron in its outermost shell to a chlorine atom, which has an unfilled orbital in *its* outermost shell. A sodium ion (Na^+) and a chloride ion (Cl^-) are the outcome of this interaction. **(b)** In each crystal of table salt, or NaCl, many sodium and chloride ions remain together owing to the mutual attraction of opposite charges. Their interaction is a case of ionic bonding.

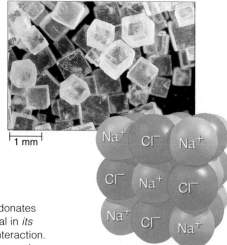

1 mm

Na^+ Cl^- Na^+

Cl^- Na^+ Cl^-

Na^+ Cl^- Na^+

b Crystals of sodium chloride (NaCl)

In a *polar* covalent bond, atoms of different elements (which have different numbers of protons) don't exert the same pull on shared electrons. The more attractive atom ends up with a slight negative charge. We say it's "electronegative." However, its effect is balanced out by the other atom, which ends up with a slight positive charge. Thus, taken together, the two atoms of a polar covalent bond have no *net* charge.

Consider a water molecule. It has two polar covalent bonds: H—O—H. The electrons are less attracted to the two hydrogens than to the oxygen, which has more

Figure 2.10 Like skydivers who briefly clasp hands to form an orderly pattern, weak attractions within and between molecules can form and break easily.

protons. The molecule is slightly negative at one end and positive at the other, but has no *net* charge overall. As you will see, its polarity attracts polar substances.

Hydrogen Bonding

The patterns of electron sharing in covalent bonds hold atoms together in specific arrangements in molecules. Some of the patterns also give rise to weak attractions and repulsions between charged functional groups of molecules, and between molecules. Like interacting skydivers, these interactions can form and break easily (Figure 2.10). Even so, they have important roles in the structure and functioning of biological molecules.

For example, a **hydrogen bond** is a weak attraction between an electronegative atom (such as an oxygen or nitrogen atom taking part in a polar covalent bond) and a hydrogen atom taking part in a second polar covalent bond. Hydrogen's slight positive charge weakly attracts the atom with the slight negative charge (Figure 2.11).

Hydrogen bonds often form between different parts of a molecule that twists and folds back on itself. Such bonds are pervasive among proteins (Section 3.7). They commonly form between two or more molecules, as when they hold the two strands of a DNA molecule together (Figure 2.11*d*). Although individually easy to break, hydrogen bonds collectively stabilize the DNA. Extensive hydrogen bonding between individual water molecules also contributes to water's life-sustaining properties, which is the topic of the next section.

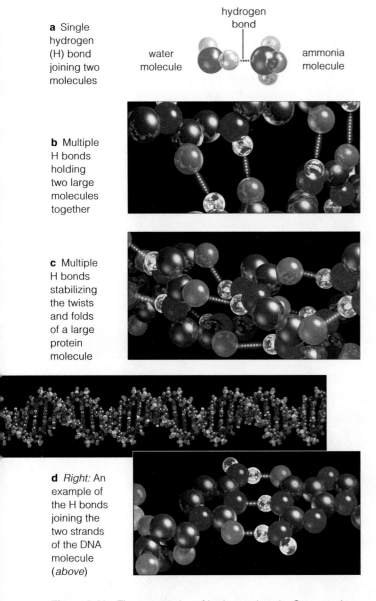

a Single hydrogen (H) bond joining two molecules

hydrogen bond

water molecule

ammonia molecule

b Multiple H bonds holding two large molecules together

c Multiple H bonds stabilizing the twists and folds of a large protein molecule

d *Right:* An example of the H bonds joining the two strands of the DNA molecule (*above*)

Figure 2.11 Three examples of hydrogen bonds. Compared to a covalent bond, a hydrogen bond is easier to break. Collectively, extensive hydrogen bonding has a major role in the structure and behavior of water, DNA, proteins, and many other substances, as you will see in later chapters.

In an ionic bond, two ions of opposite charge attract each other and stay together. Ions form when atoms gain or lose electrons and so acquire a net positive or negative charge.

In a covalent bond, atoms share a pair of electrons. When atoms share the electrons equally, the bond is nonpolar. When the sharing is not equal, the bond itself is polar—slightly positive at one end, slightly negative at the other.

In a hydrogen bond, a covalently bound atom showing a slight negative charge weakly interacts with a covalently bound hydrogen atom showing a slight positive charge.

2.5

PROPERTIES OF WATER

No sprint through basic chemistry is complete unless it leads us to the collection of molecules called water. Life originated in water. Organisms still live in it, or they cart water around with them in cells and tissue spaces. Many metabolic reactions require water as a reactant. Cell shape and internal structure depend on it. These topics will repeatedly occupy our attention in the book, so you may find it useful to become familiar with the following points about water's properties.

Polarity of the Water Molecule

A water molecule, remember, has no net charge, but the charge it does carry is unevenly distributed. Its electron arrangements and bond angles make its oxygen "end" a bit negative and its hydrogen end a bit positive, as in Figure 2.12a. Because of the polarity resulting from this charge distribution, one water molecule attracts and hydrogen-bonds with others (Figure 2.12b,c).

Sugars and other polar molecules easily hydrogen-bond with a water molecule; its polarity attracts them.

They are **hydrophilic** (water-loving) **substances**. That same polarity repels oils and other nonpolar molecules, which are **hydrophobic** (water-dreading) **substances**. Shake a bottle full of water and salad oil, then set it on a table. New hydrogen bonds soon replace those broken when you shook the bottle. As water molecules reunite, they push oil molecules aside, forcing them to cluster as oil droplets or as an oily film at the water's surface.

A thin, oily membrane separates water inside a cell from water outside it. Membrane organization, hence life itself, starts with both hydrophilic and hydrophobic interactions, as Sections 4.1 and 5.1 describe.

Water's Temperature-Stabilizing Effects

Cells consist mainly of water. They release a lot of heat energy by metabolism. If it weren't for water's hydrogen bonds, they might cook in their own juices. To see why, start with these facts: Each molecule vibrates nonstop, and its motion increases as it absorbs heat. **Temperature** is a measure of molecular motion. Compared with most

slight negative
charge at this end

The + and – balance each other; the whole molecule carries no net charge, overall

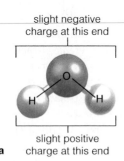

slight positive charge at this end

a

b

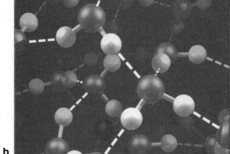

c

Figure 2.12 Water, a substance vital for life. (**a**) A water molecule's polarity. (**b**) Hydrogen bonding pattern among water molecules in liquid water. Dashed lines signify H bonds.

(**c**) The hydrogen bonding pattern of ice. Below 0°C, each water molecule hydrogen-bonds to four others in a three-dimensional lattice. Molecules are farther apart than they would be in liquid water at room temperature, when molecular motion is greater and not as many bonds form. Ice floats because it has fewer molecules than the same volume of liquid water; its lattice is less dense.

The Arctic Ocean's ice cap is melting. At current rates of melting, it will be gone in fifty years. So will polar bears. Already their seal-hunting season is shorter; the bears are thinner and producing fewer cubs.

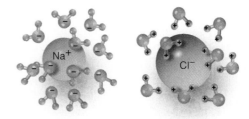

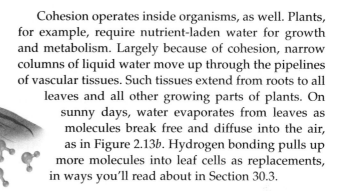

Figure 2.13 Water's cohesion. (**a**) A water strider uses its fine, water-resistant leg hairs and water's high surface tension to scoot on water. (**b**) Cohesion, and evaporation from leaves, pulls water from roots to leaves.

Figure 2.14 Two spheres of hydration.

fluids, water absorbs much more heat energy before its temperature increases measurably. Why? Much of the added energy disrupts hydrogen bonding *between* molecules instead of increasing motion inside them. Hydrogen bonds in water counter drastic temperature shifts in aquatic habitats. They do the same inside the body of a multicelled organism.

Even when liquid water's temperature is not shifting much, hydrogen bonds constantly break but form again just as fast. By contrast, large energy inputs increase molecular motion so much that hydrogen bonds stay broken, so individual molecules at the water's surface escape into air. By this process, called **evaporation**, heat energy converts liquid water to the gaseous state. The energy input overcomes the attractive forces between molecules of liquid water, molecules break free, and surface temperature of the water is lowered.

Evaporative water loss is one of the main ways in which you, and some other mammals, cool off when you sweat on hot, dry days. Sweat is about 99 percent water, and it evaporates from your skin.

Below 0°C, hydrogen bonds resist breaking and lock water molecules in the latticelike bonding pattern of ice (Figure 2.12*c*). Ice is less dense than water. During winter freezes, ice sheets may form near the surface of ponds, lakes, and streams. The ice blanket "insulates" liquid water beneath it and helps protect many fishes, frogs, and other aquatic organisms against freezing.

Water's Cohesion

Life also depends on water's cohesion. **Cohesion** means something has a capacity to resist rupturing when placed under tension—that is, stretched—as by the weight of a bug's legs (Figure 2.13*a*). Think of a lake or some other body of liquid water. Uncountable hydrogen bondings exert a continual inward pull on molecules of water at or near the surface. The bonding causes a high surface tension. You can see evidence of this when, say, flying insects splat against water and float on it.

Cohesion operates inside organisms, as well. Plants, for example, require nutrient-laden water for growth and metabolism. Largely because of cohesion, narrow columns of liquid water move up through the pipelines of vascular tissues. Such tissues extend from roots to all leaves and all other growing parts of plants. On sunny days, water evaporates from leaves as molecules break free and diffuse into the air, as in Figure 2.13*b*. Hydrogen bonding pulls up more molecules into leaf cells as replacements, in ways you'll read about in Section 30.3.

Water's Solvent Properties

Finally, water is an excellent solvent, for ions and polar molecules easily dissolve in it. A dissolved substance is known as a **solute**. In general, we say that a substance is *dissolved* after water molecules cluster around ions or molecules of it and thereby keep them dispersed in fluid. Such clusters are called "spheres of hydration." Spheres of hydration form around solutes in cellular fluid, the sap of maple trees, blood, fluid in your gut, and every other fluid associated with life.

Watch this happen when you pour some table salt (NaCl) into a cup of water. After a while, the crystals of salt separate into Na^+ and Cl^-. Each Na^+ attracts the negative end of some water molecules at the same time that Cl^- attracts the positive end of others (Figure 2.14). The spheres of hydration formed in this way keep the ions dispersed in the fluid.

A water molecule has no net charge, but it is slightly polar owing to an uneven distribution of charge. Its oxygen "end" is a bit negative and its hydrogen "end" is a bit positive.

Polarity allows water molecules to hydrogen-bond to each other and to other polar (hydrophilic) substances. Water molecules tend to repel nonpolar (hydrophobic) substances.

Water has temperature-stabilizing effects, internal cohesion, and a capacity to dissolve many substances. These properties influence the structure and functioning of organisms.

ACIDS, BASES, AND BUFFERS

A great variety of ions dissolved in the fluids inside and outside cells influence cell structure and functioning. Among the most influential are **hydrogen ions**, or H^+. Hydrogen ions are the same thing as free (unbound) protons. They have far-reaching effects largely because they are chemically active and there are so many of them.

The pH Scale

At any instant in liquid water, some water molecules are breaking apart into hydrogen ions and hydroxide ions (OH^-). This ionization of water is the basis of the **pH scale**, as in Figure 2.15. Biologists use this scale when measuring the H^+ concentration of seawater, tree sap, blood, and other fluids. Pure water (not rainwater or tapwater) always has just as many H^+ as OH^- ions. This condition, which also may occur in other fluids, signifies neutrality on the pH scale.

Neutrality is calculated to be 7. Each change by one unit from neutrality corresponds to a tenfold increase or decrease in H^+ concentration. Most commonly, the pH scale ranges from 0 (the highest H^+ concentration) to 14 (the lowest). *The greater the H^+ concentration, the lower the pH).* An easy way to sense that range is to dip your tongue in baking soda (pH 9), water (7), then lemon juice (2.3).

How Do Acids Differ From Bases?

When they dissolve in water, substances called **acids** *donate* protons (H^+) to the water solution. By contrast, substances called **bases** *accept* H^+ when dissolved in water. *Acidic* solutions, such as lemon juice, gastric fluid, and coffee, release H^+; their pH is below 7. *Basic* solutions, which include seawater, baking soda, and egg white, take up or combine with H^+. Basic solutions are also called "alkaline" fluids, with a pH above 7.

The fluid inside most human cells is about 7 on the pH scale. Most fluids bathing these cells have values between 7.3 and 7.5. This also is the case for the fluid portion of human blood. Seawater is more alkaline than the body fluids of organisms that live in it.

Most acids are weak or strong. Weak acids, such as carbonic acid (H_2CO_3), are reluctant H^+ donors. But strong acids readily give up H^+ when they dissociate in water. Examples are hydrochloric acid (HCl), nitric acid (HNO_3), and sulfuric acid (H_2SO_4).

Sniff and eat spicy food. Swallowing sends it onward to gastric fluid in your stomach. The meal stimulates cells of the stomach's lining to secrete a strong acid, HCl, that dissociates into H^+ and Cl^-. The ions increase the H^+ concentration, thereby making the gastric fluid more acidic. Increased acidity activates enzymes that digest the food's proteins and attack most bacteria that may have lurked in or on the food. When people eat too much spicy food, they may get an *acid stomach* and reach for an antacid, such as milk of magnesia. This is a base. When it dissolves, it releases magnesium ions and OH^- to neutralize the acid. OH^- combines with excess H^+ in gastric fluid, which reduces the H^+ concentration and thereby calms things down.

High concentrations of strong acids or bases also disrupt the external environment and pose dangers to life. Read the labels on bottles of ammonia, drain cleaner, or other products commonly stored in households. Many cause severe *chemical burns*. So does the sulfuric acid in car batteries.

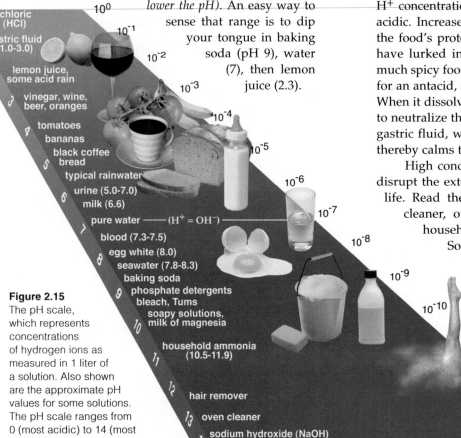

hydrochloric acid (HCl)
gastric fluid (1.0-3.0)
lemon juice, some acid rain
vinegar, wine, beer, oranges
tomatoes
bananas
black coffee
bread
typical rainwater
urine (5.0-7.0)
milk (6.6)
pure water ———— ($H^+ = OH^-$) ————
blood (7.3-7.5)
egg white (8.0)
seawater (7.8-8.3)
baking soda
phosphate detergents
bleach, Tums
soapy solutions, milk of magnesia
household ammonia (10.5-11.9)
hair remover
oven cleaner
sodium hydroxide (NaOH)

10^0
10^{-1}
10^{-2}
10^{-3}
10^{-4}
10^{-5}
10^{-6}
10^{-7}
10^{-8}
10^{-9}
10^{-10}
10^{-11}
10^{-12}
10^{-13}
10^{-14}

Figure 2.15 The pH scale, which represents concentrations of hydrogen ions as measured in 1 liter of a solution. Also shown are the approximate pH values for some solutions. The pH scale ranges from 0 (most acidic) to 14 (most basic). A change of 1 on the scale means a tenfold change in H^+ concentration.

The burning of fossil fuels—as by power plants and cars—and nitrogen fertilizers release strong acid that alter the pH of rain (Figure 2.16). Certain regions are sensitive to this *acid rain* because of their soil type and vegetation cover. The modified chemistry of habitats in such regions can harm organisms. We will return to this topic in Section 50.1.

Buffers Against Shifts in pH

Metabolic reactions are sensitive to even slight shifts in pH, for H^+ and OH^- can combine with many different molecules and alter their functions. Normally, control mechanisms respond to shifts in pH, as they do when HCl enters the stomach in response to a meal. Many of the controls involve buffer systems.

A **buffer system** is a partnership between a weak acid and the weak base that forms as the acid dissolves in water. The two work as a pair to counter slight shifts in pH. Remember, the OH^- level rises when any strong base enters a fluid. But a weak acid can neutralize part of the incoming OH^- by donating H^+. The weak acid's partner (a weak base) forms as a result of its action. Later, if a strong acid floods in, the weak base will accept H^+ from it and become the buffer system's acid.

Bear in mind, the action of a buffer system cannot make new hydrogen ions or eliminate ones that already are present. It can only bind or release them.

In all complex, multicelled organisms, diverse buffer systems operate in the internal environment—in blood and tissue fluids. For example, metabolic reactions in lungs and kidneys help control the acid–base balance of this environment, at levels suitable for life (Chapters 40 and 41). For now, simply think of what happens when the blood level of H^+ decreases and the blood is not as acidic as it should be. At such times, carbonic acid that is dissolved in blood releases H^+ and so becomes the partner base, bicarbonate:

$$H_2O + CO_2 \longrightarrow H_2CO_3 \longrightarrow HCO_3^- + H^+$$

WATER   CARBON DIOXIDE   CARBONIC ACID   BICARBONATE

When blood becomes more acidic, more H^+ becomes bound to the base, thus forming the partner acid:

$$HCO_3^- + H^+ \longrightarrow H_2CO_3 \longrightarrow H_2O + CO_2$$

BICARBONATE   CARBONIC ACID   WATER   CARBON DIOXIDE

Any buffer system can mop up only so much extra acid or base. Once its capacity is exceeded, a system's pH destabilizes quickly. Uncontrolled shifts in pH have bad effects. If blood's pH (7.3–7.5) declines even to 7,

Figure 2.16 Sulfur dioxide emissions from a coal-burning power plant. Airborne pollutants such as sulfur dioxide dissolve in water vapor to form acidic solutions. They are a major component of acid rain.

this invites a *coma*, a sometimes irreversible state of unconsciousness. An increase to 7.8 can lead to *tetany*, a potentially lethal condition in which skeletal muscles contract uncontrollably. With *acidosis*, carbon dioxide builds up in blood, too much carbonic acid forms, and blood pH plummets. With *alkalosis*, a rise in blood pH cannot be corrected. Both conditions can be lethal.

Salts

Salts are compounds that release ions *other than* H^+ and OH^- in solution. Salts and water often form when an acid and a base interact. Depending on a solution's pH, salts are able to dissolve easily. Consider how sodium chloride forms:

$$HCl \text{ (acid)} + NaOH \text{ (base)} \longrightarrow NaCl \text{ (salt)} + H_2O$$

HYDROCHLORIC ACID   SODIUM HYDROXIDE   SODIUM CHLORIDE

$$Na^+ \quad Cl^- \text{ (ionization)}$$

Many salts dissolve into ions that play key roles in cells. For example, nerve cell activity depends on ions of sodium, potassium, and calcium. Your muscles contract with the help of calcium ions. And water absorption by plant cells depends largely on potassium ions.

Hydrogen ions (H^+) and other ions dissolved in the fluids inside and outside cells affect cell function.

When dissolved in water, acidic substances release H^+, and basic (alkaline) substances accept them. Certain acid–base interactions help maintain the pH value of a fluid—that is, its H^+ concentration—by a buffering action.

A buffer system counters slight shifts in pH by releasing hydrogen ions when their concentration is too low or by combining with them when the concentration is too high.

Salts are compounds that release ions other than H^+ and OH^-, and many of those ions have key roles in cell functions.

2.7

Gold indicates text section

1. Chemistry helps us understand the nature of all the substances that make up cells, organisms, and the Earth, its waters, and the atmosphere. Each substance consists of one or more elements. Of the ninety-two naturally occurring elements, the most common in organisms are oxygen, carbon, hydrogen, and nitrogen. Organisms also have lesser amounts of many other elements, such as calcium, phosphorus, potassium, and sulfur. *CI*

2. Table 2.2 summarizes some key chemical terms that you will encounter throughout this book. *CI, 2.1–2.6*

3. Elements consist of atoms. An atom has one or more positively charged protons plus an equal number of negatively charged electrons and (except for hydrogen atoms) one or more uncharged neutrons. The protons and neutrons occupy the core region, or the atomic nucleus. Most elements have isotopes, which are two or more forms of atoms that have the same number of protons but different numbers of neutrons. *2.1*

4. Whether an atom interacts with others depends on the number and arrangement of its electrons, which occupy orbitals (volumes of space) in a series of shells around the atomic nucleus. When an atom has unfilled orbitals in its outermost shell, it tends to bond with atoms of other elements. *2.3*

5. An atom may lose or gain one or more electrons and thus become an ion, which has an overall positive or negative charge. *2.4*

6. Generally, a chemical bond is a union between the electron structures of atoms. *2.3*

   a. In an ionic bond, a positive ion and negative ion stay together because of the mutual attraction of their opposite charges. *2.4*

   b. Atoms often share one or more pairs of electrons in single, double, and triple covalent bonds. Electron sharing is equal in nonpolar covalent bonds, and it is unequal in polar covalent bonds. The interacting atoms have no net charge overall; the bond is slightly negative at one end and slightly positive at the other. *2.4*

   c. In a hydrogen bond, one covalently bonded atom (such as oxygen) that has a slight negative charge is weakly attracted to the slight positive charge of a hydrogen atom that is taking part in a different polar covalent bond. *2.4*

7. Polar covalent bonds join together three atoms in a water molecule (two hydrogens and one oxygen). The polarity of a water molecule invites hydrogen bonding between water molecules. Such hydrogen bonding is the basis of liquid water's ability to resist temperature changes more than other fluids do, display internal cohesion, and easily dissolve polar or ionic substances. These properties greatly affect the metabolic activity, structure, shape, and internal organization of cells. *2.5*

8. By the pH scale, a solution is assigned a number that reflects its H^+ concentration. A typical range is from 0 (highest concentration) to 14 (lowest). At pH 7, the H^+ and OH^- concentrations are equal. Acids release H^+ in water, and bases combine with them. Buffer systems help maintain the pH values of blood, tissue fluids, and the fluid inside cells. *2.6*

Table 2.2 *Summary of Important Players in the Chemical Basis of Life*

ELEMENT	Fundamental form of matter that occupies space, has mass, and cannot be broken apart into a different form of matter by ordinary physical or chemical means.
ATOM	Smallest unit of an element that still retains the characteristic properties of that element.
Proton (p^+)	Positively charged particle of the atomic nucleus. All atoms of an element have the same number of protons, which is the atomic number. A proton without an electron zipping around it is a hydrogen ion (H^+).
Electron (e^-)	Negatively charged particle that can occupy a volume of space (orbital) around an atomic nucleus. Electrons can be shared or transferred among atoms.
Neutron	Uncharged particle of the nucleus of all atoms except hydrogen. For a given element, the mass number is the number of protons and neutrons in the nucleus.
MOLECULE	Unit of matter in which two or more atoms of the same element, or different ones, are bonded together.
Compound	Molecule composed of two or more different elements in unvarying proportions. Water is an example.
Mixture	Intermingling of two or more elements in proportions that can and usually do vary.
ISOTOPE	One of two or more forms of an element's atoms that differ in their number of neutrons.
Radioisotope	Unstable isotope, having an unbalanced number of protons and neutrons, that emits particles and energy.
Tracer	Molecule of a substance to which a radioisotope is attached. Together with tracking devices, it is used to identify movement or destination of the substance in a metabolic pathway, the body, or some other system.
ION	Atom that has gained or lost one or more electrons, thus becoming negatively or positively charged.
SOLUTE	Any molecule or ion dissolved in some solvent.
Hydrophilic substance	Polar molecule or molecular region that can readily dissolve in water.
Hydrophilic substance	Nonpolar molecule or molecular region that strongly resists dissolving in water.
ACID	Substance that donates H^+ when dissolved in water.
BASE	Substance that accepts H^+ when dissolved in water.
SALT	Compound that releases ions other than H^+ or OH^- when dissolved in water.

Review Questions

1. What is an element? Name four elements (and their symbols) that make up more than 95 percent of the body weight of all living organisms. *CI, 2.1*

2. Define atom, isotope, and radioisotope. *2.1*

3. How many electrons can occupy each orbital around an atomic nucleus? Using the shell model, explain how the orbitals available to electrons are distributed in an atom. *2.3*

4. Define molecule, compound, and mixture. *2.3*

5. Distinguish between:
   a. ionic and hydrogen bonds *2.4*
   b. polar and nonpolar covalent bonds *2.4*
   c. hydrophilic and hydrophobic interactions *2.5*

6. If a water molecule has no net charge, then why does it attract polar molecules and repel nonpolar ones? *2.5*

7. Label the atoms in each water molecule in the sketch at right. Indicate which parts of each molecule carry a slight positive charge (+) and which carry a slight negative charge (−). *2.5*

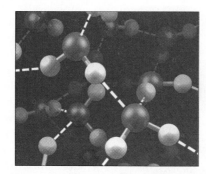

8. Define acid and base. Then describe the behavior of a weak acid in solutions having a high or low pH value. *2.6*

Self-Quiz   ANSWERS IN APPENDIX III

1. Electrons carry a _____ charge.
   a. positive   b. negative   c. zero

2. Atoms share electrons unequally in a(n) _____ bond.
   a. ionic   c. polar covalent
   b. nonpolar covalent   d. hydrogen

3. An individual water molecule shows _____ .
   a. polarity   d. spherelike hydration
   b. hydrogen-bonding capacity   e. a and b
   c. notable heat resistance   f. all of the above

4. In liquid water, spheres of hydration form around _____ .
   a. nonpolar molecules   d. solvents
   b. polar molecules   e. b and c
   c. ions   f. all of the above

5. Hydrogen ions (H^+) are _____ .
   a. the basis of pH values   d. dissolved in blood
   b. unbound protons   e. both a and e
   c. targets of certain buffers   f. all of the above

6. When dissolved in water, a(n) _____ donates H^+; however, a(n) _____ accepts H^+.

7. Match the terms with the most suitable descriptions.
   _____ trace element   a. weak acid and its partner base
   _____ buffer system   work as a pair to counter pH shifts
   _____ chemical bond   b. union between electron structures
   _____ temperature   of two atoms
                           c. less than 0.01% of body weight
                           d. measure of molecular motion
                              in some defined region

Critical Thinking

1. An ionic compound forms when calcium combines with chlorine. Referring to Figure 2.8, give the compound's formula. (*Hint:* Be sure the outermost shell of each atom is filled.)

2. David, an inquisitive three-year-old, poked his fingers into warm water inside a metal pan on the stove, then touched the pan and got a nasty burn. Explain why water in a metal pan heats up far more slowly than the pan itself.

3. When molecules absorb microwaves, which are a form of electromagnetic radiation, they move more rapidly. Explain why a microwave oven can heat foods.

4. From what you know about cohesion, devise a hypothesis to explain why water forms droplets.

5. Edward is trying to study a chemical reaction that an enzyme catalyzes (speeds up). H^+ forms in the reaction, but the enzyme is destroyed at low pH. What can he include in his reaction mix to protect the enzyme while he studies the reaction?

6. Through interactions with water and ions, a soluble protein becomes dispersed in cellular fluid. An electrically charged cushion around it makes this happen. Using the sketch at right as a guide, explain the chemical interactions through which this cushion forms. Start with the major bonds in the protein molecule.

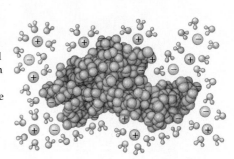

7. Reactants and products often are expressed in moles. A *mole* is a certain number of atoms or molecules of any substance, just as "a dozen" can refer to any twelve cats, roses, and so forth. Molar weight, in grams, equals the total atomic weight of all atoms making up that substance. For example, the atomic weight of carbon is 12, so one mole of carbon weighs 12 grams. A mole of oxygen (atomic weight 16) weighs 16 grams.

   State why a mole of water (H_2O) weighs 18 grams, and why a mole of glucose ($C_6H_{12}O_6$) weighs 180 grams.

Selected Key Terms

acid *2.6*	evaporation *2.5*	molecule *2.3*
atom *2.1*	hydrogen bond *2.4*	neutron *2.1*
base *2.6*	hydrogen ion (H^+) *2.6*	pH scale *2.6*
buffer system *2.6*	hydrophilic	proton *2.1*
chemical bond *2.3*	substance *2.5*	radioisotope *2.1*
cohesion *2.5*	hydrophobic	salt *2.6*
compound *2.3*	substance *2.5*	shell model *2.3*
covalent bond *2.4*	ion *2.4*	solute *2.5*
electron *2.1*	ionic bond *2.4*	temperature *2.5*
element *CI*	isotope *2.1*	tracer *2.2*
	mixture *2.3*	

Readings

Ritter, P. 1996. *Biochemistry: A Foundation.* Pacific Grove, California: Brooks/Cole. Good, easy-to-read introduction to biochemistry, with plenty of human-interest applications.

On-Line readings at Student Guide for InfoTrac:
www.brookscole.com/biology

3

CARBON COMPOUNDS IN CELLS

Carbon, Carbon, In the Sky—Are You Swinging Low and High?

Maybe you've already heard that the past decade was the warmest ever recorded. Maybe you know that a long-term rise in the lower atmosphere's temperature—a **global warming**—is under way. It is something you might want to worry about. In 2001, for example, rare thunderstorms pelted Alaska and 27 inches of rainfall in twenty-four hours washed homes away in Hawaii.

If, as predicted, the atmospheric temperature spikes even 2.5 degrees higher in this century, the sea level may rise enough to permanently flood coastal cities. Changes in rainfall patterns over broad regions may give new meaning to the expressions "crop failure" and "water wars." The already bad flooding and mudslides in the tropics and California will get worse, interiors of the continents will become drier, and deserts will expand.

And what about organisms other than ourselves? Think about the trees of those scenic forests in the high mountains of the Pacific Northwest (Figure 3.1). The trees don't do much during the winter, because the water required for growth is locked away in the form of snow and ice. The trees entered dormancy during autumn's cool, dry days; their metabolic activities idled, and growth ceased. Only water already inside them kept their living cells from dying. Only when spring arrives do the trees break dormancy. As temperatures rise and snow melts, mineral-laden water becomes available. And now photosynthetic cells take carbon dioxide from the air inside leaves and make sugars, starches, and other carbon-based compounds.

Figure 3.1 Above, a forest of conifers beneath the first snows of winter on Silver Star Mountain. The form, function, and very survival of these trees—and of people, bears, and all other organisms—start with the carbon atom and its diverse molecular partners in organic compounds. Even the gasoline and other fossil fuels that we humans have come to depend on had their beginning in carbon atoms used millions of years ago, by the growing trees of ancient forests.

Look closely and you will find the same connection between air temperature and photosynthesis all over the planet. Ultimately it affects you and all other kinds of organisms that can or can't produce their own food.

Here is something else to think about. The carbon dioxide concentration in the air declines in spring and summer when photosynthesizers take up stupendous amounts of it. It rises in autumn, when photosynthesis declines and decomposers go to work. The wastes and remains of photosynthesizers feed decomposers and fan their population growth. Collectively, decomposers release huge amounts of carbon dioxide as a metabolic by-product.

Now researchers find that plants of the northern forests and the great plains of Canada, the United States, and elsewhere are breaking dormancy earlier than they did two decades ago. Also, the atmospheric concentration of carbon dioxide is swinging more than before—as much as 20 percent more in Hawaii and a whopping 40 percent in Alaska. Global warming is probably promoting longer growing seasons, hence more uptake of carbon dioxide and the wider swings.

We don't know why the air is warming up, but a long-term rise in carbon dioxide levels in the air is accompanying it. We do know that human populations burn huge amounts of gas, coal, and other carbon-rich fossil fuels for energy. The carbon dioxide that burning releases actually may be part of the problem, as you will read in Section 48.9.

The point is this: *Carbon permeates the world of life—from the energy-requiring activities and structural organization of cells, to physical and chemical conditions that span the globe and influence ecosystems everywhere.*

With this chapter, we turn to life-giving properties that emerge from the molecular structure of carbon-rich compounds. Study the chapter well, including the summary in Section 3.10. It will serve as a foundation for understanding how diverse organisms put together such compounds, how they use them, and how the effects of all those uses ripple through the biosphere.

Key Concepts

1. Organic compounds are molecules containing carbon and at least one hydrogen atom. We define cells partly by their capacity to assemble the organic compounds called carbohydrates, lipids, proteins, and nucleic acids. These are the molecules of life.

2. Cells assemble biological molecules from pools of smaller organic compounds, including simple sugars, fatty acids, amino acids, and nucleotides.

3. Glucose and other simple sugars are carbohydrates. So are oligosaccharides, short chains of covalently bonded sugar units. The most complex carbohydrates are the polysaccharides, many of which consist of hundreds or thousands of sugar units.

4. Lipids are greasy or oily compounds that show little tendency to dissolve in water but dissolve easily in nonpolar compounds, such as other lipids. Neutral fats (triglycerides), phospholipids, waxes, and sterols are important lipids.

5. Cells use carbohydrates and lipids as building blocks and as major sources of energy.

6. Proteins show the greatest diversity in structure and function. Many serve as construction materials in cells. Many others are enzymes, a class of molecules that hugely increase the rate of specific metabolic reactions. Other proteins transport substances in cells or across membranes, contribute to cell movements, trigger changes in cell activities, and defend the body against injury and disease.

7. ATP and other kinds of nucleotides have essential roles in metabolism. DNA and RNA are strandlike nucleic acids assembled from nucleotide units. They are the basis of inheritance and reproduction.

THE MOLECULES OF LIFE—FROM STRUCTURE TO FUNCTION

What Is An Organic Compound?

Under present-day conditions on Earth, *only living cells synthesize complex carbohydrates, lipids, proteins, and nucleic acids.* These are the molecules that typify life. Different classes of biological molecules are the cell's instant energy sources, structural materials, metabolic workers, cell-to-cell signals, and libraries of hereditary information. As you will see, their three-dimensional shape influences how each functions—and that shape is influenced by how a molecule's atoms are arranged and how electric charge is distributed among them.

Molecules of life are **organic compounds**, which we define as containing the element carbon and at least one hydrogen atom. The term is a holdover from a time when chemists thought "organic" substances were the ones they got from animals and vegetables, as opposed to "inorganic" substances they got from minerals. The term persists even though scientists now make organic compounds in laboratories. It persists even though we now have reason to believe that organic compounds were present on Earth before organisms were.

The **hydrocarbons** consist only of hydrogen atoms covalently bonded to carbon. Gasoline and other fossil fuels are examples. Like other organic compounds, they have a specific number of atoms arranged in specific ways. Also, organic compounds have one or more **functional groups**, which are particular atoms or clusters of atoms covalently bonded to carbon.

In this book we adhere to a standardized color code for the main atoms of organic compounds:

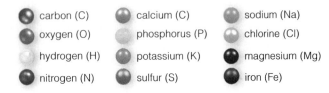

carbon (C)   calcium (C)   sodium (Na)
oxygen (O)   phosphorus (P)   chlorine (Cl)
hydrogen (H)   potassium (K)   magnesium (Mg)
nitrogen (N)   sulfur (S)   iron (Fe)

It All Starts With Carbon's Bonding Behavior

Living organisms consist mainly of oxygen, hydrogen, and carbon (Figure 2.1). The oxygen and hydrogen are mainly in the form of water. Put aside the water, and carbon makes up more than half of what's left.

Carbon's importance in life arises from its versatile bonding behavior. *Each carbon atom can covalently bond with as many as four other atoms.* Such bonds, in which two atoms share one, two, or three pairs of electrons, are relatively stable. They commonly join carbon atoms together as backbones to which hydrogen, oxygen, and other elements are attached. Such bonds are the start of the three-dimensional shapes of organic compounds.

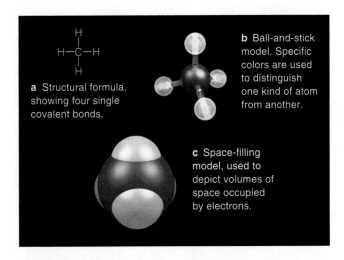

a Structural formula, showing four single covalent bonds.

b Ball-and-stick model. Specific colors are used to distinguish one kind of atom from another.

c Space-filling model, used to depict volumes of space occupied by electrons.

Figure 3.2 Molecular models for methane (CH_4), the simplest organic compound.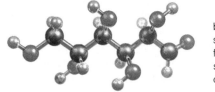

Ways of Representing Organic Compounds

Methane is the simplest organic compound of all. This colorless, odorless gas is abundant in the atmosphere, marine sediments, termite colonies, stagnant swamps, and stockyards. Its four hydrogen atoms are covalently bonded to a carbon atom (CH_4). Figure 3.2 shows ways to represent it. A ball-and-stick model is used to depict bond angles and convey how the molecule's mass is distributed (in atomic nuclei). The space-filling model is better for conveying a molecule's size and surfaces.

Now let's use the ball-and-stick model to depict an organic compound with six covalently bonded carbon atoms from which hydrogen *and* oxygen atoms project:

ball-and-stick model for the linear structure of glucose

In cells, this type of carbon backbone sometimes forms chains. But most of the time it coils back on itself, and its two ends connect to form a ring structure:

six-carbon ring structure of glucose that usually forms inside cells

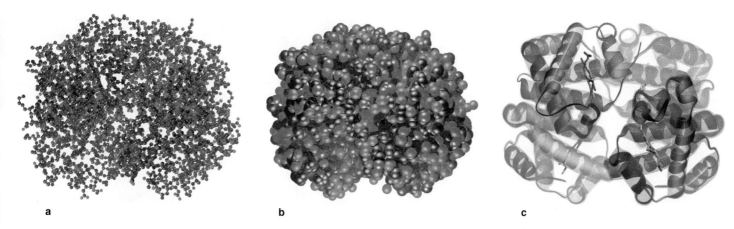

a   b   c

Figure 3.3 Visualizing the structure of hemoglobin, the oxygen-transporting molecule in red blood cells. (**a**) Ball-and-stick model, (**b**) space-filling model, and (**c**) ribbon model. Unlike the color coding for atoms, colors used for ribbon models and simple icons for complex molecules vary, depending on the context.

positively charged region

Figure 3.4 Model for the electrostatic potential energy at the surface of part of a hemoglobin molecule.

negatively charged region

We often depict carbon ring structures in simpler ways. A flat structural model may show the carbons but not other atoms bonded to it. If an icon for the ring shows no atoms at all, it is understood that one carbon atom occupies each "corner" of the ring:

simplified structural formula for a six-carbon ring

icon for a six-carbon ring

Figure 3.3 shows different ways of representing a much larger molecule: the protein hemoglobin. Like all all other vertebrates, your life depends on hemoglobin, which transports oxygen to tissue neighborhoods of all living cells in your body. The ball-and-stick and the space-filling models give you an idea of this molecule's mass and structural complexity. But neither tells you much about its oxygen-transporting function.

Now look at Figure 3.3*c*. This ribbon model shows how the hemoglobin molecule actually consists of four chains. As you will read later, each chain is a string of subunits known as amino acids. Certain regions of each chain are straight or folded, others are coiled. For now, it is enough to know that the three-dimensional shape of hemoglobin includes four pockets, each containing a small cluster of atoms known as a heme group (coded *red-orange* in this figure). Each heme group is able to bind and then release an oxygen molecule in response to differences in local tissue conditions.

More sophisticated models are now in use. Some computer models, for instance, show local differences in electric charge across molecular surfaces. Areas color-coded, say, *red* on the surface of one molecule

would be attractive to a *blue* surface on another part of the same molecule or on a different one (Figure 3.4).

Ultimately, such insights into the three-dimensional structure of molecules help us to understand how cells and multicelled organisms function. For instance, virus particles can infect a cell when they dock at specific proteins at the plasma membrane. Like Lego blocks, the proteins have ridges, clefts, and charged regions at their surface that are complementary to ridges, clefts, and charged regions of a protein at the surface of the virus particle. Design a drug molecule that matches up with a viral protein, figure out how to deliver enough copies of it into a patient, and a lot of virus particles may bind to the decoys instead of infecting body cells.

Throughout this book, you will be coming across different kinds of molecular models. In each case, the model selected gives you a glimpse into the structure and function of the molecule being described.

Carbohydrates, lipids, proteins, and nucleic acids are the main biological molecules, the organic compounds that only living cells assemble under present-day conditions in nature.

Organic compounds have diverse, three-dimensional shapes and functions that start with their carbon backbone and the bonding arrangements that arise from it.

Insights into the structure of molecules ultimately help us understand how cells, and multicelled organisms, function.

OVERVIEW OF FUNCTIONAL GROUPS

As you just read in the preceding section, functional groups are lone atoms or clusters of atoms covalently bonded to carbon atoms of organic compounds. Each has specific chemical and physical properties that are consistent from one molecule to the next. Compared to hydrocarbon regions, they are more reactive. Important features of carbohydrates, lipids, proteins, and nucleic acids arise from the number, kind, and arrangement of functional groups, such as those shown in Figure 3.5.

For example, sugars in your diet belong to a class of organic compounds called **alcohols**, which have one or more *hydroxyl* groups (—OH). Small alcohols dissolve fast because water molecules hydrogen-bond with these groups. Larger ones don't; they have long hydrocarbon

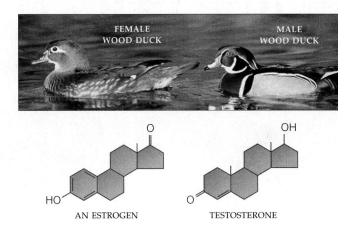

FEMALE WOOD DUCK   MALE WOOD DUCK

AN ESTROGEN   TESTOSTERONE

Figure 3.6 Observable differences in traits between the male and female wood duck (*Aix sponsa*). Two sex hormones govern the development of feather color and other traits that help males and females recognize each other and so influence reproductive success. Both hormones—testosterone and one of the estrogens—have the same carbon ring structure. They differ in the position of functional groups attached to the ring.

chains. As you will see shortly, enzyme action can split molecules and join them together at hydroxyl groups. By contrast, hydrocarbons are water insoluble. Chains of them occur in fatty acids, which is why lipids with fatty acid tails resist dissolving in water.

Carbonyl groups are highly reactive and prone to electron transfers. They are building blocks of fats and carbohydrates. *Carboxyl* groups are present in amino acids, fatty acids, and other important molecules. The *phosphate* group has oxygen atoms that form covalent bonds. It dictates ATP's energy-carrying function, and it also combines with sugars to form the backbones of DNA and RNA. As you will see, the *sulfhydryl* group, a component of the amino acid cysteine, helps stabilize the structure of many proteins.

How much can one functional group do? Consider a seemingly minor difference in the functional groups of two structurally similar sex hormones (Figure 3.6). Early on, an embryo of a wood duck, human, or any other vertebrate is neither male nor female; it just has a set of tubes and ducts that can develop either way. If an embryo starts making testosterone, that hormone will drive development of the tubes and ducts into male sex organs and, later, govern the emergence of male traits such as feather color or hairiness. Female sex organs form only in the *absence* of testosterone. Estrogens will guide the development of distinctively female traits.

Functional groups covalently bonded to carbon backbones add enormously to the structural and functional diversity of organic compounds, cells, and multicelled organisms.

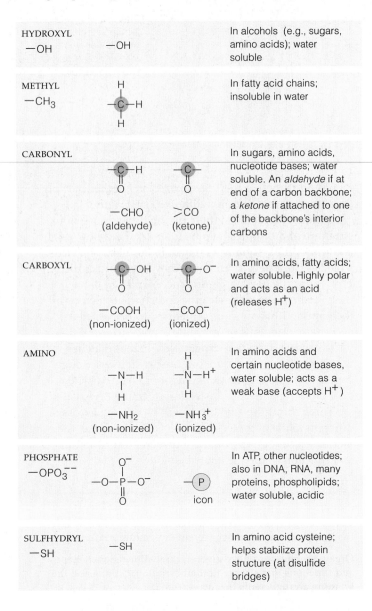

HYDROXYL —OH	—OH	In alcohols (e.g., sugars, amino acids); water soluble
METHYL —CH₃	H–C(H)(H)–	In fatty acid chains; insoluble in water
CARBONYL	—CHO (aldehyde) / >CO (ketone)	In sugars, amino acids, nucleotide bases; water soluble. An *aldehyde* if at end of a carbon backbone; a *ketone* if attached to one of the backbone's interior carbons
CARBOXYL	—COOH (non-ionized) / —COO⁻ (ionized)	In amino acids, fatty acids; water soluble. Highly polar and acts as an acid (releases H⁺)
AMINO	—NH₂ (non-ionized) / —NH₃⁺ (ionized)	In amino acids and certain nucleotide bases, water soluble; acts as a weak base (accepts H⁺)
PHOSPHATE —OPO₃⁻⁻	—O—P(O⁻)(=O)—O⁻ / (P) icon	In ATP, other nucleotides; also in DNA, RNA, many proteins, phospholipids; water soluble, acidic
SULFHYDRYL —SH	—SH	In amino acid cysteine; helps stabilize protein structure (at disulfide bridges)

Figure 3.5 Common functional groups in biological molecules, with examples of their occurrences.

HOW DO CELLS BUILD ORGANIC COMPOUNDS?

Four Families of Building Blocks

Think back on the trees of the world's great forests, as described in the chapter introduction. These producers take carbon (from carbon dioxide), water, and sunlight (for energy) and transform them first into small organic compounds. Simple sugars, fatty acids, amino acids, and nucleotides are the four major families of these small compounds. Each family includes many kinds of molecules that contain two to thirty-six carbon atoms, at the most.

Cells maintain and replenish their pools of small organic compounds, which collectively account for only about 10 percent of the organic material inside a cell. They continually withdraw some molecules as sources of energy. They withdraw others for use as individual subunits, or **monomers**, of larger molecules that they require for their structure and functioning. These larger molecules, called **polymers**, consist of three to millions of subunits that may or may not be identical. In turn, when large molecules are broken down, their released monomers may be used at once for energy, or they may rejoin the cellular pools as free molecules.

Five Categories of Reactions

So how do cells actually do the construction work? It will take more than one chapter to sketch out answers (and best guesses) to the question. At this point, simply become aware that reactions by which a cell assembles, rearranges, and splits apart organic compounds require more than energy inputs. They also require **enzymes**, a class of proteins that make metabolic reactions proceed much faster than they would on their own. Different kinds of enzymes mediate different reactions. In later chapters, you will come across specific examples of five categories of reactions:

1. *Functional-group transfer.* One molecule gives up a functional group, which another molecule accepts.

2. *Electron transfer.* One or more electrons stripped from one molecule are donated to another molecule.

3. *Rearrangement.* A juggling of its internal bonds converts one type of organic compound into another.

4. *Condensation.* Through covalent bonding, two molecules combine to form a larger molecule.

5. *Cleavage.* A molecule splits into two smaller ones.

To get a sense of these cell activities, think of what happens in a **condensation reaction**: Enzymes split off an —OH group from one molecule and an H atom from another, and a covalent bond forms between both at

dehydration synthesis

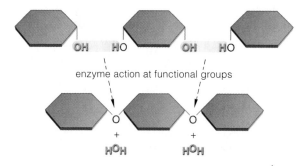

a Two condensation reactions. Enzymes remove an —OH group and H atom from two molecules, which covalently bond as a larger molecule. Two water molecules form.

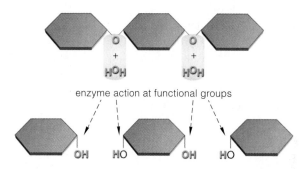

b Hydrolysis, a water-requiring cleavage reaction. Enzyme action splits a molecule into three parts, then attaches an —OH group and an H atom derived from a water molecule to each exposed site.

Figure 3.7 Examples of the types of metabolic reactions by which most biological molecules are put together, rearranged, and broken apart.

their exposed sites. The discarded atoms often form water, or H_2O (Figure 3.7a). Starch and other polymers form by way of repeated condensation reactions.

Another example: A type of cleavage reaction called **hydrolysis** is like condensation in reverse (Figure 3.7b). Enzymes split molecules at specific groups, then attach one —OH group and a hydrogen atom derived from a water molecule to the exposed sites. With hydrolysis, cells can cleave polymers into smaller molecules when these are required for building blocks or for energy.

Cells build the large molecules of life mainly from four families of small organic compounds called simple sugars, fatty acids, amino acids, and nucleotides.

Cells continually assemble, rearrange, and degrade both small and large organic compounds. They do so mainly by enzyme-mediated reactions involving the transfer of functional groups or electrons, rearrangement of internal bonds, and a combining or splitting of molecules.

CARBOHYDRATES—THE MOST ABUNDANT MOLECULES OF LIFE

Which biological molecules are most abundant? The **carbohydrates**. Most carbohydrates consist of carbon, hydrogen, and oxygen in a 1:2:1 ratio $(CH_2O)_n$. Cells use them as structural materials, transportable forms of energy, or storage forms of energy. There are three main classes, called the **monosaccharides**, **oligosaccharides**, and **polysaccharides**.

The Simple Sugars

"Saccharide" comes from a Greek word meaning sugar. A *mono*saccharide (meaning one sugar monomer) is the simplest carbohydrate. It has at least two —OH groups bonded to the carbon backbone plus an aldehyde or a ketone group. Most dissolve easily in water. Common types have a backbone of five or six carbon atoms that tends to form a ring structure when dissolved in cells or body fluids. Ribose and deoxyribose, the sugar unit of RNA and DNA, respectively, have five carbon atoms. Glucose has six (Figure 3.8*a*). Most organisms use it as their main energy source. Glucose also is a precursor

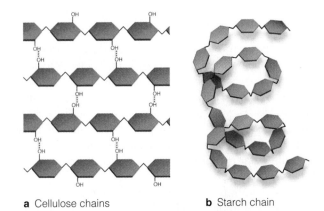

a Cellulose chains   **b** Starch chain

Figure 3.9 Bonding patterns for glucose units in cellulose and in starch. (**a**) In cellulose, bonds form between glucose chains. This pattern stabilizes the chains and allows them to become tightly bundled. (**b**) In amylose, a form of starch, bonds form between units within a chain, which coils up.

(parent molecule) of many compounds and a building block for larger carbohydrates. Vitamin C (a well-known sugar acid) and glycerol (an alcohol that has three —OH groups) are compounds made from sugar monomers.

Short-Chain Carbohydrates

Unlike the simple sugars, an *oligo*saccharide is a short chain of covalently bonded sugar monomers. (*Oligo–* means "a few.") The ones called *di*saccharides consist of only two sugar units. Lactose, sucrose, and maltose are examples. Lactose, a sugar in milk, has one glucose and one galactose unit. Sucrose, the most plentiful sugar in nature, has a glucose and a fructose unit (Figure 3.8*c*). Plants convert larger carbohydrates to sucrose, which is easily transported through leaves, stems, and roots. Table sugar is sucrose crystallized from sugarcane and sugarbeets. The backbone of some proteins and other large molecules often bears oligosaccharide side chains.

Complex Carbohydrates

The "complex" carbohydrates, or *poly*saccharides, are straight or branched chains of many sugar monomers (often hundreds or thousands) of the same or different types. Cellulose, starch, and glycogen, the most common polysaccharides, consist only of glucose—yet they have very different properties. Why? The answer starts with differences in covalent bonding patterns between their glucose units, which are joined together in chains.

In cellulose, many glucose chains stretch out side by side and hydrogen-bond to one another at —OH groups (Figure 3.9*a*). This bonding arrangement stabilizes the chains into a tightly bundled pattern that resists being digested, at least by most enzymes. Fibers of cellulose

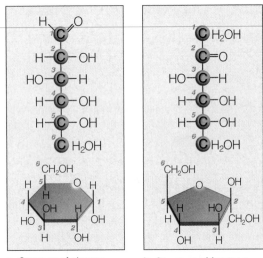

a Structure of glucose   **b** Structure of fructose

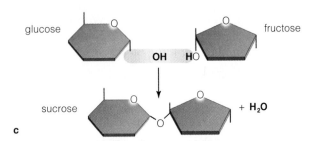

glucose   fructose

sucrose   + H_2O

c

Figure 3.8 Straight-chain and ring forms of (**a**) glucose and (**b**) fructose. For reference purposes, carbon atoms of simple sugars are numbered in sequence, starting at the end closest to the molecule's aldehyde or ketone group. (**c**) Condensation of two monosaccharides into a disaccharide.

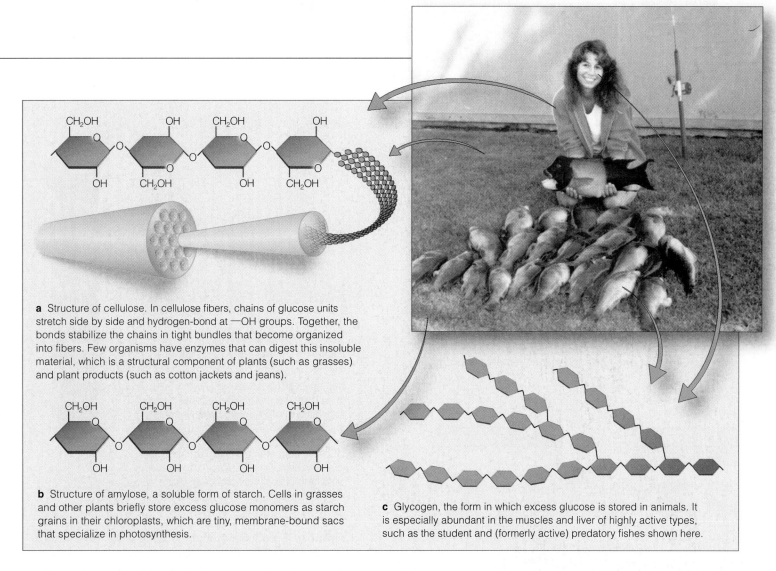

a Structure of cellulose. In cellulose fibers, chains of glucose units stretch side by side and hydrogen-bond at —OH groups. Together, the bonds stabilize the chains in tight bundles that become organized into fibers. Few organisms have enzymes that can digest this insoluble material, which is a structural component of plants (such as grasses) and plant products (such as cotton jackets and jeans).

b Structure of amylose, a soluble form of starch. Cells in grasses and other plants briefly store excess glucose monomers as starch grains in their chloroplasts, which are tiny, membrane-bound sacs that specialize in photosynthesis.

c Glycogen, the form in which excess glucose is stored in animals. It is especially abundant in the muscles and liver of highly active types, such as the student and (formerly active) predatory fishes shown here.

Figure 3.10 Molecular structure of cellulose, starch, and glycogen, and their typical locations in a few organisms. All three carbohydrates consist only of glucose units.

are a structural component of plant cell walls (Figure 3.10*a*). Like steel rods in reinforced concrete, the fibers are tough, insoluble, and resistant to weight loads and mechanical stress.

In starch, the pattern of covalent bonding puts each glucose unit at an angle relative to the next unit in line. The chain ends up coiling like a spiral staircase (Figure 3.9*b*). The coils are easily digestible. In starches that have branched chains, they are even more so. Many —OH groups project outward from the coiled chains, and this makes them readily accessible to enzymes. Plants store their photosynthetically produced sugars as large starch molecules (Figure 3.10*b*). Enzymes easily hydrolyze the starch to glucose units.

In animals, glycogen is the sugar-storage equivalent of starch in plants. Muscle and liver cells store a lot of it. When the level of sugar in blood declines, liver cells degrade glycogen, so glucose is released and enters the blood. Exercise strenuously but briefly and your muscle cells tap glycogen for a burst of energy. Figure 3.10*c* shows a few of glycogen's many branchings.

Figure 3.11 Scanning electron micrograph of a tick. Its body covering is a protective cuticle reinforced with chitin.

A different *amino* polysaccharide, chitin, is distinctive in that it has ~~nitrogen-containing~~ groups attached to its glucose monomers. Chitin is a material that strengthens external skeletons and other hard body parts of many animals, including crabs, earthworms, insects, and ticks (Figure 3.11). Chitin also is a structural material that strengthens the cell walls of many kinds of fungi.

The simple sugars (such as glucose), oligosaccharides, and polysaccharides (such as starch) are carbohydrates. Every cell requires carbohydrates as structural materials, stored forms of energy, and transportable packets of energy.

GREASY, OILY—MUST BE LIPIDS

If something is greasy or oily to the touch, you can bet it is a lipid or has lipid components. **Lipids** are nonpolar hydrocarbons. They resist dissolving in water but easily dissolve in nonpolar substances (think butter into warm cream sauce). Cells use different lipids as energy stores, as structural materials (for example, in cell membranes and surface coatings), and as signaling molecules. Let's look first at the kinds having fatty acid components—fats, phospholipids, and waxes. Then we will consider the sterols, each with a backbone of four carbon rings.

Fats and Fatty Acids

Lipids known as **fats** have one, two, or three fatty acids attached to glycerol. Each **fatty acid** has a backbone of as many as thirty-six carbon atoms, a carboxyl group (—COOH) at one end, and hydrogen atoms occupying

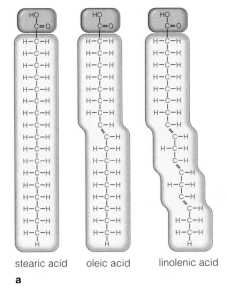

stearic acid   oleic acid   linolenic acid

a

Figure 3.12 (**a**) Structural formulas for three fatty acids. In stearic acid, the carbon backbone is fully saturated with hydrogen atoms. Oleic acid, with a double bond in its backbone, is an unsaturated fatty acid. Linolenic acid, with its three double bonds, is a polyunsaturated fatty acid. (**b**) Space-filling model for stearic acid.

b

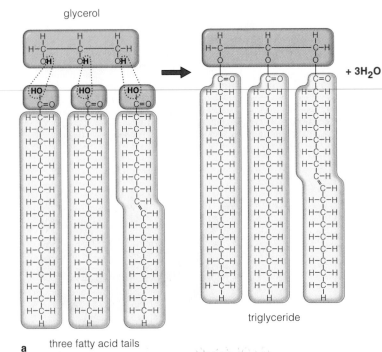

glycerol

+ 3H₂O

triglyceride

a   three fatty acid tails

b

most or all of the remaining bonding sites. Most stretch out like a flexible tail. Tails that are *unsaturated* have one or more double bonds. *Saturated* tails have single bonds only. Figure 3.12 shows examples.

Most animal fats have many saturated fatty acids, which pack together by weak interactions. They remain solid at temperatures that keep most plant fats liquid, as "vegetable oils." Similar packing interactions in plant fats aren't as stable because of rigid kinks in their fatty acid tails. That is why vegetable oils flow freely.

Butter, lard, vegetable oils, and other natural fats are mostly **triglycerides**. These "neutral" fats have three fatty acid tails attached to a glycerol unit (Figure 3.13). Triglycerides are the body's most abundant lipids and its richest energy source. Gram for gram, they yield more than twice as much energy when degraded, compared to complex carbohydrates such as starches. Quantities of triglycerides are stored as droplets in the cells of body fat (adipose tissue) in every vertebrate. A thick layer of triglycerides under the skin helps penguins and some other animals resist cold (Figure 3.13b). It insulates the body against extreme temperatures of icy habitats.

Figure 3.13 (**a**) Condensation of fatty acids and a glycerol molecule into a triglyceride. (**b**) Triglyceride-protected penguins taking the plunge.

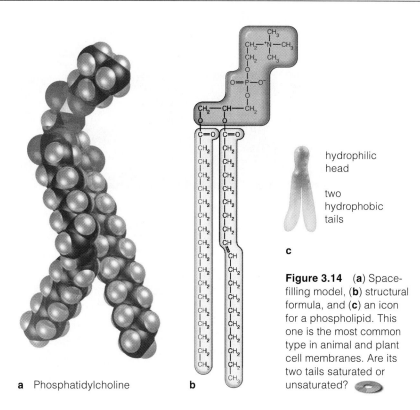

a Phosphatidylcholine   **b**

hydrophilic
head

two
hydrophobic
tails

c

Figure 3.14 (**a**) Space-filling model, (**b**) structural formula, and (**c**) an icon for a phospholipid. This one is the most common type in animal and plant cell membranes. Are its two tails saturated or unsaturated?

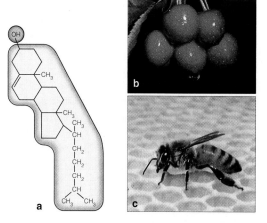

a   **b**   **c**

Figure 3.15 (**a**) Structural formula for cholesterol, the major sterol of animal tissues. Your liver produces enough cholesterol for your body. A fat-rich diet may result in high cholesterol levels in the blood, which may contribute to the clogging of small arteries. Two waxy materials: (**b**) The waxy, water-repelling cuticle of cherries. (**c**) Honeycomb, a structural material of a firm, water-repellent, waxy secretion called beeswax.

One fatty acid is a precursor of eicosanoids: local signaling molecules that include prostaglandins. As you will see, they have roles in contraction, defense, message flow in the nervous system, and other key processes.

Phospholipids

A **phospholipid** has a glycerol backbone, two fatty acid tails, and a hydrophilic "head" with a phosphate group and another polar group (Figure 3.14). Phospholipids are the main materials of cell membranes, which have two layers of lipids. Heads of one layer are dissolved in the cell's fluid interior, and heads of the other layer are dissolved in the surroundings. Sandwiched between the two are all the fatty acid tails, which are hydrophobic.

Sterols and Their Derivatives

Sterols are among the many lipids with no fatty acids. Sterols differ in the number, position, and type of their functional groups, but all have a rigid backbone of four fused-together carbon rings, as below. Eukaryotic cells have sterols in their membranes. Cholesterol is the most common type in tissues of animals. Figure 3.15a shows its structure. Also, cholesterol gets remodeled into compounds including vitamin D (necessary for good bones and teeth), steroids, and bile salts. Steroids include sex hormones of the sort shown

sterol backbone

in Figure 3.6. These hormones govern the formation of gametes and development of secondary sexual traits, such as feather color and hair distribution. Bile salts have roles in the digestion of fats in the small intestine.

Waxes

Waxes have long-chain fatty acids tightly packed and linked to long-chain alcohols or carbon rings. All have a firm consistency; all repel water. The cuticle covering aboveground plant parts consists mostly of waxes and another lipid, cutin (Figure 3.15b). It restricts water loss and fends off certain parasites. Waxy secretions protect, lubricate, and impart pliability to skin and to hair. Birds secrete waxes, fatty acids, and fats from preen glands to waterproof feathers. Bees use beeswax to construct honeycomb, which houses not only honey but new bee generations (Figure 3.15c). Tiny organisms drifting in the seas even use waxes as their main energy source.

Being largely hydrocarbon, lipids can dissolve in nonpolar substances, but they resist dissolving in water.

Triglycerides, or neutral fats, have a glycerol head and three fatty acid tails. They are the body's major energy reservoirs. Phospholipids are the main components of cell membranes.

Sterols such as cholesterol serve as membrane components and precursors of steroid hormones and other compounds. Waxes are firm yet pliable components of water-repelling and lubricating substances.

A STRING OF AMINO ACIDS: PROTEIN PRIMARY STRUCTURE

Of all large biological molecules, the **proteins** are the most diverse. The ones called enzymes make reactions happen faster than they would on their own. Structural proteins are the stuff of spider webs, butterfly wings, feathers, bone, and many other body parts and products. Some proteins transport substances across membranes and through fluids. Nutritious proteins abound in eggs and seeds. Protein hormones are signals for change in cell activities. Many proteins act as weapons against pathogens. Amazingly, cells build thousands of diverse proteins from only twenty kinds of amino acids!

Amino Acid Structure

An **amino acid** is a small organic compound consisting of an amino group (which is basic), a carboxyl group (which is an acid), a hydrogen atom, and one or more atoms called its R group. Figures 3.16 and 3.17 show the structure of some amino acids you will encounter later on in the book. Generally, their components are all bonded covalently to the same carbon atom.

Polypeptide Chain Formation

When a cell synthesizes a protein, enzymes join amino acids, one after the other, by peptide bonds. This type of covalent bond forms between the amino group ($-NH_3^+$) of one amino acid and the carboxyl group ($-COO^-$) of the next amino acid in line.

When peptide bonds join three or more amino acids, we have a **polypeptide chain**. In such chains, the carbon backbone has nitrogen atoms positioned in this regular pattern: $-N-C-C-N-C-C-$.

For each kind of protein, different amino acids are selected one at a time from the twenty kinds available.

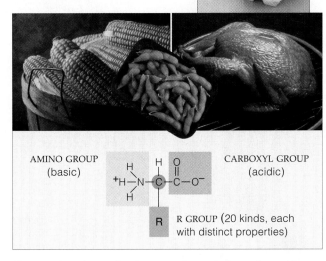

Figure 3.16 Generalized structural formula for amino acids, along with soybeans and a few other dietary sources of these small organic compounds.

Instructions encoded in the cell's DNA determine the selection. Overall, the protein's amino acid sequence is unique and is its *primary* structure (Figure 3.18).

Thousands of different kinds of proteins occur in nature. Many are *fibrous* proteins that have polypeptide chains arranged as strands or sheets. They contribute to a cell's shape and organization. *Globular* proteins have one or more chains folded in compact, rounded shapes. Most enzymes are globular proteins. So are proteins that help cells, and cell parts move.

Regardless of the type of protein, its shape and its function arise from the primary structure—that is, from information encoded in its amino acid sequence. That

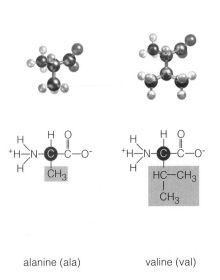
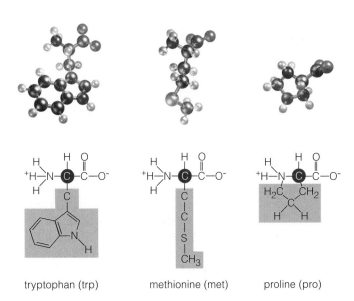

Figure 3.17 Structural formulas and ball-and-stick models for a sampling of the twenty common amino acids in cells. Appendix VI shows all twenty.

Green boxes indicate R groups, which are side chains that include functional groups. Each type of side chain makes a contribution to the chemical and physical properties of each amino acid.

alanine (ala)   valine (val)   tryptophan (trp)   methionine (met)   proline (pro)

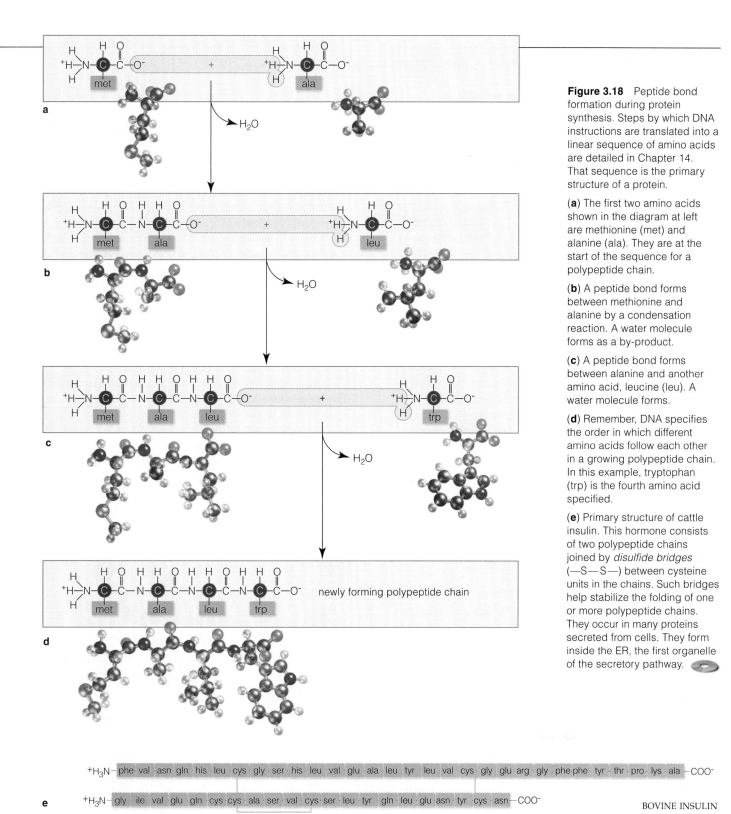

Figure 3.18 Peptide bond formation during protein synthesis. Steps by which DNA instructions are translated into a linear sequence of amino acids are detailed in Chapter 14. That sequence is the primary structure of a protein.

(**a**) The first two amino acids shown in the diagram at left are methionine (met) and alanine (ala). They are at the start of the sequence for a polypeptide chain.

(**b**) A peptide bond forms between methionine and alanine by a condensation reaction. A water molecule forms as a by-product.

(**c**) A peptide bond forms between alanine and another amino acid, leucine (leu). A water molecule forms.

(**d**) Remember, DNA specifies the order in which different amino acids follow each other in a growing polypeptide chain. In this example, tryptophan (trp) is the fourth amino acid specified.

(**e**) Primary structure of cattle insulin. This hormone consists of two polypeptide chains joined by *disulfide bridges* (—S—S—) between cysteine units in the chains. Such bridges help stabilize the folding of one or more polypeptide chains. They occur in many proteins secreted from cells. They form inside the ER, the first organelle of the secretory pathway.

newly forming polypeptide chain

BOVINE INSULIN

sequence influences which parts of a polypeptide chain will coil, bend, or interact with other chains. The type and arrangement of atoms in the coiled, stretched out, and folded parts dictate whether the protein will function as, say, an enzyme, a transporter, a receptor, or even an unlucky target for a bacterium or virus.

Each protein consists of one or more polypeptide chains of amino acids. The amino acid sequence (which kind of amino acid follows another in the chain) is unique for each kind of protein and gives rise to its unique structure, chemical behavior, and function.

HOW DOES A PROTEIN'S FINAL STRUCTURE EMERGE?

Now that you have a sense of how amino acids make a polypeptide chain, take a look at examples of the ways in which the sequence dictates a protein's final shape.

Second and Third Levels of Protein Structure

Picture the primary structure as a set of rigid playing cards joined by sticks (covalent bonds) that can swivel a bit. Each "card" is a peptide group, and some of its atoms favor bond formation with nearby atoms. Some swivels favor hydrogen bonding between amino acids along the length of a polypeptide chain. Others put R groups in positions that invite bond formation.

Certain sequences of amino acids favor a pattern of bonding that causes part of the polypeptide chain to coil and twist into a helix, a bit like a spiral staircase (Figure 3.19a). Other sequences give rise to sheetlike regions (Figure 3.19b) and to looped regions that differ in length and how they twist. These outcomes are the dominant features of a protein's *secondary* structure.

Adjacent coils or strands in a chain usually end up close together in the protein, with their side groups packed together. These regions now fold up into one to dozens of compact domains. By definition, a **domain** is a polypeptide chain, or part of it, that has become self-organized as a structurally stable, functional unit. We call domain formation the protein's *tertiary* structure.

Figure 3.19b shows adjacent, strandlike regions of one kind of polypeptide chain with looped connecting regions. Hydrogen bonds and other interactions bring about the array and help stabilize it. In this example, the regions get folded into a barrel-shaped domain.

Many proteins have coiled domains around one or more tightly packed, barrel-shaped domains. Inside the barrel's interior are hydrophobic R groups that exclude water and shield R groups of looped regions, which serve in enzyme action and other tasks. These proteins differ from ones that have coils flanking a sheetlike domain (Figure 3.20a). The sheetlike proteins are quite diverse in structure. They carry out their specific tasks at some type of crevice at the edge of the sheet.

Is the amino acid sequence for each of the world's diverse proteins totally unique? No. Some combinations of amino acids recur in most proteins. Those that give rise to barrels and helical coils are common, especially among enzymes and transporters.

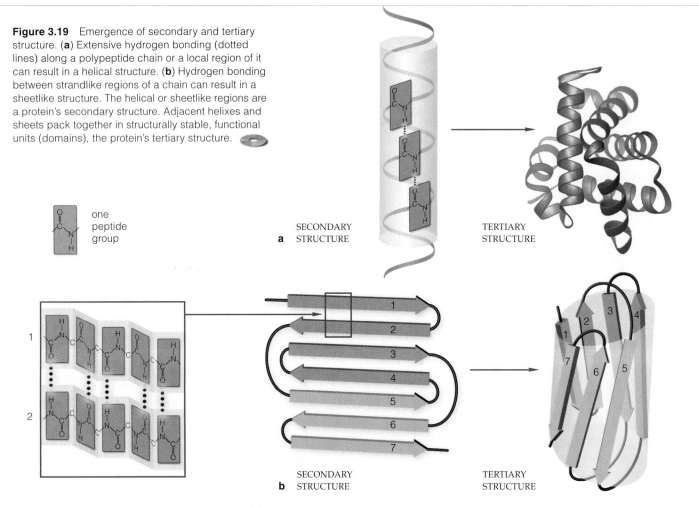

Figure 3.19 Emergence of secondary and tertiary structure. (**a**) Extensive hydrogen bonding (dotted lines) along a polypeptide chain or a local region of it can result in a helical structure. (**b**) Hydrogen bonding between strandlike regions of a chain can result in a sheetlike structure. The helical or sheetlike regions are a protein's secondary structure. Adjacent helixes and sheets pack together in structurally stable, functional units (domains), the protein's tertiary structure.

one peptide group

a SECONDARY STRUCTURE

TERTIARY STRUCTURE

b SECONDARY STRUCTURE

TERTIARY STRUCTURE

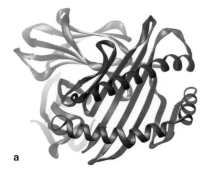

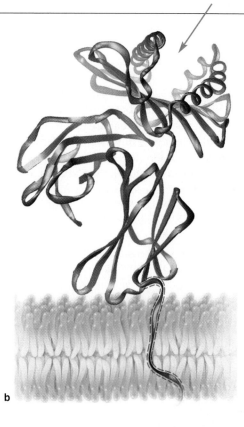

where the molecule binds and displays "enemies" (*arrow*)

one of the chains spans the plasma membrane and anchors the molecule

b

Figure 3.20 A protein that helps your body defend itself against bacteria and other foreign agents. (**a**) HLA-A2 quaternary structure, which consists of two polypeptide chains that are like jaws. (**b**) The presumed location of this protein relative to the outer surface of the plasma membrane.

Fourth Level of Protein Structure

Many final proteins consist of two or more polypeptide chains held together by hydrogen bonds and sometimes covalent bonds. Such proteins have *quaternary* structure.

Proteins called **HLAs** (human leukocyte antigens) are like this. All body cells have HLAs at their surface, but white blood cells, or leukocytes, bristle with them. HLAs function in recognizing "self"—the body's own cells—from "nonself." They bind and, later on, display fragments of nonself proteins, thus making the body's defending cells aware of an invader. We know of more than 400 slightly different versions of HLAs. Humans differ in terms of which versions occur on their cells.

Figure 3.20 shows the quaternary structure of the version designated HLA-A2. Notice, in Figure 3.20*a*, how it consists of two polypeptide chains. Part of one chain is a sheetlike domain with two helixes arching above it. These helixes are like jaws. They are able to grip a bit of a captured and digested "enemy agent" and offer it up to the body's defending cells. Receptors on the defending cells recognize and latch on to HLA molecules that are displaying foreign fragments. Such bindings trigger an immune response.

Now look at Figure 3.20*b*. Another part of the same chain continues across the plasma membrane of a body cell and ties in with fibers in the cytoplasm to anchor the whole molecule.

Protein structure doesn't stop here. Short, linear, or branched oligosaccharides get attached to many new polypeptide chains. Most proteins at the cell surface and secreted from cells are such *glyco*proteins. Lipids get attached to other proteins. One such *lipo*protein forms as proteins in blood combine with cholesterol, triglycerides, and phospholipids absorbed from the gut after a meal. Such attachments add to protein diversity.

Denaturation—How to Undo the Structure

Breaking weak bonds of a protein or any other large molecule disrupts its three-dimensional shape, an event called **denaturation**. For example, weak hydrogen bonds are sensitive to increases and decreases in temperature and pH. If the temperature or pH exceeds a protein's range of tolerance, its polypeptide chains will unwind or change shape, and the protein will lose its function. Consider the protein albumin, concentrated in the "egg white" of uncooked chicken eggs. When you cook eggs, the heat does not disrupt the strong covalent bonds of albumin's primary structure. But it destroys weaker bonds contributing to the three-dimensional shape. For some proteins, denaturation might be reversed when normal conditions are restored—but albumin isn't one of them. There is no way to uncook a cooked egg.

A protein's primary structure is the sequence of different kinds of amino acids along a polypeptide chain.

A protein has secondary structure. Local regions along the length of a polypeptide chain twist and fold into helical coils, sheetlike arrays of strands, and loops.

A protein may have tertiary structure. A polypeptide chain or parts of it becomes organized into domains: structurally stable, compact units that may have distinct functions.

A protein with quaternary structure consists of two or more polypeptide chains joined by hydrogen bonds. Covalent bonds, disulfide bridges, and other interactions stabilize it.

WHY IS PROTEIN STRUCTURE SO IMPORTANT?

Just One Wrong Amino Acid...

Earlier sections showed you elegant models for protein structure. Cells usually are good at construction tasks and turn out protein molecules that are just what their DNA specifies. But sometimes mistakes are made in the assembly process. Sometimes nature sabotages the specifications, as by ultraviolet radiation attacks. If even one amino acid is replaced by the wrong kind, the substitution may have far-reaching consequences.

Consider such a change in hemoglobin's structure. This protein has a quaternary structure; it consists of four polypeptide chains called globin. During protein synthesis, each globin molecule becomes tightly folded. The folding produces a small pocket that is chemically attractive to an iron-containing heme group, the part of hemoglobin that transports oxygen (Figure 3.21a). As you read this, each mature red blood cell in your body is transporting about a billion oxygen molecules, all bound to 250 million or so hemoglobin molecules.

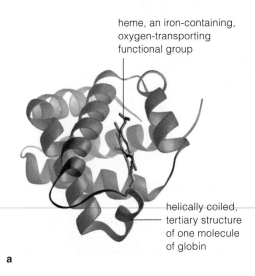

a

heme, an iron-containing, oxygen-transporting functional group

helically coiled, tertiary structure of one molecule of globin

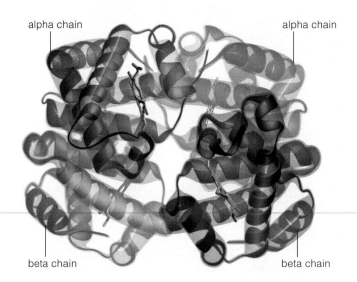

alpha chain   alpha chain

b   beta chain   beta chain

Figure 3.21 *Above:* (**a**) A globin molecule. This coiled polypeptide chain associates with heme, an iron-containing functional group that strongly attracts oxygen.

(**b**) Quaternary structure of hemoglobin, an oxygen-transporting pigment in red blood cells. It consists of four globin molecules and four heme groups. Two of the chains, designated alpha, differ a bit from the other two (beta chains) in amino acid sequence. To help you visualize all four chains, helical regions are enclosed in transparent tubes.

Right: (**c**) Normal sequence of amino acids at the start of a beta chain for hemoglobin.

(**d**) A single amino acid substitution results in the abnormal beta chain found in HbS molecules. During protein synthesis, valine was added instead of glutamate at the sixth position of the growing polypeptide chain. The ball and stick models for the original amino acid and the substitution are shown in the diagrams.

Glutamate has an overall negative charge; valine has no net charge. This difference in charge gives rise to a water-repelling, sticky patch on the HbS molecule.

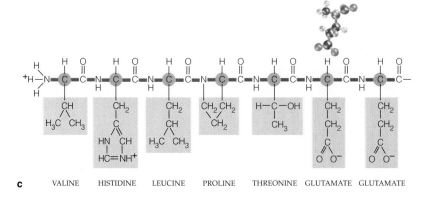

c   VALINE   HISTIDINE   LEUCINE   PROLINE   THREONINE   GLUTAMATE   GLUTAMATE

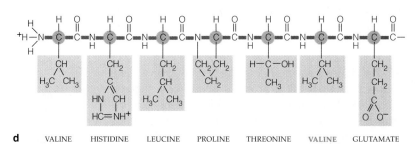

d   VALINE   HISTIDINE   LEUCINE   PROLINE   THREONINE   VALINE   GLUTAMATE

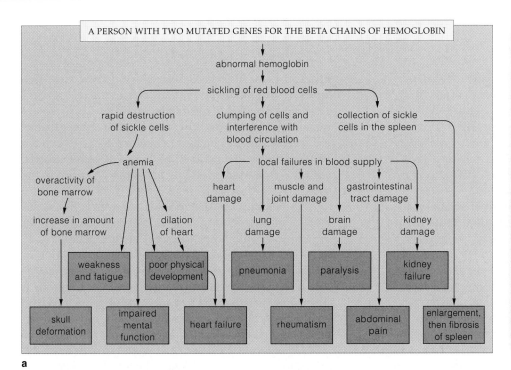

A PERSON WITH TWO MUTATED GENES FOR THE BETA CHAINS OF HEMOGLOBIN

abnormal hemoglobin

sickling of red blood cells

rapid destruction of sickle cells | clumping of cells and interference with blood circulation | collection of sickle cells in the spleen

anemia

local failures in blood supply

overactivity of bone marrow

heart damage | muscle and joint damage | gastrointestinal tract damage

increase in amount of bone marrow | dilation of heart | lung damage | brain damage | kidney damage

weakness and fatigue | poor physical development | pneumonia | paralysis | kidney failure

skull deformation | impaired mental function | heart failure | rheumatism | abdominal pain | enlargement, then fibrosis of spleen

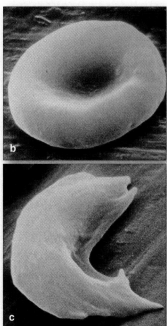

Figure 3.22 (**a**) Diagram tracking the wide range of symptoms characteristic of sickle-cell anemia, a genetic disorder. (**b**) Red blood cell from a person affected by the disorder. This scanning electron micrograph shows the surface appearance of a red blood cell from an affected individual when the blood is adequately oxygenated. (**c**) This scanning electron micrograph shows the sickle shape of a red blood cell carrying HbS when the concentration of oxygen in blood is low.

The globin in hemoglobin has two slightly different forms: alpha and beta (Figure 3.21*b*). There is a type of mutation that changes DNA's instructions for the beta chain. It specifies valine instead of glutamate at the sixth position shown in Figure 3.21*c*. Glutamate has an overall negative charge; valine, however, has no net charge. The difference puts a "sticky," water-repelling patch on beta chains. Hemoglobin that incorporates the altered chain is called HbS (instead of HbA).

Sickle-Shaped Cells and a Serious Disorder

People inherit two genes for the beta chain, one from each parent. (A "gene" is a unit of information about a heritable trait.) If one is normal but the other calls for HbS, people usually do not run into trouble. Their red blood cells can still make enough normal hemoglobin molecules to compensate for the abnormal ones. But in someone who inherited two mutated genes, red blood cells can only make HbS—the outcome being a severe genetic disorder called *sickle-cell anemia*.

Humans, like most organisms, must take in a lot of oxygen; cells use it to get energy, by aerobic respiration. Oxygen flows into our lungs, then diffuses into blood. There it binds to hemoglobin, which travels through arteries, arterioles, then the small-diameter, thin-walled capillaries that thread through all tissues. Most of the oxygen diffuses into tissues, then cells. Cellular uptake lowers its concentration in blood. Its decline is greatest at high altitudes and during strenuous activity.

Where oxygen concentrations are lowest, abnormal hemoglobin molecules are attracted to one another's sticky patches. They get stuck in rod-shaped clumps. The rods distort red blood cells into a sickle shape, as in Figure 3.22. (A sickle is a farm tool that has a long, crescent-shaped blade.) Thus distorted, cells rupture easily, and the remnants clog and rupture capillaries. Their rapid destruction leads to oxygen-starved cells in affected tissues. Clumping also causes local failures in the circulatory system's capacity to deliver oxygen and carry away carbon dioxide and other metabolic wastes. In time, ongoing expression of the mutant gene may damage tissues and organs throughout the body.

Hemoglobin, hormones, enzymes, receptors—these are the kinds of proteins necessary for your survival. Understand the structure and functions of proteins in general, and you are on your way to understanding life in its richly normal and abnormal expressions.

The molecular structure of proteins dictates how they can perform their functions. How proteins function dictates the quality of life, sometimes life or death itself.

NUCLEOTIDES AND NUCLEIC ACIDS

The Diverse Roles of Nucleotides

The small organic compounds called **nucleotides** each consist of a sugar, at least one phosphate group, and a nitrogen-containing base. The sugar is either ribose or deoxyribose. Although both sugars have a five-carbon ring structure, ribose has an oxygen atom attached to carbon 2 and deoxyribose does not. The bases have a single or double carbon ring structure. Cells put these small compounds to different uses.

The nucleotide **ATP** (adenosine triphosphate) has a string of three phosphate groups attached to its sugar component (Figure 3.23a). ATP can readily transfer a phosphate group to many other molecules inside cells, which makes the acceptor molecules energized enough to enter a reaction. ATP's phosphate-group transfers are one of the most important aspects of metabolism.

Other nucleotides have different metabolic roles. Certain kinds are subunits of **coenzymes**, or enzyme helpers, which accept hydrogen atoms and electrons stripped from molecules at one reaction site and then transfer them to different sites in the cell. Figure 3.23b shows the structure of NAD+ (short for nicotinamide adenine dinucleotide), a player in aerobic respiration. FAD (flavin adenine dinucleotide) is another kind.

Still other nucleotides serve as chemical messengers between cells and between one part of the cytoplasm and another. Later on in the book, you will be reading about one of these messengers, cAMP (cyclic adenosine monophosphate).

Last but not least, nucleotides also serve as building blocks for the larger molecules called nucleic acids.

Regarding DNA and the RNAs

Besides performing tasks for metabolic reactions, five kinds of nucleotides have vital roles in the storage and retrieval of heritable information in all cells. They are monomers for single- and double-stranded molecules classified as **nucleic acids**. In such strands, a covalent bond connects the sugar component of one nucleotide with the phosphate group of the next nucleotide in the sequence (Figure 3.24).

All cells start out life and then maintain themselves by way of instructions they inherited in some number of double-stranded molecules of deoxyribonucleic acid, or **DNA**. This nucleic acid consists of four kinds of nucleotides. Figure 3.24 gives their structural formulas. As you can see, the four differ only in their component base, which is adenine, guanine, thymine, or cytosine.

Figure 3.25 shows how hydrogen bonds between bases join the two strands together along the length of a DNA molecule. Think of each "base pair" as one rung of a ladder, and the two sugar–phosphate backbones as the ladder's two posts. The ladder twists and turns in a regular pattern, forming a double helix.

The sequence of bases in DNA encodes heritable information about how to synthesize all of the proteins that give each new cell the potential to grow, maintain itself, and reproduce. The particular bases in at least some parts of the sequence are unique to each species.

Like DNA, the **RNAs** (ribonucleic acids) consist of four kinds of nucleotide monomers. Unlike DNA, the bases are adenine, guanine, cytosine, and uracil. Also unlike DNA, RNA molecules are usually single strands

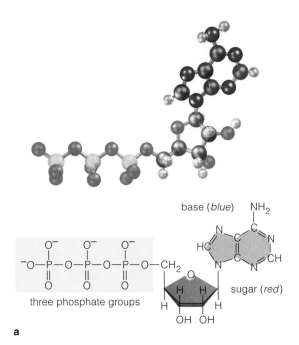

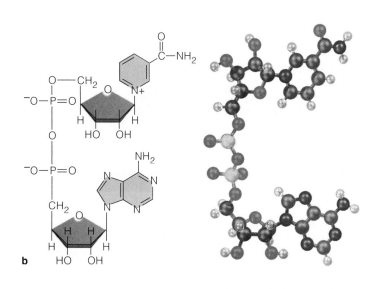

Figure 3.23 Ball-and-stick models and structural formulas for (**a**) ATP and (**b**) NAD+. Both nucleotides have central roles in cell metabolism.

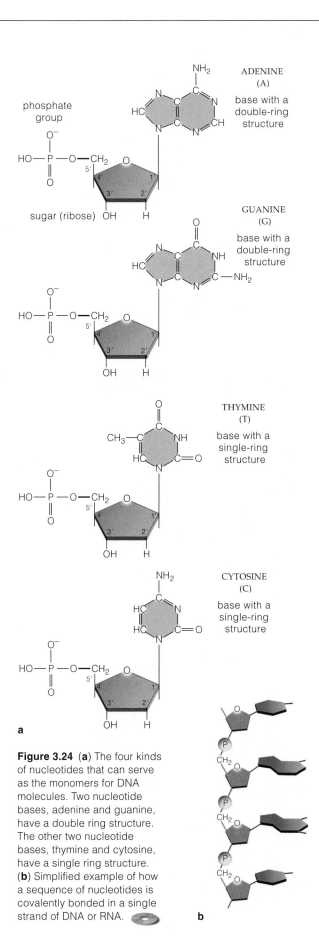

ADENINE
(A)

base with a
double-ring
structure

phosphate
group

sugar (ribose)

GUANINE
(G)

base with a
double-ring
structure

THYMINE
(T)

base with a
single-ring
structure

CYTOSINE
(C)

base with a
single-ring
structure

a

Figure 3.24 (**a**) The four kinds of nucleotides that can serve as the monomers for DNA molecules. Two nucleotide bases, adenine and guanine, have a double ring structure. The other two nucleotide bases, thymine and cytosine, have a single ring structure. (**b**) Simplified example of how a sequence of nucleotides is covalently bonded in a single strand of DNA or RNA.

b

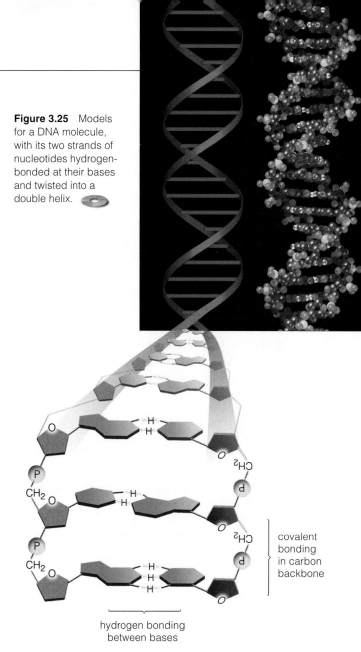

Figure 3.25 Models for a DNA molecule, with its two strands of nucleotides hydrogen-bonded at their bases and twisted into a double helix.

covalent bonding in carbon backbone

hydrogen bonding between bases

of nucleotides. Certain RNAs encode protein-building instructions, and act as a messenger to help translate the code into proteins. Other RNAs carry out the translation. Ever since cells first appeared on Earth, RNAs have been key to the processes by which genetic information is used to build proteins.

The small organic compounds called nucleotides serve as energy carriers, chemical messengers, and subunits for coenzymes and for nucleic acids. The nucleic acids consist of nucleotides joined one after another by covalent bonds.

The nucleic acid DNA consists of two nucleotide strands, joined by hydrogen bonds, and twisted as a double helix. Encoded in its nucleotide sequence is heritable information about all proteins a cell requires to survive and reproduce.

Different RNAs are single-stranded nucleic acids with roles in the processes by which a cell retrieves and uses genetic information in DNA to build proteins.

SUMMARY

Gold indicates text section

1. Organic compounds consist of carbon and at least one hydrogen atom. Their carbon atoms are often bonded covalently to form a linear or ring-shaped backbone, and functional groups are attached to the backbone. *3.1, 3.2*

2. In nature, only living cells assemble large organic compounds called biological molecules (Table 3.1). Cells build these complex carbohydrates, lipids, proteins, and nucleic acids from simple sugars, fatty acids, amino acids, nucleotides, and other small organic compounds. *3.1*

3. Cells assemble, rearrange, and break down organic compounds by different enzyme-mediated reactions, as when they transfer functional groups or electrons. *3.3*

4. Cells use simple sugars and oligosaccharides for energy and building blocks. They use complex carbohydrates as structural materials and as energy storage forms. They use the lipids (especially triglycerides) as energy storage forms, as structural components of cell membranes, and as precursors of other compounds. *3.4–3.5*

5. Proteins are the most diverse organic compounds. They function in enzyme activity, structural support, signaling, shape changes, transport, movements, and defense against disease. Nucleotides are energy carriers (especially ATP), coenzymes, and subunits for strands of the nucleic acids DNA and RNA, which interact and form the basis of inheritance and reproduction. *3.6–3.8*

6. An amino acid sequence, unique for each kind of polypeptide chain, is the start of each protein's unique structure, chemical behavior, and function. *3.6–3.9*

a. Local helical, strandlike, and looped regions form along the chain and are a protein's secondary structure. The local regions self-organize into a tertiary structure: compact, structurally stable, functional units (domains).

b. Many proteins have quaternary structure. Two or more polypeptide chains are joined and stabilized by hydrogen bonds and other interactions. *3.9*

Table 3.1 *Summary of the Main Organic Compounds in Living Things*

Category	Main Subcategories	Some Examples and Their Functions	
CARBOHYDRATES	**Monosaccharides** (simple sugars)	Glucose	Energy source
. . . contain an aldehyde or a ketone group, and one or more hydroxyl groups	**Oligosaccharides**	Sucrose (a disaccharide)	Most common form of sugar; the form transported through plants
	Polysaccharides (complex carbohydrates)	Starch, glycogen	Energy storage
		Cellulose	Structural roles
LIPIDS	**Lipids with fatty acids**		
. . . are mainly hydrocarbon; generally do not dissolve in water but do dissolve in nonpolar substances, such as other lipids	*Glycerides:* Glycerol backbone with one, two, or three fatty acid tails	Fats (e.g., butter), oils (e.g., corn oil)	Energy storage
	Phospholipids: Glycerol backbone, phosphate group, one other polar group, and (often) two fatty acids	Phosphatidylcholine	Key component of cell membranes
	Waxes: Alcohol with long-chain fatty acid tails	Waxes in cutin	Conservation of water in plants
	Lipids with no fatty acids		
	Sterols: four carbon rings; the number, position, and type of functional groups differ among sterols	Cholesterol	Component of animal cell membranes; precursor of many steroids and vitamin D
PROTEINS	**Fibrous proteins**		
. . . are one or more polypeptide chains, each with as many as several thousand covalently linked amino acids	Long strands or sheets of polypeptide chains; often tough, water-insoluble	Keratin	Structural component of hair, nails
		Collagen	Structural component of bone
	Globular proteins		
	One or more polypeptide chains folded into globular shapes; many roles in cell activities	Enzymes	Great increase in rates of reactions
		Hemoglobin	Oxygen transport
		Insulin	Control of glucose metabolism
		Antibodies	Tissue defense
NUCLEIC ACIDS (AND NUCLEOTIDES)	**Adenosine phosphates**	ATP	Energy carrier
		cAMP (Section 36.2)	Messenger in hormone regulation
. . . are chains of units (or individual units) that each consist of a five-carbon sugar, phosphate, and a nitrogen-containing base	**Nucleotide coenzymes**	NAD^+, $NADP^+$, FAD	Transfer of electrons, protons (H^+), from one reaction site to another
	Nucleic acids		
	Chains of thousands to millions of nucleotides	DNA, RNAs	Storage, transmission, translation of genetic information

Review Questions

1. Define organic compound. Name the type of chemical bond that predominates in the backbone of such a compound. *3.1*

2. Define hydrocarbon. Also define functional group. *3.1, 3.2*

3. Name the molecules of life and the families of small organic compounds from which they are built. Do they break apart most easily at their hydrocarbon portion or at functional groups? *3.2*

4. Define condensation reaction and hydrolysis. *3.3*

5. Select a carbohydrate, lipid, protein, or nucleic acid. How do its functional groups and bonds between carbon atoms in its backbone contribute to its final shape and function? *3.4–3.9*

6. Which item listed includes all of the other items listed? *3.5*
   a. triglyceride   c. wax   e. lipid
   b. fatty acid   d. sterol   f. phospholipid

7. Explain how hemoglobin's three-dimensional shape arises, starting with the primary structure of its four chains. *3.9*

Self-Quiz   ANSWERS IN APPENDIX III

1. Each carbon atom can share pairs of electrons with as many as _____ other atoms.
   a. one   b. two   c. three   d. four

2. Hydrolysis is a (an) _____ reaction.
   a. functional group transfer   d. condensation
   b. electron transfer   e. cleavage
   c. rearrangement   f. both b and d

3. _____ is a simple sugar (monosaccharide).
   a. Glucose   c. Ribose   e. both a and b
   b. Sucrose   d. Chitin   f. both a and c

4. In unsaturated fats, fatty acid tails have one or more _____ .
   a. single covalent bonds   b. double covalent bonds

5. _____ are to proteins as _____ are to nucleic acids.
   a. Sugars; lipids   c. Amino acids; hydrogen bonds
   b. Sugars; proteins   d. Amino acids; nucleotides

6. A denatured protein or DNA molecule has lost its _____ .
   a. hydrogen bonds   c. function
   b. shape   d. all of the above

7. Nucleotides occur in _____ .
   a. ATP   b. DNA   c. RNA   d. all are correct

8. Match each molecule with the most suitable description.
   _____ long sequence of amino acids   a. carbohydrate
   _____ the main energy carrier   b. phospholipid
   _____ glycerol, fatty acids, phosphate   c. protein
   _____ two strands of nucleotides   d. DNA
   _____ one or more sugar monomers   e. ATP

Critical Thinking

1. About 2,000 years ago, Plutarch suspected that the oracle at Delphi was high on a sweet-smelling gas when she made her prophecies (Figure 3.26). Recently, geologists discovered two intersecting faults beneath her temple. They analyzed sediments under the temple's foundation and found trapped molecules of methane and ethane. They also found these gases and another one, the sweet-smelling ethylene, bubbling in a natural spring near the temple. Apparently the oracle sniffed gases from oil-rich deposits, which collected in the temple's sunken floor after escaping through fissures that opened when the faults slipped.

Figure 3.26 Temple of Apollo at Delphi. To ancient Greeks, prophecies made by the temple's oracle were the voice of Apollo. A geologist now points out that the oracle made her prophecies while in a hydrocarbon-induced trance.

Reflect on the assumption of cause and effect (Section 1.5). Do you find it irritating or enlightening that science has given us a natural explanation for these "supernatural" prophecies?

2. It seems there are "good" and "bad" unsaturated fats. The double bonds of both put a bend in their fatty acid tails. But the bend in *cis* fatty acids keeps the whole tail aligned in the same direction. The bend in *trans* fatty acids makes it zigzag:

Cis fatty acid	*Trans* fatty acid

Some *trans* fatty acids occur naturally in beef. Most form by processes that solidify vegetable oils for margarine, shortening, and the like. These substances are widely used in prepared foods (such as cookies) and in french fries and other fast-food products. *Trans* fatty acids are linked to high levels of LDL, a "bad" form of cholesterol that can set the stage for a heart attack. Speculate on why your body might have an easier time dealing with *cis* fatty acids than *trans* fatty acids.

3. In the following list, identify which is the carbohydrate, the fatty acid, the amino acid, and the polypeptide:
   a. $^+NH_3$—CHR—COO$^-$ c. (glycine)$_{20}$
   b. $C_6H_{12}O_6$ d. $CH_3(CH_2)_{16}COOH$

4. A clerk in a health-food store tells you that certain "natural" vitamin C tablets extracted from rose hips are better for you than synthetic vitamin C tablets. Given your understanding of the structure of organic compounds, what would be your response? Design an experiment to test whether the vitamins differ.

Selected Key Terms

alcohol *3.2*
amino acid *3.6*
ATP *3.9*
carbohydrate *3.4*
coenzyme *3.9*
condensation reaction *3.3*
denaturation *3.7*
DNA *3.9*
domain *3.7*
enzyme *3.3*
fat *3.5*
fatty acid *3.5*
functional group *3.1*
global warming *CI*
HLA *3.7*
hydrocarbon *3.1*
hydrolysis *3.3*
lipid *3.5*
monomer *3.3*
monosaccharide *3.4*
nucleic acid *3.9*
nucleotide *3.9*
oligosaccharide *3.4*
organic compound *3.1*
phospholipid *3.5*
polymer *3.3*
polypeptide chain *3.6*
polysaccharide *3.4*
protein *3.6*
RNA *3.9*
sterol *3.5*
triglyceride *3.5*
wax *3.5*

Readings

Alberts, B., et al. 2002. *Molecular Biology of the Cell.* Fourth edition. New York: Garland. Big book, easy to read.

On-Line readings at Student Guide for InfoTrac: www.brookscole.com/biology

CELL STRUCTURE AND FUNCTION

Animalcules and Cells Fill'd With Juices

Early in the seventeenth century, a scholar by the name of Galileo Galilei put two glass lenses inside a cylinder. With this instrument he happened to look at an insect, and later he described the stunning geometric patterns of its tiny eyes. Thus Galileo, who wasn't a biologist, was one of the first to record a biological observation made through a microscope. The study of the cellular basis of life was about to begin. First in Italy, then in France and England, scholars set out to explore a world whose existence had not even been suspected.

At midcentury Robert Hooke, Curator of Instruments for the Royal Society of England, was at the forefront of those studies. When Hooke first turned a microscope to thinly sliced cork from a mature tree, he observed tiny compartments (Figure 4.1*c*). He gave them the Latin name *cellulae*, meaning small rooms—hence the origin of the biological term "cell." They actually were the interconnecting walls of dead plant cells, which is what cork is made of, but Hooke didn't think of them as being dead because neither he nor anyone else at the time knew cells could be alive. In other plant tissues, he observed cells "fill'd with juices" but didn't have a clue to what they represented.

Given the simplicity of their instruments, it is just amazing that the pioneers in microscopy observed as much as they did. Antony van Leeuwenhoek, a Dutch shopkeeper, had exceptional skill in constructing lenses and possibly the keenest vision of all (Figure 4.1*a*). By the late 1600s, he was discovering natural wonders everywhere, including "many very small animalcules, the motions of which were very pleasing to behold," in scrapings of tartar from his teeth. Elsewhere he observed protistans, sperm, and even a bacterium— an organism so small it would not be seen again for another two centuries!

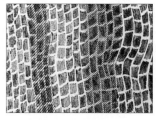
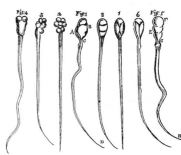

Figure 4.1 Early glimpses into the world of cells. (**a**) Antony van Leeuwenhoek, with microscope in hand. (**b**) Robert Hooke's compound microscope and (**c**) his drawing of cell walls from cork tissue. (**d**) One of van Leeuwenhoek's early sketches of sperm cells. (**e**) Cartoon evidence of the startling impact of microscopic observations on nineteenth-century London.

In the 1820s, improvements in lenses brought cells into sharper focus. Robert Brown, a botanist, noticed an opaque spot in a variety of cells and called it a nucleus. In 1838 another botanist, Matthias Schleiden, wondered if the nucleus had something to do with development. As he hypothesized, each plant cell might develop as an independent unit even though it's part of the plant.

By 1839, after years of studying animal tissues, the zoologist Theodor Schwann had this to say: Animals as well as plants consist of cells and cell products—and even though the cells are part of a whole organism, to some extent they have an individual life of their own.

A decade later a question remained: Where do cells come from? Rudolf Virchow, a physiologist, completed his own studies of a cell's growth and reproduction— that is, its division into two daughter cells. Every cell, he reasoned, comes from a cell that already exists.

And so, by the middle of the nineteenth century, microscopic analysis had yielded three generalizations, which together constitute the **cell theory**. *First, every organism is composed of one or more cells. Second, the cell is the smallest unit having the properties of life. Third, the continuity of life arises directly from the growth and division of single cells.* All three insights still hold true.

This chapter is not meant for memorization. Read it simply to gain an overview of current understandings of cell structure and function. In later chapters, you can refer back to it as a road map through the details. With its images from microscopy, this chapter and others in the book can transport you into spectacular worlds of juice-fill'd cells and animalcules.

Key Concepts

1. All organisms consist of one or more cells. The cell is the smallest unit that still retains the characteristics of life. Since the time of life's origins, each new cell has descended from a cell that is already alive. These are the three generalizations of the cell theory.

2. All cells have a plasma membrane. This outermost, double-layered membrane separates a cell's interior from its surroundings, although it selectively allows substances to cross it.

3. All cells contain cytoplasm, an organized internal region where energy conversions, protein synthesis, movements of cell parts, and other required activities proceed. In eukaryotic cells only, the DNA is enclosed in a membrane-bound nucleus. In prokaryotic cells (archaebacteria and eubacteria), the DNA simply is concentrated in part of the cell interior.

4. The plasma membrane and internal cell membranes consist mainly of lipids and proteins. The lipids are organized as two adjacent layers. This bilayer gives a membrane its basic structure and prevents water-soluble substances from freely crossing it. Proteins embedded in the bilayer or positioned at its surfaces carry out many functions, such as transporting substances across the membrane.

5. Diverse organelles, including the nucleus, divide the interior of eukaryotic cells into membrane-bound, functional compartments. Prokaryotic cells do not have comparable organelles.

6. When cells are growing, they increase faster in volume than in surface area. This physical constraint on growth influences their size and shape.

7. Microscopes modify light rays or accelerated beams of electrons in ways that allow us to form images of incredibly small specimens. They are the foundation for our current understanding of cell structure and function.

BASIC ASPECTS OF CELL STRUCTURE AND FUNCTION

Inside your body and at its moist surfaces, trillions of cells live in interdependency. In northern forests, pollen grains—four-celled structures—escape from pine trees. In scummy pondwater, a single-celled amoeba moves on its own. For humans, pines, amoebas, and all other organisms, the **cell** is the smallest unit that lives on its own or has the potential to do so (Figure 4.2). Each cell is structurally organized for metabolism. It senses and responds to its environment. And heritable instructions in its DNA give it the potential to reproduce.

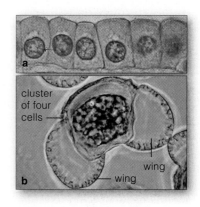

Figure 4.2 (**a**) A row of cells at the surface of a tiny tube inside a human kidney. Each cell has a nucleus (the sphere stained darker red). (**b**) A winged pollen grain. One of its cells will give rise to a sperm that may meet up with an egg and help start a new pine tree.

Structural Organization of Cells

Cells differ hugely in size, shape, and activities, as you might gather by comparing a tiny bacterium with one of your relatively giant liver cells. Yet they are alike in three respects: They all start out life with a plasma membrane, a region of DNA, and a region of cytoplasm:

1. **Plasma membrane**. This thin, outermost membrane maintains the cell as a distinct entity. By doing so, it allows metabolic events to proceed apart from random events in the environment. This membrane does not totally isolate the cell interior. Substances and signals continually move across it in highly controlled ways.

2. **Nucleus** or **nucleoid**. Depending on the species, the DNA occupies a membrane-bound sac (nucleus) in the cell or simply a region of the cell interior (nucleoid).

3. **Cytoplasm**. Cytoplasm is everything between the plasma membrane and the region of DNA. It consists of a semifluid matrix and other components, such as **ribosomes** (structures on which proteins are built).

This chapter introduces two very different kinds of cells. **Eukaryotic cells** have organelles—tiny sacs with one or more outer membranes, in the cytoplasm. One sac, the nucleus, is their defining feature. **Prokaryotic cells** don't have a nucleus; nothing intervenes between their DNA and cytoplasm. The only prokaryotic cells are archaebacteria and eubacteria. All other organisms —from amoebas to peach trees to puffball mushrooms to whales and zebras—consist of eukaryotic cells.

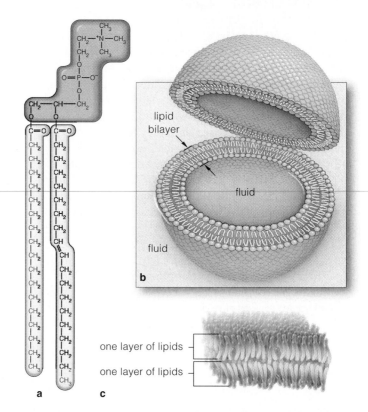

Figure 4.3 Lipid bilayer organization of cell membranes. (**a**) One of the phospholipids, the most abundant molecules of membranes. (**b,c**) These and other lipids are arranged into two layers. Their hydrophobic tails are sandwiched between the hydrophilic heads. In cells, the heads are dissolved in cytoplasm on one side of the bilayer and in extracellular fluid on the other side.

Organization of Cell Membranes

All cell membranes have the same structural framework of two sheets of lipid molecules. Figure 4.3 shows this **lipid bilayer** arrangement for the plasma membrane, the continuous, oily boundary that prevents the free passage of water-soluble substances into and out of all cells. In eukaryotic cells, other membranes divide the cytoplasm into functional zones in which substances are synthesized, processed, stockpiled, or degraded.

Diverse proteins embedded in the lipid bilayer or positioned at one of its surfaces carry out most of the membrane functions (Figure 4.4). For example, some proteins act as a channel for water-soluble substances. Others pump substances across the bilayer. Still others serve as receptors, which can latch onto hormones and other types of signaling molecules that trigger changes in cell activities. You will read more about membrane proteins in the chapter to follow.

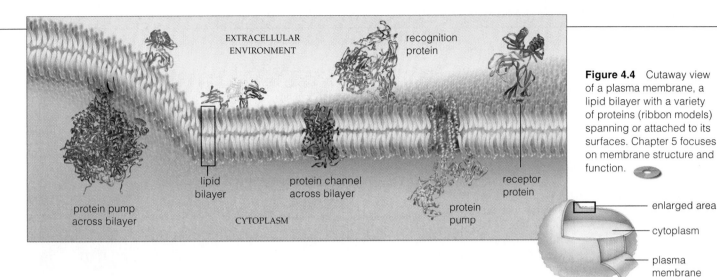

Figure 4.4 Cutaway view of a plasma membrane, a lipid bilayer with a variety of proteins (ribbon models) spanning or attached to its surfaces. Chapter 5 focuses on membrane structure and function.

EXTRACELLULAR ENVIRONMENT

recognition protein

lipid bilayer

protein channel across bilayer

receptor protein

protein pump across bilayer

protein pump

CYTOPLASM

enlarged area

cytoplasm

plasma membrane

Why Aren't All Cells Big?

You may be wondering how small cells really are. Can any be observed with the unaided human eye? Just a few, such as the "yolks" of bird eggs, cells in the red part of watermelons, and fish eggs. Eggs can get large because they are metabolically inert at maturity; most metabolically active cells are too tiny to be seen except with microscopes. To give you a sense of *how* tiny, a red blood cell is about 8 millionths of one meter across, so about 2,000 of them would fit across your thumbnail!

Why aren't all cells big? A physical relationship called the **surface-to-volume ratio** constrains increases in cell size. By this relationship, any object's volume increases with the cube of the diameter, but the surface area increases only with the square.

Apply this to a round cell. As Figure 4.5 shows, *if a cell expands in diameter during growth, then its volume will increase faster than its surface area will.* Suppose you make a round cell grow four times wider than normal. Its volume increases 64 times (4^3). But the surface area increases only 16 times (4^2). Now each unit of plasma membrane must serve four times as much cytoplasm as before. Moreover, beyond a certain point, the inward flow of nutrients and the outward flow of wastes won't be fast enough, so you'll end up with a dead cell.

A large, round cell also would have trouble moving materials *through* its cytoplasm. In small cells, random, tiny motions of molecules easily distribute materials. If a cell isn't small, you usually can expect it to be long and thin or have outfoldings and infoldings that increase its surface relative to its volume. *The smaller or narrower or more frilly-surfaced the cell, the more efficiently materials cross its surface and become distributed through the interior.*

We also see evidence of surface-to-volume constraints on body plans of multicelled species. For example, cells attach end to end in strandlike algae; each one interacts directly with its surroundings. Muscle cells are thin, but each is as long as the muscle of which it is part.

Figure 4.5 Example of the surface-to-volume ratio. This physical relationship between increases in volume and in surface area imposes restrictions on the size and the shape of cells. This includes frog eggs 2–3 mm across. They are among the largest of all animal cells. Compare Figure 4.8 in Section 4.2.

diameter (cm):	0.5	1.0	1.5
surface area (cm²):	0.79	3.14	7.07
volume (cm³):	0.06	0.52	1.77
surface-to-volume ratio:	13.17:1	6.04:1	3.99:1

All cells have an outermost plasma membrane, an internal region of cytoplasm, and an internal region of DNA. Only eukaryotic cells have a nucleus and other organelles.

A lipid bilayer, consisting of a two-layer arrangement of lipid molecules, gives a cell membrane its overall structure. Proteins embedded in the bilayer or positioned at one of its surfaces carry out diverse membrane functions.

Metabolic activity is largely a function of cell volume and surface area. As cells grow, their volume increases faster than their surface area does. The surface-to-volume ratio constrains increases in size. It also influences cell shape and the body plans of multicelled organisms.

Microscopes—Gateways to Cells

Modern microscopes are gateways to astounding worlds. Some even afford glimpses into the structure of molecules. Different kinds use wavelengths of light or accelerated electrons (Figures 4.6 and 4.7). The micrographs in Figures 4.8 and 4.9 only hint at the incredible details now being observed. A **micrograph** is simply a photograph of an image that came into view with the help of a microscope.

LIGHT MICROSCOPES   Picture a series of waves moving across an ocean. Each **wavelength** is the distance from one wave's peak to the peak of the wave behind it. Light also travels as waves from sources such as the sun and illuminated specimens. In a *compound light microscope*, two or more sets of glass lenses bend light emanating from a cell or some other specimen in ways that form an enlarged image of it (Figure 4.6a,b).

A living cell must be small or thin enough for light to pass through. It would help if cell parts differed in color and density from the surroundings, but most are nearly colorless and appear uniformly dense. To get around this problem, microscopists stain cells (expose them to dyes that react with some parts of a specimen but not others). Staining may alter and kill cells. Dead cells break down

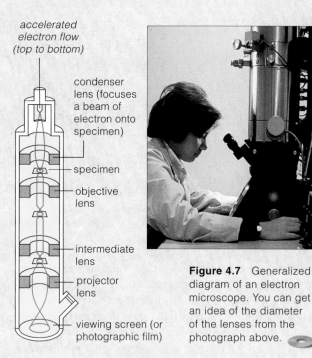

accelerated electron flow (top to bottom)

condenser lens (focuses a beam of electron onto specimen)

specimen

objective lens

intermediate lens

projector lens

viewing screen (or photographic film)

Figure 4.7   Generalized diagram of an electron microscope. You can get an idea of the diameter of the lenses from the photograph above.

rapidly, which is why technicians typically preserve or pickle a cell before staining it.

Suppose you use the best glass lens system. When you magnify the diameter of a specimen by 2,000 times or more, you discover that cell parts appear larger but are not clearer. To see why, think about the distance between the two crests of a wavelength of red or violet light. The distance is about 750 nanometers for red light and 400 nanometers for violet (wavelengths of all other colors fall in between). If a cell structure is less than one-half of a wavelength long, it will not be able to disturb enough of the rays of light streaming past to become visible.

ELECTRON MICROSCOPES   Resolution of fine details is better with the help of electrons. Electrons, recall, are particles of matter. But they also behave like waves. In electron microscopy, streams of electrons are accelerated to wavelengths of about 0.005 nanometer—about 100,000 times shorter than wavelengths of visible light. Electrons cannot pass through glass lenses, but a magnetic field can bend them from their path and focus them.

In a *transmission electron microscope*, a magnetic field is the "lens." Accelerated electrons are directed through a specimen, focused into an image, and magnified. With *scanning electron microscopes*, a narrow beam of electrons moves back and forth across a specimen to which a thin coat of metal has been applied. The metal responds by emitting electrons. A detector tied to electronic circuitry converts energy of the emitted electrons into an image of the specimen's surface on a television screen. Most of the scanning images have fantastic depth (Figure 4.9d).

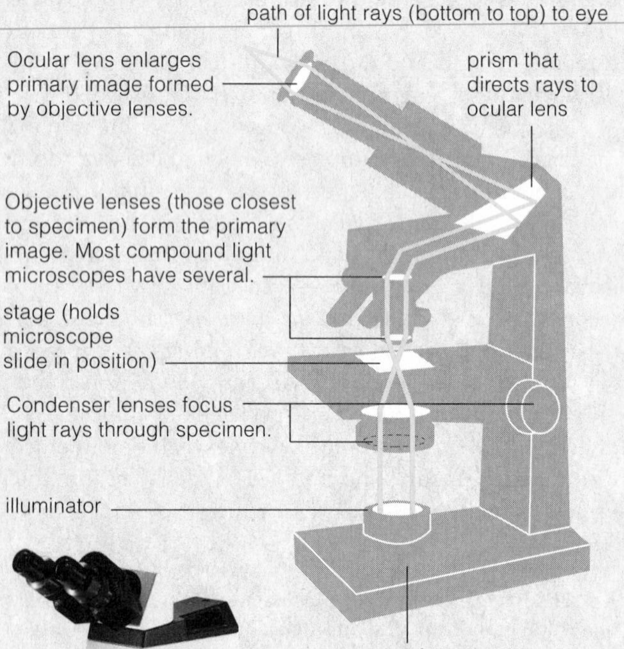

path of light rays (bottom to top) to eye

Ocular lens enlarges primary image formed by objective lenses.

prism that directs rays to ocular lens

Objective lenses (those closest to specimen) form the primary image. Most compound light microscopes have several.

stage (holds microscope slide in position)

Condenser lenses focus light rays through specimen.

illuminator

source of illumination (housed in the base of the microscope)

b

a

Figure 4.6   Diagram and exterior view of a compound light microscope.

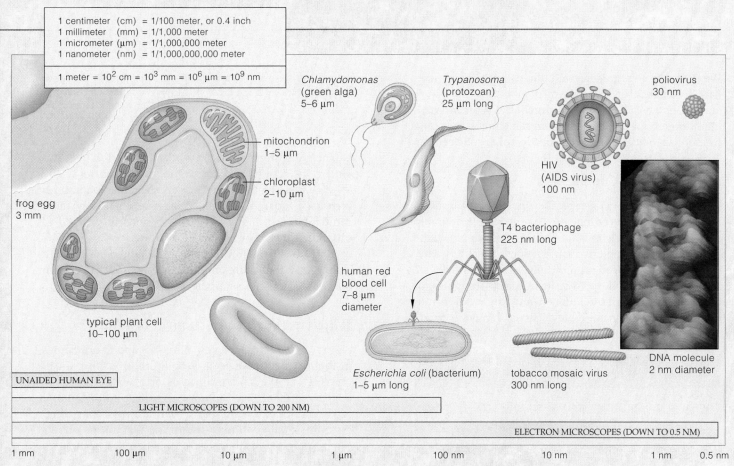

1 centimeter (cm) = 1/100 meter, or 0.4 inch
1 millimeter (mm) = 1/1,000 meter
1 micrometer (μm) = 1/1,000,000 meter
1 nanometer (nm) = 1/1,000,000,000 meter

1 meter = 10^2 cm = 10^3 mm = 10^6 μm = 10^9 nm

Chlamydomonas
(green alga)
5–6 μm

Trypanosoma
(protozoan)
25 μm long

poliovirus
30 nm

mitochondrion
1–5 μm

chloroplast
2–10 μm

HIV
(AIDS virus)
100 nm

frog egg
3 mm

T4 bacteriophage
225 nm long

human red
blood cell
7–8 μm
diameter

typical plant cell
10–100 μm

Escherichia coli (bacterium)
1–5 μm long

tobacco mosaic virus
300 nm long

DNA molecule
2 nm diameter

UNAIDED HUMAN EYE

LIGHT MICROSCOPES (DOWN TO 200 NM)

ELECTRON MICROSCOPES (DOWN TO 0.5 NM)

| 1 mm | 100 μm | 10 μm | 1 μm | 100 nm | 10 nm | 1 nm | 0.5 nm |

Figure 4.8 Units of measure used in microscopy. The photomicrograph of DNA was obtained with a *scanning tunneling microscope*, which yields magnifications up to 100 million times. The scope's needlelike probe has a single atom at its tip. When voltage is applied between the tip and an atom at a specimen's surface, electrons "tunnel" from the probe to the specimen. As the tip moves over a specimen's contours, a computer analyzes the tunneling motion and makes a three-dimensional view of the surface atoms.

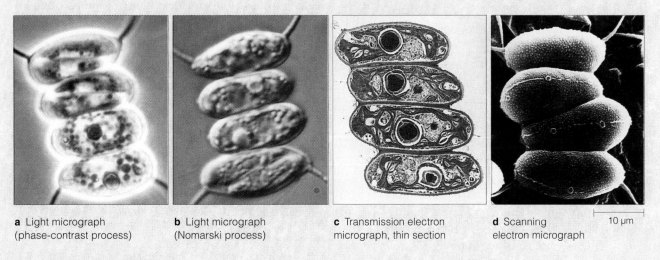

a Light micrograph
(phase-contrast process)

b Light micrograph
(Nomarski process)

c Transmission electron
micrograph, thin section

d Scanning
electron micrograph

10 μm

Figure 4.9 How different microscopes reveal different aspects of the same organism—a green alga (*Scenedesmus*). Images of all four specimens are at the same magnification. Phase-contrast and Nomarski processes (**a,b**) create optical contrasts without staining the cells. Both processes enhance the usefulness of light micrographs. As for other micrographs in the book, a horizontal bar below the micrograph, as in (**d**), provides you with a visual reference for size. A micrometer (μm) is 1/1,000,000 of a meter. Using the scale bar, can you estimate the length and width of *Scenedesmus*?

DEFINING FEATURES OF EUKARYOTIC CELLS

We turn now to organelles and other structural features that are typical of the cells of plants, animals, fungi, and protistans. We define an **organelle** as an internal, membrane-bound sac or compartment that serves one or more specialized functions inside eukaryotic cells.

Major Cellular Components

Study micrographs of a typical eukaryotic cell, such as the ones in the preceding section, and you'll probably notice that the nucleus is one of the most conspicuous features. Remember, any cell that starts out life with a nucleus is a eukaryotic cell; it houses a "true nucleus." Many other organelles and structures also are typical of these cells, although the numbers and kinds differ from one cell type to the next. Table 4.1 lists common features.

Table 4.1 *Common Features of Eukaryotic Cells*	
ORGANELLES AND THEIR MAIN FUNCTIONS	
Nucleus	Localizing the cell's DNA
Endoplasmic reticulum	Routing and modifying newly formed polypeptide chains; also, synthesizing lipids
Golgi body	Modifying polypeptide chains into mature proteins; sorting and shipping proteins and lipids for secretion or for use inside the cell
Various vesicles	Transporting or storing a variety of substances; digesting substances and structures in the cell; other functions
Mitochondria	Producing many ATP molecules in highly efficient fashion
NON-MEMBRANOUS STRUCTURES AND THEIR FUNCTIONS	
Ribosomes	Assembling polypeptide chains
Cytoskeleton	Imparting overall shape and internal organization to the cell; moving the cell and its internal structures

Think about Table 4.1 and you might find yourself asking: What is the advantage of partitioning the cell interior with such organelles? *The compartmentalization allows a large number of activities to occur simultaneously in very limited space.* Consider a photosynthetic cell in a leaf. It can put together starch molecules by one set of reactions and break them apart by another set. Yet the cell would gain nothing if the synthesis and breakdown reactions proceeded at the exact same time on the same starch molecule. Without membranes of organelles, the

Figure 4.10 *Facing page:* Generalized sketches showing some of the features of (**a**) a typical photosynthetic plant cell and (**b**) a typical animal cell.

balance of different chemical activities that helps keep eukaryotic cells alive would spiral out of control.

Organelle membranes have another function besides physically separating incompatible reactions. They allow compatible and interconnected reactions to proceed at different times. For instance, a plant's photosynthetic cells produce starch molecules in an organelle called a chloroplast, then store and later release starch for use in different reactions inside the same organelle.

Which Organelles Are Typical of Plants?

Figure 4.10*a* can start you thinking about organelles inside a typical plant cell. Bear in mind, calling a cell "typical" is like calling a cactus or a water lily or an apple tree a "typical" plant. As is the case for animal cells, variations on the basic plan are mind-boggling. With this qualification in mind, the illustration shows some common locations of organelles and structures you are likely to observe in many plant cells.

Which Organelles Are Typical of Animals?

Now start thinking about organelles of a typical animal cell, such as the one in Figure 4.10*b*. Like plant cells, it has a nucleus, mitochondria, and the other common features listed in Table 4.1. *These structural similarities correlate with functions that are necessary for survival, regardless of the cell type.* We will return to this concept throughout the book.

Comparing Figure 4.10*a* with 4.10*b* also will give you an initial idea of how plant and animal cells differ in their structure. For example, you will never observe an animal cell surrounded by a cell wall. (You might see many different kinds of fungal and protistan cells with one, however.) What other differences can you identify?

Eukaryotic cells have a number of organelles. These are internal, membrane-bound sacs and compartments with specific metabolic functions.

Organelles physically separate chemical reactions, many of which are incompatible. They also separate different reactions in time, as when certain molecules are assembled, stored, then used later in other reaction sequences.

All eukaryotic cells contain certain organelles (such as the nucleus) and structures (such as ribosomes) that perform functions essential for survival. Specialized cells also may incorporate additional kinds of organelles and structures.

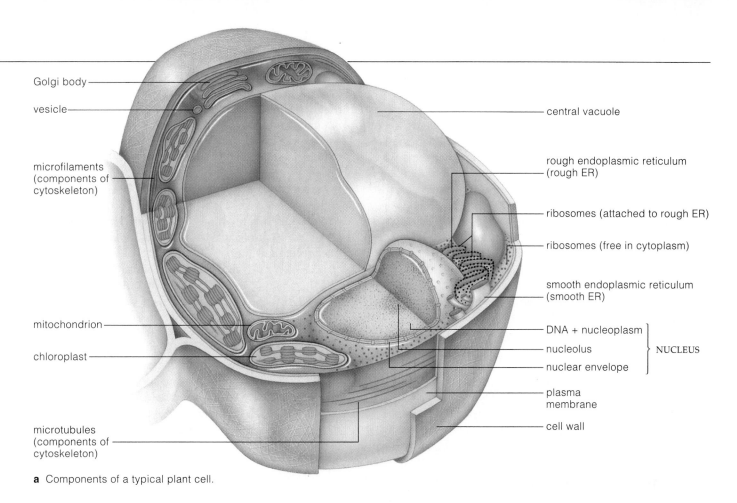

Golgi body

vesicle

microfilaments
(components of
cytoskeleton)

mitochondrion

chloroplast

microtubules
(components of
cytoskeleton)

central vacuole

rough endoplasmic reticulum
(rough ER)

ribosomes (attached to rough ER)

ribosomes (free in cytoplasm)

smooth endoplasmic reticulum
(smooth ER)

DNA + nucleoplasm

nucleolus   NUCLEUS

nuclear envelope

plasma
membrane

cell wall

a Components of a typical plant cell.

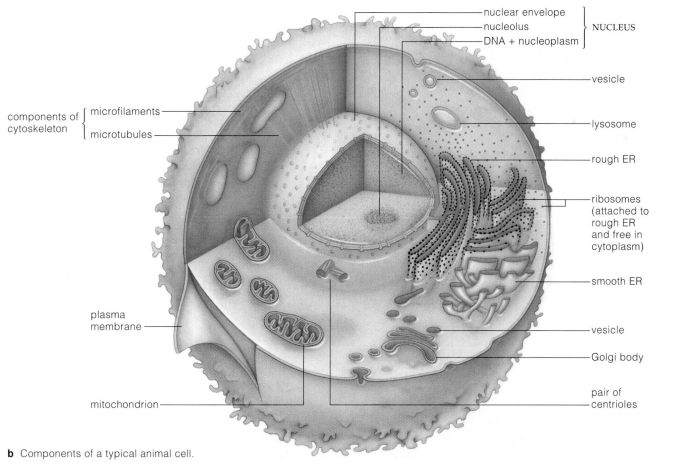

nuclear envelope

nucleolus   NUCLEUS

DNA + nucleoplasm

components of   microfilaments
cytoskeleton   microtubules

vesicle

lysosome

rough ER

ribosomes
(attached to
rough ER
and free in
cytoplasm)

smooth ER

vesicle

Golgi body

plasma
membrane

pair of
centrioles

mitochondrion

b Components of a typical animal cell.

THE NUCLEUS

Constructing, operating, and reproducing cells simply cannot be done without carbohydrates, lipids, proteins, and nucleic acids. It takes a class of proteins—enzymes—to build and use these molecules. Said another way, a cell's structure and function begin with proteins. *And instructions for building proteins are located in DNA.*

Unlike bacteria, eukaryotic cells have their genetic instructions distributed among several to many DNA molecules of different lengths. Each human body cell, for instance, has forty-six DNA molecules. End to end,

the molecules would extend outward for a meter or so. Similarly, stretch a frog cell's twenty-six DNA molecules end to end and they would extend ten meters. We have very little idea what frogs are doing with so much of it, but that's a lot of DNA!

Eukaryotic cells, recall, protect DNA in a nucleus. Figure 4.11 shows the structure of this organelle. The nucleus has two functions. *First*, it physically separates all the DNA molecules from the cytoplasm's complex metabolic machinery. (Or, as one professor put it, the nucleus keeps chromosomes in and mitochondria out.) Localizing DNA in a sac makes it easier for a parent cell to copy and sort out hereditary instructions before the cell divides. This is important, because each of the two daughter cells that form later requires a full set of DNA molecules. *Second*, the nuclear membranes serve as a boundary where cells control the movement of substances and signals to and from the cytoplasm. Table 4.2 lists the components of the nucleus.

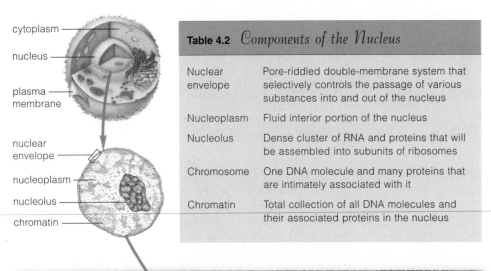

cytoplasm

nucleus

plasma membrane

nuclear envelope

nucleoplasm

nucleolus

chromatin

Table 4.2	*Components of the Nucleus*
Nuclear envelope	Pore-riddled double-membrane system that selectively controls the passage of various substances into and out of the nucleus
Nucleoplasm	Fluid interior portion of the nucleus
Nucleolus	Dense cluster of RNA and proteins that will be assembled into subunits of ribosomes
Chromosome	One DNA molecule and many proteins that are intimately associated with it
Chromatin	Total collection of all DNA molecules and their associated proteins in the nucleus

Nuclear Envelope

Unlike cells, a nucleus has two outer membranes, one wrapped around the other. Its double-membrane system is the **nuclear envelope**. It has not one but two lipid bilayers in which many molecules of proteins are embedded (Figure 4.12). The nuclear envelope surrounds the nucleoplasm, the fluid portion of a nucleus. The innermost surface contains attachment sites for strandlike proteins that anchor DNA to the envelope and also help keep it organized. On the outer surface is a profusion of ribosomes. Proteins are

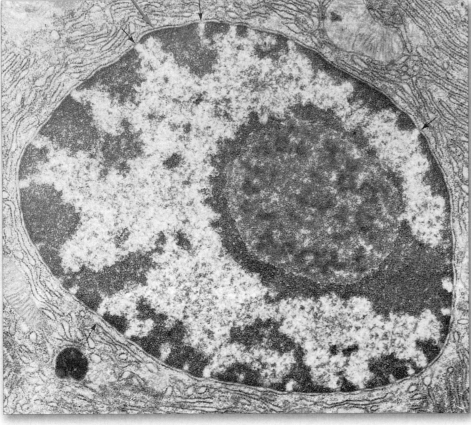

Figure 4.11 Pancreatic cell nucleus. The small arrows on this transmission electron micrograph point to pores where control systems operate to restrict or permit the passage of specific substances across the nuclear envelope.

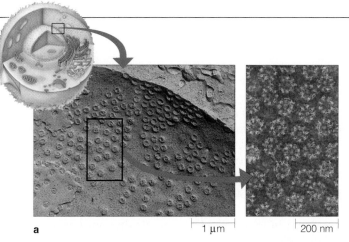

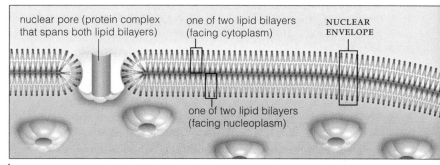

nuclear pore (protein complex that spans both lipid bilayers)

one of two lipid bilayers (facing cytoplasm)

NUCLEAR ENVELOPE

one of two lipid bilayers (facing nucleoplasm)

a | 1 μm | 200 nm   b

Figure 4.12 (**a**) Part of the outer surface of a nuclear envelope. *Left:* This specimen was fractured to show the intimate layering of its two lipid bilayers. *Right:* Nuclear pores. Each pore across the envelope is an organized cluster of membrane proteins. It permits the selective transport of substances into and out of the nucleus. (**b**) Sketch of the nuclear envelope's structure.

assembled on membrane-bound ribosomes or on other ribosomes positioned in the cytoplasm.

Like all cell membranes, a nuclear envelope's lipid bilayers prevent water-soluble substances from moving freely into and out of a nucleus. However, many pores, each a cluster of proteins, span both bilayers. Small, water-soluble molecules and ions can cross the nuclear envelope at the pores. As you will see later in the book, large molecules (including the ribosomal subunits) also cross through the pores, in highly controlled ways.

Nucleolus

Take another look at the nucleus in Figure 4.11. Inside is a somewhat spherical, dense mass of material. What is it? When eukaryotic cells grow, one or more of these masses form in the nucleus. Each is a **nucleolus** (plural, nucleoli), a construction site where certain RNAs and proteins are combined to make ribosomal subunits. The subunits move through nuclear pores to the cytoplasm. There, two subunits converge as an intact ribosome, where amino acids are assembled into proteins.

Chromosomes

When eukaryotic cells aren't dividing, their individual DNA molecules do not show up in micrographs, which look grainy (Figure 4.11). As cells get ready to divide, however, they duplicate their DNA molecules so each daughter cell will receive all of the required hereditary instructions. Then, the DNA no longer looks grainy; each molecule folds into a compact structure.

Early microscopists bestowed the name *chromatin* on the seemingly grainy substance and *chromosomes* on

the condensed structures. We now define **chromatin** as the cell's collection of DNA, together with all proteins associated with it (Table 4.2). Each **chromosome** is one DNA molecule and its associated proteins, regardless of whether it is in threadlike or condensed form:

one chromosome (one threadlike DNA molecule + proteins; not duplicated)

one chromosome (threadlike but now duplicated; two DNA molecules + proteins)

one chromosome (duplicated and also condensed)

In other words, the appearance of "the chromosome" changes over the life of a eukaryotic cell! In chapters to come, you will look at different aspects of chromosomes, so you may find it useful to remember this point.

What Happens to the Proteins Specified by DNA?

Outside the cell nucleus, new polypeptide chains for proteins are assembled on ribosomes. What happens to them? Many are stockpiled in the cytoplasm or are used at once. Others enter the endomembrane system. As described in the next section, endoplasmic reticulum, Golgi bodies, and diverse vesicles make up this system.

Thanks to DNA's instructions, many proteins take on specific, final forms in the endomembrane system. Also in this system, lipids are assembled and packaged by enzymes and other proteins (these, too, were built according to DNA's instructions). As you will see next, vesicles deliver the proteins and lipids to specific sites within the cell or to the plasma membrane, for export.

The nucleus, an organelle with two outer membranes, keeps the DNA molecules of eukaryotic cells separated from the metabolic machinery of the cytoplasm.

The localization makes it easier to organize the DNA and to copy it before a parent cell divides into daughter cells.

Pores across the nuclear envelope help control the passage of many substances between the nucleus and cytoplasm.

THE ENDOMEMBRANE SYSTEM

The **endomembrane system** is a series of functionally connected organelles in which lipids are assembled and new polypeptide chains are modified. Its products are sorted and shipped to different destinations. Figure 4.13 shows how its organelles—the ER, Golgi bodies, and vesicles—interconnect with one another.

Endoplasmic Reticulum

The functions of the endomembrane system begin with **endoplasmic reticulum**, or **ER**. In animal cells, the ER is continuous with the nuclear envelope, and it extends through the cytoplasm. Its membranes appear rough or smooth, depending on whether ribosomes are attached to the membrane facing the cytoplasm.

We typically observe *rough* ER arranged into stacks of flattened sacs with many ribosomes attached (Figure 4.14a). Every new polypeptide chain is synthesized on ribosomes. But only the newly forming chains having a built-in signal can enter the space within rough ER or become incorporated into ER membranes. (That signal is a sequence of fifteen to twenty specific amino acids.) Once the chains are in rough ER, enzymes may attach oligosaccharides and other side chains to them. Many specialized cells secrete the final proteins. Rough ER is abundant in such cells. For example, in your pancreas, ER-rich gland cells make and secrete enzymes that end up in the small intestine and help digest your meals.

Smooth ER is free of ribosomes and curves through cytoplasm like connecting pipes (Figure 4.14b). Many cells assemble most lipids inside the pipes. Smooth ER is well developed in seeds. In liver cells, certain drugs as well as toxic metabolic wastes are inactivated in it. Sarcoplasmic reticulum, a type of smooth ER in skeletal muscle cells, functions in controlling contraction.

Golgi Bodies

In **Golgi bodies**, enzymes put the finishing touches on proteins and lipids, sort them out, then package them inside vesicles for shipment to specific locations. For example, an enzyme in one Golgi region might attach a phosphate group to a new protein, thereby giving it a mailing tag to its proper destination.

Commonly, a Golgi body looks rather like a stack of pancakes; it is composed of a series of flattened, membrane-bound sacs (Figure 4.15). In functional terms, the last portion of a Golgi body corresponds to the top pancake. Here, vesicles form as patches of the membrane bulge out, then break away into the cytoplasm.

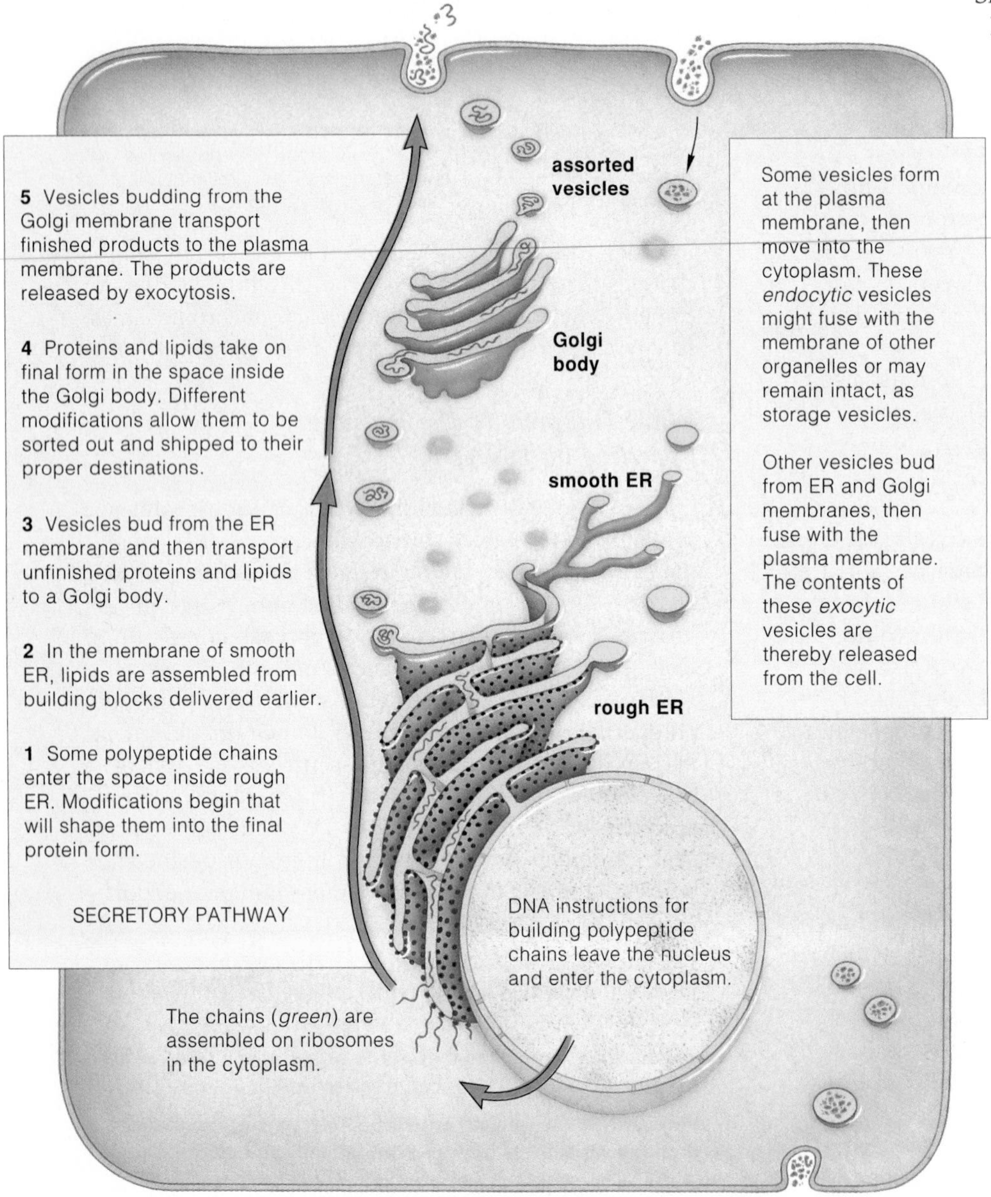

5 Vesicles budding from the Golgi membrane transport finished products to the plasma membrane. The products are released by exocytosis.

4 Proteins and lipids take on final form in the space inside the Golgi body. Different modifications allow them to be sorted out and shipped to their proper destinations.

3 Vesicles bud from the ER membrane and then transport unfinished proteins and lipids to a Golgi body.

2 In the membrane of smooth ER, lipids are assembled from building blocks delivered earlier.

1 Some polypeptide chains enter the space inside rough ER. Modifications begin that will shape them into the final protein form.

SECRETORY PATHWAY

assorted vesicles

Golgi body

smooth ER

rough ER

Some vesicles form at the plasma membrane, then move into the cytoplasm. These *endocytic* vesicles might fuse with the membrane of other organelles or may remain intact, as storage vesicles.

Other vesicles bud from ER and Golgi membranes, then fuse with the plasma membrane. The contents of these *exocytic* vesicles are thereby released from the cell.

DNA instructions for building polypeptide chains leave the nucleus and enter the cytoplasm.

The chains (*green*) are assembled on ribosomes in the cytoplasm.

Figure 4.13 Endomembrane system, a membrane system in the cytoplasm that synthesizes, modifies, packages, and ships proteins and lipids. *Green* arrows highlight a secretory pathway by which some proteins and lipids are packaged and released from many types of cells, including gland cells that secrete mucus, sweat, and digestive enzymes.

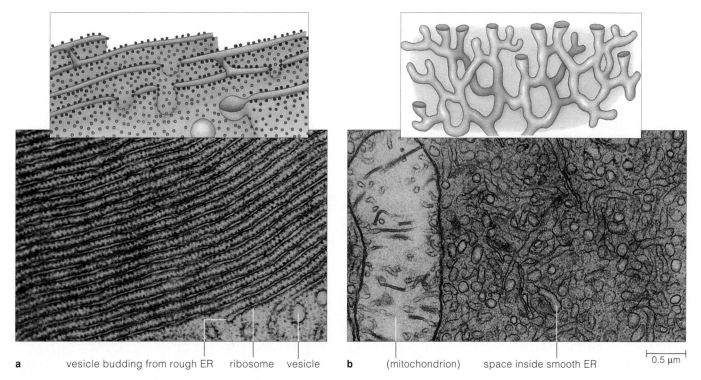

a   vesicle budding from rough ER   ribosome   vesicle   b   (mitochondrion)   space inside smooth ER   0.5 μm

Figure 4.14 Transmission electron micrographs and sketches of endoplasmic reticulum. (**a**) Many ribosomes dot the flattened surfaces of rough ER that face the cytoplasm. (**b**) This section reveals the diameters of many interconnected, pipelike regions of smooth ER.

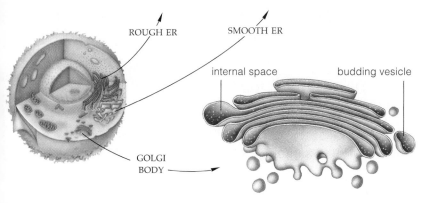

ROUGH ER   SMOOTH ER

internal space   budding vesicle

GOLGI BODY

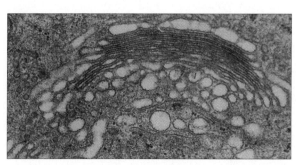

Figure 4.15 Sketch and micrograph of a Golgi body from an animal cell.   0.25 μm

A Variety of Vesicles

Vesicles are tiny, membranous sacs that move through the cytoplasm or take up positions in it. The lysosome, a common type, buds from Golgi membranes of animal cells and some fungal cells. **Lysosomes** are organelles of intracellular digestion. They hold enzymes that digest complex carbohydrates, proteins, nucleic acids, and some lipids. Often they fuse with vesicles formed earlier from patches of plasma membrane that surrounded bacteria, molecules, and other items that docked at membrane receptors. Lysosomes also digest entire cells and cell parts. For example, when a tadpole is developing into an adult frog, its tail gradually disappears. Lysosomal enzymes are responding to developmental signals and are helping to destroy cells that make up the tail.

Peroxisomes, another type, are sacs of enzymes that break down fatty acids and amino acids. A product of the reactions, hydrogen peroxide, is toxic, as the Chapter 5 introduction describes. But enzyme action converts it to water and oxygen or uses it to break down alcohol and other toxins. Drink alcohol, and the peroxisomes of liver and kidney cells will degrade nearly half of it.

In the ER and Golgi bodies of the endomembrane system, many proteins take on final form and lipids are assembled.

Lipids, proteins (such as enzymes), and other items become packaged in vesicles destined for export, storage, membrane building, intracellular digestion, and other cell activities.

MITOCHONDRIA

Recall, from Section 3.9, that ATP is an energy carrier. At many different reaction sites, it delivers energy—in the form of phosphate groups—that drives nearly all cell activities. Many ATP molecules form when organic compounds are fully broken down to carbon dioxide and water in a **mitochondrion** (plural, mitochondria).

Only eukaryotic cells contain these organelles. Figure 4.16 gives an idea of their structure. ATP-forming reactions that proceed in mitochondria extract more energy from organic compounds than can be done by any other means. They cannot run to completion without plenty of oxygen. Like all other land-dwelling vertebrates, every time you breathe in, you take in oxygen mainly for the mitochondria in cells—in your case, many trillions of cells.

Each mitochondrion has a double-membrane system. The outer one faces the cytoplasm. The inner one usually is folded repeatedly.

What's the point of such a complex system? It creates two compartments. Enzymes and other proteins of electron transport systems stockpile hydrogen ions in the outer compartment. Ions then flow to the inner compartment in controlled ways. Energy inherent in the flow drives ATP formation, as Chapter 8 describes. Oxygen binds to hydrogen and to the spent electrons, forming the end product, water. Its removal keeps the transfer systems clear for operation.

All eukaryotic cells contain one or more mitochondria. You may find only one in a single-celled yeast. You might find one thousand or more in cells that demand a lot of energy, such as those of muscles. Liver cells, too, contain a profusion of mitochondria. This alone tells you that the liver is a particularly active, energy-demanding organ.

Mitochondria are a lot like bacteria in their size and biochemistry. They have their own DNA and some ribosomes. They divide independently. Did they evolve from ancient prokaryotic cells through "endosymbiosis"? By this idea, a predatory, amoeba-like cell engulfed other cells, which escaped digestion. The engulfed cells, then their descendants, reproduced in the host. In time they became permanent, protected residents. And some of the structures and functions that they formerly needed to maintain an independent life were no longer essential and were lost. In this way the descendants evolved into mitochondria. We return to this idea of endosymbiotic origins in Section 20.4.

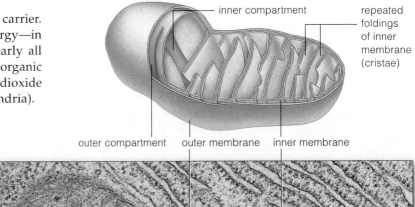

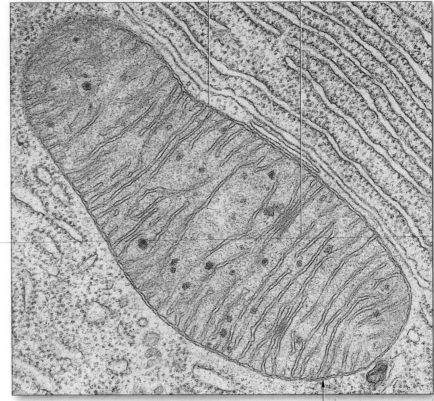

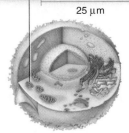

Figure 4.16 Sketch and transmission electron micrograph, thin section, of a typical mitochondrion. This organelle specializes in the production of large quantities of ATP, the main energy-carrying molecule between different reaction sites in cells. The process that produces ATP in mitochondria can't proceed without free oxygen.

The organelles called mitochondria are the ATP-producing powerhouses of all eukaryotic cells.

Energy-releasing reactions proceed at the compartmented, internal membrane system of mitochondria. The reactions, which require oxygen, produce far more ATP than can be produced by any other cellular reactions.

SPECIALIZED PLANT ORGANELLES

Chloroplasts and Other Plastids

Many plant cells contain plastids, a general category of organelles that specialize in photosynthesis or storage. Three types are common in different tissues of plants: chloroplasts, chromoplasts, and amyloplasts.

Of all eukaryotic cells, only the photosynthetic ones have **chloroplasts**. These organelles use sunlight energy to drive the formation of ATP and NADPH. These compounds are then used to assemble sugars and some other organic compounds.

Most chloroplasts have oval or disklike shapes. The semifluid interior, the stroma, is enclosed in two outer membrane layers. In the stroma is an inner membrane, called the thylakoid membrane, that forms a single sac, or compartment. That compartment is folded into interconnecting, flattened disks that are often arranged as stacks, as in Figure 4.17. Botanists call each stack a granum (plural, grana).

The first stage of photosynthesis starts where many light-trapping pigments, enzymes, and other proteins are embedded in the thylakoid membrane. It ends with formation of ATP and NADPH. The ATP and NADPH are used at sites in the stroma where sugars, then starch and other organic compounds, are built from carbon dioxide and water. Clusters of new molecules of starch (starch grains) may briefly accumulate in the stroma.

The most abundant photosynthetic pigments are chlorophylls that reflect or transmit green light. Others include carotenoids, which reflect or transmit yellow, orange, and red light. The relative abundances of such pigments contribute to the colors of plant parts.

In many ways, chloroplasts are like photosynthetic bacteria. Like mitochondria, they may have evolved by endosymbiosis in the manner outlined in Sections 4.6 and 20.4.

Chromoplasts lack chlorophylls, but they have an abundance of carotenoids. These pigments are the basis of red-to-yellow colors of many flowers, autumn leaves, ripening fruits, and carrots and other roots. The colors commonly attract animals that pollinate the plants or disperse seeds.

Amyloplasts lack pigments. Often they store starch grains and are abundant in cells of stems, potato tubers (underground stems), and seeds.

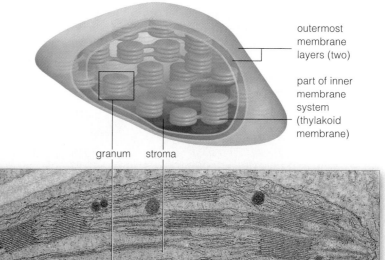

outermost membrane layers (two)

part of inner membrane system (thylakoid membrane)

granum   stroma

Figure 4.17 Generalized sketch of a chloroplast, the key defining feature of every photosynthetic eukaryotic cell. The transmission electron micrograph shows a thin section of a chloroplast from corn (*Zea mays*).

0.5 µm

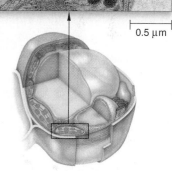

Central Vacuole

Many mature, living plant cells have a **central vacuole** (Figure 4.10*a*). This fluid-filled organelle stores amino acids, sugars, ions, and toxic wastes, and helps the cell grow. A central vacuole expands during growth. Fluid pressure builds up inside the cell and forces its pliable wall to enlarge. The cell surface area increases, and so does the rate at which water and other substances can be absorbed across the plasma membrane.

In most cases, the central vacuole increases so much in volume that it takes up 50 to 90 percent of the cell's interior. The cytoplasm ends up as a very narrow zone between the central vacuole and plasma membrane.

Photosynthetic eukaryotic cells have chloroplasts and other plastids that function in food production and storage.

Many plant cells have a central vacuole. When this storage vacuole enlarges during growth, cells are forced to enlarge, and this increases the surface area available for absorption.

SUMMARY OF TYPICAL FEATURES OF EUKARYOTIC CELLS

Before leaving the organelles, take a moment to study Figures 4.18 and 4.19 to put them together into a big picture of cell structure and function.

Figure 4.18 Transmission electron micrograph of a photosynthetic cell from a blade of Timothy grass (*Phleum pratense*), in cross-section. The sketches highlight key organelles.

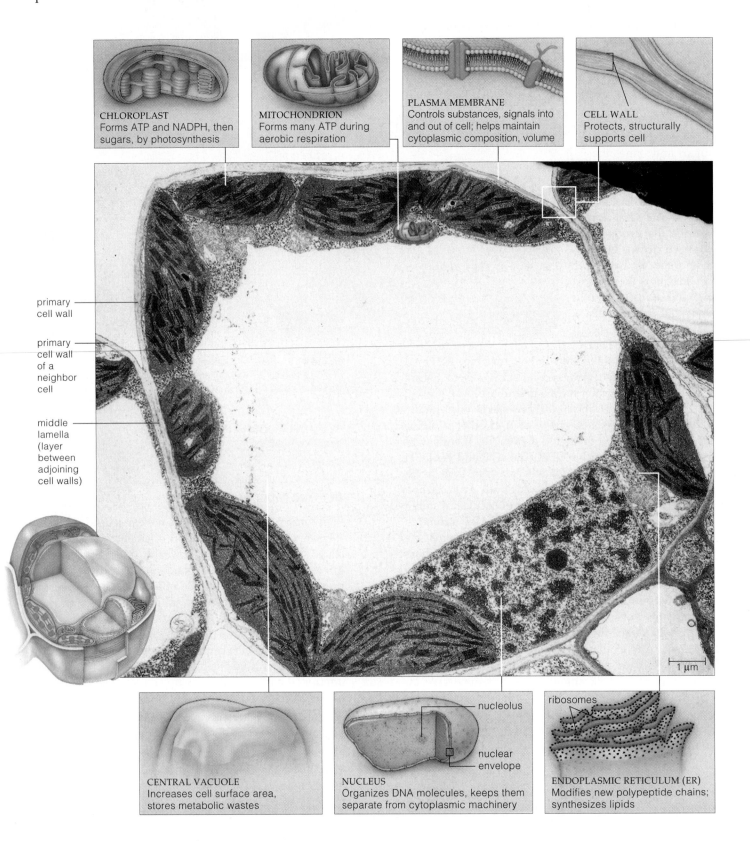

CHLOROPLAST
Forms ATP and NADPH, then sugars, by photosynthesis

MITOCHONDRION
Forms many ATP during aerobic respiration

PLASMA MEMBRANE
Controls substances, signals into and out of cell; helps maintain cytoplasmic composition, volume

CELL WALL
Protects, structurally supports cell

primary cell wall

primary cell wall of a neighbor cell

middle lamella (layer between adjoining cell walls)

1 μm

CENTRAL VACUOLE
Increases cell surface area, stores metabolic wastes

NUCLEUS
Organizes DNA molecules, keeps them separate from cytoplasmic machinery

nucleolus

nuclear envelope

ribosomes

ENDOPLASMIC RETICULUM (ER)
Modifies new polypeptide chains; synthesizes lipids

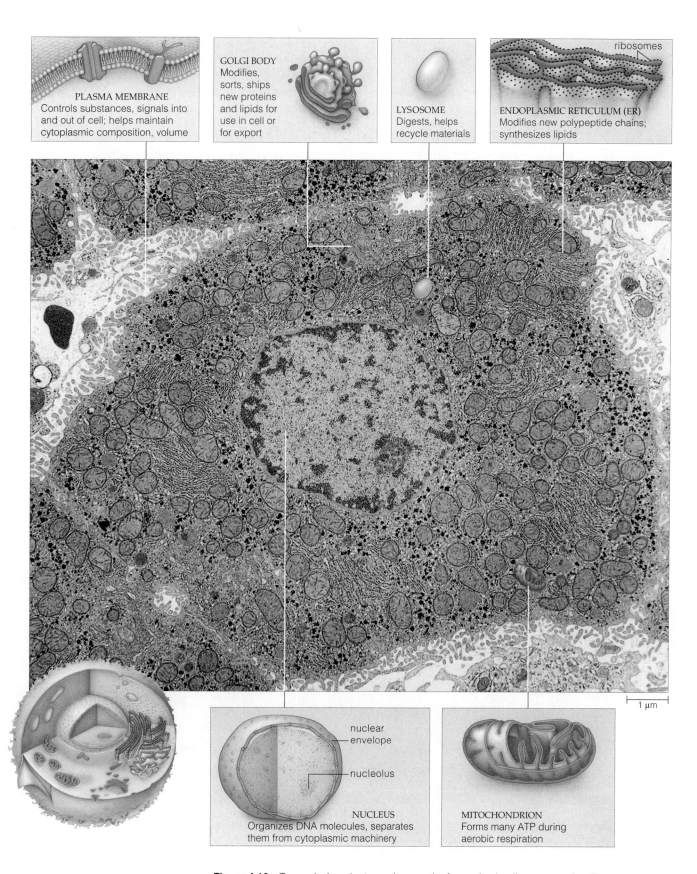

PLASMA MEMBRANE
Controls substances, signals into and out of cell; helps maintain cytoplasmic composition, volume

GOLGI BODY
Modifies, sorts, ships new proteins and lipids for use in cell or for export

LYSOSOME
Digests, helps recycle materials

ribosomes

ENDOPLASMIC RETICULUM (ER)
Modifies new polypeptide chains; synthesizes lipids

1 µm

nuclear envelope

nucleolus

NUCLEUS
Organizes DNA molecules, separates them from cytoplasmic machinery

MITOCHONDRION
Forms many ATP during aerobic respiration

Figure 4.19 Transmission electron micrograph of an animal cell, cross-section. The sketches highlight key organelles. The specimen is a cell from the liver of a rat.

EVEN YOUR CELLS HAVE A SKELETON

What pops into mind when you hear the word *skeleton*? An organized system of bones that includes a skull and rib cage? That is only one kind of skeleton in nature. If you use the general meaning of the word—*something that forms a structural framework*—then eukaryotic cells, too, have a skeleton. A complex, interconnected system of protein filaments extends between their nucleus and plasma membrane. Different parts of this system, the **cytoskeleton**, reinforce, organize, and move internal cell parts and often function in motility. Many parts are permanent, and others form only at certain times in the life of a cell. Figure 4.20 shows examples from one of the main types of cells making up your body.

Microtubules and **microfilaments** are two classes of cytoskeletal elements with roles in most eukaryotic cell movements. Certain animal cells also have ropelike **intermediate filaments** that impart strength. Although prokaryotic cells do not contain a cytoskeleton, some do have reinforcing, actin-like protein filaments.

Microtubules—The Big Ones

Microtubules function mainly in internal organization and in sustained, directional movements that put cell structures and organelles in new locations. Many of them become assembled in a spindle apparatus, which moves chromosomes in specific ways before the cell divides. Other microtubules help shunt chloroplasts, vesicles, and other organelles about.

A microtubule is a straight, hollow cylinder which, at twenty-five nanometers across, is the largest of all cytoskeletal elements in plant and animal cells (Figure 4.21a). Cells build it from tubulin monomers. Tubulin consists of two chemically distinct polypeptide chains, each folded into a globular shape. When the cylinder is being assembled, all of its monomers are oriented in the same direction. This assembly pattern puts slightly different chemical properties at opposite ends of the cylinder. Like bricks being stacked up into a new wall, monomers join the cylinder's *plus* (fast-growing) end, which at first grows freely through the cytoplasm.

In animal cells, microtubules usually grow out in all directions from the centrosome, a type of microtubule organizing center (MTOC). Microtubules form at these sites of dense material, and their *minus* (slow-growing) ends remain anchored in them. But microtubules are inherently unstable, and they can abruptly fall apart.

Cells control which microtubules disassemble and which ones remain intact by selectively capping some of them with other proteins. For instance, caps might be added only to microtubules lengthening in a specific direction—say, the ones that happen to be reinforcing the advancing end of a white blood cell on the prowl. Microtubules are not needed at the trailing end of the cell; they are not capped, and so they fall apart.

Microtubules are targets in the ongoing chemical warfare between plants and the animals that eat them. For instance, the autumn crocus (*Colchicum autumnale*) in Figure 4.22a makes colchicine. This poison blocks the assembly and promotes disassembly of microtubules in animal cells. It does not hurt the plants, which have an evolved insensitivity to it. (Colchicine cannot bind well to the type of tubulin monomers that plants produce.) Similarly, taxol is a poison made by the western yew

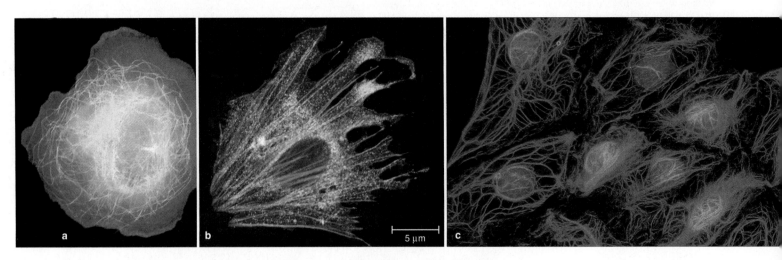

Figure 4.20 Cells stained to reveal their cytoskeleton. (**a**) Microtubules (*gold*) and distribution of actin filaments (*red*) in a pancreatic epithelial cell. This cell secretes bicarbonate, which functions in digestion by neutralizing stomach acid. The *blue* region is DNA. (**b**) Actin filaments (*red*) and accessory proteins (*green* and *blue*) help new epithelial cells such as this one migrate to proper locations in tissues while human embryos are developing. (**c**) Keratin intermediate filaments (*red*) of kangaroo rat cells in culture. The *blue*-stained organelle in each cell is the nucleus.

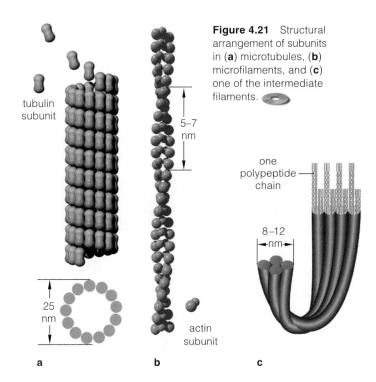

Figure 4.21 Structural arrangement of subunits in (**a**) microtubules, (**b**) microfilaments, and (**c**) one of the intermediate filaments.

tubulin subunit

5–7 nm

25 nm

one polypeptide chain

8–12 nm

actin subunit

a   b   c

Figure 4.22 Two sources of microtubule poisons: (**a**) Autumn crocus, *Colchicum autumnale*. (**b**) Western yew, *Taxus brevifolia*.

(*Taxus brevifolia*), shown in Figure 4.22*b*. Taxol also can block cell division in animal cells. Doctors now use a synthetic version of taxol to stop the uncontrolled cell divisions that give rise to some kinds of tumors.

Microfilaments—The Thin Ones

Microfilaments, at five to seven nanometers across, are the thinnest cytoskeletal elements. Each is composed of two helically twisted polypeptide chains. The chains are assembled from actin monomers (Figure 4.21*b*).

Like microtubules, microfilaments can be assembled and disassembled in controlled ways. They are often arranged as bundles or networks. For instance, the **cell cortex** is a mesh of cytoskeletal elements beneath the plasma membrane. It has bundles, crosslinked arrays, and gel-like arrays of microfilaments. Some reinforce a cell's shape. Some reconfigure the surface, as when the bundle around an animal cell's midsection contracts to pinch the cell in two. Some anchor membrane proteins. Others serve in muscle contraction. When crosslinked ones loosen in some large cells, *cytoplasmic streaming* occurs. Local gel-like regions become more fluid, and they flow vigorously. The flow redistributes substances and cell components through the cell interior.

Myosin and Other Accessory Proteins

The genes for tubulin and actin apparently have been highly conserved over time; all eukaryotic cells make similar forms of these monomers for microtubules and microfilaments. In spite of the uniformity, both kinds of cytoskeletal elements play diverse roles. How? Other proteins associate with them to execute different tasks. For example, as you will see, kinesin and myosin are two kinds of *motor* proteins. They use ATP energy to move on cytoskeletal tracks and help move some cell component to new locations. The *crosslinking* proteins splice together microfilaments as the cell cortex.

Intermediate Filaments

Intermediate filaments are the most stable elements of some cytoskeletons. They are between eight and twelve nanometers wide (Figure 4.21*c*). The six known groups strengthen and help maintain the shape of cells or cell parts. For example, lamins help form a basketlike mesh that reinforces the nucleus. They anchor many adjacent actin and myosin filaments into units of contraction inside muscle cells. Desmins and vimentins help hold the contractile units in position. Different cytokeratins reinforce cells that make nails, claws, horns, and hairs.

In certain animal cells only, intermediate filaments are present in the cytoplasm. Because they have only one or at most two kinds of these filaments, researchers can use them to identify the type of cell. Such typing is a useful tool in diagnosing the tissue origin of different forms of cancer cells. Intermediate filaments also may reinforce the nuclear envelope of all eukaryotic cells.

A cytoskeleton is the basis of the shape, internal structure, and movement of eukaryotic cells. It consists of the same kinds of protein filaments in all eukaryotic cells. Accessory proteins associate with these filaments and extend their range of functions.

Microtubules, the largest cytoskeletal element, are cylinders made of tubulin monomers. They help spatially organize the cell interior and have roles in moving cell components.

Microfilaments, the thinnest cytoskeletal element, consist of two polypeptide chains of actin monomers, helically coiled together. They form flexible, linear bundles and networks that reinforce or restructure the cell surface.

Intermediate filaments help strengthen and maintain the shape of cells or cell parts. Some are present only in the cytoplasm of certain types of animal cells. Others probably help reinforce the nuclear envelope of all eukaryotic cells.

HOW DO CELLS MOVE?

Chugging Along With Motor Proteins

Think of the bustle at a train station during the busiest holiday season, and you get an idea of what is going on inside a cell. Instead of people, cells move vesicles, chromosomes, and other components from one place or position to another. For tracks, they use microtubules and microfilaments. Instead of train engines, they use **motor proteins** which, when repeatedly energized by ATP hydrolysis, move cell components in a sustained, directional fashion. Generally, motor proteins called *kinesins* and *dyneins* move along microtubules; *myosins* move along microfilaments (Figures 4.23 and 4.24).

Consider the kinesins, a diverse group of freight engines. Some use microtubules to move chromosomes during cell division. Like dyneins, some can make one microtubule slide over another. Others line up ER and Golgi bodies with microtubules that extend from the centrosome to the cell periphery. Still others chug along tracks in nerve cells that extend from your spine to your toes. A number of these engines are organized in series, with each moving its freight partway along the track before transferring it to the next engine in line. It takes them a day or two to move vesicles filled with signaling molecules to the end of the line. In plants, kinesins can put chloroplasts in new light-intercepting positions as the angle of the sun changes overhead.

How do they do it? Here we leave the train, so to speak, in favor of a recent model for kinesin "walking." A kinesin molecule has two helical domains that can bind a specific structure or organelle at one end. The other end has two globular domains, analogous to feet, that can bind to a microtubular track (Figure 4.24). One foot moves forward when ATP energizes it. Then the other foot does the same and moves ahead, past the first foot. Through repeats of this molecular action, kinesin chugs along the microtubule, dragging freight with it. Typically it finishes a 1,000-nanometer journey in a few seconds, then lets go of the microtubule.

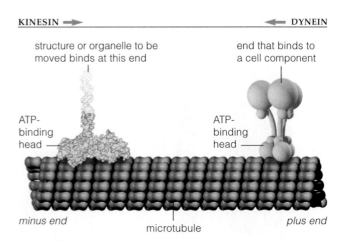

Figure 4.23 Kinesin, one of the motor proteins. Most kinesins move toward the plus end of a microtubule (or microfilament). Most dyneins move toward the minus end. If a microtubule is anchored near the center of a cell, kinesins move components away from it; dyneins move them toward it. These proteins are not drawn to scale. Each "foot" is about the size of one tubulin subunit.

Myosins work in much the same way. They move structures on microfilaments or slide one microfilament over another. For example, each muscle cell is divided lengthwise into many contractile units (Chapter 37). Each unit has microfilaments arranged in parallel with bundles of myosin. Myosin is shaped like a golf club, with a shaft and head. A head energized by ATP binds a microfilament, thrusts it in some direction, and lets go. Repeated power strokes by many myosin molecules make microfilaments slide toward each unit's center. A muscle cell shortens (contracts) as all the units shorten.

Cilia, Flagella, and False Feet

Besides moving internal components, many cells move their body or some part of it through the environment. Some types do this with **flagella** (singular, flagellum)

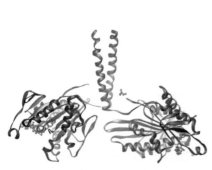

Figure 4.24 (**a**) Ribbon model for kinesin. (**b**) One model for how the kinesin molecule's two heads repeatedly move along a microtubule, driven by phosphate-group transfers from ATP.

a

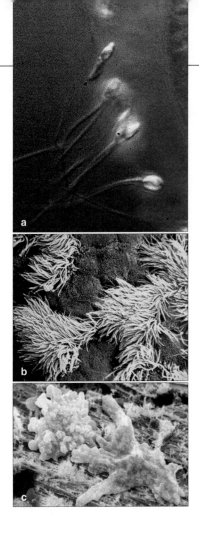

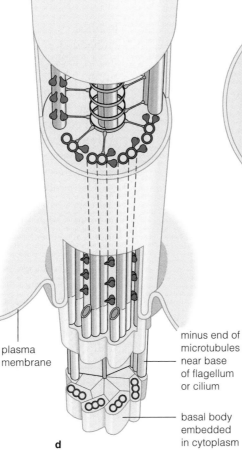

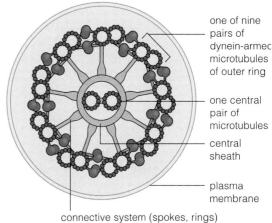

one of nine pairs of dynein-armed microtubules of outer ring

one central pair of microtubules

central sheath

plasma membrane

connective system (spokes, rings)

plasma membrane

minus end of microtubules near base of flagellum or cilium

basal body embedded in cytoplasm

d

Figure 4.25 A few motile structures. (**a**) Human sperm. Each sperm has a flagellum that has one function: moving its DNA-packed head to an egg. (**b**) From the lining of an airway to the lungs, the free surface of ciliated cells (*yellow*) and mucus-secreting cells (*brown*). (**c**) Pseudopods of a macrophage (*light blue*) about to contact another cell. (**d**) Internal organization of flagella and cilia. Both structures have a 9 + 2 array: a ring of nine pairs of microtubules around two pairs at the core. All are connected by spokes and linking elements that restrict the range of sliding.

or **cilia** (cilium). Compared to cilia, flagella usually are longer and less profuse on a cell. Sperm and other free-living, motile cells use them as whiplike tails to swim, as in Figure 4.25*a*. In many multicelled species, ciliated cells stir air or fluid. Many thousands line the airways in your chest (Figure 4.25*b*). Their coordinated motion directs mucus-trapped particles away from the lungs.

Both motile structures have an internal, cylindrical array of nine pairs of microtubules plus a central pair. Protein spokes and links connect to and stabilize this "9+2 array," which arises from a barrel-shaped MTOC known as a **centriole**. The centriole remains positioned below the finished array as a **basal body** (Figure 4.25*d*).

Both flagella and cilia beat by a sliding mechanism that begins at the minus end of microtubules. Inside an unbent flagellum (or cilium), all microtubule doublets extend the same distance into the tip. When the motile structure bends, doublets on the side bending the most are being displaced farthest from the tip.

The motor protein dynein forms many short arms down the length of each microtubule pair in the outer ring. When energized by ATP, all the arms bind the pair in front of them, tilt in a short, downward stroke, then let go. As the bound pair slides down, its arms

bind and force the doublet in front of it to slide down a bit also, and so on around the ring. The microtubules cannot slide far. The system of spokes and links stops them, but it has enough slack built in to permit each microtubule to *bend* a bit. Thus a sliding motion caused by motor proteins is converted to a bending motion.

A final example: Amoebas, macrophages, and many other free-living cells form **pseudopods** ("false feet"). These temporary, irregular lobes project from the body and are used for locomotion and prey capture (Figure 4.25*c*). Microfilaments rapidly elongate in the lobe and push it forward. Motor proteins may chug along these tracks, dragging the plasma membrane with them. The animal cell in Figure 4.20*b* also was moving forward with the help of sheetlike extensions of microfilaments.

Cell contractions and migrations, chromosome movements, and all other forms of cell motility arise at organized arrays of microtubules, microfilaments, and accessory proteins.

When energized by ATP, motor proteins move steadily, in specific directions, along tracks of microtubules and microfilaments. They deliver specific cell components that are bound to them to new locations or into new positions.

CELL SURFACE SPECIALIZATIONS

Our survey of eukaryotic cells concludes with a look at cell walls and some other specialized surface structures. Many are architectural marvels constructed from cell secretions. Others are clustered membrane proteins that structurally and functionally connect neighboring cells.

Eukaryotic Cell Walls

Single-celled eukaryotic species are directly exposed to their surroundings. Many have a **cell wall**, a structural component that wraps continuously around the plasma membrane. A cell wall protects and physically supports its owner. The wall is porous, so water and solutes can easily move to and from the plasma membrane. A great variety of protistans, as in Figure 4.26, have a cell wall. So do plant cells and many types of fungal cells.

For instance, young cells in actively growing parts of plants secrete gluelike polysaccharides (such as pectin), glycoproteins, and cellulose. The

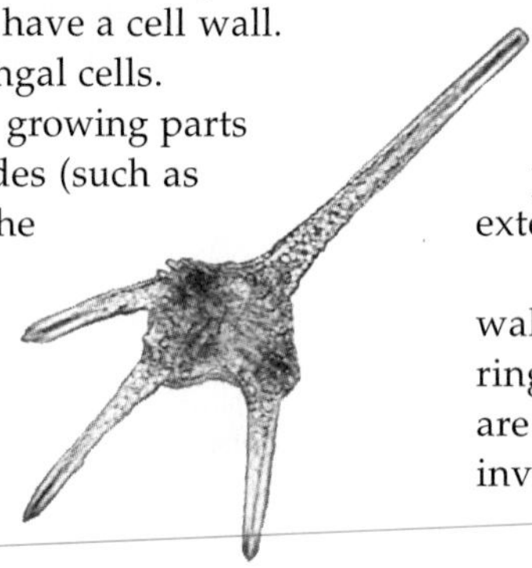

Figure 4.26 From a freshwater habitat, one of the single-celled, walled protistans (*Ceratium*, a dinoflagellate).

cellulose molecules form ropelike strands that become embedded in the gluey matrix. All secretions combined form the cell's **primary wall** (Figure 4.27). Being sticky, primary walls cement abutting cells together. They are thin and pliable, which permits the cell to continue to enlarge under the pressure of incoming water.

At cell surfaces exposed to air, waxes and other cell secretions accumulate. The deposits form a cuticle. This semitransparent, protective surface cover helps restrict evaporative water loss from plants (Figure 4.28a).

Plant cells that produce only a thin wall retain the capacity to divide or to change shape as they grow and develop. Other plant cells stop enlarging as they mature and start secreting material on the primary wall's inner surface. The deposits form a rigid, **secondary wall** that reinforces cell shape (Figure 4.27e). Whereas cellulose makes up less than 25 percent of the cell's primary wall, the secondary wall deposits are more extensive and contribute more to its structural support.

In woody plants, up to 25 percent of the secondary wall is made of lignin. Lignin consists of a six-carbon ring to which a three-carbon chain and an oxygen atom are attached. It makes plant parts more waterproof, less inviting to plant-attacking organisms, and stronger.

Figure 4.27 Examples of cell walls in (**a**) flax plants. (**b**) Primary cell wall of young cells in flower petals. Cell secretions form the middle lamella, a layer between the walls of adjoining cells. The layer is thickest in adjoining corners. Plasmodesmata, membrane-lined channels across the adjacent walls, connect the cytoplasm of neighboring cells. (**c**,**d**) Part of the lustrous fibers in a cell from a flax stem. We make linen from such fibers, which are three times stronger than cotton fibers. (**e**) In flax fibers, as in many other types of plant cells, more layers become deposited inside the primary wall. The layers stiffen the wall and help maintain its shape. Later, the cell dies, leaving the stiffened walls behind. This also happens in water-conducting pipelines that thread through most plant tissues. Interconnected, stiffened walls of dead cells form the tubes.

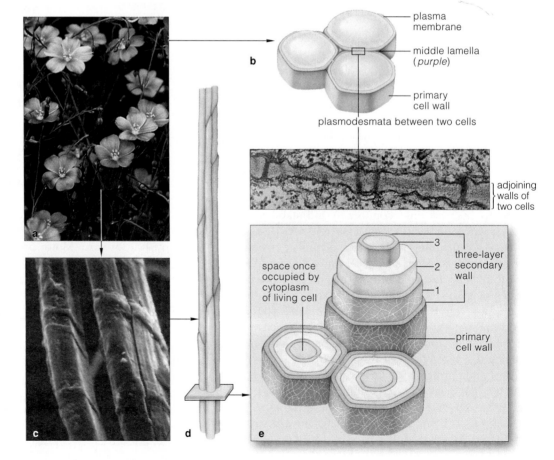

plasma membrane

middle lamella (*purple*)

primary cell wall

plasmodesmata between two cells

adjoining walls of two cells

space once occupied by cytoplasm of living cell

three-layer secondary wall

primary cell wall

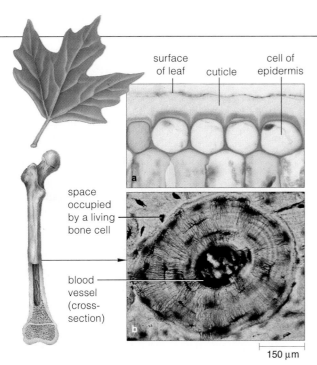

surface of leaf | cuticle | cell of epidermis

a

space occupied by a living bone cell

blood vessel (cross-section)

b

150 µm

Figure 4.28 (**a**) Section through a plant cuticle, an outer surface layer of cell secretions. (**b**) Cross-section through compact bone tissue, stained for microscopy.

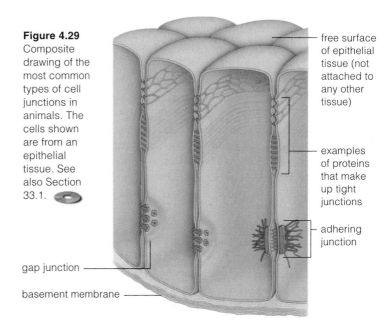

Figure 4.29 Composite drawing of the most common types of cell junctions in animals. The cells shown are from an epithelial tissue. See also Section 33.1.

free surface of epithelial tissue (not attached to any other tissue)

examples of proteins that make up tight junctions

adhering junction

gap junction

basement membrane

Matrixes Between Animal Cells

Animal cells have no walls. But intervening between many of these cells are matrixes made of cell secretions and materials taken up from the surroundings. Think of cartilage at the knobby ends of your leg bones. Cartilage consists of scattered cells plus collagen or elastin fibers formed from their secretions, all embedded in a ground substance of polysaccharides (Section 33.2). Or think of bone tissue, in which an extensive matrix separates the living bone cells (Figure 4.28*b* and Section 37.5).

Cell Junctions

Even when a wall or some other structure imprisons a cell in its own secretions, the only contact that cell has with the outside world is through its plasma membrane. In multicelled species, membrane components project into adjoining cells and into the surrounding medium. Among the components are **cell junctions** where a cell sends and receives signals and materials, and where it recognizes and cements itself to cells of the same type.

In plants, for instance, many tiny channels cross the adjacent primary walls of living cells and interconnect their cytoplasm. Figure 4.27*b* shows a few. Each channel is a plasmodesma (plural, plasmodesmata). The plasma membranes of adjoining cells merged during growth. Now they line the channels, so there is an uninterrupted flow of substances between the cells. That is how living cells in the plant body engage in quick exchanges.

In most animal tissues, three categories of cell-to-cell junctions are common (Figure 4.29). *Tight* junctions link

the cells of most body tissues, including epithelia that line outer surfaces, internal cavities, and organs. They seal abutting cells so water-soluble substances can't pass between them. That is why gastric fluid cannot leak across the stomach's lining and damage internal tissues. *Adhering* junctions join cells in tissues of the skin, heart, and other organs subjected to stretching. Gap junctions link the cytoplasm of neighboring cells. They are open channels for a rapid flow of signals and substances.

Cell Communication

Also at the plasma membrane are diverse receptors for signals from other cells. **Cell communication** refers to ways in which one cell signals another to change its activities. For example, certain hormones dock at body cells as a signal to start dividing. That external signal is essential for growth and development of all complex organisms. Throughout this book you will come across specific examples of cell communication. Chapters 15, 28, 32, 34, 36, 39, and 43 especially provide details of the kinds of signaling mechanisms used to coordinate activities of all living cells in the multicelled body.

A variety of protistan, plant, and fungal cells have a porous but protective wall that surrounds the plasma membrane. The cells themselves secrete wall-forming materials.

The combined secretions of many cells form the cuticle that covers the surface of plants and certain animals, as well as the extracellular matrixes of animal tissues.

In multicelled organisms, diverse proteins at the plasma membrane function as physical links, cytoplasmic bridges, and receptors for communication signals between cells.

4.12

PROKARYOTIC CELLS

We turn now to prokaryotic cells. Unlike the cells you've considered so far, a nucleus does not enclose their DNA. *Prokaryotic* means "before the nucleus." Biologists chose the word to remind us that this type of cell appeared on Earth before the nucleus evolved in the forerunners of eukaryotic cells.

These are the smallest known cells. Most are not much more than a micrometer wide; even rod-shaped species are only a few micrometers long. In structural terms, they are the simplest cells to think about. Most have a semirigid or rigid cell wall that surrounds the plasma membrane, supports the cell, and imparts shape to it (Figure 4.30*a*). As you will see later, their wall differs from walls of eukaryotic cells. Dissolved substances can move freely to and from the plasma membrane because the wall is permeable. Commonly, sticky polysaccharides enclose a wall and help cells attach to interesting surfaces, such as river rocks, teeth, and the vagina. Many pathogenic (disease-causing) eubacteria have a capsule made of thick, jellylike polysaccharides. It surrounds and protects the wall.

As in eukaryotic cells, the plasma membrane of prokaryotic cells mediates the flow of substances to and from the cytoplasm. It, too, has proteins that serve as channels, transporters, and receptors for diverse signals and substances. It, too, incorporates built-in machinery for metabolic reactions, such as the breakdown of energy-rich compounds. In most of the photosynthetic prokaryotic cells, organized arrays of proteins in the plasma membrane harness light energy, then convert it to chemical energy of ATP.

Prokaryotic cells are so small they have only a tiny volume of cytoplasm. But they have many ribosomes on which polypeptide chains are built. The cytoplasm is distinct from that of eukaryotic cells; it is continuous with an irregularly shaped region of DNA. Membranes don't enclose this region, which is the nucleoid (Figure 4.30*b*). A circular DNA molecule, also called a bacterial chromosome, occupies the region. Given their tiny size and internal simplicity, the cells don't need a complex cytoskeleton, but actin-like protein filaments do form a scaffold that helps shape the sturdier cell wall.

Projecting out from the surface of many prokaryotic cells are one or more bacterial flagella. These threadlike motile structures differ from eukaryotic flagella; they do not have a 9 + 2 array of microtubules. Bacterial flagella help a cell move rapidly through fluid environments. Other surface projections include pili (singular, pilus), the main protein filaments that help many types of cells attach to various surfaces, even to one another.

We recognize two major groups of prokaryotic cells: Archaebacteria and Eubacteria (Sections 19.7 and 21.3).

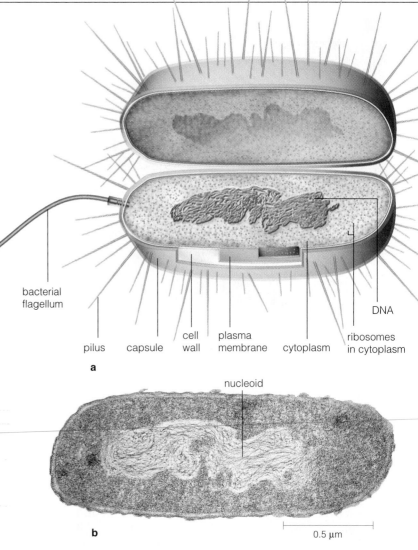

bacterial flagellum

pilus   capsule   cell wall   plasma membrane   cytoplasm   ribosomes in cytoplasm

DNA

a

nucleoid

b   0.5 µm

Figure 4.30 (**a**) Generalized sketch of a typical prokaryotic body plan. (**b**) Micrograph of the bacterium *Escherichia coli*.

The gut of humans and other mammals holds a huge population of a normally harmless strain of *E. coli*. Some harmful strains can contaminate meat sold commercially. One strain also can taint cider. Cooking meat thoroughly or pasteurizing cider kills these pathogens. Where this was not done, people who ate the meat or drank the cider became sick. Some died.

Facing page: (**c**) Researchers manipulated this *E. coli* cell to release its single, circular molecule of DNA. (**d**) Cells of various bacterial species are shaped like balls, rods, or corkscrews. Ball-shaped cells of *Nostoc*, a photosynthetic eubacterium, stick together in a thick, jellylike sheath of their own secretions. Chapter 21 gives other examples. (**e**) Like this *Pseudomonas marginalis* cell, many species have one or more bacterial flagella that propel the cell body in fluid environments.

Together, the species in these groupings are the most metabolically diverse organisms of all. Different kinds have managed to exploit energy and raw materials in just about every kind of environment, ranging from hot springs and snow fields to hot, dry deserts. In addition, ancient prokaryotic cells gave rise to all the protistans,

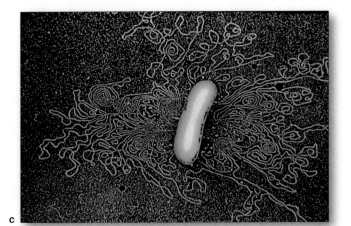

c

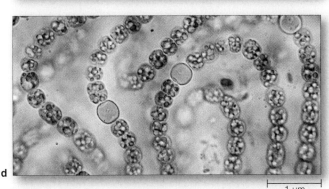

d

1 μm

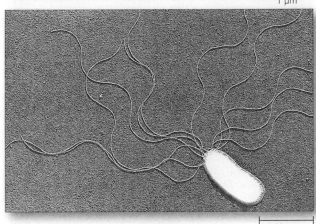

e

10 μm

plants, fungi, and animals ever to appear on Earth. The evolution, structure, and functions of these remarkable cells are topics of later chapters.

Archaebacteria and eubacteria are prokaryotic cells; their DNA is not housed inside a nucleus. Most species have a cell wall around the plasma membrane. Generally, they do not have organelles comparable to those of eukaryotic cells.

These are the simplest cells, but as a group they show the most metabolic diversity. Their metabolic activities proceed at the plasma membrane and within the cytoplasm.

SUMMARY   **Gold** indicates text section

1. Three generalizations constitute the cell theory: **CI**
   a. All living things are composed of one or more cells.
   b. The cell is the smallest entity that still retains the properties of life. That is, it lives independently or has a heritable, built-in capacity to do so.
   c. New cells arise only from cells that already exist.

2. All cells start out life with a plasma membrane, a region of DNA, and cytoplasm. The plasma membrane (a thin, outer membrane) maintains a cell as a distinct entity and lets metabolic activities proceed apart from random events in the environment. Many substances and signals cross it in controlled ways. The cytoplasm consists of ribosomes, structural elements, fluids, and (only in eukaryotic cells) organelles between the region of DNA and the plasma membrane. *4.1*

3. Membranes of lipids and proteins are essential for cell structure and function. The lipids form two layers, with their hydrophobic tails sandwiched between their hydrophilic heads (which are dissolved in fluid). This bilayer gives the membrane its basic structure. Water-soluble substances cannot cross it. Proteins embedded in the bilayer or attached to it have diverse roles. Many are channels or pumps where water-soluble substances cross the bilayer, others are receptors, and so on. *4.1*

4. The surface-to-volume ratio constrains cell size, cell shape, and the body plans of multicelled organisms. *4.1*

5. Modern microscopes harness wavelengths of light or focused beams of electrons to reveal cell structures. *4.2*

6. Membranes divide the cytoplasm of eukaryotic cells into functional compartments (organelles). Prokaryotic cells do not have comparable organelles. *4.3, 4.12*

7. Organelle membranes separate metabolic reactions inside the cytoplasm and let different kinds proceed in orderly fashion. Many similar reactions proceed at the plasma membrane of prokaryotic cells. *4.3*
   a. The nuclear envelope functionally separates DNA from the metabolic machinery of the cytoplasm. *4.4*
   b. The endomembrane system includes the ER, Golgi bodies, and vesicles. Many new proteins are modified into final form and lipids are assembled in this system. Finished products are packaged and then shipped off to destinations inside or outside the cell. *4.5*
   c. Mitochondria specialize in producing numerous ATP molecules by oxygen-requiring reactions. *4.6*
   d. Chloroplasts specialize in capturing energy from the sun and using it to build organic compounds. *4.7*

8. The cytoskeleton of eukaryotic cells functions in cell shape, internal organization, and movement. Many cells have walls and other surface specializations. *4.9, 4.10*

9. Table 4.3 on the next page summarizes the defining features of prokaryotic and eukaryotic cells.

Table 4.3 *Summary of Typical Components of Prokaryotic and Eukaryotic Cells*

Cell Component	Function	PROKARYOTIC Archaebacteria, Eubacteria	EUKARYOTIC Protistans	Fungi	Plants	Animals
Cell wall	Protection, structural support	✔*	✔*	✔	✔	None
Plasma membrane	Control of substances moving into and out of cell	✔	✔	✔	✔	✔
Nucleus	Physical separation and organization of DNA	None	✔	✔	✔	✔
DNA	Encoding of hereditary information	✔	✔	✔	✔	✔
RNA	Transcription, translation of DNA messages into polypeptide chains of specific proteins	✔	✔	✔	✔	✔
Nucleolus	Assembly of subunits of ribosomes	None	✔	✔	✔	✔
Ribosome	Protein synthesis	✔	✔	✔	✔	✔
Endoplasmic reticulum (ER)	Initial modification of many of the newly forming polypeptide chains of proteins; lipid synthesis	None	✔	✔	✔	✔
Golgi body	Final modification of proteins, lipids; sorting and packaging them for use inside cell or for export	None	✔	✔	✔	✔
Lysosome	Intracellular digestion	None	✔	✔*	✔*	✔
Mitochondrion	ATP formation	**	✔	✔	✔	✔
Photosynthetic pigments	Light–energy conversion	✔*	✔*	None	✔	None
Chloroplast	Photosynthesis; some starch storage	None	✔*	None	✔	None
Central vacuole	Increasing cell surface area; storage	None	None	✔*	✔	None
Bacterial flagellum	Locomotion through fluid surroundings	✔*	None	None	None	None
Flagellum or cilium with 9 + 2 microtubular array	Locomotion through or motion within fluid surroundings	None	✔*	✔*	✔*	✔
Complex cytoskeleton	Cell shape; internal organization; basis of cell movement and, in many cells, locomotion	Rudimentary***	✔*	✔*	✔*	✔

  * Known to be present in cells of at least some groups.
  ** Many groups use oxygen-requiring (aerobic) pathways of ATP formation, but mitochondria are not involved.
*** Protein filaments form a simple scaffold that helps support cell wall in at least some species.

Review Questions

1. State the three key points of the cell theory. *CI*

2. Suppose you wish to observe the three-dimensional surface of an insect's eye. Would you see more details with the aid of a compound light microscope, transmission electron microscope, or scanning electron microscope? *4.2*

3. Label the organelles in the two diagrams below of a plant cell and an animal cell. What are the main differences between the two kinds of cells? In what ways are they similar? *4.3*

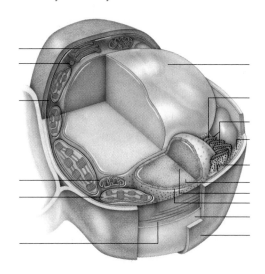

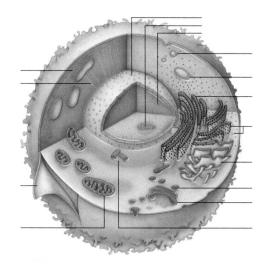

4. Describe three features that all cells have in common. After reviewing Table 4.3, write a paragraph on the key differences between eukaryotic and prokaryotic cells. *4.1, 4.3, 4.12*

5. Briefly characterize the structure and function of the cell nucleus, nuclear envelope, and nucleolus. *4.4*

6. Distinguish between chromosome and chromatin. Do chromosomes always maintain the same appearance during the life of a cell? *4.4*

7. Name three kinds of organelles that are components of the cell's endomembrane system. Briefly describe the function of each component. *4.5*

8. Is this statement true or false: Plant cells have chloroplasts, but not mitochondria. Explain your answer. *4.3, 4.6, 4.7*

9. What are the functions of the central vacuole? *4.7*

10. Define cytoskeleton. How does it aid in cell functions? *4.9*

11. What gives rise to the 9 + 2 array of cilia and flagella? *4.10*

12. Which of the following kinds of organisms typically have walled cells: bacteria, protistans, fungi, plants, animals? *4.11*

13. State the function and describe the general characteristics of cell walls. For example, are the walls permeable? Why or why not? *4.11*

14. As they grow and develop, many kinds of plant cells form a secondary wall. Is this wall deposited inside or outside the surface of the primary wall that formed earlier? *4.11*

15. In multicelled organisms, coordinated interactions depend on physical linkages and chemical communications between cells. What types of junctions are present between neighboring animal cells? Plant cells? *4.11*

16. Label the parts of the bacterial cell diagram shown at the upper right. *4.12*

Self-Quiz   ANSWERS IN APPENDIX III

1. Cell membranes consist mainly of a _____ .
   a. carbohydrate bilayer and proteins
   b. protein bilayer and phospholipids
   c. lipid bilayer and proteins
   d. none of the above

2. Organelles _____ .
   a. are membrane-bound compartments
   b. are typical of eukaryotic cells, not prokaryotic cells
   c. separate chemical reactions in time and space
   d. all of the above are features of the organelles

3. Cells of many protistans, plants, and fungi, but not animals, commonly have _____ .
   a. mitochondria   c. ribosomes
   b. a plasma membrane   d. a cell wall

4. Is this statement true or false: The plasma membrane is the outermost component of all cells. Explain your answer.

5. Unlike eukaryotic cells, prokaryotic cells _____ .
   a. lack a plasma membrane   c. do not have a nucleus
   b. have RNA, not DNA   d. all of the above

6. Match each cell component with its function.
   ____ mitochondrion   a. synthesis of polypeptide chains
   ____ chloroplast   b. initial modification of new
   ____ ribosome   polypeptide chains
   ____ rough ER   c. final modification of proteins;
   ____ Golgi body   sorting, shipping tasks
                           d. photosynthesis
                           e. formation of many ATP

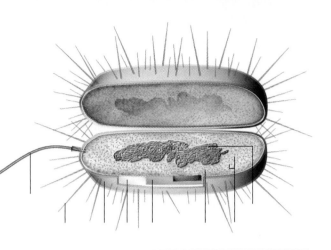

Critical Thinking

1. In *The Immunity Syndrome*, a classic Star Trek episode, a stupendous amoeba engulfs a starship. Spock uses a shuttle-craft to plant explosives inside the amoeba to blow it up before it reproduces and destroys the universe. What is wrong with this scenario? Similarly, why is it likely that you will never encounter even a two-ton amoeba on a sidewalk?

2. Your professor shows you an electron micrograph of a cell with large numbers of mitochondria and Golgi bodies. You notice that this particular cell also contains a great deal of rough endoplasmic reticulum. What kinds of cellular activities would require such an abundance of the three kinds of organelles?

3. The 1 in 30,000 individuals affected by a very rare disorder, *immotile cilia syndrome*, cannot clear mucus from airways that lead to their lungs. Bacteria thrive in the mucus that collects in the airways, and their metabolic by-products damage tissues. Chronic inflammation follows and contributes to the damage.

   Researchers using video microscopy discovered that the cilia of affected people beat poorly or not at all, even when their surface appearance is normal. Speculate on which components of the cilium may be contributing to the disorder. How do you suppose they became abnormal?

Selected Key Terms

basal body *4.10*	endomembrane	nuclear envelope *4.4*
cell *4.1*	system *4.5*	nucleoid *4.1*
cell communication	ER (endoplasmic	nucleolus *4.4*
4.11	reticulum) *4.5*	nucleus *4.1, 4.4*
cell cortex *4.9*	eukaryotic cell *4.1*	organelle *4.3*
cell junction *4.11*	flagellum *4.10*	peroxisome *4.5*
cell theory *CI*	Golgi body *4.5*	plasma membrane *4.1*
cell wall *4.11*	intermediate	primary wall *4.11*
central vacuole *4.7*	filament *4.9*	prokaryotic cell *4.1, 4.12*
centriole *4.10*	lipid bilayer *4.1*	pseudopod *4.10*
chloroplast *4.7*	lysosome *4.5*	ribosome *4.1*
chromatin *4.4*	microfilament *4.9*	secondary wall *4.11*
chromosome *4.4*	micrograph *4.2*	surface-to-volume
cilium *4.10*	microtubule *4.9*	ratio *4.1*
cytoplasm *4.1*	mitochondrion *4.6*	vesicle *4.5*
cytoskeleton *4.9*	motor protein *4.10*	wavelength *4.2*

Readings

Alberts, B., et al., 2002. *Molecular Biology of the Cell.* Fourth edition. New York: Garland. Clear look at the cytoskeleton.

Wolfe, S., 1993. *Molecular and Cellular Biology.* Belmont, California: Brooks–Cole. Good introduction to the kinds of experiments and research that are revealing details of cell structure and function.

On-Line readings at Student Guide for InfoTrac: www.brookscole.com/biology

A CLOSER LOOK AT CELL MEMBRANES

One Bad Transporter and Cystic Fibrosis

As small as it may be, a cell is a living thing engaged in risky business. Think of the cells in your body and how they must respond to something as ordinary as water. At all times, they must move water and solutes in one direction or the other across their plasma membrane. The membrane crossings must be exceedingly selective, for they help keep conditions favorable for survival of

individual cells and, ultimately, you. If all goes well, cells take in and send out the right amounts—not too little, not too much. But who is to say that life always goes well?

CFTR is one type of protein channel across the plasma membrane of epithelial cells. Sheets of these cells line sweat glands, airways and sinuses, and ducts in the digestive and reproductive systems. Chloride ions move through them, and water follows to form a thin film on the free surface of the linings. Mucus, which lubricates tissues and helps prevent infection, slides freely on the watery film.

Sometimes the gene coding for CFTR is mutated. Not enough chloride and water cross to the free surface of the linings, so the film does not form. In its absence, mucus gets dried out, thick, and difficult to displace.

For example, thickened mucus clogs ducts from the pancreas, so digestive enzymes cannot reach the small intestine where most food is digested and absorbed. Weight loss follows. Sweat glands secrete too much salt and alter the body's water–salt balance, which affects the heart and other organs. Males become sterile.

What happens in airways to the lungs? Ciliated cells normally sweep away bacteria and other particles that become trapped in the mucus (Section 4.10). Not now. **Biofilms** form. These consist of microbial populations anchored to the lining by stiff, sticky polysaccharides of their own making. Biofilms strongly resist the body's defenses and antibiotics. *Pseudomonas aeruginosa* is the most efficient of the colonizers. It causes low-grade infections that may persist for years. Most patients die because their lungs fail. Most expect to live no longer than thirty years. At present there is no cure.

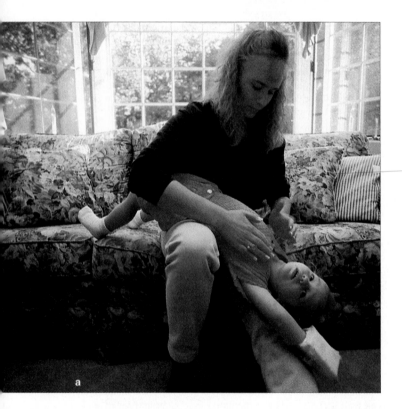

a

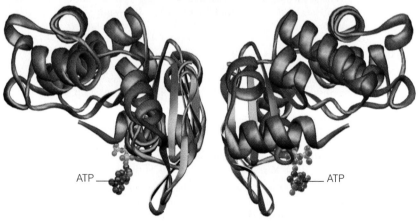

ATP   ATP

b

Figure 5.1 (**a**) Child affected by cystic fibrosis, or CF, who each day endures chest and back thumps, and repositionings to dislodge the thick mucus that collects in airways to the lungs. The symptoms can vary from one affected individual to the next, partly because the abnormal protein that causes CF has mutated in hundreds of different ways. Environmental factors and a person's genetic makeup also affect the outcome.

(**b**) Partial ribbon model for one of the ABC transporters, which include CFTR. The two parts shown are the ATP-driven motors that widen an ion channel across the plasma membrane. Other molecular details of CFTR, including the structure of its two membrane-spanning domains, are being worked out.

The mutant CFTR protein that sets these events in motion is the molecular basis of *cystic fibrosis* (CF), now the most common fatal genetic disorder in the United States. More than 10 million people carry the mutant gene. However, CF develops only when they have inherited two copies of it, one from each parent. This happens in about 1 of every 3,300 live births.

CFTR belongs to the family of **ABC transporters**, membrane proteins in all prokaryotic and eukaryotic species. Some, like CFTR, are channels through which hydrophobic substances cross the membrane. Others pump substances across. By their action, some types even control what other membrane proteins are doing.

In all but 10 percent of CF patients, loss of a single amino acid during protein synthesis causes the fatal disorder. Before a new CFTR protein is shipped to the plasma membrane, it is supposed to be modified in the endomembrane system you read about in Chapter 4. Copies of the mutant protein do indeed enter the ER, but enzymes destroy 99 percent of them before they even reach Golgi bodies. Thus few chloride channels reach their normal destinations.

CFTR may also contribute to sinus problems of an estimated 30 million people in the United States alone. *Sinusitis* is a chronic inflammation of the lining of cavities in the skull, around the nose. For one study at Johns Hopkins University, researchers took blood samples from 147 sinusitis patients and from a control group of 147 unaffected people. They focused on only sixteen of the 500+ known mutated forms of the CFTR gene. It turns out that ten of the sinusitis patients inherited a single copy of one of the sixteen mutated CFTR genes, the only ones considered for this study.

By this example, you can see how problems among a huge number of individuals can arise when cells are denied working copies of just one of the proteins that span the plasma membrane. Think about it, and you also may come to appreciate even more the precision with which normal cells use their proteins.

For eukaryotic cells, that precision extends also to the membranes of those internal compartments called organelles. Aerobic respiration, photosynthesis, and many other processes require the selective movement of substances across internal boundary layers, too. Gain insight into the structure and functioning of cell membranes, and you gain insight into survival at life's most fundamental level.

Key Concepts

1. All cell membranes consist primarily of lipids, organized as a double layer. The lipid bilayer gives the membrane its primary structure and prevents haphazard movement of water-soluble substances across it. Diverse proteins embedded in the bilayer or associated with one of its surfaces carry out most membrane functions.

2. The plasma membrane incorporates many receptor proteins, which receive chemical signals that trigger changes in cell activities. It incorporates recognition proteins, which identify a cell as being of a certain type. Spanning all cell membranes are transport proteins through which specific water-soluble substances can cross the bilayer.

3. A concentration gradient is a difference in the number per unit volume of molecules (or ions) of a substance between two regions. The molecules tend to show a net movement down such a gradient, to the region where they are less concentrated. This behavior is called diffusion.

4. Metabolism depends on concentration gradients that drive the directional movements of substances. Cells have mechanisms for increasing or decreasing water and solute concentrations across the plasma membrane and internal cell membranes.

5. With passive transport, a solute crosses a membrane simply by diffusing through a channel in the interior of a transport protein. With active transport, a different kind of transport protein pumps a solute across the membrane, against its concentration gradient. Active transport requires an energy input, typically from ATP.

6. By a molecular behavior called osmosis, water diffuses across any selectively permeable membrane to a region where its concentration is lower.

7. Larger packets of material move across the plasma membrane by processes of endocytosis and exocytosis.

MEMBRANE STRUCTURE AND FUNCTION

Revisiting the Lipid Bilayer

Imagine you are peering at a living cell with the aid of a microscope. What do you suppose would happen if you punctured it with an extremely fine needle? You may be surprised to find that its cytoplasm won't ooze out. Its plasma membrane will flow over the puncture site and seal it! Like the cytoplasmic and extracellular fluids bathing it, a cell membrane is not a solid, static wall; it has a fluid quality.

How does a fluid membrane remain distinct from its fluid surroundings? Think back on what you know about phospholipids, its most abundant components. Each **phospholipid**, recall, has a phosphate-containing head and two fatty acid tails attached to a glycerol backbone (Figure 5.2*a*). The head is hydrophilic, which means it readily dissolves in water. Its two tails are hydrophobic; water repels them. When you immerse many phospholipid molecules in water, they interact with water molecules and with one another until they spontaneously cluster as a sheet or film at the water's surface. Their interactions might even force them to become positioned as two layers, with all fatty acid tails sandwiched between all hydrophilic heads. This **lipid bilayer** arrangement is the structural basis of cell membranes (Section 4.1 and Figure 5.2*c*).

A lipid bilayer arrangement minimizes the number of hydrophobic groups exposed to water. A puncture in a cell membrane invites sealing behavior because it is energetically unfavorable; too many hydrophobic groups become exposed to the surroundings.

Few cells get jabbed by fine needles. But the self-sealing behavior of membrane phospholipids is good for more than damage control. Among other things, it helps vesicles form. When, say, a patch of ER or Golgi membrane is budding off at one end, its phospholipids are being repelled by water in the cytoplasm around it. The water "pushes" the phospholipids together, the bud rounds off into a vesicle, and the rupture seals.

What Is the Fluid Mosaic Model?

Figure 5.3 shows a small patch of cell membrane that corresponds to the **fluid mosaic model**. By this model, a cell membrane has a mixed composition—a *mosaic*—of phospholipids, glycolipids, sterols, and proteins. The phospholipid heads differ, and their tails differ in length and saturation. (Unsaturated fatty acids have one or more double bonds in their backbone and fully saturated ones have none.) Glycolipids are structurally similar to them, but their heads also have one or more sugars attached to them. In animal cell membranes, cholesterol is the most abundant sterol (Figure 5.2*b*). Plant cell membranes incorporate phytosterols.

Also by this model, a membrane is *fluid* as a result of the motions and interactions of its component parts. Most of the phospholipids are free to drift sideways. They also spin about their long axis and flex their tails, which keeps adjacent molecules from packing together into a solid layer. The short or kinked (unsaturated) hydrophobic tails contribute to the membrane fluidity. As you will read later, some membrane proteins also are free to drift laterally through the fluid bilayer.

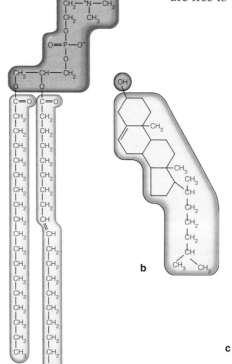

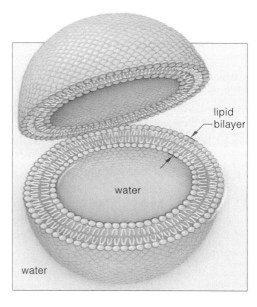

Figure 5.2 **(a)** Structural formula for phosphatidylcholine. This is a phospholipid and one of the most common components of animal cell membranes. *Orange* indicates its hydrophilic head; *yellow* indicates its hydrophobic tails.

(b) Structural formula for the main sterol in animal tissues: cholesterol. Phytosterols are its equivalent in plant tissues.

(c) Spontaneous organization of lipid molecules into two layers (a bilayer structure). When immersed in liquid water, their hydrophobic tails become sandwiched between their hydrophobic heads, which dissolve in the water.

lipid bilayer

water

water

a

b

c

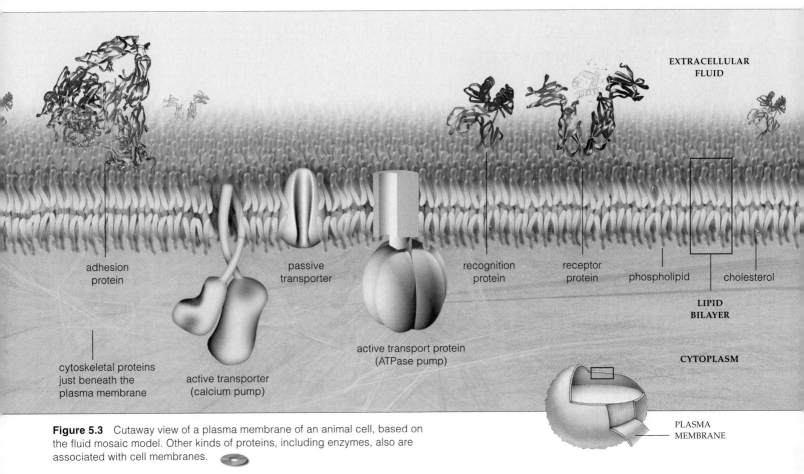

Figure 5.3 Cutaway view of a plasma membrane of an animal cell, based on the fluid mosaic model. Other kinds of proteins, including enzymes, also are associated with cell membranes.

The fluid mosaic model is a good starting point for exploring membranes. But bear in mind, membranes differ in which molecules they incorporate and how their molecules are arranged. Even the two surfaces of the same bilayer differ. For example, oligosaccharides and other carbohydrates are attached to many protein and lipid components of a plasma membrane, but only on its outward-facing surface (Figure 5.3). The number and kinds of carbohydrates differ from one species to the next. They differ even among different cells of the same individual.

How do membrane surfaces become so distinct? Their characteristics start inside the ER, where many polypeptide chains are modified before being shipped to their final destinations (Section 4.5). Proteins that will become part of the plasma membrane are shipped off in vesicles that fuse with a Golgi body. There they are further modified, then shipped off in other vesicles that fuse with the plasma membrane. Figure 5.4 shows how fusion releases the proteins to the outside face of the membrane.

The next section introduces the kinds of membrane proteins you will be encountering in many chapters. Here again, you will see how the function of a protein correlates with its molecular structure.

A cell membrane includes two layers composed mainly of lipids, phospholipids especially. This lipid bilayer is the structural foundation for the membrane and also serves as a barrier to water-soluble substances.

Hydrophobic parts of membrane lipids are sandwiched between the hydrophilic parts, which are dissolved either in cytoplasmic fluid or in extracellular fluid.

Many different proteins having different functions are embedded in the membrane or positioned at its surfaces.

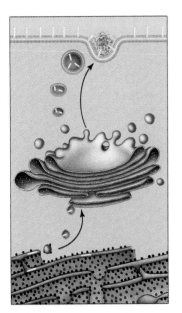

Figure 5.4 Source of the asymmetrical distribution of proteins, carbohydrates, and lipids in membranes. Proteins that function at the cell surface start out as new polypeptide chains that pass through the ER and Golgi bodies. There, they become modified. Many are then shipped to the plasma membrane inside vesicles. A vesicle's membrane fuses with the plasma membrane. As it does, proteins inside automatically become oriented correctly in the plasma membrane.

A GALLERY OF MEMBRANE PROTEINS

Where Are the Proteins Positioned?

Cells make contact with their surroundings through lipids and proteins of their plasma membrane. Here, they receive and send out a variety of substances and signals. For multicelled organisms, this is where cells that defend the body encounter evidence of an attack. This is where cells adhere to one another in tissues or to the extracellular matrix. Proteins have roles in most of these tasks, and here again we see how molecular structure translates into specific functional roles.

Integral proteins interact with hydrophobic parts of the bilayer's phospholipids, and they are not easy to remove. Most span the bilayer, with their hydrophilic domains extending past both of its surfaces. *Peripheral* proteins are positioned at the membrane surface, not in the bilayer. Weak interactions, including hydrogen bonds, allow them to associate with integral proteins and with polar heads of the membrane lipids.

What Are Their Functions?

Figure 5.5 introduces the main categories of membrane proteins and their functions. All span the bilayer. In multicelled organisms, **adhesion proteins** help cells of the same type locate, stick together, and remain in the proper tissues. After tissues form, some adhesion sites become part of cell junctions, as Section 4.11 describes.

The **communication proteins** form channels that match up across the plasma membranes of two cells; they let signals and substances flow rapidly between their cytoplasm. **Receptor proteins** bind extracellular substances, such as hormones, that trigger changes in cell activities. For example, enzymes that crank up machinery for cell growth and division are switched on when a specific hormone binds with receptors for it. Different cells have many different combinations of receptors. **Recognition proteins** are like molecular fingerprints; they identify each cell as belonging to a particular tissue or individual. **Transport proteins** passively let solutes cross the membrane through their interior or they actively pump them through.

Other proteins, including enzymes, are embedded in cell membranes. As you will see, some enzymes also are associated with other kinds of proteins and are necessary for their functioning.

Membrane proteins include transporters, which passively or actively move water-soluble substances across the lipid bilayer. They also include diverse enzymes.

The plasma membrane, especially of multicelled species, has many receptors as well as proteins that function in cell-to-cell adhesion, cell communication, and self-recognition.

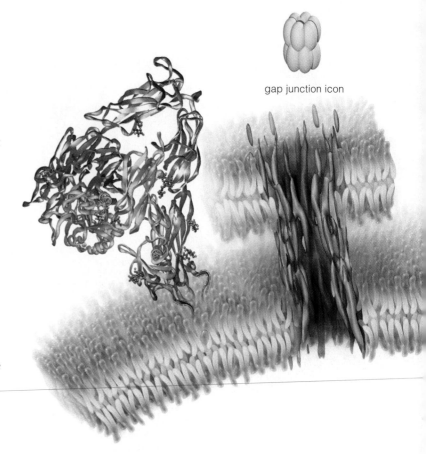

gap junction icon

ADHESION PROTEINS

These proteins are embedded within the plasma membrane. They help one cell adhere to another or to a protein, such as collagen, that is part of an extracellular matrix.

Integrins, including the one shown here, also relay signals across the membrane. Other examples: The cadherins of one cell bind with identical cadherins in adjoining cells. The selectins join together cells by adhering to specific polysaccharides. They are abundant in endothelium, the special lining of blood vessels and the heart.

CELL-TO-CELL COMMUNICATION PROTEINS

These proteins match up with identical proteins in the plasma membrane of an adjoining cell and form a channel, which directly connects their cytoplasm. Chemical and electrical signals flow through the channel.

This type, a cardiac gap junction, is abundant in your heart muscle cells. Signals flow so fast from cell to cell that they all contract together, as a functional unit. This model shows only the parts of the gap junction assembly with greatest density; details are being worked out.

Figure 5.5 Major categories and functions of membrane proteins. You will encounter specific examples of these proteins in this book, especially those with the simple icons above them.

recognition protein icon

passive transporter icon

active transporter icon
(P type)

active transporter icon
(F type)

RECEPTOR PROTEINS

Receptors embedded in a membrane are docks for diverse hormones and other signals. They cause target cells to change their activities. Targets are any cells with receptors for the signaling molecule.

A signal might make a cell synthesize a certain protein, speed or block a metabolic reaction, secrete some substance, or get ready to divide.

Shown above, a receptor for somatotropin (also called growth hormone). Copies of this receptor occur on most cells of the vertebrate body. A somatotropin molecule (*gray*) bound to it triggers cell divisions that bring about increases in size during growth and development.

RECOGNITION PROTEINS

Each plasma membrane has glycoproteins (and glycolipids) with sugar side chains projecting above it. These identify a cell as *nonself* (foreign) or *self* (belonging to one's own body or a tissue).

Some function in tissue defense. Remember the HLA proteins (Section 3.7)? Foreign fragments bound to HLA incite immune cells to battle. Still other recognition proteins help cells of the same type find and attach to one another as tissues are forming.

Shown here, one of the glycophorins, an identity tag on red blood cells. *See also* Section 11.4 on ABO blood groups.

PASSIVE TRANSPORTERS

Passive transporters are channels that allow specific solutes to move through them without requiring an energy input. The solute simply diffuses across the membrane, following concentration or electric gradients (Section 5.4).

Diverse animal cells and plant vacuoles have plenty of open channels for water (aquaporins). Other channels change shape when the solute passes through them. The example shown is GluT1, which binds glucose and helps it cross a red blood cell's plasma membrane. A chloride–bicarbonate cotransporter passively lets chloride and bicarbonate ions cross a membrane in opposite directions at the same time.

Other kinds are *ion-selective channels* with molecular gates. Some gates open or close fast when a small molecule binds to them or when the distribution of electric charge across a membrane changes. Nerve and muscle cells have gated channels for sodium, calcium, potassium, and chloride ions.

ACTIVE TRANSPORTERS

Active transport proteins pump a specific solute across the membrane to the side where it is more concentrated and less likely to move on its own. Some pumps are cotransporters that let one kind of solute flow passively "downhill" while pumping a different kind "uphill."

ATP-dependent transporters are known as ATPase pumps. Most *P type* ATPases consist of one polypeptide chain. The calcium pump shown here (*blue*) is one example. The sodium–potassium pump is another.

V type ATPases pump hydrogen ions (H$^+$). They activate hydrolytic enzymes in fungal, plant, and animal cells.

F type ATPases in prokaryotic, plant, and animal cells pump H$^+$ across membranes. They can also run in reverse, with a flow of H$^+$ through them driving the synthesis of ATP. Hence their more precise name, the ATP synthases (Chapters 7 and 8).

Do Membrane Proteins Stay Put?

Some time ago, researchers figured out how to split a plasma membrane right down the middle of its bilayer. They found that proteins were not spread like a coat on the bilayer, as some had hypothesized, but rather that a great many proteins were embedded in it:

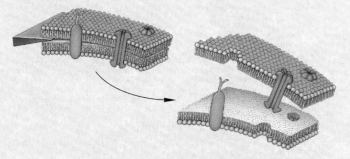

Were those proteins rigidly positioned in the membrane? No one knew until researchers designed an ingenious experiment. They induced an isolated human cell and an isolated mouse cell to fuse. The plasma membranes of the cells from the two species merged to form a single, continuous membrane in the new, hybrid cell. Most of the membrane proteins had become mixed together in less than an hour (Figure 5.6).

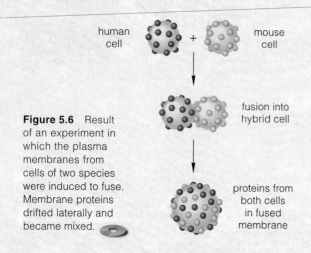

Figure 5.6 Result of an experiment in which the plasma membranes from cells of two species were induced to fuse. Membrane proteins drifted laterally and became mixed.

human cell + mouse cell

fusion into hybrid cell

proteins from both cells in fused membrane

We now know that many membrane proteins are free to move laterally through the lipid bilayer, but others do indeed stay put. Some kinds cluster together and do not move relative to one another. The protein receptors for acetylcholine, which function in information flow through nervous systems, are like this. Other kinds are tethered to cytoskeletal elements that limit their lateral movements. That glycophorin shown in Section 5.2 is attached to a mesh of crosslinked spectrin molecules. So is a type of transport protein that exchanges chloride ions for bicarbonate ions across the plasma membrane.

THINK DIFFUSION

What determines whether a substance will move one way or another across the lipid bilayer or through a membrane protein? Part of the answer has to do with something called diffusion. Think about the water at a bilayer's surfaces. Plenty of substances are dissolved in it, but the kinds and amounts are not the same near the two surfaces. The membrane itself set up the difference and is now maintaining it. How? Each cell membrane shows **selective permeability**: Its molecular structure lets some substances but not others cross it in certain ways, at certain times.

As one example, the lipid chains of the bilayer are largely nonpolar. As a result, molecular oxygen, carbon dioxide, and other small, nonpolar molecules can freely cross it. Although water molecules are polar, some slip through gaps that open up when the hydrocarbon tails flex and bend (Figure 5.7).

By contrast, the bilayer is not permeable to large, polar molecules such as glucose, nor is it permeable to ions (Figure 5.7). Membrane proteins help them across. Sometimes water molecules in which these substances are dissolved cross with them.

The membrane barriers and crossings are essential, because metabolism depends on the cell's capacity to increase, decrease, and maintain concentrations of the molecules and ions for specific reactions.

What Is A Concentration Gradient?

Now picture molecules or ions of some substance near a cell membrane. They move constantly and randomly collide and bounce off one another. They collide more frequently in places where they are more concentrated. When the concentration in one region is not the same as in an adjoining region, this condition is a *gradient*. Pulling this all together, a **concentration gradient** is a difference in the number per unit volume of ions or molecules of a substance between adjoining regions.

In the absence of other forces, any substance moves from a region where it is more concentrated to a region where it is less concentrated. *At biological temperatures, the thermal energy of the molecules drives this movement.* Although these molecules collide randomly and career back and forth millions of times per second, their *net* movement is away from the place where they are most concentrated.

Diffusion is the name for the net movement of like molecules or ions down a concentration gradient. It is a factor in how substances move across membranes and through cytoplasmic fluid. In multicelled species, it also moves substances between regions and between the body and its environment. For instance, if oxygen builds up in leaf cells, it may diffuse into air in the leaf, then into air outside where its concentration is lower.

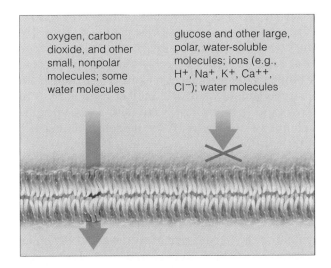

Figure 5.7 Selective permeability of cell membranes. Small, nonpolar molecules and some water molecules cross the lipid bilayer. Ions and large, polar, water-soluble molecules and water that is dissolving them cannot cross it. Transport proteins actively or passively help them across.

oxygen, carbon dioxide, and other small, nonpolar molecules; some water molecules

glucose and other large, polar, water-soluble molecules; ions (e.g., H+, Na+, K+, Ca++, Cl−); water molecules

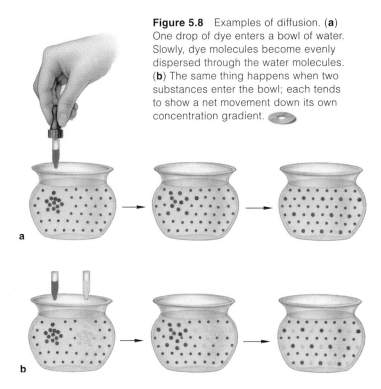

Figure 5.8 Examples of diffusion. (**a**) One drop of dye enters a bowl of water. Slowly, dye molecules become evenly dispersed through the water molecules. (**b**) The same thing happens when two substances enter the bowl; each tends to show a net movement down its own concentration gradient.

a

b

Like other substances, oxygen tends to diffuse in a direction established by its *own* concentration gradient, not gradients of other substances dissolved in the same fluid. You can observe the outcome of this tendency by squeezing a drop of dye into water. The dye molecules diffuse to the region where they are less concentrated. At the same time, water molecules move to the region where *they* are less concentrated (Figure 5.8).

What Determines Diffusion Rates?

Several factors influence the rates of movement down a concentration gradient. Among these are the gradient's steepness, molecular size, temperature, and electric or pressure gradients that may be present.

Diffusion is faster when gradients are steep. Why? Far more molecules move out from a region of greater concentration compared to the number moving in. As the gradient decreases, so does the difference in how many are moving either way. The individual molecules would still be in motion if the gradient were to vanish. But the total number moving one way or the other in a given interval would stay much the same. When the net distribution of molecules is nearly uniform in two adjoining regions, we call this *dynamic equilibrium.*

What about temperature? More heat energy (that is, higher temperature) causes molecules to move faster and collide more often. That is why diffusion is more rapid than in a cooler adjoining region. What about molecular size? Generally, smaller molecules flow down their concentration gradient faster than large ones do.

An electric gradient also can influence the rate and direction of diffusion. An **electric gradient** is simply a difference in electric charge between adjoining regions. Example: Many ions are dissolved in the fluids bathing a membrane, each contributing to the electric charge of one fluid or the other. Opposite charges attract. So the fluid with the more negative charge, overall, exerts the greatest pull on positively charged substances, such as sodium ions. Many events, including information flow through nervous systems, involve the combined force of electric and concentration gradients.

Finally, as you will see shortly, diffusion also may be affected by a **pressure gradient**, which is a difference in the pressure being exerted in adjoining regions.

Molecules or ions of a substance constantly collide because of their inherent energy of motion. The collisions result in diffusion, a net outward movement of a substance from one region into an adjoining region where it is less concentrated.

A concentration gradient is a form of energy. It can drive the directional movement of a substance across a cell membrane. The steepness of the gradient, temperature, molecular size, electric gradients, and pressure gradients influence the rates of diffusion.

Metabolic reactions rely on chemical energy inherent in concentrated amounts of molecules and ions. Cells have mechanisms to increase or decrease those concentrations across the plasma membrane and internal cell membranes.

TYPES OF CROSSING MECHANISMS

Before getting into the actual mechanisms that move substances across membranes, study the overview in Figure 5.9. These are the mechanisms that help supply cells and organelles with raw materials and get rid of wastes. Collectively, they help maintain the volume and pH of cells or organelles within functional ranges.

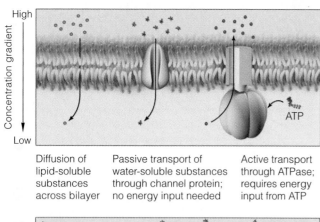

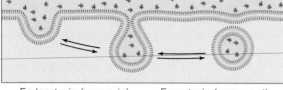

Figure 5.9 Membrane crossing mechanisms.

Again, small, nonpolar molecules such as oxygen diffuse across a lipid bilayer. Polar molecules and ions cross by diffusing through transport proteins that span the bilayer. The passive transporters simply permit the substance to follow its concentration gradient across. This is called *passive transport*, or "facilitated" diffusion.

ATPase pumps engage in *active transport*. They, too, let polar substances cross the membrane through their interior, but the net direction of movement is against the concentration gradient. Unlike diffusion across the bilayer or passive transport, active transport requires an input of energy to counter a concentration gradient.

Other mechanisms move substances in bulk across the plasma membrane. *Exocytosis* involves fusion of the plasma membrane and a membrane-bound vesicle that formed inside the cytoplasm. *Endocytosis* involves an inward sinking of a patch of plasma membrane, which seals back on itself to form a vesicle in the cytoplasm.

Different mechanisms work with and against concentration gradients to move substances across the lipid bilayer itself or through proteins that span the bilayer.

HOW DO THE TRANSPORTERS WORK?

All transporters let water-soluble molecules and ions diffuse through some kind of channel or tunnel inside them. When the solute enters a channel opening at one side of the membrane, it weakly binds to the protein. The protein's shape changes, making the opening close behind the bound solute and another open in front of it. This exposes the solute to the fluid that bathes the membrane's other surface. There, the solute binding site reverts to its former state and releases the cargo.

Passive Transport

Passive transport is the name for unassisted diffusion of a specific solute through a transport protein. Which way it moves across the membrane depends only on its concentration gradient, electric gradient, or both.

Energetically speaking, the membrane crossing costs only what the cell has already spent to produce and maintain the gradients. Passive transport itself adds no more to the energy cost. The *net* direction of movement

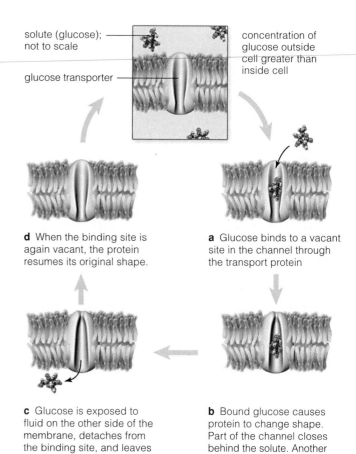

d When the binding site is again vacant, the protein resumes its original shape.

a Glucose binds to a vacant site in the channel through the transport protein

c Glucose is exposed to fluid on the other side of the membrane, detaches from the binding site, and leaves the channel.

b Bound glucose causes protein to change shape. Part of the channel closes behind the solute. Another part opens in front of it.

Figure 5.10 Passive transport, using a glucose transporter as the model. Glucose moves both ways across a membrane. But the *net* movement will be down the concentration gradient until concentrations are equal on both sides of the membrane.

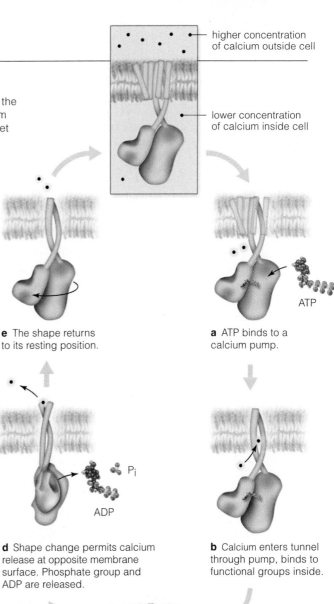

Figure 5.11 Active transport by a calcium pump. This partial model shows the channel across the membrane. ATP transfers a phosphate group to a calcium pump. Reversible changes in this transporter's shape can result in greater net movement of solute particles *against* the concentration gradient.

higher concentration of calcium outside cell

lower concentration of calcium inside cell

e The shape returns to its resting position.

a ATP binds to a calcium pump.

d Shape change permits calcium release at opposite membrane surface. Phosphate group and ADP are released.

ADP

P$_i$

b Calcium enters tunnel through pump, binds to functional groups inside.

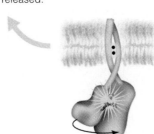

c ATP transfers a phosphate group to pump. Energy input will cause pump's shape to change.

in a given interval depends on how many molecules or ions of a specific solute are randomly contracting the transporters (Figure 5.10). Encounters are simply more frequent on the side of the membrane where the solute concentration is greatest. So the solute's *net* movement is in the direction where it is less concentrated.

If nothing else were going on, the passive transport would continue until concentrations were equal across the membrane. But other processes affect the outcome.

For instance, the glucose transporters (Figure 5.10) passively move glucose from blood into the cell, where it is used for biosynthesis and instant energy. When the blood glucose level is high, cells maintain the gradient even when uptake is rapid. How? As fast as glucose molecules are diffusing into cells, others are entering metabolic reactions. In other words, when the cells use glucose, they are maintaining a concentration gradient that favors uptake of *more* glucose.

Active Transport

Only in a dead cell have solute concentrations become equal on both sides of membranes. Living cells never stop expending energy to pump solutes into and out from their interior. In **active transport**, energy-driven protein motors help move a specific solute across the cell membrane, *against* its concentration gradient.

Reflect on the active transporters in Figure 5.5. As one of them binds ATP, a channel through its interior opens up. Only a particular kind of solute can enter the channel and bind reversibly to molecular groups that line it. Binding causes the transporter to accept a phosphate group from the ATP. That phosphate-group transfer changes the transporter's shape. And with that change, the solute is released and enters the fluid on the other side of the cell membrane.

Figure 5.11 shows one **calcium pump**. This active transporter helps keep the concentration of calcium in a cell at least a thousand times lower than it is outside. A different protein, the **sodium–potassium pump,** is a cotransporter. When ATP activates it, this pump binds sodium ions (Na$^+$) from one side of the membrane and releases them on the other side. The release facilitates binding of potassium ions (K$^+$) at a different site inside the tunnel through the transporter, which reverts to its original shape after K$^+$ is released to the other side.

Through operation of such active transport systems, concentration and electric gradients can be maintained across membranes. The gradients are vital to many cell activities and physiological processes, including muscle contraction and nerve cell (neuron) function. We return to the mechanisms of active transport in later chapters.

All transport proteins bind solutes on one side of a cell membrane and reversibly change shape. The charge shunts the solute through some type of tunnel in their interior.

In passive transport, a solute diffuses through a transporter, and its net movement is down its concentration gradient.

In active transport, the net diffusion of one type of solute is uphill, against its concentration gradient. The transporter must be activated by an energy input from ATP to counter the energy inherent in the gradient.

WHICH WAY WILL WATER MOVE?

By far, more water diffuses across cell membranes than any other substance, so the main factors that influence its directional movement deserve special attention.

Osmosis

Think about something as simple as a running faucet or as stupendous as Niagara Falls. Either way, moving water is providing you with an example of **bulk flow** —the mass movement of one or more substances in response to pressure, gravity, or some other external force. Bulk flow accounts for some water movement in plants and animals. A beating heart generates the fluid pressure that pumps blood (mostly water) through the body. Sap flows in the conducting tissues that thread through maple trees, and this, too, is bulk flow.

What about the movement of water into and out of cells or organelles? If the concentration of water is not equal across a cell membrane, **osmosis** tends to occur. Water molecules will tend to diffuse down the water concentration gradient and thus cross that membrane, which is selectively permeable.

The concentrations of solutes at the two sides of selectively permeable membranes influence osmosis. This type of membrane lets the

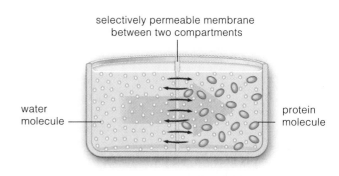

Figure 5.12 Demonstration of how a solute concentration gradient influences osmotic movement. Start with a container divided by a membrane that water but not proteins can cross. Pour water into the left compartment. Pour the same volume of a protein-rich solution into the right compartment. There, proteins occupy some of the space. The net diffusion of water in this example is from left to right (large *gray* arrow).

small, polar water molecules cross but restricts passage of large, polar molecules. The side with the most solute particles ends up with the lower concentration of water (Figure 5.12). To illustrate this, dissolve some glucose in water. Compared to an equivalent volume of water, the glucose solution has fewer water molecules. Why? Because each glucose molecule is occupying space that was formerly occupied by water molecules.

It is primarily the *total number* of molecules or ions, not the type of solute, that dictates the concentration of water. Dissolve an amount of an amino acid or urea in a liter of water, and the water concentration changes about as much as it did in the glucose solution. Add some sodium chloride (NaCl) to a liter of water, and it dissociates into equal numbers of sodium ions and chloride ions. This results in twice as many particles of solute as there were in the glucose solution, and so the concentration of water will decrease proportionately.

Effects of Tonicity

Suppose you decide to test the statement that water tends to move toward a region where solutes are *more* concentrated. You make three sacs from a membrane that water but not sucrose can cross (Figure 5.13). You

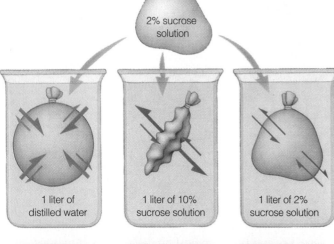

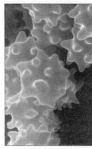

2% sucrose solution

1 liter of distilled water | 1 liter of 10% sucrose solution | 1 liter of 2% sucrose solution

a

HYPOTONIC CONDITIONS

Water diffuses into red blood cells, which swell up

HYPERTONIC CONDITIONS

Water diffuses out of the cells, which shrink

ISOTONIC CONDITIONS

No net movement of water, no change in cell size or shape

b

Figure 5.13 Effect of tonicity on water movement. (**a**) Arrow widths indicate the direction and the relative amounts of flow. (**b**) The micrograph by each sketch shows shapes that a human red blood cell assumes when you put it in fluids of higher, lower, or equal solute concentrations. Solutions inside and outside this type of cell normally are balanced. A red blood cell has no built-in mechanisms to help it adjust to drastic changes in solute levels in its fluid surroundings.

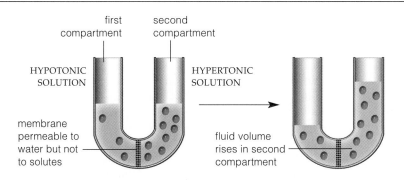

first compartment | second compartment

HYPOTONIC SOLUTION | HYPERTONIC SOLUTION

membrane permeable to water but not to solutes

fluid volume rises in second compartment

Figure 5.14 Increase in fluid volume owing to osmosis. In time, the net diffusion across a membrane separating two compartments is equal, but the fluid volume in compartment 2 is greater because the membrane is impermeable to solutes.

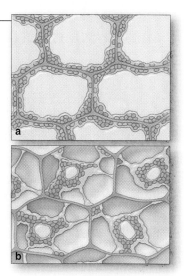

Figure 5.15 (**a**) Young plant cells. (**b**) Example of what plasmolysis, an osmotically induced loss of internal fluid pressure, does to the cells. The cytoplasm and central vacuole both shrink, and the plasma membrane moves away from the cell wall.

fill each sac with a solution that is 2 percent sucrose. You immerse one sac in a liter of distilled water (which has no solutes). You immerse another in a solution that is 10 percent sucrose. You immerse the third sac in a solution that is 2 percent sucrose. Each time, tonicity dictates the extent and direction of water movement.

Tonicity refers to the relative solute concentrations of two fluids. When two fluids on opposing sides of a membrane differ in solute concentration, the **hypotonic solution** is the one with fewer solutes. The one that has more is the **hypertonic solution**. Water tends to diffuse from hypotonic to hypertonic fluids. **Isotonic solutions** have the same solute concentrations, so water shows no net osmotic movement from one to the other.

Normally, the fluids inside and outside your cells are isotonic. When any tissue fluid becomes drastically hypotonic, too much water diffuses into cells, and they burst. When fluid becomes too hypertonic, an outward diffusion of water shrivels them.

Most cells have built-in mechanisms that adjust to shifts in tonicity. Red blood cells do not. (Figure 5.13 shows what happened to them in an observational test of tonicity differences.) That's why severely dehydrated patients get infusions of a solution isotonic with blood.

Effects of Fluid Pressure

Animal cells avoid bursting by engaging in selective transport of solutes across the plasma membrane. Cells of plants and many protistans, fungi, and bacteria do the same with the help of pressure on their cell walls.

Pressure differences as well as solute concentrations influence the osmotic movement of water. Take a look at Figure 5.14. It shows how water continues to move across a membrane between a hypotonic solution and a hypertonic solution until the solute concentration is the same on both sides. As you can see, the *volume* of the formerly hypertonic solution has increased (because its solutes cannot diffuse out).

Hydrostatic pressure is pressure that any volume of fluid exerts against a wall, membrane, or some other structure enclosing it. (In plants, this is called the *turgor* pressure.) The greater the fluid's solute concentration, the greater will be the hydrostatic pressure it exerts.

Living cells cannot increase in volume indefinitely (Section 4.1). At some point, fluid pressure that builds up inside the cell counters water's inward diffusion. That point is the **osmotic pressure**, the amount of force preventing further increase in a solution's volume.

Think of a young plant cell, with its pliable primary wall. Water diffuses into the cell as it grows. This puts more fluid pressure on its wall and makes it expand, so the cell volume can increase. Further expansion of the wall (and the cell) ends when internal fluid pressure develops enough to counterbalance the water uptake.

Plant cells are vulnerable to water losses, which can occur when soil dries out or becomes too salty. Water stops diffusing in and starts diffusing out, so internal fluid pressure falls. An osmotically induced shrinkage of cytoplasm is called plasmolysis (Figure 5.15). Plants adjust somewhat to the loss of pressure, as when they actively take up potassium ions against a concentration gradient by the mechanisms outlined in Section 5.6.

As you will read in Chapters 38 and 42, hydrostatic and osmotic pressure also influence the distribution of water in the tissue fluids and cells of animals.

Osmosis is a net diffusion of water between two solutions that differ in solute concentration and that are separated by a selectively permeable membrane. The greater the number of molecules and ions dissolved in a solution, the lower its water concentration will be.

Water tends to move osmotically to regions of greater solute concentration (from hypotonic to hypertonic solutions). There is no net diffusion between isotonic solutions.

The fluid pressure that a solution exerts against a membrane or wall also influences the osmotic movement of water.

MEMBRANE TRAFFIC TO AND FROM THE CELL SURFACE

Exocytosis and Endocytosis

Transport proteins can move only small molecules and ions into or out of cells. When it comes to taking in or expelling large molecules or particles, cells use vesicles that form through exocytosis and endocytosis.

By **exocytosis**, a vesicle moves to the cell surface, and the protein-studded lipid bilayer of its membrane fuses with the plasma membrane. While this exocytic vesicle is losing its identity, its contents are released to the surroundings (Figure 5.16a). In three pathways of **endocytosis**, a cell takes in substances near its surface.

A small patch of plasma membrane balloons inward and pinches off. The result, an endocytic vesicle, then transports its contents to some organelle or stores them in the cytoplasm (Figures 4.13 and 5.16b).

By *receptor-mediated* endocytosis, the first pathway, membrane receptors chemically recognize and bind to a specific substance—for example, a hormone, vitamin, or mineral. The receptors become concentrated in tiny pits that form in the plasma membrane (Figure 5.17). Each pit looks like a woven basket on its cytoplasmic side. The basket consists of protein filaments (clathrin) interlocked into stable, geometric patterns. When the pit sinks into the cytoplasm, the basket closes back on itself and becomes the vesicle's structural framework.

The second pathway, *bulk-phase* endocytosis, is less selective. An endocytic vesicle forms around a small volume of extracellular fluid regardless of what kinds of substances happen to be dissolved in it. Bulk-phase endocytosis operates at a fairly constant rate in nearly all eukaryotic cells. By continually pulling patches of plasma membrane into the cytoplasm, this pathway compensates for membrane that steadily arrives from the cytoplasm in the form of exocytic vesicles.

The third pathway is **phagocytosis** (meaning cell eating). By this active form of endocytosis, a cell engulfs microbes, particles, and cellular debris. Amoebas and some other protistans get food this way. In multicelled species, macrophages and some other white blood cells phagocytize pathogenic viruses or bacteria, cancerous body cells, and other threats to health.

Phagocytosis, too, is mediated by receptors. First, a target binds with receptors bristling from a phagocytic

Figure 5.16 (**a**) Exocytosis and (**b**) endocytosis.

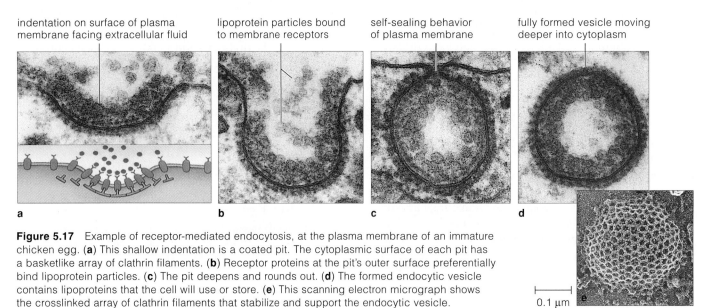

indentation on surface of plasma membrane facing extracellular fluid

lipoprotein particles bound to membrane receptors

self-sealing behavior of plasma membrane

fully formed vesicle moving deeper into cytoplasm

Figure 5.17 Example of receptor-mediated endocytosis, at the plasma membrane of an immature chicken egg. (**a**) This shallow indentation is a coated pit. The cytoplasmic surface of each pit has a basketlike array of clathrin filaments. (**b**) Receptor proteins at the pit's outer surface preferentially bind lipoprotein particles. (**c**) The pit deepens and rounds out. (**d**) The formed endocytic vesicle contains lipoproteins that the cell will use or store. (**e**) This scanning electron micrograph shows the crosslinked array of clathrin filaments that stabilize and support the endocytic vesicle.

0.1 μm

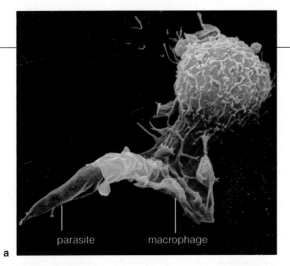

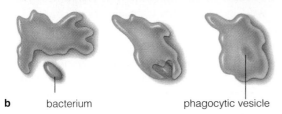

Figure 5.18 Phagocytosis. (**a**) Scanning electron micrograph showing a macrophage in the act of engulfing *Leishmania mexicana*. This parasitic protozoan causes a potentially fatal disease called leishmaniasis. Bites from infected sandflies transmit the parasite to humans. (**b**) Diagram of phagocytosis. Lobes of this amoeba's cytoplasm extend outward and surround its target. The plasma membrane of the extensions fuses, forming a phagocytic vesicle. This type of endocytic vesicle moves deeper into the cytoplasm and fuses with lysosomes. Its contents are digested. The molecular bits, and the vesicle's membrane components, are recycled elsewhere (Figure 5.19).

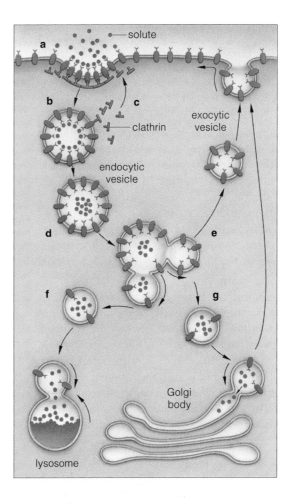

a Molecules get concentrated inside coated pits of plasma membrane.

b Endocytic vesicles form from the pits.

c Vesicles lose molecules of clathrin, which return to plasma membrane.

d Enclosed molecules are sorted and often released from receptors.

e Many sorted molecules are cycled back to the plasma membrane.

f,g Many other sorted molecules are delivered to lysosomes and stay there or are degraded. Still others are routed to spaces in the nuclear envelope and inside ER membranes, and others to Golgi bodies.

Figure 5.19 Cycling of membrane lipids and proteins. This example starts with receptor-mediated endocytosis. The plasma membrane gives up small patches of itself to endocytic vesicles that form from coated pits. It gets membrane back from exocytic vesicles that budded from ER membranes and Golgi bodies. The membrane initially used for endocytic vesicles will cycle receptor proteins and lipids back to the plasma membrane.

cell's membrane. This binding triggers synthesis and the crosslinking of microfilaments into a network just beneath the plasma membrane. Driven by ATP motors, these microfilaments contract, which squeezes some cytoplasm toward the margins of the cell, thus forming the lobes called pseudopods (Figure 5.18). Pseudopods flow over the target and fuse at their tips. The result is one phagocytic vesicle, which sinks into the cytoplasm. There it fuses with lysosomes. Inside these organelles of intracellular digestion, trapped items are digested to fragments and smaller, reusable molecules.

Membrane Cycling

As long as a cell stays alive, exocytosis and endocytosis continually replace and withdraw patches of its plasma membrane. And they apparently do so at rates that can maintain the plasma membrane's total surface area.

As one example, neurons release neurotransmitters in bursts of exocytosis. Each neurotransmitter is a type of signaling molecule released from one cell that acts on neighboring cells. Cell biologists John Heuser and T. S. Reese documented an intense burst of endocytosis

abruptly after a major episode of exocytosis in neurons, and that counterbalanced it. Figure 5.19 provides more examples of ways in which cells cycle their membrane lipids and proteins.

Whereas transport proteins in a cell membrane deal only with ions and small molecules, exocytosis and endocytosis move large packets of materials across a plasma membrane.

By exocytosis, a cytoplasmic vesicle fuses with the plasma membrane, so that its contents are released outside the cell. By endocytosis, a small patch of the plasma membrane sinks inward and seals back on itself, forming a vesicle inside the cytoplasm. Membrane receptors often mediate this process.

SUMMARY

Membrane Structure and Function

1. The plasma membrane is a structural and functional boundary between the cytoplasm and the surroundings of all cells. In eukaryotic cells, organelle membranes subdivide the fluid portion of the cytoplasm into many functionally diverse compartments. **5.1**

2. A cell membrane consists of two water-impermeable layers of lipids (phospholipids especially) and proteins associated with the layers, as shown in Figure 5.20. **5.1**

   a. Fatty acid tails and other hydrophobic parts of the lipids are sandwiched between hydrophilic heads.

   b. Many different kinds of proteins are embedded in the lipid bilayer or positioned at one of its two surfaces. The proteins carry out most membrane functions.

3. These are the key features of the fluid mosaic model of membrane structure: **5.1**

   a. A cell membrane shows fluid behavior, mainly because its lipid components twist, move laterally, and flex hydrocarbon tails. Also, some of these lipids have ring structures and many have kinked (unsaturated) or short fatty acid tails, all of which disrupt what might otherwise be tight packing within the bilayer.

   b. A membrane is a mosaic, or composite, of lipids and proteins. The proteins are embedded in the bilayer and positioned at its surface. Its two layers differ in the number, kind, and arrangement of lipids and proteins.

4. Each cell membrane incoporates transport proteins and proteins that structurally reinforce it. The plasma membrane also has adhesion proteins, communication proteins, recognition proteins, and diverse receptors. Differences in the number and types of proteins affect responsiveness to substances at the membrane, as well as cell metabolism, pH, and volume. **5.2**

   a. Transport proteins help water-soluble substances cross all membranes by passing through their interior, which opens to both sides of the membrane.

   b. Receptor proteins bind extracellular substances, and binding triggers alterations in metabolic activities.

Recognition proteins are like molecular fingerprints; they identify cells as being of a given type. Adhesion proteins help cells of tissues adhere to one another and to proteins of the extracellular matrix. Communication junctions span the plasma membranes of adjacent cells to transfer substances and signals rapidly from the cytoplasm of one to the other.

Movement of Substances Into and Out of Cells

1. Molecules or ions of a substance tend to move from a region of higher to lower concentration. Diffusion is movement in response to a concentration gradient. **5.4**

   a. Diffusion rates are influenced by the steepness of the concentration gradient, temperature, and molecular size, as well as by gradients in electrical charge and pressure that may occur between two regions.

   b. Cells have built-in mechanisms that work with or against gradients to move solutes across membranes.

   c. Metabolism requires chemical energy inherent in concentration and electric gradients across cell membranes.

2. Oxygen, carbon dioxide, and other small nonpolar molecules diffuse across a membrane's lipid bilayer. Ions and large, polar molecules such as glucose cross it with the passive or active help of transport proteins. Water moves through proteins and across the bilayer. **5.4**

3. A transport protein moves water-soluble substances across a membrane by reversible changes in its own shape. Passive transport doesn't require energy inputs; solutes diffuse through a channel inside the protein's interior. Active transport requires an energy input from ATP. The protein, an ATPase pump, moves the solute against its concentration gradient. **5.6**

4. Osmosis is defined as the diffusion of water across a selectively permeable membrane in response to a water concentration gradient, pressure gradient, or both. **5.7**

5. By exocytosis, a cytoplasmic vesicle moves into the plasma membrane. Its membrane fuses with it, and its contents are automatically released to the outside. 5.8

6. By endocytosis, a patch of plasma membrane forms a vesicle. It does so by receptor recognition of specific solutes, by the indiscriminate uptake of solutes dissolved in the extracellular fluid, or by phagocytosis ("cell eating," an engulfment of cells as well as substances). **5.8**

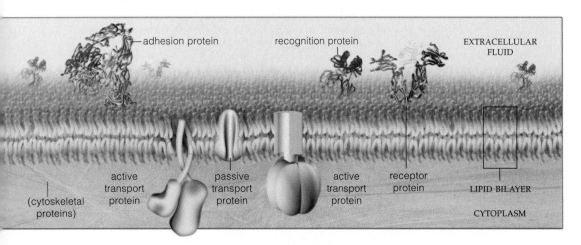

Figure 5.20 Summary of major types of proteins associated with the plasma membrane.

Review Questions

1. Describe the fluid mosaic model of cell membranes. What imparts fluidity to the membrane? What makes it a mosaic? *5.1*

2. State the functions of transport proteins, receptor proteins, recognition proteins, and adhesion proteins. *5.2*

3. Define diffusion. Does diffusion occur in response to a solute concentration gradient, an electric gradient, a pressure gradient, or some combination of these? *5.4*

4. If all transport proteins work by changing shape, then how do passive transporters differ from active transporters? *5.6*

5. Define osmosis. *5.7*

6. Define hypertonic, hypotonic, and isotonic solutions. Does each term refer to a property inherent in a given solution? Or are the terms used only when comparing solutions? *5.7*

7. Is the white blood cell in Figure 5.21 disposing of a worn-out red blood cell by endocytosis, phagocytosis, or both? *5.8*

Self-Quiz   ANSWERS IN APPENDIX III

1. Cell membranes consist mainly of a _____ .
   a. carbohydrate bilayer and proteins
   b. protein bilayer and phospholipids
   c. lipid bilayer and proteins

2. In a lipid bilayer, _____ of lipid molecules are sandwiched between _____ .
   a. hydrophilic tails; hydrophobic heads
   b. hydrophilic heads; hydrophilic tails
   c. hydrophobic tails; hydrophilic heads
   d. hydrophobic heads; hydrophilic tails

3. Most membrane functions are carried out by _____ .
   a. proteins   c. nucleic acids
   b. phospholipids   d. hormones

4. Plasma membranes incorporate _____ .
   a. transport proteins   c. recognition proteins
   b. adhesion proteins   d. all of the above

5. Immerse a living cell in a hypotonic solution, and water will tend to _____ .
   a. move into the cell   c. show no net movement
   b. move out of the cell   d. move in by endocytosis

6. _____ can readily diffuse across a lipid bilayer.
   a. Glucose   c. Carbon dioxide
   b. Oxygen   d. b and c

7. Sodium ions cross a membrane through transport proteins that receive an energy boost. This is an example of _____ .
   a. passive transport   c. facilitated diffusion
   b. active transport   d. a and c

8. Vesicle formation occurs in _____ .
   a. membrane cycling   c. endocytosis and exocytosis
   b. phagocytosis   d. all of the above

Critical Thinking

1. Certain species of bacteria thrive in environments where the temperatures approach the boiling point of water—for example, in the steam-venting fissures of volcanoes and in hot springs of Yellowstone National Park. Assume that the lipid bilayer of the bacterial cell membranes consists mainly of phospholipids. What features might the fatty acid tails of the phospholipids have that help stabilize the membranes at such extreme temperatures?

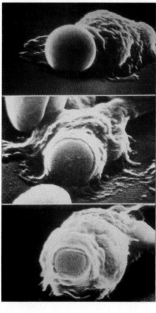

Figure 5.21   Go ahead, name the mystery membrane mechanism.

Figure 5.22   *Paramecium* contractile vacuoles.

2. Water moves osmotically into *Paramecium*, a single-celled protistan of aquatic habitats. If unchecked, the influx would bloat the cell and rupture its plasma membrane. An energy-requiring mechanism involving contractile vacuoles expels the excess (Figure 5.22). Water enters tubelike extensions of this organelle and collects in a central space in the vacuole. When full, the vacuole contracts and squirts excess water out of a pore that opens to the outside. Are the fluid surroundings hypotonic, hypertonic, or isotonic relative to *Paramecium*'s cytoplasm?

3. Many cultivated fields in California are heavily irrigated. Over the years, most of the water has evaporated from the soil, leaving behind all of the solutes in the irrigation water. What kinds of problems will the altered soil conditions cause plants?

4. Imagine you're a juvenile shrimp living in an *estuary*, where fresh water draining from the land mixes with saltwater from the sea. Many people own homes around a lake and want boat access to the sea. They ask their city for permission to build a canal to your estuary. If they succeed, what may happen to you?

Selected Key Terms

ABC transporter *CI*	hypotonic solution *5.7*
active transport *5.6*	isotonic solution *5.7*
adhesion protein *5.2*	lipid bilayer *5.1*
biofilm *CI*	osmosis *5.7*
bulk flow *5.7*	osmotic pressure *5.7*
calcium pump *5.6*	passive transport *5.6*
communication protein *5.2*	phagocytosis *5.8*
concentration gradient *5.4*	phospholipid *5.1*
diffusion *5.4*	pressure gradient *5.4*
electric gradient *5.4*	receptor protein *5.2*
endocytosis *5.8*	recognition protein *5.2*
exocytosis *5.8*	selective permeability *5.4*
fluid mosaic model *5.1*	sodium–potassium pump *5.6*
hydrostatic pressure *5.7*	transport protein *5.2*
hypertonic solution *5.7*	

Readings

Nelson, D., and M. Cox. 2000. *Lehninger's Principles of Biochemistry.* Third edition. New York: Worth.

6

GROUND RULES OF METABOLISM

Growing Old With Molecular Mayhem

Somewhere in those slender strands of DNA inside your cells are snippets of instructions for making two kinds of enzymes: superoxide dismutase and catalase (Figure 6.1). Both are *antioxidants*. Like vitamins E and C, they help keep you from growing old before your time by cleaning house, so to speak. Working as a team, they neutralize many potentially toxic wastes through a reaction that uses oxygen (O_2) to strip hydrogen atoms from molecules. Many organisms ranging from bacteria to roundworms to plants also use these two enzymes.

Stripping hydrogen atoms from molecules releases electrons, and O_2 normally picks them up. Sometimes it picks up only one. That one electron is not enough to complete the reaction. But it's enough to give the oxygen a negative charge (O_2^-).

Like other unbound molecular fragments with the wrong number of electrons, O_2^- is a **free radical**. Free radicals escape from a variety of enzyme-catalyzed reactions, including the digestion of fats and amino acids. They slip away from electron transfer chains. And they form when x-rays and other kinds of ionizing radiation bombard water and other molecules.

Free radicals are highly reactive. When they dock with a molecule, they alter its structure and function. They even attack molecules that aren't open to just any reaction. Such molecules include DNA, cell membrane lipids, and membrane receptors that, when activated, tell cells it is time to die (Section 15.6).

Enter superoxide dismutase. Under its biochemical prodding, two rogue oxygen molecules combine with hydrogen ions. Hydrogen peroxide (H_2O_2) and O_2 are the outcome. Hydrogen peroxide, a normal by-product of some reactions, is lethal to cells when it accumulates. Enter catalase. Under its prodding, two molecules of hydrogen peroxide react and split into water and oxygen: $2H_2O_2 \longrightarrow 2H_2O + O_2$. This crucial reaction normally occurs before hydrogen peroxide does major damage. And it occurs often. One catalase molecule can degrade 40 million hydrogen peroxide molecules per second.

As we age, our cells make copies of enzymes in ever diminishing numbers, in crippled form, or both. When this happens to superoxide dismutase and catalase, free radicals and hydrogen peroxide accumulate. Like loose cannons, they career through cells and blast away at

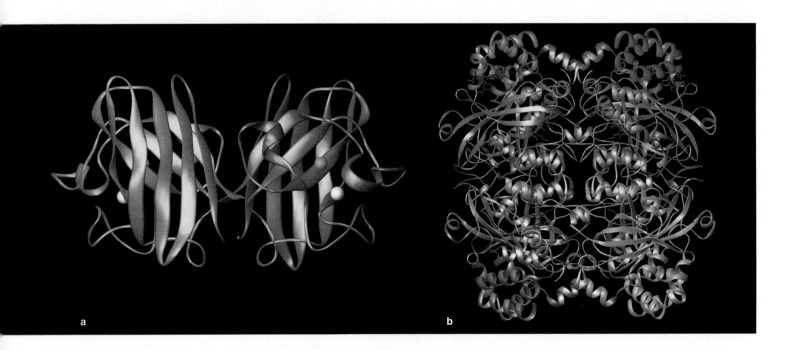

a   b

Figure 6.1 Ribbon models for (**a**) superoxide dismutase and (**b**) catalase. Both enzymes help keep free radicals in check. We find superoxide dismutase molecules in the cytoplasm, nucleus, and peroxisomes, organelles of intracellular digestion in plants and animals. One form is secreted to the extracellular matrix in all mammalian tissues, especially the heart, lungs, pancreas, and placenta. Catalase is concentrated in peroxisomes but is also in the cytoplasm and mitochondrial membranes.

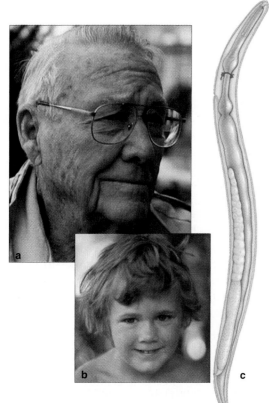

Figure 6.2 (**a**) Owner of skin that has a spattering of age spots, evidence of free radicals on the loose. At one time he, like the boy in (**b**), had many smoothly functioning molecules of superoxide dismutase and catalase. (**c**) The worm of choice for aging studies, *Caenorhabditis elegans*.

the structural integrity of proteins, DNA, lipids, and other molecules. Cells suffer or die.

Those brown "age spots" on an older person's skin are evidence of free radical assaults (Figure 6.2*a*). Each spot is a mass of brownish-black pigments that build up in cells when free radicals take over—all for the want of specific enzymes. Worse yet, free radicals also may usher in heart disease and other disorders.

Knowing this, Simon Melov, Bernard Malfroy, and others experimented with tiny roundworms that don't normally live more than a month (Figure 6.2*c*). The worm's life span increased when synthetic superoxide dismutase and catalase were added to their diet. In some cases it even doubled! The researchers also fed enzymes to worms that were genetically engineered to succumb prematurely to free radicals. Those worms lived just as long as the unaltered ones.

With this example, we turn to **metabolism**, the cell's capacity to acquire energy and use it to build, degrade, store, and release substances in controlled ways. Each extra day in that worm's life reminds us that cells must continually take in energy and use it to drive every one of their activities. At times, the activities may seem remote from your interests. But they help define who you are and who you will become, age spots and all.

Key Concepts

1. Cells engage in metabolism, or chemical work. They use energy to stockpile, build, rearrange, and break apart substances. Cells use energy for mechanical work, as when they move flagella. They also channel energy into electrochemical work, as when they move ions into or out of an organelle compartment.

2. Energy flows in one direction, from usable to less usable forms. Organisms maintain their complex organization by being resupplied with energy lost from someplace else.

3. All organisms secure energy from outside sources. Sunlight energy is the ultimate source for the web of life. Another source is chemical bond energy of molecules in the physical environment or in one species that serves as food for another.

4. ATP, the main carrier of energy in cells, couples reactions that release energy with other reactions that require it. ATP primes molecules to react by transferring a phosphate group to them.

5. Many aspects of metabolism involve electron transfers, or oxidation–reduction reactions. Major transfers occur at electron transfer chains in both photosynthesis and aerobic respiration.

6. On their own, chemical reactions proceed too slowly to sustain life. Enzymes greatly increase the reaction rates in cells. Often they have helpers, such as NAD^+, that assist in the reaction or that transfer electrons and hydrogen released in the reactions to other sites.

7. Enzymes and other kinds of molecules commonly take part in orderly sequences of reactions called metabolic pathways. The coordinated operation of these pathways maintains, increases, and decreases the relative amounts of substances in cells.

8. By controlling a key step of a metabolic pathway, cells can bring about rapid shifts in their activities.

ENERGY AND THE UNDERLYING ORGANIZATION OF LIFE

Defining Energy

If you've ever watched a house cat stalk a mouse, you know it "freezes" its position to avoid detection before springing at its unsuspecting prey. Like everything else in the universe that's stationary, the cat has a store of **potential energy**—a capacity to do work, simply owing to its position in space and the arrangement of its parts. When that cat springs, some of its potential energy is transformed into **kinetic energy**, the energy of motion.

Energy on the move is doing work when it imparts motion to other things. In skeletal muscle cells within the cat, ATP gave up some of its potential energy to molecules of contractile units and set them in motion. The combined motions in many muscle cells resulted in the movement of whole muscles. The transfer of energy from ATP also resulted in the release of another form of energy called **heat**, or *thermal* energy.

The potential energy of molecules has its own name: **chemical energy**. It is measurable, as in kilocalories. A kilocalorie is the same thing as 1,000 calories, which is the amount of energy it takes to heat 1,000 grams of water from 14.5°C to 15.5°C at standardized pressure.

What Can Cells Do With Energy?

All organisms have specific adaptations for securing energy from their environment. Some capture energy from the sun, and others extract energy from inorganic or organic substances in the environment. Regardless of the source, energy inputs are coupled to thousands of energy-requiring processes in cells. Cells use energy for *chemical* work—to stockpile, build, rearrange, and break apart substances. They channel it into *mechanical* work—to move flagella and other cell structures and (in multicelled species) the whole body or portions of it. They channel it into *electrochemical* work—to move charged substances into or out of the cytoplasm or an organelle compartment.

How Much Energy Is Available?

Like single cells, we cannot create energy from scratch; we must get it from someplace else. Why? According to the **first law of thermodynamics**, the total amount of energy in the universe remains constant. More energy cannot be created; existing energy cannot vanish. It can only be converted from one form to some other form.

Think about what the first law means. The universe has only so much energy, distributed in various forms. One form may be converted to another, as when corn plants absorb energy from the sun and convert it to the chemical energy of starch. After you eat and digest corn, your cells extract energy from starch and convert it to other forms, such as mechanical energy for movement.

With each metabolic conversion, a bit of the energy escapes to the surroundings, as heat. Even when you "do nothing," your body gives off about as much heat as a 100-watt lightbulb because of energy conversions in your cells. Released energy is transferred to atoms and molecules making up the air, and the conversion of thermal to kinetic energy "heats up" the surroundings (Figure 6.3). The kinetic energy increases the number of ongoing, random collisions among molecules in the air. And with each collision, a bit more energy is released as heat. However, none of the energy ever vanishes.

The One-Way Flow of Energy

In cells, energy available for conversion resides mainly in the arrangement of atoms and covalent bonds in complex organic compounds, such as glucose, starch, glycogen, and fatty acids. Many of these bonds are said to have a high energy content. When the compounds enter metabolic reactions, the bonds break or become rearranged. During the molecular commotion, some heat energy is lost to the surroundings. In general, cells can't recapture energy lost as heat.

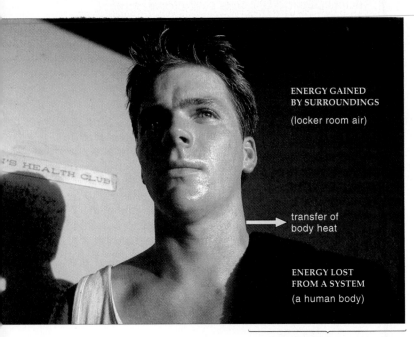

ENERGY GAINED
BY SURROUNDINGS
(locker room air)

transfer of
body heat

ENERGY LOST
FROM A SYSTEM
(a human body)

NET ENERGY CHANGE = 0

Figure 6.3 Example of how the total energy content of any system, *together with its surroundings*, remains constant.

"System" means all matter in a specific region, such as a human body, a plant, a DNA molecule, or a galaxy. "Surroundings" can be a small region in contact with the system or as vast as the whole universe. The system shown (a human male) is giving off heat to the surroundings (a locker room) by evaporative water loss from sweat. What one region loses, the other region gains, so the total energy content of both does not change.

Figure 6.4 Example of the one-way flow of energy into the world of life that compensates for the one-way flow of energy out of it. The sun continually loses energy, much of it in the form of wavelengths of light (Section 7.2). Living cells intercept some of the energy and convert it to useful forms, stored in bonds of organic compounds. Each time a metabolic reaction proceeds in cells, stored energy is released—and some inevitably is lost to the surroundings, mostly as heat.

The lower photograph shows green, water-dwelling, photosynthetic cells (*Volvox*). They live in tiny, spherical colonies. The orange cells serve in reproduction. They form new colonies inside the parent sphere.

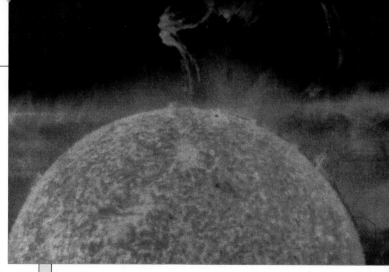

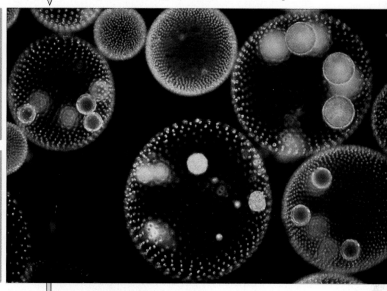

ENERGY LOST
one-way flow of energy from sun to Earth's environment

ENERGY GAINED
one-way flow of energy from environment to organisms

Producer organisms harness sun's energy, use it to build organic compounds from simple raw materials available in their environment.

All organisms tap potential energy stored in organic compounds to drive energy conversions that keep them alive. Some energy is lost with each conversion.

ENERGY LOST
one-way flow of energy from organisms back to the environment

For example, your cells release usable energy from glucose when they break all of its covalent bonds. After many steps, six molecules of carbon dioxide and six of water remain. Compared with glucose, these leftovers are more stable. But chemical energy in the arrangement of their atoms and bonds is much less than the overall chemical energy of glucose. Why? *Some energy was lost at each step leading to their formation.* In other words, glucose is a better source of usable energy.

What about the heat that was transferred from cells to their surroundings when carbon dioxide formed? That heat just isn't useful. Cells can't convert it to other forms, so they can't use it to do work.

Bad news for cells of the remote future: The amount of "low-quality" energy in the universe is increasing. No energy conversion can ever be 100 percent efficient; even highly efficient ones lose heat. From our standpoint, the total amount of energy in the universe is spontaneously flowing from usable forms to unusable forms. Billions of years from now, energy may not be available for conversions; all of it may be dissipated in heat.

Without energy inputs to maintain it, any organized system tends to become more and more disorganized over time. **Entropy** is a measure of the degree of a system's disorder. Think of the Egyptian pyramids—once highly organized, presently crumbling, and many thousands of years from now, dust. It seems the ultimate destination of pyramids and everything else in the universe is a state of maximum entropy. And that, basically, is the point to remember about the **second law of thermodynamics**.

Can life be one glorious pocket of resistance to the depressing flow toward maximum entropy? After all, in every new organism, new bonds form and hold atoms together in precise arrays. So molecules become more organized and have a richer store of energy, not poorer!

Yet a simple example will show that the second law does indeed apply to life on Earth. The primary energy source for life is the sun—which has been releasing energy since it first formed. Plants can capture sunlight energy, convert it to other forms, then lose energy to other organisms that feed, directly or indirectly, on the plants. At each energy transfer from the sun onward, some energy is lost as heat and joins the universal pool. *Overall, energy still flows in one direction.* The world of life maintains its amazing degree of organization only because it is being resupplied with energy that's being lost from someplace else (Figure 6.4).

The amount of energy in the universe remains constant. Energy can undergo conversion from one form to another, but it cannot be created out of nothing or destroyed.

From our perspective, all energy in the universe is flowing spontaneously from usable to unusable forms. A steady, one-way flow of sunlight energy into the interconnected web of life compensates for the steady flow of energy leaving it.

ENERGY INPUTS, OUTPUTS, AND CELLULAR WORK

Cells and Energy Hills

When cells convert one form of energy to another, the amount of potential energy available to them changes. The greater the amount a cell taps into, the larger the energy change, and the more work can be done.

Imagine a Martian watching the NASA Rover scoot around her planet. She decides to push it to the top of a rocky hill, which requires tapping into potential energy stored in her muscle cells (Figure 6.5). Once the Rover is at the top, it has enough potential energy (owing to its position) to roll to the base on its own. The higher it is relative to its final position at the hill's base, the greater the energy change. This same rule applies to the cellular world, in which temperature and pressure remain fairly constant: *Energy changes in living cells tend to proceed spontaneously in the direction that results in a decrease in usable energy.*

Remember glucose ($C_6H_{12}O_6$)? It's made of carbon dioxide ($6CO_2$) and water ($12H_2O$). All three substances have chemical bond energy, but glucose has more than the other two combined. This means making glucose from them isn't something that happens on its own. It

is as if carbon dioxide and water are at the base of an "energy hill." On their own, they simply don't have enough energy for an uphill run to make glucose.

In photosynthetic cells, energy inputs from the sun drive reactions that make glucose, the result being a net increase in usable energy (Figure 6.5a). We say the reaction sequence is *endergonic* (meaning energy in).

Cells also run the reactions in reverse, from glucose (at the top of the energy hill) to carbon dioxide and water (at the base). Energetically, the downhill run is favored; it will proceed on its own and end with a net loss in energy. Such reactions are *exergonic*, meaning energy out (Figure 6.5b).

ATP Couples Energy Inputs With Outputs

Cells stay alive by *coupling* energy inputs to energy outputs, primarily with **ATP** (adenosine triphosphate). In this nucleotide, covalent bonds join the five-carbon sugar ribose, the nucleotide base adenine, and three phosphate groups (Figure 6.6). Hundreds of different enzymes can split off the outer phosphate group and join it to another molecule, which thus becomes primed

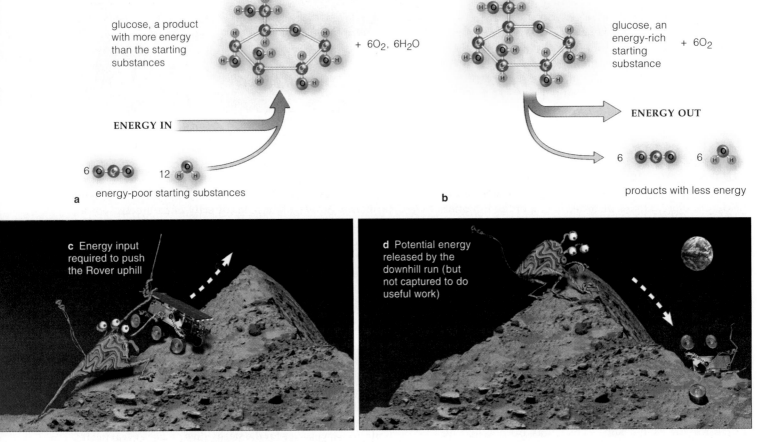

glucose, a product with more energy than the starting substances

$+ 6O_2, 6H_2O$

ENERGY IN

6   12

a energy-poor starting substances

glucose, an energy-rich starting substance

$+ 6O_2$

ENERGY OUT

6   6

b

products with less energy

c Energy input required to push the Rover uphill

d Potential energy released by the downhill run (but not captured to do useful work)

Figure 6.5 Energy changes involved in (**a,b**) chemical work and (**c,d**) mechanical work.

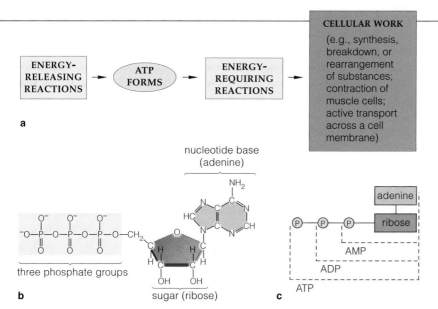

ENERGY-RELEASING REACTIONS → ATP FORMS → ENERGY-REQUIRING REACTIONS → CELLULAR WORK (e.g., synthesis, breakdown, or rearrangement of substances; contraction of muscle cells; active transport across a cell membrane)

a

nucleotide base (adenine)

NH₂

three phosphate groups

sugar (ribose)

b

adenine

ribose

AMP

ADP

ATP

c

Figure 6.6 (**a**) A key energy relationship in all living cells. ATP, an energy carrier, couples energy-releasing reactions with hundreds of diverse energy-requiring ones. (**b**) Structural formula for ATP. (**c**) Successive phosphate-group transfers turn ATP into ADP (adenosine diphosphate), then AMP (adenosine monophosphate).

Ca⁺⁺   Ca⁺⁺   Ca⁺⁺   Ca⁺⁺   Ca⁺⁺
   Ca⁺⁺   Ca⁺⁺   Ca⁺⁺   Ca⁺⁺

Ca⁺⁺

ADP

Pᵢ

energy input   energy output

ATP

Figure 6.7 Example of cellular work initiated when ATP gives up energy to a molecule involved in a specific metabolic reaction. At the plasma membrane, ATP transfers one of its phosphate groups to an active transport protein. Remember Section 5.6? When activated, this transporter changes shape and pumps calcium ions across the membrane and out of the cell, against their concentration gradient.

to enter a reaction. Such a phosphate-group transfer is called a **phosphorylation**.

As you know, covalent bonds are stable interactions between atoms. Why, then, is ATP so ready to give up its phosphates? The phosphates are arranged in a way that puts a number of negative charges close together. The like charges repel each other, which destabilizes the molecule's tail. Getting rid of a phosphate group helps stabilize the molecule and results in more stable products of lower energy. The products are adenosine diphosphate, or **ADP**, and a free inorganic phosphate atom, which we abbreviate as P_i (Figure 6.7). In other words, hydrolysis of ATP is an energetically favorable reaction that releases a large quantity of usable energy.

Think of ATP in cells as being like the currency in an economy. Cells earn it by investing in reactions that release energy. They spend it in reactions that require energy to drive hundreds of activities. That's why we often use a cartoon coin to symbolize ATP.

Because ATP is the main energy carrier for so many reactions, you might speculate that cells should have a way to renew it, and you would be right. When ATP gives up its terminal phosphate group to a molecule, it becomes ADP. This is converted back to ATP when an enzyme attaches inorganic phosphate (or a phosphate group derived from some other molecule) to it.

Regenerating ATP this way is an important aspect of metabolism. It is called the **ATP/ADP cycle**:

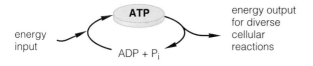

energy input → ATP → ADP + P_i → energy output for diverse cellular reactions

Some Electron Transfers Drive ATP Formation

Energy changes in cells involve transfers of electrons between molecules. These electron transfers are often called **oxidation–reduction reactions**. One molecule is said to be oxidized (it loses one or more electrons) as another molecule is reduced (it gains the electrons).

In photosynthetic organisms—the primary *providers* of electrons—sunlight energy splits water molecules, or H_2O. The oxygen escapes. But electrons and hydrogens are transferred in ways that drive the formation of ATP as well as energy-rich molecules from carbon dioxide. Glucose can be formed this way.

All cells *use* electrons to get at the energy stored in complex molecules. For instance, in aerobic respiration, an energy input jump-starts the release of electrons from glucose. As glucose is degraded to carbon dioxide and water, electron transfers along the way help drive ATP formation. Water forms when the spent electrons are transferred to free oxygen. Section 6.4 takes a closer look at the nature of these electron transfers.

On their own, energy changes in cells spontaneously run in the direction that results in a decrease in usable energy.

ATP, an energy carrier in all living cells, couples energy-releasing reactions with energy-requiring ones. It does so by phosphate-group transfers, which release enough usable energy to activate molecules—to prime them to react.

Cells obtain energy with the help of oxidation–reduction reactions, which are electron transfers from one substance to another. Such reactions are central to the formation of ATP during photosynthesis and aerobic respiration.

CELLS JUGGLE SUBSTANCES AS WELL AS ENERGY

How cells get energy is only one aspect of metabolism. Another is the accumulation, conversion, and disposal of substances by energy-driven reactions.

Participants in Metabolic Reactions

Participants of metabolic reactions go by these names: **Reactants** are the substances that enter a reaction. Any substance that forms during a reaction (or a sequence of reactions) is an **intermediate**. Substances left at the end of a reaction are the **products**. The **energy carriers** activate enzymes and other molecules by phosphate-group transfers. ATP is the main energy carrier. For now, think of enzymes as proteins that speed specific reactions. (We now know a few RNAs also function as enzymes.) The **cofactors** are metal ions and coenzymes, such as NAD^+ and some other organic compounds. They assist enzymes by picking up electrons, atoms, or functional groups from a reaction site and giving them up at different sites. **Transport proteins**, recall, help substances across cell membranes. Controls over these proteins affect metabolism by adjusting concentrations of the substances necessary for specific reactions.

Table 6.1 summarizes these participants. Learn their names; you will be encountering them later on.

What Are Metabolic Pathways?

The concentrations of thousands of substances change continually within living cells. Most of the substances enter or leave by orderly, enzyme-mediated sequences called **metabolic pathways**. Often, small molecules are used to build larger molecules of higher bond energies,

such as complex carbohydrates, complex lipids, and proteins. These are *biosynthetic* (or anabolic) pathways. They cannot proceed without energy inputs. The main biosynthetic pathway is photosynthesis.

Degradative (or catabolic) pathways are exergonic, overall. They break down large molecules to smaller products with lower bond energies. Aerobic respiration is the main degradative pathway. It completely breaks down glucose to carbon dioxide and water, with the release of a considerable amount of usable energy.

The linear pathways advance in a straight line, from substrates to end products. The cyclic pathways occur in a circle; the final step is the conversion of an intermediate back to the reactant. In other words, the cycle regenerates its own point of entry. The branched pathways send a reactant or some intermediate down two or more reaction sequences (Figure 6.8).

Are the Reactions Reversible?

Don't let Figure 6.8 lead you to think that metabolic reactions always proceed in one direction. They might start out in the "forward" direction, from reactants to products. Most can also run in reverse, with products being converted back to reactants. Whenever you see a chemical equation with opposing arrows, this signifies the reaction is reversible. Each arrow means *yields*:

$$A + B \rightleftarrows C$$

STARTING SUBSTANCES   PRODUCT

Which way such a reaction runs depends partly on the energy content of the participants. It also depends on the reactant-to-product ratio. When the energy level

Table 6.1	*Participants in Metabolic Reactions*
REACTANT	Substance that enters a metabolic reaction or pathway; also called an enzyme's substrate
INTERMEDIATE	Substance formed between reactants and end products of a reaction or pathway
PRODUCT	Substance left at end of reaction or pathway
ENZYME	Usually a protein that enhances reaction rates; a few RNAs also do this
COFACTOR	Coenzyme (such as NAD^+) or metal ion; assists enzymes or taxis electrons, hydrogen, or functional groups between reaction sites
ENERGY CARRIER	Mainly ATP in cells; couples energy-releasing reactions with energy-requiring ones
TRANSPORT PROTEIN	Protein that passively assists substances across a cell membrane or actively pumps them across

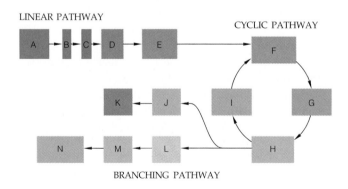

Figure 6.8 Types of reaction sequences—linear, cyclic, and branched—for different metabolic pathways. An arrow signifies an enzyme-mediated step. Here, a linear pathway converts reactant A to product F. In a cyclic pathway, F is a reactant for a series of steps that, as one outcome, regenerate F so the cycle can turn again. One of the cyclic pathway's intermediates, H, becomes a reactant for a branching pathway that splits it and converts it two products, N and K.

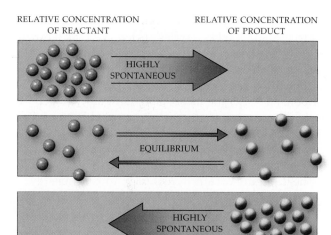

RELATIVE CONCENTRATION OF REACTANT   RELATIVE CONCENTRATION OF PRODUCT

HIGHLY SPONTANEOUS

EQUILIBRIUM

HIGHLY SPONTANEOUS

Figure 6.9 Chemical equilibrium. When the concentration of reactant molecules is high, a reaction runs most strongly in the forward direction, to products. When the concentration of product molecules is high, it runs most strongly in reverse. At equilibrium, the rates of the forward and reverse reactions are the same.

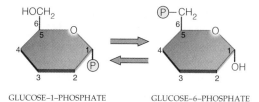

GLUCOSE–1–PHOSPHATE   GLUCOSE–6–PHOSPHATE

Figure 6.10 A reversible reaction. Glucose is primed to enter reactions when a phosphate group becomes attached to it. When there is a high concentration of glucose–1–phosphate, the reaction tends to run in the forward direction. With a high glucose–6–phosphate concentration, it runs in reverse.

The 1 and 6 of these names simply identify which carbon atom of the glucose ring has a phosphate group attached to it.

and concentration of reactant molecules are high, this is an energetically favorable state. That is, the reaction tends to run spontaneously and strongly in the forward direction. When the product concentration gets high enough, however, more molecules or ions of product are available to revert spontaneously to reactants.

Any reversible reaction tends to run spontaneously toward **chemical equilibrium**: the time when it runs at about the same pace in both directions (Figure 6.9). The *amounts* of reactant and product molecules usually differ at that time. It's like a party with people drifting into and out of two rooms. The number in each room stays the same—say, 30 in one and 10 in the other—even as individuals move from one to the other.

Each reversible reaction has a specific equilibrium ratio. For instance, glucose–1–phosphate forms from glucose–6–phosphate. (As Figure 6.10 shows, both are glucose, but they have a phosphate group attached to a different carbon atom.) In this case, the forward and reverse reactions run at the same rate when there are nineteen molecules of glucose–6–phosphate for every glucose–1–phosphate molecule. The equilibrium ratio in this case is 19:1.

Why bother to think about this? *Each cell can bring about big changes in its activities by controlling a few steps of reversible metabolic pathways.*

For instance, when your cells need a very quick fix of energy, they can rapidly break down glucose to two molecules of pyruvate. They can break it down by a sequence of nine enzyme-mediated steps of a pathway called glycolysis (Section 8.2). When glucose supplies are too low, cells can reverse this pathway and swiftly build glucose from pyruvate and other substances. Six of the reaction steps are reversible; the other three are bypassed. An input of ATP energy drives the bypass reactions in the energetically unfavorable direction.

What would happen if your cells did not have this reverse pathway? During times of starvation, when the blood glucose concentration becomes dangerously low, your cells would not be able to build replacements fast enough to counter the stress.

No Vanishing Atoms at the End of the Run

One other point: When reactions run in the forward or reverse direction, they rearrange atoms, but they never destroy them. By the **law of conservation of mass**, the total mass of all of the substances that enter a reaction equals the total mass of all of the products. When you study any chemical equation, count up the atoms of each reactant and product molecule. There should be as many atoms of each element to the right of the arrow as to the left. When you write equations for metabolic reactions, they must balance this way.

Cells can increase, maintain, and decrease concentrations of substances by coordinating thousands of reactions.

The reactions start with reactants and end with products. Substances formed in between are intermediates. Energy carriers (ATP especially), enzymes, cofactors, and transport proteins are key players in metabolic reactions.

Metabolic pathways are orderly, enzyme-mediated reaction sequences. Biosynthetic (anabolic) pathways build large molecules of higher bond energies from smaller molecules. The degradative (catabolic) pathways break down large molecules to smaller products with lower bond energies.

Cells can rapidly shift their metabolism in major ways by controlling key steps of reversible pathways.

ELECTRON TRANSFER CHAINS IN THE MAIN METABOLIC PATHWAYS

During both photosynthesis and aerobic respiration, certain electron transfers are made at membrane-bound chains of enzymes. Become familiar now with the way these chains work, and you will have a fine head start toward understanding both metabolic processes.

Photosynthetic cells make ATP, the source of energy for making glucose from CO_2 and H_2O. All cells that engage in aerobic respiration break down glucose to make a lot of ATP. It takes a series of steps to build this molecule, and it takes a series of steps to break it down.

Why? Glucose, a reduced, energy-rich molecule, is not as stable as CO_2 and H_2O. (The atmosphere holds so much oxygen that CO_2 is the most stable form of carbon, and H_2O the most stable form of hydrogen.) If you toss a cupful of glucose into a wood-burning fire, its carbon atoms and hydrogen atoms will let go of one another at once and combine with oxygen in the air. However, all of the energy released will be lost as heat (Figure 6.11a). By contrast, in cells, energy is released more efficiently by a series of oxidation–reduction reactions. First, glucose is primed to react (Figure 6.10). Then electrons and hydrogen atoms are stripped from it and transferred to coenzymes.

$NADP^+$ is an electron-transferring coenzyme in photosynthesis. NAD^+ and FAD are counterparts in aerobic respiration. We write their reduced forms as NADPH, NADH, and $FADH_2$, respectively. Each functions to transfer hydrogen as well as electrons *to* electron transfer chains. You may hear someone refer to the chains as electron transport systems, but enzymes, not transport proteins, are involved.

Electron transfer chains consist of specific kinds of enzymes and other molecules that are organized to accept and give up electrons, one after the other, at a cell membrane between two compartments. The electrons are at a higher energy level when they enter the chain than when they leave. Think of them as being moved down an energy staircase and losing a bit of energy at certain steps (Figure 6.11b).

Your focus on these stepwise transfers should not be the electrons themselves but rather on the attraction that electrons hold for hydrogen. Free hydrogen ions (H^+) associate with electrons entering the transfer chains. At certain transfer steps, components of the chain shunt these ions into the compartment on the other side of the membrane. There, the ions accumulate until H^+ concentration and electric gradients across the membrane have become established. The force of the gradients propels H^+ out of the compartment, through transport proteins that span the membrane. This ion flow drives ATP formation.

In short, the electron transfer chains contribute energy, in the form of H^+ concentration and electric gradients, for making ATP at membrane transport proteins.

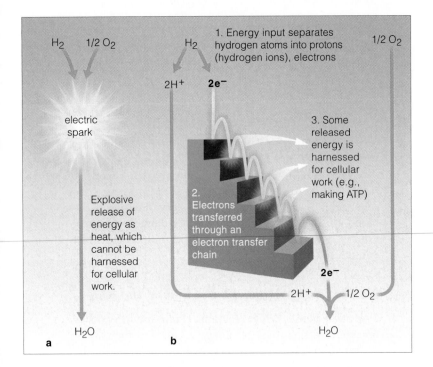

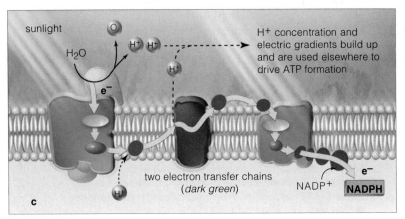

Figure 6.11 Uncontrolled versus controlled energy release. (**a**) Oxygen and hydrogen exposed to an electric spark react and release energy all at once. (**b**) Electron transfer chains let the same reaction occur in small, manageable steps that access released energy. (**c**) Electron-donating complexes and electron transfer chains in a chloroplast. The chains accept electrons that were excited to a higher energy level by sunlight and released from water. A coenzyme, $NADP^+$, accepts electrons at the end of the second chain.

Photosynthesis and aerobic respiration rely on electron transfers to secure energy for ATP formation.

Electron transfer chains associated with cell membranes contribute energy, inherent in H^+ concentration and electric gradients, to make ATP at other membrane sites.

ENZYMES HELP WITH ENERGY HILLS

What would happen if you left a cupful of glucose out in the open? Not much. Even though its conversion to CO_2 and H_2O is energetically favored in the presence of oxygen, years would pass before you could expect to see evidence of it. How long does the conversion take in your body? A few seconds. *Enzymes make the difference.*

Without enzymes, you would quickly cease to exist. Reactions would not occur fast enough for you to process food, build new cells and get rid of worn-out ones, keep your brain working, contract muscles, and do everything else you have to do to stay alive.

By definition, **enzymes** are catalytic molecules; they speed the rate at which a specific reaction approaches chemical equilibrium. An enzyme alters only the rate of a reaction, not its outcome.

Except for a few RNAs, enzymes are proteins. Like other proteins, their function depends on their primary structure and final, three-dimensional structure, for the reasons described earlier in Sections 3.6 through 3.8.

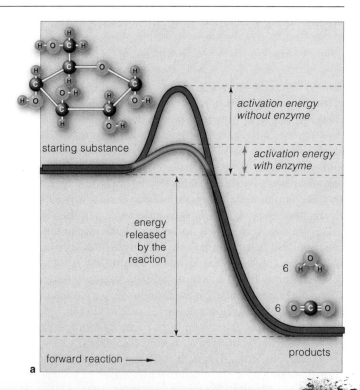

Figure 6.12 Activation energy. (**a**) A certain amount of energy is required to get a metabolic reaction going. That amount, the activation energy, allows the reaction to spontaneously proceed to the end products. (**b**) An enzyme enhances the reaction rate by lowering the amount of activation energy. Refer back to the analogy in Figure 6.5. Enzymes make the energy hill smaller; they lower the energy barrier.

All enzymes share four features. First, enzymes do not make anything happen that could not happen on its own, but they make it happen hundreds to millions of times faster. Second, reactions do not permanently alter or use up enzymes; the same enzyme molecule may work repeatedly. Third, the same type of enzyme usually catalyzes the forward and reverse directions of a reversible reaction. Fourth, an enzyme is picky about its substrates. Its only substrates are specific substances that the enzyme can chemically recognize, bind, and modify in precise ways. For instance, thrombin is one of the enzymes needed to clot blood. It only recognizes and cleaves a side-by-side arrangement of arginine and glycine—two amino acids—in a certain protein. This converts the protein to one of the clotting factors.

To get a sense of how enzymes work, you have to know about activation energy. **Activation energy** is the minimum amount of collision energy required to get a reaction going. The actual amount is not the same for all reactions. And the collision may be spontaneous or enzymes may promote it.

Activation energy is an energy barrier—something like the hill depicted in Figure 6.12. That barrier must be surmounted one way or another before the reaction will proceed. An enzyme lowers the energy barrier, in ways that will be described next.

Enzymes enhance reaction rates by lowering the amount of activation energy required to get the reaction going.

HOW DO ENZYMES LOWER ENERGY HILLS?

The Active Site

How do enzymes lower the energy hill and so increase rates of reactions? *They present a microenvironment that is energetically more favorable for reaction, compared to the environment at large.* Each contains one or more **active sites**: pockets or crevices where substrates are bound and where specific reactions are catalyzed. Reactants for an enzyme—its **substrates**—have a surface region that is complementary in its size, shape, solubility, and charge to the active site. This complementary fit is why an enzyme can selectively identify its substrate among the thousands of substances in cells. Figure 6.13 depicts one enzyme meeting up with its substrate.

Sometimes functional groups of the enzyme alone carry out a reaction. However, one or more cofactors often help out. Cofactors, again, include metal ions and coenzymes, which are complex organic compounds that may or may not have a vitamin component.

Metal ions, coenzymes, or both are often bound so tightly to an active site that they are *prosthetic* groups, as functionally important as an artificial limb is to an amputee. The "heme" in hemoglobin is an example.

Transition at the Top of the Hill

Think back on the key categories of enzyme-mediated reactions introduced in Chapter 3. During functional group transfers, a molecule gives up a functional group and another molecule accepts it. In electron transfers, one or more electrons are stripped from a molecule and donated to another. In rearrangements, the juggling of internal bonds converts one molecule to another. In condensation, two molecules become covalently bound to form a larger molecule. Finally, in cleavage, a larger molecule splits into two smaller ones.

Thus, when we talk about the amount of activation energy required to convert each kind of substrate to its products, *we really are talking about the energy it takes to align reactive chemical groups, to briefly destabilize electric charges, and to rearrange, create, and break bonds.*

These events can bring a substrate to its **transition state**, the point when a reaction can easily run in either direction, to product or back to a reactant. A substrate is bound most tightly to an enzyme in this state.

How Enzymes Work

During an enzyme-mediated reaction, various bonds form between the substrate and the enzyme, cofactor, or both. Many of these bonds are weak. But energy is released as each forms, and it promotes the formation of an enzyme–substrate complex. Energy released from all of the weak interactions is the **binding energy**. Its release in the transition state is partial payment on the cost of the reaction—the activation energy.

Binding energy can speed a reaction rate by several mechanisms. Depending on the enzyme, the following four mechanisms work alone or in combination with one another to bring about the transition state.

Helping substrates get together. Substrate molecules rarely react if their concentrations are low. Binding at an active site is like a localized boost in concentration. The boost can increase the rate by 10,000 to 10,000,000 times, depending on the particular reaction.

Orienting substrates in positions favoring reaction. On their own, substrates collide from random directions. Weak but extensive bonds at an active site put reactive groups on precise collision courses far more frequently.

Shutting out water. Sometimes substrates are bound so tightly that water molecules are partially or wholly squeezed out of the active site. For some reactions, the resulting nonpolar microenvironment can lower the activation energy by as much as 500,000 times. This is what happens when an enzyme attaches a carboxyl group ($-COO^-$) to a molecule.

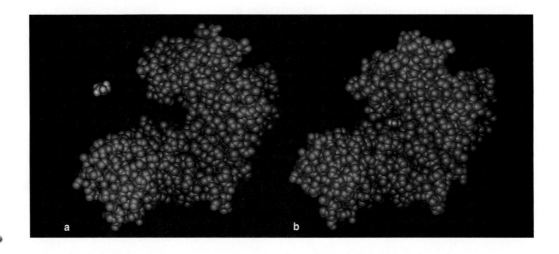

Figure 6.13 Space-filling model for hexokinase, a major enzyme. Hexokinase attaches a phosphate group to glucose. (**a**) The glucose molecule, color-coded *red*, is heading for the enzyme's active site, which is inside a cavity in the enzyme (*green*). (**b**) As glucose enters the cavity, parts of the enzyme briefly close around it.

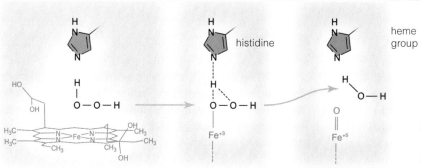

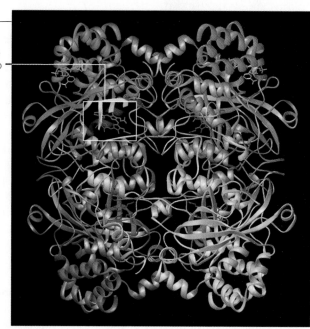

a One molecule of hydrogen peroxide (H_2O_2) enters one of the four cavities of a catalase molecule, where it will become the substrate for the reaction. The heme group enclosed in the cavity is shown in *red*.

b One of the amino acids (histidine) projecting into the cavity attracts one of the substrate's H atoms, leaving an O atom free to bind to the heme group's iron (Fe).

c Binding to the iron causes the substrate's O—O bond to stretch and break. Water (H_2O) forms when the —OH fragment attracts an H atom away from the histidine. (The iron releases the oxygen in a second reaction with another H_2O_2 molecule. Oxygen and another H_2O molecule form.)

Figure 6.14 A closer look at catalase, an enzyme that breaks down hydrogen peroxide to water and oxygen. In cells, four catalase chains associate with each other as a quaternary structure. Each folded chain forms a cavity around one heme group. An iron atom in the heme group assists in catalysis. The electrical properties of the enzyme-bound heme ensure that harmful oxygen radicals are not produced as a by-product of hydrogen peroxide breakdown.

Inducing changes in enzyme shape. In most cases, the weak interactions between an enzyme and its substrate induce change in the enzyme's shape. By the **induced-fit model**, functional groups on the substrate's surface are *not quite* complementary to functional groups in the active site. As the enzyme aligns structurally with the substrate, the substrate itself bends, which pushes it toward the transition state. Precise complementarity would not work. If a substrate were to bind too tightly to the active site, the enzyme–substrate complex would become too stable, and reaction would be impeded.

About Those Cofactors

What roles do cofactors play when a substrate has been suitably aligned in an active site? You can get an idea by considering cofactors that transfer electrons.

For instance, unstable, charged intermediates form in many reactions. Left alone, they would swiftly revert to the form of the substrate. Electron (and hydrogen) transfers to and from the substrate or intermediate can tip the reaction toward products rather than substrates.

Cofactors mediate most of these reactions. Example: The metal ions bind so tightly with aspartate, cysteine, histidine, or glutamate that they become built into the structure of many enzymes. They readily give up and accept electrons. When they interact with a substrate or intermediate, they shift electron arrangements in ways that promote product formation. This is what goes on

at the heme in catalase. Heme has a complex organic ring structure, with iron (Fe^{++}) at its center. Figure 6.14 shows how this metal ion works. Nearly one-third of all known enzymes use one or more metal ions.

Why Are Enzymes So Big?

Most enzymes, like the one in Figure 6.13, are very big molecules compared to their substrates. Why is this so? Rapid, repeated catalysis demands a stable chemical environment. An enzyme's polypeptide chains must be long enough to fold repeatedly in particular directions, and not just to afford structural stability. Folding also must put specific amino acids and functional groups in locations and orientations that favor interaction with the water that bathes the enzyme's outer surface—and with a substrate molecule that contacts the active site.

The activation energy for a reaction is the energy it takes to align reactive chemical groups, briefly destabilize electric charges, and rearrange, create, and break bonds.

Activation energy drives substrates to the transition state, the point when a reaction can spontaneously run either to product or back to substrate.

Increasing substrate concentrations, orienting substrates, excluding water, and changing the enzyme and substrate shapes are the main mechanisms that lower the activation energy at an active site.

ENZYMES DON'T WORK IN A VACUUM

You probably don't get much done when you feel too hot or cold or out of sorts because you ate too many sour plums or salty potato chips. When the cupboard is bare, you focus on food. Maybe you call a friend to go shopping with you, and when you drive too fast to the grocery store, police tend to slow you down. In such ways, you have a lot in common with enzymes. They, too, respond to shifts in temperature, pH, and salinity, and to the relative abundances of particular substances. Many depend on helpers for specific tasks. And all normal enzymes respond to metabolic police.

How Is Enzyme Activity Controlled?

Each cell controls its enzyme activity. By coordinating control mechanisms, it maintains, lowers, or increases the concentrations of substances. Controls adjust how fast enzyme molecules are built and become available, or they work on enzymes that are already synthesized.

For example, with *allosteric* control, some molecule binds to an enzyme at a site other than the active site. (*Allo–* means different, *steric* means structure, or state.) The active site changes shape in a way that allows or prevents enzyme action. Figure 6.15 shows models for the binding, which is reversible. Study these models, then picture a bacterium synthesizing tryptophan and other amino acids used to build proteins. Soon, enough proteins are built. But tryptophan synthesis won't stop until the concentration of tryptophan molecules causes **feedback inhibition**. With this feedback mechanism, a change caused by an activity *shuts down the activity*.

In this pathway, a feedback loop starts and ends at an allosteric enzyme. When tryptophan accumulates, unused molecules bind with the allosteric site. Binding shuts down the enzyme, hence the rest of the pathway.

What if tryptophan is scarce when the demand for it steps up? Enzyme molecules are not inactivated, so tryptophan will be built. Such feedback loops quickly adjust concentrations of many substances (Figure 6.16).

In humans and other multicelled organisms, control of enzyme activity is just amazing. Cells not only work to stay alive, they work with other cells in ways that benefit the whole body! For instance, this vast cellular work relies on hormones, a type of signaling molecule. Specialized cells release hormones. Any cell that has receptors for a particular hormone can respond to it. Its program for building a protein or some other activity changes. The hormone trips cell controls into action—and the activity of enzymes increases or slows down.

Do Temperature and pH Affect Enzymes?

Temperature, recall, is a measure of molecular motion. A rise in temperature boosts reaction rates by making substrates collide more often with active sites. Yet past some temperature (which differs among enzymes), the increased motion disrupts the weak bonds holding the enzyme in its three-dimensional shape. Substrates no longer bind to the active site, so reaction rates decline sharply, as Figure 6.17a shows. When temperatures rise above or fall below the range of tolerance, metabolism is disrupted. That's what happens with extremely high fevers. People usually do not survive when the body's internal temperature reaches 44°C (112°F).

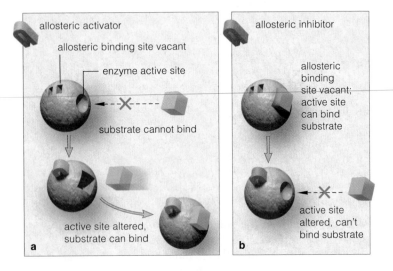

Figure 6.15 Simple sketches of allosteric control of enzyme activity. (**a**) An active site is unblocked when an activator protein binds to a vacant allosteric site. (**b**) An active site is blocked when an inhibitor protein binds to a vacant allosteric site.

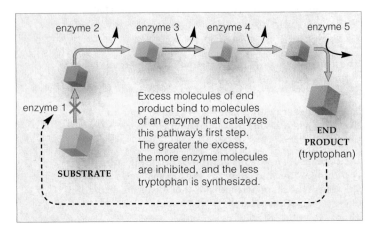

Excess molecules of end product bind to molecules of an enzyme that catalyzes this pathway's first step. The greater the excess, the more enzyme molecules are inhibited, and the less tryptophan is synthesized.

Figure 6.16 Example of feedback inhibition of a metabolic pathway. Five kinds of enzymes act in sequence. They convert a substrate to the product, tryptophan.

a

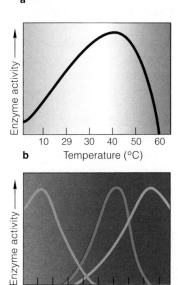

b

Temperature (°C)

10 29 30 40 50 60

c

pH

2 3 4 5 6 7 8 9 10

Figure 6.17 Enzymes and the environment. (**a**) Siamese cats show observable effects of differences in temperature. Epidermal cells that give rise to their fur produce a brownish-black pigment, melanin. More melanin is produced in fur on ears and paws than elsewhere on the cat body. A heat-sensitive enzyme that helps control melanin production is less active in warmer parts of the body, which end up with lighter fur. (**b**) How one enzyme's activity shifts as the temperature rises. (**c**) How three kinds of enzymes respond to a rise in pH.

pH values that extend beyond an enzyme's range of tolerance also take a toll (Figure 6.17b). Most enzymes work best when pH is between 6 and 8. Trypsin, for instance, is active in the small intestine, where pH is 8 or so. Pepsin, a protein-digesting enzyme, is one of the exceptions. It functions in gastric fluid, a highly acidic liquid (pH of about 1–2) that denatures most enzymes.

In addition, most enzymes do not work well when fluids are saltier than usual. High salinity (extremely high ion concentrations) disrupts the interactions that help hold enzymes in their three-dimensional shapes.

Control mechanisms govern the synthesis of new enzymes and stimulate or inhibit the activity of existing enzymes. By controlling enzymes, cells control the concentrations and kinds of substances available to them.

Enzymes function best when the cellular environment stays within limited ranges of temperature, pH, and salinity. The actual ranges differ from one type of enzyme to the next.

Beer, Enzymes, and Your Liver

That catalase molecule you've been reading about is also a foot soldier against alcoholic attacks on the body. Think of what happens after someone drinks 12 ounces of beer, 6 ounces of wine, or 1.5 ounces of eighty-proof vodka. All contain the same amount of ethanol, or ethyl alcohol (C_2H_6O). This sugar has water-soluble and fat-soluble components. The body absorbs about 20 percent from the stomach and about 80 percent from the small intestine. The bloodstream swiftly transports more than 90 percent of it to the liver, which breaks it down to a nontoxic molecule called acetate (acetic acid). If ethanol were to accumulate to high levels, cells would die and tissues would be damaged. A healthy person's liver detoxifies ethanol before this happens.

The liver can detoxify only so much in a given hour. Its alcohol-metabolizing enzymes set the reaction rate, which is slower than the rate of absorption. Alcohol dehydrogenases in liver cells are key enzymes in the detoxification reactions. They catalyze the conversion of ethyl alcohol to acetaldehyde, which is toxic at high levels. Another enzyme, acetaldehyde dehydrogenase, swiftly breaks down this intermediate to acetate and other nontoxic forms. Also, when the concentration of alcohol in the blood is high, one of the cytochromes assists in the detoxification. So does catalase.

Given the liver's central role in alcohol metabolism, habitually heavy drinkers gamble with alcohol-induced liver diseases. Over time, the capacity to tolerate alcohol diminishes because there are fewer and fewer liver cells —hence fewer enzymes—available for detoxification. In *alcoholic hepatitis*, inflammation and destruction of liver tissue is widespread. Another disease, *alcoholic cirrhosis* permanently scars the liver and eventually destroys its functioning (Figure 6.18).

Such a loss is devastating. The liver is the largest gland in the human body, and its functioning impacts all other organs. As you will see later in the book, it has major roles in digesting and absorbing fats from food. It is central to controlling the synthesis and uptake of carbohydrates, lipids, and proteins by body cells. It is a blood reservoir that is adjusted to help maintain the volume of blood through the body. It helps return excess fluid that collects in tissues to the general circulation. And its enzymes inactivate many toxic compounds besides acetaldehyde. Think about it.

Figure 6.18 *Left*: Normal liver tissue. *Right*: What alcohol-induced disease can do to it.

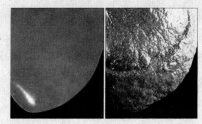

LIGHT UP THE NIGHT—AND THE LAB

BIOLUMINESCENT ORGANISMS   In warm waters of the tropical seas, above gardens and fields on summer nights, you may be lucky enough to witness organisms lighting up the night sky with their collective flashings. Many species flash with orange, yellow, yellow-green, or blue light, as in Figure 6.19.

Flashers emit light when enzymes called luciferases convert chemical energy to light energy. Figure 6.19*c* shows the molecular structure of the firefly luciferase. The reactions start when, in the presence of oxygen, an ATP molecule transfers a phosphate group to a highly fluorescent substance: luciferin. This destabilizes the molecule, which makes it enter reactions that involve electron transfers. At one reaction step, some energy is released as *fluorescent* light. This type of light is emitted when any kind of destabilized molecule reverts to a stable configuration. When organisms flash with such light, this is called **bioluminescence**.

RESEARCH APPLICATIONS   It is possible to think about organisms from a "gee-whiz-ain't-nature-grand" point of view. Or you can come up with novel ways to think about what they do and how they do it. The latter way of thinking puts you squarely in the camp of biologists.

And so we have biologists who thought to borrow light from fireflies and use it to make bioluminescent gene transfers. Copies of genes for bioluminescence can now be inserted into a variety of organisms, such as bacteria, plants, and mice (Figure 6.20).

Here is a practical application of their work. Each year, 3 million people die from a lung disease caused by *Mycobacterium tuberculosis*. No single antibiotic is effective against all of the different bacterial strains, so

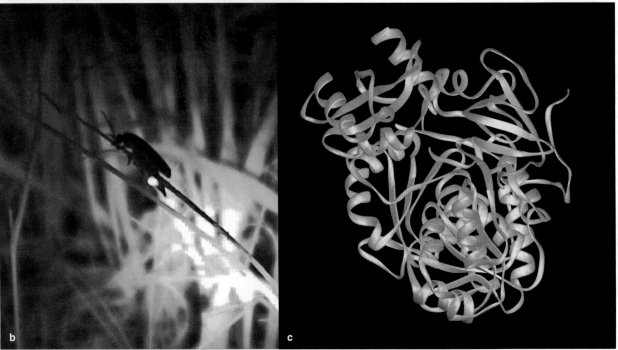

Figure 6.19   (**a**) Stirring up bioluminescent marine organisms near Vieques Island, Puerto Rico. As many as 5,000 free-living cells called dinoflagellates may be in each liter of water, and each flashes with blue light when agitated. (**b**) North American firefly (*Photinus pyralis*) emitting a flash from its light organ. Peroxisomes in this organ are packed with luciferase molecules. Firefly flashes help potential mates find each other in the dark. (**c**) Ribbon model for firefly luciferase. This enzyme catalyzes the reaction that releases light. Its single polypeptide chain is folded into multiple domains.

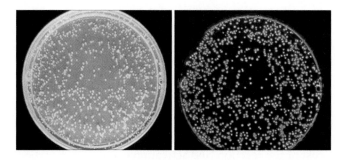

Figure 6.20 Colonies of bioluminescent bacteria in daylight (*left*) and glowing in a culture dish in the dark (*right*).

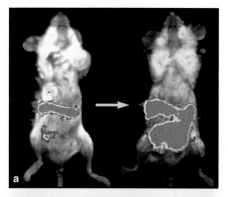

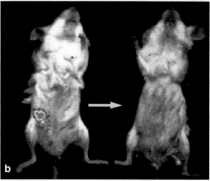

Figure 6.21 Using bioluminescent bacterial cells to chart the location of infectious bacteria inside living laboratory mice and their spread through body tissues. (**a**) False-color images in this pair of photographs show how the infection spread in a control group that had not been given a dose of antibiotics. (**b**) This pair shows how antibiotics had killed most of the infectious bacterial cells.

an infected person can't get effective treatment until the strain that is causing the infection is identified. A fast way to do this is to expose a sample of bacterial cells taken from a patient to luciferase genes. These genes will usually slip into the bacterial DNA of some cells. Clinicians isolate the cells, then expose colonies of the descendants to different antibiotics. When one antibiotic does *not* work, the colonies glow. Their cells have made gene products—including luciferase. An antibiotic works when cells of the colonies don't glow; the antibiotic killed them.

Christopher and Pamela Contag, two postdoctoral students at Stanford University, wanted to light up bacteria that cause *Salmonella* infections in laboratory mice. Why? Researchers of viral or bacterial diseases typically have to infect dozens to hundreds of mice for experiments. The only option has been to kill infected mice and examine tissues to find out whether infection occurred—a costly, tediously painstaking practice that also requires dispatching the experimental animals.

The Contags approached David Benaron, a medical imaging researcher at Stanford, with this hypothesis: If live, infectious bacteria were made bioluminescent, then flashes will shine through the tissues of infected animals. In a preliminary test of this novel idea, the researchers put glowing *Salmonella* cells into a thawed chicken breast from a market. It glowed from inside.

The Contags transferred bioluminescence genes into three strains of *Salmonella*. Then they injected the strains into mice of three experimental groups and used a digital imaging camera to track the infection in each group. The first strain was weak; the mice were able to fight off the infection in less than six days and did not glow. The second strain was not as weak but could not spread through the mouse body; it remained localized. The third strain was dangerous. It spread rapidly through the entire mouse gut—which glowed.

Thus bioluminescent gene transfer, combined with imaging of enzyme activity, can be used to track the course of infection and to evaluate the effectiveness of drugs in living organisms (Figure 6.21). It might have uses in gene therapy, where copies of functional genes replace defective or cancer-causing ones in patients.

Why use bioluminescent organisms to conclude a chapter? They give us visible signs of metabolism—of the cell's capacity to acquire energy and use it to build, break apart, store, and release substances in controlled ways. Each flash reminds us that living cells are taking in energy-rich solutes, building membranes, storing things, replenishing enzymes, and checking out their DNA. A constant supply of energy drives all of these activities. And the flashes can remind us, too, of how modern-day biologists are busy unlocking the secrets of metabolism and putting them to use in major ways.

Bioluminescence is one outcome of enzyme-mediated reactions that release energy as fluorescent light.

Biologists have transferred genes for bioluminesence, the luciferases, into a variety of organisms. These gene transfers have research applications and practical uses.

1. Cells store, break down, and dispose of substances by acquiring and using energy and raw materials from outside sources. Metabolism, the sum of these energy-driven activities, underlies their survival. **CI**

2. Two laws of thermodynamics affect life. *First*, energy undergoes conversion from one form to another, but its total amount neither increases nor decreases; the total amount in the universe holds constant. *Second*, energy spontaneously flows in one direction, from usable forms to forms that are less and less usable. **6.1**

3. All matter has potential energy, as measured by the capacity to do work, owing to its position in space and the arrangement of its parts. Cells use potential energy to do mechanical, chemical, and electrical work. **6.1**

4. Without energy, a cell (like all organized systems) gets disorganized. It loses chemical potential energy in each metabolic reaction, mainly as heat. It stays alive (organized) by balancing energy outputs and inputs. The sun is life's primary energy source. Plants and other photosynthesizers convert sunlight energy to chemical bond energy of organic compounds. Plants, as well as organisms that feed on plants and each other, use energy in organic compounds to do cellular work. **6.1, 6.2**

5. Cells conserve energy by coupling energy-releasing reactions with energy-requiring ones. ATP is the main coupling agent; it primes molecules to react by giving up a phosphate group to them. **6.2**

6. Cells increase, maintain, and lower concentrations of substances by coordinating thousands of reactions. They rapidly shift rates of metabolism by controlling a few steps of reversible pathways. **6.3, 6.6**

7. Metabolic reactions start with reactants and end with products. Substances that form in between are intermediates. Energy carriers (mainly ATP), enzymes, cofactors, and transport proteins are key players. **6.3**

8. Orderly, enzyme-mediated reaction sequences are called metabolic pathways. The biosynthetic (anabolic) ones build large molecules with higher bond energies from smaller molecules. Photosynthesis is an example. Degradative (catabolic) pathways break down large molecules to small products with lower bond energies. Aerobic respiration is an example. **6.2, 6.3**

9. In photosynthesis and aerobic respiration, electron transfers (oxidation–reduction reactions) are used in forming ATP. At certain membranes, electron transfer chains release energy, inherent in H^+ concentration and electric gradients, to make ATP at other sites. **6.2, 6.4**

10. Enzymes are catalysts; they greatly enhance rates of specific reactions. Pockets or cavities in their surface offer energetically favorable microenvironments for the reaction; these are the active sites. **6.4, 6.5**

a. Nearly all enzymes are protein molecules; some RNAs also are catalytic.

b. The activation energy for a reaction is the energy it takes to align reactive chemical groups, destabilize electric charges, and then rearrange, create, and break bonds. It drives substrates to the transition state, which is the point at which a reaction can spontaneously run either to product or back to a reactant.

c. As bonds briefly form between the enzyme and substrate during the transition state, energy is released. This binding energy is a partial payment on the energy cost of the reaction (the activation energy).

d. Enzymes lower the activation energy by boosting concentrations of substrates in the active site, orienting substrates in positions that favor reaction, shutting out most or all of the water from the active site, and often changing their shape.

11. Sometimes functional groups of the enzyme carry out a reaction. More often, one or more cofactors help out by functional group transfers or electron transfers. Cofactors are metal ions and coenzymes, which are complex organic compounds with or without a vitamin component. When bound tightly to an active site, they are called prosthetic groups. Heme is one example. It is one of nearly a third of all known enzymes that include at least one metal ion. **6.6**

12. Different controls over enzyme action influence the kinds and amounts of substances available at any time in the cell. Enzyme action is best within a limited range of temperature, pH, and salinity. **6.7**

13. Bioluminescence is one result of enzyme-mediated reactions that release energy as fluorescent light. The transfer of genes for bioluminescence into organisms has research and practical applications. **6.9**

Review Questions

1. Define free radical. How does it relate to aging? *CI*

2. State the first and second laws of thermodynamics. Does life violate the second law? *6.1*

3. Define and give a few examples of potential energy. *6.1*

4. Give examples of a change in potential energy involved in (a) mechanical work and (b) chemical work. *6.2*

5. Make a simple diagram of the ATP molecule. Highlight which parts can be transferred to another molecule. *6.2*

6. What is an oxidation–reduction reaction? *6.2*

7. Define the following terms: metabolic pathway, substrate (or reactant), intermediate, and end product. Is photosynthesis a biosynthetic (anabolic) pathway overall? *6.3*

8. What is an electron transfer chain, and how does it work? *6.4*

9. Describe four key features of enzymes. *6.5*

10. Define activation energy, then state four ways in which enzymes may lower it. *6.5, 6.6*

11. Define feedback inhibition as it relates to the activity of an allosteric enzyme. *6.7*

Self-Quiz   ANSWERS IN APPENDIX III

1. _____ is life's primary source of energy.
   a. Food   c. Sunlight
   b. Water   d. ATP

2. If we liken chemical equilibrium to the bottom of an energy hill, then an _____ reaction is an uphill run.
   a. endergonic   c. ATP-assisted
   b. exergonic   d. both a and c

3. Which statement is *not* correct? A metabolic pathway _____ .
   a. has an orderly sequence of reaction steps
   b. is mediated by one enzyme, which initiates the reactions
   c. may be biosynthetic or degradative, overall
   d. all of the above

4. Which is *not* true of chemical equilibrium?
   a. Product and reactant concentrations are always equal.
   b. The rates of the forward and reverse reactions are the same.
   c. There is no further net change in product and reactant concentrations.

5. Electron transfer chains involve _____ .
   a. enzymes and cofactors   c. cell membranes
   b. electron transfers   d. all of the above

6. Enzymes are _____ .
   a. enhancers of reaction rates   d. not influenced by salinity
   b. influenced by temperature   e. a through c
   c. influenced by pH   f. all of the above

7. All enzymes incorporate a(n) _____ .
   a. active site   c. metal ion
   b. coenzyme   d. all of the above

8. NAD$^+$, FAD, and NADP$^+$ are _____ .
   a. coenzymes   c. allosteric enzymes
   b. metal ions   d. both a and b

9. Match each substance with the most suitable description.
   ___ coenzyme or metal ion   a. reactant or substrate
   ___ adjusts gradients at membrane b. enzyme
   ___ substance entering a reaction c. cofactor
   ___ substance formed while   d. intermediate
       a reaction is proceeding   e. product
   ___ substance at end of reaction f. energy carrier
   ___ enhances reaction rate   g. transport protein
   ___ mainly ATP

Critical Thinking

1. Cyanide, a toxic compound, can bind to an enzyme that is a component of electron transfer chains. The outcome is cyanide *poisoning*. Binding prevents the enzyme from donating electrons to a nearby acceptor molecule in the system. What effect will this have on ATP formation? From what you know of ATP's function, what effect will this have on a person's health?

2. *Pyrococcus furiosus* thrives at 100°C, the boiling point of water. This species of bacterium was discovered growing in a volcanic vent in Italy. Enzymes isolated from *P. furiosus* cells do not function well below 100°C. What is it about the structure of these enzymes that allows them to remain stable and active at such high temperatures? (Hint: Review Section 3.7, which summarizes the interactions that maintain protein structure.)

3. In cells, superoxide dismutase (Figure 6.1 and *above right*) has a quaternary structure; it consists of two polypeptide chains. In each chain, a strandlike domain is arrayed as a barrel around a copper ion and zinc ion (coded *orange* and *white*).

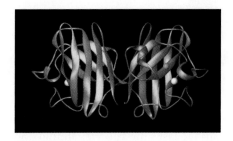

Which part of the barrel is probably hydrophobic? Which part is hydrophilic? Do you suppose substrates bind inside or outside the barrels? Do the metal ions have a role in catalysis?

4. AZT, or azidothymidine, is a drug used to alleviate symptoms of *AIDS*, or acquired immunodeficiency syndrome. AZT is very similar in molecular structure to thymidine, one of the DNA nucleotides. AZT also can fit into the active site of an enzyme produced by the virus that causes AIDS. Infection puts the viral enzyme and viral genetic material (RNA) into a cell. The cell wrongly uses the RNA as a template (structural pattern) for joining nucleotides to form viral DNA. The viral enzyme takes part in the assembly reactions. Propose a model to explain how AZT might inhibit replication of the virus inside cells.

5. The bacterium *Clostridium botulinum* is an obligate anaerobe, meaning it dies quickly in the presence of oxygen. It lives in oxygen-free pockets in soil, and it can enter a metabolically inactive, resting state by forming spores. The spores may end up on the surfaces of garden vegetables. When the picked vegetables are being canned, the spores must be destroyed. If they are not destroyed, *C. botulinum* may grow and produce botulinum, a toxin. In improperly cooked and tainted canned foods, the toxin is active and can cause a type of food poisoning called *botulism*.

This bacterium is not able to produce superoxide dismutase or catalase. Develop a hypothesis to explain how the absence of these enzymes is related to its anaerobic life-style.

Selected Key Terms

Readings

Adams, S. 22 October 1994. "No Way Back." *New Scientist*. A refreshing look at the second law of thermodynamics.

Branden, C., and J. Tooze, 1999. *Introduction to Protein Structure*. Second edition. New York: Garland.

7 HOW CELLS ACQUIRE ENERGY

Sunlight and Survival

Think about the last time you were hungry and craved a bit of apple, lettuce, chicken, pizza, or any other food. Where did it come from? For the answers, look past the refrigerator, the market or restaurant, or even the farm. Look to individual plants, the starting point for nearly all edibles you put into your mouth. Plants use environmental sources of energy and raw materials to make glucose and other organic compounds. Organic compounds, recall, are built on a framework of carbon atoms. So the questions become these:

1. *Where does the carbon come from in the first place?*

2. *Where does the energy come from to drive the synthesis of carbon-based compounds?*

Answers vary, because they depend on an organism's mode of nutrition. see p. 128

Plants generally are "self-nourishing" organisms, or **autotrophs**. Their carbon source is carbon dioxide (CO_2), a gaseous compound in the air and dissolved in water. Plants, some bacteria, and many protistans are **photoautotrophs**, meaning they use sunlight for **photosynthesis**. By this process, energy from the sun drives formation of ATP and the coenzyme NADPH. ATP delivers energy to sites where glucose and other carbohydrates are built. NADPH delivers electrons and hydrogen building blocks to the same sites.

Many other organisms are **heterotrophs**, meaning they must feed on autotrophs, one another, and organic wastes. (*Hetero*– means other, as in "being nourished by other organisms.") That is how most bacteria, many protistans, and all fungi and animals stay alive. Unlike plants, heterotrophs can't feed themselves with energy and raw materials from the physical environment.

How do we know such things? We didn't have a clue until the mid-seventeenth century. Before then, most people assumed that plants got raw materials to make food from soil. By 1882 a few chemists had an inkling that plants use sunlight, water, and something in the air to make food. A botanist, T. Engelmann, wondered: What parts of sunlight do plants favor?

As Engelmann knew, free oxygen is released during photosynthesis. He also knew that some bacteria need oxygen for aerobic respiration, as most organisms do. He hypothesized: If the bacteria require oxygen, then we can expect them to gather in places where the most photosynthesis is going on. He put a water droplet containing bacterial cells on a microscope slide with a photosynthesizer, the green alga *Spirogyra*. He used a crystal prism to break up a beam of sunlight and cast a spectrum of colors across the slide.

Bacteria gathered mostly where violet and red light fell on the green alga. Algal cells released more oxygen in the part illuminated by light of those colors—the very best light for photosynthesis (Figure 7.1).

Such observations showed us how photosynthesis works. They also helped reveal one of nature's great patterns, *for photosynthesis is the main pathway by which carbon and energy enter the web of life.* Once a cell makes or takes up organic compounds, it uses or stores them. *All* autotrophs and heterotrophs store energy in organic compounds and release it by other processes, such as aerobic respiration. Figure 7.2 previews the chemical links between photosynthesis and aerobic respiration, the focus of this chapter and the next.

Such processes might seem far removed from your daily interests. But remember, the food that nourishes you and most other organisms can't be produced or used without them. We will return to this point in later chapters. It provides perspective on many issues, such as nutrition, dieting, agriculture, human population growth, genetic engineering, and the impact of pollution on our sources of food—hence on our survival.

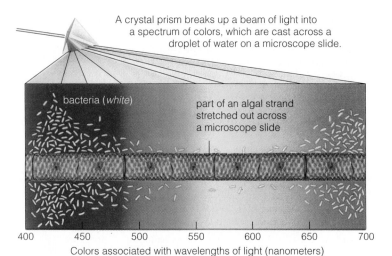

A crystal prism breaks up a beam of light into a spectrum of colors, which are cast across a droplet of water on a microscope slide.

bacteria (*white*)

part of an algal strand stretched out across a microscope slide

400   450   500   550   600   650   700

Colors associated with wavelengths of light (nanometers)

Figure 7.1 Results from T. Engelmann's observational test that correlated portions of visible light with photosynthesis in *Spirogyra*, a strandlike green alga. Many oxygen-requiring bacterial cells moved to the colors where algal cells released the most oxygen, which is a by-product of their photosynthetic activity.

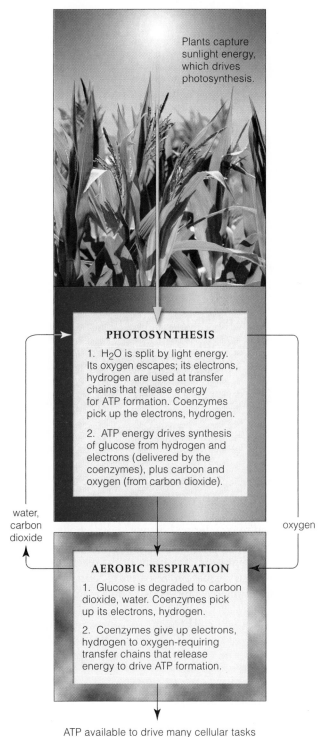

PHOTOSYNTHESIS

Plants capture sunlight energy, which drives photosynthesis.

1. H_2O is split by light energy. Its oxygen escapes; its electrons, hydrogen are used at transfer chains that release energy for ATP formation. Coenzymes pick up the electrons, hydrogen.

2. ATP energy drives synthesis of glucose from hydrogen and electrons (delivered by the coenzymes), plus carbon and oxygen (from carbon dioxide).

water, carbon dioxide

oxygen

AEROBIC RESPIRATION

1. Glucose is degraded to carbon dioxide, water. Coenzymes pick up its electrons, hydrogen.

2. Coenzymes give up electrons, hydrogen to oxygen-requiring transfer chains that release energy to drive ATP formation.

ATP available to drive many cellular tasks

Figure 7.2 Links between photosynthesis—the main energy-requiring process in the world of life—and aerobic respiration, the main energy-releasing process.

Key Concepts

1. Plants, some bacteria, and many protistans use sunlight energy, carbon dioxide, and water to make glucose and other carbohydrates, which have a backbone of carbon atoms. The metabolic process by which they do this is called photosynthesis.

2. Photosynthesis is the main process by which carbon and energy become available to the web of life.

3. In the first stage of photosynthesis, sunlight energy is trapped and converted to chemical bond energy in ATP molecules. Typically, water molecules are split, their electrons and hydrogen atoms are picked up by the coenzyme $NADP^+$ to form NADPH, and their oxygen atoms are released as a by-product.

4. In the second stage of photosynthesis, ATP delivers energy to the sites of biosynthesis. NADPH delivers electrons and hydrogen, which combine with carbon and oxygen to form glucose. Carbon dioxide provides the carbon and oxygen for the synthesis reactions.

5. In the photosynthetic cells of plants and in many protistans, both stages of the reactions proceed inside organelles called chloroplasts.

6. Often we summarize photosynthesis this way:

$$12H_2O + 6CO_2 \xrightarrow{\text{LIGHT ENERGY}} 6O_2 + C_6H_{12}O_6 + 6H_2O$$

WATER   CARBON DIOXIDE   OXYGEN   GLUCOSE   WATER

PHOTOSYNTHESIS—AN OVERVIEW

Sit outdoors on a warm, cloudless day and you'll get hot but you'll never build your own food. How do plants do it? Think of it! They have to plug into the sun's energy, but not so much that it might cook them. They have to do magic and turn the sun's energy into ATP's chemical bond energy. They have to get carbon, oxygen, and hydrogen from somewhere and combine them into carbohydrates. As you might imagine, these multiple tasks can't all be done in the same place. Different functions, different structures.

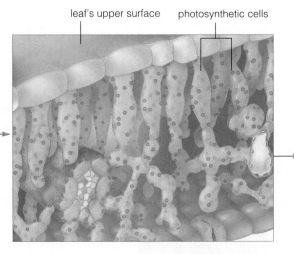

leaf's upper surface   photosynthetic cells

a Sow thistle (*Sonchus*) leaf

b Cutaway of a small section from the leaf. Its upper and lower surfaces enclose many photosynthetic cells.

c One of the photosynthetic cells, with green chloroplasts.

Where the Reactions Take Place

Plants and protistans called algae carry out the tasks in **chloroplasts**: organelles specialized for photosynthesis (Section 4.7). Figure 7.3 shows the structure of one type. All chloroplasts have two outermost membranes that surround a mostly fluid interior, the **stroma**. Another membrane in the stroma is often highly folded into what looks like stacked sacs connected by channels (Figure 7.3*d,e*). We refer to the stacks as thylakoids. But the **thylakoid membrane** makes up only one functional compartment, no matter how it is folded.

The capture and conversion of sunlight energy to chemical energy takes place at thylakoids. With these *light-dependent* reactions, plants get the electrons and hydrogen for making food. Amazingly, sunlight splits water molecules. Water's oxygen diffuses away. But its electrons flow through electron transfer chains at the thylakoid membrane. (Remember Section 6.4?) When electrons flow through the systems, hydrogen moves inside the thylakoid compartment. ATP forms when it flows out. Waste not, want not: The helpful coenzyme $NADP^+$ picks up the electrons and hydrogen.

Making glucose doesn't require light. These *light-independent* reactions proceed elsewhere, in the stroma. This is where ATP gives up energy, the coenzyme gives up electrons and hydrogen, and carbon dioxide (CO_2) is dismantled for its carbon and oxygen atoms.

Overall, you can think about all the reactants and products just described in terms of a simple equation:

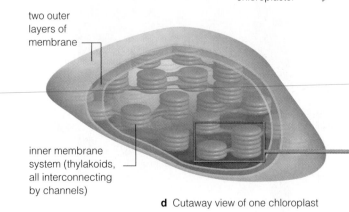

two outer layers of membrane

inner membrane system (thylakoids, all interconnecting by channels)

d Cutaway view of one chloroplast

Figure 7.3 Zooming in on sites of photosynthesis inside a leaf from a typical plant.

In case you're wondering how we know what we know: Remember the description of tracers in Section 2.2? By attaching radioisotopes to atoms of the reactants shown in the summary equation, researchers found out where each atom of the reactants ends up in the products:

REACTANTS: $12H_2O$ $6CO_2$

PRODUCTS: $6O_2$ $C_6H_{12}O_6$ $6H_2O$

But Things Don't <u>Really</u> End With Glucose

Notice how we show glucose as the end product of the reactions. We do so to keep the chemical bookkeeping simple. But you'll find little glucose in a chloroplast.

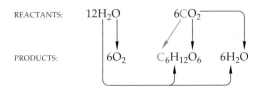

$$12H_2O + 6CO_2 \xrightarrow{\text{LIGHT ENERGY}} 6O_2 + C_6H_{12}O_6 + 6H_2O$$

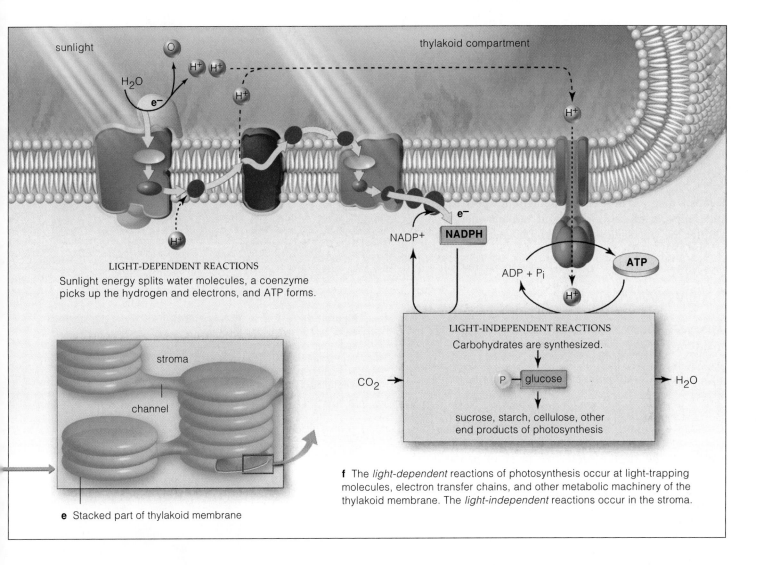

sunlight

O

H_2O

e^-

H^+ H^+

H^+

thylakoid compartment

H^+

LIGHT-DEPENDENT REACTIONS
Sunlight energy splits water molecules, a coenzyme
picks up the hydrogen and electrons, and ATP forms.

H^+

e^-

NADP$^+$ **NADPH**

ADP + P$_i$

ATP

H^+

stroma

channel

LIGHT-INDEPENDENT REACTIONS
Carbohydrates are synthesized.

CO_2 →

P — glucose

→ H_2O

sucrose, starch, cellulose, other
end products of photosynthesis

f The *light-dependent* reactions of photosynthesis occur at light-trapping
molecules, electron transfer chains, and other metabolic machinery of the
thylakoid membrane. The *light-independent* reactions occur in the stroma.

e Stacked part of thylakoid membrane

Each new glucose molecule has an attached phosphate group; it's primed to react with something else. Almost always, it quickly enters reactions that make sucrose, starch, or cellulose. These are the main end products of photosynthesis even though we "stop" with glucose.

The sketch below is a simple version of Figure 7.3. We repeat it in sections that follow to help reinforce your grasp of how the details fit into the big picture:

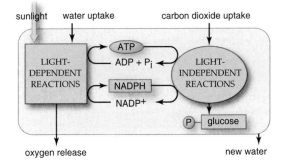

sunlight   water uptake   carbon dioxide uptake

ATP
ADP + P$_i$

LIGHT-
DEPENDENT
REACTIONS

NADPH

NADP$^+$

LIGHT-
INDEPENDENT
REACTIONS

P — glucose

oxygen release   new water

We turn next to the details. Before we do, think about this: Two thousand chloroplasts, lined single file, would be no wider than a dime. Imagine all of the chloroplasts in just one corn or rice plant—each a tiny factory for producing sugars and starch—and you may get an idea of the magnitude of the metabolic events required to feed you and every other organism on this planet.

Chloroplasts, the organelles of photosynthesis in plants and many protistans, specialize in food production.

In the first stage of photosynthesis, sunlight energy drives ATP and NADPH formation, and oxygen is released. In chloroplasts, this stage occurs at the thylakoid membrane system that weaves through the chloroplast's stroma.

The second stage occurs in the stroma. Energy from ATP drives glucose formation. For these synthesis reactions, carbon dioxide provides the carbon and oxygen atoms, and NADPH provides the electrons and hydrogen atoms.

SUNLIGHT AS AN ENERGY SOURCE

Properties of Light

The sun is the start of a one-way flow of energy through almost all ecosystems (Figure 7.4). On average, the amount adds up to 7,000 kilocalories per square meter for the whole planet. Photoautotrophs intercept about 1 percent of it.

Radiant energy from the sun undulates across space in a manner analogous to the waves crossing a sea. The horizontal distance between the crests of every two successive waves is defined as a **wavelength**. Figure 7.5*a* is a diagram of the **electromagnetic spectrum**, the full range of all wavelengths of radiant energy. The shorter the wavelength, the higher its energy:

 low-energy wavelength

high-energy wavelength

All wavelengths of visible light combined appear white to us. We and many other organisms see the individual wavelengths as different colors.

Collectively, different photoautotrophs absorb the wavelengths between 380 and 750 nanometers. Shorter wavelengths, including ultraviolet (UV) radiation, are energetic enough to break chemical bonds of organic compounds; they can kill cells. Hundreds of millions of years ago, before the protective layer of ozone (O_3) formed in the atmosphere, the lethal UV bombardment kept photoautotrophs below the surface of the seas.

How do we know these things? Direct a ray of white light through a crystal prism, as Engelmann did, and you see how the prism bends different wavelengths by different degrees. Or look at a rainbow (Figure 7.5*b*). Rainbows form as the sun's rays pass through water droplets in moisture-rich air. The droplets bend longer wavelengths (yellow to red) more than they bend the shorter (violet to blue) ones, which separates them into nearly circular arcs of color. We usually see only part of the circle, a "bow" (Figure 7.5*b*). But blue is always on the inside of the arc, and red on the outside.

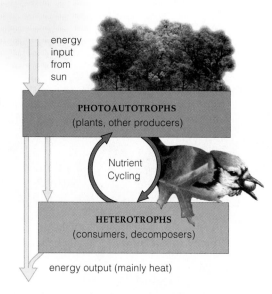

Figure 7.4 Model for how photoautotrophs are the basis of a one-way flow of energy through nearly all ecosystems, or webs of life, starting with energy inputs from the sun.

Figure 7.5 (a) The electromagnetic spectrum. Visible light is only one of many forms of electromagnetic radiation in the spectrum. **(b)** I can make a rainbow! Observational test of the properties of light. B. Mantoni, a student, created a circular rainbow from a bright flash of light in its center. He pointed his camera at a mirror. The light from the camera's flash unit was reflected off the mirror back to the camera lens, which bent the individual wavelengths to form this circular rainbow.

b

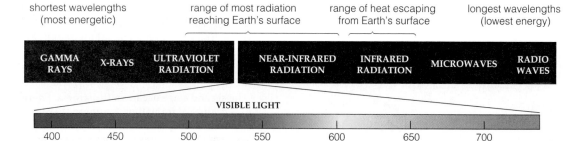

shortest wavelengths (most energetic)   range of most radiation reaching Earth's surface   range of heat escaping from Earth's surface   longest wavelengths (lowest energy)

| GAMMA RAYS | X-RAYS | ULTRAVIOLET RADIATION | NEAR-INFRARED RADIATION | INFRARED RADIATION | MICROWAVES | RADIO WAVES |

VISIBLE LIGHT

400   450   500   550   600   650   700

Wavelengths of light (nanometers)

a

Figure 7.6 (**a**) Absorption spectra that show the efficiency with which two chlorophylls, *a* and *b*, respond to wavelengths. (**b**) Absorption spectra for beta-carotene, which is one of the carotenoids, and for one of the phycobilins.

(**c**) Combined energy absorption efficiency, as indicated by the dashed line, of chlorophylls *a* and *b*, the carotenoids, and the phycobilins.

The background colors indicate the range of wavelengths being absorbed. The graph lines indicate the extent to which each of these is absorbed by a particular pigment. The color of the line itself indicates the main color that the pigment is transmitting. Thus the lines are green for chlorophylls, yellow-orange for the carotenoids, and purple and blue for the phycobilins.

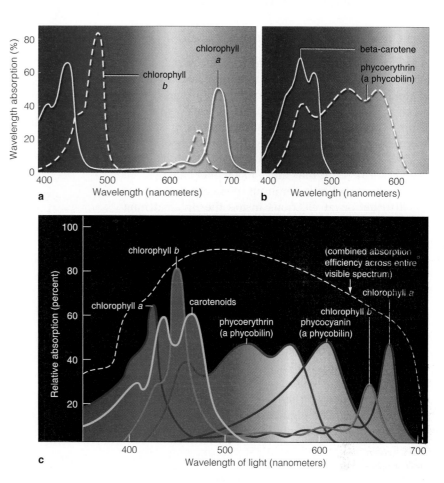

Although light travels in waves, the energy itself has a particle-like quality. When it is absorbed, the energy of visible light can be measured as if it were organized in packets, which we call **photons**. Each type of photon has a fixed amount of energy. The types having the most energy travel as the shortest wavelengths, which correspond to blue-violet light. Photons that have the least amount of energy travel as long wavelengths, and these correspond to red light.

Pigments—Molecular Bridge From Sunlight to Photosynthesis

Engelmann used a prism as part of an experiment to identify which wavelengths drive photosynthesis in a green alga (Figure 7.1). But molecular biology was far in the future, and he was not able to identify the actual bridge between sunlight and photosynthetic activity.

Pigments are the molecular bridge. **Pigments** are a class of molecules that absorb wavelengths of light, and organisms put them to many uses. Most pigments absorb only some wavelengths and transmit the rest. A few absorb so many wavelengths, they appear dark or black. Some phycobilins in algae and the melanins in animals are like this.

An **absorption spectrum** is a diagram that shows how effectively a pigment molecule absorbs different wavelengths in the spectrum of visible light. Consider the chlorophylls, the main pigments of photosynthesis. As Figure 7.6*a* indicates, the chlorophylls absorb all wavelengths except very little of the yellow-green and green ones, which they mainly transmit. That is why plant parts with an abundance of chlorophylls appear green to us.

Other, *accessory* pigments of different colors also are present in photoautotrophs (Figure 7.6*b*,c). They differ in kind and amounts among species. The dashed line in Figure 7.6*c* indicates how their assistance extends the range of wavelengths that can drive photosynthesis. As you can see, the combined absorption efficiency is high.

Radiation from the sun travels in waves, which differ in length and energy content. Shorter wavelengths have the most energy, and long wavelengths have less.

We perceive wavelengths of visible light as different colors and measure their energy content in packets called photons.

Chlorophylls and certain other pigments absorb specific wavelengths of visible light. They are the molecular bridge between the sun's energy and photosynthetic activity.

THE RAINBOW CATCHERS

The Chemical Basis of Color

How do pigment molecules impart color to plants and other organisms? Each kind has a light-catching array of atoms, often joined by alternating single and double bonds (Figure 7.7). In such arrays, electrons distributed around atomic nuclei can absorb photons of specific energies, which correspond to different colors of light.

Recall, from Section 2.3, that when electrons of an atom absorb energy, they move to a higher energy level. In a pigment molecule, an input of energy destabilizes the distribution of electrons inside the light-catching array. The excited electrons rapidly return to a lower energy level, the electron distribution stabilizes, and some energy is released in the form of light. When any destabilized molecule emits light as it reverts to a more stable configuration, this is called a **fluorescence**.

Excitation occurs only when the amount of energy of a photon matches the amount required to boost an electron to a higher energy level. Suppose a sunbeam strikes a pigment. If photons corresponding to red and orange wavelengths match the amount required for the boost, the pigment will absorb them. In this case, the photons corresponding to blue and violet wavelengths are a mismatch. The pigment molecule cannot absorb them. Because it transmits (reflects) these photons, it appears blue or violet to us.

On the Variety of Photosynthetic Pigments

Most pigments respond to only part of the rainbow of visible light. If acquiring energy is so vital, why doesn't each kind of photosynthetic pigment go after the whole rainbow? In other words, why isn't each one black?

If early photoautotrophs evolved in the seas, then so did their pigments. Ultraviolet and red wavelengths do not penetrate water as deeply as green and blue wavelengths do. Possibly natural selection favored the evolution of different pigments at different depths. Many relatives of the red alga in Figure 7.8a live deep in the sea, and some of them are nearly black. Green

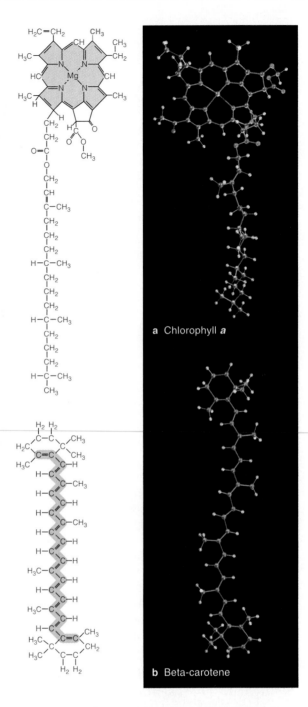

a Chlorophyll *a*

b Beta-carotene

Figure 7.7 Structural formulas and ball-and-stick models for (**a**) chlorophyll *a* and (**b**) beta-carotene. The light-catching array of both pigments is tinted the color of the light it transmits. The backbone, a pure hydrocarbon, dissolves in the lipid bilayer of photosynthetic cell membranes. Chlorophylls *a* and *b* differ only in one functional group at the position shaded *red* ($-CH_3$ for chlorophyll *a* and $-COO^-$ for chlorophyll *b*).

Figure 7.8 (**a**) A red alga from a tropical reef. (**b**) A green alga (*Codium*) from shallow coastal waters.

Figure 7.9 Leaf color. In intense-green leaves, photosynthetic cells continually make chlorophyll, which masks the accessory pigments. In autumn, chlorophyll synthesis lags behind its breakdown in many species. Other pigments are unmasked, and more colors show through. Water-soluble anthocyanins accumulate in leaf cells. They appear red when plant fluids are slightly acidic, blue when they are basic (alkaline), or colors in between in fluids of intermediate pH. Soil conditions contribute to pH values.

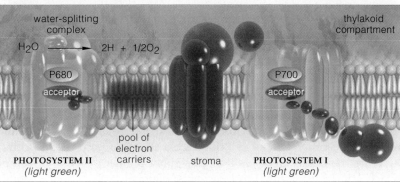

Figure 7.10 Arrangement of the two types of photosystems, II and I, in a chloroplast's thylakoid membranes. Components of neighboring electron transfer chains are coded *dark green*. One of these chains extends from photosystem II to photosystem I. The second chain gives up the electrons to the coenzyme NADP⁺.

algae, including the one shown in Figure 7.8b, live in shallow water. Their chlorophyll molecules absorb red wavelengths, and accessory pigments harvest different ones. Some accessory pigments also function as shields against ultraviolet radiation.

Today, **chlorophylls** are the primary pigments in all but one marginal group of photoautotrophs. Remember, photosynthetic pigments respond best to red and blue-to-violet light. Chlorophyll *a* absorbs both (Figure 7.7a). As one of several accessory pigments in chloroplasts, chlorophyll *b* absorbs blue and red-orange wavelengths missed by chlorophyll *a*. Plants, green algae, and a few bacterial photoautotrophs have this pigment.

All photoautotrophs also contain **carotenoids**. These accessory pigments absorb blue-violet and blue-green wavelengths that chlorophylls miss. They reflect red, orange, and yellow ones. Their backbone usually has a carbon ring at each end. Many, including beta-carotene, are fat-soluble, pure hydrocarbons (Figure 7.7b). The xanthophylls incorporate oxygen; others have proteins attached and appear purple, violet, blue, green, brown, or black. You can expect an abundance of carotenoids in flowers, fruits, and vegetables in this color range.

Carotenoids are less abundant than chlorophylls in green leaves, but in many plants they become visible in autumn (Figure 7.9). Each year, tourists spend about a billion dollars just to watch the three-week demise of chlorophyll in the deciduous trees of New England.

Anthocyanins and **phycobilins** also are accessory pigments. Many eukaryotic species make an abundance of deep red to purple anthocyanins (think red algae,

purple mushroom caps, and blueberries, for example). The blue phycobilins are the signature pigments of red algae and cyanobacteria.

Where Are Photosynthetic Pigments Located?

Some bacteria of ancient lineages have light-trapping pigments embedded in their plasma membrane. The archaebacterium *Halobacterium halobium* has a unique purple pigment (bacteriorhodopsin), not chlorophylls. Cyanobacteria have chloropyll *a*, and other prokaryotes have bacteriochlorophylls that serve the same function but differ slightly in their structure. Various accessory pigments also are embedded in the plasma membrane of photosynthetic eubacteria.

This chapter's focus, remember, is on the thylakoid membrane system inside chloroplasts. Its pigments are organized in clusters known as **photosystems**. In some plant species, the membrane has many thousands of these clusters. Each photosystem consists of proteins and 200 to 300 pigment molecules. As Figure 7.10 shows, it is located next to its functional partner—an electron transfer chain. How these components interact is the topic of the next two sections.

The chlorophylls are the main photosynthetic pigments in photoautotrophs. Carotenoids and other accessory pigments enhance their light-harvesting function.

Photosystems (pigment clusters) are the functional partners of electron transfer chains in the thylakoid membrane.

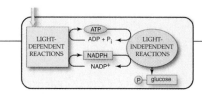

THE LIGHT-DEPENDENT REACTIONS

We turn now to the key events of the **light-dependent reactions**, the first stage of photosynthesis. *First*, many photosystems harvest sunlight energy, which also splits water molecules into oxygen, hydrogen, and electrons. *Second*, the energy input is converted to the chemical bond energy of ATP. *Third*, the coenzyme NADP$^+$ picks up the electrons, and hydrogen accompanies them.

What Happens to the Absorbed Energy?

When a photosynthetic pigment absorbs photons, some of its electrons are boosted to a higher energy level. As electrons return to a lower energy level, recall, they emit the extra energy as fluorescent light.

If nothing else were around to intercept the emitted energy, all of it would escape as light and heat. But a photosystem's pigment clusters prevent this. Most of the pigments harvest energy from photons. Rather than releasing excitation energy as a fluorescent afterglow, a harvester randomly transfers it to a neighbor pigment, which randomly passes it to a neighbor pigment, and so on. Figure 7.11 shows this "random walk" of energy.

Some energy is lost as heat with each transfer. Soon, the energy remaining corresponds to a wavelength that only the photosystem's **reaction center** can trap. That center is a special chlorophyll *a* molecule. It passes the energy, in the form of excited electrons, to an acceptor molecule positioned near an electron transfer chain.

An **electron transfer chain** is an organized array of enzymes, coenzymes, and other proteins embedded in or anchored to a cell membrane (Section 6.4). Electrons are transferred step-by-step through the array, and a bit of energy is released at each step. In chloroplasts, much of the energy drives ATP formation.

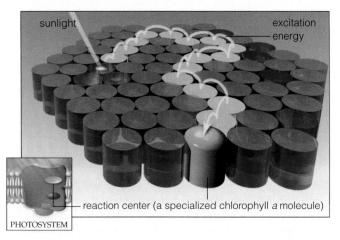

Figure 7.11 A sampling of pigments of a photosystem that harvest photon energy. Notice the random flow of energy among them. The energy of excitation quickly reaches a reaction center which, when suitably activated, releases electrons for photosynthesis.

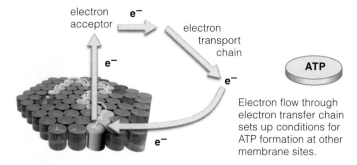

Figure 7.12 Cyclic pathway of ATP formation.

Electron flow through electron transfer chain sets up conditions for ATP formation at other membrane sites.

Cyclic and Noncyclic Electron Flow

In ancient bacteria, photosynthesis evolved as a way to form ATP, not to make organic compounds. Inorganic compounds provided hydrogen and electrons for the synthesis reactions. Sunlight energy drove electrons in a cycle: from a photosystem called type I, to a transfer chain, and back to the photosystem (Figure 7.12). This *cyclic* pathway has a reaction center designated P700, and it still runs in all photoautotrophs.

Electron flow through one photosystem does not pack enough punch to make NADPH. But more than 2 billion years ago, more energy became available for electron transfers, as if two batteries became hooked together. A different photosystem—type II—evolved. When types II and I operate together, they acquire enough energy to get electrons from water molecules.

The more recent path of electron flow is *noncyclic*. Electrons flow from water to photosystem II, through a transfer chain to photosystem I, then another transfer chain. From there, NADP$^+$ accepts two electrons and a hydrogen ion (H$^+$) to become NADPH (Figure 7.13*a*).

Let's give the electron flow a closer look. Sunlight striking photosystem II starts **photolysis**. This reaction splits water molecules into oxygen, electrons, and H$^+$ (unbound hydrogen). New electrons from water replace the electrons released from the photosystem's reaction center (P680). ATP still forms. The mechanism is a bit complicated, so we reserve that story for Section 7.5. For now, follow the electrons leaving the first transfer chain. They still have some extra energy. They get more from sunlight at photosystem I. This energy boost puts them at a higher energy level, which lets them enter the second chain. They lose only a small bit of energy before NADP$^+$ picks them up at the end of the second chain.

Today, both the cyclic and noncyclic pathways of electron flow operate in all photosynthetic organisms. Which one dominates at a given time depends on the organism's metabolic demands for ATP and NADPH.

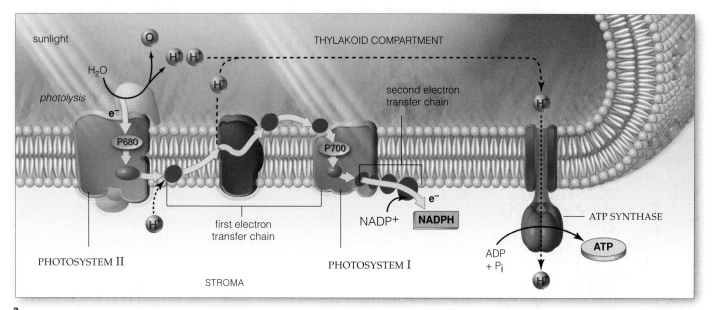

a

Figure 7.13 (**a**) Filling in details for Figure 7.3—how ATP and NADPH form in the noncyclic pathway of photosynthesis. *Yellow* arrows show electron flow. Electrons released from water molecules that were split by photolysis move through two photosystems (*light green*) and two electron transfer chains (*dark green*). The joint operation of two photosystems boosts electrons to an energy level high enough to drive NADPH formation.

As you will read shortly, hydrogen ions also move across the thylakoid membrane in ways that lead to ATP formation at proteins called ATP synthases.

(**b**) Diagram the energy changes in the noncyclic pathway, and you will end up with this "Z scheme."

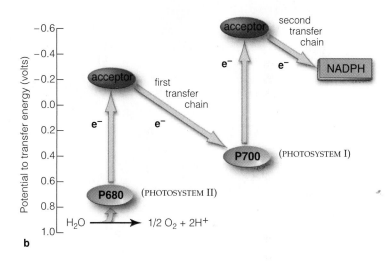

b

The Legacy—A New Atmosphere

On sunny days you can see bubbles of oxygen escaping from aquatic plants (Figure 7.14). When the noncyclic pathway first evolved, the oxygen it released dissolved in seawater, wet mud, and similar prokaryotic habitats. By about 1.5 billion years ago, stupendous quantities of dissolved oxygen were escaping to what had been an oxygen-free atmosphere. The atmosphere changed forever, and so did the world of life. The global rise in free oxygen now favored the chemical evolution of aerobic respiration. This became the most efficient metabolic pathway for releasing usable energy from organic compounds. From an evolutionary perspective, the emergence of the noncyclic pathway allowed you and every other animal to be around today, breathing the oxygen that helps keep your cells alive.

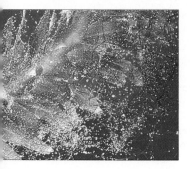

Figure 7.14 Visible evidence of photosynthesis—bubbles of oxygen escaping from the leaves of *Elodea*, a plant that lives in freshwater habitats.

In the light-dependent reactions, sunlight energy drives the release of electrons from photosystems in thylakoid membranes. Electron flow through nearby electron transfer chains results in the formation of ATP, NADPH, or both.

ATP alone forms in a cyclic pathway that uses one type of photosystem. Both ATP and NADPH form by a noncyclic pathway in which electrons flow from water, through two types of photosystems, and finally to NADP⁺.

All photosynthetic species employ one or both pathways in response to the cell's changing needs for ATP and NADPH.

Oxygen, a by-product of the noncyclic pathway, changed the early atmosphere and made aerobic respiration possible.

CASE STUDY: A CONTROLLED RELEASE OF ENERGY

Spill some sugar into a campfire and you'll release its energy all at once, but not in a form that does any good. Events in the chloroplast control the release of energy initially provided by sunlight, and they direct it into doing useful cellular work.

Inputs of light energy cause enzymes to repeatedly split water molecules. The released oxygen combines to form O_2. In this form it diffuses out of the chloroplast and out of the cell. The released hydrogen does not depart. It gets stockpiled as hydrogen ions (H^+) inside the compartment formed by the thylakoid membrane (Figure 7.15a).

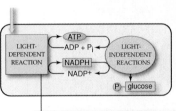

of the chain shunt the ions across the membrane, as in Figure 7.15b.

By a combination of photolysis and electron transport, H^+ is now more concentrated in the thylakoid compartment than in the stroma. Now, this unequal distribution of positively charged ions means there is a difference in charge across the membrane. H^+ concentration and electric gradients have become established.

The combined force of the two gradients propels H^+ into the stroma, through ATP synthases that span the thylakoid membrane. You read about this type of transport protein in Section 5.2. ATP synthases have built-in machinery that responds to the ion flow. They catalyze the attachment of unbound phosphate (P_i) to

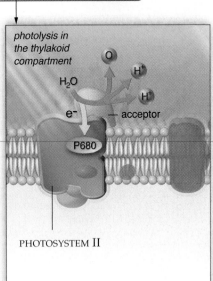

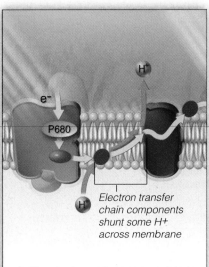

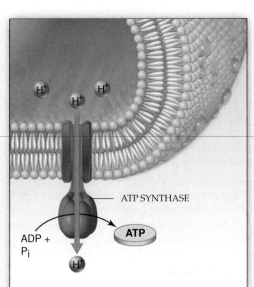

a Hydrogen ions released by photolysis (splitting of water molecules) accumulate inside the thylakoid compartment. (O_2 forms when two oxygen atoms combine. It diffuses out of the chloroplast, then out of the photosynthetic cell.)

b More hydrogen ions accumulate in the compartment when components of electron transfer chains accept excited electrons (from photolysis). Components of the chain also pick up hydrogen ions in the stroma and shunt them across the membrane.

c Ions follow the concentration and electric gradients across the thylakoid membrane. They flow to the stroma, through the interior of ATP synthases that span the membrane. The flow drives the formation of ATP from the cell's pool of ADP and P_i.

Figure 7.15 From photosynthesis, an example of how energy is released and directed to cellular tasks. Sunlight energy at photosystem II sets in motion events that build concentration and electric gradients, which are tapped to make ATP. It also leads to the formation of NADPH.

At the same time, an acceptor molecule picks up the electrons and transfers them to an electron transfer chain positioned next to it in the thylakoid membrane. The membrane has many, many of these chains. While they are all accepting electrons, they also are picking up H^+ from the stroma; these ions are attracted to the negative charge of the electrons. Certain components

a molecule of ADP that is present in the stroma. ATP forms by this phosphate-group transfer (Figure 7.15c).

The sequence of events just described is called the chemiosmotic model for ATP formation in chloroplasts. As you will read in the next chapter, the same model applies to ATP formation in mitochondria.

In chloroplasts, ongoing photolysis and transfers of electrons in transport systems set up H^+ concentration and electric gradients across the thylakoid membrane. Ion flow from the thylakoid compartment to the stroma drives ATP formation.

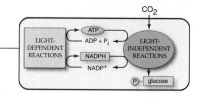

THE LIGHT-INDEPENDENT REACTIONS

The **light-independent reactions** are the "synthesis" part of photosynthesis. They occur as a cyclic pathway called the **Calvin–Benson cycle**. ATP energy drives the reactions. NADPH delivers hydrogen and electrons to build sugars. Carbon dioxide (CO_2), present in the air (or water) around photosynthetic cells, provides the carbon and oxygen. We say these reactions are light-independent because they do not depend directly on sunlight. They proceed just as well in the dark, as long as ATP and NADPH are available.

How Do Plants Capture Carbon?

Suppose a CO_2 molecule diffuses into air spaces inside a leaf, then into a photosynthetic cell, and then into a chloroplast's stroma. An enzyme attaches its carbon atom to **RuBP** (ribulose bisphosphate), a compound having a backbone of five carbon atoms. **Rubisco** (for RuBP carboxylase) is the enzyme's name. The result is an unstable six-carbon intermediate that splits into two molecules of **PGA** (phosphoglycerate). PGA is a stable organic compound with a three-carbon backbone.

Incorporating a carbon atom from CO_2 into a stable organic compound is known as **carbon fixation**. Quite simply, food cannot be produced without this first step of the cyclic light-independent reactions.

How Do Plants Build Glucose?

The cycle yields phosphorylated glucose (Section 6.3). It also regenerates the RuBP. For our purposes, we can simply focus on the carbon atoms of the substrates, intermediates, and end products, as in Figure 7.16.

PGA accepts one phosphate group from ATP, plus hydrogen and electrons from NADPH. The resulting intermediate is **PGAL** (for phosphoglyceraldehyde). To build *one* six-carbon sugar phosphate, carbon atoms from six CO_2 must be fixed, and twelve PGAL must form. Most of the PGAL gets rearranged into RuBP, which is used to fix more carbon. Two PGAL combine, thus forming glucose with a phosphate group attached to its six-carbon backbone.

When phosphorylated this way, glucose is primed to enter reactions. Plants use it as a building block for their main carbohydrates, such as sucrose, cellulose, and starch. *Synthesis of these organic compounds by other pathways marks the end of the light-independent reactions.*

With six turns of the cycle, enough RuBP molecules form to replace the ones used in carbon fixation. The ADP, NADP+, and released phosphate diffuse through the stroma, back to the light-dependent reaction sites. They may be converted again to NADPH and ATP.

During daylight hours, photosynthetic cells convert phosphorylated glucose to sucrose or starch. Sucrose is

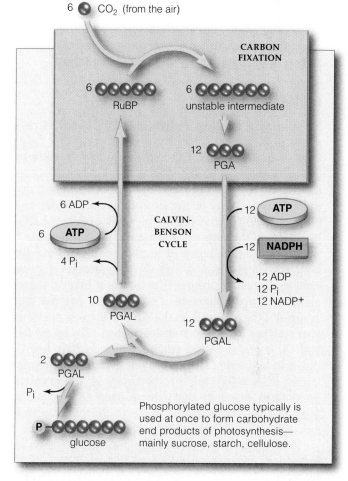

Figure 7.16 Light-independent reactions of photosynthesis. *Brown* circles signify carbon atoms of key molecules. All intermediates have one or two phosphate groups attached; for clarity, we show only the phosphate group on glucose. Many water molecules that formed in the light-dependent reactions enter this pathway; six remain at the end. Appendix VI, Figure C, shows details.

the most readily transportable carbohydrate in plants. Starch is the most common storage form. The cells also convert excess PGAL to starch, which they briefly store as starch grains inside the stroma. After the sun goes down, they convert starch to sucrose for export to other living cells in leaves, stems, and roots.

The products and intermediates of photosynthesis end up as energy sources and as building blocks for all lipids, amino acids, and other organic compounds that plants require for growth, survival, and reproduction.

In light-independent reactions of the Calvin–Benson cycle, glucose is made with hydrogen from NADPH, and carbon and oxygen from CO_2. The reactions regenerate ADP, NADP+, and the RuBP that fixes carbon for this cycle.

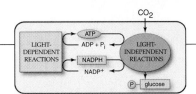

FIXING CARBON—SO NEAR, YET SO FAR

If sunlight intensity, air temperature, rainfall, and soil composition never varied, photosynthesis might be the same in all plants. But environments differ. And details of photosynthesis differ, as you can sense by comparing carbon-fixing adaptations to stressful conditions.

C4 Versus C3 Plants

Plant growth requires carbon, but CO_2 is not always plentiful in leaves. A leaf has a waxy cover that helps plants conserve water. CO_2 diffuses in and O_2 diffuses out at **stomata** (singular, stoma), tiny openings across the leaf surface (Figure 7.17). Stomata close on hot, dry days. Closure of all of the stomata restricts water loss but stops CO_2 from diffusing in. Cells use up the CO_2 inside, and O_2 builds up because it can't diffuse out.

When cells are engaged in photosynthesis, oxygen accumulates in a leaf. A high O_2 concentration triggers *photorespiration*, a process that lowers a plant's sugar-making capacity. Remember rubisco, the enzyme that attaches CO_2 to RuBP in the Calvin–Benson cycle? It also attaches *oxygen* to it instead when CO_2 levels fall and O_2 levels rise. The resulting intermediate breaks down to only one PGA and to one glycolate molecule. Unlike PGA, which helps form sugars, glycolate enters breakdown reactions that release fixed carbon in the form of CO_2—which is back where the plant started but with only half the sugars to show for it.

PGA is the first intermediate of **C3 plants** such as basswood, bluegrass, and beans (Figure 7.18a). "C3" refers to its *three* carbons. In **C4 plants**, such as corn, the first to form is *four*-carbon oxaloacetate. The C4 plants also close stomata on hot, dry days. But the CO_2 level doesn't decline as much because these plants fix carbon twice, in two types of photosynthetic cells.

Mesophyll cells use PEP, a three-carbon compound, to fix CO_2. The enzyme catalyzing this step will not use oxygen, regardless of the CO_2/O_2 ratio. Oxaloacetate forms, then malate, which diffuses to adjoining *bundle-sheath cells* (Figure 7.18b). There, pyruvate forms. CO_2 is released, then enters the Calvin-Benson cycle. With the help of ATP, the pyruvate regenerates PEP for use in the C4 pathway in the mesophyll cells.

With this pathway, C4 plants can get by with small stomata, lose less water, and make more glucose than C3 plants can when days are hot, bright, and dry.

Experiments show that photorespiration lowers the photosynthetic efficiency of many C3 plants. Example: Grow hothouse tomatoes at CO_2 levels high enough to block photorespiration and growth rates dramatically increase. So why hasn't natural selection eliminated such a wasteful process? Look to rubisco. This enzyme evolved long ago, when the atmosphere had little O_2 and an abundance of CO_2. Maybe some gene coding

a Leaves of basswood (*Tilia*), a typical C3 plant.

upper epidermis
palisade mesophyll
spongy mesophyll
lower epidermis

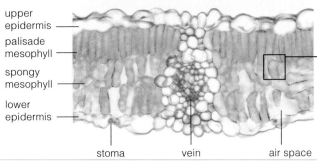

stoma   vein   air space

b Leaves of corn (*Zea mays*), a typical C4 plant

upper epidermis

mesophyll cell

bundle-sheath cell

lower epidermis

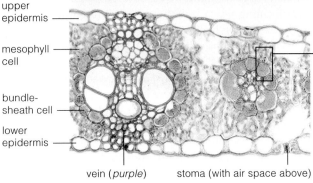

vein (*purple*)   stoma (with air space above)

Figure 7.17 Comparison of the internal leaf structure, in cross-section, for a C3 plant and a C4 plant.

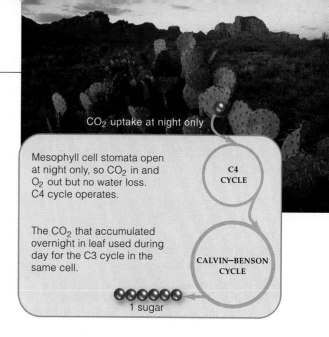

Figure 7.19 Prickly pear (*Opuntia*), a CAM plant. These plants open stomata and fix carbon at night. They include orchids, pineapples, and many succulents besides cacti.

CO₂ uptake at night only

Mesophyll cell stomata open at night only, so CO₂ in and O₂ out but no water loss. C4 cycle operates.

C4 CYCLE

The CO₂ that accumulated overnight in leaf used during day for the C3 cycle in the same cell.

CALVIN–BENSON CYCLE

1 sugar

Stomata closed, CO₂ can't get in and O₂ can't get out.

Low CO₂, high O₂ in leaf. Rubisco fixes oxygen, not carbon, in mesophyll cells.

RuBP → 6 PGA + 6 glycolate

CALVIN–BENSON CYCLE

5 PGAL   6 PGAL

1 PGAL

CO₂ + water

Twelve turns of the cycle, not just six, to make one 6-carbon sugar.

a C3 carbon fixation with low CO₂ / high O₂ levels in leaf

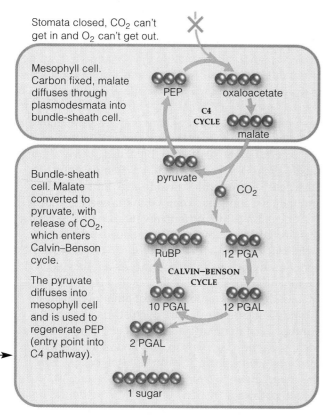

Stomata closed, CO₂ can't get in and O₂ can't get out.

Mesophyll cell. Carbon fixed, malate diffuses through plasmodesmata into bundle-sheath cell.

PEP   oxaloacetate

C4 CYCLE   malate

pyruvate

Bundle-sheath cell. Malate converted to pyruvate, with release of CO₂, which enters Calvin–Benson cycle.

The pyruvate diffuses into mesophyll cell and is used to regenerate PEP (entry point into C4 pathway).

CO₂

RuBP   12 PGA

CALVIN–BENSON CYCLE

10 PGAL   12 PGAL

2 PGAL

1 sugar

b C4 carbon fixation with low CO₂ / high O₂ levels in leaf

Figure 7.18 Two ways to fix carbon in hot, dry weather, when there is too little CO₂ and too much O₂ in leaves. (**a**) The Calvin–Benson (C3) cycle is common in evergreens and many nonwoody plants of temperate zones, such as basswood and bluegrass. (**b**) The C4 cycle is common in grasses, corn, and other plants that evolved in the tropics and that fix CO₂ twice.

for its structure can't mutate without bad effects on its primary role—carbon-fixing activity.

Over the past 50 to 60 million years, the C4 cycle evolved independently in many lineages. Before then, atmospheric CO₂ levels were higher, and so C3 plants had the selective advantage in hot climates. Which cycle will be the most adaptive in the future? The CO₂ levels have been rising for decades and may double in the next fifty years. If so, photorespiration will again lose out—and many vital crop plants may benefit.

CAM Plants

We see a carbon-fixing adaptation to desert conditions in cacti. A cactus is one of the *succulents*; it has juicy, water-storing tissues and thick surface layers that limit water loss. It cannot open stomata on hot days without losing water. It opens them and fixes CO₂ *at night*.

Many plants are adapted this way. They are called **CAM plants** (short for Crassulacean *Acid* Metabolism). At night their mesophyll cells use a C4 cycle much like that shown in Figure 7.19. They store malate and other products until the next day, when their stomata close. Then malate releases CO₂, which enters reactions of the Calvin–Benson (C3) cycle in the same cells. In this way, photosynthesis proceeds without water loss.

Many plants die in prolonged droughts, but some CAM plants survive by closing stomata even at night. They keep fixing the CO₂ from aerobic respiration. Not much forms. It is enough to maintain low metabolic rates, which can only allow slow growth. Try growing cacti in Seattle or other places with a mild climate, and they will compete poorly with C3 and C4 plants.

Compared to C3 plants, the C4 plants and CAM plants have modified ways of fixing carbon for photosynthesis. The modifications counter stresses imposed by hot, dry conditions in their environments.

AUTOTROPHS, HUMANS, AND THE BIOSPHERE

We conclude this chapter with a story to reinforce how photosynthesizers and other autotrophs fit in the world of living things. It is a story of mind-boggling numbers of single-celled and multicelled species on land and in the sunlit waters of the Earth.

In spring, you sense renewed growth of autotrophs on land when leaves unfurl and lawns and fields turn green. You may not be aware that uncountable numbers of single-celled autotrophs also drift through surface waters of the world ocean. You can't see them without a microscope; a row of 7 million cells of one aquatic species would be less than a quarter-inch long. In some regions, a cupful of seawater might hold 24 million cells of one species. And that wouldn't include cells of all the other aquatic species suspended in the cup.

Most of the drifters are photoautotrophic prokaryotic cells and protistans. Together they are the "pastures of the seas," producers that ultimately feed most other marine organisms. The pastures "bloom" in the spring, when nutrient inputs sustain rapid reproduction. At that time seawater becomes warmer and enriched with nutrients that winter currents churn up from the deep.

Until NASA gathered data from space satellites, we had no idea of their numbers and distribution. Figure 7.20a gives visual evidence of their activities one winter in surface waters of the North Atlantic Ocean. Figure 7.20b gives evidence of a springtime bloom stretching from North Carolina all the way past Spain!

Collectively, the cells help shape the global climate, for they deal with staggering numbers of reactant and product molecules. For instance, they sponge up nearly half the carbon dioxide that we humans release each year, as when we burn fossil fuels or burn vast tracts of forests to clear land for farming. Without the aquatic photoautotrophs, atmospheric carbon dioxide would accumulate more rapidly and possibly contribute to global warming. Remember the Chapter 3 introduction? If the atmosphere warms by only a few degrees, then sea levels will rise and all lowlands near the coasts of continents and islands will become submerged.

Although such global change is a real possibility, tons of industrial wastes, sewage, and fertilizers in runoff from croplands still drain into the ocean each day. The pollutants seriously alter the chemical composition of seawater. How long will the marine photoautotrophs be able to function in this chemical brew? The answer will have impact on your own life in more ways than one.

Other autotrophs affect your life in ways you might not expect. On the seafloor, in hot springs, in coal mine wastes are prokaryotic cells called **chemoautotrophs**. Such cells secure carbon dioxide as a carbon source, and they use inorganic compounds as energy sources.

For example, **hydrothermal vents** are fissures in the seafloor where molten rock superheats water. Certain archaebacteria use hydrogen sulfide in the mineral-rich water for hydrogen and electrons. As another example, chemoautotrophs in soil obtain energy from nitrogen-containing wastes and remains of other organisms.

The chemoautotrophs are microscopically small, but they exist in monumental numbers. They affect the cycling of nitrogen, phosphorus, and other elements through the biosphere. We consider their impact on the environment in later chapters. In this unit, we turn next to pathways by which cells release chemical bond energy of glucose and other biological molecules—the chemical legacy of diverse autotrophs everywhere.

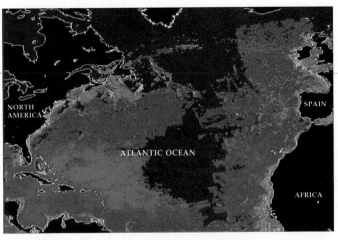

a Photosynthetic activity in winter.

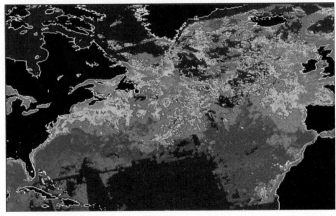

b Photosynthetic activity in spring.

Figure 7.20 Two satellite images that help convey the magnitude of photosynthetic activity during springtime in the North Atlantic portion of the world ocean. In these color-enhanced images, *red-orange* shows where chlorophyll is most concentrated.

Diverse photoautotrophs and chemoautotrophs produce the food that sustains you and all other heterotrophs. Mind-boggling numbers of these producers live in the sunlit waters of the Earth as well as on land.

SUMMARY   *Gold* indicates text section

1. Cell structure and functions are based on organic compounds, the synthesis of which depends on sources of carbon and energy. Plants and other autotrophs get carbon from carbon dioxide and energy from the sun or inorganic compounds. Animals and other heterotrophs aren't self-nourishing; they get carbon and energy from organic compounds synthesized by autotrophs. *CI*

2. Photosynthesis is the main process by which carbon and energy enter the web of life. Its two stages are the light-dependent and light-independent reactions. This equation and Figure 7.21 summarize the process: *7.1*

$$12H_2O + 6CO_2 \xrightarrow{\text{LIGHT ENERGY}} 6O_2 + C_6H_{12}O_6 + 6H_2O$$

WATER   CARBON   OXYGEN   GLUCOSE   WATER
           DIOXIDE

3. Chloroplasts are organelles of photosynthesis in algae and plants. Two outer membranes enclose its semifluid interior (stroma). A third membrane in the stroma forms a single compartment; often it folds back on itself in disk-shaped stacks (thylakoids). The light-dependent reactions take place at this membrane. But the light-independent reactions take place in the stroma. *7.1*

4. The light-dependent reactions start at photosystems that each include 200 to 300 pigments. *7.2–7.5*

    a. Chlorophyll *a*, the main photosynthetic pigment, absorbs all wavelengths of visible light but only a little of the yellow-green and green ones. Accessory pigments (e.g., carotenoids) absorb other wavelengths.

    b. A cyclic pathway forms ATP alone. Photosystem I gives up excited electrons to an electron transfer chain, which sends the "spent" ones back to the photosystem.

    c. The noncyclic pathway forms ATP and NADPH. Water molecules are split into hydrogen, electrons, and oxygen. Excited electrons enter photosystem II, then a transfer chain, photosystem I, then a second transfer chain. NADP⁺ picks up the electrons and hydrogen to form NADPH. By 1.5 billion years ago, oxygen released by this pathway had accumulated in the atmosphere. It ultimately made possible aerobic respiration.

5. ATP energy drives the light-independent reactions (by phosphate-group transfers). Hydrogen and electrons from NADPH, plus carbon and oxygen (from carbon dioxide), are used as building blocks for glucose. Most glucose is used at once to synthesize starch, cellulose, and other end products of photosynthesis. *7.6*

    a. The enzyme rubisco affixes carbon from CO_2 to the sugar RuBP, the start of the the Calvin–Benson cycle. Two PGA form. These are converted to two PGAL with ATP energy and hydrogen and electrons from NADPH.

    b. For every six carbon atoms that enter the cycle, twelve PGAL form. Two are used to form a six-carbon sugar phosphate. The rest are used to regenerate RuBP.

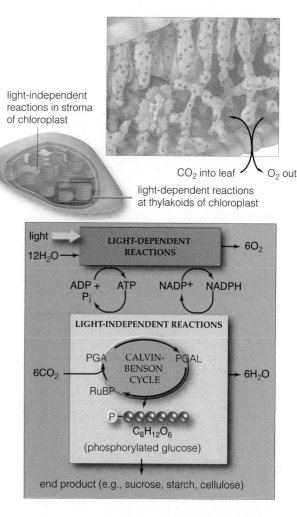

Figure 7.21 Summary of photosynthesis. In the light-dependent reactions, sunlight energy is converted to chemical bond energy of ATP. In a noncyclic pathway, water molecules are split by photolysis. NADP⁺ conserves the released electrons and hydrogen (as NADPH). Oxygen is released as a by-product.

ATP energy drives the light-independent reactions. Hydrogen and electrons from NADPH, plus carbon and oxygen from carbon dioxide, are used to form glucose. In the Calvin–Benson (C3) cycle, the RuBP on which the cycle turns is regenerated, and new molecules of water form. *Each turn* of the cycle requires one CO_2, three ATP, and two NADPH. Because each glucose molecule has a backbone of six carbon atoms, its formation requires six turns of the cycle.

6. Rubisco, a carbon-fixing enzyme, evolved when air had far more CO_2 and less O_2. Today, when CO_2 is used up and O_2 builds up in leaves, rubisco attaches oxygen (not carbon) to RuBP. This process, photorespiration, is wasteful: one PGA and glycolate form. Glycolate cannot be used to form sugars and is later degraded. *7.6*

7. Photorespiration predominates in C3 plants such as sunflowers under hot, dry conditions. Stomata close, so O_2 from photosynthesis builds up in leaves to levels higher than CO_2 levels. Corn and other C4 plants raise the CO_2 level by fixing carbon twice, in two cell types. CAM plants run the C3 cycle during the day when stomata are closed, using CO_2 fixed by the C4 cycle at night, when stomata are open. *7.7*

Review Questions

1. A cat eats a bird, which earlier ate a caterpillar that chewed on a weed. Which organisms are autotrophs? Which are the heterotrophs? *CI*

2. Summarize the photosynthesis reactions as an equation. Name where each stage takes place inside a chloroplast. *7.1*

3. What is the function of ATP in photosynthesis? What is the function of NADPH? *7.1*

4. Which of the following pigments are most visible in a maple leaf in summer? Which become the most visible in autumn? *7.3*
    a. chlorophylls   c. anthocyanins
    b. phycobilins   d. carotenoids

5. How does chlorophyll *a* differ in function from accessory pigments during the light-dependent reactions? *7.3, 7.4*

6. With respect to the light-dependent reactions, how do the cyclic and noncyclic pathways of electron flow differ? *7.4*

7. What substance does *not* take part in the Calvin–Benson cycle: ATP, NADPH, RuBP, carotenoids, O_2, CO_2, or enzymes? *7.6*

8. Fill in the blanks for the diagram at right. Which substances are the original sources of the carbon atoms and hydrogen atoms used in the synthesis of glucose in the Calvin–Benson cycle? *CI, 7.6*

9. On hot, dry days, oxygen from photosynthesis accumulates in leaves. Explain why, and explain what happens in C3 plants versus C4 plants. *7.7*

10. Are photoautotrophs the only "self feeders"? If not, give examples and where you might find them. *CI, 7.8*

1. Photosynthetic autotrophs use _____ from the air as a carbon source and _____ as their energy source.

2. In plants, light-*dependent* reactions proceed at the _____ .
    a. cytoplasm   c. stroma
    b. plasma membrane   d. thylakoid membrane

3. In the light-*dependent* reactions, _____ .
    a. carbon dioxide is fixed   c. CO_2 accepts electrons
    b. ATP and NADPH form   d. sugar phosphates form

4. The light-*independent* reactions proceed in the _____ .
    a. cytoplasm   b. plasma membrane   c. stroma

5. When a photosystem absorbs light, _____ .
    a. sugar phosphates are produced
    b. electrons are transferred to ATP
    c. RuBP accepts electrons
    d. light-dependent reactions begin

6. Identify which of the following substances accumulates inside the thylakoid compartment of chloroplasts during the light-dependent reactions:
    a. glucose   c. chlorophyll   e. hydrogen ions
    b. carotenoids   d. fatty acids

7. The Calvin–Benson cycle starts when _____ .
    a. light is available
    b. light is not available
    c. carbon dioxide is attached to RuBP
    d. electrons leave a photosystem

8. ATP phosphorylates _____ in the light-independent reactions.
    a. RuBP   c. PGA
    b. $NADP^+$ d. PGAL

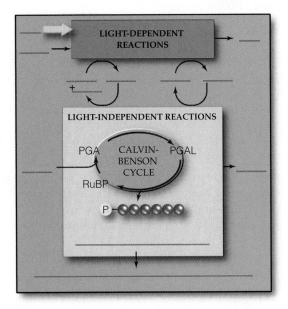

9. Match each event with its most suitable description.
    ____ photon absorption   a. rubisco required
    ____ NADPH formation   b. ATP, NADPH required
    ____ CO_2 fixation   c. electrons cycled back
    ____ PGAL formation   to photosystem I
    ____ ATP formation only   d. energy matches amount
                                     required to excite electrons
                                  e. photolysis required

Critical Thinking

1. Suppose a garden in your neighborhood is filled with red, white, and blue petunias. Explain each of the three floral colors in terms of which wavelengths of light the flower is absorbing and which wavelengths it is transmitting or reflecting.

2. About 200 years ago, Jan Baptista van Helmont performed an experiment on the nature of photosynthesis. He wanted to know where growing plants acquire the raw materials necessary to increase in size. For his experiment, he planted a tree seedling weighing 5 pounds in a barrel filled with 200 pounds of soil. He watered the tree regularly.

   Five years passed. Van Helmont again weighed the tree and the soil. The tree weighed 169 pounds, 3 ounces. The soil weighed 199 pounds, 14 ounces. Because the tree weight had increased so much and soil weight had decreased so little, he concluded the tree had gained weight after absorbing the water he had added to the barrel.

   Given what you know about the composition of biological molecules, why was van Helmont's conclusion misguided? Reflect on the current model of photosynthesis and then give a more plausible explanation of his results.

3. During the late 1980s a graduate student, Cindy Lee Van Dover, was puzzling over an eyeless shrimp (Figure 7.22). Divers discovered it deep in the Atlantic Ocean, near the base of a *hydrothermal vent*. At these fissures in the seafloor, molten rock rises and then mixes with cold seawater. Mineral-rich water, superheated by 650 degrees, spews into the perpetually

Figure 7.22 Preserved specimen of an eyeless shrimp recovered at a hydrothermal vent.

Figure 7.23 Where do the colors of butter, lemon butterfly fishes, morpho butterflies, scarlet macaws, and green pythons come from?

dark surroundings. Chemoautotrophic bacteria are the basis of thriving ecosystems near the vents (Sections 7.9 and 49.11).

When Van Dover saw videotapes of shrimps on the vent, she noticed a pair of bright strips on their back. By studying lab specimens, she found that the strips are connected to a nerve. Could eyeless shrimps have a novel sensory organ?

Steven Chamberlain, a neuroscientist, confirmed that the strips are a sensory organ with light-absorbing pigmented cells (photoreceptors). Later, Ete Szuts, a pigment expert, isolated the pigment. Its absorption spectrum is the same as that for rhodopsin, a visual pigment in eyes like yours.

Van Dover had a hunch: If the strips are light sensitive, then we should expect to find light on the seafloor. On later dives, special cameras confirmed her prediction. Like coils in a toaster, hydrothermal vents release heat (infrared radiation). They also release faint radiation at the low end of the visible spectrum— light up to nineteen times brighter than infrared.

A sunlit tree typically absorbs a billion billion photons per square inch per second. Photoautotrophic bacteria that live 72 meters (240 feet) below the Black Sea's surface encounter a trillion photons per square inch per second. *Just as many photons are available to hydrothermal vent organisms.*

Van Dover casually asked a colleague, "Hey, what if there's enough light for photosynthesis?" The response was, "What a stupid idea."

Euan Nisbet, who studies ancient environments, thought about that. The first cells arose about 3.8 billion years ago. As Nisbet and Van Dover later hypothesized, What if those cells arose at hydrothermal vents? What if they used inorganic compounds such as hydrogen sulfide (for their hydrogen and electrons) and carbon dioxide (for carbon)?

If so, survival probably depended on being able to move away from light at vents and thereby avoid being boiled alive. Millions of years later, some bacteria that may have been their descendants were evolving near the ocean's surface. In time they used their light-detecting machinery to absorb sunlight. They had become adapted to a new energy source—they were photosynthetic.

Did light-sensing machinery of deep-sea bacteria become modified for shallow-water photosynthesis? Some observations support this intriguing hypothesis. A few clues: The absorption spectra for chlorophyll in evolutionarily ancient photosynthetic bacteria correspond to light measured at the vents. Also, photosynthetic machinery incorporates iron, sulfur, manganese, and other minerals—all of which are abundant at hydrothermal vents.

If the earliest photoautotrophs evolved deep in the seas, their pigments evolved in the seas, also. Speculate on how natural selection may have favored the evolution of different pigments at different depths, starting at hydrothermal vents.

4. There are only about eight classes of pigment molecules, but this limited group gets around in the world. For example, animals synthesize the brownish-black melanin and some other pigments, but not carotenoids. These form in photoautotrophs and move up through food webs, as when tiny aquatic snails graze on green algae and then flamingos eat the snails.

Flamingos modify those ingested carotenoids in plenty of ways. For instance, their cells split beta-carotene molecules to form two molecules of vitamin A. This vitamin is the precursor of retinol, a visual pigment that transduces light into electric signals in the flamingo's eyes. Beta-carotene molecules also become dissolved in fat reservoirs under the skin. From there they are taken up by cells that give rise to bright pink feathers.

Pick one of the animals (or the one that produced the butter) shown in Figure 7.23. Do some research into its life cycle and its diet. Use your research to identify some possible sources for the pigments that impart the colors shown.

5. Krishna is using radioactively labeled carbon atoms ($^{14}CO_2$) in the laboratory. The photosynthesizing plants he is studying take them up. Identify the compound in which the labeled carbon will appear first: NADPH, PGAL, pyruvate, or PGA.

Selected Key Terms

absorption spectrum 7.2	light-independent reactions 7.6
anthocyanin 7.3	PGA 7.6
autotroph CI	PGAL 7.6
C3 plant 7.7	photoautotroph CI
C4 plant 7.7	photolysis 7.4
Calvin–Benson cycle 7.6	photon 7.2
CAM plant 7.7	photosynthesis CI
carbon fixation 7.6	photosystem 7.3
carotenoid 7.3	phycobilin 7.3
chemoautotroph 7.8	pigment 7.2
chlorophyll 7.3	reaction center 7.4
chloroplast 7.1	rubisco 7.6
electromagnetic spectrum 7.2	RuBP 7.6
electron transfer chain 7.4	stoma (stomata) 7.7
fluorescence 7.3	stroma 7.1
heterotroph CI	thylakoid membrane 7.1
hydrothermal vent 7.8	wavelength 7.2
light-dependent reactions 7.4	

Readings

Bazzaz, F., and E. Jajer. January 1992. "Plant Life in a CO_2-Rich World." *Scientific American*, 266:68–74.

Zimmer, C. November 1996. "The Light at the Bottom of the Sea." *Discover*, 63–73.

8

HOW CELLS RELEASE STORED ENERGY

The Killers Are Coming! The Killers Are Coming!

Thanks to selective breeding experiments gone wrong, descendants of "killer" bees that flew out of South America a few decades ago buzzed across the border between Mexico and Texas. By 1995, they had invaded 13,287 square kilometers of southern California and were busily setting up colonies. By 1998, when nectar-rich flowers bloomed profusely in the deserts after heavy El Niño storms, they moved even farther west and north than biologists had predicted.

When provoked, the bees are terrifying. To give an example, when a construction worker started a tractor a few hundred yards from a hive, thousands of agitated bees flew into action. They entered a nearby subway station, where they stung passengers on the platform and in the trains. They killed one person and injured a hundred more.

Where did the bees come from? In the 1950s, some queen bees were shipped over from Africa to Brazil for breeding experiments. Why? Honeybees are serious business. They are the source of nutritious honey. Also, they are rented to commercial orchards, where their collective pollinating activities may significantly enhance fruit production. Imagine yourself enclosing an orchard tree in a cage that keeps out all pollinators. Without pollinators to spread pollen about, less than 1 percent of that tree's flowers will set fruit. Now imagine putting a hive of honeybees inside the cage with the tree. As many as 40 percent of the flowers will set fruit.

Compared to their relatives in Africa, the bees of Brazil are sluggish pollinators and honey producers. By cross-breeding the two strains, researchers thought they might come up with a strain of mild-mannered but zippier bees. So they put some local bees and imported ones together inside netted enclosures, complete with artificial hives. Then they let nature take its course.

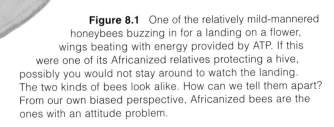

Figure 8.1 One of the relatively mild-mannered honeybees buzzing in for a landing on a flower, wings beating with energy provided by ATP. If this were one of its Africanized relatives protecting a hive, possibly you would not stay around to watch the landing. The two kinds of bees look alike. How can we tell them apart? From our own biased perspective, Africanized bees are the ones with an attitude problem.

Twenty-six African queen bees escaped. That was bad enough. Then beekeepers got wind of preliminary experimental results. After learning that the first few generations of offspring were more energetic but not overly aggressive, they imported hundreds of African queens and encouraged them to mate with the locals. And they set off a genetic time bomb.

Before long, African bees became established in commercial hives—and in wild bee populations. Their traits became dominant. The "Africanized" bees do everything other bees do, but they do more of it faster. Their eggs develop faster into adults. Adults fly faster, outcompete other bees for nectar, and even die sooner.

When something disturbs their hives or swarms, Africanized bees become extremely agitated. They can stay that way for up to eight hours. Whereas a mild-mannered honeybee might chase an intruding animal fifty yards or so, an Africanized bee squadron will chase it a quarter of a mile. If they catch up with it, they can collectively sting it to death.

Doing things faster means having a continuous supply of energy and efficient ways of using it. An Africanized bee's stomach can hold thirty milligrams of sugar-rich nectar—which is enough fuel to fly sixty kilometers. That's more than thirty-five miles! Besides this, compared to other kinds of bees, an Africanized bee's flight muscle cells have larger mitochondria. These organelles specialize in releasing a great deal of energy from sugars and other organic compounds, then converting it to the energy of ATP.

Whenever they tap into the stored energy of organic compounds, Africanized bees reveal their biochemical connection with other organisms. Study a primrose or puppy, a mold growing on stale bread, an amoeba in pondwater, or a bacterium living on your skin, and you will discover that their energy-releasing pathways differ in some details. But all of the pathways require characteristic starting materials. They yield predictable products and by-products. And they yield the universal energy currency of life—ATP.

All through the biosphere, organisms use energy and raw materials in similar ways. *At the biochemical level, we find undeniable unity among all forms of life.* We return to this idea at the chapter's end.

Key Concepts

1. For all organisms, cells release energy stored in glucose and other organic compounds, then use it in ATP production. The energy-releasing metabolic pathways differ from one another. But all of the main ones start with the breakdown of glucose to pyruvate.

2. The initial breakdown reactions, glycolysis, can proceed in the presence of oxygen or in its absence. Said another way, it is the first stage of either aerobic or anaerobic energy-releasing pathways.

3. The types of pathways called fermentation and anaerobic electron transfer do not use oxygen as an electron receptor. They proceed only in the cytoplasm, and none yields more than a small amount of ATP for each glucose molecule metabolized.

4. Another pathway, aerobic respiration, also starts in the cytoplasm. But it runs to completion in organelles called mitochondria. Compared to the other pathways, it releases far more energy from glucose.

5. Aerobic respiration consists of three stages. First, pyruvate forms from glucose (during glycolysis). Second, pyruvate is broken down to carbon dioxide, and coenzymes pick up the released electrons and hydrogen for delivery to electron transfer chains. Third, electron transfer chains help set up conditions that favor ATP formation. Free oxygen accepts the electrons at the end of the line and combines with hydrogen to form water.

6. Over evolutionary time, photosynthesis and aerobic respiration became connected on a global scale. An oxygen-rich atmosphere evolved through long-term photosynthetic activity. It sustains aerobic respiration. In turn, the by-products of this pathway—carbon dioxide and water—serve as raw materials when the photosynthesizers make organic compounds:

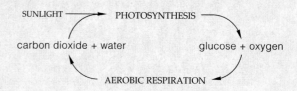

HOW DO CELLS MAKE ATP?

Organisms stay alive by taking in energy. Plants and all other organisms that engage in photosynthesis get energy from the sun. Animals get energy secondhand, thirdhand, and so on, by eating plants and one another. Regardless of its source, energy must be in a form that can drive thousands of life-sustaining reactions. Energy that becomes converted into the chemical bond energy of adenosine triphosphate—ATP—serves that function.

Plants make ATP during photosynthesis and use it to make glucose and other carbohydrates. *But they and every other organism also produce ATP by breaking down glucose and other carbohydrates, lipids, and proteins.* The reactions release electrons as well as hydrogens from intermediates. Both are delivered to electron transfer chains, as described in Section 6.4, and the outcome is central to the energy-releasing pathways.

Comparison of the Main Types of Energy-Releasing Pathways

Energy-releasing pathways originated long before our oxygen-rich atmosphere evolved by 1 billion years ago, when conditions were different. So the pathways must have been *anaerobic*; they did not use oxygen. Many prokaryotes and protistans still live in places where oxygen is absent or not always available. To make ATP, they use anaerobic reactions (fermentation or anaerobic electron transfer). Some of your own cells use anaerobic routes for short periods when they are not receiving enough oxygen. But **aerobic respiration**, an oxygen-dependent pathway of ATP formation, is the dominant pathway. Each breath you take provides your actively respiring cells with a fresh supply of oxygen.

Make note of this point: *The main energy-releasing pathways all start with the same reactions in the cytoplasm.* During this initial stage of reactions, called **glycolysis**, enzymes cleave and rearrange a glucose molecule into two molecules of **pyruvate**, an organic compound with a backbone of three carbon atoms. Once glycolysis is over, the energy-releasing pathways differ. Most importantly, only the aerobic pathway continues in a mitochondrion (Figure 8.2). There, oxygen serves as the final acceptor

a

of electrons used during the reactions. The anaerobic pathways start and end in the cytoplasm. A substance other than oxygen is the final electron acceptor.

As you examine the energy-releasing pathways in sections to follow, keep in mind that the reaction steps do not proceed by themselves. Enzymes catalyze each step, and intermediates formed at one step function as substrates for the next enzyme in the pathway.

Overview of Aerobic Respiration

Of all energy-releasing pathways, aerobic respiration gets the most ATP for each glucose molecule. Whereas anaerobic routes have a net yield of two ATP, aerobic respiration commonly yields thirty-six or more. If you were a bacterium, you would not require much ATP. Being far larger, more complex, and highly active, you depend on the aerobic pathway's high yield. When a molecule of glucose is used as the starting material, aerobic respiration can be summarized this way:

$$C_6H_{12}O_6 \; + \; 6O_2 \longrightarrow 6CO_2 \; + \; 6H_2O$$

GLUCOSE   OXYGEN   CARBON DIOXIDE   WATER

However, as you can see, the summary equation only tells us what the substances are at the start and finish of the pathway. In between are three reaction stages.

Figure 8.2 Where the aerobic and anaerobic pathways of ATP formation start and finish.

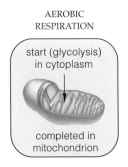

AEROBIC RESPIRATION

start (glycolysis) in cytoplasm

completed in mitochondrion

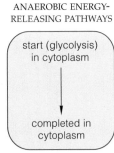

ANAEROBIC ENERGY-RELEASING PATHWAYS

start (glycolysis) in cytoplasm

completed in cytoplasm

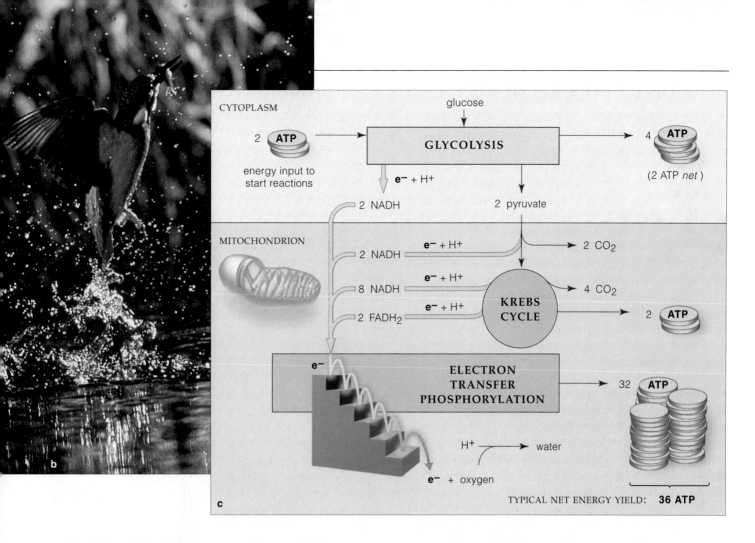

Figure 8.3 Overview of the three stages of aerobic respiration. From start to finish, the typical net energy yield from a glucose molecule is thirty-six ATP. (**a,b**) Only this pathway delivers enough ATP to build and maintain redwoods and all other large, multicelled organisms. It alone delivers enough ATP for highly active animals, including bees, humans, and kingfishers.

(**c**) In the first stage (glycolysis), enzymes partially break down glucose to pyruvate. In the second stage—mainly the Krebs cycle—enzymes degrade pyruvate to carbon dioxide. NAD^+ and FAD pick up electrons and hydrogen stripped from intermediates during both stages. In the final stage, electron transfer phosphorylation, the reduced coenzymes (NADH and $FADH_2$) give up electrons and hydrogen to electron transfer chains. Energy released during the flow of electrons through the chains drives ATP formation. Oxygen accepts the electrons at the end of the third stage.

Use Figure 8.3 while we outline the reactions. Again, glycolysis is the first stage. The second stage is mainly a cyclic pathway, the **Krebs cycle**. Enzymes break down pyruvate to carbon dioxide and water, thereby releasing electrons and hydrogen.

NAD^+ (for nicotinamide adenine dinucleotide) and **FAD** (flavin adenine dinucleotide) serve in glycolysis and the Krebs cycle. Remember, these coenzymes assist enzymes by accepting electrons and hydrogen derived from the intermediates. Unbound hydrogen atoms are naked protons, or hydrogen ions (H^+). When the two coenzymes are carrying electrons and hydrogen, they are abbreviated NADH and $FADH_2$.

Few ATP form during glycolysis or the Krebs cycle. The big energy harvest comes in the third stage, after the coenzymes give up the electrons and hydrogen to electron transfer chains. The chains are the machinery of **electron transfer phosphorylation**. They set up H^+ concentration and electric gradients, which drive ATP formation at nearby membrane proteins. It is in this final stage that so many ATP molecules are produced. As it ends, oxygen inside the mitochondrion accepts the "spent" electrons from the last component of each transport system. Oxygen picks up hydrogen at the same time and thereby forms water.

Nearly all metabolic reactions run on energy released from glucose and other organic compounds, then converted to chemical bond energy of ATP. The main energy-releasing pathways start in the cytoplasm with glycolysis, a stage of reactions that break down glucose to pyruvate.

The most common anaerobic pathways, which include the fermentation routes, end in the cytoplasm. Each has a net energy yield of two ATP.

Aerobic respiration, an oxygen-dependent pathway, runs to completion in mitochondria. From start (glycolysis) to finish, it typically has a net energy yield of thirty-six ATP.

GLYCOLYSIS: FIRST STAGE OF ENERGY-RELEASING PATHWAYS

Let's track what happens to a glucose molecule in the first stage of aerobic respiration. Remember, the same things happen to glucose in the anaerobic pathways.

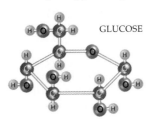

GLUCOSE

As you know, glucose is one of the simple sugars (Section 3.2). Each molecule consists of six carbon, twelve hydrogen, and six oxygen atoms, all covalently bonded together. Its carbons are the backbone. In the cytoplasm, glucose or some other simple sugar is partly broken down during glycolysis to two molecules of pyruvate, a three-carbon compound:

glucose $\longrightarrow$ glucose–6–phosphate $\longrightarrow$ 2 pyruvate

The first steps of glycolysis are *energy-requiring*. As Figure 8.4 indicates, they proceed only when two ATP molecules each transfer a phosphate group to glucose. Such phosphate-group transfers, remember, are known as phosphorylations. In this case, they raise the energy content of glucose to a level high enough to allow entry into the *energy-releasing* steps of glycolysis.

The first energy-releasing step cleaves the activated glucose into two molecules. We can call each of them PGAL (phosphoglyceraldehyde). Each PGAL becomes converted to an unstable intermediate that gives up a phosphate group to ADP, and so ATP forms. The next intermediate in the sequence does the same thing.

And so a total of four ATP form by **substrate-level phosphorylation**. We define this metabolic event as the direct transfer of a phosphate group from a substrate of a reaction to some other molecule—in this case, ADP. Remember, though, two ATP were invested to jump-start the reactions. So the *net* energy yield is two ATP.

Meanwhile, the coenzyme NAD^+ picks up electrons and hydrogens liberated from each PGAL molecule, thus becoming NADH. When NADH delivers its cargo to a different reaction site, it reverts to NAD^+. Said another way, like other coenzymes, NAD^+ is reusable.

In sum, glycolysis converts energy stored in glucose to a transportable form of energy, in ATP. NAD^+ picks up electrons and hydrogen stripped from glucose. These have roles in the next stage of reactions. So do the end products of glycolysis—the two molecules of pyruvate.

Glycolysis is a stage of reactions that partially break down glucose or some other carbohydrate to two molecules of pyruvate, thereby releasing energy.

Two NADH and four ATP form. However, when we subtract the two ATP required to start the reactions, the *net* energy yield of glycolysis is two ATP per glucose molecule.

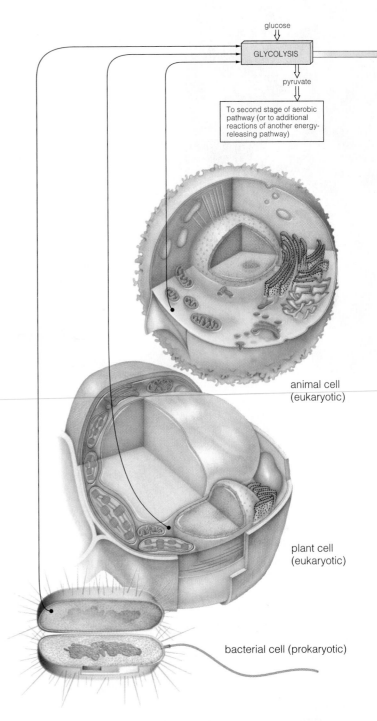

glucose

GLYCOLYSIS

pyruvate

To second stage of aerobic pathway (or to additional reactions of another energy-releasing pathway)

animal cell (eukaryotic)

plant cell (eukaryotic)

bacterial cell (prokaryotic)

Figure 8.4 Glycolysis, the first stage of the main energy-releasing pathways. All prokaryotic and eukaryotic cells use glycolysis, which occurs in the cytoplasm. In this example, glucose is the starting material. Two pyruvate, two NADH, and four ATP form. Because cells invest two ATP to start glycolysis, the *net* energy yield is two ATP. Appendix V (Figure A) gives more details.

Depending on the type of cell and environmental conditions, the pyruvate may enter the second set of reactions of the aerobic pathway, including the Krebs cycle. Or it may be used in other reactions, such as those of fermentation.

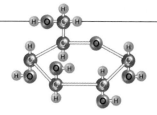

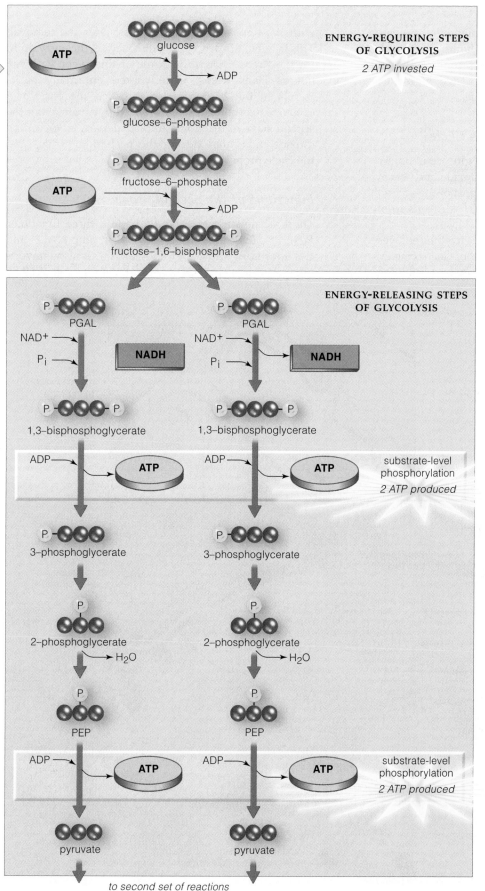

ENERGY-REQUIRING STEPS OF GLYCOLYSIS

2 ATP invested

glucose

ATP → ADP

P – glucose–6–phosphate

P – fructose–6–phosphate

ATP → ADP

P – fructose–1,6–bisphosphate – P

ENERGY-RELEASING STEPS OF GLYCOLYSIS

P – PGAL

NAD^+ →
P_i → **NADH**

P – PGAL

NAD^+ →
P_i → **NADH**

P – 1,3–bisphosphoglycerate – P

P – 1,3–bisphosphoglycerate – P

ADP → ATP

ADP → ATP

substrate-level phosphorylation
2 ATP produced

P – 3–phosphoglycerate

P – 3–phosphoglycerate

P – 2–phosphoglycerate → H_2O

P – 2–phosphoglycerate → H_2O

P – PEP

P – PEP

ADP → ATP

ADP → ATP

substrate-level phosphorylation
2 ATP produced

pyruvate

pyruvate

to second set of reactions

a This diagram tracks the six carbon atoms of the glucose molecule shown above. Glycolysis starts with an energy investment of two ATP.

b One ATP makes a phosphate-group transfer to glucose, and some atoms in glucose are rearranged in response.

c A phosphate-group transfer from the second ATP causes rearrangements that form fructose–1,6–bisphosphate. This intermediate can be split easily.

d It splits at once into two molecules, each with a three-carbon backbone. We can call these two PGAL.

e Two NADH form when each PGAL gives up two electrons and a hydrogen atom to NAD^+.

f One ATP forms as each PGAL also combines with inorganic phosphate (P_i) and transfers a phosphate group to ADP.

g *Thus, two ATP formed by the direct transfer of a phosphate group from two intermediates of the reactions.* The original energy investment of two ATP has been paid off.

h In the next two reactions, the two intermediates each release a hydrogen atom and an —OH group, which then combine to form water.

i Two 3–phosphoenolpyruvate (PEP) molecules result. Each PEP transfers a phosphate group to ADP.

j *Once again, two ATP have formed by substrate-level phosphorylation.*

k In sum, the net energy yield from glycolysis is two ATP for each glucose molecule entering the reactions. Two molecules of pyruvate, the end product, may enter the next set of reactions in an energy-releasing pathway.

SECOND STAGE OF THE AEROBIC PATHWAY

Suppose two pyruvate molecules formed by glycolysis leave the cytoplasm and enter a **mitochondrion** (plural, mitochondria). In this organelle, the second and third stages of the aerobic pathway are completed. Figure 8.5 shows its structure and functional zones.

Preparatory Steps and the Krebs Cycle

It is during the second stage that the glucose is finally broken down completely to carbon dioxide and water. Two ATP form, but the most striking part of the second stage is the transfer of electrons and hydrogens from intermediates to many coenzymes.

In a few preparatory reactions, an enzyme removes a carbon from each pyruvate molecule. Coenzyme A, an enzyme helper, becomes **acetyl–CoA** by combining

with the two-carbon fragment left after the removal. This fragment is passed to **oxaloacetate**, the entry point for the Krebs cycle. (The name of this cyclic pathway honors Hans Krebs, who worked out many of its details in the 1930s. It also is called the citric acid cycle.) *Six* carbons, three from each pyruvate, enter the second stage of reactions. And *six* depart, in six carbon dioxide molecules, during the preparatory steps and the cycle proper (Figure 8.6).

Functions of the Second Stage

Think of the second stage as having three functions. First, it loads electrons and hydrogen onto NAD^+ and FAD, resulting in NADH and $FADH_2$. It also converts organic carbons into CO_2. Second, it forms two ATP by

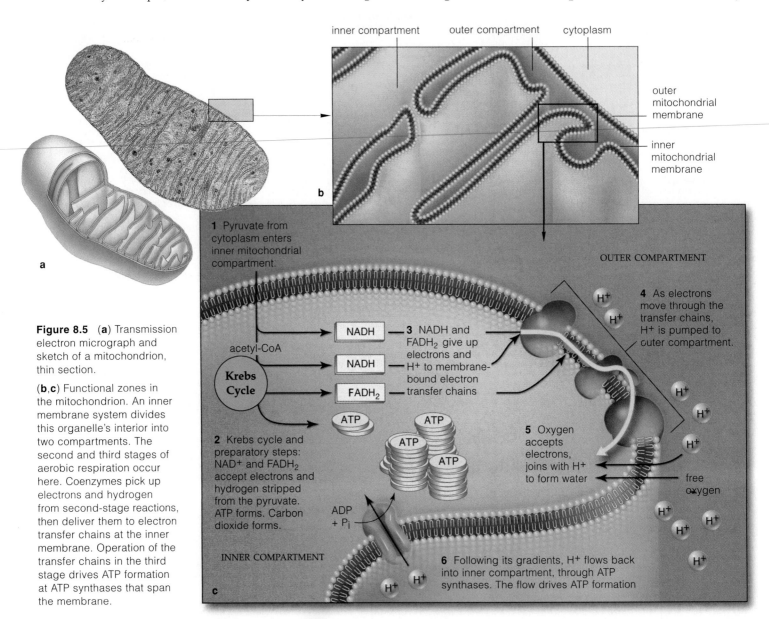

Figure 8.5 (**a**) Transmission electron micrograph and sketch of a mitochondrion, thin section.

(**b,c**) Functional zones in the mitochondrion. An inner membrane system divides this organelle's interior into two compartments. The second and third stages of aerobic respiration occur here. Coenzymes pick up electrons and hydrogen from second-stage reactions, then deliver them to electron transfer chains at the inner membrane. Operation of the transfer chains in the third stage drives ATP formation at ATP synthases that span the membrane.

inner compartment   outer compartment   cytoplasm

outer mitochondrial membrane

inner mitochondrial membrane

1 Pyruvate from cytoplasm enters inner mitochondrial compartment.

OUTER COMPARTMENT

acetyl-CoA

Krebs Cycle

NADH

NADH

$FADH_2$

ATP

3 NADH and $FADH_2$ give up electrons and H^+ to membrane-bound electron transfer chains

4 As electrons move through the transfer chains, H^+ is pumped to outer compartment.

ATP   ATP   ATP   ATP

2 Krebs cycle and preparatory steps: NAD^+ and $FADH_2$ accept electrons and hydrogen stripped from the pyruvate. ATP forms. Carbon dioxide forms.

5 Oxygen accepts electrons, joins with H^+ to form water

free oxygen

ADP + P_i

INNER COMPARTMENT

6 Following its gradients, H^+ flows back into inner compartment, through ATP synthases. The flow drives ATP formation

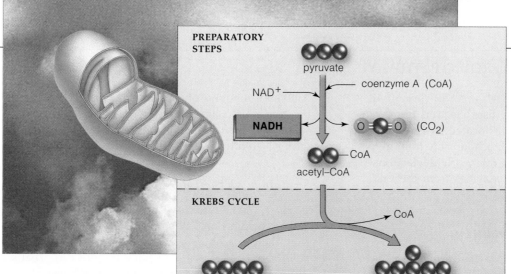

PREPARATORY STEPS

pyruvate

NAD⁺ → coenzyme A (CoA)

NADH

$O = C = O$ (CO_2)

CoA

acetyl–CoA

a Pyruvate from glycolysis enters a mitochondrion. A carbon atom is released, as CO_2. A coenzyme binds with the two-carbon fragment, becoming acetyl–CoA. NADH forms as NAD⁺ picks up hydrogen and electrons. *Thus, one NADH forms during a few steps preceding the Krebs cycle.*

KREBS CYCLE

CoA

oxaloacetate

citrate

b The four-carbon oxaloacetate is the entry point into the Krebs cycle. Acetyl–CoA transfers two carbons to it, forming citrate, a six-carbon compound. Citrate becomes rearranged into another intermediate.

g The final reactions regenerate oxaloacetate. NAD⁺ picks up hydrogen and electrons, forming NADH. *Thus, for each turn of the Krebs cycle, three NADH and one FADH₂ form.*

NADH

NAD⁺

malate

isocitrate

NAD⁺

NADH

$O = C = O$

c Another carbon atom is released, as CO_2. NADH forms when NAD⁺ picks up hydrogen and electrons.

$H_2O →$

f FAD picks up electrons and hydrogen, forming FADH₂.

fumarate

α-ketoglutarate

FADH₂

FAD

NAD⁺ ← CoA

NADH

$O = C = O$

d Another carbon atom is released as CO_2, another NADH forms, and a coenzyme A molecule binds to the intermediate. *At this point, for each turn of the cycle, three carbon atoms have been released.* This balances out the three carbons that entered the mitochondrion (in pyruvate).

succinate

CoA

succinyl–CoA

ATP

ADP + phosphate group

e A phosphate group replaces the coenzyme A and is attached to ADP. *Thus, for each turn of the Krebs cycle, one ATP forms by substrate-level phosphorylation.*

Figure 8.6 Second stage of aerobic respiration: the Krebs cycle and reaction steps that precede it. For each three-carbon pyruvate molecule, three CO_2, one ATP, four NADH, and one FADH₂ molecules form in the Krebs cycle and the preparatory steps before it. The steps shown occur *twice* for each glucose molecule that entered glycolysis. Why? The glucose was degraded earlier to *two* pyruvate molecules.

---

substrate-level phosphorylations. Third, it rearranges the Krebs cycle intermediates into oxaloacetate. This is important; cells have only so much oxaloacetate, which must be regenerated to keep the reactions going.

The two ATP do not add much to the small yield from glycolysis. But the reactions load ten coenzymes with electrons and hydrogen. So far, then, the final stage of the aerobic pathway will get all of these coenzymes:

Glycolysis:	2 NADH
Pyruvate conversion before Krebs cycle:	2 NADH
Krebs cycle:	2 FADH₂ + 6 NADH
Coenzymes sent to third stage:	2 FADH₂ + 10 NADH

Overall, these are the key points to remember about the second stage of aerobic respiration:

In the second stage of aerobic respiration, two pyruvate molecules from glycolysis enter a mitochondrion.

All of pyruvate's carbon atoms (originally from glucose) are released in the form of carbon dioxide. Two ATP form. The oxaloacetate that is the entry point for the second-stage cyclic reactions is regenerated.

Ten coenzymes are reduced by electrons (and hydrogen) released during pyruvate breakdown. With two coenzymes that formed in glycolysis, they will deliver electrons and hydrogen to sites of the final stage of the aerobic pathway.

THIRD STAGE OF THE AEROBIC PATHWAY

In the aerobic pathway's third stage, ATP formation goes into high gear. This stage uses electron transfer chains and ATP synthases, both located in the inner membrane that divides the mitochondrion into two compartments (Figure 8.7). Both interact with electrons and hydrogen, delivered by coenzymes from the first two stages of the aerobic pathway.

Electron Transfer Phosphorylation

When electrons enter the transfer chains, hydrogen is released and so becomes ionized (H^+). As the electrons pass through the chains, hydrogen ions inside the inner compartment are shuttled to the outer compartment. These repeated shuttlings set up H^+ concentration and electric gradients across the inner membrane.

The only way ions can follow the gradient and flow back to the inner compartment is through the interior of ATP synthases (Figure 8.7). Their flow through these transport proteins drives the formation of ATP from ADP and unbound phosphate. Free oxygen helps keep the transport chains clear for operation. It withdraws spent electrons at the end of the chains and combines with H^+ to form water, a by-product.

Summary of the Energy Harvest

Thirty-two ATP typically form during the third stage of aerobic respiration. When you add these to the yield from the preceding stages, the net harvest is thirty-six ATP from one glucose molecule (Figure 8.8). That's a lot! Anaerobic pathways may use up eighteen glucose molecules to get the same net yield.

Thirty-six ATP is a typical yield only. The amount depends on the type of cell and prevailing conditions. For instance, the yield is lower when an intermediate is pulled away from the reactions and used elsewhere.

Also, NADH from the cytoplasm can't even enter a mitochondrion. It only delivers electrons and hydrogen to it; transport proteins in the membrane shuttle these across. NAD^+ or FAD already inside accepts them and delivers the electrons to transport systems. But FAD drops them off at a *lower* step in a transport system "staircase," so its deliveries produce less ATP.

A final point. Glucose, remember, has more stored energy (in more covalent bonds) compared to products of its full breakdown (carbon dioxide, water). About 686 kilocalories of energy are released as each mole of

Figure 8.7 Electron transfer phosphorylation, the third and final stage of aerobic respiration. The reactions occur at electron transfer chains and at ATP synthases, a type of transport protein, in the inner mitochondrial membrane. Each electron transfer chain consists of specific enzymes, cytochromes, and other proteins. The membrane divides the mitochondrion's interior into two compartments.

NADH and $FADH_2$ give up electrons and hydrogen to the transfer chains. When electrons are transferred *through* the chains, unbound hydrogen (H^+) is shuttled across the membrane, to the outer compartment:

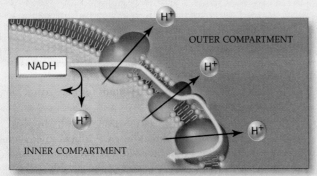

The H^+ concentration is now greater in the outer compartment. Concentration and electric gradients across the membrane have been set up. H^+ follows these gradients, through the interior of ATP synthases. Energy released by the flow drives the formation of ATP from ADP and unbound phosphate (P_i). Hence the name, electron transfer *phosphorylation*:

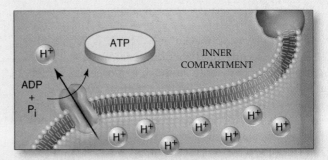

Do these events sound familiar? They should. ATP forms in much the same way in chloroplasts. By the *chemiosmotic* model, H^+ concentration and electric gradients across a cell membrane drive ATP formation. In this case, H^+ flows in the opposite direction compared to the flow in chloroplasts.

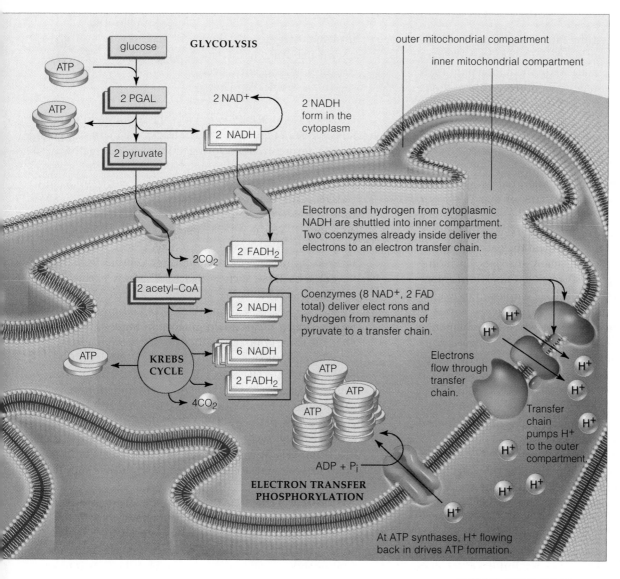

GLYCOLYSIS

glucose

ATP

ATP

2 PGAL

2 pyruvate

2 NAD⁺

2 NADH form in the cytoplasm

2 NADH

outer mitochondrial compartment

inner mitochondrial compartment

2CO₂

2 FADH₂

2 acetyl–CoA

Electrons and hydrogen from cytoplasmic NADH are shuttled into inner compartment. Two coenzymes already inside deliver the electrons to an electron transfer chain.

2 NADH

Coenzymes (8 NAD⁺, 2 FAD total) deliver electrons and hydrogen from remnants of pyruvate to a transfer chain.

ATP

KREBS CYCLE

6 NADH

2 FADH₂

4CO₂

Electrons flow through transfer chain.

ATP
ATP
ATP

ADP + Pᵢ

H⁺ H⁺ H⁺ H⁺ H⁺ H⁺ H⁺ H⁺

Transfer chain pumps H⁺ to the outer compartment.

ELECTRON TRANSFER PHOSPHORYLATION

At ATP synthases, H⁺ flowing back in drives ATP formation.

2 ATP

a In glycolysis, 2 ATP used; 4 ATP form by *substrate-level phosphorylation.* So *net* yield is 2 ATP.

2 ATP

b In Krebs cycle of second stage, 2 ATP form by *substrate-level phosphorylation.*

4 ATP

c In third stage, NADH from glycolysis used to form 4 ATP by *electron transfer phosphorylation.*

28 ATP

d In third stage, NADH and FADH₂ from second stage used to make 28 ATP by *electron transfer phosphorylation.*

TYPICAL NET ENERGY YIELD   36 ATP

Figure 8.8 Summary of the net energy harvest from aerobic respiration. Thirty-six ATP per glucose molecule is common. The actual yield varies. Shifting concentrations of reactants, intermediates, and end products affect it. So do membrane crossing mechanisms, which vary among different cell types.

Cells differ in how they use the NADH from glycolysis, which can't enter mitochondria. These NADH give up electrons and hydrogen to transport proteins in the outer mitochondrial membrane, which shuttle them across. NAD⁺ or FAD already inside accepts them, forming NADH or FADH₂.

Any NADH inside a mitochondrion delivers electrons to the highest entry point into a transfer chain. When it does, enough H⁺ is pumped across the inner membrane to make *three* ATP. FADH₂ delivers them to a lower entry point. Fewer hydrogen ions are pumped, so only *two* ATP can form.

In liver, heart, and kidney cells, for example, electrons and hydrogen from glycolysis enter the highest entry point of transfer chains, so the energy harvest is thirty-eight ATP. More commonly, as in skeletal muscle and brain cells, they are transferred to FAD, so the harvest is thirty-six ATP.

glucose is degraded to these more stable end products. (Section 2.7, *Critical Thinking* question 7, gives a simple definition of molar weight.) Much of the freed energy escapes as heat, but about 39 percent is stored in ATP.

In the final stage of the aerobic pathway, many coenzymes deliver electrons to transfer chains in a mitochondrion's inner membrane. The membrane forms two compartments.

Coenzymes give up hydrogen and electrons to the transfer chains. Electrons passing through these chains cause the unbound hydrogen (H⁺) to be shuttled into the inner compartment. The resulting H⁺ concentration and electric gradients drive ATP formation as H⁺ flows back across the membrane. Oxygen is the final electron acceptor.

Again, from start (glycolysis in the cytoplasm) to finish (in mitochondria), the pathway commonly has a net yield of thirty-six ATP for every glucose molecule metabolized.

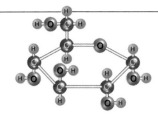

8.5

ANAEROBIC ROUTES OF ATP FORMATION

So far, we have tracked the fate of a glucose molecule through the pathway of aerobic respiration. We turn now to its use as a substrate for fermentation pathways. Remember, these are anaerobic pathways; they do *not* use oxygen as the final acceptor of the electrons that ultimately drive the ATP-forming machinery.

Fermentation Pathways

Diverse organisms use fermentation pathways. Many are prokaryotic cells and protistans of marshes, bogs, mud, deep-sea sediments, the animal gut, canned food, sewage treatment ponds, and other oxygen-free places. Some fermenters actually will die if they are exposed to oxygen. Bacteria that cause many diseases, including botulism and tetanus, are like this. Other fermenters, including the acidophilus bacteria employed by yogurt manufacturers, are indifferent to oxygen's presence. Still others use oxygen, but they switch to fermentation when oxygen becomes scarce.

Glycolysis serves as the first stage of fermentation pathways, as it does for aerobic respiration. Glycolysis also requires enzymes that catalyze the breakdown of glucose and rearrangement of the fragments into two pyruvate molecules. Here again, two NADH form, and the net energy yield is two ATP.

However, fermentation reactions do not completely break down glucose to carbon dioxide and water. Also, they produce no more ATP beyond the tiny yield from glycolysis. *Fermentation's final steps simply regenerate NAD+, the coenzyme that assists the breakdown reactions.*

Fermentation yields enough energy to sustain many single-celled anaerobic organisms. It even helps some aerobic cells through times of stress. But it isn't enough to sustain large, multicelled organisms. And that is one reason why you will never see an anaerobic elephant.

LACTATE FERMENTATION   With these points in mind, take a look at Figure 8.9, which tracks the main steps of **lactate fermentation**. During this anaerobic pathway, the pyruvate molecules from glycolysis—the first stage of its reactions—accept hydrogen and electrons from NADH. This transfer regenerates NAD+. At the same time, it converts each pyruvate to lactate, a three-carbon compound. You may hear people call this compound lactic acid, which is the non-ionized form. Lactate, its ionized form, is far more common in cellular fluids.

Some bacteria, such as *Lactobacillus*, rely exclusively on this anaerobic pathway. Their activities often spoil food. Even so, some fermenters have commercial uses, as when they break down glucose in huge vats where cheeses, sauerkraut, and other products are made.

Also, in humans, rabbits, and many other animals, some types of cells use this anaerobic pathway for a quick fix of ATP. When your own demands for energy are intense but brief—say, during a short race—muscle cells use this pathway. They can't do so for long; they would throw away too much of glucose's energy for too little ATP. Muscles fatigue and lose their ability to contract when these cells deplete their glucose supply.

ALCOHOLIC FERMENTATION   In the anaerobic route called **alcoholic fermentation**, enzymes convert each pyruvate molecule from glycolysis to an intermediate form: acetaldehyde. The NADH transfers electrons and hydrogen to acetaldehyde and thereby converts it to an alcoholic end product—ethanol (Figure 8.10).

Some single-celled fungi called yeasts are famous for their use of this pathway. One type, *Saccharomyces cerevisiae*, makes bread dough rise. Bakers mix the yeast with sugar, then blend both into dough. Fermenting yeast cells release carbon dioxide. Bubbles of this gas expand the dough (make it rise). Oven heat forces the gas out of the dough, and a porous product remains.

Beer and wine producers use yeasts on a large scale. Vintners use both wild yeasts and cultivated strains of *S. ellipsoideus*. Both yeasts are active until the alcohol concentration in wine vats exceeds 14 percent. (Wild yeasts die when the concentration exceeds 4 percent.) That is why some birds can get drunk on naturally fermented berries. That is why landscapers don't plant prodigious berry-producing shrubbery near highways;

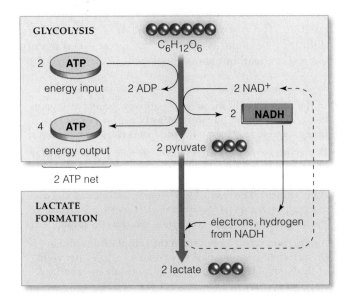

Figure 8.9   Lactate fermentation. In this anaerobic pathway, electrons end up in lactate, the reaction product.

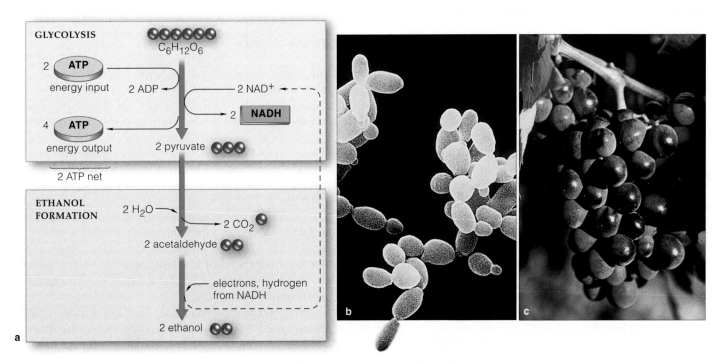

GLYCOLYSIS

$C_6H_{12}O_6$

2 ATP — energy input

2 ADP

2 NAD^+

2 **NADH**

4 ATP — energy output

2 ATP net

2 pyruvate

ETHANOL FORMATION

2 H_2O

2 CO_2

2 acetaldehyde

electrons, hydrogen from NADH

2 ethanol

a

Figure 8.10 (**a**) Alcoholic fermentation, one of the anaerobic pathways. Acetaldehyde, an intermediate of the reactions, is the final acceptor of electrons. Ethanol is the end product. Yeasts, single-celled organisms, use this pathway. (**b**) Activity of one species of *Saccharomyces* makes bread dough rise. (**c**) Another species lives on sugar-rich tissues of ripe grapes.

drunk birds doodle into windshields (Figure 8.11). Wild turkeys similarly have been known to get tipsy when they gobble fermenting apples in untended orchards.

Anaerobic Electron Transfers

Especially among prokaryotic cells, we see less common energy-releasing pathways, some of which are topics of later chapters in this book. For example, many species have key roles in the global cycling of sulfur, nitrogen, and other crucial elements. Collectively, their metabolic activities influence nutrient availability for organisms in ecosystems everywhere.

Some archaebacteria and eubacteria use **anaerobic electron transfers**. Electrons from organic compounds flow through transfer chains in the plasma membrane and H^+ flows out through ATP synthases to form ATP. Some inorganic compound often is the final electron acceptor. The net energy yield is always small.

Even as you read this, some anaerobic species that live in waterlogged soil are stripping electrons from a variety of compounds. They dump electrons on sulfate. Hydrogen sulfide, a putrid-smelling gas, is the result. Sulfate-reducing species also thrive in aquatic habitats that are enriched with decomposed organic material. They even live on the ocean floor, near hydrothermal vents. As described in Section 49.11, they are part of some unique communities.

Figure 8.11 Robin feasting on the fermented berries of a pyracantha bush.

In fermentation pathways, an organic substance that forms during the reactions serves as the final acceptor of electrons from glycolysis. The reactions regenerate NAD^+, which is required to keep the pathway operational.

By anaerobic electron transfers, an inorganic substance (but not oxygen) usually serves as the final electron acceptor.

Fermentation pathways typically have a net energy yield of two ATP for each glucose molecule metabolized. The net yield is similarly small for anaerobic electron transfers.

ALTERNATIVE ENERGY SOURCES IN THE HUMAN BODY

So far, you have looked at what happens after a lone glucose molecule enters an energy-releasing pathway. Now you can start thinking about what cells do when they have too many or too few glucose molecules.

Carbohydrate Breakdown in Perspective

THE FATE OF GLUCOSE AT MEALTIME   While you or any other mammal is eating, glucose and other small organic molecules are being absorbed across the gut lining, then blood transports them through the body. A rise in the glucose level in blood prompts the pancreas to release insulin. This hormone stimulates cells to take up glucose faster. Cells convert the incoming glucose to glucose–6–phosphate and trap it inside the cytoplasm. (When phosphorylated, glucose cannot be transported back out, across the plasma membrane.) Look again at Figure 8.4, and you see that glucose–6–phosphate is the first activated intermediate of glycolysis.

If your glucose intake exceeds cellular demands for energy, ATP-producing machinery goes into high gear. Unless a cell is rapidly using its ATP, the cytoplasmic concentration of ATP can rise to high levels. When that happens, the glucose–6–phosphate gets diverted into a biosynthesis pathway. This pathway assembles glucose units into glycogen, a storage polysaccharide (Section 3.4). The pathway is especially favored in muscle and liver cells, which maintain the largest glycogen stores.

THE FATE OF GLUCOSE BETWEEN MEALS   When you are not eating, glucose is not entering the bloodstream, and its level in the blood declines. If the decline were not countered, that would be bad news for the brain, your body's glucose hog. At any time, your brain is taking up more than two-thirds of the freely circulating glucose, because its many hundreds of millions of cells use this sugar alone as their preferred energy source.

The pancreas responds to the decline by secreting glucagon. This hormone prompts liver cells to convert glycogen back to glucose and send it back to the blood. Only liver cells do this; muscle cells won't give it up. The blood glucose level rises, and brain cells keep on functioning. Thus, *hormones control whether your body's cells use free glucose as an energy source or tuck it away.*

A word of caution: Don't let the preceding examples lead you to believe cells squirrel away large amounts of glycogen. In adult humans, glycogen makes up merely 1 percent or so of the body's total energy reserves, the energy equivalent of two cups of cooked pasta. Unless you eat on a regular basis, you will deplete the liver's small glycogen stores in less than twelve hours.

Of the total energy reserves in, say, a typical adult who eats well, 78 percent (about 10,000 kilocalories) is concentrated in body fat and 21 percent in proteins.

Energy From Fats

How does the body access its huge reservoir of fats? A fat molecule, recall, has a glycerol head and one, two, or three fatty acid tails. Most fats get stored in your body as triglycerides, with three tails each. Triglycerides accumulate in fat cells of adipose tissue, which forms at buttocks and other strategic places beneath the skin.

When blood glucose levels decline, triglycerides are tapped as an energy alternative. Enzymes in fat cells cleave the bonds between the glycerol and fatty acids, which enter the blood. Enzymes in the liver convert the glycerol to PGAL, an intermediate of glycolysis. Nearly all cells take up these circulating fatty acids. Enzymes cleave the backbone of the fatty acids. The fragments are converted to acetyl–CoA, which can enter the Krebs cycle (Figures 8.6 and 8.12).

Compared to glucose, a fatty acid tail has far more carbon-bound hydrogen atoms, so it yields far more ATP. Between meals or during sustained exercise, fatty acid conversions supply about one half of the ATP that muscle, liver, and kidney cells require.

What happens if you eat too many carbohydrates? Exceed the glycogen-storing capacity of your liver and muscle cells, and the excess gets converted to fats. *Too much glucose ends up as excess fat.* In the United States, 25 percent of the people have a combination of genes that lets them eat as much as they like without gaining weight. A diet too rich in carbohydrates keeps the other 75 percent fat. Their insulin levels stay elevated, which "tells" the body to store fat, not use it for energy. We return to this topic in Section 36.7 and Chapter 41.

Energy From Proteins

Eat more proteins than your body requires to grow and maintain itself, and its cells won't store them. Enzymes split dietary proteins into amino acid units. Then they remove the amino group ($-NH_3^+$) from each unit, and ammonia (NH_3) forms. What happens to the leftover carbon backbones? The outcome varies, depending on conditions of the moment. Maybe they'll get converted to fats or carbohydrates. Or maybe they'll enter the Krebs cycle (Figure 8.12), where coenzymes pick up the hydrogen and electrons stripped from carbon atoms. The ammonia that forms undergoes conversions and becomes urea. This nitrogen-containing waste product would be toxic if it accumulated to high concentrations. Normally your body excretes urea, in urine.

As this brief discussion makes clear, maintaining and accessing the body's energy reserves is complicated business. Hormonal controls over the disposition of glucose are special only because glucose is the fuel of choice for the all-important brain. However, as you will

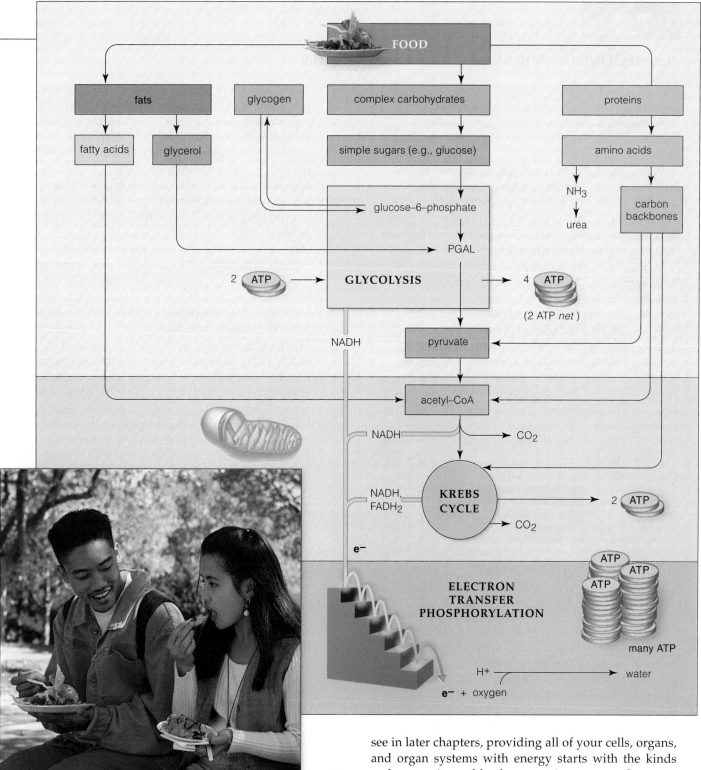

FOOD

fats → fatty acids, glycerol

glycogen

complex carbohydrates → simple sugars (e.g., glucose)

proteins → amino acids → NH₃ → urea; carbon backbones

glucose–6–phosphate

PGAL

2 ATP → **GLYCOLYSIS** → 4 ATP (2 ATP *net*)

NADH

pyruvate

acetyl–CoA

NADH → CO₂

NADH, FADH₂

KREBS CYCLE → 2 ATP

CO₂

e⁻

ELECTRON TRANSFER PHOSPHORYLATION

ATP ATP ATP ATP

many ATP

H+ → water

e⁻ + oxygen

Figure 8.12 Alternative energy sources in the human body. The diagram shows reaction sites where a variety of organic compounds can enter the stages of aerobic respiration.

Complex carbohydrates, fats, and proteins cannot enter the aerobic pathway directly. In humans and other mammals, the digestive system, and individual cells, must first break apart these molecules into simpler, degradable subunits.

see in later chapters, providing all of your cells, organs, and organ systems with energy starts with the kinds and proportions of food you put in your mouth.

This concludes our look at aerobic respiration and other energy-releasing pathways. The section to follow may help you get a sense of how they fit into the larger picture of life's evolution and interconnectedness.

In humans and other mammals, the entrance of glucose or other organic compounds into an energy-releasing pathway depends on the kinds and proportions of carbohydrates, fats, and proteins in the diet as well as on the type of cell.

PERSPECTIVE ON THE MOLECULAR UNITY OF LIFE

In this unit you read about photosynthesis and aerobic respiration—the main pathways by which cells trap, store, and release energy. What you may not know is that the two pathways became linked, on a grand scale, over evolutionary time.

When life originated long ago, the atmosphere had little free oxygen. The first single-celled organisms probably made ATP by reactions similar to glycolysis, so fermentation pathways probably dominated. More than a billion years passed before the oxygen-evolving pathway of photosynthesis emerged.

Gradually, oxygen accumulated in the atmosphere. Some cells could use it to accept electrons, perhaps as a chance outcome of mutated proteins in their electron transfer chains. In time, certain descendants of those early aerobic cells abandoned photosynthesis entirely. Among them were the forerunners of animals and all other organisms that engage in aerobic respiration.

With aerobic respiration, a flow of carbon, hydrogen, and oxygen through the metabolic pathways of living organisms came full circle. For the final products of this aerobic pathway—carbon dioxide and water—are the same materials necessary to build organic compounds in photosynthesis:

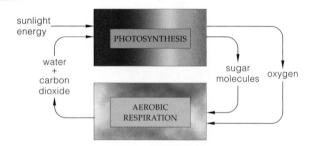

Perhaps you have difficulty seeing the connection between yourself—a highly intelligent being—and such remote-sounding events as energy flow and the cycling of carbon, hydrogen, and oxygen. Is this really the stuff of humanity?

Think back, for a moment, on the structure of a water molecule. Two hydrogen atoms sharing electrons with oxygen may not seem close to your daily life. Yet, through that sharing, water molecules show polarity and hydrogen-bond with one another. Their chemical behavior is a beginning for the organization of lifeless matter that leads to the organization of all living things.

For now you can visualize other diverse molecules interspersed through water. The nonpolar kinds resist interaction with water; polar kinds dissolve in it. On their own, the phospholipids among them assemble into a two-layered film. Such lipid bilayers, recall, are the framework of cell membranes, hence all cells. From the beginning, the cell has been the basic *living* unit.

The essence of life is not some mysterious force. It is molecular organization and metabolic control. With a membrane to contain them, reactions *can* be controlled. With mechanisms built into their membranes, cells respond to energy changes and shifting concentrations of solutes in the environment. Response mechanisms operate by "telling" proteins—enzymes—when and what to build or tear down.

And it is not some mysterious force that creates the proteins. DNA, the double-stranded treasurehouse of inheritance, has the chemical structure—*the chemical message*—that allows molecule to reproduce molecule, one generation after the next. In your own body, DNA strands tell trillions of cells how countless molecules must be built or torn apart for their stored energy.

So yes, carbon, hydrogen, oxygen, and other atoms of organic molecules are the stuff of you, and us, and all of life. But it takes more than molecules to complete the picture. Life continues as long as an unbroken flow of energy sustains its grand organization. Molecules are assembled into cells, cells into organisms, organisms into communities, and so on up through the biosphere. It takes energy inputs from the sun to maintain these levels of organization. And energy flows through time in one direction—from organized to less organized forms. Only as long as energy continues to flow into the web of life can life continue in all its rich expressions.

And so life is no more *and no less* than a marvelously complex system for prolonging order. Sustained by energy transfusions from the sun, life continues by its capacity for self-reproduction. With the hereditary instructions contained in DNA, energy and materials are organized, generation after generation. Even with the death of individuals, life elsewhere is prolonged. With each death, molecules are released and may be cycled once more, as raw materials for new generations.

With this flow of energy and cycling of material through time, each birth is affirmation of our ongoing capacity for organization, each death a renewal.

The diversity of life, and its continuity through time, arises from the unity of life at the molecular level.

SUMMARY

1. Phosphate-group transfers from ATP are central to metabolism. Autotrophic cells alone tap energy from the environment to make ATP for building carbohydrates. *All* cells make ATP by pathways that release chemical energy from organic compounds, such as glucose. *8.1*

2. After glucose enters an energy-releasing pathway, enzymes derive electrons and hydrogen from reaction intermediates formed along the way. Coenzymes pick these up and deliver them to other reaction sites, where the pathway ends. NAD^+ is the main coenzyme; FAD is also used in the aerobic route. Reduced forms of these coenzymes are designated NADH and $FADH_2$. *8.1–8.2*

3. All of the main energy-releasing pathways start with glycolysis, which begins and ends in the cytoplasm and can proceed in the presence or absence of oxygen. *8.2*

   a. In glycolysis, enzymes break down glucose to two pyruvate molecules. Two NADH and four ATP form.

   b. The net energy yield is two ATP (because two ATP had to be invested up front to get the reactions going).

4. Aerobic respiration continues on through two more stages: (1) the Krebs cycle and a few steps preceding it, and (2) electron transfer phosphorylation. These stages occur only in mitochondria of eukaryotic cells. *8.2, 8.3*

5. The second stage of the aerobic pathway starts when an enzyme strips a carbon atom from each pyruvate. Coenzyme A binds the two-carbon fragment, forming acetyl–CoA, and transfers this to oxaloacetate, the entry point of the Krebs cycle. These cyclic reactions and a few steps before them load ten coenzymes with electrons and hydrogen (eight NADH, plus two $FADH_2$). Two ATP form. Three carbon dioxide molecules are released for each pyruvate that entered this second stage. *8.3*

6. The third stage of the aerobic pathway proceeds at a membrane dividing the mitochondrion's interior into two compartments. Electron transfer chains and ATP synthases are embedded in this inner membrane. *8.4*

   a. The chains accept electrons and hydrogen from coenzymes from the first two stages, and establish H^+ concentration and electric gradients.

   b. H^+ follows the gradients and flows from the outer to the inner compartment. It does so through the interior of ATP synthases. Energy released by the ion flow drives formation of ATP from ADP and unbound phosphate.

   c. Oxygen picks up the spent electrons at the end of the transfer chain and combines with hydrogen ions to form water. That is, oxygen is the final acceptor of the electrons that initially resided in glucose.

7. Aerobic respiration has a typical net energy yield of thirty-six ATP for each glucose molecule metabolized. Yields vary by cell type and cellular conditions. *8.4*

8. Fermentation and anaerobic electron transport start with glycolysis but are anaerobic, start to finish. *8.5*

   a. Lactate fermentation has a net energy yield of two ATP, which form in glycolysis. The remaining reaction regenerates NAD^+. The two NADH from glycolysis transfer electrons and hydrogen to two pyruvate from glycolysis. Two lactate molecules are the products.

   b. Alcoholic fermentation has a net energy yield of two ATP from glycolysis, and its remaining reactions serve to regenerate NAD^+. Enzymes convert pyruvate from glycolysis to acetaldehyde, and carbon dioxide is released. NADH from glycolysis transfers electrons and hydrogen to the two acetaldehyde molecules, thus forming two ethanol molecules, the products.

   c. Certain bacteria use anaerobic electron transfer. Electrons stripped from some organic compound flow through transfer chains in the plasma membrane, and H^+ flows out through ATP synthases to form ATP. An inorganic compound in the environment is often the final electron acceptor.

9. In humans and other mammals, simple sugars such as glucose from carbohydrates, glycerol and fatty acids from fats, and carbon backbones of amino acids from proteins can enter ATP-producing pathways. *8.6*

10. Life shows great unity at the molecular level. *8.7*

Review Questions

1. Is this true or false: Aerobic respiration occurs in animals but not plants, which make ATP only by photosynthesis. *8.1*

2. Using the diagram below of the aerobic pathway, fill in the blanks with the number of molecules of pyruvate, coenzymes, and end products. Write in the net ATP formed in each stage, then the net ATP formed from start (glycolysis) to finish. *8.1*

3. Is glycolysis energy-*requiring* or energy-*releasing*? Or do both kinds of reactions occur during glycolysis? *8.2*

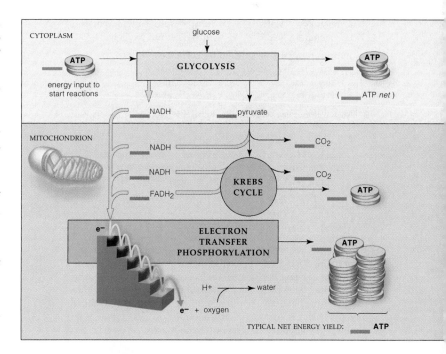

4. In what respect does *electron transfer* phosphorylation differ from *substrate-level* phosphorylation? 8.1, 8.2

5. Sketch the double-membrane system of the mitochondrion and show where electron transfer chains and ATP synthases are located. 8.3

6. Name the compound that is the entry point for the Krebs cycle, and state whether it directly accepts the pyruvate from glycolysis. For each glucose molecule, how many carbon atoms enter the Krebs cycle? How many depart from it, and in what molecular form? 8.3

7. Is this statement true or false: Muscle cells cannot contract at all when deprived of oxygen. If true, explain why. If false, name the alternative(s) available to them. 8.5

Self-Quiz ANSWERS IN APPENDIX III

1. Glycolysis starts and ends in the _____ .
   a. nucleus   c. plasma membrane
   b. mitochondrion   d. cytoplasm

2. Which of the following does *not* form during glycolysis?
   a. NADH   c. $FADH_2$
   b. pyruvate   d. ATP

3. Aerobic respiration is completed in the _____ .
   a. nucleus   c. plasma membrane
   b. mitochondrion   d. cytoplasm

4. In the last stage of aerobic respiration, _____ is the final acceptor of electrons that originally resided in glucose.
   a. water   c. oxygen
   b. hydrogen   d. NADH

5. _____ engage in lactate fermentation.
   a. *Lactobacillus* cells   c. Sulfate-reducing bacteria
   b. Muscle cells   d. a and b

6. In alcoholic fermentation, _____ is the final acceptor of the electrons stripped from glucose.
   a. oxygen   c. acetaldehyde
   b. pyruvate   d. sulfate

7. The fermentation pathways produce no more ATP beyond the small yield from glycolysis, but the remaining reactions _____ .

   a. regenerate ADP   c. dump electrons on an inorganic
   b. regenerate NAD^+ substance (not oxygen)

8. In certain organisms and under certain conditions, _____ can be used as an energy alternative to glucose.
   a. fatty acids   c. amino acids
   b. glycerol   d. all of the above

9. Match the event with its most suitable metabolic description.
   ____ glycolysis   a. ATP, NADH, $FADH_2$, CO_2,
   ____ fermentation   and water form
   ____ Krebs cycle   b. glucose to two pyruvate
   ____ electron transfer   c. NAD^+ regenerated, two ATP net
        phosphorylation   d. H^+ flows through ATP synthases

Critical Thinking

1. Diana suspects that a visit to her family doctor is in order. After eating carbohydrate-rich food, she always experiences sensations of being intoxicated and becomes nearly incapacitated, as if she had been drinking alcohol. She even wakes up with a hangover the next day. Having completed a course in freshman biology, Diana has an idea that something is affecting the way her body is metabolizing glucose. Explain why.

2. The cells of your body absolutely do not use nucleic acids as alternative energy sources. Suggest why.

3. The human body's energy needs and its programs for growth depend on balancing the levels of amino acids in blood with proteins in cells. Cells of the liver, kidneys, and intestinal lining are especially important in this balancing act. When the levels of amino acids in blood decline, lysozymes in cells can rapidly digest some of their proteins (structural and contractile proteins are spared, except in cases of malnutrition). The amino acids released this way enter the blood and thereby help maintain the required levels.

Suppose you embark on a body-building program. You already eat plenty of carbohydrates. However, a nutritionist recommends that you follow a protein-rich diet that includes protein supplements. Speculate on how extra dietary proteins will be put to use, and in which tissues.

4. Each year, Canada geese (*above*) lift off in precise formation from their northern breeding grounds. They head south to spend the winter months in warmer climates, then make the return trip in spring. As is the case for other migratory birds, their flight muscle cells are efficient at using fatty acids as an energy source. (Remember, the carbon backbone of fatty acids can be cleaved into small fragments that can be converted to acetyl–CoA for entry into the Krebs cycle.)

Suppose a lesser Canada goose from Alaska's Point Barrow has been steadily flapping along for three thousand kilometers and is nearing Klamath Falls, Oregon. It looks down and notices a rabbit sprinting like the wind from a coyote with a taste for rabbit. With a stunning burst of speed, the rabbit reaches the safety of its burrow.

Which energy-releasing pathway predominated in muscle cells in the rabbit's legs? Why was the Canada goose relying on a different pathway for most of its journey? And why wouldn't the pathway of choice in goose flight muscle cells be much good for a rabbit making a mad dash from its enemy?

5. Reflect on this chapter's introduction and question 4 above. Which energy-releasing pathway is predominating in agitated Africanized bees chasing a farmer through a cornfield?

Selected Key Terms

acetyl–CoA *8.3*   Krebs cycle *8.1*
aerobic respiration *8.1*   lactate fermentation *8.5*
alcoholic fermentation *8.5*   mitochondrion *8.3*
anaerobic electron transfer *8.5* NAD^+ *8.1*
electron transfer   oxaloacetate *8.3*
   phosphorylation *8.1*   pyruvate *8.1*
FAD *8.1*   substrate-level
glycolysis *8.1*   phosphorylation *8.2*

Readings

DeMauro, S. October 1998. "Hooked on Mitochondria." *Quest.* Volume 5, Number 5. A brief look at mitochondrial disorders.

Wolfe, S. 1995. *An Introduction to Molecular and Cellular Biology.* Belmont, California: Wadsworth. Exceptional reference text.

On-Line readings at Student Guide for InfoTrac:
www.brookscole.com/biology

II Principles of Inheritance

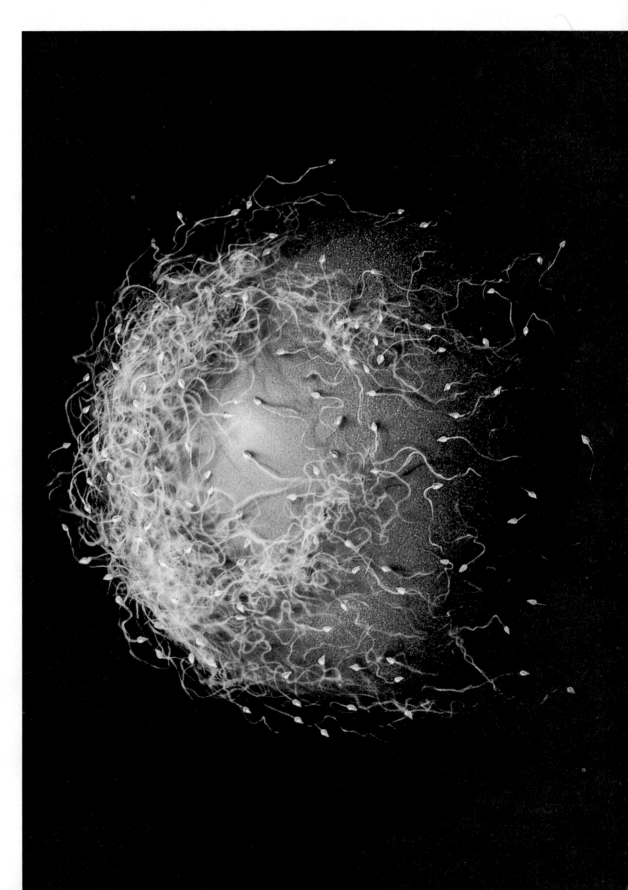

Human sperm, one of which will penetrate this mature egg and so set the stage for the development of a new individual in the image of its parents. This exquisite art is based on a scanning electron micrograph.

9

CELL DIVISION AND MITOSIS

From Cell to Silver Salmon

Five o'clock, and the first rays from the sun dance over the wild Alagnak River of the Alaskan tundra. It is September, and life is ending and beginning in the cold waters. Thousands of mature silvers—coho salmon— have returned from the open ocean to spawn in their shallow home waters. The females are tinged with red, the color of spawners, and they are dying.

This morning you watch a female salmon releasing translucent eggs into a shallow depression that her fins made in the gravel riverbed (Figure 9.1). A male releases a cloud of sperm, which fertilize the eggs. Trout and other predators eat most of the eggs, but some survive and give rise to a new generation.

Within three years the pea-sized eggs have become sexually mature salmon, fashioned from billions of cells. In time, on some September morning, some of their cells will serve as sperm or eggs and take part in an ongoing story of birth, growth, death, and rebirth.

Your body, too, emerged through *cell divisions*, as it did for salmon and all other multicelled species. Inside your mother, a fertilized egg divided in two, then the two became four, and so on until billions of cells were growing, developing in specialized ways, and dividing at different times to form different parts. Your body now has more than 65 trillion living cells. Many are still dividing.

Understanding cell division—and, ultimately, how new individuals are put together in the image of their parents—begins with answers to three questions. *First,* what instructions are necessary for inheritance? *Second,* how are those instructions duplicated for distribution into daughter cells? *Third,* by what mechanisms are the duplicated instructions parceled out to daughter cells? We'll need more than one chapter to consider the nature of cell reproduction and other mechanisms of inheritance. Even so, the points made early in this chapter can help you keep the overall picture in focus.

Begin with the word **reproduction**. In biology, this means that parents produce a new generation of cells or multicelled individuals like themselves. The process starts in cells programmed to divide. The ground rule for division is this: *Parent cells must give their daughter*

sexually mature
female salmon

Figure 9.1 The last of one generation and the first of the next in Alaska's Alagnak River— from spawning coho salmon (*Onchorhynchus kisutch*) to eggs, to fingerlings.

cells specific hereditary instructions, encoded in DNA, and enough metabolic machinery to start up their own operation.

DNA, recall, contains instructions for synthesizing proteins. Some proteins are structural materials. Many are enzymes that speed the assembly of specific organic compounds, such as the lipids that cells use as building blocks and sources of energy. Unless a daughter cell receives the necessary instructions for making proteins, it simply will not be able to grow or function properly.

Also, the cytoplasm of a parent cell already contains enzymes, organelles, and other operating machinery. When a daughter cell inherits what looks like a blob of cytoplasm, it really is getting start-up machinery—which will keep that cell operating until it can use its inherited DNA for growing and developing on its own.

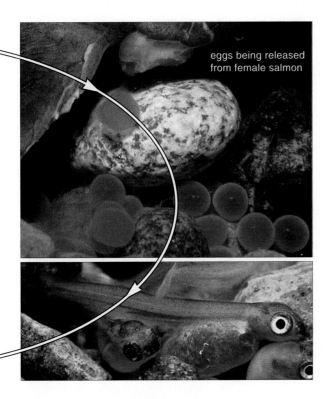

eggs being released from female salmon

Fingerlings—young fishes growing, from fertilized eggs, by way of mitotic cell divisions

Key Concepts

1. The continuity of life depends on reproduction. By this process, parents produce a new generation of cells or multicelled individuals like themselves. Cell division is the bridge between generations.

2. When a cell divides, its two daughter cells must each receive a required number of DNA molecules and some cytoplasm. For eukaryotic cells, a division mechanism called mitosis sorts out the DNA into two new nuclei. A separate mechanism divides the cytoplasm.

3. The cell cycle starts each time a daughter cell forms. It ends when the cell completes its own division. One turn of the cycle proceeds through interphase, mitosis, and cytoplasmic division. A cell spends most of its life in interphase. That is when its mass and the number of its components increase, and that is when the DNA is duplicated.

4. Proteins with structural and functional roles are attached to eukaryotic DNA molecules. Each DNA molecule with its attached proteins is a chromosome.

5. Members of a species have a characteristic number of chromosomes in cells. Their chromosomes differ from one another in length and shape, and they carry different portions of the hereditary instructions.

6. Body cells of many organisms have a diploid chromosome number; they contain two of each type of chromosome characteristic of the species.

7. Mitosis keeps the chromosome number constant from one cell generation to the next. So if a parent cell is diploid, its daughter cells also will be diploid.

8. Mitotic cell division is the basis of growth and tissue repair in multicelled eukaryotes. It also is the means by which single-celled eukaryotes and many multicelled eukaryotes reproduce asexually.

DIVIDING CELLS: THE BRIDGE BETWEEN GENERATIONS

Overview of Division Mechanisms

Hereditary instructions of plants, animals, and all other eukaryotic organisms are distributed among a number of DNA molecules. Before the cells of such organisms reproduce, they must undergo *nuclear* division. **Mitosis** and **meiosis** are two nuclear division mechanisms. Both sort out and then package a parent cell's DNA into new nuclei for their forthcoming daughter cells. A separate mechanism splits the cytoplasm into daughter cells.

Multicelled organisms grow, replace dead or worn-out cells, and repair tissues by way of mitosis and the cytoplasmic division of body cells. We call the cells that make up the body **somatic cells**. Also, many protistans, fungi, plants, and certain animals reproduce asexually by mitotic cell division (Table 9.1).

By contrast, meiosis occurs only in **germ cells**, a cell lineage set aside for the formation of gametes (such as sperm and eggs) and sexual reproduction. As you will read in the next chapter, meiosis has much in common with mitosis, but the end result is different.

What about the prokaryotic cells—the archaebacteria and eubacteria? They reproduce asexually by an entirely different mechanism called prokaryotic fission. We will consider prokaryotic fission later, in Section 21.2.

Some Key Points About Chromosomes

Remember the introduction to chromosomes in Section 4.6? You read that a eukaryotic DNA molecule, together with proteins attached to it, is a **chromosome**. Before a cell enters nuclear division, it duplicates every one of its chromosomes. Each chromosome and its copy stay attached to each other until late in mitosis, as **sister chromatids**. Figure 9.2 is a simple way to think about unduplicated and duplicated chromosomes.

Except in some laboratories, you'll never see naked eukaryotic DNA. Its chromosomal proteins are always bound tightly to it. This is so even when chromosomes are stretched out in threadlike form in a nondividing cell. You can actually see these proteins at extreme magnification; they look like beads on a string (Figure 9.3). The chromosomal proteins are **histones**. Some types of histones are like a spool for DNA. A bit of the DNA molecule winds twice around each one, as Figure 9.3*d* indicates. Each histone–DNA spool is one unit of organization called a **nucleosome**. Nucleosomes can be loosened up in controlled ways to give enzymes access to specific DNA regions (Section 15.1).

Early in mitosis (and in meiosis), the proteins and DNA interact in ways that make the chromosome coil back on itself repeatedly, into a tightly condensed form (Figure 9.3). What's the point of so much condensation? It probably helps keep the chromosomes from getting tangled up when they are being moved and sorted into parcels for daughter cells.

As each type of duplicated chromosome condenses, a pronounced constriction appears in the same region along its length (Figure 9.3*a*). This constricted region is the **centromere**. Disk-shaped structures also show up here; they are docking sites for microtubules with roles in nuclear division. The centromere's location differs among the different types of chromosomes in a cell.

Mitosis and the Chromosome Number

Each species has a characteristic **chromosome number**, the sum total of chromosomes in cells of a given type. Human somatic cells have 46, those of gorillas have 48, and those of pea plants have 14.

Table 9.1	Cell Division Mechanisms
Mechanisms	**Functions**
MITOSIS, CYTOPLASMIC DIVISION	In all multicelled eukaryotes, the basis of (1) increases in body size during growth, (2) replacement of dead or worn-out cells, and (3) repair of damaged tissues. Also, the basis of *asexual* reproduction in single-celled and many multicelled eukaryotes.
MEIOSIS, CYTOPLASMIC DIVISION	In single-celled and multicelled eukaryotes, the basis of gamete formation and sexual reproduction.
PROKARYOTIC FISSION	In bacterial cells only, the basis of asexual reproduction.

a One unduplicated chromosome   **b** One chromosome (duplicated)

one chromatid — one chromatid — two sister chromatids

Figure 9.2 A simple way to visualize a chromosome in the unduplicated state and the duplicated state. Chromosomes are duplicated before nuclear division.

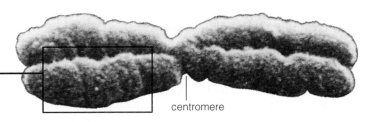

centromere

a A duplicated human chromosome at metaphase, when it is most condensed.

Figure 9.3 One model of levels of organization in a human chromosome at its most condensed form.

b At times when a chromosome is most condensed, the chromosomal proteins interact, which packages loops of already coiled DNA into a "supercoiled" array.

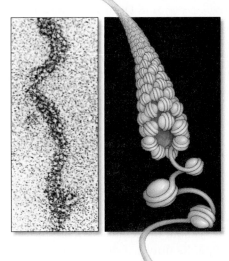

c At a deeper level of structural organization, the chromosomal proteins and DNA are organized as a cylindrical fiber.

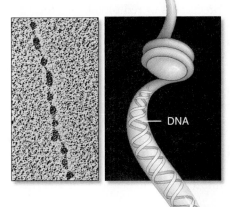

d Immerse a chromosome in saltwater and it loosens up to a beads-on-a-string organization. The "string" is one DNA molecule. Each "bead" is a nucleosome.

DNA

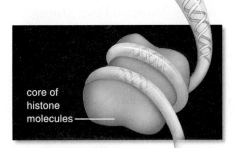

core of histone molecules

e A nucleosome consists of part of a DNA molecule looped twice around a core of histones.

Actually, your 46 chromosomes are like volumes of two sets of books. Each set is numbered, from 1 to 23. For example, you have two "volumes" of chromosome 22—that is, *a pair of them*. Except for certain pairings of sex chromosomes, both have the same length and shape. They carry the same portion of hereditary instructions for the same traits. Think of them as two sets of books on how to build a house. Your father gave you one set. Your mother had her own ideas about wiring, storage, plumbing, and so on; she gave you an alternate edition. Her set covers the same topics, but her instructions are slightly different for many of them.

We say the chromosome number is **diploid**, or 2*n*, if a cell has two of each type of chromosome characteristic of the species. The body cells of humans, gorillas, pea plants, and a great many other organisms are like this. By contrast, as Chapter 10 describes, eggs and sperm of such organisms have a *haploid* chromosome number (*n*). This means that they contain only one of each type of chromosome characteristic of the species.

With mitosis, a diploid parent cell can produce two diploid daughter cells. This doesn't mean each merely gets forty-six or forty-eight or fourteen chromosomes. If only the total mattered, one cell might get, say, two pairs of chromosome 22 and no pairs whatsoever of chromosome 9. Neither would be able to function like the parent cell *without two of each type of chromosome*.

A eukaryotic chromosome consists of one DNA molecule and many proteins that structurally organize the DNA.

A chromosome number is the sum total of chromosomes in a body cell or germ cell of a particular organism. Cells with a diploid chromosome number have two of each type of chromosome, usually from two parents.

Mitosis is a nuclear division mechanism that sorts out a required number of chromosomes for each daughter cell. A separate mechanism divides the cytoplasm.

Mitosis keeps the chromosome number constant, division after division, from one cell generation to the next. Thus, if a parent cell is diploid, its daughter cells will be diploid.

THE CELL CYCLE

Let's start thinking about cell reproduction in terms of an orderly sequence of events called a **cell cycle**. It starts every time a new daughter cell forms by mitosis and cytoplasmic division and ends when that cell finishes its own division (Figure 9.4). Thus, *mitosis, cytoplasmic division, and interphase constitute one turn of the cell cycle.*

The Wonder of Interphase

Interphase is the portion of the cell cycle when a cell increases in mass, roughly doubles the number of its cytoplasmic components, and duplicates its DNA. For most types of cells, interphase is the longest portion of the cycle. Biologists divide it into three parts:

G1   Interval ("Gap") of cell growth and functioning before the onset of DNA replication

S   Time of "Synthesis" (DNA replication)

G2   Second interval (Gap), after DNA replication, when the cell prepares for division

G1, S, and G2 are no more than code names for some amazing events. Consider what cells do with their DNA. If you could coax the DNA molecules from just one of your somatic cells to stretch in a single line, one after another, that line would extend past the fingertips of your outstretched arms. If you could do the same with salamander DNA, a single line of it would extend ten meters! The wonder is, enzymes and other proteins in cells can selectively access, activate, and silence DNA's instructions. They also make base-by-base copies of the DNA molecules. They do most of this in interphase.

G1, S, and G2 of interphase have distinct patterns of biosynthesis. Most of your cells remain in G1, when they build nearly all of the proteins, carbohydrates, and lipids they use or export. Cells destined to divide enter S, when they copy their DNA as well as the histones and other proteins associated with it. During G2, these cells produce proteins that will drive mitosis to completion.

Once S begins, events normally proceed at about the same rate in all cells of a species and continue through mitosis. Given this observation, you may be wondering whether the cell cycle has built-in molecular brakes. It does. Apply the brakes that are supposed to work in G1, and the cycle stalls in G1. Lift the brakes, and the cell cycle then runs to completion. Said another way, *control mechanisms govern the rate of cell division.*

Imagine a car losing its brakes just as it starts down a steep mountain road. As you will read later on in the book, that is how cancer starts. Controls over division are lost, and the cell cycle cannot stop turning.

The cell cycle lasts about the same length of time for cells of the same type. Its duration differs among cells of different types. Examples: All neurons (nerve cells) in your brain are arrested at G1 of interphase, and usually won't divide again. Each second, 2 million to 3 million precursors of red blood cells form to replace worn-out ones circulating in your body. Early in a sea urchin's development, the number of cells doubles every two hours.

Adverse conditions often disrupt the cell cycle. When deprived of a vital nutrient, for instance, the free-living cells called amoebas do not leave interphase. Even so, if any cell proceeds past a certain point in interphase, the cycle normally continues regardless of outside conditions because of built-in controls over its duration.

Mitosis Proceeds Through Four Stages

A cell that's making the transition from interphase to mitosis has stopped constructing new cell parts. Its DNA has already been replicated. Major changes now proceed smoothly through four stages: **prophase**, **metaphase**, **anaphase**, and **telophase**.

Figure 9.5 shows these stages in a plant cell. Notice that all the chromosomes are changing positions. They aren't doing so on their own. A **spindle apparatus**, of

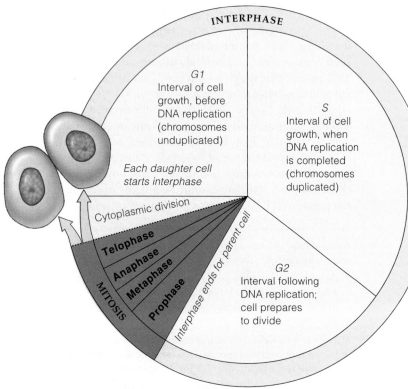

Figure 9.4 Eukaryotic cell cycle, generalized. The length of each interval differs among different cell types.

A CELL AT
INTERPHASE:

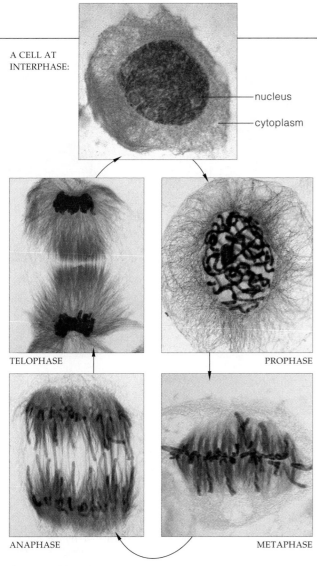

nucleus

cytoplasm

TELOPHASE

PROPHASE

ANAPHASE

METAPHASE

Figure 9.5 Mitosis in a cell from the African blood lily, *Haemanthus.* The chromosomes are stained *blue* and the many microtubules, *red.* Before reading further, take a moment to become familiar with the labels on the micrographs.

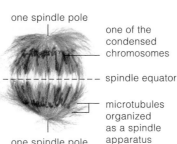

one spindle pole

one of the condensed chromosomes

spindle equator

microtubules organized as a spindle apparatus

one spindle pole

two sets of microtubules, moves them. Each set extends from one of the spindle's two poles (its end points), and each overlaps the other at the spindle equator, midway between the poles. This "bipolar" spindle will establish the final chromosome destinations before mitosis ends and the cell divides in two.

How important are spindle microtubules? One clue comes from plants of the genus *Colchicum.* These plants make colchicine, a poison that evolved against browsing animals. It blocks microtubule assembly and promotes their disassembly. This poison is a favorite of researchers who study cancer and other expressions of cell division.

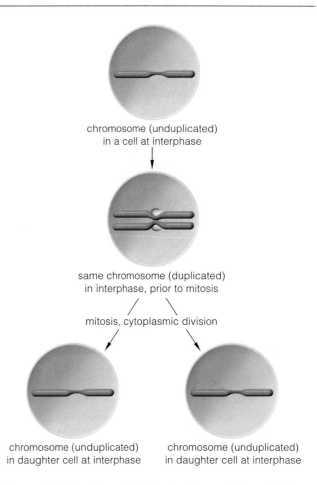

chromosome (unduplicated) in a cell at interphase

same chromosome (duplicated) in interphase, prior to mitosis

mitosis, cytoplasmic division

chromosome (unduplicated) in daughter cell at interphase

chromosome (unduplicated) in daughter cell at interphase

Figure 9.6 Simple way to think about how mitosis maintains the chromosome number, one generation after the next. For clarity, we track only one chromosome.

Another microtubule poison is mentioned in Section 9.6, *Critical Thinking* question 3. Spindles in cells fall apart within seconds or minutes after exposure to it.

Before turning the page to consider the mechanism of mitosis, hold on to these thoughts: Chromosomes are duplicated *before* mitosis. A spindle apparatus moves the sister chromatids of duplicated chromosomes apart *during* mitosis. As Figure 9.6 shows, that is how mitosis maintains the chromosome number through turn after turn of the cell cycle.

Interphase, mitosis, and cytoplasmic division constitute one turn of the cell cycle.

In interphase, a new cell increases its mass, roughly doubles the number of its cytoplasmic components, and duplicates its chromosomes. The cycle ends after the cell undergoes mitosis and then divides. Molecular mechanisms control the rate of cell division.

Mitosis has four consecutive stages: prophase, metaphase, anaphase, and telophase. Its microtubular spindle moves chromosomes that were duplicated earlier, in interphase.

MITOSIS

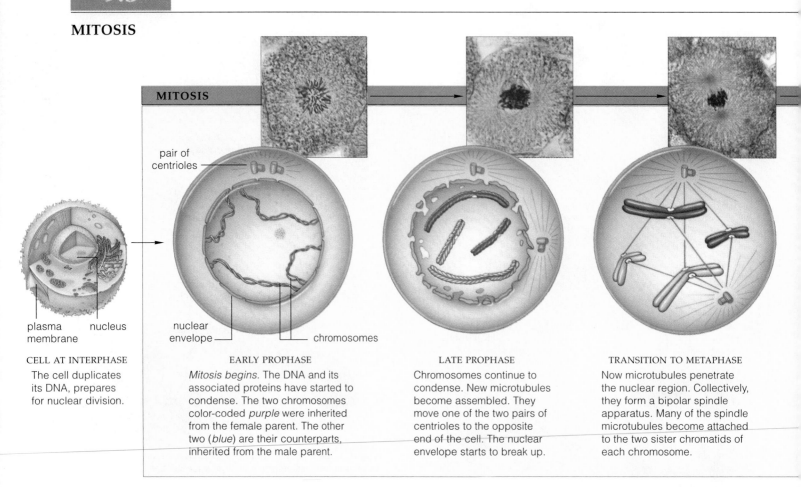

MITOSIS

pair of centrioles

plasma membrane   nucleus

nuclear envelope   chromosomes

CELL AT INTERPHASE
The cell duplicates its DNA, prepares for nuclear division.

EARLY PROPHASE
Mitosis begins. The DNA and its associated proteins have started to condense. The two chromosomes color-coded *purple* were inherited from the female parent. The other two (*blue*) are their counterparts, inherited from the male parent.

LATE PROPHASE
Chromosomes continue to condense. New microtubules become assembled. They move one of the two pairs of centrioles to the opposite end of the cell. The nuclear envelope starts to break up.

TRANSITION TO METAPHASE
Now microtubules penetrate the nuclear region. Collectively, they form a bipolar spindle apparatus. Many of the spindle microtubules become attached to the two sister chromatids of each chromosome.

Prophase: Mitosis Begins

We know a cell is in prophase, the first stage of mitosis, when its chromosomes are visible in light microscopes as threadlike forms. ("Mitosis" is from the Greek *mitos*, for thread.) Each chromosome was duplicated earlier, in interphase; each is two sister chromatids joined at the centromere. In early prophase, all of these chromatids twist and fold. By late prophase, they will be condensed into thicker, compact, rod-shaped forms.

Meanwhile, in the cytoplasm, most microtubules of the cytoskeleton are breaking down to tubulin subunits (Section 4.9). The subunits reassemble near the nucleus as microtubules of the spindle. While new microtubules are assembling, the nuclear envelope physically prevents them from interacting with the chromosomes inside the nucleus. However, the nuclear envelope starts to break up as prophase draws to a close (Figure 9.7).

Many cells have two barrel-shaped **centrioles**. Each centriole started duplicating itself during interphase, so there are two pairs of them when prophase is under way. Microtubules start moving one pair to the opposite pole of the newly forming spindle. Centrioles, recall, give rise to flagella or cilia. If you observe them in cells of an organism, you can bet that flagellated cells (such as sperm) or ciliated cells develop during its life cycle.

Figure 9.7 Mitosis in a generalized animal cell. By this nuclear division mechanism, each daughter cell ends up with the same chromosome number as the parent cell. For clarity, we track only two pairs of chromosomes from a diploid (*2n*) cell. The picture is almost always more complicated, as you may sense from the above micrographs of mitosis in a whitefish cell.

Transition to Metaphase

So much happens between prophase and metaphase that we often call this transitional time "prometaphase." The nuclear envelope breaks up completely into many small, flattened vesicles, so the chromosomes are now free to interact with microtubules. Some microtubules that have been lengthening from the two spindle poles harness each chromosome and pull on it. The two-way pulling orients it with respect to the poles. However, other microtubules extend from both spindle poles as two sets that overlap at the spindle's midpoint. They push the poles apart by interacting and ratcheting past each other. The push–pull forces become balanced when the chromosomes reach the spindle's midpoint.

When all of the duplicated chromosomes are aligned midway between the poles of a completed spindle, we call this metaphase (*meta*– means "midway between"). The alignment is crucial for the next stage of mitosis.

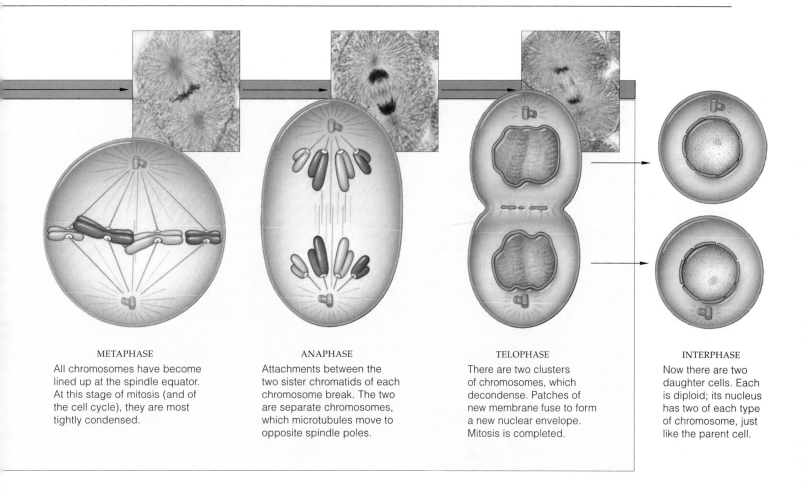

METAPHASE

All chromosomes have become lined up at the spindle equator. At this stage of mitosis (and of the cell cycle), they are most tightly condensed.

ANAPHASE

Attachments between the two sister chromatids of each chromosome break. The two are separate chromosomes, which microtubules move to opposite spindle poles.

TELOPHASE

There are two clusters of chromosomes, which decondense. Patches of new membrane fuse to form a new nuclear envelope. Mitosis is completed.

INTERPHASE

Now there are two daughter cells. Each is diploid; its nucleus has two of each type of chromosome, just like the parent cell.

From Anaphase Through Telophase

At anaphase, the sister chromatids of each chromosome separate from each other and move to opposite spindle poles. Two mechanisms bring this about.

First, microtubules that latched on to the centromere region of each chromosome shorten (they disassemble) and *pull* the chromosome to a spindle pole. Think of the centromere as a train chugging along a railroad track, except the track falls apart after the train passes over it. Activated motor proteins are the engine. They project in orderly arrays from the microtubules and drive the chromosome onward by sliding motions (Section 4.10).

Second, the spindle itself elongates as overlapping microtubules continue to ratchet past one another and *push* the spindle's two poles even farther apart. These microtubules, too, incorporate motor proteins, and they actively slide past one another where they overlap.

Once each chromatid is separated from its sister, we recognize it as a separate chromosome.

Telophase gets under way as soon as each of the two clusters of chromosomes arrives at a spindle pole. The chromosomes, no longer harnessed to the microtubules, return to threadlike form. Vesicles derived from the old nuclear envelope fuse and form patches of membrane around the chromosomes. Patch joins with patch, and soon a new nuclear envelope separates both clusters of chromosomes from the cytoplasm. If the parent cell was diploid, each cluster contains two chromosomes of each type. With mitosis, remember, each new nucleus has the same chromosome number as the parent nucleus. Once two nuclei form, telophase is over—and so is mitosis.

Prior to mitosis, each chromosome in a cell's nucleus is duplicated, so that it consists of two sister chromatids.

In prophase, microtubules assemble outside the nucleus and start to form a bipolar spindle. The nuclear envelope breaks down, so the microtubules are now free to harness and move the duplicated chromosomes.

At metaphase, all chromosomes are aligned and oriented, with respect to the poles, at the spindle equator.

At anaphase, microtubules move sister chromatids of each chromosome apart, to opposite spindle poles. Some pull on them; others push the spindle poles apart, increasing the distance between the former sister chromatids.

At telophase, there are two clusters of chromosomes, and a new nuclear envelope forms around each cluster.

Thus mitosis forms two daughter nuclei. Each has the same chromosome number as the parent cell's nucleus.

DIVISION OF THE CYTOPLASM

The cytoplasm usually divides at some time between late anaphase and the end of telophase. As you might well conclude by comparing Figure 9.8 with Figure 9.9, the actual mechanism of **cytoplasmic division** (or, as it is often called, cytokinesis) differs among organisms.

Cell Plate Formation in Plants

As described in Section 4.11, most plant cells are walled, which means their cytoplasm can't be pinched in two. Cytoplasmic division of such cells involves **cell plate formation**, as shown in Figure 9.8. By this mechanism, vesicles packed with wall-building materials fuse with one another and with remnants from the microtubular spindle. Together, they form a disklike structure—a cell plate. At this location, deposits of cellulose accumulate. In time, the cellulose deposits are thick enough to form

light micrograph of a cell plate forming in a dividing plant cell

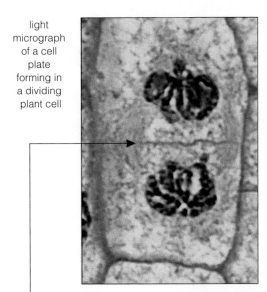

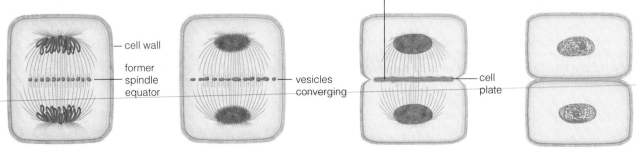

— cell wall

former spindle equator

— vesicles converging

— cell plate

a As mitosis ends, vesicles converge at the spindle equator. They contain cementing materials and structural materials for a new primary cell wall.

b A cell plate starts forming as membranes of the vesicles fuse. Materials inside the vesicles get sandwiched between two new membranes that elongate along the plane of the cell plate.

c Cellulose is deposited on the inside of the "sandwich." (In time, deposits will form two cell walls. Other deposits will form a middle lamella and cement the walls together; refer to Section 4.11.)

d A cell plate grows at its margins until it fuses with the parent cell's plasma membrane. During growth, new plant cells expand. Their primary wall is still thin, and new material is deposited on it.

Figure 9.8 Cytoplasmic division of a plant cell, as brought about by cell plate formation.

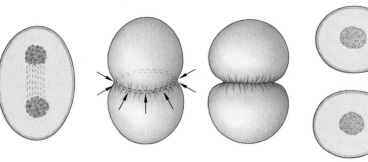

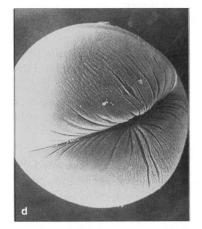

a Mitosis is over, and the spindle is now disassembling.

b At the former spindle equator, a ring of microfilaments attached to the plasma membrane contracts. As its diameter shrinks, it pulls the cell surface inward.

c Contractions continue until the ring cuts the cell in two.

Figure 9.9 (**a**–**c**) Cytoplasmic division of an animal cell. (**d**) Scanning electron micrograph of the cleavage furrow at the plane of the former spindle's equator. Beneath it, a band of microfilaments attached to the plasma membrane contracts and pulls the surface inward. The furrow deepens until the cell is cut in two.

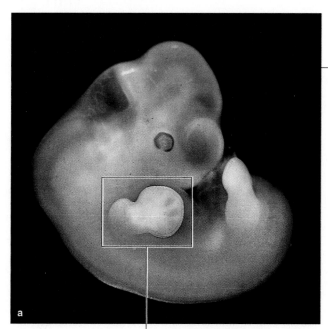

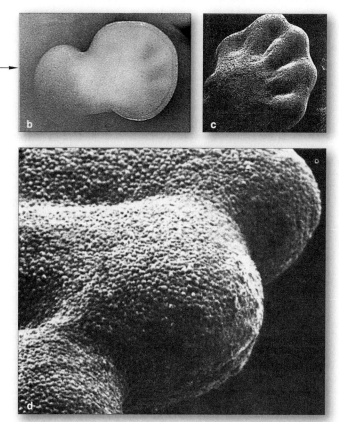

future arm and hand of embryo, five weeks old

a crosswall. That new crosswall bridges the cytoplasm and divides the parent cell into two daughter cells.

Cytoplasmic Division of Animal Cells

Unlike plant cells, an animal cell isn't confined within a cell wall, and its cytoplasm typically pinches in two. Look at Figure 9.9, a surface view of a newly fertilized animal egg. An indentation is forming about midway between the two poles of the egg. The indentation in its plasma membrane is a **cleavage furrow**. It's the first visible sign that the cytoplasm inside an animal cell is dividing. The furrow will extend all around the cell and continue to deepen along the plane of the former spindle's midpoint until the cell is cut in two.

A band of microfilaments beneath the cell's plasma membrane generates force for a cut. These cytoskeletal elements are organized to interact and slide past one another (Section 4.10). As they do, they pull the plasma membrane inward until there are two daughter cells. Each has a nucleus, cytoplasm, and plasma membrane.

Perspective on Mitotic Cell Division

This concludes our introduction to mitotic cell division. Look now at your hands and try to visualize the cells making up your palms, thumbs, and fingers. Imagine the divisions that produced all the generations of cells that preceded them as you were developing early on, inside your mother (Figure 9.10). And be grateful for the astonishing precision of the mechanisms that led to their formation at certain times, in certain numbers, for the alternatives can be terrible indeed. Why? Good

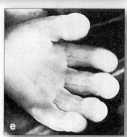

Figure 9.10 (**a–e**) Transformation from a paddlelike structure into the fingers of a human hand by mitosis, cytoplasmic divisions, and other processes necessary for embryonic development. In (**d**), you can see many individual cells at this high magnification.

health—and survival itself—depends absolutely on the proper timing and the completion of cell cycle events, including mitosis. Certain genetic disorders arise from mistakes in the duplication or distribution of even one chromosome. Also, when normal controls that prevent cells from dividing are lost, unchecked cell divisions may destroy surrounding tissues and, ultimately, the organism. Section 9.5 describes a landmark case of such losses, which we will explore further in Section 15.6.

After mitosis, a separate mechanism cuts the cytoplasm into two daughter cells, each with a daughter nucleus.

Cytoplasmic division in plants often involves the formation of a cell plate and a crosswall between the adjoining, new plasma membranes of daughter cells.

Cleavage is a form of cytoplasmic division in animals. Rings of microfilaments around a parent cell's midsection slide past one another in a way that pinches the cytoplasm in two.

Henrietta's Immortal Cells

Each human starts out as a single fertilized egg. By the time of birth, mitotic cell divisions and other processes have resulted in a human body of about a trillion cells. Even in adults, billions of cells still divide. For example, cells of the stomach's lining divide every day. Liver cells usually do not divide, but if part of the liver becomes injured or diseased, repeated cell divisions will yield more new cells until the damaged part is finally replaced.

In 1951, George and Margaret Gey of Johns Hopkins University were trying to develop a way to keep human cells dividing *outside* the body. With such isolated cells, these researchers and others could investigate basic life processes. They also could conduct studies of cancer and other diseases without having to experiment directly on patients and risk human lives. The Geys used normal and diseased human cells, which local physicians had sent them. But they couldn't stop the descendants of those precious cells from dying out within a few weeks.

Mary Kubicek, a laboratory assistant, worked with the Geys in their efforts to start a self-perpetuating lineage of cultured human cells. She was about to give up after dozens of failed attempts. Even so, in 1951 she decided to prepare one more sample of cancer cells for culture. She gave the sample the code name **HeLa cells**, for the first two letters of the patient's first and last names.

The HeLa cells began to divide. In four days there were so many cells that Kubicek subdivided them into more culture tubes. Sadly, cancer cells in the patient were dividing as rapidly. Six months after she was diagnosed with cancer, tumor cells had spread to tissues throughout her body. Eight months after the diagnosis, Henrietta Lacks, a young woman from Baltimore, was dead.

Although Henrietta passed away, some of her cells lived on in the Geys' laboratory as the first successful human cell culture. In time, HeLa cells were shipped to research laboratories all over the world. Some even traveled into space for experiments on the *Discoverer XVII* satellite. Each year hundreds of research projects depend on them. Henrietta was only thirty-one when runaway cell divisions killed her. Now, decades later, her legacy is helping humans everywhere, through cellular descendants that are still dividing day after day.

Figure 9.11 Dividing HeLa cells, a legacy of Henrietta Lacks, who was a casualty of cancer. Her cellular contribution to science is still helping others every day.

SUMMARY   *Gold* indicates text section

1. Through specific division mechanisms, a parent cell provides each of its daughter cells with the hereditary instructions (DNA) and cytoplasmic machinery required to start up its own operation. *CI, 9.1*

   a. In eukaryotic cells, the nucleus divides by mitosis or meiosis. Cytoplasmic division typically follows.

   b. Prokaryotic cells divide by prokaryotic fission.

2. Each eukaryotic chromosome is one DNA molecule with numerous proteins attached. The chromosomes in a cell differ in length, shape, and which portion of the hereditary instructions they carry. *9.1*

   a. We define chromosome number as the sum total of chromosomes in cells of a given type. Cells having a diploid chromosome number (2*n*) contain two of each kind of chromosome.

   b. Mitosis divides the nucleus into two equivalent nuclei, each with the same chromosome number as the parent cell. It maintains the chromosome number from one cell generation to the next.

   c. Mitosis is the basis of growth, tissue repair, and cell replacements among multicelled eukaryotes. It is the basis of asexual reproduction in many single-celled eukaryotes. (Meiosis occurs only in germ cells.)

3. A duplicated chromosome has two DNA molecules attached at the centromere. For as long as the two stay connected to each other, they are sister chromatids. *9.1*

4. A cell cycle starts when a new cell forms. It proceeds through interphase and ends when the cell reproduces by mitosis and cytoplasmic division. In interphase, a cell carries out its functions. When it is to divide again, the cell increases in mass, roughly doubles the number of its cytoplasmic components, then duplicates each of its chromosomes in preparation for division. *9.2*

5. Mitosis has four continuous stages: *9.3*

   a. Prophase. The duplicated, threadlike chromosomes start to condense. A spindle starts to form. The nuclear envelope starts to break up; its remnants form vesicles during the *transition* to metaphase (or prometaphase). Some microtubules from both poles of the developing spindle push the poles apart. Others directly attach to one of two sister chromatids of each chromosome.

   b. Metaphase. *At* metaphase, all chromosomes have become aligned at the spindle equator.

   c. Anaphase. Microtubules pull sister chromatids of each chromosome away from each other, to opposite spindle poles. Now each type of parental chromosome is represented by a daughter chromosome at both poles.

   d. Telophase. Chromosomes decondense to threadlike form. A new nuclear envelope forms around them. Each nucleus has the parental chromosome number.

6. Separate mechanisms divide the cytoplasm near the end of nuclear division or afterward (in plants, by cell plate formation; in animals, by cleavage). *9.4*

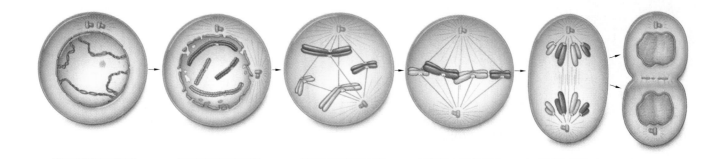

_____ _____ _____ _____ _____ _____

Review Questions

1. Define mitosis and meiosis, two mechanisms that operate in eukaryotic cells. Does either one divide the cytoplasm? *9.1*

2. Define somatic cell and germ cell. *9.1*

3. What are chromosomal proteins? What are histones, and how do they interact with a DNA molecule? *9.1*

4. What is a chromosome called when it is in the unduplicated state? In the duplicated state (with two sister chromatids)? *9.1*

5. Describe the microtubular spindle and its functions. What role do motor proteins play in its operation? *9.2, 9.3*

6. Using the diagram above as a guide, name and describe the key features of the stages of mitosis. *9.3*

7. Briefly explain how cytoplasmic division differs between a typical plant cell and a typical animal cell. *9.4*

Self-Quiz   ANSWERS IN APPENDIX III

1. Mitosis and cytoplasmic division function in _____ .
   a. asexual reproduction of single-celled eukaryotes
   b. growth, tissue repair, and sometimes asexual reproduction in many multicelled eukaryotes
   c. gamete formation in prokaryotes
   d. both a and b

2. A duplicated chromosome has _____ chromatid(s).
   a. one   b. two   c. three   d. four

3. In a chromosome, a _____ is a constricted region with attachment sites for microtubules.
   a. chromatid   b. cell plate   c. centromere   d. cleavage

4. A somatic cell having two of each type of chromosome has a(n) _____ chromosome number.
   a. diploid   b. haploid   c. tetraploid   d. abnormal

5. Interphase is the part of the cell cycle when _____ .
   a. a cell ceases to function
   b. a germ cell forms its spindle apparatus
   c. a cell grows and duplicates its DNA
   d. mitosis proceeds

6. After mitosis, the chromosome number of a daughter cell is _____ the parent cell's.
   a. the same as   c. rearranged compared to
   b. one-half   d. doubled compared to

7. Only _____ is not a stage of mitosis.
   a. prophase   b. interphase   c. metaphase   d. anaphase

8. Match each stage with the events listed.
   _____ metaphase   a. sister chromatids move apart
   _____ prophase   b. chromosomes start to condense
   _____ telophase   c. chromosomes decondense and daughter nuclei form
   _____ anaphase   d. all duplicated chromosomes are aligned at the spindle equator

Critical Thinking

1. Suppose you have a way to measure the amount of DNA in a single cell during the cell cycle. You first measure the amount at the G1 phase. At what points during the rest of the cell cycle would you predict changes in the amount of DNA per cell?

2. The cervix is part of the uterus, a chamber in which embryos develop. The *Pap smear* is a screening procedure that can detect *cervical cancer* in its earliest stages. Treatments range from freezing precancerous cells or killing them with a laser beam to removal of the uterus (a hysterectomy). The treatments are 90+ percent effective when this cancer is detected early. Survival chances plummet to less than 9 percent after the cancer spreads.

   Most cervical cancers develop slowly. Unsafe sex increases the risk. A key risk factor is infection by human papillomaviruses that cause genital warts (Section 44.15). In 93 percent of all cases, viral genes coding for the tumor-inducing proteins had become inserted into the DNA of previously normal cervical cells.

   Not all women request Pap smears. Many wrongly believe the procedure is costly. Many don't recognize the importance of abstinence or "safe" sex. Others simply don't want to think about whether they have cancer. Knowing what you've learned so far about the cell cycle and cancer, what would you say to a woman who falls into one or more of these groups?

3. Pacific yews (*Taxus brevifolius*) face extinction. People started stripping its bark and killing trees when they heard that *taxol,* a chemical extract from the bark, may fight breast and ovarian cancer. (Synthesizing taxol in laboratories may save the species.) Taxol is a poison that prevents microtubule disassembly. What does this tell you about its potential as an anticancer drug?

4. X rays and gamma rays emitted from some radioisotopes chemically damage DNA, especially in cells engaged in DNA replication. High-level exposure can result in *radiation poisoning.* Hair loss and damage to the lining of the stomach and intestines are two early symptoms. Speculate why. Also speculate on why highly focused radiation therapy is used against some cancers.

Selected Key Terms

anaphase *9.2*	cytoplasmic	mitosis *9.1*
cell cycle *9.2*	division *9.4*	nucleosome *9.1*
cell plate	diploid (chromosome	prophase *9.2*
formation *9.4*	number) *9.1*	reproduction *CI*
centriole *9.3*	germ cell *9.1*	sister
centromere *9.1*	HeLa cell *9.5*	chromatid *9.1*
chromosome *9.1*	histone *9.1*	somatic cell *9.1*
chromosome	interphase *9.2*	spindle
number *9.1*	meiosis *9.1*	apparatus *9.2*
cleavage furrow *9.4*	metaphase *9.2*	telophase *9.2*

Readings

Murray, A., and M. Kirschner. March 1991. "What Controls the Cell Cycle?" *Scientific American* 264(3): 56–63.

MEIOSIS

Octopus Sex and Other Stories

The couple clearly are interested in each other. First he caresses her with one tentacle, then with another—and another and another. She reciprocates with a hug here, a squeeze there. This goes on for hours. Finally the male reaches under his mantle, a fold of tissue that drapes around most of his body. He removes a packet of sperm from a reproductive organ and inserts it into an egg chamber beneath the female's mantle. For every sperm that fertilizes an egg, a new octopus may develop.

Unlike the one-to-one coupling between a male and a female octopus, sex for the slipper limpet is a group enterprise. Slipper limpets are marine animals, relatives of the familiar land snails. Before becoming transformed into a sexually mature adult, a slipper limpet must go through a free-living stage of development called a larva. When a limpet larva is about to undergo its programmed transformation, it settles on a rock or pebble or shell. If it settles down all by itself, it will become a female. If another larva settles on the first limpet and proceeds to develop, that second limpet will function right off as a male. However, the second one will switch gears and develop into a female if a third limpet develops into a male on top of *it*. Later, that third limpet will also become a female if still another limpet develops into a male on top of it—and so on amongst ten or more limpets!

Slipper limpets typically live in such piles, with the bottom one always being the oldest female and the uppermost one being the youngest male (Figure 10.1*a*). Until they switch their sex, male limpets release sperm, which

Figure 10.1 Examples of variations in reproductive modes of eukaryotic organisms. (**a**) Slipper limpets, busily perpetuating the species by group participation in sexual reproduction. The tiny crab in the foreground is merely a passerby. (**b**) Live birth of an aphid, a type of insect that reproduces sexually in autumn but can switch to an asexual mode in summer.

fertilize a female's eggs. Fertilized eggs develop into immature larvae, which further develop into females or males that can later become females—and so it goes, from one sexually flexible generation to the next.

Limpets are not alone in having unusual variations in their mode of reproduction. For example, sexual reproduction is common in many life cycles, but so are asexual episodes based on mitotic cell divisions. Orchids, dandelions, and many other plants reproduce very well with or without sex. Aquatic animals called flatworms can engage in sex or split their small body into two roughly equivalent parts, each of which grows and develops into a new flatworm.

And what about those aphids! In summer, nearly all aphids are females, which produce more females from *unfertilized* egg cells (Figure 10.1*b*). Only when autumn approaches do male aphids develop and do their part in the sexual phase of the life cycle. Even then, females that manage to survive over the winter can do without the opposite sex. Come summer, they begin another round of producing offspring all by themselves.

These examples only hint at the immense variation in reproductive modes among eukaryotic organisms. And yet, despite the variation, *sexual* reproduction dominates nearly all of the life cycles. And it always involves the same events. Briefly, cells set aside for sexual reproduction duplicate their chromosomes before they divide. **Germ cells**, immature reproductive cells that develop in male and female animals, are a fine example. They undergo meiosis and cytoplasmic division. Cellular descendants of germ cells mature and become **gametes**, or sex cells. When a male and female gamete manage to get together, they form the first cell of a new individual by way of fertilization.

With this chapter, we turn to the kinds of cells that serve as the bridge between generations of organisms. Specialized phases of reproduction and development, including asexual episodes, loop out from the basic life cycle of many eukaryotic species. Regardless of the specialized details, all of the life cycles turn on three events: *meiosis, the formation of gametes, and fertilization*. These three interconnected events are the hallmarks of sexual reproduction. As you will see in many chapters throughout the book, they have contributed to the diversity of life.

Key Concepts

1. Sexual reproduction proceeds through three key events: meiosis, gamete formation, and fertilization. Sperm and eggs are familiar gametes.

2. Meiosis, a nuclear division mechanism, occurs only in cells set aside for sexual reproduction. The immature germ cells of male and female animals are examples. Meiosis sorts out the chromosomes of a germ cell into four new nuclei. After meiosis is completed, gametes form by way of cytoplasmic division and other events.

3. Cells with a diploid chromosome number contain two of each type of chromosome characteristic of the species. The two function as a pair during meiosis. Commonly, one chromosome of the pair is maternal, with hereditary instructions from a female parent. The other is paternal, with the same categories of hereditary instructions from a male parent.

4. Meiosis divides the chromosome number by half for each forthcoming gamete. Thus, if both parents have a diploid chromosome number ($2n$), the gametes that form will be haploid (n). Later, a union of two gametes at fertilization will restore the diploid number in the new individual ($n + n = 2n$).

5. Each pair of chromosomes swaps segments during meiosis and exchanges hereditary information about certain traits. Also, meiosis randomly assigns one of each pair of chromosomes to a forthcoming gamete; *which* one of the pair ends up in a given gamete is a matter of chance. Hereditary instructions are further shuffled at fertilization. All three reproductive events lead to variations in traits among offspring.

6. In most plants, spore formation and other events intervene between meiosis and gamete formation.

COMPARING SEXUAL WITH ASEXUAL REPRODUCTION

When an orchid, flatworm, or aphid reproduces all by itself, what sort of offspring does it get? By the process of **asexual reproduction**, one parent alone produces offspring, and each offspring inherits the same number and kinds of genes as its parent. **Genes** are particular stretches of chromosomes—that is, of DNA molecules. Taken together, the genes for each species contain all the heritable bits of information necessary to make new individuals. Rare mutations aside, this means asexually produced individuals can only be *clones*, or genetically identical copies of the parent.

Inheritance gets much more interesting with **sexual reproduction**. This process involves meiosis, formation of gametes, and fertilization (union of the nuclei of two gametes). In most sexual reproducers, such as humans, the first cell of a new individual contains *pairs of genes* on pairs of homologous chromosomes. Typically, one of each pair is maternal and the other paternal in origin.

If instructions encoded in every pair of genes were identical down to the last detail, sexual reproduction would produce clones, also. Just imagine—you, every single person you know, the entire human population might be a clone, with everybody looking alike. But the two genes of a pair might *not* be identical. Why not? A gene's molecular structure can change; that is what we mean by mutation. Depending on their structure, two genes that happen to be paired in a person's cells may "say" slightly different things about a trait. Each unique molecular form of the same gene is called an **allele**.

Such tiny differences affect thousands of traits. For example, whether your chin has a dimple depends on which pair of alleles you inherited at one chromosome location. One kind of allele at that location says "put a dimple in the chin." Another kind says "no dimple." This leads us to a key reason why members of sexually reproducing species don't all look alike. *Through sexual reproduction, offspring inherit new combinations of alleles, which lead to variations in the details of their traits.*

This chapter gets into the cellular basis of sexual reproduction. More importantly, it starts you thinking about far-reaching effects of gene shufflings at different stages of the process. The process introduces variations in traits among offspring that are typically acted upon by agents of natural selection. Thus, *variation in traits is a foundation for evolutionary change.*

Asexual reproduction produces genetically identical copies of the parent. Sexual reproduction introduces variations in the details of traits among offspring.

Sexual reproduction dominates the life cycles of eukaryotic species. Meiosis, formation of gametes, and fertilization are the basic events of this process.

HOW MEIOSIS HALVES THE CHROMOSOME NUMBER

Think "Homologues"

Think back to the preceding chapter and its focus on mitotic cell division. Unlike mitosis, **meiosis** divides chromosomes into separate parcels not once but *twice* prior to cell division. Unlike mitosis, it is the first step leading to the formation of gametes. Gametes, recall, are sex cells such as sperm or eggs. In most multicelled eukaryotic organisms, gametes develop from cells that arise in specialized reproductive structures or organs. Figure 10.2 shows examples of where gametes form.

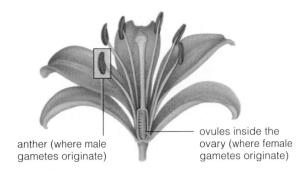

anther (where male gametes originate)

ovules inside the ovary (where female gametes originate)

a Flowering plant

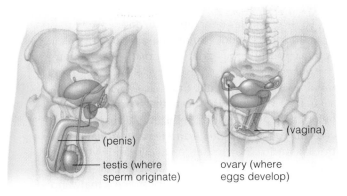

(penis)

testis (where sperm originate)

(vagina)

ovary (where eggs develop)

b Human male   **c** Human female

Figure 10.2 Examples of gamete-producing structures.

As you know, the **chromosome number** is the sum total of chromosomes in cells of a given type (Section 9.1). Germ cells start out with the same chromosome number as somatic cells (the rest of the body's cells). If a cell has a **diploid number** (2*n*), it has a *pair* of each type of chromosome, often from two parents. Except for a pairing of nonidentical sex chromosomes, each pair has the same length, shape, and assortment of genes. And they line up with each other at meiosis. We call them **homologous chromosomes** (*hom*– means alike).

As you can probably deduce from Figure 10.3, your own germ cells have 23 + 23 homologous chromosomes.

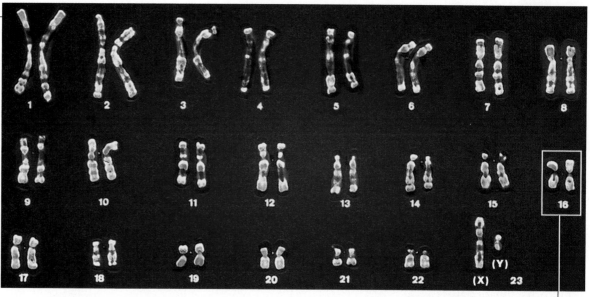

Figure 10.3 From a diploid cell of a human male, twenty-three pairs of homologous chromosomes. The last two are a pair of sex chromosomes (XY or XX). An XY pair differs in length, shape, and which genes they carry. But the two still pair up as homologues during meiosis.

one pair of homologous chromosomes

After meiosis, 23 chromosomes—one of each type—end up in gametes. That is, meiosis halves the chromosome number, so gametes have a **haploid number** (*n*).

Two Divisions, Not One

Meiosis is like mitosis in some ways, but the outcome is different. As in mitosis, a germ cell duplicates its DNA in interphase. The two DNA molecules and their associated proteins remain attached at the centromere, the notably constricted region along their length. For as long as they remain attached, we call them **sister chromatids**:

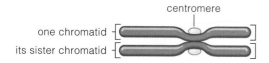

one chromosome in the duplicated state

As in mitosis, the microtubules of a spindle apparatus move the chromosomes in prescribed directions.

With meiosis alone, however, *chromosomes go through two consecutive divisions that end with the formation of four haploid nuclei*. There is no interphase between the nuclear divisions, which we call meiosis I and meiosis II:

	MEIOSIS I		MEIOSIS II
interphase (*DNA replication before meiosis I*)	PROPHASE I	no interphase (*no DNA replication before meiosis II*)	PROPHASE II
	METAPHASE I		METAPHASE II
	ANAPHASE I		ANAPHASE II
	TELOPHASE I		TELOPHASE II

Each duplicated chromosome becomes aligned with its partner—*homologue to homologue*—during meiosis I.

After the two chromosomes of each pair have lined up with each other, they are moved apart:

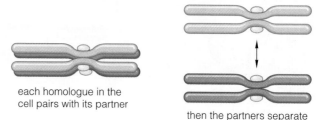

each homologue in the cell pairs with its partner

then the partners separate

The cytoplasm typically starts to divide at some point after each homologue is separated from its partner. The two daughter cells formed this way are haploid; each has *one* of each type of chromosome. But don't forget, the chromosomes are still in the duplicated state.

Next, during meiosis II, *the two sister chromatids of each chromosome are separated from each other*:

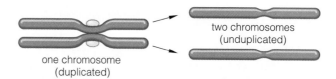

one chromosome (duplicated)

two chromosomes (unduplicated)

Each sister chromatid is now a chromosome in its own right. Four nuclei now form. Often the cytoplasm divides once more. The final outcome is four haploid cells.

Figure 10.4, on the next two pages, gives you a closer look at key events of meiosis and their consequences.

Meiosis is a type of nuclear division mechanism. It reduces the chromosome number of a parental cell by half—to the haploid number (*n*)—in daughter cells.

Meiosis, the first step leading to gamete formation, proceeds only in cells that are set aside for sexual reproduction.

VISUAL TOUR OF THE STAGES OF MEIOSIS

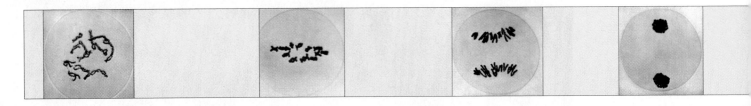

MEIOSIS I

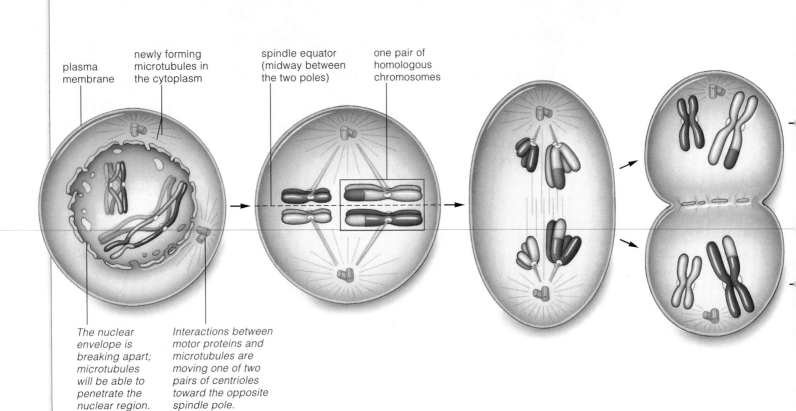

plasma membrane

newly forming microtubules in the cytoplasm

spindle equator (midway between the two poles)

one pair of homologous chromosomes

The nuclear envelope is breaking apart; microtubules will be able to penetrate the nuclear region.

Interactions between motor proteins and microtubules are moving one of two pairs of centrioles toward the opposite spindle pole.

PROPHASE I

Each duplicated chromosome is in threadlike form, but now starts to condense. It pairs with its homologue, and the two typically swap segments. The swapping, called crossing over, is indicated by the break in color on the pair of larger chromosomes. Some of the microtubules of a newly forming spindle become attached to each chromosome's centromere.

METAPHASE I

As in mitosis, motor proteins attached to microtubules move the chromosomes and move the spindle poles apart. They tug the chromosomes into position midway between the spindle poles. Thus the spindle becomes fully formed, owing to dynamic interactions among the motor proteins, microtubules, and the chromosomes themselves.

ANAPHASE I

Microtubules extending from the poles and overlapping at the spindle equator *lengthen* and push the poles apart. Other microtubules extending from the poles to the chromosomes *shorten*, thereby pulling each chromosome away from its homologous partner. These motions move the homologous partners to opposite poles.

TELOPHASE I

The cytoplasm of the germ cell divides at some point. There are now two haploid (*n*) cells. Each cell has one of each type of chromosome that was present in the parent (2*n*) cell. *However, all chromosomes are still in the duplicated state.*

Figure 10.4 Sketches of meiosis in a generalized animal cell. This nuclear division mechanism reduces the chromosome number in immature reproductive cells by half (to the haploid number) for forthcoming gametes. To keep things simple, we track only two pairs of homologous chromosomes. Maternal chromosomes are shaded *purple* and paternal chromosomes *blue*. The light micrographs above show corresponding stages in the formation of pollen grains in a lily (*Lilium regale*).

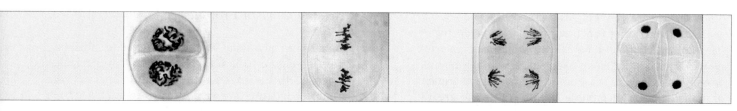

MEIOSIS II

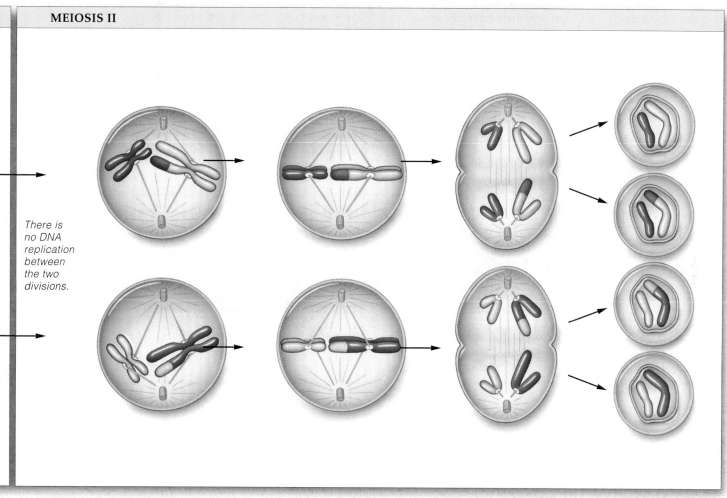

There is no DNA replication between the two divisions.

PROPHASE II

Microtubules have already moved one member of the centriole pair to the opposite pole of the spindle in each of the two daughter cells. Now, during prophase II, microtubules attach to the chromosomes, and motor proteins drive the movement of chromosomes toward the spindle's equator.

METAPHASE II

In each daughter cell, interactions among motor proteins, spindle microtubules, and each duplicated chromosome have moved all of the chromosomes so that they are positioned at the spindle equator, midway between the two poles.

ANAPHASE II

The attachment between the two chromatids of each chromosome breaks. Each of the former "sister chromatids" is now a chromosome in its own right. Motor proteins drive the movement of the newly separated chromosomes to opposite poles of the spindle.

TELOPHASE II

By the time telophase II is over, there will be four daughter nuclei. When cytoplasmic division is completed, each new, daughter cell will have a haploid chromosome number (*n*). All of the chromosomes will now be in the unduplicated state.

Of the four haploid cells that form by way of meiosis and cytoplasmic divisions, one or all may develop into gametes and function in sexual reproduction. In plants, cells that form after meiosis is over may develop into spores, which take part in a stage of the life cycle that precedes gamete formation. (The telophase II micrograph shows spores that will develop into pollen grains.)

A CLOSER LOOK AT KEY EVENTS OF MEIOSIS I

Study the overview in Sections 10.2 and 10.3, and you can sense the overriding function of meiosis: *a reduction of the chromosome number by half for forthcoming gametes.* However, two other major events occur during meiosis: crossing over at prophase I and the random alignment of homologues at metaphase I. Both contribute greatly to the adaptive advantage of sexual reproduction.

The advantage, recall, is the production of offspring with new combinations of alleles. Those combinations are translated into a new generation of individuals that differ in the details of some number of traits.

Crossing Over in Prophase I

Prophase I of meiosis is a time of major gene shufflings. Reflect on Figure 10.5a, which shows two chromosomes condensed to threadlike form. All chromosomes in a germ cell condense this way. As they do, each is drawn close to its homologue. Molecular interactions stitch homologues together point by point along their length, with little space between. The intimate, parallel array favors **crossing over**, a molecular interaction between two of the *non*sister chromatids of a pair of homologous

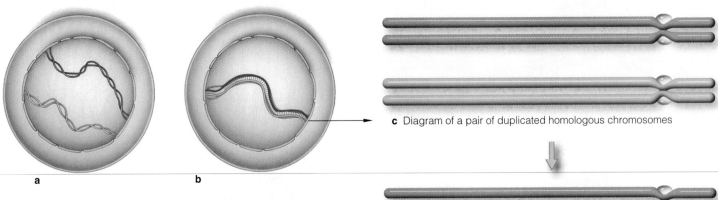

a   b

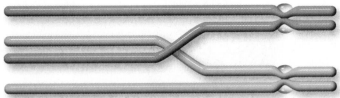

c Diagram of a pair of duplicated homologous chromosomes

Figure 10.5 Key events of prophase I, the first stage of meiosis. For clarity, this diagram shows only one pair of homologous chromosomes and one crossover event. (Typically, more than one crossover occurs.) *Blue* signifies the paternal chromosome; *purple* signifies its maternal homologue.

(**a**) Both chromosomes were duplicated earlier, in interphase. Early in prophase I, the sister chromatids of each duplicated chromosome are in thin, threadlike form. They are positioned so closely together that they look like a single thread.

(**b**) Each chromosome becomes zippered to its homologue, so all four chromatids are intimately aligned. When the two sex chromosomes have different forms (such as X paired with Y) they still get zippered together, although only in a small region at the ends.

(**c,d**) We show the pair of chromosomes as if they were already condensed, then teased apart to give you a sense of what goes on. Bear in mind, their double-stranded DNA molecules are still tightly aligned at this stage. The intimate contact allows one crossover (and usually more) to happen at intervals along the length of nonsister chromatids.

(**e**) Nonsister chromatids exchange segments. As prophase I ends, the chromosomes continue to condense into thicker, rodlike forms. Then they unzipper from each other except at places where they physically cross each other. Such places are called chiasmata (singular, chiasma, meaning "cross"). The chiasmata migrate toward the chromosome ends. They are evidence of crossovers at various places in the chromosomes.

(**f**) What is the function of crossing over? It breaks up old combinations of alleles and puts new ones together in pairs of homologous chromosomes.

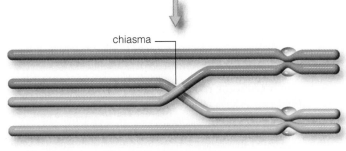

d Crossover between nonsister chromatids of the two chromosomes

chiasma

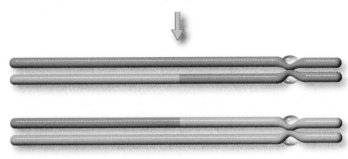

e Nonsister chromatids exchange segments

f Homologues have new combinations of alleles

chromosomes. Nonsister chromatids break at the same places along their length. At these break points, they exchange corresponding segments—that is, genes.

Gene swapping would be pointless if each type of gene never varied. But remember, a gene can come in slightly different forms: alleles. You can bet that some number of the alleles on one chromosome will *not* be identical to their partner alleles on the homologue. Each crossover is a chance to swap slightly different versions of hereditary instructions for particular traits.

We will look at the mechanism of crossing over in later chapters. For now, simply remember this: *Crossing over leads to recombinations among genes of homologous chromosomes, hence to variation in traits among offspring.*

Metaphase I Alignments

Major shufflings of whole chromosomes begin during the transition from prophase I to metaphase I, the second stage of meiosis. Suppose the shufflings are happening right now in one of your germ cells. Crossovers have already made genetic mosaics of the chromosomes, but put this aside in order to simplify tracking. Just call the twenty-three chromosomes you inherited from your mother the *maternal* chromosomes and the twenty-three homologues from your father the *paternal* chromosomes.

Spindle microtubules have already oriented one chromosome of each pair toward one spindle pole and its homologue toward the other (refer to Section 9.3). They are busily moving all of the chromosomes, which soon will become positioned at the spindle's equator.

Will all maternal chromosomes be directed to one spindle pole and all paternal chromosomes directed to the other? Maybe, but probably not. Remember, initial contacts between microtubules and chromosomes are random. Because of the random grabs, the positioning of maternal or paternal chromosomes at the spindle's equator at metaphase I follows no particular pattern. Carry this thought one step further. *Either one* of each pair of homologous chromosomes can end up at either spindle pole after they move apart at anaphase I.

Think about the possibilities when you are tracking merely three pairs of homologues. As you can see from Figure 10.6, by metaphase I, three pairs of homologues may be arranged in any one of four possible positions. Here, eight combinations (2^3) of maternal and paternal chromosomes are possible for forthcoming gametes.

Of course, a human germ cell has twenty-three pairs of homologous chromosomes, not just three. So every time a human germ cell gives rise to sperm or eggs, we can expect a grand total of *8,388,608* (or 2^{23}) possible combinations of maternal and paternal chromosomes!

Moreover, in each sperm or egg, many hundreds of alleles inherited from the mother might not "say" the

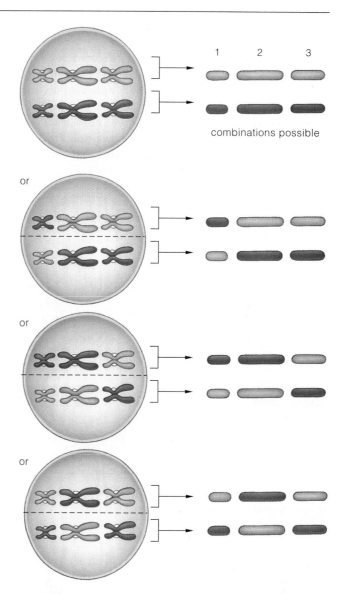

Figure 10.6 Possible outcomes for the random alignment of three pairs of homologous chromosomes at metaphase I. We label three types of chromosomes 1, 2, and 3. The maternal chromosomes are *purple*; paternal ones are *blue*. With merely four possible alignments, eight combinations of maternal and paternal chromosomes are possible in gametes.

exact same thing about hundreds of different traits as the alleles inherited from the father. Are you beginning to get an idea of why such fascinating combinations of traits show up even in the same family?

Crossing over is an interaction between a pair of homologous chromosomes. It breaks up old combinations of alleles and puts new ones together during prophase I of meiosis.

The random attachment and subsequent positioning of each pair of maternal and paternal chromosomes during metaphase I lead to different combinations of maternal and paternal traits in each new generation.

FROM GAMETES TO OFFSPRING

The gametes that form following meiosis are not all the same in their details. For example, human sperm have one tail, opossum sperm have two, and roundworm sperm have none. Crayfish sperm look like pinwheels. Most eggs are microscopic in size, yet an ostrich egg tucked inside its shell is as large as a baseball. From its appearance alone, you might not believe that a plant's gamete is even remotely like an animal's.

Later chapters explain how gametes form in the life cycles of specific organisms, including humans. Figure 10.7 and the rest of this section may help you keep the details in perspective.

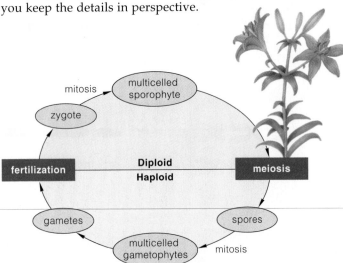

a Generalized life cycle for most kinds of plants

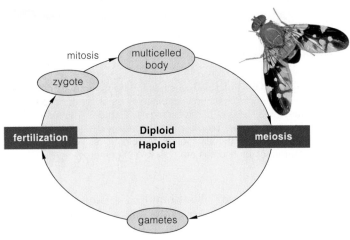

b Generalized life cycle for animals

Figure 10.7 Generalized life cycles for (**a**) most plants and (**b**) animals. The zygote is the first cell that forms when the nuclei of two gametes fuse at fertilization.

For plants, a sporophyte (spore-producing body) develops, by way of mitotic cell divisions, from the zygote. After meiosis, gametophytes (gamete-producing bodies) form. A lily plant is a sporophyte. Gametophytes form in parts of its flowers.

Chapters 21 through 26, 32, 43, and 44 have specific examples of life cycles for representative organisms.

Gamete Formation in Plants

For pine trees, apple trees, roses, dandelions, corn, and nearly all other familiar plants, certain events intervene between meiosis and the time that gametes develop and mature. Among other things, spores form.

Spores are haploid resting cells, often walled, that are good at resisting drought, cold, and other adverse environmental conditions. When favorable conditions return, spores germinate (resume growth) and develop into a haploid body or structure that produces gametes. So *gamete*-producing bodies and *spore*-producing bodies develop during the life cycle of most kinds of plants. Figure 10.7*a* is a generalized diagram of these events.

Gamete Formation in Animals

In male animals, gametes form by a process known as spermatogenesis. A diploid germ cell grows in size in the male's reproductive system. It becomes a primary spermatocyte. This large immature cell enters meiosis and cytoplasmic divisions. Four haploid cells result and develop into spermatids (Figure 10.8). These immature cells change in form and develop a tail. Each becomes a **sperm**, a common type of mature male gamete.

In female animals, gametes form by a process called oogenesis. In human females, for instance, a diploid germ becomes an **oocyte**, or immature egg. Unlike sperm, an oocyte stockpiles many cytoplasmic components, and its four daughter cells differ in size and function (Figure 10.9). As an oocyte divides after meiosis I, one daughter cell (a secondary oocyte) gets nearly all the cytoplasm. The other cell, the first polar body, is small. Later, both cells undergo meiosis II and cytoplasmic division. One daughter cell of the secondary oocyte develops into a second polar body. The other gets most of the cytoplasm and develops into a gamete. The mature female gamete is called an ovum (plural, ova) or, more often, an **egg**.

And so we have one egg and three polar bodies. The polar bodies don't function as gametes and aren't rich in nutrients or cytoplasm. In time they degenerate. But the very fact that they formed means the egg has a suitable (haploid) chromosome number. Also, by getting most of the cytoplasm, the egg has enough start-up machinery to support the new individual right after fertilization.

More Shufflings at Fertilization

The chromosome number characteristic of the parents is restored at **fertilization**, a time when a female and male gamete unite and their haploid nuclei fuse. Fertilization would double the chromosome number for each new generation if meiosis didn't precede it. Such doublings would disrupt the hereditary instructions, usually for

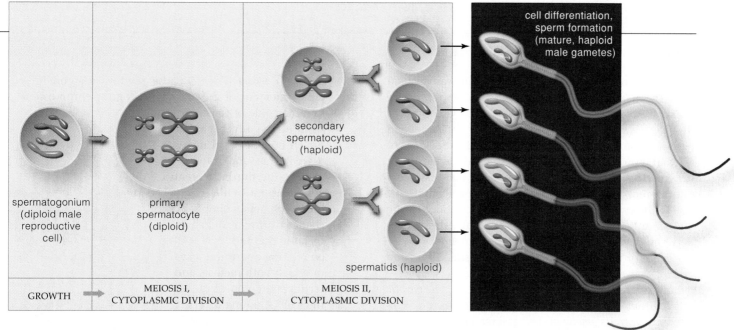

Figure 10.8 Generalized sketch of sperm formation in male animals. Figure 44.2 shows a specific example (how sperm form in human males).

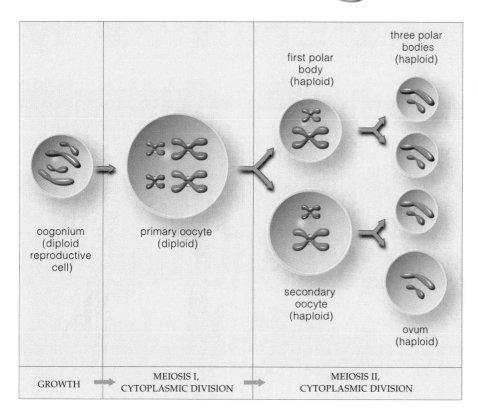

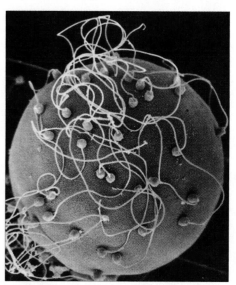

Figure 10.9 Egg formation in female animals. The eggs are far larger than sperm, as the micrograph of sea urchin gametes (*above*) suggests. Also, the three polar bodies are much smaller than the egg. Figure 44.4 gives a specific example (how human eggs form).

the worse. Why? The instructions work as an intricate, fine-tuned package in each individual.

Fertilization also adds to variation among offspring. Reflect upon the possibilities for humans alone. During prophase I, an average of two or three crossovers take place in each human chromosome. Even without these crossovers, the random positioning of pairs of paternal and maternal chromosomes at metaphase I results in one of millions of possible chromosome combinations in each gamete. And of all the male and female gametes that are produced, *which* two actually get together is a matter of chance. The sheer number of combinations that can exist at fertilization is staggering!

Cumulatively, crossing over, the distribution of random mixes of homologous chromosomes into gametes, and fertilization contribute to variation in the traits of offspring.

MEIOSIS AND MITOSIS COMPARED

In this unit our focus has been on two nuclear division mechanisms. Single-celled eukaryotic species reproduce asexually by way of mitosis, followed by cytoplasmic division. Many multicelled eukaryotic species depend on mitosis and cytoplasmic division during episodes of asexual reproduction in their life cycle. All depend on it for growth and tissue repair. By contrast, meiosis occurs only in reproductive cells, such as germ cells that give rise to the gametes used in sexual reproduction. Figure 10.10 summarizes the main similarities and differences between the two nuclear division mechanisms.

The end results of the two mechanisms differ in a crucial way. *Mitotic cell division only produces clones—genetically identical copies of a parent cell. But meiotic cell division, in conjunction with fertilization, promotes variation in traits among offspring.* First, crossing over at prophase I of meiosis puts new combinations of alleles in chromosomes. Second, the random assignment of either member of a pair of homologous chromosomes to either pole of the spindle at metaphase I affects gametes, which end up with mixes of maternal and paternal alleles. And third, different combinations of alleles are brought together simply by chance during fertilization. In later chapters, you will be reading about the ways in which both meiosis and fertilization contribute to the truly stunning diversity and evolution of sexually reproducing organisms.

A *somatic cell* with a diploid chromosome number (2n) is at interphase. Before mitotic division begins, its DNA is replicated (all chromosomes are duplicated).

Figure 10.10 Summary of mitosis and meiosis. Both diagrams use a diploid (2n) animal cell as the example. They are arranged to help you compare similarities and differences between the division mechanisms. The maternal chromosomes are coded *purple* and the paternal chromosomes *blue*.

MEIOSIS I

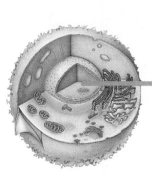

A *germ cell* with a diploid chromosome number (2n) is at interphase. Before mitotic division begins, its DNA is replicated (all chromosomes are duplicated).

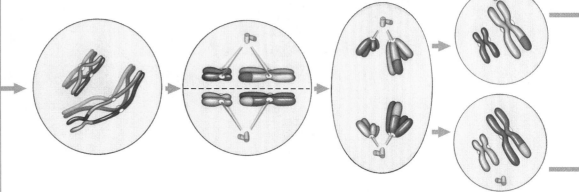

PROPHASE I

Each duplicated chromosome (consisting of two sister chromatids) condenses to threadlike form, then rodlike form. *Crossing over* occurs. Each chromosome unzips from its homologue. Each gets attached to the spindle in transition to metaphase.

METAPHASE I

All chromosomes are now positioned at the spindle's equator.

ANAPHASE I

Each chromosome is separated from its homologue. They are moved to opposite poles of the spindle.

TELOPHASE I

When the cytoplasm divides, there are two cells. Each has a haploid (n) number of chromosomes, but these are still in the duplicated state.

MITOSIS

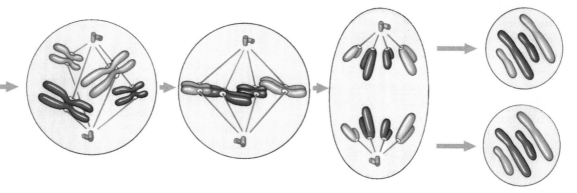

PROPHASE

Each duplicated chromosome (consisting of two sister chromatids) condenses from threadlike form to rodlike form. Each gets attached to the spindle during the transition to metaphase.

METAPHASE

All chromosomes are now positioned at the spindle's equator.

ANAPHASE

Sister chromatids of each chromosome are separated from each other. These new, daughter chromosomes are moved to opposite poles of the spindle.

TELOPHASE

When the cytoplasm divides, there are two cells. Each is diploid (2n)—*it has the same chromosome number as the parent cell.*

MEIOSIS II

There is no DNA replication between the two divisions.

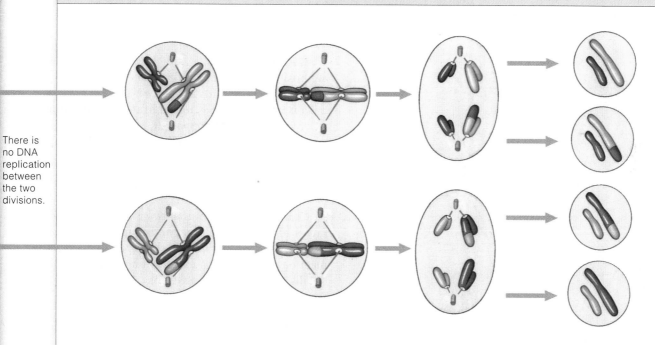

PROPHASE II

Before prophase II, the two centrioles in each new cell were moved apart and a new spindle formed. Now, each chromosome becomes attached to the spindle and starts moving toward its equator.

METAPHASE II

All chromosomes are now positioned at the spindle's equator.

ANAPHASE II

Sister chromatids of each chromosome are separated from each other. These new, daughter chromosomes are moved to opposite poles of the spindle.

TELOPHASE II

Four daughter nuclei form. When the cytoplasm divides, each new cell is haploid (n). *The original chromosome number has been reduced by half.* One or all of these cells may become gametes.

1. Eukaryotic life cycles often have asexual as well as sexual phases. Asexual reproduction results in clones. Sexual reproduction (by meiosis, gamete formation, and fertilization) leads to variation in traits. *C1, 10.1*

   a. Meiosis, a nuclear division mechanism, reduces the chromosome number of a parent germ cell by half. It precedes the formation of haploid gametes, such as sperm in males and eggs in females.

   b. At fertilization, a sperm and an egg nuclei fuse, which restores the chromosome number (Figure 10.11).

2. A germ cell with a diploid chromosome number (2*n*) has *two* of each type of chromosome characteristic of its species. Commonly, one of each pair of chromosomes is maternal and the other is paternal. *CI, 10.1*

3. Each pair of maternal and paternal chromosomes has homology; the two are alike. Except for a pairing of nonidentical sex chromosomes (e.g., X with Y), the two have the same length, shape, and gene sequence. They interact during meiosis. *10.2*

4. Chromosomes become duplicated during interphase. Each consists of two DNA molecules that stay attached (as sister chromatids) during meiosis. *10.2*

5. Meiosis consists of two consecutive divisions. Both require a microtubular spindle apparatus. *10.2–10.3*

   a. During meiosis I, spindle microtubules attach to the centromere region of each duplicated chromosome. Motor proteins attached to the microtubules separate it from its partner, the homologous chromosome.

   b. In meiosis II, similar interactions move the sister chromatids of each chromosome away from each other.

6. Meiosis I, the first nuclear division, is characterized by the following events and outcomes: *10.2–10.4*

   a. Crossing over occurs in prophase I. Two nonsister chromatids of each pair of homologous chromosomes break at corresponding sites and exchange segments. This puts new combinations of alleles together. Alleles (slightly different molecular forms of the same gene) specify different versions of the same trait.

   b. Different combinations of alleles lead to variation in the details of a given trait among offspring.

   c. Also in prophase I, a microtubular spindle forms outside the nucleus, and the nuclear envelope starts to break up. In cells with duplicated pairs of centrioles, one pair starts moving to the opposite spindle pole.

   d. All of the pairs of homologous chromosomes have become positioned at the spindle equator at metaphase I. Both the maternal chromosome and its homologue have become oriented at random, toward either pole.

   e. In anaphase I, spindle microtubules interact with each duplicated chromosome and move it away from its homologue, to opposite spindle poles.

7. Meiosis II (second nuclear division) is characterized by these events and outcomes: *10.3*

   a. At metaphase II, all the duplicated chromosomes are positioned at the spindle equator.

   b. Sister chromatids are moved apart in anaphase II. Each is now a separate, unduplicated chromosome.

   c. By the end of telophase II, four nuclei—each with a haploid chromosome number (*n*)—have been formed.

8. When the cytoplasm divides, there are four haploid cells. One or all may serve as gametes (or as plant spores that give rise to gamete-producing bodies). *10.3, 10.5*

9. Crossing over, the chance allocation of different mixes of pairs of maternal and paternal chromosomes to different gametes, and the chance of any two gametes meeting at fertilization all contribute to the variation in details of traits among offspring. *10.5, 10.6*

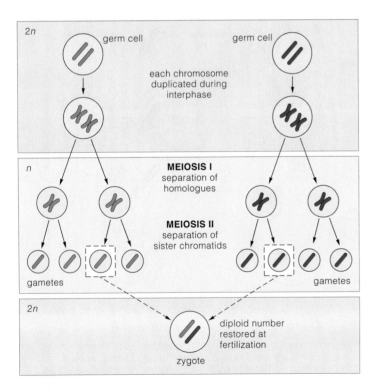

Figure 10.11 Summary of changes in chromosome number at different stages of sexual reproduction, using two diploid (2*n*) germ cells as the example. Meiosis reduces the chromosome number by half (*n*). Union of haploid nuclei of two gametes at fertilization restores the diploid number.

Review Questions

1. The diploid chromosome numbers for the somatic cells of a few organisms are listed at *right*. How many chromosomes will end up in the gametes of each organism? *10.2*

Fruit fly, *Drosophila melanogaster*	8
Garden pea, *Pisum sativum*	14
Corn, *Zea mays*	20
Frog, *Rana pipiens*	26
Earthworm, *Lumbricus terrestris*	36
Human, *Homo sapiens*	46
Chimpanzee, *Pan troglodytes*	48
Amoeba, *Amoeba*	50
Horsetail, *Equisetum*	216

2. A diploid germ cell has four pairs of homologous chromosomes, designated AA, BB, CC, and DD. Which of the chromosomes will be present in gametes? *10.2, 10.3*

3. Look at the chromosomes in the germ cell in the diagram at *right*. Is this cell at anaphase I or anaphase II? *10.3*

The cell is at anaphase ___ rather than anaphase ___ . I know this because:

_____
_____
_____

4. Define meiosis and describe its stages. In what respects is meiosis *not* like mitosis? *10.2, 10.6*

5. Actor Michael Douglas (Figure 10.12*a*) inherited a gene from each parent that influences the chin dimple trait. One form of the gene called for a dimple and the other didn't, but one is all it takes for this particular trait. Figure 10.12*b* shows what the chin of Mr. Douglas might have looked like if he had inherited two ordinary forms of the gene instead. What is the name for the alternative forms of the same gene? *10.1, 10.4*

6. Outline the main steps by which gametes form in plants. Do the same for gamete formation in animals. *10.5*

7. Genetically speaking, what is the key difference between the outcomes of sexual and asexual reproduction? *10.2, 10.6*

Self-Quiz   ANSWERS IN APPENDIX III

1. Sexual reproduction requires _____ .
   a. meiosis   c. fertilization
   b. gamete formation d. all of the above

2. An animal cell having two rather than one of each type of chromosome has a _____ chromosome number.
   a. diploid   c. normal gamete
   b. haploid   d. both b and c

3. Generally, a pair of homologous chromosomes _____ .
   a. carries the same genes   c. interacts at meiosis
   b. has the same length, shape   d. all of the above

4. Meiosis _____ the parental chromosome number.
   a. doubles   c. maintains
   b. reduces   d. corrupts

5. Meiosis is a division mechanism that produces _____ .
   a. two cells   c. eight cells
   b. two nuclei   d. four nuclei

6. Before the onset of meiosis, all chromosomes are _____ .
   a. condensed   c. duplicated
   b. released from protein d. both b and c

7. Duplicated chromosomes move away from their homologue and end up at the opposite spindle pole during _____ .
   a. prophase I   c. anaphase I
   b. prophase II   d. anaphase II

8. Sister chromatids of each duplicated chromosome move apart and end up at opposite spindle poles during _____ .
   a. prophase I   c. anaphase I
   b. prophase II   d. anaphase II

9. Match each term with its description.
   ____ chromosome   a. different molecular forms
          number   of the same gene
   ____ alleles   b. none between meiosis I and II
   ____ metaphase I   c. pairs of homologous
   ____ interphase   chromosomes are now aligned
                          at the spindle equator
                       d. sum total of chromosomes
                          in all cells of a given type

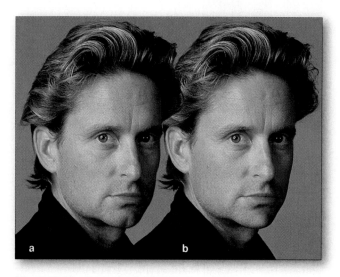

Figure 10.12 Example of the chin dimple trait (actually a fissure in the chin surface).

Critical Thinking

1. Explain why you can expect meiosis rather than mitosis to give rise to genetic differences between parent cells and their daughter cells.

2. An organism has four pairs of homologous chromosomes (AA, BB, CC, and DD). If it self-fertilizes, what chromosome combinations can you expect to see in its offspring?

3. Assume that you can measure the amount of DNA in the nucleus of a primary oocyte, then in the nucleus of a primary spermatocyte. Each gives you a mass *m*. What mass of DNA would you expect to find inside the nucleus of each mature gamete (egg and sperm) that forms after meiosis? What mass of DNA would you expect to find (1) in the nucleus of the zygote formed at fertilization and (2) in that zygote's nucleus after the first DNA duplication?

4. As you know, aphids can reproduce asexually and sexually. Females reproduce by themselves and give birth to live females in summer. They also reproduce sexually, and lay eggs that survive in summer. They also reproduce sexually, and lay eggs that survive the winter and hatch in spring. Aphids happen to be tasty to various predators. Speculate on how their reproductive flexibility might be an adaptation to agents of predation.

Selected Key Terms

allele *10.1*   germ cell *CI*
asexual reproduction *10.1*   haploid number *10.2*
chromosome number *10.2*   homologous chromosome *10.2*
crossing over *10.4*   meiosis *10.2*
diploid number *10.2*   oocyte *10.5*
egg (ovum) *10.5*   sexual reproduction *10.1*
fertilization *10.5*   sister chromatid *10.2*
gamete *CI*   sperm *10.5*
gene *10.1*   spore *10.5*

Readings

Klug, W., and M. Cummings. 2000. *Concepts of Genetics.* Sixth edition. New York: Macmillan.

Ridley, M., June 2001. "Sex, Errors, and the Genome." *Natural History.*

11

OBSERVABLE PATTERNS OF INHERITANCE

A Smorgasbord of Ears and Other Traits

Basketball ace Charles Barkley has them. So does actor Tom Cruise. Actress Joan Chen doesn't, and neither did a monk named Gregor Mendel. To see how you fit in with these folks, use a mirror to check out your ears. Is the fleshy lobe at the base of each ear attached to the side of your head? If so, you and Barkley and Cruise have something in common. Or is the fleshy lobe not attached, so that you can flap it back and forth? If so, you are like Chen and Mendel (Figure 11.1).

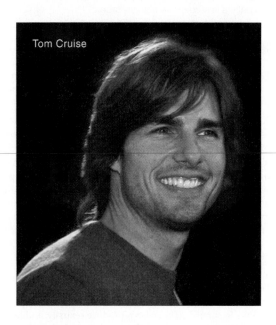
Tom Cruise

Charles Barkley

Figure 11.1 Attached and detached earlobes of a few representative humans. This sampling provides observable evidence of a trait governed by genes, which occur in different molecular forms in the human population. Which version of the trait do you have? It depends on which molecular forms of certain genes you inherited from your mother and father. As Gregor Mendel perceived, observable traits can be used to identify patterns of inheritance from one generation to the next.

Whether a person is born with attached or detached earlobes depends on which genes he or she inherited. Genes come in slightly different molecular forms—alleles. Some of them have information about detached lobes. The information is put to use as a human body is developing inside the mother. It has to do with a death warrant, a molecular signal sent to all cells positioned between the head and the forming lobes. That signal, as well as the receptors for it on target cells, must be properly synthesized at the proper time. If not, cells don't die and earlobes don't detach.

We all have genes for thousands of traits, including earlobes, cheeks, lashes, and eyeballs. Most traits vary in their details from one person to the next. Remember, we inherit pairs of genes on pairs of chromosomes. In some pairs, one allele has strong effects and overwhelms the other allele's contribution. The outgunned allele is said to be recessive to the dominant one. If you have

detached earlobes, dimpled cheeks, long lashes, or large eyeballs, you have some number of dominant alleles that affect the trait in predictable ways.

When both alleles of a pair are recessive, nothing masks their effect on a trait. You get *attached* earlobes with certain pairs of recessive alleles, *flat* feet with other pairs, a *straight* nose with other pairs, and so on.

How did we discover such remarkable things about our genes? It started with Mendel. By analyzing garden

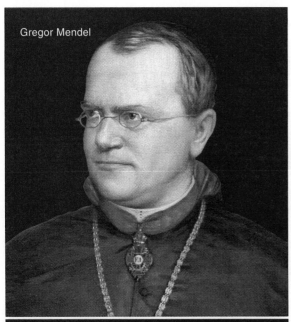

Gregor Mendel

Joan Chen

pea plants generation after generation, Mendel found indirect but *observable* evidence of how parents bestow units of hereditary information—genes—on offspring. This chapter focuses on both the methods and the results of Mendel's experiments. They remain a classic example of how a scientific approach can pry open important secrets about the natural world. And to this day, they serve as the foundation for modern genetics.

Key Concepts

1. Genes are units of information that influence the expression of heritable traits. Alleles, which are slightly different molecular forms of a gene, may affect traits in different ways.

2. Genes have specific locations on the chromosomes of a species. Humans, pea plants, and other organisms with a diploid chromosome number inherit *pairs* of genes, at equivalent locations on pairs of homologous chromosomes.

3. When the two members of each pair of homologous chromosomes are separated from each other during meiosis, their pairs of genes are moved apart, also, and end up in separate gametes. Gregor Mendel found indirect evidence of this gene segregation when he crossbred pea plants showing different versions of the same trait, such as purple versus white flowers.

4. Each pair of homologous chromosomes in a germ cell is sorted out for distribution into one gamete or another independently of how the other pairs of homologous chromosomes are assorted. Mendel discovered indirect evidence of this when he tracked many plants having observable differences in two traits, such as flower color and plant height.

5. Nonidentical alleles affect the forms of traits that Mendel happened to study. One allele is said to be dominant, in that its effect on a trait masks the effect of a recessive allele paired with it.

6. Not all traits have clearly dominant or recessive forms. One allele of a pair may be fully or partially dominant over its partner or codominant with it. Two or more gene pairs often influence the same trait, and some single genes influence many traits.

7. The environment introduces variation in traits.

MENDEL'S INSIGHT INTO INHERITANCE PATTERNS

More than a century ago, people wondered about the basis of inheritance. It was common knowledge that sperm and eggs both transmit information about traits to offspring. But few suspected that the information is organized in units (genes). Instead, the idea was that a father's blob of information "blended" with the mother's blob, like cream into coffee, at fertilization.

However, carried to its logical conclusion, blending would slowly dilute a population's shared pool of hereditary information until there was only a single intermediate version left of each trait. If that were so, then why did, say, freckles keep showing up among children of nonfreckled parents through the generations? Why weren't all the descendants of a herd of white stallions and black mares gray? The blending theory could scarcely explain the obvious variation in traits that people could observe with their own eyes. Nevertheless, few disputed the theory.

Yet it could not be reconciled with Charles Darwin's theory of natural selection. According to a key premise of Darwin's theory, individuals of a population show variation in heritable traits. Variations that improve the chance of surviving and reproducing show up with greater frequency than those that do not in subsequent generations. Less advantageous variations may persist among fewer individual or may vanish. It is not that separate versions of a given trait are "blended out" of the population. Rather, *each version of a trait may persist in a population, at frequencies that can change over time.*

Just before Darwin presented his theory, someone was gathering evidence that eventually would support his key premise. A monk, Gregor Mendel, had already guessed that sperm and eggs carry distinct "units" of information about heritable traits. By carefully analyzing traits of pea plants generation after generation, Mendel found indirect but *observable* evidence of how parents transmit genes to offspring.

Mendel's Experimental Approach

Mendel spent most of his adult life in a monastery in Brno, a city near Vienna that has since become part of the Czech Republic. However, Mendel was not a man of narrow interests who accidentally stumbled onto principles of great import. The monastery of St. Thomas was close to European capitals that were the centers of scientific inquiry.

Having been raised on a farm, Mendel was aware of agricultural principles and their applications. He kept abreast of the breeding experiments and developments described in the available literature. He was a member of the regional agricultural society. He won awards for developing improved varieties of vegetables and fruits. Shortly after entering the monastery, he took courses in mathematics, physics, and botany at the University of Vienna. Few scholars of his time showed interest in both plant breeding and mathematics.

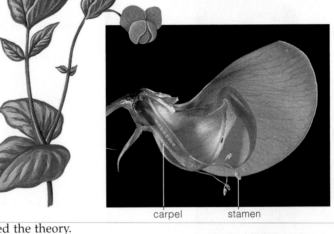

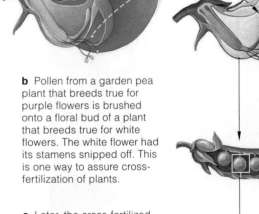

carpel   stamen

Figure 11.2 Garden pea plant (*Pisum sativum*), the organism Mendel chose for experiments on his ideas about inheritance.

a Garden pea flower. The section shows the location of its stamens and carpel. Sperm-producing pollen grains form in stamens. Eggs develop, fertilization takes place, and seeds mature inside carpels.

b Pollen from a garden pea plant that breeds true for purple flowers is brushed onto a floral bud of a plant that breeds true for white flowers. The white flower had its stamens snipped off. This is one way to assure cross-fertilization of plants.

c Later, the cross-fertilized plant produces seeds (in pea pods). Each seed is allowed to grow and develop into a new plant.

d The flower color of each new plant can be used as visible evidence of patterns in how hereditary material might be transmitted to it from each parent plant.

Shortly after his university training, Mendel began experimenting with the garden pea plant, *Pisum sativum* (Figure 11.2). This plant is self-fertilizing. Its male and female gametes (call them sperm and eggs) develop in different parts of the same flower, where fertilization occurs. Individual pea plants generally breed true for certain traits. In other words, successive generations are just like the parents in one or more traits, as when all offspring grown from seeds of self-fertilized, white-flowered parent plants have white flowers.

Pea plants may be cross-fertilized by transferring pollen from one plant's flower to the flower of another plant. Mendel knew he could open the flower buds of a plant that bred true for a trait, such as white flowers, and snip out its stamens. (Stamens bear pollen grains in which sperm develop.) He could brush those buds with pollen from a plant that bred true for a *different* version of the same trait—say, purple flowers. As he hypothesized, such clearly observable differences would help him track a given trait through many generations. If there were patterns to the trait's inheritance, *then the patterns might tell him something about heredity itself.*

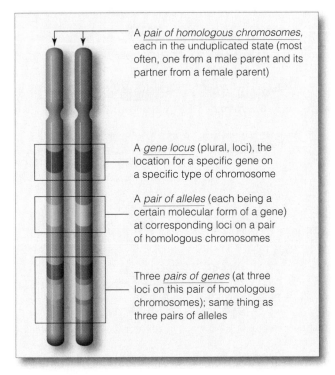

A *pair of homologous chromosomes*, each in the unduplicated state (most often, one from a male parent and its partner from a female parent)

A *gene locus* (plural, loci), the location for a specific gene on a specific type of chromosome

A *pair of alleles* (each being a certain molecular form of a gene) at corresponding loci on a pair of homologous chromosomes

Three *pairs of genes* (at three loci on this pair of homologous chromosomes); same thing as three pairs of alleles

Figure 11.3 A few genetic terms illustrated. Diploid organisms have pairs of genes, on pairs of homologous chromosomes. For example, you inherited one chromosome of each pair from your mother and the other, homologous chromosome from your father.

Most genes come in slightly different molecular forms, called alleles. Different alleles specify different versions of the same trait. An allele at one location on a chromosome may or may not be identical to its partner on the homologous chromosome.

Some Terms Used in Genetics

Having read the chapter on meiosis, you already have insight into the mechanisms of sexual reproduction, which is more than Mendel had. Neither he nor anyone else of his era knew about chromosomes. So he could not have known that a chromosome number is reduced by half in gametes, then restored when gametes meet at fertilization. Even so, Mendel sensed what was going on. As we follow his thinking, let's simplify the story by substituting a few of the modern terms used in studies of inheritance (see also Figure 11.3):

1. **Genes** are units of information about specific traits, and they are passed from parents to offspring. Each gene has a specific location (locus) on a chromosome.

2. Cells with a diploid chromosome number ($2n$) have pairs of genes, on pairs of homologous chromosomes.

3. Mutation can alter a gene's molecular structure. The alteration may change the gene's information about a trait (as when the gene for flower color specifies purple and a mutated version specifies white). All the different molecular forms of the same gene are called **alleles**.

4. When offspring inherit a pair of *identical* alleles for a trait, generation after generation, they represent a **true-breeding lineage**. Offspring of a cross between two different true-breeding individuals are **hybrids**, which have inherited *nonidentical* alleles for a trait.

5. When both alleles of a pair are identical, this is a *homozygous* condition. When the two are not identical, this is a *heterozygous* condition.

6. An allele is *dominant* when its effect on a trait masks that of any *recessive* allele paired with it. Mendel used capital letters for dominant alleles and lowercase letters for recessive ones. *A* and *a* are examples.

7. Putting this all together, a **homozygous dominant** individual has a pair of dominant alleles (*AA*) for the trait being studied. A **homozygous recessive** individual has a pair of recessive alleles (*aa*). And a **heterozygous** individual has a pair of nonidentical alleles (*Aa*).

8. Two terms help keep the distinction clear between genes and the traits they specify. **Genotype** refers to the particular alleles an individual carries. **Phenotype** refers to an individual's observable traits.

9. When tracking the inheritance of traits through generations of offspring, these abbreviations apply:

P	parental generation
F$_1$	first-generation offspring
F$_2$	second-generation offspring

MENDEL'S THEORY OF SEGREGATION

Mendel had an idea. In every generation, a plant might inherit two "units" (genes) of information about a trait, one from each parent. He used **monohybrid crosses** to test his idea. For such crosses, two parents that breed true for different forms of a trait produce F_1 offspring that are heterozygous ($AA \times aa \longrightarrow Aa$). The actual experiment is an intercross between two of the identical F_1 heterozygotes—the "monohybrids" ($Aa \times Aa$).

Predicting Outcomes of Monohybrid Crosses

Mendel tracked many traits over two generations. For one set of experiments, he crossed true-breeding purple-flowered plants and true-breeding white-flowered ones. All plants of the next generation had purple flowers. Mendel let those F_1 plants self-fertilize. Later on, some of the F_2 offspring produced white flowers!

If Mendel's hypothesis were correct—if each plant inherited two units of information about flower color—then the "purple" unit had to be dominant. Why? It masked the unit for "white" in the F_1 plants.

Let's rephrase his thinking. Germ cells of pea plants are diploid, with pairs of homologous chromosomes. Assume one parent is homozygous dominant (AA) and the other is homozygous recessive (aa) for flower color. Following meiosis, a sperm or egg carries one allele for flower color (Figure 11.4). So when a sperm fertilizes an egg, only one outcome is possible: $A + a \longrightarrow Aa$.

Mendel knew about sampling error (Section 1.5). He crossed a great many plants and tracked thousands of offspring. He also counted and recorded the number of dominant and recessive plants. On average, three of every four F_2 plants had the dominant phenotype, and one had the recessive phenotype (Figure 11.5).

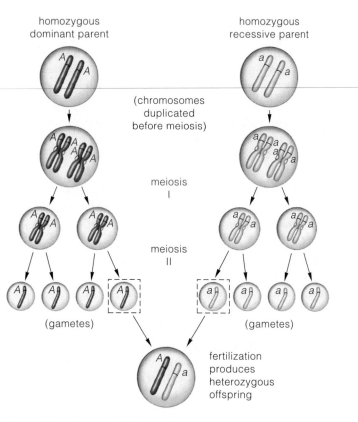

Figure 11.4 Monohybrid cross, showing how one gene of a pair segregates from the other gene. Two parents that breed true for two versions of a trait produce only heterozygous offspring.

Figure 11.5 *Right:* Numerical results from Mendel's monohybrid cross experiments with the garden pea plant (*P. sativum*). The numbers are his counts of the F_2 plants that carried dominant or recessive hereditary "units" (alleles) for the trait. On average, the dominant-to-recessive ratio was 3:1.

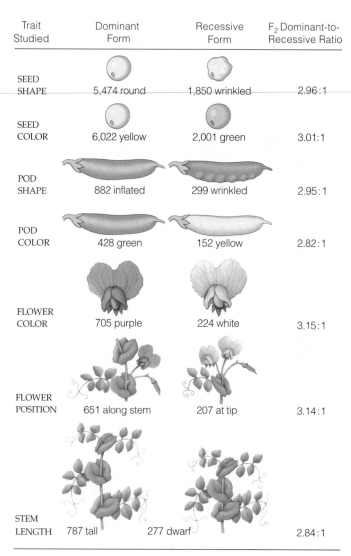

Trait Studied	Dominant Form	Recessive Form	F_2 Dominant-to-Recessive Ratio
SEED SHAPE	5,474 round	1,850 wrinkled	2.96:1
SEED COLOR	6,022 yellow	2,001 green	3.01:1
POD SHAPE	882 inflated	299 wrinkled	2.95:1
POD COLOR	428 green	152 yellow	2.82:1
FLOWER COLOR	705 purple	224 white	3.15:1
FLOWER POSITION	651 along stem	207 at tip	3.14:1
STEM LENGTH	787 tall	277 dwarf	2.84:1
		Average ratio for all traits studied:	**3:1**

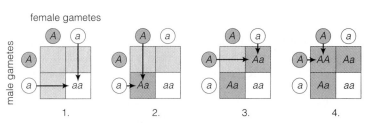

female gametes

male gametes

1.   2.   3.   4.

Figure 11.6 Punnett-square method of predicting the probable outcome of a genetic cross. Circles represent gametes. *Italic* letters on gametes represent dominant or recessive alleles. In the squares are the different genotypes possible among offspring. In this case, gametes are from a self-fertilizing heterozygous (*Aa*) plant.

Figure 11.7 *Right*: Results from one of Mendel's monohybrid crosses. On average, the dominant-to-recessive ratio among the second-generation (F$_2$) plants was 3:1.

To Mendel, the ratio suggested that fertilization is a chance event with a number of possible outcomes. And he had an understanding of probability, which applies to chance events *and therefore could help him predict the possible outcomes of genetic crosses.* **Probability** simply means this: The chance that each outcome of a given event will occur is proportional to the number of ways in which the event can be reached.

The **Punnett-square method**, explained in Figure 11.6 and applied in Figure 11.7, may help you visualize the possibilities. As you can see, if half of a plant's sperm (or eggs) were *a* and half were *A*, then four outcomes would be possible each time a sperm fertilized an egg:

POSSIBLE EVENT:	PROBABLE OUTCOME:
sperm *A* meets egg *A*	1/4 *AA* offspring
sperm *A* meets egg *a*	1/4 *Aa*
sperm *a* meets egg *A*	1/4 *Aa*
sperm *a* meets egg *a*	1/4 *aa*

By this prediction, an F$_2$ plant has three chances in four of getting at least one dominant allele (purple flowers). It has one chance in four of getting two recessive alleles (white flowers). That is a probable phenotypic ratio of three purple to one white, or 3:1.

Mendel's observed ratios were not *exactly* 3:1. You can see this for yourself by looking at the numerical results listed in Figure 11.5. Why did Mendel put aside the deviations? To understand why, flip a coin several times. As we all know, a coin is just as likely to end up heads as tails. But often a coin ends up heads, or tails, several times in a row. So if you flip the coin only a few times, the observed ratio might differ greatly from the predicted ratio of 1:1. Flip the coin many, many times, and you are more likely to come close to the predicted ratio. Mendel understood the rules of probability—and observed a large number of offspring. Almost certainly, this kept him from being confused by minor deviations from the predicted results of the experimental crosses.

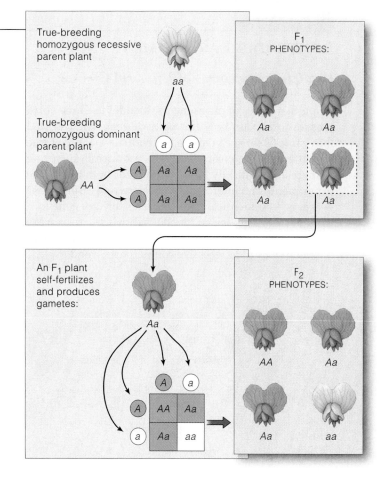

Testcrosses

By running **testcrosses**, Mendel gained support for his prediction. In this type of experimental test, an organism shows dominance for a specified trait but its genotype is unknown, so it is crossed to a known homozygous recessive individual. Test results may reveal whether the organism is homozygous dominant or heterozygous.

Mendel tested one prediction that purple-flowered F$_1$ offspring were heterozygous by crossing them with true-breeding, white-flowered plants. If they were all homozygous dominant, then all the testcross offspring would show the dominant form of the trait. If they were heterozygous, then there would be about as many dominant as recessive plants. Sure enough, when old enough to flower, about half of the testcross offspring had purple flowers (*Aa*) and half had white (*aa*). Can you construct two Punnett squares to show the possible outcomes of this testcross?

The results from Mendel's monohybrid crosses and testcrosses became the basis of a theory of **segregation**, which we state here in modern terms:

MENDEL'S THEORY OF SEGREGATION Diploid cells have pairs of genes, on pairs of homologous chromosomes. The two genes of each pair are separated from each other during meiosis, so they end up in different gametes.

INDEPENDENT ASSORTMENT

Predicting Outcomes of Dihybrid Crosses

In another series of experiments, Mendel used **dihybrid crosses** to explain how *two* pairs of genes are assorted into gametes. In such crosses, individuals that breed true for different versions of *two* traits produce F$_1$ offspring that are all identically heterozygous for both traits. The experiment is an intercross between two F$_1$ "dihybrids" —that is, two identical heterozygotes for two gene loci.

Let's diagram one of Mendel's dihybrid crosses. We can use *A* for flower color and *B* for height as dominant alleles and, as their recessive counterparts, *a* and *b*:

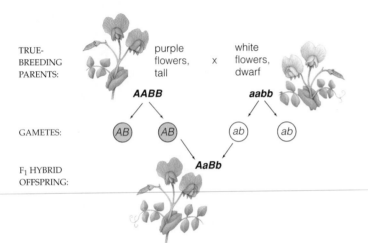

TRUE-
BREEDING
PARENTS: purple flowers, tall x white flowers, dwarf

AABB   **aabb**

GAMETES: AB   AB   ab   ab

F$_1$ HYBRID
OFFSPRING: **AaBb**

As Mendel would have predicted, the F$_1$ offspring from this cross are all purple-flowered and tall (*AaBb*).

When those F$_1$ plants reproduce, how will the two gene pairs be assorted into gametes? The answer partly depends on which chromosomes carry the pairs. Assume that one pair of homologous chromosomes carry the *Aa* alleles and a different pair carry the *Bb* alleles. Now think of how all chromosomes become positioned at the spindle equator during metaphase I of meiosis (Figures 10.6 and 11.8). The chromosome with the *A* allele might be positioned to move to either spindle pole (then into one of four gametes). Its homologue with the *a* allele might also move to either spindle pole. And the same can happen to the homologous chromosomes that carry the *B* and *b* alleles. After meiosis and gamete formation, then, four combinations of alleles are possible in sperm or eggs: 1/4 *AB*, 1/4 *Ab*, 1/4 *aB*, and 1/4 *ab*.

Given the alternative alignments of chromosomes at metaphase I, several allelic combinations are possible at fertilization. Simple multiplication (four kinds of sperm times four kinds of eggs) tells us sixteen combinations of gametes are possible in the F$_2$ offspring of a dihybrid cross. Use the Punnett-square method to diagram the probabilities (Figure 11.9). Now add up all the possible phenotypes and you get 9/16 tall purple-flowered, 3/16 dwarf purple-flowered, 3/16 tall white-flowered, and 1/16 dwarf white-flowered plants. That is a probable phenotypic ratio of 9:3:3:1. Results from one dihybrid cross that Mendel described were close to this ratio.

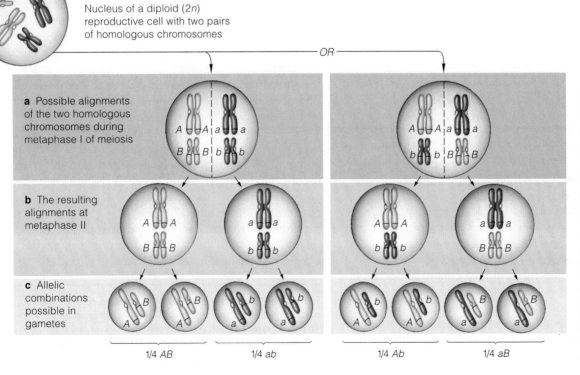

Nucleus of a diploid (2*n*) reproductive cell with two pairs of homologous chromosomes

OR

a Possible alignments of the two homologous chromosomes during metaphase I of meiosis

b The resulting alignments at metaphase II

c Allelic combinations possible in gametes

1/4 *AB*   1/4 *ab*   1/4 *Ab*   1/4 *aB*

Figure 11.8 One case of independent assortment. This example tracks two pairs of homologous chromosomes. An allele at a given locus on a chromosome may or may not be identical with its partner allele on the homologous chromosome. Either chromosome of a pair may become attached to either pole of the spindle during meiosis. As you can see, when just two pairs are being tracked, two different metaphase I lineups are possible.

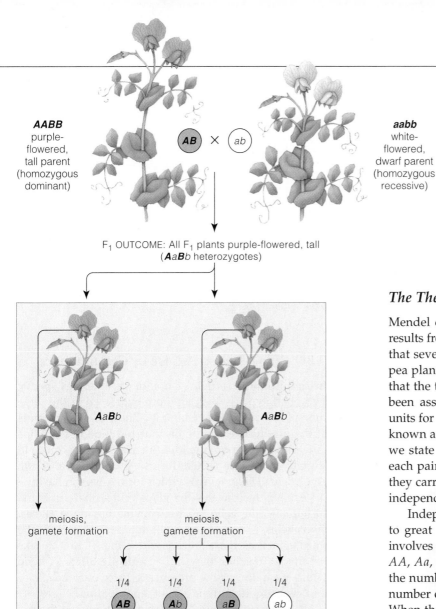

AABB
purple-
flowered,
tall parent
(homozygous
dominant)

AB × ab

aabb
white-
flowered,
dwarf parent
(homozygous
recessive)

F$_1$ OUTCOME: All F$_1$ plants purple-flowered, tall
(**A**a**B**b heterozygotes)

AaBb

AaBb

meiosis,
gamete formation

meiosis,
gamete formation

	1/4 AB	1/4 Ab	1/4 aB	1/4 ab
1/4 AB	1/16 **AABB**	1/16 **AABb**	1/16 **AaBB**	1/16 **AaBb**
1/4 Ab	1/16 **AABb**	1/16 **AAbb**	1/16 **AaBb**	1/16 **Aabb**
1/4 aB	1/16 **AaBB**	1/16 **AaBb**	1/16 **aaBB**	1/16 **aaBb**
1/4 ab	1/16 **AaBb**	1/16 **Aabb**	1/16 **aaBb**	1/16 *aabb*

Possible outcomes of cross-fertilization

ADDING UP THE F$_2$ COMBINATIONS POSSIBLE:

☐ 9/16 or 9 purple-flowered, tall

☐ 3/16 or 3 purple-flowered, dwarf

☐ 3/16 or 3 white-flowered, tall

☐ 1/16 or 1 white-flowered, dwarf

Figure 11.9 Results from Mendel's dihybrid cross between parent plants that bred true for different versions of two traits: flower color and plant height. *A* and *a* represent the dominant and recessive alleles for flower color. *B* and *b* signify the dominant and recessive alleles for plant height. As the Punnett square indicates, the probabilities of certain combinations of phenotypes among the F$_2$ offspring occur in a 9:3:3:1 ratio, on average.

The Theory in Modern Form

Mendel could do no more than analyze the numerical results from his dihybrid crosses, because he didn't know that seven pairs of homologous chromosomes carry the pea plant's "units" of inheritance. It just seemed to him that the two units for the first trait he was tracking had been assorted into gametes independently of the two units for the other trait. In time his interpretation became known as the theory of **independent assortment**, which we state here in modern terms: By the end of meiosis, each pair of homologous chromosomes—and the genes they carry—have been sorted for shipment into gametes independently of how the other pairs were sorted out.

Independent assortment and hybrid intercrosses lead to great genetic variation. In a monohybrid cross that involves only one gene pair, three genotypes are possible: *AA*, *Aa*, and *aa*. We can represent this as 3^n, where *n* is the number of gene pairs. With more pairs of genes, the number of possible combinations increases dramatically. When the parents differ in ten gene pairs, almost 60,000 genotypes are possible. When they differ in twenty gene pairs, the number approaches 3.5 billion!

In 1865, Mendel presented his ideas to the Brünn Natural History Society. His ideas had little impact. The next year he published a paper, and apparently it was read by few and understood by no one. In 1871 he became abbot of the monastery, and his experiments gradually gave way to administrative tasks. He died in 1884, never to know his experiments would become the starting point for the development of modern genetics.

Mendel's segregation theory still stands. Hereditary material is indeed organized in units (genes) that retain their identity and are segregated for distribution into gametes. But the theory of independent assortment does not apply to *all* gene combinations, as you will see in the next chapter.

MENDEL'S THEORY OF INDEPENDENT ASSORTMENT By the end of meiosis, genes on pairs of homologous chromosomes have been sorted out for distribution into one gamete or another independently of gene pairs of other chromosomes.

DOMINANCE RELATIONS

For the most part, Mendel studied traits having clearly dominant or recessive forms. As the remaining sections of this chapter will make clear, however, the expression of other traits is not as straightforward.

Incomplete Dominance

With **incomplete dominance**, one allele of a pair is not fully dominant over its partner, so the phenotype of the heterozygote is *somewhere in between* the phenotypes of the two homozygotes. Example: Cross a true-breeding red snapdragon and a true-breeding white one. All F_1 offspring will have pink flowers. Cross two F_1 plants and you can expect red, pink, and white snapdragons in a predictable ratio (Figure 11.10). What causes this inheritance pattern? Red snapdragons have two alleles that allow them to make an abundance of red pigment molecules. White snapdragons have two mutant alleles that make them pigment-free. Pink snapdragons are heterozygous; the red allele they carry specifies enough pigment to make flowers pink, but not red.

homozygous parent   ×   homozygous parent

All F_1 offspring are heterozygous for flower color:

Cross two of the F_1 plants, and the F_2 offspring will show three phenotypes in a 1:2:1 ratio:

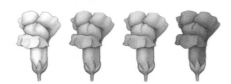

Figure 11.10 Visible evidence of incomplete dominance in heterozygous (pink) snapdragons, in which an allele for red pigment is paired with a "white" allele.

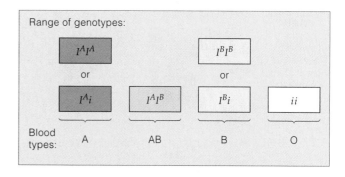

Range of genotypes:

$I^A I^A$		$I^B I^B$	
or		or	
$I^A i$	$I^A I^B$	$I^B i$	ii

Blood types: A   AB   B   O

Figure 11.11 Possible allelic combinations for ABO blood typing.

ABO Blood Types: A Case of Codominance

In **codominance**, a pair of nonidentical alleles specify two phenotypes, which are both expressed at the same time in heterozygotes. Example: One of the glycolipids at the plasma membrane of your red blood cells helps give these cells a unique identity. However, it comes in slightly different molecular forms. An analytical method, *ABO blood typing*, reveals which form a person has.

An enzyme dictates the glycolipid's final structure. In humans, the gene for that enzyme has three alleles. Two alleles, I^A and I^B, are codominant when paired. The third, *i*, is recessive; a pairing with either I^A or I^B masks its effect. Together, they are a **multiple allele system**, which we define as the presence of three or more alleles of a gene among individuals of a population.

Before each glycolipid became positioned at the cell surface, it was modified in the endomembrane system (Section 4.5). An oligosaccharide chain was attached to a lipid, then a sugar was attached to the chain. Alleles I^A and I^B specify different versions of the enzyme that attaches the sugar. The enzymes attach *different* sugars, which gives the glycolipid different identities: A or B.

Which alleles do you have? With either $I^A I^A$ or $I^A i$, you have type A blood. With $I^B I^B$ or $I^B i$, your blood is type B. With codominant alleles $I^A I^B$, it's AB—meaning you have both versions of the sugar-attaching enzyme. But if you are homozygous recessive (*ii*), the molecules never did get the final sugar attached to them. Your blood type isn't A or B; that is what type "O" means. Figure 11.11 summarizes the possibilities.

With *transfusions*, the blood of two people mixes. The recipient's immune system perceives any incompatible red blood cells as "nonself." In such cases, it acts against them and may cause death (Section 38.4).

One allele may be fully dominant, incompletely dominant, or codominant with its partner on the homologous chromosome.

MULTIPLE EFFECTS OF SINGLE GENES

Expression of the alleles at just a single location on a chromosome may have positive or negative effects on two or more traits. This phenotypic outcome of a single gene's activity is known as **pleiotropy** (after the Greek *pleio–*, meaning more, and *–tropic*, meaning to change).

Consider the gene for fibrillin 1 (Figure 11.12). This protein occurs in the extracellular matrix of connective tissues—the most abundant and widely distributed of all tissues in the vertebrate body. By itself or together with another protein (elastin), it becomes assembled into long, thin strands that are especially abundant in the heart, skin, blood vessels, tendons, and around the eyes. The effects of fibrillin may come into play when embryos are developing. They may be important in the formation of the elastic fibers of connective tissues.

Sometimes the gene for fibrillin mutates. When that happens, many connective tissues weaken throughout the body. The result is the genetic disorder known as *Marfan syndrome*. Marfan syndrome affects 1 in 10,000 people throughout the world. It affects both women and men of any ethnicity.

Table 11.1 lists some of the key symptoms of Marfan syndrome. That single gene mutation has often severe repercussions upon the skeleton, cardiovascular system (the heart and blood vessels), lungs, eyes, and skin. For example, the large blood vessel that transports blood away from the heart, the aorta, may have a weakened wall region, which has been known to rupture during strenuous exercise (Figure 11.12). Until recent medical advances, most patients died of such cardiovascular complications before they were fifty years old.

FLO HYMAN

Figure 11.12 (**a**) Flo Hyman, captain of the U.S. volleyball team that won an Olympic silver medal in 1984. Two years later, during a game in Japan, she slid silently to the floor and died. A dime-size weak spot in the wall of her aorta, the main artery transporting blood away from the heart, had abruptly burst. (**b**) Ribbon model for two calcium-binding domains that are part of the protein fibrillin. A mutation in this region is thought to give rise to Marfan syndrome. Besides Hyman, at least two college basketball stars have died abruptly as a result of this syndrome.

More than 220 fibrillin 1 gene mutations have been identified. One adversely affects the biosynthesis of the protein, its secretion from cells, and its deposition into the extracellular matrix.

Think about what happens to the aorta. First, the structure and activity of all smooth muscle cells in the blood vessel wall change. Then the cells infiltrate and multiply in the epithelial lining of the wall. Calcium deposits build up in the lining, and the wall becomes enflamed. Elastic fibers are split into fragments, which leads to a thin, weakened wall.

The alleles at a single gene location may have positive or negative effects on two or more traits.

The effects may not be simultaneous. Rather, they may have repercussions over time. The gene may lead to an alteration in one trait. That change may alter another trait, and so on.

Table 11.1 *Pleiotropic Effects Often Associated With Marfan Syndrome*

SKELETON Affected people typically lanky, loose jointed, with limbs, fingers, toes often disproportionately long; teeth often crowded; protruding or sunken breastbone; abnormally curved spine; joint pain; flat feet.

HEART AND BLOOD VESSELS Abnormal, often leaky valve between heart chambers; wall of aorta weakened, stretched, extremely vulnerable to tearing or rupturing.

LUNGS Tiny airs sacs where gases are exchanged sometimes become stretched or swollen, so risk of lung collapse.

NERVOUS SYSTEM Protective connective tissue around brain, spinal cord weakens, stretches, presses on lower backbone, causing discomfort or pain in abdominal region.

EYES Lens displacement in more than 50 percent of those affected (higher or lower, shifted to one side because fibers that hold it in place are damaged); nearsightedness common; glaucoma (high fluid pressure inside eyeball) or cataracts (cloudy lens) may develop early.

SKIN Stretch marks common.

INTERACTIONS BETWEEN GENE PAIRS

Often a trait results from interactions among products of two or more gene pairs. For example, two alleles of a gene may mask expression of another gene's alleles, so some expected phenotypes may not appear at all. Such interactions between the product of pairs of genes are called **epistasis** (meaning the act of stopping).

Hair Color in Mammals

Epistasis is common among the gene pairs responsible for skin or fur color in mammals. Consider the black, brown, or yellow fur of Labrador retrievers (Figure 11.13). The different colors arise from variations in the amount and distribution of melanin, a brownish-black pigment. A variety of enzymes and other products of many gene pairs affect different steps in the production of melanin and its deposition in certain body regions.

The alleles of one gene specify an enzyme required to produce melanin. Expression of allele B (black) has a more pronounced effect and is dominant to b (brown). Alleles of a different gene control the extent to which molecules of melanin will be deposited in a retriever's hairs. Allele E permits full deposition. Two recessive alleles (ee) reduce deposition, and fur will be yellow.

In some individuals, those two gene pairs are not able to interact, owing to a certain allelic combination at still another gene locus. There, a gene (C) calls for tyrosinase, the first of several enzymes in a melanin-producing pathway. An individual bearing one or two dominant alleles (CC or Cc) can make the functional enzyme. An individual bearing two recessive alleles (cc) cannot. When the biosynthetic pathway for melanin production gets blocked, then *albinism*—the absence of melanin—is the resulting phenotype (Figure 11.14).

a BLACK LABRADOR

b YELLOW LABRADOR

c CHOCOLATE LABRADOR

Figure 11.13 The heritable basis of coat color among Labrador retrievers. The trait arises through interactions among the alleles of two pairs of genes.

One kind of gene is involved in melanin production. Allele B (black) of this gene is dominant to allele b (brown). A different kind of gene influences the deposition of melanin pigment in individual hairs. Allele E of this gene promotes melanin deposition, but a pairing of recessive alleles (ee) of the gene blocks deposition, and a yellow coat results.

F_1 offspring of a dihybrid cross produce F_2 offspring in a 9:3:4 ratio, as the Punnett-square diagram at *right* indicates.

The yellow Labrador in (**b**) probably has genotype BBee, because it can produce melanin but cannot deposit pigment in hairs. After thinking about this Punnett square, can you guess why?

HOMOZYGOUS PARENTS: $BBEE \times bbee$

F_1 PUPPIES: BbEe

ALLELIC COMBINATIONS POSSIBLE AMONG F_2 PUPPIES:

	BE	Be	bE	be
BE	BBEE	BBEe	BbEE	BbEe
Be	BBEe	BBee	BbEe	Bbee
bE	BbEE	BbEe	bbEE	bbEe
be	BbEe	Bbee	bbEe	bbee

RESULTING PHENOTYPES:

☐ 9/16 or 9 black

☐ 3/16 or 3 brown

☐ 4/16 or 4 yellow

Figure 11.14 A rare albino rattlesnake. Like other animals that can't produce melanin, its body surface is white, overall, and its eyes are pink. In birds and mammals, surface coloration arises largely from pigments in feathers, fur, or skin. In fishes, amphibians, and reptiles, it depends on color-bearing cells. Some of these cells contain melanin or yellow-to-red pigments. Others contain crystals that reflect light and thus alter the surface coloration.

The mutation affecting melanin production in the snake shown here had no effect on the production of its yellow-to-red pigments and light-reflecting crystals. That's why the snake's skin appears to be iridescent yellow as well as white. Its eyes look pink because melanin is absent from a tissue layer in each eyeball. Without melanin to absorb it, red light is reflected from blood vessels in the eyes.

Figure 11.15 Interaction between two genes that affect the same trait in domestic chicken breeds. The initial cross is between a Wyandotte (with a rose comb, **b**, on the crest of its head) and brahma (pea comb, **c**). With complete dominance at the locus for pea comb and at the locus for rose comb, products of the two gene pairs interact and give rise to a walnut comb (**a**). With full recessiveness at both gene loci, products of the genes interact and give rise to a single comb (**d**).

Comb Shape in Poultry

In some cases, interaction between two gene pairs results in a phenotype that neither pair can produce by itself. Geneticists W. Bateson and R. Punnett identified two interacting gene pairs (R and P) that affect comb shape in chickens. Allelic combinations of rr at one gene locus and pp at the other locus result in the least common phenotype, the single comb. The presence of dominant alleles (R, P, or both) results in varied phenotypes.

Take a look at Figure 11.15. The diagram shows the combinations of alleles that interact to specify the rose, pea, and walnut combs shown in the photographs.

Gene interactions affect phenotype, as when alleles of one gene mask the expression of another gene, and when some expected phenotypes may not appear at all.

HOW CAN WE EXPLAIN LESS PREDICTABLE VARIATIONS?

Regarding the Unexpected Phenotype

As Mendel demonstrated, the phenotypic effects of one or two pairs of certain genes show up in predictable ratios when you track them from one generation to the next. Besides this, interactions among two or more gene pairs also can produce phenotypes in predictable ratios, as the example of Labrador coat color demonstrated in Section 11.6.

However, even if you were to track a single gene over the generations, you might find that the resulting phenotypes were not quite what you had expected.

Consider *camptodactyly*, a rare genetic abnormality that affects the shape and the movement of fingers. People who carry the mutant allele for this heritable trait have immobile, bent fingers on both hands. Other people have immobile, bent fingers on the left or right hand only. The fingers of others who carry the mutant allele aren't affected in any obvious way.

What causes such odd variation? Remember, most organic compounds are synthesized by a sequence of metabolic steps. *And different enzymes, each the product of a gene, regulate different steps.* Maybe one gene mutated in one of a number of possible ways. Maybe the gene product blocks the pathway or causes it to run nonstop or not long enough. Maybe poor nutrition or another factor that is variable in the individual's environment affects a crucial enzyme in the pathway. These are the kinds of variable factors that often introduce far less predictable variations in the phenotypes resulting from gene expression.

Continuous Variation in Populations

Generally, individuals of a population display a range of small differences in most traits. This characteristic of populations is called **continuous variation**. It's mainly an outcome of the number of genes affecting a trait and the number of environmental factors influencing their expression. Usually, the greater the number of genes and environmental factors, the more continuous will be the expected distribution of all the versions of that trait.

Look in a mirror at your eye color. The colored part is the iris, a doughnut-shaped, pigmented structure just beneath the cornea. Its color is the cumulative outcome of a number of gene products. Some products take part in the stepwise production and distribution of melanin, the same light-absorbing pigment that influences coat color in mammals. Dark eyes that seem to be almost black have dense deposits of melanin molecules inside the iris. These molecules absorb most of the incoming light. Melanin deposits aren't as extensive in brown eyes, and some unabsorbed light is being reflected out. Light brown or hazel eyes have even less (Figure 11.16).

Green, gray, or blue eyes don't have green, gray, or blue pigments. Their iris incorporates some amount of melanin, but not much. As a result, many or most of the blue wavelengths of light that do enter the eyeball are reflected out.

How might you describe the continuous variation of some trait within a group, such as the students in Figure 11.17a? The students range from very short to very tall, with average heights much more common than either extreme. You might start out by dividing the full range of different phenotypes into measurable categories. Next, you can count all of the individual students in each category. Doing so will give you the relative frequencies of all the phenotypes, distributed across the range of measurable values.

The bar chart in Figure 11.17c plots the proportion of students in each category against the range of the measured phenotypes. Here, the shortest vertical bars represent categories with the least number of students. The tallest bar represents the category with the greatest number of students. Finally, draw a graph line around all the bars and you end up with a bell-shaped curve.

Figure 11.16 A small sampling from the range of continuous variation in the color of human eyes. The products of different pairs of genes interact in producing and then distributing the pigment melanin. Among other things, melanin helps color the eye's iris. Different combinations of alleles result in small differences in eye color. Thus, the frequency distribution for the eye-color trait is continuous over a range from black to light blue.

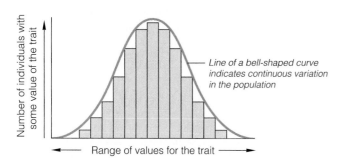

5'3" 5'4" 5'5" 5'6" 5'7" 5'8" 5'9" 5'10" 5'11" 6'0" 6'1" 6'2" 6'3" 6'4" 6'5"
Height (feet/inches)

a Two examples of continuous variation: Biology students (males, *above*; females, *right*) organized according to height.

Line of a bell-shaped curve indicates continuous variation in the population

Number of individuals with some value of the trait

Range of values for the trait

b Idealized bell-shaped curve for a population that displays continuous variation in a trait.

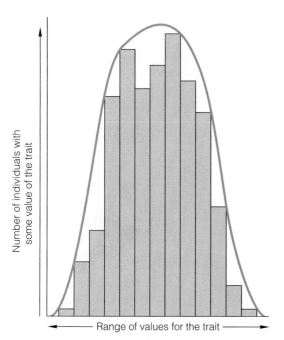

Number of individuals with some value of the trait

Range of values for the trait

c Specific bell curve corresponding to the distribution of a trait (height) among the females in the far-right photograph in (**a**).

4'11" 5'0" 5'1" 5'2" 5'3" 5'4" 5'5" 5'6" 5'7" 5'8" 5'9" 5'10" 5'11"
Height (feet/inches)

Figure 11.17 Continuous variation in body height, a trait that is one of the characteristics of the human population.

(**a**) Jon Reiskind and Greg Pryor decided to show the frequency distribution for height among biology students at the University of Florida. They divided students into two groups: male and female. For each group, they divided the range of possible heights, measured the students, and assigned each to the appropriate category.

(**b**) A bar graph is commonly used to depict continuous variation in a population. In such graphs, the proportion of individuals in each category is plotted against the range of measured phenotypes. The curved line above this particular set of bars is an idealized example of the kind of "bell-shaped" curve that emerges for populations showing continuous variation in a trait. The bell-shaped curve in (**c**) is a specific example of this type of diagram.

Such "bell curves" are typical of populations that show continuous variation in a trait.

Enzymes and other products of genes regulate each step of most metabolic pathways. Mutations, gene interactions, and environmental conditions may affect one or more of the steps. The outcome is variation in phenotypes.

For most traits, individuals of a population show continuous variation—that is, a range of small differences.

The greater the number of genes and environmental factors that can influence a trait, the more continuous will be the expected distribution of all versions of that trait.

ENVIRONMENTAL EFFECTS ON PHENOTYPE

We have mentioned, in passing, that the environment often contributes to variable gene expression among a population's individuals. Before leaving this chapter, consider a few cases of phenotypic variations that arise from environmental effects on genotype.

Possibly you have thought about the fur color of a Himalayan rabbit or Siamese cat. These mammals both have dark fur in some body regions and lighter fur in others. Let's just focus on Himalayan rabbits. They are homozygous for the c^h allele of the gene for tyrosinase. Tyrosinase is one of the enzymes necessary to make melanin. The c^h allele specifies a heat-sensitive form of the enzyme. And that enzyme is active only when the air temperature around the body is below about 33°C.

When cells that give rise to this rabbit's hairs grow under warmer conditions, they cannot make melanin, so the hairs appear light. This happens in body regions that are massive enough to conserve a fair amount of metabolic heat. Ears and other slender extremities are cooler because they tend to lose metabolic heat faster.

Figure 11.18 gives an experiment that demonstrated an environmental effect on this allele.

We also can identify environmental effects on genes that govern plant phenotypes. Consider an experiment with yarrow plants. Yarrow plants are able to grow from cuttings. This makes them a good experimental organism. Why? Cuttings from the same plant will all have the same genotype, so experimenters can discount genes as a basis for differences that appear among them.

In this case, three yarrow cuttings were grown at three elevations. The two plants grown at the lowest and highest elevation fared best; the one at the medium elevation grew poorly, as Figure 11.19 shows.

But remember the sampling error trap (Chapter 1)? The researchers did the same growth experiments for *many* yarrow plants. They saw no consistent pattern in phenotypic variation; it was too great. For instance, a cutting from one plant developed best at the medium elevation. The conclusion? Different yarrow genotypes react differently across a range of environments.

Icepack is strapped onto a hair-free patch.

New hair growing in patch exposed to cold is black.

Figure 11.18 Observable effect of differences in environmental conditions on the expression of genes in animals. A Himalayan rabbit normally has black hair only on its long ears, nose, tail, and lower leg limbs. For one experiment, a patch of a rabbit's white fur was plucked clean, then an icepack was secured over the hairless patch. Where the colder temperature had been maintained, the hairs that grew back were black.

Himalayan rabbits are homozygous for an allele of the gene for tyrosinase, an enzyme required to make melanin. As described in the text, this allele specifies a heat-sensitive form of the enzyme, which functions only when air temperature is below about 33°C.

a Mature cutting at high elevation (3,050 meters)

b Mature cutting at medium elevation (1,400 meters)

c Mature cutting at low elevation (30 meters above sea level)

Figure 11.19 Experiment demonstrating the effect of environmental conditions on gene expression in a yarrow plant (*Achillea millefolium*). Cuttings from the same parent plant were grown in the same soil batch but at three different elevations.

Figure 11.20 Environmental effect on gene expression in *Hydrangea macrophylla*, a common garden plant. Different plants that carry the same alleles may have floral colors ranging from pink to blue. This case of color variation arises because of differences in the acidity of soil in which a plant happens to be growing.

Figure 11.21 A San Diego State University student exhibiting the tongue-rolling trait to a tongue-roll-challenged student. Once thought to be genetically determined, this trait is mostly learned in the individual's environment.

As another example, plant a hydrangea in a garden and it may make pink blossoms instead of the expected blue ones, depending on its environment (Figure 11.20). Genes for its floral color are affected by soil acidity.

As one more example, for years, tongue-rolling was viewed as a genetically determined trait (Figure 11.21). It now appears to be mostly learned by individuals.

And so we conclude this chapter, which has dealt with heritable and environmental factors that give rise to variations in phenotype. What is the take-home lesson? Simply this: An individual's phenotype is an outcome of complex interactions among its genes, enzymes and other gene products, and environmental factors.

Individuals of most populations or species show complex variation for many traits. The variation arises not only from gene mutations and cumulative gene interactions. It arises also in response to variations in environmental conditions.

SUMMARY *Gold* indicates text section

1. A gene is a unit of information about a heritable trait. Alleles of a gene are different molecular versions of that information. Through experimental crosses with pea plants, Mendel gathered indirect evidence that diploid organisms have pairs of genes, and that genes retain their identity when transmitted to offspring. *CI, 11.2*

2. For a particular trait being studied, an individual who has inherited two dominant alleles (AA) is said to be homozygous dominant. A homozygous recessive individual has inherited two recessive alleles (aa), and a heterozygote has two nonidentical alleles (Aa). *11.1*

3. An individual's specific combination of alleles is its genotype. Its observable traits are its phenotype. *11.1*

4. A hybrid is an individual offspring from any cross between parents of different genotypes. In monohybrid crosses, two individuals that breed true for different versions of the same trait produce F_1 offspring that are identically heterozygous at one gene pair. *11.1, 11.2*

5. Mendel's monohybrid crosses between garden pea plants gave indirect evidence that some forms of a gene may be dominant over other, recessive forms. *11.2*

6. All F_1 offspring of the parental cross $AA \times aa$ were Aa. Crosses between F_1 monohybrids resulted in these combinations of alleles in F_2 offspring: *11.2*

	A	*a*
A	*AA*	*Aa*
a	*Aa*	*aa*

AA (dominant)
Aa (dominant)
Aa (dominant)
aa (recessive)
} the expected phenotypic ratio of 3:1

7. Results from Mendel's monohybrid crosses led to the formulation of a theory of segregation. In modern terms, diploid organisms have pairs of genes, on pairs of homologous chromosomes. The genes of each pair segregate from each other at meiosis, so each gamete formed ends up with one or the other gene. *11.2*

8. For dihybrid crosses, individuals that breed true for different versions of *two* traits produce F_1 offspring that are all identical heterozygotes for both genes. The experiment is an intercross between two F_1 dihybrids. Phenotypes of the F_2 offspring from Mendel's dihybrid crosses were close to a 9:3:3:1 ratio: *11.3*

> 9 dominant for both traits
> 3 dominant for *A*, recessive for *b*
> 3 dominant for *B*, recessive for *a*
> 1 recessive for both traits

9. Mendel's dihybrid crosses led to his formulation of the theory of independent assortment. In modern terms, by the end of meiosis, the gene pairs of two homologous chromosomes have been sorted out for distribution into one gamete or another, independently of how the gene pairs of other chromosomes were sorted out. *11.3*

10. Certain factors influence gene expression. *11.4–11.8*

    a. In some cases, one allele of a pair is incompletely dominant or codominant. *11.4*

    b. Products of pairs of genes often interact in ways that influence the same trait. *11.5–11.8*

    c. One gene may have positive or negative effects on two or more traits, a condition called pleiotropy. *11.5*

    d. Environmental conditions to which an individual is subjected may affect gene expression. *11.8*

Review Questions

1. Distinguish between these terms: *11.1*
    a. gene and allele
    b. dominant allele and recessive allele
    c. homozygote and heterozygote
    d. genotype and phenotype

2. What is a true-breeding lineage? A hybrid? *11.1*

3. Distinguish between monohybrid cross, dihybrid cross, and testcross. *11.2, 11.3*

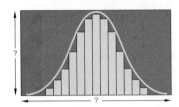

4. Do segregation and independent assortment occur during mitosis, meiosis, or both? *11.2, 11.3*

5. In the bell-shaped curve at left, a diagram of continuous variation in a population, what do the bars and the curved line represent? *11.7*

Self-Quiz   ANSWERS IN APPENDIX III

1. Alleles are _____ .
    a. different molecular forms of a gene
    b. different phenotypes
    c. self-fertilizing, true-breeding homozygotes

2. A heterozygote has a _____ for the trait being studied.
    a. pair of identical alleles
    b. pair of nonidentical alleles
    c. haploid condition, in genetic terms
    d. a and c

3. The observable traits of an organism are its _____ .
    a. phenotype   c. genotype
    b. sociobiology   d. pedigree

4. Second-generation offspring from a cross are the _____ .
    a. F_1 generation   c. hybrid generation
    b. F_2 generation   d. none of the above

5. F_1 offspring of the monohybrid cross $AA \times aa$ are _____ .
    a. all AA c. all Aa
    b. all aa d. 1/2 AA and 1/2 aa

6. Refer to question 5. Assuming complete dominance, the F_2 generation will show a phenotypic ratio of _____ .
    a. 3:1   b. 9:1   c. 1:2:1   d. 9:3:3:1

7. Crosses between F_1 pea plants resulting from the cross $AABB \times aabb$ lead to F_2 phenotypic ratios close to _____ .
    a. 1:2:1 b. 3:1   c. 1:1:1:1   d. 9:3:3:1

8. Match each example with the most suitable description.
    _____ dihybrid cross   a. *bb*
    _____ monohybrid cross   b. $AABB \times aabb$
    _____ homozygous condition   c. *Aa*
    _____ heterozygous condition   d. $Aa \times Aa$

Critical Thinking—Genetics Problems
ANSWERS IN APPENDIX IV

1. One gene has alleles *A* and *a*. Another has alleles *B* and *b*. For each genotype listed, what type(s) of gametes will be produced? Assume independent assortment occurs before gametes form.
    a. *AABB*   c. *Aabb*
    b. *AaBB*   d. *AaBb*

2. Refer to Problem 1. What will be the genotypes of offspring from the following matings? Indicate the frequencies of each genotype among them.

    a. *AABB* × *aaBB*   c. *AaBb* × *aabb*
    b. *AaBB* × *AABB*   d. *AaBb* × *AaBb*

3. In one experiment, Mendel crossed a pea plant that bred true for green pods with one that bred true for yellow pods. All the F_1 plants had green pods. Which form of the trait (green or yellow pods) is recessive? Explain how you arrived at your conclusion.

4. Return to Problem 1. Assume you now study a third gene having alleles *C* and *c*. For each genotype listed, what type(s) of gametes will be produced?
    a. *AABB CC*   c. *Aa BB Cc*
    b. *Aa BB cc*   d. *Aa Bb Cc*

5. Mendel crossed a true-breeding tall, purple-flowered pea plant with a true-breeding dwarf, white-flowered plant. All F_1 plants were tall and had purple flowers. If an F_1 plant self-fertilizes, then what is the probability that a randomly selected F_2 offspring will be heterozygous for the genes specifying height and flower color?

6. *DNA fingerprinting* is a method of identifying individuals by locating unique base sequences in their DNA molecules (Section 16.3). Before researchers refined the method, attorneys often relied on the ABO blood-typing system to settle disputes over paternity. Suppose that you, as a geneticist, are asked to testify during a paternity case in which the mother has type A blood, the child has type O blood, and the alleged father has type B blood. How would you respond to the following statements?

    a. Attorney of the alleged father: "The mother's blood is type A, so the child's type O blood must have come from the father. My client has type B blood; he could not be the father."

    b. Mother's attorney: "Because further tests prove this man is heterozygous, he must be the father."

7. Suppose you identify a new gene in mice. One of its alleles specifies white fur. A second allele specifies brown fur. You want to determine whether the relationship between the two alleles is one of simple dominance or incomplete dominance. What sorts of genetic crosses would give you the answer? On what types of observations would you base your conclusions?

8. Your sister moves away and gives you her purebred Labrador retriever, a female named Dandelion. Suppose you decide to breed Dandelion and sell puppies to help pay for your college tuition. Then you discover that two of her four brothers and sisters show *hip dysplasia*, a heritable disorder arising from a number of gene interactions. If Dandelion mates with a male Labrador known to be free of the harmful alleles, can you guarantee to a buyer that puppies will not develop the disorder? Explain your answer.

9. A dominant allele *W* confers black fur on guinea pigs. A guinea pig that is homozygous recessive (*ww*) has white fur. Fred would like to know whether his pet black-furred guinea pig is homozygous dominant (*WW*) or heterozygous (*Ww*). How might he determine his pet's genotype?

10. Red-flowering snapdragons are homozygous for allele R^1. White-flowering snapdragons are homozygous for a different allele (R^2). Heterozygous plants (R^1R^2) bear pink flowers. What

phenotypes should appear among first-generation offspring of the crosses listed? What are the expected proportions for each phenotype?

a. $R^1R^1 \times R^1R^2$ c. $R^1R^2 \times R^1R^2$
b. $R^1R^1 \times R^2R^2$ d. $R^1R^2 \times R^2R^2$

Notice, in Problem 10, that in cases of incomplete dominance it is inappropriate to refer to either allele of a pair as dominant or recessive. When the phenotype of a heterozygous individual is halfway between those of the two homozygotes, then there is no dominance. Such alleles are usually designated by superscript numerals, as shown here, rather than by uppercase letters for dominance and lowercase letters for recessiveness.

11. Two pairs of genes affect comb type in chickens (Figure 11.15). When both genes are recessive, a chicken has a single comb. A dominant allele of one gene, P, gives rise to a pea comb. Yet a dominant allele of the other (R) gives rise to a rose comb. An epistatic interaction occurs when a chicken has at least one of both dominants, $P\_\_ R\_\_$, which gives rise to a walnut comb. Predict the ratios resulting from a cross between two walnut-combed chickens that are heterozygous for both genes ($PpRr$).

12. As described in Section 3.8, a single mutant allele gives rise to an abnormal form of hemoglobin (Hb^S instead of Hb^A). Homozygotes (Hb^SHb^S) develop sickle-cell anemia. But the heterozygotes (Hb^AHb^S) show few outward symptoms.

Suppose a woman's mother is homozygous for the Hb^A allele. She marries a male who is heterozygous for the allele, and they plan to have children. For *each* of her pregnancies, state the probability that this couple will have a child who is:

a. homozygous for the Hb^S allele
b. homozygous for the Hb^A allele
c. heterozygous Hb^AHb^S

13. Certain dominant alleles are so vital for normal development that an individual who is homozygous recessive for a mutant recessive form of the allele is unable to survive. Such recessive, *lethal alleles* can be perpetuated by heterozygotes.

Consider the Manx allele (M^L) in cats. Homozygous cats (M^LM^L) die when they are still embryos inside the mother cat. In heterozygotes (M^LM), the spine develops abnormally, and the cats end up with no tail whatsoever (Figure 11.22).

Two M^LM cats mate. Among their *surviving* progeny, what is the probability that any one kitten will be heterozygous?

14. A recessive allele c is responsible for *albinism*, an inability to produce or deposit melanin in tissues. Humans and some other organisms can have this phenotype (Figure 11.23). In each of the following cases, what are the possible genotypes of the father, of the mother, and of their children?

Figure 11.22 Manx cat, which has no tail.

Figure 11.23 An albino male in India.

a. Both parents have normal phenotypes; some of their children are albino and others are unaffected.

b. Both parents are albino and have only albino children.

c. The woman is unaffected, the man is albino, and they have one albino child and three unaffected children.

15. Kernel color in wheat plants is determined by two pairs of genes. Alleles of one pair show incomplete dominance over alleles of the other pair.

For the gene pair at one locus on the chromosome, allele A^1 imparts one dose of red color to the kernel, whereas allele A^2 does not. At the second locus, allele B^1 gives one dose of red color to the kernel, whereas allele B^2 does not. One kernel with genotype $A^1A^1B^1B^1$ is dark red. A different kernel with genotype $A^2A^2B^2B^2$ is white. All other genotypes have kernel colors in between the two extremes.

a. Suppose you cross a plant grown from a dark red kernel with a plant grown from a white kernel. What genotypes and what phenotypes would you expect among the offspring?

b. If a plant with genotype $A^1A^2B^1B^2$ self-fertilizes, what genotypes and what phenotypes would be expected among the offspring? In what proportions?

Selected Key Terms

allele *11.1*
codominance *11.4*
continuous variation *11.7*
dihybrid cross *11.3*
epistasis *11.6*
F_1 *11.1*
F_2 *11.1*
gene *11.1*
genotype *11.1*
heterozygous *11.1*
homozygous dominant *11.1*
homozygous recessive *11.1*

hybrid offspring *11.1*
incomplete dominance *11.4*
independent assortment *11.3*
monohybrid cross *11.2*
multiple allele system *11.4*
phenotype *11.1*
pleiotropy *11.5*
probability *11.2*
Punnett-square method *11.2*
segregation *11.2*
testcross *11.2*
true-breeding lineage *11.1*

Readings

Fairbanks, D. J., and W. R. Andersen. 1999. *Genetics: The Continuity of Life.* Monterey, California: Brooks-Cole.

Orel, V. 1996. *Gregor Mendel: The First Geneticist.* New York: Oxford University Press.

On-Line readings at Student Guide for InfoTrac:
www.brookscole.com/biology

12

HUMAN GENETICS

The Philadelphia Story

Positioned at strategic locations in chromosomes are genes concerned with the cell cycle—that is, with cell growth and division. Some specify enzymes and other proteins that perform these tasks. Other genes control whether, when, and how fast the tasks are completed. When something disrupts the controls, cell growth and division can spiral out of control and lead to cancer.

The first abnormal chromosome tied to cancer was named the *Philadelphia chromosome* after the city where someone discovered it. This chromosome shows up in cells of some people affected by a type of leukemia.

The disorder starts with stem cells in bone marrow. All stem cells are unspecialized and retain the capacity for mitotic cell division. A portion of their descendants divide and become specialized. With leukemias, there are too many descendants—far too many of the white blood cells in charge of housekeeping and defense.

Leukemic cells may crowd out the stem cells that give rise to red blood cells and platelets. Anemia and internal bleeding follow. Leukemic cells also infiltrate blood and the liver, lymph nodes, spleen, and other organs, where they interfere with basic functions. Left to itself, the cancerous transformation kills the patient.

No one knew about the Philadelphia chromosome until microscopists learned how to identify its physical appearance. Chromosomes, recall, are most condensed at metaphase of mitosis. Then, their size, length, and centromere location are easiest to identify. A **karyotype** is a preparation of metaphase chromosomes based on their defining features. In this chapter, you will learn how to make a karyotype diagram from a photograph of metaphase chromosomes, just as microscopists do. Such diagrams yield useful information when they are compared with a standard karyotype for a species.

The Philadelphia chromosome shows up clearly with *spectral* karyotyping. This newer research and diagnostic tool artificially colors chromosomes, and the colors are clues to their structure (Figure 12.1).

As it turns out, the Philadelphia chromosome is physically longer than its normal counterpart, human chromosome 9. The extra length is actually a piece of chromosome 22! What happened? By chance, both chromosomes broke inside a stem cell. Each broken piece was reattached, but on the wrong chromosome. At the broken end of chromosome 9, a gene with a role in cell division fused with the control region of a gene at the broken end of chromosome 22. In its mutant fusion form, that gene is expressed far more than it should be. Uncontrolled divisions of white blood cells are the outcome (Figure 12.2).

The Philadelphia story is a glimpse into the world of modern genetics research. Reading it invites you to think about how far you have come in this unit of the book. You first looked at cell division, the starting point of inheritance. You looked at how chromosomes and the genes they carry are shuffled during meiosis, then at fertilization. You also mulled over Mendel's insights into inheritance and some exceptions to his conclusions. Here, you have a current example of what we know about the chromosomal basis of inheritance.

How did we get to that understanding? The answer requires a bit of history, which picks up from where we left Mendel.

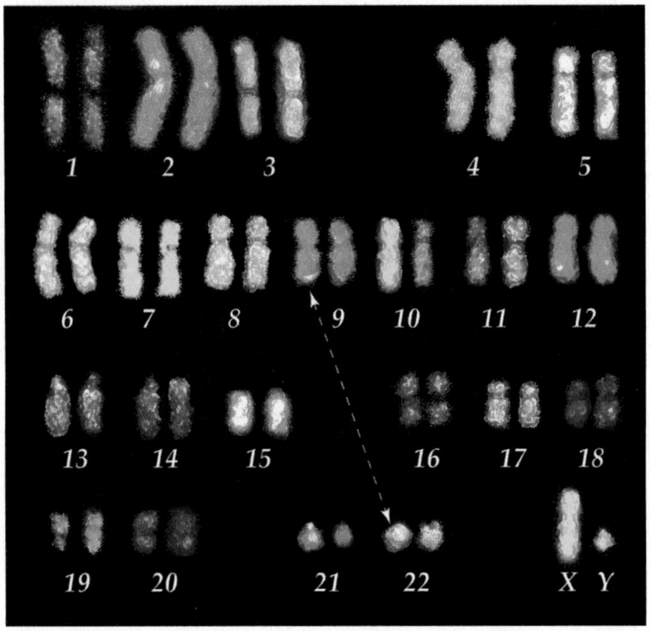

Figure 12.1 Revealing the origin of a killer—how reciprocal translocation between human chromosome 9 and chromosome 22 might appear with the help of spectral karyotyping, a new imaging technique. This particular translocation, the Philadelphia chromosome, results in a gene abnormality that gives rise to chronic myelogenous leukemia (CML), a type of cancer. Gleevec, a new oral drug, inactivates the abnormal protein product of the gene, thus arresting the dangerous cell proliferation. In preliminary tests, the drug put fifty of fifty-three CML patients into remission. So far, its side effects have been mild, compared to severe side effects of chemotherapy.

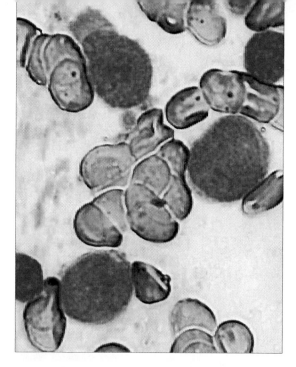

Figure 12.2 Blood typical of chronic myelogenous leukemia. Abnormal, immature white blood cells are starting to crowd out the normal cell types.

In 1884 Mendel had just passed away, and his paper on pea plants had been gathering dust in a hundred libraries for nearly two decades. Then the resolving power of microscopes improved and rekindled interest in the hereditary material. Walther Flemming had seen threadlike bodies—chromosomes—in dividing cells. Could chromosomes be the hereditary material?

Microscopists soon realized that each gamete has half the number of chromosomes of a fertilized egg. In 1887, August Weismann hypothesized that a special division process halves the chromosome number before gametes form. Sure enough, meiosis was discovered that same year. Weismann promoted another hypothesis: If the halved chromosome number is restored at fertilization, then half of the chromosomes in our cells must come from the father and half from the mother. His view was hotly debated. It prompted a flurry of experimental crosses—just like the ones Mendel had carried out.

Finally, in 1900, researchers came across Mendel's paper while checking literature related to their own genetic crosses. To their surprise, their experimental results confirmed what Mendel's results had already suggested: Diploid cells generally have two copies of each gene, and the two copies segregate from each other before gametes form.

In the decades to follow, researchers learned more about chromosomes. You'll come across a few high points of their work as we continue with our look at inheritance. As the Philadelphia story tells you, the methods of analysis are not remote from your interests. An inherited collection of information in chromosomal DNA gives rise to traits that, for better or worse, define each organism, young and old alike.

Key Concepts

1. The cells of humans and many other sexually reproducing species contain pairs of homologous chromosomes that interact during meiosis. Typically, one chromosome of each pair is maternal in origin, and its homologue is paternal in origin.

2. Each gene has its own specific location, or locus, in a particular chromosome.

3. The molecular form of a gene occupying a given locus may be slightly different from one chromosome to the next. All of the different molecular forms of a gene are called alleles.

4. The combination of alleles along the length of a chromosome does not necessarily remain intact through meiosis and gamete formation. By the event called crossing over, some alleles along its length swap places with their partner on the homologous chromosome. Alleles that swap places may or may not be identical.

5. Allelic recombinations contribute to variations in the phenotypes of offspring.

6. A chromosome may change structurally, as when a segment of it is deleted, duplicated, inverted, or moved to a different location. Also, the chromosome number of an individual's cells may change as a result of an improper separation of duplicated chromosomes during meiosis or mitosis.

7. Chromosome structure and chromosome number rarely change. When changes do occur, they may result in genetic abnormalities or genetic disorders.

CHROMOSOMES AND INHERITANCE

Genes and Their Chromosome Locations

Earlier chapters described the structure of chromosomes and what happens to them during meiosis. To refresh your memory and get a general sense of where you are going from here, take a moment to read this list:

1. **Genes** are units of information about heritable traits. The genes of eukaryotic cells are distributed among chromosomes. Each gene has its own location —a gene locus—in one type of chromosome.

2. Any cell with a diploid chromosome number (2*n*) has inherited pairs of **homologous chromosomes**. All but one pair are identical in length, shape, and gene sequence. The single exception is a pairing of nonidentical sex chromosomes, such as X with Y. The two members of a pair of homologous chromosomes interact and segregate from each other during meiosis.

3. A gene at one locus may have the same form or a slightly different one compared to its partner gene on the homologous chromosome. When considering a population as a whole, *which* forms are inherited usually varies from one individual to the next.

4. All the different molecular forms of a gene that are possible at a given locus are called **alleles**. New alleles arise through mutation.

5. A *wild-type* allele is the most common form of a gene, either in a natural population or in a standard, laboratory-bred strain of a species. Any one of the less common forms of a gene is a *mutant* allele.

6. Genes on the same chromosome are physically tied together. The farther apart two of the genes are, the more vulnerable they are to **crossing over**. By this event, homologous chromosomes exchange corresponding segments (Figure 12.3*a,b*).

7. Crossing over results in **genetic recombination**— combinations of alleles in chromosomes that were not present in the parental cell (Figure 12.3*c*).

8. **Independent assortment** refers to the random alignment of each pair of homologous chromosomes at metaphase I of meiosis. It results in nonparental combinations of alleles in gametes and offspring.

9. On rare occasions, the structure of chromosomes changes abnormally during mitosis or meiosis. So does the parental chromosome number.

Autosomes and Sex Chromosomes

In all but one case, a pair of homologous chromosomes normally are alike in length, shape, and gene sequence. Microscopists discovered the exception in the late 1800s with analytical methods such as karyotyping (Section 12.2). A distinctive chromosome is present in females *or* males of many species, but not both. As an example,

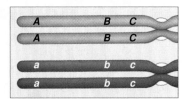

a A pair of duplicated homologous chromosomes (two sister chromatids each). This example has nonidentical alleles at three gene loci (*A* with *a*, *B* with *b*, and *C* with *c*).

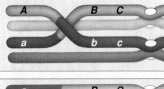

b In prophase I of meiosis, a crossover event occurs: two nonsister chromatids exchange corresponding segments.

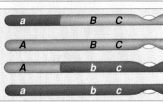

c This is the outcome of the crossover: genetic recombination between nonsister chromatids (which are shown after meiosis as unduplicated, separate chromosomes).

Figure 12.3 Review of crossing over. As shown earlier in Section 10.4, this event occurs in prophase I of meiosis.

a human male diploid cell has one **X chromosome** and one **Y chromosome** (XY). A human female diploid cell has two X chromosomes (XX). This inheritance pattern is common among mammals, fruit flies, and many other species. We see a different pattern in butterflies, moths, some birds, and certain fishes. For them, inheriting two identical sex chromosomes results in a male. Inheriting two nonidentical sex chromosomes results in a female.

The human X and Y chromosomes differ physically; one is shorter than the other (Figure 12.4*f*). They also differ in which genes they carry. Even so, they are able to synapse (become zippered together briefly) in a small region along their length. That bit of zippering allows them to function as homologues during meiosis.

The human X and Y chromosomes fall into the more general category of **sex chromosomes**. The term refers to types of chromosomes that, in certain combinations, determine a new individual's sex—whether a male or a female will develop. All other chromosomes in cells are the same in both sexes; they are **autosomes**.

Diploid cells have pairs of genes, on pairs of homologous chromosomes. At each gene locus, the alleles (alternative forms of a gene) may be either identical or nonidentical.

Crossing over and other events during meiosis give offspring new combinations of alleles and parental chromosomes.

Abnormal events during meiosis or mitosis can change the structure and number of chromosomes.

Autosomes are the pairs of chromosomes that are the same in males and females of a species. One other pair, the sex chromosomes, govern the sex of a new individual.

Karyotyping Made Easy

Karyotype diagrams help answer questions about an individual's chromosomes. Chromosomes are the most condensed and easiest to identify in cells going through metaphase of mitosis. Technicians don't count on finding a cell that happens to be dividing in the body when they go looking for it. They culture cells in vitro (literally, "in glass"). They put a sample of cells, usually from blood, into a glass container. The container holds a solution that stimulates cell growth and mitotic cell divisions.

Adding colchicine to a culture medium arrests cell division at metaphase. This extract of *Colchicum* plants also blocks spindle formation (Section 4.9). If a spindle can't form, sister chromatids of duplicated chromosomes won't be able to separate at anaphase. By using suitable colchicine concentrations and exposure times, technicians can stockpile metaphase cells. This improves the odds of finding candidates for karyotype diagrams.

Following colchicine treatment, the culture medium is transferred to tubes of a centrifuge (Figure 12.4a). Cells have greater mass and density than the solution bathing them. The spinning force moves them farthest from the center of rotation, to the bottom of the attached tubes.

The cells are transferred to a saline solution. When immersed in this hypotonic fluid, they swell (by osmosis) and move apart. The metaphase chromosomes also move apart. The cells are ready to be mounted on a microscope slide, fixed as by air-drying, and stained.

Chromosomes take up some stains uniformly along their length, which allows identification of chromosome size and shape. By other staining procedures, horizontal bands show up along the length of the chromosomes of certain species. If researchers direct a ray of ultraviolet light at the chromosomes treated with special fluorescent stains, the bands will fluoresce, as in Figure 10.3.

Next, the chromosomes are photographed through the microscope, and the image is enlarged. The photographed chromosomes are cut apart and arranged by size, shape, and length of their arms. All the pairs of homologous chromosomes are horizontally aligned by centromeres, as in Figure 12.4f. Today, such images of chromosomes are being cut and pasted electronically, on computers.

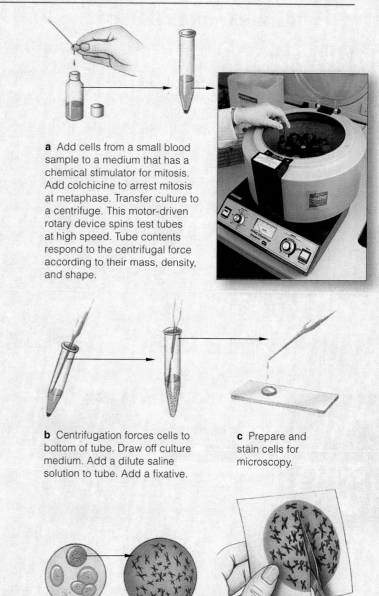

a Add cells from a small blood sample to a medium that has a chemical stimulator for mitosis. Add colchicine to arrest mitosis at metaphase. Transfer culture to a centrifuge. This motor-driven rotary device spins test tubes at high speed. Tube contents respond to the centrifugal force according to their mass, density, and shape.

b Centrifugation forces cells to bottom of tube. Draw off culture medium. Add a dilute saline solution to tube. Add a fixative.

c Prepare and stain cells for microscopy.

d Put cells on a microscope slide. Observe.

e Photograph one cell through microscope. Enlarge image of its chromosomes. Cut the image apart. Arrange chromosomes as a set.

Figure 12.4 (**a–e**) How to prepare karyotypes. (**f**) Human karyotype. Human somatic cells hold 22 pairs of autosomes and 1 pair of sex chromosomes (XX or XY). That is a diploid number of 46. These are metaphase chromosomes; each is in the duplicated state, consisting of two sister chromatids joined at the centromere.

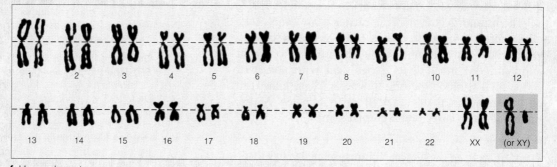

f Human karyotype

SEX DETERMINATION IN HUMANS

Genetic analysis of human cells has yielded evidence that every normal egg produced by a female has one X chromosome. Half the sperm cells produced by a male carry an X chromosome and half carry a Y.

If an X-bearing sperm fertilizes an X-bearing egg, the new individual will develop into a female. If the sperm happens to carry a Y chromosome, the individual will develop into a male (Figure 12.5).

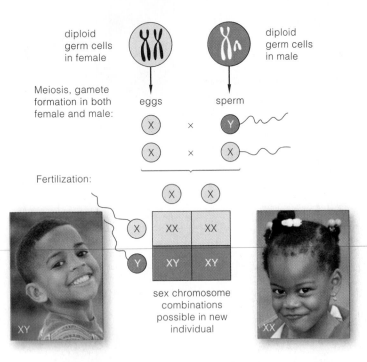

Figure 12.5 Pattern of sex determination in humans.

The human Y chromosome carries only 330 genes. However, one of them is the master gene for male sex determination. Its expression leads to the formation of testes, the primary male reproductive organs (Figure 12.6). In that gene's absence, ovaries form. Ovaries are the primary female reproductive organs. Testes and ovaries both produce important sex hormones, which influence the development of particular sexual traits.

The human X chromosome carries 2,062 genes. Like other chromosomes, it carries some genes associated with sexual traits, such as the distribution of body fat and hair. But most of its genes deal with *nonsexual* traits, such as blood-clotting functions. These genes can be expressed in males as well as in females. Males, remember, also carry one X chromosome.

A certain gene on the human Y chromosome dictates that a new individual will develop into a male. In the absence of the Y chromosome (and the gene), a female develops.

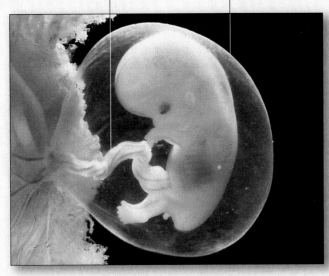

umbilical cord (lifeline between the embryo and the mother's tissues)

amnion (a protective, fluid-filled sac surrounding and cushioning the embryo)

a A human embryo, eight weeks old and about an inch long. The mass at left is part of the placenta, an organ that forms from maternal and embryonic tissues.

Figure 12.6 Boys, girls, and the Y chromosome.

For about the first four weeks of its existence, a human embryo has neither male nor female traits, regardless of whether it is XY or XX. Then ducts and other internal structures that can develop either way start forming.

(a–c) In an XX embryo, ovaries (the primary female reproductive organs) start to form *in the absence of a Y chromosome*. By contrast, in an XY embryo, testes (the primary male reproductive organs) start to form during the next four to six weeks. Apparently, a gene region on the Y chromosome governs a fork in the developmental road that can lead to maleness.

The newly forming testes start to produce testosterone and other sex hormones. These hormones are crucial for the development of a male reproductive system. By contrast, in an XX embryo, newly forming ovaries start to produce different kinds of sex hormones, particularly estrogens. Estrogens are crucial for the development of a female reproductive system.

The master gene for male sex determination is named *SRY* (short for the *Sex-determining Region* of the *Y* chromosome). The same gene has been identified in DNA from male humans, chimpanzees, mice, rabbits, pigs, horses, cattle, and tigers, among others. None of the females tested had the gene. Tests with mice indicate that the gene region becomes active about the time that testes are starting to develop.

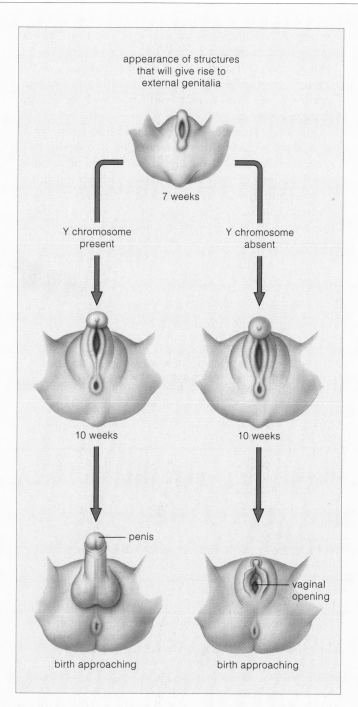

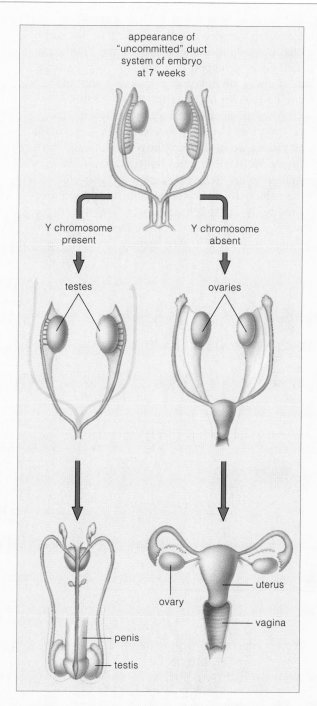

b External appearance of reproductive organs forming in human embryos.

The *SRY* gene encodes certain regulatory proteins. As you will read in Chapter 15, regulatory proteins influence the activity of genes and their products. The *SRY* gene product regulates a cascade of reactions necessary for male sex determination.

c A duct system that forms early in the human embryo. It may develop into male *or* female primary reproductive organs, depending on whether the *SRY* gene is present or absent.

WHAT MENDEL DIDN'T KNOW: CROSSOVERS AND RECOMBINATIONS

By the early 1900s, researchers had an inkling that each gene has a specific location on a chromosome. Through hybridization experiments with mutant fruit flies, of a type called *Drosophila melanogaster*, Thomas H. Morgan and his coworkers helped confirm this. For instance, they found one gene governing eye color and another governing body color on this fruit fly's X chromosome. Figure 12.7 describes one series of their experiments.

Some *Drosophila* experiments dealt with two mutant genes on the X chromosome: *w* for white eyes and *y* for yellow body. Were these genes linked? That is, do they stay together on an X chromosome in meiosis and end up in the same gamete? Experimental results pointed to the possibility of "sex-linked genes." In time, many genes were assigned to sex chromosomes, and they are now called *X-linked* and *Y-linked* genes.

Researchers eventually identified a large number of genes on each type of chromosome—that is, a **linkage group**. *D. melanogaster*, for example, has four linkage groups that correspond to its four pairs of homologous chromosomes. Indian corn (*Zea mays*) has ten linkage groups, corresponding to its ten pairs of homologous chromosomes. And humans have twenty-three linkage groups and so on.

If all linked genes stayed together through meiosis, then there would be no recombination of linked genes. That is, you could always predict the ratio of possible combinations of phenotypes among, say, F₂ offspring of dihybrid crosses. (Here you might wish to review Section 10.4 and Figure 11.3.) Yet certain results from the *Drosophila* experiments did not match expectations. Some genes were not "tightly linked."

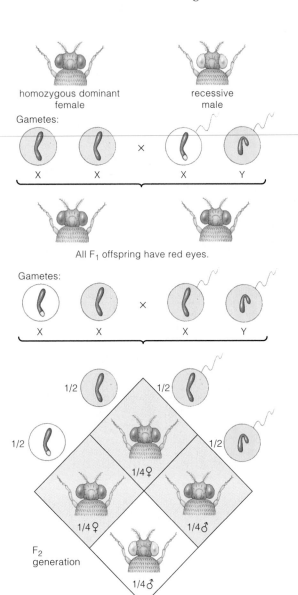

Figure 12.7 X-linked genes—clues to inheritance patterns.

Thomas Morgan, an embryologist, discovered a genetic basis for a relationship between sex determination and some of the *nonsexual* traits. For instance, human males and females have blood-clotting factors, but hemophilia, a rare blood-clotting disorder, occurs most often in males. This sex-linked outcome was not like anything Mendel saw in his hybrid crosses of pea plants. For pea plants, it made no difference which parent carried a recessive allele.

Morgan was studying eye color and other nonsexual traits of *Drosophila melanogaster*. This fruit fly is a great choice for laboratory experiments. It can live in small bottles on agar, cornmeal, molasses, and yeast. A female lays hundreds of eggs in a few days and offspring reproduce in less than two weeks. In a single year, Morgan could track traits through nearly thirty generations of thousands of flies.

At first all of the flies were wild type for eye color; they had brick-red eyes. Then Morgan got lucky; mutation in a gene controlling eye color presented him with a white-eyed male. He established true-breeding strains of white-eyed males and females for **reciprocal crosses**. (In the first of such paired crosses, one parent displays the trait of interest. In the second cross, the other parent displays it.) He let white-eyed males mate with homozygous red-eyed females. All F₁ offspring had red eyes, and some F₂ males had white eyes (*see diagram*).

Then Morgan let true-breeding red-eyed males mate with white-eyed females. The results surprised him. Half of the F₁ offspring turned out to be red-eyed females, and half were white-eyed males. And of the F₂ offspring, 1/4 were red-eyed females, 1/4 white-eyed females, 1/4 red-eyed males, and 1/4 white-eyed males.

The results implied a relationship between an eye-color gene and sex determination. Probably the gene locus was on a sex chromosome. Which one? Because females (XX) could be white-eyed, the recessive allele had to be on one of their X chromosomes. What if white-eyed males (XY) had the recessive allele on their X chromosome and their Y chromosome had no corresponding eye-color allele? In that case, they would have white eyes. They would have no dominant allele to mask the effect of the recessive one, as the diagram shows.

And so Morgan's idea of an X-linked gene dovetailed with Mendel's concept of segregation. By proposing that a specific gene is located on an X chromosome but not on the Y, Morgan explained his reciprocal crosses. His experimental results matched predicted outcomes.

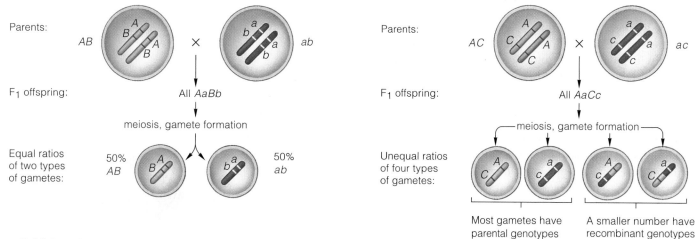

a Full linkage between two genes (no crossovers): half of the gametes have one parental genotype and half have the other.

b Incomplete linkage; crossing over affected the outcome.

Figure 12.8 Examples of gene linkages that affect a dihybrid cross. In (**a**), linkage between two gene loci is complete. In (**b**), it is incomplete.

Example: In one set of experiments, true-breeding mutant females (white eyes, yellow body) were crossed with wild-type males (red eyes, gray body). As the researchers expected, 50 percent of the F_1 offspring had one or the other parental phenotype. However, of 2,205 F_2 offspring, 129 were recombinants. That's 1.3 percent.

As it turns out, some alleles tend to remain together more often than others through meiosis. They are closer together on the same chromosome and less vulnerable to crossovers! Imagine any two genes at two different locations on the same chromosome. *The probability that a crossover will disrupt their linkage is proportional to the distance that separates the two loci.* Suppose genes A and B are twice as far apart as two other genes, C and D:

We would expect crossing over to disrupt the linkage between A and B much more often.

Two genes are very closely linked when the distance between them is small. Their allelic combinations nearly always end up inside the same gamete. Linkage is more vulnerable to crossover when the distance between them is greater (Figure 12.8). When two gene loci are very far apart, crossing over is so frequent that the genes assort independently of each other into gametes.

Unlike fruit flies, humans do not lend themselves to experimental crosses. Even so, all gene linkages have been identified by tracking the phenotypes in certain families over the generations.

For example, recessive alleles at two gene loci on the human X chromosome give rise to *color blindness* and *hemophilia*, described shortly. One female carried both alleles but was symptom-free. Her father was, too, so she must have inherited a normal X chromosome from him (remember, males have only one X and one Y). The X chromosome inherited from her mother must have carried both mutant alleles. She gave birth to six sons. Three developed color blindness and hemophilia; two were unaffected. Here was phenotypic evidence of no recombination. However, her sixth son started life as a fertilized, recombinant egg; he was color-blind only.

Tracking many families affected by these disorders revealed that recombination frequencies of these two genes are low because they are very closely linked at one end of the X chromosome. More generally, studies showed that crossovers are not rare. For humans and most other eukaryotic species, meiosis cannot even be completed properly unless every pair of homologous chromosomes takes part in at least one crossover.

The human linkage map was completed during the 1990s. Today, linkage studies are being enhanced by gene sequencing, which reveals the physical distances between genes. (Take a look at www.ncbi.nlm.nih.gov.) These genetic and physical maps are both part of the Human Genome Initiative, a topic of Chapter 16.

Genes at different loci on the same chromosome belong to the same linkage group and do not assort independently during meiosis.

Crossing over between homologous chromosomes disrupts gene linkages and results in nonparental combinations of genes in chromosomes.

The farther apart two genes are on the same chromosome, the greater will be the frequency of crossing over and genetic recombination between them.

HUMAN GENETIC ANALYSIS

Some organisms, including pea plants and fruit flies, are ideal for genetic analysis. They grow and reproduce rapidly in small spaces, under controlled conditions. It does not take very long to track a trait through many generations. Humans are another story. We live under variable conditions in diverse environments. We select our own mates and reproduce if and when we want to. Humans live as long as the geneticists who study them, so tracking traits through generations is tedious. Most human families are not large, so there are not enough offspring for easy inferences about inheritance.

Constructing Pedigrees

To get around the problems associated with analyzing human inheritance, geneticists put together pedigrees. A **pedigree** is a chart of the genetic connections among individuals. Standardized methods, definitions, and symbols for representing individuals are utilized when constructing it (Figure 12.9a).

When geneticists analyze pedigrees, they rely on their knowledge of probability and Mendelian inheritance patterns, which may yield clues to the genetic basis for a trait. For example, clues might suggest that an allele responsible for a certain disorder is dominant or recessive or that it is located on a certain autosome or sex chromosome.

Gathering a great many family pedigrees increases the numerical base for analysis. When any trait follows a simple Mendelian inheritance pattern, a geneticist has greater confidence for predicting the probability of its occurrence among children of prospective parents. We will return to this topic later.

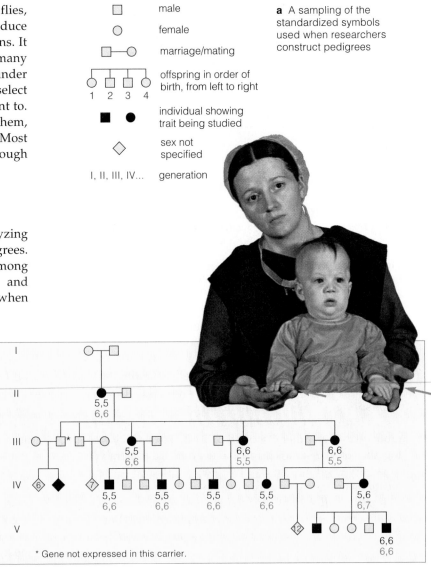

a A sampling of the standardized symbols used when researchers construct pedigrees

□ male
○ female
□—○ marriage/mating
offspring in order of birth, from left to right
1 2 3 4
■ ● individual showing trait being studied
◇ sex not specified
I, II, III, IV... generation

* Gene not expressed in this carrier.

b Pedigree for a family in which polydactyly recurs as one symptom of Ellis–van Creveld syndrome (Section 17.11)

Figure 12.9 (**a**) Some standardized symbols used in constructing pedigree diagrams. (**b**) Example of a pedigree for *polydactyly*. With this condition, an individual has extra fingers, extra toes, or both. Expression of the gene for this trait varies among individuals. *Black* numerals signify the known number of fingers on each hand; *blue* numerals signify the number of toes on each foot.

(**c**) From human genetic researcher Nancy Wexler, a pedigree for *Huntington disease*, by which the human nervous system progressively degenerates. Wexler and her team pieced together an extended family tree for nearly 10,000 Venezuelans. Analysis of affected and unaffected individuals revealed that a dominant allele on human chromosome 4 is the genetic culprit. Wexler has a special interest in Huntington disease; she herself has a 50 percent chance of developing it.

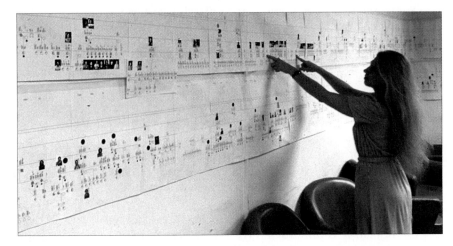

c Nancy Wexler and a pedigree she constructed to track Huntington disease

Table 12.1 Examples of Human Genetic Disorders and Genetic Abnormalities

Disorder or Abnormality*	Main Consequences	Disorder or Abnormality*	Main Consequences
AUTOSOMAL RECESSIVE INHERITANCE		**X-LINKED RECESSIVE INHERITANCE**	
Albinism 11.6; 11.9 CT	Absence of pigmentation	Androgen insensitivity syndrome 36.2	XY individual but having some female traits; sterility
Blue offspring 17.12 CT	Bright blue skin coloration	Color blindness 12.4, 12.6, 35.9	Inability to distinguish among some or all colors
Cystic fibrosis 5 CI	Excessive glandular secretions leading to tissue, organ damage	Fragile X syndrome 12.6	Mental impairment
Ellis–van Creveld syndrome 12.5, 17.11	Extra fingers, toes, short limbs	Hemophilia 12.4, 12.6, 12.12 CT	Impaired blood-clotting ability
Fanconi anemia 12.11	Physical abnormalities, bone marrow failure	Muscular dystrophies 12.12 CT, 15.7 CT, 37.10	Progressive loss of muscle function
Galactosemia 12.6	Brain, liver, eye damage	X-linked anhidrotic dysplasia 15.4	Mosaic skin (patches with or without sweat glands); other effects
Phenylketonuria (PKU) 12.10	Mental impairment		
Sickle-cell anemia 3.8. 12.11, 14.4,17.9, 38.3	Adverse pleiotropic effects on organs throughout body	**CHANGES IN CHROMOSOME NUMBER**	
		Down syndrome 12.9, 44.4	Mental impairment; heart defects
AUTOSOMAL DOMINANT INHERITANCE		Turner syndrome 12.10	Sterility; abnormal ovaries, abnormal sexual traits
Achondroplasia 12.6	One form of dwarfism		
Camptodactyly 11.7	Rigid, bent fingers	Klinefelter syndrome 12.10	Sterility; mild mental impairment
Familial hypercholesterolemia 16 CI	High cholesterol levels in blood; eventually clogged arteries	XXX syndrome 12.10	Minimal abnormalities
Huntington disease 12.5, 12.6	Nervous system degenerates progressively, irreversibly	XYY condition 12.10	Mild mental impairment or no effect
Marfan syndrome 11.5, 12.12 CT	Abnormal or no connective tissue	**CHANGES IN CHROMOSOME STRUCTURE**	
Polydactyly 12.5	Extra fingers, toes, or both	Chronic myelogenous leukemia 12 CI	Overproduction of white blood cells in bone marrow; organ malfunctions
Progeria 12.7	Drastic premature aging	Cri-du-chat syndrome 12.8	Mental impairment; abnormally shaped larynx
Neurofibromatosis 14.5	Tumors of nervous system, skin		

*Italic numbers indicate sections in which a disorder is described. CI signifies Chapter Introduction. CT signifies an end-of-chapter Critical Thinking question.

Regarding Human Genetic Disorders

Table 12.1 is a list of some heritable traits that have been studied in detail. A few are abnormalities, or deviations from the average condition. Said another way, a **genetic abnormality** is nothing more than a rare or uncommon version of a trait, as when a person is born with six toes on each foot instead of five. Whether an individual or society at large views an abnormal trait as disfiguring or merely interesting is subjective. As the classic novel *The Hunchback of Notre Dame* suggests, there is nothing inherently life-threatening or even ugly about it.

By comparison, a **genetic disorder** is an inherited condition that sooner or later will cause mild to severe medical problems. A **syndrome** is a recognized set of symptoms that characterize a given disorder.

Because alleles underlying severe genetic disorders put people at great risk, they are rare in populations. Why, then, don't they disappear entirely? First, we can expect rare mutations to introduce new copies of the alleles into the population. Second, in heterozygotes, a harmful allele is paired with a normal one that may cover its functions, so it still can be passed to offspring.

You may hear someone refer to a genetic disorder as a disease, but the terms aren't always interchangeable. A disease also is an abnormal alteration in the way the body functions, and it, too, is characterized by a set of symptoms. But **disease** is illness caused by infectious, dietary, or environmental factors, not by inheritance of mutant genes. That said, it is still appropriate to call an illness a *genetic* disease if factors alter previously workable genes in a way that disrupts body functions.

With these qualifications in mind, we turn next to examples of inheritance in the human population. As you'll see, some show simple Mendelian patterns. Many traits have been traced to a dominant or recessive allele on an autosome or X chromosome. Others arise from changes in the structure or number of chromosomes.

For many genes, pedigree analysis might reveal simple Mendelian inheritance patterns that will allow inferences about the probability of their transmission to children.

A genetic abnormality is a rare or less common version of an inherited trait. A genetic disorder is an inherited condition that results in mild to severe medical problems.

EXAMPLES OF INHERITANCE PATTERNS

Autosomal Recessive Inheritance

For some traits, inheritance patterns reveal two clues that point to a recessive allele on an autosome. First, if both parents are heterozygous, any child of theirs will have a 50 percent chance of being heterozygous and a 25 percent chance of being homozygous recessive, as Figure 12.10a indicates. Second, if the parents are both homozygous recessive, any child of theirs will be, also.

On average, 1 in 100,000 newborns is homozygous for a recessive allele that causes *galactosemia*. It cannot make functional molecules of an enzyme that prevents a product of lactose breakdown from accumulating to toxic levels. Lactose normally is converted to glucose and galactose, then to glucose–1–phosphate (which is broken down by glycolysis or converted to glycogen). The full conversion is blocked in galactosemics:

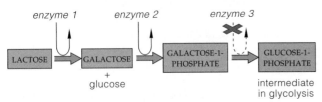

Galactose in blood rises to levels that can be detected in urine. Excessive amounts damage the eyes, liver, and brain, and cause malnutrition, diarrhea, and vomiting. Untreated galactosemics often die in childhood. When they are quickly put on a restricted diet that excludes dairy products, they can grow up symptom-free.

Autosomal Dominant Inheritance

Two clues point to an autosomal dominant allele for a trait. First, the trait typically appears each generation; usually the allele is expressed, even in heterozygotes. Second, if one parent is heterozygous and the other is homozygous recessive, any child of theirs will have a 50 percent chance of being heterozygous (Figure 12.10b).

A few dominant alleles persist in populations even though they cause severe genetic disorders. Some persist by spontaneous mutations. For others, expression of the dominant allele may not interfere with reproduction, or affected people reproduce before symptoms are severe.

For example, *Huntington disease* is characterized by progressive involuntary movements and deterioration of the nervous system; death is inevitable (Figure 12.9). Symptoms may not start to show up until an affected individual is past age thirty. Most people have already reproduced by then. Affected individuals usually die in their forties or fifties, sometimes before they realize they transmitted the mutant allele to their children.

As another example, *achondroplasia* affects about 1 in 10,000 people. Commonly, the homozygous dominant

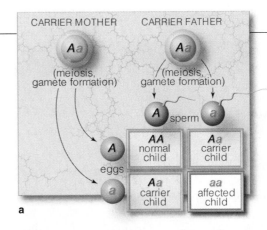

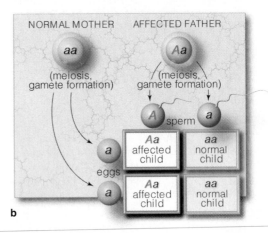

Figure 12.10 (**a**) One pattern for autosomal recessive inheritance. In this case, both parents are heterozygous carriers of the recessive allele (coded *red*). (**b**) A pattern for autosomal dominant inheritance. This dominant allele (*red*) is fully expressed in the carriers.

Figure 12.11 Infanta Margarita Teresa of the Spanish court and maids, including the achondroplasic woman (*right*).

condition leads to stillbirth, yet heterozygotes are able to reproduce. While heterozygotes are young, the cartilage of the skeleton forms improperly. At maturity, affected people have abnormally short arms and legs relative to other body parts (Figure 12.11). Adult achondroplasics are less than 4 feet, 4 inches tall. Often the dominant allele has no other phenotypic effects.

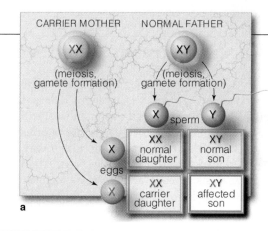

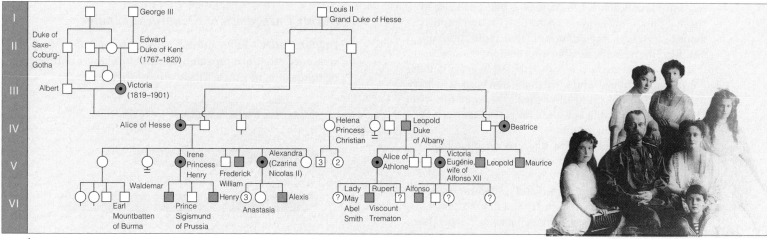

Figure 12.12 (**a**) One pattern for X-linked inheritance. In this example, assume the mother carries the recessive allele on one of her X chromosomes (*red*).

(**b**) Partial pedigree for Queen Victoria's descendants, showing carriers and affected males who carried the X allele for hemophilia A. Of the Russian royal family members in the photograph, the mother was a carrier; Crown Prince Alexis was hemophilic.

Figure 12.13 Fragile X chromosome from a cultured cell. The arrow points to the fragile site.

X-Linked Recessive Inheritance

Distinctive clues often show up when a recessive allele on an X chromosome causes a genetic disorder. First, males show the recessive phenotype more often than females. A dominant allele on the other X chromosome can mask the allele in females. The allele is not masked in males, who have one X chromosome (Figure 12.12*a*). Second, a son can't inherit the recessive allele from his father. A daughter can. If she is heterozygous, there is a 50 percent chance each son of hers will inherit the allele.

Color blindness, an inability to distinguish between some or all colors, is a common X-linked recessive trait. For instance, in red–green color blindness, an individual lacks some or all of the sensory receptors that normally respond to visible light of red and green wavelengths.

Hemophilia A, a blood-clotting disorder, is a case of X-linked recessive inheritance. Normally a blood-clotting mechanism quickly stops bleeding from minor injuries. Clotting requires several gene products, some on the X chromosome. If one of the X-linked genes is mutated in a human male, bleeding will be prolonged. About 1 in 7,000 males is affected by hemophilia A. Clotting time is close to normal in heterozygous females.

The frequency of hemophilia A was high in royal families of nineteenth-century Europe, in which close relatives often married. Queen Victoria of England was a notable carrier (Figure 12.12*b*). At one time eighteen of her sixty-nine descendants housed the recessive allele.

Fragile X syndrome, an X-linked recessive disorder that causes mental retardation, affects 1 in 1,500 males in the United States. In cultured cells, an X chromosome carrying the mutant allele is constricted near the end of its long arm (Figure 12.13). This constriction is called a fragile site because the end of the chromosome tends to break away. The term is misleading, for the end breaks away only in cultured cells, not in cells in the body.

A mutant gene, not breakage, causes the syndrome. It specifies a protein required for normal development of brain cells. Within that gene, a segment of DNA is repeated several times. Certain mutations result in the addition of many more repeats in the DNA; hence their name, expansion mutations. The outcome is a mutant allele that cannot function properly. Because that allele is recessive, males who inherit it and females who are homozygous for it have fragile X syndrome. Expansion mutations are now known to be the cause of Huntington disease and some other genetic disorders.

Genetic analyses of family pedigrees have revealed simple Mendelian inheritance patterns for certain traits, as well as for many genetic disorders that arise from expression of specific alleles on an autosome or X chromosome.

Progeria—Too Young to Be Old

Imagine being ten years old with a mind trapped inside a body that is rapidly getting a bit more shriveled, more frail—old—with each passing day. You are just barely tall enough to peer over the top of the kitchen counter, and you weigh less than thirty-five pounds. Already you are bald and have a crinkled nose. Possibly you have a few more years to live. Would you, like Mickey Hayes and Fransie Geringer, still laugh with your friends?

Of every 8 million newborn humans, one is destined to grow old far too soon. On one of its autosomes, that rare individual carries a mutant gene that gives rise to *Hutchinson–Gilford progeria syndrome*. Through hundreds, thousands, then many billions of DNA replications and mitotic cell divisions, terrible information encoded in that gene was systematically distributed to every cell in the growing embryo, and later in the newborn. Its legacy will be accelerated aging and a greatly reduced life span. The photograph of Mickey and Fransie in Figure 12.14 shows some of the symptoms.

The mutation causes gross disruptions in interactions among genes that bring about the body's growth and development. Observable symptoms start before age two. Skin that should be plump and resilient starts to thin. Skeletal muscles weaken. Tissues in limb bones that should lengthen and grow stronger start to soften. Hair loss is pronounced; extremely premature baldness is inevitable. There are no documented cases of progeria running in families, so a gene must undergo random, spontaneous mutation. Probably the gene is dominant over a normal partner on the homologous chromosome.

Most progeriacs can expect to die in their early teens as a result of strokes or heart attacks. These final insults are brought on by a hardening of the walls of arteries, a condition typical of advanced age. When Mickey turned eighteen, he was the oldest living progeriac. Fransie was seventeen when he died.

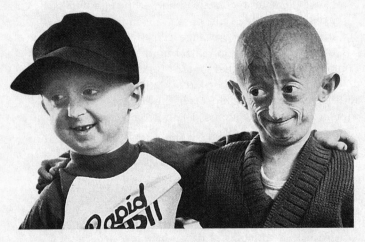

Figure 12.14 Two boys who met at a gathering of progeriacs at Disneyland, California, when they were not yet ten years old.

12.8

CHANGES IN CHROMOSOME STRUCTURE

On rare occasions, the physical structure of one or more chromosomes changes. The result is a genetic disorder or abnormality. Such changes occur spontaneously in nature. They also are induced in research laboratories by exposure to chemicals or by irradiation. Either way, changes may be detected by microscopic examination and karyotype analysis of cells at mitosis or meiosis. Let's now review four kinds of structural change. As you will see, some have severe or lethal consequences.

Major Categories of Structural Change

DUPLICATION Even normal chromosomes have gene sequences that are repeated several to many hundreds or thousands of times. These are **duplications**:

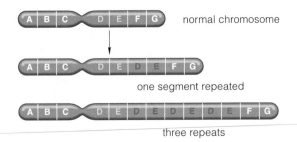

normal chromosome

one segment repeated

three repeats

INVERSION With an **inversion**, a linear stretch of DNA within the chromosome becomes oriented in the reverse direction, with no molecular loss:

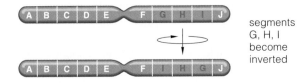

segments G, H, I become inverted

TRANSLOCATION As the chapter introduction showed, the Philadelphia chromosome that has been linked to a form of leukemia results from **translocation**, whereby a broken part of a chromosome becomes attached to a *non*homologous chromosome. Most translocations are reciprocal (both chromosomes exchange broken parts):

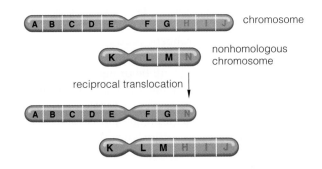

chromosome

nonhomologous chromosome

reciprocal translocation

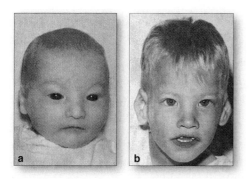

Figure 12.15 (a) Male infant who later developed cri-du-chat syndrome. Ears are low on the side of the head relative to the eyes. (b) The same boy, four years later. By this age, affected humans stop making mewing sounds typical of the syndrome.

DELETION Viral attacks, irradiation (ionizing radiation especially), chemical assaults, and other environmental agents may cause a **deletion**, the loss of some segment of a chromosome:

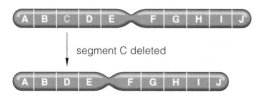

Most deletions are lethal or cause serious disorders in mammals, for they disrupt normal gene interactions that underlie a program of growth, development, and maintenance activities. For example, one deletion from human chromosome 5 results in mental impairment and the development of an abnormally shaped larynx. When affected infants cry, they produce sounds rather like a cat's meow. Hence *cri-du-chat* (cat-cry), the name of this disorder. Figure 12.15 shows an affected child.

Does Chromosome Structure Evolve?

Some changes in chromosome structure tend not to be conserved; they are selected against over evolutionary time. Even so, among many species, we see interesting signs of past changes. For example, duplications are common. Many have not harmed their bearers. Over billions of years, neutral ones have been accumulating and are now built into the DNA of all species.

Duplicates of gene sequences with neutral effects might have an adaptive advantage. Having two of the same gene could free up one of them for potentially beneficial mutations. The other gene would still issue the required product. The duplicated gene region could become slightly modified, then the products of the two genes could function in slightly different or novel ways.

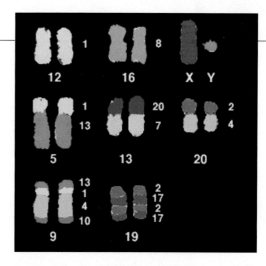

Figure 12.16 Spectral karyotype of duplicated chromosomes of the gibbon, one of the apes. The colors identify regions of gibbon chromosomes that are structurally identical with human chromosomes (compare Figure 12.1).

Top row: Chromosomes 12, 16, X, and Y are the same in both primates. *Second row:* Translocations are present in gibbon chromosomes 5, 13, and 20, and they correspond to regions of human chromosomes 1, 13, 20, 7, 2, and 4.

Third row: Gibbon chromosome 9 corresponds to regions of several human chromosomes. Duplications in gibbon chromosome 19 occur in human chromosomes 2 and 17.

Apparently, several kinds of duplicated, modified genes were pivotal in evolution. Think back on those gene regions for the polypeptide chains of hemoglobin (Section 3.8). In humans and other primates, the regions have multiple nucleotide sequences that are strikingly similar. The sequences specify whole families of chains, each with a slight structural difference. The structural differences translate into slightly different efficiencies in hemoglobin's capacity to bind and transport oxygen under a range of cellular conditions.

It also appears that certain duplications, inversions, and translocations helped put the primate ancestors of humans on a unique evolutionary road. They seem to have contributed to divergences that led to the modern apes and humans. Of the twenty-three pairs of human chromosomes, eighteen are nearly identical with their counterparts in chimpanzees and gorillas. The other five pairs differ at inverted and translocated regions. You can observe the dramatic similarities for yourself by comparing chromosomes of a human with those of a gibbon, one of the apes (Figure 12.16).

On rare occasions, a segment of a chromosome may become duplicated, inverted, moved to a new location, or deleted.

Most chromosome changes are harmful or lethal when they alter gene interactions that underlie growth, development, and maintenance activities.

Over evolutionary time, many changes have been conserved; they confer adaptive advantages or have had neutral effects.

CHANGES IN CHROMOSOME NUMBER

Occasionally, abnormal events occur before or during cell division, then gametes and new individuals end up with the wrong chromosome number. The consequences range from minor to lethal physical changes.

Categories and Mechanisms of Change

With **aneuploidy**, individuals usually have one extra or one less chromosome. This condition is a major cause of human reproductive failure. Possibly it affects about one-half of all fertilized eggs. As autopsies reveal, most *miscarriages* (spontaneous aborting of embryos before pregnancy reaches full term) are aneuploids.

With **polyploidy**, individuals have three or more of each type of chromosome. About one-half of all species of flowering plants are polyploid (Section 18.3). Often, researchers can induce polyploidy in undifferentiated plant cells by exposing them to colchicine (Section 12.2). Some species of insects, fishes, and other animals are polyploids. However, polyploidy is lethal for humans. All but about 1 percent of human polyploids die before birth, and the rare newborns die soon after birth.

Chromosome numbers can change during mitotic or meiotic cell divisions. Suppose a cell cycle proceeds through DNA duplication and mitosis but is arrested before the cytoplasm divides. The cell is now *tetra*ploid, with four of each type of chromosome. Suppose one or more pairs of chromosomes fail to separate in mitosis or meiosis, an event called **nondisjunction**. Some or all of the forthcoming cells will have too many or too few chromosomes, as in the Figure 12.17 example.

The chromosome number also may change during fertilization. Visualize a normal gamete that unites by chance with an $n + 1$ gamete (one extra chromosome).

The new individual will be "trisomic" ($2n + 1$); it will have three of one type of chromosome and two of every other type. And what if an $n - 1$ gamete unites with a normal n gamete? Then, the new individual will turn out to be "monosomic" ($2n - 1$).

Nearly all inherited changes in chromosome number arise by nondisjunction at meiosis or when gametes are forming. Mitotic cell divisions perpetuate the mistake as an embryo is developing. Let's start with one of the changes in the number of autosomes. The next section looks mainly at altered numbers of sex chromosomes.

Case Study: Down Syndrome

A trisomic 21 newborn, with three chromosomes 21, can be expected to develop *Down syndrome*, the most frequent change in human chromosome number. This autosomal disorder occurs once in every 800 to 1,000 births. It affects more than 350,000 people in the United States alone. Figure 12.18 shows a typical trisomic 21 karyotype and a few affected individuals.

Nondisjunction during meiosis accounts for about 95 percent of all Down syndrome cases. Translocation and mosaicism account for the remainder. With **mosaicism**, nondisjunction occurs in one of the embryonic cells that form after fertilization. But only the descendants of that altered cell inherit three chromosomes 21; other cells in the body have the normal chromosome number. Translocation may occur while gametes are forming or after fertilization. Part of chromosome 21 breaks away and then becomes attached to a different chromosome in the gamete or embryonic cell.

At this time we do not know why nondisjunction occurs. We know its incidence rises with advancing age

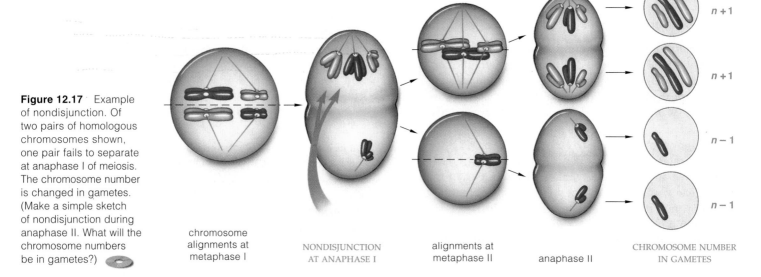

Figure 12.17 Example of nondisjunction. Of two pairs of homologous chromosomes shown, one pair fails to separate at anaphase I of meiosis. The chromosome number is changed in gametes. (Make a simple sketch of nondisjunction during anaphase II. What will the chromosome numbers be in gametes?)

chromosome alignments at metaphase I

NONDISJUNCTION AT ANAPHASE I

alignments at metaphase II

anaphase II

CHROMOSOME NUMBER IN GAMETES

$n + 1$

$n + 1$

$n - 1$

$n - 1$

Figure 12.18 Karyotype revealing the trisomic 21 condition of a young male. Both the young girl and the teenager shown here were lively participants in the Special Olympics held annually in San Mateo, California.

of the mother (Figure 12.19). Nondisjunction also may originate with the father, but not nearly as frequently. Whatever causes it, we see the same outcome. While a new individual was developing, mitotic cell divisions put an extra chromosome 21 (or an extra portion of it) in some or all of its new cells.

Characteristics associated with the disorder include upwardly slanted eyes, a skin fold that starts at the inner corner of the eye, a deep crease across each palm and foot sole, one (not two) horizontal furrows across the fifth finger, a tongue that is large relative to the oral cavity, somewhat flattened facial features, and poor muscle tone. Affected children often have heart defects and respiratory and digestive problems, most of which are now treatable. Adults often develop Alzheimer's disease. With good medical care, trisomics 21 now have a life expectancy of fifty-five years, on average.

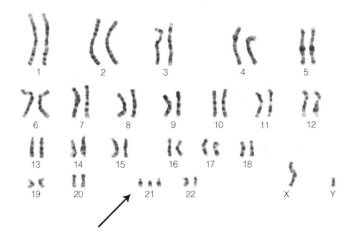

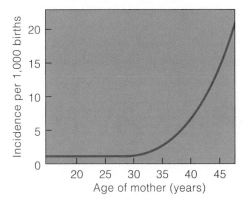

Figure 12.19 Relationship between the frequency of Down syndrome and mother's age at the time of childbirth. Results shown here are from a study of 1,119 affected children who were born in Victoria, Australia, between 1942 and 1957. The risk of giving birth to a trisomic 21 individual increases with the mother's age. This may seem counterintuitive, for about 80 percent of trisomics 21 are born to women not yet 35 years old. However, these women are in age categories with the highest fertility rates; they simply have more babies.

Not all trisomic 21 individuals develop all of the symptoms of Down syndrome. Some newborns show only a few. Besides this, some of the so-called defining features also appear in the population at large.

That said, most trisomics 21 display moderate to severe mental impairment. Heart defects are common. The skeleton develops abnormally, so older children have shortened body parts, loose joints, and poorly aligned bones of the hips, fingers, and toes. Muscles and muscle reflexes are weaker than normal. Speech and other motor skills develop slowly.

Trisomy 21 is one of many conditions that may be detected through prenatal diagnosis, as described in Section 12.11. With medical intervention and special training that begins early on, trisomic 21 individuals can take part in normal activities. As a group, they tend to be cheerful and affectionate people, and they derive great pleasure from socializing (Figure 12.18).

Nondisjunction prior to cell division in reproductive cells or early embryonic cells results in an altered number of autosomes or sex chromosomes. The change affects the course of development and the resulting phenotypes.

CASE STUDIES: CHANGES IN THE NUMBER OF SEX CHROMOSOMES

Besides causing most of the alterations in the number of autosomes, nondisjunction also causes most alterations in the number of X and Y chromosomes. The frequency of such changes is 1 in 400 live births. Most often they lead to difficulties in learning and motor functioning, including speech, although problems may be so subtle that the underlying cause is not even diagnosed.

Changes in sex chromosome number are only a bit less common than they are for the autosomal changes, partly because the phenotypes of affected individuals usually are not severely altered. Most of these changes can be diagnosed before birth by procedures described in the next section.

Female Sex Chromosome Abnormalities

TURNER SYNDROME Inheriting an X chromosome and no corresponding X or Y chromosome gives rise to *Turner syndrome*, which affects 1 in 2,500 to 10,000 or so newborn girls. Figure 12.20 shows one XO karyotype. Nondisjunction that originates with the father accounts for 75 percent of all the cases. We see fewer people with Turner syndrome, compared to other sex chromosome abnormalities. The likely reason is at least 98 percent of all XO zygotes spontaneously abort early in pregnancy. From one study, we know that about 20 percent of all spontaneously aborted embryos in which chromosome abnormalities were detected were XO.

Despite the near lethality, XO survivors are not as disadvantaged as other aneuploids. They grow up well proportioned, albeit short—4 feet, 8 inches tall, on the average. Their childhood behavior is typically normal.

Most Turner females don't have functional ovaries, can't produce normal amounts of sex hormones, and are infertile. The reduction of sex hormones adversely affects development of secondary sexual traits, such as breast enlargement. These females do produce a few immature eggs, but by the time they are two years old,

the eggs are destroyed. Possibly as a consequence of their arrested sexual development and small stature, XO teenagers often are passive and easily intimidated by peers. Hormone replacement therapy or corrective surgery, or both, may mitigate some of the symptoms that cause these individuals distress.

XXX SYNDROME A few females inherit three, four, or five X chromosomes. The XXX condition occurs at a frequency of about 1 in 1,000 live births. Most adults are an inch or so taller and more slender than average. They are fertile. Except for slight learning difficulties, these females have a normal appearance and tend to fall within the normal range of social behavior.

Male Sex Chromosome Abnormalities

KLINEFELTER SYNDROME One in 500 to 2,000 males has inherited one Y and two or more X chromosomes, mainly as an outcome of nondisjunction. They have an XXY or, more rarely, XXXY, XXXXY, or XY/XXY mosaic genotype. About 67 percent of the affected individuals inherited the extra chromosome from their mother.

The symptoms of the resulting *Klinefelter syndrome* develop after the onset of puberty. For example, XXY males tend to be taller than average and overweight. Most have a normal outward appearance. However, the testes and prostate gland usually are much smaller than average; the penis and scrotum are not.

In affected individuals, testosterone production is lower than normal and estrogen production higher. This has feminizing effects. For example, sperm counts are low. Hair on the face and elsewhere is sparse, the voice is pitched high, and the breasts are somewhat enlarged. Testosterone injections from puberty onward can reverse the feminized traits, although not the low fertility. Even without such reversal, XXY phenotypes are not that different. Many affected individuals were

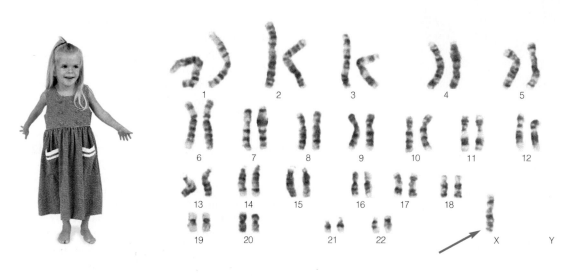

Figure 12.20
Example of an XO karyotype from a female affected by Turner syndrome.

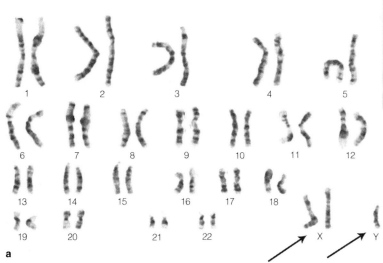

a

not even aware of their chromosomal abnormality until they were tested for infertility.

Although many XXY males fall within the normal range of intelligence, some have problems with speech and short-term memory. Those who are not diagnosed and given guidance early on often have difficulties of the sort highlighted in Figure 12.21.

XYY CONDITION   About 1 in 500 to 1,000 males has one X and two Y chromosomes. Those with this *XYY condition* tend to be taller than average. Some show mild mental impairment, but most are phenotypically normal. XYY males were once thought to be genetically predisposed to become criminals. The conclusion was erroneous. It was based on small numbers of cases in narrowly selected groups, including prison inmates. The investigators often knew who the XYY males were, which may have biased their evaluations.

Also, there were no **double-blind studies**, by which different investigators gather data independently of one another and match them up only after both sets of data are completed. In this case, the same investigators gathered karyotypes *and* personal histories. Fanning the stereotype was a sensationalized report in 1968 that a mass-murderer of young nurses was XYY. He wasn't.

In 1976 a Danish geneticist reported on a large-scale study. It was based on the records of 4,139 tall males, twenty-six years old, who had reported to their draft board. Besides giving results of physical examinations and intelligence testing, the records provided clues to social and economic status, educational background, and any criminal convictions. Only twelve of the males were XYY, which meant there were more than 4,000 males in the control group. The only significant finding was that tall, mentally impaired males who engage in criminal activity are just more likely to get caught—irrespective of karyotype.

Figure 12.21   (**a**) Karyotype of a male with Klinefelter syndrome. (**b**) One person's story: Stefan as a baby, and later in life. Until his teenage years, he was shy and reserved, yet would become enraged for no apparent reason. His parents sensed something was wrong. Psychologists and doctors could not pinpoint the problem. They assumed he had learning disabilities that affected comprehension, auditory processing, memory, and abstract thinking. A psychologist told him he was stupid, lazy, and would be lucky to graduate from high school.

Stefan did graduate from high school and went on to college, where he obtained a BS degree in business administration and also in sports management. He never discussed his learning disabilities. Rather, he took pride in doing the work on his own and not being treated differently.

It was not until Stefan was 25 years old that he was diagnosed with Klinefelter syndrome. During a routine physical examination, a physician noticed that Stefan's testes were smaller than normal. Subsequently, laboratory tests and karyotyping revealed his 46XY/47XXY mosaic condition.

That same year, Stefan began a job as a software engineer. It was tough, but he felt he would not have been nearly as successful if he hadn't learned to work on his own. Having a full-time position helped him open doors to volunteer work with the Klinefelter syndrome network. During his volunteer work, he met his future fiancée, whose son also has the syndrome.

Recent research suggests that as many as 64 percent to 85 percent of all XXY, XXX, and XYY children have not even been properly diagnosed. Some dismiss them unfairly as being underachievers without having a clue to the genetic basis for their learning disabilities.

***

Sex chromosome abnormalities are most often caused by nondisjunction during meiosis. They typically cause subtle difficulties with learning, speech, and other motor skills.

These abnormalities are only slightly less common than autosomal abnormalities, partly because they are rarely lethal and usually have much less severe effects.

Prospects in Human Genetics

With the arrival of their newborn, parents typically ask, "Is our baby normal?" Quite naturally, they want their baby to be free of genetic disorders, and most babies are. But what are the options when they are not?

We do not approach heritable disorders and diseases the same way. We attack diseases with antibiotics, surgery, and other weapons. But how do we attack a heritable "enemy" that can be transmitted to offspring? Should we institute regional, national, or global programs to identify people who might be carrying harmful alleles? Do we tell them they are "defective" and run a risk of bestowing a disorder on their children? Who decides which alleles are harmful? Should society bear the cost of treating genetic disorders before and after birth? If so, should society also have a say in whether an affected embryo will be born at all, or whether it should be aborted? An **abortion** is the expulsion of a pre-term embryo or fetus from the uterus.

Such questions are only the tip of an ethical iceberg. We don't have answers that are universally acceptable.

PHENOTYPIC TREATMENTS　Often, symptoms of genetic disorders can be minimized or suppressed by dietary controls, adjustments to environmental conditions, and surgical intervention or hormone replacement therapy.

For example, dietary control works in *phenylketonuria*, or PKU. A certain gene specifies an enzyme that converts one amino acid to another—phenylalanine to tyrosine. If an individual is homozygous recessive for a mutated form of the gene, the first of these amino acids accumulates inside the body. If excess amounts are diverted into other pathways, then phenylpyruvate and other compounds may form. High levels of phenylpyruvate in the blood can impair the functioning of the brain. When affected people restrict their intake of phenylalanine, they are not required to dispose of excess amounts, so they can lead normal lives. Among other things, they can avoid soft drinks and other food products that are sweetened with aspartame, a compound that contains phenylalanine.

Environmental adjustments can counter or minimize symptoms of some disorders, as when albinos avoid direct sunlight. Surgery can repair conditions such as a *cleft lip*, an opening in the upper lip that did not seal, as it should have, during embryonic development. This condition usually arises by interactions among multiple genes from both parents and environmental factors.

GENETIC SCREENING　Through large-scale screening programs in the general population, affected persons or carriers of a harmful allele often can be detected early enough to start preventive measures before symptoms develop. For example, most hospitals in the United States routinely screen newborns for PKU, so today it is less common to see people with symptoms of this disorder.

GENETIC COUNSELING　If a first child or close relative has a severe heritable problem, prospective parents may worry about their next child. They may request help in evaluating their options from a qualified professional counselor. *Genetic counseling* often includes diagnosis of parental genotypes, detailed pedigrees, and genetic testing for hundreds of known metabolic disorders. Geneticists, too, help predict risks for genetic disorders. Counselors must remind prospective parents that the same risk usually applies to each pregnancy.

PRENATAL DIAGNOSIS　Methods of *prenatal diagnosis* are used to determine the sex of embryos or fetuses and more than a hundred genetic conditions. (*Prenatal* means before birth. The term *embryo* applies until eight weeks after fertilization, after which *fetus* is appropriate.)

Suppose a woman who is forty-five years old becomes pregnant and worries about Down syndrome. Usually between 8 and 12 weeks after pregnancy, she might ask for prenatal diagnosis by *amniocentesis* (Figure 12.22). With this procedure, a clinician withdraws a tiny sample of fluid inside the amnion, a membranous sac enclosing the fetus. Some cells that the fetus shed are suspended in the sample. The cells are cultured and analyzed.

Chorionic villi sampling (CVS) is a different diagnostic procedure. A clinician withdraws cells from the chorion, a fluid-filled, membranous sac that surrounds the amnion.

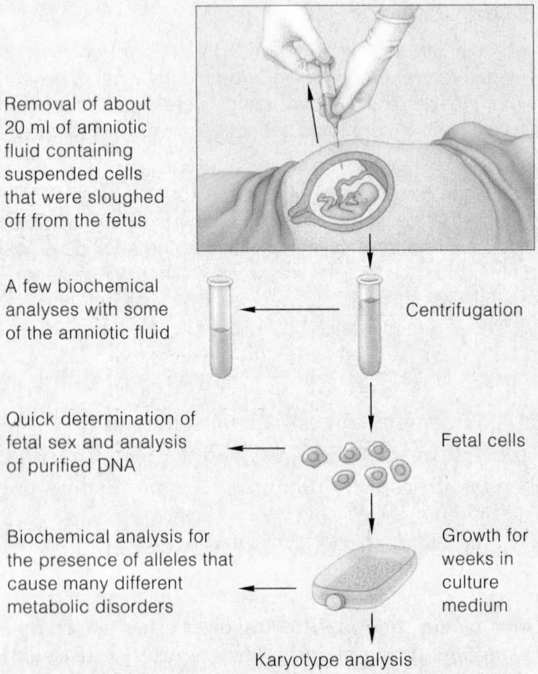

Removal of about 20 ml of amniotic fluid containing suspended cells that were sloughed off from the fetus

A few biochemical analyses with some of the amniotic fluid

Centrifugation

Quick determination of fetal sex and analysis of purified DNA

Fetal cells

Biochemical analysis for the presence of alleles that cause many different metabolic disorders

Growth for weeks in culture medium

Karyotype analysis

Figure 12.22　Amniocentesis, a prenatal diagnostic tool.

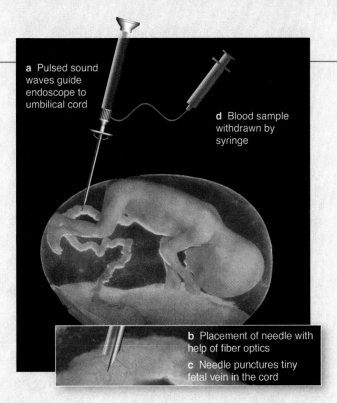

a Pulsed sound waves guide endoscope to umbilical cord

d Blood sample withdrawn by syringe

b Placement of needle with help of fiber optics

c Needle punctures tiny fetal vein in the cord

Figure 12.23 Fetoscopy for prenatal diagnosis.

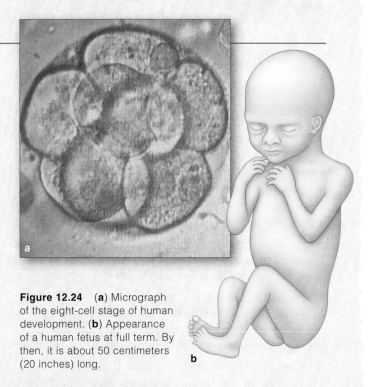

Figure 12.24 (**a**) Micrograph of the eight-cell stage of human development. (**b**) Appearance of a human fetus at full term. By then, it is about 50 centimeters (20 inches) long.

CVS can be performed weeks before amniocentesis. It yields results as early as the eighth week of pregnancy.

Direct visualization of a developing fetus is possible with *fetoscopy*. An endoscope, a fiberoptic device, uses pulsed sound waves to scan the uterus and visually locate particular parts of the fetus, umbilical cord, or placenta (Figure 12.23). Fetoscopy has been used to diagnose blood cell disorders, such as sickle-cell anemia and hemophilia.

All three procedures may cause infection or puncture the fetus. If a punctured amnion doesn't reseal itself fast, too much amniotic fluid leaks out, which harms the fetus. Amniocentesis increases the risk of miscarriage by 1 to 2 percent. CVS has a 0.3 percent risk that a future child will have missing or underdeveloped fingers and toes. Fetoscopy raises risk of a miscarriage by 2 to 10 percent.

Parents-to-be probably should seek counseling from a doctor to help them weigh the risks and benefits of such procedures in terms of their own circumstances. They may wish to ask about the small overall risk of 3 percent that any child will have some kind of birth defect. They may ask about the severity of a genetic disorder that a child might be at risk of developing. And they might consider how old the woman is at the time of pregnancy.

REGARDING ABORTION What happens when prenatal diagnosis does reveal a serious problem? Do prospective parents opt for induced abortion? We can only say here that they must weigh their awareness of the severity of the genetic disorder against ethical and religious beliefs. Worse, they must play out their personal tragedy on a larger stage, dominated by a nationwide battle between fiercely vocal "pro-life" and "pro-choice" factions. We return to this volatile issue in Sections 44.14 and 44.16.

PREIMPLANTATION DIAGNOSIS Another procedure, *preimplantation diagnosis*, relies on **in-vitro fertilization**. In vitro, recall, means "in glass." Sperm and eggs from prospective parents are placed in an enriched medium in a glass or plastic petri dish. One or more eggs may get fertilized. In two days, mitotic cell divisions may convert one of these into a ball of eight cells (Figure 12.24*a*).

According to one view, the tiny, free-floating ball is a *pre*-pregnancy stage. Like the unfertilized eggs discarded monthly from a woman, it isn't attached to the uterus. Its cells all have the same genes and are not yet committed to giving rise to specialized cells of a heart, lungs, and other organs. Doctors take one of the undifferentiated cells and analyze its genes. If that cell has no detectable genetic defects, the ball is inserted into the uterus.

Some couples who are at risk of passing on muscular dystrophy, cystic fibrosis, and other disorders have opted for the procedure. Many *"test-tube" babies* have been born in good health and are free of the mutant alleles.

In 2000, genetic screening of a dozen fertilized eggs revealed an embryo free of a certain mutated gene. Both prospective parents carried the gene, and their first-born child, Molly, had developed *Fanconi anemia*. She had no thumbs, deformed arms, an incomplete brain, and the prospect of dying soon from leukemia. When her new brother was born, blood stem cells from the umbilical cord were transfused into her. The cells took hold. At this time Molly's bone marrow is producing red blood cells and platelets that are countering the anemia. The first embryo genetically selected to save a life is doing just that.

12.12

SUMMARY
Gold indicates text section

1. Genes, the units of instruction for heritable traits, are arranged one after the other along chromosomes. Each gene has its own specific location, or locus, on one type of chromosome. Different molecular forms of that gene, called alleles, may occupy the locus. *12.1*

2. Human somatic cells are diploid (2*n*). They contain twenty-three pairs of homologous chromosomes that interact during meiosis. Each pair has the same length, shape, and genes (except for an XY pairing). *12.1*

3. An allele on one chromosome may or may not be identical to the allele at the equivalent locus on the homologous chromosome. *12.1*

4. Human females have two X chromosomes. Males have one X paired with one Y. All other chromosomes are autosomes (the same in both females and males). A gene on the Y chromosome determines sex. *12.1, 12.3*

5. Karyotypes, used in genetic analysis, are constructed to compare an individual's chromosomes (based upon their defining structural features) against standardized preparations of metaphase chromosomes. *CI, 12.2*

6. Genes on the same chromosome represent a linkage group. However, crossing over (breakage and exchange of segments between homologues) disrupts linkages. The farther apart two gene loci are along the length of a single chromosome, the greater will be the frequency of crossovers between them. *12.4*

7. Pedigrees are charts of genetic connections through lines of descent. Pedigrees provide clues to inheritance of a trait. *12.5*

8. Mendelian patterns of inheritance are characteristic of certain dominant or recessive alleles on autosomes or on the X chromosome. *12.5, 12.6*

9. A chromosome's structure may be altered on rare occasions. A segment may be deleted, inverted, moved to a new location (translocated), or duplicated. *12.8*

10. The chromosome number can change. Gametes and offspring may get one more or one less chromosome than the parents (aneuploidy). They may get three or more of each type of chromosome (polyploidy). Nondisjunction during meiosis accounts for most of these changes in chromosome number. *12.9*

11. Changes in chromosome structure or number often result in genetic abnormalities or disorders. Phenotypic treatments, genetic screening, genetic counseling, and prenatal diagnosis are social responses. *12.8–12.10*

12. Crossing over adds to potentially adaptive variation in traits in a population. Most but not all changes in chromosome number or structure are harmful or lethal. Over evolutionary time, some have become established in chromosomes of all species. *12.8*

Review Questions

1. What is a gene? What are alleles? *12.1*

2. Distinguish between: *12.1, 12.2*
   a. homologous and nonhomologous chromosomes
   b. sex chromosomes and autosomes
   c. karyotype and karyotype diagram

3. Define genetic recombination, and describe how crossing over can bring it about. *12.4*

4. Define pedigree. Explain the difference between genetic abnormality and genetic disorder, using examples. *12.5*

5. Contrast a typical pattern of autosomal recessive inheritance with that of autosomal dominant inheritance. *12.6*

6. Describe two clues that often show up when a recessive allele on an X chromosome causes a genetic disorder. *12.6*

7. Distinguish among a chromosomal deletion, duplication, inversion, and translocation. *12.8*

8. Define aneuploidy and polyploidy. Make a simple sketch of an example of nondisjunction. *12.9*

Self-Quiz   ANSWERS IN APPENDIX III

1. _____ align and segregate during _____ .
   a. Homologues; mitosis
   b. Genes on nonhomologous chromosomes; meiosis
   c. Homologues; meiosis
   d. Genes on one chromosome; mitosis

2. The probability of a crossover occurring between two genes on the same chromosome is _____ .
   a. unrelated to the distance between them
   b. increased if they are closer together on the chromosome
   c. increased if they are farther apart on the chromosome

3. Genetic disorders are caused by _____ .
   a. altered chromosome number   c. mutation
   b. altered chromosome structure   d. all of the above

4. A recognized set of symptoms that characterize a specific disorder is a _____ .
   a. syndrome   b. disease   c. pedigree

5. Chromosome structure can be altered by a _____ .
   a. deletion   c. inversion   e. all of the above
   b. duplication   d. translocation

6. Nondisjunction is the name for _____ .
   a. crossing over in mitosis
   b. segregation in meiosis
   c. failure of chromosomes to separate during meiosis
   d. multiple independent assortments

7. A gamete affected by nondisjunction would have _____ .
   a. a change from the normal chromosome number
   b. one extra or one missing chromosome
   c. the potential for a genetic disorder
   d. all of the above

8. Match the chromosome terms appropriately.
   ____ crossing over   a. number and defining features of an
   ____ deletion   individual's metaphase chromosomes
   ____ nondisjunction   b. chromosome segment moves to a
   ____ translocation   nonhomologous chromosome
   ____ karyotype   c. disrupts gene linkages at meiosis
   ____ linkage group   d. causes gametes to have abnormal
                                     chromosome numbers
                              e. loss of a chromosome segment
                              f. all genes on a given chromosome

Figure 12.25 Mutant fruit fly with vestigial wings.

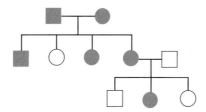

Figure 12.26 Go ahead, identify the mystery pedigree.

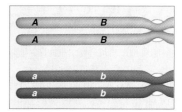

Figure 12.27 Two of your linked genes, on a pair of homologous chromosomes.

Critical Thinking—Genetics Problems

ANSWERS IN APPENDIX IV

1. Human females are XX and males are XY.
   a. Does a male inherit the X from his mother or his father?
   b. With respect to X-linked alleles, how many different types of gametes can a male produce?
   c. If a female is homozygous for an X-linked allele, how many types of gametes can she produce with respect to that allele?
   d. If a female is heterozygous for an X-linked allele, how many types of gametes can she produce with respect to that allele?

2. The wild-type allele of a gene locus governing wing length in *D. melanogaster* results in normal long wings. Figure 12.25 shows how homozygosity for a mutant recessive allele results in formation of vestigial (short) wings. Suppose you expose a homozygous dominant, long-winged fly to x-rays, then cross it with a homozygous recessive, vestigial-winged fly. The eggs develop into adults. Most are heterozygous, with long wings. A few have vestigial wings. What might explain these results?

3. *Marfan syndrome* is a genetic disorder with pleiotropic effects (Section 11.15). Affected people are often very tall and thin, with double-jointed fingers. The spine curves abnormally. Arms, fingers, and lower limbs are disproportionately long. Eye lenses dislocate easily. Thin tissue flaps in the heart that act like valves to direct blood flow may billow the wrong way when the heart contracts. The heart beats irregularly. The main artery from the heart is fragile. Its diameter widens; its wall may tear.

   Genetic analysis shows the disorder follows a pattern of autosomal dominant inheritance. What is the chance any child will inherit the allele if one parent is heterozygous for it?

4. In the Figure 12.26 pedigree, does the phenotype indicated by *red* circles and squares follow a Mendelian inheritance pattern that is autosomal dominant, autosomal recessive, or X-linked?

5. One kind of *muscular dystrophy*, a genetic disorder, is due to a recessive X-linked allele. Usually, symptoms start in childhood. Over time, a slowly progressing loss of muscle function leads to death, usually by age twenty or so. Unlike color blindness, this disorder is nearly always restricted to males. Suggest why.

6. Suppose you carry two linked genes with alleles *Aa* and *Bb*, respectively, as in Figure 12.27. If the crossover frequency between these two genes is zero, what genotypes would be expected among the gametes you produce, and with what frequencies?

7. Individuals showing *Down syndrome* usually have an extra chromosome 21, so their body cells contain 47 chromosomes.
   a. At which stages of meiosis I and II could a mistake occur that could result in the altered chromosome number?
   b. In a few cases, 46 chromosomes are present, including two chromosomes 21 with a normal appearance and a longer-than-normal chromosome 14. Explain how this chromosome abnormality may arise.

8. In the human population, mutation of two different genes on the X chromosome causes two types of X-linked *hemophilia* (types A and B). In a few known cases, a woman is heterozygous for both mutant alleles (one on each of her two X chromosomes). All her sons should have either hemophilia A or B. Yet, on very rare occasions, such a woman gives birth to a son who does not have hemophilia, and his one X chromosome does not have either mutant allele. Explain how such an X chromosome could arise.

9. Think back on the chapter definitions of genetic disorder and genetic abnormality. Then think about how subjective we are in terms of our acceptance of a condition that is out of the ordinary. Example: Would you find a cleft lip as acceptable, even attractive, on someone not as gorgeous as Joaquin Phoenix, shown at right? You will not find an answer to this question in Appendix IV. We present it simply as a way to invite critical thinking about what we as a society consider to be "ideal" phenotypes.

Selected Key Terms

abortion *12.11*	genetic abnormality *12.5*	linkage group *12.4*
allele *12.1*	genetic disorder *12.5*	mosaicism *12.9*
aneuploidy *12.9*	genetic	nondisjunction *12.9*
autosome *12.1*	recombination *12.1*	pedigree *12.5*
crossing over *12.1*	homologous	polyploidy *12.9*
deletion *12.8*	chromosome *12.1*	reciprocal cross *12.4*
disease *12.5*	independent	sex chromosome *12.1*
double-blind	assortment *12.1*	syndrome *12.5*
study *12.10*	inversion *12.8*	translocation *12.8*
duplication *12.8*	in-vitro fertilization *12.11*	X chromosome *12.1*
gene *12.1*	karyotype *CI*	Y chromosome *12.1*

Readings

Fairbanks, D., and W. R. Andersen. 1999. *Genetics: The Continuity of Life*. Monterey, California: Brooks-Cole.

13

DNA STRUCTURE AND FUNCTION

Cardboard Atoms and Bent Wire Bonds

One might have wondered, in the spring of 1868, why Johann Friedrich Miescher was collecting cells from the pus of open wounds and, later, from the sperm of a fish. Miescher, a physician, wanted to identify the chemical composition of the nucleus. These particular cells have very little cytoplasm, which makes it easier to isolate the nuclear material for analysis.

Miescher finally succeeded in isolating an organic compound having the properties of an acid. Unlike most substances in cells, it contained a notable amount of phosphorus. Miescher called the substance nuclein. He had discovered what came to be known many years later as **deoxyribonucleic acid**, or **DNA**.

The discovery did not cause even a ripple through the scientific community. At the time, no one really knew much about the physical basis of inheritance— that is, *which chemical substance encodes the instructions for reproducing parental traits in offspring*. Few even suspected that the cell nucleus might hold the answer. For a time, researchers generally believed hereditary instructions had to be encoded in the structure of some unknown class of proteins. After all, heritable traits are spectacularly diverse. Surely the molecules encoding information about those traits were structurally diverse also. Proteins are put together from potentially limitless combinations of twenty different amino acids, so the thinking was that they could function as the sentences (genes) in each cell's book of inheritance.

By the early 1950s, however, the results of a few ingenious experiments clearly indicated that DNA is the substance of inheritance. Moreover, in 1951, Linus Pauling did something no one had done before. Through his training in biochemistry, a talent for model building, and a few great educated guesses, Pauling deduced the three-dimensional structure of the protein collagen. His discovery was truly electrifying. If someone could pry open the secrets of proteins, then why not assume the same might be done for DNA? And if the structural details of the DNA molecule were worked out, then wouldn't they provide clues to its biological functions? *Who would go down in history as having discovered the very secrets of inheritance?*

Scientists around the world started scrambling after that ultimate prize. Among them were James Watson, a young postdoctoral student from Indiana University, and Francis Crick, an unflappably exuberant researcher at Cambridge University. Exactly how could DNA, a molecule that consists of only four kinds of subunits, hold genetic information? Watson and Crick spent long hours arguing over everything they had read about the size, shape, and bonding requirements of the subunits of DNA. They fiddled with cardboard cutouts of the subunits. They badgered chemists to help them identify any potential bonds they might have overlooked. Then they assembled models from bits of metal connected by wire "bonds" bent at seemingly suitable angles.

In 1953, Watson and Crick put together a model that fit all the pertinent biochemical rules and all the facts about DNA they had gleaned from other sources. They had discovered the structure of DNA (Figures 13.1 and 13.2). And the breathtaking simplicity of that structure enabled them to solve another long-standing riddle— *how the world of life can show such unity at the molecular level and yet show such spectacular diversity at the level of whole organisms.*

Figure 13.1 James Watson and Francis Crick posing in 1953 by their newly unveiled structural model of DNA.

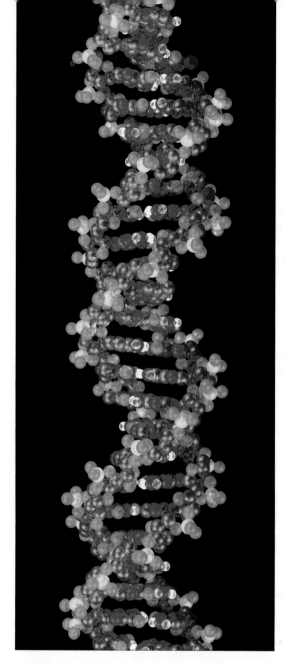

Figure 13.2 A recent space-filling model for DNA, corresponding to the prototype that Watson and Crick put together decades ago.

Key Concepts

1. In all living cells, DNA molecules are storehouses of information about heritable traits.

2. In a DNA molecule, two strands of nucleotides twist together, like a spiral stairway. Each strand of the molecule consists of four kinds of nucleotides that are the same except for one component—a nitrogen-containing base. The four bases are adenine, guanine, thymine, and cytosine.

3. Many nucleotides are arranged one after another in each strand of the DNA molecule. In at least some regions, the order in which one kind of nucleotide follows another is unique for each species. Hereditary information is encoded in the particular sequence of nucleotide bases.

4. Hydrogen bonds connect bases of one strand of the DNA molecule to bases of the other strand. As a rule, adenine pairs (hydrogen-bonds) with thymine, and guanine with cytosine.

5. Before any cell divides, its DNA is replicated with the assistance of enzymes and other proteins. Each double-stranded DNA molecule starts unwinding. As it does so, a new, complementary strand is assembled bit by bit on the exposed bases of each parent strand according to the base-pairing rule stated above.

See p. 10

With this chapter, we turn to investigations that led to our current understanding of DNA. The story is more than a march through details of its structure and function. *It also is revealing of how ideas are generated in science.* On the one hand, having a shot at fame and fortune quickens the pulse of men and women in any profession, and scientists are no exception. On the other hand, science proceeds as a community effort, with individuals sharing not only what they can explain but also what they do not understand. Even when an experiment fails to produce the anticipated results, it may turn up information that others can use or lead to questions that others can answer. Unexpected results, too, may be clues to something important about the natural world.

DISCOVERY OF DNA FUNCTION

Early and Puzzling Clues

The year was 1928. Frederick Griffith, an army medical officer, was attempting to develop a vaccine against *Streptococcus pneumoniae*, a bacterium that is one cause of the lung disease pneumonia. (When introduced into the body, vaccines mobilize internal defenses against a real attack. Many vaccines are preparations of killed or weakened bacterial cells.) Griffith never did develop a vaccine. But his work unexpectedly opened a door to the molecular world of heredity.

Griffith isolated and cultured two different strains of the bacterium. He noticed that colonies of one strain had a rough surface appearance, but those of the other strain appeared smooth. He designated the two strains R and S and used them in a series of four experiments:

1. Laboratory mice were injected with live R cells. The mice did not develop pneumonia, as Figure 13.3 shows. *The R strain was harmless.*

2. Other mice were injected with live S cells. The mice died. Blood samples taken from them teemed with live S cells. *The S strain was pathogenic* (disease-causing).

3. S cells were killed by exposure to high temperature. Mice injected with these cells did not die.

4. Live R cells were mixed with heat-killed S cells and injected into mice. The mice died—and blood samples from them teemed with *live* S cells!

What was going on in the fourth experiment? Maybe heat-killed S cells in the mixture weren't really dead. But if that were true, then mice injected with heat-killed S cells alone (experiment 3) would have died. Maybe harmless R cells in the mixture had mutated into a killer form. But if that were true, then mice injected with the R cells alone (experiment 1) would have died.

The simplest explanation was as follows: *Heat killed the S cells but did not destroy their hereditary material—including the part that specified "how to cause infection."* Somehow, that material had been transferred from the dead S cells to living R cells, which put it to use.

Further experiments showed the harmless cells had indeed picked up information on causing infections and were permanently transformed into pathogens. After a few hundreds of generations, descendants of those transformed bacterial cells were still infectious!

The unexpected results of Griffith's experiments intrigued Oswald Avery and his fellow biochemists. Later, they also transformed harmless bacterial cells with *extracts* of killed pathogenic cells. Finally in 1944, after rigorous chemical analyses, they felt confident in reporting that the hereditary substance in their extracts probably was DNA—not proteins, as was then widely believed. To give experimental evidence for their conclusion, they reported that they had added certain protein-digesting enzymes to some extracts, but cells exposed to those extracts were transformed anyway. To other extracts, they had added an enzyme that digests DNA but not proteins. Doing so blocked hereditary transformation.

Despite these impressive experimental results, many biochemists refused to give up on the proteins. Avery's findings, many said, probably applied only to bacteria.

Confirmation of DNA Function

By the early 1950s molecular detectives, including Max Delbrück, Alfred Hershey, Martha Chase, and Salvador Luria, were using viruses as experimental subjects. The viruses they had selected, called **bacteriophages**, infect *Escherichia coli* and other bacteria.

Viruses are biochemically simple infectious agents. They aren't alive, but they hold hereditary information about building more new virus particles. At some point after a virus infects a host cell, viral enzymes take over the cell's metabolic machinery, which starts churning out substances necessary to make new virus particles.

genetic material
viral coat
sheath
base plate
tail fiber

a

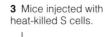
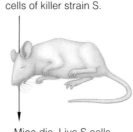

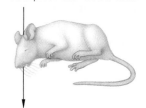

1 Mice injected with live cells of harmless strain R.

2 Mice injected with live cells of killer strain S.

3 Mice injected with heat-killed S cells.

4 Mice injected with live R cells *plus* heat-killed S cells.

Figure 13.3
Results of Griffith's experiments with a harmless and a pathogenic strain of *Streptococcus pneumoniae*, as described in the text above.

Mice do not die. No live R cells in their blood.

Mice die. Live S cells in their blood.

Mice do not die. No live S cells in their blood.

Mice die. Live S cells in their blood.

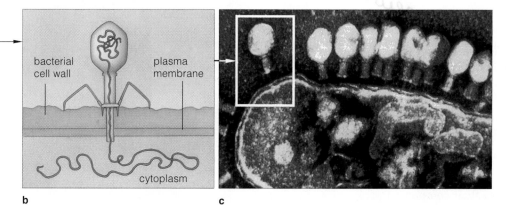

b

c

Figure 13.4 (a,b) *Far left*: Structural organization of a T4 bacteriophage. The diagram shows the genetic material of this type of virus being injected into the cytoplasm of a host cell. The genetic material of this bacteriophage is DNA (the *blue*, threadlike strand). (c) Micrograph of T4 virus particles that infected the bacterium *Escherichia coli*, which thereby became an unwilling host.

In the figure, the diagram labels read: bacterial cell wall, plasma membrane, cytoplasm.

Figure 13.5 Examples of the landmark experiments that pointed to DNA as the hereditary substance. In the 1940s, Alfred Hershey and Martha Chase were studying the biochemical basis of inheritance. They knew that certain bacteriophages consist of proteins and DNA. *Did the proteins, DNA, or both contain viral genetic information?*

To find a possible answer, Hershey and Chase designed two experiments. They started with two known biochemical facts: First, bacteriophage proteins incorporate sulfur (S) but not phosphorus (P). Second, by contrast, bacteriophage DNA incorporates phosphorus but not sulfur.

(a) In one experiment, some bacterial cells were grown on a culture medium that included a radioisotope of sulfur, ^{35}S. When bacterial cells synthesized proteins, they had to take up the radioisotope—which the researchers used as a tracer. (Here you may wish to review Section 2.2.)

After the cells were labeled with the tracer, bacteriophages were allowed to infect them. As the infection ran its course, the host cells synthesized viral proteins. These proteins also became labeled with ^{35}S. So did the new generation of virus particles.

Labeled bacteriophages were allowed to infect a new batch of unlabeled bacteria that were suspended in a fluid culture medium. Afterward, Hershey and Chase whirred the fluid in a kitchen blender. Whirring dislodged the viral protein coats from the cells, so the particles became suspended in the fluid medium. Chemical analysis revealed the presence of labeled protein in the fluid. There was very little labeled protein in the preparation of bacterial cells.

(b) For the second experiment, Hershey and Chase cultured more bacterial cells. The phosphorus available to the cells for synthesizing DNA included the radioisotope ^{32}P. Later, bacteriophages were allowed to infect the cells.

As predicted, viral DNA synthesized inside infected cells became labeled. So did a new generation of virus particles. The labeled particles were allowed to infect bacteria that were suspended in a fluid medium. They were dislodged from the host cells.

Analysis showed the vast majority of labeled viral DNA stayed inside host cells, where its hereditary instructions had to be used to make more virus particles. Here was evidence that DNA is the genetic material of this type of virus.

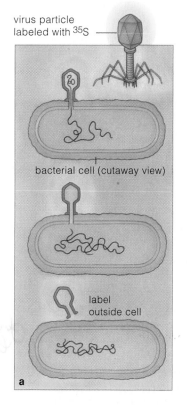

virus particle labeled with ^{35}S

bacterial cell (cutaway view)

label outside cell

a

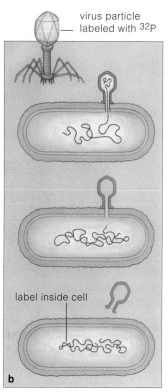

virus particle labeled with ^{32}P

label inside cell

b

By 1952, researchers knew that some bacteriophages consist only of DNA and a protein coat. Also, electron micrographs revealed that the main part of the viruses remains *outside* the cells they are infecting (Figure 13.4). Possibly, such viruses were injecting genetic material alone *into* host cells. If that were true, was the material DNA, protein, or both? Through many experiments, researchers accumulated strong evidence that DNA, not proteins, serves as the molecule of inheritance. Figure 13.5 describes two of these landmark experiments.

Information for producing the heritable traits of single-celled and multicelled organisms is encoded in DNA.

DNA STRUCTURE

What Are the Components of DNA?

pentose

Long before the bacteriophage studies were under way, biochemists knew that DNA contains only four types of nucleotides—the building blocks of nucleic acids. Each **nucleotide** consists of a five-carbon sugar (which, in DNA, is deoxyribose), a phosphate group, and one of the following nitrogen-containing bases:

adenine	guanine	thymine	cytosine
A	G	T	C

All four types of nucleotides have their component parts organized the same way (Figure 13.6). But **T** and **C** are pyrimidines, which are single-ring structures. **A** and **G** are purines, which are larger, bulkier molecules; they have double-ring structures.

By 1949, Erwin Chargaff, a biochemist, had shared with the scientific community two crucial insights into the composition of DNA. First, the amount of adenine relative to guanine differs from one species to the next. However, the amount of thymine is equal to that of adenine, and amount of cytosine is equal to the amount of guanine. We may show this as:

$$A = T \quad \text{and} \quad G = C$$

The proportions of those four kinds of nucleotides relative to one another were tantalizing clues. In some way, the proportions almost certainly were related to the arrangement of the nucleotides in a DNA molecule.

The first convincing evidence of that arrangement emerged from Maurice Wilkins's research laboratory in England. Rosalind Franklin, one of Wilkins's colleagues, had obtained especially good **x-ray diffraction images** of DNA fibers. (Maybe for the reasons sketched out in Section 13.3, Franklin's contribution has only recently been acknowledged.) X-ray diffraction images can be made by directing a beam of x-rays at a molecule. The molecule scatters the beam in patterns that are captured on film. The pattern consists only of dots and streaks; it alone does not reveal molecular structure. However, researchers can use photographic images of the patterns to calculate the positions of the molecule's atoms.

DNA does not readily lend itself to x-ray diffraction. But researchers can rapidly spin a suspension of DNA

All chromosomes in a cell contain DNA. What does DNA contain? Four kinds of nucleotides, A, G, T, and C. Here are the structural formulas for those nucleotides:

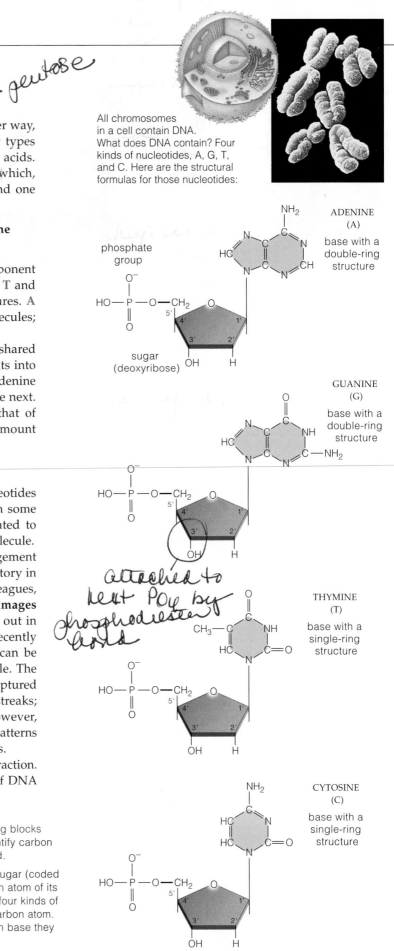

attached to next PO₄ by phosphodiester bond

Figure 13.6 Four kinds of nucleotides used as building blocks for DNA. Small numerals on the structural formulas identify carbon atoms to which other parts of the molecule are attached.

Each nucleotide in a DNA molecule has a five-carbon sugar (coded *red*) and a phosphate group attached to the fifth carbon atom of its carbon ring structure. Each nucleotide also has one of four kinds of nitrogen-containing bases (*blue*) attached to the first carbon atom. The four kinds of nucleotides in DNA differ only in which base they have: adenine, guanine, thymine, or cytosine.

Figure 13.7 Composite of different models for a DNA double helix. The two sugar–phosphate backbones run in *opposing* directions. Think of the sugar units (deoxyribose) of one strand as being upside down.

By comparing the numerals used to identify each carbon atom of the deoxyribose molecule (1', 2', 3', and so on), you see that one strand runs in the 5' → 3' direction and the other runs in the 3' → 5' direction.

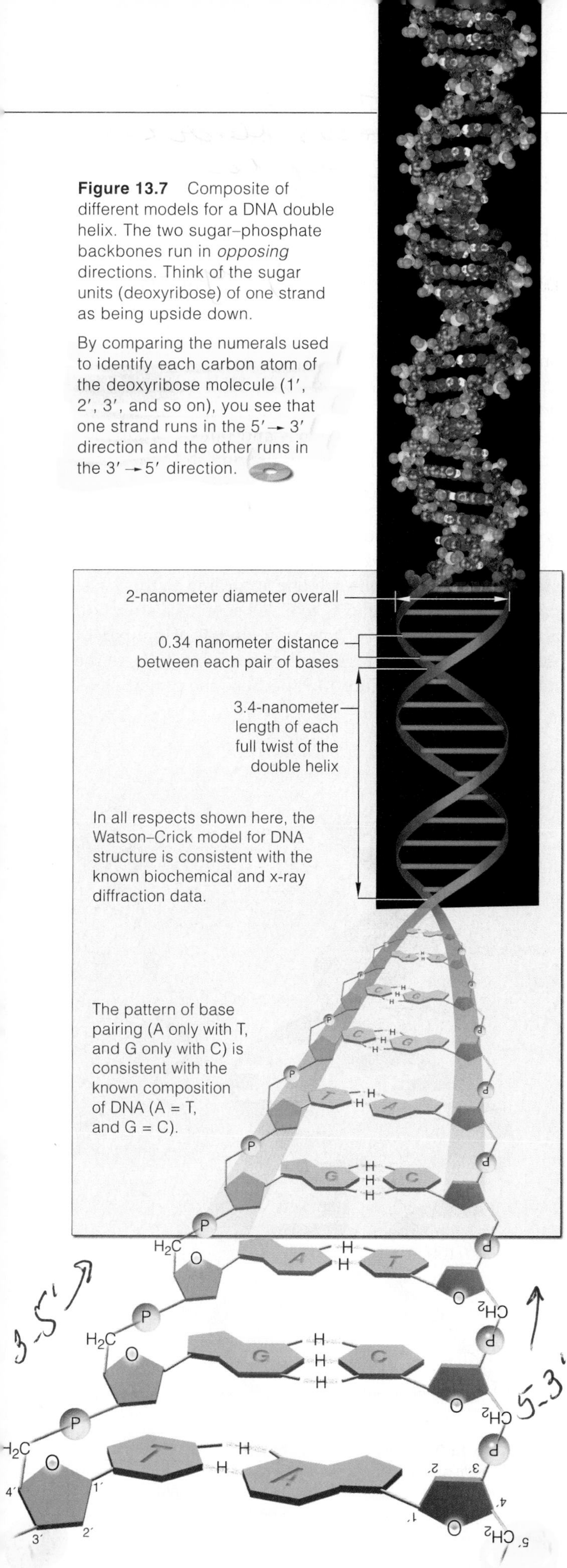

2-nanometer diameter overall

0.34 nanometer distance between each pair of bases

3.4-nanometer length of each full twist of the double helix

In all respects shown here, the Watson–Crick model for DNA structure is consistent with the known biochemical and x-ray diffraction data.

The pattern of base pairing (A only with T, and G only with C) is consistent with the known composition of DNA (A = T, and G = C).

molecules, spool them onto a rod, and gently pull them into gossamer fibers, like cotton candy. If the atoms in DNA were arranged in a regular order, x-rays directed at a fiber should scatter in a regular pattern that could be captured on film. As calculations based on Franklin's images strongly indicated, a DNA molecule is long and thin, with a 2-nanometer diameter. Some molecular configuration repeats every 0.34 nanometer along its length, and another every 3.4 nanometers.

Could the sequence of nucleotide bases be twisting, like a circular stairway? Certainly Pauling thought so. After all, he had discovered a helical shape in collagen. He and everyone else—including Wilkins, Watson, and Crick—were thinking "helix." Watson later wrote, "We thought, why not try it on DNA? We were worried that *Pauling* would say, why not try it on DNA? Certainly he was a very clever man. He was a hero of mine. But we beat him at his own game. I still can't figure out why."

Pauling, it turned out, made a big chemical mistake. His model had hydrogen bonds at phosphate groups holding DNA's structure together. That does happen in highly acidic solutions. It doesn't happen in cells.

Patterns of Base Pairing

As Watson and Crick perceived, DNA consists of *two* strands of nucleotides, held together at their bases by hydrogen bonds (Figure 13.7). These bonds form when the two strands run in opposing directions and twist to form a double helix. Two kinds of base pairings form along the length of the molecule: **A—T** and **G—C**. This bonding pattern allows variation in the order of bases. For example, even a tiny stretch of DNA from a rose, a gorilla, a human, or any other organism might be:

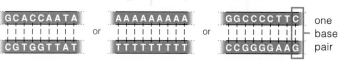

All DNA molecules show the same bonding pattern, but each species has unique base sequences in its DNA. *This molecular constancy and variation among species is the foundation for the unity and diversity of life.*

Intriguingly, recent computer simulations show that if you want to pack a string into the least space, coil it into a helix. Could this space-saving advantage also be a factor in nature? Did natural selection favor it in the molecular evolution of the DNA double helix? Maybe.

The pattern of base pairing between the two strands in DNA is constant for all species—A with T, and G with C. Each species also has some unique differences in *which* base pair follows the next along the length of its DNA molecules.

Rosalind's Story

In 1951, Rosalind Franklin arrived at King's Laboratory of Cambridge University with impressive credentials. Earlier, in Paris, she had refined existing procedures for x-ray diffraction while studying the structure of coal. She also had devised a new mathematical approach to interpreting x-ray diffraction images and had built three-dimensional models of molecules, as Pauling had done. Now she had been asked to run an x-ray crystallography laboratory, which she would create with state-of-the-art equipment. Her assignment? Investigate the structure of DNA.

No one bothered to tell her that, down the hall, Maurice Wilkins was already working on the puzzle. Even the graduate student assigned to assist her failed to mention it. And no one bothered to tell Wilkins about Franklin's assignment, so he assumed she was a technician hired to do his x-ray crystallography work because he did not know how to do it himself. And so began a poisonous clash. To Franklin, Wilkins seemed inexplicably prickly. To Wilkins, Franklin displayed an appalling lack of the deference that technicians usually show to researchers.

Wilkins had a prized cache of crystalline DNA fibers —each having parallel arrays of hundreds of millions of DNA molecules—which he gave to his "technician."

Five months later, Franklin gave a talk on what she had learned so far. DNA, she said, may have two, three, or four parallel chains twisted in a helix, with phosphate groups projecting outward. She had measured DNA's density and assigned DNA fibers to 1 of 230 categories of crystals, based on the symmetry of their parallel chains.

With his background in crystallography, Crick would have recognized the significance of that symmetry *if* he had been present. (To wit, *paired* chains running in opposite directions would look the same even if flipped 180°. Two paired chains? No. DNA's density ruled that out. But *one pair* of chains? Yes!) Watson was in the audience, but he didn't comprehend what Franklin was talking about.

Later, Franklin created an outstanding x-ray diffraction image of wet DNA fibers that fairly screamed *Helix!* She also worked out DNA's length and diameter. But she had been working with dry fibers for so long she didn't dwell on her new data. Wilkins did. In 1953, he allowed Watson to see Franklin's exceptional x-ray diffraction image and reminded him of what she had reported fourteen months earlier. And when Watson and Crick finally did focus on her data, they had the final bits of information necessary to start building a DNA model—one that had two helically twisted chains running in opposing directions.

Figure 13.8 Researcher Rosalind Franklin.

DNA REPLICATION AND REPAIR

How Is a DNA Molecule Duplicated?

The discovery of DNA structure was a turning point in studies of inheritance. Until then, no one could explain **DNA replication**, or how the molecule of inheritance is duplicated before the cell divides. Once Watson and Crick had assembled their model, Crick understood at once how this might be done.

As he knew, enzymes can easily break the hydrogen bonds between the two nucleotide strands of a DNA molecule. When these enzymes and other proteins act on the molecule, one strand can unwind from the other, thereby exposing stretches of nucleotide bases. Cells have stockpiles of free nucleotides, and these can pair with the exposed bases.

Each parent strand remains intact, and a companion strand is assembled on each one according to this base-pairing rule: **A** to **T**, and **G** to **C**. As soon as a stretch of a new, partner strand forms on a stretch of the parent strand, the two twist together into a double helix, in the manner shown in Figure 13.9. Because the parent DNA

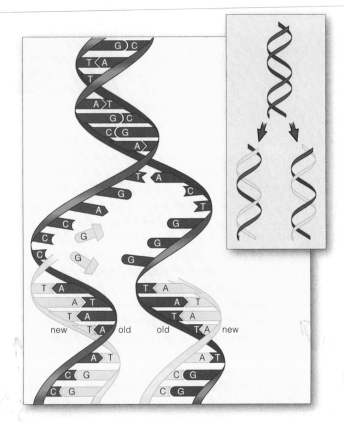

Figure 13.9 Overview of the semiconservative nature of DNA replication. The original two-stranded DNA molecule is shown in *blue*. Each parent strand remains intact. A new strand (*yellow*) is assembled on each one.

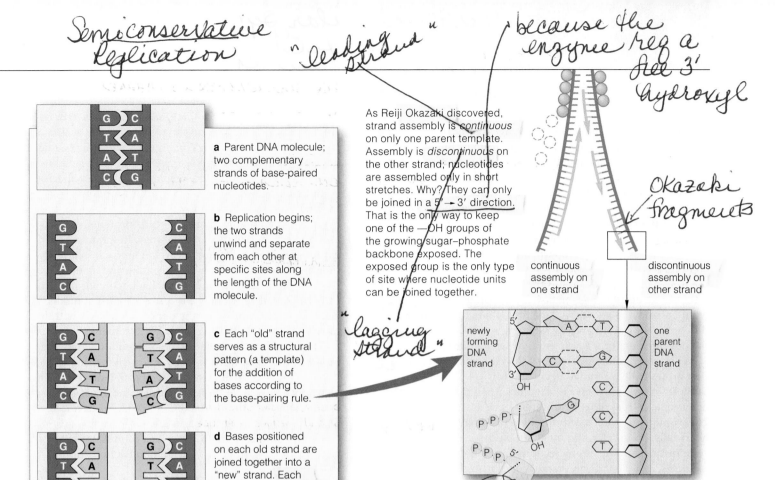

Handwritten annotations (top):

Semiconservative Replication   "leading strand"   because the enzyme req a free 3' hydroxyl

Okazaki fragments

a Parent DNA molecule; two complementary strands of base-paired nucleotides.

b Replication begins; the two strands unwind and separate from each other at specific sites along the length of the DNA molecule.

c Each "old" strand serves as a structural pattern (a template) for the addition of bases according to the base-pairing rule.

d Bases positioned on each old strand are joined together into a "new" strand. Each half-old, half-new DNA molecule is just like the parent molecule.

As Reiji Okazaki discovered, strand assembly is *continuous* on only one parent template. Assembly is *discontinuous* on the other strand; nucleotides are assembled only in short stretches. Why? They can only be joined in a 5' → 3' direction. That is the only way to keep one of the —OH groups of the growing sugar–phosphate backbone exposed. The exposed group is the only type of site where nucleotide units can be joined together.

"lagging strand"

continuous assembly on one strand

discontinuous assembly on other strand

newly forming DNA strand

one parent DNA strand

Figure 13.10 Closer look at DNA replication: base additions to new nucleotide strands being assembled on a parent template.

Handwritten (below figure): unwinding proteins   DNA gyrase

See p. 256 for PCR

strand is conserved during the replication process, half of every double-stranded DNA molecule is "old" and half is "new" (Figure 13.9). That is why biologists refer to the process as *semiconservative* replication.

DNA replication uses a team of molecular workers. In response to cellular signals, the replication enzymes become active along the length of the DNA molecule. Together with other proteins, some enzymes unwind the strands in both directions and prevent them from rewinding. Enzyme action jump-starts the unwinding but isn't necessary to unzip hydrogen bonds between the strands; hydrogen bonds are individually weak.

Now enzymes called **DNA polymerases** attach short stretches of free nucleotides to the unwound portions of a parent template (Figure 13.10). The free nucleotides themselves actually drive the strand assembly. Each has three phosphate groups. DNA polymerase splits off two of them, releasing energy that drives the attachments.

DNA ligases fill in tiny gaps between the new short stretches to form a continuous strand. Then enzymes wind the template strand and complementary strand together to form a DNA double helix.

As you will read in Section 15.1, some replication enzymes have uses in recombinant DNA technology.

Handwritten (bottom): between Okazaki fragments

Monitoring and Fixing the DNA

Cells have trouble replicating DNA with structurally broken or altered strands. **DNA repair** processes have evolved that minimize damage. DNA ligases fix some breaks in strands. Specialized DNA polymerases can fix mismatched base pairs or replace mutated bases with undamaged ones, as described in Section 14.4. During replication, some even extend a growing strand past a lesion. This confers a survival advantage on the cell, which commits suicide (by issuing signals for its own death) when damage arrests replication. But the special polymerases do their bypass trick on *undamaged* DNA as well. Over time, such bypasses allow spontaneous mutations to accumulate. They can give rise to genetic disorders when other repair systems aren't operating.

DNA is replicated prior to cell division. Enzymes unwind its two strands. Each strand remains intact throughout the process—it is conserved—and enzymes assemble a new, complementary strand on each one.

Enzymes involved in replication also repair the DNA where base-pairing errors have crept into the nucleotide sequence.

Cloning Mammals—A Question of Reprogramming DNA

Imagine the possibility of **cloning**—making a genetically identical copy—of yourself. Is the image that far-fetched? Consider this: Researchers have been cloning complex animals for more than a decade. For example, some use in vitro fertilization methods to grow cattle embryos in petri dishes. After a fertilized egg starts dividing, they split the early cluster of cells. The two clusters develop as two identical-twin cattle embryos, get implanted in surrogate mothers, and are born as cloned calves (Figure 13.11a).

A researcher who clones farm animals derived from embryonic cells has to wait for the clones to grow up to see if they display a desired trait. Using a differentiated cell from an adult would be faster, for a prized genotype would already be known. At one time, though, tricking a differentiated cell into reprogramming its DNA to direct the development of a whole embryo seemed impossible.

What does "differentiated" mean? When an embryo first grows from a fertilized egg, all of its cells have the same DNA and are pretty much alike. Then different embryonic cells start using different parts of their DNA. Their unique selections commit them to being liver cells, heart cells, brain cells, and other specialists in structure, composition, and function (Sections 15.3 and 43.4).

In 1997 in Scotland, Ian Wilmut coaxed a differentiated sheep cell to become the "uncommitted" first cell of an embryo. His group had slipped nuclei from differentiated cells into unfertilized eggs from which the nucleus had been removed (compare Figure 13.12). Of hundreds of modifed eggs, one developed into a whole animal. The cloned lamb, named Dolly, grew into a healthy adult and gave birth to a lamb of her own (Figure 13.11b).

In science, extraordinary claims call for extraordinary proof—in this case, successful repeats of the experiment at a time when no one thought mammals could ever be cloned. Researchers around the world have since cloned sheep and mice, cows, pigs, and goats. Some of the mice have been cloned through six generations. Researchers have now successfully cloned some endangered species.

But cloning processes may introduce random errors in gene expression. As you will read in Chapter 43, the cytoplasm of a mammalian egg holds proteins, mRNAs, and other components that have roles in guiding gene expression, starting with the early embryo. It may take months or years for an egg that is maturing in an ovary to stockpile required components in specific locations in the cytoplasm. Yet cloning processes use differentiated eggs that must reprogram the donor cell's DNA within minutes or hours after it is inserted into them.

That may be why fewer than 3 percent of all cloning efforts result in healthy animals. Even then, some of the "successful" clones develop medical problems, including heart and lung abnormalities, sudden and gross obesity, and a damaged immune system. In 2002, for example, Dolly developed arthritis at an unusually early age.

Being mammals, are humans also candidates for cloning? We return to this question in Section 16.10.

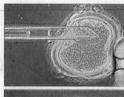

a A pipette holds an egg in place while suction is used to pull its nucleus into a fine, hollow needle.

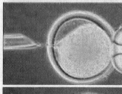

b Only the cytoplasm remains inside the plasma membrane of the egg.

c A skin cell from an animal to be cloned is transferred into the egg.

d Electric shock triggers fusion of the skin cell with the egg cytoplasm.

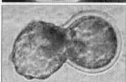

e The recipient egg starts to divide within a few hours. Seven days later, the cloned first cell of the embryo gets transferred into a host animal.

Figure 13.11 (a) Student with two genetically identical Holsteins, prized milk producers, obtained by embryos split during in vitro fertilization. (b) Dolly, a cloned sheep. She started life as a differentiated cell that was extracted from an adult ewe, then induced to start mitotic cell divisions. She is shown with her first lamb. She is able to breed normally and reproduce the old-fashioned way.

Figure 13.12 One type of nuclear transfer process.

13.6

SUMMARY

Gold indicates text section

1. For all living cells, hereditary information is encoded in DNA (deoxyribonucleic acid). *CI*

2. DNA consists of nucleotide subunits. Each of these has a five-carbon sugar (deoxyribose), one phosphate group, and one of four kinds of nitrogen-containing bases (adenine, thymine, guanine, or cytosine). *13.2*

3. A DNA molecule consists of two nucleotide strands twisted together as a double helix. Bases of one strand pair (hydrogen-bond) with bases of the other. *13.2*

4. The bases of the two strands in a DNA double helix pair in constant fashion. Adenine pairs with thymine (**A** to **T**), and guanine with cytosine (**G** to **C**). *Which* base pair follows the next (**A–T**, **T–A**, **G–C**, or **C–G**) varies along the length of the strands. *13.2*

5. Overall, the DNA of one species includes a number of unique stretches of base pairs that set it apart from the DNA of all other species. *13.2*

6. During DNA replication, enzymes unwind the two strands of a double helix and assemble a new strand of complementary sequence on each parent strand. Two double-stranded molecules result. One strand of each molecule is old (is conserved); the other is new. *13.4*

7. Repair systems fix damaged DNA strands during replication. Special DNA polymerases bypass lesions, which allows mutations to accumulate in DNA. *13.4*

Review Questions

1. Name the three molecular parts of a nucleotide in DNA. Also name the four different bases in these nucleotides. *13.2*

2. What kind of bond joins two DNA strands in a double helix? Which nucleotide base-pairs with adenine? With guanine? *13.2*

3. Explain how DNA molecules can show both constancy and variation from one species to the next. *13.2*

Self-Quiz   ANSWERS IN APPENDIX III

1. Which is *not* a nucleotide base in DNA?
   a. adenine   c. uracil   e. cytosine
   b. guanine   d. thymine   f. All are in DNA.

2. What are the base-pairing rules for DNA?
   a. A–G, T–C   c. A–U, C–G
   b. A–C, T–G   d. A–T, G–C

3. One species' DNA differs from others in its _____ .
   a. sugars   c. base sequence
   b. phosphates   d. all of the above

4. When DNA replication begins, _____ .
   a. the two DNA strands unwind from each other
   b. the two DNA strands condense for base transfers
   c. two DNA molecules bond
   d. old strands move to find new strands

5. DNA replication requires _____ .
   a. free nucleotides   c. many enzymes
   b. new hydrogen bonds   d. all of the above

6. Cell differentiation involves _____ .
   a. cloning   c. selective gene expression
   b. nuclear transfers d. both b and c

7. Match the DNA terms appropriately.
   ____ DNA polymerase   a. two nucleotide strands that
   ____ constancy in   are twisted together
        base pairing   b. A with T, G with C
   ____ replication   c. hereditary material duplicated
   ____ DNA double helix   d. replication enzyme

Critical Thinking

1. Chargaff's data suggested that adenine pairs with thymine, and guanine pairs with cytosine. What other data available to Watson and Crick suggested that adenine–guanine and cytosine–thymine pairs normally do not form?

2. One of Matthew Meselson and Frank Stahl's experiments supported the semiconservative model of DNA replication. The researchers made "heavy" DNA by growing *Escherichia coli* in a medium enriched with ^{15}N, a heavy isotope of nitrogen. They prepared "light" DNA by growing *E. coli* in the presence of ^{14}N, the more common isotope. An available technique helped them identify which replicated molecules were heavy, light, or hybrid (one heavy strand, one light). Use two pencils of different colors, one for heavy strands and one for light. Starting with a DNA molecule having two heavy strands, sketch daughter molecules that would form after replication in a ^{14}N-containing medium. Sketch the four DNA molecules that would form if the daughter molecules were replicated a second time in the ^{14}N medium.

3. Mutations, permanent changes in base sequences of genes, are the original source of genetic variation—the raw material of evolution. Yet how can mutations accumulate, given that cells have repair systems that can rapidly fix structurally altered or discontinuous DNA strands during replication?

4. As Section 4.12 indicates, a pathogenic strain of *E. coli* has acquired an ability to produce a dangerous toxin that causes medical problems and fatalities. This is especially the case for young children who have ingested undercooked, contaminated beef. Develop hypotheses to explain how a normally harmless bacterium such as *E. coli* can become a pathogen.

5. In 1999, scientists discovered a woolly mammoth that had been frozen in glacial ice for the past 20,000 years. They thawed it very carefully so they could use its DNA to clone a woolly mammoth. It turns out there wasn't enough material to work with. But they plan to try again the next time a frozen woolly mammoth comes along. Consider Section 13.5, then speculate on the pros and cons of cloning an extinct animal.

Selected Key Terms

adenine (A) *13.2*
bacteriophage *13.1*
cloning *13.5*
cytosine (C) *13.2*
deoxyribonucleic acid (DNA) *CI*
DNA ligase *13.4*
DNA polymerase *13.4*

DNA repair *13.4*
DNA replication *13.4*
guanine (G) *13.2*
nucleotide *13.2*
thymine (T) *13.2*
x-ray diffraction
   image *13.2*

Readings

Watson, J. 1978. *The Double Helix.* New York: Atheneum. Highly personal view of scientists and their methods, interwoven into an account of how DNA structure was discovered.

FROM DNA TO PROTEINS

Beyond Byssus

Picture a mussel, of the sort shown in Figure 14.1. Hard-shelled but soft of body, it is using its muscular foot to probe a wave-scoured rock. At any moment, pounding waves can whack the mussel into the water, hurl it repeatedly against the rock with shell-shattering force, and so offer up a gooey lunch for gulls.

By chance, the mussel's foot comes across a crevice in the rock. The foot moves, broomlike, and sweeps the crevice clean. It presses down, forcing air out from underneath it, then arches up. The result is a vacuum-sealed chamber, much like the one that forms when a plumber's rubber plunger is being squished down and up to unclog a drain. Into this vacuum chamber the mussel spews a fluid that's made of keratin and other proteins. The fluid bubbles into a sticky foam. Now, by curling its foot into a tubelike shape and pumping the foam through it, the mussel produces sticky threads about as wide as a human whisker. It varnishes these threads with another type of protein and ends up with an adhesive. With this adhesive, which we call byssus, the mussel anchors itself to the rock.

Byssus is the world's premier underwater adhesive. Nothing humans have manufactured comes close to it; water degrades or deforms synthetic adhesives. Byssus fascinates biochemists, dentists, and surgeons looking for better ways to do tissue grafts and rejoin severed nerves. Genetic engineers insert mussel DNA into yeast cells. These cells, which reproduce in huge numbers, are

Figure 14.1 Mussels (*Mytilus californianus*) busily demonstrating the importance of proteins for survival. When mussels come across a suitable anchoring site, they use their muscular foot like a plumber's plunger and create a vacuum chamber. In this chamber they manufacture the world's best underwater adhesive from a mix of proteins. The adhesive anchors the mussels to rocks in their wave-swept habitat.

3 processes
nuc acid

rep - DNA syn
transcription - RNA syn
translation - protein syn

"factories" for translating mussel genes into useful quantities of proteins. This exciting work, like the mussel's own byssus-building efforts, starts with one of life's universal precepts: *Every protein is synthesized in accordance with instructions in DNA.*

You are about to trace the steps leading from DNA to proteins. Many enzymes are players in this pathway. So is another kind of nucleic acid besides DNA. The same steps produce *all* proteins, from mussel-inspired adhesives to the keratin in your hair and fingernails to the insect-digesting enzymes of a Venus flytrap.

Start out by thinking of each cell's DNA as a book of protein-building instructions. The alphabet used to create the book is simple enough: A, T, G, and C (for the nucleotide bases adenine, thymine, guanine, and cytosine). How do you get from that alphabet to a protein? The answer starts with DNA's structure.

DNA, recall, is a double-stranded molecule. Which kind of nucleotide base follows the next along the length of a strand—the **base sequence**—differs from one kind of organism to the next. The two strands unwind entirely from each other when DNA is being replicated. However, at other times in a cell's life, the two strands unwind only in certain regions to expose certain base sequences—genes. Most genes contain instructions for building proteins.

It takes two steps, **transcription** and **translation**, to carry out a gene's protein-building instructions. In eukaryotic cells, transcription proceeds in the nucleus. A newly exposed base sequence in DNA serves as a structural pattern—a template—for assembling a strand of **ribonucleic acid** (RNA) from the cell's pool of free nucleotides. Sooner or later the RNA moves into the cytoplasm, where translation proceeds. At this second step, RNA directs the assembly of amino acids into polypeptide chains. The newly formed chains become folded into the three-dimensional shapes of proteins.

In short, DNA guides the synthesis of RNA, then RNA guides the synthesis of proteins:

$$\text{DNA} \xrightarrow{\textit{transcription}} \text{RNA} \xrightarrow{\textit{translation}} \textbf{PROTEIN}$$

The newly synthesized proteins will play structural and functional roles in cells. Some even will have roles in synthesizing more DNA, RNA, and proteins.

Key Concepts

1. Organisms cannot stay alive without enzymes and other proteins. Proteins consist of polypeptide chains, which consist of amino acids. The sequence of amino acids corresponds to a gene, which is a sequence of nucleotide bases in a DNA molecule.

2. The path leading from genes to proteins consists of two steps, called transcription and translation.

3. During transcription, the double-stranded DNA molecule is unwound at a gene region, and then an RNA molecule is assembled on the exposed bases of one of the strands.

4. Translation uses three classes of RNA molecules: messenger RNA, transfer RNA, and ribosomal RNA.

5. During translation, amino acids are joined together sequentially into a polypeptide chain, in a sequence specified by messenger RNA. Transfer RNA delivers the amino acids one at a time to the construction site. Ribosomal RNA catalyzes the chain-building reaction.

6. With few exceptions, the genetic "code words" by which DNA's instructions are translated into proteins are the same in all species.

7. A mutation is a permanent alteration in a gene's base sequence. Mutations are the original source of genetic variation in populations.

8. Mutations introduce changes in protein structure, protein function, or both. The changes may lead to small or large differences in traits among individuals of a population.

HOW IS RNA TRANSCRIBED FROM DNA?

The Three Classes of RNA

Before turning to the details of protein synthesis, be clear on one point. The chapter introduction may have left you with the impression that synthesis of proteins requires only one class of RNA molecules. Actually, it requires three. Transcription of most genes produces **messenger RNA**, or **mRNA**—the only class of RNA that carries *protein-building* instructions. Transcription of some other genes produces **ribosomal RNA**, or **rRNA**, a major component of ribosomes. Ribosomes, recall, are the structural units upon which polypeptide chains are assembled. Transcription of still other genes produces **transfer RNA**, or **tRNA**, which delivers amino acids one by one to a ribosome in the order specified by mRNA.

The Nature of Transcription

An RNA molecule is almost but not quite like a single strand of DNA. RNA, too, consists of only four types of nucleotides. Each nucleotide has a five-carbon sugar, ribose (not DNA's deoxyribose), a phosphate group, and a base. Three types of bases—adenine, cytosine, and guanine—are the same in RNA and DNA. But in RNA, the fourth type of base is **uracil**, not thymine (Figure 14.2). Like thymine, uracil can pair with adenine. This means a new RNA strand can be put together on a DNA region according to base-pairing rules (Figure 14.3).

Transcription resembles DNA replication in another way. Enzymes add nucleotides to a growing RNA strand one at a time, in the 5' → 3' direction. Section 13.4 has a simple explanation of strand assembly.

But transcription *differs* from DNA replication in three key respects. First, only a selected stretch of one DNA strand rather than the whole molecule is used as the template. Second, instead of DNA polymerases, the type of enzyme known as **RNA polymerase** catalyzes nucleotide additions to a growing RNA strand. Third, at the end of transcription there is a single, free strand of RNA nucleotides, not a double helix.

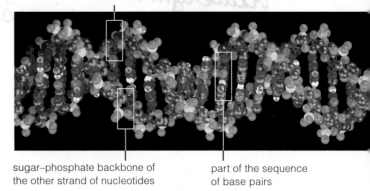

sugar–phosphate backbone of one strand of nucleotides in a DNA double helix

sugar–phosphate backbone of the other strand of nucleotides

part of the sequence of base pairs

a Location of nucleotide bases in DNA

Transcription is initiated at a **promoter**. This base sequence in DNA signals the start of a gene. Proteins position an RNA polymerase on DNA and help start transcription at the promoter. The enzyme moves along the DNA strand, joining one nucleotide after another (Figure 14.4). When it reaches a certain point in the gene, the new RNA molecule is released as a free transcript.

Finishing Touches on mRNA Transcripts

In eukaryotic cells, each new molecule of mRNA is not in its final form. This "pre-mRNA" must be modified before its protein-building instructions can be put to use. Just as a dressmaker might snip off some threads or add bows on a dress before it leaves the shop, so do eukaryotic cells tailor their pre-mRNA.

For example, enzymes attach a cap to the 5' end of pre-mRNA. The cap, a nucleotide, incorporates a methyl group and phosphate groups. Enzymes also attach a tail of about 100 to 300 nucleotides to the 3' end of the pre-mRNA transcript. The new tail becomes wound up with proteins. Later on, in the cytoplasm, the cap will help bind the mRNA to a ribosome. Enzymes will also slowly destroy the wound-up tail from the tip on back.

Figure 14.2 Structural formula for one of the four types of RNA nucleotides. The three others have a different base (adenine, guanine, or cytosine instead of the uracil shown here). Compare Section 13.2, which shows DNA's four nucleotides. Notice how the sugars of DNA and RNA differ at one group only (*yellow*).

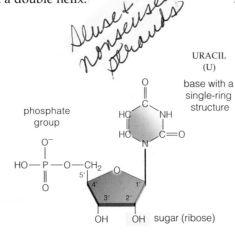

URACIL (U)

base with a single-ring structure

phosphate group

sugar (ribose)

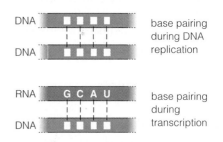

DNA

DNA

base pairing during DNA replication

RNA — G C A U

DNA

base pairing during transcription

Figure 14.3 An example of base pairing of RNA with DNA during transcription, compared to base pairing during DNA replication.

transcription   translation
DNA ——————————→ RNA ——————————→ PROTEIN

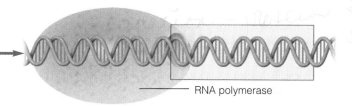

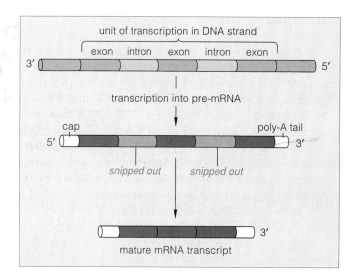

b RNA polymerase initiates transcription at a promoter region in the DNA. It will recognize the base sequence located downstream from that site as a template for linking together the nucleotides adenine, cytosine, guanine, and uracil into a strand of RNA.

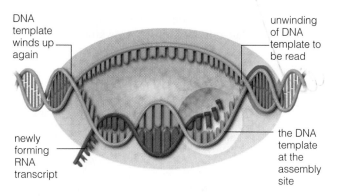

DNA template winds up again

unwinding of DNA template to be read

newly forming RNA transcript

the DNA template at the assembly site

c All through transcription, the DNA double helix becomes unwound in front of the RNA polymerase. Short lengths of the newly forming RNA strand briefly wind up with its DNA template strand. New stretches of RNA unwind from the template (and the two DNA strands wind up again).

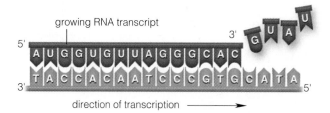

growing RNA transcript

5' A U G G U G U U A G G G C A C   3'   G U A U

3' T A C C A C A A T C C C G T G C A T A   5'

direction of transcription ——————→

d What happened at the assembly site? RNA polymerase catalyzed the base-pairing of RNA nucleotides, one after another, with exposed bases on the DNA template strand.

A U G G U G U U A G G G C A C G U A U

e At the end of the gene region, the last stretch of the new mRNA transcript is unwound and released from the DNA.

Figure 14.4 The process of gene transcription, by which an RNA molecule is assembled on a DNA template. The diagram in (**a**) shows a gene region in part of a DNA double helix. In this region, the base sequence of one of the two nucleotide strands (not both) is about to be used as a template for transcription of a molecule of RNA, in the manner shown in (**b**) through (**e**).

unit of transcription in DNA strand

exon intron exon intron exon

3'   5'

transcription into pre-mRNA

cap   poly-A tail
5'   3'

snipped out   *snipped out*

3'

mature mRNA transcript

Figure 14.5 Transcription and modification of new mRNA in the nucleus of eukaryotic cells. Its cap is a nucleotide with functional groups attached. Its tail is a sequence of adenine nucleotides (hence the name, poly-A tail).

Such tails "pace" the access of enzymes to the mRNA. That controlled access determines how long an mRNA molecule will last. It helps keeps the protein-building messages intact for as long as the cell requires them.

Besides these alterations, the pre-mRNA itself gets modified. Most eukaryotic genes contain one or more **introns**, base sequences that must be removed before a pre-mRNA molecule can be translated. These introns intervene between **exons**, the parts that are still in the mRNA when it's translated into protein. As Figure 14.5 shows, introns are transcribed along with the exons but are snipped out before the mRNA leaves the nucleus in mature form. (Think of it this way: *Ex*ons are *ex*ported from the nucleus and *in*trons stay *in* the nucleus, where they are degraded.)

Many introns are actually sites where instructions for building a protein can be snipped apart and spliced together in more than one way. With this alternative splicing, different cells in the body use the same gene for making different versions of a pre-mRNA transcript, and so the resulting proteins differ slightly in form and function. We return to this topic in Section 15.3.

eg collagen in different conn tissue (handwritten)

During gene transcription, a sequence of exposed bases in one of the two strands of a DNA molecule serves as the template upon which RNA polymerase assembles a single strand of RNA. In this case, adenine base-pairs with uracil, and cytosine with guanine.

Before leaving the nucleus, each new mRNA transcript, or pre-mRNA, undergoes modification into final form.

Handwritten notes at top: "Gene = sequence of triplets that codes for 1 protein (if no quat structure) or 1 polypeptide"

DECIPHERING THE mRNA TRANSCRIPTS

What Is the Genetic Code?

Like a strand of DNA, an mRNA molecule is a linear sequence of nucleotides. What are the protein-building "words" encoded in that sequence? Each is a certain number of nucleotides that codes for an amino acid.

Ribosomes "read" nucleotide bases *three at a time*, as triplets. Base triplets in an mRNA strand were given this name: **codons**. Figure 14.6 will give you an idea of how the order of different codons in an mRNA strand dictates the order in which particular amino acids will be added to a growing polypeptide chain.

Count the codons listed in Figure 14.7, and you see that there are sixty-four kinds. Notice how most of the twenty kinds of amino acids correspond to more than one codon. Glutamate corresponds to the code words GAA *and* GAG, for example. AUG codes for the amino acid methionine and is the start point for translation of the mRNA transcripts. Said another way, the "three-bases-at-a-time" selections start at a particular AUG in the transcript's nucleotide sequence. Methionine is the first amino acid in new polypeptide chains. Codons UAA, UAG, and UGA do not correspond to an amino

Handwritten note in margin: "one gene = 1000 bp"

acid. They serve as STOP signals that prevent further additions of amino acids to a new polypeptide chain.

The set of sixty-four different codons is the **genetic code**. It is the basis of protein synthesis in all organisms.

Structure and Function of tRNA and rRNA

In a cell's cytoplasm are pools of free amino acids and free tRNA molecules. The tRNAs each have a molecular "hook," an attachment site for amino acids. They also have an **anticodon**, a nucleotide triplet that can base-pair with a codon (Figure 14.8). When tRNAs bind to the codons, they automatically position their attached amino acids in the order specified by mRNA.

A cell has a cytoplasmic pool of sixty-four kinds of codons, but it is able to utilize fewer kinds of tRNAs.

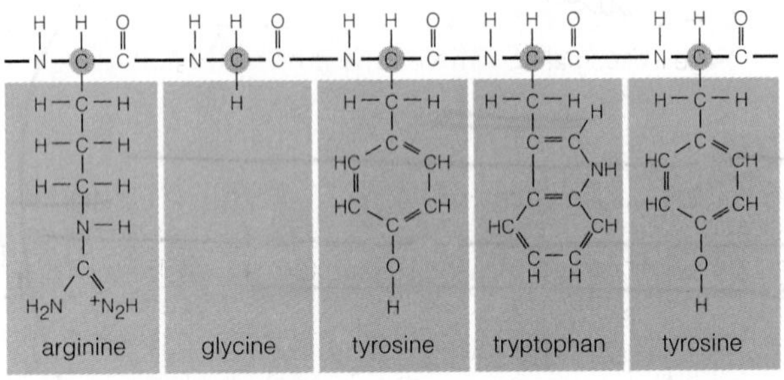

Handwritten note: "like velcro"

a Part of the amino acid sequence that was put together when mRNA was translated into a polypeptide chain

b Part of the mRNA strand that was transcribed from DNA

c The base sequence of a gene region in DNA

Figure 14.6 The correspondence between genes and proteins, as deduced by Marshall Nirenberg, Philip Leder, Severo Ochoa, and Gobind Korana. (**a**) A sequence of amino acids, part of a protein's polypeptide chain. (**b**) An mRNA transcript. Every three nucleotide bases, equaling one codon, calls for one of the chain's amino acids. (**c**) Exposed bases on one strand of a DNA double helix that is unwound during transcription. It is the template for assembling an mRNA strand. Referring to Figure 14.7, can you fill in the blank codon for tryptophan in the mRNA strand in (**b**)?

first base	second base				third base
	U	C	A	G	
U	phenylalanine	serine	tyrosine	cysteine	U
	phenylalanine	serine	tyrosine	cysteine	C
	leucine	serine	STOP	STOP	A
	leucine	serine	STOP	tryptophan	G
C	leucine	proline	histidine	arginine	U
	leucine	proline	histidine	arginine	C
	leucine	proline	glutamine	arginine	A
	leucine	proline	glutamine	arginine	G
A	isoleucine	threonine	asparagine	serine	U
	isoleucine	threonine	asparagine	serine	C
	isoleucine	threonine	lysine	arginine	A
	methionine (or START)	threonine	lysine	arginine	G
G	valine	alanine	aspartate	glycine	U
	valine	alanine	aspartate	glycine	C
	valine	alanine	glutamate	glycine	A
	valine	alanine	glutamate	glycine	G

Figure 14.7 The genetic code. Codons in mRNA are nucleotide bases "read" in blocks of three. Sixty-one of the base triplets correspond to specific amino acids. Three others are signals to stop translation. The *left* vertical column lists choices for the first of three nucleotides in an mRNA codon. The top horizontal row lists choices for the second codon. The *right* vertical column lists choices for the third. Example: reading from left to right, the triplet UGG corresponds to tryptophan. Both UUU and UUC correspond to phenylalanine.

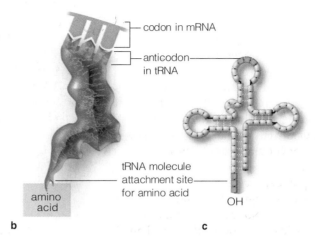

a

b
amino acid

tRNA molecule attachment site for amino acid

codon in mRNA

anticodon in tRNA

OH

c

Figure 14.8 (**a**) Stick model for one type of tRNA molecule. (**b**) An icon for tRNA that you will come across in illustrations to follow. The "hook" at the lower end of this icon represents a binding site for a specific amino acid. (**c**) Structural features common to all tRNAs.

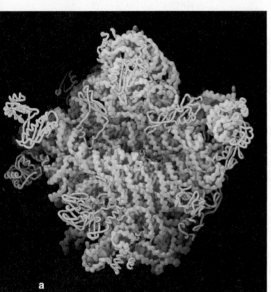

a

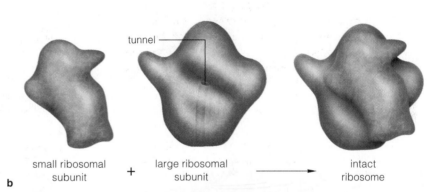

tunnel

small ribosomal subunit + large ribosomal subunit → intact ribosome

b

Figure 14.9 (**a**) Ribbon model for the large subunit of a bacterial ribosome. It consists of two rRNA molecules (*gray*) and thirty-one structural proteins (*gold*), which stabilize the structure. At one end of a tunnel inside this subunit, rRNA catalyzes polypeptide chain assembly. This is an ancient, highly conserved molecular structure. Its role is so vital that the corresponding subunit of eukaryotic ribosomes, which is bigger, is probably similar in structure and function. (**b**) Model for the small and large subunits of a ribosome.

How do tRNAs match up with more than one type of codon? According to base-pairing rules, adenine must pair with uracil, and cytosine with guanine. For codon–anticodon interactions, however, the rules loosen up for the third base of the codon. To give one example, AUU, AUC, and AUA specify isoleucine. All three of these codons can pair with a single type of tRNA that carries isoleucine. Such freedom in codon–anticodon pairing at a base is known as the "wobble effect."

Before anticodons interact with codons of an mRNA strand, that strand must bind to a ribosome. As shown in Figure 14.9, each ribosome has two subunits. These are assembled in the nucleus from rRNA and structural proteins, which stabilize the ribosome's structure. The enzyme action of rRNA drives protein synthesis.

At some point the subunits are shipped separately to the cytoplasm. There, intact, functional ribosomes are put together, each from two subunits, when messages encoded in mRNA are to be translated.

The nucleotide sequence of both DNA and mRNA encodes protein-building instructions. The genetic code is a set of sixty-four base triplets, which are nucleotide bases read in blocks of three. A codon is a base triplet in mRNA.

Different combinations of codons specify the amino acid sequence of different polypeptide chains, start to finish.

mRNAs are the only molecules that carry protein-building instructions from DNA into the cytoplasm.

tRNAs deliver amino acids to ribosomes, where they base-pair with codons in the order specified by mRNA.

Ribosomes, composed of rRNA and proteins, are structures upon which amino acids are assembled into polypeptide chains. An rRNA catalyzes chain assembly.

HOW IS mRNA TRANSLATED?

Stages of Translation

The protein-building code built into mRNA transcripts of DNA becomes translated at intact ribosomes in the cytoplasm. Translation proceeds through three stages: initiation, elongation, and termination.

During the stage called *initiation*, an initiator tRNA (the only one that can start transcription) and an mRNA transcript are both loaded onto a ribosome. First, the initiator tRNA binds with the small ribosomal subunit.

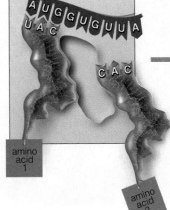

binding site for mRNA

P (first binding site for tRNA)   *A* (second binding site for tRNA)

d Simplified model for binding sites at one end of the tunnel through the large ribosomal subunit, as shown in Figure 14.9. One site is for an mRNA transcript. Two others are for tRNAs that deliver amino acids to the intact ribosome.

e The initiator tRNA has become positioned in the first tRNA binding site (called *P*) on the ribosomal platform. Its anticodon matches up with the START codon (AUG) of the mRNA, which also has become positioned in *its* binding site. Another tRNA is about to move into the platform's second tRNA binding site (called *A*). It is one that can bind with the codon following the START codon.

amino acid 1

amino acid 2

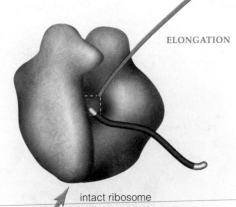

ELONGATION

c As the final step of the initiation stage, a large ribosomal subunit joins with the small one. Once this initiation complex has formed, chain *elongation*—the second stage of translation—can get under way.

intact ribosome

INITIATION

b *Initiation*, the first stage of translating the mRNA transcript, is about to begin. An initiator tRNA (one that can start this stage) is loaded onto a platform of a small ribosomal subunit. The small subunit/tRNA complex attaches to the 5′ end of the mRNA. It moves along the mRNA and "scans" it for an AUG START codon.

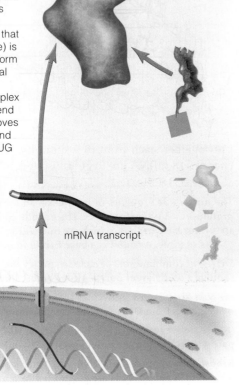

a A mature mRNA transcript leaves the nucleus by passing through pores across the nuclear envelope. Thus it enters the cytoplasm, which contains pools of many free amino acids, tRNAs, and ribosomal subunits.

mRNA transcript

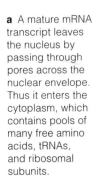

Figure 14.10 Translation, the second step of protein synthesis.

The START codon for the transcript, AUG, matches up with that tRNA's anticodon. Second, a large ribosomal subunit binds with the small subunit. When joined this way, the ribosome, mRNA, and tRNA are an initiation complex (Figure 14.10*b*). The next stage can begin.

In the *elongation* stage of translation, a polypeptide chain is assembled as the mRNA passes between the two ribosomal subunits, a bit like a thread being moved through the eye of a needle. Part of the rRNA molecule located at the center of the large ribosomal subunit has unusual acidity. This region functions as an enzyme. It catalyzes the joining of individual amino acids, and does so in the sequence that is dictated by the codon sequence in the mRNA molecule.

Figure 14.10*f–i* shows how a peptide bond forms between the most recently attached amino acid and the next one delivered to the intact ribosome while the polypeptide chain is growing. Here, you might wish to look once more at Section 3.6 (Figure 3.18), which has a sketch and description of peptide bond formation.

During the last stage of translation, *termination*, a STOP codon in the mRNA moves onto the platform. No tRNA has a corresponding anticodon. Proteins known as release factors bind to the ribosome. They trigger enzyme activity that detaches the mRNA *and* the chain from the ribosome (Figure 14.10*j–l*).

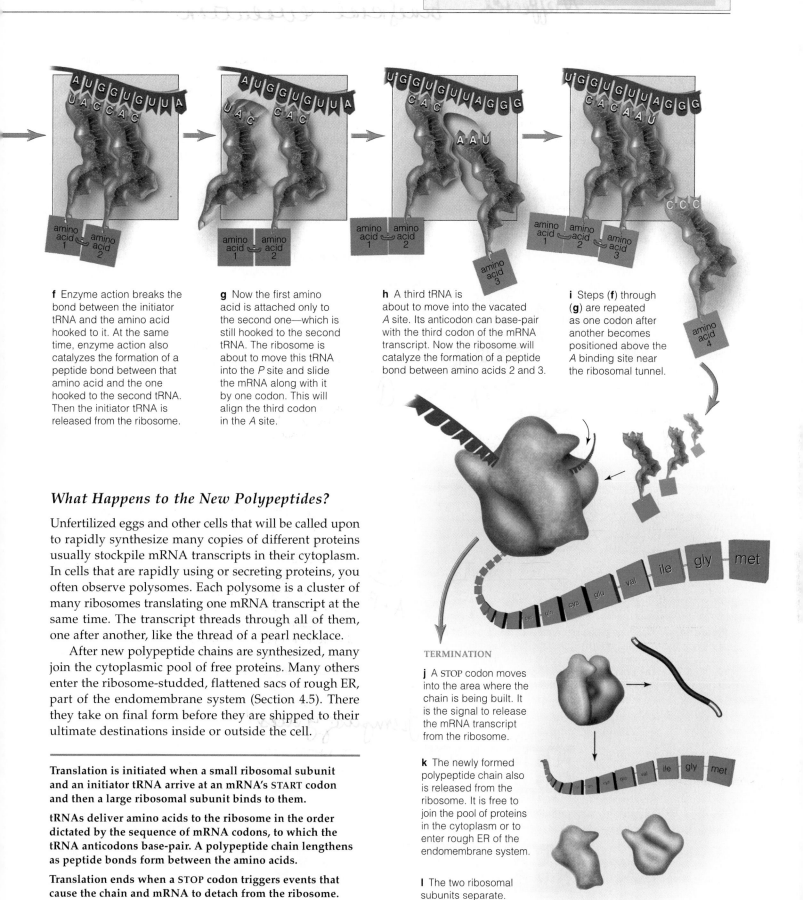

f Enzyme action breaks the bond between the initiator tRNA and the amino acid hooked to it. At the same time, enzyme action also catalyzes the formation of a peptide bond between that amino acid and the one hooked to the second tRNA. Then the initiator tRNA is released from the ribosome.

g Now the first amino acid is attached only to the second one—which is still hooked to the second tRNA. The ribosome is about to move this tRNA into the *P* site and slide the mRNA along with it by one codon. This will align the third codon in the *A* site.

h A third tRNA is about to move into the vacated *A* site. Its anticodon can base-pair with the third codon of the mRNA transcript. Now the ribosome will catalyze the formation of a peptide bond between amino acids 2 and 3.

i Steps (**f**) through (**g**) are repeated as one codon after another becomes positioned above the *A* binding site near the ribosomal tunnel.

What Happens to the New Polypeptides?

Unfertilized eggs and other cells that will be called upon to rapidly synthesize many copies of different proteins usually stockpile mRNA transcripts in their cytoplasm. In cells that are rapidly using or secreting proteins, you often observe polysomes. Each polysome is a cluster of many ribosomes translating one mRNA transcript at the same time. The transcript threads through all of them, one after another, like the thread of a pearl necklace.

After new polypeptide chains are synthesized, many join the cytoplasmic pool of free proteins. Many others enter the ribosome-studded, flattened sacs of rough ER, part of the endomembrane system (Section 4.5). There they take on final form before they are shipped to their ultimate destinations inside or outside the cell.

Translation is initiated when a small ribosomal subunit and an initiator tRNA arrive at an mRNA's START codon and then a large ribosomal subunit binds to them.

tRNAs deliver amino acids to the ribosome in the order dictated by the sequence of mRNA codons, to which the tRNA anticodons base-pair. A polypeptide chain lengthens as peptide bonds form between the amino acids.

Translation ends when a STOP codon triggers events that cause the chain and mRNA to detach from the ribosome.

TERMINATION

j A STOP codon moves into the area where the chain is being built. It is the signal to release the mRNA transcript from the ribosome.

k The newly formed polypeptide chain also is released from the ribosome. It is free to join the pool of proteins in the cytoplasm or to enter rough ER of the endomembrane system.

l The two ribosomal subunits separate.

IF progeny affected (handwritten)

harmful - noticeable (handwritten)
harmless - no noticeable effect (handwritten)
beneficial - evolution (handwritten)

DO MUTATIONS AFFECT PROTEIN SYNTHESIS?

Whenever a cell puts its genetic code into action, it is making precisely those proteins that form its structure and carry out its functions. If something changes a gene's code words, the resulting protein may change, also. If the protein is central to cell architecture or metabolism, we can expect the outcome to be an abnormal cell.

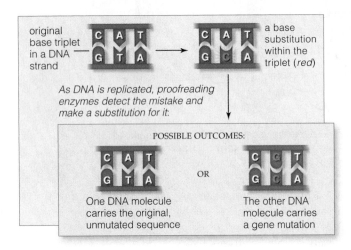

original base triplet in a DNA strand

a base substitution within the triplet (*red*)

As DNA is replicated, proofreading enzymes detect the mistake and make a substitution for it:

POSSIBLE OUTCOMES:

OR

One DNA molecule carries the original, unmutated sequence

The other DNA molecule carries a gene mutation

a Example of a base-pair substitution

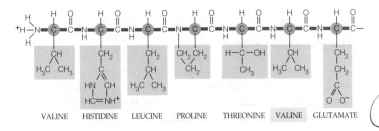

VALINE   HISTIDINE   LEUCINE   PROLINE   THREONINE   VALINE   GLUTAMATE

b Outcome of the base-pair substitution

Figure 14.11 Common types of mutations. (**a**) One example of a base-pair substitution. (**b**) This base-pair substitution is a type of molecular change that caused a single amino acid to be replaced in the beta chains of hemoglobin (valine instead of glutamate). Sickle-cell anemia is the result.

Gene sequences do change. Sometimes one base gets substituted for another in the nucleotide sequence. At other times, an extra base is inserted into the sequence or a base is lost from it. Such small-scale changes in the nucleotide sequence of genes in the DNA molecule are **gene mutations**. There is some leeway here; remember, more than one codon may specify the same amino acid. For instance, if UCU were changed to UCC, it probably wouldn't have dire effects because both codons specify serine. However, many mutations give rise to proteins with altered or lost functions.

Common Gene Mutations and Their Sources

Figure 14.11 shows a common gene mutation. One base (adenine) was wrongly paired with another (cytosine) as DNA was replicated. Specialized DNA polymerases fix such errors in growing DNA strands (Section 13.4). But some keep assembling a new strand right past an error, and such a bypass can establish a mutation in the DNA molecule. This particular mutation is a **base-pair substitution**. Its outcome? One amino acid may replace another during protein synthesis. That is what happens in people who carry *Hb^S*, the mutant allele that causes sickle-cell anemia (Section 3.8).

① (handwritten)

Figure 14.12 shows a different mutation. One *extra* base was inserted into a gene region. Remember, DNA polymerases read nucleotide sequences in blocks of three. The insertion shifted the "three-bases-at-a-time" reading frame by one base; hence the term *frameshift* mutation. The altered gene has a different message, so an altered version of the protein will be synthesized. Frameshift mutations fall in broader categories of mutation called **insertions** and **deletions**. In such cases, one to several base pairs are inserted into DNA or deleted from it.

② (handwritten)
A + B (handwritten)

As another example, **transposons**—or transposable elements—may bring about mutation when they jump around in the genome. Barbara McClintock discovered that these segments of DNA move spontaneously from

"*jumping genes*" (handwritten)

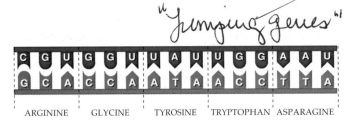
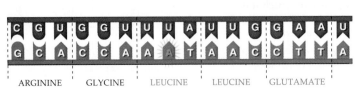

mRNA transcribed from the DNA

PART OF PARENTAL DNA TEMPLATE

resulting amino acid sequence

ARGININE   GLYCINE   TYROSINE   TRYPTOPHAN   ASPARAGINE

altered message in mRNA

A BASE INSERTION (*RED*) IN DNA

the altered amino acid sequence

ARGININE   GLYCINE   LEUCINE   LEUCINE   GLUTAMATE

Figure 14.12 Example of an insertion, a mutation in which an extra base gets inserted into a gene region of DNA. This insertion has caused a *frameshift*; it has changed the reading frame for base triplets in the DNA and in the mRNA transcript of that region. As a result, the wrong amino acids will be called up when the mRNA transcript becomes translated into protein.

[handwritten annotation at top: Mutations can be harmful, harmless, or actually beneficial → adaptive trait that helps an individual survive & reprod! under prevailing environmental conditions]

Figure 14.13 Barbara McClintock, who won a Nobel Prize for her research that showed some DNA segments slip into and out of different locations in DNA molecules. The segments are transposons. In her hands is an ear of Indian corn (*Zea mays*). The curiously nonuniform coloration of its kernels sent her on the road to discovery.

Each corn kernel is a seed that can grow into a new corn plant. All of its cells have the same pigment-coding genes. Yet some kernels are colorless or spottily colored. In the ancestor of the plant from which this ear of corn was plucked, a gene in a germ cell left its position in one DNA molecule, invaded a different DNA molecule, and shut down a pigment-encoding gene.

The plant inherited the mutation. As cell divisions proceeded in the growing plant, none of the mutated cell's descendants could synthesize pigment molecules. Wherever they were located, the kernel tissue was colorless. Later, in some cells, the movable DNA slipped out of the pigment-encoding gene. All descendants of *those* cells produced pigment—and colored kernel tissue.

one location to another in the same DNA molecule or to a different one. Often they inactivate genes into which they become inserted. Their unpredictability can cause interesting variations in traits. You can read about two examples in Figure 14.13 and *Critical Thinking* question 3 at this chapter's end.

Causes of Gene Mutations

[handwritten: mutagen that causes cancer = carcinogen]

Many mutations arise spontaneously when DNA is being replicated. This shouldn't be surprising, given the swift pace of replication and the huge pools of free nucleotides concentrated near a growing DNA strand. Specialized DNA polymerases repair most of the mistakes, but they *[handwritten: happen naturally]* bypass a small number with predictable frequency.

Each gene has a characteristic **mutation rate**. This is the probability that it will mutate spontaneously during a specified interval, such as each DNA replication cycle. (The *rate* isn't the same as the *frequency* of a mutation—the number of times it is found in some population, as in the 500,000 people resulting from 1 million gametes.)

Not all mutations are spontaneous. Many result after exposure to mutagens, or mutation-causing agents in the environment. Two classes of radiation are mutagenic. High-energy wavelengths of **ionizing radiation**, such as x-rays, can damage DNA directly. They also damage it *[handwritten: gamma rays ↓ radioisotopes]* indirectly by the action of free radicals that form when they ionize water and other molecules. Repair enzymes may not restore the altered base sequences. Ionizing radiation that deeply penetrates living tissue leaves a trail of free radicals. X-ray doses used for dental work and internal medicine diagnoses are extremely low to minimize mutation. However, x-rays have a cumulative effect, so repeated exposure even to low levels over the years can cause problems.

Nonionizing radiation simply boosts electrons to a higher energy level. But DNA easily absorbs one form, ultraviolet (UV) light. Two of its nucleotides, cytosine

and thymine, are highly vulnerable to excitation that can alter their base-pairing properties. The next chapter's introduction describes one possible mutation.

Natural and synthetic chemicals accelerate rates of *[handwritten: eg mustard gas]* spontaneous mutations. For instance, substances called **alkylating agents** transfer methyl and ethyl groups to reactive sites in DNA's bases or phosphate groups. At the alkylated sites, DNA is more susceptible to base-pair alterations that invite mutation. Many **carcinogens**, or cancer-causing agents, operate by alkylating DNA.

[handwritten: Mutations are the source of variations in heritable traits]

The Proof Is In the Protein

[handwritten: P. 10]

Spontaneous mutations are rare in terms of a human life. The rate for eukaryotes in general ranges between 10^{-4} and 10^{-6} per gene per generation. If one arises in a somatic cell, its good or bad effects won't endure, for it cannot be passed on to offspring. If the mutation arises in a germ cell or gamete, however, it may well enter the evolutionary arena. The same can happen if a mutation arises in an asexually reproducing organism or cell.

In all such cases, nature's test is this: *A protein that is specified by a heritable mutation may have harmful, neutral, or beneficial effects on the individual's ability to function in the prevailing environment.* Also, gene mutations have had powerful evolutionary consequences—and that will be a major theme of the next unit of the book.

A gene mutation is an alteration in one to several bases in the nucleotide sequence of DNA. The most common are base-pair substitutions, base insertions, and base deletions.

Each gene has a spontaneous and characteristic mutation rate, which may be accelerated by exposure to harmful radiation and certain chemicals in the environment.

A protein specified by a mutated gene may have harmful, neutral, or beneficial effects on the ability of an individual to function in the prevailing environment.

SUMMARY

Gold indicates text section

1. Single cells and multicelled organisms cannot stay alive unless they build enzymes and other proteins. A protein consists of one or more polypeptide chains. Each chain is a linear sequence of amino acids. *CI*

   a. The amino acid sequence of a polypeptide chain corresponds to a gene region in one strand of the DNA double helix. That region is a sequence of nucleotide bases. DNA's bases are adenine, thymine, guanine, and cytosine (A, T, G, and C).

   b. Transcription and translation are two steps in the path from genes to proteins (Figure 14.14):

$$\textbf{DNA} \xrightarrow{\textit{transcription}} \textbf{RNA} \xrightarrow{\textit{translation}} \textbf{PROTEIN}$$

2. The path requires three classes of RNA molecules, or ribonucleic acids:

   a. Messenger RNA (mRNA) is the only class of RNA that carries a protein-building message. *14.1, 14.2*

   b. Ribosomal RNA (rRNA) and structural proteins that stabilize it are the components of ribosomes. All polypeptide chains are assembled on ribosomes. *14.2*

   c. Transfer RNA (tRNA) binds free amino acids in the cytoplasm and gives them up at a ribosome, in the sequence dictated by a sequential message in mRNA. Different kinds bind different amino acids. *14.2, 14.3*

3. In transcription, DNA is unwound at a gene region. Exposed bases on one strand function as a template for assembling an RNA strand from the cell's pool of free nucleotides. The RNA-to-DNA rules for base-pairing are that guanine pairs with cytosine, and *uracil*—not thymine—pairs with adenine: *14.1*

   a. Different RNAs are assembled on different genes.

   b. In eukaryotic cells, the mRNA transcripts become modified into final form before being shipped from the nucleus. We call this transcript processing.

4. In translation, mRNA, tRNAs, and rRNA interact to build polypeptide chains. Afterward, the chains twist, fold, and may be additionally modified into a protein's final, three-dimensional shape. *CI, 14.3*

   a. Translation follows a genetic code. The code is a set of sixty-four base triplets; each is a series of three nucleotide bases. *Triplet* refers to the way the bases are "read" three at a time during translation at a ribosome.

   b. A base triplet in mRNA is a codon. An anticodon is a complementary triplet in a tRNA molecule. Some combination of codons specifies what the amino acid sequence of a polypeptide chain will be, start to finish.

5. Translation proceeds through three stages: *14.3*

   a. Initiation. One small ribosomal subunit and one initiator tRNA bind with the mRNA and move along it until they encounter an AUG START codon. The small subunit binds with a large ribosomal subunit.

   b. Chain elongation. tRNAs deliver amino acids to an intact ribosome. Their anticodons base-pair with the mRNA codons. Part of the rRNA of the large ribosomal subunit catalyzes peptide bond formation between every two amino acids, forming a new polypeptide chain.

   c. Chain termination. An mRNA STOP codon moves onto the ribosomal platform, making the polypeptide chain and the mRNA detach from the ribosome.

6. Gene mutations are heritable, small-scale changes in the base sequence of DNA. Many arise spontaneously while DNA is being replicated or after it is exposed to ultraviolet or ionizing radiation, alkylating agents, or some other mutagen in the environment. *14.4*

TRANSCRIPTION   *Assembly of RNA on unwound gene regions of DNA molecule*

mRNA   **rRNA**   **tRNA**

Pre-mRNA
transcript
processing

protein
subunits

mature mRNA
transcripts

ribosomal
subunits

mature
tRNA

TRANSLATION

Convergence
of RNAs

cytoplasmic
pools of
amino
acids,
ribosomal
subunits,
and tRNAs

At an intact
ribosome,
synthesis of
a polypeptide
chain at the
binding sites
for mRNA
and tRNAs

ile   gly   met

val

glu

cys

FINAL PROTEIN

For use in cell or for export

Figure 14.14 Summary of the flow of genetic information from DNA to proteins in eukaryotic cells. DNA is transcribed into RNA in the nucleus; RNA is translated in the cytoplasm. Prokaryotic cells don't have a nucleus; both transcription and translation proceed in their cytoplasm.

Review Questions

1. Are the polypeptide chains of proteins assembled on DNA? If so, state how. If not, state how they are assembled, and on which molecules. *CI, 14.3*

2. Name the three classes of RNA and state the function of each class in protein synthesis. *14.1, 14.2*

3. The pre-mRNA transcripts of eukaryotic cells contain both introns and exons. Are the introns or exons snipped out before the transcript leaves the nucleus? *14.1*

4. Distinguish between codon and anticodon. *14.2*

5. Name the three stages of translation. Briefly describe the key events of each stage. *14.3*

6. Define gene mutation. Give three examples of agents that cause mutations. *14.4*

7. Do all mutations arise spontaneously? Are environmental agents always the trigger for mutation? *14.4*

8. Define and then state the possible outcomes of the following types of mutation: base-pair substitution, base insertion, and insertion of a transposon at a new location in the DNA. *14.4*

Self-Quiz   ANSWERS IN APPENDIX III

1. DNA contains many different genes that are transcribed into different _____ .
   a. proteins   c. mRNAs, tRNAs, and rRNAs
   b. mRNAs only   d. all are correct

2. An RNA molecule is _____ .
   a. a double helix   c. always double-stranded
   b. usually single-stranded   d. usually double-stranded

3. An mRNA molecule is produced by _____ .
   a. replication   c. transcription
   b. duplication   d. translation

4. Each codon calls for a specific _____ .
   a. protein   c. amino acid
   b. polypeptide   d. carbohydrate

5. Referring to Figure 14.7, use the genetic code to translate the mRNA sequence UAUCGCACCUCAGGAGACUAG. Notice that the first codon in the frame is UAU. Which amino acid sequence is being specified?

   a. TYR—ARG—THR—SER—GLY—ASP—

   b. TYR—ARG—THR—SER—GLY

   c. TYR—ARG—TYR—SER—GLY—ASP—

6. Anticodons pair with _____ .
   a. mRNA codons   c. tRNA anticodons
   b. DNA codons   d. amino acids

7. Match the terms with the most suitable description.
   ____ alkylating   a. parts of mature mRNA transcript
        agent   b. base triplet coding for amino acid
   ____ chain   c. second stage of translation
        elongation   d. base triplet that pairs with codon
   ____ exons   e. one environmental agent that
   ____ genetic code   induces mutation in DNA
   ____ anticodon   f. set of sixty-four codons for mRNA
   ____ intron   g. parts removed from a pre-mRNA
   ____ codon   transcript

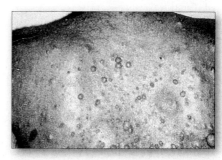

Figure 14.15 Soft skin tumors on a person with neurofibromatosis, an autosomal dominant disorder.

Critical Thinking

1. Sandra discovered a tRNA with a mutation in DNA that encodes the anticodon 3'-AUU instead of 3'-AAU. In cells with the mutated tRNA, what will be the effect on protein synthesis?

2. A DNA polymerase made an error during the replication of an important gene region of DNA. None of the DNA repair enzymes detected or repaired the damage. A portion of the DNA strand with the error is shown here:

```
...AATTCC ACTCCTATGG
...TTAAGG TGAGGATACC
```

After the DNA molecule is replicated and two daughter cells have formed, one cell is carrying a mutation and the other cell is normal. Develop a hypothesis to explain this observation.

3. *Neurofibromatosis* is a human autosomal dominant disorder caused by mutations in the *NF1* gene. It is characterized by soft, fibrous tumors in the peripheral nervous system and skin as well as abnormalities in muscles, bones, and internal organs (Figure 14.15).

Because the gene is dominant, an affected child usually has an affected parent. Yet in 1991, scientists reported on a boy who had neurofibromatosis yet whose parents did not. When they examined both copies of his *NF1* gene, they found the copy he had inherited from his father contained a transposon. Neither the father nor the mother had a transposon in any of the copies of their own *NF1* genes. Explain the cause of neurofibromatosis in the boy and how it arose.

Selected Key Terms

alkylating agent *14.4*   mRNA (messenger RNA) *14.1*
anticodon *14.2*   mutation rate *14.4*
base sequence *CI*   nonionizing radiation *14.4*
base-pair substitution *14.4*   promoter *14.1*
carcinogen *14.4*   ribonucleic acid (RNA) *CI*
codon *14.2*   RNA polymerase *14.1*
deletion (of base) *14.4*   rRNA (ribosomal RNA) *14.1*
exon *14.1*   transcription *CI*
gene mutation *14.4*   translation *CI*
genetic code *14.2*   transposon *14.4*
insertion (of base) *14.4*   tRNA (transfer RNA) *14.1*
intron *14.1*   uracil *14.1*
ionizing radiation *14.4*

Readings

Crick, F. 1988. *What Mad Pursuit: A Personal View of Scientific Discovery.* New York: HarperCollins. Crick's autobiography.

Friedberg, E., R. Wagner, and M. Radman. 31 May 2002. "Specialized DNA Polymerases, Cellular Survival, and the Genesis of Mutations." *Science* 296:1627–1630.

CONTROLS OVER GENES

When DNA Can't Be Fixed

Not long ago, Laurie Campbell turned eighteen and happened to notice a black mole on her skin. It was an odd, encrusted lump with a ragged border. No fool, she quickly made an appointment with her family doctor, who ordered a biopsy. The mole turned out to be a *malignant melanoma*, the deadliest form of skin cancer.

Moles and other tumors are **neoplasms**, abnormal masses of cells that ignore controls over growth and division. Benign neoplasms stay put. Malignant ones are **cancers**, with cells that can break away, invade other tissues, and give rise to more abnormal masses. Laurie was lucky. She detected a cancer in its earliest stage before it could spread through her body. Now she regularly checks out other skin moles. She is aware of having become a statistic. In 2001, there were more than a million cases of skin cancer in the United States and 7,700 deaths from malignant melanoma.

Laurie is smart. She plotted the position of each mole on her body. Once a month, she uses that body map as a guide for a quick, thorough self-examination. Figure 15.1 shows examples of what she looks for. Laurie also schedules a medical examination every six months.

Ultraviolet wavelengths in the sun's rays, tanning lamps, and other sources of nonionizing radiation can cause skin cancer. For instance, they promote covalent bonding between two neighboring thymine bases in a DNA strand. The result is an abnormal, bulky structure in a DNA molecule —a thymine dimer.

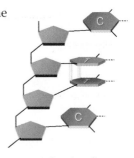

thymine dimer

Normally, at least seven gene products interact as a DNA repair mechanism to remove the bulky lesion. When mutation in one or more of those genes disrupts the mechanism, thymine dimers can accumulate in skin cells. They may trigger cancers by upsetting the normal gene controls over cell growth and division.

You are at risk if you habitually irritate moles, as by shaving or wearing abrasive clothing. You are at risk if your skin is chronically chapped, cracked, or sore. You are at risk if there is a history of cancer in your family or if you have endured radiation therapy. Like Laurie,

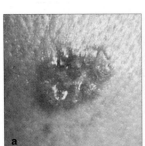
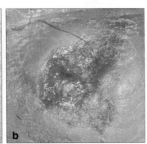
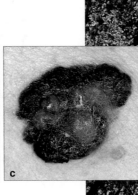

Figure 15.1 Examples of what can happen when repair enzymes can't fix damaged DNA. (**a**) *Basal cell carcinoma*, the most common form of skin cancer. This slow-growing, raised lump may be uncolored, reddish-brown, or black. (**b**) *Squamous cell carcinoma*, the second most common skin cancer. The pink growths, firm to the touch, spread rapidly under the surface of skin exposed to the sun. (**c**) *Malignant melanoma* spreads fastest. The malignant cells form dark, encrusted lumps. They may itch like an insect bite or bleed easily. *Right:* Laurie Campbell avoiding the sun—and melanoma.

you are at risk simply if you burn easily. Tanning or otherwise staying out in the sun without protection is ill advised; damaged DNA is a reality for all of us.

Why start a chapter with such awful prospects? Doing so might help focus your attention on how lucky individuals are when gene controls operate as they should. **Gene controls** are molecular mechanisms that govern when and how fast specific genes will be transcribed and translated, and whether gene products will be switched on or silenced.

In all prokaryotic and eukaryotic cells, many of the controls make transcription rates rise or fall in response to concentrations of nutrients and other substances. For example, genes that specify the enzymes required to metabolize sugars are transcribed in controlled ways. Some controls adjust the rates of transcription according to how much sugar is available. Others activate, inactivate, and degrade the gene products, including the sugar-digesting enzyme molecules that have finished their task.

In eukaryotic cells, still other controls guide mRNA transcript processing, transport of mature RNAs from the nucleus, and how fast RNAs are translated in the cytoplasm. Other controls guide the modification of new polypeptide chains in the endomembrane system.

Controls also guide the contribution that eukaryotic cells will make to a long-term program of growth and development. This is especially so for large, complex, multicelled species. As part of the program, different cell lineages activate and suppress fractions of their genes in different ways. Certain genes are expressed once, some of the time, or not at all. The result? Most cells become specialized in structure, composition, and function. We call this process and its outcome cell differentiation.

Explaining control of gene activity is like trying to explain a full symphony orchestra to someone who has never seen one or heard it perform. You have to identify all of the many separate parts before their interactions start to make sense! Gene controls weave through the story line of many chapters throughout the book. For that reason, take some time now to become acquainted with just a few of the molecular players and their amazing functions.

Key Concepts

1. In cells, a variety of controls govern when, how, and to what extent genes are expressed. The control elements operate in response to preprogrammed schedules of development. They also operate in response to changing chemical conditions and to reception of external signals.

2. Control is exerted by way of regulatory proteins and other molecules that operate before, during, or after gene transcription. The control elements interact with DNA, with RNA that has been transcribed from DNA, or with the resulting polypeptide chains or the final proteins.

3. Prokaryotic cells depend on rapid responses to short-term changes in nutrient availability and other aspects of their surrounding environment. They commonly use regulatory proteins that help make quick adjustments in gene transcription rates, which compensate for the changes.

4. Eukaryotic cells also use controls over short-term shifts in diet and levels of activity. In complex multicelled species, they rely as well on long-term controls over an intricate program of growth and development.

5. Controls over eukaryotic cells come into play when new cells contact one another in developing tissues. They also come into play when these cells start interacting with their neighbors by way of hormones and other signaling molecules.

6. The cells of a multicelled organism inherit the same genes, but different cell types activate or suppress many of the genes in different ways. Their controlled, selective use of their genes leads to the synthesis of the proteins that give each type of cell its distinctive structure, function, and products.

TYPES OF CONTROL MECHANISMS

Different mechanisms control gene expression through interactions with DNA, RNA, and the new polypeptide chains or final proteins. Some mechanisms respond to rising or falling concentrations of a nutrient or some other substance. Others respond to external signaling molecules that call for change.

The control agents include **regulatory proteins** that intervene before, during, or after gene transcription or translation. They also include signaling molecules such as hormones, which initiate change in some cell activity when they dock at suitable receptors.

By **negative control**, regulatory proteins slow down or curtail gene activity. By **positive control**, regulatory proteins promote or enhance gene activities. Control mechanisms for transcription involve noncoding base sequences in DNA that do not specify proteins. For example, **promoters** are base sequences that signal the start of a gene. As another example, the **enhancers** are binding sites for some activator proteins.

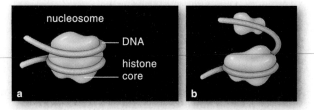

Figure 15.2 How DNA–histone packing in nucleosomes may be loosened to make gene regions available for transcription. Attaching an acetyl group to a histone makes it loosen its grip on DNA wound around it. Enzymes that attach or detach acetyl groups are known to be associated with transcription.

Control also is exerted with chemical modification. For instance, regions of newly replicated DNA can be shut down by **methylation**, the attachment of a methyl group ($-CH_3$) to nucleotide bases. Inactivated genes are often heavily methylated. And certain genes are activated by demethylation. Also, access to genes is partly controlled through **acetylation**: attaching acetyl groups from the histones that structurally organize the DNA (Section 9.1 and Figure 15.2).

When, how, and to what extent a gene is expressed depends on the type of cell and its functions, on the cell's chemical environment, and on signals for change.

Gene expression is controlled by regulatory proteins that interact with one another, with control elements built into the DNA, with RNA, and with newly synthesized proteins.

Control also is exerted through chemical modifications that inactivate or activate specific gene regions or the histone proteins that organize the DNA.

BACTERIAL CONTROL OF TRANSCRIPTION

Think about the dot of the letter "i." About a thousand bacterial cells would stretch side by side across the dot. Just imagine, each of those microscopic specks depends as much on gene controls as you do! When nutrients are plentiful and other environmental conditions also favor growth, the cells swiftly grow and divide. Gene controls promote the rapid synthesis of enzymes that catalyze nutrient digestion and other growth-related activities. Translation starts even before the RNA transcripts are finished. Bacteria, recall, have no nuclear envelope that keeps DNA away from ribosomes in the cytoplasm.

Often, prokaryotic genes for enzymes of a metabolic pathway are clustered together in the same sequence as the reaction steps. All genes in such a sequence may be transcribed as one continuous mRNA transcript.

With this bit of background, consider how one kind of prokaryote adjusts transcription rates downward or upward, depending on the availability of nutrients.

Negative Control of the Lactose Operon

Escherichia coli, an enteric bacterium, lives on sugars and other ingested nutrients in the mammalian gut. While mammals are infants, they live on milk only. Milk does not contain glucose, the sugar that *E. coli* cells prefer. It contains lactose, a different sugar. Once mammals are weaned, their milk intake typically declines or stops.

E. coli cells still take advantage of lactose when it is available. They activate a set of three adjoining genes coding for lactose-metabolizing enzymes. A promoter precedes the genes in the bacterial DNA, and operators are positioned on both sides of it.

An **operator** is a binding site for a repressor protein that can prevent gene transcription. This arrangement, in which a promoter and set of operators control more than one bacterial gene, is an **operon** (Figure 15.3).

In the absence of lactose, the repressor molecule binds to a set of operators. Binding causes the part of the DNA with the promoter to loop outward, as shown in Figure 15.3c. When looped this way, the promoter is inaccessible to RNA polymerase. Therefore, the operon genes can't be transcribed when they are not required.

When lactose *is* present, *E. coli* cells convert some of it to allolactose. This sugar binds to the repressor and alters its shape. In altered form, the repressor cannot bind to the operators. The looped DNA unwinds, RNA polymerase can start transcription, and so the lactose-degrading enzymes are produced when required.

Positive Control of the Lactose Operon

E. coli cells pay far more attention to glucose than to lactose. They transcribe genes for its breakdown faster, and continuously. Even when lactose is in the gut, the

inducer/
effector

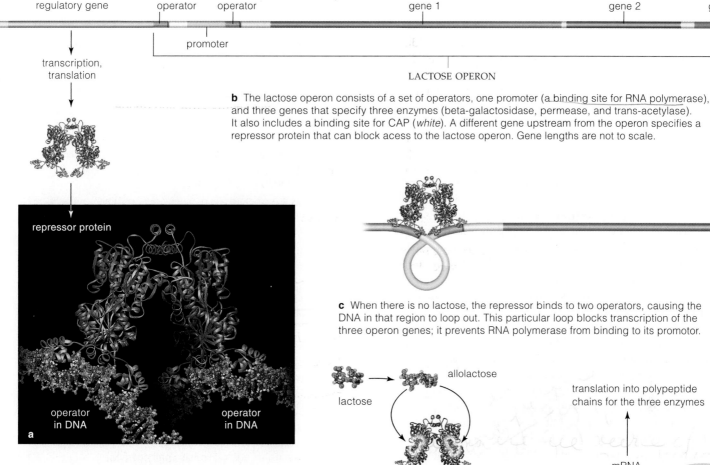

regulatory gene | operator | operator | gene 1 | gene 2 | gene 3

promoter

transcription, translation

LACTOSE OPERON

b The lactose operon consists of a set of operators, one promoter (a binding site for RNA polymerase), and three genes that specify three enzymes (beta-galactosidase, permease, and trans-acetylase). It also includes a binding site for CAP (*white*). A different gene upstream from the operon specifies a repressor protein that can block acess to the lactose operon. Gene lengths are not to scale.

repressor protein

operator in DNA

operator in DNA

a

c When there is no lactose, the repressor binds to two operators, causing the DNA in that region to loop out. This particular loop blocks transcription of the three operon genes; it prevents RNA polymerase from binding to its promotor.

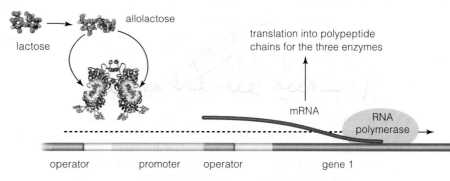

allolactose

lactose

translation into polypeptide chains for the three enzymes

mRNA

RNA polymerase

operator | promoter | operator | gene 1

Figure 15.3 **(a)** Computer model for the lac repressor protein, bound to two operators of a prokaryotic DNA molecule. **(b–d)** Negative control of the lactose operon. The operon's first gene codes for an enzyme that splits the disaccharide lactose into glucose and galactose. The second enzyme codes for an enzyme that helps transport lactose into cells. The third helps metabolize certain sugars.

d When lactose is present, some is converted to a form that can bind to the repressor protein and alter its shape. The altered repressor cannot bind to operators. RNA polymerase can transcribe the operon genes.

lactose operon is not used much—unless there is no glucose. These conditions call for an **activator** protein called CAP. This activator exerts positive control over the lactose operon by making a promoter more inviting to RNA polymerase. But CAP can't issue the invitation until it is bound to a chemical messenger, cAMP (short for cyclic adenosine monophosphate). When cAMP and the activator are complexed together and bound to the promoter, they make it far easier for RNA polymerase to start transcribing genes.

When glucose is plentiful, ATP forms by glycolysis, but an enzyme needed to synthesize cAMP is inhibited. That enzyme is released from inhibition when glucose is scarce and lactose is available. cAMP accumulates, CAP–cAMP complexes form, and the lactose operon genes are rapidly transcribed. The gene products allow lactose to be used as an alternative energy source.

Humans are born with a gene for lactase, a lactose-digesting enzyme made by intestinal cells. Before three years pass, lactase levels start to decline in genetically predisposed people. *Lactose intolerance* begins, because lactose can't be degraded. It moves into the colon and fans the growth of resident bacterial populations. One gaseous by-product of their metabolism accumulates and distends the colon, causing pain. Short fatty acid chains released by the reactions lead to diarrhea, which is often severe. People can avoid symptoms by drinking milk that has predigested lactose or by taking lactose-digesting enzymes before ingesting dairy products.

Transcription rates of bacterial genes for nutrient-digesting enzymes are quickly adjusted downward and upward by control systems that respond to nutrient availability.

GENE CONTROLS IN EUKARYOTIC CELLS

Like bacteria, eukaryotic cells control short-term shifts in diet and in levels of activity. If those cells happen to be among hundreds or trillions of cells in a multicelled organism, long-term controls also enter the picture, for gene activities change during development.

Cell Differentiation and Selective Gene Expression

Later in the book, you will be reading about controls over development, particularly in Chapters 32, 43, and 44. For now, tentatively accept this basic premise: All of the cells in your body started out life with the same genes, because every one arose by mitotic cell divisions from one fertilized egg. Many of those inherited genes specify proteins that are essential for the structure and everyday functioning of every cell. That is why those genes are controlled in ways that promote ongoing, low levels of transcription.

Even so, *nearly all of your cells became specialized in composition, structure, and function*. This process of **cell differentiation** proceeds during the development of all multicelled species. It arises as embryonic cells and the cell lineages descended from them activate a fraction of their genes in selective ways.

For example, nearly all of your body cells use genes that specify the enzymes of glycolysis on an ongoing basis. Yet only your immature red blood cells activate genes for hemoglobin. Your liver cells activate genes required to synthesize enzymes that neutralize certain toxins, but they are the only ones that do. When your eyes first formed, only certain cells accessed the genes necessary for synthesis of crystallin. No other cells can activate the genes for this protein, which helped make transparent fibers of the lens in each eye.

Controls Before and After Transcription

In all large, complex eukaryotic organisms, many of the genes that govern housekeeping tasks are continuously transcribed at low levels. In the case of other genes, transcription rates are adjusted up and down. Why? Tissue fluids of such organisms are the body's internal environment. Individual cells of the body continually deliver or secrete substances into that environment, and withdraw substances from it. The ongoing inputs and outputs cause slight shifts in the concentrations of nutrients, signaling molecules, metabolic products, and other solutes. Most often, transcription rates for those genes rise or fall by small degrees in response.

As the examples in Figure 15.4a indicate, some gene sequences are repeatedly duplicated or rearranged in genetically programmed fashion prior to transcription. Besides this, programmed chemical modifications often

a CONTROLS RELATED TO TRANSCRIPTION. At any time, most genes of a multicelled organism are shut down permanently or temporarily. Genes necessary for a cell's everyday tasks are under positive controls that promote ongoing, low levels of transcription. These controls help provide a cell with enough enzymes and other proteins to carry out its most basic functions. Transcription of many other genes shifts. In this case, controls work to assure chemical responsiveness even when concentrations of specific substances rise or fall only slightly.

Also, even before some genes are transcribed, parts of them are amplified, rearranged, or chemically modified in permanent or reversible ways. These changes are not mutations. They are programmed events that affect how the gene will be expressed, if at all.

1. *Gene amplification*. Immature amphibian eggs and glandular cells of some insect larvae copy the same genes again and again when they require enormous numbers of the products. Sometiems, multiple rounds of DNA replication produce hundreds or thousands of copies prior to transcription (Sections 15.4 and 15.5).

2. *DNA rearrangements*. Some genes have many base sequences in the DNA, and they are put together in different ways to make different product molecules. This happens when B lymphocytes, a special class of white blood cells, are forming (Section 39.5). Different cells transcribe and translate the uniquely sequences into different versions of antibodies. These protein weapons act against specific pathogens .

3. *Chemical modification.* Histones and other proteins interact with eukaryotic DNA in organized ways. The DNA–protein packaging, plus chemical modifications to particular sequences, influences gene expression. Normally, just a small fraction of a cell's genes are available for transcription. A dramatic shutdown occurs in female mammals. As Section 15.4 describes, this event is called X chromosome inactivation.

Figure 15.4 Examples of the levels of control over gene expression in eukaryotes.

shut down many genes. So does the orderly packaging of DNA by histones and other chromosomal proteins, as you may realize after reflecting on the organization of eukaryotic chromosomes (Section 9.1).

Controls also come into play after genes have been transcribed. As Figure 15.4b–d indicates, many controls govern RNA transcript processing, transport of mature RNAs from the nucleus, and rates of translation in the cytoplasm. Other controls deal with the modification of new polypeptide chains. Others deal with activating, inhibiting, and breaking down existing proteins.

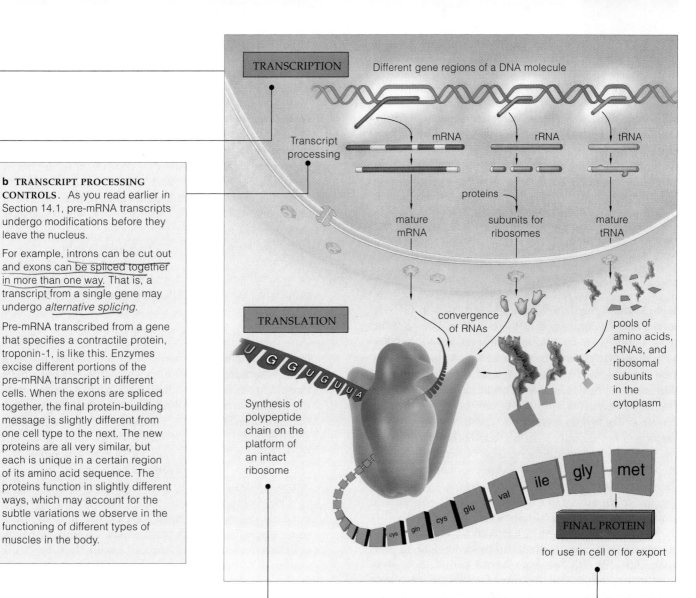

TRANSCRIPTION

Different gene regions of a DNA molecule

Transcript processing

mRNA · rRNA · tRNA

proteins

mature mRNA · subunits for ribosomes · mature tRNA

b TRANSCRIPT PROCESSING CONTROLS. As you read earlier in Section 14.1, pre-mRNA transcripts undergo modifications before they leave the nucleus.

For example, introns can be cut out and exons can be spliced together in more than one way. That is, a transcript from a single gene may undergo *alternative splicing*.

Pre-mRNA transcribed from a gene that specifies a contractile protein, troponin-1, is like this. Enzymes excise different portions of the pre-mRNA transcript in different cells. When the exons are spliced together, the final protein-building message is slightly different from one cell type to the next. The new proteins are all very similar, but each is unique in a certain region of its amino acid sequence. The proteins function in slightly different ways, which may account for the subtle variations we observe in the functioning of different types of muscles in the body.

TRANSLATION

convergence of RNAs

pools of amino acids, tRNAs, and ribosomal subunits in the cytoplasm

Synthesis of polypeptide chain on the platform of an intact ribosome

U G G U G U U A

cys · gln · cys · glu · val · ile · gly · met

FINAL PROTEIN

for use in cell or for export

c CONTROLS OVER TRANSLATION. Controls govern when, how fast, and how often an mRNA transcript is translated. Sections 36.2 and 43.3 provide elegant examples.

A transcript's stability affects how many protein molecules can be produced from it. Enzymes digest transcripts from the poly-A tail on up (Section 14.1). The tail's length and its attached proteins affect how fast it is digested. Also, after leaving the nucleus, some transcripts are temporarily or permanently inactivated. Example: In unfertilized eggs, many transcripts are inactivated and stored in the cytoplasm. These "masked messengers" will not be available for translation until after fertilization, when great numbers of protein molecules will be required for the early cell divisions of the new individual.

d CONTROLS FOLLOWING TRANSLATION. Many newly formed polypeptide chains enter the endomembrane system (Section 4.5). They undergo modification, as when enzymes attach specific oligosaccharides or phosphate groups to them.

Diverse control mechanisms govern the activation, inhibition, and stability of enzymes and other molecules used in protein synthesis. Allosteric control of tryptophan synthesis is an example (Section 6.7).

Consider enzymes, the proteins that catalyze nearly all metabolic reactions. Besides selectively transcribing and translating the genes for enzymes, diverse control systems activate and inhibit the molecules of enzymes that have already been synthesized in the cell.

Just imagine the coordination necessary to govern which of the cell's thousands of types of enzymes are to be stockpiled, deployed, or degraded in a specified interval. *That coordination helps govern all short-term and long-term aspects of cell structure and function.*

In multicelled species, a variety of gene controls guide the moment-by-moment activities that maintain cells. Other controls guide intricate, long-term patterns of the body's growth and development.

Cells of complex organisms inherit the same genes, yet most become specialized in composition, structure, and function. This process of cell differentiation arises when different populations of cells activate and suppress their genes in highly selective, unique ways.

15.4

TYPES OF CONTROL MECHANISMS

By some estimates, cells of complex organisms rarely use more than 5 to 10 percent of their genes at a given time. One way or another, control mechanisms are keeping most of the genes inactivated. *Which genes are active depends on the type of organism, the stage of growth and development it is passing through, and the controls that are operating at different stages.* You will be coming across some elegant cases of this in later chapters. For now, two examples will make the point.

Homeotic Genes and Body Plans

Some kinds of regulatory proteins bind with promoters, enhancers, and one another to control transcription of specific genes. As an example, most eukaryotic species have **homeotic genes**. These are a class of master genes; they interact with one another and also with control elements to bring about the formation of tissues and organs in accordance with the basic body plan. The master genes are transcribed in specific order in local tissue regions. Gradients in the concentrations of gene products result. Depending on their position relative to these gradients, other genes are transcribed or remain silent. In animal embryos, for instance, different genes respond to a gradient in ways that lead to the formation of the body's anterior–posterior axis (Section 43.5).

Homeotic genes were discovered by way of single mutations in *Drosophila* that transformed one body part into a different one. For example, the *antennapedia* gene is actively transcribed in regions of the embryo that will become the thorax, complete with legs. Its transcription normally is restricted in other regions, such as the one destined to become the head. Figure 15.5*a* shows what happens when it is wrongly activated in an embryo's head cells. Similarly, a homeotic gene in corn plants controls leaf vein formation. Its mutation gives rise to veins that are twisted instead of organized in flat planes.

Homeotic genes code for homeodomains, which are regulatory proteins that contain a sequence of sixty or so amino acids. That sequence, a "homeobox," can bind to short control sequences in promoters and enhancers (Figure 15.5*b*). We know of more than a hundred kinds of homeodomain proteins that control transcription by common mechanisms. They occur in all eukaryotes. Many are even interchangeable among distantly related organisms such as yeasts and humans, which suggests they evolved among the most ancient eukaryotic cells. In many cases, the homeobox sequences differ only in *conservative* amino acid substitutions; even when one amino acid has replaced another, it has similar chemical properties. We return to this topic in Chapter 43.

X Chromosome Inactivation

Female humans and female calico cats have something in common at the cellular level. Although both have two X chromosomes in their diploid cells, one is in its threadlike form; the other one is scrunched up even during interphase. The scrunching isn't a chromosome abnormality. It is a programmed shutdown of all but about three dozen genes on *one* of two homologous X chromosomes. This **X chromosome inactivation** occurs in the diploid cells of all female placental mammals.

One X chromosome is inactivated when the females are early embryos, no more than a tiny ball of dividing cells. The outcome is random; *either* chromosome may be condensed this way. One cell might shut down the maternal X chromosome, another cell next to it might shut down the paternal X chromosome or the maternal X chromosome, and so on. The condensed one looks like a dense spot inside the interphase nucleus (Figure 15.6*a*). Researchers named a condensed X chromosome a **Barr body** after Murray Barr, its discoverer.

Once that first random selection is made in a cell, however, all the descendant cells make the exact same selection as they go on dividing to form tissues. When fully developed, then, *each female mammal bears patches of tissue where genes of the maternal X chromosome are*

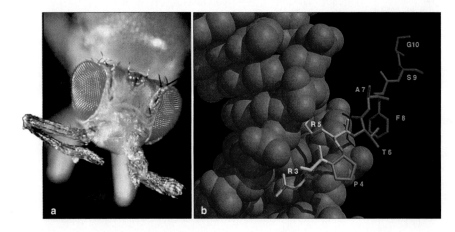

Figure 15.5 (**a**) Experimental evidence that genes control the development of body parts. In normal *Drosophila* larvae, activation of genes in a certain group of cells gives rise to the head's antennae. A larva with a mutant form of the *antennapedia* gene develops into an adult fly with legs instead of antennae on its head. This is one of the genes controlled by homeodomains, a type of regulatory protein. (**b**) Model for a homeodomain binding to a transcriptional control sequence in a DNA molecule.

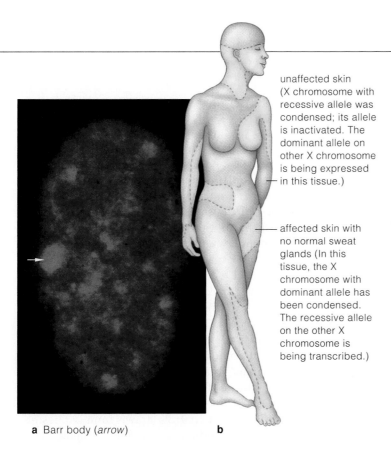

unaffected skin (X chromosome with recessive allele was condensed; its allele is inactivated. The dominant allele on other X chromosome is being expressed in this tissue.)

affected skin with no normal sweat glands (In this tissue, the X chromosome with dominant allele has been condensed. The recessive allele on the other X chromosome is being transcribed.)

a Barr body (*arrow*)

b

Figure 15.6 (**a**) Micrograph of an inactivated X chromosome, called a Barr body, as it appears in a human female's somatic cell during interphase. The X chromosome is not condensed this way in human male cells. (**b**) Anhidrotic ectodermal dysplasia, a mosaic pattern of gene expression. The condition arises rarely as a result of random X chromosome inactivation.

Figure 15.7 Why is this female calico cat "calico"? In her cells, one X chromosome carries a dominant allele for the brownish-black pigment melanin. The allele on her other X chromosome specifies yellow fur. At an early stage of the cat's embryonic development, one of the two X chromosomes was inactivated at random in each cell that had formed by then.

In all descendants of those cells, the same chromosome also became inactivated, which left them with only one functional allele for the coat-color trait. We see patches of different colors, depending on which allele was inactivated in cells that formed a given tissue region. (The white patches result from a gene interaction involving the "spotting gene," which blocks melanin synthesis entirely.)

being expressed—and patches where genes of the paternal X chromosome are being expressed. She is a "mosaic" for the X chromosomes! As you know, a pair of alleles on two homologous chromosomes might or might not be identical. When they are not, skin and other tissues of female mammals may have different features from one tissue patch to the next. Mary Lyon, a geneticist, was the discoverer of this **mosaic tissue effect** of random X chromosome inactivation.

We see the mosaic tissue effect in human females who are heterozygous for a rare recessive allele that blocks sweat gland formation. Their skin is a mosaic of tissues with and without sweat glands. This is just one symptom of *anhidrotic ectodermal dysplasia*. Where sweat glands are absent, the mutant allele is on the active X chromosome (Figure 15.6*b*). You can see the same effect in female calico cats, which are heterozygous for a coat color allele on their X chromosomes (Figure 15.7).

What's the point of the shutdown? Remember, male and female mammals differ in their sex chromosomes (XY versus XX). However, a subset of genes, mainly on the X chromosome's long arm, must be expressed at the *same* levels in males and females. Otherwise, the individual will not develop properly. Inactivating one

of the X chromosomes in XX embryos is called **dosage compensation**. It is a control mechanism that balances gene expression between sexes, starting with the early stages of development.

The *XIST* gene governs X chromosome inactivation. Early in the development of a female embryo, the *XIST* gene on each X chromosome is transcribed. Its product, a type of RNA molecule, accumulates like paint on each chromosome to inactivate almost all of the other genes. Next, *XIST* gene transcription ends on one of the two X chromosomes as it becomes the target of methylation. The addition of methyl groups to DNA usually blocks access to genes. Most of the additions take place where there are two adjacent cytosine–guanine pairs. Pairings like this are clustered in control elements in the DNA.

Which genes are transcribed at a given time depends on the type of organism, the stage of its growth and development, and the kinds of controls operating at different stages.

Homeotic gene expression during development is a case in point. Its products map out the basic body plan. Another case is X chromosome inactivation, a control mechanism that balances gene expression between sexes.

EXAMPLES OF SIGNALING MECHANISMS

The story that opened this chapter introduced the idea that a great variety of signals influence gene activity. The following examples from animals and plants will give you a sense of their effects at the molecular level.

Hormonal Signals

Hormones, a major category of signaling molecules, can stimulate or inhibit gene activity in target cells. Any cell with receptors for a given hormone is a target. Animal cells secrete hormones into tissue fluid. Most of these molecules are picked up by the bloodstream, which distributes them to cells some distance away.

Some hormones bind to membrane receptors at the target cell surface. Others enter the cell and promote transcription of genes by binding to specific activator proteins. In turn, base sequences in the DNA molecule called enhancers can bind the activators. One way or another, an activator–enhancer complex ends up right next to the promoter for the targeted gene. Because RNA polymerase can bind avidly with it, the complex facilitates rapid gene transcription.

Consider the effect of **ecdysone**, a hormone with key roles in many insect life cycles. Immature forms called larvae grow rapidly during part of the cycles and continually feed on organic matter, such as leaves. Preparing food for digestion requires copious amounts of saliva. In their salivary gland cells, DNA had been replicated again and again. The multiple copies of DNA molecules stayed together in parallel array, forming a **polytene chromosome**. When the ecdysone binds to its receptor on the salivary gland cells, its signal triggers rapid transcription of the many copies of genes in the DNA. Gene regions that respond to this hormonal signal puff out while being transcribed, as shown in Figure 15.8. After

this, translation of mRNA transcripts made from the genes produces the protein components of saliva.

In vertebrates, certain hormones have widespread effects on gene expression because many types of cells have receptors for them. As one example, the pituitary gland secretes somatotropin, or growth hormone. This hormonal signal stimulates synthesis of all the proteins required for cell division and, ultimately, the body's growth. Most cells have receptors for somatotropin.

Other vertebrate hormones only signal specific cells at specific times. Prolactin, secreted from the pituitary gland, is like this. A few days after a female mammal gives birth, prolactin can be detected in her blood. This hormone activates genes in mammary gland cells that have receptors for it. Those genes have responsibility for milk production. Liver cells and heart cells have the same genes but do not have the receptors necessary to respond to signals from prolactin.

We will return to this topic later on. Particularly in Chapter 32, 36, 43, and 44, you will come across elegant examples of hormonal controls drawn from studies of plant and animal reproduction and development.

Sunlight as a Signal

Plant a few seeds from a corn or bean plant in a pot that holds moist, nutrient-rich soil. Next, let the seeds germinate, but keep them in total darkness. After eight days have passed, they will develop into seedlings that are spindly and pale, because they have no chlorophyll (Figure 15.9). Now expose those seedlings to one burst of dim light from a flashlight. Within ten minutes, the seedlings will start to convert their stockpiles of certain molecules to the activated forms of chlorophylls, the light-trapping pigment molecules that donate electrons for the reactions of photosynthesis.

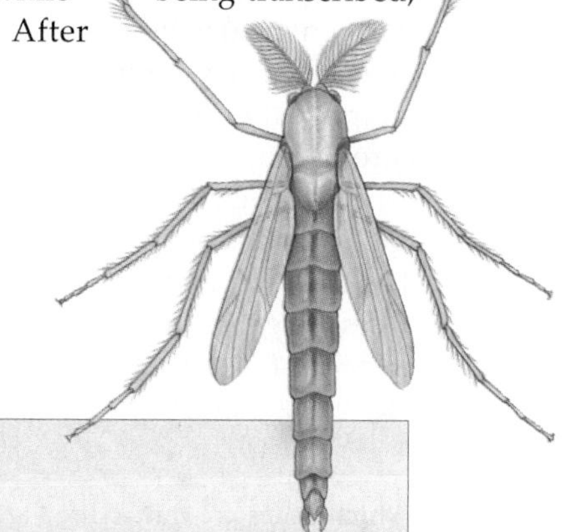

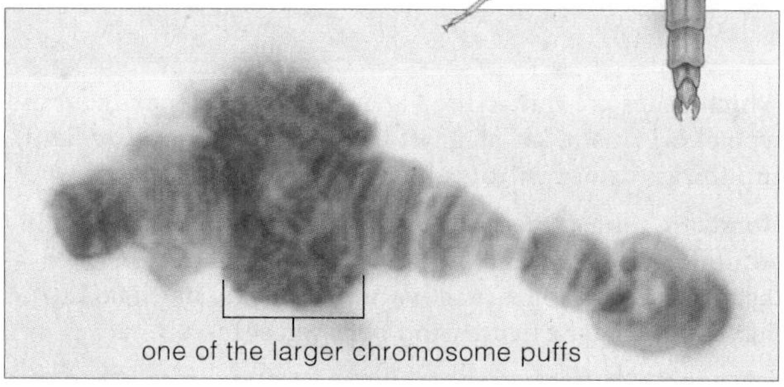

Figure 15.8 Visible evidence of transcription in a polytene chromosome from a larva of a midge (*Chironomus*). Midges are flies. Tiny, short-lived, winged adults often swarm together, which helps them find mates fast. They look like mosquitoes but don't bite.

Most midge larvae develop in aquatic habitats. To sustain their rapid growth, they feed continually, as on decaying organic material. They require a lot of saliva, and they must continuously transcribe genes for saliva's protein components. Those genes have undergone amplification.

Ecdysone, a hormone, serves as a regulatory protein that helps promote their transcription. Midge chromosomes loosen and puff out in regions where the genes are being transcribed in response to the hormonal signal. Puffs are largest and most diffuse where transcription is most intense. Staining techniques reveal banding patterns in the chromosomes, as in the micrograph.

one of the larger chromosome puffs

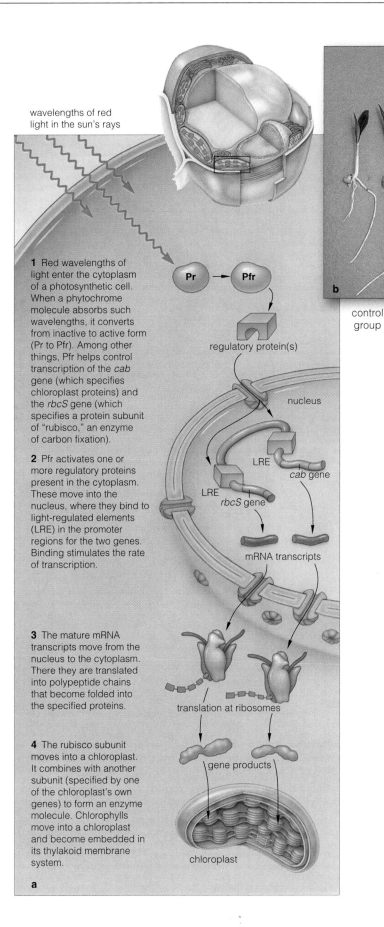

1 Red wavelengths of light enter the cytoplasm of a photosynthetic cell. When a phytochrome molecule absorbs such wavelengths, it converts from inactive to active form (Pr to Pfr). Among other things, Pfr helps control transcription of the *cab* gene (which specifies chloroplast proteins) and the *rbcS* gene (which specifies a protein subunit of "rubisco," an enzyme of carbon fixation).

2 Pfr activates one or more regulatory proteins present in the cytoplasm. These move into the nucleus, where they bind to light-regulated elements (LRE) in the promoter regions for the two genes. Binding stimulates the rate of transcription.

3 The mature mRNA transcripts move from the nucleus to the cytoplasm. There they are translated into polypeptide chains that become folded into the specified proteins.

4 The rubisco subunit moves into a chloroplast. It combines with another subunit (specified by one of the chloroplast's own genes) to form an enzyme molecule. Chlorophylls move into a chloroplast and become embedded in its thylakoid membrane system.

wavelengths of red light in the sun's rays

Pr → Pfr

regulatory protein(s)

nucleus

LRE
cab gene

LRE
rbcS gene

mRNA transcripts

translation at ribosomes

gene products

chloroplast

a

b

control group   experimental group

Figure 15.9 (**a**) Sunlight as a signal for gene expression. One model for the mechanism by which phytochrome might help to control transcription of genes in plants. Red wavelengths can convert the phytochrome molecule from inactive form (here designated Pr) to active form (Pfr). In this form, the phytochrome can serve as a regulator of transcription.

(**b**) Experiment showing the effect of an absence of light on corn seedlings. The two seedlings at *left*, the control group, were grown inside a sunlit greenhouse. The other two seedlings, the experimental group, were grown in total darkness for eight days. The dark-grown plants were not able to convert stockpiled precursors of chlorophyll molecules to active form, and they never did green up.

Phytochrome, a blue-green pigment, is a signaling molecule that helps plants adapt over the short term to changes in light conditions. Section 32.4 takes a close look at this molecule. For now, simply be aware that it alternates between active and inactive forms. At sunset, at night, or in shade, far-red wavelengths predominate. At such times, the phytochrome in cells is inactive. It is activated at sunrise, when red wavelengths dominate the sky. Also, the amount of red or far-red wavelengths that a plant intercepts varies from day to night and as seasons change.

Such variations act as controls over phytochrome activity. That activity influences transcription of certain genes at certain times of day and certain times of year. These genes specify a number of enzymes and other proteins that help seeds to germinate, stems to lengthen and then branch, and leaves to grow. The proteins also help flowers, fruits, and seeds to form.

Elaine Tobin and her coworkers at the University of California, Los Angeles, performed experiments that provided evidence in favor of phytochrome control. Using dark-grown seedlings of duckweed (*Lemna*), they discovered a marked increase in the number of certain mRNA transcripts after a one-minute exposure to red light. Exposure had enhanced transcription of the genes for proteins that bind chlorophylls and for rubisco, an enzyme that mediates carbon fixation (Sections 7.6 and 7.7). In the absence of the proteins, chloroplasts do not develop properly and they don't turn green.

Hormones and other signaling molecules, including diverse kinds that respond to environmental stimuli, have profound influence on gene expression.

Lost Controls and Cancer

THE CELL CYCLE REVISITED   Every second, millions of cells in your skin, bone marrow, gut lining, liver, and elsewhere divide and replace worn-out, dead, and dying predecessors. They don't divide willy-nilly. Mechanisms control the expression of genes that specify enzymes and other proteins required for cell growth, DNA replication, chromosome movements, and cytoplasmic division. And they control when the division machinery is put to rest.

The cell cycle has built-in checkpoints, where proteins monitor chromosome structure and other aspects of the preceding phase of the cycle. They also sense whether or not conditions favor division. The proteins are products of **checkpoint genes**, which are the basis of mechanisms that advance, delay, or block the cell cycle.

For example, the protein products called **kinases** are a class of enzymes that phosphorylate molecules. Some signal that DNA replication is finished and call on the cell to make the transition from interphase to mitosis (Section 9.2). The products called **growth factors** invite transcription of genes with roles in the body's growth. Example: one of these, epidermal growth factor (EGF), activates tyrosine kinase when it binds to target cells in epithelial tissues. Binding is a signal to start mitosis. The product of checkpoint gene *p53* can put the brakes on division when chromosomes are damaged.

ONCOGENES   Many cancers are known to arise through mutations in one or more checkpoint genes. A mutated gene that has the potential to induce cancer falls into the broader category of **oncogenes**.

Oncogenes can form following insertion of viral DNA into a cell's DNA. They also can form after mutagens alter the structure of DNA. As you read in Section 14.4, ultraviolet radiation and nonionizing radiation, such as

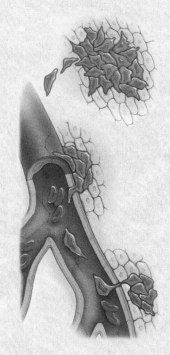

a Cancer cells break away from their home tissue.

b The metastasizing cells become attached to the wall of a blood vessel or lymph vessel. They secrete digestive enzymes onto it. Then they cross the wall at the breach.

c Cancer cells creep or tumble along inside blood vessels, then leave the bloodstream the same way they got in. They start new tumors in new tissues.

Figure 15.11 Steps in metastasis.

x-rays, are mutagens. So are many natural and synthetic compounds, including asbestos and certain substances in tobacco smoke.

And remember those chromosome alterations and gene mutations described in Sections 12.8 and 14.4? Some cancers arise when base substitutions or deletions alter a gene or one of the control elements that deal with its transcription. Other cancers may arise by translocations or by transposons that destabilize gene controls.

CHARACTERISTICS OF CANCER   What happens to a cell when it undergoes cancerous transformation? At the very least, all cancer cells display four characteristics. First, *the plasma membrane and cytoplasm change profoundly.* That membrane becomes more permeable, its proteins become lost or altered, and abnormal proteins form. The cytoskeleton shrinks severely, becomes disorganized, or both (Figure 15.10). Enzyme action shifts, as in amplified reliance on glycolysis. Second, *cancer cells grow and divide abnormally.* Controls against overcrowding in tissues are lost; cell populations reach high densities. The number of small blood vessels that service the growing cell mass increases abnormally. Third, *cancer cells have a weakened capacity for adhesion.* Recognition proteins become lost or altered, so cells cannot stay anchored in proper tissues. They break away and may establish colonies in other, distant tissues. **Metastasis** (meh-TA-stuh-SIS) is the name for the process of abnormal cell migration and tissue invasion (Figure 15.11). Fourth, *cancer cells usually are lethal.* Unless they are eradicated, their uncontrollable divisions put the individual on a painful road to death.

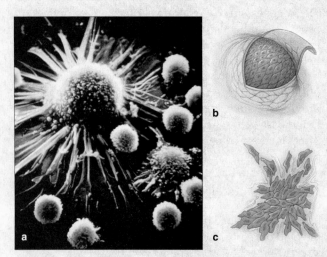

Figure 15.10   (**a**) A patrol of white blood cells surrounding a cancer cell. (**b**) Benign tumor. (**c**) Malignant tumor.

Signal to die docks at receptor.

Signal causes activation of ICE-like proteases.

Figure 15.12 Artist's representation of weapons of cell death being unleashed inside a cell. Normally, cells self-destruct when they finish their functions or become altered in ways that could threaten the body as a whole. When controls over programmed cell death are lost, an altered cell may start dividing abnormally and give rise to cancer.

Cells of common skin moles and other noncancerous, *benign* neoplasms grow abnormally but slowly, and they retain the surface recognition proteins that hold them in their home tissue. Unless benign neoplasms become too large or irritating, doctors usually leave them alone.

Abnormally growing and dividing cells of a *malignant* neoplasm have destructive physical and metabolic effects on the surrounding tissues. These are grossly disfigured, metastasizing cancer cells (Figure 15.11c). They break loose from home tissues, enter lymph or blood vessels, then slip out and invade other tissues where they do not belong. There, they may start growing as new masses.

Each year in the developed countries alone, 15 to 20 percent of all deaths result from cancer. Cancer isn't just a human problem. Researchers have observed cancers in most of the animal species they have studied to date.

HERE'S TO SUICIDAL CELLS! We conclude with a case study of a gene and its cancerous transformation.

The first cell of a new multicelled individual contains marching orders that will guide its descendants along a program of growth, development, and reproduction, and often death. As part of that program, many cells heed calls to self-destruct when they finish their prescribed function. If they become altered in ways that might pose a threat to the body as a whole, as by infection or cancer, they can execute themselves. **Apoptosis** (app-uh-TOE-sis) is the name for this form of cell death. It starts with molecular signals that activate and unleash lethal weapons of self-destruction, which were stockpiled earlier in the cell.

Protein-cleaving enzymes called **ICE-like proteases** are such weapons. Think of them as sheathed Ninja knives. When popped open, they chop apart structural proteins, including the building blocks of cytoskeletal elements and nucleosomes that organize the DNA (Figure 15.12).

A body cell in the act of suicide shrinks away from its neighbors. Its cytoplasm seems to roil as its surface repeatedly bubbles outward and inward. No longer are its chromosomes extended through the nucleoplasm; they bunch up near the nuclear envelope. The nucleus, then the cell, breaks apart. Suicidal cells or remnants of them are swiftly engulfed by phagocytic white blood cells, which patrol and protect the body's tissues.

The timing of cell death is predictable in some cells, such as keratinocytes. These cells form densely packed sheets of dead cells that are continually sloughed off and replaced at the surface of skin. They have a three-week life span, more or less. Keratinocytes and other body cells, even kinds that are supposed to last for a lifetime, can be induced to die ahead of schedule. All it takes is sensitivity to the signals that can activate ICE-like proteases and other enzymes of death.

Control genes suppress or trigger programmed cell death. One of these, *bcl-2*, helps keep normal body cells from dying before their time. The knives stay sheathed in cancer cells, which are supposed to—but do not—commit suicide on cue. Quite possibly, they lost normal receptors that would permit contact with their signaling neighbors. Maybe they are receiving abnormal signals about when to grow, divide, or cease dividing. In some cancers, for instance, the suppressor gene *bcl-2* has been tampered with or shut down. Apoptosis is no longer in the cards.

Cancer is a multistep process involving more than one oncogene. But researchers have already identified many of the mutated forms of checkpoint genes that contribute to it. Remember how one form of leukemia, described in Chapter 12's introduction, is responding to the molecular targeting of one gene's product? More anticancer drugs and gene therapies should soon follow.

SUMMARY   **Gold** indicates text section

1. Cells are equipped with controls that govern gene expression; that is, which gene products appear, when, and in what amounts. *When* control mechanisms come into play depends on cell type, on prevailing chemical conditions, and on signals from other cell types that can change a target cell's activities. *CI*

    a. In all prokaryotic and eukaryotic cells, many of the controls adjust transcription rates in response to concentrations of nutrients and other substances.

    b. In eukaryotic cells of complex organisms, controls contribute to a program of growth and development. They allow a fraction of the same genes to be expressed selectively in different lineages of cells. Selective gene expression is the beginning of diverse specializations in cell structure, composition, and function. We call this process cell differentiation.

2. Regulatory proteins (e.g., activators and inhibitors of transcription), hormones, and some DNA sequences are typical control agents. *15.1–15.4*

    a. Diverse control agents interact with one another, with operators and other control elements built into DNA, with RNA, and with gene products.

    b. Gene expression also is controlled with chemical modifications that inactivate and activate gene regions. X chromosome inactivation is an example.

    c. Control elements called promoters are the binding sites in DNA that signal the start of a gene. Enhancers are binding sites in DNA for some activator proteins.

3. In all cell types, two of the most common types of control systems inhibit or enhance gene transcription. In negative control systems, a regulatory protein binds at a specific DNA sequence to *block* transcription of one or more genes. In positive control systems, a regulatory protein binds to DNA to *promote* transcription. *15.1*

4. Most prokaryotic cells do not have great structural complexity and do not undergo complex development. Most of their gene control systems adjust transcription rates in response to nutrient availability. Control of operons (groupings of functionally related prokaryotic genes and control elements) is an example. *15.2*

5. Compared to prokaryotic cells, eukaryotic cells use more complex gene controls. Gene expression changes in response to external conditions and is subject to long-term controls over growth and development. *15.3–15.5*

    a. X chromosome inactivation in XX embryos of all female mammals is an outcome of interactions between a product of a gene (*XIST*) and control elements in one X chromosome. This interaction is an example of dosage compensation, a control mechanism that maintains a crucial balance in gene expression between the sexes.

    b. Hormones and phytochromes are two examples of signaling molecules that have roles in selective gene expression. Both can affect transcription.

6. Checkpoint gene products advance, delay, or block the cell cycle. They include the protein kinases. Some checkpoint gene mutations lead to cancer. A mutated form of such a gene is an oncogene. *15.6*

7. Cancer involves loss of normal controls over the cell cycle and mechanisms of programmed cell death. In all cancer cells, the structure and function of the plasma membrane and cytoplasm are severely compromised. Cancer cells grow and divide abnormally. Mechanisms for self-destruction are suppressed. Cancer cells have a weakened capacity for adhesion to their home tissue. Unless eradicated, cancers typically can be lethal. *15.6*

Review Questions

1. A plant, fungus, and animal consist of diverse cell types. How might the diversity arise, given that body cells in each of these organisms inherit the same set of genetic instructions? As part of your answer, define cell differentiation and the general way that selective gene expression brings it about. *CI, 15.3*

2. In what fundamental way do negative and positive controls of transcription differ? Is the effect of one or the other form of control (or both) reversible? *15.1*

3. Distinguish between: *15.1, 15.2*
    a. repressor protein and activator protein
    b. promoter and operator
    c. methylation and acetylation

4. Describe one type of control of transcription for the lactose operon in *E. coli*, a prokaryotic cell. *15.2*

5. Using the sketch on the facing page, define three types of gene controls and note the levels at which they take effect. Does the sketch work for both prokaryotic and eukaryotic cells? Why or why not? *15.3*

6. What is a Barr body? Does it appear in the cells of human males, human females, or both? Explain your answer. *15.4*

7. Briefly describe the general characteristics of normal cells. Then explain the difference between a benign tumor and a malignant tumor. *15.6*

Self-Quiz   ANSWERS IN APPENDIX III

1. Cell differentiation _____ .
    a. occurs in all complex multicelled organisms
    b. requires unique genes in different cells
    c. involves selective gene expression
    d. both a and c
    e. all of the above

2. The expression of a given gene depends on the _____ .
    a. type of cell and its functions   c. environmental signals
    b. chemical conditions   d. all of the above

3. Regulatory proteins interact with _____ .
    a. DNA   c. gene products
    b. RNA   d. all of the above

4. A base sequence signaling the start of a gene is a(n) _____ .
    a. promoter   c. enhancer
    b. operator   d. activator protein

5. In prokaryotic cells but not eukaryotic cells, a(n) _____ is a type of base sequence that precedes genes of an operon.
    a. lactose molecule   c. operator
    b. promoter   d. both b and c

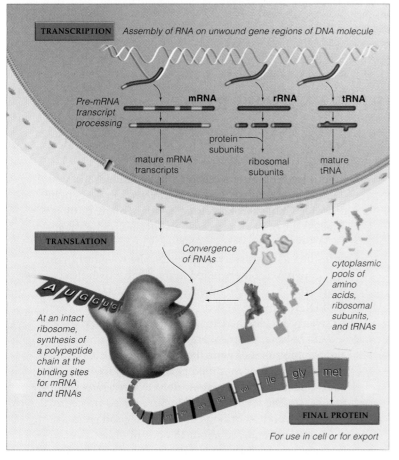

TRANSCRIPTION *Assembly of RNA on unwound gene regions of DNA molecule*

Pre-mRNA
transcript
processing

mRNA   **rRNA**   **tRNA**

protein
subunits

mature mRNA
transcripts

ribosomal
subunits

mature
tRNA

TRANSLATION

*Convergence
of RNAs*

*cytoplasmic
pools of
amino
acids,
ribosomal
subunits,
and tRNAs*

A U G G U

*At an intact
ribosome,
synthesis of
a polypeptide
chain at the
binding sites
for mRNA
and tRNAs*

gln glu cys glu val ile gly met

FINAL PROTEIN

For use in cell or for export

---

transcription of the lactose operon when cells of this bacterial strain are subjected to these conditions:

   a. Lactose and glucose are available.
   b. Lactose is available but glucose is not.
   c. Both lactose and glucose are absent.

2. *Duchenne muscular dystrophy*, a genetic disorder, affects boys almost exclusively. Early in childhood, muscles begin to atrophy (waste away) in affected individuals, who typically die in their teens or early twenties (Section 37.10). Muscle biopsies of a few women who carry an allele that is associated with the disorder reveal some body regions of atrophied muscle tissue. Yet muscles adjacent to a region of atrophy are normal or even larger and more chemically active, as if to compensate for the weakness of the adjoining region. Give a brief explanation of these observations.

3. Unlike most rodents, guinea pigs are well developed at the time of birth. Within a few days, they can eat grass, vegetables, and other plant material. Suppose a breeder decides to separate the baby guinea pigs from their mothers after three weeks. He wants to keep the males and females in different cages. But it is difficult to identify the sex of young guinea pigs. Suggest a test that the breeder can perform to identify their sex.

4. Individuals affected by *pituitary dwarfism* cannot synthesize somatotropin (also called growth hormone). Children with this genetic abnormality will be below the range of normal height. Develop a hypothesis to explain why therapy that is based on somatotropin injections is effective.

5. The closer a mammalian species is to humans in its genetic makeup, the more useful information it may yield in laboratory studies of the mechanisms of cancer. Do you support the use of any mammal for cancer research? Why or why not?

---

6. An operon most typically governs _____ .
  a. bacterial genes   c. genes of all types
  b. a eukaryotic gene   d. DNA replication

7. Eukaryotic genes guide _____ .
  a. fast short-term activities   c. development
  b. overall growth   d. all of the above

8. X chromosome inactivation is an example of _____ .
  a. a chromosome abnormality   c. dysfunctional calico cats
  b. dosage compensation   d. dysfunctional women

9. Hormones may _____ gene transcription in target cells.
  a. promote   c. participate in
  b. inhibit   d. both a and b

10. Apoptosis is _____ .
  a. cell division after severe tissue damage
  b. programmed cell death by suicide
  c. a popping sound in mutated toes

11. ICE-like proteases are _____ .
  a. structural proteins   c. environmental signals
  b. lethal weapons   d. low-temperature enzymes

12. Match the terms with their most suitable descriptions.
  ____ phytochrome   a. helps animal body plan develop
  ____ Barr body   b. mutated form of any gene that
  ____ oncogene   can induce cancer
  ____ homeotic gene   c. hormone with key role in
  ____ ecdysone   insect life cycles
               d. helps plants adapt to daily
                 and seasonal changes in light
               e. inactivated X chromosome

Critical Thinking

1. Geraldo isolated a strain of *E. coli* in which a mutation has affected the capacity of CAP to bind to a region of the lactose operon, as it would do normally. State how the mutation affects

---

Selected Key Terms

acetylation *15.1*
activator *15.2*
apoptosis *15.6*
Barr body *15.4*
cancer *CI*
cell differentiation *15.3*
checkpoint gene *15.6*
dosage compensation *15.4*
ecdysone *15.5*
enhancer *15.1*
gene control *CI*
growth factor *15.6*
homeotic gene *15.4*
hormone *15.5*
ICE-like protease *15.6*

kinase *15.6*
metastasis *15.6*
methylation *15.1*
mosaic tissue effect *15.4*
negative control system *15.1*
neoplasm *CI*
oncogene *15.6*
operator *15.2*
operon *15.2*
phytochrome *15.5*
polytene chromosome *15.5*
positive control system *15.1*
promoter *15.1*
regulatory protein *15.1*
X chromosome inactivation *15.4*

Readings

Duke, R., D. Ojcius, and J. Ding-E Young. December 1996. "Cell Suicide in Health and Disease." *Scientific American*, 80–87.

Murray, A., and M. Kirschner. March 1991. "What Controls the Cell Cycle?" *Scientific American*, 264(3): 56–63.

Tijan, R. February 1995. "Molecular Machines That Control Genes." *Scientific American*, 54–61.

Travis, J. 5 August 2000. "Silence of the Xs." *Science News* 158: 92–94. A look at mechanism of X chromosome inactivation. Raises questions about "junk DNA" (base sequences that have accumulated in DNA yet have no known function).

On-Line readings at Student Guide for InfoTrac:
www.brookscole.com/biology

16

RECOMBINANT DNA AND GENETIC ENGINEERING

Mom, Dad, and Clogged Arteries

Butter! Bacon! Eggs! Ice cream! Cheesecake! Possibly you think of such foods as enticing, off-limits, or both. After all, who among us doesn't know about animal fats and the dreaded cholesterol?

Soon after you feast on such fatty foods, cholesterol enters the bloodstream. Cholesterol is important. It is a structural component of animal cell membranes, and without membranes, there would be no cells. Cells also remodel cholesterol into a variety of molecules, such as the vitamin D necessary for the development of good bones and teeth. Normally, however, the liver itself synthesizes enough cholesterol for your cells.

Some proteins circulating in blood combine with cholesterol and other substances to form lipoprotein particles. *High*-density lipoproteins, or *HDLs*, transport cholesterol to the liver, where it's metabolized. Usually, *low*-density lipoproteins, or *LDLs*, end up inside cells that store or use cholesterol. But sometimes too many LDLs form. The excess infiltrates the elastic walls of arteries and helps form atherosclerotic plaques (Figure 16.1). These abnormal masses interfere with blood flow. If they clog one of the tiny coronary arteries that deliver blood to the heart, a heart attack may result.

How well you handle dietary cholesterol depends on which alleles you got from your parents. For example, one gene specifies a protein receptor for LDLs. If you inherited two "good" alleles and don't go overboard on the fatty foods, your blood level of cholesterol may stay low to moderate and your arteries may never clog up.

Inherit two copies of a certain mutated allele, and you will develop *familial hypercholesterolemia*, a rare genetic disorder. Cholesterol levels get so high, many of those affected die of heart attacks in childhood or their teens.

In 1992 a woman from Quebec, Canada, became a milestone in the history of genetics. She was thirty years old. Like two brothers felled in their early twenties by heart attacks, she had inherited the mutant allele for the LDL receptor. She survived a heart attack at sixteen. At twenty-six, she had coronary bypass surgery.

At the time, people were hotly debating the risks and the promises of **gene therapy**—the transfer of one or more normal or modified genes into an individual's body cells to correct a genetic defect or boost resistance to disease. The woman opted for an untried procedure to give her body working copies of the good gene.

Surgeons removed about 15 percent of her liver. Researchers put cells from it in a nutrient-rich medium to promote cell growth and division. *And they spliced the functional allele for the LDL receptor into the genetic material of a harmless virus.* The modified virus infected the cultured liver cells, thereby inserting copies of the good gene into them. When the liver cells went on to reproduce, they made copies of the good gene, also.

About a billion modified cells were infused into the woman's portal vein, a blood vessel that leads directly to the liver. Some cells took up residence in the liver and started to make the missing cholesterol receptor. Two years later, a fraction of the woman's liver cells were behaving normally and sponging up cholesterol from blood. Her LDL blood levels declined nearly 20 percent. There was no sign of the arterial clogging that nearly killed her. Her cholesterol levels were still higher than normal. Yet the intervention demonstrated the safety and feasibility of gene therapy for some patients.

As you might gather from this pioneering clinical work, recombinant DNA technology has huge potential for medicine, agriculture, and industry. Think of it! For thousands of years, we humans have been changing

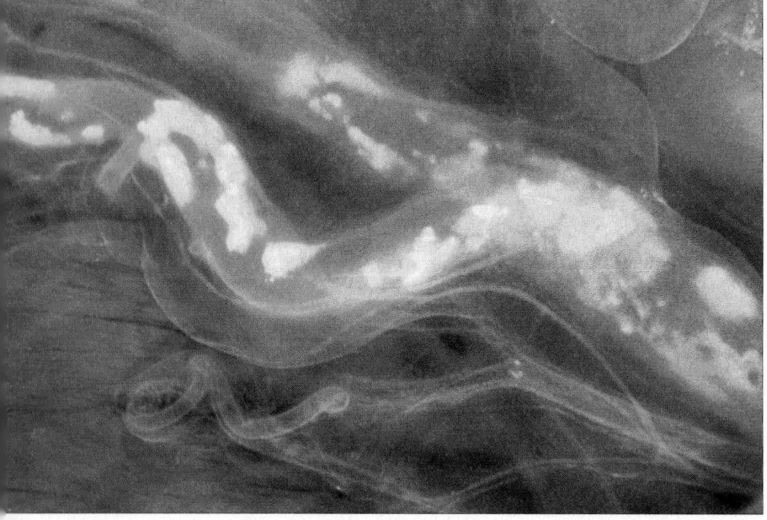

Figure 16.1 Life-threatening plaques (*bright yellow, white*) in one of the coronary arteries. These small arteries deliver blood to the heart. Abnormally high levels of cholesterol contribute to the formation of plaques that may clog them. Gene therapies based on recombinant DNA technology have the potential to counter many such threats to health.

Figure 16.2 One outcome of genetic manipulation by way of artificial selection practices: a large kernel from a modern strain of corn next to the tiny kernels of an ancestral species, which was recovered from a prehistoric cave in Mexico.

genetically based traits. By artificial selection practices, we produced new plants and new breeds of cattle, cats, dogs, and birds from wild ancestral stocks. We were selective agents for meatier turkeys, sweeter oranges, seedless watermelons, spectacular ornamental roses, and big juicy corn kernels (Figure 16.2). We conjured up hybrids such as the mule (horse × donkey) and plants that bear tangelos (tangerine × grapefruit).

Of course, we have to remember we're newcomers on the evolutionary stage. During the 3.8 billion years before we even made our entrance, nature conducted uncountable numbers of genetic experiments. Nature's tools have included mutation, crossing over, and other events that introduce changes in genetic messages. The countless changes gave rise to life's rich diversity.

But the striking thing about human-directed change is that the pace has picked up. Researchers analyze genes with **recombinant DNA technology**. They cut and recombine DNA from different species and insert it into bacterial, yeast, or mammalian cells. The cells replicate their DNA and divide rapidly. They copy the foreign DNA as if it were their own and churn out useful quantities of recombinant DNA molecules. The technology also is the basis of **genetic engineering**. By this process, genes are isolated, modified, and inserted into the same organism or into a different one. Protein products of the modified genes may cover the function of their missing or malfunctioning counterparts.

The new technology does not come without risks. With this chapter, we consider its basic aspects. At the chapter's end, we also address some ecological, social, and ethical questions related to its application.

Key Concepts

1. Genetic experiments have been proceeding for billions of years, through gene mutations, crossing over, genetic recombination, and other natural events.

2. Humans are now purposefully bringing about rapid genetic changes by way of recombinant DNA technology. Such manipulations are called genetic engineering.

3. With this technology, researchers isolate, cut, and splice together gene regions from different species. Then they greatly amplify the number of copies of genes that interest them. The genes, and sometimes the proteins they specify, are produced in quantities that are large enough to use for research and for practical applications.

4. Three activities are at the heart of recombinant DNA technology. First, procedures based on specific types of enzymes are used to cut DNA molecules into fragments. Second, the fragments are inserted into cloning tools, such as plasmids. Third, the fragments containing the genes of interest are identified, then copied rapidly and repeatedly.

5. Genetic engineering involves isolating, modifying, and inserting genes back into the same organism or into a different one. The goal is to beneficially modify traits that the genes influence. Human gene therapy, which focuses on controlling or curing genetic disorders, is an example.

6. The new technology raises social, legal, ecological, and ethical questions regarding its benefits and risks.

A TOOLKIT FOR MAKING RECOMBINANT DNA

Restriction Enzymes

In the 1950s, the scientific community was agog over the discovery of DNA's structure. The excitement gave way to frustration. No one could figure out the sequence of nucleotides, the order of genes and gene regions along a chromosome. Robert Holley and his colleagues did sequence a small tRNA. They used digestive enzymes that broke the molecule into fragments small enough to be chemically characterized. But DNA molecules? These are far longer than RNAs. No one knew how to cut one into fragments long enough to have unique, and thus analyzable, sequences. Digestive enzymes can cut DNA, but not in any particular order. There was no telling how the fragments had been arranged in the molecule.

Then Hamilton Smith discovered that *Haemophilus influenzae* has an enzyme that makes cuts at specific sites in foreign DNA inserted into it by a bacteriophage. Such enzymes, which can restrict bacteriophage growth in a host cell, are now called **restriction enzymes**.

In time, several hundred strains of bacteria offered up a toolkit of restriction enzymes that recognize and cut specific sequences of four to eight bases in DNA. Table 16.1 lists a few. They are among many that make staggered cuts, which leave single-stranded "tails" on the end of a DNA fragment. Tails made by *Taq*I are two bases long (CG). *Eco*RI makes tails four bases long (AATT).

How many cuts do restriction enzymes make? That depends in part on the molecule. *Not*I only recognizes a rare eight-base sequence in the DNA of mammals, so when it makes its cuts, most of the fragments are tens of thousands of base pairs long. That is long enough for studying the **genome**—all of the DNA in a haploid number of chromosomes for a species. For the human species, the genome is about 3.2 billion base pairs long.

Table 16.1	*Examples of Restriction Enzymes*	
Bacterial Source	Enzyme's Abbreviation	Its Specific Cut
Thermus aquaticus	*Taq*I	5' T C G A 3' / 3' A G C T 5'
Escherichia coli	*Eco*RI	5' G A A T T C 3' / 3' C T T A A G 5'
Nocardia otitidus caviarum	*Not*I	5' G C G G C C G C 3' / 3' C G C C G G C G 5'

Modification Enzymes

DNA fragments with staggered cuts have sticky ends. "Sticky" means a restriction fragment's single-stranded tail can base-pair with a complementary tail of any other DNA fragment or molecule cut by the same restriction enzyme. *Mix DNA fragments cut by the same restriction enzyme, and the sticky ends of any two fragments having complementary base sequences will base-pair and form a recombinant DNA molecule:*

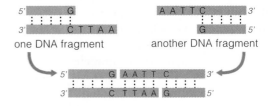

Notice the nicks where such fragments base-pair. **DNA ligase**, a modification enzyme, can seal these nicks:

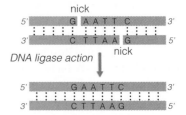

Cloning Vectors for Amplifying DNA

Restriction and modification enzymes make it possible to insert foreign DNA into bacterial cells. As you know, each bacterial cell has only one chromosome, a circular DNA molecule. Many also inherit plasmids. A **plasmid** is a very small circle of extra DNA that has just a few genes and that gets replicated along with the bacterial chromosome (Figure 16.3). Bacteria usually can survive without plasmids, but some of the genes offer benefits, as when they confer resistance to antibiotics.

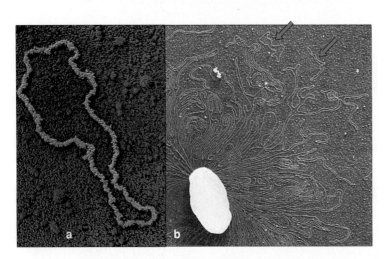

Figure 16.3 (**a**) A single plasmid. (**b**) Plasmids (*arrows*) from a ruptured *Escherichia coli* cell.

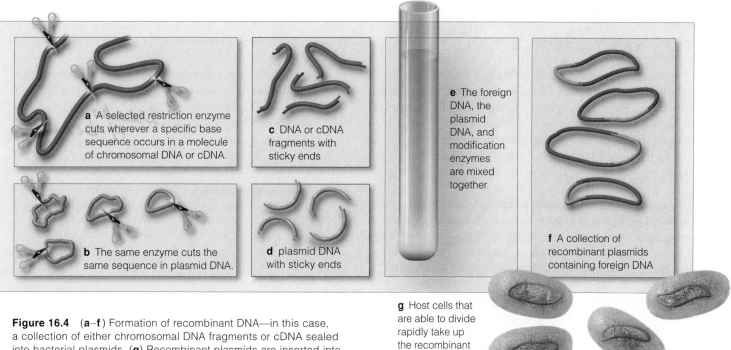

a A selected restriction enzyme cuts wherever a specific base sequence occurs in a molecule of chromosomal DNA or cDNA.

b The same enzyme cuts the same sequence in plasmid DNA.

c DNA or cDNA fragments with sticky ends

d plasmid DNA with sticky ends

e The foreign DNA, the plasmid DNA, and modification enzymes are mixed together

f A collection of recombinant plasmids containing foreign DNA

g Host cells that are able to divide rapidly take up the recombinant plasmids

Figure 16.4 (**a–f**) Formation of recombinant DNA—in this case, a collection of either chromosomal DNA fragments or cDNA sealed into bacterial plasmids. (**g**) Recombinant plasmids are inserted into host cells that can rapidly amplify the foreign DNA of interest.

Under favorable conditions, bacteria divide rapidly and often—every thirty minutes, for some species—so that huge populations of genetically identical cells form. Before cell division, replication enzymes duplicate the bacterial chromosome. They also replicate the plasmid, sometimes repeatedly, so a cell can hold many identical copies of foreign DNA. In research laboratories, foreign DNA typically is inserted into a plasmid for replication. The outcome is called a **DNA clone**, because bacterial cells have made many identical, "cloned" copies of it.

A modified plasmid that accepts foreign DNA is a **cloning vector**. It can insert foreign DNA into a host bacterium, yeast, or some other cell that can be the start of a "cloning factory." It may give rise to a population of rapidly dividing descendant cells, all with identical copies of the foreign DNA (Figure 16.4).

Reverse Transcriptase To Make cDNA

Researchers analyze genes to unlock secrets about gene products and how they are put to use. Yet even when a host cell takes up a gene, it may not be able to make the protein. For instance, most eukaryotic genes have introns (noncoding sequences). mRNA transcripts of the genes can't be translated until the introns are snipped out and coding regions (exons) spliced together. Bacterial cells can't snip out the introns, so they often can't translate human genes into proteins.

Researchers get around this problem with **cDNA**, a DNA strand "copied" from a mature mRNA transcript. **Reverse transcriptase** catalyzes transcription in reverse. This enzyme from RNA viruses builds a complementary DNA strand on an mRNA transcript (Figure 16.5). Then

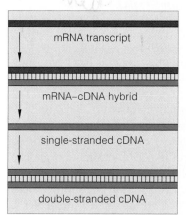

mRNA transcript

mRNA–cDNA hybrid

single-stranded cDNA

double-stranded cDNA

a Reverse transcriptase catalyzes the assembly of a single DNA strand on a mature mRNA transcript. The result is an mRNA–cDNA hybrid molecule.

b Enzymes remove the mRNA and use the cDNA strand as a template to assemble another DNA strand. The result is double-stranded cDNA, "copied" from an mRNA template.

Figure 16.5 Formation of cDNA from an mRNA transcript.

other enzymes remove the RNA from the mRNA–cDNA molecule and substitute a complementary DNA strand. The outcome is double-stranded cDNA.

Double-stranded cDNA can be further modified, as by attaching signals for transcription and translation. The modified cDNA can be inserted into a plasmid for amplification. Bacterial cells exposed to the recombinant plasmids often take them up and then use the cDNA instructions for synthesizing a protein of interest.

Restriction enzymes and modification enzymes cut apart chromosomal DNA or cDNA and splice it into plasmids and other cloning vectors. Recombinant plasmids can be taken up by bacteria or other cells that divide rapidly, thus making multiple, identical quantities of the foreign DNA.

16.2

PCR—A FASTER WAY TO AMPLIFY DNA

The polymerase chain reaction, widely known as **PCR**, is another way to amplify fragments of chromosomal DNA or cDNA. These copy-making reactions proceed with astounding speed in test tubes, not in bacterial cloning factories. Primers get them going.

What Are Primers?

Primers are synthetic, short nucleotide sequences, ten to thirty or so nucleotides long, that can base-pair with complementary sequences in DNA. The workhorses of DNA replication—the DNA polymerases—chemically recognize primers as START tags. Following a computer program, machines synthesize a primer one step at a time. How do researchers decide on the order of these nucleotides? First, they must identify short nucleotide sequences located just before and just after the DNA region from a cell that interests them. Then they build primers that have the complementary sequences.

What Are the Reaction Steps? *thermostable*

For PCR, the enzyme of choice is a DNA polymerase extracted from *Thermus aquaticus*, a bacterium that lives in superheated water of hot springs. It has been found in water heaters, also. This enzyme is not destroyed at the elevated temperatures required to unwind a DNA double helix. Most DNA polymerases are denatured and permanently lose their activity at such temperatures.

Researchers mix together primers, the molecules of DNA polymerase and cellular DNA from an organism, and free nucleotides. Next, they expose the mixture to precise cycles of temperature. At the beginning of each temperature cycle, the two strands of all molecules of DNA in the mixture unwind from each other.

Primers become positioned on exposed nucleotides at targeted sites according to base-pairing rules (Figure 16.6). Each round of reactions doubles the number of DNA molecules amplified from the targeted site. Thus, if there are 10 such molecules in a test tube, there soon will be 20, then 40, 80, 160, 320, 640, 1,280, and so on. Any targeted region from a single DNA molecule can be rapidly amplified into billions of molecules.

In short, *PCR amplifies samples that contain even tiny amounts of DNA.* At this writing, it is the amplification procedure of choice in thousands of laboratories all over the world. As you'll see in the next section, such samples can be obtained even from a single hair follicle or drop of blood left at the scene of a crime.

PCR is a method of amplifying chromosomal DNA or cDNA inside test tubes. Compared to cloning methods, PCR is far more rapid.

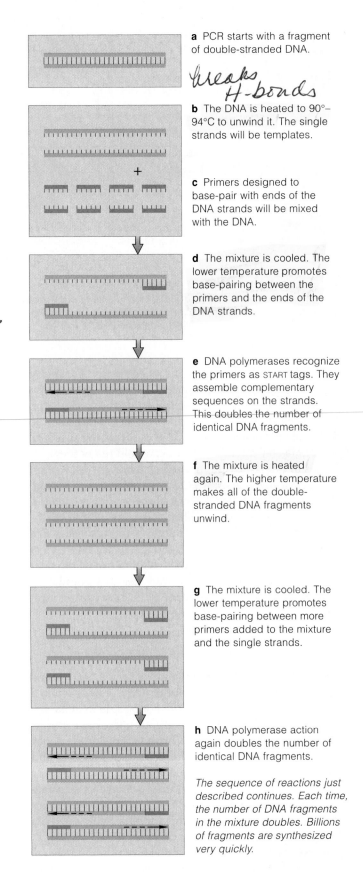

a PCR starts with a fragment of double-stranded DNA.

breaks H-bonds

b The DNA is heated to 90°–94°C to unwind it. The single strands will be templates.

c Primers designed to base-pair with ends of the DNA strands will be mixed with the DNA.

d The mixture is cooled. The lower temperature promotes base-pairing between the primers and the ends of the DNA strands.

e DNA polymerases recognize the primers as START tags. They assemble complementary sequences on the strands. This doubles the number of identical DNA fragments.

f The mixture is heated again. The higher temperature makes all of the double-stranded DNA fragments unwind.

g The mixture is cooled. The lower temperature promotes base-pairing between more primers added to the mixture and the single strands.

h DNA polymerase action again doubles the number of identical DNA fragments.

The sequence of reactions just described continues. Each time, the number of DNA fragments in the mixture doubles. Billions of fragments are synthesized very quickly.

Figure 16.6 The polymerase chain reaction (PCR).

DNA Fingerprints

Except for identical twins, no two people have exactly the same base sequence in their DNA. By detecting the differences in sequences, scientists can distinguish one person from another. As you know, each human has a unique set of fingerprints, a marker of identity. Like all other sexually reproducing species, each human also has a **DNA fingerprint**, a unique array of DNA sequences that were inherited from parents in a Mendelian pattern. DNA fingerprints are so accurate that they can reveal differences even between full siblings.

More than 99 percent of the DNA in all humans is exactly the same. However, DNA fingerprinting focuses only on the portion that tends to be highly variable from one person to the next. Throughout the human genome are **tandem repeats**—short DNA sequences present in many copies, one after the other, in a chromosome.

For example, one individual's DNA might have the five bases TTTTC repeated four times in one location, yet another individual might have them repeated fifteen times in the same location. One might have five repeats of the bases CGG, and another might have fifty of them.

Repetitive DNA sequences have increased or decreased in number over many generations. They slipped into the DNA as spontaneous changes while the DNA was being replicated. In these regions, the mutation rate is relatively high (Section 14.4).

Researchers detect such differences at tandem-repeat sites with **gel electrophoresis**. This laboratory technique uses an electric field to force molecules through a viscous gel. In this case, it separates DNA fragments according to length. Size alone dictates how far each fragment will move through the gel. Tandem repeats of different sizes migrate at different rates.

A gel is immersed in a buffered solution, then DNA fragments from individuals are added to the gel. When an electric current is applied to the solution, one end of the gel takes on a negative charge, and the other end a positive charge. With their negatively charged phosphate groups, DNA fragments migrate through the gel toward the positively charged pole. They do so at different rates and separate into bands according to length. The smaller the fragment is, the faster it migrates. After a set time, researchers identify fragments of different lengths by staining the gel or specifically highlighting fragments that contain tandem repeats.

DNA fingerprints help forensic scientists identify criminals, victims, and innocent suspects. A few drops of blood, semen, or cells from a hair follicle at a crime scene or on a suspect's clothing often yield enough to do the trick. For example, Figure 16.7 shows some tandem-repeat DNA fragments separated by gel electrophoresis. They are DNA fingerprints from seven people and from blood collected at a crime scene. Notice how much all but two of the DNA fingerprints differ. Can you identify

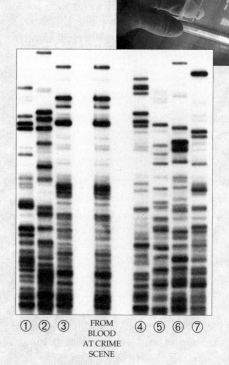

Figure 16.7 *Left:* Comparison of DNA fingerprints from a bloodstain discovered at a crime scene and from blood samples of seven suspects (the circled numbers). Given what you have learned about gel electrophoresis, point out which of the seven is a match. *Above:* The gel shown in this photograph had been stained earlier with a substance that can make DNA fragments fluoresce when placed under ultraviolet light.

① ② ③   FROM BLOOD AT CRIME SCENE   ④ ⑤ ⑥ ⑦

which pattern exactly matches the pattern from the blood collected at the crime scene?

DNA fingerprint analysis even confirmed that human bones exhumed from a shallow pit in Siberia belonged to five members of the Russian imperial family, all shot to death in secrecy in 1918.

When DNA fingerprinting was first used as evidence in court, attorneys challenged conclusions based on it. But DNA fingerprinting is now firmly established as an accurate, unambiguous procedure. Today it is routinely submitted as evidence in disputes over paternity. It is widely used to convict the guilty and exonerate innocent suspects. To give an early example, in 1998, an eleven-year-old girl was raped, stabbed, and strangled to death in Elisabethfehn, Germany. In the largest mass screening yet conducted, the police requested blood samples from 16,400 males in and around the vicinity. A single DNA fingerprint from a thirty-year-old mechanic matched the evidence, and he confessed to the crime.

The variation in tandem repeats also can be detected as restriction fragment length polymorphisms, or RFLPs. (These are DNA fragments of different sizes, cleaved by restriction enzymes.) In the case of tandem repeats, a restriction enzyme cleaves the DNA flanking the repeat. Alternatively, researchers might use PCR to amplify the tandem-repeat region. Either way, differences in the size of the fragments, which reveal genetic differences, can be detected with electrophoresis.

HOW IS DNA SEQUENCED?

In 1995, researchers accomplished what was little more than a dream a few decades ago. They determined the full DNA sequence for a species: *Haemophilus influenzae*, a bacterium that causes human respiratory infections. Since then, the genomes of other species have been fully sequenced. A draft sequence of the human genome has now been completed.

The molecular sleuths are using **automated DNA sequencing**. This laboratory method gives the sequence of cloned DNA or PCR-amplified DNA in a few hours.

Researchers use the four standard nucleotides (T, C, A, and G) for automated DNA sequencing. They also use four modified versions, which we can represent as T*, C*, A*, and G*. Each modified version is labeled with a molecule that fluoresces a certain color when it passes through a laser beam. Every time one of these modified nucleotides gets incorporated into a growing DNA strand, it arrests DNA synthesis.

Before the reactions, researchers mix the eight kinds of nucleotides together. They add millions of copies of the DNA to be sequenced along with a type of primer and molecules of DNA polymerase. Then they separate the DNA into single strands, and the reactions begin.

The primer binds with its complementary sequence on one of the strands. DNA polymerase synthesizes a new DNA strand starting at one end of the primer. One by one, it adds nucleotides in the order dictated by the exposed sequence in the template strand. Each time, one of the standard nucleotides *or* one of the modified versions may be attached.

Suppose that DNA polymerase encounters a T in a DNA template strand. It will catalyze the base-pairing of either A or A* to it. If A is added to the new strand, replication will continue. But if A* is added, replication will stop; the modified nucleotide will *block* addition of any more nucleotides to that strand. The same thing happens at each nucleotide in a template strand. When a standard nucleotide is attached, replication proceeds; when a modified version is attached, replication stops.

Remember, the starting mixture contained millions of identical copies of the DNA sequence. Because either a standard or a modified nucleotide could be added at every exposed base, the new strands end at different locations in the sequence. The mixture now contains millions of copies of tagged fragments having different lengths. These can now be separated by length into *sets* of fragments. And each set corresponds to only one of the nucleotides in the entire base sequence.

The automated DNA sequencer is a machine that separates the sets of fragments by gel electrophoresis. The set having the shortest fragments migrates fastest through the gel and reaches the end of it first. The last set to reach the end has the longest fragments. Because the fragments have a modified nucleotide at the 3' end,

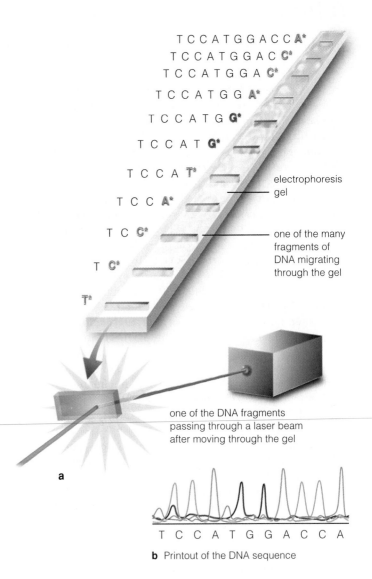

electrophoresis gel

one of the many fragments of DNA migrating through the gel

one of the DNA fragments passing through a laser beam after moving through the gel

a

T C C A T G G A C C A

b Printout of the DNA sequence

Figure 16.8 Automated DNA sequencing. (**a**) DNA fragments from an organism's genome become labeled at their end with a modified nucleotide that fluoresces a certain color. (**b**) Printout of the DNA sequence used in this example. Each peak indicates absorbance by a particular labeled nucleotide.

each set fluoresces a certain color as it passes through a laser beam (Figure 16.8*a*). The automated sequencer detects the color and indicates which nucleotide is on the end of the fragments in each set. It assembles the information from all nucleotides in the sample, and in this way it reveals the entire DNA sequence.

Figure 16.8*b* is a printout from an automated DNA sequencer. Each peak along that tracing represents the detection of a particular color as the sets of fragments reached the end of the gel.

With automated DNA sequencing, the order of nucleotides in a cloned or amplified DNA fragment can be determined.

FROM HAYSTACKS TO NEEDLES—ISOLATING GENES OF INTEREST

Any genome consists of thousands of genes. *E. coli* has 4,279, for instance, and humans apparently have about 30,000. What if you wanted to learn about or modify the structure of any one of those genes? First, you'd have to isolate that gene from all others in the genome.

There are several ways to do this. If part of the gene sequence is already known, you can design primers to amplify the entire gene or part of it by PCR. Often, though, researchers must isolate and clone a gene. First they make a **gene library**. This is a mixed collection of bacteria that contain different cloned DNA fragments. One is the gene of interest. A *genomic* library contains cloned DNA fragments from an entire genome. A *cDNA* library contains DNA derived from mRNA. Often it is the most useful because it is free of introns. But the gene is still hidden like a needle in a haystack. How can you isolate it? One way is to use a nucleic acid probe.

What Are Probes?

A **probe** is a very short stretch of DNA labeled with a radioisotope so that it can be distinguished from other DNA molecules in a given sample. Part of the probe must be able to base-pair with some portion of the gene of interest. Any base-pairing that takes place between sequences of DNA (or RNA) from different sources is called **nucleic acid hybridization**.

How do you acquire a suitable probe? Sometimes part of the gene or a closely related gene has already been cloned, in which case it can be used as the probe. If the gene's structure is a mystery, you still may be able to work backward from the amino acid sequence of its protein product, assuming that protein is already available. By using the genetic code as a guide (Section 14.2), you could build a DNA probe that is more or less similar to the gene of interest.

Screening for Genes

Once you have a gene library and a suitable probe, you are ready to hunt down the gene. Figure 16.9 shows the steps of one isolation method. The first step is to take bacterial cells of the library and spread them apart on the surface of a gelled growth medium in a petri plate. When spread out sufficiently, individual cells undergo division. Each cell starts a colony of genetically identical cells. The bacterial colonies appear as hundreds of tiny white spots on the surface of a culture medium.

After colonies appear, you lay a nylon filter on top of the colonies. Some cells stick to the filter at locations that mirror the locations of the original colonies. You use solutions to rupture the cells, and the DNA so released sticks to the filter. You denature the DNA to single strands and then add the probes. The probes hybridize

a Bacterial colonies, each derived from a single cell, grow on a culture plate. Each colony is about 1 millimeter across.

b A nitrocellulose or nylon filter is placed on the plate. Some cells of each colony adhere to it. The filter mirrors how the colonies are distributed on the culture plate.

c The filter is lifted off and put into a solution. The cells stuck to it rupture; the cellular DNA sticks to the filter.

d Also, the DNA is denatured to single strands at each site. A radioactively labeled probe is added to the filter. It binds to DNA fragments that have a complementary base sequence.

e The probe's location is identified by exposing the filter to x-ray film. The image that forms on the film reveals the colony that has the gene of interest.

Figure 16.9 Use of a probe to identify bacterial colonies that have taken up a gene library.

only with DNA from the colony that took up the gene of interest. If you expose the probe-hybridized DNA to x-ray film, the pattern formed by the radioactivity will identify that colony. With this information you can now culture cells from that colony alone, knowing that it will be the only one that can replicate the cloned gene.

Probes may be used to identify one particular gene among many in gene libraries. Bacterial colonies that have taken up the library can be cultured to isolate the gene.

USING THE GENETIC SCRIPTS

As researchers decoded the genetic scripts of species, they opened the door to astonishing possibilities. For example, genetically engineered bacteria now produce medically valued proteins. Huge bacterial populations produce useful quantities of the desired gene products in stainless steel vats. *Diabetics* who must receive insulin injections every day are among the beneficiaries. At one time, medical supplies of this pancreatic hormone were extracted only from pigs and cattle. Later on, synthetic genes for human insulin were transferred into *E. coli* cells. The cells were the start of populations that became the first large-scale, cost-effective bacterial factory for proteins. Besides insulin, human somatotropin, blood-clotting factors, hemoglobin, interferon, and a variety of drugs and vaccines are also manufactured with the help of genetically engineered bacteria.

Other kinds of modified bacteria hold potential for industry and for cleaning up environmental messes. For instance, as you know, many microorganisms can break down organic wastes and help cycle nutrients through ecosystems. Certain modified bacteria can break down crude oil into less harmful compounds. When sprayed onto oil spills, as from a shipwrecked supertanker, they might help avert an environmental disaster. Others are genetically engineered to sponge up excess phosphates or heavy metals from the environment.

In addition, bacterial species that contain plasmids offer benefits for basic research, agriculture, and gene therapy. For instance, deciphering the messages encoded in bacterial genes helps us reconstruct the evolutionary history of life. Sections 19.6 and 19.7 will give you an idea of how comparisons of DNA or RNA from different organisms can reveal evolutionary secrets. The next two sections give a few more examples of benefits.

What about the "bad" bunch—the pathogenic fungi, bacteria, and viruses? Natural selection favors mutated genes that improve a pathogen's chances of evading a host organism's natural defenses. Mutation is frequent among pathogens that reproduce rapidly. So designing new antibiotics and other defenses against new gene products is a constant challenge. But if we learn about the genes, we may have advance warning of the "plan of attack." The story of HIV, a rapidly mutating virus that causes AIDS, gives insight into the magnitude of the problem (Section 39.10).

Knowing about genes allows us to genetically engineer beneficial microorganisms for uses in medicine, industry, agriculture, and environmental remediation.

Knowing about genes can help us reconstruct evolutionary histories of species.

Knowing about genes may help us design effective, timely counterattacks against rapidly mutating pathogens.

DESIGNER PLANTS

Regenerating Plants From Cultured Cells

Many years ago, Frederick Steward and his coworkers cultured cells from carrot plants. They induced some of them to develop into small embryos. As you will read in Section 31.7, some embryos grew into whole plants. Today, researchers routinely regenerate crop plants and many other plant species from cultured cells. They use ingenious methods to pinpoint a gene in a culture that may contain even millions of cells. Suppose they add a pathogen's toxin to a culture medium. Suppose a few cells happen to carry a gene that confers resistance to the toxin. Those cells will be the only ones to survive.

Whole plants regenerated from preselected cultured cells can be hybridized with other varieties of plants. In this way, desired genes might be transferred to a plant lineage to improve herbicide resistance, pest resistance, and other traits that can benefit crops and gardens.

Consider this larger aspect of the research: The food supply for most of the human population is vulnerable. Generally, farmers prefer genetically similar varieties of high-yield plants and have discarded the more diverse, older varieties. But genetic uniformity makes food crops vulnerable to many kinds of pathogenic fungi, viruses, and bacteria. That is why botanists comb the world for seeds of the wild ancestors of potatoes, corn, and other crop plants. They send their prizes—seeds with genes of a plant's lineage—to **seed banks**. These safe storage facilities are designed to preserve genetic diversity.

The problem is huge. Example: In 1970 a new fungal strain of *Southern corn leaf blight* destroyed much of the United States corn crop. All of the plants carried the same gene that conferred susceptibility to the disease. Ever since that devastating epidemic, seed companies have been much more attentive to offering genetically diverse corn seeds. And where does the diversity come from? Plant breeders call on the seed banks.

How Are Genes Transferred Into Plants?

The Ti (Tumor-inducing) plasmid from *Agrobacterium tumefaciens* is one vehicle for inserting new or modified genes into plants. *A. tumefaciens* infects many kinds of flowering plants (Figure 16.10). Some Ti plasmid genes

crown gall tumor

Figure 16.10
Crown gall tumor on a woody plant, an abnormal tissue growth triggered by a gene in the Ti plasmid.

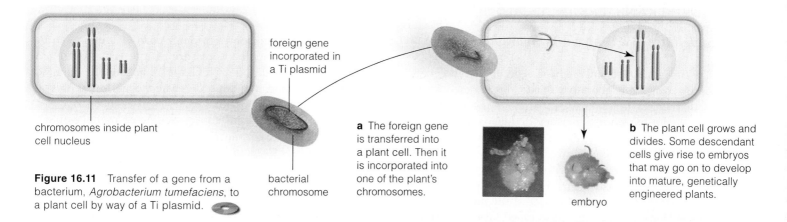

Figure 16.11 Transfer of a gene from a bacterium, *Agrobacterium tumefaciens*, to a plant cell by way of a Ti plasmid.

chromosomes inside plant cell nucleus

foreign gene incorporated in a Ti plasmid

bacterial chromosome

a The foreign gene is transferred into a plant cell. Then it is incorporated into one of the plant's chromosomes.

b The plant cell grows and divides. Some descendant cells give rise to embryos that may go on to develop into mature, genetically engineered plants.

embryo

invade a plant's DNA, then induce the formation of abnormal tissue masses. We call the masses crown gall tumors. Before introducing genes from a plasmid into plant cells, researchers remove the tumor-inducing genes and then insert a desired gene into it. They place plant cells in a culture of the modified bacteria. Some cells may take up the gene, and whole plants may be regenerated from the cellular descendants, as in Figure 16.11. Expression of foreign genes in plants sometimes is dramatic, as in Figure 16.12.

In nature, *A. tumefaciens* infects plants called dicots, which include beans, peas, potatoes and other vital crops. Geneticists have modified the bacterium so that it also delivers genes into monocots that are vital food crops, including wheat, corn, and rice. Some researchers use electric shocks or chemicals to deliver modified genes into plant cells. Some also blast microscopic particles coated with DNA into them.

Despite many obstacles, improved varieties of crop plants have been engineered or are in the works. For example, genetically engineered cotton plants display resistance to a herbicide (Figure 16.13). Farmers spray the herbicide in fields of modified cotton plants to kill weeds. The plants are not affected.

Also on the horizon are engineered plants that can be factories for pharmaceuticals. A few years ago, for example, tobacco plants that were engineered to produce hemoglobin and some other proteins were planted in a test field in North Carolina. Afterward, ecologists found

Figure 16.12 A modified plant that glows in the dark. A firefly gene was inserted into the plant's DNA and is being expressed. (Refer to Section 6.9). The gene's product is luciferase, an enzyme with a role in bioluminescence.

Figure 16.13 (**a**) Control plant (*left*) and three genetically engineered aspen seedlings. Vincent Chiang and coworkers suppressed a regulatory gene involved in a lignin biosynthetic pathway. The modified plants synthesized normal lignin, but not as much. Lignin production decreased by as much as 45 percent—yet cellulose production increased 15 percent. Root, stem, and leaf growth were greatly enhanced. Plant structure did not suffer. Wood harvested from such trees might make it easier to manufacture paper and some clean-burning fuels, such as ethanol. (Lignin, a tough polymer, strengthens the secondary cell walls of plants. Before paper can be made from wood, the lignin must be chemically extracted.)

(**b**) *Left:* Cotton plant, used as the control. *Right:* Genetically engineered cotton plant with a gene for herbicide resistance. Both plants were sprayed with a weedkiller that currently has widespread application in cotton fields.

no trace of foreign genes or proteins in the soil or in other plants or animals in the neighborhood. Another example: In a Stanford University laboratory, mustard plants synthesized biodegradable plastic beads that are suitable for use in manufacturing plastics.

Genetically diverse plant species are essential to protecting our vulnerable food supply. Genetic engineers can design plants with new beneficial traits.

GENE TRANSFERS IN ANIMALS

Supermice and Biotech Barnyards

The first mammals enlisted for experiments in genetic engineering were laboratory mice. Consider an example of this work. R. Hammer, R. Palmiter, and R. Brinster corrected a hormone deficiency that leads to dwarfism in mice. Insufficient levels of somatotropin (or growth hormone) cause the abnormality. The researchers used a microneedle to inject the gene for rat somatotropin into fertilized mouse eggs, which they implanted in an adult female. The gene was successfully integrated into mouse DNA, and the eggs developed into mice. Young mice in which that foreign gene was expressed grew 1-1/2 times larger than their dwarf littermates. In other experiments, researchers transferred the gene for human somatotropin into a mouse embryo. The gene became integrated into mouse DNA, and the modified embryo grew up to be a "supermouse" (Figure 16.14).

Today, human gene transfers are being attempted in research into the molecular basis of genetic disorders. "Biotech barnyards" have animals that compete with bacterial factories as genetically engineered sources of proteins. Goats produce CFTR protein (to treat cystic fibrosis) and TPA (to counter effects of a heart attack). Cattle may soon be producing human collagen, which can help repair cartilage, bone, and skin.

Remember the first case of cloning a mammal from an adult cell? A nucleus extracted from a mammary gland cell of a ewe was inserted into an enucleated egg. Signals from the egg cytoplasm sparked development of an embryo, which was implanted into a surrogate mother. This resulted in *Dolly* (Section 13.5).

For years, researchers had been cloning animals from embryonic tissue, but not adults that already displayed some sought-after trait. Now, however, if they could just refine the steps of their cloning processes, they might be able to maintain some genotype indefinitely.

Not long after Dolly's debut, genetically engineered clones of mice, cattle, and other animals were produced. Example: Steve Stice and his colleagues produced designer cattle. They started with a culture of cells from cattle. They induced the development of six genetically identical calves. They also engineered targeted changes in the cloning cell lineage.

Stice would like to genetically engineer cattle resistant to bovine spongiform encephalopathy (mad cow disease, Section 21.8). His method also has been used to put the human serum albumin gene

into the chromosomes of dairy cows. Albumin can be used to control blood pressure. At present, this protein must be separated from large quantities of donated human blood. It would be easier to get quantities of the protein from bountiful supplies of milk.

Given how rapidly the technologies are developing, is the genetic engineering and cloning of humans not far behind? We return to this thought in Section 16.10.

Mapping and Using the Human Genome

In the late 1980s, biologists were in an uproar over a costly proposal before the National Institutes of Health (NIH) to map the entire human genome. Many argued that benefits for medicine and pure research would be incalculable. Others said the mapping couldn't be done and would divert funding from "worthy" endeavors. But then the molecular biologist Leroy Hood invented automated DNA sequencing, and the race was on.

By 1990, the Human Genome Initiative was under way. The international effort came with a projected price tag of 3 billion dollars—about 92 cents for each base pair in the heritable script for human life. By 1991, not even 2,000 genes had been sequenced. The pace picked up after Craig Venter realized even a bit of cDNA could be used as a molecular hook to drag the whole sequence of its parent gene out of a cDNA library. He named the hooks ESTs (for *Expressed Sequence Tags*).

Later, Venter and Hamilton Smith used a software program, the TIGR Assembler, to decipher the first full genome of an organism—the bacterium *H. influenzae*. In 2000, researchers of the Human Genome Initiative along with Venter's team jointly announced that they had completed a rough draft of the human genome. Its extensive noncoding portions are now being analyzed.

Today, studies of the genome of humans and other organisms are now grouped together as a new research

Figure 16.14 Evidence of a successful gene transfer. Two ten-week-old mouse littermates. *Left:* This one weighs 29 grams. *Right:* This one weighs 44 grams. It grew from a fertilized egg into which a gene for human somatotropin had been inserted.

field: **genomics**. The branch called *structural* genomics is concerned with actual mapping and sequencing of genomes of individuals. The branch called *comparative* genomics is more concerned with possible evolutionary relationships of groups of organisms. The similarities and differences being discovered among the different genomes are analyzed.

Comparative genomics has practical applications as well as potential for research. The starting premise is that the genomes of all existing organisms are derived from common ancestral ones. For example, pathogens share some conserved genes with their human hosts, although their ancestors diverged long ago from the lineage that led to humans. By comparing the shared gene sequences, how those sequences are organized, and where they have come to differ, it may be possible to advance understanding of where immune defenses against pathogens are strongest and where they are the most vulnerable. In such ways, it is helpful to think of genomes as the keys to ancient molecular locks that became more and more complicated over time.

Genomics obviously has potential for human gene therapy. However, now that the human genome is fully sequenced, it still is not easy to manipulate it. Genetic researchers have to insert a modified gene into a host cell of a particular tissue. They have to be sure the gene gets inserted at a suitable site in a given chromosome. They must also make sure that the targeted type of cell will synthesize the specified protein at suitable times, in suitable amounts.

Today, most experimenters employ stripped-down viruses as vectors that put genes into cultured human cells. They know cells are able to incorporate foreign genes into their DNA. But viral genetic material can undergo rearrangements, deletions, and other changes that can shut down or disrupt gene expression. What about developing synthetic, streamlined versions of human chromosomes? Maybe the machinery for DNA replication and protein synthesis will work on them.

Some gene therapies simply put modified cells into a tissue. This may help, even if the cells only make 10 to 15 percent of a required protein. But no one can yet predict where the genes will end up. The danger is that the insertion will disrupt the function of other genes, including those controlling cell growth and division. One-for-one gene swaps by homologous recombination are possible. Oliver Smithies, one of the best at using this process, can put genes right where they should go, but only once every 100,000 or so tries.

Although the technical details are still being worked out, modified genes are being transferred into cells of humans and other mammals during experimental and clinical trials.

SAFETY ISSUES

Many years have passed since foreign DNA was first transferred into a plasmid. That gene transfer ignited a debate that will continue well into the next century. The issue is this: *Do potential benefits of gene modifications and gene transfers outweigh potential dangers?*

Genetically engineered bacteria are "designed" so they cannot survive except in the laboratory. As added precautions, "fail-safe" genes are built into the foreign DNA in case they escape. These genes are silent unless the captives are exposed to environmental conditions—whereupon the genes get activated, with lethal results for their owner. Say the package includes a *hok* gene next to a promoter of the lactose operon (Section 14.2). Sugars are plentiful in the environment. If they were to activate the *hok* gene, the protein product of that gene would destroy membrane function and the wayward cell.

What about a worst-case scenario? Remember how retroviruses are used to insert genes into cultured cells? If they escape from laboratory isolation, what might be the consequences? For instance, check out Section 47.9.

And what about genetically engineered plants and animals released into the environment? For example, Steven Lindlow thought about how frost destroys many crops. Knowing that a surface protein of a bacterium promotes formation of ice crystals, he excised the "ice-forming" gene from bacterial cells. As he hypothesized, spraying "ice-minus bacteria" on strawberry plants in an isolated field prior to a frost would help plants resist freezing. He actually had deleted a *harmful* gene from a species, yet a bitter legal battle ensued. The courts ruled in his favor. His coworkers sprayed a strawberry patch; nothing bad happened.

Then there was one potato plant designed to kill the insects that attack it. It also was too toxic for people to eat. Or think of how crop plants compete poorly with weeds for nutrients. Many have been designed to resist weedkillers so farmers can spray for weeds and not worry about killing their crops. Some of the herbicides do have toxic effects on more than their targets. If crop plants offer herbicide resistance, will farmers be less or more apt to spread them about?

And what if engineered plants or animals transfer modified genes to organisms in the wild? Think of how the advantage in acquiring resources would tilt toward vigorous weeds blessed with herbicide-resistant genes. Such possibilities are why standards for rigorous and extended safety tests are in place *before* the modified organisms enter the environment.

For more on safety issues, turn to *Critical Thinking* question 1 at the chapter's end.

Rigorous safety tests are carried out before genetically modified organisms are released into the environment.

BIOTECHNOLOGY IN A BRAVE NEW WORLD

Before you leave this unit of the book, reflect on what you have learned so far. You started out by examining cell division mechanisms, the means by which parents pass on their DNA to each new generation. You then moved on to the chromosomal and molecular basis of inheritance. You continued with a glimpse into the gene controls that guide the continuation of life from one generation to the next. The sequence you followed parallels the history of genetics research.

And now, with this chapter, you have arrived at a point in time when molecular geneticists hold keys to the kingdom of inheritance. They are now unlocking and changing genomes at a stunning pace. They are making **DNA microarrays**, or gene chips, each with thousands of DNA sequences from a genome stamped onto a glass plate the size of a business card. Within days instead of years, researchers can now find out which genes are silent and which are being expressed in a tissue, pinpoint mutations, track host–microbe interactions, diagnose genetic diseases, and see how drugs or therapies influence gene expression.

Each gene chip is bathed in a solution containing labeled probes. These probes are RNA transcripts that were extracted from the cells being studied—say, cancer cells from a patient. The probes bind only to complementary base sequences on the gene chip and make them glow in fluorescent light. Analysis of the glowing spots on the chip can reveal which of the thousands of genes in the cells are inactive and which are switched on. This kind of molecular knowledge is about to revolutionize how we diagnose, treat, and prevent the many diverse forms of cancer.

And we as a society are working our way through bioethical aspects of such astonishing research even as it is swirling past us.

Who Gets Well?

Think about the potential of gene therapy. The historic case mentioned at the start of this chapter was the first proof that medical applications of genetic engineering might alleviate suffering and save lives. More proof comes from two male infants with a severe combined immune deficiency called *SCID-X1*. With this genetic disorder, having two copies of a certain mutant allele disables the immune system. The infants were living as "bubble boys," in germ-free isolation tents, because they could not fight infections. Then geneticists used a viral vector to insert copies of the nonmutated gene stem cells from the boys' bone marrow. They infused the genetically modified cells into the marrow. Months later, both boys left their isolation tents. For the past year, at least, their immune system is working as it should. The boys are healthy and living at home.

We already have identified more than 15,500 genetic disorders. Should we dismiss them simply because they are rare in the population at large? Whose lives will they touch? In any given year, genetic disorders affect 3 to 5 percent of all newborns and underlie 20 to 30 percent of all infant deaths. They account for 50 percent of all people who are mentally impaired and close to 25 percent of all hospital admissions. And what of the age-related disorders that await all of us? Critics who would like to put a lid on genetic research would do well to put a face on those among us who are badly and painfully bent or broken.

Who Gets Enhanced?

To many of us, human gene therapy to correct genetic disorders seems like a socially acceptable goal. Now take this idea one step further. Is it socially desirable or even acceptable to change some genes of a normal human (or sperm or egg) to alter or enhance traits?

The idea of being able to select desirable human traits is called *eugenic engineering*. Yet who decides which forms of a trait are most "desirable"? Would it be desirable to engineer taller or blue-eyed or fair-skinned boys and girls? Would it be okay to engineer "superhumans" with amazing strength or intelligence?

For a survey conducted not long ago, more than 40 percent of the Americans interviewed said that it would be okay to use gene therapy to make smarter or better looking babies. One poll of British parents found 18 percent willing to use genetic enhancement to prevent children from being aggressive and 10 percent willing to keep them from growing up to be homosexual.

Through gene transfers, J. Z. Tsien and others have produced mice with enhanced memory and learning abilities. Maybe their work heralds help for those who have Alzheimer's disease or dementia, perhaps even for those who just want to have more brain power.

Geneticist Dean Hamer even predicts we will soon be tinkering with genes in order to change forms of behavior held to be socially undesirable. Violent forms of aggression and drug addictions come to mind.

Send in the Clones? Don't Bother, They're Here

Each year, about 75,000 people are on waiting lists for an organ transplant from a human donor, but donors are in short supply. There is talk of harvesting organs from pigs, which function very much like organs from humans do. Transferring an organ from one species to another is called **xenotransplantation**.

But the human immune system battles anything that it recognizes as "nonself," and it would reject an unmodified pig organ at once. At the free surface of

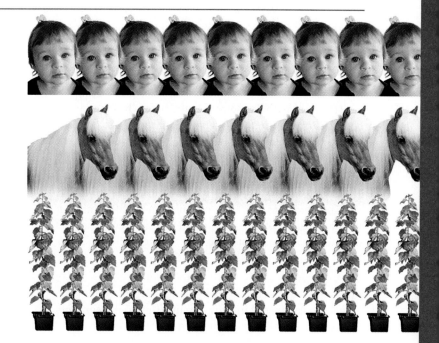

cells of a pig organ's blood vessels is a certain sugar component of a glycoprotein. Antibodies circulating in human blood would bind at once to the sugar. Binding would doom the transplant. It would trigger a cascade of reactions that would, within hours, cause massive coagulation inside the vessels (Sections 11.4 and 34.4). Potent drugs can suppress the immune response, but they make the organ recipient vulnerable to infections.

Pig DNA contains two copies of *Ggta1*, the gene for alpha-1,3-galactosyltransferase. This enzyme catalyzes a key step in the biosynthesis of the sugar that human antibodies latch on to. Knowing this, biotechnologists worked to knock the two copies of the gene out of pig DNA. Recently, they did succeed in excising one of the copies. They went on to transfer "knockout cells" (those with the excised gene) into enucleated pig eggs, in the manner described in Section 13.5.

Some eggs developed into embryos, which were implanted in a host sow. The embryos developed into a clone of piglets, each without one copy of the *Ggta1* gene. Two piglets died shortly afterward; another died a few weeks later. The survivors appear to be healthy even though one piglet has a defective eye and small ear flaps. Biotechnologists plan to breed the cloned pigs by conventional methods to produce offspring in which both copies of the *Ggta1* gene are knocked out. (Remember your Mendelian genetics? If both parents lack one copy of the *Ggta1* gene, there is one chance in four that one of their offspring will have no copies.)

A potential problem for xenotransplantation is that pigs carry a virus similar to the one that causes AIDS in humans. For the experiments just described, the researchers used miniature pigs, which apparently cannot transmit the virus to human cells.

Pigs are mammals. So are humans. Is human cloning next? Are there people willing to fund research that might yield a clone of a child, either "as is" or with genetically engineered "enhancements"? Will a human female be able to reproduce copies of herself without involving men? Will men be able to pay to have their DNA inserted into a cell stripped of its nucleus, have the cell implanted in a surrogate mother, and make a little repeat of themselves?

Lee Silver, a biologist at Princeton University, has suggested that anyone who thinks human cloning will move slowly is naive. One couple has already pledged up to 500 million dollars to a controversial company to clone their dead infant. The company says it has lots of potential customers and surrogate mothers lined up.

At this writing, several countries have now banned human cloning, but the bans leave room for cloning technology with uses in basic research. Nonscientists and scientists alike are actively debating whether any form of human cloning should be allowed.

Weighing the Benefits and Risks

Some say that the DNA of any organism must never be altered. Put aside the fact that nature itself has been altering DNA ever since the origin of life. The concern is that we as a species simply do not have the wisdom to bring about beneficial changes without causing irreparable harm to ourselves or to the environment.

To be sure, when it comes to altering human genes, one is reminded of our very human tendency to leap before we look. And yet, when it comes to restricting genetic modifications of any sort, one also is reminded of another old saying: "If God had wanted us to fly, he would have given us wings." Something about the human experience did give us a capacity to imagine wings of our own making—and that capacity carried us to the frontiers of space.

Where are we going from here? To gain perspective on our future, spend some time reading about our past. Ours is a history of survival in the face of challenges, threats, bumblings, and sometimes disasters on a grand scale. It is also a story of our connectedness with the environment and with one another.

The basic questions now confronting you are these: Should we be more cautious, believing the risk takers may go too far? And what do we as a species stand to lose if risks are not taken? There are no simple answers. But weigh the options when you consider arguments for and against genetically engineered food, stem cell research, and other issues in biotechnology, as you will be doing later in this book.

Our ability to manipulate genomes is outpacing attempts to think through some of its bioethical implications.

SUMMARY   **Gold** indicates text section

1. Uncountable numbers of gene mutations and other forms of genetic "experiments" have been proceeding in nature for at least 3 billion years. *CI*

2. Through artificial selection practices, humans have been manipulating the genetic character of a great many species for at least 11,000 years. Recombinant DNA technology enormously expands our capacity to genetically modify organisms. *CI*

3. In genetic engineering, specific genes are modified and inserted into the same organism or a different one. In gene therapy, copies of normal or modified genes are inserted into individuals to correct a genetic defect or boost resistance to disease. *CI, 16.9, 16.10*

4. A genome is all the DNA in the haploid chromosome number for a species. By recombinant DNA technology, a genome can be cut into fragments, then the fragments can be amplified (copied over and over again) to make useful quantities that permit analysis of the nucleotide sequence of the genome or a specific portion of it. *16.1*

5. Researchers work with chromosomal DNA cut by restriction enzymes or with cDNA. (A cDNA strand is transcribed by the enzyme reverse transcriptase from a mature mRNA transcript). Chromosomal DNA is better for questions about DNA regions that control gene expression or that contain introns. cDNA is better for questions about the amino acid sequence of the protein of interest. *16.1*

6. The use of plasmids or other cloning vectors is one way to amplify DNA fragments. Many bacteria contain plasmids: small circles of DNA with a few genes in addition to those of the bacterial chromosome. *16.1*

   a. Certain restriction enzymes make staggered cuts that leave the fragments with single-stranded tails. Such tails base-pair with complementary tails of any other DNA cut by the same enzyme, such as plasmid DNA.

   b. DNA ligase, a modification enzyme, seals base-pairing sites between plasmid DNA and foreign DNA. A plasmid modified to accept foreign DNA is a cloning vector; it can deliver foreign DNA into a bacterium or some other cell that can start a population of rapidly dividing descendant cells. All of the descendants have identical copies of the foreign DNA. Collectively, all of the identical copies are a DNA clone.

7. Currently, the polymerase chain reaction (PCR) is the fastest way to amplify fragments of chromosomal DNA or cDNA. Bacterial factories are not needed; the reactions occur in test tubes. The reactions use a supply of nucleotide building blocks and primers: synthetic, short nucleotide sequences that will base-pair with any complementary DNA sequence and that enzymes (DNA polymerases) recognize to start replication. Each round of replication doubles the number of fragments. *16.2*

8. For sexually reproducing species, no two individuals have exactly the same DNA base sequence (except for identical twins). When restriction enzymes are used to cut an individual's DNA, the result is a unique array of restriction fragments called a DNA fingerprint. *16.3*

   a. The DNA in all humans is more than 99 percent identical except at tandem repeats (short stretches of repeated base sequences, such as TTTC). The number and combination of tandem repeats is unique in each individual and can be detected by gel electrophoresis.

   b. DNA fingerprinting has uses in forensic science, as in resolving crimes and paternity suits.

9. Automated DNA sequencing can rapidly reveal the base sequence of cloned DNA or PCR-amplified DNA fragments. It labels fragments with one of four modified nucleotides, then separates them according to length by gel electrophoresis. Each label fluoresces a certain color under a laser beam. A machine reads each as it peels off the gel and assembles the whole sequence. *16.4*

10. A gene library is a mixed collection of bacterial cells that took up different cloned DNA or cDNA fragments. A gene may be isolated from the library with use of a probe, a very short stretch of radioactively labeled DNA that is known or suspected to be similar to or identical with part of that gene and can base-pair with it. A base pairing between nucleotide sequences from different sources is called nucleic acid hybridization. *16.5*

11. Generally, recombinant DNA technology and genetic engineering have potential for research and applications in medicine and agriculture, in the home and industry. As with any new technology, the potential benefits must be weighed against potential risks, including ecological and social repercussions. *CI, 16.9–16.10*

Review Questions

1. Distinguish between recombinant DNA technology and genetic engineering. *CI*

2. In the following diagram, which restriction enzyme made the cuts (indicated in *red*) in part of a DNA molecule from two different organisms? *16.1*

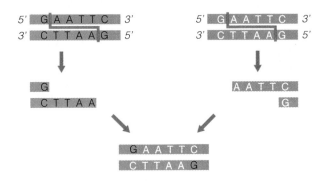

3. Distinguish these terms from one another: *16.1*
   a. chromosomal (genomic) DNA and cDNA
   b. cloning vector and DNA clone

4. Define PCR. Can fragments of chromosomal DNA, cDNA, or both be amplified by PCR? *16.2*

5. Define DNA fingerprinting. Briefly describe which portions of the DNA are used in DNA fingerprinting. *16.3*

6. Outline the steps of automated DNA sequencing. *16.4*

7. Define cDNA library, then briefly explain how a gene can be isolated from it. Define probe and nucleic acid hybridization as part of your answer. *16.5*

8. Give three examples of applications that can be derived from knowledge of an organism's genome. *CI, 16.6–16.8*

9. Name one of the ways in which modified genes have been inserted into mammalian cells. *16.8*

10. Define gene therapy. Once the human genome has been fully sequenced, why will it be difficult to manipulate its genes to advantage? *CI, 16.8*

Self-Quiz   ANSWERS IN APPENDIX III

1. _____ is the transfer of normal genes into body cells to correct a genetic defect.
   a. Reverse transcription   c. Gene mutation
   b. Nucleic acid hybridization   d. Gene therapy

2. DNA fragments result when _____ cut DNA molecules at specific sites.
   a. DNA polymerases   c. restriction enzymes
   b. DNA probes   d. RFLPs

3. Fill in the blank: _____ are small circles of bacterial DNA that are separate from the circular bacterial chromosome.

4. Foreign DNA that was inserted into a plasmid and then replicated many times in a population of bacteria is a _____ .
   a. DNA clone   c. DNA probe
   b. gene library d. gene map

5. By reverse transcription, _____ is assembled on _____ .
   a. mRNA; DNA   c. DNA; enzymes
   b. cDNA; mRNA   d. DNA; agar

6. PCR stands for _____ .
   a. polymerase chain reaction
   b. polyploid chromosome restrictions
   c. polygraphed criminal rating
   d. politically correct research

7. By gel electrophoresis, fragments of a gene library can be separated according to _____ .
   a. shape   b. length   c. species

8. Automated DNA sequencing relies on _____ .
   a. supplies of standard and labeled nucleotides
   b. primers and DNA polymerases
   c. gel electrophoresis and a laser beam
   d. all of the above

9. Match the terms with the most suitable description.
   ____ DNA fingerprint   a. selecting "desirable" traits
   ____ Ti plasmid   b. deciphering 3.2 billion base pairs
   ____ nature's genetic   of 23 human chromosomes
        experiments   c. used in some gene transfers
   ____ nucleic acid   d. unique array of DNA fragments
        hybridization   inherited in Mendelian pattern
   ____ Human Genome   from each of two parents
        Initiative   e. base pairing of nucleotide
   ____ eugenic   sequences from different
        engineering   DNA or RNA sources
                           f. mutations, crossovers

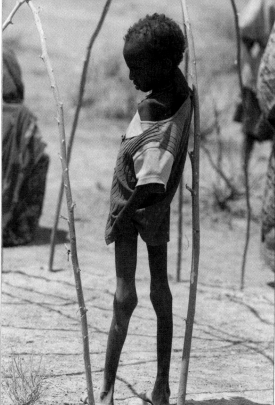

Figure 16.15 *Above:* Activists ripping some genetically modified crop plants from an experimental field in Great Britain. *Below:* Malnourished child in southeastern Ethiopia.

Critical Thinking

1. Avoiding genetically engineered food is probably impossible in the United States. At least 45 percent of cotton, 38 percent of soybean, and 25 percent of corn crops have been engineered to withstand weedkillers or to make their own pesticides. For years, modified corn and soybeans have found their way into breakfast cereals, tofu, soy sauce, vegetable oils, beer, soft drinks, and other food products. They are fed to farm animals.

By contrast, public resistance to genetically engineered food is high in Europe, especially Britain. Many people, including the Prince of Wales, speak out against what the tabloids call "Frankenfood." Protesters routinely vandalize crops (Figure 16.15). Worries abound that such foods may be more toxic, have lower nutritional value, and promote natural selection for antibiotic resistance. Some people worry that designer plants will cross-pollinate with wild plants to produce "superweeds."

Biotechnologists envision a new Green Revolution. They argue that designer plants can hold down food production costs, reduce dependence on pesticides and herbicides, enhance crop yields, and offer improved flavor, nutritional value, even salt tolerance and drought tolerance.

Yet the chorus of critics in Europe may provoke a trade war with the United States. The issue is not small potatoes, so to speak. In 1998, the value of agricultural exports reached about 50 billion dollars. Flattery, threats, and bullying are rampant on both sides of the Atlantic. Restrictions on genetic engineering will have profound impact on United States agriculture, and inevitably the impact will trickle down to what you eat and how much you pay for it.

All of which invites you to read up on scientific research related to this issue and form your own opinions. The alternatives are to be swayed either by media hype (the term Frankenfood, for instance) or by biased reports from groups (such as chemical manufacturers) with their own agendas.

Possibly start with Christopher Bond's article "Politics, Misinformation, and Biotechnology" in the 18 February 2000 *Science* (287:1201). Bond argues that biotechnology attempts to solve real-world problems of sickness, hunger, and dwindling resources; and that "hysteria and unworkable propositions advanced by those who can afford to take their next meal for granted have little currency among those who are hungry." For example, think about the body masses of the individuals in the two photographs in Figure 16.15.

2. Lunardi's Market put out a bin of tomatoes having splendid vine-ripened redness, flavor, and texture. The sign posted above the bin identified them as genetically engineered produce. Most shoppers selected unmodified tomatoes in the adjacent bin even though those tomatoes were pale pink, mealy-textured, and tasteless. Which ones would you pick? Why?

3. Ryan, a forensic scientist, obtained a very small bit of DNA from material at a crime scene. In order to examine the sample by DNA fingerprinting, he must amplify the sample by PCR, the polymerase chain reaction. By his estimate, there are 50,000 copies of the DNA in his sample. Calculate the number of copies Ryan will have after fifteen cycles of PCR.

4. A game warden in Africa confiscated eight ivory tusks from elephants. Some tissue is still attached to the tusks. Now she must determine whether the tusks were taken illegally from northern populations of endangered elephants or from other populations of elephants to the south that can be hunted legally. How can she use DNA fingerprinting to find the answer?

5. The Human Genome Initiative is undergoing completion, and knowledge about a number of the newly discovered genes is already being used to detect genetic disorders. Ask yourself: What will be done with genetic information about individuals? Will insurance companies and potential employers request it? At this writing, many women have already refused to take advantage of genetic screening for a gene associated with the development of breast cancer. Should medical records about people participating in genetic research and genetic clinical services be made available to other individuals? If not, how could such information be protected?

6. Animal rights activists are up in arms over the possibility of raising pigs as organ sources for xenotransplantation. Are you for or against this possibility? Why or why not?

7. What if it were possible to create life in test tubes? This is the question behind attempts to model and eventually create *minimal organisms*, which we define as living cells having the smallest possible set of genes that are necessary to survive and reproduce.

As recent experiments by Craig Venter and Claire Fraser revealed, *Mycoplasma genitalium*, a bacterium with 517 genes (and 2,209 transposons) might be a candidate. By disabling the genes one at a time in the laboratory, they discovered that this bacterium may contain no more than 265–350 essential protein-coding genes.

What if the genes were to be synthesized one at a time and inserted into an engineered cell consisting only of a plasma membrane and cytoplasm? Would the cell come to life?

The possibility that it might prompted Venter and Fraser to seek advice from a panel of bioethicists and theologians. As Arthur Caplan, a bioethicist at the University of Pennsylvania reported, no one on the panel objected to synthetic life research. They felt that much good might come of it, provided scientists didn't claim to have found "the secret of life."

The 10 December 1999 issue of *Science* includes an essay from the panel and an article on *M. genitalium* research. Read both, then write down your thoughts about creating life in a test tube.

Selected Key Terms

automated DNA sequencing *16.4*	genomics *16.8*
cDNA *16.1*	nucleic acid hybridization *16.5*
cloning vector *16.1*	PCR *16.2*
DNA clone *16.1*	plasmid *16.1*
DNA fingerprint *16.3*	primer *16.2*
DNA ligase *16.1*	probe (nucleic acid) *16.5*
DNA microarray *16.10*	recombinant DNA technology *CI*
gel electrophoresis *16.3*	restriction enzyme *16.1*
gene library *16.5*	reverse transcriptase *16.1*
gene therapy *CI*	seed bank *16.7*
genetic engineering *CI*	tandem repeats *16.3*
genome *16.1*	xenotransplantation *16.10*

Readings

Cho, M., et al. 10 December 1999. "Ethical Considerations in Synthesizing a Minimal Genome." *Science* 286: 2087–2090.

"The Human Genome." *Science* 291: 5507. The entire 16 February 2001 issue consolidates current understandings and questions.

Lai, L., et al. Published online 3 January 2002. "Production of 1,3-Galactosyltransferase Knockout Pigs by Nuclear Transfer Cloning," *Science*. For interested and fearless students who want to get right into cutting-edge research.

Pennisi, E. June 2000. "Finally, The Book of Life and Instructions for Navigating It." *Science* 288: 5475.

Watson, J. D., et al. 1992. *Recombinant DNA*. Second edition. New York: Scientific American Books.

On-Line readings at Student Guide for InfoTrac:
www.brookscole.com/biology

III Principles of Evolution

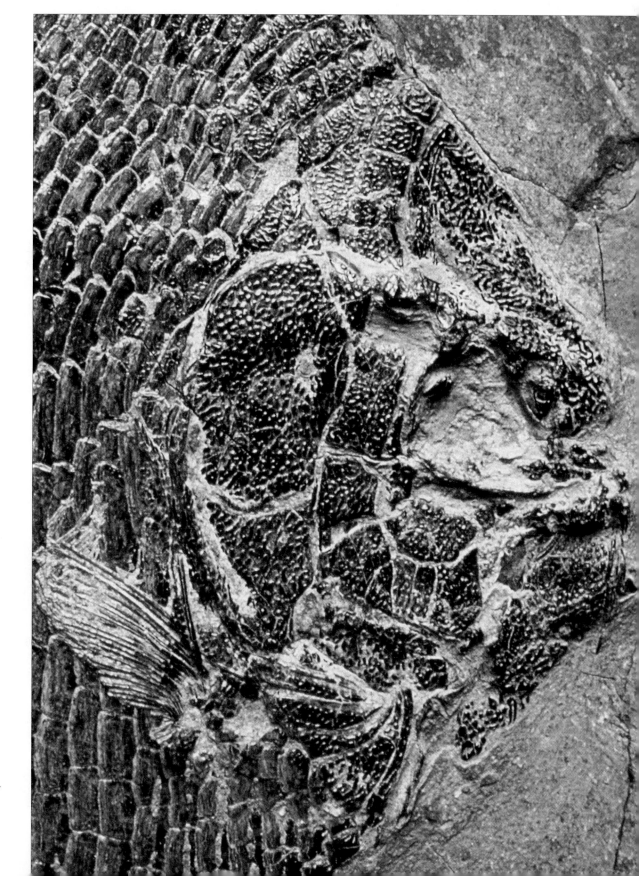

Millions of years ago, a bony fish died, and sediments gradually buried it. Today its fossilized remains are studied as one more piece of the evolutionary puzzle.

See pp 10–11

MICROEVOLUTION

Designer Dogs

We humans have tinkered rather ruthlessly with the modern descendants of a long and distinguished lineage. The lineage originated some 40 million years ago with the appearance of tree-dwelling carnivores that looked rather like small weasels. Their descendants evolved along separate branchings that now include weasels, badgers, otters, skunks, bears, pandas, raccoons —and dogs.

About 50,000 years ago, we began domesticating wild dogs. No doubt the advantages of doing so were important. Times were tough in the days before police protection and supermarkets. Dogs welcomed to the hearth could guard people and their possessions. They could corner, kill, eat, and thus dispose of big rats and other unwelcome vermin.

By 14,000 years ago, we started to develop different varieties (breeds) through artificial selection. Individual dogs having desirable forms of traits were selected from each new litter and, later, encouraged to breed. Those having undesired forms of traits were passed over.

After favoring the pick of the litter over hundreds or thousands of generations, we ended up with sheep-herding border collies, badger-hunting dachshunds, wily retrievers, and sled-pulling huskies. And at some point we began to delight in the odd, extraordinary dog. Imagine! In practically no time at all, evolutionarily speaking, we picked our way through the pool of variant dog alleles and came up with such extremes as Great Danes and chihuahuas (Figure 17.1).

Sometimes our canine designs exceeded the limits of biological common sense. For example, how long would a tiny, finicky-eating, nearly hairless, nearly defenseless chihuahua last in the wild? Not long. What about the English bulldog, bred for a very short snout and a compressed face? Long ago, breeders thought these particular traits would allow the dogs to get a better grip on the nose of a bull. (Why they wanted dogs to bite bulls is a story in itself.) So now the roof of the bulldog mouth is ridiculously wide and often flabby, so bulldogs have trouble breathing. Sometimes they get so short of air they pass out.

Through our centuries-old fascination with artificial selection, we produced thousands of varieties of crop plants, ornamental plants, cats, cattle, and birds as well as dogs. With the currently available technologies of genetic engineering, we are now mixing the genes of many different species and producing incredible new varieties, including tobacco plants that produce useful quantities of hemoglobin, and mustard plants that produce plastic.

So, when you hear someone wonder about whether "evolution" takes place, remind yourself that **evolution** simply means genetic change in a line of descent over the generations. Selective breeding practices provide abundant, tangible evidence that heritable changes do, indeed, occur. The actual mechanisms that bring about change are the focus of this chapter. Later chapters will explain their role in the evolution of new species from parent species.

Figure 17.1 Two designer dogs. About 50,000 years ago, humans began domesticating wild dogs. From that ancestral stock, artificial selection resulted in startling diversity among rather closely related breeds, such as the Great Dane (*legs, left*) and the chihuahua (*possibly fearful of being stepped on, right*).

Key Concepts

1. Starting in the fifteenth century, explorers found puzzling differences in the world distribution of species. Anatomists identified similarities and differences in body structure and patterning among embryos and adult forms of major groups of animals. In sequences of sedimentary rocks, geologists found sequences of fossils. The findings were suggestive of biological evolution, or heritable changes in lines of descent over time.

2. Charles Darwin came of age at a time when many scholars were attempting to reconcile the startling findings from global explorations, comparative morphology, and geology with prevailing cultural beliefs. He, and Alfred Wallace, came up with a theory of evolution by natural selection to explain the findings. Their evolutionary theory and the theories of others start with variation in traits.

3. All individuals of a population generally have the same number and kinds of genes. Those genes give rise to a diverse array of traits that characterize the population.

4. Mutations may result in two or more slightly different molecular forms of a gene—alleles—that influence a trait in different ways. Individuals of a population vary in the details of a trait when they inherit different combinations of alleles.

5. Any allele may become more or less common in a population relative to other kinds at a gene locus, or it may disappear. _Microevolution_ refers to changes in the allele frequencies of a population over time.

6. The frequencies of alleles change as an outcome of mutation, gene flow, genetic drift, and natural selection. Mutation alone gives rise to new alleles. Gene flow, genetic drift, and natural selection shuffle existing alleles into, through, or out of populations.

EARLY BELIEFS, CONFOUNDING DISCOVERIES

The Great Chain of Being

Our story begins more than two thousand years ago, when the seeds of biological inquiry were starting to take hold among the ancient Greeks. At the time, popular belief held that supernatural beings intervened directly and often in human affairs. As an example, people "knew" that angry gods inflicted epilepsy, an ailment known as the sacred disease. And yet, from one of the physicians of the school of Hippocrates, these thoughts come down to us:

It seems to me the disease called sacred . . . has a natural cause, as other diseases have. Men think it divine merely because they do not understand it. But if they called everything divine that they did not understand, there would be no end of divine things! . . . If you watch these fellows treating the disease, you see them use all kinds of incantations and magic—but they are also careful in regulating diet. Now if food makes the disease better or worse, how can they say it is the gods who do this? . . . It does not really matter whether you call such things divine or not. In Nature, all things are alike in this, in that they can be traced to preceding causes.

—Sacred Disease (400 B.C.)

He had perceived the link between cause and effect in nature—a key premise of modern science (Section 1.5).

Aristotle was foremost among the early naturalists, and he described the world around him in great detail. He had no reference books or instruments to guide him, for biological science in the Western world began with the great thinkers of this age. Yet here was a man who was no mere collector of random tidbits of information. In his thoughtful descriptions of nature we see evidence of a mind perceiving connections between observations and attempting to explain the order of things.

Aristotle believed (as did others) that each kind of organism was distinct from all the rest. Nevertheless, he wondered about organisms with traits that seemed to blur such distinctions. For example, some sponges look very much like plants but do not make their own food, as plants do. They capture and digest it, as animals do. In time, Aristotle came to view nature as a continuum of organization, from lifeless matter through complex forms of plant and animal life.

By the fourteenth century, Aristotle's idea had been transformed into a rigid view of life. A Chain of Being was seen to extend from the "lowest" forms of life to humans, then on up to spiritual beings. Each kind of being, or "species" as it was called, was a separate link in the great chain. All links were designed and forged at the same time, at the same center of creation, and had not changed since then. Scholars thought that once they discovered, named, and described all of the links, the meaning of life would be revealed to them.

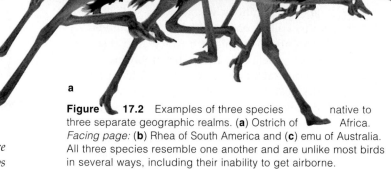

Figure 17.2 Examples of three species native to three separate geographic realms. (**a**) Ostrich of Africa. *Facing page:* (**b**) Rhea of South America and (**c**) emu of Australia. All three species resemble one another and are unlike most birds in several ways, including their inability to get airborne.

Questions From Biogeography

Until the fifteenth century, naturalists were not aware that the world is a great deal bigger than Europe, so the task of locating and describing all species seemed manageable. Then globe-spanning explorations began. Naturalists were soon overwhelmed by descriptions of tens of thousands of plants and animals that explorers were discovering in Asia, Africa, the Pacific Islands, and the New World.

In 1590, the naturalist Thomas Moufet attempted to sort through the bewildering array. He simply gave up and wrote such gems as this description of locusts and grasshoppers: "Some are green, some black, some blue. Some fly with one pair of wings; others with more; those that have no wings they leap; those that cannot fly or leap they walk . . . Some there are that sing, others are silent." It was not a work of subtle distinctions.

Even so, a few scholars began to examine the world distribution of organisms, a discipline now known as **biogeography** (Section 19.3 and Chapter 49). They soon realized that many plants and animals are unique to isolated places, such as remote oceanic islands. They also were aware that certain species separated by great distances resemble one another (Figure 17.2).

How did so many species get from one center of creation to oceanic islands and other isolated, remote locations? And what do the similarities and differences among them mean?

Questions From Comparative Morphology

By the eighteenth century, many scholars were engaged in **comparative morphology**, the systematic study of similarities and differences in the body plans between major groups, such as different kinds of vertebrates. Think of the bones of a human arm, whale flipper, and

bat wing. They differ in size, shape, and function. Yet all have similar locations in the body. They consist of the same tissues, arranged in the same overall patterns. They develop in similar ways in embryos. Comparative morphologists who deduced all of this wondered: Why are some animals that are so different in some features so much alike in others?

By one hypothesis, basic body plans were so perfect there was no need to come up with a new one for each organism at the time of their creation. Yet if that were so, then how could there be parts with no function? For example, some snakes have bones that correspond to a pelvic girdle—a set of bones to which hind legs attach (Figure 17.3). Snakes don't have legs. Why the bones? Humans have bones that correspond to a few tailbones of many other mammals, but they don't have a tail. Why, then, do they have parts of one? (After reading Chapters 19 and 26, you may have a good idea.)

Questions About Fossils

From the late 1600s on, geologists added to the rising confusion. They began mapping layers of rocks at sites where erosion or quarrying had cut deep into the earth. They found similar layers around the world. Such beds consist of distinct, multiple layers of sedimentary rock of the sort shown in Section 19.3. Most agreed that sand and other sediments had been deposited at different times and had slowly compacted into layers, with the deepest layers being the oldest. And it dawned on them that they could correlate certain **fossils**, or stone-hard evidence of life in ancient times, with specific layers.

For example, deep layers contained fossils of simple marine organisms. Some fossils in the overlying layers were similar but structurally more complex. The fossils in layers above these closely resembled living marine organisms. What did increasing structural complexity among fossils of a given type represent? Were these *sequences* of fossils, separated in time? Another puzzle: many fossils were unique in some traits yet similar to

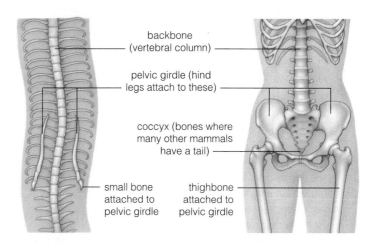

Figure 17.3 *Left:* Python bones corresponding to the pelvic girdle of other vertebrates, including humans (*right*). Small "hind limbs" protrude through the skin on the underside of the snake.

certain existing species in other traits! Could species so similar yet so far apart in time be *related* in some way?

Taken as a whole, the findings from biogeography, comparative morphology, and geology did not fit with prevailing beliefs. Georges-Louis Leclerc de Buffon and a few others started to formulate novel hypotheses. If dispersal of all species from a center of creation was not possible, given the vast oceans and other barriers, *then perhaps species had originated in more than one place.* And if they were not created in a perfect state—and fossil sequences and the presence of "useless" body parts in certain organisms suggested they were not—*then perhaps species had been modified over time.* Awareness of change in lines of descent—evolution—was in the wind.

Awareness of biological evolution emerged over centuries, through the cumulative observations of many naturalists, biogeographers, comparative anatomists, and geologists.

A FLURRY OF NEW THEORIES

Squeezing New Evidence Into Old Beliefs

In the nineteenth century, naturalists tried to reconcile the growing evidence of change in lines of descent with a traditional conceptual framework that did not allow for change. Foremost among them was Georges Cuvier, a respected anatomist. For years he had compared body plans of fossils and living organisms. He acknowledged the abrupt changes in the fossil record, corresponding to discontinuities between some layers of sedimentary beds. Was the record evidence of changing populations of organisms that lived in those ancient environments? Cuvier thought so. And he was right, as you will see from the evolutionary story in Chapters 19 and 20.

Based on that inference, Cuvier came up with his own explanation. For an intellectual framework, he started with a prevailing theory of **catastrophism**. By this theory, stupendous floods, earthquakes, and other catastrophic events had occurred on rare occasions in the past, but they were divinely invoked; otherwise the Earth was just an unchanging stage for the human drama. According to Cuvier, there was but one time of creation that populated the world with every species. A global catastrophe destroyed many of them. The survivors repopulated the world. These were not *new* species; naturalists simply hadn't yet found fossils of them that would date to the time of creation. Later on, more species were destroyed, then repopulation by the survivors followed, and so on, as recorded by fossils.

Many scholars accepted his theory, but others kept at the puzzle. One hypothesis—inheritance of *acquired* characteristics—was pushed by Jean Lamarck. During each individual's life, thought Lamarck, environmental pressures and internal "needs" bring about permanent changes in body form and functioning, then offspring inherit the necessary changes. And so life, created long ago in a simple state, gradually improved. The force for change was a drive toward perfection, up the Chain of Being. The drive was centered in nerves that directed an unknown "fluida" to body parts in need of change.

Apply his hypothesis to modern giraffes. Say they had a short-necked ancestor. Pressed by a need to find food, it kept stretching its neck to browse upon leaves beyond the reach of other animals. Stretching directed fluida to its neck, which lengthened permanently. The longer neck was inherited by offspring, which stretched their necks, also. So generations of animals desiring to reach ever loftier leaves led to the modern giraffe.

As Lamarck correctly inferred, the environment *is* a factor in evolution. His hypothesis, however, like others proposed at the time, has not been supported by tests carried out since then. There is no evidence that the environment modifies traits of all existing individuals in a way that can be passed on to offspring.

Figure 17.4 (**a**) Charles Darwin and (**b**) a blue-footed booby, one of many species he observed during his five-year voyage around the world on the *Beagle*. (**c**) A replica of the ship, sailing off South America's coast. On this trip, Darwin also ventured into the Andes. He observed fossils of marine organisms in rock layers 3.6 kilometers above sea level. (**d–f**) The Galápagos Islands are isolated in the ocean far to the west of Ecuador. They arose through volcanic action about 5 million years ago, so species could not have originated there. Winds or ocean currents must have carried them to the new islands.

Voyage of the Beagle

In 1831, in the midst of this confusion, Charles Darwin was twenty-two years old and wondering what to do with his life. Ever since he was eight, he had wanted to hunt, fish, collect shells, or simply watch insects and birds—anything but sit in school. Later, at his father's insistence, he did try to study medicine in college. The crude, painful procedures used on patients at that time sickened him. His by-now exasperated father urged him to become a clergyman, so he packed for Cambridge. His grades were good enough to earn him a degree in theology. But he spent most of his time among faculty members with leanings toward natural history.

John Henslow, a botanist, perceived Darwin's real interests. He arranged for Darwin to function as ship's naturalist aboard H.M.S. *Beagle*. The *Beagle* was about to embark on a five-year voyage that would take Darwin around the world (Figure 17.4). Abruptly, the young man who hated schoolwork and had no formal training as a naturalist started to work with enthusiasm.

The *Beagle* sailed first to South America to complete work on mapping the long coastline. During the Atlantic

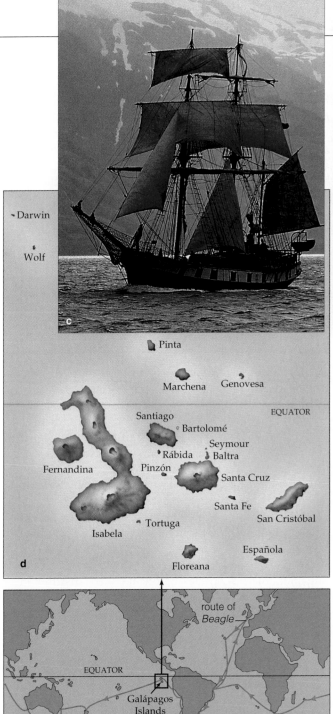

- Darwin
- Wolf
- Pinta
- Marchena
- Genovesa
- Santiago
- Bartolomé
- EQUATOR
- Seymour
- Rábida
- Baltra
- Fernandina
- Pinzón
- Santa Cruz
- Santa Fe
- San Cristóbal
- Tortuga
- Isabela
- Española
- Floreana

d

route of *Beagle*

EQUATOR

Galápagos Islands

e

crossing, Darwin collected and studied marine life. He read Henslow's parting gift, the first volume of Charles Lyell's *Principles of Geology*. During stops at the coast and at islands, he saw diverse species in environments ranging from sandy shores to high mountains. And he started circling the question of evolving life, which was now on the minds of many respected individuals.

Darwin started mulling over a rather radical theory that Lyell was advancing in his book. Lyell and other geologists were arguing that catastrophes had no more

effect on Earth history than did subtle processes of change. They had thought about how long it takes for rains, the pounding surf, and other forces of nature to sculpt the landscape. For years geologists had chipped away at layers of sandstones, limestones, and other rocks, which form after sediments erode from the land and accumulate in the beds of rivers and seas. They thought about how sedimentary beds often consist of a number of stacked layers. If, they hypothesized, the deposition took place as gradually in the past as it did in their own era, then surely it took many millions of years—not a few thousand—for such thick stacks to form. They even managed to incorporate earthquakes and other infrequent events into their view of Earth history. After all, major floods, more than a hundred great earthquakes, and twenty or so volcanic eruptions typically occur each year, so catastrophes aren't unusual.

Their view of gradual, uniformly repetitive change became the theory of **uniformity**. It directly challenged prevailing views of the age of the Earth.

The theory bothered scholars who firmly believed the Earth was only about 6,000 years old. They thought people had recorded everything that happened during those thousands of years, and in all that time no one ever mentioned seeing a species evolve. Yet, by Lyell's calculations, it must have taken millions of years to mold the present landscape. *Wasn't that enough time for species to evolve in diverse ways?* Later, Darwin thought so. But exactly *how* did they evolve? He would end up devoting the rest of his life to that burning question.

Prevailing beliefs can influence how we interpret clues to natural processes and their observable outcomes.

Darwin's observations during a global voyage helped him think about natural processes in a novel way.

DARWIN'S THEORY TAKES FORM

Old Bones and Armadillos

After Darwin returned to England in 1836, he talked with other naturalists about possible evidence that life evolves. By carefully studying all of the notes from his journey, he came up with some possibilities.

In Argentina, for example, he had observed fossils of glyptodonts, which are now extinct. Of all animals on Earth, only living armadillos are like glyptodonts (Figure 17.5). And of all places on Earth, armadillos live only in the same places where glyptodonts once lived. If the two kinds of animals had been created at the same time, lived in the same place, and were so much alike, why is only one still alive? Wouldn't it be reasonable to assume glyptodonts were early ancestors of armadillos? Many of their shared traits might have been retained through countless generations. Other traits might have been modified in the armadillo branch of a family tree. Descent with modification—it seemed possible. What, then, could be the driving force for evolution?

A Key Insight—Variation in Traits

While Darwin assessed his notes, an influential essay by Thomas Malthus, a clergyman and economist, made him look closely at a topic of social interest. Malthus had correlated population size with famine, disease, and war. Humans, he claimed, run out of food, living space, and other resources because they reproduce too much. The larger a population gets, the more people there are to reproduce in each generation. Population size burgeons, resources dwindle, and the struggle to live intensifies. Many people starve, get sick, and engage in war and other forms of competition for remaining resources.

After Darwin reflected on his personal observations, he suspected that any population has the capacity to produce more individuals than the environment is able to support. To give one example, a single sea star can release 2,500,000 eggs per year, but the seas obviously do not fill up with sea stars. Nearly all of the eggs and larvae of each generation end up inside the bellies of predators. Many of the survivors starve or succumb to disease or some other environmental assault.

Assume that the environment restricts the number of reproducing individuals. Which individuals will be the winners and losers? What influences the outcome? Darwin thought about the populations he had observed during his voyage. As he recalled, individuals were not alike down to the last detail. They varied in size, color, and other traits. *It dawned on him that variations in traits might affect an individual's ability to secure resources—and to survive and reproduce in particular environments.*

Did the Galápagos Islands show evidence of this? Between these volcanic islands and the South American coastline are 900 kilometers of open ocean. The islands offer diverse habitats along their rocky shores, deserts, and mountain flanks. Nearly all of their inhabitants live nowhere else—although they resemble species on the mainland. Were these remote islands colonized by species that flew, floated, or were blown over from the mainland? If so, then in the different island habitats, island-hopping descendants of the colonizers might have undergone modifications, over time, as adaptive responses to local conditions.

As Darwin later learned from other naturalists in England, thirteen species of finches were distributed throughout the Galápagos. He himself had collected specimens of some of the birds, and now he attempted to correlate the variations in their traits with specific environmental challenges.

Imagine yourself in his place. You notice the finches of one population have a large, strong bill suitable for cracking seeds (Figure 17.6). Yet a few individuals with a slightly stronger bill crack seeds that are too tough for their neighbors. If most of the seeds being produced in a particular habitat in a given interval have hard coats, then a strong-billed bird will have a competitive

Figure 17.5 (a) Pleistocene glyptodont, about as big as a Volkswagon "beetle." Glyptodonts are extinct. They share unusual traits and a restricted distribution with existing armadillos (b), even though these animals are widely separated in time. To Darwin, they were a clue that helped him develop a theory of evolution by natural selection. (c) One of many fossils of glyptodont armor.

Figure 17.6 Four species of finches that live on the Galápagos Islands. (**a**) *Geospiza conirostris* and (**b**) *G. scandens*, both with a bill adapted for eating cactus flowers and fruits. Other finches have thick, strong bills that crush cactus seeds. (**c**) *Certhidea olivacea*, a tree-dwelling finch, resembles warblers in song and behavior. It uses its slender beak to probe for insects. (**d**) *Camarhynchus pallidus* feeds on wood-boring insects such as termites. It has learned to break cactus spines and twigs to suitable lengths, and then hold the "tools" and use them to probe bark for insects hidden from its view.

Figure 17.7 Alfred Wallace, who studied in Malaysia. Wallace worked out a theory of natural selection long after Darwin did but was first to report it. He quickly circulated a brief letter that described his views about the process, making it "his" theory.

edge. It will have a better chance than the other birds of surviving and producing offspring. Assuming the trait is heritable, the same advantage will be bestowed on that bird's strong-billed descendants.

Take these thoughts one step further. If factors in the environment continue to "select" the most adaptive version of a trait, then the population will become one of mostly strong-billed birds. *And a population is evolving if forms of heritable traits are changing over the generations.*

Recall, from Section 1.4, that Darwin offered pigeon breeding and other *artificial* selection practices as a way

to explain *natural* selection of traits in the wild. When breeders favor pigeons with, say, black-feathered tails, they encourage black-tailed pigeons of each generation to mate, but not white-tailed ones. It was an easy way to show how selection could lead to an increase in the frequency of one form of a trait in a captive population.

Anticipating that his view would be controversial, Darwin waited to announce it and searched for flaws in his reasoning. He waited too long. More than a decade after he wrote but did not publish his theory, another respected naturalist—Alfred Wallace—sent him a letter (Figure 17.7). Wallace had developed the same theory but was the first to circulate a short letter about it! Like Wallace, most scholars thought Darwin should get most of the credit and prevailed on him to formally present a paper at the same time Wallace presented his. The next year, in 1859, Darwin published his detailed evidence in support of the theory, *On the Origin of Species*.

Although you may have heard that Darwin's book fanned an intellectual firestorm, the idea that diversity is the product of evolution was accepted almost at once by most naturalists. But Darwin's specific explanation, of gradual evolution by natural selection, was fiercely debated. Nearly seventy years passed before advances in a new field, genetics, led to widespread acceptance of his explanation. Until that happened, people generally associated Darwin's name mainly with the premise that life evolves—something others had suggested before him.

DARWIN'S THEORY OF EVOLUTION BY NATURAL SELECTION. A population can evolve (change over time) when individuals differ in one or more heritable traits that are responsible for differences in the ability to survive and reproduce.

INDIVIDUALS DON'T EVOLVE—POPULATIONS DO

Examples of Variation in Populations

As Charles Darwin perceived, *individuals do not evolve; populations do.* By definition, a **population** is a group of individuals of the same species that are occupying a given area. To understand how a population evolves, start with variation in the features of its individuals.

Certain features characterize every population. All members share the same body plan, as when jays have two wings, feathers, three toes forward and one toe back, and so on. These are *morphological* traits (*morpho–*, meaning form). Cells and body parts of all members work much the same way during metabolism, growth, and reproduction. These are *physiological* traits, relating to how the body functions in its environment. And the members respond the same way to basic stimuli, as when babies instinctively imitate adult facial expressions. These are *behavioral* traits.

Especially for sexually reproducing species, details of most traits vary among individuals. Pigeon feathers or snail shells vary in patterns or colors within a population (Section 1.4 and Figure 17.8). Some individuals of a frog population may be more sensitive to winter cold or better at attracting a mate than others. Humans differ in the distribution, color, texture, and amount of hair. These examples merely hint at the stunning variation in populations; almost every trait of every species is variable.

Many traits, such as those of Gregor Mendel's pea plants, vary in *qualitatively different* ways. They come in two or more distinct forms (or morphs) in a population, as when feathers are yellow or white. Such qualitative variation is known as **polymorphism**. Other traits, such as eye color and height, show continuous, *quantitatively different* variation. That is, individuals of a population show small, incremental differences in traits that may be quantified, as Section 11.7 describes.

The "Gene Pool"

Information about heritable traits occurs in genes, which are specific regions of DNA molecules. In general, all individuals of a population have the same number and kinds of genes. We say "in general" because males and females of sexually reproducing populations differ in a number of genes on the sex chromosomes.

Think of all the genes in the entire population as a **gene pool**—a pool of genetic resources that, in theory at least, is shared by all members of a population and then their offspring, the next generation. Each kind of gene in that pool is most often present in two or more slightly different molecular forms, called **alleles**.

Figure 17.8 *Facing page:* From different Caribbean islands, variation in shell color and banding patterns in populations of the same snail species. Different individuals carry different alleles for the genes that specify most traits. The shells differ because their owners had different combinations of alleles at particular gene locations along their chromosomes.

Individuals have different combinations of alleles. This leads to variations in phenotype—differences in the details of traits. For example, whether your hair is black, brown, red, or blond depends on which alleles of certain genes you inherited from your two parents. And don't forget: *Offspring inherit genes, not phenotypes.* Environmental conditions often modify expression of a gene (Section 11.8). But phenotypic variation resulting from their effects lasts no longer than the individual.

Which alleles end up in a given gamete and later in the new individual? Five events shape the outcome, as described in earlier chapters and summarized here:

Sources of gene variation

1° 1. Gene mutation (produces <u>new</u> alleles)

2° 2. Crossing over at meiosis I (results in novel combinations of alleles in chromosomes)

3. Independent assortment at meiosis I (puts mixes of maternal and paternal chromosomes in gametes)

4. Fertilization (combines alleles from two parents)

5. Change in chromosome number or structure (leads to the loss, duplication, or repositioning of genes)

Of all the events just listed, only mutation *creates* new alleles. The other four only shuffle *existing* alleles into different combinations—but what a shuffle! Consider this: Each human gamete inherits one of 10^{600} possible combinations of alleles. Not even 10^{10} humans are alive today. Unless you are an identical twin, it is extremely unlikely that any other person with your precise genetic makeup has ever lived, or ever will.

Stability and Change in Allele Frequencies

Imagine yourself in a big garden in summer, watching butterflies flitting about. They look the same except in wing color. A few have white wings; more have blue. Perhaps, you muse, the "blue" allele is more common than the "white" allele. By using genetic analysis, you could identify the **allele frequencies**, the abundance of each kind of allele in the population. You could track the rate of genetic change over the generations.

Suppose that you start with the "Hardy–Weinberg rule," as given in Section 17.5, as a theoretical reference point for measuring patterns of change. At that point, called **genetic equilibrium**, frequencies of alleles at a given gene locus remain stable one generation after the

you know, gene interactions underlie the phenotypes of complex organisms. A mutation that has severe effects on phenotype typically leads to death; it is a **lethal mutation**.

By comparison, a **neutral mutation** does not help *or* harm an individual. Natural selection can neither increase nor decrease the frequency of neutral mutations in the population, for these do not have any discernible effect on the individual's chances of surviving and reproducing. For example, if you carry a mutant gene that resulted in attached earlobes instead of detached ones, this alone should not stop you from surviving and reproducing just as well as anybody else.

Every so often, a mutation bestows an advantage. For instance, a product of a mutant gene that affects growth might make a corn plant grow larger or faster and so give it the best access to sunlight and nutrients. Or maybe a neutral mutation turns out to be beneficial after conditions in the environment change. Even if the advantage is small, chance events or natural selection might preserve the mutant gene in the DNA and favor its representation in the next generation.

Mutations are so rare, they usually have little or no immediate effect on the population's allele frequencies. But beneficial mutations, and neutral ones, have been accumulating in lineages for billions of years. Through all that time, they have represented the raw material for evolutionary change—the basis for the staggering range of biological diversity, past and present.

In the evolutionary view, then, the reason you don't look like a bacterium or an avocado or an earthworm or even neighbors down the street began with different mutations that originated at different times in the past, in different lines of descent.

p 281

next. The population is *not* evolving in terms of that gene, for five conditions are being met. First, no gene mutations have occurred. Second, that population is very large. Third, it is isolated from other populations of the species. Fourth, the gene has no effect at all on survival or reproduction. Fifth, all mating is random.

Rarely, if ever, do all five conditions prevail at the same time in nature. Gene mutation is infrequent but inevitable. Three processes—*natural selection, gene flow, and genetic drift*—may drive a population away from genetic equilibrium within even one generation. The term **microevolution** refers to small-scale changes in allele frequencies brought about by mutation, natural selection, gene flow, and genetic drift.

p 287 + 294

p 288

Mutations Revisited

Mutations, remember, are heritable changes in DNA that usually give rise to altered gene products. They are the original source of alleles. We cannot predict exactly when or in which individual they will appear. Yet each gene has its own **mutation rate**, which is the probability of its mutating during or in between DNA replications. On average, the rate is between 10^{-5} and 10^{-6} per gene locus per gamete, each generation. Thus, in a single reproductive season, about 1 in 100,000 to 1,000,000 gametes has a new mutation at a given locus.

Mutations may give rise to structural, functional, or behavioral alterations that can diminish an individual's chances of surviving and reproducing. Even one small biochemical change can have devastating effects.

For instance, skin, bones, tendons, and many other vertebrate organs cannot develop without collagen, a structural protein. If the gene specifying the molecular form of collagen mutates, drastic changes in the lungs, arteries, skeleton, and other body parts may ensue. As

in a species, gen variation betw pop.s

Adaptive traits (p 10)

Certain morphological, physiological, and behavioral traits characterize a population. The traits differ in their details from one individual to the next.

Differences in the combinations of alleles that individuals carry give rise to variations in phenotype. By phenotypic variation, we mean differences in the details of structural, functional, and behavioral traits that the individuals of a population have in common.

In sexually reproducing species, a population's individuals share a pool of genetic resources—that is, a gene pool.

Mutation alone creates *new* alleles. Natural selection, gene flow, and genetic drift change *allele frequencies* in a gene pool. The evolutionary story starts with these changes.

When Is a Population *Not* Evolving?

Remember those earlobes we talked about in Chapter 11? Ask yourself: Is the number of individuals with *detached* earlobes staying the same, one generation after the next, in the whole human population? And what about *attached* earlobes? How can we know whether or not a population is evolving with respect to earlobes or any other trait?

Almost a hundred years ago, a mathematician and a doctor came up with a way to find the answer. Working independently, they thought about what it would take to maintain an idealized population's allele frequencies. They came up with what we now call the **Hardy–Weinberg rule**, in their honor. They started with a formula that represents this idea: In any population at genetic equilibrium, the proportions of genotypes at a gene locus for which there are two kinds of alleles are

$$p^2\,AA + 2pq\,Aa + q^2\,aa = 1$$

where p is the frequency of allele A, and q is the frequency of allele a. They applied this rule to the formula: *The allele frequencies will not change through successive generations if there is no mutation, if the population is infinitely large and is isolated from other populations of the same species, if mating is random with respect to the alleles, and if all individuals survive and reproduce equally.* Stated differently, the rule presents a hypothesis for determining whether any of the listed assumptions currently apply to a population.

Suppose you want to test whether a population's allele frequencies stay the same from one generation to the next in the absence of evolutionary forces. You decide to track a pair of alleles through a population of butterflies. These sexually reproducing organisms have pairs of genes, on pairs of homologous chromosomes. You start by assuming the population is presently at genetic equilibrium. The pair of alleles you are interested in affect wing color. Allele A specifies dark-blue wings. Allele a is associated with white wings. And the heterozygous (Aa) condition results in medium-blue wings.

In the population as a whole, the frequencies of A and a must add up to 1. For example, if A occupies half of all the loci for this gene in the population, then a must occupy the other half ($0.5 + 0.5 = 1$). If A occupies 90 percent of all the loci, then a must occupy 10 percent ($0.9 + 0.1 = 1$). No matter what the proportions of the two kinds of alleles,

$$p + q = 1$$

During meiosis in germ cells, the alleles of each pair segregate and end up in separate gametes. Thus, p is also the proportion of gametes with the A allele, and q is the proportion with the a allele. To find the frequencies of the three genotypes (AA, Aa, and aa) that are possible in the next generation, make a Punnett square:

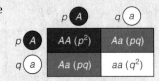

	p A	q a
p A	$AA\ (p^2)$	$Aa\ (pq)$
q a	$Aa\ (pq)$	$aa\ (q^2)$

The frequencies of the genotypes add up to 1:

$$p^2 + 2pq + q^2 = 1$$

To see whether the allele frequencies and genotypic frequencies will stay the same through the generations, you work through an example. You find the population has 1,000 butterflies, each of which produces two gametes:

490 AA individuals produce 980 A gametes
420 Aa individuals produce 420 A and 420 a gametes
90 aa individuals produce 180 a gametes

You notice the frequency of A among the 2,000 gametes is

$$(980 + 420)/2{,}000 = 0.7$$

Also,

$$q = (420 + 180)/2{,}000 = 0.3$$

At fertilization, the gametes combine at random and give rise to the next generation, as given in the Punnett square. Assuming the population remains constant at 1,000 individuals, you now have

$$p^2\ AA = 0.7 \times 0.7 = 0.49 \qquad (490\ AA\ \text{individuals})$$
$$2pq\ Aa = 2 \times 0.7 \times 0.3 = 0.42 \qquad (420\ Aa\ \text{individuals})$$
$$q^2\ aa = 0.3 \times 0.3 = 0.09 \qquad (90\ aa\ \text{individuals})$$

and

$$p^2 + 2pq + q^2 = 0.49 + 0.42 + 0.09 = 1$$

The allele frequencies have not changed:

$$A = \frac{2 \times 490 + 420}{2{,}000\ \text{alleles}} = \frac{1{,}400}{2{,}000} = 0.7 = p$$

$$a = \frac{2 \times 90 + 420}{2{,}000\ \text{alleles}} = \frac{600}{2{,}000} = 0.3 = q$$

You notice that the genotype frequencies haven't changed, either. And as long as the five assumptions of the Hardy–Weinberg rule are being met, frequencies should stay the same through the generations. To test this, you calculate allele frequencies in the gametes of the *next* generation:

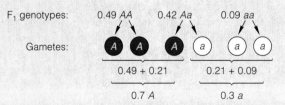

| F_1 genotypes: | 0.49 AA | 0.42 Aa | 0.09 aa |

Gametes:	A A	A a	a a
	0.49 + 0.21	0.21 + 0.09
	0.7 A	0.3 a

which is back where you started from. Because the allele frequencies for dark-blue, medium-blue, and white wings are the same as they were in the original gametes, they will yield the same phenotypic frequencies you observed in the second generation.

You could do similar calculations for other wing colors. You could go on with your calculations until you run out of paper (or patience). As long as the five assumptions continue to hold, however, the allele frequencies and the range of values for the wing-color trait will not change, as you can see from Figure 17.9.

Therefore, when genotypes and phenotypes do *not* show up in the proportions you predicted on the basis of the Hardy–Weinberg rule, this tells you that one or more conditions of the rule are being violated. And the hunt can begin for the specific evolutionary force, or forces, driving the change.

STARTING POPULATION

490 *AA* butterflies
Dark-blue wings

420 *Aa* butterflies
Medium-blue wings

90 *aa* butterflies
White wings

THE NEXT GENERATION

490 *AA* butterflies

420 *Aa* butterflies

90 *aa* butterflies

NO CHANGE

THE NEXT GENERATION

490 *AA* butterflies

420 *Aa* butterflies

90 *aa* butterflies

NO CHANGE

Figure 17.9 A hypothetical population of butterflies at genetic equilibrium.

NATURAL SELECTION REVISITED

We now turn from our idealized population that never changes to the real-world processes of change. Of these processes, natural selection might account for most of the morphological and physiological changes that have accrued throughout the history of life.

Darwin, recall, was able to explain natural selection after correlating his understanding of inheritance with certain features of populations and their environments. Before we consider the modes of natural selection, let's restate his correlations in modern terms:

1. *Observation:* All populations in nature have the reproductive capacity to increase in numbers through the generations.

2. *Observation:* No population is able to increase indefinitely, for its individuals will run out of food, living space, and other resources that sustain it.

3. *Inference:* Sooner or later, the individuals of a population will end up competing for resources.

4. *Observation:* All of the individuals generally have the same genes, which specify the same assortment of traits. Collectively, their genes represent a pool of heritable information.

5. *Observation:* Most, if not all, kinds of genes occur in slightly different molecular forms (alleles), which give rise to differences in phenotypic details.

6. *Inferences:* Some phenotypes are better than others at helping the individual compete for resources, hence to survive and reproduce. Alleles for those phenotypes therefore increase in the population, and other alleles decrease. In time the genetic change leads to increased **fitness**—an increase in adaptation to the environment.

7. *Conclusions:* **Natural selection** is the outcome of variations in shared traits that affect which individuals of a population survive and reproduce each generation. This microevolutionary process results in adaptation to the environment (Section 28.2).

Evolutionary biologists have been documenting the results of natural selection in thousands of field studies of populations of all kinds of organisms. They find that this microevolutionary process has different results. As you will see in sections to follow, sometimes the result is a shift in the range of values for a given trait in some direction. At other times, the result may be stabilization or disruption of an existing range of values.

As Darwin perceived, natural selection is the outcome of variations in shared traits that influence which individuals of a population survive and reproduce in each generation. Natural selection can lead to increased fitness—that is, to an increase in adaptation to the environment.

3 Kinds of Natural Selection (handwritten)

DIRECTIONAL CHANGE IN THE RANGE OF VARIATION

What Is Directional Selection?

In cases of **directional selection**, allele frequencies that give rise to a range of variation in phenotype tend to shift in a consistent direction. The shift is a response to directional change in the environment or to one or more novel environmental conditions. A new mutation that proves beneficial also sets it in motion. Either way, the forms at one end of the phenotypic range become more common than the midrange forms (Figure 17.10). Let's look at a few documented cases of this outcome.

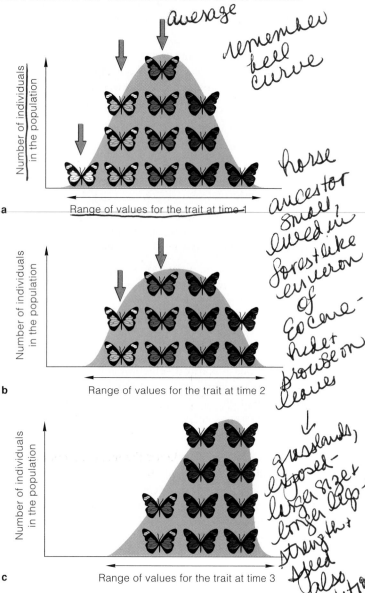

average (handwritten)
remember bell curve (handwritten)
horse ancestor small lived in forest-like environ of Eocene hidet browse on leaves (handwritten)
grasslands, exposed — larger size + longer legs? strength + speed also dentition change (handwritten)

a
Range of values for the trait at time 1

b
Range of values for the trait at time 2

c
Range of values for the trait at time 3

Figure 17.10 Directional selection, using phenotypic variation within a population of butterflies as the example. A bell-shaped curve (*green*) represents the range of continuous variation in wing color. The most common form (*medium blue*) is between extreme forms of the trait (*white* at one end of the curve, *deep blue* at the other). *Orange* arrows signify which forms are being selected against over time.

The Case of the Peppered Moths

In England, biologists tracked directional selection in about a hundred moth species, including the peppered moth (*Biston betularia*). Peppered moths feed and mate at night. During the day, they rest motionless on birches and other trees. Their wings and body have a mottled pattern in shades that range from light gray to nearly black. Their behavior, coloration, and wing patterning camouflage them from moth-eating birds. Like all birds, the ones that prey on moths hunt during the day.

The industrial revolution started in England in the 1850s. Outpourings of sooty smoke altered conditions in many parts of the surrounding countryside. Before then, light moths were the most common form, and a dark form was rare. Also before conditions changed, light-gray speckled lichens grew thickly on tree trunks. Lichens can camouflage light moths that rest on them, but not dark ones (Figure 17.11*a*).

Lichens are sensitive to air pollution. Between 1848 and 1898, soot and other factory pollutants started to kill the lichens and darken tree trunks. In the modified environment, the less common moth form was better camouflaged (Figure 17.11*b*). Researchers came up with a hypothesis: If the original conditions favored light moths, then the *changed* conditions favored dark ones.

In the 1950s, H. B. Kettlewell used a *mark–release–recapture method* to test that prediction. He bred both moth forms in captivity and marked hundreds of them so they could be easily identified after being released in the wild. He released them near highly industrialized areas around Birmingham and in an unpolluted area of Dorset. Later on, more dark moths were recaptured in the polluted area and more light ones in the pollution-free area (Table 17.1). Observers also were stationed in blinds near groups of moths that had been tethered to trees. They directly observed birds capturing more of the light moths around Birmingham and more of the dark moths around Dorset. Here was strong evidence that directional selection was operating.

Strict pollution controls went into effect in 1952. Lichens made a comeback. Tree trunks became free of soot, for the most part. As you might have predicted, phenotypes shifted in the reverse direction as a result of natural selection. Where pollution has decreased, the frequency of dark moths has been decreasing as well.

Pesticide Resistance

Widespread use of chemical pesticides in agriculture also results in directional selection. Initial applications kill most of the insects, worms, or other pests, but some individuals usually manage to survive. Some aspect of their structure, physiology, or behavior helps them resist

Figure 17.11 Peppered moths (*Biston betularia*). The light form of the moth is less visible to birds on lichen-covered tree trunks (**a**). The dark form is less visible after pollution kills the lichens and darkens the trunks (**b**).

Table 17.1 *Marked Peppered Moths Recaptured in Polluted and Nonpolluted Areas*		
	Near Birmingham (pollution high)	Near Dorset (pollution low)
LIGHT-GRAY MOTHS		
Released	64	393
Recaptured	16 (25%)	54 (13.7%)
DARK-GRAY MOTHS		
Released	154	406
Recaptured	82 (53%)	19 (4.7%)

Data after H. B. Kettlewell.

the chemical effects. When the resistance has a heritable basis, it becomes more common in each new generation. These chemicals are agents of selection that favor the most resistant forms! Today, 450 different species resist one or more pesticides. Worse, the pesticides also kill natural predators of the pests. When freed from natural constraints, the populations of resistant pests burgeon, and crop damage is greater than ever. This outcome of directional selection is called *pest resurgence*.

Maybe pesticide use will decline in fields of plants that are genetically engineered to resist pests. Even modified plants will not win the coevolutionary arms race, but they might help keep food supplies one step ahead of the pests. This is a contentious issue; many consumers are leery of genetically engineered food, as you read in Section 16.11 (*Critical Thinking* question 1).

What about using *biological controls*? By this practice, natural enemies of pests, including parasitic wasps and predatory beetles, are raised in commercial insectaries. Large populations are released at preselected sites. The practice has an advantage in that a control species can coevolve with the pests. Farmers must replace the ones that migrate from the fields or are destroyed at harvest time. They pass on the cost to consumers.

Antibiotic Resistance

When your grandparents were young, tuberculosis, pneumonia, and scarlet fever caused one-fourth of the annual deaths in the United States. Since the 1940s, we have been relying on natural and synthetic antibiotics to fight such bacterial diseases. **Antibiotics**, remember, are toxins that some microorganisms in soil release to destroy bacterial competitors for nutrients (Section 1.4). Streptomycins, for example, prevent protein synthesis in target cells. Penicillins disrupt formation of covalent bonds that hold bacterial cell walls together. Penicillin derivatives weaken the wall until it ruptures.

Antibiotics must be prescribed with restraint and care. Why? Besides executing their intended function, they often disrupt resident bacterial populations that compete for nutrients in the intestines. They can disrupt bacterial and yeast populations in the vagina, as well. Imbalances follow and lead to secondary infections.

Also, antibiotics have been overprescribed in the human population. All too frequently, people demand them for simple infections that they often can fight on their own. As a consequence, antibiotics have lost their punch. Over time, they killed the most susceptible cells of target populations. However, they also favored their replacement by much more resistant cells. Each year, millions of people around the world die from cholera, tuberculosis, and other bacterial diseases. Vancomycin, once held in reserve as "the antibiotic of last resort," is no longer effective against some pathogenic strains of enteric (gut-dwelling) bacteria. In 1996, the World Health Organization announced that, in the dangerous race for supremacy, the pathogens are sprinting ahead.

With directional selection, allele frequencies underlying a range of variation tend to shift in a consistent direction in response to directional change in the environment.

SELECTION AGAINST OR IN FAVOR OF EXTREME PHENOTYPES

As you have seen, natural selection can bring about a directional shift in a population's range of phenotypic variation. Depending on the prevailing environmental conditions, the process also may favor either the most common or the most extreme phenotypes in that range.

Stabilizing Selection

In **stabilizing selection**, intermediate forms of a trait in a population are favored, so alleles for extreme forms are eliminated (Figure 17.12). This mode of selection tends to counter mutation, gene flow, and genetic drift, and it preserves the most common phenotypes.

The gallmaking fly *Eurosta solidaginis* is evidence of stabilizing selection. Its larvae bore into stems of a tall goldenrod (*Solidago altissima*) and feed on tissues until they metamorphose into adults. In response, the tissue cells multiply rapidly and encase the invaders in a gall, a tumorous mass. We know from genetic analyses that flies with different phenotypes cause galls of different sizes—large, small, and in between—to form.

A wasp parasitizes the larvae (Figure 17.13). It can penetrate small galls, so flies that cause small galls to form are at risk. Also, the downy woodpecker and some other birds eat the larvae. They can peck through large galls, so the flies that cause large galls to form are at risk. Think it through, and you see that the flies that cause *intermediate-sized* galls to form are favored. Their larvae are less vulnerable to wasps *and* to birds.

And so parasitic wasps work against one extreme phenotype and predatory birds work against the other. In this fly population, the intermediate phenotype has the highest survival rate and fitness.

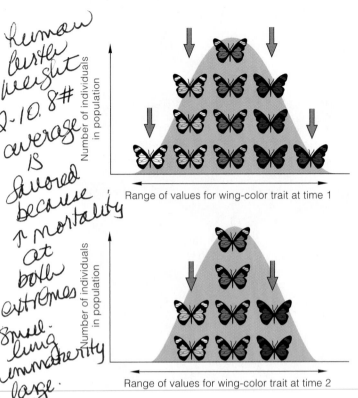

Figure 17.12 Stabilizing selection, using phenotypic variation within a population of butterflies as the example.

Range of values for wing-color trait at time 1

Range of values for wing-color trait at time 2

Range of values for wing-color trait at time 3

c Parasitic wasp (an agent of selection)

a Gall-making fly (producer of tasty larvae)

b Gall on a goldenrod stem that houses fly larvae

d Downy woodpecker (an agent of selection)

Figure 17.13 Example of stabilizing selection. Larvae of the fly *Eurosta solidaginis* (**a**) induce formation of a type of tumor called a gall (**b**) on goldenrod stems. (**c**) A parasitic wasp (*Eurytoma gigantea*) has an egg-laying device that penetrates only the thin wall of small galls. Its eggs develop into larvae, then the wasp larvae eat fly larvae. (**d**) Downy woodpeckers (*Dendrocopus pubescens*) and other birds can use their bill to chisel into large-size galls. They, too, eat the larvae.

Warren Abrahamson and his coworkers monitored twenty *Eurosta* populations in Pennsylvania. They found that larvae inside small and large galls have low relative fitnesses. They also found that larvae in intermediate-size galls have relatively high fitnesses. The fly's natural enemies target larvae in small and large galls; they stabilize the fly's range of phenotypic variation.

a

Range of values for wing-color trait at time 1

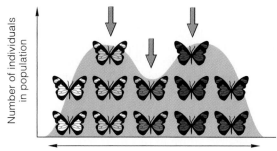

b

Range of values for wing-color trait at time 2

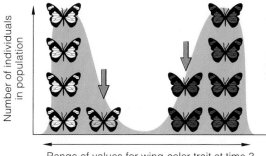

c

Range of values for wing-color trait at time 3

Figure 17.14 Disruptive selection, using phenotypic variation within a population of butterflies as the example.

③
Disruptive Selection

In **disruptive selection**, forms at both ends of a range of variation are favored and intermediate forms are selected against (Figure 17.14). Thomas Smith found an example of this in a rain forest in Cameroon, West Africa. Smith had read about unusual variation in populations of the black-bellied seedcracker (*Pyrenestes ostrinus*). These African finches have large or small bills—but no sizes in between. The pattern holds for females and males through the geographic range. (This is just remarkable; imagine every person in Texas being four feet *or* six feet tall, with no intermediates.) If the bill pattern is unrelated to gender or geography, what causes it?

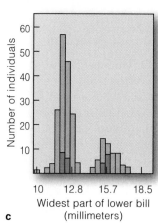

Figure 17.15 Disruptive selection among African finches. Feeding trials showed that birds with large bills are better users of hard seeds. Small-billed birds are better at using soft seeds, not hard ones. (**a,b**) Two individuals displaying small and large bill sizes.

(**c**) Graph of survival of juvenile birds during the dry season, when competition for resources is most intense. Individuals with very small, very large, or intermediate-size bills can't feed efficiently on either type of seed; they survive poorly. For this graph, *tan* bars indicate the number of nestlings; *orange* bars indicate the survivors among them. The findings are based on measurements of 2,700 netted individuals.

Smith hypothesized: Seed-cracking ability directly affects finch survival. If only two bill sizes persist, then disruptive selection may be eliminating birds that have intermediate-size bills. What selection pressures could be at work on feeding performance? Cameroon's swamp forests flood during the wet season; lightning-sparked fires burn during the dry season. Two species of sedge (fire-resistant, grasslike plants) dominate these forests. One sedge has hard seeds and the other has soft. When finches reproduce, hard *and* soft seeds are abundant.

All birds prefer soft seeds for as long as they can get them. However, the birds with small bills are better at cracking soft seeds; birds with large bills are better at cracking hard ones. Soft seeds and other food supplies dwindle as the dry season peaks. That is when small-billed birds, and the youngest ones, are at a competitive disadvantage. Many don't survive (Figure 17.15).

Smith also performed experimental crosses between large- and small-billed birds. All offspring had large *or* small bills. Along with other data, the crosses suggest that two alleles at one autosomal gene locus control bill size and feeding performance.

With stabilizing selection, intermediate phenotypes are favored and extreme phenotypes at both ends of the range of variation are eliminated.

With disruptive selection, intermediate forms are selected against; extreme forms in the range of variation are favored.

MAINTAINING VARIABILITY IN A POPULATION

Sexual Selection

Individuals of most sexually reproducing species have a distinctively male or female phenotype. We call this sexual dimorphism (*dimorphos,* "having two forms"). How does this condition arise, and what maintains it? Natural selection is taking the form of **sexual selection**. The traits being favored are advantageous, with respect to survival and reproduction, simply because males or females prefer them. Through nonrandom mating, the alleles for preferred traits prevail over the generations.

Sexual dimorphism is particularly striking among many mammals and birds, as in Figure 17.16. The males of many species are larger and have flashier coloration and patterning. They are often more aggressive than the females. Remember those male bighorn sheep butting

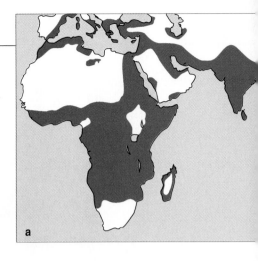

Figure 17.17 (**a**) Distribution of malaria cases in Africa, Asia, and the Middle East in the 1920s, before the start of mosquito control programs. (**b**) The distribution and frequency of people with the sickle-cell trait. Note the close correlation between the maps.

heads (Section 1.3)? Fighting wastes time and energy, and it may cause serious injuries. Why, then, do alleles that contribute to aggressive behavior still persist in a population? An increased chance of mating may offset the costs. Male bighorn sheep fight only to control areas where receptive females gather during a winter rutting season. Winners mate often, with a number of females. Losers won't challenge a stronger, larger male.

Often, females are the agents of selection. They exert direct control over reproductive success by choosing their mates. We return to this topic in Chapter 46.

Maintaining Two or More Alleles

Balancing selection refers to all forms of selection that maintain two or more alleles for a trait in a population. When this type of genetic variation persists over time, we call it **balanced polymorphism** (after *polymorphos,* "having many forms"). A population shows balanced polymorphism when nonidentical alleles for some trait are being maintained at frequencies above 1 percent. Allele frequencies might shift slightly, but over time, they often return to the same values. Smith's African finches are a good example of the effect of balancing selection through the generations.

Sickle-Cell Anemia—Lesser of Two Evils?

Balanced polymorphism may result when conditions favor heterozygotes, which carry nonidentical alleles for the trait. Compared to homozygotes, which carry identical alleles for the trait, they have higher fitness.

Let's look at the environmental pressures that favor an Hb^A/Hb^S pairing in humans. Hb^S specifies a mutant form of hemoglobin, an oxygen-transporting protein in blood. Homozygotes (Hb^S/Hb^S) develop *sickle-cell anemia,* a genetic disorder that has serious pleiotropic effects on phenotype (Section 3.8). The frequency of Hb^S is high in tropical and subtropical regions of Asia and Africa. Often, Hb^S/Hb^S homozygotes die in their early teens or early twenties. Yet in those same regions, the heterozygotes (Hb^A/Hb^S) make up nearly one-third

Figure 17.16 An outcome of sexual selection. This male bird of paradise (*Paradisaea raggiana*) is engaged in a flashy courtship display. He caught the eye (and, perhaps, the sexual interest) of the smaller, less colorful female. Males of this species compete fiercely for females, the selective agents. (Why do you suppose drab-colored females have been favored?)

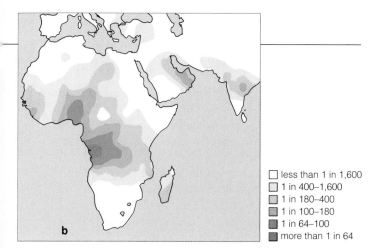

less than 1 in 1,600
1 in 400–1,600
1 in 180–400
1 in 100–180
1 in 64–100
more than 1 in 64

b

of the human population! Why is this combination of alleles maintained at such high frequency?

The balancing act, an outcome of natural selection, is most pronounced in areas with the highest incidence of *malaria* (Figure 17.17 and Section 22.6). A mosquito transmits *Plasmodium*, the parasitic agent of disease, to humans. The parasite multiplies in the liver, and later in red blood cells. The cells rupture and so release new parasites during severe, recurring bouts of infection.

Who is more likely to survive recurring infections? Hb^A/Hb^S heterozygotes. The allelic combination gives them two forms of hemoglobin, with interesting results. They produce enough *nonmutated* molecules to support body functions—and the *altered* molecules distort red blood cells in a way that slows down circulation. The slowdown in blood flow is a factor in getting through recurrences because it hampers the parasite's ability to travel and rapidly infect new cells. The distorted cells are later disposed of as they trickle through the spleen.

So the persistence of the "harmful" Hb^S allele is a matter of relative evils. Natural selection has favored one allelic combination, Hb^A/Hb^S, because its bearers show greater fitness in places where malaria is most prevalent. In these environments, the combination has a higher fitness than either Hb^S/Hb^S or Hb^A/Hb^A.

Malaria has been a selective force for over 2,000 years in tropical and subtropical habitats of Asia and the Middle East. Although sickle-cell anemia occurs at high frequencies in all these regions, its symptoms are far less severe than they are in Central Africa—where the Hb^S allele became established much later in time. It seems likely that other gene products are mediating some pleiotropic effects of the Hb^S allele; in some way they minimize symptoms of the disorder.

With sexual selection, some version of a trait simply gives the individual an advantage in reproductive success. Sexual dimorphism is one outcome of sexual selection.

Balanced polymorphism is a state in which natural selection is maintaining two or more alleles through the generations at frequencies greater than 1 percent.

GENE FLOW

Individuals of the same species don't necessarily stay put. A population loses alleles whenever an individual permanently leaves it, an act called *emigration*. It gains alleles when new individuals permanently move in, an act called *immigration*. This **gene flow**, a physical flow of alleles between populations, tends to counter genetic differences that we expect to arise through mutation, natural selection, and genetic drift. And it helps keep separated populations genetically similar.

Think of the acorns that blue jays disperse when they store nuts for the winter. Each fall the jays may make hundreds of round trips from acorn-bearing oak trees to bury acorns in the soil of their home territories, which may be up to a mile away (Figure 17.18). The alleles flowing in with "immigrant acorns" help reduce genetic differences that might otherwise arise among neighboring stands of oak trees. Gene flow apparently is operating among peppered moths, also, for the form *not* camouflaged by the prevailing background color is being maintained at frequencies higher than expected.

Figure 17.18 Gene flow among oak populations, courtesy of feathered travel agents. Blue jays hoard acorns in their home territory, but they might shop at nut-bearing trees up to a mile away. Some acorns contribute to the allele pool of an oak population some distance away from the parent tree.

Or think of the millions of people from politically explosive, economically bankrupt countries who seek a more stable homeland. The scale of their movements is unprecedented, but not unique. Throughout human history, immigrations may have minimized many of the genetic differences that otherwise would have built up among geographically separated groups of people.

Gene flow is the physical movement of alleles into and out of a population, through immigration and emigration.

GENETIC DRIFT

Chance Events and Population Size

Genetic drift is a random change in allele frequencies over the generations, brought about by chance alone. The magnitude of its effect on genetic diversity and on the range of phenotypes relates to population size. Its impact tends to be minor or insignificant in very large populations but significant in small ones.

Sampling error, a rule of probability, helps explain the difference. By this rule, the fewer times a chance event occurs, the greater will be the variance from the expected outcome of that occurrence. You saw a simple demonstration of this in Section 1.5.

Also think back on the coin-flipping example given earlier, in Section 11.2. Each time you flip a coin, there is a 50 percent chance it will turn up heads or tails. With, say, only ten flips, the odds are low that the coin will turn up heads 5 times and tails 5 times. But with a thousand flips, you are more likely to get 500 heads and 500 tails (or close to it). Similarly, sampling error applies each time random mating and fertilization take place in a population.

Bear in mind, genetic drift has nothing to do with how a population got small in the first place. *Genetic drift simply increases the chance that an allele will become more or less prevalent when the number of individuals in a population is small.*

Figure 17.19 is a computer simulation of the effect of genetic drift in two populations: one large, the other small. The outcomes parallel the findings of an actual experiment with beetles (*Tribolium*) that mate with one another at random. Researchers grouped 1,320 beetles into twelve populations of 10 beetles and twelve of 100. Which beetles ended up in which group was a matter of chance. At the outset, the frequency of a wild-type allele (call it *A*) was 0.5. The researchers tracked allele *A* for twenty generations. Every time, they randomly removed some offspring to maintain each population's original size. At the end of the experiment, *A* was not the only allele left in the large groups, but it was fixed in seven of the small groups. **Fixation** means that only one kind of allele remains at a particular locus in the population; all individuals are homozygous for it.

Thus, *in the absence of other forces, random change in allele frequencies leads to the homozygous condition and a loss of genetic diversity over the generations.* This happens in all populations; it just happens faster in small ones. Once alleles inherited from an original population are fixed, their frequencies will not change again unless mutation or gene flow introduce new alleles.

Bottlenecks and the Founder Effect

Genetic drift is pronounced when very few individuals rebuild a population or found a new one. This happens after a **bottleneck**, a severe reduction in population size brought about by intense pressure or by some calamity. Say contagious disease, habitat loss, hunting, or an erupting volcano nearly wipes out a population. Even if a moderate number do survive the bottleneck, allele frequencies will have been altered at random.

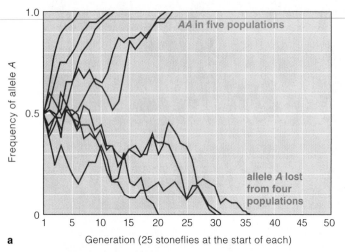

a Generation (25 stoneflies at the start of each)

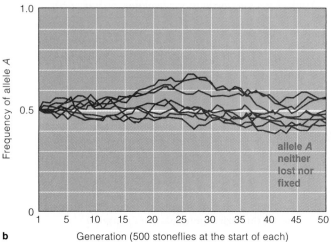

b Generation (500 stoneflies at the start of each)

Figure 17.19 Computer simulation of the effect of genetic drift on one allele's frequency in small populations and in large populations. Equal fitness is assumed for three simulations (*AA* = 1, *Aa* = 1, and *aa* = 1).

(a) The size of nine populations of a species (stoneflies, in this case) was maintained at 25 breeding individuals in each generation, through fifty generations. The five graph lines reaching the top of the diagram tell you that allele *A* became fixed in five of the small populations. The four lines plummeting off the bottom of the diagram tell you it was lost from four of them. As you can see, *alleles can be fixed or lost even in the absence of selection.*

(b) The size of nine other populations was maintained at 500 individuals in each generation, through fifty generations. Allele *A* did not become fixed in any of these large populations. The magnitude of genetic drift was much less in every generation than in the small populations.

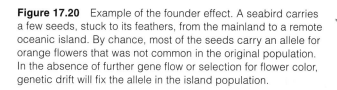

phenotypes of original population

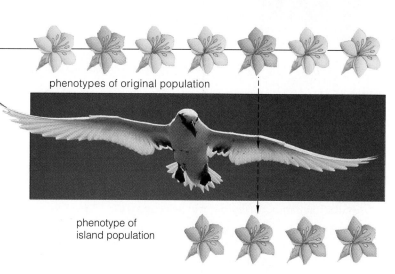

phenotype of
island population

Figure 17.20 Example of the founder effect. A seabird carries a few seeds, stuck to its feathers, from the mainland to a remote oceanic island. By chance, most of the seeds carry an allele for orange flowers that was not common in the original population. In the absence of further gene flow or selection for flower color, genetic drift will fix the allele in the island population.

Example: In the 1890s hunters killed all but twenty of a large population of northern elephant seals. The population eventually recovered. After its size reached 30,000, electrophoretic analysis of a sample of twenty-four genes revealed no variation in the population. A number of alleles had been lost during the bottleneck.

The genetic outcome can be similarly dicey after a few individuals leave a population and establish a new one elsewhere, as ancestors of Darwin's finches did on the Galápagos Islands. This form of bottlenecking is called a **founder effect**. By chance, allele frequencies of the founders may not be the same as frequencies of the original population. In the absence of further gene flow, genetic drift will operate on the small population. As you may deduce, the founder effect is pronounced on isolated islands (Figure 17.20).

Genetic Drift and Inbred Populations

Inbreeding refers to nonrandom mating among closely related individuals, which have many identical alleles in common. Inbreeding is a form of genetic drift in a small population—that is, within the group of relatives that are preferentially interbreeding. Like genetic drift, inbreeding leads to the homozygous condition. It also can lower fitness when the alleles that are increasing in frequency are recessive and have harmful effects.

Most human societies forbid or discourage incest (inbreeding between parents and children or between siblings). But inbreeding among other close relatives is common in small communities that are geographically or culturally isolated from a larger population. The Old

Order Amish of Pennsylvania, for instance, are a highly inbred group having distinct genotypes. One outcome of their inbreeding is the high frequency of a recessive allele that causes *Ellis–van Creveld syndrome*. Affected individuals have extra fingers or toes and short limbs. Section 12.5 shows one affected individual. The allele probably was rare when the small number of founders immigrated to Pennsylvania. Today, though, about 1 in 8 are heterozygous and 1 in 200 are homozygous for it.

Bottlenecks and inbreeding are a bad combination for endangered species, the populations of which have become smaller and vulnerable to extinction. Cheetahs, for instance, apparently survived a drastic bottleneck in the nineteenth century.

Inbreeding among survivors and their descendants resulted in strikingly similar alleles among the 20,000 existing cheetahs (Figure 17.21). One, a mutated allele, affects fertility. Most male cheetahs have a low sperm count, and 70 percent of the sperm are abnormal. Other shared alleles result in far lower resistance to disease. Therefore, infections that are seldom life-threatening to other cat species can be devastating to cheetahs. In one outbreak of *feline infectious peritonitis* in a wild animal park, the viral pathogen had very little effect on captive lions, but it killed many cheetahs. For them, infection triggered uncontrollable inflammation. Fluid filled the body cavity that houses the heart and other organs, and the cats died in agony. There is no vaccine.

The Florida panther also is highly inbred. Pressure from hunting and urban sprawl has helped put this cat on the endangered species list. Only seventy remain.

Genetic drift is the random change in allele frequencies over the generations, brought about by chance alone. The magnitude of its effect is greatest in small populations, such as the ones that make it through a bottleneck.

Barring mutation, selection, and gene flow, the chance losses and increases of the various alleles at a given locus lead to the homozygous condition and a loss of genetic diversity.

Figure 17.21 A few of the remaining cheetahs, which have some bad alleles that made it through a severe bottleneck.

SUMMARY   *Gold* indicates text section

1. Awareness of evolution—changes in lines of descent over time—emerged through comparisons of the body structure and patterning for major groups of animals (comparative morphology), questions about the world distribution of plants and animals (biogeography), and observations of fossils in sedimentary rock layers. As first proposed by Charles Darwin and Alfred Wallace, evolution can occur by natural selection. *CI, 17.1–17.3*

2. Individuals of a population generally have the same number and kind of genes. But genes come in different allelic forms, and this leads to variations in traits. *17.4*

3. A population is evolving when some forms of a trait (and the underlying alleles) are becoming more or less common relative to the other kinds one generation after the next. This happens through mutation, gene flow, genetic drift, and natural selection (Table 17.2). *17.4*

4. Gene mutations (heritable changes in DNA) are the only source of *new* alleles. New combinations of existing alleles arise by crossing over, independent assortment at meiosis, and mixing of alleles at fertilization. *17.4*

5. Genetic equilibrium, a state in which a population is not evolving, is a baseline to measure change. Here is the Hardy–Weinberg rule: Genetic equilibrium occurs only if there is no mutation, the population is very large and isolated from other populations of the same species, there is no selection, mating is random, and all members survive and reproduce equally. *17.4–17.5*

6. Natural selection is the difference in survival and reproduction among individuals of a population that differ in details of heritable traits. It leads to increased fitness, or adaptation to the environment. *17.6–17.9*

   a. Pressures may shift a range of variation for a trait in one direction (*directional* selection), favor extremes (*disruptive* selection), or eliminate extremes and favor intermediate forms (*stabilizing* selection).

   b. Selection can result in balanced polymorphism. A population is in this state when nonidentical alleles for a given trait are being maintained over the generations at frequencies greater than 1 percent.

   c. Sexual selection, by females or males, can lead to forms of traits that favor reproductive success. Sexual dimorphism (persistence in the phenotypic differences between males and females) is one outcome.

7. Gene flow is a change in allele frequencies brought about by the physical movement of alleles into and out of a population (by immigration and emigration). *17.10*

8. Genetic drift is a change in allele frequencies over the generations due to chance alone. Its effect is greater in small populations than in large ones. *17.11*

9. Drift has greatest impact following a bottleneck (a severe reduction, then a recovery, in population size). Alleles making it through a bottleneck may give rise to

Table 17.2	*Summary of Microevolutionary Processes*
MUTATION	A heritable change in DNA
NATURAL SELECTION	Change or stabilization of allele frequencies owing to differences in survival and reproduction among variant individuals of a population
GENETIC DRIFT	Random fluctuation in allele frequencies over time due to chance occurrences alone
GENE FLOW	Change in allele frequencies as individuals leave or enter a population

differences in the range of phenotypes compared to the original population. In bottlenecking, one type of founder effect, a few emigrants from one population establish a small subpopulation in a new environment. *17.11*

Review Questions

1. Define biogeography and comparative anatomy. How did studies in both disciplines contradict the idea that species have remained unchanged since the time of their creation? *17.1*

2. Define evolution. Define evolution by natural selection. Can an individual evolve? *CI, 17.3, 17.4*

3. Name three broad categories of heritable traits that help characterize a population. *17.4*

4. Explain the difference between continuous variation and polymorphism. *17.4*

5. How do lethal mutations and neutral mutations differ? *17.4*

6. Define genetic equilibrium. Which occurrences can drive allele frequencies away from genetic equilibrium? *17.4*

7. Define fitness, with respect to phenotype. *17.6*

8. Identify the mode of selection (stabilizing, directional, or disruptive) for the diagrams on the facing page. *17.7, 17.8*

9. Define bottleneck and the founder effect. Are these cases of genetic drift, or do they merely set the stage for it? *17.11*

Self-Quiz   ANSWERS IN APPENDIX III

1. Biologists define evolution as _____ .
   a. the origin of a species
   b. heritable change in a line of descent over generations
   c. inheritance of characteristics acquired by the individual

2. Darwin saw that populations of Galápagos finches _____ .
   a. show variation in traits
   b. resemble birds in South America
   c. are adapted to different island habitats
   d. all of the above

3. Individuals don't evolve; _____ do.

4. Genetic variation gives rise to variation in _____ traits.
   a. morphological   c. behavioral
   b. physiological   d. all of the above

5. Sickle-cell anemia first appeared in Asia, the Middle East, and Africa. The causative allele entered the U.S. population when people were forcibly brought over from Africa prior to the Civil War. In microevolutionary terms, this is a case of _____ .
   a. mutation   c. gene flow
   b. genetic drift   d. natural selection

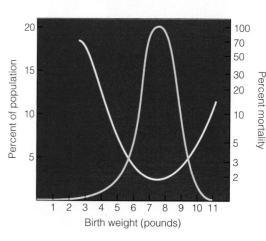

Figure 17.22
The weight distribution for 13,730 human newborns (the *yellow* curve) correlated with mortality rate (*white* curve).

6. Natural selection may occur when there are _____ .
   a. differences in the adaptiveness of forms of traits to prevailing environmental conditions
   b. differences in survival and reproduction among individuals that differ in one or more traits
   c. both a and b

7. Directional selection _____ .
   a. eliminates uncommon forms of alleles
   b. shifts allele frequencies in a steady, consistent direction
   c. favors intermediate forms of a trait
   d. works against adaptive traits

8. Disruptive selection _____ .
   a. eliminates uncommon forms of alleles
   b. shifts allele frequencies in a steady, consistent direction
   c. doesn't favor intermediate forms of a trait
   d. both b and d

9. Match the evolution concepts.
   _____ gene flow
   _____ natural selection
   _____ mutation
   _____ genetic drift
   a. source of new alleles
   b. changes in a population's allele frequencies due to chance alone
   c. allele frequencies change owing to immigration, emigration, or both
   d. differences in survival and reproduction among variant individuals

Critical Thinking

1. Prospects for human newborns of very high or very low birth weight are not good. Being born too small increases the risk of stillbirth or early infant death (Figure 17.22). Also, pre-term instead of full-term pregnancies increase the risks. Are these outcomes of directional, disruptive, or stabilizing selection?

2. For centuries *tuberculosis*, or TB, killed people throughout the world. A bacterium (*Mycobacterium tuberculosis*) causes this contagious lung disease. TB declined steadily in the United States over the years, but caseloads are now rising. The AIDS epidemic, an influx of immigrants from regions where TB is still common, and severe overcrowding in tenements and homeless shelters contribute to the increase. Antibiotics once cured TB in six to nine months. But antibiotic-resistant strains of *M. tuberculosis* have evolved. Two or even three different antibiotics are used simultaneously to block different metabolic pathways of the bacterium. How might this more aggressive approach help get around the problem of antibiotic resistance?

3. A few families in a remote region of Kentucky show a high frequency of *blue offspring*, an autosomal recessive disorder.

Skin of affected individuals appears bright blue. Homozygous recessives lack the enzyme diaphorase. The enzyme catalyzes reactions that maintain hemoglobin in its normal molecular form. Without it, a blue form of hemoglobin accumulates in blood. Skin and blood capillaries are transparent, so pigment colors in blood show through and contribute to skin color. This is why the skin of nonmutated individuals has a pinkish cast—and why blue skin is bluish. Formulate a hypothesis to explain why the blue-offspring trait recurs among a cluster of families but is rare in the human population at large.

4. In 1996, after reflecting on evidence of evolution that has been accumulating for more than a hundred years, Pope John Paul II acknowledged that the theory of evolution "has been progressively accepted by researchers, following a series of discoveries in various fields of knowledge. The convergence, neither sought nor fabricated, of the results of work that was conducted independently is in itself a significant argument in favor of this theory." His words infuriated individuals who believe in a strict, literal interpretation of the biblical account of creation. Some of these individuals have been demanding that biology instructors treat evolution as "just a theory" and give equal time to the biblical interpretation.

Refer to Section 1.5 on the nature of scientific inquiry. Do you equate scientific and religious explanations of life's origin and history as alternative theories? Why or why not?

Selected Key Terms

allele *17.4*	gene pool *17.4*
allele frequency *17.4*	genetic drift *17.11*
antibiotic *17.7*	genetic equilibrium *17.4*
balanced polymorphism *17.9*	Hardy–Weinberg rule *17.5*
biogeography *17.1*	inbreeding *17.11*
bottleneck *17.11*	lethal mutation *17.4*
catastrophism *17.2*	microevolution *17.4*
comparative morphology *17.1*	mutation rate *17.4*
directional selection *17.7*	natural selection *17.6*
disruptive selection *17.8*	neutral mutation *17.4*
evolution *CI*	polymorphism *17.4*
fitness *17.6*	population *17.4*
fixation *17.11*	sampling error *17.11*
fossil *17.1*	sexual selection *17.9*
founder effect *17.11*	stabilizing selection *17.8*
gene flow *17.10*	uniformity (theory of) *17.2*

Readings

Darwin, C. 1957. *Voyage of the Beagle*. New York: Dutton.

Grant, P., and R. Grant. 26 April 2002. "Unpredictable Evolution in a 30-Year Study of Darwin's Finches." *Science*, 707–711. Conceptual bridge from micro- to macroevolution.

SPECIATION

The Case of the Road-Killed Snails

If you happen to be a snail living in a garden in Bryan, Texas, it doesn't take much to keep your genes away from snails in a backyard across the street. By day, the sunbaked asphalt would be about as inviting to a snail as a desert would be to a catfish. Besides, day or night, a street-traversing snail is vulnerable to cars, trucks, skateboards, and bicycles (Figure 18.1*a*). That strip of asphalt is a formidable barrier to the flow of genes between populations. For snails, that is.

Whether any physical barrier deters gene flow depends in large part on the organism's mode of locomotion or dispersal. It also depends on how fast and how long an organism *can* move in response to environmental factors or its own hormonal signals.

A snail is not swift, and it does not roam very far from its home population. Compare it to the wild black duck, leg-banded in Virginia in 1969, that turned up eight years later in Korea. Compare it to the wandering albatross, one of the supreme barrier busters. After lifting off from Kerguelen Island in the Indian Ocean, one of these birds soared westward past the southern tip of Africa, across the Atlantic, and around South America's Cape Horn. After traveling thirteen thousand kilometers, it landed in Chile. With a wingspan of 3.65 meters (12 feet) and a lightweight body, that bird was able to exploit the great prevailing winds of the Southern Hemisphere.

And yet, in 1859, some snails did cross an ocean. Humans had transported garden-variety snails (*Helix aspersa*) from France, then released them in California. The idea was that the snails would multiply and meet culinary demands for that French delicacy *escargots aux fines herbes*. It was bad enough that the importers brought over a less tasty species by mistake. Worse, the exotic species exceeded expectations and became an absolute nuisance in

Figure 18.1 (**a**) A snail (*Helix aspersa*) encountering a major barrier to gene flow. It only appears to be reading the warning label. (**b**) Results from a study of neighboring populations of snails, all descended from founders that ended up in a small town in Texas in the 1930s. For each population, a circle represents relative abundances of three alleles (color-coded *green*, *orange*, and *blue*) for an enzyme, leucine aminopeptidase. Genetic variation is greater between populations living on opposite sides of Twenty-Second Street. For example, notice the higher frequencies of the allele coded *orange* in the block to the west of the street.

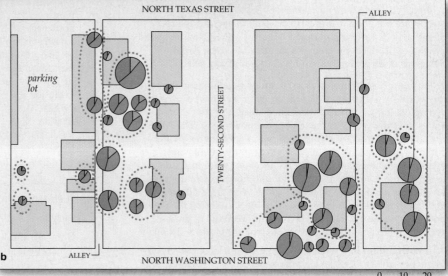

gardens and commercial nurseries through much of the southwestern United States.

By the 1930s, *H. aspersa* had hitched rides to Bryan, Texas, possibly as eggs in the soil of plant containers. They founded small colonies in patches of vegetation. Forty years later Robert Selander, now of Pennsylvania State University, was on his hands and knees with a few graduate students, scouring the patches on two adjacent city blocks. Why? He was interested in the effect of genetic drift on introduced populations. He and his students collected every snail—2,218 of them—from fourteen local colonies. For each colony, they figured out the allele frequencies for five different genes.

The analysis revealed some genetic variation among colonies on the same block—and major differences in allele frequencies *between* different blocks. Figure 18.1*b* shows an example for one of the genes studied.

What if the genetic differences between populations continue to increase, as by natural selection or genetic drift? Will the time come when snails from opposite sides of a street can no longer interbreed successfully even if they manage to get together? In other words, *will they become members of separate species?* Or will stabilizing selection work against significant genetic divergence by eliminating extreme phenotypes from neighboring populations? After all, how far can a workable package of *H. aspersa* genes evolve in such close-together colonies that encounter such similar environmental pressures in a small town in Texas?

And with these questions, we arrive at the topic of **speciation**—of changes in allele frequencies that are significant enough to mark the formation of descendant species from ancestral ones. Charles Darwin proposed long ago that natural selection might lead to speciation. But he wasn't sure how this actually happens. No one was around to watch species originate in the distant past. No one lives long enough to know whether many current populations are at intermediate stages leading to speciation. Even so, evidence is accumulating that supports certain speciation models.

As you read through this chapter, it is important for you to recognize that speciation is not the same thing as natural selection. Evolutionary biologists now view speciation as a *potential outcome* of natural selection and of other microevolutionary processes that may work in concert with it.

Key Concepts

1. A species consists of one or more populations of individuals that have the potential to interbreed under natural conditions and produce fertile offspring, and that are reproductively isolated from other such populations. This definition is restricted to sexually reproducing organisms.

2. Populations of the same species have a shared genetic history, they are maintaining genetic contact over time, and they are evolving independently of other species.

3. Speciation is the actual process by which daughter species evolve from a parent species.

4. By one model, speciation starts when a geographic barrier arises between populations or subpopulations of a species. Thereafter, mutation, natural selection, and genetic drift operate independently in each population. As a result, the populations genetically diverge from one another. The divergence may be irreversible.

5. Diverging populations may start to differ in certain alleles that affect their morphological, physiological, or behavioral traits associated with reproduction. When the genetic differences affecting reproduction become great enough, the populations can no longer interbreed even if they end up coexisting in the same area later on. Speciation is completed.

6. The model of speciation just described, which applies to geographically separated populations, is known as allopatric speciation. This might be the way most species originate in nature. Sympatric speciation and parapatric speciation might be less prevalent.

7. By the sympatric speciation model, species form within the home range of their parent species. By the parapatric speciation model, adjacent populations become distinct species while still maintaining contact along a common border between their home ranges.

8. The timing, rate, and direction of speciation differ among the branches of a given lineage. They also differ from one lineage to the next. The extinction of some number of species is inevitable for all lineages.

ON THE ROAD TO SPECIATION

What Is a Species?

"If it looks like a duck, walks like a duck, and quacks like a duck, probably it's a duck." Let's use this familiar saying as a starting point for defining what is and what is not a species. **Species** is a Latin word. It simply means "kind," as in "a particular kind of duck."

Early naturalists defined species mostly in terms of morphological traits. But common sense told them to consider other factors. Example: Sometimes individuals of a species look very different only because they grew and developed under different environmental conditions. The two plants in Figure 18.2 are a case in point.

In fact, individuals of most species vary greatly in morphological details. So perhaps we should also look for a basic function that unites populations of a species and isolates them from populations of other species.

Reproduction is such a basic, defining function. It is actually central to the **biological species concept**. Ernst Mayr, an evolutionary biologist, phrased the concept this way: Species are groups of interbreeding natural populations that are reproductively isolated from other such groups. In his view, it doesn't matter how much phenotypes vary. The populations belong to the same species as long as their members have the same form, physiology, and behaviors that let them interbreed and produce fertile offspring. Being able to contribute to a shared gene pool is qualification for membership.

Mayr's species concept does not apply to asexually reproducing organisms, such as bacteria. However, it's a useful guide for research into factors that define the vast majority of species, which do reproduce sexually.

If we subscribe to Mayr's concept, then speciation is attainment of reproductive isolation. Bear in mind, this does not mean that reproductive isolation evolves purposefully to promote the formation of a species or to maintain its separate identity. Rather, *any structural, functional, or behavioral difference that favors reproductive isolation is simply a by-product of genetic change.*

Gene flow alone counters genetic changes between populations of the same species. **Gene flow**, recall, is a movement of alleles into and out of populations by immigration and emigration. It exerts a homogenizing effect even when two populations are geographically separate as long as some genes are exchanged between them. Thus, gene flow is a microevolutionary process that helps maintain a common reservoir of alleles.

What if some geographic barrier arises and blocks intermingling of genes between some populations or subpopulations of a species? Genetic divergence will follow. Through **genetic divergence**, the gene pools of genetically separated populations accumulate distinct differences. How? Gene mutation, natural selection, and genetic drift modify each one independently of

Figure 18.2 Pronounced morphological differences between two plants of the same species. Shown are mature leaves of arrowheads (*Sagittaria sagittifolia*) growing (**a**) on land and (**b**) in water. The difference is attributable to adaptive responses to different environmental conditions, not to genetic differences.

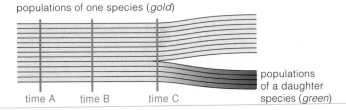

populations of one species (*gold*)

time A   time B   time C

populations of a daughter species (*green*)

Figure 18.3 Simplified diagram of genetic divergence. Each horizontal band represents a different population.

others. Figure 18.3 is a simple way to think about the process. Each of its horizontal bands represents one population. At times A and B, gene flow kept them in the same species. But a geographic barrier isolated some of them at time C and allowed their divergence.

Most species arose by gradual genetic divergence. Yet, as you will see, many flowering plants and other species arose far more rapidly. Said another way, *the process of speciation can vary in its duration.*

Depending on how populations interact and their patterns of distribution, *speciation also may vary in its details.* The preceding paragraphs started you thinking about what is probably the main route, which is called allopatric speciation. Later in the chapter, you will be taking a closer look at this route and two others.

Reproductive Isolating Mechanisms

Let's now define **reproductive isolating mechanisms** as any heritable feature of body form, functioning, or behavior that prevents interbreeding between one or more genetically divergent populations. Some prevent successful mating or pollination between individuals of the divergent populations, so hybrid zygotes can't form. Let's look first at some *prezygotic* mechanisms.

Figure 18.4 (**a**) Behavioral isolation, represented by a sampling of courtship displays that precede sex between a male and female albatross. Individuals of their species recognize tactile, visual, and acoustical components of the displays.

(**b**) A case of temporal isolation. Adult *Magicicada septendecim*. This periodical cicada matures underground and emerges to reproduce every 17 years. Often its populations overlap the habitats of a sibling species (*M. tredecim*), which reproduces every 13 years. Adults live only a few weeks. Birds consider them a taste thrill and may frenziedly gorge on them to the point of vomiting, which makes one suspect why periodical cicadas stay underground for so long.

(**c**) Mechanical isolation, as demonstrated by the precise fit between a zebra orchid and a wasp species, one of its few pollinators.

Behavioral isolation: Behavioral differences bar gene flow between related species in the same vicinity. For instance, before male and female birds copulate, they often participate in courtship rituals (Figure 18.4*a*). A female is genetically prewired to recognize distinctive singing, head bobbing, wing spreading, or prancing of a male of her species as an overture to sex. Females of different species usually ignore his behavior.

Temporal isolation: Individuals of populations that are diverging could still interbreed, but not when they are reproducing at different times. Periodical cicadas (*Magicicada*) of the eastern United States are separated in time. These winged insects mature underground and feed on juicy roots. Three species differ in size, color, and song. Every 17 years they emerge and reproduce (Figure 18.4*b*). Each has a "sibling species," which is about the same, morphologically. Its sibling species emerges every 13 years. Only once every 221 years do the two release gametes in the same year!

Mechanical isolation: Incompatibility between body parts is a way to keep potential mates or pollinators mechanically isolated (Figure 18.4*c*). Two sage species keep their pollen to themselves by attracting different insect pollinators. The pollen-bearing stamens of one extend from a nectar cup, above petals that form a big platform for big pollinators. Small bees that land on it typically do not brush against the stamens and pick up pollen. By contrast, the landing platform of the other species is too small to hold big pollinators.

Ecological isolation: Populations adapted to different microenvironments in a single habitat may be isolated ecologically. In the seasonally dry foothills of the Sierra Nevada are two species of manzanita shrubs. One lives in open forests of conifers at elevations between 600 and 1,850 meters, and the other at elevations between 750 and 3,350 meters. Where their ranges overlap, the two rarely hybridize. Like other manzanitas, they have

built-in, physiological mechanisms that enhance water conservation during the dry season. But one species does well only in far more sheltered sites, where water stress is not intense. The other is adapted to drier, more exposed sites on rocky hillsides. These ecological differences make cross-pollination unlikely.

Gametic mortality: Gametes of different species may have evolved incompatibilities at the molecular level. For example, when the pollen of one species lands on a plant of a different species, it usually doesn't recognize the molecular signals that are supposed to trigger its growth down through the plant's tissues, to the egg.

Postzygotic isolating mechanisms may act while an embryo is developing. Unsuitable interactions among some of its genes or gene products cause early death, sterility, or hybrids of low fitness. The hybrid offspring are often weak; survival rates are low. Some hybrids are sturdy but sterile. Mules, the offspring of a female horse and male donkey, are like this.

THE BIOLOGICAL SPECIES CONCEPT. **A species is one or more populations of individuals that (1) are interbreeding under natural conditions and producing fertile offspring, and (2) are reproductively isolated from other such populations.**

Speciation is the process by which a daughter species forms from a population or subpopulation of a parent species. The process varies in its details and in the length of time it takes to complete reproductive isolation.

By what may be the most common speciation mechanism, a geographic barrier separates populations of a species. Mutation, natural selection, and genetic drift independently operate in each population, so differences build up in the gene pools. This outcome is called genetic divergence.

Reproductive isolating mechanisms may evolve simply as by-products of the genetic changes. They are heritable traits that, one way or another, prevent interbreeding.

SPECIATION IN GEOGRAPHICALLY ISOLATED POPULATIONS

Allopatric Speciation Defined

If genetic changes required for a species to originate can result from *physical separation* between populations, then allopatry may be the main speciation route. By the **allopatric speciation** model, a physical barrier arises between populations or subpopulations of a species and stops gene flow. (*Allo–* means different; *patria* can be taken to mean homeland.) Reproductive isolating mechanisms evolve in the populations. And speciation is complete when interbreeding among the populations will not occur even if changing circumstances put them back together in the same area.

Whether a geographic barrier proves to be effective at blocking gene flow depends on an organism's means of travel, how fast it travels, and whether it is forced or behaviorally inclined to disperse. Reflect on those garden snails and the wandering albatross described at the start of this chapter.

The Pace of Geographic Isolation

A measurable distance often separates the populations of a species, and gene flow among them is more of an intermittent trickle than a steady stream. Also, barriers can rapidly arise and shut off the trickles. In the 1800s, a huge earthquake buckled part of the Midwest. It altered the course of the Mississippi River, which cut through habitats of insects that couldn't swim or fly. It stopped gene flow between insects on opposite shores.

The fossil record indicates that geographic isolation also occurs over great spans of time. This happened after huge glaciers advanced down into North America and Europe during ice ages and cut off populations of plants and animals from one another. In time, glaciers retreated, and descendants of the separate populations met up. Although related by descent, some were no longer reproductively compatible; they had evolved into new species. Genetic divergence was not as great between other separated populations; descendants are

still interbreeding. Reproductive isolation in their case was incomplete, so speciation did not occur.

As another example, the Earth's crust is fractured into gigantic plates, somewhat like a cracked eggshell. Slow but colossal movements of the plates have altered the configurations of land masses (Section 19.3). When Central America formed, part of an ancient ocean basin was uplifted and became the Isthmus of Panama. This new geographic barrier divided marine populations and invited allopatric speciation.

In the 1980s, John Graves compared four enzymes from the muscle cells of two related species of isthmus fishes (Figure 18.5). Both species are strong swimmers. As Graves knew, temperature has different effects on different enzymes. Seawater on the Pacific side of the isthmus is cooler by about 2°–3°C than on the Atlantic side. It also varies seasonally. Graves discovered that the four "Pacific" enzymes he studied function better at lower temperatures than the equivalent "Atlantic" enzymes. He then subjected the four pairs of enzyme molecules to gel electrophoresis (Section 16.3). In this case, the laboratory method revealed slight differences in electric charge between two of the pairs. Those two differ slightly in amino acid sequences.

Graves drew two tentative conclusions: Selection pressures that were imposed by small differences in environmental conditions may have been the impetus for divergences in the molecular structure of certain enzymes. In addition, alternative molecular forms of the enzymes studied display detectable differences in catalytic activity. They might be indicators that closely related populations of fishes on opposite sides of the isthmus are diverging from one another.

How might you interpret results from this study of geographical isolation? In the populations on each side of the isthmus, a gradual accumulation of neutral or adaptive gene mutations might be a microevolutionary "foot in the door." The genetic divergences among the related fish populations might be evidence of allopatric speciation in progress.

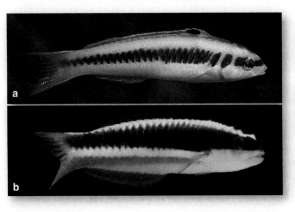

ISTHMUS OF PANAMA

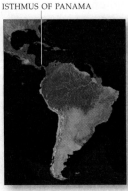

Figure 18.5 (**a**) Blue-headed wrasse (*Thalassoma bifasciatum*) from the Atlantic Ocean near the Isthmus of Panama. (**b**) Cortez rainbow wrasse (*T. lucasanum*) from the Pacific Ocean near the isthmus. The two species may be related by descent from a shared ancestral population that split when geologic forces created the isthmus. You might see individuals that differ in body coloration and patterning from these specimens. Such variation is common among reef fishes.

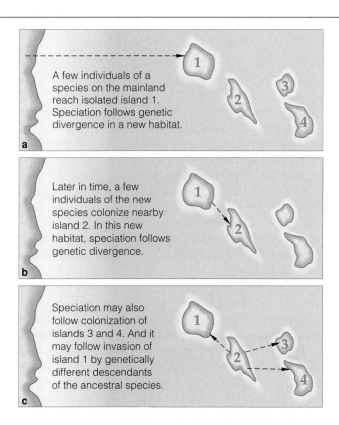

a A few individuals of a species on the mainland reach isolated island 1. Speciation follows genetic divergence in a new habitat.

b Later in time, a few individuals of the new species colonize nearby island 2. In this new habitat, speciation follows genetic divergence.

c Speciation may also follow colonization of islands 3 and 4. And it may follow invasion of island 1 by genetically different descendants of the ancestral species.

Figure 18.6 Sketch of allopatric speciation on an isolated archipelago. (Can you envision other possibilities?)

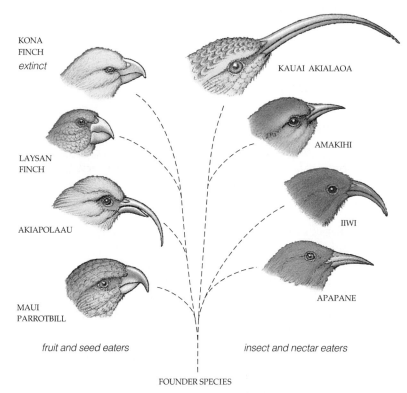

KONA FINCH *extinct*

KAUAI AKIALAOA

LAYSAN FINCH

AMAKIHI

AKIAPOLAAU

IIWI

MAUI PARROTBILL

APAPANE

fruit and seed eaters   *insect and nectar eaters*

FOUNDER SPECIES

Figure 18.7 A few Hawaiian honeycreepers, a fine example of how a new arrival in species-poor habitats on an isolated archipelago can be the start of a flurry of allopatric speciation.

Allopatric Speciation on Archipelagos

An **archipelago** is an island chain some distance away from a continent. Some archipelagos are so close to the mainland that gene flow is more or less unimpeded, and there is little if any speciation. The Florida Keys are like this. Other archipelagos are isolated enough to serve as laboratories for the study of evolution (Figure 18.6). Foremost among these are the Hawaiian Archipelago, nearly 4,000 kilometers from the California coast, and the Galápagos Islands, about 900 kilometers from the Ecuadorian coast. The islands of both chains are merely the tops of great volcanoes, some still active, that rose from the deep seafloor. When each volcano first broke the surface of the sea, its fiery surface was devoid of life.

Remember those Galápagos finches (Section 17.3)? A few finches from the mainland apparently evolved in isolation on one Galápagos island. Later, some of their descendants reached other islands in the chain. In those new and unoccupied habitats—even in different parts of the same habitats—conditions varied, and the island hoppers were subject to a variety of selection pressures. In time, genetic divergences within and between the islands paved the way for new episodes of allopatric speciation. Later on, some island-hopping new species even invaded the island of their ancestors. The distances between islands are enough to foster divergences, but not enough to stop occasional colonizations.

Bursts of speciation in the Hawaiian Archipelago have been quite dramatic. They are a premier example of adaptive radiation in a species-poor environment, as described in Section 18.4. The youngest island, Hawaii, formed less than a million years ago. Here alone we see diverse habitats, ranging from lava beds to rain forests to alpine grasslands and high, snow-capped volcanoes. When the ancestors of Hawaiian honeycreepers arrived, they found a veritable buffet of fruits, seeds, nectars, and tasty insects—and not many competitors for them. The near absence of competition from other bird species fanned allopatric speciations in vacant adaptive zones. Figure 18.7 hints at the resulting variation that arose among different species of Hawaiian honeycreepers. Today, these and thousands of other species of animals and plants that evolved in the archipelago are found nowhere else. As another example of the potential for speciation, the Hawaiian Islands make up less than 2 percent of the world's land mass. Yet they are home to 40 percent of all fruit fly (*Drosophila*) species.

ALLOPATRIC SPECIATION. **Some type of physical barrier intervenes between populations or subpopulations of a species and prevents gene flow among them. By doing so, it favors genetic divergence and speciation.**

uncommon

MODELS FOR OTHER SPECIATION ROUTES

Geographic isolation may not always be required for reproductive isolation. In the models for sympatric and parapatric speciation, populations diverge genetically even while they are in contact with one another.

Sympatric Speciation

By the model for **sympatric speciation**, a species may form *within* the home range of an existing species, in the absence of a physical barrier. (*Sym*– means together with, as in "together with others in the homeland.")

EVIDENCE FROM CICHLIDS IN AFRICA   In 1994, Ulrich Schliewen found circumstantial evidence of sympatric speciation. Schliewen and his coworkers studied fishes (cichlids) that live together in two lakes in Cameroon, West Africa. Each lake basin is the collapsed cone of a small volcano (Figure 18.8). Yet eleven cichlid species coexist in one lake. Nine coexist in the other.

The researchers analyzed mitochondrial DNA from all species in one lake. They did the same for all species in the other lake as the basis for comparison. In their nucleotide sequences, the species of each lake are like each other and *not* like related species in nearby lakes and rivers. For example, nine species in one lake are the only cichlids having a unique base-pair substitution in the gene that specifies the protein cytochrome *b*.

Physical and chemical conditions are too uniform in the lakes to foster geographic separation. For instance, shorelines are so uniform that even small topographic barriers are absent. Thus, allopatric populations could

not have evolved, even on a small scale. The lakes are isolated except for tiny creeks trickling in from higher elevations; there has been no gene flow from outside. The lakes must have been colonized before connections with a nearby river system were severed. Also, cichlids are not sluggish swimmers. Besides being very mobile, individuals of different species often make contact in the small lakes; they must be living in sympatry.

Because clusters of species are more closely related to one another than to species anywhere else, we may assume they share a common ancestor. Because of their restricted distribution, we may assume they formed in the same lake. And yet nothing about the lakes isolates different populations. Therefore, the ancestors of each cichlid species were never isolated from one another while they were evolving in separate directions.

Within each lake, species do show small degrees of *ecological* separation; differences in feeding preferences put them in different places. Some cichlid species feed in open waters and others at the lake bottom. Even so, they all *breed* close to the bottom, in sympatry. It may be that small-scale ecological separation was enough to influence sexual selection among potential mates. Over the generations, it may have fostered the reproductive isolation that can lead to speciation.

SPECIATION BY WAY OF POLYPLOIDY   Quite possibly, sympatric speciation has been a prevalent evolutionary event among flowering plants. About half of all known flowering plant species are polyploid. In **polyploidy**, somatic cells of individuals have three or more of each type of chromosome characteristic of the species. Such changes arise when chromosomes separate improperly during meiosis or mitosis. They also arise when a germ cell replicates its DNA but fails to divide, then goes on to function as a gamete (Section 12.9).

Speciation may have been rapid for many flowering plants that engage in self-fertilization or some asexual reproductive mode. Suppose they produced polyploid offspring. If the extra chromosomes paired with each other at meiosis, maybe the extra set of genes did no harm. Common bread wheat is one of the species that might have arisen by polyploidy. In this case, though, cross-fertilization also was involved (Figure 18.9).

Allen Orr has evidence that polyploid animals are rare because of a failed **dosage compensation**. By this normal event, genes on sex chromosomes are expressed at the same levels in females *and* males. For instance, by X chromosome inactivation in female mammals, genes of only one X chromosome are expressed in males *and* females (Section 15.4). Dosage compensation changes with polyploidy; extra genes are expressed, with bad or fatal effects. The mechanism is absent in plants, so their chromosome doublings are not as problematic.

Figure 18.8   A crater lake in Cameroon, West Africa. Nine kinds of cichlids apparently arose by way of sympatric speciation in this small, isolated crater lake. There is not even small-scale geographic isolation. Populations of cichlids separate by feeding preferences, but all breed near the lake bottom, in sympatry.

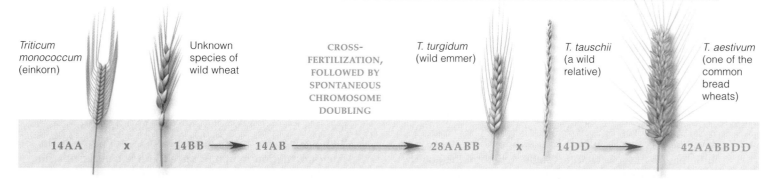

Triticum monococcum (einkorn)   Unknown species of wild wheat   CROSS-FERTILIZATION, FOLLOWED BY SPONTANEOUS CHROMOSOME DOUBLING   *T. turgidum* (wild emmer)   *T. tauschii* (a wild relative)   *T. aestivum* (one of the common bread wheats)

14AA   x   14BB → 14AB ————→ 28AABB   x   14DD → 42AABBDD

a By 11,000 years ago, human groups were cultivating wild wheats. Einkorn (*Triticum monococcum*) is still around. It has a diploid chromosome number of 14 (two sets of 7 chromosomes, or 14AA). Einkorn probably hybridized with another wild wheat species with the same chromosome number.

b About 8,000 years ago, polyploidy arose in a population of AB plants. Such plants may have evolved from sterile, self-fertilizing AB hybrid wheat. Wild emmer (*T. turgidum*) plants are tetraploid (AABB), with a chromosome number of 28 (two sets of 14). They are fertile; their A chromosomes pair with each other at meiosis, and the B chromosomes pair with each other.

c An AABB plant probably hybridized with *T. tauschii*, a wild relative of wild emmer. Its diploid chromosome number must have been 14 (two sets of 7 DD). The hybrid descendants now include common bread wheats. One of these, *T. aestivum*, has a chromosome number of 42 (six sets of 7 AABBDD).

Figure 18.9 Presumed sympatric speciation in wheat by polyploidy and hybridizations. Wheat grains 11,000 years old have been found in the Near East. Diploid wild wheats still grow there.

BULLOCK'S ORIOLE   BALTIMORE ORIOLE

BULLOCK'S ORIOLE   BALTIMORE ORIOLE

HYBRID ZONE

Figure 18.10 A possible case of parapatric speciation? Hybrids arise along a common border of the geographic ranges of two species of orioles. To the east are Baltimore orioles; to the west are Bullock's orioles. Hybridization was once so common that these two birds were considered subspecies. In 1997, the American Ornithologists' Union said they are separate species.

Parapatric Speciation

By the model for **parapatric speciation**, neighboring populations become distinct species while maintaining contact along a common border. (*Para–* means near, as in "near another homeland.") Interbreeding individuals produce hybrid offspring in the region, which is called a **hybrid zone**. Evidence that might support parapatric speciation is sketchy. Why? It's often difficult to know whether or not geographically separated populations are only subspecies. *Subspecies* refers to geographically distinct populations of the same species.

For example, Bullock's orioles differ from Baltimore orioles in the color patterning of most of their feathers. Male orioles have different territorial songs and mate with their own kind, usually. As K. Corbin determined, the birds also differ in allele frequencies for several enzymes. Their geographic ranges differ but overlap in the American Midwest (Figure 18.10). Interbreeding was once common in the hybrid zone but is becoming less frequent. Is it an example of parapatric speciation in progress? Maybe. But it might also be an example of "secondary contact." That is, gene flow may be going on between previously isolated subspecies that diverged only recently from a common ancestor.

SYMPATRIC SPECIATION. Daughter species arise from a group of individuals within an existing population. This seems to have been a common speciation event among the polyploid species of flowering plants.

PARAPATRIC SPECIATION. Adjacent populations evolve into distinct species even while maintaining contact along their common border.

PATTERNS OF SPECIATION

Branching and Unbranched Evolution

All species—past and present—are related by descent. They share genetic connections through lineages that extend back in time to the molecular origin of the first prototypic cells, some 3.8 billion years ago. Subsequent chapters focus on evidence that supports this view. In anticipation of those chapters, let's start thinking about ways to interpret the large-scale histories of species.

The fossil record provides evidence of two patterns of evolutionary change in lineages. One is branched, the other unbranched. The first is called **cladogenesis** (from the Greek *klados*, meaning branch; and *genesis*, meaning origin). This is the pattern by which a lineage splits, with populations becoming genetically isolated and then diverging in different evolutionary directions. This is the pattern of speciation described earlier.

By the second pattern, **anagenesis**, changes in allele frequencies and in morphology accumulate within an unbranched line of descent. (In this context, *ana*– means renewed.) Directional changes are confined within one lineage, and gene flow never does cease among all of its populations. At some point in time, allele frequencies and morphology have changed so much that we assign a separate name to the descendants. Such changes are occurring in the moth populations you read about earlier.

eg endoSymbiosis

Evolutionary Trees and Rates of Change

Evolutionary trees summarize information about the continuity of relationship among species. Figure 18.11 is a simple way to start thinking about how tree diagrams are constructed. Each *branch* represents a single line of descent from a common ancestor. And each *branch point* represents a time of genetic divergence and speciation, as brought about by microevolutionary processes. Such trees also can convey whether the speciation process was gradual or rapid.

If a branch is just slightly angled, this means that the species originated by many small morphological changes over long time spans (Figure 18.11a). Such change is the main premise of the **gradual model of speciation**, which fits well with numerous fossil sequences. Example: In many layers of sedimentary rock we find vertical sequences of intricately perforated shells of foraminiferans, a common type of protistan. The observable characteristics of foraminiferan fossils in the sequences offer evidence of slow change.

Alternatively, evolutionary tree diagrams for some lineages are constructed with short horizontal branches that make an abrupt 90-degree turn, as in Figure 18.11b. The **punctuation model of speciation** is consistent with such diagrams. By this model, most of the changes in morphology are compressed into a brief period when populations are starting to diverge—say, within only hundreds or thousands of years. Bottlenecks, founder effects, strong directional selection, or a combination of these foster rapid speciation. The daughter species recover fast from the adaptive wrenching, then change little during the next 2 million to 6 million years or so. The proponents of this model argue that reproductive cohesion has indeed prevailed for about 99 percent of the history of most lineages. They cite strong evidence of abrupt change in many parts of the fossil record.

Changes in lines of descent apparently have been gradual, abrupt, or both. Species originated at different times, and they differ in how long they have persisted. Remember, some lineages have endured without much change, often for millions of years. Other lineages have branched bushily and sometimes spectacularly in times of adaptive radiation.

Adaptive Radiations

An **adaptive radiation** is a burst of divergences from a single lineage that give rise to many new species, each adapted to an unoccupied or a new habitat or to using novel resources. Figure 18.12 is an example. In the past, species of a lineage often radiated into vacant **adaptive zones**. Think of adaptive zones as ways of life, such as "burrowing in seafloor sediments" or "catching winged

PRESENT

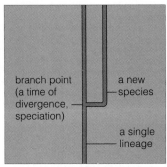

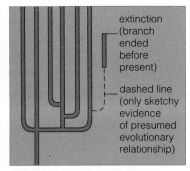

Geologic time

←— Change in form —→

Figure 18.11 How to read evolutionary tree diagrams.

a Branching at a slight angle; speciation after gradual changes in traits over geologic time.

b Horizontal branching; rapid changes near time of speciation. A long vertical branch signifies little change after speciation.

c Many branchings of the same lineage at or near the same point in geologic time; adaptive radiation occurred.

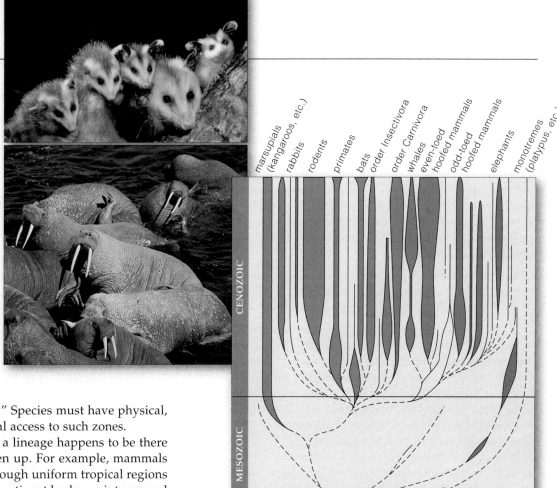

Figure 18.12 Example of adaptive radiation, as shown by an evolutionary tree diagram. This one started about 65 million years ago, at the start of a geologic era called the Cenozoic, and it led to mammals as different as the opossum and walrus. Variations in the branch widths correspond to the number of groups in the lineage, as represented at different points in time. The wider the branch, the greater the species diversity.

insects in the air at night." Species must have physical, evolutionary, or ecological access to such zones.

Physical access means a lineage happens to be there when adaptive zones open up. For example, mammals were once distributed through uniform tropical regions of a huge continent. The continent broke up into several land masses that gradually drifted apart. Habitats and resources changed in many different ways on different masses and set the stage for independent radiations.

Evolutionary access means that a modification to an existing structure or function will permit a lineage to exploit the environment in new or more efficient ways. Such modifications are **key innovations**. Example: The forelimbs of certain five-toed vertebrates evolved into wings. And that innovation opened up novel adaptive zones for the ancestors of birds and bats (Section 19.4).

Ecological access means that the lineage can enter vacated adaptive zones or displace the resident species. Chapters 20 and 47 offer examples.

Extinctions—End of the Line

Extinction is the irrevocable loss of a species. The fossil record shows twenty or more **mass extinctions**—large, catastrophic losses of entire families or other major groups. For example, 250 million years ago, 95 percent of all known species were abruptly lost. In other times, only some groups vanished. By studying the statistical distribution of extinctions, David Raup found that most were clustered in time even though they differed in size. Like George Simpson before him, he realized that times of reduced diversity follow extinction events, then new species arise and fill vacant adaptive zones.

Apparently, luck has a lot to do with it. When one asteroid struck the Earth and wiped out dinosaurs, for

example, mammals were among the survivors that radiated into the vacated adaptive zones. Also, species have tended to be hit hard by long-term changes in the global climate, especially the species adapted to warm, tropical conditions. Unlike species adapted to cold, they had nowhere to go. Asteroids, drifting continents, climatic change—as you will see in later chapters, these are some of the major factors that have contributed to a general pattern of extinctions. In Chapter 27, you also will read about a mass extinction that is currently under way, thanks largely to us.

Lineages have changed gradually, abruptly, or both. Their member species originated at different times and have differed in how long they have persisted.

Adaptive radiations are bursts of divergences from a single lineage that gave rise to many new species, each adapted to a vacant or new habitat or to using a novel resource.

Repeated and often large extinctions occurred in the past. After times of reduced diversity, new species formed and occupied new or vacated adaptive zones.

Together, the persistence, branchings, and extinctions of species account for the range of biodiversity at any point in geologic time.

18.5

Gold indicates text section

1. A species is a single kind of organism, recognized partly in terms of its morphology. *18.1*

2. By the biological species concept, a species is one or more populations of individuals that are interbreeding and producing fertile offspring under natural conditions and that are reproductively isolated from other such populations. The concept identifies a species mainly in terms of the portion of alleles that promote or maintain reproductive isolation. Also, it applies only to sexually reproducing organisms. *18.1*

3. The populations of a species have a shared genetic history, are maintaining genetic contact over time, and are evolving independently of other species. *18.1*

4. Speciation is a process by which descendant species form from a population or subpopulation of ancestral species; attainment of reproductive isolation. *CI, 18.1*

   a. Example: Genetic divergence proceeds after some geographic barrier stops gene flow among populations of the same species. Differences accumulate in the gene pools of the isolated populations as microevolutionary processes operate independently in each one.

   b. Microevolutionary processes—mutation, natural selection, and genetic drift—may by chance give rise to reproductive isolating mechanisms. Such mechanisms prevent interbreeding between populations and thus promote irreversible genetic differences between them.

   c. Speciation also can occur instantaneously, as by polyploidy or other changes in chromosome number.

5. Prezygotic isolating mechanisms prevent mating or pollination between individuals of populations. They include differences in reproductive timing or behavior, incompatibilities in reproductive structures or gametes, and occupation of different microenvironments in the same area. Postzygotic mechanisms lead to early death, sterility, or unfit hybrid offspring. They take effect after fertilization, while the embryo develops. *18.1*

6. Here are three current models for speciation:

   a. Allopatric speciation. Geographic barriers prevent gene flow between the populations of a species, which genetically diverge in such a way that interbreeding will not occur in nature even if their individuals make contact with each other later on. Allopatric speciation might be the most prevalent speciation route. *18.2*

   b. Sympatric speciation. Reproductive isolation of individuals in the same home range leads to speciation. Speciation by polyploidy is an example. *18.3*

   c. Parapatric speciation. Adjacent populations give rise to a new species while maintaining contact along a border between their home ranges. *18.3*

7. Lineages differ in their time, rates, and direction of speciation. Speciation can proceed gradually, rapidly, or both. Extensive branching of a single lineage (a burst of divergences from it) in the same geologic time span is an adaptive radiation. The lineage may be physically in place when new adaptive zones (ways of life) open up. It may access zones by key innovations. It may enter vacated zones or displace resident species. *18.4*

8. Evolutionary tree diagrams indicate the relationships among groups of species. Each tree branch represents a line of descent (lineage). Branch points are speciation events, as brought about by natural selection, genetic drift, and other microevolutionary processes. *18.4*

9. Extinction is the irrevocable loss of a species. The fossil record shows twenty or more mass extinctions: large, catastrophic events in which entire families or other major groups were abruptly lost. *18.4*

10. Persistences, branchings, and extinctions of species account for the range of biodiversity (Table 18.1). *18.4*

Table 18.1 *Summary of Processes and Patterns of Evolution*

MICROEVOLUTIONARY PROCESSES

Mutation	Original source of alleles	Stability or change in a species is the outcome of balances or imbalances among all of these processes, the effects of which are influenced by population size and by the prevailing environmental conditions.
Gene flow	Preserves species cohesion	
Genetic drift	Erodes species cohesion	
Natural selection	Preserves or erodes species cohesion, depending on environmental pressures	

MACROEVOLUTIONARY PROCESSES

Genetic persistence	Basis of the unity of life. The biochemical and molecular basis of inheritance extends from the origin of first cells through all subsequent lines of descent.
Genetic divergence	Basis of life's diversity, as brought about by adaptive shifts, branchings, and radiations. Rates and times of change varied within and between lineages.
Genetic disconnect	Extinction. End of the line for a species. Mass extinctions are catastrophic events in which major groups are lost abruptly and simultaneously.

Review Questions

1. How does the biological species concept differ from a definition of species based on morphological traits alone? *18.1*

2. Define speciation and describe a speciation model. *CI, 18.1–18.3*

3. Give examples of reproductive isolating mechanisms. *18.1*

4. With respect to evolutionary tree diagrams, describe what each of the following features represents: *18.4*
   a. a single line
   b. soft-angled branching of line
   c. horizontal branching
   d. vertical continuation of branch
   e. many branchings of a line
   f. dashed line
   g. branch ending before present

5. Define and give an example of an adaptive radiation. *18.4*

6. Define extinction. What kinds of events might bring about extinctions? *18.4*

Self-Quiz ANSWERS IN APPENDIX III

1. Sexually reproducing individuals of a species ______.
   a. can interbreed under natural conditions
   b. can produce fertile offspring
   c. have a shared genetic history
   d. all of the above

2. Reproductive isolating mechanisms ______.
   a. prevent interbreeding
   b. prevent gene flow
   c. reinforce genetic divergence
   d. all of the above

3. When potential mates occupy overlapping ranges but reproduce at different times, this is a case of ______ isolation.
   a. postzygotic
   b. mechanical
   c. temporal
   d. gametic

4. In an evolutionary tree diagram, a branch point represents a ______, and a branch that ends represents ______.
   a. single species; incomplete data on lineage
   b. single species; extinction
   c. time of divergence; extinction
   d. time of divergence; speciation complete

5. An evolutionary tree diagram with horizontal branches that abruptly become vertical is consistent with the ______.
   a. gradual model of speciation
   b. punctuation model of speciation
   c. idea of small changes in form over long spans of time
   d. both a and c

6. Match each term with its most suitable description.

____ cladogenesis	a. burst of genetic divergences from a single lineage
____ anagenesis	b. catastrophic disappearance of major groups of organisms
____ adaptive radiation	c. branching lineages
____ extinction event	d. a species lost from a lineage
____ mass extinction	e. genetic and morphological change in unbranched lineage

Critical Thinking

1. You notice several duck species in the same lake habitat. All the females of the various species look quite similar to one another. But the males differ in the patterns and colors of their feathers. Speculate on which forms of reproductive isolation may be keeping each species distinct. How does the appearance of the male ducks provide a clue to the answer?

2. One family of mammals includes true horses (*Equus*), zebras (*Hippotrigris* and *Dolichohippus*), and donkeys and asses (*Asinus*). Zebroids are hybrid offspring of wild zebras and domesticated horses that were confined to the same pasture (Figure 18.13). The unnatural confinement breached the reproductive barriers between the two lineages. The barriers have been in place since a genetic divergence got under way more than 3 million years ago. What does the breach suggest about the genetic changes required to attain irreversible reproductive isolation in nature?

3. Suppose a small population that was founded by only a few individuals gets cut off from gene exchange with the main population of their species. By Mayr's view of the *founder effect*, allele frequencies for a few gene products shift as an outcome of genetic drift, and the shift triggers a number of changes in other genes that are affected by those alleles. If that is the case, then genetic changes in newly founded populations may be so great that speciation is rapid. It actually may be too rapid for us to identify it in the fossil record. Does this view correspond to the gradual model or the punctuated model of speciation?

Figure 18.13 A mixed herd of zebroids and horses.

4. A key innovation, recall, is some modification in structure or function that permits a species to exploit the environment in a more efficient or novel way, compared to the ancestral species. Identify a key innovation of an existing species, such as humans. Then describe how that innovation might be the basis of an adaptive radiation in environments of the distant future.

5. Richard Lenski uses bacterial populations in culture tubes to develop model systems for studying evolution. He is fond of saying that such a population is the equivalent of the entire human population. But he can replicate it several times over, that bacteria produce several generations in a day, and that he can store them in the deep freeze, then bring them back to active form, unaltered, to directly compare ancestors and descendants. Are bacterial models relevant to evolutionary studies of sexually reproducing organisms? Before you answer, read a short article by P. Raine and M. Travisano entitled "Adaptive Radiation in a Heterogenous Environment" (*Nature*, 2 July 1998, 69–72).

Selected Key Terms

adaptive radiation *18.4*
adaptive zone *18.4*
allopatric speciation *18.2*
anagenesis *18.4*
archipelago *18.2*
biological species concept *18.1*
cladogenesis *18.4*
dosage compensation *18.3*
evolutionary tree *18.4*
extinction *18.4*
gene flow *18.1*
genetic divergence *18.1*
gradual model, speciation *18.4*
hybrid zone *18.3*
key innovation *18.4*
mass extinction *18.4*
parapatric speciation *18.3*
polyploidy *18.3*
punctuation model, speciation *18.4*
reproductive isolating mechanism *18.1*
speciation *CI*
species *18.1*
sympatric speciation *18.3*

Readings

Futuyma, D. 1998. *Evolutionary Biology*. Third edition. Sunderland, Massachusetts: Sinauer.

Mayr, E. 1976. *Evolution and the Diversity of Life*. Cambridge, Massachusetts: Belknap Press of Harvard University Press.

On-Line readings at Student Guide for InfoTrac: www.brookscole.com/biology

THE MACROEVOLUTIONARY PUZZLE

Measuring Time

How do you measure time? Is your comfort level with the past limited to your own generation? Maybe you can still relate to a few centuries of human history. But *geologic* time? Comprehending the very distant past requires an enormous intellectual leap, from the familiar to the unknown.

In science, this leap through time starts with the premise that any aspect of the natural world, past as well as present, has one or more underlying causes. It is an attempt to identify causes by studying physical and chemical aspects of the Earth, by analyzing fossils, and by comparing the morphology and biochemistry of organisms. It is a willingness to test hypotheses with experiments, models, and novel technologies. This shift from experience to inference—from the known to what can only be surmised—has given us astounding glimpses into the past. Consider this case in point.

Asteroids are rocky, metallic bodies, a few meters to 1,000 kilometers across, that are hurtling through space (Figure 19.1). When our solar system's planets formed, their gravitational pull swept most asteroids from the sky. At least 6,000 still orbit the sun in a belt between Mars and Jupiter. Dozens of others cross the Earth's orbit. The chance of impact is analogous to Russian roulette on a cosmic scale.

Many past catastrophic impacts altered the course of evolution. One of them wiped out the last of the dinosaurs, as indicated partly by a layer of iridium distributed throughout the world. Iridium is rare on Earth but not in asteroids, and that thin layer dates precisely to what is called the K–T boundary. Past that abrupt transition between the Mesozoic and Cenozoic eras, there are no more fossils of dinosaurs, anywhere.

Consider, next, that it has been only about 10,000 years since the first modern humans (*Homo sapiens*) walked the Earth. They were preceded by dozens of humanlike forms that evolved in Africa over the past 5 million years. Like us, those forms were hominids. So why are we the only ones left?

We have evidence of twenty impacts during that time span. Unlike today's large human populations, which are dispersed all over the world, the early hominids lived in small bands in Africa. What if most were cosmic casualties and our ancestors were just

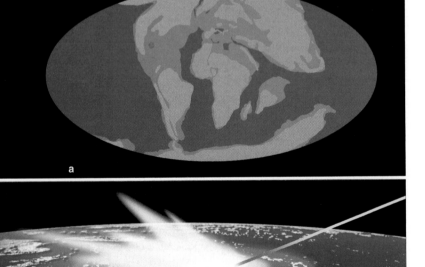

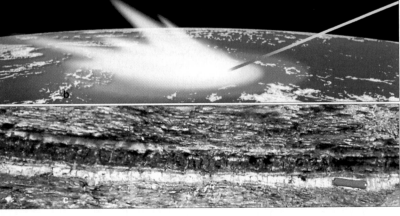

Figure 19.1 (**a**) Distribution of land masses at the Cretaceous–Tertiary (K–T) boundary between the Mesozoic and Cenozoic eras. *Light blue* indicates shallow seas. *Red* indicates where a huge asteroid hit the Earth. (**b**) Reconstruction of the angle of impact and of material ejected skyward. (**c**) A sample of the worldwide, iridium-rich layer that dates precisely to the K–T boundary. The pocket knife provides the scale. (**d**) An asteroid in space. This one would fit halfway between Baltimore and Washington, D.C. (**e**) What might have happened in the last few minutes of the Cretaceous, a topic of Sections 20.6 and 20.7.

plain lucky? Think about it. We know one object from space hit the southern Pacific Ocean, west of what is now Chile, about 2.3 million years ago. It was about 2 kilometers (1.2 miles) in diameter. Suppose it had collided with the Earth just a few hours earlier. Given the Earth's rotation, southern Africa would have been hit, our early ancestors would have been incinerated, and we would not be around today to ask about them.

Now that we know what to look for, we are seeing more and more craters in satellite images of the Earth. One of them, in Iraq, is 3.2 kilometers across. Energy released by that impact, which occurred less than 4,000 years ago, would have been equivalent to the detonation of hundreds of nuclear weapons.

And this brings us back to the present. In 1908, during an annual meteor shower, a streak appeared in the Siberian sky followed by an explosion above a remote area called Tunguska. The object was only thirty to sixty meters across. It was not a hard impact. Yet the energy released exceeded the Hiroshima blast. Trees up to 30 kilometers away were seared on one side, and they were all flattened like matchsticks. One eyewitness 100 kilometers away said the northern sky appeared to be on fire, and he felt a great heat as if his shirt were burning. Air pressure disturbances were recorded as far away as England.

More recently, in the early 1990s, the orbits of four asteroids took them halfway between the Earth and the Moon. In October 2028, a large asteroid will sweep past, uncomfortably close to the Earth or the Moon.

What we have here are cases of cause and effect in nature. And if we can figure out what an asteroid impact will do to us, we can figure out how impacts affected life in the past. We *can* comprehend life long before our own. This chapter introduces some of the tools and evidence used to interpret patterns, trends, and rates of change among lineages. Along the way you will read about their underlying causes, including good and bad cosmic luck that began even before the distant time of the origin of life.

Key Concepts

1. In the evolutionary view, all species that have ever lived are related by descent—some closely and others remotely so.

2. Macroevolution is the name for the major patterns, trends, and rates of change among lines of descent, or lineages, during the history of life. The fossil record, the geologic record, and radiometric dating of rocks yield evidence of macroevolution. So do morphological comparisons of existing groups of species.

3. The fossil record has enough detail to reveal major patterns of descent with modification. The record will never be complete in all of its details, but gaps are to be expected. As we know now, parts were lost owing to major movements of the Earth's crust and other geologic events. Also, there is a low probability of any organism becoming fossilized, and even of finding it.

4. Biogeography is the study of the distribution of species through time and through the environment. A theory about crustal movements helps explain some puzzling aspects of the global distribution of life in the past.

5. Comparisons of body forms among major groups may reveal evolutionary relationships. Adults and embryos of different lineages often show similarities in one or more body parts that reflect descent from a common ancestor.

6. Biochemical comparisons within and between major lineages provide strong evidence of macroevolution.

7. Patterns in the diversity of life are being revealed through taxonomy, phylogenetic reconstruction, and classification. Taxonomy is concerned with identifying and naming new species. Phylogenetic reconstruction infers evolutionary connections by certain analytical methods. Classification is concerned with organizing information about species into retrieval systems.

Evidence four evolution :
1. Paleontology
2. Biogeography
3. Embryology
4. Morphology

FOSSILS—EVIDENCE OF ANCIENT LIFE

About 500 years ago, Leonardo da Vinci was brooding about seashells entombed in rocks of northern Italy's high mountains, hundreds of kilometers from the sea. How did they get there? According to the traditional explanation, a stupendous flood had occurred, and the floodwaters had deposited the shells in the mountains (Figure 19.2). But many of the shells were thin, fragile, and intact. Surely they would have been battered to bits if they had been swept across such great distances and all the way up into the mountains.

Leonardo also brooded about the rocks. They were stacked like cake layers, and some contained shells but others had none. Then he remembered how large rivers swell with spring floodwaters and deposit silt in the sea. Were the layers deposited long ago, in separate intervals? If so, then shells in the mountains could be evidence of layered communities of organisms that once lived in the seas! Leonardo didn't announce his novel idea, which might have been considered heresy at the time.

By the 1700s, **fossils** were accepted as the buried remains and impressions of organisms that had lived in the past. (*Fossil* comes from a Latin word for something that was "dug up.") They were still being interpreted through the prism of prevailing cultural beliefs, as when a Swiss naturalist excitedly unveiled the remains of a giant salamander and announced that they were the skeleton of a man who had drowned in the great flood. By midcentury, however, scholars began to question the interpretations.

Extensive mining, quarrying, and canal excavations were under way. The diggers were finding similar rock layers and similar fossil sequences in distant places, such as the cliffs on both sides of the English Channel. The layers appeared to record the passing of geologic time. Thus, the ordered array of fossils within the rock layers might be a historical record of life—*a fossil record*.

Fossilization

Figure 19.3 shows a few fossils. Most fossils discovered so far are bones, teeth, shells, seeds, spores, and other hard parts. Unless organisms are buried quickly, their soft parts decompose or scavengers make short work of them. Fossilized feces (coprolites) hold residues of species that were eaten in ancient environments. Also, imprints of leaves, stems, tracks, burrows, and other *trace* fossils give indirect evidence of past life.

Fossilization is a very slow process that starts when an organism, or traces of it, becomes buried in volcanic ash or sediments. Sooner or later, water infiltrates the organic remains, which become infused with dissolved metal ions and other inorganic compounds. More and more sediments gradually accumulate above the burial site. They exert ever increasing pressure on the remains. Over great spans of time, the pressure and the chemical changes transform those remains to stony hardness.

Preservation is favored when organisms are buried rapidly in the absence of oxygen. Gentle entombment by volcanic ash or anaerobic mud is best. Preservation also is favored when a burial site remains undisturbed. Most often, however, erosion and other geologic insults crush, deform, break, or scatter the fossils. This is one reason why fossils are relatively rare.

Fossils in Sedimentary Rock Layers

Stratified (stacked) layers of sedimentary rock are rich sources of fossils. They formed long ago by a gradual deposition of volcanic ash, silt, and other materials. Thus sand and silt piled up when rivers transported them from land to the sea, just as Leonardo thought. The sand became compressed into sandstone, and silt into shale. Deposits were not continuous. For example, sea levels changed when a cooling trend caused ice ages. Tremendous amounts of water became locked up in glaciers, rivers dried up, and depositions ceased in some places. Later on, when the climate warmed and glaciers melted, depositions resumed.

We call the formation of sedimentary rock layers **stratification**. We can expect that the deepest layers were the first to form, and layers closest to the surface, the last. We can expect that most layers formed horizontally, because particles tend to settle in response to gravity. You may see tilted

Figure 19.2 From the Sistine Chapel in Italy, Michelangelo's painting of the onset of a catastrophic flood.

Figure 19.3 (**a**) A fossil hunter's dream: a complete skeleton of a bat that lived 50 million years ago. Erosion and other forces left few ancient burial sites undisturbed, so intact fossils are rare. Even the jumbled parts of the ducklike birds in (**b**) are a good find. It will take hours of preparation and analysis to identify the species. (**c**) Fossilized parts of the oldest known land plant (*Cooksonia*). Its stems were not even seven centimeters tall. (**d**) The skeletal remains of an ichthyosaur. This marine reptile lived 200 million years ago. Section 20.6 includes art that shows what it looked like.

or ruptured layers, as along a road cut into the side of a mountain. Most are evidence of geologic disturbances after the time of deposition.

Understand how rock layers form, and you realize that fossils in a particular layer are from a similar age in Earth history. Specifically, *the older the layer, the older the fossils*. Given that rock layers formed in sequence, then their fossil assemblages are unique to sequential ages. That is why fossils can be used to assign relative dates to the record of the rocks. The next section will show how the fossil layers have been used as reference points in constructing the geologic time scale.

Interpreting the Fossil Record

We have fossils for more than 250,000 known species. Judging from the current range of diversity, there must have been many, many millions more species. We never will recover fossils for most of them, so the record of past life is incomplete. Why is this so?

The odds are against finding evidence of an extinct, ancient species. At least one of its individuals had to die and be gently buried before something decomposed it or ate it. The burial site had to escape obliteration by erosion, lava flows, and other geologic events. And the fossil had to end up in a place where someone could actually find it, as in a sedimentary layer exposed by a river cutting down through a canyon. And fossils do not form in many habitats, such as mountaintops.

Also, most species of ancient communities did not lend themselves to preservation. For example, unlike bony fishes and hard-shelled mollusks, jellyfishes and soft-bodied worms don't show up as much in the fossil record. Probably they were just as common, or more so.

Also think about population density and body size. A population of plants might release millions of spores in a growing season. The earliest humans lived in small groups and produced few offspring. What are the odds of finding even one fossilized human bone compared to finding spores of plants that lived at the same time?

Finally, think about a line of descent, or **lineage**, on a remote volcanic island that ended up sinking into the sea. Or think about one lineage that lasted only briefly and another that endured for billions of years. Which is more likely to be represented in the fossil record?

Fossils are physical evidence of organisms that lived in the distant past. Stratified layers of sedimentary rock that are rich in fossils are a historical record of life. The deepest layers generally hold the oldest fossils.

The fossil record is an incomplete record of life in the past. Major geologic changes have obliterated much of it. Also, our sampling of the record is slanted toward species that had large bodies, hard parts, dense populations, broad distribution, and persistence through time.

Even so, the fossil record is now substantial enough to reconstruct the sequence of changes in the history of life.

Dating Pieces of the Puzzle

ORIGIN OF THE GEOLOGIC TIME SCALE Probably for as long as they have been digging up rocks, people have been finding fossilized leaves, shells, and other evidence of past life. *How do we know how old the fossils are?*

We didn't have a clue until long after early geologists thought about fossils in stratified rock layers. As they hypothesized, newly formed rocks accumulate on top of older ones, so the oldest fossils are in the deepest layers. With this in mind, they constructed a chronology of Earth history—a **geologic time scale**—by counting backwards through layer upon layer of sedimentary rock.

Through comparative studies, they discovered four abrupt transitions in fossil sequences from around the world. They decided to use the transitions as boundaries between four great intervals in time. The first interval,

corresponding to the oldest known fossils, was named the Proterozoic. They named later intervals the Paleozoic, the Mesozoic, and the "modern" era, the Cenozoic.

Figure 19.4 is a current version of the geologic time scale. We correlate it with **macroevolution**—with major patterns, trends, and rates of change among lineages. Example: At the Proterozoic–Paleozoic boundary, all modern animal phyla emerged in adaptive radiations (the Cambrian explosion). At the Paleozoic–Mesozoic boundary, nearly all families perished. The Mesozoic–Cenozoic boundary marks a mass extinction of dinosaurs and other reptiles; a radiation of mammals followed.

The Proterozoic turned out to be an immense span. It, too, was divided into great intervals. Life originated in one of those intervals, the Archean eon.

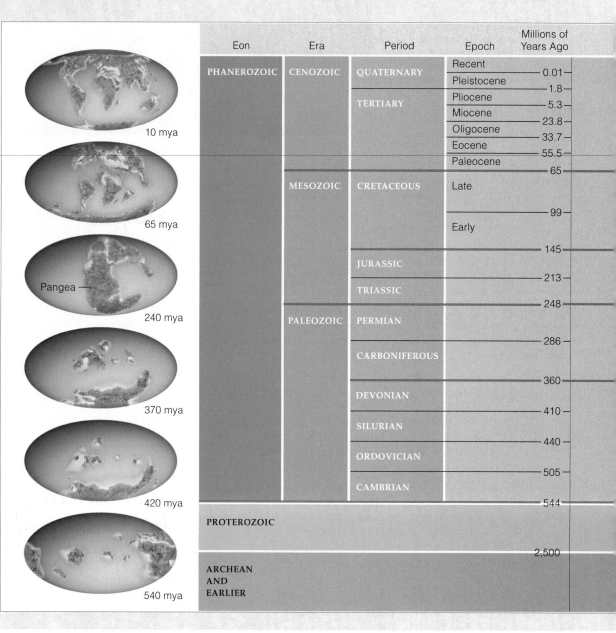

Figure 19.4 Geologic time scale. The major boundaries mark times of the greatest mass extinctions. Radiometric dating methods allowed researchers to assign absolute dates. Life arose in the Archean. The maps show changes in the distribution of land masses and oceans over time. The next chapter correlates the geologic changes with the history of life, as previewed in the far-right column.

The time spans are not to the same scale. If they were, the Archean and Proterozoic portions would run off the page. Think of the spans as minutes on a clock that runs from midnight to noon. If we say that life originated at midnight, the Paleozoic began at 10:04 A.M., the Mesozoic at 11:09 A.M., and the Cenozoic at 11:47 A.M. The Recent epoch of the Cenozoic started in the last 0.1 second before noon.

10 mya

65 mya

Pangea

240 mya

370 mya

420 mya

540 mya

Eon	Era	Period	Epoch	Millions of Years Ago
PHANEROZOIC	CENOZOIC	QUATERNARY	Recent	
			Pleistocene	0.01
				1.8
		TERTIARY	Pliocene	5.3
			Miocene	
				23.8
			Oligocene	33.7
			Eocene	
				55.5
			Paleocene	
				65
	MESOZOIC	CRETACEOUS	Late	
				99
			Early	
				145
		JURASSIC		213
		TRIASSIC		
				248
	PALEOZOIC	PERMIAN		286
		CARBONIFEROUS		
				360
		DEVONIAN		410
		SILURIAN		
				440
		ORDOVICIAN		505
		CAMBRIAN		
				544
PROTEROZOIC				2,500
ARCHEAN AND EARLIER				

RADIOMETRIC DATING The geologic time scale only had relative ages until the discovery of radioactive decay. Absolute dates have been assigned to it by **radiometric dating**. This method is used to measure the proportions of (1) a radioisotope in a mineral trapped long ago in a new rock and (2) a daughter isotope that formed from it in the same rock. Remember, the number of protons and neutrons in the atomic nucleus of a radioactive element is *unstable* (Section 2.1). The nucleus spontaneously decays —it gives up energy and one or more particles of itself— until it reaches a more stable configuration.

Half-life is the time it takes for half of a given quantity of a radioisotope to decay into a less unstable, daughter isotope (Figure 19.5). The clock-like decay rate is constant; changes in pressure, temperature, or chemical state can't

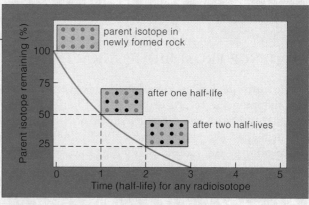

Figure 19.5 Decay of radioisotopes at a fixed rate to more stable form. The half-life of each kind is the time it takes for 50 percent of a sample to decay. After two half-lives, 25 percent of the sample has decayed, and so on.

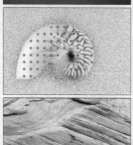
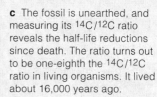

a A living mollusk took up trace amounts of ^{14}C carbon and a lot more ^{12}C, the more stable and most common carbon isotope.

b The mollusk died, was buried in sand, and was fossilized. From its death onward, the proportion of ^{14}C to ^{12}C started declining by radioactive decay. ^{14}C has a half-life of 5,370 years. Half was gone from the fossil in 5,370 years, half of what remained was gone in another 5,370 years, and so on.

c The fossil is unearthed, and measuring its $^{14}C/^{12}C$ ratio reveals the half-life reductions since death. The ratio turns out to be one-eighth the $^{14}C/^{12}C$ ratio in living organisms. It lived about 16,000 years ago.

Figure 19.6 Dating a fossil with a radioisotope. Carbon 14 (^{14}C) forms in the atmosphere, where it combines with oxygen as carbon dioxide. Along with greater quantities of the more stable carbon isotopes, trace amounts of carbon 14 enter food webs via photosynthesis. All organisms incorporate it.

Major Geologic and Biological Events
That Occurred Millions of Years Ago (mya)

1.8 mya to present. Major glaciations. Modern humans evolve. The most recent *extinction crisis* is under way.

65–1.8 mya. Major crustal movements, collisions, mountain building. Tropics, subtropics extend poleward. When climate cools, dry woodlands, grasslands emerge. *Adaptive radiations* of flowering plants, insects, birds, mammals.

65 mya. Asteroid impact; *mass extinction* of all dinosaurs and many marine organisms.

99–65 mya. Pangea breakup continues, inland seas form. *Adaptive radiations* of marine invertebrates, fishes, insects, and dinosaurs. Origin of angiosperms (flowering plants).

145–99 mya. Pangea starts to break up. Marine communities flourish. *Adaptive radiations* of dinosaurs.

145 mya. Asteroid impact? *Mass extinction* of many species in seas, some on land. Mammals, some dinosaurs survive.

248–213 mya. *Adaptive radiations* of marine invertebrates, fishes, dinosaurs. Gymnosperms dominant land plants. Origin of mammals.

Mass extinction. Ninety percent of all known families lost.

286–248 mya. Supercontinent Pangea and world ocean form. On land, *adaptive radiations* of reptiles and gymnosperms.

360–286 mya. Recurring ice ages. On land, *adaptive radiations* of insects, amphibians. Spore-bearing plants dominate; cone-bearing gymnosperms present. Origin of reptiles.

360 mya. *Mass extinction* of many marine invertebrates, most fishes.

410–360 mya. Major crustal movements. Ice ages. *Mass extinction* of many marine species. Vast swamps form. Origin of vascular plants. *Adaptive radiation* of fishes continues. Origin of amphibians.

440–410 mya. Major crustal movements. *Adaptive radiations* of marine invertebrates, early fishes.

505–440 mya. All land masses near equator. Simple marine communities flourish until origin of animals with hard parts.

544–505 mya. Supercontinent breaks up. Ice age. *Mass extinction.*

2,500–544 mya. Oxygen accumulates in atmosphere Origin of aerobic metabolism. Origin of eukaryotic cells. Divergences lead to eukaryotic cells, then protistans, fungi, plants, animals.

3,800–2,500 mya. Origin of photosynthetic prokaryotic cells.

4,600–3,800 mya. Origin of Earth's crust, first atmosphere, first seas. Chemical, molecular evolution leads to origin of life (from proto-cells to anaerobic prokaryotic cells).

---

alter it. Example: Uranium 238, a mineral in most volcanic rocks, decays into thorium 234. The still-unstable daughter isotope decays into something else, and so on through a series of intermediate daughter isotopes to the final, most stable configuration for this series: lead 206. Researchers using uranium 238, with its half-life of 4.5 billion years, found that the Earth formed more than 4.6 billion years ago. The most recent fossils can be dated using carbon 14, with its half-life of 5,370 years (Figure 19.6). *< 50,000 years old*

Radiometric dating works for volcanic rocks or ashes, but most fossils are in sedimentary rocks. The only way to date most fossil-containing rocks is to determine their position relative to volcanic rocks in the same area. The dating method has an error factor of less than 10 percent.

EVIDENCE FROM BIOGEOGRAPHY

While early geologists were busily discovering fossils and mapping the record of the rocks, Charles Darwin was so taken with one of their theories of Earth history that it helped shape his view of life. By the **theory of uniformity**, recall, mountain building and erosion had repeatedly worked over the Earth's surface in precisely the same ways through time (Section 17.2). Great stacks of sedimentary layers, as in Figure 19.7, were evidence of this. Yet the more geologists clinked their hammers against the rocks, the more they realized that repetitive change was only part of the picture. Like life, the Earth had changed irreversibly; it evolved.

An Outrageous Hypothesis

For instance, the Atlantic coasts of South America and Africa seemed to "fit" like jigsaw puzzle pieces. Were all continents once joined as a "supercontinent," which later broke up? Did the fragments drift apart? This idea came to be known as continental drift. Geologists also wondered if mountain ranges that parallel many coasts had formed when some of the huge fragments collided.

Models for the proposed supercontinent abounded. Alfred Wegener gave the name **Pangea** to his version. In constructing his model, he used glacial deposits as clues to infer past climate zones. Other clues came from the world distribution of fossils and existing species.

Figure 19.7 A splendid slice through time: the Grand Canyon of the American Southwest, once part of an ocean basin. Its sedimentary rock layers formed slowly, over hundreds of millions of years. Later, geologic forces lifted the ancient stacked layers above sea level. Later still, the erosive force of rivers carved the canyon walls and exposed the layers.

Most scholars could not accept the hypothesis. They knew the continents are thinner and not as dense as the mantle, a rocky region below them. To them, continents plowing on their own across the mantle seemed about as plausible as a paper towel plowing across water in a bathtub. They preferred the theory of uniformity.

Then scientists made an amazing discovery. When iron-rich rocks first form, their internal structure takes on a north–south orientation in response to the Earth's magnetism. Yet the orientations in ancient rocks were *not* aligned with the magnetic poles. Using 200-million-year-old rocks from North America and Europe as a magnetic guide, scientists made sketches in which rocks were rotated into their inferred, original north–south alignment. The alignments dovetailed only when they moved North America and western Europe together.

More puzzles! Deep-sea probes revealed evidence of seafloor spreading: Molten rock erupts at mid-oceanic ridges in the seafloor. It flows laterally in both directions from the ridges, then hardens to form new crust. The spreading new crust forces the older crust down into deep trenches elsewhere in the seafloor. The ridges and trenches are the edges of huge, relatively thin plates, somewhat like the pieces of a cracked eggshell (Figure 19.8). The crustal plates move exceedingly slowly, and the movement rafts continents to new positions.

What did the findings mean? Continental drift had to be just one part of a broader explanation of crustal movements, a **plate tectonics theory**. Researchers soon found ways to demonstrate the predictive power of the new theory. For instance, in Africa, India, Australia, and South America, the same succession of coal seams, glacial deposits, and basalt held fossils of a seed fern (*Glossopteris*, Figure 19.9) and a mammal-like reptile (the therapsid *Lystrosaurus*, Section 20.6). The plant's seeds were too heavy and the animal was too small for geographic dispersal across the open ocean. But what if both had originated on **Gondwana**—another proposed supercontinent—which was even older than Pangea?

A geologist predicted that fossils of *Glossopteris* and *Lystrosaurus* would also be discovered in Antarctica in a series of glacial deposits, coal seams, and basalt, like that in the southern continents. Sure enough, Antarctic explorers found the same series and fossils—evidence favoring the prediction and the plate tectonics theory.

Drifting Continents, Changing Seas

Look closely at Figure 19.9. In the remote past, crustal movements put huge land masses on collision courses. In time the masses converged to form supercontinents. Later, these split apart at deep rifts and formed new ocean basins. Gondwana was an early supercontinent. It drifted south from the tropics, across the south polar

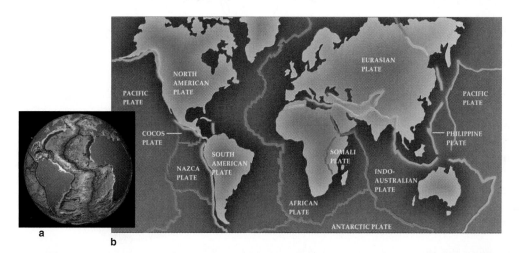

a

b

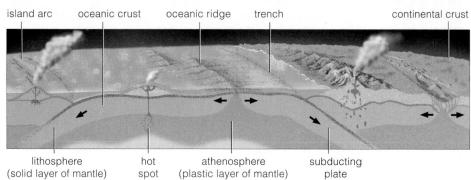

island arc   oceanic crust   oceanic ridge   trench   continental crust

lithosphere   hot   athenosphere   subducting
(solid layer of mantle)   spot   (plastic layer of mantle)   plate

c

Figure 19.8 Some forces of geologic change. (**a,b**) Present configuration of the Earth's crustal plates. These rigid plates slowly split apart, drift, and collide. The globe in (**a**) shows the newest crust (*red*, less than 10 million years old) and the oldest (*blue*, 180 million years old).

(**c**) Huge plumes of molten material drive the movement. Plumes well up from the interior, spread laterally beneath the crust, and rupture it at deep mid-oceanic ridges. At the ridges, molten material seeps out, cools, and gradually forces the seafloor away from the rupture. Seafloor spreading displaces plates from the ridges.

(**b**) Often, the leading edge of one plate plows under an adjoining plate and uplifts it. The *blue* lines show where this is now happening. The Cascades, Andes, and other mountain ranges paralleling the coasts of continents formed this way.

Long ago, superplumes violently ruptured the crust over "hot spots" in the mantle. The Hawaiian Archipelago has formed this way. Continents also rupture. Deep rifting and splitting are now occurring in Missouri, at Lake Baikal in Russia, and in eastern Africa.

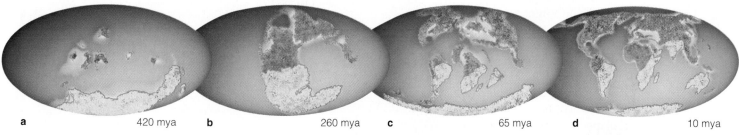

a   420 mya   **b**   260 mya   **c**   65 mya   **d**   10 mya

Figure 19.9 Reconstructions of drifting continents. (**a**) Gondwana (*yellow*), 420 million years ago. (**b**) All land masses collided to form Pangea. About 260 million years ago, seed ferns (*Glossopteris*) and other plants lived on the Gondwana portion of this supercontinent but nowhere else. (**c**) Positions of fragments from Pangea breakup 65 million years ago. (**d**) Their position 10 million years ago.

Glossopteris fossil

region, and then north until it crunched into other land masses to form a single world continent—Pangea. That supercontinent extended from pole to pole; one world ocean lapped its coastlines. All the while, slow erosive forces of water and wind resculpted the surface of the land. As if this weren't enough, asteroids and meteorites bombarded the crust. The impacts and their aftermath had long-term effects on global temperature and climate.

All of those changes in the land, oceans, and atmosphere profoundly influenced the evolution of life. Imagine early life flourishing in the warm, shallow waters along the shorelines of continents. The shorelines vanished when continents collided. Such changes were devastating for many lineages. Yet, even as old habitats vanished, new ones opened for the survivors—and evolution took off in new directions. As you will see in the unit to follow, the histories of the Earth and life are inseparable.

Over the past 3.8 billion years, changes in the Earth's crust, the atmosphere, and the oceans profoundly affected the evolution of life.

A180 vestigial structures - fully dev in group of organisms, but reduced or nonfunctional in related groups

EVIDENCE FROM COMPARATIVE MORPHOLOGY

Evolution, remember, simply means heritable changes in lines of descent. **Comparative morphology** provides good evidence of descent with modification. This field of inquiry focuses on the body form and structures of groups of organisms, such as vertebrates and flowering plants. Often it reveals a similarity in one or more body parts that has a genetic basis, that reflect inheritance from a common ancestor. Such body parts are known as **homologous structures**. (*Homo*– means the same.) The similarities are evident even when different kinds of organisms use the body parts for different functions.

Morphological Divergence

Recall, from Chapter 18, that populations of the same species diverge genetically when the gene flow between them ceases. Over time, populations typically diverge in the morphological traits that help characterize their species. Such change from the body form of a common ancestor is a major pattern of macroevolution. We call this **morphological divergence**. (*Morpho*– is Greek for body form.)

Even when two related species diverge considerably with respect to morphology, they remain alike in other ways. Remember, they started out with the same body form and the same structures, then they evolved in different directions by different modifications to the shared plan. Look carefully, and you might be able to identify underlying similarities.

An example: All vertebrates on land are descended from the first amphibians. Divergences led to reptiles, then to birds and mammals. We know about the "stem" reptile (the one ancestral to all other reptiles and to birds and mammals). We have the fossilized, five-toed limb bones of a species that crouched quite low to the ground (Figure 19.10a). Its descendants expanded into many new habitats on land; a few even returned to the seas when the environment changed.

That five-toed limb was a key innovation for the pterosaurs, birds, and bats. It was evolutionary clay that became molded into different kinds of limbs with different functions. In lineages leading to penguins and porpoises, it was modified into flippers that assisted in swimming. In a lineage leading to the modern horse, it

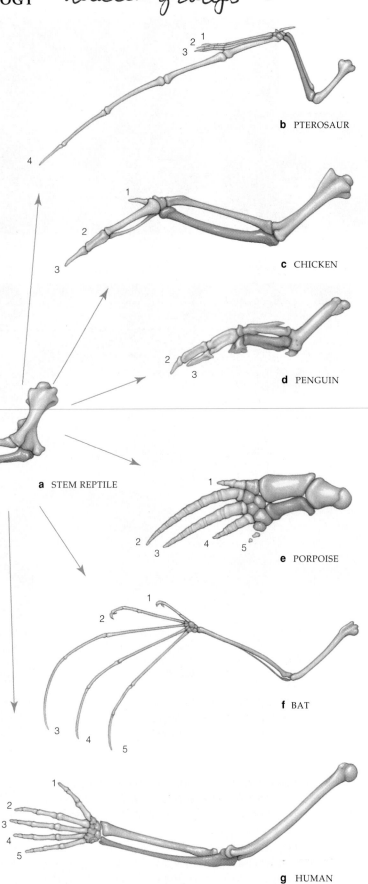

b PTEROSAUR

c CHICKEN

d PENGUIN

a STEM REPTILE

e PORPOISE

f BAT

g HUMAN

Figure 19.10 Morphological divergence among vertebrate forelimbs, starting with the stem reptile. Similarities in the number and position of skeletal elements were preserved as diverse forms evolved. Some of the bony elements were lost during the course of evolution (compare the numbers 1 through 5). The drawings are not to the same scale.

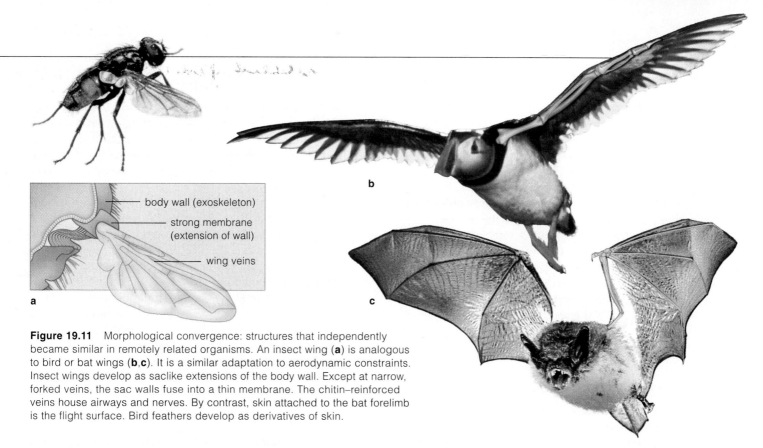

Figure 19.11 Morphological convergence: structures that independently became similar in remotely related organisms. An insect wing (**a**) is analogous to bird or bat wings (**b**,**c**). It is a similar adaptation to aerodynamic constraints. Insect wings develop as saclike extensions of the body wall. Except at narrow, forked veins, the sac walls fuse into a thin membrane. The chitin–reinforced veins house airways and nerves. By contrast, skin attached to the bat forelimb is the flight surface. Bird feathers develop as derivatives of skin.

In figure: body wall (exoskeleton); strong membrane (extension of wall); wing veins

was modified into long, one-toed limbs suitable for fast running. Among moles, it became a stubby shape that is good for burrowing into the earth. Among elephants, it became strong and pillarlike, suitable for supporting great weight. The five-toed limb became modified into the human arm and hand. The human hand also has a thumb positioned in opposition to four fingers; it is the basis of grasping and precision movements.

Even though the forelimbs of vertebrates are not all the same in size, shape, or function from one group to the next, there is clear resemblance in the structure and positioning of the bony elements. There is resemblance, too, in the nerves, blood vessels, and muscles present in the forelimbs. Comparisons of vertebrate embryos also reveal strong resemblance in how these structures develop (Chapter 43). Such similarities in body parts point to common ancestry.

Morphological Convergence

Body parts with similar forms or functions in different lineages aren't always homologous structures. They may have evolved *independently* in lineages that aren't closely related. If comparable body parts were subjected to similar environmental pressures, there may have been selection for similar modifications, and so they ended up resembling each other. **Morphological convergence** refers to the independent evolution of body structures that did become similar in remotely related organisms.

For instance, you just saw how the forelimbs of birds and bats are homologous structures. Are their wings also homologous? No. The bird flight surface evolved

as a sweep of feathers, derived from skin. The forelimb structurally supports it. The flight surface of bats is a thin membrane, an extension of the skin itself, that is attached to and also reinforced by bony elements of the forelimb (Figure 19.11).

The insect wing, too, resembles bird and bat wings in its function. But it shows no underlying homology. The insect wing develops as an extension of an outer body wall reinforced with chitin. It has no underlying bony elements to support it, as Figure 19.11 shows.

The convergent resemblance is evidence that bats, birds, and insects—three separate lineages—adapted independently to the aerodynamic constraints on flight. Their wings are **analogous structures**. They were not derived from comparable body parts; the convergent resemblance is only indicative of similar functions. The Greek *analogos* means similar to one another.

With morphological *divergence*, comparable body parts became modified in different ways in different lines of descent from a common ancestor.

Such divergences resulted in homologous structures. Even if they differ in size, shape, or function, these body parts have an underlying similarity owing to shared ancestry.

In morphological *convergence*, dissimilar body parts evolved independently in lineages that are not closely related but that responded to similar environmental pressures.

Analogous structures are the outcome of morphological convergence. Such body parts resemble one another because they are adaptations to similar environmental constraints, not because of close shared ancestry.

EVIDENCE FROM PATTERNS OF DEVELOPMENT *embryology*

Comparing the patterns by which groups of plants or animals undergo growth and development yields good evidence of evolution. Remember, the body of each new plant and animal follows a long-term, inherited pattern of development (Sections 15.1 and 15.3). Later chapters will show how the body develops in stages. For now, simply note that it cannot develop properly unless each stage is successfully completed before the next begins.

Many constraints on evolution come into play as an embryo is developing. If gene mutations or changes in chromosome number or structure arise, they tend to be selected against if they disrupt an essential stage in the embryo's development. But every so often, change has a neutral or beneficial effect. Let's look at examples of how it might shift a developmental step in a way that natural selection favors.

Developmental Program of Larkspurs

Flowering plants offer evidence of evolution through changes in rates of development. Consider *Delphinium decorum*, a larkspur. A ring of petals guides honeybees to the nectar-storing tube of its flowers (Figure 19.12). At the flower's center, outward-bulging reproductive structures give bees something to hang on to as they gather nectar and pollinate the flower. *D. nudicaule*, a larkspur of more recent origin, has compact flowers and no landing platform. This is not a problem for its pollinators—hummingbirds—which can hover in front of the flower as they gather nectar from it.

At maturity, *D. nudicaule* flowers strongly resemble buds of *D. decorum* (Figure 19.12). That resemblance is probably a result of some change in a pattern of gene expression; something slowed down the rate of floral development in *D. nudicaule*. Think about the plant in which the first slowdown appeared. It could produce only compact flowers that discouraged bees—but not hummingbirds. Now dependent on novel pollinators, the plant was reproductively isolated from *D. decorum*. In time it gave rise to *D. nudicaule*, a distinct species.

Developmental Program of Vertebrates

Now consider the diverse vertebrates, which range from fishes to the amphibians, reptiles, birds, and mammals. By comparing how their embryos develop, researchers have found compelling evidence of their evolutionary connection with one another.

The life cycles of all vertebrates proceed through six stages of development, and body parts that form in early embryonic stages are similar. Figure 19.13a shows a tail, the start of a backbone, and limb buds that have formed in the same pattern in pig and human embryos.

The early embryos of vertebrates strongly resemble one another because they inherited the same ancient plan for development. Tissues form as cells divide and start to interact in prescribed ways. Then the gut, heart, bones, skeletal muscles, and other parts grow and develop in ordered spatial patterns (Chapter 43).

During the evolution of all vertebrates, most DNA alterations that disrupted pivotal steps of development probably had lethal effects on later stages. At least the similarity among early embryos of the different groups suggests that such disruptions were selected against.

How, then, did adults of different groups get to be so different? Some differences probably came about by

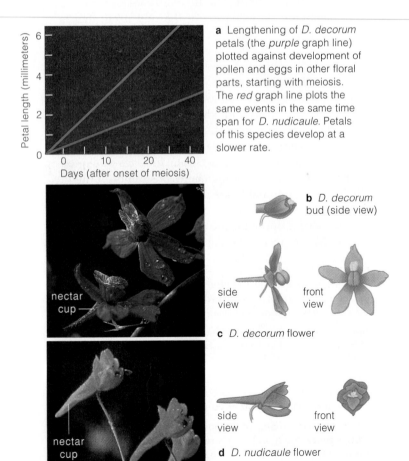

a Lengthening of *D. decorum* petals (the *purple* graph line) plotted against development of pollen and eggs in other floral parts, starting with meiosis. The *red* graph line plots the same events in the same time span for *D. nudicaule*. Petals of this species develop at a slower rate.

Petal length (millimeters)

Days (after onset of meiosis)

nectar cup

b *D. decorum* bud (side view)

side view · front view

c *D. decorum* flower

nectar cup

side view · front view

d *D. nudicaule* flower

Figure 19.12 From comparative morphology, evidence of an evolutionary relationship between two larkspurs (*Delphinium*). The time required for flowers to develop and mature is similar in both species. The *rate* of development is slower for most floral structures of *D. nudicaule*. Changes in petal shape are not nearly as great as they are for the other species, *D. decorum*.

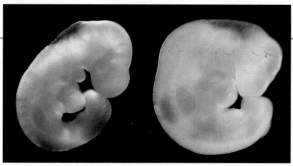

a EARLY MOUSE EMBRYO EARLY HUMAN EMBRYO

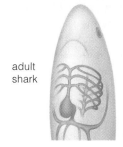

adult shark

early human embryo (three millimeters in length)

b

Figure 19.13 From comparative embryology, some evidence of evolutionary relationship among vertebrates.

(**a**) Adult vertebrates are diverse, yet their embryos are quite similar at very early stages. Diversity arises as embryos start developing differently at later stages. (**b**) Fishlike structures still form in early embryos of reptiles, birds, and mammals. In fish embryos, a two-chambered heart (*orange*), some veins (*blue*), and parts of arteries (aortic arches, *red*) develop and persist in adult fishes. The same structures form in early human embryos but do not persist as such in adults.

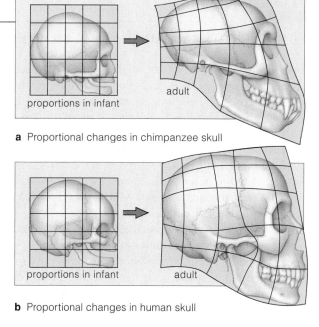

a Proportional changes in chimpanzee skull

b Proportional changes in human skull

Figure 19.14 Pronounced morphological differences between two lines of descent, possibly an outcome of changes in timing of developmental steps. Compare the proportional changes in the skull bones of a human and a chimpanzee. The skulls are quite similar in their infants.

The infant skulls are depicted as paintings on an unstretched blue rubber sheet divided into a grid. Stretching the sheet deforms the grid. For the adult skulls, differences in the size and shape of corresponding grids reflect differences in the patterns of growth.

heritable changes in the onset, rate, or completion time of developmental steps. Such changes could have made the relative sizes of tissues or organs larger or smaller. They would modify the body's structure, and in some cases adult forms might retain some juvenile features.

The "rubber grid" in Figure 19.14 indicates how a change in growth rates might underlie the pronounced proportional differences in skull bones of two primates: chimpanzees and humans. From infants to adults, the grid for humans stays fairly intact in the cranial region that encloses the brain. For chimps, facial bone growth outstrips the growth of cranial bones; the grid becomes greatly modified. Adult skulls for the two lineages end up with significant proportional differences.

Roy Britten, Wanda Reynolds, and other molecular biologists hypothesize that transposons, not single gene mutations, gave rise to most of the variation among lineages. As you read in Chapter 15.4, **transposons** are DNA segments that can move spontaneously to new locations in a genome. And remember enhancers and promoters? These short DNA sequences help switch genes on and off (Sections 15.1 and 15.3). Transposons have their *own* enhancers and promoters. Thus, when they insert themselves next to a gene at a new location, transposons may have powerful regulatory effects on it.

Only primates carry the *Alu* transposon. They have for at least 30 million years. The Alu enhancer binds estrogen, thyroid hormones, and other hormones that control the timing of many developmental events. Was such a transposon pivotal in primate evolution?

Think about this: Between 6 million and 4 million years ago, the forerunners of humans and chimpanzees diverged from the same ancestral stock. More than 98 percent of human DNA is identical with chimpanzee DNA. How can the remainder account for the large morphological differences between the two primates? Humans happen to carry about a million copies of the Alu transposon, which make up more than 5 percent of our genome. At least some of those copies might have fanned or restricted gene expression in certain tissues. Reflect again on the proportional changes in developing skull bones between humans and chimpanzees. Among early ancestors of humans, something did change the duration of bone growth in a heritable way. Ever since, the rapid growth characteristic of the chimpanzee skull bones has been reduced in humans.

Similarities in patterns of development may be clues to evolutionary relationship among plant and animal lineages.

Heritable changes that alter key steps in a developmental program may be enough to bring about major differences in the adult forms of related lineages. Transposons as well as single gene mutations may bring about such changes.

EVIDENCE FROM COMPARATIVE BIOCHEMISTRY — same macromolecules

All species have a mix of ancestral and novel traits. The kinds and numbers of traits they do or don't share are clues to how closely they are related. This is also the case for biochemical traits. Each species, recall, inherits a DNA base sequence, which encodes instructions for making the RNAs and proteins required to produce an individual. But mutations accumulated in the diverging lineages that led to new species. Thus, *we can expect the DNA, RNA, and proteins of closely related species to show more similarities than those of distantly related species.*

Morphological studies alone suggest that monkeys, humans, chimpanzees, and other primates are related. Differences in the amino acid sequence of a protein such as hemoglobin, which all three primates produce, would be evidence of this. Additional evidence might come from comparing their DNA base sequences. Such comparisons give quantitative measures of relatedness between two species that cannot be distinguished on the basis of form alone. For example, automated DNA sequencing can reveal the sequence of cloned DNA or PCR-amplified DNA in a few hours (Section 16.4). For many species, the DNA base sequences and the amino acid sequences for a number of proteins already have been worked out and are available on the Internet.

Protein Comparisons

Suppose the amino acid sequences of a gene's product are the same or nearly so in two species. The absence of mutation implies the species are close relatives. If the sequences differ greatly, many mutations must have accumulated in them. A long time must have passed since the species shared a common ancestor.

One highly conserved gene specifies cytochrome *c*, a protein component of the electron transfer chains in species ranging from aerobic bacteria to humans. The protein's primary structure in humans consists of 104 amino acids. Figure 19.15 shows how the amino acid sequences for cytochrome *c* from a fungus, a plant, and an animal show striking similarity. Now think about this: The *entire* cytochrome *c* sequence is identical in humans and chimpanzees. It differs by 1 amino acid in rhesus monkeys, 18 in chickens, 19 in turtles, and 56 in yeasts.

On the basis of this biochemical information, would you assume that humans are more closely related to a chimpanzee or a rhesus monkey? A chicken or a turtle?

Nucleic Acid Comparisons

Typically, structural alterations that resulted from gene mutations are dispersed through nucleotide sequences of DNA and RNA molecules. Some number of unique alterations have accumulated in each lineage.

Nucleic acid hybridization refers to a base-pairing between DNA or RNA sequences from two different sources. Section 16.5 describes how researchers induce a DNA molecule from different species to unwind, then to recombine as hybrid molecules. The amount of heat required to pull a hybrid molecule apart is a measure of the similarity between its two strands. It takes more heat energy to disrupt the hybridized DNA of closely related species. DNA–DNA hybridization was one of the first molecular tools used to measure evolutionary distance. For example, it was used to construct a family tree for pandas and bears (Figure 19.16).

Today, automated DNA sequencing provides faster, more accurate results (Section 16.4). This method is used to compare nuclear DNA, mitochondrial DNA (mDNA), and ribosomal RNA (rRNA). The restriction fragments cleaved from DNA are compared after being separated by gel electrophoresis. For eukaryotic species, mDNA is especially useful. Its base sequence is rather short, and mDNA mutates ten times faster than nuclear DNA. rRNA lends itself to comparisons of prokaryotic species. Why? Genes for rRNA are so vital to all cells, they haven't mutated much, even in remotely related species. As you will see, they have been used to work out early divergences in life's history.

Computer programs based on DNA sequencing are used to construct evolutionary trees, which often have reinforced morphological studies and the fossil record. But gene transfers between different species can slant the results. For example, it appears that gene swapping was rampant among early prokaryotes. Certain traits of prokaryotic species especially may be an outcome of DNA recombinations, so careful analysis is required.

Figure 19.15 Primary structure of the protein cytochrome *c*, a component of electron transfer systems in cells. The three amino acid sequences are from a yeast (*top row*), wheat (*middle row*), and primate (*bottom row*). The sequences are highly conserved, even in these evolutionarily distant lineages. *Gold* shows where they are identical. The probability that this pronounced molecular resemblance resulted by chance alone is extremely low.

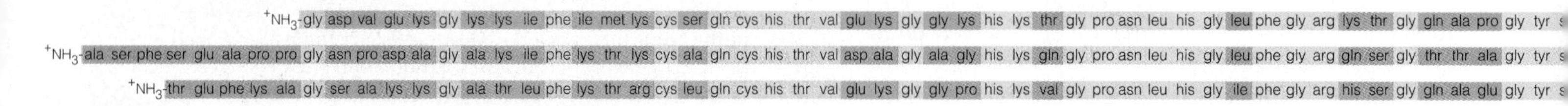

$^+NH_3$-gly asp val glu lys gly lys lys ile phe ile met lys cys ser gln cys his thr val glu lys gly gly lys his lys thr gly pro asn leu his gly leu phe gly arg lys thr gly gln ala pro gly tyr s

$^+NH_3$-ala ser phe ser glu ala pro pro gly asn pro asp ala gly ala lys ile phe lys thr lys cys ala gln cys his thr val asp ala gly ala gly his lys gln gly pro asn leu his gly leu phe gly arg gln ser gly thr thr ala gly tyr s

$^+NH_3$-thr glu phe lys ala gly ser ala lys lys gly ala thr leu phe lys thr arg cys leu gln cys his thr val glu lys gly gly pro his lys val gly pro asn leu his gly ile phe gly arg his ser gly gln ala glu tyr s

RACCOON   RED PANDA   GIANT PANDA

SPECTACLED   SLOTH SUN   BLACK   POLAR   BROWN
BEAR   BEAR   BEAR   BEAR   BEAR   BEAR

DIVERGENCE
15–20 million years ago

DIVERGENCE
approximately
40 million years ago

Figure 19.16 DNA–DNA hybridization studies of the red panda, giant panda, and bears.

Molecular Clocks

Molecular comparisons may also identify the timing of divergences when certain assumptions are made about the constancy of evolutionary change.

Think about our own species. When compared to our close primate relatives, we have some unique traits, the result of mutations that accumulated after divergence from the ancestral primate stock. We also share many genes with other species. For example, the last shared ancestor of certain yeasts and humans lived hundreds of millions of years ago. Less than 10 percent of the known yeast gene products have changed significantly in overall structure; the rest are highly conserved.

Even so, neutral mutations have introduced slight structural differences in the highly conserved genes of different lineages. **Neutral mutations**, recall, have little or no effect on survival or reproduction. They can slip past agents of selection and accumulate in the DNA.

By some calculations, neutral mutations in highly conserved genes accumulated at a regular rate. Think of the accumulation of neutral mutations in a lineage as a series of predictable ticks of a **molecular clock**. Turn the hands of the clock back, so the total number of ticks will "unwind" down through the great geologic intervals of the past. Where the last tick stops, that is roughly the time of origin for the lineage.

How are molecular clocks calibrated in time? The number of differences between the base sequences or

The giant panda, a plant-eating mammal, has a meat-eating animal's gut. It can't digest tough cellulose fibers of plants as efficiently as hoofed plant-eating mammals, which have multiple stomach chambers. The chambers hold huge populations of bacteria that make cellulose-digesting enzymes (Section 41.1). Pandas eat only bamboo which, pound for pound, has fewer nutrients than meat. They must grasp and eat a lot of bamboo to get enough proteins, fats, and carbohydrates. Their rounded, short-toed paws can't grab bamboo leaves. They use a bony digit like a thumb in opposition to the five toes on a paw to grip leaves.

A giant panda looks like a bear but resembles a red panda, which lives in the same general area in China and eats bamboo. Bears and the red panda don't have thumbs. How are all three related? Hybridization studies between single strands of their DNA gave an answer. Identifying mismatched nucleotide bases revealed the evolutionary distances among them, and supported an earlier hypothesis. The last shared ancestor of all three lived more than 40 million years ago. A genetic divergence took place among its descendants. One branching led to modern raccoons and to the red panda. Another led to bears. Between 20 million and 15 million years ago, a split from the bear lineage gave rise to ancestors of the giant panda. So, the giant panda appears to be much more closely related to bears than to the red panda.

degree of similarity as measure of relatedness

amino acid sequences between species is plotted against a series of branch points inferred from the fossil record. Such a graph reflects relative divergence times among species, even among phyla and kingdoms.

Biochemical similarity is greatest among the most closely related species and weakest among the most remote.

ala asn lys asn lys gly ile ile trp gly glu asp thr leu met glu tyr leu glu asn pro lys lys tyr ile pro gly thr lys met ile phe val gly ile lys lys lys glu glu arg ala asp leu ile ala tyr leu lys lys ala thr asn glu-COO⁻

ala asn lys asn lys ala val glu trp glu glu asn thr leu tyr asp tyr leu leu asn pro lys lys tyr ile pro gly thr lys met val phe pro gly leu lys lys pro gln asp arg ala asp leu ile ala tyr leu lys lys ala thr ser ser-COO⁻

ala asn ile lys lys asn val leu trp asp glu asn asn met ser glu tyr leu thr asn pro lys lys tyr ile pro gly thr lys met ala phe gly gly leu lys lys glu lys asp arg asn asp leu ile thr tyr leu lys lys ala cys glu-COO⁻

see p. 8

HOW DO WE INTERPRET THE EVIDENCE?

Identifying, Naming, and Classifying Species

Taxonomy is a field of biology that attempts to identify, name, and classify species. Taxonomists face hurdles. Available information about species can be interpreted differently and arranged in different ways.

Let's start with a system of naming species, which nearly all biologists accept. Corn plants, vanilla orchids, houseflies, humans—these and all other species have a scientific name that gives people throughout the world a way to know they're talking about the same organism (Figure 19.17). Taxonomists assign names on the basis of a system developed by Carolus Linnaeus, a respected early naturalist. By his **binomial system**, each species is assigned a two-part Latin name.

The first part of the species name is generic; it is descriptive of similar species that are thought to be the same type of organism and are grouped together. Such a grouping is a **genus** (plural, genera). The second part is the **specific epithet**. Together with the generic name, it refers to one kind of organism only.

There is but one *Ursus maritimus* (polar bear). *Ursus arctos* is the brown bear; *Ursus americanus* is the black bear. The first letter of such generic names is capitalized but the specific epithet is not. The specific epithet is not used without the full or abbreviated generic name that precedes it, because it also can be the second name of a species in a different group. Example: *U. americanus* means black bear, *Homarus americanus* means Atlantic lobster. *Bufo americanus* means American toad. Hence, one should never order *americanus* for dinner unless one is willing to take what one gets.

Groupings of Species—The Higher Taxa

Classification systems are organized ways to retrieve information about species. There was a time when just about everything in nature was classified as animal, vegetable, or mineral. Linnaeus devised a two-kingdom system, with all of life divided into animals or plants according to similarities and differences. His system lasted more than two centuries, but it had nowhere to put the tens of thousands of bacterial species.

Later, biologists used more inclusive groupings to reflect relationships. Groupings of species are called the **higher taxa** (singular, taxon). Family, order, class, phylum, and kingdom are examples. Taxonomists now use **phylogeny** to assign species to ever higher taxa. The term refers to evolutionary relationships among species, from the most ancestral forms through genetic divergences that led to all descendant species.

We find clues to phylogenies in the fossil record, the geologic record, morphology, and biochemistry. Yet how do we interpret them? *Classical* taxonomy uses degrees of morphological divergence to construct evolutionary tree diagrams. In *cladistic* taxonomy, the branch points in the trees are the measure (*clad*– means branch). Only species sharing derived traits are grouped past a branch point for the last common ancestor. A **derived trait** is a novel feature that evolved once and is shared *only* by descendants of the ancestor in which it first evolved.

Evolutionary tree diagrams called **cladograms** group taxa on the basis of derived traits. *All* descendants from an ancestral species in which the trait first evolved are called a **monophyletic group** (meaning "single tribe").

Example: Sharks, crocodiles, birds, and mammals have a heart. Sharks do not have lungs, but crocodiles, birds, and mammals do. Birds and crocodiles have a gizzard, but mammals do not. Only mammals have fur; only birds have feathers. We can use these traits to put together a cladogram:

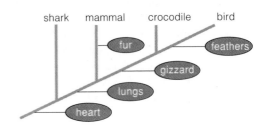

Section 19.8 shows you how to make these evolutionary trees. Such trees do not directly convey "who came from whom." They only suggest distances between groups by the position of higher to lower branch points between them and the last shared ancestor.

KINGDOM	Plantae	Plantae	Animalia	Animalia
PHYLUM	Anthophyta	Anthophyta	Arthropoda	Chordata
CLASS	Monocotyledonae	Monocotyledonae	Insecta	Mammalia
ORDER	Poales	Asparagales	Diptera	Primates
FAMILY	Poaceae	Orchidaceae	Muscidae	Hominidae
GENUS	*Zea*	*Vanilla*	*Musca*	*Homo*
SPECIES	*Z. mays*	*V. planifolia*	*M. domestica*	*H. sapiens*
COMMON NAME	corn	vanilla orchid	housefly	human

Figure 19.17 Taxonomic classification of four organisms. Each is assigned to ever more inclusive categories (higher taxa), from species to kingdom.

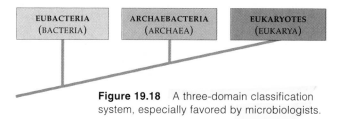

Figure 19.18 A three-domain classification system, especially favored by microbiologists.

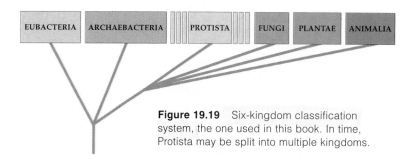

Figure 19.19 Six-kingdom classification system, the one used in this book. In time, Protista may be split into multiple kingdoms.

Classification Systems

Robert Whittaker proposed one of the first phylogenetic systems. He assigned species to one of five kingdoms. He used morphological similarities and differences, modes of nutrition, cell structure, and developmental features as clues to their evolutionary connections.

Whittaker's system places all the prokaryotic cells in kingdom **Monera**. Collectively, these single cells show more metabolic diversity than eukaryotes. They include many thousands of producers (photoautotrophs plus chemoautotrophs), and pathogens and decomposers (heterotrophs). Kingdom **Protista** has single-celled and multicelled eukaryotes with more internal complexity than prokaryotes, but not as much fungi, plants, and animals. It includes photoautotrophs and heterotrophs, such as diverse pathogens and parasites.

Kingdom **Fungi** holds multicelled, heterotrophic eukaryotes that depend on extracellular digestion and absorption. Different kinds are decomposers (nutrient cyclers), pathogens, and parasites. Kingdom **Plantae** holds the multicelled producers (photoautotrophs) with vascular tissues. More than 295,000 kinds are known. Kingdom **Animalia** contains eukaryotic, multicelled, heterotrophic species. We know of more than a million kinds, including herbivores, carnivores, and parasites that range from the microscopic to giants.

Whittaker's system enjoyed wide acceptance until recently. For years, biologists knew that the chemical composition, plasma membrane, and wall of single-celled prokaryotes called archaebacteria have unique features. The differences did not seem to be significant enough to pull them from kingdom Monera. Then the microbiologist Carl Woese and other researchers at the University of Illinois employed new gene sequencing methods (Chapter 16). After comparing the rRNA base sequences from a variety of organisms, they proposed dividing prokaryotic cells into two major taxa. As they argued, archaebacteria differ from the eubacteria (true bacteria) as much as they do from eukaryotic cells.

Strong evidence favoring their conclusion came in 1996. Carol Bult and her colleagues sequenced all 1.7 million base pairs of DNA taken from the bacterium *Methanococcus jannaschii*. Then Woese and Bult worked to decipher the sequence. More than half the genes had never been seen before. Some are closer to the genes of

humans and other eukaryotes than to eubacterial genes. Does the biochemical evidence point to three very early branchings in the history of life? Probably.

THREE-DOMAIN SYSTEM Evidence of this primordial branching has given rise to a **three-domain system**: Eubacteria (bacteria), Archaebacteria (archeans), and Eukarya, which includes the protistans, plants, fungi, and animals (Figure 19.18). Most systematists favor the system. Those most reluctant to adopt it argue it does not give suitable weight to the far greater biodiversity of eukaryotes. Also, recent sequencing studies imply that early prokaryotes must have swapped genes a lot, hence were an "ancestral community." Lineages that led from that community are still being deciphered.

A SIX-KINGDOM SYSTEM At this point, we don't have a universally accepted scheme. However, consensus is growing among biologists to replace the five-kingdom system with one that better reflects life's evolutionary history. In a **six-kingdom classification system**, the first great divergence after the origin of life gave rise to the forerunners of eubacteria and archaebacteria. Another divergence led to all single-celled eukaryotes, some of which gave rise to all multicelled forms. In this system, **Archaebacteria** and **Eubacteria** are kingdoms, just like protistans, fungi, plants, and animals (Figure 19.19). The evolutionary tree diagram shown in Section 19.8 correlates this system with the three-domain scheme.

Each species is assigned a two-part scientific name (genus and specific epithet). It is classified by ever more inclusive groupings of species, the higher taxa.

Classification systems organize and simplify the retrieval of information about species. Phylogenetic systems attempt to reflect evolutionary relationships among species.

Reconstructing the evolutionary history of a given lineage must be based on detailed understanding of the fossil record, morphology, life-styles, and habitats of its representatives, and on biochemical comparisons with other groups.

Recent evidence, especially from comparative biochemistry, favors a six-kingdom classification system: Archaebacteria, Eubacteria, Protista, Fungi, Plantae, and Animalia.

Constructing a Cladogram

To see how phylogenetic reconstruction works, construct and interpret a cladogram. Select seven vertebates from different groups, based on morphological, physiological, and behavioral traits (or characters) that clearly differ among them. The selected organisms, the *in-group*, have shared traits that suggest they may be related. Also select a different organism as a reference point for evaluating evolutionary distances within the in-group.

Suppose you select a lamprey, trout, lungfish, turtle, cat, gorilla, and human. To keep things simple, you focus on the presence ($+$) or absence ($-$) of six traits:

Taxon	Traits (Characters)					
	Jaws	Limbs	Hair	Lungs	Tail	Shell
Lamprey	$-$	$-$	$-$	$-$	$+$	$-$
Turtle	$+$	$+$	$-$	$+$	$+$	$+$
Cat	$+$	$+$	$+$	$+$	$+$	$-$
Gorilla	$+$	$+$	$+$	$+$	$-$	$-$
Lungfish	$+$	$-$	$-$	$+$	$+$	$-$
Trout	$+$	$-$	$-$	$-$	$+$	$-$
Human	$+$	$+$	$+$	$+$	$-$	$-$

After reading Chapter 26, you decide the lamprey is only distantly related to the other vertebrates listed. For instance, it lacks jaws and paired appendages. It can be the *outgroup*, the organism with the fewest derived traits. Any deviation from an outgroup is a derived trait, which implies morphological divergence and branching in the evolutionary tree. In the table below, each zero (0) across the columns of traits for each vertebrate indicates an ancestral condition. Each one (1) means the vertebrate shows the derived trait:

Taxon	Traits (Characters)					
	Jaws	Limbs	Hair	Lungs	Tail	Shell
Lamprey	0	0	0	0	0	0
Turtle	1	1	0	1	0	1
Cat	1	1	1	1	0	0
Gorilla	1	1	1	1	1	0
Lungfish	1	0	0	1	0	0
Trout	1	0	0	0	0	0
Human	1	1	1	1	1	0

Now you look for derived traits that the selected groups do or do not share. For example, the human and gorilla share five derived traits, and the human and the cat share only four. You can now make a simple cladogram. (In practice, systematists often use many traits to examine many taxa. Often it takes a computer to find the pattern that is best supported by so much information.

Figure 19.20*a* has information based on the jaw trait. All vertebrates *other than the outgroup* have jaws. The jaw

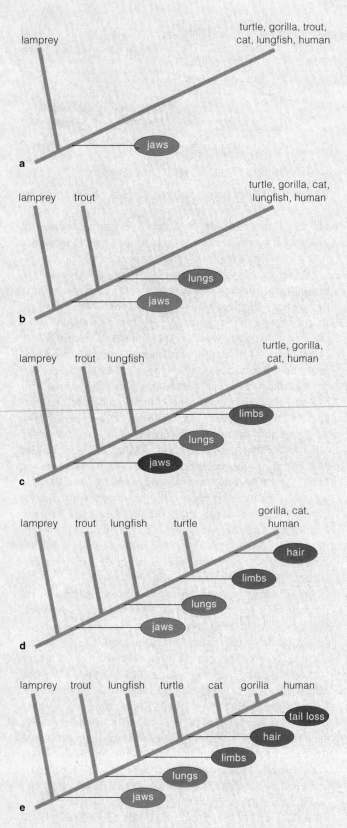

Figure 19.20 Step-by-step construction of a cladogram. This evolutionary tree diagram groups taxa by derived traits.

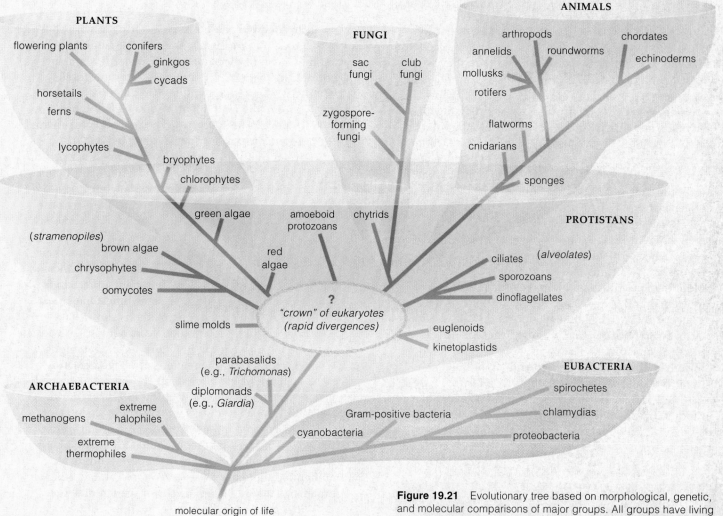

PLANTS

flowering plants   conifers
                ginkgos
                cycads
horsetails
ferns
lycophytes
           bryophytes
           chlorophytes

FUNGI

sac fungi   club fungi

zygospore-forming fungi

ANIMALS

arthropods
annelids   roundworms
mollusks
rotifers
       chordates
       echinoderms
flatworms
cnidarians
      sponges

green algae   amoeboid protozoans   chytrids

PROTISTANS

(*stramenopiles*)

brown algae
chrysophytes
oomycotes
      red algae

slime molds

?
*"crown" of eukaryotes
(rapid divergences)*

ciliates   (*alveolates*)
sporozoans
dinoflagellates
euglenoids
kinetoplastids

parabasalids
(e.g., *Trichomonas*)

diplomonads
(e.g., *Giardia*)

EUBACTERIA

spirochetes
chlamydias
proteobacteria

ARCHAEBACTERIA

methanogens   extreme halophiles
extreme thermophiles

Gram-positive bacteria

cyanobacteria

molecular origin of life

Figure 19.21 Evolutionary tree based on morphological, genetic, and molecular comparisons of major groups. All groups have living representatives. The *blue*, *tan*, and *green* branchings correspond to the three-domain system.

is a derived trait. Figure 19.20*b* offers information about lungs. All of the vertebrates except the lamprey and trout have lungs. The lungfish, human, gorilla, turtle, and cat seem be a monophyletic group, with a shared ancestor. (They share two derived traits with the lamprey and one with the trout.) Figure 19.20*e* shows how a cladogram develops as you keep adding information to it.

How do you "read" the final cladogram? Remember, it suggests relative relatedness, not ancestry. A human is more closely related to a gorilla than to a cat. The gorilla isn't the *ancestor* of humans (it is a modern organism), but both share a more recent common ancestor than either does with the cat. The cat, human, and gorilla as a group are more closely related than they are to the turtle. These relationships are easy to deduce. Less obviously, a human is more closely related to a lungfish than to a trout. Follow its lines or branches from the human and lungfish back to the intersection (or node) where they meet. Next, trace the lungfish and trout branches to their intersection, which is

closer to the base of the tree than is the lungfish–human intersection. The higher up an intersection is on a tree, the more derived traits are shared. The lower it is, the fewer traits they share with other groups under study.

Another way to think about this is to regard the axis of the cladogram as a time bar, but one without absolute dates. *The lower the position of the branch point between two groups on a cladogram, the more distant is the most recent common ancestor of the taxa being compared.*

Because they don't have absolute dates, cladograms don't correlate with geologic time. Following the branches back to intersections shows which groups have the most recent common ancestor. Also, modern, *existing* species are neither ancestors nor descendants of one another.

Comparative biochemistry also helps us construct cladograms. Figure 19.21, an evolutionary tree of life, is based largely on comparisons of ribosomal RNA and protein-coding genes of major groups. This tree correlates six recognized kingdoms with three-domain systems.

Interpreting and Misinterpreting the Past

ARCHAEOPTERYX When Charles Darwin formally presented his theory of evolution by natural selection, scholars knew that if they subscribed to it, they would have to accept tentatively that all species are related by descent to ever more ancient species. Darwin saw with his own eyes that artificial selection by pigeon breeders and others could mold traits of a population in no time at all. It was possible, he argued, for natural selection to bring about the evolution of one species into a separate species, with one or more traits that were uniquely its own, over hundreds or thousands of generations.

However, if new species evolved from older ones, then where were the transitional forms in the fossil record, the so-called "missing links"? Where were the fossils with traits that bridged major groups? More than a century would pass before fossils, and molecular and genetic analyses, would yield some answers. In Darwin's time, the presumed absence of transitional forms was an obstacle to acceptance of his theory. Ironically, a fossil of just such a transitional form had already been unearthed by workmen at a limestone quarry near Solnhofen, Germany.

The fossil, about the size of a pigeon, looked like a theropod, a small meat-eating dinosaur. It, too, had a long, slender, bony tail, three long, clawed fingers on each forelimb, and a heavy jaw fringed with short, spiky teeth (Figure 19.22). Later, diggers unearthed another fossil of the same type. Later still, someone noticed the feathers. If these were dinosaur fossils, what were they doing with *feathers*? Upon careful examination, the feathers proved to be like those of modern birds. The specimen type was given the name *Archaeopteryx* (meaning "ancient winged one").

Between 1860 and 1988, six *Archaeopteryx* specimens and a fossilized feather were found. Anti-evolutionists tried to dismiss them as forgeries. Someone, they said, had pressed modern bird bones and feathers against wet plaster. The dried, imprinted plaster casts merely looked like fossils. Microscopic examination confirmed that the fossils are real. Further confirmation of their antiquity came from the remains of obviously ancient species of worms, jellyfishes, and many other kinds of organisms preserved in the same limestone layers.

Today, radiometric dating methods show that *Archaeopteryx* lived 150 million years ago. When you examine the photograph in Figure 19.22, you might wonder: How could the remains of winged creatures be so well preserved after so much time? It happened that *Archaeopteryx* lived in tropical forests bordering a lagoon. The lagoon was large, warm, and stagnant. Coral reefs barred inputs of oxygenated water from the sea, so it was not a favorable habitat for scavenging animals that might have feasted upon—and thereby obliterated the remains of—*Archaeopteryx* and other organisms that fell from the sky or drifted offshore.

With each tropical storm surge, fine sediments swept over the reefs. They gently buried the carcasses littered at the bottom of the lagoon. In time, the soft, muddy sediments became compacted and hardened. They became a fine limestone tomb for more than 600 species, including *Archaeopteryx*.

A WHALE OF A STORY Fossils of ancient whales give us another example of transitional forms between major groups. For some time, evolutionists accepted that the ancestors of whales were tetrapods that had

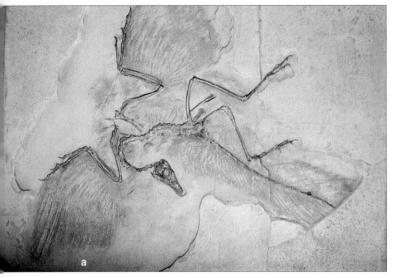

a

b   *DROMAEOSAURUS*   *ARCHAEOPTERYX*

Figure 19.22 (**a**) One of the *Archaeopteryx* fossils and (**b**) comparison of its form with that of a duck-size, two-legged dinosaur. Birds and dinosaurs arose from early reptiles. *Archaeopteryx*, like dinosaurs, had a long, slender bony tail, a heavy jaw with serrated teeth, and three long fingers. Yet it had feathers. Such characters suggest it was on or close to the lineage that gave rise to birds. *Dromaeosaurus*, the small dinosaur shown, had no feathers; a fossil of what may be a close relative apparently does.

Figure 19.23 (**a**) Reconstruction, based on fossils found in Pakistan, of *Rodhocetus*. This cetacean lived 47 million years ago, along the shores of the Tethys Sea. Its ankle bones point to a close evolutionary connection between early whales and hoofed land mammals.

(**b**) Reconstruction of *Basilosaurus*, an ancient whale. Newly recovered fossilized hind limbs have ankle bones even though this whale was fully aquatic. Unlike *Rodhocetus*, it did not use hindlimbs to support its body weight. (**c**) Beluga whale and a young observer in a modern-day aquarium. Living whales have no vestiges of ankle bones.

a

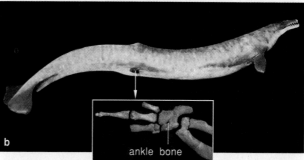

ankle bone

b

c

walked on land, then took up life in the water about 55 million years ago. Fossils revealed gradual changes in skeletal features that made an aquatic life possible. But which lineage of four-legged mammals gave rise to the earliest whales? The answer recently came from Philip Gingerich, graduate student Iyad Zalmout, and their coworkers. While digging in Pakistan, they found fossils of early aquatic whales. Intact, sheep-like ankle bones *and* archaic whale skull bones were present in the same fossilized skeletons (Figure 19.23).

Before this, some researchers contended that early whales were morphologically similar to mesonychians, now-extinct mammals that looked like hyenas with hooves. Others said DNA, immunological, and genetic comparisons suggested that cetaceans—which include whales, dolphins, and porpoises—descended from the artiodactyls. All of these hooved mammals have an even number of toes (two or four) on each foot. They include hippos, camels, pigs, deer, sheep, and cows. The form of the ankle bones of fossilized, early whales from Pakistan is the same as the unique ankle bones of extinct as well as modern artiodactyls. And all of the artiodactyls are running animals. Modern cetaceans, including the beluga whale in Figure19.23*c*, no longer have even a remnant of an ankle bone.

These fossil finds dovetail with morphological and molecular data. The convergence is sound evidence of an evolutionary link between certain aquatic mammals and a major group of mammals on land.

No one was around to witness such transitions in the history of life. But the fossils are real, just as the morphology, biochemistry, and molecular makeup of living organisms are real. Radiometric dating assigns fossils to their place in time. *Life's evolution is not "just a theory."* Remember Section 1.5? In science, a theory differs from speculation because its predictive power has been tested in the natural world many times, in many different ways that might disprove it. If a theory stands, there is a very high probability it is not wrong.

Evolutionists argue all the time among themselves. They argue over how best to interpret the evidence of life's evolution. They debate over which mechanisms and events underlie its long history. At the same time, they do not ignore the evidence of evolution, which is there for us to accumulate and interpret.

Interpreting the evolutionary history of life scientifically requires a huge intellectual shift, from direct observation to inferences based on the fossil record, morphological comparisons, and molecular studies.

19.10

SUMMARY   *Gold* indicates text section

1. Evidence of evolution comes from the fossil record, geologic record, and morphological and biochemical comparisons of major groups of organisms. *19.1–19.9*

2. Fossils are recognizable, physical evidence of life in the distant past. They start to form after an organism or traces of it become buried in volcanic ash or sediments. Minerals infuse organic remains, sediments accumulate slowly above them; increasing pressure and chemical changes transform them to stony hardness. *19.1, 19.2*

   a. Fossils are present in layers of sedimentary rock. The deepest layers accumulate first; they are the oldest. Thus, the older the layers, the older the fossils.

   b. The fossil record's completeness varies in terms of the species represented, where they lived, and the stability of burial sites since fossilization occurred.

3. All species, past and present, are related by way of descent from their common ancestors, starting with the origin of the first living cells. Macroevolution refers to patterns, trends, and rates of change among groups of species over long time spans. Clues to the past hint at the continuity of relationship in nature. *CI, 19.2, 19.9*

4. Scientists use abrupt transitions in the sequence of fossil assemblages as boundary markers in a geologic time scale. Life arose in the Archean eon. Fossils date from the Archean eon and the Proterozoic, Paleozoic, Mesozoic, and Cenozoic eras. *19.2*

5. Like life, the Earth has evolved. Some of its features change repetitively, as when mountains rise and erode slowly. Other features have changed irreversibly in ways that are explained by plate tectonics theory: *19.3*

   a. The crust is fractured into huge, thin, rigid plates that gradually split apart, drift, and collide with one another. Plates raft the land masses along with them.

   b. Molten rock wells up in plumes from the interior of the Earth. At mid-oceanic ridges, it seeps out, cools, hardens, and laterally displaces older portions of the crust. Seafloor spreading forces older crust into deep trenches. Mountain ranges paralleling coasts formed as one plate thrust under another plate and uplifted it.

   c. Large-scale, long-term changes in the Earth's land masses have changed the ocean and atmosphere, and have profoundly influenced the evolution of life.

6. Comparative morphology often reveals similarities that reflect evolutionary relationship. *19.4*

   a. In cases of morphological divergence, comparable body parts were modified in different ways in different lines of descent from a common ancestor. Modifications resulted in homologous structures: body parts that have underlying similarities owing to shared ancestry, even if they differ in size, shape, or function.

   b. In cases of morphological convergence, dissimilar body parts evolve independently in different lineages but in response to similar environmental pressures. The modifications resulted in analogous structures: body parts that resemble one another through adaptation to similar environmental pressures, not by shared ancestry.

7. Similarities in patterns of development may be clues to evolutionary relationship. Changes in DNA that alter the onset, rate, or time of developmental steps probably brought about most morphological differences between related lineages. For example, transposons altered gene expression in humans, relative to chimpanzees. *19.5*

8. Comparative biochemistry reveals similarities and differences among species at the molecular level. *19.6*

   a. Nucleic acid hybridization, a base-pairing between DNA or RNA from two sources, is a rough measure of evolutionary distance between them. Automated gene sequencing is a faster way to compare nuclear DNA, mitochondrial DNA, and ribosomal RNA. Computer programs use the results to construct family trees.

   b. Conserved genes have an accumulation of neutral mutations which, like predictable ticks of a molecular clock, are calibrated in time. The number of differences between base sequences or amino acid sequences from different species is plotted against a series of branch points, inferred from fossil records, to identify relative divergence times among groups.

9. Taxonomists attempt to identify, name, and classify species. By the Linnaean binomial system, each kind of organism is assigned a two-part Latin name. The first part (genus) identifies morphologically similar species derived from a common ancestor. In combination with the species epithet, it identifies the species. *19.7*

10. Classification systems are organized ways to retrieve information on species. Its higher taxa are ever more inclusive groups from species to genera, families, orders, classes, phyla, to kingdoms. Phylogenetic systems show inferred evolutionary relationships. *19.7–19.9*

11. A three-domain system recognizes archaebacteria, eubacteria, and eukaryotes as three major branchings in the tree of life. Archaebacteria, Eubacteria, Protista, Plantae, Fungi, and Animalia are the groupings of the six-kingdom classification system. *19.7, 19.8*

Review Questions

1. Explain why evolutionary biologists expect to find gaps in the fossil record. *19.1*

2. Distinguish macroevolution from microevolution. *19.2*

3. Name three eras of the geologic time scale. *19.2*

4. Did life originate in the Archean or Proterozoic eon? *19.2*

5. Define radiometric dating. What does half-life mean? *19.2*

6. Give an example of how immense crustal movements have influenced the evolution of life. *19.3*

7. Define and give examples of the difference between: *19.4*
   a. homologous and analogous structures
   b. morphological divergence and convergence

8. Comparative morphology refers to comparisons of body form and structures, embryonic and adult, for major lineages. Describe an example of such a comparison. *19.4*

9. Name a protein specified by a gene that has been highly conserved in organisms ranging from bacteria to humans. *19.6*

10. Why do evolutionary biologists apply heat energy to hybrid molecules that contain DNA from two species? *19.6*

11. What type of mutation is the basis of a molecular clock? What does the last tick of a molecular clock signify? *19.6*

12. How are groups of organisms organized in the three-domain system compared to the six-kingdom system? *19.7*

Self-Quiz   ANSWERS IN APPENDIX III

1. Morphological convergences may lead to _____ .
   a. analogous structures   c. divergent structures
   b. homologous structures   d. both a and c

2. Heritable changes in DNA underlying morphological differences between lineages _____ .
   a. may have been caused mostly by transposons
   b. affected the onset, rate, and time of development steps
   c. both a and b

3. A classification system that is _____ is based on presumed evolutionary relationship.
   a. epigenetic   c. phylogenetic
   b. credited to Linnaeus   d. both b and c

4. *Pinus banksiana, Pinus strobus,* and *Pinus radiata* are _____ .
   a. three families of pine trees
   b. three different names for the same organism
   c. three species belonging to the same genus
   d. both a and c

5. Increasingly inclusive taxa range from _____ to _____ .
   a. kingdom; species   c. genera; kingdom
   b. kingdom; genera   d. species; kingdom

6. Match these terms suitably.
   ____ phylogeny   a. accumulation of neutral mutations
   ____ fossil   b. evidence of life in distant past
   ____ stratification   c. similar body parts in different
   ____ homologous   lineages owing to common descent
        structure   d. e.g., insect wing and bird wing
   ____ molecular clock e. evolutionary relationship among
   ____ analogous   species, ancestors to descendants
        structure   f. layers of sedimentary rock

Critical Thinking

1. At the end of your backbone are several small, fused bones, called the coccyx (Section 17.1, Figure 17.3). Could the coccyx be a vestigial structure—all that's left of a tail that was a feature of the evolutionarily distant vertebrate (and primate) ancestors of humans? Or is it the start of a newly evolving structure? Make an educated guess, then describe some ways in which you might test whether your guess is plausible.

2. Comparative biochemistry helps us estimate evolutionary relationship and approximate times for divergences from ancestral stocks. Base sequence comparisons and amino acid comparisons yield good estimates. Reflect on the genetic code (Section 14.3), then suggest why it may be a useful measure of mutations, mutation rates, and biochemical relatedness.

3. Shannon thinks there are too many kingdoms and sees no good reason to make another one for something as small as archaebacteria. "Keep them with the other prokaryotes!" she

Figure 19.24   A human and a whale inspecting each other.

says. Taxonomists would call her a "lumper." By contrast, Andrew is a "splitter." He sees no good reason to withhold kingdom status from archaebacteria simply because they are part of a microscopic world that not many people know about. Which may be the more useful: more or fewer boundaries between groups of organisms? Explain your answer.

4. When walking along a path cut into the side of a steep mountain, you pass rocky layers that are fractured and folded back on themselves in bizarre patterns. You look closely and discover a fossilized shell in the "highest" layer in the series. How would you use the plate tectonics theory to help you decide whether the fossil is of recent or ancient origin?

5. See for yourself the kind of evidence that links modern whales (Figures 19.23 and 19.24) with hippos. Search for examples of fossilized skeletons, using the genus names in Section 19.9 as an entry point into the literature and the web.

Selected Key Terms

analogous structure *19.4*	molecular clock *19.6*
Animalia *19.7*	Monera *19.7*
Archaebacteria *19.7*	monophyletic group *19.7*
asteroid *CI*	morphological convergence *19.4*
binomial system *19.7*	morphological divergence *19.4*
cladogram *19.7*	neutral mutation *19.6*
classification system *19.7*	nucleic acid hybridization *19.6*
comparative morphology *19.4*	Pangea *19.3*
derived trait *19.7*	phylogeny *19.7*
Eubacteria *19.7*	Plantae *19.7*
fossil *19.1*	plate tectonics theory *19.3*
fossilization *19.1*	Protista *19.7*
Fungi *19.7*	radiometric dating *19.2*
genus *19.7*	six-kingdom
geologic time scale *19.2*	classification system *19.7*
Gondwana *19.3*	specific epithet *19.7*
half-life *19.2*	stratification *19.1*
higher taxon (taxa) *19.7*	taxonomy *19.7*
homologous structure *19.4*	theory of uniformity *19.3*
lineage *19.1*	three domain system *19.7*
macroevolution *19.2*	transposon *19.5*

Readings

Dott, R., Jr., and R. Batten. 1998. *Evolution of the Earth*. Fourth edition. New York: McGraw-Hill.

Ochert, A. December 1999. "Transposons." *Discover*, 59–66.

Ridley, M. 1999. *Genome: The Autobiography of a Species in 23 Chapters*. New York: HarperCollins.

THE ORIGIN AND EVOLUTION OF LIFE

In the Beginning . . .

Some clear evening, watch the moon as it rises from the horizon and think of the 380,000 kilometers between it and you. *Five billion trillion times* farther away from you are galaxies—systems of stars—at the boundary of the known universe. Wavelengths that travel through space move faster than anything else— millions of meters per second—yet long wavelengths that originated from faraway galaxies many billions of years ago are only now reaching the Earth.

By all known measures, all of the near and distant galaxies in the vast space of the universe are moving away from one another, which means the universe must be expanding. One prevailing view of how the colossal expansion came about may account for every bit of matter in every living thing.

Think about how you rewind a videotape on a VCR. Then imagine "rewinding" the universe. As you do this, the galaxies start moving back together. After 12 to 15

Figure 20.1 Part of the great Eagle nebula, a hotbed of star formation 7,000 light-years from Earth, in the constellation Serpens. (The Latin *nebula* means mist.) Not shown in this image are a few huge, young stars above the pillars. For the past few million years, intense ultraviolet radiation from the stars has been eroding the less dense surface of the pillars. Immense globs of denser gases and dust that have resisted erosion are visible at the surface. Each pillar is wider than our solar system—more than 10 billion miles across! New stars are hatching from the protruding globs; some shine brightly on the tips of gaseous streamers.

billion years of rewinding, all galaxies, all matter, and all of space are compressed into a hot, dense volume about the size of the sun. You have arrived at time zero.

That incredibly hot, dense state lasted only for an instant. What happened next is known as the **big bang**, a stupendous, nearly instantaneous distribution of matter and energy throughout the universe. About a minute later, temperatures dropped a billion degrees. Fusion reactions created most of the light elements, including helium, which are still the most abundant elements in the universe. Radio telescopes can detect cooled, diluted background radiation—a relic of the big bang—left over from the beginning of time.

Over the next billion years, uncountable numbers of gaseous particles collided and condensed under gravity's force to become the first stars. When the stars were massive enough, nuclear reactions ignited inside them and gave off tremendous light and heat. Massive stars continued to contract, and many became dense enough to promote the formation of heavier elements.

All stars have a life history, from birth to an often spectacularly explosive death. In what might be called the original stardust memories, the heavier elements released during the explosions became swept up in the gravitational contraction of new stars. They became the raw materials for the formation of even heavier elements. Even as you read this page, the Hubble space telescope is revealing astounding glimpses of ongoing star-forming activity, as in the dust clouds of Orion, Serpens, and other constellations (Figure 20.1).

Now imagine a time long ago, when explosions of dying stars ripped through our galaxy and left behind a dense cloud of dust and gas that extended trillions of kilometers in space. As the cloud cooled, countless bits of matter gravitated toward one another. By 4.6 billion years ago, the cloud had flattened into a slowly rotating disk. At the dense, hot center of that disk, the shining star of our solar system—the sun—was born.

The rest of this chapter is a sweeping slice through time, one that cuts back to the formation of the Earth and life's chemical origins. It is the bridge to the next unit, which will take us along lines of descent that led to the present range of biodiversity.

The story is not complete. Even so, the available evidence from many avenues of research points to a principle that can help us organize bits of information about the past: *Life is a magnificent continuation of the physical and chemical evolution of the universe, of galaxies and stars, and of the planet Earth.*

Key Concepts

1. We have evidence that life originated somewhere around 3.8 billion years ago. The origin and subsequent evolution of life have been correlated with the physical and chemical evolution of the universe, the stars, and the planet Earth.

2. All inorganic and organic compounds required for self-replication, assembly of membranes, and metabolism—for the structure and functions of all living cells—could have formed spontaneously under conditions that apparently prevailed on the early Earth.

3. The history of life, from its chemical beginnings to the present, spans five intervals of geologic time. It extends through two great eons—the Archean and Proterozoic—and the Paleozoic, Mesozoic, and Cenozoic eras.

4. Not long after life originated, divergences led to two great prokaryotic lineages called the archaebacteria and eubacteria. Soon afterward, the ancestors of eukaryotes diverged from the branch leading to archaebacteria.

5. Archaebacteria and eubacteria were the dominant forms of life during the Archean and Proterozoic eons. Eukaryotic cells originated late in the Proterozoic, and they soon became spectacularly diverse. A theory of endosymbiosis helps explain the profusion of specialized organelles that evolved in eukaryotic cells.

6. The fossil record reveals a history of persistences, extinctions, and radiations of different lineages.

7. Throughout life's history, asteroid impacts, drifting and colliding continents, and other environmental assaults have had profound impact on the direction of evolution.

The Scottish physicist James Clerk Maxwell (1831-79), an early authority on Saturn's rings, had, as cosmologists should, a poetic bent:

At quite uncertain times and places,
the atoms left their heavenly path,
And by fortuitous embraces,
Engendered all that being hath.

The phrase "fortuitous embraces," although lovely, is not explanatory. Knowledge, tickled from the heavens, is the business of a small band of possible explainers — the people of JPL and NASA, government at its best.

CONDITIONS ON THE EARLY EARTH

Origin of the Earth

Figure 20.1 shows part of one of the vast clouds in the universe. The clouds are mostly hydrogen gas, along with water, iron, silicates, hydrogen cyanide, ammonia, methane, formaldehyde, and other small inorganic and organic substances. The contracting cloud that became our solar system probably was similar in composition. The cloud's edges cooled between 4.6 and 4.5 billion years ago. Electrostatic attractions and gravity's pull caused mineral grains and ice orbiting the new sun to start clumping together (Figure 20.2). In time, larger, faster clumps collided and shattered. Some grew more massive by sweeping up asteroids, meteorites, and other rocky remnants of collisions. They evolved into planets.

While the Earth formed, heat generated by asteroid impacts, internal compression, and radioactive decay of minerals melted much of its rocky interior. Molten nickel, iron, and other heavy materials moved into the interior; lighter ones floated to the surface. The process resulted in a crust, mantle, and core. The **crust** became an outer zone of basalt, granite, and other low-density rocks. It rested on a zone of intermediate-density rocks, the **mantle**. In turn, the mantle enveloped an immense core of high-density, partially molten nickel and iron.

Four billion years ago, the Earth was a thin-crusted inferno (Figure 20.3a). Less than 200 million years later, life appeared on its surface! We have no record of its origin, probably because movements in the mantle and crust, volcanic activity, and erosion obliterated all traces of it. Still, we can put together a plausible explanation of how life originated by considering three questions:

First: Can we identify physical and chemical conditions that prevailed on the Earth when life originated?

Second: Do the known principles of physics, chemistry, and evolution support or disprove the hypothesis that large organic molecules formed spontaneously and evolved into molecular systems with the fundamental properties of life?

Third: Can we design experiments to test the hypothesis that living systems emerged through chemical evolution?

The First Atmosphere

Hot gases blanketed the Earth when the first patches of crust formed. We suspect that this first atmosphere was a mix of gaseous hydrogen (H_2), nitrogen (N_2), carbon monoxide (CO), and carbon dioxide (CO_2). Did it hold gaseous oxygen (O_2)? Probably not. Rocks subjected to intense heat, as happens during volcanic eruptions, do release oxygen, but not much. Also, free oxygen would have reacted at once with other elements. Remember, oxygen has an electron vacancy in its outermost shell, and it tends to bond with other atoms (Section 2.3).

If the early atmosphere had not been relatively free of oxygen, organic compounds necessary to assemble cells would not have been able to form on their own, spontaneously. Any oxygen would have attacked their structure and disrupted their functioning.

What about water? Although dense clouds cloaked the early Earth, any water falling on the molten surface must have evaporated at once. In time, the crust cooled and grew solid. For millions of years, rainfall and runoff eroded mineral salts from rocks. The salt-laden waters collected in crustal depressions, forming the first seas.

If liquid water had not accumulated, membranes—which take on their bilayer organization only in water—could not have formed. No membrane, no cell. Life at its most basic level *is* the cell, which has a capacity to survive and reproduce on its own.

Synthesis of Organic Compounds

A cell is made of proteins, complex carbohydrates and lipids, and nucleic acids. Existing cells assemble these molecules from smaller organic compounds: the simple sugars, fatty acids, amino acids, and nucleotides. Energy from the environment drives these synthesis reactions. Were small organic compounds present on the early Earth? Were there sources of energy that spontaneously drove their assembly into the large molecules of life?

Mars, meteorites, the Earth's moon, and the Earth formed at the same time, from the same cosmic cloud. Rocks from Mars, meteorites, and the moon all hold precursors of biological molecules, so these precursors must have been on the early Earth. *Energy from the sun, lightning, or heat vented from the crust could have been enough to drive their combination into organic molecules.*

Stanley Miller conducted the first experimental test of that prediction. He mixed water, methane, hydrogen, and ammonia in a reaction chamber (Figure 20.3b). He

Figure 20.2 Representation of the cloud of dust, gases, and clumps of rock and ice around the early sun.

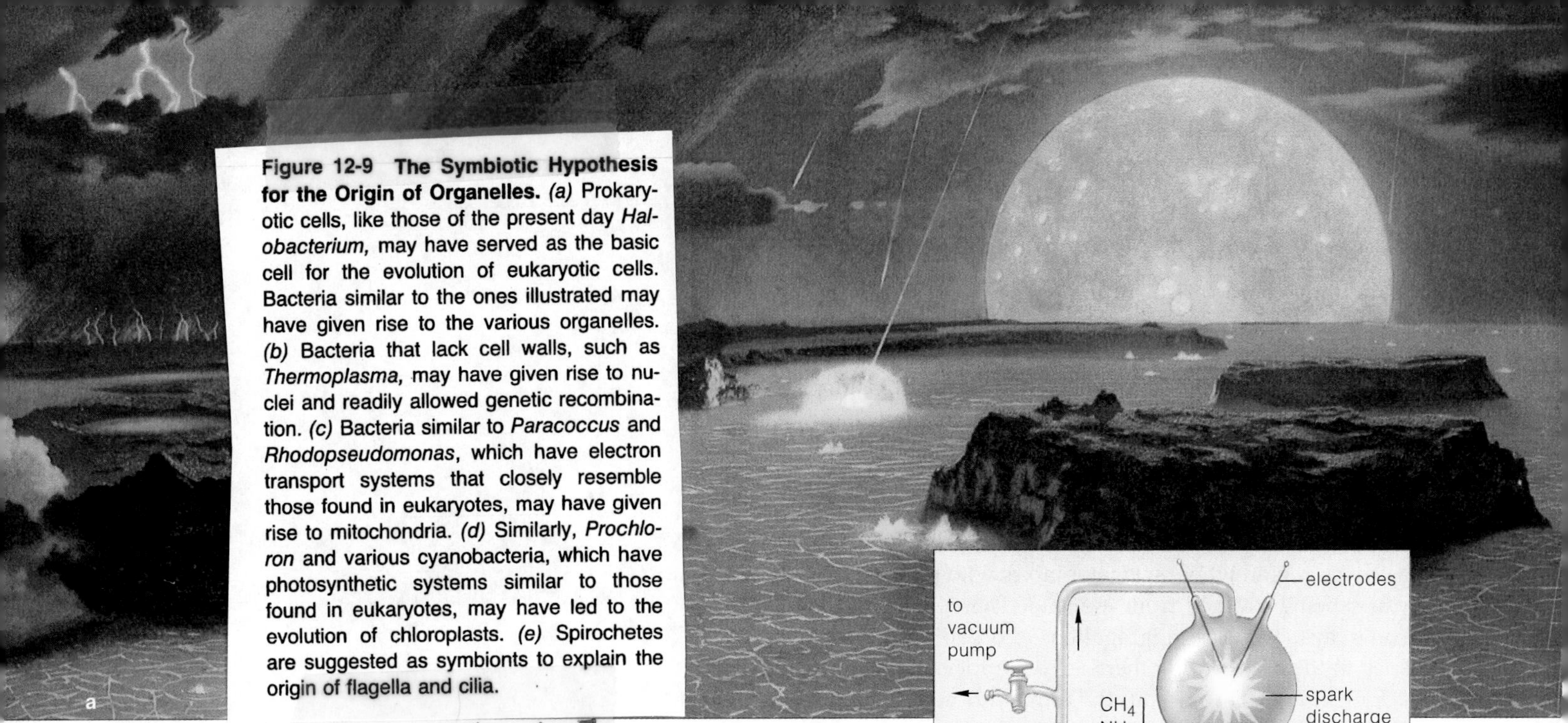

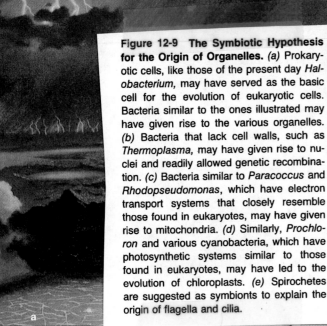

Figure 12-9 The Symbiotic Hypothesis for the Origin of Organelles. (a) Prokaryotic cells, like those of the present day *Halobacterium*, may have served as the basic cell for the evolution of eukaryotic cells. Bacteria similar to the ones illustrated may have given rise to the various organelles. **(b)** Bacteria that lack cell walls, such as *Thermoplasma*, may have given rise to nuclei and readily allowed genetic recombination. **(c)** Bacteria similar to *Paracoccus* and *Rhodopseudomonas*, which have electron transport systems that closely resemble those found in eukaryotes, may have given rise to mitochondria. **(d)** Similarly, *Prochloron* and various cyanobacteria, which have photosynthetic systems similar to those found in eukaryotes, may have led to the evolution of chloroplasts. **(e)** Spirochetes are suggested as symbionts to explain the origin of flagella and cilia.

pp 66-67

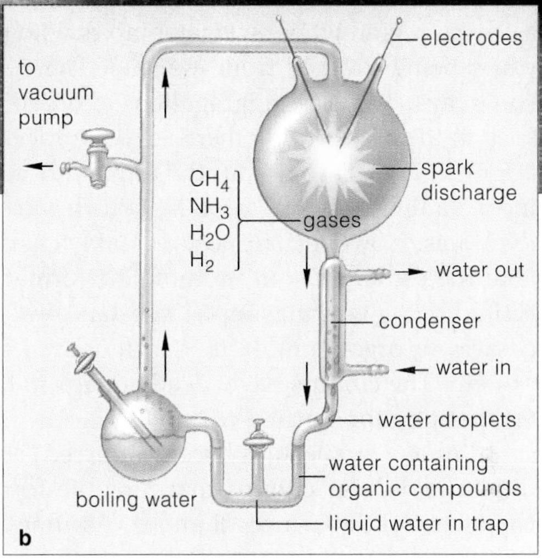

Figure 20.3 **(a)** Representation of the Earth during its formation, when the moon's orbit was closer than it is today. If the Earth had condensed into a planet of smaller diameter, its gravitational mass would not have been great enough to hold on to an atmosphere. If it had settled into an orbit closer to the sun, water would have evaporated from its hot surface. If the Earth's orbit had been more distant from the sun, its surface would have been colder, and water would have been locked up as ice. Without liquid water, life as we know it never would have originated on Earth.

(b) Stanley Miller's experimental apparatus, used to study the synthesis of organic compounds under conditions that presumably existed on the early Earth. The condenser cools circulating steam so that water droplets form.

kept the mixture circulating and exposed it to a spark discharge to simulate lightning. Amino acids and other small organic compounds formed in less than a week.

In other experiments that simulated conditions on the early Earth, glucose, ribose, deoxyribose, and other sugars formed spontaneously from formaldehyde, and adenine from hydrogen cyanide. Ribose and adenine are part of ATP, NAD^+, and other vital nucleotides.

However, if *complex* organic compounds had formed directly in seawater, could they have lasted long? The spontaneous direction of the reactions would have been toward hydrolysis, not condensation, in water. So how did more lasting bonds form?

By one hypothesis, clay in the rhythmically drained muck of tidal flats and estuaries served as templates (structural patterns) for the spontaneous assembly of proteins and other complex compounds. Clay consists of thin, stacked layers of aluminosilicates, and its metal ions attract amino acids. Expose clay to amino acids, warm it with the sun's rays, then alternately moisten it and dry it out. Condensation reactions will occur and form proteins and other complex organic compounds.

By another hypothesis, complex organic compounds formed spontaneously near hydrothermal vents, where archaebacteria are thriving today (Section 7.9, *Critical*

Thinking question 3). Experimental tests by Sidney Fox and others show that when amino acids are heated and placed in water, they spontaneously order themselves into small protein-like molecules he calls proteinoids.

However the first proteins formed, their molecular structure dictated how they could interact with other compounds. Suppose some proteins functioned as weak enzymes and hastened bond formation between amino acids. An enzyme-directed synthesis of proteins would have had selective advantage. Protein configurations that promoted such reactions would win the chemical competition for available amino acids. They would have been favored in another way—for proteins have the capacity to bind metal ions and other metabolic agents.

As you will see, the evolution of metabolism was based on such chemical modifications. For now, simply reflect on the possibility that selection was at work before cells appeared, favoring the chemical evolution of enzymes and other complex organic compounds.

Many different experiments offer indirect evidence that the complex organic molecules characteristic of life could have formed under the physical and chemical conditions that prevailed on the early Earth.

EMERGENCE OF THE FIRST LIVING CELLS

Origin of Agents of Metabolism

A defining characteristic of life is *metabolism*. The word refers to all the reactions by which cells harness energy and use it to drive their activities, such as biosynthesis. During the first 600 million years or so of Earth history, enzymes, ATP, and other organic compounds may have assembled spontaneously, perhaps in the same places. If so, the close association naturally promoted chemical interactions and the start of metabolic pathways.

Imagine an ancient estuary rich in clay deposits. It is a coastal region where seawater mixes with mineral-rich water being drained from the land. There, beneath the sun's rays, countless aggregations of organic molecules stick to the clay. At first there are quantities of an amino acid; call it *D*. Throughout the estuary, *D* molecules get incorporated into new proteins—until the supply of *D* dwindles. However, suppose a weakly catalytic protein is in the estuary. It can promote the formation of *D* by acting on a plentiful, simpler substance—*C*. By chance, clumps of organic molecules include that enzyme-like protein. The clumps have an advantage in the chemical competition for starting materials.

In time, *C* molecules become scarce. The advantage tilts to molecular clumps that promote formation of *C* from simpler substances *B* and *A*. Assume *B* and *A* are carbon dioxide and water. The atmosphere and the seas contain essentially unlimited amounts of both. Thus chemical selection has favored a synthetic pathway:

$$A + B \longrightarrow C \longrightarrow D$$

Some clumps prove better than others at absorbing and using energy. Which molecules could bestow such an advantage? Think of the energy-trapping pathway that now dominates the world of life: photosynthesis. It starts at pigments called chlorophylls. A porphyrin ring is the part of a chlorophyll molecule that absorbs light and gives up electrons (Figure 20.4). Porphyrins also occur in cytochromes, components of the electron transfer systems in all photosynthetic and aerobically respiring cells. They are assembled spontaneously from formaldehyde—one of the molecular legacies of cosmic clouds. Was porphyrin an agent of electron transfers in some of the early metabolic pathways? Perhaps.

Origin of Self-Replicating Systems

Another defining characteristic of life is a capacity for reproduction, which now starts with protein-building instructions in DNA. Molecules of DNA are fairly stable and are easily replicated before each cell division. As you know from earlier chapters, molecules of RNA, enzymes, and other factors are necessary to translate the DNA instructions into proteins.

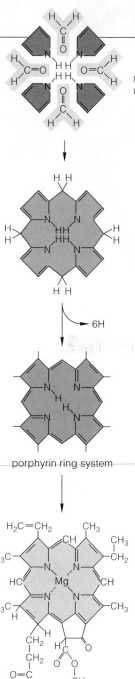

four formaldehyde molecules with four pyrrole rings

Figure 20.4 One hypothetical sequence by which formaldehyde, an organic compound, underwent chemical evolution into porphyrin. Formaldehyde was present when the Earth formed. Porphyrin is the light-trapping and electron-donating component of chlorophyll molecules. It also is a component of cytochrome, which is a protein component of electron transfer systems that are part of many metabolic pathways.

6H

porphyrin ring system

chlorophyll *a*

Most existing enzymes get assistance from small organic molecules or metal ions called coenzymes. Intriguingly, some categories of coenzymes have a structure identical to that of RNA nucleotides. Another clue: Mix together and then heat up precursors of ribonucleotides and short chains of phosphate groups, and they self-assemble into single strands of RNA. On the early Earth, energy from the sun's rays or from geothermal events might have driven the spontaneous formation of RNA from such starting materials.

Very simple self-replicating systems of RNA, enzymes, and coenzymes have been produced in some laboratories. Did RNA become the information-storing templates upon which simple proteins became synthesized? Maybe. We now know that an rRNA in the ribosome catalyzes the translation of mRNA into proteins (Section 14.2). A ribosome is like irregularly shaped pieces of a puzzle, all joined together. The pieces include many proteins that stabilize the rRNA just about everywhere except at an active site where the large ribosomal subunit contacts the smaller subunit. Ribosomes, remember, are highly conserved structures. The ones in eukaryotic cells are not that different

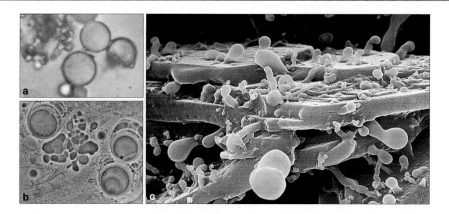

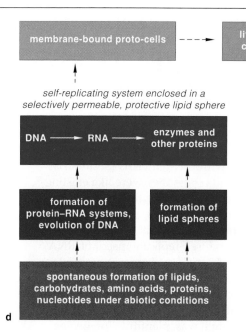

Figure 20.5 Microscopic spheres of (**a**) proteins and (**b**) lipids, self-assembled under abiotic conditions. (**c**) Nanobes, which may resemble proto-cells. Australian researchers found them in rocks 3.8 kilometers (3 miles) below the Earth's surface, where temperatures reach 1700°C (338°F). They are only 20 to 15 nanometers across. Yet they have a cell membrane that encloses DNA and other organic compounds. And they grow. (**d**) A possible sequence of events that led to the first self-replicating systems, to proto-cells, then to the first living cells.

from those in prokaryotic cells. We might take this to mean rRNA's catalytic behavior is a metabolic function that emerged early in Earth history.

Did an **RNA world**, in which RNA was a template for protein synthesis, precede the origin of DNA? As you know, DNA and RNA nucleotides are very similar. Their sugar component differs by one functional group. Three of their four bases are identical. And RNA's uracil differs from DNA's thymine by one functional group.

So what is the point of having DNA? Compared to RNA, this double-stranded molecule can persist as long nucleotide chains in far more stable fashion. Computer analyses also reveal that helical coiling of a strandlike molecule is the best way to pack the most subunits in the smallest space. DNA is just a better way to *package* more and more information about building proteins.

Until we identify the chemical ancestors of RNA and DNA, the story of life's origin will be incomplete. Filling in the details will require imaginative sleuthing. For instance, researchers ran a computer program that incorporated information about natural energy sources and simple inorganic compounds of the sort thought to have been present on the early Earth. They asked their advanced computer to subject the chosen compounds to random chemical competition and natural selection as might have occurred untold billions of times in the past. They ran their program again and again. Always the outcome was the same: *Simple precursors invariably evolved—and they spontaneously organized themselves as interacting systems of large, complex molecules.*

Origin of the First Plasma Membranes

Experimental tests are more revealing of the origin of the plasma membrane—the outermost membrane of all

living cells. This cellular component consists of a lipid bilayer that incorporates diverse proteins, which carry out different functions (Section 5.1). It controls which substances move into and out of the cell. Without such control, cells can neither exist nor reproduce.

Probably molecular evolution led to **proto-cells**, or simple membrane sacs that surrounded and protected information-storing templates and metabolism from the environment. We know simple membrane sacs can form spontaneously. In one experiment, heated amino acids formed protein-like chains, which were placed in hot water. After cooling, the chains assembled into small, stable spheres (Figure 20.5*a*). Like cell membranes, the spheres were selectively permeable to substances. In other experiments, spheres picked up lipid molecules from the surroundings, and a lipid–protein film formed at their surface. Also, "nanobes" rather like proto-cells have been discovered deep in the Earth (Figure 20.5*c*).

In still other experiments, by David Deamer and his coworkers, fatty acids and glycerol combined to form long-tail lipid molecules under laboratory conditions that simulated conditions in evaporating tidepools. The lipids self-assembled into tiny, water-filled sacs, many of which were like cell membranes (Figure 20.5*b*).

In short, there are gaps in the story of life's origins. But there also is experimental evidence that chemical evolution led to the organic molecules and structures characteristic of life. Figure 20.5*d* shows likely steps in the chemical evolution that preceded the first cells.

Although the story is not yet complete, many laboratory experiments and computer simulations indirectly show that chemical and molecular evolution could have given rise to proto-cells.

20.3

ORIGIN OF PROKARYOTIC AND EUKARYOTIC CELLS

The first living cells originated in the **Archean** eon, which lasted from 3.8 billion to 2.5 billion years ago. Those cells emerged as molecular extensions of the evolving universe, of our solar system and the Earth. Maybe they originated at tidal flats or hydrothermal vents (Section 7.9). Fossils show they were like existing bacteria. Specifically, they all were **prokaryotic cells**, with no nucleus. Maybe they were only membrane-bound, self-replicating sacs of RNA, DNA, and other organic molecules. Free oxygen was absent, so they secured energy through anaerobic pathways—fermentation, probably. Energy was plentiful, for geologic processes had already enriched the seas with organic compounds. So "food" was available, there were no predators, and cell structures were free from oxygen attacks.

Figure 20.6 Evolutionary tree of life showing possible connections among major lineages and the origins of some major eukaryotic organelles.

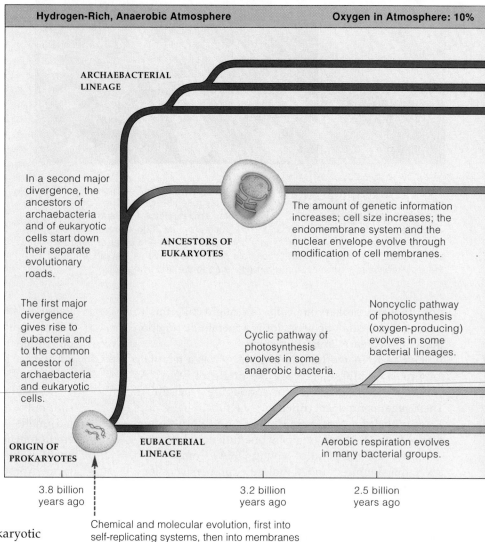

Hydrogen-Rich, Anaerobic Atmosphere   **Oxygen in Atmosphere: 10%**

ARCHAEBACTERIAL LINEAGE

In a second major divergence, the ancestors of archaebacteria and of eukaryotic cells start down their separate evolutionary roads.

The first major divergence gives rise to eubacteria and to the common ancestor of archaebacteria and eukaryotic cells.

ANCESTORS OF EUKARYOTES

The amount of genetic information increases; cell size increases; the endomembrane system and the nuclear envelope evolve through modification of cell membranes.

Cyclic pathway of photosynthesis evolves in some anaerobic bacteria.

Noncyclic pathway of photosynthesis (oxygen-producing) evolves in some bacterial lineages.

ORIGIN OF PROKARYOTES   EUBACTERIAL LINEAGE   Aerobic respiration evolves in many bacterial groups.

3.8 billion years ago   3.2 billion years ago   2.5 billion years ago

Chemical and molecular evolution, first into self-replicating systems, then into membranes of proto-cells by 3.8 billion years ago

Some populations of those first prokaryotic cells apparently diverged in two major directions shortly after the time of origin. One lineage gave rise to the **eubacteria**. The other lineage gave rise to the common ancestor of **archaebacteria** and **eukaryotic cells** (Figure 20.6).

Between 3.5 and 3.2 billion years ago, light-trapping pigments, systems of electron transfers, and other metabolic machinery evolved in some anaerobic eubacteria. With these innovations, the cells became the earliest practitioners of the cyclic pathway of photosynthesis. (You read about this ATP-forming pathway in Section 7.3.) Sunlight, an unlimited source of energy, had been tapped. For nearly 2 billion years, photosynthetic descendants of those cells dominated the living world. Their tiny but numerous populations formed large mats in which sediments collected. The mats built up, one atop the other. Calcium deposits hardened and preserved the mats, which came to be called **stromatolites** (Figure 20.7).

Figure 20.7 Stromatolites exposed at low tide in western Australia's Shark Bay. These mounds started forming 2,000 years ago. They are structurally identical with stromatolites that formed more than 3 billion years ago.

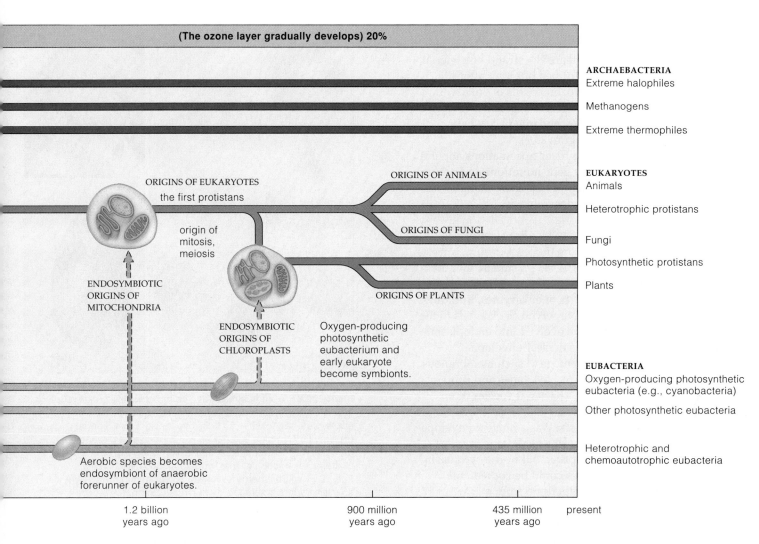

(The ozone layer gradually develops) 20%

ARCHAEBACTERIA
Extreme halophiles

Methanogens

Extreme thermophiles

ORIGINS OF EUKARYOTES
the first protistans

ORIGINS OF ANIMALS

EUKARYOTES
Animals

Heterotrophic protistans

origin of
mitosis,
meiosis

ORIGINS OF FUNGI

Fungi

Photosynthetic protistans

ENDOSYMBIOTIC
ORIGINS OF
MITOCHONDRIA

ORIGINS OF PLANTS

Plants

ENDOSYMBIOTIC
ORIGINS OF
CHLOROPLASTS

Oxygen-producing
photosynthetic
eubacterium and
early eukaryote
become symbionts.

EUBACTERIA
Oxygen-producing photosynthetic
eubacteria (e.g., cyanobacteria)

Other photosynthetic eubacteria

Aerobic species becomes
endosymbiont of anaerobic
forerunner of eukaryotes.

Heterotrophic and
chemoautotrophic eubacteria

| 1.2 billion years ago | 900 million years ago | 435 million years ago | present |

By the dawn of the **Proterozoic** eon, 2.5 billion years ago, photosynthetic machinery had become altered in some eubacterial species, and the noncyclic pathway of photosynthesis emerged. Oxygen, one of the pathway's by-products, started to accumulate. In time, this had two irreversible effects. First, *an oxygen-rich atmosphere stopped the further chemical origin of living cells*. Except in a few anaerobic habitats, complex organic compounds could no longer form spontaneously and resist attack. Second, *aerobic respiration became the dominant energy-releasing pathway*. In many prokaryotic lineages, selection favored metabolic equipment that neutralized oxygen by using it as an electron acceptor! This key innovation contributed to the rise of multicelled eukaryotes and their invasion of far-flung environments (Section 7.3).

Eukaryotic cells evolved in the Proterozoic, possibly before 1.2 billion years ago. We have fossils, 900 million years old, of complex algae, fungi, and plant spores. As you know, organelles are the premier defining features of eukaryotic cells. *Where did they come from?* The next section describes a few plausible hypotheses.

Rodinia, the first supercontinent, formed 1.1 billion years ago. By about 800 million years ago, stromatolites along its shorelines were declining dramatically. Were newly evolved, bacteria-eating *animals* using them as a concentrated source of food? Fossilized embryos—cell clusters no wider than a pin—give hints that the first animals were soft-bodied forerunners of the modern sponges and marine worms. Before 570 million years ago, in precambrian times, some of their descendants started the first adaptive radiation of animals.

The first living cells evolved by about 3.8 billion years ago, during the Archean eon. All were prokaryotic. Most, if not all, probably made ATP by fermentation routes.

Early on, ancestors of archaebacteria and eukaryotic cells diverged from the lineage that led to modern eubacteria.

Oxygen-releasing photosynthetic bacteria evolved. In time, the oxygen-enriched atmosphere put an end to the further spontaneous chemical origin of life. That atmosphere was a key selection pressure in the evolution of eukaryotic cells.

see
p. 66
&
Test
bank
p. 22
p. ___

20.4 CONNECTIONS

Where Did Organelles Come From?

Thanks to Andrew Knoll, William Schopf, Jr., and other globe-hopping microfossil hunters, we have evidence of early life. One fossil treasure is a strand of bacterial cells 3.5 billion years old, formed not long after life originated. We have Proterozoic fossils of eukaryotic cells with a few membrane-bound organelles (Figure 20.8). Today, most of the descendant species have a profusion of organelles (Figure 20.9).

Where did organelles come from? Speculations abound. Some probably evolved by gene mutations and natural selection. For others, though, researchers make a case for evolution by way of endosymbiosis.

ORIGIN OF THE NUCLEUS AND ER Prokaryotic cells do not have an abundance of organelles. But some species have interesting infoldings of the plasma membrane that incorporate enzymes and other metabolic agents (Figure 20.10). In forerunners of eukaryotes, infoldings into the cytoplasm may have served as channels to the surface. And they may have evolved into endoplasmic reticulum (ER) and into the nuclear envelope.

What would be the advantage of such membranous enclosures? Maybe they protected genes and protein products from "foreigners." Bacterial species often transfer plasmid DNA among themselves. So do simple eukaryotic cells called yeasts. Was a nuclear envelope favored because it got the cell's genes, and enymes of replication and transcription, out of the cytoplasm? If so, vital genetic messages could be copied and read, free of metabolic competition from a potentially unmanageable hodgepodge of foreign genes. Similarly, ER channels may have protected a cell's proteins and other organic compounds from metabolically hungry "guests"—foreign cells that in time became permanent residents in a host cell, as described next.

THEORY OF ENDOSYMBIOSIS Accidental partnerships among a variety of prokaryotic species probably arose countless times in lineages that gave rise to eukaryotic cells. Maybe some guests evolved into mitochondria,

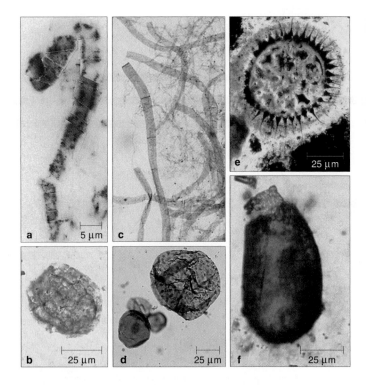

Figure 20.8 From Australia, (**a**) a strand of walled prokaryotic cells 3.5 billion years old and (**b**) one of the oldest known eukaryotes—a protistan 1.4 billion years old. From Siberia, (**c**) a multicelled alga 900 million to 1 billion years old and (**d**) eukaryotic microplankton 850 million years old. (**e**) From China, a eukaryotic cell that lived about 560 million years ago. (**f**) From Spitsbergen, Norway, a protistan that was alive 75 million years before the dawn of the Cambrian.

chloroplasts, and other organelles. This is a story of endosymbiosis, as developed in greatest detail by Lynn Margulis. (*Endo–* means within; *symbiosis* means living together.) In cases of **endosymbiosis**, one species (the guest) spends its entire life cycle *inside* another species (the host). The interaction benefits one or both.

By one hypothesis, eukaryotic cells evolved through endosymbiosis long after the noncyclic pathway of

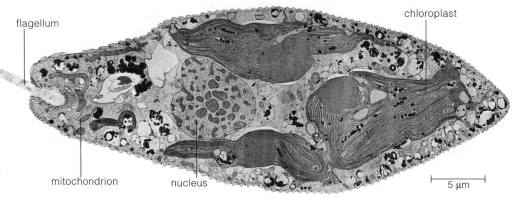

Figure 20.9 A fine example of diverse organelles—hallmarks of eukaryotic cells: *Euglena*, a single-celled protistan, sliced lengthwise. It also has a long flagellum, which could not completely fit in the image area at this magnification.

flagellum

chloroplast

mitochondrion nucleus 5 µm

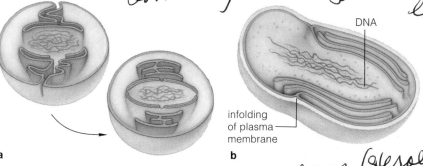

Figure 20.10 (**a**) Model for the origin of endoplasmic reticulum and the nuclear envelope. In the prokaryotic ancestors of eukaryotic cells, infoldings of the plasma membrane gave rise to both cell components. (**b**) Such infoldings are present in the cytoplasm of many kinds of existing bacteria, including *Nitrobacter*, sketched here in cutaway view.

a

b

DNA

infolding of plasma membrane

photosynthesis emerged and oxygen accumulated to notable levels in the atmosphere. In some bacterial groups, one type of electron transfer system in the plasma membrane had evolved. It now incorporated "extra" cytochromes that donated electrons to oxygen. With this key innovation, the bacteria extracted energy from organic compounds by aerobic respiration.

Possibly before 1.2 billion years ago, forerunners of eukaryotes were engulfing aerobic bacteria. Maybe they resembled existing soft-bodied, amoeboid cells that weakly tolerate free oxygen. Maybe they trapped food by sending out cytoplasmic extensions and used endocytic vesicles to get the food into the cytoplasm.

A key point of the theory is that the aerobic bacteria resisted digestion and thrived in the protected, nutrient-rich environment. In time, they were releasing excess ATP. Their hosts came to depend on that ATP for their growth, increased activity, and assembly of hard parts and other structures. The guests benefited, too. They no longer had to seek food or build components for metabolic work the hosts did for them. The anaerobic and aerobic cells became incapable of independent life. The guests had evolved into mitochondria, supreme suppliers of ATP.

EVIDENCE OF ENDOSYMBIOSIS Nature continues to tinker with many endosymbionts. Consider the cell in Figure 20.11. Its mitochondria are like bacteria in size and structure. Each replicates its own DNA and divides independently of the host cell's division. The inner mitochondrial membrane is like a bacterial plasma membrane. In its DNA and mRNA, a few genetic code words with unique meanings are translated into a few proteins that are required for mitochondrial tasks. The "mitochondrial code" has a few unique codons that do not occur in the genetic code of living cells.

Chloroplasts, too, might be modified descendants of oxygen-evolving, photosynthetic bacteria. Perhaps predatory aerobic bacteria engulfed photosynthetic cells. Maybe the cells escaped digestion, absorbed nutrients from the host's cytoplasm, and continued to function. By providing aerobically respiring hosts with oxygen, they would have promoted endosymbiosis.

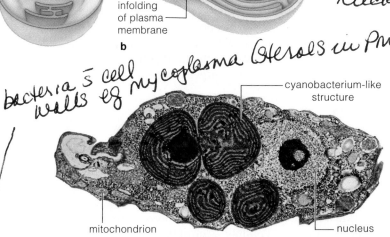

cyanobacterium-like structure

mitochondrion

nucleus

Figure 20.11 *Cyanophora paradoxa*, a protistan. Its mitochondria resemble bacteria. Its photosynthetic structures look like spherical cyanobacteria (which are photosynthetic) without the cell wall.

Chloroplasts do resemble some eubacteria in their metabolism and overall nucleic acid sequence. They differ from one another in shape and light-absorbing pigments, like photosynthetic bacteria do. Their DNA is self-replicating, and they, too, divide independently of cell division. Did chloroplasts originate a number of times in different lineages? Maybe.

Adding to the intrigue are ciliated protistans and marine slugs that "enslave" chloroplasts! The slugs eat algae but retain the algal chloroplasts in their gut cells. The chloroplasts draw nutrients from host tissues. They photosynthesize and release oxygen for weeks.

However they arose, new kinds of cells did appear on the evolutionary stage. They had endomembranes, a nucleus, and mitochondria, chloroplasts, or both. They were eukaryotic cells, the first **protistans**. With their efficient metabolic strategies, the early protistans underwent rapid divergences and adaptive radiations. In no time at all, evolutionarily speaking, some of their descendants gave rise to all plants, fungi, and animals, as sketched out earlier in Figure 20.6.

The nucleus and ER may have evolved through infoldings of the plasma membrane. Mitochondria, chloroplasts, and other organelles might have evolved by endosymbiosis. Such theories may help explain how you and all other eukaryotes evolved from simple prokaryotic beginnings.

LIFE IN THE PALEOZOIC ERA

We divide the **Paleozoic** into the Cambrian, Ordovician, Silurian, Devonian, Carboniferous, and Permian periods. The supercontinent Rodinia had split apart before this era. Some of its fragments now straddled the equator, and warm, shallow seas lapped their margins. Global conditions restricted any pronounced seasonal changes in winds, ocean currents, and the upward churning of nutrient deposits on the seafloor. Nutrient supplies in equatorial seas were stable but limited. Polar regions were inhospitably icy; and to the south, other fragments had collided to form a new supercontinent, Gondwana:

GONDWANA

Most major animal phyla had evolved earlier in the precambrian seas. Possibly some of their ancestors were among the **Ediacarans**, peculiar organisms shaped like fronds, disks, and blobs that nearly defy classification (Figure 20.12*a,b*). Like Ediacarans, the early Cambrian animals had highly flattened bodies with a fine surface-to-volume ratio for taking up nutrients (Figure 20.12*c*). Most thrived on or in seafloor sediments, where dead organisms and organic debris settled. They ranged from sponges to simple vertebrates, and they were diverse.

This was a time when new predators and prey with armorlike shells, spines, mouths, and amazing feeding structures rapidly evolved. We find fossilized animals with punctures, missing chunks, and healed wounds in sedimentary beds. Things were starting to get lively!

What caused the "Cambrian explosion" of diversity? Asteroid impacts? After tapering off about 3 billion years ago, the rate of impacts apparently did increase abruptly around this time. Were the genes governing early growth and development less intertwined than they are now? If so, selection against novel traits may have been less severe. What about the warm waters of new equatorial seas? They must have been adaptive zones, with novel opportunities for getting food.

Early in the Ordovician, widespread flooding of the land masses gave rise to broad, tropical seaways. The emergence of the vast marine environments promoted adaptive radiations. Many new reef organisms, such as the swift, shelled predators called nautiloids, evolved. Among their descendants are the chambered nautiluses (Section 27.3). Trilobites, one of the most common but unswift invertebrates, nearly vanished (Figure 20.12*d*).

Huge glaciers formed across the southern portion of Gondwana at the end of the Ordovician—one of the coldest times in Earth history. Enormous volumes of water were locked up in ice sheets, and this drained shallow seas everywhere. More than 70 percent of all marine groups—including the reef builders, trilobites, placoderms, and jawless fishes—became extinct at the Ordovician–Silurian boundary.

Gondwana drifted north during Silurian times. The sea level rose as great ice sheets melted, and adaptive radiations occurred in the seas. Large coral reefs formed, algae flourished, and the first jawed fishes appeared. Small plants took hold in wet lowlands (Figure 20.13*b,c*).

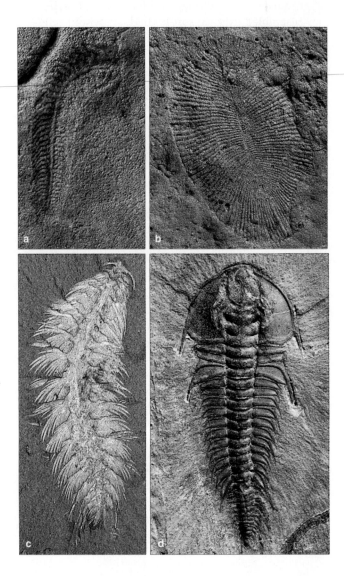

Figure 20.12 Representatives from the precambrian and Cambrian seas. Two Ediacarans, about 600 million years old: (**a**) *Spriggina* and (**b**) *Dickensonia*. The oldest known Ediacarans lived 610 million years ago and the most recent in Cambrian times, 510 million years ago. (**c**) From British Columbia's Burgess Shale, a fossilized marine worm. (**d**) A beautifully preserved fossil of one of the earliest trilobites.

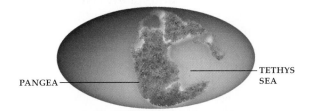

Figure 20.13 (**a**) Life in the Silurian seas. Some of the shelled animals (nautiloids) were as long as a kayak. (**b**) At the close of the Silurian and on into the Devonian, forerunners of modern ferns and club mosses dominated swamps and marshes. (**c**) Fossils of a Devonian plant (*Psilophyton*) that may have been one of the earliest ancestors of the conifers and other seed-bearing plants. (**d**) Reptiles (*Dimetrodon*) of a hotter, drier time, the Permian. Some fossils of these carnivores have been found in Texas. Giant club mosses and horsetails had declined. Conifers, cycads, and other gymnosperms replaced them during the Permian.

So did fungi and invertebrates, such as spiders, insects, and segmented worms. The Devonian was a pivotal time in vertebrate evolution. Armor-plated fishes with massive jaws diversified in the seas and in fresh water. Fishes that gave rise to amphibians invaded the land. They had lobed fins—the forerunners of legs and other limbs—and simple lungs. Lobed fins and lungs were key innovations that would prove to be advantageous for life out of water, in dry land habitats (Section 26.6).

Warm-water species especially were hit hard by a Devonian mass extinction. Global cooling, a decline in sea level, a meteorite impact, or some combination of these might have been the cause. Afterward, plants and insects embarked on adaptive radiations on land.

In the Carboniferous, land masses were submerged and drained many times. Organic debris accumulated, became compacted, and was converted to coal (Section 23.4). Insects, amphibians, and early reptiles flourished in vast swamp forests of the Permian (Figure 20.13*d*). The early ancestors of cycads, ginkgos, conifers, and other seed-producing plants dominated the forests.

Late in the Permian, 95 percent of all known species were lost in the greatest of all mass extinctions. New evidence suggests that "The Great Dying" followed a hit by a comet or asteroid. Whatever the trigger, there was a colossal outpouring of lava, a severe disruption of the global carbon cycle, and a major glaciation.

Pangea had already been forming long before then. By the time the Permian drew to a close, it extended from pole to pole. One world ocean, including the vast Tethys Sea, lapped its margins:

PANGEA —   — TETHYS
                    SEA

And just as the Paleozoic mass extinctions had done, the new distribution of oceans, land masses, and land elevations redirected the course of life's evolution.

Early in the Paleozoic era, organisms of all six kingdoms were flourishing in the seas. By the end of the era, many lineages had successfully invaded the land, including wet lowlands of the new supercontinent Pangea.

LIFE IN THE MESOZOIC ERA

Speciation on a Grand Scale

We divide the **Mesozoic** into the Triassic, Jurassic, and Cretaceous periods. It lasted about 180 million years. In the middle of the Jurassic, the supercontinent Pangea started to break up. Its fragments slowly drifted apart. The fossil record tells us those geographically isolated fragments favored divergences and speciation:

This was an era of spectacular expansion in the range of global diversity. Invertebrates and fishes underwent adaptive radiations in the seas. Conifers and other seed-bearing **gymnosperms,** insects, and reptiles became the visibly dominant lineages on land. Flowering plants—**angiosperms**—emerged during the late Jurassic or the early Cretaceous. In less than 40 million years, they would displace the conifers and related plants in most environments (Figure 20.14 and Section 23.7).

Rise of the Ruling Reptiles

Early in the Triassic, the first **dinosaurs** evolved from a reptilian lineage. They were not much bigger than a turkey. Possibly most species had high metabolic rates, and maybe they were warm-blooded. Many sprinted about on two legs. But the Triassic dinosaurs weren't the dominant land animals. The rulers were *Lystrosaurus* and other plant-eating, mammal-like reptiles that were too big to be bothered by most predators (Figure 20.15).

Adaptive zones opened up for dinosaurs following a catastrophe that left a string of five craters in what is now France, Quebec, Manitoba, and North Dakota. One crater is as big as Rhode Island. About 213 million years ago, fragments from an asteroid or a comet fell in a row as the Earth spun beneath them, just as huge fragments from comet Shoemaker–Levy 9 fell in a row on Jupiter in 1994. The blast waves, global firestorm, lava flows, and earthquakes must have been stupendous. Most animals lucky enough to survive this episode of mass extinction were smaller, had higher rates of metabolism, and were less vulnerable than others to drastic changes in air temperature. These attributes also characterized survivors of later mass extinctions.

Descendants of the surviving dinosaurs became the ruling reptiles; they endured for 140 million years. Some species reached monstrous proportions. Among them were the ultrasaurs, taller than a four-story building.

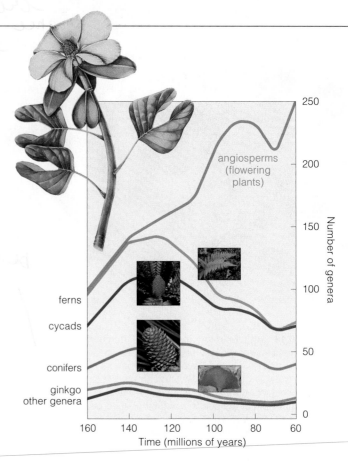

Figure 20.14 Range of diversity among vascular plants of the Jurassic and Cretaceous. Conifers and other gymnosperms were dominant. They started to decline even before flowering plants began a great adaptive radiation that continued into the Cenozoic. *Upper left:* From the Cretaceous, a floral shoot of *Archaeanthus linnenbergeri.* In many traits, this now-extinct flowering plant resembled living magnolias.

Many dinosaurs perished in another mass extinction at the end of the Jurassic, then in a pulse of extinctions during the Cretaceous. Immense outpourings of molten rock may have triggered changes in the global climate. Or maybe asteroids or comets inflicted the blows. Not all lineages survived. Yet some did recover, and new ones evolved. Duckbilled dinosaurs appeared in forests and swamps. Tanklike *Triceratops* and other plant eaters flourished in open habitats. They were prey for the agile and swift *Velociraptor* of motion picture fame.

About 120 million years ago, global temperatures skyrocketed 25 degrees. By one theory, plumes spread out beneath the crust and "greased" the crustal plates into moving twice as fast. A superplume or a rash of them broke through the crust. In what is now the South Atlantic, the crust opened like a zipper. Basalt and lava poured from fissures; volcanoes spewed nutrient-rich ashes. The plumes released great quantities of carbon dioxide, one of the key "greenhouse" gases that absorb some heat radiating from the Earth before it escapes to outer space. The nutrient-enriched planet warmed, and

Figure 20.15 Life in the Mesozoic. (**a**) *Lystrosaurus*, a therapsid (mammal-like reptile) of the Triassic. This tusked herbivore, about 1 meter (3 feet) long, fed on the fibrous plants of dry floodplains. As Section 19.3 describes, its fossils are distributed on widely separated continents and are evidence in favor of the plate tectonics theory.

(**b**) *Temnodontosaurus*, an ichthyosaur that hunted large squid, ammonites, and other prey in the warm, shallow seaways of the early Jurassic. Fossils measuring 9 meters (30 feet) long have been found in England and Germany.

(**c**) From the Cretaceous, cycads, ferns, *A. linnenbergeri* and other flowering plants, and *Deinonychus*, an agile dinosaur with keen eyesight, serrated teeth, and powerful jaws. The short forelimbs of this bipedal carnivore had sharp claws, including a large one shaped like a sickle; hence its name, which means "terrible claw." Fossils 3 meters (9 feet) long have been found in Texas.

it remained warm for 20 million years. Photosynthetic organisms flourished on land and in the shallow seas. Their remains were slowly buried and converted into the world's vast oil reserves.

About 65 million years ago, the last dinosaurs and many marine organisms vanished in a mass extinction. As described in the next section, their disappearance apparently coincided with a direct hit by an asteroid the size of Mount Everest. Imperceptibly, over a great span of time, crustal movements carried the impact site to what is now the Yucatán peninsula.

The Mesozoic was a time of major adaptive radiations and of a mass extinction in which the last dinosaurs and many marine organisms disappeared.

Horrendous End to Dominance

How often have people, puffed up with self-importance, set out to conquer neighbors, lands, and the seas? Think about it. Then think about the dinosaurs. Were they good at reigning supreme? No question about it. Their lineage dominated the land for 140 million years. In the end, did it matter? Not a bit. Sixty-five million years ago, at the Cretaceous–Tertiary (K–T) boundary, the remaining members of their excellent lineage perished. Why? Bad luck.

By analyzing iridium levels in soils, gravity maps, and other evidence, Walter Alvarez and Luis Alvarez hypothesized that an asteroid impact caused the K–T mass extinction. Later on, researchers identified the impact site. Massive movements in the crust transported it to its present location in the Gulf of Mexico (Figure 19.1). The impact crater is 9.6 kilometers deep and 300 kilometers across, wider than Connecticut. This crater, along with other evidence, strongly supports what is now called the **K–T asteroid impact theory**.

To make a crater that big, the asteroid had to hit the Earth at 160,000 kilometers per hour. It blasted at least 200,000 cubic kilometers of dense gases and debris into the sky. The crust heaved violently. Monstrous waves 120 meters high raced across the ocean. They obliterated life on islands, then slammed into the continents.

Some thought atmospheric debris blocked sunlight for months, which caused plants and other producers on land to wither and die, and many animals to starve to death. However, the volume of debris blasted aloft would not have been large enough to have had such consequences. So what did happen? It was a mystery.

Then comet Shoemaker–Levy 9 slammed into Jupiter. Particles blasted into the Jovian atmosphere triggered an intense heating of an area larger than the Earth. That event supports a **global broiling hypothesis**, proposed first by H. J. Melosh and his colleagues. Briefly stated, energy released at the K–T impact site was equivalent to detonating 100 million nuclear bombs. Trillions of tons of vaporized debris rose in a colossal fireball, then rapidly condensed into particles the size of sand grains. Seconds later, a cooler fireball of steam, carbon dioxide, and unmelted rock formed. When the debris plummeted to the Earth, it raised the atmosphere's temperature by thousands of degrees. The sky must have been ten times hotter than it gets above Death Valley in summer. In one horrific glowing hour, nearly all plants erupted in flames and all animals out in the open were broiled alive.

Remember the premise of cause and effect? Christine Janis put it this way: In a world dominated by dinosaurs, mammals survived by being small, efficient at exploiting food, and flexible in their behavior. They became artful dodgers, sidelined until the luck of the dinosaurs ran out. They originated before the dinosaurs, and they outlasted them because of a chance, catastrophic event.

LIFE IN THE CENOZOIC ERA

The breakup of Pangea triggered events that continued into the present era, the **Cenozoic**. At the dawn of the Cenozoic, major land masses were on collision courses:

Coastlines fractured. The Cascades, Andes, Himalayas, and Alps rose through volcanic activity, uplifting, and other events at crustal rifts and plate boundaries. These geologic changes brought about major shifts in global climates that influenced the further evolution of life.

Paleocene climates were warmer and wet. Tropical forests, subtropical forests, and woodlands extended farther north and south than today, even into polar regions. Mammals had evolved long ago, but when the dinosaurs vanished, they began a spectacular adaptive radiation. Many different carnivores, omnivores, and herbivores emerged, with specialized teeth that could shear, shred, grind, mash, or nip food (Section 26.10). The first types did not get bigger; they remained small forest dwellers throughout the Paleocene (Figure 20.16).

Figure 20.16 Reconstructions of a few early Paleocene species. Diverse mammals lived in dense forests of sequoia and other plants in what is now Wyoming. On the ground, raccoonlike *Chriacus* faces a tree-climbing rodent (*Ptilodus*). Higher up is *Peradectes*, a marsupial.

Figure 20.17 (**a**) From the late Eocene and on into the early Miocene, *Indricotherium*, the "giraffe rhinoceros." At 15 tons and 5.5 meters high at the shoulder, it was the largest mammal we know about. It browsed on the canopies of trees in the woodlands.

(**b**) Saber-tooth cat (*Smilodon*) in a dry woodland of the Pleistocene. *Smilodon* fossils have been recovered from the pitch pools in Rancho LaBrea, California.

The long-term warming trend continued into the Eocene epoch. Subtropical forests extended north even into polar regions and supported a far richer variety of mammals. Many modern mammalian species had their start in the early Eocene. Among them were primates, bats, rodents, the even-toed hoofed mammals (hippos, camels and other artiodactyls) and odd-toed mammals, including the tiny, earliest horse (*Hyracotherium*), and rhinos. Scavenging carnivores about the size of weasels spread throughout the northern land masses. Although tree-dwelling insect eaters dwindled, ground dwellers flourished. Among them were the glyptodonts (Section 17.3) and stocky, hoofed mammals to the south.

By the late Eocene, Australia split from Antarctica, and a passageway opened up between the Arctic and North Atlantic oceans. Deep ocean circulation patterns changed, the icing of Antarctica began, and northern land masses became cooler and drier. Tropical forests gave way to woodlands and dry grasslands. Seasons were pronounced, and as plant growth patterns shifted, most of the early mammals became extinct.

From the Oligocene on through the Pliocene, many grazing and browsing animals evolved in woodlands and grasslands. Among them were gazelle-like camels, the "giraffe rhinoceros," and the saber-tooth cats that ambushed them (Figure 20.17).

The climate cooled again in the Miocene as great mountain ranges arose and altered air circulation and rainfall patterns. Grasslands and specialized browsers emerged. Starting in the Pliocene, about 1.8 million years ago, a seesawing of glaciations and interglacials began. Species richness declined in emerging prairies, steppes, pampas, and deserts that are with us today. Early humans faced these challenges (Chapter 26).

Currently, the distribution of land masses favors biodiversity in the tropical forests of South America, Madagascar, and Southeast Asia, and tropical marine ecosystems of the Pacific archipelagos. Yet we are in the midst of a new mass extinction. About 50,000 years ago, nomadic humans started decimating the migrating herds of wild animals of the Northern Hemisphere. In a few thousand years, major groups of mammals were extinct. The pace of extinction has picked up as humans hunt for food, fur, feathers, and fun, as natural habitats are converted for farm animals and crops. We consider the current extinction crisis in Chapters 27 and 50.

Crustal movements and mountain building during the Cenozoic triggered major shifts in climate. Conditions favored a great adaptive radiation of mammals, first in tropical forests, then through woodlands and grasslands.

SUMMARY   *Gold* indicates text section

1. The evolutionary story of life begins with the "big bang," a model for the origin of the universe. By this model, all matter and space were once compressed in a fleeting state of incredible heat and density. Time began with the near-instantaneous distribution of matter and energy through the known universe, which has been expanding ever since. Helium and other light elements formed just after the big bang. Heavier elements were created by the formation, evolution, and death of stars. All elements of the solar system, the Earth, and life are products of this physical and chemical evolution. *CI*

2. Four billion years ago, the Earth had a high-density core, an intermediate-density mantle, and a thin, highly unstable crust of low-density rocks. It seems likely that gaseous hydrogen, nitrogen, carbon monoxide, and carbon dioxide made up most of the first atmosphere. Under conditions that prevailed at the time, neither free oxygen nor liquid water could accumulate at the Earth's surface. *20.1*

3. For hundreds of millions of years after the crust cooled, runoff from rains put dissolved mineral salts and other compounds in crustal depressions, where the early seas formed. Life as we know it would not have originated without that salty liquid water. *20.1*

4. Many studies and experiments have yielded indirect evidence that life originated under the conditions that existed on the early Earth. *20.1, 20.2*

a. The composition of cosmic clouds and rocks from planets and the Earth's moon indicate that precursors of complex molecules associated with life were present. In laboratory tests simulating the primordial conditions, including the absence of free oxygen, these precursors spontaneously assembled into sugars (such as glucose), amino acids, and other organic compounds.

b. Chemical principles and computer simulations suggest that metabolic pathways evolved by chemical competition for limited supplies of organic molecules that had accumulated by natural geologic processes.

c. In laboratory studies, self-replicating systems of RNA, enzymes, and coenzymes have been synthesized. Double-stranded, helically coiled DNA probably evolved as the most efficient way of packing the most protein-building instructions in the smallest space. Also, its long nucleotide chains are more stable than RNA chains. In laboratory simulations of conditions thought to have prevailed on the early Earth, lipids and lipid–protein membranes having some of the properties of current cell membranes have formed spontaneously.

5. Since its origin about 3.8 billion years ago, life has been influenced by changes in the Earth's atmosphere, crust, and oceans. Forces of change included asteroid impacts, volcanism, and crustal movements. The forces altered the distribution of land masses and the seas (Figure 20.18) as well as regional and global climates. Another major force of change: activities of organisms, especially the oxygen-releasing photosynthesizers and, far more recently, the human species. *20.2–20.8*

6. The history of life spans these intervals: *20.2–20.8*
   a. Archean: 3.8 billion to 2.5 billion years ago
   b. Proterozoic: 2.5 billion to 544 million years ago
   c. Paleozoic: 544 to 248 million years ago
   d. Mesozoic: 248 to 65 million years ago
   e. Cenozoic: 65 million years ago to the present

Middle Miocene. Polar regions again iced over, as in Cambrian. All land masses are assuming their current distribution

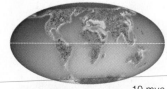

10 mya

Cretaceous into Tertiary. Extinction of dinosaurs; rise of mammals

65 mya

Permian into Triassic. Swamp forests (eventual coal source); seed plants evolve

Pangea —

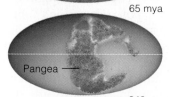

240 mya

Devonian. Jawed fishes evolve, diversify; ancestors of amphibians invade land

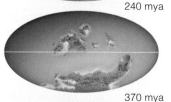

370 mya

Silurian. Sea level rises, diverse marine life; plants, invertebrates invade land

420 mya

Cambrian. Fragments of first supercontinent (Rodinia). Explosion of diversity in equatorial seas; icy polar regions

540 mya

Figure 20.18 Summary of changes in the distribution of land masses and the seas, correlated with life's evolution. Compare this with Figure 20.6, which also correlates the evolution of life with changes in atmospheric concentrations of oxygen.

7. The first living cells arose in the Archean and were prokaryotic. The first divergence led to eubacteria and to the ancestor of archaebacteria and eukaryotes. Some eubacteria used a cyclic photosynthesis pathway. *20.3*

8. A noncyclic photosynthetic pathway evolved in the Proterozoic in some eubacteria. Oxygen, the pathway's by-product, accumulated in the atmosphere; it blocked the further spontaneous formation of organic molecules. The spontaneous origin of life was over. *20.3*

9. Abundant free oxygen was a selective pressure that favored aerobic respiration, a key step in the origin of eukaryotes. It helped form an ozone layer. This shield against UV radiation allowed some aquatic lineages to move into low wetlands. *20.3, 20.5*

10. The nucleus and endoplasmic reticulum may have evolved through infoldings and modifications of the plasma membrane. Mitochondria, chloroplasts, and other organelles may have evolved by endosymbiosis between aerobic and and anaerobic species of bacteria. All eukaryotes, from protistans to plants, fungi, and animals, may have evolved in such ways from simple prokaryotic beginnings. *20.4*

11. By the early Paleozoic, diverse organisms of all six lineages were established in the seas. Before it ended, land was invaded. Ever since, there have been pulses of mass extinctions and adaptive radiations. *20.5–20.8*

12. Dinosaurs were the dominant land animals for 140 million years. An asteroid impact wiped out the last of them at the Cretaceous–Tertiary boundary. *20.6, 20.7*

13. During the Cenozoic, mammals became dominant lineages by default; they were poised to radiate into adaptive zones vacated by the dinosaurs. *20.8*

Eon	Era	Period	Epoch	Millions of Years Ago	Major Geologic and Biological Events That Occurred Millions of Years Ago (mya)
PHANEROZOIC	CENOZOIC	QUATERNARY	Recent	0.01–	1.8 mya to present. Major glaciations. Modern humans evolve. The most recent *extinction crisis* is under way.
			Pleistocene	1.8	
		TERTIARY	Pliocene	5.3	65–1.8 mya. Major crustal movements, collisions, mountain building. Tropics, subtropics extend poleward. When climate cools, dry woodlands, grasslands emerge. *Adaptive radiations* of flowering plants, insects, birds, mammals.
			Miocene	23.8	
			Oligocene	33.7	
			Eocene	55.5	
			Paleocene	65	65 mya. Asteroid impact; *mass extinction* of all dinosaurs and many marine organisms.
	MESOZOIC	CRETACEOUS	Late		99–65 mya. Pangea breakup continues, inland seas form. *Adaptive radiations* of marine invertebrates, fishes, insects, and dinosaurs. Origin of angiosperms (flowering plants).
				99	145–99 mya. Pangea star ts to break up. Marine communities flourish. *Adaptive radiations* of dinosaurs.
			Early	145	145 mya. Asteroid impact? *Mass extinction* of many species in seas, some on land. Mammals, some dinosaurs survive.
		JURASSIC		213	248–213 mya. *Adaptive radiations* of marine invertebrates, fishes, dinosaurs. Gymnosperms dominant land plants. Origin of mammals.
		TRIASSIC		248	*Mass extinction.* Ninety percent of all known families lost.
	PALEOZOIC	PERMIAN		286	286–248 mya. Supercontinent Pangea and world ocean form. On land, *adaptive radiations* of reptiles and gymnosperms.
		CARBONIFEROUS		360	360–286 mya. Recurring ice ages. On land, *adaptive radiations* of insects, amphibians. Spore-bearing plants dominate; cone-bearing gymnosperms present. Origin of reptiles.
		DEVONIAN			360 mya. *Mass extinction* of many marine invertebrates, most fishes.
				410	410–360 mya. Major crustal movements. Ice ages. *Mass extinction* of many marine species. Vast swamps form. Origin of vascular plants. *Adaptive radiation* of fishes continues. Origin of amphibians.
		SILURIAN		440	440–410 mya. Major crustal movements. *Adaptive radiations* of marine invertebrates, early fishes.
		ORDOVICIAN		505	505–440 mya. All land masses near equator. Simple marine communities flourish until origin of animals with hard parts.
		CAMBRIAN		544	544–505 mya. Supercontinent breaks up. Ice age. *Mass extinction.*
PROTEROZOIC				2,500	2,500–544 mya. Oxygen accumulates in atmosphere. Origin of aerobic metabolism. Origin of eukaryotic cells. Divergences lead to eukaryotic cells, then protistans, fungi, plants, animals.
ARCHEAN AND EARLIER					3,800–2,500 mya. Origin of photosynthetic prokaryotic cells. 4,600–3,800 mya. Origin of Earth's crust, first atmosphere, first seas. Chemical, molecular evolution leads to origin of life (from proto-cells to anaerobic prokaryotic cells).

Review Questions

1. Compare chemical and physical conditions that are thought to have prevailed on the Earth 4 billion years ago with current conditions. *20.1*

2. Describe some experimental evidence for the (a) spontaneous origin of large organic molecules, (b) self-assembly of proteins, and (c) formation of organic membranes and spheres. *20.1, 20.2*

3. Summarize the key points of the theory of endosymbiotic origins for mitochondria and chloroplasts. Cite evidence that favors this theory. *20.4*

4. Describe the prevailing conditions that probably favored the Cambrian "explosion" of diversity among marine animals, as evidenced by the fossil record. *20.5*

5. When did plants, fungi, and insects invade the land? What kind of vertebrates first invaded the land, and when? *20.5*

6. What were global conditions like when the gymnosperms and dinosaurs originated? *20.6*

7. Briefly explain how an asteroid impact, followed at once by a disastrous episode of "global broiling," may have caused the mass extinction at the K–T boundary. *20.7*

8. Would you expect the Paleozoic, Mesozoic, or Cenozoic to be called "the age of mammals"? As part of your answer, explain some of the major differences between global conditions during each era. *20.5–20.8*

Self-Quiz   ANSWERS IN APPENDIX III

1. Through study of the geologic record, we know that the evolution of life has been profoundly influenced by _____ .
   a. tectonic movements of the Earth's crust
   b. bombardment of the Earth by celestial objects
   c. profound shifts in land masses, shorelines, and oceans
   d. physical and chemical evolution of the Earth
   e. all of the above

2. _____ was the first to obtain indirect evidence that organic molecules could have been formed on the early Earth.
   a. Darwin   c. Fox
   b. Miller   d. Margulis

3. An abundance of _____ was conspicuously absent from the Earth's atmosphere 4 billion years ago.
   a. hydrogen   c. carbon monoxide
   b. nitrogen   d. free oxygen

4. Life as we know it originated by _____ .
   a. 4.6 billion years ago
   b. 2.8 million years ago
   c. 3.8 billion years ago
   d. 3.8 million years ago

5. Which of the following statements is false?
   a. The first living cells were prokaryotic.
   b. The cyclic pathway of photosynthesis first appeared in some eubacterial species.
   c. Oxygen began accumulating in the atmosphere after the noncyclic pathway of photosynthesis evolved.
   d. In the Proterozoic, the rise in atmospheric oxygen enhanced the spontaneous formation of organic molecules.
   e. All are correct.

6. The first eukaryotic cells emerged during the _____ .
   a. Paleozoic   d. Proterozoic
   b. Mesozoic   e. Cenozoic
   c. Archean

7. Match the geologic time interval with the events listed.
   _____ Archean   a. major radiations of dinosaurs, origin of flowering plants and mammals
   _____ Proterozoic
   _____ Paleozoic   b. chemical evolution, origin of life
   _____ Mesozoic   c. radiations of flowering plants, insects, birds, mammals; humans emerge
   _____ Cenozoic
                        d. free oxygen present; origin of aerobic metabolism, protistans, fungi, animals
                        e. rise of early plants on land, origin of amphibians, and origin of reptiles

Critical Thinking

1. Briefly explain, in terms of hydrophilic and hydrophobic interactions, how proto-cells might have formed in water from aggregations of lipids, proteins, and nucleic acids.

2. Reflect on Figure 20.18 and on this: The Atlantic Ocean is widening, and the Pacific Ocean and Indian Ocean are closing in on each other. Many millions of years from now, continents will collide and form a second Pangea. Write a short essay on what conditions might be like on that future supercontinent and what types of species might survive on it. (*Hint:* Check out the Paleomap Project animations on the Internet.)

3. According to one estimate, there is a chance that about 10^{20} planets have formed in the universe that are capable of sustaining life—but there is only one chance at intelligent life per planet. Given your knowledge of molecular biology and evolutionary processes, do you find this estimate plausible? If so, speculate on why the odds are so low.

4. We know of a number of large asteroids that may intersect Earth's orbit in the distant future. There probably are a number we don't know about. Would you use this as an excuse not to worry about polluting the environment, not to take care of your physical health (as by avoiding drugs), and not to care about our cultural evolution? Why or why not?

Selected Key Terms

angiosperm *20.6*
archaebacterium *20.3*
Archean *20.3*
big bang *CI*
Cenozoic *20.8*
crust, of Earth *20.1*
dinosaur *20.6*
Ediacaran *20.5*
endosymbiosis theory *20.4*
eubacterium *20.3*
eukaryotic cell *20.3*
global broiling hypothesis *20.7*
gymnosperm *20.6*
K–T asteroid impact theory *20.7*
mantle, of Earth *20.1*
Mesozoic *20.6*
Paleozoic *20.5*
prokaryotic cell *20.3*
Proterozoic *20.3*
protistan *20.4*
proto-cell *20.2*
RNA world *20.2*
stromatolite *20.3*

Readings

Gould, S. J., editor. 1993. *The Book of Life.* New York: Norton. Beautifully illustrated survey of life's evolutionary past.

Harder, B. 23 March 2002. "Water for the Rock." *Science News* 161: 184–186. Where did Earth's water come from?

Impey, C., and W. Hartman. 2000. *The Universe Revealed.* Belmont, California: Wadsworth. Paperback.

Prothero, D., and R. Dott, Jr. 2002. *Evolution of the Earth.* Sixth edition. New York: McGraw-Hill.

Wright, K. March 1997. "When Life Was Odd." *Discover* 18(3).

On-Line readings at Student Guide for InfoTrac: www.brookscole.com/biology

IV Evolution and Biodiversity

Patterns of diversity in nature, represented by plants of exquisite colors and by fungi that are intertwined with other organisms in structures called lichens. Some of the lichens are branching; flattened ones encrust this chunk of granite.

21

PROKARYOTES AND VIRUSES

The Unseen Multitudes

Did a friend ever mention that you are nearly 1/1,000 of a mile tall? Probably not. What would be the point of measuring people in units as big as miles? Even so, we think that way, in reverse, whenever we measure microorganisms. For the most part, **microorganisms** are prokaryotic cells and single-celled protistans too small to be seen without the aid of a microscope.

The bacterial cells in Figure 21.1 are a case in point. To measure them, you would have to divide one meter into a thousand units, or millimeters. Next, you would have to divide one of the millimeters into a thousand smaller units, or micrometers. To give you a sense of how small that is, a single millimeter would be about as small as the dot of this "i." And a *thousand* bacteria would fit side by side on top of the dot!

Viruses are smaller still, as you might deduce after looking at Figure 21.2. We measure them in nanometers (billionths of a meter). However, viruses are not alive. We consider them in this chapter because one kind or another infects almost every living thing.

Prokaryotic cells are the smallest organisms, but they vastly outnumber the individuals in all other kingdoms combined. Their reproductive potential is staggering. Under ideal conditions, cells of some species divide about every twenty minutes. If that reproduction rate were to hold constant, a single cell would have nearly a billion descendants in ten hours!

So why don't the unseen multitudes take over the world? Sooner or later, their burgeoning populations use up available nutrients and pollute the surroundings with their own metabolic wastes. They alter the very conditions that initially favored their reproduction. Besides, certain kinds of viruses attack just about every prokaryotic species and help keep population sizes in check. So do seasonal changes in living conditions.

Of course, you probably do not find much comfort in this when you serve as host for one of the pathogenic types. **Pathogens** are infectious, disease-causing agents that invade target organisms and multiply inside or on them. Disease follows when the metabolic activities

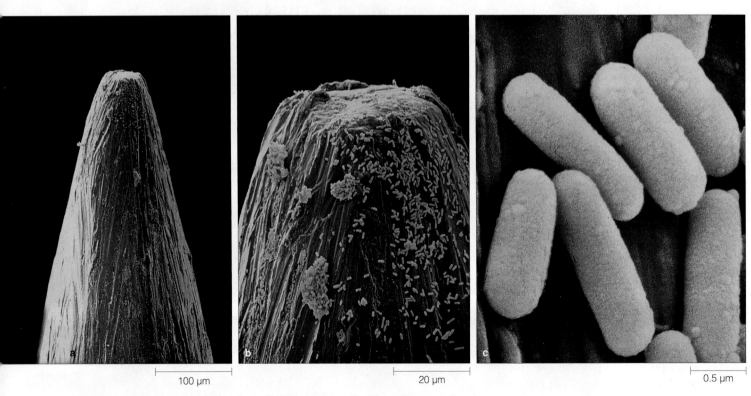

100 µm   20 µm   0.5 µm

Figure 21.1 (**a–c**) How small are bacteria? Shown here, *Bacillus* cells peppering the tip of a pin. These prokaryotic cells in (**c**) are magnified more than 16,600 times.

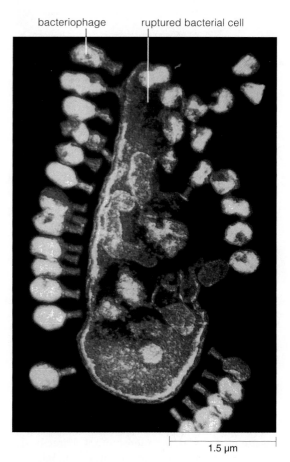

bacteriophage   ruptured bacterial cell

1.5 µm

Figure 21.2 How small are the viruses? Shown here, bacteriophage particles, each about 225 nanometers tall, that infected a bacterial cell. The cell has ruptured open.

of pathogenic cells damage body tissues and interfere with their normal functioning.

Certain pathogens can indeed make you suffer, but they should not give every microorganism a bad name. For example, think back on the uncountable numbers of photosynthetic bacterial cells in the seas (Section 7.8). Collectively, they help provide food and oxygen for entire communities and have major roles in the global cycling of carbon. Or think about the kinds of bacterial species that feed on organic debris. Together with other decomposers, they help cycle nutrients that sustain entire communities.

From the human perspective, microorganisms are good or bad, even dangerous. Basically, however, they are simply surviving and reproducing like the rest of us, in ways that are the topics of this chapter.

Key Concepts

1. In structural terms, the simplest forms of life are the prokaryotic cells called archaebacteria and eubacteria. Most cells are microscopically small. Viruses are smaller still.

2. Prokaryotic cells do not have a profusion of internal, membrane-bound organelles, as eukaryotic cells do. Collectively, these cells show great metabolic diversity, and many kinds display complex behavior.

3. Most of these cells reproduce through prokaryotic fission. This cell division mechanism follows DNA replication. It results in two genetically equivalent daughter cells.

4. Prokaryotic cells were the first living organisms on Earth. Not long after they originated, they diverged into three lineages that gave rise to eubacteria, to the ancestors of archaebacteria, and to eukaryotic cells.

5. A virus, a noncellular infectious particle, consists of nucleic acid (either DNA or RNA), a protein coat, and sometimes an outer envelope. It cannot replicate without pirating metabolic machinery of a specific host cell.

6. Nearly all viral multiplication cycles require five steps: Attachment to a host cell, penetration of its plasma membrane, replication of viral DNA or RNA and synthesis of viral proteins, assembly of new viral particles, and then release from the infected cell.

7. Most of us tend to judge microorganisms through the prism of human interests. Yet their lineages are the most ancient, their adaptations are diverse, and they are simply surviving and reproducing like the rest of us.

see p. 76

21.1

CHARACTERISTICS OF PROKARYOTIC CELLS

Of all organisms, prokaryotic cells are the most far-flung and abundant. Thousands of kinds live in diverse places, such as deserts, hot springs, glaciers, and seas. Some have been carrying on for millions of years 2,780 meters (9,121 feet) below the Earth's surface! Billions live in a handful of rich soil. The ones in your gut and on your skin outnumber your own cells. Prokaryotic cells have the longest evolutionary history. Trace any lineage back far enough and you find prokaryotic ancestors. From *Escherichia coli* to amoebas, elephants, clams, and coast redwoods, all organisms interconnect, regardless of size, numbers, and evolutionary distance. Table 21.1 and Figure 21.3 introduce features that help characterize the remarkable prokaryotes.

Splendid Metabolic Diversity

Every organism takes in energy and carbon, which are essential for their nutrition. Compared to other species, however, prokaryotic cells show the greatest diversity in their means of securing resources.

Like plants, *photoautotrophic* species build organic compounds by photosynthesis; they are light-driven "self-feeders." They tap sunlight for energy and use carbon dioxide as a carbon source. The photosynthetic machinery is part of their plasma membrane. Some of the photosynthesizers use electrons and hydrogen from water molecules for the synthesis reactions, and they release oxygen as a waste product. Others are strictly anaerobic; they die in the presence of oxygen. They get electrons and hydrogen from inorganic compounds, such as gaseous hydrogen and hydrogen sulfide.

The *chemoautotrophs* are self-feeders that use carbon dioxide as their carbon source. Some oxidize organic compounds for energy, which is then used in synthesis reactions. Other species obtain the required energy by oxidizing inorganic substances, such as iron, gaseous hydrogen, sulfur, and nitrogen compounds.

Photoheterotrophs are not self-feeders. They capture sunlight energy for photosynthesis. They get carbon from various organic compounds, such as fatty acids and carbohydrates, that other organisms produced.

Chemoheterotrophic prokaryotic cells are parasites or saprobes, not self-feeders. Parasitic types draw glucose and other nutrients from a living host. Saprobic types get nutrients by digesting organic products, wastes, or remains of other organisms.

Sizes and Shapes

By now you have a general sense of the microscopically small sizes of prokaryotic cells. Their width and length are between 0.5 and 1 micrometers, on the average. A

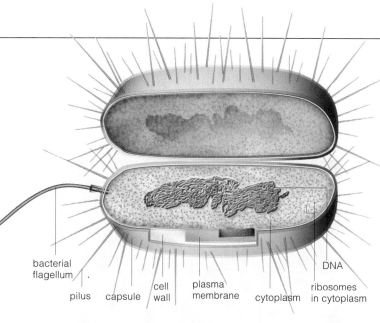

few species are as large as 500 micrometers. Three basic shapes are common among them. A spherical shape is a **coccus** (plural, cocci; from a word meaning berries). A rod shape is a **bacillus** (plural, bacilli, meaning small staffs). A cell body having one or more twists is called a **spirillum** (plural, spirilla):

coccus   bacillus   spirillum

Don't let the simple categories fool you. Cocci may also be oval or flattened. Bacilli may be skinny (like straws) or tapered (like cigars). Surface extensions make some species look like stars. Square species live in salt ponds in Egypt. Also, after daughter cells divide, they may stick together in chains, sheets, and other aggregations, as in Figure 21.4a. Some spiral species are curved like a comma, and others are flexible or like stiff corkscrews.

Structural Features

Prokaryotic means these cells were around before the evolution of the nucleated cell (*pro-*, before; and *karyon*, nucleus). Few have membrane-bounded compartments

Figure 21.3 Generalized body plan of a prokaryotic cell.

Table 21.1 *Characteristics of Prokaryotic Cells*

1. No membrane-bound nucleus.
2. Generally a single chromosome (a circular DNA molecule); many species also contain plasmids.
3. Cell wall present in most species.
4. Reproduction mainly by prokaryotic fission.
5. Collectively, great metabolic diversity among species.

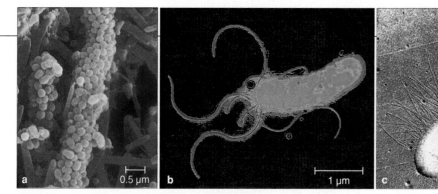

Figure 21.4 (a) Surface view of numerous bacilli and cocci attached to a human tooth. (b) *Helicobacter pylori* cell, with its tuft of flagella. This pathogen can colonize the stomach lining and trigger inflammation. If untreated, an infection can lead to gastritis, peptic ulcers, and possibly stomach cancer. *H. pylori* may contaminate water and food, especially unpasteurized milk. Currently, a combination of antibiotics and an antacid rids the body of the pathogen, after which ulcers heal. (c) Example of pili, filamentous structures that project from the surface of many bacterial cells. This *Escherichia coli* cell is dividing in two.

Figure 21.5 Scanning electron micrographs of prokaryotic cell walls. (a) Dividing *Bacillus subtilis* cell. Its wall has a smooth surface texture. (b) A dividing *E. coli* cell. When such Gram-negative bacteria are prepared for electron microscopy, the cell wall wrinkles this way.

of any sort for isolating metabolic events; reactions take place in the cytoplasm or at the plasma membrane. For example, protein synthesis proceeds at ribosomes that are distributed through the cytoplasm or attached to the side of the plasma membrane facing the interior. The structural simplicity does not mean prokaryotic cells are inferior to eukaryotic cells. They are tiny and reproduce quite fast, and so they do not require great internal complexity.

Usually, the plasma membrane is surrounded by a **cell wall**. This semirigid, permeable structure helps the cell maintain shape and resist rupturing when internal fluid pressure rises (Figure 21.5; compare Section 5.7). Eubacterial cell walls are composed of peptidoglycan molecules. In such molecules, peptide groups crosslink many polysaccharide strands to one another.

When doctors must diagnose an infectious disease, they may use a staining reaction called the **Gram stain**, which can help identify many bacterial species. First, a sample of the unknown cells is exposed to purple dye, then iodine, then an alcohol wash and a counterstain. Cells of *Gram-positive* species stay purple-colored. Cells of *Gram-negative* species lose color after the wash, but the counterstain turns them pink (Figure 21.6).

A sticky mesh, or **glycocalyx**, often encloses the cell wall. It consists of polysaccharides, polypeptides, or both. When highly organized and attached firmly to the wall, it forms a capsule. When less organized and loosely attached to the wall, it forms a slime layer. The mesh helps a prokaryotic cell attach to teeth, mucous membranes, rocks in streambeds, and other interesting surfaces. It also helps some encapsulated species avoid being engulfed by phagocytic, infection-fighting cells of their host organism.

Figure 21.6 Gram staining. Cocci and rods smeared on a slide are stained with a purple dye (such as crystal violet), washed off, then stained with iodine. All cells are now purple. The slide is washed with alcohol, which renders the Gram-negative cells colorless. Now the slide is counterstained (with safranin), washed, and dried. Gram-positive cells (*Staphylococcus aureus* in this case) stay purple. The counterstain, however, colors the Gram-negative cells (*E. coli*) light pink.

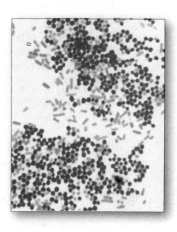

■ stain with purple dye
□ stain with iodine
□ wash with alcohol
■ counterstain with safranin

Some species have one or more motile structures called **bacterial flagella** (Figure 21.4b). These don't have the same structure as a eukaryotic flagellum, and they do not operate the same way. Rather they rotate like a propeller. Many species also have **pili** (singular, pilus). These short, filamentous proteins project above the cell wall, as in Figure 21.4c. Some pili help cells adhere to surfaces. Others help them attach to one another as a prelude to conjugation, an interaction that is described in the next section.

Prokaryotic cells generally are microscopic in size. Nearly all have a wall around the plasma membrane. Many have a capsule or slime layer around the cell wall.

PROKARYOTIC GROWTH AND REPRODUCTION

The Nature of Growth

Between divisions, prokaryotic cells grow by increases in component parts. We measure the growth of a large, multicelled organism in terms of increases in size, but doing this for a microscopically small cell would be a bit pointless. Instead, we measure prokaryotic growth as an increase in the number of cells in a population. Under ideal conditions, each cell divides in two, the division of two cells results in four, four result in eight, and so forth. Many types can divide every half hour; a few do so every ten or twenty minutes. Such rates of increase lead to large population sizes in short order.

In nature, a few species live in habitats that would restrict the growth of other prokaryotic species. We find a few clinging to life in extreme environments, such as Antarctica, the Negev Desert, and deep in the Earth (Section 20.2). They endure in or on rocks and, on rare occasions, they reproduce. A few live in highly acidic wastewater from mining operations.

K-12, a strain of *E. coli* originally isolated from the human gut, has been cultivated for such a long time in the laboratory that it no longer is able to grow when reintroduced into its natural habitat. As an outcome of microevolutionary processes, it has become adapted to the conditions in its artificial laboratory environment.

Prokaryotic Fission

A prokaryotic cell nearly doubles in size, then divides in two. Each daughter cell inherits a single **bacterial chromosome**—a circularized, double-stranded DNA molecule that has only a few proteins attached to it. In some species, the daughter cell merely buds from the parent cell. Most often, however, a cell reproduces by a division mechanism called **prokaryotic fission**.

Prokaryotic fission starts as a parent cell replicates its DNA (Figure 21.7). It now has two DNA molecules, attached to the plasma membrane at adjacent sites. The cell also synthesizes lipid and protein molecules, which become incorporated into the membrane between the two attachment sites. Membrane growth moves the two DNA molecules apart. New wall material is deposited above the membrane. The membrane and wall grow through the cell midsection and divide the cytoplasm. The result is two genetically equivalent daughter cells.

Especially in microbiology, you may hear someone refer to this division mechanism as *binary* fission, but such usage can cause confusion. The same term applies to one form of asexual reproduction among flatworms and some other multicelled animals. It refers to growth by mitotic cell divisions, then division of the whole body into two parts of the same or different sizes.

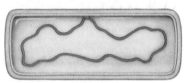

a The bacterial chromosome is attached to the plasma membrane before DNA replication.

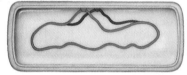

b Replication starts and proceeds in two directions from some point in the bacterial chromosome.

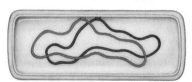

c The DNA copy is attached at a membrane site near the attachment site of the parent DNA molecule.

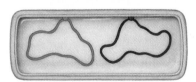

d The two DNA molecules are moved apart by membrane growth between the two attachment sites.

e New membrane and new wall material are added transversely, through the cell's midsection.

f The ongoing, orderly deposition of membrane and wall material at the midsection cuts the cell in two.

Figure 21.7 Reproduction by prokaryotic fission, a cell division mechanism. The micrograph shows the cytoplasmic division of *Bacillus cereus*, as brought about by the formation of new membrane and wall material. This cell has been magnified 13,000 times its actual size.

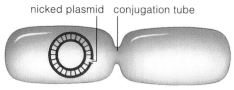

nicked plasmid   conjugation tube

a A conjugation tube has already formed between a donor and a recipient cell. An enzyme has nicked the donor's plasmid.

b DNA replication starts on the nicked plasmid. The displaced DNA strand moves through the tube and enters the recipient cell.

c In the recipient cell, replication starts on the transferred DNA.

d The cells separate from each other; the plasmids circularize.

Figure 21.8 Simplified sketch of conjugation between two prokaryotic cells. For clarity, the size of this plasmid has been greatly increased and the chromosome is not shown.

Conjugation Between Cells

A new prokaryotic cell also may inherit one or more plasmids. A **plasmid**, recall, is a small, self-replicating circle of extra DNA with a few genes (Section 16.1). The F (Fertility) plasmid includes genes that confer the means to engage in a form of **conjugation**. In this case, a donor cell transfers plasmid DNA to a recipient cell. The transfers occur among many bacterial species, such as *Salmonella*, *Streptococcus*, and *E. coli*—even between *E. coli* and yeast cells in the laboratory. The F plasmid carries genetic instructions for synthesizing a structure called a sex pilus. Sex pili at the surface of a donor cell can hook onto a recipient cell and pull it right next to the donor. Shortly after the two cells make contact, a conjugation tube develops between them. After this, plasmid DNA is transferred through the tube, in the manner shown in Figure 21.8.

Eubacteria and archaebacteria reproduce by a cell division mechanism, prokaryotic fission, that follows replication of DNA. Each daughter cell inherits one DNA molecule (one chromosome). Many species also transfer plasmid DNA.

PROKARYOTIC CLASSIFICATION

Just a few decades ago, reconstructing the evolutionary history of prokaryotes seemed to be an impossible task. Except for stromatolites (Section 20.3), the most ancient groups are not well represented in the fossil record. Most are not represented at all.

Given their elusive histories, the many thousands of known species traditionally have been classified by **numerical taxonomy**. By this practice, the traits of an unidentified cell are compared with those of a known prokaryotic group. Traits typically include cell shape, motility, staining attributes of the cell wall, nutritional requirements, metabolic patterns, and the presence or absence of endospores. The greater the total number of traits that a cell has in common with the known group, the closer is the inferred relatedness.

Since the 1970s, nucleic acid hybridization studies, gene sequencing, and other methods of comparative biochemistry have been revealing compelling evidence of bacterial phylogenies (Section 19.7). Notably useful are comparisons of ribosomal RNAs, which are crucial for protein synthesis. rRNAs can change extensively in certain base sequences without loss of function. Many small changes that have accumulated in the rRNAs of different lineages can be directly measured.

| EUBACTERIA (BACTERIA) | ARCHAEBACTERIA (ARCHAEA) | EUKARYOTES (EUKARYA) |

Figure 21.9 The three-domain scheme, an evolutionary tree diagram based on evidence from comparative biochemistry.

Surprisingly, biochemical analyses are uniting some groups that did not appear to be related on the basis of other tests. For this chapter, the important insight is that a key divergence began shortly after prokaryotic cells originated. One branch led to **eubacteria**, the most common prokaryotic cells. (Here, *eu–* means "typical.") The other branch led to **archaebacteria** and to the first eukaryotic cells. Although earlier chapters considered these evolutionary connections, take a quick look now at Figure 21.9 before starting the next sections.

Prokaryotic cells are classified by numerical taxonomy: the total percentage of observable traits they have in common with a known prokaryotic group. They also are classified more directly by comparisons at the biochemical level.

All prokaryotic cells are assigned to one of two lineages, the eubacteria and the archaebacteria.

MAJOR PROKARYOTIC GROUPS

By now, you've probably sensed that *species* is the basic unit in prokaryotic classification schemes. Even so, the definition that fits sexually reproducing species does not fit prokaryotes, which do not form reproductively isolated populations of interbreeding individuals. Each prokaryotic cell generally does its own thing. Also, these tiny cells do not display spectacular variation in traits. The variations that do show up are determined by relatively few genes. If two cells under study show only minor differences, one of them might be classified as a **strain**, not a new species.

These are a just few of the difficulties in classifying prokaryotic cells. Until evolutionary relationships are sorted out, we will continue grouping them mainly by numerical taxonomy, as described earlier. Table 21.2 lists major groupings that we touch upon in this book. Appendix I lists their habitats and characteristics.

[handwritten: pp 539-540 Stanier]

Table 21.2	*Archaebacteria and Eubacteria*
Some Major Groups	Representatives
ARCHAEBACTERIA	
Extreme thermophiles	*Sulfolobus, Thermoplasma*
Methanogens	*Methanobacterium*
Extreme halophiles	*Halobacterium, Halococcus*
EUBACTERIA	
Photoautotrophs:	
Cyanobacteria, green sulfur bacteria, and purple sulfur bacteria	*Anabaena, Nostoc, Rhodopseudomonas, Chloroflexus*
Photoheterotrophs:	
Purple nonsulfur and green nonsulfur bacteria	*Rhodospirillum, Chlorobium*
Chemoautotrophs:	
Nitrifying, sulfur-oxidizing, and iron-oxidizing bacteria	*Nitrosomonas, Nitrobacter, Thiobacillus*
Chemoheterotrophs:	
Spirochetes	*Spirochaeta, Treponema*
Gram-negative aerobic rods and cocci	*Pseudomonas, Neisseria, Rhizobium, Agrobacterium*
Gram-negative facultative anaerobic rods	*Salmonella, Escherichia, Proteus, Photobacterium*
Rickettsias and chlamydias	*Rickettsia, Chlamydia*
Myxobacteria	*Myxococcus*
Gram-positive cocci	*Staphylococcus, Streptococcus, Deinococcus*
Endospore-forming rods and cocci	*Bacillus, Clostridium*
Gram-positive nonsporulating rods	*Lactobacillus, Listeria*
Actinomycetes	*Actinomyces, Streptomyces*

[handwritten annotations: "Colors due to bact Chlorophylls + Carotenoids", "some", "Some Rhodopseudomonas"]

ARCHAEBACTERIA

We divide the kingdom of archaebacteria into three major groups: the extreme thermophiles, methanogens, and extreme halophiles. In many ways, archaebacteria are unique in their composition, structure, metabolism, and nucleic acid sequences. In some respects they differ as much from eubacteria as they do from eukaryotes. Many investigators believe that existing archaebacteria, which can withstand conditions as hostile as those on the early Earth, resemble the first living cells. Hence the name of this group (*archae*– means "beginning").

EXTREME THERMOPHILES Geothermally heated soils, sulfur-rich hot springs, and wastes from coal mines are habitats of the "heat lovers," or **extreme thermophiles** (Figure 21.10). These archaebacteria are the most heat-tolerant prokaryotes known. Some populations grow at above-boiling temperatures; all do best at temperatures above 80°C! Nearly all are strict anaerobes that require sulfur as an electron acceptor or electron donor.

Solfolobus, the first extreme thermophile discovered, grows in acidic hot springs. So does *Thermus aquaticus*, which biotechnologists utilize as a source of extremely heat-stable DNA polymerases (Section 16.2).

Certain extreme thermophiles live in shallow water around volcanoes. We find some species off the coast of Italy, where geothermally heated seawater spews from openings in the sediments. Other species are the basis of food webs in sediments around hydrothermal vents, where the surrounding water can exceed 110°C. They use hydrogen sulfide escaping from these vents as an electron source. Their existence at the vents is cited as evidence that life may have originated on the seafloor.

Figure 21.10 Representative habitat for extreme thermophiles: hot, sulfur-rich water in Emerald Pool, Yellowstone National Park. Microbes with an abundance of carotenoids impart a yellow-orange color to the rim of this hot spring.

Figure 21.11 Some methanogens.
(**a**) *Methanococcus jannaschii*, a heat-loving methane producer. (**b**) A dense population of *Methanosarcina* cells. Each cell has a thick polysaccharide wall. (**c**) Representative habitat for methanogens. The stomach chambers of cattle and other ruminants house methanogen populations. Cattle belch —a lot—and release quantities of methane. You may have noticed the resulting exceptionally pungent air around stockyards.

Figure 21.12 Representative habitats for extreme halophiles. (**a**) Great Salt Lake, Utah. When light intensity and salinity are high and pH and nutrient levels low, halophiles make purple pigments and a green alga (*Dunaliella salina*) makes beta-carotene pigments. Collectively, the pigments turn the water reddish-purple. (**b**) The commercial seawater evaporating ponds next to San Francisco Bay. *Halobacterium* and *D. salina* thrive under the hypersaline conditions, where salt levels exceed that of seawater (for example, 300 versus 35 grams per liter).

METHANOGENS The **methanogens** (methane makers) live in oxygen-free habitats, such as swamps, the gut of termites and mammals, and stockyards (Figure 21.11). They are strict anaerobes; free oxygen kills them.

Methanogens produce ATP by anaerobic electron transfer, a pathway described in Section 8.5. Usually, they get electrons from hydrogen gas (H_2), but some species obtain them from ethanol and other alcohols. Nearly all use carbon dioxide as their carbon source and as a final electron acceptor for the reactions, which end with the formation of methane (CH_4).

As a group, methanogens produce about 2 billion tons of methane per year. Most of the methane escapes from wetlands, termites, ruminants, landfills, and rice paddies. Collectively, this metabolic by-product affects the levels of carbon dioxide in the atmosphere and the cycling of carbon through ecosystems.

Long ago, methane accumulated and formed huge deposits on the seafloor. Geologists recently discovered a deposit of 35 billion tons 400 kilometers off the South Carolina coast. If oceanic circulation patterns change in the future, as they have in the past, then all of that gas might move abruptly and explosively to the surface. The rapid, simultaneous release of so much methane would drastically alter the atmosphere and the global climate, hence life through the biosphere (Section 48.9).

EXTREME HALOPHILES There are many salty habitats around the world, but **extreme halophiles** (salt lovers) thrive in exceptionally salty ones. The Great Salt Lake, the Dead Sea, and seawater evaporation ponds are like this (Figure 21.12).

Extreme halophiles spoil commercial sea salt, salted fish, and animal hides. Most produce ATP by aerobic pathways. When oxygen levels are low, they switch to photosynthesis as an ATP-forming pathway. They have a unique light-absorbing pigment, bacteriorhodopsin, in their plasma membrane. Absorption of light energy triggers a metabolic process that leads to an increase in an H^+ gradient across their plasma membrane. ATP forms when these ions follow their gradient and flow through the interior of transporter proteins, which span the membrane (compare Section 7.5).

Like the first cells on Earth, archaebacteria now live in extremely inhospitable habitats. In some respects, they differ as much from eubacteria as from eukaryotic cells.

EUBACTERIA—THE TRUE BACTERIA

We recognize more than 400 genera of prokaryotes. By far, most species are true bacteria (*eu–*, true). Unlike the archaebacteria, they have fatty acids in their plasma membrane. Nearly all have peptidoglycan in their cell wall. Evolutionary histories are still being worked out, so we still rely primarily on taxonomic classification. Here we will focus on modes of nutrition to give you a sense of their biodiversity.

A Sampling of Biodiversity

PHOTOAUTOTROPHIC BACTERIA Cyanobacteria, once known as blue-green algae, are the classic example of photoautotrophic eubacteria. They are also among the most common photoautotrophs. All cyanobacteria are aerobic cells that engage in photosynthesis. Most types live in ponds and other freshwater habitats. They may grow as mucus-sheathed chains of cells, which often form thick, dense, slimy mats at the surface of nutrient-enriched water (Figure 21.13*a*).

Anabaena and other types also convert nitrogen gas (N_2) to ammonia for use in biosynthesis. As nitrogen compounds dwindle, some of the cells develop into **heterocysts**. These cells make a nitrogen-fixing enzyme (Figure 21.13*b*,*c*). They synthesize and share nitrogen compounds with photosynthetic cells; in return, they receive carbohydrates. Shared substances move freely across cytoplasmic junctions between cells.

Anaerobic photoautotrophs, such as green bacteria, get electrons from hydrogen sulfide and hydrogen gas, not from water. They may resemble anaerobic bacteria in which the cyclic pathway of photosynthesis evolved.

CHEMOAUTOTROPHIC BACTERIA Most of the species in this group influence the global cycling of nitrogen, sulfur, and other nutrients. For example, nitrogen is an essential building block for amino acids and proteins. Without it, there would be no life. Nitrifying bacteria in soil obtain electrons from ammonia. Plants use the end product, nitrate, as a nitrogen source.

CHEMOHETEROTROPHIC BACTERIA Nearly all known bacterial species are in this category. Many, such as the pseudomonads, are decomposers; their enzymes break down organic compounds and even pesticides in soil. Other "beneficial" species include *Lactobacillus* (used to make pickles, sauerkraut, buttermilk, and yogurt) and diverse actinomycetes (sources of antibiotics). *E. coli* makes vitamin K and compounds that help us digest fat, it helps newborn mammals digest milk, and its activities help stop many food-borne pathogens from colonizing the gut. Sugarcane and corn benefit from a symbiont, the nitrogen-fixing spirochete *Azospirillum*. The plants use some of the nitrogen initially fixed by

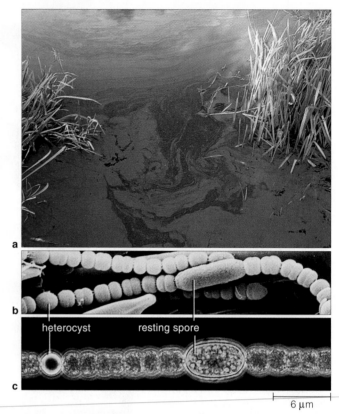

Figure 21.13 A common photoautotrophic bacterium. (**a**) Huge populations of cyanobacteria form dense mats near the surface of a nutrient-enriched pond. (**b**,**c**) Some of the cells develop into resting spores when conditions restrict growth. A nitrogen-fixing heterocyst is also shown. Section 4.12 gives other examples of eubacteria.

the bacterium and give up some sugars to it. *Rhizobium* is a fine symbiont with peas, beans, and other legumes. *Deinococcus radiodurans* resists high radiation doses that kill other organisms. One genetically engineered *E. coli* strain holds promise for bioremediation. Gene products from this strain allow it to break down toxic wastes, such as extremely nasty mercury compounds.

Also in this category are most pathogenic bacteria. We admire pseudomonads as decomposers in soil, not as contaminants of soap, antiseptics, and other carbon-rich goods. They can be a major problem because they transfer plasmids with antibiotic-resistance genes.

Some *E. coli* strains cause a form of diarrhea that is the main cause of infant death in developing countries. *Clostridium botulinum* can taint fermented grain as well as food in improperly sterilized or sealed cans and jars. Its toxins cause *botulism*, a form of poisoning that can disrupt breathing and lead to death. *C. tetani*, one of its relatives, causes the disease *tetanus* (Section 37.11).

Like many other bacteria that typically live in soil, *C. tetani* can form an **endospore**. This resting structure forms inside the cell, around one copy of the bacterial chromosome and part of the cytoplasm (Figure 21.14).

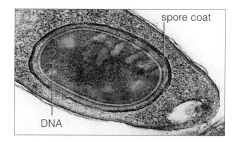

Figure 21.14 Endospore developing inside a cell of *Clostridium tetani*, one of the dangerous pathogens.

spore coat

DNA

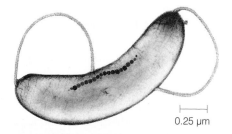

Figure 21.16 Magnetotactic bacterium. Inside the cytoplasm is a chain of magnetite particles that acts like a compass as the cell moves.

0.25 μm

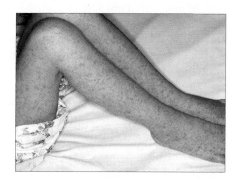

Figure 21.15 Rash typical of Rocky Mountain spotted fever, a disease caused by *Rickettsia rickettsii*. Bites by ticks that are infected transmit the bacterium to humans.

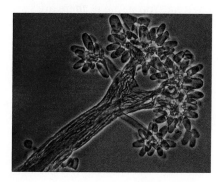

Figure 21.17 Migrating cells of myxobacteria (*Chondromyces crocatus*). Cells of this species aggregate and move about as a single unit. Some cells differentiate into spore-bearing structures.

Endospores form when a depletion of nitrogen or some other nutrient arrests cell growth. They are released as free spores when the plasma membrane ruptures. They are very resistant to heat, drying out, irradiation, acids, disinfectants, and boiling water. Endospores can stay dormant, in some cases for many decades. If favorable conditions return, each becomes metabolically active and may then develop into a single bacterial cell.

Many chemoautotrophic eubacteria taxi from host to host in the gut of insects. For example, bites from infected ticks transmit *Borrelia burgdorferi* from deer and some other wild animals to humans, who develop *Lyme disease*. A "bull's-eye" rash often develops around the bite (Section 25.18). Severe headaches, backaches, chills, and fatigue follow. Without prompt treatment, symptoms get worse. Another tick-traveling pathogen, *Rickettsia rickettsii*, causes *Rocky Mountain spotted fever*. After the bacterium enters a host through a tick bite, it penetrates the cytoplasm and the nucleus of host cells. Three to twelve days after this, a high fever and severe headache develop. Three to five days after that, a rash forms on the extremities (Figure 21.15). Diarrhea and gastrointestinal cramps are common. If untreated, the symptoms can persist for more than two weeks.

Regarding the "Simple" Bacteria

Bacteria are small. Their insides are not elaborate. *But bacteria are not simple.* A brief look at their behavior will reinforce this point. Bacteria move toward nutrient-rich regions. Aerobes move toward oxygen, and anaerobes avoid it. Photosynthetic types move toward light but move away from light if it is too intense. Many species tumble away from toxins. Such behaviors often start when shifts in chemical conditions or light intensity activate special membrane receptors. Such changes lead to alterations in metabolic activities within the cell, and the outcomes may be an adjustment in the direction of the cell's movement.

Magnetotactic bacteria contain a chain of magnetite particles that serves as a tiny compass (Figure 21.16). The compass helps them sense which way is north and also down. These bacteria swim toward the bottom of a body of water, where oxygen concentrations are lower and therefore more suitable for their growth.

Some types even show *collective* behavior, as when millions of *Myxococcus xanthus* cells form a "predatory" colony. These cells secrete enzymes that digest "prey," such as cyanobacteria, that become stuck to the colony. Then the cells absorb the breakdown products. What's more, the cells migrate, change direction, and move as a single unit toward what may be food!

Many myxobacteria colonies form **fruiting bodies** (spore-bearing structures). Under suitable conditions, some cells in the colony differentiate and form a slime stalk, others form branchings, and others form clusters of spores (Figure 21.17). The spores disperse when a cluster bursts open; each may give rise to a new colony. As you will see in the next chapter, certain eukaryotic species also form spore-bearing structures.

True bacteria are the most common and diverse prokaryotic cells. They are adapted to nearly all environments.

THE VIRUSES

Defining Characteristics

In ancient Rome, *virus* meant "poison" or "venomous secretion." In the late 1800s, this rather nasty word was bestowed on newly discovered pathogens, smaller than the bacteria being studied by Louis Pasteur and others. Many viruses deserve the name. They attack humans, cats, cattle, birds, insects, plants, fungi, protistans, and bacteria. You name it, there are viruses that infect it.

Today we define a **virus** as a noncellular infectious agent that has two characteristics. First, a viral particle consists of a protein coat wrapped around a nucleic acid core—that is, around its genetic material. Second, the virus cannot reproduce itself. It can be reproduced only after its genetic material enters the host cell and directs that cell's biosynthetic machinery into making many copies of itself.

The genetic material of a virus is DNA *or* RNA. The coat consists of one or more types of protein subunits organized into a rodlike or a polyhedral (many-sided) shape, as in Figure 21.18. The coat protects the genetic material during the journey to a new host cell. It also contains proteins that can bind with specific receptors on host cells. An envelope, made mostly of membrane remnants from a previously infected cell, encloses the coat of some viruses. The envelope bristles with spikes of glycoproteins. Coats of complex viruses have sheaths, tail fibers, and other accessory structures attached.

The vertebrate immune system detects certain viral proteins. The problem is, genes for many viral proteins mutate at high frequencies, so a virus often can elude the immune fighters. For example, people susceptible to lung infections get new "flu shots" each year because envelope spikes on influenza viruses keep changing.

Examples of Viruses

Each kind of virus can multiply only in certain hosts. It cannot be studied easily unless the investigator cultures living host cells. This is why much of our knowledge of viruses comes from **bacteriophages**, a group of viruses that infect bacterial cells. Unlike cells of humans and other complex, multicelled species, bacterial cells can be cultured easily and rapidly. This is also why bacteria and bacteriophages were used in early experiments to determine the function of DNA (Section 13.1). They are still used as research tools in genetic engineering.

Table 21.3 lists some major groups of animal viruses. These viruses contain double- or single-stranded DNA or RNA, which is replicated in various ways. Animal viruses range in size from parvoviruses (18 nanometers) to brick-shaped poxviruses (350 nanometers). Many kinds cause diseases, such as the common cold, certain cancers, warts, herpes, and influenza (Figure 21.19*a,b*). One, HIV, is the trigger for *AIDS*. By attacking certain white blood cells, HIV weakens the immune system's ability to fight infections that might not otherwise be life-threatening. Researchers who attempt to develop drugs against HIV and other diseases use HeLa cells and other cell lineages for initial experiments (Section

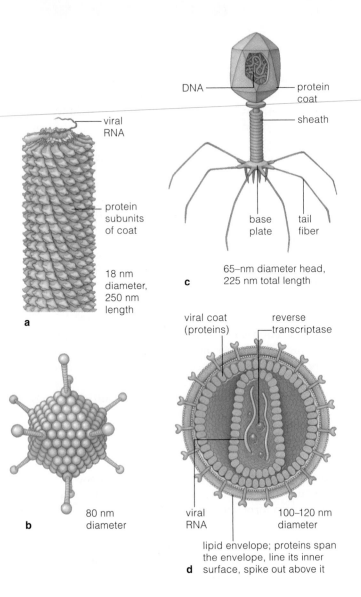

DNA — protein coat

— sheath

— viral RNA

— protein subunits of coat

base plate — tail fiber

18 nm diameter, 250 nm length

a

c 65–nm diameter head, 225 nm total length

viral coat (proteins) — reverse transcriptase

80 nm diameter

b

viral RNA — 100–120 nm diameter

d lipid envelope; proteins span the envelope, line its inner surface, spike out above it

Figure 21.18 Body plans of viruses. (**a**) *Helical* viruses, such as this tobacco mosaic virus, have a rod-shaped coat of protein subunits coiled helically around nucleic acid. (**b**) This adenovirus and other *polyhedral* viruses have a many-sided coat. (**c**) T-even bacteriophages and other *complex* viruses incorporate accessory parts attached to the coat. (**d**) Membrane encases the coat of the *enveloped* viruses. HIV, shown here, is an example.

Table 21.3 Classification of Some of the Major Animal Viruses

DNA VIRUSES	Some Diseases/Consequences
Parvoviruses	Gastroenteritis; roseola (fever, rash) in small children; aggravation of symptoms of sickle-cell anemia
Adenoviruses	Respiratory infections (fever, cough, sore throat, rash), diarrhea in infants, conjunctivitis (inflamed, pebbly eye membranes); some cause tumors
Papovaviruses	Benign and malignant warts
Orthopoxviruses	Smallpox, cowpox, monkeypox
Herpesviruses:	
H. simplex type I	Oral herpes, cold sores
H. simplex type II	Genital herpes (Section 44.15)
Varicella–zoster	Chicken pox, shingles
Epstein–Barr	Infectious mononucleosis; cancers of skin, liver, cervix, pharynx; Burkitt's lymphoma (malignant tumor of jaw, face)
Cytomegalovirus	Hearing loss, mental impairment
Hepadnavirus	Hepatitis B (severe liver infection)

RNA VIRUSES	Some Diseases/Consequences
Picornaviruses:	
Enteroviruses	Polio, hemorrhagic eye disease, hepatitis A (infectious hepatitis)
Rhinoviruses	Common cold
Hepatitis A virus	Inflammation of liver, kidneys, spleen
Togaviruses	Forms of encephalitis (inflammation in the brain), rubella
Flaviviruses	Yellow fever (fever, chills, jaundice), dengue (fever, severe muscle pain), St. Louis encephalitis
Coronaviruses	Upper respiratory infections, colds
Rhabdoviruses	Rabies, other animal diseases
Filoviruses	Hemorrhagic fevers, as by *Ebola* virus (Section 21.9)
Paramyxoviruses	Measles, mumps, respiratory ailments
Orthomyxoviruses	Influenza
Bunyaviruses	
Bunyamwera virus	California encephalitis
Phlebovirus	Hemorrhagic fever, encephalitis
Hantavirus	Hemorrhagic fever, kidney failure
Arenaviruses	Hemorrhagic fevers
Retroviruses:	
HTLV-I, HTLV-II*	Adult T-cell leukemia
HIV	AIDS
Reoviruses	Respiratory and intestinal infections

* Human T-cell leukemia virus.

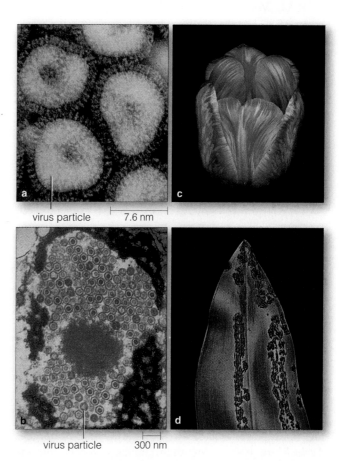

virus particle — 7.6 nm

virus particle — 300 nm

Figure 21.19 Some viruses and their effects. (**a**) Particles of a DNA virus that causes a herpes infection in humans. (**b**) Particles of an enveloped RNA virus that causes influenza in humans. Spikes project from the lipid envelope. (**c**) Streaking of a tulip blossom. A harmless virus infected pigment-forming cells in the colorless parts. (**d**) An orchid leaf infected by a rhabdovirus.

9.5). Later they must use laboratory animals, and then human volunteers, to test a new drug for toxicity and effectiveness. Why? A functioning immune system is necessary to test responses to the drugs.

Plant viruses must breach plant cell walls to cause diseases. They typically hitch rides on the piercing or sucking devices of insects that feed on plant juices. Some RNA viruses infect tobacco plants (the tobacco mosaic virus), barley, potatoes, and other major crop plants. Certain DNA viruses infect such valuable crops as cauliflower and corn. Figure 21.19c,d shows visible effects of two viral infections.

A virus is a noncellular infectious particle that consists of nucleic acid enclosed in a protein coat and sometimes an outer envelope. It cannot multiply without pirating the metabolic machinery of a specific type of host cell.

Diverse viruses infect organisms of all kingdoms.

VIRAL MULTIPLICATION CYCLES

What Happens During Viral Infections?

Once they infect host cells, different viruses multiply in a variety of ways. Even so, nearly all multiplication cycles proceed through five basic steps, as outlined here:

1. *Attachment.* A virus attaches to a host cell by molecular groups that can chemically recognize and lock on to specific molecular groups at the cell surface.

2. *Penetration.* Either the whole virus or its genetic material alone penetrates the cell's cytoplasm.

3. *Replication and synthesis.* In an act of molecular piracy, the viral DNA or RNA directs the host cell into producing many copies of the viral nucleic acids and viral proteins, including enzymes.

4. *Assembly.* The viral nucleic acids and viral proteins become organized as new infectious particles.

5. *Release.* By one mechanism or another, new virus particles are released from the cell.

Consider the lytic and lysogenic pathways that are common among bacteriophage multiplication cycles. In a **lytic pathway**, steps 1 through 4 proceed rapidly, and new particles are released when a host cell undergoes **lysis.** Here, lysis means that damage to a cell's plasma membrane, wall, or both is allowing the cytoplasm to dribble out. In this case, new virus particles dribble out as the cell dies. Late into most lytic pathways, the host cell synthesizes a viral enzyme that causes the damage and swift destruction (Figure 21.20).

In a **lysogenic pathway**, a latent period extends the cycle. The virus does not kill its host outright. Instead, a viral enzyme cuts a host chromosome, then integrates viral genes into it. When an infected host cell prepares to divide, it replicates the recombinant molecule. As a result of that single instance of genetic recombination, miniature time bombs are passed on to all of the cell's descendants. Later on, a molecular signal or some other stimulus may reactivate the cycle.

Latency is part of the multiplication cycles of many viruses, not just bacteriophages. Type I *Herpes simplex*, the cause of *cold sores* (fever blisters), is a case in point. Almost everyone harbors this virus. It remains latent in facial tissues inside a ganglion (a cluster of neuron cell bodies). Sunburn and other stress factors can reactivate the virus. When this happens, the virus particles move down to the neuron endings, near the skin. There they infect epithelial cells and cause painful skin eruptions.

The multiplication cycle of the RNA viruses has an interesting twist to it. In a host cell's cytoplasm, their RNA serves as a template for synthesizing either DNA or mRNA. For example, HIV, a retrovirus, carts its own

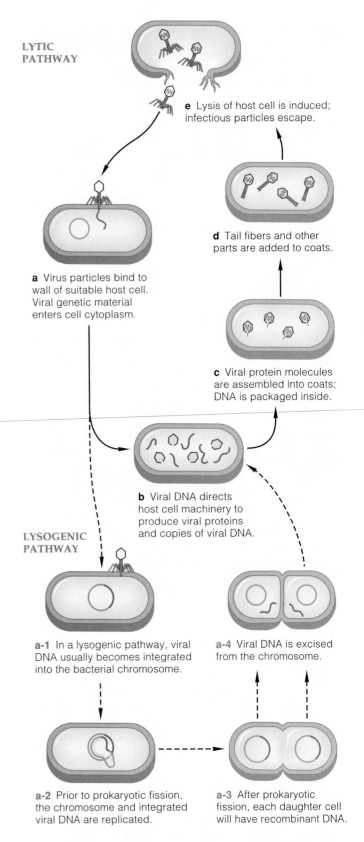

LYTIC PATHWAY

e Lysis of host cell is induced; infectious particles escape.

d Tail fibers and other parts are added to coats.

a Virus particles bind to wall of suitable host cell. Viral genetic material enters cell cytoplasm.

c Viral protein molecules are assembled into coats; DNA is packaged inside.

b Viral DNA directs host cell machinery to produce viral proteins and copies of viral DNA.

LYSOGENIC PATHWAY

a-1 In a lysogenic pathway, viral DNA usually becomes integrated into the bacterial chromosome.

a-4 Viral DNA is excised from the chromosome.

a-2 Prior to prokaryotic fission, the chromosome and integrated viral DNA are replicated.

a-3 After prokaryotic fission, each daughter cell will have recombinant DNA.

Figure 21.20 Generalized multiplication cycle for some of the bacteriophages. Virus particles may be produced and released by a lytic pathway. For certain viruses, the lytic pathway may expand to include a lysogenic pathway.

Figure 21.21 Multiplication cycle of one type of enveloped DNA virus, here infecting a generalized animal cell.

enzymes into cells. These assemble DNA on viral RNA by reverse transcription (Sections 16.1 and 39.10).

Like other enveloped viruses, the herpes viruses can enter a host cell by their own version of endocytosis, then leave it by budding from the plasma membrane. Figure 21.21 shows how they accomplish this.

What About Viroids and Prions?

Viroids, tightly folded strands or circles of RNA, are smaller than anything in viruses. They, too, cause infections even though they do not have protein-coding genes. They have no protein coat, but the tight folding might help protect them from the host's enzymes. These bits of "naked" RNA cause many diseases in plants. They destroy many millions of dollars' worth of potatoes, citrus, and other crop plants every year.

Viroids actually resemble introns, the noncoding portions of eukaryotic DNA you read about in Section 14.1. They might be self-splicing introns that escaped from DNA molecules. Or they might be remnants left over from an ancient RNA world (Section 20.2).

Prions are largely—if not entirely—abnormal, less soluble forms of proteins necessary for the operation of neurons, the communication cells of nervous systems. Prions catalyze the conversion of the normal proteins into more prions. They coagulate as massive deposits inside the brain and cause fatal degenerative diseases. Have you heard of *kuru* and *Creutzfeldt–Jakob diseases* (CJD) in humans? Both diseases progressively destroy muscle coordination and brain function. Each year, 1 in 1 million people develops CJD. Symptoms include the loss of vision and speech, rapid mental deterioration, spastic paralysis, and inevitable death; there is no cure. Scrapie, a prion-linked disease of sheep, is so named because infected sheep rub against trees or posts until they scrape off most of their wool.

Prions also cause the fatal *mad cow disease*, or BSE (bovine spongiform encephalopathy). Infected animals stagger about, drool heavily, and show other symptoms. Spongy holes and amyloid deposits form in the brain. BSE causes a variant CJD in humans. Before 2000, about 750,000 infected cattle entered the human food chain. Ground-up tissues of sheep that died of scrapie had been used as a supplement in cattle feed; contaminated tissue as small as a peppercorn can lead to infection.

Scrapie has been around for two centuries in Britain but posed no risk to humans before now. Did BSE arise by some mutation that crossed the species barrier from

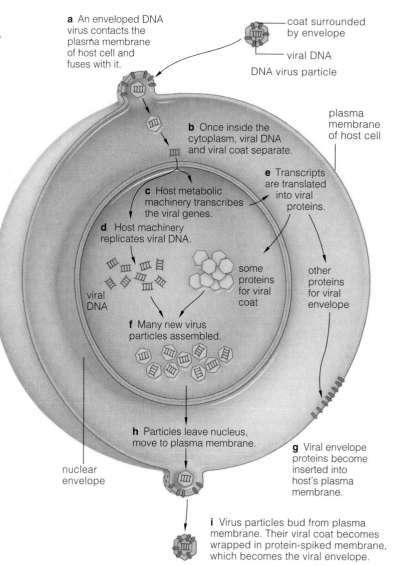

a An enveloped DNA virus contacts the plasma membrane of host cell and fuses with it.

coat surrounded by envelope

viral DNA

DNA virus particle

b Once inside the cytoplasm, viral DNA and viral coat separate.

plasma membrane of host cell

e Transcripts are translated into viral proteins.

c Host metabolic machinery transcribes the viral genes.

d Host machinery replicates viral DNA.

viral DNA

some proteins for viral coat

other proteins for viral envelope

f Many new virus particles assembled.

h Particles leave nucleus, move to plasma membrane.

nuclear envelope

g Viral envelope proteins become inserted into host's plasma membrane.

i Virus particles bud from plasma membrane. Their viral coat becomes wrapped in protein-spiked membrane, which becomes the viral envelope.

j The finished particle is equipped to infect a new potential host cell.

sheep to cattle? Maybe. Today, comprehensive safeguard measures are in place, but consumer confidence has plummeted. Export markets dried up; the British cattle industry has not recovered. Depending on assumptions about the incubation period and other variables, the number of deaths caused by this outbreak may range between 500 and 1,000, according to recent estimates.

Viral multiplication cycles include five steps: Attachment to a suitable host cell, cell penetration, viral DNA or RNA replication and synthesis of viral proteins, assembly of new viral particles, and release from the infected cell.

Some bacteriophage replication cycles follow a rapid, lytic pathway and an extended, lysogenic pathway. The replication cycles of RNA viruses involve the use of viral RNA as a template for synthesizing DNA or mRNA.

EVOLUTION AND INFECTIOUS DISEASES

THE NATURE OF DISEASE   Just by being human, you are a potential host for diverse pathogenic bacteria, viruses, fungi, protozoans, and parasitic worms. When a pathogen invades your body, it multiplies in cells and tissues; we call this an **infection**. Its outcome, **disease**, occurs when defenses cannot be mobilized fast enough to keep the pathogen's activities from interfering with body functions. In *contagious* diseases, pathogens must directly contact a new host. They travel in secretions of infected people, as in explosively wet sneezes.

During an **epidemic**, a disease spreads fast through part of a population for a limited time, then subsides. If an epidemic breaks out in several countries at the same time, this is a **pandemic**. AIDS is pandemic. *Sporadic* diseases, such as whooping cough, occur irregularly and affect very few people. *Endemic* diseases pop up more or less continually but do not spread far in large populations. Tuberculosis is like this. So is impetigo, a highly contagious bacterial infection that often spreads no further than, say, a single day-care center.

Now consider disease in terms of the pathogen's prospects for survival. *A pathogen stays around only for as long as it has access to outside sources of energy and raw materials.* To a microscopic organism or virus, a human is a treasurehouse of both. With bountiful resources, a pathogen can multiply or replicate itself to amazing population sizes. Evolutionarily speaking, the ones that leave the most descendants win.

Two barriers prevent pathogens from evolving to a position of world dominance. First, any species with a history of being attacked by a specific pathogen has coevolved with it and has built-in defenses against it. An example is the focused response of the vertebrate immune system. Second, if a pathogen kills too quickly, it might disappear along with the individual host. This is one reason why most pathogens have less-than-fatal effects. After all, infected individuals who live longer spread more germs and thus contribute to a pathogen's reproductive or replicative success.

Usually, the individual will die only if it becomes host to overwhelming numbers of a pathogen, if it is a novel host with no coevolved defenses, or if a mutant pathogenic strain has emerged and has breached the current defenses.

Being equipped with an evolutionary perspective, you probably can perceive the connection on your own. *The greater the population density of host individuals, the greater the kinds and frequencies of infectious diseases transmitted among them.* This brings us to the bad news.

EMERGING PATHOGENS   Thanks to planes, trains, and automobiles, people travel often and in droves around the world. Among their exotic destinations are virgin tropical forests and other remote regions. There, a human body had been an infrequent or nonexistent opportunity for pathogens. Now, strange and often dangerous pathogens are opportunistic about the novel, two-legged packages of nutrients entering their habitats. Within hours, they may infect travelers, who often carry pathogens back home.

Some of the **emerging pathogens** have been around for a long time and are only now taking advantage of the presence of novel human hosts. Others are newly mutated strains of existing pathogens.

Consider *Ebola*, one of several viruses that cause a deadly, hemorrhagic fever (Figure 21.22a). It may have coevolved with monkeys in Africa's tropical forests. By 1976, it was infecting humans. It kills between 70 and 90 percent of its victims. No vaccine or treatment is available for the disease, which starts with high fever and flu-like aches. Within a few days, nausea, vomiting, and diarrhea begin. Blood vessels are destroyed. Blood seeps from the circulatory system into the surrounding tissues and out through all the body's orifices. The liver and kidneys may rapidly turn to mush. Patients often become deranged and die of circulatory shock. With each *Ebola* epidemic, agencies around the world are mobilized. Quarantines limit the spread of the disease.

Or consider the *monkeypox* virus, a relative of the smallpox virus. After infecting a host, this DNA virus replicates in bone marrow, the lymph nodes, and the spleen. The bloodstream transports new viral particles to the skin, where they cause large, hardened, pus-filled sores (Figure 21.22b). When the virus overwhelms the immune system, secondary infections result in death.

Like smallpox, monkeypox kills as many as one in ten. Unlike smallpox, which is highly contagious, it was not much of a threat before now. Its few victims were mainly children who trapped, skinned, and ate infected monkeys in African forests. Virus particles slipped into the body through cuts and cracked skin on the hands.

A mutant monkeypox virus entered villages. More than *Ebola* and other attention-grabbing pathogens, it

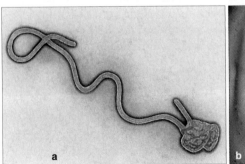
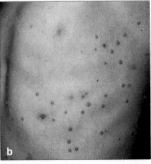

Figure 21.22   (**a**) *Ebola* virus particle. (**b**) Skin sores on a human infected with the monkeypox virus.

Table 21.4 *The Eight Deadliest Infectious Diseases*			
Disease	Main Agents	Estimated New Cases per Year	Estimated Deaths per Year
Acute respiratory infections*	Bacteria, viruses	1 billion	4.7 million
Diarrheas**	Bacteria, viruses, protozoans	1.8 billion	3.1 million
Tuberculosis	Bacteria	9 million	3.1 million
Malaria	Sporozoans	110 million	2.5–2.7 million
AIDS	Virus (HIV)	5.6 million	2.6 million
Measles	Viruses	200 million	1 million
Hepatitis B	Virus	200 million	1 million
Tetanus	Bacteria	1 million	500,000

* Includes pneumonia, influenza, and whooping cough.
** Includes amoebic dysentery, cryptosporidiosis, and gastroenteritis.

Figure 21.23 On a kitchen knife's "smooth" blade, bits of food and *Pseudomonas*. The cell's surface pili attached to the blade. Think about it next time you use any unwashed kitchen utensil that contacted raw beef, poultry, or seafood.

has the potential to cause pandemics. It easily replicates in humans and resists breakdown in the environment. Evolutionary biologist Paul Ewald says it has the right starting material for making a very nasty pathogen.

The smallpox vaccine works against monkeypox virus. But the World Health Organization essentially eradicated smallpox decades ago. There are very few stores of the vaccine, although with new bioterrorism threats on the world stage, more may be manufactured.

DRUG-RESISTANT STRAINS There is an old saying that when you attack nature, it will come back at you with a pitchfork. At this point in the book, we have looked at antibiotic resistance in several contexts. Here we correlate it with the upsurge in infectious diseases.

Example: Preschoolers have an immune system that is still developing. As the number of working mothers has skyrocketed, so has the number of preschoolers enrolled in day-care centers. Each center is a crowded population of hosts who are vulnerable to pathogens, including *Streptococcus pneumoniae*. This bacterium causes pneumonia, meningitis, and chronic middle-ear infections (Section 35.10, *Critical Thinking* question 4). Each year, it kills 40,000 to 50,000 people of all ages. Once, drug-resistant strains were almost unheard of. Now they are becoming the rule, not the exception.

Most of our 6.2 billion selves live in crowded cities. During any specified interval, as many as 50 million of us are on the move within and between countries in search of a better life. Is it any wonder that pathogenic agents of cholera, tuberculosis, and other diseases are spreading globally, with a vengeance (Table 21.4)?

PATHOGENS IN FOOD SUPPLIES Food poisoned by pathogens affects more than 80 million people in the United States each year. About 9,000 people a year die from food poisoning, mainly the very young or old, or those with compromised immune systems.

Often, people who say they have "the 24-hour flu" are more likely to be infected by *Salmonella enteriditis* and pathogenic *E. coli* strains. Both may be present in undercooked beef or poultry, water, unpasteurized milk or cider, vegetables grown in manure-fertilized fields, and ripened fruit picked up from the ground.

For example, in 1966, *S. enteriditis* slipped into some ingredients for a popular ice cream. More than 220,000 people ended up with stomach cramps, diarrhea, and other symptoms of the disease *salmonella*. A few years earlier, hundreds were sickened and three children died after eating restaurant hamburgers contaminated with *E. coli*. Unpasteurized apple juice sickened others.

Pathogens in the kitchen may cause at least half of all cases of food poisoning. Electron microscopes reveal microbes on wood or plastic cutting boards, even sleek, stainless steel knives (Figure 21.23).

Carlos Enriquez and some of his colleagues at the University of Arizona sampled 75 dishrags and 325 sponges in some homes. They discovered *Salmonella*, *E. coli*, *Pseudomonas*, and *Staphylococcus* colonies in most of them. Bacteria can live as long as two weeks in a wet sponge. Use sponges to wipe down a kitchen, and you can spread bacteria all over. Antibacterial soaps, detergents, and weak solutions of household bleach get rid of them. You can sanitize sponges by putting them through a dishwasher cycle.

Do you think food poisoning is of small concern? The annual cost of treating known infections ranges between 5 billion and 22 billion dollars.

In evolutionary terms, like any other kind of organism, the pathogen that leaves the most descendants wins.

21.10

1. Archaebacteria and the true bacteria are prokaryotic cells; they do not have a nucleus. Some have membrane infoldings and other structures in the cytoplasm. None has a profusion of organelles. *21.1*

   a. Three bacterial cell shapes are common: cocci (spheres), bacilli (rods), and spirilla (spirals).

   b. Nearly all eubacteria have a cell wall consisting of peptidoglycans. It protects the plasma membrane and resists rupturing. The composition and structure of the cell wall help identify particular species.

   c. A sticky mesh of polysaccharides (a glycocalyx) may surround the wall as a capsule or slime layer. It helps the bacterium attach to substrates and sometimes resists a host's infection-fighting mechanisms.

   d. Some species have one or more bacterial flagella, structures that rotate like a propeller and function in motility. Many have pili, filamentous proteins that help cells adhere to a surface or that facilitate conjugation.

2. As a group, prokaryotes show metabolic diversity, as in modes of acquiring energy and carbon. *21.1*

   a. *Photoautotrophs* use sunlight and carbon dioxide during photosynthesis. They include cyanobacteria and other oxygen-releasing species of eubacteria. They also include green nonsulfur and purple nonsulfur bacteria that do not produce oxygen; these get electrons from sulfur and other inorganic compounds, not from water.

   b. *Photoheterotrophs* use sunlight energy and organic compounds as sources of carbon. Some archaebacteria and eubacteria are like this.

   c. *Chemoautotrophs*, including the nitrifying bacteria, use carbon dioxide but not sunlight. Different kinds get energy by stripping electrons from a variety of organic or inorganic substances.

   d. *Chemoheterotrophs* are parasites (which get carbon and energy from living hosts) or saprobes (which feed on the organic products, wastes, or remains of other organisms). Most known prokaryotic species are like this. They include major decomposers and pathogens.

3. Prokaryotic fission is a cell division mechanism used only by prokaryotic cells. It involves replication of the bacterial chromosome and division of a parent cell into two genetically equivalent daughter cells. *21.2*

4. Many species have plasmids, small circles of DNA that are replicated independently of the single, circular bacterial chromosome. Plasmids may be transferred to daughter cells, to cells of the same species, or to entirely different species by conjugation. *21.2*

5. A major divergence occurred soon after the origin of life. One lineage gave rise to eubacteria, the other to common ancestors of archaebacteria and eukaryotes. Archaebacteria (methanogens, extreme halophiles, and extreme thermophiles) live in extreme environments, like those in which life originated. They are unique in wall structure and other features, and they share some features with eukaryotic cells. By far, the most common existing prokaryotic cells are eubacteria. *21.3–21.5*

6. Prokaryotes make behavioral responses to stimuli, as when they move toward nutrients. *21.6*

7. Viruses are nonliving, noncellular agents that infect particular species. *21.7*

   a. Each virus particle consists of a core of DNA or RNA and a protein coat that sometimes is enclosed in a lipid envelope. Many glycoproteins project like spikes from the envelopes. The coats of complex viruses have sheaths, tail fibers, and other accessory structures.

   b. A virus particle multiplies only after its genetic material enters a host cell and directs synthesis of the molecules necessary to produce new virus particles.

8. Nearly all viral multiplication cycles have five steps: attachment to a suitable host cell, cell penetration, viral DNA or RNA replication followed by protein synthesis, assembly of new viral particles, and release. *21.8*

9. Two pathways are common in multiplication cycles of bacteriophages (bacteria-infecting viruses). In a lytic pathway, multiplication is fast; new viral particles are released by lysis. In a lysogenic pathway, the infection enters a latent period. A host cell is not killed outright, and the viral nucleic acid may undergo recombination with a host cell chromosome. *21.8*

10. Multiplication cycles of viruses are diverse. They are rapid or they enter a latent phase. Penetration and release of most enveloped types occur by endocytosis and budding. DNA viruses spend part of the cycle in the nucleus of a host cell. RNA viruses complete the cycle in the cytoplasm. The viral RNA is the template for mRNA synthesis and for protein synthesis. *21.8*

Review Questions

1. Describe the basic features of prokaryotic cells. *21.1*

2. Define and make a sketch of what happens to the DNA during prokaryotic fission. *21.2*

3. Compared to your own bodily growth, how is bacterial growth measured? *21.2*

4. With respect to prokaryotic classification, what are some of the pitfalls of numerical taxonomy? How are comparisons of rRNA sequences assisting classification efforts? *21.3*

5. Name a few of the photoautotrophic, chemoautotrophic, and chemoheterotrophic eubacteria. *21.4–21.6*

6. Describe some eubacteria that are likely to give humans the most trouble, medically speaking. *21.6, 21.9*

7. What is a virus? List the five steps of viral multiplication cycles. *21.7, 21.8*

8. Distinguish between:
   a. microorganism and pathogen *CI*
   b. infection and disease *21.9*
   c. epidemic and pandemic *21.9*

9. Label the components of the viruses shown at *right. 21.7*

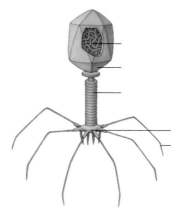 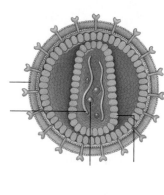

Self-Quiz ANSWERS IN APPENDIX III

1. Nondividing prokaryotic cells have _____ chromosome(s) and may have extra circles of _____ called plasmids.
   a. one; RNA
   c. one; DNA
   b. two; RNA
   d. two; DNA

2. _____ live in habitats much like those of the early Earth.
   a. Cyanobacteria
   c. Archaebacteria
   b. Eubacteria
   d. Protozoans

3. Which of the following is not grouped with archaebacteria?
   a. halophiles
   c. thermophiles
   b. cyanobacteria
   d. methanogens

4. Archaebacteria and eubacteria reproduce by _____ .
   a. mitosis
   c. prokaryotic fission
   b. meiosis
   d. longitudinal fission

5. Eubacterial cell walls are composed of _____ ; and a sticky mesh of polysaccharides, a _____ , often surrounds the wall.
   a. peptidoglycan; plasma membrane
   b. cellulose; glycocalyx
   c. cellulose; plasma membrane
   d. peptidoglycan; glycocalyx

6. Most prokaryotic cells are _____ , and they include major decomposers and pathogens.
   a. photoautotrophs
   c. chemoautotrophs
   b. photoheterotrophs
   d. chemoheterotrophs

7. Viruses are _____ .
   a. the simplest living organisms
   d. both a and b
   b. infectious particles
   e. both b and c
   c. nonliving

8. Viruses have a _____ and a _____ .
   a. DNA core; carbohydrate coat
   b. DNA or RNA core; plasma membrane
   c. DNA-containing nucleus; lipid envelope
   d. DNA or RNA core; protein coat

9. Match the terms with their most suitable description.
   ____ archaebacteria   a. infectious small protein
   ____ eubacteria   b. nonliving infectious particle with
   ____ virus   nucleic acid core and protein coat
   ____ plasmid   c. at home in stockyards
   ____ prokaryotic   d. methanogens, extreme halophiles,
        fission   extreme thermophiles
   ____ methanogen   e. most common prokaryotic cells
   ____ prion   f. small circle of bacterial DNA
                          g. bacterial cell division mechanism

Critical Thinking

1. *Salmonella* bacteria cause a form of food poisoning. They often live in poultry and eggs, but not in newly hatched chicks until chicks eat bacteria-laden feces of healthy adult chickens. Harmless bacteria ingested this way colonize the surface of intestinal cells, leaving no place for *Salmonella* to take hold. Some farmers raise thousands of chicks in confined quarters with no adult chickens. Should they consider feeding the chicks a known mixture of bacteria from a lab or a mixture of unknown bacteria from healthy adult chickens? Devise an experiment to test which approach may be more effective.

2. Reflect on mad cow disease (Section 21.8). The FDA wants to prohibit all protein supplements for sheep, cattle, and other ruminants. Investigate this practice in the United States.

3. Some strains of picornaviruses cause the highly contagious foot and mouth disease. The hosts include cattle, water buffalo, sheep, and pigs. Symptoms include extensive lesions and major tissue erosion, lameness, weight loss, milk loss, and often death. The viral strain causing a recent epidemic was first identified in Asia, then in Europe and elsewhere. It devastated farmers still recovering from an outbreak of BSE. There is no vaccine; at this writing, more than a million animals have been destroyed to stop the spread of the disease. How is the virus influencing travel and the global economy? Hint: Search for hoof and mouth disease on the Internet, starting with http://www.cdc.gov/.

4. Think about Figure 21.1. What does it tell you about the risks that drug abusers take when they share unsterilized needles? What does it tell you about medical uses of needles in developing countries where hygiene is often marginal?

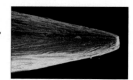

Selected Key Terms

archaebacterium *21.3*
bacillus (bacilli) *21.1*
bacterial chromosome *21.2*
bacterial flagellum *21.1*
bacteriophage *21.7*
cell wall *21.1*
coccus (cocci) *21.1*
conjugation *21.2*
disease *21.9*
emerging pathogen *21.9*
endospore *21.6*
epidemic *21.9*
eubacterium *21.3*
extreme halophile *21.5*
extreme thermophile *21.5*
fruiting body *21.6*
glycocalyx *21.1*
Gram stain *21.1*
heterocyst *21.6*

infection *21.9*
lysis *21.8*
lysogenic pathway *21.8*
lytic pathway *21.8*
methanogen *21.5*
microorganism *CI*
numerical taxonomy *21.3*
pandemic *21.9*
pathogen *CI*
pilus (pili) *21.1*
plasmid *21.2*
prion *21.8*
prokaryotic cell *21.1*
prokaryotic fission *21.2*
spirillum (spirilla) *21.1*
strain (bacterial) *21.4*
viroid *21.8*
virus *21.7*

Readings

Madigan, M., J. Martinko, and J. Parker. 2003. *Brock Biology of Microorganisms.* Tenth edition. Englewood Cliffs, New Jersey: Prentice-Hall. Exceptionally lucid, well-illustrated survey.

Daszak, P., A. Cunningham, and A. Hyatt. 21 January 2000. "Emerging Infectious Diseases of Wildlife—Threats to Biodiversity and Human Health." *Science,* 287:443–449.

Hively, W. May 1997. "Looking for Life in All the Wrong Places." *Discover,* 76–85. Account of microbes that live in the most extreme environments on and in the earth.

22

PROTISTANS

Confounding Critters at the Crossroads

More than 2 billion years ago, in tidal flats and soils, in estuaries, lagoons, lakes, and streams, prokaryotic cells were inconspicuously changing the world. Ever since the time of life's origin, Earth's atmosphere had been free of oxygen, and anaerobic species had reigned supreme. Their realm was about to shrink, drastically.

An oxygen-releasing pathway of photosynthesis was operating in vast prokaryotic populations. At first free oxygen combined with iron in rocks. The iron-rich deposits gradually rusted out—oxidized. Free oxygen started to accumulate; having little to combine with, it slowly became more concentrated in water, then in air.

The oxygen-enriched atmosphere was a selection pressure of global dimensions. Anaerobic species that couldn't neutralize the reactive, potentially lethal gas were restricted to black mud, stagnant water, and other anaerobic habitats. Others became oxygen tolerant.

It must have been a short evolutionary step from neutralizing oxygen to using it metabolically. Why? Aerobic species emerged in most prokaryotic groups. That key innovation in metabolism started rampant competition for resources. In the presence of so much oxygen, energy-rich organic compounds no longer could accumulate by geochemical processes. Organic compounds formed by living cells became the premier source of carbon and energy. New ways of acquiring and using organic compounds evolved. So did novel partnerships, predators, and parasites. And some of those evolutionary experiments led to protistans.

Sections 20.3 and 20.4 introduced the story of how protistans probably have endosymbiotic origins. Here we pick up the story with a look at the diversity of their descendants. Of all existing species, **protistans** are most like the earliest eukaryotes yet differ from the

prokaryotes. As defining characteristics, eukaryotic cells have a nucleus, large ribosomes, mitochondria, ER, and Golgi bodies. Their chromosomes consist of DNA molecules to which histones and other proteins are attached. They have microtubules as cytoskeletal elements. Many have chloroplasts. They divide by way of mitosis, meiosis, or both.

Of course, these also are defining characteristics of plant, fungal, and animal cells. In fact, until very recently, protistans were defined largely by what they are *not* (as in, not bacteria and not multicelled plants or fungi or animals). Confoundingly diverse species were essentially dumped into a separate kingdom at an evolutionary crossroads between prokaryotes and "higher" organisms. However, as this chapter suggests, the phylogenetic threads are now being untangled, and the "protistan" story is starting to make sense.

Key Concepts

1. For some time, single-celled eukaryotic organisms have been set aside in the catch-all kingdom Protista. Such organisms are relatively easy to distinguish from prokaryotes (archaebacteria and eubacteria) but have been difficult to classify with respect to plants, fungi, animals, and one another. They differ enormously from one another in morphology and life-styles.

2. Diverse spore-forming parasitic and predatory molds as well as free-living single-celled decomposers, predators, grazers, and parasites have been grouped together as protistans. So have richly varied groups of free-living, colonial, and multicelled species of photoautotrophs. Far from being insignificant, many of the photoautotrophs are the basis of food webs in nearly all aquatic habitats.

3. Molecular and biochemical comparisons, together with the fossil record and morphological studies, are clarifying the evolutionary picture. Protistans are now being grouped phylogenetically, and classification systems are being reconstructed as a result.

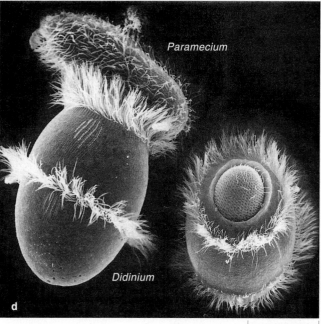

50 μm

Figure 22.1 Sampling of protistan diversity. (**a**) Underwater view of a kelp forest. A few *Macrocystis* species thrive in coastal waters of North and South America, New Zealand, Tasmania, and near most sub-antarctic islands. (**b**) Foraminiferan. These cells usually host symbionts, such as the tiny golden algal cells visible here. (**c**) Three spore-bearing structures of *Physarum*, a plasmodial slime mold, on a rotting log. (**d**) Mealtime for *Didinium*, a ciliate with a big mouth. *Paramecium*, a different ciliate, is poised at the mouth (*left*) and swallowed (*right*).

AN EMERGING EVOLUTIONARY ROAD MAP

A Phylogenetic Approach

Grouped in the traditional kingdom Protista are diverse photosynthetic types, ranging from microscopic cells to huge seaweeds. Others are decomposers, like some bacteria and fungi. Still others are parasites or even predators. The vast majority are single-celled species, but nearly every lineage also has multicelled forms.

As evolutionary relationships come into sharper focus, the traditional groups are shifting. Acceptance of the major shifts takes time, as it probably did when research warranted the split of a two-kingdom system in an earlier time. Many educators have been waiting for broader consensus on the new taxonomies before introducing them to their students.

However, Walter Judd at the University of Florida convinced us to move in the direction now favored by nearly all systematists. Here is a summary of his thoughts on a phylogenetic system of classification that recognizes *monophyletic* groups, as in Figure 22.2.

Such groups, recall, share a common ancestor and derived traits that are present in no other group.

"Kingdom Protista" is not a monophyletic group. The grouping exists in our minds, not in nature, and it makes our understanding of evolutionary relationships and change more difficult.

For example, chlorophytes—green algae and their closest relatives—were assigned to it. But they share many derived characters with plants and may be more closely related to them than to other protistan groups. Among these characters are certain molecular traits, two "whiplash" flagella at their anterior end, as well as chloroplasts that have a double outer membrane, starch grains, and chlorophylls *a* and *b*.

As another example, some classification schemes group protistans mainly in terms of their modes of nutrition. But this approach can obscure evolutionary changes. The **euglenoids** (Euglenophyta) are a classic example of an evolutionary puzzle. These free-living, flagellated cells abound in freshwater and in stagnant

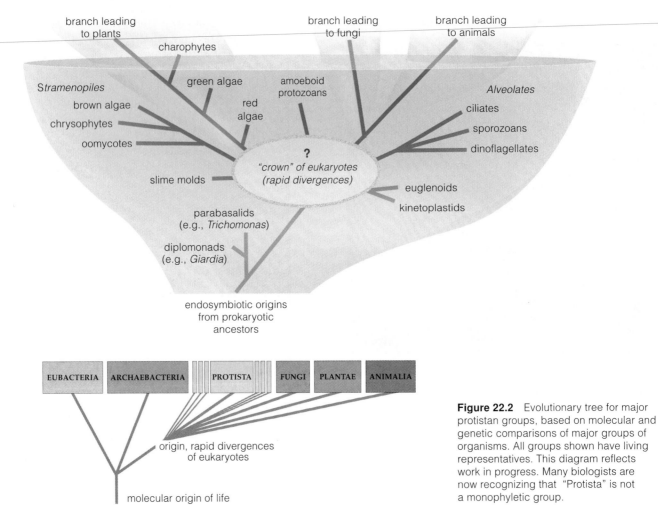

Figure 22.2 Evolutionary tree for major protistan groups, based on molecular and genetic comparisons of major groups of organisms. All groups shown have living representatives. This diagram reflects work in progress. Many biologists are now recognizing that "Protista" is not a monophyletic group.

ponds and lakes. Most of the 1,000 known species are photoautotrophs. The rest are heterotrophs that subsist on organic compounds dissolved in the water. Among their organelles are chloroplasts with chlorophylls *a* and *b*, and carotenoids—the same pigments found in chloroplasts of green algae and plants. But these cells also have two flagella and a contractile vacuole, like the flagellated protozoans described next (Figure 22.3).

Also like some of the protozoans, euglenoids have a **pellicle**. This flexible body covering has many spiral strips of a translucent, protein-rich material. Some of their pigments form an eyespot, which partly shields a light-sensitive receptor that helps direct and keep the cell in places most suitable for photosynthesis.

Generally, "self-feeders" make their own vitamins, which are required for growth. But all photoautotrophic euglenoids must get vitamin B_{12} from the surroundings; most cannot make vitamin B_1, either.

If groupings should reflect phylogenetic history, we have an explicit, operational criterion for recognizing them. Are euglenoids a lineage related to flagellated protozoans? Probably yes, based on their pellicle and molecular characters. Are they on the same branch as green algae? No. The traits they share with green algae arose through endosymbiosis, with the algal "guests" evolving into euglenoid chloroplasts.

And are the euglenoids a monophyletic group? Yes, as evidenced by unique traits—a storage carbohydrate that they alone make and the type of eyespot that is present only among the euglenoids.

What's In a Name?

Once a monophyletic group is identified, it must be assigned a rank as required by nomenclature codes. Here is where things get difficult, because ranks only have relative meaning. The plants are a monophyletic grouping—a clade—within Eukaryotes. As you will see in the next chapter, within the plant clade are a number of subclades, including mosses, liverworts, and vascular plants. But these in turn have subclades. For instance, vascular plants encompass several major monophyletic groups, such as ferns, conifers, cycads, and flowering plants. Within the flowering plants are other subclades, such as monocots, magnoliids, and eudicots—and so it goes.

Sets within sets. The important point is that ranking should reflect the *set* relationships among groups. If the eukaryotes are a kingdom, then green plants will be ranked as a phylum, and vascular plants a class. If eukaryotes are a domain, then green plants could be ranked as a kingdom, and vascular plants a phylum. You can see that absolute ranks are arbitrary. Even so, a group's rank should hold meaning relative to the

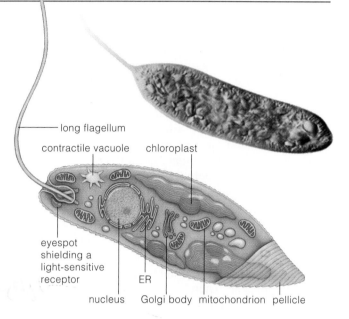

Figure 22.3 Body plan and light micrograph of *Euglena*. See also Section 20.4.

long flagellum
contractile vacuole
chloroplast
eyespot shielding a light-sensitive receptor
ER
nucleus
Golgi body
mitochondrion
pellicle

more inclusive and less inclusive monophyletic groups that surround it (Section 19.8).

Assigning names to nongroups such as protistans distorts our understanding of the evolutionary pattern of life. Recognizing various monophyletic groups as a series of sets within sets reflects a phylogenetic pattern. And the breakup of Protista into multiple groups in no way impacts the recognition of groups along other branches of the tree of life.

Some biologists have proposed that rankings be abolished and that we just name monophyletic groups. In Judd's own teaching, he has entirely left out ranks and finds students do not miss them. He shows them an evolutionary tree and talks about monophyletic groups such as eukaryotes, green plants, flowering plants, monocots, animals, deuterostomes, chordates, tetrapods, and amniotes. His students understand the characteristics of the group, how the members of the group go about living, and how patterns of character change have occurred in their evolutionary history.

Whether a group is called a phylum or class makes no difference when focusing on interesting biological questions. Judd chooses to spend time on issues more important to him than taxonomic rank—and Judd is a respected taxonomist! His problem with "Protista" is not that it is recognized at the rank of kingdom, but that it is recognized at all.

The rankings of classification schemes are in our minds, not in nature. We use them as passwords for information retrieval. However, the best passwords reflect rather than obscure evolutionary relationships.

ANCIENT LINEAGES OF FLAGELLATED PROTOZOANS

Those euglenoids you just read about are among the evolutionary branchings that are traditionally referred to as flagellated protozoans. All are heterotrophic cells, defined in part by the presence of one or more flagella. They include **kinetoplastids**, such as trypanosomes.

In many areas, trypanosomes parasitize humans. *Trypanosoma brucei* causes *African sleeping sickness*, a severe disease of the nervous system (Figure 22.4). The tsetse fly is its vector between hosts. In South America and Mexico, *T. cruzi* causes *Chagas disease*. Bugs pick up the parasite when feeding upon infected humans and other animals. The parasite multiplies inside the insect gut, then may be excreted onto a host. Scratched skin invites infection. The liver and spleen enlarge, eyelids and the face swell up, then the brain and heart become seriously damaged. There is no treatment or cure.

Even more ancient groups of flagellates are called **parabasilids** and **diplomonads**. These are free-living predatory and parasitic cells. One group has flagella *and* pseudopods, which points to an evolutionary link with the amoebas as well. Free-living types abound in freshwater or marine habitats. Parasitic types inhabit the moist tissues of certain plants and animals.

Trichomonads are parabasilids. Many are parasitic. *Trichomonas vaginalis*, a worldwide nuisance, can infect human hosts during sexual intercourse (Figure 22.5a). Unless trichomonad infections are treated promptly, they damage the urinary and reproductive tracts.

Diplomonads include *Giardia lamblia*, an internal parasite of humans, foraging cattle, and wild animals

[handwritten annotations: "1 flagella", "4 or more flagella"]

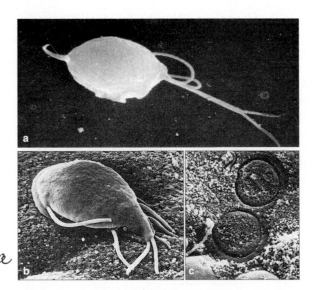

Figure 22.5 (a) Scanning electron micrograph of the motile feeding stage of *Trichomonas vaginalis*. This pathogen causes trichomoniasis, a sexually transmitted disease. (**b,c**) Scanning electron micrographs of *Giardia lamblia*, an animal-like flagellate that causes intestinal disturbances. The cell adheres to epithelium by means of its sucking disk. After it detaches, its disk often leaves a distinctive impression on the epithelial surface.

such as beavers. Cells (trophozoites) survive outside the body in cysts, usually in feces-contaminated water. A **cyst**, a body covering of cell secretions, helps resist stressful conditions. Ingesting *G. lamblia* cysts invites infection. In the stomach, the cysts release trophozoites, which become attached to the intestinal lining (Figure 22.5b,c). After reproducing, new parasites move into the large intestine (colon), then are expelled in feces. They infect new hosts who ingest contaminated water or food. The parasite also may be transmitted by way of anal intercourse with an infected partner.

G. lamblia often causes only mild intestinal upsets. But it also causes a severe form of gastroenteritis called *giardiasis*. Disease symptoms include bloating, nausea, intestinal cramps, and explosive, foul-smelling, watery diarrhea. The illness may recur over months or years.

More than 20 percent of the human population may be infected at a given time. Giardiasis is prevalent in the developing countries, in overcrowded regions with poor sanitation and water quality. In the United States, waterborne outbreaks occur mainly in mountains from coast to coast. For safety, hikers and campers should boil any water drawn from wilderness sources, such as remote streams, before drinking it.

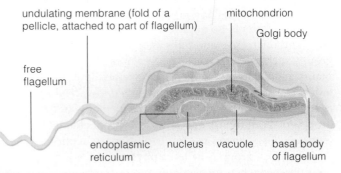

undulating membrane (fold of a pellicle, attached to part of flagellum)

mitochondrion

Golgi body

free flagellum

endoplasmic reticulum

nucleus

vacuole

basal body of flagellum

Figure 22.4 Light micrograph and body plan of *Trypanosoma brucei*. This animal-like flagellate causes African sleeping sickness.

Throughout the world, internal parasites of some flagellate lineages cause serious human diseases.

AMOEBOID PROTOZOANS

pseudopod

At some stage in their evolution, **amoeboid protozoans** (Sarcodina) lost their permanent motile structures. They now move by cytoplasmic streaming and pseudopod formation (Sections 4.9 and 4.10). **Pseudopods,** or "false feet," are dynamic, reversible cytoplasmic extensions of the cell (Figure 22.6a). The group includes the naked amoebas, foraminiferans, heliozoans, and radiolarians.

Rhizopods

Naked amoebas and foraminiferans belong to a group called **rhizopods.** Their cytoskeletal elements change continually. These structurally support the soft body of naked amoebas, which never show any symmetry. You find these protistans in damp soil, saltwater, and fresh water. Most types are free-living phagocytes that engulf other protozoans and bacteria. Some live in the gut of invertebrates or vertebrates and normally do no harm. A few are opportunistic parasites; when conditions fan their population growth, intestinal problems follow. The same thing happens after some free-living types are ingested.

Most foraminiferans live on the seafloor. They have a perforated external shell of secretions hardened with calcium carbonate (Figure 22.6b). Prey-trapping, thin pseudopods covered with mucus extend through the perforations. All but 1 percent of the named species are extinct. Countless fossilized foraminiferan shells were compressed over geologic time into large formations of sedimentary rock (Figure 22.6f). We mine the calcified remains and use them for many commercial products, such as cement and blackboard chalk.

Actinopods

Actinopods ("ray feet") include the radiolarians and heliozoans. Their name refers to the numerous slender, reinforced pseudopods that radiate from the body.

Like foraminiferans, radiolarians are richly evident in the fossil record; their ornate, silica-hardened parts do not break down easily (Figure 22.6e). Their stiffened pseudopods and thin skeletal parts project outward from a capsule of cytoplasm. Nearly all radiolarians drift with ocean currents as components of **plankton.** The word refers to aquatic communities of drifting or motile organisms, mostly microscopic. A few species of radiolarians form colonies. Secretions from individual cells cement neighbors together.

Pseudopods of heliozoans ("sun animals") radiate like the sun's rays (Figure 22.6c). Most species are free-floating or bottom dwellers of freshwater habitats. As with radiolarians, a membrane or capsule divides the cell into two biochemically distinct zones, one with the nucleus and the other with so many digestive vacuoles

copepod

Figure 22.6 (**a**) *Amoeba proteus,* one of the naked amoebas. Compared with most amoeboid protozoans, it has stubbier pseudopods. *A. proteus* is a favorite for laboratory experiments in biology classes. (**b**) One foraminiferan. Prey, including this copepod, get trapped in its sticky pseudopods. A microtubular core reinforces each pseudopod. (**c**) A living heliozoan. Silica deposits reinforce its pseudopods.

A look at how tiny cells can make grand structures. Shells of (**d**) a foraminiferan and (**e**) a radiolarian. More than 200 million years ago, fossilization and then compression of countless shells like these formed the white cliffs of Dover, England (**f**).

that the cell looks frothy. Vacuoles impart buoyancy to the cell and help keep it suspended in water.

Amoeboid protozoans are soft-bodied single cells, some with hardened skeletal elements. All form pseudopods for use in motility and prey capture. Free-living and gut-dwelling species, some parasitic, belong to this group.

THE CILIATES

Molecular studies reveal that the **ciliates** (Ciliophora) are related to sporozoans and dinoflagellates. All are **alveolates**, the only group with tiny, membrane-bound sacs (alveoli, singular alveolus) underneath their outer membranes. The sacs may stabilize the cell surface.

Many ciliates have a profusion of cilia at the surface. You read about these fine motile structures in Section 4.10. Free-swimming species use them as they move through freshwater and marine habitats. They prey on bacteria, algae, and one another, as Figure 22.1*d* so aptly suggests. Their cilia beat in such a synchronized pattern over the body surface that they call to mind a soft wind through a field of tall grasses.

Paramecium is typical of the group (Figure 22.7*a*). A fully grown cell is 150 to 200 micrometers long. Outer

membranes form a pellicle. This body covering may be rigid or flexible. A gullet starts at an oral depression at the body surface. Inside it, some cilia sweep food-laden water into the cell's body. Food becomes enclosed in enzyme-filled vesicles and is digested. Like amoeboid protozoans, *Paramecium* has a higher internal solute concentration compared to its surroundings. Both must constantly counter water's tendency to diffuse into the cell (by osmosis) by using **contractile vacuoles**. Thin, tubular extensions from the center of these organelles collect excess water that moved into the cell. A filled vacuole contracts and forces water through a pore to the outside (Section 5.9, *Critical Thinking* question 2).

Also, like the other protistan groups, ciliates show diversity among their life-styles. For example, about 65

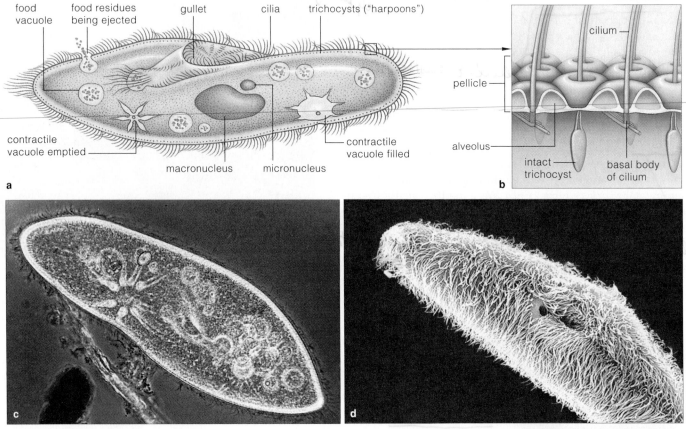

Figure 22.7 Representatives of two groups of ciliates.

Paramecia. (**a**) Generalized body plan for the genus *Paramecium*. (**b**) Components of the pellicle. Trichocysts, which span the pellicle, are organelles that discharge threads when the cell is irritated. Trichocysts might be a defense against predators. (**c**) Phase-contrast light micrograph of a living paramecium. (**d**) Scanning electron micrograph of the arrays of cilia at the surface of these single-celled predators.

(**e**) Hypotrich. This free-living species lives in the Bahamas. The hypotrichs are the most animal-like of the ciliates. They run around on leglike tufts of cilia. Some have a "head" end with modified sensory cilia.

leg-like tufts of cilia

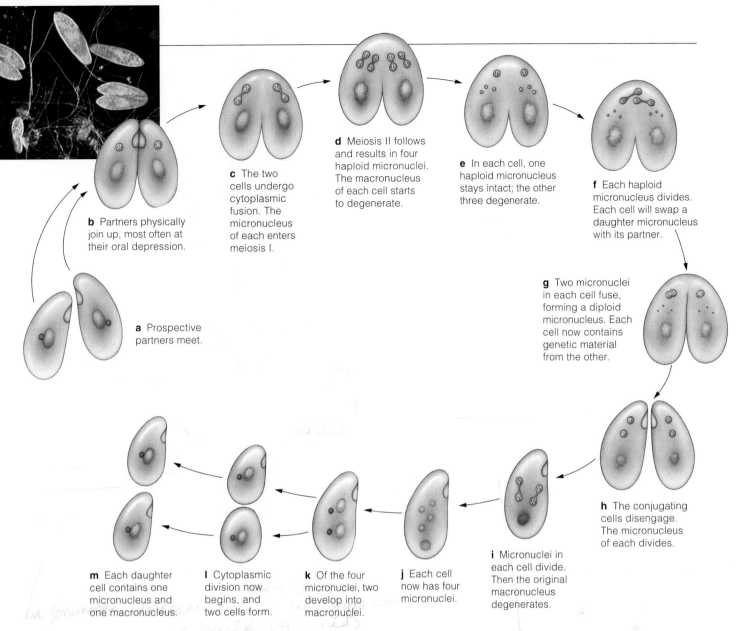

b Partners physically join up, most often at their oral depression.

a Prospective partners meet.

c The two cells undergo cytoplasmic fusion. The micronucleus of each enters meiosis I.

d Meiosis II follows and results in four haploid micronuclei. The macronucleus of each cell starts to degenerate.

e In each cell, one haploid micronucleus stays intact; the other three degenerate.

f Each haploid micronucleus divides. Each cell will swap a daughter micronucleus with its partner.

g Two micronuclei in each cell fuse, forming a diploid micronucleus. Each cell now contains genetic material from the other.

h The conjugating cells disengage. The micronucleus of each divides.

i Micronuclei in each cell divide. Then the original macronucleus degenerates.

j Each cell now has four micronuclei.

k Of the four micronuclei, two develop into macronuclei.

l Cytoplasmic division now begins, and two cells form.

m Each daughter cell contains one micronucleus and one macronucleus.

Figure 22.8 Generalized diagram of protozoan conjugation, an unusual form of sexual reproduction, as demonstrated by two ciliates that first encounter each other in (**a**).

percent of the ciliate species are free-living and motile. Others permanently or temporarily attach themselves to some substrate, often by a stalk extending from the cell body. Some form colonies. About 30 percent are symbionts with other organisms. *Balantidium coli* is the largest ciliate that parasitizes humans. It causes severe diarrhea and other intestinal problems.

Like most protozoans, ciliates reproduce sexually and asexually. An asexual process called **binary fission** divides the body in two parts. (The division plane is random for amoebas, longitudinal for flagellates, and transverse for ciliates.) Binary fission gets interesting among ciliates because each cell commonly has two types of nuclei. A small, diploid *micro*nucleus divides by mitosis. And the large *macro*nucleus lengthens and splits in two a bit sloppily; some of the DNA may spill

out. Most often, sexual reproduction occurs by a unique form of **conjugation**, as shown in Figure 22.8. In brief, the partner ciliates repeatedly divide their micronuclei, swap two daughter micronuclei, and allow two others to fuse and form a diploid macronucleus to replace the one that disappears. And you probably thought sex among the single-celled critters was simple!

Most ciliates bear a profusion of cilia. These are motile structures used for swimming and for beating food into an oral cavity. Each ciliate usually has two types of nuclei, which are distributed to daughter cells in unique ways during asexual and sexual reproduction.

Like other protistans, the ciliates show great diversity in life-styles and diverse variations on the basic body plan.

THE SPOROZOANS

Sporozoan is an informal name for parasitic alveolates that complete part of the life cycle *inside specific cells of host organisms.* These parasites form sporozoites, a type of motile infective stage. Some types form cysts. At one end of the cell body is a complex of distinct structures that function in penetrating host cells. The sporozoans are often grouped as phylum Apicomplexa.

Many sporozoans cause serious human diseases. For example, *cryptosporidiosis* is a waterborne disease caused by *Cryptosporidium.* The parasite invades epithelial cells of the small intestine or the respiratory system (Figure 22.9). Two to ten days after the infection, people suffer stomach cramps, watery diarrhea, and a slight fever. Symptoms may recur, often months or years later in people having a weak immune system. Infected people who do not show symptoms can still infect others.

The parasite can survive in lakes, rivers, streams, swimming pools, jacuzzis, chlorinated drinking water, and ice. (*Cryptosporidium,* unlike most pathogens, does not succumb to chlorine.) It contaminates most water supplies. The only way to get rid of it is to pass water through a reverse osmosis filter or bring it to a rolling boil for one minute. It should not be present in distilled water, and apparently it cannot live in hot coffee or tea.

Another sporozoan, *Pneumocystis carinii,* thrives in lungs of humans and many domestic and wild animals. Malnutrition or a weakened immune system give it the opportunity to threaten its host. For example, it causes a deadly form of pneumonia in about two-thirds of all AIDS cases. A resistant cyst forms during its life cycle. Sporozoites released during an asexual phase develop into a stage that lives in interstitial tissues of the lungs. Tiny air sacs inside the lungs fill with foamy material teeming with parasites. Fever, coughing, shallow rapid breathing, and bluish skin around the mouth and eyes follow. Untreated patients die from asphyxia; they stop breathing. Even with treatment, death rates are high.

[handwritten annotation: A FUNGUS]

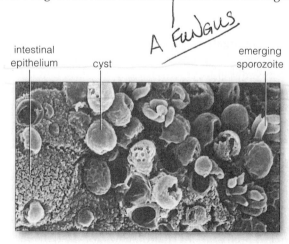

intestinal epithelium   cyst   emerging sporozoite

Figure 22.9 *Cryptosporidium* in intestinal epithelium.

Figure 22.10 Stand-off between a well-informed mother-to-be and a possible reservoir for the sporozoan *Toxoplasma.*

The sporozoan *Toxoplasma* uses domestic and wild cats as definitive hosts. Its intermediate hosts are other domestic and wild animals, and humans. Its cysts may be present in raw or undercooked meat. Cockroaches and flies also can move cysts from cat feces onto food. The disease *toxoplasmosis* has flu-like symptoms. It isn't common, but small epidemics do occur. The disease is dangerous for immune-compromised people and for embryos of pregnant women. It can cause miscarriages or birth defects if the parasite infects the unborn child. Any cat, no matter how coddled, may pass cysts. That is why all pregnant women should avoid stray cats and never clean children's sandboxes, take care of housecat "accidents," or empty a litterbox (Figure 22.10).

The next section focuses on the sporozoan agent of malaria, *Plasmodium.* Why all the attention on the most notorious flagellates, ciliates, and sporozoans? After all, less than two dozen species cause serious diseases in humans. But hundreds of millions of us are infected every year. We have no effective vaccines against them.

Sporozoans are internal parasites that produce infective, motile stages. Many species form encysted stages.

Malaria and the Night-Feeding Mosquitoes

Each year in Africa alone, about a million people die of *malaria*, a long-lasting disease that four different parasitic species of *Plasmodium* can cause. This type of sporozoan currently has infected more than 100 million people.

Only female mosquitoes of the genus *Anopheles* can transmit the parasites to human hosts. Their bite delivers sporozoites (an infective, motile stage) to the host's blood. The bloodstream transports sporozoites to the liver, where they undergo multiple fission. Some of the resulting cells, called merozoites, asexually reproduce inside red blood cells, which they kill. Others develop into male and female gametocytes, which mature into gametes (Figure 22.11).

Symptoms of malaria begin after infected cells abruptly rupture and release merozoites, metabolic wastes, and cellular debris into the individual's bloodstream. Shaking, chills, a burning fever, and drenching sweats are classic symptoms. After one episode, symptoms subside for a few weeks or months. Infected individuals might even feel normal, but relapses inevitably recur. In time, anemia and gross enlargement of the liver and spleen may result.

Amazingly, the *Plasmodium* life cycle is sensitive to the body temperature and oxygen levels inside humans and mosquitoes. The gametocytes cannot mature in humans, who are warm-bodied and have little free oxygen in their blood (most is bound to hemoglobin). They are induced to mature inside the mosquito, which has a lower body temperature and which incidentally slurps in oxygen from the air along with gametocyte-containing blood. Mature gametes fuse to form zygotes. These repeatedly divide and form many sporozoites, which migrate to the female mosquito's salivary glands and await the next bite.

Malaria has been around for a long time. People were describing its symptoms more than 2,000 years ago. It got its name in the seventeenth century, when Italians made a connection between the disease and noxious gases from swamps near Rome, where mosquitoes flourished (*mal*, bad; *aria*, air). By severely incapacitating so many people, malaria contributed to the decline of the ancient Greek and Roman empires. Much later in time, it incapacitated many soldiers during the United States Civil War, World War II, and the Korean and Vietnam conflicts.

Historically, malaria has been most prevalent in the tropical and subtropical parts of Africa. Now, the cases reported in North America and elsewhere are increasing dramatically, owing to hordes of globe-hopping travelers and unprecedented levels of human immigration.

Travelers who intend to visit countries with high rates of malaria are advised to use antimalarial drugs such as chloroquine. But certain strains of *Plasmodium* are now resistant to the drugs, and a vaccine has been difficult to develop. *Vaccines* are preparations that induce the body to build up resistance to a specific pathogen. Experimental vaccines for malaria are not equally effective against all the stages that develop during sporozoan life cycles. In general terms, this is a problem for any vaccine that researchers hope to develop against most parasites having complex life cycles.

At this writing, a vaccine to stop the parasite from invading red blood cells is undergoing clinical trials. So far, it proved 34 percent effective in 30 adult male volunteers who were given four doses. A single dose protected 34 percent of another volunteer group.

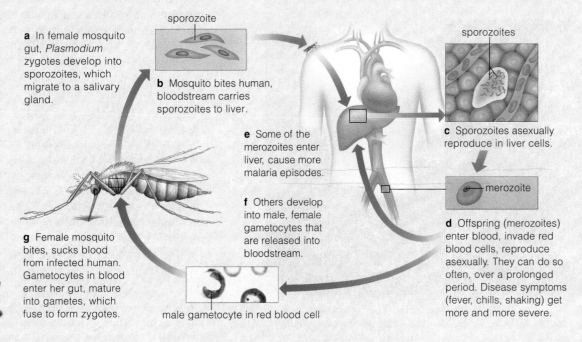

a In female mosquito gut, *Plasmodium* zygotes develop into sporozoites, which migrate to a salivary gland.

sporozoite

b Mosquito bites human, bloodstream carries sporozoites to liver.

sporozoites

c Sporozoites asexually reproduce in liver cells.

merozoite

d Offspring (merozoites) enter blood, invade red blood cells, reproduce asexually. They can do so often, over a prolonged period. Disease symptoms (fever, chills, shaking) get more and more severe.

e Some of the merozoites enter liver, cause more malaria episodes.

f Others develop into male, female gametocytes that are released into bloodstream.

male gametocyte in red blood cell

g Female mosquito bites, sucks blood from infected human. Gametocytes in blood enter her gut, mature into gametes, which fuse to form zygotes.

Figure 22.11 Life cycle of one species of *Plasmodium*, a type of sporozoan that causes malaria.

THE CELL FROM HELL AND OTHER DINOFLAGELLATES

One branch of the alveolates includes more than 1,200 species of **dinoflagellates** (Pyrrhophyta). Most species are single photosynthetic cells. Collectively, they are key producers of marine phytoplankton. A few types are symbionts with corals. Some are bioluminescent and make warm seawater shimmer at night (Section 6.8).

Each dinoflagellate has two flagella, one occupying a circular groove around the cell body (Figure 22.12). Plates of cellulose form a shell-like structure around the body. When the cell reproduces, its cellulose plates are divided between two daughter cells. These cells are yellow-green, green, blue, brown, or red, depending on their pigments and their endosymbiotic history.

From time to time, certain species of dinoflagellates (and some of the chrysophytes) undergo stupendous increases in population size, an event called an **algal bloom**. During some dinoflagellate blooms, the water turns rust-red or brown. When their cell concentrations briefly reach 6 to 8 million cells per liter of water along a coast, we call this a **red tide** (Figure 22.13).

A few species produce toxins that kill fish and other animals. In one year, for example, 150,000 grebes and

Figure 22.13
Red tide near California's central coast.

5,000 brown pelicans died at the Salton Sea in southern California. People who eat seafood tainted with some of the toxins can suffer brain damage.

Since 1991, algal blooms have killed billions of fish near North Carolina, Virginia, and Maryland. The agent of death, *Pfiesteria piscicida,* is a dinoflagellate known as "the cell from hell." The *P. piscicida* life cycle has at least twenty-four flagellated, encysted, and amoeboid stages (Figure 22.12). In estuary sediments, cysts form if conditions are harsh. Flagellated stages attach to prey and suck out juices. When a large school of menhaden or other oily fishes linger to feed, their excretions incite encysted cells to emerge and release toxins. The toxins prevent fishes from escaping (by slowing them down), then eat away patches of fish skin. *P. piscicida* feeds on sloughed epidermis, blood, and tissue juices oozing from the open sores. And then the flagellates give rise to amoeboids, which feast on the dead fish. All of these changes may unfold within a matter of hours.

Nitrogen, phosphorus, and other nutrients in runoff from fields and in raw sewage cause algal blooms all over the world. Blooms often develop in warm, shallow water that has become enriched with nutrients. JoAnn Burkholder, a North Carolina State University botanist, correlated *Pfiesteria* blooms with hundreds of millions of gallons of raw sewage from hog and chicken farms. More than 16 million hogs live in 3,500 industrial-scale hog farms in eastern North Carolina. Untreated wastes in holding lagoons often spill into rivers that drain into the poorly flushed Chesapeake Bay estuaries and into rivers in Alabama, Delaware, Florida, and Virginia. For instance, after one heavy rain, a spill from a hog farm exceeded by three times the volume of oil spilled from the tanker *Valdez* (Section 50.8). In 1997 alone, *Pfiesteria* inflicted a 60 million dollar loss on fishing and tourism industries. The magnitude of the loss finally prodded health, agricultural, and environmental agencies into coordinating research to address the problem.

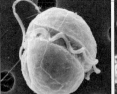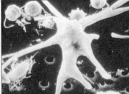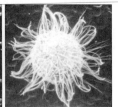

a   flagellated stage   amoeboid stage   encysted stage

Figure 22.12   (**a**) Stages in the life cycle of the dinoflagellate *Pfiesteria piscicida.* (**b**) The kind of damage inflicted by this so-called cell from hell. (**c**) Part of a fish kill resulting from a dinoflagellate bloom.

Dinoflagellates help form the food base for marine communities. Human activities that fan their explosive population growth can trigger enormous damage.

OOMYCOTES—ANCIENT STRAMENOPILES

We turn now to representatives of a major evolutionary branching, the **stramenopiles**, that may soon be given kingdom status. Some of its most ancient lineages are flagellates. Even multicelled species have flagellated spores. Stramenopiles are unique in that one of their two flagella has thin filaments projecting from it, like tinsel. The cells also have four outer membanes, which hints at endosymbiotic encounters in the past.

Stramenopiles range from colorless flagellates and water molds to groups that are mainly photosynthetic; the diatoms and the yellow-green, golden, and brown algae. Let's start here with the water molds and their relatives, the downy mildews and white rusts. All are oomycotes. The name, which means egg fungus, refers to a large egg cell that forms by way of meiosis inside a chamber called an oogonium (Figure 22.14a).

Most of the 580 species of **water molds** are saprobic decomposers of aquatic habitats. They draw nutrients from dead organisms and organic debris. One of the parasitic types, *Saprolegnia* (Figure 22.14b), makes short work of aquarium fish, but it usually attacks already damaged tissues. None has chloroplasts.

Other oomycotes, the **downy mildews**, are major pathogens. *Plasmopara viticola* is one of them. It rapidly molds grapevines and fruits. This species has become a recurring threat to vineyard productivity.

Another pathogen, *Phytophthora infestans*, lives up to its name—plant destroyer. More than a century ago, Irish peasants survived mostly on potatoes. Between

Figure 22.15 (**a**) Effects of an attack by *Phytophthora ramorum*, an oomycote that causes sudden oak death. Many thousands of trees in California's coastal forests are dying from the disease. Symptoms include oozing from bark, as shown in (**b**).

1845 and 1860, the growing seasons became cool and damp. Year after year, *P. infestans* spores spread along thin films of water on the plants. *Late blight*, a rotting of plant parts, was epidemic. In one fifteen-year period, a third of the population of Ireland starved to death, died during an outbreak of typhoid fever (a secondary effect), or fled to the United States and elsewhere.

A far more aggressive strain has emerged in Russia (Figure 22.14c). It recently evolved, by way of sexual reproduction, and it can withstand harsh winters. The destruction of this major food crop may cause famines and social upheavals in many parts of Russia.

An alarmingly opportunistic strain, *P. ramorum*, can kill healthy trees in one growing season. It has attacked Australian eucalyptus trees. In California, it is causing a *sudden oak death* epidemic. It has destroyed tens of thousands of trees and has jumped to novel hosts, such as madrone (*Arbutus*) and the coast redwoods (*Sequoia sempervirens*). The telltale sign of infection is a blackish oozing from cankers on tree trunks (Figure 22.15b).

antheridium (emptied)

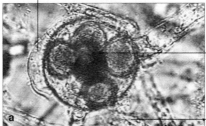

egg inside oogonium

hypha (one filament of mycelium)

Figure 22.14
Oomycotes.
(**a**) *Saprolegnia*, a parasitic type.
(**b**) One outcome of a *Saprolegnia* attack on tissues of an aquarium fish. (**c**) Potato demolished by a particularly virulent strain of *Phytophtora infestans* that recently evolved in Russia.

Stramenopiles are a major evolutionary group that includes flagellates and species that produce flagellated spores. The oomycotes are one of its ancient lineages. These are saprobic decomposers of aquatic habitats. Some types are serious parasites and pathogens.

PHOTOSYNTHETIC STRAMENOPILES—CHRYSOPHYTES AND BROWN ALGAE

We turn now to the stramenopiles that are primarily photosynthetic groups. Like oomycotes, most produce two flagella, one with unique fine hairs. They all have chloroplasts with four outer membranes, and they all have chlorophylls *a* and *c*. They synthesize oils, fats, and other storage products.

Chrysophytes

Most of the **chrysophytes** (Chrysophyta) are free-living photosynthetic cells that contain chlorophylls a, c_1, and c_2. The golden algae, yellow-green algae, diatoms, and coccolithophores are in this group.

Silica scales or other hard parts cover cells of many of the 500 species of **golden algae**, as in Figure 22.16*a*. Chlorophylls of these cells are masked by fucoxanthin, a golden-brown carotenoid pigment. Except for their chloroplasts, some amoeboid species closely resemble the true amoebas. **Yellow-green algae** do not contain fucoxanthin. The 600 or so known species are common components of phytoplankton in aquatic habitats. Most species are not motile, but they all produce flagellated gametes. Figure 22.16*b* shows an example.

The 5,600 existing species of **diatoms** have a silica "shell," which has two perforated parts that overlap like a pillbox (Figure 22.16*c*). For 100 million years, finely crushed shells of 35,000 extinct species accumulated at the bottom of lakes and seas. Many sediments contain the deposits, which are quarried for use in insulation, abrasives, and filters. More than 270,000 metric tons are quarried annually near Lompoc, California.

Calcium carbonate plates protect **coccolithophores** (Figure 22.16*d*). Most of the 500 or so existing species are single-celled photoautotrophs of marine habitats, especially in the tropics. Accumulations of plates in the past helped form ocean sediments, as well as chalk and limestone deposits, including Dover's white cliffs. During algal blooms, mucus secretions around so many cells can clog fish gills. Also, one of their metabolic wastes, dimethyl sulfide, is noxious enough to make migratory fishes deviate from their normal routes.

Brown Algae

Walk along a rocky shore at low tide and you might come across olive-green and brown seaweeds (Figure 22.17). They are among the 1,500 species called **brown algae** (Phaeophyta). Most thrive in cool or temperate seawater, from the intertidal zone through the open ocean. Masses of one type of brown alga, *Sargassum*, are the basis of an enormous floating ecosystem in the Sargasso Sea, between the Azores and the Bahamas.

Different species appear olive-green, golden, or dark brown, depending on the pigments. Like chrysophytes, brown algae contain carotenoids such as fucoxanthin, as well as chlorophylls a, c_1, and c_2. They range in size from microscopic filaments like *Ectocarpus* to the giant kelps, which are twenty to thirty meters long. Their life cycles are diverse, and have sexual and asexual phases. Gametophytes (gamete-producing bodies) develop and alternate with the development of sporophytes (spore-producing bodies) during the life cycle.

Macrocystis, *Laminaria*, and the other giant kelps are the largest, most complex protistans. Their multicelled sporophytes have stipes (stemlike parts), blades (leaflike parts), and holdfasts (anchoring structures). Hollow,

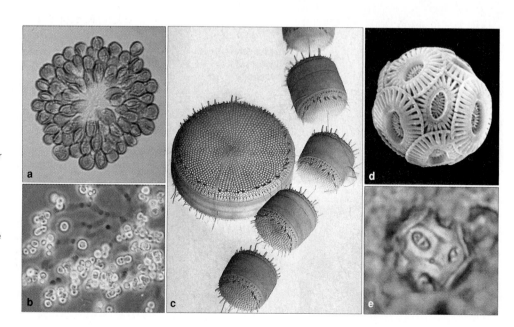

Figure 22.16 A few chrysophytes.

(**a**) *Synura*, a golden alga with a strong, fishy odor. It forms colonies in phytoplankton. (**b**) *Mischococcus*, a yellow-green alga that also forms colonies in phytoplankton.

(**c**) Diatom "shells," which fit together like a pillbox. The four smaller shells shown here formed by cell division. New walls form inside a parent cell wall, which means that diatoms become smaller with each division. When too small, a spore forms inside each shell. Spores germinate, grow, and become a full-size diatom.

(**d**) Coccolithophore "shell." (**e**) A fossilized Jurassic coccolithophore, about 11 microns in diameter.

Figure 22.17 Representative brown algae. (**a**) Body plan of *Macrocystis*. Figure 22.1*a* is an example of how these giant kelps can form dense underwater forests. (**b**,**c**) *Postelsia*, a brown alga thriving along coasts exposed to heavy surf from Vancouver Island down to central California. More than a hundred deeply grooved blades top a highly resilient stipe, which is fastened to rocky substrates by a mound of anchoring structures. The spores never disperse far from the parent sporophyte. At low tide, they simply drip onto rocks from the grooved blades.

gas-filled bladders impart buoyancy to the stipes and blades and keep them upright in the water. The stipes contain tubelike arrays of elongated cells. These tubes swiftly carry dissolved sugars and other products of photosynthesis to living cells throughout the kelp body. Vascular plants also transport sugars through similar kinds of tubes. However, this is a case of convergent evolution only; these two groups of large, multicelled organisms are not closely related.

Giant kelp beds function as productive ecosystems. Think of them as underwater forests, homes in which great numbers of different bacteria and protistans, as well as fishes and other animals, carry out their lives. A kelp forest's dimensions are variable. They change with shifts in ocean currents. They also change according to the abundances of sea urchins (which feed on kelp debris) and sea otters (which feed on sea urchins).

For instance, in the late 1950s, a warm ocean current displaced cooler currents off the California coast. With this temperature shift, *Macrocystis* did not fare well, and the extensive beds off La Jolla and Palos Verdes almost disappeared. Without organic remains for sea urchins to feed on, these invertebrates fed directly on kelps instead. Quantities of quicklime were dumped in the water to reduce the number of sea urchins, and kelps made a comeback. So did fishes, lobster, abalone, and other species that make the kelp beds their home.

Macrocystis and some other species are harvested commercially. Extracts from them become ingredients in ice cream, pudding, jelly beans, salad dressings, beer, canned and frozen foods, cough syrups, toothpaste, cosmetics, floor polish, and paper. Alginic acid from the cell walls of some species is used to make algins, which are added to various products as thickening, emulsifying, and suspension agents. In the Far East especially, people harvest kelps as sources of food and mineral salts, and as a fertilizer for crops.

The mostly photosynthetic stramenopiles include the chrysophytes and brown algae. They share distinctive traits, including chloroplasts with four membranes. They produce chlorophylls *a* and *b*, and storage fats and oils. Most have two flagella, one bearing tinsel-like filaments.

GREEN ALGAE AND THEIR CLOSEST RELATIVES

Of all protistans, **green algae** (Chlorophyta) and their close relatives (Zygophyta and Charophyta) are the most like plants in their structure and biochemistry. Indeed, many botanists are now including them with plants! All green algae are photosynthetic. Again, as in plants, their chlorophylls are of the molecular types designated *a* and *b*. Like plants, green algae store their carbohydrates as starch grains inside chloroplasts. The cell walls of some species are composed of cellulose, pectins, and other polysaccharides typical of plants. They, too, usually have two anterior whiplash flagella.

Figure 22.18 has a sampling of the more than 7,000 known species. They include single-celled, sheetlike, tubular, filamentous, and colonial forms. You won't be able to see many without the aid of a microscope. Most, including the *Micrasterias* cell shown in Figure 22.18*d*, live in freshwater. *Micrasterias* is one of thousands of desmid species, which are important food producers in nutrient-poor ponds and peat bogs. Green algae also grow at the ocean surface, in marine sediments, just below the surface of soil, and on rocks, tree bark, other organisms, and snow. Some are symbionts with fungi, protozoans, and a few marine animals. A colonial form (*Volvox*) is a hollow, whirling sphere of 500 to 60,000 flagellated cells. Those white, powdery beaches in the tropics are largely the work of uncountable numbers of *Halimeda* cells, which formed calcified walls, then died and disintegrated (Figure 22.18*f*). Someday, green algae

Figure 22.18 Representative green algae. (**a**) One of the marine species (*Codium fragilis*) with a pronounced branching form.

Many green algae are microscopic, *C. magnum* is taller than you are. In 1956 another species of *Codium*, from Washington State, was accidentally introduced into the Connecticut River where it empties into the Atlantic. Puget Sound populations of this alga had been kept in check by herbivores that evolved with them. In its new habitat, there were no native *Codium* species or herbivores that were adapted to grazing on it. The introduced alga spread along the coast as far north as Maine and as far south as North Carolina in a little more than two decades.

(**b**) Sea lettuce (*Ulva*) grows in estuaries and attaches to kelps in the seas. Reproductive cells form within and are released from margins of the sporophyte and gametophyte, both of which have the form shown in the diagram.

(**c**) *Volvox*, a colony of interdependent cells that resemble free-living, flagellated *Chlamydomonas* cells. This colony is rupturing; released daughter cells will found new colonies.

(**d**) *Micrasterias*, a single-celled freshwater green alga.

(**e**) From a marine habitat, *Acetabularia*, fancifully called the mermaid's wineglass. Each individual, a multinucleate cell mass, has a rootlike structure, stalk, and cap in which gametes form.

(**f**) *Halimeda incrassata*. Its large, branched cells form by repeated nuclear division. The cell walls form only during reproductive phases.

(**g**) *Udotea cyathiformis*, one of 100+ microscopic *Udotea* species in tropical and subtropical waters. Many are calcified.

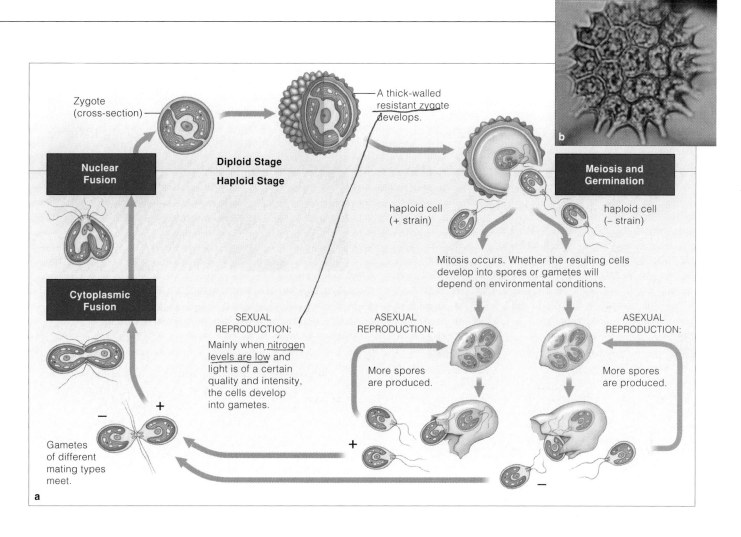

Figure 22.19 (**a**) Life cycle of a species of *Chlamydomonas*, a common green algae of freshwater habitats. This single-celled species reproduces asexually most of the time and sexually under certain environmental conditions. (**b**) Light micrograph of one of the zygotes.

Labels in figure (a):

Zygote (cross-section)

A thick-walled resistant zygote develops.

Nuclear Fusion

Diploid Stage

Haploid Stage

Meiosis and Germination

haploid cell (+ strain)

haploid cell (– strain)

Cytoplasmic Fusion

Mitosis occurs. Whether the resulting cells develop into spores or gametes will depend on environmental conditions.

SEXUAL REPRODUCTION:

Mainly when nitrogen levels are low and light is of a certain quality and intensity, the cells develop into gametes.

ASEXUAL REPRODUCTION:

More spores are produced.

ASEXUAL REPRODUCTION:

More spores are produced.

Gametes of different mating types meet.

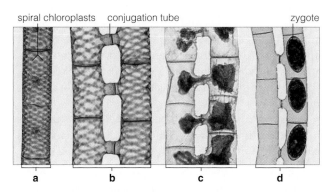

spiral chloroplasts   conjugation tube   zygote

a   b   c   d

Figure 22.20 One mode of sexual reproduction in *Spirogyra*, or watersilk. (**a**) This zygophyte has spiral, ribbonlike chloroplasts. (**b**) A conjugation tube forms between cells of adjacent haploid filaments of different mating strains. (**c,d**) The cellular contents of one strain pass through the tubes into cells of the other strain, where zygotes form. Zygotes develop thick walls. They undergo meiosis when they germinate and give rise to haploid filaments.

might even accompany astronauts on long missions in outer space. Green algae can grow in very small spaces, they release oxygen as a by-product of photosynthesis (the crew can't live without oxygen supplies), and they can take up carbon dioxide exhaled by the aerobically respiring crew.

Green algae employ diverse modes of reproduction. *Chlamydomonas* provides a classic example. This single-celled alga is twenty-five or so micrometers wide. It is able to reproduce sexually. Most of the time, however, it engages in asexual reproduction, with up to sixteen daughter cells forming by mitotic cell division within the confines of the parent cell wall. Daughter cells may live at home for a while, but sooner or later they leave by secreting enzymes that digest what's left of their parent. Figure 22.19 shows the life cycle of one species. To give a final example, Figure 22.20 shows a relative (*Spirogyra*, a zygophyte), reproducing sexually.

The green algae show great diversity in size, morphology, life-styles, and habitats. The structure and biochemistry of some groups indicate they are close to or within the boundary of the plant kingdom.

22.11

RED ALGAE

Of 4,100 known species of **red algae** (Rhodophyta), nearly all are marine; only 200 live in freshwater. Red algae are most abundant in warm currents and tropical seas, often at surprising depths (265 meters below the surface) in clear water. A few occur in phytoplankton. Some encrusting types contribute to the formation of coral reefs and coral banks (Section 27.3).

Figure 22.22 A red alga (*Bonnemaisonia hamifera*). Such a growth pattern (branching, filamentous) is the most common. Section 7.3 shows a red alga with sheetlike growth.

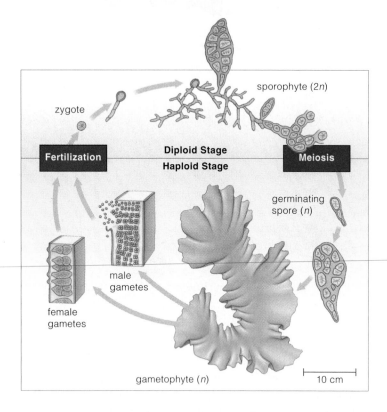

Figure 22.21 Life cycle of *Porphyra*. From 1623 to the 1950s, Japanese fishermen cultivated and harvested a species of this red alga in early fall. The rest of the year, it seemed to vanish.

Kathleen Drew-Baker studied the sheetlike form of *P. umbilicus* in the laboratory. She observed gametes forming in packets interspersed between vegetative cells near the sheet margins; the sheet was a gametophyte. She also observed gametes in a petri dish. Zygotes that formed after fertilization grew on bits of shell inside the dish before developing into a tiny branching, filamentous form. This was how *Porphyra* spent most of the year!

People already recognized the filamentous form as common pinkish growths on shells. But it differed so much from the sheetlike form that it was viewed as a separate species. It is the alga's diploid sporophyte (spore-producing body).

Drew-Baker's discovery of the alternating stages in the alga's life history, and the realization that its diploid stage could be grown on shells or other calcium-rich surfaces, revolutionized the nori industry. Within a few years, researchers worked out the *P. tenera* life cycle, the harvested species. By 1960, *nori cultivation* was a billion-dollar industry.

Red algae actually are red, green, purple, or black because accessory pigments mask their chlorophyll *a*. Most of the pigments are phycobilins. These are good at absorbing green and blue-green wavelengths that penetrate deep waters. The chlorophylls are efficient at absorbing red and blue wavelengths, but these may not penetrate the water far. Chloroplasts of red algae resemble cyanobacteria, which suggests endosymbiotic origins. Some species have cell walls hardened with calcium carbonate. None has flagella.

The life cycle of most species includes multicelled stages, but these lack tissues or organs (Figure 22.21). Reproductive modes are diverse, with complex asexual and sexual phases. Figure 22.22 shows an example.

Red algae have a flexible, slippery texture owing to mucous material in their cell walls. Agar is made from extracts of wall material from a few species. The inert, gelatinous substance is used as a moisture-preserving agent in baked goods and cosmetics, as a setting agent for jellies, and as culture gels. We also shape it into soft capsules for delivering drugs and food supplements into the body. Carrageenan, extracted from *Eucheuma*, is a stabilizer in paints, dairy products, and many other emulsions.

Humans find different species of *Porphyra* tasty as well as nutritious. You may know them from sushi bars as *nori*, a wrapping for rice and fish (Figure 22.21).

Red algae are photosynthetic, with phycobilins and other accessory pigments that mask their chlorophyll *a*. Usually, multicelled stages develop during the life cycles.

Most species of red algae are aquatic. They show great diversity in size, morphology, life-styles, reproductive modes, and habitats.

SLIME MOLDS

As a final example of protistan diversity, **slime molds** are free-living, amoebalike cells for part of the life cycle. Like amoebas, *cellular* slime molds (Acrasiomycota) and *plasmodial* slime molds (Myxomycota) are predators. Their cells crawl on rotting plant parts and eat bacteria, yeasts, spores, and organic compounds. If nutrients are scarce, starving amoebas gather into a slimy mass that

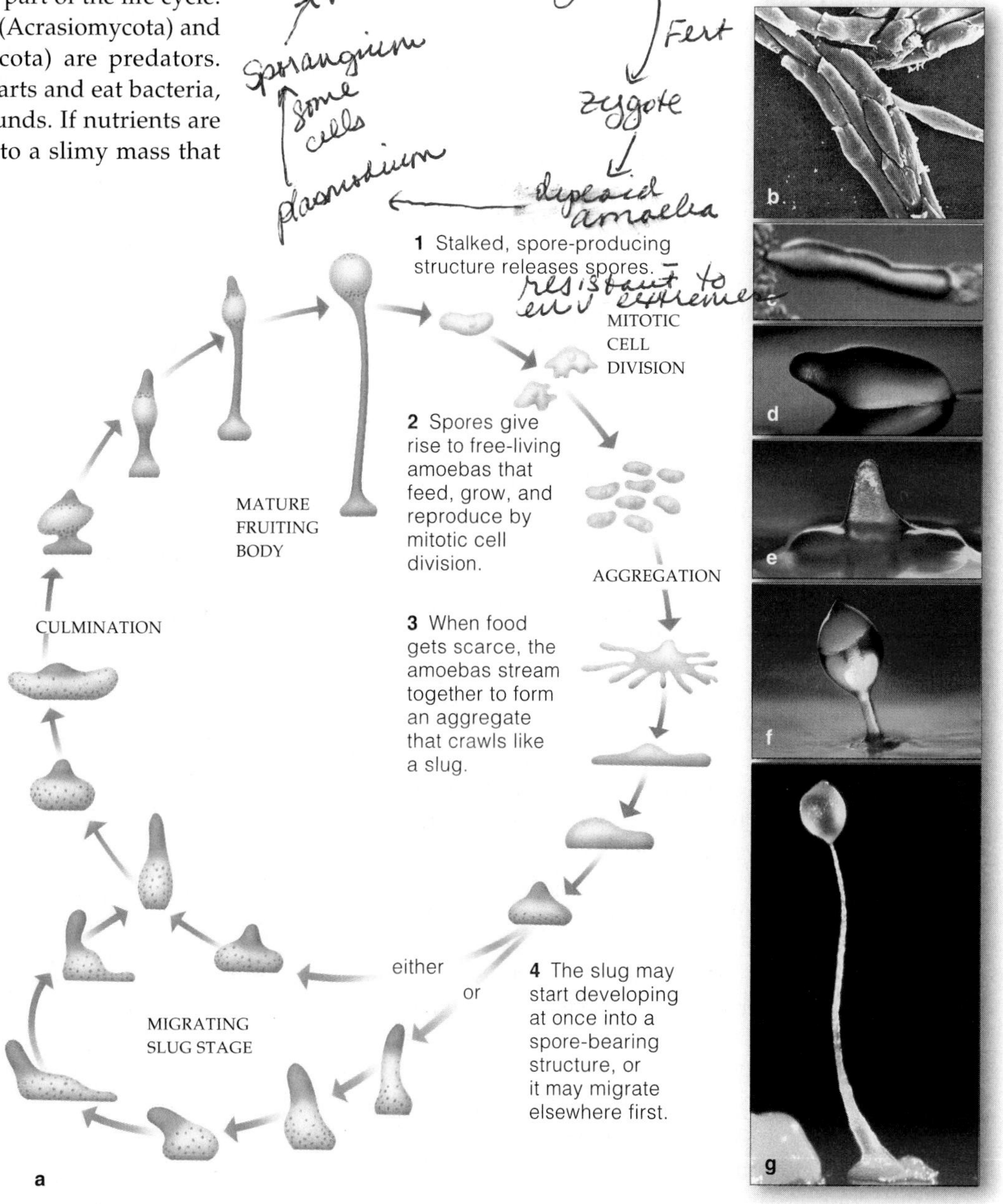

Figure 22.23 A cellular slime mold, *Dictyostelium discoideum*. (**a**) Its life cycle includes a spore-producing stage. Spores give rise to free-living amoebas, which grow and divide until food (soil bacteria) dwindles.

When starved of nutrients, amoebas secrete cyclic AMP, a signaling molecule that makes them stream toward one another. Their plasma membrane becomes sticky, so they adhere to one another. A cellulose sheath forms around the cell mass, which starts crawling like a slug (**b–d**). Some slugs may incorporate 100,000 amoebas.

In a migrating slug, the amoebas differentiate into prestalk (*red*), prespore (*white*), and anteriorlike cells (*brown dots*). Prestalk cells secrete ammonia in amounts that vary with temperature and light intensity. The slug moves fastest in response to intermediate levels of ammonia (not too little, not too much). Those levels correspond to warm, moist conditions (not too cold or hot, not soppy or dry).

The prestalk and prespore cells differentiate and form a stalked, spore-bearing structure (**e–g**). Anteriorlike cells sort into two groups and may help elevate nonmotile spores for dispersal from the top of the spore-bearing structure.

MATURE FRUITING BODY

CULMINATION

MIGRATING SLUG STAGE

1 Stalked, spore-producing structure releases spores.

MITOTIC CELL DIVISION

2 Spores give rise to free-living amoebas that feed, grow, and reproduce by mitotic cell division.

AGGREGATION

3 When food gets scarce, the amoebas stream together to form an aggregate that crawls like a slug.

either or

4 The slug may start developing at once into a spore-bearing structure, or it may migrate elsewhere first.

may migrate to a better place and form a spore-bearing structure. Collective contractions of single cells move the mass. Amoebalike cells differentiate, and some form a spore-bearing structure. Released spores germinate on warm, damp surfaces. Later on, each gives rise to an amoebalike cell. Sexual reproduction also occurs.

Dictyostelium discoideum, often used in research into development, is one of 70 kinds of cellular slime molds (Figure 22.23). Figure 22.1*c* shows some spore-bearing structures of one of the 500 species of plasmodial slime molds. The plasma membrane of these amoebalike cells

breaks down when they get together. Cytoplasm flows freely from cell to cell and distributes nutrients and oxygen throughout the mass. A streaming mass of this sort (the plasmodium) often occupies about 0.3 square meter and migrates if food runs out. You may see one crossing a lawn or a road, or even climbing a tree.

During a slime mold life cycle, amoeboid cells aggregate to form a migrating mass. Cells in the mass differentiate, then form reproductive structures and spores or gametes.

SUMMARY

1. Traditionally, protistans have been classified as the structurally simplest eukaryotes. They have a nucleus, mitochondria, ER, and ribosomes larger than those of prokaryotes. They have microtubules and two or more chromosomes (DNA complexed with many histones and other proteins). Like other eukaryotic cells, they engage in mitosis and meiosis. Many have chloroplasts and other plastids. *CI*

2. Molecular and biochemical studies, together with the fossil record and morphological comparisons, show that "protistans" are not a monophyletic group. Rather, they are many monophyletic lineages, the evolutionary relationships of which are being worked out. *22.1*

3. Phylogenies of major groups of "protozoans" have been deciphered: *22.1–22.7*

   a. The euglenoids, kinetoplastids, parabasalids, and diplomonads are closely related heterotrophic cells that are defined in part by having one to several flagella. Some parasites, such as *Trypanosoma brucei* and *Giardia lamblia*, cause serious human diseases. *22.1, 22.3*

   b. As they evolved, amoeboid protozoans lost their permanent motile structures and now use cytoplasmic streaming and pseudopods. These soft-bodied cells are amoebas (no hard parts) or foraminiferans, heliozoans, and radiolarians (silica-hardened parts). *22.4*

   c. Ciliates, sporozoans, and dinoflagellates are now recognized as a phylogenetic group called alveolates. They alone have tiny, membrane-bound sacs (alveoli) beneath the outer membranes. Many are parasites (e.g., *Plasmodium* agents of malaria). *22.4–22.7*

   d. Most flagellates and ciliates reproduce asexually or sexually. Often binary fission, an asexual process, divides the body into two parts. The division plane is random for amoebas, longitudinal for flagellates, and transverse for ciliates. Some bud from a parent. The life cycles of many have encysted stages. *22.4–22.7*

4. Water molds, chrysophytes, and brown algae are ranked as stramenopiles. They are defined in part by unique flagella, one tinselly (with many thin filaments projecting from it). Their four-membraned chloroplasts hint at endosymbiotic origins.

   a. Water molds are saprobic, colorless decomposers. They secrete enzymes that digest organic matter, then absorb breakdown products. *22.8*

   b. Most chrysophytes are free-living photosynthetic cells with chlorophylls a, c_1, and c_2, and distinctive accessory pigments. Example: golden algae and brown algae contain fucoxanthin. Brown algal species range from the microscopic to giant kelps. *22.9*

5. Green algae and closely related groups are the most like plants. Many botanists actually are including them in the plant kingdom. Their chloroplasts, too, contain chlorophylls a and b, and store carbohydrates in the form of starch grains. Some have cellulose, pectins, and other polysaccharides in their cell wall, like plant cells do. *22.10*

6. Red algae are a separate lineage of photosynthetic cells that may be related to cyanobacteria. Phycobilins are accessory pigments in both. Most red algal species have multicelled stages in the life cycle. *22.11*

7. Slime molds are free-living predatory, amoebalike cells for part of their life cycle. They also aggregate to form a migrating mass in which cells differentiate and form reproductive structures. *Dictyostelium discoideum* is a well-studied cellular slime mold. *22.12*

8. It has taken us more than one chapter to consider prokaryotes and protistans—the simplest eukaryotes. Table 22.1 summarizes the similarities and differences among the prokaryotes and eukaryotes.

Review Questions

1. Outline the general characteristics of protistans. Compare the derived traits of a few major lineages to explain why "the protistans" is not a monophyletic group. *22.1*

2. Name one or more defining traits for these related groups of heterotrophic, flagellated cells: *22.1, 22.2*
   | a. euglenoids | c. kinetoplastids |
   | b. parabasalids | d. diplomonads |

3. Amoebas, foraminiferans, radiolarians, and heliozoans all have pseudopods. How do amoebas differ from the others? *22.3*

4. List the names of three alveolate groups. What is a key defining trait that they share? State one of the defining traits of the alveolates. *22.4–22.6*

5. Select a parasitic flagellated protozoan or alveolate, then briefly explain how it adversely affects human affairs, such as crop yields and human health. *22.2, 22.5, 22.6*

6. What is a key defining trait of the stramenopile groups? List some characteristics of one of the oomycotes, chrysophytes, or brown algae that belong to this major group. *22.8, 22.9*

7. List some of the characteristics that imply green algae and related forms are the closest relatives of plants and might even belong in the plant kingdom. *22.10*

8. To which prokaryotic group do the red algae appear to be phylogenetically connected, by endosymbiosis? *22.11*

9. Is this statement true or false: In all slime molds, amoeboid cells aggregate, and the plasma membrane of every cell breaks down, thus forming a streaming mass. *22.12*

Self-Quiz   ANSWERS IN APPENDIX III

1. Trypanosomes cause which disease(s)?
a. toxoplasmosis	d. African sleeping sickness
b. Chagas disease	e. malaria
c. amoebic dysentery	f. both b and d

2. Amoebas, foraminiferans, and radiolarians have _____ .
   a. pseudopods   b. pellicles   c. cilia   d. all of the above

3. Alveolates include _____ .
   | a. ciliates | c. dinoflagellates |
   | b. sporozoans | d. all of the above |

Table 22.1 Comparison of Prokaryotes With Eukaryotes

	Prokaryotes	Eukaryotes
Organisms represented:	Archaebacteria, Eubacteria	Protistans, fungi, plants, and animals
Ancestry:	Two major lineages that evolved more than 3.5 billion years ago	Equally ancient prokaryotic ancestors gave rise to forerunners of eukaryotes, which evolved more than 1.2 billion years ago
Level of organization:	Single-celled	Protistans, single-celled or multicelled. Nearly all others multicelled; division of labor among differentiated cells, tissues, and often organs
Typical cell size:	Small (1–10 micrometers)	Large (10–100 micrometers)
Cell wall:	Most with no distinctive wall	Cellulose or chitin; none in animal cells
Membrane-bound organelles:	Rarely; no nucleus, no mitochondria	Typically profuse; nucleus present; most with mitochondria
Modes of metabolism:	Both anaerobic and aerobic	Aerobic modes predominate
Genetic material:	One chromosome; plasmids in some	Chromosomes of DNA plus many associated proteins in a nucleus
Mode of cell division:	Prokaryotic fission, mostly; some reproduce by budding	Nuclear division (mitosis, meiosis, or both) associated with one of various modes of cytoplasmic division

4. _____ have contractile vacuoles.
   - a. Slime molds
   - b. Amoeboid protozoans
   - c. Ciliates
   - d. both b and c

5. The "cell from hell" is a _____ .
   - a. parasitic protozoan
   - b. dinoflagellate
   - c. water mold
   - d. pathogenic red alga

6. Algin is used in ice cream, pudding, salad dressing, jelly beans, beer, cough syrup, toothpaste, cosmetics, and other products. Certain _____ are sources of algin.
   - a. green algae
   - b. brown algae
   - c. red algae
   - d. dinoflagellates

7. Stramenopiles (water molds, chrysophytes, and brown algae) share a unique derived trait, in that _____ project from one of their two flagella.
   - a. tinselly filaments
   - b. brown cilia
   - c. forked ends
   - d. cysts

8. Most water molds are _____ decomposers.
   - a. parasitic
   - b. saprobic
   - c. autotrophic
   - d. chemosynthetic

9. _____ are free-living, amoebalike cells that crawl on rotting plant parts and engulf bacteria, spores, and organic compounds.
   - a. Water molds
   - b. Amoeboid protozoans
   - c. Sporozoans
   - d. Slime molds

10. Match the terms with the most suitable descriptions.
   - ____ binary fission
   - ____ red tide
   - ____ water mold
   - ____ *Plasmodium*
   - ____ plankton
   - ____ green algae
   - a. dinoflagellate bloom along coast
   - b. decomposer; recycles nutrients
   - c. agent of malaria
   - d. closest relatives of plants
   - e. asexual reproduction mode
   - f. aquatic community, drifters or weak swimmers, microscopic

Critical Thinking

1. Suppose you vacation in a developing country. Sanitation and standards of personal hygiene are poor. Having read about parasitic protozoans in water and damp soil, what would you consider safe to drink? Which foods may be best to avoid or which food preparation methods might make them safe to eat?

2. As you read in this chapter, input from heavily fertilized cropland or raw sewage fans algal blooms. Think about it.

Do you accept massive kills of aquatic species, birds, and other forms of wildlife as an unfortunate but necessary side effect of human life?

If you find the environmental cost unacceptable, how would you stop the pollution? And assuming that you could stop it, what sorts of measures would you take to help feed the huge numbers of individuals that make up the human population? Bear in mind that our population is heavily dependent on high-yield (and very heavily fertilized) crops. How do you suggest we dispose of the accumulation of fecal matter and other wastes of more than 6.2 billion people?

Selected Key Terms

actinopod 22.3
algal bloom 22.7
alveolate 22.4
amoeboid protozoan 22.3
binary fission 22.4
brown alga 22.9
chrysophyte 22.9
ciliate 22.4
coccolithophore 22.9
conjugation (protozoan) 22.4
contractile vacuole 22.4
cyst 22.2
diatom 22.9
dinoflagellate 22.7
diplomonad 22.2
downy mildew 22.8
euglenoid 22.1

golden alga 22.9
green alga 22.10
kinetoplastid 22.2
parabasilid 22.2
pellicle 22.1
plankton 22.3
protistan CI
pseudopod 22.3
red alga 22.11
red tide 22.7
rhizopod 22.3
slime mold 22.12
sporozoan 22.5
stramenopile 22.8
water mold 22.8
yellow-green alga 22.9

Readings

Margulis, L. 1993. *Symbiosis in Cell Evolution*. Second edition. New York: Freeman. Paperback.

Margulis, L., and K. Schwartz. 1992. *Five Kingdoms*. Second edition. New York: Freeman. Paperback.

Satchell, M. 28 July 1997. "The Cell From Hell." *U.S. News and World Report*, 26–28.

Scagel. R. et al. 1984. *Plants: An Evolutionary Survey*. Belmont, California: Wadsworth.

On-Line readings at Student Guide for InfoTrac: www.brookscole.com/biology

PLANTS

Pioneers In a New World

Seven hundred million years ago, no shorebirds stirred and noisily announced the dawn of a new day. There were no crabs to clack their claws together and skitter off to burrows. The only sounds were the rhythmic, muffled thuds of waves in the distance, at the outer limits of another low tide. More than 3 billion years before, life had originated in the waters of the Earth. Now, quietly, the invasion of the land was under way.

Why did it happen then? Astronomical numbers of photosynthetic cells had come and gone, and oxygen-producing types had changed the atmosphere. High above the Earth, the sun's energy had converted much of the oxygen into a dense ozone layer—a shield against lethal doses of ultraviolet radiation. Until then, life had remained beneath the surface of water and mud.

Were cyanobacteria the first to adapt to intertidal zones, where mud dried out with each retreating tide? Were they the first in the shallow, freshwater streams flowing down to the coasts? Probably. Fossils suggest that later in time, some green algae and fungi made the journey together. Ancient species of green algae that lived near the water's edge or made it onto land gave rise to plants. And through millions of years thereafter, descendant plants became the basis of communities in coastal lowlands, near the snow line of high mountains, and in just about all places in between (Figure 23.1).

We have tantalizing fossils of those first pioneers. We are learning more about them through comparative studies of many modern species. Today, as in the late precambrian, cyanobacteria and green algae grow as mats in nearshore waters and freshwater streams, of the sort shown in Figure 23.1a. After a volcano erupts or a glacier slowly retreats, cyanobacteria are the first to colonize the barren rocks. Symbiotic interactions between green algae and fungi follow. Their organic wastes and remains accumulate and create pockets of soil. Mosses and other plants soon become established in the newly forming soil and further enrich it.

With this chapter we turn to the plants. Nearly all species are multicelled photoautotrophs. They build organic compounds by absorbing energy from the sun, carbon dioxide from the air, and minerals from water. These metabolic wizards also split water molecules. In doing so, they get stupendous numbers of the electrons and hydrogen atoms necessary for their growth into multicellular forms as tall as redwoods, as vast as an aspen forest that is one interconnected clone.

We know of at least 295,000 kinds of existing plants. Be glad their ancient ancestors left the water. Without those pioneers in a new world, we humans and other land-dwelling animals never would have made it onto the evolutionary stage.

Figure 23.1 (**a**) Filaments of a green alga in a shallow stream. More than 400 million years ago, green algae that may have been ancestral to all plants lived in similar streams that meandered down to the shores of ancient continents. Some land-dwelling descendants of ancestral forms: (**b**) Moss growing on rocks. (**c**) Ponderosa pine. (**d**) Flowers of one of the most prized of all flowering plants—orchids—which are native to tropical rain forests. With this chapter, we turn to the beginning—and end of the line—of some ancient lineages.

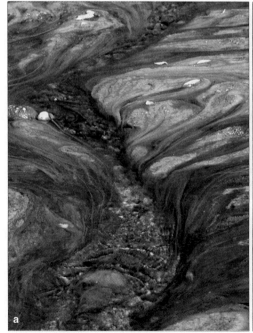

1. With rare exceptions, the plant kingdom consists of multicelled photoautotrophs. From earlier chapters, you know that chlorophylls *a* and *b* are the main photosynthetic pigments in plants. In this respect they are like green algae, their closest relatives.

2. Unlike their algal ancestors, which were adapted to aquatic habitats, most existing plants live on land.

3. Many aspects of plant structure are correlated with the interception of sunlight, absorption of water and mineral ions, and conservation of water. Lignin-reinforced tissues help support upright growth. Root systems mine the soil for water and ions, and internal tissues conduct water and solutes to all of the living cells in belowground and aboveground parts.

4. Reproductive adaptations help plants withstand dry periods. During the life cycle, a sporophyte develops roots, stems, and leaves, and forms gametes. It retains its gametes, supplying them with water and nutrients. It also disperses each new generation in ways that are responsive to the prevailing conditions in its habitat.

5. Early genetic divergences gave rise to bryophytes, then seedless vascular plants, and then seed-bearing vascular plants. Of these categories, the seed producers were the most successful at radiating into drier, and higher, environments.

6. Seed-bearing vascular plants called gymnosperms include the cycads, ginkgos, gnetophytes, and conifers. The angiosperms, another group of vascular plants, bear flowers as well as seeds. Today, three classes of flowering plants are recognized. They are informally called the magnoliids, monocots, and eudicots.

[handwritten annotations at top: "Non vascular - bryophytes / Vascular / Seedless / Seed bearing / No flowers / Flowers" and "in addition to cellulose" and "water proofing"]

TRENDS IN PLANT EVOLUTION

Today the plant kingdom consists of at least 295,000 kinds of photoautotrophs and a few heterotrophs. *Yet within all of that diversity, we find recurring structures that correlate with present and past functions.*

Vascular plants, with internal tissue systems that conduct water and solutes through roots, stems, and leaves, are the most diverse. Fewer than 19,000 species are *non*vascular plants called **bryophytes**. We group these as liverworts, hornworts, and mosses.

Whisk ferns, lycophytes, horsetails, and ferns are *seedless* vascular plants. Cycads, ginkgos, gnetophytes, and conifers are a group of *seed-bearing* vascular plants called **gymnosperms**. The **angiosperms**, another group of vascular plants, bear flowers and seeds. Magnoliids, eudicots, and monocots are the three major lineages of flowering plants.

Like bacterial and protistan photoautotrophs, the plants are producers for ecosystems. Their ancestors evolved in water more than 700 million years ago. About 260 million years later, simple stalked plants were growing in muddy sediments along coasts and streams. Then the evolutionary pace picked up; it took only 60 million years or so for plants to radiate into diverse land habitats, through long-term changes in structure, function, and modes of reproduction.

Evolution of Roots, Stems, and Leaves

Simple underground structures started to form when plants colonized the land, and they evolved into root systems in vascular plant lineages. Most **root systems** have many underground absorptive structures which, taken together, have a large surface area. They rapidly take up soil water and dissolved mineral ions. In many species, the root system also anchors the plant.

Aboveground, **shoot systems** evolved. Their stems and leaves absorb energy from the sun and carbon dioxide from the air. Stems grew taller and branched

only after plants developed the biochemical capacity to synthesize and deposit **lignin**, an organic compound, in cell walls. Collectively, lignin-strengthened cell walls structurally support stems, which grow in patterns that increase the total light-intercepting surface of leaves.

In many lineages, cellular pipelines that conduct water and solutes contributed to the evolution of roots, stems, and leaves. They evolved as components of two vascular tissues, xylem and phloem. **Xylem** distributes water and dissolved ions in plants; **phloem** distributes dissolved sugars and other photosynthetic products.

Life on land also depended on water conservation, which had not been a problem in most aquatic habitats. Shoots became protected by a **cuticle**, a waxy coat that helps conserve water on hot, dry days. Also, **stomata** (singular, stoma), tiny openings across the surfaces of leaves and some stems, evolved. They control carbon dioxide absorption and restrict evaporative water loss. Later chapters describe these tissue specializations.

From Haploid to Diploid Dominance

Life cycles changed as early plants radiated into higher, drier places. Think of algal gametes, which can meet up only in liquid water. Remember **gametophytes**, the gamete-producing bodies described in Section 10.5? They dominate the *haploid* (n) phase of algal life cycles. The diploid ($2n$) phase simply is the zygote.

Now look at Figure 23.2. *The diploid phase dominates most plant life cycles.* After a diploid zygote forms at fertilization, mitotic cell divisions and cell enlargements change it into a **sporophyte**, a multicelled, diploid plant body (for example, a pine tree). After some of its cells undergo meiosis, haploid resting cells—**spores**—form. (Sporophyte means spore-producing body.) Mitotic cell divisions of spores produce the gametophytes.

The shift to diploid dominance was an adaptation to land. Most habitats on land show seasonal changes in the availability of free water and nutrients. Natural selection probably favored the further development of root systems. It also must have favored mycorrhizae in which young roots interact with fungal symbionts; the interaction helps plants take up water and mineral ions even during dry seasons (Section 24.4).

Unlike algae and bryophytes, vascular plants have sporophytes that are structurally more complex and larger than their gametophytes. Especially among the

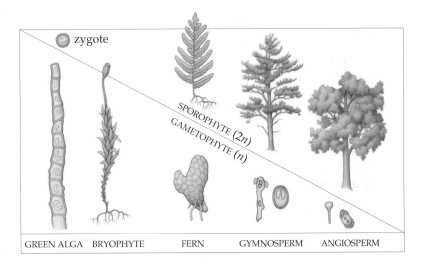

zygote

SPOROPHYTE (2n)
GAMETOPHYTE (n)

GREEN ALGA   BRYOPHYTE   FERN   GYMNOSPERM   ANGIOSPERM

Figure 23.2 Evolutionary trend among plants, from gametophyte (haploid) dominance to sporophyte (diploid) dominance in the life cycle. These representative species range from a green alga (*Ulothrix*) to a flowering plant. The trend occurred as early plants were colonizing habitats on land. See also Section 10.5.

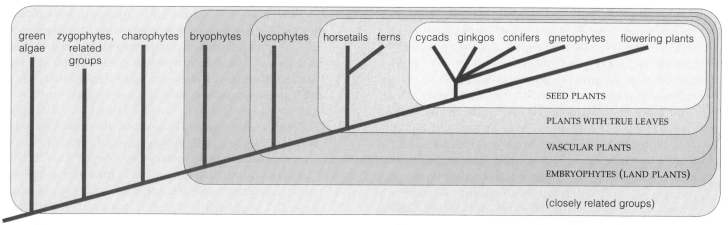

Figure 23.3 Evolutionary tree diagram for plants. Notice the nested monophyletic groups.

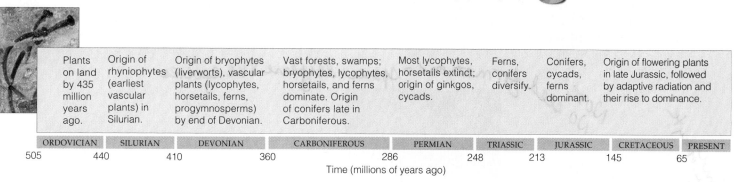

| Plants on land by 435 million years ago. | Origin of rhyniophytes (earliest vascular plants) in Silurian. | Origin of bryophytes (liverworts), vascular plants (lycophytes, horsetails, ferns, progymnosperms) by end of Devonian. | Vast forests, swamps; bryophytes, lycophytes, horsetails, and ferns dominate. Origin of conifers late in Carboniferous. | Most lycophytes, horsetails extinct; origin of ginkgos, cycads. | Ferns, conifers diversify. | Conifers, cycads, ferns dominant. | Origin of flowering plants in late Jurassic, followed by adaptive radiation and their rise to dominance. |

ORDOVICIAN	SILURIAN	DEVONIAN	CARBONIFEROUS	PERMIAN	TRIASSIC	JURASSIC	CRETACEOUS	PRESENT
505	440	410	360	286	248	213	145	65

Time (millions of years ago)

Figure 23.4 Milestones in plant evolution. The photograph shows fossils of *Cooksonia*, one of the first vascular plants.

gymnosperms and, later, angiosperms, the sporophyte became the dominant phase of the life cycle. It could retain, nourish, and protect developing gametophytes and the young sporophytes. What was the advantage? *Fertilization and dispersal of each new generation could be timed with arrival of suitable environmental conditions.*

Evolution of Pollen and Seeds

Like some seedless species, seed-bearing plants produce two types of spores. We call this condition *hetero*spory, as opposed to *homo*spory (one type of spore). In both gymnosperms and angiosperms, macrospores give rise to female gametophytes in which eggs form and get fertilized. Microspores (smaller spores) are the start of **pollen grains**, cellular structures that become mature, sperm-bearing male gametophytes. Pollen grains reach eggs via air currents, insects, birds, and so on; they do not require free-standing water to reach them. In this respect they differ greatly from algae. The evolution of pollen grains contributed to the successful radiation of seed-bearing plants into high and dry habitats.

In drier habitats, seed production had advantages. The female gametophytes of seed-bearing plants form within sporophyte tissues. Each **seed** has an embryo sporophyte, nutritious tissues, and an outer coat. It helps the embryo sporophyte survive through seasons that do not favor growth. It was no coincidence that seed plants rose to dominance during Permian times, when shifts in climate were extreme.

Before turning to the spectrum of diversity among plants, review Figures 23.3 and 23.4. You can use them as maps of the branching evolutionary roads.

The plant kingdom includes multicelled, photosynthetic species called bryophytes, seedless vascular plants, and seed-bearing vascular plants. Most of these live on land.

Most plant lineages became structurally adapted to life on land. They have root and shoot systems, a waxy cuticle, stomata, vascular tissues, and lignin-reinforced tissues.

Sporophytes with well-developed roots, stems, and leaves came to dominate the life cycle of most land plants. Parts of these complex sporophytes nourish and protect fertilized eggs and embryos until conditions favor their growth.

Some plants started to produce two types of spores, not one. This led to the evolution of male gametes well adapted for dispersal without liquid water and to the evolution of seeds.

THE BRYOPHYTES

Modern bryophytes include 18,600 species of **mosses**, **liverworts**, and **hornworts**. These nonvascular plants are mostly well adapted to grow in fully or seasonally moist habitats (Figures 23.1 and 23.5). Yet you will find certain mosses growing slowly in deserts, even on the windswept plateaus of Antarctica. Mosses particularly are sensitive to air pollution. In regions where the air quality is poor, mosses are few or absent.

Bryophytes are less than twenty centimeters (eight inches) tall. Their leaflike, stemlike, and rootlike parts have no xylem or phloem; hence the name nonvascular plants. Like some algae and lichens, they can dry out, then revive after absorbing water. Most have rhizoids: elongated cells or threadlike absorptive structures that also attach the gametophytes to the soil.

Bryophytes are the simplest plants to display three features that evolved in early land plants. *First*, a cuticle prevents water loss from aboveground parts. *Second*, a cellular jacket around the parts that produce sperm and

eggs holds in moisture. *Third*, their gametophytes are notably larger than the sporophytes, and they dominate the life cycle. Embryo sporophytes start development inside some gametophyte tissues. Even when they are mature, they are not dispersed. They remain *attached* to the gamete-producing body and draw some nutritional support from it.

With 10,000 species, mosses are the most common bryophytes. The gametophytes of some species grow in clusters and form low, cushiony mounds (Figure 23.1*b*). Those of others commonly grow in branched, feathery patterns on tree trunks and branches in humid climates. Eggs and sperm develop in tiny, jacketed vessels at the shoot tips of gametophytes. Sperm reach the eggs by moving through moisture that clings to plants. Zygotes give rise to sporophytes, each composed of a stalk and a jacketed structure in which spores develop.

Figure 23.6 shows one of 350 kinds of peat mosses (*Sphagnum*). Whereas most bryophytes grow slowly, the

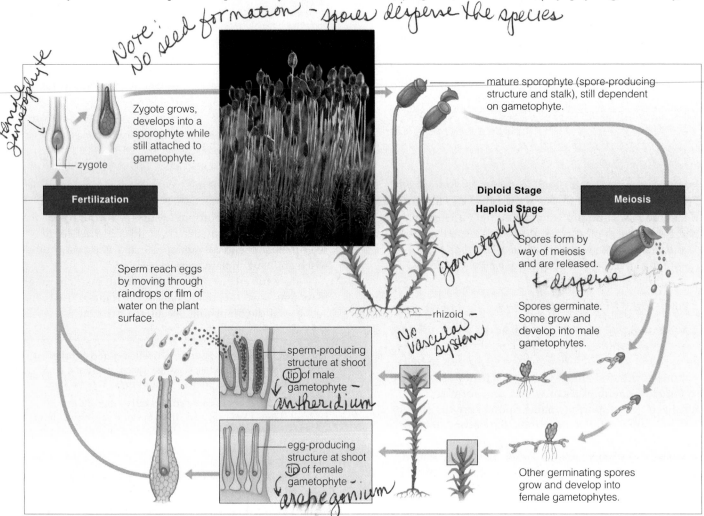

Figure 23.5 Life cycle of a moss (*Polytrichum*), a bryophyte. The moss sporophyte remains attached to the gametophyte, which provides it with nutrients and water.

Figure 23.6 (**a**) Peat moss (*Sphagnum*). A few sporophytes, the brown, jacketed structures on white stalks, are attached to a gametophyte. (**b**) Peat bog in Ireland. This family is cutting out blocks of peat and stacking them to dry as a home fuel source. Enough peat also is harvested in Ireland to generate electricity in peat-burning power plants.

sporophyte

gametophyte

a

a

female gametophyte   male gametophyte

thallus (leaflike part)   gemma

b

Figure 23.7 *Marchantia*, one of the liverworts. (**a**) Like other liverworts, this nonvascular plant reproduces sexually. Unlike the other types, it forms male and female reproductive parts on different plants. Shown here, a sexually reproducing population. (**b**) *Marchantia* also reproduces asexually by way of gemmae, multicelled vegetative bodies that develop in tiny cups on the plant body. Gemmae grow and develop into individual plants after splashing raindrops transport them to suitable sites.

peat mosses grow fast enough to yield twelve metric tons of organic matter per hectare annually, or about twice as much as corn plants yield. Nova Scotia, for example, is exporting 200,000 bales annually to Japan and the United States. Peat soaks up five times as much water as cotton does, owing to the large, dead cells in their leaflike parts. The acids that peat mosses produce hamper growth of bacterial and fungal decomposers. Because of their high absorbency and good antiseptic properties, peat was used as an emergency poultice on the wounds of soldiers during World War I.

The remains of peat mosses slowly accumulate into compressed, exceedingly moist mats called **peat bogs**. In cold and temperate regions, peat bogs cover an area equal to one-half of the United States. Only the most acid-tolerant plants, including cranberries, blueberries, larch, and Venus flytraps, can grow in the bogs, which can be as acidic as vinegar.

Nearly all peat harvested and dried in Ireland and elsewhere is burned to generate electricity in power plants. Compared to coal burning, peat fires generate fewer pollutants. Every so often, peat harvesters come across exceptionally well-preserved bodies of humans who lived about 2,000 to 3,000 years ago. The highly acidic bogs, which kept the bodies from decomposing, apparently were sites of ceremonial human sacrifices.

So as not to dwell on the macabre, let us leave this section with the liverworts and their interesting ways of reproducing, as described in Figure 23.7.

Bryophytes are nonvascular plants with flagellated sperm that require liquid water to reach and fertilize the eggs.

In these plants, a sporophyte develops within gametophyte tissues. It remains attached to the gametophyte and receives some nutritional support from it.

EXISTING SEEDLESS VASCULAR PLANTS

Figure 23.4 shows a fossil of an early seedless vascular plants. Descendants of certain lineages are still with us; we call them **whisk ferns**, **lycophytes**, **horsetails**, and **ferns**. Like their ancestors, they differ from bryophytes in three key respects. The sporophyte does not remain attached to a gametophyte, it has true vascular tissues, and it is the larger, longer lived phase of the life cycle.

Most seedless vascular plants reside in wet, humid places. Their gametophytes lack vascular tissues. The only way flagellated sperm can reach eggs is to swim through water droplets. A few species in dry habitats reproduce sexually only during brief, seasonal pulses of heavy rains. In one sense, whisk ferns, lycophytes, horsetails, and ferns are the "amphibians" of the plant kingdom. None has fully escaped the aquatic habitats of their ancestors.

Lycophytes

About 350 million years ago, lycophytes (Lycophyta) included tree-sized members of swamp forests. About 1,100 far tinier species exist today. The most familiar are club mosses, members of communities in the Arctic, the tropics, and regions between. Sporophytes of most club mosses have leaves and a branching rhizome that gives rise to vascularized roots and stems.

Sporophytes of *Lycopodium* are informally known as ground pines, partly because they resemble miniature Christmas trees complete with "cones" (Figure 23.8*a*). These are not the same as pine cones. But they, too, are reproductive structures formed from modified leaves called sporophylls. These leaves bear hollow chambers in which spores form. Each chamber is one sporangium (plural, sporangia). In this case, sporophylls form tight clusters on the ends of stems. Conelike reproductive structures made of modified leaves are characteristic of many groups. Each is a one **strobilus** (plural, strobili). *Selaginella* is a lycophyte group that produces two kinds of spores in the same strobilus; it is heterosporous.

Spores form by meiosis. After their dispersal, they germinate and become small, free-living gametophytes.

Whisk Ferns

Whisk ferns (Psilophyta) are not actually ferns. Florist suppliers in Hawaii, Louisiana, Florida, Puerto Rico, and other tropical and subtropical areas cultivate these plants, which look a bit like whisk brooms. One kind, *Psilotum*, is a unique vascular plant. It probably evolved from an ancestral species that formed roots, but now its sporophytes have none. They have **rhizomes**: short, branched, mainly horizontal absorptive stems that grow underground. Small outgrowths on the photosynthetic branched stems are probably reduced leaves. All of the plants have xylem and phloem (Figure 23.8*b*).

Horsetails

Tree-sized sphenophytes (Sphenophyta) flourished in ancient swamp forests. Twenty-five or so smaller species

Figure 23.8 (**a**) Sporophyte of a lycophyte (*Lycopodium*). (**b**) Sporophytes of a whisk fern (*Psilotum*), a seedless vascular plant. Tips of stubby branchlets bear pumpkin-shaped spore-producing structures. (**c**) Vegetative stem of *Equisetum*. It loosely resembles a horsetail. (**d**) Nonphotosynthetic, fertile stems of *Equisetum*. At their tip is a strobilus, a cluster of spore-producing structures. (**e**) Closer look at a fertile stem. Each petal-shaped part holds many spores, formed by way of meiosis.

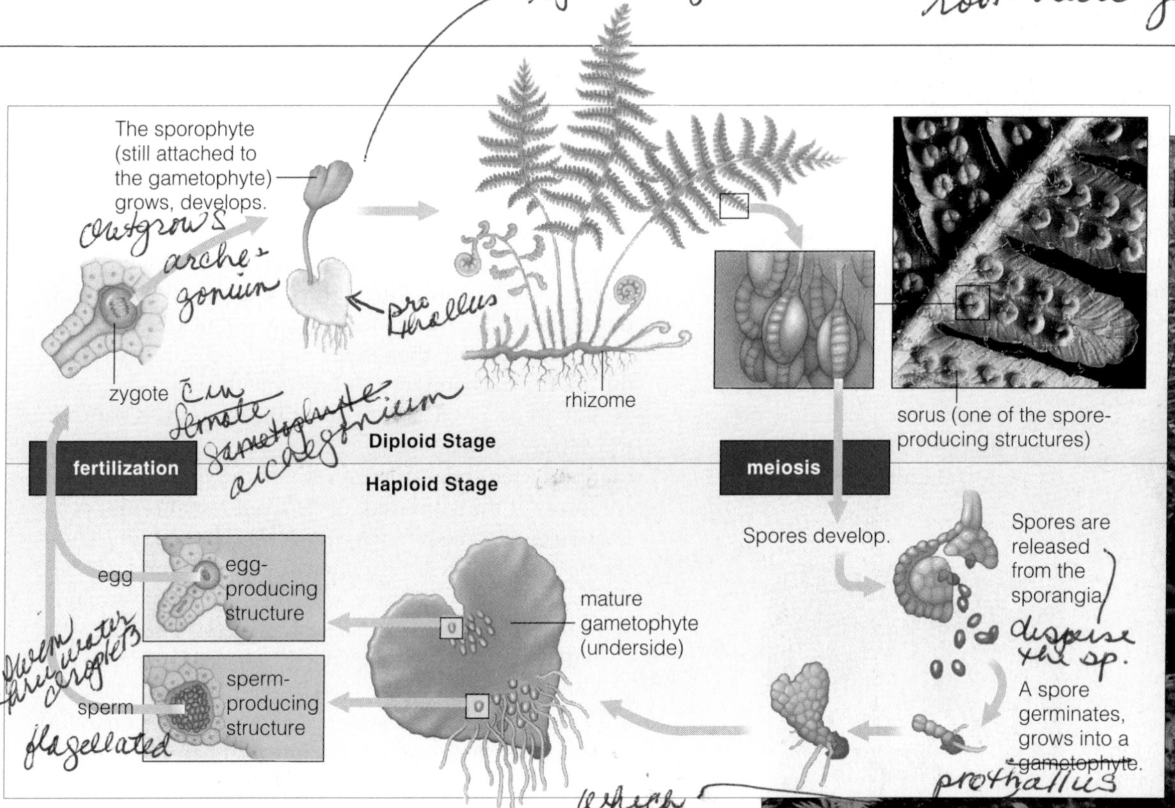

Figure labels (within figure):

The sporophyte (still attached to the gametophyte) grows, develops.

sporophyte develops a rhizome from which the root-bearing fronds develop [handwritten]

zygote *in female gametophyte archegonium* [handwritten]

outgrows archegonium [handwritten]

prothallus [handwritten]

rhizome

sorus (one of the spore-producing structures)

fertilization

Diploid Stage

Haploid Stage

meiosis

egg

egg-producing structure

between thin water droplets [handwritten]

sperm

sperm-producing structure

flagellated [handwritten]

mature gametophyte (underside)

Spores develop.

Spores are released from the sporangia

disperse the sp. [handwritten]

A spore germinates, grows into a gametophyte.

prothallus [handwritten]

Figure 23.9 Life cycle of a fern. The photograph shows ferns *which bears archegonia + antheridia on underside* [handwritten] growing in a moist habitat in Indiana. Ferns with finely divided fronds are in the foreground.

of one genus, *Equisetum*, made it to the present. These are the horsetails, and their body plan has not changed much over the past 300 million years.

Horsetails thrive in streambank muds, vacant lots, roadsides, and other disrupted habitats. Figure 23.8c–e shows the vegetative photosynthetic stems and fertile stems of one species. Its spores give rise to free-living gametophytes 1 millimeter to 1 centimeter across. The sporophytes of most horsetails have rhizomes, hollow photosynthetic stems, and scale-shaped leaves. Stems contain a ringlike array of xylem and phloem strands. Ribs reinforced with silica structurally support the stems and give them a texture like sandpaper. Pioneers of the American West, who did not have many places to store and wash towels, gathered horsetails on their westward journeys and used them as pot scrubbers.

Ferns

With 12,000 or so species, the ferns (Pterophyta) are the largest and the most diverse group of seedless vascular plants. All but about 380 are native to the tropics, but you find them in gardens and in homes throughout the world. Their size range is stunning. The leaves of some floating types are less than 1 centimeter across; some of the tropical tree ferns are 25 meters (82 feet) tall. One climbing fern has a modified leaf stalk 30 meters long.

Most ferns have vascularized rhizomes that give rise to roots and leaves. Exceptions include tropical tree ferns and epiphytes. (*Epiphyte* refers to any aerial plant that grows attached to tree trunks or branches.) While they develop, young fern leaves (fronds) are coiled in a way that resembles a fiddlehead. When mature, these fronds commonly are divided into leaflets.

You may have noticed rust-colored patches on the lower surface of many fern fronds. Each patch is one cluster of sporangia, of a type known as a sorus (plural, sori). At dispersal time, sporangia snap open with such force that spores catapult through the air. Germinating spores develops into a small gametophyte. One green, heart-shaped gametophyte is shown in Figure 23.9.

Seedless vascular plants (whisk ferns, lycophytes, horsetails, and ferns) have sporophytes adapted to conditions on land. Yet they have not entirely escaped their aquatic ancestry. Films of water must be available at the time of sexual reproduction if flagellated sperm are to reach the eggs.

Ancient Carbon Treasures

About three hundred million years ago, in the middle of the Carboniferous, mild climates prevailed and swamp forests carpeted the wet lowlands of the continents. The absence of pronounced seasonal swings in temperature favored plant growth through much of the year. The plants having lignin-reinforced tissues and well-developed root and shoot systems had the competitive edge under these growth conditions. Some evolved into giants. Massively stemmed lycophyte trees—the giant club mosses—topped out at nearly forty meters (Figure 23.10).

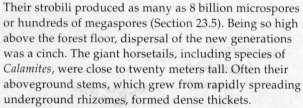

Lepidodendron

Their strobili produced as many as 8 billion microspores or hundreds of megaspores (Section 23.5). Being so high above the forest floor, dispersal of the new generations was a cinch. The giant horsetails, including species of *Calamites*, were close to twenty meters tall. Often their aboveground stems, which grew from rapidly spreading underground rhizomes, formed dense thickets.

As it happened, the sea level rose and fell fifty times during the Carboniferous. Each time the sea receded, the swamp forests flourished. When the sea moved back in, forest trees became submerged and buried in sediments that protected them from decay. Gradually the sediments compressed the saturated, undecayed remains into peat. Each time more sediments accumulated, they subjected the peat to increased heat and pressure that made it even more compact. In this way, compressed organic remains were transformed into great seams of **coal** (Figure 23.10).

With its high percentage of carbon, coal is energy rich and is one of our premier "fossil fuels." It took a fantastic amount of photosynthesis, burial, and compaction to form each major seam of coal in the Earth. It has taken us only a few centuries to deplete much of the world's known coal deposits. Often you will hear about annual production rates for coal or some other fossil fuel. But how much do we really produce each year? None. We simply *extract* it from the Earth. Coal is a nonrenewable source of energy.

stem of a giant lycophyte (*Lepidodendron*)

seed fern (*Medullosa*); probably related to the progymnosperms, which may have been among the earliest seed-bearing plants

stem of a giant horsetail (*Calamites*)

Figure 23.10 Reconstruction of a Carboniferous forest. The boxed inset shows part of a seam of coal.

THE RISE OF SEED-BEARING PLANTS

Seed-bearing plants arose about 360 million years ago, as the Devonian period gave way to the Carboniferous. In terms of diversity, numbers, and distribution, they would become the most successful groups of the plant kingdom. Seed ferns, gymnosperms, and (much later) angiosperms became dominant. They all differed from seedless vascular plants in three important respects.

First, seed-bearing plants produce pollen grains: sperm-bearing male gametophytes. Remember, these plants produce two types of spores. Their **microspores** give rise to pollen grains. Unlike the spores of seedless vascular plants, they do not have a "tetrad scar," which marks the cleavage planes between four spores that form during meiotic cell division (Figure 23.11).

Like a suitcase, a pollen grain is a means of getting its contents (the sperm) to the eggs, even during times of prolonged drought. Seedless vascular plants do not have such an advantage; without predictable rains and moisture, their sperm simply cannot reach the eggs, and this has adverse effects on reproductive success. By contrast, pollen grains of gymnosperms simply drift with air currents. Those of angiosperms also are loaded onto insects, birds, bats, and other animals that carry them to eggs. **Pollination** is the name for the arrival of pollen grains on the female reproductive structures. By this process, seed-bearing plants escaped dependence on free water for fertilization.

Second, seed-bearing plants produce **megaspores** as well as microspores. These develop within **ovules**, the female reproductive structures which, at maturity, are seeds (Figure 23.12). Each ovule consists of a female gametophyte (with egg cell), nutrient-rich tissue, and a jacket of cell layers which, recall, develops into the seed coat. One zygote will form inside the ovule when a sperm reaches and then fertilizes the egg. An embryo sporophyte will develop, and when the time comes to depart from the parent plant, the coat around the seed will help protect the embryo during its journey.

Third, compared with seedless vascular plants, the gymnosperms enjoyed water-conserving traits, such as a thicker cuticle and stomata recessed below the leaf surface. These traits gave gymnosperms a competitive advantage in drier, cooler climates. Such conditions were ushered in as the Carboniferous gave way to the Permian. Before then, **seed ferns** of the type shown in Figure 23.10 rose to dominance, and they prevailed for about 70 million years. The seed ferns probably have evolutionary links with **progymnosperms**, which were among the earliest plants to produce seedlike structures or seeds. When the global climate became cooler and drier, swamplands disappeared. So did the seed ferns. New kinds of seed-bearers—the cycads, conifers, and other kinds of gymnosperms—evolved and replaced them.

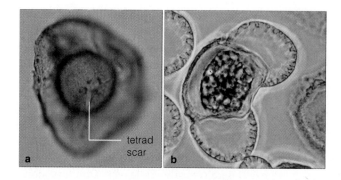

Figure 23.11 (**a**) From the Devonian-Carboniferous boundary, a fossilized spore of a lycophyte. Its tetrad scar is typical of the spores of seedless plants. (**b**) A pine pollen grain (*Pinus*). Like the pollen of other seed-bearing plants, it has no tetrad scar.

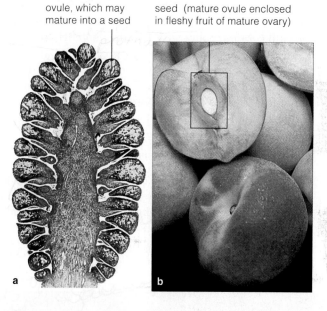

Figure 23.12 (**a**) Longitudinal section through a seed-bearing pine cone. These seeds are exposed on the cone scales; they are not protected by any sporophyte tissues. (**b**) A mature ovule, or seed, of a peach (*Prunus*), one of the angiosperms. It is protected by a seed coat, which is enclosed within the fleshy, edible tissue of the fruit.

In time, gymnosperms were the primary producers that directly or indirectly sustained most consumers, including the dinosaurs. What some folks call the Age of Dinosaurs, botanists call the Age of Cycads.

Seed-bearing plants rely on pollen grains, ovules that mature into seeds, and tissue adaptations to dry conditions.

GYMNOSPERMS—PLANTS WITH "NAKED" SEEDS

With a bit of history behind us, we turn now to a survey of some of the existing gymnosperms. Unlike seeds of flowering plants, which are enclosed inside a chamber called an ovary, gymnosperm seeds are perched, in an exposed way, on a spore-producing structure. (*Gymnos* means naked; *sperma* is taken to mean seed.)

Conifers

Conifers (Coniferophyta) are woody trees and shrubs that have needlelike or scalelike leaves. Conifers alone produce true **cones**. These reproductive structures are clusters of papery or woody scales that bear exposed ovules on their upper surface. The most abundant trees of the Northern Hemisphere (pines, as in Figure 23.1c) and the tallest (coast redwoods) are conifers. The oldest trees of all are bristlecone pines (Figure 23.13). One 4,725-year-old tree sprouted when the Egyptians were building the Great Sphinx. Other well-known conifers are the firs, yews, spruces, junipers, larches, cypresses, the bald cypress, podocarps, and the dawn redwood.

Most conifers shed some leaves all year long and remain leafy, or *evergreen*. A few are *deciduous*, meaning they shed all leaves in the fall. Their cones are clusters of modified leaves around spore-producing structures. Figure 23.14 gives examples.

Lesser Known Gymnosperms

CYCADS   About 100 species of **cycads** (Cycadophyta) made it to the present. Figure 23.14 shows examples of their pollen-bearing and seed-bearing cones, which form on separate plants. Beetles and air currents can transfer pollen from "male" plants to "female" plants. At first

Figure 23.13 Bristlecone pines (*Pinus longaeva*) growing, very slowly, near the timberline in the Sierra Nevada.

glance, you might mistake cycad leaves for those of a palm tree, but palms are flowering plants. Most of the cycads now inhabit tropical and subtropical areas. One species (*Zamia*) grows wild in Florida and is planted as an ornamental. Elsewhere, cycad seeds and the cycad trunks are ground into a flour that is toxic until its alkaloids are removed. Many cycad species are now vulnerable to extinction.

Figure 23.14 Gymnosperm reproductive structures. (**a**) Male pine strobili releasing pollen. (**b**) Female pine cone at the time of pollination. (**c**) From a juniper (*Juniperus*), cones with a berrylike appearance. These cones are made of fused-together, fleshy scales. (**d**) Pollen-bearing strobilus of a "male" cycad (*Zamia*). (**e**) Seed-bearing strobilus of a "female" cycad. Of all existing gymnosperms, cycads have the largest seed-bearing strobili. Some grow as long as one meter and weigh more than fifteen kilograms.

Figure 23.15 *Ginkgo biloba.* (**a**) Its fleshy-coated seeds and distinctive leaf shape. (**b**) Pollen-bearing strobili. (**c**) A fossilized ginkgo leaf that formed at the Cretaceous–Tertiary boundary. Today, 65 million years later, the leaf structure has changed little, if at all. (**d**) Ginkgo sporophyte in summer.

Figure 23.16 (**a**) Sporophyte of *Ephedra viridins* sporophyte. Closer look at (**b**) a pollen-bearing cone and (**c**) a seed-bearing cone of *Ephedra.* (**d**) Sporophyte of *Welwitschia mirabilis,* and (**e**) a close-up of some of its seed-bearing cones.

GINKGOS The **ginkgos** (Ginkgophyta) were a diverse group in dinosaur times. The only surviving species is the maidenhair tree, *Ginkgo biloba.* Like a few species of larch and some other gymnosperms, these plants are deciduous. Several thousand years ago, ginkgo trees were widely planted around temples in China. Then the natural populations nearly became extinct, even though ginkgos seem hardier than many other trees. Perhaps they became tempting as a source of firewood.

Today, male ginkgo trees are widely planted. They have attractive, fan-shaped leaves and are resistant to insects, disease, and air pollutants. Figure 23.15 shows the thick, fleshy seeds of female trees, which are the size of plums. The trees are not favorites of gardeners. When stepped on, the seeds give off quite a stench.

GNETOPHYTES At present, there are three genera of woody plants known as the **gnetophytes** (Gnetophyta). Trees and leathery leafed vines of *Gnetum* thrive in the humid tropics. The shrubby *Ephedra* lives in California deserts and some other arid regions (Figure 23.16*a–c*). Photosynthesis proceeds in its green stems.

Welwitschia mirabilis grows in hot deserts of south and west Africa. Most of this plant's sporophyte is a deep-reaching taproot. The exposed portion is a woody disk-shaped stem. On the stem are cones and one or two strap-shaped leaves. These leaves split lengthwise repeatedly as the plant ages (Figure 23.16*d,e*).

Conifers, cycads, ginkgos, and gnetophytes are the major groups of modern-day gymnosperms.

Like their Permian ancestors, gymnosperms are adapted to dry climates. They bear seeds on the exposed surfaces of cones and other spore-producing structures.

A CLOSER LOOK AT THE CONIFERS

Before we leave the gymnosperms, let's use conifers as an example of reproductive strategies. Depending on the species, a conifer's life cycle lasts a year or more.

Pine Life Cycle

Most of us recognize a pine tree sporophyte when we see one (Figures 23.1 and 23.17). Female cones of these trees are strobili in which the megaspores develop into female gametophytes. In the male strobili, microspores form, and these will develop into pollen grains. Each spring, millions of pollen grains drift away from the male strobili. Pollination is completed when some land on the ovules of female cones. After each germinates, a tubular structure forms from it. A germinating pollen grain—the sperm-bearing male gametophyte—grows toward the egg in a female gametophyte. For the pines, fertilization occurs within a year after pollination.

Let's look inside an ovule. Once fertilized, an egg gives rise to an embryo sporophyte. Outer layers of a jacket around the embryo and the female gametophyte mature into a hard coat. The seed coat will protect the

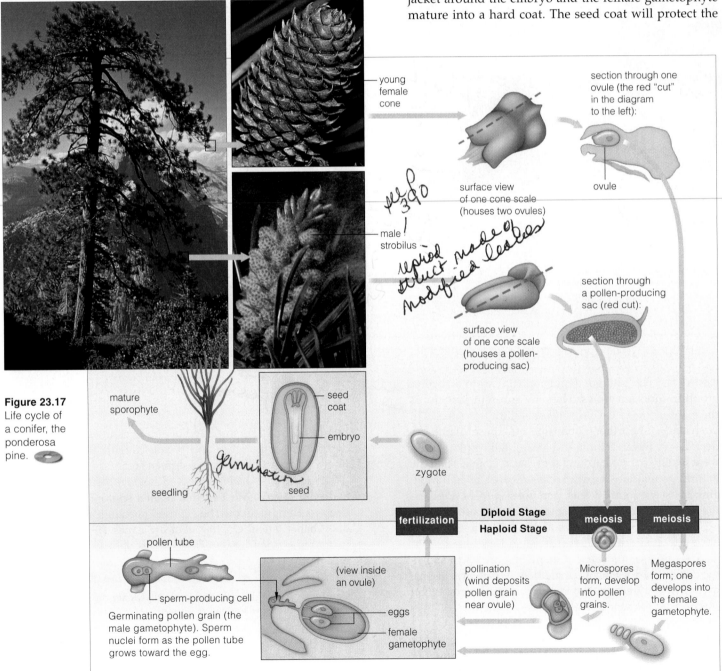

Figure 23.17 Life cycle of a conifer, the ponderosa pine.

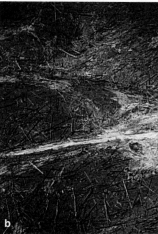

Figure 23.18 Sampling of the deforestation under way around the world. Only deforested tracts in North America are shown; Section 50.4 focuses on the rapidly disappearing tropical rain forests.

From America's heartland, logged-over acreage in Arkansas (**a**). From the eastern seaboard, one of the denuded bits of North Carolina (**b**). Clear-cut peaks in Washington (**c**) and Alaska (**d**).

These are not isolated examples. In the early 1980s, 400 million board feet of timber were being cut in the Olympic Peninsula of Washington every year. In Arkansas, about one-third of the Ouachita National Forest has been clear-cut. Its once-diverse forest communities were replaced by "tree farms" of a single species of pine. Throughout the world, huge tracts of land deforested years ago still show no signs of recovery.

embryo sporophyte after its dispersal from the parent plant. Nutrients inside will help it through the critical time of germination, before its roots and shoots become fully functional.

Deforestation and the Conifers

Conifers dominated many land habitats during the Mesozoic, but their slow reproductive pace put them at a competitive disadvantage when the flowering plants began their stunning adaptive radiation (Section 20.6). Coniferous forests still predominate in the far north, at higher elevations, and in some parts of the Southern Hemisphere. However, existing conifers face more than competition with flowering plants for resources. Now they are vulnerable to **deforestation**: the removal of all trees from large tracts, as by clear-cutting (Figure 23.18). Conifers just have the bad luck to be premier sources of lumber, paper, and other wood products required in human societies. We return to this topic in Chapter 50.

In places where flowering plants flourish, conifers are at a competitive disadvantage, partly because they take so long to reproduce. Rampant deforestation isn't helping them one bit, either.

ANGIOSPERMS—FLOWERING, SEED-BEARING PLANTS

Characteristics of Flowering Plants

Only angiosperms produce the specialized reproductive structures called **flowers** (Figure 23.19). *Angeion,* which means vessel, refers to the female reproductive parts at the center of a flower. The enlarged base of the "vessel" is the floral ovary, where ovules and seeds develop.

As you will read in Chapter 31, nearly all flowering plants coevolved with **pollinators**—insects, bats, birds, and other animals that withdraw nectar or pollen from a flower and transfer pollen to its female reproductive parts. Recruiting animals as assistants in reproduction contributed to the success of flowering plants, which have dominated the land for 100 million years.

At least 260,000 species of flowering plants thrive in diverse habitats. They range in size from duckweeds (a millimeter or so long) to towering *Eucalyptus* (some of these trees are more than 100 meters tall). A few rare species, including mistletoes and Indian pipe, are not even photosynthetic. They withdraw nutrients directly from other plants or from mycorrhizae.

Magnoliids, **monocots**, and **eudicots** (true dicots) are the three major groups of flowering plants. More ancient groups include the water lilies (Figure 23.19).

Magnoliids include magnolias, avocados, nutmeg, sassafras, and black pepper plants. Among the 170,000 dicots are most herbaceous (nonwoody) plants, such as cabbages and daisies; most flowering shrubs and trees, such as maple, oak, and apple trees; and cacti. Among the 80,000 or so monocots are orchids, palms, lilies, and grasses, such as rye, sugarcane, corn, rice, and wheat. Many other highly valued crop plants are monocots. Appendix I lists other examples.

Representative Life Cycle—A Monocot

The next unit deals with the structure and function of flowering plants. For now, simply start thinking about how a large sporophyte dominates life cycles. It retains and nourishes gametophytes, and disperses its sperm-bearing pollen grains. Endosperm, a nutritious tissue, surrounds the embryo sporophytes inside the seeds of

Figure 23.19 (**a**) Evolutionary tree diagram for flowering plant groups. Their reproductive structures—flowers—have roles in pollination and in seed formation. (**b**) A hummingbird pollinator withdrawing nectar from a passion flower (*Passiflora*). (**c**) Sacred lotus (*Nelumbo nucifera*), an aquatic species. The radial pattern of this flower is typical of ancient lineages. (**d**) More recently evolved lineages such as the pansies (*Viola*) have a bilateral pattern, with roughly equivalent left and right parts. (**e**) Indian pipe (*Monotropa uniflora*), a nonphotosynthetic species, draws nutrients from the mycorrhizae of some photosynthetic plants. (**f**) Dwarf mistletoe (*Arceuthobium*). It parasitizes plants directly and stunts the growth of forest trees in the western United States.

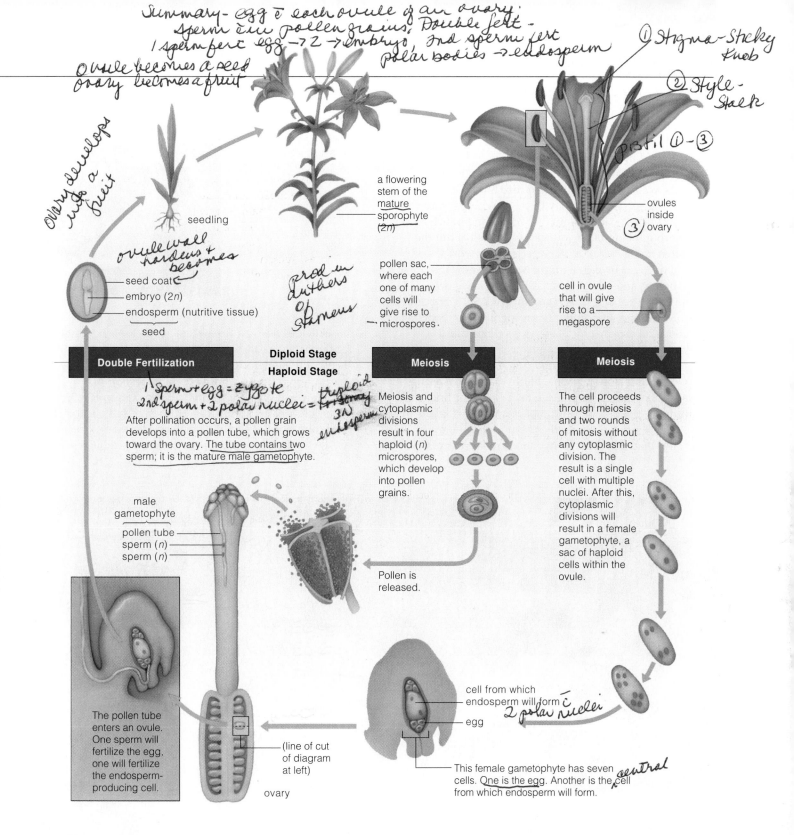

Handwritten annotations:

Summary- egg ē each ovule of an ovary.
Sperm in pollen grains. Double fert -
1 sperm fert egg → 2 → embryo. 2nd sperm fert
polar bodies → endosperm

Ovule becomes a seed
ovary becomes a fruit

① Stigma-Sticky Knob
② Style- Stalk
Pistil ①-③

Ovary develops into a fruit

ovule wall hardens + becomes

prod in anthers of stamens

1 Sperm + egg = zygote
2nd sperm + 2 polar nuclei = triploid 3n endosperm

③ ovules inside ovary

2 polar nuclei

central cell

Printed labels and text:

a flowering stem of the mature sporophyte (2n)

seedling

seed coat
embryo (2n)
endosperm (nutritive tissue)
seed

pollen sac, where each one of many cells will give rise to microspores.

cell in ovule that will give rise to a megaspore

Double Fertilization

Diploid Stage
Haploid Stage

Meiosis

Meiosis

After pollination occurs, a pollen grain develops into a pollen tube, which grows toward the ovary. The tube contains two sperm; it is the mature male gametophyte.

Meiosis and cytoplasmic divisions result in four haploid (n) microspores, which develop into pollen grains.

The cell proceeds through meiosis and two rounds of mitosis without any cytoplasmic division. The result is a single cell with multiple nuclei. After this, cytoplasmic divisions will result in a female gametophyte, a sac of haploid cells within the ovule.

male gametophyte
pollen tube
sperm (n)
sperm (n)

Pollen is released.

The pollen tube enters an ovule. One sperm will fertilize the egg, one will fertilize the endosperm-producing cell.

(line of cut of diagram at left)

ovary

cell from which endosperm will form c̄

egg

This female gametophyte has seven cells. One is the egg. Another is the cell from which endosperm will form.

Figure 23.20 Representative flowering plant life cycle. This example is for a lily (*Lilium*), one of the monocots.

"Double" fertilization is a distinctive feature of flowering plant life cycles. A male gametophyte delivers two sperm to an ovule. One sperm fertilizes the egg, and the other fertilizes a cell that gives rise to endosperm, a tissue that will nourish the forthcoming embryo. In most flowering plants, endosperm cells have three nuclei; endosperm cells in lilies have five. Sections 31.1 and 31.3 detail the flowering plant life cycles, using a true dicot as the example.

flowering plants. As seeds develop, ovaries (along with other structures) mature into fruits. Fruits protect and help disperse embryos. Figure 23.20 is a representative monocot life cycle. Section 31.3 has a eudicot life cycle.

Angiosperms are the most successful plants, in terms of diversity, numbers, and distribution. They alone produce flowers. Most species coevolved with animal pollinators.

SEED PLANTS AND PEOPLE

Which plants give taste thrills and which kill? Starting with trials and errors of the earliest human species, we have acquired intimate knowledge of plants. By 500,000 years ago, *Homo erectus* clans in China were stashing pine nuts, walnuts, hazelnuts, and rose hips in caves, and roasting seeds. At least by then, grown-ups were teaching children which plants are edible and which are toxic. Through language, youngsters learned about the plants that had evolved in their parts of the world.

By about 11,000 years ago, we were domesticating wheat, barley, and other plants, this being a way to get reliable quantities of food. Of an estimated 3,000 species that different human populations recognized as food, only about 200 became *the* major crops (Figure 23.21).

Plant lore still threads through our lives. We learned to use fast-growing, soft-wooded conifers for lumber and paper—and slow-growing, hard-wooded flowering plants such as cherry, maple, and mahogany for fine furniture. We make twine and rope from century plant leaves (*Agave*), cords and textiles from Manila hemp fibers, and thatched roofs from palm fronds and grass.

Figure 23.22 A bride tossing her bouquet to single women who attended her wedding, a ritual that supposedly reveals who will be the next to marry. (A sixty-eight-year-old optimist caught this one.)

Insecticides derived from Mexican cockroach plants kill cockroaches, fleas, lice, and flies. Extracts of neem tree leaves kill nematodes, insects, and mites but not the natural predators of those common pests.

Flowers grace homes and customs (Figure 23.22). Their oils impart scents to perfumes; oils of eucalyptus and camphor have medicinal uses. Digitalin extracted from foxglove (*Digitalis purpurea*) plants stabilizes the

Figure 23.21 Edible treasures from flowering plants. (**a**) A sampling of fruits, some of which we call "vegetables." All fruits function in seed dispersal. (**b**) From the American Midwest, mechanized harvesting under way in a field of common bread wheat, *Triticum*. (**c**) Triticale, a popular hybrid grain. Its parental stocks are wheat and rye (*Secale*). It combines the high yield of wheat with rye's tolerance of harsh climates.

(**d**) Indonesians gathering tender shoots of tea plants, evergreen shrubs related to camellias. Plants growing on hillsides in moist, cool regions yield leaves with the best flavors. Only the terminal bud and two or three of the youngest leaves are picked for the finest teas.

(**e**) From Hawaii, a field of sugarcane (*Saccharum officinarum*). Wild stock of this cultivated species may have evolved in New Guinea. Sap extracted from its cut stems is boiled down to make sucrose crystals (table sugar) and syrups.

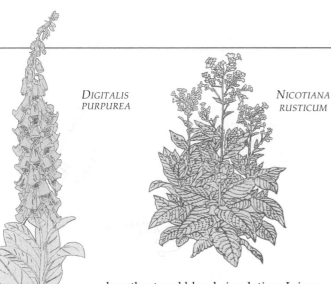

DIGITALIS PURPUREA

NICOTIANA RUSTICUM

HYOSCYAMUS

CANNABIS SATIVA

heartbeat and blood circulation. Juices from *Aloe vera* leaves soothe sun-damaged skin. Alkaloids from periwinkle leaves can even slow the growth of some cancer cells.

Have we, in all this time, also learned how to abuse plants? You bet. Through much of recorded history, people also figured out how to use leaves in harmful ways. The ancient Mayans cultivated tobacco plants (*Nicotiana tabacum* and *N. rusticum*) and introduced

European explorers to tobacco smoking. Mayan priests thought that the smoke rising from pipes carried their priestly thoughts upward, to the gods. People continue to smoke, chew, and tuck in their mouth tobacco plant leaves and are candidates for the hundreds of thousands of annual deaths from lung, mouth, and throat cancers. Heavy smoking of *Cannabis sativa*, source of marijuana and other mind-altering substances, is linked to low sperm counts. Cocaine, derived from coca leaves, has medicinal uses. It also is abused by millions around the world who are addicted to its mind-altering properties, with devastating social and economic effects.

ATROPA BELLADONNA

And what about that henbane (*Hyoscyamus niger*)? What about that belladonna? Their toxic alkaloids have been tapped for the occasional murder as well as for medicine. Remember Hamlet's father? As he slept, he was sneakily dispatched by someone who poured a solution of henbane into his ear. Remember Juliet's heartbroken, suicidal Romeo? After sipping a potion of nightshade, he quickly dropped dead.

Those self-proclaimed witches of the Middle Ages used henbane and atropine from nightshade during some of their suspect rituals. They used sticks to apply atropine solutions to their body. By doing so, they induced sensations of weightlessness. During their atropine-induced sprees, they quite mistakenly believed that they were flying off to rendezvous with demons. Hence those Halloween cartoons of witches flying hither and yon on their broomsticks.

Richly diverse uses and sometimes abuses of seed plants thread through human history.

SUMMARY

1. Ancient green algae apparently gave rise to plants. Plants invaded land by 435 million years ago. Nearly all species are multicelled photoautotrophs. Table 23.1 summarizes and compares major groups. *CI, 23.1*

2. Table 23.2 summarizes key trends in plant evolution as identified by comparisons among lineages: *23.1*

   a. Structural adaptations to dry habitats, including stomata, a cuticle, and vascular tissues (xylem, phloem).

   b. A shift from haploid to diploid dominance during the life cycle. Complex sporophytes evolved; they hold on to, nourish, and protect spores and gametophytes.

   c. A shift to two spore types, not one (homospory to heterospory). For gymnosperms and flowering plants, this led to the evolution of pollen grains and seeds.

3. Nearly all vascular plants live on land. Their cuticle and stomata conserve water. Root systems get nutrients from soil. The upright and branched growth patterns

---

Table 23.1 *Comparison of Major Plant Groups*

Nonvascular land plants. Fertilization requires free water. Haploid dominance. Cuticle, stomata present in some.

BRYOPHYTES	18,600 species. Moist, humid habitats.

Seedless vascular plants. Fertilization requires free water. Diploid dominance. Cuticle, stomata present.

LYCOPHYTES	1,100 species with simple leaves. Mostly wet or shady habitats.
WHISK FERNS	7 species, sporophytes with no obvious roots or leaves. *Psilotum.*
HORSETAILS	25 species of single genus. Swamps, disturbed habitats.
FERNS	12,000 species. Wet, humid habitats in mostly tropical, temperate regions.

Gymnosperms—vascular plants with "naked seeds." Free water not required for fertilization. Diploid dominance. Cuticle, stomata present.

CONIFERS	550 species, mostly evergreen, woody trees and shrubs having pollen- and seed-bearing cones. Widespread distribution.
CYCADS	185 slow-growing tropical, subtropical species.
GINKGO	1 species, a tree with fleshy-coated seeds.
GNETOPHYTES	70 species. Limited to some deserts, tropics.

Angiosperms—vascular plants with flower, protected seeds. Free water not required for fertilization. Diploid dominance. Cuticle, stomata present.

FLOWERING PLANTS

Monocots	80,000 species. Floral parts often arranged in threes or in multiples of three; one seed leaf; parallel leaf veins common.
Eudicots	At least 170,000 species. Floral parts often arranged in fours, fives, or multiples of these; two seed leaves; net-veined leaves common.

Magnoliids and basal groups

Table 23.2 *Evolutionary Trends Among Plants*

Bryophytes	Ferns	Gymnosperms	Angiosperms
Nonvascular ⟶ Vascular —————————————⟶			
Haploid dominance ⟶ Diploid dominance —————————⟶			
Spores of one type ⟶ Spores of two types ———————⟶			
Motile gametes ——————⟶ Nonmotile gametes* —⟶			
Seedless ——————————⟶ Seeds ——————⟶			

* Require pollination by wind, insects, animals, etc.

of shoot systems intercept sunlight and carbon dioxide. Tissues enclose and protect spores and gametes. *23.1*

4. Mosses, liverworts, and hornworts are bryophytes. These nonvascular plants have no complex xylem and phloem. Their flagellated sperm can reach the eggs only by swimming through films or droplets of water. *23.2*

5. Whisk ferns, lycophytes, horsetails, and ferns are all seedless vascular plants. Like the bryophytes, they too require free water for fertilization. *23.3*

6. Gymnosperms and flowering plants (angiosperms) are seed-bearing vascular plants. They produce pollen grains (mature male gametophytes) that develop from microspores, and female gametophytes (with egg cells) that develop from megaspores. *23.5–23.8*

   a. Megaspores form in ovules: female reproductive structures consisting of a female gametophyte, nutritive tissue, and a jacket of cell layers. Part of the outer jacket develops into a seed coat. A seed is a mature ovule.

   b. The evolution of pollen grains freed these plants from dependence on water for fertilization. Their seeds are efficient means of dispersing new generations, even during hostile conditions. Pollen grains and seeds were key adaptations in the move to high and dry habitats.

7. Angiosperms produce flowers. Most coevolved with pollinators (e.g., insects) that move the pollen to female reproductive parts. Fruit encases most seeds and aids in their dispersal. *23.8*

Review Questions

1. List a few structural and reproductive modifications that helped plants invade and diversify in habitats on land. *23.1*

2. Does the haploid phase or diploid phase dominate the life cycles of most plants? *23.1*

3. Name representatives of the following groups of plants and then compare their main characteristics: (*refer to Table 23.1*)

   a. bryophytes and seedless vascular plants *23.2, 23.3*
   b. gymnosperms and angiosperms *23.1, 23.6–23.8*

4. Distinguish between:

   a. root system and shoot system *23.1*
   b. xylem and phloem *23.1*
   c. sporophyte and gametophyte *23.1*
   d. ovule and seed *23.1, 23.5*
   e. microspore and megaspore *23.5*

Figure 23.23 From forests to urban housing developments—where many conifers end up.

Self-Quiz ANSWERS IN APPENDIX III

1. Which of the following statements is *not* true?
   a. Gymnosperms are the simplest vascular plants.
   b. Bryophytes are nonvascular plants.
   c. Lycophytes and angiosperms are both vascular plants.
   d. Only angiosperms produce flowers.

2. Which does *not* apply to gymnosperms and angiosperms?
   a. vascular tissues   c. single spore type
   b. diploid dominance   d. all of the above

3. Of all land plants, bryophytes alone have independent _____ and attached, dependent _____ .
   a. sporophytes; gametophytes
   b. gametophytes; sporophytes
   c. rhizoids; zygotes
   d. rhizoids; stalked sporangia

4. Whisk ferns, lycophytes, horsetails, and ferns are classified as _____ plants.
   a. multicelled aquatic   c. seedless vascular
   b. nonvascular seed   d. seed-bearing vascular

5. A seed is _____ .
   a. a female gametophyte   c. a mature pollen tube
   b. a mature ovule   d. an immature embryo

6. Match the terms appropriately.
   _____ gymnosperm
   _____ sporophyte
   _____ lycophyte
   _____ ovule
   _____ bryophyte
   _____ gametophyte
   _____ stomata
   _____ angiosperm
   a. gamete-producing body
   b. help control water loss
   c. "naked" seeds
   d. only plant that produces flowers
   e. spore-producing body
   f. nonvascular land plant
   g. seedless vascular plant
   h. seed originates from it

Critical Thinking

1. Figure 23.23 shows a forest in the Nahmint Valley of British Columbia, before and after logging. It also shows wood frames of homes that are in the process of being built. Reflect on these photographs and Figure 23.18. To stop the loggers, would you chain yourself to a tree in an old-growth forest scheduled for clear-cutting? If your answer is yes, would you also give up the chance of owning a wood-frame home (as most homes are, in developed countries)? What about forest products, including newspapers, toilet tissue, and fireplace wood?

2. With respect to question 1, multiply each of your answers by 6.2 billion (there are more than that many people today) and describe what might happen when, inevitably, we run out of trees. Also describe what you might consider to be some of the pros and cons of tree farms—say, of a single species of pine.

3. Elliot Meyerowitz of the California Institute of Technology has studied the genetic basis of flower formation in *Arabidopsis thaliana*. By inducing mutations in seeds of this small weed, he discovered three genes (*A*, *B*, and *C*) that interact in different parts of a flower. Gene interactions lead to the formation of different structures—sepals, petals, stamens, and carpels—from the same mass of undifferentiated tissue. Compare Section 28.5. Using what you know about gene control, suggest ways in which the *A*, *B*, and *C* genes may be controlling flower development.

4. Genes nearly identical to the *A*, *B*, and *C* genes of *A. thaliana* also have been isolated from snapdragons and other flowering plants. Ancestors of these plants evolved by 150 million years ago. They quickly rose to dominance in nearly all land habitats.
   Review the general introduction to adaptive radiation in Section 18.4. Then speculate on how the spectacular and rapid adaptive radiation of flowering plants came about.

Selected Key Terms

angiosperm 23.1	hornwort 23.2	pollinator 23.8
bryophyte 23.1	horsetail 23.3	progymnosperm 23.5
coal 23.4	lignin 23.1	rhizome 23.3
cone 23.6	liverwort 23.2	root system 23.1
conifer 23.6	lycophyte 23.3	seed 23.1
cuticle 23.1	magnoliid 23.8	seed fern 23.5
cycad 23.6	megaspore 23.5	shoot system 23.1
deforestation 23.7	microspore 23.5	spore 23.1
eudicot 23.8	monocot 23.8	sporophyte 23.1
fern 23.3	moss 23.2	stoma (stomata) 23.1
flower 23.8	ovule 23.5	strobilus 23.3
gametophyte 23.1	peat bog 23.2	vascular plant 23.1
ginkgo 23.6	phloem 23.1	whisk fern 23.3
gnetophyte 23.6	pollen grain 23.1	xylem 23.1
gymnosperm 23.1	pollination 23.5	

Readings

Gray, J., and W. Shear. September–October 1992. "Early Life on Land." *American Scientist* 80:444–456.

Moore, R., W. D. Clark, and D. S. Vodopich. 1998. *Botany.* Second edition. New York: WCB/McGraw Hill.

FUNGI

Ode to the Fungus Among Us

When push comes to shove, plants need certain fungi more than they need us. (Actually, they don't need us at all.) Fungi were there, as symbionts, when plants first invaded the land. **Symbiosis**, recall, refers to species that live together and closely interact. In cases of **mutualism**, the interaction benefits both partners or does one of them no harm. Lichens and mycorrhizae are like this.

A **lichen** is a vegetative body in which a fungus is intertwined with one or more photosynthetic organisms. From the Antarctic to the Arctic, lichens live in habitats that are just too hostile to support most organisms.

Lichens absorb minerals from substrates. They make antibiotics against bacteria that can decompose them. They make toxins against invertebrate larvae that graze upon them. Coincidentally, their metabolic products enrich soil or help form new soil from bedrock. This is what happens when lichens colonize barren sites, such as bedrock exposed by retreating glaciers. Conditions improve, and other species move in and replace the pioneers. This is probably what happened when plants first invaded land. Cyanobacteria-containing lichens even help maintain ecosystems. They capture nitrogen and convert it to a form that plants use. *Lobaria* alone secures 20 percent of the nitrogen used by trees of old-growth forests in the Pacific Northwest (Figure 24.1).

Lichens also give early warning of deteriorating environmental conditions. How? They absorb toxins but can't get rid of them. From studies in New York City and in England, we know that when lichens die around human habitats, air pollution is getting bad.

a SULFUR SHELF FUNGUS *Polyporus*

Figure 24.2 Fungal species from southeastern Virginia. This sampling hints at the rich diversity within the kingdom Fungi.

Some fungi and young tree roots also are mutualists. Their interaction is a **mycorrhiza** (plural, mycorrhizae), which means "fungus-root." Underground parts of the fungus grow through the soil and afford a huge surface area for absorption. The fungus swiftly takes up many ions of phosphorus and other minerals when these are abundant, and it releases ions to the plant when they are scarce. And what does the plant give the fungus in return? It gives up some sugars. The loss is a trade-off; many plants can't grow well without mycorrhizae.

Many fungi also help plants by being **decomposers**. Figure 24.2 shows just a few of these beneficial species. Like all other decomposers, they break down organic compounds in their surroundings. But few organisms besides fungi digest dinner while it's still on the table. As fungi grow into or on organic matter, their enzyme secretions digest it into bits that their cells absorb. We call this mode of nutrition **extracellular digestion and absorption**. Plants benefit when they take up some of the released nutrients. And plants, remember, are the primary producers of nearly all ecosystems.

Keep this global perspective in mind. Why? As you will see, some fungi do cause diseases in humans, pets

Figure 24.1 *Lobaria oregana*, one of the premier lichens.

PURPLE CORAL FUNGUS *Clavaria*

c RUBBER CUP FUNGUS *Sarcosoma*

BIG LAUGHING MUSHROOM *Gymnophilus*

TRUMPET CHANTERELLE *Craterellus* **f** SCARLET HOOD *Hygrophorus*

and farm animals, ornamental plants, and crop plants. Some are notorious spoilers of food supplies. Others, however, help us manufacture substances ranging from antibiotics to cheeses. We show a tendency to assign "value" to fungi and other organisms in terms of their direct effect on our lives. There is nothing wrong with battling dangerous species and admiring the beneficial ones. Yet we should not lose sight of the greater roles of fungi or any other kind of organism in nature.

Key Concepts

1. Fungi are heterotrophs. Together with heterotrophic bacteria, they are the biosphere's decomposers. The saprobic types get nutrients from nonliving organic matter. Parasitic types get them from tissues of living hosts.

2. Fungi secrete enzymes that digest food outside their body, then the fungal cells absorb breakdown products. Their metabolic activities release carbon dioxide to the atmosphere and return many nutrients to the soil, where they become available to plants and other producers.

3. Most fungi are multicelled. A mycelium, the food-absorbing portion of a fungal body, develops during their life cycle. Each mycelium is a mesh of hyphae. Hyphae are filaments that elongate at their tips, with the nuclei inside dividing by mitosis.

4. In many species, some modified hyphae become interwoven into a reproductive structure, as typified by a mushroom. Fungal spores develop in or on such structures. After germinating, a spore may grow and develop into a new mycelium.

5. Many fungi are symbionts. Lichens consist of fungi that are partnered with algae and other organisms. Mycorrhizae are mutually beneficial associations of fungi with young roots of land plants. Metabolic activities of cells making up fungal hyphae provide the plants with nutrients. The plants provide the fungi with carbohydrates.

6. We tend to assign value to plants and fungi in terms of their direct effect on our lives. Our battles with the "bad" ones and reliance on the "good" ones should start from a solid understanding of their long-established roles in nature.

CHARACTERISTICS OF FUNGI

Mode of Nutrition

Fungi are heterotrophs. This means they require organic compounds that other organisms synthesize. Most are **saprobes**, which take nutrients from nonliving organic matter and cause its decay. Others are **parasites**, which extract nutrients from tissues of a living host. When the cells of any species grow in or on organic matter, they secrete digestive enzymes and then absorb breakdown products. This mode of extracellular digestion is good for plants, because plants readily absorb a portion of the released nutrients and carbon dioxide by-products. Without fungi and heterotrophic bacteria, communities would gradually become buried in their own garbage, nutrients would not be cycled, and life would be over.

Major Groups

For many of us, "fungi" are mushrooms sold in grocery stores. But commercial mushrooms are fungal body parts of only a few species of a highly diverse group. Figure 24.2 shows a few of the 56,000 fungal species we know about. There may be at least a million more we don't know about! The fossil record suggests that fungi evolved by 900 million years ago. Some accompanied simple plants onto land 435 million years ago. Three major lineages were well established about 100 million years after that. They are classified as the **zygomycetes** (Zygomycota), **sac fungi** (Ascomycota), and **club fungi** (Basidiomycota). Others are chytrids (Section 47.6). The puzzling kinds known as "imperfect fungi" are lumped together but aren't a formal taxonomic group. The vast majority of species in all these groups are multicelled.

P854

Key Features of Fungal Life Cycles

Fungi reproduce asexually quite often. They reproduce sexually, too, when the opportunity presents itself. They form stupendous numbers of nonmotile spores. **Spores** are reproductive cells and multicelled structures, often walled, that germinate after dispersal from the parent. In multicelled species, a spore gives rise to a **mycelium** (plural, mycelia). This mesh of branched filaments has a good surface-to-volume ratio and grows quickly over or into organic matter. Each filament in a mycelium is a **hypha** (plural, hyphae). Hyphal cells commonly have chitin-reinforced walls. Their cytoplasm interconnects, so nutrients flow unimpeded through the mycelium.

Fungi are major decomposers that engage in extracellular digestion and absorption of organic matter. Most species are saprobic, and some are parasitic. The multicelled types form absorptive mycelia and spore-producing structures.

CONSIDER THE CLUB FUNGI

A Sampling of Spectacular Diversity

Fungal life cycles and life-styles show dizzying variety. The most we can do here is to sample a few species, starting with club fungi. The 25,000 or so club fungi include mushrooms, shelf fungi, coral fungi, puffballs, and stinkhorns. Figures 24.2 through 24.4 are examples. Some saprobic species are important decomposers of litter on and in soil. Other species are symbionts with young roots of trees. The fungal rusts and smuts often destroy entire crops of wheat, corn, and other valued plants. Cultivation of the common mushroom (*Agaricus brunnescens*) is a multimillion-dollar business; this is a mushroom of grocery-store and pizza-topping fame.

Figure 24.3 Club fungi. (**a**) The red coral fungus *Ramaria*. (**b**) Fly agaric mushroom (*Amanita muscaria*). This species induces hallucinations. In ancient societies of Central America, Russia, and India, it was used in rituals to induce trances.

(**c**) California's *A. ocreata*. Eat this one and you might die. In outward appearance, it resembles its relative *A. phalloides*, the death cap mushroom. Taste as little as five milligrams of its toxin, and you will start to vomit and suffer diarrhea eight to twenty-four hours later. Then your liver and kidneys will degenerate and you may die within a few days.

Except for the rubber cup fungus, all the species in Figure 24.2 also are club fungi.

Figure 24.4 Generalized life cycle that applies to many club fungi. When hyphal cells of two compatible mating strains make contact, their cytoplasm fuses but the nuclei do not. Cell divisions result in a dikaryotic mycelium in which each cell has two nuclei. When conditions are favorable, mushrooms form. Club-shaped, spore-bearing structures (basidia) form on gills, the mushroom cap's leaflike inner surface. Inside each structure, the two nuclei fuse. The result is a diploid zygote. With zygote formation, the cycle starts again.

The scanning electron micrograph below shows part of a mycelium, the underground portion of a club fungus that absorbs water and dissolved nutrients.

[handwritten notes:] Fruiting bodies form. FB of a mushroom is has a stalk & a cap. ↓ basidiocarp

After nuclear fusion, the basidium (now 2n) produces and bears haploid spores at the four tips of the cell

Diploid Stage

nuclear fusion

Haploid Stage

meiosis

Basidia, each with two nuclei (n + n), form on gill margins

spore (n)

at gill margin

gill

Spores are released.

Each germinating spore gives rise to a hypha that grows and becomes a branching mycelium.

cap

stalk

After cytoplasmic fusion, a dikaryotic (n + n) mycelium gives rise to spore-bearing bodies (e.g., mushrooms).

cytoplasmic fusion

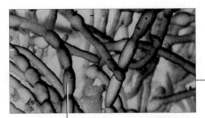

one hyphal cell among the many hyphae that make up a mycelium

Yet some of its relatives produce toxins that can kill you or any other organism that nibbles on them.

Have you ever wondered which organisms are the oldest and the largest? The honey mushroom, *Armillaria ostoyae*, might be the winner. In Oregon, one individual has been spreading through forest soil for 2,400 years. It now extends through 2,200 acres to an average depth of 3 feet. Imagine 1,665 football fields side by side and you get an idea of how big that is. So far, scientists have not identified any larger organism. *A. ostoyae* definitely is not one of the symbionts. Its hyphae penetrate and clog tree roots, and ultimately kill the tree.

Example of a Fungal Life Cycle

Let's use *A. brunnescens*, the common mushroom, as an example of a fungal life cycle. Like many other club fungi, it forms **mushrooms**. These reproductive bodies are short-lived, aboveground parts; the mycelium is buried in soil or decaying wood. Each mushroom has a stalk and a cap with fine tissue sheets (gills) suspended from it. Spores form on club-shaped structures (basidia, singular basidium) on the gills. The spores, of a type called **basidiospores**, are dispersed from the mushroom. When they land on a suitable site, they may germinate and give rise to a haploid mycelium.

If a hyphal cell of one strain of club fungus meets up with that of a compatible mating strain, they may undergo cytoplasmic fusion. But their nuclei won't fuse immediately. The fused portion may start a *dikaryotic* mycelium, in which hyphal cells have one nucleus of each mating type (Figure 24.4). A mycelium may grow extensively, then form mushrooms when nutrients and moisture favor reproduction. Each basidium that forms in a mushroom is dikaryotic at first. Then its two nuclei fuse, the result being a zygote. Very quickly, the zygote undergoes meiosis and haploid spores develop on top of small stalks. Air currents disperse them.

When you notice mushrooms or any other fungus growing outdoors, think twice before nibbling on them. Unlike the common mushroom just described, many are toxic (Figure 24.3). Mushrooms gathered in the wild must be accurately identified as edible before being popped into the mouth. As the saying goes, there are old mushroom hunters and bold mushroom hunters— but no old, bold mushroom hunters.

The most familiar mushrooms are reproductive structures of club fungi. A mushroom's gills display club-shaped, spore-bearing structures on their surface. Although some mushrooms are edible, others are toxic enough to kill.

SPORES AND MORE SPORES

A fungus has a thing about spores. It produces sexual spores, asexual spores, or both, depending on contact with a suitable hypha, food availability, and how cool or damp conditions are. Its spores are usually small and dry, and air currents disperse them. Each spore that germinates can be the start of a hypha and a mycelium. Stalked reproductive structures may develop on many of the hyphae and produce asexual spores. After these spores germinate, each may be the start of still *another* extensive mycelium. In no time at all, that one fungus and staggering numbers of its descendants are busily decomposing organic stuff or pirating nutrients from a host! Look at what can happen to a slice of stale bread:

Each major group of fungi has unique sexual spores. The club fungi form basidiospores, zygomycetes form zygospores, and sac fungi form ascospores.

Producers of Zygospores

Consider the zygomycetes. Parasitic species feed on insects. Most saprobic types live in soil, decaying plant or animal material, and stored food. You just saw what *Rhizopus stolonifer*, the black bread mold, does to bread. When it reproduces sexually, a diploid zygote forms. This **zygospore** is a thick-walled sexual spore enclosed in a thin, clear covering (Figure 24.5). It undergoes meiosis and gives rise to a specialized hypha that bears a spore sac, which is called a sporangium. A number of spores form in the sporangium, and each may give rise to a new mycelium. Stalked hyphae grow from such mycelia. Asexual spores form in a sporangium perched on top of each stalk.

Producers of Ascospores

We know of more than 30,000 kinds of sac fungi. The vast majority are multicelled. Most form sexual spores called **ascospores** in sac-shaped cells. This unique type of cell is an ascus (plural, asci). Reproductive structures

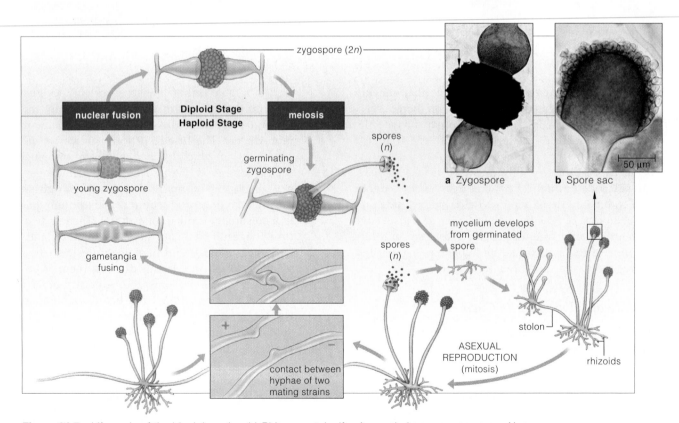

Figure 24.5 Life cycle of the black bread mold *Rhizopus stolonifer*. Asexual phases are common. Also, different mating strains (+ and −) reproduce sexually. Either way, haploid spores form and give rise to mycelia. Chemical attraction between a + hypha and a − hypha makes them fuse. Two gamete-producing structures (gametangia) form, each with several haploid nuclei. The nuclei pair up. Each pair fuses and forms a zygote. Some zygotes disintegrate. Others become thick-walled zygospores and may be dormant for several months. Meiosis occurs as the zygospore germinates, and asexual spores form.

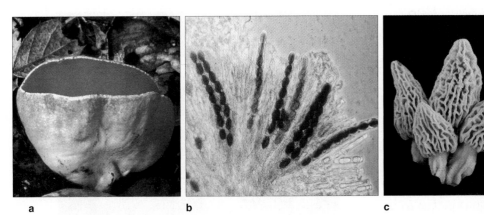

a　　　　b　　　　c　　　　d　　　　e

Figure 24.6 Sac fungi. (**a**) *Sarcoscypha coccinia*, the scarlet cup fungus. (**b**) Saclike structures on the cup's inner surface produce sexual spores (ascospores) by meiosis. (**c**) One of the morels (*Morchella esculenta*). This edible species has a poisonous relative. (**d**) From *Eupenicillium*, chains of asexual spores of a type called conidiospores. These drift away from the chains, like dust, after even the slightest jiggling. "Conidia" means dust. (**e**) Cells of *Candida albicans*, agent of "yeast" infections of the vagina, mouth, intestines, and skin. The cell at the lower left is budding.

[handwritten: Yellow + green molds on food — color is conidia — Bluish streaks in blue cheese are patches of conidiospores]

of tightly interwoven hyphae enclose the asci. Different kinds are shaped like flasks, globes, and shallow cups (Figures 24.2*c* and 24.6).

In this groups are truffles and morels (Figure 24.6*c*). Truffles are underground symbionts with roots of oak and hazelnut trees. Pigs and dogs are trained to snuffle out truffles in the woods. In France, truffles are now being cultivated on the roots of inoculated seedlings.

Aspergillus makes citric acid, used in candies and soft drinks, and ferments soybeans for soy sauce. Some *Penicillium* species "flavor" Camembert and Roquefort cheeses; others make penicillin antibiotics. Most food-spoiling red, bluish-green, and brown fungal molds are a conidial stage of multicelled sac fungi (Figure 24.6*d*). Salmon-colored *Neurospora sitophila* forms many spores and is extremely difficult to eradicate in bakeries and research laboratories. One of its relatives, *N. crassa*, is an important organism in genetic research.

Sac fungi also include about 500 species of single-celled yeasts (although still other yeasts are classified as club fungi). Yeasts reproduce sexually when two cells fuse and become a spore-producing sac. Some types live in the nectar of flowers and on fruits and leaves. Bakers and vintners put fermenting by-products of vast populations of yeasts to use.

For example, the carbon dioxide by-products of *Saccharomyces cerevisiae* leaven bread. The commercial production of wine and beer depends on its ethanol end product. Many yeast strains with useful properties have been developed through artificial selection and genetic engineering. Then again, *Candida albicans*, a notorious relative of "good" yeasts, causes vexing infections in humans (Figure 24.6*e*).

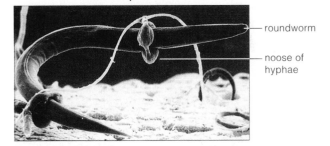

roundworm

noose of hyphae

Figure 24.7 *Arthrobtrys dactyloides*, an imperfect fungus. This predatory species forms a nooselike ring that swells rapidly with turgor pressure when stimulated. The "hole" in the noose shrinks and captures a worm, into which hyphae grow.

Elusive Spores of the Imperfect Fungi

Imperfect fungi are set aside in a taxonomic holding station not because they are somehow defective but mainly because no one has yet discovered what kind of sexual spores they produce (if any). Figure 24.7 shows one of these species, a puzzling predatory fungus, that is awaiting formal classification. Investigators recently reunited the previously orphaned *Aspergillus, Candida,* and *Penicillium* with their kin—other sac fungi.

Through their exuberant and rapid production of asexual and sexual spores, fungi take quick advantage of available organic matter, whether it has been discarded or is part of a living or dead organism. Their penchant for making spores is central to their success as decomposers and parasites.

THE SYMBIONTS REVISITED

Recall, from the introduction, that symbiosis refers to species that live together in close ecological association. Often one is a parasite's victim, not a partner. In cases of mutualism, interaction benefits both partners or does one of them no harm. Here are more detailed examples.

Lichens

In the single vegetative body called a lichen, a fungus is intertwined with one or more photosynthetic species. The fungal part is the *myco*biont. The photosynthetic part is the *photo*biont. Of about 13,500 known types of lichens, nearly half incorporate sac fungi. Only 100 or so species serve as photobionts, and most often these are green algae and cyanobacteria.

A lichen forms after the tip of a fungal hypha binds with a suitable host cell. Both lose their wall, and their cytoplasm fuses or the hypha induces the host cell to cup around it. The mycobiont and the photobiont grow and multiply together. The lichen commonly has distinct layers. The overall pattern of growth may be leaflike, flattened, pendulous, or erect (Figures 24.1 and 24.8).

Lichens typically colonize sites that are hostile for most organisms, including sunbaked or frozen rocks, fence posts, gravestones, and plants, even the tops of

c

dispersal fragment
(cells of mycobiont
and of photobiont)

cortex (outer
layer of
mycobiont)

photobionts

medulla (inner
layer of loosely
woven hyphae)

cortex

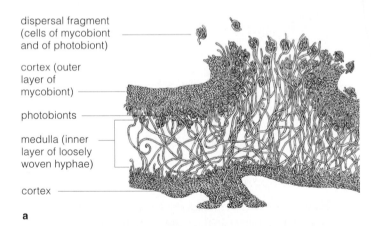

a

b

Figure 24.8 (**a**) Sketch of a stratified lichen, cross-section. (**b**) Encrusting lichens on a rock. (**c**) Leaflike lichen on a birch tree. (**d**) *Usnea*, a pendant lichen commonly called old man's beard. (**e**) *Cladonia rangiferina*, an erect, branching lichen.

d

e

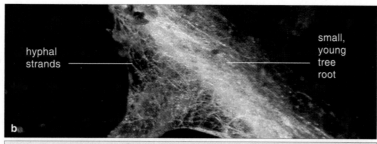

hyphal
strands

small,
young
tree
root

Figure 24.9 (**a**) Lodgepole pine (*Pinus contorta*) seedling, longitudinal section. Notice the extent of the mycorrhiza compared to the shoot system, which is only about four centimeters tall. (**b**) The mycorrhiza of a hemlock tree. (**c**) Effects of the presence or absence of mycorrhizae on plant growth. The juniper seedlings at the left are six months old. They were grown in sterilized, phosphorus-poor soil with a mycorrhizal fungus. The seedlings at the right were grown without the fungus.

giant Douglas firs. Almost always, the fungus is the largest component. Cyanobacteria typically reside in a separate structure inside or outside the main body. The fungus gets a long-term source of nutrients, which it absorbs from photobiont cells. Nutrient withdrawals affect the photobiont's growth a bit, but the lichen may help shelter it. If more than one fungus is present in the lichen, it might be a mycobiont, a parasite, or even an opportunist that is using the lichen as a substrate.

Mycorrhizae

Fungi, recall, also are mutualists with young tree roots, as mycorrhizae (Figure 24.9*a*). Without mycorrhizae, plants cannot grow as efficiently. In *ecto*mycorrhizae, hyphae form a dense net around living cells in roots but do not penetrate them (Figure 24.9*b*). Other hyphae form a velvety wrapping around roots as the mycelium radiates through soil. Ectomycorrhizae are common in temperate forests. They help the trees survive seasonal shifts in temperature and rainfall. About 5,000 fungal species, mostly club fungi, enter into such associations.

The more common *endo*mycorrhizae form in about 80 percent of all vascular plants. These fungal hyphae penetrate plant cells, as they do in lichens. Fewer than 200 species of zygomycetes serve as the fungal partner.

Their hyphae branch extensively, forming tree-shaped absorptive structures in plant cells. Hyphae also extend for several centimeters into the soil. Chapter 30 offers a closer look at these beneficial species.

As Fungi Go, So Go the Forests

Since the early 1900s, collectors have recorded data on wild mushroom populations in European forests. As the records tell us, the number and kinds of fungi are declining at alarming rates. Mushroom gatherers can't be the cause, because inedible as well as edible species are vanishing. However, the decline does correlate with rising air pollution. Vehicle exhaust, smoke from coal burning, and emissions from nitrogen fertilizers pump ozone, nitrogen oxides, and sulfur oxides into the air. Normally, as a tree ages, one species of mycorrhizal fungus gives way to another, in predictable patterns. When fungi die, trees lose this vital support system, and they become vulnerable to severe frost and drought. Are the North American forests at risk, also? Conditions there are deteriorating in comparable ways.

Both lichens and mycorrhizae are symbiotic associations between fungi and other organisms, with mutual benefits.

A Look at the Unloved Few

You know you are a serious student of biology when you view organisms objectively in terms of their place in nature, not in terms of their impact on humans in general and you in particular. As a student you salute saprobic fungi as vital decomposers and praise parasitic fungi that help keep populations of harmful insects and weeds in check. The true test is when you open the fridge to get a bowl of strawberries and discover a fungus beat you to them. The true test is when a fungus starts feeding on warm, damp tissues between your toes and turns skin scaly, reddened, and cracked (Figure 24.10*a*).

Which home gardeners wax poetic about black spot or powdery mildew on roses? Which farmers happily hand over millions of dollars a year to sac fungi that attack corn, wheat, peaches, and apples (Figure 24.10*b*)? Who rejoices that a certain sac fungus, *Cryphonectria parasitica*, blitzed the chestnut trees in eastern North America?

Who willingly inhales airborne spores of *Ajellomyces capsulatus*? After landing on soil, these dimorphic beasties form mycelia. When they alight in moist lung tissues, they form yeastlike cells that cause *histoplasmosis*, a respiratory disease. The body's macrophages normally eliminate the threat, but debris from the battle results in calcified lung tissue. Heavy exposure to spores invites pneumonia.

And household molds! Thank them for sinus, ear, and lung infections, hearing losses, memory losses, and asthma attacks, boosted 300 percent in the past twenty years. The worst culprits are *Stachybotrys*, *Memnoliella*, *Cladosporium*, and certain *Aspergillus* and *Penicillium* species.

Some fungi have even tweaked human history. One notorious species, *Claviceps purpurea*, parasitizes rye and other cereal grains (Figure 24.10*c*). Give it credit; we use some of its by-products (alkaloids) to treat migraines and to shrink the uterus after childbirth to stop hemorrhages. But these alkaloids can be toxic. Eat a lot of bread made with tainted rye flour and you end up with *ergotism*. The symptoms include vomiting, diarrhea, hallucinations, hysteria, and convulsions. Untreated, the disease turns limbs gangrenous and brings on death.

Ergotism epidemics were common in Europe in the Middle Ages, when rye was a major crop. They thwarted Peter the Great, the Russian czar who became obsessed with conquering ports along the Black Sea for his nearly landlocked empire. Soldiers laying siege to the ports ate mostly rye bread and fed rye to their horses. The soldiers went into convulsions and the horses into "blind staggers." Ergotism outbreaks may have been used as an excuse to launch witch-hunts in the early American colonies.

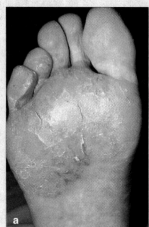

Figure 24.10 Love those fungi! (**a**) On one young fellow, athlete's foot, courtesy of *Epidermophyton floccosum*. (**b**) Apple scab, the trademark of *Venturia inaequalis*. (**c**) Rye plant infected by that historically significant fungus, *Claviceps purpurea*.

Table 24.1 *Some Pathogenic and Toxic Fungi**

ZYGOMYCETES	
Rhizopus	Food spoilage
ASCOMYCETES	
Ajellomyces capsulatus	Histoplasmosis
Aspergillus (some)	Aspergilloses (allergic reactions; sinus, ear, and lung infections; *A. flavus* toxin linked to cancers)
Candida albicans	Infection of mucous membranes
Claviceps purpurea	Ergot of rye, ergotism
Coccidioides immitis	Valley fever
Cryphonectria parasitica	Chestnut blight
Microsporum, Trichophyton, Epidermophyton	Various species cause ringworms of scalp, body, nails, beard, athlete's foot
Monilinia fructicola	Brown rot of peaches, other stone fruits
Ophiostoma ulmi	Dutch elm disease
Venturia inaequalis	Apple scab
Verticillium	Plant wilt
BASIDIOMYCETES	
Amanita (some species)	Severe mushroom poisoning
Puccinia graminis	Black stem wheat rust
Tilletia indica	Smut of cereal grains
Ustilago maydis	Smut of corn

* After C. Alexopoulos, C. Mims, and M. Blackwell. 1996.

SUMMARY

1. Fungi are heterotrophs and major decomposers. The saprobes feed on nonliving organic matter; parasites feed on tissues of living organisms. Some fungal species are symbiotic partners with other organisms. Cells of all species secrete digestive enzymes that break down food to small molecules, which the cells absorb. *24.1*

2. Nearly all fungi are multicelled. The food-absorbing part, the mycelium, is a mesh of filaments (hyphae). The aboveground reproductive parts (e.g., mushrooms) form from tightly interwoven hyphae. *24.1, 24.2*

3. Major groups are zygomycetes, ascomycetes (sac fungi), and basidiomycetes (club fungi). Each produces distinctive sexual and asexual spores. When a sexual phase can't be detected or is absent from the life cycle, a fungus is assigned to an informal category called the imperfect fungi. *24.1–24.3*

4. Lichens are mutualistic associations of fungi with photosynthesizers (e.g., green algae and cyanobacteria). Mycorrhizae are mutualistic associations of a fungus and young roots. Fungal hyphae give up nutrients to the plant and get back carbohydrates. *CI, 24.4*

Review Questions

1. Describe the fungal mode of nutrition, and explain how the structure of mycelia facilitates this mode. *24.1*

2. How does a lichen differ from a mycorrhiza? *CI, 24.4*

3. List conditions that influence fungal spore formation. *24.3*

4. What makes fungi such successful decomposers? *24.3*

Self-Quiz   ANSWERS IN APPENDIX III

1. A mycorrhiza is a _____ .
   a. fungal disease of the foot
   b. fungus–plant relationship
   c. parasitic water mold
   d. fungus of barnyards

2. Parasitic fungi obtain nutrients from _____ .
   a. tissues of living hosts
   b. nonliving organic matter
   c. only living animals
   d. none of the above

3. Saprobic fungi derive nutrients from _____ .
   a. nonliving organic matter
   b. living plants
   c. living animals
   d. both b and c

4. New mycelia form after _____ germinate.
   a. hyphae   b. mycelia   c. spores   d. mushrooms

5. A mushroom is _____ .
   a. the food-absorbing part of a fungal body
   b. the part of the fungal body not constructed of hyphae
   c. a reproductive structure
   d. a nonessential part of the fungus

6. Match the terms appropriately.
   ____ zygomycete   a. mushrooms, shelf fungi
   ____ conidia   b. type of sexual spore
   ____ hypha   c. *Penicillium's* chains of asexual spores
   ____ club fungi   d. each filament in a mycelium
   ____ ascospore   e. black bread mold
   ____ sac fungi   f. truffles, morels, some yeasts

Figure 24.11   Reproductive structures of *Pilobolus*, a name from a Greek word for "hat-thrower." The "hats" actually are spore sacs.

Critical Thinking

1. *Pilobolus* is a type of fungus that commonly dines on horse dung. Each morning, stalked reproductive hyphae emerge from irregularly spaced piles of dung. By early afternoon, they have dispersed spores to sunlit grasses where horses feed. The spores pass through the horse gut unharmed and exit with their own pile of dung. At the tip of each stalked hypha is a dark-walled, spore-containing sac (Figure 24.11). Just below the sac, the stalk is differentiated into a vesicle, swollen with a fluid-filled central vacuole. At the base of the vesicle is a ring of light-sensitive, pigmented cytoplasm. The stalk bends as it grows until its wall is parallel with the sun's rays and light strikes all of the ring. When that happens, turgor pressure builds up inside the central vacuole until the vesicle ruptures. The forceful blast can propel spore sacs two meters away—which is amazing, considering that the stalk is less than ten millimeters tall. Reflect on the examples of fungi discussed in this chapter. Would you say *Pilobolus* is a zygomycete, a club fungus, or a sac fungus?

2. Renee sees in the laboratory that the fungus *Trichoderma* grows well in distilled water. It continues to do so even after she rigorously treats the water and glassware to remove all traces of organic carbon. This fungus is not a photoautotroph. Suggest a metabolic life-style that lets it grow under these conditions.

3. *Trichoderma* is being tested as a natural pest control agent. Laboratory experiments demonstrated that some strains of this fungus combat other fungi that cause plant diseases. Some even promote seed germination and plant growth. During one set of twenty trials, workers increased lettuce yields by 54 percent. What concerns must be addressed before *Trichoderma* can be released into the environment for commercial applications?

Selected Key Terms

ascospore *24.3*	fungus *24.1*	sac fungus
basidiospore *24.2*	hypha *24.1*	(ascomycetes) *24.1*
club fungus	lichen *CI*	saprobe *24.1*
(basidiomycetes) *24.1*	mushroom *24.2*	spore (fungal) *24.1*
decomposer *CI*	mutualism *CI*	symbiosis *CI*
extracellular digestion	mycelium *24.1*	zygomycetes *24.1*
and absorption *CI*	mycorrhiza *CI*	zygospore *24.3*
	parasite *24.1*	

Readings

Moore-Landecker, E. 1996. *Fundamentals of the Fungi*. Fourth edition. Englewood Cliffs, New Jersey: Prentice-Hall.

ANIMALS: THE INVERTEBRATES

Madeleine's Limbs

In August of 1994, about 900 million years after the first animals appeared on Earth, Madeleine made *her* entrance. As they are wont to do, her grandmothers and aunts made a count on the sly—arms, legs, ears, and eyes, two of each; fully formed mouth and nose—just to be sure these were present and accounted for.

One grandmother, having been too long in the company of biologists, experienced an epiphany as she witnessed Madeleine's birth. In that profound instant she sensed ancestral connections emerging from the distant past and, through her, into the future.

Madeleine's body plan did not emerge out of thin air. Thirty-five thousand years ago, people just like us were having children just like Madeleine. If we are interpreting the past correctly, then six million years ago, individuals on the evolutionary road to humans resembled her in some respects but not others. Sixty million years ago, *their* primate ancestors were giving birth precariously, up in the trees. And 250 million years ago, the mammalian ancestors of those primates were giving birth—and so on back in time to the very first animals, which had no limbs or eyes or noses at all.

We have few clues to what the first animals looked like. By the dawn of the Cambrian, however, they had given rise to all major groups of invertebrates—animals without backbones—and even to Madeleine's backboned but limbless ancestors.

We know this from the fossil record. Part of that record tells us about a community that flourished 530 million years ago in a submerged basin between a steep reef and the coast of an early continent. Protected from ocean currents, sediments piled up against the reef. About 500 feet below the surface, the water was oxygenated and clear. Small, well-developed animals lived in, on, or above the dimly lit, muddy sediments (Figure 25.1).

Like castles built from wet sand along a seashore, their living quarters were unstable. Part of the bank of sediments slumped abruptly. An underwater avalanche buried the animals, so scavengers could not remove traces of the dead. Over time, muddy silt continued to rain down; pressure and chemical change transformed it into finely stratified shale. Soft parts of the flattened animals became shimmering, mineralized films.

Sixty-five million years ago, part of the seafloor was plowing under the North American crustal plate, and western Canada's mountain ranges were rising. By 1909, the fossils were high in the eastern mountains of British Columbia. In that year a fossil hunter tripped on a chunk of shale, which split into fine layers—and so the Burgess Shale story came to light.

In this chapter and the next, you will be comparing major groups of animals. Comparisons give insight into evolutionary relationships among them and help us construct family trees, as in Figure 25.2. Don't assume that structurally simple animals of the most ancient lineages are somehow evolutionarily stunted or "primitive." As you will see, they, too, are exquisitely adapted to their environment.

Figure 25.1 Reconstruction of a few Cambrian animals, based on fossils of the Burgess Shale, British Columbia. Section 20.5 shows two of the fossils.

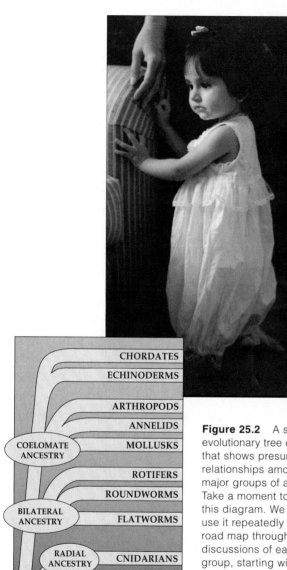

Figure 25.2 A simple evolutionary tree diagram that shows presumed relationships among major groups of animals. Take a moment to study this diagram. We will use it repeatedly as a road map through our discussions of each group, starting with invertebrates. In the chapter to follow, we will continue with the chordates—including young Madeleine.

CHORDATES

ECHINODERMS

ARTHROPODS

ANNELIDS

COELOMATE ANCESTRY — MOLLUSKS

ROTIFERS

ROUNDWORMS

BILATERAL ANCESTRY — FLATWORMS

RADIAL ANCESTRY — CNIDARIANS

MULTICELLED ANCESTRY — SPONGES

SINGLE-CELLED, PROTISTAN-LIKE ANCESTORS

As you poke through branches of the animal family tree, keep the greater evolutionary story in mind. At each branch point, microevolutionary processes gave rise to workable changes in body plans. Madeleine's uniquely human traits, and yours, emerged through modification of traits that evolved earlier in countless generations of vertebrates and, before them, in ancient invertebrate forms.

Key Concepts

1. All animals are multicelled, aerobic heterotrophs that ingest or parasitize other organisms. Nearly all kinds have tissues, organs, and organ systems, and most are motile during at least part of their life cycle. Animals reproduce sexually, and many also reproduce asexually. Their embryos develop through a series of continuous stages.

2. Animals originated late in the precambrian. Their descendants include the more than 2 million modern species we have identified so far. More than 1,950,000 of these species are invertebrates, or animals without a backbone. Fewer than 50,000 are vertebrates—animals with a backbone.

3. Comparisons of the body plans of existing animals, in conjunction with the fossil record, reveal several evolutionary trends. The most revealing aspects of evolving body plans are the animal's type of symmetry, gut, and cavity (if any) between the gut and body wall; whether it has a distinct head end; and whether the body is divided into a series of segments.

4. Placozoans and sponges are structurally simple animals with no body symmetry. Both are at the cellular level of body construction. Cnidarians show radial symmetry, and they are at the tissue level of body construction.

5. Flatworms, roundworms, rotifers, and nearly all other animals more complex than cnidarians show bilateral symmetry. They have tissues, organs, and organ systems.

6. Not long after flatworms evolved, divergences began that would lead to two major lineages. One evolutionary branching gave rise to the mollusks, annelids, and arthropods. The other branching gave rise to the echinoderms and chordates.

7. By biological measures, including diversity, sheer numbers, and distribution, the arthropods have been the most successful animal group. And on land, the insects have been the most successful arthropods.

OVERVIEW OF THE ANIMAL KINGDOM

General Characteristics of Animals

What, exactly, are **animals**? We can only define them by a list of characteristics, not with a sentence or two. *First*, animals are multicelled. In most cases their body cells form tissues that become arranged as organs and organ systems. The body cells of nearly all species have a diploid chromosome number. *Second*, all animals are heterotrophs that get carbon and energy by ingesting other organisms or by absorbing nutrients from them. *Third*, animals require oxygen for aerobic respiration. *Fourth*, animals reproduce sexually and, in many cases, asexually. *Fifth*, most animals are motile during at least part of the life cycle. *Sixth*, life cycles include stages of embryonic development. Briefly, mitotic cell divisions transform an animal zygote into a multicelled embryo. The embryonic cells give rise to primary tissue layers: **ectoderm**, **endoderm**, and, in most species, **mesoderm**. The layers in turn give rise to all tissues and organs of the adult, as described in Sections 33.6 and 43.2.

Variations in Body Plans

Mammals, birds, reptiles, amphibians, and fishes are the most familiar animals. All are **vertebrates**; they contain a backbone. Yet, of more than 2 million known species of animals, fewer than 50,000 are vertebrates! What we call the **invertebrates** are animals with many defining features, but a backbone isn't one of them.

Animals are grouped into more than thirty phyla. Table 25.1 lists the groups surveyed in this book. Their shared traits arose very early, before divergences from a common ancestor gave rise to diverse lineages. Later, morphological differences accumulated among them, and they took off in different directions. How might we get a conceptual handle on the modern descendants—on animals as different as flatworms, toads, spiders, hummingbirds, humans, and giraffes? We can compare similarities and differences with respect to five basic features. These are body symmetry, cephalization, type of gut, type of body cavity, and segmentation.

BODY SYMMETRY AND CEPHALIZATION With very few exceptions, animals are radial or bilateral. Those with **radial symmetry** have body parts arranged regularly around a central axis, like spokes of a bike wheel. So a cut down the center of a hydra (Figure 25.3a) divides it into equal halves; another cut at right angles to the first

Phylum	Some Representatives	Existing Species
PLACOZOA (*Trichoplax*)	Simplest animal; like a tiny plate, but with layers of cells	1
PORIFERA (poriferans)	Sponges	8,000
CNIDARIA (cnidarians)	Hydrozoans, jellyfishes, corals, sea anemones	11,000
PLATYHELMINTHES (flatworms)	Turbellarians, flukes, tapeworms	15,000
NEMATODA (roundworms)	Pinworms, hookworms	20,000
ROTIFERA (rotifers)	Crown of cilia, tiny body yet great internal complexity	2,000
MOLLUSCA (mollusks)	Snails, slugs, clams, squids, octopuses	110,000
ANNELIDA (segmented worms)	Leeches, earthworms, polychaetes	15,000
ARTHROPODA (arthropods)	Crustaceans, spiders, insects	1,000,000+
ECHINODERMATA (echinoderms)	Sea stars, sea urchins	6,000
CHORDATA (chordates)	Invertebrate chordates: Tunicates, lancelets	2,100
	Vertebrates:	
	Fishes	21,000
	Amphibians	4,900
	Reptiles	7,000
	Birds	8,600
	Mammals	4,500

Table 25.1 *Animal Phyla Surveyed in This Book*

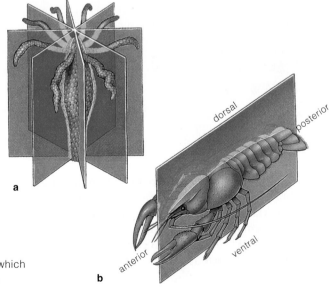

Figure 25.3 Examples of body symmetry among animals. (**a**) A hydra, which is radially symmetrical, and (**b**) a bilaterally symmetrical crayfish.

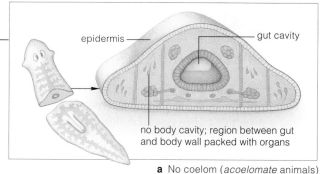

a No coelom (*acoelomate* animals)

epidermis — — gut cavity

no body cavity; region between gut and body wall packed with organs

Figure 25.4 Type of body cavity (if any) in animals.

epidermis — — gut cavity

unlined body cavity (pseudocoel) around gut

b Pseudocoel (*pseudocoelomate* animals)

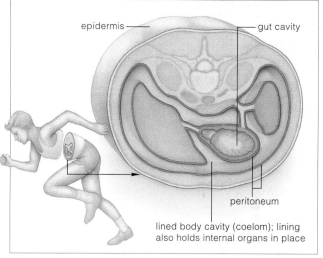

epidermis — — gut cavity

peritoneum

lined body cavity (coelom); lining also holds internal organs in place

c Coelom (*coelomate* animals)

divides it into equal quarters. Radial animals live in water. Their body plan is adapted to intercepting food drifting or swimming toward them from any direction.

Animals with **bilateral symmetry** have a body axis passing from an *anterior* end (front) to a *posterior* end (back). The body is separated into right and left sides along this main axis, and has a *dorsal* surface (backside) and *ventral* surface (underside). Figure 25.3*b* shows the body plan, which originated among forward-creeping species. In such animals, the forward end encountered food and other stimuli first, so there was selection for **cephalization**. By this evolutionary process, nerve cells and sensory parts became concentrated in the head. Bilateral body plans and cephalization evolved jointly, with heritable changes in one bringing about changes in the other. Paired brain regions and paired nerves, sensory structures, and muscles were the outcomes.

TYPE OF GUT   A **gut** is a sac projecting into the body or part of a tube through the body. In it, food is digested, then absorbed into the internal environment. A saclike gut has a single opening (mouth) for taking in food and disposing of residues. A tubular gut, with openings at two ends (mouth and anus), is a "complete" digestive system. It includes specialized regions that break up, digest, absorb, store, and get rid of material. As these more efficient digestive systems evolved, they helped make increases in body size and activity possible.

BODY CAVITIES   In between the gut and the body wall of most bilateral animals is a body cavity (Figure 25.4). One type of cavity, the **coelom**, has a peritoneum. This is a tissue lining that also encloses organs in the coelom and helps hold them in place. For example, you have a coelom. A sheetlike muscle, the diaphragm, divides it into two smaller cavities. Your heart and lungs occupy the upper (thoracic) cavity; your stomach, intestines, and other organs occupy the lower (abdominal) cavity.

The coelom was a key innovation in the evolution of larger, complex animals from small ancestral forms. Why? It favored increases in size and activity because it cushioned and protected the internal organs, and let them move independently of the body wall. Some of the less complex invertebrates don't even have a body cavity. Tissues fill the region between the gut and the body wall. Some have a pseudocoel, or "false coelom." This, too, is a body cavity, but it has no peritoneum.

SEGMENTATION   Segmented animals have a repeating series of body units that may or may not be similar to one another. The many segments of earthworms have a similar outward appearance. Insect segments are fused

into three units (head, thorax, and abdomen) and differ greatly from one another. Especially among the insects, richly diverse head parts, legs, wings, and many other appendages evolved from less specialized segments.

Animals are multicelled, aerobically respiring heterotrophs that ingest other organisms or absorb nutrients from them. Nearly all animals have tissues, organs, and organ systems. Most have a diploid chromosome number.

Animals reproduce sexually and, in many cases, asexually. They go through a period of embryonic development, and most are motile during at least part of the life cycle.

Animal body plans differ in body symmetry, cephalization, type of gut, type of body cavity, and segmentation.

PUZZLES ABOUT ORIGINS

Judging from genetic evidence and radiometric dating of fossilized tracks, burrows, and microscopic embryos, animals originated between 1.2 billion and 670 million years ago, during precambrian times. *Where did they come from?* They probably evolved from some protistan lineages, but we don't know which ones (Section 20.3).

By one hypothesis, the forerunners of animals were ciliates, like *Paramecium*, with multiple nuclei in a one-celled body. Supposedly as they evolved, each nucleus was compartmentalized inside individual cells of the multicelled body. But we do not know of any existing animal that develops by compartmentalization.

By another hypothesis, multicelled animals arose from spherical colonies of a number of flagellated cells, maybe like the *Volvox* colony shown in Section 22.10. In time, as a result of mutation, some cells in the colony became modified in ways that enhanced reproduction and other specialized tasks. And so began the division of labor that characterizes multicellularity.

Suppose colonies became flattened over time and started creeping on the seafloor. The creeping life-style could have favored the evolution of cell layers, such as those of *Trichoplax adhaerens*. This is the only known **placozoan** (Placozoa, after *plax*, plate; and *zoon*, animal). *T. adhaerens* is a soft-bodied marine animal shaped like pita bread. It has no symmetry, no mouth, and only two layers of several thousand cells. It briefly humps up as its body glides over food (Figure 25.5). Gland cells in the lower layer secrete digestive enzymes onto food, then individual cells absorb the breakdown products. Reproduction may be asexual (by budding or fission) or sexual, by means not understood. In sum, structurally and functionally, *Trichoplax* is as simple as animals get.

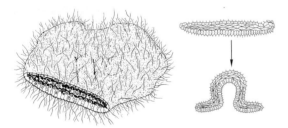

Figure 25.5 Cutaway views of *Trichoplax adhaerens*, an animal with a two-layer body measuring about three millimeters across.

Possibly the question of origins requires more than one answer. It may be that different lineages descended from more than one group of protistan-like ancestors.

Multicelled animals arose from protistan-like ancestors that may have resembled ciliates, colonial flagellates, or both.

SPONGES—SUCCESS IN SIMPLICITY

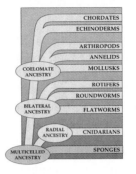

Sponges (Porifera) are animals with no symmetry, no tissues, and no organs. Yet the sponges are one of nature's success stories. They have been abundant ever since the precambrian, particularly in coral reefs and coastal waters. Of 8,000 or so known species, only 100 or so live in fresh water.

Sponges are homes for many other animals, such as worms and shrimps. Some are big enough to sit in; others are as tiny as fingernails. Their shape is flattened, sprawling, lobed, compact, tubelike, or vaselike (Figures 25.6 and 25.7).

The sponge body has no symmetry. Flattened cells line its outer surface and inner cavities. But the linings aren't much more specialized than those cell layers of *Trichoplax*, and they aren't like tissues of other animals. Amoeboid cells live in a gelatin-like substance between the two linings and play several roles (Figure 25.7b).

Commonly, tough protein fibers and sharp, glasslike spicules of calcium carbonate or silica stiffen the body. The skeletal elements may be a reason why sponges as a group have endured so long. Cleveland Hickman put it this way: Most predators discover that sampling any sponge is about as pleasant as eating a mouthful of glass splinters embedded in fibrous gelatin. Besides, chemically speaking, many sponges stink.

Water flows into a sponge's body through many microscopic pores and chambers, then out through one or more large openings. It does so when thousands or millions of phagocytic **collar cells** beat their flagella. The cells are part of the inner lining (Figure 25.7c). The "collars" are food-trapping structures. Bacteria and other edibles get trapped in them, then are engulfed. Some of the food is transferred to the amoebalike cells for further breakdown, storage, and distribution.

Figure 25.6 A sprawling, red-orange sponge, one of many types that encrust underwater ledges in temperate seas.

Figure 25.7 (**a**, **b**) Body plan of a simple sponge. The outer lining, mostly flattened cells, also has some contractile cells around a large opening at the top. Slow contraction directs water flow through the body. In a gelatin-like matrix between the inner and outer linings, some amoeba-like cells secrete materials for spicules and fibers. Others digest and transport food. They retain the capacity to divide. Their descendants can differentiate into any other type of sponge cell. The amoeba-like cells also have roles in asexual processes, such as gemmule formation.

(**c**) Many flagellated, phagocytic cells line the body's inner canals and chambers. Each has a collar of food-trapping structures called microvilli. Fine filaments connect microvilli to each other, forming a "sieve" that strains food particles from water. Cells at the collar's base engulf trapped food.

(**d**) Venus's flower basket (*Euplectella*). Silica spicules of this marine sponge are fused in a rigid network. A thin layer of cells stretches over them. A tuft of spicules anchors the body. (**e**) Basket sponge releasing a cloud of sperm into the water.

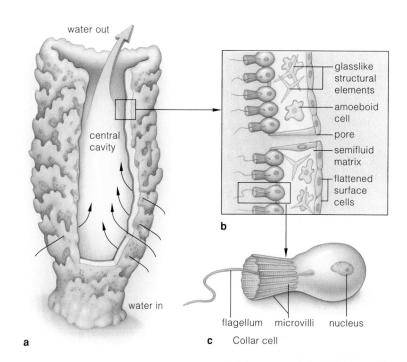

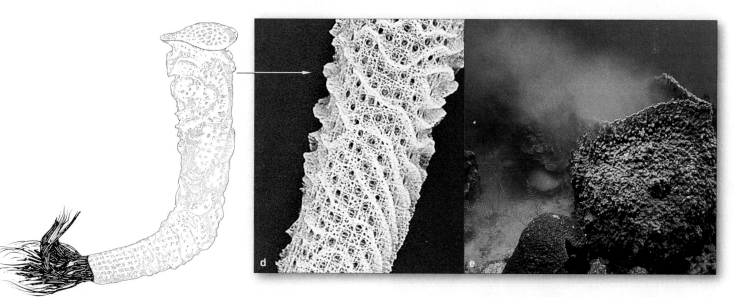

Most sponges reproduce sexually. They release the sperm into the water (Figure 25.7*e*). But they retain the fertilized eggs and the developing embryos. A young sponge develops into a microscopic, swimming larva. A **larva** (plural, larvae) is a type of sexually immature stage that develops into an adult, the sexually mature form of the species. As you will later see, larval stages develop during the life cycle of many animals.

Some kinds of sponges also reproduce asexually by fragmentation. Small fragments break away from the parent and grow into new sponges. Most freshwater species also reproduce asexually by way of gemmules.

These are clusters of sponge cells, some of which form a hard covering around others. The clusters inside are protected from extreme cold or drying out. Later, when favorable conditions return, the gemmules germinate and establish a new colony of sponges.

Sponges have no symmetry, tissues, or organs; they are at the cellular level of construction.

Even so, sponges have successfully endured through time, possibly because most predators find their spicule-rich and often stinky bodies unappetizing.

CNIDARIANS—TISSUES EMERGE

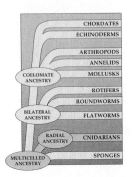

Cnidarians (Cnidaria) are radial, tentacled animals, most of which live in the seas. Jellyfishes belong to the group called scyphozoans. Anthozoans (sea anemones and their kin) and hydrozoans (such as *Hydra*) also are in the phylum. Of the roughly 11,000 species, we find fewer than 50 in fresh water. All species are at the tissue level of organization. They alone make **nematocysts**, capsules with tubular threads that can be discharged. The thread tips ooze sticky prey-trapping substances, sport prey-piercing barbs, or deliver toxins that can inflict painful stings (Figure 25.8). Hence the name Cnidaria, after the Greek word for nettle.

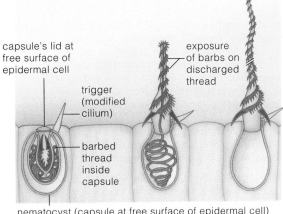

nematocyst (capsule at free surface of epidermal cell)

Figure 25.8 Nematocyst before and after prey (not shown) touched its trigger and made the capsule "leakier" to water. As water diffused in, turgor pressure built up and forced the thread to turn inside out. Barbs at the tip pierced the prey's body.

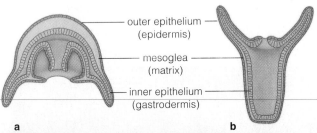

a

b

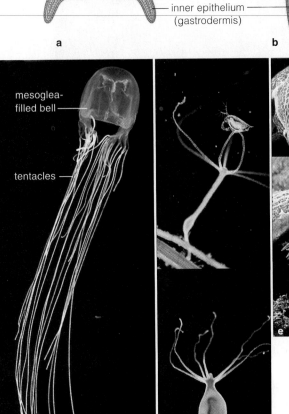

Cnidarian Body Plans and Life Cycles

The **medusa** (plural, medusae) and **polyp** are the most common cnidarian body plans. Both have a saclike gut, as in Figure 25.9. Medusae float and look like bells or saucers. The mouth, centered under the bell, often has tentacles and other extensions used in prey capture and feeding. Tubelike polyps have a tentacle-ringed mouth at one end; the other end attaches to a substrate.

Trichoplax draped over food and digesting it is like a gut on the run. But the cnidarian gut is a permanent food-processing chamber. Its sheetlike inner lining, the gastrodermis, has glandular cells that secrete digestive enzymes. Epidermis lines body surfaces (Figure 25.9a). Each lining is an **epithelium** (plural, epithelia), a tissue with one free surface bathed in a body fluid or exposed to the outside. All animals more complex than sponges have this tissue. In cnidarians, it incorporates a "nerve net" of interacting **nerve cells**. These receive signals from receptors and then signal **contractile cells** to carry out responses. Contractile cells shorten and go back to their original length. The nerve net, a simple nervous system, only controls changes in movement and shape.

Between the outer and inner linings is a layer of gelatinous secreted material called mesoglea ("middle jelly"). In a jellyfish, the volume of mesoglea imparts

Figure 25.9 Common cnidarian body plans: (**a**) medusa and (**b**) polyp, midsection. Sections 27.3, 34.5, and 37.3 provide other examples. (**c**) Sea wasp (*Chironix*), a jellyfish with tentacles up to fifteen meters long. Its toxin can kill you within minutes. (**d**) Hydrozoan polyp (*Hydra*) capturing and then digesting prey. (**e**) Sea anemone using its hydrostatic skeleton to escape from a sea star. First it closes its mouth, then contractile cells generate force against water inside the gut. The sea anemone's body changes shape and thrashes about, generally away from the predator.

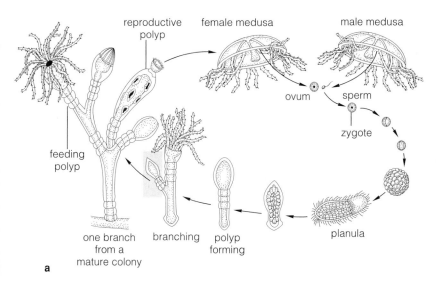

female medusa | male medusa

reproductive
polyp

feeding
polyp

ovum | sperm

zygote

one branch
from a
mature colony | branching | polyp
forming

planula

a

gas-filled medusa | tentacle of one polyp

b

c

three polyps (reproductive,
feeding, and prey-capturing)
attached one after the other

interconnected
skeletons of polyps
of a colonial coral

Figure 25.10 (**a**) Life cycle of a hydrozoan (*Obelia*) that has medusa and polyp stages. Established colonies have thousands of feeding polyps. (**b**) Portuguese man-of-war (*Physalia*). (**c**) A colonial coral.

buoyancy and serves as a firm yet deformable skeleton against which contractile cells act. Visualize many such cells in the jellyfish bell. Their contraction narrows the bell, forcing water to jet out from beneath it. The bell returns to its resting position, cells contract again, and the jellyfish moves. What about polyps, most of which have little mesoglea? Contractile cells act against water in their gut cavity. Any fluid-filled cavity or cell mass against which contractile cells can act is a **hydrostatic skeleton**. The cavity's volume or mass doesn't change; it is simply shunted about to change the body's shape (Figure 25.9*e*). As you will read in Chapter 37, nearly all animals have some type of skeletal system against which the force of contraction is applied.

Medusae, polyps, or both grow and develop during cnidarian life cycles. *Obelia* is a good example (Figure 25.10*a*). Its medusa is a sexual stage. Embedded within its epithelium are **gonads**, which are primary (gamete-producing) reproductive organs. *Obelia* gonads release gametes simply by rupturing. After zygotes form, they develop into **planulas**, a type of swimming or creeping larva that usually has ciliated epidermal cells. In time, a mouth opens at one end of a planula, which develops into a polyp or medusa that starts the cycle again.

Examples of Cnidarian Diversity

Some cnidarians form colonies and are representatives of this phylum's diversity. You may know about one of the colonial hydrozoans, the Portuguese man-of-war (*Physalia*). Its nematocyst toxin is dangerous to bathers, fishermen, and fish prey. *Physalia* lives mainly in warm waters, but currents also move it to the cooler Atlantic

coastal waters of North America and Europe. A blue, gas-filled float grows from the planula's body. It keeps the colony near the water's surface, where winds move it (Figure 25.10*b*). Beneath the float, groups of polyps and medusae interact as "teams" for feeding, defense, reproduction, and other specialized tasks.

Another example: Colonial anthozoans such as the reef-forming corals interconnect by external skeletons (Figure 25.10*c*). Reefs consist mainly of accumulated, compressed skeletons. The reef-forming corals do best in warm, clear seas, where they take up nutrients that tidal action stirs up. Dinoflagellates, a photosynthetic protistan, are mutualists with corals. They supply their hosts with oxygen, recycle mineral wastes, and adjust seawater pH in a way that helps calcium deposits build up their host's skeleton. In return, the corals provide these photosynthesizers with a relatively safe habitat, nicely exposed to the sun's rays as well as to dissolved carbon dioxide and minerals. Their interaction cycles scarce nutrients quickly, directly, and efficiently.

You won't come across coral reefs in shallow, still water, which can get hot enough to kill corals. Also, corals go into osmotic shock after heavy rains freshen seawater. That is why you won't see reefs where a river drains into the sea or near rainfall-drenched land.

Cnidarians are radial animals equipped with tentacles, a saclike gut, epithelia, a simple nerve net, and a hydrostatic skeleton. They alone produce nematocysts.

Cnidarians are at the tissue level of construction; compared with sponges, they have layers of cells interacting in more coordinated fashion to carry out specific tasks.

ACOELOMATE ANIMALS—AND THE SIMPLEST ORGAN SYSTEMS

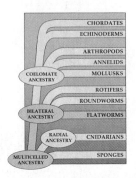

Moving beyond the cnidarians in our survey, we find animals that range from flatworms to humans. Most are bilateral, and all contain organs. Each **organ** is a structural unit of two or more tissues that are arrayed in a specific pattern and that perform a common task. Most organs interact with others. By definition, an **organ system** is composed of two or more organs interacting chemically, physically, or both in ways that contribute to the survival of the whole organism.

Flatworms are the simplest animals at the organ-system level of construction. As you will see, a few of the parasitic types have the capacity to make our lives miserable. Parasites, recall, reside in or on living hosts and feed on their tissues. Most do not kill hosts, at least not until after they have reproduced. The *definitive* host harbors the mature stage of a parasite's life cycle. One or more *intermediate* hosts harbor immature stages.

Characteristics of Flatworms

Among the 15,000 or so known species of **flatworms** (phylum Platyhelminthes) are turbellarians, flukes, and tapeworms. Most of these bilateral, cephalized animals have simple organ systems inside a flattened body, as in Figure 25.11. The digestive system is a pharynx and a saclike gut, often with branchings. A **pharynx** simply is a muscular tube; flatworms use it for feeding.

Species differ in their reproductive systems. Most of their individuals are **hermaphrodites**, with female *and* male gonads, including a penis (a sperm-delivery structure). Flatworms can still reproduce sexually. One also may fertilize itself and make a clone of genetically identical offspring.

Major Groups of Flatworms

TURBELLARIANS Most turbellarians (Turbellaria) live in the seas; only the planarians and a few others live in fresh water. They eat tiny animals or suck juices from dead or damaged ones. Each planarian can reproduce asexually by *transverse* fission: It divides in half at its midsection, then each half regenerates missing parts. Like you, a planarian adjusts the composition and the volume of its body fluids. Its water-regulating system has one or more branched tubes called protonephridia (singular, protonephridium). These tubes extend from pores at the body surface to bulb-shaped flame cells in tissues. When excess water diffuses into the flame cells, a tuft of cilia "flickering" in the bulb drives the water through the tubes to the outside (Figure 25.11*b*).

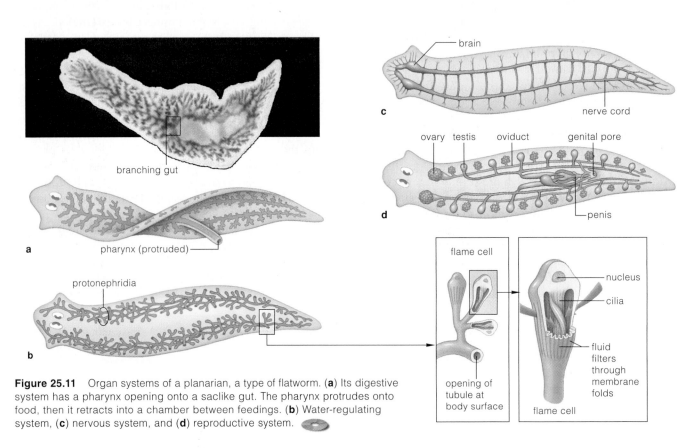

Figure 25.11 Organ systems of a planarian, a type of flatworm. (**a**) Its digestive system has a pharynx opening onto a saclike gut. The pharynx protrudes onto food, then it retracts into a chamber between feedings. (**b**) Water-regulating system, (**c**) nervous system, and (**d**) reproductive system.

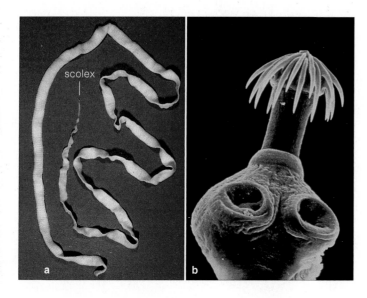

Figure 25.12 (a) Sheep tapeworm. (b) Scolex of a tapeworm that uses a shorebird as its primary host.

FLUKES Flukes (Trematoda) are parasitic flatworms. Their life cycle has sexual and many asexual phases, and one to four hosts. Almost always, a snail or a clam is the initial host for larval or juvenile stages, and some vertebrate is the definitive host. Section 25.7 provides a detailed look at a blood fluke (*Schistosoma*). Each year, schistosomes parasitize 200 million of us.

TAPEWORMS The tapeworms (Cestoda) are parasites of vertebrate intestines. Ancestral tapeworms probably had a gut, then lost it during their evolution in the host intestines, which are habitats rich in predigested food. Species now attach to the intestinal wall with a scolex, a structure with suckers, hooks, or both (Figure 25.12).

Proglottids bud behind a scolex. Each is a new unit of the tapeworm body. The units are hermaphroditic; they mate and transfer sperm to one another. Older proglottids, the ones farthest from the scolex, store the fertilized eggs. They break off from younger ones, then leave the body in feces. Later on, the eggs may reach an intermediate host. Section 25.7 shows how proglottids form in the life cycle of a representative tapeworm.

In some respects, the simplest turbellarians, larval flukes, and larval tapeworms resemble the planulas of cnidarians. The resemblance inspires speculation that bilateral animals evolved from planula-like ancestors. They may have done so by increased cephalization and the development of tissues derived from mesoderm.

Flatworms are among the simplest bilateral, cephalized, acoelomate animals with organ systems.

ROUNDWORMS

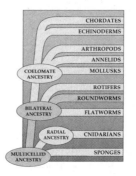

Roundworms (Nematoda) thrive nearly everywhere. They may be the most abundant animals alive. Shallow saltwater or freshwater sediments often hold one million roundworms per square meter. In a handful of rich topsoil, maybe thousands of scavenging types make fast work of rotting plant parts and dead earthworms. We know of about 20,000 species, but there may be at least a hundred times more than this.

A roundworm's cylindrical body is usually tapered at both ends and has bilateral features. A cuticle covers it. Animal **cuticles** are tough, protective, often flexible body coverings. A roundworm is the simplest animal with a complete digestive system. Between the gut and body wall is a false coelom, most of which is packed with reproductive organs. Cells in all tissues absorb nutrients from coelomic fluid and give up wastes to it.

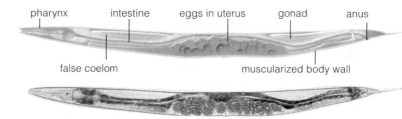

Figure 25.13 Body plan and micrograph of the roundworm *Caenorhabditis elegans*.

Parasitic species harm humans, cats, dogs, cows, sheep, soybeans, potatoes, and other hosts. But most are free-living and harmless or beneficial, as when their populations cycle nutrients for communities. One species—*Caenorhabditis elegans*—has endeared itself to many who study inheritance, development, and aging (Figure 25.13). Why? It has the same general body plan and tissues as complex animals (cephalized, bilateral, a complete digestive system). It is transparent and small (only about 1 millimeter long), so its development can be followed with micrographs of sections cut thinly, in series, along the body length. Also, its generation time is short. And hermaphroditic individuals make clones.

C. elegans was the first multicelled organism to have its genome fully sequenced. Of its 19,900 genes, only 3,000 are essential for normal structure and function.

Roundworms are cylindrical, bilateral, cephalized animals with a false coelom and complete digestive system. Many are parasites, but most cycle nutrients in communities.

A Rogue's Gallery of Worms

A number of parasitic flatworms and roundworms call the human body home. One of these, the Southeast Asian blood fluke *Schistosoma japonicum*, causes *schistosomiasis*. It requires a human definitive host, standing water for its swimming larvae, and an aquatic snail as an intermediate host. Figure 25.14 shows its life cycle. White blood cells of infected humans attack the masses of fluke eggs. Clumps of grainy debris form in tissues. In time, the liver, spleen, bladder, and kidneys deteriorate.

Some tapeworms, too, parasitize humans. Different kinds use pigs, freshwater fish, or cattle as intermediate hosts. Humans get infected by eating raw, improperly pickled, or insufficiently cooked pork, fish, or beef that contains tapeworm larvae (Figure 25.15).

The guinea worm (a parasitic roundworm) makes thin, serpentlike ridges in human skin. For thousands of years, healers have extracted the "serpents" from infected people by slowly winding them out on a stick. Or consider the roundworms called pinworms and hookworms. *Enterobius vermicularis*, a parasitic pinworm of temperate regions, lives in the human large intestine. At night, the females migrate to the anal region and lay eggs. Their activity makes skin itch, and when individuals scratch in response they transfer eggs to their hands, then to other objects. Newly laid eggs contain embryonic pinworms. In less than a few hours, they develop into juveniles, ready to hatch if another human inadvertently ingests them.

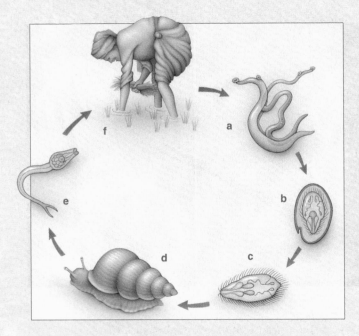

Figure 25.14 Life cycle of *Schistosoma japonicum*. (**a**) This blood fluke grows, matures, and mates in human hosts. (**b,c**) Fertilized eggs exit in feces, then hatch as ciliated, swimming larvae. (**d**) Larvae burrow into an aquatic snail and multiply asexually. (**e,f**) Fork-tailed, swimming larvae develop, leave the snail, then bore into skin of a human host. They migrate to thin-walled intestinal veins, and the cycle begins anew.

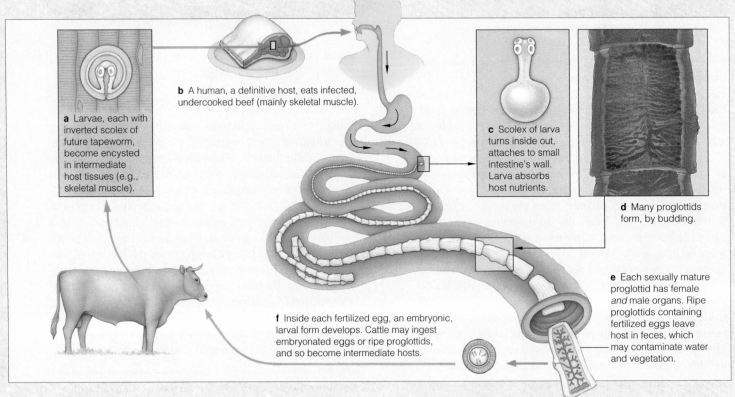

a Larvae, each with inverted scolex of future tapeworm, become encysted in intermediate host tissues (e.g., skeletal muscle).

b A human, a definitive host, eats infected, undercooked beef (mainly skeletal muscle).

c Scolex of larva turns inside out, attaches to small intestine's wall. Larva absorbs host nutrients.

d Many proglottids form, by budding.

e Each sexually mature proglottid has female *and* male organs. Ripe proglottids containing fertilized eggs leave host in feces, which may contaminate water and vegetation.

f Inside each fertilized egg, an embryonic, larval form develops. Cattle may ingest embryonated eggs or ripe proglottids, and so become intermediate hosts.

Figure 25.15 Life cycle of a beef tapeworm, *Taenia saginata*.

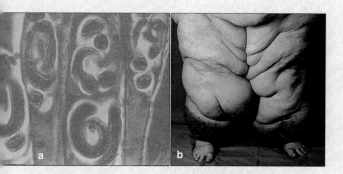

Figure 25.16 (**a**) Juveniles of the roundworm *Trichinella spiralis* in the muscle tissue of a host animal. (**b**) Severe edema caused by the roundworm *Wuchereria bancrofti*.

Adult hookworms live in a host's small intestine. Using toothlike devices or sharp ridges around their mouth, they cut into the intestinal wall, feed on blood and other tissues, and so compete with their host for nutrients. Adult females, about a centimeter long, can release a thousand eggs daily. After leaving the body via feces, the eggs hatch into juveniles. When a juvenile hookworm contacts bare human skin, it cuts its way inside. It travels the bloodstream to the lungs, works its way into air spaces, and moves up the windpipe. When the host swallows, the parasite is transported to the stomach and then to the small intestine. There, it may mature and live for several years.

The roundworm *Trichinella spiralis* causes trichonosis, a disease with painful, sometimes fatal symptoms. People become infected mainly by eating undercooked meat from infected pigs or certain game animals. Encysted juveniles are hard to detect when inspecting fresh meat. Adults live in the lining of the small intestine, where the females release juveniles (see Figure 25.16*a*). The juveniles infiltrate blood vessels, then muscles, where they become encysted. That is, they secrete a covering around themselves and enter a resting stage.

Figure 25.16*b* shows a result of prolonged, repeated infections by another roundworm, *Wuchereria bancrofti*. Adult worms get lodged in lymph nodes, organs that filter lymph: excess tissue fluid. Lymph normally flows back to blood, but the worm obstructs the flow. When an obstruction causes fluid to back up and accumulate in tissue spaces, the legs and other body regions enlarge grossly. We call this condition *elephantiasis*.

A mosquito is *Wuchereria*'s intermediate host. This parasite's females produce active young that travel at night in the bloodstream. When a mosquito sucks blood from an infected person, juveniles enter the insect's tissues. They move close to the insect's sucking device and enter a new host when blood is drawn again.

ROTIFERS

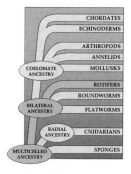

Like the roundworms, the **rotifers** (Rotifera) are bilateral, cephalized animals with a false coelom. All but 5 percent live in fresh water, such as lakes, ponds, and droplets on plants. One liter of pond water typically holds 40 to 500 rotifers; 5,000 have been recorded on a few occasions. They all are predators of bacteria and microscopic algae.

Most rotifers are less than one millimeter long, yet rarely do we see so many organs in so little space. A pharynx, an esophagus, digestive glands, a stomach, protonephridia, and most often an intestine and anus are packed inside. Nerve cell bodies clustered in the head integrate body activities. Certain rotifers have "eyes," clusters of absorptive pigments.

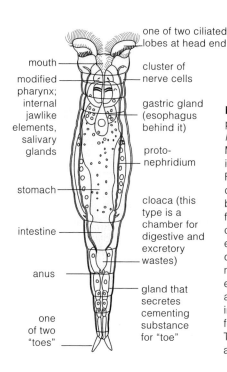

Figure 25.17 Body plan of the rotifer *Philodina roseola*. Males are unknown in many species. Females produce diploid eggs that become diploid females. Some also can make haploid eggs that develop directly into haploid males. If a haploid egg is fertilized by a male, it develops into a female. Males form occasionally. They are dwarfed, and short-lived.

Labels for Figure 25.17:
- one of two ciliated lobes at head end
- mouth
- modified pharynx; internal jawlike elements, salivary glands
- cluster of nerve cells
- gastric gland (esophagus behind it)
- proto-nephridium
- stomach
- cloaca (this type is a chamber for digestive and excretory wastes)
- intestine
- anus
- gland that secretes cementing substance for "toe"
- one of two "toes"

As Figure 25.17 shows, two "toes" exude substances that attach free-living species to substrates at feeding time. A feature unique to rotifers is a crown of cilia at the head end that assists in swimming and in wafting food toward the mouth. Its rhythmic motion reminded some early microscopists of a turning wheel; hence the phylum's name (rotifer means wheel-bearer).

The rotifers are bilateral, cephalized animals with ciliated lobes at their head end and a false coelom that is packed with many kinds of specialized organs.

A MAJOR DIVERGENCE

Late in the precambrian, bilateral animals not much more complex than modern flatworms evolved. Soon afterward, they gave rise to two lineages of coelomate animals (Figure 25.2). We call these two great lineages **protostomes** and **deuterostomes**. Mollusks, annelids, and arthropods are protostomes. The echinoderms and chordates are deuterostomes.

Different mutations accumulated in these separate lineages. Some altered how an embryo develops from a fertilized egg. For instance, mitotic cell divisions cut an egg's cytoplasm along prescribed planes to form a ball of cells. Section 43.3 describes what goes on. For now, simply know that in protostomes, the pattern is *spiral* cleavage: earliest cuts occur at oblique angles relative to the main body axis. In deuterostomes, the pattern is *radial* cleavage: certain early cuts are parallel with the original body axis and others are perpendicular to it:

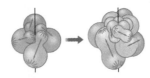

Early protostome embryo. Its four cells are undergoing cleavages *oblique to* the original body axis:

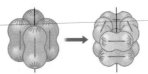

Early deuterostome embryo. Its four cells are undergoing cleavages *parallel with* and *perpendicular to* the original body axis:

As other examples, the first opening that forms in an early protostome embryo becomes a mouth; an anus forms elsewhere. The first opening in a deuterostome embryo becomes the anus and the second becomes the mouth. Also, a protostome's coelom forms from spaces that open in mesoderm, but a deuterostome's coelom forms from outpouchings of the gut wall:

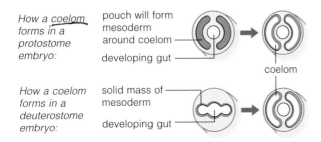

How a coelom forms in a protostome embryo: pouch will form mesoderm around coelom — developing gut

coelom

How a coelom forms in a deuterostome embryo: solid mass of mesoderm — developing gut

Such modifications to the embryonic stages of the two kinds of animals led to key differences in body plans.

Soon after coelomate animals evolved in Cambrian times, two great lineages—the protostomes and deuterostomes—evolved through mutations that affected how their embryos develop. This led to major differences in body plans.

A SAMPLING OF MOLLUSKS

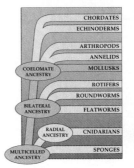

From our explorations in gardens and books, most of us would have no trouble recognizing a land snail when we see one. Yet few know much about its relatives in one of the largest of all animal groups, phylum Mollusca. There are more than 110,000 known species.

As the name implies, **mollusks** have fleshy soft bodies (*molluscus*, a Latin word, means soft). These bilateral animals have a small coelom. Many have a shell or a reduced version of one. The shell is made of calcium carbonate and proteins secreted from cells of a tissue that drapes like a skirt over the body mass. This is the **mantle**, a tissue unique to mollusks. Respiratory organs (ctenidia) are a type of gill having thin-walled leaflets. Most mollusks have a fleshy foot. Many have a radula, a tonguelike, toothed organ that rasps away at food destined for the gut. The mollusks with a well-developed head have eyes and tentacles, but not all have a head. Figure 25.18 has a generalized body plan.

Beyond these generalizations, there are no "typical" mollusks. They range in size from tiny snails that live in treetops to huge predators of the seas. In this section and the next, we sample just four classes: the chitons, gastropods, bivalves, and cephalopods.

The largest class, the gastropods ("belly foots"), has 90,000 species of slugs and snails. It includes limpets, periwinkles, nudibranchs, and garden snails. The name

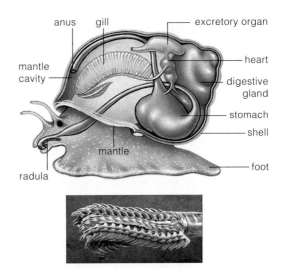

Figure 25.18 Body plan of an aquatic snail, a gastropod. The scanning electron micrograph shows one kind of radula, a feeding device of snails and some other mollusks. When a radula rhythmically protracts and retracts, it rasps food and draws it toward the gut on the retraction stroke.

Figure 25.19 A sampling of molluscan diversity. (**a**) Land snail, a gastropod. (**b**) Sea slugs, a type of gastropod called a nudibranch. Different gastropods creep, swim, or float as they graze on, prey on, or parasitize other organisms. (**c**) Chiton from the intertidal zone of Monterey Bay. Its broad foot clings to substrates.

(**d**) Octopus. Its eyes resemble yours but form differently (Section 35.6). Like other cephalopods, it has a well-developed nervous system. (**e**) Scallop. The light-sensitive "eyes" (small dark dots) of this bivalve fringe its shell's two halves. Iridescent mother-of-pearl lines the shells of many bivalves, including a few pearl producers. (**f**) Squid (*Dosidiscus*) and diver inspecting each other.

(**g**) External shell, longitudinal section, of a chambered nautilus. This cephalopod, the only existing one with an external shell, can swiftly change its buoyancy. It sinks fast as a jet propulsion siphon sucks gas-filled fluid into the shell's chambers and rises as it shoots fluid out. The nautilus constructs and lives in the newest, largest chamber, which is visible in this cut shell. Some shells grow wider than a dinner plate.

refers to the way a mollusk's foot spreads out when it crawls. Many species have a spirally coiled or conical shell. Coiling compacts the organs into a mass that is balanced above the body, like a backpack. Some species have lost the shell or have only a remnant of one.

Chitons are slow-moving or sedentary grazers with a dorsal shell divided into eight plates (Figure 25.19*c*). The bivalves—animals having a "two-valved shell"— include clams, scallops, oysters, and mussels (Figure 25.19*e*). Some bivalves are only a millimeter across. A few giant clams are over a meter across and weigh 225 kilograms (close to 500 pounds). Humans have eaten one type of bivalve or another since prehistoric times.

Cephalopods are active predators of the seas. In this class are the swiftest invertebrates (squids), the largest (the giant squid), and the smartest (octopuses, Figure 25.19*d*). Show an octopus an object with a distinctive shape and then give it a mild electric shock, and it will thereafter avoid that object. With respect to memory and the capacity to learn, octopuses and some of the squids are the world's most complex invertebrates.

Mollusks are bilateral, soft-bodied, coelomate animals that differ tremendously in body size, body details, and life-styles.

EVOLUTIONARY EXPERIMENTS WITH MOLLUSCAN BODY PLANS

Maybe it was their fleshy, soft bodies—so forgiving of chance evolutionary changes in morphology—that gave the ancestors of mollusks the potential to diversify in so many ways and to radiate into so many habitats. Let's explore this idea by using a few characteristics of our representative mollusks.

Twisting and Detwisting of Soft Bodies

Notice, in Figure 25.20, how evolution put an unusual twist in the soft snail body. Its anus dumps wastes near the mouth! As a gastropod embryo develops, a cavity between its mantle and shell twists counterclockwise by 180°. So does nearly all of the visceral mass (the gut, heart, gills, and other internal organs). This process is called **torsion**, and it occurs only in gastropods.

Such a drastic rearrangement of body parts could have come about by mutations that affected retractor muscles, which attach a gastropod embryo or larva to its shell. The muscles on the body's right side develop before those on the left. As they grow toward the front of the body, they drag the mantle cavity and organs with them. It takes them a few hours at most to do this.

By one hypothesis, torsion proved to be adaptive, for the head could withdraw into the mantle cavity in times of danger. However, in some 1985 experiments by J. Pennington and F. Chia, predators ate just as many torsioned larvae as "pre-torsioned" ones. Was torsion a bad evolutionary experiment? By putting the gills, anus,

and kidneys right above the mouth, it certainly created a potential problem in terms of hygiene. At the least, the uptake of discharged wastes in ancestral torsioned species must have been distasteful.

In fact, we find evolutionary compensations for this state of affairs. Most gastropods now have enough cilia in this region to create currents that sweep the wastes away. Also, torsion is not as pronounced as it once was in some lineages. Nudibranchs and sea slugs (Figure 25.20c) have even undergone detorsion. The soft larval body twists, but then it untwists. At some point in their evolution, these mollusks also lost most of the mantle cavity and all of their ctenidia. Most species have other outgrowths that function in gas exchange.

Hiding Out, One Way or Another

If you were small, edible, and soft of body, an external shell would be a distinct advantage, as it is for chitons, clams, and limpets (Figure 25.20d). A chiton disturbed by predators or exposed by a receding tide hunkers under its shell. Muscles in its foot pull the body mass down as the mantle edge around the shell's rim presses like a suction cup against a rock. The eight-plated shell is flexible. Pluck a chiton from a rock, and it rolls up in a ball until it can unroll and get reattached elsewhere.

Besides building a shell, protection also can be had by hiding in sand, sediments, or mud. A bivalve's head is not much to speak of, but its foot is usually large and

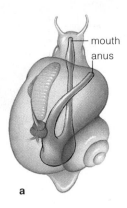

mouth
anus

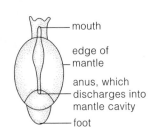

BEFORE TORSION

mouth
edge of mantle
anus, which discharges into mantle cavity
foot

a

c

Figure 25.20 (**a**) Dorsal view of an aquatic snail. (**b**) Torsion, a process that occurs during the development of gastropods only. (**c**) Sea slug *Aplysia*, also called a sea hare. Two flaps above its dorsal surface are foot extensions that undulate and ventilate the mantle cavity. Like many other gastropods, *Aplysia* is hermaphroditic. It can function as a male, a female, or both. (**d**) Ventral view of a blue limpet crawling on glass.

AFTER TORSION

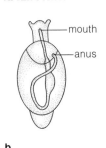

mouth
anus

b

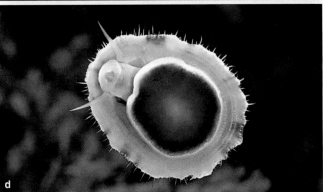

d

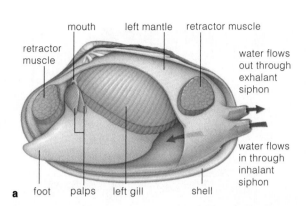

Figure 25.21 (**a**) Body plan of a clam, with half of its shell removed. In nearly all bivalves, gills serve in collecting food and in respiration. As water moves through the mantle cavity, mucus on the gills traps food bits. Cilia move the mucus and food to palps, where final sorting takes place before suitable bits are driven to the mouth. (**b**) A scallop escaping from a sea star by clapping its valves and producing a propulsive water jet.

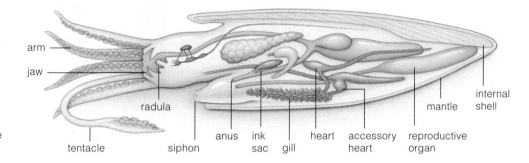

Figure 25.22 Generalized body plan of a cuttlefish, one of the cephalopods.

specialized for burrowing. Bivalves that dig in sand or mud have paired siphons. These are extensions of the mantle edges, fused into tubes (Figure 25.21a). Water is drawn into the mantle cavity through one siphon and leaves through the other, carrying wastes with it. One wonders what preyed on the ancestors of geoducks of the Pacific Northwest, which have siphons more than a meter long.

On the Cephalopod Need for Speed

Some 500 million years ago the cephalopods, with their buoyant, chambered shells, were the supreme predators in Ordovician seas (Section 20.5). And yet, of a lineage having more than 7,000 ancestral species, the shell of all but one of the existing descendant species is reduced or gone (Figure 25.22).

What happened? This evolutionary trend coincided with an adaptive radiation of the bony fishes—which preyed on cephalopods or were strong competitors for the same prey. During what may have been a long-term race for speed and wits, cephalopods lost their thick external shell, and they became streamlined and highly active. Of all mollusks, they now have the largest brain relative to body size and display the most complex behavior. Nerves swiftly deliver signals from the brain to muscles that carry out responses to food or danger. Except for the chambered nautilus, they also discharge dark fluid from an ink sac, maybe to confuse predators.

Jet propulsion became the name of the game. All cephalopods force a jet of water from the mantle cavity and a funnel-shaped siphon. Water is drawn into the cavity as mantle muscles relax. Next, muscles contract at the same time that the free edge of the mantle closes down on the head. The rapid squeeze shoots the water out as a jet through the siphon. The brain controls the siphon's activity and the direction of escape or pursuit.

Supporting all the activity are efficient respiratory and circulatory systems. The cephalopods are the only mollusks with a closed circulatory system. The heart pumps blood to two gills, then two accessory (booster) hearts speed blood flow, oxygen uptake, and carbon dioxide removal—for muscle cells especially.

Lively stories emerge when evolutionary theory is used to interpret the fossil record and the range of existing species diversity, as we have done for the mollusks.

ANNELIDS—SEGMENTS GALORE

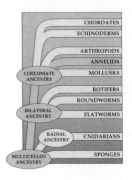

Maybe you've seen earthworms after a downpour. They wriggle out of their flooded burrows to avoid drowning. Earthworms are among the 15,000 or so species of bilateral, segmented animals we call **annelids** (Annelida, meaning ringed forms). Their relatives are the leeches and the less familiar but far more diverse polychaetes (Figures 25.23 and 25.24). Their body actually has repeating segments, not "rings" as the phylum name implies. The segmentation is highly pronounced. Except in leeches, nearly all segments have pairs or clusters of chitin-reinforced bristles on each side of the body. These are also called setae or chaetae, which are just formal names for bristles. When pushed into soil, the bristles provide traction for crawling and burrowing. Earthworms, which are oligochaetes, have a few bristles on their segments; the marine polychaete worms typically have many (*oligo–*, few; *poly–*, many).

Advantages of Segmentation

The segmented body apparently had great evolutionary potential for some annelid lineages. Most segments of an earthworm are still similar, but leeches have suckers at both ends, and polychaetes have an elaborate head and fleshy-lobed appendages called parapods ("closely resembling feet"). Existing species offer us glimpses of the kinds of modifications in body plans that permitted increases in size, more complex internal organs, and segments adapted for specialized tasks.

Annelid Adaptations—A Case Study

Let us use earthworms as our representative annelid. Partitions divide their body into a series of coelomic chambers, each with repeats of muscles, blood vessels,

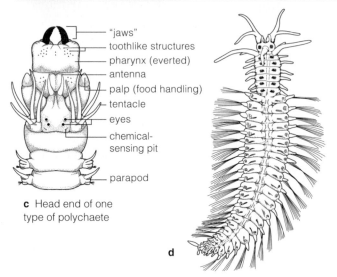

c Head end of one type of polychaete

"jaws"
toothlike structures
pharynx (everted)
antenna
palp (food handling)
tentacle
eyes
chemical-sensing pit
parapod

Figure 25.23 Leeches. Most leeches have sharp jaws and a blood-sucking device. This one is shown before and after gorging on human blood. For at least 2,000 years, *Hirudo medicinalis*, a freshwater leech, has been employed as a blood-letting tool to "cure" problems ranging from nosebleeds to obesity.

Some surgeons still use leeches, but more selectively, as when they draw off pooled blood after a severed ear, lip, or fingertip is reattached. A patient's body can't do this on its own until the severed blood circulation routes are reestablished.

before feeding

after feeding

Figure 25.24 (a) A familiar annelid—one of several species of earthworms. (b–d) Polychaetes, which give us a sense of the dizzying variety of modifications that have evolved in this group, starting with a segmented, coelomic body plan.

The polychaete shown in (b) is one of the tube-dwellers. Featherlike structures at its head end are coated with mucus. After the mucus has trapped bacteria and other bits of food, coordinated beating of cilia sweeps them to the mouth. Most polychaetes live in marine habitats. They actually are one of the most common types of animals along coasts. Many are predators or scavengers; others dine on algae.

Figure 25.25 Earthworm body plan. (**a**) Midbody, transverse section. (**b**) A nephridium, one of many functional units that help maintain the volume and the composition of body fluids. (**c**) Portion of the closed circulatory system. The system is linked in its functioning with nephridia. (**d**) Part of the digestive system, near the worm's head end. (**e**) Part of the nervous system.

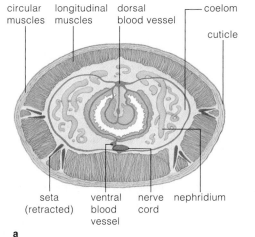

circular muscles · longitudinal muscles · dorsal blood vessel · coelom · cuticle

seta (retracted) · ventral blood vessel · nerve cord · nephridium

a

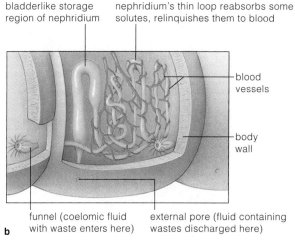

bladderlike storage region of nephridium · nephridium's thin loop reabsorbs some solutes, relinquishes them to blood

blood vessels

body wall

funnel (coelomic fluid with waste enters here) · external pore (fluid containing wastes discharged here)

b

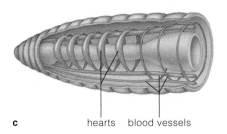

c hearts · blood vessels

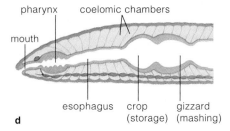

pharynx · coelomic chambers

mouth

esophagus · crop (storage) · gizzard (mashing)

d

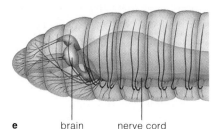

e brain · nerve cord

branching nerves, and other organs. The gut extends through all chambers, from mouth to anus. As in all annelids, an earthworm's body surface is enclosed in a flexible cuticle of secreted material (Figure 25.25). The permeable cuticle is good for gas exchange but not for conserving water. That is one reason annelids cannot venture from aquatic habitats or moist habitats on land.

Earthworms are scavengers. They ingest moistened soil rich in detritus (decomposing organic bits, such as leaf litter). One worm can eat the equivalent of its own weight each day. When many worms burrow and feed, they collectively aerate the soil and lift nutrients to the surface, to the benefit of many plants.

Like other annelids, earthworms have a hydrostatic skeleton. Their muscles work against fluid-cushioned coelomic chambers. In each segment's wall is a layer of circular muscles (Figure 25.25a). Longitudinal muscles bridge several segments. As they contract, the circular muscles relax. The segments shorten and fatten, and their bristles anchor them by plunging into soil. The pattern then reverses. The segments lengthen and their bristles retract, thereby freeing them for movement. The whole worm moves forward by muscle contractions in one segment after another, down the length of its body.

An earthworm burrows by closing off the coelomic fluid inside its anterior segments and then contracting circular muscles to compress the fluid and extend the longitudinal muscles. Segments closer to the back have their bristles plunged into the wall of the burrow; these anchor the worm as its head is being pushed forward.

Annelids contain a system of **nephridia** (singular, nephridium), a series of units that control the volume and composition of body fluids. Usually, each unit has a funnel that collects excess fluid from one coelomic chamber and drains it into small tubes in the chamber behind it (Figure 25.25b). The tubes lead to a bladder, which delivers fluid to a pore at the surface in the body wall of the next coelomic chamber.

The worm's head end has a rudimentary **brain**. This cluster of nerve cell bodies integrates sensory input and commands for muscle responses for the whole body. Paired **nerve cords**, each a bundle of long extensions of nerve cell bodies, lead out from the brain. These are pathways for fast communication. In each segment, the paired cords broaden into a ganglion (plural, ganglia), a cluster of nerve cell bodies that controls local activity.

Finally, as is the case for most annelids, earthworms have a closed circulatory system. Multiple hearts and the muscularized walls of blood vessels contract and keep blood circulating in one direction. Smaller vessels service the gut, nerve cord, and body wall.

Annelids are bilateral, coelomate, segmented worms that have complex organ systems. Some show the great degree of specialization possible with a segmented body plan.

ARTHROPODS—THE MOST SUCCESSFUL ORGANISMS ON EARTH

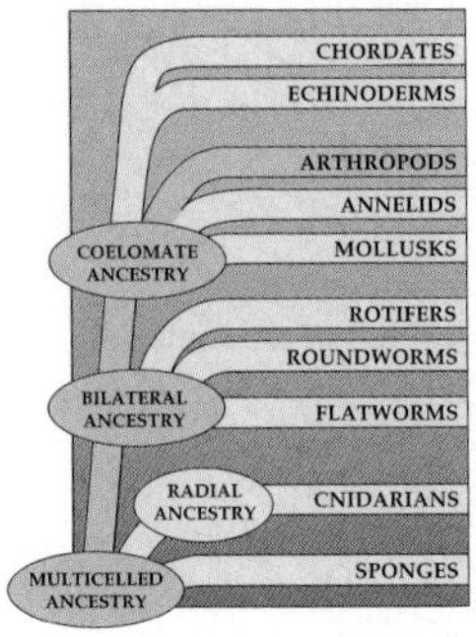

Arthropod Diversity

Evolutionarily speaking, "success" means having the most offspring, the greatest number of species, and the most habitats; fending off threats and competition efficiently; and having the capacity to exploit the greatest amounts and kinds of food. These are the features that come to mind when we attempt to characterize **arthropods** (Arthropoda). We have already identified more than a million species—mostly insects—and researchers are discovering new ones weekly.

Of four major lineages, trilobites (Section 20.5) are extinct. The other three are chelicerates (spiders and their relatives), crustaceans (such as crabs and shrimps), and uniramians (centipedes, millipedes, and insects).

Adaptations of Insects and Other Arthropods

Six important adaptations contributed to the success of arthropods in general and to the insects in particular: a hardened exoskeleton, jointed appendages, fused and modified segments, specialized respiratory structures, efficient nervous system and sensory organs, and often a division of labor in the life cycle.

HARDENED EXOSKELETONS   Arthropods have a cuticle of chitin, proteins, and surface waxes that are often impregnated with calcium carbonate deposits. It acts as a rigid, protective external skeleton—an **exoskeleton**. Such cuticles might have evolved as defenses against predation. They took on added functions when some arthropods first invaded the land. They support a body deprived of water's buoyancy, and their waxy surface limits evaporative water loss. Hard cuticles restrict size increases, but arthropods grow in spurts, by **molting**. At certain stages of the life cycle, they secrete a soft new cuticle under the old one, which they shed (Figure 25.26). The body mass increases by the uptake of air or water, and by repeated, rapid cell division before the new cuticle can harden.

JOINTED APPENDAGES   If an arthropod had a uniformly hardened cuticle, it wouldn't move much. But the arthropod cuticle is thinner at *joints,* where two body parts

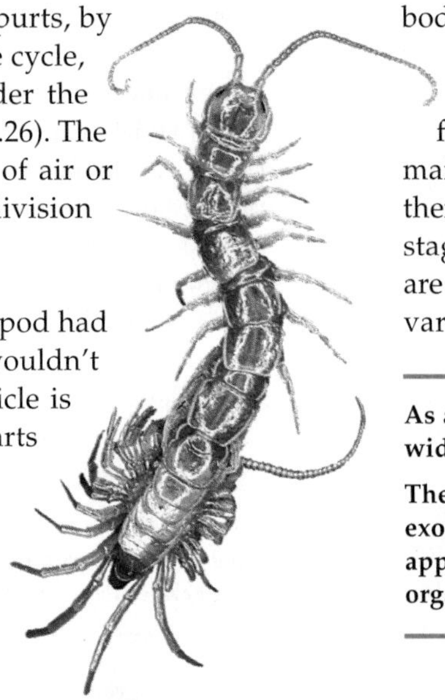

Figure 25.26   Molting, demonstrated by a red-orange centipede wriggling out of its old exoskeleton.

abut. Muscles associated with joints make the thinner cuticle bend in certain directions and move body parts. The jointed exoskeleton proved to be a key innovation; it led to appendages as diverse as wings, antennae, and legs (*arthropod* means "jointed foot").

FUSED AND MODIFIED SEGMENTS   The first arthropods were segmented, like the annelid stock that presumably gave rise to them. In most existing descendants, serial repeats of the body wall and organs are masked; many fused-together segments are modified to perform more specialized functions. For example, in the ancestors of insects, different segments fused to form three regions—head, thorax, and abdomen—which morphologically diverged from one another in remarkable ways.

RESPIRATORY STRUCTURES   Many aquatic arthropods depend on gills for gas exchange. Air-conducting tubes evolved among insects and other land dwellers. Insect tracheas have pores on the body surface. They branch into tubes that deliver oxygen directly to tissues. They support flight and other energy-consuming activities.

SPECIALIZED SENSORY STRUCTURES   Intricate eyes and other sensory organs contributed to arthropod success. Example: Species having a wide angle of vision can get visual information coming from many directions.

DIVISION OF LABOR   Moths, butterflies, beetles, flies, and many other species divide the job of surviving and reproducing among different stages of development. For many species, the new individual is a *juvenile*—a miniaturized version of the adult that simply changes in size and proportion until reaching sexual maturity. Other species show **metamorphosis**, meaning the body form changes from embryo to adult. Under hormonal commands, their size increases, tissues reorganize, and body parts are remodeled. Sections 25.15, 25.17, and 36.8 have examples of this transitional time. Immature stages such as caterpillars typically specialize in feeding and growing in size. Adult stages specialize mainly in dispersal and reproduction. For such species, then, the life cycle turns on a *division of labor:* different stages of development become specialized in ways that are well adapted to environmental conditions—seasonal variation in food resources, water supplies, and so on.

As a group, the arthropods are exceptionally abundant and widespread, and they have enormously different life-styles.

Their success arises largely from their hardened, jointed exoskeletons; fused, modified body segments; specialized appendages; specialized respiratory, nervous, and sensory organs; and often a division of labor in the life cycle.

A LOOK AT SPIDERS AND THEIR KIN

The chelicerates originated in shallow seas early in the Paleozoic. The only surviving marine species are a few mites, sea spiders, and horseshoe crabs (Figure 25.27). The familiar chelicerates on land—spiders, scorpions, ticks, and chigger mites—are all classified as arachnids. We might say this about the whole group: Never have so many been loved by so few.

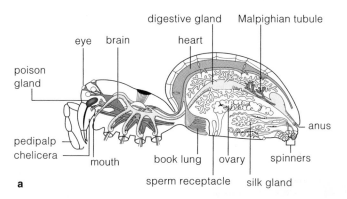

Figure 25.27 Horseshoe crab, (**a**) ventral and (**b**) dorsal views. Five pairs of legs are one of the defining features of this chelicerate. They are hidden beneath the hard, shieldlike cover.

Figure 25.28 (**a**) A spider's internal organization. (**b**) Wolf spider. Like most spiders, this one helps keep populations of insects in check. Its bite is harmless to humans. Section 25.18 describes two of its dangerous relatives.

Bites of the blood-sucking ticks that parasitize deer, mice, and other vertebrates can cause maddening itches and often serious diseases. For example, bites of some ticks transmit the bacterial agents of Rocky Mountain spotted fever or Lyme disease to humans (Sections 21.6 and 25.18). Most mites are free-living scavengers.

Spiders and scorpions are efficient predators; they sting or bite and may subdue prey with venom. When most kinds sting or bite people, the reaction might be painful but only rarely does it lead to serious problems. Collectively, the spiders especially are beneficial in that they prey upon great numbers of pestiferous insects.

Arachnids have segments fused into a forebody and hindbody. The forebody's jointed appendages include four pairs of legs, a pair of pedipalps with primarily sensory functions, and a pair of chelicerae that inflict wounds and discharge venom. The appendages of the

hindbody spin out silk threads for webs and for egg cases. Most webs are netlike. One type of spider spins a vertical thread with a ball of sticky material at the end. It uses a leg to swing the ball at insects passing by!

Inside the body is an *open* circulatory system, with a heart that pumps blood into body tissues, then receives blood through small openings in its wall. Some of the slowly circulating blood travels through moist folds of book lungs. These respiratory organs, which resemble loose pages of a book, greatly increase the surface area that is available for gas exchange with the air. Figure 25.28*a* shows the arrangement of book lungs and other major organs inside the spider body.

The spiders, scorpions, and their relatives have a variety of appendages specialized for predatory or parasitic life-styles.

A LOOK AT THE CRUSTACEANS

Shrimps, lobsters, crabs, barnacles, pillbugs, and other crustaceans got their name because they have a hard yet flexible "crust" (an external skeleton)—but so do nearly all arthropods. Only some of the 35,000 species live in fresh water or on land. The vast majority live in marine habitats, where they are so abundant they have been dubbed the insects of the seas. Lobsters and crabs are "giants" of this subphylum; most crustaceans are less than a few centimeters long. All have major roles in food webs, and humans harvest many edible types.

The simplest crustaceans may resemble their annelid ancestors in having pairs of similar appendages along most of their body. Among the more complex lineages,

unspecialized appendages evolved into many diverse structures of the sort shown in Figure 25.29. The strong claws of lobsters and crabs, for example, are used to collect food, intimidate other animals, and sometimes dig burrows. Feathery appendages of barnacles comb microscopic bits of food from the water.

Many crustaceans have sixteen to twenty segments; some have more than sixty. In crabs, lobsters, and some other crustaceans, the dorsal cuticle extends back from the head as a shieldlike cover over some or all of the segments. The head has two pairs of antennae, a pair of jawlike appendages (mandibles), and two other pairs of food-handling appendages. Crayfish, crabs, lobsters,

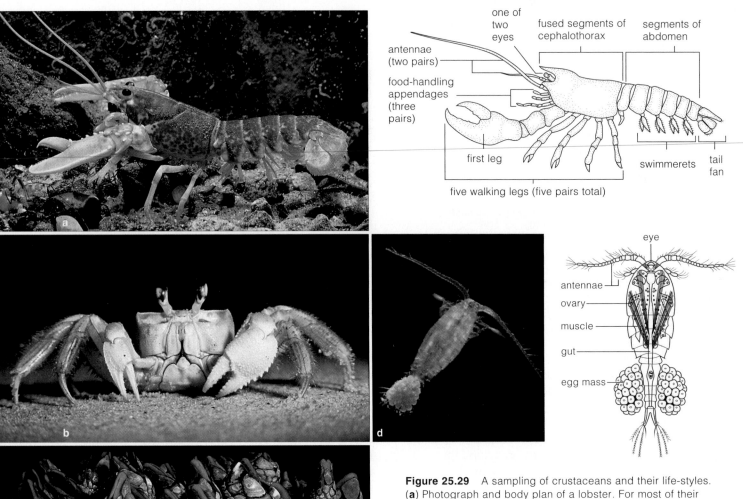

Figure 25.29 A sampling of crustaceans and their life-styles. (**a**) Photograph and body plan of a lobster. For most of their lives, lobsters are secretive. (**b**) Crab. Guess why crabs skitter actively and openly across sand and rocks at night but not in the day. (**c**) Stalked barnacles. Adults cement themselves to one spot. You might mistake them for mollusks, but as soon as they open their hinged shell to filter-feed, you see jointed appendages, the hallmark of arthropods. (**d**) Photograph and body plan of a copepod. Copepods are free-living filter feeders, predators, or parasites. This female has a pair of long antennae. She carries her eggs with her.

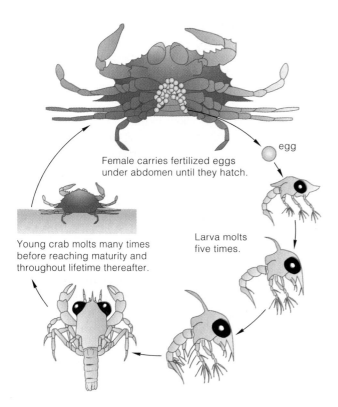

Figure 25.30 Life cycle of a crab. The larval and juvenile stages molt repeatedly and grow in size.

Female carries fertilized eggs under abdomen until they hatch.

egg

Larva molts five times.

Young crab molts many times before reaching maturity and throughout lifetime thereafter.

shrimps, and their various relatives are equipped with five pairs of walking legs.

Of all the arthropods, only barnacles have a calcified "shell." This modified external skeleton protects them from predators, drying winds, battering surf, and strong currents. Adult barnacles cement themselves to rocks, wharf pilings, and similar surfaces (Figure 25.29c). A few kinds attach themselves only to the skin of whales.

Figure 25.29d shows a copepod. Copepods are less than two millimeters long and are the most numerous animals in aquatic habitats. They also are abundant on land. About 1,500 kinds parasitize various invertebrates and fishes. But the majority—8,000 species—consume phytoplankton, the "pastures" of aquatic habitats. Some prey on larval or small adult invertebrates, fish eggs, and fish larvae, which they grab with a pair of food-handling appendages. In turn, the copepods are food for different invertebrates, fishes, and baleen whales.

Like other arthropods, crustaceans repeatedly molt and shed the exoskeleton during their life cycle. Figure 25.30 shows a crab's larval stages and increases in size.

Crustaceans differ greatly in their number and kind of appendages. As is the case for arthropods in general, they repeatedly replace their external skeleton by molting.

HOW MANY LEGS?

The **millipedes** and **centipedes** have a long, segmented body with many legs. Of course, millipedes don't have "a thousand," as their name implies. Most have about 100 pairs of legs, although one impressive individual grew 752. Centipedes have between 15 and 177 pairs of legs, not a nicely rounded number of "one hundred."

As a millipede develops, its pairs of segments fuse, and each segment in its cylindrical body ends up with two pairs of legs (Figure 25.31a). Millipedes scavenge decaying plant material in soil and forest litter.

Figure 25.31 (a) Millipede. (b) A Southeast Asian centipede.

Centipedes have a flattened body, and all but two segments have a pair of walking legs. All species are fast-moving, aggressive predators outfitted with fangs and venom glands. They prey on insects, earthworms, and snails. The one in Figure 25.31b can subdue small lizards, toads, and frogs. A house centipede (*Scutigera*) often hides in buildings and hunts cockroaches, flies, and other pests. Although helpful in this respect, its vaguely terrifying body keeps it from being welcomed.

Mild-mannered, scavenging millipedes and aggressive, predatory centipedes do not lend themselves to leg counts as they walk by.

A LOOK AT INSECT DIVERSITY

As a group, insects share the adaptations listed earlier in Section 25.13. Here we expand the list. Insects have a head, a thorax, and an abdomen. The head has paired sensory antennae and mouthparts for biting, chewing, sucking, or puncturing (Figure 25.32). Three pairs of legs and usually two pairs of wings project from the thorax. Insects are the *only* winged invertebrates. Their abdomen has egg-laying devices and other reproductive structures. They have a foregut, midgut (where digestion occurs), and hindgut (where water is reabsorbed). They get rid of wastes through **Malpighian tubules**, small tubes that connect with the midgut. When proteins are digested, the nitrogen-containing wastes diffuse from blood into the tubules and are converted into harmless crystals of uric acid. These crystals are eliminated with feces. This system lets land-dwelling insects dispose of potentially toxic wastes without losing precious water.

We've already catalogued more than 800,000 species of insects. If we use sheer numbers and distribution as the yardstick, the most successful ones are small in size and have a staggering reproductive capacity. You may find some species growing and reproducing in great numbers on a single plant that would be an appetizer for another animal. By one estimate, if all offspring of one female fly survived and reproduced for six more generations, she would have over 5 trillion descendants!

Also, the most successful insects are winged. They move among food sources that are too widely scattered to be exploited by other kinds of animals. The capacity for flight contributed greatly to their success on land.

Finally, insect life cycles commonly proceed through stages that allow exploitation of different resources at different times. As an insect embryo develops, organs required for feeding and other vital activities form and become functional. Before an insect becomes an adult (the sexually mature form), it goes through immature, post-embryonic stages. **Nymphs** and **pupae**, as well as larvae, are such stages. Like human infants, nymphs of some insects have the form of miniature adults. Unlike children, these forms molt (Section 36.8). Other insects develop through post-embryonic stages of reactivated growth, tissue reorganization, and remodeling of body parts. Metamorphosis, remember, is the name for this resumption of growth and the transformation of post-embryonic stages into an adult. Transformation is more drastic in some species than in others (Figure 25.33).

The features that contribute to insect success also make them our most aggressive competitors. Insects

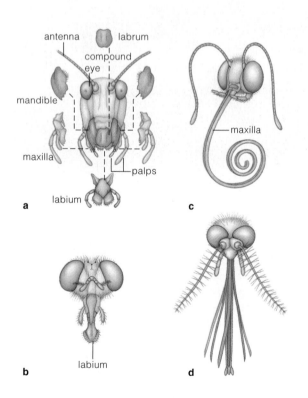

Figure 25.32 Examples of insect appendages. Head parts typical of (**a**) grasshoppers, a chewing insect; (**b**) flies, which sponge up nutrients; (**c**) butterflies, which siphon nectar; and (**d**) mosquitoes, with piercing and sucking appendages.

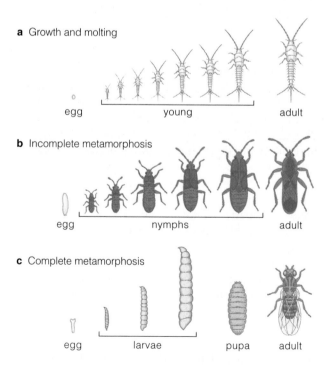

Figure 25.33 Examples of post-embryonic development. (**a**) Young silverfish, adults in miniature, change little except in size and proportion as they mature into adults. (**b**) True bugs show *incomplete* metamorphosis, gradual, partial change from the immature form until the last molt. (**c**) Fruit flies show *complete* metamorphosis. Tissues of immature forms are destroyed and replaced before the adult emerges.

Figure 25.34 Representative insects. (**a**) Duck louse (order Mallophaga). It eats bits of feathers and skin. (**b**) European earwig (order Dermaptera), one of the common household pests. (**c**) Flea (order Siphonaptera), with strong legs for jumping onto and off animal hosts. (**d**) Mediterranean fruit fly (order Diptera). Its larvae destroy citrus fruit and other crops.

(**e**) Stinkbugs (order Hemiptera), newly hatched. (**f**) At center, a honeybee (order Hymenoptera) attracting its hive mates by its dancing, as described in Section 46.4. (**g**) Ladybird beetles (order Coleoptera) swarming. These beetles are commercially raised and released as biological controls of aphids and other pests. Also in this order, *Dyastes*, a scarab beetle (**h**). With over 300,000 species, Coleoptera is the largest order of the animal kingdom. (**i**) Swallowtail butterfly and (**j**) luna moth (order Lepidoptera). Microscopic scales cover the wings and body of most butterflies and moths. (**k**) One of the dragonflies (order Odonata). It swiftly captures and eats insects in midflight.

destroy crops, stored food, wool, paper, and timber. As some stealthily draw blood from us and from our pets, they commonly transmit pathogens. Yet many pollinate flowering plants, including highly valued crop plants. And many "good" insects attack or parasitize the ones we would rather do without. Figures 25.33 and 25.34 show representatives of major orders of insects.

As a group, insects show immense variation on the basic arthropod body plan. Many have wings, and their life cycles have stages that allow exploitation of different and often widely scattered food sources. Many species produce great numbers of small individuals that pass through immature, post-embryonic stages before the adult form emerges.

Unwelcome Arthropods

Given that there are now more than 6 billion of us on the planet, very, very few of us encounter the truly nasty arachnids—certain spiders, ticks, and scorpions that can directly or indirectly inflict memorable pain. All things considered, humans collectively do more harm to more species than, say, the spiders, most of which do good by popping off uncountable numbers of insect pests. However, the "good" ones have some "bad" relatives, and it doesn't hurt to know them when we see them.

HARMFUL SPIDERS   All spiders of the genus *Loxosceles* are poisonous to humans and other mammals, but we seldom meet most of them. One exception, the brown recluse (*L. reclusa*), favors hiding in warm places. These include houses, especially near refrigerator motors and clothes dryers, and in the folds of clothing or linens in closets. You can identify it by the violin-shaped marking on its cephalothorax (Figure 25.35a). A bite ulcerates the skin and heals poorly. Some bitten people have required grafts, and a few have died. *L. laeta*, a related species in Chile, Argentina, and Peru, is similarly dangerous.

About five people die each year in the United States as a result of black widow bites (Figure 25.35b)—orders of magnitude less than deaths attributed to, say, drug overdoses or accidents involving SUVs. Even so, besides being painful, this spider's neurotoxin causes muscles to stiffen, so medical attention should be prompt.

MIGHTY MITES   Of all arachnids, mites are among the smallest, most diverse, and most widely distributed. Of an estimated 500,000 species, a few parasitic mites—ticks especially—are harmful. Figure 25.36 shows a deer tick (*Ixodes*), a vector for *Lyme disease* in the northeastern and

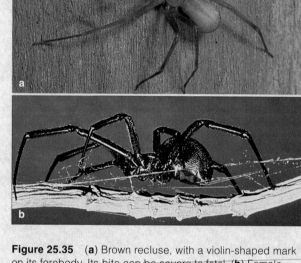

Figure 25.35 (**a**) Brown recluse, with a violin-shaped mark on its forebody. Its bite can be severe to fatal. (**b**) Female black widow, with a red, hourglass-shaped marking on the underside of her shiny black abdomen. Males are smaller and do not bite.

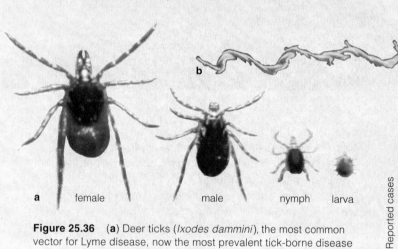

a   female   male   nymph   larva

Figure 25.36 (**a**) Deer ticks (*Ixodes dammini*), the most common vector for Lyme disease, now the most prevalent tick-borne disease in the United States. The female is three millimeters long. All stages of the life cycle can transmit the disease agent, the spirochete *Borrelia burgdorferi* (**b**), to humans. (**c**) A bull's-eye rash, an extreme reaction to a tick bite. (**d**) Graph of the rising number of reported cases of Lyme disease in the United States.

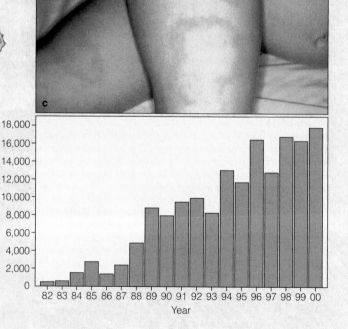

Figure 25.37 *Centruroides sculpturatus*, a scorpion from Arizona that is one of the most dangerous species known.

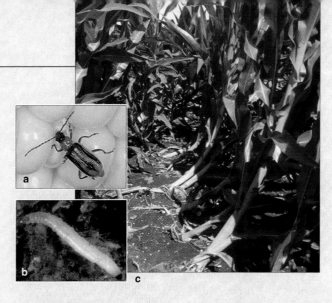

Figure 25.38 (**a**) Adult western corn rootworm (*Diabrotica virgifera*). (**b**) Larvae on corn roots. (**c**) Damaged corn plants.

north-central United States. (Different ticks spread the disease in the western and southeastern states.) Usually the tick feeds on white-tail deer, mice, other mammals, and birds. *Borrelia burgdorferi*, the agent of Lyme disease, is transmitted to humans mainly by the nymphal stages, probably because these are no bigger than a pinhead and escape detection. Detection is critical, because ticks normally transmit the bacterium after they have been feeding two days or more. Adult ticks are easier to spot, so they are more likely to be removed quickly.

Ticks don't fly or jump. They crawl onto grasses and shrubs, then onto animals that brush past, then typically into the scalp, groin, or other hairy, hidden body parts. Look for them after walks in the wild. Different types can transmit pathogens that cause serious diseases, such as Rocky Mountain spotted fever (Section 21.6), scrub typhus, tularemia, babesiasis, and encephalitis. A few can even kill directly. Their toxin interferes with nerves in a way that leads to progressive paralysis, from the lower extremities upward. The tick must be removed before the bitten person stops breathing.

Ending on a less scary but irritating note, think of the house dust mites. They don't kill you, but the allergies they provoke in people can be irritating to the extreme.

SCORPION COUNTRY   Oversize, prey-seizing pincers, a venom-dispensing stinger at the tip of a narrowed, jointed abdomen—what's not to love about scorpions? Actually, scorpions seldom dispense venom unless their prey—other small arthropods, the occasional small lizard or mouse—puts up a fight. In the wild they hunt at night and rest during the day in burrows or under logs, stones, and bark. They can remain hidden without eating and still survive for as long as a year.

We find scorpions in cold regions, including Canada and the southern Andes and southern Alps. Some live in moist forests. Most live in hot, dry regions, where they

are active all year. The ones that can hurt or kill people aren't the fiercest looking, and there aren't very many of them. *Centruroides sculpturatus*, an Arizona scorpion, is the most dangerous species in the United States (Figure 25.37). Its close relatives in Mexico have a reputation for causing fatalities, especially among small children.

When in scorpion country, it is not a good idea to walk about barefoot at night or to put on shoes without first turning them upside down and shaking them vigorously.

BEETLES EVERYWHERE   First in America's Midwest and now in the Balkans, *Diabrotica virgifera* is making the rounds in cornfields as a representative of one of our major competitors for food. Figure 25.38 shows what it looks like and the damage it causes. Its larvae feed on corn roots until the severely damaged plants keel over. Each year, about a billion dollars' worth of crops are lost to these pests alone.

The astronomical losses compel farmers to spread about 30 million pounds of pesticides annually through cornfields—about half the total for all row crops in the United States. As you might expect, many rootworms now show pesticide tolerance. Inventive researchers are targeting adults, which are fond of a bitter plant juice called curcurbitacin. This juice evolved as a natural pesticide in cucumbers, melons, squashes, and other plants. Rootworms apparently appropriated the chemical as a defense against bird predators. (Birds learn to avoid rootworms after a taste trial that ends in vomiting.)

Robert Schroder and others concocted a bait for the pests. In one test, they laced watermelon juice with red dye 28, a photoactive chemical. Adult rootworms gorged themselves, turned deep red, and after five minutes of sun exposure, dropped dead. The sun's rays triggered oxidation reactions that destroyed the insect tissues. Beneficial insects avoid the dyed juice. In this ongoing contest, chalk one up for the researchers.

THE PUZZLING ECHINODERMS

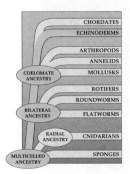

We arrive, finally, at the second lineage of coelomate animals, the deuterostomes. The **echinoderms** (Echinodermata) are invertebrate members of this lineage. Species in this phylum include the feather stars, sea urchins, sea cucumbers, and brittle stars (Figure 25.39), as well as sea lilies (crinoids), sand dollars, and sea biscuits. Sea lilies look a bit like stalked plants and flourished in Silurian times (Figure 20.5). About 13,000 echinoderm species are known from the fossil record, but most of these are extinct. Nearly all of the 6,000 or so existing species live in marine habitats.

An echinoderm body wall bears a number of spines, spicules, or plates made rigid with calcium carbonate. (Echinodermata means spiny-skinned.) The structures serve defensive functions, as you might suspect if you have ever stepped barefoot on a sea urchin. Its spines trigger painful swelling when they break off under the skin. Most echinoderms have a well-developed internal skeleton, which is composed of calcium carbonate and other substances secreted from specialized cells. Oddly, the adults are radial with some bilateral features. Most species even have bilateral larvae. Did the echinoderms evolve from bilateral ancestors, in which some radial features appeared at a later time? Maybe.

Adult echinoderms have no brain. However, their decentralized nervous system allows them to respond

Figure 25.39 Representative echinoderms. (**a**) A bed of sea urchins, which move about on spines and tube feet. (**b**) Feather star, with finely branched food-gathering appendages. (**c**) Sea cucumber, with rows of tube feet along its body. (**d**) Brittle star. Its arms (rays) make rapid, snakelike movements.

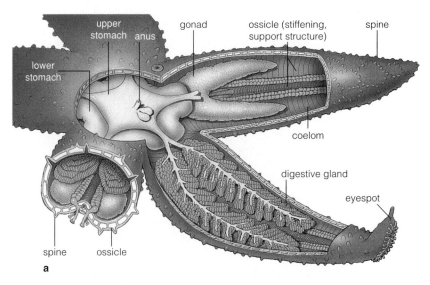

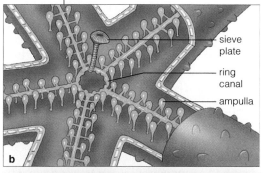

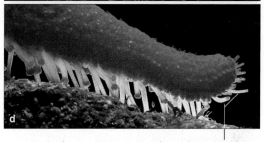

tube feet on the ventral surface of a sea star arm

Figure 25.40 (**a**) Radial body plan of a sea star. Its water-vascular system, combined with many tube feet, is the basis of locomotion. (**b**–**d**) Five-armed sea star, with closer views of tube feet.

to information about food, predators, and so forth that is coming from different directions. For instance, any arm of a sea star that touches the shell of a tasty mussel can become the leader, directing the rest of the body to move in a direction suitable for capturing its prey.

Figures 25.39 and 25.40 show examples of tube feet. These fluid-filled, muscular structures have suckerlike adhesive disks. Sea stars use their tube feet for walking, burrowing, clinging to rocks, or gripping a clam or snail about to become a meal. Tube feet are components of a **water-vascular system** unique to echinoderms. In sea stars, that system includes a main canal in each arm. Short side canals extend from them and deliver water to the tube feet. Each tube foot contains a fluid-filled, muscular structure shaped rather like the rubber bulb on a medicine dropper. As the bulb contracts, it forces fluid into the foot and causes it to lengthen.

Tube feet change shape constantly as muscle action redistributes fluid through the water-vascular system. Hundreds of tube feet may move at a time. After being released, each one swings forward, reattaches to the substrate, then swings backward and is released before swinging forward again. Their motions are splendidly coordinated, so sea stars glide rather than lurch along.

On their ventral surface, sea stars have a formidable feeding apparatus, such as the one shown here:

Some eager sea stars simply swallow their prey whole. Others push part of their stomach outside the mouth and around their prey, then start digesting their meal even before swallowing it. Sea stars get rid of coarse, undigested residues through the mouth even though they do have a small anus.

With their curious traits, echinoderms are a suitable point of departure for this chapter. Even though broad trends have been identified in animal evolution, keep in mind that there are confounding exceptions to the perceived macroevolutionary patterns.

Echinoderms are coelomate animals with spines, spicules, or plates in the body wall. From the evolutionary perspective, they are a puzzling mix of bilateral and radial features.

SUMMARY
Gold indicates text section

1. Animals are multicelled, aerobic heterotrophs that ingest or parasitize other organisms. Most kinds have diploid body cells organized into tissues, organs, and organ systems. Animals reproduce sexually and often asexually. They undergo embryonic development. Most are motile during at least part of the life cycle. *25.1*

2. Animals range from structurally simple placozoans and sponges to vertebrates. *25.1*

   a. By comparing major animal phyla and integrating the information with the fossil record, biologists have identified major evolutionary trends among them.

   b. Revealing aspects of body plans are the type of symmetry, gut, and cavity (if any) between the gut and body wall; whether there is a head end; and whether the body is divided into segments (Figure 25.41).

3. *Trichoplax*, a placozoan, may be the simplest living animal known. It is composed of little more than two layers of cells with a fluid matrix in between. *25.2*

4. A sponge body has no symmetry. It is at the cellular level of construction. Although it consists of several kinds of cells, these are not organized like the epithelia and other tissues seen in complex animals. *25.3*

5. Cnidarians include the jellyfishes, sea anemones, and hydras. They have radial symmetry, and they are at the tissue level of construction. Cnidarians alone produce nematocysts, capsules with dischargeable threads that are used mainly in prey capture. *25.4*

6. Most animals more complex than cnidarians have bilateral symmetry, tissues, organs, and organ systems. Their gut is saclike, as in flatworms, or complete, with anus and mouth. Those more complex than flatworms have a cavity between the gut and body wall (a coelom, with its peritoneum, or a false coelom). *25.5, 25.6*

7. Two major lineages diverged shortly after flatworms evolved. One (protostomes) gave rise to the mollusks, annelids, and arthropods. The other (deuterostomes) gave rise to echinoderms and chordates. *25.9*

8. All mollusks have a fleshy, soft body and a mantle. Most have a shell or a remnant of one. They differ in size, body details, and life-styles. *25.10, 25.11*

9. Annelids (earthworms, polychaetes, leeches) have complex organs in a series of coelomic chambers. Their segmented body also is a hydrostatic skeleton. *25.12*

10. Collectively, arthropods are the most successful of all groups in diversity, numbers, distribution, defenses, and capacity to exploit food resources. *25.13p–25.18*

    a. All arthropods, including arachnids, crustaceans, and insects, have hardened, jointed exoskeletons. They also have modified segments, specialized appendages, specialized respiratory, nervous, and sensory organs, and (in insects only) wings.

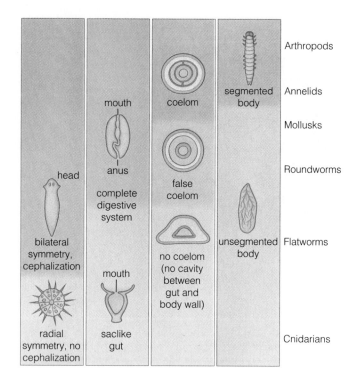

Figure 25.41 Summary of key trends in the evolution of animals, as identified by comparing body plans of major phyla. Not all of these features evolved in every group.

   b. Arthropods develop by growth in size and molting or develop through a series of immature stages, such as larvae and nymphs. Many metamorphose; the tissues of immature forms undergo major reorganization, and body parts are remodeled before the adult emerges.

11. Echinoderms have spines, spicules, or plates in their body wall. Evolutionarily, the phylum is puzzling. The larvae of most species form bilateral features, but they go on to develop into basically radial adults. *25.19*

Review Questions

1. List the six main features that characterize animals. *25.1*

2. When attempting to discern evolutionary relationships among major groups of animals, which aspects of their body plans provide the most useful clues? *25.1*

3. What is a coelom? Why was it important in the evolution of certain animal lineages? *25.1*

4. Name some animals with a saclike gut. Evolutionarily, what advantages does a complete gut afford? *25.1, 25.4–25.6*

5. Choose a species of insect that lives in your neighborhood and describe some of the observable adaptations that underlie its success. *25.13, 25.14, 25.16–25.18*

Self-Quiz   ANSWERS IN APPENDIX III

1. Which is *not* a general characteristic of the animal kingdom?
   a. multicellularity; most have tissues, many form organs
   b. exclusive reliance on sexual reproduction
   c. motility at some stage of the life cycle
   d. embryonic development during the life cycle

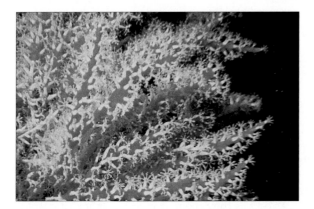

Figure 25.42 A soft, branching coral (*Telesto*).

2. Between the gut and body wall of most animals is a _____.
   a. pharynx   c. coelom
   b. pseudocoelom   d. archenteron

3. Jellyfishes, sea anemones, and their relatives have _____ symmetry, and their cells form _____ .
   a. radial; mesoderm   c. radial; tissues
   b. bilateral; tissues   d. bilateral; mesoderm

4. Most animals that are more complex than cnidarians have _____ symmetry, and _____ forms in their embryos.
   a. radial; mesoderm   c. bilateral; mesoderm
   b. bilateral; endoderm   d. radial; endoderm

5. Which phylum contains members that are notorious for causing serious diseases in humans?
   a. cnidarians   c. segmented worms
   b. flatworms   d. chordates

6. _____ have a coelom and pronounced segmentation, and a dizzying variety of modifications to the segments.
   a. arthropods   c. sponges   e. sea stars
   b. annelids   d. snails and clams   f. vertebrates

7. _____ are the most evolutionarily successful animals.
   a. arthropods   c. sponges   e. sea stars
   b. annelids   d. snails and clams   f. vertebrates

8. Match the terms with the appropriate groups.
   ____ sponges   a. spiny-skinned
   ____ cnidarians   b. vertebrates and kin
   ____ flatworms   c. flukes and tapeworms
   ____ roundworms   d. no tissue organization
   ____ rotifers   e. no males for some
   ____ mollusks   f. nematocysts, radial symmetry
   ____ annelids   g. hookworms, elephantiasis
   ____ arthropods   h. jointed exoskeleton
   ____ echinoderms   i. "belly-foots" and kin
   ____ chordates   j. segmented worms

Critical Thinking

1. A carnivorous sponge was found in an underwater cave in the Mediterranean Sea. Unlike other sponges, it has no pores or canals. Hooklike spicules on surface branchings trap shrimp and other animals. The branchings envelop prey, which cells digest. Which body parts had to evolve for such a feeding strategy?

2. Tapeworms are hermaphroditic. What selective advantages might this feature offer, and in what kinds of environments?

3. People who eat raw oysters or clams harvested from sewage-polluted waters develop mild to severe gastrointestinal ailments. Think about the feeding modes of these mollusks and develop a hypothesis to explain why mollusk eaters can get sick.

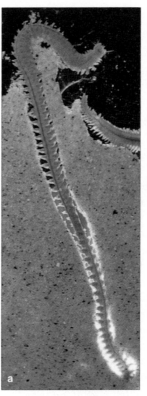

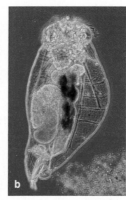

Figure 25.43 Name the mystery animals.

4. You are diving in the calm, warm waters behind a large tropical reef. You see something that looks like a plant with many tentacles (Figure 25.42). This is a branching, soft coral with many polyps. You will not see it growing on reef surfaces exposed to the open sea. Formulate two hypotheses that might explain its distribution.

5. The marine worm in Figure 25.43*a* has a segmented body. Most segments are similar; on their sides are bristles that the worm uses to dig burrows in sea sediments. The tiny animal in Figure 25.43*b* is packed with organs, and its head end has a crown of cilia. The females of most species can reproduce without help from males. To which groups do these animals belong?

Selected Key Terms

animal *25.1*
annelid *25.12*
arthropod *25.13*
bilateral
  symmetry *25.1*
brain *25.12*
centipede *25.16*
cephalization *25.1*
cnidarian *25.4*
coelom *25.1*
collar cell *25.3*
contractile cell *25.4*
cuticle *25.6*
deuterostome *25.9*
echinoderm *25.19*
ectoderm *25.1*
endoderm *25.1*
epithelium *25.4*
exoskeleton *25.13*

flatworm *25.5*
gonad *25.4*
gut *25.1*
hermaphrodite *25.5*
hydrostatic
  skeleton *25.4*
invertebrate *25.1*
larva *25.3*
Malpighian
  tubule *25.17*
mantle *25.10*
medusa *25.4*
mesoderm *25.1*
metamorphosis *25.13*
millipede *25.16*
mollusk *25.10*
molting *25.13*
nematocyst *25.4*
nephridium *25.12*
nerve cell *25.4*

nerve cord *25.12*
nymph *25.17*
organ *25.5*
organ system *25.5*
pharynx *25.5*
placozoan *25.2*
planula *25.4*
polyp *25.4*
proglottid *25.5*
protostome *25.9*
pupa *25.17*
radial symmetry *25.1*
rotifer *25.8*
roundworm *25.6*
sponge *25.3*
torsion *25.11*
vertebrate *25.1*
water-vascular
  system *25.19*

Readings

Pearse, V., et al. 1987. *Living Invertebrates*. Palo Alto, California: Blackwell.

Pechenik, J. 1995. *Biology of Invertebrates*. Third edition. Dubuque: W. C. Brown.

Raloff, J. July 10, 1999. "The Bitter End: Enticing Agricultural Pests to Their Last Repast." *Science News* 156:24–26.

ANIMALS: THE VERTEBRATES

So You Think the Platypus Is a Stretch?

Growing up, we learn about cats, dogs, horses, and cows, and these familiar vertebrates shape our notions of what animals are supposed to look like. More than once, calling something an animal has been viewed as a bit of a stretch. Consider:

In 1798, the first platypus specimen delivered to the British Museum was nearly dismissed as a prank, a sly stitching of a duck bill and webbed feet onto a pelt of some animal about half the size of a housecat. It was no joke. The duck-billed platypus (*Ornithorhynchus anatinus*) is a representative of a mammalian lineage that arose 150 million years ago, after the breakup of the supercontinent Pangea. Its fleshy bill does look ducklike, and its broad, flat, furry tail resembles that of another mammal, the beaver (Figure 26.1*a*). Like turtles and birds, the platypus has a cloaca, which is an enlarged duct through which feces, excretions from kidneys, *and* gametes pass. The female platypus has mammary glands but lays shelled eggs, as turtles and birds do. The young hatch pink and unfinished, as embryonic stages too helpless to fend for themselves.

Yet, as odd as they seem, this combination of traits is adaptive in remote streams and lagoons of Australia and Tasmania, the natural habitat of the platypus. At night this mammal dives underwater. Besides storing fat, its thick tail functions as a rudder in water. Nostrils, eyes, and ears are shut, but 800,000 sensory receptors in the bill detect prey (by oscillations in water pressure or electric fields), even flicks of a shrimp tail. The flattened bill scoops up snails, shrimps, worms, mussels, and insect larvae. Horny pads on jaws quickly grind shelled or hard-cuticled prey. Dense fur offers warmth at night and all day in a cool burrow, which the platypus dug with those strong claws projecting from its hind feet.

And now consider making a really big intellectual stretch past this particular oddity. A tiny bag of an animal has ancestors close to the start of a lineage that led to platypuses, you, and all other vertebrates.

The salps in Figure 26.1*b* are 12 millimeters (about 0.5 inch) long. Their ancestors probably were among the first chordates, an evolutionary branch of bilateral, coelomate animals (Figure 26.2). Encircling each bag-shaped body are bands of muscles. As these circular muscles contract and relax, they pump seawater from the top of the body out through the other end. The tiny, recurring jets of oxygenated, tidbit-bearing water function in respiration, feeding, and locomotion.

Like other chordates, the salp *Thalia democratica* has a hollow nerve cord. While it is growing up, one end of the cord develops into a ganglion, a cluster of nerve cell bodies that functions as a rudimentary brain. An "eye" develops on top of it. The *T. democratica* brain is organized a bit like neural structures in sea star larvae (Section 45.5). Sea stars, recall, are echinoderms. The salp brain also resembles a small part of the vertebrate

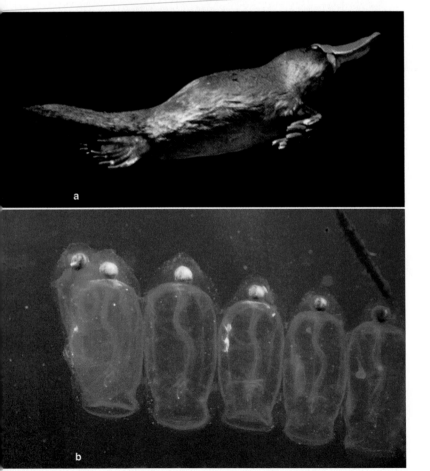

Figure 26.1 From the somewhat unfamiliar to one of our most remote chordate relatives: (**a**) Platypus underwater. (**b**) Salps (*Thalia*). We find these urochordates in marine plankton, usually in warm and temperate seas but also in cold, deep water. They swim independently or in loose chains as shown here. Salps can reproduce asexually or sexually every few days or weeks. Under favorable conditions, population sizes burgeon, and the density of *Thalia* chains in the water can be eye-popping.

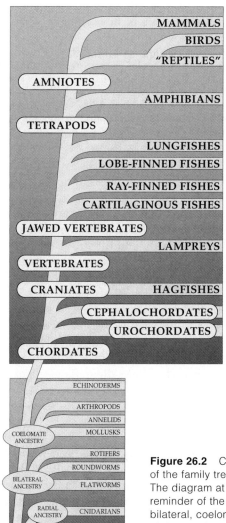

MAMMALS

BIRDS

"REPTILES"

AMNIOTES

AMPHIBIANS

TETRAPODS

LUNGFISHES

LOBE-FINNED FISHES

RAY-FINNED FISHES

CARTILAGINOUS FISHES

JAWED VERTEBRATES

LAMPREYS

VERTEBRATES

CRANIATES HAGFISHES

CEPHALOCHORDATES

UROCHORDATES

CHORDATES

ECHINODERMS

ARTHROPODS

ANNELIDS

COELOMATE ANCESTRY MOLLUSKS

ROTIFERS

ROUNDWORMS

BILATERAL ANCESTRY FLATWORMS

RADIAL ANCESTRY CNIDARIANS

MULTICELLED ANCESTRY

SINGLE-CELLED, PROTISTAN-LIKE ANCESTORS

Figure 26.2 Chordate branch of the family tree for animals. The diagram at the *left* is a reminder of the multicelled, bilateral, coelomate ancestry of the vertebrates and their chordate relatives.

Key Concepts

1. Vertebrates and some invertebrates are chordates. Five features develop in chordate embryos and set them apart from other animals: a supporting rod for the body (notochord), a dorsal nerve cord, a pharynx, gill slits in the pharynx wall, and a tail extending past the anus. Some or all of the features persist in adults.

2. Living vertebrates are informally known as jawless fishes, cartilaginous fishes, bony fishes, amphibians, reptiles, birds, and mammals.

3. Four evolutionary trends occurred among certain vertebrate lineages. A vertebral column replaced the notochord as a partner of muscles used in locomotion. Jaws evolved, sparking the evolution of novel sensory organs and brain expansions. On land, lungs replaced gills, and more efficient blood circulation enhanced lung function. Fleshy fins having skeletal supports evolved into the limbs of amphibians, reptiles, birds, and mammals.

4. The primate branch of the mammalian lineage includes the prosimians, tarsioids, and anthropoids (monkeys, apes, and humans). Apes and humans are hominoids. Humans and some extinct species with a mosaic of apelike and humanlike traits are hominids.

5. Starting in the Miocene, a long-term cooling trend led to seasonal changes in habitats and food sources. As food sources became scarcer, early hominoids spread out through Africa and, later, entered Europe and southern Asia. One lineage gave rise to hominids.

6. Unlike earlier hominids, *Homo erectus* and *H. sapiens* showed exceptional behavioral flexibility and creative experimentation with the environment, as when they used fire. These characteristics helped them survive dispersal into novel and often harsh environments.

brain that processes visual signals and controls a few motor neurons. These traits hint at an evolutionary link with echinoderms *and* vertebrates.

Through numerous morphological and biochemical comparisons, researchers are refining the phylogenetic connections among animals. In this chapter we sample the major lineages of vertebrates and the characteristics they share with other chordates. The evolutionary tree diagram in Figure 26.2 is a road map through these groups, which range from fishes to mammals. The chapter concludes with a brief history of the human species, beginning with its mammalian and primate ancestors. As you consider the groups, remember this key concept: *Each kind of animal is a mosaic of traits, many conserved from remote ancestors and others unique to its branch on the animal family tree.*

THE CHORDATE HERITAGE

The preceding chapter left off with echinoderms, one of the most ancient lineages of the deuterostome branch of the animal family tree. Dominating this branch are the more recently evolved **chordates** (Chordata). These mostly coelomate, bilateral animals have four unique features that are evident in chordate embryos. In many species, all four features persist into adulthood.

First, a **notochord**, a long rod of stiffened tissue (not cartilage or bone), helps support the body. *Second*, the nervous system of chordate embryos develops from a dorsal tube, the **nerve cord**, which runs parallel to the notochord and the gut. The nerve cord's anterior end increases in mass and becomes modified into a brain. *Third*, chordate embryos have slits in the wall of their **pharynx**, a type of muscularized tube that functions in feeding, respiration, or both. *Fourth*, a tail forms in embryos and extends past the anus.

Only about 2,100 species of chordates are grouped as urochordates and cephalochordates. Like sea stars, these are "invertebrates," with no backbone. Far more chordates (about 48,000 species) are **vertebrates**. These animals have a backbone of cartilage or bone, and a brain in a chamber of cartilage or bone. Appendix I has an expanded classification scheme for vertebrates. Unit VI has details of their body plans and functions. Here we start with their closest invertebrate relatives.

Urochordates

The salp described earlier is one of the **urochordates**, baglike chordates having a caudal notochord. (*Caudal* means at the posterior or tail end of the body.) Also in this group are the tunicates. Their name refers to the gelatinous or leathery "tunic" adults secrete around themselves. Most are no more than a few centimeters long and range from intertidal zones to the deep ocean.

Some are solitary, others colonial. Adults usually stay attached to substrates. Sea squirts, the most common types, spurt water through one of their siphons when something irritates them.

Sea squirt larvae are bilateral swimmers. Their firm, flexible notochord (a series of cells) acts like a torsion bar. It interacts with bands of muscles just under the epidermis. When muscles on one side of the tail or the other contract, the notochord bends, then springs back as muscles relax. A strong, side-to-side motion propels the larva forward. Most fishes use muscles and their backbone for the same kind of motion. The notochord is lost as larvae undergo metamorphosis, a time when drastic tissue remodeling and reorganization transform the larva into an adult (Section 25.13 and Figure 26.3).

Tunicates are **filter feeders**: they direct a current of water through part of the body and filter food from it. Tunicates have no coelom; a saclike chamber replaced it as they evolved. Seawater flows through a siphon into the chamber and then past **gill slits**: openings in the pharynx wall. It flows out through another siphon. This pharynx also is a respiratory organ. Dissolved oxygen is less concentrated in blood vessels adjoining the pharynx than in seawater. So it diffuses from water into blood. Carbon dioxide diffuses down its gradient, from blood into water leaving the pharynx.

Cephalochordates

The fish-shaped **cephalochordates** have a head with a simple brain that develops from the anterior end of the nerve cord (*cephalo–*, head). Their nerve cord develops as it does in vertebrates. Their brain, too, is dense with nerve cells (neurons); it often controls reflex responses to light and to other stimuli. These animals burrow into shallow sediments and swim fast, but not far.

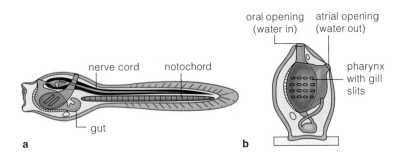

Figure 26.3 (**a**) Generalized larval form of a sea squirt, a tunicate. Each new larva briefly swims about. Metamorphosis starts when its head end attaches to a substrate. Its tail, notochord, and most of the nervous system are resorbed (recycled and used to form new tissues). Many slits form in the pharynx wall. Organs rotate until openings through which water flows into and out from the pharynx are pointing away from the substrate, as in (**b**). (**c**) Adult sea squirts (*Rhopalaea crassa*).

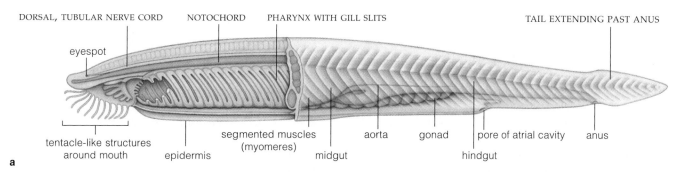

DORSAL, TUBULAR NERVE CORD   NOTOCHORD   PHARYNX WITH GILL SLITS   TAIL EXTENDING PAST ANUS

eyespot

tentacle-like structures
around mouth   epidermis   segmented muscles
(myomeres)   midgut   aorta   gonad   pore of atrial cavity   hindgut   anus

a

Figure 26.4 (**a**,**b**) Body plan and burrowing behavior of lancelets. (**c**) Reconstruction of one of the earliest known craniates, which resembles lancelets in some respects and the vertebrates in others. A fossil of one craniate (*Myllokunmingia*) recovered in China is about 530 million years old. It dates from the Cambrian explosion of animal diversity.

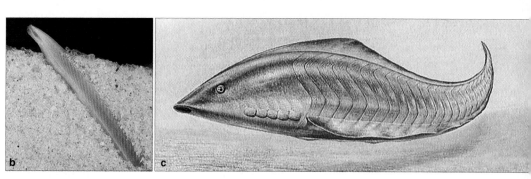

Cephalochordates are between thirty and seventy millimeters long. Their body tapers at both ends; hence their common name, lancelets. Their segmented body shows the four chordate characters (Figure 26.4a,b).

Like vertebrates, lancelets have segmented muscles, or myomeres, and a notochord. The contractile units in muscle cells run parallel with the body's long axis. Contractile force is directed against the notochord and brings about side-to-side swimming motions.

Like tunicates, lancelets are filter-feeders. Cilia that line as many as 200 gill slits beat in coordination and move water into the pharynx, where food gets caught in mucus. Trapped food is delivered to the rest of the gut. Their pharynx has a large food-trapping surface area relative to the body length. It is enough to sustain this filter-feeding animal (Figure 26.4a). As you will see, the pharynx turned out to be an organ with amazing evolutionary possibilities for the vertebrates.

Early Craniates

All fishes, amphibians, reptiles, birds, and mammals alive today are **craniates**. They have a brain inside a cranium, which is a chamber of cartilage or bone. The first craniates arose before 530 million years ago. Like lancelets, they had a notochord and segmented muscles (Figure 26.4c). They also had fins, which indicates they were active swimmers. They resembled the larvae of lampreys, which are modern fishes without jaws.

Thirty million years later, the earliest jawless fishes (agnathans) had evolved. **Ostracoderms** were among

these craniates. Protecting their body were armorlike plates of bony tissue and dentin, a hardened tissue that still persists in vertebrate teeth. Perhaps their armor was good against pincers of giant sea scorpions. It was not good against **jaws**: hinged, bony feeding structures. Jaws evolved in some craniates, and when these species began an adaptive radiation, ostracoderms vanished.

Placoderms were among the earliest fishes that had vertebrae as well as jaws. Their jaws were expansions of the first gill-supporting structures, sometimes with razor-sharp edges. Armor plates protected their head, but not much else. Before, chordate feeding strategies had been limited to filtering, sucking, and rasping food. Placoderms evolved into predators, some formidable and huge. When those predatory fishes started biting or tearing off large chunks of prey, they triggered an evolutionary contest involving offensive and defensive adaptations that has continued to the present. We pick up that part of the vertebrate story in the next section.

Chordate embryos alone have this combination of traits: a notochord, a tubular dorsal nerve cord, a pharynx with slits in its wall, and a tail extending past the anus.

Tunicate larvae and lancelets use a notochord and muscles for fishlike swimming motions.

Tunicates and lancelets are filter feeders. They strain food from water at their pharynx, a muscular tube with fine slits in its wall. Tunicates also use the pharynx in gas exchange.

The first vertebrates arose during the Cambrian, along with all other major animal phyla.

TRENDS IN VERTEBRATE EVOLUTION

The emergence of vertebrates profoundly changed the course of animal evolution. In contrast to their nearly brainless filter-feeding relatives, vertebrates have **bone tissue**, which consists of secretions from specialized cells hardened with mineral deposits. A series of bony segments called **vertebrae** (singular, vertebra) replaced the notochord as the functional partner of segmented muscles. That flexible column, part of an endoskeleton (internal skeleton), was less cumbersome than external armor plates. The union of bony segments and muscle segments enhanced maneuverability, and hard bones invited more forceful contractions. The outcome? *Agile, fast-moving fishes began their dominance of the seas.*

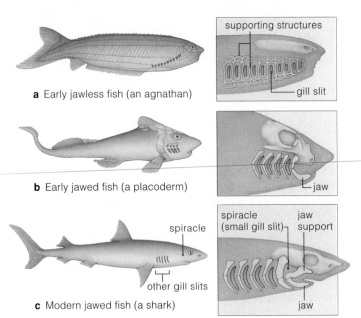

a Early jawless fish (an agnathan)

b Early jawed fish (a placoderm)

c Modern jawed fish (a shark)

supporting structures

gill slit

jaw

spiracle
(small gill slit)   jaw support

jaw

spiracle

other gill slits

Figure 26.5 Comparison of the gill-supporting structures in jawless fishes and jawed fishes. Cartilage reinforced the mouth's rim in early jawed vertebrates. Modern jawed fishes draw water in through a gill slit between the jaws and a supporting element. (**a**) Gill supports are just under the skin. (**b,c**) The gill supports are internal relative to the surface of the gills. (**d**) The human standing at left will give you an idea of the size of one ancient jawed fish, the placoderm *Dunkleosteus*.

In a related trend, jaws evolved from the first of a series of structural supports for the gill slits (Figure 26.5). Jaws were a key innovation, an evolutionary foot in the door for novel feeding possibilities. They started a coevolutionary race for defenses and new ways to overcome them. For example, selective agents favored better eyes for detecting predators and bigger brains that could plan fast getaways. *A trend toward complex sensory organs and nervous systems started among ancient fishes and continued among vertebrates on land.*

Another trend began when paired fins evolved. **Fins** are appendages that help propel, stabilize, and guide the body through water (compare Figure 26.9). In some lineages, fleshy ventral fins had skeletal supports that were forerunners of limbs. *Paired, fleshy fins were a starting point for all legs, arms, and wings that evolved among amphibians, reptiles, birds, and mammals.*

Another trend started with a change in respiration. In lancelets, oxygen and carbon dioxide simply diffuse across the skin. In most early vertebrates, gills evolved. **Gills** are respiratory organs with a moist, thin, folded surface. They are richly endowed with blood vessels and represent a large surface area for gas exchange. For instance, five to seven pairs of a shark's gill slits are continuous with gills that extend from the pharynx to the body's surface. When a shark opens its mouth and closes external gill openings, the pharynx expands. Oxygen diffuses *from* water in the mouth into gills as carbon dioxide diffuses *into* the water. Muscles now constrict the pharynx and force oxygen-depleted, carbon-dioxide-enriched water out through gill slits.

Among larger and more active fishes, gills became more efficient. But gills can't work out of water; their thin surfaces stick together without a flow of water to keep them moist. In some fishes ancestral to the land vertebrates, pouches developed from the gut wall. The pouches evolved into **lungs**—internally moistened sacs for gas exchange. In a related trend, modifications to the heart enhanced the pumping of oxygen and carbon dioxide through the body. *Ancestors of land vertebrates relied less on gills and more on lungs. And more efficient circulatory systems accompanied the evolution of lungs.*

The earliest jawless fishes became extinct when the placoderms and other jawed fishes diversified. These fishes, too, became extinct during the Carboniferous as faster, brainier predators replaced them in the seas. In sections to follow, you will read about the jawless and jawed descendants of groups that, through adaptation, key innovations, and plain luck, made it to the present.

The vertebral column, jaws, paired fins, and lungs were pivotal developments in the evolution of lineages that gave rise to fishes, amphibians, reptiles, birds, and mammals.

EXISTING JAWLESS FISHES

The hagfishes and lampreys may be descended from jawless fishes like the ostracoderms, which were adapted to freshwater habitats. We still find some lampreys in fresh water. Others, including hagfishes, became adapted to life in the seas.

Hagfishes

Of all existing chordates, only the hagfishes have a partial cranium and a cartilage backbone in a cylindrical body (Figure 26.6). They lack jaws and paired fins. Most species are less than one meter long. They burrow into sediments on the seafloor and prey on small invertebrates, mainly polychaetes. They also scavenge injured or dead fish, often eating from the inside out. When hagfishes are on the defensive, they secrete up to a gallon of extremely sticky mucus around themselves, as in Figure 26.6*b*. Fishermen who accidentally snag hagfishes typically find this behavior disgusting. Evolutionarily, however, slime has served this soft-bodied and otherwise tastily vulnerable lineage quite well. And it has been doing so for far longer than humans have been around.

A hagfish is nearly blind, but it has well-developed means of smelling and feeling out potential food. Its four pairs of sensory tentacles are arrayed around the mouth. Instead of jaws, the hagfish has two pairs of rasping appendages on a tonguelike structure. Its rates of metabolism are not enough to sustain a highly active life style. On the other hand, hagfishes can wait out an absence of food for as long as seven months. They produce and release large eggs relative to their body size. A thick-shelled hagfish egg is approximately 2.5 centimeters (1 inch) long.

Lampreys

Like hagfishes, lampreys are an evolutionarily ancient lineage of jawless fishes. Unlike hagfishes, lampreys are parasites more than scavengers. They attach to prey with a suckerlike oral disk and rasp away at flesh with horny mouthparts (Figure 26.7).

Some nonparasitic species live in freshwater brooks and rivers that have rocky, gravelly beds. Usually they are no more than 15.5 to 18 centimeters long. Where the water runs swiftly, these typically secretive species will hollow out depressions in the gravel and lay eggs.

Sea lampreys evolved in the Atlantic Ocean. These are big fish; they can grow to more than 14 centimeters (1.5 feet) long. They latch onto another big fish, then suck out its juices and tissues. In the twelve to twenty months of its adult life, each lamprey can kill as much

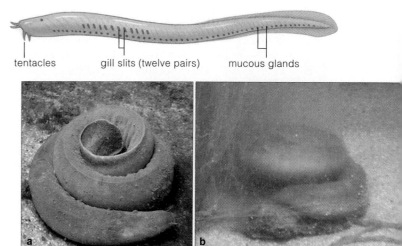

tentacles   gill slits (twelve pairs)   mucous glands

Figure 26.6 Hagfish body plan. The photographs show one hagfish before and after it secretes slimy mucus from its pores.

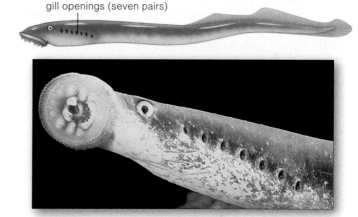

gill openings (seven pairs)

Figure 26.7 Lamprey body plan. The photograph shows its oral disk pressed against aquarium glass.

as 18 kilograms (40 pounds) of fish. Fishermen often observe several lampreys attached to the same host.

Sea lampreys are aggressive fishes. Where they have invaded freshwater habitats, they have threatened the native fish species. For instance, in the early 1800s, they invaded the Great Lakes of North America. Probably they entered the Hudson River, then canals that were built for commerce. By 1946, they were established in all the Great Lakes. Rainbow trout, lake trout, salmon, whitefish, chubs, catfishes, and other natives have no coevolved defenses against them. Sometimes only one in seven survives the parasitic attack.

Hagfishes and lampreys get along without jaws; they latch onto prey with their efficient, specialized mouthparts.

EXISTING JAWED FISHES

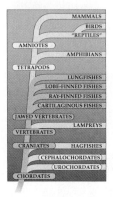

Unless you study underwater life, you may not know that fishes are the world's dominant vertebrates. Their numbers exceed those of all other vertebrate groups combined. And they show far more diversity; we already know of at least 21,000 kinds of bony fishes alone.

The form and behavior of a fish tell us about the challenges it faces in water. Water is 800 times denser than air, and it resists fast motion through it. In response to this constraint, many marine fishes are streamlined for pursuit or escape. Sharks are like this (Figure 26.8*a*). Their long, trim body reduces friction, and tail muscles are organized for propulsive force and forward motion. But some bottom-dwelling fishes such as rays have a flattened body (Figure 26.8*b*). Their shape is good for hiding, not for high-speed runs.

A trout suspended in shallow water shows another adaptation to water's density. Like many fishes, it can maintain neutral buoyancy with a **swim bladder**. This adjustable flotation device exchanges gases with blood inside the body. When a trout gulps air at the water's surface, it is adjusting the volume of its swim bladder.

Figure 26.8 Cartilaginous fishes: (**a**) Galápagos shark. (**b**) Manta ray. The two projections on its head unfurl and waft plankton to the broad mouth between them. Mantas get huge but are gentle with human divers. (**c**) Chimaera.

Fishes With a Skeleton of Cartilage

Cartilaginous fishes (Chondrichthyes) include about 850 species of skates, sharks, and chimaeras (Figure 26.8). These marine predators have prominent fins, a skeleton of cartilage, and five to seven gill slits on both sides. Most have a few scales or many rows of them. Scales are small, bony plates that typically protect the body surface without weighing it down.

At fifteen meters head to tail, some sharks are among the largest living vertebrates. Sharks continually shed and replace their hard, sharp teeth, which are modified scales. They use their jaws to grab and rip off chunks of prey. Attacks on humans have given the entire group a bad reputation. Yet sharks have hunted invertebrates, fishes, and marine mammals for many millions of years. Surfboards with human legs dangling over the sides are a relatively new temptation for some of them.

The mainly bottom-dwelling skates and rays have flattened teeth that crush shelled prey. The largest, the manta ray, is up to six meters across. Stingrays have a venom gland in their tail that may help deter predators (Figure 26.8*b*). The tail or fins of other rays have electric organs that stun prey with up to 200 volts of electricity.

The thirty or so species of chimaeras feed mostly on mollusks. With their bulky body and slender tail, they look a bit like a rat; hence the common name, ratfishes. They have a venom gland in front of the dorsal fin.

Fishes With a Skeleton of Bone

All but 4 percent of the existing fish species are jawed and have a bony endoskeleton. These fishes are the most numerous and diverse vertebrates. The three lineages are known as ray-finned fishes, lobe-finned fishes, and lungfishes, and they originated more than 400 million years ago. The body plans of these fishes differ greatly. Marine predators commonly have a torpedo shape, a flexible body, and strong tail fins used in swift pursuit. Many reef dwellers are small finned and box shaped; they can easily navigate narrow spaces. Those with an elongated, flexible body, such as the moray eel, wriggle through muddy sediments and crevices that conceal them. The cryptic body shape of sea horses and many bottom-dwelling species can hide them from predators, prey, or both. Figure 26.9 shows examples.

Ray-finned fishes have fin supports derived from skin. Most have maneuverable fins and thin, flexible scales that don't hamper complex swimming motions. Their ancestors had sac-shaped outpouchings from the wall of the esophagus, a tube to the gut. These evolved into lunglike sacs that supplemented the gills in gas exchange. In most species, the sacs evolved not into lungs but into swim bladders.

One group of ray-finned fishes includes sturgeons and paddlefishes of the Mississippi River basin. The most abundant group, teleosts, include salmon, tuna,

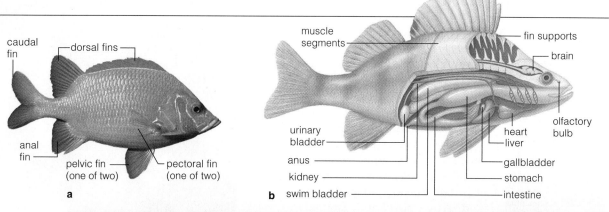

a

Figure 26.9 Bony fishes.

Examples of ray-finned fishes: (**a**) Soldierfish, a teleost. (**b**) Internal organs of a perch, also a teleost. (**c**) Sea horse, which anchors itself to substrates with its tail. (**d**) Long-nose gar.

(**e**) Living coelacanth (*Latimeria*), of the only surviving lineage of the lobe-finned fishes.

(**f**) Australian lungfish.

f

rockfish, catfish, perch, minnows, moray eels, flying fish, sculpins, blennies, scorpionfish, and pikes.

Coelacanths are the only living **lobe-finned fishes**. Fleshy body extensions form part of their ventral fins, which have some skeletal elements inside. Lobe-finned fishes have lunglike sacs, but these do not function in gas exchange. Respiration is restricted to gills.

Lungfishes have gills and one or a pair of "lungs" that are modified gut wall outpouchings. These don't function in buoyancy; they take in oxygen and rid the body of carbon dioxide. Lungfishes must surface and gulp air; they drown if kept underwater. Three kinds currently live in Africa, South America, and Australia (Figure 26.9*f*). In dry seasons, when streams dwindle, lungfishes encase themselves in slimy mud that keeps them from drying out until the next rainy season.

Do such fishes share a common ancestor with four-legged walkers—**tetrapods**? Probably. Like tetrapods, a lungfish has a separate circuit for blood to the lungs. Its skullbones are arranged the same way. Even its tooth enamel is the same. The "limbs" of lobe-finned fishes, lungfishes, and tetrapods are about the same in their size, position, and structure. The lungfish even uses its limbs to move itself forward on underwater substrates.

Limb joints, digits, a fishlike caudal fin, and other traits of several fossils indicate that walking actually originated in water, not on land. In the middle to late Devonian, some fishes were using limbs and digits for swimming and crawling. As you will see next, this key innovation contributed to the tetrapod move onto land.

Ray-finned fishes are now the most diverse and abundant vertebrates. The lobe-finned fishes have fleshy ventral fins reinforced with skeletal parts. The lungfishes have simple lunglike sacs that supplement respiration by gills.

Walking probably originated in water during the Devonian, among the aquatic forerunners of tetrapods on land.

THE RISE OF AMPHIBIANS

Amphibians are vertebrates with a mostly bony endoskeleton and either four legs or a four-legged aquatic ancestor (Figure 26.10). Their body plan and their mode of reproduction are somewhere between "fishes" and "reptiles."

What event favored the move of certain aquatic tetrapods onto land? As we now know, asteroids hit the Earth five times during the Devonian. One of the last impacts coincided with a mass extinction of marine life, maybe by obliterating much of the oxygen at the sea surface and in swampy habitats near the coasts. The tetrapods with lungs had the advantage; they could get enough oxygen by gulping in air. At the close of the Devonian, some were semiaquatic and spending time on land.

It would not have taken much of a genetic change for the transition from lobed fins to limbs. Remember the enhancers that control gene transcription (Section 15.5?) One governs genes involved in the formation of digits on limb bones. As described in Sections 19.5 and 43.5, even a single mutation in one of these so-called master genes can lead to a big change in morphology.

Life in the new, drier habitats was both dangerous and promising. Temperatures shifted far more on land than in the water, air didn't support the body as well as water did, and water was not always plentiful. But air is richer in oxygen. Amphibian lungs continued to be modified in ways that enhanced oxygen uptake. Also, circulatory systems became more efficient at rapidly moving oxygen to cells. Both modifications increased the capacity for aerobic respiration, the pathway that makes enough ATP to sustain more active life-styles.

New sensory information also challenged the early amphibians. Swamp forests supported vast numbers of edible insects and other invertebrate prey. Animals having better vision, hearing, and balance—the senses most advantageous on land—were favored. And brain regions concerned with interpreting and responding to sensory input expanded.

All of the frogs, toads, salamanders, and caecilians alive today are descended from those first amphibians. None has escaped the water entirely. Even when they use gills or lungs as a respiratory surface, amphibians also exchange oxygen and carbon dioxide across their thin skin. However, respiratory surfaces must be kept moist, and skin dries easily in air.

Some existing amphibians spend their entire life in water. Others species lay eggs in water or they produce aquatic larvae. The species adapted to land either lay eggs in moist places or, in a few cases, protect embryos during development inside the moist adult body.

inside lobed fin, bony or cartilaginous structures (*orange*) undergoing modification

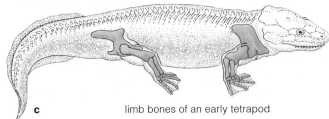

limb bones of an early tetrapod

Figure 26.10 (**a**) Devonian tetrapods. *Acanthostega*, submerged, and *Ichthyostega*. Their skull, caudal tail, and fins were fishlike. Unlike fishes, they had four limbs with digits, and a short neck. (**b,c**) Proposed evolution of skeletal elements in fins into the limb bones of aquatic tetrapods and early amphibians.

Frogs and Toads

Frogs and toads are the most familiar amphibians, and with close to 4,000 species, they also have been the most successful (Figure 26.11*a,b*). They use long hindlimbs and powerful muscles to catapult into the air or barrel through water. Most frogs capture prey by flipping a sticky-tipped tongue out of their mouth. An adult eats about any animal it can stuff into its mouth. Frog skin has mucous glands, poison glands, and antibiotics that afford protection against diverse pathogens in aquatic habitats. This is the case for amphibians in general. The body surface of the poisonous species often has bright warning coloration. In Unit VI, we will take a look at how frogs are put together and how they function.

a fish swimming

a salamander walking

Figure 26.11 Amphibians. (**a**) A frog, splendidly jumping. (**b**) American toad. (**c**) Terrestrial stage in the life cycle of a red-spotted salamander. (**d**) Fish versus salamander locomotion. (**e**) Caecilian. How do you suppose this long, thin amphibian burrows through soil? (Reflect on Section 25.12.)

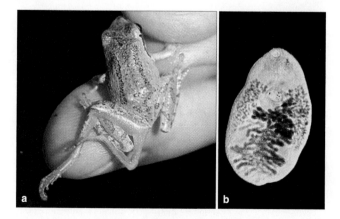

Figure 26.12 (**a**) Deformed frog with extra legs, an outcome of a parasitic infection. Stanley Sessions of Hartwick College and Stanford graduate Pieter Johnson identified a parasitic infection that affects frog limbs. (**b**) Trematodes (*Ribeiroia*) are the culprits. They burrow into tadpole limb buds and physically or chemically alter cells. Infected tadpoles grow extra legs or none at all. With heavy exposure to *Ribeiroia*, the number of tadpoles that successfully complete metamorphosis declines. Trematode cysts have been found in weirdly legged frogs and salamanders in California, Oregon, Arizona, and New York.

Frog population sizes are plummeting. Increases in parasitic attacks, predation, ultraviolet radiation, habitat losses, and chemical pollution are factors in this (Figure 26.12, also Sections 36.8 and 47.6). Even muddy hiking boots can import parasites into remote frog habitats.

Salamanders

About 380 species of salamanders and their kin, the newts, live in north temperate zones and tropical parts of Central and South America. Most are less than fifteen centimeters long. The forelimbs and hindlimbs are about the same size, and most project at right angles from the body. As they walk, all salamanders bend from side to side, like fishes and early amphibians (Figure 26.11*d*). Probably the first tetrapods on land also walked this way. Larval and adult salamanders are carnivorous. Adults of some species retain a few larval features. For example, an adult Mexican axolotl retains the larva's external gills. Also, its teeth and bones stop developing at an early stage. In addition, as in some other species, axolotl larvae are sexually precocious; they can breed.

Caecilians

As some amphibians evolved, they lost their limbs and their vision, but not their prey-capturing jaws. They gave rise to worm-shaped caecilians (Figure 26.11*e*). Most of the 160 or so species burrow through moist soil, using senses of touch and smell to pursue insects and earthworms. The few aquatic types use electrical cues.

In body form and behavior, amphibians show resemblances to aquatic tetrapods and reptiles. Most species have not fully escaped dependency on aquatic or moist habitats to complete their life cycle.

THE RISE OF AMNIOTES

Divergences from amphibians late in the Carboniferous gave rise to the **amniotes**. These were the first vertebrates to form eggs with four internal membranes that conserve water and cushion an embryo, and metabolically support it. The early amniotes laid leathery or calcified eggs (Figure 26.13). Many of their living descendants still do. These include mammals, turtles, lizards, snakes, tautaras, crocodiles, and birds. They are the only tetrapods that can reproduce successfully away from aquatic habitats; their embryos can develop to an advanced stage before being hatched or born in dry habitats. The structure and formation of amniote eggs is a topic of Chapter 44.

Three other adaptations helped free amniotes from aquatic habitats. All amniotes have toughened, dry, or scaly skin that restricts water loss (Figure 26.13c). As is the case for some amphibians, the eggs are fertilized internally after a copulatory organ deposits sperm in a female's body. Also, amniotes have a pair of kidneys that are good at conserving water in controlled ways.

Synapsids and *sauropsids* are the major groups of amniotes. Synapsids are mammals and early mammal-like reptiles (such as *Lystrosaurus*, Section 20.6).

Sauropsids include what we conventionally call the **"reptiles,"** but birds also belong to this evolutionary group. Compared to amphibians, even the early reptiles chased insects and other vertebrates with far greater cunning and speed. With their well-muscled jawbones and formidable teeth, they were able to seize and apply sustained, crushing force on prey. Their limbs typically were better at supporting the body's trunk on land.

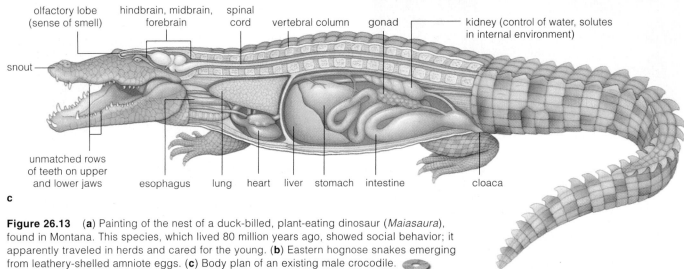

Figure 26.13 (**a**) Painting of the nest of a duck-billed, plant-eating dinosaur (*Maiasaura*), found in Montana. This species, which lived 80 million years ago, showed social behavior; it apparently traveled in herds and cared for the young. (**b**) Eastern hognose snakes emerging from leathery-shelled amniote eggs. (**c**) Body plan of an existing male crocodile.

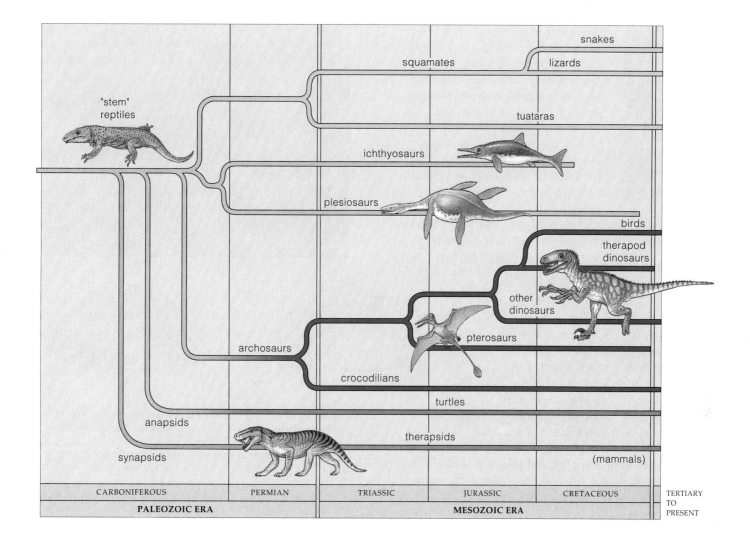

Labels in figure: snakes, squamates, lizards, "stem" reptiles, tuataras, ichthyosaurs, plesiosaurs, birds, therapod dinosaurs, other dinosaurs, archosaurs, pterosaurs, crocodilians, turtles, anapsids, therapsids, synapsids, (mammals)

CARBONIFEROUS | PERMIAN | TRIASSIC | JURASSIC | CRETACEOUS | TERTIARY TO PRESENT

PALEOZOIC ERA | **MESOZOIC ERA**

Figure 26.14 Evolutionary history of the amniotes.

For many species, the nervous system became more complex. A reptile's brain is small compared to the rest of the body mass, but it governs forms of behavior unknown among amphibians. The cerebral cortex, the brain's most recent addition, originated with reptiles.

Crocodilians (Figure 26.13c) were the first animals with a muscular, four-chambered heart fully separated into two halves. Blood enters the first chamber of each half, and the second pumps it out. The separation lets oxygen-rich blood travel from lungs to the rest of the body, and oxygen-poor blood from the body to lungs, in two separated circuits (Section 38.1). Gas exchange across skin, so vital for amphibians, was abandoned by reptiles, nearly all of which have well-developed lungs.

Adaptive radiations of early reptiles led to fabulous diversity. In the Triassic, the group called the dinosaurs evolved. For the next 125 million years, they were the dominant land vertebrates. The fossilized nests of one

kind, *Maiasaura*, hold eggs and juveniles a few months old, at most (Figure 26.13a). These fossils suggest that at least some dinosaurs showed parental behavior; they took care of their young through an extended period of dependency. The K–T asteroid impact wiped out the last of the dinosaurs, or at least what is conventionally thought of as dinosaurs (Section 20.7). We now know that a specialized group of feathered dinosaurs lived through that mass extinction. We call them birds. As Figure 26.14 indicates, crocodilians, turtles, tuataras, snakes, lizards, and birds are reptilian groups that made it to the present, along with mammals.

Amniotes were the first vertebrates to escape dependency on free-standing water through major modifications in their organ systems. Amniotes also produce eggs, often shelled or leathery, with internal membranes that conserve water, and cushion and metabolically support the embryo inside.

Major groups are synapsids (mammals and mammal-like reptiles) and sauropsids, which include reptiles and birds.

A SAMPLING OF EXISTING REPTILES

The name reptile is derived from the Latin *repto*, meaning to creep. Possibly, some early reptiles did creep slowly in muddy swamps and on dry land. But others race, lumber and slither or swim about, like the mosasaurs did long ago. Some (the long-gone pterosaurs) even flew about. The group is just so diverse that calling an animal a reptile is simply another way of saying it has basic amniote traits but not the derived traits of birds and mammals. It is better to view "reptiles" as an evolutionary grade, as we do for "fishes." They are not a monophyletic group.

Turtles

The 250 existing species of turtles live inside a shell that is attached to the skeleton. When most are threatened, they pull their head and limbs inside (Figure 26.15*a,b*). It is a body plan that works well; it has been around since Triassic times. Only among sea turtles and other highly mobile types has the shell become reduced in size.

Instead of teeth, turtles have tough, horny plates that are suitable for gripping and chewing food. They have strong jaws and often a fierce disposition that helps deter predators. But all turtles lay eggs on land, then leave them. Predators eat most of the eggs, so few new turtles hatch. Sea turtles are slow-moving on land, and have been hunted to the brink of extinction.

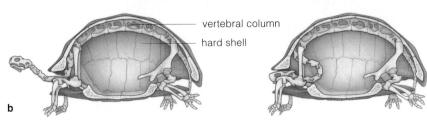

Figure 26.15 A few reptiles. (**a**) Galápagos tortoise. (**b**) Turtle shell and skeleton with head withdrawn and extended. *Facing page:* (**c**) Basilik lizard, which can run across water. By stomping its feet, it creates air pockets on the water surface that keep it afloat. (**d**) Rattlesnake in mid-strike. (**e**) Tuatara (*Sphenodon*). (**f**) Spectacled caiman, a crocodilian. Its peglike upper and lower teeth do not match up. The crocodilian body and life-styles have not changed much for nearly 200 million years. Their future is not rosy. Housing developments encroach on many of their habitats. Their belly skin is in demand for wallets, shoes, and handbags.

Lizards

Lizards and snakes, which make up 95 percent or so of living reptiles, are distant relatives of dinosaurs. Most are small, but the Komodo monitor lizard is big enough to capture a young water buffalo. The longest modern snake would span ten yards of a football field.

Most of the 3,750 kinds of lizards are insect eaters of deserts and tropical forests. They include aggressive, sticky-toed geckos and iguanas (see the photograph in the Unit VI introduction). Most lizards use their small, peglike teeth to snag prey. Chameleons rely on accurate flicks of their very sticky prey-capturing tongue, which is longer than their body (Section 33.4).

Being small themselves, most lizards are prey for many other animals. Some attempt to startle predators and intimidate rivals by flaring their throat fan; others outrun them (Figure 26.15*c*). Many give up their tail as a predator grabs them. The detached tail wriggles for a bit and may be distracting enough to permit a getaway.

Snakes

During the early Cretaceous, short-legged, long-bodied lizards gave rise to elongated, limbless snakes. Most of the 2,300 existing species of snakes slither in S-shaped waves, much like salamanders do. "Sidewinders" make J-shaped movements and leave distinctive trails across loose sand and sediments. Some species still have bony remnants of ancestral hindlimbs (Section 17.1).

All snakes are carnivores. They have flexible skull bones and jaws; some swallow prey wider than they are. Pythons and boas coil around prey and suffocate it into submission. Fanged types, including coral snakes and rattlesnakes (a pit viper) bite and subdue prey with venom (Figure 26.15*d*). Snakes usually do not act aggressively toward people. Even so, rattlesnakes and other poisonous types bite 8,000 or so people annually in the United States and kill about 12 of them. In India, king cobras and other snakes bite 200,000 or so people and kill about 9,000 of them.

venom
gland

hollow
fang

Even the most feared snakes are vulnerable during their life cycle; birds and other predators relish snake eggs. The female snakes store sperm and lay several clutches of fertilized eggs at intervals after they mate, which improves the odds that at least some will hatch.

Tuataras

Besides having reptilian traits, tuataras are like modern amphibians in some aspects of their brain and in their way of walking. The two existing species live on small, windswept islands near New Zealand (Figure 26.15*e*). Their body plan has not changed much for the past 140 million years. Like lizards, tuataras have a third "eye" under the skin, with a retina, a lens, and nerves to the brain. It only registers changes in daylength and light intensity. Does it have roles in hormonal controls over reproduction, as described in Section 36.8? Maybe. The tuataras, like turtles, may live longer than sixty years. They engage in sex only after they are twenty years old.

Crocodilians

Modern crocodiles and alligators, the closest relatives of birds and dinosaurs, live in or near water. Among them are the largest living reptiles. Crocodiles and alligators have powerful jaws, a long snout, and sharp teeth, as in Figure 26.15*f*. The feared "man-eaters" of southern Asia and Nile crocodiles weigh as much as 1,000 kilograms. They drag a mammal or bird into water, tear it apart by violently spinning about, then gulp down the chunks. Like the other reptiles and birds, crocodilians adjust body temperature with behavioral and physiological mechanisms. They are like birds in displaying complex social behaviors, as when parents guard nests and assist hatchlings into water. This trait and others suggest that crocodilians and birds share a common ancestor.

Existing crocodilians are the closest relatives of dinosaurs and birds. Turtles, like tuataras, have changed little in body plan for millions of years. Lizards and snakes represent about 95 percent of all living reptiles.

BIRDS

The capacity for flight evolved in four groups—insects, pterosaurs (extinct), birds, and bats. Of these groups, feathers are a trait unique to **birds** and some of the extinct dinosaurs. *Feathers* are lightweight structures derived from skin that are used for flight and insulation. There are about 28 orders of birds and close to 9,000 named species. They vary in size, proportions, coloration, and capacity for flight. One of the smallest known birds weighs 2.25 grams (0.08 ounce). The largest living bird, the ostrich, weighs up to 150 kilograms (330 pounds). It cannot fly, but it is an impressive sprinter (Section 17.1). Many birds, such as warblers, parrots, and other perching types, differ markedly in feather coloration and territorial behavior. Bird song and other social behaviors are topics of later chapters. Figures 26.16 through 26.18 highlight just a few basic features of this diverse group of vertebrates.

The first birds evolved during an adaptive radiation of reptiles in the Mesozoic. They apparently diverged from a lineage of small theropod dinosaurs, which were carnivores that ran about on two legs. Feathers evolved as highly modifed reptilian scales. *Archaeopteryx* was in or near that lineage. As you read in Section 19.9, it had reptilian *and* avian traits, including feathers. Birds still share numerous traits with their closest relatives, the dinosaurs and crocodilians. For example, they still have scales on their legs and have some of the same internal structures. Most engage in parental behavior. And all species lay amniote eggs. In these eggs, an amnion, chorion, allantois, and yolk sac surround the embryo (Figure 26.16). All four "extraembryonic" membranes have roles in how the embryo develops. You will come across examples of amniote eggs in Chapters 43 and 44.

Feathers help a bird's body conserve metabolically generated heat when the outside temperature declines. Also, elastic sacs connected to bird lungs help dissipate excess heat as they force warmed air out of the body.

Bird flight involves more than feathers. It involves bones with a lightweight, honeycombed structure and efficient modes of respiration and circulation. High rates of metabolism sustain flight, and those rates depend on a strong flow of oxygen through the body. The elastic sacs that connect to the lungs greatly enhance oxygen uptake (Section 40.3). Like mammals, birds also have a large, durable, four-chambered heart. The heart pumps oxygen-enriched blood to the lungs and to the rest of the body along separate circuits, as it probably did in the reptilian ancestors of birds.

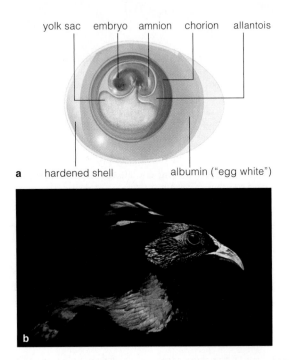

Figure 26.16 Two of the key defining characteristics of birds. (**a**) Amniote egg. (**b**) Feathers. The striking plumage of this male monal pheasant, a native of India, is a result of sexual selection. This bird is an endangered species. As is the case for many other bird species, its bright, jewel-colored feathers end up adorning humans—in this case, on the caps of native tribespeople.

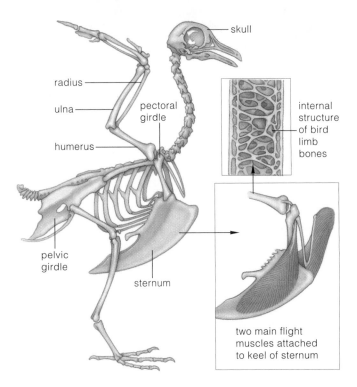

Figure 26.17 Body plan of a typical bird. Flight muscles attach to a large, keeled breastbone (sternum).

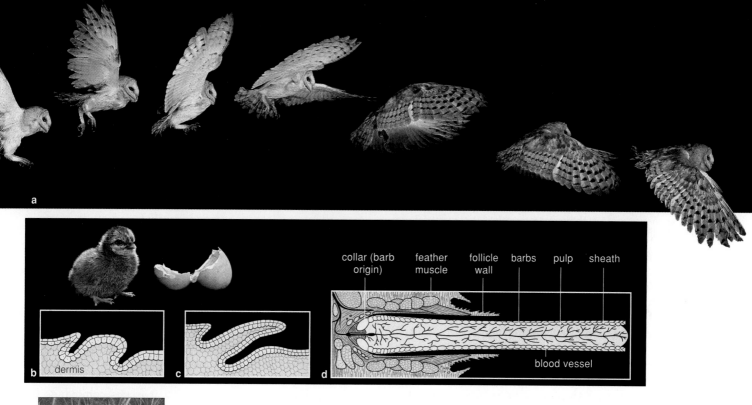

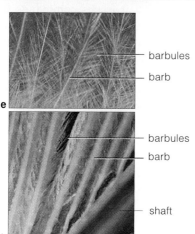

collar (barb origin)   feather muscle   follicle wall   barbs   pulp   sheath

dermis

blood vessel

barbules

barb

barbules

barb

shaft

Figure 26.18 Bird flight. (**a**) Of all living animals, only birds and bats fly by *flapping* their wings. A bird wing is an intricate system of lightweight bones and feathers. Section 19.4 includes a look at the evolution of bird forelimbs.

How a typical feather grows and develops from a region of actively dividing epidermal cells in a chick embryo. (**b**) One of many buds on an early chick embryo's surface. (**c**) Bud growing into a cone-shaped structure. Its base sinks into the skin. The depression around it will become a feather follicle. (**d**) A layer of horny cells at the cone's surface differentiates into a sheath. An epidermal layer under the sheath is the start of a feather. The underlying dermis contains many blood vessels. It becomes the pulp, which nourishes the growing feather but does not contribute to its structure.

(**e**) Down feathers, which function in insulation. (**f**) Structure of flight feathers, which gain strength from their hollow central shaft and interlocked lattice of barbs and barbules.

Flight also demands an airstream, low weight, and a powerful downstroke that can provide lift—a force at right angles to the airstream. A bird wing, a modified forelimb, is composed of feathers and lightweight bones attached to powerful muscles. Its bones do not weigh much, owing to profuse air cavities in the bone tissue (Figures 26.17 and 26.18). For example, the skeleton of a frigate bird, with its seven-foot wingspan, weighs a mere four ounces. That is less than the feathers weigh!

The flight muscles attach to an enlarged breastbone (sternum) and to the upper limb bones adjacent to it (Figure 26.17). Muscle contraction creates the powerful downstroke required for flight. Wings, with their long flight feathers, serve as airfoils. Usually, a bird spreads out long feathers on a downstroke and thus increases the size of the surface pushing against the air. On the upstroke, a bird folds its feathers somewhat, so each wing presents the least possible resistance to air.

Flight is astonishing among many migratory birds. **Migration** is a recurring pattern of movement between two or more regions in response to environmental rhythms. Seasonal change in daylength is one cue that influences internal timing mechanisms, or biological clocks. It causes physiological and behavioral changes in individuals. Such changes induce migratory birds to make round trips between distant regions that differ in climate. For example, Canada geese spend the summer nesting at marshes and lakes in the northern United States and in Canada (see Section 8.8, *Critical Thinking* question 4). The wintering grounds are in New Mexico and other parts of the southern United States.

Of all existing animals, birds alone have feathers, which they use in flight, in heat conservation, and in socially significant communication displays.

THE RISE OF MAMMALS

Mammalian Traits

The **mammals** are vertebrates with hair and mammary glands (hence the name of the class Mammalia, from the Latin *mamma*, meaning breast). Of all species, they alone share these traits. The females feed their young with milk, a nutritious secretion from mammary glands, the ducts of which open upon the body's ventral or anterior surface (Figure 26.19*a*). Mammals also are diverse. For instance, in size alone, existing species range from the 1.5-gram Kitti's hog-nosed bat to 100-ton whales. Species differ greatly in the amount, distribution, and type of hair.

A few aquatic mammals, including the whales, lost most of their hair after their land-dwelling ancestors returned to the seas. But look closely and you see that even the whale snout has some "whiskers"—modified hairs that serve sensory functions, just as they do in dogs and cats. More typically, mammals have a furry coat of underhair (a dense, soft, insulative layer that traps heat) and coarser, longer guard hairs that protect the insulative layer from wear and tear (Figure 26.19*b*). When wet, the guard hairs of platypuses, otters, and other aquatic mammals become matted down like a protective blanket, and the underhair stays dry.

Mammals also are distinctive in that most care for the young for an extended period, and adults serve as models for their behavior. Young mammals are born with a capacity to learn and to repeat behaviors that have survival value. It is among mammals that we find the most stunning shows of behavioral flexibility—a capacity to expand on basic activities with novel forms of behavior—although the trait is far more pronounced in some species than in others. Behavioral flexibility coevolved with expansion of the brain, especially the cerebral cortex. Remember, this outermost layer of the forebrain receives, processes, and stores information from sensory structures, and it issues commands for complex responses. We find the most highly developed cerebral cortex among the mammals named primates. We will soon turn to their story.

Unlike the other amniotes, which typically swallow prey whole, most mammals secure, cut, and sometimes chew food before swallowing it. They also differ from them in **dentition** (that is, in the type, number, and size of teeth). Mammals have four distinctive types of upper and lower teeth that match up and work together to crush, grind, or cut food (Figure 26.19*c*). Their incisors (flat chisels or cones) nip or cut food. Horses and other grazing mammals have large incisors. Canines, with piercing points, are longest in meat-eating mammals. Premolars and molars (cheek teeth) are a platform with surface bumps, or cusps; they crush, grind, and shear food. If a mammal has large, flat-surfaced cheek teeth, you can bet its ancestors evolved in places where tough, fibrous plants were abundant foods. As you will see later on in the chapter, fossilized jaws and teeth from the early primates as well as from species that apparently were on the road to modern humans offer clues to their life-styles.

molars   premolars canines incisors

c

Figure 26.19 Three distinctly mammalian traits. (**a**) A human baby busily demonstrating a key defining feature: It derives nourishment from mammary glands. (**b**) A pair of juvenile raccoons displaying their fur coat. (**c**) Unlike the teeth of their amniote ancestors, the upper and lower rows of mammalian teeth match up. (Compare Figure 26.15*f*.)

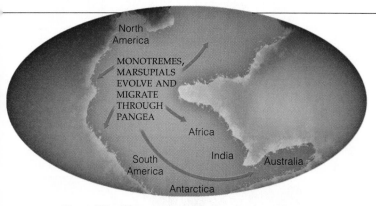

a About 150 million years ago, during the Jurassic

North
America

MONOTREMES,
MARSUPIALS
EVOLVE AND
MIGRATE
THROUGH
PANGEA

Africa

South
America   India   Australia

Antarctica

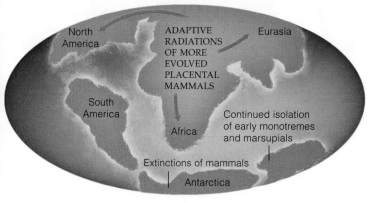

b Between 100 and 85 million years ago, during the Cretaceous

PLACENTAL
MAMMALS
EVOLVE;
ADAPTIVE
RADIATIONS
BEGIN

Isolation of
the early
monotremes,
marsupials on
this land mass

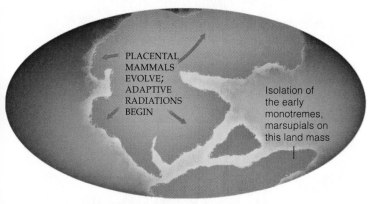

c About 20 million years ago, during the Miocene

North
America

ADAPTIVE
RADIATIONS
OF MORE
EVOLVED
PLACENTAL
MAMMALS

Eurasia

South
America

Africa

Continued isolation
of early monotremes
and marsupials

Extinctions of mammals

Antarctica

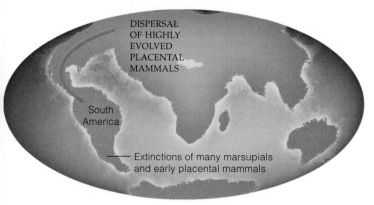

d About 5 million years ago, during the Pliocene

DISPERSAL
OF HIGHLY
EVOLVED
PLACENTAL
MAMMALS

South
America

Extinctions of many marsupials
and early placental mammals

Figure 26.20 Adaptive radiations of mammals.

Mammalian Origins and Radiations

More than 200 million years ago, during the Triassic, a genetic divergence from small, hairless synapsids gave rise to the therapsids, the early ancestors of mammals (Figure 26.14). By Jurassic times, diverse plant-eating and meat-eating mammals called therians had evolved.

Early therians were the size of a mouse, and they were endowed with hair and major changes in the jaws, teeth, and body form. For instance, their four limbs were positioned upright under the body's trunk. This skeletal arrangement made it easier to walk erect, but a trunk higher from the ground was not as stable. At that time the cerebellum, a brain region dealing with the body's balance and spatial positioning, started expanding.

The therians coexisted with dinosaurs through the Cretaceous. When the last of the dinosaurs vanished, diverse adaptive zones opened up for three lineages of those previously inconspicuous mammals (Sections 18.4 and 20.8). New opportunities, differences in traits, and key geologic events put those lineages—the **monotremes** (egg-laying mammals), **marsupials** (pouched mammals), and **eutherians** (placental mammals)—on very different paths to the present. Compared with monotremes and marsupials, both of which retained many archaic traits, placental mammals had the competitive edge. They had higher metabolic rates, more precise ways of regulating body temperature, and a new way of nourishing their developing embryos. You will read about those traits in later chapters. For now, it is enough to know that the placental mammals radiated into many new adaptive zones throughout the world, at the expense of their less competitive relatives.

By the late Jurassic, ancestors of monotremes and marsupials were in southern Pangea (Figure 26.20*a*). After Pangea split apart, those on the huge, drifting fragment that would become Australia were isolated from the ancestors of placental mammals, which were evolving on other continents (Figure 26.20*b,c*).

On the fragment that would become South America, monotremes were replaced by marsupials and early placental mammals. Then a land bridge rejoined South and North America in the Pliocene, and highly evolved placental mammals radiated southward, then rapidly replaced many of those previously isolated mammals (Figure 26.20*d*). Only opossums and a few other species successfully invaded lands in the other direction.

Mammals alone have hair and mammary glands. They have distinctive dentition, a highly developed nervous system, and a notable capacity for behavioral flexibility.

Much of mammalian history was a matter of luck, of species with particular traits being in the right or wrong places on a changing geologic stage at particular times.

PORTFOLIO OF EXISTING MAMMALS

Evolutionarily distant, geographically isolated lineages often evolved in similar ways in similar habitats and came to resemble each other in form and function. We call this **convergent evolution** (Section 19.4). With their interesting history, the three lineages of mammals offer classic examples, as Table 26.1 suggests.

The only living monotremes are two species of spiny anteaters and the duck-billed platypus (Figure 26.21a). Figure 26.21b shows the spiny anteater from Australia; the other lives in the mountains of New Guinea. These small, burrowing mammals feed mostly on ants. Like porcupines, the spiny anteaters bristle with protective spines (modified hairs). As is the case for platypuses, females lay eggs. Unlike platypuses, they do not dig out nests. They incubate a single egg, and their hatchling suckles and completes its development in a skin pouch that forms *temporarily* (by muscle contractions) on the mother's ventral surface.

Most of the 260 existing species of marsupials are native to Australia and nearby islands; a few live in the Americas (Figure 26.21c,d). The tiny, blind, and hairless newborns suckle and finish developing in a *permanent* pouch on the mother's ventral surface. The Tasmanian devil is the largest carnivorous marsupial. Its habit of baring fangs as a threat display (Figure 26.21d), together with hair-raising screeches, coughs, and snarls, give it an undeserved bad reputation. It is just a scavenger, a bit famous for its rowdy communal feeds at carcasses.

Figure 26.21 Monotremes: (**a**) Platypus with offspring and (**b**) spiny anteater (*Tachyglossus*). Two marsupials: (**c**) Adult koala (*Phascolarctos cinereus*). Its ancestors evolved millions of years ago, when the climate turned drier and drought-resistant plants evolved. Koalas fed only on eucalyptus leaves. When Europeans arrived, they cleared eucalyptus forests for farmland. Millions of the slow-moving koalas were shot for their pelts. In the 1930s they became a protected species but their numbers declined because nothing protected the eucalyptus trees. This still is the case in much of their home range. (**d**) Young Tasmanian devil.

Table 26.1 Convergences Among Mammalian Groups		
Life-Style	Home	Mammalian Family
Aquatic invertebrate eater	North America Central America Australia	Water shrew (Soricidae) Water mouse (Cricedidae) Platypus (Ornithorhynchidae)
Land-dwelling carnivore	North America Australia	Wolf (Canidae) Tasmanian wolf (Thylacinedae)
Land-dwelling anteater	South America Africa Australia	Giant anteater (Myrmecophagidae) Aardvark (Orycteropodidae) Spiny anteater (Tachyglossidae)
Ground-dwelling leaf, tuber eater	North America South America Eurasia	Pocket gopher (Geomyidae) Tuco-tuco (Ctenomyidae) Mole rat (Spalacidae)
Tree-dwelling leaf eater	South America Africa Madagascar Australia	Howler monkey (Cebidae) Colobus monkey (Cercopithecidae) Woolly lemur (Indriidae) Koala (Phascolarctidae)
Tree-dwelling nut, seed eater	Southeast Asia Africa Australia	Flying squirrel (Sciuridae) Flying squirrel (Anomaluridae) Flying squirrel (Phalangeridae)

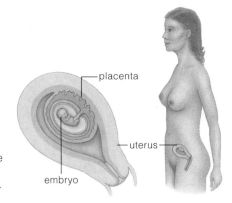

Figure 26.22 Location of the placenta in a human female.

placenta

embryo

uterus

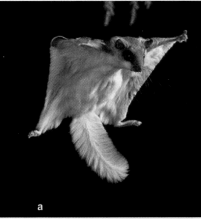

Figure 26.23 Placental mammals. (**a**) A flying squirrel, really just a glider. The only flying mammals are bats (**b**); this one is a grounded Kitti's hog-nosed bat. (**c**) A camel is adapted to traversing hot deserts. (**d**) Manatees live in water and eat seaweed. (**e**) Harp seals swim in frigid waters after prey and sunbathe on ice. (**f**) Thick, insulative fur camouflages an arctic fox from prey. In summer, the light-brown fur blends with golden-brown grasses. In winter, the fur turns white and blends with the snow-covered ground.

All other modern descendants of therians are called placental mammals. The **placenta** is a spongy tissue of maternal and fetal membranes (Figure 26.22). It forms inside a pregnant female's uterus, a chamber where an embryo develops in relative freedom and protection from harsh outside conditions. The embryo gets oxygen and nutrients across the placenta, which also carries away its metabolic wastes.

We will take a closer look at the placenta's structure and function in Section 44.9. For now the point is that, in general, placental mammals have a developmental advantage over their closest relatives. They grow faster in the uterus than marsupials do in their pouch, and

many species are fully formed (and less vulnerable to predators) at birth. Figure 26.23 shows a few species; Appendix I lists the major groups.

The body form and function, behavior, and ecology of the mammals will occupy our attention later in the book. Chapter 27, for example, will provide insight into why many existing mammalian species are threatened by a mass extinction that humans are bringing about.

Convergent evolution occurred among the families of all three lineages of mammals that now occupy similar habitats in different regions of the world.

TRENDS IN PRIMATE EVOLUTION

Since the start of this chapter, you have traveled about 570 million years on branching evolutionary roads. You moved from tiny baglike animals to craniates, then to the jawless and jawed fishes having a backbone. By the Devonian's end some 360 million years ago, you were moving past early tetrapods. These had skullbones, jaws, a backbone, and lungs—*and* they were walking on four limbs that evolved from fleshy lobed fins. You moved past early tetrapods to amniotes—reptiles and mammals. From this point on, you will be traveling on roads that led to primates, then humans.

The mammals called **primates** include prosimians, tarsioids, and anthropoids. The first prosimians were *arboreal*, or tree-dwelling. They dominated northern forests for millions of years, then monkeys and apes evolved and nearly displaced them. The only living tarsioids, small tarsiers of Southeast Asia, are between prosimians and anthropoids in their characters (Figure 26.24*a*). Monkeys, apes, and humans are **anthropoids**. The spider monkeys (Figure 26.24*b*) are typical ceboids (New World monkeys). The baboons, langurs, and macaques are cercopithecoids (Old World monkeys).

In biochemistry and body form, apes are closer to humans than to monkeys. Apes include the gibbon, siamang, orangutan, gorilla, chimpanzee, and bonobo. All apes and humans, and their extinct ancestors, are **hominoids**. All of the humanlike and human species, past and present, are further classified as **hominids**.

Most primates live in tropical or subtropical forests, woodlands, or savannas (grasslands with scattered trees). The vast majority are tree dwellers. Yet no one feature sets them apart from other mammals, and each lineage has its defining traits. Five trends define one that led to humans. They were set in motion as primates adapted to life up in the trees. *First*, there was less reliance on the sense of smell and more on daytime vision. *Second*, skeletal modifications promoted **bipedalism**—upright walking—which freed the hands for novel tasks. *Third*, bone and muscle changes led to refinements in hand movements. *Fourth*, teeth became less specialized. *Fifth*, changes in the brain became interlocked with changes in behavior and the evolution of culture. **Culture** is the sum of behavior patterns of a social group, passed on to generations through learning and symbolic behavior. In brief, *"uniquely" human traits emerged by modification of traits that had evolved earlier, in ancestral forms.*

Enhanced daytime vision. Early primates observed the world through one eye on each side of their head. Later ones had forward-directed eyes, an arrangement that is better for sampling shapes and movement in three dimensions. Through other modifications, their eyes responded more exquisitely to variations in color and light intensity (dim to bright light). Being able to interpret and respond swiftly to novel, diverse stimuli proved advantageous for life in the trees.

Upright walking. How a primate walks depends on its arm length and the shape and positioning of its shoulder blades, pelvic girdle, and backbone. With arm and leg bones about the same length, a monkey can climb, leap, and run on four legs, but not two. A gorilla walks on two legs and on the knuckles of its two long arms. Humans and bonobos are bipedal—they can routinely stride and run about on two legs. Compared to an ape or a monkey, humans have a shorter, more flexible backbone that has an S-shaped curve (Figure 26.24). Skeletal change favoring bipedalism was a key innovation that evolved in ancestors of hominids.

Power grip and precision grip. How did we get our versatile hands? Early mammals spread their toes apart to support the body as they walked or ran on four legs. Primates still spread fingers or toes. Many cup their fingers, as when monkeys lift food to the mouth. Among ancient tree-dwelling primates, modifications to handbones allowed them to wrap fingers around objects (*prehensile* movements) and to touch a thumb to a fingertip (*opposable* movements). In time, the hands became freed from load-bearing functions. Later, when hominids evolved, so did power and precision grips:

power grip   precision grip

These hand positions gave early humans a capacity to make and use tools. They were a foundation for unique technologies and cultural development.

Teeth for all occasions. Before hominids evolved, modifications in jaws and teeth accompanied a shift from eating insects to fruits and leaves, then a mixed diet. Later, rectangular jaws and long canines came to be defining features of monkeys and apes. Along the road that led to modern humans, a bow-shaped jaw and smaller teeth of about the same length evolved.

Brains, behavior, and culture. Long-term shifts in reproductive and social behavior accompanied the shift to an arboreal life-style. In many lineages, parents put more effort in fewer offspring. Maternal care became intense and offspring started to require longer periods of dependency and learning (Figure 26.25).

Expansion of brain regions became interlocked with selection for more complex behavior. Novel behaviors promoted new neural connections in brain regions that

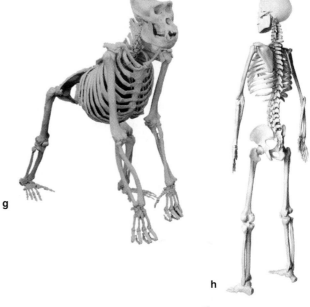

Figure 26.24 Representative primates. (**a**) A tarsier, small enough to sit on one of your hands, is a vertical climber and leaper. (**b**) Spider monkey, an agile four-legged climber, leaper, and runner. (**c**) Gibbon, a brachiator, with body and limbs adapted to swing arm over arm through trees. (**d**) Male gorilla, a knuckle-walker. (**e**) Bonobos, our closest primate relatives. Like us, they walk upright.

Comparison of the skeletal organization and stance of (**f**) monkeys, (**g**) gorillas, and (**h**) humans. The images are not shown at the same scale. A monkey has long, thin limbs relative to its torso, and its long, thin digits are better at gripping than supporting body weight. Like chimpanzees, gorillas use their forelimbs to climb and help support their body weight. Most often, gorillas walk on all fours. Humans are two-legged walkers. These differences in locomotion arose through modifications of the basic skeletal plan of mammals.

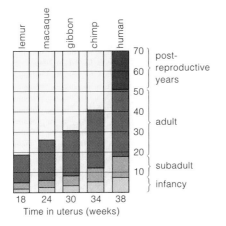

Figure 26.25 Trend, among primates, toward longer life spans and greater dependency of offspring on adults.

dealt with sensory inputs, a brain with more intricate wiring favored more novel behavior, and so on. We see evidence of such interlocking in the parallel evolution of the hominid brain and culture. Culture evolves over time by enrichment in learning and symbolic behavior, especially language. A capacity for language arose in ancestral human species. It required modifications in the capacity of the bony chamber that houses the brain, as well as greater complexity in the brain itself.

These key adaptations evolved along the pathway from arboreal primates to modern humans: complex, forward-directed vision; bipedalism; refined hand movements; generalized dentition; and an interlocked elaboration of brain regions, behavior, and culture.

FROM EARLY PRIMATES TO HOMINIDS

Origins and Early Divergences

The mammals called primates evolved about 60 million years ago, in tropical Paleocene forests. Like the small rodents and tree shrews they resembled (Figure 26.26), they had huge appetites and foraged at night for eggs, insects, seeds, and buds under the trees. They had a long snout and a good sense of smell, suitable for snuffling food and predators. They clawed their way up stems and branches, although not with much speed or grace. In the Eocene (about 55 to 37 million years ago), some primates had moved into the trees. They had a shorter snout, enhanced daytime vision, a larger brain, and far better ways to grasp objects. *How did these traits evolve?*

Consider the trees. Trees offered abundant food and safety from ground-dwelling predators. They also were habitats of uncompromising selection. Picture an Eocene morning with dappled leaves swaying in the breeze, colorful fruit, and predatory birds. An odor-sensitive, long snout would not have been of much use, because air currents tend to disperse odors. But a brain able to evaluate motion, depth, shape, and color would have been favored. So would a brain that quickly estimated body weight, distance, wind speed, and the suitability of destinations; adjustments had to be quick. Skeletal changes that increased fitness also would have helped. For instance, eye sockets facing forward instead of on the sides of the skull would enhance depth perception.

By 36 million years ago, tree-dwelling anthropoids had evolved. They were on or close to the lineage that led to monkeys and apes. One had forward-directed eyes, a snoutless, flattened face, and an upper jaw with shovel-shaped front teeth. It must have used its hands to grab food. Some early anthropoids lived in swamps infested with predatory reptiles. Was that one reason why it became imperative to think fast, grip strongly, and avoid adventures on the ground? Maybe.

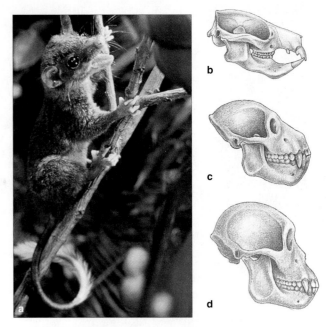

Figure 26.26 (**a**) Tree shrew of Indonesia. The skull shape and teeth of a few early primates: (**b**) From the Paleocene, *Plesiadapis* was as tiny as a tree shrew and had rodentlike teeth. (**c**) The monkey-sized *Aegyptopithecus*, an Oligocene anthropoid, predates a divergence that led to Old World apes and monkeys. (**d**) The apelike dryopiths lived in the Miocene. Some dryopiths were as large as a chimpanzee.

Between 23 and 5 million years ago, in the Miocene, apelike forms—*the first hominoids*—evolved and spread through Africa, Asia, and Europe. At that time, shifts in land masses and ocean circulation caused a long-term change in climate (Figure 26.27). Africa became cooler, drier, and more seasonal. Tropical forests, with their edible soft fruits, leaves, and abundant insects, started giving way to open woodlands and later to grasslands. Food had become drier, harder, and difficult to find. Hominoids that had evolved in lush forests had two options: Move into new adaptive zones or die out. Not all made the transition; most became extinct. But one was the common ancestor of two enduring lineages that arose by 7 million years ago. One gave rise to the great apes, and the other to the first *hominids*.

The First Hominids

Sahelanthropus tchadensis was one species that evolved in Central Africa about 6 or 7 million years ago, during the time when the ancestors of humans were becoming distinct from the apes (Figures 26.28). Was it an ape or a hominid? The braincase of one fossil is no bigger than that of a chimpanzee. Yet, like bigger brained hominids that would follow it much later in time, the fossil also has a shorter, flattened face, a pronounced brow ridge, and smaller canines. Apparently it lived by an ancient

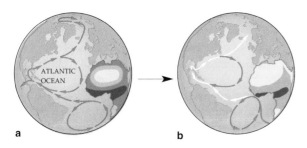

Figure 26.27 Model for a long-term shift in global climate. (**a**) Before the Isthmus of Panama formed, the salinity of surface ocean currents were similar around the world. Circulation patterns kept Arctic waters warm. (**b**) After the isthmus formed, North Atlantic surface currents grew saltier and heavier. They sank before reaching the Arctic. Water stayed colder in polar regions. The Arctic ice cap formed, ushering in a long-term trend toward a cooler, drier climate in Africa.

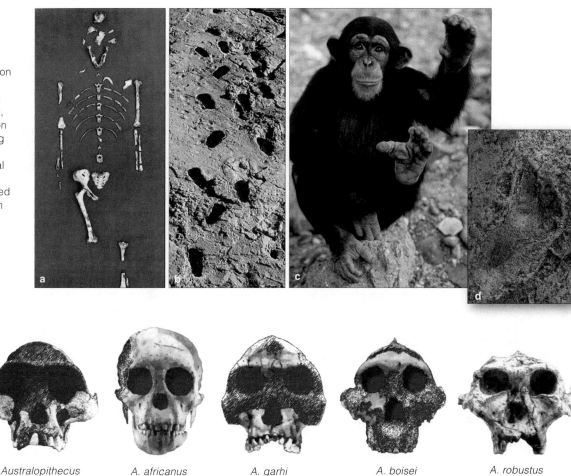

Figure 26.28 (**a**) Remains of "Lucy" (*Australopithecus afarensis*), who lived 3.2 million years ago. (**b**) At Laetoli in Tanzania, Mary Leakey found these footprints, made in soft, damp, volcanic ash 3.7 million years ago. (**c,d**) The arch, big toe, and heel marks of such footprints are signs of bipedal hominids. Unlike apes, early hominids didn't have a splayed big toe, as the chimpanzee in (**c**) obligingly demonstrates.

| *Sahelanthropus tchadensis* 7–6 million years | *Australopithecus afarensis* 3.6–2.9 million years | *A. africanus* 3.2–2.3 million years | *A. garhi* 2.5 million years (first tool user?) | *A. boisei* 2.3–1.4 million years (huge molars) | *A. robustus* 1.9–1.5 million years |

Figure 26.29 Representative fossils of African hominids that lived between 7 and 1.4 million years ago.

lake, in a region that was a mosaic of woodlands and grasslands. As more fossils are found, they may tell us whether *S. tchadensis* was one of our earliest ancestors.

The Miocene through the Pliocene was a "bushy" time of evolution for hominids in central, eastern, and southern Africa. By this we mean that many diverse forms evolved during that interval, and we still do not know how they are related. Most are informally called the **australopiths**, or "southern apes." *Australopithecus anamensis*, *A. afarensis*, and *A. africanus* were gracile, or slightly built (Figure 26.29). *A. boisei* and *A. robustus* were robust, or muscular and heavily built. Like apes, australopiths had a large face, protruding jaws, and a small skull (and brain) size. Yet they differed in key respects from hominoids. Their thick-enameled molars could grind harder foods, for example. And hominids walked upright. We know this from their fossilized hip bones and limb bones, and from footprints. In one case, about 3.7 million years ago *A. afarensis* walked across newly fallen volcanic ash during a light rain, which turned that ash to quick-drying cement (Figure 26.28*b*).

When bipedal hominids started to evolve in the late Miocene, they were still adapted to forested habitats. In the trees, their hands became good at gripping objects strongly and precisely. Their descendants left the trees for life on the ground and did not become specialized in running fast on all fours. They became fully upright and used their manipulative skills to advantage. They kept their hands free to hold offspring and probably to carry precious food during their foraging expeditions.

Primates evolved from small, rodentlike mammals that moved into the trees 60 million years ago.

The first hominoids (apelike forms) evolved and radiated through Africa, Europe, and southern Asia. A long-term, global cooling trend may have ushered in new selection pressures for modification in their form and behavior.

By 7 million years ago, divergences had given rise to the first hominids. Many diverse forms evolved during the late Miocene and on through the Pliocene. All were humanlike in a crucial respect: They walked upright.

EMERGENCE OF EARLY HUMANS

What might the fossilized fragments of early hominids tell us about our own origins? The record is still too sketchy for us to know how those diverse forms were related to one another, let alone which ones may have been the ancestors of humans. Besides, which traits do characterize **humans**—members of the genus *Homo*?

Well, what about brains? The modern human brain is the basis of fine analytical and verbal skills, complex social behavior, and technological innovation. It sets us apart from apes, which have a far smaller skull volume and brain size (Figure 26.30). Yet this feature alone can't tell us when certain hominids made the leap to becoming human. Why? Their brain size fell within the range of the apes. They did make simple tools, but so do chimps and some birds. We have no clues to their social behavior.

We are left to speculate on physical traits, indicated by many fossils—a skeleton adapted for bipedalism; manual dexterity; larger brain volume; a smaller face; and smaller, more thickly enameled teeth. These traits, which apparently originated during the late Miocene, are signs of what may have been the earliest humans— *Homo habilis*, a name that means "handy man."

Between 2.4 and 1.6 million years ago, early forms of *Homo* were living in dry woodlands adjoining the savannas of eastern and southern Africa (Figure 26.31). Fossilized teeth tell us they could have been eating hard-shelled nuts, dry seeds, soft fruits, leaves, and

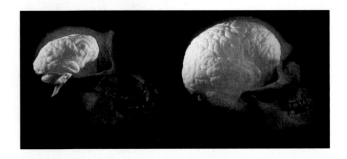

Figure 26.30 Comparison of brain size for a chimpanzee (*left*) and a modern human (*right*).

insects—all seasonal foods. Most likely, *H. habilis* had to think ahead, to plan when to gather and store foods that would help it survive the cold dry season.

H. habilis shared its habitat with predators such as saber-tooth cats. The cats' teeth could impale prey and shear off flesh but couldn't crush open marrow bones. Carcasses with meat shreds clinging to bones afforded nutrients in nutrient-stingy places. *H. habilis* was not a full-time carnivore. But it may have enriched its diet opportunistically, by scavenging carcasses.

Fossil hunters have found many stone tools dating to the time of *H. habilis*. They cannot say for sure that *H. habilis* was the only species that made them. Perhaps australopiths as well as *H. habilis* used sticks and other

Homo rudolfensis
2.4–1.8
million years

H. habilis
1.9–1.6
million years

Figure 26.31 *Left:* Reconstruction of *Homo habilis* in an East African woodland. Two australopiths are in the distance. *Above:* Two fossils of early humans.

Figure 26.32 From Olduvai Gorge in Africa, a sampling of stone tools. *Left to right:* crude chopper, more refined chopper, hand ax, and cleaver.

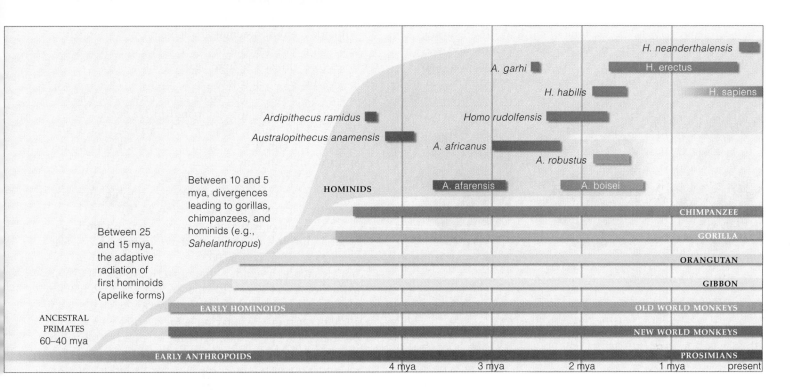

Figure 26.33 Summary of the presumed evolutionary branchings in the primate family tree, including lineages on or near the evolutionary road that led to modern humans.

perishable materials for tools before then, as modern apes do, but we have no way of knowing.

Maybe ancestors of modern humans started down a toolmaking road by picking up rocks to crack marrow bones. Maybe they scraped flesh from bones with sharp flakes split naturally from rocks. However it happened, at some point they started *shaping* stone implements. Paleontologist Mary Leakey was the first to discover evidence of stone toolmaking in sedimentary layers at Africa's Olduvai Gorge. In the deepest layers are the oldest tools—crudely chipped pebbles (Figure 26.32). Early humans might have used them to smash food, dig for roots, and poke insects from bark. More recent layers have more complex tools.

At such sites we find fossils of an early form of *Homo* that was twice as brainy as australopiths and obviously ate well. There apparently was no selection pressure for more creativity in securing food resources; stone tools did not change much for the next 500,000 years.

Before turning the page, take a look at Figure 26.33 to get a sense of where we have been on the primate evolutionary road, and where we go from here.

Compared to australopiths, early humans had a larger brain, smaller face, and smaller, thickly enameled teeth. Theirs was a stable adaptive zone; the design of their stone tools did not change much for half a million years.

EMERGENCE OF MODERN HUMANS

Ancestors of modern humans stayed put in Africa until about 2 million years ago. Then a divergence gave rise to *Homo erectus*—a species related to modern humans (Figures 26.34 and 26.35). Its name means upright man. Although its forerunners also were upright walkers, *H. erectus* populations did the name justice. They trekked out of Africa, turned left into Europe, and right into Asia. Some walked to China; fossils from the former Soviet republic of Georgia and Southeast Asia are 1.8 million and 1.6 million years old. *H. erectus* survived as ice sheets advanced—more than once—into northern Europe, southern Asia, and North America.

Whatever pressures triggered the far-flung travels, this was a time of physical changes, as in skull size and leg length. It also was a time of cultural lift-off for the human lineage. *H. erectus* had a larger brain and was a more creative toolmaker. Its social organization and communication skills must have been well developed.

| *H. erectus*
2 million–53,000? years | *H. neanderthalensis*
200,000–30,000 years | *H. sapiens*
100,000–? |

Figure 26.34 Fossils of early modern human forms.

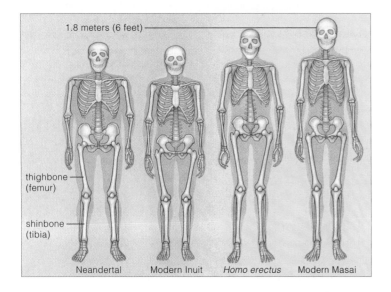

1.8 meters (6 feet)

thighbone (femur)

shinbone (tibia)

Neandertal   Modern Inuit   *Homo erectus*   Modern Masai

Figure 26.35 Body build correlated with climate. Humans adapted to cold climates have a heat-conserving body: stockier, shorter legs, compared with humans adapted to hot climates.

Figure 26.36 Cave paintings at Lascaux, France.

How else can we explain its successful dispersal? From southern Africa to England, different populations used the same kinds of hand axes and other tools to pound, scrape, shred, chop, cut, and whittle materials. They withstood environmental challenges by building fires and using furs for clothing. Clear evidence of fire use dates from an ice age in the early Pleistocene.

As the Middle Eastern fossils indicate, *Homo sapiens* had evolved by 100,000 years ago. Early *H. sapiens* had smaller teeth and jaws compared to *H. erectus* (Figure 26.34). Many individuals also had a novel feature—a chin. Facial bones were smaller, the skull was higher and rounder, and the brain had a larger volume. And these humans may have developed complex language. Their origin and details of their geographic dispersal are hotly debated, however (Section 26.15).

One group of early humans, the massively built and large-brained Neandertals, lived in Europe and the Near East from 200,000 to 30,000 years ago. Some were the first to adapt to colder climates (Figure 26.35). Their extinction coincided with the arrival of anatomically modern humans in those areas between 40,000 and 30,000 years ago. We have no evidence that they warred or interbred with the new arrivals. We don't know what happened to them. We do know Neandertal DNA has unique sequences, and these may not be represented in the gene pools of modern European populations.

From 40,000 years ago to today, human evolution has been almost entirely cultural, not biological—and so we leave our story. A point to remember: Humans spread rapidly through the world by devising *cultural* means to deal with a broader range of environments. Compared to their predecessors, they developed rich cultures (Figure 26.36). Hunters and gatherers persist in parts of the world, but other groups have moved from "stone-age" technology to the age of "high tech," which attests to the remarkable behavioral plasticity and depth of human adaptations.

Cultural evolution has outpaced the biological evolution of the only remaining human species, *H. sapiens*. Today, humans everywhere rely on cultural innovation to adapt rapidly to a broad range of environmental challenges.

Out of Africa—Once, Twice, Or . . .

If researchers are interpreting the fossil record of human evolution correctly, then Africa was the cradle for us all. At this writing, at least, no one has found any human fossils older than 2 million years *except* in Africa. There, *H. erectus* coexisted for a time with *H. habilis* before some populations dispersed to the cooler grasslands, forests, and mountains of Europe and Asia. They apparently left Africa in waves, between about 2 million and 500,000 years ago. Judging from a few fossils from Java, isolated *H. erectus* populations may have endured until some time between 53,000 and 37,000 years ago.

But when did *H. sapiens* originate? *Here we find a good example of how the same body of evidence can be interpreted in different ways.* The **multiregional model** and **African emergence model** are interpretations of the distribution of early and modern humans. Both models are based on measurements of genetic distances among existing human populations. Biochemical and immunological studies do show the greatest genetic distance is between *H. sapiens* populations native to Africa and all other human populations. The next greatest distance separates Southeast Asian and Australian populations from others (Figure 26.37). Both also correlate *H. sapiens* fossils to specific times in the past (Figure 26.38). Even so:

By the multiregional model, *H. erectus* spread through many regions by about 1 million years ago. The separate populations evolved in regionally distinct ways because they faced different selection pressures. Subpopulations ("races") of *H. sapiens* evolved from them but did not evolve into separate species. Why? Gene flow continued among them, even to the present. For example, when Alexander the Great's armies swept eastward, they also contributed "blue-eye genes" from Greeks to the allele pool of generally brown-eyed subpopulations in Africa, the Near East, and Asia.

The African emergence model does not dispute fossil evidence that *H. erectus* populations evolved in different ways in different places. However, it holds that modern

humans originated in sub-Saharan Africa between 200,000 and 100,000 years ago. Later, their populations moved out of Africa into regions along the routes shown in Figure 26.38. In each region where they settled, they replaced archaic *H. erectus* populations that preceded them. Only then were regional phenotypic differences superimposed on the *H. sapiens* body plan.

In support of this model, the oldest known *H. sapiens* fossils are from Africa. Also, in Zaire, a finely wrought barbed-bone harpoon and other exquisite tools suggest that African populations were as skilled at making tools as *Homo* populations known earlier from Europe. Also, in 1998, researchers at the University of Texas and in China announced findings from the Chinese Human Genome Diversity Project. Detailed analysis of gene patterns from forty-three ethnic groups in Asia suggest that modern humans moved from Central Asia, along the coast of India, then on into Southeast Asia and southern China. Populations later moved north and northwest into China, then into Siberia, and then on down into the Americas.

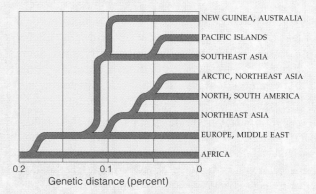

Figure 26.37 One proposed family tree for populations of modern humans (*Homo sapiens*) native to different regions. The tree is based on nucleic acid hybridization studies of many genes (including those for mitochondrial DNA and the ABO blood group) and immunological comparisons.

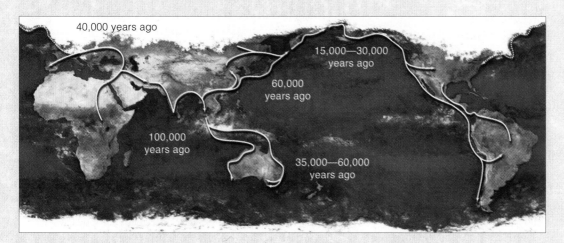

Figure 26.38 Estimated times when populations of early *H. sapiens* were colonizing different regions of the world, based on radiometric dating of fossils. The presumed dispersal routes (*white* lines) seem to support the African emergence model.

SUMMARY   *Gold* indicates text section

1. Generally, chordate embryos have a notochord, a dorsal hollow nerve cord, a pharynx with gill slits (or hints of these), and a tail extending past the anus. Some or all of these traits persist in adults. Chordates with a backbone are vertebrates. The tunicates (including sea squirts) and lancelets are invertebrate chordates. *26.1*

2. Conventional vertebrate groups are jawless fishes, jawed armored fishes (extinct), cartilaginous fishes, bony fishes, amphibians, reptiles, birds, and mammals. These groups are in the process of being renamed to better reflect evolutionary relationships. *26.1*

3. The earliest vertebrates, jawless fishes that arose in the Cambrian, included ostracoderms. Lampreys and hagfishes are their modern descendants. Jawed fishes also arose in the Cambrian. A once-dominant lineage, placoderms, became extinct. Other lineages gave rise to jawed fishes having a cartilaginous endoskeleton or a bony endoskeleton. The ray-finned fishes, lobe-finned fishes, and lungfishes are in this group. *26.1–26.4*

4. Four evolutionary trends are associated with certain vertebrate lineages: *26.2, 26.4*

   a. A vertebral column supplanted the notochord as a structural element against which muscles act. This development led to fast-moving predatory animals.

   b. Jaws evolved from gill-supporting elements. This led to increased predator–prey competition; it favored more efficient nervous systems and sensory organs.

   c. In early lobe-finned fishes, paired fins evolved into fleshy lobes with inner structural elements. Such lobes were forerunners of paired limbs of tetrapods.

   d. In early lungfishes, lunglike sacs assisted gills in respiration. Such organs proved adaptive when some Devonian tetrapods ventured onto land. In a related development, the circulatory system became far more efficient at distributing oxygen through the body.

5. In body plan and reproductive mode, amphibians are between fishes and the amniotes called "reptiles." Amphibians were the first tetrapods to live on land. But most life cycles still have aquatic stages. *26.5*

6. Amniotes were the first vertebrates to form eggs with four internal membranes that conserve water for the embryo, cushion it, and metabolically support it. Early amniotes laid leathery or calcified eggs; many still do. Amniotes include mammals, turtles, lizards, snakes, tuataras, crocodiles, and birds. All reproduce successfully away from aquatic habitats. *26.6*

7. All of the amniotes have well-developed circulatory, respiratory, nervous, and sensory systems. *26.6–26.11*

   a. Birds have feathers, which are used in flight, heat conservation, and social displays. *26.8*

   b. Mammals have milk-secreting mammary glands, insulative hair or thick skin, distinctive dentition, and a highly developed cerebral cortex. Most adults nurture offspring through extended dependency. *26.9–26.10*

   c. Mammals evolved from small, hairless synapsids in the Triassic. After the extinction of the last dinosaurs, and after the isolation caused by Pangea's breakup, three lineages radiated into new adaptive zones: egg-laying mammals (monotremes), pouched mammals (marsupials), and placental mammals (eutherians).

   d. Early placental mammals had higher metabolic rates, more precise control of body temperature, and an efficient way to nourish embryos. Many outcompeted monotremes and marsupials. Some species of the three lineages also converged morphologically.

8. Primates include prosimians (lemurs and related forms), tarsioids, and anthropoids (monkeys, apes, and humans). Apes and humans alone are hominoids. The australopiths and anatomically modern humans (*Homo sapiens*) and others of their lineage, including *H. habilis* to *H. erectus*, are classified as hominids. *26.11–26.13*

   a. The first primates were tiny, rodentlike mammals that evolved by 60 million years ago in tropical forests.

   b. Hominoids (apelike forms) evolved in Africa in the Miocene, between 23 and 7 million years ago. Some gave rise to hominids by 6 or 7 million years ago.

   c. The origin and early evolution of hominids might relate to a long-term cooling trend when land masses and ocean circulation patterns shifted. Tropical forests gave way to dense woodlands, then grasslands.

9. Modern humans emerged by modification of traits starting when tree-dwelling ancestors relied less on the sense of smell and more on enhanced daytime vision. Manipulative skills increased when hands were freed from load-bearing functions. Miocene ancestors shifted to bipedalism, omnivorous eating habits, and increases in brain complexity and behavior. *26.11*

10. *H. habilis*, the earliest known *Homo* species, evolved by 2.5 million years ago. It was one of the first stone toolmakers. *Homo erectus*, ancestral to modern humans, evolved by 2 million years ago. *H. erectus* populations radiated out of Africa, into Asia and Europe. We have 100,000-year-old fossils of modern humans (*H. sapiens*). About 40,000 years ago, cultural evolution outstripped biological evolution of the human form. *26.13–26.14*

Review Questions

1. Which traits distinguish chordates from other animals? *26.1*

2. List the four major trends in vertebrate evolution. Which living vertebrates are the most abundant? *26.2, 26.4*

3. Name the three major lineages of fishes that have a bony endoskeleton. Which uses lunglike sacs for gas exchange? *26.4*

4. List these groups in evolutionary order, from most ancient to most recent: amniotes, cephalochordates, tetrapods, jawed vertebrates, urochordates, vertebrates, and tetrapods. *CI–26.6*

5. Is this statement true or false: Skin, gills, and lungs are used for gas exchange among amphibians. *26.5*

6. Birds, crocodilians, and dinosaurs share what traits? *26.6–26.8*

7. List some characteristics that distinguish each of the three mammalian lineages from other amniotes. *26.9, 26.10*

8. Define hominoid and hominid. *26.11*

9. Briefly describe some of the conserved physical traits that connect anatomically modern humans with their mammalian ancestors, then their primate ancestors. *26.9, 26.11, 26.13*

Self-Quiz   ANSWERS IN APPENDIX III

1. Only _____ have a notochord, a tubular dorsal nerve cord, a pharynx with slits in the wall, and a tail extending past the anus.
   a. echinoderms
   b. tunicates and lancelets
   c. vertebrates
   d. both b and c
   e. all of the above

2. Gills function in _____ .
   a. respiration
   b. circulation
   c. food trapping
   d. water regulation
   e. both a and c

3. A shift from a reliance on _____ to reliance on _____ was pivotal in the evolution of all vertebrates.
   a. the notochord; a backbone
   b. filter feeding; jaws
   c. gills; lungs
   d. all of the above

4. It now appears that the first walkers were _____ .
   a. some Devonian fishes
   b. Devonian amphibians
   c. Cretaceous amphibians
   d. Silurian reptiles

5. The evolutionary road to humans included _____ .
   a. craniates
   b. jawed vertebrates
   c. tetrapods
   d. amniotes
   e. a, b, and d
   f. all of the above

6. Generally, the only amphibian groups that entirely escaped dependency on free-standing water are _____ .
   a. salamanders
   b. toads
   c. caecilians
   d. none of the above

7. Reptiles moved fully onto land owing to _____ .
   a. tough skin
   b. internal fertilization
   c. good kidneys
   d. amniote eggs
   e. none of the above
   f. all of the above

8. A four-chambered heart evolved first in _____ .
   a. bony fishes
   b. amphibians
   c. birds
   d. mammals
   e. crocodilians
   f. both c and d

9. _____ have highly efficient circulatory and respiratory systems, and a complex nervous system and sensory organs.
   a. Reptiles
   b. Birds
   c. Mammals
   d. all of the above

10. Various mammals _____ .
   a. hatch
   b. complete their embryonic development in pouches
   c. complete their embryonic development in the uterus
   d. both b and c
   e. all of the above

11. Match the organisms with the appropriate features.
   _____ jawless fishes
   _____ cartilaginous fishes
   _____ bony fishes
   _____ amphibians
   _____ reptiles
   _____ birds
   _____ mammals
   a. complex cerebral cortex, thick skin or hair
   b. respiration by skin and lungs
   c. include coelacanths
   d. include hagfishes
   e. include sharks and rays
   f. complex social behavior, feathers
   g. first with amniote eggs

Figure 26.39 What do you suppose this frilled lizard is doing if not announcing an attitude problem?

Critical Thinking

1. Describe the factors that might contribute to the collapse of native fish populations in a lake after the introduction of a novel predator, such as the lamprey.

2. The frilled lizard in Figure 26.39 is flaring a ruff of neck skin. Research the literature and report on hypotheses about the possible adaptive value of this behavior.

3. Think about the flight muscles of birds and their demands for oxygen and ATP energy. What type of organelle would you expect to be profuse in these muscles? Explain your reasoning.

4. Kathie and Gary, two amateur fossil hunters, unearthed the complete fossil of a mammal. How can they determine whether it was a herbivore, carnivore, or omnivore?

5. In Australia, many species of marsupials are competing with recently introduced placental mammals (such as rabbits) for resources—but they are not winning. Explain how it is that placental mammals that did not even evolve in the Australian habitats show greater fitness than the native mammals.

Selected Key Terms

Vertebrates

amniote *26.6*
amphibian *26.5*
bird *26.8*
bone tissue *26.2*
cartilaginous fish *26.4*
cephalochordate *26.1*
chordate *26.1*
convergent evolution *26.10*
craniate *26.1*
dentition *26.9*
eutherian *26.9*
filter feeder *26.1*
fin *26.2*
gill *26.2*
gill slit *26.1*

jaw *26.1*
lobe-finned fish *26.4*
lung *26.2*
lungfish *26.4*
mammal *26.9*
marsupial *26.9*
migration *26.8*
monotreme *26.9*
nerve cord *26.1*
notochord *26.1*
ostracoderm *26.1*
pharynx *26.1*
placenta *26.10*
placoderm *26.1*
primate *26.11*
ray-finned fish *26.4*
"reptile" *26.6*
swim bladder *26.4*

tetrapod *26.4*
urochordate *26.1*
vertebra *26.2*
vertebrate *26.1*

Human Evolution

African emergence model *26.15*
anthropoid *26.11*
australopith *26.12*
bipedalism *26.11*
culture *26.11*
hominid *26.11*
hominoid *26.11*
human *26.13*
multiregional model *26.15*

Readings

Romer, A. S., and T. S. Parsons. 1986. *The Vertebrate Body.* Sixth edition. Philadelphia: Saunders. The classic.

Tattersall, I. 1998. *Becoming Human: Human Evolution and Human Uniqueness.* Pennsylvania: Harvest Books.

On-Line readings at Student Guide for InfoTrac:
www.brookscole.com/biology   473

BIODIVERSITY IN PERSPECTIVE

The Human Touch

In 1722, on Easter morning, a European explorer landed on a small volcanic island and found a few hundred skittish, hungry Polynesians living in caves. The island had no trees whatsoever, only dry, withered grasses and scorched, shrubby plants. There were about 200 massive stone statues near the coast and 700 unfinished, abandoned ones in inland quarries (Figure 27.1). Some weighed fifty tons. Without trees for wood and fibrous plants to make ropes, how were they erected? There were no wheeled carts or draft animals. How did the statues get from the quarries to the coast?

Two years later James Cook visited the island. When islanders paddled out to his ships, he noted that they didn't have the knowledge or materials to keep their canoes from leaking; they spent half the time bailing water. He saw only four canoes on the whole island. Nearly all of the statues had been tipped over, often onto spikes that shattered the faces on impact.

Later, researchers solved the mystery of the statues. Easter Island, as it came to be called, is no bigger than 165 square kilometers (64 square miles). Voyagers from the Marquesas discovered this eastern outpost of Polynesia around A.D. 350, possibly after being blown off course during storms. The place was a paradise. Its fertile volcanic soil supported dense palm forests, hauhau trees, toromino shrubs, and lush grasses. The new arrivals built large, buoyant, oceangoing canoes from long, straight palms strengthened with rope made of fibers from hauhau trees. They used toromino wood as fuel to cook fishes and dolphins; they cleared forests to plant taro, bananas, sugarcane, and sweet potatoes.

By 1400, between 10,000 and 15,000 descendants were living on the island. Society had become highly structured, given the need to cultivate, harvest, and distribute food for so many. Crop yields had declined; precious soil nutrients had been carried off by each harvest and by ongoing, neglected erosion. Edible species vanished from the waters around the island, so fishermen had to build larger canoes to sail out ever farther, into the open and more dangerous sea.

Figure 27.1 On Easter Island, a few of the massive stone tributes to the gods. Apparently islanders erected the statues as a plea for divine intervention after their once-large population destroyed the biodiversity of their tropical paradise. The plea didn't have any effect whatsoever on reversing the losses on land and in the surrounding sea. The human population didn't recover, either.

Survival was at stake. Those in power appealed to the gods. They directed the community to carve divine images of unprecedented size and power, and to move newly carved statues over miles of rough terrain to the coast. As Jo Anne Van Tilburg recently demonstrated, they probably lashed the statues to canoe-shaped rigs, and then rolled them along a horizontal "ladder" of greased logs. She got the idea after observing how existing islanders haul canoes from water onto land.

By now, islanders had all of the arable land under cultivation. They had eaten all the native birds, and no new birds came to nest. They were raising and eating rats, the descendants of hitchhikers on the first canoes. They coveted palm seeds, now a scarce delicacy.

Wars broke out over dwindling food and space. By about 1550 no one was venturing offshore to harpoon fishes or dolphins. They couldn't build canoes because the once-rich forests were gone. They had cut down all of the palms. Hauhau trees were extinct; islanders had used every last one as firewood for cooking.

The islanders turned to the only remaining source of animal protein. They started to hunt and eat one another.

Central authority crumbled, and gangs replaced it. As gang wars raged, those on the rampage burned the remaining grasses to destroy hideouts. The rapidly dwindling population retreated to caves and launched raids against perceived enemies. Winners ate the losers and tipped over the statues. Even if the survivors had wanted to leave the island, they no longer had a way to do so. What could they have been thinking when they chopped down the last palm?

And so, from archeological and historical records, we know Easter Island initially sustained a society that flourished amid abundant resources, then fell apart abruptly. In life's greater evolutionary story, its loss of biodiversity might not even warrant a footnote. After all, compare the loss to the millions of species that appeared and became extinct in the distant past!

Yet there is a lesson here. The near-total destruction that 15,000 Polynesians wrought on biodiversity was confined to a small bit of land in the vastness of the Pacific Ocean. What are the requirements and whims of *6.2 billion* people doing to global biodiversity today? Is the loss of species on Easter Island an isolated case? Or should we take what happened there as a warning of a worldwide extinction crisis of our own making, one that is even now under way? Let's take a look.

Key Concepts

1. The current range of global biodiversity is largely an outcome of an overall pattern of abrupt extinction events and slow recoveries. It takes tens of millions of years to reach the level of biodiversity that prevailed before a mass extinction.

2. The loss of individual species as well as extinctions of major groups has often had complicated causes, the understanding of which requires scientific analysis.

3. We classify species that are living in a restricted geographic region and are extremely vulnerable to extinction as endangered. Extinction rates have been rising in the past four decades as a result of habitat loss, habitat fragmentation, species introductions, overharvesting, and the illegal wildlife trade.

4. The rapidly growing human population is expected to reach 9 billion by 2050. Its collective demands for food, materials, and living space are now threatening biodiversity around the world and are the cause of a new extinction crisis.

5. Conservation biology is an attempt to preserve diversity. It includes systematic surveys of the full range of biodiversity, analysis of its evolutionary and ecological origins, and identification of methods to maintain and use biodiversity for the benefit of the human population.

6. Systematic surveys focus initially on identifying hot spots, or limited areas where many species are in danger of extinction owing to human activities. At the next research level, broader or multiple hot spots are researched. Data are combined into a global picture of ecoregions where biodiversity is most threatened.

ON MASS EXTINCTIONS AND SLOW RECOVERIES

Based on many lines of evidence accumulated over the past few centuries, an estimated 99 percent of all species that have ever lived are extinct. Even so, the full range of biodiversity is greater now than it has ever been at any time in the past.

Reflect on the evolutionary stories of the preceding chapters in this unit. For at least the first 2 billion years of life's history, single-celled prokaryotes dominated the evolutionary stage. They did so until the Cambrian period, when atmospheric oxygen started to approach current levels. At that time, some of the single-celled eukaryotes began genetic divergences and intricate species interactions, which led to the origin of diverse protistans, plants, fungi, and animals.

You saw how mass extinctions slashed biodiversity on land and in the seas. Each global episode spurred evolutionary change and radiations into the newly vacated adaptive zones. However, recovery to the same level of biodiversity has been exceedingly slow, requiring 20 to 100 million years. Figure 27.2a reviews the pattern of five major extinctions and recoveries.

That pattern is only a composite of what happened to the major taxa. Lineages, remember, differ in their time of origin, the extent to which the member species diverged, and how long they persisted. If you consider ongoing survival and reproduction to be the measures of success, then each became a loser or a winner when environmental conditions changed in drastic or novel ways. To appreciate this point, reflect on Figure 27.2b. It shows the evolutionary history of representative lineages. Many histories were combined to give us the overall pattern of Figure 27.2a.

Era	Period	MASS EXTINCTION UNDER WAY
CENOZOIC	QUATERNARY —— 1.8 mya TERTIARY	With high population growth rates and cultural practices (e.g., agriculture, deforestation), humans become major agents of extinction.
	—— 65	*MASS EXTINCTION*
MESOZOIC	CRETACEOUS —— 145 JURASSIC —— 213 TRIASSIC	Slow recovery after Permian extinction, then adaptive radiations of some marine groups and plants and animals on land. Asteroid impact at K–T boundary, 85% of all species disappear from land and seas.
	—— 248	*MASS EXTINCTION*
PALEOZOIC	PERMIAN —— 286 CARBONIFEROUS	Pangea forms; Land area exceeds ocean surface area for first time. Asteroid impact? Major glaciation, colossal lava outpourings, 90%–95% of all species lost.
	—— 360	*MASS EXTINCTION*
	DEVONIAN —— 410 SILURIAN	More than 70% of marine groups lost. Reef builders, trilobites, jawless fishes, and placoderms severely affected. Meteorite impact, sea level decline, global cooling?
	—— 440	*MASS EXTINCTION*
	ORDOVICIAN —— 505 CAMBRIAN	Second most devastating extinction in seas; nearly 100 families of marine invertebrates lost.
	—— 544	*MASS EXTINCTION*
	(precambrian)	Massive glaciation; 79% of all species lost, including most marine microorganisms.

a

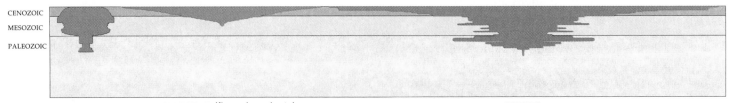

b

Figure 27.2 (**a**) Review of the five greatest mass extinctions and subsequent slow recoveries in the past. More mass extinctions occurred than are shown here. Compare Figure 19.4, in Section 19.2. (**b**) Within the framework of this generalized graph, patterns of extinction and recoveries differed significantly for different major taxa, as these selected examples indicate.

a

Figure 27.3 Two extinct and one threatened species. (**a**) Charles Knight's magnificent painting of *Tylosaurus*, one of the mosasaurs. This powerful marine lizard flourished in shallow, nearshore waters of the Cretaceous seaways. Indonesia's Komodo dragon, which grows 10 feet (3 meters) long, is a living relative. We think the Komodo dragon is a giant, yet imagine meeting up with a mosasaur. Some were 40 feet (12 meters) long.

(**b**) The dodo (*Raphus cucullatus*), a large, flightless bird that evolved on Mauritius, then vanished more than 300 years ago. Certain trees (*Calvaria major*) also evolved on that island, as did tortoises. Did the tree coevolve with dodos, tortoises, or both? Either may have fed on the tree's large fruits. For example, the large dodo gizzard could have partially digested the thick-walled coat of seeds in the fruits, just enough to help seeds germinate after leaving the gut. Either way, after the dodo became extinct, the seeds stopped germinating. Only thirteen trees were still standing by the mid-1970s. By some estimates they are more than 300 years old. Each year they produce seeds, but these apparently cannot break out of the seed coats without help. Botanists now employ turkeys or gem polishers as substitute grinders.

b

What causes mass extinctions? The answer may not be obvious. For example, the K–T asteroid impact did deliver the coup de grace for some lineages, including the dinosaurs and mosasaurs (Figure 27.3*a*). But other factors must have been at work as well. Why? Many dinosaurs had already become extinct during the 10 million years that preceded the impact. As another example, insect lineages persisted right through it.

Maybe tectonic change was a contributing factor. Pangea started breaking up long before the impact, and deep, cold currents from the south had moved into warmer, equatorial seas. Many marine groups perished. Oceanic changes affected the atmosphere, the climate, and through them, life on land. The fossil and geologic records combined show that the K–T asteroid did indeed abruptly eliminate some lineages. But they also show that circumstances were bringing about mass extinctions long before the asteroid hit.

Now extend this thought about hidden causes of extinctions to individual species. Long ago Dutch sailors clubbed to death every last dodo, a flightless bird that lived only on Mauritius. Then a type of tree native to the island stopped reproducing. It was not until the 1970s that a hypothesis emerged: If the tree depended intimately on a coevolved species that became extinct, then the tree would be vulnerable to extinction, also. Was its partner the dodo? Maybe (Figure 27.3*b*).

Biodiversity is greater now than it has ever been in the past.

The current range of global biodiversity is an outcome of an overall pattern of mass extinctions and slow recoveries in the history of life. Within that pattern, lineages differ in which member species persisted or became extinct.

The loss of individual species, as well as mass extinctions, may have obvious or complicated causes.

THE NEWLY ENDANGERED SPECIES

No biodiversity-shattering asteroids have hit the Earth for 65 million years. *Yet the sixth major extinction event is under way.* Throughout the world, human activities are driving many species to extinction. Consider the mammalian lineage, which has outlasted the dinosaurs. About 2 million years ago, early humans started to hunt mammals in earnest. About 11,000 years ago, they were encroaching on habitats in a big way. With agriculture, domesticated species were favored at the expense of wild stock. And think of the once-vast herds of bison, which nearly disappeared in the 1800s. As adventurers moved westward, they shot too many bison for sport.

Only 4,500 or so species of mammals made it to the present. Of these, 300 (including most whales, wild cats, otters, and primates other than humans) have the bad luck to be enrolled in the endangered species club. An **endangered species** is any endemic species extremely vulnerable to extinction. *Endemic* means it originated in one geographic region and is found nowhere else.

We are one species among millions. Yet, primarily as an outcome of our rapid population growth during the past forty years, *we are threatening other species with habitat losses, species introductions, overharvesting, and illegal wildlife trading* (Figures 27.4 through 27.6). If the pace of extinction continues, recovery to former levels of biodiversity may take millions of years.

Habitat Loss and Fragmentation

Edward O. Wilson defines **habitat loss** as the physical reduction in suitable places to live, as well as the loss

Figure 27.5 Confiscated products made from endangered species. The illegal wildlife trade threatens 600+ species, yet all but 10 percent goes undetected. Some black market prices: Live large saguaro cactus ($5,000–15,000), bighorn sheep head ($10,000–60,000), polar bear ($6,000), grizzly ($5,000), bald eagle ($2,500), peregrine falcon ($10,000), live chimpanzee ($50,000), live mountain gorilla ($150,000), Bengal tiger hide ($100,000), and rhinoceros horn ($28,600 per kilogram).

of suitable habitat as an outcome of chemical pollution. Habitat loss is one of the significant threats to more than 90 percent of endemic species facing extinction.

Biodiversity is greatest in the tropics. There, habitats on land are assaulted daily by deforestation, a topic you will read about in Sections 50.4 and 50.5. Also under attack are grasslands, wetlands, and aquatic habitats

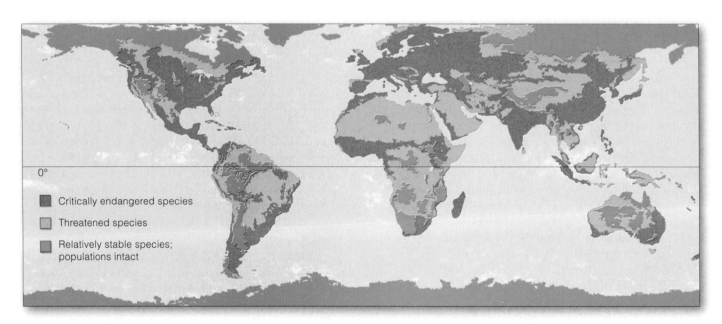

- Critically endangered species
- Threatened species
- Relatively stable species; populations intact

Figure 27.4 Threats to biodiversity: major regions of the world in which species are critically endangered, threatened, or relatively stable, as projected from 1998 through 2018.

Figure 27.6 (**a**) A minke whale kill. Despite a worldwide ban, minke whales are still harvested, even inside ocean sanctuaries.

(**b**) Graph of whale harvesting. It reflects declining numbers of great whales, which are at the brink of extinction.

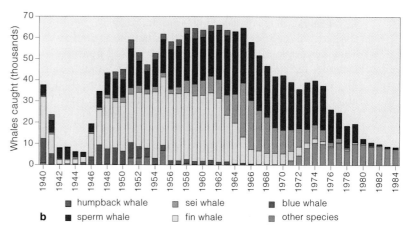

Whales caught (thousands)

■ humpback whale   ■ sei whale   ■ blue whale
■ sperm whale   □ fin whale   ■ other species

b

(Section 27.3). Other examples: 98 percent of tallgrass prairies, 50 percent of wetlands, and 85–95 percent of old-growth forests of the United States alone are gone.

In **habitat fragmentation**, a habitat is chopped into patches having a separate periphery. This is damaging for biodiversity. Species at the habitat's periphery are more vulnerable to temperature changes, winds, fires, predators, and pathogens. Also, patches may not be large enough to support population sizes required for successful breeding. And they may not have enough food or other resources to sustain a species.

ISLAND BIOGEOGRAPHY AND HABITAT ISLANDS   Section 47.10 describes factors shaping biodiversity on islands. Islands are a fraction of the Earth's surface, yet about half of the plants and animals driven to extinction since 1600 were island natives that had nowhere else to go. R. MacArthur and Wilson used island biogeography as a model to estimate the number of current and future extinctions in places surrounded by logging, mining, urbanization, and other disruptions. National parks, tropical forests, reserves, and lakes are like this. Think of them as **habitat islands** in seas of unsuitable habitat. Applying this model to an island habitat, a 50 percent loss will drive about 10 percent of endemic species to extinction. A 90 percent loss will drive about 50 percent of endemic species to extinction.

INDICATOR SPECIES   Conservation biologists often study **indicator species**, which offer warning of changes in habitat and impending widespread loss of biodiversity. Birds are one example. Different types live in all major land regions and climate zones, they respond quickly to changes in habitats, and they are relatively easy to track.

Consider songbirds that migrate annually between tropical forests and summer breeding grounds in North America. In one nine-year period, biologists surveyed 64 species. Population sizes of 44 species fell owing to deforestation in the winter habitats and fragmentation

of summer habitats. Also, intrusion of farms, freeways, and suburbs chopped forests into patches. This made it easier for skunks, opossums, squirrels, raccoons, rats, and jays to feast on eggs and juveniles of migratory songbirds. About 69 percent of the 9,600 known species of birds are now confronted with habitat loss.

Other Contributing Factors

Whether by accident or deliberate importation, a species sometimes moves from its home range and successfully insinuates itself into a different habitat. Each habitat has only so many resources, and its occupants compete for their share. Introduced species have adaptations that might make them highly competitive, so they often displace one or more endemic species. You will read about these *exotic* species in Sections 47.8 and 47.9. For now, it is enough to know that species introductions are a major factor in almost 70 percent of the cases where endemic species are being driven to extinction.

In a sad commentary on human nature, the more rare a wild animal becomes, the more its value soars in the black market (Figure 27.5).

Overharvesting also is reducing biodiversity. Whales are an example (Figure 27.6). Humans have killed them for food, lubricating oil, fuel, cosmetics, fertilizer, even corset stays. Substitutes exist for all whale products. Yet fishermen of several nations still ignore a moratorium on slaughtering large whales, and they routinely cross boundaries of sanctuaries set aside for recoveries.

An endangered species is any endemic species extremely vulnerable to extinction.

For the past forty years, human activities have raised rates of extinction through habitat losses, species introductions, overharvesting, and illegal wildlife trading.

Habitat loss is the physical reduction in suitable living spaces and habitat closure by pollution.

Case Study: The Once and Future Reefs

Coral reefs are wave-resistant formations consisting of the accumulated remains of countless marine organisms. They develop mainly in the clear, warm waters between latitudes 25° north and 25° south (Figures 27.7 and 27.8). Hardened coral parts have formed each reef's spine. Red algae such as *Corallina* contributed mineral-hardened cell walls to it. Secretions from other organisms helped cement things together.

The massive, pocketed spine is home to living corals and a staggering number of other species. For example, Australia's Great Barrier Reef supports 500 coral species, 3,000 fish species, 1,000 molluscan species, and 40 kinds of sea snakes. Figure 27.8c hints at the wealth of warning colors, tentacles, and stealthy behavior. These are a few of the signs of danger and fierce competition for resources among species packed together in limited space.

Dinoflagellates, which are photosynthetic, often live as symbionts in tissues of reef-building corals (Section 25.4). The dinoflagellates find protection in the tissues. They provide the coral polyp with oxygen and recycle its mineral wastes. When stressed, coral polyps expel the symbionts. When stressed for more than a few months, they die; only bleached hard parts remain (Figure 27.8b).

Abnormal, widespread bleaching in the Caribbean and tropical Pacific began in the 1980s. So did increases in sea surface temperature, which may be a key stress factor. Is the damage one outcome of global warming (Chapter 3)? If so, as marine biologists Lucy Bunkley–Williams and Ernest Williams suggest, the future looks grim for reefs. They may be devastated within three decades.

Also, humans can directly destroy reefs, as by raw sewage discharges into nearshore waters of populated islands. Massive oil spills, as occurred during the Persian Gulf war, have calamitous effects. So do commercial dredging operations and mining for coral rock.

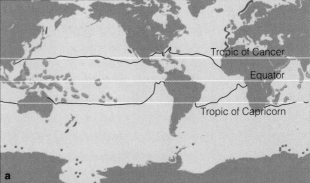

a

b

Figure 27.8 (**a**) Distribution map for coral reefs (*orange*) and coral banks (*green*). Nearly all reef-building corals are limited to regions with warm sea surface temperatures, here enclosed in dark lines. Beyond latitudes 25° north and south, solitary and colonial corals (*red*) form coral banks in temperate seas as well as in cold seas above the continental shelves.

(**b**) Coral bleaching in 1998 at Australia's Great Barrier Reef, which parallels Queenland's coast for 2,500 kilometers. This is the largest example of biological architecture, but it is not one continuous reef. It is a string of thousands of reefs, some of which are 150 kilometers (95 miles) across.

(**c**) The facing page shows a small sampling of biodiversity from representative coral reefs.

a

b

c

d

Figure 27.7 Three types of coral reef formations. *Fringing* reefs form near land when rainfall and runoff are light, as on the downwind side of the most recently formed volcanic islands. Many reefs in the Hawaiian Islands and Moorea (**a**) are like this. *Barrier* reefs form parallel with the shore of continents and volcanic islands, as in Bora Bora (**b**). Behind them are calm lagoons a few meters to sixty meters deep. Ring-shaped *atolls* are coral reefs and coral debris. They fully or partly enclose a shallow lagoon that a channel often links to the open ocean (**c**). Biodiversity isn't great in shallow water, which can get too hot for corals to survive. (**d**) Coralline alga, one of the reef builders.

Where commercial fishermen from Japan, Indonesia, and Kenya move in, reef life is vanishing. No simple nets for these fellows; they drop dynamite in the water. Fish hiding in the coral are blasted out and float dead to the surface. Also, sodium cyanide squirted into the water stuns fish, which float to the surface. Some survivors end up as tropical fish in pet stores. Endangered species are transported to exotic restaurants, where they are killed and served up as exorbitantly priced status symbols. On small, native-owned islands, fishing rights are traded for paltry sums. The fishermen destroy the reefs, which then no longer sustain the small human populations that have depended on them for survival.

Reef biodiversity is in danger around the world, from Australia and Southeast Asia to the Hawaiian Islands, Galápagos Islands, Gulf of Panama, Florida, and Kenya. To give a final example of this, the biodiversity on the coral reef off Florida's Key Largo has been reduced by 33 percent since 1970.

LIONFISH

CORAL REEF IN THE RED SEA

MORAY EEL

CHAMBERED NAUTILUS

SEA ANEMONE

c   DAISY CORAL

PILLAR CORAL

CROWN-OF-THORNS SEA STAR

Rachel's Warning

In 1951 Rachel Carson (Figure 27.9), an oceanographer and marine biologist, wrote a book about human impact on the ocean. *The Sea Around Us* was translated into thirty-two languages and sold more than 2 million copies. Seven years later, officials happened to spray mosquito-controlling DDT near the home and private bird sanctuary of one of her friends, who became agitated over the subsequent agonizing deaths of several birds. That friend begged Carson to find someone who might study the effects of pesticides on birds and other wildlife.

Carson discovered that independent, critical research on the subject was almost nonexistent. She conducted an extensive survey of the literature and then methodically developed information about the harmful effects of the widespread use of pesticides. In 1962, she published her findings. *Silent Spring*, the book's title, was an allusion to the silencing of "robins, catbirds, doves, jays, wrens, and scores of other bird voices" following pesticide exposure.

Many scientists, politicians, and policymakers read *Silent Spring*, and the public embraced it. Manufacturers of chemicals viewed Carson as a threat to their booming pesticide sales, and they swiftly mounted a campaign to discredit her. Critics and industry scientists claimed the book was full of inaccuracies, made selective and biased use of research findings, and failed to balance the picture with an account of the benefits of pesticides. Some even claimed that, as a woman, Carson simply was incapable of understanding such a highly technical subject. Others charged she was a hysterical woman and a radical nature lover trying to scare the public in order to sell books.

During those intense attacks, Carson knew she had terminal cancer. Yet she strongly defended her research and successfully countered her critics. She died in 1964—eighteen months after publication of *Silent Spring*—without knowing that her efforts would be the impetus for what is now known as the environmental movement in the United States. The new field of conservation biology is one outgrowth of that movement.

Figure 27.9 Rachel Carson, who helped awaken public interest in human impact on nature. She died without knowing she helped start the environmental movement.

CONSERVATION BIOLOGY

All countries can count their wealth in three forms: material, cultural, and biological. Until recently, many undervalued their biological wealth—*their biodiversity*, which is a source of food, medicine, and other goods. No one knows how much countries around the world have already lost.

For instance, in the 1970s, a Mexican college student poking around in Jalisco discovered *Zea diploperennis*, a wild species of maize. Unlike domesticated corn, it resists disease and lives for more than one growing season. Gene transfers from the wild species to *Z. mays* have the potential to enormously boost production of corn for hungry people in Mexico and elsewhere. That species was growing in a mountain habitat no larger than ten hectares (twenty-five acres). The student saved it in the nick of time. A week later, humans started to clear the habitat with machetes and controlled burns. They would have pushed *Z. diploperennis* into oblivion.

There is a monumental problem. In the next fifty years, human population size may reach 9 billion, with most growth proceeding in the developing countries. Each individual must have raw materials, energy, and living space. How many species will they crowd out? The countries with the greatest monetary wealth have the least biological wealth and are using most of the world's natural resources (Section 45.9). The countries with the least monetary wealth and the most biological wealth have the fastest growing populations.

Awareness of the impending extinction crisis gave rise to **conservation biology**. We define this field of pure and applied research as (1) a systematic survey of the full range of biological diversity, (2) an attempt to decipher the evolutionary and ecological origins of diversity, and (3) an attempt to identify methods that might maintain and use biodiversity for the good of the human population. Its goal is to conserve and use, in a sustainable way, as much biodiversity as is feasible.

The Role of Systematics

Realistically, we simply don't have enough time, money, or people to complete a global survey. Possibly 99 percent of existing species have not even been described, and we don't know where most of them live. Thus initial work has focused on identifying **hot spots**. These are habitats of many species that are found nowhere else and are in greatest danger of extinction through human activities.

At the first survey level, researchers target a limited area, such as an isolated valley. A complete survey is out of the question, so they inventory birds, mammals, fishes, butterflies, and other indicator species for the habitat. At the next level, broader areas that are major or multiple hot spots are systematically explored. The widely separated forests of Mexico are an example of a

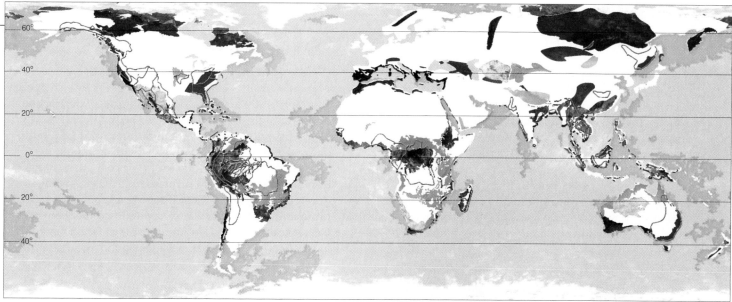

Figure 27.10 One map of the most vulnerable regions of land and seas, compiled by the World Wildlife Fund.

■ Tropical forest	■ Mediterranean shrub
■ Temperate forest	▨ Desert
■ Northern coniferous forest	■ Arctic tundra
▨ Tropical grassland and savanna	■ Mangrove swamp
▨ Temperate grassland	▨ Marine ecoregion
▨ Mountain grassland	▨ Freshwater ecoregion

major hot spot. Research stations are set up at different latitudes and elevations across the region.

At the highest survey level, hot spot inventories are combined with existing information on biodiversity of ecoregions. An **ecoregion** is a broad land or ocean region defined by climate, geography, and producer species. For example, Figure 27.10 is a work in progress by the World Wildlife Fund. Currently this organization has identified 238 regions crucial to global biodiversity: 142 land, 53 fresh water, and 43 marine. Initially, it selected 25 for immediate conservation efforts. Such efforts on behalf of biodiversity will be refined over the coming decades.

Bioeconomic Analysis

The systematic expansion of species inventories serves as the baseline for economic analysis—for assigning future value to ecoregions. This is a novel concept in human history. To give examples, when humans first ventured eastward from Southeast Asia, they literally ate their way across the Pacific, through New Zealand, Easter Island, and the Hawaiian Islands. The flightless birds and other endemic species with no evolutionary experience with humans did not stand a chance. A wave of extinctions also accompanied humans as they expanded through the Americas. What counted in the past, as counts today, is food on the table for self and family. It counted for the impoverished truck driver who said, after shooting one of the two last imperial woodpeckers, "It was a great piece of meat."

Such thinking by economically pressed individuals might be countered by showing them that sustaining biodiversity has more economic value than destroying it. Should governments—and individuals—protect a habitat, withdraw resources in sustainable fashion, or

obliterate it? We might analyze what each threatened species may offer. For example, some may be sources of medicines and other chemical products. Developing countries do not have laboratories for chemical analysis. Large pharmaceutical companies do. One of the largest has been paying the National Institute of Biodiversity in Costa Rica to collect and identify promising species and send in chemical samples extracted from them. If natural chemicals end up being marketed, Costa Rica will get a share of the royalties, which are earmarked for conservation programs.

Sustainable Development

Ultimately, biodiversity will be better protected when species can be used for the long-term good of local economies. Identifying and implementing uses is easier said than done; technical problems and social barriers are huge. Yet there are successes, as you will read next.

Conservation biology entails a systematic survey of the full range of biodiversity, analysis of its evolutionary and ecological origins, and identification of methods to maintain and use biodiversity for the benefit of humans.

Researchers identify local hot spots (limited areas with many endemic species in greatest danger of extinction by human activities), then broader or multiple hot spots, which they combine into a global picture of biodiversity.

To counter economic demands of the human population, the future economic value of biodiversity must be determined.

RECONCILING BIODIVERSITY WITH HUMAN DEMANDS

When people locked in poverty are confronted with the choice of saving endangered species or themselves, which will they choose? Deprived of education and resources that people in developed countries take for granted, they clear tropical forests. They hunt animals and dig up plants in "protected" parks and reserves. They try to raise crops and herds on marginal land. Instead of simply asking them to stop, we might assist them in identifying ways in which they might earn a living from their biological wealth—that is, from the biodiversity in threatened habitats.

A Case for Strip Logging

Section 50.4 describes the staggering pace of destruction in the once-great tropical rain forests, the richest homes of biodiversity on land. For now, simply think about one concept of how to minimize the destruction.

Besides providing wood for local economies, trees of tropical rain forests yield diverse, exotic woods prized by developed countries. What if they could be logged in a profitable, sustainable way that preserved biodiversity? Gary Hartshorn was the first to propose **strip logging** in portions of forests that are sloped and have a number of streams (Figure 27.11). The idea is to clear a narrow corridor that parallels contours of the land, and use the upper part of it as a road to haul away logs. After a few years, saplings start to grow in the corridor. Another corridor is cleared above the roadbed. Vital nutrients that runoff leaches from the exposed soil trickle into the first corridor. There they are taken up by saplings, which benefit from all the nutrient input by growing more rapidly. Later, a third corridor is cut above the second one —and so on in a profitable cycle of logging, which the habitat sustains over time.

Ranching and the Riparian Zones

Reflect upon the Figure 27.10 map, and you sense at once that the biodiversity crisis is not confined to Third World countries. Also take a look at Figure 27.12, which shows a riparian zone

uncut forest

cut 1 year ago

dirt road

cut 3–5 years ago

cut 6–10 years ago

uncut forest

stream in watershed

Figure 27.11 Strip logging, a practice that might sustain biodiversity even as it enhances the regeneration of economically valued forest trees.

Figure 27.12 Riparian zone restoration. This example comes from Arizona's San Pedro River, shown before and after restoration efforts.

in the western United States. Each **riparian zone** is a relatively narrow corridor of vegetation along a stream or river. Its plants afford a major line of defense against flood damage by sponging up water during the spring runoffs and summer storms. Shade cast by the canopy of taller shrubs and trees helps conserve water during droughts. Riparian zones also offer food, shelter, and shade for wildlife, particularly in arid and semiarid regions. For example, in the western United States, 67 to 75 percent of the endemic species must spend all or part of their life cycle in riparian zones. They include more than 136 species of songbirds, some of which nest only in riparian vegetation.

Cattle raised in the American West provide beef for much of the human population. Compared with wild ungulates, cattle drink a lot more water, so they tend to congregate at rivers and streams. There, they trample and feed until grasses and herbaceous shrubs are gone. It takes only a few head of cattle to destroy riparian vegetation. All but 10 percent of the riparian vegetation of Arizona and New Mexico has already disappeared, mainly into the stomachs of grazing cattle.

Restricted access and development of watering sites for cattle away from streams can help conserve riparian zones. So can rotation of livestock and the provision of supplemental feed at different grazing areas. Cattle ranchers resist implementing such measures. Putting in more fencing, for instance, is costly.

Jack Turnell is one rancher who knows almost as much about riparian biodiversity as he knows about cattle. He runs cattle on a 32,000-hectare (80,000-acre) ranch south of Cody, Wyoming, and on 16,000 hectares of Forest Service land for which he has grazing rights. Ten years after he took over the ranch, he decided to raise cattle in an unconventional way by systematically rotating them away from riparian areas. Turnell also crossed Hereford and Angus cattle with a breed from France that does not consume as much water. He made most of his ranching decisions in consultation with specialists in range and wildlife management.

Willows and other plants are sustaining diversity in the restored zones. And Turnell's ranching approach is profitable; grasses maintained in the riparian zones help put more meat on his cattle. His may be only a small step toward sustainable ranching, but it appears to be a step in the right direction.

Throughout much of the world, the unprecedented rate of human population growth is driving the current mass extinction event.

Protecting biodiversity depends on finding ways for people to make a living from it without destroying it. Can such a balance be struck with many billions of people?

SUMMARY *Gold* indicates text section

1. Global biodiversity is greater now than ever, but the current rate of species losses is high enough to suggest that an extinction crisis is under way. *27.1*

2. After mass extinctions, it has taken 20 million to 100 million years for biodiversity to recover to the previous level. Taxa have different evolutionary histories. Some were lost in mass extinctions. Others passed through the events relatively unscathed. Extinctions of individual species, as well as mass extinctions, may have obvious or hidden, complicated causes. *27.1*

3. Human population size, expected to reach 9 billion within fifty years, is causing the sixth major extinction crisis. Human population growth is most rapid in areas with the richest, most vulnerable biodiversity. *27.2*

4. Habitat losses, habitat fragmentation, introduction of species into novel habitats, overharvesting, and illegal wildlife trading threaten endemic species. An endemic species originated in and is limited to one geographic region. An endangered species is defined as an endemic species extremely vulnerable to extinction. *27.2*

5. Habitat loss refers to physical reduction or chemical pollution of suitable places for species to live. Habitat fragmentation is carving a habitat into isolated patches. It puts species at risk, as by splitting up populations to sizes that cannot promote successful breeding. *27.2*

6. Island biogeography models help predict the number of current and future extinctions. A habitat island is a habitat in a sea of possibly destructive human activities, such as logging. Generally, destruction of 50 percent of an island habitat (or habitat island) will drive one-tenth of its species to extinction. Destruction of 90 percent will drive one-half to extinction. *27.2*

7. Indicator species are birds and other easily tracked species that can provide warning of changes in habitats and impending widespread loss of biodiversity. *27.2*

8. Humans are directly destroying coral reefs and other regions, as when commercial fishermen dynamite reefs to harvest fish. They may be destroying them indirectly, as by contributing to global warming (with concurrent rises in sea surface temperatures and sea levels). *27.3*

9. Conservation biology is a field of pure and applied research. Its goal is to conserve and use biodiversity in sustainable ways. This goal involves (a) a systematic survey at three ever more inclusive levels of the full range of biodiversity, (b) an analysis of biodiversity's origins in evolutionary and ecological terms, and (c) identification of methods that might maintain and use biodiversity for the benefit of the human population, which may otherwise destroy it. *27.5*

10. The systematic survey of conservation biology is now proceeding at three levels: (a) Local hot spots (for

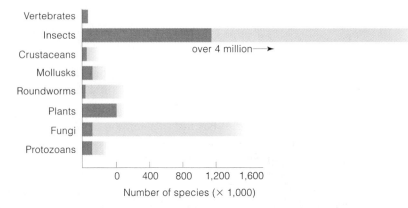

Figure 27.13 Current species diversity for a few major taxa. *Red* indicates the number of named species; *gold* indicates the estimated number of species not yet discovered.

example, an isolated valley) are identified and indicator species inventoried. Hot spots are habitats that have many species in great danger of extinction because of human activities. (b) Major hot spots or multiple ones are inventoried. Then research stations are set up across broader areas to gather data by latitude and elevation. (c) Data gathered at the first two levels are combined with data on ecoregions; these are the most vulnerable broad regions of land and seas throughout the world. Data gathered for such maps will be refined into an increasingly detailed picture of global biodiversity. **27.5**

11. Protecting biodiversity depends on finding ways for people to make a living from it without destroying it. To counter growing economic demands, biodiversity's future economic value must be determined. Methods by which local economies can exploit that biodiversity in sustainable ways must be developed. **27.5**

Review Questions

1. How many major mass extinctions have occurred, including the one that is now under way? 27.1

2. Define endangered, endemic, and indicator species. 27.2

3. Distinguish between island habitat and habitat island. 27.2

4. State the goal of conservation biology and briefly describe its three-pronged approach to achieving that goal. 27.5

5. Define hot spot. Why are conservation biologists focusing on hot spots rather than quickly completing a global survey? 27.5

Self-Quiz   ANSWERS IN APPENDIX III

1. Following mass extinctions in the past, recovery to the same level of biodiversity has taken many _____ of years.
   a. hundreds   b. millions   c. billions

2. Species may be driven to extinction by _____ .
   a. obvious factors   c. human activities
   b. complicated, hidden factors   d. all of the above

3. The goal(s) of conservation biology is (are) to _____ .
   a. conduct a three-level, systematic survey of all biodiversity
   b. analyze biodiversity's evolutionary and ecological origins
   c. identify ways to maintain and use biodiversity for people
   d. all of the above

4. Strip logging _____ .
   a. can sustain forests   c. destroys wild habitats
   b. is profitable   d. both b and c

Critical Thinking

1. Figure 27.13 compares the current estimated number of species for some major taxa. Given what you learned about taxa from chapters in this unit, which are most vulnerable in the current extinction crisis? Which are likely to pass through it with much of their biodiversity intact? List some reasons why (for example, compare the distribution of member species).

2. Visit or study a riparian zone near where you live. Imagine visiting it five years from now, then ten years after that. Given its location, what kinds of changes do you predict for it?

3. Mentally transport yourself to a tropical rain forest of South America. Imagine you don't have a job. There are no jobs in sight, not even in overcrowded cities some distance away. You have no money or contacts to get you anywhere else. And yet you are the sole supporter of a large family. Today a stranger approaches you. He tells you he will pay good money if you can capture alive a certain brilliantly feathered parrot in the forest. You know the parrot is rare; it is seldom seen. However, you have an idea of where it lives. What will you do?

4. As an alternative to burial on land, ashes of cremated people are now being mixed with pH-neutral concrete and used to build *artificial reefs*. Each perforated, concrete ball weighs 400 to 4,000 pounds and is designed to last 500 years. So far, 100,000 of these reefs help marine life in 1,500 locations (Figure 27.14). Proponents say that the burial costs less, does not waste land, and is ecologically friendly. Your thoughts?

Figure 27.14
One of the new realities: casket in the ground or a concrete building block for an artificial reef?

5. In his review of this chapter, ecologist Robin Tyser offered these comments: Students might find the current biodiversity crisis too overwhelming to contemplate. But people are making a difference. For example, they helped reestablish viable bald eagle populations in the continental United States and wolves in northern Wisconsin. Daniel Janzen is working to recreate a dry forest ecosystem in Costa Rica. There is much good in the world. Do a library search or computer research and report on one or two success stories you may find inspiring.

Selected Key Terms

conservation biology 27.5	endangered species 27.2	hot spot 27.5
coral reef 27.3	habitat fragmentation 27.2	indicator species 27.2
ecoregion 27.5	habitat island 27.2	riparian zone 27.6
	habitat loss 27.2	strip logging 27.6

Readings

Wilson, Edward O. 1992. *The Diversity of Life*. Cambridge, Massachusetts: Belknap Press.

On-Line readings at Student Guide for InfoTrac:
www.brookscole.com/biology

Principles of Anatomy and Physiology

Two bottlenose dolphins leaping from the sea, to which they are most exquisitely adapted, into air—where they cannot survive for long. This transient bridging of two very different environments invites you to ask: Are there fundamental constraints on how any organism is put together and how it functions, regardless of where it lives? And can we identify patterns in the diverse responses to recurring challenges?

HOW PLANTS AND ANIMALS WORK

On High-Flying Geese and Edelweiss

Each year, Mount Everest beckons irresistibly to some humans simply because it is the highest place on Earth (Figure 28.1). Being the highest place on Earth, it has the thinnest air. Climbers must condition themselves for months before their assault on the summit. They make short-term changes in how they breathe and how many oxygen-transporting red blood cells they produce. Even then, breathing near the summit will be agonizing. A few climbers won't make it back.

If oxygen is so scarce, then how does the barheaded goose (*Anser indicus*) regularly wing its way back and forth over the highest peaks of the Himalayas? How does its geographically separate relative, the Andean goose (*Cloephaga melanoptera*), live comfortably at 6,000 meters above sea level in the Andes of South America?

The type of hemoglobin synthesized by their red blood cells binds oxygen more strongly than human hemoglobin is able to do. Yet their hemoglobin is the same as ours, except for one amino acid substitution. Which type of amino acid was substituted differs between the two kinds of geese. But both types disrupt the same weak interaction in two of four polypeptide chains that make up the hemoglobin molecule. The

outcome, a greater oxygen-binding capacity, helps keep both birds flying high in different parts of the world.

Look about and you also find plants that survive brutal conditions at 4,100 meters (13,500 feet) above sea level, higher than the timberline of the Himalayan range. One species, *Leontopodium alpinum*, is the more ancient and equally endangered relative of Europe's edelweiss (Figure 28.2). It is anchored in pockets of soil in tiny fissures in rocks, where its roots are protected from soil contractions and expansions each time the below-freezing winter weather alternates with spring thaws. Its aboveground parts form short, tight cushions close to the rocks, and its sturdy stems resist the strong mountain winds. Its flowers appear woolly, so thickly are they covered with fine epidermal hairs. The hairs protect floral reproductive structures from ultraviolet radiation, which is intense at high altitudes. They also help the plant resist losing moisture to winds.

These examples can start you thinking about how plants and animals are put together and how their body parts function in particular environments. Because each human, barheaded goose, and alpine plant shares the same general characteristics with others of its species,

Figure 28.1 A climber near the summit of Mount Everest, where the simple act of taking in enough oxygen becomes torture. Yet the barheaded goose (Figure 28.2*a*) flies over the Himalaya range every year. It migrates back and forth between lakes, marshes, and rivers in high mountain valleys to warmer wintering grounds in India.

Figure 28.2 (**a**) Barheaded goose (*Anser indicus*). (**b**) *Leontopodium alpinum*, a low-growing plant that evolved above the timberline in the Himalayas.

you might infer that structural traits have a genetic basis and are vulnerable to mutation. You also might infer that long-standing adaptations are outcomes of natural selection and offer a better fit with prevailing environmental conditions. You might infer that the individuals of at least some species besides humans can adapt briefly to more extreme conditions than they are used to. You may also find the examples reinforce what you already suspect—*that the structure of a body part typically correlates with a present or past function.*

Such thoughts are your passport to the world of anatomy and physiology.

Anatomy is the study of an organism's form—that is, its morphology—from its molecular foundation to its physical organization as a whole. **Physiology** is the study of patterns and processes by which an organism survives and reproduces in the environment. It deals with how structures are put to use and the nature of metabolic and behavioral adjustments to changing conditions. It also deals with how, and to what extent, physiological processes can be controlled.

This chapter introduces some important concepts. Study them well, for you will apply them in the next two units of the book. They can help deepen your sense of how plants and animals function under stressful as well as favorable environmental conditions.

Key Concepts

1. Anatomy is the study of body form at many levels of structural organization, from molecules through tissues, organs and, for most animals, organ systems.

2. Physiology is the study of how the body functions in response to predictable and unexpected aspects of the environment.

3. The structure of most body parts correlates with current or past functions. Typical examples are tissues that cover the plant or animal body or line its internal parts; structures such as woody stems that function as scaffolds for upright growth; and thick, strong leg bones with load-bearing functions.

4. Most aspects of body form and function are long-term adaptations. They are heritable features that evolved in the past and have continued to be effective under prevailing conditions. Individuals of many species also adapt relatively quickly, over the short term, to stressful conditions.

5. For multicelled organisms, the extracellular fluid that bathes living cells is an internal environment. Cells, tissues, organs, and organ systems typically contribute to maintaining the internal environment within a range that individual cells can tolerate. This concept helps us understand the functions and interactions of body parts.

6. Homeostasis is the name for stable operating conditions in the internal environment. Negative and positive feedback mechanisms are among the controls that work to maintain these conditions.

7. Cells of tissues and organs communicate with one another by secreting diverse hormones and other signaling molecules into extracellular fluid, and selectively responding to signals from other cells.

LEVELS OF STRUCTURAL ORGANIZATION

From Cells to Multicelled Organisms

In most plants and animals, cells, tissues, organs, and organ systems split up the work and contribute to the survival of the body as a whole. A separate lineage of cells gives rise to the body parts that will function in reproduction. The body shows a **division of labor**.

A **tissue** is a community of cells and intercellular substances, all interacting in one or more tasks. For example, wood tissue and bony tissue both function in structural support. An **organ** consists of two or more tissues, organized in specific proportions and patterns, that perform one or more related tasks. A leaf adapted for photosynthesis, an eye that responds to light in the surroundings—these are examples of organs. An **organ system** is composed of different organs that interact physically, chemically, or both during common tasks. A plant's shoot system, with organs of photosynthesis and reproduction, is like this (Figure 28.3). And so is the digestive system of an animal, which takes in food, breaks it down, absorbs the breakdown products, and gets rid of the residues and wastes.

Growth Versus Development

The structural organization of a plant or animal emerges during growth and development. Be sure you understand at the outset that these two terms are not the same. Generally, **growth** of a multicelled organism means its cells increase in number, size, and volume. **Development** refers to the successive stages in the formation of specialized tissues, organs, and organ systems. In other words, we measure growth in *quantitative* terms, and we measure development in *qualitative* terms.

Structural Organization Has a History

The structural organization of each tissue, organ, and organ system has an evolutionary history. Think back on how plants invaded the land, and you have clues to how their structures reflect specific functions. Plants that left aquatic habitats found plenty of sunlight and carbon dioxide for photosynthesis, oxygen for aerobic respiration, and mineral ions. As they dispersed farther away from their aquatic cradle, however, they faced a new challenge: how to keep from drying out in air.

Think about that when you look at micrographs of cellular pipelines in the roots, stems, and leaves of a typical plant (Figure 28.3). They conduct tiny streams of water from soil to the leaves. Stomata, small gaps across the leaf epidermis, open and close in ways that help conserve water (Section 7.7). Root systems grow

reproductive organ
(tomato flower)

stem tissues, cross-section, for support, storage, water distribution, food distrbution

SHOOT SYSTEM
(aboveground parts)

ROOT SYSTEM
(belowground parts, mostly)

Figure 28.3 Morphology of a tomato plant (*Lycopersicon*). Different cell types make up various vascular tissues that conduct either water and dissolved minerals or organic substances. They thread through other tissues that make up most of the plant. A different tissue covers surfaces of the root and shoot systems.

root water-conducting cells, longitudinal section

Figure 28.4 Some of the structures that function in human respiration. Cells carry out specialized tasks, as when ciliated cells sweep bacteria and other airborne particles out of airways to and from two organs, a pair of lungs. Tissues arranged as tubes (blood capillaries), a fluid connective tissue called blood, and epithelial tissue in the form of thin air sacs function in gas exchange inside the lungs. Other organs, including the airways, are components of this organ system.

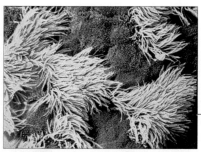

individual cells, ciliated and mucus-secreting, that line respiratory airways

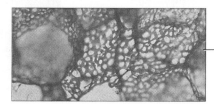

lung tissue (tiny air sacs) laced with blood capillaries—one-cell-thick tubular structures that hold blood, which is a fluid connective tissue

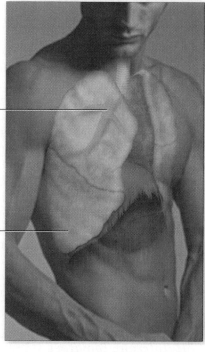

organs (lungs), part of an organ system (respiratory tract) of a whole organism

outward to places where water and minerals are more concentrated in soil. Inside the stems, cells with lignin-reinforced walls collectively support upright growth.

Similarly, respiratory systems of land animals are adaptations to life in air. Gases move into and out of the body by diffusing across a moist surface. This is not a problem for aquatic organisms. However, moist surfaces dry out in air. Land animals have moist sacs inside their body for gas exchange (Figure 28.4). And this brings us to the *internal* environment.

The Body's Internal Environment

To stay alive, plant and animal cells must be bathed in a fluid that offers nutrients and carries away metabolic wastes. In this they are no different from free-living cells. But plants and animals consist of thousands to trillions of cells. Each cell must draw nutrients from and dump wastes into the fluid bathing them.

Body fluids *not* inside cells—extracellular fluids—are an **internal environment**. Functionally, changes in the composition and volume of this environment affect cell activities. The type and number of ions are vital. They must be kept at concentrations compatible with metabolism. It makes no difference whether the plant or animal is simple or complex. *Its components require a stable fluid environment for all of its living cells.* As you will see, this concept is absolutely central to learning how plants and animals work.

How Do Parts Contribute to the Whole?

The next two units describe how each plant or animal carries out these functions: It provides its cells with a stable fluid environment. It makes or gets nutrients and other raw materials, distributes them, and disposes of wastes. It has ways to protect itself against injury and attack. It also has the capacity to reproduce, and often it helps nourish and protect its offspring during their early growth and development.

The big picture is this: Each living plant or animal cell engages in metabolic activities that ensure its own survival. But the structural organization and collective activities of cells in tissues, organs, and organ systems sustain the whole body. They work to keep operating conditions in the internal environment within tolerable limits for individual cells—a state called **homeostasis**.

The structural organization of a plant or animal emerges during growth and development and is correlated with basic functions.

Cells, tissues, and organs require, and collectively help maintain, a favorable internal environment. The internal environment consists of all body fluids not inside cells.

The basic functions include maintaining a stable internal environment, securing substances and distributing them through the body, disposing of wastes, protecting the body, reproducing, and often nurturing offspring.

THE NATURE OF ADAPTATION

Defining Adaptation

"Adaptation" is one of those words that have different meanings in different contexts. An individual plant or animal often quickly adjusts its form, function, and behavior. Junipers that germinated in inhospitably windy places are stunted compared to more sheltered individuals of the same species. A sudden clap of thunder may make you lurch the first time you hear it, but you may get used to the sound over time and eventually ignore it. These are *short-term* adaptations to the environment, because they last only as long as the individual does.

Over the long term, an **adaptation** is some heritable aspect of form, function, behavior, or development that improves the odds for surviving and reproducing in a given environment. It is an *outcome* of microevolution —natural selection especially—an enhancement of the fit between the individual and prevailing conditions.

Salt-Tolerant Tomatoes

As a simple example of a long-term adaptation, we can compare the way different tomato species respond to saltwater. Tomatoes evolved in Ecuador, Peru, and the Galápagos Islands. The commercial tomato in grocery stores, *Lycopersicon esculentum*, has eight close relatives in the wild. Mix together ten grams of table salt and sixty milliliters of water, then pour it into a soil-filled container in which an *L. esculentum* plant is growing. Within thirty minutes, the plant will wilt severely (Figure 28.5*a*). Even if soil has only 2,500 parts per million of salt, this species will grow poorly.

Yet the Galápagos tomato (*L. cheesmanii*) survives and reproduces in seawater-washed soils. Researchers

Figure 28.5 (**a**) Severe and rapid wilting of a commercial tomato plant (*Lycopersicon esculentum*) after a drink of salty water. (**b**) Galápagos tomato plant, *L. cheesmanii*.

showed that its salt tolerance is a long-term, heritable adaptation. Gene transfers from the wild species into the commercial species yielded a small, edible tomato. It tolerates irrigation with water that is two parts fresh and one part salty. The hybrid is garnering interest in places where fresh water is scarce and where salts have accumulated in fields used for agriculture.

It may take modification of only a few traits to get salt-tolerant plants. For instance, researchers discovered that revving up a single gene for a sodium/hydrogen ion transporter lets tomato plants use salty water and still bear edible fruits. The fruits are slightly more salty, but the leaves store most of the excess salts.

No Polar Bears in the Desert

You can safely assume that a polar bear (*Ursus maritimis*) is well adapted to the Arctic environment, and that its form and function would be a total flop in a desert (Figure 28.6). You might be able to make some educated guesses about

Figure 28.6 Which adaptations of a polar bear (*Ursus maritimis*) won't help in a desert? Which ones help an oryx (*Oryx beisa*)? For each animal, make a preliminary list of possible structural and functional adaptations relative to the environment. Later, after you finish Unit VI, see how you can expand the list.

why that is so. However, detailed knowledge of its anatomy and physiology might make you view it—or any other animal or plant—with respect. How does a polar bear maintain its internal temperature when it sleeps on ice? How can its muscles function in frigid water? How often must it eat? How does it find food? Conversely, how can an oryx walk about all day in the blistering heat of an African desert? How does it get enough water when there is no water to drink? You will find some answers, or at least ideas about how to look for them, in the next two units of this book.

Adaptation to What?

From the examples just given, you may be thinking it is fairly easy to identify a direct relationship between an adaptation and some aspect of the environment. But the environment in which a trait originally evolved may be very different from the one prevailing now.

Consider the llama, a native of the cloud-piercing peaks of the Andes range paralleling the western coast of South America (Figure 28.7). It can live comfortably 4,800+ meters (16,000 feet) above sea level. Compared to humans living at lower elevations, its lungs have far more air sacs and blood vessels, which formed as it was growing up. A llama heart has larger chambers, so it pumps larger volumes of blood. It does not have to stimulate production of extra blood cells, as people do when they move permanently from low to higher elevations. (The extra cells make the blood "stickier," so the heart has to pump harder.) However, the most publicized adaptation is this: Llama hemoglobin is better than ours at latching onto oxygen. It can pick up oxygen inside the lungs much more efficiently.

Superficially, at least, the oxygen-binding affinity of llama hemoglobin appears to be an adaptation to thin air at high altitudes. But is it? Apparently not.

Llamas belong to the same family as dromedary camels. They share camelid ancestors that evolved in the Eocene grasslands and deserts of North America. Later on, a genetic divergence occurred. The ancestors of camels entered Asia's low-elevation grasslands and deserts by way of a land bridge that later submerged owing to a long-term rise in sea level. The ancestors of llamas moved in a different direction—down the Isthmus of Panama, and on into South America.

Intriguingly, the dromedary camel's hemoglobin also has a high oxygen-binding capacity. So if the trait arose in a shared ancestor, then in what respect has it been adaptive at *low* elevations when it is adaptive at *high* elevations? In this case we can rule out convergent evolution. Why? These animals are very close kin, and their most recent ancestors faced different challenges in environments with different oxygen concentrations.

Figure 28.7 Adaptation to what? A trait is an adaptation to a specific environment. Hemoglobin of llamas, which live at high altitudes, has a high oxygen-binding affinity. But so does hemoglobin of camels, which live at lower elevations.

Who knows why the trait was originally favored? Eocene climates were alternately warm and cool, and hemoglobin's oxygen-binding capacity does go down as temperatures go up. Did it prove adaptive during a long-term shift in the air temperature? Or did the trait have neutral effects at first? For example, what if the mutant gene for the trait became fixed in some ancestral population simply by chance?

Or what if the nonmutated allele interacted closely with another gene that could not be dispensed with? For example, mechanisms that control certain stages of animal development can't be significantly modified without causing major, and usually lethal, disruptions to the basic body plan. As you will read in Section 43.5, molecular details of most genes that code for these mechanisms have been conserved through time.

Use all of these "what-ifs" as a reminder to think critically about the connections between an organism's form and function. Identifying those connections takes a lot of guesswork, research, and experimental tests.

A long-term adaptation is a heritable aspect of form, function, behavior, or development that contributes to the fit between the individual and its environment.

An adaptive trait improves the odds of surviving and reproducing, or at least it did so under conditions that prevailed when genes encoding the trait first evolved.

Observable traits are not always easy to correlate with specific conditions in the organism's environment.

MECHANISMS OF HOMEOSTASIS IN ANIMALS

In preparation for your trek through the next two units, take time now to get a firm grasp of what homeostasis means to survival. Animal physiologists were the first to identify this state and the mechanisms that maintain it, so let's start with a couple of examples of animals.

Like other adult humans, your body contains many trillions of living cells. Each cell must draw nutrients from and dump wastes into the same fifteen liters of fluid, which is less than sixteen quarts. Again, fluid not inside the cells of your body is the extracellular fluid. Much of this is **interstitial fluid**; it fills spaces between cells and tissues. The rest is **plasma**, the fluid portion of blood. Interstitial fluid exchanges many substances with the cells it bathes and also with blood.

Homeostasis, again, is a state in which the body's internal environment is being maintained within some range that its living cells can tolerate. In nearly all animals, three kinds of components interact to maintain it. They are sensory receptors, integrators, and effectors.

Sensory receptors are cells or cell parts that detect forms of energy, such as pressure. Any specific form of energy that a receptor has detected is a **stimulus**. When one chimp kisses another, pressure on its lips changes. Receptors in the lip tissues translate the stimulus into signals that flow to the brain (Figure 28.8).

The brain is an example of an **integrator**, a central command post that pulls together information about stimuli and issues signals to muscles, glands, or both. Muscles and glands—the body's **effectors**—carry out suitable responses. One response to a kiss is flushing with pleasure and kissing back. Of course, a kiss can't continue indefinitely. Doing so would prevent eating, breathing deeply, and various other activities needed to maintain operating conditions in the body.

How does the brain reverse physiological changes induced by the kiss? Receptors can only provide it with information about how things *are* operating. The brain also evaluates information about how things *should be* operating relative to "set points." One example of a set point is a certain concentration of carbon dioxide in the blood coursing through a certain artery. When carbon dioxide's concentration deviates sharply from the set point, the brain initiates actions that will return it to an effective operating range. It does so by way of signals that cause specific effectors in different body regions to increase or decrease certain activities.

Negative Feedback

Feedback mechanisms are important controls that help keep physical and chemical aspects of the body within tolerable ranges. A major type is the **negative feedback mechanism**: An activity is initiated and changes some condition, and when the condition is altered enough, it triggers a response that reverses the change. Figure 28.9 shows one model for such control.

Think of a furnace with a thermostat. A thermostat senses the surrounding air's temperature relative to a preset point on a thermometer built into the furnace's control system. When the temperature falls below the preset point, it sends signals to a switching mechanism that turns on the furnace. When the air becomes heated enough to match the prescribed level, the thermostat signals the mechanism, which shuts off the furnace.

Similarly, feedback mechanisms help keep the core body temperature of chimpanzees, humans, huskies, and many other animals near 37°C (98.6°F) even in hot or cold weather. Thinks of a husky running around on a hot summer day. Its body heats up. Receptors trigger events that slow down the whole dog *and* its cells. The husky searches for shade and flops down under a tree. Moisture from its respiratory system evaporates from the tongue and carries away some body heat with it, as

STIMULUS (input into the system)

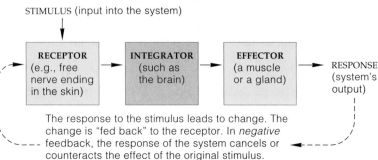

| RECEPTOR (e.g., free nerve ending in the skin) | → | INTEGRATOR (such as the brain) | → | EFFECTOR (a muscle or a gland) | → RESPONSE (system's output) |

The response to the stimulus leads to change. The change is "fed back" to the receptor. In *negative* feedback, the response of the system cancels or counteracts the effect of the original stimulus.

Figure 28.8 A kiss. It cannot continue indefinitely without disrupting oxygen flow, food intake, and other physiological events that maintain the internal environment. Negative feedback sets in at the organ level. It reverses the changes brought about by the pressure on the lips.

The husky is overactive on a hot, dry day and its body surface temperature rises.

| RECEPTORS in skin and elsewhere detect the temperature change. | → | An INTEGRATOR (the hypothalamus, a brain region) compares input from the receptors against a set point. | → | Some EFFECTORS (pituitary gland and thyroid gland) trigger widespread adjustments. |

RESPONSE

Temperature of circulating blood starts decreasing.

Many EFFECTORS carry out specific responses:

SKELETAL MUSCLES	SMOOTH MUSCLE IN BLOOD VESSELS	SALIVARY GLANDS	ADRENAL GLANDS
Husky rests, starts to pant (behavioral changes).	Blood carrying metabolically generated heat shunted to skin, some heat lost to surroundings.	Secretions from glands increase; evaporation from tongue. Both have a cooling effect, especially on the brain.	Output drops, husky is less stimulated.

Activity of the body in general slows down (behavioral change).

The overall slowdown in activities results in less metabolically generated heat.

Figure 28.9 Homeostatic controls over the internal temperature of a husky's body. The *blue* arrows indicate the main control pathways. The dashed line shows how a feedback loop is completed.

the Figure 28.9 indicates. These control mechanisms and others counter the overheating by curbing the activities that naturally generate metabolic heat and by giving up the body's excess heat to the surrounding air.

Positive Feedback

In some cases, **positive feedback mechanisms** operate. These controls initiate a chain of events that *intensify* change from an original condition, and after a limited time, the intensification reverses the change. Positive feedback is associated with instability in a system. For example, during sexual intercourse, chemical signals from the nervous system of a human female might induce her to make intense physiological responses to stimulation from her partner. Responses may stimulate changes in her partner that stimulate the female more, and so on until she reaches an explosive, climax level of excitation. Now the affected parts of the body return to normal, and homeostasis prevails.

As another example, at the time of childbirth, the fetus exerts pressure on the wall of its mother's uterus. The pressure stimulates the production and secretion of oxytocin, a hormone. Oxytocin causes muscle cells in the wall to contract. Contractions exert pressure on the fetus, which exerts more pressure on the wall, and so on until the fetus is expelled from her body.

What we have been describing is a general pattern of detecting, evaluating, and responding to a continual flow of information about the animal's internal and external environments. During all of the activity, organ systems operate together in astoundingly coordinated fashion. In time you should find yourself asking these questions about their operation:

1. Which physical or chemical aspects of the internal environment are organ systems working to maintain as conditions change?

2. By what means are organ systems kept informed of the various changes?

3. By what means do they process the information?

4. What mechanisms are set in motion in response?

As you will read in Unit VI, the organ systems of most animals are under neural and endocrine control.

Homeostatic control mechanisms help maintain physical and chemical aspects of the body's internal environment within ranges that are most favorable for cell activities.

DOES THE CONCEPT OF HOMEOSTASIS APPLY TO PLANTS?

Plants differ from animals in some important respects. In young plants, for instance, new tissues arise only at the very tips of actively growing roots and shoots. In animal embryos, tissues and organs form all through the body. Also, plants do not respond to stimuli with a centralized integrator, such as a brain. This means that direct comparisons between plants and animals are not always possible.

But plants do have decentralized mechanisms that work to maintain the internal environment and ensure survival for the body as a whole. It is a bit of a leap, but the concept of homeostasis can be applied to them, too. The next two examples will make the point.

Walling Off Threats

Unlike people, trees consist mostly of dead and dying cells. Also unlike people, trees cannot run away from attacks. And when a pathogen infiltrates their tissues, trees cannot unleash an immune system in response, because trees have none. Some trees live in habitats too harsh or too remote for most pathogens, so they have been able to grow, albeit slowly, for thousands of years. Remember the bristlecone pines?

Most trees growing elsewhere can wall off threats to their internal environment by building a fortress of thickened cell walls around wounds. At the same time, they deploy phenols and other compounds that are toxic to invaders. For example, some cells of conifers and other trees secrete resin. A heavy resin flow can saturate and protect bark and wood near the attack site. It also may seep into the soil and litter above the roots. Taken together, plant responses to an attack are called **compartmentalization**.

So potent are some toxins that they also kill cells of the tree itself. As compartments form around injured or infected or poisoned sites, the tree lays down new tissues over them. This works well when the tree acts strongly or is not under massive attack. Drill holes in a tree species that makes a strong response and it walls off the wounds fast. Drill holes in a tree species that makes a moderate response, and it decays lengthwise and somewhat laterally. Drill holes in a species that is a weak compartmentalizer, though, and it will massively decay (Figure 28.10). Even strong compartmentalizers live only so long. If they form too many walls, they shut off the vital flow of water and solutes to living cells.

Bob Tiplady, a plant health specialist, has reported on compartmentalization in response to attack by fungi of the genus *Armillaria*. These fungi destroy shrubs as well as trees in forests and under cultivation throughout the world. They are saprobic decomposers that help to cycle nutrients back to plants. However, they become opportunistic pathogens among trees already stressed by wounds, aging, drought, insects, and air pollution. They infiltrate roots and draw nutrients from narrow zones of actively growing tissues beneath the bark.

In temperate climates, aboveground symptoms of a fungal infection include a gradual decline in growth. Leaves are undersized or yellow, and they often drop prematurely. In summer, the youngest branches may die back or brown quickly and die.

When compartmentalization is strong, an infection can be stopped. Even then, localized decay continues, as in Figure 28.10. An infected tree may die abruptly or slowly, one limb at a time. For instance, if soil holds ample water, the tree might stay green for up to two years after an *Armarillia* infection has decayed most of its roots. Dry spells add to the stress and invite a quick death. That is one reason fungus-infected trees seem to give up abruptly during a prolonged drought.

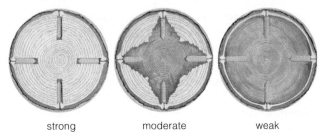

strong   moderate   weak

a Compartmentalization responses

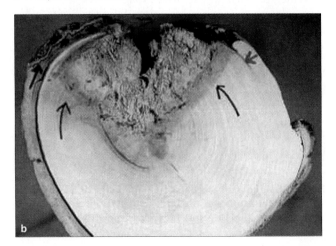

Figure 28.10 (**a**) Pattern of drilling into the stems of three tree species for an experiment to test the effectiveness of compartmentalization. From left to right, the decay patterns (*green*) for three species that are strong, moderate, and weak compartmentalizers. (**b**) Compartmentalized wood rot. Toxic secretions by this tree (*Populus tremuloides*) weakly countered an *Armillaria* attack. A barrier became established (*red* arrow), but the fungus breached it and spread into the wood.

1 A.M.

6 A.M.

NOON

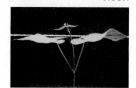

3 P.M.

10 P.M.

c   MIDNIGHT

Sand, Wind, and the Yellow Bush Lupine

Anyone who has tiptoed barefoot across sand near the coast on a hot, dry day has a tangible clue to why few plants grow in it. One of the exceptions is the yellow bush lupine, *Lupinus arboreus* (Figure 28.11*a*).

L. arboreus is a colonizer of soil exposed by fires or abandoned after being cleared for agriculture. It also becomes established along windswept, sandy shores of the Pacific Ocean. Like its relatives in the legume family, it houses nitrogen-fixing symbionts in its roots, which gives it a competitive edge in nitrogen-deficient soils (Chapters 24 and 30).

A big environmental challenge near the beach is the scarcity of fresh water. Leaves of a yellow bush lupine are structurally adapted for water conservation. Each has a surprisingly thin cuticle. But a dense array of fine epidermal hairs projects above the cuticle, especially on the lower leaf surface. Like the hairs on edelweiss, they trap moisture escaping from stomata, and they might also reflect heat. The trapped, moist air slows evaporation and helps keep water inside the leaf.

These leaves make homeostatic responses to the environment. They fold along their length, like the two parts of a clam shell, and resist the moisture-sucking force of the wind. Each folded, hairy leaf is better at holding moisture escaping from stomata (Figure 28.11*b*).

Leaf folding by *L. arboreus* is a controlled response to changing conditions. When winds are strong and the potential for water loss is high, its leaves fold tightly. We find the least-folded leaves near the plant's center or on the side most sheltered from wind. Folding also is a response to heat as well as to wind. When the air temperature is highest during the day, leaves fold at an angle that helps reflect the sun's rays away from their surface. This response minimizes heat absorption.

About Rhythmic Leaf Folding

Just in case you think leaf folding couldn't possibly be a coordinated response, look at Figure 28.11*c*. Like some other plants, this one holds its leaves horizontally in

Figure 28.11 (**a**,**b**) The yellow bush lupine, *Lupinus arboreus*, an exotic species in this habitat by a sandy shore. Introduced into northern California in the early 1900s, it is taking over sand dunes; it outcompetes the native species, to the enormous consternation of conservationists.

(**c**) Observational test of the rhythmic movements of the leaves of a young bean plant (*Phaseolus*). The investigator, physiologist Frank Salisbury, kept this plant in full darkness for twenty-four hours. Its leaves continued to fold and unfold independently of sunrise (6 A.M.) and sunset (6 P.M.).

the day and folds them closer to its stem at night. Keep this plant in full sun or darkness for a few days and it will continue to move its leaves into and out of the "sleep" position, independently of sunrise and sunset. This folding response might help reduce heat loss at night, when air becomes cooler, and so maintain the plant's internal temperature within a tolerable range.

Rhythmic leaf movements are just one example of **circadian rhythm**, a biological activity that is repeated in cycles, each lasting for close to twenty-four hours. Circadian means "about a day." As you will see later, in Chapter 32, a pigment molecule called phytochrome may be part of homeostatic controls over leaf folding.

Homeostatic mechanisms are at work in plants, although they are not governed from central command posts as they are in most animals. Compartmentalization and rhythmic leaf movements are two examples of responses to specific environmental challenges.

COMMUNICATION AMONG CELLS, TISSUES, AND ORGANS

Signal Reception, Transduction, and Response

Reflect on an earlier overview of how adjoining cells communicate, as through plasmodesmata in plants and gap junctions in animals (Section 4.11). Also think back on how free-living *Dictyostelium* cells issue signals to get together and then differentiate into a spore-bearing structure. These amoeboid cells do so in response to dwindling supplies of food—an environmental cue for change (Section 22.12). In large multicelled organisms, cells in one tissue or organ also signal cells in different tissues or organs that are often quite a distance away. They do so in response to changes in the internal and external environments. Their local and long-distance signals call for local and regional changes in metabolic activities, gene expression, growth, and development.

The molecular mechanisms by which cells "talk" to one another evolved early in the history of life, among prokaryotic species. Many persist in the most complex eukaryotic organisms. In many cases, they involve three events—*activation of a receptor, as by reversible binding of a signaling molecule, then transduction of the signal into a molecular form that can operate inside the cell, and then the functional response.*

Most receptors are membrane proteins of the sort shown in Section 5.2. When activated, many change shape, which starts signal transduction. One enzyme often activates many molecules of a different enzyme, which activates many molecules of another kind, and so on in cascading reactions that amplify the signal. In some pathways, small, easily diffusible molecules that are already inside the cytoplasm help broadcast the message through the cell; they are second messengers.

In the next two units, you will come across diverse cases of signal reception, transduction, and response. For now, two simple examples will give you a sense of the kinds of events they set in motion.

Communication in the Plant Body

Have you ever noticed green nubbins along the branch of a young tree? Each holds a community of cells, a meristem, much like tissues of stem cells in your body. They can divide again and again, and the descendants give rise to new tissues and organs. In plants, growth and development of new parts are under the control of genes, signaling molecules, and environmental cues. Incoming light, gravity, seasonal temperature changes, and how long it stays dark at night are such cues.

You might be surprised that growth, development, and survival of a plant require extensive coordination among cells, just as they do in animals. Plant cells, too, communicate with others that are often relatively far away in the plant body. **Hormones** are the main signals for cell plant communication. They initiate changes in cell activities when they dock at suitable receptors.

But hormones do not work in a vacuum. Expression of specific genes in some cells can change a hormone's level in plant tissues or mediate cell sensitivity to it. Expression of many genes is activated or inhibited in response to environmental cues. Chapter 32 describes the main control mechanisms in plants. Here, consider a few genes that control how to make floral organs. These *organ-identity* genes are analogous to homeotic genes, which guide the patterned formation of organs in animals (Sections 43.5 and 43.6).

All flowers are variations on the same pattern of growth and development. Focus on the common wall cress, *Arabidopsis thaliana* (Figure 28.12). It is a valued organism for genetics experiments. This plant is small and has a short generation time; forty-two days after seeds germinate, the new plants have made seeds.

Genetic analysis of *A. thaliana* mutants support an **ABC model** for flowering—that three groups of genes designated *A*, *B*, and *C* are master switches for floral

petal

carpel

stamen

sepal

Figure 28.12 Observable evidence that meristematic cells at a shoot tip of *Arabidopsis thaliana* were reprogrammed to make a flower instead of more leaves. In (**a**), notice the individual cells making up tissues of the rudimentary structures.

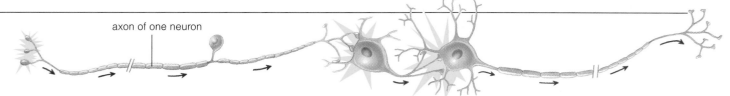

a Reception of environmental signal (pressure) by a sensory neuron in a toe.

b Signal transduced; gates on ion channels across neural membrane open. Rapid ion flow changes the electric and concentration gradients across membrane. Flow reversals self-propagate along length of membrane.

c Signal transduced again; ion flow triggers release of signaling molecules from neuron. This is the response to the signal. Molecules diffuse to another neuron and bind with receptors on it. Binding triggers same transductions in this neuron, and the next, and so on through communication lines from receptor cells to the brain or spinal cord, and on to effectors.

Figure 28.13 (**a**–**c**) Example of signal reception, transduction, and response in the vertebrate nervous system. A myelin fatty wrapping around the long extensions of these three cells functions in signal propagation. (**d**) Artist's interpretation of the damage caused by an autoimmune attack on myelin sheaths around motor axons. Such attacks cause many disorders, including multiple sclerosis.

development. These groups encode factors governing gene transcription for the products that make sepals, petals, and other structures. They reprogram the mass of rapidly dividing meristematic cells that, until now, were the source of leaves. The descendant cell lineages become arranged in whorls around the shoot's tip, which is a genetically guided floral pattern.

Sepals form as cells in the outer whorls of the floral meristem divide and differentiate. These are the only cells that express *A* genes. Petals form from cells in the whorl inside them, the only cells to express both *A* and *B* genes. *B* and *C* gene interactions switch on activities that result in pollen-producing reproductive structures, the stamens. Cells of the innermost whorl switch on *C* genes only. These cells give rise to a single carpel, the reproductive structure where eggs, then seeds, form.

Another master gene, designated *Leafy*, flips on the ABC switches. And what flips on *Leafy*? At this writing, most evidence is pointing to a steroid hormone.

Communication in the Animal Body

We see elegant cases of signal reception, transduction, and response in cells of the vertebrate nervous system. Chapters 34 and 35 will describe how communication lines of cells called neurons extend through the body. Parts of each neuron are signal input zones; each has a patch of receptor-rich membrane. Other parts, called axons, are typically long extensions with output zones. Chemical signals are released from output zones and diffuse to receptors on a nearby cell (Figure 28.13).

How is a signal transduced? The receptors are on gated ion channels across the plasma membrane. When they bind a signaling molecule, the gates open, and ions flow across the membrane. The distribution of charge across the membrane changes abruptly, which causes ions to flow back across the membrane at an adjacent

patch, and so on in a self-propagating way. When the disturbance reaches the output zone, it triggers release of signaling molecules that act on the next cell.

Many axons have a sheath of **myelin**, a lipid-rich membrane. (Cells called oligodendrocytes make these sheaths by wrapping around the axons like jellyrolls.) A myelin sheath insulates an axon in a way that makes messages travel a lot faster to an output zone. Once in place, it prevents a neuron from sprouting more axons, so it also affects how the developing brain gets wired. It takes twenty years or more before the human brain becomes fully myelinated.

Sometimes the myelin sheaths get damaged. One outcome is *multiple sclerosis*—a gradual loss of brain and spinal cord function. An autoimmune response is made against a myelin component that may have been modified following a viral infection (Figure 28.13*d*).

Plant and animal cells communicate with one another by secreting signaling molecules into extracellular fluid, and selectively responding to signals from other cells.

Cell communication involves mechanisms for receiving signals, transducing them, and bringing about change in metabolism or gene expression in target cells.

Signal transduction commonly involves shape changes in receptors and other membrane proteins as well as activation of enzymes and second messengers inside the cytoplasm.

RECURRING CHALLENGES TO SURVIVAL

Plants and animals have such spectacularly diverse body plans that we sometimes forget how much they have in common. How often do we even think of the connections between, say, Tina Turner and a tulip (Figure 28.14)? Yet the connections are there, and they start with what individual cells in each multicelled body are doing. Nearly all of the cells are working in the best interest of the whole body—*but the whole body is also working in the best interest of individual cells.*

Constraints on Gas Exchange

Tina's cells, and the tulip's, depend on gas exchanges across their plasma membrane. Like most multicelled heterotrophs, Tina has a fine way to get oxygen to the plasma membrane of each one of her aerobically respiring cells and take away their carbon dioxide wastes. Like most autotrophs, a tulip plant can get carbon dioxide to its photosynthetic cells and remove their oxygen by-products. It also delivers oxygen to aerobically respiring cells throughout the body, as in roots. Both organisms have specific body structures that help individual cells by acquiring, transporting, exchanging, and getting rid of gases.

These structures evolved in response to certain environmental parameters. Remember **diffusion**, the net movement of like molecules or ions of a substance from a region where they are more concentrated to a region where they are less concentrated? Once gases reach a cell's plasma membrane, they diffuse across it. Which way? That depends on the direction of the concentration gradient for each one. And which body parts help keep each gradient pointed in the right direction? The questions will lead you to the stomata of leaves (Chapter 30) and to the respiratory and circulatory systems of animals (Chapters 38 and 40).

Figure 28.14 What do these organisms have in common besides their good looks?

Requirements for Internal Transport

As you know, metabolic reactions are amazingly fast. If it takes too long for substances to diffuse through the body or to and from the surface, the body will shut down. That is one of the reasons cells and multicelled organisms have the sizes and shapes that they do. As they grow, their volume expands in three dimensions (say, length, width, and depth), but the surface increases in two dimensions only. And that is the essence of the **surface-to-volume ratio**. If a body were to become one huge unbroken mass, it would not have enough surface area for fast, efficient exchanges with the environment.

When a body or body part is thin, as it is for the lily pads and flatworm shown at right, substances can easily diffuse between its individual cells and the environment. But the farther individual cells are from an exchange point with the environment, the more they depend on systems of rapid internal transport.

Most plants and animals have vascular tissues, which transport substances to and from cells. In plants, soil water and mineral ions are distributed to aboveground parts through xylem. The photosynthetic products made in leaves are distributed through phloem. Each leaf vein shown in Figure 28.15*a* contains long strands of xylem and phloem.

In large animals, too, vascular tissues extend from an interface with the environment to individual cells. Each time Tina belts out a song, she must swiftly move oxygen into her lungs and carbon dioxide out of them. Part of her vascular system threads intricately through lung tissue, where gases are exchanged. From there, large-diameter vessels swiftly transport oxygen to all other regions, then branch into tiny capillaries (Figure 28.15*b*). Blood flow slows here, so diffusion proceeds at a suitable pace between blood, interstitial fluid, and cells. The capillaries branch into transport tubes that deliver carbon dioxide to the lungs, where it is exhaled and where oxygen is again picked up (Chapter 40).

As in plants, that vascular system also transports nutrients and water. Unlike plants, the one in animals interconnects with other organ systems that have roles in maintaining a more complex internal environment. One of these, the immune system, uses the vascular system as highways for moving white blood cells and chemical weapons to tissues that are under siege. Another organ system specializes in regulating the concentration and volume of the internal environment.

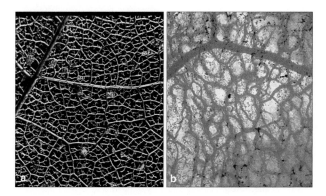

Figure 28.15 Transport lines to and from cells: (**a**) Veins visible in a decaying leaf, and (**b**) human veins and capillaries.

Figure 28.16 Protecting body tissues from predation: (**a**) Cactus spines. (**b**) Quills of a porcupine (*Erethizon dorsatum*).

Maintaining a Solute–Water Balance

Plants and animals continually gain and lose water and solutes. They regularly produce metabolic wastes. Given all the inputs and outputs, how is the volume and composition of their internal environment kept within a tolerable range? Plants and animals differ hugely in this respect. Yet we still can find common responses among them at the molecular level.

Substances tend to follow concentration gradients as they move into and out of the body. They do so as they move to and from one body compartment into another. At such interfaces, we find sheetlike tissues in which cells are engaged in **active transport**, the pumping of specific substances *against* the gradients.

In roots, active transport mechanisms help control which solutes move into the plant. In leaves, they help control water loss and gas exchange by making stomata open only at certain times. In animals, we find such mechanisms in kidneys and many other organs. As you will see again and again in the next two units, *active and passive transport mechanisms help maintain metabolism and the internal environment by continually adjusting the kinds, amounts, and directional movements of substances.*

Requirements for Integration and Control

We could introduce you to many other structural and functional similarities between plants and animals. But let's sign off with the cells of certain tissues, which release the signaling molecules that coordinate and integrate activities of the body as a whole. Sections 28.4 and 28.5 gave a few examples. These are the signals that respond to changes in the internal and external environments. Different kinds guide events by which the body grows, develops, maintains, and often prepares for reproduction—events that will be our focus in chapters to come.

On Variations in Resources and Threats

Beyond the common challenges are resources and dangers that differ among habitats. A **habitat** is the place where individuals of a species normally live.

What are the physical and chemical characteristics of the habitat? Is water plentiful, with suitable kinds and amounts of solutes? Is the habitat rich or poor in nutrients? Is it sunlit, shady, or dark? Is it warm or cool, hot or icy, windy or calm? How much do outside temperatures change from day to night? Are seasonal changes slight or pronounced?

And what about the biotic (living) components of the habitat? What kinds of producers, predators, prey, pathogens, and parasites live there? Is competition for resources and reproductive partners fierce among others of the species? These are the kinds of variables that promote diversity in anatomy and physiology.

Even in all that diversity, we often can find similar responses to similar environmental challenges. Think about the sharp spines of a cactus and sharp quills of a porcupine (Figure 28.16). These specialized structures function effectively as deterrents to most predators. Both are derived from specialized epidermal cells. In both cases, vascular tissues and other body parts helped nurture those cells. And by contributing to the formation of defensive structures at the body surface, individual epidermal cells help protect the whole body against an environmental threat.

Plant and animal cells function in ways that help ensure survival of the body as a whole. At the same time, tissues and organs that make up the body function in ways that help ensure survival of individual living cells.

The connection between each cell and the body as a whole is evident in the requirements for—and contributions to—gas exchange, nutrition, internal transport, stability in the internal environment, and defense.

SUMMARY
Gold indicates text section

1. Anatomy is the study of body form. Physiology is the study of body functioning in the environment. The structure of most body parts correlates with current or past functions. *CI, 28.1*

2. A plant or animal's structural organization emerges during growth and development. Each cell in its body performs metabolic functions that ensure its survival. Cells are organized in tissues, organs, and often organ systems, which function in coordinated ways to ensure survival of the whole organism. *28.1, 28.6*

3. Long-term adaptations are specific heritable traits that evolved in the past and continue to be effective, or at least neutral, under prevailing conditions. The traits are specific aspects of the form, function, behavior, or development of the body. *28.2*

4. Tissues, organs, and organ systems work together to maintain a stable internal environment (extracellular fluid) needed for individual cell survival. At homeostasis, conditions in the internal environment are balanced at levels most favorable for cell activities. *28.1, 28.3, 28.4*

5. Feedback controls help maintain internal operating conditions. Example: By negative feedback, a change in some condition, such as body temperature, triggers a response that results in reversal of the change. *28.3*

6. Cells of multicelled organisms communicate directly with one another. They communicate with cells some distance away, in different tissues and organs. They secrete signaling molecules into extracellular fluid and selectively respond to signals themselves. *28.5*

7. Cell communication involves reception of signals, transduction of them, and a response, such as a change in metabolism, in gene expression, or in development. Signal transduction often involves shape changes in receptors and other membrane proteins. It affects the activity of enzymes and often of second messengers in the cytoplasm of the target cells. *28.5*

Review Questions

1. Define tissue, organ, and organ system. *28.1*

2. Distinguish between growth and development. *28.1*

3. What does the term internal environment mean? *28.1*

4. Define long-term adaptation. Is it a process of evolution or an outcome of microevolutionary processes? *28.2*

5. Define homeostasis and give an example of how certain mechanisms work to maintain it. *28.3, 28.4, 28.6*

6. Briefly define the receptors, integrators, and effectors in the animal body and state how they interact. *28.3*

7. Do plants have decentralized homeostatic mechanisms? *28.4*

8. Describe a form of communication between two or more plant or animal tissues. *28.5*

Self-Quiz
ANSWERS IN APPENDIX III

1. An increase in the number, size, and volume of plant cells or animal cells is called _____ .
   a. growth   c. differentiation
   b. development   d. all of the above

2. The internal environment consists of _____ .
   a. all body fluids   c. all body fluids outside cells
   b. all fluids in cells d. interstitial fluid

3. As basic functions, a plant or animal must _____ .
   a. maintain a stable internal environment
   b. get and distribute water and solutes through the body
   c. dispose of wastes
   d. defend the body
   e. all of the above

4. Cell communication always involves signal _____ .
   a. reception   c. response
   b. transduction   d. all of the above

5. Match the terms with their most suitable description.
   ____ physiology   a. study of how body parts function
   ____ circadian rhythm b. signaling molecule
   ____ homeostasis   c. 24-hour or so cyclic activity
   ____ hormone   d. stable internal environment
   ____ negative feedback e. activity changes some condition, change causes its own reversal

Critical Thinking

1. Reflect on prevailing conditions in a desert in New Mexico or Arizona; on the floor of a shady, moist forest in Georgia or Oregon; and in Alaska's arctic tundra. Consider the kinds of plants and animals living in each environment. Now "design" a plant or animal that might function even more efficiently in each place. Be sure to consider how its basic requirements, such as the acquisition of water and nutrients, will be met.

2. Refer to Section 28.2. Set up class teams and see which team can come up with the longest list of the most plausible answers to the questions asked in the caption to Figure 28.6.

Selected Key Terms

ABC model *28.5*	integrator *28.3*
active transport *28.6*	internal environment *28.1*
adaptation, long-term *28.2*	interstitial fluid *28.3*
anatomy *CI*	myelin *28.5*
circadian rhythm *28.4*	negative feedback *28.3*
compartmentalization *28.4*	organ *28.1*
diffusion *28.6*	organ system *28.1*
development *28.1*	physiology *CI*
division of labor *28.1*	plasma *28.3*
effector *28.3*	positive feedback *28.3*
growth *28.1*	sensory receptor *28.3*
habitat *28.6*	stimulus *28.3*
homeostasis *28.1*	surface-to-volume ratio *28.6*
hormone *28.5*	tissue *28.1*

Readings

Hochachka, P. and G. Somero. 2002. *Biochemical Adaptation: Mechanism and Process in Physiological Evolution.* New York: Oxford University Press. Integrative look at how cellular systems are adapted to environmental stresses.

On-Line readings at Student Guide for InfoTrac: www.brookscole.com/biology

V Plant Structure and Function

The sacred lotus, *Nelumbo nucifera*, busily doing what its ancestors did for well over 100 million years—flowering spectacularly during the reproductive phase of its life cycle.

PLANT TISSUES

Plants Versus the Volcano

On a clear spring day in 1980, in a richly forested region of the Cascade Range of southwestern Washington, Mount Saint Helens erupted and 500 million metric tons of ash blew skyward. Within minutes, shock waves from the blast had flattened or incinerated hundreds of thousands of tall, mature trees growing near the mountain's northern flanks.

Rivers of hot volcanic ash surged down the slopes at rates exceeding forty-four meters per second. They turned into nightmarish rivers of cementlike mud when intense heat from the blast melted and released more than 75 billion liters of water that had been locked up in the mountain's snowfields and glacial ice.

In one mind-numbing moment, about 40,500 hectares—100,000 acres—of magnificent forests dominated by hemlock and Douglas fir had been transformed into barren sweeps of land (Figure 29.1a,b). In the aftermath of the violent eruption, we gained stunning insight into what the world must have looked like long ago, before the first plants started to colonize habitats on land.

Yet it did not take long for existing plants to move back into habitats that their ancestors had claimed. In less than a year's time, the seeds of a variety of flowering plants, including fireweed and blackberry, sprouted near the grayed trunks of fallen trees around Mount Saint Helens.

Before ten years had passed, young willows and alders took hold near riverbanks, and low shrubs cloaked the land (Figure 29.1c). In time they afforded pockets of shade, which favored germination of seeds of the slower growing but ultimately dominant species, the hemlocks and Douglas fir (Figure 29.1d). In less than a century, the forest will be as it once was.

Figure 29.1 (**a**,**b**) Grim reminder of what the world would be like without plants: aftermath of the violent eruption of Mount Saint Helens in 1980. Nothing remained of the forest that surrounded this Cascade volcano. (**c**) In less than a decade, seed-bearing vascular plants were making a comeback. (**d**) Twelve years after the eruption, the seedlings of a dominant species, Douglas fir (*Pseudotsuga menziesii*), were starting to reclaim the land.

With this example, we open a unit dedicated to the seed-bearing vascular plants, with emphasis on the flowering types. In terms of distribution and diversity, they are the most successful plants on Earth.

This first chapter is an overview of plant tissues and body plans. Next, Chapter 30 explains how the seed-bearing plants absorb and distribute water and mineral ions, conserve water, and distribute organic substances among roots, stems, and leaves. Chapters 31 and 32 take a look at their patterns of growth, development, and reproduction. As you will see, their structure and physiology (that is, the functioning of the plant body) help them survive sometimes hostile conditions on land—even momentary takeovers by volcanoes.

Key Concepts

1. Angiosperms (flowering plants) and, to a lesser extent, gymnosperms are groups that now dominate the plant kingdom. All are seed-bearing vascular plants. They have complex aboveground shoot systems of stems, leaves, and reproductive parts. Most species have complex root systems that grow downward and outward through soil.

2. We find three major tissue systems in seed-bearing vascular plants. A ground tissue system makes up the bulk of the plant body. A vascular tissue system distributes water, dissolved mineral ions, and the products of photosynthesis. A dermal tissue system covers and protects plant surfaces.

3. Simple plant tissues—parenchyma, collenchyma, and sclerenchyma—are each composed of no more than one type of cell.

4. Complex plant tissues incorporate two or more types of cells. Xylem and phloem, which are vascular tissues, are like this. So are the dermal tissues called epidermis and periderm.

5. Plants lengthen and thicken only through mitotic cell divisions and the accompanying cell growth at meristems. At these localized regions of the plant body, rapid divisions of undifferentiated cells give rise to all specialized cell lineages that form mature tissues.

6. Each growing season, shoots and roots lengthen. The lengthening, called primary growth, originates only at apical meristems in shoot and root tips.

7. For many plant species, shoots and roots also thicken during the growing season. Typically, lateral meristems inside shoots and roots give rise to an increase in diameter, which is called secondary growth. Wood is one outcome of secondary growth.

OVERVIEW OF THE PLANT BODY

Earlier, in Chapter 23, we surveyed representatives of the 295,000 known species of plants. Even that sprint through diversity revealed why no one species can be used as a typical example of plant body plans. When we hear the word "plant," however, we usually think of well-known species of seed-bearing vascular plants—gymnosperms (including pine trees) and angiosperms (flower-producing plants, such as roses, corn, cactuses, and elms). With 260,000 species, angiosperms dominate the plant kingdom. Its major groups are **magnoliids, eudicots** (true dicots) and **monocots** (Section 23.8). We focus here on the true dicots and monocots.

Shoots and Roots

Many flowering plants have a body plan similar to that shown in Figure 29.2. Aboveground are the **shoots:** stems, leaves, flowers (reproductive shoots), and other structures. Stems offer structural support for upright growth, and some of its tissues also conduct water and solutes. Upright growth gives photosynthetic cells in young stems and leaves favorable exposure to sunlight. **Roots** are specialized structures that most often grow downward and outward through soil. An underground root system absorbs water and dissolved minerals, and it typically anchors the aboveground parts. A root also stores food, then releases it as required for its own cells and for distribution to living cells aboveground.

Three Plant Tissue Systems

Stems, branches, leaves, and roots all consist of three major tissue systems (Figure 29.2). The **ground tissue system** serves basic functions, such as food and water storage. The **vascular tissue system** has two different tissues that distribute water and solutes. The **dermal tissue system** covers and protects plant surfaces.

Some tissues in each system are simple, in that they have only one type of cell. Parenchyma, collenchyma, and sclerenchyma are in this category. Other tissues are complex, with organized arrays of two or more types of cells. Xylem, phloem, and epidermis are like this.

The next sections describe the tissue organization of shoots and roots. You may find it easier to follow them by studying Figure 29.3. It reviews different ways in which botanists cut tissue specimens from plants.

Figure 29.2 Body plan for a tomato plant (*Lycopersicon*). Its vascular tissues (*purple*) conduct water, dissolved minerals, and organic substances. They thread through ground tissues that make up most of this plant. Dermal tissue (epidermis in this case) covers surfaces of the root and shoot systems.

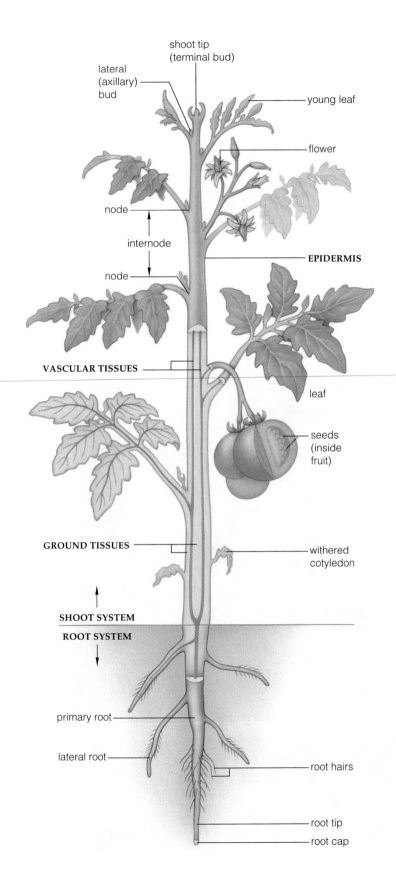

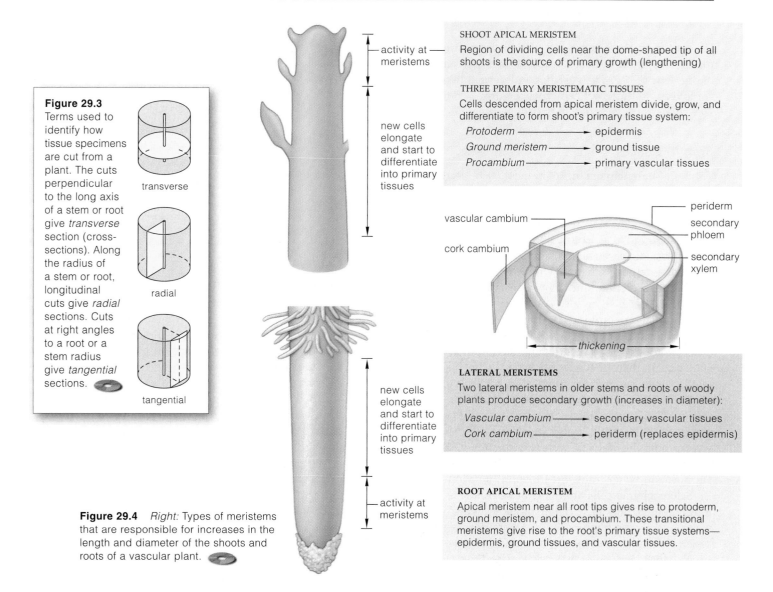

Figure 29.3 Terms used to identify how tissue specimens are cut from a plant. The cuts perpendicular to the long axis of a stem or root give *transverse* section (cross-sections). Along the radius of a stem or root, longitudinal cuts give *radial* sections. Cuts at right angles to a root or a stem radius give *tangential* sections.

transverse

radial

tangential

activity at meristems

new cells elongate and start to differentiate into primary tissues

new cells elongate and start to differentiate into primary tissues

activity at meristems

Figure 29.4 *Right:* Types of meristems that are responsible for increases in the length and diameter of the shoots and roots of a vascular plant.

SHOOT APICAL MERISTEM
Region of dividing cells near the dome-shaped tip of all shoots is the source of primary growth (lengthening)

THREE PRIMARY MERISTEMATIC TISSUES
Cells descended from apical meristem divide, grow, and differentiate to form shoot's primary tissue system:

Protoderm ──────────→ epidermis
Ground meristem ──────→ ground tissue
Procambium ──────────→ primary vascular tissues

vascular cambium

cork cambium

periderm

secondary phloem

secondary xylem

thickening

LATERAL MERISTEMS
Two lateral meristems in older stems and roots of woody plants produce secondary growth (increases in diameter):

Vascular cambium ──────→ secondary vascular tissues
Cork cambium ──────────→ periderm (replaces epidermis)

ROOT APICAL MERISTEM
Apical meristem near all root tips gives rise to protoderm, ground meristem, and procambium. These transitional meristems give rise to the root's primary tissue systems—epidermis, ground tissues, and vascular tissues.

Where Do Plant Tissues Originate?

Different plant tissues become active at different times during the growing season. Most growth proceeds at **meristems**, which are localized regions of dividing cells (Figure 29.4). In other regions, cellular descendants of meristems are maturing or have already matured.

Apical meristems, in the tips of shoots and roots, are where plant parts start to lengthen. Populations formed here develop into protoderm, ground meristem, and procambium. These are immature forms of the primary tissues—epidermis, ground tissue, and vascular tissues respectively. Taken as a whole, the *lengthening* of stems and roots represents the plant's primary growth.

Also during a growing season, the older stems and roots of many plants thicken. Increases in girth start with *lateral* meristems, each a cylindrical array of cells that forms in stems and roots. One lateral meristem, the **vascular cambium**, produces secondary vascular

tissues. The other, **cork cambium**, produces a sturdier covering (periderm) that replaces epidermis. Taken as a whole, the *thickening* of stems and roots represents secondary growth.

Vascular plants have stems that support upright growth and conduct substances, leaves that function in photosynthesis, shoots specialized for reproduction, and other structures. They also have roots that absorb water and solutes. Roots often anchor aboveground parts and store food.

A ground tissue system makes up most of the young plant body. A vascular tissue system distributes water, dissolved ions, and photosynthetic products through it. A dermal tissue system covers and protects plant surfaces.

Shoots and roots lengthen (put on primary growth) when their apical and primary meristems are active. In many plants, older stems and roots also thicken (add secondary growth) when lateral meristems called vascular cambium and cork cambium are active.

TYPES OF PLANT TISSUES

We turn now to an overview of the organization and functions of plant tissues. Simple tissues are composed of one type of cell. The vascular and dermal tissues are complex, with different cell types. Figures 29.5 through 29.9 show examples from these tissue categories.

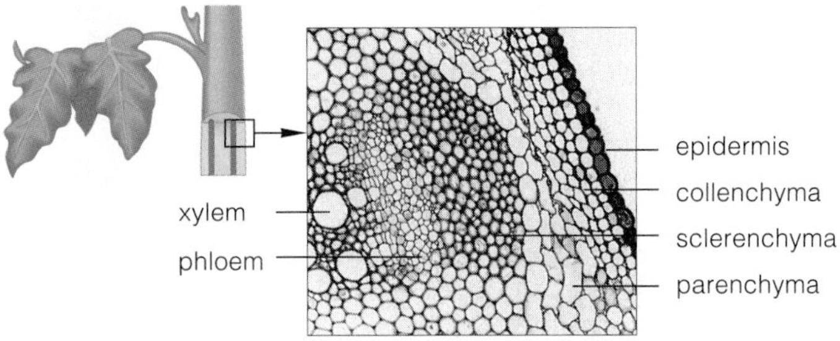

Simple Tissues

Tissues of **parenchyma** make up most of the soft, moist, primary growth of roots, stems, leaves, flowers, and fruits. Most parenchyma cells are pliable, thin-walled, and many-sided. Mature cells are alive and can still divide, often to heal wounds. In leaves, mesophyll is a photosynthetic parenchyma, and air spaces between its cells enhance gas exchange. Parenchyma also has roles in storage, secretion, and other specialized tasks. The vascular tissue systems also contain parenchyma cells.

Collenchyma provides flexible support for primary tissues. Patches or cylinders of its living, usually long cells often support a lengthening stem and form leaf stalk ribs. In their unevenly thickened cell walls, pectin (a pliable polysaccharide) glues cellulose fibrils together.

Most cells in **sclerenchyma** have lignin-impregnated thick walls. Lignin stiffens these walls and gives them compressive strength. It also resists fungal attacks and waterproofs the walls of water-conducting cells. Land plants could not have evolved without its mechanical support and water transport functions (Section 23.1).

Sclerenchyma cells are fibers or sclereids. *Fibers* are long, tapered cells in vascular tissue systems of some stems and leaves (Figure 29.7a). Fibers flex, twist, and resist stretching. We use certain kinds to make cloth, rope, paper, and other valued commodities. *Sclereids* are stubbier cells. Think of a hard seed coat, coconut shell, or peach pit or a pear's gritty texture; sclereids are the source of such features (Figure 29.7b).

Figure 29.5 Locations of simple tissues and complex tissues in one kind of plant stem, transverse section.

epidermis
collenchyma
sclerenchyma
parenchyma

xylem
phloem

Complex Tissues

VASCULAR TISSUES Two vascular tissues—xylem and phloem—distribute substances through plants. Fibers and parenchyma often sheath their conducting cells.

Xylem conducts water and dissolved mineral ions. It also helps mechanically support plants. Figure 29.8a,b shows examples of its conducting cells. The cells, *vessel members* and *tracheids*, are dead at maturity, and their lignified walls interconnect. Collectively, the cell walls form water-conducting pipelines and strengthen plant parts. Water flows into and out of the adjoining cells through numerous pits in the cell walls.

Phloem conducts sugars and other solutes. Its main conducting cells, called *sieve-tube members*, are alive at maturity (Figure 29.8c). They form tubes that connect at openings in their side walls and end walls. Sugars made by photosynthetic cells in leaves are loaded into sieve-tube members with the help of specialized, living parenchyma cells called *companion cells*. Sugars moving through the phloem pipelines are unloaded in regions where cells are growing or storing food. The chapter to follow describes this process.

Figure 29.6 Three examples of simple tissue from the stem of a sunflower plant (*Helianthus*), cross-section. Parenchyma makes up the bulk of the plant body. Collenchyma and sclerenchyma help support and also strengthen plant parts.

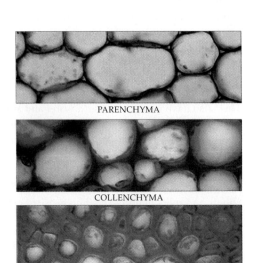

PARENCHYMA

COLLENCHYMA

SCLERENCHYMA

thick, lignified secondary wall

b

SCLEREIDS

FIBERS

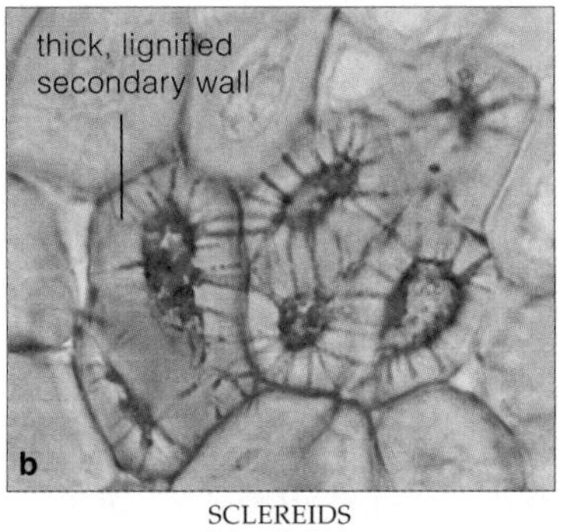

Figure 29.7 Two different kinds of sclerenchyma. (**a**) Strong fibers from flax stems. Compare Figure 4.27a,c. (**b**) From a pear, the sclereids called stone cells, cross-section.

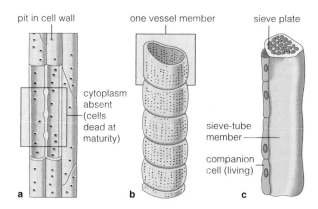

pit in cell wall   one vessel member   sieve plate

cytoplasm absent (cells dead at maturity)

sieve-tube member

companion cell (living)

a   b   c

Figure 29.8 From xylem, portions of (**a**) tracheids and (**b**) a vessel. Pipelines made of such cells conduct water and dissolved ions. (**c**) One type of phloem cell. Long tubes of many cells conduct sugars and other organic compounds.

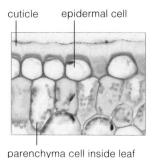

cuticle   epidermal cell

parenchyma cell inside leaf

Figure 29.9 Light micrograph of a section through the upper surface of a kaffir lily leaf. The plant cuticle is made of secretions from epidermal cells. Inside the leaf are many photosynthetic parenchyma cells.

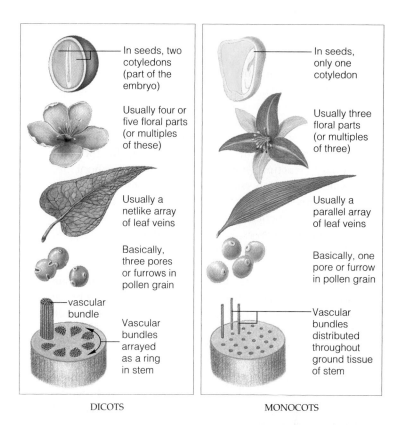

In seeds, two cotyledons (part of the embryo)

Usually four or five floral parts (or multiples of these)

Usually a netlike array of leaf veins

Basically, three pores or furrows in pollen grain

vascular bundle

Vascular bundles arrayed as a ring in stem

In seeds, only one cotyledon

Usually three floral parts (or multiples of three)

Usually a parallel array of leaf veins

Basically, one pore or furrow in pollen grain

Vascular bundles distributed throughout ground tissue of stem

DICOTS   MONOCOTS

Figure 29.10 Comparison of some defining features of the true dicots (eudicots) and monocots. Both classes of flowering plants consist of the same simple and complex tissues, but their body parts show some differences in structural organization.

DERMAL TISSUES A dermal tissue system, **epidermis**, covers surfaces of primary plant parts. In most plants, it is mainly a single layer of unspecialized cells. Waxes and cutin, a fatty substance, coat outward-facing cell walls. The surface coating is a **cuticle**, which helps the plant conserve water and in some cases resists attacks by microorganisms (Figure 29.9).

Stem and leaf epidermis contains many specialized cells. For instance, pairs of guard cells change shape in response to changing conditions. As they do, a gap—or **stoma** (plural, stomata)—closes or opens between them. The next chapter looks at how stomata work as control points for the movement of water vapor, oxygen, and carbon dioxide across the epidermis. Periderm replaces epidermis in stems and roots with secondary growth. As you will see, it includes dead cork cells with walls heavily impregnated with suberin, a fatty substance.

Dicots and Monocots—Same Tissues, Different Features

True dicots (eudicots) include most trees and shrubs other than conifers, such as maples, roses, cacti, and beans. Palms, lilies, orchids, ryegrass, wheat, corn, and sugarcane are familiar monocots. Dicots and monocots are similar in structure and function but differ in some distinctive ways. Dicot seeds have two cotyledons and monocot seeds have only one. Cotyledons are leaflike structures commonly known as seed leaves. They form in seeds as part of a plant embryo, and they store or absorb food for it. After a seed germinates, cotyledons wither, and new leaves grow and start to make food. Figure 29.10 shows other differences between dicots and monocots.

Most of the plant body (ground tissue system) consists of parenchyma, collenchyma, and sclerenchyma. Each of these simple tissues is composed of only one type of cell.

Xylem and phloem are vascular tissues. In xylem, pipelines made of tracheids and vessel members conduct water and dissolved ions. In phloem, sieve-tube members interact with companion cells to distribute organic compounds.

Of two dermal tissues, epidermis covers the surfaces of the primary plant body. Periderm replaces epidermis on plant parts with extensive secondary growth.

Dicots and monocots consist of the same tissues, but each has some of the tissues organized in distinctive ways.

PRIMARY STRUCTURE OF SHOOTS

How Do Stems and Leaves Form?

Next time you or a friend eats a bundle of bean sprouts or alfalfa sprouts, pull one aside to look at its structure. That seedling started forming while it was still inside a seed coat. It already has a primary root and shoot. In the primary shoot's tip, apical meristem and its descendant tissues are laying out orderly frameworks for the stem primary structure (Figure 29.11). Below the shoot tip, tissues become specialized as cells divide at different rates in prescribed directions, and differentiate in size, shape, and function. These tissues make up distinctive stem regions, leaves, and lateral (axillary) buds from which lateral shoots develop. Lateral shoots give rise to side branches and reproductive structures.

Briefly, as a typical shoot lengthens, bulges of tissue develop along the sides of apical meristem. Each bulge is one immature leaf (Figures 29.11 and 29.12). While growth continues, the stem lengthens between tier after tier of new leaves. Each part of the stem where one or more leaves are attached is a node. As shown in Figure 29.2, the stem region between two successive nodes is an internode. Buds develop in the upper angle where leaves attach to the stem. Each **bud** is an undeveloped shoot of mostly meristematic tissue, often protected by bud scales (modified leaves). Buds give rise to stems, leaves, and flowers in ways described in Section 32.5.

Internal Structure of Stems

While the primary plant body of a monocot or dicot is forming, the ground, vascular, and dermal tissues of its stems become organized in distinctive ways. Most often, primary xylem and phloem develop inside the same sheath of cells, as **vascular bundles**. The bundles are multistranded cords threading lengthwise through the ground tissue system of primary and lateral shoots.

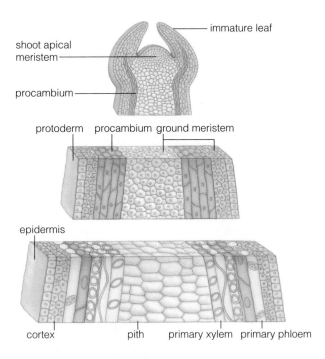

Figure 29.11 Successive stages in primary growth, starting with activity at shoot apical meristem of a typical dicot and continuing at primary meristem tissues derived from it. Notice the progressive differentiation of most tissue regions.

They commonly develop in two genetically dictated patterns. In most dicot stems, long bundles form a ring that divides the ground tissue into a cortex and pith (Figure 29.13a). A stem's **cortex** is the region between the vascular bundles and the epidermis. Its **pith** is the center of the stem, inside the ring of vascular bundles. Ground tissue of the plant's roots becomes similarly divided, into root cortex and pith. A different pattern is common inside the stems of most monocots and some

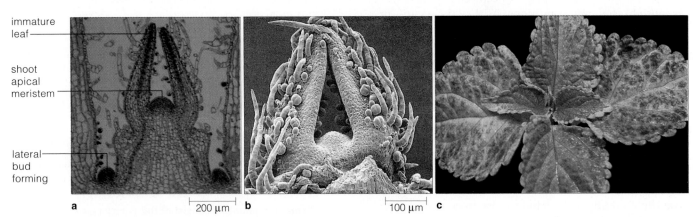

Figure 29.12 (**a**) Light micrograph of a *Coleus* shoot tip, cut longitudinally through its center. (**b**) Scanning electron micrograph of its surface. (**c**) New *Coleus* leaves.

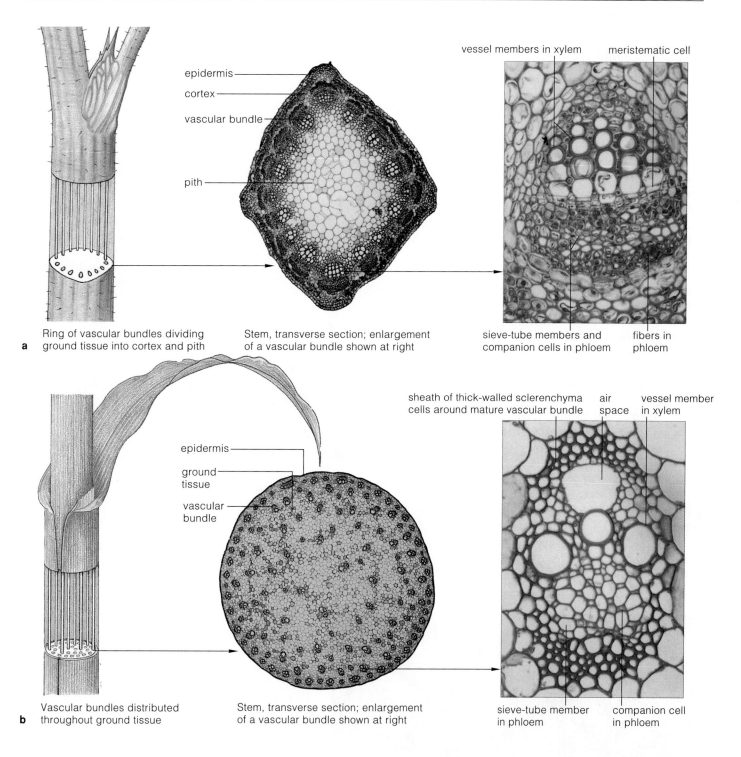

a Ring of vascular bundles dividing ground tissue into cortex and pith

epidermis

cortex

vascular bundle

pith

Stem, transverse section; enlargement of a vascular bundle shown at right

vessel members in xylem

meristematic cell

sieve-tube members and companion cells in phloem

fibers in phloem

b Vascular bundles distributed throughout ground tissue

epidermis

ground tissue

vascular bundle

Stem, transverse section; enlargement of a vascular bundle shown at right

sheath of thick-walled sclerenchyma cells around mature vascular bundle

air space

vessel member in xylem

sieve-tube member in phloem

companion cell in phloem

Figure 29.13 Internal organization of cells and tissues inside the stems from a dicot and a monocot. (**a**) Part of a stem from alfalfa (*Medicago*), a dicot. In many species of dicots and conifers, the vascular bundles develop in a more or less ringlike array in the ground tissue system, as shown here. The portion of the ground tissue between the ring and the surface of the stem is the cortex. The portion enclosed within the ring is the pith. (**b**) Part of a stem from corn (*Zea mays*), a monocot. In most monocots and some nonwoody dicots, vascular bundles are scattered through the ground tissue, as shown.

dicots. The long vascular bundles inside the stem are scattered throughout its ground tissue (Figure 29.13*b*). How substances are conducted through such vascular systems is a topic of the next chapter.

Shoot apical meristems give rise to the primary plant body, which develops a distinctive internal structure (as in the pattern in which its vascular bundles are arranged).

A CLOSER LOOK AT LEAVES

Similarities and Differences Among Leaves

Every **leaf** that forms is a metabolic factory equipped with many photosynthetic cells. Yet leaves vary greatly in size, shape, surface details, and internal structure. A duckweed leaf is no more than 1 millimeter (0.04 inch) across; leaves of one palm (*Attalea*) are 12 meters (40 feet) across. Different leaves are shaped like needles, blades, spikes, cups, tubes, and feathers. They differ hugely in color, odor, and edibility; many form toxins. Leaves of birches and other species of *deciduous* plants wither and drop away from stems with the approach of winter. Leaves of camellias and other *evergreen* plants also drop, but not all at the same time.

A typical leaf has a flat blade, as in Figures 29.14*a* and 29.15. It has a stalk (petiole) that attaches it to the stem. *Simple* leaves are undivided, but many are lobed. *Compound* leaves have blades divided into leaflets, all oriented in the same plane. Leaves of most monocots, such as ryegrass and corn, are flat surfaced like a knife blade. The blade's base encircles and sheathes the stem.

Leaves of most species are thin, with a high surface-to-volume ratio. Their flat surface grows and orients itself perpendicular to light. A leaf often projects from a stem in patterns that minimize shading of other leaves. For example, a clover leaf stalk is attached to the stem at right angles to its neighbors (Figure 29.15*c*).

Such leaf adaptations intercept as much of a plant's energy source —sunlight—as possible. They also help oxygen diffuse out and carbon dioxide diffuse in. And when a leaf

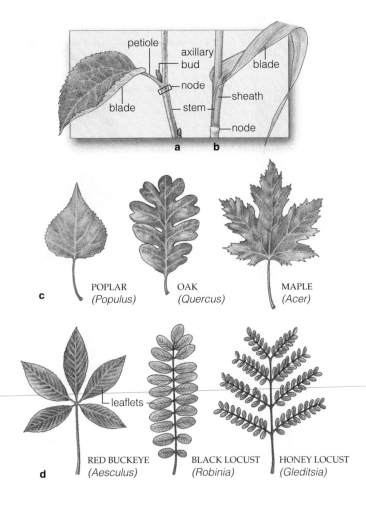

Figure 29.14 Common leaf forms of (**a**) dicots and (**b**) monocots. Examples of (**c**) simple leaves and (**d**) compound leaves.

POPLAR (*Populus*) OAK (*Quercus*) MAPLE (*Acer*)

RED BUCKEYE (*Aesculus*) BLACK LOCUST (*Robinia*) HONEY LOCUST (*Gleditsia*)

is thick, you can assume it belongs to a succulent or some other plant of arid habitats and serves in water storage as well as photosynthesis. Another example: The leaves of many desert plants orient themselves parallel with the sun's rays to reduce heat absorption to tolerable levels.

Leaf Fine Structure

In its fine structure, too, each leaf is adapted to intercept energy from the sun's rays and promote gas exchange. In addition, many have distinctive surface specializations.

Figure 29.15 (**a**) Decaying dicot leaf, with its netlike veins. (**b**) Parallel veins of a monocot leaf (*Agapanthus*). (**c**) Leaf orientation of a four-leaf clover (*Trifolium*).

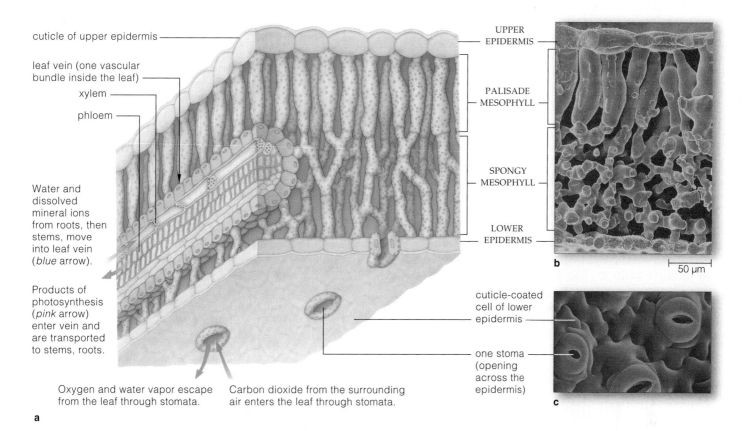

cuticle of upper epidermis

leaf vein (one vascular bundle inside the leaf)

xylem

phloem

Water and dissolved mineral ions from roots, then stems, move into leaf vein (*blue* arrow).

Products of photosynthesis (*pink* arrow) enter vein and are transported to stems, roots.

Oxygen and water vapor escape from the leaf through stomata.

Carbon dioxide from the surrounding air enters the leaf through stomata.

UPPER EPIDERMIS

PALISADE MESOPHYLL

SPONGY MESOPHYLL

LOWER EPIDERMIS

b

50 µm

cuticle-coated cell of lower epidermis

one stoma (opening across the epidermis)

c

a

Figure 29.16 (**a**) Diagram of leaf structure for many kinds of flowering plants. (**b**) Scanning electron micrograph of the tissue organization of a leaf from the kidney bean plant (*Phaseolus*). Notice the compact organization of the epidermal cells. (**c**) Stomata, each a tiny opening across the epidermis, appear when paired guard cells are in their plumped configuration. See also Figure 7.17.

LEAF EPIDERMIS Epidermis covers every leaf surface exposed to the air. It may be smooth, sticky, or slimy, with "hairs," scales, spikes, hooks, glands, and other surface specialties. The Chapter 30 introduction has two examples. A cuticle covers the sheetlike, compact array of epidermal cells; it restricts the loss of precious water (Figures 29.9 and 29.16). Most leaves have many more stomata on their lower surface. In arid or cold habitats, the stomata are often located at depressions in the leaf surface, together with thickly coated epidermal hairs. Both leaf adaptations help conserve water.

MESOPHYLL—PHOTOSYNTHETIC GROUND TISSUE As you read earlier, **mesophyll** is a type of parenchyma specializing in photosynthesis. Inside a leaf, most of its cells are exposed to air spaces (Figure 29.16). Carbon dioxide reaches cells by diffusing into the leaf through stomata and through the air inside; oxygen diffuses the opposite way. Adjoining cells exchange substances fast at plasmodesmata. These junctions freely interconnect the cytoplasm of both cells, as Section 4.11 describes.

Leaves oriented perpendicular to the sun have two mesophyll regions. Attached to the upper epidermis is

palisade mesophyll—columnar parenchymal cells with more chloroplasts and more photosynthetic potential, compared to cells of the *spongy* mesophyll layer below them (Figure 29.16). Monocot leaves grow vertically and intercept light from all directions. As you might expect, their mesophyll is not organized as two layers.

VEINS—THE LEAF'S VASCULAR BUNDLES Leaf **veins** are vascular bundles, usually strengthened with fibers. Their continuous strands of xylem rapidly move water and dissolved nutrients to all mesophyll cells, and the continuous strands of phloem carry the photosynthetic products—especially sugars—away from them. In most dicots, the veins branch lacily into a number of minor veins embedded inside mesophyll. In most monocots, the veins are more or less similar in length, and they run parallel with the leaf's long axis (Figure 29.15).

A leaf's structure is adapted for sunlight interception, gas exchange, and distribution of water, dissolved nutrients, and photosynthetic products. Leaves of each species have a characteristic size, shape, and often surface specializations.

PRIMARY STRUCTURE OF ROOTS

Taproot and Fibrous Root Systems

When a seed germinates, the first part to poke through the seed coat is a **primary root** (Figure 29.17). In nearly all dicot seedlings, it increases in girth while it grows downward. Later on, **lateral roots** form in the primary root's tissues, perpendicular to its axis, and then erupt through epidermis. Youngest lateral roots are closest to the root tip. A **taproot system** is a primary root with its lateral branchings. Dandelions, carrots, and oak trees are examples of plants with this system (Figure 29.17c).

By comparison, the primary root of most monocots, such as grasses, is short-lived. In its place, adventitious roots arise from the stem, and then lateral roots branch from these. (*Adventitious* means the structures form at an unusual location.) The lateral roots are all more or less alike in diameter and length. Together, roots that form this way are a **fibrous root system** (Figure 29.17d).

Internal Structure of Roots

In Figure 29.17a, notice the meristems inside the tip of one root. Many cellular descendants of these meristems divide, enlarge, elongate, and become cells of primary tissue systems. Notice also the root cap, a dome-shaped mass of cells at the tip. The apical meristem produces the cap, which in turn protects the meristem.

Protoderm gives rise to root epidermis, the plant's absorptive interface with the soil. Some epidermal cells send out extensions called **root hairs**. Collectively, root hairs enormously increase the surface area available for taking up water and dissolved nutrients. Only first-time or foolish gardeners would yank a plant from the ground when transplanting it. Such yanking would tear off too much of the highly fragile absorptive surface.

The apical meristem also gives rise to the ground tissue system and to the **vascular cylinder**. A vascular

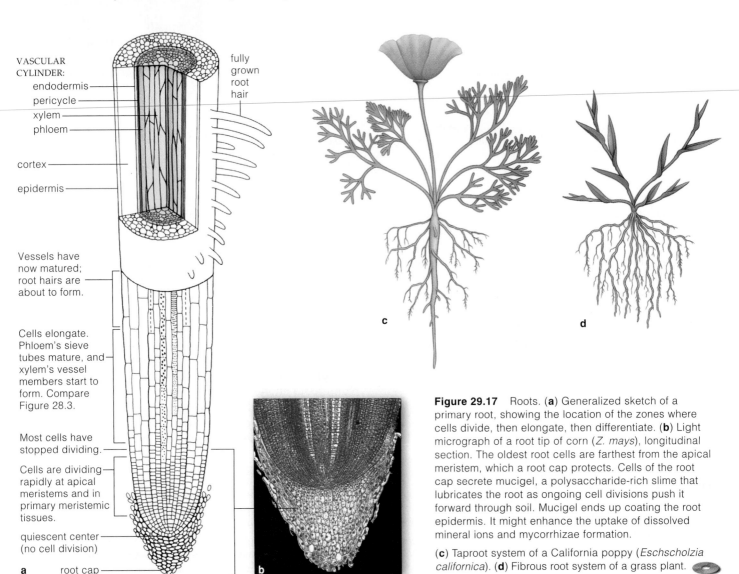

VASCULAR CYLINDER:
endodermis
pericycle
xylem
phloem

cortex

epidermis

fully grown root hair

Vessels have now matured; root hairs are about to form.

Cells elongate. Phloem's sieve tubes mature, and xylem's vessel members start to form. Compare Figure 28.3.

Most cells have stopped dividing.

Cells are dividing rapidly at apical meristems and in primary meristemic tissues.

quiescent center (no cell division)

a   root cap

c

d

Figure 29.17 Roots. (**a**) Generalized sketch of a primary root, showing the location of the zones where cells divide, then elongate, then differentiate. (**b**) Light micrograph of a root tip of corn (*Z. mays*), longitudinal section. The oldest root cells are farthest from the apical meristem, which a root cap protects. Cells of the root cap secrete mucigel, a polysaccharide-rich slime that lubricates the root as ongoing cell divisions push it forward through soil. Mucigel ends up coating the root epidermis. It might enhance the uptake of dissolved mineral ions and mycorrhizae formation.

(**c**) Taproot system of a California poppy (*Eschscholzia californica*). (**d**) Fibrous root system of a grass plant.

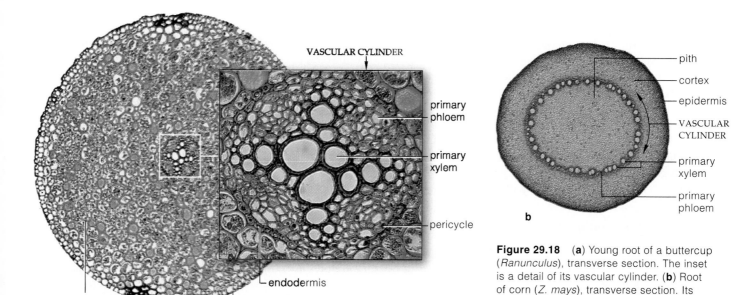

VASCULAR CYLINDER

primary
phloem

primary
xylem

pericycle

endodermis

epidermis

root cortex

a

pith

cortex

epidermis

VASCULAR
CYLINDER

primary
xylem

primary
phloem

b

Figure 29.18 (**a**) Young root of a buttercup (*Ranunculus*), transverse section. The inset is a detail of its vascular cylinder. (**b**) Root of corn (*Z. mays*), transverse section. Its vascular cylinder divides the ground tissue into two zones—cortex and pith.

cylinder consists of primary xylem and phloem and one or more layers of cells called the pericycle. Figure 29.18*a* shows a vascular cylinder at the center of a dicot root's cortex. Figure 29.18*b* shows how the vascular cylinder divides the ground tissue system of one type of monocot into cortex and pith regions. With either pattern, there are plenty of air spaces between cells of the ground tissue system, so oxygen can easily diffuse

through them. Like other cells in the plant, all living root cells depend on oxygen for aerobic respiration.

When water enters a root, it moves from cell to cell until it reaches the endodermis, a layer of cells around the vascular cylinder. Where endodermal cells abut, their walls are waterproofed, so incoming water is forced to pass through their cytoplasm. As described in Chapter 30, this arrangement controls the movement of water and dissolved substances into the vascular cylinder.

The pericycle is just inside the endodermis. Some of its cells divide repeatedly and form lateral roots, which erupt through the cortex and epidermis (Figure 29.19).

Regarding the Sidewalk-Buckling, Record-Breaking Root Systems

Unless tree roots start to buckle a sidewalk or choke off a sewer line, most of us do not pay much attention to flowering plant root systems. Roots mine the soil for water and minerals, and most reach a depth of 2 to 5 meters. In hot deserts, where free water is scarce, one hardy mesquite shrub sent its roots down 53.4 meters (175 feet) near a stream bed. Some cacti have shallow roots radiating outward for 15 meters. Someone once measured the roots of a young rye plant that had been growing for four months in 6 liters of soil water. If the surface area of that root system were laid out as one sheet, it would occupy more than 600 square meters!

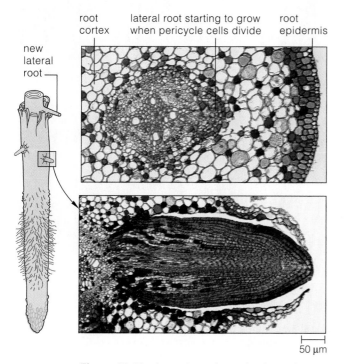

root
cortex

lateral root starting to grow
when pericycle cells divide

root
epidermis

new
lateral
root

50 μm

Figure 29.19 Lateral root formation in a primary root from a willow tree (*Salix*), transverse section.

Roots provide a plant with a tremendous surface area for absorbing water and solutes. Taproot systems consist of a primary root and lateral branchings. Fibrous root systems consist of adventitious roots that replace the primary root.

ACCUMULATED SECONDARY GROWTH—THE WOODY PLANTS

Flowering plant life cycles extend from germination to seed formation, then death. **Annuals** complete their life cycle in a single growing season, and they typically are "nonwoody," or herbaceous. Examples are marigolds and alfalfa. **Biennials** such as carrots can live for two consecutive growing seasons. Their roots, stems, and leaves form the first season; flowers form, seeds form, and the plant dies the next season. **Perennials** continue vegetative growth and seed formation year after year. And roots and stems thicken in many of them.

Woody and Nonwoody Plants Compared

Like all gymnosperms, some monocots, many eudicots, and magnoliids add secondary growth in two or more growing seasons; they are *woody* plants. Early in life, their stems and roots are like those of nonwoody plants. Differences emerge after their lateral meristems become active and start producing large amounts of secondary vascular tissues, especially the secondary xylem. Again, periderm replaces epidermis on roots and stems that continue to add secondary growth.

The differences are especially pronounced in some of the perennial plants in which the vascular cambium

has become reactivated each growing season, often for hundreds or thousands of years. Ongoing meristematic activity has resulted in giants. To give an example, at last measure, the massive trunk of one coast redwood (*Sequoia sempervirens*) towered more than 110 meters above the forest floor. By one estimate, that redwood's accumulated secondary growth may weigh close to 100 metric tons. Another example: The tree with the largest girth is one of the chestnuts (*Castanea*) that is growing in Sicily. To walk completely around the base of it, you would have to pace off 58 meters.

What Happens at the Vascular Cambium?

Massive stems and roots originate with the vascular cambium. Look at Figure 29.20. Each spring, primary growth resumes at this stem's buds; secondary growth is added *inside* it. Fully formed vascular cambium in a stem is like a cylinder, one or a few cells thick. Some of its cells (fusiform initials) give rise to secondary xylem and phloem that extend *lengthwise* through the stem. Other cells of the vascular cambium (ray initials) give rise to *horizontal* rays of parenchyma, in a pattern a bit like a sliced pie. Through these vascular tissues, water

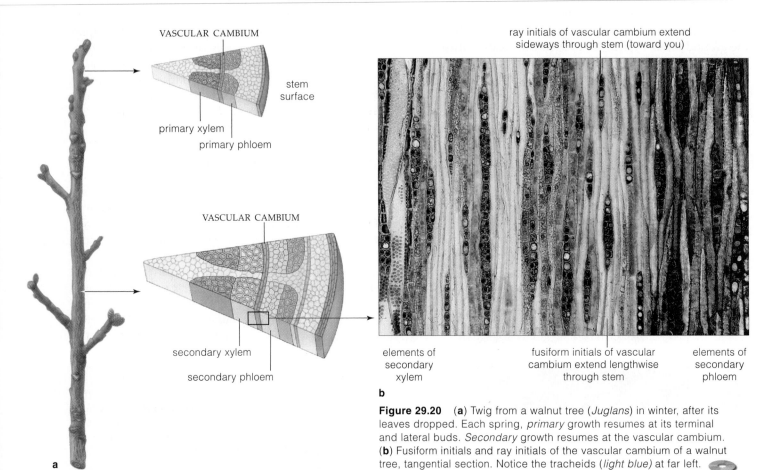

Figure 29.20 (**a**) Twig from a walnut tree (*Juglans*) in winter, after its leaves dropped. Each spring, *primary* growth resumes at its terminal and lateral buds. *Secondary* growth resumes at the vascular cambium. (**b**) Fusiform initials and ray initials of the vascular cambium of a walnut tree, tangential section. Notice the tracheids (*light blue*) at far left.

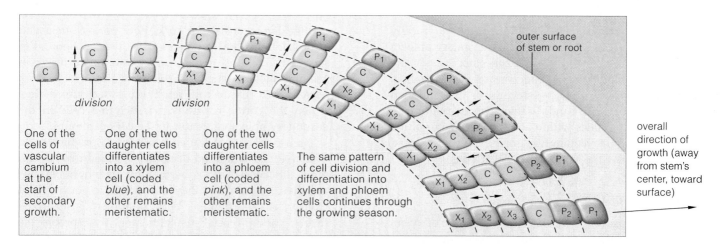

Figure 29.21 Pattern of activity at vascular cambium. Reading left to right, ongoing cell divisions enlarge the inner core of secondary xylem and displace vascular cambium toward the stem or root surface.

Within the figure:

One of the cells of vascular cambium at the start of secondary growth.

division

One of the two daughter cells differentiates into a xylem cell (coded *blue*), and the other remains meristematic.

division

One of the two daughter cells differentiates into a phloem cell (coded *pink*), and the other remains meristematic.

The same pattern of cell division and differentiation into xylem and phloem cells continues through the growing season.

outer surface of stem or root

overall direction of growth (away from stem's center, toward surface)

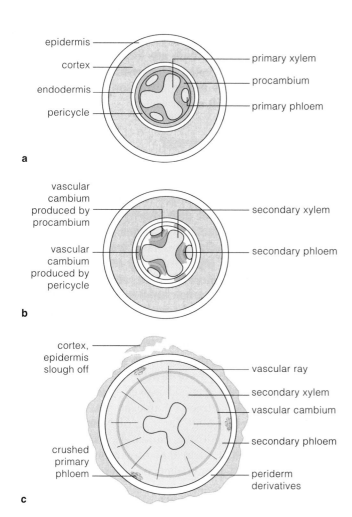

Labels for figure 29.22a:
epidermis
cortex
endodermis
pericycle
primary xylem
procambium
primary phloem

a

Labels for figure 29.22b:
vascular cambium produced by procambium
vascular cambium produced by pericycle
secondary xylem
secondary phloem

b

Labels for figure 29.22c:
cortex, epidermis slough off
crushed primary phloem
vascular ray
secondary xylem
vascular cambium
secondary phloem
periderm derivatives

c

Figure 29.22 Secondary growth in one type of woody root.
(**a**) This is how tissues are organized as primary growth ends.
(**b,c**) A thin cylinder of vascular cambium forms and gives rise to secondary xylem and phloem. Cell divisions are parallel with the vascular cambium. The cortex ruptures as the root thickens.

and solutes travel up, down, and sideways through the enlarging woody stem.

Figure 29.21 shows the growth pattern at vascular cambium. Secondary xylem forms on this meristematic tissue's *inner* face. Secondary phloem forms on its *outer* face. As the inner core of xylem gradually thickens, it displaces meristematic cells toward the surface of the stem. Its cells also maintain a ring of vascular cambium by dividing sideways, in a widening circle.

We have been focusing on how stems thicken, but bear in mind that secondary xylem and phloem also form at vascular cambium in the plant's roots. Figure 29.22 shows one of the patterns of secondary growth at vascular cambium in the root of a typical plant.

What are the selective advantages of woody stems and roots? Remember, like all other organisms, plants compete for resources. Plants having taller stems or broader canopies that defy the pull of gravity intercept more energy streaming in from the sun. With a greater supply of energy for photosynthesis, they have the metabolic means to form large root and shoot systems. With larger systems, they can be more competitive in acquiring resources—and ultimately to be successful, in reproductive terms, in particular habitats.

In woody plants, secondary vascular tissues form at a ring of vascular cambium inside older stems and roots. Wood is an accumulation of secondary xylem especially.

With their sturdier tissues, woody plants defy gravity and grow taller and broader. Where competition for sunlight is intense, the ones that intercept the most sunlight win. Other factors being equal, having more energy to drive photosynthesis provides advantages in terms of metabolic capabilities, growth, and reproductive success.

A CLOSER LOOK AT WOOD AND BARK

Where secondary xylem (wood) is extensive, it typically makes up about 90 percent of a tree. Secondary phloem is restricted to a relatively thin zone just outside the vascular cambium. This phloem consists of thin-walled, living parenchyma cells and sieve tubes that are often interspersed between bands of thick-walled, reinforcing fibers. Only the tubes within about a centimeter of the vascular cambium remain functional. The rest are dead and help protect the living cells beneath them.

Formation of Bark

As seasons pass and a tree ages, its inner core of xylem continues its outward expansion. The resulting pressure is directed toward the stem or root surface. Eventually it ruptures the cortex and the outer part of secondary phloem. Some cortex and epidermis split away. A new surface cover, the periderm, forms from cork cambium. Together, the periderm and secondary phloem constitute **bark**. In other words, bark is composed of all tissues outside the vascular cambium (Figures 29.23 and 29.24).

Periderm consists of cork, new parenchyma, and cork cambium that produces these tissues. Soon after vascular cambium forms, cork cambium forms from the outermost parenchyma cells of the stem or the root cortex. Such cells, recall, retain the capacity to divide. When the cortex ruptures, parenchyma cells in the secondary phloem give rise to the cork cambium.

Cell divisions at the cork cambium produce **cork**. This tissue is composed of densely packed rows of cell walls, thickened with suberin. Only its innermost cells are alive, because only they have access to nourishment from xylem and phloem. With its numerous suberized layers, cork protects, insulates, and waterproofs the stem or root surface. Cork also forms over wounded tissues. When leaves are about to drop, cork forms at the place where petioles attach to stems.

Like all living plant cells, the cells in woody stems and roots require oxygen for aerobic respiration and give off carbon dioxide wastes. So how do these gases get across the suberized, corky surface of bark? They do so through lenticels, which are localized areas where the packing of cork cells is loosened up a bit. Those dark spots you may have noticed on a wine bottle's cork are all that is left of lenticels.

Heartwood and Sapwood

Wood's appearance and function change as a stem or root grows older. The core becomes **heartwood**. This is a dry tissue that is no longer transporting water and solutes. It helps the tree defy gravity and is a dumping ground for various metabolic wastes, including resins, tannins, gums, and oils. Eventually, wastes clog and fill in the oldest xylem pipelines. They commonly darken heartwood, make it more aromatic, and strengthen it.

In the early 1900s, when lumbermen were active in California's groves of old-growth redwoods, someone cut a tunnel through heartwood of a few of the biggest trees, the better to drive an automobile through them.

Sapwood is all of the secondary growth in between the vascular cambium and heartwood (Figure 29.24a). Unlike heartwood, sapwood is wet, usually pale, and not as strong. Maple trees are an example. Each spring, from early March through early April, New Englanders insert metal tubes into the sapwood of sugar maples. Sap, a sugar-rich fluid in the secondary xylem, drips through the tubes, into buckets positioned below.

Figure 29.23 The thick, fire-resistant bark of a champion of secondary growth—a coast redwood (*Sequoia sempervirens*).

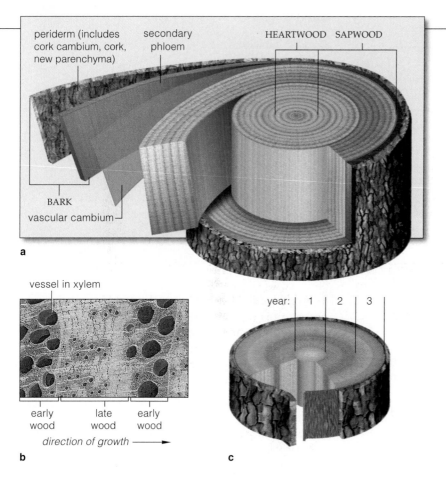

vessel in xylem

early wood | late wood | early wood

direction of growth ——→

b

c

Figure 29.24 (**a**) Stem having extensive secondary growth. (**b**) Scanning electron micrograph of early and late wood cut from red oak (*Quercus rubra*). (**c**) Radial cut through a stem with three annual rings. The first year, the stem put on primary and some secondary growth; the next two years, it added secondary growth.

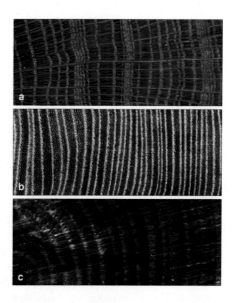

Figure 29.25 Growth layers, or tree rings, of (**a**) oak and (**b**) elm, two hardwood dicots that are durable and strong. (**c**) Pine growth layers. Pine is a softwood. It is lightweight, resists warping, and grows faster than the hardwoods. It is commercially farmed as a source of relatively inexpensive lumber.

Each growth layer shown corresponds to one growing season. (The elm sequence spans the years 1911 to 1950.) Differences in the widths of growth layers correspond to shifts in climate, including water availability. Count the rings and you have clues to a tree's age and to climates and life in the past.

Early Wood, Late Wood, and Tree Rings

Vascular cambium becomes inactive for part of the year in regions having cool winters or prolonged dry spells. *Early* wood, formed at the start of the growing season, has large-diameter, thin-walled xylem cells. By contrast, *late* wood, which forms in dry summers, has xylem cells with smaller diameters and thicker walls. Cut a transverse section from a trunk and you see alternating bands of early and late wood, which reflect the light differently. These visible differences are **growth rings** or, informally, "tree rings" (Figures 29.24 and 29.25).

Seasonal change is predictable in temperate regions, and trees growing there usually add one growth ring per year. In deserts, thunderstorms rumble through at different times of year, and trees respond by adding more than one growth ring in the same season. In the tropics, seasonal change is almost nonexistent. That is why growth rings are not a feature of tropical trees.

Oak, hickory, and other dicot trees that evolved in temperate and tropical regions are all **hardwoods**; they have vessels, tracheids, and fibers in their xylem. Pines, redwoods, and other conifers are **softwood** trees; their

xylem contains tracheids and rays of parenchyma, but no vessels or fibers. Lacking fibers, the trees are weaker and less dense than the hardwoods (Figure 29.25).

Limits to Secondary Growth

Some trees, including bristlecone pines and redwoods, gradually put on secondary growth for centuries. But most die far sooner from old age and environmental insults. Compartmentalization, that response to attack you read about in Section 28.4, eventually shuts off the flow of water and solutes through the vascular system.

Bark consists of all living and nonliving tissues outside the vascular cambium—that is, secondary phloem and periderm.

Periderm consists of cork (the outermost covering of woody stems and roots), cork cambium, and new parenchyma.

Wood may be classified by its location and functions (as in heartwood versus sapwood) and by the type of plant (many dicots produce hardwood, and conifers produce softwood).

SUMMARY

1. Seed-bearing vascular plants include gymnosperms and angiosperms (the flowering plants). Their shoots (stems, leaves, and other structures) and roots consist of dermal, ground, and vascular tissue systems. *29.1*

2. All plant growth originates at meristems, localized regions of cells that retain the capacity to divide. *29.1*

   a. Primary growth (lengthening of stems and roots) originates at apical meristems in root and shoot tips.

   b. In many plants, secondary growth (increases in diameter) originates inside stems and roots, at lateral meristems called vascular cambium and cork cambium.

3. Parenchyma, sclerenchyma, and collenchyma are the simple tissues, with only one cell type (Table 29.1). *29.2*

   a. Parenchyma cells, alive and metabolically active at maturity, make up the bulk of ground tissue systems. They function in a variety of tasks. The ones making up mesophyll, for example, are photosynthetic.

   b. Collenchyma supports growing plant parts. With its thick, lignified cell walls, sclerenchyma functions in mechanical support.

4. Complex tissues include vascular tissues (xylem and phloem) and dermal tissues (epidermis and periderm). Each has two or more cell types (Table 29.1). *29.2*

   a. Vascular tissues distribute water and dissolved substances throughout a plant. Vascular bundles (each having xylem and phloem clustered together inside a cellular sheath) thread through the ground tissue.

   b. Water-conducting cells of xylem are not alive at maturity. Their lignified, pitted walls interconnect and serve as pipelines for water and dissolved minerals.

   c. Phloem's conducting cells are alive at maturity. The cytoplasm of adjoining cells interconnects across perforated end and side walls. In leaves, sugars and other photosynthetic products are loaded into the cells; often companion cells assist in this. The unloading takes place wherever cells are growing or storing food.

   d. Epidermis covers and protects the outer surfaces of primary plant parts. Periderm replaces epidermis on plants showing extensive secondary growth. *29.2, 29.6*

5. Stems support upright growth and conduct water and solutes through their vascular bundles. Monocot stems often have vascular bundles distributed through ground tissue. Most dicot stems have a ring of bundles dividing the ground tissue into cortex and pith. *29.3*

6. Leaves have veins and mesophyll (photosynthetic parenchyma) between the upper and lower epidermis. Air spaces around the photosynthetic cells enhance gas exchange. Water vapor and gases cross the epidermis through numerous tiny openings called stomata. *29.4*

7. Roots absorb water and dissolved mineral ions for distribution to aboveground parts. Most anchor plants and store food. Some help support shoots. *29.1, 29.5*

Table 29.1	*Summary of Flowering Plant Tissues and Their Components*
SIMPLE TISSUES	
Parenchyma	Parenchyma cells
Collenchyma	Collenchyma cells
Sclerenchyma	Fibers or sclereids
COMPLEX TISSUES	
Xylem	Conducting cells (tracheids, vessel members); parenchyma cells; sclerenchyma cells
Phloem	Conducting cells (sieve-tube members); parenchyma cells; sclerenchyma cells
Epidermis	Undifferentiated cells; also guard cells and other specialized cells
Periderm	Cork; cork cambium; new parenchyma

8. Wood (secondary xylem) is classified by location and function (as in heartwood or sapwood) and plant type (hardwood of many dicots, softwood of conifers). Bark consists of secondary phloem and periderm. *29.6, 29.7*

Review Questions

1. List some functions of roots and shoots. *29.1*

2. Name and define the basic functions of a flowering plant's three main tissue systems. *29.1*

3. Describe the differences between:
   a. apical, transitional, and lateral meristems *29.1*
   b. parenchyma and sclerenchyma *29.2*
   c. xylem and phloem *29.2*
   d. epidermis and periderm *29.2, 29.7*

4. Which of the two stem sections below is typical of most dicots? Which is typical of most monocots? Label the main tissue regions of both sections. *29.2*

5. In Figure 29.26, is the plant with the yellow flower a dicot or a monocot? What about the plant with the purple flower? *29.2*

Figure 29.26 Flower of (**a**) St. John's wort (*Hypericum*) and (**b**) an iris (*Iris*).

1. Roots and shoots lengthen through activity at _____ .
   a. apical meristems   c. vascular cambium
   b. lateral meristems   d. cork cambium

2. Older roots and stems thicken through activity at _____ .
   a. apical meristems   c. vascular cambium
   b. cork cambium   d. both b and c

3. Soft, moist plant parts consist mostly of _____ cells.
   a. parenchyma   c. collenchyma
   b. sclerenchyma   d. epidermal

4. Xylem and phloem are _____ tissues.
   a. ground   b. vascular   c. dermal   d. both b and c

5. _____ conducts water and ions; _____ conducts food.
   a. Phloem; xylem   c. Xylem; phloem
   b. Cambium; phloem   d. Xylem; cambium

6. Buds give rise to _____ .
   a. leaves   c. stems
   b. flowers   d. all of the above

7. Mesophyll consists of _____ .
   a. waxes and cutin   c. photosynthetic cells
   b. lignified cell walls d. cork but not bark

8. In early wood, cells have _____ diameters, _____ walls.
   a. small; thick   c. large; thick
   b. small; thin   d. large; thin

9. Match the plant parts with the most suitable description.
   ____ apical meristem   a. masses of xylem
   ____ lateral meristem   b. source of primary growth
   ____ xylem, phloem   c. corky surface covering
   ____ periderm   d. source of secondary growth
   ____ vascular cylinder   e. distribution of water, food
   ____ wood   f. central column in roots

Critical Thinking

1. Sylvia lives in Santa Barbara, where droughts are common and a long-term abundance of water is not. She replaced most of her garden with drought-tolerant plants and cut back the size of the lawn. The lawn does not get a light sprinkling every day. Sylvia waters it only twice a week in the evening, after the sun goes down. Then the lawn gets a good soak to a depth of several inches. Why is her strategy good for lawn grasses?

2. *Girdling* means making a continuous cut right through the vertical phloem all the way around a tree trunk. Without the phloem, food from leaves cannot reach roots. If the roots die, so, in time, will the tree. In northern California, anti-logging activists and loggers have been in conflict. Not too long ago, a redwood given the name "Luna" became a symbol for the activists. One night an anti-activist buzz-sawed around most of Luna's trunk. The partially girdled tree hasn't died yet. Do some research and then speculate on why someone wanted to destroy the symbolic tree in the first place.

3. Oscar and Lucinda meet on a trip through a tropical rain forest and fall in love. In the exuberance of the moment, he carves their initials into the bark of a small tree. They never do get together, though. Ten years later, the still-heartbroken Oscar searches for the tree. Given what you know about primary and secondary growth, will he find the carved initials higher relative to ground level? If he goes berserk and cuts down the tree, what kind of growth rings will he see?

4. Environmental conditions apparently worked against the early immigrants from England who tried to settle in North

severe drought conditions 1587–1589   Jamestown drought 1606–1612

Figure 29.27 *Left:* Map of Virginia's Tidewater region. *Right:* Growth layers of a bald cypress that was living when the first of the English colonizers were in North America.

America. Some attempted to establish a colony on Roanoke Island in 1560 or so, in Virginia's Tidewater region. The colony lasted twenty-seven years, then it vanished from the historical record. Historians thought the colony failed because the people were poor planners. However, tree growth layers tell a very different story.

Scientists extracted a wood core from a bald cypress in the region (they didn't harm the tree). Its growth layers revealed that the colonizers were in the wrong place at the wrong time. They were smack in the middle of the worst prolonged drought to hit the mid-Atlantic seaboard over the past eight hundred years. Between the years 1587 and 1589, drought conditions were especially severe. That is precisely when the Roanoke Island colonizers disappeared (Figure 29.27).

Securing and growing food must have been challenging enough. But the colonizers of this region also had to drink brackish water. During the prolonged drought, salts became more concentrated than ever in their water supply. Either the colonizers dispersed elsewhere or the salts poisoned them.

If the secrets locked inside trees intrigue you, do some research into a field of study called dendroclimatology. For example, find out what growth layers reveal about fluctuations in the climate where you live. See if you can correlate it with human events at the time the changes were occurring.

Selected Key Terms

annual 29.6	hardwood 29.7	sapwood 29.7
bark 29.7	heartwood 29.7	sclerenchyma 29.2
biennial 29.6	lateral root 29.5	shoot 29.1
bud 29.3	leaf 29.4	softwood 29.7
collenchyma 29.2	magnoliid 29.1	stoma
cork 29.7	meristem 29.1	(stomata) 29.2
cork cambium 29.1	mesophyll 29.4	taproot system 29.5
cortex 29.3	monocot 29.1	vascular bundle 29.3
cuticle (plant) 29.2	parenchyma 29.2	vascular
dermal tissue system 29.1	perennial 29.6	cambium 29.1
epidermis 29.2	periderm 29.7	vascular
eudicot (true dicot) 29.1	phloem 29.2	cylinder 29.5
fibrous root system 29.5	pith 29.3	vascular tissue
ground tissue	primary root 29.5	system 29.1
system 29.1	root 29.1	vein (leaf) 29.4
growth ring 29.7	root hair 29.5	xylem 29.2

Readings

Raven, P., et al. 1999. *Biology of Plants*. Sixth edition. New York: Freeman.

PLANT NUTRITION AND TRANSPORT

Flies for Dinner

How often do we think that plants actually do anything impressive? Being mobile, intelligent, and emotional, we humans tend to be fascinated more with ourselves than with immobile, expressionless plants. Yet plants don't just stand around soaking up sunlight. Consider the Venus flytrap (*Dionaea muscipula*), a flowering plant native to the bogs of North and South Carolina. Its two-lobed, spine-fringed leaves open and close like a steel trap (Figure 30.1*a–d*). Like all other plants, it cannot grow properly without nitrogen and other nutrients, which happen to be scarce in the soil of bogs. However, plenty of insects fly in from places around the bogs.

Sticky sugars ooze from epidermal glands onto the surface of the flytrap's leaf. The sugars entice insects to land. As they do, they brush against hairlike structures that project from the leaf surface. These are triggers for the trap. When an insect touches two hairs at the same time or the same hair twice in rapid succession, the two lobes of the leaf snap shut. Now digestive juices pour out from cells of the leaf. They pool around the insect, dissolve it, and so release nutrients from it. In this way the Venus flytrap makes its own nutrient-rich water, which it proceeds to absorb!

The Venus flytrap is only one of several species of **carnivorous plants**. We call them this even though it takes a leap of the imagination to put their mode of nutrient acquisition—a form of extracellular digestion and absorption—in the same category as the chompings of lions, dogs, and other meat eaters. Besides, not all carnivorous plants actively spring traps. Some types have fluid-filled traps into which prey slip, slide, or fall and then simply drown (Figure 30.2).

All carnivorous plants evolved in habitats where nitrogen and other nutrients are hard to come by. For instance, you may also come across plants with bizarre nutrient-acquiring habits in shallow freshwater lakes and streams, which contain only dilute concentrations of dissolved minerals.

Given the variety and numbers of insects and other animals that attack plants, you can just imagine how endearing the carnivorous plants are to botanists. With their plucky modes of nutrition, these plants also are a fine way to start thinking about **plant physiology**—the study of adaptations by which plants function in their environment. As you already know, nearly all plants are photoautotrophs that use energy from sunlight to drive the synthesis of organic compounds from water, carbon dioxide, and some minerals. Like people, they do not

VENUS FLYTRAP, OPEN FOR DINNER

Figure 30.1 Do plants take nutrition seriously? You bet. (**a**) A Venus flytrap (*Dionaea muscipula*), a carnivorous plant. It makes up for scarce nutrients in its habitat by absorbing them from animals that land on its leaves. (**b**) A fly stuck in sugary goo on a lobed leaf. (**c**) It brushes against hairlike triggers projecting from the leaf; the base of one is shown here. (**d**) An activated leaf snaps shut in just half a second. How? Mesophyll cells right below the epidermis are compressed when the trap is open. Spring the trap and turgor pressure makes the cells decompress abruptly. Whoosh!

base of epidermal hair   epidermal gland

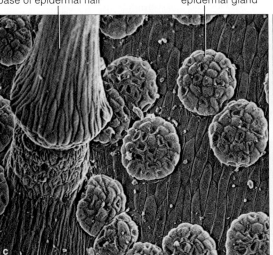

Figure 30.2 Cobra lily (*Darlingtonia californica*). Its leaves form a "pitcher" that is partly filled with digestive juices. Insects lured in by irresistible odors often cannot find the way back out; light shining through the pitcher's patterned dome confuses them. They just wander around and down, adhering to downward-pointing leaf hairs—which are slickened with wax above the potent vat.

have unlimited supplies of the resources necessary to nourish themselves. Of every 1 million molecules of air, only 350 are carbon dioxide. Unlike the soggy habitats of Venus flytraps, most soils are frequently dry. And nowhere except in overfertilized gardens does soil water hold lavish amounts of dissolved minerals. As you continue with these chapters on vascular plants, keep this point in mind: *Many aspects of plant structure and function are responses to low concentrations of vital environmental resources.*

Key Concepts

1. Many aspects of a plant's structure and function are adaptive responses to low concentrations of water, minerals, and other environmental resources.

2. A plant's root system takes up water from soil and also mines the soil for nutrients. For many land plants, mycorrhizae and bacterial symbionts assist in nutrient uptake. In a given habitat, soil properties affect water and nutrient availability for plants.

3. A plant's cuticle and its many stomata function in the conservation of water, a scarce resource in most habitats on land. Stomata are passageways across the epidermis of leaves and, to a lesser extent, stems. When open, stomata permit gas exchange. When closed, they help control water loss.

4. Stomata open during the day, when photosynthesis proceeds. Carbon dioxide diffuses into leaves, oxygen diffuses out—and water loss is rapid. Most plant species conserve water by closing stomata at night.

5. In flowering plants and other vascular plants, the flow of water and solutes through xylem and phloem functionally connects all living cells of roots, stems, and leaves. Xylem serves in the uptake and distribution of water and dissolved mineral ions. Phloem functions in distributing photosynthetically produced sugars and other organic compounds.

6. Water absorbed from soil moves on up through xylem and into leaves. By the process of transpiration, dry air around leaves promotes evaporation through stomata. The force of evaporation pulls continuous columns of water molecules that are hydrogen-bonded to one another from roots to aboveground parts.

7. By the energy-requiring process of translocation, sucrose and other organic compounds are distributed throughout the plant. Organic compounds produced by photosynthetic cells in leaves are loaded into conducting cells of phloem. They are unloaded at the plant's actively growing regions or at storage regions.

PLANT NUTRIENTS AND THEIR AVAILABILITY IN SOILS

Nutrients Required for Plant Growth

So far, we've mentioned nutrients in passing. But exactly what are they? A **nutrient** is any element essential for an organism because no other element can indirectly or directly fulfill its metabolic role. Essential elements for plants are the oxygen, carbon, and hydrogen required for photosynthesis. Plants also need least thirteen other elements (Table 30.1). These usually are dissolved in soil water in ionic forms that reversibly bind with clay. Ions of calcium (Ca^{++}) and potassium (K^+) are examples. Plants easily exchange hydrogen ions for these weakly bound elements.

Nine essential elements are *macro*nutrients. Normally they are required in amounts above 0.5 percent of the plant's dry weight (weighed after all the water has been removed from the plant). The other elements listed are *micro*nutrients; they make up traces (usually a few parts per million) of the dry weight. Even the trace amounts obtained from soil are essential for normal growth.

Properties of Soil

Soil consists of mineral particles mixed with variable amounts of decomposing organic material, or **humus**. Weathering of hard rocks yields these minerals. Dead organisms and organic litter (fallen leaves, feces, and

so on) make up the humus. Water and oxygen occupy spaces between the particles and organic bits.

Soils differ in the proportions of mineral particles and how much these are compacted. The three main particle sizes are sand, silt, and clay. Letting beach sand dribble between your fingers can give you a sense that sand particles are large (0.05 to 2 millimeters across). Rubbing silt between your fingers won't tell you much. You won't be able to distinguish among the individual particles, which are only about 0.002 to 0.05 millimeter across. Clay particles are the finest of all.

How suitable is a given soil for plant growth? Is it gummy when wet because it does not have enough air spaces? Does it form hard clods when dry? The answer depends partly on the relative proportions of sand, silt, and clay. The more clay, the finer the soil's texture.

Each clay particle consists of thin, stacked layers of aluminosilicates with negatively charged ions at their surfaces. Clay attracts and binds (adsorbs) positively charged mineral ions dissolved in water that trickles through soil as well as the water molecules themselves. Ions and water cling reversibly to clay. This chemical behavior is vital for all plants. With its high adsorption capacity, clay latches on to many nutrients for plants even as water percolates on past and drains away.

Too much clay is bad for plants. Tightly packed clay particles exclude air spaces, so root cells are deprived

Table 30.1 *Essential Elements and Plant Functioning*

MACRONUTRIENT	Some Functions	Some Deficiency Symptoms	MICRONUTRIENT	Some Functions	Some Deficiency Symptoms
Carbon Hydrogen Oxygen	Raw materials for second stage of photosynthesis	None; all are abundantly available (water, carbon dioxide are sources)	Chlorine	Role in root and shoot growth and photolysis	Wilting; chlorosis; some leaves die
Nitrogen	Protein, nucleic acid, coenzyme, chlorophyll component	Stunted growth; light-green older leaves; older leaves yellow and die (these are symptoms that define a condition called chlorosis)	Iron	Roles in chlorophyll synthesis and in electron transport	Chlorosis; yellow and green striping in leaves of grass species
Potassium	Activation of enzymes; contributes to water–solute balances that influence osmosis*	Reduced growth; curled, mottled, or spotted older leaves; burned leaf edges; weakened plant	Boron	Roles in germination, flowering, fruiting, cell division, nitrogen metabolism	Terminal buds, lateral branches die; leaves thicken, curl, become brittle
Calcium	Component in control of many cell functions; cementing of cell walls	Terminal buds wither, die; deformed leaves; poor root growth	Manganese	Chlorophyll synthesis; coenzyme action	Dark veins, but leaves whiten and fall off
Magnesium	Chlorophyll component; activation of enzymes	Chlorosis; drooped leaves	Zinc	Role in forming auxin, chloroplasts, starch; enzyme component	Chlorosis; mottled or bronzed leaves; root abnormalities
Phosphorus	Phospholipid, nucleic acid, ATP component	Purplish veins; stunted growth; fewer seeds, fruits	Copper	Component of several enzymes	Chlorosis; dead spots in leaves; stunted growth
Sulfur	Component of most proteins, two vitamins	Light-green or yellowed leaves; reduced growth	Molybdenum	Part of enzyme used in nitrogen metabolism	Pale green, rolled or cupped leaves

* All mineral elements contribute to water–solute balances; potassium is notable because there is so much of it.

Figure 30.3 (**a**) Some of the soil horizons that developed in one habitat in Africa. (**b**) Profile of a heavily leached soil. Such soils are common in cool, moist coniferous forests. Breakdown of pine needle litter makes the soil water highly acidic, so nutrients are easily leached from surface layers. Acid-resistant materials such as quartz remain and give the layers closest to the surface an ash-gray appearance. Iron and aluminum oxides stain deeper layers.

(**c**) Erosion forms gullies that channel runoff from the land. As gullies deepen and widen, erosion becomes more rapid. When topsoil is depleted, productivity declines, and fertilizers typically are trucked in to replace lost nutrients.

O HORIZON
Fallen leaves and other organic material littering the surface of mineral soil

A HORIZON
Topsoil, which contains some percentage of decomposed organic material and which is variably deep (only a few centimeters deep in deserts, but elsewhere extending as far as thirty centimeters below the soil surface)

B HORIZON
Compared with the A horizon, larger soil particles, not much organic material, but greater accumulation of minerals; extends thirty to sixty centimeters below soil surface

C HORIZON
No organic material, but partially weathered fragments and grains of rock from which soil forms; extends to underlying bedrock

BEDROCK

of oxygen for aerobic respiration. Packing also retards water penetration into the soil. Runoff, and nutrient loss, is severe in heavy clay soils. The best soils are **loams**, which have roughly the same proportions of sand, silt, and clay.

Humus in soil also affects plant growth. Generally, humus has abundant negatively charged organic acids. It weakly binds and retains dissolved mineral ions of opposite charge. Humus also has a high capacity to absorb and swell with water, then shrink as water is gradually released. Its alternating swelling and shrinking aerates the soil. As decomposers gradually work it over, humus releases nutrients that plants can take up.

In general, the soils that contain 10 to 20 percent humus are most favorable for plant growth. The worst soils of all are less than 10 percent humus or more than 90 percent humus, the latter being a feature of swamps and bogs.

Soils are classified by *profile* properties, their layered characteristics. Soils in different places are in different stages of development. Figure 30.3 has two examples. **Topsoil**, the uppermost soil layer, is the A horizon. This is the most essential layer for plant growth, and its depth is variable from one habitat to the next.

Leaching and Erosion

Leaching is the removal of nutrients from soil as water percolates through it. It is heaviest in sandy soils, which are not as good as clay at binding nutrients. **Erosion** is a movement of land under the force of wind, running water, and ice (Figure 30.3*b,c*). For example, erosion from all the farmlands drained by the Mississippi River puts about 25 billion metric tons of topsoil into the Gulf

of Mexico each year. Whether by leaching or erosion, the loss of nutrients from soil is bad for plants and for all organisms that depend on plants for survival.

Nutrients are essential elements. No other element can substitute for their direct or indirect roles in the metabolic activities that sustain growth and keep organisms alive.

The mineral component of soil includes particles ranging from large-grained sand to silt and fine-grained clay. These particles, clay especially, reversibly bind water molecules and dissolved mineral ions and thereby make them more accessible for uptake by plant roots.

Soil also contains humus, which is a reservoir of organic material, rich in organic acids, in different stages of decay. Most plants grow best in soils having equal proportions of sand, silt, and clay, as well as 10 to 20 percent humus.

HOW DO ROOTS ABSORB WATER AND MINERAL IONS?

In terms of energy outlays, mining the soil for mineral ions and water molecules clinging to clay particles is an expensive task. Plants spend considerable energy on building extensive root systems. Wherever the soil's texture and composition change, new roots must form to replace old ones and infiltrate different regions. It isn't that roots "explore" soil for resources. Rather, the patches of soil where the concentrations of water and mineral ions are greater stimulate outward growth.

Absorption Routes

Think back on the preceding chapter's discussion of a typical root's structure (Section 29.5). Water molecules in soil are only weakly bound to clay particles, so they readily cross the root epidermis and continue into the **vascular cylinder**, a column of vascular tissue. There, a cylindrical layer of cells, an **endodermis**, wraps around the column. A band of waxy deposits—the **Casparian strip**—is embedded in abutting endodermal cell walls (Figure 30.4). Water molecules can't penetrate it. They infiltrate only the unwaxed wall regions, pass through cells, then cross unwaxed wall regions on the opposite side. This is the only way water and solutes move into the vascular cylinder. Like all cells, endodermal cells have many transport proteins embedded in the plasma membrane. The proteins let some solutes but not others cross it (Section 5.7). *The transport proteins of endodermal cells are control points where plants adjust the quantity and types of solutes absorbed from soil water.*

Roots of many plants also have an **exodermis**, a cell layer just beneath their surface (Figure 30.4a). Walls of exodermal cells commonly have a Casparian strip that functions like the one next to a root vascular cylinder.

Specialized Absorptive Structures

ROOT HAIRS Vascular plants require great amounts of water. Roots of a mature corn plant absorb more than three liters of water daily. They could not do so without **root hairs**. Recall, from the preceding chapter, that root hairs are slender extensions of specialized epidermal cells. They greatly increase the surface area available for absorption (Section 29.5 and Figure 30.5). When a plant is putting on primary growth, its system of roots may develop millions or billions of root hairs.

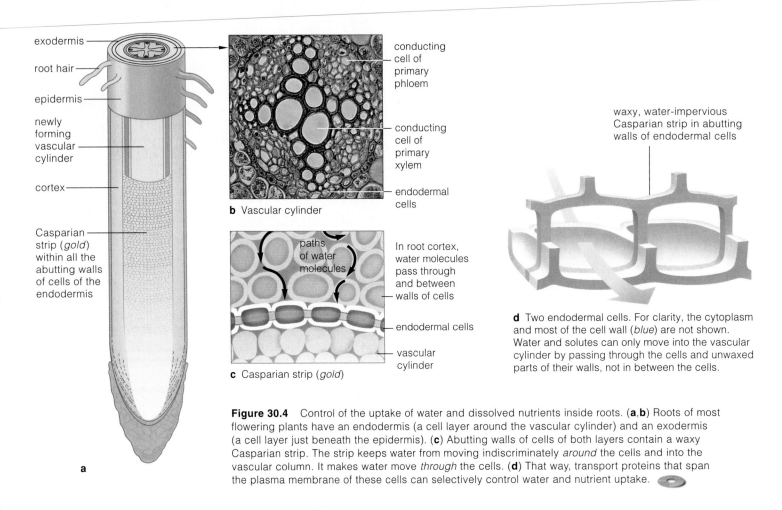

b Vascular cylinder

c Casparian strip (*gold*)

d Two endodermal cells. For clarity, the cytoplasm and most of the cell wall (*blue*) are not shown. Water and solutes can only move into the vascular cylinder by passing through the cells and unwaxed parts of their walls, not in between the cells.

Figure 30.4 Control of the uptake of water and dissolved nutrients inside roots. (**a,b**) Roots of most flowering plants have an endodermis (a cell layer around the vascular cylinder) and an exodermis (a cell layer just beneath the epidermis). (**c**) Abutting walls of cells of both layers contain a waxy Casparian strip. The strip keeps water from moving indiscriminately *around* the cells and into the vascular column. It makes water move *through* the cells. (**d**) That way, transport proteins that span the plasma membrane of these cells can selectively control water and nutrient uptake.

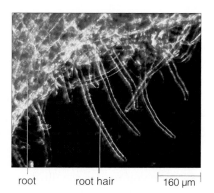

Figure 30.5 An example of root hairs. These thin extensions of the young root's epidermal cells specialize in absorption of both water and dissolved ions.

root   root hair |—— 160 μm ——|

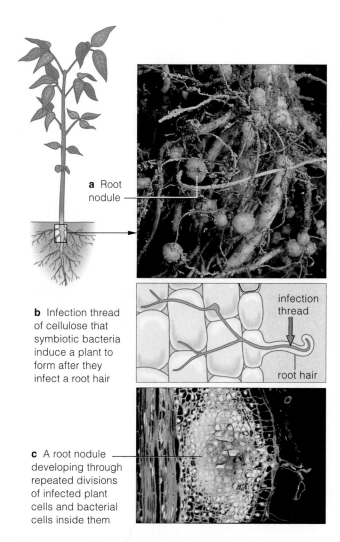

a Root nodule

b Infection thread of cellulose that symbiotic bacteria induce a plant to form after they infect a root hair

infection thread

root hair

c A root nodule developing through repeated divisions of infected plant cells and bacterial cells inside them

Figure 30.6 (a) Nutrient uptake at root nodules of legumes that are mutualists with nitrogen-fixing bacteria (*Rhizobium* and *Bradyrhizobium*). When infected by the bacteria, root hair cells form a thread of cellulose deposits. Bacteria use the thread as a highway to invade plant cells in the root cortex.

(**b,c**) Infected plant cells and bacterial cells inside them divide repeatedly. Together they form a swollen mass that becomes a root nodule. The bacteria start fixing nitrogen when membranes of plant cells surround them. The plant takes up some of the nitrogen; the bacteria take up some photosynthetic compounds.

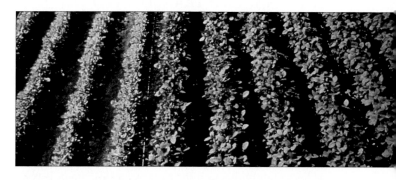

Figure 30.7 Demonstration of the effect of root nodules on plant growth. At *left*, rows of soybean plants growing in nitrogen-poor soil. At *right*, growing in the same soil, plants in these rows were inoculated with *Rhizobium* cells and developed root nodules.

ROOT NODULES Certain bacteria and fungi help many plants absorb dissolved nutrients and get something in return. **Mutualism** is the name for such a two-way flow of benefits between species (Chapter 24). Think about how nitrogen deficiency limits growth of plants. Gaseous nitrogen ($N\equiv N$, or N_2) is plentiful in air. But plants don't have the metabolic machinery for **nitrogen fixation**. By this process, enzymes split all N_2 covalent bonds and attach both atoms to organic compounds. To obtain high crop yields, farmers apply nitrogen-rich fertilizers or encourage the growth of nitrogen-fixing bacteria in the soil. These bacteria convert N_2 to forms that they can utilize. In this case, string beans, peas, alfalfa, clover, and other legumes have an advantage. Nitrogen-fixing bacteria are symbionts in their roots, in localized swellings called **root nodules** (Figures 30.6 and 30.7). Bacterial cells withdraw some of the organic compounds, originally produced by photosynthesis in the leaves. In return, the plants absorb some nitrogen that the bacterial cells secured.

MYCORRHIZAE Also think back on the **mycorrhizae** (singular, mycorrhiza). As described in Chapter 24, a mycorrhiza is a symbiotic interaction between a young root and a fungus. Hence the name, meaning "fungus-root." Fungal filaments—hyphae—form a velvety cover around roots or penetrate the root cells. Collectively, hyphae have a large surface area that absorbs mineral ions from a larger volume of soil than the roots can do. The fungus can absorb sugars and nitrogen-containing compounds from root cells. The root cells obtain some scarce minerals that the fungus is better able to absorb.

Gymnosperms and flowering plant roots control the uptake of water and dissolved nutrients at the vascular cylinder's endodermis and at a similar layer near the root surface.

Root hairs, root nodules, and mycorrhizae greatly enhance the uptake of water and dissolved nutrients.

HOW IS WATER TRANSPORTED THROUGH PLANTS?

Transpiration Defined

By now, you have a sense of how the distribution of water and dissolved mineral ions to all living cells is central to plant growth and functioning. Let's turn now to a model for how water actually moves from a plant's roots to its stems, then into leaves.

Plants, recall, use only a fraction of the water they absorb for growth and metabolism. Most of that water is lost, mainly through the numerous stomata in leaves. Evaporation of water molecules from leaves, stems, and other plant parts is a process called **transpiration**.

Cohesion–Tension Theory of Water Transport

This brings up an interesting question. Assuming that plants lose most of the absorbed water from their leaves, how does water actually get *to* the leaves? What gets it to the uppermost leaves of plants, including redwoods and other trees that may be more than 100 meters tall?

In a plant's vascular tissues, water moves through a complex tissue called **xylem**. Recall, from Section 29.2, that the water-conducting cells of xylem are **tracheids** and **vessel members**. Figure 30.8 provides examples. These cells are dead at maturity, and only their lignin-impregnated walls are left. This means the conducting cells in xylem can't be actively pulling water "uphill."

Some time ago, the botanist Henry Dixon came up with a useful way to explain water transport in plants. By his **cohesion–tension theory**, the water inside xylem is pulled upward by air's drying power, which creates continuous negative pressures—that is, tensions. These tensions extend all the way from leaves to roots. Figure 30.9 illustrates Dixon's theory. Think about these three points as you review the illustration:

First, air's drying power causes transpiration: the evaporation of water from all plant parts exposed to air—but most notably at stomata. Transpiration puts the water confined in xylem's waterproofed conducting tubes into a state of tension. And that tension extends from veins inside leaves, down through the stems, and on into young roots where water is being absorbed.

Second, the unbroken, fluid columns of water show *cohesion*; they resist rupturing while they are pulled up under *tension*. (Here you may wish to reflect on Section 2.5.) The collective strength of all the hydrogen bonds between water molecules in the narrow, tubular xylem cells imparts this cohesion.

Third, for as long as molecules of water continue to escape from a plant, the continuous tension inside the

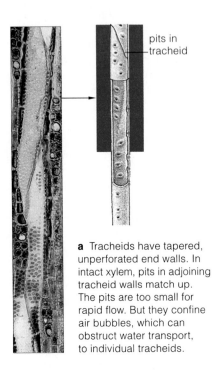

a Tracheids have tapered, unperforated end walls. In intact xylem, pits in adjoining tracheid walls match up. The pits are too small for rapid flow. But they confine air bubbles, which can obstruct water transport, to individual tracheids.

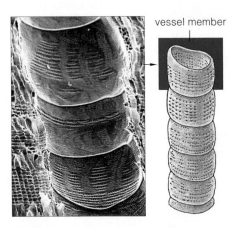

b Close-up of three of the adjoining members that make up a vessel. The thick, finely perforated walls of these dead cells connect one after another to form vessels, another type of water-conducting tube in xylem. The walls of all these tubes contain an abundance of lignin, which strengthens them and also makes them waterproof.

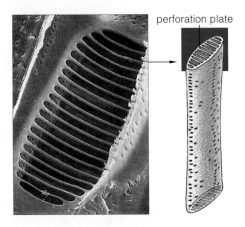

c Perforation plate at the end wall of one type of vessel member. The perforated ends permit water and air bubbles to flow unimpeded through the conducting tube. This might be one reason why natural selection favored the retention of tracheids and vessel members in the same plants.

Figure 30.8 Scanning electron micrographs and sketches of a few types of tracheids and vessel members from xylem. These water-conducting tubes are made of the walls of cells, which are dead at maturity. The cell walls still remain interconnected, and so form the tubes. Tracheids probably evolved before vessel members. Both occur in nearly all vascular plants.

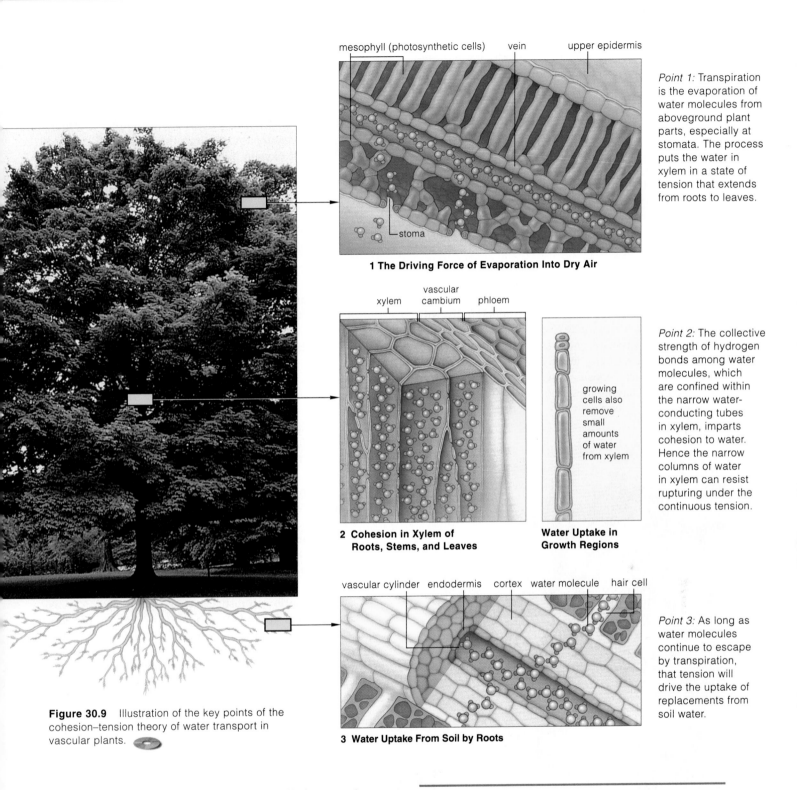

mesophyll (photosynthetic cells)   vein   upper epidermis

stoma

1 The Driving Force of Evaporation Into Dry Air

Point 1: Transpiration is the evaporation of water molecules from aboveground plant parts, especially at stomata. The process puts the water in xylem in a state of tension that extends from roots to leaves.

xylem   vascular cambium   phloem

growing cells also remove small amounts of water from xylem

2 Cohesion in Xylem of Roots, Stems, and Leaves

Water Uptake in Growth Regions

Point 2: The collective strength of hydrogen bonds among water molecules, which are confined within the narrow water-conducting tubes in xylem, imparts cohesion to water. Hence the narrow columns of water in xylem can resist rupturing under the continuous tension.

vascular cylinder   endodermis   cortex   water molecule   hair cell

3 Water Uptake From Soil by Roots

Point 3: As long as water molecules continue to escape by transpiration, that tension will drive the uptake of replacements from soil water.

Figure 30.9 Illustration of the key points of the cohesion–tension theory of water transport in vascular plants.

xylem permits more molecules to be pulled upward from the roots, and therefore to replace them.

Hydrogen bonds are strong enough to hold water molecules together inside the xylem. However, they are not strong enough to prevent the water molecules from breaking away from one another during transpiration and then escaping from leaves, through stomata.

Transpiration is the process of evaporation from plant parts.

By the cohesion–tension theory of water transport, this process is the key source of tensions in water in xylem. The tensions extend from leaves to roots, and they allow columns of water molecules that are hydrogen-bonded to one another to be pulled upward through the plant body.

HOW DO STEMS AND LEAVES CONSERVE WATER?

At least 90 percent of the water transported from roots to a leaf goes right out by evaporating into the air. Cells use only about 2 percent for photosynthesis, membrane functions, and other activities, but that amount must be maintained. So how do plants know how much water they have? They sense **turgor pressure**—the pressure against a cell wall that arises from the movement of water into that cell. When a plant's soft parts are erect, you know that as much water is moving into its cells as is moving out. When the cells lose water, the plant wilts; its soft parts droop and water-dependent events are disrupted (Sections 5.7 and 28.2). Yet plants are not entirely at the mercy of changes in water availability. They have a cuticle, and they have stomata.

The Water-Conserving Cuticle

Even mildly water-stressed plants would rapidly wilt and die without their **cuticle** (Figure 30.10). Epidermal cells secrete this translucent, water-impermeable layer, which coats cell wall regions exposed to the air. At the cuticle surface are deposits of waxes, which are water-insoluble lipids having long fatty-acid tails. The cuticle itself consists of waxes embedded in **cutin**, an insoluble lipid polymer. Beneath all the waxes, cellulose fibers weave through the cutin. A layer of polysaccharides (pectins) often helps bind the cuticle to the cell walls.

A cuticle does not bar the passage of light rays into photosynthetic parts of the plant. It does restrict water loss. It also restricts the *inward* diffusion of the carbon dioxide necessary for photosynthesis and the *outward* diffusion of oxygen by-products. Recall, from Section 7.7, that a buildup of oxygen in the air spaces inside a leaf has bad effects on the rate of photosynthesis.

Controlled Water Loss at Stomata

How do carbon dioxide and oxygen get past the cuticle-covered, water-conserving epidermis? They cross it in controlled fashion at **stomata** (singular, stoma, a Greek word meaning mouth). Gas exchange and evaporation of water occur mainly at stomata.

Stomata usually are open in the day to help support photosynthesis. Water is lost, but given sufficient soil moisture, roots replace it. At night, stomata are closed. Water is conserved, and carbon dioxide accumulates in leaves as cells engage in aerobic respiration. Like you, plants depend on this energy-releasing pathway to get enough ATP to drive most of their metabolic reactions.

Stomata close mainly in response to water loss. Each stoma is defined by a pair of specialized parenchyma cells called **guard cells** (Figure 30.11). When the pair swell with water, the turgor pressure makes both bend, so a gap (stoma) forms between them. When they lose water and turgor pressure drops, the two cells collapse against each other, so the gap closes.

When a plant is water stressed, its stomata close in response to a signal from the hormone called abscisic acid (ABA). Remember the Section 28.5 introduction to signal reception, transduction, and response? ABA is such a signal. It binds to receptors on each guard cell's plasma membrane. Some experiments indicate that the binding causes the opening of gated channels across the membrane. Calcium ions (Ca^{++}) flow into the cells, where they may act as a second messenger. They cause other channels to open that let chloride ions (Cl^-) and a negatively charged organic compound, malate, flow rapidly from the cytoplasm to the extracellular matrix.

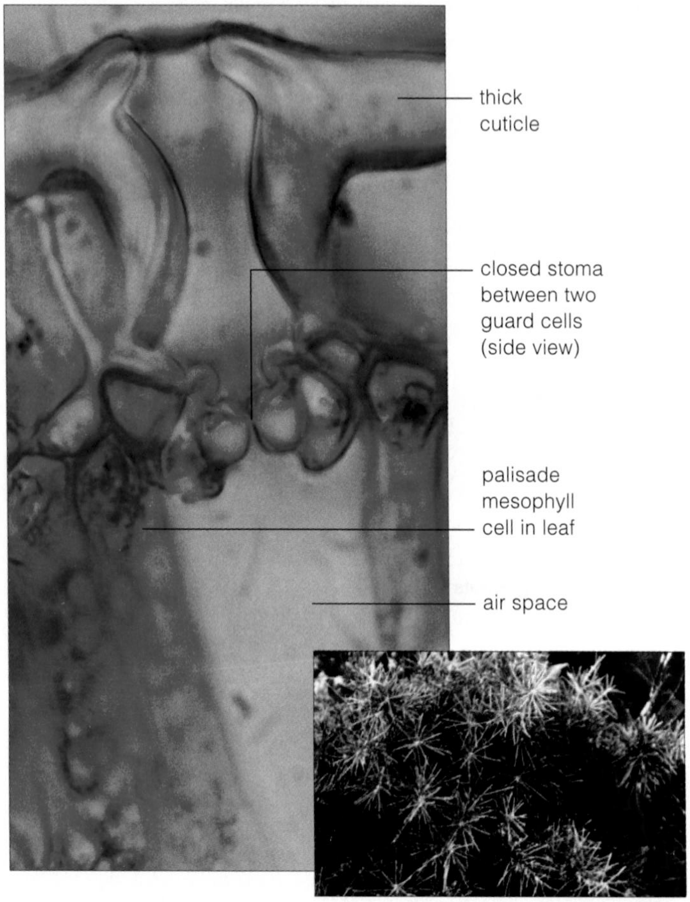

thick cuticle

closed stoma between two guard cells (side view)

palisade mesophyll cell in leaf

air space

Figure 30.10 From Australia, cuticle of the circular, needle-like leaves of *Hakea gibbosa*, known as the rock needlebush, longitudinal section. Water, carbon dioxide, and oxygen cross the cuticle at stomata. Like other plants in seasonally dry habitats, it has a very thick cuticle that helps restrict water loss. (Aquatic plants have thin cuticles or none at all.)

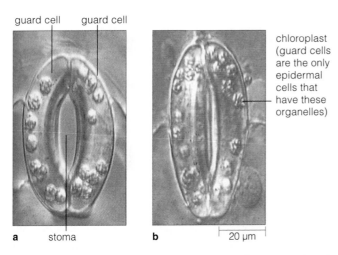

guard cell   guard cell

chloroplast (guard cells are the only epidermal cells that have these organelles)

a   stoma   b   20 μm

Figure 30.11 Stomatal action. Whether a stoma is open or closed at any given time depends on the shape of two guard cells that define this small opening across a cuticle-covered leaf epidermis. (**a**) This stoma is open. Water entered collapsed guard cells, which swelled under turgor pressure and moved apart, thus forming the stoma. (**b**) This stoma is closed. Water left the swollen guard cells, which collapsed against each other and closed the stoma.

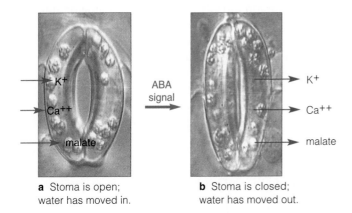

K^+ ABA signal K^+

Ca^{++} Ca^{++}

malate   malate

a Stoma is open;   **b** Stoma is closed;
water has moved in.   water has moved out.

Figure 30.12 Hormonal control of stomatal closure. (**a**) When a stoma is open, high solute concentrations in the cytoplasm of both guard cells have raised the turgor pressure, keeping the cells plumped up. (**b**) In a water-stressed plant, the hormone abscisic acid binds to receptors on the guard cell plasma membrane. It activates a signal transduction pathway that lowers solute concentrations inside the cells, which lowers the turgor pressure and closes the stoma.

The flow creates an electric and concentration gradient for potassium ions (K^+) across the membrane.

With the loss of all these ions, the cytoplasm in the guard cells has a higher water potential (Section 5.7). Now water follows its gradient and moves out of the cells. The turgor pressure falls, and so the stoma closes. Experiments indicate that the stoma opens when ions move in reverse, into the guard cells (Figure 30.12).

Environmental signals also affect whether stomata are opened or closed. They include the concentration of carbon dioxide inside leaves, the incoming light, and

Figure 30.13 (**a**) Stomata at holly leaf surface, set against a backdrop of smog in Central Europe. (**b**) What they look like when a holly plant is growing in industrialized regions. Gritty airborne pollutants clog its stomata and prevent much of the sun's rays from reaching photosynthetic cells in the leaf.

temperature. For example, photosynthesis starts after the sun comes up. As the morning progresses, carbon dioxide levels fall in photosynthetic cells, including the guard cells (Figure 30.12a). The decline helps trigger the active transport of K^+ into these cells. So do blue wavelengths of light, which penetrate the atmosphere better as the sun arcs higher in the sky. Water follows potassium ions into the cells. The inward movement increases the turgor pressure and the stoma opens.

Carbon dioxide levels in the cells rise when the sun sets and photosynthetic activity ceases. Potassium, then water, move out of guard cells. And so stomata close.

CAM plants, including most cacti, conserve water differently. They open stomata at night, when they fix carbon dioxide by the metabolic C4 pathway described in Section 7.7. The next day, when their stomata close, CAM plants use carbon dioxide in photosynthesis.

As this section makes clear, plant survival depends on stomatal function. Think about it when you are out and about on smog-shrouded days (Figure 30.13).

Water-dependent events in plants are severely disrupted when water loss exceeds uptake at roots for extended periods. Wilting is one observable outcome.

Transpiration and gas exchange occur mainly at stomata. These numerous small openings span the waxy cuticle, which covers all plant epidermal surfaces exposed to air.

Plants open and close stomata at different times to control water loss, carbon dioxide uptake, and oxygen disposal, all of which affect rates of photosynthesis and plant growth.

HOW ARE ORGANIC COMPOUNDS DISTRIBUTED THROUGH PLANTS?

Whereas xylem distributes water and minerals through the plant, the vascular tissue called **phloem** distributes organic products of photosynthesis. Like xylem, phloem consists of many conducting tubes, fibers, and strands of parenchyma cells. Remember Sections 29.3 and 29.6? They showed phloem's distribution patterns in dicots and monocots. Unlike xylem, phloem has **sieve tubes** through which organic compounds flow swiftly. These tubes consist of *living* cells called sieve-tube members. The cells are positioned side by side and end to end in vascular tissues. Their abutting end walls, called sieve plates, have many pores (Figure 30.14*a*).

Adjoining the long sieve tubes are **companion cells** (Figure 30.14*b*). Like the sieve-tube members, these are differentiated parenchyma cells that help load organic compounds into sieve tubes.

Let's see what happens to sucrose and other organic products of photosynthesis in phloem. Leaf cells use some of the products for their own activities. The rest travels to roots, stems, buds, flowers, and fruits. In most cells, carbohydrates are stored as starch in plastids. Proteins and fats, built from carbohydrates and amino acids distributed to living cells, often are stored in seeds. Avocados and some other fruits also accumulate fats.

Starch molecules are too large for transport across the plasma membrane of the cells that make them and are too insoluble for transport to other regions of the

Figure 30.15 A honeydew droplet that is exuding from the end of an aphid gut. The tubular mouthpart of this small insect penetrated a conducting tube in phloem. The sugary fluid was under high pressure in phloem and was forced out through the gut's terminal opening.

plant. Overall, proteins are too large and fats are too insoluble for transport from storage sites. *But the cells can convert storage forms of organic compounds to smaller solutes that are more easily transported by the phloem.* For instance, the cells degrade starch to glucose monomers. When one of these monomers combines with fructose, the result is sucrose—an easily transportable sugar.

Simple experiments with aphids show that sucrose is the main carbohydrate transported inside phloem. These insects were anesthetized by exposure to high levels of carbon dioxide as they were feeding on juices in conducting tubes of phloem. The body of the aphids was detached from their mouthparts—which were still embedded in the sieve tubes. For most of the plants studied, sucrose was the most abundant carbohydrate in the fluid that was being forced out of the tubes.

Translocation

Translocation is the technical name for the transport of sucrose and all other organic compounds through the phloem of a vascular plant. High pressure drives this process. Often the pressure is five times higher than in automobile tires. Aphids demonstrate the magnitude of it when they force their mouthparts into sieve tubes and feed upon the dissolved sugars. The high pressure can force fluid through an aphid gut and out the other end, as "honeydew" (Figure 30.15). Park a car under some trees being attacked by aphids, and it might get spattered by sticky honeydew droplets, thanks to the high fluid pressure in phloem.

Pressure Flow Theory

Phloem translocates the photosynthetic products along decreasing pressure and solute concentration gradients. The *source* of the flow is any region of the plant where organic compounds are being loaded into sieve tubes. Common sources are mesophylls—the photosynthetic tissues in leaves. The flow ends at a *sink*, which is any plant region where the products are being unloaded, used, or stored. Example: While flowers and fruits are growing during development, they are sink regions.

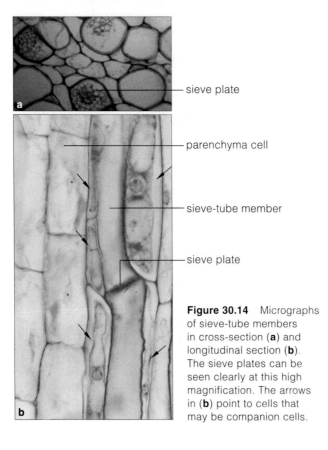

sieve plate

parenchyma cell

sieve-tube member

sieve plate

Figure 30.14 Micrographs of sieve-tube members in cross-section (**a**) and longitudinal section (**b**). The sieve plates can be seen clearly at this high magnification. The arrows in (**b**) point to cells that may be companion cells.

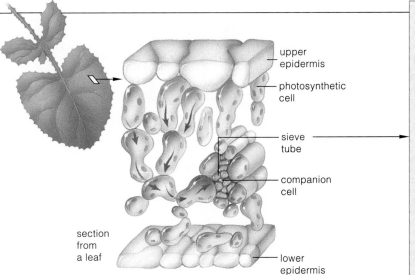

upper
epidermis

photosynthetic
cell

sieve
tube

companion
cell

section
from
a leaf

lower
epidermis

a *Loading at a source.* Photosynthetic cells in leaves are a common source of organic compounds that must be distributed through a plant. Small, soluble forms of these compounds move from the cells into phloem (a leaf vein).

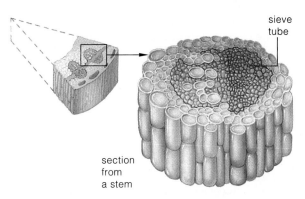

sieve
tube

section
from
a stem

b *Translocation along a distribution path.* Fluid pressure is greatest inside sieve tubes at the source. It pushes the solute-rich fluid to a sink, which is any region where cells are growing or storing food. There, the pressure is lower because cells are withdrawing solutes from the tubes.

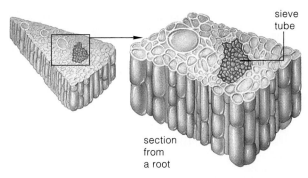

sieve
tube

section
from
a root

c *Unloading at the sink.* Solutes are unloaded from sieve tubes into cells at the sink; water follows. Translocation continues as long as solute concentration gradients and a pressure gradient exist between the source and the sink.

Figure 30.16 Translocation of organic compounds in sow thistle (*Sonchus*). Review Section 7.6 to get an idea of how translocation relates to photosynthesis in vascular plants.

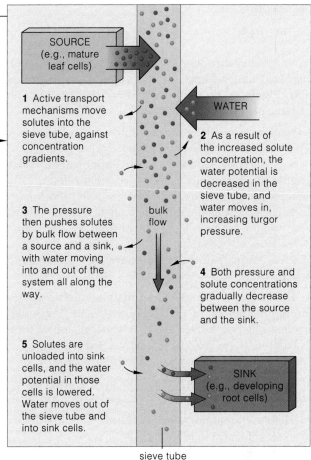

SOURCE
(e.g., mature
leaf cells)

1 Active transport mechanisms move solutes into the sieve tube, against concentration gradients.

WATER

2 As a result of the increased solute concentration, the water potential is decreased in the sieve tube, and water moves in, increasing turgor pressure.

3 The pressure then pushes solutes by bulk flow between a source and a sink, with water moving into and out of the system all along the way.

bulk
flow

4 Both pressure and solute concentrations gradually decrease between the source and the sink.

5 Solutes are unloaded into sink cells, and the water potential in those cells is lowered. Water moves out of the sieve tube and into sink cells.

SINK
(e.g., developing
root cells)

sieve tube

Why do organic compounds flow from a source to a sink? According to the **pressure flow theory**, internal pressure builds up at the source end of the sieve tube system and *pushes* the solute-rich solution on toward any sink, where solutes are being removed.

Experimental evidence for the pressure flow theory was obtained from studies of sow thistle (*Sonchus*). Use Figure 30.16 to track what happens after sucrose moves from photosynthetic cells into small veins in the leaves of this plant. Companion cells in veins spend energy to load sucrose into adjacent sieve-tube members. As the sucrose concentration increases in the tubes, water also moves in, by osmosis. More internal fluid pressure is exerted against the sieve-tube walls. When the pressure increases, it pushes the sucrose-laden fluid out of the leaf, into the stem, and on to the sink.

Plants store organic compounds in the form of starch, fats, and proteins. They convert the storage forms to sucrose and other small units that are soluble and easily translocated.

Translocation is the distribution of organic compounds to different plant regions. It depends on concentration and pressure gradients in the sieve-tube system of phloem.

Gradients exist as long as companion cells load compounds into the sieve tubes at sources, such as mature leaves, and as long as compounds are removed at sinks, such as roots.

SUMMARY　　　　　　　*Gold* indicates text section

1. Vascular plants depend on the distribution of water, dissolved mineral ions, and organic compounds to its cells. Figure 30.17 summarizes how that distribution sustains plant growth and survival. *CI, 30.1–30.3*

2. Root systems efficiently take up water and nutrients, which commonly are scarce in soil. *30.1, 30.2*

   a. Collectively, a vascular plant's root hairs (slender extensions of specialized root epidermal cells) greatly increase the surface area available for absorption.

   b. Bacterial and fungal symbionts help many plants take up mineral ions. They benefit from the interaction by using some products of photosynthesis. Examples of these mutualistic interactions are root nodules and mycorrhizae.

3. Plants distribute water and dissolved mineral ions through water-conducting tubes of xylem, a vascular tissue. Cells called tracheids and vessel members form the tubes but are dead at maturity. Their waterproofed walls interconnect as narrow pipelines. *30.3*

4. Plants lose water by transpiration, or evaporation of water from leaves and other parts exposed to air. *30.3*

5. Here are the points of the cohesion–tension theory of water transport in plants: *30.3*

   a. Inside xylem's water-conducting cells, continuous negative pressures (tensions) extend from leaves to the roots. Transpiration causes the tension.

   b. When water molecules escape from the leaves, replacements are pulled into the leaf under tension.

   c. The collective strength of numerous hydrogen bonds between water molecules imparts cohesion that allows water molecules to be pulled up as continuous fluid columns. The "cohesion" part of the theory refers to the capacity of the hydrogen bonds between water molecules to resist rupturing when put under tension.

6. Most plants have a waxy, water-impermeable cuticle covering their aboveground parts. They lose water and take up carbon dioxide at stomata. A pair of guard cells defines each of these microscopically small openings across the epidermis of leaves and stems. Guard cells are specialized parenchyma cells. *30.4*

7. Stomata open and close at different times. Control of their action balances water conservation with carbon dioxide uptake and the release of oxygen. *30.4*

   a. In many kinds of plants, stomata open during the day. Such plants lose water but take in carbon dioxide for photosynthesis. Stomata close at night; thus, water and the carbon dioxide released by aerobically respiring cells are conserved for the next day.

   b. CAM plants open stomata and fix carbon dioxide by the C4 pathway at night. During the day, they close their stomata and use the carbon fixed the night before for photosynthesis.

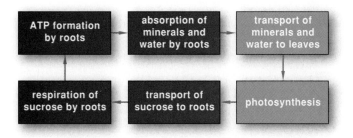

Figure 30.17 Summary of the interdependent processes that sustain the growth of vascular plants. All living cells in plants require oxygen, carbon, hydrogen, and at least thirteen mineral ions. They all produce ATP, which drives metabolic activities.

8. Plants distribute organic compounds through sieve tubes of phloem, a vascular tissue. Translocation is the distribution of organic compounds in phloem. *30.5*

9. According to the pressure flow theory, translocation is driven by differences in solute concentrations and pressure between source and sink regions. A source is any site where organic compounds are being loaded into sieve tubes (e.g., mature leaves). A sink is any site where compounds are being unloaded from sieve tubes. Roots are examples. Solute concentration gradients and pressure gradients exist for as long as companion cells expend energy to load solutes into the sieve tubes and for as long as solutes are removed at the sink. *30.5*

Review Questions

1. Define soil, then distinguish between: *30.1*
   a. humus and loam
   b. leaching and erosion
   c. macronutrient and micronutrient (for plants)

2. Define nutrient. What are signs that a plant is deficient in one of the essential nutrients listed in Table 30.1? *30.1*

3. What is the function of the Casparian strip in roots? *30.2*

4. Using Dixon's model, explain how water moves from soil upward through tall plants. *30.3*

5. Describe the structure and function of a plant cuticle. *30.4*

6. Which type of ion influences stomatal action? *30.4*

7. Explain translocation according to the pressure flow theory described in this chapter. *30.5*

Self-Quiz　　ANSWERS IN APPENDIX III

1. Carbon, hydrogen, oxygen, nitrogen, and potassium are examples of _____ for plants.
   a. macronutrients　　　d. essential elements
   b. micronutrients　　　e. both a and d
   c. trace elements

2. A _____ strip in endodermal cell walls forces water and solutes to move through root cells, not around them.
   a. cutin　　b. lignin　　c. Casparian　　d. cellulose

3. The nutrition of some plants depends on a root–fungus association known as a _____ .
   a. root nodule　　　c. root hair
   b. mycorrhiza　　　d. root hypha

4. The nutrition of some plants depends on a root–bacterium association known as a _____ .
   a. root nodule   c. root hair
   b. mycorrhiza   d. root hypha

5. Water can be pulled up through a plant by the cumulative strength of _____ between water molecules.

6. Water evaporation from plant parts is called _____ .
   a. translocation   c. transpiration
   b. expiration   d. tension

7. Water transport from roots to leaves is explained by _____ .
   a. the pressure flow theory
   b. differences in source and sink solute concentrations
   c. the pumping force of xylem vessels
   d. the cohesion–tension theory

8. In daytime, most plants lose _____ and take up _____ .
   a. water; carbon dioxide   c. oxygen; water
   b. water; oxygen   d. carbon dioxide; water

9. At night, most plants conserve _____ , and _____ accumulates.
   a. carbon dioxide; oxygen c. oxygen; water
   b. water; oxygen   d. water; carbon dioxide

10. In phloem, organic compounds flow through _____ .
   a. collenchyma cells   c. vessels
   b. sieve tubes   d. tracheids

11. Match the concepts of plant nutrition and transport.
   ____ stomata   a. evaporation from plant parts
   ____ nutrient   b. response to scarce soil nutrients
   ____ sink   c. balancing water loss with
   ____ root system   carbon dioxide requirements
   ____ hydrogen   d. cohesion in water transport
        bonds   e. sugars unloaded from sieve tubes
   ____ transpiration f. organic compounds distributed
   ____ translocation   through the plant body
                  g. element with roles in metabolism
                  that no other element can fulfill
                  for an organism

Critical Thinking

1. Home gardeners, like farmers, must ensure that their plants have access to nitrogen from either nitrogen-fixing bacteria or fertilizer. Insufficient nitrogen stunts plant growth; leaves turn yellow and die. Which major classes of biological molecules incorporate nitrogen? How would a low nitrogen level in plants affect biosynthesis and cause symptoms of nitrogen deficiency?

2. When moving a plant from one place to another, it helps to include some native soil around the roots. Explain why, given what you know about mycorrhizae and root hairs.

3. In the sketch at *right*, label the actual stoma. Now think about how Henry discovered a way to keep all of a plant's stomata open at all times. He also figured out how to keep those of another plant closed all the time. Both plants died. Explain why.

4. Allen is studying the rate of transpiration from tomato plant leaves. He notices that several environmental factors, including wind and relative humidity, affect the rate. Explain why.

5. You have just returned home from a three-day vacation. Your plants tell you, by their severe wilting, that you forgot to water them before you departed. Being aware of the cohesion–tension theory of water transport, explain what happened to them.

6. Not having a green thumb, your friend Stephanie decides to grow zucchini plants, which are very forgiving of poor soils

Figure 30.18 Middle fork of the Salmon River, Idaho.

and amateur gardeners. She plants too many seeds and ends up with far too many plants. Among them you happen to notice a stunted plant and decide to find out what happened to it. After many experiments, you decide that its leaves are producing a mutated, malfunctioning form of an enzyme that is necessary for the formation of sucrose. Knowing what you do about the pressure flow theory, explain why plant growth was hindered.

7. At the middle fork of the Salmon River in Idaho, clear water from a wilderness area (visible at *right* in Figure 30.18) converges on brown-colored water (visible at *left* and *center*). The brown water is enriched with silt from a wilderness area that was disturbed by cattle ranching and some other human activities. Knowing what you do about the nature of topsoil, what is probably happening to plant growth in the disturbed habitat? Knowing how fishes acquire oxygen for aerobic respiration, how might the silt be affecting their survival?

Selected Key Terms

CAM plant *30.4*	leaching *30.1*	root nodule *30.2*
carnivorous plant *CI*	loam *30.1*	sieve tube *30.5*
Casparian strip *30.2*	mutualism *30.2*	soil *30.1*
cohesion–tension	mycorrhiza *30.2*	stoma (stomata) *30.4*
theory *30.3*	nitrogen	topsoil *30.1*
companion cell *30.5*	fixation *30.2*	tracheid *30.3*
cuticle *30.4*	nutrient *30.1*	translocation *30.5*
cutin *30.4*	phloem *30.5*	transpiration *30.3*
endodermis *30.2*	plant	turgor pressure *30.4*
erosion *30.1*	physiology *CI*	vascular cylinder *30.2*
exodermis *30.2*	pressure flow	vessel member *30.3*
guard cell *30.4*	theory *30.5*	xylem *30.3*
humus *30.1*	root hair *30.2*	

Readings

Hausenbuiller, R. 1985. *Soil Science: Principles and Practices.* Third edition. Dubuque, Iowa: W. C. Brown.

Hopkins. W. G. 1995. *Introduction to Plant Physiology.* New York: Wiley.

Raven, R., R. Evert, and S. Eichhorn. 1999. *Biology of Plants.* Sixth edition. New York: Freeman.

Salisbury, F., and C. Ross. 1992. *Plant Physiology.* Fourth edition. Belmont, California: Wadsworth.

On-Line readings at Student Guide for InfoTrac:
www.brookscole.com/biology

PLANT REPRODUCTION

A Coevolutionary Tale

We find flowering plants almost everywhere, from icy tundra to deserts to oceanic islands. What accounts for their distribution and diversity? Consider the **flower**, a specialized reproductive shoot. About 435 million years ago, when plants first invaded the land, insects that ate decaying plant parts and spores probably weren't far behind. The plant smorgasbord seems to have favored natural selection of winged insects with an amazing array of sucking, piercing, and chewing mouthparts.

By 390 million years ago, in humid coastal forests, seed-bearing plants were making pollen grains. These tiny, sperm-bearing packages can travel to eggs, which develop in ovules in female plant parts. Pollen is rich in nutrients. At first, air currents may have dispersed it to the ovules. Then hungry insects made the connection between "plant parts with pollen" and "food." Plants lost some pollen to insects but gained a reproductive advantage. How? Unlike air currents, pollen-dusted insects clambering over plants could deliver pollen right to the ovules.

Plant structures evolved in novel or modified ways that were more enticing to insects, so pollen dispersal became more efficient. Insects that became specialized in detecting and gathering pollen from a specific source gained a competitive edge over insects that spent more time and energy on random searches for food.

What we are describing is a case of **coevolution**. The word refers to two or more species jointly evolving as an outcome of close ecological interactions. A heritable change in one species affects selection pressure operating between them, so the other species evolves, too.

In this case, the more enticing plants enjoyed more home deliveries, produced more seeds, and *improved their chances of reproductive success*. And so, over time, plants with distinctive flowers, fragrances, and sugar-rich nectar coevolved with pollinators.

A **pollinator** is any agent that transfers pollen from male to female reproductive parts of flowers of the same plant species. Besides insects, pollinating agents include air currents, water currents, bats, birds, and other animals (Figures 31.1 and 31.2).

You can correlate many floral features with specific mutualists. For instance, a flower's reproductive parts are positioned so that pollinating agents will brush past them. Many parts are positioned above nectar-filled floral tubes the same length as the feeding device of a preferred pollinator. Red and yellow flowers attract birds, which have great daytime vision but a poor sense of smell. As you might suspect, plants that birds visit don't divert metabolic resources to making fragrances. Red flowers don't attract beetles. Neither do flowers with nectar cups large and deep enough for beetles to

Figure 31.1 Example of adaptations uniting a flowering plant with its pollinator. The giant saguaro of Arizona's Sonoran Desert has large, showy white flowers at the tips of its spiny arms. Insects and birds visit the flowers by day, and bats visit by night. The plant offers these animals nectar. And the animals transport pollen grains, which stick to their body, from one cactus plant to another.

How we see it

How bees see it

Figure 31.2 (**a**) Bahama woodstar sipping nectar from a hibiscus blossom. Like other hummingbirds, it forages for nectar in midflight. Its long, narrow bill coevolved with long, narrow floral tubes. (**b,c**) Shine ultraviolet light on a gold-petaled marsh marigold to reveal its bee-attracting pattern.

drown in. Like flies, beetles pollinate flowers that smell like rotten meat, moist dung, or decaying litter on the forest floor—where beetles first evolved.

Daisies and other fragrant flowers with distinctive patterns, shapes, and red or orange components attract butterflies, which forage by day. Nectar-sipping bats and most moths forage by night. They pollinate intensely sweet-smelling flowers with white or pale petals that are more visible than colored petals in the dark. Long, thin mouthparts of moths and butterflies reach nectar in narrow floral tubes or floral spurs. The Madagascar hawkmoth uncoils a mouthpart the same length as a narrow floral spur of an orchid, *Angraecum sesquipedale*. It is 22 centimeters (more than 8–1/2 inches) long!

Flowers with sweet odors and yellow, blue, or purple parts attract bees. Their ultraviolet-light-absorbing pigments form patterns that say, "Nectar here!" Unlike us, bees see ultraviolet light and find nectar guides alluring (Figure 31.2*b,c*). Unlike beetles, they also have long mouthparts that extend into floral tubes.

In short, flowers contribute to reproductive success of the plants that bear them. This chapter focuses on the reproductive modes and development of these plants.

Key Concepts

1. Sexual reproduction is the premier reproductive mode of flowering plant life cycles. It involves the formation of spores and gametes, both of which develop inside the specialized reproductive shoots called flowers.

2. Microspores form in a flower's male reproductive parts and develop into pollen grains, the sperm-bearing male gametophytes. Animals, air currents, and other pollinating agents transfer pollen grains to the female reproductive parts of flowers.

3. Megaspores develop in ovules, structures that form on the inner ovary wall of the flower's female reproductive parts. Within each ovule, a mature female gametophyte forms from a megaspore. One of its cells is the egg.

4. After sperm fertilize the eggs, ovules mature into seeds. Each seed consists of an embryo sporophyte and various tissues that function in its nutrition and protection.

5. While seeds are developing, tissues of the ovary and sometimes neighboring tissues mature into fruits, which function in seed dispersal. Air currents, water currents, and animals function as dispersal agents.

6. Many species of flowering plants also reproduce asexually by mechanisms of vegetative growth, parthenogenesis, and tissue culture propagation.

REPRODUCTIVE STRUCTURES OF FLOWERING PLANTS

Think Sporophyte and Gametophyte

Like us, flowering plants engage in sex. They produce, nourish, and protect sperm and eggs in reproductive systems. Like human females, they nourish and protect developing embryos. They put out floral invitations to third parties—pollinators that get sperm to eggs. Long before we thought of it, they were using perfumes and colors to improve the odds for sexual success.

When you hear the word "plant," you may think of something like a cherry tree (Figure 31.3a). The tree is a typical **sporophyte**, a vegetative body that grows, by mitotic cell divisions, from a fertilized egg. During the life cycle, this sporophyte bears flowers—floral shoots specialized for reproduction. In flowers, haploid spores give rise to haploid bodies called **gametophytes** (Section 10.5). Sperm develop inside male gametophytes, and the eggs develop inside female gametophytes. Fusion of a haploid sperm with a haploid egg at fertilization results in a cell with two sets of genetic instructions.

Components of Flowers

Flowers form at floral shoots of the primary plant body. As they form, they differentiate into sepals and petals (nonfertile parts) and stamens and carpels (fertile parts). Figures 31.3b and 31.4 show how the parts are arranged. All connect to a receptacle, the modified base of a floral shoot. Peel open a rosebud. Sepals, the outer whorl of leaflike parts, surround leaflike petals, which surround the fertile parts. This is a common pattern. Collectively, sepals are a flower's calyx, and petals are its corolla.

Like leaves, sepals and petals have ground tissues, vascular tissues, and epidermis. Why are many flowers sweet smelling? Some epidermal cells of petals produce fragrant oils. What gives the petals their shimmer and color? Cells of the ground tissue contain pigments, such as carotenoids (yellow to red-orange) and anthocyanins (red to blue), plus small, light-refracting crystals. What does a flower's fragrance or its coloration, patterning, or arrangement of petals do? It attracts pollinators.

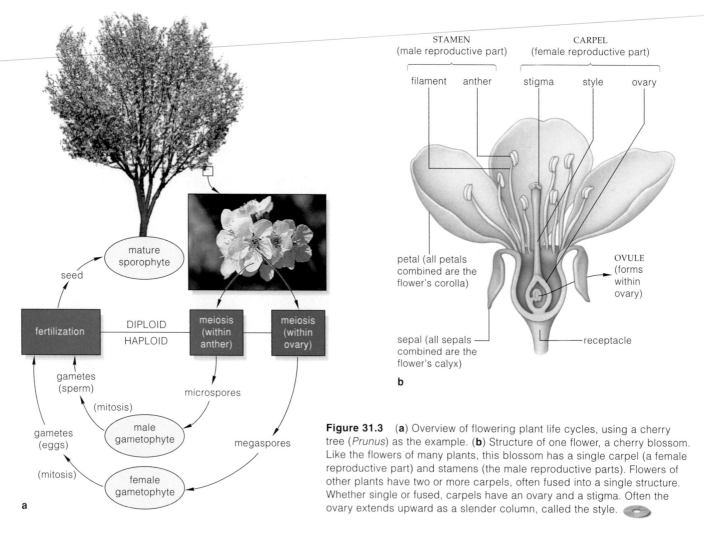

Figure 31.3 (**a**) Overview of flowering plant life cycles, using a cherry tree (*Prunus*) as the example. (**b**) Structure of one flower, a cherry blossom. Like the flowers of many plants, this blossom has a single carpel (a female reproductive part) and stamens (the male reproductive parts). Flowers of other plants have two or more carpels, often fused into a single structure. Whether single or fused, carpels have an ovary and a stigma. Often the ovary extends upward as a slender column, called the style.

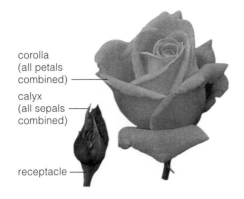

Figure 31.4 Location of floral parts in a prized cultivated plant, the rose (*Rosa*).

corolla (all petals combined)

calyx (all sepals combined)

receptacle

Where Pollen and Eggs Develop

Nearly all **stamens**, a flower's male reproductive parts, consist of an anther and a one-veined stalk (filament). An anther is divided into pollen sacs, the chambers in which walled, haploid spores form and give rise to the male gametophytes called **pollen grains** (Figure 31.5).

A flower's female reproductive parts are located at its center. You may have heard someone call these parts pistils, but their more recent name is **carpels**.

Many flowers have one carpel. Others have more, often fused as a compound structure. The lower part of single or fused carpels—the **ovary**—is where the eggs develop, fertilization takes place, and seeds mature. The more formal name for flowering plants, angiosperm, refers to the carpel (from the Greek *angeion*, meaning vessel, and *sperma*, seed). The upper portion of a carpel is the stigma. This sticky or hairy surface tissue captures pollen grains and favors their germination. Commonly the stigma is elevated upon a style, a slender extension of the upper ovary wall (Figure 31.3*b*).

Not every flower has stamens and carpels. Some species have *perfect* flowers, with both female and male parts. Others have *imperfect* flowers; they have female or male parts. In certain species such as oaks, the same individual has male and female flowers. In willows and other species, they are on separate, individual plants.

Sexual reproduction is the dominant reproductive mode of flowering plant life cycles. Such cycles alternate between the production of sporophytes (spore-producing bodies) and gametophytes (gamete-producing bodies).

Flowering plant sporophytes form from a fertilized egg by mitotic cell divisions and cell growth. A sporophyte is all of the plant body except for its male and female gametophytes.

Gametophytes arise from haploid spores, which form in stamens and carpels—the male and female reproductive parts of the flower.

Each male gametophyte gives rise to a sperm-producing pollen grain. Each female gametophyte develops inside a carpel's ovary; one of its cells is an egg. Also inside the ovary, fertilization takes place and seeds mature.

Pollen Sets Me Sneezing

In the year 1835 Sidney Smith wrote, "I am suffering from my old complaint, the hay-fever . . . that sets me sneezing; and if I begin sneezing at twelve, I don't leave off till two o'clock, and am heard distinctly in Taunton [six miles distant] when the wind sets that way."

Sidney suffered what is now called *allergic rhinitis*. This is the name for hypersensitivity to a normally harmless substance. Such hypersensitivity is common.

The pollen of ragweed, shown in Figure 31.5, brought on Sidney's dreaded hay fever, but it wasn't out to get him. Every spring and summer, flowering plants release pollen grains. In many millions of people, white blood cells respond by mounting an immune response against some proteins that project from the surface of pollen grain walls. The pollen grains from different plants have differences in their wall proteins. The immune system of a person who is hypersensitive might chemically respond to some of these but not others. The misdirected response results in a profusely runny nose, reddened and itchy eyelids, congestion, and bouts of sneezing.

Hay fever is a genetic abnormality; it runs in families. Some individuals simply are genetically predisposed to overreact to some kinds of pollen. Infections, emotional stress, and changes in temperature may also trigger overreactions to pollen.

Figure 31.5 Pollen grains of (**a**) grass, (**b**) chickweed, and (**c**) ragweed plants. Pollen grains of most families of plants differ in size, wall sculpturing, and number of wall pores.

31.3

A NEW GENERATION BEGINS

From Microspores to Pollen Grains

We turn now to pollen grain formation. While anthers are growing, four masses of spore-producing cells form by mitotic cell divisions. Walls develop around them. Each anther now has four chambers, called pollen sacs (Figure 31.6a). Haploid **microspores** form when cells in the sacs undergo meiosis and cytoplasmic division. They go on to develop an elaborately sculpted wall. The walled microspores divide once or twice, by mitosis, and then form pollen grains. These enter a period of arrested growth and in time will be released from the anther. Components of the pollen grain wall will help protect them from decomposers in their surroundings.

As soon as many types of pollen grains form, they produce sperm nuclei, which are the male gametes of flowering plants. Other types don't do this until after they reach a carpel and start growing toward its ovule. In short, each pollen grain is a mature *or* an immature male gametophyte, depending on the plant species.

From Megaspores to Eggs

Meanwhile, one or more masses of cells are forming on the inner wall of the flower's ovary. Each is the start of an **ovule**, a structure that houses a female gametophyte and that may become a seed. As each cell mass grows, a tissue forms within it, and one or two protective layers called **integuments** form around it. Inside the mass, a cell divides by meiosis, and four haploid spores form. Spores formed in flowering plant ovaries are generally larger than microspores and are called **megaspores**.

Commonly, all megaspores but one disintegrate. That one undergoes mitosis three times without cytoplasmic division. At first, it has eight nuclei (Figure 31.6b). Each nucleus migrates to a prescribed location, and then the cytoplasm divides. The result, a seven-celled embryo sac, is a female gametophyte. One cell has two nuclei. This endosperm mother cell will help form the **endosperm**, a nutritive tissue for the embryo. Another cell is the egg.

From Pollination to Fertilization

Flowering plants release pollen in spring. You are very aware of this reproductive event if you are one of the millions of people who experience hay fever, an allergic reaction to wall proteins of pollen grains (Section 31.2).

Specifically, **pollination** is a transfer of pollen grains to a receptive stigma. Air or water currents, birds, bats, insects, and many other agents of pollination carry out such transfers, as you read in the chapter introduction.

Once a pollen grain lands on a receptive stigma, it germinates. For this event, germination means a pollen grain resumes growth and then develops into a tubular

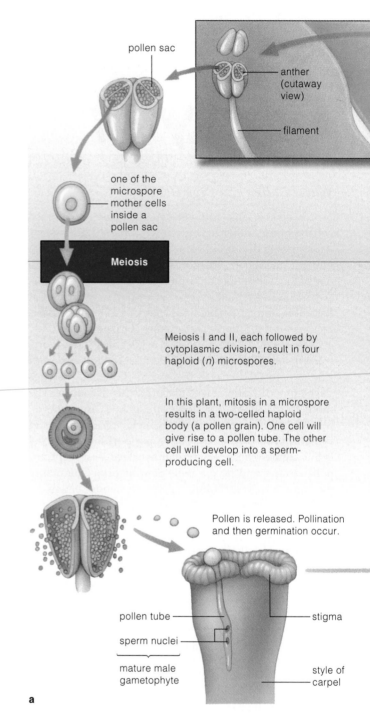

pollen sac

anther (cutaway view)

filament

one of the microspore mother cells inside a pollen sac

Meiosis

Meiosis I and II, each followed by cytoplasmic division, result in four haploid (*n*) microspores.

In this plant, mitosis in a microspore results in a two-celled haploid body (a pollen grain). One cell will give rise to a pollen tube. The other cell will develop into a sperm-producing cell.

Pollen is released. Pollination and then germination occur.

pollen tube

sperm nuclei

stigma

mature male gametophyte

style of carpel

a

Figure 31.6 Life cycle of cherry (*Prunus*), one of the flowering plants classified as a eudicot (true dicot). (**a**) How pollen grains develop and germinate. (**b**) Events in this ovary's ovule.

structure. This "pollen tube" grows down through the tissues of the ovary and carries the sperm nuclei with it (Figure 31.6b). Chemical and molecular cues guide the growth of the pollen tube through the ovary's tissues,

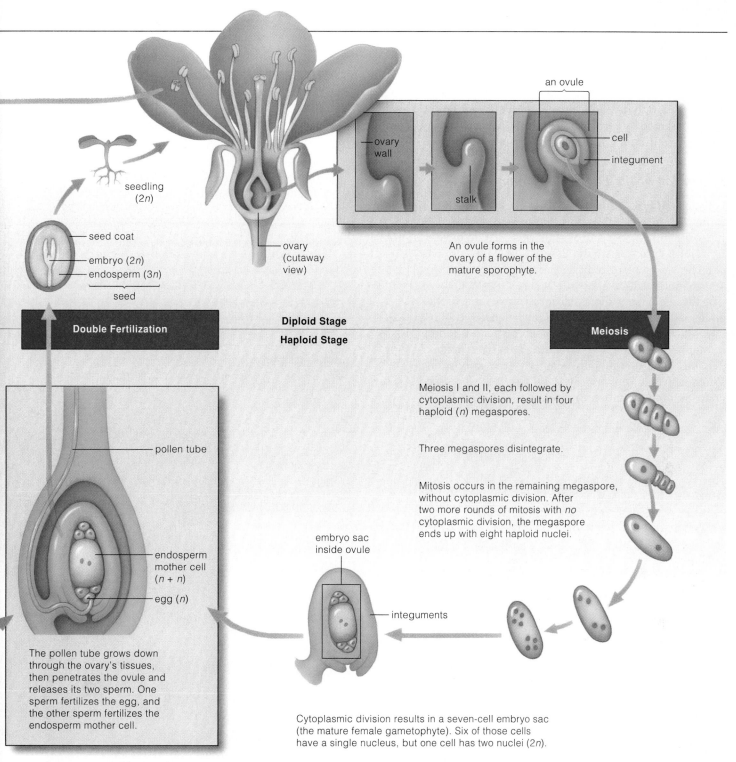

seedling
(2n)

seed coat

embryo (2n)

endosperm (3n)

seed

ovary
(cutaway
view)

an ovule

ovary
wall

stalk

cell

integument

An ovule forms in the
ovary of a flower of the
mature sporophyte.

Double Fertilization

Diploid Stage

Haploid Stage

Meiosis

Meiosis I and II, each followed by
cytoplasmic division, result in four
haploid (n) megaspores.

Three megaspores disintegrate.

Mitosis occurs in the remaining megaspore,
without cytoplasmic division. After
two more rounds of mitosis with *no*
cytoplasmic division, the megaspore
ends up with eight haploid nuclei.

pollen tube

endosperm
mother cell
(n + n)

egg (n)

embryo sac
inside ovule

integuments

The pollen tube grows down
through the ovary's tissues,
then penetrates the ovule and
releases its two sperm. One
sperm fertilizes the egg, and
the other sperm fertilizes the
endosperm mother cell.

Cytoplasmic division results in a seven-cell embryo sac
(the mature female gametophyte). Six of those cells
have a single nucleus, but one cell has two nuclei (2n).

b

toward the egg chamber and sexual destiny. When the
pollen tube reaches an ovule, it penetrates the embryo
sac, and its tip ruptures to release two sperm.

"Fertilization" generally means fusion of a sperm
nucleus with an egg nucleus. But **double fertilization**
occurs in flowering plants. In species having a diploid
chromosome number, one sperm nucleus fuses with an
egg nucleus to form a diploid (2n) zygote. Meanwhile,
the other sperm nucleus fuses with both nuclei of the

endosperm mother cell. The resulting cell has a triploid
(3n) nucleus and will give rise to the nutritive tissue.

In flowering plants, sperm cells form within pollen grains,
the male gametophytes. Ovules form inside ovaries, and
female gametophytes (with egg cells) develop within them.

After pollination and double fertilization, an embryo and
nutritive tissue form in the ovule, which becomes a seed.

FROM ZYGOTES TO SEEDS AND FRUITS

After fertilization, the newly formed zygote embarks on a course of mitotic cell divisions that lead to a mature embryo sporophyte. It develops as part of an ovule and is accompanied by formation of a **fruit**: a mature ovary, with or without other, neighboring tissues.

Formation of the Embryo Sporophyte

Let's look at how a shepherd's purse (*Capsella*) embryo forms. By the time it reaches the stage in Figure 31.7e, two **cotyledons**, or seed leaves, have started to develop from lobes of meristematic tissue. Cotyledons are part of all flowering plant embryos. Dicot embryos have two, and monocot embryos have one. The *Capsella* embryo, like those of many dicots, absorbs nutrients from the endosperm and stores them in its cotyledons. By contrast, in corn, wheat, and most other monocots, endosperm is not tapped until a seed germinates. Digestive enzymes get stockpiled in a monocot embryo's thin cotyledons. When activated, the enzymes will help transfer stored endosperm to the growing seedling.

Table 31.1 *Categories and Examples of Fruits*

SIMPLE FRUITS From one ovary of one flower.

Dry fruit

Dehiscent. Fruit wall splits along definite seams to release seeds. Legume (e.g., pea, bean), poppy, larkspur, mustard

Indehiscent. Fruit wall does not split on seams to release seeds. Acorn, grains (e.g., corn), sunflower, carrot, maple

Fleshy fruit

Berry. Compound or simple ovary, many seeds. Tomato, grape, banana

Pepo. Ovary wall has hard rind. Cucumber, watermelon

Hesperidium. Ovary wall has leathery rind. Orange, lemon

Drupe. One or two seeds. Thin skin, part of flesh around a seed enclosed in hardened ovarian tissue. Peach, cherry, apricot, almond, olive

AGGREGATE FRUITS

Many ovaries of one flower, all attached to the same receptacle. Many seeds. Raspberry, blackberry (not really "berries")

MULTIPLE FRUITS

Combined from ovaries of many flowers. Pineapple, fig

ACCESSORY FRUITS

Most tissues of the flesh are not derived from the ovary; e.g., mainly from the receptacle. Pome (apple, pear), strawberry

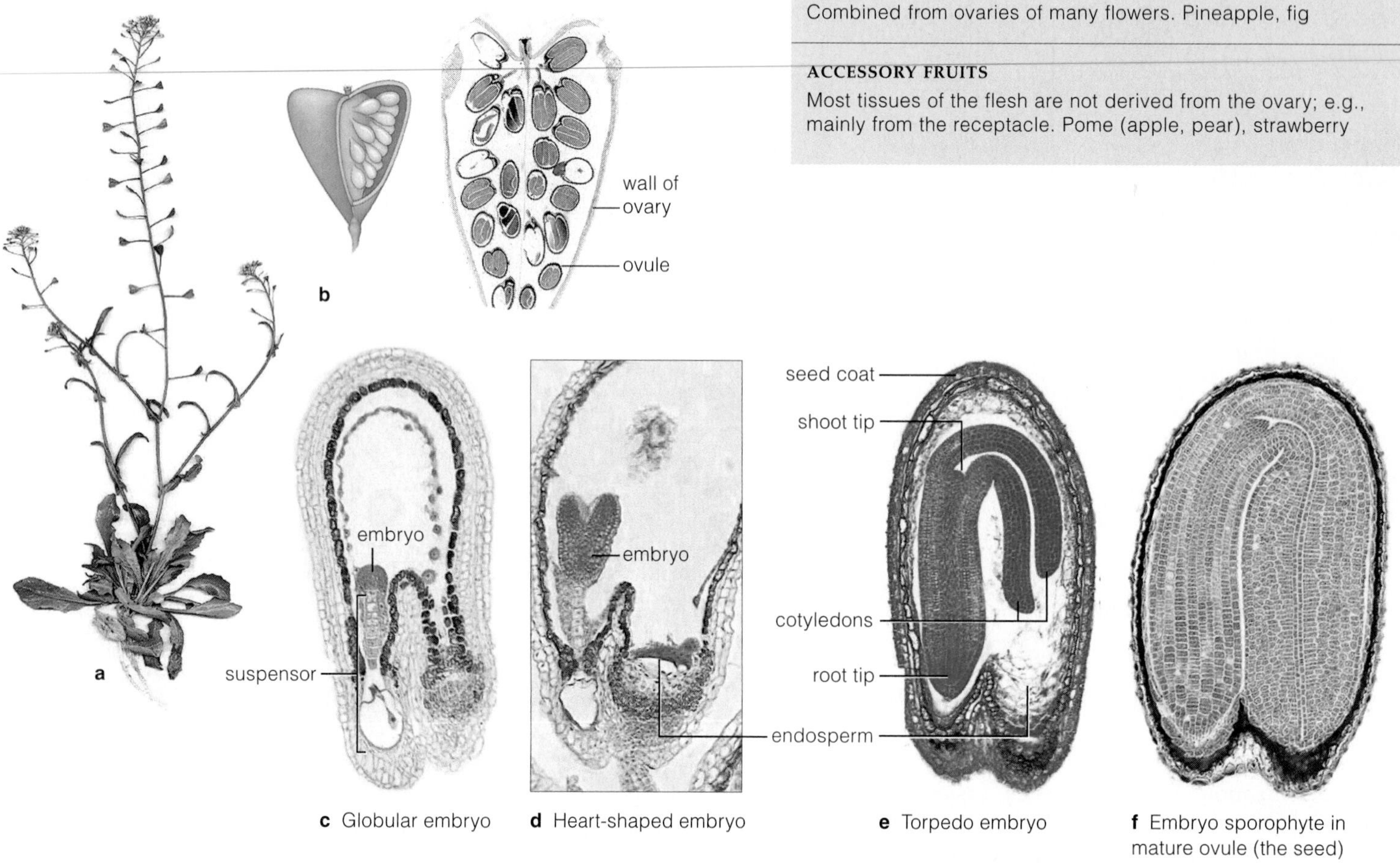

Figure 31.7 Dicot seed development. (**a**) Mature sporophyte of shepherd's purse (*Capsella*). (**b**) Seeds in ovary. (**c,d**) The suspensor transfers nutrients to the early embryo sporophyte from the parent plant. The embryo is well developed in (**e**). It is mature in (**f**). Micrographs are not to the same scale.

fruit wall — cotyledons

Figure 31.8 From flowers to fruits—three examples. (**a–d**) Strawberry (*Fragaria*), a fragrant accessory fruit. Most of the fleshy tissues arise from the flower's receptacle. Numerous fruits are perched on the surface of the mature receptacle; each has a hard fruit wall around the embryo sporophyte.

(**e**) Pineapple (*Ananas*), a multiple fruit. Native Americans cultivated pineapples. To them, it was a symbol of hospitality. They used the juice as a base for an alcoholic beverage. Christopher Columbus thought the fruit looked like pine cones, hence the name.

(**f–i**) Fruit formation on an apple (*Malus*) tree. When the petals drop, this is a sign that the eggs have become fertilized.

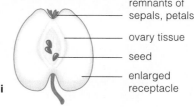

remnants of
sepals, petals

ovary tissue

seed

enlarged
receptacle

i

Seeds and Fruit Formation

From the time a zygote forms inside an ovule until it has become a mature embryo sporophyte, the parent plant transfers nutrients to ovule tissues. Food reserves accumulate in endosperm or in cotyledons. The ovule eventually separates from the wall of the ovary, and its integuments thicken and harden into a seed coat. The embryo, food reserves, and coat are a self-contained package—a **seed**, which we define as a mature ovule.

When seeds form, other floral parts change and start to form fruits. We put fruits into four categories: *Simple* (dry or fleshy, derived from one ovary), *aggregate* (from many separate ovaies of one flower), *multiple* (many separate ovaries attached to the same receptacle of one flower), and *accessory* (most tissues not derived from the ovary). Table 31.1 lists examples.

At maturity, acorns and some other simple dry fruits are intact; pea pods unzip to release seeds. Peaches, as in Section 23.5, are simple fleshy fruits. Strawberries are an aggregate fruit that has many seeds on a fleshy, expanded receptacle (Figure 31.8*a–d*). Pineapples are a cone-shaped multiple fruit. They form when ovaries of many separate flowers fuse together into a fleshy mass as they enlarge. A waxy, hardened rind covers the fruit (Figure 31.8*e*). Apples, an accessory fruit, are mainly an enlarged receptacle and calyx (Figure 31.8*f–i*).

Fleshy fruits have three tissue divisions, although sometimes the three are not immediately visible. The *endo*carp is the innermost portion around the seed or seeds. *Meso*carp is the fleshy portion. *Exo*carp is the skin. Together, the three fruit regions are a **pericarp**. In acorns and other dry fruits, the pericarp is typically thin. In true berries, such as tomatoes and grapes, the pericarp is thin skinned and relatively soft at maturity.

A mature ovule, which encases an embryo sporophyte and food reserves inside a protective coat, is a seed.

A mature ovary, with or without additional floral parts that have become incorporated into it, is a fruit.

Fruits are classified by whether they are dry or fleshy, are derived from one or more ovaries, and incorporate other tissues besides those of the ovary.

Fleshy fruits have three regions. Innermost is the endocarp, which surrounds the seeds. The mesocarp is the fleshy part, and the exocarp is the skin. Together, the three regions are known as the pericarp.

DISPERSAL OF FRUITS AND SEEDS

The fruits you read about in the preceding section have a common function: *seed dispersal*. To gain insight into their structure, think about how they coevolved with particular dispersing agents—currents of air or water, or animals passing by.

Consider the wind-dispersed fruit of maples (*Acer*), as shown in Figure 31.9a. The pericarp of a maple fruit extends out like wings. When the fruit breaks in half and drops from a tree, it interacts with air currents and spins sideways. The wings whirl the fruit far enough away that the embryo sporophytes inside seeds will not have to compete with the parent plant for water, minerals, and sunlight.

Brisk winds have been known to transport fruits as far as ten kilometers from a tree. Air currents easily lift the tiny fruit of dandelions by their outward-pluming "parachute." And air currents readily disperse orchid seeds, which are as fine as dust particles.

Other fruits taxi to new locations on or in animals, including many birds, mammals, and insects. Some can adhere to feathers, feet, or fur with hooks, spines, hairs,

and sticky surfaces. Fruits of cocklebur, bur clover, and bedstraw are like this. Embryo sporophytes tucked in the seed coats of fleshy fruits survive being eaten and assaulted by the digestive enzymes in an animal's gut. Besides digesting the fruit's flesh, these enzymes digest some of the seed coat. The digested portions make it easier for the embryo to break through a hard coat after the seeds are expelled from the animal body in feces and start to germinate. Germination is a topic that will be described in the next chapter.

Some water-dispersed fruits have heavy wax coats, and others, including sedges, have sacs of air that help them float. With its thick, water-repelling pericarp, the coconut palm fruit is adapted to travel across the open ocean. If it doesn't wash onto a beach soon enough, saltwater might penetrate the pericarp and kill the embryo. Coconuts can drift hundreds of kilometers before this happens.

a　　wing　seed (in carpel)　　**b**

Figure 31.9 (**a**) Winged seeds of maple (*Acer*). (**b**) A few students taking a chocolate ice cream break and indirectly contributing to the long-term reproductive success of cacao (*Theobroma cacao*). (**c**) Cacao fruit, or pod. Each contains up to forty seeds ("beans"). Seeds are processed into cocoa butter and essences of chocolate products. Interactions among at least a thousand compounds in chocolate exert compelling (some say addictive) effects on the human brain. The average American buys 8 to 10 pounds of chocolate annually. The average Swiss citizen craves a whopping 22 pounds. With that kind of demand, *T. cacao* seeds are cherished and "dispersed" in orderly arrays in plantations, where the trees that grow from them are carefully tended.

Humans are grand agents of dispersal. In the past, explorers transported seeds all over the world, although most places now control imports. Cacao (Figure 31.9c), oranges, corn, and other plants were domesticated and encouraged to reproduce in new habitats, in far greater numbers than they did on their own. And in evolutionary terms, reproductive success is what life is all about.

Seeds and fruits are structurally adapted for dispersal by air currents, water currents, and many kinds of animals.

Why So Many Flowers and So Few Fruits?

In the Sonoran Desert of Arizona (Figure 31.10a), the giant saguaro produces as many as a hundred flowers at the tips of its huge, spiny arms. Remember the large, showy flowers in Figure 31.1? Each flower blooms for only twenty-four hours. Again, insects visit the flowers by day, and bats by night. The animals benefit by getting nectar, and the giant saguaro benefits because the animals deliver their dusting of pollen grains to other giant saguaros.

The flowers that do get pollinated, courtesy of insects or bats, now begin the task of producing seeds and fruits. Petals wilt, then they wither as the many egg cells inside the ovary become fertilized. Each egg-containing ovule expands and matures into a seed. During the same interval, ovaries develop into fruits (Figure 31.10). Weeks later, the plum-sized fruits split open, exposing a bright red interior and dark seeds.

White-winged doves feast on the ripe fruits. When they fly away, each might carry hundreds of seeds in their gut. A few seeds may elude the gut's digestive enzymes and later end up on the ground. There, each seed will have a tiny chance of growing into a giant saguaro.

commonly produce far more flowers than they do mature fruits. They seem to be producing flowers excessively and giving up chances to make seeds and leave descendants. Doing so is not a reproductive adaptation that you would expect, based on Darwinian evolutionary theory.

Well, *suppose that some of the plants simply did not receive enough pollen to have all their egg cells fertilized.* After all, unfertilized flowers cannot produce seeds and fruits. This hypothesis has been experimentally tested for some plant species. Researchers carefully brushed pollen on every single flower. In many instances, the plants still failed to produce fruit for every flower!

Alternatively, *suppose the presumed "excess" flowers are formed strictly to produce pollen for export to other plants.*

Figure 31.10 (**a**) Giant saguaro growing in Arizona's Sonoran Desert. (**b**) Two fruits. The one above set, and the one below it failed to set. (**c**) A red, maturing fruit.

One of the puzzling aspects of this annual event is the frequency with which saguaro flowers fail to give rise to fruit (Figure 31.10b, for example). Why would a cactus invest so much energy constructing a hundred flowers if only thirty or so of them will set fruit? Compare one of these "inefficient" plants with a species that produces a hundred flowers, all of which develop into fruit. Which do you think will leave more descendants?

Common sense might lead you to conclude, "The more fruits, the better." Yet saguaros—like many other plants—

Pollen grains are small. Energetically speaking, they are inexpensive to produce compared to large, calorie-rich, seed-containing fruits. Thus, for a fairly small investment, a plant might reap a large reward in offspring that carry its genes.

The idea that some kinds of plants actually set aside flowers exclusively for pollen export is only now being tested for saguaros. Perhaps you might like to design and carry out experiments that will help clear up the mystery of the "excess" flowers of saguaros and similar plants.

ASEXUAL REPRODUCTION OF FLOWERING PLANTS

Asexual Reproduction in Nature

The features of sexual reproduction that we considered in earlier sections dominate flowering plant life cycles. Bear in mind, many species also reproduce asexually by modes of **vegetative growth**, as listed in Table 31.2. In essence, new roots and shoots grow right out from extensions or fragments of parent plants. Such asexual reproduction proceeds by way of mitosis. This means a new generation of offspring is genetically identical to the parent; it is a clone.

One "forest" of quaking aspen (*Populus tremuloides*) provides us with an impressive example of vegetative reproduction. The leaves of this flowering plant species tremble even in the slightest breeze, hence the name. Figure 31.11 is a panoramic view of the shoot systems of one individual. The root system of the parent plant gave rise to adventitious shoots, which became separate shoot systems. Barring rare mutations, individuals at the north end of this vast clone are genetically identical to individuals at the south end. Water travels from roots near a lake all the way to shoot systems in much drier soil. Dissolved ions travel in the opposite direction.

As long as environmental conditions favor growth and regeneration, such clones are about as close as one can get to being immortal. No one knows how old the aspen clones are. The oldest known clone is a ring of creosote bushes (*Larrea divaricata*) growing in the Mojave Desert. It has been around for the past 11,700 years.

The reproductive possibilities are amazing. Watch strawberry plants send out horizontal, aboveground stems (runners), then watch the new roots and shoots develop at every other node. The oranges you eat may be descended from a single tree in southern California that reproduced by **parthenogenesis**. With this type of process, an embryo develops from an unfertilized egg.

Parthenogenesis can be stimulated when pollen has contacted a stigma even though a pollen tube has not grown through the style. Maybe certain hormones from the stigma or pollen grains diffuse to the unfertilized egg and trigger formation of an embryo. That embryo is 2*n* as a result of fusion of the products of mitotic cell division in the egg. A 2*n* cell outside the gametophyte also may be stimulated to develop into an embryo.

Induced Propagation

Most houseplants, woody ornamentals, and orchard trees are clones. People typically propagate them from cuttings or fragments of shoot systems. For example, with suitable encouragement, a severed African violet leaf may form a callus from which adventitious roots develop. A callus is one type of meristem. Remember, a meristem is a localized region of plant cells that retain the potential for mitotic cell division.

Or consider a twig or bud from one plant, grafted onto a different variety of some closely related species. Vintners in France, for instance, graft prized grapevines onto disease-resistant root stock from America.

Frederick Steward and his colleagues were pioneers in **tissue culture propagation**. They cultured small bits of phloem from the differentiated roots of carrot plants (*Daucus carota*) in rotating flasks. They used a liquid growth medium that contained sucrose, minerals, and vitamins. The liquid also contained coconut milk, which Steward knew was rich in then-unidentified growth-inducing substances. As the flasks rotated, individual cells that were torn away from the tissue bits divided and formed multicelled clumps, which sometimes gave rise to new roots (Figure 31.12a). Steward's experiments were among the first to demonstrate that some cells of

Table 31.2 *Asexual Reproductive Modes of Flowering Plants*

Mechanism	Examples	Characteristics
VEGETATIVE REPRODUCTION ON MODIFIED STEMS		
1. Runner	Strawberry	New plants arise at nodes along aboveground horizontal stems.
2. Rhizome	Bermuda grass	New plants arise at nodes of underground horizontal stems.
3. Corm	Gladiolus	New plants arise from axillary buds on short, thick, vertical underground stems.
4. Tuber	Potato	New shoots arise from axillary buds (tubers are the enlarged tips of slender underground rhizomes).
5. Bulb	Onion, lily	New bulbs arise from axillary buds on short underground stems.
PARTHENOGENESIS		
	Orange, rose	An embryo develops without nuclear or cellular fusion (for example, from an unfertilized haploid egg or by developing adventitiously, from tissue surrounding the embryo sac).
VEGETATIVE PROPAGATION		
	Jade plant, African violet	A new plant develops from tissue or structure (a leaf, for instance) that drops from the parent plant or is separated from it.
TISSUE CULTURE PROPAGATION		
	Orchid, lily, wheat, rice, corn, tulip	A new plant is induced to arise from a parent plant cell that has not become irreversibly differentiated.

Figure 31.11 A mere portion of Pando the Trembling Giant, so named by Michael Grant and his coworkers at the University of Colorado, who studied its genetic makeup. This "forest" in Utah is actually a single asexually reproducing male organism, of a type called quaking aspen (*Populus tremuloides*). Its root system extends through about 106 hectares (about 262 acres) of soil and functionally supports its 47,000 genetically identical shoots; that is, the trees. By one estimate, this clone weighs more than 5,915,000 metric tons.

Figure 31.12 (**a**) Cultured cells from a carrot plant. At the time this photograph was taken, roots and shoots of embryonic plants were already forming. (**b**) Young orchid plants, developed from cultured meristems and young leaf primordia. Orchids are one of the most highly prized cultivated plants. Before the meristems were cloned, they were difficult to hybridize. From the time seeds form, it can take seven years or more until a new plant bears flowers. In natural habitats, the seeds normally will not germinate unless they interact with a specific fungus.

specialized tissues still house the genetic instructions required to produce an entire individual.

Researchers now use shoot tips and other parts of individual plants for tissue culture propagations. The techniques prove useful when an advantageous mutant arises. Such a mutant may show resistance to a disease that is crippling to wild-type plants of the same species. Tissue culture propagation can result in hundreds, even thousands of identical plants from merely one mutant specimen. The techniques are already being employed in efforts to improve major food crops, including corn, wheat, rice, and soybeans. They also are being used to increase production of hybrid orchids, lilies, and other prized ornamental plants (Figure 31.12*b*).

Besides reproducing sexually, flowering plants engage in asexual reproduction, as by vegetative growth.

SUMMARY   *Gold* indicates text section

1. Sexual reproduction is the main reproductive mode of flowering plant life cycles. Diploid sporophytes, or spore-producing plant bodies, develop, as do haploid gametophytes (gamete-producing bodies). *31.1*

   a. The sporophyte is a multicelled vegetative body with roots, stems, leaves, and, at some point, flowers. Flowers of many species coevolved with animals that feed on nectar or pollen and also serve as pollinators. Gametophytes form in male and female floral parts.

   b. Many flowering plants also reproduce asexually. They can do this naturally (as by runners, rhizomes, and bulbs) and artificially (as by cuttings and grafting).

2. Flowers typically have sepals, petals, and one or more stamens, which are male reproductive structures. They also have carpels, the female reproductive structures. Most or all of the reproductive structures are attached to a receptacle, the modified end of a floral shoot. *31.1*

   a. Anthers of stamens contain pollen sacs in which cells divide by meiosis. A wall develops around each resulting haploid cell (a microspore), which develops into a sperm-bearing pollen grain (male gametophyte).

   b. A carpel (or two or more carpels fused together) has an ovary where the eggs develop, fertilization takes place, and seeds mature. A stigma is a sticky or hairy surface tissue above the lower portion of the ovary. It captures pollen grains and promotes their germination.

3. Inside carpels, ovules form on the inner ovary wall. Each consists of a female gametophyte with an egg cell, an endosperm mother cell, a surrounding tissue, and one or two protective layers called integuments. *31.3*

   a. At maturity, each ovule is a seed; its integuments form the seed coat.

   b. The ovary, and sometimes other tissues, matures into a fruit, which is a seed-containing structure.

4. Female gametophytes typically form this way: *31.3*

   a. Four haploid megaspores form after meiosis. All but one usually disintegrate.

   b. The one remaining megaspore undergoes mitosis three times but not cytoplasmic division. Its multiple nuclei move to prescribed positions in the cytoplasm.

   c. With cytoplasmic division, a female gametophyte (in this case, a seven-cell, eight-nuclei body) develops. One cell is an egg. The cell with two nuclei (endosperm mother cell) will help give rise to endosperm, which is a nutritive tissue for the forthcoming embryo.

5. Pollination refers to the arrival of pollen grains on a suitable stigma. Once it has been transferred, a pollen grain germinates. It becomes a pollen tube that grows down through tissues of the ovary, carrying the two sperm nuclei with it. *31.3*

6. At double fertilization, one sperm nucleus fuses with one egg nucleus, thereby forming a diploid (2n) zygote.

The other sperm nucleus fuses with both nuclei of the endosperm mother cell to form a cell that will give rise to nutritive tissue in the seed. *31.3*

7. After double fertilization, the endosperm forms, the ovule expands, and the seed and fruit mature. *31.4*

8. Fruits function to protect and disperse seeds, which germinate after dispersal from the parent plant. As an example, fruits can attract animals and other dispersal agents; and lightweight fruits can be dispersed by wind. Some have hooks and such that attach to animals. *31.5*

9. Many flowering plants can reproduce asexually by vegetative growth. For example, new shoot systems may arise by mitotic divisions at nodes or buds along modified stems of a parent plant, by parthenogenesis or by vegetative propagation—from fragments or parts severed from a parent plant. *31.7*

Review Questions

1. Define flower and define pollinator. What kinds of animals are attracted to flowers with red and orange components? What kinds respond to the flowers that smell like decaying organic material? What are some of the ways in which night-foraging animals find flowers in the dark? *CI*

2. Label the floral parts. Explain the role each part plays in the reproduction of flowering plants. *31.1*

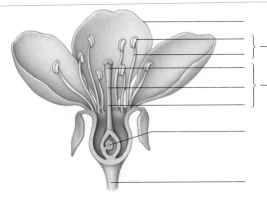

3. Distinguish between these terms:
   a. Sporophyte and gametophyte *31.1*
   b. Stamen and carpel *31.1*
   c. Ovule and ovary *31.1, 31.3*
   d. Microspore and megaspore *31.3*
   e. Pollination and fertilization *CI, 31.3*
   f. Pollen grain and pollen tube *31.1, 31.3*

4. Describe the steps by which an embryo sac, a type of female gametophyte, forms in a dicot such as cherry (*Prunus*). *31.3*

5. Define the difference between a seed and a fruit. *31.4*

6. Do food reserves accumulate in endosperm, cotyledons, or both? If both, in what ways do these structures differ? *31.4*

7. Name an example of a simple dry fruit, simple fleshy fruit, accessory fruit, aggregate fruit, and multiple fruit. *31.4*

8. Name the three regions of a fleshy fruit. What is the name for all three regions combined? *31.4*

9. Define and give an example of an asexual reproductive mode used by a flowering plant. *31.7*

Self-Quiz   ANSWERS IN APPENDIX III

1. The flowers of many species coevolved with insects, birds, and other agents that function as _____ .

2. The _____ , which bears flowers, roots, stems, and leaves, dominates the life cycle of flowering plants.
    a. sporophyte
    b. gametophyte
    c. sporangium and its derivatives
    d. gametangium and its derivatives

3. Male gametophytes of flowering plants produce _____ , and the female gametophytes produce _____ .
    a. megaspores; eggs   c. eggs; sperm
    b. sperm; microspores   d. sperm; eggs

4. A _____ is a closed vessel that contains an ovary in which eggs develop, fertilization occurs, and seeds mature.
    a. pollen sac   c. receptacle
    b. carpel   d. sepal

5. Seeds are mature _____ ; fruits are mature _____ .
    a. ovaries; ovules   c. ovules; ovaries
    b. ovules; stamens   d. stamens; ovaries

6. After meiosis within pollen sacs, haploid _____ form.
    a. megaspores   c. stamens
    b. microspores   d. sporophytes

7. Following meiosis in ovules, _____ megaspores form.
    a. two   c. six
    b. four   d. eight

8. The seed coat forms from which structure(s)?
    a. ovule integuments   c. endosperm
    b. ovary   d. residues of sepals

9. Cotyledons develop as part of all flowering plant _____ .
    a. seeds   c. fruits
    b. embryos   d. ovaries

10. The outermost region of a fleshy fruit is the _____ .
    a. pericarp   c. mesocarp
    b. endocarp   d. exocarp

11. Development of a new plant from a tissue or structure that drops or is separated from the parent plant is called _____ .
    a. parthenogenesis
    b. exocytosis
    c. vegetative propagation
    d. nodal growth

12. Match the terms with the most suitable description.
    ____ double   a. formation of zygote and first
        fertilization   cell of endosperm
    ____ ovule   b. outcome of two species
    ____ mature female   interacting in close ecological
        gametophyte   fashion over geologic time
    ____ asexual   c. contains a female gametophyte
        reproduction   and has potential to be a seed
    ____ coevolution   d. an embryo sac, commonly with
                                seven cells (one has two nuclei)
                             e. mitotic cell division at bud or
                                node produces a new plant

Critical Thinking

1. Wanting to impress her many friends with her sophisticated botanical knowledge, Dixie Bee is preparing a plate of tropical fruits for a party and cuts open a papaya (*Carica papaya*) for the first time. Inside are great numbers of slime-covered seeds, surrounded by soft flesh and soft skin (Figure 31.13). Knowing

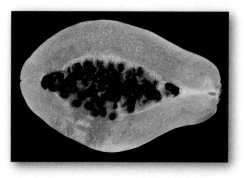

Figure 31.13 Seeds in the fruit of a papaya (*Carica papaya*).

that her friends will ask what sort of fruit this is and which parts are the endocarp, mesocarp, and exocarp, she panics, runs to her biology book, and opens it to Section 31.4. What does she find out? What will she tell them when they ask what kinds of agents disperse the seeds of such a fruit from the parent plant?

2. Observe several kinds of flowers growing in the area where you live. Given the coevolutionary links between flowering plants and their pollinators, describe what sorts of pollination agents your floral neighbors might depend upon.

3. Elaine, a plant physiologist, succeeded in cloning genes for pest resistance into petunia cells. How can she use tissue culture propagation to produce many petunia plants with those genes?

4. Before cherries, apples, peaches, and many other fruits ripen and the seeds inside them mature, their flesh is bitter or sour. Only later does it become tasty to animals that assist in seed dispersal. Develop a hypothesis of how this feature improves the odds for the plant's reproductive success.

Selected Key Terms

carpel *31.1*	ovule *31.3*
coevolution *CI*	parthenogenesis *31.7*
cotyledon *31.4*	pericarp *31.4*
double fertilization *31.3*	pollen grain *31.1*
endosperm *31.3*	pollination *31.3*
flower *CI*	pollinator *CI*
fruit *31.4*	seed *31.4*
gametophyte *31.1*	sporophyte *31.1*
integument *31.3*	stamen *31.1*
megaspore *31.3*	tissue culture
microspore *31.3*	propagation *31.7*
ovary *31.1*	vegetative growth *31.7*

Readings

Grant, M. October 1993. "The Trembling Giant." *Discover* 14(10): 82–89.

Proctor, M., and P. Yeo. 1973. *The Pollination of Flowers.* New York: Taplinger.

Raven, P., R. Evert, and S. Eichhorn. 1999. *Biology of Plants.* Sixth edition. New York: Freeman.

Rost, T., M. Barbour, C. R. Stocking, and T. Murphy. 1998. *Plant Biology.* Belmont, California: Wadsworth.

Stern, K. 2000. *Introductory Plant Biology.* Eighth edition. New York: McGraw-Hill.

PLANT GROWTH AND DEVELOPMENT

Foolish Seedlings, Gorgeous Grapes

A few years before the American stock market grew feverishly and then collapsed catastrophically in 1929, a researcher in Japan came across a substance that caused runaway growth and subsequent collapse of rice plants. Ewiti Kurosawa was studying what the Japanese call *bakane*, or the "foolish seedling" effect on rice plants. Stems of rice seedlings that had become infected with *Gibberella fujikuroi*, a fungus, grew twice as long as the stems of uninfected plants. The elongated stems were weak and spindly. Eventually they fell over, and the infected plants died. Kurosawa soon discovered that he could trigger the disease by applying extracts of the fungus to plants. Many years later, other researchers purified the disease-causing substance from fungal extracts. The substance was named gibberellin.

Gibberellin, as we now know, is one of the premier plant hormones. **Hormones**, remember, are signaling molecules secreted by some cells that travel to target cells, where they stimulate or inhibit gene activity (Section 15.5). Any cell that bears molecular receptors for a given hormone is its target. And the targets may be in the same tissue or some distance away.

Researchers have isolated more than eighty different forms of gibberellin from the seeds of flowering plants as well as from fungi. The changes that gibberellins trigger make young cells elongate, which makes stems lengthen (Figure 32.1). In nature, gibberellins help seeds and buds break dormancy and resume growth in spring.

Expose a cabbage plant to a suitable concentration of a gibberellin and its growth may well astound you (Figure 32.2). Applications of gibberellins make celery stalks longer and crispier, and they prevent the skin of navel oranges in orchard groves from ripening too quickly. Walk past those plump seedless grapes in the produce bins of grocery stores and observe how fleshy fruits of the grape plant (*Vitis*) grow in bunches along stems. Gibberellin applications made young cells elongate and stems lengthen between the internodes. This opened up more space between individual grapes, which thereby grew larger. Air could circulate better between grapes, which made it harder for pathogenic, grape-loving fungi to take hold and do damage.

Together, gibberellin and other plant hormones orchestrate plant growth and development. They also

Figure 32.1 Demonstrations of hormonal effects on plant growth and development. (**a**) Seedless grapes radiating market appeal. Gibberellin caused the stems to lengthen, which improved air circulation around grapes and gave them more room to grow. Grapes got larger and weighed more, making growers happy (grapes are sold by weight). (**b**) This young California poppy (*Eschscholzia californica*) was left alone. (**c**) Gibberellin was applied to this young poppy plant.

Figure 32.2 Foolish cabbages! To the left, in front of the ladder, are two untreated cabbages that were the controls. To the right, three cabbage plants treated with gibberellins

Key Concepts

1. From the time a plant seed germinates, a number of hormones influence growth and development.

2. Hormones are signaling molecules between cells. One cell type produces and then secretes a particular kind of hormone, which stimulates or inhibits gene activity in other cell types that take up molecules of the hormone. The changes in gene activity have predictable effects, as when they trigger the mitotic cell divisions and other processes that make stems grow longer.

3. The known plant hormones are auxins, gibberellins, cytokinins, abscisic acid, and ethylene. We also have indirect evidence of other kinds of hormones, which have not yet been identified.

4. Plant hormones govern predictable patterns of development, including the extent and direction of cell growth and differentiation in particular plant parts.

5. Plant hormones help adjust patterns of growth and development in response to environmental rhythms, including seasonal shifts in daylength and temperature. In addition, they help adjust the patterns of response to environmental circumstances in which an individual plant finds itself, such as the amount of sunlight or shade, moisture, and so on at a given site.

6. Commonly, two or more kinds of plant hormones must interact with one another to bring about specific effects on growth and development.

respond to cues from rhythmic changes of the seasons, as when days grow longer and warmer in spring after the short, cold nights of winter. Those grapes, cabbage leaves, or celery stalks that find their way into your mouth are the culmination of a plant's own exquisitely controlled programs of growth and development.

Here we continue with a story that started in the preceding chapter. That chapter traced the formation and development of a flowering plant zygote into a mature embryo, housed in a protective seed coat. At some point after its dispersal from a parent plant, the embryo is transformed into a seedling, which in turn grows and develops into a mature sporophyte. In time, the sporophyte typically forms flowers, then seeds and fruits. Depending on the species, it drops old leaves throughout the year or all at once, in autumn.

This part of the story surveys the heritable, internal mechanisms that govern plant development, and the environmental cues that turn such mechanisms on or off at different times, in different seasons.

PATTERNS OF EARLY GROWTH AND DEVELOPMENT—AN OVERVIEW

How Do Seeds Germinate?

Let's first consider the overall patterns of growth and development for dicots and monocots, starting with the events that take place inside a seed. Figure 32.3 shows the embryo sporophyte inside a grain of corn. (A grain, remember, is a seed-containing dry fruit.) The growth of the embryo idles before or after the seed is dispersed from the parent plant. Later, if all goes well, the seed germinates. **Germination** is the process by which some immature stage in the life cycle of a species resumes growth after a period of arrested development.

Germination depends on environmental factors, such as soil temperature, moisture, and oxygen level, and the number of daylight hours. The factors vary with the seasons. For instance, mature seeds don't hold enough water for cell expansion or metabolism. In many land habitats, ample water is available on a seasonal basis, so seed germination coincides with the spring rains. By the process of **imbibition**, water molecules move into a seed, being attracted mainly to hydrophilic groups of proteins stored in endosperm or cotyledons. As more water moves in, the seed swells and its coat ruptures.

Figure 32.4 (**a**,**b**) Pattern of growth and development of a dicot, the common bean plant (*Phaseolus vulgaris*). When this seed germinates, the embryo in it resumes growth. A hypocotyl (a hook-shaped part of the shoot below the cotyledons) forces a channel through soil, so the food-storing cotyledons are pulled up without being torn apart. At the surface, sunlight makes the hook straighten. Photosynthetic cells make food for several days in the cotyledons, which then wither and fall off. Foliage leaves take over photosynthesis. Flowers form in buds at nodes. (**c**) Cotyledons growing out of a seed coat.

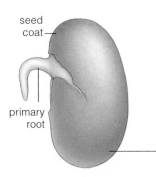

a Bean seedling at the close of germination.

Figure 32.5 Pattern of growth and development of a monocot, corn (*Zea mays*). (**a**,**b**) A corn grain germinates. As the seedling grows through soil, a thin sheath (coleoptile) protects the new leaves. In corn seedlings, adventitious roots develop at the base of the coleoptile. (**c**) Coleoptile and primary root of a corn seedling. (**d**) The coleoptile and first foliage leaf of two seedlings poking out from the soil surface.

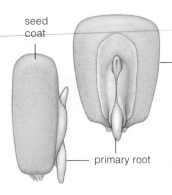

a Corn grain at the close of germination.

Once the seed coat splits, more oxygen reaches the embryo, and aerobic respiration moves into high gear. The embryo's meristematic cells now divide rapidly. The root meristem is usually activated first. Its descendants divide, elongate, and give rise to the first, primary root of a seedling sporophyte. Germination is over when the seedling's primary root breaks through the seed coat.

Genetic Programs, Environmental Cues

Figures 32.4 and 32.5 show patterns of germination, growth, and development for dicots and monocots. The patterns have a heritable basis, being dictated by genes. All cells in a plant arise from the same cell (a zygote), so they generally inherit the same genes. But unequal divisions between daughter cells and their positions in

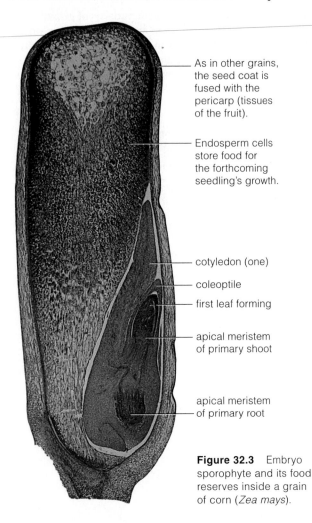

As in other grains, the seed coat is fused with the pericarp (tissues of the fruit).

Endosperm cells store food for the forthcoming seedling's growth.

cotyledon (one)

coleoptile

first leaf forming

apical meristem of primary shoot

apical meristem of primary root

Figure 32.3 Embryo sporophyte and its food reserves inside a grain of corn (*Zea mays*).

one foliage leaf (this type is divided into three leaflets)

primary leaf

primary leaf

point at which cotyledons were attached

withered cotyledon

primary leaf

cotyledons (two)

hypocotyl

branch roots

primary root

root nodule

b Parts that form during a bean plant's early growth and development

coleoptile

branch root

primary root

b Parts that form during a corn plant's early growth and development

first foliage leaf

first internode of stem

adventitious root

branch root

primary root

prop roots that form on corn seedlings and that afford additional support for the rapidly growing stem

first foliage leaf

coleoptile

a developing plant lead to differences in metabolic mechanisms and output. Cell activities start to differ with the onset of selective gene expression (Chapter 15). For example, the genes that govern the synthesis of growth-stimulating hormones are activated in some cells but not others. As you will see shortly, such events seal the developmental fate of different cell lineages.

Bear in mind, plants commonly adjust prescribed growth patterns in response to unusual environmental pressures. Suppose a seed germinates in a vacant lot and a heavy paper bag blows on top of it. The primary

shoot will quickly bend and grow out from under the bag, in the direction of sunlight impinging on it. Interactions among enzymes, hormones, and other gene products in its cells result in this growth response.

Plant growth involves cell divisions and cell enlargements. Plant development requires cell differentiation, as brought about by selective gene expression.

Interactions among genes, hormones, and the environment govern how each individual plant grows and develops.

WHAT THE MAJOR PLANT HORMONES DO

When a plant undergoes **growth**, the number, size, and volume of its cells increase. The cells form specialized tissues, organs, and organ systems through successive stages during **development**. These cells communicate with one another by secreting signaling molecules into extracellular fluid and selectively responding to signals from other cells. Hormones are foremost among the signals. As you know from Chapters 15 and 28, after a signal is received by a target cell, it gets transduced in a way that causes a specific change in metabolism or gene expression. Signal transduction often involves changes in the shape of the receptor, then activation of enzymes and of second messengers in the cytoplasm.

Growth proceeds by cell divisions and enlargements at meristems, as described in Chapter 29. About half of the resulting daughter cells do not increase in size, but they retain their capacity to divide. Their descendants divide in genetically prescribed planes and expand in prescribed directions to give rise to parts with specific shapes and functions. Figure 32.6 shows an example.

Young cells enlarge as they take up water and fluid pressure, or *turgor* pressure, acts on their primary wall. Just as a soft balloon inflates easily, a soft-walled cell expands under turgor pressure. Its wall thickens with polysaccharide additions, and more cytoplasm forms.

Hormones affect the selective gene expression that underlies patterns of plant growth and development, starting with seed germination. Table 32.1 lists the five

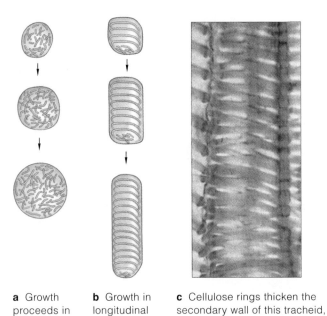

a Growth proceeds in all directions

b Growth in longitudinal direction only

c Cellulose rings thicken the secondary wall of this tracheid, a conducting tube of xylem

Figure 32.6 How cells get their shape. Microtubules in the cytoplasm inherited from a parent cell are already arranged in patterns that govern how cellulose microfibrils will be oriented in the cell wall. (**a**) On exposure to ethylene, their orientation becomes more random. Because a primary wall is elastic all over, the cell expands in all directions and assumes a spherical shape. (**b,c**) On exposure to gibberelins, microtubules assume a transverse orientation and the cellulose microfibrils become similarly arranged. Such a cell can only lengthen.

Table 32.1 *Overview of Plant Signaling Molecules and Their Effects*

Signaling Molecule	Source and Mode of Transport	Stimulatory or Inhibitory Effects
HORMONES		
Gibberellins	Synthesized in apical meristem of bud, leaf, roots, seeds. May travel in xylem and phloem	Makes stems lengthen by stimulating cell division and elongation; often contributes to flowering; helps end bud, seed dormancy
Auxins	Synthesized mainly in apical meristem of bud, leaf, seed. Diffuses from cells in one direction	Stimulates apical dominance, tropisms, vascular cambium divisions, vascular tissue development, fruit formation. Inhibits leaf, fruit abscission
Cytokinins	Synthesized mainly in root tip. Travels from roots to shoots in xylem	Stimulates cell division, leaf expansion. Inhibitory effect on leaf aging. Its application can release buds from apical dominance
Ethylene (a gas)	Synthesized in most parts undergoing ripening, aging, or stress. Diffusion in all directions	Stimulates fruit ripening, abscission of leaf, flower, fruit
Abscisic acid (ABA)	Synthesized in mature leaf in response to water stress; also in stems, unripened fruit	Stimulates stomatal closure, development of embryo sporophytes, distribution of photosynthetic products to seeds, product storage and protein synthesis in seeds. May influence dormancy in some species
GROWTH REGULATORS		
Brassinolides	Steroid	Influences cell division, elongation; required for normal growth; protects against pathogens, stress
Jasmonates	Volatile compound derived from fatty acid	Influences seed germination, root growth, protein storage, defense
Salicylic acid	Phenolic compound structurally like aspirin	Influences tissue defense responses to pathogens
Systemin	Peptide; synthesized in damaged tissues	Influences defense responses in damaged tissues

Figure 32.7 Examples of the effects of auxin. (**a**) *Left:* A cutting from a gardenia plant four weeks after auxin was applied to its base. *Right:* An untreated cutting used as a control.

(**b**) Experiments demonstrating how IAA in a coleoptile tip causes elongation of cells below it. (1) First, cut off the tip of an oat coleoptile. The cut stump will not elongate as much as a normal oat coleoptile (2) used as the control. (3) Place a tiny block of agar under the cut tip and leave it for several hours. IAA diffuses into the agar. (4) Now put the agar on top of another de-tipped coleoptile. Its elongation will proceed about as rapidly as in an intact coleoptile growing next to it (5).

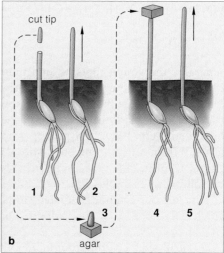

major classes of hormones. They are the gibberellins, auxins, cytokinins, abscisic acid, and ethylene.

Gibberellins stimulate cell division and elongation. As described earlier, they make stems lengthen. They help seeds break dormancy in spring by making cells of the primary root lengthen. In at least some cases, they influence flowering and the development of fruit.

Auxins control cell elongations that lengthen shoots and coleoptiles. A **coleoptile** is a thin sheath that keeps some primary shoots from shredding as growth pushes them up through soil (Figure 32.7*a*). In addition, auxins help vascular tissues and vascular cambium develop. They help keep leaves, flowers, and fruits from falling prematurely. Also, auxin produced in seeds stimulates growth of the ovary wall during fruit formation.

Auxins form in the apical meristems of shoots and coleoptiles. They travel a long distance toward the base of the plant by diffusing through parenchyma cells in vascular tissues. Auxin also causes *apical dominance*, an inhibition of lateral bud growth. Gardeners routinely pinch off the shoot tips to stop auxin from diffusing through stems, so lateral buds are free to branch out.

IAA (indoleacetic acid) is the most pervasive auxin in nature. Orchardists use it to thin flowers in spring so trees yield fewer but larger fruits. They also use it to prevent premature fruit drop so all fruit in the orchard can be picked at the same time, which cuts labor costs.

2,4-D, a synthetic auxin, is a widely used **herbicide** (Section 32.6). It does not seem to harm humans when properly handled. But mixing it with equal parts of a related compound, 2,4,5-T, yields *Agent Orange*. This herbicide was used to clear forested war zones during the Vietnam conflict. Later, experiments linked dioxin, a carcinogen and trace contaminant of 2,4,5-T, to birth defects, miscarriages, leukemias, and liver disorders. 2,4,5-T is now banned in the United States.

Cytokinins stimulate rapid cell division (the name refers to cytoplasmic division). They are abundant in root and shoot meristems and in maturing fruits. They oppose auxin's effects by promoting growth of lateral buds. At least in tissue cultures, cytokinins and auxins work together in promoting rapid cell divisions. Also, cytokinins keep leaves from aging before their time. They are used for basic research and to prolong the shelf life of cut flowers and other horticultural goods.

Abscisic acid (ABA) helps plants adapt to seasonal changes, as by inducing bud dormancy and inhibiting cell growth and premature seed germination. ABA also contributes to stomatal closure when a plant is water stressed (Section 30.4). Growers typically apply ABA to nursery stock before shipping it, because plants are not as vulnerable to injury when they are dormant.

Ethylene, the only gaseous plant hormone, induces fruit ripening, leaf drop, and other aging responses. Ancient Chinese knew to burn incense to ripen fruit faster. By the early 1900s, growers were ripening citrus fruits by storing them in sheds with kerosene stoves. Food distributors now use ethylene to ripen tomatoes and other green fruit after they ship them to grocery stores. Fruit picked green doesn't bruise or deteriorate as fast. Ethylene exposure brightens citrus rinds before the fruit is displayed in the market.

Besides the hormones just described, plants also produce growth regulators of the sort listed in Table 32.1. There are more than these, including some as-yet unknown signal or signals that induce flowering.

Plant hormones and growth regulators are required for normal plant growth and development.

Hormones are signaling molecules secreted by specific cells that influence gene activity in target cells, which may be nearby or some distance away in the multicelled body.

The main categories of plant hormones are gibberellins, auxins, cytokinins, abscisic acid, and ethylene.

ADJUSTING THE DIRECTION AND RATES OF GROWTH

What Are Tropisms?

Generally, the young roots of land plants grow down through soil, and the shoots grow upright through air. Both also can adjust the direction of growth in response to environmental stimuli, as when a new shoot turns toward sunlight. When any root or shoot turns toward or away from an environmental stimulus, we call this a plant tropism (after the Greek *trope*, for turning). As the following examples illustrate, these responses are an outcome of hormone-mediated shifts in the rates at which different cells in the plant grow and elongate.

GRAVITROPISM   The first root to break through a seed coat always curves downward, and the coleoptiles and new stems curve upward. Figure 32.8*a* has an example. These are growth responses to the Earth's gravitational force; they are forms of **gravitropism**.

Figure 32.8*b* shows how to track this response for a seedling turned on its side in a dark room. The stem curves up even in the absence of light. How? In that horizontally oriented stem, cell elongations slow down greatly on the side facing up, but elongations on the lower side increase fast. Different rates of elongation on opposing sides of the stem are enough to make it bend upward. Something apparently makes cells that are on the "bottom" of a stem turned on its side *more* sensitive to a hormone, and those on top *less* so.

Auxin, together with a growth-inhibiting hormone in roots, may trigger such responses. Turn a young root on its side and remove its root cap, and it will *not* curve down. Put the cap back on, and the root curves down. Elongating root cells won't stop growing if you remove the root cap; if anything, cells will grow faster. Suppose a growth inhibitor in root cap cells gets *redistributed* inside a root turned on its side. If gravity causes the inhibitor to leave the cap and accumulate in cells on the root's lower side, the cells won't elongate as much as cells on the upper side. The root will curve down.

Gravity-sensing mechanisms of plants are based on **statoliths**, which generally are clusters of particles in a number of different cells. In plants, they are clusters of unbound starch grains in modified plastids. Figure 32.9 gives one example. These plastids are made denser by the statoliths. They collect near the bottom of root cells in response to gravity and settle downward until they rest in the lowest cytoplasmic region. Statolith redistribution may cause an *auxin* redistribution inside the cells, which initiates the gravitropic response.

PHOTOTROPISM   When stems or leaves adjust the rate and direction of growth in response to light, they are showing **phototropism**. These adjustments favor the light-dependent reactions of photosynthesis, which do not proceed in the dark. Plants that reorient themselves to maximize light interception have an advantage.

Charles Darwin pondered on phototropism after it dawned on him that a coleoptile was growing toward light on one side of its tip. But it wasn't until the 1920s that Fritz Went, a graduate student in Holland, linked phototropism with a growth-promoting substance. He was the person who named the substance auxin (after the Greek *auxein*, "to increase"). Went demonstrated that auxin moves from the tip of a coleoptile into cells

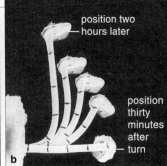

Figure 32.8 Observational tests of gravitropic responses by plants. (**a**) Responses of a corn primary root and shoot that are growing in their normal orientation (*left*) and when turned upside down (*right*). (**b**) One way to measure a gravitropic response. Force a sunflower seedling to grow in a darkened room for five days. Then turn it on its side, mark the shoot at 0.5-centimeter intervals, and watch what happens.

Figure 32.9 Observational test of the gravitropic responses by young roots. (**a**) Normal orientation of plastids in root cap cells of corn. Statoliths in these cells settle downward. (**b**) Turn the root sideways. Five to ten minutes later, plastids have settled to the new "bottom" of the cells. A gravity-sensing mechanism using statoliths may be sensitive to auxin redistribution in a root tip. A difference in auxin concentrations may cause cells on the "top" of a root turned sideways to elongate faster than cells on the bottom. Different elongation rates will make the root tip curve downward.

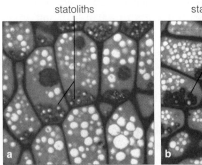
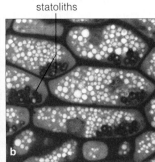

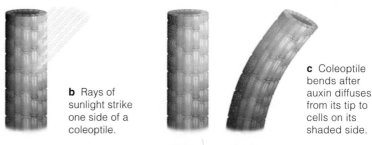

b Rays of sunlight strike one side of a coleoptile.

c Coleoptile bends after auxin diffuses from its tip to cells on its shaded side.

Figure 32.10 (a) Phototropism by tomato seedlings. (b,c) Hormone-mediated differences in the rates of cell elongation induce coleoptiles and stems to bend toward light.

less exposed to light and makes them elongate faster than cells on the illuminated side. The difference in rates of growth results in bending toward light.

You can observe a phototropic response by putting seedlings of sun-loving plants in a dark room next to a window through which rays of sunlight are streaming. Tomato seedlings will do. They will start curving in the direction of the most light (Figure 32.10).

Blue wavelengths of light are known to induce the strongest phototropic response. These wavelengths can be absorbed by **flavoprotein**. This is a yellow pigment, and it may be the receptor that transduces light energy into the phototropic bending mechanism.

THIGMOTROPISM Plants also shift their direction of growth when they contact solid objects. This response is called **thigmotropism** (after the Greek *thigma*, which means "touch"). Auxin and ethylene may have roles in the response. Vines, which are stems too slender or soft to grow upright without support, make such a contact response. So do tendrils, which are modified leaves and stems that wrap around objects and help support the plant (Figure 32.11). You can observe what happens as they grow against a stem of another plant. Within minutes, cells on the contact side stop elongating and the vine or tendril starts curling around the stem, often more than once. Afterward, the cells on both sides will resume growth at the same rate.

Responses to Mechanical Stress

Mechanical stress, as inflicted by prevailing winds and grazing animals, can inhibit stem elongation and plant growth. You can see such effects on trees growing near the snowline of windswept mountains; they are stubby compared to trees of the same species that are growing at lower elevations. Similarly, plants grown outdoors commonly have shorter stems than plants grown in a

Figure 32.11 Passion flower (*Passiflora*) tendril busily twisting thigmotropically.

Figure 32.12 Effect of mechanical stress on tomato plants. (**a**) This plant, the control, grew in a greenhouse. (**b**) Each day for twenty-eight days, this plant was mechanically shaken for thirty seconds. (**c**) This one had two shakings each day.

greenhouse. You can observe this response to stress by shaking a plant daily for a brief period. Doing so will inhibit growth of the whole plant (Figure 32.12).

Plants adjust the direction and rate of their growth in response to environmental stimuli.

HOW DO PLANTS KNOW WHEN TO FLOWER?

All flowers are variations on the same pattern of growth and development. In Section 28.5, you read about the expression of genes that control floral development in *Arabidopsis thaliana*. A hormonal signal activates these genes. But what activates the hormone-secreting cells?

An Alarm Button Called Phytochrome

All organisms have internal mechanisms that preset the time for recurring changes in biochemical events. Some of the internal timing mechanisms—**biological clocks**—trigger shifts in daily activities. Section 28.5 gave one example, the rhythmic leaf movements of a bean plant. That leaf movement is a *circadian* rhythm, a biological activity repeated in cycles, each lasting for close to twenty-four hours. Experiments by Ruth Satter, Richard Crain, and their colleagues at the University of Connecticut showed that phytochrome has a role in leaf movements. Satter, a pioneer in the study of timing mechanisms, was one of the first to correlate plant rhythms with "hands of a biological clock."

Biological clocks also help bring about seasonal adjustments in basic patterns of growth, development, and reproduction—including flower formation.

Some biological clocks have an alarm button called **phytochrome**. This blue-green pigment is part of the switching mechanisms that promote or inhibit growth for a variety of plant parts. Phytochrome is a receptor for red and far-red wavelengths of light. At sunrise, red wavelengths dominate the sky. They are the signal that causes phytochrome to change shape to Pfr, its active molecular form (Figure 32.13). The signal is transduced in one of two ways. Either cells are induced to take up free calcium ions (Ca++) or organelles are induced to

release them. The response starts as ions combine with calcium-binding proteins in cells. At sunset, at night, or in the shade, the signal transduction pathway is reversed. At such times, wavelengths are primarily far-red. They cause phytochrome to revert to its inactive molecular form (Pr).

Photoperiodism is a biological response to change in the length of daylight relative to darkness in a cycle of twenty-four hours. Phytochrome's active form, Pfr, sets in motion events that result in the transcription of genes. The gene products include signaling molecules that deal with seed germination, shoot elongation and branching, expansion of leaves, flower, fruit, and seed formation, and dormancy (Figure 32.14).

Flowering—A Case of Photoperiodism

Different flowering plants start diverting more energy to forming flowers at different times of year. Tulips flower in spring, for example, and chrysanthemums in autumn. It seems probable that the secretion of one or more flower-inhibiting and flower-inducing hormones depends on timed phytochrome responses to cues from the environment. Despite intensive searches, however, no one has yet identified the elusive hormone(s).

It is a puzzle. Tulips, spinach, and other **long-day** plants flower in the spring, when daylength exceeds some critical value. Chrysanthemums, poinsettias, and cockleburs are among the **short-day** plants. They flower in late summer or early autumn, when daylength is shorter than a critical value. **Day-neutral** plants simply flower when they are mature enough to do so.

Actually, these are not very good names, because the environmental cue is *night length*, not daylength.

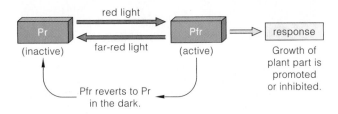

Figure 32.13 Interconversion of the phytochrome molecule from active form (Pfr) to inactive form (Pr). This blue-green pigment is part of a switching mechanism that promotes or inhibits the growth of a variety of plant parts.

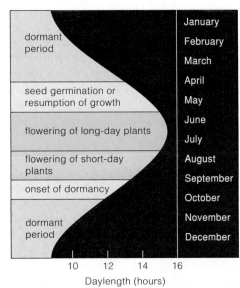

Figure 32.14 Plant growth and development as correlated with the number of hours of light available each day. The number changes with the passing of the seasons. The data shown reflect photoperiodic responses of plants that grow in temperate regions of North America. In such regions, rainfall and temperature shift with the seasons.

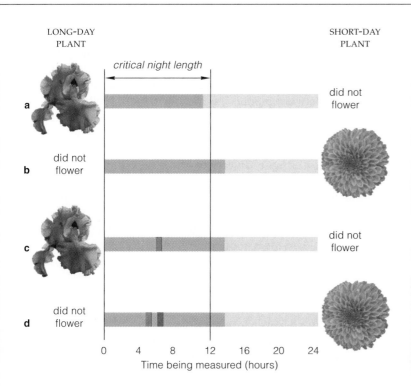

LONG-DAY PLANT

SHORT-DAY PLANT

critical night length

a did not flower

b did not flower

c did not flower

d did not flower

Time being measured (hours)
0 4 8 12 16 20 24

Figure 32.15 Experiments showing that short-day plants flower by measuring night length. Each horizontal bar signifies 24 hours. *Yellow* bars signify daylight; *blue* bars signify night. (**a**) Long-day plants flower when the night is *shorter* than a length that is critical for flowering. (**b**) Short-day plants flower when the night is *longer* than a critical value. (**c**) When an intense red flash interrupts a long night, both kinds of plants respond as if it were a short night. (**d**) A short pulse of far-red light after the red flash cancels the disruptive effect of the red flash.

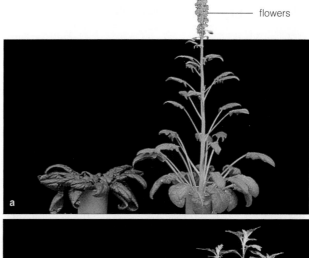

flowers

a

b

Figure 32.16 Experiment involving flowering responses of (**a**) spinach, a long-day plant, and (**b**) chrysanthemum, a short-day plant. In both photographs, the plant at *left* grew under short-day conditions. The plant at *right* grew under long-day conditions.

Figures 32.15 and 32.16 show the responses of long-day and short-day plants to a range of light conditions. They all respond to the wavelengths that predominate at dawn and at dusk. When experimenters interrupt a critical dark period with a pulse of red light, they reset the clocks. "Short-day" plants flower only when nights are longer than a critical value. And "long-day" plants flower when the night is shorter than a critical value.

And so spinach plants won't flower and form seeds unless they are exposed to ten hours of darkness for two weeks. That's why you would not want to start, say, a spinach seed farm in the tropics, where that set of cues is not available. Chrysanthemum growers stall the flowering process by exposing plants to a flash of light at night to break up one long night into two short ones. That is why they can offer chrysanthemums in spring as well as autumn. Cockleburs normally flower after only one night longer than 8–1/2 hours. When artificial light interrupts their dark period for even a minute, they won't flower. And why didn't poinsettias planted along California's interstate highways put out flowers? Headlights from vehicles zipping past during the night inhibited the flowering response.

The unidentified hormone(s) that are thought to be central to flowering might be produced in leaf cells and transported to new floral buds. Experiments suggest this is so. For example, trim all but one leaf away from a cocklebur plant. Next, cover that leaf entirely with black paper for 8–1/2 hours. The plant will proceed to produce flowers. However, cut off that one cocklebur leaf right after the dark period is completed, and the plant will not produce flowers.

Like other organisms, flowering plants have biological clocks, which are internal time-keeping mechanisms.

Phytochrome, a blue-green pigment, is part of a switching mechanism for phototrophic responses to light of red and far-red wavelengths. Its active form, Pfr, might trigger the secretion of one or more hormones that induce and inhibit flowering at different times of year.

The main environmental cue for flowering is the length of night—that is, the hours of darkness, which vary with the seasons. Different kinds of plants form flowers at different times of year, depending on their phytochrome mechanism.

LIFE CYCLES END, AND TURN AGAIN

Senescence

While leaves and fruits are growing, cells inside them produce auxin (IAA), which moves into stems. There it interacts with cytokinins and gibberellins to maintain growth. When autumn approaches and the length of daylight decreases, plants start to withdraw nutrients from leaves, stems, and roots, then distribute them to flowers, fruits, and seeds. Deciduous plants shed leaves as a growing season ends. They channel nutrients to storage sites in twigs, stems, and roots before their leaves die and drop. The dropping of flowers, leaves, fruits, and other plant parts is known as **abscission**. An abscission zone is made of thin-walled parenchyma cells at the base of a petiole or some other part about to drop from the plant. Figure 32.17 shows one example.

Senescence is the sum total of processes that lead to death of a plant or some of its parts. The recurring cue is a decrease in daylength. But other factors, such as drought, wounds, and nutrient deficiencies, also bring it about. The cue causes a decline in IAA production in leaves and fruits. A different signal that the plant itself may produce stimulates abscission zone cells to make ethylene. Cells enlarge, deposit suberin in their walls, and produce enzymes that digest cellulose and pectin in middle lamellae. A middle lamella is the cementing layer between plant cell walls (Section 4.11). While cells continue to enlarge and their walls become digested, they separate from one another. Eventually the leaf or other plant part above the abscission zone drops away.

Interrupt the diversion of nutrients into flowers, seeds, or fruits, and you stop aging of a plant's leaves, stems, and roots. For example, if you remove each new flower or seed pod from a plant, its leaves and stems will remain vigorous and green much longer (Figure 32.18). Gardeners routinely remove flower buds from many kinds of plants to maintain vegetative growth.

Entering Dormancy

As autumn approaches and days grow shorter, many perennials and biennials start to shut down growth. They do so even when temperatures are still mild, the sky is bright, and water is plentiful. When a plant stops growing under conditions that seem (to us) suitable for growth, it has entered a state of **dormancy** in which its metabolic activities idle. Ordinarily, the plant's buds will not resume growth until there is a convergence of precise environmental cues in early spring.

Short days, long, cold nights, and dry soil that is deficient in nitrogen are strong cues for dormancy. Researchers tested this by interrupting the long dark period of some Douglas firs with a short period of red light. The plants responded as if nights were shorter and days longer. They continued to grow taller (Figure 32.19). In this experiment, conversion of Pr to Pfr by red light during the dark period prevented dormancy.

tissues of stem   cells of abscission zone

Figure 32.17 Light micrograph of an abscission zone in a red maple (*Acer*), in longitudinal section. The zone forms at the base of leaf petioles.

control (pods not removed)   experimental plant (pods removed)

Figure 32.18 The observable results from an experiment in which seed pods were removed from a soybean plant. Removal delayed its senescence.

Figure 32.19 Experiment to test the effect of the relative length of day and night on Douglas firs. The fir at left was exposed to 12-hour light/12-hour dark cycles for one year. Its buds stayed dormant; "daylength" was too short. The fir at right was exposed to 20-hour light/4-hour dark cycles. It continued growing. The fir in the middle was exposed to 12-hour light/11-hour dark cycles with 1 hour of light interrupting the middle of the dark period. The interruption prevented bud dormancy; it caused Pfr formation at a sensitive time in the normal day–night cycle.

Potted plant grown inside a greenhouse did not flower.

Branch exposed to cold outside air flowered.

Figure 32.20 Localized effect of cold temperature on dormant buds of a lilac (*Syringa*).

For this experiment, one branch of the lilac plant was positioned so that it protruded from a greenhouse during a cold winter. The rest of the plant was kept inside and exposed only to warmer temperatures. Only buds on the branch exposed to low outside temperatures resumed growth in spring.

In nature, buds might enter dormancy because less Pfr can form when daylength shortens in late summer.

The requirement for multiple cues for dormancy has adaptive value. For example, if temperature were the only cue, then warm autumn weather might make plants flower and seeds germinate—and winter frost would kill them. By contrast, with artificial selection, growers have developed seeds that germinate promptly in greenhouses any time of year.

Breaking Dormancy

Temperatures as well as daylength change seasonally in most places, and dormancy-breaking mechanisms work between fall and spring. The temperatures often become milder, and rains and nutrients again become available. With the return of favorable conditions, life cycles turn once more. Seeds germinate; buds resume growth and give rise to new leaves, then to flowers.

Breaking dormancy probably requires gibberellins and abscisic acid. Often it does not occur without prior exposure to low temperatures at specific times of year. Temperatures required to break dormancy vary among plants. For example, Delicious apple trees growing in Utah require 1,230 hours near 43°F (6°C). Apricot trees grown there require 720 hours. Generally, trees in the southern United States require less cold exposure than the trees growing in northern states and Canada. If you

live in Colorado and order a young peach tree from a Georgia nursery, the tree might start spring growth too soon and be killed by late frost or heavy snow.

Vernalization

Flowering, too, is often a response to seasonal changes in temperatures. To give one example, unless buds of some biennials and perennials become exposed to low winter temperatures, flowers will not form on their stems in spring. The low-temperature stimulation of flowering is called **vernalization** (from *vernalis*, which means "to make springlike"). Figure 32.20 shows how you can gather experimental evidence of this effect.

As long ago as 1915, the plant physiologist Gustav Gassner studied flowering of some cereal plants after exposing their seeds to controlled temperatures. As an example, he germinated seeds of winter rye (*Secale cereale*) at near-freezing temperatures. Plants flowered the next summer even when they were planted in late spring. Vernalization is common in agriculture.

Multiple cues from the environment influence hormonal secretions that stimulate or inhibit processes of growth and development during the life cycle of plants. These cues include changes in daylength, temperature, moisture, and nutrient availability.

GROWING CROPS AND A CHEMICAL ARMS RACE

In this unit, you glimpsed the nature of plant structure and function. Think about that the next time you eat strawberries, corn, beans, or anything else grown on fertilized cropland. How did the plants you eat for nutrients get enough of those nutrients? How did they manage to compete with weeds and avoid pests that ruin or gobble up nearly half of what we try to grow?

Most plants aren't entirely defenseless. They evolved under selection pressure of attacks by insects and other organisms and often repel attackers with natural toxins. A **toxin** is an organic compound, a normal metabolic product of one species, but its chemical effects harm or kill a different organism that contacts it. Humans, too, encounter traces of natural toxins, even in such familiar foods as hot peppers, potatoes, figs, celery, rhubarb, and alfalfa sprouts. Still, we do not die in droves, so we must have chemical defenses against these toxins.

a *2,4-D* (2,4-dichlorophenoxyacetic acid), a synthetic auxin widely used as a herbicide. Enzymes of weeds and microbes cannot easily degrade it, compared to natural auxins.

b *Atrazine*, the bestselling herbicide, kills weeds weeds in a few days, as do glyphosate (Roundup), alachlor, (Lasso), and daminozide (Alar). It now appears that atrazine causes abnormal sexual development in frogs, even in trace amounts below the level allowed in drinking water.

c Dichlorodiphenyltrichloroethane, or *DDT*. It takes two to fifteen years for this nerve cell poison to break down. Chlordane, another type of insecticide, also persists for a long time in the environment.

d *Malathion*. Like other organophosphates, it is cheap, breaks down faster than chlorinated hydrocarbons, and is more toxic. Organophosphates represent half of all insecticides used in the United States. Some are now banned for crops; application of others must end at least three weeks before harvest. Farmers who contest this policy want the Environmental Protection Agency to consider economic and trade issues as well as human health.

Figure 32.21 A few pesticides, some more toxic than others.

Figure 32.22 One of the crop dusters that intervene in the competition for nutrients between crop plants and pests, including weeds.

Just a few thousand years ago, farmers used sulfur, lead, arsenic, and mercury to protect crop plants from insects. They freely dispensed the highly toxic metals until the late 1920s, when someone figured out they were poisoning people. Traces of the toxic metals still turn up in contaminated lands.

The farmers also used organic compounds extracted from leaves, flowers, and roots as natural pesticides. In 1945, scientists started to make synthetic toxins and to identify mechanisms by which toxins attack pests. The *herbicides*, including the synthetic auxin mentioned in Figure 32.21a, kill weeds by disrupting metabolism and growth. *Insecticides* clog the airways of a target insect, disrupt its nerves and muscles, or prevent its reproduction. *Fungicides* work against harmful fungi, including the mold that makes aflatoxin, one of the deadliest poisons. By 1995, people in the United States were spraying or spreading more than 1.25 billion pounds of toxins through fields, gardens, homes, and industrial and commercial sites (Figure 32.22).

Some pesticides also kill birds and other predators that help control pest population sizes. Besides this, targeted pests have been developing resistance to the chemical arsenal, for reasons given earlier (Chapter 1). Pesticides cannot be released haphazardly, for they can be inhaled, ingested with food, or absorbed through skin. Different types are active for weeks or years. Some trigger rashes, headaches, hives, asthma, and joint pain in millions of people. Some trigger life-threatening allergic reactions in abnormally sensitive people.

DDT and other long-lasting pesticides are currently banned in the United States. Even rapidly degradable ones are subjected to rigorous application standards and ongoing safety tests.

And how we protect crop plants is just one aspect of what must be done to support a human population that now exceeds 6.1 billion. Should essential crop plants be genetically engineered? Can we, and should we, make them better at fixing nitrogen or tolerating salts? What about bruise resistance? Built-in pesticide resistance? Here you might reflect once more on Sections 28.2, 16.7, and the first *Critical Thinking* question in 16.11.

How well plants grow and develop ultimately affects the growth and development of the human population.

SUMMARY

Gold indicates text section

1. This chapter started with what happens after seeds are dispersed from parent plants. Inside each seed, the embryo sporophyte has been dormant. It germinates by imbibition—it absorbs water, resumes growth, and breaks through its seed coat. Now control mechanisms govern the growth and development of the new plant. The seedling increases in volume and mass. Its tissues and organs develop. Later, fruits and new seeds form, then older leaves drop. *32.1*

2. How individual plants grow and develop depends on interactions among their genes, their hormones, and cues from the environment. *32.2*

   a. A plant's genes govern the synthesis of enzymes and other proteins necessary for metabolism, hence for all cell activities. How and when each type of enzyme functions depend on hormonal action.

   b. Hormones are a category of signaling molecules. After being produced and secreted by some cells, they travel to target cells in different parts of the plant body and stimulate or inhibit gene activity. As is the case for other kinds of organisms, any cell that bears receptors for a particular hormone is its target.

   c. The prescribed growth patterns are influenced by environmental cues. Often, they become adjusted in response to unusual environmental pressures.

3. Plant hormones bring about predictable patterns of growth and development. They also trigger responses to environmental rhythms (such as seasonal changes in daylength and temperature) and variations in shade, sunlight, and other circumstances. *32.1, 32.2*

4. Five major categories of plant hormones have been identified. Probably there are others. *32.2–32.5*

   a. Gibberellins promote stem elongation, they help seeds and buds break dormancy in spring, and they may help induce the flowering process.

   b. Auxins promote coleoptile and stem elongation. They have roles in phototropism and gravitropism.

   c. Cytokinins stimulate cell division, promote leaf expansion, and retard leaf aging.

   d. Abscisic acid promotes bud and seed dormancy, and it limits water loss by promoting stomatal closure.

   e. Ethylene promotes fruit ripening and abscission.

5. Plant parts make tropic responses to light, gravity, and other environmental conditions. Hormones induce a difference in the rate and direction of growth on two sides of the part, which causes it to turn or move. *32.3*

   a. With gravitropism, roots grow downward and stems grow upright in response to the Earth's gravity. Some gravity-sensing mechanisms in plants are based on statoliths (clusters of particles in cells).

   b. With phototropism, stems and leaves adjust rates and directions of growth in response to light. A yellow pigment (flavoprotein) might be involved; it absorbs blue wavelengths that trigger the strongest response.

   c. With thigmotropism, plants adjust their direction of growth in response to contact with solid objects.

6. Plants respond to mechanical stress, as when strong winds inhibit stem elongation and plant growth. *32.3*

7. A biological clock is any internal, time-measuring mechanism that has a biochemical basis. *32.4*

   a. Circadian rhythms are biological activities that recur in cycles that each last about twenty-four hours. Rhythmic movements of leaves are an example.

   b. Photoperiodism is a biological response to change in the relative length of daylight and darkness in the twenty-four-hour cycle. Photoperiodism is seasonal in plants. Phytochrome, a blue-green pigment, is part of a switching mechanism for a clock. It helps promote or inhibit germination, stem elongation, leaf expansion, stem branching, and flower, fruit, and seed formation.

   c. Long-day plants flower during spring or summer, when there are more hours of daylight than darkness. Short-day plants flower when daylength is less. Day-neutral plants flower regardless of daylength.

8. Senescence is the sum of processes leading to the death of a plant or plant structure. *32.5*

9. Dormancy is a state in which a perennial or biennial stops growing even when conditions appear suitable for continued growth. A decrease in Pfr levels might trigger dormancy. Breaking dormancy might involve exposure to certain temperatures and hormonal action, including gibberellins and abscisic acid. *32.5*

Review Questions

1. Explain the process by which seeds germinate. *32.1*

2. Briefly describe how cells of a new plant grow, enlarge, and take on specific shapes. *32.1*

3. List the five known types of plant hormones and describe the known functions of each. *32.2*

4. List a few plant growth regulators and their functions. *32.2*

5. Define plant tropism. What is the difference between phototropism and photoperiodism? *32.3, 32.4*

6. What is phytochrome, and what role does it play in the flowering process? *32.4*

7. Explain the differences between long-day, short-day, and day-neutral plants. *32.4*

8. Define dormancy and senescence. Give examples. *32.5*

Self-Quiz   ANSWERS IN APPENDIX III

1. Seed germination is over when the _____ .
   a. embryo sporophyte absorbs water
   b. embryo sporophyte resumes growth
   c. primary root pokes out of the seed coat
   d. cotyledons unfurl

Figure 32.23 Field of sunflowers (*Helianthus*) that are busily demonstrating solar tracking.

2. Which of the following statements is false?
   a. Auxins and gibberellins promote stem elongation.
   b. Cytokinins promote cell division but retard leaf aging.
   c. Abscisic acid promotes water loss and dormancy.
   d. Ethylene promotes fruit ripening and abscission.

3. Plant hormones _____ .
   a. interact with one another
   b. are influenced by environmental cues
   c. are active in plant embryos within seeds
   d. are active in adult plants
   e. all of the above

4. Plant growth depends on _____ .
   a. cell division   c. hormones
   b. cell enlargement   d. all of the above

5. _____ are the strongest stimulus for phototropism.
   a. Red wavelengths   c. Green wavelengths
   b. Far-red wavelengths   d. Blue wavelengths

6. Light of _____ wavelengths makes phytochrome switch from inactive to active form; light of _____ wavelengths has the opposite effect.
   a. red; far-red   c. far-red; red
   b. red; blue   d. far-red; blue

7. The flowering process is a _____ response.
   a. phototropic   c. photoperiodic
   b. gravitropic   d. thigmotropic

8. Abscission occurs during _____ .
   a. seed germination   c. senescence
   b. flowering   d. dormancy

9. Senescence involves a decrease in _____ in leaves and fruits and an increase in _____ at abscission zones.
   a. IAA; ethylene   c. Pfr; gibberellin
   b. ethylene; IAA   d. gibberellin; abscisic acid

10. Match the plant reproduction and development terms.
    ____ vernalization   a. water moves into seeds
    ____ senescence   b. unequal growth following
    ____ imbibition   contact with solid objects
    ____ thigmotropism   c. lateral bud formation inhibited
    ____ apical   d. low-temperature stimulation
         dominance   of the flowering process
                             e. all processes leading to death
                                of plant or plant part

Critical Thinking

1. Given what you know about the growth of plants (Chapter 29), would you expect hormones to influence primary growth only? What about secondary growth in, say, a redwood tree?

2. Plant growth depends on photosynthesis, which depends on inputs of light energy from the sun. How, then, can seedlings that were germinated in a dark room grow taller than different seedlings that germinated in the sun?

3. *Solar tracking* refers to the observation that many plants are able to maintain the flat blades of their leaves at right angles to the sun throughout the day. Figure 32.23 gives an example. This tropic response maximizes the harvesting of the sun's rays by leaves. Suggest the name of one type of molecule that might be involved in the response.

4. Belgian scientists isolated a mutated gene in wall cress (*Arabidopsis thaliana*) that produces excess amounts of auxin. Predict what some of the resulting phenotypic traits might be.

5. Remember Section 28.5? All flowers are variations on a basic pattern of growth and development. Two groups led by Elliot Meyerowitze and Detlef Weigel recently identified the genes responsible for that plan in *Arabidopsis*. The master gene (*leafy*) activates other genes that contribute to the formation of sepals (gene *A*), petals (gene *B*), and reproductive structures (gene *C*). Discovering these genes and their interactions has been called the botanical equivalent of isolating master genes of *Drosophila* development. Speculate on what kind of internal and external signals switch on the leafy gene in the first place.

6. Cattle typically are given somatotropin, an animal hormone that makes them grow bigger (the added weight means greater profits). There is a major concern that such hormones may have unforeseen side effects on beef-eating humans. Would you think plant hormones applied to crop plants can affect humans also? Why or why not?

Selected Key Terms

abscisic acid *32.2*	ethylene *32.2*	photoperiodism *32.4*
abscission *32.5*	flavoprotein *32.3*	phototropism *32.3*
auxin *32.2*	germination *32.1*	phytochrome *32.4*
biological clock *32.4*	gibberellin *CI*	senescence *32.5*
coleoptile *32.2*	gravitropism *32.3*	short-day
cytokinin *32.2*	growth *32.2*	plant *32.4*
day-neutral	herbicide *32.2*	statolith *32.3*
plant *32.4*	hormone *CI*	thigmotropism *32.3*
development *32.2*	imbibition *32.1*	toxin *32.6*
dormancy *32.5*	long-day plant *32.4*	vernalization *32.5*

Readings

Meyerowitz, E. M. November 1994. "Genetics of Flower Development." *Scientific American* 271: 56–65.

Raven, P. , R. Evert, and S. Eichhorn. 1999. *Biology of Plants.* Sixth edition. New York: Freeman/Worth.

Rost, T., M. Barbour, C. Stocking, and T. Murphy. 1998. *Plant Biology.* Belmont, California: Wadsworth. Paperback.

Salisbury, F., and C. Ross. 1992. *Plant Physiology.* Fourth edition. Belmont, California: Wadsworth.

On-Line readings at Student Guide for InfoTrac:
www.brookscole.com/biology

VI Animal Structure and Function

How many and what kinds of body parts does it take to function as a lizard in a tropical forest? Make a list of what comes to mind as you start reading Unit VI, then see how resplendent the list can become at the unit's end.

ANIMAL TISSUES AND ORGAN SYSTEMS

Meerkats, Humans, It's All the Same

After a cold night in Africa's Kalahari Desert, animals small enough to fit inside a coat pocket emerge stiffly from their burrows. These "meerkats" are a type of mongoose. They stand on their hind legs and face east, exposing their chilled bodies to the warm rays of the morning sun (Figure 33.1). Meerkats don't know it, but sunning behavior helps their enzymes. If their body's internal temperature were to fall below some tolerable range, the action of countless enzyme molecules in their cells would falter and metabolism would suffer.

Once meerkats warm up, they fan out from their burrows and search for food. Into the meerkat gut go insects and the occasional lizard. These are pummeled, dissolved, and digested into glucose and other nutritious bits small enough to move across the gut wall, into the blood, and on to the body's cells. Aerobic machinery in the cells cracks apart molecules of glucose and other organic compounds and releases energy. A respiratory system supplies the machinery with oxygen and takes away carbon dioxide leftovers.

All of this activity changes the composition and volume of the **internal environment**—blood and the interstitial fluid (tissue fluid) that bathes all living cells of any complex animal. Drastic changes in these fluids would kill the cells, but a urinary system works to keep

that from happening. Governing this system and the others are the nervous and endocrine systems. They interact as a central command post to mobilize the body as a whole for everything from simple housekeeping tasks to heart-thumping flights from predators.

And so meerkats start us thinking more closely about this unit's topics: how the animal body and its parts are physically put together (its *anatomy*) and how it functions in the environment (its *physiology*). This chapter is an introduction to the animal tissues and organ systems we will consider. It also reminds us of the concept of **homeostasis**. As you read earlier, in Section 28.3, homeostasis is a state in which operating conditions in the internal environment stay within a range that the body's cells can tolerate. When all goes well, that state is being maintained by the coordinated activities of cells, tissues, organs, and organ systems.

Amazingly, all complex animals are constructed of only four basic types of tissues. These are epithelial, connective, muscle, and nervous tissues. Each **tissue**, remember, is a community of cells and intercellular substances that perform one or more tasks, such as contraction by muscle tissue. Each **organ** consists of different tissues organized in the proportions and patterns necessary to carry out specific tasks. Thus

the vertebrate heart has predictable proportions and arrangements of epithelial, connective, muscle, and nervous tissues. Each **organ system** consists of two or more organs interacting physically, chemically, or both in a common task. We see this in the interconnected arteries and other vessels that transport blood through the body under the driving force of a beating heart.

By the end of this unit, you may have an abiding appreciation for the sheer magnitude of the division of labor and integration among cells, tissues, organs, and organ systems. And whether you look at a flatworm or salmon, a meerkat or human, keep reminding yourself that each animal is structurally and physiologically adapted to perform four overriding tasks:

1. *Maintain conditions in the internal environment within ranges that living cells can tolerate.*

2. *Acquire water, nutrients, and other raw materials, distribute them through the body, and dispose of wastes and residues.*

3. *Afford protection against injury or attack from viruses, bacteria, and other agents of disease.*

4. *Reproduce, then often help nourish and protect new individuals during their early growth and development.*

Figure 33.1 In the Kalahari Desert, gray meerkats (*Suricata suricatta*) face the sun's warming rays, as they do each morning. This simple behavior helps them maintain internal body temperature. How animals function in their environment is the focus of this unit.

Key Concepts

1. In most animals, cells interact at three levels of organization—in tissues, organs, and organ systems.

2. Most animals are constructed of four types of tissues. We call them epithelial, connective, muscle, and nervous tissues.

3. Epithelial tissues line the body's surface and its internal cavities and tubes. Different types have protective and secretory functions.

4. A variety of connective tissues bind, support, strengthen, protect, and insulate other tissues. They include soft connective tissues, cartilage, bone, blood, and adipose tissue.

5. Muscle tissues contract (shorten) and return to their resting position. By interacting with some type of skeleton, they move the body and its parts.

6. Nervous tissue consists of neurons, the basic units of communication in nervous systems, and of diverse neuroglial cells that structurally and functionally support the neurons.

7. All vertebrates have the same general assortment of organ systems. The systems arise from just three kinds of primary tissues, called ectoderm, mesoderm, and endoderm, that start forming in early embryos.

8. Each tissue, organ, and organ system has a specialized function. Different kinds contribute to maintaining stable operating conditions in the internal environment, acquiring and distributing substances, disposing of wastes, and protecting the body. Some kinds function in reproduction.

EPITHELIAL TISSUE

General Characteristics

Epithelium (plural, epithelia) is a sheetlike tissue with a free surface facing the outside environment or a body fluid. *Simple* epithelium, with one layer of cells, serves as a lining for body cavities, ducts, and tubes. *Stratified* epithelium, such as the epidermis of your skin, has two or more cell layers, and it often functions in protection. Figure 33.2 has examples of this type of animal tissue.

All cells in epithelium are positioned close together, with little intervening material. They make, absorb, and secrete (release) substances. As is the case for cells in nearly all animal tissues, specialized junctions serve as structural and functional links between them.

Cell-to-Cell Contacts

In epithelium and other tissues, we see three classes of cell junctions. **Tight junctions** prevent substances from leaking across the tissue. **Adhering junctions** are like spot welds; they cement neighboring cells together. **Gap junctions** are open channels between the cytoplasm of abutting cells. They facilitate the rapid transfer of ions and small molecules from one cell to its neighbors.

Consider tight junctions. The cells in some epithelia have parallel rows of proteins that form tight seals and fuse each cell with its neighbors. The junctions prevent most substances from leaking across the tissue's free surface. Substances reach tissues below only by passing through the cytoplasm of epithelial cells (Figure 33.3).

free surface of epithelium

simple squamous epithelium

basement membrane

connective tissue

a

Figure 33.2 (**a**) Some basic characteristics of epithelium. Epithelium has a free surface exposed to a body fluid or the outside environment. Between the opposite surface and the underlying connective tissue is a basement membrane. All epithelial cells rest on this layer of extracellular material. The sketch below the athlete shows this arrangement for simple epithelium, with one layer of cells. The light micrograph below her arm is a section of the upper portion of stratified epithelium, which has more than one layer of cells. Its cells become flattened out near the free surface.

(**b**) Light micrographs and sketches of three simple epithelia, showing three common shapes of cells for this tissue.

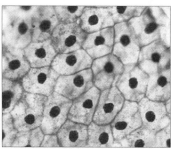

b

TYPE: Simple squamous
DESCRIPTION: Friction-reducing slick, single layer of flattened cells
COMMON LOCATIONS: Lining of blood and lymph vessels, heart; air sacs of lungs; peritoneum
FUNCTION: Diffusion; filtration; secretion of lubricants

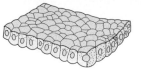

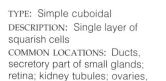

TYPE: Simple cuboidal
DESCRIPTION: Single layer of squarish cells
COMMON LOCATIONS: Ducts, secretory part of small glands; retina; kidney tubules; ovaries, testes; bronchioles
FUNCTIONS: Secretion, absorption

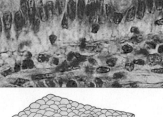

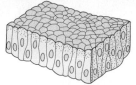

TYPE: Simple columnar
DESCRIPTION: Single layer of tall cells; free surface may have cilia, mucus-secreting glandular cells, microvilli
COMMON LOCATIONS: Glands, ducts; gut; parts of uterus; small bronchi
FUNCTION: Secretion; absorption; ciliated types move substances

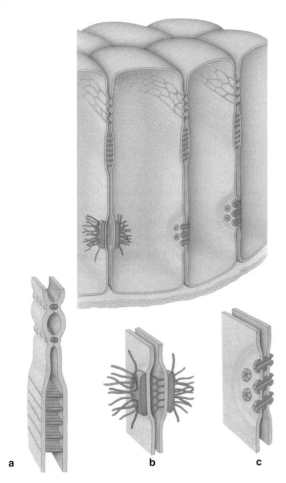

a   b   c

TIGHT JUNCTION	ADHERING JUNCTION	GAP JUNCTION
Strands (rows of proteins) running parallel with the free surface of the tissue; they block leaking between adjoining cells.	Adjoining cells adhere at a mass of proteins. The mass is anchored beneath their plasma membrane by many intermediate filaments of the cytoskeleton.	Cylindrical arrays of proteins span the plasma membrane of adjoining cells. They pair up as open channels for signals between cells.

Figure 33.3 Examples of cell junctions.

Built-in controls make the plasma membrane of these cells selectively permeable. They allow some substances but not others to move across, through the interior of transport proteins (Sections 4.11, 5.2, and 5.6).

Example: Gastric fluid inside the stomach is highly acidic. If it were to leak across the stomach's epithelial lining, it would digest the proteins of your own body instead of those brought in with meals. Actually, that happens in people who have peptic ulcers. Many tight junctions in such linings function as a leakproof barrier between all cells near their free surface.

Adhering junctions (spot welds or collars) act as a unit to hold cells together in all animal tissues. They are profuse in the skin's surface layer and other tissues that are subjected to ongoing abrasion (Figure 33.3). The gap junctions let small molecules and ions diffuse

Figure 33.4 Section through the glandular epithelium of a frog. This frog, of the genus *Dendrobates*, makes one of the most lethal glandular secretions known. (Natives of one tribe in Colombia use its exocrine gland secretion to poison tips of darts, which they shoot through blowguns.) Pigment-rich cells branching through this epithelium impart color to the skin. The striking coloration of all poisonous frogs evolved as a clear warning signal to predators. In essence, it tells them "Don't even think about it."

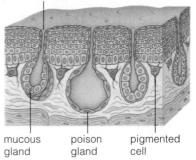

pore that opens at skin surface

mucous gland   poison gland   pigmented cell

swiftly among cells. They enhance chemical or electrical *communication* between certain cells that must rapidly coordinate their activities. They are profuse in the heart, skeletal muscles, and other highly active organs.

Glandular Epithelium and Glands

In the structurally simple invertebrates, epithelia have **gland cells**. These cells secrete products, unrelated to their own metabolism, that are to be used elsewhere. In more complex animals, such cells occur in glandular epithelium and in glands, which are secretory organs derived from epithelia. Note that *secretion* isn't the same as *excretion*, the concentration and removal of metabolic wastes or excess substances of no use to the body.

Exocrine glands secrete mucus, saliva, earwax, oil, milk, digestive enzymes, and other cell products. Most exocrine products are released onto the free surface of epithelium through ducts or tubes, as in Figure 33.4.

By contrast, **endocrine glands** have no ducts. Their products are hormones, which they secrete directly into fluid bathing the gland. Molecules of animal hormones typically enter the bloodstream, which distributes them to target cells elsewhere in the body (Chapter 36).

Epithelia are sheetlike tissues lining the body's surface and its cavities, ducts, and tubes. Epithelia have one free surface facing a body fluid or the outside environment. Junctions structurally and functionally connect the adjoining cells.

Glands are secretory organs derived from epithelium.

CONNECTIVE TISSUE

Of all tissues in complex animals, connective tissues are the most abundant and widely distributed. They range from soft connective tissues to the specialized types called cartilage, bone tissue, adipose tissue, and blood (Table 33.1). In all connective tissues except blood, cells secrete fibers of structural proteins: collagen or elastin. (This is the collagen that plastic surgeons use to plump wrinkled skin and lips.) These cells also secrete modified polysaccharides, which accumulate between cells and fibers as the connective tissue's "ground substance."

Soft Connective Tissues

Generally, these tissues have the same components but different proportions of them. **Loose connective tissue** contains fibers and fibroblasts (cells that produce and secrete the fibers), all loosely arranged in a semifluid ground substance (Figure 33.5a). The tissue often acts as a framework for epithelium. White blood cells patrol it and mount early counterattacks against pathogens—as when bacteria invade skin at an abrasion or breach the digestive, respiratory, or urinary tract lining and enter the internal environment.

Dense, irregular connective tissue has fibroblasts and many fibers (mostly collagen-containing ones) that are oriented every which way. This tissue is present in skin and forms protective capsules around organs that

Table 33.1 *Categories of Connective Tissues*	
SOFT	SPECIALIZED
Loose connective tissue	Cartilage
Dense, irregular connective tissue	Bone tissue
Dense, regular connective tissue	Adipose tissue
(ligaments, tendons)	Blood

do not stretch much (Figure 33.5b). In **dense, regular connective tissue**, the fibroblasts occur in rows between many parallel bundles of fibers. Tendons, which attach skeletal muscle to bone, have this tissue. Its bundles of collagen fibers help tendons resist being torn (Figure 33.5c). Dense, regular connective tissue also is present in elastic ligaments, which attach one bone to another. Its elastic fibers facilitate movement around joints.

Specialized Connective Tissues

Like rubber, **cartilage** is a pliable yet solid intercellular material that resists compression. Its cells make and secrete cartilage, then get imprisoned in cavities inside their secretions (Figure 33.5d). In vertebrate embryos, bone forms on cartilage deposits and in time replaces most of them. The cartilage left imparts shape to the nose, outer ear, and some other body parts. Cartilage

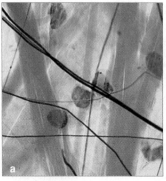
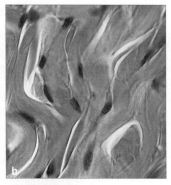
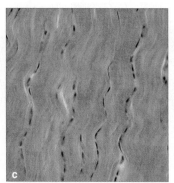
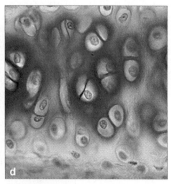

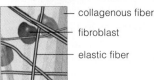
— collagenous fiber
— fibroblast
— elastic fiber

— collagenous fibers

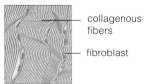
— collagenous fibers
— fibroblast

— ground substance with very fine collagen fibers
— cartilage cell (chondrocyte)

TYPE: Loose connective tissue
DESCRIPTION: Fibers, fibroblasts, other cells loosely arranged in extensive ground substance
COMMON LOCATIONS: Beneath skin and most epithelia
FUNCTION: Elasticity, diffusion

TYPE: Dense, irregular connective tissue
DESCRIPTION: Collagen fibers, fibroblasts occupy most of the ground substance
COMMON LOCATIONS: In skin and in capsules around some organs
FUNCTION: Structural support

TYPE: Dense, regular connective tissue
DESCRIPTION: Collagen fibers bundled in parallel, long rows of fibroblasts, little ground substance
COMMON LOCATIONS: Tendons, ligaments
FUNCTION: Strength, elasticity

TYPE: Cartilage
DESCRIPTION: Chondrocytes inside pliable, solid ground substance
COMMON LOCATIONS: Nose, ends of long bones, airways, skeleton of cartilaginous fish, vertebrate embryo
FUNCTION: Support, flexion, low-friction surface for joint movements

Figure 33.5 Examples of soft connective tissues and specialized connective tissues.

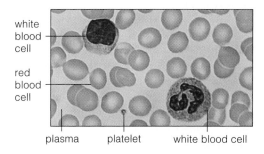

Figure 33.7 Some components of human blood. This tissue's straw-colored, liquid matrix (plasma) is primarily water in which many nutrients, diverse proteins, oxygen, carbon dioxide, ions, and other substances are dissolved.

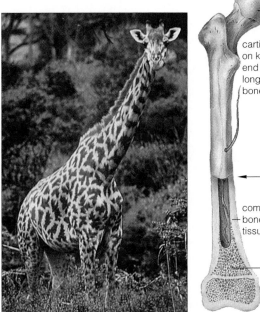

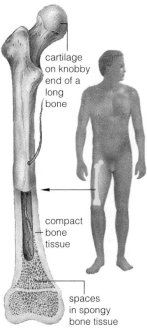

Figure 33.6 Cartilage and bone tissue. Spongy bone tissue has needlelike hard parts with spaces between. Compact bone tissue is more dense. Bone, a load-bearing tissue, resists compression. Over time, it favored increases in size for many land vertebrates, including this giraffe. It gives large animals selective advantages. They can ignore most predators and roam farther for food and water. They also can gain or lose heat more slowly than small animals because they have a lower surface-to-volume ratio.

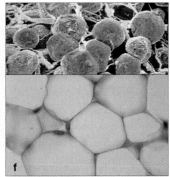

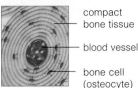

TYPE: Bone tissue

DESCRIPTION: Collagen fibers, osteocytes occupying extensive calcium-hardened ground substance

LOCATION: Bones of all vertebrate skeletons

FUNCTION: Movement, support, protection

TYPE: Adipose tissue

DESCRIPTION: Large, tightly packed fat cells occupying most of ground substance

COMMON LOCATIONS: Under skin, around the heart and kidneys

FUNCTION: Energy storage, insulation, padding

also protects and cushions the joints between bones of the limbs, the vertebral column, and elsewhere.

Bone tissue is mineral hardened; its collagen fibers and ground substance are strengthened with calcium salts (Figures 33.5e and 33.6). This is the main tissue of bones. In vertebrate skeletons, different bones support and protect softer tissues and organs. Limb bones, such as the long bones of your two legs, have weight-bearing functions. They interact with skeletal muscles attached to them to bring about movement. Parts of some bones also are production sites for blood cells.

When many cells convert excess carbohydrates and lipids to fats and other energy storage forms, a few fat droplets form inside their cytoplasm. But fat droplets nearly fill the cells in **adipose tissue**, which specializes in fat storage and serves as an insulating layer, mainly beneath skin (Figure 33.5f). Fats move rapidly to and from the cells by way of fine blood vessels threading through this tissue.

Because **blood** is derived primarily from connective tissue, many biologists classify it as a connective tissue. Blood serves transport functions. Circulating within its *plasma*, a fluid medium, are great numbers of red blood cells, white blood cells, and platelets (Figure 33.7). Red blood cells efficiently deliver oxygen to metabolically active tissues and then carry carbon dioxide and other wastes away from them. Plasma is largely water, but it has a great number of different kinds of proteins, ions, and other substances dissolved in it. We will be taking a closer look at this complex tissue in Section 38.2.

Diverse types of connective tissues bind together, support, strengthen, protect, and insulate other tissues in the body.

Soft connective tissues consist of protein fibers as well as a variety of cells arranged in a ground substance.

Cartilage, bone, blood, and adipose tissue are specialized connective tissues. Cartilage and bone are both structural materials. Blood is a fluid connective tissue with transport functions. Adipose tissue is a reservoir of stored energy.

MUSCLE TISSUE

In muscle tissues, cells *contract*—forcefully shorten in response to stimulation from the outside—and then relax and passively lengthen. These tissues have many cells arranged in parallel arrays. Layers of muscles and muscular organs contract in a coordinated fashion. The three types of tissues are skeletal, smooth, and cardiac muscle tissues.

Skeletal muscle tissue is the main tissue of muscles attached to bones. It functions in maintaining posture

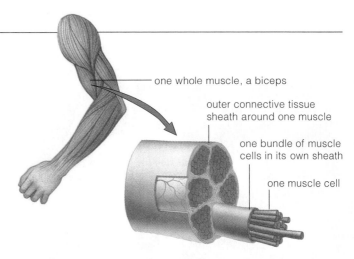

one whole muscle, a biceps

outer connective tissue sheath around one muscle

one bundle of muscle cells in its own sheath

one muscle cell

Figure 33.9 Location and general arrangement of the muscle cells in a typical skeletal muscle. Its many muscle cells are bundled together in parallel to direct the force of contraction against a bone.

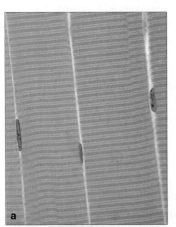

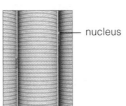

nucleus

TYPE: Skeletal muscle

DESCRIPTION: Bundles of cylindrical, long, striated contractile cells; many mitochondria; often reflex-activated but can be consciously controlled

LOCATIONS: Partner of skeletal bones, against which it exerts great force

FUNCTION: Locomotion, posture; head, limb movements

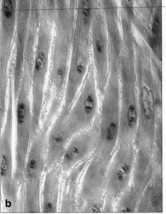

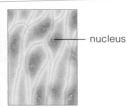

nucleus

TYPE: Smooth muscle

DESCRIPTION: Contractile cells tapered at both ends; not striated

LOCATIONS: Wall of arteries, sphincters, stomach, intestines, urinary bladder, many other soft internal organs

FUNCTION: Controlled constriction; motility (as in gut); arterial blood flow

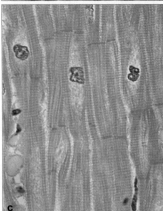

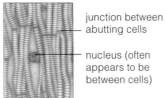

junction between abutting cells

nucleus (often appears to be between cells)

TYPE: Cardiac muscle

DESCRIPTION: Unevenly striated, fused-together cylindrical cells that contract as a unit owing to signals at gap junctions between them

LOCATIONS: Heart wall

FUNCTION: Pump blood forcefully through circulatory system

Figure 33.8 Characteristics and examples of skeletal muscle tissue, smooth muscle tissue, and cardiac muscle tissue.

and in moving the body and its assorted parts. Many transverse stripes (or striations) of its long, cylindrical skeletal muscle cells represent many repeating units of the contractile proteins actin and myosin (Figures 33.8*a* and 33.9). Rapidly contracting muscle cells use a great deal of energy, which their abundant mitochondria and enzymes of glycolysis provide. Chapter 37 includes a look at this tissue's structure and functioning.

Smooth muscle tissue is in the wall of the stomach, lungs, and other soft internal organs of vertebrates. Its contractile cells incorporate many mitochondria. Unlike skeletal muscle, contractile proteins in its tapered cells are not arranged in a series of repeating units, so this tissue is not striated (Figure 33.8*b*). Its contractions are slower than in skeletal muscle but they can be sustained longer. They function in gut motility, urinary bladder emptying, sphincter closure, blood flow in arteries, and other internal organ activities. Smooth muscle action is said to be "involuntary" because we usually can't make it contract just by thinking about it. (We can do so with skeletal muscle; its action is said to be "voluntary.")

Cardiac muscle tissue is a contractile tissue present only in the heart (Figure 33.8*c*). Its muscle cells contract as a unit owing to signals at gap junctions where their plasma membranes fuse together. When one cell gets a signal to contract, others are stimulated to contract, too. This tissue's cells hold more mitochondria, so it is not as evenly striated as skeletal muscle tissue. Section 38.6 deals with this tissue's structure and function.

Muscle tissue, which can contract (shorten) in response to stimulation, helps move the body and specific body parts.

Skeletal muscle is the only muscle tissue attached to bones. Smooth muscle tissue occurs in many soft internal organs. Cardiac muscle alone makes up the contractile walls of the heart. Connective tissue sheathes all three types of muscle.

NERVOUS TISSUE

Of all tissues, **nervous tissue** exerts the greatest control over the body's responsiveness to changing conditions. Its **neurons** are excitable cells that form communication lines in most nervous systems. A variety of cells called **neuroglia** protect and support the neurons structurally and metabolically. Neuroglia makes up more than half of the volume of your own nervous system.

Like other excitable cells, the neuron responds in a specific way to adequate stimulation. It propagates an electric disturbance along its plasma membrane to its endings, or output zone. There, the disturbance causes events that may trigger the stimulation or inhibition of adjacent neurons and other cells.

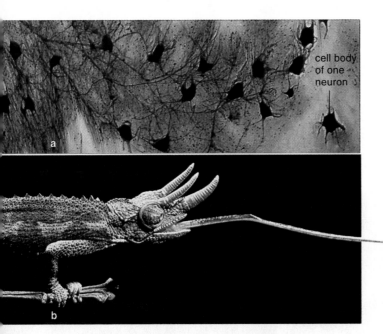

cell body of one neuron

Figure 33.10 (**a**) Motor neurons, which relay signals from the brain or spinal cord to muscles and glands. Diverse neurons interact to detect and process information about internal and external conditions, and initiate responses. (**b**) Without neurons, this chameleon could not detect an edible insect, calculate its distance, and command a long, sticky, prey-capturing tongue to uncoil and hit its mark with stunning speed.

For example, more than a hundred billion neurons are organized as communication lines throughout your body. Some detect specific changes in environmental conditions. Others coordinate immediate and long-term responses to change. The type shown in Figure 33.10*a* delivers signals from the brain to muscles and glands. How neurons function is a topic of later chapters.

Neurons are the basic units of communication in nervous tissue. Different kinds detect specific stimuli, integrate information, and issue or relay commands for response.

Frontiers in Tissue Research

As you know by now, a tissue is more than the sum of its cells. When each new animal grows and develops, its cells interact and become organized in prescribed ways to give rise to the body's diverse tissues. Also, cells of each new tissue synthesize specific gene products that are vital for normal body functioning.

For decades, medical researchers attempted to grow artificial tissues in quantity in the laboratory. And now, *lab-grown epidermis* is a reality. It is used to speed up the healing of extensive burns, ulcers, cancerous lesions, and blistering disorders. One company makes it by mixing cells from foreskins (discarded from circumcised male infants) and proteins (from cattle tendons). The mixture is added to rows of small, shallow dishes that hold a culture medium enriched with nutrients and growth factors. The cells multiply to form spherical, paper-thin grafts. A bit of foreskin no larger than a postage stamp contains enough undifferentiated cells to make 200,000 grafts, each with a five-day shelf life.

Surgeons apply such grafts to a wound and bind it with a gauze dressing. During the next few weeks, the graft's cells interact biochemically and structurally with a patient's cells to regenerate damaged or missing tissue. Use of such lab-grown skin is less costly and less risky than surgery to get a patch of the patient's skin.

On the horizon are *designer organs*: capsules holding preselected groups of living cells that synthesize specific hormones, enzymes, growth factors, and other substances. The idea is to surgically snip a bit of epithelium from a patient and use it to encapsulate the preselected cells. The capsule would be derived from the patient's own epithelial cells, so it wouldn't be chemically recognized as "foreign" and attacked by the immune system. That is what happens when the body of a patient rejects organ or tissue implants. Such rejection can have severe medical consequences.

Biotechnologists are close to understanding how to synthesize molecular cues that will help designer organs stick to suitable sites in the body. Once attached, they might become integral parts of normal body functioning.

Ultimately, the goal of this research is to put together packages of cells able to make life-saving substances that are absent in patients who suffer from severe genetic disorders or chronic diseases. For example, imagine the potential for people with type 1 *diabetes mellitus*. This metabolic disorder results in an elevated concentration of glucose in the blood. Affected people produce little if any insulin, the hormone that signals cells to take up glucose from the blood. Unless they get insulin injections on a regular basis, they die. However, if a customized, insulin-secreting organ could be successfully installed inside the body of such individuals, daily injections of insulin may become a thing of the past.

ORGAN SYSTEMS

Overview of Major Organ Systems

Figure 33.11 is an overview of eleven organ systems of a typical vertebrate, an adult human. Figure 33.12 lists some terms used when describing positions of various organs. It also shows the major body cavities in which a number of important organs are located.

Each organ system contributes to the survival of all living cells in the animal body. You might think this is stretching things. For example, how could muscles and bones be helping each small, individual cell stay alive? Yet interactions between the skeletal system and the muscular system let us move, say, toward sources of nutrients and water. Parts of these organ systems help keep blood circulating to cells, as when contractions of leg muscles help move blood in veins back to the heart. The circulatory system rapidly transports oxygen and other substances dissolved in blood to cells, and moves metabolic products and wastes away from them. The respiratory system swiftly delivers oxygen from the air to the circulatory system and takes up carbon dioxide wastes from it, skeletal muscles assist the respiratory system—and so it goes, throughout the entire body.

Figure 33.12 (**a**) Major cavities in the human body. (**b,c**) Directional terms and planes of symmetry for the vertebrate body. Notice how the *midsagittal* plane divides the body into right and left halves. Most vertebrates, such as fishes and rabbits, move with their main body axis parallel with the Earth's surface. For them, *dorsal* pertains to their back or upper surface, and *ventral* pertains to the opposite, lower surface.

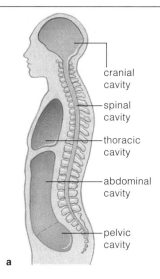

- cranial cavity
- spinal cavity
- thoracic cavity
- abdominal cavity
- pelvic cavity

a

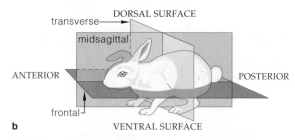

transverse — DORSAL SURFACE
midsagittal
ANTERIOR ___ POSTERIOR
frontal
VENTRAL SURFACE
b

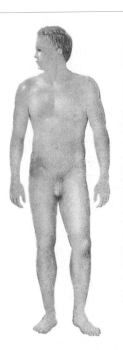

INTEGUMENTARY SYSTEM	MUSCULAR SYSTEM	SKELETAL SYSTEM	NERVOUS SYSTEM	ENDOCRINE SYSTEM	CIRCULATORY SYSTEM
Protects body from injury, dehydration, and some pathogens; controls its temperature; excretes certain wastes; receives some external stimuli.	Moves body and its internal parts; maintains posture; generates heat by increases in metabolic activity.	Supports and protects body parts; provides muscle attachment sites; produces red blood cells; stores calcium, phosphorus.	Detects external and internal stimuli; controls and coordinates the responses to stimuli; integrates all organ system activities.	Hormonally controls body functioning; works with nervous system to integrate short-term and long-term activities.	Rapidly transports many materials to and from cells; helps stabilize internal pH and temperature.

Figure 33.11 Overview of human organ systems and their functions.

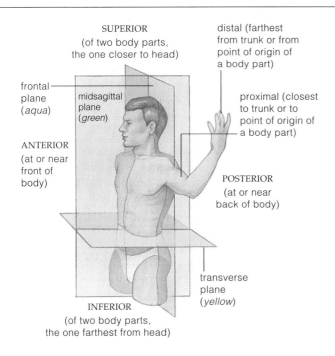

SUPERIOR
(of two body parts,
the one closer to head)

distal (farthest
from trunk or from
point of origin of
a body part)

frontal
plane
(*aqua*)

midsagittal
plane
(*green*)

proximal (closest
to trunk or to
point of origin of
a body part)

ANTERIOR
(at or near
front of
body)

POSTERIOR
(at or near
back of body)

transverse
plane
(*yellow*)

INFERIOR
(of two body parts,
the one farthest from head)

c Unlike quadrupedal animals, humans walk upright, with their main body axis perpendicular to the ground. *Anterior* refers to the front of an upright walker; it corresponds to ventral, as in (**b**). *Posterior* refers to the back; it corresponds to dorsal.

Tissue and Organ Formation

Where do tissues of organ systems come from? They start with sperm and eggs that arise from *germ* cells, or immature reproductive cells. (All other cells in the body are *somatic*, after a Greek word for body.) A zygote forms after a sperm fertilizes an egg, then mitotic cell divisions produce an early embryo. In vertebrates, cells become arranged as three primary tissues—ectoderm, mesoderm, and endoderm, the embryonic forerunners of all tissues in the adult. **Ectoderm** gives rise to the skin's outer layer and to the nervous system's tissues. **Mesoderm** gives rise to muscles and bones and to most of the circulatory, reproductive, and urinary systems. **Endoderm** gives rise to the lining of the digestive tract and to organs derived from it (Section 43.2).

In general, all vertebrates have the same kinds of organ systems. Each organ system serves specialized functions, such as gas exchange, blood circulation, and locomotion.

Vertebrate tissues, organs, and organ systems arise from three primary tissues that form in the early embryo. The primary tissues are ectoderm, mesoderm, and endoderm.

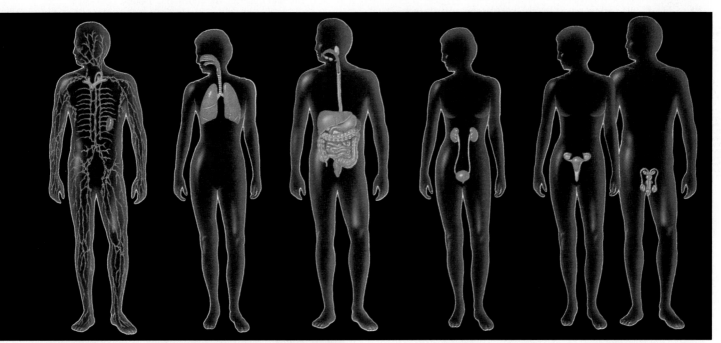

LYMPHATIC SYSTEM	RESPIRATORY SYSTEM	DIGESTIVE SYSTEM	URINARY SYSTEM	REPRODUCTIVE SYSTEM
Collects and returns some tissue fluid to the bloodstream; defends the body against infection and tissue damage.	Rapidly delivers oxygen to the tissue fluid that bathes all living cells; removes carbon dioxide wastes of cells; helps regulate pH.	Ingests food and water; mechanically, chemically breaks down food, and absorbs small molecules into internal environment; eliminates food residues.	Maintains the volume and composition of internal environment; excretes excess fluid and blood-borne wastes.	*Female:* Produces eggs; after fertilization, affords a protected, nutritive environment for the development of new individuals. *Male:* Produces and transfers sperm to the female. Hormones of both systems also influence other organ systems.

SUMMARY

Gold indicates text section

1. Tissues, organs, and organ systems work together to maintain a stable internal environment (extracellular fluid) needed for individual cell survival. At homeostasis, conditions in the internal environment are balanced at levels most favorable for cell activities. *CI*

2. A tissue is an aggregation of cells and intercellular substances that perform a common task. An organ is a structural unit of different tissues combined in definite proportions and patterns that allow them to perform a common task. An organ system has two or more organs interacting chemically, physically, or both in ways that contribute to the survival of the body as a whole. *CI*

3. Cell-to-cell junctions occur in most animal tissues. Tight junctions prevent substances from leaking across the tissue. Adhering junctions cement neighboring cells together. Gap junctions are open channels between the cytoplasm of abutting cells. They afford rapid transfer of ions and small molecules between cells. *33.1*

4. Epithelial tissues cover external body surfaces and line internal cavities and tubes. *33.1*

  a. Epithelium has a free surface exposed to some body fluid or to the outside environment. Between the opposite surface and the underlying connective tissue is a basement membrane on which epithelial cells rest.

  b. Gland cells secrete (release) products, unrelated to their own metabolism, that are to be used elsewhere. The cells occur in glandular epithelium and in glands, which are secretory organs derived from epithelia.

  c. Exocrine glands secrete mucus, saliva, earwax, oil, milk, digestive enzymes, and other cell products. Most exocrine products are released onto the free surface of epithelium through ducts or tubes.

  d. Endocrine glands are ductless. Their products are hormones, which they secrete directly into interstitial fluid that bathes the gland. From there, the hormones usually enter the bloodstream, which distributes them to target cells elsewhere in the body.

5. Different connective tissues bind together, support, strengthen, protect, and insulate other tissues. Most have fibers of structural proteins (especially collagen), fibroblasts, and other cells in a ground substance. *33.2*

  a. Loose connective tissue, with a semifluid ground substance, is present under skin and most epithelia.

  b. Dense, irregular connective tissue contains mostly collagen fibers and fibroblasts. It is present in skin and forms protective capsules around a number of organs.

  c. Dense, regular connective tissue, such as that in tendons, contains parallel bundles of collagen fibers. It protects and structurally supports organs.

  d. Cartilage is a solid, pliable intercellular material with structural and cushioning roles. Bone, the weight-bearing tissue of the vertebrate skeleton, interacts with skeletal muscle to bring about movement.

  e. Blood, a specialized connective tissue, consists of plasma, cellular components, and dissolved substances. Adipose tissue, another specialized connective tissue, is a reservoir of energy; it consists mainly of fat cells.

6. Muscle tissues contract (shorten), then return to the resting position. They help move the body or parts of it. The three types are skeletal muscle, smooth muscle, and cardiac muscle tissue. *33.3*

7. Nervous tissue intercepts and integrates information about internal and external conditions, and governs the body's responses to change. Neurons are the units of communication in nervous systems. *33.4*

Review Questions

1. What makes up the internal environment of animals? What does homeostasis mean with respect to that environment? *CI*

2. Name four basic tasks that the animal body performs. *CI*

3. Define animal tissue, organ, and organ system. List and define the functions of the eleven major organ systems of the human body. *CI, 33.6*

4. Describe the characteristics of epithelial tissue in general. Then describe the various types of epithelial tissues in terms of specific characteristics and functions. *33.1*

5. List the major types of connective tissues; add the names and characteristics of their specific types. *33.2*

6. Identify and describe the following tissues: *33.1–33.3*

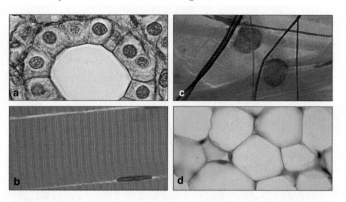

7. Identify this category of tissue and its characteristics: *33.4*

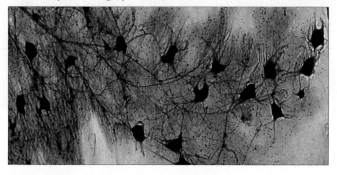

8. What type of cell serves as the basic unit of communication in nervous systems? *33.4*

9. Name the three kinds of primary tissues that form in early vertebrate embryos. *33.6*

Self-Quiz ANSWERS IN APPENDIX III

1. _____ tissues have closely linked cells and a free surface.
   a. Epithelial   c. Nervous
   b. Connective   d. Muscle

2. _____ function in cell-to-cell communication.
   a. Tight junctions   c. Gap junctions
   b. Adhering junctions d. all of the above

3. In most animals, glands are located in _____ tissue.
   a. epithelial   c. muscle
   b. connective   d. nervous

4. In most _____, cells secrete fibers of collagen and elastin.
   a. epithelial tissues c. muscle tissues
   b. connective tissues d. nervous tissues

5. _____ has a semifluid ground substance and occurs under most epithelia.
   a. Dense, irregular   c. Dense, regular
      connective tissue   connective tissue
   b. Loose connective tissue d. Cartilage

6. _____, a specialized connective tissue, is mostly plasma with cellular components and various dissolved substances.
   a. Irregular connective tissue c. Cartilage
   b. Blood   d. Bone

7. After you eat too many carbohydrates and proteins, your body converts the excess to storage fats, which accumulate in _____.
   a. connective tissue proper c. adipose tissue
   b. dense connective tissue   d. both b and c

8. In your own body, _____ can shorten (contract).
   a. epithelial tissue   c. muscle tissue
   b. connective tissue   d. nervous tissue

9. Only _____ muscle tissue has a striated appearance.
   a. skeletal   c. cardiac
   b. smooth   d. a and c are correct

10. Components of _____ detect and coordinate information about changes and control responses to those changes.
   a. epithelial tissue   c. muscle tissue
   b. connective tissue   d. nervous tissue

11. Cells of complex animals _____.
   a. survive by their own metabolic activities
   b. contribute to the survival of the whole animal
   c. help maintain extracellular fluid
   d. all of the above

12. In vertebrates, _____ is the embryonic source of tissues in the adult.
   a. ectoderm   c. mesoderm
   b. endoderm   d. all are correct

13. Match the terms with the most suitable description.
   ____ exocrine gland   a. strong, pliable; like rubber
   ____ endocrine gland   b. secretion through duct
   ____ cartilage   c. stable internal environment
   ____ homeostasis   d. contracts, not striated
   ____ smooth muscle   e. cements neighboring
   ____ neuroglia   cells together
   ____ adhering junction f. supports neurons
                          g. ductless secretion

Critical Thinking

1. *Anhidrotic ectodermal dysplasia* is a disorder associated with a recessive allele on the mammalian X chromosome (Section 15.4).

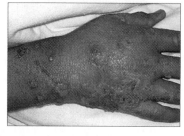

Figure 33.13 Following even mild exposure to sunlight, the type of blistering and ulcers that form on the skin of a person affected by porphyria.

As one symptom of this genetic disorder, affected males and females have no sweat glands in tissues where the recessive allele is expressed. What type of tissue are we talking about?

2. Adipose tissue and blood are often said to be "atypical" connective tissues. Compared with other connective tissues, which of their features are *not* typical?

3. Reflect on the nature of adaptation, as presented in Section 28.2. Jot down a few thoughts on the functions of animal hairs, which are derived from specialized cells of an epidermis. Next, think about plant hairs, derived from specialized epidermal cells. Especially in deserts and other dry habitats, many plants have so many surface hairs that they look almost furry. Now ask: Are animal hairs and plant hairs both adaptive responses to the same environmental pressures? (A few clues: Check Sections 26.10 and 28.4, as well as the Chapter 30 introduction.)

4. *Porphyria*, a genetic disorder, shows up in about 1 in 25,000 individuals. Affected people lack certain enzymes that are part of a metabolic pathway leading to the formation of heme, which is the iron-containing group of hemoglobin. An accumulation of porphyrins (intermediates of the pathway) results in jarring symptoms, especially after sunlight exposure. Lesions and scars form on the skin (Figure 33.13). Hair grows thickly on the face and hands. Gums retreat from teeth, and this makes the canines take on a fanglike appearance. Symptoms worsen on exposure to assorted substances, including garlic and alcohol. Affected individuals avoid sunlight and aggravating substances. They also may get injections of heme from normal red blood cells.

Porphyria may have been the source of vampire stories that were being told even before the Middle Ages. Superstitious folk didn't have a clue to the basis of the condition, but molecular biologists do. Research and report on its possible causes.

Selected Key Terms

adhering junction *33.1*	gap junction *33.1*
adipose tissue *33.2*	gland cell *33.1*
blood *33.2*	homeostasis *CI*
bone tissue *33.2*	internal environment *CI*
cardiac muscle tissue *33.3*	loose connective tissue *33.2*
cartilage *33.2*	mesoderm *33.6*
dense, irregular	nervous tissue *33.4*
connective tissue *33.2*	neuroglia *33.4*
dense, regular	neuron *33.4*
connective tissue *33.2*	organ *CI*
ectoderm *33.6*	organ system *CI*
endocrine gland *33.1*	skeletal muscle tissue *33.3*
endoderm *33.6*	smooth muscle tissue *33.3*
epithelium *33.1*	tight junction *33.1*
exocrine gland *33.1*	tissue *CI*

Readings

Bloom, W., and D. W. Fawcett. 1995. *A Textbook of Histology*. Twelfth edition. Philadelphia: Saunders.

Telford, I., and C. Bridgman. 1995. *Introduction to Functional Histology*. Second edition. New York: HarperCollins.

Wright, C. November 1999. "Ready-to-Wear Flesh." *Discover*.

On-Line readings at Student Guide for InfoTrac:
www.brookscole.com/biology   577

34

INTEGRATION AND CONTROL: NERVOUS SYSTEMS

Why Crack the System?

Suppose your biology instructor asks you to volunteer for an experiment. You'll get a microchip implanted in your brain. It will make you feel really good. But it may mess up your health, lop ten years off your life, and destroy part of your brain. Your behavior will change for the worse. You may have trouble finishing school, keeping a job, and having a normal family life.

The longer the chip is implanted, the less you will want to give it up. You won't get paid. You will pay the experimenter—first at bargain rates, then a little more each week. The chip is illegal. Get caught using it, and you and the experimenter go to jail.

Sometimes Jim Kalat, a professor at North Carolina State University, proposes this experiment, which of course is hypothetical. Hardly any students volunteer. Then he substitutes *drug* for microchip and *dealer* for experimenter, and a surprising number of students come forward! Like 30 million other Americans, the "volunteers" seem ready to use drugs that may rewire and corrupt their nervous system permanently.

The self-destruction shows up in unexpected places. For instance, each year 300,000 or so newborns already are addicted to crack, thanks to their addicted mothers. *Crack*, a cheap form of cocaine, relentlessly stimulates

brain regions that govern the sense of pleasure. Crack dampens normal urges to eat and sleep, and it raises blood pressure. Elation and sexual desire intensify. In time, brain cells that produce the stimulatory chemicals can't keep up with the abnormal demand. The chemical vacuum makes crack users frantic and then profoundly depressed. Only crack makes them feel good again.

Addicted babies quiver with "the shakes" and respond to the world with chronic irritation. They are abnormally small. As they developed in their mother, their tissues did not get enough oxygen and nutrients. Why? As one side effect, crack causes blood vessels to constrict, so maternal blood vessels cannot deliver enough required nutrients to the developing fetus.

Crack babies can't respond to rocking and to other normally soothing forms of stimulation. A year or more may pass before they even recognize their own mother. If untreated, they are likely to be emotionally unstable children, prone to aggressive outbursts and stony silences. A damaged nervous system is their legacy.

Think about it. The **nervous system** evolved as a way to sense and respond to changing conditions inside and outside the body, with increasing precision. Awareness of sounds and sights, of odors, of hunger

Figure 34.1 Owner of an evolutionary treasure—a complex brain, the foundation for our memory and reasoning, and our future. *Facing page:* Communication lines typical of vertebrate nervous systems.

and passion, fear and rage—all begin with the flow of information along communication lines of the nervous system. Are the lines silent until they receive outside signals, much as telephone lines wait to carry calls from all over the country? No. For example, before you were born, excitable cells called neurons started to organize themselves into gridworks in newly forming tissues. They began to chatter among themselves. All through your life, in moments of danger or quiet reflection, excitement or sleep, their chattering has continued and will continue for as long as you do.

As in all vertebrates, your body has three classes of neurons. **Sensory neurons** respond to stimuli and relay information about them to the spinal cord and brain. A **stimulus** is a form of energy, such as light, that specific receptors detect. In the spinal cord and brain, **interneurons** receive and process sensory input, then influence activities of other neurons. **Motor neurons** relay information away from the brain and spinal cord to the body's effectors—muscles and glands—which carry out the specified responses (Figure 34.1). More than half the volume of your nervous system consists of cells, collectively called neuroglia, that metabolically assist, protect, and structurally support the neurons.

Our initial focus in this chapter will be on the structure and function of neurons. Later on, we will consider how neurons interact in nervous systems.

Key Concepts

1. Neurons are the basic units of communication in nearly all nervous systems. Collectively, many neurons interact to detect and integrate information about external and internal conditions, then select or control muscles and glands in ways that result in suitable responses.

2. Neurons are a type of excitable cell, which means a stimulus can disturb the distribution of electric charge across their plasma membrane. When strong enough, the disturbances form the basis of messages that travel from a neuron's input zone to its output zone, next to a neighboring cell.

3. Information flow through the nervous system depends on the moment-by-moment integration of excitatory and inhibitory signals that act upon each neuron in a given pathway.

4. The simplest nervous systems are the nerve nets of radial animals, such as hydras and sea anemones. Nervous systems of most animals show pronounced cephalization and bilateral symmetry.

5. Vertebrate nervous systems are functionally divided into central and peripheral regions. The brain and spinal cord make up the central nervous system. Paired nerves that thread through the remainder of the body are the main components of the peripheral nervous system.

6. The somatic nerves of the peripheral nervous system deal with skeletal muscles. The autonomic nerves deal with the heart, lungs, and other soft internal organs, most of which are enclosed within body cavities.

7. The vertebrate brain has three functional divisions, the hindbrain, midbrain, and forebrain. The brain's most ancient tissues deal with reflex control over breathing, blood circulation, and other functions that are essential for staying alive.

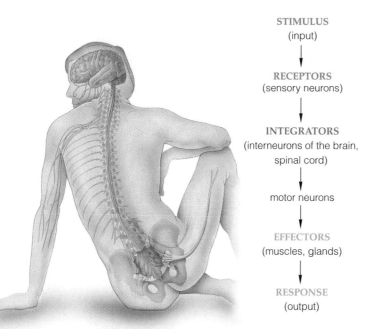

STIMULUS
(input)

↓

RECEPTORS
(sensory neurons)

↓

INTEGRATORS
(interneurons of the brain,
spinal cord)

↓

motor neurons

↓

EFFECTORS
(muscles, glands)

↓

RESPONSE
(output)

NEURONS—THE COMMUNICATION SPECIALISTS

Functional Zones of a Neuron

A neuron has a nucleated cell body with cytoplasmic extensions that differ in number and length. Typically, the cell body and slender extensions called **dendrites** are *input* zones for information. At a nearby patch of plasma membrane, the *trigger* zone, the input may give rise to signals that travel along an **axon**, a slender and often long extension that is a neuron's *conducting* zone. The axon endings are the neuron's *output* zones. There, information is sent to other cells (Figure 34.2).

What Is a "Resting" Neuron?

In the body's communication lines, each neuron is only a single unit. It accepts and sends on information about stimuli. And where does that capacity to accept and respond to information originate? It all starts with a difference in charge across the plasma membrane.

When a neuron is "at rest," or not being stimulated, the cytoplasmic fluid next to the plasma membrane is negatively charged with respect to the interstitial fluid just outside. We measure such charges in millivolts. Think of the amount of energy inherent in maintaining the steady voltage difference across a neural membrane as the **resting membrane potential**. For many neurons, that amount is about −70 millivolts.

When a patch of membrane in the neuron's trigger zone is weakly stimulated, the voltage difference at the patch changes only slightly, if at all. But a strong signal might trigger an **action potential**. We may define this as a brief reversal in the voltage difference across the plasma membrane, with the cytoplasmic fluid next to the membrane becoming positive relative to the fluid outside. That initial reversal sets in motion a series of fleeting reversals, just like itself. And these move away from the trigger zone, down to an axon's output zone.

Gradients Required for Action Potentials

How does the neuron keep itself ready for an action potential? It does so by working with and against ion gradients across its plasma membrane. Potassium ions (K^+), sodium ions (Na^+), and other charged substances can't cross the membrane's lipid bilayer. They can only move through channels inside transport proteins that span the bilayer—where some control can be exerted over which ones cross at a given time.

As Figure 34.3 shows, certain passive transporters have open channels. They never shut, so they let certain ions leak through them all the time. (Different passive transporters have channels with molecular gates that open only when the resting neuron receives adequate stimulation at an input zone.)

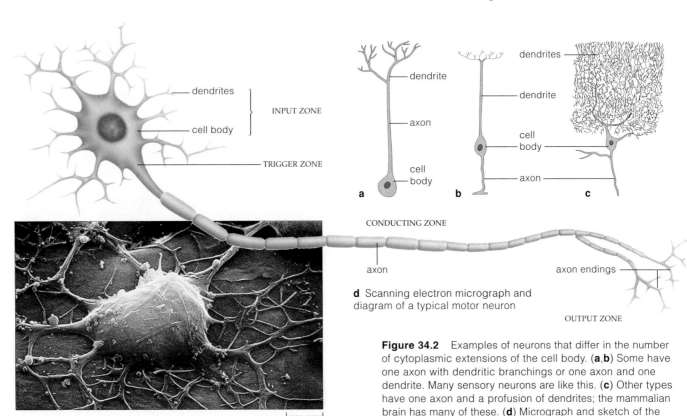

d Scanning electron micrograph and diagram of a typical motor neuron

10 μm

Figure 34.2 Examples of neurons that differ in the number of cytoplasmic extensions of the cell body. (**a**,**b**) Some have one axon with dendritic branchings or one axon and one dendrite. Many sensory neurons are like this. (**c**) Other types have one axon and a profusion of dendrites; the mammalian brain has many of these. (**d**) Micrograph and sketch of the functional zones of a motor neuron.

Figure 34.3 How ions move across the plasma membrane of a neuron. Spanning the lipid bilayer are selective ion channels, as described in Section 5.2. Sodium–potassium pumps that also span the bilayer actively move ions across.

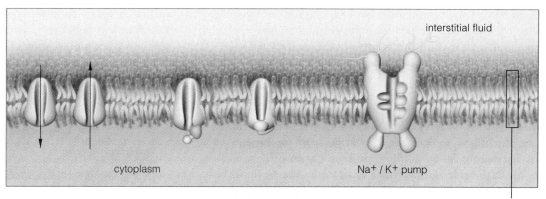

interstitial fluid

cytoplasm

Na⁺ / K⁺ pump

a Passive transporters with open channels let ions continually leak across the membrane.

b Other passive transporters have voltage-sensitive gated channels that open and shut during and between action potentials. They assist the diffusion of Na⁺ and K⁺ across the membrane, down their concentration gradients.

c Active transporters pump Na⁺ and K⁺ across the membrane, against their concentration gradients. They counter ion leaks and restore resting membrane conditions.

lipid bilayer of neural membrane

Suppose there are 15 sodium ions inside a motor neuron for every 150 outside. Suppose there are 150 potassium ions inside for every 5 outside. The two ion concentration gradients across the neural membrane can be shown this way:

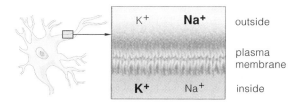

K⁺   **Na⁺**   outside

plasma membrane

K⁺   Na⁺   inside

Each set of large-to-small letters signifies which side of the membrane has the greater concentration of the ion.

Gated channels are closed in a motor neuron at rest, so the Na⁺ can't follow its gradient and diffuse across the membrane. Some K⁺ is following its gradient and leaking out through a few passive transporters. The leaks make the cytoplasm inside the neuron slightly more negative, so some K⁺ is attracted back in. There is no more net movement of K⁺ when the inward pull of electric charge balances its outward-directed diffusion. The difference in charge across the neural membrane at this time is the resting membrane potential.

Ah, but a tiny fraction of the K⁺ that leaked outside is still outside. Also, a tiny fraction of Na⁺ is leaking in through a few open channels (Figure 34.4). Do the leaks mean that the gradients will gradually disappear? No. **Sodium–potassium pumps** maintain those gradients. They also restore them after they are reversed during an action potential. The pumps are active transporter proteins that span the plasma membrane (Sections 5.3 and 5.7). When a phosphate-group transfer from ATP

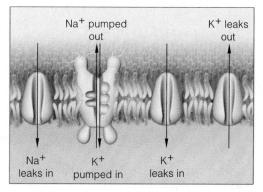

interstitial fluid next to membrane

Na⁺ pumped out

K⁺ leaks out

plasma membrane

cytoplasm just inside membrane

Na⁺ leaks in

K⁺ pumped in

K⁺ leaks in

Figure 34.4 Pumping and leaking processes that dictate the distribution of sodium and potassium ions across the plasma membrane of a neuron at rest. The arrow widths indicate the magnitude of the movements. Notice how the total inward and outward movements for each kind of ion are balanced.

activates them, they pump K⁺ into the neuron and Na⁺ out at the same time, *against* their gradients.

With this bit of information on the gradients across the neural membrane, we are ready to look at how an action potential arises at a trigger zone and propagates itself, undiminished, to an output zone.

An undisturbed neuron maintains ion concentration and electric gradients across its plasma membrane. We measure this as a steady voltage difference, and energy inherent in it is the resting membrane potential. An abrupt, short-lived reversal in that voltage difference is an action potential.

Sodium–potassium pumps restore and maintain the resting membrane potential in between action potentials.

HOW ARE ACTION POTENTIALS TRIGGERED AND PROPAGATED?

Action potential propagation isn't hard to follow if you already know something about the gradients across the neural membrane. And so we now build on Section 34.1.

Approaching Threshold

When you weakly stimulate a neuron at its input zone, you disturb the ion balance across its membrane, but not much. Imagine putting a bit of pressure on the skin of a snoozing cat by gently tapping a toe on it. Tissues beneath the skin surface have receptor endings—input zones of sensory neurons. Patches of plasma membrane at these endings deform under pressure and let some ions flow across. The flow slightly changes the voltage difference across the membrane. In this case, pressure has produced a graded, local signal.

influx of ions, the cytoplasmic side of the membrane becomes less negative. This causes more gates to open and more sodium to enter. The ever increasing, inward flow of sodium is a case of **positive feedback**, whereby an event intensifies as a result of its own occurrence:

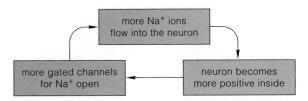

At threshold, opening of sodium gates no longer depends on the strength of the stimulus. The positive-feedback cycle is under way, and the inward-rushing sodium itself is enough to open the gated channels.

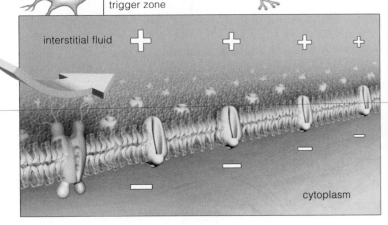

a Membrane at rest (inside negative with respect to the outside). An electrical disturbance (*yellow* arrow) spreads from an input zone to an adjacent trigger region of the membrane, which has a great number of gated sodium channels.

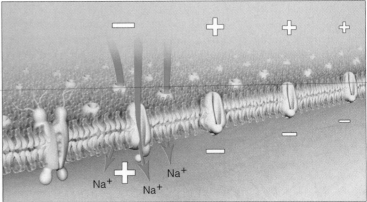

b A strong disturbance initiates an action potential. Sodium gates open. The sodium inflow decreases the negativity inside the neuron. The change causes more gates to open, and so on until threshold is reached and the voltage difference across the membrane reverses.

Figure 34.5 Propagation of an action potential along the axon of a motor neuron.

Graded means that signals arising at an input zone vary in magnitude. They are small to large, depending on the stimulus intensity or duration. *Local* means these signals do not spread far from the site of stimulation. Why? It takes certain kinds of ion channels to propagate a signal, and input zones simply don't have them.

When a stimulus is intense or long-lasting, graded signals spread from the input zone into an adjoining trigger zone. This patch of membrane is richly endowed with voltage-sensitive gated channels for sodium ions. *And this is where a certain amount of change in the voltage difference across the plasma membrane triggers an action potential.* The amount is the neuron's threshold level.

When these gates open, positively charged sodium ions flow into the neuron, as in Figure 34.5. With the

An All-or-Nothing Spike

Figure 34.6 shows a recording of the voltage difference across the plasma membrane before, during, and after an action potential. Notice how the membrane potential peaks once threshold is reached. All action potentials in a neuron spike to the same level above threshold as an *all-or-nothing* event. Once a positive-feedback cycle starts, nothing stops full spiking. Unless threshold is reached, the membrane disturbance subsides when the stimulation ends, and an action potential won't occur.

Each spike lasts for only a millisecond or so. Why? At the patch of membrane where the charge reversed, gated sodium channels close and shut off the sodium inflow. And about halfway into the reversal, potassium

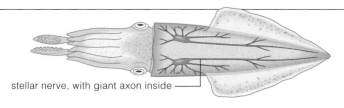

stellar nerve, with giant axon inside

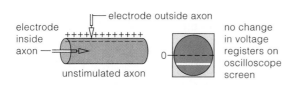

electrode outside axon

electrode inside axon

unstimulated axon

0

no change in voltage registers on oscilloscope screen

a

Figure 34.6 Action potentials. (**a**) When researchers started to study neural function, a squid (*Loligo*) yielded evidence of action potential spiking. This squid's "giant" axons are large enough to slip electrodes inside. (**b**) Researchers put electrodes inside and outside the axon, then stimulated it. The electrodes detected the voltage change, which showed up as deflections in a beam of light across the screen of an oscilloscope connected to the electrodes. (**c**) Typical waveform (*yellow* line) for an action potential on an oscilloscope screen.

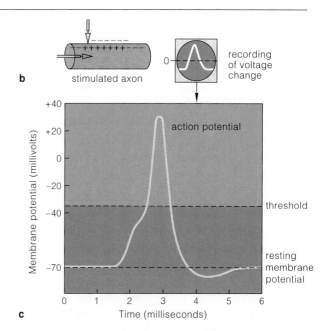

b stimulated axon

recording of voltage change

c

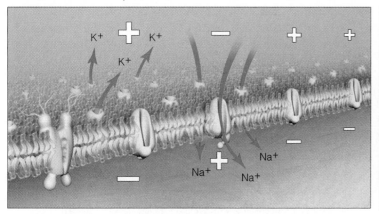

c With the reversal, sodium gates shut and potassium gates open (*red* arrows). Potassium follows its gradient out of the neuron. Voltage is restored. The disturbance triggers an action potential at the adjacent site, and so on, away from the point of stimulation.

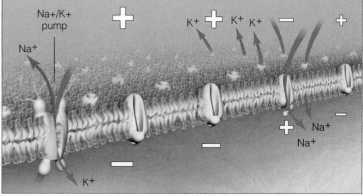

d Following each action potential, the inside of the plasma membrane becomes negative once again. However, the sodium and potassium concentration gradients are not yet fully restored. Active transport at sodium–potassium pumps restores them.

channels opened, so potassium flows out. This restores the voltage difference at the patch but not the original gradients. Sodium–potassium pumps actively transport sodium back outside and potassium inside. Once this is done, most potassium gates close and sodium gates are in their original position—ready to be opened with the arrival of a suitable signal at the membrane patch.

The Direction of Propagation

During an action potential, the inward rush of sodium ions affects the charge distribution across the adjacent membrane patch, where an equivalent number of gated channels open. Gated channels open in the *next* patch, and the next, and so on. This positive feedback event is self-propagating and does not diminish in magnitude. You might be wondering: Do action potentials spread

back to the trigger zone? No. For a brief period after the inward rushing of sodium ions, the voltage-gated channels remain insensitive to stimulation, so sodium ions cannot move through them. This is one reason why action potentials do not spread back to the patch of membrane where they were initiated. It is why they propagate themselves away from it.

Ions cross the neural membrane through transport proteins that serve as gated or open channels. At a suitably disturbed trigger zone, sodium gates open in an all-or-nothing way, and the inward-rushing sodium causes an action potential. Sodium–potassium pumps restore the original ion gradients.

Sodium gates across the membrane are briefly inactivated after an action potential, which is one reason why an action potential is self-propagating away from a trigger zone.

34.3

CHEMICAL SYNAPSES

When action potentials reach a neuron's output zone, they usually do not proceed farther than this. But their arrival may induce the neuron to release one or more **neurotransmitters**, which are signaling molecules that diffuse across chemical synapses. A **chemical synapse** is a narrow cleft between a neuron's output zone and an input zone of a neighboring cell (Figure 34.7). Some clefts are between two neurons. Other clefts intervene between a neuron and a muscle cell or gland cell.

At its output zone, a *pre*synaptic neuron has vesicles filled with neurotransmitter. It also has gated channels for calcium ions, which are more concentrated outside the resting membrane. The gates open when an action potential arrives, and ions diffuse into the neuron. The influx induces the synaptic vesicles to fuse with and become part of the plasma membrane. Thus they dump neurotransmitter molecules into the synaptic cleft.

Neurotransmitter molecules diffuse across the cleft. The *post*synaptic cell's membrane has protein receptors that bind to specific neurotransmitters. Binding changes the receptor shape, so a passage opens up through the protein's interior. Ions now cross the plasma membrane by diffusing through the passageway (Figure 34.7c).

How does the postsynaptic cell respond? It depends on the type and concentration of neurotransmitter, the types of receptors, and the number and type of voltage-gated channels in that cell's membrane. Such factors

influence whether a given neurotransmitter will have an *excitatory* effect and help drive the postsynaptic cell's membrane toward the threshold of an action potential. The factors influence whether it will have an *inhibitory* effect and drive the membrane away from threshold.

Consider the neurotransmitter **acetylcholine** (ACh). It has both excitatory and inhibitory effects on the brain, spinal cord, glands, and muscles. For example, it acts at chemical synapses between a motor neuron and muscle cell, as in Figure 34.8. ACh is released from the motor neuron, diffuses across the cleft, and binds to that cell's receptors for it. ACh has excitatory effects on muscle cells and may trigger action potentials, which in turn may initiate muscle contraction (Section 37.8).

A Smorgasbord of Signals

Acetylcholine is only one of a veritable smorgasbord of signals that neurons deliver to target cells. For example, serotonin's targets in brain regions help control sensory perception, sleeping, body temperature, and emotions. Norepinephrine's targets influence emotions, dreaming, and waking up. Dopamine, too, works in brain regions that deal with emotions. GABA (gamma aminobutyric acid) is the most common inhibitory signal working in the brain. Valium and other antianxiety drugs work by enhancing GABA's effects.

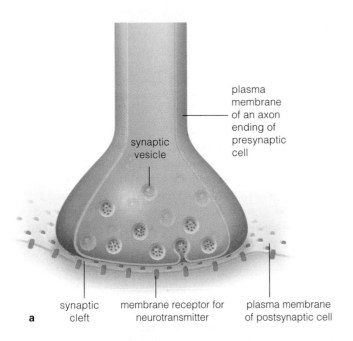

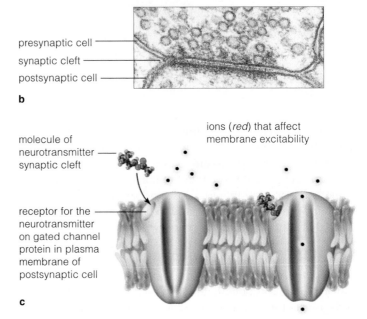

Figure 34.7 Example of a chemical synapse. (**a,b**) A thin cleft separates a presynaptic cell from a postsynaptic cell. Arrival of an action potential at an axon ending triggers release of molecules of neurotransmitter from the presynaptic cell. (**c**) The molecules diffuse across the cleft. They bind to receptors that are part of gated proteins of the postsynaptic cell membrane. The gates open. Ions flow in and trigger a graded potential at the membrane site.

Neuromodulators are signaling molecules that reduce or magnify a neurotransmitter's effects on adjacent or distant neurons. For example, substance P induces pain perception, and endorphins (natural painkillers) inhibit the release of this substance from sensory nerves. Also, neuromodulators may influence memory and learning, sexual activity, body temperature, and emotions.

Synaptic Integration

Anywhere from 1,000 to 10,000 lines of communication synapse on a typical neuron in your brain, which holds at least 100 billion neurons. As long as you are alive, those neurons hum with messages about doing what it takes to be a human. At any moment, a great number of excitatory and inhibitory signals may wash over input zones of a postsynaptic cell. Some drive its membrane closer to threshold; others maintain the resting level or drive it away from threshold. Said another way, signals compete for control of the neuron's membrane.

All synaptic signals are graded potentials. The type we call an EPSP (for excitatory postsynaptic potential) has a *depolarizing* effect. This simply means it drives a membrane closer to threshold. An IPSP (for inhibitory postsynaptic potential) may have a *hyperpolarizing* effect and drive a membrane away from threshold, or it may help maintain the membrane at its resting level.

With **synaptic integration**, competing signals from more than one presynaptic cell reach the input zone of a neuron at the same time and are summed. By this process, two or more signals arriving at a neuron may be dampened, suppressed, reinforced, or sent onward to other cells in the body. Figure 34.9 is a composite of an EPSP recording, an IPSP recording, and a recording

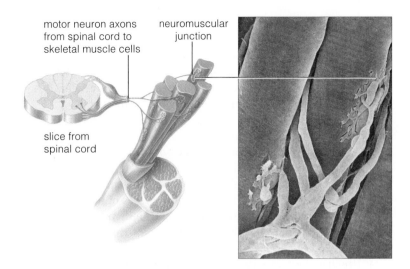

Figure 34.8 Neuromuscular junctions. Each junction is an area of chemical synapsing between the axon endings of a motor neuron and a muscle cell.

of their summation. Integration also occurs when the postsynaptic cell is being bombarded repeatedly with neurotransmitter molecules from a single presynaptic cell that is responding to a series of action potentials.

How Is Neurotransmitter Removed From the Synaptic Cleft?

Signaling depends on the prompt, controlled removal of neurotransmitter molecules from synaptic clefts. Some molecules simply diffuse out of the cleft. Enzymes in the cleft get rid of others, as when acetylcholinesterase breaks apart ACh. Transport proteins in the membrane of presynaptic cells also actively pump molecules back inside or into neighboring neuroglial cells.

What happens if neurotransmitter accumulates in the cleft? As one example, cocaine blocks the uptake of dopamine. Molecules of this neurotransmitter linger in synaptic clefts and keep on stimulating target cells. At first the abnormal stimulation invites euphoria (intense pleasure). Its later effects are severe (Section 34.13).

Neurotransmitters are signaling molecules that bridge a synaptic cleft. The cleft is a thin gap between two neurons or between a neuron and a muscle cell or gland cell.

Neurotransmitters have excitatory or inhibitory effects on different kinds of target cells. Synaptic integration is the moment-by-moment combining of excitatory signals and inhibitory signals acting on the same postsynaptic cell.

Summation allows messages flowing through the nervous system to be reinforced or downplayed, sent onward or suppressed. It is vital for normal body functioning.

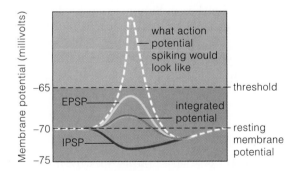

Figure 34.9 Example of synaptic integration. The *yellow* line shows how an EPSP of a certain magnitude would register on an oscilloscope screen if it were acting alone. The *purple* line shows the effect of an IPSP if *it* were acting alone. The *red* line shows what occurs when these two signals arrive at a postsynaptic cell membrane at the same time. In this case, threshold is not reached when they are integrated. So an action potential (the *white* dashed line) cannot be initiated in the cell.

PATHS OF INFORMATION FLOW

Blocks and Cables of Neurons

By synaptic integration, messages arriving at a neuron may be reinforced and sent on to its neighbors. What determines the direction in which a message travels? It depends on how neurons are organized in the body.

For example, your brain's staggering numbers of neurons engage in something like block parties. Regional blocks of hundreds or thousands of them get excitatory and inhibitory signals. They integrate signals entering their block, then send out new ones in response. In the regions of *divergent* circuits, the processes of neurons in one block fan out to form connections with other blocks. In regions of *convergent* circuits, signals from many are relayed to just a few. In regions of *reverberating* circuits, neurons synapse back on themselves and repeat signals like gossip that just won't go away. Such circuits make your eye muscles twitch rhythmically as you sleep.

In the cablelike **nerves**, long axons of many sensory neurons, motor neurons, or both are bundled in parallel inside connective tissue (Figure 34.10). They form long-distance communication lines between the brain and spinal cord, and the rest of the body. Each long axon has a myelin sheath, an electric insulator that enhances the rate of action potential propagation. The sheath is a series of neuroglial cells (Schwann cells) wrapped like jelly rolls around the axon. It hampers ion movements

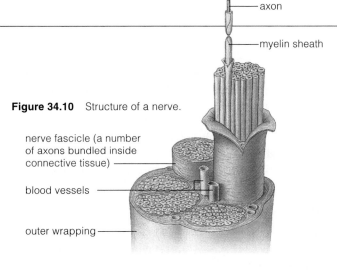

Figure 34.10 Structure of a nerve.

nerve fascicle (a number of axons bundled inside connective tissue)

blood vessels

outer wrapping

axon

myelin sheath

across the membrane. But spaced along the axon are unsheathed nodes with voltage-sensitive, gated sodium channels (Figure 34.11). Ion disturbances tend to flow until reaching a node, where they initiate a new action potential. By jumping node to node, action potentials travel 120 meters per second in large sheathed axons.

Multiple sclerosis, recall, is an autoimmune disorder. White blood cells wrongly attack nerves in the spinal cord (Section 28.5). Myelin sheaths, then axons, become chronically inflamed. Some people are predisposed to the disorder; viral infection may trigger it. Disrupted information flow leads to muscle weakening, fatigue, pain, uncontrollable movements, and other symptoms. Up to 500,000 people in the United States are affected.

Reflex Arcs

A reflex arc is an example of information flow through a nervous system. **Reflexes** are automatic movements in response to stimuli. In the simplest arcs, sensory neurons synapse directly on motor neurons.

The *stretch reflex* is an example. In this case, a muscle contracts after gravity or some other load stretches it. Suppose you hold out a bowl and keep it stationary as someone drops some peaches into it. The weight added to the bowl makes your hand drop. This causes a muscle in your arm, the biceps, to stretch. Stretching activates receptor endings of the sensory organs called muscle

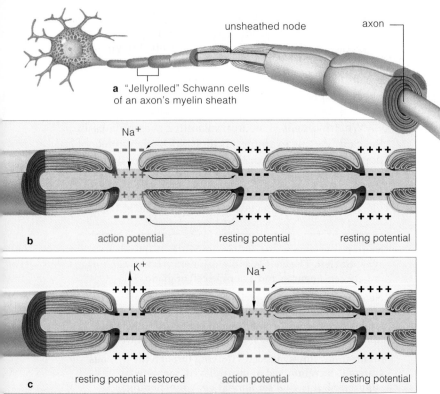

unsheathed node

axon

a "Jellyrolled" Schwann cells of an axon's myelin sheath

Na⁺

++++   ++++

action potential   resting potential   resting potential

b

K⁺   Na⁺

++++   ++++

resting potential restored   action potential   resting potential

c

Figure 34.11 Action potential propagation in a sheathed neuron. (**a**) A myelin sheath blocks ion flow across the neural membrane. Ions cross only at unsheathed nodes. Each node has dense arrays of gated sodium channels. (**b**) A disturbance caused by an action potential spreads along the axon. When it reaches a node, sodium gates open in the positive feedback cycle that starts a new action potential. (**c**) The disturbance spreads swiftly to the next node, where it triggers a new action potential, and so on down the neuron.

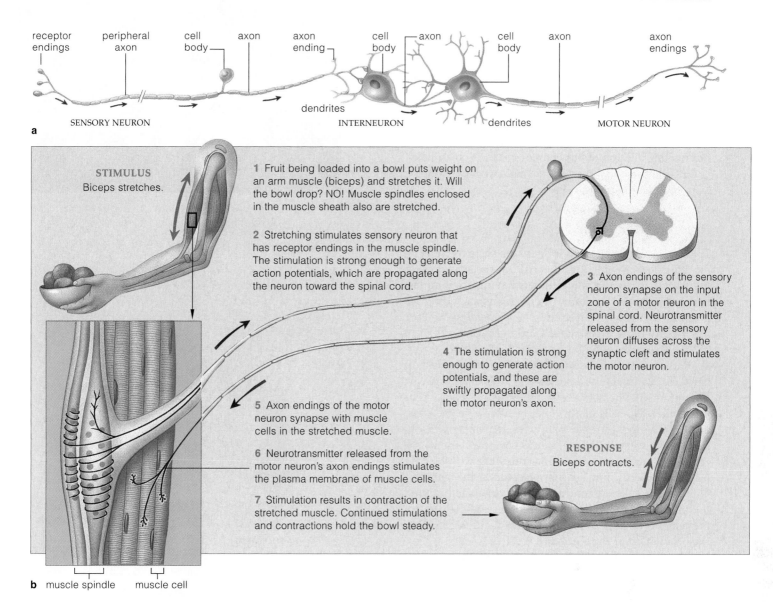

a

receptor endings | peripheral axon | cell body | axon | axon ending | cell body | axon | cell body | axon | axon endings

dendrites

SENSORY NEURON

INTERNEURON

dendrites

MOTOR NEURON

STIMULUS
Biceps stretches.

1 Fruit being loaded into a bowl puts weight on an arm muscle (biceps) and stretches it. Will the bowl drop? NO! Muscle spindles enclosed in the muscle sheath also are stretched.

2 Stretching stimulates sensory neuron that has receptor endings in the muscle spindle. The stimulation is strong enough to generate action potentials, which are propagated along the neuron toward the spinal cord.

3 Axon endings of the sensory neuron synapse on the input zone of a motor neuron in the spinal cord. Neurotransmitter released from the sensory neuron diffuses across the synaptic cleft and stimulates the motor neuron.

4 The stimulation is strong enough to generate action potentials, and these are swiftly propagated along the motor neuron's axon.

5 Axon endings of the motor neuron synapse with muscle cells in the stretched muscle.

6 Neurotransmitter released from the motor neuron's axon endings stimulates the plasma membrane of muscle cells.

7 Stimulation results in contraction of the stretched muscle. Continued stimulations and contractions hold the bowl steady.

RESPONSE
Biceps contracts.

b muscle spindle   muscle cell

Figure 34.12 (**a**) General direction of information flow in nervous systems. Sensory neurons relay signals *into* the spinal cord and brain, where they synapse with interneurons. Interneurons *within* the spinal cord and brain integrate signals. Many synapse with motor neurons, which carry signals *away* from the spinal cord and brain. (**b**) Stretch reflex. In skeletal muscle, stretch-sensitive receptors of sensory neurons (one is shown here) are located in muscle spindles. Stretching generates action potentials, which reach sensory axon endings in the spinal cord. These synapse with a motor neuron that carries signals to contract, from the spinal cord back to the stretched muscle.

spindles. The endings, enclosed in a sheath running parallel with the muscle, are the input zones of sensory neurons. In the spinal cord, axons of these neurons synapse with motor neurons. The axons of these motor neuron axons lead back to the stretched muscle (Figure 34.12). At the axon endings, action potentials trigger the release of ACh, which initiates contraction by cells making up the muscle (Section 37.8). As long as ACh molecules continue to be released, the motor neurons continue to be excited. The stimulation helps keep your hand steadily positioned against the force of gravity.

In nearly all reflex pathways, sensory neurons also interact with interneurons, which activate or suppress all motor neurons required for a coordinated response.

Vertebrates have interneurons organized in information-processing blocks. Cablelike nerves that have long axons of sensory neurons, motor neurons, or both connect the brain and spinal cord with the rest of the body.

Reflex arcs, in which sensory neurons synapse directly on motor neurons, are the simplest paths of information flow.

INVERTEBRATE NERVOUS SYSTEMS

We turn now from the messages sent through nervous systems to the systems themselves. At this point in the book, you know all animals except sponges have some type of nervous system in which nerve cells, such as neurons, are oriented relative to one another in signal-conducting and information-processing highways. At a minimum, the cells making up the communication lines receive information about changing conditions outside and inside the body, then they elicit suitable responses from muscle and gland cells.

Regarding the Nerve Net

To appreciate the diversity of nervous systems, start by recalling that animals first evolved in the seas. It is in the seas that we still find animals having the simplest nervous systems. They are the cnidarians, including sea anemones. These invertebrates have *radial* symmetry. Their body parts are arranged around a central axis, a bit like the spokes of a bike wheel (Section 26.1).

Radial animals rely on a **nerve net**, a loose mesh of nerve cells intimately associated with epithelial tissue (Figure 34.13). The nerve cells interact with sensory cells and contractile cells along reflex pathways in the same epithelial tissue. In reflex pathways, remember, sensory stimulation triggers simple, stereotyped movements.

In all cnidarians, one pathway dealing with feeding behavior extends from sensory receptors in the tentacles, along nerve cells, to contractile cells organized around the mouth. In jellyfishes, other reflexes are the basis of movements used in swimming slowly and in changing the orientation of the soft body.

The nerve net itself extends through the animal's body, but information flow through it is not focused. It simply commands the body wall to slowly contract and expand or tentacles to move through water. Its owners will never dazzle you with bursts of speed or precision acrobatics. Either they are weak swimmers or they are sedentary types that spend most of their life attached to substrates. Yet in their watery world, the nerve net is adaptive. It is equally responsive to signals about food or danger that might come from any direction.

On the Importance of Having a Head

Flatworms are the simplest animals having a bilateral nervous system (Figure 34.14). *Bilateral* symmetry means having equivalent body parts on the left and right sides of the body's midsagittal plane. (Imagine sliding down a staircase banister that turns into a razor and you may never forget the location of the midsagittal plane.) Both sides have the same array of muscles that function in moving the body forward. Both have the same array of nerves to control the muscles, and so on.

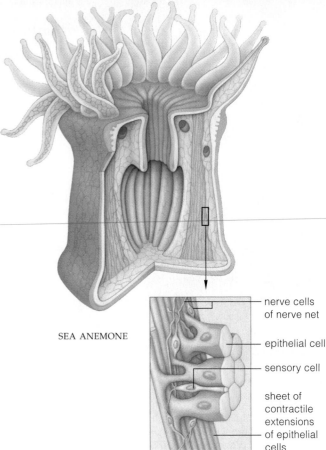

SEA ANEMONE

nerve cells of nerve net

epithelial cell

sensory cell

sheet of contractile extensions of epithelial cells

Figure 34.13 Nerve net of a sea anemone, a cnidarian. Its nerve cells interact with sensory and contractile cells. Both cell types are embedded in epithelium between the outer epidermis and a jellylike midlayer (mesoglea) of the body wall.

The flatworm's ladderlike nervous system has two cordlike nerves. The two run longitudinally through the body and have many side branches. In their head end, some flatworms have two ganglia. A *ganglion* (plural, ganglia) is a cluster of nerve cell bodies that is a local integrating center. Ganglia inside the head of a flatworm coordinate signals coming from paired sensory organs (two eyespots, for example), and they provide some control over the nerves.

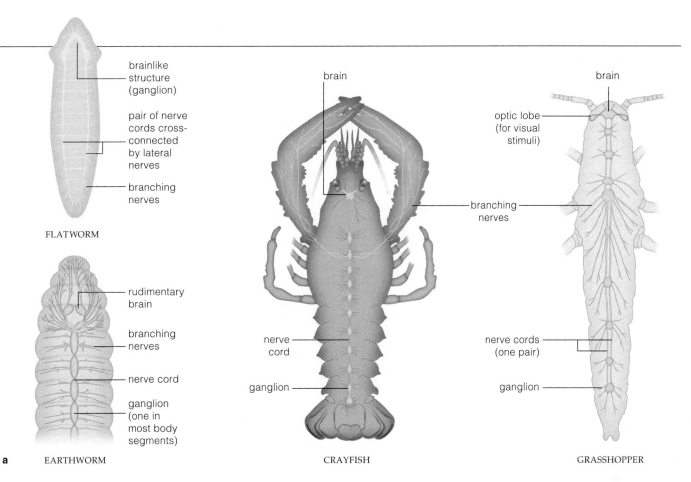

FLATWORM

brainlike structure (ganglion)

pair of nerve cords cross-connected by lateral nerves

branching nerves

rudimentary brain

branching nerves

nerve cord

ganglion (one in most body segments)

a   EARTHWORM

brain

branching nerves

nerve cord

ganglion

CRAYFISH

brain

optic lobe (for visual stimuli)

branching nerves

nerve cords (one pair)

ganglion

GRASSHOPPER

Figure 34.14 (**a**) Bilateral nervous systems of a few invertebrates. The sketches are not to the same scale. (**b**) Generalized sketch of a planula and a micrograph of the planula stage that develops in the life cycle of the jellyfish *Aurelia*.

Did bilateral nervous systems evolve from nerve nets? Maybe. In nearly all animals more complex than flatworms, we find local nerve nets—plexuses—such as the one in your intestinal wall. Intriguingly, a planula (a self-feeding larval stage) develops in some cnidarian life cycles. Like flatworms, it has a flattened body and uses cilia to swim or crawl about (Figure 34.14*b*).

Visualize a planula on the Cambrian seafloor. A mutation blocks its metamorphosis into the adult yet doesn't stop reproductive organs from maturing. This does happen in some larvae (Section 26.5). The crawling planula kept on crawling but now had the capacity to reproduce. Its offspring inherited the mutant gene for forward mobility. They crawled longer and entered new parts of their habitat—food-rich, food-poor, and maybe dangerous. The mutation had survival value, because it led to more rapid, more effective responses to diverse stimuli. So a concentration of sensory cells in the body's leading end—not the trailing end—was favored.

Regardless of how it happened, cephalization (the formation of a head) and bilateral symmetry developed in most invertebrate lineages (Chapter 26). Patterns of cephalization and bilateral symmetry also are evident in the paired sensory structures, brain centers, nerves, and skeletal muscles of all vertebrates, including you.

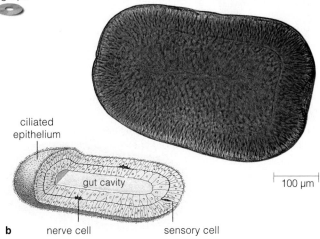

ciliated epithelium

gut cavity

100 μm

b   nerve cell   sensory cell

All animals except sponges have a nervous system, in which nerve cells are oriented relative to one another in signal-conducting and information-processing highways.

Radial animals have a nerve net, a diffuse mesh of nerve cells that take part in simple reflex pathways involving sensory cells and contractile cells of an epithelial tissue.

Cephalized, bilateral animals have a nervous system with a brain or ganglia at the head end, as well as paired nerves and paired sensory structures.

VERTEBRATE NERVOUS SYSTEMS—AN OVERVIEW

Evolution of the Spinal Cord and Brain

Hundreds of millions of years ago, the earliest fishlike vertebrates were evolving (Section 26.3). A column of bony segments was taking over the functions of their notochord, a long rod of stiffened tissue that worked with their segmented muscles to bring about movement. Above the notochord, the hollow, tubular nerve cord was undergoing modification. It was the forerunner of the spinal cord and brain. These developments were the basis of different life-styles. Not long afterward, genetic divergences from lineages of filter-feeding, scavenging fishes gave rise to swift, jawed predators of the seas.

At first, simple reflex pathways prevailed. Sensory neurons synapsed directly with motor neurons, which directly signaled muscles to contract. No other neurons

helped refine the responses to stimuli. However, in the world of fast-moving vertebrates, predators or prey that had more effective ways of assessing and responding to food and danger had a competitive edge.

The senses of smell, hearing, and balance became keener among the vertebrates that invaded land. Bones and muscles evolved in ways that allowed specialized movements. Also, the brain became variably thickened with nervous tissue that could integrate rich sensory information and issue orders for complex responses.

The oldest regions of the vertebrate brain still deal with reflex coordination of respiration and other vital functions. But many more interneurons now synapse on the sensory and motor neurons of ancient pathways and with one another in the newer brain regions. In the most complex vertebrates, interneurons receive, store, retrieve, and compare information about experiences. They weigh possible responses. And they give our own species the capacity to reason, remember, and learn.

The nerve cord persists in all vertebrate embryos. We call it the **neural tube**. As an embryo grows and develops, it expands into a brain and spinal cord, but to different degrees among different vertebrate groups (Figure 34.15). A vertebral column encloses the spinal cord. Adjacent tissues give rise to nerves in all body regions; these connect with the spinal cord and brain.

The System's Functional Divisions

Figure 34.16 hints at the expansions of nervous tissue in the human nervous system. It shows major paired nerves of this bilateral system. What it cannot show are the 100 billion interneurons in the brain alone. Humans do have the most intricately wired nervous system in

FOREBRAIN. Receives, integrates sensory information from nose, eyes, and ears; in land-dwelling vertebrates, contains the highest integrating centers

MIDBRAIN. Coordinates reflex responses to sight, sounds

HINDBRAIN. Reflex control of respiration, blood circulation, other basic tasks; in complex vertebrates, coordination of sensory input, motor dexterity, and possibly mental dexterity

(start of spinal cord)

a Expansion of the dorsal, hollow nerve cord into more complex, functionally distinct regions in certain lineages.

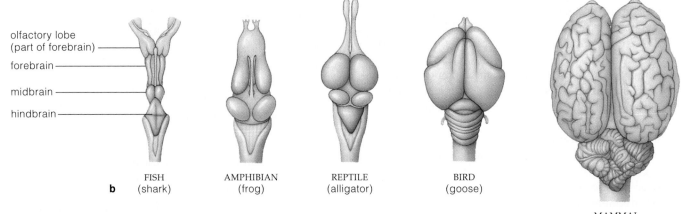

olfactory lobe (part of forebrain)
forebrain
midbrain
hindbrain

b

FISH (shark)   AMPHIBIAN (frog)   REPTILE (alligator)   BIRD (goose)   MAMMAL (horse)

Figure 34.15 Evolutionary trend toward an expanded, more complex brain. The trend became apparent after morphological comparisons were made of the brains of some vertebrates. These dorsal views are not to the same scale.

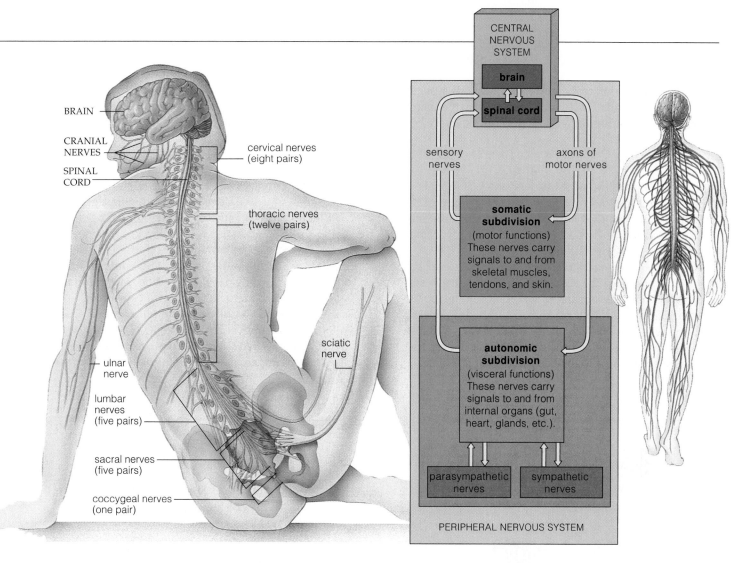

Figure 34.16 Brain, spinal cord, and some major peripheral nerves of the human nervous system. The system also has twelve pairs of cranial nerves that originate from the brain. Other vertebrates have a similar nervous system.

Figure 34.17 Functional divisions of the human nervous system. The central nervous system is coded *blue*, somatic nerves *green*, and autonomic nerves *red*. Sensory nerves leading to the central nervous system are also called *afferent* (meaning "to bring to"). Motor neurons leading away from the central nervous system to muscles and glands are *efferent* ("to carry outward").

the animal world. Yet similar patterns of organization and information flow occur among other vertebrates.

Investigators typically approach the complexity of the vertebrate nervous system by functionally dividing it into central and peripheral regions (Figure 34.17). All of the interneurons are confined to the **central nervous system**, which consists of the spinal cord and brain. The **peripheral nervous system** consists mainly of nerves that extend through the rest of the body. Nerves carry signals into and out of the central nervous system.

Inside the brain and spinal cord, the communication lines are called tracts, not nerves. The tracts making up *white* matter have axons with glistening white myelin sheaths; they function in rapid signal transmission. By contrast, *gray* matter consists of unmyelinated axons, dendrites, and cell bodies of neurons, plus neuroglial cells. Neuroglia, remember, protects or structurally and

functionally supports neurons. It makes up more than half the volume of vertebrate nervous systems.

The coevolution of nervous, sensory, and motor systems made more complex life-styles possible among vertebrates.

The vertebrate nervous system has become so intricately wired that the structure and functions of its central and peripheral regions are described separately.

The central nervous system consists of the brain and spinal cord. The peripheral nervous system consists of nerves that thread through the rest of the body and carry signals into and out of the central region.

Myelinated axons make up the white matter of tracts inside the spinal cord and the brain. Neuroglia, unmyelinated axons, dendrites, and cell bodies of neurons make up the gray matter of those tracts.

WHAT ARE THE MAJOR EXPRESSWAYS?

Let's now take a look at the peripheral nervous system and the spinal cord. The two interconnect as the main expressways for information flow through the body.

Peripheral Nervous System

SOMATIC AND AUTONOMIC SUBDIVISIONS In humans, the peripheral nervous system includes thirty-one pairs of *spinal* nerves, which connect with the spinal cord. The system also includes twelve pairs of *cranial* nerves that connect directly with the brain.

Cranial and spinal nerves are further classified by function. Those carrying signals about moving the head, trunk, and limbs are **somatic nerves**. Their sensory axons deliver information from receptors in the skin, skeletal muscles, and tendons into the central nervous system. Their motor axons deliver the commands issued by the brain and spinal cord to the body's skeletal muscles.

By contrast, spinal and cranial nerves dealing with smooth muscle, cardiac (heart) muscle, and glands are **autonomic nerves**. They deal with the viscera—internal organs and structures. Autonomic nerves carry signals to and from these organs and structures.

PARASYMPATHETIC AND SYMPATHETIC NERVES Figure 34.18 shows the two categories of autonomic nerves. We call them parasympathetic and sympathetic. Normally they work antagonistically, with signals from one kind opposing signals from the other. However, both carry excitatory and inhibitory signals to the internal organs. Often their signals arrive at the same time at muscle or gland cells and compete for control over them. In such cases, synaptic integration at the cellular level leads to minor adjustments in the organ's level of activity.

Parasympathetic nerves dominate when the body is not receiving much outside stimulation. They tend to

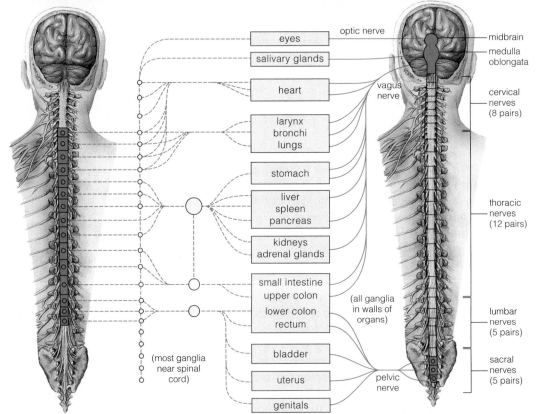

SYMPATHETIC OUTFLOW FROM THE SPINAL CORD	*Examples of Responses*
	Heart rate increases
	Pupils of eyes dilate (widen, let in more light)
	Glandular secretions in airways to lungs decrease
	Salivary gland secretions thicken
	Stomach and intestinal movements slow down
	Sphincters (rings of muscle) contract

PARASYMPATHETIC OUTFLOW FROM THE SPINAL CORD AND BRAIN	*Examples of Responses*
	Heart rate decreases
	Pupils of eyes constrict (keep more light out)
	Glandular secretions in airways to lungs increase
	Salivary gland secretions become dilute
	Stomach and intestinal movements increase
	Sphincters (rings of muscle) relax

Figure 34.18 Autonomic nervous system. This diagram shows the major sympathetic nerves and parasympathetic nerves leading from the central nervous system to some major organs. Remember, there are *pairs* of both kinds of nerves, servicing both right and left halves of the body. The ganglia are clusters of the cell bodies of neurons that have their axons bundled together in nerves.

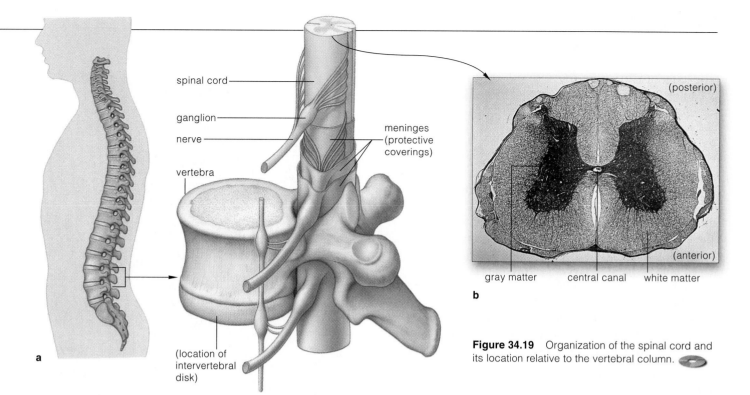

spinal cord

ganglion

nerve

vertebra

meninges (protective coverings)

(location of intervertebral disk)

a

(posterior)

(anterior)

gray matter   central canal   white matter

b

Figure 34.19 Organization of the spinal cord and its location relative to the vertebral column.

slow down the body overall and divert energy to basic "housekeeping" tasks, such as digestion.

Sympathetic nerves dominate in times of sharpened awareness, stress, excitement, or danger. Housekeeping tasks tend to be shelved. And the body prepares to fight or escape if threatened, or to frolic (as in play behavior or sexual behavior).

For example, signals from sympathetic nerves usually make the heart beat a bit faster even as signals from parasympathetic nerves make it beat a bit slower. The rate depends on integration of these opposing signals. If something scares you, parasympathetic input to the heart drops. Sympathetic signals make the heart beat faster by triggering epinephrine secretions. They also make you breathe faster and start to sweat. In this state of intense arousal, you are primed to fight or play hard, or get away fast. Hence the term *fight–flight response.*

Suppose that the stimulus for a fight–flight response ends. Sympathetic activity may decrease abruptly, and parasympathetic activity may rise suddenly. You might observe this "rebound effect" after someone has been instantly mobilized to rush onto a road to save a child from an oncoming car. The person might well faint as soon as the child has been swept out of danger.

Spinal Cord

By definition, the **spinal cord** is a vital expressway for signals between the peripheral nervous system and the brain. Also, sensory and motor neurons make direct reflex connections in the spinal cord, as they do for the stretch reflex. The spinal cord threads through a canal made of bones of the vertebral column (Figure 34.19a).

The bones, and ligaments attached to them, protect the cord. So do meninges, three tubelike coverings around the spinal cord and brain. The coverings are tough but vulnerable. *Meningitis*, an often-fatal disease, is brought on by certain viral and bacterial infections. Symptoms include severe headaches, fever, a stiff neck, and nausea.

Signals swiftly travel up and down the spinal cord in bundles of myelinated axons. These axons, which make up the cord's white matter, surround the gray matter (Figure 34.19b). Gray matter, again, consists of dendrites and cell bodies of neurons, and neuroglial cells. It plays a key role in controlling reflexes for limb movements, as when you dance or wave your arms about, and for organ activity, such as bladder emptying.

Experiments tell us about such pathways. Examples: Between a frog's spinal cord and brain are circuits that command bent legs to straighten. Sever the circuits at the base of the brain, and the legs become paralyzed— but only for about a minute. Extensor reflex pathways in the spinal cord recover quickly and have the frog hopping about in no time. There is little or no recovery after similar damage in humans and other primates— the vertebrates with the greatest cephalization.

Nerves of the peripheral nervous system connect the brain and spinal cord with the rest of the body.

The somatic nerves of the peripheral nervous system deal with skeletal muscle movements. Its autonomic nerves deal with smooth muscle, cardiac muscle, and glands.

The spinal cord is a vital expressway for signals between the brain and peripheral nerves. Some of its interneurons also exert direct control over certain reflex pathways.

THE VERTEBRATE BRAIN

The anterior end of the spinal cord is continuous with the **brain**, which is the body's master control center. The brain receives, integrates, stores, retrieves, and issues information. It coordinates responses to sensory input by adjusting activities through the body. Like the spinal cord, the brain is protected by bones and membranes.

Reflect on Figures 34.15*a* and 34.20. In the vertebrate embryo, three successive portions of the neural tube give rise to the forebrain, midbrain, and hindbrain. The nervous tissue that evolved first in all three portions is the **brain stem**. This tissue is still identifiable in the adult brain, and it still has many basic reflex centers. Over time, expanded layers of gray matter developed from the brain stem. Biologists correlate these recent additions with an increasing reliance on three major sensory organs—the nose, ears, and eyes. The forebrain and possibly the hindbrain region called the cerebellum have the newest additions of gray matter.

Hindbrain

The medulla oblongata, cerebellum, and pons are parts of the hindbrain. In the **medulla oblongata** are reflex centers for a number of vital tasks, such as respiration and blood circulation. This brain region coordinates motor responses with certain complex reflexes, such as coughing. It also influences sleep centers in the brain.

The **cerebellum** integrates sensory signals from the eyes, ears, and muscle spindles with motor signals from the forebrain. It helps control motor skills. More recent expansions of the cerebellum in humans may contribute to language and some other forms of mental dexterity.

Bands of many axons extend from both sides of the cerebellum to the **pons** (meaning "bridge"). Although lodged inside the brain stem, the pons is a major traffic center for information passing between the cerebellum and the higher integrating centers of the forebrain.

Midbrain

The midbrain coordinates the reflex responses to sights and sounds. The midbrain's roof, the *tectum*, is a more recent addition of gray matter. In amphibians as well as fishes, the tectum coordinates nearly all sensory input and initiates motor responses. Its importance is evident from experiments that surgically deprive a frog of its highest brain center but leave the tectum intact. The frog will still be able perform almost all of the tasks that frogs normally perform.

In most vertebrates (not mammals), the midbrain has a pair of optic lobes, brain centers that deal with sensory input from eyes. Mammalian eyes deserted the tectum, so to speak. They formed important functional connections with centers in the forebrain. The tectum

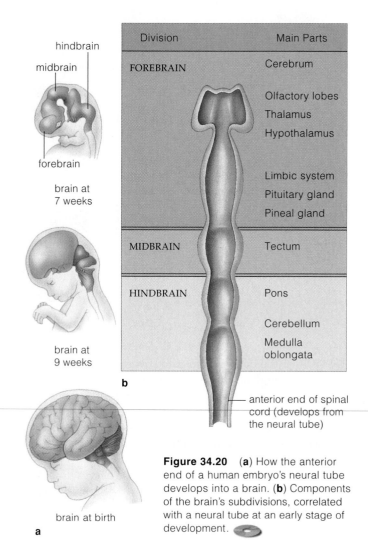

Figure 34.20 (**a**) How the anterior end of a human embryo's neural tube develops into a brain. (**b**) Components of the brain's subdivisions, correlated with a neural tube at an early stage of development.

of mammals became a reflex center that quickly relays sensory signals to those higher integrating centers.

Evolution of the Forebrain

For much of vertebrate history, chemical odors that slowly diffused through aquatic habitats were the most important clues to survival. An olfactory lobe dealing primarily with odors from predators, prey, and mates was a key forebrain structure (Figure 34.20*b*). So were a pair of tissue outgrowths from the brain stem, where olfactory input and responses to it were integrated. In time the outgrowths expanded greatly, especially after certain vertebrates invaded the land. These outgrowths became the two hemispheres of the **cerebrum**.

Another forebrain region, the **thalamus**, evolved as a coordinating center for sensory input and as a relay station for input to the cerebrum. Below the thalamus, the **hypothalamus** evolved into the premier center for

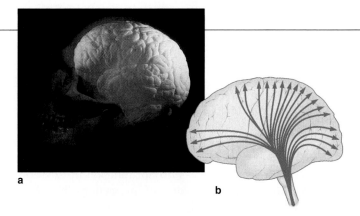

a

b

Figure 34.21 (**a**) Model showing the surface of the cerebral cortex of the human brain. (**b**) Communication pathways of the reticular formation, an evolutionarily ancient, diffuse network of neurons. The formation extends from the anterior end of the spinal cord on up to the highest integrative centers of the cerebral cortex.

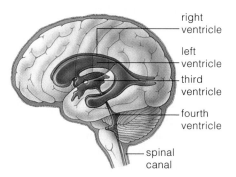

right
ventricle

left
ventricle

third
ventricle

fourth
ventricle

spinal
canal

Figure 34.22 Cerebrospinal fluid (*blue*). This extracellular fluid fills the four interconnected cavities in the brain and in the spinal cord's central canal.

homeostatic control over the internal environment. It also became central to the behaviors related to internal organ activities, such as thirst, hunger, and sex, and to emotional expression, such as sweating with fear.

Reticular Formation

You may be thinking that the brain is tidily subdivided into three main regions. This is not the case. An ancient network of interneurons extends from the uppermost part of the spinal cord, through the brain stem, and on into higher integrative centers of the cerebral cortex (Figure 34.21). This major network of interneurons, the **reticular formation**, persists as a low-level pathway to motor centers of the medulla oblongata and spinal cord. It also activates centers in the cerebral cortex and helps govern many activities of the nervous system.

Part of the reticular formation promotes chemical changes that affect the states of consciousness, such as sleeping and waking. Serotonin is a neurotransmitter from one of its sleep centers. It inhibits other neurons that arouse the brain and maintain wakefulness. High serotonin levels cause drowsiness and sleep. Its effects are inhibited by substances released from another brain center that bring about wakefulness.

We can track the states of consciousness with EEGs, which are recordings of summed electrical activity that appear as wave forms. The pattern for someone who is meditating (relaxed, eyes closed) is an alpha rhythm. Wave trains become larger, slower, and more erratic during the transition to sleep, when sensory input is low and the mind is more or less idling. People who are awakened from slow-wave sleep usually say they were not dreaming; they often seemed to be mulling over recent, ordinary events. Punctuating slow-wave sleep is the REM pattern of rapid eye movements (the eyes jerk under closed lids), irregular breathing, faster heartbeat, twitching fingers, and usually vivid dreams.

Protection at the Blood–Brain Barrier

The hollow neural tube that forms in embryos persists in adults, as a continuous system of fluid-filled cavities and canals. Inside the system is cerebrospinal fluid, a clear extracellular fluid that cushions the brain and the spinal cord against jarring movements (Figure 34.22).

A mechanism called the **blood–brain barrier** protects the brain and spinal cord by exerting some control over which solutes enter the cerebrospinal fluid. No other portion of extracellular fluid has solute concentrations kept within such narrow limits. Even normal changes in fluid composition that accompany, say, eating and exercising are opposed here. Why? Some blood-borne hormones and amino acids can change the functioning of neurons. Also, shifting levels of certain ions (such as K^+) can skew the threshold for action potentials.

The barrier works at the plasma membrane of cells making up the wall of blood capillaries that service the brain. Tight junctions fuse all the abutting cell walls in most brain regions—so water-soluble substances must move *through* these cells to reach the brain. Membrane transport proteins let glucose and other vital nutrients as well as some ions move into and out of cells. They bar some toxins and metabolic wastes, such as urea. The barrier does not keep out fat-soluble substances, such as oxygen, carbon dioxide, anesthetics, alcohol, caffeine, and nicotine. It is nonexistent around the brain stem's vomiting center and the hypothalamus (guess why).

The vertebrate brain develops from a hollow neural tube, which persists in adults as a system of cavities and canals filled with cerebrospinal fluid. The fluid cushions nervous tissue from sudden, jarring movements. The nervous tissue is subdivided into the hindbrain, forebrain, and midbrain.

The brain stem is the most ancient nervous tissue in all three regions. It affords reflex control over basic functions necessary for survival. The highest integrative centers are in the forebrain, especially in the cerebral cortex.

THE HUMAN CEREBRUM

Functional Divisions of the Cerebral Cortex

Figure 34.23 shows components of the human brain. In an average-sized person, the brain weighs 1,300 grams (3 pounds). Over half of its volume is neuroglia.

A fissure divides the human cerebrum into left and right cerebral hemispheres. Each half has an outer layer of gray matter, the **cerebral cortex**. Below the cortex is white matter (axons), which has patches of gray matter called basal nuclei. The left half deals primarily with analytical skills, speech, and mathematics. It typically dominates the right hemisphere, which has more to do with visual–spatial relationships and with music. Each hemisphere responds to sensory input mainly from the opposite side of the body. For example, signals about pressure on the right arm travel to the left hemisphere. Signals flow back and forth along a connecting band of nerve tracts, the corpus callosum, and help coordinate activities of the two hemispheres.

Each hemisphere has four subdivisions, the frontal, occipital, temporal, and parietal lobes, that receive and process different signals. EEGs and PET scans (Section 2.2) are used to trace electrical activity in each lobe.

Taken as a group, humans are exceptionally good at comprehending, communicating, remembering, and voluntarily acting upon information. This brain power helps define our humanness. It arises in a layer of gray matter in the cerebral cortex, and it governs conscious behavior. We functionally divide this layer into *motor* areas (control of voluntary motor activity), *sensory* areas (perception of what sensations mean), and *association* areas (integration of information preceding conscious action). The cortical areas do not function in isolation. Consciousness is an outcome of interactions that occur throughout the cortex (Figures 34.24 and 34.25).

MOTOR AREAS The entire body is spatially mapped out in the primary motor cortex of each hemisphere's frontal lobe. This area controls coordinated movements of skeletal muscles. Thumb, finger, and tongue muscles get much of the area's attention. This gives you an idea of how much control is required for voluntary hand movements and verbal expression (Figure 34.24).

The frontal lobe also contains the premotor cortex, Broca's area, and the frontal eye field. Learned patterns of motor skills are the domain of the premotor cortex. Play a piano concerto, pound on a computer keyboard, dribble a basketball—such repetitive movements are signs that this region is coordinating the simultaneous, sequential movements of a number of muscle groups. Broca's area (most often in the left hemisphere) and a corresponding area in the right hemisphere contain the

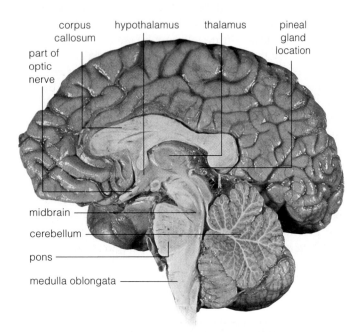

Figure 34.23 Right cerebral hemisphere of a human brain, sagittal view. Not visible is the reticular formation, which extends between the upper spinal cord and cerebrum. Each hemisphere has a layer of gray matter, the cerebral cortex, about 2–4 millimeters (1/8 inch) thick. Interneuron cell bodies and dendrites, unmyelinated axons, neuroglia, and blood vessels make up the gray matter.

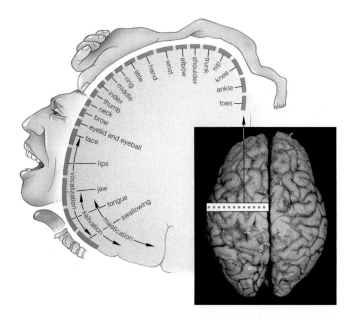

Figure 34.24 Diagram of a slice through the primary motor cortex of the left cerebral hemisphere of the human brain. Sizes of different body parts draped over different parts of the slice are distorted to show which receive the most precise control. The photograph is a dorsal view of both cerebral hemispheres.

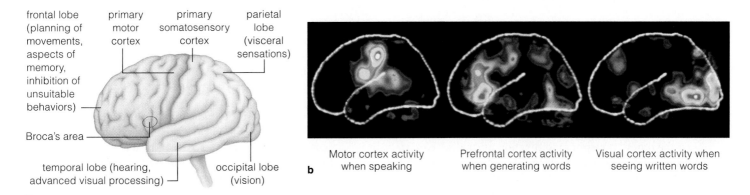

frontal lobe
(planning of
movements,
aspects of
memory,
inhibition of
unsuitable
behaviors)

primary
motor
cortex

primary
somatosensory
cortex

parietal
lobe
(visceral
sensations)

Broca's area

temporal lobe (hearing,
advanced visual processing)

occipital lobe
(vision)

a

Motor cortex activity
when speaking

Prefrontal cortex activity
when generating words

Visual cortex activity when
seeing written words

b

Figure 34.25 (**a**) Primary receiving and integrating centers of the human cerebral cortex. Primary cortical areas receive signals from receptors on the body's periphery. Association areas coordinate and process sensory input from different receptors. (**b**) Three PET scans identifying which brain regions were active while an individual performed three tasks: speaking, generating words, and observing written words.

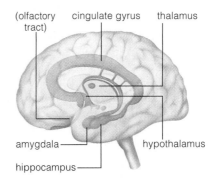

(olfactory tract)   cingulate gyrus   thalamus

amygdala

hypothalamus

hippocampus

Figure 34.26 Limbic system components.

controls over tongue, throat, and lip muscles that are used in speech. The controls are activated when we are about to speak, even when we plan voluntary motor activities other than speaking. The frontal eye field, above Broca's area, controls voluntary eye movements.

SENSORY AREAS Different parts of the cortex include sensory areas. As you will read in the next chapter, the body is spatially mapped out in the parietal lobe, in its primary somatosensory cortex. This is the key receiving center for sensory input from skin and joints. Also in this lobe, a primary cortical area deals with perception of taste. At the back of the occipital lobe, the primary visual cortex receives signals from the eyes. Perception of sounds and odors arises in primary cortical areas in each temporal lobe.

ASSOCIATION AREAS These areas occur throughout the cortex (but not in the primary motor and sensory areas). Each integrates and responds to many inputs. For instance, the visual association area surrounding the primary visual cortex helps us recognize something we observe by comparing it with visual memories. The most complex association area, the prefrontal cortex, is the basis of complex learning, intellect, and personality. Without it, we would be incapable of abstract thought, judgment, planning, and having concern for others.

Connections With the Limbic System

Encircling the upper brain stem is the **limbic system**, which controls emotions and has roles in memory. It includes the hypothalamus, amygdala, cingulate gyrus, hippocampus, and parts of the thalamus (Figure 34.26).

The hypothalamus is a clearinghouse for emotions and visceral activity. The amygdala is crucial for emotional stability, interpreting social cues and, together with the hippocampus, converting stimuli to long-term memory. The cingulate (belt-shaped) gyrus affects the will to act.

The limbic system is evolutionarily related to the olfactory lobes and still deals with the sense of smell. That is one reason why you may feel warm and fuzzy when your brain recalls the cologne of a special person.

By its connections with the prefrontal cortex and other brain centers, the limbic system correlates organ activities with self-gratifying behavior, such as eating and sex. That is why the limbic system is called our emotional–visceral brain. It can make you feel your stomach or heart is on fire with passion or indigestion. However, reasoning in the cerebral cortex can override or dampen rage, hatred, or other "gut reactions."

The cerebrum of humans and other complex vertebrates is divided into two hemispheres. Each half receives, processes, and coordinates responses to sensory input from primarily the opposite side of the body.

The cerebral cortex (each hemisphere's outermost layer of gray matter) contains motor, sensory, and association areas. Interactions among the areas govern conscious behavior. The cerebral cortex also interacts with the limbic system, which governs emotions and contributes to memory.

Sperry's Split-Brain Experiments

Some time ago, the neurosurgeon Roger Sperry and his colleagues demonstrated some intriguing differences in perception between the two cerebral hemispheres of epileptics. Severe *epilepsy* is characterized by seizures, sometimes as often as every half hour. The seizures are analogous to an electrical storm in the brain. Sperry asked: Would *cutting* the corpus callosum of epileptics confine the electrical storm to one hemisphere, leaving at least the other hemisphere to function normally? Earlier studies of laboratory animals and humans whose corpus callosum had been damaged suggested this might be so.

Sperry performed the surgery on some patients. The electrical storms did subside in frequency and intensity. Cutting the neural bridge ended what must have been a positive feedback loop of ever intensifying electrical disturbances between the two hemispheres. The "split-brain" patients were able to lead what seemed, on the surface, entirely normal lives.

But then Sperry devised some elegant experiments to test whether their conscious experience was indeed "normal." Given that the corpus callosum contains 200 million axons, surely *something* was different. Something was. "The surgery," Sperry later reported, "left these people with two separate minds, that is, two spheres of consciousness. What is experienced in the right hemisphere seems to be entirely outside the realm of awareness of the left."

Sperry presented the two hemispheres of split-brain patients with two different portions of the same visual stimulus. Researchers knew at the time that the visual connections to and from one hemisphere are mainly concerned with the opposite half of the visual field, as shown in Figure 34.27. Sperry projected words—say, COWBOY—onto a screen so that COW fell in the left half of the visual field, and BOY fell in the right (Figure 34.28).

The subjects of this experiment reported *seeing* the word BOY. The left hemisphere, which controls language, perceived only the letters BOY. However, when asked to write the perceived word with the left hand—a hand that was deliberately blocked from a subject's view—the subject wrote COW. The right hemisphere "knew" the other half of the word (COW) and had directed the left hand's motor response. But it could not tell the left hemisphere what was going on because of the severed corpus callosum. The subject knew that a word was being written, but could not say what it was!

Thus Sperry showed that signals across the corpus callosum coordinate the functioning of the two cerebral hemispheres, each of which had responded to visual signals from the opposite side of the body.

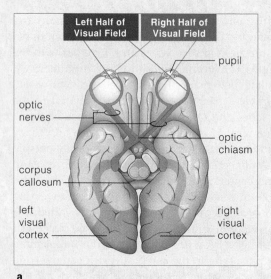

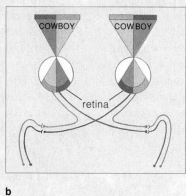

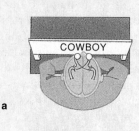

Figure 34.27 **(a)** Sensory pathway by which sensory input about visual stimuli reaches the visual cortex of the human brain.

(b) Each eye gathers visual information at the retina, a thin layer of densely packed photoreceptors at the back of the eyeball (Section 35.7). Light from the *left* half of the visual field strikes receptors on the right side of both retinas. Parts of two optic nerves carry signals from the receptors to the right cerebral hemisphere. Light from the *right* half of the visual field strikes receptors on the left side of both retinas. Parts of the optic nerves carry signals from them to the left hemisphere.

Figure 34.28 One example of the response of a split-brain patient to visual stimuli. As described in the text, this type of experiment demonstrated the importance of the corpus callosum in coordinating activities between the two cerebral hemispheres.

HOW ARE MEMORIES TUCKED AWAY?

Memory refers to a brain's capacity to store and retrieve information about past sensory experience. Without it, learning and adaptive modifications of behavior would be impossible. Information is stored in stages. In *short-term* storage, neural excitation lasts a few seconds to a few hours. This stage is limited to a few bits of sensory information—numbers, words of a sentence, and so on. In *long-term* storage, seemingly unlimited amounts of information get tucked away more or less permanently, as shown in Figure 34.29.

Only part of the sensory input reaching the cerebral cortex is selected for the short-term memory bins where information is processed for relevance. If irrelevant, it's forgotten. Otherwise it is consolidated in long-term bins. Emotional states influence the outcome. So does having time to repeat or rehearse the input.

Our brain processes facts and skills separately. *Facts*, soon forgotten or stored in long-term bins, include faces, names, dates, words, smells, and other bits of explicit information, plus the circumstance in which they were learned. That is why you may associate, say, the smell of sun-warmed watermelon with a long-ago great picnic at the beach. By contrast, *skills* are gained by practicing specific motor activities. A skill such as slam-dunking a basketball or playing a violin concerto is best recalled by performing it, not by rehashing the circumstances under which the skill was initially learned.

Separate memory circuits handle different kinds of input. A circuit leading to fact memory (Figure 34.30*a*) starts with inputs at the sensory cortex that flow to the two structures of the limbic system, the amygdala and hippocampus. The amygdala acts as the gatekeeper; it

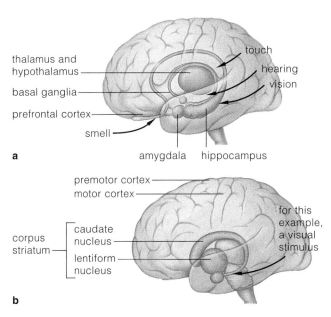

a

b

Figure 34.30 Diagrams of possible circuits involved in (**a**) fact memory and (**b**) skill memory.

connects the sensory cortex with parts of the thalamus and hypothalamus that govern emotional states. The hippocampus mediates learning and spatial relations. Information flows on to the prefrontal cortex. There, multiple banks of fact memories are retrieved and used to stimulate or inhibit other parts of the brain. The new input also flows to basal ganglia, structures that send it back to the cortex in a feedback loop. The loop reinforces the input until it is consolidated in long-term storage.

Skill memory also starts at the sensory cortex, but this circuit routes sensory input to the corpus striatum, which promotes motor responses (Figure 34.30*b*). Motor skills involve muscle conditioning. Thus, as you might suspect, the circuit extends to the cerebellum, the brain region that coordinates motor activity.

Amnesia is a loss of memory. Its severity depends on whether the hippocampus, amygdala, or both have been damaged, as by a severe head blow. But the loss doesn't affect the ability to learn new skills. With *Parkinson's disease*, basal ganglia are destroyed and learning ability is lost, yet skill memory is retained. *Alzheimer's disease* usually starts late in life and involves structural changes in the cerebral cortex and hippocampus. Often, affected people can recall long-known facts, such as their Social Security number. But they have trouble remembering what has just happened to them. In time they become confused, depressed, and incoherent.

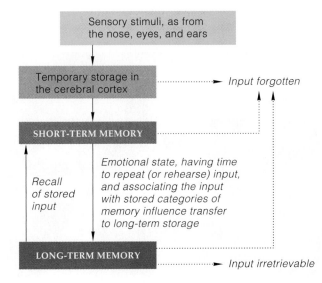

Figure 34.29 Stages of memory processing, starting with the temporary storage of sensory inputs in the cerebral cortex.

Memory, the storage and retrieval of sensory information, arises from circuits between the cerebral cortex and the limbic system, thalamus, and hypothalamus. Sensory input is processed for short-term and long-term storage.

REFLECTIONS ON THE NOT-QUITE-COMPLETE TEEN BRAIN

Part of this chapter is designed to help you gain deeper understanding of the structure and function of the human brain. Did the idea of learning about the prefrontal cortex, amygdala, and other parts of the brain drive you screaming from the classroom? Were you guessing that never in a million years would knowledge of brain structures have anything to do with how you choose to function in your daily life? Guess again.

To give an example, reflect on your teenage years. As a group, teenagers often go through mood swings and episodes of rage, narcissism, insolence, defiance of authority, zero concentration, and occasionally self-mutilation, as with tattooing and body-piercing. If you sometimes puzzled even yourself, rest assured that until recently, many teenage behaviors also were puzzling to the experts.

MY LIMBIC SYSTEM MADE ME DO IT   Conventional wisdom was that the brain is fully developed before puberty ends, with all 100 billion neurons hardwired. As it turns out, all of the necessary connections are not completed before the early twenties!

Like arms, legs, and other body parts, different parts of the brain are on different developmental schedules. For instance, the prefrontal cortex is one of the last parts to be completed. As you know, this major coordinating center keeps tabs on other brain regions, including the limbic system. The prefrontal cortex mediates decision making, sifts through and makes sense of ambiguous information, and reinforces or dampens raw emotions generated in the limbic system. However, even while the prefrontal cortex is still under construction, the limbic system is developing fast! In effect, a teenager is like a car with a revved-up engine and no brakepads.

Neuropsychologist Deborah Yurgelum-Todd and graduate student Abigail Baird tested the hypothesis that teenagers do not have the neural wiring necessary to exercise sound judgment and control emotion. They showed standardized photographs of human faces that were expressing fear. Fear is a primal emotion, a "gut reaction." They asked fifteen adults and fifteen teens to state which emotion the photographs expressed. All the adults identified fear. All but four teenagers gave one or more wrong answers; they said the facial expressions were conveying anger or discomfort.

Then the researchers used *MRI* (magnetic resonance imaging) to compare adult and teenage brains. MRI is a method of forming images of organ structure or activity every few seconds. Yurgelum-Todd and Baird wanted to see which brain parts were being activated in response to the photographs. In the adult brain, an image of fear elicited activity in the limbic system *and* in the prefrontal cortex. MRIs of the teenage brain also revealed a strong limbic system response to the same image—but very little activity in the prefrontal cortex.

From such results, Yurgelum-Todd put forth this hypothesis: Teenagers are still learning to interpret the content of the social world, and their prefrontal cortex is not helping enough with the interpretations.

For years, Jay Giedd and coworkers at the National Institutes of Health have been using MRI to map the brain structure of nearly a thousand children ranging from three to eighteen years old. They discovered that interneurons of the prefrontal cortex begin sprouting rapidly to form new connections around ages nine or ten. By the time the individual is twelve or so, most connections are starting to wither. It is as if the brain undergoes a pruning process to remove synapses that turn out to be useless over the long haul.

Figure 34.31 Teenagers busily demonstrating the primacy of a limbic system revved up by surges of neurotransmitters and sex hormones.

In the meantime, teenagers may not be able to work their way through an unpruned forest of synapses with much proficiency. If so, then this might explain why they have trouble organizing several tasks or tracking several thoughts at once. It may be that they do not yet have the brainpower to quickly call up memories and emotions necessary for making such decisions.

A CHEMICAL BREW MADE ME DO IT Other factors are at work here. When the prefrontal cortex is still far from finishing its development, neurotransmitters and sex hormones are busy at work in the teenage brain. For example, Giedd's team found that the amygdala swells during puberty as an outcome of a surge of testosterone. The swelling is more notable in boys, but it also happens in girls. (The female body uses testosterone as a precursor for synthesizing estrogen.) The amygdala, recall, helps control emotional states—especially anger and fear. Its sudden growth spurt might be why both sexes become more aggressive and irritable during the teen years.

As neurobiologist Sara Leibowitz has pointed out, starting at puberty, most girls also find themselves gaining weight. This is expected; the female body must have a certain percentage of fat if it is to mature sexually. Why does this happen? At this time the hypothalamus steps up production and secretion of an appetite-stimulating hormone.

And what about teen zombies in the classroom? Mary Carskadon argues that they just need more sleep, 9–1/4 hours per night, to be specific. Why? Perhaps because most of the somatotropin and other hormones required for growth and sexual maturation are released while the body sleeps. The brain has a biological clock, in the form of melatonin secretion,

that governs circadian rhythms; it dictates when you wake up and fall asleep. In one study, Carskadon's team asked students to fall asleep during the day. Many students must have been monumentally sleep deprived; they did so within three or four minutes.

Also, memory and learning get a boost during REM sleep, when chemical levels in brain centers are adjusted and short-term memory banks are emptied for the new day. Students deprived of REM sleep are depressed and irritable. They can't retrieve memories quickly, their judgment is compromised, and they do not perform well on tests to measure reaction times. Sleep-deprived teens get the most C's and D's on tests. Well-rested teens get the most A's and B's.

In most adolescents, the blood level of serotonin briefly declines, and impulsive behavior is associated with the decline. Also, without a finished prefrontal cortex to act as traffic cop, teenagers are far more open than adults to invading the brain's pleasure center. Interneurons in that center release dopamine, a neurotransmitter governing arousal and motivation. Novel behaviors, especially those with an element of risk or danger, stimulate the center. So do cocaine and some other drugs, of the sort described in the next section. Sneak out late for a rock concert? Sure. Snort cocaine? Why not?

If you are not yet in your twenties, don't jump to the conclusion that you don't have to nurture your brain (as by reading) or make choices (as in behaving responsibly instead of impulsively). How someone decides to exercise the brain helps shape the forming neural circuits, which in turn profoundly influence behavior and success later in life.

Here's the connection. Each of us possesses a body of great complexity. Its architecture, its functioning are legacies of millions of years of evolution. Its nervous system is unparalleled in the living world. One of its most astonishing products, language, is an encoding of the shared experiences of groups of individuals in time and space. Through the evolution of the nervous system, the sense of history was born, and the sense of destiny. Perhaps one of the sorriest consequences of indifference toward brain function is the implicit denial of this legacy—the denial of self when we do not ask, and cease to care.

Through the evolution of the nervous system, we can ask how we came to be, and where we are headed from here.

How we ultimately function as individuals depends largely on whether we decide to nurture, ignore, or abuse our nervous system even while it is forming. And it is still forming well into the second decade of our life.

Drugging the Brain

Broadly speaking, a drug is a substance introduced into the body to provoke a specific physiological response. Some drugs mediate illness or stress. Some initially fan the pleasure that is associated with sex and other forms of self-gratifying behavior.

Many drugs are habit-forming. Even if the body can function well without them, a person continues to use such drugs for real or imagined relief. Often the body develops tolerance of such drugs; it takes larger or more frequent doses to produce the same effect. Habituation and tolerance are signs of **drug addiction**, a chemical dependence on a drug (Table 34.1). *The drug has rewired the brain and assumed an "essential" biochemical role in the body.* Abruptly deprive addicts of the drug, and they can expect physical pain and mental anguish. The body goes through biochemical upheaval. Add to this the increased potential for serious diseases. Many addicts who crave fast fixes share unsterilized needles. In so doing, they put themselves and others at risk of developing AIDS, hepatitis B, and hepatitis C. They may lose skin, muscles, adipose tissue, and eventually their life to *necrotizing fasciitis*. This disease is an outcome of infection by the so-called flesh-eating bacterium *Clostridium perfringens*.

STIMULANTS Stimulants make you more alert, then depress you. *Caffeine* in coffee, tea, chocolate, and many soft drinks is one. Low doses act at the cerebral cortex to increase alertness. Higher doses act at the medulla oblongata to make you clumsy and mentally incoherent. Another stimulant, the *nicotine* in tobacco, mimics ACh. It has widespread effects by directly stimulating a variety of sensory receptors. Nicotine addiction has staggering health, social, and financial costs, which you will read about in Section 40.7.

An estimated 1.2 to 3.7 million Americans are *cocaine* abusers. This stimulant fans pleasure by blocking the reabsorption of norepinephrine, dopamine, and other neurotransmitters from synaptic clefts (Figure 34.32). These signaling molecules accumulate and incessantly stimulate postsynaptic cells. Heart rate, blood pressure, and sexual appetite rise. In time the molecules diffuse away, but neurons can't synthesize replacements fast enough. The sense of pleasure is lost as hypersensitized postsynaptic cells demand stimulation. With long-term, heavy use of cocaine, pleasure becomes impossible.

Granular cocaine is inhaled (snorted). Abusers burn crack cocaine and inhale the smoke. As the start of this chapter implied, crack is extremely addictive. Its highs are higher and its crashes more devastating. The social and economic tolls are extreme. In the first half of 2001, 100,000 people ended up in emergency rooms for cocaine-related disorders. There is no antidote for overdoses that induce seizures, respiratory failure, or heart failure.

Amphetamines induce massive release of dopamine and norepinephrine. Addicts often smoke, snort, inject, or gulp a form called *crank*. Its effects range from euphoria and sexual arousal to a pounding heart, agitation, dry mouth, tremors, and paranoia. Crank kills the appetite (people have become addicts after using it to lose weight). In time, dopamine and norepinephrine synthesis declines; the brain depends more on artificial stimulation.

The long-term outcomes? Malnutrition, psychosis, depression, memory loss, and damage to the brain, heart, lungs, and liver. Crank is made cheaply in home kitchens from such caustic ingredients as drain cleaners, and abuse is epidemic. Figure 34.33 gives just one addict's story.

Another synthetic amphetamine is MDMA (or *Ecstasy*, Adam, or XTC). Users say it enhances sex, trust, and

---

Table 34.1 *Warning Signs of Drug Addiction**
1. Tolerance—it takes increasing amounts of the drug to produce the same effect.
2. Habituation—it takes continued drug use over time to maintain the self-perception of functioning normally.
3. Inability to stop or curtail use of the drug, even if there is persistent desire to do so.
4. Concealment—not wanting others to know of the drug use.
5. Extreme or dangerous behavior to get and use a drug, as by stealing, asking more than one doctor for prescriptions, or jeopardizing employment by drug use at work.
6. Deterioration of professional and personal relationships.
7. Anger and defensive behavior when someone suggests there may be a problem.
8. Preference for drug use over previous customary activities.

* Three or more of these signs may be cause for concern.

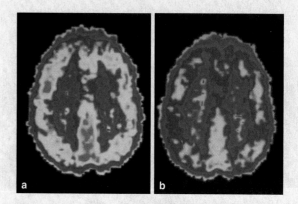

Figure 34.32 (**a**) Normal brain activity revealed by a PET scan. (**b**) PET scan of a comparable section that reveals cocaine's effect. *Red* indicates greatest activity, followed by *yellow, green,* then *blue* for the least activity.

tranquility. They don't say as much about the confusion, depression, severe anxiety, blurred vision, nausea, and hypertension. They may not care that its long-term use permanently damages the brain and liver. By the way, destroy dopamine-secreting neurons and you end up with tremors, lack of coordination, and often paralysis of the sort seen in patients with Parkinson's disease.

Ecstasy is one of many *club drugs* used at raves, dance clubs, and bars. Especially when mixed with alcohol, these drugs can kill you. Sexual predators are sneaking colorless, tasteless, and odorless club drugs into the drinks of individuals they wish to assault.

DEPRESSANTS, HYPNOTICS These addictive drugs affect nerves; some act at the reticular formation and thalamus. Responses vary. You might sink into emotional relief, drowsiness, sleep, coma, or death, depending on the dose and your physiological and emotional states. Low doses have the most effect on inhibitory synapses. Initially, users feel excited or euphoric. Increased doses suppress excitatory synapses and lead to depression. Depressants and hypnotics amplify each other, as when alcohol plus barbiturates heightens depression.

Alcohol, or ethyl alcohol, acts at the plasma membrane to alter cell function. Like nicotine and cocaine, it is lipid soluble and swiftly crosses the blood–brain barrier. Some people mistakenly think alcohol is a harmless stimulant because of the initial "high" it produces. But it is a potent drug that has put binge drinkers into comas and coffins. Even low alcohol intake leads to diminished judgment, disorientation, and uncoordinated movements. Long-term addiction can lead to liver failure (Section 6.9).

ANALGESICS When severe stress leads to physical or emotional pain, the brain produces natural *analgesics*, or pain relievers. Two kinds, endorphins and enkephalins, influence many parts of the nervous system. *Codeine*, *heroin*, and other narcotic analgesics sedate the body, relieve pain, and are among the most addictive drugs. Deprivation of heroin after massive doses is followed by hyperactivity, anxiety, fever, chills, severe vomiting, cramping, and diarrhea.

PSYCHEDELICS, HALLUCINOGENS These drugs skew sensory perception by interfering with the action of acetylcholine, norepinephrine, or serotonin. Lysergic acid diethylamide, or *LSD*, alters serotonin's roles in inducing sleep, controlling the body's core temperature, and mediating sensory perception. Even in small doses, LSD warps perceptions, as when some users "perceived" they could fly and "flew" off buildings.

The hallucinogen *marijuana* is made from crushed leaves, flowers, and stems of the plant *Cannabis*. In low

Figure 34.33 Portrait of a five-year methamphetamine abuser. One night, Steve Wade downed a fifth of 151-proof rum, then injected meth five times and snorted meth ten times. His body temperature skyrocketed to 110 degrees, hearing in one ear shut down, and his cognitive skills and reasoning dissolved.

Five years later, he still shakes too much to shave with a razor. Either individual objects or the whole room won't stop spinning. Steve forgets words in midsentence. Speech therapists help him practice simple expressions, such as "How?" "What?" and "Are you okay?" A poster on his wall has this insight: "They say meth won't kill you. But you'll wish it had." Today, Steve is in his early thirties and irrevocably disabled.

doses it is like a depressant. It slows down but does not impair motor activity; it relaxes the body and elicits mild euphoria. It also causes disorientation, anxiety bordering on panic, delusions, and hallucinations. Like alcohol, it interferes with the performance of complex tasks, such as driving a car. In one study, pilots showed a marked deterioration in instrument-flying ability for more than two hours after smoking marijuana. In time, marijuana impairs the immune system and mental functioning.

According to Alan Leshner, director of the National Institute on Drug Abuse, people don't intend to get lung cancer when they first start to smoke, or get clogged arteries when they eat fatty foods, or become addicts when they first use drugs. Yet the brain, behavior, and experience interact and modify one another, and prolonged drug abuse inevitably leads to addiction.

Leshner also points to recent evidence that all major addictive drugs stimulate the release of dopamine, the neurotransmitter directly involved in the sense of pleasure. For example, most people who smoke think this is just a habit. Actually, they smoke because of the dopamine spike when nicotine hits neurons in the brain. All addicts specifically and uncontrollably crave the dopamine spike, and the vast majority cannot stop the craving. Prolonged drug use has rewired the brain in abnormal ways, with abnormal outcomes. Just ask Steve Wade (Figure 34.33).

SUMMARY   *Gold* indicates text section

1. Nervous systems sense and interpret specific aspects of the environment and issue commands for responses to them. Most communication lines consist of sensory neurons, interneurons, and motor neurons. *CI, 34.4*

   a. Sensory neurons are receptors that detect stimuli, which are specific forms of energy, such as light.

   b. Interneurons are integrators in the brain and spinal cord. They receive and interpret signals from sensory receptors, then issue commands for suitable responses.

   c. Motor neurons carry commands away from the brain and spinal cord to the body's effectors. Effectors are muscle and gland cells that carry out responses.

Table 34.2 *Summary of the Central Nervous System**

FOREBRAIN	Cerebrum	Localizes and processes sensory inputs; initiates and controls skeletal muscle activity. Governs memory, emotions, and abstract thought in the most complex vertebrates
	Olfactory lobe	Relays sensory input from the nose to olfactory centers of cerebrum
	Thalamus	Has relay stations for conducting sensory signals to and from cerebral cortex; has role in memory
	Hypothalamus	With the pituitary gland, a homeostatic control center that adjusts the volume, composition, and temperature of internal environment. Governs behaviors affecting organ functions (e.g., thirst and hunger) and expression of emotion
	Limbic system	A complex of brain structures that governs emotions; has roles in memory
	Pituitary gland (Chapter 36)	With hypothalamus, provides endocrine control of metabolism, growth, and development
	Pineal gland (Chapter 36)	Helps control some circadian rhythms; also has role in mammalian reproductive physiology
MIDBRAIN	Tectum	In fishes and amphibians, its centers coordinate sensory input (as from optic lobes) and motor responses. In mammals, its centers (mainly reflex centers) swiftly relay sensory input to forebrain
HINDBRAIN	Pons	Is a "bridge" of tracts between the cerebrum and cerebellum; its other tracts connect spinal cord with forebrain. Works with the medulla oblongata to control the rate and depth of respiration
	Cerebellum	Coordinates motor activity for moving limbs and maintaining posture, and for spatial orientation
	Medulla oblongata	Its tracts relay signals between the spinal cord and pons; its reflex centers help control heart rate, adjustments in blood vessel diameter, respiratory rate, vomiting, coughing, other vital functions
SPINAL CORD		Makes reflex connections for limb movements. Many of its tracts carry signals between the brain and the peripheral nervous system

* The reticular formation extends from the spinal cord to the cerebral cortex.

2. A neuron's dendrites and cell body are input zones. If arriving signals spread to a trigger zone (e.g., the start of an axon), they may trigger an action potential that can be propagated to an output zone (axon endings). *34.1*

   a. Transport proteins pepper the plasma membrane from the trigger zone to axon endings of a neuron. They serve as gated or open channels for the passage of ions.

   b. The controlled flow of ions across the membrane is the basis of an action potential: an abrupt, short-lived reversal in the voltage difference across the membrane in response to adequate stimulation. *34.1, 34.2*

   c. Sodium–potassium pumps counter small ion leaks and help maintain ion gradients across the membranes. They also restore the gradients after an action potential.

3. At output zones, action potentials trigger the release of neurotransmitter molecules, which diffuse across the chemical synapse between a neuron and another neuron, a muscle cell, or a gland cell. They may excite the post-synaptic cell membrane (drive it closer to threshold) or inhibit it (drive it away from threshold). *34.3*

4. Integration is the moment-by-moment summation of excitatory and inhibitory signals reaching synapses on a neuron. It is a way to play down, suppress, reinforce, or send on information through a nervous system. *34.4*

5. The simplest nervous systems are nerve nets, such as those of cnidarians and other radial animals. Their meshwork of nerve cells forms reflex connections with contractile and sensory cells of the epithelium. Most animals have a bilateral, cephalized nervous system with a brain or ganglia at their anterior end. These animals have cordlike nerves: axons of sensory neurons, motor neurons, or both bundled inside a sheath. *34.5*

6. The vertebrate central nervous system is composed of a spinal cord and brain (Table 34.2). The peripheral nervous system is mostly nerves that connect all body regions with the spinal cord and brain. *34.6–34.9*

   a. Skeletal muscles are serviced by somatic nerves. Soft internal organs (viscera) are serviced by autonomic nerves (parasympathetic and sympathetic) that work in opposition continuously to adjust organ activities.

   b. Parasympathetic nerve outflow causes energy to be diverted to basic housekeeping tasks if stimulation is low. Sympathetic nerve outflow stimulates activities during times of heightened awareness or danger.

Review Questions

1. Describe sensory neuron, interneuron, and motor neuron in terms of their structure and functions. Label this motor neuron's functional zones: *CI, 34.1, 34.4*

2. Define resting membrane potential, graded potential, and action potential. *34.1, 34.2*

3. A neuron at rest is controlling the ion distribution across its plasma membrane. Identify two kinds of ions. Do they leak across the membrane, are they pumped across, or both? *34.1*

4. With respect to action potentials, explain threshold level, all-or-nothing spikes, and self-propagation. *34.2*

5. Define chemical synapse and neurotransmitter. Choose an example of a neurotransmitter and state where it acts. *34.3*

6. Define synaptic integration. *34.3*

7. What is a myelin sheath? Do all neurons have one? *34.4*

8. Define reflex, then give an example of a reflex arc. *34.4*

9. Contrast the nervous system of a radial animal with that of a bilateral animal. *34.5, 34.6*

10. Distinguish between the following:
    a. brain and brain stem *34.8*
    b. central and peripheral nervous system *34.6*
    c. cranial and spinal nerves *34.7*
    d. somatic and autonomic nerves *34.7*
    e. parasympathetic and sympathetic nerves *34.7*

11. Define cerebrospinal fluid and the blood–brain barrier. *34.8*

12. Explain the limbic system in terms of some of its component parts and their functions. *34.9*

Self-Quiz   ANSWERS IN APPENDIX III

1. Action potentials occur when _____ .
   a. a neuron receives adequate stimulation
   b. sodium gates open in an ever accelerating way
   c. sodium–potassium pumps kick into action
   d. both a and b

2. The resting membrane potential is maintained by _____ .
   a. ion leaks   c. neurotransmitters
   b. ion pumps   d. both a and b

3. Neurotransmitters diffuse across a _____ .
   a. chemical synapse   c. myelin sheath
   b. membrane pump   d. both a and b

4. A nerve may consist of bundled-together axons of _____ .
   a. sensory neurons   c. sensory and motor neurons
   b. motor neurons   d. all of the above

5. Is this statement true or false: White matter and gray matter are components of the spinal cord alone.

6. Match the terms with their most suitable description.
   ____ synaptic   a. at input zone of excitable cell
        integration   b. summation of all signals arriving
   ____ muscle spindle   at any neuron at the same time
   ____ graded, local   c. arises at trigger zone
        potential   d. stretch-sensitive receptor
   ____ action potential

7. Match the component with its main functions.
   ____ spinal cord   a. motor, sensory, association areas
   ____ medulla   for most complex integration
        oblongata   b. homeostatic control of internal
   ____ hypothalamus   environment, internal organs
   ____ limbic system   c. our "emotional brain"
   ____ cerebral cortex   d. reflex control of respiration, blood
                             circulation, other basic activities
                          e. expressway between brain and
                             peripheral nervous system; also,
                             direct reflex connections

Critical Thinking

1. In human newborns and premature babies, the blood–brain barrier is not fully developed. Explain why this might be reason enough to pay careful attention to their diet.

2. When Jennifer was six years old, a man lost control of his car and hit a tree in front of her house. She ran over to him and screamed when she saw blood from a head wound dripping on a bouquet of red roses. Thirty-five years later, someone gave her a bottle of *Tea Rose* perfume. When she sniffed the perfume, she became frightened and extremely anxious. A few minutes later she also had a vivid recollection of the accident. Explain this incident in terms of what you learned about memory.

3. Eric typically drinks one cup of coffee nearly every hour, all day long. By midafternoon he has trouble concentrating on his studies, and he feels tired and more than a little clumsy. Drinking another cup of coffee doesn't make him more alert. Explain how caffeine in coffee might produce such symptoms.

4. *Epilepsy* is a neurological disorder characterized by brief, recurring episodes of sensory and motor malfunctioning. The episodes, or seizures, may arise when reverberating circuits in the brain are wrongly activated. Muscles contract involuntarily and lights, sounds, and odors are often sensed even when receptors in the eyes, ears, and nose aren't being stimulated.

   The drug valproic acid can eliminate or lessen the severity of the seizures by stimulating the body's synthesis of GABA. Why would stepping up GABA production help?

5. Think back on Sections 34.12 and 34.13. As a class project, do a web search for accounts of drug addicts. Report on how an individual became addicted, whether he or she knew about the long-term consequences, and on the outcomes. Compare reports in class. Do common elements show up in all of these stories?

Selected Key Terms

Neural Function

acetylcholine (ACh) *34.3*
action potential *34.1*
axon *34.1*
chemical synapse *34.3*
dendrite *34.1*
drug addiction *34.13*
interneuron *CI*
motor neuron *CI*
nerve *34.4*
neurotransmitter *34.3*
positive feedback *34.2*
reflex *34.4*
resting membrane potential *34.1*
sensory neuron *CI*
sodium–potassium pump *34.1*
stimulus *CI*
synaptic integration *34.3*

Nervous Systems

autonomic nerve *34.7*
blood–brain barrier *34.8*

brain *34.8*
brain stem *34.8*
central nervous system *34.6*
cerebellum *34.8*
cerebral cortex *34.9*
cerebrum *34.8*
hypothalamus *34.8*
limbic system *34.9*
medulla oblongata *34.8*
memory *34.11*
nerve net *34.5*
nervous system *CI*
neural tube *34.6*
parasympathetic nerve *34.7*
peripheral nervous
 system *34.6*
pons *34.8*
reticular formation *34.8*
somatic nerve *34.7*
spinal cord *34.7*
sympathetic nerve *34.7*
thalamus *34.8*

Readings

Romer, A., and T. Parsons. 1986. *The Vertebrate Body*. Sixth edition. Philadelphia: Saunders. It's been around a while but has fine insights into the evolution of vertebrate nervous systems.

SENSORY RECEPTION

Different Strokes for Different Folks

You might be reluctant to pet a python or scratch a bat behind its ears. But you have to give them credit for being vertebrates with special traits (Figure 35.1). In rows of pits above and below a python's mouth are thermoreceptors that detect infrared energy—in this case, the body heat of small, night-foraging mammals, the prey of choice. When stimulated, these receptors send messages to the brain, which processes them and issues commands to muscle cells. In no time at all, the snake aims and executes a stunningly accurate strike.

A motionless, edible frog would not stimulate the receptors; the snake would slither past it. Why? A frog has cool skin and blends with colors of its habitat. The python does not have receptors or a neural program that can respond to these traits.

Or consider the bats. Nearly all of these mammalian species sleep during the day and spread their webbed wings at dusk. Different kinds take to the air in search of nectar, fruit, frogs, or insects. Many sensory receptors in their eyes, nose, ears, mouth, and skin are not all that different from yours. Others are very different. They help even the tiny-eyed, nearly blind species navigate and capture flying insects swiftly in the dark!

How do they do it? Bats happen to be masters of **echolocation**. They emit calls, as the bat in Figure 35.1*b* is doing. When sound waves of the calls bounce off insects, trees, and other objects, acoustical receptors inside the bat's ears detect the echoes and send signals about them to the bat brain.

As an echolocating bat flies, it emits a steady stream of about ten clicking sounds per second. You can't hear these clicks. They are ultrasounds, of an intensity beyond the range of sound waves that receptors in human ears can detect. When a bat hears a pattern of distant echoes from, say, an airborne mosquito, it increases the rate of ultrasonic clicks to as many as 200 per second. That is faster than a machine gun can fire bullets. In the few milliseconds of silence between the clicks, sensory receptors detect the echoes and deliver messages about them to the brain. The bat brain rapidly constructs a "map" of the sounds, which the bat follows during its maneuvers through the night world.

With this chapter we turn to **sensory systems**, the means by which animals receive signals from inside and outside the body, then decode them in ways that give rise to awareness of sounds, sights, odors, and other sensations. Information about different stimuli also is integrated to give rise to compound sensations. For instance, "wetness" is not a single stimulus. Our perception of it arises from simultaneous inputs that concern pressure, touch, and temperature. Two points to keep in mind: **Perception** is an understanding of what a stimulus means. It isn't the same as **sensation**, which is simply conscious awareness of a stimulus.

Figure 35.1 Examples of sensory receptors. (**a**) Thermoreceptors in pits above and below a python's mouth detect body heat, or infrared energy, of nearby prey. (**b**) Some bats listen to echoes of their own high-frequency sounds. Their brain constructs a sound map from the echoes that bounce back from prey and other objects in the surroundings. Such maps help this night-flying bat capture insects in midair without the help of eyes.

The system components are sensory neurons, nerve pathways, and specific brain regions. These receptors are the front doors of the nervous system. We define each type in terms of a **stimulus**, which is a specific form of energy detected by a sensory receptor:

Mechanoreceptors detect forms of mechanical energy (changes in pressure, position, or acceleration).

Thermoreceptors detect heat energy.

Pain receptors (nociceptors) detect tissue damage.

Chemoreceptors detect chemical energy of specific substances dissolved in the fluid surrounding them.

Osmoreceptors detect changes in solute concentration (water volume) in the surrounding fluid.

Photoreceptors detect visible and ultraviolet light.

Depending upon the kinds and numbers of their sensory receptors, animals sample the environment in different ways and differ in their awareness of it. Unlike bees, you have no receptors for ultraviolet light and do not see many flowers the way they do. Unlike many bats, you and bees have no receptors for ultrasound. Unlike pythons, you, bees, and bats have no receptors for detecting warm-blooded prey in the dark. In this chapter, you will encounter processes and molecular structures responsible for these fascinating differences.

b

Key Concepts

1. Sensory systems are portions of the nervous system. Each consists of specific types of sensory receptors, nerve pathways from receptors to the brain, and brain regions that receive and process sensory information.

2. A stimulus is a form of energy that has activated a specific type of sensory receptor, which is either a sensory neuron or a specialized cell adjacent to it. Photoreceptors detect light energy, thermoreceptors detect heat energy, and so on.

3. A sensation is conscious awareness of change in some aspect of the external or internal environment. It begins when sensory receptors detect a stimulus. The stimulus energy becomes converted to a graded, local signal that may contribute to initiating an action potential.

4. Information about a stimulus becomes encoded in the number and the frequency of action potentials sent to the brain along particular nerve pathways. Specific brain regions translate information into a sensation.

5. The somatic sensations include touch, pressure, pain, temperature, and muscle sense.

6. Taste, smell, hearing, and vision are special senses.

OVERVIEW OF SENSORY PATHWAYS

Take a look at the young gymnast in Figure 35.2*a*. He is precisely maintaining his body's position in space in response to input from sensory receptors. That input has reached his spinal cord and his brain by way of sensory pathways. In this case, some pathways involve muscle spindles of the sort you read about earlier, in Section 34.4. Muscle spindles are just one of the kinds of sensory receptors listed in Table 35.1.

Sensory receptors convert the energy of a stimulus into action potentials, the basis of messages that travel through the nervous system. **Action potentials**, recall, are brief reversals in the resting membrane potential of an excitable cell. In receptor endings, they start after a stimulus disturbs a patch of plasma membrane called the trigger zone (Sections 34.1 through 34.3). A weak disturbance may give rise to a local, graded potential. This type of signal, remember, does not spread far from the point of stimulation. And unlike action potentials, its magnitude can vary.

When a stimulus is intense or when it is repeated rapidly enough for a summation of local signals, action potentials may result. Each action potential propagates itself as a sequence of equivalent reversals away from the trigger zone. It travels from the receptor endings to axon endings of sensory neurons, which lead into the spinal cord or brain (Figure 35.2*b*). There, the endings release neurotransmitter molecules. These may trigger action potentials in the interneurons or motor neurons adjacent to them. And certain interneurons are part of an information pathway to specific brain regions.

Figure 35.2 Example of a how we make use of sensory pathways—in this case, from receptors in many muscles throughout the body to the spinal cord and brain.

Action potentials traveling along a sensory neuron are not like a wailing ambulance siren. *They do not vary in amplitude.* How, then, does the brain assess a given stimulus? It assesses (1) *which* nerve pathways happen to be delivering the action potentials, (2) the *frequency* of the action potentials traveling along each axon that is part of the pathway, and (3) the *number* of axons the stimulus recruited.

Table 35.1 *Main Categories of Sensory Receptors*

Category	Examples	Stimulus
MECHANORECEPTORS		
Touch, pressure	Certain free nerve endings and Pacinian corpuscles in skin	Mechanical pressure against body surface
Baroreceptor	Carotid sinus (Section 39.7)	Pressure changes in the fluid that bathes them
Stretch	Muscle spindle in skeletal muscle	Stretching
Auditory	Hair cells in organ inside ear	Vibrations (sound or ultrasound waves)
Balance	Hair cells in organ inside ear	Movement of some body fluid in confined space
THERMORECEPTORS	Certain free nerve endings	Change in temperature (heating, cooling)
PAIN RECEPTORS (NOCICEPTORS)*	Certain free nerve endings	Tissue damage (e.g., distortions, burns)
CHEMORECEPTORS		
Internal chemical sense	Carotid bodies in blood vessel wall	Substances (O_2, CO_2, etc.) dissolved in extracellular fluid
Taste	Taste receptors of tongue	Substances dissolved in saliva, etc.
Smell	Olfactory receptors of nose	Odors in air, water
OSMORECEPTORS	Hypothalamic osmoreceptors (Section 42.2)	Change in solute concentration (water volume) of the fluid that bathes them
PHOTORECEPTORS		
Visual	Rods, cones of eye	Wavelengths of light

* Extremely intense stimulation of any sensory receptor also may be perceived as pain.

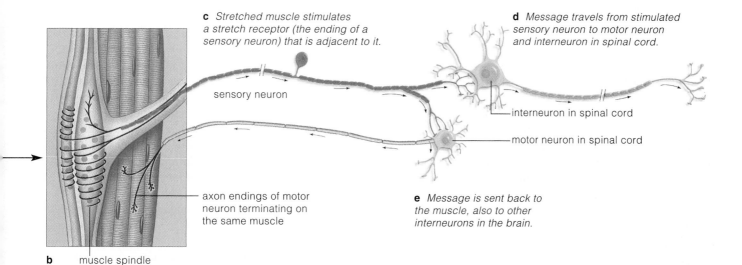

c *Stretched muscle stimulates a stretch receptor (the ending of a sensory neuron) that is adjacent to it.*

sensory neuron

d *Message travels from stimulated sensory neuron to motor neuron and interneuron in spinal cord.*

interneuron in spinal cord

motor neuron in spinal cord

axon endings of motor neuron terminating on the same muscle

e *Message is sent back to the muscle, also to other interneurons in the brain.*

b muscle spindle

First, an animal's brain is prewired in a genetically programmed way to interpret action potentials only in certain ways. That is why you "see stars" when one of your eyes gets poked, even in a darkened room. Many photoreceptors in the assaulted eye were mechanically disturbed enough to give rise to messages that traveled along one of two optic nerves into the brain. Your brain always interprets any signals arriving from an optic nerve as "light."

Second, when the stimulus is strong, receptors fire action potentials more frequently than they do when a stimulus is weak. The same receptor might detect the sound of a throaty whisper or a wild screech. The brain senses the difference from variations in the frequency of the signals being sent to it.

Third, a stronger stimulus can recruit more sensory receptors than a weaker stimulus is able to do. Gently tap a small patch of skin on one of your arms and you activate a few receptors. Press harder on the same spot and you activate more receptors in a larger area. The increased disturbance translates into action potentials in many different sensory axons. The brain interprets that combined activity as an increase in the stimulus intensity. Figure 35.3 is an example of this effect.

In some cases, the frequency of action potentials decreases or stops entirely even when the stimulus is being maintained at constant strength. Any decrease in the response to a stimulus is a **sensory adaptation**. For example, after you put on clothing, your awareness of the pressure it exerts against your skin ceases. Inside your skin, some mechanoreceptors adapt rapidly to the sustained stimulation; they are of the type that can only signal a change in the stimulus—its onset or removal. By contrast, other receptors adapt slowly or not at all; they help the brain monitor specific stimuli all the time. The stretch receptor shown in Figure 35.2*b* is like this. Many of these sensory receptors continually inform the

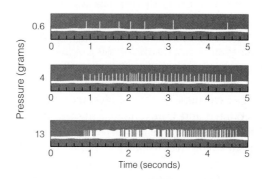

Figure 35.3 Recordings of action potentials from a pressure receptor with endings in a human hand. They correspond to variations in stimulus strength. A thin rod was pressed against skin with the amount of pressure indicated to the left of each diagram. Vertical bars above each thick horizontal line record individual action potentials. Increases in their frequency correspond to increases in the stimulus strength.

brain about changes in the length of muscles, and that information helps you maintain balance and posture.

In sections to follow, we will turn to some specific examples of the sensory receptors listed in Table 35.1. As you will see, the types that are present at more than one location in the body contribute to what we call **somatic sensations**. Other types of sensory receptors are restricted to particular locations, such as inside the eyes or ears. They contribute to the **special senses**.

A sensory system has sensory receptors for specific stimuli, nerve pathways that conduct information from receptors to the brain, and brain regions that receive the information.

The brain assesses a given stimulus based on which nerve pathways are carrying action potentials, the frequency of action potentials traveling along each axon of that pathway, and the number of axons recruited into action.

SOMATIC SENSATIONS

Somatic sensations begin with receptors in the body's surface tissues, skeletal muscles, and walls of internal organs. These receptors are most highly developed in birds and mammals; amphibians and fishes have only a few. The receptor inputs travel to the spinal cord and into the **somatosensory cortex**, part of the outermost gray matter of the cerebral hemispheres (Section 34.9). The interneurons in this brain region are organized like maps for individual parts of the body's surface. Map regions having the largest areas correspond to the body parts that have the greatest sensory acuity and that require the most intricate control. Such parts include the fingers, thumbs, and lips (Figure 35.4).

Receptors Near the Body Surface

You and other mammals discern sensations of touch, pressure, cold, warmth, and pain near the body surface. Regions with the greatest number of sensory receptors, such as the fingertips and tip of the tongue, are most sensitive to stimulation. Less sensitive regions, such as the back of the hand, do not have nearly as many.

Free nerve endings are the simplest receptors. These thinly myelinated and unmyelinated (naked) branched endings of sensory neurons are distributed in skin and internal tissues. Different types are mechanoreceptors, thermoreceptors, and pain receptors. All adapt slowly to stimulation. One subpopulation gives rise to a sense of prickling pain, as when you jab a finger with a pin. Another contributes to itching or warming sensations that are provoked by chemicals, including histamine. Two thermoreceptive types have peak sensitivities that are higher and lower than normal body temperature,

respectively. One mechanoreceptive type coils around hair follicles and detects hair movements (Figure 35.5).

Encapsulated receptors are common near the body surface. The Meissner's corpuscle adapts very slowly to low-frequency vibrations. It is abundant in fingertips, lips, eyelids, nipples, and genitals. The bulb of Krause, an encapsulated thermoreceptor, is activated at 20°C (68°F) or lower. Below 10°C, it contributes to painful freezing sensations. Ruffini endings are slowly adapting encapsulated types. These respond to steady touching and pressure, and to temperatures above 45°C (113°F).

The Pacinian corpuscle, also encapsulated, is widely distributed deep in the skin, where it responds to fine textures. It also resides near freely movable joints and in some internal organs. Onionlike layers of membrane alternating with fluid-filled spaces enclose this sensory ending (Figure 35.5). They help it detect rapid pressure changes associated with touch and vibrations.

Muscle Sense

That gymnast shown in Figure 35.2a is sensing limb motions and his body's position in space with the help of many mechanoreceptors in skeletal muscle, joints, tendons, ligaments, and skin. The stretch receptors of muscle spindles are among them. These sensory organs run parallel with skeletal muscle cells. Their responses to stimulation depend on how much and how fast the muscle stretches.

What Is Pain?

Pain is the perception of injury to some body region. Sensations of *somatic* pain start with signals from pain receptors in skin, skeletal muscles, joints, and tendons. Sensations of *visceral* pain are associated with internal organs. They start with detection of excessive chemical stimulation, muscle spasms, muscle fatigue, excessive distension of digestive tract regions, inadequate blood flow to organs, and other abnormal conditions.

Superficial somatic pain arises at or near the skin surface. It is sharp or prickling and often does not last long. Deep somatic pain, which arises farther down in skin, muscles, or joints, is more diffuse and lasts longer. This is also the case for visceral pain. We characterize both as burning or gnawing pain, or a dull ache.

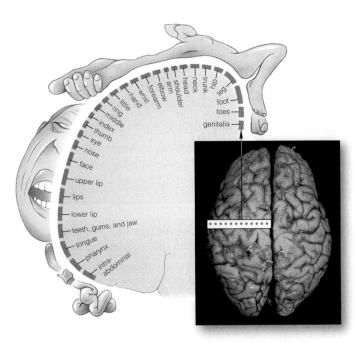

Figure 35.4 Differences in the representation of different body parts in the primary somatosensory cortex. This region is a strip of cerebral cortex, a little wider than 2.5 centimeters (about an inch), from the top of the human head to above the ear.

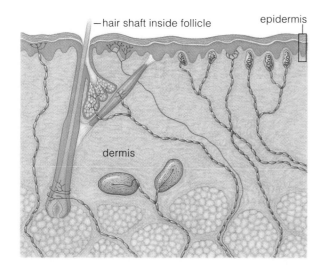

—hair shaft inside follicle

epidermis

dermis

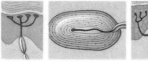

free nerve
endings

Pacinian
corpuscle

Ruffini
endings

bulb of
Krause

Meissner's
corpuscle

Figure 35.5   A sampling of receptors in human skin.

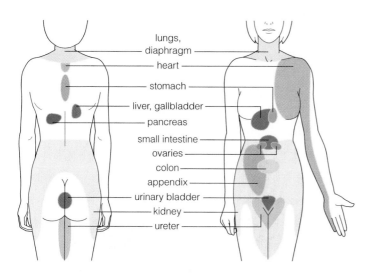

lungs,
diaphragm
heart
stomach
liver, gallbladder
pancreas
small intestine
ovaries
colon
appendix
urinary bladder
kidney
ureter

Figure 35.6   Referred pain. Receptors in some internal organs detect painful stimuli. Instead of localizing the pain at the organs, the brain projects the sensation to the skin areas indicated.

When cells are damaged, they release chemicals that activate neighboring pain receptors. The most potent, the bradykinins, open the floodgates for an outpouring of histamine, prostaglandins, and other participants in the inflammatory response (Section 39.3). They might also bind with and activate pain receptor endings.

Signals from pain receptors reach interneurons in the spinal cord and induce them to release substance P. This neurotransmitter causes signals to be relayed along two pathways. Some signals reach the thalamus and the sensory cortex, which quickly analyzes the intensity and type of pain. Other signals alert the brain stem and limbic system, which mediates arousal and emotional responses to pain. Now signals from the hypothalamus and midbrain stimulate the release of endorphins and enkephalins, natural opiates that reduce pain perception.

Morphine, hypnosis, and natural childbirth techniques may also stimulate release of these natural opiates.

Pain tolerance depends on emotional state, cultural factors, and possibly how old we are (tolerance might increase with aging). Extremely intense or long-lasting stimulation may lead to *hyperalgesia*, or amplification of pain. Someone with a severe sunburn experiences this amplified effect when stepping into a hot shower, for example.

Referred Pain

Responses to pain depend on the ability of the brain to identify the affected tissue and to project the sensation back to it. For example, get smacked in the face with a snowball and you "feel" the contact on facial skin. However, sensations of pain from some internal organs may be wrongly projected to part of the skin. We all have this capacity for *referred pain*, whereby visceral sensations are perceived as somatic sensations. As one example, people suffering a heart attack may perceive pain radiating from the chest wall, from the neck, and along the left shoulder and arm (Figure 35.6).

How can the brain be so mistaken? The answer lies in the nervous system's construction. Sensory inputs from skin and certain internal organs enter the spinal cord in the same segments of the cord. Because skin receives more painful stimuli than internal organs do, its signals travel more often along the somatic pathway to the brain. The brain might interpret most sensory input as arriving from skin, the more common source.

Referred pain is not the same as the *phantom pain* reported by amputees. They often sense the presence of a missing body part, as if it were still there. In some undetermined way, severed sensory nerves continue to respond to the amputation. The brain projects the pain back to the missing part, past the healed region.

Diverse sensory receptors near the body surface and in internal organs detect touch, pressure, temperature, pain, motion, and positional changes of body parts. Their input flows through the spinal cord to the somatosensory cortex and other brain regions where somatic sensations arise.

CHEMICAL SENSES

The special senses of smell and taste are both *chemical senses*. Their sensory pathways start at chemoreceptors, which become activated when they bind molecules of a substance that is dissolved in the fluid bathing them. The receptors wear out and are replaced with new ones on an ongoing basis. In both pathways, sensory input travels from chemoreceptors through the thalamus and on to the cerebral cortex. There, perceptions of certain stimuli take shape and undergo fine-tuning. Sensory input also travels to the limbic system, which integrates it with emotional states and stored memories (Sections 34.9 and 34.11).

Olfactory receptors detect water-soluble or volatile (easily vaporized) substances. A human nose has about 5 million; a bloodhound nose has more than 200 million. The receptor axons lead into one of two olfactory bulbs. In these small brain structures, the axons synapse with groups of cells that sort out the components of a given scent and relay the information along an olfactory tract for further processing in the cerebrum (Figure 35.7).

Pheromones, signaling molecules secreted from the exocrine glands of one individual, influence the social behavior of other individuals of its species, including potential sex partners (Chapter 46). Bombykol is one of

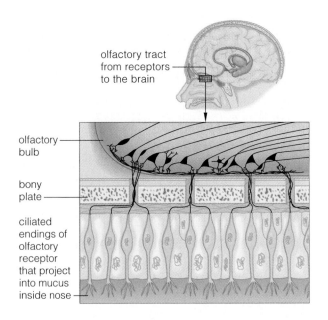

olfactory tract from receptors to the brain

olfactory bulb

bony plate

ciliated endings of olfactory receptor that project into mucus inside nose

Figure 35.7 Sensory pathway from the sensory endings of olfactory receptors in the nose to the cerebral cortex and limbic system. Receptor axons pass through holes in a bony plate between the nasal lining and the brain. In the earliest vertebrates, an olfactory bulb and olfactory lobe dominated the forebrain (Section 34.8).

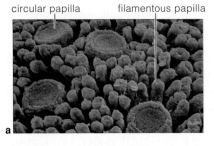

circular papilla   filamentous papilla

a

Figure 35.8 Taste receptors in the human tongue. Circular papillae ring the epithelial tissue that contains taste buds. Filamentous papillae do not contribute to taste; their movements direct chemical-laden fluid past the receptors.

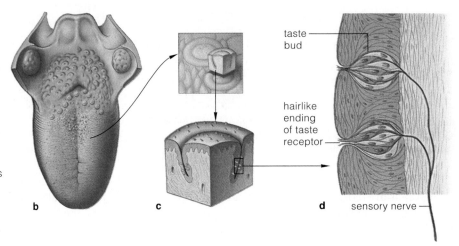

taste bud

hairlike ending of taste receptor

b   c   d   sensory nerve

the sex pheromones. Olfactory receptors of a male silk moth hit by only one bombykol molecule per second fire action potentials. They help a male find a female in the dark, even if she is more than a kilometer upwind. Human pheromones may influence sexual preferences, menstrual cycles, and menopause. The receptors and olfactory pathway for them are not yet identified.

Different animals taste with their mouth, antennae, legs, tentacles, or fins, depending on where their **taste receptors** are. In your mouth, throat, and especially the upper surface of your tongue, these chemoreceptors are located in about 10,000 sensory organs called taste buds

(Figure 35.8). In a taste bud, fluids in the mouth move through a pore to receptor cells. Thousands of perceived tastes are some combination of five primary sensations: sweet (as elicited by glucose and other simple sugars), sour (acids), salty (NaCl and other salts), bitter (plant toxins, including alkaloids), and *umami* (glutamate, the brothy or savory taste typical of aged cheese and meat).

The senses of smell and taste both start at chemoreceptors. Both involve sensory pathways that lead to processing regions in the cerebral cortex and in the limbic system.

SENSE OF BALANCE

All animals assess and respond to displacements from *equilibrium*, at which the body is balanced in relation to gravitation, acceleration, and other forces that affect its positions and its movements. Even jellyfishes right themselves after tilting upside-down. The first *organs* of equilibrium evolved among fishes, amphibians, and reptiles and had little, if anything, to do with hearing. You have a pair of them, one in each of your inner ears.

A **vestibular apparatus** in each ear consists of two sacs—the utricle and saccule—and three semicircular canals (Figure 35.9a). Fluid fills the sacs and canals, which interconnect as one continuous system. It moves when the head rotates, and the fluid pressure activates sensory receptors. Other receptors are activated when the body starts and stops moving in a straight line.

Figure 35.9 (**a**) Human vestibular apparatus. Organs of (**b**) static equilibrium and (**c**) dynamic equilibrium. In both organs, hair cells project into a membrane above them. They are stimulated when the head moves or stops moving. The membrane is displaced, and the hair cells bend.

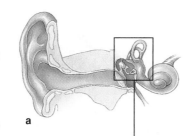

a

Inside the bulging base of a semicircular canal is a crista, an organ of *dynamic* equilibrium. It has a cupula, a gelatinous mass with hair cells projecting into it (Figure 35.9c). **Hair cells** are a type of mechanoreceptor. Rotate the head horizontally, vertically, or diagonally, and you displace fluid in the canal oriented in the same direction. Fluid pressure bends the cupula. It also bends the hair cells and deforms their plasma membrane; thus action potentials arise. These hair cells synapse with sensory neurons. Axons of these neurons converge, and they form a vestibular nerve to the brain.

In amphibians, birds, and mammals, input from the hair cells is integrated with input from the skin, joints, and tendons. By controlling eye muscles, integration keeps the eyes focused as the head rotates. It also helps maintain awareness and memories of body positions and motion in space, as exemplified by Sarah Hughes.

A stroke, inner ear infection, loose particles in the semicircular canals, or conflicting sensory inputs may

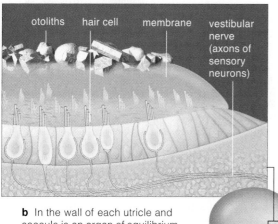

otoliths hair cell membrane vestibular nerve (axons of sensory neurons)

b In the wall of each utricle and saccule is an organ of equilibrium, a thick, weighted membrane and hair cells in a patch of epithelium.

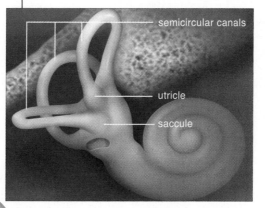

semicircular canals

utricle

saccule

VESTIBULAR APPARATUS

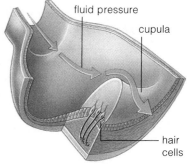

fluid pressure

cupula

hair cells

c In the bulging base of each semicircular canal is a crista, an organ of dynamic equilibrium. It consists of hair cells that project into a gel-like mass above them.

Inside each utricle and saccule is an organ of *static* equilibrium that is activated by linear acceleration or deceleration. The organ is a thick membrane resting on hair cells that project upward from the floor of the sac. Embedded in the membrane are **otoliths**, a tiny mass of calcium carbonate crystals that make the membrane heavier than the fluid in the sac (Figure 35.9b). When the body starts or stops moving, the fluid accelerates past the weighted membrane, which slides in response. The sliding bends the hair cells and activates them.

cause *vertigo*, an illusion of spinning in space. Vertigo may occur when you stand still high above the ground, say, on the top floor of a skyscraper and look down. The vestibular system reports that you are motionless, but your eyes report that your body is floating in space.

Organs of equilibrium help keep the body balanced in relation to gravity, velocity, acceleration, and other forces that influence its position and movement.

SENSE OF HEARING

Properties of Sound

Many arthropods and most vertebrates have a sense of **hearing**, or sound perception. Sounds are waves of compressed air, a form of mechanical energy. Each time you clap your hands, you force molecules outward and create a low-pressure state in the region they vacated. Such pressure variations are depicted as wave forms in which *amplitude* corresponds to loudness, or intensity.

Typically, amplitude is measured in decibels. Every ten decibels signifies a tenfold increase in the intensity above the faintest sounds humans hear. The *frequency* of a sound is the number of wave cycles per second. Each cycle extends from the start of one wave peak to the start of the next. The more cycles per second, the higher the frequency and perceived pitch (Figure 35.10).

Unlike a tuning fork's pure tone, most sounds are combinations of waves of different frequencies. Their timbre or quality arises from various combinations of overtones. This property can help you recognize, say, voices of people you know when talking on the phone.

Evolution of the Vertebrate Ear

After vertebrates invaded the land, the sense of hearing became far more important than it had been in aquatic habitats. Parts of the **inner ear** that deal with the sense of balance did not change much. Other parts gradually expanded into structures that receive and process the sounds traveling through air.

Thus the **middle ear** evolved. Its structures amplify and transmit air waves to the inner ear. In reptiles, a shallow depression formed on each side of the head and evolved into an eardrum. This thin membrane rapidly vibrates in response to air waves. Behind the eardrum of all existing crocodiles, birds, and mammals is an air-filled cavity and small bones, which transmit vibrations to the inner ear. Forerunners of these bones structurally supported gill pouches in early fishes, then

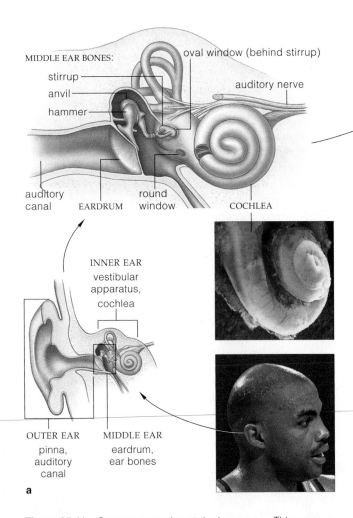

Figure 35.11 Sensory reception at the human ear. This sensory organ collects, amplifies, and sorts out sound waves (acoustical stimuli). The external flaps of the outer ear collect sound waves, which move through an auditory canal to the eardrum. The eardrum is also called the tympanic membrane.

became part of the jaw joint. In short, some of the bones that once supported gas exchange became modified for feeding functions, and later still they became modified for hearing among reptiles, birds, and mammals.

Among many mammalian species, a pair of sound-collecting **external ears** also became well developed. Usually, each external ear has a pinna: a skin-covered, sound-collecting, folded flap of cartilage that projects from the side of the head. Each external ear also has a deep auditory canal leading to the middle ear.

Figure 35.11 shows the outer, middle, and inner ear of humans. Sounds collected at the external ear move down the auditory canal to an eardrum, a membrane that vibrates with the incoming sound waves. Bones of the middle ear pick up the vibrations. They amplify a stimulus by transmitting the force of pressure waves to a smaller surface, the oval window. That "window" is

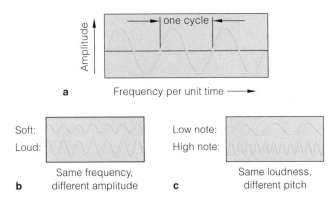

Figure 35.10 The wavelike properties of sound.

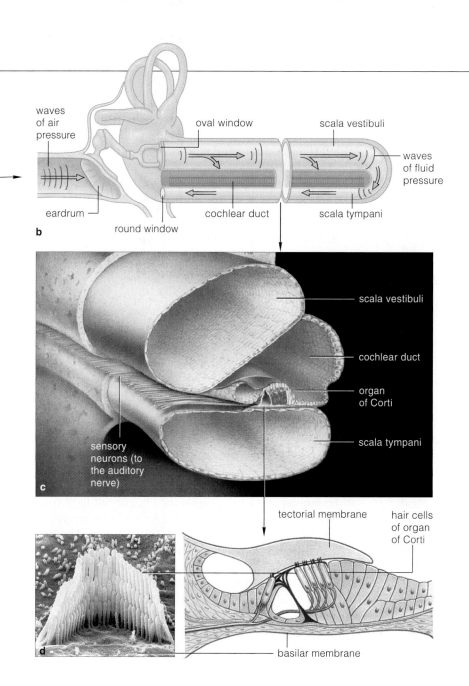

waves
of air
pressure

oval window

scala vestibuli

waves
of fluid
pressure

eardrum

b

round window

cochlear duct

scala tympani

scala vestibuli

cochlear duct

organ
of Corti

scala tympani

sensory
neurons (to
the auditory
nerve)

c

tectorial membrane

hair cells
of organ
of Corti

d

basilar membrane

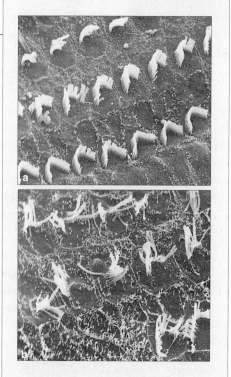

Figure 35.12 Results of an experiment on the effect of intense sound on the inner ear. (**a**) From a guinea pig ear, three rows of hair cells that project into the tectorial membrane in the organ of Corti. (**b**) Hair cells inside the same organ after twenty-four hours of exposure to noise levels comparable to extremely loud music.

To give you a sense of what "loud" is, a ticking watch measures 10 decibels (100 times louder than the threshold of hearing for humans). Normal conversation is about 60 decibels (about a million times louder), a food blender operating at high speed is about 90 decibels (a billion times louder), and an amplified rock concert is about 120 decibels (a trillion times louder).

an elastic membrane in front of the **cochlea**, a pea-sized part of the inner ear that sorts out pressure waves.

Take a look at Figure 35.11*b*, an "uncoiled" diagram of the cochlea. Pressure waves make the oval window bow in and out, which produces fluid pressure waves in two ducts (scala vestibuli and scala tympani). These waves reach another membrane, the round window. The round window passively moves in and out, which compensates for the pressure differences.

The cochlear duct, the third duct of the inner ear, sorts out the pressure waves. Its *basilar* membrane is narrow and stiff, then it broadens and becomes more flexible deeper in the coil. Thus, along its length, the basilar membrane can vibrate differently to sounds of different frequencies. The higher pitched sounds are detected as vibrations early in the coil. And sounds of lower frequency cause vibrations deeper in the coil.

Attached to the inner basilar membrane is the organ of Corti, with vibration-sensitive **acoustical receptors** (Figure 35.11*c,d*). These mechanoreceptors are hair cells that have specialized cilia embedded in an overhanging *tectorial* membrane. The cilia bend as pressure waves displace the basilar membrane. This bending transduces the mechanical energy of pressure waves into action potentials, which travel along an auditory nerve into the brain. These delicate structures are damaged by chronic exposure to intense sounds, such as amplified music and jet planes taking off (Figure 35.12).

Hearing in land vertebrates starts with structures that collect, amplify, and sort out sound waves. In the inner ear, the sound waves produce fluid pressure variations, which hair cells transduce into action potentials.

SENSE OF VISION

What Are the Requirements for Vision?

All organisms are sensitive to light. Although they have no photoreceptor cells, sunflowers track the sun as it arcs across the sky. Even single-celled amoebas abruptly stop moving when you shine a light on them. But we do not find photoreceptor cells until we poke about among the invertebrates. Even then, not all species that have photoreceptors "see" as you do. Many might detect a change in light intensity, as a mollusk on the seafloor might do when a fish swims above it. However, they cannot discern the size or shape of such objects.

At the minimum, the sense of **vision** requires eyes and image perception in brain centers that receive and interpret patterns of visual stimulation. Such images are pieced together from information about the shape, brightness, position, and movement of visual stimuli. **Eyes** are sensory organs that incorporate a tissue with a dense array of photoreceptors. All photoreceptors have this in common: *They incorporate pigment molecules that can absorb photon energy, which is converted to excitation energy in sensory neurons.*

Nearly all photoreceptor cells have some portion of their surface infolded to a small or large extent, which increases the surface area available for photochemical reactions. Their structure can be traced to the kind of epidermal cells that gave rise to them. The *rhabdomeric*

photoreceptors are derived from cells with microvilli. They predominate in flatworms, mollusks, annelids, arthropods, and echinoderms. By contrast, the *ciliary* photoreceptors have a photoreceptive surface derived from the membrane of a cilium. We find these among the cnidarians, some flatworms, and all vertebrates.

A Sampling of Invertebrate Eyes

SIMPLE EYES Earthworms and some other animals that lack eyes still respond to light, because photoreceptors are dispersed through their integument. The simplest eye is an ocellus (plural, ocelli). It is a patch or cuplike depression of the integument in which photoreceptors and pigmented cells are concentrated. Figure 35.13*a* shows one of these. Such eyes help the animal use light as a cue to orient the body, detect a predator's shadow, or adjust biological clocks.

COMPLEX EYES Vision first evolved in fast-moving predators, which had to discriminate among prey and other objects in their rapidly changing visual field. A **visual field**, recall, is the part of the outside world that an animal sees (Section 34.10). At the least, different photoreceptors must sample the intensity of light from different parts of the visual field. The brain interprets the intensities as contrasting parts of a visual image. Image perception differs among invertebrates. Except in cephalopods and a few other animals, it is not good. Why? Good images require many photoreceptors, and the eyes of most invertebrates are too small to have a lot of them. For example, the planarian's pigment cup only has about 200 photoreceptors, compared with the 70,000 per square millimeter in an octopus eye.

Image formation benefits from a **lens**, a transparent body that bends all light rays from a given point in the visual field so they converge onto photoreceptors (see Section 4.2). Abalones have a spherical lens that bends incoming rays but not to the same points, so blurred images form (Figure 35.13*b*). Images improve if some space intervenes between the lens and photoreceptors, as in a snail eye. They get even better with a **cornea**, a transparent cover that directs light rays onto the lens, as happens in a snail or conch (Figure 35.13*c,d*).

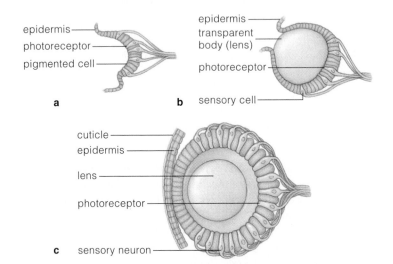

a

b   sensory cell

epidermis
photoreceptor
pigmented cell

epidermis
transparent body (lens)
photoreceptor

cuticle
epidermis
lens
photoreceptor
c   sensory neuron

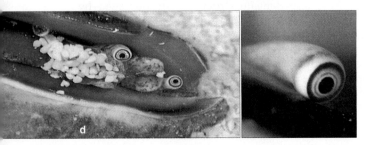

d

Figure 35.13 Organization of invertebrate eyes, longitudinal section. There are far more photoreceptors than can be shown in simple art. (**a**) Limpet ocellus, a shallow depression in the epidermis that incorporates light-sensitive receptors. (**b**) Abalone eye, with its spherical, transparent lens. (**c**) Eye of a land snail. (**d**) Well-developed eyes of a conch, which is peering into the waters of the Great Barrier Reef along the east coast of Australia.

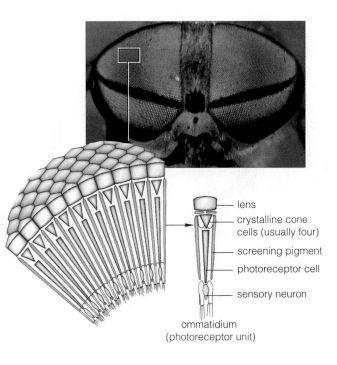

Figure 35.14 Compound eyes of a deerfly. The lens of each photosensitive unit (ommatidium) directs light onto a crystalline cone, which focuses light on photoreceptor cells.

lens
crystalline cone cells (usually four)
screening pigment
photoreceptor cell
sensory neuron

ommatidium
(photoreceptor unit)

Figure 35.15 Approximation of light reception in an insect eye. This image of a butterfly was formed after a researcher detached an insect's compound eye and took a photograph through its outer surface. It may not be what the insect "sees." Integration of signals sent to the brain from photoreceptors may produce a sharper image. This representation does suggest how the separate ommatidia *sample* the overall visual field.

Invertebrates on land are better adapted for visually discriminating objects. Most arthropods have *simple* eyes with one lens for all photoreceptors. Spiders are like this. Crustaceans and insects have complex, *compound* eyes. A **compound eye** contains many closely packed rhabdomeric units (Figure 35.14). The compound eyes of some species contain thousands of these units, of a sort called ommatidia (singular, ommatidium).

According to the mosaic theory of image formation, each unit samples only a small part of the visual field. The brain builds images based on signals about differences in light intensities, with each unit contributing a small bit to the formation of a visual mosaic (Figure 35.15).

Of all invertebrates, octopuses and squids have the most complex eyes. Like vertebrates, they have **camera eyes**, so named because the eyeball is structured along the lines of a camera. The interior is a darkened chamber. Light enters it only through the pupil, an opening in a ring of contractile tissue called the iris (equivalent to a camera's diaphragm). Behind the pupil, a lens focuses light onto an array of photoreceptors (a camera's light-sensitive film). Axons of sensory nerves converge as an optic tract that leads to the brain (Figure 35.16).

Cephalopods and vertebrates are not closely related, so similarities in eye structure and function might be a result of convergent evolution. By one theory, a group of genes that once affected development of the nervous system were appropriated for the task of eye formation. Whether this might have happened more than once in different animal lineages is not yet known.

Vision requires eyes (sensory organs with a dense array of photoreceptors) and a capacity for image formation in the brain, based on patterns of visual stimulation.

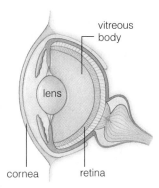

vitreous body

lens

cornea   retina

Figure 35.16 Octopus eye. The pupil constricts into slits and flares open in response to stimuli, changing light conditions, or the presence of a potential mate or enemy. Experiments show that cephalopods form distinct images. They sense the size, shape, and vertical and horizontal orientation and possibly color of objects. They are nearsighted. Even so, they can snag tiny prey three meters away in the water.

STRUCTURE AND FUNCTION OF VERTEBRATE EYES

The Eye's Structure

Vertebrate vision starts at the **retina**, a tissue of densely packed photoreceptors. It involves image formation in brain centers that can receive and interpret patterns of stimulation from different parts of each eye. Images are based upon information about the shape, brightness, position, and movement of visual stimuli. With this in mind, take a look at Figure 35.17, which diagrams the human eye. Like most vertebrate eyes, this eyeball has a three-layered wall (Table 35.2).

The eyeball's outermost layer consists of the sclera and cornea. The sclera, the dense, fibrous "white" of an eye, protects most of the eyeball. The cornea, made of transparent collagen fibers, covers the rest of it.

The middle layer has a choroid, ciliary body, iris, and pupil. The choroid is a darkly pigmented, vascularized tissue. It absorbs the light that photoreceptors have not absorbed and prevents it from scattering inside the eye. Suspended in back of the cornea is a doughnut-shaped, pigmented iris (the Latin *iris* refers to the rings of a rainbow). The dark "hole" in the center of an iris is the pupil, the entrance for light. When rays of bright light strike an eye, circular muscles embedded within the iris contract, so the pupil's diameter shrinks. In dim light, radial muscles in the iris contract, so the pupil enlarges.

The inner layer, which includes the retina, is at the back of your eyeball. In birds of prey, such as owls and hawks, it is closer to the eyeball's roof (Figure 35.18).

The eye's interior has a lens. A clear fluid called the aqueous humor bathes the lens, which is constructed of layers of transparent proteins. The vitreous body is a jellylike substance that fills a chamber behind the lens.

Figure 35.18 Here's looking at you! In owls and some other birds of prey, photoreceptors are concentrated more on top of the inner eyeball, not back. Such birds look down more than up when they fly and scan the ground for a meal. When they are on the ground, they cannot see something overhead very easily unless they turn their head almost upside-down.

Eyes with a large pupil and iris tell us something about the life-style of their owners. Animals that move about actively at night or in dimly lit habitats are less likely to stumble, bump into objects, or fall off cliffs when they intercept as much of the available light as they can. The greater the amount of incoming light, the better is the visual discrimination. A large pupil lets in

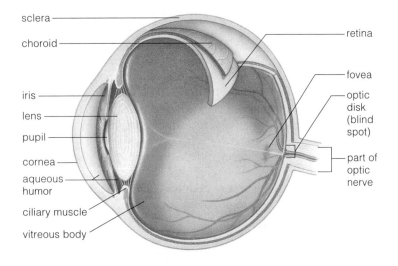

Figure 35.17 Structure of the human eye, cutaway view.

Labels: sclera, choroid, iris, lens, pupil, cornea, aqueous humor, ciliary muscle, vitreous body, retina, fovea, optic disk (blind spot), part of optic nerve

Table 35.2	Components of the Vertebrate Eye	
WALL OF EYEBALL (THREE LAYERS)		
Fibrous Tunic	*Sclera.* Protects eyeball	
	Cornea. Focuses light	
Vascular Tunic	*Choroid.* Blood vessels nutritionally support wall cells; pigments prevent light scattering	
	Ciliary body. Its muscles control lens shape; its fine fibers hold lens in upright position	
	Iris. Adjusting iris controls incoming light	
	Pupil. Serves as entrance for light	
Sensory Tunic	*Retina.* Absorbs and transduces light energy	
	Fovea. Increases visual acuity	
	Start of optic nerve. Carries signals to brain	
INTERIOR OF EYEBALL		
Lens	Focuses light on photoreceptors	
Aqueous humor	Transmits light, maintains pressure	
Vitreous body	Transmits light, supports lens and eyeball	

Figure 35.19 Pattern of retinal stimulation in the human eye. A transparent cornea positioned in front of the pupil is curved, so it changes the trajectories of light rays as they enter the eye. The pattern is upside-down and reversed left to right, compared with the stimulus pattern in the visual field.

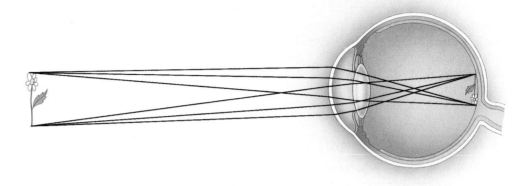

Figure 35.20 Two focusing mechanisms. A ciliary muscle encircles the lens and attaches to it. (**a**) When it relaxes, the lens flattens; the focal point moves farther back and brings distant objects into focus. (**b**) When the muscle contracts, the lens bulges; so the focal point moves closer and brings close objects into focus.

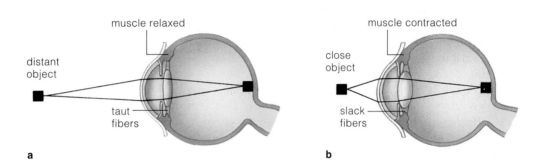

muscle relaxed

distant object

taut fibers

a

muscle contracted

close object

slack fibers

b

large quantities of light. A pupil ringed by a large iris may be dilated or constricted, which admits more or less light. This is a useful option for animals that are active at night as well as during the day.

When light rays do converge at the back of the eye, they stimulate the retina in a distinct pattern. Because the cornea and lens have a curved surface, light rays coming from a particular point in the visual field strike them at different angles, so their trajectories change. (As described in Section 4.2, rays of light bend at the boundaries between substances of different densities. Bending sends them off in new directions.) Because of their newly angled trajectories, light rays that converge at the back of the eye stimulate the retina in a pattern that is upside-down and reversed left to right, relative to the original source of the light rays. Figure 35.19 gives a simplified diagram of this outcome.

Visual Accommodation

Light rays emanating from sources at varying distances from the eye strike the cornea at different angles, which means that they could end up being focused at different distances behind it. This would do the brain no good. However, adjustments in the position or shape of the lens normally focus all incoming stimuli onto the retina. We call these lens adjustments **visual accommodation**. Without them, light rays from distant objects would be improperly focused in front of the retina, and rays from close objects would be focused behind it.

In the fish and reptilian eye, muscles move the lens forward or back, like a camera's focusing apparatus. Extending the distance between the lens and the retina moves the focal point forward; shrinking the distance will move it back. The lens shape itself is adjusted in birds and mammals. A ciliary muscle encircles their lens and is attached to it by fiberlike ligaments (Figure 35.17). When this muscle relaxes, the lens flattens and thereby moves the focal point farther back, as in Figure 35.20a. When the muscle contracts, the lens bulges and moves the focal point forward, as in Figure 35.20b.

In some cases, the lens cannot be adjusted enough to make the focal point match up precisely with the retina. For example, in some people, the eyeball is not shaped quite right, and the position of the lens is either too close or too far away from the retina. When this is the case, accommodation alone cannot bring about an exact match. Eyeglasses often correct both visual disorders, which you undoubtedly know by their familiar names: nearsightedness and farsightedness (Section 35.9).

For most vertebrate eyes, the eyeball has three layers, and it encloses a lens, an aqueous humor, and a vitreous body. Adjustments in the positioning or shape of the lens focus incoming visual stimuli onto the retina inside.

A protective sclera and a light-focusing cornea make up the outer layer. The middle layer has a vascularized, pigmented choroid and other parts that admit and control incoming light. Photoreception occurs at the retina of the inner layer.

CASE STUDY—FROM SIGNALING TO VISUAL PERCEPTION

We conclude the basics of this chapter with one of the classic examples of neuronal architecture—the sensory pathway from the retina to the brain. It shows how raw visual input is received, transmitted, and combined in ways that lead to awareness of light and shadows, of colors, of near and distant objects in the outside world.

How Is the Retina Organized?

Information flow begins as light strikes the retina. The retina's basement layer, a pigmented epithelium, covers the choroid. Resting on this layer are densely packed arrays of **rod cells** and **cone cells**, which are two classes of ciliary photoreceptors (Figure 35.21). Rod cells detect very dim light. During the night or in darkened places, they contribute to coarse perception of movement by detecting changes in light intensity across the visual field. Cone cells detect bright light. They contribute to sharp vision and color perception during the day.

Distinct layers of neurons are located above the rods and cones. The first to accept the visual information from photoreceptors are bipolar sensory neurons, then ganglion cells. Axons of these ganglion cells form the two optic nerves to the brain.

Before signals depart from the retina, they converge dramatically. The input from 125 million photoreceptors converges on only 1 million ganglion cells. Signals also flow laterally among horizontal cells and amacrine cells.

Both kinds of neurons interact to dampen or enhance signals before ganglion cells get them. Thus, *a great deal of synaptic integration and processing goes on even before visual information is sent to the brain.*

Neuronal Responses to Light

ROD CELLS   A rod cell's outer segment contains several hundred membrane disks, each peppered with about 10^8 molecules of a visual pigment called rhodopsin. The membrane stacking and the extremely high density of pigments greatly increase the likelihood of intercepting packets of light energy—that is, photons of particular wavelengths. The action potentials that result from the absorption of even a few photons can lead to conscious awareness of objects in dimly lit surroundings, as in dark rooms or late at night.

Rhodopsin effectively absorbs photons of blue-to-green wavelengths (Section 7.2). Absorption changes the pigment's shape and triggers a cascade of reactions that affect ion channels and pumps across the rod cell's plasma membrane. The voltage difference shifts across the membrane as gated sodium channels close, and a graded potential results. This graded potential reduces the ongoing release of a neurotransmitter that inhibits nearby sensory neurons. Released from inhibition, the neighbors fire off signals about the visual stimulus that will travel to the brain.

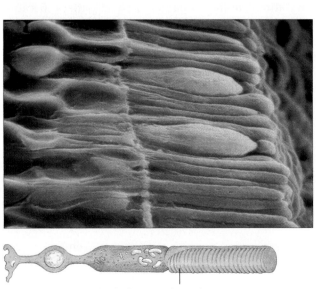

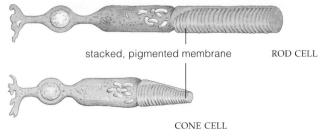

Figure 35.21 (**a**) Organization of photoreceptors and sensory cells in the mammalian retina. (**b**) Scanning electron micrograph and sketches showing the general structure of rods and cones.

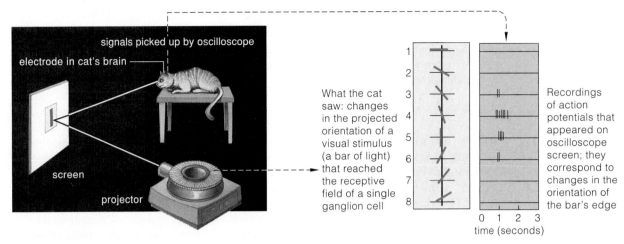

Figure 35.22 Example of experiments into the nature of receptive fields for visual stimuli. David Hubel and Torsten Wiesel implanted an electrode in an anesthetized cat's brain. They placed the cat in front of a small screen upon which different patterns of light were projected—here, a hard-edged bar. Light or shadow falling on part of the screen excited or inhibited signals sent to a single neuron in the visual cortex.

Tilting the bar at different angles produced changes in the neuron's activity. A vertical bar image produced the strongest signal (*numbered 5 in the sketch*). When the bar image tilted slightly, signals were less frequent. When it tilted past a certain angle, signals stopped.

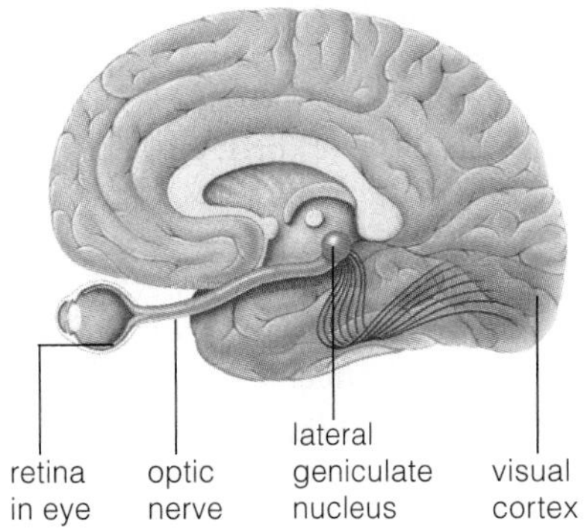

Figure 35.23
Diagram of the sensory pathway from the retina to the brain.

retina in eye   optic nerve   lateral geniculate nucleus   visual cortex

CONE CELLS The sense of color and of daytime vision starts with photon absorption by red, green, and blue cone cells, each with a different kind of visual pigment. Here again, photon absorption reduces the release of a neurotransmitter that otherwise inhibits the neurons that are adjacent to the photoreceptors. Near the center of the retina is a funnel-shaped depression called the fovea. This is the retinal area with the greatest density of photoreceptors and the greatest visual acuity. Cone cells of the fovea are the ones that discriminate the most precisely between adjacent points in space.

RESPONSES AT RECEPTIVE FIELDS Neurons in the eye collectively respond to information about light in highly organized ways. The retinal surface is organized into receptive fields, or restricted areas that influence the activity of individual sensory neurons. For example, the field for each ganglion cell is a small circle. Some of these cells respond best to a spot of light ringed by dark in the field's center. Others respond to rapidly changing light intensity, a spot of one color, or motion.

For one experiment, sensory cells generated action potentials after detecting a bar at certain orientations (Figure 35.22). But they did not respond to uniform, diffuse illumination, which is just as well. Doing so would send a confusing array of signals to the brain.

ON TO THE VISUAL CORTEX The part of the outside world that an individual actually sees is its visual field. The right side of both retinas intercepts light from the left half of the visual field; the left side intercepts light from the right half. The optic nerve leading away from each eye delivers signals about a stimulus from the *left* visual field to the right cerebral hemisphere. It delivers signals from the *right* visual field to the left hemisphere, as described in Sections 34.9 and 34.10.

Optic nerve axons end in a layered brain region, the lateral geniculate nucleus (Figure 35.23). Each layer has a map corresponding to receptive fields and deals with one aspect of a stimulus, such as form, movement, depth, color, and texture. After early processing, signals rapidly and simultaneously reach different portions of the visual cortex. There, final integration organizes the electrical activity and produces the sensation of sight.

Each eye is an outpost of the brain, collecting and analyzing information about the distance, shape, brightness, position, and movement of visual stimuli. Its sensory pathway starts at the retina and proceeds along an optic nerve to the brain.

Organization of visual signals begins at receptive fields in the retina, it is further processed in the lateral geniculate nucleus, and it is finally integrated in the visual cortex.

Disorders of the Human Eye

Of all the sensory receptors that the human brain requires to help keep you alive and functioning independently in the environment, fully two-thirds are located in your pair of eyes. They are photoreceptors, and they do more than detect light. They also let you see the world in a rainbow of colors. Eyes are one of our most important sources of information about the outside world.

Given their essential role, it is no surprise that so much attention is paid to eye disorders resulting from injuries, inherited abnormalities, diseases, and advancing age. The consequences range from relatively harmless conditions, such as nearsightedness, to total blindness. Each year, many millions of people must deal with such consequences.

COLOR BLINDNESS Occasionally, some or all of the cone cells that selectively respond to light of red, green, or blue wavelengths fail to develop in individuals. The rare individuals who have only one of three kinds of cones are totally *color-blind*. They perceive the world in shades of gray.

Consider a common genetic abnormality, *red–green color blindness*. This X-linked, recessive trait shows up most often in males. The retina of affected individuals lacks some or all of the cone cells with pigments that normally respond to light of red or green wavelengths. Most of the time, people who are affected by red–green color blindness merely have trouble distinguishing red from green in dim light. Some cannot distinguish between the two even in bright light.

FOCUSING PROBLEMS Other heritable abnormalities arise from misshapen features of the eye that affect the focusing of light. *Astigmatism*, for example, results from corneas with an uneven curvature; they cannot bend incoming light rays to the same focal point.

Nearsightedness, or myopia, often occurs when the horizontal axis of the eyeball is longer than the vertical axis. It also occurs when the ciliary muscle responsible for adjusting the lens contracts too strongly. The outcome is that images of distant objects are focused in front of the retina instead of on it (Figure 35.24).

Farsightedness, or hyperopia, is the opposite of myopia. Here, the vertical axis of the eyeball is longer than the horizontal axis (or the lens is "lazy"). As a result, close images are focused behind the retina (Figure 35.25).

Even a normal lens loses some of its natural flexibility as a person grows older. That is why people over forty years old often start wearing eyeglasses.

EYE DISEASES The structure and functioning of the eyes are vulnerable to infectious diseases. For example, *histoplasmosis*, a fungal infection of the lungs that is described in Section 24.5, can lead to retinal damage, which can cause partial or total loss of vision. To give another example, *Herpes simplex*, a virus that causes skin sores, also can infect the cornea and cause it to ulcerate.

Trachoma is an extremely contagious disease that has blinded millions, mostly in North Africa and the Middle East. The culprit is a bacterium that also is responsible for the sexually transmitted disease chlamydia (Section

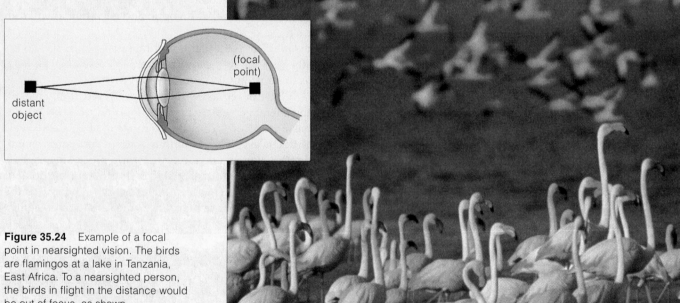

(focal point)

distant object

Figure 35.24 Example of a focal point in nearsighted vision. The birds are flamingos at a lake in Tanzania, East Africa. To a nearsighted person, the birds in flight in the distance would be out of focus, as shown.

44.15). The eyeball and the lining of the eyelids (the conjunctiva) become damaged. The damaged tissues are entry points for bacteria that can cause secondary infections. In time the cornea can become so scarred that blindness follows.

AGE-RELATED PROBLEMS   You have probably heard of *cataracts*, a gradual clouding of the lens. It is a problem associated with aging, although it also may arise from eye injury or diabetes. Possibly cataracts form when the transparent proteins that make up the eye's lens undergo structural changes. Clouding may alter the trajectory of incoming light rays. If the lens becomes opaque, light cannot enter the eye at all.

Glaucoma, another age-related problem, results when excess aqueous humor accumulates inside the eyeball. The blood vessels that service the retina collapse under the increased fluid pressure. Vision deteriorates as sensory neurons of the retina and optic nerve die off.

Although we often associate chronic glaucoma with advanced age, conditions that give rise to the disorder actually start to develop in middle age. If doctors detect the fluid pressure that tends to build up in the eye early enough, they can relieve the fluid pressure with drugs or with surgery before damage becomes severe.

EYE INJURIES   *Retinal detachment* is the eye injury that we read about most often. It may follow a physical blow to the head or illness that causes a tear in the retina. As the jellylike vitreous body oozes through the torn region, the retina is lifted away from the underlying choroid. In time the retina may peel away entirely, leaving its vital blood supply behind.

Early symptoms of a detached retina include blurred vision, flashes of light that occur even in the absence of outside visual stimulation, and loss of peripheral vision. Without medical intervention, the person may become totally blind in the damaged eye.

NEW TECHNOLOGIES   Today a variety of tools can be used to correct some eye disorders. In *corneal transplant surgery*, for example, a defective cornea can be removed, and then an artificial cornea (made of clear plastic) or a natural cornea from a donor can be stitched in its place. Within a year, the patient can be fitted with eyeglasses or contact lenses. Similarly, cataracts can be surgically corrected by removing the clouded lens and replacing it with an artificial one, although the operation is difficult and not always successful.

Severely nearsighted people sometimes opt for *radial keratotomy*, a still-controversial surgical procedure in which tiny, spokelike incisions are made around the edge of the cornea to flatten it more. When all goes well, the adjustment eliminates the need for corrective lenses. But vision sometimes is overcorrected or undercorrected, in which case the patient must undergo further surgery.

As a final example, retinal detachment may be treated with *laser coagulation*, a painless technique in which a laser beam seals off leaky blood vessels and "spot welds" the retina to the underlying choroid.

Figure 35.25   Example of a focal point in farsighted vision. To affected individuals, the flamingos standing in water in the foreground would be out of focus, as shown.

SUMMARY

Gold indicates text section

1. A stimulus is a specific form of energy that the body detects by means of sensory receptors. A sensation is an awareness that stimulation has occurred. Perception is understanding what the sensation means. *CI*

2. Sensory receptors are endings of sensory neurons or specialized cells next to them. They respond to stimuli, or specific forms of energy such as mechanical pressure and light. Animals respond to aspects of the internal or external environment only when they have receptors that are sensitive to the energy of stimuli. *CI, 35.1*

    a. Mechanoreceptors, including free nerve endings and hair cells, detect mechanical energy related to touch, pressure, motion, and changes in position.

    b. Thermoreceptors detect radiant energy (heat).

    c. Pain receptors (nociceptors) detect tissue damage.

    d. Chemoreceptors, such as olfactory receptors and taste receptors, detect chemical substances dissolved in the fluid bathing them.

    e. Osmoreceptors detect changes in water volume (hence solute concentrations) in the surrounding fluid.

    f. Photoreceptors detect light. Rods and cones of the retina in the human eye are examples.

3. A sensory system has sensory receptors for specific stimuli and nerve pathways that lead from receptors to receiving and processing centers in the brain. The brain assesses a stimulus based upon which nerve pathway is delivering signals, the frequency of signals traveling along individual axons of the pathway, and the number of axons that were recruited into action. *35.1*

4. Somatic sensations include touch, pressure, pain, temperature, and muscle sense. The receptors associated with these sensations are not localized in a single organ or tissue. Special senses include taste, smell, hearing, balance, and vision. Receptors associated with these senses typically reside in sensory organs, such as eyes, or some other particular body region. *35.2*

5. The senses of taste and smell both involve sensory pathways from chemoreceptors to processing regions in the cerebral cortex and limbic system. *35.3*

6. Organs of equilibrium in the vertebrate inner ear detect gravity, acceleration, and other forces that affect the body's positions and movements. The vertebrate sense of hearing (sound perception) is based on paired outer, middle, and inner ears that respectively collect, amplify, and sort out sound waves. *35.4, 35.5*

7. Vision requires eyes: sensory organs having a dense array of photoreceptors, as in a retina. It requires a capacity for image formation in the brain, based upon incoming patterns of visual stimulation. Vision and discrimination among objects evolved first among the earliest fast-moving, predatory animals. *35.6–35.8*

Review Questions

1. Define stimulus. *CI* (*see also Section 28.3*)

2. Name six categories of sensory receptors and the type of stimulus energy that each kind detects. *CI, 35.1*

3. What are the basic components of a sensory system? How does the brain assess the nature of a given stimulus? *35.1*

4. How do somatic sensations differ from special senses? *35.2*

5. What is pain? Describe one type of pain receptor. *35.2*

6. Name the five primary taste sensations. Which substances elicit each type of sensation? *35.3*

7. Define pheromone and give an example. *35.3*

8. Which evolved first, a sense of balance or sense of hearing? Do both involve the outer, middle, and inner ear? *35.4, 35.5*

9. How does vision differ from light sensitivity? What sensory organs and structures does vision require? *35.6*

10. Label the component parts of the human eye: *35.7*

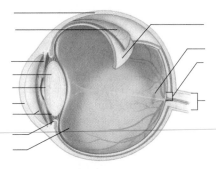

Self-Quiz   ANSWERS IN APPENDIX III

1. A _____ is a specific form of energy that is detected by a sensory receptor.

2. Conscious awareness of a stimulus is called a _____ .

3. _____ is actually understanding what particular sensations mean.

4. Each sensory system consists of _____ .
    a. nerve pathways from specific receptors to the brain
    b. sensory receptors
    c. brain regions that deal with sensory information
    d. all of the above

5. _____ may be defined as a decrease in the response to an ongoing stimulus.
    a. Perception   c. Visual accommodation
    b. Sensory adaptation   d. both b and c

6. _____ detect mechanical energy associated with changes in pressure, position, or acceleration.
    a. Chemoreceptors   c. Photoreceptors
    b. Mechanoreceptors   d. Thermoreceptors

7. Detecting light energy is the function of _____ .
    a. chemoreceptors   c. photoreceptors
    b. mechanoreceptors   d. thermoreceptors

8. Free nerve endings are _____ .
    a. mechanoreceptors   c. pain receptors
    b. thermoreceptors   d. all of the above

9. Is this statement true or false: Sensations of visceral pain are associated with internal organs.

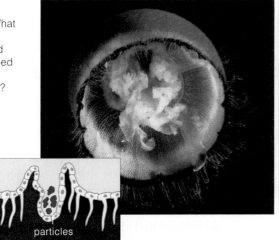

Figure 35.26 What are the mystery particles located on the bell-shaped margin of this jellyfish medusa?

particles

Figure 35.27 One way to test whether your organs of equilibrium are working.

10. We now recognize _____ primary sensations of taste.
a. two   b. three   c. four   d. five

11. In humans, the _____ ear contains organs that can detect accelerated, rotational motions of the head.
a. outer   b. middle   c. inner   d. all of these

12. The inner layer of your eyeball's wall includes the _____ .
a. retina   c. lens and choroid
b. sclera and cornea   d. both a and c

13. _____ detect bright light; _____ detect very dim light.
a. Rods; cones   b. Cones; rods

14. Match each term with the most suitable description.
_____ somatic sensations
_____ somatosensory cortex
_____ special senses
_____ variations in stimulus intensity
_____ visual field
_____ stimulus

a. the part of the outside world that an animal sees
b. e.g., vision (by way of eyes) and hearing (by way of ears)
c. form of energy that a specific sensory receptor can detect
d. encoded in the frequency and number of action potentials
e. touch, pressure, temperature, pain, and muscle sense
f. has maps of body surface

Critical Thinking

1. On the bell-shaped rim of the jellyfish shown in Figure 35.26 are tiny saclike structures that hold calcium particles next to a sensory cilium. When the bell tilts, the particles slide over the cilium. Would you assume that these structures contribute to the sense of taste, smell, balance, hearing, or vision?

2. Wayne, on standby for the last flight from San Francisco to New York, was assigned the last available seat on the plane. It was in the last row, where vibrations and noise from the jet engines are most pronounced. When Wayne got off the plane in New York, he was speaking loudly and having trouble hearing what other people were saying. The next day, things were back to normal. What happened to his hearing during the flight?

3. Juanita made an appointment with her doctor because she was experiencing recurring episodes of dizziness. Her doctor immediately asked her whether "dizziness" meant she had sensations of lightheadedness, as if she were going to faint, or whether it meant she had sensations of vertigo, a feeling that she herself or objects near her were spinning around.

Why did Juanita's doctor emphasize this clarification early in his evaluation of her condition?

4. Michael, now three years old, experiences chronic *middle-ear infections* (acute otitis media). This type of infection is common among children in day-care centers and elementary schools. Symptoms start when thick mucus forms during a throat or nose infection. Bacteria then infect the mucus that backs up into the eustachian tube, which connects the back of the throat with the middle ear. Michael felt pain, became feverish and irritable, and had trouble hearing. Then his left eardrum ruptured, and a jellylike substance dribbled out. His mother was startled and upset, but the pediatrician told her that if Michael's eardrum had not ruptured on its own, he would have had to insert a small drainage tube into it. Why did the pediatrician consider this a necessary surgical intervention?

5. Are organs of dynamic equilibrium, static equilibrium, or both activated during a roller coaster ride (Figure 35.27)?

Selected Key Terms

acoustical receptor *35.5*
action potential *35.1*
camera eye *35.6*
chemoreceptor *CI, 35.1*
cochlea *35.5*
compound eye *35.6*
cone cell *35.8*
cornea *35.6*
echolocation *CI*
encapsulated receptor *35.2*
external ear *35.5*
eye *35.6*
free nerve ending *35.2*
hair cell *35.4*
hearing *35.5*
inner ear *35.5*
lens *35.6*
mechanoreceptor *CI, 35.1*
middle ear *35.5*
olfactory receptor *35.3*
osmoreceptor *CI, 35.1*

otolith *35.4*
pain *35.2*
pain receptor *CI, 35.1*
perception *CI*
pheromone *35.3*
photoreceptor *CI, 35.1*
retina *35.7*
rod cell *35.8*
sensation *CI*
sensory adaptation *35.1*
sensory system *CI*
somatic sensation *35.1*
somatosensory cortex *35.2*
special sense *35.1*
stimulus *CI*
taste receptor *35.3*
thermoreceptor *CI, 35.1*
vestibular apparatus *35.4*
vision *35.6*
visual accommodation *35.7*
visual field *35.6*

Readings

Delcomyn, F. 1998. *Foundations of Neurobiology.* New York: Freeman.

Sherwood, L. 1997. *Human Physiology.* Third edition. Belmont, California: Wadsworth.

INTEGRATION AND CONTROL: ENDOCRINE SYSTEMS

Hormone Jamboree

In the 1960s, at her camp in a forest by the shore of Lake Tanganyika in Tanzania, the primatologist Jane Goodall let it be known that bananas were available. One of the first chimpanzees attracted to the delicious food was a female—Flo, as she came to be called (Figure 36.1). Flo brought along her offspring, an infant female and a juvenile male, and exhibited commendable parental behavior toward them.

Three years passed, and Goodall observed that Flo's earlier preoccupation with motherhood gave way to a preoccupation with sex. She also observed that male chimpanzees followed Flo to the camp by the lake and stayed for more than the bananas.

Sex, Goodall discovered, is *the* premier binding force in the social life of chimpanzees. These primates do not mate for life as eagles do, or wolves. Before the rainy season begins, mature females that are entering their fertile cycle become sexually active. Changing hormone levels drive behavioral and physical changes. Sexual organs, for instance, become swollen and vividly pink. The swellings are strong visual signals to males. They are the flags of sexual jamborees, of great gatherings of highly stimulated chimps in which any males present may have a turn at copulating with the same female.

The gathering of many flag-waving females draws together chimpanzees, which forage alone or in small family groups for most of the year. It reestablishes bonds that unite a rather fluid community.

Infants and juveniles now play with one another and with the adults. Their aggressive and submissive jostlings help map out future dominance hierarchies. Flo happened to be a high-ranking female in the social hierarchy. Through her sexual attractiveness and direct solicitations, she built strong alliances with many males.

Figure 36.1 (**a**) Primatologist Jane Goodall in Gombe National Park, near the shores of Lake Tanganyika, scouting for chimpanzees. (**b**) Flo and two of her offspring, all subjects of long-term field observations. Flo helped Goodall clarify the central role of sex—and of the hormones that orchestrate it—in the social life of these primate relatives of humans.

b

Through her high status and aggressive behavior, she helped her male offspring win confrontations with other young male chimps.

The hormone-induced swelling during the fertile period of female chimps lasts somewhere between ten and sixteen days. Yet females are fertile for only one to five days. Sex hormones induce swelling even after a female gets pregnant. We can hypothesize that sexual selection has favored prolonged swelling. Males groom a sexually attractive female more often, protect her, give her more food, and let her tag along to new foraging sites. The more that males accept a female, the higher she rises in the social hierarchy—and, evolutionarily speaking, the more her individual offspring benefit.

Through their effects, hormones help orchestrate the growth, development, and reproductive cycles of nearly all animals, from the invertebrate worms to chimpanzees and humans. They influence minute-by-minute and day-to-day metabolic functions. Through their interplays with one another and with the nervous system, hormones profoundly influence the physical appearance, well-being, and behavior of individuals. Behavior, too, affects whether individuals will survive, either on their own or as part of social groups.

This chapter focuses primarily on hormones—on their sources, targets, and interactions as well as the mechanisms involved in their secretion. If the details start to seem remote, remember that this is the stuff of life. Hormones underwrote Flo's appearance, behavior, and rise through the chimpanzee social hierarchy. Just imagine what they have been doing for you.

Key Concepts

1. Hormones and other signaling molecules have central roles in integrating the activities of individual cells in ways that benefit the whole body.

2. Only cells with molecular receptors for a specific hormone are its targets. Hormones operate by serving as signals for change in the activities of their targets.

3. Many types of hormones influence transcription of genes and protein synthesis in target cells. Other types call for alterations in existing molecules and structures in cells.

4. Some hormones work by binding with and altering membrane characteristics, such as the permeability of a plasma membrane to a particular solute. Others interact with a cell's DNA, either directly or after first binding with a receptor in the cytoplasm.

5. Some hormones help the body adjust to short-term shifts in diet and levels of activity. Others help induce long-term adjustments in cell activities that underlie the body's growth, development, and reproduction.

6. In vertebrates, the hypothalamus and pituitary gland interact in ways that coordinate the activities of a number of endocrine glands. Together, they exert control over many aspects of the body's functioning.

7. Hormones, neural signals, shifts in local chemical conditions, and environmental signals function as triggers for the secretion of particular hormones.

36.1

THE ENDOCRINE SYSTEM

Hormones and Other Signaling Molecules

Throughout their lives, cells must respond to changing conditions by taking up and releasing various chemical substances. In vertebrates, the responses of millions to many billions of cells must become integrated in ways that help keep the whole body alive and functioning.

Signaling molecules help integrate activities within and between cells, and between tissues. We call them hormones, neurotransmitters, local signaling molecules, and pheromones. Each acts on target cells. A "target" is any cell that has receptors for a signaling molecule and that may change its activities in response to it. Targets may or may not be adjacent to cells that send signals.

By definition, **animal hormones** are the secretory products of endocrine glands, endocrine cells, and some neurons that the bloodstream delivers to nonadjacent target cells. They are this chapter's focus.

Neurotransmitters, released from axon endings of neurons, act swiftly on target cells by diffusing across a tiny synaptic cleft between them. You read about them in Section 34.3. **Local signaling molecules** released by many types of cells alter conditions in local tissues. For instance, some prostaglandins restrict or enhance blood flow to certain tissues. They do this by causing smooth muscle cells in blood vessel walls to contract or relax.

Pheromones, nearly odorless secretions of certain exocrine glands, diffuse through water or air to targets outside the animal body. These hormone-like secretions act on cells of other individuals of the same species and help integrate social behavior. For example, female silk moths secrete bombykol as a sex attractant, and soldier termites secrete alarm signals when ants attack their colony (Chapter 46). Among vertebrates, a vomeronasal organ is a common pheromone detector. Even humans have one, but we don't know much about its function. Human pheromones or other chemical signals do affect sexual behavior. Do they work at the subconscious level, also? For example, do they trigger inexplicable impressions, such as instant good or bad "feelings" about somebody you have just met? Maybe.

Discovery of Hormones and Their Sources

The word "hormone" dates to the early 1900s. W. Bayliss and E. Starling were trying to find out what triggers the secretion of pancreatic juices when food is traveling through the canine gut. As they knew, acids are mixed with food in the stomach, and the pancreas secretes an alkaline solution after the acidic mixture is propelled forward into the small intestine. Was the nervous system or something else stimulating the pancreatic response?

To find an answer, Bayliss and Starling blocked the nerves—but not blood vessels—that serviced the upper

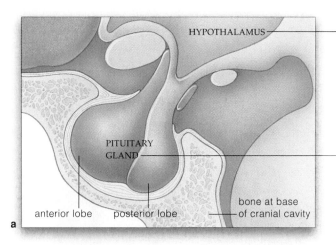

Figure 36.2 (a) A major neural–endocrine control center. The pituitary gland interacts closely with the hypothalamus, a part of the brain that also secretes some hormones. (b) *Facing page:* Overview of the key components of the human endocrine system and of the primary effects of their main hormonal secretions. The system also includes endocrine cells in many organs, such as the liver, kidneys, heart, small intestine, and skin.

small intestine of a laboratory animal. Later, acidic food entered the small intestine. The pancreas still secreted the alkaline solution. More telling, some extracts of cells from the intestinal lining—a *glandular* epithelium—also induced the response. Most likely, glandular cells were the source of the pancreas-stimulating substance.

The substance came to be called secretin. Proof of its existence and mode of action confirmed a centuries-old idea: *The bloodstream picks up internal secretions that can influence the activities of organs inside the body.* Starling coined the word "hormone" for such internal glandular secretions (after *hormon*, meaning to set in motion).

Later on, researchers identified many other kinds of hormones and their sources. Figure 36.2 is a simplified picture of the locations of the following major sources of hormones for the human body. Bear in mind, most vertebrates have the same hormone sources:

Pituitary gland
Adrenal glands (*two*)
Pancreatic islets (*numerous cell clusters*)
Thyroid gland
Parathyroid glands (*in humans, four*)
Pineal gland
Thymus gland
Gonads (*two*)
Endocrine cells of the hypothalamus, stomach, small intestine, liver, kidneys, heart, placenta, skin, adipose tissue, and other organs

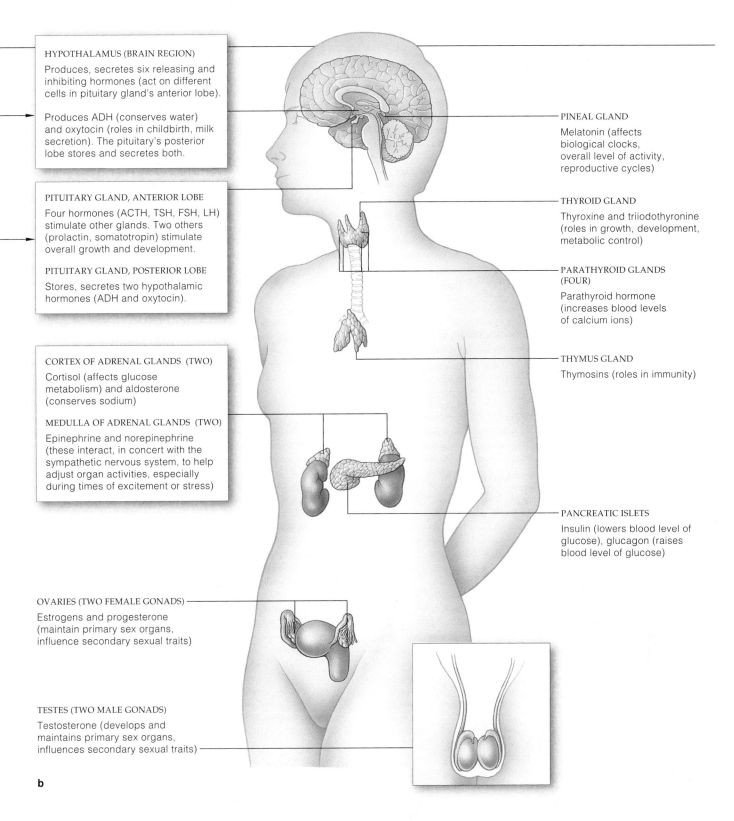

HYPOTHALAMUS (BRAIN REGION)

Produces, secretes six releasing and inhibiting hormones (act on different cells in pituitary gland's anterior lobe).

Produces ADH (conserves water) and oxytocin (roles in childbirth, milk secretion). The pituitary's posterior lobe stores and secretes both.

PITUITARY GLAND, ANTERIOR LOBE

Four hormones (ACTH, TSH, FSH, LH) stimulate other glands. Two others (prolactin, somatotropin) stimulate overall growth and development.

PITUITARY GLAND, POSTERIOR LOBE

Stores, secretes two hypothalamic hormones (ADH and oxytocin).

CORTEX OF ADRENAL GLANDS (TWO)

Cortisol (affects glucose metabolism) and aldosterone (conserves sodium)

MEDULLA OF ADRENAL GLANDS (TWO)

Epinephrine and norepinephrine (these interact, in concert with the sympathetic nervous system, to help adjust organ activities, especially during times of excitement or stress)

OVARIES (TWO FEMALE GONADS)

Estrogens and progesterone (maintain primary sex organs, influence secondary sexual traits)

TESTES (TWO MALE GONADS)

Testosterone (develops and maintains primary sex organs, influences secondary sexual traits)

PINEAL GLAND

Melatonin (affects biological clocks, overall level of activity, reproductive cycles)

THYROID GLAND

Thyroxine and triiodothyronine (roles in growth, development, metabolic control)

PARATHYROID GLANDS (FOUR)

Parathyroid hormone (increases blood levels of calcium ions)

THYMUS GLAND

Thymosins (roles in immunity)

PANCREATIC ISLETS

Insulin (lowers blood level of glucose), glucagon (raises blood level of glucose)

b

Collectively, the body's sources of hormones came to be called the **endocrine system**. The name implies there is a separate control system for the body, apart from the nervous system. (*Endon* means "within" and *krinein* is taken to mean "secrete.") Later, biochemical research and electron microscopy studies revealed that endocrine sources and the nervous system function in intricately connected ways, as you will see shortly.

Integration of cell activities depends on hormones and other signaling molecules. Each type of signaling molecule acts on target cells, which are any cells that have receptors for it and that may alter their activities in response to it.

Components of the endocrine system and certain neurons produce and secrete hormones, which the bloodstream takes up and distributes to nonadjacent target cells.

36.2

SIGNALING MECHANISMS

The Nature of Hormonal Action

Like other signaling molecules, hormones interact with protein receptors of target cells. Their interactions have diverse effects on physiological processes. Some induce a target cell to increase its uptake of glucose, calcium, or some other substance. Others stimulate or inhibit a target in ways that adjust the rates of protein synthesis or metabolism, change cell shape, or alter the structure of proteins or other components in the cytoplasm.

Remember Section 28.5? Hormonal action involves three events: (1) activation of a receptor as it reversibly binds the hormone, (2) transduction of the hormonal signal into a molecular form that can operate inside the cell, and (3) the functional response.

To a large extent, responses to hormones depend on two factors. *First,* different hormones act on different mechanisms in target cells. *Second,* not all types of cells can respond to a given signal. For example, many have receptors for cortisol, so this hormone has widespread effects through the body. By contrast, only certain cells of small tubes in the kidneys have receptors for ADH. This hormone initiates a highly directed response to a decrease in the volume of extracellular fluid.

With these points in mind, let's look briefly at the effects of two main categories of signaling molecules, the steroid and peptide hormones (Table 36.1).

Characteristics of Steroid Hormones

Steroid hormones are lipid-soluble molecules, derived from cholesterol, that are secreted by cells in adrenal glands and in primary reproductive organs, or gonads. Testosterone, one of the sex hormones, is an example.

Think of how testosterone affects secondary sexual traits associated with "maleness." These traits develop normally only if target cells have working receptors for testosterone. In *androgen insensitivity syndrome*, a genetic

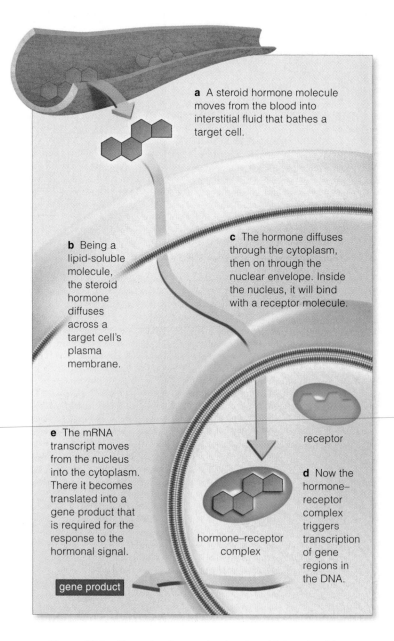

a A steroid hormone molecule moves from the blood into interstitial fluid that bathes a target cell.

b Being a lipid-soluble molecule, the steroid hormone diffuses across a target cell's plasma membrane.

c The hormone diffuses through the cytoplasm, then on through the nuclear envelope. Inside the nucleus, it will bind with a receptor molecule.

receptor

d Now the hormone–receptor complex triggers transcription of gene regions in the DNA.

hormone–receptor complex

e The mRNA transcript moves from the nucleus into the cytoplasm. There it becomes translated into a gene product that is required for the response to the hormonal signal.

gene product

Figure 36.3 Example of a mechanism by which a steroid hormone initiates changes in a target cell's activities.

disorder, receptors are defective. The affected person is male (XY), with testes that secrete testosterone. But the target cells can't respond to the hormone, so secondary sexual traits that do develop are like those of females.

How does a steroid hormone operate? Being lipid soluble, it can diffuse directly across the lipid bilayer of a cell's plasma membrane (Figure 36.3). Once inside the cytoplasm, it commonly moves into the nucleus and binds to a receptor. Some types bind with a receptor in the cytoplasm. The hormone–receptor complex moves into the nucleus and interacts with a specific region of DNA. Different kinds inhibit or stimulate transcription

Table 36.1	Two Main Categories of Hormones

Type	Examples
Steroid and steroid-like hormones	Estrogens (feminizing effects), progestins (related to pregnancy), androgens (such as testosterone; masculinizing effects), cortisol, aldosterone. Thyroid hormones and vitamin D act like steroid hormones.
Peptide hormones	
Peptides	Glucagon, ADH, oxytocin, TRH
Proteins	Insulin, somatotropin, prolactin
Glycoproteins	FSH, LH, TSH

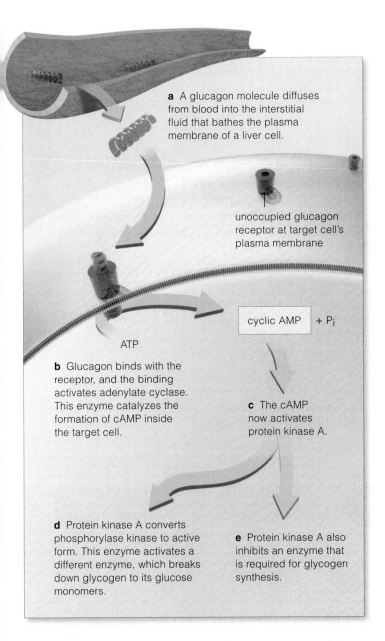

a A glucagon molecule diffuses from blood into the interstitial fluid that bathes the plasma membrane of a liver cell.

unoccupied glucagon receptor at target cell's plasma membrane

ATP

cyclic AMP $+ P_i$

b Glucagon binds with the receptor, and the binding activates adenylate cyclase. This enzyme catalyzes the formation of cAMP inside the target cell.

c The cAMP now activates protein kinase A.

d Protein kinase A converts phosphorylase kinase to active form. This enzyme activates a different enzyme, which breaks down glycogen to its glucose monomers.

e Protein kinase A also inhibits an enzyme that is required for glycogen synthesis.

Figure 36.4 Example of a mechanism by which a peptide hormone initiates changes in a target cell's activities. Here, glucagon binds at a receptor and initiates reactions inside the cell. In this case cyclic AMP, which is one type of second messenger, relays the signal into the cell interior.

Characteristics of Peptide Hormones

Traditionally, **peptide hormones** have been defined as water-soluble signaling molecules with 3 to 180 amino acids. Many peptides, polypeptides, and glycoproteins are in this category. When peptide hormones bind to receptors at the plasma membrane, specific membrane-bound enzyme systems become activated. These systems initiate reactions that lead to a cellular response.

Think of a liver cell. Glucagon, a peptide hormone, acts when the blood glucose level is low. It stimulates that cell to convert glycogen to glucose, which ends up in the blood. Glucagon binds with a plasma membrane receptor that extends into the cytoplasm. Binding causes formation of a **second messenger**—a small molecule in the cytoplasm that relays signals from hormone–receptor complexes at the plasma membrane into the cell. cAMP (cyclic adenosine monophosphate) is this messenger.

An enzyme system at the plasma membrane starts the response to glucagon binding. Adenylate cyclase, an enzyme, is activated. It causes a cascade of reactions by converting ATP to many cyclic AMP molecules (Figure 36.4). These products serve as signals to convert many molecules of another type of enzyme, a protein kinase, to its active form. The kinases activate other enzymes, and so on until a final reaction converts the glycogen stored inside the cell to glucose. In short order, a huge number of molecules execute the hormonal response.

Or think of a muscle cell. It has receptors that bind insulin, a protein hormone. As one of its functions, an insulin–receptor complex induces glucose transporters to move through the cytoplasm and insert themselves into the plasma membrane. The transporters, a type of protein, help the muscle cell take up glucose faster. The signal also activates enzymes that help store glucose.

Bear in mind, there are other hormone categories, including the catecholamines such as epinephrine. Like glucagon, epinephrine binds with membrane receptors and triggers the release of a second messenger, cyclic AMP, that assists in the cellular response.

Hormones work this way: Hormone molecules reversibly bind with a target cell's receptors, and the hormonal signal is transduced into a molecular form that elicits a functional response in the cell. Responses include adjusting specific metabolic reactions and rates of protein synthesis.

Steroid hormones interact with a cell's DNA after entering the nucleus directly or after binding with a receptor in the cytoplasm. Some act by altering membrane properties.

Peptide hormones bind to a membrane receptor. Binding itself is a signal for enzyme-mediated, intracellular events. Often a second messenger inside the cytoplasm relays the signal into the cell interior.

---

of gene regions into mRNA. Translation of such mRNA transcripts results in enzymes and other proteins that can carry out a response to the hormonal signal.

Still other steroid hormones have receptors on the plasma membranes. They modify functions of a target cell by altering the properties of the membrane itself.

Part of a group of genes that specifies receptors for steroid hormones also specifies thyroid hormone and vitamin D receptors. Thyroid hormones and vitamin D are not steroid hormones, but they behave like them.

THE HYPOTHALAMUS AND PITUITARY GLAND

From Section 34.8 you already know the **hypothalamus** is part of the forebrain's centers for homeostatic control of the internal environment, the viscera, and emotional states. Secretory neurons extend down into a slender stalk at its base, then into the lobed, pea-size **pituitary gland**. The hypothalamus and the pituitary interact as a major neural–endocrine control center.

The *posterior* lobe of the pituitary gland secretes two hormones that the hypothalamus produces. Its *anterior* lobe produces and secretes its own hormones. Most of these help control the release of hormones from other endocrine glands (Table 36.2). The pituitary of many vertebrates—*not* humans—has an intermediate lobe as well. In many cases, the third lobe secretes a hormone that governs reversible changes in skin or fur color.

Posterior Lobe Secretions

Figure 36.5*a* shows how hypothalamic neurons extend into the posterior lobe and end next to a capillary bed. The neurons produce antidiuretic hormone (ADH) and oxytocin, then store them in their axon endings. They secrete these hormones into interstitial fluid. Hormone molecules diffuse into capillaries, and the bloodstream distributes them through the body. ADH acts inside the kidneys. Its action adjusts the volume and composition of extracellular fluid in ways that help conserve water when necessary. Oxytocin has roles in reproduction. It triggers uterine contractions during childbirth and also induces milk release when offspring are being nursed.

Anterior Lobe Secretions

ANTERIOR PITUITARY HORMONES Other hypothalamic neurons release hormones next to a capillary bed in the stalk at its base. From there, the hormones flow into a *second* capillary bed, then diffuse into the anterior lobe (Figure 36.6). Their anterior lobe targets are cells that secrete these hormones:

Corticotropin	ACTH
Thyrotropin	TSH
Follicle-stimulating hormone	FSH
Luteinizing hormone	LH
Prolactin	PRL
Somatotropin (growth hormone)	STH (or GH)

All of these hormones have widespread effects on the body. ACTH and TSH orchestrate the secretions from the adrenal glands and thyroid gland, respectively, as described shortly. FSH and LH have essential roles in

Table 36.2 *Hormones Released From the Mammalian Pituitary Gland*

Pituitary Lobe	Secretions	Designation	Main Targets	Primary Actions
POSTERIOR Nervous tissue (extension of hypothalamus)	Antidiuretic hormone	ADH	Kidneys	Induces water conservation as required during control of extracellular fluid volume and solute concentrations
	Oxytocin	OCT	Mammary glands	Induces milk movement into secretory ducts
			Uterus	Induces uterine contractions during childbirth
ANTERIOR Glandular tissue, mostly	Corticotropin	ACTH	Adrenal cortex	Stimulates release of cortisol, an adrenal steroid hormone
	Thyrotropin	TSH	Thyroid gland	Stimulates release of thyroid hormones
	Follicle-stimulating hormone	FSH	Ovaries, testes	In females, stimulates estrogen secretion, egg maturation; in males, helps stimulate sperm formation
	Luteinizing hormone	LH	Ovaries, testes	In females, stimulates progesterone secretion, ovulation, corpus luteum formation; in males, stimulates testosterone secretion, sperm release
	Prolactin	PRL	Mammary glands	Stimulates and sustains milk production
	Somatotropin (or growth hormone)	STH (GH)	Most cells	Promotes growth in young; induces protein synthesis, cell division; roles in glucose, protein metabolism in adults
INTERMEDIATE* Glandular tissue, mostly	Melanocyte-stimulating hormone	MSH	Pigmented cells in skin and other integuments	Induces color changes in response to external stimuli; affects some behaviors

* Present in most vertebrates (not humans). MSH is associated with the anterior lobe in humans.

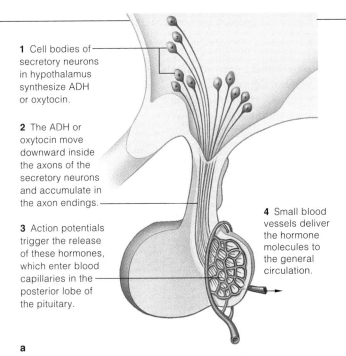

1 Cell bodies of secretory neurons in hypothalamus synthesize ADH or oxytocin.

2 The ADH or oxytocin move downward inside the axons of the secretory neurons and accumulate in the axon endings.

3 Action potentials trigger the release of these hormones, which enter blood capillaries in the posterior lobe of the pituitary.

4 Small blood vessels deliver the hormone molecules to the general circulation.

a

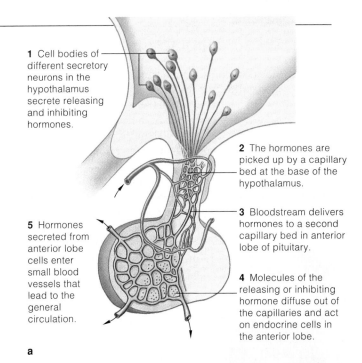

1 Cell bodies of different secretory neurons in the hypothalamus secrete releasing and inhibiting hormones.

2 The hormones are picked up by a capillary bed at the base of the hypothalamus.

3 Bloodstream delivers hormones to a second capillary bed in anterior lobe of pituitary.

5 Hormones secreted from anterior lobe cells enter small blood vessels that lead to the general circulation.

4 Molecules of the releasing or inhibiting hormone diffuse out of the capillaries and act on endocrine cells in the anterior lobe.

a

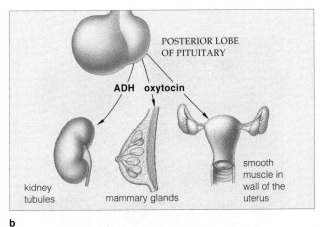

POSTERIOR LOBE OF PITUITARY

ADH   oxytocin

kidney tubules   mammary glands   smooth muscle in wall of the uterus

b

Figure 36.5 (**a**) Functional links between the hypothalamus and the posterior lobe of the pituitary gland. (**b**) Main targets of the posterior lobe secretions.

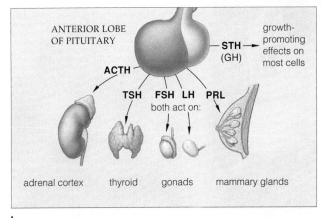

ANTERIOR LOBE OF PITUITARY

ACTH

TSH   FSH LH   PRL
both act on:

STH (GH) growth-promoting effects on most cells

adrenal cortex   thyroid   gonads   mammary glands

b

Figure 36.6 (**a**) Functional links between the hypothalamus and the anterior lobe of the pituitary gland. (**b**) Main targets of the anterior lobe secretions.

gamete formation and hormonal secretions required in sexual reproduction of animals, a topic of Chapter 44. Somatotropin affects metabolism in many tissues and triggers liver secretions that affect growth of bone and soft tissues (Table 36.2 and Figure 36.6*b*). Prolactin acts on different cell types. It is best known for stimulating and sustaining milk production in mammary glands, after other hormones prime the tissues. It also affects hormone production in the ovaries of some species.

ABOUT THE HYPOTHALAMIC TRIGGERS Most of the hypothalamic hormones acting in the anterior lobe of the pituitary are **releasers**. This means they stimulate secretion of hormones from target cells. For example, GnRH (gonadotropin-releasing hormone) brings about secretion of FSH and LH, which are gonadotropins. As

another example, TRH (thyrotropin-releasing hormone) stimulates secretion of thyrotropin. Other hypothalamic hormones acting in the anterior lobe are **inhibitors** of their targets. Somatostatin is one inhibitor. It is a signal to slow down somatotropin and thyrotropin secretion.

The hypothalamus and pituitary gland produce eight kinds of hormones and interact to control their secretion.

The posterior lobe of the pituitary stores and secretes two hypothalamic hormones, ADH and oxytocin, both of which target specific cell types.

The anterior lobe of the pituitary produces and secretes six hormones, ACTH, TSH, FSH, LH, PRL, and STH. These trigger release of other hormones from other glands, with a variety of effects throughout the body.

Abnormal Pituitary Outputs

The body does not churn out enormous quantities of hormone molecules. Two researchers, Roger Guilleman and Andrew Schally, realized this when they isolated the first known releasing hormone. In their four-year attempt to secure TRH, they dissected 500 metric tons of brains and 7 metric tons of hypothalamic tissue from sheep and ended up with only a single milligram of it.

The body depends on those small but significant amounts of hormones. It depends also on their secretion at controlled frequencies. If the controls fail, a hormone will either be undersecreted or oversecreted. The body's form or function will become altered as a result.

Consider *gigantism*, which results from the overproduction of somatotropin. Adults who are affected by this condition are proportionally like average-size people, but they are notably larger. Figure 36.7 shows examples. *Pituitary dwarfism* develops as an outcome of underproduction of somatotropin. Proportionally, the affected adults are like an average person, but they are much smaller (Figure 36.7b).

In some cases, the somatotropin output may become excessive in adults, when their long bones have stopped lengthening. *Acromegaly* follows. In time, cartilage, bone, and other connective tissues in the jaws, feet, and hands thicken abnormally. So do epithelial tissues of the skin, nose, eyelids, lips, and tongue. Figure 36.7c shows an affected female.

Another example: ADH secretion may dwindle or stop entirely when the posterior lobe of the pituitary is damaged, as by a blow to the head. This is one cause of *diabetes insipidus*. Symptoms include excessive excretion of dilute urine as well as life-threatening dehydration. Patients often respond to hormone replacement therapy involving synthetic ADH.

Figure 36.7 Examples of the outcome of abnormalities in the secretion of a hormone. (**a**) Manute Bol, an NBA center, is 7 feet 6–3/4 inches tall, an outcome of excessive secretion of somatotropin (STH) in childhood. (**b**) More examples of the effect of STH on body growth. The male at the center of this photograph is affected by gigantism, which resulted from excessive STH production during childhood. The male at the right is affected by pituitary dwarfism, an outcome of an underproduction of STH in childhood. The male at the left is of average height.

(**c**) Acromegaly, which resulted from excessive production of STH during adulthood. Before this female reached maturity, she was symptom-free.

SOURCES AND EFFECTS OF OTHER HORMONES

Table 36.3 lists hormones from endocrine sources other than the pituitary. The rest of this chapter gives a few examples of their effects and of the controls over their output. As you read, keep the following points in mind.

First, hormones often have interacting effects. One or more may oppose, add to, or prime target cells for another hormone's effects. *Second*, negative feedback mechanisms often control secretion. When a hormone's concentration increases or decreases in a body region, the change triggers events that respectively dampen or stimulate its further secretion. *Third*, a target cell may react differently at different times. Responses depend on the hormone's concentration and on the functional state of the cell's receptors. *Fourth*, environmental cues may be important mediators of hormonal secretion.

Hormone secretion and its effects depend on hormone interactions, feedback mechanisms, variations in the state of target cells, and sometimes environmental cues.

Table 36.3 *Hormone Sources Other Than the Mammalian Hypothalamus and Pituitary Gland*

Source	Secretion(s)	Main Targets	Primary Actions
ADRENAL CORTEX	Glucocorticoids (including cortisol)	Most cells	Promote breakdown of glycogen, fats, and proteins as energy sources; thus help raise blood level of glucose
	Mineralocorticoids (including aldosterone)	Kidney	Promote sodium reabsorption (sodium conservation); help control the body's salt–water balance
ADRENAL MEDULLA	Epinephrine (adrenaline)	Liver, muscle, adipose tissue	Raises blood level of sugar, fatty acids; increases heart rate and force of contraction
	Norepinephrine	Smooth muscle of blood vessels	Promotes constriction or dilation of certain blood vessels; thus helps control the flow of blood volume to different body regions
THYROID	Triiodothyronine, thyroxine	Most cells	Regulate metabolism; have roles in growth, development
	Calcitonin	Bone	Lowers calcium level in blood
PARATHYROIDS	Parathyroid hormone	Bone, kidney	Elevates calcium level in blood
GONADS			
Testes (in males)	Androgens (including testosterone)	General	Required in sperm formation, development of genitals, maintenance of sexual traits, growth, and development
Ovaries (in females)	Estrogens	General	Required for egg maturation and release; preparation of uterine lining for pregnancy and its maintenance in pregnancy; genital development; maintenance of sexual traits; growth, development
	Progesterone	Uterus, breasts	Prepares, maintains uterine lining for pregnancy; stimulates development of breast tissues
PANCREATIC ISLETS	Insulin	Liver, muscle, adipose tissue	Promotes cell uptake of glucose, thus lowers glucose level in blood
	Glucagon	Liver	Promotes glycogen breakdown, raises glucose level in blood
	Somatostatin	Insulin-secreting cells	Inhibits digestion of nutrients, hence their absorption from gut
THYMUS	Thymopoietin	T lymphocytes	Poorly understood regulatory effect on T lymphocytes
PINEAL	Melatonin	Gonads (indirectly)	Influences daily biorhythms, seasonal sexual activity
STOMACH, SMALL INTESTINE	Gastrin, secretin, etc.	Stomach, pancreas, gallbladder	Stimulate activities of stomach, pancreas, liver, gallbladder required for food digestion, absorption
LIVER	Somatomedins	Most cells	Stimulate cell growth and development
KIDNEYS	Erythropoietin	Bone marrow	Stimulates red blood cell production
	Angiotensin*	Adrenal cortex, arterioles	Helps control secretion of aldosterone (hence sodium reabsorption, and blood pressure)
	1,25-hydroxyvitamin D_6* (calcitriol)	Bone, gut	Enhances calcium reabsorption from bone and calcium absorption from gut
HEART	Atrial natriuretic hormone	Kidney, blood vessels	Increases sodium excretion; lowers blood pressure

* Kidneys produce *enzymes* that modify precursors of this substance, which enters the general circulation as an activated hormone.

FEEDBACK CONTROL OF HORMONAL SECRETIONS

What controls hormonal secretion? A look at a few of the endocrine glands listed in Table 36.3 will give you a sense of some of the control mechanisms. Briefly, the hypothalamus, pituitary, or both signal a gland to step up or slow down its secretory activity, so a hormone's concentration changes in the blood or elsewhere. With that change, a feedback mechanism trips into action.

With **negative feedback**, a rise in the concentration of a secreted hormone *inhibits* further secretion. With **positive feedback**, a rising concentration of a secreted hormone *stimulates* further secretion.

Negative Feedback From the Adrenal Cortex

Humans have a pair of adrenal glands, one above each kidney (Figure 36.2b). Some cells of the **adrenal cortex**, an adrenal gland's outer layer, secrete glucocorticoids and other hormones. Glucocorticoids raise the blood level of glucose. One type, cortisol, acts when the body is so stressed that the glucose level declines below a set point that acts as an alarm. It starts a stress response, which a negative feedback mechanism later cuts off.

Figure 36.8 shows what goes on. With a low glucose level, the hypothalamus secretes a releasing hormone, CRH, that makes the anterior pituitary secrete ACTH (corticotropin). ACTH makes cells of the adrenal cortex secrete cortisol. In response to cortisol, liver cells break

down stored glycogen, adipose cells degrade fat, and skeletal muscle cells degrade proteins. The breakdown products—glucose, fatty acids, and amino acids—enter blood. Except for brain cells, cells take up all three, not just glucose, as sources of energy (Section 8.6). And so the blood glucose level rises above the set point. The threat to homeostasis ends. Now the hypothalamus and pituitary issue signals that inhibit secretion of CRH and ACTH, thus stopping further cortisol secretion.

Chronic stress, injury, or illness causes the nervous system to keep this response going. Why? Prolonged inflammation damages tissues, and cortisol suppresses the inflammation. Hence the use of cortisol-like drugs for asthma and other chronic inflammatory disorders.

Local Feedback in the Adrenal Medulla

Some neurons reside in the **adrenal medulla**, the adrenal gland's inner region. Their secretions, epinephrine and norepinephrine, are neurotransmitters in some contexts and hormones in others. Sympathetic nerves deliver hypothalamic signals to the adrenal medulla. Suppose they call for norepinephrine's release. Soon, hormone molecules collect in the synaptic cleft between target cells and axon endings from the nerve. A local negative feedback mechanism now operates at receptors on the axon endings. Excess amounts of norepinephrine bind to the receptors and shut down its further release.

When you are excited or stressed, epinephrine and norepinephrine help adjust blood circulation, and fat and carbohydrate metabolism. They boost heart rate, make arteriole diameters widen or narrow in different regions, and dilate airways to the lungs. These highly controlled events direct more of the total blood volume —hence oxygen flow—to active heart and muscle cells that are demanding a lot of energy. These are features of the *fight–flight response* (Section 34.7).

Cases of Skewed Feedback From the Thyroid

The human **thyroid gland** is located at the base of the neck in front of the trachea, or windpipe (Figures 36.2b and 36.9a–c). Thyroxine and triiodothyronine, its main hormones, have widespread effects. In their absence, many tissues cannot develop normally. Also, metabolic rates of warm-blooded animals depend on them. The importance of feedback control of their secretion comes into sharp focus in cases of abnormal thyroid output.

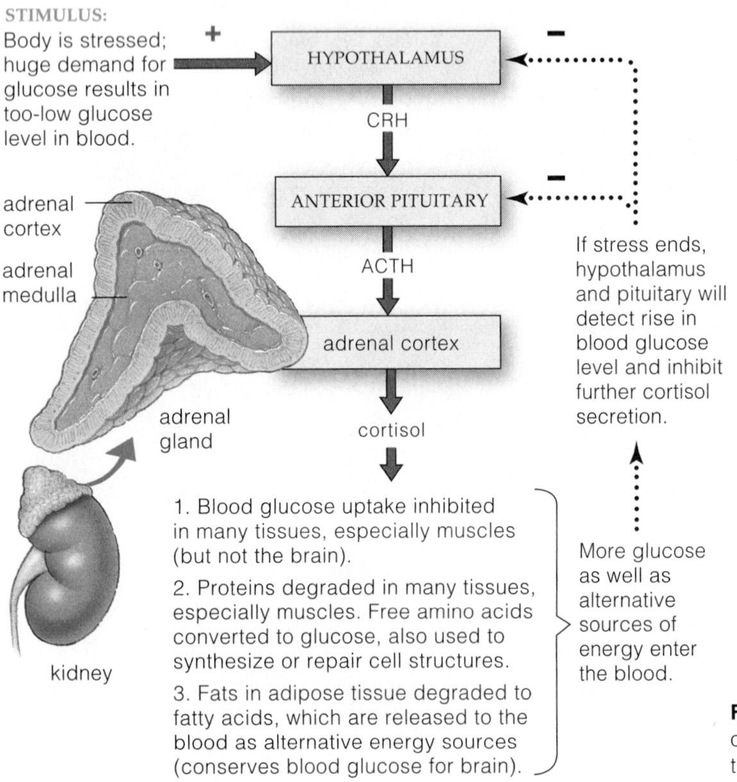

STIMULUS:
Body is stressed; huge demand for glucose results in too-low glucose level in blood.

adrenal cortex

adrenal medulla

adrenal gland

kidney

HYPOTHALAMUS

CRH

ANTERIOR PITUITARY

ACTH

adrenal cortex

cortisol

If stress ends, hypothalamus and pituitary will detect rise in blood glucose level and inhibit further cortisol secretion.

1. Blood glucose uptake inhibited in many tissues, especially muscles (but not the brain).

2. Proteins degraded in many tissues, especially muscles. Free amino acids converted to glucose, also used to synthesize or repair cell structures.

3. Fats in adipose tissue degraded to fatty acids, which are released to the blood as alternative energy sources (conserves blood glucose for brain).

More glucose as well as alternative sources of energy enter the blood.

Figure 36.8 Structure of the human adrenal gland. One gland rests on top of each kidney. The diagram shows a negative feedback loop that governs cortisol secretion from the adrenal cortex.

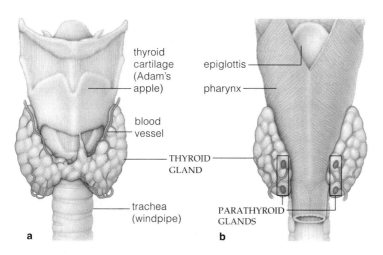

Figure 36.9 (**a**) Human thyroid gland, anterior view. (**b**) Posterior view, showing the location of four parathyroid glands. (**c**) Position of the thyroid in the body. (**d**) A mild case of goiter, displayed by Maria de Medici in the year 1625. A rounded neck was considered to be a sign of great beauty during the late Renaissance. It showed up regularly in parts of the world where iodine supplies were not sufficient for normal thyroid function.

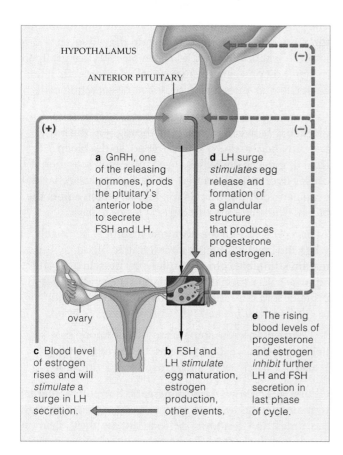

Figure 36.10 Feedback loops to the hypothalamus and the pituitary gland from the ovaries during the menstrual cycle, a recurring reproductive event. Positive feedback triggers egg release from an ovary. Negative feedback after its release prevents release of another egg until the cycle is over.

Example: Thyroid hormones cannot be synthesized without iodide, a form of iodine. Iodine-deficient diets cause one or both lobes of the thyroid gland to enlarge (Figure 36.9*d*). The enlargement, a *simple goiter*, occurs after low blood levels of thyroid hormones cause the anterior pituitary to secrete TSH (thyroid-stimulating hormone). The thyroid grows, but it cannot produce its hormones. So TSH secretion continues as a result of the ongoing, abnormal feedback loop to the pituitary.

Hypothyroidism is the clinical name for low blood levels of thyroid hormones. Affected adults are often sluggish, overweight, intolerant of cold, dry-skinned, confused, and depressed. Simple goiter is no longer common in places where people use iodized salt.

Graves' disorder and other forms of *toxic goiter* result from excessive levels of thyroid hormones in blood, a condition we call *hyperthyroidism*. Symptoms include heat intolerance, irritability, anxiety, tremors, fatigue, difficulty sleeping, protruding eyes, and an irregular, pounding heartbeat. Some of the cases are autoimmune disorders; antibodies wrongly stimulate thyroid cells (Section 39.9). Other cases arise after the thyroid has become inflamed or develops nodules or tumors.

Feedback Control of the Gonads

Gonads are *primary* reproductive organs, which make and secrete gametes and certain sex hormones. Testes (singular, testis) in males and ovaries in females are examples. Testes secrete testosterone; ovaries secrete estrogens and progesterone. These hormones influence secondary sexual traits (as they did for those chimps described earlier), and feedback controls govern their secretion. Figure 36.10 is a preview of the feedback loops from ovaries to the hypothalamus and pituitary during the menstrual cycle, a key topic of Chapter 44.

Feedback mechanisms control secretions from endocrine glands. In many cases, feedback loops to the hypothalamus, pituitary, or both govern the secretory activity.

Negative feedback slows down the further secretion of a hormone. Positive feedback enhances its further secretion.

DIRECT RESPONSES TO CHEMICAL CHANGES

The hypothalamus and pituitary gland are not the main controls over some endocrine glands or cells. Instead, chemical changes in the internal environment directly stimulate or inhibit the hormone secretions.

Secretions From Parathyroid Glands

Humans have four **parathyroid glands** on the posterior surface of the thyroid gland (Figure 36.9*b*). The glands secrete parathyroid hormone (PTH), the key regulator of calcium levels in blood. Calcium ions have roles in muscle contraction, enzyme action, blood clotting, and other activities. PTH is secreted when the calcium level is low. Its secretion slows when the calcium level rises. PTH acts on cells present in the skeleton and kidneys.

PTH induces living bone cells to secrete enzymes that digest bone tissue, and this releases calcium and other minerals into interstitial fluid. Calcium enters the blood, then small tubes in the kidneys. There, PTH enhances calcium reabsorption. It also induces some kidney cells to secrete enzymes that act on blood-borne precursors of an active form of vitamin D_3, a hormone (Section 37.2). The activated form stimulates intestinal cells to absorb much more calcium from the lumen of the gut. In children who have vitamin D deficiency, not enough calcium and phosphorus are absorbed, so rapidly growing bones develop improperly. The resulting bone abnormality, *rickets*, is characterized by bowed legs, a malformed pelvis, and in many cases a malformed skull and rib cage (Figure 36.11).

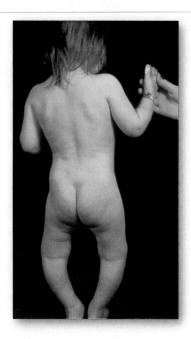

Figure 36.11 A child who is affected by rickets.

Effects of Local Signaling Molecules

Many cells detect changes in the surrounding chemical environment and alter their activity, often in ways that counteract or amplify those changes. The cells secrete various local signaling molecules, the effects of which are confined to the immediate vicinity of change. Target cells take up most signaling molecules so rapidly that few enter the general circulation.

At least sixteen types of prostaglandins, each having a twenty-carbon fatty acid structure with a five-carbon ring, occur in many tissues. Production and secretion of these local signaling molecules rise as local chemical conditions change, with potent results. Some make rings of smooth muscle in arteriole walls constrict or dilate; this helps redirect blood flow through the body. Others function in inflammation, control of stomach acid and intestinal motility, and childbirth. They also function in overall endocrine control. They might have therapeutic use for asthma, ulcers, hypertension, and heart attacks.

Growth factors are good examples of local signaling molecules. They affect transcription of genes required in development. Rita Levi-Montalcini discovered one, a nerve growth factor (NFG) that helps neurons survive and guides their direction of growth in an embryo. We know that in the presence of NFG, immature neurons can survive indefinitely in tissue culture. Without it, they die within a few days.

Secretions From Pancreatic Islets

The pancreas is a gland having exocrine and endocrine functions. Its *exocrine* cells secrete digestive enzymes into the small intestine. Its *endocrine* cells are grouped in 2 million or so small clusters called **pancreatic islets**. Each islet has three types of hormone-secreting cells:

1. *Alpha* cells in the pancreas secrete the hormone glucagon. Between meals, cells throughout the body take up and use glucose from blood, so the blood level of glucose decreases. Then, glucagon secretion causes liver cells to convert the storage polysaccharide glycogen and amino acids to glucose, which enters the blood. *Glucagon raises the level of glucose in blood.*

2. *Beta* cells secrete the hormone insulin. After meals, when the level of glucose circulating in blood is high, insulin stimulates glucose uptake by muscle cells and adipose cells especially. It promotes the synthesis of proteins and fats, and it inhibits protein conversion to glucose. *Insulin lowers the level of glucose in blood.*

3. *Delta* cells secrete somatostatin, a hormone that helps control digestion and absorption of nutrients. It also can block secretion of insulin and glucagon.

Figure 36.12 shows how pancreatic hormones interact to maintain the level of glucose in blood even though the times and amounts of food intake vary. Bear in mind, insulin is the only hormone that causes cells to take up and store glucose in forms that can be swiftly tapped when required. Its crucial role in carbohydrate, protein, and fat metabolism becomes clear when we consider people who cannot produce enough insulin or whose target cells cannot respond to it.

Figure 36.12 Some of the homeostatic controls over glucose metabolism.

Following a meal, glucose enters the blood faster than cells can take it up. The blood glucose level rises, and the chemical change stimulates pancreatic beta cells to secrete insulin. The main targets, liver and muscle cells, use glucose at once and store excess amounts as glycogen. In this way, insulin *lowers* the glucose level in blood.

Between meals, the blood glucose level decreases, and the chemical change stimulates pancreatic alpha cells to secrete glucagon. Target cells with receptors for this hormone convert glycogen back to glucose, which enters the blood. In this way, glucagon *raises* the glucose level in blood.

The nervous system helps control glucose metabolism. For example, as you read earlier, the hypothalamus orders the adrenal medulla to secrete glucocorticoids when the body is being stressed. As one effect, glycogen is broken down to glucose in the liver, and glycogen synthesis slows, in liver and muscle tissue especially.

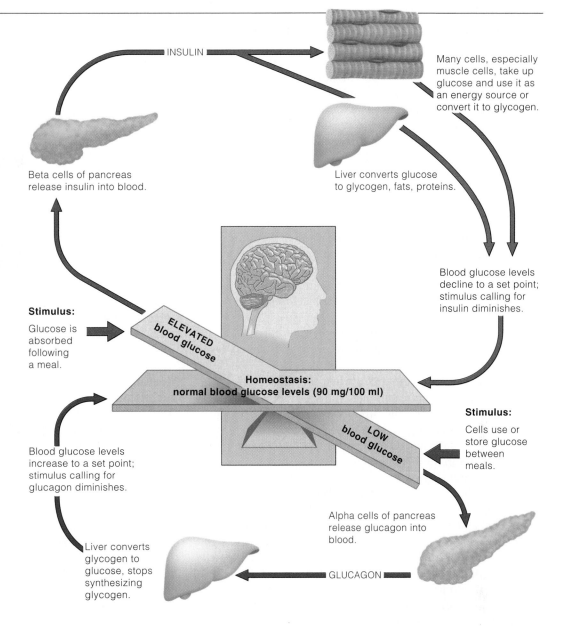

INSULIN

Beta cells of pancreas release insulin into blood.

Many cells, especially muscle cells, take up glucose and use it as an energy source or convert it to glycogen.

Liver converts glucose to glycogen, fats, proteins.

Blood glucose levels decline to a set point; stimulus calling for insulin diminishes.

Stimulus:
Glucose is absorbed following a meal.

ELEVATED blood glucose

Homeostasis: normal blood glucose levels (90 mg/100 ml)

LOW blood glucose

Stimulus:
Cells use or store glucose between meals.

Blood glucose levels increase to a set point; stimulus calling for glucagon diminishes.

Liver converts glycogen to glucose, stops synthesizing glycogen.

Alpha cells of pancreas release glucagon into blood.

GLUCAGON

For example, insulin deficiency may lead to *diabetes mellitus*. In this disorder, excess glucose accumulates in blood, then in urine. Urination becomes excessive, so the body's water–solute balance is disrupted. Affected people become dehydrated and thirsty, abnormally so. Without a steady supply of glucose, their body cells start depleting their own fats and proteins as sources of energy. Weight loss is one outcome. Another is that ketones accumulate in the blood and urine. Ketones are normal acidic products of fat breakdown. When they accumulate, they contribute to water losses and alter the body's acid–base balance. Such imbalances disrupt brain function. In extreme cases, death follows.

In *type 1 diabetes*, the body mounts an autoimmune response against its insulin-secreting beta cells. White blood cells mistakenly identify the beta cells as foreign and kill them. A combination of genetic susceptibility and environmental factors gives rise to this disorder, which is less common than other forms of diabetes but more dangerous in the short term. Usually, symptoms first appear in childhood and adolescence (the disorder is also called juvenile-onset diabetes). Type 1 diabetic patients survive with insulin injections.

In *type 2 diabetes*, insulin levels are close to or above normal, but target cells cannot respond to it. Symptoms usually develop in middle age, for beta cells of affected people produce less insulin over time. Affected people lead normal lives by controlling diet and weight. Some prescription drugs enhance insulin action or secretion.

Secretions from some endocrine glands and endocrine cells are direct homeostatic responses to chemical changes in the internal environment.

HORMONES AND THE EXTERNAL ENVIRONMENT

This last section of the chapter invites you to reflect on a key point. An individual's growth, development, and reproduction begin with genes and hormones, and so does behavior. *But certain environmental factors commonly influence gene expression and hormonal secretion, and they do so in predictable ways.* Later chapters invite analysis of specific environmental effects on animals. For now, it is enough to consider the following examples.

Daylength and the Pineal Gland

Inside the vertebrate brain is the **pineal gland** (Figure 36.2). It secretes melatonin, a hormone, in the absence of light. Thus the blood level of melatonin varies from day to night and through the seasons. In many species, the variations help control growth and development of gonads, the primary reproductive organs. How? Their melatonin is a piece of a **biological clock**, an internal timing mechanism. It helps control reproductive cycles and reproductive behavior.

Think of a hamster in winter, when there are more hours of darkness than in summer. The melatonin level in its blood is high, and this suppresses sexual activity. When daylength is longest, in summer, the blood level is low and hamster sex peaks. Similarly, in the fall and winter, melatonin indirectly suppresses the growth of a male white-throated sparrow's gonads (Figure 36.13*a*). When daylight increases in spring, stepped-up gonadal activity leads to production of hormones that influence singing behavior, as described in Section 46.1. With his distinctive song, the male sparrow defines his territory.

Does melatonin influence human behavior as well? Perhaps. Clinical observations and studies suggest that decreased melatonin levels may trigger **puberty**, the age at which reproductive organs and structures start to mature. For example, some patients who did not have a functional pineal gland entered puberty prematurely.

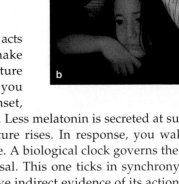

Figure 36.13 (**a**) Male white-throated sparrow, belting out a song that began, indirectly, with an environmentally induced decrease in melatonin secretion from the pineal gland. (**b**) Annie blanketing her winter blues.

Melatonin also acts on neurons that make the body temperature decline and make you drowsy after sunset, when light wanes. Less melatonin is secreted at sunrise, so body temperature rises. In response, you wake up and become active. A biological clock governs the cycle of sleep and arousal. This one ticks in synchrony with daylength. We have indirect evidence of its action from night workers who cannot get to sleep in the morning. The same thing happens to travelers from the United States to Europe who go through a few days of jet lag. Two or three hours past midnight they are wide awake; a couple of hours past noon they are ready for sleep.

Seasonal affective disorder (SAD) hits some people in winter. They become exceedingly depressed, binge on carbohydrates, and have an almost overwhelming need to sleep (Figure 36.13*b*). These "winter blues" might develop when a biological clock is out of sync with the seasonally shorter daylengths. Clinically administered melatonin worsens the seasonal symptoms. Exposure to intense light, which shuts down pineal activity, often leads to dramatic improvement.

Thyroid Function and Frog Habitats

Since 1994, in natural habitats around the world, the number of deformed frogs has been skyrocketing. We are beginning to understand why. In 1999, researchers decided to expose *Xenopus laevis* embryos to Minnesota and Vermont lakewater. Half the samples came from lakes where deformity rates were low. The other half came from "hot spots" with as many as twenty kinds of dissolved pesticides—and with high deformity rates.

When the frog embryos developed into tadpoles, the ones that had been raised in the hot-spot water had bent spines, malformed eyes, and a malformed mouth. Some did not metamorphose into adults (Figure 36.14). The other frog embryos of the sampling were normal.

Thyroid hormones orchestrate much of vertebrate development. For instance, a surge in thyroid hormone output stimulates a tadpole to metamorphose into an

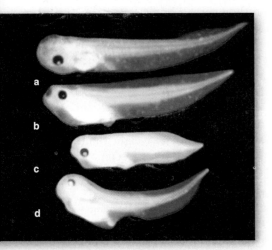

Figure 36.14 *Xenopus laevis* tadpoles showing possible environmental effects on the thyroid, hence on development. (**a**) Tadpole raised in water taken from a lake where there were few deformed frogs. (**b–d**) Three tadpoles raised in water taken from three "hot spot" lakes that had increasingly higher concentrations of dissolved chemical compounds, both natural and synthetic.

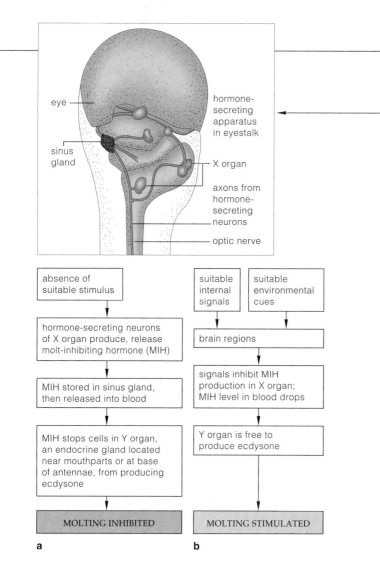

eye

hormone-
secreting
apparatus
in eyestalk

sinus
gland

X organ

axons from
hormone-
secreting
neurons

optic nerve

absence of suitable stimulus		suitable internal signals	suitable environmental cues

hormone-secreting neurons of X organ produce, release molt-inhibiting hormone (MIH)

brain regions

MIH stored in sinus gland, then released into blood

signals inhibit MIH production in X organ; MIH level in blood drops

MIH stops cells in Y organ, an endocrine gland located near mouthparts or at base of antennae, from producing ecdysone

Y organ is free to produce ecdysone

MOLTING INHIBITED

MOLTING STIMULATED

a

b

Figure 36.15 (**a**,**b**) Steps in the hormonal control of molting in crustaceans, including crabs (**c**). The steps differ a bit in insects, which do not use a molt-inhibiting hormone. Rather, stimulation of the insect brain makes certain neurons secrete ecdysiotropin. This hormone induces other neurons to make and secrete still another hormone, which targets ecdysone-producing cells in prothoracic glands. (**d**) This insect, one of the cicadas, is emerging from its old cuticle.

adult. Are conditions in hot-spot habitats interfering with thyroid functions? Maybe. Embryos that developed in hot-spot water displayed few symptoms, or none at all, when they were given extra thyroid hormones.

Other studies correlated some frog deformities with ultraviolet radiation and with infections by a parasitic protistan (Section 26.6). "Chemical soup" habitats that disrupt thyroid function are now added to the list.

Comparative Look at a Few Invertebrates

Although this chapter's focus has been on vertebrates, do not lose sight of the fact that all organisms produce signaling molecules of one sort or another. Let's look at hormonal control of **molting**, a periodic discarding and replacement of a hardened cuticle that otherwise would limit increases in body mass. Molting occurs in the life cycle of all insects, crustaceans, and other invertebrates with thick cuticles (Sections 25.13, 25.15, and 25.17).

Although details vary from group to group, molting is largely under the control of **ecdysone**. This steroid hormone is derived from cholesterol and is chemically related to many crucial vertebrate hormones. In insects and crustaceans, molting glands synthesize and store

ecdysone and then release it for distribution through the body at molting time. Hormone-secreting neurons in the brain seem to control its release. They respond to a combination of environmental cues, such as light and temperature, as well as internal signals (Section 15.5).

Figure 36.15 provides examples of the control steps, which differ in crustaceans and insects. During premolt and molting periods, coordinated interactions among ecdysone and other hormones bring about structural and physiological changes. The interactions make the old cuticle separate from the epidermis and muscles. They induce changes that dissolve inner layers of the cuticle and recycle the remnants. The interactions also trigger changes in metabolism and in the composition and volume of the internal environment. They promote cell divisions, secretions, and pigment formation, all of which go into making the new cuticle. Simultaneously, hormonal interactions control heart rate, muscle action, color changes, and other physiological processes.

Environmental cues, such as changes in light intensity from day to night and seasonal changes in daylength, influence certain hormonal secretions.

SUMMARY　　　　　　　　　　*Gold* indicates text section

1. The cells of complex animals continually exchange substances with the body's internal environment. Their withdrawals and secretions are integrated in ways that ensure cell survival through the whole body. *36.1*

2. Integration of cell activities requires the stimulatory or inhibitory effects of signaling molecules. *36.1, 36.2*

　　a. Signaling molecules are chemical secretions from a cell that adjust the behavior of other, target cells.

　　b. Any cell with molecular receptors for a signaling molecule is its target. Target cells may or may not be adjacent to the cell that sends the signal.

　　c. There are different kinds of signaling molecules. Hormones as well as neurotransmitters, local signaling molecules, and pheromones are the main kinds.

3. In target cells, hormones influence gene activation, protein synthesis, and alterations in existing enzymes, membranes, and other cellular components. Hormones exert physiological effects by interacting with specific protein receptors at the plasma membrane or inside the cytoplasm of target cells. *36.2*

　　a. Steroid hormones enter the target cell's nucleus directly, or after binding with an intracellular receptor, or by binding with plasma membrane receptors.

　　b. Being water soluble, the protein hormones do not cross the bilayer. They bind to membrane receptors at the cell surface. Their effect is exerted with the help of transport proteins as well as second messengers in the cytoplasm, some of which trigger the actual response.

4. The posterior lobe of the pituitary stores and secretes two hypothalamic hormones, ADH and oxytocin. ADH targets kidney cells, thus influencing extracellular fluid volume. Oxytocin acts on cells in mammary glands and the uterus to influence reproductive events. *36.3*

5. Hypothalamic hormones known as releasing and inhibiting hormones control secretions from a variety of cells of the anterior lobe of the pituitary gland. *36.3*

6. The anterior lobe makes and secretes six hormones: ACTH, TSH, FSH, LH, PRL, and STH. These trigger secretion from the adrenal cortex, the thyroid gland, gonads, and mammary glands. They elicit a variety of responses throughout the body. *36.3*

7. Vertebrates have other sources of hormones (e.g., the adrenal medulla; parathyroid, thymus, and pineal glands; pancreatic islets; endocrine cells in the liver, stomach, small intestine, and heart). *36.1, 36.5–36.8*

8. Interactions among hormones, feedback mechanisms, the number and kind of target cell receptors, variations in the state of target cells, and often environmental cues influence hormone secretion and actions. *36.5–36.8*

9. The secretion of local signaling molecules, such as prostaglandins, is a direct response to a change in the localized chemical environment. *36.7*

　　a. In general, secretion of hormones such as insulin and parathyroid hormone can change rapidly when the extracellular concentration of some substance must be homeostatically controlled.

　　b. Hormones such as somatotropin have prolonged, slow, and often irreversible effects, as on development.

10. Environmental cues, such as the change in sunlight intensity from day to night and the seasonal changes in daylength, influence some hormonal secretions. *36.8*

Review Questions

1. Name the endocrine glands typical of most vertebrates and state where each is located in the human body. *36.1*

2. Distinguish among hormones, neurotransmitters, local signaling molecules, and pheromones. *36.1*

3. A hormone molecule binds with a receptor on a plasma membrane. It does not enter the cell. Binding activates a second messenger in the cell, which triggers an amplified response to the hormonal signal. State whether the molecule is a steroid hormone or a peptide hormone. *36.2*

4. Which secretions of the posterior lobe of the pituitary gland have the targets indicated? (Fill in the blanks.) *36.3*

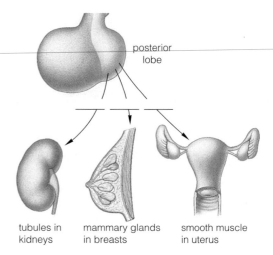

posterior lobe

tubules in kidneys　　mammary glands in breasts　　smooth muscle in uterus

5. Which secretions of the anterior lobe of the pituitary gland have the targets indicated? (Fill in the blanks.) *36.3*

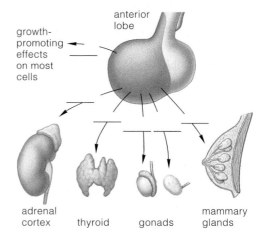

anterior lobe

growth-promoting effects on most cells

adrenal cortex　　thyroid　　gonads　　mammary glands

1. _____ are molecules released from a signaling cell that have effects on target cells.
   a. Hormones
   b. Neurotransmitters
   c. Pheromones
   d. Local signaling molecules
   e. both a and b
   f. a through d

2. Hormones are products of _____ .
   a. endocrine glands
   b. some neurons
   c. exocrine cells
   d. a and b
   e. a and c
   f. a, b, and c

3. Second messengers include _____ .
   a. steroid hormones
   b. protein hormones
   c. cyclic AMP
   d. both a and b

4. ADH and oxytocin are hypothalamic hormones secreted from the _____ lobe of the pituitary gland.
   a. anterior
   b. posterior
   c. intermediate
   d. secondary

5. GnRH is a(n) _____ secreted by hypothalamic neurons.
   a. releasing hormone
   b. inhibiting hormone
   c. corticotropin
   d. somatotropin

6. Which do *not* stimulate hormone secretions?
   a. neural signals
   b. local chemical changes
   c. hormonal signals
   d. environmental cues
   e. All of the above can stimulate secretion.

7. _____ lowers blood sugar levels; _____ raises it.
   a. Glucagon; insulin
   b. Insulin; glucagon
   c. Gastrin; insulin
   d. Gastrin; glucagon

8. The pituitary detects a rising hormone concentration in blood and inhibits the gland secreting the hormone. This is a _____ feedback loop.
   a. positive
   b. negative
   c. long-term
   d. b and c

9. Match the hormone source with the closest description.
   _____ adrenal medulla
   _____ thyroid gland
   _____ parathyroids
   _____ pancreatic islets
   _____ pineal gland
   _____ prostaglandin
   a. affected by daylength
   b. potent local effects
   c. raise blood calcium level
   d. epinephrine source
   e. insulin, glucagon
   f. hormones require iodide

Critical Thinking

1. The zebra offspring being nursed in Figure 36.16 is too young to nourish itself by eating grasses. Its source of nutrients is its mother's milk. Outline how secretions from the hypothalamus and the pituitary gland influence milk production and secretion.

2. In winter, with its fewer daylight hours than in the summer, Maxine became exceedingly depressed, craved carbohydrate-rich foods, and stopped exercising regularly. She put on a great deal of weight. Her doctor diagnosed her condition as *seasonal affective disorder* (SAD), the winter blues. Maxine was advised to purchase a cluster of intense, broad-spectrum lights and sit near them for at least an hour each day. The treatment quickly lifted the cloud of depression. Use your knowledge of the secretory activity of the pineal gland to explain why Maxine's symptoms appeared and why the prescribed therapy worked.

3. Suzannah is affected by *type 1 insulin-dependent diabetes*. One day, after injecting herself with too much insulin, she starts to shake and feels confused. Her doctor recommends a glucagon injection. What caused her symptoms? How would an injection of glucagon help?

Figure 36.16 Female zebra nursing her offspring.

4. In one application of recombinant DNA technology, growth hormone—somatotropin—is now commercially available for treating pituitary dwarfism. Although it's illegal to do so, some athletes use somatotropin instead of anabolic steroids. (Refer to Section 37.12, *Critical Thinking* question 4.) Why? Somatotropin can't be detected by drug test procedures currently employed in sports medicine. Explain how athletes might believe this hormone can improve their performance.

5. *Osteoporosis* is a condition in which loss of calcium results in thin, brittle bones. Combined with other treatments, vitamin D_3 injections are sometimes recommended. Explain why.

Selected Key Terms

adrenal cortex 36.6
adrenal medulla 36.6
biological clock 36.8
ecdysone 36.8
endocrine system 36.1
gonad 36.6
hormone, animal 36.1
hypothalamus 36.3
inhibitor (hypothalamic) 36.3
local signaling
   molecule 36.1
molting 36.8
negative feedback 36.6

neurotransmitter 36.1
pancreatic islet 36.7
parathyroid gland 36.7
peptide hormone 36.2
pheromone 36.1
pineal gland 36.8
pituitary gland 36.3
positive feedback 36.6
puberty 36.8
releaser (hypothalamic) 36.3
second messenger 36.2
steroid hormone 36.2
thyroid gland 36.6

Readings

Goodall, J. 1986. *The Chimpanzees of Gombe.* Cambridge, Massachusetts: Belknap Press of Harvard University Press.

Goodman, H. 1994. *Basic Medical Endocrinology.* Second edition. New York: Raven Press.

Hadley, M. 1995. *Endocrinology.* Fourth edition. Englewood Cliffs, New Jersey: Prentice-Hall.

Raloff, J. 2 October 1999. "Thyroid Linked to Some Frog Defects." *Science News* 156:212.

Sherwood, L. 2001. *Human Physiology.* Fourth edition. Belmont, California: Wadsworth.

PROTECTION, SUPPORT, AND MOVEMENT

Of Men, Women, and Polar Huskies

In 1989 Will Steger and his dog-sled team walked on ice for seven months, enduring temperatures of −113°F and a blizzard that lasted for more than seven weeks. They crossed Antarctica, all 6,023 kilometers (3,741 miles) of it. In 1995 this legendary polar explorer set out with four men, two women, and thirty-three sled dogs to cross 3,220 kilometers of the Arctic Ocean in one season. Ice blankets this northernmost ocean in winter and thins treacherously during the spring thaw. The sled dogs stayed with the team for two-thirds of the journey. Steger flew them out only when the team had to cross too much melting ice and switched to using canoes.

To Steger's mind, polar huskies were the heroes of the crossings, the members of the team that worked hardest and pulled all the weight (Figure 37.1). The traits of this mixed breed were modified through years of artificial selection among Canadian and Greenland huskies (bred for size and strength), Siberian huskies (bred for intelligence), and Alaskan racing dogs (bred for spirit and endurance).

A husky's leg bones are sturdy yet lightweight. Its forelegs move freely, thanks to a deep but not too broad rib cage. You won't see the massive muscles of its hind legs in sprinting greyhounds or cheetahs. They are the muscles of a load-pulling, long-distance runner. The husky has tough, calloused foot pads—cushions against sharp ice and frozen rock. Like many other mammals, it has a fur coat with underhair: a dense, soft, insulative layer that traps heat. Its coarser, longer, slightly oily guard hairs protect the underhair from wear and tear. On winter nights, a husky gets comfortable and covers its nose with its furry tail, oblivious of drifting snow.

Steger and his teammates, Victor Boyarsky, Julie Hanson, Martin Hignell, Paul Pregont, and Takako Takano, did not even approach the polar husky's stamina and built-in protection against the elements. Long before the polar crossings, the team followed a regimen of diet and exercise to put their arm and leg muscles in peak condition. Human legs are adapted for long-distance walking, not load-pulling motion.

Figure 37.1 In Ely, Minnesota, Will Steger and his polar huskies warming up for an Arctic crossing.

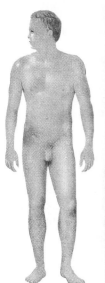

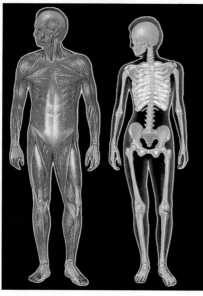

Figure 37.2 From left to right, overview of the human integumentary, muscle, and skeletal systems.

And our skin can't withstand bitter cold. Lacking the fur coat of mammals that evolved in polar climates, team members used clothes that could insulate and protect them from cold without restricting movement. From this perspective, it was human ingenuity that let humans keep company with polar huskies, which are supremely adapted for the challenges of life on ice.

With this chapter, we turn to three systems that give the body of most animals its superficial features, shape, and capacity to move. We call them the integumentary system, muscle system, and skeletal system. Figure 37.2 shows examples from a representative vertebrate.

Animals ranging from worms to humans have an outer covering, or **integument** (after the Latin *integere*, meaning to cover). As you saw in Chapter 25, most integuments are tough barriers against outside threats. For example, the roundworms and insects, crabs, and other arthropods have a chitin-hardened cuticle. In this chapter the story picks up with the evolution of vertebrate skin, including its specialized structures.

Also, regardless of the species, movement of the body or parts of it requires contractile cells and some medium or structure against which contractile force is applied. Among different animal groups, hydrostatic skeletons, exoskeletons, and endoskeletons receive the applied force, as you will see shortly.

Key Concepts

1. Nearly all animals have an integument, some type of skeleton, and muscles. An integument is the outer covering of the animal body. Vertebrate skin is a prime example.

2. Skin protects the body from abrasion, ultraviolet radiation, bacterial attack, and other environmental assaults. It also contributes to overall body function, as when it helps control moisture loss.

3. Three categories of skeletal systems are common in the animal kingdom. We call them hydrostatic skeletons, exoskeletons, and endoskeletons. Each has body fluids or structural elements, such as bones, against which a contractile force can be applied.

4. Bones are collagen-rich, mineralized organs. They function in movement, protection and support of soft organs, and mineral storage. Blood cells form only in certain bones. Ligaments or cartilage keep adjacent bones together by bridging the joint between them. Skeletal muscles are attached to bones at tendons.

5. Many responses to changes in external and internal conditions involve muscles that move the body or parts of it. In response to suitable stimulation, the cells of muscle tissue contract, or shorten.

6. Smooth muscle and cardiac muscle are responsible for the motions of internal organs. Skeletal muscle helps move the body's limbs and other structural elements and maintain their spatial positions.

7. In each skeletal and cardiac muscle cell, many threadlike structures called myofibrils are divided into sarcomeres. The sarcomere is the basic unit of contraction. It has parallel arrays of actin and myosin filaments. ATP-driven interactions between the actin and myosin shorten the sarcomeres of a muscle and collectively account for its contraction.

37.1

EVOLUTION OF VERTEBRATE SKIN

Let's go back in time to the Cambrian seas, when jawed predatory fishes were evolving. Some prey species had heavy armor-plated integuments. Although protective, armor plates do not lend themselves to moving fast or making hairpin turns. Among many of the bony fishes, a thinner, layered integument—skin—evolved.

Vertebrate skin consists of an outer **epidermis** (one or more epithelial sheets) and an underlying **dermis** of dense connective tissue. The scales of most bony fishes originate with differentiated cells inside the dermis. Species with no scales, such as catfish, commonly have a thick, mucus-coated skin as a deterrent to predation.

We can identify two evolutionary trends among the vertebrates that moved onto dry land: keratinization and the emergence of recessed glands with ducts to the skin's surface. First in amphibians and then in reptiles, birds, and mammals, more and more keratinocytes and melanocytes developed in the outer epidermal layers. **Keratinocytes** are cells that produce the water-resistant protein keratin. **Melanocytes** are the cells that produce and then give up melanin to keratinocytes. Melanin, a brownish-black pigment, is one of the body's barriers to harmful ultraviolet radiation (UV) from the sun.

Let's focus now on human skin, the largest organ of your body. It can stretch, conserve body water, and fix small cuts or burns. Skin cells also make a precursor for the vitamin D hormone that is important in calcium metabolism. When the nervous system directs blood flow and metabolic heat to and from skin, this helps adjust body temperature. Sensory receptor endings in skin help the brain assess events on the outside. And some of its cells function in defense (Section 37.2).

The epidermis is a stratified epithelium having an abundance of cell junctions and no extracellular matrix

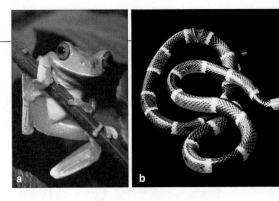

Figure 37.4 (**a**) Tree frog and (**b**) coral snake. Their skin colors arise from melanocytes and chromatophores. Some chromatophores make yellow, orange, and red pigments. Others make colorless crystals that reflect or refract light, which we perceive as iridescent blue to gold or white.

(Section 33.1). Continual mitotic cell divisions rapidly push epidermal cells from deeper layers to the sheet's free surface. Because of wear and tear at the surface, along with pressure exerted by the growing cell mass, the older cells are flattened and dead by the time they reach the outer layers (Figure 37.3). The oldest cells are continually abraded off or flake away.

The dermis, mainly a dense connective tissue, has stretch-resisting elastin fibers and supportive collagen fibers. Blood and lymph vessels, and sensory receptors, thread through it. The hypodermis, a continuous layer beneath skin, anchors the dermis to structures below it (Figure 37.3). It consists of loose connective tissue and adipose tissue that has insulative and cushioning roles.

Skin coloration ranges from very dark to very light owing to differences in the distribution and metabolic activity of melanocytes. In pale skin, little melanin is produced, so the pigment hemoglobin inside red blood cells is not masked. The skin often looks pink because hemoglobin's color shows through thin-walled blood

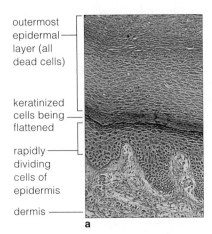

outermost epidermal layer (all dead cells)

keratinized cells being flattened

rapidly dividing cells of epidermis

dermis

a

Figure 37.3 (**a**) Section through human skin. Such thickened skin is typical of the palms and soles. (**b**) Sketch of the skin structure typical of body parts with hair, oil glands, and sweat glands.

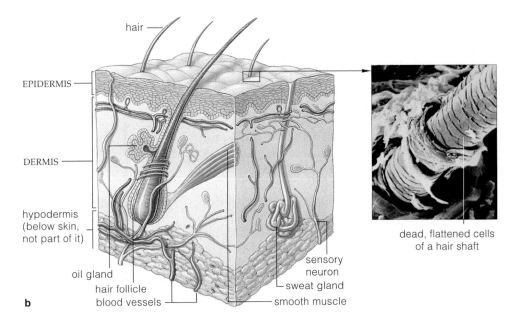

hair

EPIDERMIS

DERMIS

hypodermis (below skin, not part of it)

oil gland

hair follicle

blood vessels

sensory neuron

sweat gland

smooth muscle

b

dead, flattened cells of a hair shaft

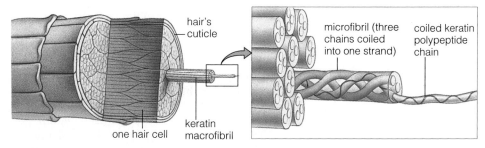

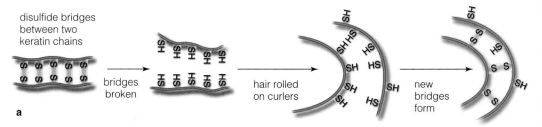

Figure 37.5 Structure of hair. Dead, flattened hair cells form a tubelike cuticle around a hair shaft. These cells arise from modified skin cells and synthesize polypeptide chains of the protein keratin. Disulfide bridges link three chains into fine fibers, which are bundled into larger, cable-like fibers. The fibers nearly fill the cells, which in time die off.

Figure 37.6 (a) Curly or straight hair? That depends on disulfide bridges between keratin chains. Bridges break when exposed to chemicals. Their keratin chains are held in new positions, either by flattening out hair or rolling it around curlers. Exposure to a different chemical causes new disulfide bridges to form between different sulfur-bearing amino acids. The displaced bonding locks the chains in new positions. That is how, many women, including actress Nicole Kidman (b,c), straighten their naturally curly hair.

vessels and the epidermis. Carotene, a yellow-orange pigment, also contributes to skin color. Unlike humans, other vertebrates have many chromatophores that color their skin (Figure 37.4).

Human skin has a number of exocrine glands. As in other mammals, ducts extend from the skin surface to mammary glands, the main mass of which is recessed below the dermis. Another example: Fluid secreted by sweat glands is 99 percent water, with dissolved salts, traces of ammonia, vitamin C, and other substances. Sympathetic nerves control 2.5 million sweat glands in human skin. Secretions from those in the palms, soles, forehead, and armpits help lower body temperature. Secretions from others increase during stress, pain, and sexual foreplay, and prior to menstruation.

Except on the palms and soles, human skin contains oil glands (or sebaceous glands). These lubricate and soften hair and skin. Their secretions kill many surface bacteria. When bacteria do infect oil gland ducts, they can cause acne, an inflammation of skin.

Each **hair**, a flexible structure of mostly keratinized cells, has a root embedded in skin and a shaft above its surface (Figure 37.3). Cells divide near the root's base, are pushed upward, then flatten and die. Flattened cells of the shaft's outer layer overlap, like roof shingles. If mechanically abused, the cells frizz as "split ends." On average, the human scalp includes about 100,000 hairs,

although genes, nutrition, and hormones influence hair growth and density. Protein deficiency causes hair to thin (amino acids are necessary for keratin synthesis). So do high fever, emotional stress, and excess vitamin A intake. When the body produces abnormal amounts of testosterone, *hirsutism*, or excessive hairiness, might be one result. This hormone influences patterns of hair growth and other secondary sexual traits.

And remember how protein structure and function pervade the world of life? Keratin is one of the fibrous proteins used in constructing scales, horns, hooves, beaks, claws, nails (flattened claws), and that uniquely mammalian structure, hair. Figure 37.5 indicates how three polypeptide chains form fibrils that are bundled into cablelike fibers. The keratinocytes become packed with the fibers, die off, flatten, and then form a cuticle around a hair shaft. Disulfide bridges hold together the three keratin chains in each fibril. Figure 37.6 shows what happens when hair is curled or straightened.

The vertebrate integument, skin, is composed of an outer epidermis (one or more sheets of epithelium) and an inner dermis. Skin is a protective, pliable covering.

Skin's keratinized, melanin-shielded epidermal cells help the body conserve water, minimize damage by ultraviolet radiation, and resist mechanical stress.

Sunlight and Skin

THE VITAMIN CONNECTION   Sit outside in the sun and you give some of your skin's epidermal cells a chance to make cholecalciferol, a precursor of vitamin D. **Vitamin D** is a generic name for steroid-like compounds that help the body absorb calcium from food. Expose one type of skin cell to sunlight and it produces vitamin D from a precursor related to cholesterol, then secretes it. This is a hormone-like action, which means skin acts like an endocrine gland when exposed to sunlight. The vitamin D reaches the kidneys, where it is converted to an active form (vitamin D_3) that will help absorptive cells of the intestinal lining take up dietary calcium.

As you know, the sun's rays also contain ultraviolet (UV) wavelengths, which are harmful. Dark skin is able to block UV radiation while still making vitamin D. It also blocks the breakdown of a B vitamin, **folate**. Among other things, folate happens to be essential for proper embryonic development, hence for reproduction.

Humans first evolved beneath the intense sun of the African savanna. Skin alone supplied them with enough vitamin D. What happened when some populations left the tropics? What happened after they moved to higher altitudes or into caves and bundled up in clothing? Then, dietary sources of vitamin D became more important. And selection pressure for embryo-protecting dark skin probably eased up. Nina Jablonksi and George Chaplin compared the extent to which skin reflects rather than absorbs light with satellite data on annual UV levels. The researchers conducted studies in more than fifty countries. They found a match between increasingly lower UV levels and increasingly lighter skin.

SUNTANS AND SHOE-LEATHER SKIN   When exposed to UV radiation, the melanocytes in skin are stimulated to make melanin. Production accelerates slowly, then peaks about ten days after the initial exposure. Before then, unprotected skin can become mildly to severely burned.

Ongoing exposure produces the "tan" that so many light-skinned people covet (Figure 37.7). Dark-skinned people have better initial protection from UV radiation. Yet even in naturally dark skin, prolonged UV exposure causes elastin fibers in connective tissue of the dermis to clump together. Skin loses its resiliency as a result. In time, it starts looking like old shoe leather.

Skin will age anyway. As we grow older, epidermal cells divide less often. Skin thins and is more vulnerable to injury; glandular secretions that once kept it soft and moistened dwindle. Collagen and elastin fibers in the dermis break down and become sparser, so skin loses its elasticity and wrinkles deepen. People accelerate the aging process and look older faster in several ways, as by continually exposing skin to dry winds or tobacco smoke—as well as by excessive tanning.

SUNLIGHT AND THE FRONT LINE OF DEFENSE   Besides having cells that make melanin and keratin, skin has cells that help protect the body against pathogens and against cancer. The two guardians of the epidermis are Langerhans and Granstein cells.

Langerhans cells are phagocytes that develop in bone marrow, then take up stations in skin. After engulfing virus particles or bacteria, they display molecular alarm signals at their plasma membrane. The signals mobilize the immune system. UV radiation damages these cells. This may be why sunburns commonly trigger *cold sores*, small, painful blisters that announce the recurrence of a *Herpes simplex* infection. Nearly everyone harbors the *H. simplex* virus. It hides in the face, inside a ganglion. (A ganglion, recall, is a cluster of neuron cell bodies.) Sunburns and other stress factors can reactivate the virus. Virus particles move down the neurons to the axonal endings in skin, where they infect epithelial cells and cause the eruptions.

When UV radiation damages the Langerhans cells, it weakens the body's first lines of defense against injury and attack. It may activate proto-oncogenes and initiate the cancerous transformation of skin cells. (You read about this in the Chapter 15 introduction and Section 15.4.) Skin cancers grow rapidly and spread to adjacent lymph nodes unless they are surgically removed.

By interacting with white blood cells, **Granstein cells** put the brakes on immune responses in skin. They issue suppressor signals and thereby help keep the responses from spiraling out of control. Their roles are not fully understood, but researchers know that these cells are less vulnerable than Langerhans cells to the damaging effects of ultraviolet radiation.

Figure 37.7   Demonstration of how shoe-leather skin forms.

TYPES OF SKELETONS

Operating Principles for Skeletons

Nearly all animals move by activating, contracting, and relaxing muscle cells. But muscle cells alone can't move the body. *Muscles require the presence of some medium or structural element against which the force of contraction is applied.* A skeletal system fulfills this requirement.

Three types of animal skeletons are common. With a **hydrostatic skeleton**, muscles work against an internal body fluid and redistribute it within a confined space. The confined fluid resists compression, like a waterbed. With an **exoskeleton**, rigid, *external* body parts such as a cuticle or a shell receive the applied force of muscle contraction. With an **endoskeleton**, *internal* body parts receive the applied force of muscle contraction.

Examples From the Invertebrates

We see hydrostatic skeletons among the soft-bodied invertebrates, such as a sea anemone with a soft body and saclike gut (Figure 37.8). When longitudinal muscles in its body wall contract (shorten) and the radial ones relax (lengthen), its body is squat. When longitudinal muscles relax and radial muscles contract (thus forcing fluid out of the gut), the body lengthens and becomes upright. An earthworm, too, has a hydrostatic skeleton. It uses the coelomic chambers of its segmented body (Section 25.12). Contracting and relaxing the fluid-filled segments in sequence move the body forward. The body thrashes forward and back by alternating contractions of muscles on one side and then the other.

Figure 37.9 Jumping spiders. When these spiders jump, their hind legs extend hydraulically as blood surges into them under high pressure.

Recall how an arthropod has a hinged exoskeleton? Its hard parts are moved like levers by sets of muscles attached to them (Section 19.4). This is especially the case for a winged insect's cuticle, which extends over certain body segments and the gaps between segments. Being pliable at the gap, it functions like a hinge when contracting muscles raise a wing or the body part to which that wing is attached. With this arrangement, small contractions bring about large movements:

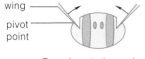

wing
pivot point

Dorsal-ventral muscles contract; exoskeleton pops out, wings move up.

Exoskeleton pops back, muscles relax, wings move down.

A final example: Like other arthropods, a jumping spider uses a hinged exoskeleton for movements. But it also uses body fluids to transmit force when leaping at prey (Figure 37.9). Spiders have an open circulatory system, and their heart pumps blood into body tissues. By contracting its muscles, a spider makes blood surge into hind leg spines. (It's like squeezing a rubber glove partly filled with water so that the limp fingers become erect.) And it is an effective use of hydraulic pressure. *Hydraulic* refers to the fluid pressure within a tube.

resting position, which typically is assumed at low tide when currents are not delivering food

feeding position

Figure 37.8 Hydrostatic skeleton of the sea anemone. Contractile cells in the body wall run longitudinal to the main body axis and radially around the gut. *Left:* Radial cells are relaxed; longitudinal ones are contracted. *Right:* The radial cells are contracted and the longitudinal ones are relaxed. The body extends upward, to its feeding position.

Animal skeletons have structural elements or body fluids against which the force of contraction may be applied.

EVOLUTION OF THE VERTEBRATE SKELETON

From notochord to vertebral column, from supports for gills to jaws, from structural elements inside lobed fins to limbs—Section 26.2 introduced these pivotal evolutionary trends among vertebrates that invaded land. Many other modifications helped those lineages make the transition to life in a new medium—air.

Reflect on the structural responses to body weight deprived of water's buoyancy. For example, the pelvic and pectoral girdles of fishes function as a stable base for moving fins, which propel, guide, and stabilize the body in water. Among early four-legged vertebrates, the girdles transferred body weight to the limbs and had more surface area to which limb muscles could become attached. Limbs that became positioned closer to the body mass helped hold it upright and assisted in the thrusts necessary for locomotion on land. The cage of hardened bones attached to the backbone protected the heart, lungs, and other soft internal organs from collapsing under the weight of the body.

Reflect now on the sweep through time in Figure 37.10, which moves from the cartilaginous skeleton of a fish to the bony endoskeletons that evolved among vertebrates on land. And use it to identify conserved features in your own skeleton of 206 bones. Notice, in Figure 37.11, the pectoral girdle (at the shoulders), pelvic girdle (at the hips), and the paired arms, hands, legs, and feet. This is the *appendicular* portion of the human skeleton, the legacy of some ancient tetrapod. Notice the flat shoulder blades and slender collarbone of the pectoral girdles. It's easy to think "fish out of water" when someone falls hard on an outstretched arm and dislocates a shoulder blade or fractures a collarbone, the bone broken most often.

What of the *axial* portion of the human skeleton? Its jaws and other skull bones, twelve pairs of ribs, breastbone, and twenty-six vertebrae are legacies of early craniates and jawed vertebrates. Those **vertebrae** (singular, vertebra), bony segments of a backbone, extend from the base of the skull to the pelvic girdle. What does your vertebral column do now? Its bony parts offer attachment sites for paired muscles and form a canal for the spinal cord. And that column transmits your torso's weight to the lower limbs of a two-legged walker (Sections 26.11 and 26.12).

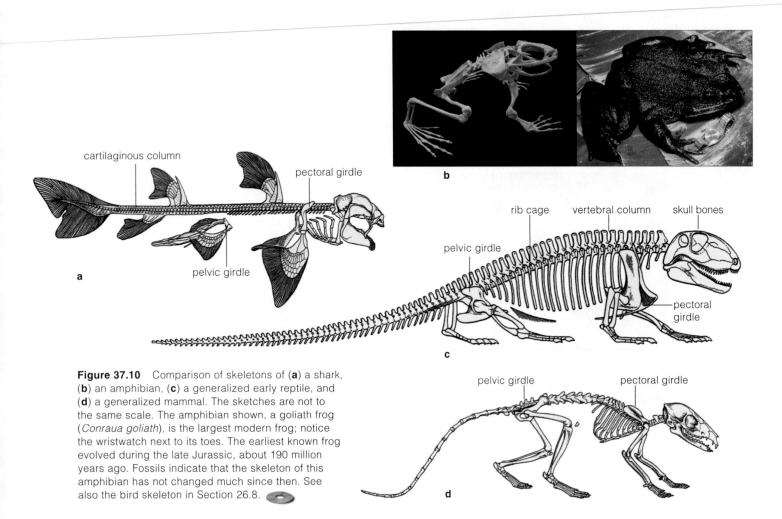

Figure 37.10 Comparison of skeletons of (**a**) a shark, (**b**) an amphibian, (**c**) a generalized early reptile, and (**d**) a generalized mammal. The sketches are not to the same scale. The amphibian shown, a goliath frog (*Conraua goliath*), is the largest modern frog; notice the wristwatch next to its toes. The earliest known frog evolved during the late Jurassic, about 190 million years ago. Fossils indicate that the skeleton of this amphibian has not changed much since then. See also the bird skeleton in Section 26.8.

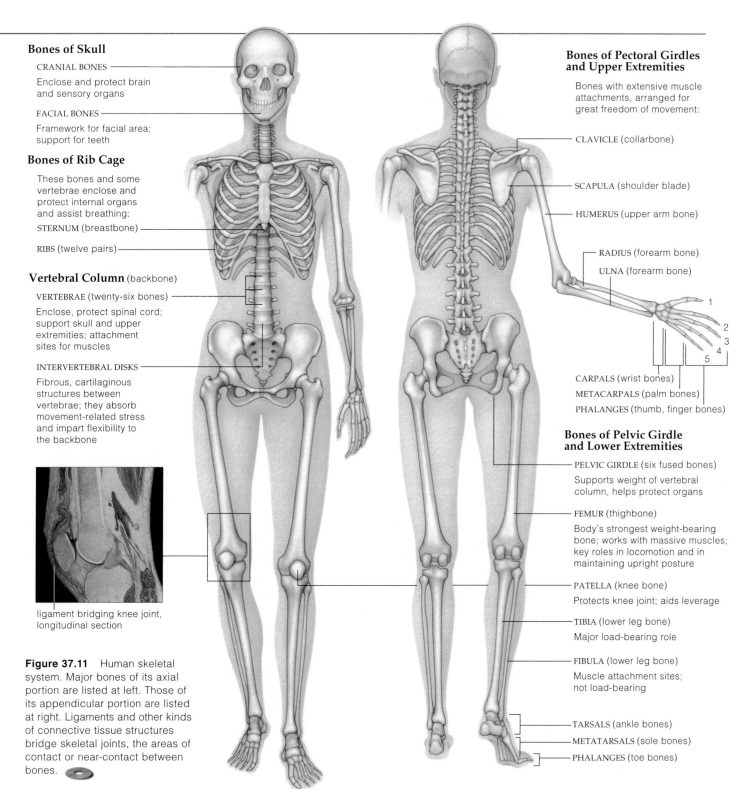

Bones of Skull

CRANIAL BONES

Enclose and protect brain and sensory organs

FACIAL BONES

Framework for facial area; support for teeth

Bones of Rib Cage

These bones and some vertebrae enclose and protect internal organs and assist breathing:

STERNUM (breastbone)

RIBS (twelve pairs)

Vertebral Column (backbone)

VERTEBRAE (twenty-six bones)

Enclose, protect spinal cord; support skull and upper extremities; attachment sites for muscles

INTERVERTEBRAL DISKS

Fibrous, cartilaginous structures between vertebrae; they absorb movement-related stress and impart flexibility to the backbone

ligament bridging knee joint, longitudinal section

Figure 37.11 Human skeletal system. Major bones of its axial portion are listed at left. Those of its appendicular portion are listed at right. Ligaments and other kinds of connective tissue structures bridge skeletal joints, the areas of contact or near-contact between bones.

Bones of Pectoral Girdles and Upper Extremities

Bones with extensive muscle attachments, arranged for great freedom of movement:

CLAVICLE (collarbone)

SCAPULA (shoulder blade)

HUMERUS (upper arm bone)

RADIUS (forearm bone)

ULNA (forearm bone)

CARPALS (wrist bones)

METACARPALS (palm bones)

PHALANGES (thumb, finger bones)

Bones of Pelvic Girdle and Lower Extremities

PELVIC GIRDLE (six fused bones)

Supports weight of vertebral column, helps protect organs

FEMUR (thighbone)

Body's strongest weight-bearing bone; works with massive muscles; key roles in locomotion and in maintaining upright posture

PATELLA (knee bone)

Protects knee joint; aids leverage

TIBIA (lower leg bone)

Major load-bearing role

FIBULA (lower leg bone)

Muscle attachment sites; not load-bearing

TARSALS (ankle bones)

METATARSALS (sole bones)

PHALANGES (toe bones)

Our hominid ancestors started walking upright more than 4 million years ago, and the change resulted in a pronounced S-shaped curve in the backbone. In between the bony segments are **intervertebral disks**, cartilaginous shock absorbers and flex points. But the bones and disks are stacked against gravity. A severe or rapid shock can force a disk to slip out of place or rupture—become *herniated*—and cause chronic pain.

No longer aquatic, no longer tetrapods, we must live with the costs as well as the benefits of getting around in the world on two legs.

The skeleton of land vertebrates is a collection of earlier adaptations to life in water and later modifications to a life deprived of water's buoyancy.

A CLOSER LOOK AT BONES AND JOINTS

By definition, **bones** are complex organs that function in movement, support, protection, mineral storage, and formation of blood cells (Table 37.1). Bones that support and anchor skeletal muscles help maintain or change the positions of body parts. Some form hard compartments that enclose and protect the brain, the lungs, and other internal organs. Bones are reservoirs for calcium and phosphorus ions. Continual deposits and withdrawals of ions from bone help maintain blood levels of calcium and phosphorus, thus supporting metabolic activities. Some bones (not all) are sites of blood cell formation.

Human bones range in size from middle ear bones smaller than lentils to clublike thighbones, or femurs (Figure 37.12). They have long, flat, short or cubelike, or irregular shapes. All consist of connective tissues—bone tissue especially—and epithelia. Bone tissue consists of mature bone cells, called **osteocytes**, as well as collagen fibers in a calcium-hardened ground substance.

The shaft and both ends of thighbones have *compact* bone tissue, which resists mechanical shock. This tissue develops as Haversian systems: many thin, cylindrical, dense layers around interconnecting canals for blood vessels and nerves that service living bone cells. Also in the bone ends and shaft is *spongy* bone tissue, which imparts strength but does not weigh much. Abundant spaces make it look spongy but its flattened parts are firm. **Red marrow**, a major site of blood cell formation, fills spaces in some bones, such as the breastbone. The cavities in most mature bones hold **yellow marrow**, a largely fatty region that converts to red marrow and makes more red blood cells in times of severe blood loss.

Bone Formation and Remodeling

Bone tissue first forms in an embryo, where many bones are constructed on cartilage models. On these models, **osteoblasts**—the bone-forming cells—secrete organic substances that become mineralized. Cartilage in the bone shaft breaks down and a marrow cavity opens up

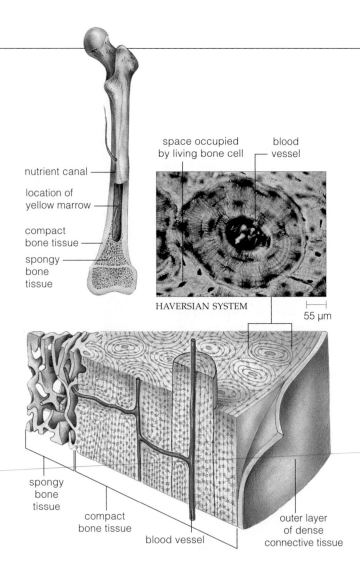

Figure 37.12 (**a**) Human thighbone structure, and (**b**) its spongy and compact bone tissue.

Labels: nutrient canal; location of yellow marrow; compact bone tissue; spongy bone tissue; space occupied by living bone cell; blood vessel; HAVERSIAN SYSTEM; 55 µm; spongy bone tissue; compact bone tissue; blood vessel; outer layer of dense connective tissue

Table 37.1 *Functions of Bone*
1. *Movement.* Bones interact with skeletal muscle to change or maintain the position of the body and its parts.
2. *Support.* Bones support and anchor muscles.
3. *Protection.* Many bones form hard compartments that enclose and protect soft internal organs.
4. *Mineral storage.* Bones are a reservoir for calcium and phosphorus, the deposits and withdrawals of which help maintain ion concentrations in body fluids.
5. *Blood cell formation.* Only certain bones contain regions where blood cells are produced.

(Figure 37.13). Once osteoblasts are imprisoned in their own secretions, we call them osteocytes.

The total bone mass in healthy young adults doesn't change much, yet mineral ions and osteocytes are being simultaneously removed and replaced all the time. The mineral deposits and removals help maintain blood levels of calcium and phosphorus, and they also adjust bone strength. Two cell types interact in this process, which we call **bone remodeling**. Osteoblasts deposit bone, and **osteoclasts** secrete enzymes that digest bone's organic matrix. Liberated ions enter interstitial fluid, from which they can be absorbed by the bloodstream.

Where Bones Meet—Skeletal Joints

Joints, areas of contact or near-contact between bones, have a bridge of connective tissue. Short connecting fibers bridge *fibrous* joints, and cartilage straps bridge *cartilaginous* joints. **Ligaments**, or long straps of dense connective tissue, bridge *synovial* joints (Figure 37.11).

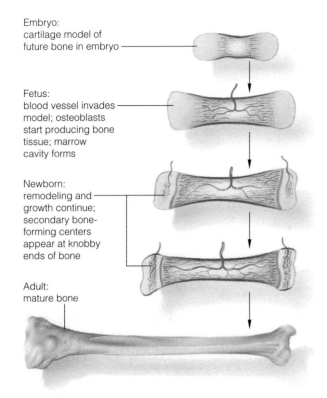

Embryo:
cartilage model of
future bone in embryo

Fetus:
blood vessel invades
model; osteoblasts
start producing bone
tissue; marrow
cavity forms

Newborn:
remodeling and
growth continue;
secondary bone-
forming centers
appear at knobby
ends of bone

Adult:
mature bone

Figure 37.13 Long bone formation, starting with osteoblast activity in a cartilage model (here, already formed in the embryo). These bone-forming cells are active first in the shaft region, then at the knobby ends. In time, cartilage is left only at the ends.

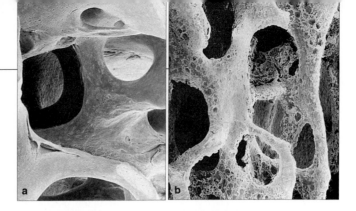

Figure 37.14 Bone affected by osteoporosis. (**a**) Section through normal bone tissue; mineral deposits continually replace withdrawals. (**b**) After the onset of osteoporosis, mineral replacements lag behind withdrawals. The tissue erodes; bones become hollow and brittle.

Some fibrous joints connect the flat skull bones of a fetus. At childbirth, the loose connections allow bones to slide over each other a bit and keep the skull from fracturing. The newborn's skull still has fibrous joints and membranous areas called soft spots, or fontanels. By childhood, the fibrous tissue hardens and the skull bones are fused together as a single unit.

Cartilaginous joints bridge the vertebrae, ribs, and breastbone, and they allow slight movements. Synovial joints move freely; ligaments stabilize them. The knee joint is one of these. Cartilage cushions the abutting bones and helps absorb shocks. A flexible capsule of dense connective tissue surrounds the area of contact. A membrane lines the capsule's interior, and its cells secrete a fluid that lubricates the joint.

Like other joints, knee joints are vulnerable to stress. They let you swing, bend, and turn the long bones below them. As you run, they absorb the force of your weight. Abruptly stretch or twist a knee joint too far and you *strain* it. Tear its ligaments or tendons and you *sprain* it. Move the wrong way and you may dislocate attached bones. In collision sports such as football, a blow to the knee often severs ligaments. These must be reattached within ten days. Why? Phagocytes in a lubricating fluid inside the joint usually clean up after everyday wear and tear. They can turn shredded ligaments to mush.

Joint inflammation and degenerative disorders are collectively called "arthritis." In *osteoarthritis*, cartilage at knees and other freely movable joints wears off as a person ages. Most often, the joints in the fingers, knees, hips, and vertebral column are affected. In *rheumatoid arthritis*, synovial membranes in joints become inflamed and thicken, cartilage degenerates, and bone deposits accumulate. This is an autoimmune response (Chapter 39). It has a genetic basis, but bacterial or viral infection may trigger it. Rheumatoid arthritis can begin at any age, but symptoms usually emerge before age fifty.

Bones and the Blood Level of Calcium

Calcium is vital for neural function, contraction, and other activities, and its level in blood is one of the most tightly controlled aspects of metabolism. Bones and teeth store all but 1 percent or so of the calcium in the human body. Negative feedback loops from blood to glands govern the rates of calcium release and uptake. The thyroid gland releases calcitonin when the calcium level rises. This hormone weakly suppresses osteoclast action and slows calcium's release into blood. When the calcium level declines, parathyroid glands release PTH (parathyroid hormone). The PTH promotes calcium release from bones and its reabsorption from kidneys, enhances osteoclast action, and helps activate vitamin D, which stimulates absorption of dietary calcium.

In a bone placed under compression by exercise or by injury, the osteoblasts deposit more minerals than the osteoclasts withdraw. The bone becomes denser and stronger. As any person ages, however, the backbone and other bones decrease in mass, a condition called *osteoporosis* (Figure 37.14). Physical inactivity, declining activity of bone-forming cells, calcium losses, excessive protein intake, and deficient secretion of sex hormones contribute to the disorder.

Bones are collagen-rich, mineralized organs that function in movement, support, protection, storage of calcium and other minerals, and blood cell formation.

SKELETAL–MUSCULAR SYSTEMS

How Do Muscles and Bones Interact?

Skeletal muscles are the functional partners of bones. Each skeletal muscle contains bundles of hundreds to many thousands of muscle cells, which look like long, striped fibers. In muscle tissue, remember, muscle cells contract (shorten) in response to adequate stimulation. They lengthen in response to gravity and other loads. When you run, dance, breathe, squint, or scribble notes, many muscle cells are working to move your body or change the positions of some of its parts.

Dense connective tissue bundles many muscle cells together and extends beyond them, forming a cordlike or straplike **tendon**. Each tendon attaches a muscle to bone (Figure 37.15). Most of these attachment sites are like a gearshift in a car. *They form a lever system in which a rigid rod is attached to a fixed point and can move about it.* Muscles connect to bones (rigid rods) near a joint (fixed point). When they contract, they transmit force to bones and so make them move. Tendons often rub against bones, as at the knees and wrists. They slide inside fluid-filled sheaths, which helps reduce friction.

Skeletal muscles interact with one another, also. Some work in pairs or groups and so promote a movement. Others work in opposition; the action of one opposes or reverses another muscle's action. Figure 37.16 shows how opposing muscle groups work to move frog legs.

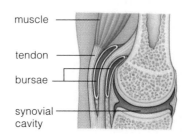

Figure 37.15 Tendon. Bursae are positioned between tendons and bones (or some other structure). A bursa is a flattened sac filled with synovial fluid. They help reduce friction during movements.

muscle
tendon
bursae
synovial cavity

Also look at Figure 37.17. Extend your right arm forward, then place your left hand over the biceps in your upper right arm and slowly bend the elbow. Feel the biceps contract? Even when a biceps contracts only a bit, it causes a large motion of the bone connected to it. This is the case for most leverlike arrangements.

Bear in mind, only *skeletal* muscle is the functional partner of bone. As mentioned earlier, smooth muscle is mainly a component of soft internal organs, such as the stomach (Section 33.3). Cardiac muscle is present only in the heart wall. We will look at the structure and

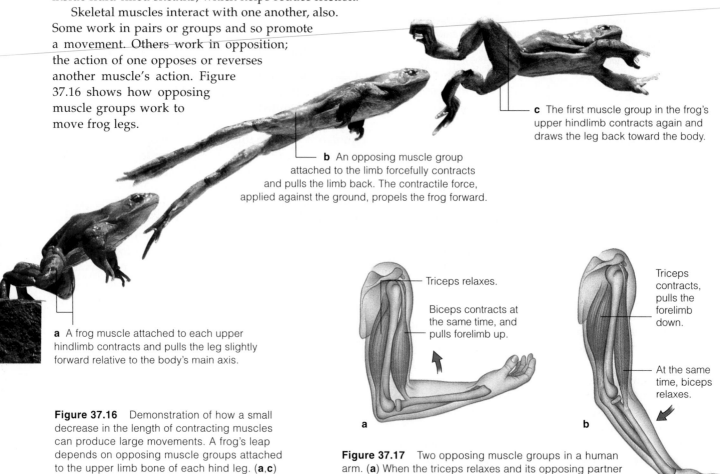

c The first muscle group in the frog's upper hindlimb contracts again and draws the leg back toward the body.

b An opposing muscle group attached to the limb forcefully contracts and pulls the limb back. The contractile force, applied against the ground, propels the frog forward.

a A frog muscle attached to each upper hindlimb contracts and pulls the leg slightly forward relative to the body's main axis.

Triceps relaxes.

Biceps contracts at the same time, and pulls forelimb up.

Triceps contracts, pulls the forelimb down.

At the same time, biceps relaxes.

a   b

Figure 37.16 Demonstration of how a small decrease in the length of contracting muscles can produce large movements. A frog's leap depends on opposing muscle groups attached to the upper limb bone of each hind leg. (**a**,**c**) One muscle group pulls the limb forward a bit, toward the body's midline. (**b**) Another pulls the limb back and away from the body.

Figure 37.17 Two opposing muscle groups in a human arm. (**a**) When the triceps relaxes and its opposing partner (biceps) contracts, the elbow joint flexes and the forearm bends upward. (**b**) When the triceps contracts and the biceps relaxes, the forearm straightens out.

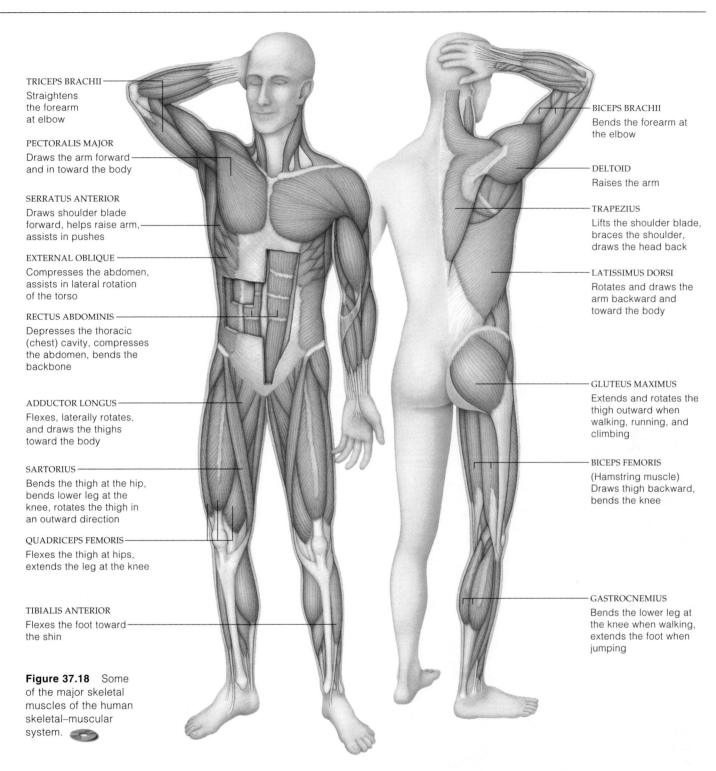

TRICEPS BRACHII
Straightens the forearm at elbow

PECTORALIS MAJOR
Draws the arm forward and in toward the body

SERRATUS ANTERIOR
Draws shoulder blade forward, helps raise arm, assists in pushes

EXTERNAL OBLIQUE
Compresses the abdomen, assists in lateral rotation of the torso

RECTUS ABDOMINIS
Depresses the thoracic (chest) cavity, compresses the abdomen, bends the backbone

ADDUCTOR LONGUS
Flexes, laterally rotates, and draws the thighs toward the body

SARTORIUS
Bends the thigh at the hip, bends lower leg at the knee, rotates the thigh in an outward direction

QUADRICEPS FEMORIS
Flexes the thigh at hips, extends the leg at the knee

TIBIALIS ANTERIOR
Flexes the foot toward the shin

BICEPS BRACHII
Bends the forearm at the elbow

DELTOID
Raises the arm

TRAPEZIUS
Lifts the shoulder blade, braces the shoulder, draws the head back

LATISSIMUS DORSI
Rotates and draws the arm backward and toward the body

GLUTEUS MAXIMUS
Extends and rotates the thigh outward when walking, running, and climbing

BICEPS FEMORIS
(Hamstring muscle) Draws thigh backward, bends the knee

GASTROCNEMIUS
Bends the lower leg at the knee when walking, extends the foot when jumping

Figure 37.18 Some of the major skeletal muscles of the human skeletal–muscular system.

functioning of smooth muscle as well as cardiac muscle in later chapters of this unit.

Human Skeletal–Muscular System

The human body has more than 600 skeletal muscles, some superficial, others deep in the body wall. Some, such as facial muscles, attach to the skin. The trunk has muscles of the thorax, backbone, abdominal wall, and pelvic cavity. Other groups of muscles attach to upper and lower limb bones. Figure 37.18 shows a few of the main skeletal muscles and lists their functions. We turn next to the mechanisms underlying their contraction.

Only skeletal muscles transmit contractile force to bones. Many work as a group to promote a movement; many others work to oppose or reverse the action of others.

HOW DOES SKELETAL MUSCLE CONTRACT?

Consider the dancer in Figure 37.19. Her bones move in a given direction when skeletal muscles attached to them shorten. A muscle shortens because its cells are shortening. And when each muscle cell shortens, the individual units of contraction inside it are shortening. The basic units of contraction, **sarcomeres**, are highly organized arrays of actin, myosin, and other proteins.

The organization starts with muscle cells that run parallel with a muscle's long axis. Each cell has many **myofibrils**: threadlike, cross-banded cell structures that also run in parallel. Cardiac muscle cells have the same cross-bands. When stained, these bands alternate in a light–dark pattern that makes the cells look striped, or striated (Figure 37.19b). Transecting each light band is a mesh of cytoskeletal elements—a Z band (or Z disk).

Two Z bands flank every sarcomere and anchor its many components (Figure 37.19c). Desmins and other intermediate filaments around each Z band help keep neighboring sarcomeres laterally aligned.

A sarcomere has thick filaments made of **myosin**, a motor protein with a club-shaped head. The sarcomere also has thin filaments, each made of two tightly coiled strands of **actin** monomers. Tropomyosin, troponin, and other regulator proteins are bound tightly along the length of the thin filaments (Figure 37.19e).

All filaments run parallel with the myofibril. Thick ones have their myosin heads just a few nanometers away from the thin ones. Anchored to each Z band is one of two sets of actin filaments that does not extend to the sarcomere's center. Thin elastic filaments, of the

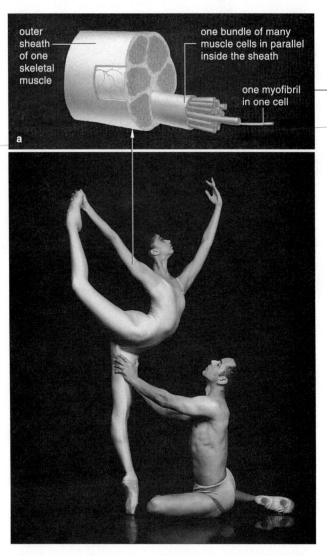

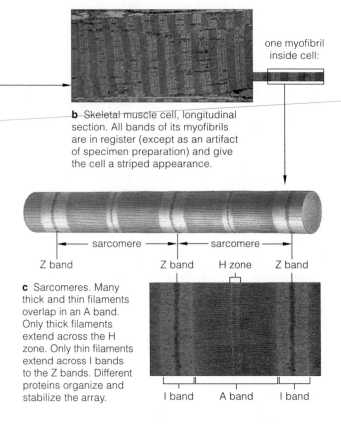

b Skeletal muscle cell, longitudinal section. All bands of its myofibrils are in register (except as an artifact of specimen preparation) and give the cell a striped appearance.

c Sarcomeres. Many thick and thin filaments overlap in an A band. Only thick filaments extend across the H zone. Only thin filaments extend across I bands to the Z bands. Different proteins organize and stabilize the array.

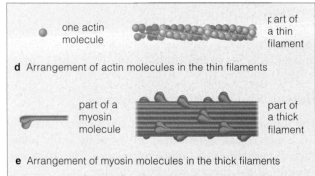

d Arrangement of actin molecules in the thin filaments

e Arrangement of myosin molecules in the thick filaments

Figure 37.19 From the Dance Theatre of Harlem, an example of exquisite control of skeletal muscle movements. (**a–e**) This sequence shows skeletal muscle components from a biceps down to molecules with contractile properties.

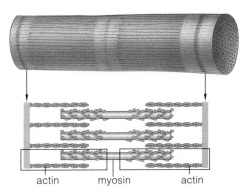

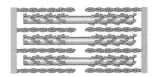

actin   myosin   actin

a   Sarcomere between contractions

b   Same sarcomere, contracted

Figure 37.20 (**a**,**b**) Arrangement of the actin and myosin filaments that interact to reduce the width of a sarcomere. (**c–g**) Sliding-filament model for the contraction of a sarcomere in a myofibril of a muscle cell. For clarity, we show the action of only one myosin head. Many myosin heads, repeatedly making many cross-bridges and power strokes, are required to shorten each sarcomere.

Z band   Z band

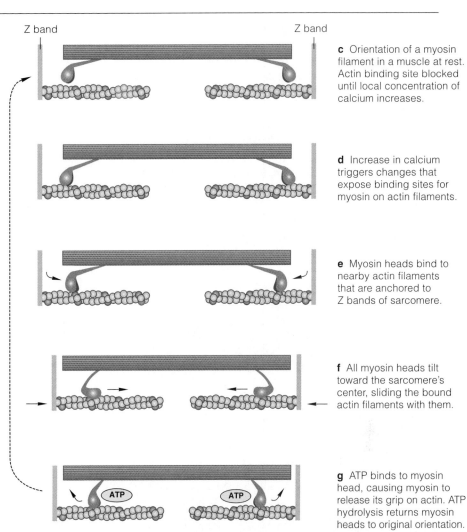

c Orientation of a myosin filament in a muscle at rest. Actin binding site blocked until local concentration of calcium increases.

d Increase in calcium triggers changes that expose binding sites for myosin on actin filaments.

e Myosin heads bind to nearby actin filaments that are anchored to Z bands of sarcomere.

f All myosin heads tilt toward the sarcomere's center, sliding the bound actin filaments with them.

g ATP binds to myosin head, causing myosin to release its grip on actin. ATP hydrolysis returns myosin heads to original orientation.

protein **titin**, also are anchored to the Z band and run parallel with the rest. They keep the myosin filaments centered in a contracting sarcomere and also account for a relaxed muscle's passive resistance to stretching.

Thus the muscle bundles, muscle cells, myofibrils, and thick, thin, and elastic filaments all have the same parallel orientation. What's the point of this? *It focuses the force of contraction on a bone in a particular direction.*

How do sarcomeres shorten and contract a muscle? By the **sliding-filament model** for contraction, all the myosin filaments remain stationary, and they use short, ATP-driven power strokes to slide the two sets of actin filaments over them, toward the sarcomere's center. Both sets of actin filaments slide over the myosin. This is roughly analogous to two pocket doors closing, but pulling both walls to which they are attached along with them. Thus the width of the sarcomere shrinks.

Each myosin head binds repeatedly to sites on an actin filament. The head is one of the ATPases, a type of enzyme that binds and hydrolyzes ATP (Section 5.2). Each binding between the two is called a **cross-bridge**.

As you will see in the next section, the myosin head forms a cross-bridge to the actin when a change in the

local concentration of calcium ions exposes a binding site for it. Binding causes the head to tilt and pull the actin filament toward the center of the sarcomere. Next, ATP binds to the myosin head and breaks its grip on the actin. The ATP becomes hydrolyzed, and the energy released makes the head revert to its original position (Figure 37.20*c–g*). Depending on the calcium levels, the head may attach to another binding site, tilt in another power stroke, and so on. A single contraction of each sarcomere requires hundreds upon hundreds of myosin heads performing a series of short power strokes down the length of their neighbors, the actin filaments.

A skeletal muscle shortens through combined decreases in the length of its numerous sarcomeres. Sarcomeres are the basic units of skeletal and cardiac muscle contraction.

The parallel orientation of a skeletal muscle's components directs the force of contraction toward a bone that must be pulled in some direction.

By energy-driven interactions between myosin and actin filaments, the many sarcomeres of a muscle cell shorten and collectively account for its contraction.

37.8

WHAT CONTROLS CONTRACTION?

Contracting skeletal muscles can move the body and parts of it at certain times, in certain ways. They do so in response to commands from the nervous system. The commands reach them by way of motor neurons and stimulate or inhibit contraction of sarcomeres.

Like all other cells, a muscle cell shows a difference in electric charge across its plasma membrane. That is, the cytoplasm just beneath the membrane is a bit more negative than the interstitial fluid outside it. But only in muscle cells, neurons, and other *excitable* cells does that difference reverse abruptly and briefly. An **action potential** is the name for the abrupt reversal in charge

across a plasma membrane. It arises when charged ions cross the membrane in an ever accelerating way. Such a disturbance, remember, spreads away from the point of stimulation without diminishing (Section 34.2).

Suppose that action potentials arise in a muscle cell. They spread rapidly away from the stimulation point, then along tubular extensions of the plasma membrane (Figure 37.21). The tubes have anchor points for the actin filaments of the myofibril's sarcomeres. They also interconnect with a system of membranous chambers that wraps lacily around the myofibrils. That system, the **sarcoplasmic reticulum**, takes up, stores, and then releases calcium ions in controlled ways.

Arrival of action potentials causes an outward flow of calcium ions from the chambers. The released ions diffuse into the myofibrils and reach actin filaments. In a resting muscle cell, the actin binding sites are blocked and myosin cannot form cross-bridges with them. The calcium ions clear the binding sites, so contraction can proceed. After contraction, calcium ions are actively transported back into the membrane storage system.

What blocks cross-bridge binding sites in a muscle at rest? A tropomyosin filament as well as troponins are located in or near grooves at the surface of the actin filaments (Figure 37.22). When the calcium level is low, these proteins are joined so tightly that the tropomyosin is forced slightly outside the groove. In that position, it blocks the cross-bridge binding site. However, when signals from the nervous system trigger an inflow of calcium into the sarcomere, enough calcium ions bind with the troponin to make its shape change. When that

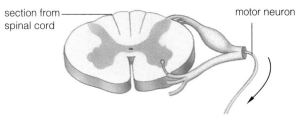

a Messages generated in nervous system trigger action potentials in a motor neuron extending from the spinal cord to a skeletal muscle.

b Acetylcholine (Ach) released from axon endings of motor neuron triggers action potentials in muscle cells.

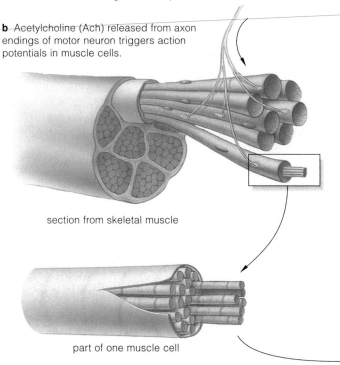

c Action potentials spread along muscle cell plasma membrane and reach the sarcoplasmic reticulum.

Figure 37.21 Pathway by which the nervous system stimulates or inhibits skeletal muscle contraction. The plasma membrane of each muscle cell encloses myofibrils. Tubular extensions of it, the T tubules, have anchor points for actin filaments at the Z bands of all sarcomeres inside.

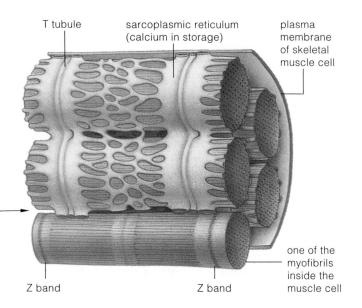

d Action potentials trigger release of calcium ions from sarcoplasmic reticulum threading among myofibrils.

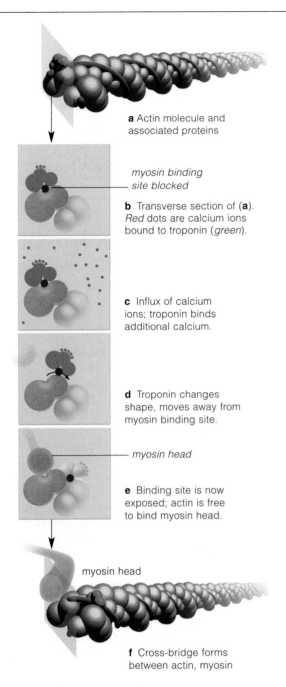

a Actin molecule and associated proteins

myosin binding site blocked

b Transverse section of (**a**). *Red dots are calcium ions bound to troponin (green).*

c Influx of calcium ions; troponin binds additional calcium.

d Troponin changes shape, moves away from myosin binding site.

— *myosin head*

e Binding site is now exposed; actin is free to bind myosin head.

myosin head

f Cross-bridge forms between actin, myosin

Figure 37.22 Actin and tropomyosin filaments, and troponins, in a skeletal muscle cell. When troponin binds free calcium, it changes shape and makes tropomyosin move. A binding site for a myosin head becomes exposed. This action occurs in series at many binding sites along the actin molecule.

happens, the troponin has a different molecular grip on the tropomyosin filament, which is now free to move into the groove and expose the binding site.

Certain commands from the nervous system initiate action potentials in muscle cells. These action potentials are signals for cross-bridge formation, hence for contraction.

ENERGY FOR CONTRACTION

All cells require ATP, but only in muscle cells does the demand skyrocket in so short a time. When a muscle cell at rest is called upon to contract, phosphate donations from ATP must occur twenty to a hundred times faster. But a cell has only a small supply of ATP at the start of contractile activity. At such times, it forms ATP by a quick reaction. An enzyme simply transfers phosphate from **creatine phosphate**, an organic compound, to ADP. A cell has about five times as much creatine phosphate as ATP, so this reaction is good for a few contractions. And that is enough to buy time for a relatively slower ATP-forming pathway to kick in (Figure 37.23).

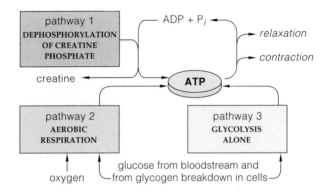

Figure 37.23 Three metabolic pathways by which ATP forms in muscle cells in response to demands of physical exercise.

During prolonged, moderate exercise, the oxygen-requiring reactions of aerobic respiration typically can supply most of the ATP needed for contraction. Suppose a muscle cell taps its store of glycogen for glucose (the starting substrate) in the first five to ten minutes. That cell depends on glucose and fatty acid deliveries from blood for the next half hour or so of sustained activity. For contractile activity longer than this, fatty acids are the main fuel source, as Section 8.6 describes.

What happens when intensive exercise exceeds the capacity of the respiratory and circulatory systems to deliver the oxygen for aerobic respiration? Glycolysis alone contributes more of the total ATP being formed. Remember, by this set of anaerobic reactions, glucose is not fully broken down, so the net ATP yield is small. But muscle cells use this metabolic route for as long as their glycogen stores can be tapped for glucose before fatigue sets in, as you will see in the next section.

After intense exercise, deep, rapid breathing helps repay the body's **oxygen debt**, incurred when ATP use by muscles exceeded the aerobic pathway's deliveries.

During exercise, ATP availability in muscle cells affects whether contraction will proceed, and for how long.

PROPERTIES OF WHOLE MUSCLES

Muscle Tension

Whether a muscle actually shortens during cross-bridge formation depends on the external forces acting on it. Collectively, the cross-bridges exert **muscle tension**. By definition, this is a mechanical force that a contracting muscle exerts on an object, such as a bone. Opposing it is a load, either the weight of an object or gravity's pull on the muscle. Only when muscle tension exceeds the load does a stimulated muscle shorten.

Isotonically contracting muscles shorten and move a load (Figure 37.24). An *isometrically* contracting muscle develops tension but doesn't shorten. With *lengthening* contraction, a muscle lengthens when an external load is greater than its tension for the period of contraction. This happens to leg muscles as you walk down stairs.

A muscle's tension relates to the formation of cross-bridges in its cells and to the number of cells recruited into action. Consider the **motor unit**: a motor neuron and all of the muscle cells that form junctions with its endings. By experimentally stimulating the motor unit with an electrical impulse, we can induce an action potential and then make a recording of an isometric contraction. It takes 50 milliseconds for the tension to increase, then it peaks and declines. This response is a **muscle twitch** (Figure 37.25a). Its duration depends on the load and on the cell type. For example, fast-acting muscle cells rely on glycolysis—not efficient, but fast—and use up ATP faster than slow-acting cells do.

If a new stimulus is applied before a response ends, the muscle will twitch again. **Tetanus** is a sustained contraction that results from the repeated stimulation of a motor unit, so twitches mechanically run together. (In *tetanus*, a disease by the same name, toxins disrupt muscle relaxation, as described in Section 37.11.) Figure 37.25c shows a recording of tetanic contraction.

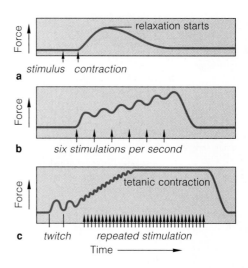

Figure 37.25 Three recordings of twitches in muscles artificially stimulated in three different ways. (**a**) A single twitch. (**b**) Summation of twitches following six stimulations per second. (**c**) Tetanic contraction following twenty stimulations per second.

What Is Muscle Fatigue?

When a muscle is kept in a state of tetanic contraction by continuing high-frequency stimulation, the result is **muscle fatigue**. We define it as a decline in a muscle's capacity to generate force; that is, a decline in tension.

After a few minutes of rest, a fatigued muscle will contract again in response to stimulation. The extent of recovery depends largely on how long and how often it was stimulated before. Muscles associated with brief, intense exercise (such as weight lifting) fatigue fast but recover fast. The muscles associated with prolonged, moderate exercise fatigue slowly. They take longer to recover, often up to twenty-four hours. The molecular mechanisms causing muscle fatigue are unknown, but glycogen depletion is a factor.

Muscle fatigue, together with a lack of stretching, probably causes "charley horses." These muscle cramps are involuntary, large, and often painful contractions that can persist for fifteen minutes or more. Although any skeletal muscle can cramp, the "hamstrings" that span the back of the thigh and calf, as well as the foot's plantar muscles, are commonly affected. How can you end a cramp? Try massaging the muscle, gently stretch it, and keep it stretched. A heating pad might help.

What Are Muscular Dystrophies?

Muscular dystrophies are a class of genetic disorders in which muscles progressively weaken and degenerate. Different forms affect skeletal muscle and, to a lesser

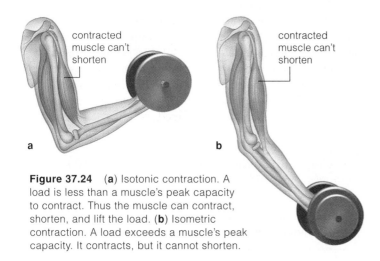

Figure 37.24 (**a**) Isotonic contraction. A load is less than a muscle's peak capacity to contract. Thus the muscle can contract, shorten, and lift the load. (**b**) Isometric contraction. A load exceeds a muscle's peak capacity. It contracts, but it cannot shorten.

extent, cardiac and smooth muscles. Among children, the most common form is called Duchenne muscular dystrophy. Myotonic muscular dystrophy is the most common among adults.

The disorder originates with mutation of a gene on the X chromosome. The gene specifies dystrophin, one of a group of proteins of the T-tubule extensions of the muscle cell's plasma membrane. This complex connects the membrane to actin filaments at the Z bands. Unless actin is firmly bound to the long dystrophin molecules at these bands, the sarcomeres cannot contract. In some muscular dystrophies, mutation of that gene results in reduced amounts of dystrophin or none at all. Being X-linked, the disorder develops most often in males, at a frequency of about 1 in 3,500. There are no cures.

Muscles, Exercise, and Aging

A muscle's properties depend on how often, how long, and how intensely it is put to use. With regular **exercise** (that is, increased levels of contractile activity), muscle cells do not increase in number. Rather, they increase in size and metabolic activity, and become more resistant to fatigue. Think of *aerobic exercise*—not intense, but long in duration. Aerobic exercise increases the number of mitochondria in both fast and slow muscle cells, and increases the blood capillaries that service them. Such physiological changes improve endurance. By contrast, *strength training* (intense, short-duration exercise such as weight lifting) makes fast-acting muscle cells form more myofibrils and enzymes for glycolysis. Strong, bulging muscles may develop, although these muscles do not have much endurance. They fatigue rapidly.

Muscle tension declines in adult humans after age thirty or forty. Older people might exercise just as long and intensely as younger ones, but their muscles cannot adapt (change) in response to the same extent. Even so, some adaptation can be beneficial. Aerobic exercise can improve blood circulation. And, as it turns out, modest strength training slows the loss of muscle tissue that is an inevitable part of the aging process.

As a beneficial side effect, aerobic exercise among the middle-aged and elderly can lift major depression as well as drugs do. Studies indicate that it also may improve memory and the capacity to simultaneously plan and organize complex tasks. Its effects correlate with activity in the frontal and prefrontal cortex. In short, exercise is good for the brain.

Collectively, the cross-bridges that form in sarcomeres of muscle cells exert a mechanical force in response to an external force acting on the muscle. That force is muscle tension. Properties of muscles vary with age and activity.

A Bad Case of Tetanic Contraction

What happens if something disrupts information flow in nervous systems? Ask *Clostridium botulinum*, an anaerobic, endospore-forming bacterium that normally lives in soil. It can cause the disease *botulism*. Its endospores may be present in improperly preserved, canned, or stored food, and these resting structures can germinate in anyone who eats the food. One metabolic product of the bacterium is toxic. It enters neurons that synapse with muscle cells and blocks the release of acetylcholine (ACh). Without ACh, muscles cannot contract. They become more and more flaccid and paralyzed. Unless a poisoned individual receives antitoxins within ten days, breathing becomes impossible and the heart stops beating.

A related bacterium, *C. tetani*, lives in the gut of horses, cattle, and other grazing animals, even many people. Its endospores survive in soil, especially manure-rich ones. They persist for years if sunlight and oxygen do not reach them. They resist disinfectants, heat, and boiling water. When they enter the body through a deep puncture or a deep cut, they germinate in anaerobic, dead tissues. The bacteria do not spread from dead tissues. They produce a toxin that the blood or nerves deliver to the spinal cord and brain. In the spinal cord, the toxin acts on interneurons that help control motor neurons. It blocks the release of inhibitory neurotransmitters (GABA and glycine) and frees the motor neurons from normal inhibitory control. Symptoms of the disease *tetanus* are about to begin.

After four to ten days, the overstimulated muscles stiffen and go into spasm. They cannot be released from contraction, and prolonged, spastic paralysis follows. Fists and jaws may remain clenched (the disease is also called lockjaw). The back may arch severely and permanently. Once respiratory and cardiac muscles become paralyzed, death nearly always follows.

Vaccines were not available for soldiers of early wars, when dead cavalry horses and manure littered battlefields (Figure 37.26). Today, vaccines have all but eradicated tetanus in the United States.

Figure 37.26 Painting of a young victim of a contaminated battle wound, as he lay dying in a military hospital.

SUMMARY   *Gold* indicates text section

1. Most animals have an integumentary system, which covers the body's surface. Examples are insect cuticles and vertebrate skin. Skin protects against abrasion, UV radiation, dehydration, and many pathogens. It helps control body temperature; blood flow to skin dissipates heat. Its sensory receptors detect external stimuli. Skin exposed to sunlight has an endocrine function; it helps produce vitamin D, a hormone-like substance required for absorption of calcium from food. *CI, 37.1, 37.2*

2. Skin consists of two regions: an outer epidermis and underlying dermis, where rapid divisions replace cells continually being shed or abraded away. Keratinocytes (keratin producers) are the most abundant cells. Others are melanocytes (melanin producers), and Langerhans cells and Granstein cells (both types defend the body against pathogens and cancer cells). *37.1, 37.2*

3. Movement of the animal body or parts of it requires contractile cells and some medium or structure against which contractile force can be applied. *37.3*

   a. With hydrostatic skeletons, such as that of a sea anemone, body fluids accept the contractile force and, as a result, are redistributed within a confined space.

   b. With exoskeletons, such as that of an insect, rigid external body parts accept the force of contraction.

   c. With endoskeletons, rigid internal structures such as bones receive the applied force of contraction.

4. Bones are organs with osteocytes (living bone cells) in a mineralized, collagen-rich ground substance. They have roles in moving, protecting, and supporting body parts; storing minerals; and, in bones that contain red and yellow marrow, forming blood cells. *37.5*

5. A skeletal joint, a gap between bones, is bridged by connective tissues. Short fibers bridge fibrous joints. Cartilage bridges cartilaginous joints. Ligaments bridge synovial joints. *37.5*

6. Tendons attach skeletal muscles to bones. Skeletal muscles and bones interact as a system of levers, with rigid rods (bones) moving at fixed points (joints). Many muscles work together or in opposition to bring about movement or positional changes in body parts. *37.6*

7. Smooth, cardiac, and skeletal muscle cells contract (shorten) in response to adequate stimulation. Many soft internal organs incorporate smooth muscle. Only the heart incorporates cardiac muscle. Skeletal muscle is the functional partner of bones. *37.6*

8. Inside each skeletal muscle cell are many myofibrils arranged parallel with the long axis. These threadlike structures contain filaments of actin and myosin, also organized in parallel. Every myofibril is transversely divided into sarcomeres, the basic units of contraction. The parallel orientation of all parts directs the contractile force at a bone to be pulled in some direction. *37.7*

9. In response to stimulation from the nervous system, skeletal muscles shorten by decreases in the length of all of its sarcomeres. Key points for the sliding-filament model of muscle contraction follow: *37.7, 37.8*

   a. Action potentials cause the release of calcium ions from a membrane system (sarcoplasmic reticulum) that threads around the cell's myofibrils. Calcium diffuses into sarcomeres, and binds to and pulls aside accessory proteins on actin filaments. Binding sites for myosin are now exposed so cross-bridges can form.

   b. Each cross-bridge is a brief attachment between a myosin head and an actin binding site. Cross-bridges form during repeated, ATP-driven power strokes. The repeated strokes make actin filaments slide past myosin filaments, and collectively they shorten the sarcomere.

10. Muscle cells obtain the ATP for contraction in three ways: *Dephosphorylation of creatine phosphate* is a direct, fast pathway good for a few seconds of contraction. *Aerobic respiration* is a route that predominates during prolonged but moderate exercise. *Glycolysis* takes over when intense exercise exceeds the body's capacity to deliver oxygen to muscle cells. *37.9*

11. Muscle tension is a mechanical force caused by cross-bridge formation. The load (gravity or weight) is an opposing force. Stimulated muscles shorten if tension exceeds the load and lengthen if tension is less than it. Exercise and aging affect muscle properties. *37.10*

Review Questions

1. List the functions of skin. Describe the distinctive features of its regions. Is the hypodermis part of skin? *CI, 37.1*

2. Identify four cell types in vertebrate skin and briefly describe their functions. *37.1, 37.2*

3. Distinguish between:
   a. hydrostatic skeleton, exoskeleton, and endoskeleton *37.3*
   b. vertebra and intervertebral disk *37.4*
   c. ligament and tendon *37.5, 37.6*

4. What are the functions of bones? What is a joint? *37.5*

5. Which hormones help control calcium ion concentrations in blood? Name their general effects on bone tissue turnover. *37.5*

6. In what respect are tendons like a car's gearshift? *37.6*

7. Review Figure 37.19. On your own, sketch and label the fine structure of a muscle, down to one of its individual myofibrils. Identify the basic unit of contraction in the myofibrils. *37.7*

8. What role does calcium play in control of contraction? What role does ATP play, and by what routes does it form? *37.8, 37.9*

Self-Quiz   ANSWERS IN APPENDIX III

1. Which is *not* a function of skin?
   a. make vitamin D   c. initiate movement
   b. conserve water   d. help control temperature

2. _____ are shock pads and flex points.
   a. Vertebrae   c. Marrow cavities
   b. Femurs   d. Intervertebral disks

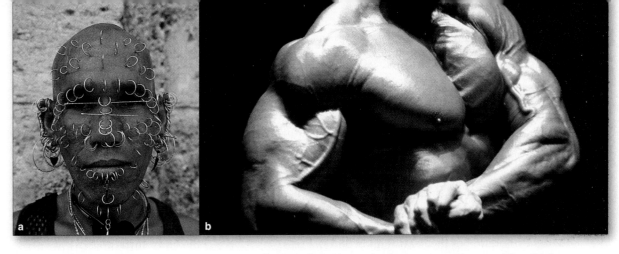

Figure 37.27 (**a**) A demonstration of the curious invasions of the integument that have no immediately obvious adaptive value. (**b**) Extremely pumped-up muscles of a human male.

3. Blood cells form in _____ .
   a. red marrow
   b. all bones
   c. certain bones only
   d. a and c

4. The _____ is the basic unit of contraction.
   a. myofibril
   b. sarcomere
   c. muscle fiber
   d. myosin filament

5. Muscle contraction requires _____ .
   a. calcium ions
   b. ATP
   c. action potential arrival
   d. all of the above

6. ATP for muscle contraction can be formed by _____ .
   a. aerobic respiration
   b. glycolysis
   c. creatine phosphate breakdown
   d. all of the above

7. Match the M words with their defining feature.
   _____ muscle
   _____ muscle twitch
   _____ muscle tension
   _____ melanin
   _____ myosin
   _____ marrow
   _____ metacarpals
   _____ myofibrils
   _____ muscle fatigue

   a. actin's partner
   b. all in the hands
   c. blood cell production
   d. decline in tension
   e. brownish-black pigment
   f. motor unit response
   g. force exerted by cross-bridges
   h. bundles of muscle cells in connective tissue sheaths
   i. threadlike parts in muscle cell

the cost of securing and using resources to grow and maintain a massive body is offset by reproductive rewards. Now think of the social rewards for millions of athletes who use anabolic steroids to increase muscle mass and strength (Figure 37.27*b*). *Anabolic steroids*, synthetic hormones, mimic testosterone. This sex hormone governs secondary sexual traits and also increases aggressive behavior, which is often associated with maleness.

By stimulating protein synthesis, anabolic steroids induce rapid gains in muscle mass and strength during weight-training and exercise programs. However, at high concentrations, they trigger a sharp decline in testosterone production, resulting in acne, baldness, shrinking testes, infertility, and maybe early heart disease in men. Even occasional use may damage kidneys or set the stage for cancers of the liver, testes, and prostate gland. Among women, anabolic steroids deepen the voice and produce pronounced facial hair. Breasts may shrink and the menstrual cycle may become irregular.

Not every steroid user develops severe side effects. More common are mental difficulties, called *'roid rage* or *body-builder's psychosis*. Users become irritable and increasingly aggressive. Some men have become uncontrollably aggressive, manic, and delusional. One steroid user accelerated his car to high speed and deliberately drove it into a tree.

Would you say that natural selection is favoring or working against anabolic steroid use? This is not a trick question.

Critical Thinking

1. The nose, lips, tongue, navel, and private parts are common targets for *body piercing*, cutting holes into the body so jewelry and other objects can be threaded through them (Figure 37.27*a*). *Tattooing*, or using permanent dyes to create patterns in skin, is another fad. Besides being painful, both skin invasions invite bacterial infections, chronic viral hepatitis, AIDS, and other nasty diseases if piercers and tattooers reuse unsterilized needles, dye, razors, gloves, swabs, or trays. Months may pass before problems arise, so the cause-and-effect connection isn't always obvious. If you dismiss the risks and think unregulated tissue invasions are okay, how could you make sure the equipment used is sterile?

2. For young women, the recommended daily allowance (RDA) of calcium is 800 milligrams/day. During pregnancy, the RDA is 1,200 milligrams/day. Why the increase? What might happen to a pregnant woman's bones without the larger amount?

3. Compared to most people, Joe and other long-distance runners have a greater number of muscle cells with more mitochondria. Sprinters have muscle cells that contain more of the enzymes necessary for glycolysis but fewer mitochondria. Think about how these two forms of exercise differ. Then explain why the muscle cells differ between the two kinds of runners.

4. Extreme sexual dimorphism in body size involves differences in muscle mass (compare Section 17.9). This may be a measure of how much male mammals invest in fighting capacity. Maybe

Selected Key Terms

Integument
actin 37.7
dermis 37.1
epidermis 37.1
folate 37.2
Granstein cell 37.2
hair 37.1
integument CI
keratinocyte 37.1
Langerhans cell 37.2
melanocyte 37.1
skin, vertebrate 37.1
vitamin D 37.2

Skeletal-Muscle Function
action potential 37.8
bone 37.5
bone remodeling 37.5

creatine
  phosphate 37.9
cross-bridge 37.7
endoskeleton 37.3
exercise 37.10
exoskeleton 37.3
hydrostatic
  skeleton 37.3
intervertebral
  disk 37.4
joint 37.5
ligament 37.5
motor unit 37.10
muscle fatigue 37.10
muscle tension 37.10
muscle twitch 37.10
myofibril 37.7
myosin 37.7

osteoblast 37.5
osteoclast 37.5
osteocyte 37.5
oxygen debt 37.9
red marrow 37.5
sarcomere 37.7
sarcoplasmic
  reticulum 37.8
skeletal muscle 37.6
sliding-filament
  model 37.7
tendon 37.6
tetanus 37.10
titin 37.7
vertebra
  (vertebrae) 37.4
yellow marrow 37.5

Readings

Sherwood, L. 1997. *Human Physiology*. Third edition. Belmont, California: Wadsworth.

Travis, J. 15 January 2000. "Boning Up." *Science News* 157: 41–43.

On-Line readings at Student Guide for InfoTrac:
www.brookscole.com/biology

CIRCULATION

Heartworks

For Dr. Augustus Waller, Jimmie the bulldog was no ordinary pooch. Connected to wires and with his feet soaked in buckets of salty water, Jimmie was on a four-footed exploration of the heart (Figure 38.1a). Press your fingers to your chest a few inches left of center, between the fifth and sixth ribs, and feel the repetitive thumpings of your heart. The same rhythms intrigued Waller and other nineteenth-century physiologists.

They wondered: Is there a pattern of electrical currents that corresponds to each heartbeat? Could they find out by devising a painless way to record currents at the body surface?

That's where Jimmie and the buckets of salty water came in. Saltwater happens to be an efficient conductor of electricity. In Waller's experiment, it picked up faint signals from Jimmie's beating heart through the skin of his legs and conducted them to a crude monitoring device. With that device, Waller made one of the first recordings of heart activity (Figure 38.1c). Today we call such a recording an electrocardiogram, or ECG.

A graph of your heart's normal electrical activity would look very much the same. The pattern emerged a few weeks after you started growing, by mitotic cell divisions, from a fertilized egg in your mother. Early on, some embryonic cells differentiated into cardiac muscle cells and started to contract on their own. One small patch of cells was the first to start contracting. They set the pace for the other cells and have been functioning as your heart's pacemaker ever since.

If all goes well, that patch of cardiac muscle cells will continue to contract as it should until the day you die. It is a natural pacemaker; it sets the baseline rate at which blood is pumped out of the heart, through blood vessels, then back to the heart. The rate, about seventy beats every minute, is moderate. But signals from the nervous and endocrine systems continually adjust it. When you run, skeletal muscle cells demand more

a   BUCKET   BUCKET

b   JIMMIE   DR. WALLER

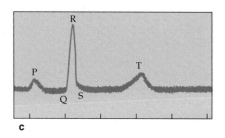

c

Figure 38.1 A bit of history in the making. (**a**) Jimmie the bulldog taking part in a painless experiment. (**b**) The physiologist Augustus Waller and his beloved pet bulldog sharing a moment in Waller's study after the experiment, which yielded one of the world's first electrocardiograms (**c**).

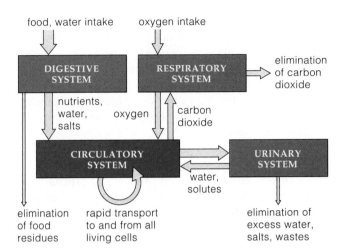

food, water intake   oxygen intake

DIGESTIVE
SYSTEM

RESPIRATORY
SYSTEM

elimination
of carbon
dioxide

nutrients,
water,
salts

oxygen

carbon
dioxide

CIRCULATORY
SYSTEM

URINARY
SYSTEM

water,
solutes

elimination
of food
residues

rapid transport
to and from all
living cells

elimination of
excess water,
salts, wastes

Figure 38.2 Diagram of functional connections between the circulatory, respiratory, digestive, and urinary systems, which interact in transporting substances to and from all living cells in the animal body. Their integrated activities help maintain favorable operating conditions in the internal environment.

blood-borne oxygen and glucose than they do when you sleep. At such times of intense exercise, your heart starts pounding more than twice as fast, and this helps deliver sufficient blood to muscle tissues.

We've come a long way from Waller and Jimmie in monitoring the heart. Sensors now detect faint signals characteristic of an impending heart attack. Internists use computers to analyze a patient's beating heart, and ultrasound probes to get images of it on video screens. Cardiologists routinely substitute battery-powered pacemakers for faltering natural ones.

With this chapter, we turn to the **circulatory system**, the means by which substances move rapidly to and from the interstitial fluid that bathes all living cells in nearly all animals. The system continually accepts the oxygen, nutrients, and other substances that an animal secures, through its respiratory and digestive systems, from the environment (Figure 38.2). It also picks up carbon dioxide and other wastes from cells and gives them up to the respiratory and urinary systems for disposal. Its operation is central to maintaining the operating conditions of the internal environment within a tolerable range. And that is the state we call homeostasis.

Key Concepts

1. All cells survive by exchanging substances with their surroundings. In vertebrates and many other animals, substances rapidly move to and from cells by way of a closed circulatory system.

2. Blood, a fluid connective tissue, is the transport medium of circulatory systems. It transports oxygen, carbon dioxide, plasma proteins, vitamins, hormones, lipids, and many other solutes. Blood also transports metabolically generated heat.

3. In birds and mammals, a four-chambered, muscular heart pumps blood through two separate circuits of blood vessels, both of which lead back to the heart. One circuit flows through the lungs, where it picks up oxygen that cells require, primarily for aerobic respiration. The other circuit delivers oxygenated blood to all body regions, from which it picks up carbon dioxide released by the aerobic pathway.

4. Arteries and veins are both large-diameter, low-resistance blood vessels. Arteries rapidly transport oxygenated blood away from the heart. Veins transport oxygen-poor blood to the heart and serve as blood volume reservoirs.

5. The flow volume through each organ is regulated at arterioles. In response to control signals, the diameter of these blood vessels widens in some body regions and narrows in others. The coordinated responses of arterioles through the body divert more of the flow volume to organs that are most active at a given time.

6. Small-diameter, thin-walled capillaries make up diffusion zones. As a volume of blood spreads out through a capillary bed, it slows and allows time for the exchanges between blood and interstitial fluid.

7. Pressure forces some fluid out of capillaries, into interstitial fluid. Capillaries reabsorb some of it, and the lymphatic system returns the remainder to the blood. Lymphatic organs cleanse blood of infectious agents and other threats.

EVOLUTION OF CIRCULATORY SYSTEMS

From Structure to Function

Imagine that an earthquake has closed off highways around your neighborhood. Grocery trucks can't enter and waste-disposal trucks can't leave, so food supplies dwindle and garbage piles up. Cells would face similar predicaments if your body's highways were disrupted. The highways are part of a circulatory system, which swiftly moves substances to and from cells. The system helps maintain a favorable neighborhood, so to speak, and this is vital work. Your differentiated cells are just too specialized to fend for themselves. They interact to maintain the volume, composition, and temperature of **interstitial fluid**, a tissue fluid bathing them. **Blood**, a circulating connective tissue, interacts with interstitial fluid. Its continuous deliveries and pickups help keep conditions tolerable for enzymes and other molecules that carry out cell activities. Interstitial fluid and blood are the body's "internal environment."

Blood flows inside blood vessels—tubes that differ from one another in their wall thickness and diameter. A muscular pump, the **heart**, generates pressure that keeps blood flowing. As is the case for many animals, your circulatory system is *closed*; it confines the blood inside continuously connected walls of a heart and blood vessels. This is not the case in *open* circulatory systems of most mollusks and arthropods, such as insects. In an open system, blood flows out from the vessels into sinuses, or spaces inside body tissues. Blood mingles with the tissue fluids. Then it moves back into the heart by way of openings in the blood vessels or the heart wall (Figure 38.3a,b).

Think about a closed system's overall "design." The total *volume* of blood is continuously pumped out from the heart, then flows through blood vessels back to the heart. Flow *velocity* (speed) is highest in large-diameter transport vessels. It slows in specific body regions, where there must be enough time for blood to exchange substances with the interstitial fluid and cells. The required slowdown proceeds at **capillary beds**. There, blood spreads out through a great many small-diameter blood vessels called **capillaries**.

During any interval, the same volume of blood is moving forward through capillary beds as elsewhere in the system. But it is doing so at a more leisurely pace because of the great total cross-sectional area of the capillaries. The simple analogy given in Figure 38.3e may help you grasp this concept.

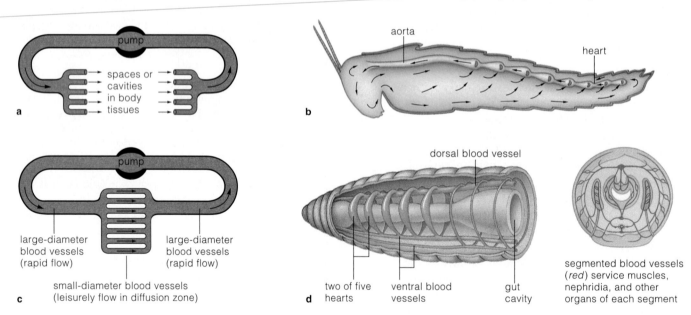

Figure 38.3 Open and closed circulatory systems. (**a,b**) In a grasshopper's open system, a heart (not like yours) pumps blood through a vessel, the aorta. Blood moves into tissue spaces and mingles with fluid bathing cells. It reenters the heart at openings in the heart wall. (**c,d**) An earthworm's closed system confines blood in several pairs of muscular hearts near the head end and inside blood vessels.

(**e**) Relationship between flow velocity and total cross-sectional area in any closed circulatory system. Visualize two fast rivers flowing into and out from a lake. The flow *rate* is the same in all three places; an identical volume of water moves from point *1* to point *3* during the same interval. Flow *velocity* decreases within the lake. Why? The volume spreads out through a larger cross-sectional area and moves forward a shorter distance during the same length of time.

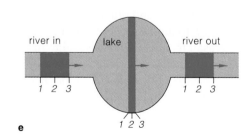

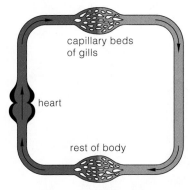

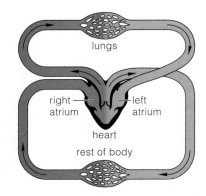

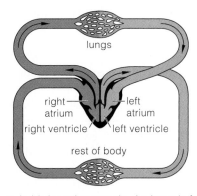

a In fishes, a two-chambered heart (atrium, ventricle) pumps blood in one circuit. Blood picks up oxygen in gills, delivers it to rest of body. Oxygen-poor blood flows back to heart.

b In amphibians, a heart pumps blood through two partially separate circuits. Blood flows to lungs, picks up oxygen, returns to heart. But it mixes with oxygen-poor blood still in the heart, flows to rest of body, returns to heart.

c In birds and mammals, the heart is fully partitioned into two halves. Blood circulates in two circuits: from the heart's right half to lungs and back, then from the heart's left half to oxygen-requiring tissues and back.

Figure 38.4 Comparison of the closed circulatory system of four vertebrate groups.

Evolution of Vertebrate Circulation

Although humans and other existing vertebrates have a closed circulatory system, the pump and plumbing for fishes, amphibians, birds, and mammals differ in their details. The differences evolved over hundreds of millions of years and arose during the move of some vertebrate lineages onto land. Remember Section 26.3? Fishes—the first vertebrates—had gills. Gills, like all respiratory structures, have a thin, moist surface that oxygen and carbon dioxide can diffuse across. Later, in the ancestors of all land vertebrates, lungs evolved that supplemented gas exchange. Being *internally moistened* sacs, lungs had advantages for the move onto dry land. Also advantageous were concurrent modifications in circulatory systems, which pick up oxygen from lungs and transport carbon dioxide wastes back to them.

Consider this: In fishes, blood flows in *one* circuit (Figure 38.4*a*). Pressure generated by a two-chambered heart forces it from gill capillary beds into the largest artery, through capillary beds of organs, and back to the heart. Extensive gill capillaries offer so much resistance to flow that pressure drops considerably before blood enters the main artery. Blood delivery is enough for the activity level of most fishes. It would not be enough for the more active life-styles of most land vertebrates.

As amphibians were evolving, their heart became partitioned into right and left halves—only partly so, but enough to direct blood flow through *two* partially separated circuits (Figure 38.4*b*). The partitioning continued in reptiles called crocodilians. It became complete in the two side-by-side pumps of birds and mammals (Figure 38.4*c*). Their heart's *right* half pumps oxygen-poor blood to lungs, where the blood picks up oxygen and gives up carbon dioxide. That freshly

oxygenated blood flows to the heart's left half. This route is the **pulmonary circuit**.

In the **systemic circuit**, the heart's *left* half pumps freshly oxygenated blood to every tissue and organ where oxygen is used and carbon dioxide forms. Then the oxygen-poor blood flows to the heart's right half.

The double circuit is a rapid and efficient mode of blood delivery. It supports the high levels of activity typical of vertebrates whose ancestors evolved on land.

Links With the Lymphatic System

The heart's pumping puts pressure on blood flowing through the circulatory system. Partly because of this, small amounts of water, nutrients, and a few proteins dissolved in blood are forced from capillaries into the interstitial fluid. Excess fluid and reclaimable solutes move back into capillaries or enter drainage vessels, which return them to the circulatory system. These vessels are part of the **lymphatic system**. As you will see, other parts help cleanse the fluid.

Circulatory systems function in moving substances rapidly to and from all living cells in a multicelled body. A closed system generally confines the transport medium—blood—within its walls. In an open system, blood also intermingles with tissue fluids, then moves back inside the walls.

Blood flows fastest in large-diameter vessels between the vertebrate heart and capillary beds, where the flow velocity drops enough for efficient exchanges with interstitial fluid.

Vertebrate closed systems coevolved with the respiratory and lymphatic systems. In fishes, blood flows in one circuit. In birds and mammals, it flows in two, through a partitioned heart that works as two side-by-side pumps. The double circuit supports the high levels of activity typical of most vertebrates that evolved on land.

CHARACTERISTICS OF BLOOD

Functions of Blood

Blood is a connective tissue with multiple functions. It transports oxygen, nutrients, and other solutes to cells. It carries away metabolic wastes and secretions such as hormones. It helps stabilize internal pH and serves as a highway for phagocytic cells that fight infections and scavenge debris in tissues. In birds and mammals, blood helps equalize body temperature by moving excess heat from regions of high metabolic activity (such as skeletal muscles) to the skin, where heat can be dissipated.

2 µm

8 µm
average
diameter

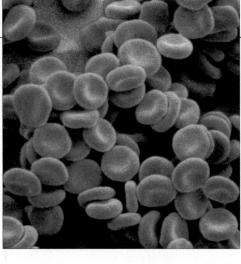

Figure 38.6 Size and shape of red blood cells.

Blood Volume and Composition

The volume of blood depends on body size and on the concentrations of water and solutes. Blood volume for average-size adult humans is about 6 to 8 percent of the total body weight. That amounts to about four or five quarts. As is the case for all vertebrates, human blood is a viscous fluid that is thicker than water and slower flowing. Its components are **plasma**, **red blood cells**, **white blood cells**, and **platelets**. Normally, the plasma accounts for 50 to 60 percent of the total blood volume.

PLASMA Prevent a blood sample in a test tube from clotting, and it separates into a red, cellular portion and the plasma, a straw-colored liquid, which floats on the cellular portion (Figure 38.5). Plasma is mostly water, and it serves as a transport medium for blood cells and platelets. Plasma also functions as a solvent for ions and molecules, including hundreds of different plasma proteins. Some of the plasma proteins transport lipids

and fat-soluble vitamins through the body. Others have roles in blood clotting or in defense against pathogens. Collectively, the concentration of the plasma proteins influences the fluid volume of blood, for it affects the movement of water between the blood and interstitial fluid. Glucose and other simple sugars as well as lipids, amino acids, vitamins, and hormones are dissolved in plasma. So are oxygen, carbon dioxide, and nitrogen.

RED BLOOD CELLS Erythrocytes, or red blood cells, are biconcave disks (Figure 38.6). They transport oxygen from lungs to all aerobically respiring cells, and some carbon dioxide wastes from them. When oxygen first diffuses into blood, it binds to hemoglobin, the iron-containing pigment that makes red blood cells look red. Section 3.8 shows its structure. Oxygenated blood is bright red. Oxygen-poor blood is darker red but looks blue through blood vessel walls near the body surface.

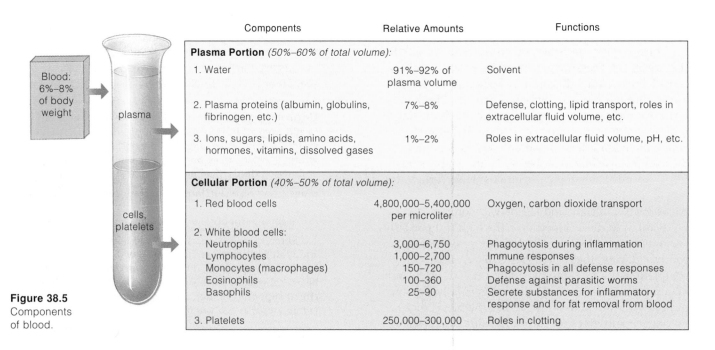

	Components	Relative Amounts	Functions
Plasma Portion *(50%–60% of total volume):*			
	1. Water	91%–92% of plasma volume	Solvent
	2. Plasma proteins (albumin, globulins, fibrinogen, etc.)	7%–8%	Defense, clotting, lipid transport, roles in extracellular fluid volume, etc.
	3. Ions, sugars, lipids, amino acids, hormones, vitamins, dissolved gases	1%–2%	Roles in extracellular fluid volume, pH, etc.
Cellular Portion *(40%–50% of total volume):*			
	1. Red blood cells	4,800,000–5,400,000 per microliter	Oxygen, carbon dioxide transport
	2. White blood cells:		
	Neutrophils	3,000–6,750	Phagocytosis during inflammation
	Lymphocytes	1,000–2,700	Immune responses
	Monocytes (macrophages)	150–720	Phagocytosis in all defense responses
	Eosinophils	100–360	Defense against parasitic worms
	Basophils	25–90	Secrete substances for inflammatory response and for fat removal from blood
	3. Platelets	250,000–300,000	Roles in clotting

Blood:
6%–8%
of body
weight

plasma

cells,
platelets

Figure 38.5
Components
of blood.

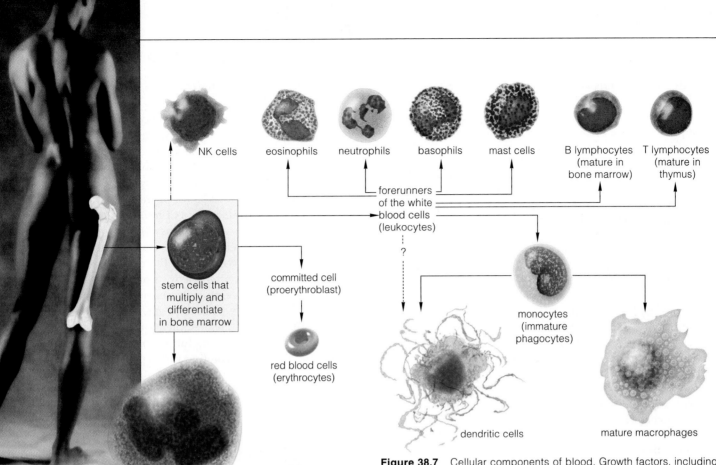

NK cells   eosinophils   neutrophils   basophils   mast cells   B lymphocytes (mature in bone marrow)   T lymphocytes (mature in thymus)

forerunners of the white blood cells (leukocytes)

stem cells that multiply and differentiate in bone marrow

committed cell (proerythroblast)

red blood cells (erythrocytes)

monocytes (immature phagocytes)

dendritic cells   mature macrophages

megakaryocytes   platelets

Figure 38.7 Cellular components of blood. Growth factors, including erythropoietin and interleukin 3, stimulate growth and differentiation in the cell lineages. The next chapter describes white blood cell functions. Section 38.10 describes the role of platelets in blood clotting.

In bone marrow, red blood cells develop from stem cells (Figure 38.7). Generally speaking, **stem cells** stay unspecialized and retain their capacity for mitotic cell division. Their daughter cells divide, also, but only a portion go on to differentiate into specialized types.

Mature red blood cells no longer have their nucleus, nor do they require it. They have enough hemoglobin, enzymes, and other proteins to function for about 120 days. The body's own phagocytes engulf the oldest red blood cells and those already dead. Normally, ongoing replacements keep the cell count fairly stable.

A **cell count** is a measure of the number of cells of a given type in a microliter of blood. For example, the average number of red blood cells per cubic centimeter is 5.4 million in males and 4.8 million in females.

WHITE BLOOD CELLS Leukocytes, or white blood cells, arise from stem cells in bone marrow. They function in daily housekeeping and defense. Some types patrol tissues. They target or engulf damaged or dead cells and anything chemically recognized as foreign to the body. Many others are massed together in lymph nodes and the spleen, which are organs of the lymphatic system. There they divide to produce armies of cells that battle specific viruses, bacteria, and other threats to health.

Figure 38.7 shows the main categories of white blood cells, which differ in size, nuclear shape, and staining traits. Their numbers change, depending on whether an individual is active, healthy, or under siege. You'll read about them in the next chapter. For now, a few examples will give you a sense of what they do. Neutrophils are fast-acting phagocytes. Like basophils, they have roles in inflammatory responses to invasions and tissue damage. All immune responses against specific threats use two classes of lymphocytes: B and T cells. Macrophages and dendritic cells can call up immune responses. NK cells destroy virus-infected and cancerous body cells.

PLATELETS Some stem cells in bone marrow give rise to megakaryocytes. These "giant" cells shed cytoplasmic fragments wrapped in a bit of plasma membrane. The membrane-bound fragments are what we call platelets. Each platelet lasts only five to nine days, but hundreds of thousands are always circulating in blood. They can release substances that initiate blood clotting.

Vertebrate blood has major roles in transporting substances, defense, clotting, and maintaining the volume, composition, and temperature of the internal environment.

Blood Disorders

The body continually replaces blood cells, for good reason. Besides aging and dying off regularly, blood cells typically encounter a variety of pathogens that use them as places to complete their life cycle. Besides this, sometimes blood cells malfunction as a result of gene mutations.

RED BLOOD CELL DISORDERS Consider the **anemias**. These disorders result from too few red blood cells or deformed ones. Oxygen levels in blood cannot be kept high enough to support normal metabolism. Shortness of breath, fatigue, and chills follow. *Hemorrhagic* anemias result from sudden blood loss, as from a severe wound; *chronic* anemias result from low red blood cell production or ongoing but slight blood loss, as from a bleeding ulcer.

Some bacteria and protozoans replicate in blood cells. They escape by lysis and kill the cells, thus causing some *hemolytic* anemias. Insufficient iron in the diet results in *iron deficiency* anemia; red blood cells can't make enough normal hemoglobin without iron. B_{12} *deficiency* anemia is a potential hazard for strict vegetarians and alcoholics. Red blood cells form but cannot divide without vitamin B_{12}, which you can get from meats, poultry, and fish.

As described elsewhere in the book, *sickle-cell anemia* arises from a mutation that alters hemoglobin. Another mutation partially or fully blocks synthesis of the globin chains of hemoglobin; it causes *thalassemias*. Too few red blood cells form, and those that do are thin and fragile.

Polycythemias—symptoms of far too many red blood cells—make blood flow sluggish. Some bone marrow cancers lead to this condition. So can *blood doping*. Some athletes about to compete in strenuous events withdraw and store their own red blood cells, then reinject them a few days prior to competition. The withdrawal triggers red blood cell formation as the body attempts to replace "lost" cells, so when withdrawn cells are put back in the body, the cell count bumps up. The idea is to increase oxygen-carrying capacity and endurance. Blood doping leads to temporary high blood pressure and also alters blood's viscosity. Some believe blood doping works, but others call it unethical. It is banned from the Olympics.

WHITE BLOOD CELL DISORDERS Possibly you've heard of *infectious mononucleosis*. An Epstein–Barr virus causes this highly contagious disease, which results from the formation of too many monocytes and lymphocytes. After a few weeks of fatigue, aches, low-grade fever, and a chronic sore throat, patients usually recover.

Recovery is chancy for *leukemias*. These cancers suppress or impair white blood cell formation in bone marrow (Chapter 12). Radiation therapy and chemotherapy kill cancer cells, but side effects are often severe. Remissions (symptom-free periods of a chronic illness) may last for months or years. New gene therapies are putting some leukemias into remission with only mild side effects.

BLOOD TRANSFUSION AND TYPING

Concerning Agglutination

A decline in blood volume or blood cell counts triggers countermeasures. If the volume were to decrease more than 30 percent, circulatory shock and death may occur.

Blood from donors can be transfused into patients affected by a blood disorder or severe blood loss. Such **blood transfusions** can't be hit-or-miss. Why not? *Red blood cells of a potential donor and recipient may not have the same kind of recognition proteins at their surface.* Some of the proteins are "self" markers. They identify cells as belonging to one's own body. If the donor's cells have the "wrong" marker, a recipient's immune system will recognize them as foreign, with serious consequences.

Figure 38.8 shows what happens when blood from incompatible donors and recipients intermingle. In a defense response called **agglutination**, proteins called antibodies that are circulating in the plasma act against

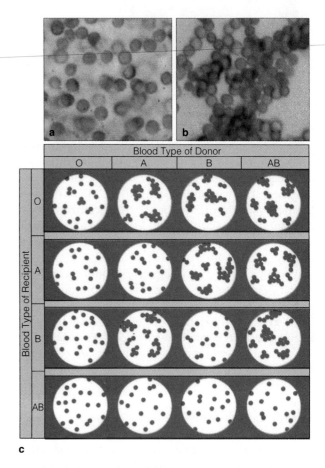

Figure 38.8 Light micrographs showing (**a**) an absence of agglutination in a mixture of two different yet compatible blood types and (**b**) agglutination in a mixture of incompatible types. (**c**) Responses in blood types A, B, AB, and O when mixed with samples of the same and different types.

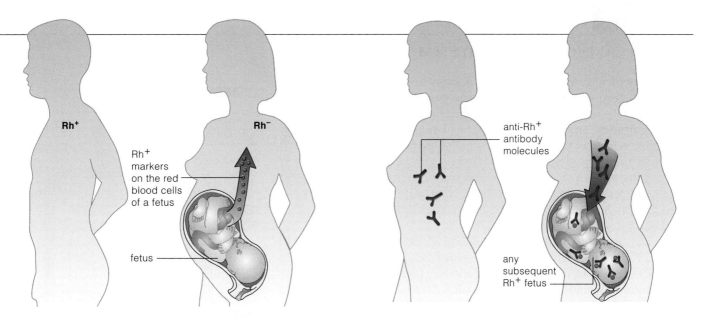

a A forthcoming child of an Rh⁻ woman and Rh⁺ man inherits the gene for the Rh⁺ marker. During childbirth, some of its cells bearing the marker may leak into the maternal bloodstream.

b The foreign marker stimulates antibody formation. If the same woman becomes pregnant again and if her second fetus (or any other) inherits the gene for the marker, her circulating anti-Rh⁺ antibodies will act against it.

Figure 38.9 Antibody production in response to Rh⁺ markers on red blood cells of a fetus.

the foreign cells and cause them to clump together. As you'll see in the next chapter, antibodies bind to specific foreign markers and target any cell or particle bearing them for destruction by the immune system. Also, the large number of foreign cells transfused into a patient end up in numerous clumps that can clog small blood vessels and damage tissues. Without treatment, death may follow. The same thing can happen during certain pregnancies when antibodies diffuse from a mother's circulatory system into that of her unborn child.

With knowledge of surface markers and antibodies, scientists analyze the self markers on a person's red blood cells. Blood typing is based on these analyses.

ABO Blood Typing

Molecular variations in one kind of self marker on red blood cells are analyzed with **ABO blood typing**. The genetic basis of such variation is described in Section 11.4. People with one form of the marker are said to have type A blood; those with another form have type B blood. Those with both forms of the marker on their red blood cells have type AB blood. Others do not have either form of the marker; they have type O blood.

If you are type A, your antibodies ignore A markers but will act against B markers. If you are type B, your antibodies will ignore B markers but will act against A markers. If you are type AB, your antibodies ignore both forms of the marker, so you can tolerate donations of type A, B, AB, or O blood. If you are type O, you have antibodies against both forms of the marker, so your options are limited to type O donations.

Rh Blood Typing

Rh blood typing is based on the presence or absence of the Rh marker (first identified in blood of *Rh*esus monkeys). If you are type Rh⁺, your blood cells bear this marker. If you are type Rh⁻, they do not. Usually people don't have antibodies against Rh markers. But an Rh⁻ recipient of transfused Rh⁺ blood will make antibodies against them, and these stay in the blood.

If an Rh⁻ woman is impregnated by an Rh⁺ man, there is a chance the fetus will be Rh⁺. Some fetal red blood cells usually leak into a woman's blood during childbirth and her body makes antibodies against Rh (Figure 38.9). If she gets pregnant again, Rh antibodies will enter the bloodstream of her new fetus. If the fetus is Rh⁺, then her antibodies will cause its red blood cells to swell, rupture, and release hemoglobin.

Erythroblastosis fetalis follows an immune reaction of an Rh⁻ mother against her Rh⁺ fetus. Too many cells are killed, and the fetus dies. If diagnosed before birth, a newborn survives when its blood is slowly replaced with transfusions that are free of Rh antibodies. Today, a known Rh⁻ woman can be treated right after her first pregnancy with an anti-Rh gamma globulin (RhoGam) that will protect her next fetus. The drug will inactivate any Rh⁺ fetal blood cells that are circulating through her bloodstream before she can become sensitized and start producing potentially dangerous antibodies.

To avoid symptoms of blood incompatibilities, red blood cells should be typed before transfusions or pregnancies.

HUMAN CARDIOVASCULAR SYSTEM

"Cardiovascular" comes from the Greek *kardia* (heart) and Latin *vasculum* (vessel). In a human cardiovascular system, a muscular heart pumps the blood into large-diameter **arteries**. The blood flows into small, muscular **arterioles**, which branch into even smaller diameter capillaries (introduced earlier). Blood flows continually from capillaries into small **venules**, and from there to large-diameter **veins** that return it to the heart.

As in most vertebrates, a partition separates a human heart into a double pump, which drives blood through two cardiovascular circuits (Figure 38.10). Each circuit has its own arteries, arterioles, capillaries, venules, and veins. The *pulmonary* circuit, a short loop, oxygenates blood. It leads from the heart's right half to capillary beds in both lungs, then returns to the heart's left half. The *systemic* circuit, a longer loop, starts at the heart's left half. Its main artery, the **aorta**, accepts oxygenated blood, which flows through arterioles, capillary beds in all regions, then veins that deliver the oxygen-poor blood to the heart's right half. Figure 38.11 identifies key blood vessels of both circuits and their functions.

For most of the systemic branchings, a given volume of blood flows through only one capillary bed. The branch to the intestines is one of the exceptions. Blood picks up glucose and other absorbed substances from one bed, then moves through another capillary bed in the liver. This organ has many crucial roles, including metabolizing substances (Section 6.8). The second bed gives the liver time to process absorbed nutrients.

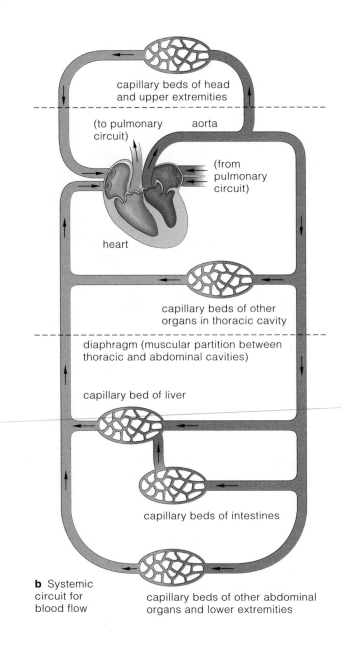

b Systemic circuit for blood flow

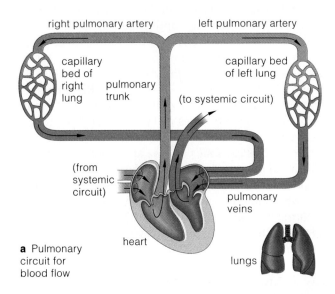

a Pulmonary circuit for blood flow

Figure 38.10 (**a**,**b**) Pulmonary and systemic circuits for blood flow through the human cardiovascular system. Blood vessels carrying oxygenated blood are color-coded *red*. Those carrying oxygen-poor blood are color-coded *blue*. (**c**) Distribution of the heart's output in a person at rest. The lungs receive all blood pumped out of the heart's right half. Organs serviced by the systemic circuit receive the portions indicated from the heart's left half. Control mechanisms adjust the flow distribution when required.

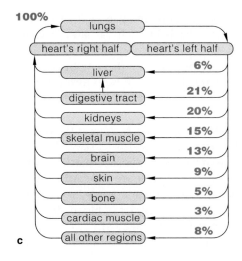

100%

lungs	
heart's right half / heart's left half	
liver	6%
digestive tract	21%
kidneys	20%
skeletal muscle	15%
brain	13%
skin	9%
bone	5%
cardiac muscle	3%
all other regions	8%

c

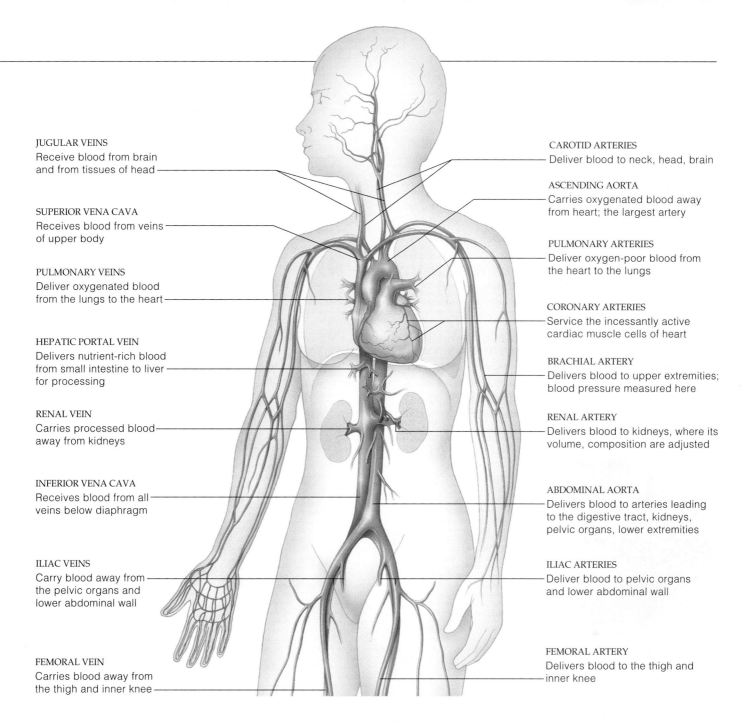

JUGULAR VEINS
Receive blood from brain and from tissues of head

SUPERIOR VENA CAVA
Receives blood from veins of upper body

PULMONARY VEINS
Deliver oxygenated blood from the lungs to the heart

HEPATIC PORTAL VEIN
Delivers nutrient-rich blood from small intestine to liver for processing

RENAL VEIN
Carries processed blood away from kidneys

INFERIOR VENA CAVA
Receives blood from all veins below diaphragm

ILIAC VEINS
Carry blood away from the pelvic organs and lower abdominal wall

FEMORAL VEIN
Carries blood away from the thigh and inner knee

CAROTID ARTERIES
Deliver blood to neck, head, brain

ASCENDING AORTA
Carries oxygenated blood away from heart; the largest artery

PULMONARY ARTERIES
Deliver oxygen-poor blood from the heart to the lungs

CORONARY ARTERIES
Service the incessantly active cardiac muscle cells of heart

BRACHIAL ARTERY
Delivers blood to upper extremities; blood pressure measured here

RENAL ARTERY
Delivers blood to kidneys, where its volume, composition are adjusted

ABDOMINAL AORTA
Delivers blood to arteries leading to the digestive tract, kidneys, pelvic organs, lower extremities

ILIAC ARTERIES
Deliver blood to pelvic organs and lower abdominal wall

FEMORAL ARTERY
Delivers blood to the thigh and inner knee

Figure 38.11 Location and functions of major blood vessels of the human cardiovascular system.

Figure 38.10*c* shows you how the two blood circuits distribute the heart's output to different organs. The percentages listed are for a person at rest. As you will see, change in levels of physical activity, plus shifting internal and external conditions, lead to adjustments in the flow distribution by the systemic route. Example: Blood flow to and from a skeletal muscle varies greatly, depending on what that muscle is being called upon to do. It varies to and from skin. When the body gets too cold, blood flow—hence metabolic heat—gets diverted away from the skin's extensive capillary beds. When the body is too hot, blood flow to these beds increases and heat radiates away from the skin's surface. Only the brain is exempt; it cannot tolerate flow variations.

The human cardiovascular system consists of two separate circuits, pulmonary and systemic, for blood flow.

The pulmonary circuit is a short loop from the heart's right half, through both lungs, to the heart's left half. Oxygen-poor blood flowing into the circuit rapidly picks up oxygen and gives up carbon dioxide at capillary beds in the lungs.

The longer systemic circuit starts at the heart's left half and aorta. It delivers oxygen and accepts carbon dioxide at capillary beds of all metabolically active regions. Then its veins deliver oxygen-poor blood to the heart's right half.

THE HEART IS A LONELY PUMPER

A human heart beats 2.5 billion times during a seventy-year life span, so you know it must be a durable pump. Its structure speaks of its durability. The pericardium, a double sac of tough connective tissue, protects and anchors the heart to nearby structures (*peri*, around). A fluid between its layers lubricates the heart during its perpetual wringing motions. The inner layer is actually part of the heart wall (Figure 38.12). The myocardium, the bulk of the wall, has cardiac muscle cells tethered to elastin and collagen fibers. The fibers are so densely crisscrossed, they serve as a "skeleton" against which contractile force can be applied. Cardiac muscle cells need so much oxygen, they have a separate, *coronary* circulation; coronary arteries branch off the aorta and lead to the heart's own capillary bed. The inner layer of the heart wall is composed of connective tissue and endothelium. This sheet of single cells is the type of epithelium found only in the heart and blood vessels.

Each half of the heart has two chambers: an atrium (plural, atria) and a ventricle. Blood flows into the atria, then the ventricles, then out through the great arteries: the aorta and pulmonary trunk. Between each atrium and ventricle is an AV valve (short for atrioventricular). Between each ventricle and the artery leading out from it is a semilunar valve. Both types are *one-way* valves. They are alternately forced open and then shut to help keep the blood moving in the forward direction only.

Cardiac Cycle

Each time the heart beats, its four chambers go through phases of systole (contraction) and diastole (relaxation). The sequence of contraction and relaxation is a **cardiac cycle**. First, relaxed atria fill with blood. Increasing fluid pressure forces both AV valves open. Blood flows into the ventricles, which fill when the atria contract (Figure 38.13). As the filled ventricles contract, the rising fluid pressure forces the AV valves shut. It rises so sharply above the pressure in the great arteries that it forces the semilunar valves open—and blood leaves the heart. The

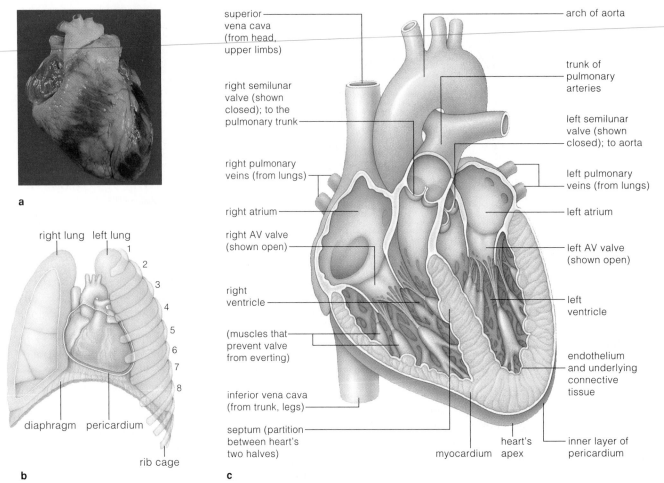

a

b

right lung left lung

diaphragm pericardium

rib cage

c

superior vena cava (from head, upper limbs)

right semilunar valve (shown closed); to the pulmonary trunk

right pulmonary veins (from lungs)

right atrium

right AV valve (shown open)

right ventricle

(muscles that prevent valve from everting)

inferior vena cava (from trunk, legs)

septum (partition between heart's two halves)

arch of aorta

trunk of pulmonary arteries

left semilunar valve (shown closed); to aorta

left pulmonary veins (from lungs)

left atrium

left AV valve (shown open)

left ventricle

endothelium and underlying connective tissue

myocardium heart's apex inner layer of pericardium

Figure 38.12 (**a**) Photograph of the human heart and (**b**) its location in the thoracic cavity. (**c**) Cutaway view of the heart, showing its wall and internal organization.

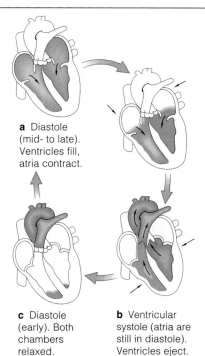

a Diastole (mid- to late). Ventricles fill, atria contract.

c Diastole (early). Both chambers relaxed.

b Ventricular systole (atria are still in diastole). Ventricles eject.

Figure 38.13 Blood flow in a cardiac cycle. Blood and heart movements generate a "lub-dup" sound at the chest wall. At each "lub," AV valves are closing as ventricles contract. At each "dup," semilunar valves are closing as ventricles relax.

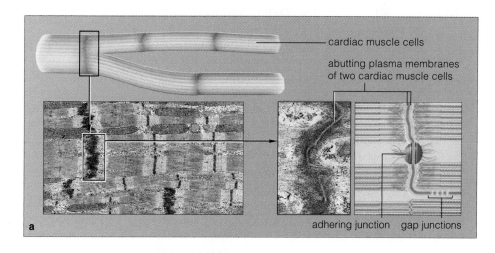

cardiac muscle cells

abutting plasma membranes of two cardiac muscle cells

adhering junction   gap junctions

a

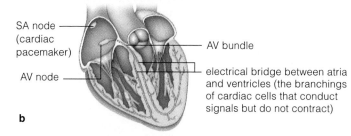

SA node (cardiac pacemaker)

AV node

AV bundle

electrical bridge between atria and ventricles (the branchings of cardiac cells that conduct signals but do not contract)

b

Figure 38.14 (**a**) Micrograph of abutting ends of cardiac muscle cells, held together by tight junctions and adhering junctions. Gap junctions connecting their cytoplasm afford low electrical resistance and enhance the rapid spread of excitation from cell to cell. (**b**) Location of noncontracting cardiac muscle cells of the cardiac conduction system.

ventricles relax, semilunar valves close, and the already filling atria are ready to repeat the cycle. And so, atrial contraction simply helps fill the ventricles. *Contraction of the ventricles is the driving force for blood circulation.*

Cardiac Muscle Contraction

Like skeletal muscle, cardiac muscle is striated (striped). In response to action potentials, the sarcomeres of its cells contract in the manner predicted by the sliding-filament model (Section 37.7). This tissue, too, requires ATP for contraction, which a great many mitochondria in the myocardium provide. But cardiac muscle cells are structurally unique. They are branching, short, and connected at their end regions. Gap junctions span both plasma membranes of adjoining cells. They let action potentials spread swiftly through cardiac tissue in waves of excitation that wash over the heart (Figure 38.14).

In addition, about 1 percent of the cardiac muscle cells *do not* contract. They function instead as a **cardiac conduction system**. About seventy times each minute, these specialized cells initiate and propagate waves of excitation that travel in a rhythmic sequence from atria to ventricles. Their synchronized excitation directs the heart's efficient pumping. Each wave starts at the SA (sinoatrial) node, a cluster of cell bodies in the wall of the right atrium (Figure 38.14). It passes through the wall to another cluster of cell bodies, the AV node. *This is the only electrical bridge between the atria and ventricles, which are insulated everywhere else by connective tissue.* After the AV node, conducting cells are arranged as a bundle in the partition between the heart's two halves. These cells branch, and branches deliver the excitatory wave up the ventricle walls. The ventricles contract in response with a twisting movement, upward from the heart's apex, that ejects blood into the great arteries.

The SA node fires action potentials faster than the rest of the system, so it acts as the **cardiac pacemaker**. Its spontaneous, rhythmic signals are the basis for the normal rate of heartbeat. The nervous system can only *adjust* the rate and strength of contractions dictated by the pacemaker. Even if all nerves leading to a heart are severed, the heart will keep on beating!

The heart's construction reflects its role as a durable pump. Under the spontaneous, rhythmic signals from the cardiac pacemaker, its branching, abutting cardiac muscle cells contract in synchrony, almost as if they were a single unit.

The heart has four chambers (each half has one atrium and one ventricle). Contraction of the ventricles is the driving force for blood circulation away from the heart.

BLOOD PRESSURE IN THE CARDIOVASCULAR SYSTEM

As you have seen, blood flows to and from the human heart through arteries, arterioles, capillaries, venules, and then veins. Figure 38.15*a–d* shows the structure of these blood vessels. Arteries are the main transporters of oxygenated blood through the body. Arterioles are sites of control over the volume of blood flow through every organ. Blood capillaries and, to a lesser extent, venules are the diffusion zones of the cardiovascular system. Veins function as blood volume reservoirs and transporters of oxygen-poor blood back to the heart.

Two factors greatly influence the flow rate through each type of blood vessel. First, the flow rate is directly proportional to the pressure gradient between the start and end of the vessel. Second, the flow rate is inversely proportional to the vessel's resistance to flow.

Blood pressure is fluid pressure imparted to blood by heart contractions. In the pulmonary and systemic circuits, the *difference* in pressure between two points determines the flow rate. The beating heart establishes high pressure at the start of a circuit. Flowing blood rubs against the vessel walls, and the friction impedes flow. The major factor determining resistance to flow is blood vessel radius. Simply put, resistance increases as the tubes narrow down. Even a twofold decrease in radius increases the resistance sixteenfold.

Mainly because of frictional losses after the heart pumps out blood, pressure differs along the circuits. It is highest in contracting ventricles and still high at the start of arteries. It continues to drop and is lowest in relaxed atria (Figure 38.16).

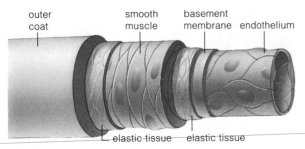

a ARTERY

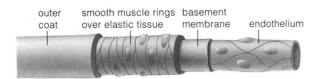

b ARTERIOLE

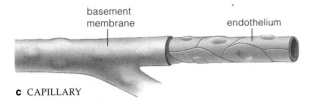

c CAPILLARY

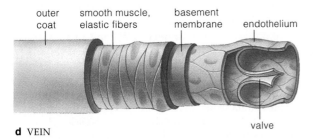

d VEIN

Figure 38.15 Structure of blood vessels. The basement membrane around each vessel's endothelium is a noncellular layer, rich with proteins and polysaccharides. Small venules resemble capillaries in structure and function.

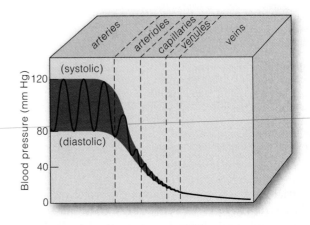

Figure 38.16 Plot of the decline in fluid pressure for a volume of blood moving through the systemic circuit.

Arterial Blood Pressure

With their large diameter and low resistance to flow, arteries are fast transporters of oxygenated blood. They also are pressure reservoirs that smooth out pulsations in pressure caused by each cardiac cycle. Their thick, muscular, elastic wall bulges from the large volume of blood that ventricular contraction forces into them. The wall recoils, and this forces the blood forward through the circuit while the heart is relaxing.

Suppose you want to measure your blood pressure daily. Typically, you'd take a reading from the brachial artery of an upper arm while resting (Figure 38.17). For one cardiac cycle, *systolic* pressure is the peak pressure that contracting ventricles exert against the artery wall. *Diastolic* pressure is the lowest arterial blood pressure of a cardiac cycle, reached when ventricles are relaxing. You can estimate the mean arterial pressure as *diastolic pressure + 1/3 pulse pressure*. (The difference between the highest and lowest readings is the pulse pressure.)

Figure 38.17 Measuring blood pressure.

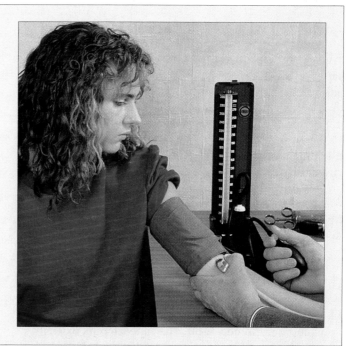

A hollow cuff attached to a pressure gauge is wrapped around an upper arm. The cuff is inflated with air to a pressure above the highest pressure of the cardiac cycle—at systole, when the ventricles contract. Above this pressure, you cannot hear any sounds through a stethoscope positioned below the cuff and above the artery, for no blood is flowing through the vessel.

Air in the cuff is slowly released, so some blood flows into the artery. The turbulent flow causes soft tapping sounds. When the tapping starts, the gauge's value is the systolic pressure, or about 120 mm mercury (Hg) in young adults at rest. This value means the measured pressure would force mercury to move upward 120 millimeters in a narrow glass column.

More air is released from the cuff. Just after the sounds grow dull and muffled, blood is flowing continuously, so the turbulence and tapping end. The silence corresponds to diastolic pressure at the end of a cardiac cycle, just before the heart pumps out blood. The reading is usually about 80 mm Hg. In this case, pulse pressure (the difference between the highest and lowest pressure readings) is 120 − 80, or 40 mm Hg. Assuming you're an adult in good health, this average resting value stays fairly constant over a few weeks or months.

How Can Arterioles Resist Flow?

Track a volume of blood through the systemic circuit and you notice the greatest pressure drop at arterioles (Figure 38.16), which present the most flow resistance. This control mechanism allows time for more or less of the total flow volume to be diverted to specific organs. Control signals act at smooth muscle ringing arteriole walls. Some cause smooth muscle cells to relax, which results in **vasodilation**—an enlargement, or dilation, of blood vessel diameter. Others cause smooth muscle cells to contract, which results in **vasoconstriction**—a shrinking of blood vessel diameter.

The nervous and endocrine systems can control the arteriole diameter. For example, sympathetic nerves end on smooth muscle cells of many arterioles. Increased activity in these nerves initiates action potentials that stimulate contraction and widespread vasoconstriction. Epinephrine and angiotensin are two of the signaling molecules that trigger changes in arteriole diameter.

Arteriole diameter also is adjusted when changes in metabolism shift localized concentrations of substances in a tissue. Such local chemical changes are "selfish" in that they invite or divert blood flow to meet a tissue's own metabolic needs. As you run, skeletal muscle cells use up oxygen, and levels of carbon dioxide, H^+, K^+, and other solutes rise. The conditions make arterioles in the vicinity dilate. More blood flows through active muscles to deliver more raw materials and carry away cell products and metabolic wastes. When the muscles relax and demand fewer blood deliveries, the oxygen level increases and makes the arterioles constrict.

Controlling Mean Arterial Blood Pressure

Mean arterial pressure depends on cardiac output and on total resistance through the vascular system. Cardiac output is adjusted by controls over the rate and strength of heartbeats. The total resistance is adjusted mainly by coordinating vasoconstriction at arterioles. Of course, organs and tissues vary in their demands for blood's pickups and deliveries, so the balance between cardiac output and total resistance is juggled all the time.

A **baroreceptor reflex** is the key short-term control over arterial pressure. It starts as baroreceptors detect changes in arterial mean pressure and pulse pressure. The most vital of these mechanoreceptors are located in carotid arteries (which deliver blood to the brain) and in the arch of the aorta (which delivers blood to the rest of the body). Baroreceptors continually generate signals that reach a control center in the part of the hindbrain called the medulla oblongata. In response, the control center adjusts its own signals, which then travel along parasympathetic and sympathetic nerves to the blood vessels and heart. For instance, when the mean arterial pressure rises, the center commands the heart to beat more slowly and contract less forcefully.

Long-term controls over blood pressure are exerted at kidneys, which adjust the volume and composition of the blood. And that is a topic of another chapter.

Mean arterial pressure is an outcome of controls over the heart's output and over the total resistance to blood flow through the vascular system, as exerted mainly at arterioles.

FROM CAPILLARY BEDS BACK TO THE HEART

Capillary Function

THE NATURE OF THE EXCHANGES   Capillary beds, again, serve as diffusion zones for exchanges between blood and interstitial fluid. Living cells in any body tissue die quickly when deprived of these exchanges. Brain cells die within four minutes.

Between 10 billion and 40 billion capillaries service the human body. Collectively, these components of the circulatory system offer a tremendous surface area for gas exchanges. They extend into nearly all tissues. At least one threads as close as 0.001 centimeter to each living cell. Proximity to cells is vital, because shorter distances mean faster diffusion rates. Think about this: It would take years for oxygen to diffuse on its own from your lungs all the way down your legs. By then your toes, and the rest of you, would long be dead.

Red blood cells (8 micrometers wide) squeeze single file through capillaries, which are 3 to 7 micrometers wide and deform only a bit. The squeeze puts oxygen-transporting red blood cells, and substances dissolved in plasma, in direct contact with the exchange area—the capillary wall—or only a short distance away from it.

And remember, the capillaries collectively present a greater cross-sectional area than the arterioles leading into them, so the flow velocity cannot be as great in the capillary beds. The slowdown means there is plenty of time for interstitial fluid to exchange substances with the 5 percent or so of the total volume of blood moving forward through these beds during a specified interval. Here you may wish to reflect on Figure 38.3*e*.

Each capillary is a single sheet of endothelial cells in the form of a tube (Figure 38.15*c*). The cells abut one another, but the "fit" varies in different body regions. In most cases, narrow, water-filled clefts occur between endothelial cells, so some of the fluid in blood leaks out into interstital fluid. By contrast, in capillaries that service the brain, tight junctions join the cells together and clefts are nonexistent. Substances cannot "leak" between these cells but must move through them. Thus the junctions are a functional part of the blood–brain barrier described in Section 34.8.

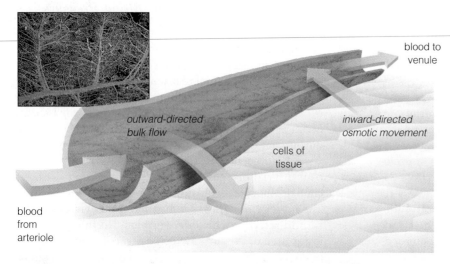

blood to venule

outward-directed bulk flow

inward-directed osmotic movement

cells of tissue

blood from arteriole

Figure 38.18   Bulk flow in an idealized capillary bed. The fluid movement plays no significant role in diffusion of solutes. However, it is important in maintaining the distribution of extracellular fluid between blood and interstitial fluid.

The fluid movements across a capillary wall result from the two opposing effects of ultrafiltration and reabsorption.

At a capillary's arteriole end, the difference between blood pressure and interstitial fluid pressure forces some plasma, but very few plasma proteins, to leave the capillary. Plasma moves by bulk flow through clefts between endothelial cells of the capillary wall. Ultrafiltration is the bulk flow of fluid *out* of the capillary.

Reabsorption is an osmotic movement of some interstitial fluid *into* the capillary as a result of a difference in the water concentration between interstitial fluid and plasma. Plasma, with its dissolved protein components, has a greater solute concentration and therefore a lower water concentration.

Reabsorption near the end of a capillary bed tends to balance ultrafiltration at the beginning. Normally there is only a small *net* filtration of fluid, which the lymphatic system returns to the blood.

ARTERIOLE END OF CAPILLARY BED	
Outward-Directed Pressure:	
Hydrostatic pressure of blood in capillary:	35 mm Hg
Osmosis due to interstitial proteins:	28 mm Hg
Inward-Directed Pressure:	
Hydrostatic pressure of interstitial fluid:	0
Osmosis due to plasma proteins:	3 mm Hg
Net Ultrafiltration Pressure:	
(35 – 0) – (28 – 3) = 10 mm Hg	
ULTRAFILTRATION FAVORED	

VENULE END OF CAPILLARY BED	
Outward-Directed Pressure:	
Hydrostatic pressure of blood in capillary:	15 mm Hg
Osmosis due to insterstitial proteins:	28 mm Hg
Inward-Directed Pressure:	
Hydrostatic pressure of interstitial fluid:	0
Osmosis due to plasma proteins:	3 mm Hg
Net Reabsorption Pressure:	
(15 – 0) – (28 – 3) = – 10 mm Hg	
REABSORPTION FAVORED	

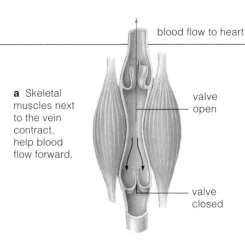

blood flow to heart

a Skeletal muscles next to the vein contract, help blood flow forward.

valve open

valve closed

b Skeletal muscles relax, and valves in vein shut. Backflow prevented.

valve closed

valve closed

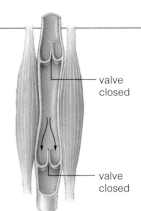

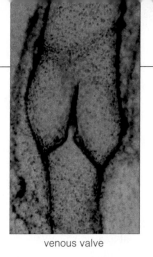

venous valve

Oxygen, carbon dioxide, and most other small, lipid-soluble substances also move across the capillary wall by diffusing through the lipid portion of an endothelial cell's plasma membrane and on through its cytoplasm. Certain proteins leave and enter blood by endocytosis or exocytosis. Small, water-soluble substances such as ions enter and leave at the clefts between endothelial cells. So do the white blood cells, as you will see in the chapter to follow.

MECHANISMS OF EXCHANGE Diffusion is not the only way substances are exchanged across a capillary wall. When fluid pressures acting on the wall are imbalanced, there may be bulk flow one way or the other across it. Bulk flow, recall, is a movement of water and solutes in the same direction in response to fluid pressure.

As Figure 38.18 shows, water and solutes in blood and interstitial fluid affect the flow direction. At the start of the capillary bed, the outward-directed blood pressure is greater than the inward-directed osmosis brought about by plasma proteins. In a process called **ultrafiltration**, a small amount of protein-free plasma is pushed out in bulk at clefts in the capillary wall. The balance shifts farther on. Whereas blood pressure has steadily declined, the osmotic pressure has stayed the same. When inward-directed osmosis exceeds outward pressure, tissue fluid returns to the capillary through clefts between cells. This process is called **reabsorption**.

The normal outcome is a tiny *net* outward flow of fluid from a capillary bed, which the lymphatic system returns to blood. This bulk flow helps maintain a fluid balance between interstitial fluid and blood, which is vital. Blood pressure requires an adequate volume of blood. If it plummets, as by hemorrhaging, interstitial fluid may be reabsorbed to help counter the loss.

High blood pressure can cause excess ultrafiltration. If too much fluid collects in interstitial spaces, this is *edema*. Some edema occurs during physical exercise, when arterioles dilate in many tissues. It also results from obstructions in veins or from heart failure. Edema is extreme during *elephantiasis*. This, disease, recall, is brought on by roundworm infection and a subsequent obstruction of lymphatic vessels (Section 25.7).

Figure 38.19 Vein function. Valves in medium-sized veins prevent backflow of blood. Bulging of adjacent contracting skeletal muscles helps raise fluid pressure in a vein.

Venous Pressure

What happens to the flow velocity after it continues on past the vast cross-sectional area of the capillary beds? Remember, the capillaries merge into venules, or "little veins." These in turn merge into large-diameter veins. Once again the total cross-sectional area is reduced, and flow velocity increases as blood returns to the heart.

In functional terms, venules are a bit like capillaries. Some solutes diffuse across their wall, which isn't much thicker than the thin capillary wall. Some control over capillary pressure also is exerted at these vessels.

Veins are large-diameter, low-resistance transport tubes to the heart (Figure 38.15*d*). Their valves prevent backflow. When gravity beckons, venous flow reverses direction and pushes the valves shut. The vein wall can bulge greatly under pressure, more so than an arterial wall. Thus veins are reservoirs for variable volumes of blood. Taken together, the veins of an adult human can hold up to 50 to 60 percent of the total blood volume.

The vein wall contains some smooth muscle. When blood must circulate faster, as during physical exercise, the smooth muscle contracts. The wall stiffens and the vein bulges less, so that venous pressure increases and drives more blood back to the heart. Also, when limbs move, skeletal muscles bulge against the veins in their vicinity. They help raise the venous pressure and drive blood back to the heart (Figure 38.19). Rapid breathing also contributes to increased venous pressure. Inhaled air pushes down on internal organs and thereby alters the pressure gradient between the heart and veins.

Capillary beds are diffusion zones for exchanges between blood and interstitial fluid. Here, too, bulk flow contributes to the fluid balance between blood and interstitial fluid.

Venules overlap somewhat with capillaries in function.

Veins are highly distensible blood volume reservoirs, and they help adjust flow volume back to the heart.

Cardiovascular Disorders

Each year, cardiovascular disorders disrupt the lives of 40 million people and kill a million of them. All affect blood circulation; they are the leading cause of death in the United States. The two most common are *hypertension*, or sustained high blood pressure, and *atherosclerosis*, a progressive thickening of the arterial wall that narrows the lumen. Both damage or destroy heart muscle, this being a *heart attack*. Both cause *strokes*, or brain damage.

Warning signs of a heart attack are pain or a sensation of squeezing behind the breastbone, numbness or pain in the left arm, sweating, and nausea. Women, more often than men, also experience neck and back pain, tiredness, shortness of breath, anxiety, a sense of indigestion, a fast heartbeat, and low blood pressure.

RISK FACTORS   Researchers correlate these factors with an increased risk of developing cardiovascular disorders:

1. Smoking (Section 40.7)
2. Genetic predisposition to heart attack
3. High level of cholesterol in the blood
4. High blood pressure
5. Obesity (Section 41.10)
6. Lack of regular exercise
7. Diabetes mellitus (Section 36.7)
8. Age (the older you get, the greater the risk)
9. Gender (until age fifty, males are at much greater risk than females)

Exercising regularly, eating properly, and not smoking help lower the risk. With too little exercise and too much food, adipose tissue masses increase in the body, more capillaries develop to service them, and the heart has to work harder to pump blood through the increasingly divided vascular circuits. In people who smoke, nicotine in tobacco makes the adrenal glands secrete epinephrine. This hormone causes blood vessel diameters to constrict, thus boosting heart rate and blood pressure. Also, carbon monoxide in cigarette smoke competes with oxygen for binding sites on hemoglobin. As a result, the heart has to pump harder to deliver enough oxygen to cells. Smoking also predisposes people to atherosclerosis even if their blood level of cholesterol is normal.

The next paragraphs describe some of the damage inflicted by cardiovascular disorders.

HYPERTENSION   Gradual increases in flow resistance through small arteries lead to hypertension. Over time, blood pressure stays above 140/90 even during rest. Heredity may be a factor; the disorder tends to run in families. Diet is a factor for some people; high salt intake raises blood pressure and the heart's workload. In time the heart may enlarge and fail to pump blood effectively. High blood pressure may contribute to a "hardening" of arterial walls, which hampers the delivery of oxygen to the brain, heart, and other vital organs.

Hypertension is called a "silent killer" because affected individuals may show no outward symptoms. Even when they know their blood pressure is high, some people tend to resist helpful medication, changes in diet, and regular exercise. Of 23 million hypertensive Americans, most don't seek treatment. About 180,000 die each year.

ATHEROSCLEROSIS   With *arteriosclerosis*, arteries thicken and lose their elasticity. With atherosclerosis, the condition worsens as cholesterol and other lipids build up in the arterial wall, thus narrowing the lumen (Figure 38.20).

Recall, from the Chapter 15 introduction, that the liver produces enough cholesterol to satisfy the body's needs. A cholesterol-rich diet typically increases the cholesterol concentration in blood. When cholesterol circulates in blood, it is bound to proteins as *low-density* lipoproteins, or **LDLs**. Cells throughout the body have LDL receptors, and they take up LDL—with its cholesterol cargo—for use in metabolism. They do not take up all of it. Some of the excess cholesterol becomes attached to proteins as

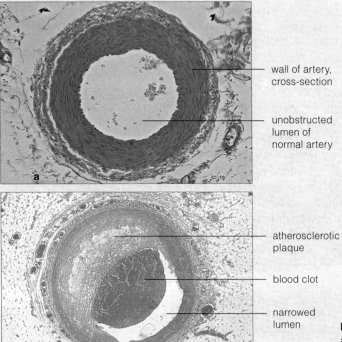

wall of artery, cross-section

unobstructed lumen of normal artery

atherosclerotic plaque

blood clot

narrowed lumen

Figure 38.20 Sections from (**a**) a normal artery and (**b**) an artery with a lumen narrowed by an atherosclerotic plaque. A blood clot almost clogs this one.

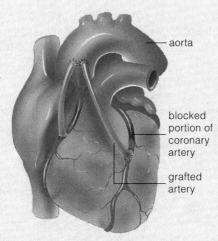

Figure 38.21
Two coronary bypasses (color-coded *green*), which extend from the aorta and past two clogged parts of the coronary arteries.

aorta

blocked portion of coronary artery

grafted artery

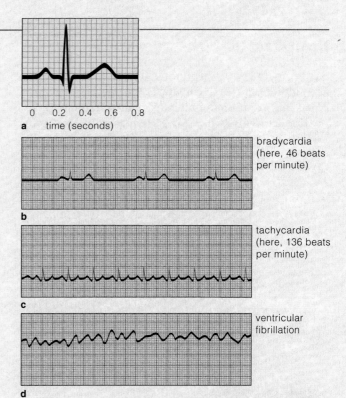

a time (seconds)

bradycardia (here, 46 beats per minute)

b

tachycardia (here, 136 beats per minute)

c

ventricular fibrillation

d

Figure 38.22 (**a**) ECG of a single, normal human heartbeat. (**b–d**) Three examples of recordings of arrhythmias.

high-density lipoproteins, or **HDLs**. The blood transports HDLs back to the liver, where they are metabolized.

For a variety of reasons, some people just can't sweep enough LDL from their blood. As the blood level of LDL rises, so does the risk of atherosclerosis. LDLs infiltrate the wall of arteries. In the wall, abnormal smooth muscle cells multiply, and connective tissue components increase in mass. Cholesterol accumulates in the endothelial cells and in clefts between them. Calcium deposits actually form microscopic slivers of bone on top of the lipids! Then a fibrous net forms over the whole mass. This *atherosclerotic plaque* sticks out into the arterial lumen (Figure 38.20*b*).

The bony slivers shred the endothelium. Platelets gather at the damaged site and initiate clot formation. Conditions worsen as fatty globules in the plaque become oxidized. The globules change shape, and many end up resembling molecules at the surface of common bacteria. One of the bacteria is a type that triggers the formation of bonelike calcium deposits in the lungs. And so the call goes out to bacteria-fighting monocytes. An inflammatory response gets under way, and certain chemicals released during the response activate the genes for bone formation. Normally, this is good; it helps wall off the invaders and keeps the infection from spreading. It is bad news for arteries.

Enlarging plaques and clots narrow or block arteries. Blood flow to tissues that an artery services dwindles or stops. A clot that stays in place is a *thrombus*. When it gets dislodged and travels the bloodstream, it is an *embolus*. Coronary arteries, which have very narrow diameters, are especially vulnerable to clogging. When they narrow to one-quarter of their former diameter, the outcome ranges from *angina pectoris*, or mild chest pains, to a heart attack.

Physicians can use *stress electrocardiograms* to diagnose atherosclerosis in coronary arteries. These are recordings of electrical activity of the cardiac cycle made as a person exercises on a treadmill. They also diagnose the condition

by *angiography*. By this procedure, an opaque dye injected into blood reveals blockages of blood flow in x-ray film.

Surgery may alleviate severe blockage. With *coronary bypass surgery*, a section of an artery from the chest is stitched to the aorta and to the coronary artery below the narrowed or blocked region, as shown in Figure 38.21. With *laser angioplasty*, laser beams directed at the plaques vaporize them. With *balloon angioplasty*, a small balloon inflated within a blocked artery flattens the plaques.

ARRHYTHMIAS ECGs detect irregular heart rhythms, or *arrhythmias* (Figure 38.22). Arrhythmias aren't always abnormalities. For example, endurance athletes may show *bradycardia*, a below-average resting cardiac rate. As an adaptation to ongoing strenuous exercise, their nervous system has adjusted the cardiac pacemaker's rate of contraction downward. Exercise or stress often causes 100+ heartbeats a minute, or *tachycardia*.

Atrial fibrillation, an abnormal, irregular contraction of the atria, affects over 10 percent of the elderly and young people with heart diseases. Coronary occlusions or another disorder may cause irregular rhythms that rapidly lead to a dangerous condition called *ventricular fibrillation*. Cardiac muscle contracts haphazardly in parts of the ventricles, so blood pumping falters. Within seconds, a person loses consciousness, which may be a sign of impending death. A strong electric shock to the chest sometimes restores normal cardiac function.

HEMOSTASIS

Small blood vessels are vulnerable to ruptures, cuts, and similar injuries. **Hemostasis** is a process that stops blood loss from them and constructs a framework for repairs. It involves blood vessel spasm, platelet plug formation, and blood coagulation (Figure 38.23).

STIMULUS: Blood vessel damaged

RESPONSE

FIRST PHASE	SECOND PHASE	THIRD PHASE
Vascular spasm for 1/2 hour constricts vessel at site, slows blood loss.	Platelets aggregate, stick together within 15 seconds to plug site.	Clot formation starts after 30 seconds or more.

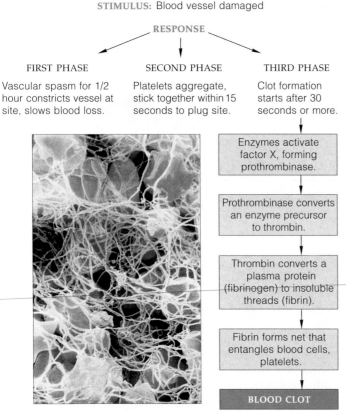

Enzymes activate factor X, forming prothrombinase.

Prothrombinase converts an enzyme precursor to thrombin.

Thrombin converts a plasma protein (fibrinogen) to insoluble threads (fibrin).

Fibrin forms net that entangles blood cells, platelets.

BLOOD CLOT

Figure 38.23 Hemostasis.

In the vascular phase of the process, smooth muscle located in the damaged wall contracts in an automatic response called a spasm. Vasoconstriction may stanch the blood flow as a temporary fix. In the platelet phase, platelets clump together to temporarily fill the breach. They release substances that help prolong the spasm and attract more platelets. In the coagulation phase, plasma proteins convert blood to a gel and form a clot. First, rod-shaped fibrinogens react to collagen fibers exposed by the damage. They stick together as long, insoluble threads that adhere to the exposed collagen, forming a net that traps blood cells and platelets. The entire mass is a clot. Finally, the clot retracts and forms a compact mass, thus sealing the breach in the wall.

> The body routinely repairs damage to small blood vessels and prevents blood loss. The repair process involves blood vessel spasm, platelet plug formation, and coagulation.

LYMPHATIC SYSTEM

We conclude this chapter with a brief section on the manner in which the lymphatic system supplements blood circulation. Think of this section as a bridge to the next chapter, on immunity. Why? The lymphatic system also helps defend the body against injury and attack. This important system is composed of drainage vessels, lymphoid organs, and lymphoid tissues. When tissue fluid moves into the vessels, we call this lymph.

Lymph Vascular System

The portion of the lymphatic system called the **lymph vascular system** consists of many tubes that collect and deliver water and solutes from the interstitial fluid to ducts of the circulatory system. Its main components are lymph capillaries and lymph vessels (Figure 38.24).

The lymph vascular system serves three functions. First, its vessels are drainage channels for water and plasma proteins that have leaked out from the blood at capillary beds and must be delivered back to the blood circulation. Second, the system also takes up fats that the body has absorbed from the small intestine and delivers them to the general circulation, in the manner described in Section 41.5. Third, it delivers pathogens, foreign cells and material, and cellular debris from the body's tissues to the lymph vascular system's highly organized disposal centers, the lymph nodes.

The lymph vascular system starts at capillary beds, where fluid enters the lymph capillaries. The capillaries have no obvious entrance; water and solutes move into clefts between cells that form flaplike valves. As you can see from Figure 38.24b, these are areas where the endothelial cells overlap. Lymph capillaries merge and become lymph vessels, which have a larger diameter. Lymph vessels contain some smooth muscle in their wall. They also have valves in the lumen that prevent backflow. These vessels converge into collecting ducts that drain into veins in the lower neck.

Lymphoid Organs and Tissues

The other portion of the lymphatic system has roles in the body's defense responses to injury and attack. We call them the *lymphoid* organs and tissues. The major types are the lymph nodes, spleen, and thymus, as well as the tonsils, and patches of tissue in the wall of the small intestine and appendix.

Lymph nodes are strategically located at intervals along lymph vessels (Figure 38.24c). Before entering blood, the lymph trickles through at least one node for filtration. Masses of lymphocytes take up stations in the nodes after forming in bone marrow. When they recognize cellular debris or an invader, they multiply rapidly and form large armies to destroy it.

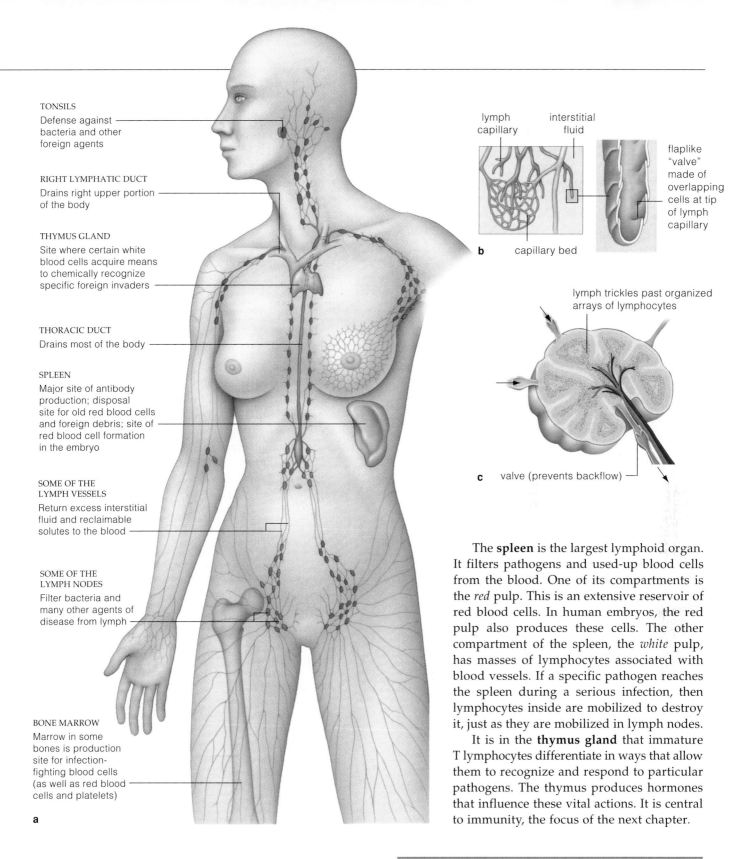

TONSILS
Defense against bacteria and other foreign agents

RIGHT LYMPHATIC DUCT
Drains right upper portion of the body

THYMUS GLAND
Site where certain white blood cells acquire means to chemically recognize specific foreign invaders

THORACIC DUCT
Drains most of the body

SPLEEN
Major site of antibody production; disposal site for old red blood cells and foreign debris; site of red blood cell formation in the embryo

SOME OF THE LYMPH VESSELS
Return excess interstitial fluid and reclaimable solutes to the blood

SOME OF THE LYMPH NODES
Filter bacteria and many other agents of disease from lymph

BONE MARROW
Marrow in some bones is production site for infection-fighting blood cells (as well as red blood cells and platelets)

a

lymph capillary   interstitial fluid

flaplike "valve" made of overlapping cells at tip of lymph capillary

b   capillary bed

lymph trickles past organized arrays of lymphocytes

c   valve (prevents backflow)

The **spleen** is the largest lymphoid organ. It filters pathogens and used-up blood cells from the blood. One of its compartments is the *red* pulp. This is an extensive reservoir of red blood cells. In human embryos, the red pulp also produces these cells. The other compartment of the spleen, the *white* pulp, has masses of lymphocytes associated with blood vessels. If a specific pathogen reaches the spleen during a serious infection, then lymphocytes inside are mobilized to destroy it, just as they are mobilized in lymph nodes.

It is in the **thymus gland** that immature T lymphocytes differentiate in ways that allow them to recognize and respond to particular pathogens. The thymus produces hormones that influence these vital actions. It is central to immunity, the focus of the next chapter.

Figure 38.24 (**a**) Components of the human lymphatic system and their functions. The small *green* ovals represent some of the major lymph nodes. Not shown are patches of lymphoid tissue in the small intestine and in the appendix. (**b**) Diagram of lymph capillaries at the start of a drainage network, the lymph vascular system. (**c**) Cutaway view of a lymph node. Its inner compartments are packed with arrays of infection-fighting white blood cells.

The lymph vascular portion of the lymphatic system returns water and solutes from tissue fluid to blood, and it delivers fats and foreign material to lymph nodes.

The lymphatic system also includes lymph nodes and other lymphoid organs that have specific roles in defending the body against tissue damage and infectious diseases.

SUMMARY   **Gold** indicates text section

1. The closed circulatory system of humans and other vertebrates consists of a heart (muscular pump), many blood vessels (arteries, arterioles, capillaries, venules, and veins), and blood. It functions in the rapid internal transport of substances to and from cells. In birds and mammals, blood flows in two separate circuits. One is pulmonary and the other is systemic. *38.1, 38.5*

2. Blood helps maintain favorable conditions for cells. This fluid connective tissue consists of plasma, red and white blood cells, and platelets. It delivers oxygen and other substances to interstitial fluid around cells. It also picks up cell products and wastes from that fluid. *38.2*

   a. Plasma, the liquid portion of blood, is a transport medium for blood cells and platelets. Plasma is a solvent for plasma proteins, simple sugars, lipids, amino acids, mineral ions, vitamins, hormones, and several gases.

   b. Red blood cells transport oxygen from the lungs to all body regions. They are packed with hemoglobin, an iron-containing pigment that binds reversibly with oxygen. They also transport some carbon dioxide from interstitial fluid to the lungs.

   c. Phagocytic white blood cells cleanse tissues by engulfing dead cells, cellular debris, and anything else recognized as not being part of the body. White blood cells called lymphocytes form great armies that destroy specific bacteria, viruses, and other disease agents.

3. Blood flows in these two circuits in humans: *38.5*

   a. A pulmonary circuit loops between the heart and lungs. Oxygen-poor blood from systemic veins enters the heart's *right* atrium, is pumped through pulmonary arteries to the two lungs, picks up oxygen, then flows through pulmonary veins to the heart's left atrium.

   b. A systemic circuit loops between the heart and all tissues. Oxygenated blood in the *left* atrium flows into the left ventricle, is pumped into the aorta, then is distributed to capillary beds. There, the blood gives up oxygen and picks up carbon dioxide. Systemic veins return the blood to the heart's right atrium.

4. The human heart, a double pump, beats incessantly. Each half has two chambers: atrium and ventricle. Blood flows into the atria, ventricles, then the great arteries. Ventricular contraction drives blood circulation. One-way valves enforce the forward-directed flow. *38.6*

5. The cardiac conduction system is the basis for the heart's rhythmic, spontaneous contractions. *38.6*

   a. One percent of the cardiac muscle cells are not contractile. They are specialized to initiate and distribute action potentials, independently of the nervous system. The nervous system only adjusts the rate and strength of the basic contractions; it does not initiate them.

   b. The system's SA node fires action potentials the fastest and is the cardiac pacemaker. Waves of excitation starting here wash over the atria, then down the heart's partition, then up the ventricles. The ventricles contract in a wringing motion that ejects blood from the heart.

6. Blood pressure is highest in contracting ventricles. It drops in arteries, then in arterioles, capillaries, venules, and veins. It is lowest in relaxed atria. *38.7, 38.8*

   a. Arteries, rapid-transport vessels and a reservoir for pressure, smooth out pressure changes that result from heartbeats and thereby smooth out blood flow.

   b. Arterioles are major sites of control over the flow volume through different body regions.

   c. Capillary beds are diffusion zones where blood and interstitial fluid exchange substances.

   d. Venules overlap capillaries in function. Veins are rapid-transport vessels and a blood volume reservoir that is tapped to adjust flow volume back to the heart.

7. Hemostasis is a process of stopping blood loss from small blood vessels that have been injured. *38.10*

8. The lymphatic system has these functions: *38.11*

   a. Its vascular portion takes up water and plasma proteins that seep out of blood capillaries, then returns them to circulating blood. It transports absorbed fats and delivers pathogens and foreign material to disposal centers. Lymph capillaries and vessels are components.

   b. Its lymphoid organs and tissues have production centers for lymphocytes. Some are battlegrounds where organized arrays of lymphocytes fight disease agents.

Review Questions

1. State the functions of the circulatory and lymphatic systems. Distinguish between blood and interstitial fluid. *CI, 38.1*

2. Distinguish between systemic and pulmonary circuits. *38.1*

3. Describe the cellular components of blood. Then describe the plasma portion of blood. *38.2*

4. First distinguish between the functions of the human heart's atria and ventricles. Label the components of the heart diagram at right. *38.6*

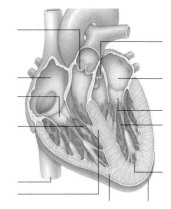

5. State the functions of arteries, arterioles, capillaries, veins, and venules, plus lymph vessels. Identify the four types of blood vessels shown below. *38.7, 38.8*

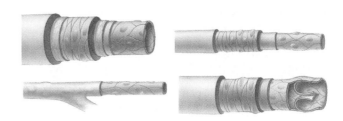

Self-Quiz ANSWERS IN APPENDIX III

1. Cells directly exchange substances with _____ .
   a. blood vessels   c. interstitial fluid
   b. lymph vessels   d. both a and b

2. Which are *not* components of blood?
   a. plasma
   b. blood cells and platelets
   c. gases and other dissolved substances
   d. All of the above are components of blood.

3. The _____ produces red blood cells, which transport _____ and some _____ .
   a. liver; oxygen; mineral ions
   b. liver; oxygen; carbon dioxide
   c. pancreas; oxygen; hormones
   d. bone marrow; oxygen; carbon dioxide

4. The _____ produces white blood cells, which function in _____ and _____ .
   a. liver; oxygen transport; defense
   b. lymph nodes; oxygen transport; pH stabilization
   c. bone marrow; housekeeping; defense
   d. bone marrow; pH stabilization; defense

5. In the pulmonary circuit, the heart's _____ half pumps blood to lungs, then _____ blood flows to the heart.
   a. right; oxygen-poor   c. right; oxygen-rich
   b. left; oxygen-poor   d. left; oxygen-rich

6. In the systemic circuit, the heart's _____ half pumps _____ blood to all body regions.
   a. right; oxygen-poor   c. right; oxygen-rich
   b. left; oxygen-poor   d. left; oxygen-rich

7. Blood pressure is high in _____ and lowest in _____ .
   a. arteries; veins   c. arteries; ventricles
   b. arteries; relaxed atria d. arterioles; veins

8. _____ contraction drives blood circulation; and blood pressure is highest in contracting _____ .
   a. Atrial; ventricles   c. Ventricular; arteries
   b. Atrial; atria   d. Ventricular; ventricles

9. Which is *not* a function of the lymphatic system?
   a. delivers disease agents to disposal centers
   b. contributes to immunity
   c. delivers oxygen to cells
   d. returns water and plasma proteins to blood

10. Match the type of blood vessel with its major function.
    ____ arteries   a. diffusion
    ____ arterioles   b. control of blood volume distribution
    ____ capillaries   c. transport, blood volume reservoirs
    ____ venules   d. overlap capillary function
    ____ veins   e. transport, pressure reservoirs

11. Match the components with their most suitable description.
    ____ capillary beds   a. two atria, two ventricles
    ____ lymph vascular system b. bathes body's living cells
    ____ human heart chambers c. driving force for blood
    ____ blood   d. zones of diffusion
    ____ heart contractions   e. starts at capillary beds
    ____ interstitial fluid   f. fluid connective tissue

Critical Thinking

1. Shirelle, who is using a light microscope to examine a human tissue specimen, sees red blood cells moving single file through thin-walled tubes. She makes a photomicrograph (Figure 38.25). What type of blood vessel has she captured on film?

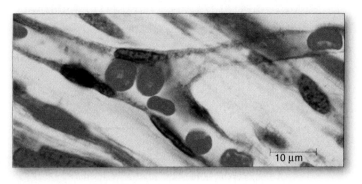

Figure 38.25 Light micrograph of a branching blood vessel.

2. In individuals who have weak or leaky valves in their veins, fluid pressure associated with the backflow of blood causes venous walls below the valves to bulge outward. In time, the walls become stretched and flabbily distorted, a condition called *varicose veins*. Some people are genetically predisposed to develop the condition, but the cumulative mechanical stress associated with prolonged standing, pregnancy, and aging can contribute to it. With chronic varicosity, the legs themselves become swollen. Speculate on why this swelling might occur. Also speculate on why veins close to the leg surface are more susceptible to varicosity than those deeper in the leg tissues.

3. Infection by *Streptococcus pyogenes*, a hemolytic bacterium, may trigger an inflammation that ultimately damages valves in the heart. The disease symptoms of *rheumatic fever* follow. Explain how this disease must affect the heart's functioning and what kinds of symptoms would arise as a consequence.

Selected Key Terms

ABO blood typing *38.4*
agglutination *38.4*
anemia *38.3*
aorta *38.5*
arteriole *38.5*
artery *38.5*
baroreceptor reflex *38.7*
blood *38.1*
blood pressure *38.7*
blood transfusion *38.4*
capillary (blood) *38.1*
capillary bed *38.1*
cardiac conduction system *38.6*
cardiac cycle *38.6*
cardiac pacemaker *38.6*
cell count *38.2*
circulatory system *CI*
HDL *38.9*
heart *38.1*
hemostasis *38.10*
interstitial fluid *38.1*

LDL *38.9*
lymph node *38.11*
lymph vascular system *38.11*
lymphatic system *38.1, 38.11*
plasma *38.2*
platelet *38.2*
pulmonary circuit *38.1, 38.5*
reabsorption *38.8*
red blood cell *38.2*
Rh blood typing *38.4*
spleen *38.11*
stem cell *38.2*
systemic circuit *38.1, 38.5*
thymus gland *38.11*
ultrafiltration *38.8*
vasoconstriction *38.7*
vasodilation *38.7*
vein *38.5*
venule *38.5*
white blood cell *38.2*

Readings

Amir, M. September–October 1996. "Secrets of the Heart." *The Sciences*. Effects of exercise on heart function.

Baraga, M. 3 May 1996. "Finding New Drugs to Treat Stroke." *Science*, 274.

Randall, D., Burggren, W., French, K. 1997. *Eckert Animal Physiology*. Fourth edition. New York: Freeman.

On-Line readings at Student Guide for InfoTrac:
www.brookscole.com/biology

IMMUNITY

Russian Roulette, Immunological Style

Until about a century ago, smallpox epidemics swept repeatedly through the world's cities. Some outbreaks were so severe that only half of the stricken survived. Survivors had permanent scars on the face, neck, arms, and shoulders. Yet seldom did they contract the disease again; they were said to be "immune" to smallpox.

No one knew what caused smallpox, but the idea of acquiring immunity was dreadfully appealing. In twelfth-century China, people in good health gambled with deliberate infections. They sought out individuals who had survived mild cases of smallpox and were mildly scarred. They removed bits of crusts from the scars, ground them up, and inhaled the powder.

By the seventeenth century Mary Montagu, wife of the English ambassador to Turkey, was championing inoculation. She went so far as to poke bits of smallpox scabs into her children's skin. Others soaked threads in fluid from smallpox sores, then poked them into skin. Some people survived the chancy practices, and they

acquired immunity to smallpox. Yet many developed raging infections. As if the odds weren't dangerous enough, the crude inoculation practices also made the body vulnerable to several other infectious diseases.

While this immunological version of Russian roulette was going on, Edward Jenner was growing up in the countryside of England. At the time, it was common knowledge that people who contracted cowpox never got smallpox. (Cowpox is a mild disease that can be transmitted from cattle to humans.) No one thought

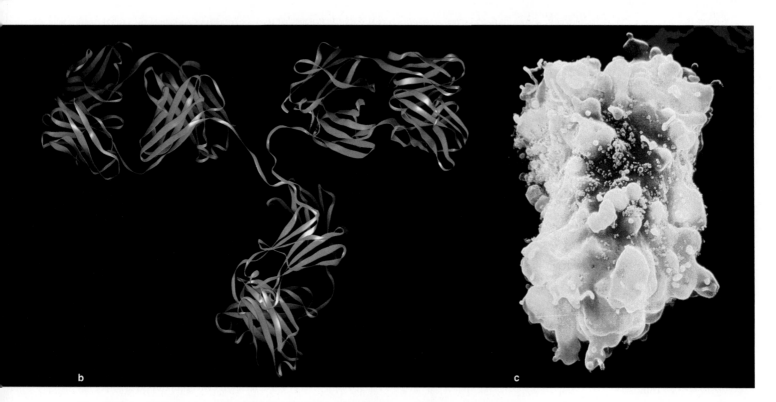

Figure 39.1 (**a**) Statue honoring Edward Jenner's work to develop a vaccine against smallpox, one of the most dreaded diseases in human history. (**b**) Computer model for an antibody molecule, a type of weapon that Jenner's procedure mobilized. (**c**) White blood cell being attacked by HIV (*blue particles*), the virus that causes AIDS. Immunologists are working to develop effective weapons against this modern-day scourge.

much about this until 1796, when Jenner, by now a physician, injected material from a cowpox sore into the arm of a healthy boy. Six weeks later, after the reaction subsided, Jenner injected some material from *smallpox* sores into the boy (Figure 39.1*a*). He had hypothesized that the earlier injection might provoke immunity to smallpox. Fortunately he was right; the boy remained free of smallpox. Jenner had developed an effective immunization procedure against a specific pathogen.

The French mocked Jenner's procedure by calling it **vaccination**. The word literally means "encowment." Much later Louis Pasteur, an influential French chemist, developed similar immunization procedures for other diseases. In acknowledgment of Jenner's pioneering work, Pasteur also called his procedures vaccinations, and only then did the word become respectable.

By Pasteur's time, improved light microscopes had revealed bacteria, fungal spores, and other previously invisible forms of life. As Pasteur himself discovered, microorganisms abound in ordinary air. Did some cause diseases? Probably. Could they drift onto food or drink and spoil it? He demonstrated that they could and did.

Pasteur found a way to kill most of the suspected pathogens (disease agents) in food or beverages. As he and others knew, boiling killed the suspects. He also knew that boiling changes the properties of wine—or beer or milk, for that matter. He devised a way to heat beverages to a temperature low enough not to ruin them but high enough to kill most of the pathogens suspended in them. We still rely on his antimicrobial method, which was named pasteurization in his honor.

In the late 1870s Robert Koch, a German physician, linked a specific pathogen to a specific disease: anthrax. In one experiment, Koch injected blood from animals with anthrax into healthy animals. Later, the blood of the recipient animals teemed with cells of the bacterium *Bacillus anthracis*—and the animals developed anthrax. More convincing, injections of *B. anthracis* cells cultured outside the animal body also caused the disease!

Thus, by the beginning of the twentieth century, the promise of understanding the basis of infectious disease and immunity loomed large—and the battles against those diseases were about to begin in earnest. Since that time, spectacular advances in microscopy, biochemistry, and molecular biology have increased our understanding of the body's defenses. Today we have greater insight into responses to tissue damage in general and immune responses to specific pathogens or tumor cells. The responses are the focus of this chapter.

Key Concepts

1. The vertebrate body has physical, chemical, and cellular defenses against pathogenic microorganisms, malignant tumor cells, and other agents capable of destroying tissues and the individual itself.

2. In the early stages of tissue invasion and damage, white blood cells and certain proteins in the plasma portion of blood escape from capillaries. They execute a rapid counterattack in response to a general alarm, not to the presence of a specific pathogen, so we call this a nonspecific inflammatory response. Phagocytic white blood cells ingest invading agents and clean up tissue debris. Plasma proteins promote phagocytosis, and some also destroy invaders directly.

3. If the invasion persists, certain white blood cells make immune responses. The cells can chemically recognize distinct configurations on molecules that are abnormal or foreign to the body, such as those on bacterial cells and viruses. A foreign or abnormal molecule that triggers an immune response is called an antigen.

4. In one type of immune response, activated B cells produce and secrete enormous quantities of antibodies. Antibodies are molecules that bind to a specific antigen, thus tagging it for destruction.

5. In another type of immune response, cytotoxic T cells directly destroy body cells already infected by a variety of intracellular pathogens. Cytotoxic cells also can destroy some tumor cells.

THREE LINES OF DEFENSE

You continually cross paths with astoundingly diverse **pathogens**—the viruses, bacteria, fungi, protozoans, parasitic worms, and other agents that cause diseases. Having coevolved with most of them, you and other vertebrates have three lines of defense, so you need not lose sleep over this. Most pathogens cannot breach the body surface. If they do, they face chemical weapons and white blood cells, or leukocytes, that attack anything perceived as foreign. Other white blood cells zero in on specific targets. Table 39.1 lists the three lines of defense.

Surface Barriers to Invasion

Most often, pathogens can't get past the linings of body surfaces, including skin. Think of skin as a habitat of low moisture, low pH, and thick layers of dead cells. Normally harmless bacteria tolerate these conditions. Few pathogens can compete with dense populations of established types unless conditions change. Repeatedly enclose your toes in warm, damp shoes, for instance, and you may be inviting certain fungi to penetrate the sodden, weakened tissues and cause *athlete's foot.*

Similarly helpful bacteria live on the mucous lining of the gut or vagina. Think of *Lactobacillus* populations in the vagina. Lactate, their acidic fermentation product,

helps maintain a low pH that most bacteria and fungi can't tolerate. Think of the branching, tubular airways into the lungs. When air rushes in, it flings airborne bacterial cells into the sticky, mucus-coated tube walls. That mucus contains protective substances, especially a type of **lysozyme**. Enzymes in this category digest cell walls of bacteria and thus invite their death. As a final touch, broomlike cilia lining the airways sweep away trapped and enzymatically destroyed pathogens.

There are more defenses. Tears, saliva, and gastric fluid contain lysozyme and other protective substances. Tears give eyes a sterile wash. Urine's flushing action and low pH help remove pathogens from the urinary tract. Diarrhea flushes pathogens from the gut, which is why stopping diarrhea can prolong infection. (Even so, severe cases lead to dehydration, especially in children. Oral hydration therapy uses a glucose–salt solution to reverse such dangerous losses of water and ions.)

Nonspecific and Specific Responses

All animals react to tissue damage. Even small aquatic invertebrates have phagocytic cells and antimicrobial substances, including lysozymes. But the more complex the animal, the more complex are the systems to defend it. Think back on the evolution of the circulatory and lymphatic systems of vertebrates that invaded the land (Section 38.1). As the circulation of body fluids became more efficient, so did the means to defend the body. Phagocytic cells and plasma proteins could be swiftly transported to tissues under attack and could intercept pathogens trickling along the vascular highways.

Plasma proteins evolved. Some promote rapid clot formation after tissue damage; others destroy invading pathogens or target them for phagocytosis. In addition, highly focused responses to specific dangers evolved.

In short, *internal defenses against a tremendous variety of pathogens are in place even before damage occurs.* Ready and waiting are white blood cells and plasma proteins. They make *nonspecific* responses to damaged tissue in general, not to a particular pathogen. Other white blood cells may recognize unique molecular configurations of a *specific* pathogen. The resulting immune response will run its course whether or not tissues are damaged.

Table 39.1 *The Vertebrate Body's Three Lines of Defense Against Pathogens*

BARRIERS AT BODY SURFACES (*nonspecific* targets)

Intact skin; mucous membranes at other body surfaces

Infection-fighting chemicals (e.g., lysozyme in tears, saliva)

Normally harmless bacterial inhabitants of skin and other body surfaces that can outcompete pathogenic visitors

Flushing effect of tears, saliva, urination, and diarrhea

NONSPECIFIC RESPONSES (*nonspecific* targets)

Inflammation:

1. Fast-acting white blood cells (neutrophils, eosinophils, and basophils)
2. Macrophages (also take part in immune responses)
3. Complement proteins, blood-clotting proteins, and other infection-fighting substances

Organs with pathogen-killing functions, such as lymph nodes

Some cytotoxic cells (e.g., NK cells) with a range of targets

IMMUNE RESPONSES (*specific* targets only)

T cells and B cells; dendritic cells, macrophages alert them

Communication signals (e.g., interleukins and interferons) and chemical weapons (e.g., antibodies, perforins)

Intact skin, mucous membranes, antimicrobial secretions, and other barriers at the body surface constitute the first line of defense against invasion and tissue damage.

Inflammation and other internal, nonspecific responses to invasion are the second line of defense.

Immune responses against specific invaders, as executed by armies of white blood cells and their chemical weapons, are the third line of defense.

COMPLEMENT PROTEINS

A set of plasma proteins has roles in nonspecific *and* specific defenses. Collectively, they are the **complement system**. About twenty kinds circulate in the blood, in inactive form. If even a few molecules of one kind are activated, they trigger cascading reactions that activate many molecules of another complement protein. Each molecule activates many molecules of another kind of protein at the next reaction step, and so on. Deploying huge numbers of molecules has the following effects.

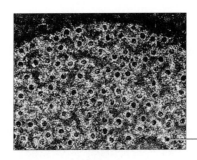

Figure 39.2 Micrograph of the surface of a pathogenic cell. Membrane attack complexes made the holes.

— hole in membrane

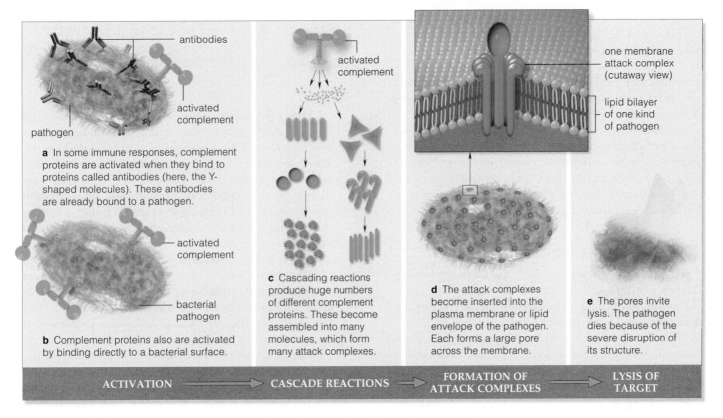

a In some immune responses, complement proteins are activated when they bind to proteins called antibodies (here, the Y-shaped molecules). These antibodies are already bound to a pathogen.

b Complement proteins also are activated by binding directly to a bacterial surface.

c Cascading reactions produce huge numbers of different complement proteins. These become assembled into many molecules, which form many attack complexes.

d The attack complexes become inserted into the plasma membrane or lipid envelope of the pathogen. Each forms a large pore across the membrane.

e The pores invite lysis. The pathogen dies because of the severe disruption of its structure.

ACTIVATION ⟹ CASCADE REACTIONS ⟹ FORMATION OF ATTACK COMPLEXES ⟹ LYSIS OF TARGET

Figure 39.3 Formation of membrane attack complexes. One reaction pathway starts as complement proteins bind to bacterial surfaces. Another operates in immune responses to specific invaders. Both pathways result in membrane pore complexes that induce lysis in pathogens.

Some complement proteins form attack complexes: molecular structures with an inner channel (Figures 39.2 and 39.3). The complexes are inserted into the plasma membrane of many pathogens and initiate lysis. They also are inserted into the wall of certain bacterial cells. **Lysis**, remember, is the gross structural disruption of a plasma membrane or cell wall; it results in cell death.

Certain activated proteins promote inflammation, a nonspecific defense response described next. By cascades of synthesis reactions, they form concentration gradients in irritated or damaged tissues. Phagocytic white blood cells follow the gradients to invasion sites. In addition, complement proteins bind to the surface of many types

of invaders. Together, they form a complement "coat" on the invaders that promotes phagocytosis.

In such ways, the complement proteins target many bacteria, parasitic protistans, and enveloped viruses.

The complement system, a set of about twenty kinds of plasma proteins circulating in blood, takes part in cascades of reactions that defend the body against many bacteria, some parasitic protistans, and enveloped viruses.

Complement proteins take part in nonspecific defenses. They also take part in immune responses, as when some are activated by antibody molecules bound to pathogens.

INFLAMMATION

The Roles of Phagocytes and Their Kin

Certain types of white blood cells take part in an initial response to a damaged tissue. White blood cells, recall, arise from stem cells in bone marrow. Many of the cells circulate in blood and lymph. A great many take up stations in lymph nodes as well as in the spleen, liver, kidneys, lungs, and brain. Here you may wish to scan Sections 38.2 and 38.11 once more; they introduce the components of blood and the lymphatic system.

Like SWAT teams, three kinds of white blood cells react swiftly to danger in general but are not adapted for sustained battles. **Neutrophils**, the most abundant kind, phagocytize bacteria. They ingest, kill, and digest bacterial cells into molecular bits. **Eosinophils** secrete enzymes that put holes in parasitic worms. **Basophils** and **mast cells** make and secrete chemical substances that help keep inflammation going after it starts.

Although slower to act, the white blood cells called **macrophages** are "big eaters." Figure 39.4 shows one of these phagocytic cells. Macrophages engulf and digest just about any foreign agent. They also help clean up damaged tissues. Immature macrophages circulating in blood are classified as monocytes.

The Inflammatory Response

An inflammatory response develops when something damages or kills cells of any tissue region. Infections, punctures, burns, and other assaults are the triggers. By a mechanism known as **acute inflammation**, the fast-acting phagocytes and complement proteins, as well as other plasma proteins, escape from the bloodstream at capillary beds in the affected tissue. Localized signs that they have entered interstitial fluid and that acute inflammation is proceeding include redness, warmth, swelling, and pain, as listed in Table 39.2.

Table 39.2	*Localized Signs of Inflammation and Their Causes*
Redness	Arteriolar vasodilation increases blood flow to the affected tissue.
Warmth	Arteriolar vasodilation delivers more blood, carrying more metabolic heat, to the tissue.
Swelling	Chemical signals increase capillary permeability; plasma proteins escape and disrupt fluid balance across capillary walls; localized edema results.
Pain	Nociceptors (pain receptors) are stimulated by increased fluid pressure, local chemical signals.

Mast cells reside in connective tissues. In response to tissue damage, they make and release **histamine** and other local signaling molecules into interstitial fluid. These secretions trigger vasodilation of arterioles that snake through the tissue. Vasodilation, remember, is an increase in the diameter of a blood vessel when smooth muscle in its wall is made to relax. So blood engorges the arterioles—which reddens the tissue and warms it up with blood-borne metabolic heat.

The released histamine also increases permeability of the thin-walled capillaries in the tissue. It induces endothelial cells making up the capillary wall to pull apart farther at the narrow clefts between them. Thus the capillaries become "leaky" to plasma proteins that normally do not leave the blood. When some proteins leak out, osmotic pressure increases in the surrounding interstitial fluid. Also, blood pressure is higher because of the increased flow of blood to the tissue. As a result, ultrafiltration rises and reabsorption declines across the capillary wall. Localized edema is the outcome of this

Figure 39.4
(**a**) White blood cell squeezing out of a blood capillary at a cleft between endothelial cells. (**b**) Macrophage, caught in the act of engulfing a yeast cell.

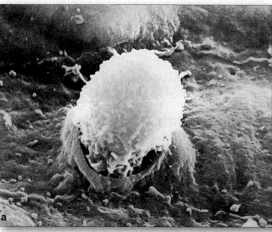
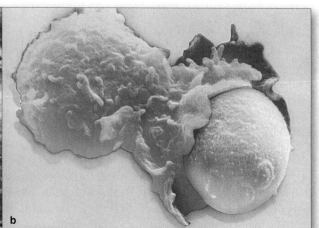

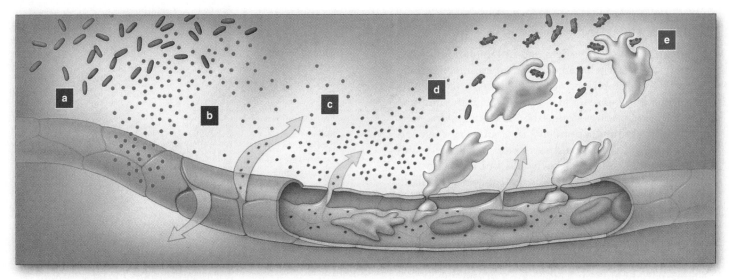

a Bacteria invade a tissue and directly kill cells or release metabolic products that damage tissue.

b Mast cells in tissue release histamine, which then triggers arteriole vasodilation (hence redness and warmth) as well as increased capillary permeability.

c Fluid and plasma proteins leak out of capillaries; localized edema (tissue swelling) and pain result.

d Complement proteins attack bacteria. Clotting factors wall off inflamed area.

e Neutrophils, macrophages, and other phagocytes engulf invaders and debris. Macrophage secretions attract even more phagocytes, directly kill invaders, and call for fever and for T and B cell proliferation.

Figure 39.5 Acute inflammation in response to a bacterial invasion. The response involves delivering phagocytes and plasma proteins to the tissue. Collectively, these components of blood inactivate, destroy, or wall off the invaders, remove chemicals and cellular debris, and prepare the tissue for repair. These are their functions in all inflammatory responses.

shift in the fluid balance across the wall. (Here you may wish to review Section 38.8.) The tissue swells with fluid, and certain free nerve endings threading through it give rise to sensations of pain. Voluntary movements aggravate the pain, so an affected person tends to avoid movements and, in doing so, promotes tissue repair.

Within hours of the first physiological responses to the damage, neutrophils are squeezing across capillary walls. They swiftly go to work. Monocytes arrive later, differentiate into macrophages, and engage in sustained action (Figure 39.5). While macrophages are engulfing the invaders and debris, they secrete a number of local signaling molecules that act as chemical mediators.

Chemical mediators called *chemotaxins* attract more phagocytes. *Interleukins* stimulate the formation of B cell and T cell armies, as described shortly. *Lactoferrin* directly kills bacteria. *Endogenous pyrogen* triggers the release of certain prostaglandins, which have the effect of raising the set point on the hypothalamic thermostat that controls the body's core temperature. A **fever** is a core temperature that reaches the higher set point.

A fever of about 39°C (100°F) is not a bad thing. It increases body temperature to a level that is "too hot" for the functioning of most pathogens. It also promotes an increase in the host's defense activities. *Interleukin-1* induces drowsiness, which reduces the body's demands for energy. And so more energy may be diverted to the

tasks of defense and tissue repair. Macrophages take part in the cleanup and repair operations.

Besides complement proteins, the plasma proteins leaking into the tissue include clotting factors, which are activated by chemicals that the phagocytes secrete. Fibrin threads form, stick to exposed collagen fibers in the damaged tissue, and trap blood cells and platelets to form a clot (Section 38.10). By walling off inflamed areas, clots can prevent or delay the spread of invaders and toxic chemicals into surrounding tissues. After the inflammation subsides, anticlotting factors, which also escaped from the capillaries, dissolve the clots.

An inflammatory response develops in a local tissue when cells are damaged or killed, as by infection. It proceeds during both nonspecific and specific defenses of tissues.

Mast cells in damaged or invaded tissues secrete histamine, which causes arterioles to vasodilate and increases capillary permeability to fluid and to plasma proteins. The tissue warms and reddens because of the localized vasodilation. Edema results from the fluid imbalance across the capillary wall. The tissue swelling causes pain.

The response involves phagocytes such as macrophages, which engulf invaders and debris and secrete chemical mediators. It requires plasma proteins, such as complement proteins that target invaders for destruction, as well as clotting factors that wall off the inflamed tissue.

OVERVIEW OF THE IMMUNE SYSTEM

Defining Features

Sometimes physical barriers and inflammation are not enough to overwhelm an invader, so an infection may become well established. Then, the body calls upon its third line of defense—the B and T cells of the immune system. Four features define the **immune system**: self/nonself recognition, specificity, diversity, and memory. *Self/nonself recognition* and *specificity* mean that B and T cells ignore the body's own cells but attack specific foreign agents. *Diversity* means that, collectively, B and T cell receptors have the potential to respond to a billion specific threats. *Memory* means a portion of the B and T cells formed during a primary response to a foreign agent is set aside for a future battle with that agent.

Self/nonself recognition starts at unique molecular configurations that give every kind of cell, virus, and substance a unique identity. An individual's own cells bear self-marker configurations. That individual's B and T cells chemically recognize self markers and normally don't attack them. If they come across nonself markers of specific foreign agents or abnormal configurations on altered body cells, they divide again and again, and form huge populations. Any molecular configuration that incites formation of lymphocyte armies and is their target is an **antigen**. The most important antigens are proteins at the surface of pathogens and tumor cells.

All cells resulting from the divisions are sensitized to the same antigen. But some subpopulations become *effectors:* fully differentiated types that kill the enemy. Others become *memory* cells. Instead of joining the first response against a specific antigen, they enter a resting phase. But they remember that antigen. If it shows up again, they will join a larger, faster response to it.

In short, immune responses involve these events: recognition of antigen, repeated cell divisions that form huge populations of lymphocytes, and differentiation into subpopulations of effector and memory cells that have receptors for one kind of antigen.

The Key Defenders

Interactions among antigen-presenting cells, activated helper T cells, B cells, and cytotoxic T cells are central to all immune responses. Here, you can start thinking about how some of them process and display antigen, and how others recognize and attack antigen bearers.

MHC markers are named after genes that code for them: the *Major Histocompatibility Complex*. They are recognition proteins at the plasma membrane of body cells. Remember Sections 3.7 and 5.2? When paired with antigen, a cell's MHC markers represent a call to arms.

Any cell that can (1) process and display antigen in association with MHC markers and (2) activate T cells

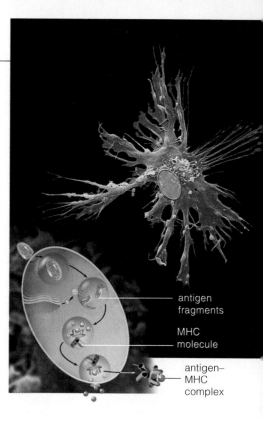

Figure 39.6
A dendritic cell —one of the major antigen-presenting cells. This class of cells patrols blood, internal organs, and skin. (The Langerhans cells described in Section 37.2 are dendritic.) They ingest, process, and then display antigen with MHC molecules.

antigen fragments

MHC molecule

antigen– MHC complex

is an **antigen-presenting cell**. Macrophages, B cells, and dendritic cells (Figure 39.6) are examples. They all ingest antigen, then the endocytic vesicle that forms around it fuses with a lysosome. The lysosome's potent enzymes cleave the antigen molecules into fragments. Some of the fragments associate with MHC markers and form **antigen–MHC complexes**, which move to the plasma membrane. There they are displayed.

Helper T cells bind to antigen–MHC complexes and secrete signals that promote immune responses. The signals call for mitotic cell divisions of any B and T cell sensitive to the antigen, and for cell differentiation into effector and memory subpopulations (Figure 39.7).

Effector **cytotoxic T cells** and **natural killer** (NK) **cells** carry out *cell-mediated* responses against infected body cells and tumor cells. All recognize antigen–MHC complexes, provided the MHC marker has a coreceptor called CD8 bound to it. They release chemical weapons that form pores in the plasma membrane of targets.

Only antigen-sensitized **B cells** carry out *antibody-mediated* responses. They alone make and secrete many antigen-binding receptor molecules called **antibodies**. When macrophages, NK cells, or neutrophils contact a foreign agent with antibody bound to it, they destroy it.

The Main Targets

Any cell with antigen–MHC complexes at its surface turns on cytotoxic cells with receptors for that antigen. This includes body cells infected with an intracellular pathogen; they, too, display bits of antigen with MHC markers at their surface. Tumor cells and cells of organ transplants also are targets. Besides this, antibodies can

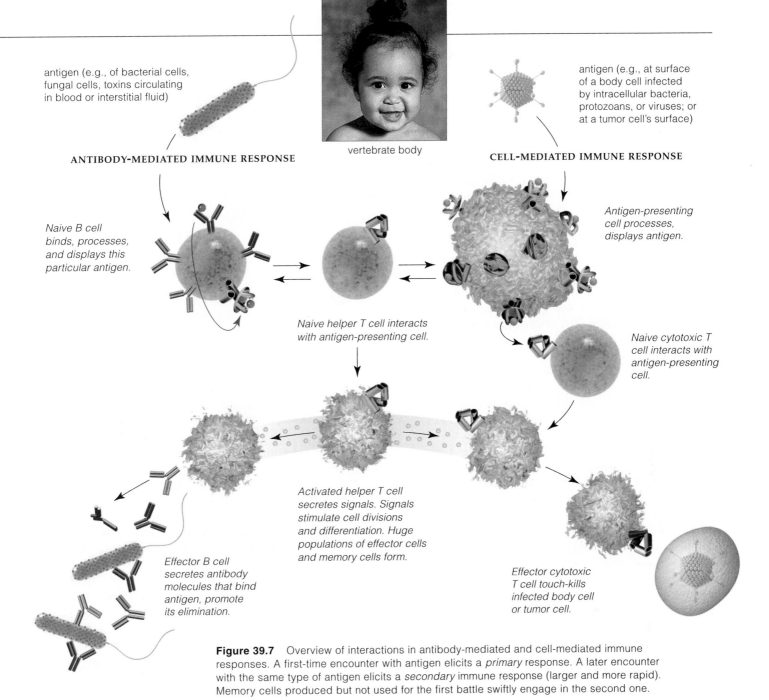

antigen (e.g., of bacterial cells, fungal cells, toxins circulating in blood or interstitial fluid)

vertebrate body

antigen (e.g., at surface of a body cell infected by intracellular bacteria, protozoans, or viruses; or at a tumor cell's surface)

ANTIBODY-MEDIATED IMMUNE RESPONSE

CELL-MEDIATED IMMUNE RESPONSE

Naive B cell binds, processes, and displays this particular antigen.

Antigen-presenting cell processes, displays antigen.

Naive helper T cell interacts with antigen-presenting cell.

Naive cytotoxic T cell interacts with antigen-presenting cell.

Activated helper T cell secretes signals. Signals stimulate cell divisions and differentiation. Huge populations of effector cells and memory cells form.

Effector B cell secretes antibody molecules that bind antigen, promote its elimination.

Effector cytotoxic T cell touch-kills infected body cell or tumor cell.

Figure 39.7 Overview of interactions in antibody-mediated and cell-mediated immune responses. A first-time encounter with antigen elicits a *primary* response. A later encounter with the same type of antigen elicits a *secondary* immune response (larger and more rapid). Memory cells produced but not used for the first battle swiftly engage in the second one.

directly inactivate antigen-bearing cells and toxins in the blood or interstitial fluid. Antibodies also promote inflammation and activate complement proteins.

Control of Immune Responses

Antigen provokes an immune response—and removal of antigen stops it. For example, by the time the tide of battle turns, effector cells and their secretions already have killed most antigen-bearing agents in the body. With less antigen, the response declines, then ends. Also, **suppressor T cells** release signals that help end both cell-mediated and antibody-mediated responses.

The immune system is defined by self/nonself recognition, specificity, diversity, and memory. Collectively, T and B cells bear receptors for millions of specific antigens: foreign molecular configurations. Recognition of antigen, not self markers on the body's own cells, triggers immune responses.

Huge T and B cell armies form by repeated mitotic cell divisions after antigen recognition. These differentiate into subpopulations of effector and memory cells, all of which are sensitized to that one kind of antigen.

Antigen-sensitized cytotoxic cells execute cell-mediated responses. Antigen-sensitized B cells execute antibody-mediated responses.

HOW LYMPHOCYTES FORM AND DO BATTLE

B and T Cell Formation

Your body is continually exposed to a mind-boggling array of antigens. However, antigen receptors of all of its B and T cell populations combined show staggering diversity—enough to recognize a billion or so different antigens. How does such diversity arise? *In each new T and B cell, gene sequences with instructions for building the antigen-binding parts of receptor molecules get shuffled, extensively and randomly, before they are expressed.* Figure 39.8 shows what happens before the antigen receptors called antibodies are synthesized. Antigen receptors are mostly proteins—polypeptide chains. While each chain is forming, its antigen-binding region folds up into a unique array of bumps and grooves having a certain distribution of charge. The only antigen in the world that will bind to that region will have complementary grooves, bumps, and distribution of charge.

As you read earlier, B cells arise from stem cells in bone marrow. Each acquires unique antigen receptors before leaving it. As a B cell matures, it starts making many copies of a unique, typically Y-shaped antibody molecule. These move to the plasma membrane, where the tail of the Y becomes embedded in the lipid bilayer and its two arms project above it. Soon, the cell bristles with antigen receptors—bound antibodies. It is now a "naive" B cell, meaning it has not yet met the antigen it is genetically programmed to detect. Its membrane-bound antibodies will not recognize any antigen–MHC complexes. They will recognize antigen alone.

T cells also arise from stem cells in bone marrow. Then they migrate to the thymus gland, and there they mature and acquire unique antigen-binding receptors called **TCRs** (short for *T-Cell Receptors*). Bristling with

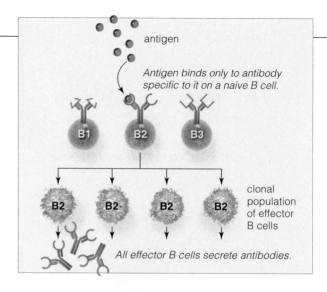

Figure 39.9 antigen

Antigen binds only to antibody specific to it on a naive B cell.

clonal population of effector B cells

All effector B cells secrete antibodies.

Figure 39.9 Clonal selection of a B cell with the one antibody able to combine with a specific antigen. Only antigen-selected B cells and T cells give rise to a clonal population of immunologically identical cells.

antigen receptors, naive T cells leave the thymus and enter the general circulation. Their TCRs will ignore free antigen; they will ignore unadorned MHC markers at the surface of the body's own cells. Their TCRs will recognize antigen–MHC complexes only.

And so each B cell or T cell bears only one type of antigen receptor. By the theory of *clonal selection*, of all body cells, antigen "chooses" (binds to) only the B or T cell displaying the receptor specific for it (Figure 39.9). Descendants of the activated cell form an enormous population of genetically identical cells—a clone—all with the same antigen receptor.

The clonal selection theory also explains how an individual acquires immunological memory of the first encounter with antigen. The term refers to the body's

Figure 39.8 Generation of antigen receptor diversity, using the antibody molecule as an example. Antibodies are proteins. Protein-building instructions are encoded in genes. In chromosomes with genes for antibodies, long DNA stretches contain different versions of the segments coding for variable regions of an antibody molecule. (Compare Figure 39.13*a*.)

Different V and J segments code for the variable region. A recombination event occurs in this region when each B cell is maturing. Any V segment may be joined to any of the J segments. Afterward, DNA intervening between them is excised. The new sequence is attached to a C (*Constant*) segment. This completes a rearranged antibody gene. That gene will be present in every descendant of the B cell.

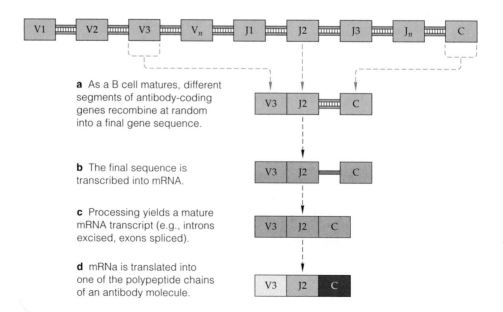

a As a B cell matures, different segments of antibody-coding genes recombine at random into a final gene sequence.

b The final sequence is transcribed into mRNA.

c Processing yields a mature mRNA transcript (e.g., introns excised, exons spliced).

d mRNa is translated into one of the polypeptide chains of an antibody molecule.

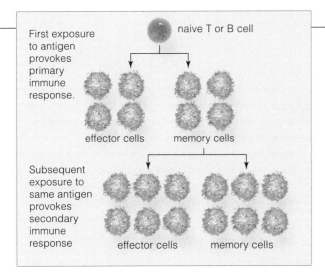

First exposure to antigen provokes primary immune response.

naive T or B cell

effector cells   memory cells

Subsequent exposure to same antigen provokes secondary immune response

effector cells   memory cells

Figure 39.10 Immunological memory. Not all B and T cells are activated during a primary immune response to antigen. A large number continue to circulate as memory cells, which become activated during a secondary immune response.

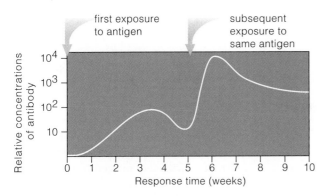

first exposure to antigen   subsequent exposure to same antigen

Relative concentrations of antibody

10^4
10^3
10^2
10

0 1 2 3 4 5 6 7 8 9 10
Response time (weeks)

Figure 39.11 Example of the differences in magnitude and duration between a primary and a secondary immune response to the same antigen. The primary response peaked twenty-four days after it started. The secondary response peaked after seven days (the span shown between weeks five and six). Antibody concentration during the secondary response was 100 times greater, 10^4 compared to 10^2.

capacity to make a secondary immune response to any subsequent encounter with the same type of antigen that provoked the primary response (Figure 39.10).

Memory cells formed during a primary immune response don't engage in that battle. They circulate for years or for decades. Compared to a primary response, antigen is intercepted sooner by battalions that already formed and are patrolling tissues. Thus populations of effector cells form faster, so the infection ends before an individual gets sick. Far greater numbers of memory B and T cells form in a secondary response (Figure 39.11). In evolutionary terms, advance preparations against further encounters with the same pathogen bestow a survival advantage on the individual.

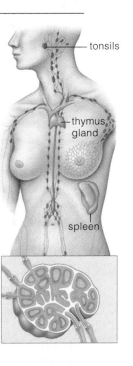

tonsils

thymus gland

spleen

Figure 39.12 Lymph nodes, which contain organized arrays of antigen-presenting cells and lymphocytes. Compare Section 38.11.

Lymphocyte Battlegrounds

Where is antigen actually captured and dealt with? In skin, in mucous membranes, and in internal organs, the macrophages and dendritic cells that form an early line of defense pick up antigen by phagocytosis or endocytosis. Then they all migrate into regional lymph nodes (Section 38.11). That is where macrophages and dendritic cells present antigen to naive B and T cells.

The antigen in interstitial fluid enters the lymph vascular system. Because lymph drains into the blood, there is always the chance that antigen will be circulated all through the body. However, lymph nodes do trap most of it. In addition, antigen that does manage to enter the blood filters on through the spleen (Figure 39.12). Every day, more circulating lymphocytes move through the spleen than through all lymph nodes combined.

In lymph nodes, antigen first percolates through the region occupied by B cells, macrophages, and dendritic cells. There it is captured, processed, and presented to helper T cells, thereby activating them. Other dendritic cells that bear receptors for antibody and complement may activate B cells in the nodes. Deeper in the node, T cells start multiplying and maturing into effector and memory cells. So do B cells, which mature and start to secrete antibodies. Quantities of antibodies and newly formed lymphocytes leave the node. Also, antigen has been prodding many lymphocytes to slip out of nearby venules and enter the node. Hence those swollen nodes we notice from time to time under the jawline.

By recombination of segments drawn at random from the receptor-encoding regions of DNA, each B cell or T cell receives a gene sequence for just one of a billion possible kinds of antigen receptors.

Immunological specificity means the clonal descendants of an antigen-"chosen" cell bind only to the selected antigen.

Immunological memory means B and T cells set aside during a primary response can make a faster and greater secondary response to another encounter with the same antigen.

Macrophages, dendritic cells, B cells, and T cells mount immune responses in lymphoid organs. They are organized for efficient interactions in the lymph nodes and spleen.

ANTIBODY-MEDIATED RESPONSE

B cells without helper T cells would be like a symphony without a conductor. (The same goes for cytotoxic T cells.) When activated, helper T cells unleash antibody-mediated and cell-mediated responses. Activation is tightly controlled to put immune fighters on suitable courses of action. Limiting how they recognize antigen is a key control. Helper T cells recognize antigen *only* when a fragment of it appears with an MHC molecule at the surface of a B cell, macrophage, or dendritic cell. Also, after helper T cells interact with antigen–MHC complexes, a signal from that antigen-presenting cell stimulates them to divide. The resulting clone of helper T cells drives the immune responses.

The Roles of Antibodies

Let's now follow a naive B cell bristling with bound antibodies. All of its antibody molecules are alike; they are specific for only one antigen. Figure 39.13 shows a typical antibody's antigen-binding sites. When antigen binds with some of the molecules, it cross-links them together. The cross-linking activates receptor-mediated endocytosis, and so the antigen is moved into the cell (Section 5.8). There, antigen–MHC complexes form and move to the B cell surface, where they are displayed.

Suppose a responsive helper T cell meets up with a B cell's antigen–MHC complex. The two cells exchange a costimulatory signal, then disengage. When the B cell again encounters unprocessed antigen, its antibodies bind to it. The binding, in conjunction with interleukins secreted from nearby activated helper T cells, drives the B cell into mitosis. Its clonal descendants differentiate

into B effector and B memory cells (Figure 39.14). The effectors (also called plasma cells) produce and secrete huge numbers of antibodies. When freely circulating antibody binds to it, antigen gets tagged for destruction, as by phagocytes and complement proteins.

Remember, the main targets of antibody-mediated responses are extracellular pathogens and toxins freely circulating in extracellular fluid. *Antibodies cannot bind to pathogens or toxins that are hidden inside a host cell.*

Classes of Immunoglobulins

B effector cells produce five classes of antibodies that are **immunoglobulins**, or **Igs**. These protein products of gene shufflings form when B cells mature and engage in immune responses. We abbreviate them as IgM, IgD, IgG, IgA, and IgE. Each has antigen-binding sites *and* other functional sites. Secreted forms look like this:

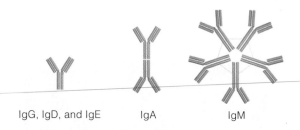

IgG, IgD, and IgE   IgA   IgM

IgM is the first to be secreted in a primary response and the first produced by newborns. IgM molecules aggregate as a structure with ten antigen binding sites. This structure more effectively binds clumped targets such as agglutinating red blood cells (Section 38.4) and

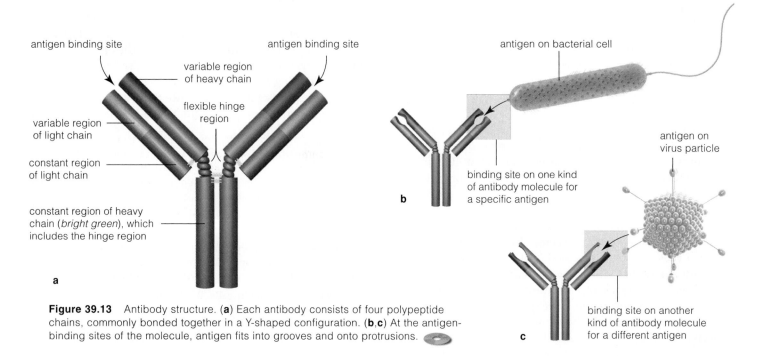

Figure 39.13 Antibody structure. (**a**) Each antibody consists of four polypeptide chains, commonly bonded together in a Y-shaped configuration. (**b,c**) At the antigen-binding sites of the molecule, antigen fits into grooves and onto protrusions.

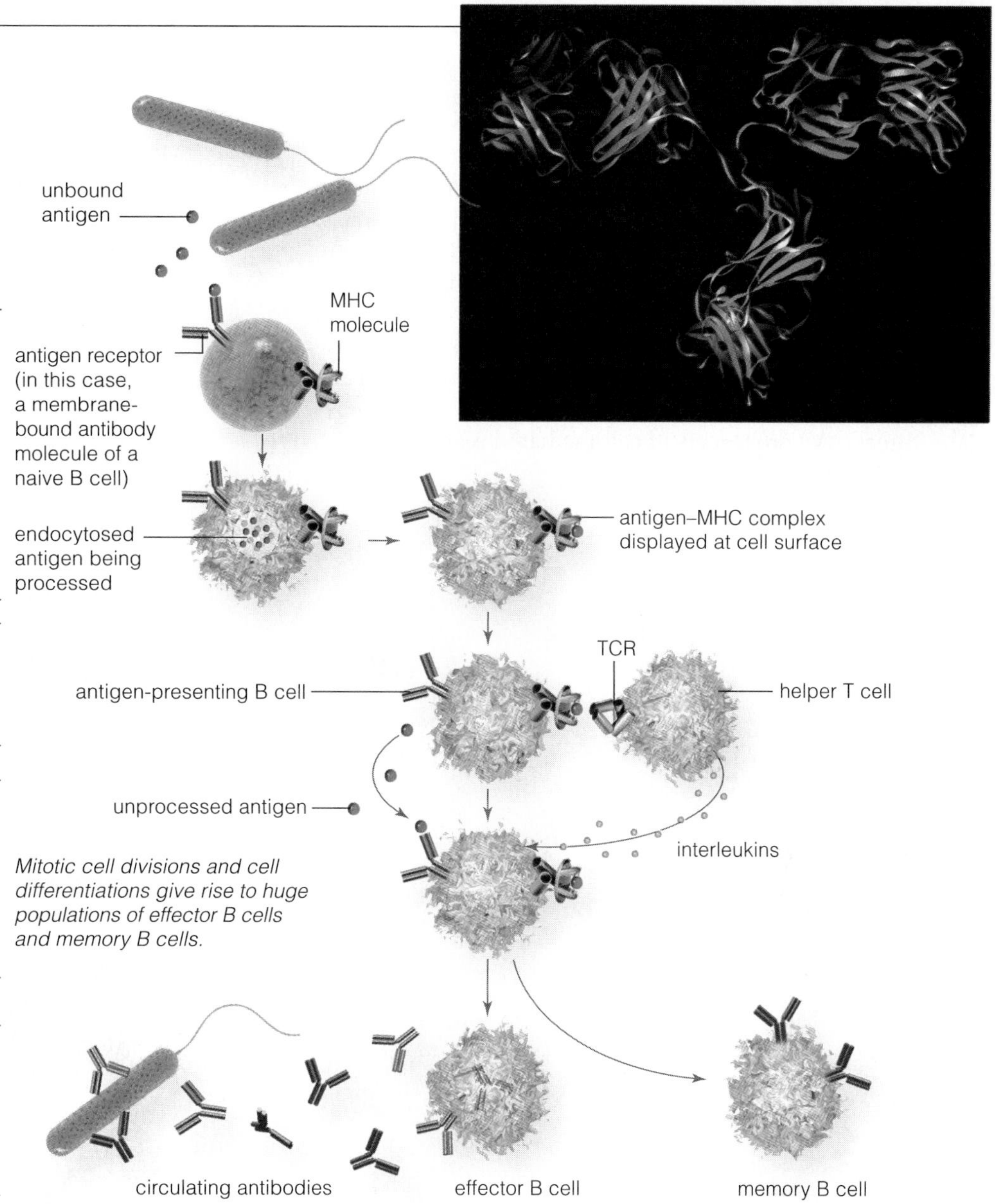

Figure 39.14 Antibody-mediated immune response. This example is a response to a bacterial invasion. The inset is a model for one type of antibody molecule.

a Each naive B cell bristles with 10 million identical antibody molecules, all specific for one antigen. When they bind it, the antigen moves into the cell in an endocytic vesicle and is digested. Some fragments bind with MHC molecules and are displayed with them at the cell surface. The B cell is now an antigen-presenting cell.

b TCRs of a helper T cell bind to the B cell's antigen–MHC complexes. Binding activates the T cell, and it stimulates the B cell to prepare for mitosis. The cells disengage.

c Unprocessed antigen binds to the same B cell. The helper T cell secretes interleukins (*blue* dots). Both events trigger repeated cell divisions and differentiations that yield armies of antibody-secreting effector and memory B cells.

d Antibody molecules secreted by effector B cells enter extracellular fluid. When they contact antigen on a bacterial cell surface, they bind it and thus tag the cell for destruction. (Compare Figures 35.3 and 35.7.)

Labels in figure: unbound antigen; MHC molecule; antigen receptor (in this case, a membrane-bound antibody molecule of a naive B cell); endocytosed antigen being processed; antigen–MHC complex displayed at cell surface; antigen-presenting B cell; TCR; helper T cell; unprocessed antigen; interleukins; Mitotic cell divisions and cell differentiations give rise to huge populations of effector B cells and memory B cells.; circulating antibodies; effector B cell; memory B cell

dense aggregations of virus particles. Along with IgM, *IgD* is the most common antibody bound at the plasma membrane of naive B cells. IgM may take part in the activation of helper T cells.

IgG makes up 80 percent or so of the blood-borne immunoglobulins. It is the most efficient one at turning on complement proteins, and it neutralizes many toxins. This long-lasting antibody easily crosses the placenta. It helps protect the developing fetus with the mother's acquired immunities. IgG secreted into early milk is also absorbed into a suckling newborn's bloodstream.

IgE triggers inflammation after attacks by parasitic worms and other pathogens. IgE antibody tails lock on to basophils and mast cells; their antigen receptors face outward. Antigen binding induces basophils and mast cells to release histamine, which fans inflammation. IgE also is involved in allergic reactions, including asthma.

IgA is the main immunoglobulin in exocrine gland secretions, including tears, saliva, and breast milk. It also is in mucus that coats surfaces of the respiratory, digestive, and reproductive tracts—the areas that are vulnerable to most infectious agents. Like IgM, it can form larger structures that can bind larger antigens. Bacteria and viruses can't bind to the cells of mucous membranes when secreted IgA is bound to them. IgA is known to be effective in fighting the agents that cause salmonella, cholera, gonorrhea, influenza, and polio.

Antibodies secreted by B effector cells bind to antigens of extracellular pathogens or toxins and tag them for disposal, as by phagocytes and complement.

The immunoglobulins IgM, IgG, IgD, IgE, and IgA are five antibody classes. They help defend against diverse threats.

CELL-MEDIATED RESPONSE

If every antigen stayed out in the open in the internal environment, maybe the antibody-mediated response would be sufficient to dispose of them. But a number of pathogens evade antibodies. They hide in body cells, drain the life out of them, and often reproduce in them. They are exposed only briefly after they slip out of one cell and before they infect others. You've already read about some of the intracellular viruses, bacteria, fungi, protozoans, and sporozoans. You know about genetically altered body cells, such as tumor cells, that pose other threats. All are targets of cell-mediated responses.

Helper T cells and cytotoxic T cells, along with NK cells, macrophages, neutrophils, eosinophils, and some other types, are the executioners. Local concentrations of interleukins, interferons, and other signals get their attention and stimulate them to divide, differentiate, and attack. Interferons also call for increases in making MHC molecules, thus enhancing the capacity of some cells to display antigen. If antibodies bind with exposed antigen, they, too, will call in some of the defenders.

Cytotoxic T cells are so sensitive to antigen–MHC complexes and to altered configurations on body cells,

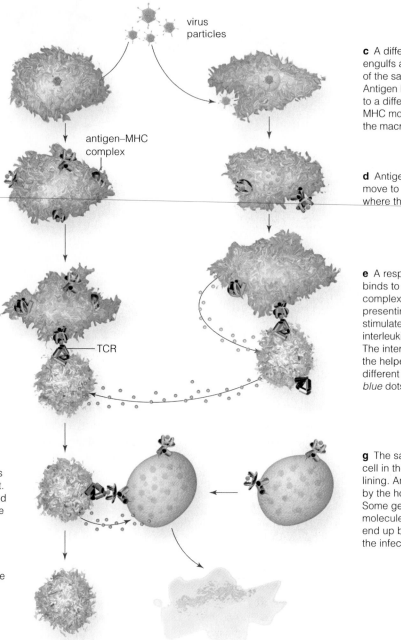

a A virus infects a macrophage. The host cell's pirated metabolic machinery makes viral proteins, which are antigenic. Enzymes of the macrophage cleave the antigens into fragments. Some fragments bind to a certain class of MHC molecules.

b Some antigen fragments are bound to class I MHC molecules. (These class I proteins occur on all nucleated cells and present antigen that originates inside infected cells.) Antigen–MHC complexes move to the cell surface and are displayed.

f The TCRs of a responsive cytotoxic T cell bind to the antigen–MHC complexes of the macrophage. Interleukins being secreted from the helper T cell stimulate the cytotoxic T cell to initiate cell divisions and differentiations. Large populations of effector and memory T cells form.

h An effector cytotoxic T cell contacts the infected cell and then touch-kills it. The effector cell releases perforins and toxic substances (*green* dots) onto the cell, which programs it for death.

i The effector disengages from the doomed cell and reconnoiters for more targets. Meanwhile, perforins make holes in the infected cell's plasma membrane. Toxins move into the cell, disrupt its organelles, and make the DNA disassemble. The cell dies.

virus particles

antigen–MHC complex

TCR

c A different macrophage engulfs and digests antigen of the same type of virus. Antigen becomes bound to a different class of MHC molecules inside the macrophage.

d Antigen–MHC complexes move to the cell surface, where they are displayed.

e A responsive helper T cell binds to the antigen–MHC complexes of the antigen-presenting cell. This binding stimulates the cell to secrete interleukins (*yellow* dots). The interleukins stimulate the helper T cell to secrete different interleukins (the *blue* dots).

g The same virus infected a cell in the respiratory tract lining. Antigen synthesized by the host cell is processed. Some gets bound to MHC molecules. The complexes end up being displayed at the infected cell's surface.

Figure 39.15 Diagram of a T cell–mediated immune response.

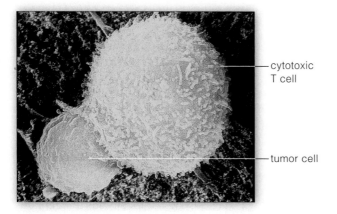

Figure 39.16 Cytotoxic T cell caught in the act of touch-killing a tumor cell.

they don't need further signals to start dividing. They also display and secrete molecules used in touch-killing infected and altered body cells, as in Figures 39.15 and 39.16. For example, perforins are protein molecules that form doughnut-shaped pores in the plasma membrane of a target cell, something like those in Figure 39.2. Cytotoxic T cells persuade target cells to commit suicide by **apoptosis**, a process described in Section 15.6. The cell's cytoplasm dribbles out, its organelles break up, and its DNA becomes fragmented. After making its lethal hit, the cytotoxic T cell then disengages from the doomed cell, and it resumes reconnoitering.

Cytotoxic T cells also contribute to the rejection of tissue and organ transplants. Parts of MHC markers on donor cells are different enough from the recipient's to be recognized as antigens. Other parts are enough alike to call for the costimulatory signal. MHC typing and matching donors to recipients minimize the risk.

What about NK cells? Antigen–MHC complexes do not activate them, so other receptors must be involved. Some of their receptors block attacks on cells with MHC markers, but apparently the inhibition is lifted for cells with altered or deficient numbers of MHC molecules. However they become activated, NK cells are mobilized when infections by some viruses, intracellular bacteria, and tumor cells lead to rapidly rising concentrations of interferons and interleukins. Being the first immune cells called into battle, they buy time while others are being activated, dividing, and differentiating.

Intracellular pathogens and altered body cells are targets of cell-mediated immune responses.

Helper T cells and cytotoxic T cells are antigen-specific defenders. NK cells, macrophages, neutrophils, eosinophils, and other cells make nonspecific responses. All cytotoxic types touch-kill infected cells and tumor cells.

Cancer and Immunotherapy

Carcinomas, sarcomas, leukemias—these chilling words refer to malignant neoplasms in skin, bone, and other tissues. They arise when viral attack, irradiation, or chemicals alter genes and cells turn cancerous (Section 15.6). The transformed cells divide repeatedly. Unless cancers can be destroyed or surgically removed, they can be lethal.

Transformed cells often bear abnormal proteins and protein fragments bound to MHC. Immune responses to them may make a tumor regress but may not be enough to kill it. Also, some tumors "hide" by releasing copies of the abnormal proteins, which saturate antigen receptors. Hidden tumors that reach a critical mass may overwhelm the immune system. The idea behind *immunotherapy* is to enlist white blood cells to destroy this threat and others.

MONOCLONAL ANTIBODIES Two decades ago, Cesar Milstein and Georges Kohler made limited amounts of antibodies that home in on tumor-specific antigens. They injected an antigen into mice, which made antibodies against it. Then they fused antibody-producing B cells from the mice with cells extracted from the B cell tumors. Clonal descendants of such hybrid cells made *monoclonal antibodies* (identical copies of antibodies). However, mice cannot produce useful amounts. Genetic engineers have now designed cows that can secrete antibodies into milk, but it may be difficult to isolate these molecules from bacteria or other pathogens that may be in the milk.

Today, genetically engineered plants can manufacture monoclonal antibodies. Cornfields might mass-produce them. Besides being cost-effective, *plantibodies* are safer than antibodies from cattle. (Few plant pathogens infect people.) The first plantibody used on human volunteers prevented infection by a bacterial agent of tooth decay.

MULTIPLYING THE TUMOR KILLERS Lymphocytes often infiltrate tumors. Researchers have removed them from a tumor and exposed them to lymphokine, an interleukin. Large populations of tumor-infiltrating lymphocytes with enhanced killing abilities have formed. These *LAK* cells (short for *Lymphokine-Activated Killers*) appear to be somewhat effective when injected back into a patient.

THERAPEUTIC VACCINES On the horizon are *therapeutic vaccines*—wake-up calls against specific tumor cells. One approach is to genetically engineer antigen so it becomes obvious to killer lymphocytes. Example: Remember those Langerhans cells (Section 37.2)? Like other dendritic cells, these phagocytes prowl on pseudopods. Once activated, they swiftly migrate to the lymph nodes and sound the alarm. Dirk Schadendorf harvested dendritic cells from melanoma patients. He cultured them with ground-up tumor cells, then injected them at intervals into patients' lymph nodes or skin. A year later, two patients had no trace of melanoma. In three others, tumors shrank more than half their size. Such "vaccines" may soon be available.

DEFENSES ENHANCED, MISDIRECTED, OR COMPROMISED

Immunization

Immunization refers to various processes that promote immunity against diseases. With *active* immunization, an antigen-containing preparation called a **vaccine** is taken orally or is injected into the body, as in Figure 39.17. An initial injection triggers a primary immune response. A subsequent injection—a booster—elicits a secondary response; more effector and memory cells form rapidly and provide long-lasting protection.

Many vaccines are manufactured from weakened or killed pathogens or use inactivated natural toxins. Still others are made of harmless viruses that have genes from three or more pathogens inserted into their DNA or RNA. After the vaccination, body cells use the novel genes to make antigens, and immunity is established.

Passive immunization protects individuals infected by agents of hepatitis B, diphtheria, tetanus, measles, and some other diseases. At-risk patients receive injections of a purified antibody. The best source is someone who has made the required antibody. Effects do not last; a patient has no antibodies or memory cells. Injections do help counter the initial attack.

A vaccine may fail or have harmful effects. In a few reported cases, vaccines have caused immunological problems; a few have caused neurological disorders. It's a good idea to assess the risks and benefits before agreeing to a procedure.

Allergies

In many people, normally harmless substances provoke inflammation, excess mucus secretion, and sometimes immune responses. Such substances are **allergens**, and hypersensitivity to them is called an **allergy**. Common allergens are pollen, many drugs and foods, dust mites, fungal spores, insect venom, perfumes, and cosmetics.

Some people are genetically predisposed to develop allergies. Infections, emotional stress, or changes in air temperature also trigger reactions that otherwise might not occur. Upon exposure to an antigen, IgE is secreted and binds to mast cells. When it then binds antigen, the mast cells secrete prostaglandins, histamine, and other substances that fan inflammation. Copious amounts of mucus are secreted, and airways constrict. Stuffed-up sinuses, labored breathing, sneezing, and a drippy nose are symptoms of allergic responses in *asthma* and *hay fever* among many millions of people (Figure 39.18).

Rarely, inflammatory reactions are life-threatening. With *anaphylactic shock*, someone allergic to wasp or bee venom might die minutes after a sting. Airways constrict massively. And fluids escape from excessively permeable capillaries and blood pressure plummets. The circulatory shock may quickly end in death.

Antihistamines (anti-inflammatory drugs) often relieve symptoms. In desensitization programs, skin tests might identify the offending allergens. Larger doses are administered slowly. Each time, the body makes more circulating IgG and memory B cells. Thus IgG, not IgE, binds with the allergen.

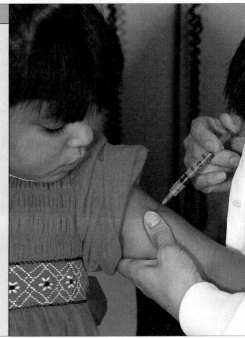

RECOMMENDED VACCINES	RECOMMENDED AGES
Hepatitis B	Birth–2 months
Hepatitis B booster	1–4 months
Hepatitis B booster	6–18 months
Hepatitis B assessment	11–12 years
DTP (Diphtheria; Tetanus; and Pertussis, or whooping cough)	2, 4, and 6 months
DTP booster	15–18 months
DTP booster	4–6 years
DT	11–16 years
HiB (*Haemophilus influenzae*)	2, 4, and 6 months
HiB booster	12–15 months
Polio	2 and 4 months
Polio booster	6–18 months
Polio booster	4–6 years
MMR (Measles, Mumps, Rubella)	12–15 months
MMR booster	4–6 years
MMR assessment	11–12 years
Pneumococcal	2, 4, and 6 months
Pneumococcal booster	12–15 months
Pneumococcal booster	1–18 years
Varicella	12–18 months
Varicella assessment	11–12 years
Hepatitis A (in selected areas)	1–12 years

Figure 39.17 From the Centers for Disease Control and Prevention, guidelines for immunizing children as of December 2002. Doctors routinely immunize infants and children. Low-cost or free vaccinations are available at many community clinics and health departments.

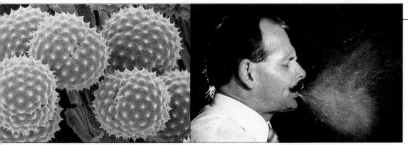

RAGWEED POLLEN . . .   . . . AND SOMETHING IT CAN PROVOKE

Figure 39.18 Example of the effects of allergens in sensitive people. Allergy sufferers who moved to deserts to escape pollen brought it with them. Half of those in Tucson, Arizona, are now sensitized to pollen from olive and mulberry trees, planted far and wide in cities and the suburbs.

Autoimmune Disorders

Sometimes self recognition fails and the immune system mounts an inappropriate attack against the body's own components. We call this **autoimmunity**. Antibodies or T cells wrongly activated can severely damage tissues.

In some cases, brakes on immune responses fail, and B and T cells become hyperactive. *Rheumatoid arthritis*, for instance, is a chronic inflammation of the joints and, often, of cardiovascular and respiratory tissues. T cells attack joints and cause an inflammatory response. Often auto-antibodies, such as IgM–IgG complexes, collect in the joints. They fan cascades of complement that invite chronic inflammation. In time, joints fill with synovial membrane cells and become immobilized.

In other cases, antibodies wrongly bind to hormone receptors and stimulate cell division. *Graves' disorder* is like this. The body makes antibodies to THS receptors in the thyroid gland. This causes increased production of thyroid hormones, which control metabolic rates and growth in many tissues. The antibodies don't respond to normal feedback controls over hormone production. Symptoms include elevated rates of metabolism, heart fibrillations, and excessive sweating (Section 36.6). The most common cause of neurological disorders, *multiple sclerosis*, is another case. It arises after autoreactive T cells trigger inflammation of myelin sheaths and also slip into cerebrospinal fluid, thereby disrupting nerves (Section 34.4). Symptoms range from mild numbness to paralysis and blindness. More than eight genes that influence immune function may heighten susceptibility to multiple sclerosis, but the activation of lymphocytes against viral infections may trigger its development.

Immune responses are stronger in women than men. And autoimmunity is far more frequent in women. We know that the estrogen receptor is involved in controls over gene expression throughout the body. Is it wrongly occupied in autoimmune responses? Maybe.

Deficient Immune Responses

Loss of immune function can have lethal consequences. *Primary* immune deficiencies—present at birth—result

Figure 39.19 A case of severe combined immunodeficiency (SCID). Ashanthi DeSilva was born without an immune system. She carries a mutated gene for ADA (adenosine deaminase). Without this enzyme, her cells could not break down adenosine, so a reaction product that's toxic to lymphocytes accumulated. High fevers, severe ear and lung infections, diarrhea, and an inability to gain weight were outcomes.

Ashanthi's parents consented to the first federally approved human gene therapy. Genetic engineers spliced the ADA gene into the genetic material of a harmless virus. The modified virus was used to deliver copies of the "good" gene into Ashanthi's bone marrow cells. Some cells incorporated the gene in their DNA and started to synthesize the missing enzyme. Ashanthi is now in her teens. Like other ADA-deficient patients who have undergone this gene therapy, she is doing well.

Researchers also take bone marrow stem cells from blood in the umbilical cord of affected newborns. (The cord connects the fetus and placenta during pregnancy and is discarded after birth.) They expose the cells to viruses that deliver copies of the ADA gene into them and to factors that stimulate mitotic cell division and growth. The cells are reinserted into newborns.

from altered genes or abnormal developmental steps. *SCID* (severe combined immunodeficiency) is like this. Figure 39.19 describes one type, called ADA deficiency. *Secondary* immune deficiencies are losses of immune function following exposure to outside agents, such as viruses. Severe immune deficiencies make individuals more vulnerable to infections by opportunistic agents, which are otherwise harmless to those in good health.

AIDS (acquired immunodeficiency syndrome) is the most common secondary immune deficiency. It almost always results in death. The next section describes how the viral agent, HIV, replicates inside lymphocytes and destroys the body's capacity to fight infections.

Immunization programs are designed to boost immunity to specific diseases.

Some heritable disorders, developmental defects, or attacks by viruses and other outside agents result in misdirected, compromised, or nonexistent immunity.

AIDS—The Immune System Compromised

CHARACTERISTICS OF AIDS   AIDS is a constellation of disorders that follow infection by HIV (short for *H*uman *I*mmunodeficiency *V*irus). The virus cripples the immune system—which makes the body extremely susceptible to usually harmless infections and rare forms of cancer. Worldwide, more than 40 million are infected (Table 39.3). At this writing 21.8 million have died, 3.8 million of them children. The death rate has declined in countries where people can afford improved, expensive treatments. It is appalling in sub-Saharan Africa. That is where all but 5 percent of the 13.2 million children orphaned by this epidemic reside. Because of the stigma associated with AIDS, the children are vulnerable to malnutrition, abuse, disease, and sexual exploitation. In seven of the region's countries, at least one in five individuals is infected.

There is no way to rid the body of known forms of the virus (HIV-I and HIV-II). *There is no cure for those already infected.* At first an infected person appears to be in good health, suffering "just a bout of the flu." Then symptoms that foreshadow AIDS emerge. Fever, many enlarged lymph nodes, fatigue, chronic weight loss, and drenching night sweats are typical. Opportunistic infections strike. They include yeast infections of the mouth, esophagus, vagina, and elsewhere, and a form of pneumonia caused by *Pneumocystis carinii*. Bruises often develop on the legs and feet especially (Figure 39.20). The bruises are signs of Kaposi's sarcoma, a cancer of blood vessel endothelium.

HOW HIV REPLICATES   HIV infects antigen-presenting macrophages and helper T cells, which are also called CD4 lymphocytes. HIV is a retrovirus. Its lipid envelope is a bit of plasma membrane acquired as it budded from an infected cell. Various proteins spike from the envelope, span it, or line its inner surface. Inside the envelope, viral coat proteins enclose two RNA strands and several copies of reverse transcriptase, a retroviral enzyme.

Once it is inside an effector or memory T cell, the viral enzyme uses the viral RNA as a template to make DNA, which is inserted into a host chromosome. In some cells, the inserted genes remain silent but are activated in a later round of infection. Transcription yields copies of viral RNA. Some of the transcripts are translated into viral proteins. Others get enclosed in viral coat proteins as new particles are put together; they will be the viral hereditary material. The particles are released by budding from the plasma membrane (Figure 39.21). These start

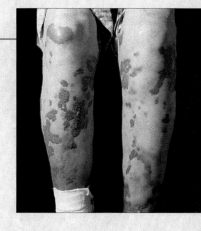

Figure 39.20 Lesions that are a sign of Kaposi's sarcoma.

a new round of infection in other cells. Each time, more antigen-presenting cells and helper T cells are impaired or killed.

A TITANIC STRUGGLE BEGINS   Infection marks the onset of a titanic battle between HIV and the immune system. In response to HIV antigenic proteins, B cells synthesize antibodies. Armies of helper T and cytotoxic T cells form. But during some infection phases, the virus infects about 2 billion helper T cells and makes 100 million to 1 billion virus particles each day. (The viral load can be measured directly.) Every two days, about half of the particles are destroyed and half of the helper T cells lost in battle are replaced. Huge reservoirs of HIV and masses of infected T cells accumulate in the lymph nodes.

Gradually, the number of virus particles in the general circulation rises and the battle tilts. The body produces fewer and fewer replacement helper T cells. It may take a decade or more, but with the erosion of the helper T cell count, the body inevitably loses its capacity to mount effective immune responses.

Some viruses, including the measles virus, produce far more virus particles in a given day, but the immune system usually wins out. Other viruses, including herpes viruses, can be in the body for a lifetime, but the immune system keeps them in check. With HIV, the system almost always loses; infections and tumors kill the individual.

HOW HIV IS TRANSMITTED   Like any other virus that infects humans, HIV requires a medium that allows it to leave one host, survive in the outside environment, then enter another host. It's transmitted when some body fluid of an infected person enters tissues of another person. In the United States, transmission initially occurred most often between homosexual males, as by anal intercourse, then among drug abusers who share blood-contaminated syringes and needles. It also spread in the heterosexual population, increasingly by vaginal intercourse.

Infected mothers can transmit HIV to offspring during vaginal birth and breast-feeding. Also, blood supplies contaminated before 1985 accounted for some AIDS cases, although careful screening procedures have been in place since then. Tissue transplants resulted in a few infections. Health care providers in some developing countries have inadvertently spread HIV by contaminated transfusions and reuse of unsterile needles and syringes.

The molecular structure of HIV does not remain stable outside the human body. This is why the virus must be

Table 39.3 AIDS Cases*	
Sub-Saharan Africa:	28,100,000
South/Southeast Asia:	6,100,000
Latin America:	1,400,000
East Asia and Pacific:	1,000,000
Central Asia/East Europe:	1,000,000
North America:	940,000
Western Europe:	560,000
Middle East/North Africa:	440,000
Caribbean Islands:	420,000
Australia/New Zealand:	15,000

* Global estimates as of December 2001.

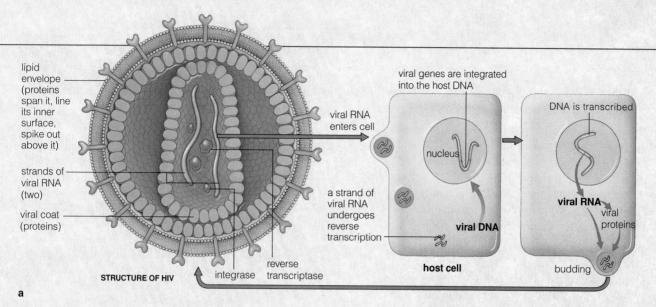

lipid envelope (proteins span it, line its inner surface, spike out above it)

strands of viral RNA (two)

viral coat (proteins)

STRUCTURE OF HIV

integrase

reverse transcriptase

viral RNA enters cell

a strand of viral RNA undergoes reverse transcription

viral genes are integrated into the host DNA

nucleus

viral DNA

host cell

DNA is transcribed

viral RNA

viral proteins

budding

a

Figure 39.21 (**a**) Replication cycle of HIV, a retrovirus. (**b**) Electron micrographs of an HIV particle budding from a host cell. Compare Sections 16.1 and 21.8.

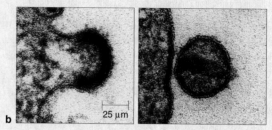

b   25 µm

directly transmitted from one host to another. It has been isolated from human blood, semen, vaginal secretions, saliva, tears, breast milk, amniotic fluid, urine, and cerebrospinal fluid. Probably it is in other fluids. Until recently, only infected blood, semen, vaginal secretions, and breast milk were thought to contain HIV particles in concentrations high enough for successful transmission. Transmission by oral sex is now known to have resulted in some cases of AIDS. HIV is not effectively transmitted by way of food, air, water, casual contact, or insect bites.

DRUG TREATMENTS   Current drugs can't cure infected people because they cannot attack HIV genes already incorporated in someone's DNA.

Given the replication mechanisms and the staggering number of replications in an infected person, the HIV genome acquires new mutations at a high rate. Natural selection operates in patients undergoing drug therapy. It favors drug resistance and fans the evolution of drug-resistant HIV populations. Also, long-term use of the currently available drugs has serious side effects.

AZT (azidothymidine) and ddI (dideoxyinosine) are among the drugs that block reverse transcription. Over time, however, they become incorporated in the DNA of the body's cells and kill them. Ritonavir and Indinavir are among the protease inhibitors, which block cleavages that convert new viral proteins into the building blocks for new virus particles. The potential side effects include nausea and vomiting, headache, diarrhea, blurred vision, and disruption of fat metabolism.

The expense of current treatments—as much as 15,000 dollars a year—effectively puts them out of reach where the AIDS epidemic is raging at its fullest, in developing countries. At present, the best option appears to be the development of a safe, effective, low-cost vaccine.

REGARDING VACCINES   Developing vaccines against HIV is a formidable challenge for several reasons. Here are just a few of them.

First, the most effective vaccines mimic a natural infection from which a person can recover. There are no recovered AIDS patients. A few rare individuals have displayed a long-term capacity to fight the infection, but whatever is going on is simply slowing the assault and does not lend itself to vaccine development.

Second, most vaccines prevent disease, not infection. HIV already integrated into an infected person's DNA cannot be eradicated easily. The best an HIV vaccine may do is help prolong life by lowering the viral load.

Third, unusually high mutation rates rapidly give rise to variations in HIV antigens. Mutated forms that evade the immune system are favored. This makes it hard for researchers to select effective antigens for vaccines.

Fourth, there is a scarcity of HIV strains with poor cell-killing abilities. Such strains are central to efforts to make antigen that can stimulate formation of cytotoxic T cell armies, the best protection against HIV.

Fifth, most vaccines are carefully tested for safety through clinical trials with laboratory animals before testing them on human volunteers. There are no suitable laboratory animals that mirror HIV infections in humans.

In short, *until researchers develop effective vaccines and treatments, checking the spread of HIV depends absolutely on persuading people to avoid or modify social behaviors that put them at risk.* We return to this topic in Section 44.15.

SUMMARY

1. Vertebrates fend off many pathogens with physical and chemical barriers at body surfaces. They also are protected by nonspecific and specific defense responses of the white blood cells listed in Table 39.4. *39.1*

   a. Nonspecific responses to irritation or damage of a tissue include inflammation and participation of organs with phagocytic functions, such as the spleen and liver.

   b. Immune responses are mounted against specific pathogens, foreign cells, or abnormal body cells.

2. Skin and mucous membranes lining body surfaces are physical barriers to infection. The chemical barriers include glandular secretions (such as lysozyme in tears, saliva, and gastric fluid) and some metabolic products of bacteria that normally reside on body surfaces. *39.1*

3. Diverse plasma proteins make up the complement system. When activated, they take part in nonspecific reactions and in immune responses to many bacteria, some parasitic protistans, and enveloped viruses. *39.2*

4. An inflammatory response develops in body tissues that have become damaged, as by infection. *39.3*

   a. Each inflammatory response starts with arteriolar vasodilation that increases the blood flow to the tissue, which reddens and becomes warmer as a result. Blood capillary permeability increases and the resulting local edema causes swelling and pain.

   b. Pathogens and dead or damaged body cells release the substances that trigger increased permeability of capillaries. White blood cells leave the blood and enter the tissue. There, they release a number of chemical mediators and engulf invaders. Also, plasma proteins enter the tissue. Complement proteins bind pathogens and induce their lysis, and they also attract phagocytes. Blood-clotting proteins wall off the damaged tissue.

5. An immune response has these features: *39.4*

   a. Self/nonself recognition. B and T cells ignore the body's cells but mount a counterattack against antigen. Antigen is a molecular configuration that lymphocytes perceive as foreign (nonself).

   b. Specificity. One antigen triggers the response.

   c. Diversity. Collectively, unique receptors on each B and T cell may detect millions of kinds of antigen.

   d. Memory. A later encounter with the same antigen triggers a secondary response (greater and more rapid).

6. Antigen-presenting cells process and bind fragments of antigen to their own MHC markers. T cell receptors can bind to displayed antigen–MHC complexes. Binding is a start signal for immune responses. *39.1, 39.4–39.7*

7. After antigen recognition, repeated cell divisions form clones of B and T cells. These differentiate into subpopulations of effector and memory cells. Chemical mediators such as interleukins secreted by white blood cells drive the responses. *39.1, 39.4–39.7*

Table 39.4 *Summary of White Blood Cells and Their Roles in Defense*

Cell Type	Main Characteristics
MACROPHAGE	Phagocyte; acts in nonspecific and specific responses; presents antigen to T cells, and cleans up and helps repair tissue damage
NEUTROPHIL	Fast-acting phagocyte; takes part in inflammation, not in sustained responses, and is most effective against bacteria
EOSINOPHIL	Secretes enzymes that attack certain parasitic worms
BASOPHIL AND MAST CELL	Secrete histamine, other substances that act on small blood vessels to produce inflammation; also contribute to allergies
DENDRITIC CELL	Processes and directly presents antigen to helper T cells
LYMPHOCYTES:	*All take part in most immune responses; following antigen recognition, form clonal populations of effector and memory cells.*
B cell	Effectors secrete five types of antibodies (IgM, IgG, IgA, and IgD known to protect a host in specialized ways; also IgE)
Helper T cell	Effectors secrete interleukins that stimulate rapid divisions and differentiation of both B cells and T cells
Cytotoxic T cell	Effectors kill infected cells, tumor cells, and foreign cells by a touch-kill mechanism
NATURAL KILLER (NK) CELL	Cytotoxic cell of undetermined affiliation; kills virus-infected cells and tumor cells by a touch-kill mechanism

8. B cells, which form and mature in bone marrow, make antibody-mediated responses. They alone make and secrete antibodies (typically Y-shaped proteins with binding sites for a specific antigen). Antibody binding neutralizes toxins, tags pathogens for destruction, or prevents pathogens from binding to body cells. *39.6*

9. Helper T cells and cytotoxic cells make cell-mediated responses. T cells arise in bone marrow but mature in the thymus, where they acquire T-cell receptors that recognize and bind antigen–MHC complexes on antigen-presenting cells. Activated cytotoxic cells directly kill virus-infected cells, tumor cells, and cells making up tissue or organ transplants. *39.7*

10. By active immunization, vaccines provoke immune responses involving production of effector and memory cells. By passive immunization, injections of purified antibodies help counter an initial infection. *39.9*

11. Allergic reactions are misguided immune responses to harmless substances. In autoimmune responses, B and T cells mount inappropriate attacks on the body's cells. Immune deficiency is a nonexistent or weakened capacity to mount a response. *39.9, 39.10*

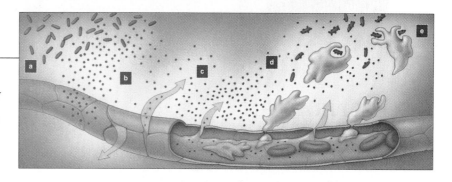

Review Questions

1. While jogging barefoot in the surf, your toes accidentally land on a jellyfish. Soon they are swollen, red, and warm to the touch. Using the diagram at right as a guide, describe events that result in these signs of inflammation. *39.3*

2. Distinguish between:
   a. neutrophil and macrophage *39.3*
   b. cytotoxic T cell and natural killer cell *39.4, 39.7*
   c. effector cell and memory cell *39.4*
   d. antigen and antibody *39.4*

3. Describe the events by which a macrophage turns into an antigen-presenting cell. *39.4, 39.6, 39.7*

4. Why is a vaccine to control AIDS so elusive? *39.10*

Self-Quiz (ANSWERS IN APPENDIX III)

1. _____ can deter many pathogens at body surfaces.
   a. Intact skin and mucous   c. Resident bacteria
      membranes   d. Urine flow
   b. Tears, saliva, gastric fluid   e. all of the above

2. Complement proteins functions in defense. They _____ .
   a. neutralize toxins   d. form pore complexes that
   b. enhance resident   cause lysis of pathogens
      bacteria   e. both a and b
   c. promote inflammation   f. both c and d

3. Macrophages are derived from _____ .
   a. basophils   c. neutrophils
   b. monocytes   d. eosinophils

4. _____ are certain molecules that lymphocytes recognize as foreign and that elicit an immune response.
   a. Interleukins   d. Antigens
   b. Antibodies   e. Histamines
   c. Immunoglobulins

5. The most important antigens are _____ .
   a. nucleotides   c. steroids
   b. triglycerides   d. proteins

6. Immunological specificity is based on _____ .
   a. antigen receptor diversity c. mast cell proliferation
   b. recombinations of   d. both a and b
      receptor genes   e. all of the above

7. Antibody-mediated responses work best against _____ .
   a. intracellular pathogens   d. both b and c
   b. extracellular pathogens   e. all of the above
   c. extracellular toxins

8. Immunoglobulin(s) _____ increase antimicrobial activity in mucus-coated surfaces of some organ systems.
   a. IgA   b. IgE   c. IgG   d. IgM   e. IgD

9. _____ would be a target of an effector cytotoxic T cell.
   a. Extracellular virus particles in blood
   b. A virus-infected body cell or tumor cell
   c. Parasitic flukes in the liver
   d. Bacterial cells in pus
   e. Pollen grains in nasal mucus

10. Match the immunity concepts.
    _____ inflammation   a. neutrophil
    _____ antibody secretion   b. effector B cell
    _____ fast-acting phagocyte c. nonspecific response
    _____ immunological   d. improper immune response
          memory   to body's own components
    _____ allergy   e. basis of secondary response
    _____ autoimmunity   f. hypersensitivity to allergen

Critical Thinking

1. As described in the chapter introduction, Edward Jenner lucked out. He performed a potentially harmful experiment on a boy who managed to survive it. What would happen if a would-be Jenner tried to do the same thing today?

2. Researchers have been attempting to develop a way to get the immune system to accept foreign tissue as "self." Speculate on some of the clinical applications of such a development.

3. Before each flu season, you get an influenza vaccination. This year you come down with "the flu" anyway. What do you suppose happened? (There are at least three explanations.)

4. In December 2001, researchers Akio Ohta and Michael Sitkovsky reported on a membrane receptor for adenosine that may be a switching mechanism for turning off an inflammatory response. Inflammation normally protects us from infection. Prolonged or misdirected inflammation can interfere with the healing of damaged tissues and may promote an autoimmune response against them. The researchers also found that caffeine blocks adenosine's action. What does this finding tell you about drinking coffee or tea when you have a fever?

5. Ellen developed *chicken pox* when she was in kindergarten. Later in life, when her children developed chicken pox, she remained healthy even though she was exposed to countless virus particles daily. Explain why.

6. Write a short essay on how immune responses contribute to stability in the internal environment. (Compare Section 28.3.)

Selected Key Terms

allergen *39.9*	immune system *39.4*
allergy *39.9*	immunization *39.9*
antibody *39.4*	immunoglobulin (Ig) *39.6*
antigen *39.4*	inflammation, acute *39.3*
antigen–MHC complex *39.4*	lysis *39.2*
antigen-presenting cell *39.4*	lysozyme *39.1*
apoptosis *39.7*	macrophage *39.3*
autoimmunity *39.9*	mast cell *39.3*
B lymphocyte (B cell) *39.4*	MHC marker *39.4*
basophil *39.3*	natural killer (NK) cell *39.4*
complement system *39.2*	neutrophil *39.3*
cytotoxic T cell *39.4*	pathogen *39.1*
eosinophil *39.3*	suppressor T cell *39.4*
fever *39.3*	T lymphocyte (T cell) *39.4*
helper T cell	TCR *39.5*
(CD4 lymphocyte) *39.4*	vaccination *CI*
histamine *39.3*	vaccine *39.9*

Readings

Goldsby, R., T. Kindt, and B. Osborne. 2000. *Kuby Immunology.* Fourth edition. New York: Freeman.

Stine, G. 2001. *AIDS Update 2001.* Upper Saddle River, New Jersey: Prentice-Hall.

On-Line readings at Student Guide for InfoTrac:
www.brookscole.com/biology

40

RESPIRATION

Of Lungs and Leatherbacks

Late at night, near a Caribbean shoreline, biologist Molly Lutcavage and her colleagues are on a turtle patrol. This time they are in luck; a female Atlantic leatherback sea turtle (*Dermochelys coriacea*) emerges from the surf to nest in the sand (Figure 40.1). All sea turtles evolved 300 million years ago, and are now on the brink of extinction. Dr. Lutcavage and others are racing to gather information about the leatherback's form, physiology, and behavior that might be used to help get it off the endangered species list.

Leatherbacks normally leave the water only to breed and lay eggs. They spend most of their lives migrating across the open ocean. For example, nine leatherbacks equipped with satellite transmitters were tracked from Mexico, across the equator, to the coasts of Chile and Peru. There, the long-term expansion of fishing fleets may have something to do with the decline of Pacific leatherbacks. Another possible factor in the decline: These turtles die after ingesting discarded plastic bags. They mistake them for jellyfishes, their main prey.

Some say the impending loss of a species is no big deal, it's only a turtle. Yet a leatherback's adaptations to life in the seas are remarkable. It almost never stops swimming, even while it sleeps. Satellite-monitored leatherbacks proved they can swim more than 12,000 kilometers (7,457 miles) annually. And leatherbacks do something else that you never would be able to do on your own. They can dive 1,000 meters (3,000 feet) below sea level.

To reach that depth, a leatherback swims and glides for nearly forty minutes without taking a breath. You, on the other hand, cannot hold your breath for more than a few minutes without passing out.

Like you and other highly active vertebrates, the leatherback relies on the large supply of energy made available through the oxygen-requiring pathway of aerobic respiration. Also like you, the leatherback has a pair of lungs that move oxygen into the body and carbon dioxide wastes out of it. But oxygen reserves in the lungs are not enough to sustain a leatherback's prolonged dive. Even at depths of 80 to 160 meters, pressure exerted by the surrounding water would make the delicate, air-filled lung surfaces collapse. Before that can happen, leatherbacks move most of the air into the tubes that lead to the lungs, where little, if any, gas exchange occurs.

However, like deep-diving marine mammals, the leatherback has an abundance of myoglobin in its heart muscle cells and in skeletal muscle cells that contract

Figure 40.1 Three individuals adapted to getting and using oxygen for aerobic respiration, each with an uncertain future. At *left*, female leatherback sea turtles (*Dermochelys coriacea*) return to the water after laying eggs in the sands of a Caribbean island. These two may be close to the end of their 300-million-year-old lineage; leatherbacks are an endangered species. *Above*, a child in Mexico City already proficient at smoking cigarettes, a behavior that ultimately will endanger her capacity to breathe.

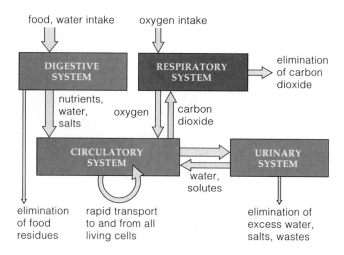

Figure 40.2 Interactions between the respiratory system and other organ systems, the foundation for homeostasis in complex animals.

slowly and resist fatiguing. Like hemoglobin, the protein myoglobin is good at binding and delivering oxygen—and it is good at storing it. Also, a leatherback has so many hemoglobin-packed red blood cells that its blood is a remarkable 21 percent oxygen by volume. That amount is typical of a human, not what you might otherwise expect to find in a "plodding" turtle. Finally, of all sea turtles, only leatherbacks can adjust how fast and deeply they breathe. Adjustments help them use oxygen efficiently when they dive and also to recover respiratory function faster when they resurface.

Few of us find ourselves diving in deep water, using tanks of compressed air to push our reliance on oxygen to the limits. Here on land, disease, smoking, and other environmental assaults push it in more ordinary ways, although the risks can be just as great.

The point is this: *Each animal has a body plan adapted to the oxygen levels of a particular habitat.* That body plan allows for **respiration**—movement of oxygen into the internal environment and movement of carbon dioxide out of it. Again, the high energy demands of animals are based on the ATP output of aerobic respiration, a metabolic pathway that releases energy from organic compounds and produces carbon dioxide leftovers.

This chapter samples a few **respiratory systems**, which function in the exchange of gases between the body and the environment. Together with other organ systems, they also contribute to homeostasis—that is, to maintaining internal operating conditions for all of the body's living cells (Figure 40.2).

Key Concepts

1. Of all organisms, multicelled animals require the most energy to drive their metabolic activities. At the cellular level, the energy comes mainly from aerobic respiration, an ATP-producing metabolic pathway that uses oxygen and releases carbon dioxide wastes.

2. By a physiological process called respiration, animals move oxygen into fluids of their internal environment and give up carbon dioxide to the external environment.

3. Oxygen diffuses into the animal body as a result of a pressure gradient. The pressure of this gas is higher in the outside air than it is in metabolically active tissues, where cells rapidly use oxygen. Carbon dioxide follows its own gradient, in the opposite direction. Its pressure is higher in tissues, where it is a by-product of metabolism, than it is in the air.

4. In most respiratory systems, oxygen and carbon dioxide diffuse across a respiratory surface, such as the thin, moist respiratory membrane inside human lungs. Blood flowing through the body's circulatory system picks up oxygen and gives up carbon dioxide at this respiratory membrane.

5. Gas exchange is most efficient when the rate of air flow matches the rate of blood flow. The nervous system brings the rates into balance by controlling the rhythmic pattern and magnitude of breathing.

THE NATURE OF RESPIRATION

The Basis of Gas Exchange

A concentration gradient, recall, is a difference in the number of molecules (or ions) of a substance between two regions. Like other substances, oxygen and carbon dioxide tend to diffuse down a concentration gradient; each shows a net outward movement from the region where molecules of it are colliding more frequently (Section 5.4). The process of respiration is based on the tendency of these two gases to follow their respective concentration gradients—or, as we say for gases, to diffuse down **pressure gradients**—that form between the animal and its surroundings.

The gases do not exert the same pressure. Pump air into a flat tire near a beach in San Diego or Miami or anywhere else at sea level. You will fill it with about 78 percent nitrogen, 21 percent oxygen, 0.04 percent carbon dioxide, and 0.96 percent of some other gases. By mercury barometer measures, atmospheric pressure at sea level is about 760 mm Hg (Figure 40.3). Oxygen exerts just part of that total pressure on the tire wall. Its "partial" pressure is greater than carbon dioxide's. And so oxygen's **partial pressure**—its contribution to the total atmospheric pressure—is 760 × 21/100, or about 160 mm Hg. When similarly measured, carbon dioxide's partial pressure is about 0.3 mm Hg.

Gases enter and leave an animal body by crossing a **respiratory surface**, a thin layer of epithelium or some other body tissue. The surface must be kept moist at all times; gaseous molecules cannot diffuse across it unless they are dissolved in fluid. What dictates the *number* of gas molecules that can move across a respiratory surface during any specified interval? According to **Fick's law**, the greater the surface area and the larger the partial pressure gradient, the faster the diffusion rate will be.

Which Factors Influence Gas Exchange?

SURFACE-TO-VOLUME RATIO   All animal body plans promote favorable rates of inward diffusion of oxygen and outward diffusion of carbon dioxide. For example, animals without respiratory organs are tiny, tubelike, or flattened, and gases diffuse directly across the body surface. These body plans meet the constraint imposed by the surface-to-volume ratio, as Section 4.1 describes. To get a sense of the ratio's effects, imagine a flatworm growing in all directions like an inflating balloon. The worm's surface area does not increase at the same rate as its volume. Once its girth exceeds even a millimeter, the diffusion distance between the body's surface and internal cells will be so great that the worm will die.

VENTILATION   Large-bodied, active animals have huge demands for gas exchange, more than diffusion alone

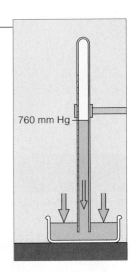

Figure 40.3 Atmospheric pressure as measured with a device called a mercury barometer. Part of the device is a glass tube in which the height of a column of mercury (Hg) increases or decreases, depending on air pressure outside the device. At sea level, the mercury rises to about 760 millimeters (29.91 inches) from the base of the tube. At this level, the pressure that the column of mercury exerts inside the tube is equal to the atmospheric pressure on the outside.

760 mm Hg

can satisfy. Diverse adaptations make the exchange rate more efficient. For example, a flap of tissue above many fish gills (respiratory organs) moves back and forth, thus stirring the surrounding water. The stirring brings more dissolved oxygen closer to the gills and moves carbon dioxide away. Among vertebrates, a circulatory system rapidly transports oxygen to cells and carbon dioxide to gills or lungs for disposal. As another example, you breathe in and out to ventilate your lungs.

TRANSPORT PIGMENTS   Rates of gas exchange get a boost with transport pigments—mainly **hemoglobin**—that help maintain the steep pressure gradients across the respiratory surface. For example, at the respiratory surface in a human lung, the oxygen concentration is high, and each hemoglobin molecule in blood weakly binds up to four oxygen molecules (Section 3.8). The circulatory system swiftly transports hemoglobin away from the respiratory surface. In oxygen-poor tissues, oxygen follows its gradient and diffuses away from the hemoglobin, thereby helping maintain the gradient that entices oxygen into blood. **Myoglobin**, a similar but smaller pigment, has a notable oxygen-storing capacity. It's abundant in heart muscle cells and skeletal muscle cells that resist fatiguing.

As the chapter introduction states, respiration is a process by which animals move oxygen into the internal environment, and move carbon dioxide wastes from aerobic respiration out of it.

Oxygen and carbon dioxide enter and leave the internal environment by diffusing across a moist respiratory surface. Like other atmospheric gases, they tend to move down their respective pressure gradients. Each type of gas exerts only part of the total pressure across a respiratory surface.

Gas exchange depends on steep partial pressure gradients between the outside and inside of the animal body. The greater the area of the respiratory surface and the larger the partial pressure gradient, the faster diffusion will proceed.

INVERTEBRATE RESPIRATION

Flatworms, earthworms, and many other invertebrates are not massive, and their life-styles do not depend on high metabolic rates (Figure 40.4a). Demands for gas exchange are simply met by **integumentary exchange**, in which gases diffuse directly across the body's surface covering (integument). This mode of respiration works only when the surface is moist, and invertebrates that rely exclusively on it are restricted to aquatic or damp habitats. (By contrast, amphibians and some other large animals use integumentary exchange also, but only as a supplement to other modes of respiration.)

Many invertebrates of aquatic habitats have moist, thin-walled respiratory organs called **gills**. Extensively folded gill walls have an increased respiratory surface area that enhances the rates of exchange between blood or some other body fluid and the surroundings. Figure 40.4b shows the much-folded gill of a sea hare (*Aplysia*). By supplementing integumentary exchange, the gill helps provide enough oxygen for this big mollusk. Some sea hares are 40 centimeters (nearly 16 inches) long.

Spiders and other invertebrates of dry habitats have a small, thick, or hardened integument that is not well

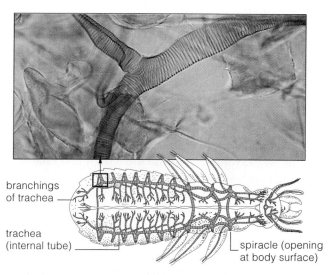

Figure 40.5 General plan of an insect tracheal system. Chitin rings reinforce many of the branching tubes of the system.

endowed with blood vessels. Their integument holds in precious water, but it is not a good respiratory surface. Such animals have an *internal* respiratory surface. For instance, most spiders have internal respiratory organs with thin, folded walls that resemble book pages; hence the name, book lungs (Section 25.14). Like most insects, millipedes, and centipedes, some other spiders have a system of internal tubes for **tracheal respiration**.

Figure 40.5 shows the tracheal system of one insect. Small openings perforate the insect's integument. Each opening, a spiracle, is the start of a tube that branches within the body. Each of the last branchings dead-ends at its fluid-filled tip, where gases diffuse directly into tissues. The tube tips are especially profuse in muscle and other tissues with high oxygen demands.

We find hemoglobin or other respiratory pigments in many invertebrates, although they are rare in insects. The respiratory pigments increase the capacity of body fluids to transport oxygen, as they do in vertebrates. In species that have a well-developed head, oxygenated blood tends to circulate first through the head end and then through the rest of the body.

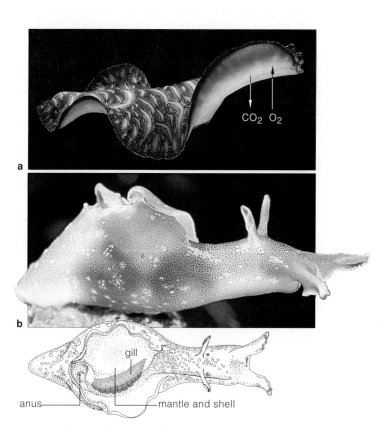

Figure 40.4 Invertebrates of aquatic habitats. (**a**) A flatworm, small enough to get along well without an oxygen-transporting circulatory system. Dissolved oxygen in such habitats reaches individual cells simply by diffusing across the body surface. (**b**) Gill of a sea hare (*Aplysia*), one of the gastropods.

Flatworms and some other invertebrates of aquatic or moist habitats do not have a massive body. By their simple mode of respiration, called integumentary exchange, oxygen and carbon dioxide diffuse directly across the body surface.

Most marine invertebrates and many freshwater types have gills of one sort or another. These respiratory organs have moist, thin, and often highly folded walls.

Most insects, millipedes, centipedes, and some spiders use tracheal respiration. Gases flow through open-ended tubes that start at the body surface and end directly in tissues.

VERTEBRATE RESPIRATION

Gills of Fishes and Amphibians

Many vertebrates use gills as respiratory organs. A few kinds of fish larvae and a few amphibians have *external* gills projecting into water. Adult fishes have a pair of *internal* gills: rows of slits or pockets at the back of the mouth that extend to the body's surface (Figure 40.6a). Whatever its form, each gill has a wall of moist, thin, vascularized epithelium.

In fishes, water flows into the mouth and pharynx, then over arrays of filaments in the gills (Figure 40.6b). Blood vessels thread through the respiratory surfaces in each filament. First, water flows past a vessel leading to the rest of the body. Blood inside has less oxygen than the water does, so oxygen diffuses into the blood. The same volume of water flows over a vessel leading into the gills. Although it already gave up some oxygen, it still holds more than the blood in this vessel does. So more oxygen diffuses into the filament. Any movement of two fluids in opposing directions is a **countercurrent flow**. With such a mechanism, a fish can extract about 80 to 90 percent of the oxygen dissolved in the water flowing past. That is more than the fish would get from a one-way flow mechanism, and at less energy cost.

Evolution of Paired Lungs

Some fishes and all amphibians, birds, and mammals have a pair of **lungs**, internal respiratory surfaces in the shape of a cavity or sac. More than 450 million years ago, lungs started to develop in some fish lineages as outpouchings of the gut wall. They must have evolved rapidly by natural selection. In oxygen-poor habitats, lungs afforded a larger surface area for gas exchange. This proved advantageous for the first tetrapods that moved onto land; air holds far more oxygen than water does (Section 26.3). Also, gills would not have worked on land. Gill filaments stick together when water does not flow through them and keep them moist.

Lungfishes of oxygen-poor habitats still have gills. They also use tiny lungs as a backup for gas exchange.

Amphibians never completed the transition to land. Salamanders especially use their skin as a respiratory surface for integumentary exchange. Frogs and toads rely more on small lungs for oxygen uptake, but most of their carbon dioxide wastes diffuse outward across the skin. Frogs also are heavy-duty breathers; they *force* air into their lungs, which they empty by contracting muscles in their body wall (Figure 40.7).

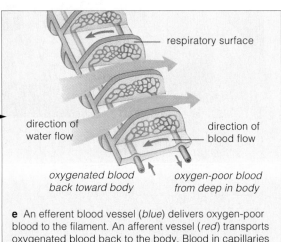

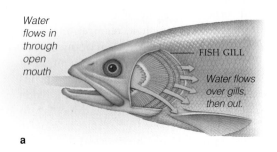

Figure 40.6 (a) One of a pair of fish gills. A bony lid, removed for this sketch, protects it.

(b) Gill ventilation. Opening the mouth and closing the lid pulls water across the gill. (c) Closing the mouth and opening the lid forces water outside.

(d) Gas exchange at a gill. Filaments in a gill have vascularized respiratory surfaces. The filament organization directs incoming water past the gas exchange surfaces.

(e) A blood vessel carries oxygen-poor blood from the body to each filament. Another vessel carries oxygenated blood into the body. Blood flows from one into the other and runs counter to the flow of water over respiratory surfaces. The countercurrent flow favors the movement of oxygen (down its partial pressure gradient) from water into the blood.

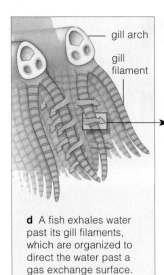

d A fish exhales water past its gill filaments, which are organized to direct the water past a gas exchange surface.

e An efferent blood vessel (*blue*) delivers oxygen-poor blood to the filament. An afferent vessel (*red*) transports oxygenated blood back to the body. Blood in capillaries connecting the two vessels flows counter to the direction of water flowing over the gas exchange surface.

  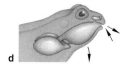 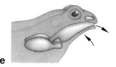

b   c   d   e

Figure 40.7 (**a**) Frog respiration. (**b**) The frog lowers the floor of its mouth and inhales air through its nostrils. (**c**) It closes its nostrils, opens the glottis, and elevates the mouth's floor. This *forces* air into the lungs. (**d**) Rhythmic ventilation assists in gas exchange. (**e**) Air is forced out when muscles in the body wall above the lungs contract and lungs elastically recoil.

a

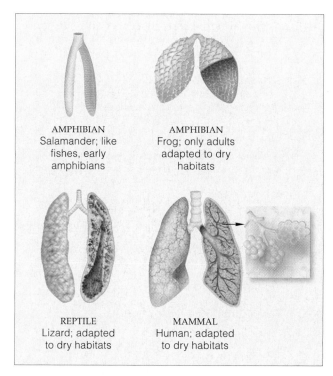

AMPHIBIAN
Salamander; like fishes, early amphibians

AMPHIBIAN
Frog; only adults adapted to dry habitats

REPTILE
Lizard; adapted to dry habitats

MAMMAL
Human; adapted to dry habitats

Figure 40.8 Generalized lung structure for four vertebrates, suggestive of an evolutionary trend from sacs for simple gas exchange to larger, more complex respiratory surfaces.

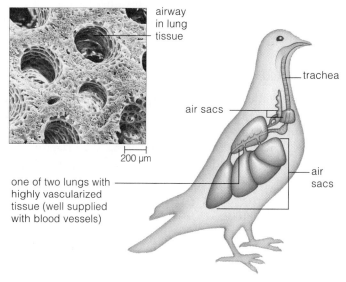

airway in lung tissue

200 μm

one of two lungs with highly vascularized tissue (well supplied with blood vessels)

trachea

air sacs

air sacs

Figure 40.9 A bird's respiratory system. Two small, inelastic lungs have a number of attached air sacs. The bird inhales and draws air into the sacs through tubes, open at both ends, that thread through vascularized lung tissue. This tissue is the respiratory surface.

The bird exhales, forcing air out of the sacs, through the tubes, and out of the trachea. So air isn't just drawn into bird lungs. It is drawn continuously through them and across respiratory surfaces. This unique ventilating system supports the high metabolic rates birds require for flight and other energy-intensive activities.

Frogs use lungs to make sounds. So do all mammals except whales. Sounds start at the entrance to a larynx. Part of a mucous membrane folds into this airway to the lungs. The folds are **vocal cords**; the gap between them is a **glottis**. Frogs force air back and forth through their glottis, and between lungs and a pair of pouches on the floor of the mouth. Air flow makes the cords vibrate. The vibrations are controlled to make different sounds.

Paired lungs are the dominant respiratory organs in reptiles, birds, and mammals (Figure 40.8). Breathing moves air by bulk flow into and out of the lungs, where capillaries thread lacily around the respiratory surface. Oxygen and carbon dioxide diffuse rapidly across the surface, following steep gradients. Oxygen enters blood capillaries and is circulated quickly through the body. Where oxygen levels are low, it diffuses into interstitial

fluid, then into cells. Carbon dioxide moves quickly in the other direction and is expelled from the lungs.

This mode of gas exchange is embellished a bit only in birds. As Figure 40.9 indicates, birds are unique in that air not only flows into and out from their lungs; it also flows *through* their lungs. With that exception in mind, we turn next to the human respiratory system, for its operating principles apply to most vertebrates.

A countercurrent flow mechanism in fish gills compensates for low oxygen levels in aquatic habitats. Internal air sacs—lungs—are more efficient in dry habitats on land.

Amphibians use integumentary exchange and force air into and out of small lungs. Ventilation of paired lungs is the major mode of respiration in reptiles, birds, and mammals.

HUMAN RESPIRATORY SYSTEM

What Are the System's Functions?

Getting oxygen from air and expelling carbon dioxide are the key functions of the human respiratory system. Breathing, a form of ventilation, alternately moves air into and out of its paired lungs. Each lung has about 300 million outpouchings. These tiny air sacs are called **alveoli** (singular, alveolus). Control mechanisms adjust breathing rates so that the inflow and outflow of air at alveoli match metabolic demands for gas exchange.

The system has other functions. Breathing is used in vocalizations, such as speech. It enhances the return of venous blood to the heart and helps dispose of excess heat and water. Breathing controls adjust the body's acid–base balance. Carbon dioxide, recall, helps form carbonic acid (H_2CO_3), which works with bicarbonate (HCO_3^-) as a buffer system. HCO_3^- accepts or releases hydrogen ions (H^+), depending on blood's pH. Deep, rapid breathing expels more carbon dioxide, so less carbonic acid forms and the body loses acid. Shallow, slow breathing has the opposite effect. Carbon dioxide builds up, more H_2CO_3 forms, and the body gains acid.

The respiratory system also has built-in mechanisms that deal with airborne foreign agents and substances

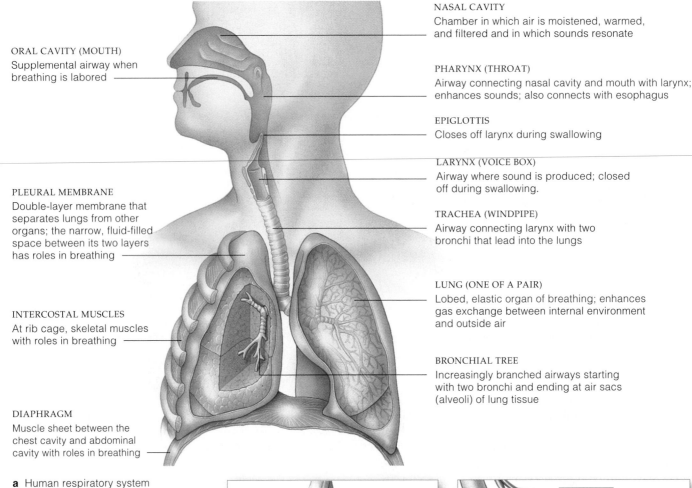

NASAL CAVITY
Chamber in which air is moistened, warmed, and filtered and in which sounds resonate

PHARYNX (THROAT)
Airway connecting nasal cavity and mouth with larynx; enhances sounds; also connects with esophagus

EPIGLOTTIS
Closes off larynx during swallowing

LARYNX (VOICE BOX)
Airway where sound is produced; closed off during swallowing.

TRACHEA (WINDPIPE)
Airway connecting larynx with two bronchi that lead into the lungs

LUNG (ONE OF A PAIR)
Lobed, elastic organ of breathing; enhances gas exchange between internal environment and outside air

BRONCHIAL TREE
Increasingly branched airways starting with two bronchi and ending at air sacs (alveoli) of lung tissue

ORAL CAVITY (MOUTH)
Supplemental airway when breathing is labored

PLEURAL MEMBRANE
Double-layer membrane that separates lungs from other organs; the narrow, fluid-filled space between its two layers has roles in breathing

INTERCOSTAL MUSCLES
At rib cage, skeletal muscles with roles in breathing

DIAPHRAGM
Muscle sheet between the chest cavity and abdominal cavity with roles in breathing

a Human respiratory system

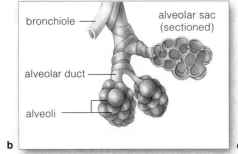

bronchiole
alveolar sac (sectioned)
alveolar duct
alveoli

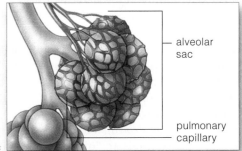

alveolar sac

pulmonary capillary

Figure 40.10 (**a**) Human respiratory system: its components and their functions. Muscles, including the diaphragm, and parts of the axial skeleton have secondary roles in respiration. (**b,c**) Location of alveoli relative to bronchioles and to lung (pulmonary) capillaries.

b

c

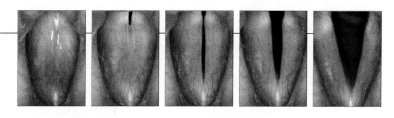

inhaled with air. Finally, it removes and inactivates or otherwise modifies numerous blood-borne substances before they can circulate through the rest of the body.

The respiratory system's role in respiration ends at alveoli. There, the circulatory system takes over the task of moving gases. Oxygen and carbon dioxide diffuse across both the alveoli and the pulmonary capillaries next to them. (The Latin *pulmo* means lung.)

From Airways Into the Lungs

It will take at least 300 million breaths to get you to age seventy-five. You might find yourself going without food for a few hours or days. But stop breathing even for five minutes and normal brain function is over.

Take a deep breath. Now look at Figure 40.10 to get an idea of where the air will travel in your respiratory system. Unless you are out of breath and panting, air has entered two nasal cavities, not your mouth. There, secretions of mucus warm and moisten it. In the cavities, ciliated epithelium and hairs filter dust and particles from air, and olfactory receptors function in the sense of smell (Section 35.3). Now air is poised at the **pharynx**, or throat. The pharynx is the entrance to the **larynx**, an airway having two paired folds of mucous membrane. The lower pair are vocal cords (Figure 40.11). When you breathe, air is forced in and out through the glottis, the gap between these cords. As in frogs (Section 40.3), air flow makes the vocal cords vibrate in ways that can be controlled to make different sounds.

Inside the folds are thick bands of elastic ligaments attached to cartilage. When the larynx muscles contract and relax, these ligaments tighten or slacken to change how much the folds are stretched. The nervous system coordinates the narrowing and widening of the glottis. For example, when it calls for increased tension in the larynx muscles, the gap between the vocal cords shrinks and you make high-pitched sounds and squeaks. The lips, teeth, tongue, and the soft roof above the tongue are enlisted to modify different sounds into patterns of vocalization, such as speech and song.

When vocal cords are inflamed as an outcome of an infection or irritation, swelling of their mucous lining interferes with their capacity to vibrate. If the swelling causes hoarseness, this condition is called *laryngitis*.

At the entrance to the larynx is an **epiglottis**. When this flap of tissue points upward, air can move into the **trachea**, or windpipe. The epiglottis points downward and shuts off the trachea's entrance when you swallow. Food and fluids enter a different tube (the esophagus) that connects the pharynx with the stomach.

The trachea branches into two airways, which lead into tissues of the lungs. Each airway is a **bronchus** (plural, bronchi). Its epithelial lining has a profusion of

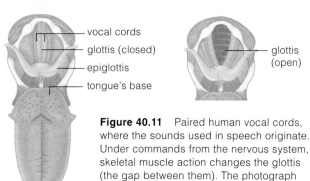

Figure 40.11 Paired human vocal cords, where the sounds used in speech originate. Under commands from the nervous system, skeletal muscle action changes the glottis (the gap between them). The photograph series gives you an idea of how the glottis changes during quiet breathing.

cilia and mucus-secreting cells that act as a barrier to infection. Bacteria and airborne particles stick to the mucus, and cilia sweep the debris-laden mucus toward the mouth. The mucus is expelled or swallowed.

Human lungs are elastic, cone-shaped organs of gas exchange. They are located inside the rib cage to the left and right of the heart and above the diaphragm. The **diaphragm** is a muscular partition between the thoracic and abdominal cavities. Lining the outer surface of the lungs and inner surface of the thoracic cavity's wall is a saclike pleural membrane. The thoracic wall and lungs press the two thin, opposing surfaces together.

A thin film of lubricating fluid cuts friction between the pleural membrane surfaces. In *pleurisy*, a respiratory ailment, the membrane becomes inflamed and swollen. Its two surfaces rub each other, so breathing is painful.

Inside each lung, air moves through finer and finer branchings of a "bronchial tree" (Figure 40.10a). These airways are **bronchioles**. Their finest branchings, the *respiratory* bronchioles, end in the cup-shaped alveoli. Alveoli usually are clustered as larger pouches called alveolar sacs, as in Figure 40.10b,c. Collectively, alveolar sacs offer a huge surface area for gas exchanges with blood. If all of your alveolar sacs were stretched out in one layer, they would cover a racquetball court floor!

Oxygen uptake and carbon dioxide removal are the major functions of the human respiratory system. Inside its pair of lungs, the circulatory system takes over the remaining tasks of respiration.

The respiratory system also has roles in moving venous blood to the heart, in vocalization, in adjusting the body's acid–base balance, in defense against harmful airborne agents or substances, in removing or modifying a number of blood-borne substances, and in the sense of smell.

CYCLIC REVERSALS IN AIR PRESSURE GRADIENTS

The Respiratory Cycle

There is a cyclic pattern to breathing, which ventilates the lungs. Each **respiratory cycle** consists of two actions: *inhalation* (a single breath of air drawn into the airways) and *exhalation* (a single breath out).

Look at Figure 40.12. *Inhalation always is an active, energy-requiring action.* When you are breathing quietly, inhalation is brought about by the contraction of the diaphragm and, to a lesser extent, external intercostal muscles. The outcome is an increase in the thoracic cavity volume. Breathe hard, and the volume increases more because neck muscles contract, thereby elevating the sternum and the first two ribs attached to them.

In a respiratory cycle, the thoracic cavity's volume increases and decreases, so pressure gradients between the air inside and outside the respiratory tract change. *Atmospheric* pressure, 760 mm Hg at sea level, is exerted by the combined weight of all atmospheric gases on all airways. Before inhalation, the *intrapulmonary* pressure in all alveoli is 760 mm Hg (Figure 40.13a).

A different gradient helps keep the lungs close to the thoracic cavity wall during the cycle even through exhalation, when lungs have a smaller volume than the thoracic cavity (Figure 40.13b,c). As the cavity expands, the lungs expand also, as a result of a pressure gradient across the lung wall.

In a person who is resting, the *intrapleural* pressure (inside the pleural sac) averages 756 mm Hg, which is lower than atmospheric pressure. Intrapleural pressure is exerted outside the lungs (inside the thoracic cavity). As it pushes inward on a lung wall, intrapulmonary pressure is pushing outward. The difference between them—4 mm Hg—is large enough to cause the lungs to stretch and fill the thoracic cavity.

In the fluid inside the pleural sac (the intrapleural fluid), the cohesiveness of water molecules also helps keep lungs close to the thoracic wall. By analogy, wet two small panes of glass and press them together. The panes easily slide back and forth but strongly resist being pulled apart. Similarly, intrapleural fluid "glues" lungs to the wall. As a result, when the thoracic cavity expands by inhalation, the lungs must expand, also.

Figure 40.12b shows what happens as you start to inhale. The diaphragm flattens and moves downward, and the rib cage is lifted upward and outward. As the

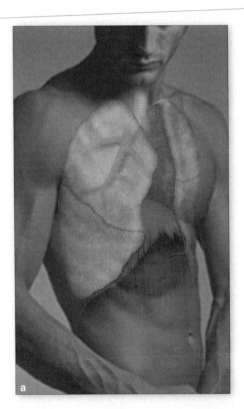

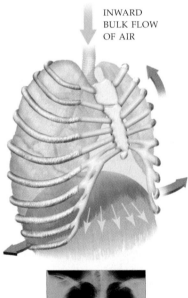

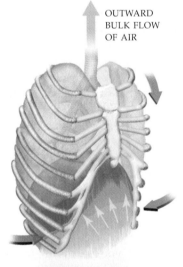

INWARD
BULK FLOW
OF AIR

OUTWARD
BULK FLOW
OF AIR

Figure 40.12 (**a**) Location of human lungs relative to the diaphragm. (**b**,**c**) Changes in the thoracic cavity's size during one respiratory cycle. The x-ray images show how the maximum inhalation possible changes the thoracic cavity volume.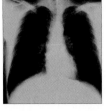

b Inhalation. The diaphragm contracts, moves down. External intercostal muscles contract and lift rib cage upward and outward. The lung volume expands.

c Exhalation. Diaphragm, external intercostal muscles return to resting positions. Rib cage moves down. Lungs recoil passively.

Figure 40.13 Changes in intrapulmonary pressure and lung volume in a respiratory cycle.

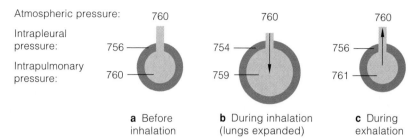

Atmospheric pressure:

Intrapleural pressure:

Intrapulmonary pressure:

a Before inhalation

b During inhalation (lungs expanded)

c During exhalation

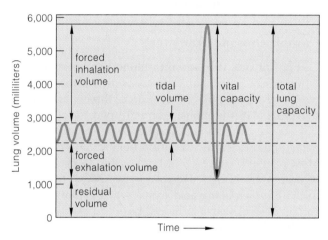

Figure 40.14 Lung volume. During quiet breathing, a tidal volume of air enters and leaves the lungs. Forced inhalation gets more air to them; forced exhalation releases some air that normally stays in them. A residual volume is trapped in partly filled alveoli even during the strongest exhalation.

thoracic cavity is expanding, the lungs expand along with it. At this time, the air pressure in all alveolar sacs combined is lower compared to atmospheric pressure. Fresh air follows this pressure gradient and flows into the airways, then into the alveoli.

The second part of the respiratory cycle, exhalation, is passive when you are breathing quietly. The muscles that brought about inhalation relax and the lungs passively recoil, with no further energy outlays. The decrease in lung volume compresses air inside the alveolar sacs. At this time, pressure inside is greater than atmospheric pressure. Air follows the gradient, flowing out from the lungs, as in Figure 40.13c.

Exhalation becomes active and energy-demanding only when you are vigorously exercising and so must expel more air. The abdominal wall muscles contract during active exhalation. Abdominal pressure increases, and it exerts an upward-directed force on the diaphragm, which is thereby pushed upward. Also, the internal intercostal muscles contract. When they do, they pull the thoracic wall inward and downward, and their action flattens the chest wall. In all such ways, the thoracic cavity's dimensions shrink. The lung volume shrinks as well when elastic tissue of the lungs passively recoils.

Lung Volumes

The lungs can hold up to 5.7 liters or so of air in adult males and 4.2 liters in adult females who are young and in good health. These are only average values; age, build, and respiratory health influence the total lung capacity. During quiet breathing, the lungs are far from being fully inflated. In general, they hold 2.7 liters at the end of one inhalation and 2.2 liters at the end of one exhalation. The lungs never do deflate completely. When air flows out and lung volume is low, walls of the smallest airways collapse and prevent further air loss.

The volume of air that can move out of the lungs in one breath after maximal inhalation is called the **vital capacity**. Humans rarely use more than half of the total vital capacity, even when taking in very deep breaths during strenuous exercise (Figure 40.14). Because the lungs are never empty, gas exchange between alveolar air and the blood proceeds even at the end of maximum exhalation. *Hence the concentrations of gases in the blood remain fairly constant throughout the respiratory cycle.*

The volume of air flowing into or out of the lungs in the respiratory cycle, the **tidal volume**, averages 0.5 liter. How much of the residual volume is available for gas exchange? Between breaths, about 1.2 liters stay in the airways; only 0.35 liter of fresh air reaches alveoli. When you breathe, say, ten times a minute, you are supplying your alveoli with (0.35 × 10) or 3.5 liters of fresh air per minute.

Breathing, which ventilates the lungs, has a cyclic pattern. The respiratory cycle consists of inhalation (one breath of air in) and exhalation (one breath of air out).

Inhalation is always an active, energy-requiring process involving contractions mainly of the diaphragm and the external intercostal muscles.

During quiet breathing, exhalation is a passive process. Muscles relax, the thoracic cavity volume decreases, and the lungs recoil elastically. Forceful exhalation is an active process that requires abdominal muscle contraction.

Breathing reverses pressure gradients between the lungs and the air outside the body.

GAS EXCHANGE AND TRANSPORT

Exchanges at the Respiratory Membrane

An alveolus is a cupped sheet of epithelial cells with a basement membrane. A pulmonary capillary consists of endothelial cells, and its basement membrane fuses with that of the alveolus. Together, the alveolar epithelium, capillary endothelium, and their basement membranes form a respiratory membrane (Figure 40.15). Gases can diffuse across this thin membrane very quickly.

Oxygen and carbon dioxide passively diffuse across the membrane in response to partial pressure gradients. An inward-directed gradient for oxygen is maintained because inhalations replenish oxygen and cells take it up continually. An outward-directed carbon dioxide gradient is maintained because cells produce carbon dioxide continually and exhalation removes it.

Oxygen Transport

Oxygen does not dissolve well in blood, so it cannot be efficiently transported on its own. In vertebrates, as in other large animals, demands for oxygen are met with a boost from hemoglobin molecules. These respiratory pigments are packed in red blood cells. A hemoglobin molecule, recall, is a compact array of four polypeptide chains and four **heme groups**, each incorporating an iron atom that binds reversibly with oxygen. *Of all the oxygen inhaled into the human body, 98.5 percent of it is bound to heme groups of hemoglobin.*

Normally, inhaled air that reaches alveoli has plenty of oxygen. The opposite is true of blood in pulmonary capillaries. Thus, in the lungs, oxygen tends to diffuse into the plasma portion of blood. Then it diffuses into red blood cells, where it binds with hemoglobin. In this way, it forms **oxyhemoglobin**, or HbO_2.

The amount of HbO_2 that forms in a given interval depends on oxygen's partial pressure. The higher the pressure, the greater will be the oxygen concentration. When plenty of oxygen molecules are available, they tend to randomly collide with the heme binding sites at a faster rate. The encounters continue until all four of the binding sites in hemoglobin are saturated.

HbO_2 molecules bind oxygen weakly and give it up where oxygen's partial pressure is lower than in the lungs. They give it up faster in tissues where blood is warmer, the pH is lower, and carbon dioxide's partial pressure is high. Such conditions prevail in contracting muscle and other metabolically active tissues.

Carbon Dioxide Transport

Carbon dioxide diffuses into blood capillaries in any tissue where its partial pressure is higher than it is in the blood flowing past. From there, three mechanisms transport it to the lungs. About 10 percent of the gas stays dissolved in blood. Another 30 percent binds with hemoglobin to form **carbamino hemoglobin** ($HbCO_2$). Most of it—60 percent—is transported in the form of bicarbonate (HCO_3^-). How do these HCO_3^- molecules form? First, carbonic acid forms when carbon dioxide combines with water in blood, but it quickly separates into bicarbonate and hydrogen ions (H^+):

$$CO_2 + H_2O \rightleftharpoons \underset{\text{CARBONIC ACID}}{H_2CO_3} \rightleftharpoons \underset{\text{BICARBONATE}}{HCO_3^- + H^+}$$

In blood plasma, that reaction converts only 1 of every 1,000 carbon dioxide molecules—which doesn't amount to much. It is a different story in red blood cells, which have the enzyme **carbonic anhydrase**. Enzyme action enhances the reaction rate 250 times! Most of the carbon dioxide not bound to hemoglobin becomes converted to carbonic acid this way. Conversion makes the blood level of carbon dioxide decline rapidly. Thus it helps maintain a gradient that promotes diffusion of carbon dioxide from interstitial fluid into the bloodstream.

What about the HCO_3^- that forms in the reactions? It tends to move out of red blood cells and into blood

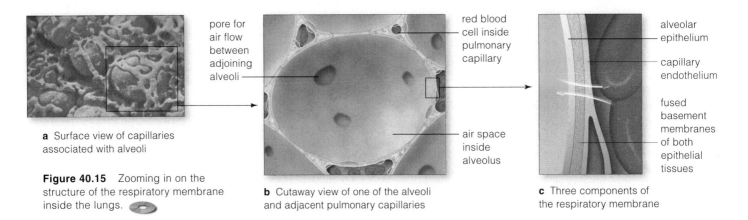

a Surface view of capillaries associated with alveoli

pore for air flow between adjoining alveoli

red blood cell inside pulmonary capillary

air space inside alveolus

alveolar epithelium

capillary endothelium

fused basement membranes of both epithelial tissues

Figure 40.15 Zooming in on the structure of the respiratory membrane inside the lungs.

b Cutaway view of one of the alveoli and adjacent pulmonary capillaries

c Three components of the respiratory membrane

plasma. What about the H^+ ions? Hemoglobin acts as a buffer for them and keeps blood from getting too acidic. A *buffer* is any molecule that combines with or releases H^+ in response to a shift in cellular pH (Section 2.6).

The reactions are reversed in alveoli, where carbon dioxide's partial pressure is lower than it is in the lung capillaries. Carbon dioxide and water form here. The gas diffuses into the alveolar sacs, then is exhaled.

At this point, reflect on Figure 40.16. It summarizes the partial pressure gradients for oxygen and carbon dioxide through the human respiratory system.

How to Match Air Flow With Blood Flow?

Gas exchange is most efficient when the rate of air flow matches the rate of blood flow. The nervous system acts to balance them by controlling the *rhythm* of breathing and its *magnitude*—that is, its rate and depth.

A respiratory center inside the medulla oblongata controls rhythmic patterns of breathing. A group of its neurons sets the pace for neurons that signal muscles involved in inhalation. The signals trigger contraction. In their absence, the muscles relax and air is exhaled. When exhalation must be active and strong, different neurons in the respiratory center issue signals to the muscles involved in exhalation.

Higher in the brain stem, in the pons, other centers smooth out the breathing rhythm set by the respiratory pacemakers. An apneustic center prolongs inhalation; a pneumotaxic center curtails it.

Controlling the magnitude of breathing depends on the partial pressures of oxygen and carbon dioxide as well as the H^+ concentration in arterial blood. Carbon dioxide has the greatest effect on ongoing adjustments. Chemoreceptors in the brain respond to H^+ increases in the cerebrospinal fluid bathing them. (Remember, H^+ is also a by-product of the reactions that kick in when blood has too much carbon dioxide.) The brain responds by signaling the diaphragm and other muscles to alter activities and so adjust the rate and depth of breathing.

Aortic bodies (in the aortic arch) and carotid bodies (at a branching of the carotid artery) notify the brain when oxygen's partial pressure falls below 60 mm Hg in arterial blood. Such life-threatening decreases can occur at very high altitudes and during severe lung diseases.

In some situations, a person can "forget" to breathe. Breathing that is briefly interrupted and then resumes spontaneously is known as *apnea*. It may stop for one or two seconds or minutes—in a few cases as often as 500 times a night. With *sudden infant death syndrome* (SIDS), a sleeping infant cannot awaken from an apneic episode. SIDS might have a heritable basis. It has been linked to mutations that cause an irregular heartbeat. Also, the risk is three times higher among infants of

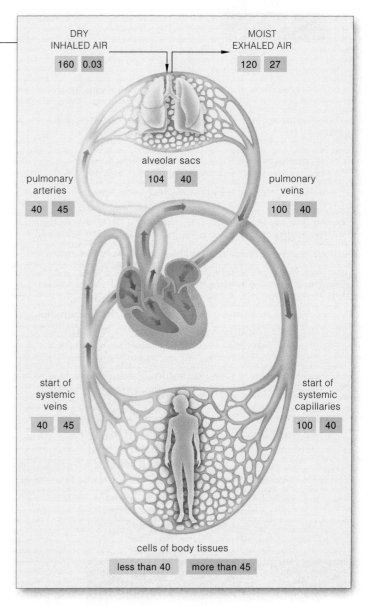

Figure 40.16 Partial pressure gradients for oxygen (*blue* boxes) and carbon dioxide (*pink* boxes) in the respiratory tract.

women who smoked cigarettes or had been exposed to secondhand smoke while pregnant. Infants who sleep on their back or sides are less vulnerable to SIDS than those who sleep on their stomach.

Driven by its partial pressure gradient, oxygen diffuses from alveolar air spaces, through interstitial fluid, and into lung capillaries. Carbon dioxide, driven by its partial pressure gradient, diffuses in the opposite direction.

Hemoglobin in red blood cells enormously enhances the oxygen-carrying capacity of blood. Most carbon dioxide is transported in blood in the form of bicarbonate, nearly all of which forms by an enzyme action in red blood cells.

Respiratory centers in the brain stem control the rhythmic pattern and magnitude of breathing, in ways that maintain appropriate levels of carbon dioxide, oxygen, and hydrogen ions in arterial blood.

When the Lungs Break Down

In large cities, in certain workplaces, even in the cloud around a cigarette smoker, airborne particles and certain gases are present in abnormally high concentrations. And they put extra workloads on the respiratory system.

BRONCHITIS   Ciliated, mucous epithelium lines the inner wall of your bronchioles (Figure 40.17). It is one of the built-in defenses that protect you from respiratory infections. Toxins in cigarette smoke and other airborne pollutants irritate the lining and may lead to *bronchitis*. With this respiratory ailment, epithelial cells along the airways become irritated and secrete excessive mucus. The mucus accumulates, and so do bacteria and particles stuck in it. Coughing brings up some of the gunk. If the source of irritation persists, so does the coughing.

Initial attacks of bronchitis are treatable. When the aggravation continues, bronchioles become chronically inflamed. Bacteria, chemical agents, or both attack the bronchiole walls. Ciliated cells in the walls are destroyed, and mucus-secreting cells multiply. Fibrous scar tissue forms and in time may narrow or obstruct the airways.

EMPHYSEMA   Thick mucus clogs the airways during persistent bronchitis. Inside the lungs, tissue-destroying bacterial enzymes attack the thin, stretchable walls of alveoli. The walls break down, and inelastic fibrous tissue forms around them. Now gas exchange proceeds at fewer alveoli, which become enlarged. In time, the lungs remain distended and inelastic, so the fine balance between air flow and blood flow is permanently compromised.

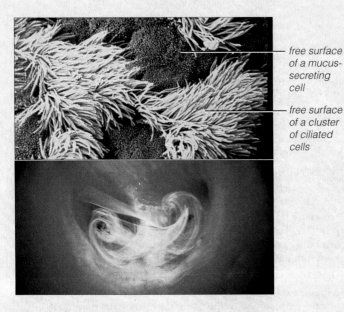

free surface of a mucus-secreting cell

free surface of a cluster of ciliated cells

Figure 40.17   Ciliated and mucus-secreting cells that line a bronchus. What do you suppose happens to them after cigarette smoke swirls in from the entrance to these two bronchi, which lead into the lungs?

Compare Figure 40.18*a* with *b* to get a sense of why it becomes hard to run, walk, and exhale. These problems are symptoms of *emphysema*, a respiratory ailment that affects over a million people in the United States alone.

A few people are genetically predisposed to developing emphysema. They do not have a workable gene for the enzyme antitrypsin, which can inhibit bacterial attack on alveoli. Poor diet and persistent or recurring colds and other respiratory infections also invite emphysema later in life. However, *smoking is the major cause of the disease*. Emphysema may develop slowly, over twenty or thirty years. When severe damage is not detected in time, the lung tissues cannot be repaired.

EFFECTS OF SMOKING   Worldwide, 3 million people die each year as a result of complications arising from tobacco smoking. The World Health Organization, the Harvard School of Health, and the World Bank estimate that deaths from smoking may surpass even those from AIDS by the year 2020. Every thirteen seconds in the United States, 1 of every 50 million smokers dies from emphysema, chronic bronchitis, or heart disease. And one nonsmoker dies of ailments brought on by breathing *secondhand smoke*—by prolonged exposure to tobacco smoke in the surrounding air. Children who breathe secondhand smoke at home and elsewhere are far more vulnerable to allergies and lung problems.

Smoking is the major cause of lung cancer. Yet every day, 3,000 to 5,000 Americans light a cigarette for the first time. Even children spend a billion dollars a year on cigarettes. Each year, direct medical costs of treating smoke-induced respiratory disorders drain 22 billion dollars from the economy.

How does cigarette smoke damage lungs? Noxious particles in the smoke from just one cigarette immobilize cilia in the bronchioles for several hours. The particles also trigger mucus secretions, which eventually clog the airways. They also kill infection-fighting macrophages in the respiratory tract. What starts out as "smoker's cough" can end in bronchitis and emphysema.

Or consider how cigarette smoke contributes to lung cancer. Inside the body, certain compounds in coal tar and in cigarette smoke become converted to highly reactive intermediates. These are the carcinogens; they provoke uncontrolled cell divisions in lung tissues. On average, 90 of every 100 smokers who develop lung cancer will die from it. If you now smoke, or are thinking about starting or quitting, you may wish to give serious thought to the information in Figure 40.18.

EFFECTS OF MARIJUANA SMOKE   In 2000, 10.7 million individuals twelve years and older in the United States smoked *pot*—marijuana (*Cannabis*)—as a way of inducing light-headed euphoria. About 1.5 million are chronic

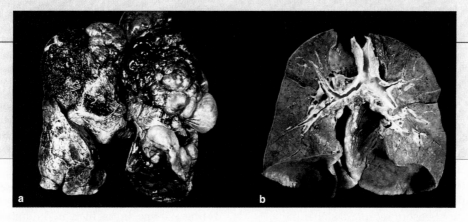

RISKS ASSOCIATED WITH SMOKING

SHORTENED LIFE EXPECTANCY: Nonsmokers live 8.3 years longer on average than those who smoke two packs daily from the midtwenties on.

CHRONIC BRONCHITIS, EMPHYSEMA: Smokers have 4–25 times more risk of dying from these diseases than do nonsmokers.

LUNG CANCER: Cigarette smoking is the major cause of lung cancer.

CANCER OF MOUTH: 3–10 times greater risk among smokers.

CANCER OF LARYNX: 2.9–17.7 times more frequent among smokers.

CANCER OF ESOPHAGUS: 2–9 times greater risk of dying from this.

CANCER OF PANCREAS: 2–5 times greater risk of dying from this.

CANCER OF BLADDER: 7–10 times greater risk for smokers.

CORONARY HEART DISEASE: Cigarette smoking is a major contributing factor.

EFFECTS ON OFFSPRING: Women who smoke during pregnancy have more stillbirths, and weight of liveborns averages less (hence, babies are more vulnerable to disease, death).

IMPAIRED IMMUNE SYSTEM FUNCTION: Increase in allergic responses, destruction of defensive cells (macrophages) in respiratory tract.

BONE HEALING: Evidence suggests that surgically cut or broken bones require up to 30 percent longer to heal in smokers, possibly because smoking depletes the body of vitamin C and reduces the amount of oxygen reaching body tissues. Reduced vitamin C and reduced oxygen interfere with production of collagen fibers, a key component of bone. Research in this area is continuing.

c

REDUCTION IN RISKS BY QUITTING

Cumulative risk reduction; after 10 to 15 years, life expectancy of ex-smokers approaches that of nonsmokers

Greater chance of improving lung function and slowing down rate of deterioration.

After 10 to 15 years, risk approaches that of nonsmokers.

After 10 to 15 years, risk is reduced to that of nonsmokers.

After 10 years, risk is reduced to that of nonsmokers.

Risk proportional to amount smoked; quitting should reduce it.

Risk proportional to amount smoked; quitting should reduce it.

Risk decreases gradually over 7 years to that of nonsmokers.

Risk drops sharply after a year; after 10 years, risk reduced to that of nonsmokers.

When smoking stops before fourth month of pregnancy, risk of stillbirth and lower birthweight eliminated.

Avoidable by not smoking.

Avoidable by not smoking.

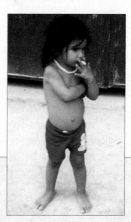

Figure 40.18 (**a**) Appearance of lung tissues from someone affected by emphysema. (**b**) Normal appearance of human lungs. (**c**) From the American Cancer Society, a list of the risks incurred by smoking and the benefits of quitting.

users who are enamored of the euphoria. However, the sense of well-being does not last long; it fades to apathy, depression, and fatigue. Users keep smoking to avoid the negative effects.

Besides causing psychological dependency, long-term use may result in chronic throat irritations, persistent coughing, bronchitis—and emphysema.

With respect to the United States, cigarette smoking is increasingly being banned from enclosed spaces, such as airline cabins and airports, restaurants, and theaters. Cigarette sales to minors are prohibited, and there are ongoing attempts to restrict billboard advertising near

elementary schools. Are tobacco companies now viewing children and women of developing countries as untapped markets? Mark Palmer, former ambassador to Hungary, thinks so and argues that it is the worst single thing we do to the rest of the world. The World Health Organization estimates that tobacco use kills 4 million people annually around the world. It estimates that deaths will rise to 10 million annually by 2030, with 70 percent of the total in developing countries. As G. H. Brundtland, head of the World Health Organization and a medical doctor puts it, tobacco remains the only legal consumer product that kills half of its regular users.

HIGH CLIMBERS AND DEEP DIVERS

Respiration at High Altitudes

Most of us live at low elevations. Of the air we breathe, one molecule in five is oxygen. When we travel 2,400 meters (about 8,000 feet) or more above sea level, there is less air pressure pushing down from above. So the gas molecules are more spread out, and the breathing game changes. At high elevations, humans can expect *hypoxia*, cellular oxygen deficiency. Sensory receptors signal the brain of the deficiency. The brain responds by sending out commands that make us hyperventilate; it makes us breathe faster and more deeply than usual.

Above 3,300 meters (10,000 feet), hyperventilating severely disrupts ion balances in cerebrospinal fluid. It may result in heart palpitations, shortness of breath, headaches, nausea, and vomiting. These are signs that the body's cells are screaming for oxygen.

Remember Section 28.2? Compared to people who live at lower elevations, llamas and other long-time residents of high mountains have lungs with far more alveoli and blood vessels, which formed as they were growing up. Their heart developed larger ventricles, so it pumps larger volumes of blood. In addition, more mitochondria formed in their muscle tissue. Llama hemoglobin has a greater affinity for oxygen. It picks up oxygen more efficiently at the lower pressures of high altitudes (Figure 40.19).

Does this mean a healthy person who grew up by the seashore cannot ever pick up roots and move to the mountains? No. By mechanisms of **acclimatization**, an individual often makes long-lasting physiological and behavioral adaptations to a new environment that is markedly different from the one left behind. At high altitudes, gradual changes in the pattern and magnitude of breathing and in the heart's output can supplant the initially acute compensations for hypoxia.

Within a few days, the reduction in oxygen delivery stimulates kidney cells to secrete more **erythropoietin**, a hormone that induces stem cells in bone marrow to divide repeatedly and give rise to red blood cells. Each second in an adult human, 2 million to 3 million red blood cells are churned out as replacements for the ones that continually die off. The stepped-up erythropoietin secretion can increase that astounding pace by six times during extreme stress. Increased numbers of circulating red blood cells improve the oxygen-delivery capacity of blood. When blood's concentration of oxygen rises enough, erythropoietin secretion slows in the kidneys.

Erythropoietin is the main hormonal mediator of red blood cell production. But human males have a larger muscle mass and greater demands for oxygen. They depend on testosterone, also. As one of its roles, this male sex hormone stimulates increases in the basic red blood cell production rate. That is why a sample of blood from males normally has a greater percentage of red blood cells compared to an equivalent sample from females. The compensatory increase in red blood cells comes at a cost. Having many more cells in the blood increases resistance to flow because the blood is more viscous. The heart must work harder to pump it.

Carbon Monoxide Poisoning

High elevations are not the exclusive culprits behind cellular oxygen deficiency. Hypoxia also can result from *carbon monoxide poisoning*. Carbon monoxide (CO), a colorless, odorless gas, occurs in exhaust fumes from gas-driven vehicles. The smoke from burning coal and wood and from tobacco contains it. CO competes with oxygen for binding sites in hemoglobin. Its binding capacity is at least 200 times greater than oxygen's! Even small amounts of this gas can tie up 50 percent of the hemoglobin in the body. That is how it can impair oxygen delivery to tissues to dangerous degrees.

Respiration in Deep Water

The higher you climb, the more air pressure decreases, but the deeper you dive, the more water pressure rises. Unless human divers use tanks of compressed air, they risk *nitrogen narcosis*, or "raptures of the deep." Why? Starting at depths of 45 meters (150 feet), the pressure causes much more gaseous nitrogen (N_2) than usual to dissolve in body tissues. Being lipid soluble, this gas

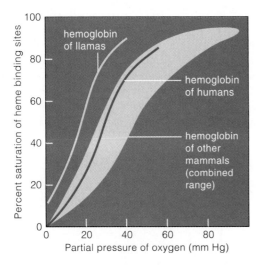

Figure 40.19 Comparison of hemoglobin's binding and releasing capacity for humans, llamas, and other animals.

easily crosses the lipid bilayer of cell membranes and disrupts metabolism. Divers become euphoric and then drowsy, as if tipsy with alcohol. Some have offered the mouthpiece of their airtanks to fishes. At lower depths, divers become clumsy, weak, then comatose as a result of nitrogen narcosis.

Returning to the surface from a deep dive requires caution. As pressure lets up, N_2 moves out of tissues and into the bloodstream. If an ascent is too rapid, N_2 enters the blood faster than the lungs can dispose of it. Nitrogen bubbles may form in the blood and tissues. The resulting pain at joints especially is known as "the bends," or decompression sickness. If bubbles obstruct blood flow to the brain, deafness, impaired vision, and muscle paralysis may be the outcome.

Humans trained to dive without oxygen tanks can stay fully submerged for about three minutes at most. Career divers, such as the ama of Korea and Japan, do only a bit better. So how do dolphins, whales, seals, and other air-breathing marine mammals routinely dive underwater, where they can't inhale oxygen and exhale carbon dioxide?

We are not talking about a quick bob under the water's surface. Weddell seals (*Leptonychotes weddelli*) of the Antarctic can stay submerged for more than an hour. They typically dive to 400 meters (1,312 feet) or less, although some have been observed at 600 meters —more than a third of a mile down. One sperm whale tracked by sonar reached 2,250 meters. And what about bottlenose dolphins (Figure 40.20)? Although they do not have to dive far to catch food, trained dolphins can reach depths of 547 meters (1,795 feet) and stay there for ten minutes.

When all marine mammals dive, highly specialized respiratory adaptations come into play. Before slipping underwater, they fill their lungs with oxygen, which binds to hemoglobin in blood and myoglobin in cells. Aerobic respiration can proceed as usual during a short dive. During prolonged dives, oxygen is preferentially distributed to the heart and central nervous system, which require a steady supply of ATP.

Basically, blood is directed away from most organs. Blood pumped from the heart travels mainly to the lungs and brain, and back to the heart. Skeletal muscles use their own myoglobin-bound oxygen as well as the hemoglobin-bound oxygen in the dense capillary beds that service them. Later on, they switch to anaerobic pathways, and lactate accumulates in tissues. When the animal surfaces, circulation to the muscles is restored.

When a sperm whale or dolphin surfaces, it blows stale air from its lungs, through a tubelike epiglottis and then a single blowhole at the body surface. Before diving again, it breathes rapidly, and elastic, extensible lungs fill fast with large volumes of air.

Figure 40.20 Bottlenose dolphins (*Tursiops truncatus*), revisited.

During a dive, special valves close its nostrils; and rings of muscle and of cartilage clamp its bronchioles. The whale's respiratory surface is not that large. But strategically located valves and local networks of blood vessels (plexuses) store and distribute the blood volume and gases in an economical fashion. A sperm whale's heart rate and metabolism slow down, and so does the rate of oxygen use and carbon dioxide formation. Also, compared to land mammals, its respiratory center is less sensitive to carbon dioxide levels.

Diving marine mammals also conserve their oxygen stores by gliding a lot. For example, when a dolphin dives eighty meters or more below the water's surface, it spends more than 78 percent of the time gliding.

During brief stays at high elevations, the human body can make acute compensatory responses to the thinner air. Over time, the body makes adaptive adjustments in breathing and cardiac output.

All diving marine mammals have specialized respiratory adaptations. When surfacing, they blow out stale air and take in oxygen that becomes bound to hemoglobin and myoglobin. They selectively distribute more oxygen to the heart and skeletal muscles during a dive. They spend much of a dive gliding, which conserves energy.

SUMMARY   **Gold** indicates text section

1. Aerobic respiration is the main metabolic pathway that yields enough energy for active life-styles. It uses oxygen and has carbon dioxide by-products. Respiration is a process by which the animal body takes in oxygen and disposes of carbon dioxide wastes. *CI, 40.1*

2. Air is a mixture of oxygen, carbon dioxide, and other gases, each exerting a partial pressure. Each gas tends to move from areas of higher to lower partial pressure. Respiratory systems make use of this tendency. *40.1*

3. Gases diffuse across a respiratory surface in animals. In human lungs, a thin respiratory membrane consists of alveolar epithelium, lung capillary epithelium, and the fused basement membranes of both. *40.1, 40.6*

4. Animals have diverse modes of respiration. *40.2, 40.3*
   a. Invertebrates with a small body mass depend only on integumentary exchange; oxygen and carbon dioxide simply diffuse across the body surface. This mode also persists in some large animals, including amphibians.
   b. Many marine and some freshwater invertebrates have gills: respiratory organs with moist, thin, and often highly folded walls. Most insects and some spiders use tracheal respiration. Gases flow through open-ended tubes leading from the body surface directly to tissues. Most spiders have book lungs, with leaflike folds.
   c. Fishes use more energy for respiration than small invertebrates or mammals. Countercurrent flow at their paired gills compensates for the low oxygen levels of aquatic habitats. Paired lungs are the dominant means of respiration in reptiles, birds, and mammals.

5. The airways of the human respiratory system consist of nasal cavities, pharynx, larynx, trachea, bronchi, and bronchioles. At the end of terminal bronchioles, alveoli are the main sites of gas exchange. *40.4*

6. Breathing ventilates the lungs. Each respiratory cycle consists of inhalation (one breath in) and exhalation (one breath out). During inhalation, the chest cavity expands, lung pressure decreases below atmospheric pressure, and air flows into the lungs. The events are reversed during normal exhalation. *40.5*

7. Driven by its partial pressure gradient, oxygen in lungs diffuses from alveolar air spaces into pulmonary capillaries, then diffuses into red blood cells and binds weakly with hemoglobin. Hemoglobin gives up oxygen at capillary beds in metabolically active tissues. Oxygen diffuses across the interstitial fluid, then into cells. *40.6*

8. Driven by its partial pressure gradient, carbon dioxide in tissues diffuses from cells, across interstitial fluid, and into the blood. Most of it reacts with water to form bicarbonate. The reactions are reversed in the lungs; carbon dioxide diffuses from the pulmonary capillaries into air spaces of alveoli, then is exhaled. *40.6*

Figure 40.21   Body surfing.

Review Questions

1. Define respiration, then define aerobic respiration. *CI, 40.1*

2. Define respiratory surface. Why must the oxygen and carbon dioxide partial pressure gradients across it be steep? *40.1*

3. What is the name of the main respiratory pigment? *40.1*

4. A few of your friends who have not taken a biology course ask you what insect lungs look like. How do you answer them (assuming your instructor is listening)? *40.2, 40.3*

5. Briefly describe how countercurrent flow through a fish gill is so efficient at taking up dissolved oxygen from water. *40.3*

6. Which animals have air sacs that ventilate the lungs: fishes, amphibians, reptiles, birds, or mammals? *40.3*

7. Distinguish between:
   a. tracheal respiration and trachea (windpipe) *40.2, 40.4*
   b. pharynx and larynx *40.4*
   c. bronchiole and bronchus *40.4*
   d. pleural sac and alveolar sac *40.4*

8. Define the functions of the human respiratory system. In the diagram below, label its components and the major bones and muscles with which it interacts during breathing. *40.4*

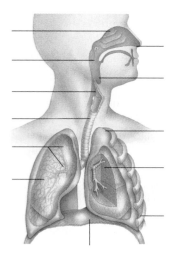

9. Explain why the human male in Figure 40.21 cannot survive on his own for very long under water. *CI, 40.8*

1. Most insects and some spiders depend on _____ .
   a. tracheal respiration   c. integumentary exchange
   b. lungs   d. both a and c

2. _____ have an abundance of respiratory pigments.
   a. Many invertebrates   c. All photoautotrophs
   b. Vertebrates   d. both a and b

3. In birds, air flows _____ .
   a. into and out of lungs   c. into air sacs
   b. through lungs   d. all of the above

4. Each human lung encloses a _____ .
   a. diaphragm   c. pleural sac
   b. bronchial tree   d. both b and c

5. In human lungs, gas exchange occurs at the _____ .
   a. two bronchi   c. alveolar sacs
   b. pleural sacs   d. both b and c

6. When you breathe quietly, inhalation is _____ and exhalation is _____ .
   a. passive; passive   c. passive; active
   b. active; active   d. active; passive

7. Hypoxia triggers increased _____ secretion.
   a. carbonic anhydrase   c. erythropoietin
   b. carbon monoxide   d. myoglobin

8. After oxygen diffuses into pulmonary capillaries, it diffuses into _____ and binds with _____ .
   a. interstitial fluid; red blood cells
   b. interstitial fluid; carbon dioxide
   c. red blood cells; hemoglobin
   d. red blood cells; carbon dioxide

9. Oxyhemoglobin (HbO_2) gives up oxygen faster where the _____ than it is in the lungs.
   a. pH is lower   c. O_2 partial pressure is higher
   b. blood is cooler   d. CO_2 partial pressure is lower

10. Sixty percent of the carbon dioxide in human blood is in the form of _____ and thirty percent in the form of _____ .
   a. carbon dioxide; carbonic acid
   b. carbonic anhydrase; bicarbonate
   c. bicarbonate; carbamino hemoglobin
   d. carbamino hemoglobin; bicarbonate

11. Match the descriptions with their components in humans.
   ____ trachea   a. airway leading into a lung
   ____ pharynx   b. gap between vocal cords
   ____ alveolus   c. fine branch of bronchial tree
   ____ hemoglobin   d. windpipe
   ____ bronchus   e. respiratory pigment
   ____ bronchiole   f. site of gas exchange
   ____ glottis   g. throat

Critical Thinking

1. Some cigarette manufacturers are urging people to "smoke responsibly." What social and biological issues are involved here? For example, what about risks to a nonsmoking spouse or to children of a smoker? To nonsmoking patrons in restaurants? To the unborn child of a pregnant smoker? In your opinion, what behavior would constitute "responsible smoking"?

2. People sometimes poison themselves with carbon monoxide by building a charcoal fire in an enclosed area. Assuming help arrives in time, what would be the best treatment: (1) placing the victim outdoors in fresh air or (2) rapidly administering pure oxygen? Explain how you arrived at your answer.

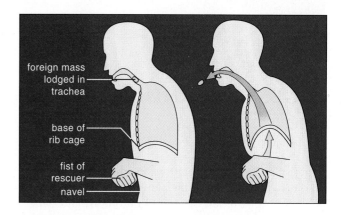

Figure 40.22 Heimlich maneuver to dislodge food stuck in the trachea. Stand behind the victim, make a fist with one hand, then position the fist, thumb-side in, against the victim's abdomen. The fist must be slightly above the navel and well below the rib cage. Now press the fist into the abdomen with a sudden upward thrust. Repeat the thrust several times if needed. The maneuver can be performed on someone who is standing, sitting, or lying down.

3. When you swallow food, contracting muscles force the epiglottis down, to its closed position. This prevents food from going down the trachea and blocking gas flow. Yet each year, several thousand people choke to death after food enters the trachea and blocks air flow for as little as four or five minutes. The *Heimlich maneuver*—an emergency procedure only—often dislodges food stuck in the trachea (Figure 40.22).

When correctly performed, the maneuver forcibly elevates the diaphragm, causing a sharp decrease in the thoracic cavity volume and an abrupt rise in alveolar pressure. Air forced up into the trachea by the increased pressure may be enough to dislodge the obstruction. Once the obstacle is dislodged, a doctor must see the person at once because an inexperienced rescuer can inadvertently cause internal injuries or crack a rib.

Reflect now on our current social climate in which lawsuits abound. Would you risk performing the Heimlich maneuver to save a relative's life? A stranger's life? Why or why not?

Selected Key Terms

acclimatization *40.8*	larynx *40.4*
alveolus (alveoli) *40.4*	lung *40.3*
bronchiole *40.4*	myoglobin *40.1*
bronchus *40.4*	oxyhemoglobin *40.6*
carbamino hemoglobin *40.6*	partial pressure *40.1*
carbonic anhydrase *40.6*	pharynx *40.4*
countercurrent flow *40.3*	pressure gradient *40.1*
diaphragm *40.4*	respiration *CI*
epiglottis *40.4*	respiratory cycle *40.5*
erythropoietin *40.8*	respiratory surface *40.1*
Fick's law *40.1*	respiratory system *CI*
gill *40.2*	tidal volume *40.5*
glottis *40.3*	trachea *40.4*
heme group *40.6*	tracheal respiration *40.2*
hemoglobin *40.1*	vital capacity *40.5*
integumentary exchange *40.2*	vocal cord *40.3*

Readings

Gorman, J. 31 March 2001. "Breathing on the Edge." *Science News*, 159: 202–204.

Sherwood, L. 1997. *Human Physiology*. Second edition. Belmont, California: Wadsworth.

On-Line readings at Student Guide for InfoTrac:
www.brookscole.com/biology

DIGESTION AND HUMAN NUTRITION

Lose It—And It Finds Its Way Back

America's fixation on Beautiful People who have nary an ounce of extra fat on their utterly perfect selves is bad enough. After all, we might rationalize our own extra ounces with the truism that beauty is only skin deep. But we can't rationalize away the diabetes, heart attacks, strokes, colon cancer, and other dangerous risks to a body packing away too much fat.

The proportion of body fat relative to total tissue mass is now supposed to be 18 to 24 percent for women less than thirty years old. For men, it is supposed to be no more than 12 to 18 percent. An estimated 34 million Americans don't even come close to the standards.

No matter how much we diet, the lost weight seems to find its way back. This physiological dilemma is an outcome of our evolutionary heritage. Like most other mammals, we have an abundance of fat-storing cells in adipose tissue. Collectively, the cells are an adaptation for survival, an energy warehouse that opens up when food is scarce. Once the fat-storing cells have formed, they are in the body to stay. Variations in food intake only affect how empty or full each cell gets.

Dieting opens the fat warehouse. The brain interprets this as *STARVATION!* and calls for a metabolic slowdown. In response, the body uses energy far more efficiently even for basic tasks, such as breathing and digesting food. It now takes less food to do the same things!

As you've probably heard, dieting is useless without long-term commitment to physical exercise. Why? Your skeletal muscles adapt to *STARVATION!* signals and burn less energy than before. Jog for four hours or play tennis for eight hours straight and you might lose a pound of fat. Meanwhile, your appetite surges. If and when you do stop dieting, your "starved" fat cells quickly refill. Unless you eat less and exercise moderately all through your life, you can't keep off extra weight (Figure 41.1).

What about gaining weight? A weight gain triggers increases in metabolism. You use 15 to 20 percent more energy than before until you lose the excess.

As if this were not enough, emotions affect weight gains and losses. As an extreme case, *anorexia nervosa* is a potentially fatal eating disorder based on a flawed assessment of body weight. Most anorexics dread being fat and hungry. Many starve themselves, overexercise, and have fears about growing up and of being sexually mature. They often have irrationally high expectations about their personal performance.

Another extreme is called *bulimia*, an out-of-control "oxlike appetite." A bulimic on an hour-long binge

Figure 41.1 A few organisms positively engaged in countering imbalances between caloric intake and energy output.

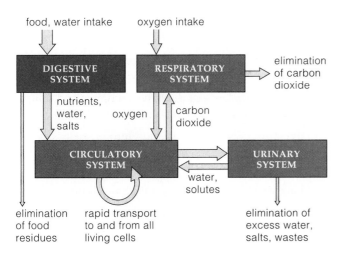

food, water intake oxygen intake

DIGESTIVE SYSTEM **RESPIRATORY SYSTEM** elimination of carbon dioxide

nutrients, water, salts oxygen carbon dioxide

CIRCULATORY SYSTEM **URINARY SYSTEM**

water, solutes

elimination of food residues rapid transport to and from all living cells elimination of excess water, salts, wastes

Figure 41.2 Functional links between the digestive, respiratory, circulatory, and urinary systems. These organ systems and others work together to supply cells with raw materials and eliminate wastes.

might take in 50,000 kilocalories—the equivalent of sixteen pounds of food—then vomit or use laxatives to get rid of it. Some follow a binge–purge routine as an "easy" way to lose weight. Others may not even like to eat but purge themselves to relieve frustration and anger. The binge–purge routines range from once per month to a few times a day. Purgings damage the gut lining. Chronic vomiting brings up gastric fluid that erodes teeth to stubs. In severe cases, the stomach can rupture and the heart and kidneys fail.

Are severe eating disorders rare? No. In the United States, an estimated 7 million females and 1 million males are anorexic or bulimic. Most are in their teens and early twenties, but the number also is increasing among preadolescents. Each year, complications kill 5 to 6 percent of the weight-obsessed individuals.

With these sobering thoughts, we start our tour of **nutrition**. The word encompasses all those processes by which an animal ingests and digests food, then absorbs the released nutrients for later conversion to the body's own carbohydrates, lipids, proteins, and nucleic acids.

The nutritional processes proceed at the **digestive system**. This body cavity or tube mechanically and chemically reduces food to particles, then to molecules that are small enough to be absorbed into the internal environment. It also eliminates unabsorbed residues. Other organ systems, especially those shown in Figure 41.2, contribute to the nutritional processes.

Key Concepts

1. In complex animals, interactions among digestive, circulatory, respiratory, and urinary systems supply the body's living cells with raw materials, dispose of wastes, and maintain the volume and composition of extracellular fluid.

2. Most digestive systems include specialized regions for food transport, processing, and storage. Different regions mechanically break apart and chemically break down food, absorb breakdown products, and eliminate the unabsorbed residues.

3. "Nutrition" refers to all processes by which an animal ingests food, digests food, and then absorbs the released nutrients, which become converted to the body's own carbohydrates, lipids, proteins, and nucleic acids.

4. A subcategory of nutritional processes is the intake of foods that are good sources of vitamins, minerals, and a number of amino acids and fatty acids that the body itself cannot produce.

5. To maintain suitable body weight and overall health, energy intake must balance energy output, largely by way of metabolic activity and regular physical exertion. Complex carbohydrates are the main source of dietary glucose. Typically, glucose is the body's main source of instant energy.

THE NATURE OF DIGESTIVE SYSTEMS

Incomplete and Complete Systems

Recall, from Chapter 25, that we do not see many organ systems until we get to the flatworms. They are among the few small invertebrates that have an **incomplete digestive system**, which means they take in food and dispose of wastes through a single opening at the body surface. The saclike, branching gut cavity of flatworms opens at the start of a tubular pharynx (Figure 41.3*a*). Food enters the sac, where it is partly digested, then is circulated to cells even as residues are sent out. Such two-way traffic does not favor regional specialization.

More complex animals have a **complete digestive system**. This is basically a tube with an opening at one end for taking in food and an opening at the other end for eliminating the unabsorbed residues (for example, a mouth and an anus). In between, the tube is subdivided into specialized regions that function in the one-way transport, processing, and storage of raw materials.

Consider a frog's complete digestive system (Figure 41.3*b*). Between its mouth and the anus are a pharynx, stomach, small intestine, and large intestine. A liver, gallbladder, and pancreas—all organs with accessory roles in digestion—connect with the small intestine. Birds, too, have a complete digestive system with a few unique regional specializations (Figure 41.3*c*).

Regardless of its complexity, a complete digestive system carries out five overall tasks:

1. **Mechanical processing and motility**: Movements that break up, mix, and propel food material.

2. **Secretion**: Release of digestive enzymes and other substances into the space inside the tube.

3. **Digestion**: Breakdown of food into particles, then into nutrient molecules small enough to be absorbed.

4. **Absorption**: Passage of digested nutrients and fluid across the tube wall and into body fluids.

5. **Elimination**: Expulsion of the undigested and unabsorbed residues from the end of the gut.

Correlations With Feeding Behavior

We can correlate any digestive system's specializations with feeding behavior. Consider the pigeon in Figure 41.3*c*. Its *food-gathering* region is a bill adapted to peck seeds off the ground. Seeds enter a mouth, a long tube (esophagus), and then a *food-processing* region, which is compactly centered in the body mass. The compactness helps balance the bird during flight. As in other seed eaters, part of the esophagus bulges out as a *crop*. This stretchable storage organ allows the bird to gulp a lot of food fast and make quick getaways from predators. Like all birds, a pigeon eats during the day and its crop fills before sunset. Soon after dark, the crop releases food, which helps sustain the bird through the night.

The first part of the pigeon stomach has a glandular lining, which secretes enzymes and other substances that function in digestion. The second part, the gizzard, is a muscular organ that grinds up food and cycles it back to the stomach. A pigeon's intestines are proportionally shorter than those in ducks or ostriches. Those birds eat plant parts rich in tough, fibrous cellulose, which requires a much longer processing time than seeds do.

And what about the digestive system of pronghorn antelope (*Antilocapra americana*)? Fall through winter, on the mountain ridges from central Canada down into northern Mexico, pronghorn antelope browse on wild

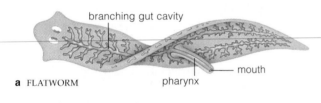

a FLATWORM
branching gut cavity
mouth
pharynx

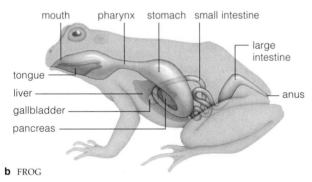

b FROG
mouth pharynx stomach small intestine
large intestine
tongue
liver
gallbladder
pancreas
anus

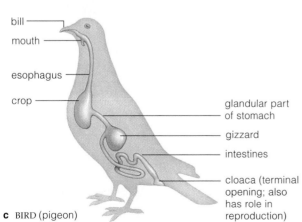

c BIRD (pigeon)
bill
mouth
esophagus
crop
glandular part of stomach
gizzard
intestines
cloaca (terminal opening; also has role in reproduction)

Figure 41.3 (**a**) Incomplete digestive system of a flatworm, which has a two-way traffic of food and undigested material through one opening. (**b,c**) Examples of a complete digestive system, a tube for one-way traffic with specialized regions and an opening at each end.

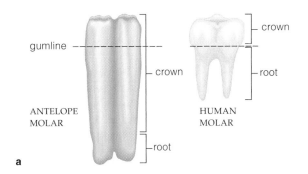

gumline

ANTELOPE
MOLAR

crown

root

crown

root

HUMAN
MOLAR

a

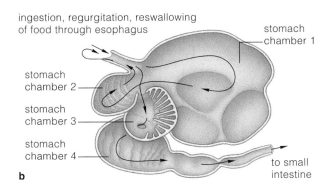

ingestion, regurgitation, reswallowing
of food through esophagus

stomach
chamber 2

stomach
chamber 3

stomach
chamber 4

b

stomach
chamber 1

to small
intestine

Figure 41.4 Regional specializations of the complete digestive system of pronghorn antelope (*Antilocapra americana*).

(**a**) Antelope molar compared with a human molar. (**b**) Multiple-chambered antelope stomach. In the first two chambers, food is mixed with fluid, kneaded, and exposed to fermentation actions of bacterial, protozoan, and fungal symbionts. Some symbionts degrade cellulose; others synthesize organic compounds, fatty acids, and vitamins. The host uses part of these substances. Kneaded food is regurgitated into the mouth, rechewed, and swallowed again. It enters the third chamber and is digested once more before entering the last stomach chamber.

Figure 41.5 Exceptionally Big Mac of the snake world.

sage. Come spring, they move to open grasslands and deserts, there to browse on new growth (Figure 41.4).

Now think of your own cheek teeth, or molars, each having a flattened crown that is a grinding platform. The crown of an antelope's molars dwarfs yours (Figure 41.4a). Why the difference? You probably don't brush your mouth against dirt as you eat. An antelope does. Abrasive bits of soil enter its mouth along with tough plant material, so the crown wears out quickly. Natural selection has favored more crown to wear down.

Antelopes are **ruminants**, a type of hoofed mammal having multiple stomach chambers in which cellulose is slowly broken down. The breakdown gets under way inside the first two of four stomach sacs (Figure 41.4b). There, bacterial symbionts release digestive enzymes to access the nutrients in cellulose. As they do, their host regurgitates and rechews the contents of the first two sacs, then swallows again. (That's what "chewing cud" means.) Pummeling the plants more than once exposes more surface area to enzymes, which gives them more time to release nutrients that the antelope (as well as the bacteria) can use. The ruminant stomach system can steadily accept food during extended feeding times and slowly liberate nutrients when the animal rests.

What about predatory and scavenging mammals? They gorge on food when they can get it and might not eat again for some time (Figure 41.5). One part of their digestive system specializes in storing a lot of food, which is digested and absorbed at a leisurely pace. As

you will see, other organs having accessory roles in digestion help ensure that there will be an adequate distribution of nutrients between meals.

An incomplete digestive system is a saclike body cavity. Food enters the system and residues leave through the same opening. A complete digestive system is a tube that has two openings and regional specializations in between.

Complete digestive systems carry out the following tasks in controlled ways: movements that break up, mix, and propel food material; secretion of digestive enzymes and other substances; breakdown of food; absorption of nutrients; and expulsion of undigested and unabsorbed residues.

VISUAL OVERVIEW OF THE HUMAN DIGESTIVE SYSTEM

Humans have a complete digestive system, a tube with many specialized organs between two openings (Figure 41.6). If the tube of an adult were fully stretched out, it would extend 6.5 to 9 meters (21 to 30 feet). From start to finish, mucus-coated epithelium lines every surface facing the lumen. (A *lumen* is the space within a tube.) The thick, moist mucus helps protect the tube wall and promotes diffusion across its inner lining.

Ingested substances advance in one direction, from the mouth through the pharynx, esophagus, then the gastrointestinal tract, or gut. The human **gut** starts at the stomach and extends through the small intestine, large intestine (colon) and rectum, to the anus. Salivary glands and a gallbladder, liver, and pancreas function as accessory organs; they secrete a variety of substances into different regions of the tube.

Major Components:

MOUTH (ORAL CAVITY)

Entrance to system; food is moistened and chewed; polysaccharide digestion starts.

PHARYNX

Entrance to tubular part of system (and to respiratory system); moves food forward by contracting sequentially.

ESOPHAGUS

Muscular, saliva-moistened tube that moves food from pharynx to stomach.

STOMACH

Muscular sac; stretches to store food taken in faster than can be processed; gastric fluid mixes with food and kills many pathogens; protein digestion starts. Secretes ghrelin, an appetite stimulator.

SMALL INTESTINE

First part (duodenum, C-shaped, about 10 inches long) receives secretions from liver, gallbladder, and pancreas.

In second part (jejunum, about 3 feet long), most nutrients are digested and absorbed.

Third part (ileum, 6–7 feet long) absorbs some nutrients; delivers unabsorbed material to large intestine.

LARGE INTESTINE (COLON)

Concentrates and stores undigested matter by absorbing mineral ions, water; about 5 feet long: divided into ascending, transverse, and descending portions.

RECTUM

Distension stimulates expulsion of feces.

ANUS

End of system; terminal opening through which feces are expelled.

Accessory Organs:

SALIVARY GLANDS

Glands (three main pairs, many minor ones) that secrete saliva, a fluid with polysaccharide-digesting enzymes, buffers, and mucus (which moistens and lubricates food).

LIVER

Secretes bile (for emulsifying fat); roles in carbohydrate, fat, and protein metabolism.

GALLBLADDER

Stores and concentrates bile that the liver secretes.

PANCREAS

Secretes enzymes that break down all major food molecules; secretes buffers against HCl from the stomach. Secretes insulin, a hormonal control of glucose metabolism.

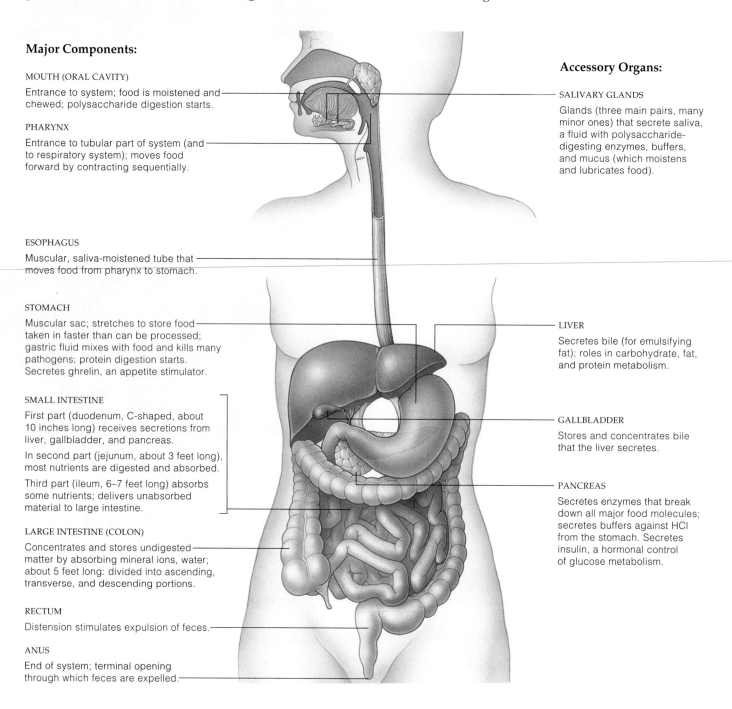

Figure 41.6 Overview of the component parts of the human digestive system and their specialized functions. Organs with accessory roles in digestion are also listed.

INTO THE MOUTH, DOWN THE TUBE

Food is chewed and polysaccharide breakdown begins in the oral cavity, or mouth. Thirty-two teeth typically project into a human adult's mouth (Figure 41.7). Each **tooth** has an enamel coat (hardened calcium deposits), dentin (a thick, bonelike layer), and an inner pulp with nerves and blood vessels. This engineering marvel can withstand years of chemical and mechanical stresses. Chisel-shaped incisors shear off chunks of food. Cone-shaped canines tear food. Premolars and molars, with broad crowns and rounded cusps, grind and crush it.

Also in the mouth is a **tongue**, an organ consisting of membrane-covered skeletal muscles that function in positioning food in the mouth, swallowing, and speech. On that tongue's surface are many circular structures with taste buds embedded in their tissues (Figure 41.8). A taste bud contains sensory receptors that respond to chemical differences in dissolved substances. The brain uses this information to give rise to our sense of taste.

Chewing mixes food with **saliva**. This fluid contains an enzyme (salivary amylase), a buffer (bicarbonate, or HCO_3^-), mucins, and water. Salivary glands beneath and in back of the tongue produce and secrete saliva through ducts to the free surface of the mouth's lining. Salivary amylase breaks down starch. The HCO_3^- helps maintain the mouth's pH when you eat acidic foods. The mucins (modified proteins) help form the mucus that binds food into a softened, lubricated ball (bolus).

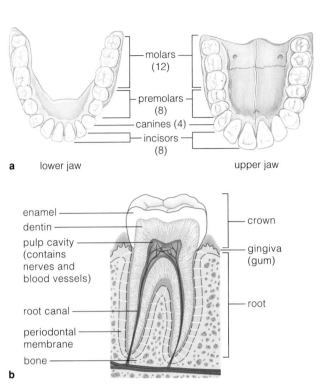

Figure 41.7 (**a**) Number and arrangement of human teeth in the upper and lower jaw. (**b**) A molar's main regions: the crown and the root. Enamel capping the crown is made of calcium deposits; it is the hardest substance in the body.

Normally harmless bacteria live on and between teeth. Daily flossing, gentle brushing, and avoiding too many sweets help keep bacterial populations in check. Without such preventive measures, conditions favor bacterial overgrowth that may lead to *caries* (tooth decay), *gingivitis* (inflamed gums), or both. Infections also can spread to the periodontal membrane that anchors teeth to the jawbone. In *periodontal disease*, infection slowly destroys the bone tissue around a tooth.

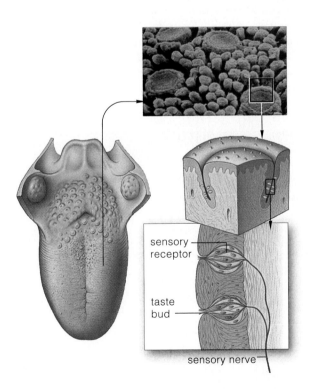

Figure 41.8 Human tongue. Circular structures at its surface house many taste buds. Filamentous structures next to them help keep food moving in the mouth.

When you swallow, contractions of tongue muscles force boluses into the **pharynx**, the tubular entrance to the esophagus *and* trachea, an airway to the lungs. To block breathing as food leaves the pharynx, a flaplike valve (the epiglottis) and the vocal cords close off the trachea. That's why you normally don't choke on food (Sections 40.4 and 40.9, *Critical Thinking* question 3).

The tubular **esophagus** connects the pharynx with the stomach. Contractions of its muscular wall propel food past a sphincter, into the stomach. A **sphincter** is a ring of smooth muscles, the contractions of which close off a passageway or an opening to the body surface.

By the action of the mouth's teeth and tongue, food gets chewed, mixed with saliva, and bound into soft, lubricated balls that will be propelled through tubes to the stomach. Enzymes in saliva start the digestion of polysaccharides.

DIGESTION IN THE STOMACH AND SMALL INTESTINE

We arrive now at the premier food-processing organs, the stomach and small intestine. Both have layers of smooth muscles, the contractions of which break apart, mix, and directionally move food. The lumen of each organ receives digestive enzymes and other secretions that help break down nutrients into fragments and then molecules small enough to be absorbed. Table 41.1 lists the names and sources of the major digestive enzymes. Carbohydrate breakdown starts in the mouth; protein breakdown starts in the stomach. Digestion of nearly all carbohydrates, lipids, proteins, and nucleic acids in food is *completed* in the small intestine.

The Stomach

The **stomach**, a muscular, stretchable sac (Figure 41.9a), has three important functions. First, it mixes and stores ingested food. Second, its secretions help dissolve and degrade the food, particularly proteins. Third, it helps control passage of food into the small intestine.

The stomach wall surface exposed to the lumen is lined with glandular epithelium. Each day, the lining's glandular cells secrete about two liters of hydrochloric acid (HCl), mucus, pepsinogens, and other substances that make up **gastric fluid**, or stomach fluid. Its potent acidity, in combination with strong contractions of the stomach wall, converts food into a thick, liquid mixture called **chyme**. Chyme's acidity kills many pathogens that enter the stomach with food. It also can cause *heartburn* when gastric fluid backs up into the esophagus.

Protein digestion starts as the high acidity alters the structure of proteins and exposes their peptide bonds. It also converts pepsinogens to pepsins: active enzymes that cleave the bonds. Protein fragments accumulate in the stomach's lumen. Meanwhile, glandular cells of the stomach lining secrete gastrin. This hormone stimulates cells of the lining that secrete HCl and pepsinogen.

Peptic ulcers form when the digestive enzymes and gastric fluid erode through the lining of the stomach or small intestine. Erosion occurs when the body doesn't secrete enough mucus and buffers or secretes too much pepsin. Heredity, chronic stress, smoking, and excessive intake of alcohol or aspirin make things worse. At least 80 percent of these ulcers start after *Helicobacter pylori* infections (Section 21.1). Antibiotic therapy cures them.

The stomach empties by waves of contraction and relaxation, of a type called peristalsis. These waves mix chyme and gather force as they approach a sphincter at the stomach's base (Figure 41.9b). The arrival of a strong contraction closes the sphincter, which squeezes most of the chyme back. All of the chyme does move into the small intestine, but only in small amounts at a time.

The Small Intestine

The small intestine has three regions: the duodenum, jejunum, and ileum (Figure 41.6). Each day, on average, 9 liters of fluid enter the first region from the stomach and through the ducts of three organs: the **pancreas**, **liver**, and **gallbladder**. At least 95 percent of the fluid is absorbed across the intestine's epithelial lining.

Enzymes secreted by cells of the intestinal lining and pancreas digest food to monosaccharides (such as glucose), monoglycerides (each having a glycerol head and fatty acid tail), free fatty acids, free amino acids, nucleotides, and nucleotide bases. Example: Like the

Table 41.1 *Major Digestive Enzymes and Their Breakdown Products*

Enzyme	Source	Where Active	Substrate	Main Breakdown Products
CARBOHYDRATE DIGESTION				
Salivary amylase	Salivary glands	Mouth, stomach	Polysaccharides	Disaccharides
Pancreatic amylase	Pancreas	Small intestine	Polysaccharides	Disaccharides
Disaccharidases	Intestinal lining	Small intestine	Disaccharides	MONOSACCHARIDES* (such as glucose)
PROTEIN DIGESTION				
Pepsins	Stomach lining	Stomach	Proteins	Protein fragments
Trypsin and chymotrypsin	Pancreas	Small intestine	Proteins	Protein fragments
Carboxypeptidase	Pancreas	Small intestine	Protein fragments	AMINO ACIDS*
Aminopeptidase	Intestinal lining	Small intestine	Protein fragments	AMINO ACIDS*
FAT DIGESTION				
Lipase	Pancreas	Small intestine	Triglycerides	FREE FATTY ACIDS, MONOGLYCERIDES*
NUCLEIC ACID DIGESTION				
Pancreatic nucleases	Pancreas	Small intestine	DNA, RNA	NUCLEOTIDES*
Intestinal nucleases	Intestinal lining	Small intestine	Nucleotides	NUCLEOTIDE BASES, MONOSACCHARIDES*

* Breakdown products small enough to be absorbed into the internal environment.

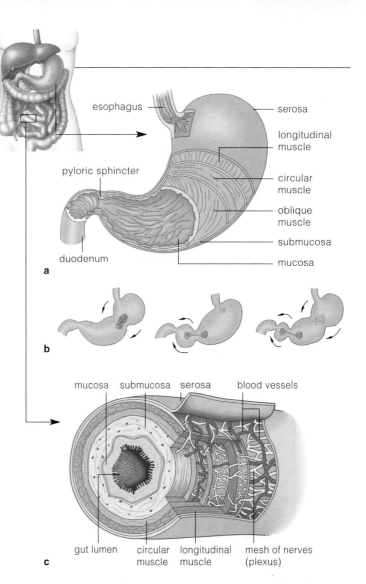

Figure 41.9 (**a**) Stomach structure. (**b**) Peristaltic wave down the stomach. (**c**) Small intestine structure. The gut wall usually has an innermost mucosa: epithelium and an underlying connective tissue layer. Next is the submucosa, a connective tissue layer with blood and lymph vessels and a mesh of nerves that locally control digestion. Next are smooth muscle layers that differ in orientation, hence in the direction of contraction. The gut wall's outermost layer of connective tissue is the serosa.

pepsin secreted from cells in the stomach lining, two pancreatic enzymes (trypsin and chymotrypsin) digest proteins to peptides. Another cleaves peptides to free amino acids. The pancreas also secretes bicarbonate, a buffer that helps neutralize HCl from the stomach.

The Role of Bile in Fat Digestion

Besides enzymes, fat digestion requires **bile**. This fluid, which the liver secretes continually, contains bile salts, bile pigments, cholesterol, and lecithin, a phospholipid. One of the pigments is iron-rich bilirubin from degraded red blood cells. After the stomach empties, a sphincter closes the main bile duct from the liver. Bile backs up into the gallbladder, which stores and concentrates it.

By a process of **emulsification**, bile salts accelerate fat digestion. Most fats in food are triglycerides. Being insoluble in water, the triglycerides tend to cluster into large fat globules in chyme. But the intestinal wall has muscle layers (Figure 41.9c). As these move and agitate chyme, the fat globules break apart into small droplets that get coated with bile salts. Bile salts carry negative charges, so the coated droplets repel each other and stay separated. The suspension of tiny fat droplets, formed by mechanical and chemical action, is the "emulsion."

Compared to large fat globules, emulsion droplets offer fat-digesting enzyme molecules a greater surface area. The enzymes can break down triglycerides more rapidly to fatty acids and monoglycerides.

Controls Over Digestion

Homeostatic controls counter changes in the internal environment. By contrast, controls over digestion act *before* food is absorbed into the internal environment. The nervous system, endocrine system, and meshes of nerves (plexuses) in the gut wall exert the control.

When food enters the stomach, it distends the wall and stimulates mechanoreceptors. Signals travel along short reflex pathways to smooth muscles and glands (longer reflex pathways also signal the brain). As one response, gut wall muscles contract and glandular cells secrete enzyme-rich fluid into the lumen or hormones into blood. The response depends partly on the chyme volume and composition. A large meal activates more mechanoreceptors in the stomach wall, so contractions

become more forceful and the stomach empties faster. High acidity or a high fat content in the small intestine triggers hormonal secretions that lead to a slowdown in stomach emptying, so chyme is not moved forward faster than it can be processed. Fear, depression, and other emotional upsets also trigger slowdowns.

What are the gastrointestinal hormones? If chyme contains amino acids and peptides, cells of the stomach lining will secrete the gastrin that induces secretion of acid into the stomach. Secretin induces the pancreas to secrete bicarbonate, and CCK (cholecystokinin) calls for secretion of pancreatic enzymes and contraction of the gallbladder. When the small intestine holds glucose and fat, GIP (glucose insulinotropic peptide) causes cells to take up more glucose by stimulating insulin secretion.

Carbohydrate breakdown starts in the mouth, and protein breakdown starts in the stomach. In the small intestine, most large organic compounds are digested to molecules small enough to be absorbed into the internal environment.

Signals from the nervous system and a nerve plexus in the gut wall, along with endocrine secretions, control digestion.

ABSORPTION FROM THE SMALL INTESTINE

Structure Speaks Volumes About Function

Only alcohol and a few other substances are absorbed from the stomach. Most nutrient absorption proceeds at the small intestine. Figure 41.10 shows the structure of the small intestine's wall. Focus first on the profuse folds of the mucosa that project into the lumen. Look closer and you see large numbers of tinier projections from each large fold. Look closer still and you see that epithelial cells at the surface of these tiny projections have a brushlike crown of great numbers of even tinier projections, all exposed to the intestinal lumen.

What is the significance of so much convolution? It has a highly favorable surface-to-volume ratio (Sections 4.1 and 28.6). Collectively, all the projections from the intestinal mucosa ENORMOUSLY increase the surface area available to interact with chyme and absorb nutrients.

Without that immense surface area, absorption would proceed hundreds of times more slowly, and that rate would not be enough to sustain human life.

Figure 41.10*c,d* shows **villi** (singular, villus), which are absorptive structures on mucosal folds. Each villus is about one millimeter long. Millions project from the mucosa. Their sheer density gives the mucosa a velvety appearance. Inside each villus is an arteriole, a venule, and a lymph vessel that function in moving substances to and from the general circulation.

Most cells in the epithelial lining of a villus have **microvilli** (singular, microvillus), which are ultrafine, threadlike projections from their free surface. Each of these cells has 1,700 or so microvilli. Hence the name, "brush border" cell. Glandular cells in the lining make and secrete digestive enzymes. Also, some phagocytic cells patrol and help protect the lining (Figure 41.10*e*).

Figure 41.10 (**a**,**b**) Structure of the mammalian small intestine. Circular folds of the intestinal mucosa are permanent. (**c**) At the free surface of each fold are many absorptive structures called villi. (**d**) Fine structure of one villus. Monosaccharides and most amino acids that cross the intestinal lining enter blood capillaries in the villus. Fats enter lymph vessels. (**e**) Epithelial cell types at the free surface of a villus. The absorptive cell has a thick crown of microvilli facing the intestinal lumen.

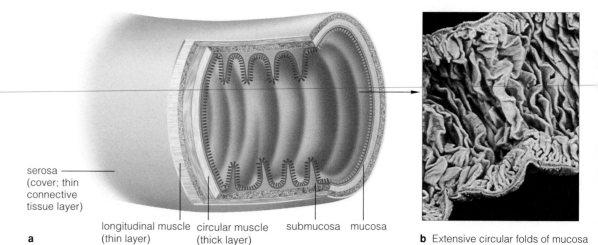

serosa (cover; thin connective tissue layer)

longitudinal muscle (thin layer) circular muscle (thick layer) submucosa mucosa

a

b Extensive circular folds of mucosa

villi (many fingerlike, epithelium-covered projections from the mucosa)

connective tissue

vesicles

artery

vein

lymph vessel

epithelium

blood capillaries

lymph vessel

microvilli at free surface of absorptive cells

cytoplasm

phagocytosis, lysozyme secretion

hormone secretion

mucus secretion (goblet cell)

absorption

c Villi on one of the folds, longitudinal section

d One villus

e Specialized cells making up the epithelium of a villus

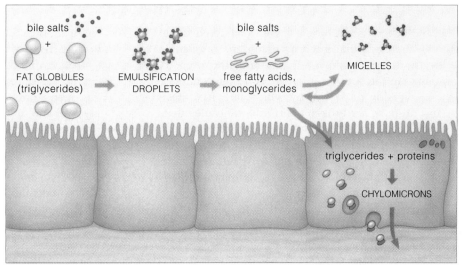

a Digestion of carbohydrates to monosaccharides, and proteins to amino acids, is completed with enzymes secreted by the pancreas and by cells of the epithelial lining.

b Monosaccharides and amino acids are actively transported across the plasma membrane of the epithelial cells, then out of the same cells and into the internal environment.

c Emulsification: The constant movement of the intestinal wall breaks up fat globules into small emulsification droplets. Bile salts prevent the fat globules from re-forming. Pancreatic enzymes digest the droplets to fatty acids and monoglycerides.

d Micelles form as bile salts combine with digestion products and phospholipids. Products can readily slip into and out of the micelles.

e Concentration of monoglycerides and fatty acids in micelles enhances gradients. This leads to diffusion of both substances across the lipid bilayer of the plasma membrane of cells making up the lining.

f Products of fat digestion reassemble into triglycerides in cells of the intestinal lining. These become coated with proteins, then are expelled (by exocytosis from the cells) into the internal environment.

Figure 41.11 Digestion and absorption in the small intestine.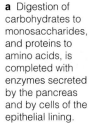

What Are the Absorption Mechanisms?

Absorption, recall, is the passage of nutrients, water, salts, and vitamins into the internal environment. The intestinal wall's large absorptive surface facilitates the process, but so does the action of its smooth muscle. In **segmentation**, rings of circular muscle contract and relax repeatedly. This creates an oscillating (back and forth) movement that constantly mixes and forces the lumen contents against the wall's absorptive surface:

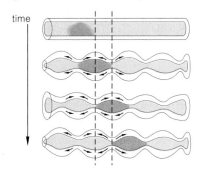

By the time the meal is halfway through the small intestine, most is broken apart and digested. Molecules of water cross the intestinal lining by osmosis. Mineral ions are selectively absorbed. Transport proteins in the plasma membrane of brush border cells actively shunt some breakdown products, including monosaccharides and amino acids, across the lining. By contrast, being lipid-soluble, fatty acids and monoglycerides diffuse right across the plasma membrane's lipid bilayer. And bile salts help them do this (Figure 41.11).

By a process called **micelle formation**, the bile salts combine with fat digestion products and phospholipids to form tiny droplets (micelles). The product molecules in the micelles continuously exchange places with those suspended in chyme. But micelles concentrate these molecules next to the intestinal lining. When they are concentrated enough, gradients promote their diffusion out of the micelles and into epithelial cells. Fatty acids and monoglycerides recombine inside the cells to form triglycerides. Then triglycerides combine with proteins into particles (chylomicrons) that leave the cells by way of exocytosis and enter the internal environment.

Once absorbed, glucose and amino acids directly enter blood vessels. Triglycerides enter lymph vessels that eventually drain into blood vessels.

With its richly folded intestinal mucosa, millions of villi, and hundreds of millions of microvilli, the small intestine has a vast surface area for absorbing nutrients.

Substances pass through the brush border cells that line the free surface of each villus by active transport, osmosis, and diffusion across the lipid bilayer of plasma membranes.

DISPOSITION OF ABSORBED ORGANIC COMPOUNDS

Earlier in the book, in Section 8.6, you considered some of the mechanisms that govern organic metabolism—specifically, the disposition of glucose and other organic compounds in the body as a whole. You saw examples of the conversion pathways by which carbohydrates, fats, and proteins are broken apart into molecules that serve as intermediates in the ATP-producing pathway of aerobic respiration. Here, Figure 41.12 rounds out the picture by showing all the major routes by which organic compounds obtained from food can be shuffled and reshuffled in the body as a whole.

energy source. There is no net breakdown of protein in muscle tissue or any other tissue during this period.

Between meals, the brain takes up two-thirds of the circulating glucose. Most body cells tap their fat and glycogen stores. Adipose tissue cells dismantle fats to glycerol and fatty acids, and release these to the blood. Liver cells degrade glycogen to release glucose, which also enters blood. Most body cells take up fatty acids as well as glucose and use them for ATP production.

Bear in mind, the liver does more than store and convert organic compounds to different forms. Example:

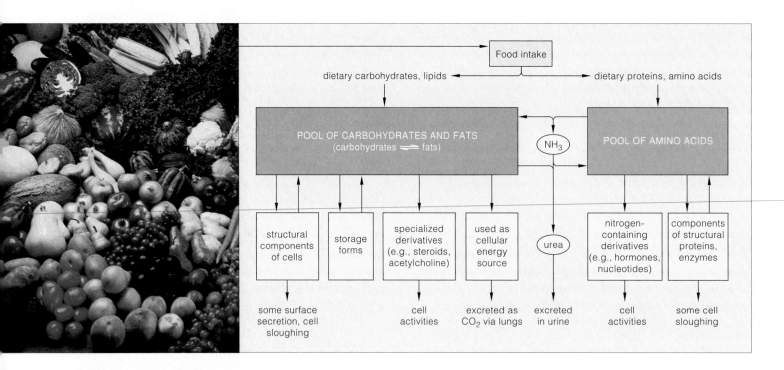

Figure 41.12 Summary of major pathways of organic metabolism. Cells continually synthesize and tear down carbohydrates, fats, and proteins. Urea forms mainly in the liver (Sections 8.6 and 42.1).

All living cells in the body continually recycle some of their carbohydrates, lipids, and proteins by breaking them apart. They use breakdown products as energy sources and building blocks. This massive molecular turnover is integrated by the nervous and endocrine systems, but a treatment of how they do this would be beyond the scope of this book.

For now, simply reflect on a few key points about organic metabolism. When you eat, your body adds to its pools of materials. Excess amounts of carbohydrates and other organic compounds absorbed from the gut are transformed mostly into fats, which become stored in adipose tissue. Some are converted to glycogen in the liver and muscle tissue. *While* compounds are being absorbed and stored, most cells use glucose as the main

It helps maintain their concentrations in blood. It also inactivates most hormone molecules, which are sent to the kidneys for excretion, in urine. The liver removes worn-out blood cells and deactivates certain toxins, including alcohol (Section 6.9). For instance, amino acid breakdown releases ammonia (NH_3), which is toxic in high concentrations. The liver converts it into the less toxic urea, which is excreted in urine.

Right after a meal, cells take up glucose absorbed from the gut and use it as a quick energy source. Excess glucose and other organic compounds are converted mainly to fats that get stored in adipose tissue. Some glucose is converted to glycogen and stored mainly in the liver and muscle tissue.

Between meals, brain cells tap two-thirds of the circulating glucose. Most body cells tap fat reservoirs as their main energy source. Fats are converted to glycerol and fatty acids, both of which enter ATP-producing pathways.

THE LARGE INTESTINE

What happens to material *not* absorbed in the small intestine? It enters the large intestine, or **colon**. The colon concentrates and stores feces: a mixture of water, undigested and unabsorbed material, and bacteria. It starts at the cecum, a cup-shaped pouch (Figures 41.6 and 41.13). It ascends on the right side of the abdominal cavity, extends across to the other side, and descends and connects with a short tube, the rectum.

Colon Functioning

Material becomes concentrated as water moves across the colon's lining. Cells of the lining actively transport sodium ions out of the lumen. Ion concentrations fall, which increases the water concentration, so water also moves out of the lumen, by osmosis. These same cells secrete mucus that lubricates feces and helps keep them from mechanically damaging the colon wall. They also secrete bicarbonate, which buffers acidic fermentation products of diverse bacteria that colonize the colon. These colonizers generally are harmless unless they breach the wall and enter the abdominal cavity.

Short, longitudinal bands of smooth muscle in the colon wall are gathered at their ends, like a series of full skirts nipped in at elastic waistbands. As they contract and relax, they move the lumen's contents back and forth against the colon wall's absorptive surface. Nerve plexuses largely control the motion, which is similar to segmentation in the small intestine but much slower.

Whereas the quick transit time through the small intestine does not favor the rapid population growth of ingested bacteria, the colon's slower motion favors it. The volume of cellulose fibers and other undigested material that cannot be decreased by absorption in the colon is called **bulk**. It adds to the volume of material and influences the transit time through the colon.

Following each new meal, secretion of gastrin and signals from the nervous system's autonomic nerves stimulate large parts of the ascending and transverse colon to contract together. Within seconds, the lumen's contents can be propelled through much of the colon's length, making way for incoming material.

The contents are stored in the last part of the colon until they distend the rectal wall enough to cause reflex action and expulsion. The nervous system controls this defecation by stimulating or inhibiting the contraction of a sphincter at the anus, the last opening of the gut.

Colon Malfunctioning

Typically, the frequency of defecation ranges from three times a day to once a week. Aging, emotional stress, a low-bulk diet, injury, or disease can result in delayed defecation, called *constipation*. The longer the delay, the

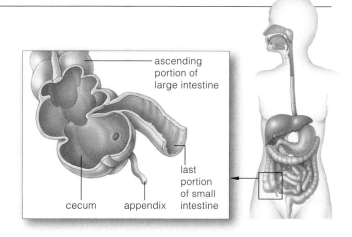

Figure 41.13 Cecum and appendix of the large intestine (colon).

more water gets absorbed, so feces harden and dry out. The abdominal discomfort is accompanied by loss of appetite, headaches, and often nausea and depression.

Hard feces may become lodged inside the **appendix**, a narrow projection from the cecum (Figure 41.13). The appendix has no known digestive functions, but, like the ileum, it holds a concentration of bacteria-fighting white blood cells. Feces that obstruct the normal blood flow and mucus secretion to this projection can cause *appendicitis*, or an inflamed appendix. Unless removed, an inflamed appendix can rupture. Then bacteria, even otherwise harmless ones, would be free to enter the abdominal cavity and cause life-threatening infections.

The colon also is vulnerable to cancer, which occurs most often among the world's wealthiest and best-fed populations. Too many of these people skip meals, eat too much and too fast when they do sit down at the table, and generally give their gut erratic workouts. Their diet tends to be rich in sugar, cholesterol, and salt, and low in bulk. Too little bulk extends the transit time of feces through the colon. The longer that irritating, potentially carcinogenic material is in contact with the colon wall, the more damage it may cause. Symptoms of *colon cancer* include a change in bowel functioning, rectal bleeding, and blood in feces. Some people appear to be genetically predisposed to develop colon cancer, but a diet too low in fiber might also be a factor.

Colon cancer is rare in rural India and in Africa, where most people can't afford to eat much more than fiber-rich whole grains. When these same people move to cities in the affluent nations and change their eating habits, the incidence of colon cancer increases.

The large intestine, or colon, functions in the absorption of water and mineral ions from the gut lumen. It also functions in the compaction of undigested residues into feces, for expulsion at the terminal opening of the gut.

HUMAN NUTRITIONAL REQUIREMENTS

Ever since the 1950s, the United States government has been trying to establish nutritional guidelines for good health. A few hundred million dollars' worth of studies later, the mainstream conclusion has been that low-fat, high-carbohydrate diets are the best. This thinking is reflected in Figure 41.4. It is a **food pyramid**, a chart of recommended amounts from different food groups. For an average-sized adult male, the daily intake is divided as 55–60 percent complex carbohydrates, 15–20 percent proteins (less for females), and 20–25 percent fats. You might call this a "fat-is-bad" diet. Its premise is that if you eat less fat, you will lose weight and live longer.

Alternative diets also abound. The *Mediterranean diet* pyramid is heaviest on grain products, then fruits and vegetables, legumes and nuts, then *olive oil* as the fat group, and then cheese and yogurt. It limits the weekly intake of fish, poultry, eggs, simple sugars, and—at its tiny top—red meat. Olive oil is as much as 40 percent of the total energy intake. This monounsaturated fat is less likely than saturated fats to raise cholesterol levels. It also is a fine antioxidant, good at eliminating free radicals. You might call this an "olive-oil-is-good" diet.

Other diets restrict carbohydrates and load up on proteins and fats. Their premise is that *carbohydrates* are the foods that make us fat, and if you eat less of them, you will lose weight and live longer. For the past three decades, the American Medical Association, the major health agencies, and most nutritionists have dismissed the "carbohydrates-are-bad" diets as threats to health.

In mid-2002, Gary Taubes reported on his interviews with leading nutritionists and endocrinologists around the country. A growing number are concluding that the low-carb diets have worked for millions of individuals, and that an "ALL-fat-is-bad-for-you" mentality may be responsible for the current obesity epidemic! **Obesity** is an excess of fat in adipose tissues, most often caused by imbalances between energy intake and output.

Who is right? We still do not have answers. Here is what we do know. The percentage of obese Americans did not change much all through the 1970s and 1980s. Coincident with the "fat-is-bad" push, it shot up in the 1980s, and now one in four Americans is obese. There are almost three times as many overweight children, and type 2 diabetes is showing up among them. Yet for

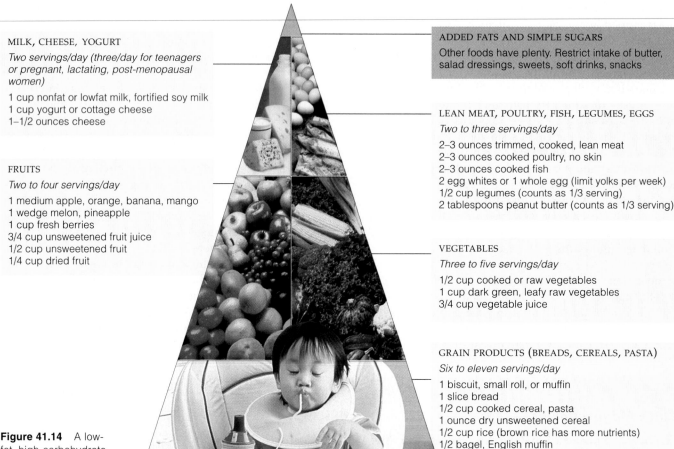

MILK, CHEESE, YOGURT
Two servings/day (three/day for teenagers or pregnant, lactating, post-menopausal women)

1 cup nonfat or lowfat milk, fortified soy milk
1 cup yogurt or cottage cheese
1–1/2 ounces cheese

FRUITS
Two to four servings/day

1 medium apple, orange, banana, mango
1 wedge melon, pineapple
1 cup fresh berries
3/4 cup unsweetened fruit juice
1/2 cup unsweetened fruit
1/4 cup dried fruit

ADDED FATS AND SIMPLE SUGARS
Other foods have plenty. Restrict intake of butter, salad dressings, sweets, soft drinks, snacks

LEAN MEAT, POULTRY, FISH, LEGUMES, EGGS
Two to three servings/day

2–3 ounces trimmed, cooked, lean meat
2–3 ounces cooked poultry, no skin
2–3 ounces cooked fish
2 egg whites or 1 whole egg (limit yolks per week)
1/2 cup legumes (counts as 1/3 serving)
2 tablespoons peanut butter (counts as 1/3 serving)

VEGETABLES
Three to five servings/day

1/2 cup cooked or raw vegetables
1 cup dark green, leafy raw vegetables
3/4 cup vegetable juice

GRAIN PRODUCTS (BREADS, CEREALS, PASTA)
Six to eleven servings/day

1 biscuit, small roll, or muffin
1 slice bread
1/2 cup cooked cereal, pasta
1 ounce dry unsweetened cereal
1/2 cup rice (brown rice has more nutrients)
1/2 bagel, English muffin

Figure 41.14 A low-fat, high-carbohydrate food pyramid, based on prevailing USDA guidelines.

Figure 41.15 Vegetarian diets. All eight essential amino acids, a small portion of the total protein intake, must be available at the same time, in certain amounts, if cells are to build their own proteins. Milk and eggs are among the *complete* proteins; they have high amounts of all eight in proportions that humans require.

Nearly all plant proteins are *incomplete*. Vegetarians must be careful to avoid protein deficiency. For instance, they might combine different foods from the three columns of incomplete proteins shown. Strict vegetarians (who also avoid dairy products and eggs) must take vitamin B_{12} and B_2 (riboflavin) supplements. In regions where animal protein is a luxury, traditional cuisines have good combinations of plant proteins, including rice/beans, chili/cornbread, tofu/rice, and lentils/wheat bread.

No Limiting Amino Acid	Low Lysine	Low Methionine, Cysteine	Low Tryptophan
Legumes: soybeans (e.g., tofu, soy milk) Grains: wheat germ Milk Cheeses (not cream) Yogurt Eggs (Meats)	Legumes: peanuts Grains: barley, rice, buckwheat, oats, corn, rye, wheat Nuts: almonds, cashews, coconut, walnuts, hazelnuts, pecans Seeds: pumpkin, sunflower	Legumes: beans (dried), black-eyed peas, garbanzos, lentils, lima beans, mung beans, peanuts Nuts: hazelnuts Fresh vegetables: asparagus, green peas, broccoli, mushrooms, parsley, potatoes, soybeans, Swiss chard	Legumes: beans (dried), garbanzos, lima beans, mung beans, peanuts Grains: cornmeal Nuts: almonds, English walnuts Fresh vegetables: corn, green peas, mushrooms, Swiss chard

the past twenty years, Americans in general have been eating less fat, cholesterol levels are declining, and the levels of "leisure exercise" are about the same.

A Carbohydrate–Insulin Connection

Coincident with the shunning of fats is a large increase in the daily consumption of refined carbohydrates—400 kilocalories more per day, mainly in the form of sugars and starch. Today the worst offenders are flour and the "wet carbohydrates"—sugar-loaded soft drinks, fruit drinks, and so-called sports drinks that now account for more than 10 percent of the average caloric intake. These refined carbohydrates have a *high-glycemic index*; within minutes of being absorbed, they cause a surge in sugar and insulin levels in blood.

The insulin surge makes cells take up sugar fast. A few hours later, the blood sugar level is lower than before, so we get hungry. But circulating insulin keeps cells from dipping into fat stores; they metabolize fat only when insulin levels are low (Section 36.7).

And so we eat more, secrete more insulin, and thus keep storing fat—mainly in the form of triglycerides. Over time, for 30 or 40 percent of the population, high triglycerides may be riskier than high cholesterol as a factor in heart disease and type 2 diabetes.

So What Do We Eat?

If insulin affects fat metabolism, then a common-sense approach would be to avoid insulin surges. High-fiber foods such as whole grains, green vegetables, legumes, and fruit to some extent take longer to digest, so blood sugar levels rise slowly instead of surging. Vegetarian diets are rich in these foods, although its practitioners must guard against protein deficiencies (Figure 41.15). Fats and proteins also prevent insulin surges.

You cannot function without fats. Cell membranes incorporate the phospholipid lecithin; and fats serve as energy reserves, cushion many internal organs, and act as insulation beneath the skin. Dietary fats help store fat-soluble vitamins. But the body can synthesize most

of its own fats from proteins and carbohydrates. The ones it can't make, such as linoleic acid, are **essential fatty acids**. Whole foods offer plenty. A teaspoonful or two each day of a polyunsaturated fat, such as olive oil or corn oil, may turn out to be enough dietary fat.

Like other animal fats, butter is a saturated fat that tends to raise the level of cholesterol in the blood. As described in earlier chapters, cholesterol is used in the synthesis of bile acids and steroid hormones, and it is a component of animal cell membranes. In some people, too much can interfere with blood circulation.

Your body also cannot function without proteins. It requires the amino acid components of dietary proteins for its own protein-building programs. Of the twenty common types, eight are **essential amino acids**. Your cells can't make them, so you must get them from food. The eight are methionine (or its metabolic equivalent cysteine), isoleucine, leucine, lysine, phenylalanine (or tyrosine), threonine, tryptophan, and valine. Most of the animal proteins are *complete*; their amino acid ratios match human nutritional needs. The plant proteins are *incomplete*, lacking one or more essential amino acids.

Eat fats and proteins, and you don't get hungry as often; they make insulin levels drop enough to trigger *ketosis*. Muscle cells use stored fat; the liver produces ketones, fats that may have evolved as a backup fuel to keep the brain working in times of famine. (This is not ketoacidosis, an abnormal accumulation of ketones that can kill diabetics.) But too much protein and too little complex carbohydrate may damage the kidneys.

What is the bottom line? It looks like the nutritional guidelines will be revised over the next few years. The jury is still out on intake levels for fats and proteins. But no reputable nutritionist or endocrinologist is arguing that refined sugars and starches are good for you.

For three decades, nutritionists have been advocating a low-fat, high-carbohydrate diet. Too many individuals have taken "carbohydrate" to include an abundance of refined sugars and starches. New studies in the works may result in major shifts in nutritional guidelines.

VITAMINS AND MINERALS

Vitamins are *organic* substances that are essential for growth and survival, for no other substances can play their metabolic roles. Most plants are able to synthesize vitamins on their own. Most kinds of animals lost the ability to do so and must get vitamins from food.

At the minimum, human cells require the thirteen vitamins listed in Table 41.2. Each vitamin has specific metabolic functions. Many reactions use several types, and the absence of one affects the functions of others.

Minerals are *inorganic* substances that are essential for growth and survival because no other substances can serve their metabolic functions. For instance, all of your cells need iron for their electron transfer chains. Red blood cells do not function at all without iron in hemoglobin. Neurons stop functioning in the absence of sodium and potassium (Table 41.3).

People who are in good health get all the vitamins and minerals they need from a balanced diet of whole

Table 41.2 *Major Vitamins: Sources, Functions, and Effects of Deficiencies or Excesses* [*]

Vitamin	Common Sources	Main Functions	Effects of Chronic Deficiency	Effects of Extreme Excess
FAT-SOLUBLE VITAMINS				
A	Its precursor comes from beta-carotene in yellow fruits, yellow or green leafy vegetables; also in fortified milk, egg yolk, fish, liver	Used in synthesis of visual pigments, bone, teeth; maintains epithelia	Dry, scaly skin; lowered resistance to infections; night blindness; permanent blindness	Malformed fetuses; hair loss; changes in skin; liver and bone damage; bone pain
D	D_3 formed in skin and in fish liver oils, egg yolk, fortified milk; converted to active form elsewhere	Promotes bone growth and mineralization; enhances calcium absorption	Bone deformities (rickets) in children; bone softening in adults	Retarded growth; kidney damage; calcium deposits in soft tissues
E	Whole grains, dark green vegetables, vegetable oils	Counters effects of free radicals; helps maintain cell membranes; blocks breakdown of vitamins A and C in gut	Lysis of red blood cells; nerve damage	Muscle weakness, fatigue, headaches, nausea
K	Enterobacteria form most of it; also in green leafy vegetables, cabbage	Blood clotting; ATP formation via electron transport	Abnormal blood clotting; severe bleeding (hemorrhaging)	Anemia; liver damage and jaundice
WATER-SOLUBLE VITAMINS				
B_1 (thiamin)	Whole grains, green leafy vegetables, legumes, lean meats, eggs	Connective tissue formation; folate utilization; coenzyme action	Water retention in tissues; tingling sensations; heart changes; poor coordination	None reported from food; possible shock reaction from repeated injections
B_2 (riboflavin)	Whole grains, poultry, fish, egg white, milk	Coenzyme action	Skin lesions	None reported
Niacin	Green leafy vegetables, potatoes, peanuts, poultry, fish, pork, beef	Coenzyme action	Contributes to pellagra (damage to skin, gut, nervous system, etc.)	Skin flushing; possible liver damage
B_6	Spinach, tomatoes, potatoes, meats	Coenzyme in amino acid metabolism	Skin, muscle, and nerve damage; anemia	Impaired coordination; numbness in feet
Pantothenic acid	In many foods (meats, yeast, egg yolk especially)	Coenzyme in glucose metabolism, fatty acid and steroid synthesis	Fatigue, tingling in hands, headaches, nausea	None reported; may cause diarrhea occasionally
Folate (folic acid)	Dark green vegetables, whole grains, yeast, lean meats; enterobacteria produce some folate	Coenzyme in nucleic acid and amino acid metabolism	A type of anemia; inflamed tongue; diarrhea; impaired growth; mental disorders	Masks vitamin B_{12} deficiency
B_{12}	Poultry, fish, red meat, dairy foods (not butter)	Coenzyme in nucleic acid metabolism	A type of anemia; impaired nerve function	None reported
Biotin	Legumes, egg yolk; colon bacteria produce some	Coenzyme in fat, glycogen formation and in amino acid metabolism	Scaly skin (dermatitis), sore tongue, depression, anemia	None reported
C (ascorbic acid)	Fruits and vegetables, especially citrus, berries, cantaloupe, cabbage, broccoli, green pepper	Collagen synthesis; possibly inhibits effects of free radicals; structural role in bone, cartilage, and teeth; used in carbohydrate metabolism	Scurvy, poor wound healing, impaired immunity	Diarrhea, other digestive upsets; may alter results of some diagnostic tests

[*] Guidelines for appropriate daily intakes are being worked out by the Food and Drug Administration.

Table 41.3 *Major Minerals: Sources, Functions, and Effects of Deficiencies or Excesses**

Mineral	Common Sources	Main Functions	Effects of Chronic Deficiency	Effects of Extreme Excess
Calcium	Dairy products, dark green vegetables, dried legumes	Bone, tooth formation; blood clotting; neural and muscle action	Stunted growth; possibly diminished bone mass (osteoporosis)	Impaired absorption of other minerals; kidney stones in susceptible people
Chloride	Table salt (usually too much in diet)	HCl formation in stomach; contributes to body's acid–base balance; neural action	Muscle cramps; impaired growth; poor appetite	Contributes to high blood pressure in susceptible people
Copper	Nuts, legumes, seafood, drinking water	Used in synthesis of melanin, hemoglobin, and some transport chain components	Anemia, changes in bone and blood vessels	Nausea, liver damage
Fluorine	Fluoridated water, tea, seafood	Bone, tooth maintenance	Tooth decay	Digestive upsets; mottled teeth and deformed skeleton in chronic cases
Iodine	Marine fish, shellfish, iodized salt, dairy products	Thyroid hormone formation	Enlarged thyroid (goiter), with metabolic disorders	Toxic goiter
Iron	Whole grains, green leafy vegetables, legumes, nuts, eggs, lean meat, molasses, dried fruit, shellfish	Formation of hemoglobin and cytochrome (transport chain component)	Iron-deficiency anemia, impaired immune function	Liver damage, shock, heart failure
Magnesium	Whole grains, legumes, nuts, dairy products	Coenzyme role in ATP–ADP cycle; roles in muscle, nerve function	Weak, sore muscles; impaired neural function	Impaired neural function
Phosphorus	Whole grains, poultry, red meat	Component of bone, teeth, nucleic acids, ATP, phospholipids	Muscular weakness; loss of minerals from bone	Impaired absorption of minerals into bone
Potassium	Diet alone provides ample amounts	Muscle and neural function; roles in protein synthesis and body's acid–base balance	Muscular weakness	Muscular weakness, paralysis, heart failure
Sodium	Table salt; diet provides ample to excessive amounts	Key role in body's salt–water balance; roles in muscle and neural function	Muscle cramps	High blood pressure in susceptible people
Sulfur	Proteins in diet	Component of body proteins	None reported	None likely
Zinc	Whole grains, legumes, nuts, meats, seafood	Component of digestive enzymes; roles in normal growth, wound healing, sperm formation, and taste and smell	Impaired growth, scaly skin, impaired immune function	Nausea, vomiting, diarrhea; impaired immune function and anemia

* The guidelines for appropriate daily intakes are being worked out by the Food and Drug Administration.

foods. Generally, vitamin and mineral supplements are necessary only for strict vegetarians, the elderly, and people suffering a chronic illness or taking medication that interferes with utilization of specific nutrients. As examples, vitamin K supplements aid calcium retention and lessen osteoporosis in elderly women. Vitamin C, vitamin E, and the beta-carotene precursor for vitamin A may lessen some aging effects and improve immune functioning by inactivating free radicals. A free radical, remember, is an atom or group of atoms that is highly reactive because it has an unpaired electron.

However, metabolism varies in its details from one person to the next, so no one should take massive doses of any vitamin or mineral supplement except under medical supervision. Also, excessive amounts of many

vitamins and minerals can harm everyone (Tables 41.2 and 41.3). Large doses of vitamins A and D are good examples. As is the case for all fat-soluble vitamins, excess amounts of these vitamins accumulate in tissues and interfere with normal metabolic activity. Similarly, sodium is present in plant and animal tissues, and it is a component of table salt. It has roles in the body's salt–water balance, muscle activity, and nerve function. However, prolonged, excessive intake of this mineral may contribute to high blood pressure in some people.

Severe shortages or self-prescribed, massive excesses of vitamins and minerals can disturb the delicate balances in body function that promote health.

Weighty Questions, Tantalizing Answers

Have you ever asked yourself *why* you want to weigh a certain amount? Are you merely fearful of "being fat"? One standard of what constitutes obesity is cultural, and it varies from one culture to the next. For example, one female student despaired over being plump until her graduate studies program in Africa, where many males found her to be one of the most desirable females on the planet. However, a lot depends on how healthy you want to be and how long you want to live. In general, thinner people live longer, maybe because they produce fewer free radicals (Chapter 6).

In the United States, obesity is now the second leading cause of death; 300,000 or so people die annually from preventable, weight-related conditions. Each year, the weight-related cases of type 2 diabetes, heart disease, hypertension, breast cancer, colon cancer, gout, gallstones, and osteoarthritis add up to a 100-billion-dollar drain on the economy. The future doesn't look any rosier: children today are 42 percent fatter than they were in 1980.

IN PURSUIT OF THE "IDEAL" WEIGHT In Figure 41.16, you see charts for estimating "ideal" weights for adults. Another indicator of obesity-related health risk is the *body mass index* (BMI), as determined by this formula:

$$BMI = \frac{\text{weight (pounds)} \times 700}{\text{height (inches)}^2}$$

You can find several web-based BMI calculators on the Internet. If your BMI value is 27 or higher, the health risk rises dramatically. Other factors that affect the risk are smoking habits, family history of heart disorders, use of sex hormones after menopause, and fat distribution. Fat stored above the belt, as with "beer bellies," can be a sign of potential health problems.

Dieting alone can't lower a BMI value. Eat less, and the body slows its metabolic rate to conserve energy. So how do you function normally over the long term while maintaining acceptable weight? *You must balance caloric intake with energy output.* For most of us, this means eating controlled portions of low-calorie, nutritious foods *and* exercising regularly. Bear in mind, we express energy stored in food as kilocalories or Calories (with a big C). A **kilocalorie** is 1,000 calories, or units of heat energy.

To figure out how many kilocalories you should take in daily to maintain a desired weight, multiply that weight (in pounds) by 10 if you are not active physically, by 15 if moderately active, and by 20 if highly active. From the value you get this way, subtract the following amount:

Age:	25–34	Subtract:	0
	35–44		100
	45–54		200
	55–64		300
	Over 65		400

For instance, if you want to weigh 120 pounds and are very active, $120 \times 20 = 2{,}400$ kilocalories. If you're thirty-five years old, then $(2{,}400 - 100)$, or 2,300 kilocalories. Such calculations give a rough estimate of caloric intake. Other factors, including height, must also be considered. An active person 5 feet, 2 inches tall doesn't need as much energy as an active 6-footer who weighs just the same.

WEIGHT GUIDELINES FOR WOMEN		**WEIGHT GUIDELINES FOR MEN**	
Starting with an ideal weight of 100 pounds for a woman who is 5 feet tall, add five additional pounds for each additional inch of height. Examples:		Starting with an ideal weight of 106 pounds for a man who is 5 feet tall, add six additional pounds for each additional inch of height. Examples:	

Height (feet)	Weight (pounds)	Height (feet)	Weight (pounds)
5' 2"	110	5' 2"	118
5' 3"	115	5' 3"	124
5' 4"	120	5' 4"	130
5' 5"	125	5' 5"	136
5' 6"	130	5' 6"	142
5' 7"	135	5' 7"	148
5' 8"	140	5' 8"	154
5' 9"	145	5' 9"	160
5' 10"	150	5' 10"	166
5' 11"	155	5' 11"	172
6'	160	6'	178

Figure 41.16 How to estimate the "ideal" weight for an adult. Values shown are consistent with a long-term Harvard study into the link between excessive weight and increased risk of cardiovascular disorders. Depending on specific factors (such as having a small, medium, or large skeletal frame), the "ideal" might vary by plus or minus 10 percent.

MY GENES MADE ME DO IT   On average, one in three Americans eats too much and has more trouble keeping off excess weight than others do. Genes have a lot to do with it. Researchers suspected as much for a long time. Studies of identical twins offered clues. (*Identical twins* are born with identical genes.) As an outcome of family problems, some of the twins had been separated at birth and were raised apart, in different households. Yet, at adulthood, the separated twins had similar body weights! People, it seemed, were born with a "set point" for body fat. Could it be that whatever set point someone inherits is the one he or she will be stuck with for life?

Many experiments confirmed this suspicion. The first started in 1950 with a seriously obese mouse (Figure 41.17). In 1995, molecular geneticists identified one of its genes that affects the set point for body fat. They named it the *ob* **gene** (guess why). Which clues led to its discovery? The nucleotide sequence in the gene region of obese mice differs from that in normal mice. The gene is active in adipose tissue, nowhere else. Its biochemical message translates into a protein released from cells and picked up by the bloodstream; the gene specifies a hormone.

Jeffrey Friedman discovered the hormone in 1994 and called it **leptin**. A nearly identical hormone was isolated in humans. Leptin is just one of a number of factors that mediate the brain's commands to suppress or whip up appetite, but it's a crucial one. By assessing the blood levels of incoming hormonal signals, an appetite control center in the hypothalamus "decides" whether the body has taken in enough food to provide enough fat for the day. If so, hypothalamic commands go out to increase metabolic rates—and to stop eating.

Expression of a mutant *ob* gene may affect the control center, making appetite skyrocket and metabolic furnaces burn less. Obese mice injected with leptin quickly shed their excess weight.

Developing therapies for obesity in humans won't be easy, because hormonal control is tricky business. Leptin also inhibits bone formation, which means its use in therapies could trigger osteoporosis. What about **ghrelin** (GRELL-in)? This newly discovered peptide hormone, secreted mainly by cells of the stomach lining, directly stimulates the appetite control center. Unlike leptin and cholecystokinin, which make you feel full, ghrelin makes you hungry and keeps you from burning fat. Your body is adapted to secrete more after you lose weight.

Leptin, ghrelin, and insulin are only three players in complex hormonal pathways involving the brain. It seems unlikely that medication to treat obesity will be developed soon. Researchers must identify all steps of the ancient pathways that control body weight. They must identify the contributions of all those components to other aspects of the body's physiology. Maybe then, effective drugs may be designed.

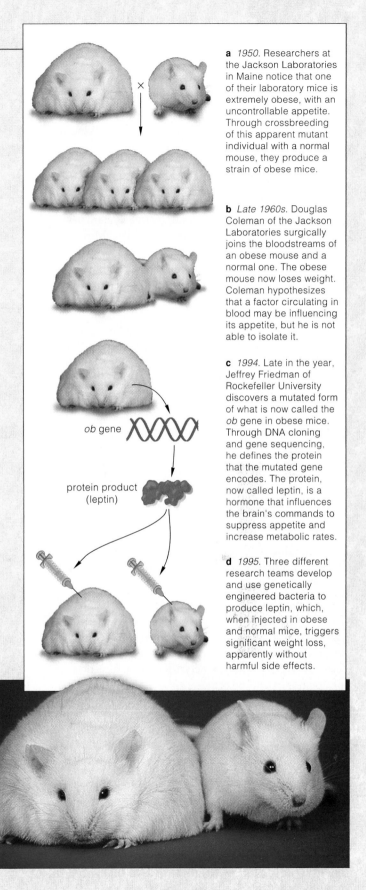

a *1950*. Researchers at the Jackson Laboratories in Maine notice that one of their laboratory mice is extremely obese, with an uncontrollable appetite. Through crossbreeding of this apparent mutant individual with a normal mouse, they produce a strain of obese mice.

b *Late 1960s*. Douglas Coleman of the Jackson Laboratories surgically joins the bloodstreams of an obese mouse and a normal one. The obese mouse now loses weight. Coleman hypothesizes that a factor circulating in blood may be influencing its appetite, but he is not able to isolate it.

c *1994*. Late in the year, Jeffrey Friedman of Rockefeller University discovers a mutated form of what is now called the *ob* gene in obese mice. Through DNA cloning and gene sequencing, he defines the protein that the mutated gene encodes. The protein, now called leptin, is a hormone that influences the brain's commands to suppress appetite and increase metabolic rates.

d *1995*. Three different research teams develop and use genetically engineered bacteria to produce leptin, which, when injected in obese and normal mice, triggers significant weight loss, apparently without harmful side effects.

ob gene

protein product (leptin)

Figure 41.17   Chronology of research developments that revealed the identity of a key factor in the genetic basis of body weight.

SUMMARY

Gold indicates text section

1. For animals, *nutrition* refers to processes by which the body takes in, digests, absorbs, and uses food. *CI*

2. Most animals have a complete digestive system—a tube that has two openings (a mouth and an anus), and regional specializations between them. Mucus-coated epithelium lines and protects every exposed surface of the tube and facilitates diffusion across its wall. *41.1*

3. These activities proceed in digestive systems: *41.1*

   a. Mechanical processing and motility: Movements that break up, mix, and propel food material.

   b. Secretion: Release of digestive enzymes and other substances from the pancreas and liver, as well as from glandular epithelium, into the gut lumen.

   c. Digestion: Breakdown of food into particles, then into nutrient molecules small enough to be absorbed.

   d. Absorption: The diffusion or transport of digested organic compounds, fluid, and ions from the gut lumen into the internal environment.

   e. Elimination: The expulsion of undigested as well as unabsorbed residues at the end of the system.

4. The human digestive system has a mouth, pharynx, esophagus, stomach, small intestine, large intestine (or colon), rectum, and anus. The salivary glands and the liver, gallbladder, and pancreas have accessory roles in the system's functions (Table 41.4). *41.2–41.4*

5. Starch digestion starts in the mouth, and protein digestion starts in the stomach. Digestion is completed and most nutrients are absorbed in the small intestine. The pancreas secretes the main digestive enzymes. Bile from the liver assists in fat digestion. *41.3, 41.4*

6. The nervous and endocrine systems, and meshes of neurons (plexuses) in the gut wall, interact to govern the digestive system. Many controls operate in response to the volume and composition of food in the gut. They cause changes in muscle activity and in secretion rates for hormones and enzymes. *41.4*

7. In absorption, cells of the intestinal lining actively transport glucose and most amino acids out of the gut lumen. Fatty acids and monoglycerides diffuse across the lipid bilayer of these cells. They are recombined in cytoplasm as triglycerides, which then are released by exocytosis, into interstitial fluid. *41.5*

8. Nutritionists advise a daily food intake in certain proportions. Example: One diet for healthy adult males of average body weight recommends 55–60 percent complex carbohydrates, 15–20 percent proteins (less for females), and 20–25 percent fats. A well-balanced diet of whole foods normally supplies all required vitamins and minerals. *41.8–41.9*

9. To maintain health and a given body weight, caloric (energy) intake must balance energy output. *CI, 41.10*

Table 41.4	*Summary of the Digestive System*
MOUTH (oral cavity)	Start of digestive system, where food is chewed, moistened; polysaccharide digestion begins
PHARYNX	Entrance to tubular parts of digestive and respiratory systems
ESOPHAGUS	Muscular tube, moistened by saliva, that moves food from pharynx to stomach
STOMACH	Sac where food mixes with gastric fluid and protein digestion begins; stretches to store food taken in faster than can be processed; gastric fluid destroys many microbes
SMALL INTESTINE	The first part (duodenum) receives secretions from the liver, gallbladder, and pancreas
	Most nutrients are digested, absorbed in second part (jejunum)
	Some nutrients absorbed in last part (ileum), which delivers unabsorbed material to colon
COLON (large intestine)	Concentrates and stores undigested matter (by absorbing mineral ions and water)
RECTUM	Distension triggers expulsion of feces
ANUS	Terminal opening of digestive system
Accessory Organs:	
SALIVARY GLANDS	Glands (three main pairs, many minor ones) that secrete saliva, a fluid with polysaccharide-digesting enzymes, buffers, and mucus (which moistens and lubricates ingested food)
PANCREAS	Secretes enzymes that digest all major food molecules and buffers against HCl from stomach
LIVER	Secretes bile (used in fat emulsification); roles in carbohydrate, fat, and protein metabolism
GALLBLADDER	Stores and concentrates bile from the liver

Review Questions

1. Define the five key tasks carried out by a complete digestive system. Then correlate some organs of such a system with the feeding behavior of a particular kind of animal. *41.1*

2. Using the sketch to the right, list the organs and accessory organs shown and the main functions of each. *Table 41.4*

3. With respect to digestive systems, define segmentation. Does segmentation occur in the stomach? Does it occur in the small intestine, colon, or both? *41.5, 41.7*

4. Using the *black* lines shown in Figure 41.18 as a guide, name the types of breakdown products small enough to be absorbed across the small intestine's lining, into the internal environment. *41.5*

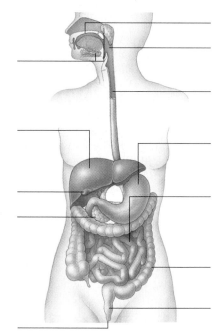

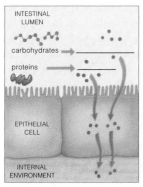 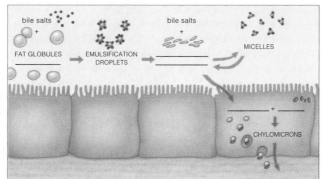

Figure 41.18 Fill in the blanks for substances that cross the lining of the small intestine.

Self-Quiz ANSWERS IN APPENDIX III

1. The _____ maintains the internal environment, supplies cells with raw materials, and disposes of metabolic wastes.
   a. digestive system   d. urinary system
   b. circulatory system   e. interaction of all systems
   c. respiratory system   listed in a through d

2. Most digestive systems have regions for _____ food.
   a. transporting   c. storing
   b. processing   d. all of the above

3. _____ secretions do not assist in digestion and absorption.
   a. Salivary gland   c. Liver
   b. Thymus gland   d. Pancreas

4. Digestion is completed and most nutrients are absorbed in the _____ .
   a. mouth   c. small intestine
   b. stomach   d. colon

5. Bile has roles in _____ digestion and absorption.
   a. carbohydrate   c. protein
   b. fat   d. amino acid

6. Monosaccharides and most amino acids are absorbed _____ .
   a. by active transport   c. at lymph vessels
   b. by diffusion   d. as fat droplets

7. Fat metabolism is influenced by _____ .
   a. insulin   c. pancreatic nucleases
   b. leptin and ghrelin   d. both a and b

8. On its own, the human body cannot produce all of the _____ it requires.
   a. vitamins and minerals   d. a through c
   b. fatty acids   e. a and c
   c. amino acids

9. Maintaining good health and normal body weight requires that _____ intake be balanced by _____ output.

10. Match the organ with its key digestive function(s).
    ____ gallbladder   a. secrete bile; many metabolic roles
    ____ stomach   b. digest, absorb most nutrients
    ____ colon   c. store, mix, dissolve food; start
    ____ pancreas   protein breakdown
    ____ salivary   d. store, concentrate bile
         gland   e. concentrate undigested matter
    ____ small   f. secrete substances that moisten food,
         intestine   start polysaccharide breakdown
    ____ liver   g. secrete digestive enzymes and
                          bicarbonate

Critical Thinking

1. A glassful of whole milk contains lactose, proteins, butterfat (mostly triglycerides), vitamins, and minerals. Explain what will happen to each component in your digestive tract.

2. As people age, the number of their body cells steadily decreases and energy needs decline. If you were planning an older person's diet, which foods would you emphasize, and why? Which ones would you deemphasize?

3. Using Section 41.10 as a reference, determine your ideal weight and then design a well-balanced program of diet and exercise that will help you achieve or maintain that weight.

4. Holiday meals often are larger and have a higher fat content than everyday ones. After stuffing themselves at an early dinner on Thanksgiving Day, Richard and other members of his family feel uncomfortably full for the rest of the afternoon. Based on what you have learned about controls over digestion, propose a biochemical explanation for their discomfort.

5. Explain some of the ways in which your digestive system helps protect you against many pathogenic bacteria that may have contaminated the kinds of food you eat. (You may wish to refer to Section 39.1, also.)

6. Angelina is chronically anorexic. Clinical tests show that she has very low blood pressure and a below-average cardiac rate (bradycardia). The heart itself has become smaller. How did anorexia bring about these life-threatening conditions?

Selected Key Terms

appendix *41.7*	liver *41.4*
bile *41.4*	micelle formation *41.5*
bulk *41.7*	microvillus (microvilli) *41.5*
chyme *41.4*	mineral *41.9*
colon *41.7*	nutrition *CI*
complete digestive system *41.1*	*ob* gene *41.10*
digestive system *CI*	obesity *41.10*
emulsification *41.4*	pancreas *41.4*
esophagus *41.3*	pharynx *41.3*
essential amino acid *41.8*	ruminant *41.1*
essential fatty acid *41.8*	saliva *41.3*
food pyramid *41.8*	segmentation *41.5*
gallbladder *41.4*	sphincter *41.3*
gastric fluid *41.4*	stomach *41.4*
ghrelin *41.10*	tongue *41.3*
gut *41.2*	tooth *41.3*
incomplete digestive system *41.1*	villus (villi) *41.5*
kilocalorie *41.10*	vitamin *41.9*
leptin *41.10*	

Readings

Kassirer, J. 17 September 1998. *New England Journal of Medicine.* Report on unproven and harmful uses of dietary supplements.

Sizer, F., E. Whitney. 2003. *Nutrition: Concepts and Controversies.* Ninth edition. Belmont: California: Wadsworth.

Taubes, G. 7 July 2002. "What If It's All Been A Big Fat Lie?" *New York Times.*

42

THE INTERNAL ENVIRONMENT

Tale of the Desert Rat

Look closely at a fish or some other marine animal, and you'll find that its body cells are exquisitely adapted to life in a salty fluid. And yet, about 375 million years ago, some lineages of animals that evolved in the seas moved onto dry land. They were able to do so partly because they brought salty fluid along with them, as an *internal* environment for their cells. Even so, it was not a simple transition. The pioneers on land and their descendants faced intense sunlight, dry winds, more pronounced swings in temperature, water of dubious salt content, and sometimes no water at all.

How did the pioneers conserve or replace the water and salts lost as a result of everyday activities? How did they manage to stay comfortably warm when their surroundings became too cold or too hot? They must have done these things. Otherwise the composition, volume, and temperature of their internal environment would have spun out of control. In short, the question is this: *How did the land-dwelling descendants of marine animals maintain internal operating conditions and so prevent cellular anarchy?*

Observe any of their existing descendants and you get a sense of what some answers might be. Suppose you focus on a tiny mammal, a kangaroo rat living in an isolated desert of New Mexico (Figure 42.1). After a brief rainy season, the sun bakes the desert sand for months. The only obvious water is imported, sloshing about in the canteens of the occasional researcher and tourist. Yet with nary a sip of free water, a kangaroo rat routinely counters threats to its internal environment.

It waits out the heat of the day inside a burrow, then forages in the cool of night for dry seeds and maybe a succulent. It is not sluggish about this. It hops rapidly and far, searching for seeds and fleeing from coyotes or snakes. All that hopping requires ATP energy and water. Seeds, chockful of energy-rich carbohydrates, provide both. Metabolic reactions that release energy from carbohydrates and other organic compounds also yield water. Each day, such "metabolic water" makes up a whopping 90 percent of a kangaroo rat's total water intake. By comparison, metabolic water is only about 12 percent of the total water intake for your body.

	KANGAROO RAT	HUMAN
WATER GAIN (milliliters)		
by ingesting solids	6.0	850
by ingesting liquids	0.0	1,400
by metabolism	54.0	350
	60.0	2,600
WATER LOSS (milliliters)		
in urine	13.5	1,500
in feces	2.6	200
by evaporation	43.9	900
	60.0	2,600

Figure 42.1 Kangaroo rat, master of water conservation in a New Mexico desert. The chart will give you an idea of how kangaroo rats and humans gain and lose water. They do so in different ways. Even so, in both cases, *water losses balance out the gains* —just as they do in all animals. How this balancing happens, and why it absolutely must happen, is this chapter's initial focus.

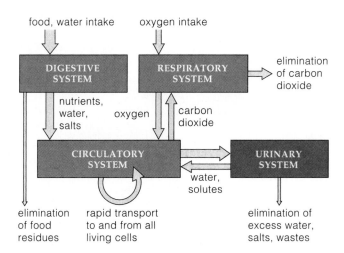

food, water intake   **oxygen intake**

DIGESTIVE SYSTEM   RESPIRATORY SYSTEM → elimination of carbon dioxide

nutrients, water, salts   oxygen   carbon dioxide

CIRCULATORY SYSTEM ⇄ URINARY SYSTEM

water, solutes

elimination of food residues   rapid transport to and from all living cells   elimination of excess water, salts, wastes

Figure 42.2 Links between the urinary system and other organ systems that contribute to homeostasis, or stability in favorable operating conditions in the animal body.

A kangaroo rat resting in a cool burrow conserves and recycles water. When it inhales, air is moistened and warmed. When it exhales, moisture condenses in its cooler nose, and some diffuses back into the body. Also, after a busy night of foraging, the kangaroo rat empties its cheek pouches of seeds. As seeds leave the mouth, they soak up water dripping from the nose. When the dripped-upon seeds are eaten, water is reclaimed.

A kangaroo rat can't lose water by perspiring; it has no sweat glands. It loses water when urinating, but its two specialized kidneys don't let it piddle away much. Kidneys filter the blood's water and solutes, including dissolved salts. They continually adjust *how much water* and *which solutes* return to the blood or depart as urine.

Overall, kangaroo rats and all other animals take in enough water and solutes to replace daily losses (Figure 42.1). How they do so will be our initial focus. Later, we will consider how mammals withstand hot, cold, and often unpredictable temperature changes on land.

As a starting point, think back on the fluids inside most animals. **Interstitial fluid** fills the spaces between cells and other tissue components. **Blood**, a fluid connective tissue, moves substances to and from all regions by way of a circulatory system. Together, interstitial fluid and blood are **extracellular fluid**. In many animals, a well-developed urinary system helps keep the volume and composition of extracellular fluid within tolerable ranges. As you will see, major organ systems interact with the urinary system in the performance of this homeostatic task (Figure 42.2).

Key Concepts

1. Animals are continually gaining and losing water and solutes, or dissolved substances. They continually produce metabolic wastes. Even with all of the inputs and outputs, the overall composition and volume of the body's extracellular fluid are kept within a range that its cells can tolerate.

2. In humans as in other vertebrates, a urinary system is crucial to balancing the intake and output of water and solutes. This system filters water and solutes from blood on an ongoing basis. It reclaims some amount of both and eliminates the rest. At different times, different amounts are reclaimed, depending on what is required to maintain extracellular fluid.

3. Kidneys are blood-filtering organs, and the urinary system of mammals has a pair of them. Packed inside each kidney are a great number of nephrons.

4. At its receiving end, each nephron cups around a tuft of blood capillaries and accepts water and solutes from it. The nephron returns most of the filtrate to the blood by giving it up to capillaries threading around the nephron's tubular parts.

5. Water and solutes not returned to the blood by way of the kidneys leave the body as a fluid called urine. During any given interval, control mechanisms make adjustments that affect whether urine is concentrated or dilute. Two hormones, ADH and aldosterone, have key roles in the adjustments.

6. The internal body temperature of animals depends on a balancing of heat produced through metabolism, heat absorbed from the environment, and heat lost to the environment.

7. The internal body temperature is maintained within a favorable range through controls over metabolic activity and adaptations in body form and behavior.

URINARY SYSTEM OF MAMMALS

The Challenge—Shifts in Extracellular Fluid

Different solid foods and fluids intermittently enter a mammal's gut. Afterward, variable amounts of absorbed water, nutrients, and other substances move into the blood, then into interstitial fluid and on into cells. Such events could easily shift the volume and composition of extracellular fluid beyond tolerable limits. But the body makes compensatory adjustments that balance out the gains and losses. Within a given time frame, it takes in as much water and solutes as it gives up.

WATER GAINS AND LOSSES To start, think about how humans and other mammals *gain* water, mainly by two processes:

> *Absorption from the gut*
> *Metabolism*

Considerable water is absorbed from solids and liquids inside the gut. As you know, water also forms as a normal by-product of many metabolic reactions. In land mammals, how much water enters the gut in the first place depends on a thirst mechanism. When they lose too much water, they seek streams, waterholes, and so on, then drink. Later in the chapter, we will look at the thirst mechanism.

Normally, a mammal *loses* water mainly by way of four physiological processes—especially the first process listed here:

> *Urinary excretion*
> *Evaporation from lungs and the skin*
> *Sweating, by some species*
> *Elimination, in feces*

Urinary excretion affords the most control over water loss. The process eliminates excess water and solutes as urine. This fluid forms in a urinary system such as the one in Figure 42.3. Water evaporates from respiratory surfaces, too. Some species lose it in sweat. A mammal in good health tends to lose very little water from the gut; most is absorbed, not eliminated in feces.

SOLUTE GAINS AND LOSSES Every mammalian body *gains* solutes mainly by the following four processes:

> *Absorption from the gut*
> *Secretion from cells*
> *Respiration*
> *Metabolism*

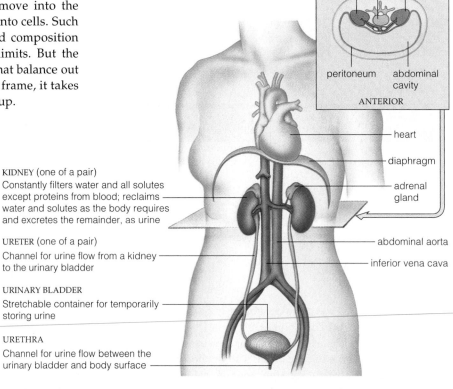

KIDNEY (one of a pair)
Constantly filters water and all solutes except proteins from blood; reclaims water and solutes as the body requires and excretes the remainder, as urine

URETER (one of a pair)
Channel for urine flow from a kidney to the urinary bladder

URINARY BLADDER
Stretchable container for temporarily storing urine

URETHRA
Channel for urine flow between the urinary bladder and body surface

Figure 42.3 Organs of the human urinary system and their functions. The two kidneys, two ureters, and urinary bladder are located *outside* the peritoneum, the membranous lining of the abdominal cavity. Compare Section 25.1.

For example, you absorb food, mineral ions, drugs, and food additives from the gut. Cell secretions and wastes, such as carbon dioxide, enter interstitial fluid and then blood. The respiratory system puts oxygen into blood; aerobically respiring cells put carbon dioxide into it.

Mammals typically *lose* solutes by three processes:

> *Urinary excretion*
> *Respiration*
> *Sweating, by some species*

The urine of mammals includes various wastes formed by the breakdown of organic compounds. For example, ammonia forms as amino groups are split from amino acids. Then **urea**, a major waste, forms in the liver when two ammonia molecules join with carbon dioxide. Also in urine are uric acid (from nucleic acid and amino acid breakdown), hemoglobin breakdown products, drugs, and food additives. Besides these solute losses, some mammals also lose mineral ions when they sweat. And all mammals lose carbon dioxide, the most abundant waste, by way of respiration.

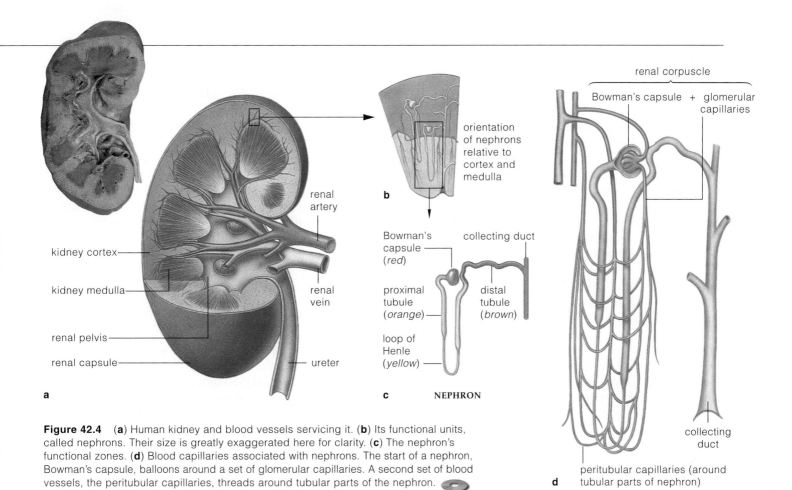

Figure 42.4 (**a**) Human kidney and blood vessels servicing it. (**b**) Its functional units, called nephrons. Their size is greatly exaggerated here for clarity. (**c**) The nephron's functional zones. (**d**) Blood capillaries associated with nephrons. The start of a nephron, Bowman's capsule, balloons around a set of glomerular capillaries. A second set of blood vessels, the peritubular capillaries, threads around tubular parts of the nephron.

Components of the Urinary System

Mammals routinely counter shifts in the composition and volume of extracellular fluid, mainly with a **urinary system**. This consists of two kidneys, two ureters, one urinary bladder, and one urethra. **Kidneys** are a pair of bean-shaped organs, about as large as a fist in an adult human (Figures 42.3 and 42.4). Each has an outer capsule of connective tissue. Blood capillaries penetrate its two inner regions, the cortex and medulla.

Kidneys filter water, mineral ions, organic wastes, and other substances from blood. Then they adjust the filtrate's composition and return all but about 1 percent to the blood. The small portion of unreclaimed water and solutes is urine. By definition, **urine** is a fluid that rids the body of water and solutes which exceed the amounts required to maintain extracellular fluid.

Urine flows from each kidney into a **ureter**, a tubular channel between it and the **urinary bladder**, a muscular sac. Urine is briefly stored here before flowing into the **urethra**, a muscular tube that opens at the body surface.

Flow from the bladder (urination) is a reflex action. When the bladder is full, the sphincter around its neck opens, smooth muscle in its balloonlike wall contracts, and so urine is forced out through the urethra. Skeletal muscle surrounds the urethra. Its contraction, which is under voluntary control, prevents urination.

Nephrons—Functional Units of Kidneys

Each human kidney has more than a million **nephrons**, slender tubules packed inside lobes that extend from the cortex of a kidney down into the medulla. *It is at the nephrons that water and solutes are filtered from blood and amounts to be reclaimed are adjusted.*

A nephron starts as a **Bowman's capsule**, where its wall cups around a set of *glomerular* capillaries. The cup and capillaries interact as one blood-filtering unit, the **renal corpuscle** (Figure 42.4*d*). Next comes a **proximal tubule** (tubular region closest to the capsule), a hairpin-shaped **loop of Henle**, and a **distal tubule** (most distant from the capsule). A **collecting duct**, the nephron's last region, is part of a duct system leading to the kidney's central cavity (renal pelvis), then into a ureter.

Blood does not give up all of its water and solutes. The unfiltered part flows into capillaries threading lacily around the nephron's tubular parts. In these *peritubular* capillaries, the blood reclaims water and solutes. Then it flows into veins and back to the general circulation.

A urinary system counters unwanted shifts in the volume and composition of extracellular fluid. In its paired kidneys, water and solutes are filtered from blood. The body reclaims most of this, but the excess leaves the kidneys as urine.

URINE FORMATION

Urine forms in nephrons by three processes: filtration, tubular reabsorption, and tubular secretion. All three processes depend on properties of cells of the nephron wall. Cells in different wall regions differ in membrane transport mechanisms and in their permeability.

Blood pressure generated by the heart's contractions drives **filtration**—the forcing of water and all solutes except proteins from the glomerular capillaries into the fluid-filled space of Bowman's capsule. The protein-free filtrate moves into the proximal tubule (Figure 42.5).

By **tubular reabsorption**, the peritubular capillaries reclaim most of the water and solutes that moved into the proximal tubule (Figure 42.6). Later, at the distal tubule, hormonal controls will help adjust the amounts conserved or excreted during a given interval.

Blood still inside peritubular capillaries holds excess ions (mainly H^+ and K^+), metabolites such as urea, and neurotransmitters, histamine, and maybe drugs or toxins. Such substances diffuse into the surrounding interstitial fluid. By **tubular secretion**, they now enter the nephron through proteins that function as channels and transporters across tubular portions of the nephron wall. Then they are excreted.

Factors Influencing Filtration

Each minute, about 1.5 liters (1–1/2 quarts) of blood flow through an adult's kidneys, and 120 milliliters of water and small solutes are filtered into nephrons. That's 180 liters of filtrate per day! A high filtration rate is possible mainly because the glomerular capillaries are 10 to 100 times more permeable to water and small solutes than other capillaries are. Also, blood pressure is high inside them. Why? Arterioles delivering blood to the nephron have a wider lumen and less flow resistance than those taking blood away. Blood dams up in the renal corpuscle and helps maintain the pressure in its capillaries.

Filtration rates also depend on flow volume to the kidneys. Neural, endocrine, and local controls maintain the flow even when blood pressure changes. Example: Dance until dawn and your nervous system commands an above-normal volume of blood to be diverted away from your kidneys and sent instead toward the energy-demanding cells in skeletal muscles and the heart. Such redistribution of blood volume involves coordinating the vasoconstriction and vasodilation of arterioles throughout the body.

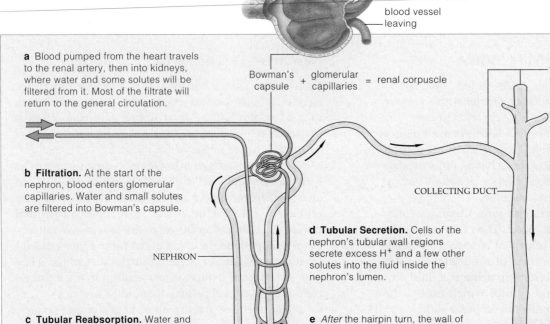

blood vessel entering

blood vessel leaving

a Blood pumped from the heart travels to the renal artery, then into kidneys, where water and some solutes will be filtered from it. Most of the filtrate will return to the general circulation.

Bowman's capsule + glomerular capillaries = renal corpuscle

f Hormonal action adjusts the urine concentration. *ADH* promotes *water* reabsorption, so the urine is concentrated. When controls inhibit ADH secretion, urine is dilute.

Aldosterone activates sodium pumps of cells in distal tubule and collecting duct walls. More sodium is reabsorbed, so urine has less of it. When controls inhibit aldosterone secretion, more sodium is excreted in urine.

b Filtration. At the start of the nephron, blood enters glomerular capillaries. Water and small solutes are filtered into Bowman's capsule.

COLLECTING DUCT

NEPHRON

d Tubular Secretion. Cells of the nephron's tubular wall regions secrete excess H^+ and a few other solutes into the fluid inside the nephron's lumen.

c Tubular Reabsorption. Water and many solutes cross the proximal tubule wall and enter interstitial fluid of the kidney cortex. Membrane transport proteins move most of the solutes across the wall. These materials then enter the peritubular capillaries.

e *After* the hairpin turn, the wall of the loop of Henle is impermeable to water. However, its cells actively pump chloride ions out of the loop; sodium passively follows. Pumping makes the interstitial fluid saltier. As a result, even more water is drawn out of collecting ducts that also run through the medulla.

g Excretion. What happens to water and solutes that were not reabsorbed or that were secreted into the tubule? They flow through a collecting duct to the renal pelvis, then are eliminated from the body by way of the urinary tract.

Figure 42.5 Urine formation.

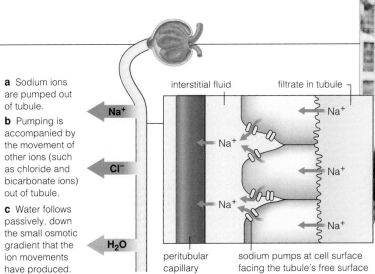

a Sodium ions are pumped out of tubule.

b Pumping is accompanied by the movement of other ions (such as chloride and bicarbonate ions) out of tubule.

c Water follows passively, down the small osmotic gradient that the ion movements have produced.

interstitial fluid

filtrate in tubule

Na+

Na+

Na+

Na+

Na+

Na+

Na+

Cl⁻

H₂O

peritubular capillary

sodium pumps at cell surface facing the tubule's free surface

Figure 42.6 Membrane pumps in the nephron wall. Sodium and water are reabsorbed here in response to, say, gulping too much water or ingesting too much salt.

Also, the walls of arterioles leading to nephrons are sensitive to pressure. These arterioles vasodilate when blood pressure falls and vasoconstrict when it rises. They help keep rates of filtration steady when pressure varies, as it does when you exercise and then rest.

Reabsorption of Water and Sodium

REABSORPTION MECHANISM Kidneys precisely adjust how much water and sodium ions the body excretes or saves. Drink too much or too little water, wolf down salty potato chips, or lose too much sodium in sweat, and responses are quick. As filtrate enters the proximal tubules, cells actively transport some sodium out of it. Other ions follow sodium into interstitial fluid. Water leaves the filtrate, too, by osmosis. The proximal tubule wall is so highly permeable that about two-thirds of the filtrate's water is reabsorbed here (Figure 42.5c,d).

Interstitial fluid is saltiest around the hairpin turn of the loop of Henle. Water moves out of the filtrate by osmosis *before* the turn. The fluid left behind gets saltier until it matches interstitial fluid. The loop wall *after* the turn is impermeable to water, but chloride is pumped out by active transport mechanisms. Interstitial fluid gets even saltier—so it attracts even more water out of the filtrate that is just entering the loop (Figure 42.5e).

The sodium removal helps dilute the fluid entering the distal tubule. Now adjustments can make the urine highly dilute, concentrated, or anywhere in between.

HORMONE-INDUCED ADJUSTMENTS When the volume of extracellular fluid drops, the hypothalamus calls for the release of antidiuretic hormone, or **ADH**, from the pituitary's posterior lobe (Section 36.3). ADH binds to receptors of distal tubules and collecting ducts, making their walls more permeable to water. More water gets reabsorbed so urine is more concentrated (Figure 42.5f). But less ADH is released when there is too much water. Then, the walls become less permeable, which means less water is reabsorbed and urine remains dilute.

Aldosterone promotes reabsorption of sodium. The extracellular fluid volume falls when too much sodium is lost. Sensory receptors in the heart and blood vessels detect the drop and stimulate renin-secreting gland cells in the wall of arterioles at glomeruli. The enzyme renin splits a plasma protein. A fragment is converted to the hormone **angiotensin II**. Among its many targets are aldosterone-secreting cells of the adrenal cortex, part of a gland on each kidney (Figure 42.3). The cells secrete more aldosterone, which stimulates cells of the distal tubules and collecting ducts to reabsorb sodium faster, so less sodium is excreted. Conversely, when there is too much sodium, aldosterone secretion declines. Less sodium is reabsorbed, so more is excreted.

THIRST BEHAVIOR A **thirst center** in the hypothalamus induces water-seeking behavior. Osmoreceptors nearby warn of decreases in blood volume and rising blood solute levels. The signals activate the thirst center *and* ADH-secreting cells in the brain. While water intake is being encouraged, urinary output is reduced.

Angiotensin II also stimulates thirst by direct effects on brain cells controlling ADH secretion, so it helps to adjust the sodium balance *and* the water balance. Thirst also is initiated when free nerve endings detect dryness in the mouth, an early sign of dehydration.

Filtration, reabsorption, and secretion in the kidneys help maintain the composition and volume of extracellular fluid.

Blood filtration rates depend on high hydrostatic pressure (generated by heart contractions) and on neural, endocrine, and local control over how much blood flows into kidneys.

The proximal tubule reabsorbs most of the filtered water and solutes. Some adjustments occur at the distal tubule: ADH promotes water conservation (concentrated urine). Aldosterone promotes sodium conservation (dilute urine).

Certain excess solutes diffuse out of peritubular capillaries and into interstitial fluid. Then, by tubular secretion, they move into the nephron for subsequent excretion.

When Kidneys Break Down

By this point in the chapter, you probably have figured out that good health depends on nephron function. Whether by illness or accident, when the nephrons of both kidneys become damaged and no longer perform their regulatory and excretory functions, we call this **renal failure**. Chronic renal failure is irreversible.

Renal failure may occur after infectious agents reach the kidneys through the bloodstream or the urethra. It may occur after someone ingests lead, arsenic, pesticides, or other toxins. Ongoing high doses of aspirin and some other drugs also can bring it about. Abnormal retention of metabolic wastes, such as the by-products of protein breakdown, results in *uremic toxicity*. Atherosclerosis, heart failure, hemorrhage, and shock diminish blood flow and disrupt filtration pressure in the kidneys.

In *glomerulonephritis*, antibody–antigen complexes get trapped in glomeruli. Unless phagocytes remove them, the complexes keep activating complement and other agents that cause widespread inflammation and tissue damage. Fortunately, this type of disease is rare.

Kidney stones form when uric acid, calcium salts, and other wastes settle out of urine and collect in the renal pelvis. These hard deposits are usually passed in urine, but they can become lodged in the ureter or the urethra. When they disrupt urine flow, they must be medically or surgically removed to prevent renal failure.

About 13 million people in the United States alone suffer from renal failure. A *kidney dialysis machine* often restores the proper solute balances. Like the kidneys, it helps maintain extracellular fluid by selectively adjusting the solutes in blood. "Dialysis" refers to an exchange of substances across an artificial membrane interposed between two solutions that differ in composition.

In *hemodialysis*, the machine is connected to an artery or a vein. Then the patient's blood is pumped on through tubes that have been submerged in a warm saline bath. The bath's mix of salts, glucose, and other substances sets up the correct concentration gradients with blood. Then the blood flows back into the body. For kidney dialysis to have optimum effect, it must be performed three times a week. Each time, the procedure takes about four hours, because the patient's blood must circulate repeatedly to improve solute concentrations in her or his body.

In *peritoneal dialysis*, fluid of suitable composition is introduced into the patient's abdominal cavity, left in there for a specific length of time, then drained out. In this case, the cavity's lining itself, the peritoneum, acts as the membrane for dialysis.

Bear in mind, kidney dialysis is used as a temporary measure in reversible kidney disorders. In chronic cases, the procedure must be used for the rest of the patient's life or until a transplant operation provides her or him with a functional kidney. With treatment and controlled diets, many patients resume fairly normal activity.

THE BODY'S ACID–BASE BALANCE

In addition to maintaining the volume and composition of extracellular fluid, kidneys help keep it from getting too acidic or too basic (alkaline). The overall **acid–base balance** of that fluid is an outcome of controls over its concentrations of H^+ and other dissolved ions. *Metabolic acidosis* hints at the importance of the balancing acts. This condition results when the kidneys cannot secrete enough H^+ to keep pace with all the H^+ that is forming during metabolism. It is life-threatening.

Buffer systems, respiration, and urinary excretion all work in concert to provide control over the acid–base balance. A buffer system, remember, consists of weak acids or bases that can reversibly bind and release ions, thus helping to minimize shifts in pH (Section 2.6).

Normally, the extracellular pH of the human body should be maintained between 7.35 and 7.45. As you know, acids lower the pH and bases raise it. A variety of acidic and basic substances enter blood after being absorbed from the gut and as by-products of normal metabolism. Cell activities typically produce an excess of acids. These dissociate into H^+ and other fragments, and pH declines. The effect is minimized when excess hydrogen ions react with buffer molecules. An example is the *bicarbonate–carbon dioxide* buffer system:

$$H^+ + HCO_3^- \rightleftharpoons H_2CO_3 \rightleftharpoons CO_2 + H_2O$$

BICARBONATE   CARBONIC ACID

In this case, the buffer system neutralizes excess H^+, and the carbon dioxide that forms during the reactions is exhaled from the lungs. Like other buffer systems in the body, however, this one has temporary effects only; it does not *eliminate* excess H^+. Only the urinary system can do so and thereby restore the buffers.

The same reactions proceed in reverse in cells of the nephron's tubular walls. HCO_3^- formed by the reverse reactions moves into interstitial fluid, then peritubular capillaries. Afterward, it enters the general circulation and buffers excess acid. The H^+ formed in the cells is secreted into the nephron and may join with HCO_3^-. The CO_2 that formed may be returned to blood, then exhaled. H^+ also may combine with phosphate ions or ammonia (NH_3), then leave the body in urine.

The kidneys work in concert with buffering systems that neutralize acids and with the respiratory system to help keep the extracellular fluid from becoming too acidic or too basic (alkaline).

A bicarbonate–carbon dioxide buffer system temporarily neutralizes excess hydrogen ions. The urinary system alone eliminates the excess ions and thus restores these buffers.

The bicarbonate–carbon dioxide buffer system is one of the key mechanisms that help maintain the acid–base balance.

ON FISH, FROGS, AND KANGAROO RATS

Now that you have a general sense of how your body maintains water and solute levels, consider what goes on in some other vertebrates, including that kangaroo rat hopping about at the start of the chapter.

Bony fishes and amphibians of freshwater habitats gain water and lose solutes (Figure 42.7a). Water moves into their internal environment by osmosis; they don't gain water by drinking it. Water diffuses across thin gill membranes in fishes and across the skin in adult amphibians. Excess water leaves as dilute urine, formed inside a pair of kidneys. In both groups of vertebrates, solute losses are balanced when more solutes come in with food and when gill or skin cells pump in sodium.

Body fluids of herring, snapper, and other marine fishes are about three times less salty than seawater. These fishes lose water by osmosis, then replace it by drinking more. They excrete ingested solutes against concentration gradients (Figure 42.7b). Fish kidneys do not have loops of Henle, so urine cannot ever become saltier than body fluids. Cells in the fish gills actively pump out most of the excess solutes present in blood.

Figure 42.7c describes the water–solute balancing act in a salmon. This type of fish spends part of its life cycle in fresh water and another part in seawater.

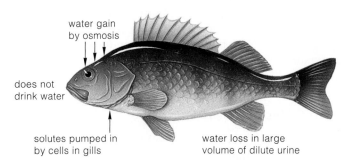

a Freshwater bony fish (body fluids far saltier than surroundings)

water gain by osmosis
does not drink water
solutes pumped in by cells in gills
water loss in large volume of dilute urine

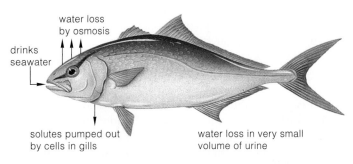

b Marine bony fish (body fluids less salty than surroundings)

water loss by osmosis
drinks seawater
solutes pumped out by cells in gills
water loss in very small volume of urine

Figure 42.7 (**a**,**b**) Water–solute balance in fishes.

(**c**) Water–solute balance by salmon. This type of fish lives in saltwater *and* fresh water. Salmon hatch in streams and later move downstream to the open sea, where they feed and mature. They return to home streams to spawn.

For most salmon, salt tolerance is an outcome of changing hormone levels that appear to be triggered by increasing daylength in spring. Prolactin, a pituitary hormone, plays a key role in sodium retention in freshwater habitats. We know this because a freshwater fish that has its pituitary gland removed will die from sodium loss—but that fish will live if prolactin is administered to it.

A salmon's salt tolerance depends on cortisol. Secretion of this steroid hormone from the adrenal cortex correlates with rises in sodium excretion, sodium–potassium pumping by cells in the salmon's gills, and absorption of ions and water from the gut. In young salmon, cortisol secretion increases prior to the seaward movement—and so does salt tolerance.

salmon avoiding grizzly while maintaining water–solute balance

c

And what about that kangaroo rat! Proportionally, the loops of Henle of its nephrons are amazingly long, compared to yours. This means a great deal of sodium is pumped out from the nephron. Therefore, the solute concentration in the interstitial fluid around the loops becomes very high. The osmotic gradient between fluid in the loops of Henle and urine is *so* steep that nearly all the water that does reach the equally long collecting ducts gets reabsorbed. The kangaroo rat gives up only a tiny volume of urine. And that urine is three to five times more concentrated than the most concentrated urine of humans.

The urinary systems of vertebrates differ in their details, such as the length of the nephron's loop of Henle. They are adapted to balance the body's gains in water and solutes with its losses of water and solutes in particular habitats.

HOW ARE CORE TEMPERATURES MAINTAINED?

We turn now to another major aspect of the internal environment, its temperature. Many physiological and behavioral responses to change help maintain a **core temperature** within the tolerance range for the body's enzymes. *Core* means the body's internal temperature, as opposed to temperatures of tissues near its surface.

Heat Gains and Heat Losses

The core temperature rises when heat that is absorbed or generated by metabolic reactions accumulates. But a warm body loses heat to cool surroundings. The core temperature stabilizes when the rate at which heat is lost balances the rate of heat gain and heat production. The heat content of any complex animal depends on a balancing between such gains and losses:

$$\text{CHANGE IN BODY HEAT} = \text{HEAT PRODUCED} + \text{HEAT GAINED} - \text{HEAT LOST}$$

Heat is gained and lost through exchanges at skin and other surfaces. Four processes—radiation, conduction, convection, and evaporation—drive the exchanges.

With **radiation**, the surface of a warm body emits heat in the form of radiant energy. For example, many of the wavelengths that the sun continually emits are in the infrared range; they are forms of heat energy. Also, metabolically active animals generate and emit heat.

With **conduction**, there is a transfer of heat between an animal and another object of differing temperature that is in direct contact with it. An animal loses heat by resting on objects cooler than it is. And it gains heat when it contacts warmer objects, as in Figure 42.8.

With **convection**, moving air or water transfers heat. Conduction plays a part in this process, for heat moves down a thermal gradient between the body and air or water next to it. Mass transfer also plays a part, with currents carrying heat away from or toward the body.

Table 42.1	*Body Temperature Versus Environmental Temperature*
General range of internal temperatures favorable for metabolism:	0°C to 40°C (32°F to 104°F)
Range of air temperatures above land surfaces:	−70°C to +85°C (−94°F to +185°F)
Range of surface temperatures of open ocean:	−2°C to +30°C (+28.4°F to +86°F)

When it is heated, air becomes less dense and moves away from the body. Even when there is no breeze, the body loses heat, for its movements create convection.

With **evaporation**, a liquid converts to gaseous form and heat is lost in the process. As described earlier, in Section 2.5, the liquid's heat content provides energy for the conversion. Evaporation from the body surface has a cooling effect, because the water molecules that are escaping carry away some energy with them.

Evaporation rates depend on humidity and on the rate of air movement. If air next to the body is already saturated with water—that is, when the local relative humidity is 100 percent—water won't evaporate. If air next to the body is hot and dry, evaporation may be the only means of countering the metabolic production of heat and the heat gains from radiation and convection.

Ectotherms, Endotherms, and In-Betweens

Animals can adjust the amount of heat lost or gained through changes in behavior and physiology, but some are better equipped than others to do so. Amphibians, snakes, lizards, and other reptiles are among animals with low metabolic rates and poor insulation (Figure 42.8). They absorb and gain heat fast, especially the small species. Core temperatures are protected mainly through heat gains from the environment, not through metabolic activities. Hence we classify these animals as **ectotherms**, which means "heat from outside."

When outside temperatures change, an ectotherm must adjust its behavior. This is *behavioral* temperature regulation. For example, that iguana shown at the start of this unit basks on warm rocks, thus gaining heat by conduction. It keeps reorienting its body to expose the most surface area to the sun's infrared radiation. It loses

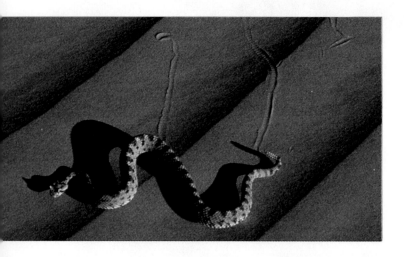

Figure 42.8 A sidewinder making J-shaped tracks across hot desert sand at dusk. As near as you can tell, how might this ectotherm be gaining and losing heat?

Figure 42.9 Short-term and long-term adaptations to environmental temperatures. (**a**) Far away from their home-range burrows, meerkats absorb radiant heat energy on a cold night in a zoo in Germany. (**b**) Pine grosbeak with fluffed feathers providing insulation against winter cold. Meerkats and this bird are both endotherms. (**c**) Dromedary camels are heterotherms; they let their core temperatures rise by as much as 14°C during the hottest hours of the day.

heat after sunset. Before its metabolic rates decrease, it crawls into a crevice or under a rock, where heat loss is not as great and it is not as vulnerable to predators.

Most birds and mammals are **endotherms** ("heat from within"). Their high metabolic rates keep them active under a wide temperature range. (Compared to a foraging lizard of the same weight, a foraging mouse uses up to thirty times *more* energy.) Metabolism, along with morphological and behavioral controls, contribute to the core temperature. Fluffed-up feathers, thick fur, fat layers, even clothing limit heat loss (Figure 42.9*a,b*).

Mammals in cold habitats often have more massive bodies than close relatives in warmer ones. For instance, the snowshoe hare of Canada (Section 47.4) has a more compact body, far shorter legs, and far shorter ears than jackrabbits of the American Southwest. Its body has a greater volume of cells, which generate metabolic heat, relative to the surface area available for heat loss. The heat dissipated from its legs and ears does not begin to match the losses from a jackrabbit's lengthy extremities.

Some birds and mammals are **heterotherms**. They maintain a fairly constant core temperature some of the time, and allow it to shift at other times. For example, a dromedary camel's temperature climbs during the hottest time of day in the desert, and then declines at night (Figure 42.9*c*). A hypothalamic mechanism raises the camel's internal thermostat, so to speak, with no adverse effects. Another example: Tiny hummingbirds have high metabolic rates. They locate and sip nectar only in the day. At night, they may shut down almost entirely. Their metabolic rates plummet, and they may

become almost as cool as their surroundings. This way, they conserve precious energy.

Camels also make behavioral responses to outside temperatures, as when they rest on surfaces that were cooled during the night. They do not pant, and they do not sweat until the core temperature rises considerably. Also, long, thick coat hairs on the camel's back send absorbed radiant heat back to the environment by way of convection and reradiation.

In general, ectotherms have an advantage in warm, humid regions, especially in the tropics. They need not spend much energy to maintain core temperatures, and more energy can be devoted to reproduction and other tasks. In terms of numbers and diversity, reptiles far exceed mammals in tropical regions. Endotherms have advantages in moderate to cold regions. For example, with their high metabolic rates, snowshoe hares, arctic foxes, and some other endotherms occupy polar habitats —where you would never find a lizard.

The internal, core temperature of an animal's body is being maintained when heat gains and heat losses are in balance.

Metabolism generates heat inside the body. Radiation, conduction, and convection can move heat down thermal gradients between the animal body and its surroundings. Evaporative heat loss carries heat away from the body.

Animals have morphological, physiological, and behavioral adaptations to environmental temperatures. Ectotherms gain heat mainly from the outside; endotherms generate most from within. Heterotherms do both at different times.

TEMPERATURE REGULATION IN MAMMALS

Control centers that maintain the core temperature of the mammalian body reside in the hypothalamus. They receive ongoing input from peripheral thermoreceptors in skin and from thermoreceptors deeper in the body (Figure 42.10). When the temperature deviates from a set point, the centers integrate complex responses that involve skeletal muscles, smooth muscle in arterioles that service skin, and, often, sweat glands. Negative feedback loops back to the hypothalamus shut off the responses when a suitable temperature is reinstated. Here you may wish to review Section 28.3.

Responses to Heat Stress

Table 42.2 lists the primary responses that mammals make to cold stress. When a mammal becomes too hot, temperature control centers in the hypothalamus call for **peripheral vasodilation**. By this response, blood vessels in the skin dilate. More blood flows to the skin, which can dissipate excess heat (Figure 42.10).

Evaporative heat loss is a response to heat stress that takes place at moist respiratory surfaces and across skin. Animals that sweat lose some water this way. For instance, humans and some other mammals have sweat glands that release water and solutes through pores at the skin's surface. An average-sized adult human has 2–1/2 million or more sweat glands. For every liter of sweat produced, 600 kilocalories of heat energy depart through evaporative heat loss.

Bear in mind, sweat that is dripping from skin does not dissipate heat. Outside temperatures must be high enough to cause water from sweat to *evaporate*. On hot, humid days, evaporation rates cannot match the rate of sweat secretion; air's high water content slows it down. During strenuous exercise, sweating may help balance heat production in skeletal muscle. Extreme sweating, as in a marathon race, forces the body to lose a vital salt—sodium chloride—and water. Large solute losses can so disrupt the extracellular fluid's composition and volume that runners may collapse and faint.

What about mammals that sweat little or not at all? Some make behavioral responses, such as licking fur or panting. "Panting" is shallow, rapid breathing that increases evaporative water loss from the respiratory tract. The body cools when water evaporates from the nasal cavity, mouth, and tongue (Section 28.3).

Sometimes peripheral blood flow and evaporative heat loss cannot counter heat stress, and *hyperthermia* occurs. This is a condition in which the core temperature rises above normal. Even a few degrees above normal can be serious for humans and other endotherms.

Fever

A fever, recall, is part of an inflammatory response to tissue injury (Section 39.3). The hypothalamus actually resets the body's thermostat, which dictates what the core temperature is supposed to be. Although mechanisms that increase metabolic heat production and cut heat loss are working, they now work at maintaining a higher temperature. The individual feels chilled at a fever's onset. When the fever "breaks," peripheral vasodilation increases and the individual sweats. As the body attempts to restore the normal core temperature, the person feels warm.

Aspirin and other anti-inflammatory drugs lower the fever, but doing so might

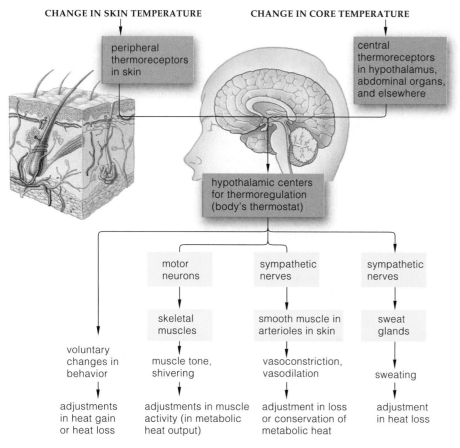

Figure 42.10 Key physiological and behavioral adjustments of a typical mammal to changes in outside temperature. This example shows the main pathways for thermoregulation in humans.

Table 42.2	*Mammalian Responses to Changes in Core Temperature*	
Stimulus	Main Responses	Outcome
Heat stress	Widespread vasodilation in skin; behavioral adjustments; in some species, sweating, panting	Dissipation of heat from body
	Decreased muscle action	Heat production decreases
Cold stress	Widespread vasoconstriction in skin; behavioral adjustments (e.g., minimizing surface parts exposed)	Conservation of body heat
	Increased muscle action; shivering; nonshivering heat production	Heat production increases

Figure 42.11 Case study: Polar bears (*Ursus maritimus*, "bear of the sea"). Unlike other bears, which hibernate, a polar bear actively moves about all through the severe Arctic winter. The outermost layer of its coat consists of coarse, hollow guard hairs that shed water when the bear shakes itself after swimming. This behavior helps keep the body from becoming chilled. The thick, soft underhair is a heat-trapping layer. Beneath the skin is a layer of brown adipose tissue, often about 11.5 centimeters (4–1/2 inches) thick, that insulates and helps generate metabolic heat.

Hypothalamic control centers adjust the core temperature of polar bears and other warmblooded mammals. The hypothalamic centers will respond to cooling of the skin and of the blood, even by only a few tenths of one degree. Surgically remove these centers and the capacity to regulate core temperature is lost.

actually prolong the healing time. Only when fevers approach dangerous levels should drugs be an option, and then only under medical supervision.

Responses to Cold Stress

Table 42.2 also lists responses that mammals make to cold stress. They are called peripheral vasoconstriction, the pilomotor response, shivering, and nonshivering heat production. Birds make the same responses, but they also have a higher set point for body temperature.

Suppose peripheral thermoreceptors detect a decline in outside temperature. They notify the hypothalamus, which commands smooth muscle in arterioles in the skin to contract. The response is **peripheral vasoconstriction**. Diameters of these arterioles constrict, which decreases the blood's delivery of heat to the body surfaces. When your fingers or toes are chilled, all but 1 percent of the blood that otherwise would flow to the skin is diverted to other regions of the body.

Also, muscle contractions make hairs (or feathers) "stand up." This **pilomotor response** creates a layer of still air next to the skin and reduces convective and radiative heat loss. Behavioral changes also minimize exposed surface areas and reduce heat loss, as when you hold both arms tightly to your body or when polar bear cubs cuddle against their mother (Figure 42.11).

A **shivering response** may occur during prolonged cold spells. Rhythmic tremors begin as skeletal muscles contract, ten to twenty times a second, when activated by signals from the hypothalamus. Shivering increases heat production by several times. However, it has a high energy cost and is not effective for long.

Prolonged or severe cold exposure also leads to a hormonal response, an increase in thyroid activity, that raises the rate of metabolism. This **nonshivering heat production** most typically involves the *brown* adipose tissue. This connective tissue is richly endowed with blood vessels and mitochondria. The controlled flow of hydrogen ions in the mitochondria is used to generate heat, not ATP.

Polar bears and other animals that are acclimatized to cold environments or hibernate have brown adipose tissue (Figure 42.11). So do human infants. Adults have little unless they are adapted to very cold conditions. For example, the tissue is present in ama, the Japanese and Korean commercial divers who spend up to six hours per day in very cold water, collecting shellfish.

Failure to defend against cold leads to *hypothermia*, a condition in which the core temperature falls below normal. In humans, a decrease by even a few degrees adversely affects brain function and leads to confusion; cooling leads to coma and death. Many mammals can recover from profound hypothermia. However, frozen cells may die unless tissues thaw under close medical supervision. The destruction of tissues as a result of localized freezing is known as *frostbite*.

Mammals counter cold stress by vasoconstriction in skin, behavioral adjustments, increased muscle activity, and shivering and nonshivering heat production.

Mammals counter heat stress by widespread peripheral vasodilation in skin and by evaporative heat loss.

SUMMARY   *Gold* indicates text section

Control of Extracellular Fluid

1. The extracellular fluid in the animal body consists of certain types and amounts of substances dissolved in water. It is distributed as interstitial fluid (in tissue spaces) and blood. Its volume and composition are maintained when an animal's daily intake and output of water as well as solutes are in balance. The following processes maintain the balance in mammals: *CI, 42.1*

   a. Water is gained by absorption from the gut and by metabolism. It is lost by urinary excretion, evaporation from lungs and skin, sweating, and elimination of feces.

   b. Solutes are gained by absorption from the gut and by secretion, respiration, and metabolism. Solutes are lost by excretion, respiration, and sweating.

   c. Losses of water and solutes are controlled mainly by adjusting the volume and composition of urine.

2. A mammal's urinary system has a pair of kidneys, a pair of ureters, a urinary bladder, and a urethra. *42.1*

3. Kidneys have many nephrons that filter blood and form urine. A nephron interacts with two sets of blood capillaries: glomerular and peritubular. *42.1*

   a. The start of a nephron is cup-shaped (Bowman's capsule). It continues as three tubular regions (proximal tubule, loop of Henle, and distal tubule, which empties into a collecting duct).

   b. Together, Bowman's capsule and the set of highly permeable glomerular capillaries within it are a blood-filtering unit: a renal corpuscle. Blood pressure forces water and small solutes from capillaries, into the fluid-filled cup. Most of the filtrate gets reabsorbed at the nephron's tubular regions and is returned to the blood. A portion is excreted as urine.

4. Urine forms in the nephron by three processes: *42.2*

   a. Filtration. Blood filtered at the glomerulus puts water and small solutes into the nephron.

   b. Reabsorption. Water and solutes to be conserved leave the nephron's tubular parts and enter capillaries that thread around them. A small volume of water and solutes remains in the nephron.

   c. Secretion. A few substances can leave peritubular capillaries and enter the nephron, for disposal in urine.

5. Two hormones, acting on cells of the distal tubule and collecting duct walls, adjust urine concentrations. ADH conserves water by enhancing reabsorption across the wall. In its absence, more water is excreted (urine is dilute). Aldosterone enhances sodium reabsorption. In its absence, sodium is excreted. Angiotensin II stimulates aldosterone secretion (sodium conservation) and thirst (water conservation by calling for more ADH). *42.2*

6. The urinary system acts in concert with buffers and with the respiratory system to maintain the acid–base balance of extracellular fluid. *42.4*

Control of Body Temperature

1. Maintaining an animal's core (internal) temperature depends on balancing metabolically produced heat and heat absorbed from and lost to the environment. *42.6*

2. Animals exchange heat with their environment by four processes: *42.6*

   a. Radiation. Emission from the body of infrared and other wavelengths. Radiant energy can be absorbed at the body surface, then converted to heat energy.

   b. Conduction. Direct transfer of heat energy from one object to another object in contact with it.

   c. Convection. Heat transfer by air or water currents; involves conduction and mass transfer of heat-bearing currents away from or toward the animal body.

   d. Evaporation. Conversion of liquid to a gas, driven by energy inherent in the heat content of the liquid. Some animals lose heat by evaporative water loss.

3. Core temperatures depend on metabolic rates and on anatomy, behavior, and physiology. *42.6*

   a. For ectotherms, core temperature depends more on heat exchange with the environment than on heat generated by metabolism.

   b. For endotherms, core temperature depends more on high metabolic rates and precise controls over heat produced and heat lost.

   c. For heterotherms, the core temperature fluctuates with environmental temperatures some of the time, and controls over heat balance operate at other times.

Review Questions

1. State the function of the urinary system in terms of gains and losses for the internal environment. State the components of the mammalian urinary system and their functions. *42.1*

2. Label the component parts of this kidney and nephron. *42.1*

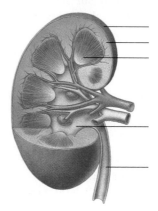 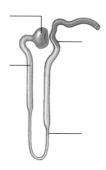

3. Define filtration, tubular reabsorption, and secretion. How does urine formation help maintain an internal environment? *42.2*

4. Which hormone or hormones promote (a) water conservation, (b) sodium conservation, and (c) thirst behavior? *42.2*

5. Name and define the physical processes by which animals gain and lose heat. What are the main physiological responses to cold stress and to heat stress in mammals? *42.6, 42.7*

Physiological Responses to Increases in Cold Stress	
36°–34°C (about 95°F)	Shivering response; rise in respiration, metabolic heat output. Peripheral vasoconstriction, more blood deeper in body. Dizziness, nausea set in.
33°–32°C (about 91°F)	Shivering response stops. Metabolic heat output drops.
31°–30°C (about 86°F)	Capacity for voluntary motion is lost. Eye and tendon reflexes inhibited. Consciousness is lost. Cardiac muscle action becomes irregular.
26°–24°C (about 77°F).	Ventricular fibrillation sets in (Section 38.9). Death follows.

Figure 42.12 Sinking of the *Titanic*, based on eyewitness accounts.

Self-Quiz   ANSWERS IN APPENDIX III

1. In mammals, water intake depends on _____ .
   a. absorption from gut   c. a thirst mechanism
   b. metabolism   d. all of the above

2. In mammals, water is lost by way of the _____ .
   a. skin   d. urinary system
   b. respiratory system   e. c and d
   c. digestive system   f. a through d

3. Water and small solutes enter nephrons during _____ .
   a. filtration   c. tubular secretion
   b. tubular reabsorption   d. both a and c

4. Kidneys send water and small solutes to blood by _____ .
   a. filtration   c. tubular secretion
   b. tubular reabsorption   d. both a and b

5. Certain excess solutes move out of peritubular capillaries that thread around the tubular parts of the nephron. These solutes are moved into the nephron during _____ .
   a. filtration   c. tubular secretion
   b. tubular reabsorption   d. both a and c

6. A nephron's reabsorption mechanism involves _____ .
   a. osmosis across nephron wall
   b. active transport of sodium across nephron wall
   c. a steep solute concentration gradient
   d. all of the above

7. _____ promotes water conservation.
   a. ADH   c. Low extracellular fluid volume
   b. Aldosterone   d. both a and c

8. _____ enhances sodium reabsorption.
   a. ADH   c. Low extracellular fluid volume
   b. Aldosterone   d. both b and c

9. Match the term with the most suitable description.
   ____ renal corpuscle   a. surrounded by saltiest fluid
   ____ urine adjustment   b. extra-long loops of Henle
   ____ loop of Henle   c. involves buffer systems
   ____ acid–base balance   d. blood-filtering unit
   ____ kangaroo rat   e. task of ADH, aldosterone

10. Match the term with the most suitable description.
    ____ ectotherm   a. heat transfer by air or water currents
    ____ endotherm   b. core temperature can fluctuate with environment or can be controlled
    ____ evaporation
    ____ heterotherm   c. emission of radiant energy
    ____ radiation   d. direct heat transfer between an object and another object in contact with it
    ____ conduction
    ____ convection   e. metabolism dictates core temperature
    f. environment dictates core temperature
    g. conversion of liquid to gas

Critical Thinking

1. Fatty tissue holds kidneys in place. Rarely, extremely rapid weight loss may cause the tissue to shrink and kidneys to slip from their normal position. If slippage puts a kink in one or both ureters and blocks urine flow, what may happen to the kidneys?

2. Drink one quart of water in one hour. What changes might you expect in your kidney function and in urine composition?

3. In 1912, the ocean liner *Titanic* left Europe on her maiden voyage to America—and a chunk of the leading edge of a glacier in Greenland broke away and floated out to sea. Late at night, off the Newfoundland coast, the iceberg and the *Titanic* made an ill-fated rendezvous. The *Titanic* was said to be unsinkable. Survival drills were neglected. There weren't enough lifeboats to hold even half the 2,200 passengers. The *Titanic* sank in about 2-1/2 hours. Rescue ships were on the scene in less than two hours, but 1,517 bodies were recovered from a calm sea. All of the dead had on life jackets; none had drowned. Name their probable cause of death, which involved the changes shown in Figure 42.12.

4. When iguanas have an infection, they rest for a prolonged period in the sun. Propose a hypothesis to explain why.

5. Out on a first date, Jon takes Geraldine's hand in a darkened theater. "*Aha!*" he thinks. "*Cold hands, warm heart!*" What does this tell us about the regulation of core temperature, let alone Jon?

Selected Key Terms

Salt–Water Balance
acid–base balance 42.4
ADH 42.2
aldosterone 42.2
angiotensin II 42.2
blood *CI*
Bowman's capsule 42.1
collecting duct 42.1
distal tubule 42.1
extracellular fluid *CI*
filtration 42.2
interstitial fluid *CI*
kidney 42.1
loop of Henle 42.1
nephron 42.1
proximal tubule 42.1

renal corpuscle 42.1
renal failure 42.3
thirst center 42.2
tubular reabsorption 42.2
tubular secretion 42.2
urea 42.1
ureter 42.1
urethra 42.1
urinary bladder 42.1
urinary excretion 42.1
urinary system 42.1
urine 42.1

Body Temperature
conduction 42.6
convection 42.6

core temperature 42.6
ectotherm 42.6
endotherm 42.6
evaporation 42.6
evaporative heat loss 42.7
heterotherm 42.6
nonshivering heat production 42.7
peripheral vasoconstriction 42.7
peripheral vasodilation 42.7
pilomotor response 42.7
radiation 42.6
shivering response 42.7

Readings

Sherwood, L. 2001. *Human Physiology*. Fourth edition. Monterey, California: Brooks/Cole.

On-Line readings at Student Guide for InfoTrac: www.brookscole.com/biology

PRINCIPLES OF REPRODUCTION AND DEVELOPMENT

From Frog to Frog and Other Mysteries

With a quavering, low-pitched call that only a female of its kind could find seductive, a male frog proclaims the onset of warm spring rains, of ponds, of sex in the night. By August the summer sun will have parched the earth, and his pond dominion will be gone. But tonight is the hour of the frog!

Through the dark, a hormone-primed female moves toward the vocal male. They meet; they dally in the behaviorally prescribed ways of their species. He clamps his forelegs above her swollen abdomen and gives her a prolonged squeeze (Figure 43.1a). Out into the water streams a ribbon of hundreds of eggs, which the male blankets with a milky cloud of sperm. Soon afterward, fertilized eggs—**zygotes**—are suspended in the water.

For the leopard frog, *Rana pipiens*, a drama begins to unfold that has been reenacted each spring, with only minor variations, for many millions of years. Within a few hours after fertilization, each zygote divides into two cells, the two divide into four, then the four into eight. In less than twenty hours after fertilization, the mitotic cell divisions have produced a ball of cells no larger than the zygote. It is an early embryonic stage, of a type known as a blastula.

The cells continue to divide, but now they start to interact by way of their surface structures and chemical secretions. At prescribed times, many change shape and migrate to prescribed positions. They all inherited the same genetic information from the zygote—yet now they start to differ in appearance and function!

Through their associations, the cells form layers of embryonic tissues and then embryonic organs. A pair of tissue regions at the embryo's surface interact with the tissues beneath them. Together they give rise to a pair of eyes. Within the embryo a heart is forming, and soon it starts an incessant, rhythmic beating. In less than a week, events have transformed the frog embryo into a swimming, algae-eating larva called a tadpole.

Several months pass. Legs form; the tail shortens and disappears. The mouth develops jaws that snap shut on insects and worms. Eventually the transformations lead to an adult frog. With luck the frog will avoid predators, disease, and other threats in the months ahead. In time

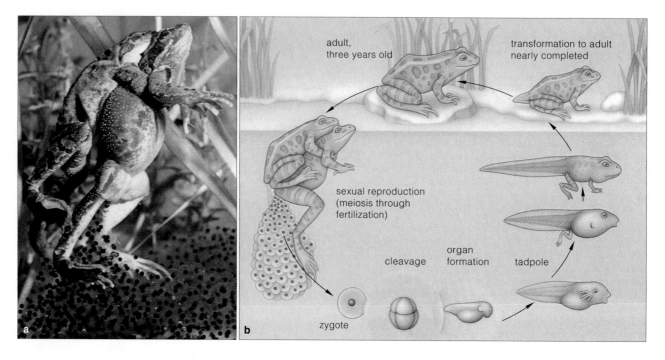

Figure 43.1 Reproduction and development of *Rana pipiens*, the leopard frog. (**a,b**) We zoom in on the life cycle as a male clasps a female in a reproductive behavior called amplexus. The female releases eggs into the water. The male releases sperm over the eggs. A zygote forms when an egg nucleus and a sperm nucleus fuse at fertilization. (**c**) Frog embryos suspended in the water. (**d**) A tadpole. (**e**) A transitional form between a tadpole and the young adult frog (**f**).

it may even call out quaveringly across a moonlit pond, and the life cycle will turn again.

Many years ago you, too, started a developmental journey when a zygote carved itself up. Three weeks into the journey, your embryonic body had the stamp of "vertebrate" on it. A mere five weeks after that, it was a recognizable human in the making!

With this chapter we turn to one of life's greatest dramas—the development of offspring in the image of sexually reproducing parents. The guiding question is this: *How does a single-celled zygote of a frog, human, or any other complex animal become transformed into all the specialized cells and structures of the adult form?* Answers will start to emerge through this chapter's survey of basic principles, then through a case study in the next chapter, on human reproduction and development.

DEVELOPING EMBRYO

Key Concepts

1. Sexual reproduction dominates the life cycle of nearly all animals, but separation into sexes has biological costs. Specialized reproductive structures must be constructed and maintained. Hormonal control mechanisms and complex forms of behavior must be attuned to the environment and to potential mates and rivals.

2. Separation into sexes affords a selective advantage. Offspring show variation in traits, which improves odds that at least some will survive and reproduce despite unexpected challenges from the environment. This reproductive advantage offsets the biological cost of the separation.

3. The life cycles of many animals proceed through six stages of embryonic development—gamete formation, then fertilization, cleavage, gastrulation, formation of organs, and growth and tissue specialization.

4. Each stage of development builds on tissues and structures that formed in the stage that preceded it.

5. In a developing embryo, the fate of each type of cell depends partly on cleavage, which distributes different regions of the fertilized egg's cytoplasm to daughter cells. It depends on interactions among cells of the embryo. These activities are the foundation for cell differentiation and morphogenesis.

6. In cell differentiation, each cell type selectively uses certain genes and synthesizes proteins not found in other types, and so becomes unique in structure and function. In morphogenesis, tissues and organs change in size, shape, and proportion. They become organized relative to one another in prescribed patterns.

7. All multicelled animals that show extensive cell differentiation undergo aging. Their cells gradually break down in structure and function, which leads to the decline of tissues, organs, and eventually the body.

THE BEGINNING: REPRODUCTIVE MODES

Sexual Versus Asexual Reproduction

In earlier chapters, you learned about the cellular basis of **sexual reproduction**. Briefly, by this reproductive mode, meiosis and gamete formation typically proceed in two prospective parents. At fertilization, a gamete from one parent fuses with a gamete from the other to form the zygote, the first cell of the new individual. You also read about **asexual reproduction**, in which a single parent organism can produce offspring by one of a variety of mechanisms. We now turn to just a few structural, behavioral, and ecological aspects of these two reproductive modes.

Picture a scuba diver accidentally kicking a sponge. A tissue fragment breaks away from the sponge body, then grows and develops by mitotic cell divisions and cell differentiation into a new sponge. Or picture one of the flatworms undergoing transverse fission while it glides along through the water. First its body constricts below the midsection. The part behind the constriction grips a substrate and starts a tug-of-war with the part in front, and then splits off a few hours later. Both parts go their separate ways, regenerate what's missing, and thus become a whole worm. Only some species do this.

In such cases of *asexual* reproduction, all offspring are genetically the same as their individual parent, or nearly so. Phenotypically they are much the same, too. We can speculate that phenotypic uniformity is useful when each individual's gene-encoded traits are highly adapted to a limited and more or less consistent set of environmental conditions. Most variations introduced into the finely tuned gene package would not do much good, and often they could do harm.

Most animals live where opportunities, resources, and danger are highly variable. For the most part, such animals reproduce sexually, meaning female and male parents bestow different mixes of alleles on offspring. As you read in Section 10.1, the resulting variation in traits improves the odds that some of the offspring, at least, should survive and reproduce even if conditions change in the environment.

Costs and Benefits of Sexual Reproduction

Separation into sexes has its costs. Cells that can serve as gametes must be set aside and nurtured. Housing and delivering gametes close to or inside a prospective mate require specialized reproductive structures. Often mating involves investing in courtship. It also requires built-in controls that synchronize the timing of gamete formation and sexual readiness in two individuals.

Just look at the question of *reproductive timing*. How do mature sperm in one individual become available exactly when the eggs mature in a different individual? Timing depends upon energy outlays for constructing, maintaining, and operating neural as well as hormonal control mechanisms in each parent. Also, parents must produce mature gametes in response to the same cues, such as a seasonal change in daylength, that mark the onset of the most appropriate time of reproduction for their species. Example: Male and female moose become sexually active only during late summer and early fall. The coordinated timing means that their offspring will be born the following spring—a time when the weather improves and food is plentiful for many months.

Locating and recognizing a likely mate of the same species is a challenge. Many invest energy in producing social signals called pheromones. They invest in visual signals such as colored and patterned feathers. Many males funnel energy into courting and keeping mates, as with complex bonding rituals and territorial defense requiring, say, claws or massive bodies (Figure 43.2).

Figure 43.2 Some biological costs associated with sexual reproduction. (**a**) Male prairie chickens deciding who gets the territory, and access to females. Reflect on the energy and raw materials directed into making (**b**) peacock feathers, (**c**) snow leopard spots, and (**d**) the body mass of a male northern elephant seal.

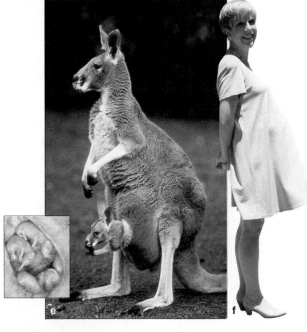

Figure 43.3 Adding to the examples in Chapters 25 and 26, another look at where the embryos of invertebrates and vertebrates develop, how they are nourished, and how (if at all) parents protect them.

Snails (**a**) and spiders (**b**) are *oviparous*. They release eggs from which the young later hatch (*ovi–*, egg; *parous*, produce). Snails are hermaphrodites and abandon their eggs. Spider eggs develop in a silk egg sac that the female anchors to a substrate or carts around with her. Females often die soon after making the sac. Some guard the sac, then cart spiderlings about for a few days and feed them.

(**c**) Birds, including these Allen's hummingbirds, are also oviparous. Fertilized eggs with large yolk reserves develop and hatch outside the mother. Unlike snails, one or both parent birds expend considerable energy feeding and caring for the young.

(**d**) Most sharks and other fishes, all lizards, and many snakes are *ovoviviparous*. Their embryos develop in their mother, receive nourishment continuously from yolk reserves, then are born live (*vivi–*, alive). Shown here, live birth of a lemon shark.

Most mammals are *viviparous*; embryos draw nutrients from maternal tissues and are born live. (**e**) In kangaroos and some other species, the embryos are born in unfinished form and finish developing in a pouch on the mother's ventral surface. Juvenile stages (joeys) continue to draw nourishment from mammary glands in the mother's pouch. (**f**) By contrast, a human female retains the fertilized egg in her body. Maternal tissues nourish the developing individual until the time of its birth, in the manner recounted in the chapter to follow.

Ensuring the survival of offspring is costly (Figure 43.3). Many invertebrates, bony fishes, and frogs simply release eggs and motile sperm into the surrounding water, as in Figure 43.1*a*. If each adult were to produce only *one* sperm or *one* egg each season, the chances for fertilization would not be good. These species invest energy in making numerous gametes, often thousands of them. As another example, nearly all land animals rely on internal fertilization, the union of sperm and egg *within* the body of a female. They invest metabolic energy to construct elaborate reproductive organs, such as a penis (by which certain males deposit sperm in a female) and a uterus (a chamber inside the females of certain species where the embryo grows and develops).

Finally, animals set aside energy in forms that can *nourish the developing individual* until it has developed enough to feed itself. For instance, nearly all animal eggs contain **yolk**. This substance, rich in proteins and lipids, nourishes embryonic stages. The eggs of some species have much more yolk than others. Sea urchins make tiny eggs with little yolk, release large numbers of them, and thus limit the biochemical investment in each one. Within twenty-four hours, each fertilized egg has developed into a freely moving, self-feeding larva. Sea stars and other predators eat most of the eggs. So

for sea urchins, bestowing as little metabolic energy as possible on each one of many individual gametes pays off, in terms of reproductive success.

By contrast, mother birds lay truly yolky eggs. Yolk nourishes the bird embryo through an extended period of development inside an eggshell that forms after the egg is fertilized. *Your* mother put huge demands on her body to protect and nourish you through nine months of development inside her, starting from the time you were a nearly yolkless egg. After becoming implanted in her uterus, you were sustained through pregnancy by physical exchanges with her tissues (Figure 43.3*f*).

As these few examples suggest, animals show great diversity in reproduction and development. However, as you will see in the sections to follow, some patterns are widespread throughout the animal kingdom, and they can serve as a framework for your reading.

Separation into male and female sexes requires special reproductive cells and structures, neural and hormonal control mechanisms, and forms of behavior.

A selective advantage—variation in traits among offspring—offsets the biological costs that are associated with the separation into sexes.

STAGES OF DEVELOPMENT—AN OVERVIEW

Embryos are a class of transitional forms between the fertilized egg and the adult. Although they all start out as a single cell, embryos of different species often look different as they grow and develop. For example, you do not look like a frog now, and you did not look like one when you and the frog were early embryos, either. However, despite such differences in appearance, it is possible to identify certain patterns in how embryos of nearly all animal species develop.

Figure 43.4 is an overview of the stages of animal development. During **gamete formation**, the first stage, eggs or sperm develop inside the reproductive organs of a parent body. **Fertilization**, the second stage, starts when a sperm's plasma membrane fuses with an egg's plasma membrane. Fertilization is over when the sperm nucleus and egg nucleus fuse and form a zygote.

The third stage, **cleavage**, is a program of mitotic cell divisions that divide the volume of egg cytoplasm into a number of **blastomeres**—smaller cells, each with its own nucleus. There is no growth during this stage. Cleavage only increases the number of cells; it does not change the original volume of the egg cytoplasm.

As cleavage draws to a close, the pace of mitotic cell division slackens. The embryo enters **gastrulation**. This fourth stage of animal development is a time of major cellular reorganization. The new cells become arranged into a gastrula with two or three primary tissues, often called germ layers. Cellular descendants of the primary tissues give rise to all tissues and organs of the adult:

1. **Ectoderm**. This is the *outermost* primary tissue layer, the one that forms first in the embryos of every animal. Ectoderm is the embryonic forerunner of the cell lineages that give rise to tissues of the nervous system and to the integument's outer layer.

2. **Endoderm**. Endoderm is the *innermost* primary tissue layer. It is the embryonic forerunner of the gut's inner lining and organs derived from the gut.

3. **Mesoderm**. This *intermediate* primary tissue layer gives rise to muscle, most of the skeleton; circulatory, reproductive, and excretory organs; and connective tissues of the gut and integument. It evolved hundreds of millions of years ago, and it was a pivotal step in the evolution of nearly all large, complex animals.

After the primary tissue layers form in embryos, they give rise to subpopulations of cells. This marks the onset of **organ formation**. The subpopulations become specialized in structure and function. Diverse tissues and organs start to form from their cellular descendants.

During the last stage of animal development, **growth and tissue specialization**, tissues and organs gradually assume their final size, shape, proportions, and functions. This stage extends into adulthood.

Now take a look at Figure 43.5. Its photographs and sketches show several stages in the embryonic development of a typical animal, a frog. Take a moment to study this

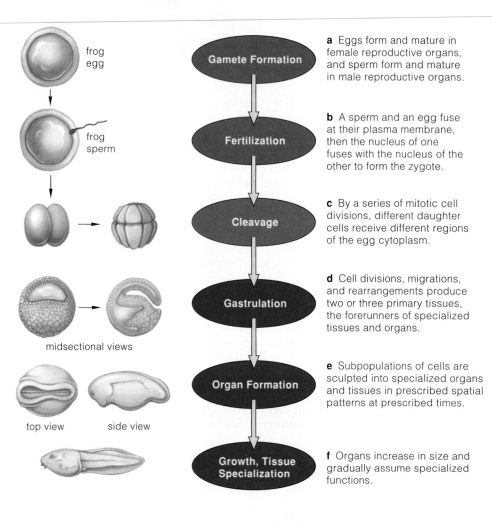

frog egg

frog sperm

midsectional views

top view   side view

Gamete Formation

a Eggs form and mature in female reproductive organs, and sperm form and mature in male reproductive organs.

Fertilization

b A sperm and an egg fuse at their plasma membrane, then the nucleus of one fuses with the nucleus of the other to form the zygote.

Cleavage

c By a series of mitotic cell divisions, different daughter cells receive different regions of the egg cytoplasm.

Gastrulation

d Cell divisions, migrations, and rearrangements produce two or three primary tissues, the forerunners of specialized tissues and organs.

Organ Formation

e Subpopulations of cells are sculpted into specialized organs and tissues in prescribed spatial patterns at prescribed times.

Growth, Tissue Specialization

f Organs increase in size and gradually assume specialized functions.

Figure 43.4 Overview of the stages of animal development. We use a few forms that appear during the frog life cycle as examples.

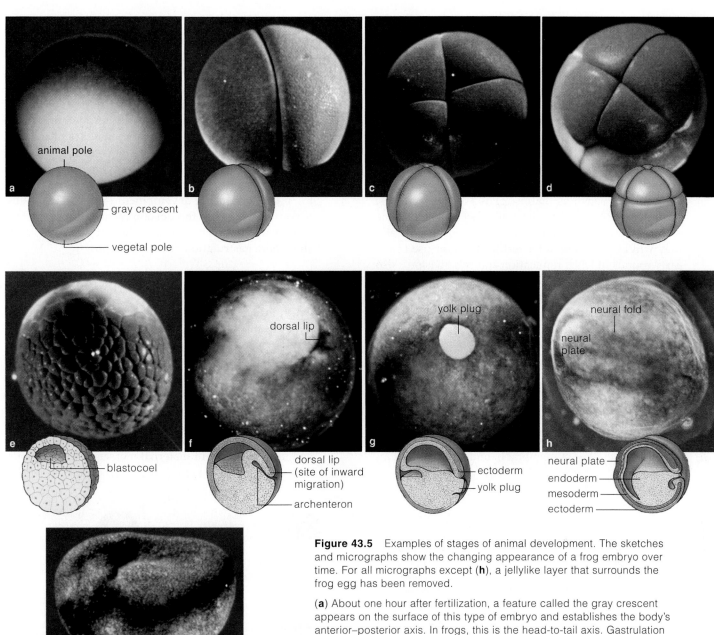

Figure 43.5 Examples of stages of animal development. The sketches and micrographs show the changing appearance of a frog embryo over time. For all micrographs except (**h**), a jellylike layer that surrounds the frog egg has been removed.

(**a**) About one hour after fertilization, a feature called the gray crescent appears on the surface of this type of embryo and establishes the body's anterior–posterior axis. In frogs, this is the head-to-tail axis. Gastrulation will start here. (**b–e**) Cleavage produces a blastula, a ball of cells with a fluid-filled cavity called a blastocoel.

(**f, g**) Cells actively migrate to new locations and are rearranged during gastrulation. (**h,i**) Primary tissue layers form, then a primitive gut cavity in which internal organs will become suspended. Cell differentiation proceeds, moving the embryo on its way to becoming a tadpole, as shown in Figure 43.1.

illustration, for it reinforces an important concept. The structures that form during one stage of development serve as the foundation for the stage that comes after it. Successful development depends on whether all of the structures develop according to normal patterns, in a prescribed sequence.

Animal development proceeds from gamete formation and fertilization through cleavage, gastrulation, then organ formation, and finally growth and tissue specialization.

Development cannot proceed properly unless each stage is successfully completed before the next begins.

EARLY MARCHING ORDERS

Information in the Egg Cytoplasm

Why don't you have an arm attached to your nose or toes growing from your navel? The patterning of body parts for all complex animals, including yourself, starts with messages present in the cytoplasm of immature eggs—or **oocytes**—even before sperm enter the picture. A **sperm**, recall, consists only of paternal DNA and a bit of equipment, including a tail, that helps it reach and penetrate an egg. Compared to a sperm, an oocyte is much larger and more complex (Section 10.5).

While an oocyte is maturing, its volume increases. Enzymes, mRNA transcripts, and other factors become stockpiled in different parts of the cytoplasm. Typically they will be activated, after fertilization, for use in the early rounds of DNA replication and cell division. Also present are tubulin molecules and factors that will help control the angle and timing of tubulin assembly into microtubules for mitotic spindles. How the cytoplasm gets cut depends on such factors and on the yolk in the cytoplasm. The amount of yolk and its distribution will influence how large the resulting blastomeres will be.

Such regionally localized aspects of the oocyte are "maternal messages." We find evidence of their effects as early as fertilization. As an example, a frog egg has pigment granules concentrated near one pole and yolk near the other. The egg's cortex, the cytoskeletal mesh just under the plasma membrane, undergoes structural reorganization when a sperm fertilizes the egg. Part of the cortex shifts toward the point of sperm entry, and it exposes a crescent-shaped area of yolky cytoplasm. This area of intermediate pigmentation, a **gray crescent**, forms near the frog egg's midsection. It establishes the body's anterior–posterior axis. You can see directional planes for this body axis and others in Section 33.6.

In itself, a gray crescent is not evidence of regional differences in maternal messages. We get such evidence from experiments and by observing embryos in which localized cytoplasmic differences are obvious enough to be tracked during development (Figure 43.6). Thus, if you were to track the development of fertilized frog eggs, you would observe that the gray crescent is the site where gastrulation normally begins.

Cleavage—The Start of Multicellularity

Once an egg is fertilized, the zygote enters cleavage. Underneath its plasma membrane, its midsection has a ring of microfilaments made of the contractile protein actin. As the microfilaments slide past one another, the ring tightens, and it pinches in the cytoplasm. The cell surface above the shrinking ring is drawn inward, as a cleavage furrow (Section 9.5). This force of contraction splits the cytoplasm into two blastomeres.

Simply by virtue of where they form, blastomeres end up with different maternal messages. This outcome

Figure 43.6 Experiments that show how a fertilized egg's cytoplasm has localized differences that help determine the fate of cells in a developing embryo. A frog egg's cortex contains granules of dark pigment concentrated near one pole. At fertilization, part of the granule-containing cortex shifts toward the point of sperm entry. This exposes lighter colored, yolky cytoplasm in a crescent-shaped gray area:

The first cleavage normally puts part of the gray crescent in both of the resulting blastomeres.

(**a**) For one experiment, the first two blastomeres that formed were physically separated from each other. Each blastomere still gave rise to a whole tadpole.

(**b**) For another experiment, a fertilized egg was manipulated so the cut through the first cleavage plane missed the gray crescent. Of two resulting blastomeres, only one received the gray crescent. It alone developed into a normal tadpole. Deprived of the maternal messages in the gray crescent, the other cell gave rise to a ball of undifferentiated cells.

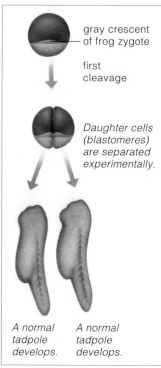

gray crescent of frog zygote

first cleavage

Daughter cells (blastomeres) are separated experimentally.

A normal tadpole develops.

A normal tadpole develops.

a EXPERIMENT 1

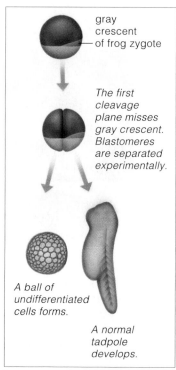

gray crescent of frog zygote

The first cleavage plane misses gray crescent. Blastomeres are separated experimentally.

A ball of undifferentiated cells forms.

A normal tadpole develops.

b EXPERIMENT 2

Figure 43.7 Comparison of early cleavage planes for (**a**) sea urchins and (**b**) mammals. Early cuts of a fertilized egg are radial in sea urchins and rotational in mammals. (**c**) To sense cleavage's impact, consider Sabra and Nina. These identical twins started out life from the same zygote. The first two blastomeres formed at cleavage, the inner cell mass, or another early stage split. The split was the start of two genetically identical, look-alike individuals.

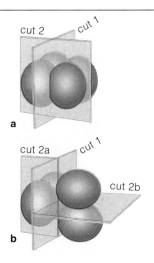

of cleavage, called **cytoplasmic localization**, helps seal the developmental fate of each cell's descendants. Its cytoplasm alone may have molecules of a protein that can activate, say, a gene coding for a certain hormone. And its descendants alone will make that hormone.

Cleavage Patterns

The simplest cleavage pattern has full cuts perpendicular to the mitotic spindle. The cuts parcel out a copy of the nucleus to each blastomere, so all the blastomeres are genetically equivalent. The cuts made during cleavage only divide the volume of cytoplasm into increasingly smaller cells; the blastomeres don't grow in size.

In most animals, a zygote's genes are silent during early cleavage; proteins and mRNAs stockpiled in the cytoplasm control how fast the cuts proceed and how the blastomeres become arranged. In mammals, certain genes must be activated first. Cleavage cannot even be completed without the protein products of those genes.

We can correlate cleavage patterns of major animal groups with the amount and distribution of yolk, which usually restricts cleavage. Cuts may go right through a nearly yolkless egg. Early cuts go only partway through an egg with yolk stockpiled at one end. That egg has polarity. Its *vegetal* pole is the yolk-rich end; its *animal* pole is closest to the nucleus. Frog eggs are like this.

Other heritable factors affect the cuts. Frog and sea urchin eggs undergo *radial* cleavage; cleavage furrows run horizontal and vertical to the animal–vegetal axis (Figure 43.7a). Successive cuts produce a **blastula**, an embryonic stage in which the blastomeres surround a fluid-filled cavity (a blastocoel). But a sea urchin egg is nearly yolkless, so the cuts produce horizontal rows of blastomeres. A frog egg's yolk impedes cuts near the vegetal pole (Figure 43.5e). More, smaller blastomeres form near its animal pole, and the blastocoel is offset. Between the time that 16 to 64 blastomeres form, any amphibian blastula looks like a mulberry. It's called a morula (after the Latin word for mulberry).

What about eggs of reptiles, birds, and most fishes? They undergo *incomplete* cleavage. The large volume of yolk restricts the early cuts to a small, caplike region near the animal pole. The result is two flattened layers of cells with a narrow cavity in between.

What about a mammal's egg? It undergoes *rotational* cleavage, with blastomeres dividing slowly at different times. The first cut is vertical through both poles. In the next cleavage, one blastomere is cut vertically and one horizontally (Figure 43.7b). The third cleavage yields a loose arrangement with space between eight cells. The cells undergo compaction; they abruptly huddle into a compact ball. Tight junctions stabilize the ones outside and seal off those inside. Gap junctions form between the inner cells and facilitate chemical communication among them. Descendants of the outer cells form a thin surface layer. Their secretions form a fluid-filled cavity in the ball, and the inner cells mass together against one side of it. We call this type of blastula a **blastocyst**. In all mammals, including humans, the inner cell mass gives rise to the embryo proper (Section 44.7).

Once in a while, the first two blastomeres, the inner cell mass, or even a later stage splits. Such a split may result in *identical twins*, which have the same genetic makeup and share a common placenta (Figure 43.7c). By contrast, *fraternal twins* arise from two oocytes that matured and then became fertilized during the same menstrual cycle. Each is serviced by its own placenta.

The egg cytoplasm contains maternal instructions in the form of regionally distributed enzymes and other proteins, mRNAs, cytoskeletal elements, yolk, and other factors.

Cleavage divides a zygote into blastomeres. Each of these cells gets part of the maternal messages distributed in egg cytoplasm. We call this outcome cytoplasmic localization.

Differences in the amount and distribution of yolk and other heritable factors give rise to different patterns of cleavage. Those patterns contribute to differences in body plans among different animal groups.

HOW DO SPECIALIZED TISSUES AND ORGANS FORM?

Nearly all animals have a gut, with tissues and organs that function in digestion and absorption of nutrients. They have surface parts that protect internal parts and detect what is going on outside. In between, most have organs, such as those dealing with structural support, movement, and blood circulation. This three-layer body plan emerges as cleavage ends and gastrulation begins. The embryo's size increases little, if any. But cells start migrating to new, predictable locations to form primary tissue layers—ectoderm, endoderm, and mesoderm.

Figure 43.8 shows steps in the formation of a fruit fly gastrula. Figure 43.9 shows the emergence of a bird embryo's anterior–posterior axis. In all vertebrates, this axis defines where a **neural tube**, the forerunner of a brain and spinal cord, will form. All such organs now start to form by cell differentiation and morphogenesis.

Cell Differentiation

All cells of an embryo have the same number and kinds of genes (they all descend from the same zygote). They all activate the genes that specify proteins necessary for cell survival, such as histones and glucose-metabolizing enzymes. From gastrulation onward, certain groups of genes are used in some cells but not in others. When a cell selectively activates genes and synthesizes proteins not found in other cell types, we call this process **cell differentiation**. You read about the molecular basis of selective gene expression in Chapter 15. Distinct cell structures, products, and functions are the outcome.

For instance, when your eye lenses developed, only some cells could activate genes for crystallins, a family of proteins that are assembled into transparent fibers for each lens. Long crystallin fibers formed in the cells, forcing them to lengthen and flatten. The differentiated cells provided each lens with unique optical properties.

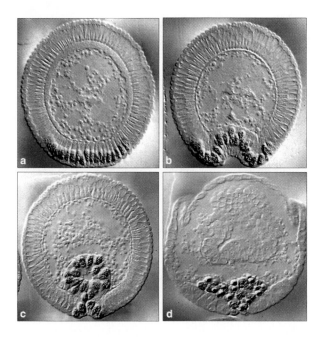

Figure 43.8 Gastrulation in a fruit fly (*Drosophila*), cross-section. The blastula (formed during cleavage) is transformed into a gastrula. Notice how some cells migrate inward.

And crystallin-producing cells are only 1 of 150 or so differentiated cell types now present in your body.

As many experiments show, nearly all cells become differentiated with no loss of genetic information. For instance, John Gurdon stripped unfertilized eggs of an African clawed frog (*Xenopus laevis*) of their nucleus. He ruptured the plasma membrane of intestinal cells from tadpoles of the same species. He left the nucleus and much of the cytoplasm of these fully differentiated cells intact and inserted them into the enucleated eggs. Some eggs gave rise to a complete frog! The intestinal

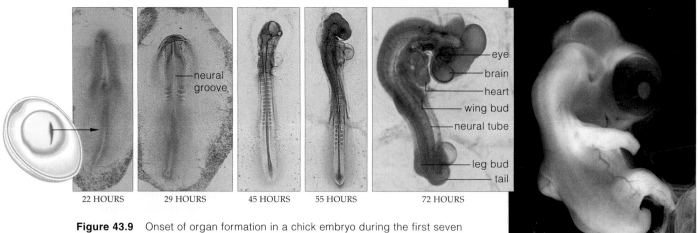

22 HOURS   29 HOURS   45 HOURS   55 HOURS   72 HOURS   168 HOURS (SEVEN DAYS OLD)

neural groove

eye
brain
heart
wing bud
neural tube
leg bud
tail

Figure 43.9 Onset of organ formation in a chick embryo during the first seven days of development. The heart begins to beat between thirty and thirty-six hours. You may have observed such embryos at the yolk surface of raw, fertilized eggs.

Figure 43.10 Examples of morphogenesis. (**a**) Cell migration. A nerve cell (*orange*) "climbs" through a developing brain to its final position, using a glial cell (*yellow*) as a highway. (**b**) How a neural tube forms. As gastrulation ends, ectoderm is a uniform sheet of cells. In some cells, microtubules lengthen. The elongating cells form a neural groove. Then rings of microfilaments at one end of certain cells constrict, so these cells become wedge shaped. Their part of the sheet folds over to form the tube.

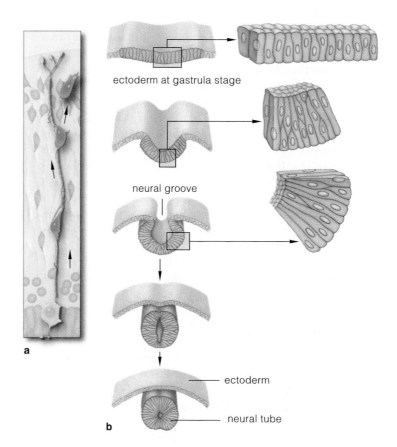

ectoderm at gastrula stage

neural groove

ectoderm

neural tube

cells still had the same number and kinds of genes as the zygote; its nucleus still had all the genes required to make all cell types that make up a frog.

Embryonic cells of humans also retain the capacity to give rise to a whole individual. That is how identical twins arise from a split blastomere (Figure 43.7).

Morphogenesis

Morphogenesis refers to a program of orderly changes in an embryo's size, shape, and proportions, the result being specialized tissues and early organs. As part of the program, cells divide, grow, migrate, and change in size. Tissues expand and fold, and the cells in some of them die in controlled ways at prescribed locations.

Think of active cell migration. *Cells send out and use pseudopods that move them along prescribed routes.* When they reach their destination, they establish contact with cells already there. For instance, forerunners of neurons interconnect this way as a nervous system is forming.

How do cells "know" where to move? They respond to adhesive cues, as when migrating nerve cells stick to adhesion proteins on glial cells but not blood vessels (Figure 43.10a). They respond to chemical gradients, too. Their migrations are coordinated by the synthesis, release, deposition, and removal of specific chemicals in the extracellular matrix. And adhesive cues tell cells when to stop. Cells will migrate to regions of strongest adhesion, but once there, further migration is impeded. Section 22.12 describes the chemical-induced migration of cells of *Dictyostelium discoideum*, a slime mold.

Also, *whole sheets of cells expand and fold inward and outward.* Microtubules lengthen and microfilament rings constrict inside cells. The assembly and disassembly of these cytoskeletal components are part of a controlled program of changes in cell shape.

Through such controlled, localized events, the size, shape, and proportions of body parts emerge. Example: Figure 43.10b shows what happens after three primary tissues form in embryos of amphibians, reptiles, birds, and mammals. At the embryo midline, ectodermal cells elongate and form a neural plate—the first sign that a region of ectoderm is on its way to becoming nervous tissue. Specific cells lengthen (as microtubules grow) or become wedge shaped (as a contracting microfilament

ring constricts them at one end). The changes in shape cause tissue flaps to fold over and meet at the midline, thus forming the neural tube (Sections 34.6 and 34.8).

As a last example, *programmed cell death helps sculpt body parts.* Embryonic cells that have a short life span execute themselves. By this form of cell death, called apoptosis, signals from some cells activate weapons of self-destruction stockpiled in other, target cells. Section 15.6 describes what happens at the molecular level. For now, think of a human embryo's hand when it is still a paddle (Section 9.4). Cartilage models of digits form inside it. Cells in thin tissue zones between digits have receptors for proteins, each a product of selective gene expression. When proteins bind to their receptors, cells commit mass suicide, and the digits separate. One rare mutation blocks apoptosis, and the digits stay webbed.

In cell differentiation, a cell selectively uses certain genes and makes certain proteins, not found in other cell types, for distinctive cell structures, products, and functions.

Morphogenesis is a program of orderly changes in body size, shape, and proportions. It results in specialized tissues and organs at prescribed locations, at prescribed times.

Morphogenesis involves cell division, active cell migration, tissue growth and foldings, changes in cell size and shape, and programmed cell death by way of apoptosis.

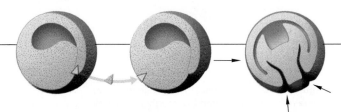

a Dorsal lip excised from donor embryo is grafted to an abnormal site in another embryo.

b Graft induces a second site of inward migration.

c Gastrula develops into a "double" tadpole. Most of its tissues originated from the host embryo.

Figure 43.11 Experimental evidence that the dorsal lip of an embryonic axolotl (a type of amphibian) controls gastrulation. When transplanted to a different site in another axolotl embryo, it organized the formation of a second set of body parts.

PATTERN FORMATION

Different cells become organized in different ways in a developing embryo. They divide, differentiate, and live or die; they migrate and stick to cells of the same type in tissues. Their descendants fill in the body details in ordered patterns. Cells of amphibian embryos make a head at one end and a tail at the other, and so on.

A cell locked in at a given place in the embryo has chemical memory; it selectively reads some genes but not others. It makes and secretes molecular signals that cause changes in the activities of its neighbors, and it responds to neighborhood signals. And its memory is passed on to all of its descendants, which go on to form tissues in places where we expect them to be. Now is the time of **embryonic induction**: Developmental fates of the embryonic cell lineages change when exposed to gene products from adjacent tissues. By these changes, specialized tissues and organs emerge from clumps of embryonic cells in predictably ordered, spatial patterns. We call their emergence **pattern formation**.

We have experimental evidence of cell memory. For instance, researchers excised the dorsal lip of a normal axolotl embryo, then grafted it into a novel location in a different axolotl embryo. Gastrulation proceeded at the recipient's own dorsal lip *and at the graft*. Siamese twins—a double embryo with two sets of body parts—developed (Figure 43.11). The dorsal lip organizes this amphibian's main axis. And it was the first embryonic signaling center discovered. Figure 43.12 gives another example of embryonic induction experiments, this one involving chick wing development.

Embryonic Signals in Pattern Formation

Cytoplasmic localization during cleavage ensures that different cells will send and receive different chemical signals. While the embryo forms, cells that started out as neighbors use the signals. So do cells that meet up by gastrulation and other morphogenetic movements.

For example, cells of the dorsal lip are the source of a morphogen. **Morphogens** are degradable molecules that diffuse as long-range signals from an embryonic signaling center. The concentration gradient helps cells assess their position and affects how they differentiate. The signal is strongest at the start of the gradient and weakens with distance. Cells in different body regions affected by the gradients receive different information, so they selectively express different genes as a result.

Other signals are short range, involving cell-to-cell contacts. Example: Changes in the expression of genes for cadherins, integrins, and other glycoproteins at the cell surface alter how cells associate with one another when the gastrula, then organs, form. When a cell starts to make a cytoskeletal protein, it may move or lengthen in some direction. If adhesion proteins are affected, a cell may cohere with neighbors, break free and migrate, or become segregated from cells of an adjoining tissue.

Also, cells respond differently to signals depending on when they receive them. Graft a bit of animal pole epithelium from an early gastrula over a rudimentary eye of an older embryo, and it gives rise to something like neural tube tissue. Leave it alone in vitro for a few hours before the graft, and it differentiates into a lens. Leave it even longer in vitro and it will not respond at all to the inductive signals from the eye tissue.

A Theory of Pattern Formation

Do the same classes of genes govern spatial patterning in all animals? It now appears so. After studies of many diverse animals, researchers have put together a **theory of pattern formation**. Here are its key points:

1. The formation of tissues and organs in ordered, spatial patterns starts with cytoplasmic localization. It continues through short-range cell-to-cell contacts and long-range signals—chemical gradients that weaken with distance from the source. Cells at the start, middle, and end of a gradient receive different information.

2. Morphogens and other inducer molecules diffuse through embryonic tissues, sequentially activating classes of **master genes**.

3. Products of **homeotic genes** and other master genes interact with control elements to map out the overall body plan, as in Section 43.6. The products appear as blocks of genes are activated and suppressed in cells along the anterior–posterior axis and dorsal–ventral axis of the embryo. Other gene products interact to fill in details of specific body parts.

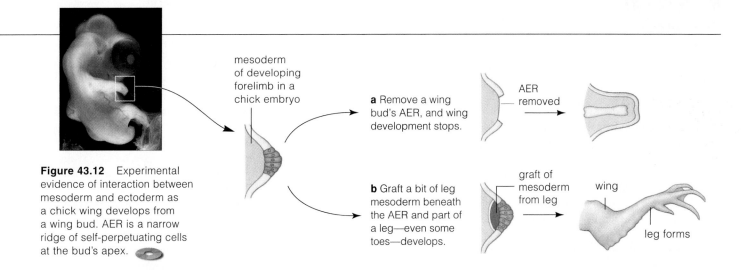

Figure 43.12 Experimental evidence of interaction between mesoderm and ectoderm as a chick wing develops from a wing bud. AER is a narrow ridge of self-perpetuating cells at the bud's apex.

mesoderm of developing forelimb in a chick embryo

a Remove a wing bud's AER, and wing development stops.

AER removed

b Graft a bit of leg mesoderm beneath the AER and part of a leg—even some toes—develops.

graft of mesoderm from leg

wing

leg forms

When organs first form, gene products and control elements activate and inhibit blocks of genes—and they do so in similar ways among all major animal groups. As the next section describes, they direct genes to map out the overall body plan. When master genes fail to do so, disaster follows, as when a heart forms at the wrong location. Once the basic body plan is locked in, the many inductions that follow have only localized effects. For example, when a lens fails to form, then only the eye will be affected.

Evolutionary Constraints on Development

How long have the master genes been operating? The basic body plan for sponges, flatworms, mollusks, flies, vertebrates, and all other major animal groups hasn't changed much for about 500 million years. *These many millions of species are variations on a few dozen plans!*

Why is this so? There are few new master genes, so variations in body plans may be more of an outcome of how control genes control each other. Example: Insect, squid, and vertebrate eyes differ in their structure, yet genes governing eye formation are nearly identical in all these groups. An *eyeless* gene controls eye formation in fruit flies (*Drosophila*). In humans, one mutation of a nearly identical gene results in eyeless babies.

Similarly, the *distalless* gene in fruit flies causes legs to grow from leg buds. The same gene controls the fate of cells that give rise to crab legs, beetle legs, butterfly wings, sea star arms, fish fins, and mouse feet. As you know, legs, fins, and wings are different structures with different functions. For instance, skeletal elements form on the outside of a beetle's legs; your legs, and fish fins, have them inside. Yet all of these structures start out as buds from the main body axis.

All those six-legged insects with us today evolved from many-legged crustaceans about 400 million years ago. And the loss of *just a single peptide* from the Ubx protein probably caused this big evolutionary change! As William McGinnis and graduate student Matthew Ronshaugen found out, that one mutation represses leg development over most of the embryonic body.

And so, rather than starting from scratch with new genes, diverse animals are using identical or similar master genes to control development. By analogy, they have old software programs in more recent computers.

For a long time, we have known about the *physical* constraints (including the surface-to-volume ratio) and *architectural* constraints (as imposed by body axes). The kinds of events described in this chapter indicate there also are *phyletic* constraints on change. These are the constraints imposed on each lineage by interactions of the master organizer genes, which operate when organs form and control induction of the basic body plan.

That might be why we have so many species and so few body plans. Once master genes had evolved and were engaged in intricate interactions, it would have been difficult to change the body's basic characteristics without destroying the embryo. Mutations have indeed added marvelous variations to animal lineages. But the basic body plans have prevailed through time; maybe it simply proved unworkable to start all over again.

Pattern formation is the ordered sculpting of embryonic cells into specialized tissues and organs. Cells differentiate through cytoplasmic localization, then through chemical memory of their position in the developing embryo.

The pattern starts with different chemical signals inside blastomeres having different maternal messages. Cell-to-cell inductive interactions fill in the details of the pattern.

Inductive interactions among classes of master genes map the basic body plan and specify where and how body parts develop. Their products are long-range and short-range beacons that help cells assess their position in the embryo. Cells respond differently to signals at different times.

Physical, architectural, and phyletic constraints limit the evolution of animal body plans. Of millions of species, we see only a few basic plans. In all major groups, master genes that guide development are similar and sometimes identical.

To Know A Fly

"Although small children have taboos against stepping on ants because such actions are said to bring on rain, there has never seemed to be a taboo against pulling off the legs or wings of flies. Most children eventually outgrow this behavior. Those who do not either come to a bad end or become biologists."

So wrote Vincent Dethier in *To Know A Fly*, a tribute to flies and their suitability for laboratory experiments. Even before he published his booklet in 1962, plenty of geneticists and developmental biologists had reached the same conclusion. *Drosophila melanogaster* was the fly of choice, and it still is. It costs almost nothing to feed or house this small insect. It reproduces fast in bottles, it has a short life cycle, and disposing of spent bodies after an experiment is a snap.

Studies of *Drosophila* at the anatomical, cytological, biochemical, and genetic levels revealed a distinctive pattern of embryonic development. Even so, the studies tell us much about embryonic development and the evolutionary history of animals in general.

For instance, mammalian eggs have relatively little yolk, so cleavage furrows cut through the entire egg. Eggs of *Drosophila* (and of most insects, fishes, reptiles, and birds) have a large yolk mass that restricts cleavage to a small part of the egg cytoplasm. In a *Drosophila* egg, the yolk mass is centrally located. It confines cleavage to the cytoplasm's periphery, an example of a *superficial* cleavage pattern (Figure 43.13*a*).

Compared to a mammalian egg, which undergoes slow cuts every twelve to twenty-four hours, superficial cleavage of a fertilized *Drosophila* egg is fast. The egg nucleus repeatedly divides but the cytoplasm does not, forming the blastoderm shown in Figure 43.13*b*. Nuclear division proceeds every eight minutes until 256 nuclei crowd together in the yolky cytoplasm. This embryonic stage is a type of cellular blastoderm. Microtubules and microfilaments in each nucleus are arrayed in ways that will guide cell formation and elongation.

Most of the nuclei migrate toward the egg periphery. First the plasma membrane folds inward around each

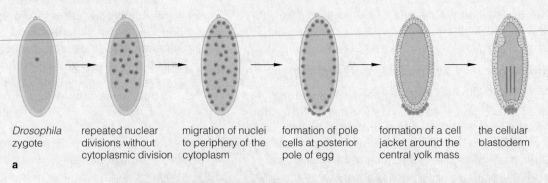

Drosophila zygote

repeated nuclear divisions without cytoplasmic division

migration of nuclei to periphery of the cytoplasm

formation of pole cells at posterior pole of egg

formation of a cell jacket around the central yolk mass

the cellular blastoderm

a

Figure 43.13 Superficial cleavage in the yolk-rich egg of *Drosophila*.

(**a**) A fertilized egg nucleus repeatedly divides. Cytoplasmic division is delayed until after the nuclei have migrated to the periphery of the cytoplasm. (**b**) Pole cells form. (**c,d**) Then other cells form; they are arranged like a jacket around the central mass of yolk. This early stage of development is a type of blastoderm.

As the blastoderm forms, polar granules migrate through the cytoplasm and then become isolated in pole cells that form at one end of the blastoderm. Pole cells are forerunners of germ cells that give rise to eggs or sperm in an adult.

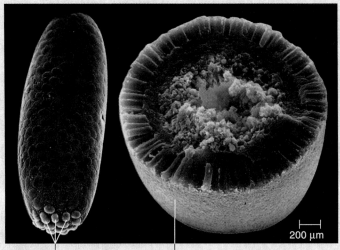

b Pole cells (forerunners of germ cells of the adult)

c Section through the blastoderm showing jacket of cells around yolk

200 µm

d Gastrulation, well under way

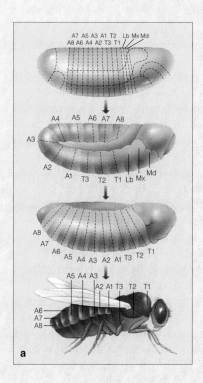

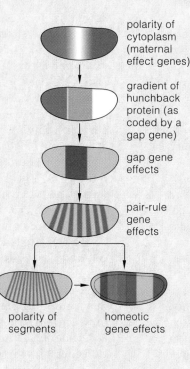

polarity of cytoplasm (maternal effect genes)

gradient of hunchback protein (as coded by a gap gene)

gap gene effects

pair-rule gene effects

polarity of segments

homeotic gene effects

a

b

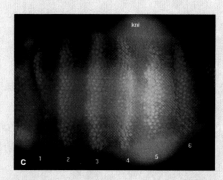

Figure 43.14 (**a**) Fate map of a *Drosophila* zygote. Dashed lines indicate the regions that normally will become different segments of the body, each having specialized parts. Each segment's fate starts with the egg's cytoplasmic localization (maternal effect genes).

(**b**) Generalized model for anterior–posterior pattern formation in a *Drosophila* embryo. Gradients of maternal effect gene products influence hunchback gene activity, which influence the activity of different gap genes, and so on to give each segment its identity. (**c**) For example, by late cleavage, spatial domains are established by gradients of one of the gap gene proteins (*green*) and by a pair-rule gene protein (*red*).

nucleus that reaches the egg's posterior pole axis. The resulting pole cells are forerunners of germ cells of the adult fly. Physical separation of the future reproductive cells from the rest of the blastoderm is one of the first events of insect development. After it occurs, the plasma membrane folds inward and forms a partition around each remaining nucleus and a bit of cytoplasm. Four hours after fertilization, about 6,000 cells are arranged like a jacket around the yolky mass.

During the late 1940s, Edward Lewis of the California Institute of Technology saw that the products of mutated genes introduce bizarre blips in the pattern formation in *Drosophila*. For instance, as a result of one mutation, a leg grew out from the head region where an antenna should have grown. You saw this result in Section 15.4.

Figure 43.14 includes a **fate map** for the surface of a *Drosophila* zygote. Such maps give you an idea of where each differentiated cell type will originate in the adult. This pattern starts with the unfertilized egg's polarity. A head end and tail end form. A series of body segments form in between. Genes specify whether each segment will grow legs, or legs and wings, and so on.

Researchers put together a model to explain how the segmented polarity of the adult arises. *Maternal effect* genes that specify regulatory proteins are transcribed, translated, or both (Figure 43.14*b*). The resulting mRNAs and proteins become localized in different regions of the egg cytoplasm. These gene products are switched on in

the zygote. They activate or suppress *gap* genes, which map out broad regions. When the product of one gap gene (hunchback) diffuses through the blastoderm, it creates a chemical gradient that influences the extent to which other gap genes are expressed. The differences in concentrations of gap gene products switch on *pair-rule* genes—the products of which accumulate in bands (stripes) corresponding to two body segments. They activate *segment polarity* genes, the products of which divide the embryo into segment-sized units. Ultimately, these interactions control expression of the *homeotic* genes that collectively govern the fate of each segment.

Although *Drosophila* has been a research favorite for more than eighty years, the pace of research has picked up since automated gene sequencing has identified its 13,500 genes. It was the second multicelled organism to have its genome sequenced fully, after *Caenorhabditis elegans*. This roundworm has 19,900 genes, and only 3,000 are essential for the body's structure and function. Section 25.6 gives some of the reasons this organism has been another favorite experimental organism for studies of genetics, development, and aging.

Researchers do not yet have a complete understanding of the multiple levels of gene interactions that govern body parts of fruit flies and roundworms. But the studies of such tiny organisms give us fascinating glimpses into the kinds of regulatory mechanisms that must come into play in other organisms as well.

WHY DO ANIMALS AGE?

Every multicelled animal that undergoes extensive cell differentiation gradually deteriorates. Through processes collectively called **aging**, its cells gradually break down in structure and function, which leads to the decline of tissues, organs, and eventually the whole body.

The Programmed Life Span Hypothesis

Do biological clocks affect aging? After all, each species has a maximum life span—20 years for dogs, 12 weeks for butterflies, 35 days for fruit flies, and so on. No human has lived past 122 years. Given the consistency within species, we can expect that genes influence the aging process. Indeed, researchers can double the life span of fruit flies by manipulating their genes.

Suppose the animal body is like a clock shop, with each type of cell, tissue, and organ ticking at its own genetically set pace. Many years ago, Paul Moorhead and Leonard Hayflick investigated such a possibility. They cultured normal human embryonic cells, which divided about fifty times, then died off. Hayflick also withdrew cultured cells that were partway through the series of divisions, then froze them for several years. After he thawed the cells and placed them in a culture medium, they proceeded to complete the in-vitro cycle of fifty doublings and died on schedule.

No cell in a human body divides more than eighty or ninety times. You may well wonder: If an internal clock ticks off their life span, then how can *cancer* cells go on dividing? The answer provides us with a clue to why *normal* cells can't beat the clock.

As you know, cells duplicate their chromosomes before they divide. Capping the chromosome ends are repetitive DNA sequences called **telomeres**. A piece of each telomere is lost during each nuclear division, and when only a nub is left, cells stop dividing and die. The exceptions are cancer cells and reproductive cells. Both make telomerase, an enzyme that can make telomeres *lengthen*. Experimentally expose some somatic cells to telomerase and they will go on dividing well past their normal life span, with no apparent bad effects.

The Cumulative Assaults Hypothesis

Another hypothesis: Over the long term, aging results from cumulative damage at the molecular and cellular levels. Both environmental assaults and spontaneous mistakes by DNA repair mechanisms are at work here.

For example, free radicals, those rogue molecular fragments described in Chapter 5, bombard DNA and all other biological molecules. This includes the DNA of mitochondria—the power plants of eukaryotic cells. Structural changes in DNA compromise the synthesis of functional enzymes and other proteins necessary for normal life processes. And free radicals are implicated in many age-related problems, such as atherosclerosis, cataracts, and Alzheimer's disease.

Problems in DNA replication and repair operations also have been implicated in aging. Researchers have correlated *Werner's syndrome*, an aging disorder, with a mutation in the gene that specifies a helicase. Such enzymes unwind nucleotide strands from each other. The mutated helicase probably does not compromise DNA replication, because affected people do not die right away. They start aging abnormally fast in their thirties and die before reaching fifty. The nonmutated gene may be crucial for repairs; people with Werner's syndrome accumulate mutations at above-normal rates. Sooner or later, damage to DNA may interfere with cell division. Like skin cells of the elderly, skin cells of Werner's patients do not divide many times.

It may be that both hypotheses have merit. Aging may be an outcome of many interconnected processes in which genes, hormones, environmental assaults, and a decline in DNA repair mechanisms come into play. Consider how living cells of all tissues depend upon exchanges of materials with extracellular fluid. Also consider how collagen is a structural element of many connective tissues. If something shuts down or mutates collagen-encoding genes, the missing or altered gene product might disrupt the flow of oxygen, nutrients, hormones, and so forth to and from cells through every connective tissue. Repercussions from such a mutation would ripple through the body.

Similarly, if mutations result in altered self markers on the body's cells, do T cells of the immune system perceive them as foreign and go on the attack? If such autoimmune responses were to become more frequent over time, they would promote greater vulnerability to disease and stress associated with old age.

In evolutionary terms, reproductive success means surviving long enough to produce offspring. Humans can reach sexual maturity in fifteen years *and* help their own children reach adulthood. We don't *need* to live longer. But we among all animals have the capacity to think about it, and we generally have decided that we like life better than the alternative. Eventually, though, even the most tenacious clingers to life must confront their own mortality. And perhaps in time we all might learn to accept the inevitable with wisdom and grace.

The human life cycle flows naturally from the time of birth, growth, and development, to production of the individual's own offspring, and on through aging to the time of death.

A biological clock ticks off the life span of all body cells, and DNA becomes increasingly damaged over time. These factors and others result in aging.

DEATH IN THE OPEN

As a leading cancer specialist, Lewis Thomas reflected with compassion on the fear of dying. Before Thomas himself died of cancer, he gave us this gift of insight:

Everything in the world dies, but we only know about it as a kind of abstraction. If you stand in a meadow, at the edge of a hillside, and look around carefully, almost everything you can catch sight of is in the process of dying, and most things will be dead long before you are. If it were not for the constant renewal and replacement going on before your eyes, the whole place would turn to stone and sand under your feet.

There are some creatures that do not seem to die at all; they simply vanish totally into their own progeny. Single cells do this. The cell becomes two, then four, and so on, and after a while the last trace is gone. It cannot be seen as death; barring mutation, the descendants are simply the first cell, living all over again. . . .

There are said to be a billion billion insects on the Earth at any moment, most of them with very short life expectancies by our standards. Someone estimated that there are 25 million assorted insects hanging in the air over every temperate square mile, in a column extending upward for thousands of feet, drifting through the layers of atmosphere like plankton. They are dying steadily, some by being eaten, some just dropping in their tracks, tons of them around the Earth, disintegrating as they die, invisibly.

Who ever sees dead birds, in anything like the huge numbers stipulated by the certainty of the death of all birds? A dead bird is an incongruity, more startling than an unexpected live bird, sure evidence to the human mind that something has gone wrong. Birds do their dying off somewhere, behind things, under things, never on the wing.

Animals seem to have an instinct for performing death alone, hidden. Even the largest, most conspicuous ones find ways to conceal themselves in time. If an elephant missteps and dies in an open place, the herd will not leave him there; the others will pick him up and carry the body from place to place, finally putting it down in some inexplicably suitable location. When elephants encounter the skeleton of an elephant in the open, they methodically take up each of the bones and distribute them, in a ponderous ceremony, over neighboring acres.

It is a natural marvel. All of the life on Earth dies all of the time, in the same volume as the new life that dazzles us each morning, each spring. All we see of this is the odd stump, the fly struggling on the porch floor of the summer house in October, the fragment on the highway. I have lived all my life with an embarrassment of squirrels in my backyard, they are all over the place, all year long, and I have never seen, anywhere, a dead squirrel.

I suppose it is just as well. If the Earth were otherwise, and all the dying were done in the open, with the dead there to be looked at, we would never have it out of our

minds. We can forget about it much of the time, or think of it as an accident to be avoided, somehow. But it does make the process of dying seem more exceptional than it really is, and harder to engage in at the times when we must ourselves engage.

In our way, we conform as best we can to the rest of nature. The obituary pages tell us of the news that we are dying away, while birth announcements in finer print, off at the side of the page, inform us of our replacements, but we get no grasp from this of the enormity of the scale. There are now billions of us on the Earth, and all must be dead, on a schedule, within this lifetime. The vast mortality, involving something over 50 million each year, takes place in relative secrecy. We can only really know of the deaths in our households, among our friends. These, detached in our minds from all the rest, we take to be unnatural events, anomalies, outrages. We speak of our own dead in low voices; struck down, we say, as though visible death can occur only for cause, by disease or violence, avoidably. We send off for flowers, grieve, make ceremonies, scatter bones, unaware of the rest of the billions on the same schedule. All of that immense mass of flesh and bone and consciousness will disappear by absorption into the Earth, without recognition by the transient survivors.

Less than half a century from now, our replacements will have more than doubled in numbers. It is hard to see how we can continue to keep the secret, with such multitudes doing the dying. We will have to give up the notion that death is a catastrophe, or detestable, or avoidable, or even strange. We will need to learn more about the cycling of life in the rest of the system, and about our connection in the process. Everything that comes alive seems to be in trade for everything that dies, cell for cell. There might be some comfort in the recognition of synchrony, in the information that we all go down together, in the best of company.

— LEWIS THOMAS, 1973

SUMMARY
Gold indicates text section

1. For animals, sexual reproduction is the dominant reproductive mode. It requires specialized reproductive structures, control mechanisms, and forms of behavior that assist fertilization and support the offspring. *43.1*

2. In normal development, each stage is successfully completed before the next stage begins. *43.2, 43.6*

3. Embryonic development typically proceeds through six consecutive stages: *43.1*

   a. Gamete formation, when oocytes (immature eggs) and sperm form in reproductive organs. Molecular and structural components are stockpiled and are localized in different parts of the oocyte. *43.2, 43.3*

   b. Fertilization, from the time a sperm penetrates an egg to the fusion of sperm and egg nuclei that results in a zygote (fertilized egg). *43.2, 43.3*

   c. Cleavage, when mitotic cell divisions transform a zygote into smaller cells (blastomeres). Cleavage does not increase the original volume of egg cytoplasm; it only increases the number of cells. It regionally divides the yolk, mRNAs, proteins, cytoskeletal elements, and other "maternal messages" among new blastomeres, an outcome called cytoplasmic localization. *43.2, 43.3*

   d. Gastrulation, when primary tissue layers (germ layers) form. Cell divisions, cell migrations, and other events lead to the formation of endoderm, ectoderm, and (in most species) mesoderm. All tissues of the adult body develop from primary tissue layers. *43.2, 43.4*

   e. Onset of organ formation. Different organs start developing by a tightly orchestrated program of cell differentiation and morphogenesis. *43.2, 43.4–43.6*

   f. Growth and tissue specialization, when organs enlarge and acquire specialized chemical and physical properties. Maturation of tissues and organs continues into post-embryonic stages. *43.2, 43.4–43.6*

4. In cell differentiation, a cell selectively uses certain genes and synthesizes proteins not found in other cell types. The outcomes are subpopulations of specialized lineages of cells that differ from one another in their structure, biochemistry, and functioning. *43.4, 43.5*

5. Morphogenesis starts at gastrulation. It changes the embryo's size, shape, and proportions. *43.4, 43.5*

6. Pattern formation means the emergence of the basic body plan and body parts in specific regions of the embryo, in orderly sequence. Cytoplasmic localization helps seal the fate of cells. Cell determination depends also on interactions among classes of master genes that specify certain products. These products are spatially organized and create chemical gradients. The gradients influence differentiation, hence the identity of each cell lineage in the embryo. *43.5, 43.6*

7. Change in the developmental fate of an embryonic cell lineage, as brought about by exposure to products released from an adjacent tissue, is called embryonic induction. *43.5, 43.6*

8. All animals showing extensive cell differentiation undergo aging, or gradual changes in structure and a decline in efficiency that typically ends in death. *43.7*

Review Questions

1. What is the main benefit of sexual reproduction compared to asexual reproduction? What are some of the biological costs of separation into sexes? *43.1*

2. Briefly describe the key events of gamete formation, fertilization, then cleavage, gastrulation, and organ formation. At which stage of development is the frog embryo in the photograph at right? *43.2*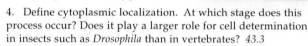

3. Does cleavage increase the volume of cytoplasm, the number of cells, or both, compared to the zygote? *43.2*

4. Define cytoplasmic localization. At which stage does this process occur? Does it play a larger role for cell determination in insects such as *Drosophila* than in vertebrates? *43.3*

5. Define blastula. What is a blastocyst? *43.3*

6. Define cell differentiation and morphogenesis. *43.4*

7. Distinguish between morphogens and master genes. *43.5*

8. Summarize the theory of pattern formation. *43.5*

9. Distinguish between the programmed life span hypothesis and the cumulative assaults hypothesis of aging. *43.7*

Self-Quiz
ANSWERS IN APPENDIX III

1. Sexual reproduction among animals is _____ .
   a. biologically costly   c. evolutionarily beneficial
   b. diverse in its details   d. all of the above

2. Development cannot proceed properly unless each stage is successfully completed before the next begins, starting with
_____ .
   a. gamete formation   d. gastrulation
   b. fertilization   e. organ formation
   c. cleavage   f. growth, tissue specialization

3. The astonishing internal complexity characteristic of most animals became possible following the evolution of _____ .
   a. ectoderm b. mesoderm c. myoderm d. endoderm

4. A cell formed during cleavage is a _____ .
   a. blastula   c. blastomere
   b. morula   d. gastrula

5. _____ distributes different maternal messages to different blastomeres.
   a. Gametogenesis   c. Morphogenesis
   b. Cleavage   d. Pattern formation

6. Primary tissue layers first appear _____ .
   a. in the egg cortex   c. in the gastrula
   b. during cleavage   d. in primary organs

7. During development, the formation of subpopulations of different cell types is the outcome of _____ .
   a. selective gene expression c. metamorphosis
   b. cell differentiation   d. a and b

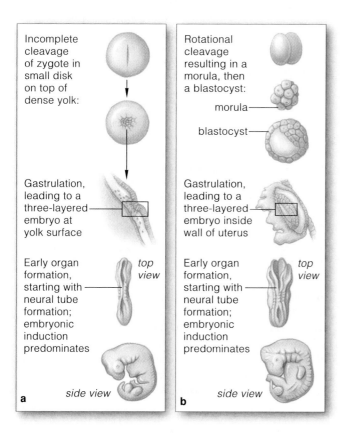

Incomplete cleavage of zygote in small disk on top of dense yolk:

Gastrulation, leading to a three-layered embryo at yolk surface

Early organ formation, starting with neural tube formation; embryonic induction predominates

top view

side view

a

Rotational cleavage resulting in a morula, then a blastocyst:

morula

blastocyst

Gastrulation, leading to a three-layered embryo inside wall of uterus

Early organ formation, starting with neural tube formation; embryonic induction predominates

top view

side view

b

Figure 43.15 Comparison of the pattern of embryonic development for (**a**) a chick and (**b**) a human.

8. Products of homeotic genes _____ .
   a. help map out the overall body plan
   b. interact with control elements
   c. appear as blocks of genes are activated and suppressed
   d. all of the above

9. Pattern formation requires _____ .
   a. master genes   c. regulatory proteins
   b. morphogens   d. all of the above

10. The orderly formation of new tissues and organs in a developing embryo involves _____ .
    a. cytoplasmic localization   c. chemical gradients
    b. cell-to-cell contacts   d. all of the above

11. The developmental fate of an embryonic cell lineage changes upon exposure to gene products from an adjacent tissue. This is a case of _____ .
    a. cleavage   c. embryonic induction
    b. cytoplasmic localization   d. apoptosis

12. *Drosophila melanogaster* cleavage is _____ .
    a. superficial   c. radial
    b. superfluous   d. rotational

13. Match the development stage with its description.
    _____ cleavage   a. egg and sperm mature in parents
    _____ gametes form   b. sperm nucleus, egg nucleus fuse
    _____ organ   c. formation of primary tissue layers
         formation   d. cytoplasmic localization helps seal
    _____ growth, tissue   fate of blastomeres
         specialization   e. organs, tissues increase in size,
    _____ gastrulation   acquire specialized properties
    _____ fertilization   f. starts when primary tissue layers
                             split into subpopulations of cells

Critical Thinking

1. In normal *Drosophila* larvae, a specific cluster of cells is the embryonic source of antennae on the head. When a larva has a certain mutant gene, it develops legs instead of antennae on its head, as in Section 15.4. Would you say that the mutated gene is first expressed during cleavage or during organ formation?

2. Before an amphibian egg enters cleavage, you divide it so that only one daughter cell gets the gray crescent. You separate the two cells, and only the one with the gray crescent gives rise to an embryo with an anterior–posterior axis, notochord, nerve cord, and back muscles. The other cell gives rise to a shapeless mass of immature gut cells and blood cells.

   Does cytoplasmic localization or embryonic induction play a greater role in these outcomes? Explain your answer.

3. Take a look at Figure 43.15. Now consider this: Despite differences in their early stages of development, the embryos of chickens and humans resemble each other after their organs start to form. Speculate on how the evolutionary constraints on vertebrate development might underlie the similarities.

4. Four hundred million years ago, six-legged insect lineages arose from crustaceans that had multiple jointed legs. Given what you know about constraints on drastic changes in an organism's morphology, why do you suppose this particular drastic change was perpetuated?

5. By the traditional view, separation of the sexes has selective advantages of the sort outlined in Section 43.1. By a highly provocative, alternative view, males and females have been waging a *"gender war,"* at the level of genes, for billions of years. Matt Ridley's *Genome* (1999, New York: HarperCollins) may challenge your thinking.

Selected Key Terms

aging *43.7*	growth and tissue
asexual reproduction *43.1*	specialization *43.2*
blastocyst *43.3*	homeotic gene *43.5*
blastomere *43.2*	master gene *43.5*
blastula *43.3*	mesoderm *43.2*
cell differentiation *43.4*	morphogen *43.5*
cleavage *43.2*	morphogenesis *43.4*
cytoplasmic localization *43.3*	neural tube *43.4*
ectoderm *43.2*	oocyte *43.3*
embryo *43.2*	organ formation *43.2*
embryonic induction *43.5*	pattern formation *43.5*
endoderm *43.2*	pattern formation, theory of *43.5*
fate map *43.6*	sexual reproduction *43.1*
fertilization *43.2*	sperm *43.3*
gamete formation *43.2*	telomere *43.7*
gastrulation *43.2*	yolk *43.1*
gray crescent *43.3*	zygote *CI*

Readings

Caldwell, M. November 1992. "How Does a Single Cell Become a Whole Body?" *Discover* 13(11): 86–93.

Gilbert, S. 2000. *Developmental Biology*. Sixth edition. Sunderland, Massachusetts: Sinauer.

McGinnis, W., and M. Kuziora. February 1994. "The Molecular Architects of Body Design." *Scientific American* 270(2): 58–66.

Nusslein-Volhard, C. August 1996. "Gradients That Organize Embryonic Development." *Scientific American*, 54–61.

HUMAN REPRODUCTION AND DEVELOPMENT

Sex and The Mammalian Heritage

Sex and romance! Exposure to the contrived linkage between the two starts early and often in Western cultures. Ad agencies use sex to sell everything from movies to underwear; and they, together with scores of romance novelists, seem indifferent to how they might be trivializing the mammalian heritage.

For instance, Victoria's Secret aside, the breasts of a female mammal function to nourish offspring that are simply too vulnerable to survive on their own. Any mammalian mother that bonds with offspring is committing herself to protect it through an extended time of dependency and learning. Extended care helps the new individual survive until it is old enough to survive on its own. We can expect that it is a result of natural selection, for such behavior increases the odds of reproductive success in the new generation.

We see this among all mammals. Less than four months after mating, a lioness will give birth to a small, blind cub. She will not mate again for the next two years; her attention will be on that one vulnerable individual (Figure 44.1). Other females in her pride will assist her in its defense—often against the males, which tend to kill cubs when they get the chance.

Nine months or so after a human female becomes pregnant, she gives birth. Besides taking longer to develop inside its mother, the new individual requires more intensive care for a far longer time. It is about twelve years from a human birth to puberty, when the primary sex organs become operational. In the course of human evolution, the males also became behavioral partners with females. They, too, became protectors and providers for the young.

Figure 44.1 Glimpses into maternal care. This complex form of behavior has become most highly developed among mammals. The offspring of these sexually reproducing species require an extended period of development, dependency, and learning.

In the preceding chapter, you became acquainted with some principles that govern the reproduction and development of animals in general. Turn now to a case study of a representative mammal and see how those principles apply.

Start with what should be an obvious assumption: In nature, the main function of sex is not recreational but rather the perpetuation of one's genes. Those genes, remember, become packaged into gametes. Gametes of human males and females start forming in a pair of primary reproductive organs, or gonads. Sperm form in **testes** (singular, testis), the gonads of males. Eggs form in **ovaries**, the gonads of females.

The primary reproductive organs also secrete sex hormones that influence reproductive functions as well as the development of **secondary sexual traits**. Such traits are features we associate with maleness and femaleness, although they do not play a direct role in reproduction. Examples are the amount and distribution of body fat, hair, and skeletal muscles.

You already know that early human embryos have neither male nor female traits (Section 12.3). Seven weeks after fertilization, however, a pair of ovaries start to develop in the embryos that did not inherit a Y chromosome, which carries the master gene for sex determination. Testes develop only in XY embryos. Ovaries and testes will be fully formed at the time of birth. It will take about a decade for them to grow to their full size and become reproductively functional. And that is where our case study picks up.

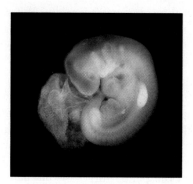

EARLY HUMAN EMBRYO

Key Concepts

1. The human reproductive system consists of a pair of primary reproductive organs, or gonads, as well as a number of accessory glands and ducts. Testes are sperm-producing male gonads, and ovaries are egg-producing female gonads.

2. In response to signals from the hypothalamus and the pituitary gland, gonads also release sex hormones. These hormones orchestrate reproductive functions and the development of secondary sexual traits.

3. Human males continually produce sperm from puberty onward. The hormones testosterone, LH, and FSH are central to the control of male reproductive functions.

4. From puberty onward, human females are fertile on a cyclic basis. Each month during their reproductive years, an egg is released from one of a pair of ovaries, and the lining of the uterus is primed for a possible pregnancy. The hormones estrogen, progesterone, FSH, and LH are central to controlling the cyclic activity.

5. As is the case for nearly all animals, embryonic development for humans starts with the formation of gametes and proceeds through fertilization, then cleavage, gastrulation, organ formation, and growth and tissue specialization.

REPRODUCTIVE SYSTEM OF HUMAN MALES

Figure 44.2 shows the primary reproductive organs of an adult human male. In this pair of testes, sperm start their formation and sex hormones are produced. Their production starts during puberty, when the secondary sexual traits emerge. Enlarging testes are often the first sign that a boy has entered puberty, and this usually happens between ages twelve and sixteen. Additional signs are a growth spurt, more hair sprouting on the face and elsewhere, and a deepening voice.

Where Sperm Form

Recall, from Section 12.3, that testes form on the wall of an XY embryo's abdominal cavity. Before birth, they descend into the scrotum, an outpouching of skin that is suspended beneath the pelvic region. At the time of birth, testes are fully formed miniatures of the adult organs. They start to produce sperm at puberty.

Figure 44.2a shows the scrotum's position in an adult. If sperm cells are to form and develop properly, the temperature inside the scrotum must remain a few degrees cooler than the core temperature of the rest of the body. A control mechanism stimulates and inhibits contraction of smooth muscles in the scrotum's wall. When air just outside the body becomes too cold, muscle contractions draw the pouch up, closer to the body mass, which is warmer. When it becomes warmer outside, the muscles relax and lower the pouch.

Packed within each testis are many small, highly coiled tubes, or **seminiferous tubules**. Sperm cells form and start to develop here in the manner described in the next section.

Where Semen Forms

Mammalian sperm travel from a testis through a series of ducts that lead to the urethra. They are not quite mature when they enter the first duct, the epididymis, which is long and coiled. Secretions from glandular cells located in the duct wall trigger events that put the finishing touches on maturing sperm cells. When they are fully mature, sperm are stored in the last stretch of each epididymis. During a male's reproductive years, mature sperm are stored in the epididymis until they are ejaculated from the body.

In a sexually aroused male, muscles in the walls of reproductive organs contract, which propels mature sperm into and through a pair of thick-walled tubes, the vasa deferentia (singular, vas deferens). Additional contractions propel the sperm even further, through a pair of ejaculatory ducts, then on through the urethra. This last tube in the series extends through the penis,

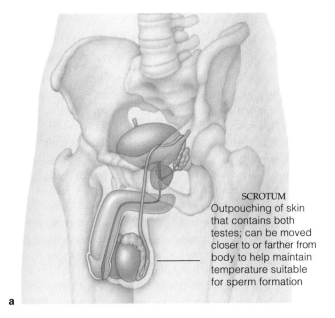

SCROTUM Outpouching of skin that contains both testes; can be moved closer to or farther from body to help maintain temperature suitable for sperm formation

a

Figure 44.2 *Above and facing page:* (**a**) Position of the human male reproductive system relative to the pelvic girdle and urinary bladder. The midsagittal section in (**b**) shows the components of the system and lists their functions, which also are summarized in Table 44.1.

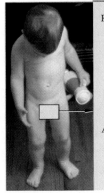

Table 44.1	*Organs and Accessory Glands of the Male Reproductive Tract*
REPRODUCTIVE ORGANS	
Testis (2)	Production of sperm, sex hormones
Epididymis (2)	Sperm maturation site and sperm storage
Vas deferens (2)	Rapid transport of sperm
Ejaculatory duct (2)	Conduction of sperm to penis
Penis	Organ of sexual intercourse
ACCESSORY GLANDS	
Seminal vesicle (2)	Secretion of large part of semen
Prostate gland	Secretion of part of semen
Bulbourethral gland (2)	Production of lubricating mucus

the male sex organ, and opens at its tip. The urethra is a duct that also functions in urinary excretion.

As sperm travel to the urethra, they are mixed with glandular secretions. The result is **semen**, a thick fluid expelled from the penis during sexual activity. While semen forms, paired seminal vesicles secrete fructose into it. Sperm use this sugar as an energy source.

Seminal vesicles also secrete prostaglandins that can cause muscles to contract. Possibly, these signaling molecules exert their effects during sexual activity. At that time, they may induce contractions in the female's

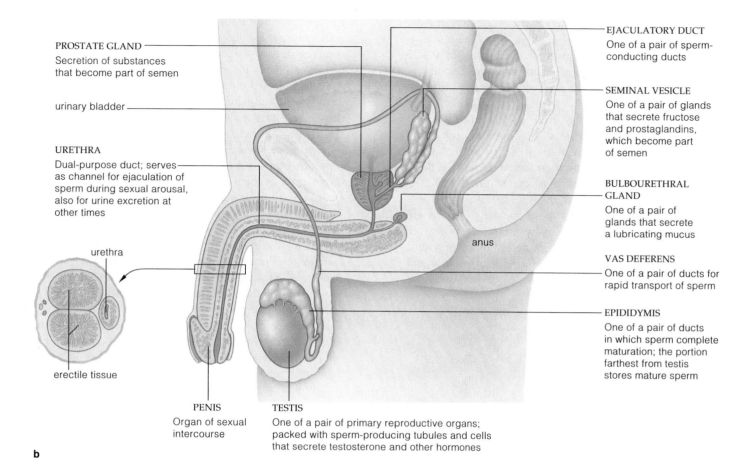

PROSTATE GLAND
Secretion of substances that become part of semen

urinary bladder

URETHRA
Dual-purpose duct; serves as channel for ejaculation of sperm during sexual arousal, also for urine excretion at other times

urethra

erectile tissue

PENIS
Organ of sexual intercourse

TESTIS
One of a pair of primary reproductive organs; packed with sperm-producing tubules and cells that secrete testosterone and other hormones

EJACULATORY DUCT
One of a pair of sperm-conducting ducts

SEMINAL VESICLE
One of a pair of glands that secrete fructose and prostaglandins, which become part of semen

BULBOURETHRAL GLAND
One of a pair of glands that secrete a lubricating mucus

VAS DEFERENS
One of a pair of ducts for rapid transport of sperm

EPIDIDYMIS
One of a pair of ducts in which sperm complete maturation; the portion farthest from testis stores mature sperm

anus

b

reproductive tract and thereby assist sperm movement through it, to the egg.

Secretions from a prostate gland may help buffer the acidic conditions in the female reproductive tract. (The pH of vaginal fluid is about 3.5–4.0, but sperm are able to swim more efficiently at pH 6.0.) Two bulbourethral glands secrete some mucus-rich fluid into the urethra when a male is sexually aroused.

Cancers of the Prostate and Testis

Until recently, cancers of the male reproductive tract did not get much media coverage. Yet *prostate cancer* is the second leading cause of death among men, exceeded only by cancers of the respiratory tract.

At this writing, an estimated 189,000 new cases may be diagnosed in 2002, and close to 32,000 men may die. The death rate for breast cancer, which is publicized to a far greater degree, is not much higher. Even so, this is a dramatic decline from the prostate cancer rates for 1989–1992, possibly as a result of widespread, earlier diagnostic screenings. In addition to prostate cancer, *testicular cancer* is a frequent cause of death among young men. About 5,000 cases are diagnosed each year.

Both cancers are painless in the early stages. When they are not detected in time, they spread silently into lymph nodes of the abdomen, chest, neck, and then the lungs. Once the cancer metastasizes, prospects are not good. Testicular cancer, for example, now kills as many as half of those stricken.

Doctors can detect prostate cancer by blood tests for increases in prostate-specific antigen (PSA) and by physical examinations. Once a month from high school onward, men should examine each testis after a warm bath or shower, when scrotal muscles are relaxed. The testis should be rolled gently between the thumb and the forefinger to check for an enlargement, hardening, or lump. Such changes may or may not cause obvious discomfort. But they must be reported so a physician may order a full examination. Treatment of testicular cancer has one of the highest success rates, provided that the cancer is caught before it starts to spread.

Human males have a pair of testes—primary reproductive organs that produce sperm and sex hormones—as well as accessory glands and ducts. The hormones influence sperm formation and the development of secondary sexual traits.

MALE REPRODUCTIVE FUNCTION

Sperm Formation

Although smaller than a golfball, each testis holds 125 meters or so of seminiferous tubules. Up to 300 wedge-shaped lobes partition the interior, and each holds two or three coiled tubules (Figures 44.3 and 44.4).

Inside a tubule wall are undifferentiated cells called spermatogonia (singular, spermatogonium). Ongoing cell divisions force them away from the wall, toward the tubule interior. During their forced departure, these cells undergo mitotic cell division. Daughter cells, the primary spermatocytes, are the ones that enter meiosis.

As Figure 44.4a shows, these nuclear divisions are accompanied by *incomplete* cytoplasmic divisions; the developing cells remain interconnected by cytoplasmic bridges. Ions and molecules required for development move freely across the bridges, so successive divisions

from every spermatogonium result in a generation of cells that mature together. These forerunners of sperm receive nourishment and signals from **Sertoli cells**, one of two cell types in seminiferous tubules.

Secondary spermatocytes form by way of meiosis I. The chromosomes of these haploid cells are still in the duplicated state; each consists of two sister chromatids. (Here you may wish to check the general overview of spermatogenesis in Section 10.5.) The sister chromatids separate from each other during meiosis II, and then daughter cells form. These are haploid spermatids that gradually develop into sperm, the male gametes.

Each mature sperm is a flagellated cell. Its head is packed with DNA and its tail has a microtubular core (Figure 44.4b). Extending over most of the head is a cap (acrosome). Enzymes inside the cap will help the sperm penetrate extracellular material around the egg prior to fertilization. Mitochondria in a midpiece behind the head supply energy for the tail's whiplike motions.

It takes nine to ten weeks for each sperm to form. From puberty onward, a human male produces sperm on a continuous basis, so many millions of cells are in different stages of development on any given day.

Hormonal Controls

Coordinated secretions of LH, FSH, testosterone, and other hormones govern reproductive function in males. Take a look at Figure 44.3b and notice the **Leydig cells** between lobes in the testes. They secrete **testosterone**, a steroid hormone essential for the growth, form, and function of the male reproductive tract. Besides having the key role in sperm formation, testosterone promotes development of male secondary sexual traits. It also stimulates sexual behavior and aggressive behavior.

LH and **FSH**, recall, are secretions of the anterior lobe of the pituitary gland (Section 36.3). Initially, both hormones were named for their effects in females. (One

seminal vesicle

vas deferens

prostate gland

bulbo-urethral gland

urethra

epididymis

SEMINIFEROUS TUBULE

penis

testis

a

wall of seminiferous tubule   Leydig cells between tubules

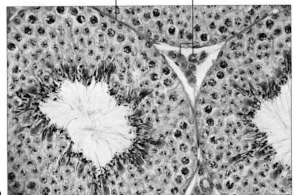

b

Figure 44.3 (a) Male reproductive tract, posterior view. Arrows show the route that sperm take before ejaculation from a sexually aroused male. (b) Light micrograph of cells in three adjacent seminiferous tubules, cross-section. Leydig cells occupy tissue spaces between the tubules.

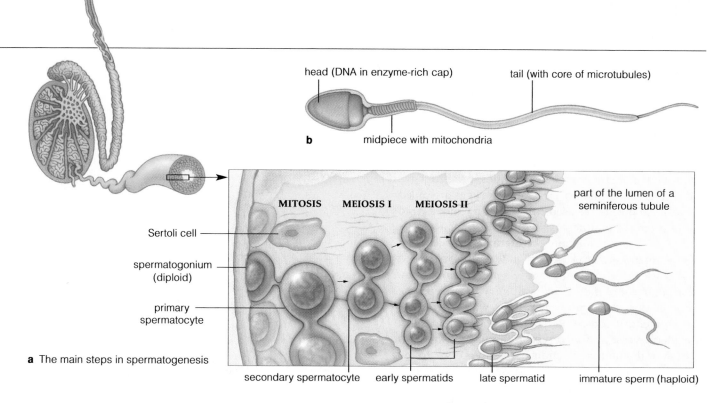

head (DNA in enzyme-rich cap)   tail (with core of microtubules)

b   midpiece with mitochondria

MITOSIS   MEIOSIS I   MEIOSIS II

part of the lumen of a
seminiferous tubule

Sertoli cell

spermatogonium
(diploid)

primary
spermatocyte

a The main steps in spermatogenesis

secondary spermatocyte   early spermatids   late spermatid   immature sperm (haploid)

Figure 44.4 (**a**) Sperm formation. The process starts with a spermatogonium, a type of diploid germ cell. Mitosis, meiosis, and incomplete cytoplasmic divisions result in a group of immature haploid cells. These differentiate and develop into mature sperm. (**b**) Structure of a mature sperm from a human male.

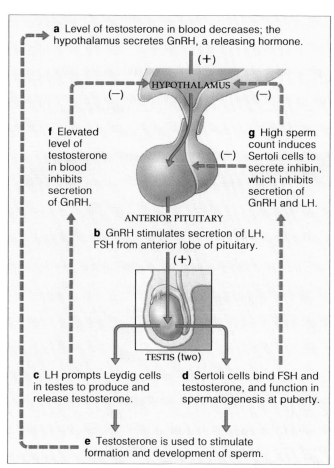

a Level of testosterone in blood decreases; the hypothalamus secretes GnRH, a releasing hormone.

(+)

HYPOTHALAMUS

(−)   (−)

f Elevated level of testosterone in blood inhibits secretion of GnRH.

g High sperm count induces Sertoli cells to secrete inhibin, which inhibits secretion of GnRH and LH.

(−)

ANTERIOR PITUITARY

b GnRH stimulates secretion of LH, FSH from anterior lobe of pituitary.

(+)

TESTIS (two)

c LH prompts Leydig cells in testes to produce and release testosterone.

d Sertoli cells bind FSH and testosterone, and function in spermatogenesis at puberty.

e Testosterone is used to stimulate formation and development of sperm.

Figure 44.5 Negative feedback loops to the hypothalamus and to the pituitary gland's anterior lobe from the testes. With these loops, excess levels of testosterone in blood slow the mechanisms leading to its production. This helps maintain the testosterone level in amounts required for sperm formation.

abbreviation stands for *Luteinizing Hormone*, the other for *Follicle-Stimulating Hormone*.) We now know their molecular structure is identical in males and females.

The hypothalamus controls secretion of LH, FSH, and testosterone, and thus controls sperm formation. When blood levels of testosterone and other factors are low, it secretes the releasing hormone **GnRH** (Figure 44.5). GnRH calls on the anterior lobe to step up the release of LH and FSH, which have targets in testes.

LH prods Leydig cells to secrete testosterone, which assists in stimulating the formation and development of sperm. Sertoli cells have receptors for FSH, which is required to jump-start spermatogenesis at puberty. We don't know whether FSH is also necessary for normal functioning of mature human testes.

Figure 44.5 also shows how feedback loops to the hypothalamus can cut back on testosterone secretion and sperm formation. An elevated testosterone level in blood slows down the release of GnRH. Also, when the sperm count is high, Sertoli cells release **inhibin**. This protein hormone acts on the hypothalamus and pituitary to cut back on the release of GnRH and FSH.

Sperm formation depends on the hormones LH, FSH, and testosterone. Negative feedback loops from the testes to the hypothalamus and pituitary gland control their secretion.

REPRODUCTIVE SYSTEM OF HUMAN FEMALES

The Reproductive Organs

We turn next to the reproductive system of the human female. Figure 44.6 shows its components, and Table 44.2 summarizes their functions. The female's *primary* reproductive organs, her pair of ovaries, produce eggs and secrete sex hormones. Her immature eggs are called **oocytes**. An oocyte released from an ovary moves into one of a pair of oviducts (also called Fallopian tubes). An oviduct is a channel into the **uterus**, a pear-shaped, hollow organ in which embryos grow and develop.

A thick layer of smooth muscle, the myometrium, makes up most of the uterine wall. As you will see, the **endometrium**, the inner lining of the wall, is vital for development. It consists of connective tissues, glands, and blood vessels. The cervix is the narrowed-down portion of the uterus. Extending from the cervix to the body surface is the vagina. This muscular tube receives sperm and functions as part of the birth canal.

At the body's surface are external genitals (vulva), including organs for sexual stimulation. Outermost are a pair of fat-padded skin folds (the labia majora). They enclose a smaller pair of skin folds (the labia minora), which are highly vascularized but have no fatty tissue. The smaller folds partly enclose the clitoris, a sex organ highly sensitive to stimulation. At the body surface, the urethra's opening is positioned about midway between the clitoris and the vaginal opening.

Overview of the Menstrual Cycle

Females of most mammalian species follow an *estrous* cycle. They are fertile and in heat (sexually receptive to males) only at certain times of year. By contrast, female primates, including humans, follow a **menstrual cycle**. They become fertile intermittently on a cyclic basis, and heat is not synchronized with the fertile periods. Said another way, female primates of reproductive age can become pregnant only at certain times during the year, but they may be receptive to sex at any time.

Briefly, an oocyte matures and is released from an ovary during each menstrual cycle. Also during the cycle, the endometrium becomes primed to receive and nourish a forthcoming embryo *if* a sperm penetrates the oocyte and fertilization occurs. However, if that oocyte does not get fertilized, then about four to six tablespoons of blood-rich fluid from the uterus will start flowing out through the vaginal canal. Such a recurring blood flow is called menstruation. It means "there is no embryo at this time," and it marks the first day of a new cycle. The uterine nest is being sloughed off and is about to be constructed once again.

The events just sketched out proceed through three phases. The cycle starts with a *follicular* phase. This is

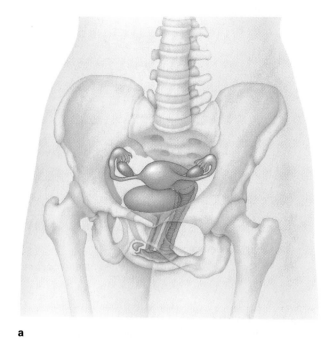

a

Figure 44.6 *Above and facing page:* (**a**) Position of the human female reproductive system relative to the pelvic girdle and urinary bladder. The midsagittal section in (**b**) shows the components of the system and lists their functions.

Table 44.2	*Female Reproductive Organs*
Ovaries	Oocyte production and maturation, sex hormone production
Oviducts	Ducts for conducting oocyte from ovary to uterus; fertilization normally occurs here
Uterus	Chamber in which new individual develops
Cervix	Secretion of mucus that enhances sperm movement into uterus and (after fertilization) reduces embryo's risk of bacterial infection
Vagina	Organ of sexual intercourse; birth canal

the time of menstruation, endometrial breakdown and rebuilding, and oocyte maturation. The next phase is limited to an oocyte's release from the ovary. We call it **ovulation**. In the cycle's *luteal* phase, the corpus luteum (an endocrine structure) forms in the ovary, and the endometrium is primed for pregnancy (Table 44.3).

All three phases are governed by feedback loops to the hypothalamus and pituitary gland from the ovaries. FSH and LH promote cyclic changes in the ovaries. As you will see, FSH and LH also stimulate the ovaries to

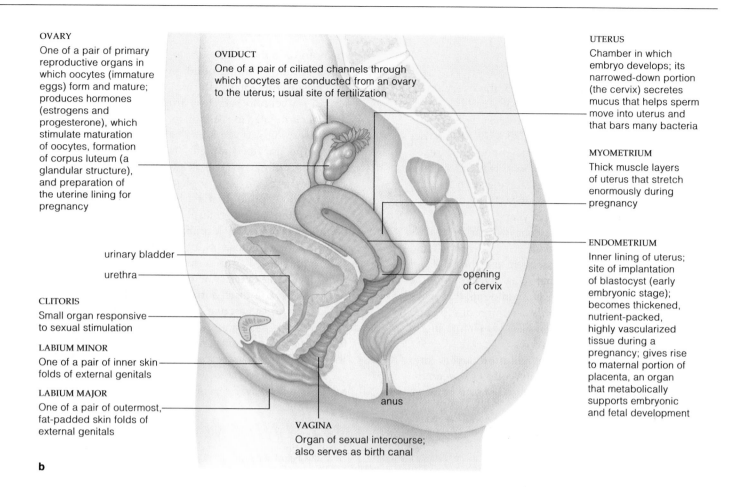

OVARY
One of a pair of primary reproductive organs in which oocytes (immature eggs) form and mature; produces hormones (estrogens and progesterone), which stimulate maturation of oocytes, formation of corpus luteum (a glandular structure), and preparation of the uterine lining for pregnancy

OVIDUCT
One of a pair of ciliated channels through which oocytes are conducted from an ovary to the uterus; usual site of fertilization

urinary bladder

urethra

CLITORIS
Small organ responsive to sexual stimulation

LABIUM MINOR
One of a pair of inner skin folds of external genitals

LABIUM MAJOR
One of a pair of outermost, fat-padded skin folds of external genitals

VAGINA
Organ of sexual intercourse; also serves as birth canal

anus

opening of cervix

UTERUS
Chamber in which embryo develops; its narrowed-down portion (the cervix) secretes mucus that helps sperm move into uterus and that bars many bacteria

MYOMETRIUM
Thick muscle layers of uterus that stretch enormously during pregnancy

ENDOMETRIUM
Inner lining of uterus; site of implantation of blastocyst (early embryonic stage); becomes thickened, nutrient-packed, highly vascularized tissue during a pregnancy; gives rise to maternal portion of placenta, an organ that metabolically supports embryonic and fetal development

b

Table 44.3	*Events of the Menstrual Cycle*	
Phase	Events	Days of Cycle*
Follicular phase	Menstruation; endometrium breaks down	1–5
	Follicle matures in ovary; endometrium rebuilds	6–13
Ovulation	Oocyte released from ovary	14
Luteal phase	Corpus luteum forms, secretes progesterone; the endometrium thickens and develops	15–28

* Assuming a 28-day cycle.

secrete sex hormones—**estrogens** and **progesterone**—to promote the cyclic changes in the endometrium.

A human female's menstrual cycles start between ages ten and sixteen. Each cycle lasts for about twenty-eight days, but this is only the average. It runs longer for some women and shorter for others. The menstrual cycles continue until a woman is in her late forties or early fifties, when her supply of eggs is dwindling and secretions of hormones slow down. This is the onset of *menopause*, the twilight of reproductive capacity.

You may have heard of *endometriosis*, a condition that arises after endometrial tissue abnormally spreads and grows outside the uterus. Endometrial scar tissue may form on ovaries or oviducts and lead to infertility. In the United States, 10 million women may be affected annually. Perhaps the condition arises when menstrual flow backs up through the oviducts and spills into the pelvic cavity. Or perhaps some embryonic cells became positioned in the wrong place before birth and were stimulated to grow when sex hormones became active at puberty. Whatever the case, estrogen still stimulates cells in the mislocated tissue. The resulting symptoms include pain during menstruation, sex, and urination.

Ovaries, the female primary reproductive organs, produce oocytes (immature eggs) and sex hormones. Endometrium lines the uterus, a chamber in which embryos develop.

Secretion of sex hormones (estrogens and progesterone) is coordinated on a cyclic basis through the reproductive years.

A menstrual cycle starts when the endometrium breaks down and blood is released. The endometrium is rebuilt and an oocyte matures. After ovulation (release of an oocyte from an ovary), the cycle ends with formation of a corpus luteum and an endometrium that is primed for pregnancy.

FEMALE REPRODUCTIVE FUNCTION

Cyclic Changes in the Ovary

Take a moment to review Section 10.5, the overview of meiosis in an oocyte. Each normal baby girl has about 2 million primary oocytes in her ovaries. By the time she is seven years old, only about 300,000 remain; her body resorbed the rest. Her primary oocytes already started meiosis, but nuclear division was arrested. It resumes in one oocyte at a time, starting with the first cycle of menstruation. About 400 to 500 oocytes will be released during her reproductive years.

Figure 44.7 shows a primary oocyte near an ovary's surface. A cell layer (of granulosa cells) surrounds and nourishes it. We call a primary oocyte and the cell layer around it a **follicle**. As the menstrual cycle starts, the hypothalamus is secreting GnRH in amounts that make the anterior pituitary step up *its* secretion of FSH and LH. The increase in hormone molecules circulating in blood causes the follicle to grow.

The oocyte increases in size, and more layers form around it. Deposits of glycoproteins collect between the cell layers and oocyte, widening the distance between them. In time they make up the **zona pellucida**. This is a noncellular coating around the oocyte.

Outside the zona pellucida, FSH and LH stimulate granulosa cells to secrete estrogens. An estrogen-rich fluid accumulates in the follicle, and the blood estrogen level rises. Eight to ten hours before its release from the ovary, the oocyte completes meiosis I. Its cytoplasm divides, forming two cells. One, the **secondary oocyte,** gets nearly all of the cytoplasm. The other cell is the first of three **polar bodies**. The meiotic parceling of chromosomes to four cells gives the secondary oocyte a haploid chromosome number—which is the number required for gametes and for sexual reproduction.

About halfway through each menstrual cycle, the pituitary gland detects the rise in estrogens in blood. It responds with a brief outpouring of LH. The surge in LH causes rapid vascular changes that make the follicle swell. It makes enzymes digest the bulging wall of the follicle, which weakens and ruptures. Fluid, along with the secondary oocyte, escapes (Figures 44.7 and 44.8).

Thus, the midcycle surge of LH triggers ovulation—the release of a secondary oocyte from the ovary.

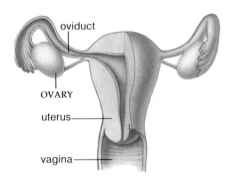

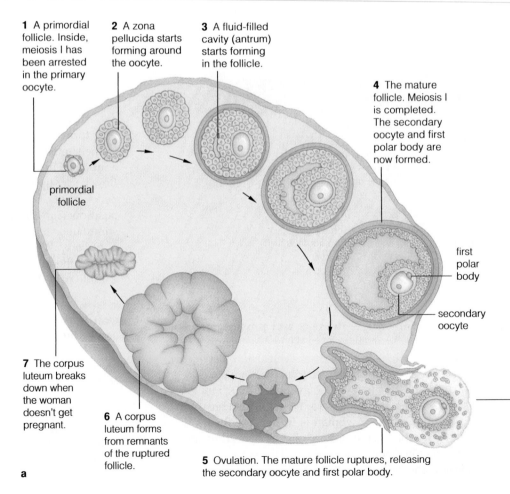

1 A primordial follicle. Inside, meiosis I has been arrested in the primary oocyte.

2 A zona pellucida starts forming around the oocyte.

3 A fluid-filled cavity (antrum) starts forming in the follicle.

4 The mature follicle. Meiosis I is completed. The secondary oocyte and first polar body are now formed.

primordial follicle

first polar body

secondary oocyte

7 The corpus luteum breaks down when the woman doesn't get pregnant.

6 A corpus luteum forms from remnants of the ruptured follicle.

5 Ovulation. The mature follicle ruptures, releasing the secondary oocyte and first polar body.

Figure 44.7 (**a**) Cyclic events in a human ovary, cross-section. A follicle stays in the same place inside an ovary all through one menstrual cycle. It does not "move around" as in this diagram, which shows the *sequence* of events.

In the cycle's first phase, a follicle grows and matures. At ovulation, the second phase, the mature follicle ruptures and releases a secondary oocyte. In the third phase, a corpus luteum forms from remnants of the follicle. If the woman doesn't get pregnant during the cycle, the corpus luteum will self-destruct.

Facing page: (**b**) A secondary oocyte being released from an ovary. It will enter an oviduct, the passageway to the uterus. (**c**) Billowing folds of an oviduct's inner surface.

oviduct
OVARY
uterus
vagina

a

Figure 44.8 Feedback control of hormone secretion in a menstrual cycle. A positive feedback loop from an ovary to the hypothalamus causes a surge in LH secretion that triggers ovulation. After release of a secondary oocyte from the ovary, negative feedback loops to the hypothalamus and pituitary inhibit FSH secretion. They block maturation of another follicle until the cycle is over.

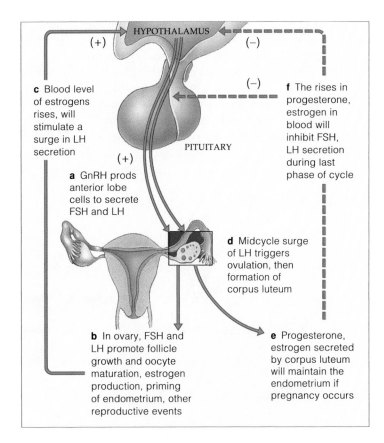

c Blood level of estrogens rises, will stimulate a surge in LH secretion

a GnRH prods anterior lobe cells to secrete FSH and LH

f The rises in progesterone, estrogen in blood will inhibit FSH, LH secretion during last phase of cycle

d Midcycle surge of LH triggers ovulation, then formation of corpus luteum

b In ovary, FSH and LH promote follicle growth and oocyte maturation, estrogen production, priming of endometrium, other reproductive events

e Progesterone, estrogen secreted by corpus luteum will maintain the endometrium if pregnancy occurs

Cyclic Changes in the Uterus

Estrogens released early in the menstrual cycle help pave the way for pregnancy. They stimulate growth of the endometrium and its glands. Before the midcycle surge of LH, follicle cells secrete some progesterone and estrogens. Blood vessels grow fast in the thickened endometrium. At ovulation, the estrogens act on tissues of the cervical canal to the vagina. The cervix starts to secrete large quantities of a thin, clear mucus that is an ideal medium for sperm to swim through.

The midcycle surge of LH that causes ovulation also causes a glandular structure, the **corpus luteum,** to form from granulosa cells left behind in the ruptured follicle. Its secretions influence the rest of the cycle.

The corpus luteum also secretes progesterone and some estrogen. Progesterone prepares the reproductive tract for the arrival of a **blastocyst,** the type of blastula formed from a fertilized mammalian egg (Section 43.3). For example, this hormone makes cervical mucus turn thick and sticky. The mucus may keep normal bacterial residents of the vagina out of the uterus. Progesterone also will maintain the endometrium during pregnancy.

A corpus luteum persists for about twelve days. All the while, the hypothalamus is calling for minimal FSH secretion, which stops other follicles from developing. If a blastocyst does not burrow into the endometrium, the corpus luteum will secrete certain prostaglandins that cause it to self-destruct in the last days of the cycle.

After this, progesterone and estrogen levels in blood decline quickly, and so the endometrium starts to break down. Deprived of oxygen and nutrients, the lining's blood vessels constrict and tissues die. Blood escapes as weakened capillary walls start to rupture. Blood as well as the sloughed endometrial tissues make up the menstrual flow, which continues for three to six days. Then the cycle begins anew, with rising estrogen levels stimulating repair and growth of the endometrium.

Eggs released late in a woman's life are at some risk of changes in chromosome number or structure when meiosis does resume. One possible outcome is that the newborn will develop Down syndrome (Section 12.9). By menopause, the supply of oocytes is dwindling, hormone secretions slow down, and in time menstrual cycles—and fertility—will be over.

During one menstrual cycle, FSH and LH stimulate growth of an ovarian follicle (a primary oocyte and the surrounding layer of cells). The first meiotic cell division results in one secondary oocyte and the first polar body.

A midcycle surge of LH triggers ovulation—the release of the secondary oocyte and the polar body from the ovary.

Early on, estrogens call for endometrial repair and growth. Estrogens and progesterone prepare the endometrium and other parts of the reproductive tract for possible pregnancy.

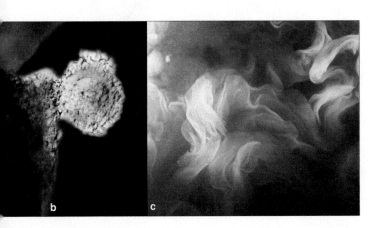

b   **c**

VISUAL SUMMARY OF THE MENSTRUAL CYCLE

By now, you probably have come to the conclusion that the menstrual cycle is not a simple tune on a biological banjo. It's a full-blown hormonal symphony!

Before continuing with your reading, take a moment to review Figure 44.9. It correlates the cyclic events in the ovary and the uterus with the changes in hormone concentrations that trigger these events. This summary illustration may leave you with a better understanding of what goes on.

Figure 44.9 Changes in the ovary and uterus, as correlated with changing hormone levels during each turn of the menstrual cycle. *Green* arrows show which hormones dominate the cycle's first phase (when the follicle matures), and then the second phase (when the corpus luteum forms).

(**a–c**) FSH and LH secretions bring about the changes in ovarian structure and function. (**d,e**) Estrogen and progesterone secretions from the ovary stimulate the changes in the endometrium.

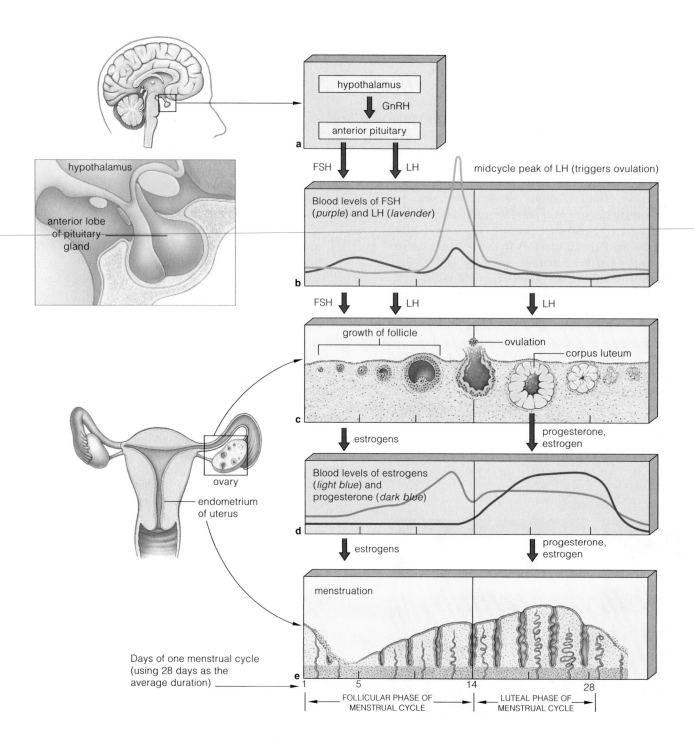

PREGNANCY HAPPENS

Sexual Intercourse

Suppose the secondary oocyte is on its way down an oviduct when a female and male are engaged in sexual intercourse, or **coitus**. The male sex act requires *erection*, whereby a normally limp penis stiffens and lengthens; and *ejaculation*, the forceful expulsion of semen into the urethra and out from the penis. As Figure 44.2 shows, the penis incorporates cylinders of spongy tissue. Many friction-activated sensory receptors are on its mushroom-shaped tip, the glans penis. In males *not* sexually aroused, the large blood vessels leading into the cylinders are vasoconstricted. In aroused males, they vasodilate and blood flows into the penis faster than it flows out, so blood collects in the spongy tissue. The resulting engorgement makes the organ stiffen and lengthen, this being helpful for penetration into the vaginal canal.

During coitus, pelvic thrusts stimulate the penis, the female's clitoris, and the vaginal wall. The mechanical stimulation induces involuntary contractions in the male reproductive tract. It forces sperm from the epididymes and the contents of seminal vesicles and prostate gland into the urethra. The substances mix and form semen, which is then ejaculated inside the vagina. A sphincter closes the bladder's neck and prevents urination during ejaculation.

Emotional intensity, hard breathing, heart pounding, and the contraction of skeletal muscles in general accompany rhythmic throbbing of pelvic muscles. During *orgasm*, the end of the sex act, strong sensations of physical release, warmth, and relaxation dominate. Similar sensations typify female orgasm. It's a common misconception that a female cannot become pregnant if she doesn't reach orgasm. Don't believe it.

Fertilization

An ejaculation can put 150 million to 350 million sperm in the vagina. If they arrive a few days before or after ovulation or any time between, fertilization can occur. Less than thirty minutes after sperm arrive, contractions move them deep into the female's reproductive tract. Just a few hundred actually reach the upper part of the oviduct, where fertilization usually takes place.

Think back on the micrograph that opens Unit II, which depicts sperm around a secondary oocyte. When

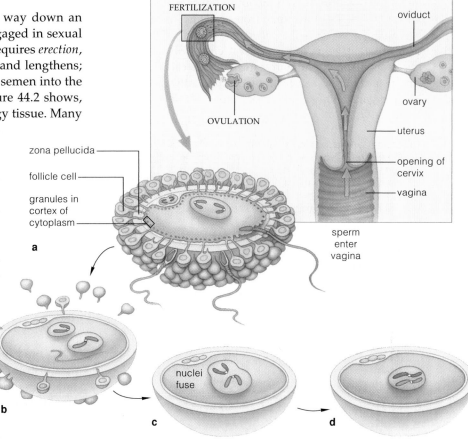

Figure 44.10 Fertilization. (**a**) Many sperm surround a secondary oocyte. Acrosomal enzymes clear a path through the zona pellucida. (**b**) When a sperm does penetrate the secondary oocyte, granules in the egg cortex release substances that make the zona pellucida impenetrable to other sperm. Penetration also stimulates meiosis II of the oocyte's nucleus. (**c**) The sperm tail degenerates; its nucleus enlarges and fuses with the oocyte nucleus. (**d**) At fusion, fertilization is over. The zygote has formed.

sperm reach an oocyte, they release digestive enzymes that clear a path through the zona pellucida. Although many sperm might get this far, usually only one fuses with the oocyte (Figure 44.10). It degenerates inside the oocyte's cytoplasm until only its nucleus and centrioles remain. Penetration induces the secondary oocyte and the first polar body to finish meiosis II.

There are now three polar bodies and one mature egg, or **ovum** (plural, ova). The sperm nucleus and egg nucleus fuse. Together, their chromosomes restore the diploid number for a brand new zygote.

The intense physiological events that accompany coitus have one function: to put sperm on a collision course with an egg. Fertilization is over with the fusion of a sperm nucleus and egg nucleus, which results in a diploid zygote.

FORMATION OF THE EARLY EMBRYO

Pregnancy lasts an average of thirty-eight weeks from the time of fertilization. It takes about two weeks for a blastocyst to form. The time span from the third to the end of the eighth week is the *embryonic* period, when the major organ systems form. When it ends, the new individual has distinctly human features and is called a **fetus**. In the *fetal* period from the start of the ninth week until birth, organs enlarge and become specialized.

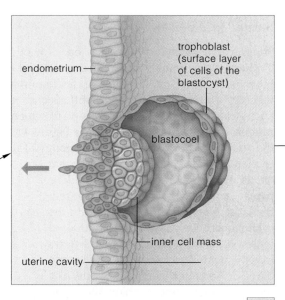

d DAY 5. A blastocoel (a fluid-filled cavity) forms in the morula as a result of secretions from the surface cells. By the thirty-two-cell stage, cells of an inner cell mass are already differentiating. They will give rise to the embryo proper. This embryonic stage is called a blastocyst.

c DAY 4. By 96 hours, there is a ball of sixteen to thirty-two cells that is shaped like a mulberry. It is a morula (after *morum*, Latin for mulberry). Cells of the surface layer will function in implantation and will give rise to a membrane, the chorion.

b DAY 3. After the third cleavage, the cells suddenly huddle together into a compacted ball, which becomes stabilized by numerous tight junctions among the outer cells. Gap junctions form among the interior cells and enhance intercellular communication.

a DAYS 1–2. Cleavage begins within 24 hours after fertilization. The first cleavage furrow extends between the two polar bodies. Subsequent cuts are rotational, so the resulting cells are not symmetrically arranged (compare Section 43.3). Until the eight-cell stage forms, the cells are loosely arranged, with considerable space between them.

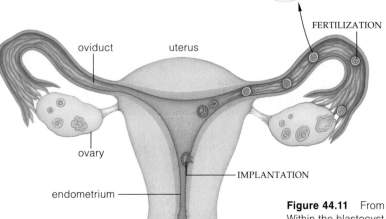

e DAYS 6–7. Surface cells of the blastocyst attach to the endometrium and start to burrow into it. Implantation is under way.

actual size

We typically call the first three months of pregnancy the *first* trimester. A *second* trimester extends from the start of the fourth month to the end of the sixth. The *third* trimester extends from the seventh month until birth. Beginning with Figure 44.11, the next series of illustrations shows characteristic features of the new individual at progressive stages of development.

Cleavage and Implantation

Three to four days after fertilization, a zygote is already undergoing cleavage as it tumbles through the oviduct. Genes are being expressed; early divisions require their products. At the eight-cell stage, the cells huddle into a compact ball. By the fifth day, a blastocyst has formed. It has a surface layer of cells (the trophoblast), a cavity filled with their secretions (blastocoel), and a cluster of cells called an inner cell mass (Figure 44.11*d*).

Six or seven days after fertilization, **implantation** is under way. By this process, the blastocyst adheres to the uterine lining, some of its cells send out projections that invade the mother's tissues, and connections start to form that will metabolically support the developing embryo through the months ahead. While the invasion

Figure 44.11 From fertilization through implantation. A blastocyst forms at cleavage. Within the blastocyst, the inner cell mass gives rise to the disk-shaped early embryo. Three extraembryonic membranes (the amnion, chorion, and yolk sac) start forming. A fourth membrane (allantois) forms after the blastocyst is implanted.

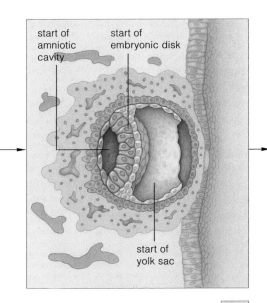

start of
amniotic
cavity

start of
embryonic disk

start of
yolk sac

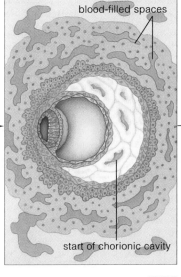

blood-filled spaces

start of chorionic cavity

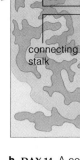

chorionic
villi

chorion

chorionic
cavity

amniotic
cavity

connecting
stalk

yolk sac

f DAYS 10–11. The yolk sac, embryonic disk, and amniotic cavity have started to form from parts of the blastocyst.

actual size

g DAY 12. Blood-filled spaces form in maternal tissue. The chorionic cavity starts to form.

actual size

h DAY 14. A connecting stalk has formed between the embryonic disk and chorion. Chorionic villi, which will be features of a placenta, start to form.

actual size

is proceeding, the inner cell mass develops into two cell layers of a flattened and somewhat circular shape. Those two layers make up the embryonic disk—and in short order they will give rise to the embryo proper.

Extraembryonic Membranes

As implantation progresses, membranes start to form outside the embryo. First a fluid-filled *amniotic* cavity opens up between the embryonic disk and part of the blastocyst's surface (Figure 44.11*f*). Then cells migrate around the wall of the cavity and form the **amnion**, a membrane that will enclose the embryo. Fluid inside the cavity will function as a buoyant cradle where the embryo can grow, move freely, and be protected from abrupt temperature changes and mechanical impacts.

While the amnion forms, other cells migrate around the inner wall of the blastocyst's first cavity. They form a lining that becomes a **yolk sac**. This extraembryonic membrane speaks of the evolutionary heritage of land vertebrates (Section 26.8). For most animals that make shelled eggs, the sac holds nutritive yolk. In humans, part of it becomes a site of blood cell formation. Part will give rise to germ cells, the forerunners of gametes.

Before a blastocyst is fully implanted, spaces open in maternal tissues and fill with blood seeping in from ruptured capillaries. Inside the blastocyst, a different cavity opens up around the amnion and yolk sac. Now fingerlike projections start to form on the cavity lining, which is the **chorion**. This new membrane will become part of the placenta, a spongy, blood-engorged tissue.

After the blastocyst is implanted, an outpouching of the yolk sac will become the third extraembryonic membrane—the **allantois**. The allantois has different functions in different animal groups. Among reptiles, birds, and some mammals, it serves in respiration and in storing metabolic wastes. In humans, blood vessels for the placenta and the urinary bladder form from it.

One more point should be made here. Cells of the blastocyst secrete a hormone, **HCG** (*H*uman *C*horionic *G*onadotropin), which stimulates the corpus luteum to keep on secreting progesterone and estrogen. In this way, the blastocyst itself prevents menstrual flow and works to avoid being sloughed off until the placenta takes over the function of secreting HCG, about eleven weeks later. By the start of the third week, the HCG can be detected in samples of the mother's blood or urine. At-home *pregnancy tests* have a treated "dip-stick" that changes color when HCG is present in urine.

A human blastocyst is composed of a surface layer of cells around a fluid-filled cavity (blastocoel) and an inner cell mass, which will give rise to the embryo proper.

Six or seven days after fertilization, the blastocyst implants itself in the endometrium. Now projections from its surface invade maternal tissues, and connections start to form that in time will metabolically support the developing embryo.

Some parts of the blastocyst give rise to an amnion, yolk sac, chorion, and allantois. These extraembryonic membranes serve different functions. Together they are vital for the structural and functional development of the embryo.

EMERGENCE OF THE VERTEBRATE BODY PLAN

By the time a woman has missed her first menstrual period, cleavage is completed and gastrulation is under way. **Gastrulation**, recall, is a stage when extensive cell divisions, migrations, and rearrangements produce the primary tissue layers (Sections 43.2 and 43.4).

Earlier, the inner cell mass had behaved *as if* it were still developing on the large ball of yolk that evolved in the reptilian ancestors of mammals. That is why it develops into a *flattened* two-layer embryonic disk, as in reptiles and birds. By now, it is surrounded by the amnion and chorion except where a stalk connects it to the chorion's inner wall. Cell divisions and migrations of one layer formed the yolk sac lining. The other layer now starts to generate the embryo proper.

For example, by the eighteenth day, the embryonic disk has two neural folds that will merge to form the neural tube (Figure 44.12b). Early in the third week, a sausage-shaped outpouching forms on the yolk sac. It is the allantois (Greek *allas*, meaning sausage). In humans, recall, it only helps form the urinary bladder and blood vessels for the placenta. Some mesoderm folds into a tube that becomes the notochord. A notochord is only a structural framework; bony segments of the vertebral column will be forming around it. Toward the end of the third week, some mesoderm gives rise to multiple, paired segments, or **somites**. These are the embryonic source of most bones, of skeletal muscles of the head and trunk, and of the dermis overlying these regions.

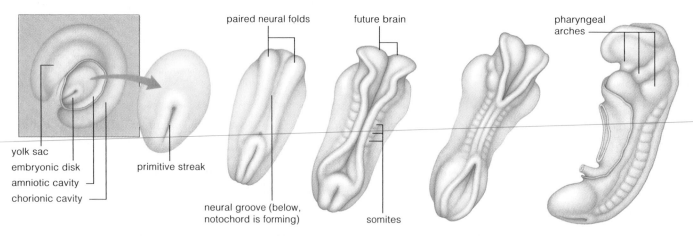

paired neural folds   future brain   pharyngeal arches

yolk sac
embryonic disk
amniotic cavity
chorionic cavity

primitive streak

neural groove (below, notochord is forming)   somites

a DAY 15. A faint band appears around a depression along the axis of the embryonic disk. This is the primitive streak, and it marks the onset of gastrulation in vertebrate embryos.

b DAYS 18–23. Organs start to form through cell divisions, cell migrations, tissue folding, and other events of morphogenesis. Neural folds will merge to form the neural tube. Somites (bumps of mesoderm) appear near the embryo's dorsal surface. They will give rise to most of the skeleton's axial portion, skeletal muscles, and much of the dermis.

c DAYS 24–25. By now, some embryonic cells have given rise to pharyngeal arches. These will contribute to the formation of the face, neck, mouth, nasal cavities, larynx, and pharynx.

Figure 44.12 Hallmarks of the embryonic period of humans and other vertebrates. A primitive streak, then a notochord form. Neural folds, somites, and pharyngeal arches form later. (**a,b**) Dorsal views, of the embryo's back. (**c**) Side view. Compare the diagrams with the Figure 44.14 photographs.

Through cell divisions and migrations, the midline of that layer thickens faintly around a depression at its surface. This "primitive streak" lengthens and thickens further the next day. It marks the onset of gastrulation (Figure 44.12a). It also defines the anterior–posterior axis of the embryo and, in time, its bilateral symmetry. Endoderm and mesoderm form from cells migrating inward at this axis. Now pattern formation begins, as described in Section 43.5. Embryonic inductions and interactions among classes of master genes map out the basic body plan characteristic of all vertebrates. And specialized tissues and organs start forming in orderly sequences, in prescribed parts of the embryo.

Pharyngeal arches start to form that will contribute to the face, neck, mouth, nose, larynx, and pharynx. Tiny spaces open up in parts of the mesoderm. In time the spaces will interconnect as the coelomic cavity.

During the third week after fertilization—when a pregnant woman has missed her first menstrual period—the basic vertebrate body plan emerges in the new individual.

A primitive streak, neural tube, somites, and pharyngeal arches form during the embryonic period of all vertebrates. Formation of the primitive streak establishes the body's anterior–posterior axis and its bilateral symmetry.

WHY IS THE PLACENTA SO IMPORTANT?

Even before the beginning of the embryonic period, extraembryonic membranes have been interacting with the uterus and sustaining the embryo's rapid growth. By the third week, tiny fingerlike projections from the chorion have grown profusely into the maternal blood that has pooled in endometrial spaces. The projections —chorionic villi—enhance the exchange of substances between the mother and the new individual. They are functional components of the placenta.

The **placenta**, a blood-engorged organ, consists of endometrial tissue and extraembryonic membranes. At full term, a placenta will make up a fourth of the inner surface of the uterus (Figure 44.13).

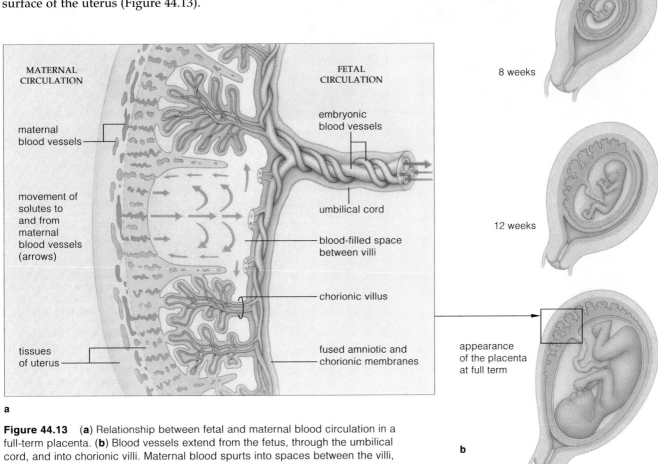

4 weeks

8 weeks

12 weeks

appearance of the placenta at full term

Figure 44.13 (**a**) Relationship between fetal and maternal blood circulation in a full-term placenta. (**b**) Blood vessels extend from the fetus, through the umbilical cord, and into chorionic villi. Maternal blood spurts into spaces between the villi, but the two bloodstreams do not intermingle. Oxygen, carbon dioxide, and other small solutes diffuse across the placental membrane surface.

A placenta is the body's way of sustaining the new individual while allowing its blood vessels to develop apart from the mother's. Oxygen and nutrients diffuse out of the maternal blood vessels, across the placenta's blood-filled spaces, then into embryonic blood vessels. (The vessels converge in the umbilical cord, the lifeline between the placenta and the new individual.) Carbon dioxide and other wastes diffuse in the other direction. The mother's lungs and kidneys quickly dispose of them.

After the third month, the placenta secretes progesterone and estrogens—and so maintains the uterine lining.

The placenta is a blood-engorged organ of endometrial and extraembryonic membranes. It permits exchanges between the mother and the new individual without intermingling their bloodstreams. Thus it sustains the individual and allows its blood vessels to develop apart from the mother's.

EMERGENCE OF DISTINCTLY HUMAN FEATURES

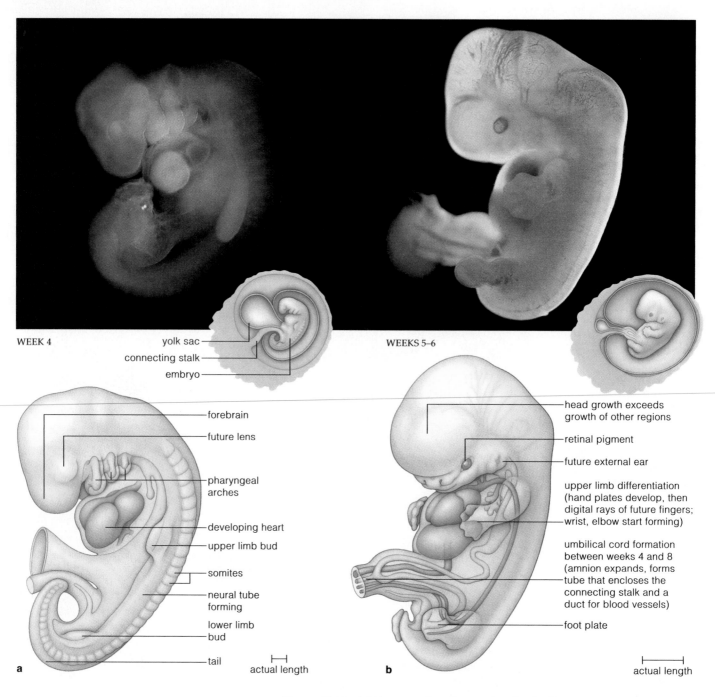

WEEK 4

yolk sac
connecting stalk
embryo

WEEKS 5–6

a
forebrain
future lens
pharyngeal arches
developing heart
upper limb bud
somites
neural tube forming
lower limb bud
tail
actual length

b
head growth exceeds growth of other regions
retinal pigment
future external ear
upper limb differentiation (hand plates develop, then digital rays of future fingers; wrist, elbow start forming)
umbilical cord formation between weeks 4 and 8 (amnion expands, forms tube that encloses the connecting stalk and a duct for blood vessels)
foot plate
actual length

When the fourth week of the embryonic period ends, the embryo is 500 times its starting size. The placenta sustained the growth spurt, but now the pace slows as details of organs fill in. Limbs form; toes and fingers are sculpted from embryonic paddles. The umbilical cord also forms, and the circulatory system becomes intricate. Growth of the all-important head now surpasses that of any other body region (Figure 44.14). The embryonic period ends after the eighth week is completed. No longer is

Figure 44.14 (a) Human embryo four weeks after fertilization. Like all vertebrates, it has a tail and pharyngeal arches. (b) Embryo at five to six weeks. (c) Embryo at the boundary between the embryonic and fetal periods. It now has human features. It floats in amniotic fluid. The chorion covers the amniotic sac but has been pulled aside here. (d) Fetus at sixteen weeks. Movements begin as nerves make functional connections with forming muscles. Legs kick, arms wave, fingers grasp, the mouth puckers. These reflex actions will be vital for life in the world outside the uterus.

the embryo merely "a vertebrate." By now, its features clearly define it as a human fetus.

During the second trimester, the fetus is moving its facial muscles. It frowns; it squints. It busily practices

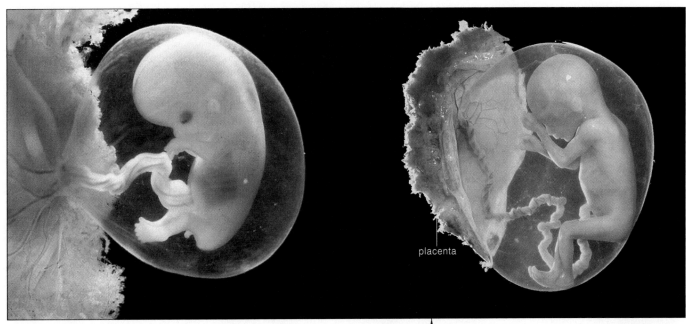

placenta

WEEK 8

final week of embryonic
period; embryo looks
distinctly human
compared to other
vertebrate embryos

upper and lower limbs well
formed; fingers and then
toes have separated

primordial tissues of
all internal, external
structures now developed

tail has become stubby

c |———— actual length ————|

WEEK 16
Length: 16 centimeters
          (6.4 inches)
Weight: 200 grams
          (7 ounces)

WEEK 29
Length: 27.5 centimeters
          (11 inches)
Weight: 1,300 grams
          (46 ounces)

WEEK 38 (full term)
Length: 50 centimeters
          (20 inches)
Weight: 3,400 grams
          (7.5 pounds)

During fetal period, length
measurement extends
from crown to heel (for
embryos, it is the longest
measurable dimension, as
from crown to rump).

d

the sucking reflex, shown in the next section. Now the mother can easily sense movements of the fetal arms and legs. When the fetus is five months old, she can hear its heartbeat through a stethoscope positioned on her abdomen. Soft, fuzzy hair (the lanugo) covers the fetal body. The skin is wrinkled, reddish, and protected from abrasion by a thick, cheeselike coating. During the sixth month, delicate eyelids and eyelashes form. During the seventh month, the eyes of the fetus open.

A fetus born prematurely (before 22 weeks) cannot survive. The situation also is grave for births before 28

weeks, mainly because the lungs have not developed enough. The risks start to drop after this. By 36 weeks, the liveborn survival rate is 95 percent. A fetus born between 36 and 38 weeks still has trouble breathing and maintaining a normal core temperature even with the best of medical care. The optimal birthing time is about 38 weeks after the estimated time of fertilization.

In the fetal period, primary tissues that formed in the early embryo become sculpted into distinctly human features.

Mother as Provider, Protector, Potential Threat

Each pregnant woman is committing much of her body's resources to a new individual. From fertilization until birth, her future child is at the mercy of her diet, health habits, and life-style (Figure 44.15).

NUTRITIONAL CONSIDERATIONS   A well-balanced diet for a pregnant woman usually provides the embryo with all required carbohydrates, lipids, and proteins. (Chapter 41 outlines human nutritional requirements.) Now her demands for vitamins and minerals rise as the placenta absorbs enough for the embryo from her blood. She must get a lot more folate (folic acid). Taking more B-complex vitamins (under medical supervision) before conception and in early pregnancy reduces the risk that her embryo will develop severe neural tube defects. Utilization of nutrients is adversely affected by smoking. In one study, pregnant women who smoked lowered the blood level of vitamin C for themselves *and* the developing fetuses,

even when their intake of that vitamin matched that of a control group.

Most fetal organs are highly vulnerable to nutritional deficiencies. For example, the brain undergoes its greatest expansion in the weeks just before and after birth. Poor nutrition then will have repercussions on intelligence and other neural functions later in life.

A pregnant woman must eat enough to gain 20 to 25 pounds, on average. If she does not, her newborn may be too underweight, at risk of postdelivery complications and, in time, impaired mental functions.

RISK OF INFECTIONS   IgG antibodies circulating in a pregnant woman's blood continually cross the placenta and protect the new individual from all but the most serious bacterial infections. Some virus-induced diseases can be dangerous in the first six weeks after fertilization, a critical time of organ formation. Suppose a pregnant

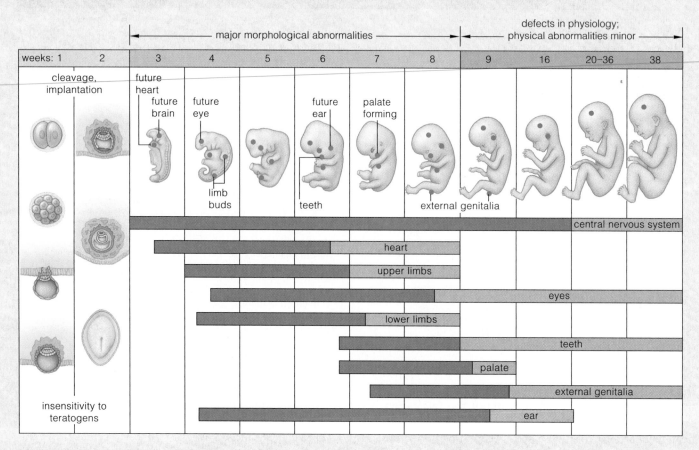

Figure 44.15   Teratogen sensitivity. *Teratogens* are drugs and other environmental factors that can induce deformities in an embryo or fetus. They usually have no effect before organs start to form. After that, they can adversely affect growth, tissue remodeling, and tissue resorption. In this graph, *dark blue* shows the highly sensitive period; *light blue* shows less severe sensitivity to teratogens.

woman contracts *rubella* (German measles) in this critical period. There is a 50 percent chance that some organs of her embryo will not form properly. For example, if she becomes infected when embryonic ears are forming, her newborn may be deaf. If she becomes infected during or after the fourth month of pregnancy, the disease will have no notable effect. A woman can avoid this risk entirely, because vaccination before pregnancy prevents rubella.

PRESCRIPTION DRUG EFFECTS   Pregnant women should not take drugs except under close medical supervision. Think of what happened when the tranquilizer *thalidomide* was routinely prescribed in Europe. Infants of women who used thalidomide in the first trimester had severely deformed arms and legs, or none at all. That drug was withdrawn from the market as soon as its connection with the deformities was apparent. However, other tranquilizers, sedatives, and barbiturates are still being prescribed. They may cause similar, although less severe, damage. Some *anti-acne drugs* increase the risk of facial and cranial deformities. Tetracycline, one of the most overprescribed antibiotics, yellows the teeth. Streptomycin causes hearing problems and may adversely affect the nervous system.

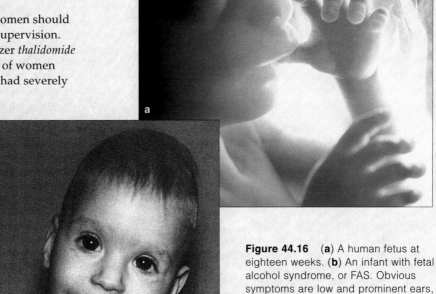

Figure 44.16   (**a**) A human fetus at eighteen weeks. (**b**) An infant with fetal alcohol syndrome, or FAS. Obvious symptoms are low and prominent ears, improperly formed cheekbones, and an abnormally wide, smooth upper lip. Growth problems and abnormalities of the nervous system can be expected. This disorder currently affects about 1 in 750 newborns in the United States.

EFFECTS OF ALCOHOL   As a fetus develops, its physiology becomes more like its mother's. Because it is so small and grows so fast, alcohol is more harmful to it. Alcohol passes freely across the placenta, so drinking it while pregnant invites *fetal alcohol syndrome* (FAS). Symptoms include a reduced brain size, a small head size, mental impairment, facial deformities, slow growth, poor coordination, and often heart defects (Figure 44.16*b*). One in 750 newborns is affected, and the symptoms are irreversible; FAS children never "catch up," physically or mentally. There may be no "safe" drinking level; any alcohol may harm the fetus. More doctors are now urging total abstinence during pregnancy.

EFFECTS OF COCAINE   Any pregnant woman who uses cocaine, especially crack, disrupts the nervous system of her future child as well as her own. Here you might wish to reflect again on the introduction to Chapter 34, which describes the kind of future that awaits the newborn of a crack addict.

EFFECTS OF TOBACCO SMOKE   Simply put, certain toxic elements in tobacco smoke can impair fetal growth and development. These elements accumulate even inside the fetuses of pregnant nonsmokers who simply are exposed to *secondhand smoke* at home or work.

Daily smoking during pregnancy leads to underweight newborns even if the woman's weight, nutritional status, and all other relevant variables match those of pregnant nonsmokers. Smoking has other effects. In Great Britain, all infants born during the same week were tracked for seven years. Infants of smokers were smaller, had twice as many heart abnormalities, and died of more postdelivery complications. At age seven, they were nearly half a year behind children of nonsmokers in average "reading age."

The mechanisms by which smoking affects a fetus are not known. Its demonstrated effects are evidence that the placenta, marvelous structure that it is, cannot prevent all assaults on the fetus that the human mind can dream up.

FROM BIRTH ONWARD

Giving Birth

Pregnancy ends thirty-eight weeks after fertilization, give or take a few weeks. The birth process is known as **labor** (or delivery). The amnion typically ruptures just before birth, so "water" (amniotic fluid) rushes from the vagina. During labor, the cervical canal dilates, and the fetus moves out from the uterus, then the vagina, and into the outside world. Strong uterine contractions are the driving force behind the expulsion.

Mild uterine contractions start in the last trimester. **Relaxin**, a hormone secreted by the corpus luteum and placenta, makes the cervical connective tissues soften and the bridges between pelvic bones looser. The fetus "drops," or shifts down; its head usually touches the cervix (Figure 44.17). Those rhythmic contractions that herald the onset of labor will increase in frequency and intensity over the next two to eighteen hours. Usually they expel the fetus within an hour after the cervix has become fully dilated.

Strong contractions detach the placenta from the uterus and expel it as the afterbirth. (Detachment is necessary; the placental hormones can block lactation.) Contractions constrict blood vessels at the ruptured attachment site and stop hemorrhaging. The umbilical cord is cut and tied at the newborn's belly. A few days after shriveling up, the stump has become the navel.

Nourishing the Newborn

Once the lifeline to the mother is severed, a newborn enters the extended time of dependency and learning that is typical of all primates. Its early survival requires an ongoing supply of milk or a nutritional equivalent. **Lactation** (milk production) occurs in mammary glands in a mother's breasts (Figure 44.18). Before pregnancy, breasts consist mainly of adipose tissue and a system of undeveloped ducts. Their size depends on how much fat they contain, not on milk-producing ability. During pregnancy, estrogen and progesterone act to develop a glandular system of milk production.

For the first few days after birth, mammary glands produce a fluid rich in proteins and lactose. **Prolactin**, a hormone that calls for synthesis of enzymes used in milk production, is secreted by pituitary's anterior lobe (Section 32.3). When a newborn suckles, the pituitary releases **oxytocin**. This hormone triggers contractions that force milk into breast tissue ducts, and it induces the uterus to shrink back to its pre-pregnancy size.

Still another hormone, **CRH** (corticotropin-releasing hormone) contributes to the timing of labor. But it also may contribute to postpartum depression, or *afterbaby blues*. CRH is secreted from the placenta. Its level in blood rises by as much as three times in pregnancy. The rise may trigger secretion of cortisol, a stress hormone,

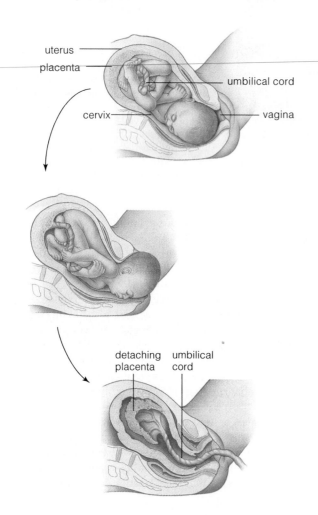

Figure 44.17 Expulsion of a human fetus and the afterbirth (the placenta, tissue fluid, and blood) during labor.

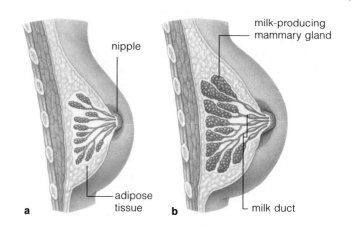

Figure 44.18 (**a**) Breast of a woman who is not pregnant. (**b**) Breast of a lactating woman.

Figure 44.19 Observable, proportional changes in the human body during prenatal and postnatal growth. Changes in overall physical appearance are slow but noticeable until the teenage years. For example, compared to the embryonic period, the head becomes proportionally smaller, the legs longer, and the trunk shorter. Correlate these drawings with the stages in Table 44.4.

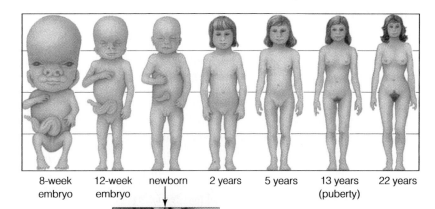

| 8-week embryo | 12-week embryo | newborn | 2 years | 5 years | 13 years (puberty) | 22 years |

to help a mother cope with extraordinary stresses that pregnancy and labor place on her. Once the placenta is expelled, her CRH levels crash to levels typical of some clinical depressions. Afterbaby blues cease when CRH secretion from the hypothalamus returns to normal.

Regarding Breast Cancer

On average, 100,000 women develop *breast cancer* each year. Obesity, and high cholesterol and estrogen levels, contribute to the cancerous transformation. Prospects

of being cured are excellent with early detection and treatment. Once a month, about a week after she has menstruated, a woman should examine her breasts. For recommended examination procedures, she can contact a physician or the American Cancer Society, which has listings in telephone directories. The society also posts information on its web site, http://www.cancer.org.

Postnatal Development

Like many species, humans start out as a miniaturized version of adults, then change in size and proportion until reaching sexual maturity. Figure 44.19 shows just a few of the proportional changes that take place as the life cycle unfolds during the prenatal ("before birth") stages and postnatal ("after birth") stages.

Postnatal growth is most rapid between the years thirteen and nineteen. Sex hormone secretions step up and bring about the development of secondary sexual traits as well as sexual maturity. Not until adulthood are the bones fully mature. Body tissues normally are maintained in top condition in early adulthood. As the years pass, tissues are harder to maintain and repair. They gradually deteriorate through aging processes of the sort described in Section 43.7.

The human life cycle flows naturally from the time of birth, growth, and development, to production of the individual's own offspring, and on through aging to the time of death.

Table 44.4 *Stages of Human Development*

PRENATAL PERIOD

Zygote	Single cell resulting from fusion of sperm nucleus and egg nucleus at fertilization.
Morula	Solid ball of cells produced by cleavages.
Blastocyst	Ball of cells with surface layer, fluid-filled cavity, and inner cell mass (the mammalian blastula).
Embryo	All developmental stages from two weeks after fertilization until end of eighth week.
Fetus	All developmental stages from ninth week to birth (about thirty-eight weeks after fertilization).

POSTNATAL PERIOD

Newborn	Individual during the first two weeks after birth.
Infant	Individual from two weeks to about fifteen months after birth.
Child	Individual from infancy to about ten or twelve years.
Pubescent	Individual at puberty, when secondary sexual traits develop; girls between ten and fifteen years, boys between twelve and sixteen years.
Adolescent	Individual from puberty until about three or four years later; physical, mental, emotional maturation.
Adult	Early adulthood (between eighteen and twenty-five years); bone formation and growth finished. Changes proceed very slowly after this.
Old age	Aging follows late in life.

Control of Human Fertility

The transformation of a zygote into an adult of intricate detail raises profound questions. *When does development begin?* As you have seen, major developmental events unfold even before fertilization. *When does life begin?* In her lifetime, a woman produces as many as 500 eggs, all of which are alive. With one ejaculation, a man releases a quarter of a billion sperm, which are alive. Before sperm and egg merge by chance and establish a new individual's genetic makeup, they are as much alive as any other form of life. It is scarcely tenable, then, to say that life begins at fertilization. *Life began more than 3 billion years ago; and each gamete, each zygote, and each sexually mature individual is but a fleeting stage in the continuation of that beginning.*

This greater perspective on life cannot diminish the meaning of conception. It is no small thing to entrust a new individual with the gift of life, wrapped in the unique evolutionary threads of our species and handed down through an immense sweep of time.

Yet how can we reconcile the marvel of individual birth with the astounding birth rate for our species? About 15,000 newborns enter the world every hour. By the time you go to bed tonight, there may be 361,381 more newborns on Earth than there were last night at that hour. In three months there may be 32,975,994 more—almost as many people as there are now in the entire state of California. *Within three months.*

Human population growth is outstripping global resources. Many millions face the horrors of starvation. Living where we do, few of us know what it means to give birth to a child, to give it the gift of life, and have no food to keep it alive.

And how can we reconcile the marvel of birth with the confusion that surrounds unwanted pregnancies? Even highly developed countries do not have adequate educational programs concerning fertility. And many people aren't inclined to exercise control. In 2000 nearly 480,000 babies were born to teenage mothers in the United States. Many parents actually encourage early boy–girl relationships without thinking through the risks of premarital intercourse and unplanned pregnancies. Advice is often condensed to a terse "Don't do it. But if you do it, be careful!" Each year, there are probably more than a million abortions among all age groups.

The motivation to engage in sex has been evolving for hundreds of million years. A few centuries of moral and ecological arguments for its suppression have not stopped that many unwanted pregnancies. Complex social factors now contribute to a population growth rate that is out of control. How will we reconcile our biological past with the need for a stabilized cultural present? Whether and how human fertility is to be controlled is a volatile issue. We return to this issue in the next chapter, in the context of principles that govern the growth and stability of all populations.

BIRTH CONTROL OPTIONS

What options are available to individuals who decide to postpone pregnancy or forgo it entirely? The most effective method of birth control is complete *abstinence*, no sexual intercourse whatsoever. Data show that it is unrealistic to expect many people to practice it.

A less reliable variation of abstinence is the *rhythm method*. The idea is to avoid intercourse in a woman's fertile period, starting a few days before ovulation and ending a few days after. A woman identifies and tracks her fertile period. She also keeps records of the length of her menstrual cycles, takes her temperature every morning when she wakes up, or both. (The body's core temperature rises by one-half to one degree just before a fertile period.) But ovulation may not be regular, and miscalculations are frequent. Also, sperm deposited in the vagina a few days before ovulation may survive until ovulation. The rhythm method *is* inexpensive; it costs nothing after you buy a thermometer. It does not require fittings and periodic medical checkups. But its practitioners run a risk of pregnancy (Figure 44.20).

Withdrawal, or removing the penis from the vagina before ejaculation, dates back at least to biblical times. But withdrawal requires very strong willpower, and the practice may fail anyway. Fluid released from the penis just before ejaculation may contain some sperm.

Douching (chemically rinsing the vagina right after intercourse) is almost useless. Sperm move out of reach of a douche ninety seconds after ejaculation.

Controlling fertility by surgical intervention is less chancy. Males may elect to have a *vasectomy*. This is a brief operation, performed under local anesthetic. The physician makes a tiny incision in the male's scrotum, then cuts and ties off each vas deferens. Afterward, sperm cannot leave the testes, so they can't be present in semen. At this writing there is no firm evidence that vasectomy disrupts hormonal functions or adversely affects sexual activity. It is reversible. But half of those who submit to the surgery can later develop antibodies against sperm and may not be able to regain fertility.

Females may opt for a *tubal ligation*. By this surgical intervention, oviducts are cauterized or cut and tied off, usually in a hospital. When the operation is performed correctly, tubal ligation is the most effective means of birth control. A few women have recurring pain in the pelvic area. The operation sometimes can be reversed.

Less drastic methods of controlling fertility involve physical or chemical barriers that prevent sperm from entering the uterus and oviducts. *Spermicidal foam* and *spermicidal jelly* are toxic to sperm. They are transferred from an applicator into the vagina before intercourse. These products aren't always reliable unless used with another device, such as a diaphragm and a condom. *IUDs* are coils inserted into the uterus by a physician. They sometimes invite pelvic inflammatory disease.

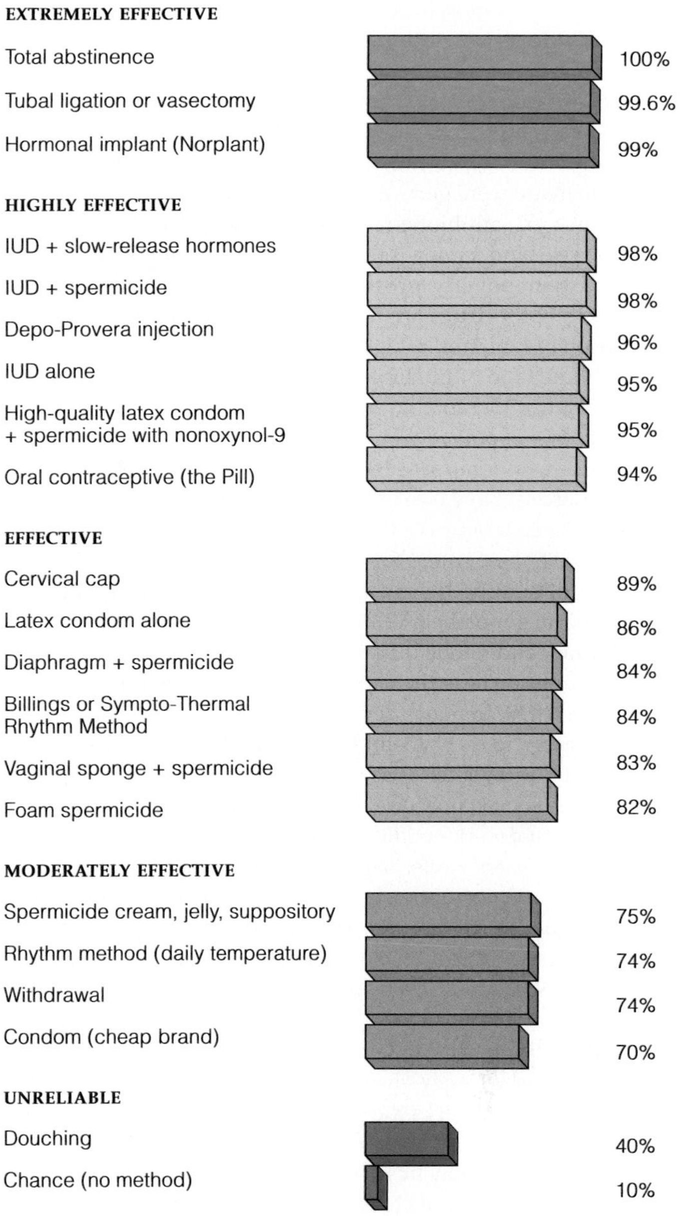

Figure 44.20 Comparison of the effectiveness of methods of contraception in the United States. The percentages also indicate the number of unplanned pregnancies per 100 couples who use only that method of birth control for a year. For example, "94% effectiveness" for oral contraceptives (the Pill) means that 6 of every 100 women will still become pregnant, on the average.

EXTREMELY EFFECTIVE

Total abstinence	100%
Tubal ligation or vasectomy	99.6%
Hormonal implant (Norplant)	99%

HIGHLY EFFECTIVE

IUD + slow-release hormones	98%
IUD + spermicide	98%
Depo-Provera injection	96%
IUD alone	95%
High-quality latex condom + spermicide with nonoxynol-9	95%
Oral contraceptive (the Pill)	94%

EFFECTIVE

Cervical cap	89%
Latex condom alone	86%
Diaphragm + spermicide	84%
Billings or Sympto-Thermal Rhythm Method	84%
Vaginal sponge + spermicide	83%
Foam spermicide	82%

MODERATELY EFFECTIVE

Spermicide cream, jelly, suppository	75%
Rhythm method (daily temperature)	74%
Withdrawal	74%
Condom (cheap brand)	70%

UNRELIABLE

Douching	40%
Chance (no method)	10%

A *diaphragm* is a flexible, dome-shaped device. It is inserted into the vagina and positioned over the cervix before intercourse. It is relatively effective when fitted initially by a doctor, used with foam or a jelly before each sexual contact, inserted correctly each time, and left in place for a prescribed length of time.

Good brands of *condoms*—thin, tight-fitting sheaths worn over the penis during intercourse—are up to 95 percent effective if used with a spermicide. Only latex condoms offer protection against sexually transmitted diseases (Section 44.15). However, condoms often tear and leak, at which time they become absolutely useless.

The *birth control pill* is an oral contraceptive made of synthetic estrogens and progestins (progesterone-like hormones). Taken daily except for the last five days of a menstrual cycle, it suppresses maturation of oocytes and ovulation. Often it corrects erratic menstrual cycles and reduces cramps. In some women it causes nausea, weight gain, tissue swelling, and headaches. With more than 50 million women taking it, "the Pill" is the most-used fertility control method. It is 94 percent effective or better when it is taken correctly. Earlier formulations were linked to blood clots, high blood pressure, and possibly cancer. Lower doses of the new formulations might lessen the risk of breast and endometrial cancer. The possibility of a correlation between the Pill and breast cancer is still under study.

Progestin injections or implants block ovulation. A *Depo-Provera* injection works for three months and is 96 percent effective. *Norplant* (six rods implanted under skin) works for five years and is 99 percent effective. Both may cause sporadic, heavy bleeding, and doctors have some trouble surgically removing Norplant rods.

A pregnancy test won't register positive until after implantation. According to one view, a woman is not pregnant until that time, so *morning-after pills* intercept pregnancy. Actually, such pills work up to seventy-two hours following unprotected sex by interfering with the hormones that control events between ovulation and implantation. One is Preven, a kit with high doses of birth control pills that suppress ovulation and block corpus luteum function. RU-486, more commonly used as an abortion pill, also is used as a morning-after pill. It can block fertilization and progesterone-dependent implantation. It also may induce abortion when taken within seven weeks of implantation (progesterone also

is vital to maintain pregnancy). It may trigger elevated blood pressure, blood clots, or breast cancer in some women. RU-486, now known as Mifeprex, is available in Europe. Although controversial, its use in the United States was granted FDA approval in 2000. By contrast, Preven had been used before as a contraceptive, and its repackaging as a morning-after pill has not been very controversial. It gained FDA approval in 1998.

Whether and how human fertility is to be controlled is a volatile issue. It centers on reconciling our biological past with seriously divergent views about our cultural present.

Sexually Transmitted Diseases

At some point in his or her life, at least one of every four individuals in the United States who engage in sexual intercourse will become infected by the pathogens that cause **sexually transmitted diseases**, or **STDs**. Tens of millions are already infected; STDs have reached epidemic proportions. Two-thirds are under age twenty-five; one-fourth are teenagers. Women and children are hardest hit. Worse yet, antibiotic-resistant strains of bacteria are on the rise, and some viral diseases simply cannot be cured.

Urban poverty, prostitution, intravenous drug abuse, and sex-for-drugs are fanning the epidemic. Yet STDs also are rampant in high schools and colleges, where students too often think, "It can't happen to me." They reject the idea that no sex—abstinence—is the only safe sex. In one poll of high school students, two-thirds of the respondents said they don't use condoms. More than 40 percent have two or more sex partners.

The economics of this health problem are staggering. In 2000, the Centers for Disease Control tallied the annual cost of treating the most prevalent STDs: *Herpes*, $759 million; gonorrhea, $1 billion; chlamydial infections, $2 billion; and pelvic inflammatory disease, $4.2 billion. This does not include the accelerating cost of treating patients with AIDS. In many developing countries, AIDS alone threatens to overwhelm health-care delivery systems and to unravel decades of economic progress.

The social consequences are sobering. Mothers bestow a chlamydial infection on 1 of every 20 newborns in the United States alone. They bestow type II *Herpes* virus on 1 in 10,000 newborns; one-half of the infected babies die, and one-fourth have severe neural defects. Every year, 1 million women develop pelvic inflammatory disease, and 100,000 to 150,000 become infertile.

AIDS   Someone can become infected by HIV, the human immunodeficiency virus, and not even know it. However, the infection marks the start of a titanic battle that the immune system almost certainly will lose (Section 39.10). At first there may be no outward symptoms. Five to ten years later, a set of chronic disorders develops. They are signs of *AIDS* (*Acquired Immune Deficiency Syndrome*). When the immune system finally is weakened, the stage is set for opportunistic infections. Normally harmless bacteria residing in the body are the first to take advantage of the lowered resistance. Dangerous pathogens take their toll. In time, an immune-compromised person simply is overwhelmed.

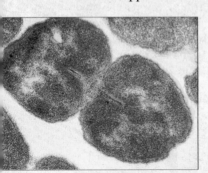

Figure 44.21   Bacterial agents of gonorrhea.

Most often, HIV spreads through anal, vaginal, and oral intercourse and by intravenous drug use. Infections usually occur through transfer of blood, semen, urine, or vaginal secretions between people. Cuts and abrasions on the penis, vagina, rectum, and oral membranes invite HIV into the internal environment of a new host.

Today there is no vaccine against HIV and no effective treatment for AIDS. If you get it, you die. *There is no cure.*

HIV was not identified until 1981, but it was present in some parts of Central Africa for at least several decades before that. In the 1970s and early 1980s, it spread to the United States and other developed countries, where most of those initially infected were male homosexuals. Today, many in the heterosexual population are infected or at risk.

Free or low-cost, confidential testing for HIV exposure is available at public health facilities and at many doctor's offices. People who are worried should know there may be a time lag between a first exposure and the first test to come out positive. It takes a few weeks to six months or more before detectable amounts of antibodies will form in response to infection. (The presence of antibodies in the blood indicates exposure to the virus.) Anyone who tests positive is capable of spreading the virus.

Public education programs attempt to stop the spread of HIV. Most health-care workers advocate safe sex, yet there is great confusion about what "safe" means. Many advocate the use of high-quality latex condoms *and* a spermicide that contains nonoxynol-9 to help prevent the transmission of the virus. Even then, there is a slight risk. As an unfortunate couple learned in 1997, no one should participate in open-mouthed kissing with someone who tests positive for HIV. Caressing is not risky *if* there are no lesions or cuts where body fluids that harbor the virus can enter the body. Pronounced lesions caused by other sexually transmitted diseases may increase susceptibility to HIV infection.

In sum, AIDS reached pandemic proportions mainly for three reasons. First, it took a while to discover that the virus can travel in semen, blood, and vaginal fluid and that *behavioral* controls can limit its spread. Second, it took time to develop ways to test symptom-free carriers, who infect others. Third, many still don't realize the medical, economic, and social consequences affect everyone.

GONORRHEA   During intercourse, *Neisseria gonorrhoeae* (Figure 44.21) can enter the body at mucous membranes of the urethra, cervix, and anal canal. This bacterium causes *gonorrhea*. An infected female may notice a slight vaginal discharge or burning sensation as she is urinating. If the infection spreads to her oviducts, it might induce severe cramps, fever, vomiting, and scarring, which may cause sterility. Males have more obvious symptoms. Within a week of infection, the penis discharges yellow pus. Urinating is more frequent and may be painful.

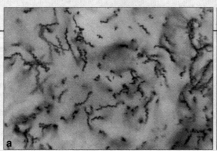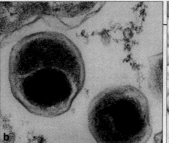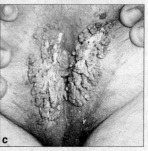

Figure 44.22 Bacterial agents of (**a**) syphilis and (**b**) chlamydia. (**c**) Genital warts on external genitalia of a young woman.

This STD is rampant even though prompt treatment quickly cures it. Why? Women experience no troubling symptoms during early stages. Also, infection does not confer immunity to the bacterium, perhaps because there are sixteen or more different strains of it. Contrary to common belief, someone can get gonorrhea over and over again. Also, oral contraceptives, used so widely, promote infection by altering vaginal pH. Populations of resident bacteria decline—and *N. gonorrhoeae* is free to move in.

SYPHILIS *Syphilis*, a dangerous STD, is caused by the spirochete *Treponema pallidum* (Figure 44.22*a*). Sex with an infected partner puts the motile, spiral bacterium on the surface of genitals or into the cervix, vagina, or oral cavity. It can enter the body through tiny epidermal cuts. One to eight weeks later, new treponemes are twisting about in a flattened, painless chancre (local ulcer). This first chancre is a symptom of the primary stage of syphilis. By then, treponemes are in the blood. Treponemes can cross the placenta of an infected pregnant woman. The result will be a miscarriage, stillbirth, or syphilitic newborn.

Chancres usually heal, but treponemes multiply in the spinal cord, brain, eyes, mucous membranes, bones, and joints. More chancres and a skin rash develop during this infectious, secondary stage of syphilis. Immune responses succeed in about 25 percent of the cases, so the symptoms do subside. Another 25 percent remain infected but are symptom free. In the rest, minor to significant lesions and scars appear in the skin and liver, bones, and other internal organs. Few treponemes form during this tertiary stage. But the host's immune system is hypersensitive to them. Chronic immune reactions may severely damage the brain and spinal cord and lead to general paralysis.

Probably because the symptoms are so alarming, more people seek early treatment for syphilis than for gonorrhea. Later stages require prolonged treatment.

PELVIC INFLAMMATORY DISEASE *Pelvic inflammatory disease* (PID) is one of the most serious complications of gonorrhea, chlamydial infections, and other STDs. It also can arise when normal bacterial residents of the vagina ascend to the pelvic region. Most commonly, the uterus, oviducts, and ovaries are affected. There is bleeding and vaginal discharge. Pain in the lower abdomen may be as severe as an acute appendicitis attack. The oviducts may become scarred. Pelvic inflammatory disease invites abnormal pregnancies as well as sterility.

GENITAL HERPES About 45 million Americans have *genital herpes*, caused by the type II *Herpes simplex* virus. Infection requires direct contact with active *Herpes* viruses or sores that contain them. Mucous membranes of the mouth and the genitals are highly susceptible to invasion. The symptoms are often mild or absent. Among infected women, small, painful blisters may appear on the vulva, cervix, urethra, or anal tissues. Among men, blisters form on the penis and anal tissues. Within three weeks, the virus enters latency, and the sores crust over and heal.

The virus is reactivated sporadically. Each time, it causes painful sores at or near the original infection site. Sexual intercourse, menstruation, emotional stress, or other infections trigger *Herpes* flare-ups. Acyclovir, an antiviral drug, decreases healing time and often decreases the pain.

CHLAMYDIAL INFECTION *Chlamydia trachomatis* (Figure 44.22*b*), a parasitic bacterium, spends part of its life cycle in genital and urinary tract cells. It can be transmitted by vaginal, oral, or anal intercourse. It causes several diseases, including *NGU* (chlamydial nongonococcal urethritis).

NGU leads to inflammation of the cervix and, in both sexes, the urethra. It can cause a burning sensation during urination. When diagnosed, the infection is easily treated. But 2 of every 3 infected females and 1 of every 2 infected males have only subtle symptoms, if any. Ignored and untreated, the infection can inflame a male's prostate gland; in females it spreads to the uterus and oviducts.

C. trachomatis is probably responsible for 50 percent of all PID cases. Pregnant infected females can expect their newborn to be premature and at risk of severe eye, ear, and lung infections. For infected females, certain *C. trachomatis* strains may substantially increase the risk of cervical cancer. Vulnerability to HIV rises for both sexes.

Chlamydial infections are now the most frequent STDs in the United States. Each year, 3 million new cases are reported, especially among teenagers and young adults.

GENITAL WARTS More than sixty types of the human papillomaviruses (HPV) are known. A few cause *genital warts*, or benign, bumplike growths (Figure 44.22*c*). HPV infection of the genitals and the anus is one of the most prevalent STDs in this country. Usually type 16 HPV does not cause obvious warts, but it may cause precancerous sores and cancers of the cervix, vagina, vulva, penis, and anus. In one Seattle study, 22 percent of female college students who were examined tested positive for the virus.

To Seek or End Pregnancy

Because of sterility or infertility, about 15 percent of all couples in the United States cannot conceive. For example, hormonal imbalances may prevent ovulation. Or a male's sperm count may be so low that fertilization would be next to impossible without medical intervention.

In vitro fertilization means conception outside the body (literally "in glass" petri dishes or test tubes). It may occur if the couple can produce normal sperm and oocytes. First the woman receives injections of a hormone that prepares her ovaries for ovulation. Afterward, the preovulatory oocyte is removed from her with a suction device. In the meantime, sperm from the male are placed in a solution that simulates fluid in the oviducts. A few hours after the sperm and suctioned oocyte make contact in a petri dish, fertilization may occur. Twelve hours later, the zygote is transferred to a solution that can sustain it through the initial cleavages. Two to four days later, the resulting ball of cells is transferred to the female's uterus.

Each attempt at in vitro fertilization costs about 8,000 dollars, and most attempts aren't successful. In 1994, each "test-tube" baby cost the nation's health-care system about 60,000 to 100,000 dollars, on average. The childless couple may believe no cost is too great. However, in an era of increased population growth and shrinking medical coverage, is the cost too great for society to bear? Court battles are being waged over this issue.

At the other extreme is **abortion**, the dislodging and removal of the blastocyst, embryo, or fetus from the uterus. At one time in the United States, abortions were forbidden by law unless the pregnancy endangered the mother's life. Later the Supreme Court ruled that the government does not have the right to forbid abortions during the early stages of pregnancy (typically up to five months). Before this ruling, there were dangerous, traumatic, and often fatal attempts to abort embryos, either by pregnant women themselves or by quacks.

From a clinical standpoint, vacuum suctioning and other methods make abortion painless for the woman, rapid, and free of complications when performed during the first trimester. Abortions performed in the second and third trimesters are extremely controversial unless the mother's life is threatened. Even so, for both medical and humanitarian reasons, the majority of people in this country generally agree that the preferred route to birth control is not through abortion. Rather, it is through sexually responsible behavior that prevents unwanted pregnancy from happening in the first place.

This textbook cannot offer you the "right" answer to a moral question because of the reasons given in Section 1.7. It *can* provide you with a serious, detailed description of how a new individual develops. Your choice of the "right" answer to the question of the morality of abortion will be just that—your choice. And it is one that can be based on objective insights into the nature of life.

SUMMARY   *Gold* indicates text section

1. Humans have a pair of primary reproductive organs (sperm-producing testes in males, and egg-producing ovaries in females), accessory ducts, and glands. Testes and ovaries also synthesize sex hormones that govern reproductive functions and secondary traits. *44.1–44.3*

   a. The hormones LH, FSH, and testosterone control sperm formation. They are part of feedback loops from the testes to the hypothalamus and anterior lobe of the pituitary gland. *44.2*

   b. The hormones estrogen, progesterone, FSH, and LH control the maturation and release of oocytes from the ovary, as well as cyclic changes in the endometrium (the inner lining of the uterus). Their secretion is part of feedback loops from ovaries to the hypothalamus and to the anterior lobe of the pituitary gland. *44.3–44.5*

2. A menstrual cycle is a recurring cycle of intermittent fertility during the reproductive years of female humans and other primates. *44.3, 44.5*

   a. Follicular phase: One of many follicles matures inside an ovary. Each follicle is an oocyte and the cell layer surrounding it. Meanwhile, the endometrium is prepared for a possible pregnancy. It breaks down each cycle if pregnancy does not occur.

   b. Ovulation: A midcycle surge of the blood level of LH triggers ovulation, the release of a secondary oocyte from the ovary.

   c. Luteal phase: After ovulation the corpus luteum, a glandular structure, develops from the remnants of the follicle. It secretes progesterone and some estrogen that prime the endometrium for fertilization. If fertilization occurs, the corpus luteum is maintained.

3. After fertilization, a blastocyst forms by cleavage and becomes implanted in the endometrium. Three primary tissue layers form that give rise to all organs. They are ectoderm, endoderm, and mesoderm. *44.7, 44.8*

4. Four extraembryonic membranes form as embryos of humans and other vertebrates develop: *44.7, 44.8*

   a. The *amnion* becomes a fluid-filled sac around the embryo, which it protects from drying out, mechanical shock, and abrupt temperature changes.

   b. The *yolk sac* stores nutritive yolk in most shelled eggs. In humans, part of the sac becomes a major site of blood formation, and some of its cells give rise to germ cells. (Germ cells later give rise to sperm and eggs.)

   c. The *chorion* is a protective membrane enclosing the embryo and the other extraembryonic membranes. It also becomes a major component of the placenta.

   d. In humans, the blood vessels for the placenta arise from the *allantois*, as does the urinary bladder.

5. The placenta, a blood-engorged organ, is composed of endometrium and extraembryonic membranes. The placenta permits embryonic blood vessels to develop

independently of the mother's but also allows oxygen, nutrients, and wastes to diffuse between them. *44.9*

6. To some extent, the placenta serves as a protective barrier for the fetus. But it cannot protect the fetus from harmful effects of the mother's nutritional deficiencies, infections, intake of prescription drugs, illegal drugs, alcohol, and cigarette smoke. *44.11*

7. During pregnancy, estrogens as well as progesterone stimulate the growth and development of mammary glands. During labor, strong uterine contractions expel the fetus and afterbirth. Nursing a newborn stimulates release of prolactin and oxytocin. These hormones, in turn, stimulate milk production and release. *44.12*

8. Whether and how to control human sexuality and fertility are major ethical issues. Sexually responsible behavior rather than abortion is generally viewed as the preferred route to birth control. *44.13, 44.14, 44.16*

Review Questions

1. Name the two primary reproductive organs of the human male and where sperm formation starts inside them. Does semen consist of sperm, glandular secretions, or both? *44.1*

2. Does sperm formation require mitosis, meiosis, or both? *44.2*

3. Study Figure 44.5. Then, on your own, sketch the feedback loops (to the hypothalamus and anterior pituitary from the testes) that govern sperm formation. Include the names of the releasing hormone as well as the hormones and cells in the testes that are involved in these loops. *44.2*

4. Name the two primary reproductive organs of the human female. What is the endometrium? *44.3*

5. Distinguish between:
   a. oocyte and ovum *44.4, 44.6*
   b. follicular phase and luteal phase of menstrual cycle *44.5*
   c. follicle and corpus luteum *44.6*
   d. ovulation and implantation *44.3, 44.6*

6. What is the menstrual cycle? Name four of the hormones that influence the cycle. Which one triggers ovulation? *44.3, 44.5, 44.7*

7. Study Figure 44.8. Then, on your own, sketch the feedback loops (to the hypothalamus and anterior pituitary from the ovaries) that govern the menstrual cycle. Include the names of the releasing hormone as well as hormones and cells in the ovarian structures involved in these loops. *44.4*

8. Describe the cyclic changes that occur in the endometrium during the menstrual cycle. What role does the corpus luteum play in the changes? *44.4, 44.5*

9. Distinguish between the embryonic period and fetal period of human development. Describe the general organization of a human blastocyst. *44.7, 44.10*

10. State the embryonic source of the amnion, yolk sac, chorion, and allantois. Also state the functions of each extraembryonic membrane with respect to the structure or functioning of the developing individual. *44.7*

11. Name some of the early organs that are the hallmark of the embryonic period of humans and other vertebrates. *44.10*

12. Describe the placenta's structure and function. Do maternal and fetal blood vessels interconnect in this organ? *44.9*

13. Name the releasing hormone with a key role in labor. Then state the role of estrogen, progesterone, prolactin, and oxytocin in ensuring that the newborn will have an ongoing supply of milk. *44.12*

14. Label the components of the human male reproductive system and state their functions: *44.1, 44.2*

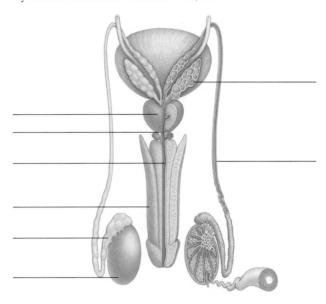

15. Label the components of the human female reproductive system and state their functions: *44.3, 44.4*

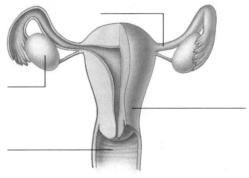

Self-Quiz   ANSWERS IN APPENDIX III

1. The _____ and the pituitary gland control secretion of sex hormones from the testes and the ovaries.

2. Reproductive function in males requires _____ .
   a. FSH   d. GnRH
   b. LH   e. inhibin
   c. testosterone   f. all of the above

3. During a menstrual cycle, a midcycle surge of _____ triggers ovulation.
   a. estrogen   b. progesterone   c. LH   d. FSH

4. During a menstrual cycle, _____ and _____ secreted by the corpus luteum prime the uterus for pregnancy.
   a. FSH; LH   c. estrogen; progesterone
   b. FSH; testosterone   d. Only estrogens prime it.

5. In implantation, a _____ burrows into the endometrium.
   a. zygote   c. blastocyst
   b. gastrula   d. morula

6. The _____ , a fluid-filled sac, surrounds and protects the embryo from mechanical shocks and keeps it from drying out.
   a. yolk sac   b. allantois   c. amnion   d. chorion

7. At full term, a placenta _____ .
   a. is composed of extraembryonic membranes
   b. directly connects maternal and fetal blood vessels
   c. keeps maternal and fetal blood vessels separated

8. (A) _____ form(s) in all vertebrate embryos.
   a. neural plate   c. pharyngeal arches   e. a through c
   b. somites   d. primitive streak   f. a through d

9. Distinctly human features emerge in the embryo by the end of the _____ week after fertilization.
   a. second   c. fourth   e. eighth
   b. third   d. fifth   f. sixteenth

10. Match each term with the most suitable description.
   _____ blastocyst   a. secondary sexual traits develop
   _____ seminiferous   b. secretions influence labor
           tubule   c. sperm start to form here
   _____ corpus   d. secretes estrogens, progesterone to
           luteum   prepare for blastocyst and maintain
   _____ secondary   endometrium during pregnancy
           oocyte   e. type of blastula formed from a
   _____ CRH, cortisol   fertilized mammalian egg
   _____ puberty   f. after fertilization, becomes mature
                              egg (ovum)

Critical Thinking

1. Suppose that, at the time a human zygote is undergoing cleavage, the first two blastomeres that form separate from each other completely. Both of the blastomeres and their cellular descendants continue to divide on schedule, on the prescribed developmental program. The result is *identical twins*, or two normal, genetically identical individuals (compare Section 43.3). By contrast, *nonidentical twins* form when two different secondary oocytes are fertilized at the same time by two different sperm. On the basis of this information, explain why nonidentical (fraternal) twins show considerable genetic variability and identical twins do not.

2. Infection by the rubella virus apparently has an inhibitory effect on mitosis. Serious birth defects result when a woman is infected during the first trimester of pregnancy, but not later. Review the developmental events that unfold during pregnancy and explain why this might be so.

3. Imagine you are an obstetrician advising a woman who has just learned she is pregnant. What instructions would you provide concerning her diet and behavior during pregnancy?

4. In the United States, teenage pregnancies as well as STD infections are rampant. Suppose the office of the Surgeon General requests your participation in a task force that will recommend practices that might reduce the incidence of both. What practices might meet with the greatest success? What kind of enthusiasm or resistance may they provoke among the teenagers in your community? Among adults? Explain why.

Selected Key Terms

Human Reproduction

coitus 44.6
endometrium 44.3
estrogen 44.3
follicle 44.4
FSH 44.2
GnRH 44.2
inhibin 44.2
in vitro fertilization 44.16
Leydig cell 44.2
LH 44.2
menstrual cycle 44.3
oocyte 44.3
ovary CI
ovulation 44.3
ovum 44.6
polar body 44.4
progesterone 44.3
secondary oocyte 44.4
secondary sexual trait CI
semen 44.1
seminiferous tubule 44.1
Sertoli cell 44.2
sexually transmitted
   disease (STD) 44.15

testis (testes) CI
testosterone 44.2
uterus 44.3

Human Development

abortion 44.16
allantois 44.7
amnion 44.7
blastocyst 44.4
chorion 44.7
corpus luteum 44.4
CRH 44.12
fetus 44.7
gastrulation 44.8
HCG 44.7
implantation 44.7
labor (birth process) 44.12
lactation 44.12
oxytocin 44.12
placenta 44.9
prolactin 44.12
relaxin 44.12
somite 44.8
yolk sac 44.7
zona pellucida 44.4

Readings

Caldwell, M. November 1992. "How Does a Single Cell Become a Whole Body?" *Discover* 13(11): 86–93.

Cohen, J., and I. Stewart. April 1994. "Our Genes Aren't Us." *Discover* 15(4): 78–84. Argument that DNA means nothing in the absence of a developmental context.

Gilbert, S. 2000. *Developmental Biology*. Sixth edition. Sunderland, Massachusetts: Sinauer.

Larsen, W. 1993. *Human Embryology*. New York: Churchill Livingston. Paperback. Recommended for the serious student, but maybe not for the fainthearted.

McGinnis, W., and M. Kuziora. February 1994. "The Molecular Architects of Body Design." *Scientific American* 270(2): 58–66.

Nilsson, L., et al. 1986. *A Child Is Born*. New York: Delacorte Press/Seymour Lawrence.

Nusslein-Volhard, C. August 1996. "Gradients That Organize Embryonic Development." *Scientific American*, 54–61.

Rathus, S., and S. Boughn. 1993. "AIDS: What Every Student Needs to Know." Dallas, Texas: Harcourt Brace. Paperback; also has sections on other sexually transmitted diseases.

Sherwood, L. 2001. *Human Physiology*. Fourth edition. Belmont, California: Brooks/Cole.

Zack, B. July 1981. "Abortion and the Limitations of Science." *Science* 213:291.

On-Line readings at Student Guide for InfoTrac:
www.brookscole.com/biology

VII Ecology and Behavior

Two organisms—a fox in the shadows cast by a snow-dusted spruce tree. What are the nature and consequences of their interactions with each other, with other kinds of organisms, and with their environment? By the end of this last unit, you may find worlds within worlds in such photographs.

45

POPULATION ECOLOGY

Tales of Nightmare Numbers

Across from Sausalito, California, the steep flanks of Angel Island rise from the waters of San Francisco Bay (Figure 45.1a). The island, set aside as a game reserve, escaped urban development but not the descendants of a few deer that well-meaning nature lovers shipped over in the early 1900s. With no natural predators to keep them in check, the few deer became many—too many for the limited food in the isolated habitat. Yet the island attracted a steady stream of picnickers from the mainland. They felt sorry for the malnourished deer and loaded picnic baskets with extra food for them.

The visitors imported so much food that scrawny deer kept on living and reproducing. In time, the herd nibbled away the native grasses, the roots of which had helped slow soil erosion on the steep hillsides. Hungry deer chewed off all the new leaves of seedlings; they killed small trees by stripping away bark and phloem. The herd was destroying the environment.

In desperation, game managers proposed using a few skilled hunters to thin the herd. They were denounced as cruel. They proposed importing a few coyotes to thin the herd naturally. Animal rights advocates said no.

As a compromise, about 200 of the 300+ deer were captured, loaded onto a boat, and shipped to suitable mainland habitats. Many were given collars with radio transmitters so that game managers could track them after the release. In less than sixty days, dogs, coyotes, bobcats, hunters, and speeding cars and trucks killed most of them. In the end, relocating each surviving deer cost taxpayers close to 3,000 dollars. The State of California refused to do it again. And no one else, anywhere, volunteered to pick up future tabs.

It's not difficult to define Angel Island's boundaries or track its inhabitants, so it's easy to draw a lesson from this tale: *A population's growth depends on environmental resources. And attempts to "beat nature" by altering the sometimes cruel outcome of limited resources only postpone the inevitable.* Does the lesson apply to other populations, in other places? Yes it does, as the next tale makes clear.

When 2002 drew to a close, there were over 6 billion people on Earth. About 3 billion already live in poverty, trying to live on the equivalent of 1 to 2 dollars a day. Next to China, India is now the most populous country (Figure 45.1b). Its population size exceeds 1 billion, and

Figure 45.1 (**a**) Angel Island, which turned out to be a good laboratory for studying population growth. (**b**) Bathers along a bank of the Ganges River in India—a small sampling of a human population that exceeds 6.2 billion. In this chapter, we turn to principles that govern the growth and sustainability of all populations.

18 million are added annually. Forty percent of India's population live in rat-infested shantytowns. They don't have enough food or fresh water and are forced to wash clothes and dishes in open sewers. The land available to raise food shrinks by 365 acres a day, on average. Why? Irrigated soil becomes too salty when it drains poorly, and there isn't enough water to flush away the salts.

Can wealthier, less densely populated nations help? After all, they use most of the world's resources. Maybe they should learn to get by more efficiently, on less. For example, people might limit their meals to cereal grains and water; give up their private cars, living quarters, air conditioners, televisions, and dishwashers; stop taking vacations and stop laundering so much; close all of the malls, restaurants, and theaters at night; and so on.

Maybe wealthier nations also should donate more surplus food than they already donate to less fortunate ones. Then again, would huge donations help or would they encourage dependency and spur more increases in population size? And what if surpluses run out?

It is a monumental dilemma. At one extreme, global redistribution of resources would help the most people survive, but at the lowest comfort level. At the other extreme, foreign aid rationed only to nations that limit population growth might encourage fewer individuals to be born to help give each a better quality of life.

The United States foreign aid program is currently based on two premises: (1) that individuals of every nation have an irrevocable right to bear children, even if unrestricted reproduction ruins the environment that must sustain them; and (2) that because human life is precious above all else, the wealthiest nations have an absolute moral obligation to save lives everywhere.

Regardless of the positions that nations take on this issue, they ultimately must come to terms with this fact: *Certain principles govern the growth and sustainability of populations over time.* These principles are the bedrock of **ecology**—the systematic study of how organisms interact with one another and with their physical and chemical environment. Ecological interactions start within and between populations, and they extend on through communities, ecosystems, and the biosphere. They are the focus of this last unit of the book.

In this chapter we look first at relationships that influence the characteristics of populations. Later, we will apply the basic principles of population growth to the past, present, and future of the human species.

Key Concepts

1. Ecological principles govern the growth and sustainability of all populations, including our own.

2. Genetic factors and a population's size, density, distribution, and the number of individuals in its various age categories influence patterns of growth.

3. When the birth rate exceeds the death rate, and when immigration and emigration are in balance, populations may show exponential growth.

4. Exponential growth is generally defined as a quantity that is increasing by a fixed percentage of the whole in a given interval. It occurs when the number of individuals approaching or already in its reproductive age category becomes ever larger.

5. A shortage of any resource that individuals require is a limiting factor on population growth.

6. Carrying capacity refers to the maximum number of individuals of a population that can be sustained indefinitely by the resources of a given environment. The number of individuals rises or falls with changes in resource availability.

7. A population may show a pattern of logistic growth. By this pattern, population density—the number of individuals in a specified area at a given point in time—is initially low. Then population size rapidly increases. It finally levels off as resource scarcity limits further increase or causes a decline in numbers.

8. Some populations show fluctuations in numbers that cannot be explained by a single growth model.

9. All populations face limits to growth, because no environment can indefinitely sustain a successively increasing number of individuals. Competition, disease, predation, and many other factors control population growth. The controls vary in their relative effects on populations of different species, and they vary over time.

CHARACTERISTICS OF POPULATIONS

By this point in the book you know that a population is a group of individuals of the same species occupying a given area. Its gene pool is the basis for characteristic ranges of morphological, physiological, and behavioral traits. When studying a population, ecologists consider its genetic make-up as well as reproductive modes and overall behavior. They also consider **demographics**, or vital statistics of the population, and that will be our focus here. The vital statistics include population size, density, distribution, and age structure.

Population size is the number of individuals that represent the population's gene pool. A population's **age structure** is the number of individuals in each of several to many age categories. For instance, all members may be grouped into *pre-reproductive*, *reproductive*, and *post-reproductive* ages. Individuals in the first category have the capacity to produce offspring when mature. Together with the actually and potentially reproducing individuals in the second category, they help make up the population's **reproductive base**.

Population density is the number of individuals in some specified area or volume of a habitat, such as the number of guppies per liter of water in a stream. A *habitat*, remember, is the type of place where a species normally lives. We characterize a habitat by its physical features, chemical features, and other species that are present. **Population distribution** is the general pattern in which a population's individuals are dispersed in a specified area.

Crude density is a measured number of individuals in a specified area. Section 45.2 outlines how ecologists make their counts, which help them track changes in population density over time. The counts do not reveal how much of a habitat is used as living space. Even areas that seem uniform, such as a long, sandy beach, are more like tapestries of light, moisture, temperature, mineral composition, and other variables. The tapestry of environmental conditions may even change from day to night or with the seasons. So a population may find one part of a habitat more suitable for occupancy than other parts, and it may do so some or all of the time.

Also, different species typically share the same area —and they compete for energy, nutrients, living space, and other resources. Most interact as predators, prey, or parasites. Later chapters describe these interactions. For now, simply note that species interactions influence population density and dispersion through a habitat.

Theoretically, populations show three distribution patterns. By these patterns, individuals are dispersed in clumps, nearly uniformly, or randomly (Figure 45.2). The individuals of most populations form aggregations at specific sites in a habitat. Why is clumping together the most common pattern of dispersion? Three possible reasons come to mind.

Figure 45.2 Three generalized patterns of population distribution.

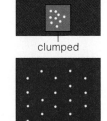
clumped

nearly uniform

random

First, each species is adapted to a limited set of ecological conditions, which usually are patchy through a habitat. For example, some parts of the habitat offer more or less shade, moisture, hiding spots, and hunting spots. Second, in the case of animals, many species gather in social groups that aid survival and reproduction, as by offering more opportunities for mating and mutual defense against predators. Third, adults of many species do not disperse their young (larvae, seeds, and other immature forms of the new generation) over very large distances. For example, sponge larvae are able to swim, but not far from their parent sponge. They clump together simply by settling down on substrates near the parents.

Where the individuals are more evenly spaced than they would be by chance alone, we call this a nearly uniform dispersal pattern. Evenly spaced fruit trees in orchards are often cited as examples. Few populations live this way in nature. You may see the pattern where habitat conditions are fairly uniform and where there is fierce competition for resources or territory.

Creosote bushes (*Larrea*) that live in arid habitats of the American Southwest may show this pattern. Large, mature plants deplete soil water all around them. Seed-eating ants and rodents forage near the plants, which offer cover from predators and the hot sun. Most seeds are devoured, which puts the seeds and seedlings at a competitive disadvantage. So seeds typically take hold only at patches where established, mature plants are weakened or have died. As one result, creosote bushes are not clumped together (Figure 45.3a). Yet neither are they dispersed at random; seeds that rodents and ants overlook and that await opportunity for life in the sun cannot travel far from the mature parent plants.

We observe random dispersion only when habitat conditions are nearly uniform, resource availability is fairly steady, and individuals of the population neither attract nor avoid one another. Example: Wolf spiders are solitary hunters on forest floors. Each generation might well be randomly spaced. However, a dispersion pattern such as this is extremely rare in nature.

Each population has its own gene pool and range of traits. It has a characteristic size, density, distribution pattern, and age structure. Environmental conditions and species interactions influence these characteristics.

Elusive Heads to Count

Does it seem like deer are just about everywhere, eating up gardens and smacking into cars on the roads? Their estimated number in the United States rose from fewer than 1 million at the turn of the century to more than 18 million today. Suppose you live in a northern California county and wonder how many deer live there. How can you go about determining their population density?

To get some idea of what your study would entail, you start with a literature search. Ecologists have baselines of approximate population densities for many organisms. For example, in some habitats, baselines show that you can expect to find as many as 5 million diatoms per cubic meter, 500 trees per hectare, 250 field mice per hectare, and so on. (A hectare is 10,000 square meters.) You come across an estimate of 4 deer per square kilometer. This tells you deer will be easier to count than, say, diatoms.

A full count is the most straightforward measurement of absolute population density. Census takers make such a count every ten years for the human population in the United States. Ecologists make counts for large organisms in small areas, such as songbirds in a forest, northern fur seals at their breeding grounds, sea stars in a tidepool, and creosote bushes in a patch of desert (Figure 45.3a). More often, they must content themselves with sampling a small part of a population and estimating total density.

You could get a map of your county and divide it into small plots, or quadrats. Quadrats are sampling areas of the same size and shape, such as rectangles, squares, and hexagons. Then you could count all the deer in several plots and extrapolate the average for the entire county. Ecologists conduct such counts for migrating herds and flocks. For example, the North American Breeding Bird Survey makes counts each year along more than 2,000 flyways for migratory waterfowl (Figure 45.3b).

Deer are among the animals that do not stay put in their habitat. So how can you be sure that the individuals you are counting in a given plot are not the same ones that you counted earlier in a different plot?

For mobile animals, ecologists sample the population density with a **capture–recapture method**. The idea is to capture individuals and mark them in some way. Deer get bright collars, squirrels get tattoos, salmon get tags, migratory birds get leg rings, and so on. Marked animals are released at time 1. Animals are captured and checked for marks at time 2. In later samplings, the proportion of marked individuals should be representative of the proportion marked in the whole population:

$$\frac{\text{Marked individuals in sampling 2}}{\text{Total captured in sampling 2}} = \frac{\text{Marked individuals in sampling 1}}{\text{Total population size}}$$

Ideally, marked and unmarked individuals of the population are captured at random, no marked animal dies during the study interval, and none of the marked animals leaves the population or gets overlooked. In the real world, though, recapturing of marked individuals might *not* be random. Squirrels that were marked after being attracted to yummy bait in boxes might now be trap-happy or trap-shy, so they might overrepresent or underrepresent the population. Instead of mailing tags of marked fish back to ecologists, some fishermen keep them as good-luck charms. Birds lose their leg rings.

Your estimate also depends on the time of year when you do a sampling. Population distribution varies over time, as in migratory responses to environmental rhythms. Few places yield abundant resources all year long, and many animals move between habitats with changes in the seasons. Canada geese and deer are like this (Figure 45.3b). In such cases, capture–recapture methods might be used more than once a year, for several years.

 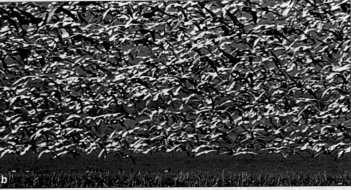

Figure 45.3 (a) Example of nearly uniform spacing: creosote bushes near Death Valley, California. (b) Snow geese, one of many species of waterfowl that migrate through 3.4 million square kilometers of air space above Canada and the United States.

POPULATION SIZE AND EXPONENTIAL GROWTH

Gains and Losses in Population Size

Populations are dynamic units of nature. Depending on the species, they may add or lose individuals every minute of every day, season, or year. Sometimes they glut portions of their habitat with individuals. Other times, individuals are scarce. Populations even drive themselves or are driven into extinction. We measure such changes in population size in terms of birth rates, death rates, and how many individuals are entering and leaving during a specified interval.

Population size increases as a result of (1) births and (2) **immigration**: the arrival of new residents from other populations of the species. Its size decreases because of (1) deaths and (2) **emigration**: individuals permanently moving out of the population. Population size for many species changes during daily or seasonal migrations. However, **migration** is a recurring round trip between two regions, and so we need not consider its transient effects in our initial study of population size.

From Zero to Exponential Growth

For our purposes, assume that immigration is balancing emigration over time, so that we can ignore the effects of both on population size. Doing so allows us to define **zero population growth** as some interval during which the number of births is balanced out by the number of deaths. The population's size is stabilized during such an interval, with no overall increase or decrease.

Births, deaths, and other variables that might affect population size can be measured in terms of **per capita** rates, or rates per individual. (*Capita* means heads, as in head counts.) Visualize 2,000 mice living in a cornfield. Twenty or so days after their eggs are fertilized, the females produce a litter, then nurse the offspring for a month or so, and get pregnant again. Collectively, if all the females give birth to 1,000 mice per month, the birth rate would be 1,000/2,000 = 0.5 per mouse per month. If 200 of the 2,000 die during that interval, the death rate would be 200/2,000 = 0.1 per mouse per month.

If we assume the birth rate and death rate remain constant, we can combine both into a single variable—the **net reproduction per individual per unit time**, or *r* for short. For our mouse population in the cornfield, *r* is 0.5 − 0.1 = 0.4 per mouse per month. This example gives us a way to represent population growth:

population growth per unit time	=	net population growth rate per individual per unit time	×	number of individuals

or, more simply, $G = rN$.

As the next month begins, 2,800 mice are scurrying about. With that net increase of 800 fertile mice, the reproductive base is larger. Assume *r* does not change. The population size expands this month, too, for a net increase of 0.4 × 2,800 = 1,120 mice. The population is now 3,920. In the months ahead, it just so happens that

			Net Monthly Increase:		New Population Size:
$G = r \times$	3,920	=	1,568	=	5,488
$r \times$	5,488	=	2,195	=	7,683
$r \times$	7,683	=	3,073	=	10,756
$r \times$	10,756	=	4,302	=	15,058
$r \times$	15,058	=	6,023	=	21,081
$r \times$	21,081	=	8,432	=	29,513
$r \times$	29,513	=	11,805	=	41,318
$r \times$	41,318	=	16,527	=	57,845
$r \times$	57,845	=	23,138	=	80,983
$r \times$	80,983	=	32,393	=	113,376
$r \times$	113,376	=	45,350	=	158,726
$r \times$	158,726	=	63,490	=	222,216
$r \times$	222,216	=	88,887	=	311,103
$r \times$	311,103	=	124,441	=	435,544
$r \times$	435,544	=	174,218	=	609,762
$r \times$	609,762	=	243,905	=	853,667
$r \times$	853,677	=	341,467	=	1,195,134

a

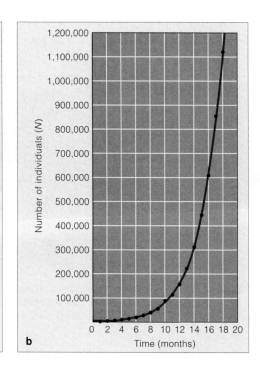

b

Figure 45.4 (**a**) Data showing net monthly increases in a population of field mice living in a cornfield. Start to finish, the numbers listed show a pattern typical of exponential growth. (**b**) Graph the data and you end up with a J-shaped growth curve.

Figure 45.5 (**a**) Effect of deaths on the rate of increase in two bacterial populations. Growth curve *1* represents a population of bacterial cells that reproduced every half hour. Growth curve *2* represents a different population. This cell's population divided every half hour, but 25 percent died between cell divisions. By comparing the two graph lines, you can see that deaths can slow the rate of increase but cannot directly prevent exponential growth.

(**b**) Do such growth curves seem far removed from your own experience? Think about the population growth of, say, the bacterial cells residing in your mouth after you present them with a smorgasbord of nutrients in candy or some other sugar-laden food.

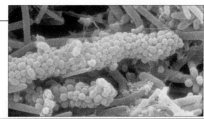

b Example of the types of bacterial cells that live in the human mouth

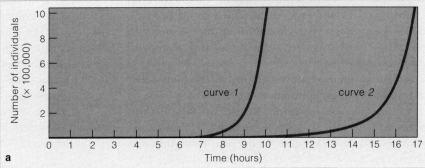

r remains constant, and a growth pattern emerges. As Figure 45.4*a* shows, *in less than two years from the time we started counting, the number of mice running around in the cornfield increased from 2,000 to more than a million!*

Plot the monthly increases against time, as in Figure 45.4*b*, and you end up with a graph line in the shape of a "J." When growth of any population over increments of time plots out as a J-shaped curve, you'll know that you are tracking exponential growth.

Exponential growth can be defined as a quantity increasing by a fixed percentage in a given interval (for example, 0.1, 1.0, and 10 percent over three successive years). So an increase in population size depends on how many individuals form the reproductive base in successive increments of time. *The larger the reproductive base, the greater will be the expansion in population size during each specified interval.* This will happen even if the per capita rate of increase does not change.

Now look at other aspects of exponential growth by supplying a single bacterium in a culture flask with all nutrients required for growth. Thirty minutes later, the one cell divides in two. Thirty minutes pass, the two cells divide, and so on every thirty minutes. Assuming no cells die between divisions, the population size will double in each interval—from 1 to 2, then 4, 8, 16, 32, and so on. The length of time it takes for a population to double in size is its **doubling time**.

The larger the population gets, the more cells there are to divide. After 9–1/2 hours (nineteen doublings), it has more than 500,000 bacterial cells. After 10 hours (twenty doublings), it has more than 1 million. Plot the doublings in size against time, and you end up with the J-shaped curve typical of unrestricted, exponential growth. Figure 45.5, curve *1*, shows this.

To examine the effect of deaths on the growth rates, start over with one bacterium. Assume 25 percent of the descendant cells die every thirty minutes. Owing to deaths, it takes about seventeen hours (not ten) for

population size to reach one million. *But deaths changed only the time scale for population growth.* You still end up with a J-shaped curve, which is curve *2* in Figure 45.5.

What Is the Biotic Potential?

Finally, visualize a population occupying a place where conditions are ideal. Every one of its individuals has adequate shelter, food, and other vital resources. No predators, pathogens, or pollutants lurk anywhere in the habitat. That population might well display its **biotic potential**, which is the maximum rate of increase per individual under ideal conditions.

Each species has a characteristic maximum rate of increase. For many bacteria, it is 100 percent every half hour or so. For humans and other large mammals, it is 0.02 to 0.05 percent per year. But the *actual* rate depends on the age at which each generation starts to reproduce, how often each individual reproduces, and how many offspring are produced. Now think about this. A human female is biologically capable of bearing twenty or more children, yet in each generation, many females do not reproduce at all. The human population has not been displaying its biotic potential. Even so, since the mid-eighteenth century, its growth has been exponential, for reasons that will soon be apparent.

During a specified interval, population size is generally an outcome of births, deaths, immigration, and emigration.

With exponential growth, population size is increasing by a fixed percentage of the whole in a given interval, because the reproductive base becomes ever larger.

Population size against time plots out as a characteristic J-shaped curve if the population is growing exponentially.

As long as the per capita birth rate remains even slightly above the per capita death rate, a population can grow exponentially.

LIMITS ON THE GROWTH OF POPULATIONS

What Are the Limiting Factors?

Most of the time, environmental circumstances prevent any population from fulfilling its biotic potential. That is why sea stars, the females of which could produce 2,500,000 eggs each year, do not fill up the oceans with sea stars. That also is why populations of humans will never fill up the planet.

In natural environments, complex interactions occur within and between populations of different species, so it is not easy to identify all the factors working to limit population growth. To get a sense of what some of the factors might be, start again with a lone bacterium in a culture flask, where you can control the variables. First you enrich the culture medium with glucose and other nutrients required for bacterial growth. Then you allow bacterial cells to reproduce for many generations.

At first, the pattern appears to be one of exponential growth. Then growth slows; and afterward, population size remains relatively stable. After that stable period, the size of the population declines quite rapidly until all of the bacterial cells have died.

What happened? When the population expanded by ever increasing amounts, it used up more and more nutrients. Dwindling nutrients were a cue to cells to stop dividing (Section 9.2). And when the existing cells exhausted the nutrient supply, they starved to death.

Any essential resource that is in short supply is a **limiting factor** on population growth. Food, minerals of certain types, refuge from predators, living quarters, and a pollution-free environment are such essentials. The number of limiting factors can be huge, and their effects can vary. Even so, one factor alone often can put the brakes on population growth.

Suppose you keep on freshening the supply of all required nutrients for that growing bacterial culture. After growing exponentially, the population still crashes. Like all other organisms, bacteria produce metabolic wastes. Wastes of that huge population were so great, they drastically altered living conditions in the culture. By its own metabolic activities, the population polluted the experimentally designed habitat and thus put a stop to any further exponential growth.

Carrying Capacity and Logistic Growth

Now visualize a small population in which individuals are dispersed through the habitat. As the population increases in size, more and more individuals must share nutrients, living quarters, and other resources. As the share available to each diminishes, fewer individuals may be born and more may die by starvation or nutrient deficiencies. Now the population's rate of growth will decrease until births are balanced or outnumbered by

deaths. Ultimately, the *sustainable* supply of resources will be the key factor determining population size. The term **carrying capacity** refers to the maximum number of individuals of a population (or species) that a given environment can sustain indefinitely.

The pattern of **logistic growth** shows how carrying capacity can affect population size. By this pattern, a small population starts growing slowly in size, then it grows rapidly, and finally its size levels off once the carrying capacity is reached. How can you represent this pattern? Start with the exponential growth, given in Section 45.3. Then add the term $(K - N)/K$, where K designates carrying capacity:

$$G = r_{max} N \frac{K - N}{K}$$

The term tells us how many individuals can be added to the population based on the proportion of resources still available. When population size is small, $(K - N)$ is close to 1. It approaches 0 when the size is close to the carrying capacity. Representing this another way,

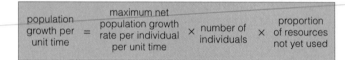

A plot of logistic growth gives an S-shaped curve (Figure 45.6). Such curves are just an approximation of what goes on in nature. For instance, a population that grows too fast can overshoot the carrying capacity. The death rate skyrockets, and the birth rate plummets. These two responses drive the number of individuals down to the carrying capacity or lower (Figure 45.7).

Density-Dependent Controls

The logistic growth equation just described deals with **density-dependent controls.** These are any factors that affect individuals of a population and that vary with population density.

For example, a small population may grow rapidly, but when it does bump up against its carrying capacity, so to speak, its size will stabilize or decline. Similarly, when other natural factors come into play, they also may drive the number of individuals in a population below the maximum sustainable level. A classic case is the increased likelihood of individuals dying under crowded living conditions. Why? Predators, parasites, and pathogens can interact more intensely with their host population and bring about a decline in density. Once population size declines, the density-dependent interactions decrease. And so the population size may increase once again.

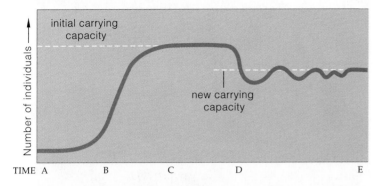

initial carrying capacity

Number of individuals →

new carrying capacity

TIME A   B   C   D   E

Figure 45.6 Idealized S-shaped curve characteristic of logistic growth. Growth slows after a phase of rapid increase (time B to C). The curve flattens out as the carrying capacity is reached (time C to D). S-shaped growth curves can show variations, as when changes in the environment lower the carrying capacity (time D to E). This happened to the human population of Ireland before 1900, when late blight, a disease caused by a water mold, destroyed the potato crops that were the mainstay of the diet (Section 22.8).

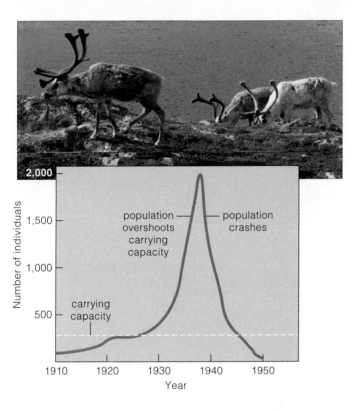

population overshoots carrying capacity

population crashes

carrying capacity

Figure 45.7 Carrying capacity and a reindeer herd. In 1910, four male and twenty-two female reindeer were introduced on St. Matthew Island in the Bering Sea. In less than thirty years, the size of the herd increased to two thousand. Individuals had to compete for dwindling vegetation, and overgrazing destroyed most of it. In 1950, herd size plummeted to eight. The growth curve reflects how the reindeer population size overshot the carrying capacity, then crashed.

Density-Independent Factors

Sometimes events result in more deaths or fewer births regardless of a population's density. For instance, each year, monarch butterflies migrate from Canada to spend the winter in Mexico's forested mountains. But logging opened up stands of the forest trees—which normally buffer temperature. In 2002, a sudden freeze, combined with deforestation, killed millions of butterflies. Both were **density-independent factors**, meaning the effects came into play independently of population density.

Similarly, heavy applications of pesticides in your backyard may also kill most insects, mice, cats, birds, and other animals. They will do this regardless of how uncrowded or dense the populations are.

Bubonic plague and *pneumonic plague*, two dangerous diseases, are examples of density-dependent controls. Both are caused by a bacterium, *Yersinia pestis*. A large reservoir of *Y. pestis* persists in rabbits, rats, and certain other small mammals. Fleas transmit it by biting new hosts. Bacterial cells reproduce rapidly in the flea gut. In time their numbers and metabolic activities disrupt digestion and cause hunger sensations. This provokes a flea into feeding more often on more of its hosts, and so the disease spreads.

A devastating episode of bubonic plague occurred in the fourteenth century. That plague swept through European cities where humans were crowded together, sanitary conditions were poor, and rats were abundant. By the time the epidemic subsided, urban populations in Europe had declined by 25 million. Both diseases are still threats to populations around the world.

Resources in short supply put limits on population growth. Together, all of the limiting factors acting on a population dictate how many individuals can be sustained.

Carrying capacity is the maximum number of individuals of a population that can be sustained indefinitely by the resources in a given environment. The number may rise or fall with changes in resource availability.

The size of a low-density population may increase slowly, go through a rapid growth phase, then level off once the carrying capacity for the population is reached. This is a logistic growth pattern.

Density-dependent controls and density-independent factors can bring about decreases in population size.

LIFE HISTORY PATTERNS

So far, we have looked at populations as if all of their members are identical throughout a given interval. For most species, however, the individuals of a population are at different stages of development. They interact in different ways with other organisms and with their environment. At different times in the life cycle, they may be adapted to using different resources, as when caterpillars eat leaves and butterflies prefer nectar. In addition, individuals at different stages might be more or less vulnerable to danger.

In short, each species has a **life history pattern**. We may define this as a set of adaptations that influence survival, fertility, and age at first reproduction. It is a set of conditions pertaining to an individual's schedule of reproduction. In this section and the next, we look at a few of the environmental variables that underlie the age-specific life history patterns.

Life Tables

Although each species has a characteristic life span, few of its individuals reach the maximum age possible. Death looms larger at some ages than at others. Also, individuals of a species tend to reproduce or emigrate during a characteristic age interval.

Age-specific patterns in populations first intrigued life insurance and health insurance companies, then ecologists. Such investigators typically track a **cohort**, a group of individuals, recorded from the time of birth until the last one dies. They also track the number of offspring born to individuals in each age interval. Life tables list the data on an age-specific death schedule. Often they are converted into much cheerier "survivorship" schedules: the number of individuals reaching some specified age (x). Table 45.1 is a typical example. Table 45.2, another example, lists data for the 1989 human population of the United States.

Dividing the population into age classes and assigning birth rates and mortality risks sometimes has practical applications. Unlike a crude census (or head count), the resulting data might be a basis for informed policy decisions about pest management, endangered species protections, social planning for human populations, and other current issues. For instance, birth and death schedules for the northern spotted owl figured in federal court rulings that put a stop to mechanized logging in the owl's habitat (old-growth forests).

Table 45.1 *Life Table for a Cohort of Annual Plants (Phlox drummondii)* *

Age Interval (days)	Survivorship (number surviving at start of interval)	Number Dying During Interval	Death Rate (number dying/number surviving)	"Birth" Rate During Interval (number of seeds produced per individual)
0–63	996	328	0.329	0
63–124	668	373	0.558	0
124–184	295	105	0.356	0
184–215	190	14	0.074	0
215–264	176	4	0.023	0
264–278	172	5	0.029	0
278–292	167	8	0.048	0
292–306	159	5	0.031	0.33
306–320	154	7	0.045	3.13
320–334	147	42	0.286	5.42
334–348	105	83	0.790	9.26
348–362	22	22	1.000	4.31
362–	0	0	0	0
		996		

* Data from W. J. Leverich and D. A. Levin. 1979. *American Naturalist* 113:881–903.

Table 45.2 *1989 Life Table for United States Human Population*

Age Interval (category for individuals between the two ages listed)	Survivorship (number alive at start of age interval, per 100,000 individuals)	Mortality (number dying during the age interval)	Life Expectancy (average lifetime remaining at start of age interval)	Reported Live Births for Total Population
0–1	100,000	896	75.3	
1–5	99,104	192	75.0	
5–10	98,912	117	71.1	
10–15	98,795	132	66.2	11,486
15–20	98,663	429	61.3	506,503
20–25	98,234	551	56.6	1,077,598
25–30	97,683	606	51.9	1,263,098
30–35	97,077	737	47.2	842,395
35–40	96,340	936	42.5	293,878
40–45	95,404	1,220	37.9	44,401
45–50	94,184	1,766	33.4	1,599
50–55	92,418	2,727	28.9	
55–60	89,691	4,334	24.7	
60–65	85,357	6,211	20.8	
65–70	79,146	8,477	17.2	
70–75	70,669	11,470	13.9	
75–80	59,199	14,598	10.9	
80–85	44,601	17,448	8.3	
85 +	27,153	27,153	6.2	
			Total:	4,040,958

Figure 45.8 Three generalized survivorship curves. **(a)** Elephants, typifying Type I populations. They have high survivorship until some age, then high mortality. **(b)** Snowy egrets, for Type II populations. They have a fairly constant death rate. **(c)** Sea star larva representing Type III populations, which show low survivorship early in life.

Patterns of Survival and Reproduction

Evolutionarily speaking, we measure the reproductive success of individuals in terms of the number of their surviving offspring. But that number varies among species, which differ in (1) how much energy and time are allocated to producing gametes, securing mates, and parenting, and in (2) the size of offspring. It seems that tradeoffs have been made in response to selection pressures, such as prevailing conditions in the habitat and the type of species interactions.

A **survivorship curve** is a graph line that emerges when ecologists plot a cohort's age-specific survival in a habitat. Each species has a characteristic survivorship curve, and three types are common in nature.

Type I curves reflect high survivorship until fairly late in life, then a large increase in deaths. Such curves are typical of elephants and other large mammals that bear only one or a few large offspring at a time, then engage in extended parental care (Figure 45.8a). As an example, a female elephant gives birth to four or five calves and devotes several years to parenting each one.

Type I curves also are typical of human populations in which individuals have access to good health care services. However, today as in the past, infant deaths cause a sharp drop at the start of the curve in regions where health care is poor. Following the drop, it levels off from childhood to early adulthood.

Type II curves reflect a fairly constant death rate at all ages. They are typical of organisms just as likely to

be killed or to die of disease at any age, such as lizards, small mammals, and large birds (Figure 45.8b).

Type III curves signify a death rate that is highest early on. We see this for species that produce many small offspring and do little, if any, parenting. Figure 45.8c shows how the curve plummets for sea stars. Sea stars release mind-boggling numbers of eggs. Their tiny larvae must rapidly eat, grow, and finish developing on their own without support, protection, or guidance from parents. Corals and other animals quickly eat most of them. Their plummeting survivorship curve is characteristic of numerous marine invertebrates, most of the insects, and many fishes, plants, and fungi.

At one time, ecologists thought selection processes were favoring *either* the early and rapid production of many small offspring *or* the late production of a few large offspring. They now realize that the two patterns are extremes at opposite ends of a range of possible life histories. In addition, both life history patterns—as well as intermediate ones—are sometimes evident in different populations of the same species, as the next section makes clear.

Tracking a cohort (a group of individuals) from birth until the last one dies reveals patterns of reproduction, death, and migration that typify the populations of a species.

Survivorship curves can reveal differences in age-specific survival among species. In some cases, such differences exist even between populations of the same species.

NATURAL SELECTION AND THE GUPPIES OF TRINIDAD

Earlier in the book, you read that those hinged, bony feeding structures called jaws evolved in Cambrian times and were a key innovation in the evolution of vertebrate predators and prey (Sections 26.1 and 26.2). No one was there to witness this early coevolutionary arms race. But the hypothesis that natural selection is a factor in vertebrate coevolution has been tested experimentally. Example:

Several years ago, drenched with sweat and with fish nets in hand, two evolutionary biologists were conducting fieldwork in the mountains of Trinidad, an island in the southern Caribbean Sea. They were after some small, live-bearing fishes that live in the shallow freshwater streams (Figure 45.9). The fishes were guppies (*Poecilia reticulata*). For eleven years the biologists, David Reznick and John Endler, studied environmental variables that affect the life history patterns of guppies.

The researchers easily distinguished male guppies from the females. The males are smaller and have bright-colored scales, which function as visual signals during mating behavior. They engage in intricate courtship maneuvers. Once they are sexually mature, they stop growing. By contrast, the females are drab

colored. They also continue to grow in size during the reproductive phase of the life cycle.

Different populations of guppies inhabit different streams in the Trinidad mountains, and sometimes different parts of the same stream. These separated populations deal with different predators, including killifishes (*Rivulus hartii*) and pike-cichlids (*Crenicichla*). A killifish is not an especially large fish. It can prey efficiently on smaller, immature guppies but not on larger adults. The larger pike-cichlid lives in other streams. It preys on larger, sexually mature guppies and tends to ignore small ones.

Starting from an understanding of natural selection, Reznick and Endler hypothesized that predation has a strong effect on guppy life history patterns. In the pike-cichlid streams, guppies mature faster and their body size is smaller at maturity, compared to guppies in the killifish-dominated streams. Also, guppies hunted by pike-cichlids reproduce earlier in life, they produce far more offspring, and they do so more frequently.

Did other variables influence the guppy life history patterns in killifish and pike-cichlid streams? To test this possibility, the researchers carefully shipped sets of guppies from each stream back to their laboratory in

GUPPY FROM A PIKE-CICHLID STREAM

PIKE-CICHLID

GUPPY FROM A KILLIFISH STREAM

KILLIFISH

Figure 45.9 (**a**) Evolutionary biologist David Reznick contemplating interactions among guppies (*Poecilia reticulata*) and their predators in a freshwater stream in Trinidad. (**b**,**c**) Two guppies, and two of the guppy eaters.

Figure 45.10 Representative guppies that inhabit streams with killifishes (**a**) and with pike-cichlids (**b**). Guppies in killifish streams tend to be larger, less bulky, and more brightly colored. The guppies in pike-cichlid streams tend to be smaller, more bulky, and duller in their color patterning. The differences between the two groups are accompanied by clear differences in their life history patterns.

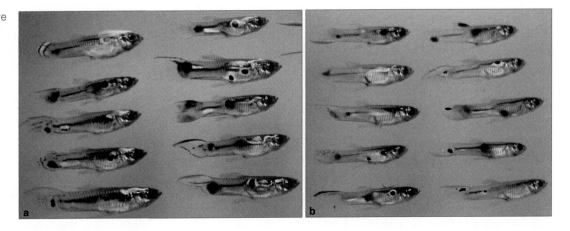

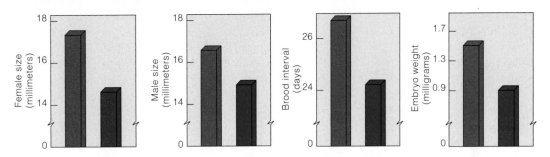

Figure 45.11 Graphs of some of the experimental evidence of natural selection among populations of guppies subjected to different predation pressures. Compared to guppies raised with killifish (the *red* bars), the guppies raised with pike-cichlids (*green* bars) differed in body size and length of time between broods. Killifish are small and prey on smaller guppies, pike-cichlids are large fish and prey on larger guppies, and so the two predators select for the differences.

the United States. They let the guppies reproduce in the absence of predators, in separate aquariums, for two generations. Physical and chemical conditions in the aquariums were identical for the experimental groups.

As it turned out, the offspring of the experimental guppy populations showed the same differences as the natural populations. The conclusion? The differences between guppies preyed upon by different predators have a genetic basis. This finding is consistent with the predictions of a theory that life history patterns evolve, due to differences in mortality schedules. Figure 45.10 shows representatives of the guppy populations.

The researchers looked into the role of predation in the evolution of differences in body size, the length of time between generations, and other aspects of guppy life history patterns. They studied many generations in the laboratory. As predicted, body sizes of the guppy lineage subjected to killifish predation over time were larger at maturity (Figure 45.11). Guppies raised with pike-cichlids tended to mature earlier.

When Reznick and Endler first visited Trinidad, they set up a field experiment by introducing guppies to a site upstream from a small waterfall. Before the

experiment, the waterfall was an effective barrier to dispersal; it had prevented guppies and all predators except killifish from moving upstream. The introduced guppies were from a population that had evolved with pike-cichlids downstream from the waterfall. That part of the stream was designated the control site.

Eleven years later, researchers revisited the stream and found that the guppy population had evolved. They based their finding on comparisons between guppies from the experimental site and the control site. Just as Reznick and Endler had predicted earlier, the neighborhood predator influenced the body size, frequency of reproduction, and other aspects of guppy life history patterns.

Later on, laboratory experiments involving two generations of guppies confirmed that the differences have a genetic basis, which is one of the requirements for the occurrence of natural selection.

Experimental tests support the hypothesis that natural selection, operating through predation, profoundly influences the life history patterns of prey populations.

HUMAN POPULATION GROWTH

The human population has now surpassed 6.2 billion. By midyear 2002, rates of natural increase for different countries ranged from 3.5 percent to below 1 percent, for an annual average of 1.3 percent. Visualize enough people to fill another New York City every month and you get the picture. Now think about this: The annual additions to that base population will result in a *larger* absolute increase each year into the foreseeable future.

Our staggering population growth continues even though 1.2 billion are already malnourished, without clean drinking water or adequate shelter. It continues even though 790 million do not have access to health care delivery systems or sewage treatment facilities. It continues primarily in already overcrowded regions. Nearly all 6.2 billion of us are living on just 10 percent of the land; 4 billion of us are crowded together within 480 kilometers (about 300 miles) of the seas.

Suppose it were possible to double the food supply to keep pace with growth. That would do little more than maintain marginal living conditions for most. The annual deaths from starvation could still be at least 10 million. Even this would come at great cost, for we are severely modifying the environment that must sustain us. Unusable cropland, loss of grasslands and forests, pollution—these are some consequences, and they do not bode well for our future.

For a while, it will be like the Red Queen's garden in Lewis Carroll's *Through the Looking Glass*, where one is forced to run as fast as one can to remain in the same place. But what happens when our population doubles again? Can you brush the doubling aside as being too far in the future to warrant concern? *It is no further removed from you than the sons and daughters of the next generation.*

How We Began Sidestepping Controls

How did we get into this predicament? For most of its history, the human population grew slowly. But during the past two centuries, increases in growth rates became astounding. There are three possible reasons for this:

> Humans steadily developed the capacity to expand into new habitats and new climate zones.
>
> Humans increased the carrying capacity in their existing habitats.
>
> Human populations sidestepped limiting factors.

The human population did all three of these things.

Reflect on the first point. The first humans evolved in woodlands, then savannas. They were vegetarians, for the most part, but they also scavenged bits of meat. Small bands of hunter–gatherers started moving out of Africa about 2 million years ago. By 40,000 years ago,

different populations of their descendants had become established in much of the world (Section 26.15).

Most species cannot expand into such a broad range of habitats. Having a notably complex brain, the early humans drew on learning and memory to figure out how to build fires, assemble shelters, make clothes and tools, and plan community hunts. Learned experiences did not die with individuals. They spread quickly from one band to another by language, which is the basis for extraordinary cultural communication. *Thus, the human population expanded into diverse, novel environments in short order, compared to the long-term geographic dispersal of other kinds of organisms.*

What about the second possibility? Starting about 11,000 years ago, many hunter–gatherer bands shifted to agriculture. Instead of following the migratory game herds, they settled in fertile valleys and other regions that favored seasonal harvesting of fruits and grains. By doing so, they developed a more dependable basis for life. A pivotal factor was the domestication of wild grasses, including species ancestral to modern wheats and rice. People harvested, stored, *and planted* seeds in one place. They domesticated animals and kept them close to home for food and for pulling plows. They dug ditches and diverted water to irrigate croplands.

Their agricultural practices increased productivity. And the larger, more dependable food supplies favored increases in population growth rates. Towns and cities developed along with the more intensive agriculture. Later in time, food supplies increased again by the use of fertilizers, herbicides, and pesticides. Transportation improved, and so did food distribution. *Thus, even at its simplest, management of food supplies through agriculture increased the carrying capacity for the human population.*

What about the third possibility—the sidestepping of limiting factors? Think about what happened when medical practices and sanitary conditions improved. Until about 300 years ago, poor hygiene, malnutrition, and contagious disease kept death rates high enough to more or less balance birth rates. Contagious diseases, a type of density-dependent control, swept unchecked through overcrowded settlements and cities that were infested with fleas and rodents. Then came plumbing and methods of sewage treatment. Over time, vaccines, antitoxins, and antibiotics were developed as weapons against many of the pathogens. Death rates dropped sharply. Births began to exceed deaths—and so, rapid population growth was under way.

In the mid-eighteenth century, people discovered how to harness energy stored in fossil fuels, starting with coal. Within a few decades, large industrialized societies began to form in western Europe and North America. Then efficient technologies developed after World War I. Large factories mass-produced tractors,

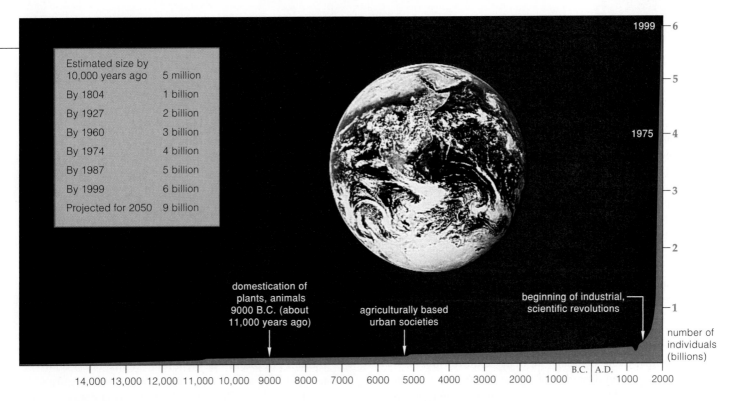

Estimated size by 10,000 years ago	5 million
By 1804	1 billion
By 1927	2 billion
By 1960	3 billion
By 1974	4 billion
By 1987	5 billion
By 1999	6 billion
Projected for 2050	9 billion

domestication of plants, animals 9000 B.C. (about 11,000 years ago)

agriculturally based urban societies

beginning of industrial, scientific revolutions

number of individuals (billions)

14,000 13,000 12,000 11,000 10,000 9000 8000 7000 6000 5000 4000 3000 2000 1000 B.C. | A.D. 1000 2000

Figure 45.12 Growth curve (*red*) for the human population. Its vertical axis shows the world population in billions. (The dip between the years 1347 and 1351 is when 60 million people died swiftly and horribly, first in Asia, then in Europe, during a bubonic plague.) Agricultural revolutions, industrialization, and health care improvements have sustained the accelerated growth pattern for the past two centuries. The *blue* box lists how long it took for the human population to increase from 5 million to 6 *billion*.

cars, and other affordable goods. Machines replaced many of the farmers who had produced the food, and fewer farmers could support a larger population.

And so, by controlling many disease agents and tapping into concentrated, existing forms of energy (fossil fuels), humans have managed to sidestep major factors that had previously limited their population growth.

Present and Future Growth

So where have the far-flung dispersals and spectacular advances in agriculture, industrialization, and health care taken us? Start with *Homo habilis* (Section 26.13). It took about 2.5 million years for human population size to reach 1 billion. As Figure 45.12 shows, it took just 123 years to reach 2 billion, another 33 to reach 3 billion, 14 more to get to 4 billion, and 13 more to get to 5 billion. And it took only 12 more years to arrive at 6 billion!

From what we know of the principles governing population growth, we can expect a decline in growth as birth rates drop or as death rates rise. Alternatively, we might expect growth to continue if technological breakthroughs increase the carrying capacity. *However, continued growth cannot be sustained indefinitely.*

Why? Continuing increases in population size can set the stage for density-dependent controls. Example: Infection by *Vibrio cholerae* results in the disease called *cholera*. This bacterium enters hosts who drink water

or eat food that is contaminated with raw sewage. It multiplies in the gut, where it produces a toxin that triggers severe diarrhea and massive fluid loss. Two to seven days afterward, infected, untreated people can die from extreme dehydration.

For centuries, *V. cholerae* has thrived and mutated in sewage-enriched rivers and estuaries (Figure 49.34). Millions are forced to bathe in polluted waters in many slums around the world. Since 1992 the largest cholera epidemic in the past century has been spreading across southern Asia. Like six previous epidemics, it began in India and spread through Africa, the Middle East, and Mediterranean countries, then reached Central America and the United States. This cholera resurgence might claim 5 million people. Existing vaccines do not work against the mutant bacterial strain causing it.

The problem is compounded by human migration on a vast scale. By some estimates, economic hardship and civil strife have put 50 million people on the move within and between countries. Will their relocations be peaceable? Will they find sustainable supplies of food, clean water, and other basic resources?

Through expansion into new habitats, cultural intervention, and technological innovation, the human population has temporarily skirted environmental resistance to growth. Its increased growth cannot be sustained indefinitely.

CONTROL THROUGH FAMILY PLANNING

Figure 45.13 presents the annual rates of population increase for major regions in 2001. The current average annual rate of increase is 1.3 percent. Add that to the current total of 6.2 billion, and world population may exceed 9 billion by 2050.

Think about the natural resources required for that many people. We will have to increase food production as well as the supplies of fresh water, energy reserves, wood, and other materials to meet basic needs, which is something that we are not doing now for almost half of the world's population. Large-scale manipulation of resources will intensify pollution; and water supplies, air quality, and food productivity will deteriorate.

More governments are recognizing that population growth, resource depletion, pollution, and the quality of life are interconnected. Most are attempting to lower long-term birth rates, as by family planning programs that help people decide how many children they will have, and when. Details vary from country to country, but all provide information on available methods of fertility control (Section 44.14).

The **total fertility rate** (TFR) is the average number of children born to women of a population during their reproductive years. TFRs are estimated on the basis of current age-specific rates. The world population TFR declined dramatically after 1950, when it averaged 6.5. That number was far above 2.1, which is considered to be the replacement level necessary to arrive at zero population growth. The average level is slightly higher than two children per couple, because some female children die before reaching reproductive age.

The birth rate slows when women bear children in their early thirties, not in mid-teens or early twenties. This delayed reproduction slows the growth rate and lowers the average number of children in families.

The world TFR is currently 2.8. In many developed countries, TFRs are at or below replacement levels, but the TFRs of developing countries are about 3.2. Even if every couple decides that they will bear no more than two children, the world population will keep growing for another sixty years! Why? A tremendous number of existing children will soon reach reproductive age.

Figure 45.14 shows some age structure diagrams for populations growing at different rates. Make note of the children who will be moving into the reproductive age category in the next fifteen years; the average range for childbearing years is 15–49. Diagrams for fast-growing populations, including Mexico, have a broad base. The human population of the United States has a relatively narrow base and is showing slow growth. Figure 45.15 tracks the 78 million *baby-boomers*. This cohort started to form in 1946 when American soldiers returned home after World War II and started raising families.

More than a third of the world population is now in the broad pre-reproductive base. This hints at the magnitude of what it will take to control world population growth.

China, for instance, supports the world's most far-reaching family planning. The government discourages premarital sex; it urges people to postpone marriage and limit families to one child. It makes available free abortions, contraceptives, and sterilization to married couples; paramedics and mobile units ensure access to these measures even in remote rural areas. Couples who pledge to have only one child receive more food, free medical care, better housing, and salary bonuses. The child will get free tuition and preferred treatment

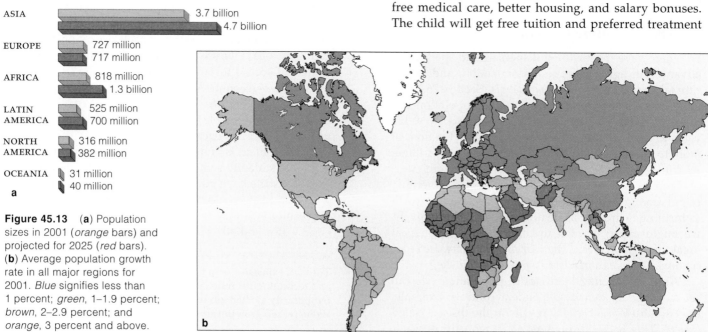

ASIA	3.7 billion	4.7 billion
EUROPE	727 million	717 million
AFRICA	818 million	1.3 billion
LATIN AMERICA	525 million	700 million
NORTH AMERICA	316 million	382 million
OCEANIA	31 million	40 million

a

Figure 45.13 (**a**) Population sizes in 2001 (*orange* bars) and projected for 2025 (*red* bars). (**b**) Average population growth rate in all major regions for 2001. *Blue* signifies less than 1 percent; *green*, 1–1.9 percent; *brown*, 2–2.9 percent; and *orange*, 3 percent and above.

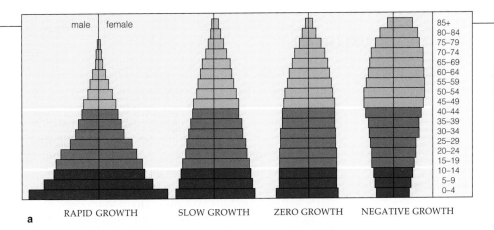

Billboard in China promoting family planning

a   RAPID GROWTH   SLOW GROWTH   ZERO GROWTH   NEGATIVE GROWTH

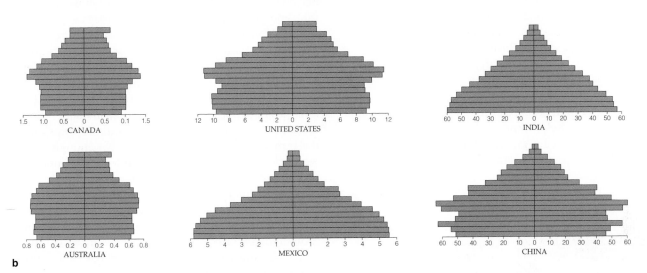

b

Figure 45.14 (**a**) General age structure diagrams for countries with rapid, slow, zero, and negative population growth rates. Pre-reproductive years are *green* bars; reproductive years, *purple;* and the post-reproductive years, *light blue*. A vertical axis divides each graph into males (*left*) and females (*right*). Bar widths correspond to the proportion of individuals in each age group. (**b**) Age structure diagrams for representative countries in 1997. Population sizes are measured in millions.

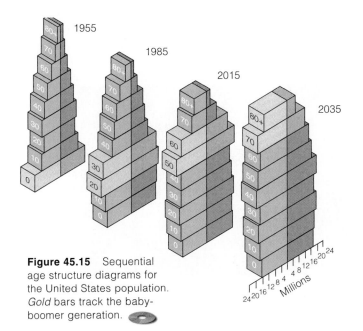

Figure 45.15 Sequential age structure diagrams for the United States population. *Gold* bars track the baby-boomer generation.

when he or she is old enough to enter the job market. Breaking the pledge means losing benefits and paying more taxes. Do you find the measures inhumane? Then what realistic alternatives would you suggest? Between 1958 and 1962 alone, 30 million Chinese may have died from starvation because of widespread famines.

Since 1972, China's TFR has declined sharply, from 5.7 to 1.8. Even so, the population time bomb has not stopped ticking. Its current population is close to 1.27 billion—and 150 million of its young females are in the pre-reproductive age category. By 2050 the population in China is projected to surpass 1.36 billion.

Family planning programs on a global scale are designed to help stabilize the size of the human population.

Even at zero population growth, human population size will continue to increase for sixty years, because its reproductive base already consists of an enormous number of individuals.

POPULATION GROWTH AND ECONOMIC DEVELOPMENT

More and more people are now aware that population growth rates and economic development interconnect. It seems economic security lifts pressure on individuals to produce a great many children to help them survive.

Demographic Transition Model

We can correlate changes in population growth with changes that often unfold in four stages of economic development. These four stages are at the heart of the **demographic transition model** (Figure 45.16).

By this model, living conditions are harshest during the *preindustrial* stage, before technology and medical advances become widespread. Birth and death rates are both high, so the population growth rate is low. Next, in the *transitional* stage, industrialization begins. Food production and health care improve. Death rates drop. But birth rates stay high; in agricultural societies, large families—hence more help in the fields—are viewed as an advantage. Populations grow fast for some time. Annual growth rates tend to average between 2.5 and 3

percent. When living conditions improve and the birth rates begin to decline, growth starts to level off.

During the *industrial* stage, when industrialization is in full swing, population growth slows dramatically. A slowdown starts mainly because people move from the country to cities, and urban couples tend to control family size. Many get caught up in the accumulation of goods and often decide the time and cost of raising more than a few children conflict with that goal.

In the *postindustrial* stage, population growth rates become negative. The birth rate falls below the death rate, and population size slowly decreases.

The United States, Canada, most of western Europe, Australia, Japan, and the nations of the former Soviet Union are in the industrial stage. Their growth rate is slowly decreasing. In Germany, Bulgaria, Hungary, and some other countries, the death rates exceed the birth rates, and the populations are getting smaller.

Mexico and other less developed countries are in the transitional stage. They don't have enough skilled workers to complete the transition to a fully industrial economy. Since 1980, their economic development has often been hindered because they must pay interest on huge debts owed to developed countries.

If population growth continues to outpace economic growth, death rates will increase. Thus many countries might get stuck in the transitional stage. Some might return to the harsh conditions of the preceding stage.

Think about how California's population increased by 12 *million* between 1970 and 1996. By July 1999, it exceeded 33 million. By 2025, it may be between 41.5 million and 51 million. For some time, California has been a symbol of the good life, drawing people from all

Figure 45.16 The demographic transition model for changes in population growth rates and in sizes, correlated with long-term changes in the economy. The model helps explain changes in western Europe and other industrialized regions.

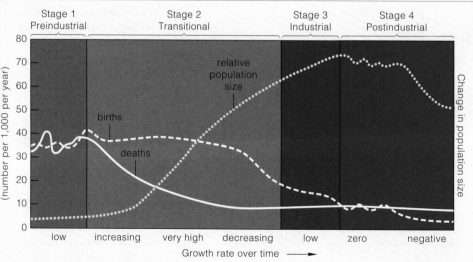

over the world. But its schools are overcrowded, funds for welfare and other social services are shrinking, and sewage treatment plants are near capacity. Shortages of water are frequent and often severe. Statewide electric power shortages and rolling blackouts occurred in 2001. The shortages, together with ongoing salinization of cropland and air pollution, are eroding the agricultural and technological bases of this state. At this writing, the economic repercussions of California's slowdown are affecting individuals throughout the nation.

Claiming that population growth affects economic health, many governments restrict immigration. Only the United States, Canada, Australia, and a few other countries now accept large annual increases. Elsewhere, difficult living conditions, harsh governmental policies, and civil strife combine to promote the exodus of more people than better-off nations can easily absorb.

A Question of Resource Consumption

This chapter opened with a brief look at the conditions endured by most people in India, with its 16 percent of the global human population. By comparison, the United States has a mere 4.7 percent. Yet other factors must be considered, because the United States and the other highly industrialized countries consume the most resources. In addition, their high rates of consumption are instigators of environmental damage within and beyond their own borders.

For example, the United States produces 21 percent of the world's goods and services. But it consumes fifty times more goods and services than the average person in India. It uses 25 percent of the world's processed minerals and a big portion of the available, nonrenewable sources of energy. People of the United States generate at least 25 percent of all pollution and trash.

By contrast, India produces only about 1 percent of all goods and services. It uses 3 percent of the available minerals and nonrenewable energy resources. It only generates about 3 percent of the pollution and trash.

Extrapolating from these numbers, G. Tyler Miller estimated that it would take 12.9 billion impoverished individuals living in India to have as much impact on the environment as 284 million people in the United States. Think about this estimate when you also reflect on the biodiversity crisis and conservation programs, as sketched out in Chapter 27.

Differences in population growth among countries correlate with levels of economic development, hence with economic security (or lack of it) of individuals. Growth rates are low in preindustrial, industrial, and post-industrial stages; they are greatest during the transition to industrialization.

SOCIAL IMPACT OF NO GROWTH

For us, as for all species, the biological implications of rapid population growth are staggering. Yet so are the social implications of what will happen when (and if) the human population decreases to the point of zero population growth—and stays there.

For instance, in a growing population, most people are in younger age brackets. If living conditions ensure constant growth over time, the age distribution should guarantee availability of a future workforce. This has social implications. Why? *It takes a large workforce to support the older age brackets.* In the United States, older, retired people often expect the federal government to provide them with subsidized medical care, low-cost housing, and many other social programs. However, as a result of improved medicine and hygiene, people in these brackets are living far longer than the elderly did when the nation's social security program was set up. The current cash benefits exceed the contributions that they made to the program when *they* were younger.

If the population ever does reach and maintain zero growth, a larger proportion of individuals will end up in the older age brackets. Even slower growth will pose problems. What happens when baby-boomers retire? Will all those individuals continue to receive goods and services when the workforce must carry more and more of the economic burden? This is not an abstract question. Put it to yourself. Just how much economic hardship are you willing to bear for the sake of your parents? For your grandparents? And how much will your children be willing or able to bear for you?

We have arrived at a major turning point, not only in our biological evolution but in our cultural evolution as well. The decisions awaiting us are among the most difficult we will ever have to make—yet it is clear that they must be made, and soon.

All species face limits to growth. We might think we are different from the rest, for our special ability to undergo rapid cultural evolution has enabled us to postpone the action of most factors that limit growth. But the crucial word is *postpone*. No amount of cultural intervention can sidestep the ultimate check of limited resources and a damaged environment.

We have sidestepped a number of the smaller laws of nature. In doing so we have become more vulnerable to those laws which cannot be repealed. Today, there may be only two options available. Either we make a global effort to limit population growth in accordance with environmental carrying capacity, or we wait until the environment does it for us.

In the final analysis, no amount of cultural intervention can repeal the ultimate laws governing population growth as imposed by the carrying capacity of the environment.

SUMMARY

Gold indicates text section

1. A population is a group of individuals of the same species occupying a given area. It has a characteristic size, density, distribution, and age structure as well as characteristic ranges of heritable traits. *45.1*

2. The growth rate for a population during a specified interval may be determined by calculating the rates of birth, death, immigration, and emigration. To simplify calculations, we may put aside effects of immigration and emigration, and combine the birth and death rates into a variable r (net reproduction per individual per unit time). Then we may represent population growth (G) as $G = rN$, where N is the number of individuals during the interval specified. *45.3*

    a. In cases of exponential growth, the reproductive base of a population increases and its size expands by a certain percentage during successive intervals. This trend plots out as a J-shaped growth curve.

    b. Any population will show exponential growth as long as its per capita birth rate stays even slightly above its per capita death rate.

    c. In logistic growth, a low-density population slowly increases in size, goes through a rapid growth phase, then levels off in size once carrying capacity is reached.

3. Carrying capacity is the name ecologists give to the maximum number of individuals in a population that can be sustained indefinitely by the resources available in their environment. *45.4*

4. The availability of sustainable resources as well as other factors that limit growth dictates population size during a specified interval. The limiting factors vary in their relative effects and vary over time, so population size also changes over time. *45.4*

5. Limiting factors such as competition for resources, disease, and predation are density-dependent. Density-independent factors, such as weather on the rampage, tend to raise the death rate or lower the birth rate more or less independently of population density. *45.4*

6. Patterns of reproduction, death, and migration vary over the life span of a species. Environmental variables help shape the life history (age-specific) patterns. *45.5*

7. The human population recently surpassed 6.2 billion. Its annual growth rate is below zero in a few developed countries. The rate is now above 3 percent in some less developed countries. In 2001 the annual growth rate for the global human population was 1.3 percent. *45.7*

8. The human population's rapid growth in the past two centuries occurred through expansion into many diverse habitats and through agricultural, medical, and technological developments that raised the carrying capacity. Ultimately, we must confront the reality of the carrying capacity and limits to growth. *45.7–45.9*

Review Questions

1. Define population size, density, and distribution. Describe a typical population's age-structure categories. *45.1*

2. Define exponential growth. Be sure to state what goes on in the age category that is a foundation for its occurrence. *45.3*

3. Define carrying capacity. Describe its effect as evidenced by a logistic growth pattern. *45.4*

4. Give examples of the limiting factors that come into play when a population of mammals (for example, rabbits, deer, or humans) reaches very high density. *45.4*

5. Define doubling time. In what year is the human population expected to surpass 9 billion? *45.3, 45.8*

6. How did earlier human populations expand steadily into new environments? How did they increase the carrying capacity in their habitats? Have we now avoided some limiting factors on population growth? Or is the avoidance an illusion? *45.7*

Self-Quiz   ANSWERS IN APPENDIX III

1. _____ is the study of how organisms interact with one another and with their physical and chemical environment.

2. A _____ is a group of individuals of the same species that occupy a certain area.

3. The rate at which a population grows or declines depends upon the rate of _____ .
    a. births   c. immigration   e. all of the above
    b. deaths   d. emigration

4. Populations grow exponentially when _____ .
    a. population size expands by ever increasing increments through successive time intervals
    b. size of low-density population increases slowly, then quickly, then levels off once carrying capacity is reached
    c. Both a and b are characteristic of exponential growth.

5. For a given species, the maximum rate of increase per individual under ideal conditions is the _____ .
    a. biotic potential   c. environmental resistance
    b. carrying capacity   d. density control

6. Resource competition, disease, and predation are _____ controls on population growth rates.
    a. density-independent   c. age-specific
    b. population-sustaining   d. density-dependent

7. A life history pattern for a population is a set of adaptations that influence the individual's _____ .
    a. survival   c. age when it starts reproducing
    b. fertility   d. all of the above

8. In 2001, the average annual growth rate for the human population at midyear was _____ percent.
    a. 0   b. 1.05   c. 1.3   d. 1.55   e. 2.7   f. 4.0

9. Match each term with its most suitable description.
    _____ carrying   a. maximum rate of increase per
          capacity   individual under ideal conditions
    _____ exponential   b. population growth plots out
          growth   as an S-shaped curve
    _____ biotic   c. the maximum number of
          potential   individuals sustainable by
    _____ limiting factor   an environment's resources
    _____ logistic growth   d. population growth plots out
                                 as a J-shaped curve
                             e. short supply of any resource
                                essential to population growth

Critical Thinking

1. If house cats that have not been neutered or spayed live up to their biotic potential, two can be the start of many kittens—12 the first year, 72 the second year, 429 the third, 2,574 the fourth, 15,416 the fifth, 92,332 the sixth, 553,019 the seventh, 3,312,280 the eighth, and 19,838,741 kittens the ninth year. Is this a case of logistic growth? Exponential growth? Irresponsible cat owners?

2. A third of the world population is below age fifteen. Describe the effect of this age distribution on the human population's future growth rate. If you conclude that this will have severe impact, what sorts of humane recommendations would you make to encourage individuals of this age group to limit family size? What are some social, economic, and environmental factors that might keep them from following the recommendations?

3. Write a short essay about a population having one of the age structures shown below. Describe what may happen to younger and older groups when individuals move into new categories.

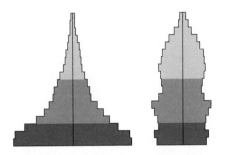

4. Figure 45.17 charts the legal immigration to the United States between 1820 and 1997. (The Immigration Reform and Control Act of 1986 accounted for the most recent dramatic increase; it granted legal status to undocumented immigrants who could prove they had lived in the country for years.) During the 1980s and 1990s, an economic downturn fanned resentment against newcomers. Many now say annual legal immigration should be restricted to 300,000–450,000 and we should crack down on undocumented immigrants. Others argue that such a policy would diminish our reputation as a land of opportunity. They also say that such a policy would discriminate against legal immigrants during crackdowns on others of the same ethnic background. Do some research, then write an essay on the

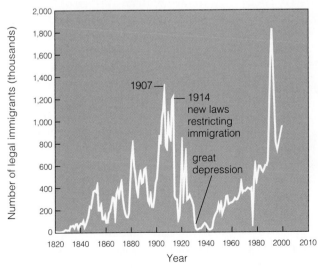

Figure 45.17 Chart of legal immigration to the United States between 1820 and 1997.

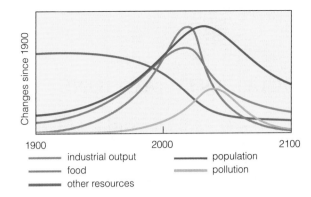

Figure 45.18 Computer-based projection of what may happen if human population size continues to skyrocket without dramatic policy changes and technological innovation. The assumptions used are that the population has already overshot the carrying capacity and current trends will continue unchanged.

pros and cons of both positions. (You may wish to start with this web site: http://www.census.gov.)

5. In his book *Living in the Environment*, G. Tyler Miller points out that the projected increase in the human population to over 9 billion by the year 2050 raises serious questions. Will there be enough food, energy, water, and other resources to sustain that many people? Will governments be able to provide adequate education, housing, medical care, and other social services for all of them? Computer models suggest that the answer is no (Figure 45.18). Yet some people claim we can adapt politically and socially to a far more crowded world, assuming harvests improve through technological innovation, every inch of arable land is cultivated, and everybody eats only grain.

There are no easy answers to these questions. If you have not yet been doing so, start following the arguments in your local newspapers, in magazines, on television, and on the Web. This will allow you to become an informed participant in a global debate that surely will have impact on your future.

Selected Key Terms

age structure *45.1*	life history pattern *45.5*
biotic potential *45.3*	limiting factor *45.4*
capture–recapture method *45.2*	logistic growth *45.4*
carrying capacity *45.4*	migration *45.3*
cohort *45.5*	per capita *45.3*
demographic transition model *45.9*	population density *45.1*
demographics *45.1*	population distribution *45.1*
density-dependent control *45.4*	population size *45.1*
density-independent factor *45.4*	*r* (net reproduction per
doubling time *45.3*	individual per unit time) *45.3*
ecology *CI*	reproductive base *45.1*
emigration *45.3*	survivorship curve *45.5*
exponential growth *45.3*	total fertility rate *45.8*
immigration *45.3*	zero population growth *45.3*

Readings

Bloom, D., and D. Canning. 18 February 2000. "The Health and Wealth of Nations." *Science* 287:1207–1209. Argues that poor health is more than a consequence of low income; that it is also one of its fundamental causes.

Miller, G. T. 2002. *Living in the Environment*. Twelfth edition. Belmont, California: Brooks/Cole.

46

SOCIAL INTERACTIONS

Deck the Nest With Sprigs of Green Stuff

In 1890, bird fanciers imported more than a hundred starlings (*Sturnus vulgaris*) from Europe and released them in New York City's Central Park. In less than a century, descendants of the introduced species had expanded their geographic range from coast to coast. They are so good at gleaning food from the agricultural fields of North America that they have outmultiplied native birds. They also have evicted great numbers of them from scarce nesting sites in the cavities of trees.

Once a male and female commandeer a tree cavity, they gather dry grass and twigs and use them to build or rebuild a nest (Figure 46.1). Theirs is no ordinary nest. They *decorate* the nest bowl with sprigs, freshly plucked.

Why do starlings do this? Do the sprigs of greenery camouflage the nest from predators? Not likely. The nests are already concealed in tree cavities. Do sprigs serve as insulative material that helps keep forthcoming eggs warm? Actually, the still-moist, green plant parts would promote heat loss, not heat conservation. Well, do the green sprigs combat parasites? Many mites parasitize birds and infest nest cavities. In short order, even a few mites produce thousands of descendants. In large numbers, mites can suck enough blood from a nestling to weaken it. They interfere with the nestling's growth, development, and survival.

Larry Clark and Russell Mason decided to test the third hypothesis. As they knew, starlings don't weave just any green plant material into the nests. Starlings choosily favor the leaves of certain plants, such as wild carrot (Figure 46.1). So the two biologists built a set of experimental nests, some with freshly cut wild carrot leaves and some without. They removed the natural nests that pairs of starlings had constructed and had already started using. Half of the nesting pairs got replacement nests decorated with sprigs of wild carrot. Replacement nests for the other pairs of starlings were sprigless.

Figure 46.2 shows the results. The number of mites in greenery-free nests was consistently greater than the number in nests decorated with greenery. At the end of one experiment, the sprig-free nests teemed with an average of 750,000 mites. Nests with sprigs contained a mere 8,000 mites.

Shoots of wild carrot happen to contain a highly aromatic steroid compound. The compound almost certainly repels herbivores and helps plants survive. By coincidence, the compound also prevents mites from maturing sexually—and so prevents mite population explosions in nests festooned with wild carrot.

Figure 46.1 A most excellent fumigator in nature—a European starling (*Sturnus vulgaris*). Starlings combat infestations of mites by decorating previously owned nests with fresh sprigs of wild carrot (*Daucus carota*), shown at left.

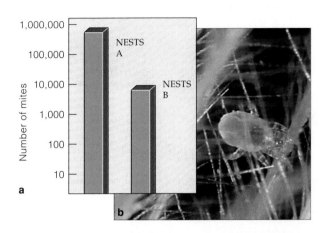

Figure 46.2 (**a**) Experimental results that point to the adaptive value of starling nest-decorating behavior. Nests designated *A* were kept free of fresh sprigs of wild carrot and other plants that contain aromatic compounds. The compounds prevent baby mites from developing into (**b**) adult mites. Other nests, designated *B*, had fresh sprigs tucked in every seven days. Head counts of mites (*Ornithonyssus sylviarum*) infesting the nests were made at the time the starling chicks left the nest, twenty-one days after this experiment was under way.

Far from being a trivial behavior, then, "decorating" a nest with aromatic greenery has adaptive value. It fumigates the nest and thus increases an individual's chance of producing healthy, surviving offspring.

And so starlings lead us into the fascinating world of research into behavior. As you will see, some of the studies focus on the adaptive value of behavioral traits to an individual's reproductive success. Others focus on the internal mechanisms that allow individuals to behave as they do.

As a starting point, recall that information encoded in genes directs the formation of tissues and organs that make up the animal body, including those of its nervous system. That system detects, processes, and integrates information about stimuli in the internal and external environments. Then it commands muscles or glands (also built according to genetic instructions) to make suitable responses. Other gene products called hormones contribute to the responses. Because genes specify the crucial substances required for constructing nervous and endocrine systems, *they are the heritable foundation for responses that animals make to stimuli*. We call these observable, coordinated responses to stimuli **animal behavior**.

Key Concepts

1. Animals make coordinated responses to stimuli, which are forms of behavior. Among individuals of an animal species, cooperative, interdependent relationships are regulated by social behavior.

2. Genes underlie behavior, as when instructions encoded in certain genes govern the development of the nervous and endocrine systems. These organ systems are the basis of detecting, interpreting, and then issuing commands for appropriate behavioral responses to stimuli.

3. Most newly born or hatched animals have neural wiring and motor systems necessary to perform some behaviors; it is not necessary to learn them through experience.

4. Most nervous systems can process and retain information about specific experiences, then use the information to adjust behavioral responses. The outcomes are called learned behaviors.

5. Behavioral mechanisms that influence the ability of the individual to pass on genes to offspring are subject to agents of natural selection.

6. Evolved modes of communication underlie social behavior. Communication signals hold clear meaning for both the sender and the receiver of signals.

7. Living in a social group has costs and benefits, as measured by an individual's ability to pass on its genes to offspring. Not every environment favors the evolution of such groups.

8. An altruistic individual helps others in ways that require it to sacrifice its own reproductive success. The evolution of altruism requires circumstances in which individuals propagate their genes indirectly, by helping relatives reproduce successfully.

BEHAVIOR'S HERITABLE BASIS

Genes and Behavior

An animal's nervous system, recall, is wired to detect, interpret, and issue commands for response to stimuli—that is, to specific aspects of the external and internal environments. One way or another, certain products of genes affect each of the steps required to assemble and operate its brain, nerve pathways, and all of its sensory receptors. If we define animal behavior as coordinated responses to stimuli, then behavior starts with genes.

Stevan Arnold found evidence of behavior's genetic basis by studying feeding preferences among coastal and inland populations of a snake species in California. Garter snakes near the coast prefer banana slugs (Figure 46.3a). Snakes living inland prefer tadpoles and small fishes. Offer them a banana slug and they ignore it.

In one set of experiments, Arnold offered captive newborn garter snakes a chunk of slug as the first meal. Coastal snake offspring usually ate the chunk and even flicked their tongue at cotton swabs drenched in essence of slug. (Snakes "smell" by tongue-flicking, which pulls chemical odors into the mouth.) However, inland snake offspring ignored the swabs and rarely ate slug meat. Here was an obvious difference between captive baby snakes that had no prior, direct experience with slugs. Arnold concluded the snakes were programmed before birth to accept or reject slugs; they certainly did not learn feeding preferences by taste trials. Maybe coastal and inland populations have allelic differences for the genes affecting how odor-detecting mechanisms form when a garter snake embryo is developing.

To test his hypothesis, Arnold crossbred coastal and inland snakes. If genes underlie the difference in food

Figure 46.3 (**a**) Banana slug, food for (**b**) an adult garter snake of coastal California. (**c**) A newborn garter snake from a coastal population, tongue-flicking at a cotton swab drenched with tissue fluids from a banana slug.

preferences, he figured that the hybrid offspring from snakes of both populations should make an *intermediate* response to slug chunks and odors. Results matched his prediction. Compared with a typical newborn inland snake, many baby snakes of mixed parentage tongue-flicked more often at the slug-scented cotton swabs—but less often than newborn coastal snakes did.

Hormones and Behavior

Gene products called hormones also guide behavior. Consider the seasonal singing behavior of a male zebra finch as one example (Figure 46.4). It arises partly as a result of seasonal differences in how much melatonin is secreted from his pineal gland (Section 36.7). The greater the secretion, the more the growth and function of his gonads are suppressed, and the less he sings.

Photoreceptors in the bird's pineal gland can absorb sunlight energy, and when they do, melatonin secretion slows. There is not enough light to slow things down in winter, which has fewer daylight hours than spring.

Figure 46.4 Hormones and the territorial song of zebra finches. (**a**) Sound spectrogram of the male's song. The spectrogram is a visual record of each note's pitch (that is, its frequency, measured in kilohertz). (**b**) Sound spectrogram of a female zebra finch that was experimentally converted to a singer. When she was a nestling, she received extra estrogen. At adulthood, she received an implant of testosterone—and started singing. Estrogen normally organizes the song system (certain parts of the brain) in developing songbird embryos. Testosterone activates the system in adult males.

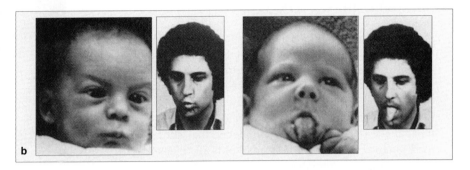

Figure 46.5 Examples of instinctive responses that human offspring make to sign stimuli. (**a**) A close-up face of an adult triggers smiling behavior in young infants. (**b**) Infants instinctively imitate adult facial expressions.

In spring, when daylength increases, less melatonin is secreted. Now the bird's gonads increase in size—and they secrete more testosterone as the breeding season starts. This sex hormone acts on cells in a **song system**, which consists of several brain structures that control muscles of a vocal organ. It induces metabolic changes that prepare the male to sing when suitably stimulated.

In most songbird embryos, differences in singing behavior between males and females also is related to sex hormones. Estrogen controls formation of the song system, which differs in its structural components and size between the males and females. Unlike humans, songbird females are XY and males are XX. And their Y chromosome has some genes that specify an estrogen-inhibiting product. Even before a male has hatched, his gonads secrete estrogen. This hormone is converted to testosterone, which stimulates the development of the masculinized brain and, in time, male sexual behavior.

Instinctive Behavior Defined

With their tongue-flicking and strikes at prey, newborn garter snakes can start us thinking about **instinctive behavior**. This term means the behavior is performed without having been learned through experience in the environment. Instead, the nervous system of a newly born or hatched animal is prewired to recognize **sign stimuli**—one or two simple, well-defined cues from the environment that can trigger a suitable response. The snake's response is a stereotyped motor program. When the snake recognizes certain sign stimuli, a **fixed action pattern** follows. It is a program of coordinated muscle activity that runs to completion independently of feedback from the environment.

Another example: When human infants are two or three weeks old, they tend to smile instinctively when an adult's face comes close to their own (Figure 46.5a). Infants make the same response to an overly simplified stimulus—a flat, face-sized mask with two dark spots where eyes would be on a human face. A mask with

Figure 46.6 A social parasite. European cuckoos lay eggs in nests of other species. This cuckoo hatchling is executing an innate behavior. It inherited knowledge of what to do without having to learn it. Even before its eyes open, it responds to the round shape of the host's eggs and shoves them out of the nest. The foster parents keep on feeding the usurper.

one "eye" won't do the trick. As Figure 46.5b suggests, older infants continue to show instinctive behavior.

And think of the cuckoo, a social parasite. Adult females lay their eggs in nests of other species. If newly hatched cuckoos eliminate the natural-born offspring, they get the undivided attention of their foster parents (Section 47.6). They are blind at birth, but when they contact an egg or another round object, they carry out a fixed action pattern. They maneuver the egg onto their back, then push it from the nest (Figure 46.6).

Animals clearly have a genetically based capacity to respond automatically to certain environmental cues. But they also have the means to process information about specific experiences, then use the information to change the responses. The modified outcome, as you will see next, is what we call learned behavior.

Genes underlie animal behavior—coordinated responses to stimuli. Certain gene products are essential in constructing and operating the nervous system, which governs behavior. Other gene products called hormones also influence the mechanisms required for particular forms of behavior.

Animals start out life neurally wired to recognize important cues and to make an instinctively suitable response, one that has not been learned through actual experience.

46.2

LEARNED BEHAVIOR

An animal processes and integrates information gained from experiences, and then uses it to vary or change responses to stimuli. We call this **learned behavior**. For example, a young toad's nervous system commands it to flip its sticky tongue instinctively at dark objects in its field of vision, which usually are edible insects. But what if the dark object is a bumblebee that stings the toad tongue? Thereafter, the toad avoids all "black-and-yellow-banded bumblebee-sized objects that sting."

Imprinting, another example, is a time-dependent form of learning triggered by exposure to sign stimuli, most often during a sensitive period when an animal is

Figure 46.7 No one can tell these imprinted baby geese that Konrad Lorenz is not Mother Goose. (Refer to Table 46.1.)

young. The imprinting of baby geese on their mother is a classic example (Table 46.1 and Figure 46.7).

Can instinct be governed only by genes and learned behavior only by the environment? No. All behavior arises from interactions between gene products *and* the "environment," which includes nutrients ingested, sensory experiences, and other inputs that can trigger changes in gene activity. Consider this: In different habitats, songbirds use variations, or dialects, of their species song. As Peter Marler found out, the males learn the full song ten to fifty days after hatching by listening to other birds sing it. As an example, a male white-crown sparrow's nervous system is prewired to recognize that song; his learning mechanism is primed to select and respond to acoustical input. However, his rendition of the species song is influenced by what he actually hears during the sensitive period.

The male birds learn parts of a song by picking up cues when other males sing. Marler raised white-crown nestlings in soundproof chambers so they could not hear adult males. The captives sang when mature, but not with the detailed structure of a typical adult song. Marler exposed other isolated captives to recordings of white-crown sparrows *and* song sparrows. At maturity, the captives sang just the white-crown song. They even mimicked the species dialect that they heard. Marler conducted many such experiments, which support the hypothesis that birdsong requires *a genetically based capacity to learn* from acoustical cues.

This isn't the whole story. In another experiment, Marler let young, hand-reared white-crowns interact with a "social tutor" of a different species, as opposed to listening to taped songs. The males tended to learn the tutor's song. Their primed learning mechanism had responded to *social experience* as well as acoustical cues.

In instinctive behavior, animals make complex, stereotyped responses to specific, often simple environmental cues. In learned behavior, responses may vary or change as a result of individual experiences in the environment.

Whether instinctive or learned, behavior develops through interactions between genes and environmental inputs.

Table 46.1 *A Few Categories of Learned Behavior*

IMPRINTING This time-dependent form of learning involves exposure to sign stimuli, most often early in development. For instance, in response to a moving object and probably to certain sounds, baby geese imprint on the mother and follow her during a short, sensitive period after hatching. They are neurally wired to learn crucial information—the identity of the individual that will protect them in the months ahead. Usually that individual is the mother or father. Imprinting occurs among many animals.

Konrad Lorenz, an early ethologist, must have presented the baby geese in Figure 46.7 with sign stimuli that made them form an attachment to him. As another outcome of imprinting, birds also direct sexual attention to members of the species they had been sexually imprinted upon when young.

CLASSICAL CONDITIONING Ivan Pavlov's early experiments with dogs are an example of classical conditioning. Just before eating, dogs will salivate. Pavlov's dogs were conditioned to salivate even in the absence of food. They did so in response to the sound of a bell or a flash of light that they initially associated with getting food. In this case, the dogs learned to associate an automatic, unconditioned response with a novel stimulus that does not normally trigger the response.

OPERANT CONDITIONING An animal learns to associate a voluntary activity with its consequences, as when a toad learns to avoid stinging insects after its first attempt to eat them.

HABITUATION An animal learns by experience *not* to respond to a situation if the response has neither positive nor negative consequences. Thus pigeons and other birds living in cities learn not to flee from people who pose no threat to them.

SPATIAL (LATENT) LEARNING By inspecting its environment, an animal acquires a mental map of a particular region, often by learning the position of local landmarks. Chickadees, for example, can store information about the position of dozens, if not hundreds, of places where they have stashed food.

INSIGHT LEARNING An animal abruptly solves some problem without trial-and-error attempts at the solution. Chimpanzees often exhibit insight learning in captivity when they suddenly solve a novel problem that their captors devise for them. Some chimps abruptly stacked and stood on several boxes *and* used a stick to reach bananas suspended out of their reach.

THE ADAPTIVE VALUE OF BEHAVIOR

Given that forms of behavior have a genetic foundation, they can be subject to evolution by natural selection. **Natural selection**, recall, is a result of differences in reproductive success among individuals that vary in heritable traits. Alleles underlying the most adaptive versions of a trait tend to increase in frequency in a population, whereas alternative alleles do not. (Alleles are different molecular forms of the same gene.) Over time, genetic changes that yield greater reproductive success for individuals spread through the population.

By using the theory of evolution by natural selection as a point of departure, we should be able to develop and test possible explanations of why a behavior has endured. We should be able to discern how some actions bestow reproductive benefits that offset reproductive costs (disadvantages) associated with them. If a behavior is adaptive, it must promote the *individual's* production of offspring. Keep this thought in mind while you read through the following terms, which you will encounter repeatedly in the remainder of this chapter:

1. **Reproductive success**: The number of surviving offspring that the individual produces.

2. **Adaptive behavior**: Any behavior that promotes propagation of an individual's genes; its frequency is maintained or increases in successive generations.

3. **Social behavior**: Interdependent interactions among individuals of the species.

4. **Selfish behavior**: Within a population, any form of behavior that increases an individual's chance to produce or protect offspring of its own regardless of the consequences for the group as a whole.

5. **Altruistic behavior**: Self-sacrificing behavior. The individual behaves in a way that helps others but lowers its own chance of producing offspring.

When biologists speak of selfish or altruistic behavior, they don't mean an individual is consciously aware of what it's doing or of the behavior's reproductive goal. A lion does not have to know that eating zebras will be good for its reproductive success. Its nervous system simply calls for *HUNTING BEHAVIOR!* when the lion is hungry and sees a zebra. Hunting behavior persists in the lion population because genes responsible for the underlying brain mechanisms are persisting, also.

As a case in point, Norwegian lemmings disperse from a population when density skyrockets and food is scarce. Many die during the exodus. Are they helping the species by committing suicide to get rid of "excess" individuals? Or do some just happen to die as a result of starvation, predation, or accidental drowning while scurrying to new locations where they may reproduce?

One of Gary Larson's instructive cartoons depicts a population of lemmings plunging over a cliff above

water, presumably in the act of mass suicide. But one lemming has an inflated inner tube around its waist! If many lemmings were suicidally altruistic, then only the "selfish" ones would reproduce. Over time, genes underlying their altruism would disappear from the population. Perhaps a more useful hypothesis would be this: Individual lemmings disperse to less crowded places where each will have a better chance to survive and reproduce, and some simply die during the move.

Another case: In northern forests, ravens scavenge carcasses of deer, elk, or moose, which are few and far between. When one of these large birds comes across a carcass—even in winter, when food is scarce—it often calls loudly and attracts a crowd of similarly hungry ravens. This calling behavior puzzled Bernd Heinrich. On the surface, it seemed to work against the caller's interests. Wouldn't a quiet raven eat more, increase its chance of surviving, and also leave more descendants compared to an "unselfish" vocalizer? If its behavior is an outcome of natural selection, then the biological cost in terms of lost calories and nutrients must be offset by some reproductive benefit for individual callers. What could be the benefit?

Maybe a lone bird picking at a carcass is exposed to predators that lie in wait for it. If that were so, then other ravens attracted by the calls could help keep an eye out for danger. But ravens are big, agile birds with very few known enemies. Heinrich stealthily watched ravens feeding alone and feeding in pairs at a carcass. He never saw a predator attack any of them.

Then he realized that territory may have something to do with it. A **territory** is an area that one or more individuals defend against competitors. After hauling a cow carcass into a Maine forest, Heinrich observed that single ravens or paired ravens don't vocally advertise food. Maybe the silent ravens were adults that had already staked out a large territory, which happened to include the spot where he put the carcass.

A pair of ravens would gain nothing by attracting others to their territory. But what if the Maine forest had been *subdivided* into territories and was defended by powerful adults? A wandering young bird would be lucky to eat at all in an aggressive pair's territory. But recruiting a gang of other, nonterritorial ravens might overwhelm the resident pair's defensive behavior.

As it turns out, only the wandering ravens advertise carcasses. Their calling behavior is basically selfish and adaptive. It gives them a shot at otherwise off-limits food and thereby promotes their reproductive success.

Behavioral biologists generally find it profitable to look for evidence of natural selection of the individual's traits rather than something that benefits the species as a whole.

COMMUNICATION SIGNALS

Competing for food, defending territory, alerting others to danger, advertising sexual readiness, forming bonds with a mate, caring for the offspring—such *intraspecific* interactions depend on forms of communication among animals, as in the case of those vocalizing ravens.

The Nature of Communication Signals

Intraspecific interactions involve mixes of instinctive and learned behaviors by which individuals send and respond to information-laden cues, encoded in stimuli. These are **communication signals**, with unambiguous meaning for the species. They include specific odors, colors, patterning, sounds, postures, and movements.

Communication signals sent by one individual, the **signaler**, are meant to change the behavior of others of the species. Those responding are the **signal receivers**. A signal evolves or persists when it tends to help the reproductive success of both the sender *and* receiver. When a signal proves disadvantageous to either party, then natural selection will favor individuals that do not send the signal or do not respond to it.

Remember **pheromones**? They are chemical signals between individuals of the same species. As you know, chemical odors from food and danger were the main stimuli that early animals had to deal with. As animals evolved, most started to rely on *signaling* pheromones to induce a receiver to respond fast. Some, including chemical alarm signals, call for aggressive or defensive behaviors. Others are sex attractants. Bombykol molecules released from female silk moths are like this (Section 36.1). The *priming* pheromones bring about generalized physiological responses. For instance, a volatile odor in the urine of certain male mice triggers and also enhances estrus in female mice.

Figure 46.8 A dog using a play bow to solicit a romp.

Acoustical signals also abound in nature, as when male songbirds sing to secure territory and attract a mate. At night, tungara frogs issue distinctive "whine-chuck" calls to attract females and warn rival males.

Some signals never vary. Zebra ears laid back flat against the head convey hostility, but ears pointing up convey its absence. Other signals convey the intensity of the signaler's message. A zebra that has its ears laid back is not too riled up when its mouth is open just a bit, also. But when its mouth is gaping, watch out. That combination is a **composite signal**: a communication signal with information encoded in two cues (or more).

Signals often take on different meaning in different contexts. A lion emits a spine-tingling roar to keep in touch with its pride *or* to threaten rivals. Also, a signal can convey information about signals to follow. Dogs and wolves solicit play behavior with a play bow, as shown in Figure 46.8. Without the bow, a receiver of this signal could construe the behavioral patterns that follow as aggressive, sexual, or even exploratory—not as playful.

Examples of Communication Displays

The play bow is a **communication display**—a pattern of behavior that is a social signal. Another common pattern, the **threat display**, unambiguously announces that a signaler is prepared to attack a signal receiver. When a rival for a receptive female confronts him, a dominant male baboon will roll his eyes upward and "yawn" to expose his formidable canines (Figure 46.9). The signaler often benefits when the rival backs down, for he retains access to the female without putting up a

Figure 46.9 *Above:* Exposed canines, part of a male baboon's threat display. *Below:* Part of a courtship display, with visual, tactile, and acoustical signals, that precedes copulation. Here a male albatross spreads his wings, an information-laden cue for the female. See also Section 18.1.

Figure 46.10 Dances of the honeybees, as examples of tactile displays. (**a**) Honeybees that visit food sources close to their hive perform a *round* dance on the honeycomb. Worker bees that maintained contact with the forager through the dance go out and search for food near the hive.

(**b**) Bees trained to visit feeding stations over 100 meters from the hive do a *waggle* dance. During the dance, a bee makes a straight run and waggles its abdomen. The waggle dancers also vary the dance speed to convey more information about the distance of a food source. For example, when a site is 150 meters away, the dance is executed much faster, with more waggles per straight run, compared with a dance concerning a food source that is 500 meters away.

(**c**) As Karl von Frisch discovered, a *straight* run's orientation varies, depending on the direction in which food is located. When he put a dish of honey on a direct line between a hive and the sun, foragers that located it returned to the hive and oriented their straight runs right up the honeycomb. When he put the honey at right angles to a line between the hive and the sun, foragers made straight runs 90 degrees to vertical. Thus, a honeybee "recruited" into foraging orients its flight *with respect to the sun and the hive*. It wastes less time and energy during its food-gathering expedition.

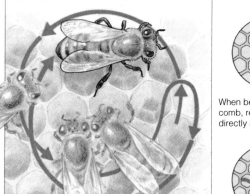

a

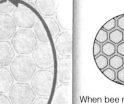

b

When bee moves straight down comb, recruits fly to source directly away from the sun.

When bee moves to right of vertical, recruits fly at 90° angle to right of the sun.

When bee moves straight up comb, recruits fly straight toward the sun.

c

fight. The signal receiver benefits because he avoids a serious beating, infected wounds, and possibly death.

Such displays are *ritualized*, with intended changes in the function of common behavior patterns. Normal movements may be exaggerated yet simplified; postures may be frozen. Feathers, manes, claws, and other body parts are often conspicuously enlarged, patterned, and colored. Ritualization is often developed to an amazing degree in **courtship displays** between potential mates. For instance, food-related displays are common among many birds. A male bird may emit calls while bowing low, as if to peck the ground for food. When a female comes running, he may spread his wings, as if to focus her attention on the ground before him (Figure 46.9). Here, one behavior (searching for food) has changed to attract a female, which sometimes leads to copulation. Consider also those bioluminescent flashes of fireflies described earlier, in Section 6.8. The male firefly uses a light-generating organ to emit a bright, flashing signal in courtship displays. A receptive female of his species may quickly answer him with a flash. Both insects may flash back and forth until they meet up.

Similarly, in **tactile displays**, a signaler touches the receiver in ritualized ways. After discovering a source of pollen or nectar, a foraging honeybee returns to its colony, a hive, and performs a complex dance. It moves about in a circle, jostling in the dark with a crowd of workers. Other bees may follow and maintain physical contact with the dancer. Honeybees that do so acquire information about the general location, distance, and direction of pollen or nectar (Figure 46.10).

Illegitimate Signalers and Receivers

Sometimes the wrong parties intercept communication signals. Termites will respond with defensive behavior if they catch a whiff of a scent from an invading ant. That scent is meant to identify the ant as a member of an ant colony, and it elicits cooperative behavior from other ants. The scent does have evolved functions, but announcing an invasion of a termite colony is not one of them. A termite also might detect the scent and kill the ant. In that case, it is an **illegitimate receiver** of a signal between individuals of a species.

We even see **illegitimate signalers**. Certain assassin bugs hook dead and drained bodies of termite prey on their dorsal surface to acquire the odor of the victims. By deceptively signaling that they "belong" to a termite colony, they can more easily hunt termite prey. Another illegitimate signaler is the female of certain predatory fireflies. If one notices a male firefly's flash, she flashes back. If she can lure him into attack range, he becomes her meal—an evolutionary cost of having an otherwise useful response to a come-hither signal.

A communication signal between individuals of the same species is an action having a net beneficial effect on both the signaler and receiver. Natural selection tends to favor communication signals that promote reproductive success.

MATES, PARENTS, AND INDIVIDUAL REPRODUCTIVE SUCCESS

For reasons we need not explore here, most people find mating and parenting behaviors of animals fascinating. How useful is selection theory in helping us interpret such behavior? Let's take a look.

Sexual Selection Theory and Mating Behavior

Competition among members of one sex for access to mates is common. So is choosiness in selecting a mate. Recall, from Section 17.9, that such activities are forms of a microevolutionary process called **sexual selection**. Sexual selection favors traits that give the individual a competitive edge in attracting and holding on to mates.

But *whose* reproductive success is it—the male's or female's? Males, recall, produce large numbers of tiny sperm, and females produce far larger but fewer eggs. For the male, success generally depends on how many eggs he fertilizes. For the female, it depends more on how many eggs she produces or how many offspring she can care for. In most cases, *the key factor influencing a female's sexual preference is the quality of the mate—not the quantity of partners*. Hangingflies, sage grouse, and bison offer examples of how most females dictate the rules of male competition (Figures 46.11 through 46.13). They also show how males employ tactics that might help them fertilize as many eggs as possible.

Female hangingflies (*Harpobittacus apicalis*) choose males that offer superior material goods. And so you may observe some males capturing and killing a moth or some other insect. After they do so, they release a sex pheromone to attract females to their "nuptial gift,"

as in Figure 46.11*a*. A female will select the male that displays a large calorie-rich offering. Only *after* she has been eating a gift for five minutes or so does she begin to accept sperm from her partner, which she stores in a special internal compartment. She allows the male to continue inseminating her, but only as long as the food holds out. Before twenty minutes are up, she can break off the mating at any point. If she does, she might well mate with another male and accept his sperm. Doing so will dilute the reproductive success of her first partner.

In western regions of North and South Dakota, we find sage grouse (*Centrocercus urophasianus*) dispersed among stands of sagebrush. During spring courtship or summer nesting, they never roam far from protective cover. The males make no attempt to secure and hold on to a large territory. During the breeding season, they congregate in the type of communal display ground called a **lek**. Each male stakes out a few square meters as his territory. The females are attracted to the lek—not to feed or nest, but rather to watch the males. With tail feathers erect and large neck pouches puffed out, a male emits booming calls and stamps about rather like a wind-up toy on his display ground (Figure 46.11*b*). Females tend to select and mate with one male only. Then they go off to nest by themselves in sagebrush, unassisted by a sexual partner. Many often choose the same male, so most of the males never mate.

The females of some species cluster in defendable groups when they are sexually receptive. Where you come across such a group, you are likely to observe males competing for access to the clusters. Competition

Figure 46.11 (**a**) Male hangingfly dangling a moth as a nuptial gift for a future mate. Females of some hangingfly species choose sexual partners on the basis of the size of prey offered to them. (**b**) Male sage grouse. Its ornamental feathers are visual signals that function in a courtship display—a dance performed at a lek. Males vigorously defend a small patch of this communal mating area. The females (the smaller brown birds) observe the prancing males before choosing the one they will mate with.

Figure 46.12 (**a**) Sexual competition between male bison, which fight for access to a cluster of females. (**b**) Subordinate member of a wolfpack. The ritualized groveling of this gray wolf is a signal of submission to the dominant male, the only breeding male of the pack.

for ready-made harems favors highly combative male lions, sheep, elk, elephant seals, and bison, just to name a few animals (Figure 46.12*a*).

Another example: The females of a wolf pack court males. Why? Unlike most male mammals, which don't invest as much as females do in raising offspring, male wolves help pups by giving them food, defending the feeding territory, and driving off infanticidal intruders. Unlike herbivores, which forage as individuals, wolves share captured prey with others of their social group. Therefore, a female's reproductive success increases if she monopolizes the benefits that males offer.

Most wolf packs have one dominant male (Figure 46.12*b*) that breeds with a dominant female. The others are nonbreeding, altruistic brothers and sisters, aunts and uncles. They all hunt and bring food to individuals that stay in the den and guard the young. Nonbreeding females may ovulate and court males, but they will fail to reproduce in the presence of a dominant female.

Costs and Benefits of Parenting

What happens after mating, when the offspring arrive? Until the offspring have developed enough to survive on their own, parents of some species care for them.

Example: Adult Caspian terns (Figure 46.13) will incubate the eggs, shelter the nestlings and feed them, then accompany and protect them after they start to fly. Such parental behavior comes at a reproductive cost. It drains time and energy that might otherwise be given

Figure 46.13 Male and female Caspian terns protecting their vulnerable chicks. Parental care has costs as well as benefits.

to improving their own chances of living to reproduce another time. Yet for many species, parenting improves the likelihood that the current generation of offspring will survive. The benefit of devoting time and energy to immediate reproductive success can outweigh the cost of reduced reproductive success at a later time.

Selection theory helps explain some aspects of mating behavior, for sexual selection favors behavioral traits that give the individual a competitive edge in reproductive success. This theory also is used to explain why adults of some species are parental, whereas others are not.

COSTS AND BENEFITS OF LIVING IN SOCIAL GROUPS

Survey the animal kingdom and you observe a range of social groups. Individuals of some species spend most of their lives alone or in small family groups. Others live in groups of thousands of related individuals. Termite and honeybee societies are like this. Still others live in social units consisting mainly of nonrelatives. Human populations are an example. Evolutionary biologists often attempt to identify costs and benefits of sociality in terms of individual reproductive success, measured by contributions to the next generation's gene pool.

Cooperative Predator Avoidance

Some groups act cooperatively against a predator and reduce the net risk to any one individual. In a flock or herd, vulnerable individuals can scan for predators, join forces in a counterattack, or engage in more effective defensive behavior (Figures 46.14 and 46.15).

Birgitta Sillén-Tullberg, a biologist, found evidence of such benefits. She was studying Australian sawfly caterpillars that live in clumps on tree branches. Figure 46.15 shows one clump. When something disturbs the caterpillars, they collectively rear up from the branch and writhe about, all the while regurgitating partially digested eucalyptus leaves. These leaves happen to be heavily impregnated with chemical compounds that are toxic to most kinds of predatory animals—most notably the songbirds that prey on the caterpillars.

Sillén-Tullberg hypothesized that individual sawfly caterpillars benefit from the coordinated act of repelling predatory birds. She used her hypothesis to predict that birds are more likely to eat a solitary caterpillar. To test her prediction, she offered young, hand-reared Great Tits (*Parus major*) the chance to eat sawfly caterpillars.

Figure 46.15 Example of social defensive behavior. Australian sawfly caterpillars form clumps in branches and collectively regurgitate a fluid (yellow blobs) that is toxic to most predators. Once the danger passes, the caterpillars swallow the fluid.

She offered the caterpillars one by one or in a group of twenty for a standard number of presentations. Ten birds that were offered one individual at a time ate an average of 5.6 caterpillars. But ten birds that were each offered a clump of caterpillars ate an average of 4.1. As predicted, individuals were somewhat safer in a group.

The Selfish Herd

Simply by their physical position within a group, some individuals form a living shield against predation on the others. They all belong to a **selfish herd**, a simple society held together (not consciously) by reproductive self-interest. The selfish-herd hypothesis has been tested for male bluegill sunfishes, which build adjacent nests on lake bottoms. The females deposit eggs where males have used their fins to hollow out depressions in mud.

If a colony of bluegill males is a selfish herd, then we can predict competition for the "safe" sites—at the center of the colony. Compared to eggs at the periphery, eggs in nests at the center are less likely to be eaten by snails and largemouth bass. Competition does indeed exist. The largest, most powerful males tend to claim the central locations. Other, smaller males assemble around them and bear the brunt of predatory attacks. Even so, they are better off in the group than on their own, fending off a bass single-handedly, so to speak.

Dominance Hierarchies

In selfish herds, personal benefits of living with others simply seem to outweigh the costs. By contrast, some individuals may *help* others survive and reproduce at their own expense. Their helpful behavior might be a cost of belonging to the group. Or maybe they do not give up their chance to reproduce entirely.

We see evidence of both possibilities in **dominance hierarchies**. In some social groups, certain individuals are subordinate to others. With a threat signal from a dominant individual of a baboon troop, subordinates will relinquish safe sleeping sites, choice food bits, and sexually receptive females. Each baboon knows its own

Figure 46.14 Cooperative defensive behavior among musk oxen (*Ovibos moschatus*). Adults form a circle, often around the young, and face outward. The ring of horns may deter predators.

Figure 46.16 Some benefits and costs of social behavior, illustrated.

(**a**) A baboon belonging to a troop is safer from leopards than a loner out in the open. This one is making its last stand with a threat display (it didn't work).

(**b**) Imagine how easily pathogens and parasites might spread through this colony of royal penguins living on Macquarie Island, between New Zealand and Antarctica.

status in the troop. Dominance hierarchies are highly developed among chimpanzees, also, as you saw in the introduction to Chapter 36.

Why do subordinates stay in the group when they must give up valuable resources? They do so to avoid injury and to remain in the group. It may be impossible to survive alone—a solitary baboon out in the open surely quickens the pulse of the first leopard to spy it (Figure 46.16*a*). Challenging a strong individual may invite injuries or death. Self-sacrificing behavior might give a subordinate a chance to reproduce if it lives long enough or if a predator or old age removes dominant peers. Some patient, subordinate wolves and baboons do indeed move up the social ladder that way.

Regarding the Costs

If social behavior is so advantageous, *then why are there so few social species among most groups of animals?* One answer is that, in some environments, costs outweigh benefits. Most obviously, when more individuals of a species live together, competition for food increases. Prairie dogs, royal penguins, herring gulls, and cliff swallows are among the many kinds of animals that live in enormous colonies (Figure 46.16*b*). All must compete for a share of the same ecological pie.

Crowded social groups also invite parasites and contagious diseases that are readily transmitted from host to host. We observe this when plagues spread like wildfire through densely crowded human populations.

Another cost is the risk of being killed or exploited by others in the group. When given the opportunity, breeding pairs of herring gulls quickly cannibalize the eggs and chicks of neighbors. Lions live in prides, yet lions living alone often could eat more meat per day. Group living costs lionesses in terms of food intake and reproductive success. Why do they stick together? Maybe they can better defend their territories against rival groups. Also, aggressive males show infanticidal behavior. Such males almost always kill cubs of a lone lioness, but two or more lionesses occasionally save some of them.

The individuals of a social group gain benefits, as through cooperative defenses or shielding effects against predators.

Individuals that belong to a social group pay costs in terms of increased competition for food, living quarters, mates, and other limited resources. They also are more vulnerable to contagious diseases and parasitic infections.

Individuals also may risk being exploited or having their offspring killed by others in the social group.

THE EVOLUTION OF ALTRUISM

A subordinate animal giving way to a dominant one can be acting in its own interest. But true self-sacrificing altruism has evolved among many vertebrate groups; it is most extreme in termite, honeybee, and ant societies.

Termite Colonies

In eucalyptus forests in Queensland, Australia, termites construct foraging tunnels for their colony along dead tree trunks. Chip into a tunnel and small, nearly white insects will bang their head against the tunnel wall and scurry away. Head banging is an acoustical signal. It results in vibrations that invite golden soldier termites to make a defensive stand (Figure 46.17).

Each soldier termite has a swollen, eyeless, tapering head with a long, pointed "nose." Disturb one and it will shoot thin jets of silvery goo out of its nose. The silvery strands release volatile compounds that attract more soldiers that collectively battle the danger (which typically is an invasion by ants). Soldiers will give up their life in defense of the colony.

Pale workers build the tunnels, which lead to places where they safely gather cellulose fibers from wood. They cart fibers to their underground nest, where they cultivate an edible fungus. Fungal hyphae grow into the fibers and absorb nutrients from them. Termites eat portions of the fungus and chew wood into small bits they can swallow. (Unlike some termite species, they do not synthesize cellulose-digesting enzymes; symbionts living in their gut do this for them.)

Soldier and worker termites are sterile. They engage in self-sacrificing behavior that helps others survive. A single queen and one or more kings serve as "parents" for the entire society.

Honeybee Colonies

Consider next a queen bee, the only fertile female in a honeybee colony, or hive (Figure 46.18a). Her sterile worker daughters feed her and relay her pheromones through the hive, and these influence the activities of all members. The queen is larger than the worker bees, partly because of the enormous egg-producing ovaries in her abdomen.

Only at certain times of year do the stingless drones develop and mature (Figure 46.18b). The drones do not work for the colony. Instead, they leave and attempt to mate with queens of other hives. If successful, they may perpetuate part of their family's genes.

The hive contains 30,000 to 50,000 workers (Figure 46.18c). They feed larvae, clean and maintain the hive, and construct honeycomb from waxy secretions. The workers store honey and pollen in honeycomb, which houses bee generations from eggs, through a series of larval stages, to pupae, to the adults (Figure 46.18d).

Adult workers live for about six weeks in spring and summer. Workers also engage in scent-fanning. Fanned air passes over a honeybee's exposed scent gland. As pheromones waft away from the scent gland, they help bees become oriented relative to the hive's entrance on

Figure 46.17 Soldier termites defending their social colony by guarding a break in a foraging tunnel, which worker termites constructed on a dead tree trunk. The members of this caste shoot out strands of glue from their pointed "nose." The glue entangles invading ants and other intruders.

Figure 46.18 Life within a honeybee colony. (**a**) A queen bee and her court of sterile worker daughters, the only female of the colony that reproduces. (**b**) A stingless drone. (**c**) One of the 30,000 to 50,000 worker bees in the hive. (**d**) Worker bees store honey or pollen in the honeycomb, which houses new bee generations. Young workers feed the larvae. (**e**) Worker bees transferring food to one another. (**f**) Worker females at the hive entrance serve as guards.

d   early larval stage   nearly mature pupa   **f**

their way to or from foraging expeditions. When the foragers return to the hive after locating a rich source of nectar or pollen, they start a dance—a tactile display that recruits more workers into taking off for the source (Figure 46.10).

In other cooperative behaviors, workers transfer food to one another, and worker females at the hive entrance act as guards (Figure 46.18*e,f*). The guard bees readily sacrifice themselves to repel intruders.

Indirect Selection

None of the altruistic individuals of the termite colony and the honeybee hive can contribute genes to the next generation. So how are genes that underlie the altruistic behavior perpetuated? By W. Hamilton's theory of **inclusive fitness**, genes that are associated with caring for one's *relatives* may be favored in some situations.

A sexually reproducing, diploid parent caring for offspring is not helping exact genetic copies of itself. Each gamete that it produces, hence each offspring, inherits one-half of its genes. Other individuals of the social group that have the same ancestors also share genes with parents. Two siblings (brothers, sisters) are as genetically similar as a parent is to one of its offspring. Nephews and nieces have inherited about one-fourth of their uncle's genes.

Suppose that selective altruism is an extension of parenting. Suppose an uncle helps his niece survive long enough to reproduce. He has made an *indirect* genetic contribution to the next generation, through the genes he shares with his niece. Altruism costs him; he may lose his own chance to reproduce. But if the cost is less than the benefit, his actions will propagate his genes and favor the spread of his kind of altruism in the species. If an uncle saves two nieces, this is equivalent to producing one surviving daughter.

And so we see that nonbreeding workers in insect societies may indirectly promote their "self-sacrifice"

genes through exceptional altruistic behavior directed toward their relatives. Generally, all the individuals in termite, honeybee, and ant colonies are members of an extended family. Sterile workers of the family support their siblings. And some of those siblings will be the future kings and queens of that colony. When a guard bee drives her stinger into a bear or raccoon, she dies, but her siblings in the hive might well perpetuate some of her genes.

Altruistic behavior is at its most extreme among certain insect societies, such as honeybee and termite colonies.

Altruistic behavior may persist when individuals pass on genes indirectly, by helping relatives survive and reproduce.

By the theory of inclusive fitness, genes associated with altruistic behavior that is directed toward relatives may spread through a population in certain situations.

Why Sacrifice Yourself?

Sterility and extreme self-sacrifice are rare in social groups of vertebrates. The known exceptions include the naked mole-rats (*Heterocephalus glaber*). These almost hairless mammals live in arid regions of eastern Africa inside cooperatively excavated burrows. They always live in clans—small social units—of anywhere from 25 to 300 individuals (Figure 46.19).

One reproducing female dominates the clan. She mates with one to three males. Other, nonbreeding members live to protect and care for the "queen" and "king" (or kings) and their offspring. Nonreproducing "diggers" excavate extensive subterranean tunnels and chambers that serve as living rooms or waste-disposal centers. They locate and cut up large tubers growing underground and deliver the bits to the queen, her retinue of males, and her offspring.

Digger mole-rats also deliver food to other helpers that loaf about, shoulder to shoulder and belly to back, with the reproductive royals. The "loafers" actually spring to action when a snake or some other enemy threatens the clan. Collectively, and at great personal risk, they chase away or attack and kill the predator.

Using indirect selection theory as our guide, we can formulate a hypothesis to account for the altruism of the mole-rats that never do breed.

If helpful behavior is genetically advantageous to some individuals in a naked mole-rat clan (*the hypothesis*), then helpers will be related to the reproducing members of the clan that benefit from their altruistic behavior (*the prediction*).

To *test* the prediction, we require information about genetic relationships among the members of the naked mole-rat colony. One way to obtain it would be to know which animals were offspring of which others. However,

constructing a family tree for an entire colony would be most laborious and difficult. Why? There are a number of breeding males. In addition, queen mole-rats die and are replaced.

H. Kern Reeve and his colleagues decided on a more practical approach. They relied upon *DNA fingerprinting*, a method of establishing degrees of genetic relatedness among individuals. As described in Section 16.3, this laboratory method starts with the formation of restriction fragments of DNA molecules. Then technicians construct a visual record of different sets of fragments of DNA from different individuals. Essentially, a set of identical twins would have identical DNA fingerprints (their DNA would have the same base sequence). Individuals with the same father and mother would have similar DNA. And so they would have similar although not identical DNA fingerprints. On average, the DNA fingerprints of genetically unrelated individuals should show far more differences than the DNA fingerprints of siblings or some other relatives.

When Reeve constructed DNA fingerprints for naked mole-rats, he discovered that all individuals from the same clan are *very* close relatives. He also found out that they are very different genetically from members of other clans. The findings indicate that each naked mole-rat clan is highly inbred as a result of generations of brother–sister, mother–son, and father–daughter matings. As you know from Section 17.11, inbreeding among individuals of a population results in extremely reduced genetic variability.

Therefore, a self-sacrificing naked mole-rat is helping to perpetuate a high proportion of the alleles it carries. As it turns out, the genotypes of helpers and the helped might be as much as 90 percent identical!

Figure 46.19 A peek at a few naked mole-rats (*Heterocephalus glaber*).

AN EVOLUTIONARY VIEW OF HUMAN SOCIAL BEHAVIOR

If we are comfortable with analyzing the evolutionary basis of the behavior of termites, naked mole-rats, and other animals, would it not be rewarding to analyze such a basis of human behavior, too? Many resist the idea. Apparently, they believe that attempts to identify the adaptive value of a human trait are attempts to define its moral or social advantage. However, there is a clear difference between trying to explain something in terms of its evolutionary history and attempting to justify it. "Adaptive" does not mean "morally right." To biologists, adaptive only means valuable in terms of transmission of the individual's genes.

Consider an example from human societies. **Adoption** refers to the acceptance of offspring of other individuals as one's own. It appears to be a form of altruistic behavior. If we wish to investigate this behavior, we can use evolutionary theory to come up with a hypothesis about it.

This does not mean that we are attempting to judge whether adoption is a moral behavior or even a socially desirable behavior. "Morality" and "desirability" are issues about which a biologist has no more to say than anybody else.

Start with a premise that all adaptations have costs and benefits. One cost is that certain adaptive behaviors might be *redirected* under rare or unusual circumstances. In many species, adults that have lost offspring adopt a substitute. Some cardinals have fed goldfish. A whale tried to lift a log out of water as if it were a distressed infant that needed help to reach the water's surface. Emperor penguins will fight to adopt orphans, as in Figure 46.20.

As John Alcock suggests, such examples inspire us to construct hypotheses about adoptions in our own societies. Adoptive behavior of humans may occur "by mistake" when parental behavior becomes focused on unrelated children. Suppose the "maladaptive error" arises through a frustrated desire to bear children, the *hypothesis*. If so, then we can expect that couples who lost an only child or are sterile and can't have children

Figure 46.20 Two adult emperor penguins competing to adopt an orphan, which penguins accept as substitute offspring.

of their own should be especially prone to adopt strangers (the *prediction*). This prediction has not been formally tested.

Now consider this. For most of their evolutionary history, humans have lived in small groups and, later, small villages. They probably had few opportunities or showed little inclination to adopt strangers. Even so, they were likely to accept the child of a deceased relative. But how would adoptive parents gain genetic representation in the next generation? If we use the inclusive fitness theory to explain the adoptive behavior of humans, we would predict that most adoptions that occur in traditional societies will involve adults taking on nephews and nieces or other relatives.

This prediction is testable, and behaviorist Joan Silk actually tested it in traditional societies. Her results showed that people will adopt related children far more often than nonrelated children.

Especially in a large, industrialized society that has welfare agencies as well as other mechanisms of adoption assistance, individuals do become parents of nonrelatives. These large societies are one kind of evolutionary environment. Parenting mechanisms evolved in the past, and their redirection toward nonrelatives says more about our evolutionary history than it does about transmission of one's genes.

Evolutionary hypotheses about the adaptive value of behavior can be tested. With tests, we gain insight into the evolution of human behavior.

It is possible to test evolutionary hypotheses regarding the adaptive value of different forms of human behavior.

There might be greater acceptance of such testing when more people come to understand that adaptive behavior and socially desirable behavior are separate issues.

In biology, "adaptive" means only that a specified trait of an individual has proved beneficial in the transmission of genes responsible for the trait.

SUMMARY
Gold indicates text section

1. Animal behavior (coordinated responses to stimuli) starts with genes that specify products required for the development and operation of nervous, endocrine, and skeletal–muscular systems. *CI, 46.1*

2. A behavior performed without having been learned by experience is instinctive, a prewired response to one or two simple, well-defined environmental cues (sign stimuli) that trigger a suitable response, such as a fixed action pattern. Experiences may lead to variations or changes in responses (learned behavior). Learning is a result of genetic and environmental inputs. *46.1, 46.2*

3. Behavior with a genetic basis is subject to evolution by natural selection. It evolved as a result of individual differences in reproductive success in past generations, when reproductive benefits of a behavior exceeded its reproductive costs (or disadvantages). *46.3*

4. Social groups require cooperative interdependency among individuals of a species. Sociality is promoted by communication signals, cues sent by one individual (a signaler) that can change the behavior of another of the same species (a signal receiver). *46.4*

5. Chemical, visual, acoustical, and tactile signals are components of communication displays. *46.4*

    a. Pheromones function in chemical communication. Signaling pheromones, such as sex attractants and alarm signals, can induce immediate change in the behavior of a receiver. Priming pheromones elicit a generalized physiological response by the receiver.

    b. Visual signals (observable actions or cues) are key components of courtship displays and threat displays.

    c. Acoustical signals are sounds that have precise, species-specific information (e.g., frog mating calls).

    d. Tactile signals are specific forms of physical contact between a signaler and a receiver.

6. Members of the same species often create obstacles to one another's reproductive success, as by selective mate choice and by competing for access to mates. *46.5*

7. Social groups have costs, including competition for limited resources and greater exposure to contagious diseases and parasites. Benefits outweigh the costs in some cases, as when predation pressure is strong. *46.6*

8. In self-sacrificing (altruistic) behavior, individuals give up chances to reproduce while helping others of their group. In most social groups, individuals do not sacrifice reproductive chances to help others. *46.7*

    a. A dominance hierarchy is a social ranking in which some individuals give way to others of their group. Often, dominant individuals force subordinates to relinquish food or some other resource.

    b. When subordinates give in to dominant members in their group, they may receive compensatory benefits of group living, such as safety from predators. In some species, subordinates may reproduce if they live long enough and dominant ones slip in the hierarchy or die.

9. By one theory of inclusive fitness, genes associated with caring for relatives (not one's direct descendants but individuals that bear a fraction of the same genes) are favored in some cases. Indirect selection can lead to extreme altruism, as among workers of some species of social insects and naked mole-rats. The workers usually do not reproduce. Altruism helps reproducing relatives survive, so altruistic individuals pass on, "by proxy," genes underlying development of this behavior. *46.7*

10. Testing the adaptive value of human behavior starts with understanding that "adaptive" only means a trait is helpful in the transmission of an individual's genes; socially desirable behavior is a separate issue. *46.9*

Review Questions

1. Explain how genes and their products, including hormones, influence mechanisms required for forms of behavior. *46.1*

2. Define these terms: instinctive behavior, sign stimulus, fixed action pattern, and learned behavior. *46.1, 46.2*

3. Contrast altruistic behavior with selfish behavior. Why does either form of behavior persist in a population? *46.3*

4. Describe some characteristics of communication signals. Then give an example of a communication display. *46.4*

5. Describe some feeding behavior or mating behavior in terms of natural (individual) selection. *46.4, 46.5*

6. List some benefits and costs of sociality. *46.6, 46.7*

Self-Quiz   ANSWERS IN APPENDIX III

1. "Starlings minimize nest mites by festooning nests with wild carrot," said Clark and Mason. Their statement was _____ .
    a. an untested hypothesis   c. a test of a hypothesis
    b. a prediction   d. a proximate conclusion

2. Genes affect the behavior of individuals by _____ .
    a. influencing the development of nervous systems
    b. affecting the kinds of hormones in individuals
    c. governing the development of muscles and skeletons
    d. all of the above

3. Generally, living in a social group costs the individual, in terms of _____ .
    a. competition for food, other resources
    b. vulnerability to contagious diseases
    c. competition for mates
    d. all of the above

4. A statement that overcrowding causes lemmings to disperse to areas that are more favorable for reproduction _____ .
    a. is consistent with Darwinian evolutionary theory
    b. is based on a theory of evolution by group selection
    c. is supported by the finding that most animals behave altruistically during their lives

5. The genetic similarity between an uncle and his nephew is _____ .
    a. the same as between a parent and its offspring
    b. greater than that between two full siblings
    c. dependent on how many other nephews the uncle has
    d. less than that between a mother and her daughter

Figure 46.21 Caterpillar under siege.

6. Match the terms with their most suitable description.

_____ fixed action pattern	a. time-dependent form of learning requiring exposure to key stimulus
_____ altruism	b. genes plus actual experience
_____ basis of instinctive *and* learned behavior	c. stereotyped motor program that runs to completion independently of feedback from environment
_____ imprinting	d. assisting another individual at one's own expense

Critical Thinking

1. A large caterpillar is crawling along a branch in a tropical forest. You poke it, and one of its responses is to let go of the branch and puff up anterior segments of its body (Figure 46.21). Propose an internal mechanism that may contribute to this defensive behavior. Also propose how the behavior may have adaptive value. How would you test your two hypotheses?

2. A cheetah scent-marks plants in its territory by releasing certain chemicals from exocrine glands. What evidence would you require to demonstrate that the cheetah's action is an evolved communication signal?

3. You observe mallard ducks in a small pond. Then you notice a rooster wading out to the ducks with amorous intent (Figure 46.22). What probably happened to the rooster early in life?

4. Develop an adaptive value hypothesis for this observation: Male lions kill offspring of females they acquire after chasing away previous pride holders that had mated with the females. How would you test your infanticide hypothesis?

Figure 46.22 Behaviorally confused rooster.

Figure 46.23 A black heron, possibly on a fish-enticing expedition.

5. What are the likely evolutionary costs and benefits to a male hangingfly of offering a nuptial gift to its potential mate? How might a female's choosiness depend on the costs and benefits?

6. While traveling through Africa, you see a black heron in a lake with its wings held over its head like an umbrella (Figure 46.23). You might suppose that, like other herons, it is looking for fish. But most other herons keep their wings folded when hunting. What is the adaptive value of this behavior? Is this heron shading the water and are minnows attracted to the shade because it looks like shelter? How would you test this idea?

7. In 2002, Svante Paabo correlated a key gene mutation with the origin of *language*. All mammals have the gene *FOXP2*, which has a 715 base-pair sequence. The human and mouse gene differ by 3 base pairs, and the human and chimpanzee gene by only 2 base pairs. This mutation, along with others, altered brain regions that control muscles in the face, throat, and vocal cords. It apparently arose about 200,000 years ago and swiftly replaced the nonmutated gene in the populations immediately ancestral to modern *Homo sapiens*. Using Sections 26.11 through 26.13, speculate on its impact on social behavior and learning.

Selected Key Terms

adaptive behavior *46.3*
adoption *46.9*
altruistic behavior *46.3*
animal behavior *CI*
communication display *46.4*
communication signal *46.4*
composite signal *46.4*
courtship display *46.4*
dominance hierarchy *46.6*
fixed action pattern *46.1*
illegitimate receiver *46.4*
illegitimate signaler *46.4*
imprinting *46.2*
inclusive fitness *46.7*
instinctive behavior *46.1*
learned behavior *46.2*
lek *46.5*
natural selection *46.3*
pheromone *46.4*
reproductive success *46.3*
selfish behavior *46.3*
selfish herd *46.6*
sexual selection *46.5*
sign stimulus *46.1*
signal receiver *46.4*
signaler *46.4*
social behavior *46.3*
song system *46.1*
tactile display *46.4*
territory *46.3*
threat display *46.4*

Readings

Alcock, J. 2001. *Animal Behavior: An Evolutionary Approach.* Seventh edition. Sunderland, Massachusetts: Sinauer.

Frisch, K. von. 1961. *The Dancing Bees.* New York: Harcourt Brace Jovanovich. A classic on the behavior of honeybees.

Wilson, E. O. 1975. *Sociobiology: The New Synthesis.* Cambridge, Massachusetts: Harvard University Press. Another classic.

On-Line readings at Student Guide for InfoTrac:

COMMUNITY INTERACTIONS

No Pigeon Is An Island

Flying through the rain forests of New Guinea is an extraordinary pigeon with cobalt blue feathers and long, lacy plumes on its head (Figure 47.1). It is about as big as a turkey, and it flaps so slowly and noisily that its flight sounds like an idling truck. Like at least twelve species of smaller pigeons in the same forest, this one perches on branches to eat fruit.

How is it possible that *twelve* species of large and small fruit-eating pigeons live in the space of the same forest? Wouldn't competition for food leave one species the winner? Actually, each species in that rain forest lives, grows, and reproduces in a characteristic way, as defined by its relationships with other organisms and with the surroundings.

Figure 47.1 *Above*, a cobalt blue, turkey-sized Victoria crowned pigeon (*Goura*) and *right*, one of twelve other, less ponderous pigeons living in the same tropical rain forest of New Guinea. In this habitat, each species has its own niche.

Big pigeons perch on the sturdiest branches when they eat. And they eat big fruit. Smaller pigeons, with their smaller bills, can't peck open big fruit. They eat small fruit hanging from slender branches that are not sturdy enough to support the weight of a turkey-sized pigeon. Species of trees in the forest differ with respect to fruit size and the diameter of fruit-bearing branches. They attract pigeons with different traits. In such ways, the twelve pigeon species *partition* the fruit supply.

How do individual trees benefit from enticing the pigeons to eat? Seeds in their fruits have tough coats, which resist the action of digestive enzymes in a pigeon gut. During the time it takes for ingested seeds to travel through the gut, pigeons fly about and dispense seed-containing droppings in more than one place. In this way, they tend to disperse seeds some distance from the parent plants. Later, when seedlings grow, the odds are better that at least some of them won't have to compete with their parents for sunlight, water, and mineral ions. The seeds that drop close to home can't compete in any significant way with the resource-gathering capacity of mature trees, which already have well-developed, extensive roots and leafy crowns.

Within the same forest, leaf-eating, fruit-munching, and bud-nipping insects interact in certain ways with other organisms and with their surroundings. So do nectar-drinking insects, birds, and bats that pollinate flowers. And so do bacteria, fungi and beetles, worms, and other invertebrates that decompose organic debris on the forest floor. Some nutrients released during the decomposition process get cycled back to the trees.

Like humans, then, no pigeon is an island, isolated from the rest of the living world. The twelve species of New Guinea pigeons eat fruit of different sizes. They disperse seeds from different sorts of trees. Dispersal influences where new trees will grow and where the decomposers will flourish. Ultimately, tree distribution and decomposition activities influence how the entire forest community is organized.

Directly or indirectly, interactions among coexisting populations organize the community to which they belong. With this chapter, we now turn to **community structure**—the patterns of species abundances, species interactions, and local biodiversity for a given community.

Key Concepts

1. A habitat is the type of place where individuals of a species normally live. A community is an association of all the populations of species that occupy the same habitat.

2. Every species in the community has its own niche, defined as the sum of all activities and relationships in which its individuals engage as they secure and use the resources required for their survival and reproduction.

3. Community structure starts with adaptive traits that offer individuals of each species the capacity to respond to the physical and chemical features of their habitat, and to the levels and patterns of resource availability over time.

4. Interactions among species influence the structure of the community. They include mutually beneficial interactions, competition, predation, and parasitism.

5. Community structure depends on the geographic location and size of the habitat, the rates at which the member species arrive and disappear, and the history of physical disturbances to the habitat.

6. The first species to occupy a particular type of habitat are replaced by others, which are replaced by still others, and so on in an expected sequence. This process, known as primary succession, produces a climax community. A climax community is a stable, self-perpetuating array of species in balance with one another and with the environment.

7. Different stages of succession often prevail within the same habitat. The variation arises as a result of local differences in soils and other conditions of the habitat, recurring disturbances such as seasonal fires, and chance events.

WHICH FACTORS SHAPE COMMUNITY STRUCTURE?

Think of a clownfish darting above a coral reef, a maple tree on a Vermont hillside, or a mole making a burrow in soft earth. The type of place where you will normally observe a clownfish, maple, or mole is its **habitat**. The habitat of an organism is characterized by physical and chemical features, such as its temperature and salinity, and by the array of other species living in it. Directly or indirectly, the populations of all species in the habitat associate with one another as a **community**.

Five factors shape community structure. *First*, the interactions among climate and topography dictate the habitat's temperatures, rainfall, type of soil, and other conditions. *Second*, the kinds and amounts of food and other resources that become available through the year affect which species can live there. *Third*, individuals of each species have adaptive traits that allow them to survive and exploit specific resources in the habitat. *Fourth*, species in the habitat interact by competition, predation, and mutually beneficial actions. *Fifth*, the overall pattern of population sizes affects community structure. This pattern reflects the history of changing population sizes, the arrivals and disappearances of species, and physical disturbances to the habitat.

Together, these factors help determine the number of species at several "feeding levels," starting with the producers and continuing through levels of consumers. They influence population sizes. They also help dictate the overall number of species. For example, high solar radiation, warm temperatures, and high humidity in tropical habitats favor growth of many kinds of plants, which support many kinds of animals. Conditions in arctic habitats do not favor great numbers of species.

Chapters to follow deal with energy flow through feeding levels and with geographic factors influencing community structure. Here we begin with interactions among species, using the niche concept as our guide.

The Niche

All organisms of a community share the same habitat —the same "address." But within that community, each kind has a distinct "profession." This means it differs from the rest in terms of the sum of its activities and relationships as it goes about securing and using the resources necessary for its survival and reproduction. This is its **niche**.

For each species, a *fundamental* niche is the one that might prevail in the absence of competition and other factors that could constrain its acquisition and use of resources. However, as you will see, such constraining factors do come into play in communities. They tend to bring about a more constrained, *realized* niche that shifts in large and small ways over time, as individuals of the species respond to a mosaic of changes.

Table 47.1	Types of Two-Species Interactions*	
Type of Interaction	Direct Effect on Species 1	Direct Effect on Species 2
Commensalism	+	0
Mutualism	+	+
Interspecific competition	−	−
Predation	+	−
Parasitism	+	−

* 0 means no direct effect on population growth;
+ means positive effect; − means negative effect.

Categories of Species Interactions

Dozens to hundreds of species interact in diverse ways even in simple communities. In spite of this diversity, we can identify five categories of interactions that have different effects on population growth (Table 47.1).

Species often interact indirectly with others in their habitat. For instance, the Canadian lynx indirectly helps plants by preying on and thus decreasing the number of snowshoe hares that eat plants. The plants indirectly help the Canadian lynx by fattening hares and also by providing cover for ambushes.

Commensalism directly helps one species but does not affect the other much, if at all. For instance, some birds use tree branches for roosting sites. The trees get nothing but are not harmed. With **mutualism**, benefits flow both ways between the interacting species. Don't think of this as cozy cooperation. Benefits flow from a two-way exploitation. With **interspecific competition**, disadvantages flow both ways between species. Finally, **predation** and **parasitism** are interactions that directly benefit one species (either the predator or the parasite) and directly hurt the other (the prey or the host).

Commensalism, mutualism, and parasitism are cases of **symbiosis** ("living together"). For at least part of the life cycle, individuals of two or more species interact with neutral, positive, or negative effects on each other.

A habitat is the type of place where individuals of a species normally live. A community consists of all populations that reside in the habitat.

Community structure arises from the habitat's physical and chemical features, resource availability over time, adaptive traits of its members, how the members interact, and the history of the habitat and its occupants. Species may have neutral, positive, or negative effects on one another.

A niche is the sum of all activities and relationships in which individuals of a species engage as they secure and use the resources necessary to survive and reproduce.

MUTUALISM

Mutualistic interactions in which positive benefits flow both ways abound in nature. When trees provide New Guinea pigeons with food and when the pigeons help disperse seeds from the trees to new germination sites, both function as mutualists. Many other plants and animals enter into such interactions. For example, most flowering plants, and the insects, birds, bats, and other animals that pollinate them, are mutualists. A few vivid examples are given in the Chapter 31 introduction.

of absorptive structures interact in two ways. Either the fungal hyphae penetrate root cells or they grow as a dense, velvety mat around them. The fungus is good at absorbing mineral ions from soil, and the plant comes to rely on pirating some of them. In turn, the fungus pilfers some of the sugar molecules the plant makes by way of photosynthesis. The fungus also depends on the plant for its reproductive success. Its spore production stops when photosynthesis ceases.

Figure 47.2 One mutualistic interaction on a rocky slope of Colorado's high desert.

(**a**) Different kinds of flowering plants of the genus *Yucca* are each pollinated exclusively by one species of yucca moth (**b**). This insect cannot complete its life cycle with any other plant.

The adult stage of the moth life cycle coincides with blossoming of yucca flowers. By using her specialized mouthparts, a female moth gathers sticky pollen and rolls it into a ball. Then she wings her way to another flower. She pierces the wall of the flower's ovary, where seeds form and develop, and lays her eggs inside. As she crawls out of the flower, she pushes a ball of pollen onto a pollen-receiving surface.

Pollen grains germinate and grow down through tissues of the ovary. They deliver sperm to the flower's eggs. The seeds develop after fertilization. Meanwhile, the moth eggs develop to the larval stage. (**c**) When larvae emerge, they eat a few seeds, then gnaw their way out of the ovary. The seeds that moth larvae do not eat give rise to new yucca plants.

Some forms of mutualism are *obligatory*. That is, the individuals of one species cannot grow and reproduce in the absence of intimate dependency with individuals of another species during the life cycle. This is the case for interactions between yucca plants and yucca moths. Each plant of this genus (*Yucca*) can be pollinated only by one species of the yucca moth genus (*Tegeticula*). In addition, the larval stages of the moth grow only in the yucca plant; they eat only yucca seeds (Figure 47.2).

We also find cases of obligatory mutualism between many fungi and plants. Mycorrhizae, remember, are intimate ecological interactions between fungal hyphae and young roots (Chapters 24 and 31). The two kinds

And what about the apparent endosymbiotic origin of eukaryotes? Long ago, phagocytic bacteria may have engulfed aerobic bacterial cells. These resisted digestion, tapped host nutrients, and reproduced independently. In time the hosts came to rely on ATP produced by the guests—which became mitochondria and chloroplasts. If those prokaryotic cells had not evolved as such close mutualists, you and all other eukaryotic species would not even be around today (Section 20.4).

In cases of mutualism, each of the participating species reaps benefits from the interaction.

COMPETITIVE INTERACTIONS

Where you find limited supplies of energy, nutrients, living space, and other natural resources, you observe organisms competing for a share of them. *Intraspecific* competition develops between individuals of the same population or species. As you probably deduced from the preceding chapter, their interactions can be fierce. *Interspecific* competition occurs between populations of different species and usually isn't as intense. Why not? *Requirements of two species may be similar, but they never are as close as they are for individuals of the same species.*

ALPINE CHIPMUNK   LODGEPOLE CHIPMUNK   YELLOW PINE CHIPMUNK   LEAST CHIPMUNK

Figure 47.3 Example of competition in nature. On the eastern slopes of the Sierra Nevada, competition confines some species of chipmunks to specific habitats.

The alpine habitat is at the highest elevation. Below it are the lodgepole pine, piñon pine, and then sagebrush habitats. The least chipmunk lives at the base of the mountains in sagebrush. Its adaptations would allow it to move up into the piñon pine habitat, but the aggressively competitive behavior of yellow pine chipmunks in that habitat won't let it. Food preferences keep the yellow pine chipmunk out of the sagebrush habitat.

Consider just two forms of competiton. *Exploitative* competition occurs when individuals have equal access to a required but limited resource. If one individual is better than another at exploiting it, competition can reduce the supply of that shared resource. By analogy, when you and a friend use straws to share a small milkshake, you may not get much if your friend uses a jumbo straw. With *interference* competition, access to some resource—regardless of whether it is abundant or scarce—is controlled by individuals who let others use only some of the resource or none at all. Thus, even if you shared a ten-gallon milkshake, you still would get less if your friend pinched your straw.

Interference abounds in nature. Chipmunks exclude other species of chipmunks from their habitats (Figure 47.3). A strangler fig tree wraps around other trees as a framework for its growth and eventually it kills them. Spring through summer, the *broadtailed* hummingbird male chases other males and females of its species from blossoms in his territory in the Rockies. In August, *rufous* hummingbirds of the Pacific Northwest migrate through the Rockies on their way to wintering grounds in Mexico. Rufous males are more aggressive, stronger competitors for the shared resource—food. They evict the male broadtails from broadtail territories.

Competitive Exclusion

Let's turn now to exploitative interactions. Any two species differ in their adaptations for getting food and avoiding enemies. In some way, one usually competes more effectively for scarce resources. The two are less likely to coexist in the same habitat when they utilize resources in very similar ways. G. Gause demonstrated this years ago by growing two species of *Paramecium* separately, then together (Figure 47.4a–c). Both species

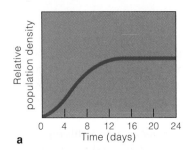

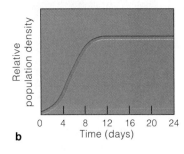

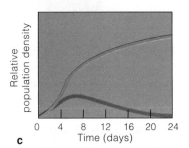

Figure 47.4 Results of competitive exclusion between two protistan species that compete for the same food. (**a**) *Paramecium caudatum* and (**b**) *P. aurelia* were grown in separate culture tubes and established stable populations. The S-shaped graph curves indicate logistic growth and stability. (**c**) Then populations were grown together. *P. aurelia* (*red* curve) drove the other species toward extinction (*blue* curve in **c**). This experiment and others suggest that two species cannot coexist indefinitely in the same habitat *when they require identical resources*. If their requirements do not overlap much, one might influence the population growth rate of the other, but they may still coexist.

used the same food (bacteria) and competed intensely for it. As experimental tests show, two species that use identical resources cannot coexist indefinitely. Many other experiments have supported this concept, which ecologists now call **competitive exclusion**.

Gause also studied two other *Paramecium* species that didn't overlap much in requirements. When grown together, one species tended to feed on bacterial cells suspended in culture tube liquid. The other ate yeast cells at the bottom of the test tube. Population growth rates slowed for both of the species—but the overlap in resource use wasn't enough for one species to fully exclude the other. The two continued to coexist.

Field experiments also reveal effects of competition. For example, N. Hairston studied salamanders in their native habitat in the Balsam Mountains and the Great Smoky Mountains. One kind, *Plethodon glutinosus,* lives at lower elevations than its relative *P. jordani*, but their home ranges do overlap in some areas (Figure 47.5).

For one experiment, Hairston removed one species or the other from test plots in the overlap areas. He left some of those plots untouched as controls. Five years later, nothing had changed in the control plots; the two species were still coexisting. By contrast, salamander populations in the test plots were growing in size. The plots cleared of *P. jordani* now had a greater proportion of *P. glutinosus*. And the plots cleared of *P. glutinosus* had a greater proportion of *P. jordani*.

Hairston concluded that, where populations of the two salamander species coexist in nature, competitive interactions suppress the growth rate of both of them.

Resource Partitioning

Think back on those twelve fruit-eating pigeon species in the same New Guinea forest. They all use the same resource: fruit. Yet they overlap only slightly in their use of it, for each is adapted to eat fruits of a particular size. They are a fine example of **resource partitioning**—the *subdividing* of some category of similar resources that lets competing species coexist.

Another competitive situation arose among three species of annual plants in plowed, abandoned fields. Like other plants, all three require sunlight, water, and dissolved mineral ions. Yet each species is adapted to exploiting a different part of the habitat (Figure 47.6). Drought-tolerant foxtail grasses have a shallow, fibrous root system that quickly absorbs rainwater. They grow where moisture in soil varies from day to day. Mallow plants, with a taproot system, grow in deeper soil that is moist early in the growing season but drier later. Smartweed's taproot system branches in topsoil and soil below the roots of the other species. It grows where the soil is continuously moist.

Figure 47.5 Two species of salamanders that were found to coexist in the same habitat: (**a**) *Plethodon glutinosus* complex and (**b**) *P. jordani.*

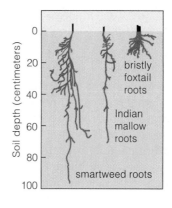

Figure 47.6 Partitioning of resources among three annual plants in a plowed, abandoned field. They differ in adaptations for securing soil water and mineral ions.

Soil depth (centimeters): 0, 20, 40, 60, 80, 100

bristly foxtail roots

Indian mallow roots

smartweed roots

In sum, species might coexist in the same habitat even if their niches do overlap. Many species compete when experimentally grown together but not in their natural habitats, where they tend to use different foods or other resources. Hairston's salamanders compete, yet they coexist at suppressed population sizes. New Guinea pigeons coexist through resource partitioning. Birds and people overlap in their need for oxygen, but atmospheric oxygen is so abundant that there really is no need to compete for it.

In some competitive situations, all individuals have equal access to a resource that they all require, but some are better than others at exploiting it. In other competitive situations, some individuals control access to a shared resource.

The more two species in the same habitat differ in which resources they use, the more likely they may coexist.

Two competing species also may coexist by sharing the same resource in different ways or at different times.

PREDATOR–PREY INTERACTIONS

Coevolution of Predators and Prey

Predators are consumers that feed on living organisms —their **prey**—and typically cause their prey's death. Diverse carnivores and omnivores fall in this category.

Many adaptations of predators (and parasites) and their victims arose by **coevolution**. This term refers to the joint evolution of two or more species that exert selection pressure on each other as an outcome of close ecological interaction. Suppose a prey individual has a new, heritable means of defense, which in time spreads through the prey population. Some predators are better than others at countering this defense. They tend to eat more, and are more likely to survive and reproduce, as are their descendants. These predators exert selection pressure that favors better prey defenses, and so on.

Coevolution has been documented among marine snails and the crabs that eat them. Over time, the snail shell has become thicker. And the crab claw has become more massive, the better to crush the thicker "armor."

Models for Predator–Prey Interactions

Under certain circumstances, predators may limit prey population size. The extent of this limitation depends on a number of factors. One of the most important of these is how individual predators respond to increases or decreases in prey density. There are three general patterns of functional responses, as in Figure 47.7a.

By the type I model, each individual predator will consume a constant number of prey individuals over time, regardless of prey abundance. The assumptions of the model are that prey become ever more abundant, that predators never get full, and that predators do not have to divert time to eating and digesting the prey that they have already caught.

By the type II model, the consumption of prey by each predator increases, but not as fast as increases in prey density. By the type III model, a predator response is lowest when prey density is at its lowest level and predation pressure lessens. This response often happens when predators switch to alternative prey. It is highest at intermediate prey densities, then levels off.

Predators that show type I and type II functional responses can limit prey at a stable equilibrium point.

Other factors besides individual predator response to prey density are at work. For instance, predator and prey reproductive rates influence the interaction. So do hiding places for prey and the presence of alternative prey or predator species. Still another factor: Shifts in environmental conditions can cause predator and prey densities to move between two stabilized levels. At the lowest level, predation strongly depresses prey density. At the highest level, predation is absent and the prey population is nearing the carrying capacity. **Carrying capacity**, recall, is the maximum number of individuals that the resources in a given environment can maintain indefinitely (Section 45.4).

The Canadian Lynx and Snowshoe Hare

In some cases, predator and prey populations oscillate. Consider the ten-year oscillation in populations of one predator—the Canadian lynx—and the showshoe hare that is its main prey (Figure 47.8). To identify the basis of this cyclic pattern, Charles Krebs and his coworkers tracked the densities of hare populations for ten years in Alaska's Yukon River Valley. They set up 1-square-kilometer control plots. They set up experimental plots that kept out predatory mammals (by electric fences), had additional food, or had fertilizer to increase plant growth. In addition, the team captured and released

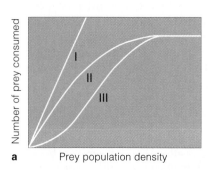
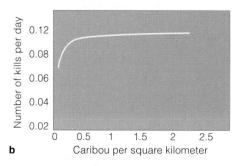

Figure 47.7 (a) Three alternative models for responses of predators to prey density. Type I: Prey consumption rises linearly as prey density rises. Type II: Prey consumption is initially high, then levels off as predators become satiated. Type III: When prey density is low, it takes longer to hunt prey, so the predator response is low. (b) A type II response in nature. For one winter month in Alaska, B. W. Dale and his coworkers observed four wolfpacks (*Canis lupus*) feeding on caribou (*Rangifer taranuds*). The interaction fits the type II model for the functional response of predators to prey density.

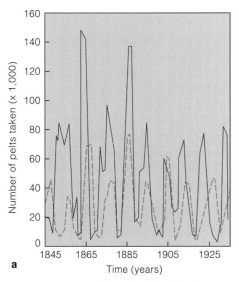

Figure 47.8 (**a**) Correspondence between abundances of Canadian lynx (*dashed* line) and snowshoe hares (*solid* line), based on counts of pelts sold by trappers to Hudson's Bay Company during a ninety-year period.

(**b**) This photograph reinforces Charles Krebs's observation that predation causes mass paranoia among snowshoe hares, which continually look over their shoulders during the declining phase of each cycle. (**c**) This photograph supports the Krebs hypothesis that there is a three-level interaction going on, one that involves plants.

The graph is a good test of whether you tend to readily accept someone else's conclusions without questioning their basis in science. Remember those sections in Chapter 1 that introduced the nature of scientific methods?

What other factors may have had impact on the cycle? Did the weather vary, with more severe winters imposing greater demand for hares (to keep lynx warmer) and higher death rates? Did the lynx compete with other predators, such as owls? Did the predators turn to alternative prey during low points of the hare cycle? When fur prices rose in Europe, did the trapping increase? When the pelt supply outstripped the demand, did trapping decline?

over 1,000 hares, lynx, squirrels, and coyotes after they put radio collars on each one of them.

In predator-free plots, the hare density doubled. In plots with extra food, it tripled. In experimenal plots with fewer predators *and* extra food, the hare density increased by elevenfold.

Nevertheless, the experimental manipulations only delayed the cyclic declines in population density. They did not stop them. Why? The fence could not keep out owls and other raptors. So predators ate 83 percent of the collared hares, only 9 percent of which starved to death. Krebs concluded that a simple predator–prey or plant–herbivore model cannot explain the results of his

experiments in the Yukon River Valley. Rather, in the case of Canadian lynx and snowshoe hare cycle, more variables are operating in a *three-level* interaction.

Predators and prey engaged in long-term interactions exert coevolutionary pressure on one another.

Stability in predator–prey interactions depends on carrying capacity and population densities, prey refuge, predator efficiency, and sometimes alternative prey.

A three-level model involving plants, herbivores, and carnivores helps explain the cyclic pattern of Canadian lynx and snowshoe hare abundances.

AN EVOLUTIONARY ARMS RACE

Populations of other species are part of any organism's environment, and the ones that interact as predators and prey exert selection pressure on each other. One must defend itself and the other must overcome the defenses. This is the basis of an evolutionary arms race that has resulted in amazing adaptations.

CAMOUFLAGE   Some prey species gain protection from **camouflage**; they hide in the open. Such species have adaptations in their form, patterning, color, and behavior that help them blend with the surroundings and escape detection. Figure 47.9 has classic examples, including *Lithops,* a kind of desert plant that resembles a rock. Only during a brief rainy season does *Lithops* flower. That is when other plants grow profusely and distract herbivores from *Lithops*—and when free water instead of juicy plant tissues is available to quench an animal's thirst.

WARNING COLORATION   Many prey species taste bad, are toxic, or inflict pain on attackers. Toxic types often have **warning coloration**, or conspicuous patterns and colors that predators learn to recognize as *AVOID ME!* signals. Maybe a young, inexperienced bird will spear a yellow-banded wasp or an orange-patterned monarch butterfly once. It will soon learn to associate the strong colors and patterning with a painful sting or with the vomiting of foul-tasting butterfly toxins.

Dangerous or repugnant species make little or no attempt to conceal themselves. Skunks are like this. So are frogs of the genus *Dendrobates*; they are among the most vivid and poisonous organisms (Section 33.1).

MIMICRY   Many prey species look a lot like dangerous, unpalatable, or hard-to-catch species. **Mimicry** is the name for such close resemblances in form, behavior, or both between one species that is the *model* for deception and another species that is its *mimic.* Figure 47.10 shows how tasty but weaponless mimics physically resemble species that predators have learned to ignore. In "speed" mimicry, sluggish prey species look like swift-moving species that predators have given up trying to catch.

MOMENT-OF-TRUTH DEFENSES   When luck runs out, survival of prey that are cornered or under attack may turn on a last-chance trick. Suppose a leopard manages to run down a tasty baboon, as shown in Figure 46.6. By turning abruptly and displaying its formidable canines, the baboon may startle and confuse this predator long enough for a getaway. Other cornered animals spew

Figure 47.9   Prey organisms demonstrating the fine art of camouflage. (**a**) What bird??? When a predator approaches its nest, the least bittern stretches its neck (which is colored like the surrounding withered reeds), thrusts its beak upward, and sways gently like reeds in the wind. (**b**) An unappetizing bird dropping? No. This edible caterpillar's body coloration and the stiff body positions that it assumes help camouflage it (hide it in the open) from predatory birds. (**c**) Find the plants (*Lithops*) hiding in the open from herbivores by their stonelike form, pattern, and coloring.

c The dangerous model   **d** One of its edible mimics . . .   **e** Another edible mimic

Figure 47.10 Examples of mimicry among insects. Many predators avoid prey that taste awful, secrete toxins, or inflict painful bites or stings. Such prey commonly display warning signals (bright coloration, bold markings, or both). Often they do not even bother to hide. Other prey species unrelated to dangerous or unpalatable ones have evolved striking behavioral and morphological resemblances to them.

The inedible butterfly in (**a**) is a model for an edible mimic, *Dismorphia* (**b**). The yellowjacket in (**c**) stings aggressively. It may be the model for nonstinging, edible wasps (**d**) and beetles (**e**) that have a similar appearance.

Figure 47.11 Examples of adaptive responses of predators to prey defenses. (**a**) Some beetles spray noxious chemicals at attackers, which deters them some of the time. (**b**) At other times, grasshopper mice plunge the chemical-spraying tail end of their beetle prey into the ground and feast on the head end. (**c**) Find the scorpionfish, a venomous predator with camouflaging fleshy flaps, multiple colors, and profuse spines. (**d**) Where do the pink flowers end and the pink praying mantis begin?

irritating chemical repellents or toxins. Earwigs, stink beetles, and skunks produce disgusting odors. Some of the beetles take aim and let loose with noxious sprays.

Also, the foliage and seeds of many plants contain bitter or hard-to-digest repellents. For example, make the mistake of nibbling on the visually luscious yellow petals of a buttercup (*Ranunculus*), and you will inflict a chemical burn on the lining of your mouth.

PREDATOR RESPONSES TO PREY In evolutionary arms races, predators counter prey defenses with their own adaptations. Among their countermeasures are stealth, camouflage, and clever ways of avoiding repellents. Consider edible beetles that direct sprays of noxious chemicals at their attackers. Grasshopper mice grab

the beetle and plunge its sprayer end into the ground, then eat the unprotected head (Figure 47.11*a,b*).

Chameleons can hold themselves motionless for extended intervals. Prey might not even notice them until the amazingly swift chameleon tongue zaps them (Section 33.4). And it is no accident that the stealthiest predators blend with their backgrounds. Think about the snow-white polar bears camouflaged against snow, golden tigers crouched in tall-stalked, golden grasses, and pastel predatory insects lurking in pastel flowers. And hope you never step barefoot on a scorpionfish concealed on the seafloor.

Predators and prey exert selection pressure on each other.

PARASITE–HOST INTERACTIONS

Evolution of Parasitism

Parasites live in or on other living organisms—their **hosts**—and feed upon specific host tissues for part of their life cycle. The hosts may or may not die from the interaction. In this ecological interaction, one organism benefits by draining nutrients from another, which is harmed. This also is true of predation, but you would not expect to come across predators that exploit another organism as food *and* habitat.

Parasites have pervasive influences in populations. By draining host nutrients, they alter how much energy and how many nutrients the population is withdrawing from the habitat. Weakened hosts are more vulnerable to predation and less attractive to potential mates. Some infections result in sterility. Some may alter the ratio of host males to females. In such ways, parasitic infections lower the birth rate, raise the death rate, and influence intraspecific and interspecific competition.

Sometimes the gradual drain of nutrients during a parasitic infection indirectly causes death, for the host becomes so weakened that it succumbs to secondary infections. In evolutionary terms, however, killing a host is not very good for a parasite's reproductive success. An infection of longer duration obviously will give the parasite more time to produce more offspring. We may therefore speculate that natural selection tends to favor parasite and host adaptations that promote some level of mutual tolerance and less-than-fatal effects.

Usually, only two types of interactions kill the host. Either the parasite attacks a novel host, which has no coevolved defenses against it, or the host is supporting too many active, individual parasites at the same time.

Kinds of Parasites

Previous chapters introduced you to diverse parasitic species. They include *ecto*parasites (which live on a host body surface) and *endo*parasites (which live within the host's body). Some parasites live on or in one host for their entire life cycle, and others are free-living some of the time and residents of different hosts at different stages of their life cycle. Many enlist insects and other arthropods as taxis from one host organism to another.

Many viruses, bacteria, protozoans, and sporozoans are *micro*parasites; they are microscopic in size and fast reproducers. *Macro*parasites include diverse flatworms and roundworms (Figure 47.12). Fleas, ticks, mites, and lice are other invertebrates in this group. Figure 47.13 is a micrograph of one of the chytrids, a parasitic fungus that is decimating the amphibians of tropical forests. Sections 21.9, 22.2 through 22.6, 24.5, 25.7, and 25.18 have more detailed cases of parasites in the bacterial, protistan, fungal, and animal kingdoms.

Even the kingdom of plants has a few. *Holo*parasitic plants are nonphotosynthetic; they withdraw nutrients and water from young roots. Broomrapes parasitize oak or beech tree roots. *Hemi*parasites retain a capacity for photosynthesis, but they withdraw nutrients and water from host plants. Mistletoe is like this; it puts out

Figure 47.12 Adult roundworms (*Ascaris*), an endoparasite, packed inside the small intestine from a host pig.

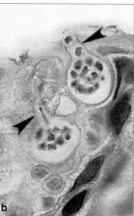

Figure 47.13 (**a**) Harlequin frog (*Atelopus varius*) from Central America. (**b**) Section through the skin of an *A. varius* individual with *chytridomycosis*, a fungal infection. Arrows point to two flask-shaped cells of the parasitic chytrid. Each contains fungal spores that are discharged to the skin surface, through the neck of the flask.

Chytridomycosis is one of the causes of the recent mass deaths of rain forest amphibians. Even in untouched forested regions (with no agriculture, deforestation, or pollution) there are long-term, catastrophic declines in many amphibian species from a large part of their habitats. This indicates that the drastic losses in populations are abnormal changes.

Figure 47.14 Example of a social parasite. European cuckoos lay eggs in nests of other species. In response to an environmental cue, a hatchling executes innate behavior. As Section 46.1 describes, even before its eyes open, it responds to the spherical shape of the host's eggs by shoving them out of the nest. The foster parents continue to raise the usurper even after it is old enough to show its true feathers, so to speak.

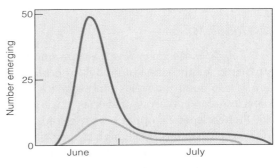

Figure 47.15 Effect of the activity of a parasitoid wasp on the emergence of adult sawflies from cocoons on the forest floor. The *brown* line represents the number emerging when wasp attacks were experimentally prevented. The *pink* line represents the number emerging after the attacks. The wasp made most of its attacks on the would-be *early* emergers from cocoons not buried as deeply in forest litter.

long extensions that invade the sapwood of host trees. Section 23.9 shows examples.

Social parasites alter the social behavior of another species to complete their life cycle. Cuckoos and North American cowbirds are examples. The adult females lay eggs in the nests of other species. Their hatchlings rid the nest of its rightful occupants or they demand, and get, most of the food (Section 46.1 and Figure 47.14).

Regarding the Parasitoids

Between predators and parasites are **parasitoids**. Larvae of these insects live inside and consume all of the soft tissues of the host, which is usually the larva or pupa of another insect. Peter Price studied the behavior of a parasitoid wasp that lays its eggs on sawfly cocoons. Sawflies reproduce once a year and lay eggs in trees. Their fertilized eggs develop into fly larvae, which feed on foliage. In time, larvae drop to the forest floor and spin cocoons after burrowing in leaf litter. Some burrow more deeply than others.

As Price noted, the first adult sawflies to emerge did so from cocoons that were the least deeply buried. Later in the season, more sawflies emerged from the more deeply buried cocoons (Figure 47.15). Parasitoid wasps tend to lay eggs on the cocoons closest to the surface of leaf litter. They exert strong selection pressure on the sawfly population, because the deep-burrowing sawfly individuals are more likely to escape detection. There are only so many cocoons near the surface, so the wasp that is able to locate cocoons deeper in the litter will be more competitive in securing food for her larvae.

In this coevolutionary contest, the host is staying ahead. Each time sawfly larvae burrow deeper, female wasps spend more time searching for them. Fewer wasp eggs are laid, so the wasp population is held in check.

Parasites as Biological Control Agents

Because parasites and parasitoids exert control over the population growth of other species, many are raised commercially and then selectively released as *biological controls*. They are often touted as a fine alternative to chemical pesticides. But less than 20 percent of existing selections qualify as effective defenses against pests.

As C. Huffaker and C. Kennett point out, effective biological control agents display five attributes. They are well adapted to a host species and to their habitat. They are good at searching for a host. Their population growth rate is high compared to that of the host. Their offspring are mobile enough to ensure dispersal. The lag time between their type III functional responses to changing numbers of the host population is minimal.

Releasing more than one kind of biological control agent in an area may trigger competition among them and lessen their overall effectiveness. Also, the shotgun approach is risky, for parasites may attack nontargeted species. In 1983, for example, F. Howarth reported that the populations of butterflies and moths native to the Hawaiian Islands are declining. Introductions of wasps that were supposed to serve as biological controls over something else are partly to blame. Later in the chapter, we return to the effect that introducing new species into a habitat has on native populations.

Like predators and their prey victims, parasites and their hosts are locked into long-term, coevolutionary contests.

Natural selection favors parasitic species that temper their demands in ways that ensure an adequate supply of hosts.

Too great a demand quickly kills the hosts and limits the duration of an attack, which sets a limit on the number of parasitic offspring that can be produced.

FORCES CONTRIBUTING TO COMMUNITY STABILITY

A Successional Model

By now you might be asking, How does a community come into being? By the classical model for **ecological succession**, a community develops in sequence, from pioneer species to an end array of species that remains stabilized. **Pioneer species**, opportunistic colonizers of vacant or vacated habitats, have high dispersal rates, grow and mature fast, and produce many offspring. In time, stronger competitor species replace the pioneers. They too, are replaced until the species array stabilizes under the prevailing habitat conditions. This persistent species composition is the **climax community**.

Primary succession is a process that begins when pioneer species colonize a barren habitat, such as a new volcanic island or land exposed when a glacier retreats (Figure 47.16). The pioneer species include lichens and plants that are small, have rapid life cycles, and are adapted to exposed sites having intense sunlight, large temperature changes, and nutrient-deficient soil. In the early years, hardy flowering plants put out many small seeds, which are quickly dispersed.

Once established, the pioneers improve conditions for other species and often set the stage for their own replacement. Also, many new arrivals are mutualists with nitrogen-fixing bacteria and outcompete the early species in nitrogen-poor habitats. Seeds of later species find shelter inside low-growing mats of the pioneers, which cannot shade out the seedlings. In time, wastes and remains accumulate, adding volume and nutrients to soil that help more species take hold. Successional species eventually crowd out the pioneers, whose spores and seeds travel as fugitives on the wind and water—destined, perhaps, for a new but temporary habitat.

In **secondary succession**, a disturbed area within a community recovers and moves again toward a climax state. The pattern is typical of abandoned fields, burned forests, and storm-battered intertidal zones. It proceeds after a falling tree opens part of an established forest's canopy. Sunlight reaches seeds and seedlings that are already on the forest floor and spurs their growth.

Do colonizers facilitate their own replacement? By a different hypothesis, early colonizers compete against any species that could replace them. And the sequence of succession actually depends on who gets there first.

The Climax-Pattern Model

At one time, some ecologists thought the same general type of community always develops in a given region because of constraints imposed by climate. However, stable communities other than "the climax community" commonly persist in a region, as when tallgrass prairie from the west extends into Indiana's deciduous forests.

Figure 47.16 (a) Primary succession in Alaska's Glacier Bay area. The glacier in this photograph has been retreating since 1794. *Facing page:* (b) As a glacier retreats, meltwater leaches nitrogen and other minerals from the newly exposed soil. Less than ten years ago, ice buried this nutrient-poor soil. The first invaders are lichens, horsetails, and seeds of fireweed and mountain avens (*Dryas*). (c) *Dryas* is a pioneer species that relies on nitrogen-fixing activities of mutualistic microbes. It grows and spreads fast over the glacial till.

(d) Within twenty years, alder, cottonwood, and willow seedlings have taken hold in drainage channels. They, too, are symbionts with nitrogen-fixing microbes. (e) Within fifty years, alders have formed mature, dense thickets in which cottonwood, hemlock, and a few evergreen spruce can grow quickly. (f) After eighty years, cottonwood and spruce trees crowd out mature alders. (g) In areas deglaciated for more than a century, dense forests of Sitka spruce and western hemlock dominate.

By the **climax-pattern model**, a community is adapted to many environmental factors—topography, climate, soil, wind, species interactions, recurring disturbances, chance events, and so on—that vary in their influence over a region. As a result, even in the same region you may see one climax stage extending into another along gradients of influential environmental conditions.

Cyclic, Nondirectional Changes

Many small-scale changes occur over and over again in patches of a habitat. Such recurring changes contribute to the internal dynamics of the community as a whole. Observe a disturbed patch of some habitat, and you might conclude that great shifts in species composition are afoot. Observe the community on a larger scale, and you might discover that the overall composition includes all of the pioneer species that are colonizing the patch, in addition to the dominant climax species.

Think of a tropical forest. It slowly develops through phases of colonization by pioneer species, increases in

small fires are prevented. And so the underbrush slowly thickens with fire-susceptible species. But dense underbrush prevents the sequoia seeds from germinating —and it also fuels hotter fires. Hotter fires damage the giants. Park rangers now set controlled fires. By periodically removing underbrush, such controlled fires promote conditions that favor the community's cyclic replacements.

Restoration Ecology

Think back on the introduction to Chapter 29, which described the regrowth after the Mount Saint Helens eruption during 1980, and you already know that secondary succession can heal disturbances on a massive scale. Such *natural* restoration of communities often takes a long time. Today we also see *active* restoration—efforts to reestablish the biodiversity in key areas adversely impacted or lost by agriculture, urban sprawl, and other human activities.

Chapter 27 included examples of this work. Here are a few more: Ecologists and volunteers are building artificial reefs along coasts, and repairing wetlands and grasslands. In Illinois, species of natural prairie communities turned up in old cemeteries and other neglected patches of land. Those species were transplanted, by hand, to a common restored site. Now this small restored patch of prairie extends over 180 hectares (445 acres). Volunteers maintain it by controlled burns, weeding by hand, and other means. Their goal is to find all of the 150–200 known species of the original community.

species diversity, and then reaches maturity. Whipping sporadically through the successional pattern, though, are heavy winds that cause treefalls. Where trees fall, gaps open in the forest canopy and more light reaches a local patch of forest floor. There, conditions favor the growth of previously suppressed small trees as well as germination of pioneers or shade-intolerant species.

Or think of the groves of sequoia trees in the Sierra Nevada of California. Some trees in this type of climax community are giants, more than 4,000 years old. Their persistence depends partly on recurring brush fires that sweep through parts of the forests. Sequoia seeds only germinate in the absence of far smaller, shade-tolerant plant species. Too much litter on the forest floor inhibits germination. Modest fires eliminate trees and shrubs that compete with young sequoias but do not damage the giant trees. Mature sequoias have thick bark that burns poorly and helps insulate their living phloem cells against modest heat damage.

At one time, fires were prevented in many sequoia groves in national and state parks—not just accidental fires from campsites and discarded cigarettes, but also natural, lightning-sparked fires. Litter builds up when

Community structure is an outcome of a balance of forces, such as predation and competition, that operate over time.

A climax community is a stable, self-perpetuating array of species in equilibrium with one another and their habitat. It develops as a sequence of species, starting with pioneers.

Similar climax stages can persist along gradients dictated by environmental factors and by species interactions. Small-scale, recurring changes help shape many communities.

Natural or deliberate ecological restoration often can repair a badly damaged climax community, provided that suitable species are available to reinstate the original biodiversity.

FORCES CONTRIBUTING TO COMMUNITY INSTABILITY

The preceding sections might lead you to believe that all communities become stabilized in predictable ways. But this is not always the case. *Community stability is an outcome of forces that have come into uneasy balance.* Resources are sustained, as long as populations do not flirt dangerously with the carrying capacity. Predators and prey coexist, as long as neither wins. Competitors have no sense of fair play. The mutualists are stingy, as when plants produce as little nectar as necessary to attract pollinators, and pollinators take as much nectar as they can for the least effort.

In the short term, disturbances can hurt the growth of some populations. You saw one example of this in Section 45.3. Also, long-term changes in climate or some other environmental variable often have destabilizing effects. If the instability is great enough, a community might change in ways that will persist even when the disturbance ends or is reversed. If some of its member species happen to be rare or do not compete well with others, they might become extinct.

How Keystone Species Tip the Balance

The uneasy balancing of forces in a community comes into sharp focus with studies of a **keystone species**, a dominant species that can dictate community structure. Consider periwinkles (*Littorina littorea*), an alga-eating snail of intertidal zones. Jane Lubchenco found that periwinkles can increase *or* decrease the diversity of algal species in different settings. In tidepools, they eat a dominant algal species (*Enteromorpha*). By doing so, they allow many other, less competitive algal species to survive. By contrast, on rocks exposed only at high tide, *Chondrus* and other red algae predominate. Periwinkles leave these tough, unpalatable species alone—and eat the competitively weak algal species that they pass up in tidepools. Thus, periwinkles increase the diversity of algae in tidepools and yet reduce it on rocks that are recurringly exposed (Figure 47.17).

Robert Paine studied the effect of keystone species through removal experiments in the intertidal zones of North America's west coast. Pounding surf, tides, and storms continually disturb intertidal zones, and living space is scarce. Paine put the sea star *Pisaster ochraceus* in control plots along with its prey: mussels (*Mytilus*), barnacles, limpets, and chitons. Then he removed all sea stars from experimental plots.

Mussels took over Paine's experimental plots. They crowded out seven other invertebrate species. Mussels, the main prey of sea stars, therefore are the strongest

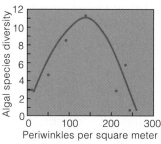

d Algal diversity in tidepools

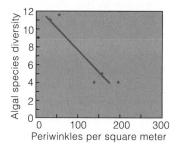

e Algal diversity on rocks that become exposed at high tide

Figure 47.17 Effect of competition and predation on community structure. (**a**) Grazing periwinkles (*Littorina littorea*) affect the number of algal species in different ways in different marine habitats. (**b**) *Chondrus* and (**c**) *Enteromorpha*, two kinds of algae in their natural habitat. (**d**) In tidepools, periwinkles graze on the dominant alga (*Enteromorpha*); less competitive algae that might otherwise be overwhelmed have an advantage. (**e**) Algal diversity is lower on rocks exposed at low tide. There, periwinkles don't graze on *Chondrus* and other kinds of red algae, the dominant species.

Table 47.2 *Detrimental Effects of Some Species Introduced Into the United States*

Species Introduced	Origin	Mode of Introduction	Outcome
Water hyacinth	South America	Intentionally introduced (1884)	Clogged waterways; other plants shaded out
Dutch elm disease: *Ophiostoma ulmi* (fungus) Bark beetle (vector)	Europe	Accidental; on infected elm timber (1930) Accidental; on unbarked elm timber (1909)	Millions of mature elms destroyed
Chestnut blight fungus	Asia	Accidental; on nursery plants (1900)	Nearly all eastern American chestnuts killed
Argentine fire ant	Argentina	In coffee shipments from Brazil? (1891)	Crop damage; native ant communities destroyed; ground-nesting birds killed
Japanese beetle	Japan	Accidental; on irises or azaleas (1911)	Over 250 plant species (e.g., citrus) defoliated
Sea lamprey	North Atlantic	Ship hulls, through canals (1860s, 1921)	Trout, whitefish destroyed in Great Lakes
European starling	Europe	Intentional release, New York City (1890)	Native songbirds displaced; crop damage; swine disease vector; huge flocks noisy, messy
House sparrow	England	Released intentionally (1853)	Crop damage; native songbirds displaced; transmission of some diseases
Green alga (*Caulerpa taxifolia* modifed strain)	Tropical seas	Probably released accidentally from saltwater aquarium (2000)	Potential for mass destruction of coastal marine ecosystems, as in Mediterranean

competitors in the absence of sea stars. Normally, sea star predation maintains the diversity of prey species because it restricts competitive exclusion by mussels. Remove all the sea stars, and the community shrinks from fifteen species to eight.

How Species Introductions Tip the Balance

Finally, consider what can happen when some residents of established communities move out from their home range and successfully take up residence elsewhere. We call such a directional movement **geographic dispersal**. It can proceed in three ways.

First, over a number of generations, a population might expand its home range by slowly moving into outlying regions that prove hospitable. Second, some individuals might be rapidly transported across great distances. This is called *jump* dispersal. It often takes individuals across regions where they could not survive on their own, as when insects travel from the mainland to Maui in a ship's cargo hold. Third, with imperceptible slowness, a population might be moved away from its home range over geologic time, as by continental drift.

The dispersal and colonization of vacant places can be amazingly rapid as well as successful. Consider one of Amy Schoener's experiments in the Bahamas. She set out plastic sponges on the barren, sandy floor of Bimini Lagoon. How fast did aquatic species take up residence in chambers of the artificial hotels? Schoener recorded occupancy by 220 species within thirty days.

Or think of the 4,500 or so nonindigenous species that successfully established themselves in the United States following jump dispersal. These are just some of the ones we know about. Some, including soybeans, rice, wheat, corn, and potatoes, have been put to use as food. But most adversely impact natural communities and farmlands (Table 47.2 and Section 47.9).

If you live in the northeastern United States, you know about imported gypsy moths. They escaped from a research facility near Boston in 1869, and today their descendants defoliate entire forests. You might know about zebra mussels, which probably entered the Great Lakes on a cargo ship's hull. They attach to and block water intake pipes for cities. The cumulative damage may exceed 5 billion dollars. *Cryphonectria parasitica*, a fungus introduced to North America almost a century ago, killed nearly all American chestnut trees.

If you live in the southeastern United States, you know about the hot sting of fire ants (*Solenopsis invicta*). Nearly a half century ago, the ants were accidentally imported from South America, possibly in the hold of a ship that docked in Mobile, Alabama. Each year, fire ants inflict enough multiple bites to make 70,000 or so people seek medical attention.

And remember those African bees released in South America? As you read in Chapter 8, their Africanized, "killer bee" descendants traveled northward, through Mexico and on into the United States. Swarms have already killed several hundred people in Latin America. At this writing, these aggressive bees have attacked 140 Americans, two of whom died from multiple stings.

Long-term shifts in climate, the rapid introduction and successful establishment of a new species, and many other disturbances can permanently alter community structure.

Exotic and Endangered Species

When you hear someone bubbling enthusiastically about an **exotic species**, you can safely bet the speaker isn't an ecologist. This is a name for a resident of an established community that was deliberately or accidentally moved from its home range and became established elsewhere. Unlike most imports, which can't take hold outside their home range, an exotic species permanently insinuates itself into a new community. Table 47.2 gives examples.

Sometimes the additions are harmless and even have beneficial effects. More often, they make native species **endangered species**, which by definition are extremely vulnerable to extinction (Section 27.2). Of all species on the rare or endangered lists or that recently became extinct, *close to 70 percent owe their precarious existence or demise to displacement by exotic species.*

THE RABBITS THAT ATE AUSTRALIA   During the 1800s, British settlers in Australia just couldn't bond with the koalas and kangaroos, so they started to import familiar animals from their homeland. In 1859, in what would be the start of a wholesale disaster, a northern Australian landowner imported and then released two dozen wild European rabbits (*Oryctolagus cuniculus*). Good food and good sport hunting, that was the idea. An ideal rabbit habitat with no natural predators—that was the reality.

Six years later, the landowner had killed 20,000 rabbits and was besieged by 20,000 more. The rabbits displaced livestock, even kangaroos. Now Australia has 200 to 300 million hippityhopping through the southern half of the country. They overgraze perennial grasses in good times and strip bark from shrubs and trees during droughts. You know where they've been; they transform grasslands and shrublands into eroded deserts (Figure 47.18). They have been shot and poisoned. Their warrens have been plowed under, fumigated, and dynamited. Even when all-out assaults reduced their population size by 70 percent,

the rapidly reproducing imports made a comeback in less than a year. Did the construction of a 2,000-mile-long fence protect western Australia? No. Rabbits made it to the other side before workers finished the fence.

In 1951, government workers introduced a myxoma virus by way of mildly infected South American rabbits, its normal hosts. This virus causes *myxomatosis*. The disease has mild effects on South American rabbits that coevolved with the virus but nearly always had lethal effects on *O. cuniculus*. Biting insects, mainly mosquitoes and fleas, quickly transmit the virus from host to host. Having no coevolved defenses against the novel virus, the European rabbits died in droves. But, as you might expect, natural selection has since favored rapid growth of populations of *O. cuniculus* resistant to the virus.

In 1991, on an uninhabited island in Spencer Gulf, Australian researchers released a population of rabbits that they had injected with a calicivirus. The rabbits died quickly and relatively painlessly from blood clots in their lungs, heart, and kidneys. In 1995, the test virus escaped from the island, possibly on insect vectors. It has been killing 80 to 95 percent of the adult rabbits in Australian regions. At this writing, researchers are now questioning whether the calicivirus should be used on a widespread scale, whether it can jump boundaries and infect animals other than rabbits (such as humans), and what the long-term consequences will be.

THE PLANTS THAT ATE GEORGIA   A vine called kudzu (*Pueraria lobata*) was deliberately imported from Japan to the United States, where it faces no serious threats from herbivores, pathogens, or competitor plants (Figure 47.19). In temperate parts of Asia, it is a well-behaved legume with a well-developed root system. It *seemed* like a good idea to use it to control erosion on hills and highway embankments in the southeastern United States. With

Figure 47.18   Part of a fence against the 200 million–300 million rabbits that are destroying Australia's vegetation.

Figure 47.19 (a) Kudzu (*Pueraria lobata*) taking over part of Lyman, South Carolina. We now find kudzu vines from East Texas to Florida and as far north as Pennsylvania. (b) *Caulerpa taxifolia* suffocating yet another marine ecosystem.

nothing to stop it, though, kudzu's shoots grew a third of a meter per day. Vines now blanket streambanks, trees, telephone poles, houses, and almost everything else in their path. Attempts to dig up or burn kudzu are futile. Grazing goats and herbicides help, but goats eat other plants, too, and herbicides contaminate water supplies. Kudzu could reach the Great Lakes by the year 2040.

On the bright side, a Japanese firm is constructing a kudzu farm and processing plant in Alabama. Asians use a starch extract from kudzu in drinks, herbal medicines, and candy. The idea is to export the starch to Asia, where the demand currently exceeds the supply. Also, kudzu may eventually help reduce logging operations. At the Georgia Institute of Technology, researchers report that kudzu might become an alternative source for paper.

THE ALGA TRIUMPHANT They just looked so perfect in saltwater aquariums, those bright-green, featherlike branches of the green alga *Caulerpa taxifolia*. Germany's Stuttgart Aquarium even developed a lovely strain and shared it with other marine institutions. Was it from the Oceanographic Museum in Monaco that the strain escaped into the wild? Some say yes, Monaco says no.

The aquarium strain of *C. taxifolia* cannot reproduce sexually. Yet. It grows by runners, just a few centimeters a day. But boat propellers and fishing nets assisted its dispersal. Since 1984, it has now blanketed more than

30,000 hectares of the seafloor along the Mediterranean coast (Figure 47.19*b*). It thrives along sandy and rocky shores or mud, and it lasts ten days when tossed in meadows. Unlike its tropical parents, this strain does well in cool water, even polluted water. It also produces a toxin that discourages herbivores. And it displaces native algae, invertebrates, and fishes. It has invaded coastal waters of Australia and the United States, where it is wiping out food webs, biodiversity, and fisheries.

In 2000, divers discovered the alga in a lagoon north of San Diego. Someone probably drained water from a home aquarium into a storm drain or the lagoon itself. As Alan Millar put it, "The Californians chose to shoot first and ask questions later." They tarped over the area to shut out sunlight, pumped chlorine into the mud to poison the alga, and used underwater welders to boil it. They identified two sea slugs, *Aplysia depilans* and *Elysia subornata*, that can eat the alga without being harmed by its toxin. However, what these two potential imports will do to coastal ecosystems is not known, so they have yet to be released as biological controls.

C. taxifolia is now banned in the United States. But it is still sold over the Internet and is still being imported, because it is easy to confuse this species with its less invasive relatives. The aquarium industry successfully lobbied against a ban on every *Caulerpa* species, even though most inspectors cannot tell one from another.

PATTERNS OF BIODIVERSITY

Biodiversity refers to the variety of organisms, habitats, and ecosystems on the Earth or in a particular location. As you might well conclude from Chapter 27 and the preceding discussions, biodiversity differs greatly from one habitat to another. Also, as ecologists have found through many field investigations, biodiversity differs in significant ways between geographic regions.

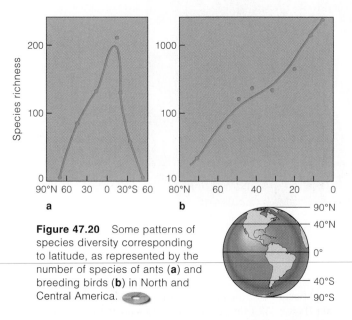

Figure 47.20 Some patterns of species diversity corresponding to latitude, as represented by the number of species of ants (**a**) and breeding birds (**b**) in North and Central America.

Figure 47.21 Aerial photograph of Surtsey, a volcanic island, at the time of its formation in the sea. Newly formed, isolated islands are natural laboratories for ecologists. The chart gives the number of colonizing species of mosses and vascular plants recorded on Surtsey between 1965 and 1973.

What Causes Mainland and Marine Patterns?

The most striking pattern of biodiversity corresponds to distance from the equator. For most groups of plants and animals, *the number of coexisting species on land and in the seas is greatest in the tropics, and it systematically declines from the equator to the poles.* Figure 47.20 shows two examples of this biodiversity pattern. What factors underlie it? Three are paramount.

First, as described in Chapter 49, tropical latitudes intercept more sunlight of consistently higher intensity, rainfall is heavier, and the growing season is longer. As one outcome, *resource availability tends to be greater and more reliable in the tropics than elsewhere.* All year long, different tree species in humid tropical forests put out new leaves, flowers, and fruit. Year in and year out, they support diverse herbivores, nectar foragers, and fruit eaters. Such specialization can't evolve to a comparable degree in temperate and arctic regions.

Second, *species diversity might be self-reinforcing.* The diversity of tree species in tropical forests is far greater than in comparable forests at higher latitudes. When more plant species compete and coexist, more herbivore species also evolve and coexist partly because no single herbivore can overcome all the chemical defenses of all the different plants. Thus, more predators and parasites tend to evolve in response to a diversity of prey and hosts. The same applies to diversity on tropical reefs.

Third, *over geologic time, speciation rates in the tropics exceeded rates of extinction.* Declines in biodiversity have occurred, but mainly during mass extinctions. Global temperatures drop after such episodes. Species adapted to cool climates tend to survive, but those adapted to the temperatures of the tropics do not. Also, millions of species in tropical forests may disappear in the next decade, a topic addressed in Chapters 27 and 50.

What Causes Island Patterns?

Some islands are laboratories for biodiversity studies. For instance, in 1965, a volcanic eruption formed a new island southwest of Iceland. Within six months, bacteria, fungi, seeds, flies, and some seabirds were established on it. One vascular plant appeared after two years, and a moss two years after that (Figure 47.21). The number of plant species rose when soils improved. Iceland was the source of all of the colonists. None originated on the island, which was named Surtsey. Yet the number

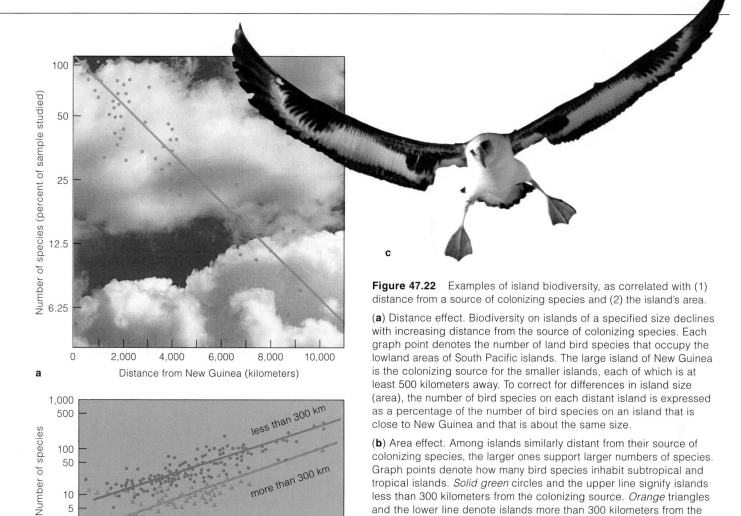

a — Distance from New Guinea (kilometers)
Number of species (percent of sample studied)

b — Area (square kilometers)
Number of species

c

Figure 47.22 Examples of island biodiversity, as correlated with (1) distance from a source of colonizing species and (2) the island's area.

(**a**) Distance effect. Biodiversity on islands of a specified size declines with increasing distance from the source of colonizing species. Each graph point denotes the number of land bird species that occupy the lowland areas of South Pacific islands. The large island of New Guinea is the colonizing source for the smaller islands, each of which is at least 500 kilometers away. To correct for differences in island size (area), the number of bird species on each distant island is expressed as a percentage of the number of bird species on an island that is close to New Guinea and that is about the same size.

(**b**) Area effect. Among islands similarly distant from their source of colonizing species, the larger ones support larger numbers of species. Graph points denote how many bird species inhabit subtropical and tropical islands. *Solid green* circles and the upper line signify islands less than 300 kilometers from the colonizing source. *Orange* triangles and the lower line denote islands more than 300 kilometers from the source areas. As you can see, this graph incorporates data on the distance effect as well as on the area effect.

(**c**) Wandering albatross, one travel agent for jump dispersals. Seabirds that island-hop long distances often have a few seeds stuck to their feathers. Seeds that successfully germinate in a new island community may give rise to a population of new immigrants.

of new species will not increase indefinitely. Why not? Studies of community patterns on islands around the world provide us with two models.

First, islands far from a source of potential colonists receive few colonizing species, and the few that arrive are adapted for long-distance dispersal (Figure 47.22). This is the **distance effect**. Second, larger islands tend to support more species than smaller islands at equivalent distances from source areas. This is the **area effect**. The larger islands tend to have more and varied habitats; most have more complex topography and extend higher above sea level. Thus they favor species diversity. Also, being bigger targets, they intercept more colonists.

Most importantly, extinctions suppress biodiversity on small islands. There, populations are small and far more vulnerable to storms, volcanic eruptions, disease, and random shifts in birth and death rates. As for any island, the number of species reflects a balance between immigration rates for new species and extinction rates for established ones. Small islands that are distant from a source of colonists have low immigration rates and high extinction rates, so they support few species once the balance has been struck for their populations.

For most groups of organisms, the number of coexisting species is highest in the tropics. The number systematically declines from the equator to the poles.

A region's biodiversity depends on many factors, such as its climate, topographical variation, possibilities for dispersal, and evolutionary history—including extinctions and the time span available for speciation.

Disturbances to habitats tend to work against competitive exclusion and therefore to favor increases in biodiversity.

Island biodiversity is a balance between immigration rates for new species and extinction rates for established species.

SUMMARY

1. A habitat is the type of place where individuals of a given species normally live—that is, their "address." A community consists of all populations of all species that occupy a habitat and that typically associate with one another, directly or indirectly. *47.1*

2. Each species has its own niche, or "profession," in the community. A niche is the sum of all activities and relationships in which its individuals engage as they secure and use the resources necessary for survival and reproduction. *47.1*

3. Mutualism, commensalism, competition, predation, and parasitism are species interactions that directly or indirectly link populations in a community. *47.1, 47.2*

4. Two species requiring the same limited resource tend to compete, as by exploiting it as fast or efficiently as possible or interfering with the other's use of it. *47.3*

   a. According to the competitive exclusion concept, when two (or more) species require identical resources, they cannot coexist indefinitely.

   b. Species are more likely to coexist if they differ in their use of resources. They may also coexist by using a shared resource in different ways or at different times.

5. Predator and prey densities often oscillate between low and high points of stability. The carrying capacity for prey, density dependencies, likely refuges for prey, predator efficiency, and often alternative prey sources influence the interactions. *47.4*

6. Predators and prey exert selection pressure on each other. After a novel, heritable trait appears and gives prey an edge, selection pressure works on the predator populations, which may evolve in response. The same occurs when a novel, heritable trait arises in a predator population. *47.4, 47.5*

   a. Evolved defenses by prey include threat displays, chemical weapons, mimicry, and camouflaging.

   b. Predators overcome defenses of prey by adaptive behavior (including stealth), camouflaging, and so on.

7. Parasites live in or on other living organisms—their hosts—and withdraw nutrients from host tissues for part of their life cycle. Hosts may or may not die as a result. Parasites and hosts coevolve; resistant hosts and only moderately harmful parasites are favored. *47.6*

8. By the classical model for ecological succession, a community develops in predictable sequence, from its pioneer species to an end array of species that persists over an entire region. *47.7*

9. A stable, self-perpetuating array of species that are in equilibrium with one another and with a particular environment is called a climax community. *47.7*

10. Similar climax stages may persist within the same region yet show variation as a result of environmental gradients and species interactions. *47.7*

11. Recurring, small-scale changes are an aspect of the internal dynamics of communities. Other changes, such as long-term shifts in climate and species introductions, may permanently alter community structure. *47.7, 47.8*

   a. Community structure reflects an uneasy balance of forces, including predation (as by keystone species) and competition, that have been operating over time.

   b. Community structure may change permanently by the introduction of species that one way or another expanded their geographic range. Dispersals may occur when a population expands its home range gradually. Jump dispersals rapidly put individuals into distant habitats. Extremely slow dispersals also have occurred over geologic time as a result of plate tectonics.

12. The number of species in a community depends on the size of a region, the colonization rate, disturbances, and extinction rates. It depends also on the levels and patterns of resource availability. *47.10*

13. Biodiversity tends to be highest in the tropics and to decline systematically toward polar regions, except during times of mass extinction. *47.10*

Review Questions

1. Distinguish between the habitat and the niche of a species. Why is it difficult to define "the human habitat"? *47.1*

2. Describe competitive exclusion. How might two species that compete for the same resource coexist? *47.3*

3. Characterize a predator and a parasite. *47.1, 47.4, 47.6*

4. Define primary and secondary succession. *47.7*

5. What is a climax community? How does the climax-pattern model help explain its structure? *47.7*

Self-Quiz   ANSWERS IN APPENDIX III

1. A habitat _____ .
   a. has distinguishing physical and chemical features
   b. is where individuals of a species normally live
   c. is occupied by various species
   d. both a and b
   e. a through c

2. A two-way flow of benefits in mutualistic interactions between species is an outcome of _____ .
   a. close cooperativeness   c. resource partitioning
   b. two-way exploitation   d. competitive coexistence

3. A niche _____ .
   a. is the sum of activities and relationships in a community by which individuals of a species secure and use resources
   b. is unvarying for a given species
   c. shifts in large and small ways
   d. both a and b
   e. both a and c

4. Two species in one habitat can coexist when they _____ .
   a. differ in their use of resources
   b. share the same resource in different ways
   c. use the same resource at different times
   d. all of the above

Figure 47.23 (**a**) A flesh fly and (**b**) one of the weevils (*Zygops*).

Figure 47.24 Water hyacinths choking a waterway in Florida.

5. A predator population and prey population _____ .
   a. always coexist at relatively stable levels
   b. may undergo cyclic or irregular changes in density
   c. cannot coexist indefinitely in the same habitat
   d. both b and c

6. Parasites _____ .
   a. tend to kill their hosts   c. feed on host tissues
   b. can kill novel hosts   d. both b and c

7. In _____ , a disturbed site in a community recovers and moves again toward the climax state.
   a. the area effect   c. primary succession
   b. the distance effect   d. secondary succession

8. The biodiversity of a given region is an outcome of _____ .
   a. climate and topography   d. both a and b
   b. possibilities for dispersal   e. a through c
   c. evolutionary history

9. Match the terms with the most suitable descriptions.
   _____ geographic dispersal   a. opportunistic colonizer of barren or disturbed places
   _____ area effect   b. dominates community structure
   _____ pioneer species   c. individuals leave home range, become successfully established elsewhere
   _____ climax community   d. more biodiversity on large islands than small at same distance from source
   _____ keystone species   e. stable, self-perpetuating array of species

Critical Thinking

1. Think of possible examples of competitive exclusion besides the ones used in this chapter, as by considering some of the animals and plants living in your own neighborhood.

2. Flesh flies (Figure 47.23*a*) have a gray and black body, red eyes, and red tail ends. They are fast fliers, and bird predators soon give up trying to catch them. Many sluggish insects, such as the weevil *Zygops rufitorquis*, resemble flesh flies (Figure 47.23*b*). This appears to be a case of _____ .

3. Frog populations all over the world declined dramatically during the 1990s. Individual frogs were being born with bizarre deformities, such as too few or too many legs (Sections 25.6 and 36.8). Should we view what is happening to frogs as a warning of drastically changing environments that will affect us, also? Are the same declines and deformities occurring in all aquatic

habitats, or are there different problems in different regions? Use the library or a computer search engine to find information. (Try these keywords: *frog deformities*.) Then write a brief list of possible answers to the preceding questions.

4. The water hyacinth (*Eichhornia crassipes*) is an aquatic plant native to South America. In the 1880s, someone fancied its blue flowers and displayed the plant at a New Orleans exhibition. Other flower fanciers took clippings and put them in ponds and streams. Unchecked by natural predators, the fast-growing plants spread through nutrient-rich waters, displaced many native species, and choked rivers and canals (Figure 47.24). They spread as far west as San Francisco. Do some research to see whether the United States now limits imports and, if so, how effective the restrictions have been.

Selected Key Terms

area effect,
   island biodiversity *47.10*
biodiversity *47.10*
camouflage *47.5*
carrying capacity *47.4*
climax community *47.7*
climax-pattern model *47.7*
coevolution *47.4*
commensalism *47.1*
community *47.1*
community structure *CI*
competitive exclusion *47.3*
distance effect,
   island biodiversity *47.10*
ecological succession *47.7*
endangered species *47.9*
exotic species *47.9*
geographic dispersal *47.8*
habitat *47.1*

host *47.6*
interspecific competition *47.1*
keystone species *47.8*
mimicry *47.5*
mutualism *47.1*
niche *47.1*
parasite *47.6*
parasitism *47.1*
parasitoid *47.6*
pioneer species *47.7*
predation *47.1*
predator *47.4*
prey *47.4*
primary succession *47.7*
resource partitioning *47.3*
secondary succession *47.7*
social parasite *47.6*
symbiosis *47.1*
warning coloration *47.5*

Readings

Krebs, C., et al. January 2000. "What Drives the 10-Year Cycle of Snowshoe Hares?" *Bioscience* 51(1): 25–35.

Ricklefs, R., and G. Miller. 2000. *Ecology*. Fourth edition. New York: Freeman. An excellent, accessible new edition.

48

ECOSYSTEMS

Crêpes for Breakfast, Pancake Ice for Dessert

Think of Antarctica and you think of ice. Mile-thick slabs of the stuff hide all but a small fraction of a vast continent whipped by fierce winds and kept frozen by murderously low temperatures, on the order of −100°F. Even so, bacteria, lichens, and mosses cling to life on patches of exposed rocky soil and on nearby islands. During the breeding season, a variety of penguins and seals form great noisy congregations, reproduce, and raise offspring. They cruise offshore or venture out in the open ocean in pursuit of krill, fishes, and squids.

The first explorers of Antarctica called it "The Last Place on Earth." For all of its harshness, the beauty of this ecosystem fired the imagination of those first travelers and others who followed them.

In 1961, thirty-eight nations signed a treaty to set aside Antarctica as a reserve for scientific research.

This was the start of scientific outposts and the trashing of Antarctica. At first, researchers discarded a few oil drums and old tires. Then prefabricated villages went up. Onto the ice and into the water went garbage, used equipment, sewage, and chemical wastes.

Antarctica even became a destination of cruise ships (Figure 48.1a). Every summer thousands of tourists, fortified by three sumptuous meals a day plus snacks, are ferried from ship to land across channels glistening with pancake ice. They trample the sparse vegetation and bob around the penguins, cameras clicking.

We do not know what tourism's long-term impact will be, but it pales when compared to other assaults. Nations looking for new sources of food turned their attention to Antarctica's krill. Word got out about potentially rich deposits of uranium, oil, and gold. By

Figure 48.1 (a) A boatload of tourists crossing a channel of "pancake ice" off the coast of Antarctica. (b) On the shore of an island near Antarctica, tourists get close to nature—maybe too close.

1988, treaty nations were poised to authorize digs and drillings. But oil spills or commercial harvesting of krill could easily destroy the fragile ecosystem. For instance, tiny, shrimplike krill are the only food source for Adélie penguins and a key food for baleen whales and other marine animals which are, in turn, food for still others.

In 1991, the treaty nations thought about all of this and imposed a fifty-year ban on mineral exploration. Research stations started to bury or incinerate wastes, treat raw sewage, and take other pollution-curbing steps. Tour operators promised to supervise tourists. Krill are not being harvested on a massive scale. Yet.

Will damage to one habitat or even one species lead to collapse of the whole? Because Antarctica is remote from the rest of the world, it is relatively easy to see how its species interconnect, and so we can be fairly sure of the answer. What about places not as sharply defined? Are they as vulnerable to disturbance or more resilient? We won't know for sure until researchers gain deeper insights into the evolutionary and ecological histories of their species, including the suspected great numbers of species we have not even discovered yet.

In the meantime, biodiversity is rapidly declining in ecosystems on land and in the seas. Competition from exotic species is one threat; the human propensity to overexploit species and destroy habitats is another. *Conservation biology* is an attempt to counter the threat by working to lower current extinction rates and design management programs that will help sustain the most vulnerable ecosystems. Chapter 27 introduced the challenges. To gain more insight into the magnitude of the task, start with a few principles of how ecosystems work. They are the topic of this chapter.

b

Key Concepts

1. An ecosystem is an association of organisms and their physical environment, interconnected by an ongoing flow of energy and a cycling of materials through it.

2. An ecosystem is an open system, with inputs, internal transfers, and outputs of both energy and nutrients.

3. Energy flows in only one direction through every ecosystem. Most commonly, energy flow begins when photosynthetic autotrophs harness sunlight energy and convert it to forms that they and other organisms of the ecosystem can use. Autotrophs are primary producer organisms for the ecosystem.

4. Energy-rich organic compounds that the primary producers synthesize become incorporated in their body tissues. Many of these compounds are stored forms of energy, and they serve as the foundation for the ecosystem's food webs. Such webs consist of a number of interconnected food chains.

5. Each chain in a food web is a linear sequence from producers through various consumers, decomposers, and detritivores.

6. Over time, most of the energy that enters a food web is lost to the environment, mainly in the form of metabolic heat. Most nutrients are cycled within food webs, but some amounts are lost to the environment.

7. The water, carbon, nitrogen, phosphorus, and other substances required for primary productivity move through biogeochemical cycles that are global in scale. Ions or molecules of the substances move slowly from environmental reservoirs, then among organisms of food webs, then back to the reservoirs.

THE NATURE OF ECOSYSTEMS

Overview of the Participants

Diverse natural systems abound on the Earth's surface. In climate, landforms, soil, vegetation, animal life, and other features, deserts differ from hardwood forests, which differ from tundra and prairies. In biodiversity and physical properties, the open seas differ from reefs, which differ from lakes. *Yet, despite the differences, such systems are alike in many aspects of structure and function.*

With few exceptions, these systems run on sunlight captured by autotrophs (self-feeders). Plants and other photosynthesizers are the most common autotrophs. They convert sunlight energy to chemical energy and use it to produce organic compounds from inorganic raw materials. By doing so, they are **primary producers** for the entire system (Figure 48.2).

All other organisms in the system are heterotrophs, not self-feeders. They extract energy from compounds that primary producers put together. **Consumers** feed on the tissues of other organisms. The consumers called *herbivores* eat plants and plant parts. Other consumers are *carnivores* that mostly consume flesh, and *parasites* that ingest tissues of living hosts but do not usually kill them. Nonliving products and the remains of producers and consumers feed **decomposers**. These heterotrophic fungi and bacteria practice extracellular digestion and absorption: they secrete enzymes that digest organic compounds in the surroundings, and their individual cells absorb some breakdown products. In this respect, decomposers differ from detritivores. **Detritivores**, such as earthworms and crabs, are heterotrophs that ingest decomposing bits of organic matter, like leaf litter.

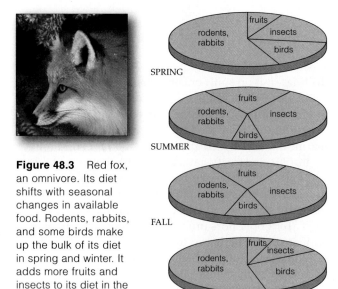

Figure 48.3 Red fox, an omnivore. Its diet shifts with seasonal changes in available food. Rodents, rabbits, and some birds make up the bulk of its diet in spring and winter. It adds more fruits and insects to its diet in the summer and fall.

Not all consumers fall into simple categories. Many are *omnivores*, which dine on assorted animals, plants, fungi, protistans, even bacteria. The red fox is like this (Figure 48.3). Also, a fox that comes across a dead bird may scavenge it. A *scavenger* is any animal that ingests dead plants, animals, or both, all or some of the time. Vultures, termites, and many beetles do this.

What are the nutrients for the system? The primary producers take up small inorganic compounds—water and carbon dioxide—from the environment as sources for hydrogen, oxygen, and carbon. They also take up phosphorus, nitrogen, and other minerals. These are the raw materials for carbohydrates, lipids, proteins, and nucleic acids. When the decomposers and detritivores work over the remains of organisms, many nutrients are released to the environment. And unless something removes them from the system, as when minerals end up dissolved in a stream that is flowing away from a meadow, autotrophs typically take them up again.

What we have just described in broad outline is an ecosystem. An **ecosystem** is an array of organisms and their physical environment, interacting through a one-way flow of energy and a cycling of materials. It is an open system, unable to sustain itself. It runs on *energy inputs*, as from the sun, and usually *nutrient inputs*, as from a creek delivering dissolved minerals into a lake. An ecosystem has *energy outputs* and *nutrient outputs*, too. Remember, energy cannot be recycled; over time, energy fixed by autotrophs is lost to the environment, mainly as metabolically generated heat. And some of the nutrients slip away. Most of this chapter explores ecosystem inputs, internal transfers, and outputs.

Figure 48.2
Simple model for ecosystems. Energy flows in one direction: into the ecosystem, through its living organisms, and then out from it. Its nutrients are cycled among autotrophs and heterotrophs. For this model, energy flow starts with autotrophs that can capture energy from the sun.

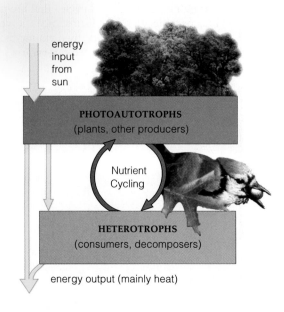

energy input from sun

PHOTOAUTOTROPHS
(plants, other producers)

Nutrient Cycling

HETEROTROPHS
(consumers, decomposers)

energy output (mainly heat)

Trophic Level	Category of Organism	Main Energy Sources	Examples of Organisms
FIFTH	**Fourth-level consumers** (heterotrophs): *Top carnivores, parasites, detritivores, decomposers*	Third-level consumers	Marsh hawks, mites, ticks, parasitic flies
FOURTH	**Third-level consumers** (heterotrophs): *Carnivores, parasites, detritivores, decomposers*	Second-level consumers	Upland sandpipers, crows, mites, ticks, parasitic flies
THIRD	**Second-level consumers** (heterotrophs): *Carnivores, parasites, detritivores, decomposers*	First-level consumers	Badgers, weasels, coyotes, spiders, garter snakes, meadow frogs
SECOND	**First-level consumers** (heterotrophs): *Herbivores, parasites, detritivores, decomposers*	Primary producers	Earthworms; moths, butterflies (adults and larvae;/ e.g., cutworms); clay-colored sparrows; voles; grasshoppers; pocket gophers; ground squirrels; saprobic fungi and bacteria
FIRST	**Primary producers** (autotrophs): *Photoautotrophs* *Chemoautotrophs*	Sunlight Inorganic substances	Grasses, composites, other flowering plants Nitrifying bacteria

Figure 48.4 Simplified picture of trophic levels in tallgrass prairie, a type of grassland in parts of the American Midwest.

Structure of Ecosystems

We can classify all organisms of an ecosystem by their functional roles in a hierarchy of feeding relationships called **trophic levels** (*troph*, nourishment). "Who eats whom?" we may ask. When organism **B** eats organism **A**, energy is transferred to **B** from **A**. All organisms at a particular trophic level are the same number of transfer steps away from some energy input into an ecosystem.

Consider organisms of a prairie ecosystem (Figure 48.4). Being first to tap energy (from the sun), plants and other producers are at the first trophic level. They are food for primary consumers, such as grasshoppers (herbivores) and earthworms (detritivores) at the next level. Primary consumers feed carnivores and parasites at the third trophic level, and so on up the hierarchy.

At each trophic level, organisms interact with the same sets of predators, prey, or both. The omnivores feed at several levels, so they are partitioned among different levels or assigned one of their own.

A **food chain** is a straight-line sequence of steps by which energy stored in autotroph tissues passes on to higher trophic levels. We have a hard time finding such simple, isolated chains in nature. Why? Species usually belong to more than one food chain, especially species at a low trophic level. It is more accurate to visualize food chains *cross-connecting* with one another, as **food webs**. And that is the topic of the next section.

Producers, consumers, decomposers, detritivores, and their physical environment make up an ecosystem. A one-way energy flow and a cycling of materials interconnect them.

A food chain, a straight-line sequence of who eats whom in an ecosystem, starts with autotrophs. A food web is a network of cross-connected food chains.

THE NATURE OF FOOD WEBS

Make a sketch of a food chain—say, from a plant in a tallgrass prairie, to a cutworm that eats its juicy parts, to a baby snake that eats the cutworm, to a sandpiper that eats the snake, and a hawk that eats the sandpiper. Your sketch should look something like Figure 48.5.

Identifying a food chain is a simple way to start thinking about who eats whom in ecosystems, but this is just part of the picture. Most often, a complex array of species competes for food, particularly at the lower trophic levels. For instance, as the food web in Figure 48.6 shows, tallgrass prairie

Figure 48.5 *Right:* Example of a simple food chain.

producers (flowering plants) feed insects and herbivorous mammals. Yet it is a hugely simplified picture of all the species interactions in this ecosystem!

How Many Energy Transfers?

Imagine separating a spider web's threads. Ecologists did this for many food webs in nature. When they compared the chains of different food webs, a pattern emerged: *In most cases, energy that producers initially captured passes through no more than four or five trophic levels.* It simply makes no difference how much energy goes into the ecosystem. Remember, transfers of energy can never be 100 percent efficient; a bit of energy is lost at each step (Chapter 6). In time, the energy that some organism would have to expend to capture another at a higher trophic level would be more than the amount of energy obtainable from it.

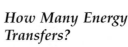

MARSH HAWK

UPLAND SANDPIPER

GARTER SNAKE

CUTWORM

PLANTS

HIGHER TROPHIC LEVELS

Complex array of carnivores, omnivores, and other consumers. Many feed at more than one trophic level all the time, seasonally, or when an opportunity presents itself

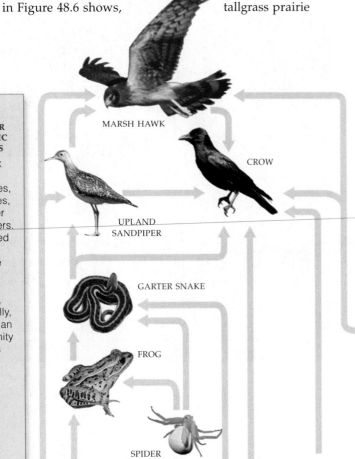

MARSH HAWK

CROW

UPLAND SANDPIPER

GARTER SNAKE

FROG

SPIDER

WEASEL  BADGER COYOTE

CLAY-COLORED SPARROW

PRAIRIE VOLE   POCKET GOPHER   GROUND SQUIRREL

SECOND TROPHIC LEVEL

Primary consumers (such as herbivores)

EARTHWORMS, INSECTS (E.G., GRASSHOPPERS, CUTWORMS)

FIRST TROPHIC LEVEL

Primary producers

GRASSES, COMPOSITES

Figure 48.6 Small, highly simplified sampling of connecting interactions in a tallgrass prairie food web.

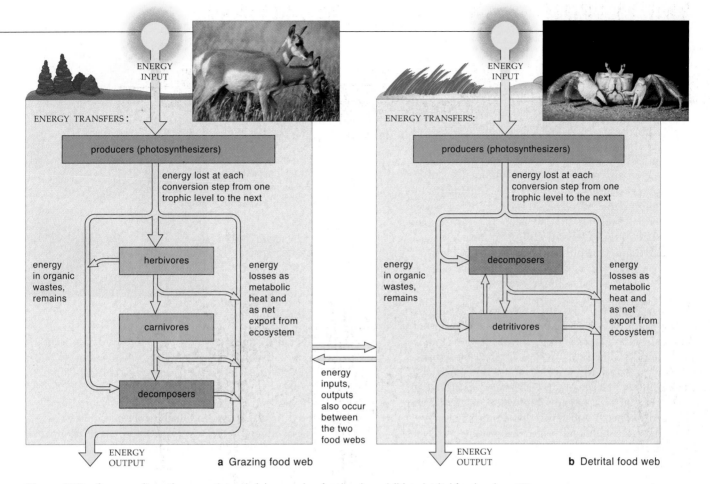

Figure 48.7 One-way flow of energy through (**a**) a grazing food web and (**b**) a detrital food web.

Even rich ecosystems with complex food webs, such as the one in Figure 48.6, do not have lengthy food chains.

Field studies and computer simulations of the food webs of marine, freshwater, and terrestrial ecosystems reveal more patterns. For instance, chains of food webs tend to be short where environmental conditions vary. Shifts in temperature, moisture, salinity, and pH are such variables. Chains are longer in stable environments, as in parts of the deep ocean. The most complex webs have the most herbivorous species, but their chains are shortest. You find such webs in grasslands. In contrast, simple food webs have more top carnivores.

Two Categories of Food Webs

In what direction does energy flow through ecosystems on land? Plants fix only a small part of the energy from the sun. They store half the energy in chemical bonds of new tissues but lose the rest as metabolic heat. Other organisms draw upon the energy that became stored in plant tissues, remains, and wastes. Then they, too, lose metabolic heat. Taken together, *all of these heat losses represent a one-way flow of energy out of the ecosystem.*

Energy from a primary source flows in one direction through two kinds of webs. In a **grazing food web**, it goes from photoautotrophs to herbivores, then through carnivores. By contrast, in a **detrital food web**, energy flows from photoautotrophs through detritivores and decomposers (Figure 48.7). Both webs cross-connect in nearly all ecosystems. For example, a connection forms when an opportunistic herring gull of a grazing food web gulps down a crab of a detrital food web.

The amount of energy moving through food webs differs from one ecosystem to the next and often varies with the seasons. In most cases, however, most of the net primary production moves through detrital food webs. You may doubt this. After all, when cattle graze heavily on pasture plants, about half of the net primary production enters a grazing food web. But cattle do not use all of the stored energy. A lot of undigested plant parts and feces become available for the decomposers and detritivores. Marshes are a similar case in point. In these aquatic ecosystems, most of the stored energy is not used until after parts of the marsh grasses die and become available for detrital food webs.

The loss of energy at each transfer in a food chain limits the number of trophic levels in ecosystems to four or five.

Tissues of living photosynthesizers are the basis of grazing food webs. Remains and wastes of photosynthesizers and consumers are the basis of detrital food webs.

Biological Magnification in Food Webs

ECOSYSTEM MODELING   We turn now to a premise that opened this chapter—that disturbances to one part of an ecosystem often can have unexpected effects on other, seemingly unrelated parts. One approach to predicting unforeseen effects of disturbances is through **ecosystem modeling**. With this method, researchers identify crucial bits of data on different ecosystem components, and then use computer programs and models to combine the bits. Results help predict outcomes of the next disturbance.

A potential danger is that investigators may not have identified all of the key relationships in the ecosystem under study and incorporated them accurately into a computer model. The most crucial fact may be one we do not yet know, as the example below makes clear.

DDT IN FOOD WEBS   DDT, recall, is a synthetic organic pesticide (Section 32.6). This fairly stable hydrocarbon is nearly insoluble in water, so you might think that it would exert its effects only where applied. But winds carry DDT in vapor form; water moves fine particles of it. Being highly soluble in fats, DDT can accumulate in tissues. Hence DDT can show **biological magnification**. By this occurrence, a substance that degrades slowly or not at all gets more and more concentrated in tissues of organisms at higher trophic levels of a food web.

Most of the DDT that becomes concentrated in all the organisms that a consumer will eat during its lifetime will end up in the consumer's own tissues. DDT and its modified forms disrupt metabolic activities and are toxic to many aquatic and terrestrial animals.

Figure 48.8   Peregrine falcon, a top carnivore of certain food webs. It almost became extinct as a result of biological magnification of DDT. A wildlife management program successfully increased its population sizes. Peregrine falcons were reintroduced into wild habitats. They also have adapted to cities. There, these raptors hunt pigeons, large populations of which are a messy nuisance.

Several decades ago, DDT started to infiltrate food webs and act on diverse organisms in ways that no one had predicted. Where it was sprayed to control Dutch elm disease, songbirds died. In small streams flowing in forests where it was sprayed to kill budworm larvae, fish died. In fields sprayed to control one kind of pest, new pests moved in. *DDT was indiscriminately killing the natural predators that keep pest populations in check!*

Then side effects of biological magnification showed up in habitats far removed from where DDT had been applied—*and much later in time.* Most vulnerable were brown pelicans, bald eagles, peregrine falcons, and other top carnivores of some food webs (Figures 48.8 and 48.9). Why? A product of DDT breakdown interferes with physiological processes. As one outcome, bird eggs developed brittle shells; many chick embryos did not even hatch. Some species faced extinction.

In the United States, DDT has been banned since the 1970s except where necessary to protect public health. Many species hit hardest have recovered, somewhat. Some birds still lay thin-shelled eggs; they pick up DDT at their winter ranges in Latin America. Even as late as 1990, a fishery near Los Angeles was closed. DDT from industrial waste discharges that had stopped twenty years before was still contaminating that ecosystem.

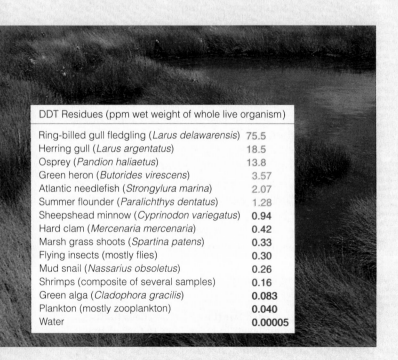

DDT Residues (ppm wet weight of whole live organism)	
Ring-billed gull fledgling (*Larus delawarensis*)	75.5
Herring gull (*Larus argentatus*)	18.5
Osprey (*Pandion haliaetus*)	13.8
Green heron (*Butorides virescens*)	3.57
Atlantic needlefish (*Strongylura marina*)	2.07
Summer flounder (*Paralichthys dentatus*)	1.28
Sheepshead minnow (*Cyprinodon variegatus*)	0.94
Hard clam (*Mercenaria mercenaria*)	0.42
Marsh grass shoots (*Spartina patens*)	0.33
Flying insects (mostly flies)	0.30
Mud snail (*Nassarius obsoletus*)	0.26
Shrimps (composite of several samples)	0.16
Green alga (*Cladophora gracilis*)	0.083
Plankton (mostly zooplankton)	0.040
Water	0.00005

Figure 48.9   Biological magnification in an estuary on the south shore of Long Island, New York, as reported in 1967 by George Woodwell, Charles Wurster, and Peter Isaacson. The researchers knew of broad correlations between the extent of DDT exposure and mortality. For instance, residues in birds known to have died from DDT poisoning were 30–295 ppm, and they were 1–26 ppm in several fish species. Some DDT concentrations measured during this study were below lethal thresholds but were still high enough to interfere with reproductive success.

STUDYING ENERGY FLOW THROUGH ECOSYSTEMS

What Is Primary Productivity?

To get an idea of how energy flow is studied, think of ecosystems on land, which typically have multicelled plants as primary producers. The rate at which the producers secure and store an amount of energy in their tissues during a specified interval is the ecosystem's **primary productivity**. How much energy actually gets stored depends on how many plants are living there and on the balance between photosynthesis (energy trapped) and aerobic respiration (energy used). All energy that gets trapped is the *gross* primary production. The *net* amount is all energy that accumulates in plants during growth and reproduction. Subtracting the energy used by plants and soil organisms from the gross primary production gives us the **net ecosystem production**.

Other factors impact net production, its seasonal patterns, and its distribution in the habitat. They do so for marine ecosystems, also (Figure 48.10). Example: Energy acquisition and storage depend in part on the body size and form of the primary producers, mineral availability, the temperature range, and the amount of sunlight and rainfall during each growing season. The harsher the conditions, the less new plant growth per season—and the lower the productivity.

What Are Ecological Pyramids?

Ecologists often will represent the trophic structure of an ecosystem in the form of an ecological pyramid. In such pyramids, the primary producers form a base for successive tiers of consumers above them.

A **biomass pyramid** depicts the weight of all of an ecosystem's organisms at each tier. Think about Silver Springs, Florida, a small aquatic ecosystem. A biomass pyramid for it, measured as grams per square meter during a specified interval, might look like this:

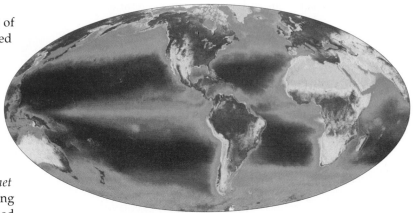

Figure 48.10 Summary of satellite data on global primary productivity from 1997 to August 2000. The most productive regions, such as rain forests, are coded *dark green*. Deserts (low productivity) are *yellow*. Productivity near the sea surface is *red* (highest) down through *orange, yellow, green,* and then *blue* (lowest). Compare the satellite maps in Section 7.8.

Some biomass pyramids are "upside-down." The smallest tier is on the bottom. A pond or the sea is like this. Its primary producers have less biomass than the consumers feeding on them. How is this possible? They consist of phytoplankton, the members of which grow and reproduce fast enough to support a much greater biomass of zooplankton. Organisms of zooplankton are bigger, grow slower, and consume less energy per unit weight. Phytoplankton fix more energy and produce more biomass than land plants (compare Section 7.8).

An **energy pyramid**, too, shows how usable energy diminishes as it flows through an ecosystem. Sunlight energy enters the pyramid base (first trophic level) and diminishes through successive levels to its tip (the top carnivores). Energy pyramids have a large energy base at the bottom and are always "right-side up." As you will see in the next section, they give a clearer picture of energy flow through successive trophic levels.

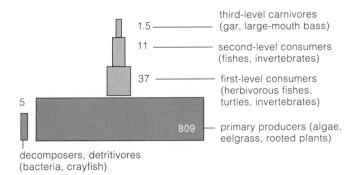

1.5 —— third-level carnivores (gar, large-mouth bass)

11 —— second-level consumers (fishes, invertebrates)

37 —— first-level consumers (herbivorous fishes, turtles, invertebrates)

5

809 —— primary producers (algae, eelgrass, rooted plants)

decomposers, detritivores (bacteria, crayfish)

Such biomass pyramids—mostly primary producers—are common. Top carnivores are few, which may be one reason why sighting them in the wild is so memorable.

Energy flows into food webs of ecosystems from an outside source, usually the sun. Energy leaves ecosystems mainly by losses of metabolic heat, which each organism generates.

Gross primary productivity is an ecosystem's total rate of photosynthesis during a specified interval. The *net* amount is the rate at which primary producers store energy in tissues in excess of their rate of aerobic respiration. Heterotrophic consumption affects the rate of energy storage.

The loss of metabolic heat and the shunting of food energy into organic wastes mean that usable energy flowing through consumer trophic levels declines at each energy transfer.

Energy Flow at Silver Springs

Imagine you are with ecologists who are bent on gathering data to construct an energy pyramid for a small freshwater spring over the course of one year. You observe them as they measure the energy that each type of individual in the spring takes in, loses as metabolic heat, stores in its body tissues, and loses in waste products. You see that they multiply the energy per individual by population size, then they calculate energy inputs and outputs. In such ways, these ecologists are able to express the flow per unit of water (or land) per unit of time.

The energy pyramid shown in Figure 48.11 summarizes the data from a long-term study of a grazing food web in an aquatic ecosystem—Silver Springs, Florida. The larger diagram in Figure 48.12 shows some of the calculations that ecologists used when they constructed the pyramid.

Given the metabolic demands of organisms and the amount of energy lost in their organic wastes, only about 6 to 16 percent of the energy entering one trophic level becomes available for organisms at the next level.

Because the efficiency of the energy transfers is so low, most ecosystems can support no more than four or five consumer trophic levels.

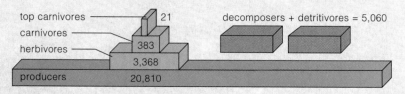

Figure 48.11 Pyramid of energy flow through Silver Springs, Florida, as measured in kilocalories/square meter/year.

Figure 48.12 Breakdown of the annual energy flow through Silver Springs, Florida, as measured in kilocalories/square meter/year.

Most of the primary producers in this small spring are aquatic plants. Most of the carnivores are insects and small fishes; the top carnivores are larger fishes. The original energy source, sunlight, is available all year long. The spring's detritivores and decomposers cycle organic compounds from the other trophic levels.

The producers trapped 1.2 percent of the incoming solar energy, and only a little more than a third of that amount became fixed in new plant biomass (4,245 + 3,368). The producers used more than 63 percent of the fixed energy for their own metabolism.

About 16 percent of the fixed energy was transferred to the herbivores, and most of this was used for metabolism or transferred to detritivores and decomposers.

Of the energy that transferred to herbivores, only 11.4 percent reached the next trophic level (carnivores). Carnivores used all but about 5.5 percent, which was transferred to top carnivores.

By the end of the specified time interval, all 20,810 kilocalories of energy that had been transferred through the system appeared as metabolically generated heat.

Bear in mind, this energy flow diagram is oversimplified, because no community is isolated from others. New individual organisms and substances continually drop from overhead leaves and branches into the springs. Also, organisms and substances are slowly lost by way of a stream that leaves the spring.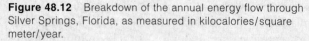

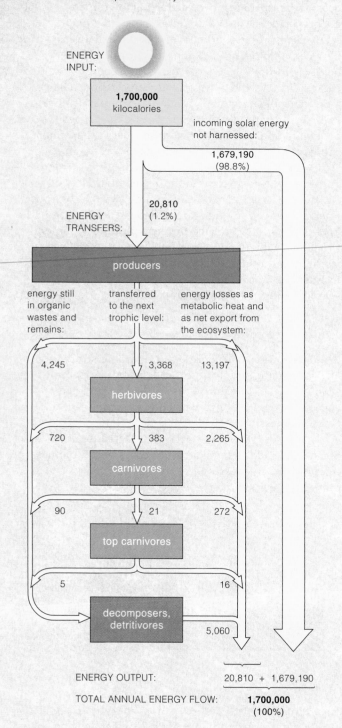

BIOGEOCHEMICAL CYCLES—AN OVERVIEW

Availability of nutrients as well as energy profoundly influences the structure of ecosystems. Photosynthetic producers need carbon, hydrogen, and oxygen, which they obtain from water and air. They require nitrogen, phosphorus, and other mineral ions. A scarcity of even one of these minerals has widespread adverse effects, for it lowers the ecosystem's primary productivity.

In a **biogeochemical cycle**, ions or molecules of a nutrient are transferred from the environment into organisms, then back to the environment, part of which functions as a vast reservoir for them. Transfer rates of nutrients into and from a reservoir are usually lower than the rates of exchange between and among organisms.

Figure 48.13 is a simple model for the relationship between most ecosystems and the geochemical part of the cycles. The model is based on four factors.

First, mineral elements that producer organisms use as nutrients typically are available in the forms of mineral ions, including ammonium (NH_4^+). *Second*, inputs from the physical environment, along with nutrient cycling activities of all of the decomposers and detritivores, maintain an ecosystem's reserves of nutrients. *Third*, the actual amount of a nutrient that is being cycled on through most major ecosystems is greater than the amount entering and leaving per year. *Fourth*, rainfall, snowfall, and the slow weathering of rocks are common sources of environmental inputs into the ecosystem's nutrient reserves. So are the combined effects of metabolic activities, such as nitrogen fixation.

Ecosystems on land also have typical outputs from the nutrient reserves. The loss of mineral ions from a river valley through ongoing erosion is an example.

There are three categories of biogeochemical cycles, based on the part of the environment that holds the largest portion of the specified ion or molecule. As you will see, in the *hydrologic* cycle, oxygen and hydrogen move in the form of water molecules. This movement is also known as the global water cycle. In *atmospheric*

cycles, a large percentage of the nutrient is in the form of an atmospheric gas. For instance, this is the case for gaseous forms of nitrogen and carbon (mainly carbon dioxide). *Sedimentary* cycles involve phosphorus and other solid nutrients that do not have gaseous forms. Solid nutrients move from land to the seafloor and return to dry land only by way of geological uplifting,

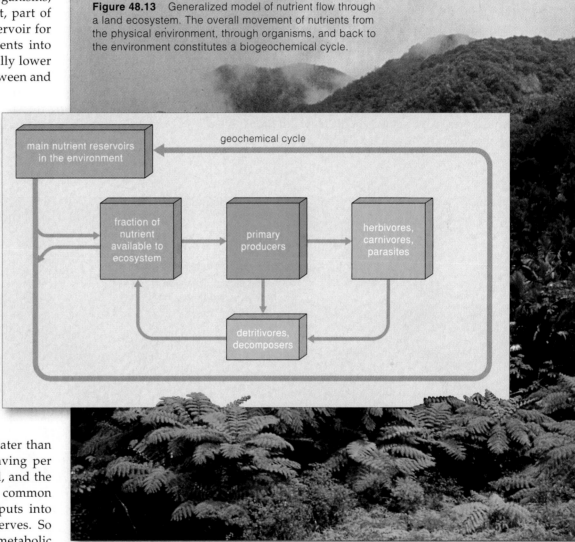

Figure 48.13 Generalized model of nutrient flow through a land ecosystem. The overall movement of nutrients from the physical environment, through organisms, and back to the environment constitutes a biogeochemical cycle.

geochemical cycle

main nutrient reservoirs in the environment

fraction of nutrient available to ecosystem

primary producers

herbivores, carnivores, parasites

detritivores, decomposers

which may take millions of years. The Earth's crust is the largest storehouse for sedimentary cycles.

Availability of nutrients profoundly influences the primary productivity on which ecosystems depend.

In a biogeochemical cycle, ions or molecules of a nutrient move slowly through the environment, then rapidly among organisms, and back to the environmental reservoir for them.

HYDROLOGIC CYCLE

Driven by ongoing inputs of solar energy, the Earth's waters move slowly, on a vast scale, from the ocean into the atmosphere, to land, and back to the ocean—the main reservoir. Water evaporating into the lower atmosphere initially stays aloft in the form of vapor, clouds, and ice crystals. It returns to the Earth in the form of precipitation—primarily rain and snow. Ocean currents and prevailing wind patterns influence the global **hydrologic cycle**, as shown in Figure 48.14.

Water is vital for all organisms. It also is a transport medium; it moves nutrients into and out of ecosystems. Its role in moving nutrients became clear in long-term studies of watersheds. A **watershed** is any region in which the precipitation becomes funneled into a single stream or river. Watersheds may be any size of interest to an investigator. For example, the Mississippi River watershed extends across about a third of the continental United States. Watersheds in the Hubbard Brook Valley of New Hampshire average 14.6 hectares (36 acres).

Most water entering a watershed seeps into soil or becomes part of surface runoff, which enters streams. Plants withdraw water and its dissolved minerals from the soil, then lose it by transpiration (Figure 48.15a).

Measuring watershed inputs and outputs has many practical applications. For instance, cities that draw upon a watershed's supply of surface water can adjust their usage in compliance with seasonal shifts in the volume of water. Measurements also reveal the extent to which vegetation cover influences the movement of nutrients through the ecosystem phase of biogeochemical cycles.

For example, you might think that surface runoff in a watershed would swiftly leach calcium ions and other minerals. But in young, undisturbed forests in Hubbard Brook watersheds, each hectare lost only 8 kilograms or so of its calcium. Also, the weathering of rocks and rainfall were replacing that lost calcium. In addition, tree roots were "mining" the soil, so calcium was being stored in a growing biomass of tree tissues.

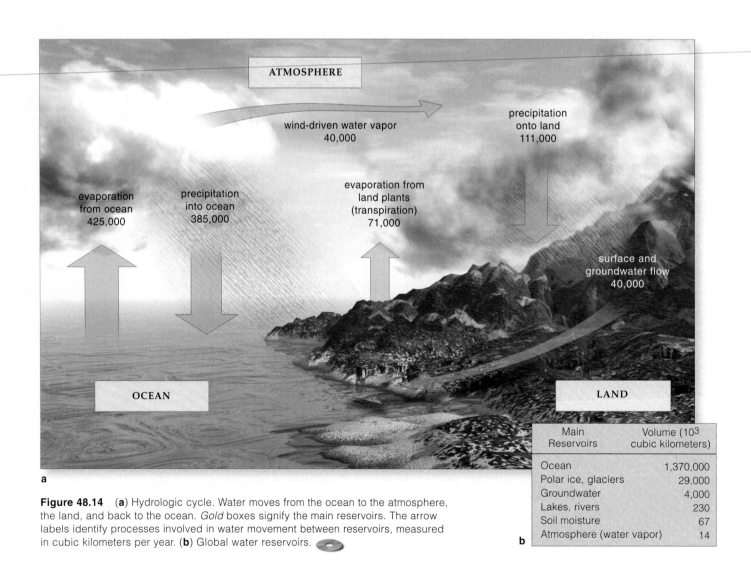

Figure 48.14 (**a**) Hydrologic cycle. Water moves from the ocean to the atmosphere, the land, and back to the ocean. *Gold* boxes signify the main reservoirs. The arrow labels identify processes involved in water movement between reservoirs, measured in cubic kilometers per year. (**b**) Global water reservoirs.

Main Reservoirs	Volume (10^3 cubic kilometers)
Ocean	1,370,000
Polar ice, glaciers	29,000
Groundwater	4,000
Lakes, rivers	230
Soil moisture	67
Atmosphere (water vapor)	14

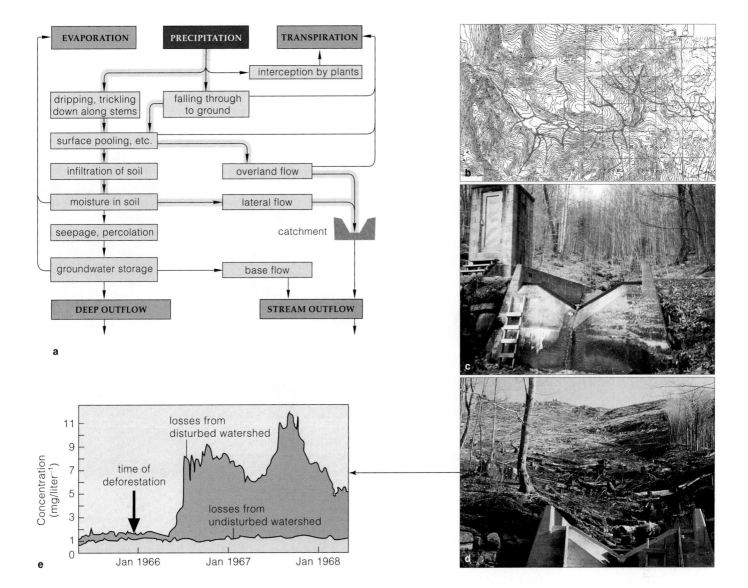

Figure 48.15 Effects of watershed disturbances. (**a**) Model for how water moves through watersheds in general. The *yellow* lines indicate the flow that researchers monitored during these experiments. The *dark blue* box is the input to the watershed; *light blue*, distribution within the watershed; and *medium blue*, outputs from it. Transpiration is the evaporation of water from leaves and other plant parts that are exposed to air. Plants tap water in the soil and in groundwater stores.

(**b**) Map of the Hubbard Brook Valley, New Hampshire. (**c**) One of the V-shaped concrete catchments where the overland flow was measured. (**d**) Researchers stripped the vegetation cover in forested plots but did not disturb the soil. Then they applied herbicides to the soil for three years to prevent regrowth. All of the surface water draining from each watershed was measured as it flowed over the catchments.

(**e**) An arrow marks the time of the experimental deforestation. Concentrations of calcium and other minerals were compared against those in water passing over a control catchment in an undisturbed area. The calcium losses were *six times* greater from the deforested area.

In experimental watersheds in the Hubbard Brook Valley, deforestation caused a shift in nutrient outputs. The results, summarized in Figure 48.15*e,* are sobering. Calcium and other nutrients cycle so slowly in nature that deforestation probably disrupts the availability of nutrients for an entire ecosystem. This is especially the case for forests that cannot regenerate themselves over the short term. Northern coniferous forests and tropical rain forests require long recovery times.

In the hydrologic cycle, water slowly moves on a global scale from the world ocean (the main reservoir), through the atmosphere, onto land, then back to the ocean.

In ecosystems on land, plants stabilize the soil and absorb dissolved minerals. By doing so, they minimize the loss of soil nutrients in runoff from land.

CARBON CYCLE

In an essential atmospheric cycle, carbon moves through the lower atmosphere and all food webs on its way to and from the ocean's water and sediments, and rocks. Its global movement is the **carbon cycle** (Figure 48.16). Sediments and rocks hold most of the carbon, followed by the ocean, then soil, the atmosphere, and biomass on land. It enters the atmosphere as cells engage in aerobic respiration, when fossil fuels burn, and when molten rock in the Earth's crust releases carbon during a volcanic eruption. Carbon dioxide (CO_2) is the most abundant form in the atmosphere. Carbon dissolved in the ocean is mainly in the forms of bicarbonate and carbonate.

Watch bubbles escape from a glass of carbonated soda left out in the sun, and you may wonder: Why doesn't all the CO_2 dissolved in the warm ocean surface waters escape to the atmosphere? Driven by winds and regional differences in water density, water makes a gigantic loop from the surface of the Pacific and Atlantic oceans on to the seafloor of the Atlantic and Antarctic seafloors. There, the dissolved CO_2 moves into deep storage reservoirs before the seawater loops back up (Figure 48.17). This loop of ocean water affects the distribution and global budget of carbon:

MAIN CARBON RESERVOIRS AND HOLDING STATIONS:

Sediments and rocks	77,000,000 (10^{15} grams)
Ocean (dissolved forms)	39,700
Soil	1,500
Atmosphere	750
Biomass on land	715

ANNUAL FLUXES IN GLOBAL DISTRIBUTION OF CARBON:

From atmosphere to plants (carbon fixation)	120
From atmosphere to ocean	107
To atmosphere from ocean	105
To atmosphere from plants	60
To atmosphere from soil	60
To atmosphere from fossil fuel burning	5
To atmosphere from net destruction of plants	2
To ocean from runoff	0.4
Burial in ocean sediments	0.1

When photosynthetic autotrophs engage in carbon dioxide fixation, they lock up billions of metric tons of carbon atoms in organic compounds each year (Sections

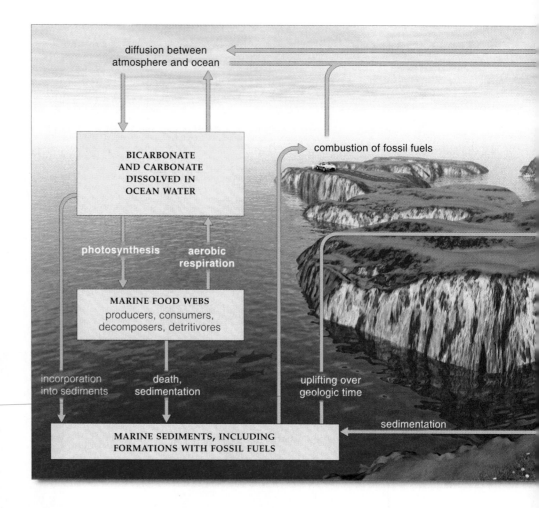

Figure 48.16 Global carbon cycle. The portion of the diagram on this page shows the movement of carbon through typical marine ecosystems. The *facing page* shows carbon's movement through ecosystems on land. The *gold* boxes indicate the main carbon reservoirs, the storage volumes of which are listed in the text.

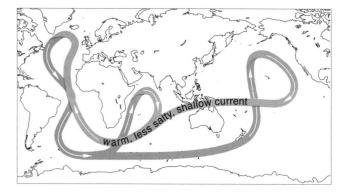

Figure 48.17 Loop of ocean water that delivers carbon dioxide to its deep ocean reservoir. It sinks in the cold, salty North Atlantic and rises in the warmer Pacific.

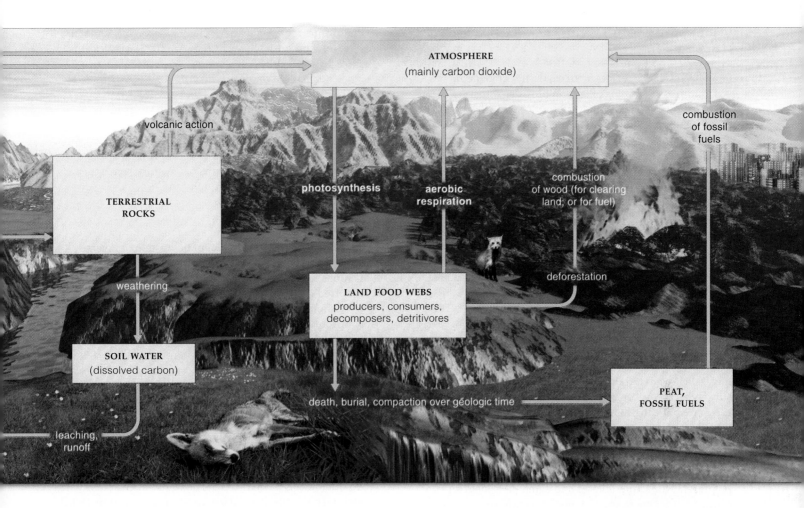

ATMOSPHERE
(mainly carbon dioxide)

volcanic action

combustion
of fossil
fuels

TERRESTRIAL
ROCKS

photosynthesis

aerobic
respiration

combustion
of wood (for clearing
land; or for fuel)

weathering

LAND FOOD WEBS
producers, consumers,
decomposers, detritivores

deforestation

SOIL WATER
(dissolved carbon)

death, burial, compaction over geologic time

PEAT,
FOSSIL FUELS

leaching,
runoff

7.8 and 48.4). Yet the average time that an ecosystem holds a given carbon atom varies greatly. For example, organic wastes and remains decompose so quickly in tropical rain forests that not much carbon accumulates at soil surfaces. In bogs, marshes, and other anaerobic sites, decomposers cannot degrade organic compounds to smaller bits, so the carbon gradually accumulates in peat and other forms of compressed organic matter.

Also, in food webs of ancient aquatic ecosystems, carbon became incorporated in staggering numbers of shells and other hard body parts. Foraminiferans and other shelled organisms died, sank through water, then were buried in sediments. The carbon remained buried for many millions of years in deep sediments until part of the seafloor was uplifted above the ocean surface by geologic forces of the sort introduced in Section 19.3. Remember those white cliffs of Dover (Section 22.3)?

As a final example, plants of the vast swamp forests of the Carboniferous also incorporated carbon atoms in their organic compounds. As you read in Section 23.4, those compounds were gradually converted into coal, petroleum, and gas reserves—which humans now tap as fossil fuels. Over the course of a few hundred years,

we have been extracting and burning fossil fuels that took millions of years to form. Here is a major problem: Fossil fuels are nonrenewable resources. They also are the energy source of choice for a huge number of the 6.2 billion people now on Earth—mainly in developed countries, which can afford them. At present, fossil fuel burning and some other human activities are releasing more carbon into the atmosphere than can be naturally cycled to the ocean's storage reservoirs. We will return to this topic in Chapter 50.

As you might deduce from the amounts listed on the preceding page, the ocean can remove only about 2 percent of the excess carbon entering the atmosphere. Most scientists now think the excess is amplifying the greenhouse effect, which means it may be contributing to global warming. The next section looks at this effect and some possible outcomes of modifications to it.

The ocean and atmosphere interact in the global cycling of carbon. Fossil fuel burning and other human activities appear to be contributing to imbalances in the global carbon budget.

From Greenhouse Gases to a Warmer Planet?

GREENHOUSE EFFECT The atmospheric concentrations of gaseous molecules play a profound role in shaping the average temperature near the surface of the Earth. That temperature has enormous effects on the global climate.

Countless molecules of carbon dioxide, water, ozone, methane, nitrous oxide, and chlorofluorocarbons are key players in interactions that dictate the global temperature. Collectively, the gases act somewhat like a pane of glass in a greenhouse—hence their name, "greenhouse gases." Wavelengths of visible light pass around them and reach the Earth's surface. But greenhouse gases impede the escape of longer, infrared wavelengths—heat—from the Earth into space. How? Gaseous molecules can absorb these wavelengths and radiate much of the absorbed energy back toward the Earth, as shown in Figure 48.18.

The constant radiation of heat energy from greenhouse gases toward Earth proceeds lockstep with the constant bombardment and absorption of wavelengths from the sun, and so heat builds up in the lower atmosphere. The **greenhouse effect** is the name for this warming action.

GLOBAL WARMING DEFINED Without the action of greenhouse gases, the Earth's surface would be cold and lifeless. However, there can be too much of a good thing. Largely as an outcome of human activities, greenhouse gases are building to atmospheric levels that are higher than they were in the past. Figure 48.19 documents the increase. As many researchers suspect, greenhouse gases may be contributing to long-term higher temperatures at the Earth's surface, an effect called **global warming**. At present, the United States and some other developed countries are generating the most greenhouse gases.

EVIDENCE OF AN INTENSIFIED GREENHOUSE EFFECT In the 1950s, researchers on Hawaii's highest volcano started measuring the atmospheric concentrations of different greenhouse gases. That remote site is almost free of local airborne contamination and is representative of conditions for the Northern Hemisphere. The Chapter 3 introduction outlined what they found. Briefly, carbon dioxide levels follow the annual cycles of plant growth. They decline in summer, when photosynthesis rates are highest. They rise in winter, when photosynthesis declines but aerobic respiration is still proceeding.

The troughs and peaks along the graph line in Figure 48.20a show the annual lows and highs. For the first time, scientists saw the integrated effects of carbon balances for land and water ecosystems of an entire hemisphere. Notice the midline of the troughs and peaks in the cycle. *And notice that the concentration is steadily increasing.*

Current data from satellites, weather stations and balloons, research ships, and supercomputer programs suggest we are past the point of being able to reverse some impacts of climate change. In 2001, the United Nations' Intergovernmental Panel on Climate Change reported that global warming and climate change are real. Human activities, including fossil fuel burning, are contributing factors. In the past decade—the warmest on record—hurricanes, floods, tornadoes, and droughts were intense. Alaska saw rare thunderstorms; Hilo, Hawaii, saw homes washed away by 27 inches of rainfall in 24 hours.

CAUSES AND EFFECTS: SOME PREDICTIONS If the lower atmosphere's temperature spikes 2.5 to 10.4 degrees higher in this century, we might see these outcomes:

Rising sea level. If the atmosphere warms by only 4°C (7°F), the sea level will rise by about 0.6 meter (2 feet). Why? Sea surface temperature will increase, and water expands when heated. Glaciers and polar ice sheets will melt faster (*Critical Thinking* question 5 in Section 48.12). The volume of water released would flood low coastal regions. High tides and storm waves will worsen.

Increased rainfall and flooding. Global warming will change regional rainfall patterns. Much of the Northern Hemisphere will get more rain. Snow will melt faster in mountain ranges, so spring runoff and summer droughts

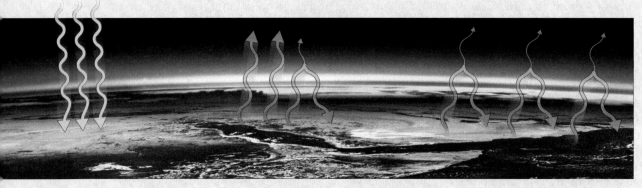

a The sun's rays penetrate the lower atmosphere, warm the Earth's surface.

b The surface radiates heat (infrared wavelengths) to the atmosphere. Some heat escapes into space. But greenhouse gases and water vapor absorb some infrared energy and radiate a portion of it back toward Earth.

c Increased concentrations of greenhouse gases trap more heat near Earth's surface. Sea surface temperature rises, more water evaporates into atmosphere. Earth's surface temperature rises.

Figure 48.18 The greenhouse effect, as executed mainly by carbon dioxide, water vapor, ozone, methane, nitrous oxide, and various chlorofluorocarbons present in the lower atmosphere.

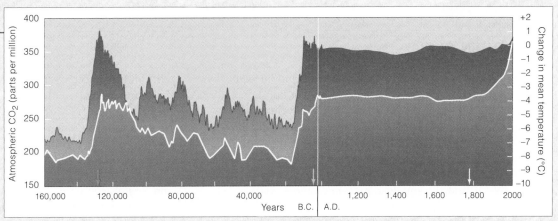

Figure 48.19 Correlation of changes in atmospheric CO₂ levels (*white* graph line) with ice ages and interglacials. The 1985 mean temperature was the baseline for tracking global changes in temperature (*red* graph line). Arrows show start of the last ice age (*red*), last interglacial (*yellow*), and the industrial revolution (*white*).

will be more severe. Already bad floods and mudslides may get even worse in the tropics and California. The interiors of the continents will become drier, and deserts will expand. Even now, severe droughts, storms, and flooding are battering the croplands and cities of the American Midwest.

Heat waves and wildfires. Heat waves will intensify in the Northern Hemisphere, and deaths from heat stroke will probably climb in cities ranging from Shanghai to Chicago. Hotter, drier summers will also invite more wildfires in Alaska, California, Florida, Brazil, and Asia.

Crop failures. Droughts will intensify in the United States, Mexico, Brazil, Africa, Europe, and Asia. Major crop losses will increase global malnutrition and famine.

Contagious disease. Warmer, wetter conditions along coasts will invite cholera, malaria, and other diseases. The West Nile virus, introduced into the United States in 1999, is expanding its range across the continent.

Water wars. By 2015, an estimated 3 billion people will run out of fresh water (Section 50.7).

Some argue that the global climate is too complex and unpredictable for us to accept data gathered so far. The majority now agree there is cause for concern, and we already see actions being taken. Chapter 50 outlines some of them, such as developing energy alternatives to fossil fuels. Are the actions none too soon? Future climate changes might turn out to be twice as violent as scientists were predicting just five years ago.

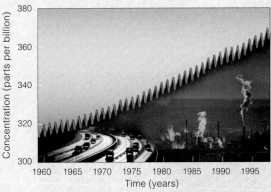

a Carbon dioxide (CO₂). Of all human activities, the burning of fossil fuels and deforestation contribute the most to increasing atmospheric levels.

c Methane (CH₄). Termite activity and anaerobic bacteria living in swamps, landfills, and the stomachs of cattle and other ruminants produce great amounts of methane as a by-product of metabolism.

b CFCs. Until restrictions were in place, CFCs were widely used in plastic foams, refrigerators, air conditioners, and industrial solvents.

d Nitrous oxide (N₂O). Denitrifying bacteria produce N₂O in metabolism. Also, fertilizers and animal wastes release enormous amounts; this is especially the case for large-scale livestock feedlots.

Figure 48.20 Recent increases in atmospheric concentrations of four greenhouse gases.

NITROGEN CYCLE

Since the time of life's origin, the atmosphere and ocean have contained nitrogen. This component of all proteins and nucleic acids moves in an atmospheric cycle called the **nitrogen cycle**. Gaseous nitrogen (N_2) makes up 80 percent or so of the atmosphere, the largest reservoir. Successively smaller reservoirs are seafloor sediments, ocean water, soil, biomass on land, atmospheric nitrous oxide (N_2O), and the smallest, marine biomass.

The atoms of N_2 are joined by triple covalent bonds ($N \equiv N$). Few organisms have the metabolic means to break them. Only certain bacteria, volcanic action, and lightning convert N_2 into forms that enter food webs.

Of all nutrients required for plant growth, nitrogen often is the scarcest. Nearly all of the nitrogen in soils was put there by nitrogen-fixing organisms. Ecosystems lose it when other bacteria "unfix" the fixed nitrogen. Land ecosystems lose more by leaching, although this is the basis of nitrogen inputs into aquatic ecosystems such as streams, lakes, and the seas (Figure 48.21).

Cycling Processes

Let's follow nitrogen atoms through the portion of the nitrogen cycle that proceeds in an ecosystem on land. They are objects of processes called nitrogen fixation, assimilation and biosynthesis, decomposition, and then ammonification.

In **nitrogen fixation**, a few kinds of bacteria convert N_2 to ammonia (NH_3). This quickly dissolves inside the cytoplasm, thus forming ammonium (NH_4^+). In aquatic ecosystems, *Anabaena, Nostoc,* and other cyanobacteria are nitrogen fixers. In many land ecosystems, *Rhizobium* and *Azotobacter* fix nitrogen. Collectively, these bacteria fix about 200 million metric tons of nitrogen each year!

Decomposition and **ammonification** are processes by which bacteria and fungi degrade nitrogenous wastes and remains of organisms. Decomposers use part of the released proteins and amino acids during metabolism. But most of the nitrogen remains in the decay products, in the form of ammonia and ammonium, which plants take up. Nitrifying bacteria, too, act on the ammonia or ammonium. In **nitrification**, they strip electrons from compounds, and nitrite (NO_2^-) is the result. Other kinds of bacteria use the nitrite during their metabolism and produce nitrate (NO_3^-), which plants then take up.

Most plants growing in nitrogen-poor soil interact as mutualists with fungi. Their roots form mycorrhizae with fungal hyphae, the collective surface area of which absorbs a lot of mineral ions from soil. Clover, beans, peas, and other legumes are mutualists with nitrogen-

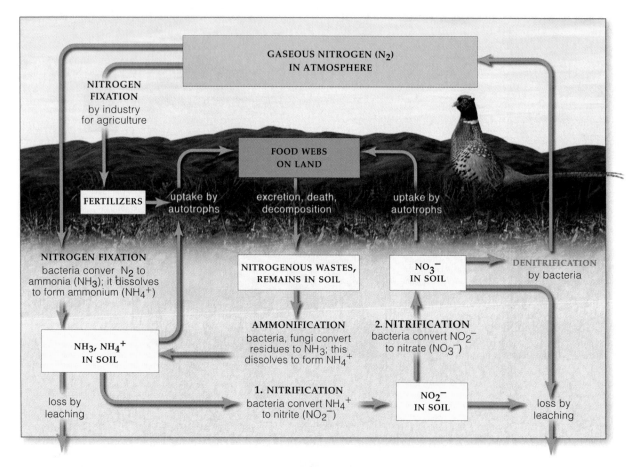

Figure 48.21 The nitrogen cycle in an ecosystem on land. Activities of nitrogen-fixing bacteria make nitrogen available to plants. Other bacterial species cycle nitrogen to plants. They break down organic wastes to ammonium and nitrates. The largest nitrogen reservoir by far is the atmosphere (mainly N_2), followed by ocean sediments, ocean water, and soil. Lesser amounts of nitrogen are in the form of nitrous oxide (N_2O) in the atmosphere and in the biomass of the seas.

fixing bacteria that form root nodules. Chapter 24 and Section 30.2 describe these interactions. How do plants use the nitrogen provided by bacterial activities? Their cells assimilate nitrogen and channel much of it into synthesizing amino acids, proteins, and nucleic acids. Plant tissues, remember, are the sole nitrogen source for animals, which directly or indirectly feed on plants.

Figure 48.22 Dead and dying spruce trees in a forest in Germany. They are cited as casualties of extremely high concentrations of nitrogen oxides and other forms of air pollution.

Nitrogen Scarcity

Ammonium, nitrite, and nitrate that form during the nitrogen cycle are vulnerable to leaching and runoff. Leaching, recall, is the removal of some soil nutrients as water percolates through it (Section 30.1). Besides losses from leaching, some nitrogen is lost to the air by **denitrification**. By this process, certain bacteria convert nitrate or nitrite to N_2 and N_2O (nitrous oxide). Usually, most denitrifying bacteria rely on aerobic respiration. In waterlogged or poorly aerated soil, they use anaerobic pathways in which nitrate, nitrite, or N_2O (not oxygen) is the final electron acceptor. Section 8.5 describes this metabolic strategy. During the reactions, fixed nitrogen is converted to N_2, much of which escapes to the air.

Also, nitrogen fixation comes at high metabolic cost to plants that are mutualists with the nitrogen fixers. In exchange for nitrogen, plants give up sugars and other photosynthetic products that require large investments of ATP and NADPH. Such plants have a competitive edge in nitrogen-poor soil. In nitrogen-rich soil, species that don't pay the metabolic price often displace them.

Human Impact on the Nitrogen Cycle

Humans are having disproportionately bad effects on the nitrogen cycle, compared to other species. Where deforestation and grassland conversion for agriculture are rampant, losses of nitrogen from soil are huge. With each forest clearing and harvest, nitrogen stored in the plants is lost. Soil erosion and leaching take away more.

Many farmers counter nitrogen losses by rotating crops, as by alternating wheat with legumes. Rotation can help keep soil stable and productive. In developed countries, farmers also spread nitrogen-rich fertilizers. They even select new strains of crop plants that have a greater capacity to take up fertilizers from soil. By such practices, the crop yields per hectare have doubled and even quadrupled over the past forty years.

But heavy fertilizer applications alter soil water and can harm plants. Remember, the surface charges of soil particles attract ions that weathering, irrigation, and fertilizers introduce to soils. By a pH-dependent process called **ion exchange**, ions dissociate from soil particles, and other ions dissolved in soil water replace them. Any change in the composition of soil water disrupts the number and kinds of ions in exchangeable form—which are the kinds that plants take up. Calcium and magnesium are the most abundant exchangeable ions. When nitrogen fertilizers are added to soil, ammonium also assumes importance. This is especially the case for ammonium sulfate $(NH_4)_2SO_4$, for all of its nitrogen is in the form of ammonium. Heavy applications of such a fertilizer greatly increase the acidity of soil water. Its many hydrogen ions compete for binding sites on soil particles. One outcome is that the levels of calcium and magnesium ions necessary for plant growth decline.

Humans introduce excess nitrogen into ecosystems in other ways, as when nitrogen-rich sewage flows into rivers, lakes, and estuaries. Section 50.7 has details on this. Fossil fuel burning by power plants and by vehicles with combustion engines also releases NO_2 and other nitrogen oxides into the air (Section 50.1). Winds carry the air pollutants to forests at high elevations and high northern latitudes, which have nitrogen-poor soils.

Nitrogen is a limiting factor on tree growth, and the forest trees are adapted to storing and cycling it. With more nitrogen, they grow faster well into summer. But the late growth is vulnerable to autumn frost and dies during the winter. In addition, the nitrogen-stimulated growth outstrips availability of phosphorus and other nutrients, so the trees soon show the effects of nutrient deficiencies. Their leaves yellow and drop, and rates of photosynthesis decline. That decline has repercussions on the growth of all plant parts. Weakened trees are far more susceptible to diseases and other environmental stresses (Figure 48.22).

The nitrogen cycle turns on processes of nitrogen fixation, assimilation and biosynthesis by all organisms, and on decomposition. Its natural cycling depends on the activity of nitrogen-fixing bacteria and on mycorrhizae.

Ecosystems lose nitrogen naturally by the activities of denitrifying bacteria, which convert nitrite or nitrate to forms that escape into the atmosphere.

Ecosystems often lose or gain too much nitrogen through human activities, in ways that compromise plant growth.

SEDIMENTARY CYCLES

A Look at the Phosphorus Cycle

We continue our survey of biogeochemical cycling with one of the key sedimentary cycles. In the **phosphorus cycle**, phosphorus passes quickly through food webs as it moves from land, to ocean sediments, and then back to land. The Earth's crust is the largest reservoir, just as it is for other minerals (Figure 48.23).

Rock formations incorporate phosphorus mainly as phosphate ions (PO_4^{3-}). By weathering and soil erosion, phosphates enter streams and rivers, then enter ocean sediments. There they slowly accumulate together with other minerals. They form insoluble deposits mainly on submerged shelves of continents. Millions of years go by. In regions where movements of crustal plates uplift part of the seafloor, phosphates become exposed on drained land surfaces. Over time, weathering and erosion release phosphates from exposed rocks. And so the cycle's geochemical phase begins again.

The ecosystem phase of the cycle is more rapid than the long-term geochemical phase. All organisms use phosphorus for synthesizing phospholipids, NADPH, ATP, nucleic acids, and other compounds. Plants take up dissolved phosphates rapidly and efficiently from soil water. Herbivores get phosphates by eating plants; carnivores get them by eating herbivores. All animals excrete phosphates as a waste in urine and feces.

Bacterial and fungal decomposers in the soil release phosphates, then plants take up dissolved forms of this mineral. By this action, plants help cycle phosphorus rapidly through an ecosystem.

Eutrophication

Sedimentary cycles, in combination with the hydrologic cycle, move most mineral elements through ecosystems on land and in water. Rainfall that evaporates from the ocean falls over land. And on its way back to the ocean, it transports silt and dissolved minerals, which primary producers use as nutrients.

Of all minerals, phosphorus is the most prevalent limiting factor in natural ecosystems around the world. Soils hold little of it, and sediments tie up most of it in aquatic ecosystems. It helps that phosphorus does not have a gaseous phase, as nitrogen does, so ecosystems lose little of their scarce supplies to the atmosphere. It helps that phosphorus resists leaching. Also, as you might predict, producers are adapted to respond rather efficiently to low phosphorus concentrations.

Some human activities are lowering the phosphorus levels of natural ecosystems. This is especially the case for tropical and subtropical regions of the developing countries. There, highly weathered soils have very little phosphorus. Enough is stored in biomass and is slowly released from decomposing organic matter to sustain undisturbed forests or grasslands. Even so, when people harvest trees and clear grassland for agriculture, they deplete the phosphorus. Crop yields start out low and soon become nonexistent. Fields are quickly abandoned, but regrowth of natural vegetation is sparse. About 1 to 2 billion hectares are already depleted of phosphorus, mostly where impoverished farmers do not engage in soil management practices and cannot afford fertilizers.

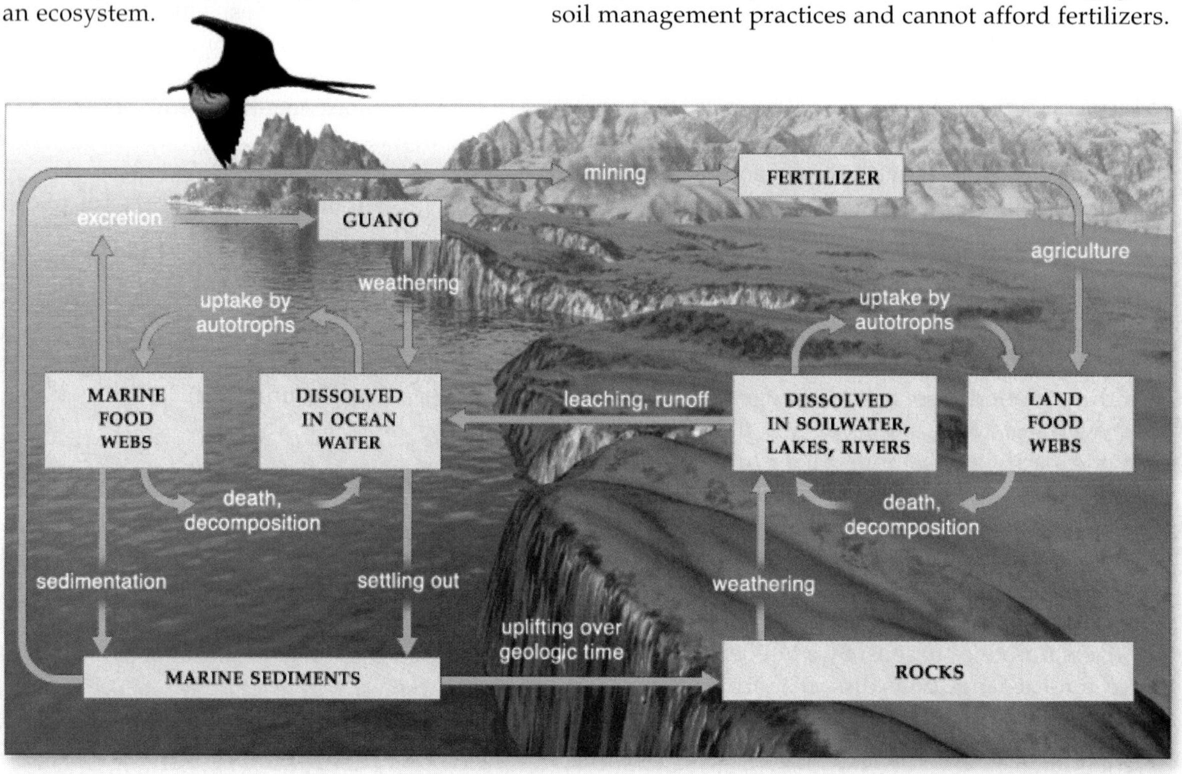

Figure 48.23
Phosphorus cycle. In this sedimentary cycle, most of the phosphorus moves in the form of phosphate ions (PO_4^{3-}) and reaches the oceans. Phytoplankton of marine food webs take up some of it, then fishes eat the plankton. Seabirds eat the fishes, and their droppings (guano) accumulate on islands. Guano is collected and used as a phosphate-rich fertilizer.

Figure 48.24 Experiment demonstrating lake eutrophication in Ontario, Canada. Researchers stretched a plastic curtain across a channel between two basins of the same lake. They added phosphorus, carbon, and nitrogen to one basin (the basin in the *background*) and carbon and nitrogen to the other (the basin in the *foreground*). Within two months, the phosphorus-enriched basin showed signs of accelerated eutrophication; a dense algal bloom turned the water green.

Developed countries have a different problem. Soils are overloaded with phosphorus after years of heavy applications of phosphorus. Without close monitoring, this mineral becomes concentrated in eroded sediments and runoff from agricultural fields. Phosphorus also is present in outflows from sewage treatment plants and from industries. It is present as well in the runoff from cleared land, even from lawns.

Dissolved phosphorus that enters streams, rivers, lakes, and estuaries can promote dense algal blooms. Like plants, photosynthetic algae cannot grow without phosphorus, nitrogen, and other mineral ions. In many freshwater ecosystems, nitrogen is not a limiting factor on algal growth because of the abundance of nitrogen-fixing bacteria. Phosphorus is. When decomposers act on the remains from algal blooms, they deplete water of oxygen, which kills fishes and other organisms.

Eutrophication is the name for nutrient enrichment of an ecosystem that is naturally low in nutrients. As you will be reading in Section 49.10, eutrophication is a natural process, but phosphorus inputs accelerate it. Field experiments, including the one shown in Figure 48.24, nicely demonstrate this outcome.

Sedimentary cycles, in combination with the hydrologic cycle, move most mineral elements, such as phosphorus, through terrestrial and aquatic ecosystems.

Agriculture, deforestation, and other human activities upset the nutrient balance of an ecosystem when they accelerate the amount of minerals entering and leaving it.

SUMMARY   *Gold* indicates text section

1. An ecosystem consists of an array of producers, consumers, detritivores, and decomposers along with their environment. It is an open system, with inputs, internal transfers, and outputs of energy and nutrients. There is a one-way flow of energy into and out from it and a cycling of materials among its organisms. *48.1*

2. Sunlight is the initial energy source for nearly all ecosystems. Photoautotrophs (the primary producers) convert that energy to other forms, such as chemical bond energy of ATP. They also assimilate nutrients that the ecosystem's heterotrophs require. *48.1*

    a. Many heterotrophs are consumers. The herbivores feed upon algae and plants. Carnivores ingest animals. Parasites withdraw nutrients from the tissues of living hosts. Omnivores use a variety of food sources.

    b. Decomposers also are heterotrophs. Most (certain fungi and bacteria) obtain energy and nutrients from organic remains and wastes. Detritivores, such as crabs and earthworms, are heterotrophs that ingest bits of decomposing or dead organisms and organic products.

3. In ecosystems, feeding relationships are structured as a hierarchy of energy transfers (trophic levels). *48.1*

    a. Primary producers are at the first trophic level; consumers are at successively higher trophic levels.

    b. Herbivores are second-level consumers; carnivores and parasites are at higher levels. The detritivores and decomposers are assigned to all trophic levels above the first level. So are the parasites. Humans and other omnivores obtain energy from more than one source, so they cannot be assigned to a single trophic level.

4. A food chain is a straight-line sequence by which energy stored in autotroph tissues is passed on through higher trophic levels. Most cross-connect as food webs. In grazing food webs, energy is transferred from the primary producers to herbivores, then carnivores. In detrital food webs, energy is transferred from primary producers through detritivores and decomposers. *48.2*

5. The amount of useful energy that flows through successive levels of consumers declines at each energy transfer in a food web because of metabolic heat loss and shunting of food energy into organic wastes. This inefficiency typically limits the number of trophic levels in a food web to no more than four or five. *48.2*

    a. In environments that show variations in salinity, temperature, and other environmental conditions, food webs tend to have short chains. In stable environments, such as parts of the deep ocean, food chains are longer.

    b. The most complex food webs have short chains; the simplest have the most top carnivores.

6. Disturbance to one aspect of an ecosystem usually has unexpected effects on other, seemingly unrelated parts. To predict consequences of a disruption, as by computer modeling, researchers attempt to identify all

the potential relationships and incorporate them into their model. *48.3*

7. Primary productivity is the rate at which the primary producers store some amount of energy in a specified interval. The net production for the ecosystem is the gross primary production minus the energy that plants and soil organisms use for themselves. *48.1, 48.4*

8. An ecosystem's primary productivity and structure require a cycling of nutrients as well as energy inputs. Water and minerals pass on through the environment, they pass among organisms, and then move back to the environment in biogeochemical cycles. *48.6*

9. Water moves through a hydrologic cycle. In land ecosystems, plants stabilize soil and minimize nutrient loss during the cycle, as by runoff. In a sedimentary cycle, Earth's crust is the main reservoir for a nutrient that has a solid form (for example, phosphorus). In atmospheric cycles, the nutrient is mainly in gaseous form. Examples: carbon in carbon dioxide, nitrogen in gaseous nitrogen. *48.6, 48.7*

10. In the carbon cycle, carbon moves through the ocean (its main reservoir), the atmosphere (as carbon dioxide, mostly), and ecosystems. Fossil fuel burning, logging, and the conversion of natural ecosystems for crops and grazing affect the global carbon budget. *48.1, 48.8*

11. Like water, ozone, CFCs, methane, and nitrous oxide, carbon dioxide is a greenhouse gas. Fossil fuel burning

and other human activities that release carbon to the atmosphere may be adding to global warming. *48.9*

12. For land ecosystems, nitrogen is a limiting factor on the total net primary productivity. Gaseous nitrogen is abundant in the atmosphere. Nitrogen-fixing bacteria convert N_2 to ammonia and nitrates, which producers take up. Mycorrhizae and root nodules, two symbiotic interactions, enhance the nitrogen uptake. *48.10*

13. Sedimentary cycles and the hydrologic cycle interact and move mineral nutrients on a global scale. *48.11*

    a. Certain human activities are depleting minerals from ecosystems, as when weathered soils of tropical forests are cleared for agriculture.

    b. Some human activities accelerate the process of eutrophication because they discharge nutrients into aquatic ecosystems and promote algal blooms.

Review Questions

1. Define an ecosystem in terms of inputs and outputs. *48.1*

2. Define primary producer, consumer, decomposer, and detritivore. Identify a few in Figure 48.25. *48.1*

3. Define and describe trophic levels. Which class of organisms is farthest from the energy input into an ecosystem? *48.1*

4. Define food chain and food web. How do grazing food webs differ from detrital food webs? *48.1, 48.2*

5. Distinguish between biomass pyramid and energy pyramid. What did the pyramids for Silver Springs reveal? *48.4, 48.5*

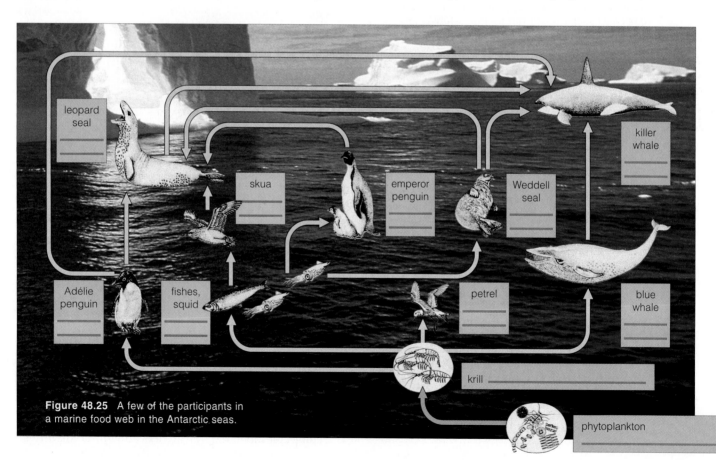

Figure 48.25 A few of the participants in a marine food web in the Antarctic seas.

leopard seal

skua

emperor penguin

Weddell seal

killer whale

Adélie penguin

fishes, squid

petrel

blue whale

krill

phytoplankton

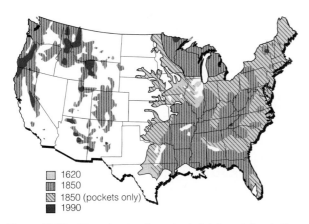

Figure 48.26 Map showing the extent of deforestation in the United States, starting from the year 1620 through 1990.

1620
1850
1850 (pockets only)
1990

6. Define three types of biogeochemical cycles and give an example of each. *48.6–48.8*

7. Define nitrogen fixation. What is the difference between nitrification and denitrification? *48.10*

Self-Quiz ANSWERS IN APPENDIX III

1. Ecosystems have _____ .
   a. energy inputs and outputs
   b. one trophic level
   c. nutrient cycling but not nutrient outputs
   d. a and b

2. Trophic levels are _____ .
   a. structured feeding relationships
   b. who eats whom in an ecosystem
   c. a hierarchy of energy transfers
   d. all of the above

3. Primary productivity on land is affected by _____ .
   a. photosynthesis and respiration by plants
   b. how many plants are neither eaten nor decomposed
   c. rainfall and temperature
   d. all of the above

4. Match the ecosystem terms with the suitable description.
   ____ producers   a. herbivores, carnivores, omnivores
   ____ consumers   b. feed on partly decomposed matter
   ____ decomposers   c. degrade organic remains, wastes
   ____ detritivores   d. photoautotrophs

Critical Thinking

1. Imagine and describe an extreme situation in which you are a participant in a food chain rather than a food web.

2. Marguerite is growing a vegetable garden in Maine. What are the variables that can affect its net primary production?

3. List as many of the agricultural products and manufactured goods as you can identify that you depend upon. Are any of them implicated in the amplified greenhouse effect?

4. Biologists Reed Noss, J. Michael Scott, and Edward LaRoe issued a report on *endangered ecosystems* in the United States. They categorize thirty natural, once-vast domains as in danger of vanishing. Deforestation, agricultural conversions, urban growth, and other human activities account for many of the

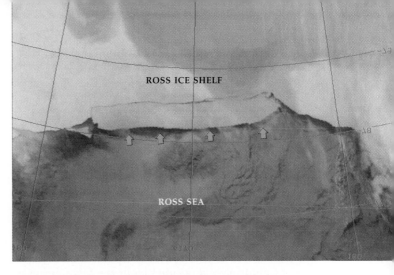

Figure 48.27 Satellite image of an iceberg roughly the same size as Connecticut (*red* arrows). This one started to break away from Antarctica in 2000 and may be the longest ever recorded.

ROSS ICE SHELF

ROSS SEA

losses. Figure 48.26 gives an example. More biodiversity has been lost at the ecosystem level than was generally recognized. Do some research on a habitat that is part of such an ecosystem. Get a sense of its decline in biodiversity, of the kinds of species that once flourished there. Would you take part in efforts to set aside land for restoration? Or do you consider such efforts too intrusive on individual rights of property ownership?

5. *Polar ice shelves* are vast, thickened sheets of ice floating on seawater. Measurements taken between 1978 and the present indicate that Antarctic ice shelves are retreating. A chunk of the Larsen Ice Shelf the size of Rhode Island broke away from Antarctica in 1995 and drifted off into the open ocean. Figure 48.27 shows another gargantuan chunk, 37 kilometers wide and 295 kilometers long, that started to split away from the Ross Ice Shelf in 2000. Also, in 1996, seawater 200 meters below the Arctic ice cap was 1°C warmer than just five years ago. Do you suppose that the retreating and fragmentation of the polar ice sheets are evidence of a long-term trend in global warming? If not, what other factors might cause such dramatic events?

Selected Key Terms

ammonification *48.10*	food web *48.1*
biogeochemical cycle *48.6*	global warming *48.9*
biological magnification *48.3*	grazing food web *48.2*
biomass pyramid *48.4*	greenhouse effect *48.9*
carbon cycle *48.8*	hydrologic cycle *48.7*
consumer *48.1*	ion exchange *48.10*
decomposer *48.1*	net ecosystem
decomposition *48.10*	production *48.4*
denitrification *48.10*	nitrification *48.10*
detrital food web *48.2*	nitrogen cycle *48.10*
detritivore *48.1*	nitrogen fixation *48.10*
ecosystem *48.1*	phosphorus cycle *48.11*
ecosystem modeling *48.3*	primary producer *48.1*
energy pyramid *48.4*	primary productivity *48.4*
eutrophication *48.11*	trophic level *48.1*
food chain *48.1*	watershed *48.7*

Readings

Brady, N., and R. Weil. 1996. *The Nature and Properties of Soils.* Eleventh edition. New Jersey: Prentice-Hall.

Krebs, C. 1994. *Ecology.* Fourth edition. New York: HarperCollins.

THE BIOSPHERE

Does a Cactus Grow in Brooklyn?

Suppose you live in the American Southwest but find yourself touring an African desert. There you notice a flowering plant with spines, tiny leaves, and fleshy, columnlike stems—just like some cactus plants back home (Figure 49.1). Or suppose you live in the coastal hills of California and decide to tour the Mediterranean coast, the southern tip of Africa, or even central Chile. There you come across many-branched, tough-leafed woody plants—very much like the many-branched, tough-leafed chaparral plants back home.

In both cases, vast geographic and evolutionary distances separate the plants. Why, then, are they alike? The question intrigues you, so you decide to compare their locations on a global map. You see that American and African desert plants live about the same distance from the equator. Chaparral plants and their distant look-alikes grow along the western and southern coasts of continents between latitudes 30° and 40°. As Charles Darwin and other naturalists did long ago, you have just stumbled onto one of many predictable patterns in nature.

What causes such patterns? With this question, we turn to **biogeography**—the study of the distribution of organisms, past and present, and of diverse processes that underlie their distribution patterns. Think back on past chance events, including the colossal breakup of Pangea more than 100 million years ago. When chunks of this supercontinent began to drift apart, most species had no choice but to go along for the ride. Travelers were dispersed to new isolated locations. There they diverged genetically from their parent populations. Among the changing populations on the fragment that became Australia were ancestors of eucalyptus trees and platypuses, of wombats and kangaroos.

Species also owe their distribution to topography, climate, and interactions with other species. With diligence, you might grow a cactus under artificial lights in a heated room in Brooklyn or another New

York City borough. Plant that cactus outside, and it will not last one winter.

As this last example reminds us, we humans tinker with species distributions. Not all of our tinkering is as harmless as growing a cactus in Brooklyn. Think of our predatory effects on the world's fisheries and the effects of our pesticide battles with insect competitors for food. Earlier chapters gave you a general picture of predation, competition, and other species interactions. This chapter considers the physical forces shaping the biosphere itself. Use it as a foundation for addressing the impact of the human species on the biosphere, the topic of the chapter to follow.

Start off with the definition of the **biosphere**: the sum of all places in which organisms live. Organisms occupy the Earth's waters—the hydrosphere, which includes the ocean, ice caps, and other forms of water, liquid and frozen. They occupy soils and sediments of the lithosphere, the Earth's outer, rocky layer. And they occupy the lower **atmosphere**, the gases and airborne particles that envelop the Earth. Air in the

Figure 49.1 Life in the biosphere—a case of morphological convergence. (**a**) *Echinocerus*, a member of the cactus family (Cactaceae), grows in hot deserts of the American Southwest. (**b**) In deserts of southwestern Africa, we find *Euphorbia*, of the spurge family (Euphorbiaceae). Although the lineages appear to be closely related, they are geographically and evolutionarily distant. Long ago, plants in both lineages put similar structures, including their leaves, to similar uses in similar habitats. Their descendants ended up resembling each other.

upper atmosphere, seventeen kilometers or so above the Earth's surface, is too thin to support life.

In this vast biosphere you find ecosystems ranging from continent-straddling forests to a rainwater pool in a small cup-shaped array of leaves. As you will see, climate profoundly influences all ecosystems except for a few at hydrothermal vents on the ocean floor.

Climate means average weather conditions, such as temperature, humidity, wind speed, cloud cover, and rainfall, over time. Many factors contribute to it. The main ones are variations in the amount of solar radiation streaming in, the Earth's daily rotation and its path around the sun, the world distribution of continents and oceans, and land mass elevations. All of these factors interact to produce prevailing winds and ocean currents that influence the global patterns of climate. The physical and chemical development of sediments and soils also depend on climate.

Climate, together with the composition of soils and sediments, influences the growth of primary producers and, through them, the distribution of ecosystems.

Key Concepts

1. Energy from the sun is the initial energy source for nearly all ecosystems on Earth. In addition, solar energy influences the global distribution of ecosystems. It does so by continually providing heat energy that warms the atmosphere and drives the Earth's weather systems.

2. Regional variations in temperature and rainfall arise through atmospheric and oceanic circulation patterns that topographic features help shape. They affect the composition of soils and sediments, which in turn affects the growth and distribution of primary producers. In such ways, the regional variations help give ecosystems their characteristics.

3. A biome is a large, regional unit of land that is characterized by the climax vegetation of ecosystems within its boundaries. Deserts and broadleaf forests are examples. Their distribution corresponds roughly with regional variations in climate, topography, and soil type.

4. The water provinces cover more than 71 percent of the Earth's surface. The oceans hold 97 percent of the free-flowing water. The remainder is distributed among inland seas, estuaries, and bodies of fresh water, including lakes, streams, and rivers.

5. Freshwater and marine ecosystems have gradients in light availability, temperature, and dissolved gases. These gradients vary daily and with the changing seasons. Primary productivity and the composition of species depend on them.

AIR CIRCULATION PATTERNS AND REGIONAL CLIMATES

Each winter Pacific Grove, California, is host to great gatherings of tourists and monarch butterflies (Figure 49.2). Similarly, caribou, Canada geese, sea turtles, and whales undertake vast migrations, moving to and from overwintering grounds. Other animals stay put. Their adaptations in body form, physiology, and behavior help them withstand seasonal change. In Canada, the United States, Japan, and Europe, flowering plants leaf out, flower, bear fruit, then drop leaves. In the ocean, astronomical numbers of photosynthetic protistans and bacteria show seasonal bursts of primary productivity. Like nearly all organisms, they are exquisitely attuned to regional climates and seasonal change.

Climate starts with incoming rays from the sun. Of the total amount of solar radiation reaching the outer atmosphere, only about half gets through to the Earth's surface. Ozone (O_3) and oxygen (O_2) molecules in the upper atmosphere absorb most of the wavelengths of ultraviolet radiation. Such wavelengths are lethal for most forms of life. Absorption is greatest at altitudes between 17 and 27 kilometers above sea level, where molecules of ozone are the most concentrated (Figure 49.3). Hence the name "ozone layer." Clouds, particles of dust, and water vapor suspended in the atmosphere absorb other wavelengths or reflect them into space.

Radiation penetrating the atmosphere warms the Earth's surface, which gives up heat by radiation and evaporation. The lower atmosphere's molecules absorb some heat and reradiate part of it toward the Earth. The effect is a bit like heat retention in a greenhouse, which lets in the sun's rays while retaining heat that the plants and soil inside are giving up (Section 48.9). Why is this greenhouse effect important? *Heat energy from the sun warms the atmosphere and ultimately drives the Earth's great weather systems.*

The sun's effect varies from one latitude to the next. Its rays are more concentrated at the equator than the poles, so air is heated more at the equator (Figure 49.4). A global pattern of air circulation starts as warm air at the equator rises and spreads north and south. The Earth rotates faster at the equator than at the poles, and its rotation under the moving air creates global belts of prevailing east and west winds. The differences in solar heating at different latitudes and the modified patterns of air circulation help define the world's major **temperature zones**, which are shown in Figure 49.5a.

The differences also help define rainfall patterns at different latitudes. Warm air holds more moisture than cool air. Air warmed at the equator picks up moisture from the seas. It rises to cooler altitudes and gives up moisture as rain, which supports luxuriant growth of equatorial forests. Being drier now, the air moves away from the equator. It becomes warmer and drier when it

Figure 49.2 Monarch butterflies. These migratory insects gather each winter in trees of California coastal regions and central Mexico. They travel hundreds of kilometers south to those places, which are cool and humid in winter. If they were to stay in their northern breeding grounds, monarchs would risk being killed by more severe climatic conditions.

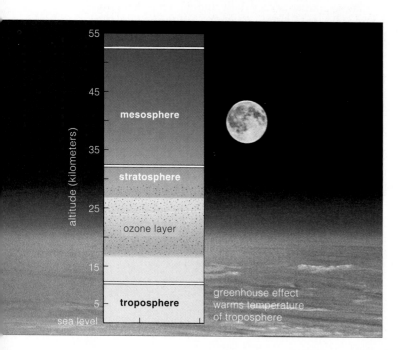

Figure 49.3 Earth's atmosphere. Most of the global air circulation proceeds in the troposphere, where temperature decreases rapidly with altitude. Most of the ultraviolet wavelengths in the sun's rays are absorbed in the upper atmosphere, mainly at the ozone layer. The ozone layer ranges between 17 and 27 kilometers above sea level.

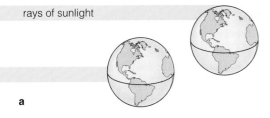

a

rays of sunlight

Figure 49.4 Global air circulation patterns arising from three interrelated factors. *First*, the sun's rays are less spread out in equatorial regions than in polar regions (**a**). Warm equatorial air rises, spreads north and south, and causes an initial pattern of air circulation (**b**).

Second, land absorbs and gives up heat faster than the ocean, and it causes air parcels above it to sink and rise faster. Air pressure is lower where air rises and is greater where air sinks. The nonuniform distribution of land and water creates regional pressure differences. This causes winds that disrupt the overall movement of air from the equator to the poles.

Third, the Earth's rotation and its curvature introduce easterly and westerly deflections in wind directions. With each full rotation, the surface turns faster beneath the air masses at the equator (where Earth's diameter is greatest) and slower beneath air masses at the poles. The difference means that a rising air mass cannot move "straight north" or "straight south." Its deflection is the source of prevailing east and west winds (**c**).

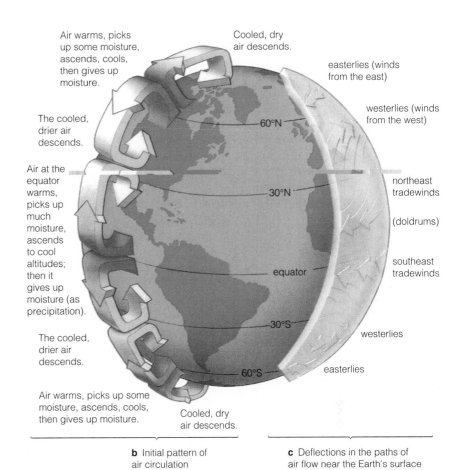

Air warms, picks up some moisture, ascends, cools, then gives up moisture.

Cooled, dry air descends.

The cooled, drier air descends.

Air at the equator warms, picks up much moisture, ascends to cool altitudes; then it gives up moisture (as precipitation).

The cooled, drier air descends.

Air warms, picks up some moisture, ascends, cools, then gives up moisture.

Cooled, dry air descends.

easterlies (winds from the east)

westerlies (winds from the west)

60°N

30°N

equator

northeast tradewinds

(doldrums)

southeast tradewinds

30°S

westerlies

60°S

easterlies

b Initial pattern of air circulation

c Deflections in the paths of air flow near the Earth's surface

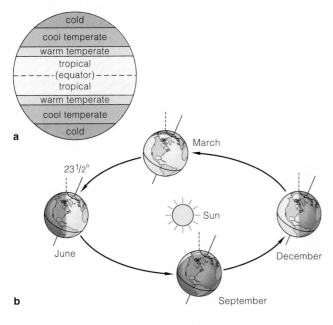

a

cold
cool temperate
warm temperate
tropical
- - - (equator) - - -
tropical
warm temperate
cool temperate
cold

23 1/2°

March

June

Sun

December

September

b

Figure 49.5 (**a**) World temperature zones. (**b**) Annual variation in incoming solar radiation. The northern end of the Earth's fixed axis tilts toward the sun in June and away from it in December, so the equator's position relative to the day–night boundary of illumination varies annually. Variations in sunlight intensity and daylength cause seasonal temperature variations.

descends at latitudes of about 30°. Deserts often form at these latitudes. Farther north and south, air again picks up moisture and ascends to higher altitudes. It creates another belt of moisture at latitudes of about 60°. Air descends in the polar regions, where low temperatures and little precipitation foster cold, dry polar deserts.

The amount of solar radiation reaching the surface also varies as the Earth rotates around the sun (Figure 49.5*b*). The variation causes changes in wind directions, daylength, and temperature. Seasonal changes are more pronounced inland, away from the ocean's moderating influence, and with distance from the equator.

In response, primary productivity alternately rises and falls on land and in the seas—and migrations shift many kinds of animals to new locations.

Latitudinal differences in the amount of solar radiation reaching the Earth produce global air circulation patterns. The Earth's rotation and its overall shape affect the patterns to produce latitudinal belts of temperature and rainfall.

Also, the Earth's annual rotation around the sun introduces seasonal changes in winds, temperatures, and rainfall.

Such factors influence the locations of different ecosystems.

THE OCEAN, LANDFORMS, AND REGIONAL CLIMATES

Ocean Currents and Their Effects

The **ocean**, a continuous body of water, covers more than 71 percent of the Earth. Driven by solar heat and wind friction, the upper 10 percent of the ocean moves in currents, which distribute nutrients through marine ecosystems and influence regional climates.

Latitudinal and seasonal variations in solar heating warm and cool that vast body of water. Where heated, the volume of water expands; where cooled, it shrinks. So sea level is about 8 centimeters, or 3 inches, higher at the equator than at either pole. The sheer volume of water in the "slope" is enough to get surface waters moving in response to gravity, most often toward the poles. Along the way, the waters warm the air parcels above them. At midlatitudes, they transfer *10 million billion* calories of heat energy per second to the air!

Rapid mass flows of water—*currents*—arise mostly from the tug of trade winds and westerlies. The Earth's rotation, its distribution of land masses, and the shapes of oceanic basins affect the direction and properties of these currents. As Figure 49.6 shows, water circulates clockwise in the Northern Hemisphere and circulates counterclockwise in the Southern Hemisphere.

Swift, deep, and narrow currents of nutrient-poor waters move parallel with the east coast of continents. As an example, 55 million cubic meters per second of warm water move northward as the Gulf Stream along the east coast of North America. Slower, shallow, and broad currents paralleling the west coast of continents move cold water toward the equator. As you will see, they move deep, nutrient-rich waters to the surface.

Why are Pacific Northwest coasts cool, foggy, and mild in summer? Winds nearing the coast give up heat to cold coastal water, which the California Current is moving toward the equator. Why are Baltimore and Boston muggy in summer? Air above the Gulf Stream gains heat and moisture, which southerly and easterly winds deliver to those cities. Why are winters milder in London and Edinburgh than in Ontario and central Canada, which are all at the same latitude? The North Atlantic Current picks up warm water from the Gulf Stream. As it flows past northwestern Europe, it gives up heat energy to the prevailing winds.

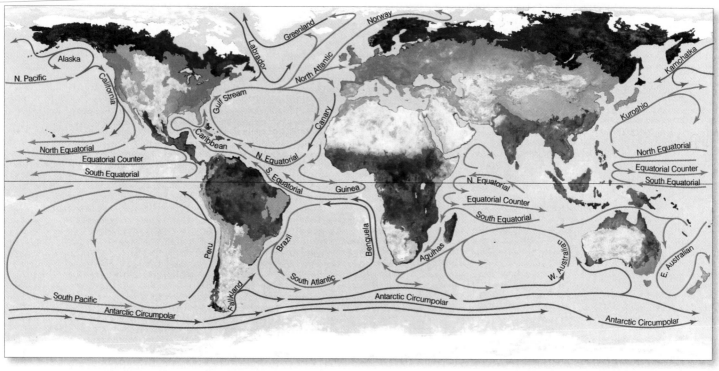

Figure 49.6 Generalized map of major climate zones correlated with surface currents and drifts of the world ocean. Warm surface currents move from the equator toward the poles. Water temperatures differ with latitude and depth, which contributes to differences in air temperature and average rainfall around the world. In which direction does a given current flow? That depends on prevailing winds, the Earth's rotation, gravity, the shapes of ocean basins, and the positions of land masses.

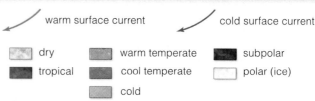

warm surface current   cold surface current

dry	warm temperate	subpolar
tropical	cool temperate	polar (ice)
cold		

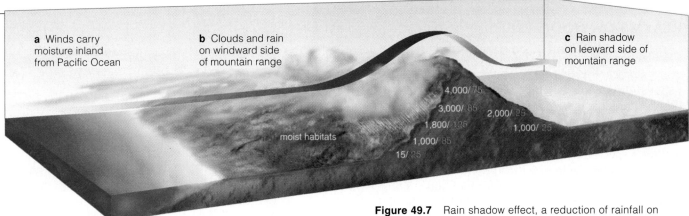

a Winds carry moisture inland from Pacific Ocean

b Clouds and rain on windward side of mountain range

c Rain shadow on leeward side of mountain range

4,000/ 75
3,000/ 85   2,000/ 25
1,800/ 125   1,000/ 25
moist habitats
1,000/ 85
15/ 25

Figure 49.7 Rain shadow effect, a reduction of rainfall on the side of high mountains away from prevailing winds. *Blue* numbers show average yearly precipitation in centimeters, as measured at different locations on both sides of the range. *White* numbers signify elevation in meters.

Regarding Rain Shadows and Monsoons

Mountains, valleys, and other aspects of topography affect regional climates. *Topography* refers to a region's physical features, such as elevation. Imagine a warm air mass picking up moisture off California's west coast. It moves inland to the Sierra Nevada, a mountain range paralleling the coast. The air cools as it rises to higher altitudes, then loses moisture as rain (Figure 49.7).

Belts of vegetation at different elevations reflect the differences in temperature and moisture. At the western base of the range are arid grasslands. Higher up, we see deciduous and evergreen species adapted to more moisture and cooler air. Higher still, a subalpine belt supports just a few evergreen species that withstand a rigorously cold habitat. Above the subalpine belt, only low, nonwoody plants and dwarfed shrubs can grow.

Air warms and can hold more water after it flows over mountain crests. As it starts to descend, it retains moisture and draws more water out of plants and soil. The result is a **rain shadow**, a semiarid or arid region of sparse rainfall on the leeward side of high mountains. (*Leeward* is the direction not facing a wind; *windward* is the direction from which wind blows.) The Himalayas, Andes, Rockies, and other great mountain ranges cause impressive rain shadows. Similarly, on the windward side of Hawaii's high volcanic peaks are lush tropical forests; arid conditions prevail on the leeward side.

Monsoons are air circulation patterns that influence the continents north or south of warm oceans. The land heats intensely, so low-pressure air parcels form above it. The low pressure draws in moisture-laden air from the ocean, as from the Bay of Bengal into Bangladesh or from the Gulf of Mexico into deserts of the American Southwest. Fantastic summer thunderstorms are one result. Converging trade winds and the equatorial sun cause intense heating and rainfall in a zone called the doldrums. The air circulating north and south creates patterns of wet and dry seasons, as in eastern Africa.

Recurring coastal breezes are like mini-monsoons. Here again, water and the adjacent land have different

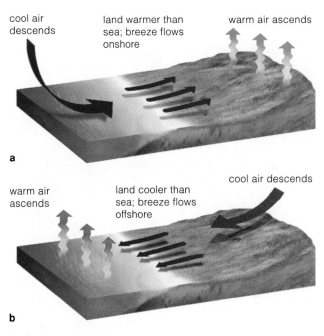

cool air descends

land warmer than sea; breeze flows onshore

warm air ascends

a

warm air ascends

land cooler than sea; breeze flows offshore

cool air descends

b

Figure 49.8 Coastal breezes, afternoon (**a**) and night (**b**).

heat capacities. In the morning, the water temperature does not rise as fast as that of land. When warmed air above the land rises, cooler marine air moves in. After sunset, the land loses heat faster than water. Then, land breezes flow in the reverse direction (Figure 49.8).

Surface ocean currents, in combination with global air circulation patterns, influence regional climates and help distribute nutrients in marine ecosystems.

Air circulation patterns, ocean currents, and landforms interact in ways that influence regional temperatures and moisture levels. Thus they also influence the distribution and dominant features of ecosystems.

49.3 REALMS OF BIODIVERSITY

Circulation patterns in the atmosphere and at the ocean surface, together with topography, give rise to regional differences in temperature and moisture. And these differences help explain why deserts, grasslands, forests, and tundra form in some places but not others. They offer insights into the distribution of ecosystems and their species, each adapted to regional conditions.

For example, the information provides clues to why many evolutionarily distant species look alike. Often, they emerged through convergent evolution, as Section 19.4 describes. Briefly, their ancestors were not closely related but faced similar selection pressures in similar yet geographically distant places. By natural selection, their body plans underwent similar modifications and ended up looking alike. The ancestors of those cactus and spurge plants in Figure 49.1 both evolved in hot,

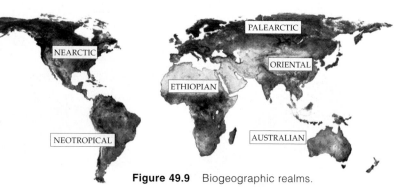

Figure 49.9 Biogeographic realms.

dry deserts, where water is scarce. Fleshy plant stems that have thick cuticles conserve precious water. And rows of sharp spines deter hungry, thirsty herbivores, which might otherwise chew on juicy plant parts.

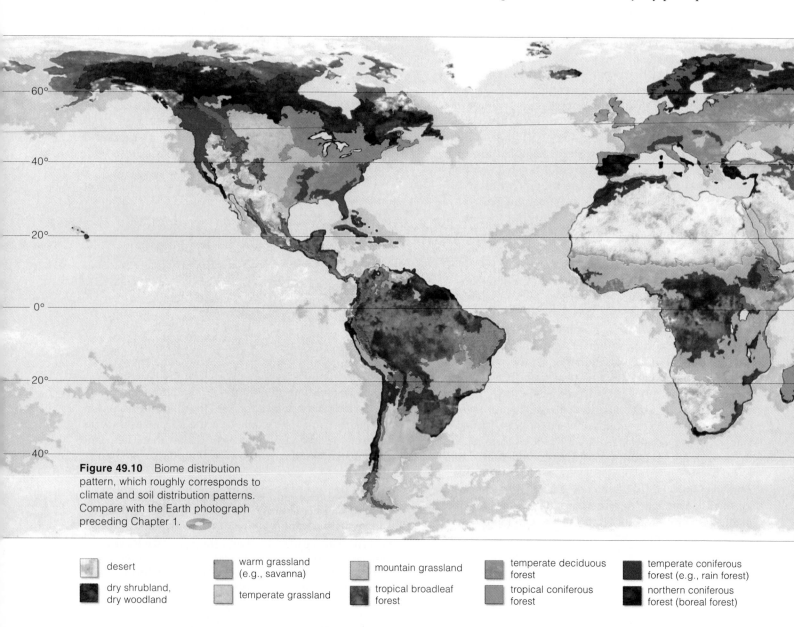

Figure 49.10 Biome distribution pattern, which roughly corresponds to climate and soil distribution patterns. Compare with the Earth photograph preceding Chapter 1.

desert	warm grassland (e.g., savanna)	mountain grassland	temperate deciduous forest	temperate coniferous forest (e.g., rain forest)
dry shrubland, dry woodland	temperate grassland	tropical broadleaf forest	tropical coniferous forest	northern coniferous forest (boreal forest)

ALPINE TUNDRA

MONTANE CONIFEROUS FOREST

DECIDUOUS FOREST

TROPICAL FOREST

High ◄───── Elevation ───── Low

TROPICAL FOREST　　TEMPERATE DECIDUOUS FOREST　　NORTHERN CONIFEROUS FOREST　　ARCTIC TUNDRA

High ◄─────────────── Moisture Availability ───────────────► Low

Figure 49.11 For North America, changes in plant form along environmental gradients. Gradients in elevation and availability of water help dictate a biome's primary productivity. Other factors, such as mean annual air temperature and soil drainage, also are at work.

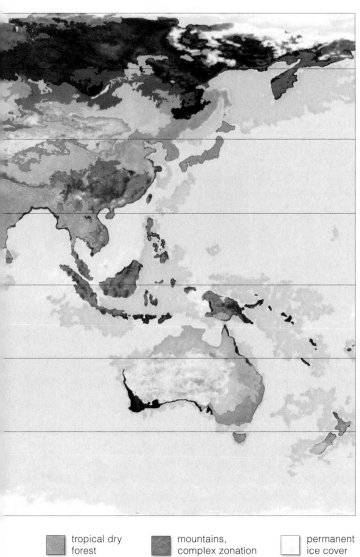

| | tropical dry forest | | mountains, complex zonation | | permanent ice cover |
| | tundra | | mangrove swamps | | marine ecoregions |

W. Sclater and, later, Alfred Wallace, bestowed the name **biogeographic realms** on six immense land areas, each with distinctive kinds and numbers of plants and animals (Figure 49.9). Biogeographic realms maintain their identity partly because of climate. They also tend to maintain a distinct identity when mountain ranges, oceans, and other physical barriers restrict gene flow, thus keeping the gene pools of each realm isolated.

A biogeographic realm may be divided into biomes. Each **biome** is a large region of land characterized by habitat conditions and community structure, including endemic species (which evolved nowhere else). Take a look at how Figure 49.10 correlates with environmental factors of the sort shown in Figure 49.11. It helps show that distinctive biomes prevail at certain latitudes and elevations. Biomes dominated by low-growing plants prevail in dry regions, at high elevations, and at high latitudes. Those dominated by tall, leafy plants prevail at tropical latitudes, temperate latitudes, and at lower elevations with warm temperatures and high rainfall. As you will see, soils also affect biome distribution.

Conservationists locate, inventory, and protect "hot spots." These are portions of biomes that are the richest in biodiversity and the most vulnerable to species loss (Section 27.5). Twenty-four of the hot spots hold more than half of all land species. In its Global 200 program, the World Wildlife Fund is also targeting **ecoregions**: large areas representative of globally important biomes *and* water provinces that are vulnerable to extinction. Figure 49.10 identifies the marine ecoregions.

The land and the seas contain realms of biodiversity. The unique identity of each biogeographic realm, biome, and ecoregion is an outcome of habitat conditions, the kinds and numbers of species, and their evolutionary history.

SOILS OF MAJOR BIOMES

Most regions of land have **soils**, which are mixtures of mineral particles and varying amounts of decomposing organic material (humus). As described in Section 30.1, the weathering of hard rocks produces coarse-grained gravel, then sand, silt, and finely grained clay. Water and air infiltrate spaces between particles. How much these particles compact together, and the proportions of each kind, differ within and between regions.

Soils have a layered structure, a *profile*, that reflects their stage of development. Figure 49.12 shows a few examples. Uppermost is topsoil, with the most humus and the most vulnerability to erosion. Topsoil may be less than a centimeter thick on steep slopes and more than one meter thick in grasslands. The lowest layer of soil consists of rocks in varying degrees of weathering.

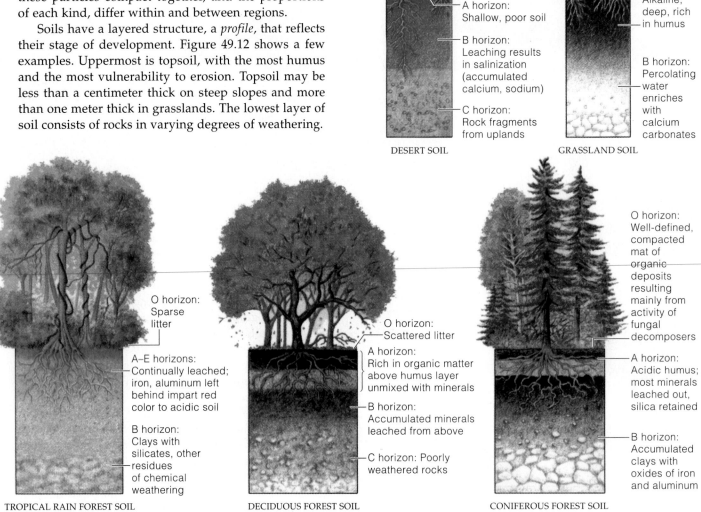

O horizon: Pebbles, little organic matter
A horizon: Shallow, poor soil
B horizon: Leaching results in salinization (accumulated calcium, sodium)
C horizon: Rock fragments from uplands
DESERT SOIL

A horizon: Alkaline, deep, rich in humus
B horizon: Percolating water enriches with calcium carbonates
GRASSLAND SOIL

O horizon: Sparse litter
A–E horizons: Continually leached; iron, aluminum left behind impart red color to acidic soil
B horizon: Clays with silicates, other residues of chemical weathering
TROPICAL RAIN FOREST SOIL

O horizon: Scattered litter
A horizon: Rich in organic matter above humus layer unmixed with minerals
B horizon: Accumulated minerals leached from above
C horizon: Poorly weathered rocks
DECIDUOUS FOREST SOIL

O horizon: Well-defined, compacted mat of organic deposits resulting mainly from activity of fungal decomposers
A horizon: Acidic humus; most minerals leached out, silica retained
B horizon: Accumulated clays with oxides of iron and aluminum
CONIFEROUS FOREST SOIL

Figure 49.12 Soil profiles from a few representative biomes. Compare Section 30.1.

The growth of most plants suffers in poorly aerated, poorly draining soils. Remember, loam topsoils have the best mix of sand, silt, and clay for agriculture. They have enough coarse particles to promote drainage and enough fine particles to retain water-soluble mineral ions that serve as nutrients for plant growth. Gravelly or sandy soils promote rapid leaching, which depletes them of water and vital minerals. Clay soils with fine, closely packed particles are poorly aerated and do not drain well. Few plants grow in waterlogged clay soils.

As farmers know, soil affects primary productivity. They grow most crops in cleared, former grasslands. Many burn or clear-cut tropical forests for agriculture, but these biomes have very little topsoil above poorly draining sublayers. Clearing exposes the topsoil, and heavy rains leach most of the nutrients (Section 50.4).

We turn now to the deserts, shrublands, woodlands, grasslands, various forests, and tundras. Bear in mind, none of these major biomes is uniform throughout. In each biome, local climates, landforms, soils, and other features favor patches of distinct communities.

Primary productivity depends on the soil profile and its proportions of sand, silt, clay, gravel, and humus.

DESERTS

Deserts form on land with less than ten centimeters or so of annual rainfall and high potential for evaporation. Such conditions prevail at latitudes of about 30° north and south. There we find great deserts of the American Southwest, of northern Chile, Australia, northern and southern Africa, and Arabia. Farther north are the high deserts of eastern Oregon, and Asia's vast Gobi and the Kyzyl-Kum east of the Caspian Sea. Rain shadows are the main reason these northern deserts are so arid.

after a rain. Annual and perennial species flower briefly but profusely after seasonal rains. Deep-rooted plants, including mesquite and cottonwood, commonly grow near the few streambeds that have a permanent underground water supply.

More than a third of the Earth's land surface is arid or semiarid, without enough rainfall to support crops. Crops in California's Imperial Valley and some other

Figure 49.13 Warm desert near Tucson, Arizona. Primary producers include creosote bushes, multistemmed ocotillos, tall saguaro cacti, and prickly pear cacti with rounded pads.

Deserts do not have lush vegetation. Rain falls in heavy, brief, infrequent pulses that swiftly erode the exposed topsoil. Humidity is so low that the sun's rays easily penetrate the air. They quickly heat the ground's surface, which radiates heat and cools quickly at night.

Although arid or semiarid conditions do not favor large, leafy plants, deserts show plenty of biodiversity. In a patch of Arizona's desert (as in Figure 49.13), you might find creosote and other deep-rooted, evergreen, woody shrubs, fleshy-stemmed, shallow-rooted cacti, tall saguaros, short prickly pears, and ocotillos, which drop leaves more than once a year and grow new ones

deserts need intensive soil management and irrigation. Without drainage programs, crop production declines from waterlogging and salt buildup. Also, many parts of the world are becoming desertlike wastelands with low biodiversity as grasslands and other productive biomes are converted for agriculture. We return to this trend, called **desertification**, in Section 50.6.

Where the potential for evaporation greatly exceeds sparse rainfall, deserts form. This condition prevails at latitudes 30° north and south and in rain shadows.

DRY SHRUBLANDS, DRY WOODLANDS, AND GRASSLANDS

Dry shrublands and **dry woodlands** prevail in western or southern coastal regions of the continents between latitudes 30° and 40°. These semiarid regions get a bit more rain than deserts, mostly during mild winters. Summers are long, hot, and arid. The dominant plants often have hardened, tough, evergreen leaves.

Dry shrublands get less than 25 to 60 centimeters of rain per year. We see them in California, South Africa, and Mediterranean regions, where they are known by such local names as fynbos and chaparral. California has 2.4 million hectares (6 million acres) of chaparral. In summer, lightning-sparked, wind-driven firestorms sweep through these biomes (Figure 49.14). The shrubs have highly flammable leaves and quickly burn to the ground. However, they are highly adapted to episodes of fire and soon resprout from their root crowns. Trees do not fare as well during firestorms. Shrubs actually "feed" the fires and have the competitive edge.

Dry woodlands dominate where annual rainfall is about 40 to 100 centimeters. The dominant trees can be tall, but they do not form a dense, continuous canopy. Eucalyptus woodlands of southwestern Australia and oak woodlands of California and Oregon are like this.

Grasslands cloak much of the interior of continents in zones between deserts and temperate forests. Warm temperatures prevail in the summer, and winters are extremely cold. Annual rainfall of 25 to 100 centimeters prevents deserts from forming, but it is not enough to support forests. Drought-tolerant primary producers survive strong winds, sparse and infrequent rainfall, and rapid evaporation. Grazing and burrowing species are the dominant animals. Their activities, combined with periodic fires, help keep shrublands and forests from encroaching on the fringes of many grasslands.

The main grasslands of North America are *shortgrass* and *tallgrass* prairie. Most form on flat or rolling land, as in Figure 49.15a. Roots of perennial plants extend profusely through topsoil. In the 1930s, the shortgrass prairie of the Great Plains was overgrazed and plowed under to grow wheat, which requires more water than the region sometimes receives. Strong winds, prolonged droughts, and unsuitable farming practices turned much of the prairie into the Dust Bowl (Section 50.6). John Steinbeck's *The Grapes of Wrath* and James Michener's *Centennial*, two historical novels, eloquently describe the impact on people living there.

The tallgrass prairie once extended west from temperate deciduous forests (Figures 48.26 and 49.15b). A variety of legumes and composites, including daisies, thrived in the continent's interior, which had richer topsoil and slightly more frequent rainfall. Farmers converted nearly all of the original tallgrass prairie for agriculture, but certain areas are now being restored (Section 47.7).

Between the tropical forests and deserts of Africa, South America, and Australia are broad belts of grasslands with a smattering of shrubs and trees. These are the **savannas** (Figure 49.15c). Rainfall averages 90 to 150 centimeters a year, and prolonged seasonal droughts are common. Where rainfall is low, fast-growing grasses dominate. Acacia and other shrubs grow in regions that get a bit more moisture. Where the rainfall is higher,

Figure 49.14 California chaparral. The dominant plants are multibranched, woody, and typically only a few meters tall. Without periodic fires, they form a nearly impenetrable vegetation cover. The boxed inset shows a firestorm racing through a chaparral-choked canyon above Malibu.

Figure 49.15 (**a**) Rolling shortgrass prairie to the east of the Rocky Mountains. Once, bison were the dominant large herbivore of North American grasslands. At one time, the abundant grasses of this biome supported 60 million of these hefty ungulates, which are hooved, plant-eating mammals.

(**b**) A rare patch of natural tallgrass prairie in eastern Kansas.

(**c**) The African savanna—a warm grassland with scattered stands of shrubs and trees. At present, more varieties and greater numbers of large ungulates live here than anywhere else. They include the migratory wildebeests shown in this photograph and giraffes, Cape buffalos, zebras, and impalas.

savannas grade into tropical woodlands with low trees and shrubs, and tall, coarse grasses. *Monsoon* grasslands form in southern Asia where heavy rains alternate with a dry season. Dense stands of tall, coarse grasses form, then die back and often burn during the dry season.

Plants of dry shrublands, dry woodlands, and grasslands are supremely adapted to surviving strong winds, grazing animals, and recurring episodes of drought and fire.

TROPICAL RAIN FORESTS AND OTHER BROADLEAF FORESTS

In forest biomes, tall trees grow close together, forming a fairly continuous canopy over a broad stretch of land. In general, there are three types of trees. Which type prevails in a region depends partly on distance from the equator. Evergreen broadleafs dominate between latitudes 20° north and south. Deciduous broadleafs are typical of moist, temperate latitudes where winters are mild. Evergreen conifers, the third type of forest tree, dominate at high, cold latitudes and in the mountains of temperate zones. We will take a look at evergreen coniferous forests in Section 49.8.

Evergreen broadleaf forests sweep across tropical zones of Africa, the East Indies and Malay Archipelago, Southeast Asia, South America, and Central America.

Annual rainfall can exceed 200 centimeters and never is less than 130 centimeters. One biome, the **tropical rain forest**, depends on regular and heavy rainfall, an annual mean temperature of 25°C, and humidity of at least 80 percent (Figure 49.16). Evergreen trees of this highly productive forest produce new leaves and shed old ones all year long. But litter does not accumulate; decomposition and mineral cycling are notably rapid, given the hot, humid climate. The soils are weathered, humus-deficient, and poor nutrient reservoirs. We will take a closer look at these forests in the next chapter.

Leaving the tropical rain forests, we enter regions where temperatures remain mild but rainfall dwindles during part of the year. This is the start of **deciduous**

Figure 49.16 Tropical rain forests, biomes of great biodiversity. Among their millions of known species are jaguars and *Rafflesia*, a leafless, foul-smelling plant with a fly-pollinated flower that grows three meters across. Bromeliads, orchids, and other epiphytes grow on trees. They obtain minerals from organic remains of leaves, insects, and other litter, which dissolve in water that collects at the base of their leaves.

SPRING

SUMMER

WINTER

AUTUMN

Figure 49.17 The changing appearance of part of a temperate deciduous forest south of Nashville, Tennessee, in spring, summer, autumn, and winter.

broadleaf forests. Trees of *tropical* deciduous forests drop some or all of their leaves in a pronounced dry season. *Monsoon* forests of India and southeastern Asia also have such trees. Farther north, in the temperate zone, rainfall is even lower. Winters are so cold, water is locked away as snow and ice. *Temperate* deciduous forests, such as those of the southeastern United States, prevail here (Figure 49.17). Decomposition is not as rapid as in the humid tropics, and many nutrients are conserved in accumulated litter on the forest floor.

Complex forests of ash, beech, birch, chestnut, elm, and deciduous oaks once stretched across northeastern North America, Europe, and East Asia. They declined drastically when farmers cleared the land. Pathogenic species introduced to North America killed nearly all chestnuts and many elms (Section 47.8). Now, maple and beech predominate in the northeast. Farther west, oak–hickory forests prevail, then oak woodlands that eventually grade into tallgrass prairie.

In forest biomes, conditions favor dense stands of tall trees that form a continuous canopy over a broad region.

CONIFEROUS FORESTS

Conifers (cone-bearing trees) are primary producers of **coniferous forests**. Most have thickly cuticled, needle-shaped leaves with recessed stomata—adaptations that help them conserve water during droughts and winter. They dominate the boreal forests, montane coniferous forests, temperate rain forests, and pine barrens.

Boreal forests stretch across northern Europe, Asia, and North America. Such forests also are called *taigas*, meaning "swamp forests." Most are in glaciated regions having cold lakes and streams (Figure 49.18*a*). It rains mostly in summer, and evaporation is low in the cool summer air. The cold, dry winters are more severe in eastern parts of these biomes than in the west, where

surrendered its rights to clear-cut one of the largest intact temperate rain forests outside of tropical regions. The tall trees, some of which are 800 years old, cloak British Columbia's Kitlope Valley.

Southern pine forests dominate the coastal plains of the south Atlantic and the Gulf states. Pine species are adapted to the dry, sandy, nutrient-poor soil and to natural fires or controlled burns. These fires do not get hot enough to damage the trees, and they open up the understory. Forests of pine, scrub oak, and wiregrass grow in New Jersey. Palmettos grow below pines and loblolly in the Deep South. During one recent Florida

ocean winds moderate the climate. Spruce and balsam fir dominate the boreal forests of North America. Pine, birch, and aspen forests form in areas that have been burned or logged. Where soil is poorly drained, highly acidic bogs dominated by peat mosses, shrubs, and stunted trees prevail. Boreal forests are less dense to the north, where they grade into arctic tundra.

In the Northern Hemisphere, montane coniferous forests extend southward through the great mountain ranges. Spruces and firs dominate in the north and at higher elevations. They give way to firs and pines in the south and at lower elevations (Figures 3.1 and 49.18*b*). Some temperate lowlands support coniferous forests. One temperate rain forest that parallels the coast from Alaska on into northern California contains some of the world's tallest trees—Sitka spruce to the far north and redwoods to the south. Logging destroyed much of this biome. One bright note: In 1994, a timber corporation

Figure 49.18 (**a**) Spruce-dominated boreal forest. (**b**) Montane coniferous forest of Yosemite Valley in California's Sierra Nevada. (**c**) Firefighter retreating from flames in a forest of pines, oaks, and palmettos by Daytona Beach, Florida.

heat wave, fires destroyed more than 320,000 acres. Smaller burns had been prohibited in counties where housing and other developments encroached on these forests. Dense, tinder-dry understory had accumulated, and it fed the fires (Figure 49.18*c*).

Coniferous forests prevail in regions in which a cold, dry season alternates with a cool, rainy season.

ARCTIC AND ALPINE TUNDRA

Tundra is derived from *tuntura*, a Finnish word for the great treeless plain between the polar ice cap and belts of boreal forests in Europe, Asia, and North America. This plain is *arctic* tundra (Figure 49.19*a,b*). Here you find extremely low temperatures; poor drainage; short growing seasons, when sunlight is nearly continuous; and poor decomposition of organic matter. The annual precipitation (rain and melting snow) varies regionally but typically is less than 25 centimeters. Lichens and short, hardy, shallow-rooted plants are a base for food webs that include voles, arctic hares, caribou, and other herbivores. Carnivores include arctic foxes, wolves, and polar bears. Many migratory birds nest here in summer.

and it accumulates in soggy masses. About 95 percent of the carbon that photoautotrophs fixed in the arctic tundra is locked up in extensive peat bogs.

A similar type of biome prevails at high elevations in mountains throughout the world. Figure 49.19*c* is one example of this *alpine* tundra. The temperatures at night are typically below freezing. It is just too cold for any trees to grow. However, unlike the arctic tundra, alpine tundra has no permafrost; the soil is thin but well drained. Grasses, heaths, and other small-leafed shrubs form low, compact cushions and mats that can withstand the buffeting of strong winds.

Figure 49.19 Arctic tundra in Russia in summer (**a**) and winter (**b**). About 4 percent of the Earth's land mass is arctic tundra, blanketed with snow for as long as nine months of the year. Most arctic tundra is in northern Russia and Canada, although we also find some in Alaska and Scandinavia. Small bands of humans have hunted, fished, and herded reindeer in these sparsely populated regions for hundreds of thousands of years. Today, mineral and fossil fuel deposits are being extracted, especially in Russia. These are fragile biomes. Plants and animals must survive extreme cold in winter. They might be vulnerable to industrial pollution. Severe climatic changes that are expected to result from increased global warming may put them at risk. (**c**) Short, compact, hardy plants typical of alpine tundra. Mount Evans, Colorado, is in the distance.

Summers last fifty to sixty days and temperatures average 3°–12°C (37°–54°F), so not much more than the surface soil thaws. Mosquitoes, flies, and other insects proliferate under these conditions. Just beneath the soil surface is a frozen layer, called **permafrost**, that never melts. It is more than 500 meters thick in some places. Permafrost prevents drainage, so the soil above it stays waterlogged. These cool, anaerobic conditions hamper nutrient cycling. Organic matter decomposes slowly,

Even in summer, shaded patches of snow persist in alpine tundra. The thin, fast-draining soil is nutrient-poor, so primary productivity is low.

At high latitudes with short, cool summers and long, cold winters, we find arctic tundra. In high mountains where seasonal changes vary with latitude but where the climate is too cold to support forests, we find alpine tundra.

FRESHWATER PROVINCES

Freshwater and saltwater provinces cover more of the Earth's surface than all biomes combined. They include the world ocean, lakes, ponds, wetlands, and coral reefs. "Typical" examples are not easy to find. Some ponds can be waded across; Lake Baikal in Siberia is over 1.7 kilometers deep. All aquatic ecosystems have gradients in light penetration, temperature, and dissolved gases, but the values differ greatly. All we can do is sample the diversity, starting with the freshwater provinces.

Lake Ecosystems

A **lake** is a body of standing fresh water produced by geologic events, as when an advancing glacier carves out the land beneath it. After the glacier retreats, water collects in the exposed basin (Figure 49.20). Over time, erosion and sedimentation alter the dimensions of the lake, which usually ends up filled or drained.

A lake has littoral, limnetic, and profundal zones (Figure 49.21). The littoral extends all around the shore to a depth at which rooted aquatic plants stop growing. Diversity is greatest in its well-lit, shallow, warm water. Beyond the littoral is the limnetic, ranging from open sunlit water to depths where light and photosynthesis are negligible. Cyanobacteria, green algae, and diatoms dominate the lake's *phyto*plankton. They feed rotifers, copepods, and other members of the *zoo*plankton. The profundal extends down through all open water below the depth at which wavelengths of light suitable for photosynthesis can penetrate. Detritus sinks through this zone to the lake's bottom. Communities of diverse bacterial decomposers live in and on bottom sediments, where they enrich the water with nutrients.

SEASONAL CHANGES IN LAKES In temperate regions having warm summers and cold winters, lakes show seasonal changes in density and temperature from the surface to the bottom. A layer of ice forms over many of them in midwinter. Water near the freezing point is the least dense and accumulates under the ice. Water at 4°C is the most dense. It collects in deeper layers that are a bit warmer than the surface layer in midwinter.

Figure 49.20 In the Canadian Rockies, the basin of Moraine Lake, formed by glacial action during the most recent ice age.

In spring, daylength increases and the air warms. Lake ice melts, the surface water warms to 4°C, and temperatures turn uniform throughout. Winds over the surface layer cause a **spring overturn**: Strong vertical movements carry dissolved oxygen near the surface to the depths, and nutrients released by decomposition move from sediments to the surface.

By midsummer a *thermocline* forms. At this midlayer of water, an abrupt temperature change blocks vertical mixing (Figure 49.22). Being warmer and not as dense, the surface water floats on the thermocline. In cooler water below, decomposers deplete dissolved oxygen. In autumn, the upper layer cools and becomes denser. It sinks, and the thermocline vanishes. During this **fall overturn**, water mixes vertically, so dissolved oxygen moves down and nutrients move up.

Primary productivity corresponds with the seasons. After a spring overturn, the longer daylengths and the cycled nutrients favor higher rates of photosynthesis. Phytoplankton and rooted aquatic plants quickly take up phosphorus, nitrogen, and other nutrients. During a growing season, the thermocline stops vertical mixing. Nutrients locked in the remains of organisms sink to deeper water, so photosynthesis slows. By late summer,

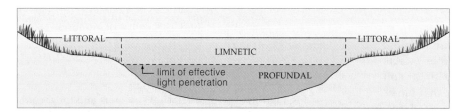

Figure 49.21 Lake zonation. The littoral extends all around the shore to a depth where aquatic plants stop growing. The profundal is all water below the depth of light penetration. Above the profundal are open, sunlit waters of the limnetic zone.

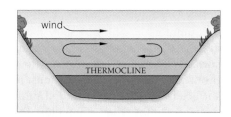

Figure 49.22 Thermal layering, which occurs in many lakes of temperate zones during the summer.

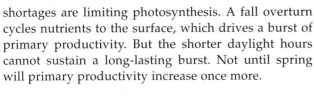

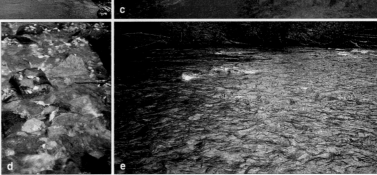

shortages are limiting photosynthesis. A fall overturn cycles nutrients to the surface, which drives a burst of primary productivity. But the shorter daylight hours cannot sustain a long-lasting burst. Not until spring will primary productivity increase once more.

TROPHIC NATURE OF LAKES   The topography, climate, and geologic history of a lake dictate the numbers and kinds of residents, how they are dispersed in the lake, and how they cycle nutrients. Soils of the surrounding region and of the lake basin contribute to the type and amount of nutrients available to support organisms.

Interplays among the climate, soil, basin shape, and metabolic activities of the lake's residents contribute to conditions that range from oligotrophy to eutrophy. The *oligotrophic* lakes are often deep, clear, and low in nutrients. Primary productivity is not great. *Eutrophic* lakes usually are shallower, nutrient rich, and high in primary productivity. Such conditions occur naturally as sediments accumulate in the lake basin. Water slowly becomes less transparent and less deep, compared to an oligotrophic lake, and aquatic communities become dominated by phytoplankton. If more sediments build up, then the final successional stage will eventually be a filled-in basin—hence no more lake.

Eutrophication, recall, refers to any processes that enrich a body of water with nutrients (Section 48.11). Human activities caused eutrophication of Seattle's Lake Washington. From 1941 to 1963, phosphate-rich sewage draining into the lake promoted cyanobacterial blooms. Slimy mats formed in summer, making the lake useless for recreation. With the abundant phosphorus, nitrogen became the limiting resource. Cyanobacteria are superior competitors for nitrogen; they can fix N_2. They became dominant. Later, sewage discharges were stopped, and by 1975, the lake neared full recovery.

Stream Ecosystems

Flowing-water ecosystems called **streams** start out as freshwater springs or seeps. They grow and merge as they flow downslope, then often combine into a river. Between a river's headwaters and end, we find three kinds of habitats: riffles, pools, and runs (Figure 49.23). *Riffles* are shallow, turbulent stretches where the water

Figure 49.23   Stream habitats in North Carolina and Virginia. (**a**) Pool. (**b**) Pool leading into a riffle. (**c**) A run, Sinking Creek. (**d**) Leaf detritus in a riffle. (**e**) Closer look at a riffle.

flows swiftly over a rough, sandy, and rocky bottom. *Pools* have deep water flowing slowly over a smooth, sandy, or muddy bottom. *Runs* are smooth-surfaced, fast-flowing stretches over bedrock or rock and sand.

A stream's average flow volume and temperature depend on rainfall, snowmelt, geography, altitude, and even the shade cast by plants. Its solute concentrations are influenced by the streambed's composition as well as by agricultural, industrial, and urban wastes.

Streams import most of the organic matter for food webs, especially in forest ecosystems. When the trees cast shade and thus hamper photosynthesis, their litter is the basis of detrital food webs. Aquatic organisms continually take up and release nutrients as the water flows downstream. Nutrients move upstream only in tissues of migratory fishes and other animals. Think of nutrients as spiraling between aquatic organisms and water as a stream flows on its one-way course, usually into a river that ends at the sea.

Ever since cities formed, streams have been sewers for industrial and municipal wastes. The wastes, along with other pollutants from poorly managed farmlands, choked many streams with sediments and chemically poisoned them. Streams, however, are resilient. They can recover impressively when pollution is controlled.

Freshwater and saltwater provinces are far more extensive than the Earth's biomes. All of the aquatic ecosystems in these water provinces have characteristic gradients in light penetration, temperature, and dissolved gases.

THE OCEAN PROVINCES

Beyond land's end are two vast provinces of the world ocean, which covers nearly three-fourths of the Earth's surface. The *benthic* province includes sediments and rocks of the ocean bottom (Figure 49.24). It begins at continental shelves and extends to deep-sea trenches. The *pelagic* province is the full volume of ocean water. Its neritic zone is all water above continental shelves, and its oceanic zone is the water of the ocean basins.

Walk along the ocean and its vastness may humble you. Its surface extends to the distant horizon, with no mountains and valleys and plains that would break it up visually into something less overwhelming. What you do not see are all of the *submerged* mountains and valleys and plains of the benthic province.

Primary Productivity in the Ocean

Photosynthesis proceeds on a stupendous scale in the ocean's upper surface, just as it does on land. Primary productivity varies seasonally, just as it does on land

(Section 7.8 and Figure 49.25). Drifting through the water are huge "pastures" of phytoplankton, the basis of food webs that include copepods, shrimplike krill, whales, squids, and fishes. Organic remains and wastes from its communities sink to the bottom of the ocean basin. There they support the detrital food webs for most benthic communities.

Near the ocean surface, 70 percent of the primary productivity might be attributable to **ultraplankton**—photosynthetic bacteria no more than 2 micrometers (40 millionths of an inch) wide. In the subtropical and tropical seas once thought to be practically devoid of producers, 0.035 gram (or 1 ounce) of water holds up to 3 million bacterial cells. In deep water too dark to support photosynthesis, food webs start with **marine snow**. These bits of organic matter drift down and become the food base for staggering biodiversity in midoceanic waters—possibly up to 10 million species. In what may be the greatest of circadian migrations, some species rise thousands of feet to feed in shallower

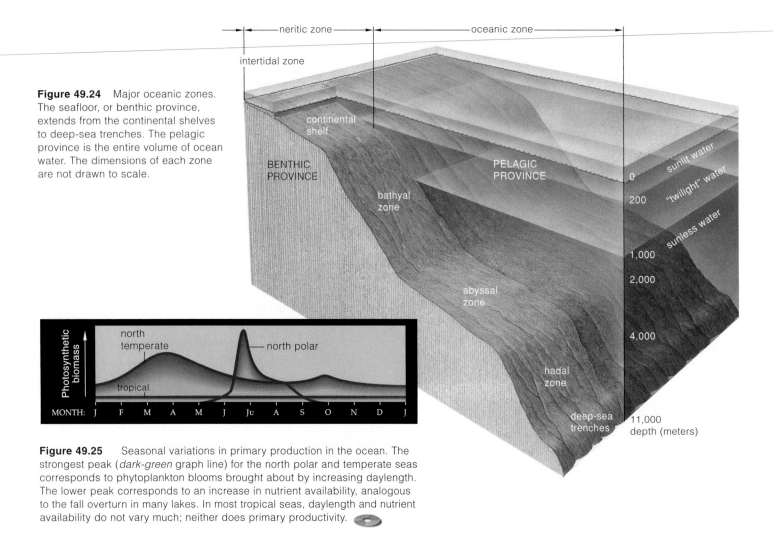

Figure 49.24 Major oceanic zones. The seafloor, or benthic province, extends from the continental shelves to deep-sea trenches. The pelagic province is the entire volume of ocean water. The dimensions of each zone are not drawn to scale.

Figure 49.25 Seasonal variations in primary production in the ocean. The strongest peak (*dark-green* graph line) for the north polar and temperate seas corresponds to phytoplankton blooms brought about by increasing daylength. The lower peak corresponds to an increase in nutrient availability, analogous to the fall overturn in many lakes. In most tropical seas, daylength and nutrient availability do not vary much; neither does primary productivity.

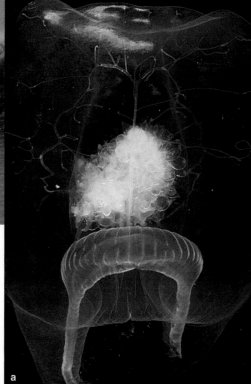

Figure 49.26 What lies beneath—a vast world of marine life that is largely unexplored. (**a**) *Praya dubia*, a relative of the Portuguese man-of-war. This bioluminescent, stinging siphonophore is one of the longest existing animals. Some measure fifty meters, from a mouthless, pulsating swimming bell to the tip of a narrow stem to which reproductive medusae, tentacles, and feeding polyps connect. *P. dubia* moves vertically in a circadian migration. (**b**) Tube worms at a hydrothermal vent on the ocean floor. (**c**) Deep-sea angler fish.

water at night, then move down the next morning. Top carnivores range from familiar types, such as sharks and giant squids, to the truly exotic (Figure 49.26*a*).

Hydrothermal Vents

In the Galápagos Rift, a volcanically active boundary between two crustal plates, we observe communities thriving near **hydrothermal vents**. Near-freezing water seeps down into these fissures in the seafloor and gets heated to extremely high temperatures. As the heated water moves up under pressure, it leaches mineral ions from rocks before being spewed out. Iron, zinc, copper sulfides, and sulfates of magnesium and calcium are dissolved in the outpourings. They settle out and form rich mineral deposits. Some of the minerals are energy sources for chemoautotrophs, which are the foundation of hydrothermal vent communities. Chemoautotrophic bacteria are the primary producers for food webs that include red tube worms, crustaceans, clams, and fishes (Figure 49.26*b*). More hydrothermal vent ecosystems occur in the South Pacific near Easter Island; the Gulf of California, about 150 miles south of the tip of Baja California, Mexico; and the Atlantic. In 1990, a team of United States and Russian scientists found one in Lake Baikal, the world's deepest lake. This lake basin seems to be splitting apart (hence the vents) and may define the beginning of a new ocean.

Did life originate in such nutrient-rich places on the seafloor? Conditions at the surface of the early Earth were about as inhospitable as you might imagine. At the least, cells living on the seafloor would have had protection from the destructive radiation bombarding the Earth before the oxygen-rich atmosphere formed. Were some descendants of those cells carted closer to the sea surface when the seafloor was uplifted during episodes of crustal crunchings? These are some of the unanswered questions that evolutionary detectives are now asking. Sections 7.9 (*Critical Thinking* question 3) and 20.1 address some of their interesting hypotheses.

Although we are most familiar with features of the land, a continuous ocean dominates most of the Earth's surface. Its primary productivity and biodiversity are staggering.

WETLANDS AND THE INTERTIDAL ZONE

Rich ecosystems develop near the coasts of continents and islands. Like freshwater ecosystems, they differ in their physical and chemical properties, including light penetration and water temperature, depth, and salinity.

Mangrove Wetlands and Estuaries

In tidal flats at tropical latitudes, we find nutrient-rich **mangrove wetlands**. "Mangrove" refers to forests in sheltered regions along tropical coasts (Figure 49.27a). Ocean waves do not reach these saltwater ecosystems, so anaerobic sediments and mud accumulate. The term also refers to a particular group of salt-tolerant species (halophytes), such as red mangrove (*Rhizophora*). Plants have shallow, spreading roots or branching prop roots that extend from the trunk (Figure 49.27b). Many have pneumatophores, root extensions that take up oxygen.

Florida's mangrove wetlands have few mangrove species, whereas those in Southeast Asia have as many as thirty or forty. Seeds of all species germinate while still attached to the parent tree. Seedlings drop to the water and float upright until currents deposit them in the mud of shallow water, where their roots take hold.

Net primary productivity of a mangrove wetland varies, depending partly on the volume and flow rate of water moving in and out with the tides. It depends on salinity and nutrient availability, which are affected by litter accumulation and turnover. But these are all rich ecosystems. Tidal circulation and nutrient inputs from streams and rivers combine to support notable biomass. Especially along the Malaysian peninsula and Gulf of Mexico, nutrient-rich detritus from mangrove wetlands supports nearby estuarine ecosystems.

An **estuary** is a partly enclosed coastal region where seawater slowly mixes with nutrient-rich fresh water from rivers, streams, and runoff (Figure 49.28). The confined conditions, slow mixing of water, and tidal

Figure 49.27 (**a**) From Florida's Everglades, a mangrove wetland. (**b**) Mangrove prop roots.

action combine to trap dissolved nutrients. Freshened water continually replenishes the nutrients, which is why estuaries can support productive ecosystems.

Primary producers are phytoplankton, salt-tolerant plants that tolerate submergence at high tide, and algae in mud and on plants. Much of the primary production enters detrital food webs in which bacterial and fungal decomposers are the first to feed. Detritus (and bacteria on and in it) feeds nematodes, snails, crabs, and fishes. Food particles suspended in the slowly moving water feed clams and other filter feeders. Estuaries nurture so many larval and juvenile stages of invertebrates and some fishes that they are often called marine nurseries. Also, many migratory birds use estuaries as rest stops.

Estuaries range from broad, shallow Chesapeake Bay, Mobile Bay, and San Francisco Bay to the narrow, deep fjords of Norway and their equivalents in Alaska and British Columbia. Texas and Florida estuaries lie

Figure 49.28 Salt marsh of a New England estuary, which interacts with the open sea in the distance. *Spartina*, a marsh grass, is the main producer. Microbe-enriched litter feeds diverse consumers that live in shallow water and muddy sediments (compare Section 48.3).

Figure 49.29 (a) One of the tidepools that are typical of rocky shores of the Pacific Northwest, with an abundance of algae and invertebrates. (b) In this coastal region, you are likely to come across pronounced vertical zonation, as shown here.

Around the world, the difference between the high and low tide marks ranges from a few centimeters in the Mediterranean Sea to more than fifteen meters in the Bay of Fundy, near Nova Scotia.

UPPER LITTORAL

MIDLITTORAL

LOWER LITTORAL

behind long sandy or muddy spits. New England's salt marshes are familiar estuaries.

Estuaries cannot be maintained without inflows of unpolluted, fresh water. Many are declining as a result of upstream diversion of fresh water combined with inputs of agricultural runoff and industrial, urban, and suburban wastes, including raw sewage.

Rocky and Sandy Coastlines

Rocky and sandy coastlines support ecosystems of the **intertidal zone**, which is not renowned for its creature comforts. Waves batter its residents. Tides alternately submerge and expose them. The higher they are, the more they dry out, freeze in winter, or bake in summer, and the less food comes their way. The lower they are, the more they must compete in limited spaces. At low tides, the birds, rats, and raccoons move in and feed on them. High tides bring the predatory fishes.

Generalizing about coastlines is not easy, for waves and tides continually resculpt them. One feature that rocky and sandy shores do share is vertical zonation.

Rocky shores often have three zones (Figure 49.29). The *upper* littoral is submerged only during the highest tide of the lunar cycle and is sparsely populated. The *mid*littoral is submerged by the highest regular tide and exposed by the lowest. Its tidepools hold red, brown, and green algae, fishes, hermit crabs, nudibranchs, and sea stars. Biodiversity is greatest in the *lower* littoral, which is exposed only by the lowest tide of the lunar cycle. In all three zones, rapid erosion prevents detritus from accumulating, so grazing food webs prevail.

Waves and currents continually rearrange stretches of loose sediments called *sandy* and *muddy* shores. Few

Figure 49.30 From Australia's Great Barrier Reef, a coral fragment washed up on Heron Island's sandy shore.

large plants grow in these unstable places, so you won't find many grazing food webs. Detrital food webs start with organic debris from land or offshore (Figure 49.30). Vertical zonation is less clear than along rocky shores. Below the low tide mark in temperate regions are sea cucumbers and blue crabs. Marine worms, crabs, and other invertebrates thrive between high and low tide marks. At night, at the high tide mark, beach hoppers and ghost crabs pop out of their burrows, seeking food.

Nutrient-rich ecosystems called mangrove swamps and estuaries develop along many coasts. Intertidal zones are not as renowned for creature comforts; waves batter them and tides alternately expose and submerge them.

EL NIÑO AND THE BIOSPHERE

Upwelling Along the Coasts

We turn now to a global event that influences primary productivity along coasts of the Northern Hemisphere and in the equatorial Pacific Ocean. Prevailing north winds parallel the western coasts and tug at the ocean surface. Wind friction starts the surface waters moving. Under the Earth's rotational force, that mass of slow-moving water is deflected westward, away from the coastlines. Cold, deep, often nutrient-rich water moves in vertically to replace it (Figure 49.31).

When cold, deep water moves up this way, we call this an **upwelling**. Upwelling occurs in the equatorial currents and along the coasts of continents, and it cools air masses above it. Thick fogbanks along California's coast are one indication that upwelling of cold water is encountering warm air above it.

In the Southern Hemisphere, commercial fishing depends on wind-induced upwelling along Peru and Chile. When prevailing coastal winds blow in from the south and southeast, they tug surface water away from shore. Cold, deeper water that the Humboldt Current delivers to the continental shelf moves to the surface. Tremendous quantities of nitrate and phosphate are pulled up, then carried north by the cold Peru Current. The nutrients sustain phytoplankton, the basis of one of the world's richest fisheries.

Oscillating Currents and Weather Patterns

Every three to seven years, the warm surface waters of the western equatorial Pacific Ocean move eastward. This massive displacement of warm water affects the prevailing wind direction. The eastward flow speeds up so much that it influences the movement of water along the coastlines of Central and South America.

Water driven into a coastline is forced downward and flows outward from the continental shelf. Near Peru's coast, the prolonged "downwelling" of nutrient-poor water displaces cooler waters of the Humboldt Current—and prevents upwelling. The current usually arrives around Christmas. Fisherman in Peru named it **El Niño** ("the little one," a reference to the baby Jesus).

The name was incorporated into a more inclusive, scientific explanation for the event: El Niño Southern Oscillation, or ENSO. Massively dislocated patterns of rainfall characterize this major climatic event, which corresponds to changes in sea surface temperatures and air circulation patterns. The "southern oscillation" is a seesawing of atmospheric pressure in the western equatorial Pacific, the largest reservoir of warm water in the world. More warm air rises in this region than anywhere else. It is the source of heavy rainfall, which releases enough heat energy to drive the world's air circulation system.

Between ENSOs, the warm reservoir and heavy rain associated with it move to the west, as in Figure 49.32a. In an ENSO, surface winds prevailing in the western equatorial Pacific pick up speed, and they "drag" the surface waters east (Figure 49.32b). As eastward water transport increases, westward transport slows. More warm water in the reservoir moves eastward, in a vast feedback loop between the ocean and atmosphere. As a result, sea surface temperatures rise, evaporation from the ocean accelerates, and air pressure declines. These events have repercussions all around the world.

Humid updrafts give birth to heavier, often violent storms and flooding along the coasts of continents. In

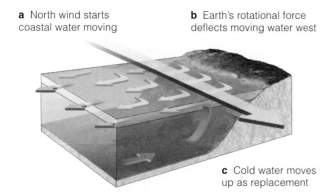

a North wind starts coastal water moving

b Earth's rotational force deflects moving water west

c Cold water moves up as replacement

Figure 49.31 Upwelling along the west coasts in the Northern Hemisphere. Prevailing north winds start surface water moving. The Earth's rotational force deflects the moving water westward. Cold, deeper, nutrient-rich water moves up to replace it. The nutrients sustain marine life, including these anchovies.

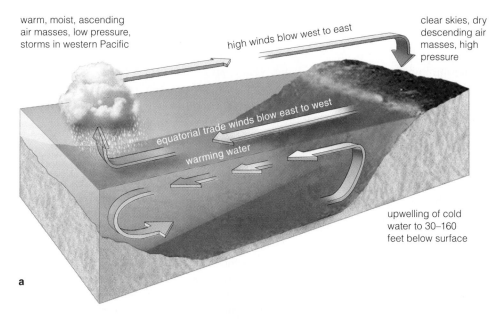

warm, moist, ascending air masses, low pressure, storms in western Pacific

high winds blow west to east

clear skies, dry descending air masses, high pressure

equatorial trade winds blow east to west

warming water

upwelling of cold water to 30–160 feet below surface

a

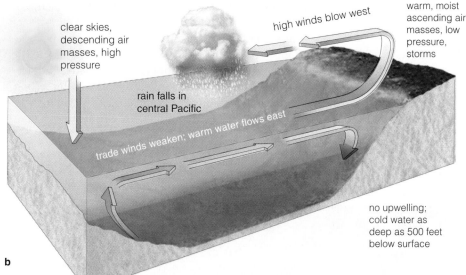

clear skies, descending air masses, high pressure

high winds blow west

warm, moist ascending air masses, low pressure, storms

rain falls in central Pacific

trade winds weaken; warm water flows east

no upwelling; cold water as deep as 500 feet below surface

b

Figure 49.32 (**a**) Typical westward flow of cold, equatorial surface water between ENSOs. (**b**) Eastward dislocation of warm ocean water during an ENSO. Some effects of the massive dislocation: (**c**) Prolonged drought and huge fires, (**d**) violent storm in Europe, and (**e**) severe ice storms.

the interior of continents, rainfall is often heavier than usual. In many regions, extended droughts prevail.

Between 1997 and 2003, for example, numbing heat waves caused hundreds of deaths, and wildfires burned up much of the natural habitats in the United States and Mexico. Monsoons were delayed in Southeast Asia, so fires coincidentally set to clear tracts of rain forest swept out of control through 4,000 square kilometers. Fires also blazed through parts of the Amazon River basin (Figure 49.32c). Australia and much of North America have been reeling from severe droughts and storms.

Intense winter storms caused massive mudslides and flooding along coasts (Figure 49.32d). One year, an ice storm struck northeastern Canada and the United States, damaging and killing millions of trees (Figure 49.32e). Ice on power lines caused power outages for millions of

people. The largest hurricane ever recorded struck the eastern Pacific; megacyclones struck the central Pacific.

On average, El Niño episodes last between six and eighteen months. In the intervening period, called **La Niña**, the weather seesaws again. Another correlation: As global warming continues, the El Niño episodes are becoming more frequent, and regional weather patterns throughout the biosphere are becoming more extreme.

A recurring westward deflection of the equatorial Pacific Ocean's surface waters leads to massive upwelling of the cold, nutrient-rich water that sustains marine ecosystems. It alternates with an eastward deflection of nutrient-poor surface waters.

The seesawing dramatically changes the global weather.

RITA IN THE TIME OF CHOLERA

We leave this chapter with a story about how one biologist solved one of nature's puzzles and helped improve human health. Rita Colwell, a microbiologist who later became Director of the National Science Foundation, drew on her knowledge of long-term adaptations of certain microbes to their environment. She was able to interpret that knowlege through her understanding of interconnections between the ocean, the atmosphere, and the land.

SEA SURFACE TEMPERATURE AND ALGAL BLOOMS For many years now, ocean-bobbing buoys and satellites in space have been gathering data on winds, currents, and sea surface temperatures. Information has been loaded into supercomputers that can simulate global weather patterns. The simulations are getting good. In 1997, scientists predicted the most powerful ENSO of the century a season in advance. Average sea surface temperatures rose nine degrees in the eastern Pacific. Warm water extended 9,660 kilometers west from Peru's coast and 320 kilometers north.

Researchers documented record-breaking effects of the 1997–1998 El Niño/La Niña rollercoaster on primary productivity in the equatorial Pacific Ocean. With the massive eastward displacement of nutrient-poor, warm water, the so-called pastures of the seas shrank almost to undetectable levels (Figure 49.33*a*). During the La Niña rebound, however, cool, nutrient-rich water welled up to the surface and was displaced westward along the equator. Satellite imagery from NASA revealed that upwelling had promoted a vast algal bloom, one that stretched across the equatorial Pacific (Figure 49.33*b*).

THE CHOLERA CONNECTION During the 1997–1998 El Niño episode, 30,000 cases of cholera were reported in Peru compared to only 60 cases from January to August in 1997. For some time, people knew that cholera was linked to water contaminated by *Vibrio cholerae* (Figure 49.34*a*). Epidemics develop after this bacterium infects humans, causes the severe diarrhea that characterizes the disease, and enters water supplies in feces. When people use contaminated water, they become infected.

What people did *not* know was where the pathogen lurked between cholera outbreaks. Year after year, no one could find it in humans or in water supplies. Then cholera would emerge simultaneously in places some distance apart—most often in coastal cities, where the urban poor draw water from rivers that enter the sea.

Colwell had been thinking about cholera for a long time. She suspected that humans did not harbor the pathogen between cholera outbreaks. Was there an

Near-absence of phytoplankton in the equatorial Pacific Ocean during the 1997–1998 El Niño event

Immense algal bloom in the equatorial Pacific Ocean during the La Niña rebound event

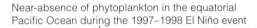
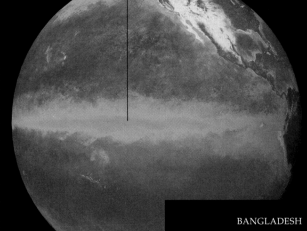

Figure 49.33 Satellite data on the primary productivity in the equatorial Pacific Ocean, using chlorophyll concentrations as the measure. (**a**) During the 1997–1998 El Niño episode, nutrient-poor water moved eastward, and photosynthetic activity was negligible. (**b**) During the subsequent La Niña episode, massive upwelling and the westward displacement of nutrient-rich water fanned this huge algal bloom. (**c**) Satellite measurements of rising sea surface temperatures in the Bay of Bengal correlated with cholera cases in the region's hospitals. Shown here, the warmest summer temperatures are coded *red*.

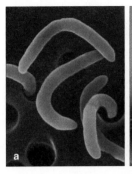

Figure 49.34 (**a**) *Vibrio cholerae*, the agent of cholera. (**b**) The copepod host in which a dormant stage of *V. cholerae* waits out environmental conditions that are unfavorable to its growth and reproduction. (**c**) Typical waterway in Bangladesh from which water samples were drawn for analysis. (**d**) In Bangladesh, Rita Colwell comparing unfiltered and filtered drinking water.

environmental reservoir for the pathogen? Maybe. But nobody had detected any cells of *V. cholerae* in water samples subjected to standard culturing. Then Colwell had a flash of insight: What if no one could find the pathogen because it changes form and enters a dormant, sporelike stage between outbreaks?

During a cholera outbreak in Louisiana, she decided to use labeled antibodies that would bind to a certain protein at the bacterium's surface. Later, antibody tests in Bangladesh pinpointed the bacterium in fifty-one of fifty-two water samples. Standard culture methods had missed it in all but seven of the samples.

V. cholerae lives in brackish water, rivers, estuaries, and seas. As Colwell knew, plankton (communities of mostly microscopic aquatic species) also thrive in these aquatic environments. She focused her search on the waters near Bangladesh, where outbreaks of cholera are endemic and seasonal.

Colwell eventually discovered a dormant stage of *V. cholerae* in copepods. These marine invertebrates graze on algae and other photoautotrophs in plankton (Figure 49.34*b*). The abundance of copepods—hence of *V. cholerae*—rises and falls with shifts in the abundance of phytoplankton, the pastures of the seas.

Colwell already knew about the seasonal variations in sea surface temperatures. Remember the old saying, Chance favors the prepared mind? In a sense, she was prepared to recognize a correlation between cholera cases and seasonal temperature peaks in the water of the Bay of Bengal. When she compared data from the 1990–1991 and 1997–1998 El Niño episodes, her correlation held. *Four to six weeks after the sea surface temperatures go up, so do the cases of cholera!*

Together with her colleague Anwarul Huq, a native of Bangladesh, Colwell is working to identify other mitigating factors. These include shifts in salinity and nutrient content that cause algal blooms, hence rises in copepod population size. Their goal is to integrate the data into a model that might be used to predict precisely where cholera will break out and to give advance warning to start filtering drinking water.

Meanwhile, she is advising Bangladeshi women to use four layers of old sari cloth as filters, which remove 99 percent of *V. cholerae* cells from the water (Figure 49.34*d*). Even though the pathogenic bacterial cells are small enough to pass through the cloth, their copepod hosts are not. The cloths can be rinsed in clean water, sun-dried, and used again and again.

Combine knowledge about life with knowledge about the physical and chemical aspects of the biosphere, and who knows what you may discover.

SUMMARY
Gold indicates text section

1. The biosphere encompasses the Earth's waters, the lower atmosphere, and the uppermost portions of its crust in which organisms live. Energy flows one way through the biosphere. Materials move through it on a grand scale to influence ecosystems everywhere. *CI*

2. The distribution of species through the biosphere is an outcome of the Earth's history, topography, climate, and interactions among species. *CI, 49.1, 49.2*

    a. "Climate" refers to average weather conditions, including temperature, humidity, wind velocity, cloud cover, and rainfall over time. It results from differences in the amount of solar radiation reaching equatorial and polar regions, the Earth's daily rotation and its annual path around the sun, the distribution of continents and seas, and the elevation of land masses.

    b. Interacting climatic factors underlie prevailing winds and ocean currents, which shape global weather patterns. Weather affects soil composition and water availability, which in turn influence the distribution and growth of primary producers in ecosystems.

3. The world's land masses are classified as six major biogeographic realms. Each is more or less isolated by oceans, mountain ranges, or desert barriers that tend to restrict gene flow between realms, so each generally maintains a characteristic array of species. *49.3*

4. Biomes (a category of major ecosystems on land) are shaped by regional variations in climate, landforms, and soils. Dominant plants reflect prevailing conditions in the biomes called deserts, dry shrublands and dry woodlands, grasslands, broadleaf forests (e.g., tropical forests), coniferous forests, and tundra. *49.3–49.9*

5. Water provinces cover more than 71 percent of the Earth's surface. They include standing fresh water (such as lakes and ponds), running fresh water (including streams), and the world ocean. All aquatic ecosystems show gradients in light penetration, water temperature, salinity, and dissolved gases. These factors vary daily and seasonally, and affect primary productivity. *49.10*

6. Wetlands, intertidal zones, rocky and sandy shores, tropical reefs, and regions of the open ocean are major marine ecosystems. Photosynthetic activity is greatest in shallow coastal waters and in regions of upwelling: an upward movement of deep, cool ocean water that often carries nutrients to the surface. *49.12, 49.13*

Review Questions

1. Define biosphere. As part of your answer, also define atmosphere, lithosphere, and hydrosphere. *CI*

2. List the major interacting factors that influence climate. *49.1*

3. List some of the ways in which air currents, ocean currents, or both may influence the region where you live. *49.1, 49.2*

4. Indicate where these biomes tend to be located in the world, and describe some of their defining features. *49.5–49.9*

   a. deserts   e. evergreen   h. alpine tundra
   b. dry shrublands   broadleaf forests   i. boreal forest
   c. dry woodlands   f. deciduous forests   j. montane
   d. grasslands   g. arctic tundra   coniferous forests

5. Define soils, then explain how the composition of regional soils affects ecosystem distribution. *49.4*

6. Describe the littoral, limnetic, and profundal zones of a large temperate lake in terms of seasonal primary productivity. *49.10*

7. Define the two major provinces of the world ocean. Is the open ocean devoid of life? *49.11*

8. Define and characterize the following types of ecosystems: mangrove wetland, estuary, and the intertidal zone. *49.12*

Self-Quiz   ANSWERS IN APPENDIX III

1. Solar radiation drives the distribution of weather systems and so influences _____ .
    a. temperature zones   c. seasonal variations
    b. rainfall distribution   d. all of the above

2. The _____ is a shield against ultraviolet wavelengths.
    a. upper atmosphere   c. ozone layer
    b. lower atmosphere   d. greenhouse effect

3. Regional variations in the global patterns of rainfall and temperature depend on _____ .
    a. global air circulation   c. topography
    b. ocean currents   d. all of the above

4. A rain shadow is a reduction in rainfall on the _____ of a mountain range.
    a. windward side   c. highest elevation
    b. leeward side   d. lowest elevation

5. Biogeographic realms are _____ .
    a. land and water provinces   c. divided into biomes
    b. six major land provinces   d. b and c

6. Biome distribution corresponds roughly with regional variations in _____ .
    a. climate   b. soils   c. topography   d. all of the above

7. Dominant plants of _____ are highly adapted to recurring episodes of lightning-sparked fires.
    a. dry shrublands   c. southern pine forests
    b. grasslands   d. all of the above

8. During _____ , deeper, often nutrient-rich water moves to the surface of a body of water.
    a. spring overturns   c. upwellings
    b. fall overturns   d. all of the above

9. Match the terms with the most suitable description.
    _____ boreal forest   a. high productivity; rapid cycling
    _____ permafrost   of nutrients (poor reservoirs)
    _____ chaparral   b. "swamp forest"
    _____ tropical   c. a type of dry shrubland
          rain forest   d. feature of arctic tundra

10. Match the terms with the most suitable description.
    _____ marine snow   a. deep, cool, often nutrient-rich
    _____ upwelling   ocean water moves upward
    _____ eutrophication   b. sediments, rocks of ocean bottom
    _____ estuary   c. partially enclosed mix of seawater
    _____ benthic   and fresh water
          province   d. nutrient enrichment of body
                     of water; reduced transparency,
                     phytoplankton blooms
                    e. basis of midocean food webs

OLIGOTROPHIC LAKE	EUTROPHIC LAKE
Deep, steeply banked	Shallow with broad littoral
Large deep-water volume relative to surface-water volume	Small deep-water volume relative to surface-water volume
Highly transparent	Limited transparency
Water blue or green	Water green to yellow- or brownish-green
Low nutrient content	High nutrient content
Oxygen abundant through all levels throughout year	Oxygen depleted in deep water during summer
Not much phytoplankton; green algae and diatoms dominant	Abundant, thick masses of phytoplankton; and cyanobacteria dominant
Abundant aerobic decomposers favored in profundal zone	Anaerobic decomposers
Low biomass in profundal	High biomass in profundal

Figure 49.35 (**a**) Crater Lake, Oregon, a collapsed volcanic cone filled with rainwater and melted snow. Like other volcanoes of the Cascade range, it started forming as a result of tectonic forces that prevailed at the dawn of the Cenozoic. (**b**) Major characteristics of oligotrophic and eutrophic lakes.

Critical Thinking

1. Kangaroos are furry, pouched mammals. Raccoons are furry placental mammals. Speculate on why Australia but not North America is the original home of kangaroos and why the reverse is true of raccoons. Use Sections 26.9 and 26.10 when devising an explanation.

2. Reflect on the world distribution of land masses and ocean water, then develop an explanation of why grassland biomes tend to form in the interior of continents, not along their coasts.

3. *Wetlands,* remember, are transitional zones between aquatic and terrestrial ecosystems in which plants are adapted to grow in periodically or permanently saturated soil. They include mangrove swamps and riparian zones (Section 27.6). Wetlands everywhere are being converted for agriculture, home building, and other human activities. Does the great ecological value of wetlands for the nation as a whole outweigh the rights of the private citizens who own many of the remaining wetlands in the United States? Should the private owners be forced to transfer title to the government? If so, who should decide the value of a parcel of land *and* pay for it? Or would such seizure of private property violate laws of the United States?

4. Observe conditions in a lake near your home or a place where you vacation. If lakes aren't part of your life, think about Oregon's Crater Lake instead (Figure 49.35*a*). Using the data in Figure 49.35*b* as a guide, would you conclude that Crater Lake is oligotrophic or eutrophic? Will it remain so over time?

5. The United States Forest Service and Geophysical Dynamics Laboratory ran supercomputer programs to predict possible outcomes of a *global warming* trend. (Here you may wish to review Section 48.9.) According to their results, by the year 2030, the United States will experience recurring, devastating forest fires. Severe storms and more frequent rainfall in the western states will accelerate erosion, especially along the coasts and in steep foothills and mountain ranges.

Increases in the deposition of sediments will have adverse effects on rivers, streams, and estuaries. Dry shrublands and woodlands may replace hardwood forests in Minnesota, Iowa, and Wisconsin, and in parts of Missouri, Illinois, Indiana, and Michigan. The breadbasket of the American Midwest will shift north into Canada. Think about where you live now and where you plan to live in the future. How would such changes affect you? Should nations get serious about finding ways to counter global warming? Or do you believe that the scientists who are concerned about this are merely alarmist and that nothing bad will happen? Explain how you have reached your conclusion.

6. Think about the ocean, which covers the majority of the Earth's surface. Its deepest regions are remote from human populations, to say the least. Now think about a modern-day problem—where to dispose of nuclear wastes and other truly hazardous materials. You will be reading about this serious problem in the next chapter. For now, simply be aware that the United States government has stockpiles of very dangerous wastes and has nowhere to put them. Would it be feasible, let alone ethical, to use the deep ocean as a "burying ground" for them? Why or why not?

Selected Key Terms

atmosphere *CI*
biogeographic realm *49.3*
biogeography *CI*
biome *49.3*
biosphere *CI*
boreal forest *49.8*
climate *CI*
coniferous forest *49.8*
deciduous broadleaf forest *49.7*
desert *49.5*
desertification *49.5*
dry shrubland *49.6*
dry woodland *49.6*
ecoregion *49.3*

El Niño *49.13*
estuary *49.12*
eutrophication *49.10*
evergreen broadleaf forest *49.7*
fall overturn *49.10*
grassland *49.6*
hydrothermal vent *49.11*
intertidal zone *49.12*
lake *49.10*
La Niña *49.13*
mangrove wetland *49.12*
marine snow *49.11*
monsoon *49.2*
ocean *49.2*

permafrost *49.9*
rain shadow *49.2*
savanna *49.6*
soil *49.4*
southern pine forest *49.8*
spring overturn *49.10*
stream *49.10*
temperature zone *49.1*
tropical rain forest *49.7*
tundra *49.9*
ultraplankton *49.11*
upwelling *49.13*

Readings

Brown, J., and A. Gibson. 1983. *Biogeography.* St. Louis: Mosby.

Dold, C. February 1999. "The Cholera Lesson." *Discover,* 71–76.

Garrison, T. 1996. *Oceanography.* Second edition. Belmont, California: Wadsworth. Accessible; read its lyrical epilogue.

Olson, D. M., and E. Dinerstein, 1998. "The Global 200: A Representation Approach to Conserving the Earth's Most Biologically Valuable Ecoregions." *Conservation Biology* 12(3).

Smith, R. 1996. *Ecology and Field Biology.* Fifth edition. New York: HarperCollins.

On-Line readings at Student Guide for InfoTrac:
www.brookscole.com/biology

50

PERSPECTIVE ON HUMANS AND THE BIOSPHERE

An Indifference of Mythic Proportions

Of all the concepts introduced in the preceding chapter, the one that should be foremost in your mind is this: *The atmosphere, ocean, and land interact in ways that help dictate living conditions throughout the biosphere.* Driven by energy streaming in continually from the sun, the interactions give rise to globe-spanning atmospheric and oceanic circulation patterns on which life ultimately depends. This final chapter turns to a related concept of equal importance. Simply put, *we have become major players in these global interactions even before we fully comprehend how they work.*

To gain perspective on what is happening, think about something we take for granted—the air around us. The composition of the present-day atmosphere is a result of geologic and metabolic events—most notably, photosynthesis—that began billions of years ago. The first humans, recall, had evolved by 2.4 million years ago. Like us, they breathed oxygen from an atmosphere of ancient origins. Like us, they were protected from the sun's ultraviolet radiation by the ozone layer, a shield in the stratosphere. The size of their population was not much to speak of, and their effect on the biosphere was trivial. About 11,000 years ago, however, agriculture began in earnest, and it laid the foundation for huge increases in population size. Starting a few centuries ago, medical and industrial revolutions expanded that foundation for growth—and human population size skyrocketed in a mere blip of evolutionary time.

Today we are extracting huge amounts of energy and resources from the environment and giving back monumental amounts of wastes. As we do so, we are destabilizing ecosystems everywhere, even though the magnitude of change might not even be recognized when measured on the scale of a lifetime (Figure 50.1).

In a few developed countries, population growth has more or less stabilized, and the resource use per individual has dropped a bit. But resource utilization levels in those places are already high. In developing

1900   1940   1954   1962

countries in Central America, Africa, and elsewhere, population sizes and resource consumption are rising fast. Yet about half of the people already are struggling to survive on less than three dollars a day.

Many of the problems sketched out in this chapter are not going to disappear tomorrow. It will take many decades, even centuries, to reverse some trends that are already in motion—and not everyone is ready to make the effort. A few enlightened individuals in Michigan or Kenya or New South Wales can commit themselves to resource conservation and to minimizing pollution. But scattered attempts will not be enough. Individuals of all nations will unite to reverse global trends only when they perceive that the dangers of *not* doing so outweigh the personal benefits of ignoring them.

Does this seem pessimistic? Think of the exhaust fumes released to the air each time you travel in a car or bus. Think of oil refineries, food-processing plants, and paper mills that supply you with goods and also release chemical wastes into the nation's waterways.

Who changes behavior first? We have no answer. We can suggest, however, that a strained biosphere can rapidly impose an answer upon us. As an individual, you might choose to cherish or brood about or ignore any aspect of the world of life. Whatever your choice, the bottom line is that you and all other organisms are in this together. Our lives interconnect to degrees that we are only now starting to comprehend.

Figure 50.1 Tracking human population growth in one small part of the world, based on historical data and satellite imaging. *Red* denotes areas of dense human settlement in and around the San Francisco Bay Area and Sacramento since 1900.

1974

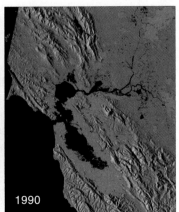
1990

Key Concepts

1. Human population growth has been skyrocketing ever since the mid-eighteenth century. At present, humans have the population size, the technology, and the cultural inclination to use energy and alter the environment at astonishing rates.

2. Pollutants are substances with which ecosystems have had no prior evolutionary experience, in terms of kinds or amounts. So adaptive mechanisms that might deal with them are not in place. In a more restrictive sense, a pollutant is any substance that accumulates in the environment in amounts that can adversely affect an individual's health, activities, and survival.

3. Many pollutants, including kinds that contribute to the formation of smog and acid rain, exert regionally harmful effects. Effects of other pollutants, including chlorofluorocarbons that attack the ozone layer in the stratosphere, are global in scale.

4. The conversion of marginally fertile lands for agriculture, highly mechanized deforestation, and other practices required to meet the demands of the growing human population are resulting in a loss of soil fertility and desertification. These practices are causing a decline in the quality and quantity of one of the most crucial of all resources—fresh water. They also are endangering biodiversity.

5. Ultimately, the world of life runs on energy inputs from the sun that drive complex interactions among the atmosphere, ocean, and land. As a population, we are intervening in these interactions in ways that may have severe consequences in the near future.

6. We as a species must come to terms with principles of energy flow and principles of resource utilization that govern all systems of life on Earth.

AIR POLLUTION—PRIME EXAMPLES

Let's start this survey of the impact of human activities by defining pollution, a central problem. **Pollutants** are substances with which ecosystems have had no prior evolutionary experience, in terms of kinds or amounts, so adaptive mechanisms that might deal with them are not in place. From the human perspective, pollutants are substances that have accumulated enough to have adverse effects on our health, activities, or survival.

Air pollutants are premier examples. As Table 50.1 shows, they include carbon dioxide, oxides of nitrogen and sulfur, and chlorofluorocarbons. Also among them are photochemical oxidants formed when the sun's rays interact with certain chemicals. The United States alone releases 700,000 metric tons of air pollutants each day. Whether these remain concentrated at the source or are dispersed depends on local climate and topography.

Table 50.1 *Major Classes of Air Pollutants*	
Carbon oxides	Carbon monoxide (CO), carbon dioxide (CO_2)
Sulfur oxides	Sulfur dioxide (SO_2), sulfur trioxide (SO_3)
Nitrogen oxides	Nitric oxide (NO), nitrogen dioxide (NO_2), nitrous oxide (N_2O)
Volatile organic compounds	Methane (CH_4), benzene (C_6H_6), chlorofluorocarbons (CFCs)
Photochemical oxidants	Ozone (O_3), peroxyacyl nitrates (PANs), hydrogen peroxide (H_2O_2)
Suspended particles	Solids (dust, soot, asbestos, lead, etc.) and droplets (sulfuric acid, oils, pesticides, etc.)

Smog

When weather conditions trap a layer of cool, dense air under a warm air layer, this is a **thermal inversion**. If pollutants are in the trapped air, winds cannot disperse them, and they might accumulate to dangerous levels. Thermal inversions have been key factors in some of the worst air pollution disasters, because they intensify an atmospheric condition called smog (Figure 50.2).

Two types of smog, gray air and brown air, form in major cities. Where winters are cold and wet, **industrial smog** develops as a gray haze over industrialized cities

that burn coal and other fossil fuels for manufacturing, heating, and generating electric power. Burning releases airborne pollutants, such as dust, smoke, soot, ashes, asbestos, oil, bits of lead and other heavy metals, and sulfur oxides. If not dispersed by winds and rain, the emissions may reach lethal concentrations. Industrial smog caused 4,000 deaths during a 1952 air pollution disaster in London. Before coal burning was restricted, New York, Pittsburgh, and Chicago also were gray-air cities. Currently, most industrial smog forms in China, India, and other developing countries as well as in the coal-dependent countries of eastern Europe.

In warm climates, **photochemical smog** forms as a brown haze over large cities. It gets concentrated where the land forms a natural basin, as it does around Los Angeles and Mexico City (Figure 50.2). The key culprit, nitric oxide, is discharged from vehicles. It reacts with oxygen in air to form nitrogen dioxide. Nitrogen dioxide exposed to sunlight reacts with hydrocarbons and forms photochemical oxidants. Most hydrocarbons are released from spilled or partly burned gasoline. Key oxidants are ozone and peroxyacyl nitrates, or **PANs**. Even traces of PANs sting eyes, irritate lungs, and damage crops.

Acid Deposition

Sulfur and nitrogen oxides are among the worst air pollutants. Coal-burning power plants, metal smelters, and factories emit most sulfur dioxides. Motor vehicles, power plants that burn gas and oil, and nitrogen-rich fertilizers produce nitrogen oxides. In dry weather, these oxides may remain airborne briefly, then fall as **dry acid deposition**. In moist air, they form nitric acid vapor, sulfuric acid droplets, and sulfate and nitrate salts. Winds may disperse them far from their source. If they fall to Earth in rain and snow, we call this wet acid deposition, or **acid rain**. Normal rainwater is pH 5 or so. Acid rain can be 10 to 100 times more

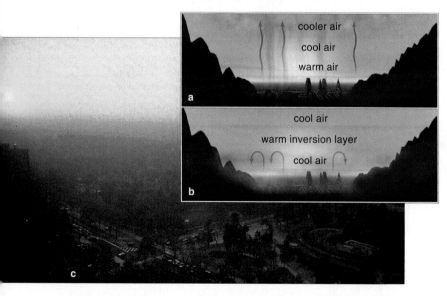

Figure 50.2 (**a**) Normal pattern of air circulation in smog-forming regions. (**b**) Air pollutants trapped under a thermal inversion layer. (**c**) Mexico City on an otherwise bright, sunny morning. Topography, staggering numbers of people and vehicles, and industry combine to make its air among the world's smoggiest. Breathing this city's air is like smoking two packs of cigarettes a day.

Figure 50.3 Average acidities of precipitation in the United States in 1998. *Red* dots pinpoint large coal-burning power and industrial plants. Prolonged exposure to air pollutants is contributing to a rapid decline of forests around the world. Weakened trees are susceptible to drought, disease, and insect pests.

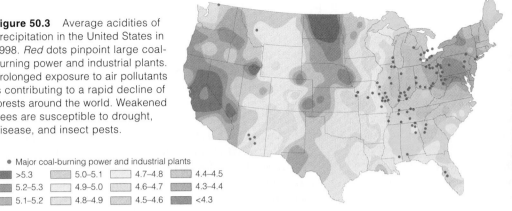

- Major coal-burning power and industrial plants

>5.3	5.0–5.1	4.7–4.8	4.4–4.5
5.2–5.3	4.9–5.0	4.6–4.7	4.3–4.4
5.1–5.2	4.8–4.9	4.5–4.6	<4.3

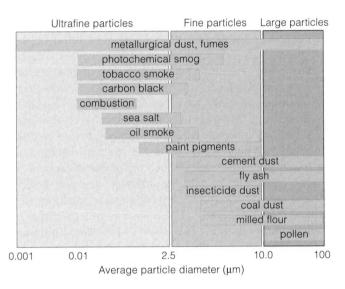

Figure 50.4 Suspended particulate matter, consisting of solids and liquid droplets small enough to remain airborne for variable intervals. Ultrafine particles reach alveoli in lungs and contribute to respiratory disorders. Carbon black is a powdered form of carbon used in paints, tires, and other goods. About 6 million tons are manufactured annually. Fly ash is a major by-product of coal combustion. More than 45 million tons are generated annually.

acidic, even as potent as lemon juice! Deposited acids eat away at metals, the marble on buildings, rubber, plastics, nylon stockings, and other materials. They also disrupt the physiology of organisms, the chemistry of ecosystems, and biodiversity.

Depending on soil type and vegetation cover, some regions are far more sensitive than others to acid rain (Figure 50.3). Highly alkaline soil can neutralize acids before they enter the streams and lakes of watersheds. Water having a high carbonate content also neutralizes acids. However, many watersheds of northern Europe, of southeastern Canada, and of regions throughout the United States have thin soils on top of solid granite. These soils cannot buffer much of the acidic inputs.

Rain in much of eastern North America is thirty to forty times more acidic than it was several decades

ago. Crop yields are declining. Fish populations have already disappeared from more than 200 lakes in New York's Adirondack Mountains. By some estimates, fish will vanish from 48,000 lakes in Ontario, Canada, in the next two decades. Pollution from industrial regions also is changing rainfall acidity enough to contribute to the decline of forest trees and the mycorrhizae that support new growth (Chapter 24).

Researchers confirmed long ago that emissions from power plants, factories, and vehicles are the key sources of air pollutants. In 1995, researchers from the Harvard School of Public Health and Brigham Young University reported this finding from a comprehensive air quality study: You will shorten your life span by a year or so if you live in cities with air having fine particles of dust, soot, smoke, or acid droplets. Smaller particles damage lung tissue. High concentrations of ultrafine particles can cause lung cancer (Figure 50.4).

At one time the world's tallest smokestack, in the Canadian province of Ontario, accounted for 1 percent by weight of the annual worldwide emissions of sulfur dioxide. But Canada gets more acid deposition from industrialized regions of the midwestern United States than it sends across its southern border. Most of the air pollutants in Scandinavian countries, the Netherlands, Austria, and Switzerland arise in industrialized parts of western and eastern Europe. *Prevailing winds—hence air pollutants—do not stop at national boundaries.*

Pollutants are substances with which ecosystems have had no prior evolutionary experience, in terms of kinds or amounts, so adaptive mechanisms are not in place to deal with them.

Accumulated pollutants can harm organisms, as when they reach levels that adversely affect human health.

Smog forms mainly as a result of fossil fuel burning in urban and industrialized regions. Airborne acidic pollutants drift down as dry particles or as components of acid rain.

Smog formation and acid deposition are two examples of air pollution in specific regions, although prevailing winds often distribute pollutants beyond regional boundaries.

OZONE THINNING—GLOBAL LEGACY OF AIR POLLUTION

Now think about the ozone layer, almost twice as high above sea level as Mount Everest, the highest place on Earth. September through mid-October, the layer thins at high latitudes above Antarctica. The seasonal **ozone thinning** is so pronounced that it was once called an "ozone hole." Since 1999, ozone thinning has peaked at 26 million square kilometers, which is about the size of North America (Figure 50.5).

Why is ozone reduction alarming? More ultraviolet radiation reaches the Earth and is triggering far more skin cancers, cataracts, and weakened immune systems. Such radiation also adversely affects photosynthesis. Substantial declines in the oxygen-releasing activity of

phytoplankton alone could change the atmosphere's composition (Section 7.8).

CFCs (chlorofluorocarbons) are the major factors in ozone reduction. These compounds of carbon, fluorine, and chlorine are odorless and invisible. You find them in refrigerators and air conditioners (as the coolants), solvents, and plastic foams. CFCs slowly escape into air and resist breakdown. When a free CFC molecule absorbs ultraviolet light, it gives up one chlorine atom. If chlorine reacts with ozone, this yields oxygen and chlorine monoxide—which reacts with free oxygen to release another chlorine atom. Each released chlorine atom breaks apart 10,000+ ozone molecules!

Chlorine monoxide levels above Antarctica are 100 to 500 times higher than at midlatitudes. Why? High-altitude ice clouds form there during winter. Winds rotate around the South Pole for most of the winter, like a dynamic fence that helps keep ice clouds from spreading into other latitudes (Figure 50.5). This also happens on a lesser scale in the Arctic. Ice crystals are surfaces upon which chlorine compounds are swiftly degraded. When air warms in the spring, the chlorine is free to destroy ozone—hence the ozone thinning.

CFCs are not the only ozone eaters. An even nastier one is methyl bromide. This fungicide persists in the atmosphere only briefly. But it is so widely used that it will account for about 15 percent of the ozone thinning in the years to come unless production stops.

A few scientists dismiss the threat of chemicals that contain chlorine and bromine. But the vast majority of those who have studied ecosystem modeling programs conclude that these chemicals pose long-term threats—not only to human health and certain crops but also to animal life in general. Substitutes are now available for most uses of CFCs; others are being developed.

Under international agreement, CFC production in developed countries has been phased out. Developing countries will phase out CFCs by 2010. Methyl bromide production will also end by then. Recent computer models suggest the area of thinning should not get any larger if international goals are met. Even assuming they are, it still will be 50 years before the ozone layer is restored to 1985 levels and another 100 to 200 years to full recovery, to pre-1950 levels. Meanwhile, you and all children and grandchildren of future generations will have to live with the effects of ozone-destroying chemicals. And if the levels of greenhouse gases also continue to rise, you can expect an increase in the size and duration of seasonal ozone depletions.

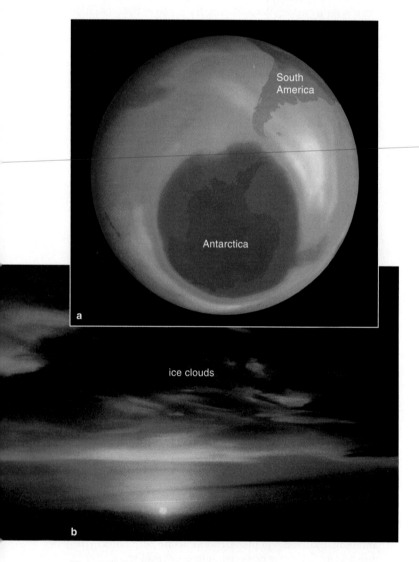

Figure 50.5 (**a**) Seasonal ozone thinning above Antarctica in the fall of 2001. The *darkest blue* coloration represents the area with the lowest ozone level, at that time the largest recorded. (**b**) The ice clouds above Antarctica have a role in ozone thinning, which occurs annually.

Air pollution can have global repercussions, as when CFCs and other compounds contribute to a thinning of the ozone layer that shields life from the sun's ultraviolet radiation.

WHERE TO PUT SOLID WASTES? WHERE TO PRODUCE FOOD?

Oh Bury Me Not, In Our Own Garbage

In natural ecosystems, one organism's wastes serve as resources for others, so by-products of existence are cycled through the system. In developing countries, many resources are scarce. People conserve what they can and discard little. In the United States and some other developed countries, most people use something once, discard it, and buy another. Each year, millions of metric tons of solid wastes are dumped, burned, and buried. Paper products make up half the total volume, which also includes 50 billion nonreturnable cans and bottles. Every week in the United States, paper made from 500,000 trees ends up in Sunday newspapers. If every reader recycled merely one of ten newspapers, 25 million trees per year could be left alone. Recycling paper would reduce the airborne pollutants released during paper manufacturing by 95 percent and require 30 to 50 percent less energy than making new paper.

Our throwaway mentality is unique in nature. We generally accept a monumental accumulation of solid wastes in human ecosystems and natural ones. Bury the garbage in landfills? What happens when the open space around cities is developed? Besides, the landfills "leak" and eventually threaten groundwater supplies. Burn the wastes in inefficient incinerators? These spew volumes of pollutants into the atmosphere, Also, they produce highly toxic ash that must be disposed of.

Recycling is affordable and feasible. At the personal level, individuals can help initiate change by refusing to buy goods that are excessively wrapped, packaged in nondegradable containers, and designed for one-time use. They can participate in curbside recycling. Such action may be more efficient than building huge resource recovery centers. These centers require such a large volume of trash to turn a profit that owners may actually encourage the throwaway mentality. Also, the toxic ash released during the recovery process must be disposed of in landfills, which eventually leak.

Converting Marginal Lands for Agriculture

The human population already uses nearly 21 percent of the Earth's land surfaces for cropland or for grazing. Another 28 percent is said to be potentially suitable for agriculture, but productivity would be so low that the conversion may not be worth the cost (Figure 50.6).

Asia and several other heavily populated regions now experience recurring, severe food shortages. Yet over 80 percent of their productive land is already under cultivation. Scientists have made valiant efforts to improve production on existing cropland. Under the banner of the **green revolution**, their research has been directed toward (1) improving the genetic character of

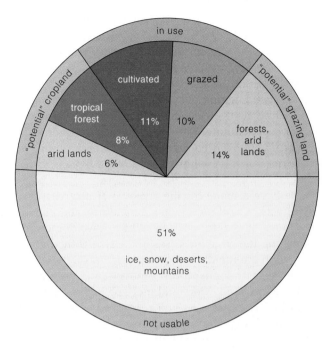

Figure 50.6 Classification of land with respect to its suitability for agriculture. Theoretically, clearing vast tracts of tropical forests and irrigating marginal lands could more than double the world's cropland. Doing so would destroy valuable forest resources, damage the environment, cause severe losses in biodiversity, and possibly cost more than it is worth.

crop plants for higher yields and (2) exporting modern agricultural practices and equipment to the developing countries. Many of these countries rely on *subsistence* agriculture, which runs on energy inputs from sunlight and human labor. They also depend heavily on *animal-assisted* agriculture, with energy inputs from oxen and other draft animals. By contrast, *mechanized* agriculture requires massive inputs of fertilizers, pesticides, and ample irrigation to sustain high-yield crops. It requires fossil fuel energy to drive farm machines. Crop yields are four times as high. But the modern practices use up 100 times more energy and minerals. Also, there are signs that limiting factors may eventually come into play to slow down further increases in crop yields.

Pressure for increased food production is greatest in parts of Central and South America, Asia, the Middle East, and Africa where human populations are rapidly expanding into marginal lands. Repercussions extend beyond national boundaries, as you will see next.

Human population growth has impact on the Earth's land. We generate huge amounts of solid wastes but reuse or recycle little. We expand into marginal lands for food production, often at high energy and environmental costs.

DEFORESTATION—AN ASSAULT ON FINITE RESOURCES

At one time, tropical forests cloaked regions that were, collectively, twice the size of Europe. For 10,000 years or more, they were home to an estimated 50 to 90 percent of all land-dwelling species. In less than four decades, human populations have destroyed more than half the forests. Most of the spectacular arrays of forest species may be lost. Logging over millions of acres every year is the equivalent of leveling eleven to thirty-four city blocks every minute. The logging extends beyond the tropics. Highly mechanized deforestation is going on in the once-vast temperate forests of the United States, Canada, Europe, Siberia, and elsewhere.

We have a name for the removal of all trees from large tracts of land for logging, agriculture, and grazing operations. It is **deforestation**. Why are we doing this? Paralleling the huge increases in the size of the human population are rapidly increasing demands for wood as fuel, lumber, and other forest products; and for grazing land and cropland. More and more people compete for dwindling resources. As you read in Chapter 27, they do so for economic profit, but also because alternative ways of life simply are not available to most people.

The world's great forests have profound influences on ecosystems. Like enormous sponges, the watersheds of forested regions absorb, hold, and gradually release water. By intervening in the downstream flow of water, they help control soil erosion, flooding, and a buildup of sediments that can clog rivers, lakes, and reservoirs. When the vegetation cover is stripped away, exposed soil becomes vulnerable to leaching of nutrients and to erosion, especially on steep slopes.

Deforestation is now greatest in Brazil, Indonesia, Colombia, and Mexico. If the clearing and destruction continue at present rates, then by 2010, only Brazil and Zaire will still contain large tropical forests. By 2035, most of their forested regions will be cleared, also.

Figures 50.7 and 50.8 give different perspectives on what is now happening in the Amazon basin of South America. Section 23.7 includes one example of heavy deforestation in North America.

In tropical regions, clearing forests for agriculture sets the stage for long-term losses in productivity. The irony is that tropical forests are one of the worst places to grow crops and raise pasture animals. In an intact forest, litter cannot accumulate, for the high temperatures and the heavy, frequent rainfall promote the rapid decomposition of organic wastes and remains. As fast as the decomposers release nutrients, the trees and other plants take them up. Deep, nutrient-rich topsoils simply cannot form.

Well before the advent of highly mechanized logging practices, people were practicing **shifting cultivation** (once referred to as slash-and-burn agriculture). They cut and burn trees, then till the ashes into the soil. The nutrient-rich ashes can sustain crops for one to several seasons. Later on, people abandon their cleared plots, for heavy leaching results in infertile soil. However, if shifting cultivation is practiced on small plots that are widely separated, a forest ecosystem

Figure 50.7 Countries that are allowing the greatest destruction of tropical forests. *Red* marks the regions where 50,000 to 170,000 square kilometers are deforested each year. *Orange* denotes "moderate" deforestation, which encompasses areas of 100 to nearly 2,000 square kilometers.

ANDES

SMOKE FROM FIRES

Figure 50.8 The vast Amazon River basin of South America in September 1988. Smoke from fires that had been deliberately set during the dry season to clear tropical forests, pasturelands, and croplands completely obscured its features. The smoke extended to the Andes Mountains near the western horizon, about 650 miles (1,046 kilometers) away. That smoke cover was the largest that astronauts had ever observed. It extended almost 175 million square kilometers (1,044,000 square miles).

The smoke plume near the center of the photograph covered an area comparable to the immense smoke cover arising from the forest fire in Yellowstone National Park in that year. During an El Niño–induced drought of 1997, fires that had been set deliberately burned out of control. A smoke cover formed again.

Extensive deforestation is not confined to equatorial regions. For instance, in the past century, 2 million acres of redwood forests along the California coast were logged over. Much of the logging in temperate forests has occurred since 1950. Many of the logs are exported to lumber mills overseas, where wages are low, then the finished lumber is often reimported.

isn't necessarily damaged extensively. But soil fertility plummets as populations expand. Why? Larger areas are cleared; plots are cleared again at shorter intervals.

Deforestation also affects the rates of evaporation, transpiration, runoff, and maybe regional patterns of rainfall. Trees release 50 to 80 percent of the water vapor above tropical forests. In logged-over regions, annual precipitation drops, and rain swiftly drains away from

Figure 50.9 Wangari Maathai, a Kenyan who organized the internationally acclaimed Green Belt Movement in 1977. The 50,000 members of this women's group are committed to establishing nurseries, raising seedlings, and planting a tree for each of the 30 million Kenyans. Together with half a million schoolchildren, they had planted more than 10 million trees by 1990. Their success inspired similar programs in more than a dozen countries in Africa. Dr. Maathai's efforts are not appreciated by her own government. Kenyan police have jailed her twice, and in 1992 they severely beat her because of her efforts on behalf of the environment.

the exposed, nutrient-poor soil. The regions get hotter and drier—and soil fertility and moisture decline. In time sparse, dry grassland or desertlike conditions may prevail instead of the formerly rich, forested biomes.

Also, tropical forests absorb much of the sunlight reaching equatorial regions. Deforested land is shinier, so to speak, and reflects more incoming energy back into space. And the combined photosynthetic activity of so many trees affects the global cycling of carbon and oxygen. Extensive tree harvesting and burning of the tree biomass releases stored carbon to the atmosphere, as carbon dioxide. In this way, deforestation might be amplifying the greenhouse effect.

Conservation biologists are attempting to reverse the trend. As three examples, a coalition of 500 groups is dedicated to preserving Brazil's remaining tropical forests. In India, women have already built and installed 300,000 inexpensive, smokeless wood stoves. Over the past decade, the stoves saved more than 182,000 metric tons' worth of trees by reducing demands for fuelwood. In Kenya, women have planted millions of trees to hold soil in place and to provide fuelwood (Figure 50.9).

Once-vast forests helped sustain rapid increases in human population growth. Recent and highly mechanized modes of deforestation are rapidly depleting these finite resources.

You and the Tropical Rain Forest

Developing nations in Latin America, Southeast Asia, and Africa have the fastest-growing populations but not enough food, fuel, and lumber. Of necessity, they turn to their forests for growth-sustaining resources. Most of the forests may disappear within our lifetime. That possibility elicits the most outcries from concerned groups in highly developed nations—which happen to use most of the world's resources, including forest products. You read about economic aspects of this issue in earlier chapters. We invite you to reflect now on a few more aspects.

For purely ethical reasons, many people condemn the destruction of so much biodiversity. Tropical rain forests have the greatest variety and numbers of insects, and the world's largest ones. They are home to the most species of birds and to plants with the largest flowers. Living in the forest canopy and understory are monkeys, tapirs, and jaguars in South America and apes, okapi, and leopards in Africa. Massive vines twist around trees. Orchids, mosses, lichens, and other organisms grow on branches, absorbing minerals that rains deliver to them. Entire communities of microbes, insects, spiders, and amphibians live, breed, and die in small pools of water that collect in furled leaves.

Their disappearance will have repercussions through human life. Only a few strains of crop plants and livestock sustain most human populations, and they are vulnerable to ever-evolving pathogens. Tissue-culture specialists and genetic engineers use genes of forest species to develop new or hybrid strains that can make our food base less vulnerable. Geneticists use them to develop more effective antibiotics and vaccines. Aspirin, the most widely used painkiller, is based on a chemical blueprint of an extract from tropical willow leaves. Many ornamental plants and spices and foods, including cocoa, cinnamon, and coffee, originated in the tropics. So did latex, gums, resins, dyes, waxes, and oils for tires, shoes, toothpaste, ice cream, shampoo, compact discs, condoms, and perfumes. Also think about this: Rampant burning of forests is releasing enough air pollutants to change the air you breathe and possibly to overheat the planet during your lifetime.

And so conservation biologists rightly decry the mass extinction, the assaults on species diversity, the depletion of much of the world's genetic reservoir. Yet something else is going on here. Too many of us grow uneasy when we pass through obliterated forests in our own country. Is it because we are losing the comfort of our heritage—a connection with our evolutionary past? Many millions of years ago, our earliest primate ancestors moved into the trees of tropical forests. Through countless generations, their nervous and sensory systems evolved and became highly responsive to information-rich, arboreal worlds.

Does our neural wiring still resonate with rustling leaves, with shafts of light and mosaic shadows? Are we innately attuned to the forests of Eden—or have time and change buried recognition of home?

Figure 50.10
Tropical rain forest in Southeast Asia.

WHO TRADES GRASSLANDS FOR DESERTS?

Long-term shifts in climate often convert grasslands to deserts. Humans do, also. **Desertification** is the name for the conversion of large tracts of natural grasslands to a more desertlike condition. It applies also when the conversion of rain-fed or irrigated croplands causes a 10 percent or greater decline in agricultural productivity. During the past fifty years, 9 million square kilometers worldwide have become desertified.

plants. They also are better at conserving water; they lose little in feces, compared to cattle.

In 1978 a biologist, David Holpcraft, began ranching antelopes, zebras, giraffes, ostriches, and other native herbivores. He also raised cattle as control groups to compare costs and meat yields on the same land. The native herds increased and provided tasty meat. Range conditions did not deteriorate; they improved. Vexing

Figure 50.11 An awe-inspiring dust storm approaching Prowers County, Colorado, in 1934.

The Great Plains of the American Midwest are dry, windy grasslands that are subjected to pronounced and recurring droughts. Extensive conversion of these grasslands to agriculture began in the 1870s. Overgrazing left the ground bare across vast tracts. In May 1934, a cloud of topsoil that blew off the land blanketed the entire eastern portion of the United States, giving the Great Plains a dubious new name—the Dust Bowl. About 3.6 million hectares (9 million acres) of cropland were destroyed. Today, without large-scale irrigation and intensive conservation farming, desertlike conditions could prevail.

Figure 50.12 Desertification in the Sahel, a region of West Africa that forms a belt between the hot, dry Sahara Desert and tropical forests. This savanna country is undergoing rapid desertification due to overgrazing, overfarming, and prolonged drought.

Today, at least 200,000 square kilometers are being converted to deserts each year. Long droughts are accelerating the process, just as one did years ago in the Great Plains (Figure 50.11). At present, however, overgrazing of livestock on marginal lands is the source of most large-scale desertification.

In Africa, for instance, there are too many cattle in the wrong places. Cattle require more water than the native wild herbivores do, so they move back and forth between grazing areas and watering holes more often. As they do, they trample grasses and compact the soil surface (Figure 50.12). By contrast, gazelles and other wild herbivores get most (if not all) of their water from

problems remained. Some local people in Africa prefer beef to game, and view cattle as the symbol of wealth in their society.

Without irrigation and conservation practices, grasslands converted for agriculture often end up as deserts.

A GLOBAL WATER CRISIS

The Earth has a lot of water, but most is too salty for human consumption and agriculture. Imagine all that water in a bathtub. Withdraw all the fresh, renewable portion from lakes, rivers, reservoirs, groundwater, and other sources, and it would barely fill a teaspoon.

Why not consider **desalinization**—removal of salt from seawater? After all, there are unlimited supplies of seawater. Desalinization processes exist; they distill seawater or force it through membranes, by reverse osmosis. Expensive fuel energy drives these processes, so they may be feasible only in Saudi Arabia and a few other countries that have small population sizes and large energy reserves. In some situations desalinization may be an alternative to running out of drinkable water. That nearly happened in Santa Barbara and some other cities in California during one prolonged drought. However, desalinization cannot solve the core problem. It may never be cost-effective for large-scale agriculture, and mountains of salts are left over.

Consequences of Heavy Irrigation

Large-scale agriculture accounts for two-thirds of the human population's use of fresh water. In many cases, irrigation water from surface sources is piped into extensive fields where water-demanding crops will not grow on their own. At its most extreme, irrigation turns some hot deserts into lush gardens, although believing these can be maintained over the long term is a bit delusional (Figure 50.13).

Irrigation itself can change the land's suitability for agriculture. Concentrations of mineral salts commonly are high in piped-in water. In regions where soil drains poorly, evaporation may cause **salinization**: a buildup of salt in soil. Salinization can stunt the growth of crop plants, eventually kill them, and decrease yields.

Land that drains poorly also becomes waterlogged. When water accumulates underground, it slowly raises the **water table**, the upper limit at which the ground is fully saturated. When the water table rises close to the ground's surface, soil gets saturated with saline water, which can damage plant roots. Properly managing the water–soil system often can reverse the waterlogging and salinization. The economic cost to do so is high.

Worse, groundwater overdrafts (the amount nature does not replenish) are high. Figure 50.14 shows areas of aquifer depletion in the United States. Overdrafts have depleted half of the great Ogallala aquifer, which supplies irrigation water for 20 percent of the United States croplands. Also, so much groundwater has been withdrawn for irrigation in California's San Joaquin Valley that the surface of the water table has dropped by as much as six meters.

Figure 50.13 Irrigated crops in the Sahara Desert, in Algeria.

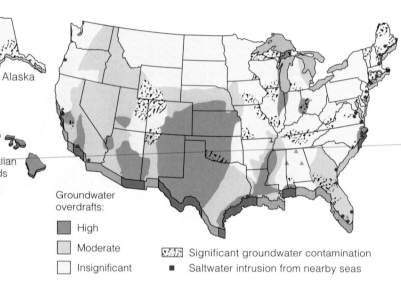

Groundwater overdrafts:
- High
- Moderate
- Insignificant

▨ Significant groundwater contamination
■ Saltwater intrusion from nearby seas

Figure 50.14 Areas of aquifer depletion, saltwater intrusion, and groundwater contamination in the United States.

Groundwater Contamination

About one-third of the world's population depends on groundwater for the water they drink. This includes about 51 percent of the United States population and 75 percent of the populations in Europe. Yet groundwater is being contaminated by toxic chemicals leached from some landfills, hazardous waste dumps, wells that store hazardous chemicals, and underground tanks used to store gasoline, oil, and solvents. Unlike flowing streams, which can cleanse themselves of biodegradable wastes, contaminated groundwater is essentially unreclaimable; it is difficult and very expensive to clean up. The only solution is to prevent toxic chemicals from reaching it.

Water Pollution

Water pollution amplifies the problem of water scarcity. Inputs of sewage, animal wastes, and toxic chemicals

Figure 50.15 An experimental wastewater treatment facility in Rhode Island. Treatment begins when sewage flows into rows of large water tanks in which water hyacinths, cattails, and other aquatic plants are growing. Decomposers in the tank degrade wastes—which contain nutrients that promote plant growth. Heat from incoming sunlight speeds the decomposition.

From these tanks, water flows through an artificial marsh of sand, gravel, and bulrushes that filter out algae and organic wastes. Then it flows into aquarium tanks, where zooplankton and snails consume microorganisms suspended in the water—and where zooplankton become food for crayfishes, tilapia, and other fishes that can be sold as bait. After ten days, the now-clear water flows into a second artificial marsh for final filtering and cleansing.

from power-generating plants and factories can make water unfit to drink. Runoff from fields puts sediments and pesticides into water, along with phosphates and other plant nutrients that promote algal blooms.

Pollutants collect in lakes, rivers, and bays before reaching the oceans. Many cities throughout the world dump untreated sewage into coastal waters. The cities along rivers and harbors maintain shipping channels by dredging the polluted muck and barging it out to sea. They also barge out sewage sludge—coarse, settled solids enriched with bacteria, viruses, and toxic metals.

In the United States, about 15,000 facilities partially treat liquid wastes from 70 percent of the population and 87,000 industries. The remaining wastes are mainly from suburban and rural populations. These are treated in lagoons or septic tanks or are directly discharged—untreated—into waterways.

There are three levels of **wastewater treatment**. In *primary* treatment, screens and settling tanks remove sludge, which is dried, burned, dumped in landfills, or treated further. Chlorine often is used to kill pathogens. It does not kill them all, and it produces carcinogens when it reacts with a number of chemicals. In *secondary* treatment, microbial populations break down organic matter after primary treatment but before chlorination. The two treatments get rid of most solids and oxygen-demanding wastes, but not all nitrogen, phosphorus, and toxins, such as pesticides and heavy metals. Water is usually chlorinated before being released.

Tertiary treatment adequately reduces pollution but is largely experimental and expensive. It is applied to only 5 percent of our nation's wastewater.

In short, most wastewater is not treated adequately. A pattern gets repeated thousands of times along our waterways. Water for drinking is drawn upstream from a city, and wastes from industry and sewage treatment are discharged downstream. It takes no great leap of the imagination to see that water pollution intensifies as rivers flow to the oceans. In Louisiana, waters drained from the central states flow toward the Gulf of Mexico. Its high pollution levels threaten public health as well as ecosystems. Water destined for drinking gets treated to remove pathogens, but treatment does not remove toxic wastes dumped by numerous factories upstream.

This rather bleak picture might be numbing to most of us, but not to biologist John Todd. He constructed experimental wastewater treatment facilities in several greenhouses and artificial lagoons (Figure 50.15). When it works properly, his solar–aquatic treatment system produces water fit to drink. Such natural alternatives cannot work for the large urban areas. But they are an attractive alternative for small towns and rural areas.

The Coming Water Wars

If current rates of human population growth and water depletion hold, the amount of fresh water available for everyone on the planet will soon be 55 to 66 percent less than it was in 1976. Already in this past decade, thirty-three nations have been engaged in conflicts over reductions in water flow, pollution, and silt buildup in major aquifers, rivers, and lakes. Among the squabblers are the United States and Mexico, Pakistan and India, and Israel and the Palestinian territories.

Remember the Persian Gulf War, mainly about oil? Unless we pull off a blue revolution equivalent to the green one, we may be in for upheavals and wars over water rights. Does this sound far-fetched? By building dams and irrigation systems at the headwaters of the Tigris and Euphrates rivers, Turkey can, in the view of one of its dam-site managers, stop the water flow into Syria and Iraq for as long as eight months "to regulate their political behavior." Regional, national, and global planning for the future is long overdue.

Water, not oil, may become the most important fluid of the twenty-first century. National, regional, and global policies for water usage and water rights have yet to be developed.

A QUESTION OF ENERGY INPUTS

Paralleling the J-shaped curve of human population growth is a dramatic rise in total and per capita energy consumption. Many people are now consuming and wasting energy on an extravagant scale. For example, in one city with a near-ideal climate, a major university built new seven- and eight-story buildings with narrow windows that cannot be opened to catch the prevailing ocean breezes. Neither the buildings nor the windows were designed or aligned to take advantage of the abundant sunlight for passive solar heating and breezes for passive cooling. Massive energy-demanding heating and cooling systems were built instead.

When you hear talk of abundant energy supplies, bear in mind that there is a large difference between the *total* and net amounts available. *Net* refers to the amount left over after subtracting the energy necessary to locate, extract, transport, store, and deliver energy to consumers. Some sources, such as direct solar energy, are renewable. Others, including coal, are not. Figure 50.16 summarizes the picture.

Fossil Fuels

Fossil fuels are the legacy of forests that disappeared many hundreds of millions of years ago, so we cannot renew them. Carbon-containing remains of the plants were buried in sediments, compacted, then chemically transformed into coal, petroleum (oil), and natural gas over a great span of time (Section 23.4).

Known reserves of petroleum will be depleted in this century, and natural gas reserves in the next, even with strict conservation. As known reserves run out in accessible areas, we turn to wilderness areas in Alaska and other fragile environments, including continental shelves. When petroleum is extracted and transported to and from remote areas, the net energy cost rises, and risks to the environment escalate. Remember the awful spectacle of the supertanker *Valdez* running aground in Alaska's Prince William Sound? Its hull broke apart and 11 million gallons of oil spilled into the pristine waters. The damage and cleanup costs were huge.

What about coal? In theory, known reserves may meet the energy needs of the human population for at least a few centuries. However, coal burning has been the primary source of air pollution. Most of the known reserves contain low-quality material that has a higher sulfur content. Unless sulfur is removed before or after this fuel is burned, sulfur dioxides are released into the atmosphere and contribute to global acid deposition. Fossil fuel burning in general releases carbon dioxide and adds to the greenhouse effect.

Extensive strip mining of coal reserves close to the surface invites problems. It reduces the land available for agriculture, grazing, and wildlife. Strip mining is occurring mainly in arid and semiarid regions where the absence of sufficient supplies of water and the poor quality of soil make restoration efforts difficult.

Nuclear Energy

NUCLEAR REACTORS In 1945, as Hiroshima burned, people recoiled in horror from the destructive force of nuclear energy. Nevertheless, by the 1950s, many were convinced that nuclear energy was an instrument of progress. Many of the industrialized but energy-poor nations, including France, now depend heavily on nuclear power.

Until recently, construction of more nuclear plants was delayed or canceled in most places. After 1970 in the United States alone, governments shelved the plans to build 117 nuclear power plants. The construction of others was stopped before they were finished. At high cost, several were converted to burn fossil fuels. What happened? People started to question the operating costs, efficiency, and environmental impact of nuclear power. And they questioned its safety.

What about safety? A nuclear plant does emit less radioactivity and carbon dioxide than coal-burning plants of the same capacity. And it emits zero sulfur dioxide. However, it has the potential for a **meltdown.** Nuclear fuel releases a lot of heat by radioactive decay. In the

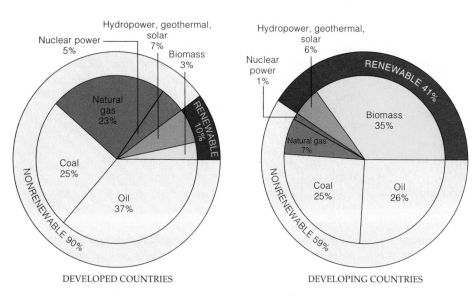

Figure 50.16 Energy consumption in countries that differ greatly in sources of energy and average per capita energy use. The values shown do not take into account energy from the sun, which is the foundation for agriculture.

a

Figure 50.17 Incident at Chernobyl. (**a**) On April 26, 1986, errors in judgment during a routine test procedure resulted in runaway reactions, explosions, and a full core meltdown at the Chernobyl power plant in Ukraine. Helicopter pilots who were supposed to drop 5,000 tons of lead, sand, clay, and other materials on the blazing core to suppress further release of radioactivity missed the target. Unimpeded and uncovered, nuclear fuel burned for nearly ten days, right on through a six-foot-thick steel and gravel barrier beneath it.

Between 185 and 250 million curies of radioactive matter may have escaped in those ten days alone. Inhaling as little as ten-millionths of a curie of plutonium can cause cancer. Thirty-one people died at once; others died of radiation poisoning in the following weeks. Inhabitants of entire villages were relocated; their former homes were bulldozed under. In time, concrete entombed the 180 tons of partially burned nuclear fuel.

(**b**) Afterward, the number of people opposed to nuclear plants rose sharply. You can get a sense of why opposition increased dramatically by studying maps of the global distribution of radioactive fallout within two weeks of the meltdown. The fallout put 300 to 400 million people at risk for leukemia and other radiation-induced disorders. Throughout Europe, hundreds of millions of dollars were lost when the fallout made crops and livestock unfit for consumption. In 1994, rainwater and air were still moving through 11,000 square meters of holes in the power plant's concrete. By 1998, the rate of thyroid abnormalities in children living immediately downwind from Chernobyl was nearly seven times as high as for those upwind; their thyroid gland collected iodine radioisotopes from the fallout.

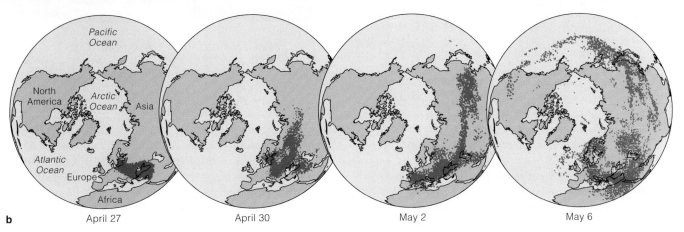

b   April 27   April 30   May 2   May 6

older plants, water circulating over the nuclear fuel absorbs heat, then turns into steam that can drive the electricity-generating turbines. Water levels may plunge around the fuel if the circulating water system leaks. If fuel heats past its melting point, the rest of the water can instantly be converted to steam. Steam formation, with other reactions, could blow the system apart and release radioactive material. An overheated core could melt right through the thick concrete slab that contains it. This happened at a nuclear plant in Chernobyl, in the Ukraine (Figure 50.17).

NUCLEAR WASTE DISPOSAL Unlike coal, nuclear fuel cannot be burned to harmless ashes. After three years or so, the fuel elements are spent, but they still contain uranium fuel as well as hundreds of new radioisotopes that formed in the reactions. The wastes are extremely radioactive and dangerous. They get extremely hot as they undergo radioactive decay, so they are plunged at once into water-filled pools. The water cools them and

keeps radioactive material from escaping. Even after being stored for several months, the isotopes remaining are lethal. Some must be kept isolated for at least 10,000 years. If a certain isotope of plutonium (^{239}Pu) is not removed, wastes must be kept isolated for a quarter of a million years! After nearly fifty years of research, scientists still cannot agree on what is the best way to store high-level radioactive wastes. Even if they could, there is no politically acceptable solution. No one wants radioactive wastes anywhere near where they live.

As if we don't have enough to worry about, after the Soviet Union's breakup, some underpaid workers of a Russian nuclear power plant sold fuel elements on the black market. The buyers? Individuals who want to make nuclear weapons and possibly deliver them into the hands of terrorist organizations.

The nuclear genie is out of the bottle, exploitable by the best and worst elements of the human population.

ALTERNATIVE ENERGY SOURCES

Less than thirty years from now, the projected human population size will be such that demands for fossil fuels will increase by 30 percent and for electricity by 265 percent. Although we can use the known energy supplies much more efficiently, we also must develop alternatives, of the sort described here.

Solar–Hydrogen Energy

Each year, incoming sunlight contains about ten times more energy than the amount in all of the known fossil fuel reserves. That is 15,000 times more energy than our population uses now. Isn't it about time that we start to collect it in earnest, in something besides crops?

For example, when exposed to sunlight, electrodes in "photovoltaic cells" produce an electric current that splits water molecules into oxygen and hydrogen gas (H_2). This can be used directly as fuel to run cars, heat and cool buildings, and generate electricity. We have had the technology to tap such **solar–hydrogen energy** since the 1940s (Figure 50.18a). Such energy is stored efficiently. It costs less to distribute H_2 than electricity, and water is the only by-product. Space satellites run on it. Here on Earth, many think fossil fuels are still "cheaper," partly as a result of government subsidies.

Unlike fossil fuels, however, sunlight and seawater are unlimited resources. The technology's potential to protect the environment is staggering. Environmental scientist G. Tyler Miller, Jr. puts it this way: "If we make the transition to an energy-efficient solar–hydrogen age, we can say good-bye to smog, oil spills, acid rain, and nuclear energy, and possibly to global warming. The reason is simple. When hydrogen burns in air, it reacts with oxygen gas to produce water vapor—not a bad thing to have coming out of tailpipes, chimneys, and smokestacks." Also, if this technology becomes cost-effective for the developing countries, at least part of the forests now being destroyed for timber and fuel might still be around for future generations.

In 1995, photovoltaic cells generated almost 4 million kilowatt-hours of net electricity in the United States (of the total electricity generation of 988.8 million). Thirty states now promote development of solar energy. In 1996, Hawaii, Iowa, and Washingon offered incentives to promote clean, renewable energy technologies.

Wind Energy

As you know, solar energy also is converted into the mechanical energy of winds. Where prevailing winds travel faster than 7.5 meters per second, we find **wind farms**. These arrays of cost-effective turbines exploit wind patterns that arise from latitudinal variations in the intensity of incoming sunlight (Figure 50.18b). One

Figure 50.18 Harnessing solar energy. (**a**) A large array of electricity-producing photovoltaic cells in panels that collect sunlight energy. (**b**) A field of turbines harvesting wind energy.

percent of California's electricity comes from wind farms. Possibly the winds of North and South Dakota alone can meet all but 20 percent of the current energy needs of the United States. Winds do not blow on a regular schedule, but the energy can be fed into utility grids as it becomes available. Wind energy also has potential for islands and other areas remote from utility grids.

What About Fusion Power?

The sun's gravitational force is enough to compress atomic nuclei to high densities; its temperatures are high enough to force atomic nuclei to fuse. We call this **fusion power**. Similar conditions do not exist on Earth, but maybe we can mimic them. Researchers confine a certain fuel (a heated gas of two isotopes of hydrogen) in magnetic fields, then bombard it with lasers. The fuel implodes, it is compressed to extremely high densities, and energy is released. The more energetic the lasers, the greater the compression, and the more the fuel will burn. The bad news is, although the amount of energy released has been steadily increasing, it will be at least fifty years before fusion reactors could be operating, and the costs will probably be high. The good news is, that's about the time fossil fuels may start running out.

Sunlight may end up sustaining the energy needs of the human population in more ways than one.

BIOLOGICAL PRINCIPLES AND THE HUMAN IMPERATIVE

Molecules, single cells, tissues, organs, organ systems, multicelled organisms, populations, communities, then ecosystems and the biosphere. These are architectural systems of life, assembled in increasingly complex ways over the past 3.8 billion years. We are latecomers to this immense biological building program. Yet within the relatively short span of 10,000 years, our activities have been changing the character of the land, ocean, and atmosphere, even the genetic character of species.

It would be presumptuous to think we alone have had profound effects on the world of life. As long ago as the Proterozoic, photosynthetic organisms were irrevocably changing the course of biological evolution by enriching the atmosphere with oxygen. During the past as well as the present, competitive adaptations led to the rise of some groups, whose dominance assured the decline of others. Change is nothing new. What *is* new is the capacity of one species to comprehend what might be going on.

We now have the population size, technology, and cultural inclination to use up energy and modify the environment at rapid rates. Where will this end? Will feedback controls operate as they do, for example, when population growth exceeds carrying capacity? In other words, will negative feedback controls come into play and keep things from getting too far out of hand?

Feedback control will not be enough, for it operates after a deviation. Our patterns of resource consumption and population growth are founded on an illusion of unlimited resources and a forgiving environment. A prolonged, global shortage of food or the passing of a critical threshold for the global climate can come too fast to be corrected; in which case the impact of the deviation may be too great to be reversed.

What about feedforward mechanisms that might serve as early warning systems? For example, when sensory receptors near the surface of skin detect a drop in outside air temperature, each sends messages to the nervous system. That system responds by triggering mechanisms that raise the body's core temperature before the body itself becomes dangerously chilled.

Extrapolating from this, if we develop feedforward control mechanisms, would it not be possible to start corrective measures before we do too much harm?

Feedforward controls alone will not work, for they operate after change is under way. Think of the DEW line—the Distant Early Warning system. It is like a vast sensory receptor for detecting missiles launched against North America. By the time it does what it is supposed to, it may be too late to stop widespread destruction.

It would be naive to assume we can ever reverse who we are at this point in evolutionary time, to de-evolve ourselves culturally and biologically into becoming less complex in the hope of averting disaster. Yet there is reason to believe we can avert disaster by using a third kind of control mechanism—a capacity to anticipate events even before they happen. We are not locked into responding only after irreversible change has begun. We have the capacity to anticipate the future—it is the essence of our visions of utopia and hell. *We all have the capacity to adapt to a future that we can partly shape.*

For instance, we can stop trying to "beat nature" and learn to work with it. Individually and collectively, we can work to develop long-term policies that take into account biotic and abiotic limits on population growth. Far from being a surrender, this would be one of the most intelligent behaviors of which we are capable.

Having a capacity to adapt and using it are not the same thing. We have already put the world of life on dangerous ground because we have not yet mobilized ourselves as a species to work toward self-control.

Our survival depends on predicting possible futures. It depends on preserving, restoring, and constructing ecosystems that fit with our definition of basic human values and available biological models. Human values can change; our expectations can and must be adapted to biological reality. *For the principles of energy flow and resource utilization, which govern the survival of all systems of life, do not change.* It is our biological and cultural imperative that we come to terms with these principles, and ask ourselves what our long-term contribution will be to the world of life.

50.11

1. Accompanying the extraordinarily rapid growth of the human population are increases in energy demands and in environmental pollution. *CI, 50.1*

2. A pollutant is any substance with which ecosystems have had no prior evolutionary experience (in terms of kinds and amounts) and thus have no mechanisms for absorbing or cycling it. Many pollutants result from human activities, and they adversely affect the health, activities, or survival of all organisms. *50.1*

3. Smog, a form of air pollution, arises in industrialized and urban regions that rely on fossil fuels. It becomes most concentrated in land basins that promote thermal inversion (a layer of cool, dense air trapped below a warm air layer). Industrial smog forms in coal-burning industrial regions having cold, wet winters. In warm climates, large cities with many fuel-burning vehicles have photochemical smog. Mainly, nitric oxide emitted from vehicles reacts with hydrocarbons in sunlight to form photochemical oxidants such as PANs. *50.1*

4. In dry weather, acidic air pollutants (e.g., oxides of nitrogen and sulfur) fall as dry acid deposition. They also dissolve in atmospheric water, then fall to Earth as wet acid deposition, or acid rain. *50.2*

5. Seasonal thinning of the ozone layer at high latitudes has become pronounced as CFCs (chlorofluorocarbons) and other air pollutants rise to the stratosphere, where they deplete ozone and allow more harmful ultraviolet radiation from the sun to reach the Earth. *50.2*

6. Land surfaces are being degraded by:
   a. Passively accepting the dumping, burning, and burial of solid wastes and not making concerted efforts to recycle or reuse materials and to reduce waste. *50.3*
   b. Engaging in rampant deforestation (destruction of vast tracts of forest biomes). *50.4*
   c. Contributing to desertification (the conversion of natural grasslands, croplands, and grazing lands, on a large scale, to desertlike conditions). *50.6*

7. Human population growth currently depends upon the expansion of agriculture, made possible by large-scale irrigation and extensive applications of fertilizers and pesticides. Global freshwater supplies are limited; even so, they are being polluted by agricultural runoff (which includes sediments, pesticides, and fertilizers), industrial wastes, and human sewage. *50.7*

8. Fossil fuel energy is nonrenewable, dwindling, and environmentally costly to extract and utilize. Nuclear energy is less polluting, but costs and risks associated with fuel containment and storing radioactive wastes are high. The rapid growth of the human population means that affordable alternatives such as wind energy and solar energy must soon be developed. *50.8, 50.9*

Review Questions

1. Define pollution and list some specific examples of water pollutants. *50.1, 50.7*

2. Distinguish among the following conditions: *50.1*
   a. industrial smog   c. dry acid deposition
   b. photochemical smog   d. wet acid deposition

3. Define CFCs and describe how they apparently contribute to seasonal thinning of the ozone layer in the stratosphere. *50.2*

4. What percent of the Earth's land is under cultivation? What percent is available for new cultivation? *50.3*

5. Define and describe possible consequences of deforestation and of desertification. *50.4, 50.5, 50.6*

6. Which human activity uses the most fresh water? *50.7*

Self-Quiz   ANSWERS IN APPENDIX III

1. Starting a few centuries ago, human population growth has been _____ .
   a. leveling off   c. accelerating
   b. growing slowly   d. not much to speak of

2. Pollutants disrupt ecosystems because _____ .
   a. their components differ from those of natural substances
   b. only humans have uses for them
   c. there are no evolved mechanisms to deal with them
   d. their only effect is on ecosystems, not humans

3. During a thermal inversion, weather conditions trap a layer of _____ air under a layer of _____ air.
   a. warm; cool   c. warm; sooty
   b. cool; warm   d. cool; sooty

4. _____ can be viewed as regional air pollution.
   a. Smog   c. Ozone layer thinning
   b. Acid rain   d. a and b

5. _____ can be viewed as air pollution with global effects.
   a. Smog   c. Ozone layer thinning
   b. Acid rain   d. b and c

6. Worldwide, two-thirds of the fresh water used annually goes to _____ .
   a. urban centers   c. treatment facilities
   b. agriculture   d. a and c

7. The upper limit at which the ground is fully saturated with water is called _____ .
   a. groundwater   c. the water table
   b. an aquifer   d. the salinization limit

8. Energy from fossil fuels is _____ ; their extraction and use come at _____ cost to the environment.
   a. renewable; low   c. renewable; high
   b. nonrenewable; low   d. nonrenewable; high

9. Nuclear energy normally pollutes _____ than fossil fuels; it poses _____ dangers than fossil fuels.
   a. less; lesser   b. more; greater   c. more; fewer   d. less; more

10. Match each term with the most suitable description.
    _____ desertification   a. possibly one of our best options
    _____ deforestation   b. soil loss, watershed damage, altered rainfall patterns follow
    _____ green
          revolution   c. attempt to improve crop production on existing land
    _____ solar–hydrogen
          energy   d. converting large tracts of natural grasslands to desertlike state
    _____ CFCs   e. invisible, odorless compounds that contribute to ozone thinning

Figure 50.19 City as ecosystem.

Critical Thinking

1. Kristen, a recent college graduate, is finding her idealism on a collision course with reality. She strongly believes people who live in the United States are obliged to make the world a level playing field for all human beings, with equality in resources, health care, education, economic security, and a pristine environment for all. Yet she also understands that the sheer size of the human population makes this impossible.

Kristen decided that she cannot be party to hard choices and actions that go against her ideals, and that she just wants nature to solve the problem for us. Comment on this true story.

2. Investigate where the water for your own city comes from and where it has been. You may find the answer illuminating.

3. Make a list of advantages you personally enjoy as a member of an affluent, industrialized society. List some drawbacks. Do the benefits outweigh the costs? This is not a trick question.

4. It has been said that economic wars, more than military wars, will determine the winners and losers among nations in the near future. The economic growth of certain nations, including some in the former Soviet Union and the Far East, has had devastating impact on the environment. Elsewhere, efforts of conservation biologists to maintain standards of environmental protection add to the cost of goods produced and put practicing nations at serious disadvantage in this global competition.

Should the United States loosen environmental laws to help ensure its economic survival? Why or why not? Should it impose pressure on indifferent nations to encourage reduction of harmful practices? Why or why not?

5. To reduce the backlog of unnamed new species and secure funding, a group of German taxonomists is offering shoppers the opportunity, for $2,500 and up, to name a species (Biopat website). Example: As a birthday present, investment banker Stan Lai's wife had a Bolivian yellow-striped brown toad named in her husband's honor (*Bufo stanlaii*, or "Stan Lai's toad").

Usually, a species name reflects its appearance or honors its discoverer. But not always; for example, one midge has the name *Dicrotendipes thanatogratus* (the "Grateful Dead" midge). The International Commission on Zoological Nomenclature in London denounces offering taxonomy to the public, which it believes will obscure science and hinder conservation efforts. What does Mrs. Lai think? "Now, every time I hear something about Bolivia, I'll pay attention. After all, I don't want to see our toad die."

Does inviting the public to pay their way into taxonomy undermine the science? Is it a good or bad way to raise public awareness of conservation biology?

6. Populations of every species utilize resources and produce wastes. Use Figure 50.19 as a starting point for a brief essay on the accumulation and uses of energy and materials, including wastes, in the human population of a large city. Contrast your description with the flow of energy and the cycling of materials that proceed in the natural population of some other organism.

7. Many psychologists believe that we spend much of our lives consciously or subconsciously searching for roots that might anchor us in bewildering or frightening times of change. Some philosophers have argued that each of us must find a mountain, river, backyard, or any other place that elicits a sense of sanctuary, of rooted connections with nature. Others believe that forming an emotional connection with nature is based on a romanticized or mystical view of it. As they argue, time would be better spent developing a scientifically sound understanding of nature and the technological means to protect and sustain its resources.

Do you agree with one or the other point of view? Or do you suspect that the two are not mutually exclusive?

Selected Key Terms

acid rain *50.1*	ozone thinning *50.2*
chlorofluorocarbon (CFC) *50.2*	PAN (peroxyacyl nitrate) *50.1*
deforestation *50.4*	photochemical smog *50.1*
desalinization *50.7*	pollutant *50.1*
desertification *50.6*	salinization *50.7*
dry acid deposition *50.1*	shifting cultivation *50.4*
fossil fuel *50.8*	solar–hydrogen energy *50.9*
fusion power *50.9*	thermal inversion *50.1*
green revolution *50.3*	wastewater treatment *50.7*
industrial smog *50.1*	water table *50.7*
meltdown *50.8*	wind farm *50.9*

Readings

Kaiser, J. 18 February 2000. "Ecologists on a Mission to Save the World." *Science* 287:1188–1192. Addresses a growing debate over how far environmental scientists should go when interpreting their findings for policy makers.

Miller, G. T., Jr. 2003. *Environmental Science*. Ninth edition. Belmont, California: Brooks/Cole.

Terrell, K. 29 April 2002. "Running on Fumes." *U.S. News & World Report*, 58–59. Fuel-cell cars need hydrogen stations.

On-Line readings at Student Guide for InfoTrac:
www.brookscole.com/biology

APPENDIX I.   CLASSIFICATION SYSTEM

This revised classification scheme is a composite of several that microbiologists, botanists, and zoologists use. The major groupings are agreed upon, more or less. However, there is not always agreement on what to name a particular grouping or where it might fit within the overall hierarchy. There are several reasons why full consensus is not possible at this time.

First, the fossil record varies in its completeness and quality (Section 19.1). As one outcome, the phylogenetic relationship of one group to other groups is sometimes open to interpretation. Today, comparative studies at the molecular level are firming up the picture, but the work is still under way.

Second, ever since the time of Linnaeus, systems of classification have been based on the perceived morphological similarities and differences among organisms. Although some original interpretations are now open to question, we are so used to thinking about organisms in certain ways that reclassification often proceeds slowly.

A few examples: Traditionally, birds and reptiles were grouped in separate classes (Reptilia and Aves); yet there are many compelling arguments for grouping the lizards and snakes in one class and the crocodilians, dinosaurs, and birds in a separate class. Some biologists have favored a six-kingdom system of classification (archaebacteria, eubacteria, protistans, plants, fungi, and animals). Many others now favor a three-domain classification system. The archaebacteria, eubacteria, and eukaryotes (alternatively, the archaea, bacteria, and eukarya) are its major groupings.

Third, researchers in microbiology, mycology, botany, zoology, and other fields of inquiry inherited a wealth of literature, based on classification systems that have been developed over time in each field of inquiry. Many simply do not wish to give up established terminology that offers access to the past.

For example, botanists and microbiologists often use *division*, and zoologists *phylum*, for taxa that actually are equivalent in hierarchies of classification. As another example, opinions are quite polarized with respect to kingdom Protista, certain members of which could easily be grouped with plants, or fungi, or animals. Indeed, the term "protozoan" is a holdover from an earlier scheme in which some single-celled organisms were ranked as simple animals.

Given the problems, why do systematists work so hard to construct hypotheses regarding the history of life? They do so because classification systems that accurately reflect evolutionary relationships are more useful, with far more predictive power for comparative studies. Such systems also serve as a good framework for studying the living world, which could otherwise be an overwhelming body of knowledge. Importantly, classifications enhance the retrieval of information that relates to living organisms.

Bear in mind, *we include this appendix on classification for your reference purposes only*. Besides being open to revision, it is by no means complete. Names in "quotes" are polyphyletic or paraphyletic groups undergoing revision. The groupings of certain animal phyla reflect common ancestry (Nematoda through Arthropoda, and Nemertea through Annelida), as emerging molecular data suggest. Also, the most recently discovered species, as from the mid-ocean province, are not listed. Many existing and extinct species of the more obscure phyla are not represented. Our strategy is to focus primarily on the organisms mentioned in the text.

PROKARYOTES AND EUKARYOTES COMPARED

As a general frame of reference, note that almost all eubacteria and archaebacteria are microscopic in size. Their DNA is concentrated in a nucleoid (a region of cytoplasm), not in a membrane-bound nucleus. All are single cells or simple associations of cells. They reproduce by prokaryotic fission or budding; they transfer genes by bacterial conjugation.

Table A lists representative types of autotrophic and heterotrophic prokaryotes. The authoritative reference, *Bergey's Manual of Systematic Bacteriology,* has called this a time of taxonomic transition. It references groups mainly by numerical taxonomy (Section 21.3) rather than by phylogeny. Our classification system does reflect evidence of evolutionary relationships for at least some bacterial groups.

The first life forms were prokaryotic. Similarities between Eubacteria and Archaebacteria have more ancient origins relative to the traits of eukaryotes.

Unlike the prokaryotes, all eukaryotic cells start out life with a DNA-enclosing nucleus and other membrane-bound organelles. Their chromosomes have many histones and other proteins attached. They include spectacularly diverse single-celled and multicelled species, which can reproduce by way of meiosis, mitosis, or both.

BACTERIA	ARCHAEA	EUKARYA			
EUBACTERIA	ARCHAEBACTERIA	PROTISTA	FUNGI	PLANTAE	ANIMALIA

DOMAIN OF EUBACTERIA (BACTERIA)

KINGDOM EUBACTERIA Gram-negative and Gram-positive prokaryotic cells. Peptidoglycan in cell wall. Collectively, great metabolic diversity; photosynthetic autotrophs, chemosynthetic autotrophs, and heterotrophs.

PHYLUM FIRMICUTES Typically Gram-positive, thick wall. Heterotrophs. *Bacillus, Staphylococcus, Streptococcus, Clostridium, Actinomycetes.*

PHYLUM GRACILICUTES Typically Gram-negative, thin wall. Autotrophs (photosynthetic and chemosynthetic) and heterotrophs. *Anabaena* and other cyanobacteria. *Escherichia, Pseudomonas, Neisseria, Myxococcus.*

PHYLUM TENERICUTES Gram-negative, wall absent. Heterotrophs (saprobes, pathogens). *Mycoplasma.*

DOMAIN OF ARCHAEBACTERIA (ARCHAEA)

KINGDOM ARCHAEBACTERIA Methanogens, extreme halophiles, extreme thermophiles. Evolutionarily closer to eukaryotic cells than to eubacteria. All strict anaerobes living in habitats as harsh as those that probably prevailed on the early Earth. Compared with other prokaryotic cells, all archaebacteria have a distinctive cell wall and unique membrane lipids, ribosomes, and RNA sequences. *Methanobacterium, Halobacterium, Sulfolobus.*

Table A   Representative Eubacteria and Archaebacteria Grouped on the Basis of Numerical Taxonomy

Some Major Groups	Main Habitats	Characteristics	Representatives
EUBACTERIA			
Photoautotrophs:			
Cyanobacteria, green sulfur bacteria, and purple sulfur bacteria	Mostly lakes, ponds; some marine, terrestrial habitats	Photosynthetic; use sunlight energy, carbon dioxide; cyanobacteria use oxygen-producing noncyclic pathway; some also use cyclic route	*Anabaena, Nostoc, Rhodopseudomonas, Chloroflexus*
Photoheterotrophs:			
Purple nonsulfur and green nonsulfur bacteria	Anaerobic, organically rich muddy soils, and sediments of aquatic habitats	Use sunlight energy; organic compounds as electron donors; some purple nonsulfur may also grow chemotrophically	*Rhodospirillum, Chlorobium*
Chemoautotrophs:			
Nitrifying, sulfur-oxidizing, and iron-oxidizing bacteria	Soil; freshwater, marine habitats	Use carbon dioxide, inorganic compounds as electron donors; influence crop yields, cycling of nutrients in ecosystems	*Nitrosomonas, Nitrobacter, Thiobacillus*
Chemoheterotrophs:			
Spirochetes	Aquatic habitats; parasites of animals	Helically coiled, motile; free-living and parasitic species; some major pathogens	*Spirochaeta, Treponema*
Gram-negative aerobic rods and cocci	Soil, aquatic habitats; parasites of animals, plants	Some major pathogens; some fix nitrogen (e.g., *Rhizobium*)	*Pseudomonas, Neisseria, Rhizobium, Agrobacterium*
Gram-negative facultative anaerobic rods	Soil, plants, animal gut	Many major pathogens; one bioluminescent (*Photobacterium*)	*Salmonella, Escherichia, Proteus, Photobacterium*
Rickettsias and chlamydias	Host cells of animals	Intracellular parasites; many pathogens	*Rickettsia, Chlamydia*
Myxobacteria	Decaying organic material; bark of living trees	Gliding, rod-shaped; aggregation and collective migration of cells	*Myxococcus*
Gram-positive cocci	Soil; skin and mucous membranes of animals	Some major pathogens	*Staphylococcus, Streptococcus*
Endospore-forming rods and cocci	Soil; animal gut	Some major pathogens	*Bacillus, Clostridium*
Gram-positive nonsporulating rods	Fermenting plant, animal material; gut, vaginal tract	Some important in dairy industry, others major contaminators of milk, cheese	*Lactobacillus, Listeria*
Actinomycetes	Soil; some aquatic habitats	Include anaerobes and strict aerobes; major producers of antibiotics	*Actinomyces, Streptomyces*
ARCHAEBACTERIA (ARCHAEA)			
Methanogens	Anaerobic sediments of lakes, swamps; animal gut	Chemosynthetic; methane producers; used in sewage treatment facilities	*Methanobacterium*
Extreme halophiles	Brines (extremely salty water)	Heterotrophic; also, unique photosynthetic pigments (bacteriorhodopsin) form in some	*Halobacterium*
Extreme thermophiles	Acidic soil, hot springs, hydrothermal vents	Heterotrophic or chemosynthetic; use inorganic substances as electron donors	*Sulfolobus, Thermoplasma*

DOMAIN OF EUKARYOTES (EUKARYA)

KINGDOM "PROTISTA"

Diverse single-celled, colonial, and multicelled eukaryotic species. Existing types are unlike prokaryotes and most like the earliest forms of eukaryotes. Autotrophs, heterotrophs, or both (Table 22.1). Reproduce sexually and asexually (by meiosis, mitosis, or both). Not a monophyletic group. The kingdom may soon be split into multiple kingdoms, and some of its groups are already being reclassified as plants or fungi.

PHYLUM "MASTIGOPHORA" Flagellated protozoans. Free-living heterotrophs; many are internal parasites. They have one to several flagella. At present, a non-monophyletic grouping of ancient lineages, including the diplomonads, parabasalids, and kinetoplastids. *Trypanosoma, Trichomonas, Giardia.*

PHYLUM EUGLENOPHYTA Euglenoids. Mostly heterotrophs, some photoautotrophs, some both depending on conditions. Most with one short, one long flagellum. Pigmented (red, green) or colorless. Related to kinetoplastids. *Euglena.*

PHYLUM SARCODINA Amoeboid protozoans. Heterotrophs, free-living or endosymbionts, some pathogens. Soft-bodied, with or without shell, pseudopods. Rhizopods (naked amoebas, foraminiferans), actinopods (radiolarians, heliozoans). *Amoeba.*

ALVEOLATES

PHYLUM CILIOPHORA Ciliated protozoans. Heterotrophs, predators or symbionts, some parasitic. All have cilia. Free-living, sessile, or motile. *Paramecium, Didinium,* hypotrichs.

PHYLUM APICOMPLEXA Heterotrophs, sporozoite-forming parasites. Complex structures at head end. Most familiar types known as sporozoans. *Cryptosporidium, Plasmodium, Toxoplasma.*

PHYLUM PYRRHOPHYTA. Dinoflagellates. Photosynthetic, mostly, but some heterotrophs. *Pfiesteria, Gymnodinium breve.*

STRAMENOPILES

PHYLUM OOMYCOTA. Water molds. Heterotrophs. Decomposers, some parasites. *Saprolegnia, Phytophthora, Plasmopara.*

PHYLUM CHRYSOPHYTA. Golden algae, yellow-green algae, diatoms. Photosynthetic. Some flagellated. *Mischococcus, Synura, Vaucheria.*

PHYLUM PHAEOPHYTA. Brown algae. Photosynthetic, nearly all endemic to temperate or marine waters. *Macrocystis, Fucus, Sargassum, Ectocarpus, Postelsia.*

GROUPS CLOSELY RELATED TO PLANTS

PHYLUM CHLOROPHYTA Green algae. Mostly photosynthetic, some parasitic. Most freshwater, some marine or terrestrial. *Chlamydomonas, Spirogyra, Ulva, Volvox, Codium, Halimeda.*

PHYLUM RHODOPHYTA Red algae. Mostly photosynthetic, some parasitic. Nearly all marine, some in freshwater habitats. *Porphyra, Bonnemaisonia, Euchema.*

PHYLUM CHAROPHYTA Stoneworts. *Chara.*

GROUPS CLOSELY RELATED TO FUNGI

PHYLUM CHYTRIDIOMYCOTA Chytrids. Heterotrophs; saprobic decomposers or parasites. *Chytridium.*

GROUPS OF SLIME MOLDS

PHYLUM ACRASIOMYCOTA Cellular slime molds. Heterotrophs with free-living, phagocytic amoeboid cells and spore-bearing stages. *Dictyostelium.*

PHYLUM MYXOMYCOTA Plasmodial slime molds. Heterotrophs with free-living, phagocytic amoeboid cells and spore-bearing stages. Aggregate into streaming mass of cells that discard their plasma membrane. *Physarum.*

KINGDOM PLANTAE

Multicelled eukaryotes. Nearly all photosynthetic autotrophs with chlorophylls *a* and *b*. Some parasitic. Nonvascular and vascular species, generally with well-developed root and shoot systems. Nearly all adapted to survive dry conditions on land; a few in aquatic habitats. Sexual reproduction predominant with spore-forming chambers and embryos in life cycle; also asexual reproduction by vegetative propagation and other mechanisms.

PHYLUM "BRYOPHYTA" Bryophytes; mosses, liverworts, hornworts. Not a monophyletic group. Seedless, nonvascular, haploid dominance. *Marchantia, Polytrichum, Sphagnum.*

VASCULAR PLANTS

PHYLUM "RHYNIOPHYTA" Earliest known vascular plants; muddy habitats. A polyphyletic group, some are primitive lycophytes. Extinct. *Cooksonia, Rhynia.*

PHYLUM LYCOPHYTA Lycophytes, club mosses. Seedless, vascular. Small leaves, branching rhizomes, vascularized roots and stems. *Lepidodendron* (extinct), *Lycopodium, Selaginella.*

PHYLUM SPHENOPHYTA Horsetails. Seedless, vascular, whorled leaves. Some stems photosynthetic, others nonphotosynthetic, spore-producing. *Calamites* (extinct), *Equisetum.*

PHYLUM CYCADOPHYTA Cycads. Gymnosperm group (vascular, bear "naked" seeds). Tropical, subtropical. Palm-shaped leaves, simple cones on male and female plants. *Zamia.*

PHYLUM PTEROPHYTA. Ferns. Large leaves, usually with sori. Largest group of seedless vascular plants (12,000 species), mainly tropical, temperate habitats. *Pteris, Trichomanes, Cyathea* (tree ferns), *Polystichum.*

PHYLUM PSILOPHYTA Whisk ferns. Seedless, vascular. No obvious roots, leaves on sporophyte, very reduced. *Psilotum.*

PHYLUM "PROGYMNOSPERMOPHYTA" The progymnosperms. Ancestral to early seed-bearing plants; extinct. *Archaeopteris.*

SEED-BEARING PLANTS (A subgroup of vascular plants)

PHYLUM "PTERIDOSPERMOPHYTA" Seed ferns. Fernlike gymnosperms; extinct. *Medullosa.*

PHYLUM CYCADOPHYTA Cycads. Group of gymnosperms (vascular, bear "naked" seeds). Tropical, subtropical. Compound leaves, simple cones on male and female plants. Plants usually palm-like. *Zamia, Cycas.*

PHYLUM GINKGOPHYTA Ginkgo (maidenhair tree). Type of gymnosperm. Seeds with fleshy outer layer. *Ginkgo.*

PHYLUM GNETOPHYTA Gnetophytes. Only gymnosperms with vessels in xylem and double fertilization (but endosperm does not form). *Ephedra, Welwitchia, Gnetum.*

PHYLUM CONIFEROPHYTA Conifers. Most common and familiar gymnosperms. Generally cone-bearing species with needle-like or scale-like leaves.

Family Pinaceae. Pines (*Pinus*), firs (*Abies*), spruces (*Picea*), hemlock (*Tsuga*), larches (*Larix*), true cedars (*Cedrus*).

Family Cupressaceae. Junipers (*Juniperus*), Cypresses (*Cupressus*), Bald cypress (*Taxodium*), redwood (*Sequoia*), bigtree (*Sequoiadendron*), dawn redwood (*Metasequoia*).

Family Taxaceae. Yews. *Taxus.*

PHYLUM ANTHOPHYTA Angiosperms (the flowering plants). Largest, most diverse group of vascular seed-bearing plants. Only organisms that produce flowers, fruits. Some families from several representative orders are listed:

BASAL FAMILIES

Family Amborellaceae. *Amborella.*
Family Nymphaeaceae. Water lilies.
Family Illiciaceae. Star anise.

<u>MAGNOLIIDS</u>

Family Magnoliaceae. Magnolias.
Family Lauraceae. Cinnamon, sassafras, avocados.
Family Piperaceae. Black pepper, white pepper.

<u>EUDICOTS</u>

Family Papaveraceae. Poppies.
Family Cactaceae. Cacti.
Family Euphorbiaceae. Spurges, poinsettia.
Family Salicaceae. Willows, poplars.
Family Fabaceae. Peas, beans, lupines, mesquite.
Family Rosaceae. Roses, apples, almonds, strawberries.
Family Moraceae. Figs, mulberries.
Family Cucurbitaceae. Gourds, melons, cucumbers, squashes.
Family Fagaceae. Oaks, chestnuts, beeches.
Family Brassicaceae. Mustards, cabbages, radishes.
Family Malvaceae. Mallows, okra, cotton, hibiscus, cocoa.
Family Sapindaceae. Soapberry, litchi, maples.
Family Ericaceae. Heaths, blueberries, azaleas.
Family Rubiaceae. Coffee.
Family Lamiaceae. Mints.
Family Solanaceae. Potatoes, eggplant, petunias.
Family Apiaceae. Parsleys, carrots, poison hemlock.
Family Asteraceae. Composites. Chrysanthemums, sunflowers, lettuces, dandelions.

<u>MONOCOTS</u>

Family Araceae. Anthuriums, calla lily, philodendrons.
Family Liliaceae. Lilies, tulips.
Family Alliaceae. Onions, garlic.
Family Iridaceae. Irises, gladioli, crocuses.
Family Orchidaceae. Orchids.
Family Arecaceae. Date palms, coconut palms.
Family Bromeliaceae. Bromeliads, pineapples, Spanish moss.
Family Cyperaceae. Sedges.
Family Poaceae. Grasses, bamboos, corn, wheat, sugarcane.
Family Zingiberaceae. Gingers.

FUNGI Nearly all multicelled eukaryotic species. Heterotrophs, mostly saprobic decomposers, some parasites. Nutrition based upon extracellular digestion of organic matter and absorption of nutrients by individual cells. Multicelled species form absorptive mycelia within substrates and structures that produce asexual spores (and sometimes sexual spores).

PHYLUM ZYGOMYCOTA Zygomycetes. Producers of zygospores (zygotes inside thick wall) by way of sexual reproduction. Bread molds, related forms. *Rhizopus, Philobolus.*

PHYLUM ASCOMYCOTA Ascomycetes. Sac fungi. Sac-shaped cells form sexual spores (ascospores). Most yeasts and molds, morels, truffles. *Saccharomycetes, Morchella, Neurospora, Sarcoscypha, Claviceps, Ophiostoma, Candida, Aspergillus, Penicillium.*

PHYLUM BASIDIOMYCOTA Basidiomycetes. Club fungi. Most diverse group. Produce basidiospores inside club-shaped structures. Mushrooms, shelf fungi, stinkhorns. *Agaricus, Amanita, Craterellus, Gymnophilus, Puccinia, Ustilago.*

"IMPERFECT FUNGI" Sexual spores absent or undetected. The group has no formal taxonomic status. If better understood, a given species might be grouped with sac fungi or club fungi. *Arthobotrys, Histoplasma, Microsporum, Verticillium.*

"LICHENS" Mutualistic interactions between fungal species and a cyanobacterium, green alga, or both. *Lobaria, Usnea, Cladonia.*

KINGDOM ANIMALIA Multicelled eukaryotes, nearly all with tissues, organs, and organ systems; show motility during at least part of their life cycle; embryos develop through a series of stages. Diverse heterotrophs, predators (herbivores, carnivores, omnivores), parasites, detritivores. Reproduce sexually and, in many species, asexually as well.

PHYLUM PORIFERA Sponges. No symmetry, tissues. *Euplectella.*

PHYLUM PLACOZOA Marine. Simplest known animal. Two cell layers, no mouth, no organs. *Trichoplax.*

PHYLUM CNIDARIA Radial symmetry, tissues, nematocysts.
    Class Hydrozoa. Hydrozoans. *Hydra, Obelia, Physalia, Prya.*
    Class Scyphozoa. Jellyfishes. *Aurelia.*
    Class Anthozoa. Sea anemones, corals. *Telesto.*

PHYLUM MESOZOA Ciliated, wormlike parasites, about the same level of complexity as *Trichoplax.*

PHYLUM PLATYHELMINTHES Flatworms. Bilateral, cephalized; simplest animals with organ systems. Saclike gut.
    Class Turbellaria. Triclads (planarians), polyclads. *Dugesia.*
    Class Trematoda. Flukes. *Clonorchis, Schistosoma.*
    Class Cestoda. Tapeworms. *Diphyllobothrium, Taenia.*

PHYLUM ROTIFERA Rotifers. *Asplancha, Philodina.*

PHYLUM NEMERTEA Ribbon worms. *Tubulanus.*

PHYLUM MOLLUSCA Mollusks.
    Class Polyplacophora. Chitons. *Cryptochiton, Tonicella.*
    Class Gastropoda. Snails (periwinkles, whelks, limpets, abalones, cowries, conches, nudibranchs, tree snails, garden snails), sea slugs, land slugs. *Aplysia, Ariolimax, Cypraea, Haliotis, Helix, Liguus, Limax, Littorina, Patella.*
    Class Bivalvia. Clams, mussels, scallops, cockles, oysters, shipworms. *Ensis, Chlamys, Mytelus, Patinopectin.*
    Class Cephalopoda. Squids, octopuses, cuttlefish, nautiluses. *Dosidiscus, Loligo, Nautilus, Octopus, Sepia.*

PHYLUM BRYOZOA Bryozoans (moss animals).

PHYLUM BRACHIOPODA Lampshells.

PHYLUM ANNELIDA Segmented worms.
    Class Polychaeta. Mostly marine worms. *Eunice, Neanthes.*
    Class Oligochaeta. Mostly freshwater and terrestrial worms, many marine. *Lumbricus* (earthworms), *Tubifex.*
    Class Hirudinea. Leeches. *Hirudo, Placobdella.*

PHYLUM NEMATODA Roundworms. *Ascaris, Caenorhabditis elegans, Necator* (hookworms), *Trichinella.*

PHYLUM TARDIGRADA Water bears.

PHYLUM ONYCHOPHORA Onychophorans. *Peripatus.*

PHYLUM ARTHROPODA.
    Subphylum Trilobita. Trilobites; extinct.
    Subphylum Chelicerata. Chelicerates. Horseshoe crabs, spiders, scorpions, ticks, mites.
    Subphylum Crustacea. Shrimps, crayfishes, lobsters, crabs, barnacles, copepods, isopods (sowbugs).
    Subphylum Uniramia.
      Superclass Myriapoda. Centipedes, millipedes.
      Superclass Insecta.
        Order Ephemeroptera. Mayflies.
        Order Odonata. Dragonflies, damselflies.
        Order Orthoptera. Grasshoppers, crickets, katydids.
        Order Dermaptera. Earwigs.

Order Blattodea. Cockroaches.

Order Mantodea. Mantids.

Order Isoptera. Termites.

Order Mallophaga. Biting lice.

Order Anoplura. Sucking lice.

Order Hemiptera. Cicadas, aphids, leafhoppers, spittlebugs, bugs.

Order Coleoptera. Beetles.

Order Diptera. Flies.

Order Mecoptera. Scorpion flies. *Harpobittacus.*

Order Siphonaptera. Fleas.

Order Lepidoptera. Butterflies, moths.

Order Hymenoptera. Wasps, bees, ants.

Order Neuroptera. Lacewings, antlions.

PHYLUM ECHINODERMATA. Echinoderms.

Class Asteroidea. Sea stars. *Asterias.*

Class Ophiuroidea. Brittle stars.

Class Echinoidea. Sea urchins, heart urchins, sand dollars.

Class Holothuroidea. Sea cucumbers.

Class Crinoidea. Feather stars, sea lilies.

Class Concentricycloidea. Sea daisies.

PHYLUM HEMICHORDATA. Acorn worms.

PHYLUM CHORDATA. Chordates.

Subphylum Urochordata. Tunicates, related forms.

Subphylum Cephalochordata. Lancelets.

CRANIATES

Superclass "Agnatha." Jawless fishes, including ostracoderms (extinct).

Class Myxini. Hagfishes.

Class Cephalaspidomorphi. Lampreys.

Subphylum Vertebrata. Jawed vertebrates.

Class "Placodermi." Jawed, heavily armored fishes; extinct.

Class Chondrichthyes. Cartilaginous fishes. (sharks, rays, skates, chimaeras).

Class "Osteichthyes." Bony fishes. Not monophyletic.

Subclass Dipnoi. Lungfishes.

Subclass Crossopterygii. Coelacanths, related forms.

Subclass Actinopterygii. Ray-finned fishes.

Order Acipenseriformes. Sturgeons, paddlefishes.

Order Salmoniformes. Salmon, trout.

Order Atheriniformes. Killifishes, guppies.

Order Gasterosteiformes. Seahorses.

Order Perciformes. Perches, wrasses, barracudas, tunas, freshwater bass, mackerels.

Order Lophiiformes. Angler fishes.

TETRAPODS (A subgroup of craniates)

Class Amphibia. Amphibians.

Order Caudata. Salamanders.

Order Anura. Frogs, toads.

Order Apoda. Apodans (caecilians).

AMNIOTES (A subgroup of tetrapods)

Class "Reptilia." Skin with scales, embryo protected and nutritionally supported by extraembryonic membranes.

Subclass Anapsida. Turtles, tortoises.

Subclass Lepidosaura. *Sphenodon*, lizards, snakes.

Subclass Archosaura. Dinosaurs (extinct), crocodiles, alligators.

Class Aves. Birds. In some of the more recent classification systems, dinosaurs, crocodilians, and birds are grouped in the same category, the archosaurs.

Order Struthioniformes. Ostriches.

Order Sphenisciformes. Penguins.

Order Procellariiformes. Albatrosses, petrels.

Order Ciconiiformes. Herons, bitterns, storks, flamingoes.

Order Anseriformes. Swans, geese, ducks.

Order Falconiformes. Eagles, hawks, vultures, falcons.

Order Galliformes. Ptarmigan, turkeys, domestic fowl.

Order Columbiformes. Pigeons, doves.

Order Strigiformes. Owls.

Order Apodiformes. Swifts, hummingbirds.

Order Passeriformes. Sparrows, jays, finches, crows, robins, starlings, wrens.

Order Piciformes. Woodpeckers, toucans.

Order Psittaciformes. Parrots, cockatoos, macaws.

Class Mammalia. Skin with hair; young nourished by milk-secreting glands of adult.

Subclass Prototheria. Egg-laying mammals (monotremes; duckbilled platypus, spiny anteaters).

Subclass Metatheria. Pouched mammals or marsupials (opossums, kangaroos, wombats, Tasmanian devil).

Subclass Eutheria. Placental mammals.

Order Edentata. Anteaters, tree sloths, armadillos.

Order Insectivora. Tree shrews, moles, hedgehogs.

Order Chiroptera. Bats.

Order Scandentia. Insectivorous tree shrews.

Order Primates.

Suborder Strepsirhini (prosimians). Lemurs, lorises.

Suborder Haplorhini (tarsioids and anthropoids).

Infraorder Tarsiiformes. Tarsiers.

Infraorder Platyrrhini (New World monkeys).

Family Cebidae. Spider monkeys, howler monkeys, capuchin.

Infraorder Catarrhini (Old World monkeys and hominoids).

Superfamily Cercopithecoidea. Baboons, macaques, langurs.

Superfamily Hominoidea. Apes and humans.

Family Hylobatidae. Gibbon.

Family "Pongidae." Chimpanzees, gorillas, orangutans.

Family Hominidae. Existing and extinct human species (*Homo*) and humanlike species, including the australopiths.

Order Lagomorpha. Rabbits, hares, pikas.

Order Rodentia. Most gnawing animals (squirrels, rats, mice, guinea pigs, porcupines, beavers, etc.).

Order Carnivora. Carnivores.

Suborder Feloidea. Cats, mongooses, hyenas.

Suborder Canoidea. Dogs, weasels, skunks, otters, raccoons, pandas, bears.

Order Pinnipedia. Seals, walruses, sea lions.

Order Proboscidea. Elephants; mammoths (extinct).

Order Sirenia. Sea cows (manatees, dugongs).

Order Perissodactyla. Odd-toed ungulates (horses, tapirs, rhinos).

Order Tubulidentata. African aardvarks.

Order Artiodactyla. Even-toed ungulates (camels, deer, bison, sheep, goats, antelopes, giraffes, etc.).

Order Cetacea. Whales, porpoises.

Metric-English Conversions

Length

English		Metric
inch	=	2.54 centimeters
foot	=	0.30 meter
yard	=	0.91 meter
mile (5,280 feet)	=	1.61 kilometer

To convert	multiply by	to obtain
inches	2.54	centimeters
feet	30.00	centimeters
centimeters	0.39	inches
millimeters	0.039	inches

Weight

English		Metric
grain	=	64.80 milligrams
ounce	=	28.35 grams
pound	=	453.60 grams
ton (short) (2,000 pounds)	=	0.91 metric ton

To convert	multiply by	to obtain
ounces	28.3	grams
pounds	453.6	grams
pounds	0.45	kilograms
grams	0.035	ounces
kilograms	2.2	pounds

Volume

English		Metric
cubic inch	=	16.39 cubic centimeters
cubic foot	=	0.03 cubic meter
cubic yard	=	0.765 cubic meters
ounce	=	0.03 liter
pint	=	0.47 liter
quart	=	0.95 liter
gallon	=	3.79 liters

To convert	multiply by	to obtain
fluid ounces	30.00	milliliters
quart	0.95	liters
milliliters	0.03	fluid ounces
liters	1.06	quarts

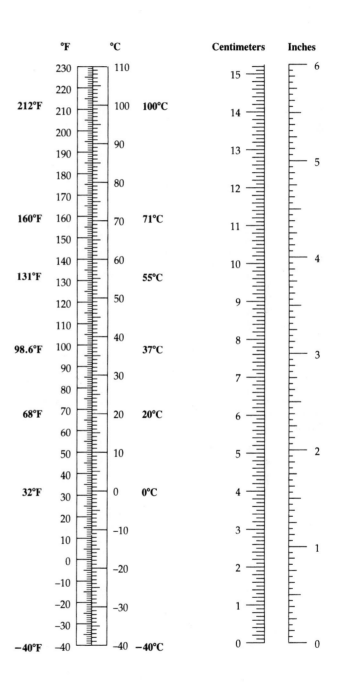

APPENDIX III. ANSWERS TO SELF-QUIZZES

CHAPTER 1

1.	Metabolism	5
2.	Homeostasis	5
3.	cell	6
4.	adaptive	10
5.	mutation	10
6.	d	4
7.	d	4, 10
8.	d	10–11
9.	d	13
10.	c	13
11.	a	13
12.	c	10
	e	11
	d	13
	b	12
	a	12

CHAPTER 2

1.	b	22
2.	c	27
3.	e	28
4.	e	29
5.	f	30–31
6.	acid, base	30
7.	c	20
	a	31
	b	25
	d	28–29

CHAPTER 3

1.	d	36
2.	e	39
3.	f	40
4.	b	42
5.	d	44, 50
6.	d	47
7.	d	50
8.	c	44
	e	50
	b	43
	d	50
	a	40

CHAPTER 4

1.	c	56
2.	d	60
3.	d	60, 74
4.	False; many cells have a wall	60, 74
5.	c	56, 76
6.	e	60, 66
	d	67
	a	56, 64
	b	64
	c	64

CHAPTER 5

1.	c	82
2.	c	82, 83
3.	a	81, 84
4.	d	84
5.	b	91
6.	d	86–87
7.	b	89
8.	d	92–93

CHAPTER 6

1.	c	99
2.	d	100
3.	b	102
4.	a	102–103
5.	d	104
6.	e	105, 108–109
7.	a	106, 107
8.	a	104

9.	c	102, 106
	g	102
	a	102
	d	102
	e	102
	b	102, 105
	f	102

CHAPTER 7

1.	carbon dioxide; light energy from sun	114
2.	d	116
3.	b	122
4.	c	116, 125
5.	d	122
6.	e	116, 124
7.	c	125
8.	c	125
9.	d	120
	e	122
	a	125
	b	125
	c	122

CHAPTER 8

1.	d	134
2.	c	138, 139
3.	b	134, 140
4.	c	134, 140
5.	d	142
6.	c	142, 143i
7.	b	142
8.	d	144, 145i
9.	b	134, 136
	c	142
	a	138–139
	d	140, 140i

CHAPTER 9

1.	d	152, 152t
2.	b	152
3.	c	152
4.	a	153
5.	c	154
6.	a	155, 157
7.	b	154
8.	d	156
	b	156
	c	157
	a	157

CHAPTER 10

1.	c	163, 164
2.	a	164
3.	d	164
4.	b	164–165
5.	d	165
6.	c	165
7.	c	166
8.	d	167
9.	d	164
	a	164
	c	166
	b	165

CHAPTER 11

1.	a	179
2.	b	179
3.	a	179
4.	b	179
5.	c	180
6.	a	181
7.	d	182
8.	b	182
	d	180
	a	179
	c	179

CHAPTER 12

1.	c	196
2.	c	201
3.	d	203, 206–208
4.	a	203
5.	e	206–207
6.	c	208
7.	d	208
8.	c	201
	e	207
	d	208
	b	206
	a	194
	f	200

CHAPTER 13

1.	c	220
2.	d	221
3.	c	221
4.	a	223
5.	d	222–223
6.	c	224
7.	d	223
	b	221, 222
	c	222
	a	221, 221i

CHAPTER 14

1.	c	228
2.	b	228
3.	c	228
4.	c	230
5.	a	230i
6.	a	230
7.	e	235
	c	232
	a	229
	f	230
	d	230
	g	229
	b	230

CHAPTER 15

1.	d	242
2.	d	239, 240
3.	d	240
4.	a	240
5.	c	240, 240i
6.	a	240
7.	d	242–243
8.	b	245
9.	d	246
10.	b	249
11.	b	249
12.	d	247
	e	244
	b	248
	a	244
	c	246

CHAPTER 16

1.	d	252
2.	c	254
3.	Plasmids	254
4.	a	255
5.	b	255
6.	a	256
7.	b	257
8.	d	258
9.	d	257
	c	260–261
	f	253
	e	259
	b	262
	a	264

CHAPTER 17

1.	b	271
2.	d	276
3.	populations	278
4.	d	278
5.	c	287
6.	c	281
7.	b	282
8.	c	285
9.	c	287
	d	281
	a	278, 279
	b	288

CHAPTER 18

1.	d	294, 295
2.	d	294, 295
3.	c	295
4.	c	300
5.	b	300
6.	c	300
	e	300
	a	300–301
	d	301
	b	301

CHAPTER 19

1.	a	313
2.	c	315
3.	c	318
4.	c	318
5.	d	318
6.	e	318
	b	306
	f	306–307
	c	312
	a	317
	d	313

CHAPTER 20

1.	e	332–341
2.	c	328–329
3.	d	328
4.	c	332, 332i
5.	d	332–333
6.	d	333
7.	b	332
	d	333
	e	336–337
	a	338–339
	c	340–341

CHAPTER 21

1.	c	350
2.	c	352–353
3.	b	354
4.	c	350
5.	d	349
6.	d	354
7.	e	356
8.	d	356
9.	d	352–353
	e	354
	b	356
	f	351
	g	350
	c	353
	a	359

CHAPTER 22

1.	f	368
2.	a	369
3.	d	366i, 370
4.	d	370
5.	b	374
6.	b	377

7.	a	375
8.	b	375
9.	d	381
10.	e	371
	a	374
	b	375
	c	372, 373
	f	368
	d	378

CHAPTER 23

1.	a	386–387
2.	c	387, 393
3.	b	388, 389
4.	c	390
5.	b	393
6.	c	394
	e	386
	g	386, 390
	h	393
	f	386, 388
	a	386
	b	386
	d	386, 398

CHAPTER 24

1.	b	404
2.	a	406
3.	a	406
4.	c	406
5.	c	407
6.	e	408
	c	409
	d	406
	a	406–407
	b	408
	f	408–409

CHAPTER 25

1.	b	416
2.	c	417
3.	c	420
4.	c	422, 423
5.	d	422–425
6.	b	430
7.	a	432
8.	d	418–419
	f	420
	c	422, 423
	g	424, 425
	e	425
	i	426
	j	430
	h	432
	a	440
	b	416, 416t

CHAPTER 26

1.	d	446
2.	a	448
3.	d	448
4.	a	451, 452
5.	f	445i, 464
6.	d	452
7.	f	454
8.	e	455
9.	d	446, 454, 458
10.	e	462–463
11.	d	449
	e	450
	c	450–451
	b	452
	g	454
	f	458–459
	a	460

APPENDIX IV.   ANSWERS TO GENETICS PROBLEMS

CHAPTER 11

1. a. *AB*

    b. *AB, aB*

    c. *Ab, ab*

    d. *AB, Ab, aB, ab*

2. a. All of the offspring will be *AaBB*.

    b. 1/4 *AABB* (25% each genotype)

    1/4 *AABb*

    1/4 *AaBB*

    1/4 *AaBb*

    c. 1/4 *AaBb* (25% each genotype)

    1/4 *Aabb*

    1/4 *aaBb*

    1/4 *aabb*

    d. 1/16 *AABB* (6.25%)

    1/8 *AaBB* (12.5%)

    1/16 *aaBB* (6.25%)

    1/8 *AABb* (12.5%)

    1/4 *AaBb* (25%)

    1/8 *aaBb* (12.5%)

    1/16 *AAbb* (6.25%)

    1/8 *Aabb* (12.5%)

    1/16 *aabb* (6.25%)

3. Yellow is recessive. Because F_1 plants have a green phenotype and must be heterozygous, green must be dominant over the recessive yellow.

4. a. *ABC*

    b. *ABc, aBc*

    c. *ABC, aBc, ABc, aBc*

    d. *ABC, aBC, AbC, abC, ABc, aBc, Abc, abc*

5. Because all F_1 plants of this dihybrid cross had to be heterozygous for both genes, then 1/4 (25%) of the F_2 plants will be heterozygous for both genes.

6. a. The mother must be heterozygous $I^A i$. The male with type B blood could have fathered the child if he were heterozygous $I^B i$.

    b. Genotype alone cannot prove the accused male is the father. Even if he happens to be heterozygous, *any* male who carries the *i* allele could be the father, including those heterozygous for type A blood ($I^A i$) or type B blood ($I^B i$) and those with type O blood (*ii*).

7. A mating between a mouse from a true-breeding, white-furred strain and a mouse from a true-breeding, brown-furred strain would provide you with the most direct evidence.

    Because true-breeding strains of organisms typically are homozygous for a trait being studied, all F_1 offspring from this mating should be heterozygous. Record the phenotype of each F_1 mouse, then let them mate with one another. Assuming only one gene locus is involved, these are possible outcomes for the F_2 offspring:

    a. All F_1 mice are brown, and their F_2 offspring segregate 3 brown : 1 white. *Conclusion*: Brown is dominant to white.

    b. All F_1 mice are white, and their F_2 offspring segregate 3 white : 1 brown. *Conclusion*: White is dominant to brown.

    c. All F_1 mice are tan, and the F_2 offspring segregate 1 brown : 2 tan : 1 white. *Conclusion*: The alleles at this locus show incomplete dominance.

8. You cannot guarantee that the puppies will not develop the disorder without more information about Dandelion's genotype. You could do so only if she is a heterozygous carrier, if the male is free of the alleles, and if the alleles are recessive.

9. Fred could use a testcross to find out if his pet's genotype is *WW* or *Ww*. He can let his black guinea pig mate with a white guinea pig having the genotype *ww*.

    If any F_1 offspring are white, then the genotype of his pet is *Ww*. If the two guinea pig parents are allowed to mate repeatedly and all the offspring of the matings are black, then there is a high probability that his pet guinea pig is *WW*.

    (If, say, ten offspring are all black, then the probability that the male is *WW* is about 99.9 percent. The greater the number of offspring, the more confident Fred can be of his conclusion.)

10. a. 1/2 red, 1/2 pink

    b. All pink

    c. 1/4 red, 1/2 pink, 1/4 white

    d. 1/2 pink, 1/2 white

11.   9/16 walnut comb

    3/16 rose comb

    3/16 pea comb

    1/16 single comb

12. Because both parents are heterozygotes ($Hb^A Hb^S$), the following are the probabilities for each child:

    a. 1/4 $Hb^S Hb^S$

    b. 1/4 $Hb^A Hb^A$

    c. 1/2 $Hb^A Hb^S$

13. A mating of two $M^L M$ cats yields:

    1/4 homozygous dominant (MM)

    1/2 heterozygous ($M^L M$)

    1/4 homozygous recessive ($M^L M^L$)

Because $M^L M^L$ is lethal, the probability that any one kitten among the survivors will be heterozygous is 2/3.

14. a. Both parents must be heterozygotes (Aa). Their children may be albino (aa) or unaffected (AA or Aa).

    b. All are aa.

    c. The albino father must be aa. They have an albino child, so the mother must be Aa. (If she were AA, they could not have an albino child.) The albino child is aa. The three unaffected children are Aa. There is a 50% chance that any child of theirs will be albino. The observed 3:1 ratio is not surprising, given the small number of offspring.

15. a. All of the offspring will have medium-red color corresponding to the genotype $A^1 A^2 B^1 B^2$.

    b. All possible genotypes could appear in the following proportions:

1/16	$A^1 A^2 B^1 B^1$	dark red
1/8	$A^1 A^1 B^1 B^2$	medium-dark red
1/16	$A^1 A^1 B^2 B^2$	medium red
1/8	$A^1 A^2 B^1 B^1$	medium-dark red
1/4	$A^1 A^2 B^1 B^2$	medium red
1/8	$A^1 A^2 B^2 B^2$	light red
1/16	$A^2 A^2 B^1 B^1$	medium red
1/8	$A^2 A^2 B^1 B^2$	light red
1/16	$A^2 A^2 B^2 B^2$	white

CHAPTER 12

1. a. Human males (XY) inherit their X chromosome only from their mother.

    b. In males, an X-linked allele will be found only on his one X chromosome. Males can produce two kinds of gametes: one kind with a Y chromosome free of the gene, and the other kind with an X chromosome bearing the X-linked allele.

    c. One. Each gamete of a woman who is homozygous for an X-linked allele will have an X chromosome that carries the allele.

    d. Two. If a female is heterozygous for an X linked allele, half of the gametes that she produces will contain one of the alleles and the other half will contain the other allele.

2. All of the offspring should be heterozygous for the gene and all should have long wings. However, because some have vestigial wings, the dominant allele might have mutated because of radiation.

3. Because Marfan syndrome is a case of autosomal dominant inheritance and because one parent bears the allele, the probability of any child inheriting the mutant allele is 50%.

4. Because the phenotype appeared in every generation shown in the diagram, this must be a pattern of autosomal dominant inheritance.

5. A daughter could develop this type of muscular dystrophy only if she were to inherit two X-linked recessive alleles—one from her father and one from her mother.

However, if a son bears the allele for the disorder on his X chromosome, then it will be expressed. He will develop the disorder, and most likely he will not father children because of his early death.

6. If no crossover occurs between the two genes, then half the chromosomes will carry alleles AB and half will carry alleles ab.

7. a. Nondisjunction could occur in anaphase I or anaphase II of meiosis.

    b. As a result of a translocation, chromosome 21 (which is small) may become attached to the end of chromosome 14. Even though the chromosome number of the new individual would be 46, its somatic cells would contain the translocated chromosome 21, in addition to two normal chromosomes 21.

8. In the mother, a crossover between the two genes at meiosis generates an X chromosome that carries neither mutant allele.

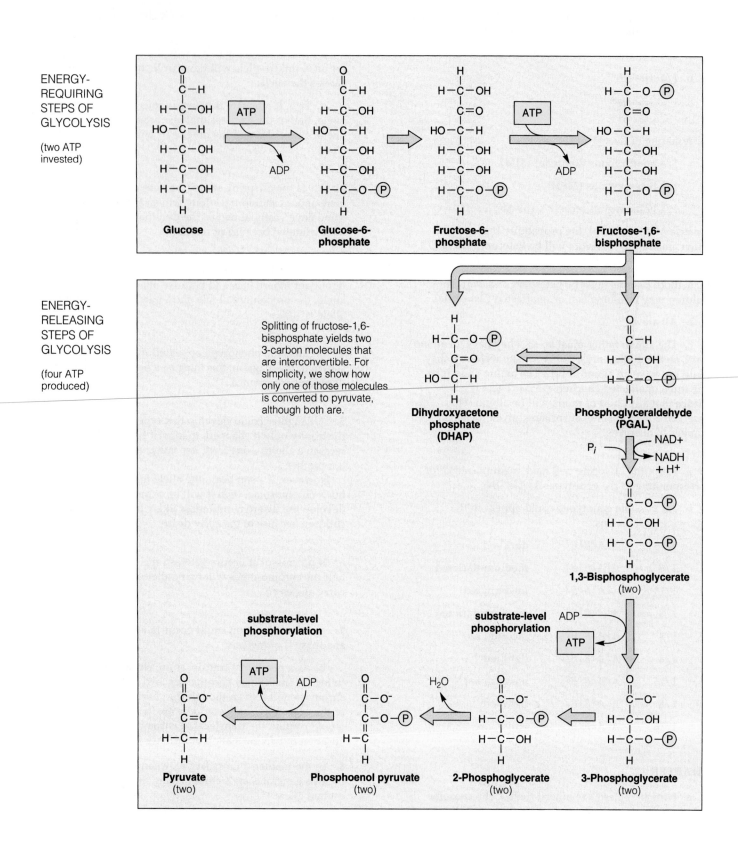

ENERGY-REQUIRING STEPS OF GLYCOLYSIS

(two ATP invested)

ENERGY-RELEASING STEPS OF GLYCOLYSIS

(four ATP produced)

Splitting of fructose-1,6-bisphosphate yields two 3-carbon molecules that are interconvertible. For simplicity, we show how only one of those molecules is converted to pyruvate, although both are.

Figure A   Glycolysis, ending with two 3-carbon pyruvate molecules for each 6-carbon glucose molecule entering the reactions. The *net* energy yield is two ATP molecules (two invested, four produced).

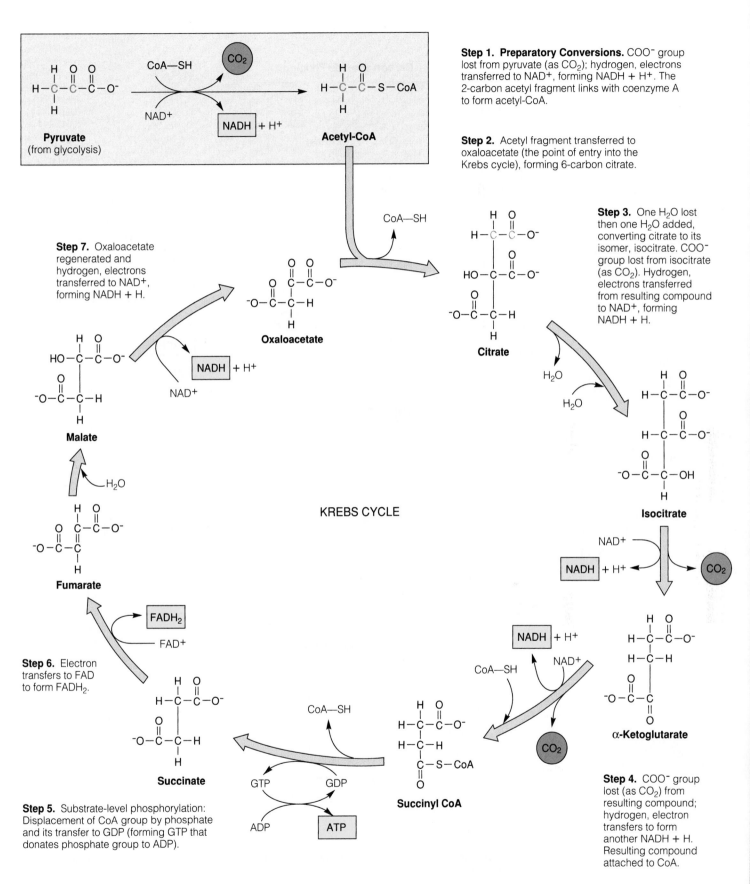

Figure B Krebs cycle, also known as the citric acid cycle. *Red* identifies carbon atoms entering the cyclic pathway (by way of acetyl-CoA) and leaving (by way of carbon dioxide). These cyclic reactions run twice for each glucose molecule that has been degraded to two pyruvate molecules.

Step 1. Preparatory Conversions. COO⁻ group lost from pyruvate (as CO_2); hydrogen, electrons transferred to NAD⁺, forming NADH + H⁺. The 2-carbon acetyl fragment links with coenzyme A to form acetyl-CoA.

Step 2. Acetyl fragment transferred to oxaloacetate (the point of entry into the Krebs cycle), forming 6-carbon citrate.

Step 3. One H_2O lost then one H_2O added, converting citrate to its isomer, isocitrate. COO⁻ group lost from isocitrate (as CO_2). Hydrogen, electrons transferred from resulting compound to NAD⁺, forming NADH + H.

Step 4. COO⁻ group lost (as CO_2) from resulting compound; hydrogen, electron transfers to form another NADH + H. Resulting compound attached to CoA.

Step 5. Substrate-level phosphorylation: Displacement of CoA group by phosphate and its transfer to GDP (forming GTP that donates phosphate group to ADP).

Step 6. Electron transfers to FAD to form $FADH_2$.

Step 7. Oxaloacetate regenerated and hydrogen, electrons transferred to NAD⁺, forming NADH + H.

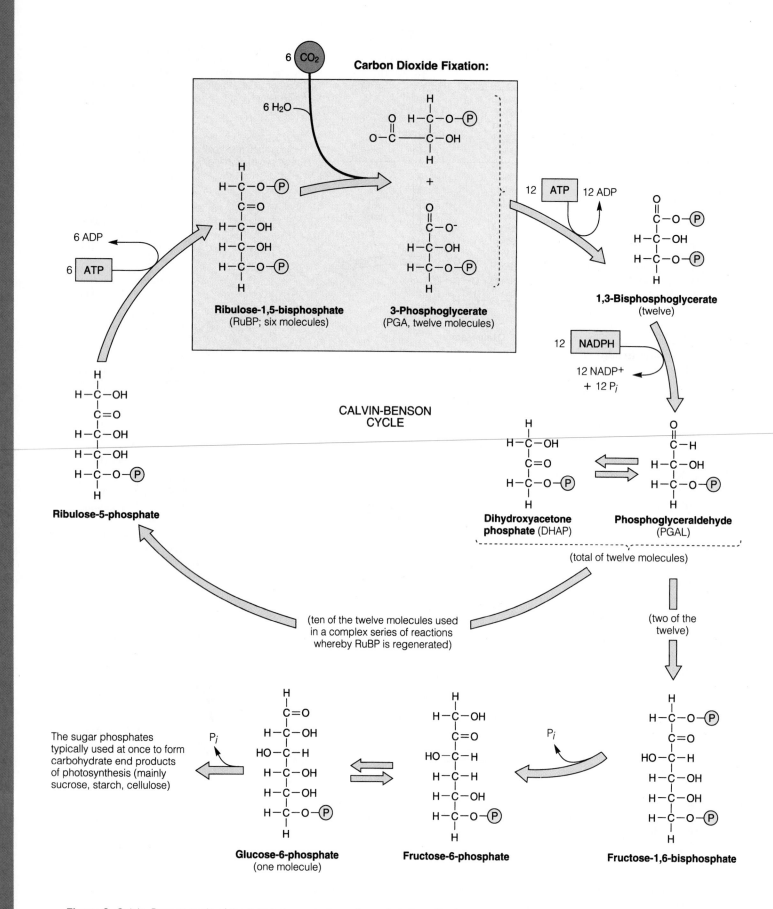

Figure C Calvin–Benson cycle of the light-independent reactions of photosynthesis.

Neutral, nonpolar side group

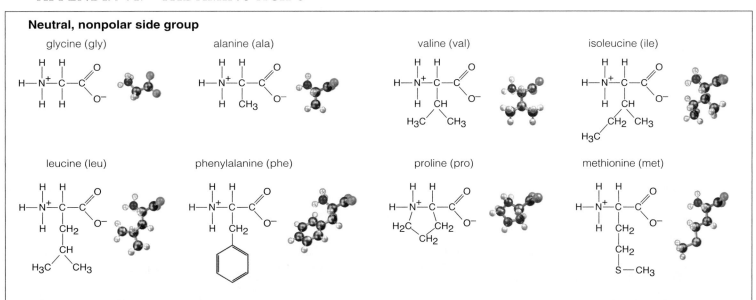

glycine (gly) alanine (ala) valine (val) isoleucine (ile)

leucine (leu) phenylalanine (phe) proline (pro) methionine (met)

Neutral, polar side group

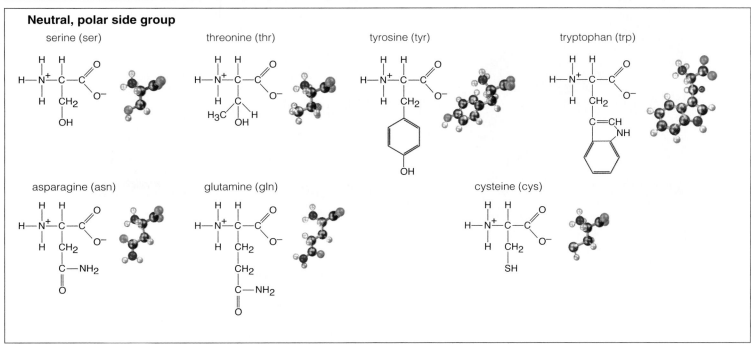

serine (ser) threonine (thr) tyrosine (tyr) tryptophan (trp)

asparagine (asn) glutamine (gln) cysteine (cys)

Acidic side group

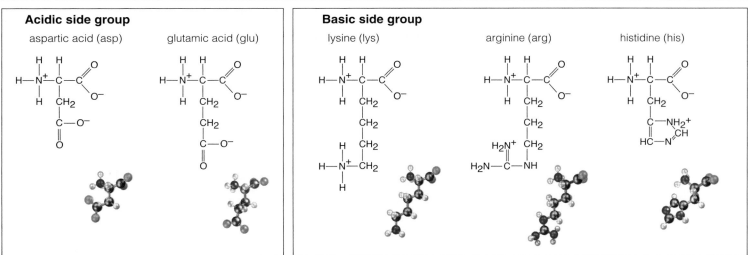

aspartic acid (asp) glutamic acid (glu)

Basic side group

lysine (lys) arginine (arg) histidine (his)

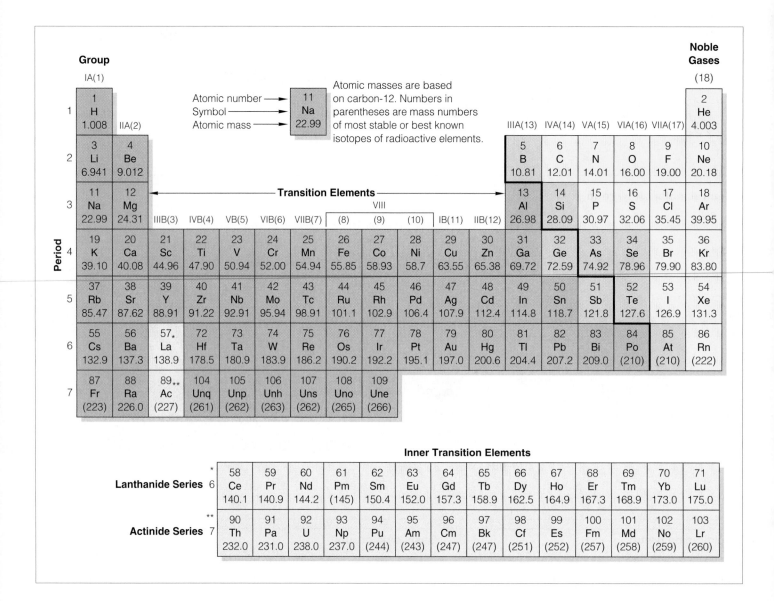

Periodic Table of the Elements

Atomic masses are based on carbon-12. Numbers in parentheses are mass numbers of most stable or best known isotopes of radioactive elements.

Key:
- Atomic number → 11
- Symbol → Na
- Atomic mass → 22.99

Main Table

Group IA(1)	IIA(2)	IIIB(3)	IVB(4)	VB(5)	VIB(6)	VIIB(7)	VIII (8)	(9)	(10)	IB(11)	IIB(12)	IIIA(13)	IVA(14)	VA(15)	VIA(16)	VIIA(17)	Noble Gases (18)
1 H 1.008																	2 He 4.003
3 Li 6.941	4 Be 9.012											5 B 10.81	6 C 12.01	7 N 14.01	8 O 16.00	9 F 19.00	10 Ne 20.18
11 Na 22.99	12 Mg 24.31											13 Al 26.98	14 Si 28.09	15 P 30.97	16 S 32.06	17 Cl 35.45	18 Ar 39.95
19 K 39.10	20 Ca 40.08	21 Sc 44.96	22 Ti 47.90	23 V 50.94	24 Cr 52.00	25 Mn 54.94	26 Fe 55.85	27 Co 58.93	28 Ni 58.7	29 Cu 63.55	30 Zn 65.38	31 Ga 69.72	32 Ge 72.59	33 As 74.92	34 Se 78.96	35 Br 79.90	36 Kr 83.80
37 Rb 85.47	38 Sr 87.62	39 Y 88.91	40 Zr 91.22	41 Nb 92.91	42 Mo 95.94	43 Tc 98.91	44 Ru 101.1	45 Rh 102.9	46 Pd 106.4	47 Ag 107.9	48 Cd 112.4	49 In 114.8	50 Sn 118.7	51 Sb 121.8	52 Te 127.6	53 I 126.9	54 Xe 131.3
55 Cs 132.9	56 Ba 137.3	57* La 138.9	72 Hf 178.5	73 Ta 180.9	74 W 183.9	75 Re 186.2	76 Os 190.2	77 Ir 192.2	78 Pt 195.1	79 Au 197.0	80 Hg 200.6	81 Tl 204.4	82 Pb 207.2	83 Bi 209.0	84 Po (210)	85 At (210)	86 Rn (222)
87 Fr (223)	88 Ra 226.0	89** Ac (227)	104 Unq (261)	105 Unp (262)	106 Unh (263)	107 Uns (262)	108 Uno (265)	109 Une (266)									

Transition Elements — span groups IIIB(3) through IIB(12), VIII covers groups (8), (9), (10).

Inner Transition Elements

Lanthanide Series (Period 6) *

58 Ce 140.1	59 Pr 140.9	60 Nd 144.2	61 Pm (145)	62 Sm 150.4	63 Eu 152.0	64 Gd 157.3	65 Tb 158.9	66 Dy 162.5	67 Ho 164.9	68 Er 167.3	69 Tm 168.9	70 Yb 173.0	71 Lu 175.0

Actinide Series (Period 7) **

90 Th 232.0	91 Pa 231.0	92 U 238.0	93 Np 237.0	94 Pu (244)	95 Am (243)	96 Cm (247)	97 Bk (247)	98 Cf (251)	99 Es (252)	100 Fm (257)	101 Md (258)	102 No (259)	103 Lr (260)

GLOSSARY

ABC model Idea that products of three groups of master genes control floral development.

ABC transporter One of a class of ATP-driven membrane pumps, each for a specific substrate (e.g., ions, sugars, amino acids).

ABO blood typing Method of characterizing an individual's blood based on proteins (A, B, or their absence) at surface of red blood cells.

abortion Premature, spontaneous or induced expulsion of the embryo or fetus from uterus.

abscisic acid Plant hormone; induces stomatal closure (conserves water); bud, seed dormancy.

abscission (ab-SIH-zhun) Dropping of leaves, flowers, fruits, or other parts from a plant.

absorption Of pigments, capture of photon energy. Of cells, uptake of substances from fluid bathing them. Of digestion, water and solute uptake into the internal environment.

absorption spectrum Range of wavelengths that one or more specified pigments can absorb.

accessory pigment Photosynthetic pigment; absorbs wavelengths missed by chlorophylls.

acclimatization Physiological and behavioral adaptation to a new habitat.

acetylation Attachment of an acetyl group to a compound, such as DNA.

acetylcholine (ACh) Neurotransmitter; acts on brain, spinal cord, glands, and muscles.

acetyl-CoA (uh-SEET-ul) Coenzyme A attached to a two-carbon fragment from pyruvate, which it transfers to oxaloacetate for the Krebs cycle.

acid [L. *acidus*, sour] Any substance that, when dissolved in water, donates hydrogen ions.

acid–base balance State in which extracellular fluid is not too acidic or too basic; outcome of controls over its concentrations of dissolved ions.

acidity Of a solution, an excess of hydrogen ions relative to hydroxyl ions.

acid rain Wet acid deposition; falling of rain (or snow) rich in sulfur and nitrogen oxides.

acoelomate (ay-SEE-luh-mate) Absence of a fluid-filled cavity between gut and body wall.

acoustical receptor Specialized, vibration-sensitive mechanoreceptor.

acoustical signal Use of sound as a form of intraspecific communication.

actin (AK-tin) Type of cytoskeletal protein.

action potential Abrupt, brief reversal in the resting membrane potential of excitable cells.

actinopod ("ray foot") Amoeboid protozoan with silica hard parts and slender, reinforced pseudopods (e.g., a radiolarian or heliozoan).

activation energy A specific minimum amount of energy required to get a specific reaction going, with or without the help of an enzyme. Reactions differ in the amount required.

activator Regulatory protein that enhances a cell activity (e.g., gene transcription).

active site Crevice in an enzyme molecule where a specific reaction is catalyzed.

active transport Pumping of a specific solute across a cell membrane against its concentration gradient, through a transport protein's interior. Requires an energy boost, typically from ATP.

adaptation [L. *adaptare*, to fit] Long-term, heritable aspect of form, function, behavior, or development that improves odds for surviving and reproducing; an outcome of microevolution (e.g., natural selection that enhances fit between individual and prevailing conditions). Of sensory neurons, a decline or end of action potentials even when a stimulus is maintained at constant strength.

adaptive behavior Behavior that contributes to the individual's reproductive success.

adaptive radiation Macroevolutionary pattern; burst of genetic divergences from a lineage that gives rise to many species, each using a novel resource or a new (or newly vacated) habitat.

adaptive trait Any aspect of form, function, or behavior that helps the individual survive and reproduce under prevailing conditions.

adaptive zone Any way of life available for organisms that are physically, ecologically, and evolutionarily equipped to live it.

adenine (AH-de-neen) A purine; a nitrogen-containing base in certain nucleotides.

ADH Antidiuretic hormone. Hypothalamic hormone; promotes water conservation.

adhering junction Mass of proteins anchored at cell membrane; helps adjoining cells adhere.

adhesion protein Plasma membrane protein that helps cells locate tissue mates and stick together.

adipose tissue A connective tissue having an abundance of fat-storing cells.

adoption Permanent acceptance of and caring for offspring of other individuals as one's own.

ADP Adenosine diphosphate (ah-DEN-uh-seen die-FOSS-fate). A nucleotide coenzyme.

adrenal cortex Outer portion of adrenal gland; secretes cortisol and aldosterone.

adrenal gland Endocrine gland; helps govern glucose metabolism and sodium reabsorption.

adrenal medulla Inner portion of adrenal gland; secretes epinephrine, norepinephrine.

aerobic respiration (air-OH-bik) [Gk. *aer*, air, + *bios*, life] Oxygen-requiring pathway of ATP formation; from glycolysis, to Krebs cycle, to electron transport phosphorylation. Typical net energy yield: 36 ATP per glucose molecule.

African emergence model Holds that *Homo sapiens* originated in Africa, then replaced archaic *Homo* populations in other regions.

age structure Number of individuals in each age category for a population.

agglutination (ah-glue-tin-AY-shun) Defense response; antibodies cause antigen to clump together, which facilitates phagocytosis.

aging Of any multicelled organism showing extensive cell differentiation, the gradual and expected deterioration of the body over time.

AIDS Acquired immune deficiency syndrome. A set of chronic disorders following infection by the human immunodeficiency virus (HIV).

alcohol Organic compound containing one or more hydroxyl groups (—OH); it dissolves readily in water. Sugars are examples.

alcoholic fermentation Anaerobic ATP-forming pathway. Pyruvate from glycolysis is degraded to acetaldehyde, which accepts electrons from NADH to form ethanol; NAD+ needed for the reactions is regenerated. Net yield: 2 ATP.

aldosterone (al-DOSS-tuh-rohn) Adrenal cortex hormone; helps control sodium reabsorption.

alga, plural **algae** Photosynthetic protistans, mainly; most are members of phytoplankton.

algal bloom Huge increases in aquatic algal population sizes after nutrient enrichments.

alkylating agent Any substance that transfers methyl or ethyl groups to a compound.

allantois (ah-LAN-twahz) [Gk. *allas*, sausage] Extraembryonic membrane used to exchange gases and store metabolic wastes of embryos of reptiles, birds, some mammals. In humans, it forms urinary bladder, placental blood vessels.

allele (uh-LEEL) One of two or more molecular forms of a gene that arise by mutation and code for different versions of the same trait.

allele frequency For a given gene locus, the relative abundances of each kind of allele among all the individuals of a population.

allergen Normally harmless substance that provokes inflammation, excess secretion of mucus, and often immune responses.

allergy Hypersensitivity to an allergen.

allopatric speciation [Gk. *allos*, different, + L. *patria*, native land] Speciation model. A physical barrier arises, separates populations or subpopulations of a species, ends gene flow, and favors divergences that end in speciation.

altruism (AL-true-IZ-um) Behavior that lowers an individual's chance of reproductive success but helps others of its species.

alveolate Protistan with many membrane-bound sacs just beneath the cell surface (e.g., a ciliate, sporozoan, or dinoflagellate).

alveolus (ahl-VEE-uh-lus), plural **alveoli** [L. *alveus*, small cavity] Cupped, thin-walled outpouching of respiratory bronchiole where lungs and blood exchange O_2 and CO_2.

amino acid (uh-MEE-no) Organic compound with an H atom, amino group, acid group, and R group, all covalently bonded to a carbon atom. Subunit of polypeptide chains.

ammonification (uh-moan-ih-fih-KAY-shun) Of nitrogen cycle, process by which soil fungi and bacteria break down nitrogenous wastes or remains to forms that plants can take up.

amnion (AM-nee-on) An extraembryonic membrane; boundary layer of a fluid-filled sac in which certain tetrapod embryos develop.

amniote A tetrapod that produces amniote eggs. Major groups are synapsids (mammals, early mammal-like reptiles) and sauropsids ("reptiles" and birds).

amniote egg Egg, often shelled, with four extraembryonic membranes (amnion, chorion, yolk sac, allantois). Pivotal adaptation in the vertebrate invasion of land.

amoeboid protozoan A predatory or parasitic protistan that moves by pseudopod formation and cytoplasmic streaming (e.g., an amoeba or a foraminiferan, heliozoan, or radiolarian).

amphibian Vertebrate with a body plan and reproductive mode between fishes and reptiles, and a mostly bony endoskeleton; one of the four-legged walkers or a descendant of one.

anaerobic electron transfer (an-uh-ROW-bik) [Gk. *an*, without, + *aer*, air] Flow of electrons through transfer chains in plasma membrane that drives ATP formation. A compound other than oxygen is the final electron acceptor.

anaerobic pathway Any metabolic pathway in which a substance other than oxygen is the final acceptor of electrons stripped from substrates.

anagenesis Speciation pattern; changes in allele frequencies and morphology accumulate within an unbranched line of descent.

analogous structures (ann-AL-uh-gus) [Gk. *analogos*, similar to one another] Body parts that once differed in evolutionarily distant lineages but converged in structure and function as responses to similar environmental pressures.

anaphase (AN-uh-faze) Of mitosis, stage when sister chromatids of each chromosome move to opposite spindle poles. Of anaphase I (meiosis), each duplicated chromosome and its homologue move to opposite poles. Of anaphase II (meiosis), sister chromatids of each chromosome move to opposite poles.

anatomy Study of body form, from molecular to organ-system levels of organization.

anemia Disorder resulting from deformed or insufficient quantity of red blood cells.

aneuploidy (AN-yoo-ploy-dee) Having one extra or one less chromosome relative to the parental chromosome number.

angiosperm (AN-gee-oh-spurm) [Gk. *angeion*, vessel, and *spermia*, seed] Flowering plant.

angiotensin II A hormone that stimulates aldosterone-secreting cells of adrenal cortex.

animal Multicelled, aerobic, motile predator or parasite; usually with tissues, organ systems; developed through embryonic stages.

animal behavior Coordinated response to stimuli involving sensory, neural, endocrine, and effector components. Has a genetic basis, can evolve, and can be modified by learning.

Animalia Kingdom of animals.

annelid An invertebrate; a segmented worm (e.g., oligochaete, leech, or polychaete).

annual Plant that lasts one growing season.

anthocyanin Blue or red accessory pigment.

anthropoid Monkey, ape, or human.

antibiotic Metabolic product of soil microbes that kills bacterial competitors for nutrients.

antibody [Gk. *anti*, against] Antigen-binding receptor made and secreted only by B cells.

anticodon Series of three nucleotide bases in tRNA; can base-pair with an mRNA codon.

antigen (AN-tih-jen) [Gk. *anti*, against, + *genos*, race, kind] Any molecular configuration that certain lymphocytes chemically recognize as nonself and that triggers an immune response.

antigen–MHC complex Antigen bound with MHC molecule at plasma membrane; activates helper T cells to start immune response.

antigen-presenting cell Cell that processes and displays antigen with MHC molecules (e.g., macrophage, B cell, dendritic cell).

aorta (ay-OR-tah) Of vertebrates, main artery of systemic circulation.

apical dominance Growth-inhibiting effect of a terminal bud on growth of lateral buds.

apical meristem (AY-pih-kul MARE-ih-stem) [L. *apex*, top, + Gk. *meristos*, divisible] Mass of dividing cells at root tips and shoot tips.

apoptosis (APP-oh-TOE-sis) Programmed cell death of body cells finished with prescribed functions or altered, as by infection or cancer.

appendicular skeleton (ap-en-DIK-yoo-lahr) Bones of limbs, hips, and shoulders.

appendix Narrow projection from the cecum.

Archaebacteria Prokaryotic domain; closer to eukaryotic cells than to eubacteria; includes methanogens, halophiles, and thermophiles; also called a kingdom.

Archean Eon in which life arose (3.8–2.5 bya).

archipelago Island chain some distance away from a continent.

area effect Idea that larger islands support more species than smaller ones at equivalent distances from sources of colonizer species.

arteriole (ar-TEER-ee-ole) Type of blood vessel between arteries and capillaries. Collectively, points of selective dilation and constriction.

artery Thick-walled, muscular, rapid-transport vessel that smooths out blood pressure pulses.

arthropod Invertebrate with hard exoskeleton, specialized segments, and jointed appendages.

artificial selection Selection of traits among a population under contrived conditions.

ascospore Sexual spore of sac fungi.

asexual reproduction Any reproductive mode by which offspring arise from a single parent and inherit the genes of that parent only.

asteroid Rocky, metallic body, a few to 1,000 kilometers across, hurtling through space.

atmosphere The volume of gases, airborne particles, and water vapor enveloping Earth.

atom Smallest unit of an element that retains the element's properties.

atomic number Of an element, the number of protons in the nucleus of each atom.

ATP Adenosine triphosphate (ah-DEN-uh-seen try-FOSS-fate). Nucleotide with adenine, ribose, and three phosphate groups; delivers energy to most energy-requiring metabolic reactions.

ATP/ADP cycle ADP forms from ATP, then ATP is regenerated from ADP, through the transfer of inorganic phosphate or a phosphate group.

ATP synthase Active transport protein that shows catalytic activity; ATP forms as H⁺ flows through a channel in the protein's interior.

australopith (OHSS-trah-low-pith) [L. *australis*, southern, + Gk. *pithekos*, ape] Early hominid.

autoimmunity Result of inappropriate attacks on normal body cells by lymphocytes.

automated DNA sequencing Fast method of sequencing cloned or PCR-amplified DNA.

autonomic nerve (AH-toe-NOM-ik) One of the nerves from central nervous system to smooth muscle, cardiac muscle, and glands of viscera.

autosome Any chromosome of a type that is the same in males and females of the species.

autotroph (AH-toe-trofe) [Gk. *autos*, self, + *trophos*, feeder] Any organism that makes its own food with an environmental energy source (e.g., sunlight) and CO_2 as its carbon source.

auxin (AWK-sin) Plant hormone; induces stem lengthening and responses to gravity and light.

axial skeleton (AX-ee-uhl) Skull, backbone, ribs, and breastbone (sternum).

axon Neuron's signal-conducting zone.

B cell B lymphocyte; only body cell that makes and displays or secretes antibody molecules.

bacillus Rod-shaped prokaryotic cell.

bacterial chromosome Circularized, double-stranded DNA molecule with few proteins.

bacteriophage (bak-TEER-ee-oh-fahj) Category of viruses that infect bacterial cells.

balanced polymorphism Form of selection in which two or more alleles for a trait are being maintained in a population over time.

bark All tissues external to vascular cambium.

Barr body Randomly condensed one of two X chromosomes in cells of female mammals.

basal body Centriole which, after giving rise to microtubules of a flagellum or cilium, stays attached to its base in cytoplasm.

base Any substance that accepts hydrogen ions (H⁺) when dissolved in water; OH⁻ forms after this. Also, an organic compound with a single or double ring structure containing nitrogen.

base-pair substitution One amino acid has replaced another during protein synthesis.

base sequence Sequential order of bases in a DNA or RNA strand.

basidiospore Sexual spore of a club fungus.

basophil Fast-acting white blood cell; secretes histamine during inflammatory response.

biennial (bi-EN-yul) Flowering plant that completes its life cycle in two growing seasons.

big bang Model for origin of universe.

bilateral symmetry Body plan with axis from anterior to posterior end, separated into right and left sides, and dorsal and ventral surface.

bile Liver secretion required for fat digestion.

binary fission Asexual reproductive mode; the body of protozoans and some other animals divides in two parts. *See* prokaryotic fission.

binding energy Energy released when weak bonds that form between a substrate, enzyme, and any cofactor break during the transition state; the partial payment on energy cost of a reaction (the activation energy).

binomial system Of taxonomy, assigning a generic and a specific name to each species.

biodiversity [Gk. *bios*, life] All species living in a specified region during a given interval.

biofilm Huge microbial populations anchored to surfaces (e.g., lung epithelia) by their own sticky, stiff polysaccharide secretions.

biogeochemical cycle Slow movement of an element from environmental reservoirs, through food webs, then back to the environment.

biogeographic realm [Gk. *bios*, life, + *geographein*, to describe the Earth's surface] One of six vast land areas with distinctive species.

biogeography Scientific study of the world distribution of species.

biological clock Internal time-measuring mechanism; adjusts activities seasonally, daily, or both in response to environmental cues.

biological magnification Ever increasing concentration of a nondegradable or slowly degradable substance in body tissues as it is passed along food chains.

biological species concept Defines a species as one or more populations of individuals that are interbreeding under natural conditions, producing fertile offspring, and are isolated reproductively from other such populations. Applies to sexually reproducing species only.

biology The scientific study of life.

bioluminescence Fluorescent light produced by an organism through an ATP-driven reaction involving enzymes (luciferases).

biomass Combined weight of all organisms at a given trophic level in an ecosystem.

biome Subdivision of a biogeographic realm.

biosphere [Gk. *bios*, life, + *sphaira*, globe] All regions of the Earth's waters, crust, and atmosphere in which organisms live.

biosynthetic pathway Metabolic pathway by which organic compounds are synthesized.

biotic potential Of population growth for a given species, the maximum rate of increase per individual under ideal conditions.

bipedalism Habitually walking upright on two feet, as by ostriches and hominids.

bird A sauropsid; only animal (besides some extinct related dinosaurs) that grows feathers.

blastocyst (BLASS-tuh-sist) Type of blastula; a surface layer of blastomeres, a cavity filled with their secretions, and an inner cell mass.

blastomere One of the small, nucleated cells that form during cleavage of animal zygotes.

blastula Early outcome of cleavage; a number of blastomeres enclose a fluid-filled cavity.

blood Fluid connective tissue of water, solutes, and formed elements (blood cells, platelets). Transports substances to and from cells, helps maintain internal environment.

blood–brain barrier Mechanism that controls which solutes enter cerebrospinal fluid.

blood pressure Fluid pressure, generated by heart contractions, that circulates blood.

bone Vertebrate organ with mineral-hardened connective tissue; roles in movement, mineral storage, protection; blood cells form in some.

bone remodeling Ongoing mineral deposits, withdrawals from bone; adjusts bone strength, maintains calcium, phosphorus levels in blood.

bone tissue Of vertebrate skeleton, a tissue of osteoblast secretions hardened with minerals.

bony fish Aquatic vertebrate with cranium and mostly bony endoskeleton.

boreal forest A swamp forest (or *taiga*) in glaciated regions with cold lakes and streams.

bottleneck Severe reduction in the size of a population, brought about by intense selection pressure or a natural calamity.

Bowman's capsule Part of a nephron; receives water and solutes being filtered from blood.

brain Of most nervous systems, integrating center that receives and processes sensory input and issues coordinated commands for responses by muscles and glands.

brain stem Most ancient nerve tissue in the vertebrate hindbrain, midbrain, and forebrain.

bronchiole Finely branched airway; part of the bronchial tree inside the lung.

bronchus, plural **bronchi** (BRONG-cuss, BRONG-kee) [Gk. *bronchos*, windpipe] Tubular airway; branches from trachea and leads into lungs.

brown alga A stramenopile, mostly marine, photoautotrophic; chlorophylls a, c_1, c_2, and carotenoids such as fucoxanthin; e.g., kelp.

bryophyte Nonvascular land plant requiring free water for fertilization. Haploid dominance in life cycle. Some with cuticle and stomata. A moss, liverwort, or hornwort.

bud Undeveloped shoot, mainly meristematic tissue. Small, protective scales often cover it.

buffer system A weak acid and the base that forms when it dissolves in water. The two work as a pair to counter slight shifts in pH.

bulk Of the vertebrate gut, the volume of undigested material in the small intestine that cannot be decreased by absorption.

bulk flow In response to a pressure gradient, movement of more than one kind of molecule in the same direction in the same medium.

C3 plant Plant that uses three-carbon PGA as the first intermediate for carbon fixation.

C4 plant Plant that uses oxaloacetate (a four-carbon compound) as the first intermediate for carbon fixation. CO_2 is fixed twice, in two cell types; helps counter photorespiration.

calcium pump Active transporter protein specific for calcium ions.

Calvin–Benson cycle Light-independent cyclic reactions of photosynthesis that form sugars, using ATP energy, NADPH, and CO_2.

CAM plant Type of plant that conserves water by opening stomata only at night, when it fixes carbon dioxide by means of a C4 pathway.

camera eye Of cephalopods and vertebrates, a camera-like eyeball with an inner darkened chamber, a single opening for light, and a retina (like film) on which visual stimuli are focused.

camouflage Coloration, form, patterning, or behavior that helps predators or prey blend with the surroundings and escape detection.

cancer Malignant tumor; mass of altered cells that divide abnormally. Potentially lethal.

capillary, blood [L. *capillus*, hair] Blood vessel with thin endothelial wall and small diameter.

capillary bed Diffusion zone, consisting of great numbers of capillaries, where blood and interstitial fluid exchange substances.

capture-recapture method For population counts; collecting, marking, releasing, then recapturing animals representative of group.

carbamino hemoglobin $HbCO_2$, the form in which 30% of the CO_2 in blood is transported.

carbohydrate Molecule of carbon, hydrogen, and oxygen mostly in a 1:2:1 ratio. Main kinds are monosaccharides, oligosaccharides, and polysaccharides. They are structural materials, energy stores, and transportable energy forms.

carbon cycle An atmospheric cycle. Carbon moves from reservoirs (sediments, rocks, the ocean), through the atmosphere (mostly as CO_2), food webs, and back to the reservoirs.

carbon fixation First of the light-independent reactions. Rubisco, an enzyme, affixes carbon (from CO_2) to RuBP or to another compound for entry into the Calvin–Benson cycle.

carbonic anhydrase Red blood cell enzyme that converts CO_2 to bicarbonate, the form in which 60% of the CO_2 in blood is transported.

carcinogen (kar-SIN-uh-jen) Any substance or agent that can trigger cancer.

cardiac conduction system (KAR-dee-ak) [Gk. *kardia*, heart, + *kyklos*, circle] Specialized cardiac muscle cells that make the heart pump by way of rhythmic, synchronized excitation.

cardiac cycle Sequence of muscle contraction and relaxation in one heartbeat.

cardiac muscle tissue A contractile tissue that is present only in the heart wall.

cardiac pacemaker Sinoatrial (SA) node; mass of self-excitatory cardiac muscle cells that are the basis of normal rate of heartbeat.

carnivore Animal that consumes mostly flesh.

carnivorous plant Plant that traps, digests, and feeds on insects and other small animals.

carotenoid (kare-OTT-en-oyd) An accessory pigment of photosynthesis; e.g., fucoxanthin.

carpel (KAR-pul) Female reproductive part of a flower. The ovary, stigma, and often a style consist of one or more carpels.

carrying capacity The maximum number of individuals in a population (or species) that a given environment can sustain indefinitely.

cartilage Connective tissue with solid, pliable intercellular material that resists compression.

cartilaginous fish Jawed fish with a cartilage endoskeleton and prominent fins (e.g., sharks).

Casparian strip Narrow, waxy, impermeable band between walls of abutting cells making up root endodermis and exodermis.

catastrophism Idea that abrupt changes in the geologic or fossil record were divinely invoked.

cDNA DNA molecule copied from a mature mRNA transcript by reverse transcription.

cell [L. *cella*, small room] Smallest living unit, it can survive and reproduce on its own, given its DNA, raw materials, and an energy source.

cell communication Mechanisms by which free-living cells of a species or the cells of a multicelled organism coordinate their activities; involves sending and receiving, transducing, and responding to signaling molecules.

cell cortex A dynamic, crosslinked mesh of cytoskeletal elements under plasma membrane.

cell count The number of cells of a given type in one microliter of blood.

cell cycle Events by which a cell increases in mass, roughly doubles its cytoplasm, duplicates its DNA, and divides in two. Extends from the time a cell forms until it completes division.

cell differentiation Key development process. Cell lineages become specialized in structure, composition, and function by activating and suppressing part of the genome selectively.

cell junction Site where cells are interacting physically, functionally, or both.

cell plate formation Process of cytoplasmic division in plant cells; after nuclear division, a cross-wall with plasma membrane on both surfaces forms and divides the cytoplasm.

cell theory All organisms consist of one or more cells, the smallest units with a capacity for independent life that no longer spontaneously arise under existing conditions on Earth.

cell wall A semirigid, permeable structure external to the plasma membrane; helps many cells retain their shape and resist rupturing.

Cenozoic The present era (65 mya to present).

centipede Long-bodied segmented arthropod; many legged, swiftly aggressive predator.

central nervous system Brain and spinal cord.

central vacuole Fluid-filled storage organelle of plant cell; its growth enhances cell surface area.

centriole (SEN-tree-ohl) Structure that gives rise to microtubules of cilia and flagella.

centromere (SEN-troh-meer) A constricted area of a chromosome that has attachment sites for spindle microtubules during nuclear division.

cephalization (SEF-ah-lah-ZAY-shun) [Gk. *kephalikos*, head] Over time, the concentration of sensory structures and nerve cells in a head.

cerebellum (ser-ah-BELL-um) Hindbrain region with reflex centers for maintaining posture and smoothing out limb movements.

cerebral cortex Thin surface layer of cerebral hemispheres; receives, integrates, and stores sensory information; coordinates responses.

cerebrospinal fluid Clear extracellular fluid, inside a system of canals and chambers, that bathes and protects the brain and spinal cord.

cerebrum (suh-REE-bruhm) Forebrain region that generally deals with olfactory input and motor responses. In mammals, it evolved into the most complex integrating center.

checkpoint gene Its protein product can help delay, advance, or block cell cycle.

chemical bond A union between the electron structures of two or more atoms or ions.

chemical energy Potential energy of molecules.

chemical equilibrium Time when a reversible reaction runs both ways at about same rate.

chemical synapse (SIN-aps) Thin cleft between a presynaptic neuron and a postsynaptic cell. Neurotransmitter molecules diffuse across it.

chemiosmotic theory Idea that electrochemical gradient drives ATP formation. H$^+$ builds up in membranous compartment, ATP forms as it flows out through ATP synthases in response to concentration and electric gradients.

chemoautotroph (KEE-moe-AH-toe-trofe) Any prokaryotic cell that synthesizes its own food using carbon dioxide as the carbon source and an inorganic substance as the energy source.

chemoreceptor Sensory receptor that detects ions or molecules dissolved in fluid bathing it.

chlorofluorocarbon (KLORE-oh-FLOOR-oh-car-bun) CFC; a compound of chlorine, fluorine, and carbon; is contributing to ozone thinning.

chlorophyll (KLOR-uh-fill) [Gk. *chloros*, green, + *phyllon*, leaf] Main photosynthetic pigment. Chlorophylls absorb all wavelengths of visible light but not much of green and yellow ones.

chloroplast (KLOR-uh-plast) The organelle of photosynthesis in plants and many protistans.

chordate Animal having a notochord, dorsal hollow nerve cord, pharynx, and gill slits in pharynx wall during at least part of life cycle.

chorion (CORE-ee-on) Type of extraembryonic membrane that becomes part of placenta. Villi form at its surface and facilitate exchange of substances between the embryo and mother.

chromatid *See* sister chromatid.

chromatin A cell's collection of DNA and all of the proteins associated with it.

chromosome (CROW-moe-some) [Gk. *chrōma*, color, + *soma*, body] Of eukaryotic cells, a DNA molecule, duplicated or unduplicated, with many associated proteins. Of prokaryotic cells, a circular DNA molecule.

chromosome number All chromosomes in a given type of cell. *See* haploidy; diploidy.

chrysophyte A category of photosynthetic protistans (e.g., golden algae, yellow-green algae, coccolithophores, diatoms).

chyme Liquified food material in the gut.

ciliate Ciliated protozoan; an alveolate that typically has profuse cilia at its surface.

cilium (SILL-ee-um), plural **cilia** Short motile or sensory structure of certain eukaryotic cells; its core is a 9 + 2 array of microtubules.

circadian rhythm (ser-KAYD-ee-un) [L. *circa*, about, + *dies*, day] Biological activity repeated in cycles, each about twenty-four hours long, independently of environmental change.

circulatory system Organ system that moves substances to and from cells, and often helps stabilize body temperature and pH. Typically consists of a heart, blood vessels, and blood.

cladogenesis [Gk. *clad-*, branch] Speciation pattern in which a lineage splits and isolated populations undergo genetic divergence.

cladogram Evolutionary tree diagram with branch points showing relative relationships. Groups closer together share a more recent common ancestor than those farther apart.

classification system A way of organizing and retrieving information about species.

cleavage Early stage of animal development. Mitotic cell divisions divide a fertilized egg into many smaller, nucleated cells; original volume of egg cytoplasm does not increase.

cleavage furrow Ringlike depression defining the cutting plane for a dividing animal cell.

climate For a specified region, the prevailing weather conditions (e.g., temperature, cloud cover, wind speed, rainfall, and humidity).

climax community Array of species that has stabilized under prevailing habitat conditions.

climax pattern model Idea that environmental factors may have different effects in a large region, so stable communities other than the climax stage may also persist in that region.

cloaca (kloe-AY-kuh) Chamber or duct in last part of gut of some animals; roles in excretion, reproduction, and sometimes respiration.

cloning Making a genetically identical copy of DNA or of an organism.

cloning vector Any plasmid, viral DNA, or some other piece of DNA that researchers use to isolate and amplify DNA of interest.

club fungus Fungus having club-shaped cells that produce and bear spores.

cnidarian (nye-DAR-ee-un) Type of radial invertebrate at the tissue level of organization; the only nematocyst producer.

coal Nonrenewable energy source that formed over 280 million years ago from submerged, undecayed, and compacted plant remains.

coccolithophore Type of chrysophyte with calcium carbonate plates.

coccus Spherically shaped prokaryotic cell.

codominance In heterozygotes, simultaneous expression of a pair of nonidentical alleles that specify different phenotypes.

codon One of 64 possible base triplets in an mRNA strand. A code word for an amino acid in a polypeptide chain; a few codons also act as START or STOP signals for translation.

coelom (SEE-lum) A peritoneum-lined cavity between the gut and body wall of most animals.

coenzyme Enzyme helper; a nucleotide that transfers electrons and H atoms stripped from substrates to a different reaction site.

coevolution Joint evolution of two closely interacting species by changes in the selection pressures operating between the two.

cofactor Metal ion, coenzyme, or other organic compound that assists an enzyme in catalysis or that transfers atoms, functional groups, or electrons to a different reaction site.

cohesion Capacity to resist rupturing when placed under tension (stretched).

cohesion theory of water transport Theory that collective cohesive strength of hydrogen bonds pulls up water molecules through xylem in response to transpiration from leaves.

cohort Group of individuals being studied.

coitus Sexual intercourse.

coleoptile Thin sheath that protects primary shoot as growth pushes it up through soil.

collar cell In sponge body wall, a flagellated cell with a "collar" of food-trapping villi.

collecting duct Last tubular region of nephron.

collenchyma (coll-ENG-kih-mah) A simple plant tissue that imparts flexible support during primary growth, as in lengthening stems.

colon (CO-lun) Large intestine.

commensalism Ecological interaction between two (or more) species in which one benefits directly and the other is affected little, if at all.

communication display Social signal, often ritualized with intended changes in functions of common patterns.

communication protein Part of a protein complex that forms an open channel between cytoplasm of adjoining cells.

communication signal Social cue encoded in stimuli such as specific body coloration, body patterning, odors, sounds, and postures.

community All populations in a habitat. Also, a group of organisms with similar life-styles.

companion cell Specialized parenchyma cell; helps load conducting cells of phloem.

comparative morphology [Gk. *morph*, form] Scientific study of comparable body parts of adults or embryonic stages of major lineages.

compartmentalization Plant responses to an attack, including secretion of resins and toxins.

competition, interspecific Type of ecological interaction in which individuals of different species compete for a share of resources.

competition, intraspecific Type of ecological interaction in which individuals of the same population compete for a share of resources.

competitive exclusion Theory that two or more species requiring identical resources cannot coexist indefinitely.

complement system Set of proteins circulating in inactive form in vertebrate blood. Different kinds promote inflammation, induce lysis of pathogens, and stimulate phagocytes during nonspecific defenses and immune responses.

complete digestive system Body tube with a mouth at one end and an anus at the other.

composite signal Communication signal with information encoded in more than one cue.

compound Molecule consisting of two or more elements in unvarying proportions (e.g., H$_2$O).

compound eye Crustacean or insect eye with multiple rodlike units, each of which samples part of the visual field.

concentration gradient A difference in the number per unit volume of molecules or ions of a substance between two regions. Molecules collide constantly and career outward to a region where they are less concentrated. All substances tend to diffuse down such gradients.

condensation reaction Covalent bonding of two molecules into a larger one; water often forms as a by-product.

conduction Exchange of heat owing to a thermal gradient between two objects.

cone Reproductive structure of certain seed-bearing plants; has clusters of scales with exposed ovules on their surface.

cone cell In a vertebrate eye, a photoreceptor that responds to intense light and contributes to sharp daytime vision and color perception.

conifer A gymnosperm; most are evergreen woody trees or shrubs with thickly cuticled needle-like or scale-like leaves; adapted to conserve water in droughts and cold winters.

coniferous forest Forest of coniferous trees.

conjugation Prokaryotic gene transfer mode; also protistan sexual reproductive mode.

connective tissue Most abundant type of animal tissue. Soft connective tissues (loose, dense, irregular, or dense regular) differ in amount, arrangement of fibroblasts, fibers, ground substance. Adipose tissue, cartilage, bone tissue, and blood are specialized types.

conservation biology International systematic survey of the full range of biodiversity, analysis of its evolutionary and ecological origins, and identification of methods to maintain and use it for the benefit of the human population.

conservation of mass, law of The total mass of all substances entering a reaction equals the total mass of all products.

consumer [L. *consumere*, to take completely] A heterotroph that feeds on cells or tissues of other organisms (e.g., herbivores, carnivores).

continuous variation Of a population, a more or less continuous range of small differences in a given trait among its individuals.

contractile cell A cell that shortens in response to stimulation and returns to a resting position.

contractile vacuole (kun-TRAK-till VAK-you-ohl) [L. *contractus*, to draw together] Organelle in some protistans; expels excess water.

control group Group used as a standard for comparison with an experimental group and, ideally, identical with it in all respects except for the one variable being studied.

convection Movement of air or water next to an object that aids conductive heat loss from it.

convergent evolution Dissimilar body parts in evolutionarily distant lineages become similar over time, as independent responses to similar kinds of environmental pressures.

coral reef Formation of accumulated hard parts of corals and other organisms in warm, clear waters between latitudes 25°N and 25°S.

core temperature Internal body temperature.

cork Periderm component; insulates, protects, and waterproofs woody stems and roots.

cork cambium Lateral meristem that replaces epidermis with cork on woody plant parts.

cornea Transparent cover over front of eye.

corpus luteum (CORE-pus LOO-tee-um) Of female mammals, a glandular structure that forms from cells of a ruptured ovarian follicle; it secretes progesterone and estrogen.

cortex A rindlike layer. In vascular plants, a ground tissue; supports parts and stores food.

cotyledon (KOT-uhl-EE-dun) Seed leaf. One develops in monocot seeds and has digestive roles; two develop in dicot seeds and store food for germination and early growth.

countercurrent flow Movement of two fluids in opposing directions, as in a fish gill.

courtship display Pattern of ritualized social behavior between potential mates. Commonly incorporates frozen postures, exaggerated yet simplified movements, and visual signals.

covalent bond (koe-VAY-lunt) [L. *con*, together, + *valere*, to be strong] Sharing of one or more electrons between atoms or groups of atoms. Nonpolar bonds share them equally. Polar bonds (slightly positive at one end, negative at the other) share them unequally.

craniate Having a brain inside a cranium. Includes all fishes, amphibians, reptiles, birds, and mammals alive today.

creatine phosphate Organic compound that transfers phosphate to ADP in a rapid, short-term, ATP-generating pathway.

CRH Corticotropin-releasing hormone. Contributes to the timing of labor.

cross-bridge formation A reversible, ATP-driven interaction between a sarcomere's actin and myosin filaments; results in a short power stroke that is the basis of muscle contraction.

crossing over At prophase I of meiosis, an interaction in which nonsister chromatids of a pair of homologous chromosomes break at corresponding sites and exchange segments; genetic recombination is the result.

crust, of Earth Outer zone of low-density rocks resting on the Earth's mantle.

culture Sum of behavior patterns of a social group, passed between generations by way of learning and symbolic behavior.

cuticle (KEW-tih-kull) Body cover. Of plants, a transparent covering of waxes and cutin on outer epidermal cell walls. Of annelids, a thin, flexible coat. Of arthropods, a lightweight exoskeleton hardened with protein and chitin.

cutin Insoluble lipid polymer that functions as the ground substance for plant cuticles.

cycad Slow-growing gymnosperm; its lineage arose in the Permian. Resembles palm trees.

cyclic AMP (SIK-lik) Type of nucleotide used in intercellular communication.

cyclic pathway of ATP formation Most ancient photosynthetic pathway. Membrane-bound photosystems give up electrons to transfer systems that return them to photosystems. Electron flow across the membrane sets up H^+ gradients that drive ATP formation.

cyst Of many microorganisms, a resting stage with thick outer layers that typically forms under adverse conditions. Of skin, abnormal, fluid-filled sac without an external opening.

cytochrome (SIGH-toe-krome) Iron-containing protein molecule of electron transfer systems.

cytokinesis (SIGH-toe-kih-NEE-sis) [Gk. *kinesis*, motion] Cytoplasmic division.

cytokinin (SIGH-toe-KYE-nin) Type of plant hormone; stimulates cell division, promotes leaf expansion, and retards leaf aging.

cytoplasm (SIGH-toe-plaz-um) All cell parts, particles, and semifluid substances between the plasma membrane and nucleus or nucleoid.

cytoplasmic division After nuclear division, a splitting of the parent cell cytoplasm that completes the formation of daughter cells.

cytoplasmic localization Parceling of a portion of maternal messages in the egg cytoplasm to each blastomere that forms during cleavage.

cytosine (SIGH-toe-seen) Pyrimidine; one of the nitrogen-containing bases in nucleotides.

cytoskeleton Inner, interconnected system of protein elements that reinforce, organize, and move the eukaryotic cell and its structures.

cytotoxic T cell T lymphocyte that touch-kills infected, cancerous, or altered body cells.

day-neutral plant Plant that flowers when mature, independently of day or night length.

deciduous broadleaf forest Its trees drop all or some of leaves in a pronounced dry season; nutrients in litter accumulate on forest floor.

decomposer [L. *dis*–, to pieces] Prokaryotic or fungal heterotroph; gets carbon and energy from products or remains of organisms. Helps cycle nutrients to producers in ecosystems.

decomposition In general, processes by which prokaryotic cells and fungi degrade organic matter (e.g., nitrogenous wastes of organisms).

deductive logic Pattern of thinking; making inferences about specific consequences or predictions that must follow from a hypothesis.

deforestation Removal of all the trees from a large tract of land (e.g., Amazon River Basin).

degradative pathway A stepwise series of metabolic reactions that break down organic compounds to products of lower energy.

deletion At cytological level, loss of a segment from a chromosome. At molecular level, loss of one to a few base pairs from a DNA molecule.

demographic transition model Model that correlates changes in population growth with four stages of economic development.

demographics A population's vital statistics.

denaturation (deh-NAY-chur-AY-shun) Loss of a molecule's three-dimensional shape as weak bonds (e.g., hydrogen bonds) are disrupted.

dendrite (DEN-drite) [Gk. *dendron*, tree] Short, slender extension from cell body of a neuron; commonly a signal input zone.

dendritic cell Type of antigen-presenting cell.

denitrification (DEE-nite-rih-fih-KAY-shun) Conversion of nitrate or nitrite by certain soil bacteria to gaseous nitrogen (N_2) and a small amount of nitrous oxide (N_2O).

dense, irregular connective tissue Animal tissue with fibroblasts, many asymmetrically positioned fibers in ground substance. In skin and in some capsules around organs.

dense, regular connective tissue Animal tissue with rows of fibroblasts between parallel bundles of fibers. In tendons, elastic ligaments.

density-dependent control Factor that limits population growth by reducing the birth rate or increasing death and dispersal rates (e.g., predation, parasitism, disease, competition).

density-independent factor Factor that causes a population's death rate to rise independently of density (e.g., a severe storm or flood).

dentition (den-TIH-shun) Collectively, the type, size, and number of an animal's teeth.

derived trait A novel feature that evolved but once and is shared only by the descendants of the ancestral species in which it evolved.

dermal tissue system Tissues that cover and protect all exposed surfaces of a plant.

dermis Skin layer beneath the epidermis that consists primarily of dense connective tissue.

desalination Removing salt from seawater.

desert Biome that forms where the potential for evaporation greatly exceeds rainfall, and where soil is thin and vegetation sparse.

desertification (dez-urt-ih-fih-KAY-shun) The conversion of grassland or irrigated or rain-fed cropland to a desertlike condition.

detrital food web Network of food chains in which energy flows mainly from plants through arrays of detritivores and decomposers.

detritivore (dih-TRY-tih-vore) Heterotroph that feeds on decomposing particles of organic matter (e.g., crab, earthworm, roundworm).

deuterostome (DUE-ter-oh-stome) [Gk. *deuteros*, second, + *stoma*, mouth] Bilateral animal in which the anus forms from first indentation in the embryo (e.g., echinoderms, chordates).

development The series of genetically guided embryonic and post-embryonic stages by which morphologically distinct, specialized body parts emerge in a new multicelled individual.

diaphragm (DIE-uh-fram) [Gk. *diaphragma*, to partition] Muscular partition between thoracic and abdominal cavities with role in breathing. Also a contraceptive device inserted into the vagina to prevent sperm from entering uterus.

diatom Type of chrysophyte; perforated silica shell parts overlap like a pillbox.

dicot (DIE-kot) [Gk. *di*, two, + *kotylēdōn*, cup-shaped vessel] Dicotyledon. Flowering plant generally characterized by embryos with two cotyledons; net-veined leaves; and floral parts arranged in fours, fives, or multiples of these.

diffusion Net movement of like molecules or ions down their concentration gradient.

digestive system Body sac or tube, often with specialized regions where food is ingested, digested, and absorbed, and where undigested residues are eliminated. Incomplete systems have one opening; complete systems have two.

dihybrid cross An intercross between two F_1 heterozygotes that are identical for two gene loci; the dihybrids are offspring of parents that bred true for different versions of two traits.

dinoflagellate Type of single-celled, flagellated, cellulose-plated protistan. Most are producers of marine phytoplankton; some cause red tides.

dinosaur One of a fabulous group of reptiles that originated in the Triassic and became the dominant land vertebrates for 125 million years.

diploidy (DIP-loyd-ee) Presence of two of each type of chromosome (i.e., pairs of homologs) in a cell nucleus at interphase. *Compare* haploidy.

diplomonad Protozoan, parasite or predator, with bilateral symmetry, two nuclei, and one to four flagella; e.g., *Giardia*.

directional selection Mode of natural selection by which allele frequencies underlying a range of phenotypic variation shift in a consistent direction, in response to directional change or to new conditions in the environment.

disaccharide (die-SAK-uh-ride) [Gk. *di*, two, + *sakcharon*, sugar] A common oligosaccharide; two covalently bound sugar monomers.

disease Outcome of infection when defenses aren't mobilized fast enough and a pathogen's activities interfere with normal body functions.

disruptive selection Mode of natural selection by which the different forms of a trait at both ends of the range of variation are favored and intermediate forms are selected against.

distal tubule Tubular part of nephron where water and sodium are selectively reabsorbed.

distance effect Idea that only species adapted for long-distance dispersal can be potential colonists of islands far from their home range.

diversity of life Sum of all variations in form, function, and behavior in all lineages, from the time of life's origin to the present.

division of labor A splitting up of tasks among cells, tissues, organs, organ systems that collectively contribute to the survival and reproduction of a multicelled organism.

DNA Deoxyribonucleic acid (dee-OX-ee-RYE-bow-new-CLAY-ik) Of cells and many viruses, the molecule of inheritance. H bonds join DNA's two helically twisted nucleotide strands, one of which has instructions (in its base sequence) for synthesizing all of the enzymes and other proteins required to build and maintain cells.

DNA clone Many identical copies of DNA that was inserted into plasmids and later amplified.

DNA fingerprint Unique array of DNA sequences inherited in a Mendelian pattern.

DNA ligase (LYE-gaze) Enzyme that seals new base-pairings during DNA replication.

DNA microarray Gene chip stamped with thousands of DNA sequences from a genome.

DNA polymerase (poe-LIM-uh-raze) Enzyme of replication and repair that assembles a new strand of DNA on a parent DNA template.

DNA repair Enzyme-mediated process that fixes small-scale alterations in a DNA strand by restoring the original base sequence.

DNA replication Any process by which a cell duplicates its DNA molecules before dividing.

domain Part or all of a polypeptide chain that forms a structurally stable, functional unit.

dominance hierarchy Social organization in which some individuals of the group have adopted a subordinate status to others.

dominant allele Of diploid cells, an allele that masks the phenotypic effect of any recessive allele paired with it.

dormancy [L. *dormire*, to sleep] A predictable time of metabolic inactivity for many spores, cysts, seeds, perennials, and some animals.

dosage compensation Any mechanism that balances gene expression between the sexes during critical early stages of development.

double-blind study Different investigators independently collect, then compare data.

double fertilization Of flowering plants only, fusion of a sperm and egg nucleus, plus fusion of another sperm nucleus with nuclei of a cell that gives rise to a nutritive tissue (endosperm).

doubling time Time it takes for a population to double in size.

downy mildew A pathogenic stramenopile.

drug addiction Dependence on a drug, which assumes an "essential" biochemical role in the body following habituation and tolerance.

dry acid deposition Airborne oxides of sulfur, nitrogen fall to Earth during dry weather.

dry shrubland Biome that forms when annual rainfall is less than 25 to 60 centimeters; short, multibranched woody shrubs dominate.

dry woodland Biome that forms when annual rainfall is about 40 to 100 centimeters; it may have many tall trees but no dense canopy.

duplication Gene sequence repeated several to many hundreds or thousands of times. Even normal chromosomes have such sequences.

ecdysone Hormone of many insect life cycles; roles in metamorphosis, molting.

echinoderm Type of invertebrate with calcified spines, needles, or plates on body wall. Radial with some bilateral features (e.g., sea stars).

echolocation Use of echoes from self-generated ultrasounds as a navigational mechanism.

ecological succession Processes by which a community develops in sequence, from pioneer species to an end array of species that remain in equilibrium over a given region.

ecology [Gk. *oikos*, home, + *logos*, reason] The scientific study of how organisms interact with one another and with their environment.

ecoregion Broad land or ocean region defined by climate, geography, and producer species.

ecosystem Array of organisms, together with their environment, interacting through a flow of energy and a cycling of materials.

ecosystem modeling Analytical method; uses computer programs and models to predict unforeseen effects of ecosystem disturbances.

ectoderm [Gk. *ecto*, outside, + *derma*, skin] The first-formed, outermost primary tissue layer of animal embryos; gives rise to nervous system tissues and integument's outer layer.

ectotherm Poorly insulated animal with low metabolic rates; maintains core temperature by absorbing environmental heat (e.g., from sun).

Ediacaran One of the species with a highly flattened body; arose in the precambrian.

effector Muscle (or gland); helps bring about movement (or chemical change) in response to neural or endocrine signals.

effector cell Differentiated antigen-specific cell that forms during immune responses.

egg Mature female gamete; an ovum.

El Niño Massive eastward movement of warm surface waters of the western equatorial Pacific; displaces cool water off South America. Recurs and disrupts climates throughout the world.

electric gradient Difference in electric charge between adjoining regions.

electromagnetic spectrum All wavelengths from radiant energy less than 10^{-5} nm long to radio waves more than 10 km long.

electron Negatively charged unit of matter, with particulate and wavelike properties, that occupies one of the orbitals around the atomic nucleus. Atoms gain, lose, or share electrons.

electron transfer Reaction in which one molecule gives up one or more electrons to another, acceptor molecule.

electron transfer chain Organized array of membrane-bound enzymes and cofactors that accept and donate electrons in series. It sets up an electrochemical gradient that makes H^+ flow across the membrane. The flow energy drives ATP formation at ATP synthases.

electron transfer phosphorylation Last stage of ATP formation by aerobic respiration; uses the hydrogen and electrons that many reduced coenzymes deliver to electron transfer chains.

element Fundamental form of matter that has mass, occupies space, and cannot be broken apart into a different form of matter, at least by ordinary physical or chemical means.

embryo (EM-bree-oh) [Gk. *en*, in, + *bryein*, to swell] Of animals, a multicelled body formed by way of cleavage, gastrulation, or other early developmental events. Of plants, a young sporophyte until time of germination.

embryonic induction In a growing embryo, release of a gene product from one tissue that affects the development of an adjacent tissue.

emerging pathogen Deadly pathogen, either a newly mutated or newly opportunistic strain.

emigration Permanent move by a resident out of a population.

emulsification Of chyme, a suspension of fat droplets coated with bile salts.

encapsulated receptor Skin mechanoreceptor; detects pressure, temperature, low vibrations.

endangered species Endemic (native) species highly vulnerable to extinction.

endergonic reaction (en-dur-GONE-ik) Type of chemical reaction having a net gain in energy.

endocrine gland Ductless gland that secretes hormones, which the bloodstream distributes.

endocrine system Integrative system of cells, tissues, and organs functionally linked to the nervous system; exerts control through its secretion of hormones and other chemicals.

endocytosis (EN-doe-sigh-TOE-sis) Cell uptake of substances via vesicle formation. Receptor-mediated endocytosis, phagocytosis, and the bulk transport of extracellular fluid are three modes of endocytosis.

endoderm [Gk. *endon*, within, + *derma*, skin] Inner primary tissue layer of animal embryos; source of inner gut lining and derived organs.

endodermis Cell layer around root vascular cylinder; influences water and solute uptake.

endomembrane system Series of organelles in which new polypeptide chains are modified and lipids are synthesized.

endometrium (EN-doe-MEET-ree-um) [Gk. *metrios*, of the womb] Inner lining of uterus.

endoplasmic reticulum or **ER** (EN-doe-PLAZ-mik reh-TIK-yoo-lum) Organelle that starts at nucleus and curves through cytoplasm. New polypeptides get side chains in rough ER (has ribosomes on its cytoplasmic side); smooth ER (has no ribosomes) is a site of lipid synthesis.

endoskeleton [Gk. *endon*, within, + *sklēros*, hard] Of chordates, an internal framework of cartilage, bone, or both; works with skeletal muscle to position, support, and move body.

endosperm (EN-doe-sperm) Nutritive tissue inside the seed of a flowering plant.

endospore Resting structure formed by some bacteria; encloses a duplicate of the bacterial chromosome and a portion of cytoplasm.

endosymbiosis Continuing physical contact between two species, one of which lives and reproduces inside the other's body.

endotherm Animal with adaptations (high metabolic rates, body form, and behavior) that help control heat gains and losses over a wide temperature range.

energy Capacity to do work.

energy carrier Molecule that delivers energy from one reaction site to another; mainly ATP.

energy pyramid Pyramidal diagram of an ecosystem's trophic structure; it shows energy losses at each transfer to another trophic level.

enhancer A short DNA base sequence that is a binding site for an activator protein.

entropy (EN-trow-pee) Measure of the degree of disorder in a system (how much energy has become disorganized, usually as dispersing heat, and is no longer available to do work). All systems need energy inputs to counter entropy.

enzyme (EN-zime) A type of protein or one of the few RNAs that catalyze reactions between substances, most often at functional groups.

eosinophil Fast-acting white blood cell; its enzyme secretions digest holes in parasitic worms during an inflammatory response.

epidemic Rapid spread, then subsidence, of a disease within a population.

epidermis Outermost tissue layer of plants and all animals above sponge level of organization.

epiglottis Flaplike structure between pharynx and larynx; its controlled positional changes direct air into trachea or food into esophagus.

epinephrine (ep-ih-NEF-rin) Adrenal hormone that influences metabolism, heart function.

epistasis (eh-PISS-tah-sis) Interaction among the products of two or more gene pairs.

epithelium (EP-ih-THEE-lee-um) Animal tissue that covers external surfaces and lines internal cavities and tubes. One surface is free and the other rests on a basement membrane.

erosion Movement of land under the force of wind, running water, and ice.

erythropoietin Hormone secreted by kidney cells; induces stem cells in bone marrow to divide and give rise to red blood cells.

esophagus (ee-SOF-uh-gus) A muscular tube just after the pharynx in the digestive tract.

essential amino acid Any amino acid that an organism cannot synthesize for itself and must obtain from food.

essential fatty acid Any fatty acid an organism cannot synthesize and must obtain from food.

estrogen (ESS-trow-jen) Female sex hormone; helps oocytes mature, primes uterine lining for pregnancy, helps maintain secondary sexual traits, affects growth and development.

estuary (EST-you-ehr-ee) Partly enclosed coast region where seawater mixes with fresh water and runoff from land, as from rivers.

ethylene (ETH-il-een) Plant hormone; promotes fruit ripening and leaf, flower, fruit abscission.

Eubacteria Domain of all prokaryotic cells except archaebacteria; also called a kingdom.

eubacterium Prokaryotic cell; has a nucleoid, but no nucleus, cytoplasm, or cell membrane; most have a cell wall, some encapsulated.

eudicot True dicot; one of three major groups of flowering plants. *See* dicot.

euglenoid Single-celled, flagellated protistan; photosynthetic but with heterotrophic feeding apparatus (it has phagocytic ancestors).

Eukarya, Eukaryotes Domain of eukaryotic cells; all protistans, plants, fungi, and animals.

eukaryotic cell (yoo-CARE-EE-oh-tic) [Gk. *eu*, good, + *karyon*, kernel] Cell having a nucleus and other membrane-bound organelles.

eutrophication Nutrient enrichment of a body of water; typically causes reduced transparency and a phytoplankton-dominated community.

evaporation [L. *e-*, out, + *vapor*, steam] The conversion of a substance from a liquid state to a gaseous state by an input of heat energy.

evaporative heat loss Response to heat stress (e.g., at respiratory surfaces and across skin).

evergreen broadleaf forest Tropical forest in regions of heavy rainfall; rapid decomposition.

evolution, biological [L. *evolutio*, unrolling] Genetic change in a line of descent. Outcome of microevolutionary events: gene mutation, natural selection, genetic drift, and gene flow.

evolutionary tree Treelike diagram; a branch point means divergence from a shared ancestor and branches signify separate lines of descent.

excitatory postsynaptic potential (EPSP) Graded potential; drives membrane toward threshold.

excretion Removal of excess water and solutes (some harmful) by urinary system or glands.

exercise Increased level of contractile activity.

exergonic reaction (EX-ur-GONE-ik) Chemical reaction that shows a net loss in energy.

exocrine gland (EK-suh-krin) [Gk. *es*, out of, + *krinein*, to separate] Glandular structure that secretes products, usually through ducts or tubes, to a free epithelial surface.

exocytosis (EK-so-sigh-TOE-sis) Release of a vesicle's contents at cell surface as it fuses with and becomes part of the plasma membrane.

exodermis Cylindrical sheet of cells inside the root epidermis of most flowering plants; helps control uptake of water and solutes.

exon One of the base sequences of an mRNA transcript that will become translated.

exoskeleton [Gk. *sklēros*, hard, stiff] External skeleton (e.g., hardened cuticle of arthropods).

exotic species Species that left its home range and became established in a new community.

experiment, scientific Test that simplifies observation in nature or in the laboratory by manipulating and controlling the conditions under which the observations are made.

exponential growth (EX-po-NEN-shul) Pattern of population growth; a quantity increasing by a fixed percentage in a given interval (e.g., 0.1, 1.0, 10 percent over three successive intervals). Plots out as a J-shaped curve.

external ear Among mammals, one of a pair of sound-collecting flaps on the sides of the head.

extinction Irrevocable loss of a species.

extracellular digestion, absorption Mode of nutrition by which certain organisms grow in or on organic matter, digest it with secreted enzymes, then absorb breakdown products.

extracellular fluid Of most animals, all fluid not in cells; plasma (blood's liquid portion) plus interstitial fluid.

extracellular matrix Deposits of secretions, other substances at free surface of tissue cells.

extreme halophile Archaebacterium of unusually saline habitats (e.g., salt lakes).

extreme thermophile Archaebacterium of very hot aquatic habitats (e.g., hot springs).

eye Sensory organ that incorporates a dense array of photoreceptors.

F_1, F_2 In genetics, first and second generations.

FAD Flavin adenine dinucleotide, a type of nucleotide coenzyme; it transfers electrons and unbound protons (H^+) between reaction sites. At such times it is abbreviated $FADH_2$.

fall overturn Vertical mix of a body of water in fall. Its upper layer cools, gets more dense, sinks. Oxygenated surface water moves down, nutrients from bottom sediments move up.

fat Lipid with a glycerol head and one, two, or three fatty acid tails. Unsaturated tails have single covalent bonds in the carbon backbone; saturated tails have one or more double bonds.

fate map Surface diagram of certain early embryos (e.g., *Drosophila*) showing where the differentiated cells of the adult originate.

fatty acid Molecule with a backbone of up to 36 carbon atoms, a carboxyl group (—COO⁻ or —COOH) at one end, and hydrogen atoms at most or all of the other bonding sites.

feedback inhibition Mechanism by which a cellular change resulting from some activity shuts down the activity that brought it about.

fermentation *See* alcoholic fermentation, lactate fermentation.

fern A seedless vascular plant of moist or wet habitats; has fronds often divided into leaflets.

fertilization [L. *fertilis*, to carry, to bear] The fusion of a sperm nucleus with the nucleus of an egg, thus forming a zygote.

fetus Animal development stage after major organ systems form until time of birth.

fever Any core temperature higher than the set point in the hypothalamic thermostat.

fibrous root system All lateral branchings of adventitious roots arising from a young stem.

Fick's law The larger the surface area and the greater the partial pressure gradient, the faster gases will diffuse across a respiratory surface.

filter feeder Animal that filters food from a current of water directed through a body part (e.g., through a sea squirt's pharynx).

filtration First step in forming urine; pressure of heart contractions filters blood (forces water and all solutes except proteins) into nephrons.

fin Of fishes generally, appendage that helps stabilize, propel, and guide body in water.

first law of thermodynamics The total amount of energy in the universe is constant. Energy cannot be created from nothing and existing energy cannot be destroyed.

fish An aquatic animal of the most ancient, diverse vertebrate lineage; one of the jawless, jawed cartilaginous, or jawed bony species.

fitness Increase in adaptation to environment, as brought about by genetic change.

fixation Loss of all but one kind of allele at a gene locus for all individuals in a population.

fixed action pattern Program of coordinated, stereotyped muscle activity that is completed independently of feedback from environment.

flagellum (fluh-JELL-um), plural **flagella** A motile structure of many free-living eukaryotic cells. Its core has a 9 + 2 array of microtubules. Typically longer than cilia.

flatworm One of the simplest invertebrates having organ systems; a turbellarian, fluke, or tapeworm. Some are notorious parasites.

flavoprotein Yellow pigment, absorbs blue wavelengths to induce phototropic response.

flower Of angiosperms only, a reproductive structure with nonfertile parts (sepals, petals) and fertile parts (stamens, carpels) attached to a receptacle (modified base of a floral shoot).

fluid mosaic model A cell membrane is a mix of lipids (organized as a bilayer) and proteins. Structural lipids make it largely impermeable to water-soluble molecules yet impart fluidity by packing variations and motions. Diverse proteins perform most membrane functions (e.g., transport, signal reception).

fluorescence A destabilized molecule emits light when reverting to more stable form.

folate [L. *folium*, leaf] Water-soluble B vitamin.

follicle (FOLL-ih-kul) Small sac, pit, or cavity, as around a hair; also a mammalian oocyte with its surrounding layer of cells.

food chain Straight-line sequence of steps by which energy stored in autotroph tissues passes on to higher trophic levels.

food pyramid Chart of a purportedly well-balanced diet; continually being refined.

food web Of ecosystems, cross-connecting food chains consisting of producers, consumers, and decomposers, detritivores, or both.

forebrain Part of vertebrate brain; includes cerebrum, olfactory lobes, and hypothalamus.

forest Biome in which tall trees grow close enough to form a fairly continuous canopy.

fossil Recognizable, physical evidence of an organism that lived in the distant past.

fossil fuel Coal, petroleum, or natural gas; a nonrenewable energy source that formed long ago from remains of swamp forests.

fossilization How fossils form. An organism or evidence of it gets buried in sediments or volcanic ash; water and dissolved inorganic compounds infiltrate it; then chemical changes and pressure from accumulating sediments above transform it to stony hardness.

founder effect A form of bottlenecking. By chance alone, a few individuals that establish a new population have allele frequencies that differ from those of the original population.

free nerve ending Simplest sensory receptor in skin, internal tissues; has unmyelinated or thinly myelinated branched endings.

free radical Any unbound molecular fragment with an unpaired electron.

fruit [L. after *frui*, to enjoy] Flowering plant's mature ovary, often with accessory structures.

fruiting body Spore-bearing structures formed by some bacteria, fungi.

FSH Follicle-stimulating hormone produced and secreted by the anterior lobe of pituitary gland; has reproductive roles in both sexes.

functional group An atom or a group of atoms that is covalently bonded to the carbon backbone of an organic compound and that influences its chemical behavior.

functional-group transfer One molecule donates a functional group to another.

Fungi Kingdom of fungi; major decomposers plus pathogens and parasites.

fungus Eukaryotic heterotroph; it secretes enzymes that digest food to bits that its cells absorb (extracellular digestion, absorption). Saprobic types use nonliving organic matter; parasitic types feed on living organisms.

fusion power Release of energy from fuel (e.g., heated gas of H isotopes) that implodes when compressed to extremely high densities.

gallbladder Organ that stores bile secreted from liver; its duct connects to small intestine.

gamete (GAM-eet) [Gk. *gametēs*, husband, and *gametē*, wife] Haploid cell, formed by meiotic cell division of a germ cell; required for sexual reproduction. Eggs and sperm are examples.

gamete formation Formation of sex cells (e.g., sperm, eggs) in reproductive tissues or organs.

gametophyte (gam-EET-oh-fite) [Gk. *phyton*, plant] Haploid gamete-producing body that forms during plant life cycles.

ganglion (GANG-lee-on), plural **ganglia** Distinct cluster of cell bodies of neurons.

gap junction Cylindrical arrays of proteins in the plasma membrane that pair up as open channels for signals between adjoining cells.

gastric fluid Highly acidic mix of secretions from the stomach's glandular epithelium (HCl, mucus, pepsinogens, etc.) that act on food.

gastrulation (gas-tru-LAY-shun) Stage of animal development; reorganization of new cells into two or three primary tissue layers.

gel electrophoresis Laboratory technique used to distinguish among molecules. Applied electric field forces them to migrate through a viscous gel and distance themselves from one another by length, size, or electric charge.

gene [German *pangan*, after Gk. *pan*, all; *genes*, to be born] Unit of information for a heritable trait, passed from parents to offspring.

gene control One of the molecular mechanisms that govern when and how fast specific genes will be transcribed and translated, and whether gene products will be activated or inactivated.

gene flow Microevolutionary process; alleles enter and leave a population as an outcome of immigration and emigration, respectively.

gene frequency Abundance of an allele with respect to others at same locus in a population.

gene library Mixed collection of bacteria that house many different cloned DNA fragments.

gene locus A gene's chromosomal location.

gene mutation A small-scale change in the nucleotide sequence of a DNA molecule.

gene pair Two alleles at the same gene locus on a pair of homologous chromosomes.

gene pool All genotypes in a population.

gene therapy Generally, a transfer of one or more normal genes into an organism to correct or lessen adverse effects of a genetic disorder.

genetic abnormality A rare or less common version of a heritable trait.

genetic code [After L. *genesis*, to be born] The correspondence between nucleotide triplets in DNA (then mRNA) and specific sequences of amino acids in a polypeptide chain; the basic language of protein synthesis in cells.

genetic disease Illness in which expression of one or more genes increases susceptibility to infection or weakens immune response to it.

genetic disorder Any inherited condition that causes mild to severe medical problems.

genetic divergence Gradual accumulation of differences in gene pools of populations or subpopulations of a species after a geographic barrier arises and separates them; thereafter, microevolution occurs independently in each.

genetic drift Change in allele frequencies over the generations due to chance alone. Its effect is most pronounced in very small populations.

genetic engineering Deliberately altering the information content of DNA molecules.

genetic equilibrium In theory, a state in which a population is not evolving. *Compare* Hardy–Weinberg rule.

genetic recombination Result of any process that puts new genetic information into a DNA molecule (e.g., by crossing over).

genome All the DNA in a haploid number of chromosomes for a given species.

genomics Study of the genome of humans and other organisms.

genotype (JEEN-oh-type) Genetic constitution of an individual; a single gene pair or the sum total of an individual's genes.

genus, plural **genera** (JEEN-US, JEN-er-ah) [L. *genus*, race or origin] A grouping of species that are more closely related to one another in their morphology, ecology, and history than to other species at the same taxonomic level.

geographic dispersal An organism moves out of its home range and becomes established in a new community, as an exotic species.

geologic time scale Time scale for the Earth's history with major subdivisions corresponding to mass extinctions. Now radiometrically dated.

germ cell Animal cell of a lineage set aside for sexual reproduction; gives rise to gametes.

germination (jur-mih-NAY-shun) Of seeds and spores, resumption of growth after dormancy, dispersal from the parent organism, or both.

ghrelin A peptide hormone that stimulates appetite control center.

gibberellin (JIB-er-ELL-un) Plant hormone; promotes elongation of stems, helps seeds and buds break dormancy, and helps flowering.

gill Respiratory organ. Most gills have a thin, moist, vascularized layer for gas exchange.

gill slit Opening in a thin-walled pharynx. Gill slits serve in food-trapping and respiration; jaws evolved from certain gill slit supports.

ginkgo Deciduous gymnosperm; its ancestors were diverse in dinosaur times.

gland Secretory cell or structure derived from epithelium and often still connected to it.

gland cell A cell that secretes products unrelated to their own metabolism for use elsewhere.

global broiling Theory that an asteroid impact caused the K–T mass extinction by creating a colossal fireball, the debris from which raised global air temperature by thousands of degrees.

global warming A long-term increase in the temperature of the Earth's lower atmosphere.

glomerular capillary Set of blood capillaries located in Bowman's capsule of a nephron.

glomerulus (glow-MARE-you-luss) [L. *glomus*, ball] Bowman's capsule, together with the glomeruli capillaries it cups around.

glottis Opening between two vocal cords.

glucagon Pancreatic hormone; stimulates glycogen, amino acid conversion to glucose.

glyceride (GLISS-er-eyed) Molecule having one, two, or three fatty acid tails attached to a glycerol backbone; one of the fats or oils.

glycerol (GLISS-er-ohl) [Gk. *glykys*, sweet, + L. *oleum*, oil] Three-carbon compound with three hydroxyl groups; component of fats and oils.

glycocalyx Sticky mesh forming a capsule or slime layer around a bacterial cell wall.

glycogen (GLY-kuh-jen) Polysaccharide of glucose monomers, highly branched; the main storage carbohydrate in animals.

glycolysis (gly-CALL-ih-sis) [Gk. *glykys*, sweet, + *lysis*, breaking apart] Breakdown of glucose or another organic compound to two pyruvate molecules. First stage of aerobic respiration, fermentation, and anaerobic electron transfer. Oxygen has no role in glycolysis, which occurs in the cytoplasm of all cells. Two NADH form. Net yield: 2 ATP per glucose molecule.

glycoprotein Protein with linear or branched oligosaccharides covalently bonded to it.

gnetophyte Woody gymnosperm, a vine or shrub; unusual in having vessels in its xylem.

GnRH Gonadotropin-releasing hormone, triggers hypothalamus to release LH and FSH.

golden alga A chrysophyte with silica scales, or other hard parts; abundant fucoxanthin.

Golgi body (GOHL-gee) Organelle of lipid assembly, polypeptide chain modification, and packaging of both in vesicles for export or for transport to locations in cytoplasm.

gonad (GO-nad) Primary reproductive organ in which animal gametes are produced.

Gondwana Paleozoic supercontinent; with other land masses, it formed Pangea.

graded potential Signal of variable magnitude at input zone of excitable cell (e.g., neuron).

gradual model, speciation Idea that species arise by many small morphological changes that accumulate over great spans of time.

Gram stain A sample of cells is exposed to purple dye, then iodine, then an alcohol wash and counterstain. Cell walls of Gram-positive species stay purple; Gram-negative turn pink.

Granstein cell Issues suppressor signals to keep immune responses (of skin) in check.

granum, plural grana In many chloroplasts, part of the thylakoid membrane organized as a stack of flattened disks.

grassland Biome that has flat or rolling land, 25–100 centimeters of annual rainfall, warm summers, and a distinct array of grazers; fires recur and regenerate dominant plant species.

gravitropism (GRAV-ih-TROPE-izm) [L. *gravis*, heavy, + Gk. *trepein*, to turn] Directional plant growth in response to gravitational force.

gray crescent Pigmented area in amphibian egg cytoplasm; sets anterior-posterior axis.

gray matter Portion of brain, spinal cord consisting of unmyelinated axons, dendrites, neuron cell bodies, and neuroglial cells.

grazing food web Network of food chains in which energy flows from plants to an array of herbivores, then to carnivores.

green alga Type of protistan biochemically, evolutionarily, and structurally most like plants (e.g., most are photoautotrophs with starch grains and chlorophylls *a*, *b* in chloroplasts).

green revolution In developing countries, use of improved crop strains, modern agricultural equipment, and practices to raise crop yields.

greenhouse effect Atmospheric gases impede escape of infrared wavelengths (heat) from Earth's sun-warmed surface, absorb them, and radiate much of the heat back toward Earth.

ground tissue system Tissues, parenchyma especially, making up most of the plant body.

growth Of multicelled species, increases in the number, size, and volume of cells. Of bacterial populations, increases in the number of cells.

growth factor Protein that plays a role in the body's growth (e.g., by promoting mitosis).

growth ring One of the alternating bands of early and late wood; a "tree ring."

growth, tissue specialization Stage of animal development when organs enlarge and assume specialized functions; continues into adulthood.

guanine Nitrogen-containing base in one of four nucleotide monomers of DNA or RNA.

guard cell One of two cells defining a stoma, an opening across leaf or stem epidermis.

gut Generally, a sac or tube from which food is absorbed into internal environment. Also, gastrointestinal tract from stomach onward.

gymnosperm (JIM-noe-sperm) [Gk. *gymnos*, naked, + *sperma*, seed] Type of vascular plant in which seeds form on exposed surfaces of reproductive structures (e.g., on cone scales).

habitat [L. *habitare*, to live in] Place where an organism or species lives; characterized by its physical and chemical features and its species.

habitat fragmentation A habitat gets chopped into patches too small to support successful breeding; populations also more vulnerable.

habitat island A region with endemic species (e.g., a forest or lake) located within a "sea" of habitats unsuitable for sustaining them.

habitat loss Reduction in suitable living space and habitat closure through chemical pollution.

hair A flexible structure rooted in skin, with a shaft above the skin's surface.

hair cell Hairlike mechanoreceptor that is activated when sufficiently bent or tilted.

half-life The time it takes for half of a given quantity of any radioisotope to decay into a different, and less unstable, daughter isotope.

haploidy (HAP-loyd-ee) Presence of only half of the parental number of chromosomes in a spore or gamete, as brought about by meiosis.

hardwood Type of strong, dense wood with many vessels, tracheids, and fibers in xylem.

Hardy–Weinberg rule Allele frequencies stay the same over the generations when there is no mutation, the population is infinitely large and isolated from other populations of the species, mating is random, and all individuals are reproducing equally and randomly.

HCG Human chorionic gonadotropin; helps maintain endometrium during the menstrual cycle and first trimester of pregnancy.

HDL A high-density lipoprotein in blood; it transports cholesterol to liver for metabolism.

hearing Perception of sound.

heart Muscular pump; its contractions keep blood circulating through the animal body.

heartwood A dry tissue at the core of aging stems and roots that no longer transports water and solutes. It helps the tree defy gravity and is a dumping ground for some metabolic wastes.

heat Thermal energy; a form of kinetic energy.

HeLa cell Cancer cell of a lineage established for research; now used in many laboratories.

helper T cell CD4 lymphocyte; has key roles in antibody- and cell-mediated immune responses. It induces responsive T and B cells to divide and differentiate rapidly into huge armies.

heme group Iron-containing functional group that reversibly binds oxygen; one associated with each of the four chains in hemoglobin.

hemoglobin (HEEM-oh-glow-bin) [Gk. *haima*, blood, + L. *globus*, ball] An iron-containing respiratory protein of red blood cells.

hemostasis (HEE-mow-STAY-sis) [Gk. *haima*, blood, + *stasis*, standing] Process that stops loss of blood from damaged blood vessel by coagulation, blood vessel spasm, platelet plug formation, and other mechanisms.

herbicide Natural or synthetic toxin that can kill plants or inhibit their growth.

herbivore [L. *herba*, grass, + *vovare*, to devour] Plant-eating animal (e.g., snail, deer, manatee).

hermaphrodite (her-MAH-froe-dyte) Individual having both male and female gonads.

heterocyst (HET-er-oh-sist) Cyanobacterial cell, self-modified, that synthesizes a nitrogen-fixing enzyme when nitrogen supplies dwindle.

heterotherm Generally, an endotherm that can lower metabolic rates; it cuts energy cost of maintaining its core temperature by slowing down its activities.

heterotroph (HET-er-oh-trofe) [Gk. *heteros*, other, + *trophos*, feeder] Organism unable to make its own organic compounds; feeds on autotrophs, other heterotrophs, organic wastes.

heterozygous condition (HET-er-oh-ZYE-guss) [Gk. *zygoun*, join together] Having a pair of nonidentical alleles at a gene locus (that is, on a pair of homologous chromosomes).

higher taxon (plural, **taxa**) One of ever more inclusive groupings that reflect relationships among species. Family, order, class, phylum, and kingdom are examples.

hindbrain Medulla oblongata, cerebellum, pons of vertebrate brain. Has reflex centers for respiration, blood circulation, and other basic functions; helps coordinate motor responses.

histamine Local signaling molecule that fans inflammation; makes arterioles vasodilate and capillaries more permeable (leaky).

histone Type of protein intimately associated with eukaryotic DNA and largely responsible for organization of eukaryotic chromosomes.

HLA One of a class of recognition proteins important in immune responses.

homeostasis (HOE-me-oh-STAY-sis) [Gk. *homo*, same, + *stasis*, standing] State in which physical and chemical aspects of internal environment (blood, interstitial fluid) are being maintained within ranges suitable for cell activities.

homeotic gene In all major animal groups, one of the master genes, the products of which interact with one another and with control elements to map out the overall body plan.

hominid [L. *homo*, man] All species on or near evolutionary road leading to modern humans.

hominoid Apes, humans, and recent ancestors.

homologous chromosome (huh-MOLL-uh-gus) [Gk. *homologia*, correspondence] Of cells with a diploid chromosome number, one of a pair of chromosomes identical in size, shape, and gene sequence, and that interact at meiosis. Nonidentical sex chromosomes (e.g., X and Y) also interact as homologues during meiosis.

homologous structures Of separate lineages, comparable body parts that show underlying similarity even when they may differ in size, shape, or function; outcome of morphological divergence from a shared ancestor.

homozygous dominant condition Having a pair of dominant alleles (*AA*) at a gene locus (on a pair of homologous chromosomes).

homozygous recessive condition Having a pair of recessive alleles (*Aa*) at a gene locus (on a pair of homologous chromosomes).

hormone [Gk. *hormon*, stir up, set in motion] Signaling molecule secreted by one cell that stimulates or inhibits activities of any cell with receptors for it. Animal hormones are picked up and transported by the bloodstream.

hornwort A bryophyte; nonvascular plant.

horsetail Seedless vascular plant; has rhizomes, scale-shaped leaves, and hollow photosynthetic stems with silica-reinforced ribs.

host Living organism exploited by a parasite. A definitive host harbors the mature stage of a parasite's life cycle. One or more intermediate hosts harbor immature stages.

hot spot Limited area where human activities are driving many species to extinction.

human A primate of the species *Homo sapiens*.

humus Decomposing organic matter in soil.

hybrid offspring Of a genetic cross, offspring having a pair of nonidentical alleles for a trait.

hybrid zone Where adjoining populations are interbreeding and producing hybrid offspring.

hydrocarbon An organic compound that has only hydrogen bonded to a carbon backbone.

hydrogen bond Weak interaction between a small, highly electronegative atom and an H atom already taking part in a polar covalent bond in the same molecule or a different one.

hydrogen ion Free (or unbound) proton; one hydrogen atom that lost its electron and now bears a positive charge (H^+).

hydrologic cycle Biogeochemical cycle driven by solar energy; water is moved through the atmosphere, on or through land, to the ocean, and back to the atmosphere.

hydrolysis (high-DRAWL-ih-sis) [L. *hydro*, water, + Gk. *lysis*, loosening] Cleavage reaction that breaks covalent bonds and splits a molecule into two or more parts. H^+ and OH^- (from a water molecule) are often attached to the newly exposed bonding sites.

hydrophilic substance [Gk. *philos*, loving] A polar molecule or molecular region that easily dissolves in water (e.g., sugars).

hydrophobic substance [Gk. *phobos*, dreading] A nonpolar molecule or molecular region that strongly resists dissolving in water (e.g., oils).

hydrostatic pressure Pressure exerted by a volume of fluid against a wall, membrane, or some other structure that encloses the fluid.

hydrostatic skeleton A fluid-filled cavity or cell mass against which contractile cells act.

hydrothermal vent A steaming fissure in the deep ocean floor; has unique ecosystems.

hypertonic solution A fluid having a greater solute concentration relative to another fluid.

hypha (HIGH-fuh), plural **hyphae** [Gk. *hyphe*, web] Fungal filament with chitin-reinforced walls; component of a mycelium.

hypothalamus [Gk. *hypo*, under, + *thalamos*, inner chamber, or *tholos*, rotunda] Forebrain center of homeostatic control over the internal environment (e.g., salt–water balance, core temperature); influences hunger, thirst, sex, other viscera-related behaviors, and emotions.

hypothesis In science, a possible explanation of a phenomenon, one that has the potential to be proved false by experimental tests.

hypotonic solution A fluid that has a lower solute concentration relative to another fluid.

ICE-like protease One enzyme of apoptosis.

illegitimate receiver Animal that intercepts signals between members of another species.

illegitimate signaler Predator that lures prey by simulating an intraspecific signal.

imbibition Water molecules move into a seed, attracted by its hydrophilic groups of proteins.

immigration One or more individuals moves from its population and takes up permanent residence in another population of its species.

immune response B cells and T cells recognize antigen and give rise to huge antigen-sensitized populations of effector and memory cells.

immune system Interacting white blood cells that defend the vertebrate body through self/nonself recognition, specificity, and memory. T and B cell antigen receptors ignore body's own cells yet collectively recognize a billion specific threats. Some B and T cells formed in a primary response are set aside as memory cells for future battles with the same antigen.

immunization Any process that promotes immunity against disease (e.g., vaccination).

immunoglobulin (Ig) One of five classes of antibodies, each with antigen-binding sites as well as other sites with specialized functions.

implantation Event in pregnancy. A blastocyst burrows into endometrium and establishes connections by which a mother will exchange substances with the embryo (and fetus) that develops from the blastocyst's inner cell mass.

imprinting Time-dependent form of learning, usually during a sensitive period for a young animal, triggered by exposure to sign stimuli.

in vitro fertilization Conception outside the body ("in glass" petri dishes or test tubes).

inbreeding Nonrandom mating among close relatives that share many identical alleles.

inclusive fitness theory Idea that genes associated with caring for relatives may be favored in some situations.

incomplete digestive system Food intake and waste output occur through a single opening.

incomplete dominance Condition in which one allele of a pair is not fully dominant; a heterozygous phenotype somewhere between both homozygous phenotypes emerges.

independent assortment theory Mendelian theory that, by the end of meiosis, each pair of homologous chromosomes (and linked genes on each one) are sorted before shipment to gametes independently of how the other pairs were sorted. Later modified to account for the disruptive effect of crossing over on linkages.

indicator species Any species that provides warning of changes in habitat and impending widespread loss of biodiversity.

induced-fit model When bound to an active site, a substrate promotes reactivity by altering an enzyme's shape; it brings about a more precise molecular fit between the two.

inductive logic Pattern of thinking; deriving a general statement from specific observations.

infection Invasion and multiplication of a pathogen in a host. Disease follows if defenses are not mobilized fast enough; the pathogen's activities interfere with normal body functions.

inflammation, acute Rapid response to tissue injury by phagocytes and certain proteins (e.g., histamine, complement, clotting factors). Signs include localized redness, heat, swelling, pain.

inheritance The transmission, from parents to offspring, of genes that specify structures and functions characteristic of the species.

inhibin Hormone that inhibits GnRH, FSH secretion from hypothalamus and pituitary.

inhibiting hormone Hypothalamic signaling molecule that suppresses a particular secretion by the anterior lobe of the pituitary gland.

inhibitor Any substance that binds with a specific molecule and blocks its function.

inhibitory postsynaptic potential (IPSP) Graded potential at an input zone of an excitable cell that drives its membrane away from threshold.

inner ear Primary organ of equilibrium and hearing; includes vestibular apparatus, cochlea.

insertion Insertion of one to a few bases into a DNA strand. Also, a movable attachment of muscle to bone.

instinctive behavior A behavior performed without having been learned by experience.

insulin Pancreatic hormone; lowers glucose level in blood (triggers protein, fat synthesis and inhibits protein conversion to glucose).

integrator A control center (e.g., brain) that receives, processes, and stores sensory input, issues commands for coordinated responses.

integument Of animals, protective body cover (e.g., skin). Of seed-bearing plants, one or more layers around an ovule; becomes a seed coat.

integumentary exchange (in-teg-you-MEN-tuh-ree) Respiration across a thin, moist, and often vascularized surface layer of animal body.

interleukin Signaling molecule between cells of immune system.

intermediate Substance that forms between the start and end of a metabolic pathway.

intermediate filament Cytoskeletal element; mechanically strengthens some animal cells.

internal environment Blood + interstitial fluid.

interneuron Neuron of brain or spinal cord.

internode The stem between two nodes.

interphase Of a cell cycle, interval between nuclear divisions when a cell increases in mass and roughly doubles the number of its cytoplasmic components. It also duplicates its chromosomes (replicates its DNA) during interphase, but *not* between meiosis I and II.

interstitial fluid (IN-ter-STISH-ul) [L. *interstitus*, to stand in the middle of something] The portion of extracellular fluid that occupies the spaces between animal cells and tissues.

intertidal zone Region between the low and high water marks of a rocky or sandy shore.

intervertebral disk Cartilaginous flex point and shock absorber between vertebrae.

intron A noncoding portion of a pre-mRNA transcript; excised before translation.

inversion Part of a chromosome that became oriented in reverse, with no molecular loss.

invertebrate Any animal without a backbone.

ion, negatively charged (EYE-on) An atom or a molecule that acquired an overall negative charge by gaining one or more electrons.

ion, positively charged Atom or molecule that acquired an overall positive charge by losing one or more electrons.

ion exchange The pH-dependent process by which ions dissociate from soil particles, and other ions dissolved in soil water replace them.

ionic bond Two ions being held together by the attraction of their opposite charge.

ionizing radiation High-energy wavelengths.

isotonic solution A fluid with the same solute concentration as another fluid being studied.

isotope (EYE-so-tope) One of two or more forms of an element's atoms that differ in the number of neutrons.

jaw Of chordates, paired, hinged cartilaginous or bony feeding structures. Many invertebrates have similar hinged feeding structures.

joint Area of contact between bones.

J-shaped curve Type of diagrammatic curve that emerges when unrestricted exponential growth of a population is plotted against time.

juvenile Of some animals, a post-embryonic stage that changes only in size and proportion to become the adult (no metamorphosis).

karyotype (CARE-ee-oh-type) Preparation of metaphase chromosomes sorted by length, centromere location, other defining features.

keratinocyte Skin cell that produces keratin, a tough, water-insoluble protein.

key innovation A structural or functional modification to the body that by chance, gives a lineage opportunity to exploit aspects of the environment in more efficient or novel ways.

keystone species A species with a dominant role in shaping community structure.

kidney One of a pair of vertebrate organs that filter ions and other substances from blood; it controls amounts returned to help maintain the internal environment.

kilocalorie 1,000 calories of heat energy; the amount needed to raise the temperature of 1 kilogram of water by 1°C. Used as the unit of measure for the caloric content of foods.

kinase Enzyme (e.g., a protein kinase) that catalyzes a phosphate-group transfer.

kinetic energy Energy of motion.

kinetochore At centromere, a mass of protein, DNA to which spindle microtubules attach.

kinetoplastid Protozoan of ancient lineage, a parasite or predator; one or two flagella, DNA mass in one mitochondrion; e.g., *Trypanosoma*.

Krebs cycle A stage of aerobic respiration, in mitochondria only; it (and a few preparatory steps) breaks down pyruvate to CO_2 and H_2O. Many coenzymes accept protons (H^+) and electrons from intermediates and deliver them to the next stage of reactions; 2 ATP form.

K–T asteroid impact theory A huge asteroid hit Earth at the K-T boundary; last dinosaurs perished during the mass extinction.

K–T boundary Of geologic time scale, end of Cretaceous and start of Tertiary period.

La Niña Cooler climatic event with global effects; occurs between El Niño events.

labor Uterine contractions that result in birth.

lactate fermentation An anaerobic pathway of ATP formation. Pyruvate from glycolysis is converted to three-carbon lactate, and NAD^+ is regenerated. Net energy yield: 2 ATP.

lactation Production and secretion of milk by hormone-primed mammary glands.

lake A body of standing fresh water in a basin formed by geologic events; characterized by light penetration, temperature, other features.

Langerhans cell Antigen-presenting cell in skin; phagocyte of virus particles, bacteria.

large intestine Colon; gut region that receives unabsorbed residues from small intestine, and concentrates and stores feces until expulsion.

larva, plural **larvae** Immature stage between an embryo and adult in many animal life cycles.

larynx (LARE-inks) Tubular airway leading to lungs. Contains vocal cords in some animals.

lateral meristem Meristem in plants that show secondary growth; vascular or cork cambium.

lateral root Outward branching from the first (primary) root of a taproot system.

LDL Low-density lipoprotein that transports cholesterol; cells take it up, but excess amounts contribute to atherosclerosis.

leaching Removal of some nutrients from soil as water percolates through it.

leaf Chlorophyll-rich plant part adapted for sunlight interception and photosynthesis.

learned behavior Lasting modification of a behavior as a result of experience or practice.

lek Communal display ground for courtship behavior by some animals, including birds.

lens A transparent body that bends light rays so that they converge onto photoreceptors.

leptin An appetite-mediating hormone.

lethal mutation Mutation with drastic effects on phenotype; usually causes death.

Leydig cell Testosterone-secreting cell in mammalian testis.

LH Luteinizing hormone. Anterior pituitary hormone; reproductive roles in males, females.

lichen (LY-kun) Symbiotic interaction between a fungus and photoautotroph.

life cycle A recurring pattern of genetically programmed events from the time individuals are produced until they themselves reproduce.

life history pattern Of a species, a set of adaptations that influence survival, fertility, and age at first reproduction.

ligament Strap of dense connective tissue that bridges a joint.

light-dependent reactions The first stage of photosynthesis. Sunlight energy is trapped and converted to chemical energy of ATP, NADPH, or both, depending on the pathway.

light-independent reactions Second stage of photosynthesis; sugar-building reactions that require phosphate-group transfers from ATP, electrons and H atoms from NADPH, and carbon from CO_2. The phosphorylated sugars from these reactions are then converted to end products (e.g., sucrose, cellulose, starch).

lignification Deposition of lignin in secondary wall of many plant cells. Stabilizes, protects, strengthens, waterproofs plant parts. A factor in evolution of vascular plants.

lignin Organic compound; three-carbon chain and oxygen atom linked to a six-carbon ring.

limbic system In cerebral hemisphere, centers that govern emotions; has roles in memory.

limiting factor Any essential resource which, in short supply, limits population growth.

lineage (LIN-ee-edge) Line of descent.

linkage group All genes on a chromosome.

lipid A mostly greasy or oily hydrocarbon; resists dissolving in water but dissolves in nonpolar substances. All cells use lipids as storage forms of energy, structural materials as in membranes, and cell products.

lipid bilayer Phospholipids, mostly, arranged in two layers; the structural basis of all cell membranes. Hydrophobic tails are sandwiched between the hydrophilic heads; the heads are dissolved in intracellular or extracellular fluid.

lipoprotein Molecule formed when proteins in blood combine with cholesterol, triglycerides, phospholipids absorbed from small intestine.

liver In vertebrates and many invertebrates, a large gland that stores, converts, and helps maintain blood levels of organic compounds; inactivates most hormone molecules that have completed their tasks; inactivates compounds that can be toxic at high concentrations.

liverwort A bryophyte; a nonvascular plant.

loam Soil with roughly the same proportions of sand, silt, and clay; best for plant growth.

lobe-finned fish Only bony fish with fleshy body extensions forming part of ventral fins, which have skeletal elements inside.

local signaling molecule A secretion by one cell type that alters chemical conditions only in localized tissue regions (e.g., prostaglandin).

logistic growth (low-JISS-tik) A population growth pattern. A low-density population slowly increases in size, enters a phase of rapid growth, then levels off in size once the carrying capacity has been reached.

long-day plant Plant that flowers in spring when daylength exceeds a critical value.

loop of Henle Hairpin-shaped, tubular part of a nephron that reabsorbs water and solutes.

loose connective tissue Animal tissue with fibers, fibroblasts loosely arrayed in semifluid ground substance.

lung Internal, sac-shaped respiratory surface that originally evolved in oxygen-poor aquatic habitats. A few fishes and amphibians, birds, reptiles, and mammals have paired lungs.

lungfish Bony fish; has gills, one or a pair of lung-like sacs that assist in respiration.

lycophyte Seedless vascular plant; needs free water to complete life cycle. Most have leaves, roots, and stems (e.g., club mosses).

lymph (LIMF) [L. *lympha*, water] Tissue fluid that has entered vessels of lymphatic system.

lymph node Lymphoid organ packed with lymphocytes; major site for immune responses.

lymph vascular system Part of lymphatic system; delivers excess tissue fluid, absorbed fats, and reclaimable solutes to blood.

lymphatic system Supplements the vertebrate circulatory system; delivers fluid, solutes from interstitial fluid to blood; its lymphoid organs have roles in body defenses.

lysis [Gk. *lysis*, a loosening] Gross damage to a plasma membrane, cell wall, or both that lets the cytoplasm leak out; causes cell death.

lysogenic pathway Latent period that extends many viral replication cycles. Viral genes get integrated into host chromosome and may stay inactivated through many host cell divisions but eventually are replicated in host progeny.

lysosome (LYE-so-sohm) Important organelle of intracellular digestion.

lysozyme Infection-fighting enzyme present in mucous membranes (e.g., of mouth, vagina).

lytic pathway Of viruses, a rapid replication pathway that ends with lysis of host cell.

macroevolution Large-scale patterns, trends, and rates of change among higher taxa.

macrophage Phagocytic white blood cell; roles in nonspecific defense and immune responses.

magnoliid One of three major flowering plant groups; includes magnolias, avocados, nutmeg.

malpighian tubule One of many small tubes that help land-dwelling insects dispose of toxic wastes without losing precious water.

mammal Amniote whose females nourish offspring with milk from mammary glands.

mangrove wetland A nutrient-rich ecosystem of tidal flats at tropical latitudes.

mantle Of mollusks, a tissue draped over the visceral mass. Of Earth, a zone of intermediate-density rocks beneath the crust.

marine snow Organic matter, drifting down from ultraplankton to midoceanic water; the basis of food webs and staggering biodiversity.

marsupial Pouched mammal.

mass extinction Catastrophic event or phase in geologic time when entire families or other major groups are irrevocably lost.

mass number Sum of all protons and neutrons in an atom's nucleus.

mast cell Basophil-like cell; secretes histamine.

master gene One of the genes whose products govern the development of an embryo's body according to an inherited plan; all major groups have similar or identical master genes.

mechanoreceptor Sensory cell or nearby cell that detects mechanical energy (changes in pressure, position, or acceleration).

medulla oblongata Hindbrain region with reflex centers for basic tasks (e.g., respiration); coordinates motor responses with complex reflexes (e.g., coughing); influences sleeping.

medusa (meh-DOO-sah) [Gk. *Medusa*, one of three sisters in Greek mythology having snake-entwined hair] Of cnidarian life cycles, a free-swimming, bell-shaped stage, often with oral lobes and tentacles extending below the bell.

megaspore Haploid spore; forms by meiosis in the ovary of seed-bearing plants; one of its cellular descendants develops into an egg.

meiosis (my-OH-sis) [Gk. *meioun*, to diminish] Two-stage nuclear division process that halves the chromosome number of a parental germ cell nucleus, to the haploid number. Basis of gamete formation (and meiospore formation).

melanin A brownish-black pigment.

melanocyte Type of skin cell that produces and gives up melanin to keratinocytes.

meltdown Accidental overheating of the core in a nuclear power plant.

memory The capacity to store and retrieve information about past sensory experience.

memory cell B or T cell formed in a primary immune response, reserved for later battles.

menstrual cycle A recurring cycle in adult human females. A secondary oocyte is released from an ovary and the uterine lining is primed for pregnancy, all under hormonal control.

meristem (MARE-ih-stem) [Gk. *meristos*, divisible] One of the localized zones where dividing cells give rise to differentiated cell lineages that form all mature plant tissues.

mesoderm (MEH-zoe-derm) [Gk. *mesos*, middle, + *derm*, skin] Primary tissue layer of all large, complex animals; gives rise to many internal organs and part of the integument.

mesophyll (MEH-zoe-fill) A photosynthetic parenchyma with an abundance of air spaces.

Mesozoic An era (240–65 mya) of spectacular expansion in the range of global diversity.

messenger RNA (mRNA) A single strand of ribonucleotides transcribed from DNA, then translated into a polypeptide chain. The only RNA encoding protein-building instructions.

metabolic pathway (MEH-tuh-BALL-ik) Orderly sequence of enzyme-mediated reactions by which cells maintain, increase, or decrease the concentrations of particular substances.

metabolism (meh-TAB-oh-lizm) [Gk. *meta*, change] All the controlled, enzyme-mediated chemical reactions by which cells acquire and use energy to synthesize, store, degrade, and eliminate substances in ways that contribute to growth, survival, and reproduction.

metamorphosis (me-tuh-MOR-foe-sis) [Gk. *meta*, change, + *morphe*, form] Major changes in body form of certain animals; hormonally controlled growth, tissue reorganization, and remodeling of body parts leads to adult form.

metaphase Of meiosis I, stage when all pairs of homologous chromosomes have become positioned at the spindle equator. Of mitosis or meiosis II, all the duplicated chromosomes are positioned at the spindle equator.

metastasis Abnormal migration of cancer cells, which may establish colonies in other tissues.

methanogen Anaerobic archaebacterium that produces methane gas as by-product.

methylation Attachment of a methyl group to an organic compound; a common gene control.

MHC marker Self-marker protein. Some are on all body cells of the individual; other kinds are unique to macrophages and lymphocytes.

micelle (my-SELL) Of fat digestion, tiny droplet of bile salts, fatty acids, and monoglycerides; role in fat absorption from small intestine.

microevolution Of a population, any change in allele frequencies resulting from mutation, genetic drift, gene flow, natural selection, or some combination of these.

microfilament Cytoskeletal element of two thin, helically twisted polypeptide chains; functions in movement, cell shape.

micrograph Photograph of an image brought into view with the aid of a microscope.

microorganism Organism, usually single celled, too small to be observed without a microscope.

microspore Walled haploid spore; becomes a pollen grain in gymnosperms or angiosperms.

microtubular spindle Bipolar array of many microtubules; forms for nuclear division and moves chromosomes apart in controlled ways.

microtubule (my-crow-TUBE-yool) Cylinder of tubulin subunits; cytoskeletal element with roles in most eukaryotic cell movements.

microtubule organizing center (MTOC) Mass of substances in cytoplasm of eukaryotic cells; number, type, and location dictate orientation and organization of cell's microtubules.

microvillus (MY-crow-VILL-us) [L. *villus*, shaggy hair] Slender extension from free surface of certain cells; arrays of many microvilli greatly increase absorptive or secretory surface area.

midbrain Part of vertebrate brain with centers for coordinating reflex responses to visual and auditory input; also relays signals to forebrain.

middle ear Structures (eardrum, ear bones) that transmit air waves to the inner ear.

migration Recurring pattern of movement between two or more regions in response to environmental rhythms (e.g., the seasons).

millipede Segmented arthropod; long body, many legs; scavenges decaying plant material.

mimicry (MIM-ik-ree) Close resemblance in form, behavior, or both between one species (the mimic) and another (its model). Serves in deception, as when an orchid mimics a female insect and attracts males, which pollinate it.

mineral Any element or inorganic compound that formed by natural geologic processes and is required for normal cell functioning.

mitochondrion (MY-toe-KON-dree-on) Double-membrane organelle of ATP formation. Only site of aerobic respiration's second and third stages. May have endosymbiotic origins.

mitosis (my-TOE-sis) [Gk. *mitos*, thread] Type of nuclear division that maintains the parental chromosome number for daughter cells. The basis of growth in size, tissue repair, and often asexual reproduction for eukaryotes.

mixture Two or more elements intermingled in proportions that can and usually do vary.

model Theoretical, detailed description or analogy that helps people visualize something that has not yet been directly observed.

molar Tooth with cusps that help crush, grind, and shear food; one of the cheek teeth.

molecular clock Model used to calculate the time of origin of one lineage relative to others; assumes that the lineage accumulates neutral mutations at predictable rates, measurable as a series of ticks back through time.

molecule Two or more atoms of the same or different elements joined by chemical bonds.

mollusk Only invertebrate with mantle draped over a soft, fleshy body; most have an external or internal shell. Diverse body plans and sizes (e.g., gastropods, bivalves, and cephalopods).

molting Periodic shedding of worn-out or too-small body structures. Permits some animals to grow in size or renew parts.

monocot (MON-oh-kot) Monocotyledon; a flowering plant with one cotyledon in seeds, floral parts usually in threes (or multiples of three), and often parallel-veined leaves.

monohybrid cross Intercross between two F_1 heterozygotes that are identical for one gene locus; offspring of two parents that breed true for different forms of a trait.

monomer Small molecule used as a subunit of polymers, such as sugar monomers of starch.

monophyletic group All descendants from an ancestral species in which a trait first evolved.

monosaccharide (MON-oh-SAK-ah-ride) [Gk. *monos*, alone, single, + *sakcharon*, sugar] One of the simple carbohydrates (e.g., glucose).

monotreme Egg-laying mammal.

monsoon Air circulation pattern that moves moisture-laden air arising from warm oceans to continents north or south of them.

morphogen Type of inducer molecule that diffuses through embryonic tissues, activating master genes in sequence; it contributes to mapping out the overall body plan.

morphogenesis (MORE-foe-JEN-ih-sis) [Gk. *morphe*, form, + *genesis*, origin] Programmed, orderly changes in body size, proportion, and shape of an animal embryo through which all specialized tissues and early organs form.

morphological convergence Macroevolutionary pattern. In response to similar environmental pressures over time, evolutionarily distant lineages evolve in similar ways and end up being alike in appearance, functions, or both.

morphological divergence Macroevolutionary pattern; genetically diverging lineages undergo change from body form of a common ancestor.

mosaicism, mosaic tissue effect Cells of same type express genes differently, so phenotypic differences emerge in same type of tissue. E.g., occurs by X chromosome inactivation in female mammals; also by nondisjunction in any cell after fertilization (only descendants of altered cell inherit the abnormal chromosome number).

moss Most common kind of bryophyte.

motor neuron Neuron that relays signals from brain or spinal cord to muscle or gland cells.

motor protein Type of protein (e.g., myosin) attached to microfilaments and microtubules; used in cell movements (e.g., contraction).

motor unit A motor neuron and all muscle cells that form junctions with its endings.

multicelled organism Organism composed of many cells with coordinated metabolic activity; most show extensive cell differentiation into tissues, organs, and organ systems.

multiple allele system Three or more slightly different molecular forms of a gene that occur among individuals of a population.

multiregional model Holds that *Homo sapiens* evolved from *H. erectus* groups that had spread through much of the world by about 1 mya and evolved into regionally distinctive "races."

muscle fatigue Decline in tension of a muscle kept in a state of tetanic contraction as a result of continuous, high-frequency stimulation.

muscle tension Mechanical force exerted by a contracting muscle; resists opposing forces (e.g., gravity or weight of an object being lifted).

muscle tissue Tissue with organized arrays of cells able to contract when stimulated.

muscle twitch Sequence of muscle contraction and relaxation in response to a brief stimulus.

mushroom Reproductive part of club fungi.

mutation [L. *mutatus*, a change, + *-ion*, act, result, or process] Heritable change in DNA's molecular structure. Original source of all new alleles and, ultimately, the diversity of life.

mutation frequency Of a population, number of times a mutation at a gene locus has arisen.

mutation rate Of a gene locus, the probability that a spontaneous mutation will occur during or between DNA replication cycles.

mutualism [L. *mutuus*, reciprocal] Symbiotic interaction that benefits both participants.

mycelium (my-SEE-lee-um), plural **mycelia** [Gk. *mykes*, fungus, mushroom, + *helos*, callus] Mesh of tiny, branching filaments (hyphae); the food-absorbing portion of most fungi.

mycorrhiza (MY-coe-RIZE-uh) "Fungus-root." A form of mutualism between fungal hyphae

and young plant roots. The plant gives up carbohydrates to the fungus, which gives up some of its absorbed mineral ions to the plant.

myelin sheath Lipid-rich membrane formed by oligodendrocytes that wrap around axons of many sensory and motor neurons; enhances long-distance propagation of action potentials.

myofibril (MY-oh-FY-brill) One of many striated threadlike structures inside a muscle cell.

myoglobin A pigment with a higher oxygen-binding capacity than hemoglobin; abundant in cardiac and skeletal muscle cells that can contract slowly and resist fatiguing.

myosin (MY-uh-sin) Motor protein; uses cytoskeletal tracks to move cell components.

NAD+ Nicotinamide adenine dinucleotide. A nucleotide coenzyme; abbreviated NADH when carrying electrons and H^+ to a reaction site.

NADP+ Nicotinamide adenine dinucleotide phosphate. A phosphorylated nucleotide coenzyme; abbreviated $NADPH_2$ when it is carrying electrons and H^+ to a reaction site.

natural killer cell Cytotoxic lymphocyte that touch-kills tumor cells and virus-infected cells.

natural selection Microevolutionary process; the outcome of differences in survival and reproduction among individuals that differ in details of heritable traits.

necrosis (neh-CROW-sis) Passive death of many cells that results from severe tissue damage.

negative control Use of regulatory proteins to trigger a slowdown in gene expression.

negative feedback mechanism A homeostatic mechanism by which a condition that changed as a result of some activity triggers a response that reverses the change.

nematocyst (NEM-at-uh-sist) [Gk. *nema*, thread + *kystis*, pouch] Cnidarian capsule that has a dischargeable, tube-shaped thread, sometimes barbed; releases a toxin or sticky substance.

neoplasm Abnormally grown mass of cells; if benign, remains in tissue. If malignant, it metastasizes; it is a cancer.

nephridium (neh-FRID-ee-um), plural **nephridia** Unit that controls composition and volume of fluid in some invertebrates (e.g., earthworms).

nephron (NEFF-ron) [Gk. *nephros*, kidney] One of the kidney's urine-forming tubules; it filters water and solutes from blood, then selectively reabsorbs adjusted amounts of both.

nerve Sheathed, cordlike bundle of the axons of sensory neurons, motor neurons, or both.

nerve cell One of the cells that receive and integrate signals from sensory receptors, and guide responses to stimuli. *See also* Neuron.

nerve cord A prominent longitudinal nerve; most animals have one, two, or three. Chordate nervous systems arise from a dorsal nerve cord.

nerve net Simple nervous system in epidermis of cnidarians and some other invertebrates; a diffuse mesh of simple, branching nerve cells interacts with contractile and sensory cells.

nervous system Integrative organ system with nerve cells interacting in signal-conducting, information-processing pathways. Detects and processes stimuli, and elicits responses from effectors (e.g., muscles and glands).

nervous tissue Connective tissue composed of neurons and often neuroglia.

net ecosystem production Of ecosystems, all the energy plants have accumulated through

their growth, reproduction in a given interval (net primary production), *minus* the energy that the plants and soil organisms have used.

neural tube The embryonic and evolutionary forerunner of brain and spinal cord.

neuroglia (NUR-oh-GLEE-uh) Collectively, cells that structurally and metabolically support neurons. They make up about half the volume of nervous tissue in vertebrates.

neuromodulator A signaling molecule; reduces or magnifies neurotransmitter effects.

neuromuscular junction Synapse between a motor neuron's axon endings and a muscle cell.

neuron (NUR-on) Type of nerve cell; basic communication unit in most nervous systems.

neurotransmitter Any of a class of signaling molecules secreted by neurons. It acts on cells next to it, then is rapidly degraded or recycled.

neutral mutation A mutation with little or no effect on phenotype, so natural selection can't change its frequency in a population.

neutron Unit in the atomic nucleus that has mass but no electric charge.

neutrophil Phagocytic white blood cell.

niche (NITCH) [L. *nidas*, nest] Sum total of all activities and relationships in which individuals of a species engage as they secure and use the resources required to survive and reproduce.

nitrification (nye-trih-fih-KAY-shun) Major nitrogen cycle process. Certain bacteria in soil break down ammonia or ammonium to nitrite, then other bacteria break down the nitrite to nitrate (a form that plants can take up).

nitrogen cycle Atmospheric cycle. Nitrogen moves from its largest reservoir (atmosphere), through the ocean, ocean sediments, soils, and food webs, then back to the atmosphere.

nitrogen fixation Major nitrogen cycle process. Certain bacteria convert gaseous nitrogen to ammonia, which dissolves in their cytoplasm to form ammonium (used in biosynthesis).

node Stem site where one or more leaves form.

noncyclic pathway of ATP formation (non-SIK-lik) [L. *non*, not, + Gk. *kylos*, circle] Light-dependent reactions of photosynthesis, using photolysis, two photosystems, two transport systems. H_2O molecules are split. Released electrons and H atoms are used in ATP and NADPH formation; the oxygen by-product is the basis of Earth's oxygen-rich atmosphere.

nondisjunction Failure of sister chromatids or a pair of homologous chromosomes to separate during meiosis or mitosis. Daughter cells end up with too many or too few chromosomes.

non-ionizing radiation Wavelengths that boost electrons to a higher energy level. DNA easily absorbs one form, ultraviolet light.

nonshivering heat production Hormone-induced increase in metabolic rate in response to prolonged or severe cold exposure.

notochord (KNOW-toe-kord) Of chordates, a rod of stiffened tissue (not cartilage or bone); forms as a supporting structure for body.

nuclear envelope Outermost portion of the cell nucleus; consists of a double membrane (two lipid bilayers and associated proteins).

nucleic acid (new-CLAY-ik) Single- or double-stranded chain of four kinds of nucleotides joined at their phosphate groups. Nucleic acids (e.g., DNA, RNA) differ in base sequences.

nucleic acid hybridization Any base-pairing between DNA or RNA from different sources.

nucleoid (NEW-KLEE-oid) Portion of bacterial cell interior in which the DNA is physically organized but not enclosed by a membrane.

nucleolus (new-KLEE-oh-lus) [L. *nucleolus*, tiny kernel] In a nondividing cell nucleus, a site for assembling protein and RNA subunits that will later join up as ribosomes in the cytoplasm.

nucleosome (NEW-klee-oh-sohm) A stretch of eukaryotic DNA looped twice around a spool of histone molecules; one of many units that give condensed chromosomes their structure.

nucleotide (NEW-klee-oh-tide) Small organic compound with deoxyribose (a five-carbon sugar), a nitrogenous base, and a phosphate group. Monomer for adenosine phosphates, nucleotide coenzymes, and nucleic acids.

nucleotide coenzyme Enzyme helper; moves electrons, hydrogen atoms from one site of catalysis to another reaction site.

nucleus (NEW-klee-us) [L. *nucleus*, a kernel] Of atoms, a central core of one or more protons and (in all but hydrogen atoms) neutrons. In a eukaryotic cell, the organelle that physically separates DNA from cytoplasmic machinery.

numerical taxonomy Study of the degree of relatedness between an unidentified organism and a known group through comparisons of traits. Used to classify prokaryotic species.

nutrient Element with a direct or indirect role in metabolism that no other element fulfills.

nutrition Processes of selectively ingesting, digesting, absorbing, and converting food into the body's own organic compounds.

nymph Immature, post-embryonic stage of some insect life cycles.

ob gene Gene affecting set point for body fat.

obesity Excess of fat in adipose tissue; caloric intake has exceeded the body's energy output.

ocean A continuous body of water that covers more than 71 percent of the Earth; its currents distribute nutrients in marine ecosystems and influence regional climates.

olfactory receptor Chemoreceptor for water-soluble or volatile substance.

oligosaccharide (oh-LIG-oh-SAC-uh-rid) Short-chain carbohydrate of two or more covalently bonded sugar monomers (e.g., disaccharides).

omnivore [L. *omnis*, all, + *vovare*, to devour] Animal that eats at more than one trophic level.

oncogene (ON-koe-jeen) Any gene having the potential to induce cancerous transformation.

oocyte Type of immature egg.

operator Very short base sequence between a promoter and bacterial genes; a binding site for a repressor that can block transcription.

operon Promoter–operator sequence serving more than one bacterial gene; part of a control that adjusts transcription rates up or down.

organ Body structure with definite form and function that consists of more than one tissue.

organ formation Developmental stage in which primary tissue layers give rise to cell lineages unique in structure and function. Descendants of those lineages give rise to all the different tissues and organs of the adult.

organ system Organs interacting chemically, physically, or both in a common task.

organelle (or-guh-NELL) Membrane-bound sac or compartment in the cytoplasm having one or more specialized metabolic functions. Most eukaryotic cells have a profusion of them.

organic compound A molecule containing carbon and at least one hydrogen atom.

osmoreceptor Sensory receptor; detects shifts in water volume (solute concentration).

osmosis (oss-MOE-sis) [Gk. *osmos*, pushing] In response to a water concentration gradient, the diffusion of water between two regions that a selectively permeable membrane separates.

osmotic pressure Pressure that operates after hydrostatic pressure develops in an enclosed region (e.g., a cell); it counters water's inward diffusion (stops further rises in fluid volume).

osteoblast Bone-forming cell; it secretes organic substances that become mineralized.

osteoclast Cell that secretes the enzymes that digest bone's organic matrix; releasing calcium and phosphorus for the bloodstream.

osteocyte A mature bone cell.

ostracoderm An early craniate; a filter-feeding bottom-dwelling jawless fish; extinct.

otolith In organ of static equilibrium, a mass of calcium carbonate crystals embedded in a membrane that is sensitive to fluid movement.

ovary (OH-vuh-ree) In most animals, a female gonad. In flowering plants, the enlarged base of one or more carpels.

oviduct (OH-vih-dukt) Duct between ovary and uterus. Formerly called a fallopian tube.

ovulation (OHV-you-LAY-shun) Release of a secondary oocyte from an ovary.

ovule (OHV-youl) [L. *ovum*, egg] Tissue mass in a plant ovary that becomes a seed; a female gametophyte with egg cell, nutritious tissue, and a jacket that will become a seed coat.

ovum (OH-vum) Mature secondary oocyte.

oxaloacetate (ox-AL-oh-ASS-ih-tate) A four-carbon compound with roles in metabolism (e.g., the point of entry into the Krebs cycle).

oxidation–reduction reaction A transfer of an electron (and often an unbound proton, or H^+) between atoms or molecules.

oxidized molecule A molecule that has lost one or more electrons.

oxygen debt Lowered O_2 level in blood when muscle cells have used up more ATP than they have formed by aerobic respiration.

oxyhemoglobin Oxygen bound to hemoglobin (HbO_2) inside red blood cells.

oxytocin A hormone of lactation; also induces uterus to shrink after pregnancy.

ozone thinning A pronounced and seasonal thinning of the atmosphere's ozone layer.

P_i Abbreviation for inorganic phosphate.

pain Perception of injury to some body region.

pain receptor A nociceptor; a sensory receptor that detects tissue damage.

Paleozoic Era from Cambrian, Ordovician, Silurian, Devonian, Carboniferous, through the Permian (544 to 248 mya).

PAN An oxidant in photochemical smog.

pancreas (PAN-cree-us) Gland with roles in digestion and organic metabolism. Secretes enzymes and bicarbonate into small intestine; also secretes insulin and glucagon.

pancreatic islet Any of the 2 million or so clusters of endocrine cells of the pancreas.

pandemic An epidemic that breaks out in several countries at the same time.

Pangea Paleozoic supercontinent upon which the first terrestrial plants and animals evolved.

parabasilid Protozoan, parasitic or predatory; four to many flagella; e.g., *Trichomonas*.

parapatric speciation Idea that neighboring populations can become distinct species while maintaining contact along a common border.

parasite [Gk. *para*, alongside, + *sitos*, food] Organism that lives in or on a host for at least part of its life cycle. It feeds on specific tissues and usually does not kill its host outright.

parasitism Symbiotic interaction in which a parasitic species benefits as it feeds on tissues of a host species, which it harms.

parasitoid Type of insect larva that grows and develops in a host organism (usually another insect), consumes its soft tissues, and kills it.

parasympathetic nerve (PARE-uh-SIM-pu-THET-ik) An autonomic nerve. Its signals slow down overall activities and divert energy to basic tasks; it also helps make small adjustments in internal organ activity by acting continually in opposition to sympathetic nerve signals.

parathyroid gland One of four small glands embedded in the thyroid; their secretions raise blood calcium levels.

parenchyma (par-ENG-kih-mah) Simple tissue that makes up the bulk of a plant; has roles in photosynthesis, storage, secretion, other tasks.

parthenogenesis (par-THEN-oh-GEN-uh-sis) An unfertilized egg gives rise to an embryo.

partial pressure Characteristic of a gas; its contribution to total atmospheric pressure.

passive transport Event in which a transport protein that spans a cell membrane passively permits a solute to diffuse through its interior. Also called facilitated diffusion.

pathogen (PATH-oh-jen) [Gk. *pathos*, suffering, + *genēs*, origin] Any virus, bacterium, fungus, protistan, or parasitic worm that can infect an organism, multiply in it, and cause disease.

pattern formation The ordered, sequential sculpting of embryonic cells into specialized animal tissues and organs; brought about by cytoplasmic localization and, later, inductive interactions among master genes, the products of which map out the basic body plan.

PCR Polymerase chain reaction. A method of enormously amplifying the quantity of DNA fragments cut by restriction enzymes.

peat bog Compressed, soggy, highly acidic mat of accumulated remains of peat mosses.

pedigree Diagram of the genetic connections among related individuals through successive generations; uses standardized symbols.

pellicle Of some protistans, a flexible body covering of protein-rich material.

peptide hormone A hormone that binds to a membrane receptor, thus activating enzyme systems that alter target cell activity. A second messenger inside the cell often relays the hormone's message.

per capita [L. *capita*, head] Of population studies, a term used when making head counts.

perception Understanding of what a given stimulus means.

perennial [L. *per-*, throughout, + *annus*, year] Plant that lives three or more growing seasons.

pericarp Three tissue divisions of a fleshy fruit; endocarp, mesocarp, and exocarp.

pericycle (PARE-ih-sigh-kul) [Gk. *peri-*, around, + *kyklos*, circle] One or more cell layers inside the endodermis; gives rise to lateral roots and contributes to secondary growth.

periderm Protective cover that replaces plant epidermis during extensive secondary growth.

peripheral nervous system (per-IF-ur-uhl) [Gk. *peripherein*, to carry around] All nerves leading into and out from the spinal cord and brain. Includes ganglia of those nerves.

peripheral vasoconstriction Diameters of arterioles constrict, decreasing blood's delivery of heat to body surface.

peripheral vasodilation Blood vessels in the skin dilate; more blood flows to the skin, which then dissipates excess body heat.

peristalsis (pare-ih-STAL-sis) Recurring waves of contraction and relaxation of muscles in the wall of a tubular or saclike organ.

peritoneum (pare-ih-tuh-NEE-um) Membrane lining of coelom.

peritubular capillary Set of blood capillaries around tubular parts of a nephron. Reabsorbs water, solutes; secretes excess H^+, other solutes.

permafrost A water-impermeable, perpetually frozen layer, up to 500 meters thick, beneath the soil surface of the arctic tundra.

peroxisome A vesicle; its enzymes digest fatty acids and amino acids to hydrogen peroxide, which is then converted to harmless products.

PGA Phosphoglycerate (FOSS-foe-GLISS-er-ate) Important intermediate of glycolysis and of the Calvin–Benson cycle.

PGAL Phosphoglyceraldehyde. Intermediate of glycolysis and of the Calvin–Benson cycle.

pH scale Measure of the concentration of free hydrogen ions (H^+) in blood, water, and other solutions. pH 0 is the most acidic, 14 the most basic, and 7, neutral.

phagocyte (FAG-uh-sight) [Gk. *phagein*, to eat, + *kytos*, hollow vessel] Cell that engulfs prey or cellular debris and foreign agents in body.

phagocytosis Of some cells, engulfment of an extracellular target by way of pseudopod formation and endocytosis.

pharynx (FARE-inks) Among invertebrates, a muscular tube to the gut. In some chordates, a gas exchange organ. In land vertebrates, a dual entrance to the esophagus and trachea.

phenotype (FEE-no-type) [Gk. *phainein*, to show + *typos*, image] Observable trait or traits of an individual that arise from gene interactions and gene–environment interactions.

pheromone (FARE-oh-moan) [Gk. *phero*, to carry, + *-mone*, as in hormone] Hormone-like, nearly odorless exocrine gland secretion. A signaling molecule between individuals of the same species that integrates social behavior.

phloem (FLOW-um) Plant vascular tissue. Live cells (sieve tubes) interconnect as conducting tubes for sugars and other solutes; companion cells help load solutes into the tubes.

phospholipid Organic compound that has a glycerol backbone, two fatty acid tails, and a hydrophilic head of two polar groups (one being phosphate). Phospholipids are the main structural component of cell membranes.

phosphorus cycle A sedimentary cycle; the movement of phosphorus (mainly phosphate ions) from land, through food webs, to ocean sediments, then back to land.

phosphorylation (FOSS-for-ih-LAY-shun) An enzyme activates a molecule by attaching inorganic phosphate to it or by mediating a phosphate-group transfer to it, as from ATP.

photoautotroph Photosynthetic autotroph; any organism that synthesizes its own organic compounds using CO_2 for carbon atoms and sunlight for energy. Nearly all plants, some protistans, and a few bacteria do this.

photolysis (foe-TALL-ih-sis) [Gk. *photos*, light, +-*lysis*, breaking apart] Reactions split water molecules using photon energy. The released electrons, hydrogen used in noncyclic pathway of photosynthesis; oxygen is a by-product.

photon One unit of energy of visible light.

photoperiodism Biological response to change in relative lengths of daylight and darkness.

photoreceptor Light-sensitive sensory cell.

photosynthesis Sunlight energy trapped, converted to chemical energy (ATP, NADPH, or both); then synthesis of sugar phosphates that are converted to sucrose, cellulose, starch, and other end products. The main pathway by which energy and carbon enter the web of life.

photosystem Cluster of many light-trapping pigments in a photosynthetic membrane.

phototropism Change in the direction of cell movement or growth in response to light (e.g., as in a stem bending toward light).

photovoltaic cell Unit in a device designed to convert sunlight energy into electricity.

phycobilin (FIE-koe-BY-lin) Type of accessory pigment; notably abundant in red algae and in cyanobacteria.

phylogeny Evolutionary relationships among species, starting with an ancestral form and including branches leading to its descendants.

physiology Study of patterns and processes by which organisms survive and reproduce.

phytochrome A light-sensitive pigment. Its controlled activation and inactivation affect plant hormone activities that govern leaf expansion, stem branching, stem lengthening, and often seed germination and flowering.

phytoplankton (FIE-toe-PLANK-tun) [Gk. *phyton*, plant, + *planktos*, wandering] Aquatic community of floating or weakly swimming photoautotrophs (e.g., "pastures of the seas").

pigment Any light-absorbing molecule.

pilomotor response Hairs or feathers become erect; creates a layer of still air next to the skin.

pilus Short, filamentous protein projecting above cell wall to help it adhere to surfaces.

pineal gland Light-sensitive endocrine gland; its melatonin secretions affect biological clocks, overall activity, and reproductive cycles.

pioneer species Any opportunistic colonizer of barren or disturbed habitats. Adapted for rapid growth and dispersal.

pith Of most dicot stems, ground tissue inside the ring of vascular bundles.

pituitary gland Endocrine gland that, with the hypothalamus, controls many physiological functions, including activity of many other endocrine glands. Its posterior lobe stores and secretes hypothalamic hormones. Its anterior lobe produces and secretes its own hormones.

placenta (plah-SEN-tuh) Blood-engorged organ inside pregnant female placental mammals; made of endometrial tissue, extraembryonic membranes. Permits exchanges between the mother and fetus without intermingling their bloodstreams; sustains the new individual and allows its blood vessels to develop separately.

placoderm An early craniate; a jawed predator with paired fins, armor plates on head; extinct.

placozoan The simplest known animal; tiny, asymmetrically soft-bodied, two cell layers.

plankton Aquatic community of primarily microscopic autotrophs and heterotrophs.

plant Generally, a multicelled photoautotroph with well-developed root and shoot systems; photosynthetic cells that include starch grains as well as chlorophylls *a* and *b*; and cellulose, pectin, and other polysaccharides in cell walls.

plant physiology Study of adaptations by which plants function in their environment.

Plantae Kingdom of plants.

planula Larva, usually with ciliated epidermis. Swims or creeps; may resemble first animals.

plasma (PLAZ-muh) Liquid portion of blood; mainly water in which ions, proteins, sugars, gases, and other substances are dissolved.

plasma membrane Outermost cell membrane; structural and functional boundary between cytoplasm and the fluid outside the cell.

plasmid A small, circular molecule of extra bacterial DNA that carries a few genes and is replicated independently of the chromosome.

plasmodesma (PLAZ-moe-DEZ-muh), plural **plasmodesmata** Plant cell junction; a channel that connects the cytoplasm of adjoining cells.

plate tectonics Theory that great slabs (plates) of the Earth's outer layer float on a hot, plastic mantle. All plates are slowly moving and have rafted continents to new positions over time.

platelet (PLAYT-let) Megakaryocyte fragment that releases substances used in clot formation.

pleiotropy (PLEE-oh-troe-pee) [Gk. *pleon*, more, + *trope*, direction] Positive or negative effects on two or more traits owing to expression of alleles at a single gene locus. Effects may or may not emerge at the same time.

polar body One of four cells that has formed by the meiotic cell division of an oocyte but that does not become the ovum.

pollen grain [L. *pollen*, fine dust] Immature or mature, sperm-bearing male gametophyte of gymnosperms and angiosperms.

pollination Arrival of a pollen grain on the landing platform (stigma) of a flower's carpel.

pollinator Agent (wind, water, an animal) that puts pollen from male on female reproductive parts of flowers of the same plant species.

pollutant Natural or synthetic substance with which an ecosystem has no prior evolutionary experience, in terms of kinds or amounts; it accumulates to disruptive or harmful levels.

polymer Large molecule of three to millions of monomers of the same or different kinds.

polymorphism (poly-MORE-fizz-um) [Gk. *polus*, many, + *morphe*, form] The persistence of two or more qualitatively different forms of a trait (morphs) in a population.

polyp (POH-lip) Vase-shaped, sedentary stage of cnidarian life cycles.

polypeptide chain An organic compound of three or more amino acids joined by peptide bonds; atoms of its backbone have this pattern: —N—C—C—N—C—C— . All proteins are composed of one or more polypeptide chains.

polyploidy (POL-ee-PLOYD-ee) Having three or more of each type of chromosome in the nucleus of a eukaryotic cell at interphase.

polysaccharide [Gk. *polus*, many, + *sakcharon*, sugar] Straight or branched chain of many covalently linked sugar units of the same or different kinds. In nature, the most common types are cellulose, starch, and glycogen.

polysome A series of ribosomes translating the same mRNA molecule at the same time.

polytene chromosome Insect chromosome consisting of many in-parallel copies of DNA.

pons A hindbrain traffic center for signals between cerebellum and forebrain centers.

population All individuals of the same species that are occupying a specified area.

population density Count of individuals of a population in a habitat (by area or volume).

population distribution Dispersal pattern for individuals of a population through a habitat.

population size The number of individuals that make up the gene pool of a population.

positive control Use of regulatory proteins to promote gene expression.

positive feedback mechanism Homeostatic control; it initiates a chain of events that intensify change from an original condition, then intensification reverses the change.

potential energy A stationary object's capacity to do work owing to its position in space or to the arrangement of its parts.

predation Ecological interaction in which a predator feeds on a prey organism.

predator [L. *prehendere*, to grasp, seize] A heterotroph that eats other living organisms (its prey), does not live in or on them (as parasites do), and may or may not kill them.

prediction Statement about what you should observe in nature if you were to go looking for a particular phenomenon; the if–then process.

pressure flow theory Organic compounds flow through phloem in response to pressure and concentration gradients between sources (e.g., leaves) and sinks (e.g., growing parts where they are being used or stored).

pressure gradient A difference in pressure between two adjoining regions.

prey Organism that another organism (e.g., a predator) captures as a food source.

primary growth Plant growth originating at root tips and shoot tips.

primary producer Type of autotroph that secures energy directly from the environment and stores some in its tissues.

primary productivity Of ecosystems, the rate at which primary producers capture and store energy in tissues during a specified interval.

primary root First root of a young seedling.

primary succession Development of climax community through a series of species that replace one another, starting with pioneers.

primary wall A thin, flexible plant cell wall of cellulose, polysaccharides, and glycoproteins; allows growing cells to divide or change shape.

primate Mammalian lineage dating from the Eocene; includes prosimians, tarsioids, and anthropoids (monkeys, apes, and humans).

primer Short nucleotide sequence designed to base-pair with any complementary DNA sequence; acts as a START tag for replication.

prion Small infectious protein that causes rare, fatal degenerative diseases of nervous system.

probability The chance that each outcome of a given event will occur is proportional to the number of ways the outcome can be reached.

probe Very short stretch of DNA labeled with a radioisotope; it is designed to base-pair with part of a gene in a DNA sample being studied.

procambium (pro-KAM-bee-um) A meristem that gives rise to primary vascular tissues.

producer Autotroph (self-feeder); it nourishes itself using sources of energy and carbon from the physical environment. Photoautotrophs and chemoautotrophs are examples.

product Substances left at end of a reaction.

progesterone (pro-JESS-tuh-rown) Of female mammals, a sex hormone secreted by ovaries and the corpus luteum.

proglottid One of many tapeworm body units that bud in sequence, behind the scolex.

progymnosperm Among the earliest plants to produce seedlike structures or seeds.

prokaryotic cell (pro-CARE-EE-oh-tic) [L. *pro*, before, + Gk. *karyon*, kernel] Archaebacterium or eubacterium; single-celled organism, most often walled; lacks the profusion of membrane-bound organelles observed in eukaryotic cells.

prokaryotic fission Cell division mechanism by which prokaryotic cells alone reproduce.

prolactin Hormone that stimulates synthesis of enzymes used in milk production.

promoter Short stretch of DNA to which RNA polymerase can bind and start transcription.

prophase Of mitosis, a stage when duplicated chromosomes start to condense, microtubules form a spindle, and the nuclear envelope starts to break up. Duplicated pairs of centrioles (if present) are moved to opposite spindle poles.

prophase I First stage of meiosis I. Each duplicated chromosome starts to condense. It pairs with its homologue; nonsister chromatids usually undergo crossing over and are attached to spindle. One pair of duplicated centrioles (if present) is moved to the opposite spindle pole.

prophase II First stage of meiosis II. Spindle microtubules attach to each chromosome and move them to spindle's equator. A centriole pair (if present) is already at each spindle pole.

protein Organic compound of one or more polypeptide chains folded and twisted into a globular or fibrous shape, overall.

Proterozoic Era in which oxygen accumulated in atmosphere; origin of aerobic metabolism and eukaryotic cells (2500 to 544 mya).

Protista Kingdom of protistans; may soon be split into multiple kingdoms.

protistan (pro-TISS-tun) [Gk. *prōtistos*, primal] Photoautotroph or heterotroph (or both) unlike bacteria; some like earliest eukaryotic cells. Has a nucleus, larger ribosomes, mitochondria, ER, Golgi bodies, chromosomes with numerous proteins, and cytoskeletal microtubules. Range in size from microscopic algae to giant kelps.

proto-cell Hypothetic cell-like stage between chemical evolution and the first living cell.

proton Positively charged particle; one or more in the nucleus of each atom. An unbound (free) proton is called a hydrogen ion (H^+).

protostome (PRO-toe-stome) [Gk. *proto*, first, + *stoma*, mouth] Lineage of coelomate, bilateral animals that includes mollusks, annelids, and arthropods. The first indentation to form in protostome embryos becomes the mouth.

protozoan A flagellated or ciliated protistan (e.g., a kinetoplastid or ciliated protozoan).

proximal tubule Nephron's tubular portion extending from Bowman's capsule.

pseudopod A dynamically extending lobe of cytoplasm used for motility or engulfment.

puberty Post-embryonic stage; gametes start to mature and secondary sexual traits emerge.

pulmonary circuit Vertebrate cardiovascular route; O_2-poor blood flows from the heart to lungs, gets oxygenated, flows back to heart.

punctuation model, speciation Idea that most morphological changes occur in a brief span when populations start to diverge; speciation is rapid, and the daughter species change little for the next 2–6 million years or so.

Punnett-square method Construction of a simple diagram as a way to predict probable outcomes of a genetic cross.

pupa, plural, **pupae** An immature, post-embryonic stage of many insect life cycles.

purine A nucleotide base with a double ring structure (e.g., adenine or guanine).

pyrimidine A nucleotide base with a single ring structure (e.g., cytosine or thymine).

pyruvate (PIE-roo-vate) Organic compound with a backbone of three carbon atoms. Two molecules form as end products of glycolysis.

r Variable in population growth equations that signifies net population growth rate. Birth and death rates are assumed to remain constant and are combined into this one variable.

radial symmetry Animal body plan having four or more roughly equivalent parts around a central axis (e.g., sea anemone).

radiation Radiant energy such as sunlight or a surface warmer than body surface temperature; exposure to it can lead to a heat gain.

radioisotope Unstable atom (uneven number of protons and neutrons). It spontaneously emits particles and energy, and so decays into a different atom over a predictable time span.

radiometric dating Method of measuring the proportions of (1) a radioisotope in a mineral trapped long ago in newly formed rock and (2) a daughter isotope that formed from it by radioactive decay in the same rock. Used to assign absolute dates to fossil-containing rocks and to the geologic time scale.

rain shadow A reduction in rainfall on the leeward side of a high mountain range that results in arid or semiarid conditions.

ray-finned fish Fish with fin supports derived from skin; usually maneuverable fins and thin, flexible scales, and swim bladders.

reabsorption In kidneys, diffusion or active transport of water and reclaimable solutes from nephrons into peritubular capillaries; amounts are controlled by ADH and aldosterone. At a capillary bed, osmotic movement of interstitial fluid into a capillary when water concentration between plasma and interstitial fluid differ.

reactant Substance that enters a reaction.

reaction center The only molecule (a special chlorophyll *a*) that can pass electrons out of a photosystem to a nearby acceptor molecule.

rearrangement, molecular Conversion of one organic compound to another through changes in its internal bonds.

receptor, molecular Membrane protein (or cytoplasmic protein) that triggers a change in cell activities after it binds signaling molecule.

receptor, sensory Sensory cell or specialized structure that can detect a stimulus.

recessive allele [L. *recedere*, to recede] Allele whose expression in heterozygotes is fully or partially masked by expression of its partner. Fully expressed in homozygous recessives.

reciprocal cross A paired cross. In the first cross, one parent displays the trait of interest. In the second, the other parent displays it.

recognition protein One of a class of plasma membrane proteins that distinguish *nonself* (foreign) from *self* (belonging to a body tissue).

recombinant DNA Any DNA molecule that incorporates one or more nonparental base sequences. Outcome of microbial gene transfer in nature or of recombinant DNA technology.

recombinant DNA technology Procedures by which DNA molecules from different species are isolated, cut up, spliced together, and then enormously amplified to useful quantities.

recombination Any enzyme-mediated reaction that inserts one DNA sequence into another. Generalized recombination uses any pair of homologous sequences between chromosomes as substrates, as occurs at crossing over. Site-specific recombination uses a short stretch of homology between bacterial and viral DNA. A different reaction inserts transposons at new, random sites in bacterial or eukaryotic genomes; no homology is required.

red alga Type of photoautotrophic protistan; most are multicelled and aquatic; an abundance of phycobilins masks their chlorophyll *a*.

red blood cell Erythrocyte. Cell that functions in the rapid transport of oxygen in blood.

red marrow Site of blood cell formation in the spongy tissue of many bones.

red tide During dinoflagellate blooms, algal concentrations reach 6 to 8 million cells per liter of water, turning it rust-red or brown.

reduced molecule A molecule to which one or more electrons have been transferred.

reflex [L. *reflectere*, to bend back] Stereotyped, simple movement in response to stimuli. In the simplest reflex arcs, it results from sensory neurons synapsing directly on motor neurons.

regulatory protein Component of mechanisms that control transcription, translation, and gene products by interacting with DNA, RNA, new polypeptide chains, or proteins (e.g., enzymes).

relaxin Hormone with role in birth process.

releaser Hypothalamic signaling molecule; enhances or slows secretion from target cells in anterior lobe of pituitary.

renal corpuscle Bowman's capsule plus the glomerular capillaries it cups around.

renal failure Nephrons of both kidneys can't perform regulatory and excretory functions.

repressor Protein that binds with an operator on bacterial DNA to block transcription.

reproduction Any process by which a parental cell or organism produces offspring. Among eukaryotes, asexual modes (e.g., binary fission, budding, vegetative propagation) and sexual modes. Prokaryotes use prokaryotic fission only. Viruses cannot reproduce themselves; host organisms execute their replication cycle.

reproductive base The number of actually and potentially reproducing individuals in a population.

reproductive isolating mechanism Heritable feature of body form, functioning, or behavior that prevents interbreeding between two or more genetically divergent populations.

reproductive success Of individuals, production of viable, fertile offspring.

reptile Carnivorous species of the amniote lineage called sauropsids, which includes birds. E.g., a crocodilian, snake, lizard, tuatara.

resource partitioning Of two or more species that compete for the same resource, a sharing of the resource in different ways or at different times, which permits them to coexist.

respiration [L. *respirare*, to breathe] For all animals, exchange of environmental oxygen with CO_2 from cells (e.g., by integumentary exchange or by way of a respiratory system).

respiratory cycle One inhalation, exhalation.

respiratory surface Thin, moist epithelium that functions in gas exchange.

respiratory system Type of organ system that exchanges gases between the body and the environment and contributes to homeostasis.

resting membrane potential Of a neuron and other excitable cells, a voltage difference across the plasma membrane that holds steady in the absence of outside stimulation.

restoration ecology Attempts to reestablish biodiversity in ecosystems severely altered by mining, agriculture, and other disturbances.

restriction enzyme One of a class of bacterial enzymes that can cut apart foreign DNA that infects a cell, as by viral attack. Important tool of recombinant DNA technology.

reticular formation A mesh of interneurons extending from the upper spinal cord, through the brain stem, and into the cerebral cortex; it is a low-level pathway of information flow.

retina Dense array of photoreceptors in an eye.

reverse transcription Synthesis of DNA on an RNA template by reverse transcriptase, a viral enzyme. Basis of RNA virus replication cycle and of cDNA synthesis in the laboratory.

Rh blood typing Method of characterizing red blood cells by a certain membrane surface protein. Rh^+ cells have it; Rh^- cells do not.

rhizoid Simple rootlike absorptive structure of some fungi and nonvascular plants.

rhizopod A group of protozoans including naked amoebas and foraminiferans. Their cytoskeletal elements change continually.

ribosomal RNA (rRNA) Type of RNA that combines with proteins to form ribosomes, on which polypeptide chains are assembled.

ribosome Two subunits of rRNA and proteins briefly joined together as a structure on which mRNA is translated into polypeptide chains.

riparian zone Narrow corridor of vegetation along a stream or river.

RNA Ribonucleic acid. Any of a class of single-stranded nucleic acids that function in transcribing and translating the genetic instructions encoded in DNA into proteins.

RNA polymerase Enzyme that catalyzes the assembly of RNA strands on DNA templates.

RNA world One model for prebiotic evolution in which RNA was the template for protein synthesis before the evolution of DNA.

rod cell Vertebrate photoreceptor sensitive to very dim light. Contributes to the coarse perception of movement across visual field.

root Plant part, typically belowground, that absorbs water and dissolved minerals, anchors aboveground parts, and often stores food.

root hair Thin extension of a young, specialized root epidermal cell. Increases root surface area for absorbing water and minerals.

root nodule Localized swelling on a root of certain legumes and other plants. Develops when nitrogen-fixing bacteria infect the plant, multiply, and become mutualists with it.

root system Underground vascular plant structures that absorb water, mineral ions.

rotifer Bilateral, cephalized animal with a false coelom, crown of cilia.

roundworm Cylindrical, bilateral, cephalized animal with a false coelom and a complete digestive system. Most cycle nutrients in communities; many are parasites.

rubisco RuBP carboxylase; an enzyme that catalyzes attachment of the carbon atom from CO_2 to RuBP and so starts the Calvin–Benson cycle of the light-independent reactions.

RuBP Ribulose bisphosphate; five-carbon organic compound; used for carbon fixation in the Calvin–Benson cycle, which regenerates it.

ruminant Hoofed, herbivorous mammal with multiple stomach chambers.

sac fungus Fungus that forms sexual spores in sac-shaped cells; e.g., truffles, morels.

salinization Salt buildup in soil through poor drainage, evaporation, and heavy irrigation.

saliva Glandular secretion that mixes with food and starts starch breakdown in mouth.

salt Compound that releases ions other than H^+ and OH^- in solution.

saltatory conduction Of myelinated neurons, a rapid form of action potential propagation. Excitation hops to nodes between jellyrolled membranes of cells making up myelin sheath.

sampling error Use of a sample or subset of a population, an event, or some other aspect of nature for an experimental group that is not large enough to be representative of the whole.

saprobe Heterotroph that obtains energy and carbon from nonliving organic matter and so causes its decay (e.g., many fungal species).

sapwood Of older stems and roots, secondary growth in between the vascular cambium and heartwood; wet, usually pale, not as strong.

sarcomere (SAR-koe-meer) One of many basic units of contraction, defined by Z lines, that subdivide a muscle cell. Shortens in response to ATP-driven interactions between its parallel arrays of actin and myosin components.

sarcoplasmic reticulum System of membrane-bound chambers; threads around muscle cells and takes up, stores, and releases the calcium ions necessary for contraction.

sauropsid One of two major amniote lineages; includes reptiles and birds.

savanna Broad belt of warm grassland with a smattering of shrubs and trees.

scale Of fishes, small, bony plates that protect the body surface without weighing it down.

Schwann cell Type of neuroglial cell that wraps around certain axons like a jellyroll. A series of these cells, separated only by small nodes, form a myelin sheath.

sclerenchyma (skler-ENG-kih-mah) Simple plant tissue that supports mature plant parts and commonly protects seeds. Most of its cells have thick, lignin-impregnated walls.

second law of thermodynamics A law of nature stating that the spontaneous direction of energy flow is from organized forms to less organized forms; with each conversion, some energy is randomly dispersed in a form (usually heat) not as useful for doing work.

second messenger Molecule within a cell that mediates a hormonal signal.

secondary oocyte Formed during meiosis; this cell gets most of the cytoplasm and a haploid number of chromosomes.

secondary sexual trait A trait associated with maleness or femaleness but with no direct role in reproduction (e.g., distribution of body hair and body fat). The primary sexual trait is the presence of male or female gonads.

secondary succession A disturbed area recovers and moves toward a climax state.

secondary wall Of older plant cells no longer growing but in need of structural support, a wall on the inner surface of the primary wall. Contains lignin in older cells of woody plants.

secretion Release of a substance from cells or glands to the external environment, to the lumen of some organ, or to interstitial fluid.

sedimentary cycle Biogeochemical cycle. An element having no gaseous phase moves from land, through food webs, to the seafloor, then returns to land through long-term uplifting.

seed Mature ovule with an embryo sporophyte inside and integuments that form a seed coat.

seed bank A safe storage facility where genes of diverse plant lineages are being preserved.

seed fern Among the earliest plants to produce seedlike structures, or seeds, in the swamplike areas during the Carboniferous.

segmentation Of animal body plans, a series of units that may or may not be similar to one another in appearance. Of tubular organs, an oscillating movement produced by rings of circular muscle in the tube wall.

segregation, theory of [L. *se-*, apart, + *grex*, herd] Mendelian theory. Sexually reproducing organisms inherit pairs of genes (on pairs of homologous chromosomes), the two genes of each pair are separated from each other at meiosis, and they end up in separate gametes.

selective gene expression Control of which gene products a cell makes or activates during a specified interval. Depends on the cell type, adjustments to changing chemical conditions, external signals, and built-in control systems.

selective permeability Of a cell membrane, a capacity to let some substances but not others cross at certain sites, at certain times owing to its bilayer structure and its transport proteins.

selfish behavior An individual protects or increases its own chance to produce offspring regardless of consequences to its social group.

selfish herd Social group held together simply by reproductive self-interest.

semen (SEE-mun) The sperm-bearing fluid expelled from a penis during male orgasm.

semiconservative replication [Gk. *hēmi*, half, + L. *conservare*, to keep] Mechanism of DNA duplication. A DNA double helix unzips and a complementary strand is assembled on exposed bases of each strand. Each conserved strand and its new partner wind up together as a double helix; a half-old, half-new molecule.

seminiferous tubule One of the small coiled tubes in testes where sperm start forming.

senescence (sen-ESS-cents) [L. *senescere*, to grow old] Processes leading to the natural death of an organism or to parts of it (e.g., leaves).

sensation Conscious awareness of a stimulus.

sensory adaptation Decrease in response to a stimulus maintained at constant strength.

sensory neuron Type of neuron that detects a stimulus and relays information about it toward an integrating center (e.g., a brain).

sensory system The "front door" of a nervous system; it detects external and internal stimuli and relays information to integrating centers that issue commands for responses.

Sertoli cell One of two cell types in seminiferous tubules; has receptors for FSH.

sessile animal (SESS-ihl) Animal that remains attached to a substrate during some stage (often the adult stage) of its life cycle.

sex chromosome A chromosome with genes that affect sexual traits. Depending on the species, somatic cells have one or two sex chromosomes of the same or different type (e.g., in mammals, XX females, XY males).

sexual dimorphism Occurrence of female and male phenotypes among the individuals of a sexually reproducing species.

sexual reproduction Production of offspring by meiosis, gamete formation, and fertilization.

sexual selection A microevolutionary process; a type of selection that favors a trait giving an individual a competitive edge in attracting or keeping a mate (favors reproductive success).

shell model Model of electron distribution in which all orbitals available to electrons of atoms occupy a nested series of shells.

shifting cultivation Cutting and burning trees in a plot of land, followed by tilling ashes into soil. Once called slash-and-burn agriculture.

shivering response Rhythmic tremors in response to cold; increases heat production.

shoot system Aboveground plant parts (e.g., stems, leaves, and flowers).

short-day plant Plant that flowers in late summer or early fall, when daylength is shorter than a critical value.

sieve-tube member One of the cells that join together as phloem's sugar-conducting tubes.

sign stimulus Simple environmental cue that triggers a response to a stimulus, which the nervous system is prewired to recognize.

signal receiver The individual responding to a communication signal.

signal reception Activation of a molecular or sensory receptor, as by reversible binding of a hormone or another signaling molecule.

signal transduction and response Conversion of an extracellular signal into a molecular or chemical form that causes a change in some activity inside the target cell.

signaler Sender of a communication signal.

sister chromatid (CROW-mah-tid) Of a duplicated chromosome, one of two DNA molecules (and associated proteins) attached at the centromere until they are separated from each other at mitosis or meiosis; each is then designated a separate chromosome.

six-kingdom classification Phylogenetic system that groups all organisms into the kingdoms Eubacteria, Archaebacteria, Protista, Fungi, Plantae, and Animalia.

skeletal muscle Organ with many muscle cells bundled inside a sheath of connective tissue.

skeletal muscle tissue Striated contractile tissue that is the functional partner of bone.

skin Of vertebrates, the integument plus structures derived from epidermal cells.

sliding filament model Model for how muscles contract, based on the ATP-driven interactions between myosin and actin filaments in each sarcomere of each muscle cell. The sarcomere shortens as actin filaments projecting from its two sides are made to slide toward its center.

slime mold Predatory protistan; amoebalike cells form fruiting bodies during life cycle.

small intestine Part of the vertebrate digestive system in which digestion is completed and most dietary nutrients are absorbed.

smog, industrial Polluted, gray-colored air in industrialized cities during cold, wet winters.

smog, photochemical Polluted, brown, smelly air that forms in warm weather in large cities that have many gas-burning vehicles.

smooth muscle tissue Nonstriated contractile tissue found in soft internal organs.

social behavior Diverse interactions among individuals of a species, which display, send, and respond to shared forms of communication that have genetic and learned components.

social parasite Species that completes life cycle by altering behavior of another species.

sodium–potassium pump Type of membrane transport protein that, when activated by ATP, selectively transports potassium ions across a membrane against its concentration gradient, and passively allows sodium ions to cross in the opposite direction.

softwood Wood with tracheids but no vessels or fibers. Weaker, less dense than hardwood.

soil Mixture of mineral particles of variable sizes and decomposing organic material; air and water occupy spaces between particles.

solar-hydrogen energy Sunlight energy captured, converted to H_2 as a fuel source.

solute (SOL-yoot) [L. *solvere*, to loosen] Any substance dissolved in a solution.

solvent Any fluid (e.g., water) in which one or more substances are dissolved.

somatic cell (so-MAT-ik) [Gk. *sōmā*, body] Any body cell that is not a germ cell. (Germ cells are the forerunners of gametes.)

somatic nervous system Nerves leading from a central nervous system to skeletal muscles.

somatic sensation Perception of touch, pain, pressure, temperature, motion, or positional changes of body parts.

somatosensory cortex Part of the outermost gray matter of the cerebral hemispheres.

somite One of many paired segments in a vertebrate embryo that give rise to most bones, skeletal muscles of head and trunk, and dermis.

song system Several brain structures that control muscles of a bird's vocal organ.

southern pine forest Dominates coastal plains of south Atlantic and Gulf states. Adapted to dry, sandy, nutrient-poor soils, and fires of relatively low heat, which can clear understory.

special sense Vision, hearing, olfaction, or another sensation that arises from a particular location, such as the eyes, ears, or nose.

speciation (spee-see-AY-shun) The formation of a daughter species from a population or subpopulation of a parent species by way of microevolutionary processes. Routes vary in their details and duration.

species (SPEE-sheez) [L. *species*, a kind] One kind of organism. Of sexually reproducing organisms, one or more natural populations in which individuals are interbreeding and are reproductively isolated from other such groups.

specific epithet A name that, combined with a genus name, designates one species alone.

sperm [Gk. *sperma*, seed] Mature male gamete.

sphere of hydration A clustering of water molecules around the molecules or ions of a substance placed in water; arises by positive and negative interactions among them.

sphincter A ring of smooth muscles that can alternatively contract and relax to close off and open a passageway to the body surface.

spinal cord Part of central nervous system, in a canal inside the vertebral column; site of direct reflex connections between sensory and motor neurons; has tracts to and from the brain.

spindle apparatus Dynamic, temporary array of microtubules that moves chromosomes in precise directions during mitosis or meiosis.

spirillum A prokaryotic cell with one or more twists in its body.

spleen A lymphoid organ that is a filtering station for blood, a reservoir of red blood cells, and a reservoir of macrophages.

sponge Animal with no body symmetry, and no tissues, and phagocytic collar cells in its body wall. Lineage dates from precambrian.

spore A reproductive or resting structure of one or a few cells, often walled or coated; used for resisting harsh conditions, dispersal, or both. May be nonsexual or sexual, as formed by way of meiosis. Some bacteria as well as sporozoans, fungi, and plants form spores.

sporophyte [Gk. *phyton*, plant] Vegetative body that produces spore-bearing structures; grows by mitotic cell divisions from a plant zygote.

sporozoan Type of parasitic protistan. Forms a motile infective stage inside specific host cells; some form cysts. One end of the body has a complex of structures used to penetrate hosts.

spring overturn Of many bodies of water (e.g., large lakes in temperate zones) in spring, a downward movement of oxygenated surface water and an upward movement of nutrient-rich water from mud and sediments below; fans primary productivity.

S-shaped curve Type of diagrammatic curve that emerges when logistic population growth is plotted against time.

stabilizing selection Mode of natural selection by which intermediate phenotypes in the range of variation are favored and extremes at both ends are eliminated.

stamen (STAY-mun) A male reproductive part of a flower (e.g., anther with stalked filament).

statolith A gravity-sensing mechanism based on clusters of particles.

STD A sexually transmitted disease.

stem cell Self-perpetuating, undifferentiated animal cell. Some of its daughter cells are self-perpetuating; others become specialized (e.g., erythrocytes from stem cells in bone marrow).

steroid hormone Lipid-soluble hormone made from cholesterol that acts on target cell DNA.

sterol (STAIR-all) Lipid with a rigid backbone of four fused carbon rings (e.g., cholesterol). Sterols differ in the number, position, and type of their functional groups.

stigma Sticky or hairy surface tissue on upper part of a carpel (or fused carpels) that captures pollen grains and favors their germination.

stimulus [L. *stimulus*, goad] A specific form of energy (e.g., pressure, light, and heat) that activates a sensory receptor able to detect it.

stoma (STOW-muh), plural **stomata** [Gk. *stoma*, mouth] A gap between two guard cells in leaf or stem epidermis. Opens or closes to control CO_2 movement into a plant and H_2O and O_2 out of it. Stomata help plants conserve water.

stomach Muscular, stretchable sac that mixes and stores ingested food, helps break it apart mechanically and chemically, and controls its expulsion (e.g., into the small intestine).

strain Compared to an organism of a known type, an organism with differences that are too minor to classify it as a separate species (e.g., *Escherichia coli* strain 018:K1:H).

stramenopile One of a major protistan group; has four outer membranes and thin tinsel-like filaments projecting from one of two flagella; e.g., a colorless water mold or downy mildew; a photosynthetic chrysophyte (golden alga, yellow-green alga, diatom, coccolithophore).

stratification Stacked layers of sedimentary rock, built up by gradual deposition of volcanic ash, silt, and other materials over time.

stream A flowing-water ecosystem that starts out as a freshwater spring or seep.

strip logging A way to minimize downside of deforestation; only a very narrow corridor that parallels contours of sloped land is cleared; the upper part is used as a road to haul away logs, then is reseeded from intact forest above it.

strobilus Of many nonflowering plants, a cluster of spore-producing structures.

stroma [Gk. *strōma*, bed] A semifluid matrix between the thylakoid membrane system and two outer membranes of a chloroplast; a zone where sucrose, starch, cellulose, and other end products of photosynthesis are assembled.

stromatolite Fossilized mats of shallow-water microbial communities, mainly cyanobacteria, from Archean to precambrian. Cell secretions blocked UV radiation but trapped sediments, and new mats grew on old ones; some are half a mile thick and hundreds of miles across.

substrate Reactant or precursor for a specific enzyme-mediated metabolic reaction.

substrate-level phosphorylation The direct, enzyme-mediated transfer of a phosphate group from a substrate to a molecule, as when an intermediate of glycolysis is made to give up a phosphate group to ADP, forming ATP.

succession, primary (suk-SESH-un) [L. *succedere*, to follow after] Ecological pattern by which a community develops in orderly progression, from the time that pioneer species colonize a barren habitat to the climax community.

succession, secondary Ecological pattern by which a disturbed area of a community recovers and moves back toward the climax state.

suppressor T cell Issues signals that help end an immune response.

surface-to-volume ratio Mathematical relation in which the volume of an object expands in three dimensions (e.g., length, width, depth) but its surface area expands in only two dimensions; a constraint on cell size and shape.

survivorship curve Plot of age-specific survival of a group of individuals in the environment, from the time of birth until the last one dies.

swim bladder Adjustable flotation sac that helps many fishes maintain neutral buoyancy in water; its volume changes as it exchanges gases with blood.

symbiosis (sim-by-OH-sis) [Gk. *sym*, together, + *bios*, life, mode of life] Individuals of one species live near, in, or on individuals of a different species for at least part of life cycle (e.g., as in commensalism, mutualism, and parasitism).

sympathetic nerve An autonomic nerve; deals mainly with increasing overall body activities at times of heightened awareness, excitement, or danger; works continually in opposition with parasympathetic nerves to make minor adjustments in internal organ activities.

sympatric speciation [Gk. *sym*, together, + *patria*, native land] A speciation event within the home range of an existing species, in the absence of a physical barrier. Such species may form instantaneously, as by polyploidy.

synapsid One of two major amniote lineages; includes mammals, mammal-like reptiles.

synaptic integration (sin-AP-tik) Moment-by-moment combining of all excitatory and inhibitory signals arriving at the trigger zone of a neuron or some other excitable cell.

syndrome A set of symptoms that may not individually be a telling clue but collectively characterize a genetic disorder or disease.

systemic circuit (sis-TEM-ik) Of vertebrates, a cardiovascular route in which oxygen-enriched blood flows from the heart through the rest of the body (where it gives up oxygen and takes up carbon dioxide), then back to the heart.

T cell T lymphocyte; a type of white blood cell vital to immune responses (e.g., helper T cells and cytotoxic T cells).

tactile display In intraspecific communication; a ritualized contact between individuals.

tandem repeat One of many short sequences of DNA, occurring one after the other, in a chromosome. Used in DNA fingerprinting.

taproot system A primary root together with all of its lateral branchings.

target cell Any cell with molecular receptors that can bind a particular signaling molecule.

taste receptor A chemoreceptor for substances dissolved in fluid bathing it.

taxonomy Field of biology that deals with identifying, naming, and classifying species.

TCR Antigen-binding receptor of T cells.

tectum Midbrain's roof. Coordinating center for most sensory inputs and initiating motor responses in fishes and amphibians. In most vertebrates (not mammals), a reflex center that relays sensory input to forebrain.

telomere A cap of repetitive DNA sequences at end of chromosome; reduced at each nuclear division. Division stops when only a bit is left.

telophase (TEE-low-faze) Of meiosis I, a stage when one member of each pair of homologous chromosomes has arrived at a spindle pole. Of mitosis and of meiosis II, the stage when chromosomes decondense into threadlike structures and two daughter nuclei form.

temperature A measure of the kinetic energy of ions or molecules in a specified region.

temperature zone Globe-spanning bands of temperature defined by latitude (e.g., cool temperate, warm temperate, equatorial).

tendon A cord or strap of dense connective tissue that attaches a muscle to bone.

territory An area that an animal is defending against competitors for mates, food, water, living space, and other resources.

test A means to determine the accuracy of a prediction, as by conducting experimental or observational tests and by developing models. Scientific tests are conducted under controlled conditions in nature or the laboratory.

testcross Experimental cross to determine whether an individual of unknown genotype that shows dominance for a trait is either homozygous dominant or heterozygous.

testis, plural **testes** A type of primary male reproductive organ (gonad); it produces male gametes and sex hormones.

testosterone (tess-TOSS-tuh-rown) A type of sex hormone necessary for the development and functioning of the male reproductive system of vertebrates.

tetanus (TET-uh-nuss) Of a muscle, a large contraction; repeated stimulation of a motor unit causes muscle twitches to run together. In a disease by the same name, a toxin prevents muscles from being released from contraction.

tetrapod Vertebrate that is a four-legged walker or a descendant of one.

thalamus (THAL-uh-muss) A forebrain region that is a coordinating center for sensory input and a relay station for signals to the cerebrum.

theory, scientific An explanation of the cause or causes of range of related phenomena. It has been rigorously tested but is still open to tests, revision, and tentative acceptance or rejection.

thermal inversion A layer of dense, cool air trapped beneath a layer of warm air; can hold air pollutants close to the ground.

thermoreceptor Sensory cell or specialized cell next to it that detects radiant energy (heat).

thigmotropism (thig-MOE-truh-pizm) [Gk. *thigm*, touch] Orientation of the direction of growth in response to physical contact with a solid object (e.g., a vine curls around a post).

thirst center Hypothalamic region, along with ADH-secreting cells; reacts to osmoreceptors for blood volume and solute levels.

threat display Ritualized intraspecific signal conveying intent to attack.

three-domain system Classification system with three primordial branchings: Eubacteria (bacteria), Archaebacteria (archeans), and Eukarya (protistans, plants, fungi, animals).

thylakoid Inner chloroplast membrane often folded as interconnected flattened sacs; forms a single compartment for hydrogen ions. Light-trapping pigments, and enzymes used to form ATP, NADPH, or both, are embedded in it.

thymine A nitrogen-containing base; one of the nucleotides in DNA (not in RNA).

thymus gland Lymphoid organ; its hormones influence T cells, which form in bone marrow but migrate to and differentiate in the thymus.

thyroid gland Endocrine gland; its hormones influence overall growth, development, and metabolic rates of warm-blooded animals.

tidal volume The volume of air flowing into or out of the lungs during the respiratory cycle.

tight junction Cell junction where strands of fibrous proteins oriented in parallel with a tissue's free surface collectively block leaks between the adjoining cells.

tissue Of multicelled organisms, a group of cells and intercellular substances that function together in one or more specialized tasks.

tissue culture propagation Inducing a tissue or organism to grow from an isolated cell of a parent tissue growing in a culture medium.

titin Thin, elastic protein that keeps myosin filaments centered in sarcomere; also enables relaxed muscles to passively resist stretching.

tongue Organ of food positioning, swallowing, speech; membrane-covered skeletal muscles.

tonicity (toe-NISS-ih-TEE) The relative solute concentrations of two fluids (e.g., cytoplasmic fluid relative to extracellular fluid).

tooth A hard, bony appendage in the mouth, consisting of inner pulp, dentin, and enamel.

topsoil Uppermost soil layer, the one most essential for plant growth.

torsion During the development of certain mollusks, a drastic twisting of the body, including the visceral mass.

total fertility rate (TFR) Average number of children born to women of a population during their reproductive years.

touch-killing Mechanism by which cytotoxic T cells directly release perforins and toxins onto a target cell and cause its destruction.

toxin Normal metabolic product of a species that can hurt or kill a different species.

trace element Any element that represents less than 0.01 percent of body weight.

tracer Substance with an attached radioisotope that researchers can track after delivering it into a cell, body, ecosystem, or some other system. Its emissions are detected as it moves through a pathway or reaches a destination.

trachea (TRAY-kee-uh), plural **tracheae** An air-conducting tube of respiratory systems. Of land vertebrates, the windpipe through which air passes between the larynx and bronchi.

tracheal respiration Of certain invertebrates (e.g., insects), respiration by way of finely branching tubes that start at openings in the integument and dead-end in body tissues.

tracheid (TRAY-kid) One of two types of cells in xylem that conduct water and mineral ions.

tract A cordlike bundle of axons of sensory neurons, motor neurons, or both inside the brain or spinal cord. Comparable to a nerve.

transcription [L. *trans*, across, + *scribere*, to write] First stage of protein synthesis. An RNA strand is assembled on exposed bases of an unwound strand of a DNA double helix. The transcript has a complementary base sequence.

transfer RNA (tRNA) An RNA that binds with and delivers amino acids to a ribosome and that pairs with an mRNA codon during the translation stage of protein synthesis.

transition state The point when a reaction can run either to product or back to reactant.

translation Stage of protein synthesis when an mRNA's base sequence becomes converted to the amino acid sequence of a new polypeptide chain by rRNA, tRNA, and mRNA interactions.

translocation Of cells, movement of a stretch of DNA to a new chromosomal location with no molecular loss. Of vascular plants, distribution of organic compounds by way of phloem.

transpiration Evaporative water loss from a plant's aboveground parts, leaves especially.

transport protein Membrane protein that passively or actively assists specific molecules or ions across a membrane's lipid bilayer. The solutes pass through a channel in its interior.

transposon DNA segment that can randomly move to different locations in a genome; may inactivate genes into which it inserts itself and cause changes in phenotype.

triglyceride (neutral fat) A neutral fat; a lipid with three fatty acid tails attached to a glycerol unit. Triglycerides are the animal body's most abundant lipid and its richest energy source.

trisomy Presence of three chromosomes of a given type in a cell, not the two characteristic of a parental diploid chromosome number.

trophic level (TROE-fik) All organisms the same number of transfer steps away from the energy input into an ecosystem.

tropical rain forest A biome characterized by regular, heavy rainfall, an annual mean temperature of 25°C, humidity of 80+ percent, and stunning but threatened biodiversity.

tropism (TROE-pizm) Of plants, a directional growth response to an environmental factor (e.g., growth toward light).

true breeding lineage Of sexually reproducing species, a lineage in which only one version of a trait appears over the generations in all parents and their offspring.

tubular reabsorption Of nephrons, a process by which peritubular capillaries reclaim water and solutes from the proximal tubule.

tubular secretion Of nephrons, excess ions, metabolites, neurotransmitters, histamine, and toxins cross the nephron wall through protein transporters and are excreted.

tumor *See* neoplasm.

tundra Biome with poor drainage, extremely low temperatures, short growing season, poor decomposition; arctic tundra is a treeless plain between the polar ice cap and belts of northern boreal forests; alpine tundra occurs at high elevations in mountains throughout the world.

tunicate A baglike, filter-feeding urochordate.

turgor pressure (TUR-gore) Type of internal fluid pressure; it acts against a cell wall when water moves into the cell by osmosis.

ultrafiltration Bulk flow of a small amount of protein-free plasma from a blood capillary when the outward-directed effect of blood pressure exceeds the inward-directed osmotic movement of interstitial fluid.

ultraplankton Photosynthetic bacteria, no more than 2μm wide, that may account for 70 percent of the ocean's primary productivity.

uniformity theory Theory that Earth's surface changes in slow, uniformly repetitive ways except for catastrophes that occur annually, such as large earthquakes and floods. Helped change Darwin's view of evolution. Has since been replaced by plate tectonics theory.

upwelling An upward movement of deep, nutrient-rich water along coasts; it replaces surface waters that move away from shore when the prevailing wind direction shifts.

uracil (YUR-uh-sill) Nitrogen-containing base of a nucleotide in RNA but not DNA. Like thymine, uracil can base-pair with adenine.

urea Waste product formed in the liver from ammonia (derived from protein breakdown) and CO_2; excreted in urine.

ureter One of a pair of tubes that conduct urine from the kidneys to the urinary bladder.

urethra Urine-conducting tube from urinary bladder to an opening at the body's surface.

urinary bladder The distensible sac in which urine is stored before being excreted.

urinary excretion Mechanism by which excess water and solutes are removed from the body through the urinary system.

urinary system Organ system that adjusts the volume and composition of blood, and thereby helps maintain extracellular fluid.

urine Fluid consisting of any excess water, wastes, and solutes; it forms in kidneys by filtration, reabsorption, and secretion.

urochordate A baglike chordate with a caudal notochord. Most of these aquatic animals have either a rudimentary brain or none at all.

uterus (YOU-tur-us) [L. *uterus*, womb] Of a female placental mammal, a muscular, pear-shaped organ in which embryos are contained and nurtured during pregnancy.

vaccination Immunization procedure against a specific pathogen.

vaccine An antigen-containing preparation designed to increase immunity to diseases.

vagina Of female mammalian reproductive system, organ that receives sperm, forms part of birth canal, and channels menstrual flow.

variable Of an experimental test, a specific aspect of an object or event that may differ over time and among individuals. A single variable is directly manipulated in an attempt to support or disprove a prediction.

vascular bundle Array of primary xylem and phloem in multistranded, sheathed cords that thread lengthwise in the ground tissue system.

vascular cambium A lateral meristem that increases stem or root diameter.

vascular cylinder Arrangement of vascular tissues as a central cylinder in roots.

vascular plant Plant with xylem, phloem, and usually well-developed roots, stems, and leaves.

vascular tissue system Xylem and phloem, the conducting tissues that distribute water and solutes through a vascular plant.

vasoconstriction The diameter of one or more blood vessels decreases in response to stimuli.

vasodilation The diameter of one or more blood vessels increases in response to stimuli.

vegetative growth A new plant grows from an extension or fragment of its parent.

vein Of a cardiovascular system, any of the large-diameter vessels that lead back to the heart. Of leaves, one of the vascular bundles that thread through photosynthetic tissues.

venule A small blood vessel that serves as a transitional conducting tube between a small-diameter capillary and a larger diameter vein.

vernalization Low-temperature stimulation of flowering.

vertebra, plural **vertebrae** One of a series of hard bones that serve as a backbone and that protect the spinal cord.

vertebrate Animal having a backbone.

vesicle (VESS-ih-kul) [L. *vesicula*, little bladder] One of various small, membrane-bound sacs in cytoplasm that function in the transport, storage, or digestion of substances.

vessel member Type of cell in xylem; dead at maturity, but its wall becomes part of a water-conducting pipeline (a vessel).

vestibular apparatus Organ of equilibrium.

vestigial (ves-TIDJ-ul) Applies to a small body part, tissue, or organ abnormally developed or degenerated and unable to function like its normal counterpart (e.g., vestigial wings of mutant fruit flies; "tail bones" of humans).

villus (VIL-us), plural **villi** Any of several fingerlike absorptive structures projecting from the free surface of an epithelium.

viroid Infectious particle of short, tightly folded strands or circles of RNA.

virus A noncellular infectious agent made of DNA or RNA, a protein coat and, in some, an outer lipid envelope; it can be replicated only after its genetic material enters a host cell and subverts the host's metabolic machinery.

viscera All soft organs inside an animal body (e.g., heart, lungs, and stomach).

vision Perception of visual stimuli. Light is focused on a dense layer of photoreceptors that sample details of a light stimulus, which are sent to the brain for image formation.

visual accommodation Adjustments in lens position or shape that focus light on a retina.

visual field The part of the outside world that an animal sees.

visual signal An observable action or cue that functions as a communication signal.

vital capacity Air volume expressed from the lungs in one breath after maximal inhalation.

vitamin Any of more than a dozen organic substances that an organism requires in small amounts for metabolism but that it generally cannot synthesize for itself.

vitamin D Calciferol, a fat-soluble vitamin; helps body absorb calcium.

vocal cord Of certain animals, one of the thickened, muscular folds of the larynx that help produce sound waves for vocalization.

warning coloration Of many toxic species and their mimics, avoidance signals (strong colors and patterns) that predators learn to recognize.

wastewater treatment Processing of liquid wastes to remove sludge, organic matter, etc.

water mold A stramenopile; mostly saprobic decomposers of aquatic habitats.

water potential Sum of two opposing forces (osmosis and turgor pressure) that can cause a directional movement of water into or out of an enclosed volume (e.g., a walled cell).

water table Upper limit at which ground in a specified region is fully saturated with water.

water–vascular system Of sea stars and sea urchins, a system of many tube feet that are deployed in synchrony for smooth locomotion.

watershed Any specified region in which all precipitation drains into one stream or river.

wavelength A wavelike form of energy in motion. The horizontal distance between the crests of every two successive waves.

wax Organic compound; long-chain fatty acids packed together and attached to long-chain alcohols or carbon rings. Waxes have a firm consistency and repel water.

whisk fern Seedless vascular plant.

white blood cell A leukocyte (e.g., eosinophil, neutrophil, macrophage, T or B cell) that, with chemical mediators, counters tissue invasion and tissue damage. Different kinds take part in nonspecific and specific defense responses.

white matter Part of brain and spinal cord; its axons, in glistening white sheaths, specialize in rapid signal transmission.

wild-type allele Allele that occurs normally or with the greatest frequency at a given gene locus among individuals of a population.

wind farm Collection of turbines used to convert mechanical energy into electricity.

wing A body part that functions in flight, as among birds, bats, and many insects.

X chromosome A type of sex chromosome. An XX mammalian embryo becomes female; an XY pairing causes it to develop into a male.

X chromosome inactivation One of the two X chromosomes in somatic cells of mammalian females condenses, which inactivates most of its genes. A dosage compensation mechanism.

X-linked gene A gene on an X chromosome.

X-linked recessive inheritance Recessive condition in which the responsible, mutated gene is on the X chromosome.

X-ray diffraction image Pattern that forms on film exposed to x-rays that have been directed at a molecule; reveals positions of atoms, not the molecular structure.

xenotransplantation The transfer of an organ from one species to another.

xylem (ZYE-lum) [Gk. *xylon*, wood] Of vascular plants, a complex tissue that conducts water and solutes through pipelines of interconnected walls of cells, which are dead at maturity.

Y chromosome Distinctive chromosome in males or females of many species, but not both (e.g., human males XY; human females, XX).

yellow alga A chrysophyte common in phytoplankton.

yellow marrow A fatty tissue in the cavities of most mature bones that produces red blood cells when blood loss from the body is severe.

Y-linked gene Gene on a Y chromosome.

yolk Protein- and lipid-rich substance that nourishes embryos in animal eggs.

yolk sac Extraembryonic membrane. In most shelled eggs, it holds nutritive yolk; in humans, part becomes a blood cell formation site, and some cells give rise to forerunners of gametes.

zero population growth Of a population, no overall increase or decrease during a specified interval; population size is stabilized.

zooplankton A community of suspended or weakly swimming heterotrophs of aquatic habitats. Most species are microscopic.

zygomycetes Parasitic or saprobic fungus that forms a thick-walled sexual spore (a diploid zygote) in a thin cover (the zygosporangium).

zygospore Sexual spore of zygomycetes.

zygote (ZYE-goat) First cell of a new individual, formed by fusion of a sperm nucleus with egg nucleus at fertilization; a fertilized egg.

ART CREDITS AND ACKNOWLEDGMENTS

This page is an extension of the copyright page. We made every effort to trace ownership of all copyrighted material and to secure permission from copyright holders. Should any question arise concerning the use of any material, we will make necessary corrections in future printings. Many thanks to the illustrators who rendered our art and to authors, publishers, and agents for granting permission to use their material.

TABLE OF CONTENTS
Page iv Above left, © Gary Head; above right, model by Dr. David B. Goodin, The Scripps Research Institute; below left, DigitalVision/Picture Quest. **Page v** Mary Osborn, Max Planck Institute for Biophysical Chemistry, Goettingen, FRG. **Page vi** Larry West/FPG. **Page vii** Above, Lisa Starr; below, © 2002 PhotoDisc, computer enhanced by Lisa Starr. **Page ix** Left to right, Lisa Starr with John McNamara, www.paleodirect.com; © Alfred Kamajian; Jean Paul Tibbles; Christopher Ralling. **Page x** Left to right, K. G. Murti/Visuals Unlimited; © Oliver Meckes/Photo Researchers, Inc.; Robert C. Simpson/Nature Stock; © Jim Christensen, Fine Art Digital Photographic Images. **Page xi** Above left to right, © 2002 PhotoDisc/Getty Images; Stanley Sessions, Hartwick College; Pieter Johnson; below left, Jane Burton/Bruce Coleman. **Page xii** Left to right, © David Parker/SPL/Photo Researchers, Inc.; © John Lotter Gurling/Tom Stack & Associates; David Cavagnaro/Peter Arnold, Inc.; Juergen Berger, Max Planck Institute for Developmental Biology, Tuebingen, Germany. **Page xiii** Left, © Cory Gray; right, © David Scharf/Peter Arnold, Inc. **Page xiv** Left to right, Eric A. Newman; © 2002 Ken Usami/PhotoDisc/Getty Images; Kenneth Garrett/National Geographic Image Collection. **Page xiv** Left to right, Bone Clones®, www.boneclones.com; Dr. John D. Cunningham/Visuals Unlimited; NSIBC/SPL/Photo Researchers, Inc. **Page xvi** Left, © micrograph Ed Reschke; right, Gunter Ziesler/Bruce Coleman. **Page xvii** Left to right, Dr. Maria Leptin, Institute of Genetics, University of Koln, Germany; © Minden Pictures; Nigel Cook/Daytona Beach News Journal/Corbis Sygma. **Page xviii** Left to right, © W. Perry Conway/CORBIS; Australian Broadcasting Company; Robert Vrijenhoek, MBARI. **Page xix** Gerry Ellis/The Wildlife Collection

INTRODUCTION NASA Space Flight Center

CHAPTER 1
1.1 John McColgan, Bureau of Land Management. **1.2** Lisa Starr. **1.3** All, Jack de Coningh. **1.4** © Y. Arthrus-Bertrand/Peter Arnold, Inc. **1.5** Left, art, Gary Head; photographs, left, Paul DeGreve/FPG; right, Norman Meyers/Bruce Coleman. **1.6** Art, Lisa Starr; photographs, (a) Walt Anderson/Visuals Unlimited; (b) Gregory Dimijian/Photo Researchers, Inc.; (c) Alan Weaving/Ardea, London. **1.7** Page 8, clockwise from above, © Lewis Trusty/Animals Animals; *Emiliania huxleyi* photograph, Vita Pariente, scanning electron micrograph taken on a Jeol T330A instrument at Texas A&M University Electron Microscopy Center; Carolina Biological Supply Company; R. Robinson/Visuals Unlimited, Inc.; © Oliver Meckes/Photo Researchers, Inc.; © James Evarts; art, Gary Head; Page 9, clockwise from above left, © John Lotter Gurling/Tom Stack & Associates; Edward S. Ross; Edward S. Ross; Robert C. Simpson/Nature Stock; © Stephen Dalton/Photo Researchers, Inc.; CNRI/SPLPhoto Researchers, Inc.; © P. Hawtin, University of Southampton/SPL/Photo Researchers, Inc.; Gary Head. **1.8** All, J. A. Bishop and L. M. Cook. **1.9** (All) Courtesy Derrell Fowler, Tecumseh, Oklahoma. **1.10** Gary Head. **1.11** All, © Gary Head. **1.12** Art, Gary Head and Preface, Inc.; photographs, Dr. Douglas Coleman, The Jackson Laboratory. **1.13** Courtesy Dr. Michael S. Donnenberg. **1.14** © JH Pete Carmichael Jr./NHPA. **Page 18** © Digital Vision/PictureQuest

Page 19 UNIT I © Cabisco/Visuals Unlimited

CHAPTER 2
2.1 (a) Art, Lisa Starr; (b) Jack Carey. **2.2** Lisa Starr. **2.3** Photograph, Gary Head; art, Lisa Starr. **2.4** (a) © John Greim/MediChrome; (b, c) Art, Raychel Ciemma; (d) Dr. Harry T. Chugani, M.D., UCLA School of Medicine. **2.5** Micrograph, Maris and Cramer. **2.6, 2.7** Lisa Starr. **2.8** Gary Head and Lisa Starr. **2.9** (a,b) Art, Lisa Starr; micrograph, © Bruce Iverson. **2.10** Vandystadt/Photo Researchers, Inc. **2.11** (a) Lisa Starr with PDB files from NYU Scientific Visualization Lab; (b–d) Lisa Starr with PDB ID:1BNA; H. R. Drew, R. M. Wing, T. Takano, C. Broka, S. Tanaka, K. Itakura, R. E. Dickerson; Structure of a B-DNA Dodecamer. Conformation and Dynamics; PNAS V. 78 2179, 1981. **2.12** (a–c) Art, Lisa Starr with PDB file from NYU Scientific Visualization Lab; (b, right) photographs © Steve Lissau/Rainbow; (c, right) © Kennan Ward/CORBIS. **2.13** (a) H. Eisenbeiss/Frank Lane Picture Agency; (b) art, Lisa Starr. **2.14** Lisa Starr. **2.15** Art, Lisa Starr; photographs, © 2000 PhotoDisc, Inc. **2.16** Michael Grecco/Picture Group. **Page 33** Left, Lisa Starr; right, Raychel Ciemma

CHAPTER 3
3.1 Above, Dave Schiefelbein; below, © Ron Sanford/Michael Agliolo/Photo Researchers. **Page 35** NASA. **Page 36** Left, Lisa Starr; Right-top, Lisa Starr with PDB file from NYU Scientific Visualization Lab; Right-bottom, Lisa Starr with PDB file from Klotho Biochemical Compounds Declarative Database. **3.2** Lisa Starr with PDB file from NYU Scientific Visualization Lab. **3.3** Lisa Starr with PDB ID: 1BBB; Silva, M. M., Rogers, P. H., Arnone, A.; A third quaternary structure of human hemoglobin A at 1.7-Å resolution; J Biol Chem 267 pp. 17248 (1992). **3.4** Lisa Starr with PDB ID: 1BBB; Silva, M. M., Rogers, P. H., Arnone, A.; A third quaternary structure of human hemoglobin A at 1.7-Å resolution; J Biol Chem 267 pp. 17248 (1992). Electrostatic potential calculated using Chime. **3.5** Art, Gary Head. **3.6** Photograph, Tim Davis/Photo Researchers, Inc.; art, Lisa Starr. **3.7, 3.8, 3.9** Lisa Starr. **3.10** Art and photograph, Lisa Starr. **3.11** © David Scharf/Peter Arnold, Inc. **3.12** (a) Art, Precision Graphics; (b) Art, Lisa Starr with PDB file courtesy of Dr. Christina A. Bailey, Department of Chemistry & Biochemistry, California Polytechnic State University, San Luis Obispo, CA. **3.13** (a) Art, Precision Graphics; (b) Clem Haagner/Ardea, London. **3.14** (a) Lisa Starr with PDB file courtesy of Dr. Christina A. Bailey, Department of Chemistry & Biochemistry, California Polytechnic State University, San Luis Obispo, CA; (b) Art, Precision Graphics; (c) Lisa Starr. **3.15** (a) Art, Precision Graphics; (b) Larry Lefever/Grant Heilman Photography, Inc.; (c) Kenneth Lorenzen. **Page 43** Lisa Starr. **3.16** All photographs © 2002 PhotoDisc/Getty Images; art, Gary Head. **3.17** Lisa Starr (bottom) with PDB files from NYU Scientific Visualization Lab. **3.18** (a–d) Lisa Starr with PDB files from NYU Scientific Visualization Lab; (e) Lisa Starr. **3.19** (a) Lisa Starr with PDB ID: 1BBB; Silva, M. M., Rogers, P. H., Arnone, A.; A third quaternary structure of human hemoglobin A at 1.7-Å resolution; J Biol Chem 267 pp. 17248 (1992); (b) Lisa Starr After: Introduction to Protein Structure 2nd ed., Branden & Tooze, Garland Publishing, Inc. **3.20** (a–b) Photo, Al Giddings/Images Unlimited; art, Lisa Starr with PDB ID: 1AKJ; Gao, G. F., Tormo, J., Gerth, U. C., Wyer, J. R., McMichael, A. J., Stuart, D. I., Bell, J. I., Jones, E. Y., Jakobsen, B. K.; Crystal structure of the complex between human CD8alpha(alpha) and HLA-A2; Nature 387 pp. 630 (1997). **3.21** (a,b) Lisa Starr with PDB ID: 1BBB; Silva, M. M., Rogers, P. H., Arnone, A.; A third quaternary structure of human hemoglobin A at 1.7-Å resolution; J Biol Chem 267 pp. 17248 (1992); (c,d) Gary Head and Lisa Starr with PDB files from New York University Scientific Visualization Center. **3.22** (a) art, Preface, Inc.; (b,c) Stanley Flegler/Visuals Unlimited. **3.23** Gary Head and Lisa Starr with PDB files from Klotho Biochemical Compounds Declarative Database. **3.24** (a) Gary Head; (b) Lisa Starr. **3.25** Lisa Starr with PDB ID:1BNA; H. R. Drew, R. M. Wing, T. Takano, C. Broka, S. Tanaka, K. Itakura, R. E. Dickerson; Structure of a B-DNA Dodecamer. Conformation and Dynamics; PNAS V. 78 2179, 1981. **3.26** © 2002 Charlie Waite/Stone/Getty Images. **Page 53** Lisa Starr

CHAPTER 4
4.1 (a) © Bettmann/Corbis; (b) Armed Forces Institute of Pathology; (c) National Library of Medicine; (e) *Monster Soup,* 1828 by William Heath, 1800–1840. **4.2** (a) Manfred Kage/Bruce Coleman; (b) George J. Wilder/Visuals Unlimited **4.3** (a) Precision Graphics; (b) Raychel Ciemma; (c) Lisa Starr. **4.4** Lisa Starr (see citation for 5.5). **4.5** Photograph, Hans Pfletschinger; art, Raychel Ciemma. **4.6** (a) Leica Microsystems, Inc., Deerfield, IL; (b) Art, Gary Head. **4.7** (a) Photograph, George Musil/Visuals Unlimited; art, Gary Head. **4.8** Art, Raychel Ciemma; micrograph, Driscoll, Youngquist, Baldeschwieler/CalTech/Science Source/Photo Researchers, Inc. **4.9** Jeremy Pickett-Heaps, School of Botany, University of Melbourne. **4.10** Raychel Ciemma and Precision Graphics. **4.11** Micrographs, Stephen Wolfe; art, Raychel Ciemma. **4.12** (a, left) Don W. Fawcett/Visuals Unlimited; (a, right) A. C. Faberge, *Cell and Tissue Research,* 151:403–415, 1974; Art, Raychel Ciemma; (b) art, Lisa Starr. **4.13** Art, Raychel Ciemma and Precision Graphics. **4.14** Micrographs (a,b) Don W. Fawcett/Visuals Unlimited; art, Raychel Ciemma. **4.15** Art left, Raychel Ciemma; right, Robert Demarest after a model by J. Kephart; micrograph, Gary Grimes. **4.16** Micrograph, Keith R. Porter; art, Raychel Ciemma. **4.17** Art above, Lisa Starr; micrograph, L. K. Shumway. **4.18** Art, Raychel Ciemma and Preface, Inc.; micrograph, M. C. Ledbetter, Brookhaven National Laboratory **4.19** Micrograph, G. L. Decker; art, Raychel Ciemma and Preface, Inc. **4.20** (a) © J. W. Shuler/Photo Researchers, Inc.; (b) Courtesy Dr. Vincenzo Cirulli, Laboratory of Developmental Biology, The Whittier Institute for Diabetes, University of California, San Diego, California; (c) Courtesy Mary Osborn, Max Planck Institute for Biophysical Chemistry, Goettingen, FRG. **4.21** Lisa Starr. **4.22** (a) John Lonsdale, www.johnlonsdale.net; (b) David C. Martin, PhD. **4.23** Lisa Starr (see citation for 4.24). **4.24** (a–b) Lisa Starr with PDB ID:3KIN; Kozielski, F., Sack, S., Marx, A., Thormahlen, M., Schonbrunn, E., Biou, V., Thompson, A., Mandelkow, E. M., Mandelkow, E.: The crystal structure of dimeric kinesin and implications for microtubule-dependent motility. Cell 91 pp. 985 (1997). **4.25** Photographs (a) © Lennart Nilsson; (b) CNRI/SPL/Photo Researchers, Inc.; art, Precision Graphics after Stephen Wolfe, *Molecular and Cellular Biology,* Wadsworth, 1993; (c) © Andrew Syred/SPL/Photo Researchers, Inc.; (d) art, Precision Graphics after Stephen L. Wolfe, *Molecular and Cellular Biology,* Wadsworth, 1993. **4.26** Ronald Hoham, Dept. of Biology, Colgate University. **4.27** Photographs, (a) Walter Hodge/ Peter Arnold, Inc.; (c) courtesy of Burlington Mills; micrograph, Biophoto Associates/Photo Researchers; art, Raychel Ciemma. **4.28** (a) George S. Ellmore; (b) Ed Reschke. **4.29** Raychel Ciemma and Lisa Starr. **4.30** (a) Art, Lisa Starr; (b) courtesy Dr. G. Cohen–Bazire; (c) K. G. Murti/Visuals Unlimited; (d) R. Calentine/Visuals Unlimited; (e) Gary Gaard, Arthur Kelman. **Page 78** Left, Raychel Ciemma; right, Lisa Starr. **Page 79** Lisa Starr

CHAPTER 5
5.1 (a) © Abraham Menashe/humanistic-photography.com; (b) Lisa Starr with PDB ID:1B0U; L.-W. Hung, I. X. Wang, K. Nikaido, P.-Q. Liu, G. F.-L. Ames, S.-H. Kim; Crystal Structure of the ATP-Binding Subunit of an Abc Transporter; Nature 396 pp. 703, 1998. **5.2** (a,b) Art, Precision Graphics; (c) art, Raychel Ciemma. **5.3** Lisa Starr (see citation for 5.5). **5.4** Lisa Starr. **5.5** Lisa Starr with (bovine ATPase) PDB ID:18HE; Menz, R. I., Walker, J. E., Leslie, A. G. W.: Structure of Bovine Mitochondrial F1-ATPase with Nucleotide Bound to All Three Catalytic Sites: Implications for the Mechanism of Rotary Catalysis. Cell (Cambridge, Mass.) 106 pp. 331 (2001); (calcium

pump) PDB ID:1EUL; Toyoshima, C., Nakasako, M., Nomura, H., Ogawa, H.: Crystal Structure of the Calcium Pump of Sarcoplasmic Reticulum at 2.6 Angstrom Resolution. Nature 405 pp. 647 (2000); (integrin) PDB ID:1JV2; Xiong, J.-P., Stehle, T., Diefenbach, B., Zhang, R., Dunker, R., Scott, D. L., Joachimiak, A., Goodman, S. L., Arnaout, M. A.: Crystal Structure of the Extracellular Segment of Integrin {Alpha}V{Beta}3 Science 294 pp. 339 (2001); (human growth hormone receptor) PDB ID:1A22; Clackson, T., Ultsch, M. H., Wells, J. A., de Vos, A. M.: Structural and functional analysis of the 1:1 growth hormone:receptor complex reveals the molecular basis for receptor affinity. J Mol Biol 277 pp. 1111 (1998); (gap junction) After: Vinzenz M. Unger, Nalin M. Kumar, Norton B. Gilula, Mark Yeager. Science 283: 1176, 1999. Three-Dimensional Structure of a Recombinant Gap Junction Membrane Channel. Computer graphics: M. Pique and M. Yeager, The Scripps Research Institute; (Glut1) PDB ID:1JA5; Zuniga, F. A., Shi, G., Haller, J. F., Rubashkin, A., Flynn, D. R., Iserovich, P., Fischbarg, J.: A Three-Dimensional Model of the Human Facilitative Glucose Transporter Glut1 J Mol Biol.Chem. 276 pp. 44970 (2001). **Page 86** Raychel Ciemma after Pinto daSilva, D. Branton, *Journal of Cell Biology,* 45:98, by permission of The Rockefeller University Press. **5.6** Gary Head and Precision Graphics. **5.7** Lisa Starr. **5.8** Raychel Ciemma and Gary Head. **5.9** Lisa Starr and Gary Head. **5.10** Lisa Starr. **5.11** Lisa Starr After: David H. MacLennan, William J. Rice and N. Michael Green, "The Mechanism of Ca2+ Transport by Sarco (Endo)plasmic Reticulum Ca2+-ATPases." JBC Volume 272, Number 46, Issue of November 14, 1997 pp. 28815-28818. **5.12** Precision Graphics. **5.13** (a) Art, Raychel Ciemma; (b) M. Sheetz, R. Painter, and S. Singer, *Journal of Cell Biology,* 70:193 (1976) by permission, The Rockefeller University Press. **5.14** Lisa Starr. **5.15** Raychel Ciemma. **5.16** Lisa Starr. **5.17** Micrographs (a–d) M. M. Perry and A. M. Gilbert; (e) Courtesy John Heuser/www.cellbio.wustl.edu. **5.18** (a) © Juergen Berger/Max Planck Institute/SPL/Photo Researchers, Inc.; (b) Art, Raychel Ciemma. **5.19** Raychel Ciemma. **5.20** Lisa Starr (see citation for 5.5). **5.21** © Prof. Marcel Bessis/SPL/Photo Researchers, Inc. **5.22** Frieder Sauer/Bruce Coleman

CHAPTER 6

6.1 Lisa Starr with (SOD) PDB ID: 1CBJ; Hough, M. A., Hasnain, S. S.: Crystallographic structures of bovine copper-zinc superoxide dismutase reveal asymmetry in two subunits: functionally important three and five coordinate copper sites captured in the same crystal. J Mol Biol 287 pp. 579 (1999); (Catalase) PDB ID: 1DGF; Putnam, C. D., Arvai, A. S., Bourne, Y., Tainer, J. A.: Active and Inhibited Human Catalase Structures: Ligand and Nadph Binding and Catalytic Mechanism J Mol Biol. 296 pp. 295 (2000). **6.2** (a,b) Photographs by Gary Head; (c) art, Raychel Ciemma. **6.3** Evan Cerasoli. **6.4** Above, NASA; below, Manfred Kage/Peter Arnold, Inc. **6.5** Lisa Starr, using photographs © 1997, 1998 from the Jet Propulsion Laboratory/California Institute of Technology and NASA, 1972. **6.6** Gary Head. **6.7** Lisa Starr (see citation for 5.11). **6.8** Precision Graphics. **6.9** Lisa Starr and Gary Head. **6.10** Gary Head. **6.11** (a,b) Lisa Starr, from B. Alberts et al, *Molecular Biology of the Cell,* 1983, Garland Publishing; (c) Lisa Starr. **6.12** (a,b) Lisa Starr using photographs © 1997, 1998 from Jet Propulsion Laboratory, California Institute of Technology and NASA, 1972. **6.13** (a,b) Thomas A. Steitz. **6.14** Lisa Starr with PDB ID: 1DGF; Putnam, C. D., Arvai, A. S., Bourne, Y., Tainer, J. A.: Active and Inhibited Human Catalase Structures: Ligand and Nadph Binding and Catalytic Mechanism J Mol Biol. 296 pp. 295 (2000). **6.15** Lisa Starr. **6.16** Lisa Starr. **6.17** (a) Douglas Faulkner/Sally Faulkner Collection; (b,c) art, Gary Head. **6.18** Left, courtesy Dr. Edward C. Klatt; right, courtesy of Downstate Medical Center, Department of Pathology, Brooklyn, NY. **6.19** (a) © Frank Borges Llosa/www.frankley.com. (b) Sara Lewis, Tufts University; (c) Art, Lisa Starr with PDB ID: 1LCI; Conti, E., Franks, N. P., Brick, P.: Crystal structure of firefly luciferase throws light on a superfamily of adenylate-forming enzymes. Structure 4 pp. 287, (1996). **6.20** Professor J. Woodland Hastings, Harvard University. **6.21** (a,b) C. Contag, *Molecular Microbiology,* November, 1985, 18 (4):593. "Photonic Detection of Bacterial Pathogens in Living Hosts." Reprinted by permission of Blackwell Science. **Page 113** Lisa Starr (see citation for 6.1)

CHAPTER 7

7.1 Art, Raychel Ciemma; micrograph, Carolina Biological Supply Company. **7.2** Art, Lisa Starr; photograph, Wernhner Krutein/PhotoVault. **Page 115** Lisa Starr. **7.3** Lisa Starr with Preface, Inc. **Page 117** Gary Head. **7.4** Gary Head and Lisa Starr with photograph from David Neal Parks. **7.5** (a) Precision Graphics; (b, above) © 2002 PhotoDisc; (b, below) Bobby Mantoni. **Page 118** Gary Head. **7.6** (a,b) Lisa Starr after Stephen L. Wolfe, *Molecular and Cellular Biology,* Wadsworth; (c) Precision Graphics and Gary Head after Govindjee. **7.7** Left, Precision Graphics; (right-a,b) Lisa Starr from PDB files from NYU Scientific Visualization Lab. **7.8** (a) Douglas Faulkner/Sally Faulkner Collection; (b) Herve Chaumeton/Agence Nature. **7.9** Larry West/FPG. **7.10** Lisa Starr. **7.11** Lisa Starr. **Page 122** Gary Head. **7.12** Lisa Starr. **7.13** Lisa Starr. **7.14** E. R. Degginger. **7.15** Lisa Starr. **7.16** Lisa Starr. **Page 125** Gary Head. **7.17** (a, above) © Bill Beatty/Visuals Unlimited; (a, below) micrograph, Bruce Iverson, computer-enhanced by Lisa Starr; (b, above) © 2001 PhotoDisc; (b, below) micrograph by Ken Wagner/Visuals Unlimited, computer-enhanced by Lisa Starr. **Page 126** Gary Head. **7.18** Gary Head. **7.19** © 2002/Jeremy Woodhouse/PhotoDisc/Getty Images; Gary Head. **7.20** (a,b) NASA. **7.21** Lisa Starr. **Page 130** Lisa Starr. **7.22** Steve Chamberlain, Syracuse University. **7.23** Clockwise from above left, Gary Head; © 2002/PhotoDisc/Getty Images; C. B. Frith and D. W. Frith/Bruce Coleman, Ltd.; © 2002/PhotoDisc/Getty Images; © Rudiger Lhenen/SPL/Photo Researchers, Inc.

CHAPTER 8

8.1 Stephen Dalton/Photo Researchers, Inc., computer-enhanced by Lisa Starr. **8.2** Raychel Ciemma and Gary Head. **8.3** (a, left) Gary Head; (a, right) © John Lotter Gurling/Tom Stack & Associates; (b) Paolo Fioratti; (c) Lisa Starr with Gary Head. **Page 136** Lisa Starr. **8.4** page 136, Raychel Ciemma and Lisa Starr; page 137, Lisa Starr and Gary Head, after Ralph Taggart. **Page 137** Lisa Starr. **8.5** (a) Art, Raychel Ciemma; micrograph, Keith R. Porter; (b,c) art, Lisa Starr. **8.6** Photo, © 2001 PhotoDisc, Inc.; art left, Raychel Ciemma; art right, Lisa Starr. **8.7** Above, Raychel Ciemma; below, Lisa Starr. **8.8** Lisa Starr with Preface, Inc. **8.9** Lisa Starr. **Page 142** Lisa Starr. **8.10** (a) Lisa Starr with Gary Head; (b) Adrian Warren/Ardea, London; (c) David M. Phillips/Visuals Unlimited. **8.11** William Grenfell/Visuals Unlimited. **8.12** Art, Lisa Starr; photograph, Gary Head. **Page 146** Art, Gary Head; photograph R. Llewellyn/SuperStock. **Page 147** Lisa Starr. **Page 148** Thomas D. Mangelsen/Images of Nature

Page 149 UNIT II © Francis Leroy, Biocosmos/Science Photo Library/Photo Researchers

CHAPTER 9

9.1 Left and right, center, Chris Huss; right above and below, Tony Dawson. **9.2** Gary Head. **9.3** Art, Lisa Starr and Raychel Ciemma; micrographs (a) C. J. Harrison et al, *Cytogenetics and Cell Genetics*, 35: 21–27 © 1983 S. Karger, A. G. Basel; (c) B. Hamkalo; (d) O. L. Miller, Jr., Steve L. McKnight. **9.4** Raychel Ciemma and Gary Head. **9.5** All micrographs courtesy Andrew S. Bajer, University of Oregon **9.6** Gary Head. **9.7** Art, Raychel Ciemma and Gary Head; micrographs, © Ed Reschke. **9.8** Art, Lisa Starr; micrograph, © R. Calentine/Visuals Unlimited. **9.9** Art, Raychel Ciemma; (d) micrograph, © D. M. Phillips/Visuals Unlimited. **9.10** (a–c, e) Lennart Nilsson from *A Child Is Born* © 1966, 1977 Dell Publishing Company, Inc.; (d) Lennart Nilsson from *Behold Man,* © 1974 by Albert Bonniers Förlag and Little, Brown & Company, Boston **9.11** Photograph right, Courtesy of the Family of Henrietta Lacks; micrograph, Dr. Pascal Madaule, France. **Page 161** Raychel Ciemma

CHAPTER 10

10.1 (a) Jane Burton/Bruce Coleman; (b) Dan Kline /Visuals Unlimited. **10.2** Raychel Ciemma. **10.3** © CNRI/SPL/Photo Researchers, Inc. **Page 165** Raychel Ciemma. **10.4** Micrographs, with thanks to the John Innes Foundation Trustees, computer-enhanced by Gary Head; art, Raychel Ciemma. **10.5** Left, Raychel Ciemma; Right, Lisa Starr. **10.6** Raychel Ciemma. **10.7** Art, Preface, Inc.; photograph, David Maitland/Seaphot Limited. **10.8** Lisa Starr. **10.9** Micrograph, David M. Phillips/Visuals Unlimited; art, Lisa Starr. **10.10** Raychel Ciemma. **10.11** Precision Graphics. **Page 175** Raychel Ciemma. **10.12** © Richard Corman/Corbis Outline

CHAPTER 11

11.1 Tom Cruise photo, ©AFP/CORBIS; Charles Barkley photo, Focus on Sports; Gregor Mendel painting, The Moravian Museum, Brno; Joan Chen photo, Fabian/Corbis Sygma. **11.2** Art, Jennifer Wardrip; photograph, Jean M. Labat/Ardea, London. **11.3** Lisa Starr. **11.4** Precision Graphics. **11.5** Raychel Ciemma and Precision Graphics. **11.6** Precision Graphics. **11.7** Raychel Ciemma with Precision Graphics. **Page 182** Raychel Ciemma and Precision Graphics. **11.8** Raychel Ciemma. **11.9** Raychel Ciemma and Precision Graphics. **11.10** Photographs, William E. Ferguson; art, Raychel Ciemma. **11.11** Art, Precision Graphics. **11.12** (a) © Bettmann/CORBIS; (b) art, Lisa Starr with PDB ID: 1EMN; Downing, A. K., Knott, V., Werner, J. M., Cardy, C. M., Campbell, I.D., Handford, P. A.: Solution structure of a pair of calcium-binding epidermal growth factor-like domains: implications for the Marfan syndrome and other genetic disorders. Cell 85 pp. 597 (1996). **11.13** Art, Preface, Inc.; photographs, (a,b) Michael Stuckey/Comstock, Inc.; (c) Bosco Broyer; photograph, Gary Head. **11.14** David Hosking. **11.15** Photographs, Ted Somes. **11.16** Above and below, Frank Cezus/FPG; Frank Cezus/FPG; © 2001 PhotoDisc, Inc.; Ted Beaudin/FPG; Stan Sholik/FPG. **11.17** (a) Photographs courtesy Ray Carson, University of Florida News and Public Affairs; art, Gary Head. **11.18** Photograph, Jane Burton/Bruce Coleman; art, D. Hennings and V. Hennings. **11.19** Photograph, © Pamela Harper/Harper Horticultural Slide Library; art, Lisa Starr, referencing Professor Otto Wilhelm Thomé, Flora von Deutschland Österreich und der Schweiz, 1885, Gera, Germany. **11.20** Left, Eric Crichton/Bruce Coleman; right, William E. Ferguson. **11.21** Evan Cerasoli. **Page 192** Gary Head. **11.22** Leslie Faltheisek. Clacritter Manx. **11.23** © Joe McDonald/Visuals Unlimited

CHAPTER 12

12.1 From "Multicolor Spectral Karyotyping of Human Chromosomes," by E. Schrock, T. Ried et al, *Science,* 26, July 1966, 273:495. Used by permission of E. Schrock and T. Reid and the American Association for the Advancement of Science, computer-enhanced by Lisa Starr. **12.2** From P. Maslak, Blast Crisis of Chronic Myelogenous Leukemia. Posted online December 5, 2001. ASH Image Bank. Copyright American Society of Hematology, used with permission. **12.3** Raychel Ciemma. **12.4** Photograph, Charles D. Winters/Photo Researchers; micrograph, Omikron/Photo Researchers; art, Raychel Ciemma. **12.5** Art, Precision Graphics and Gary Head; photographs left, © 2001 EyeWire; right, © 2001 PhotoDisc, Inc. **12.6** (a) Photograph, from Lennart Nilsson, *A Child Is Born,* © 1966, 1977 Dell Publishing Company, Inc.; (b) Robert Demarest after Patten, Carlson & others; (c) Robert Demarest with the permission of M. Cummings, *Human Heredity: Principles and Issues,* p. 126 third edition, © 1994 Brooks/Cole. All rights reserved. **12.7** Art, Raychel Ciemma and Preface, Inc.; photograph, Carolina Biological Supply Company. **12.8** Raychel Ciemma. **12.9** (a,b) Precision Graphics; (b) Dr. Victor A. McKusick; (c) Steve Uzzell. **Page 204** Precision Graphics. **12.10** Lisa Starr. **12.11** Giraudon/Art Resource, New York. **12.12** (a) Lisa Starr; (b) photograph © Bettmann/Corbis, art after V. A. McKusick, *Human Genetics,* second edition, © 1969, reprinted by permission, Prentice-Hall, Inc.,

Englewood Cliffs, N.J. **12.13** C. J. Harrison. **12.14** Eddie Adams/AP/Wide World Photos. **Page 206** Precision Graphics. **12.15** (a,b) Courtesy G. H. Valentine. **12.16** From "Multicolor Spectral Karyotyping of Human Chromosomes," by E. Schrock, T. Ried, et al, *Science,* 26, July 1966, 273:496. Used by permission of E. Schrock and T. Reid and the American Association for the Advancement of Science. **Page 207** Precision Graphics. **12.17** Raychel Ciemma. **12.18** Left, permission of Carole Lafrate; (center) courtesy of Peninsula Association for Retarded Children and Adults, San Mateo Special Olympics, Burlingame, CA; right, Courtesy Special Olympics; karyotype, © 1997, Hironao Numabe, M.D., Tokyo Medical University. **12.19** Preface, Inc. **12.20** Left, UNC Medical Illustration and Photography; right, © 1997, Hironao Numabe, M.D., Tokyo Medical University. **12.21** (a) © 1997, Hironao Numabe, M.D., Tokyo Medical University; (b, inset) Stefan Schwarz. **12.22** Raychel Ciemma. **12.23** Art, Lisa Starr; photographs, courtesy Lennart Nilsson from *A Child Is Born,* © 1966, 1977 Dell Publishing Company, Inc. **12.24** Photograph, (a) Fran Heyl Associates © Jacques Cohen, computer-enhanced by © Pix Elation; (b) Raychel Ciemma. **12.25** Carolina Biological Supply Company. **12.26** Precision Graphics. **12.27** Raychel Ciemma. **Page 215** © Mitchell Gerber/CORBIS

CHAPTER 13
13.1 A. C. Barrington Brown, © 1968 J. D. Watson. **13.2** Lisa Starr with PDB ID: 1BBB; Silva, M. M., Rogers, P. H., Arnone, A.: A third quaternary structure of human hemoglobin A at 1.7-Å resolution. J Mol Biol Chem 267 pp. 17248 (1992). **13.3** Raychel Ciemma. **13.4** (a,b) Raychel Ciemma; (c) Lee D. Simon/Science Source/Photo Researchers, Inc. **13.5** Raychel Ciemma. **13.6** Art, above, Raychel Ciemma; below, Gary Head; micrograph, Biophoto Associates/SPL/Photo Researchers, Inc. **Page 221** Preface, Inc. **13.7** Lisa Starr (see citation for 13.2). **13.9, 13.10** Precision Graphics. **13.11** (a) Daniel Fairbanks; (b) PA News Photo Library. **13.12** Courtesy of Advanced Cell Technology, Inc., Worcester, Massachusetts

CHAPTER 14
14.1 Above, Dennis Hallinan/FPG; below, © Bob Evan/Peter Arnold, Inc. **14.2, 14.3** Precision Graphics. **14.4** Lisa Starr with PDB ID: 1BBB; Silva, M. M., Rogers, P. H., Arnone, A.: A third quaternary structure of human hemoglobin A at 1.7-Å resolution. J Mol Biol Chem 267 pp. 17248 (1992). **14.5** Gary Head. **14.6, 14.7** Precision Graphics. **14.8** (a) model by Dr. David B. Goodin, The Scripps Research Institute; (b,c) Lisa Starr. **14.9** (a) Courtesy of Thomas A. Steitz from Science; (b) Lisa Starr. **14.10** Lisa Starr. **14.11** (a) Lisa Starr; (b) Precision Graphics. **14.12** Lisa Starr. **14.13** Left, Nik Kleinberg; right, Peter Starlinger. **14.14** Lisa Starr. **14.15** Courtesy of the National Neurofibromatosis Foundation

CHAPTER 15
15.1 Photos (a) Ken Greer/Visuals Unlimited; (b) Biophoto Associates/Science Source/Photo Researchers; (c) James Stevenson/SPL/Photo Researchers; right, Gary Head. **Page 238** Lisa Starr. **15.2** Raychel Ciemma and Gary Head. **15.3** Lisa Starr with PDB ID: 1CJG; Spronk, C. A. E. M., Bonvin, A. M. J. J., Radha, P. K., Melacini, G., Boelens, R., Kaptein, R.: The Solution Structure of Lac Repressor Headpiece 62 Complexed to a Symmetrical Lac Operator. *Structure (London)* 7 pp. 1483 (1999). Also PDB ID: 1LBI; Lewis, M., Chang, G., Horton, N. C., Kercher, M. A., Pace, H. C., Schumacher, M. A., Brennan, R. G., Lu, P.: Crystal structure of the lactose operon repressor and its complexes with DNA and inducer. Science 271 pp. 1247 (1996); lactose pdb files from the Hetero-Compound Information Centre - Uppsala (HIC-Up). **15.4** Raychel Ciemma and Lisa Starr. **15.5** (a) Carolina Biological Supply Company; (b) UCSF Computer Graphics Laboratory, National Institutes, NCRR Grant 01081. **15.6** (a) Dr. Karen Dyer Montomery; (b) Raychel Ciemma. **15.7** Jack Carey. **15.8** Above, Raychel Ciemma; below, W. Beerman. **15.9** (a) Raychel Ciemma and Gary Head; (b) Frank B. Salisbury. **15.10** (a) Lennart Nilsson © Boehringer Ingelheim International GmbH; (b, c) Betsy Palay/Artemis. **15.11** Betsy Palay/Artemis. **15.12** Slim Films. **Page 251** Lisa Starr

CHAPTER 16
16.1 Lewis L. Lainey. **16.2** Science VU/Visuals Unlimited. **16.3** (a) © Professor Stanley Cohen/SPL/Photo Researchers, Inc.; (b) Dr. Huntington Potter and Dr. David Dressler. **Page 254** Preface, Inc. **16.4** Lisa Starr. **16.5** Gary Head. **16.6** Gary Head. **16.7** Above, Damon Biotech, Inc.; below, Cellmark Diagnostics, Abingdon, UK. **16.8, 16.9** Lisa Starr. **16.10** Stephen Wolfe, *Molecular and Cellular Biology.* **16.11** Art, Lisa Starr; photograph, Herve Chaumeton/Agence Nature. **16.12** Keith V. Wood. **16.13** Photographs, (a) Dr. Vincent Chiang, School of Forestry and Wood Products, Michigan Technology University; (b) courtesy Calgene LLC **16.14** R. Brinster, R. E. Hammer, School of Veterinary Medicine, University of Pennsylvania. **Page 265** Above, photograph, Lisa Starr; center, © 2002 PhotoDisc; below, Dr. Vincent Chiang, School of Forestry and Wood Products, Michigan Technology University, computer-enhanced by Lisa Starr. **Page 266** Preface, Inc. **16.15** Above, © Adrian Arbib/Still Pictures; below, © AP/Wide World Photos.
Page 268 © 2002 PhotoDisc, computer-enhanced by Lisa Starr

Page 269 UNIT III © S. Stammers/SPL/Photo Researchers, Inc.

CHAPTER 17
17.1 Elliot Erwitt/Magnum Photos, Inc. **17.2** (a) Jen & Des Bartlett/Bruce Coleman; (b) Kenneth W. Fink/Photo Researchers, Inc.; (c) Dave Watts/A.N.T. Photo Library. **17.3** Raychel Ciemma. **17.4** (a) Courtesy of George P. Darwin, Darwin Museum, Down House; (b) Photo, Heather Angel, computer-enhanced by Lisa Starr; (c) Christopher Ralling; (d) art, Leonard Morgan; (e) art, Precision Graphics; (f) Dieter & Mary Plage/Survival Anglia. **17.5** (a) Charles R. Knight painting (negative CK21T), Field Museum of Natural History, Chicago; (b) © Philip Boyer/Photo Researchers, Inc.; (c) www.fossilmuseum.net. **17.6** (a) Heather Angel; (b) David Cavagnaro; (c) Dr. P. Evans/Bruce Coleman; (d) Alan Root/Bruce Coleman. **17.7** Courtesy of Down House and The Royal College of Surgeons of England **Page 278** Courtesy Derrell Fowler, Tecumseh, Oklahoma. **17.8** Alan Solem. **17.9** Raychel Ciemma and Gary Head. **17.10** Precision Graphics. **17.11** (a,b) J. A. Bishop and L. M. Cook. **Table 17.1** Data after H. B. Kettlewell. **17.12** Precision Graphics. **17.13** (a,b) Warren Abrahamson; (c) Forest W. Buchanan/Visuals Unlimited; (d) Kenneth McCrea, Warren Abrahamson; computer-enhanced by Gary Head. **17.14** Precision Graphics. **17.15** (a,b) Thomas Bates Smith; (c) Preface, Inc. **17.16** Bruce Beehler. **17.17** Precision Graphics, after Ayala and others. **17.18** Above, David Neal Parks; below, W. Carter Johnson. **17.19** (a,b) Art, Precision Graphics, after computer models developed by Jerry Coyne. **17.20** Art, Raychel Ciemma; photograph, David Cavagnaro. **17.21** Kjell Sandved/Visuals Unlimited. **Page 291** Precision Graphics. **17.22** Precision Graphics, using NIH data

CHAPTER 18
18.1 (a) Photograph, Gary Head; snail, courtesy of Larry Reed; (b) Gary Head, adapted from R. K. Selander and D. W. Kaufman, *Evolution,* 29(3), December 31, 1975. **18.2** Art, Jennifer Ward Rip. **18.3** Precision Graphics, after F. Ayala and J. Valentine, *Evolving,* Benjamin-Cummings, 1979 **18.4** (a) G. Ziesler/Zefa; (b) Alvin E. Staffan/Photo Researchers, Inc.; (c) John Alcock/Arizona State University. **18.5** (a) Fred McConnaughey/Photo Researchers, Inc.; (b) Patrice Geisel/Visuals Unlimited; right, Tom Van Sant/The Geosphere Project, Santa Monica, CA. **18.6** Preface, Inc. **18.7** Raychel Ciemma. **18.8** Above, Steve Gartlan; below, Tom Van Sant/The Geosphere Project, Santa Monica, CA. **18.9** Lisa Starr after W. Jensen and F. B. Salisbury, *Botany: An Ecological Approach,* Wadsworth, 1972. **18.10** Photographs, left © H. Clarke, VIREO/Academy of Natural Sciences; right, Robert C. Simpson/Nature Stock; art, Preface, Inc. **18.11** Precision Graphics. **18.12** Photographs above, Jack Dermid; below, Leonard Lee Rue III/FPG; art, Precision Graphics after P. Dodson, *Evolution: Process and Products,* third edition., PWS. **18.13** Jen and Des Bartlett/Bruce Coleman

CHAPTER 19
19.1 (a) Lisa Starr; (b) Lisa Starr after http://www.scotese.com/K/t.htm; (c) © David A. Kring, NASA/University of Arizona Space Imagery Center; (d) NASA Galileo Imaging Team; (e) art, Don Davis. **19.2** Vatican Museums. **19.3** (a) N. Simmons and J. Geisler, © American Museum of Natural History; (b) A. Feduccia, *The Age of Birds,* Harvard University Press, 1980; (c) H. P. Banks; (d) Jonathan Blair. **19.4** Left, globes, Lisa Starr; chart, Gary Head. **19.5** Gary Head. **19.6** (a) © 2001 PhotoDisc, Inc.; (b,c) Lisa Starr. **19.7** © Danny Lehman/CORBIS. **19.8** (a) NOAA/NGDC; (b) Leonard Morgan; (c) Gary Head. **19.9** Lisa Starr, after A. M. Ziegler, C. R. Scotese, and S. F. Barrett, "Mesozoic and Cenozoic Paleogeographic Maps," and J. Krohn and J. Sundermann (editors), *Tidal Friction and the Earth's Rotation II,* Springer–Verlag, 1983; photograph, Martin Land/Photo Researchers, Inc. **19.10** Raychel Ciemma. **19.11** (a) Art, Lisa Starr referencing Natural History Collection, Royal BC Museum; photographs, (a) Stephen Dalton/Photo Researchers, Inc., computer-enhanced by Lisa Starr; (b) Frans Lanting/Minden Pictures, computer-enhanced by Lisa Starr; (c) J. Scott Altenbach, University of New Mexico, computer-enhanced by Lisa Starr. **19.12** Art, Precision Graphics and Gary Head, after E. Guerrant, *Evolution,* 36:699–712; photographs, Gary Head. **19.13** (a) left, Sinclair Stammers/Science Photo Library/Photo Researchers, Inc.; right, Bradley R. Smith, University of Michigan; (b) After T. Storer et al, *General Zoology,* sixth edition, McGraw-Hill, 1979, reproduced by permission from McGraw-Hill, Inc. **19.14** Raychel Ciemma. **19.15** Precision Graphics. **19.16** Left to right, Kjell B. Sandved/Visuals Unlimited; Jeffrey Sylvester/FPG; Thomas D. Mangelsen/Images of Nature. **19.17** Left to right, Larry Lefever/Grant Heilman Photography, Inc.; R.I.M. Campbell/Bruce Coleman; Runk & Schoenberger/Grant Heilman Photography, Inc.; Bruce Coleman. **Page 318** Preface, Inc. **19.18, 19.19, 19.20** Gary Head. **19.21** Lisa Starr with Gary Head. **19.22** (a) P. Morris/Ardea London; (b) Raychel Ciemma. **19.23** (a) John Klausmeyer, University of Michigan Exhibit of Natural History; (b) © Bruce J. Mohn; (b, inset) Phillip Gingerich, Director, University of Michigan Museum of Paleontology; (c) © 2002 Peter Timmermans/Stone/Getty Images. **19.24** © 2002 Paolo Curto/The Image Bank/Getty Images

CHAPTER 20
20.1 Jeff Hester and Paul Scowen, Arizona State University and NASA. **20.2** Painting, William K. Hartmann. **20.3** (a) Painting, Chesley Bonestell; (b) Raychel Ciemma. **20.4** Precision Graphics. **20.5** (a) Sidney W. Fox; (b) courtesy W. Hargreaves and D. Deamer; (c) © Phillippa Uwins/University of Queensland; (d) Preface, Inc. **20.6** Raychel Ciemma and Precision Graphics. **20.7** Bill Bachmann/Photo Researchers, Inc. **20.8** (a) Stanley M. Awramik; (b–f) Andrew H. Knoll, Harvard University. **20.9** Micrograph, P. L. Walne and J. H. Arnott, *Planta,* 77:325–354, 1967. **20.10** Raychel Ciemma. **20.11** Robert K. Trench. **Page 336** Lisa Starr. **20.12** (a,b) Neville Pledge/South Australian Museum; (c, d) Dr. Chip Clark. **20.13** (a,b) Illustrations by Zdenek Burian, © Jiri Hochman and Martin Hochman; (c) Patricia G. Gensel; (d) © John Barber. **Pages 337, 338** Lisa Starr. **20.14** Megan Rohn painting, courtesy of David Dilcher; art, Gary Head; photographs clockwise from below, Runk & Schoenberger/Grant Heilman Photography, Inc.; Robert and Linda Mitchell; Ed Reschke; Lee Casebere. **20.15** (a) © John Sibbick; (b,c) © Karen Carr Studio/www.karencarr.com. **Page 340** Lisa Starr. **20.16** Painting © Ely Kish. **20.17** (a) Painting © Ely Kish; (b) Karen Carr Studio, www.karencarr.com. **20.18** Maps, Lisa Starr after A. M. Ziegler, C. R. Scotese, and S. F. Barrett, "Mesozoic and Cenozoic Paleogeographic Maps," and J. Krohn and J. Sundermann, "Paleotides

Department of Entymology, Purdue University. **25.39** (a) © Fred Bavendam/Minden Pictures; (b) Chris Huss/The Wildlife Collection; (c) Jan Haaga, Kodiak Lab, AFSC/NMFS; (d) © George Perina, www.seapix.com. **25.40** (a,b) Art, L. Calver; (c, d) photographs, Herve Chaumeton/Agence Nature. **Page 441** Jane Burton/Bruce Coleman. **25.41** Raychel Ciemma. **25.42** Walter Deas/Seaphot Limited/Planet Earth Pictures. **25.43** (a) J. Solliday/BPS; (b) Herve Chaumeton/Agence Nature

CHAPTER 26

26.1 (a) Tom McHugh/Photo Researchers, Inc.; (b) © Lawson Wood/CORBIS. **26.2** Art, Gary Head. **26.3** (a,b) Redrawn from *Living Invertebrates,* V. & J. Pearse and M. & R. Buchsbaum, The Boxwood Press, 1987. Used by permission; (c) © 2002 Gary Bell/Taxi/Getty Images. **26.4** (a) Raychel Ciemma; (b) Runk and Schoenberger/Grant Heilman Photography, Inc.; (c) © John and Bridgette Sibbick. **26.5** (a–c) Raychel Ciemma, adapted from A. S. Romer and T. S. Parsons, *The Vertebrate Body,* sixth edition, Saunders, 1986; (d, left) Photograph, Lisa Starr; (d, right) courtesy John McNamara, www.paleodirect.com, computer-enhanced by Lisa Starr. **26.6** Above, after D. H. Milne, *Marine Life and the Sea,* Wadsworth, 1995; (a,b) © Brandon D. Cole/CORBIS. **26.7** Above, after D. H. Milne, *Marine Life and the Sea,* Wadsworth, 1995; photograph below, Heather Angel. **26.8** (a) © Jonathan Bird/Oceanic Research Group, Inc.; (b) © 1999 Gido Braase/Deep Blue Productions; (c) Tom McHugh/Photo Researchers, Inc. **26.9** (a) Bill Wood/Bruce Coleman; (b) Raychel Ciemma; (c) Robert and Linda Mitchell Photography; (d) Patrice Ceisel/© 1986 John G. Shedd Aquarium; (e) © Norbert Wu/Peter Arnold, Inc.; (f) Wernher Krutein/photovault.com. **26.10** (a) © Alfred Kamajian; (b,c) Laszlo Meszoly and D. & V. Hennings. **26.11** (a) Stephen Dalton/Photo Researchers, Inc.; (b) John Serraro/Visuals Unlimited; (c) Jerry W. Nagel; (d) Leonard Morgan adapted from A. S. Romer and T. S. Parsons, *The Vertebrate Body,* sixth edition, Saunders, 1986; (e) Juan M. Renjifo/Animals Animals. **26.12** (a) Pieter Johnson; (b) Stanley Sessions, Hartwick College. **26.13** (a) © 1989 D. Braginetz; (b) Z. Leszczynski/Animals Animals; (c) Raychel Ciemma. **26.14** D. & V. Hennings and Gary Head. **26.15** (a) Mark Grantham; (b) Raychel Ciemma; (c, d) © Stephen Dalton/Photo Researchers, Inc.; (d, inset) Raychel Ciemma; (e) Heather Angel; (f) Kevin Schafer/Tom Stack & Associates. **26.16** (a) Lisa Starr; (b) Rajesh Bedi. **26.17** Raychel Ciemma. **26.18** (a) Gerard Lacz/A.N.T. Photolibrary; (b) © 2002 PhotoLink/PhotoDisc/Getty Images; art (b–d) from *The Life of Birds,* fourth edition, L. Baptista and J. C. Whelty, © 1988, Saunders. Reproduced by permission of the publisher; (e,f) Courtesy of Dr. M. Guinan, Anatomy, Physiology, and Cell Biology, School of Veterinarian Medicine, University of California at Davis. **26.19** (a) Sandy Roessler/FPG; (b) Leonard Lee Rue III/FPG; (c) art, Raychel Ciemma after M. Weiss and A. Mann, *Human Biology and Behavior,* fifth edition, Harper Collins, 1990. **26.20** Lisa Starr. **26.21** (a) Gean Phillipe Varin/Jacana/Photo Researchers, Inc.; (b) D. and V. Blagden/A.N.T. Photolibrary; (c) Corbis Images/Picture Quest; (d) Mike Jagoe/Talune Wildlife Park, Tasmania, Australia. **26.22** Raychel Ciemma. **26.23** (a) © Stephen Dalton/Photo Researchers, Inc.; (b) © Merlin D. Tuttle/Bat Conservation International; (c) © David Parker/SPL/Photo Researchers, Inc.; (d) Douglas Faulkner/Photo Researchers, Inc.; (e) Bryan and Cherry Alexander Photography; (f) Christopher Crowley. **Page 464** Art, D. and V. Hennings. **26.24** (a) Larry Burrows/Aspect Photolibrary; (c) © Art Wolfe/Photo Researchers, Inc.; (d) © Dallas Zoo, Robert Cabello; (e) courtesy of Dr. Takeshi Furuichi, Biology, Meiji–Gakuin University-Yokohama; (f) Allen Gathman, Biology Department, Southeast Missouri State University; (g) Bone Clones®, www.boneclones.com; (h) Gary Head. **26.25** Gary Head. **26.26** (a) Time, Inc., 1965. Larry Burrows Collection. **26.27** Precision Graphics after National Geographic, February 1997, page 82. **26.28** (a) Dr. Donald Johanson, Institute of Human Origins; (b) Louise M. Robbins; (c,d) Kenneth Garrett/National Geographic Image Collection. **26.29** *Sahelanthropus tchadensis,* MPFT/Corbis Sygma; other art, Lisa Starr. **26.30** Kenneth Garrett/National Geographic Image Collection. **26.31** Left, Jean Paul Tibbles; right, Lisa Starr. **26.32** Left © Elizabeth Delaney/Visuals Unlimited; all others, John Reader © 1981. **26.33** Lisa Starr. **26.34** Lisa Starr. **26.35** Raychel Ciemma. **26.36** Left, © Sandak/FPG; right, Douglas Mazonowicz/Gallery of Prehistoric Art. **26.37** Preface, Inc. **26.38** Lisa Starr; photograph, NASA. **26.39** Andrew Dennis/A.N.T. Photolibrary

CHAPTER 27

27.1 © Tom Till/Stone/Getty Images. **27.2** Art, Gary Head. **27.3** (a) Painting by Charles Knight (negative 2425), Department of Library Services, American Museum of Natural History; (b) Mansell Collection/Time, Inc. **27.4** Lisa Starr . **27.5** Steve Hillebrand, U.S. Fish & Wildlife Service **27.6** (a) © Greenpeace/Cunningham; (b) Lisa Starr, based on 1940–1984 reports from the International Whaling Commission. **27.7** (a) C. B. & D. W. Frith/Bruce Coleman; (b) Douglas Faulkner/Photo Researchers, Inc.; (c) Douglas Faulkner/Sally Faulkner Collection; (d) Sea Studios/Peter Arnold, Inc. **27.8** (a) T. Garrison, *Oceanography: An Invitation to Marine Science,* third edition, Brooks/Cole, 2000. All rights reserved; (b) © Greenpeace/Grace; (c, above left to right) Douglas Faulkner and Sally Faulkner Collection; Peter Scoones/Planet Earth Pictures; (c, center left to right) Jeff Rotman; Alex Kirstitch; Douglas Faulkner/Sally Faulkner Collection; (c, below) Douglas Faulkner/Sally Faulkner Collection. **27.9** Eric Hartmann/Magnum Photos. **27.10** Lisa Starr. **27.11** Art, Lisa Starr, with photographs © 2000 PhotoDisc, Inc. **27.12** Bureau of Land Management. **27.13** Lisa Starr. **27.14** Photograph courtesy www.eternalreefs.com

Page 487 UNIT INTRODUCTION © 2002 Stuart Westmorland/Stone/Getty Images

CHAPTER 28

28.1 Galen Rowell/Peter Arnold, Inc. **28.2** (a) © K. G. Vock/Okapia/Photo Researchers, Inc.; (b) © Patrick Johns/CORBIS. **28.3** left above, courtesy Charles Lewallen; left center, © Bruce Iverson; left below, © Bruce Iverson; right, Raychel Ciemma. **28.4** Left above, © CNRI/SPL/Photo Researchers, Inc.; left below, Dr. Roger Wagner/University of Delaware, www.udel.edu/Biology/Wags; right, art, Lisa Starr with © 2002 PhotoDisc, Inc. **28.5** (a) Gary Head; (b) John W. Merck, Jr., University of Maryland. **28.6** Left © Thomas Mangelsen; right, © Anthony Bannister/Photo Researchers, Inc. **28.7** Left, Giorgio Gwalco/Bruce Coleman; right, © David Parker/SPL/Photo Researchers, Inc. **28.8** © K & K Ammann/Taxi/Getty Images; art, Precision Graphics. **28.9** Left, art, Preface, Inc.; right, photograph, Fred Bruemmer. **28.10** (a) Gary Head with permission from Alex Shigo; (b) From *Tree Anatomy* by Dr. Alex L. Shigo, Shigo Trees and Associates; Durham, New Hampshire. **28.11** (a) www.plants.montara.com; (b) G. J. McKenzie (MGS); (c) Frank B. Salisbury. **28.12** (a) Courtesy of Hall and Bleecker; (b) Juergen Berger, Max Planck Institute for Developmental Biology, Tuebingen, Germany. **28.13** Kevin Somerville and Gary Head. **28.14** Left, © Darrell Gulin/The Image Bank/Getty Images; right, © Pat Johnson Studios Photography. **Page 500** Above, ©PhotoDisc/Getty Images; below, © Cory Gray. **28.15** (a) Heather Angel; (b) © Biophoto Associates/Photo Researchers, Inc. **28.16** (a) © Geoff Tompkinson/SPL/Photo Researchers, Inc.; (b) © John Beatty/SPL/Photo Researchers, Inc.

Page 503 UNIT V © Jim Christensen, Fine Art Digital Photographic Images

CHAPTER 29

29.1 (a) R. Barrick/USGS; (b, c) © 1980 Gary Braasch; (d) Don Johnson/Photo Nats, Inc. **29.2** Art, Raychel Ciemma. **29.3** Art, Precision Graphics. **29.4** Art, Raychel Ciemma. **29.5** micrograph, James D. Mauseth, *Plant Anatomy,* Benjamin-Cummings, 1988. **29.6** All, Biophoto Associates. **29.7** (a) D. E. Akin and I. L. Risgby, Richard B. Russel Agricultural Research Center, Agricultural Research Service, U.S. Department of Agriculture, Athens, Georgia; (b) Kingsley R. Stern. **29.8** Art, Precision Graphics. **29.9** George S. Ellmore. **29.10** Art, D. & V. Hennings. **29.11** Art, Raychel Ciemma. **29.12** (a) Robert and Linda Mitchell Photography; (b) Roland R. Dute; (c) Gary Head. **29.13** Art, D. & V. Hennings; (a, left) Ray F. Evert; (a, right) James W. Perry; (b, left) Carolina Biological Supply Company; (b, right) James W. Perry. **29.14** Art, D. & V. Hennings. **29.15** (a) Heather Angel; (b) Gary Head; (c) © 2001 PhotoDisc, Inc. **29.16** (a) Art, Raychel Ciemma; (b) C. E. Jeffree, et al., *Planta,* 172(1):20–37, 1987. Reprinted by permission of C. E. Jeffree and Springer–Verlag; (c) Dr. Jeremy Burgess/SPL/Photo Researchers, Inc. **29.17** (a) Art, after Salisbury and Ross, *Plant Physiology,* fourth edition, Wadsworth; (b) micrograph, John Limbaugh/Ripon Microslides, Inc.; (c, d) Art, Raychel Ciemma. **29.18** Micrographs, (a) Chuck Brown; (b) Carolina Biological Supply Company. **29.19** Photographs, John Limbaugh/Ripon Microslides; art after T. Rost et al., *Botany: A Brief Introduction to Plant Biology,* second edition, © 1984, John Wiley & Sons, Inc. **29.20** (a) Raychel Ciemma; (b) Alison W. Roberts, University of Rhode Island. **29.21** Art, Raychel Ciemma. **29.22** Art, Precision Graphics. **29.23** John Lotter Gurling/Tom Stack & Associates. **29.24** (a,c) Lisa Starr; (b) H. A. Core, W. A. Cote, and A. C. Day, *Wood Structure and Identification,* second edition, Syracuse University Press, 1979. **29.25** (a) © Jon Pilcher; (b) © George Bernard/SPL/Photo Researchers, Inc.; (c) © Peter Ryan/SPL/Photo Researchers, Inc. **29.26** (a,b) Edward S. Ross. **29.27** Left, NASA/USGS; right, courtesy Professor David W. Stahle, University of Arkansas

CHAPTER 30

30.1 (a,b) Robert and Linda Mitchell Photography; (c) © John N. A. Lott, *Scanning Electron Microscope Study of Green Plants,* St. Louis: C. V. Mosby Company, 1976; (d) Robert C. Simpson/Nature Stock. **30.2** David Cavagnaro/Peter Arnold, Inc. **30.3** (a) William Furgeson; (b) USDA NRCS; (c) U.S. Department of Agriculture. **30.4** (a,c,d) Art, Leonard Morgan; (b) micrograph, Chuck Brown. **30.5** Courtesy Mark Holland, Salisbury University. **30.6** Photographs, (a) Adrian P. Davies/Bruce Coleman; (c) Mark E. Dudley and Sharon R. Long; (a,b) art, Jennifer Wardrip. **30.7** NifTAL Project, University of Hawaii, Maui. **30.8** Micrographs (a) Alison W. Roberts, University of Rhode Island; (b, c) H. A. Cote, W. A. Cote and A. C. Day, *Wood Structure and Identification,* second edition, Syracuse University Press, 1979. **30.9** Left, Natural History Collections, The Ohio Historical Society; Raychel Ciemma. **30.10** Left, M. Ricketts, School of Biological Sciences, The University of Sydney, Australia; inset, Aukland Regional Council. **30.13** © Left, Don Hopey, Pittsburgh Post–Gazette, 2002, all rights reserved. Reprinted with permission; (a,b) Dr. Jeremy Burgess/SPL/Photo Researchers, Inc. **30.14** (a) Courtesy of Professor John Main, PLU; (b) © James D. Mauseth, University of Texas. **30.15** Martin Zimmerman, *Science,* 1961, 133:73–79, © AAAS. **30.16** Left, Palay/Beaubois; right, Precision Graphics. **Page 535** Palay/Beaubois. **30.18** James T. Brock

CHAPTER 31

31.1 Photographs left, courtesy Merlin D. Tuttle/Bat Conservation International; right, John Alcock, Arizona State University. **31.2** (a) Robert A. Tyrrell; (b,c) Thomas Eisner, Cornell University. **31.3** Art, Raychel Ciemma and Precision Graphics; photos, Gary Head. **31.4** John Shaw/Bruce Coleman. **31.5** (a) David M. Phillips/Visuals Unlimited; (b) © Dr. Jeremy Burgess/SPL/Photo Researchers, Inc.; (c) David Scharf/Peter Arnold, Inc. **31.6** Raychel Ciemma. **31.7** (a,c,e,f) Michael Clayton, University of Wisconsin; (b) Raychel Ciemma; (d) Dr. Charles Good, Ohio State University, Lima. **31.8** (a) Dr. Dan Legard, University of Florida GCREC, 2000; (b) Richard H. Gross; (c) © Andrew Syred/SPL/Photo Researchers, Inc.; (e) Mark Rieger; (f–h) Janet Jones. **31.9** (a) R. Carr; (b) Gary Head; (c) Rein/Zefa. **31.10** John Alcock, Arizona State University. **31.11** Russell Kaye, © 1993 The Walt Disney Co. Reprinted with permission of *Discover* Magazine. **31.12** (a) Runk & Schoenberger/Grant Heilman Photography, Inc.; (b) Kingsley R. Stern **Page 548** Raychel Ciemma. **31.13** Gary Head

CHAPTER 32

32.1 (a) Michael A. Keller/FPG; (b,c) R. Lyons/Visuals Unlimited. **32.2** Sylvan H. Wittwer/Visuals Unlimited. **32.3** © Dr. John D. Cunningham/Visuals Unlimited. **32.4** (a,b) Raychel Ciemma; (c) Herve Chaumeton/Agence Nature. **32.5** (a,b) Raychel Ciemma; (c) Barry L. Runk/Grant Heilman Photography, Inc.; (d) from Mauseth. **32.6** (a,b) Raychel Ciemma; (c) Biophot. **32.7** (a) Kingsley R. Stern; (b) Precision Graphics. **32.8** (a) Michael Clayton, University of Wisconsin; (b) John Digby and Richard Firn. **32.9** (a,b) Micrographs courtesy of Randy Moore, from "How Roots Respond to Gravity," M. L. Evans, R. Moore, and K. Hasenstein, *Scientific American,* December 1986. **32.10** (a) © Adam Hart-Davis/SPL/Photo Researchers, Inc.; (b,c) Lisa Starr. **32.11** Gary Head. **32.12** Cary Mitchell. **32.14** Precision Graphics. **32.15** Long-day plant, © Clay Perry/CORBIS; short-day plant, © Eric Chrichton/CORBIS; art, Gary Head. **32.16** (a) Jan Zeevart; (b) Ray Evert, University of Wisconsin. **32.17** Above, N. R. Lersten; below, © Peter Smithers/CORBIS. **32.18** Larry D. Nooden. **32.19** R. J. Downs. **32.20** Art, Lisa Starr; photograph, Eric Welzel/Fox Hill Nursery, Freeport, Maine. **32.21** Gary Head and Preface Graphics Inc. **32.22** Inga Spence/Tom Stack & Associates. **32.23** Grant Heilman Photography, Inc.

Page 565 UNIT VI Kevin Schafer

CHAPTER 33

33.1 David Macdonald. **33.2** (a) Photograph, Focus on Sports; micrograph, Manfred Kage/Bruce Coleman; art, Lisa Starr; (b) photographs left to right, © Ray Simons/Photo Researchers, Inc.; © Ed Reschke/Peter Arnold, Inc.; © Don W. Fawcett; art, Lisa Starr. **33.3** Raychel Ciemma and Lisa Starr. **33.4** Photograph, Gregory Dimijian/Photo Researchers, Inc.; art, Raychel Ciemma, adapted from C. P. Hickman, Jr., L. S. Roberts, and A. Larson, *Integrated Principles of Zoology,* ninth edition, W. C. Brown, 1995. **33.5** Art, Lisa Starr; photographs (a–c, e) Ed Reschke, (d) Fred Hossler/Visuals Unlimited; (f, above) Professor P. Motta/Department of Anatomy, University "La Sapienza" Rome/SPL/Photo Researchers, Inc.; (f, below) Ed Reschke. **33.6** Photograph, Roger K. Burnard; art left, Joel Ito; right, L. Calver. **33.7** Ed Reschke. **33.8** Art, Lisa Starr; micrographs (a,c) Ed Reschke; (b) © Biophoto Associates/Photo Researchers, Inc. **33.9** Robert Demarest. **33.10** (a) Lennart Nilsson from *Behold Man,* © 1974 Albert Bonniers Förlag and Little, Brown and Company, Boston; (b) Kim Taylor/Bruce Coleman. **33.11** L. Calver. **33.12** Palay/Beaubois. **Page 576** (a) Ed Reschke/Peter Arnold, Inc.; (b-d) Ed Reschke. **Page 576** From Lennart Nilsson. **33.13** Dr. Preston Maxim and Dr. Stephen Bretz, Department of Emergency Services, San Francisco General Hospital

CHAPTER 34

34.1 Left, © 2002 Darren Robb/Stone/Getty Images; right, Kevin Somerville. **34.2** Art, Robert Demarest; micrograph, Manfred Kage/Peter Arnold, Inc. **Page 581** Lisa Starr. **34.3, 34.4,** and **34.5** Lisa Starr. **34.6** Precision Graphics. **34.7** (a, c) Art, Lisa Starr; micrograph, Dr. Constantino Sotelo, *International Cell Biology,* page 83, 1977. Used by copyright permission of Rockefeller University Press. **34.8** micrograph, © Don Fawcett, Bloom and Fawcett, eleventh edition, after J. Desaki and Y. Uehara/Photo Researchers, Inc.; art, Kevin Somerville. **34.9** Gary Head. **34.10** Robert Demarest. **34.12** (a) Kevin Sommerville and Precision Graphics; (b) Robert Demarest. **34.13** Above, © 2002 Ken Usami/PhotoDisc/Getty Images; art, Raychel Ciemma. **34.14** (a) Art, Lisa Starr, after Eugene Kozloff; (b) Ron Koss, University of Alberta, Canada. **34.15** Raychel Ciemma. **34.16** Kevin Somerville. **34.17** Kevin Somerville and Precision Graphics. **34.18** Robert Demarest and Precision Graphics. **34.19** (a) Robert Demarest; (b) Manfred Cage/Peter Arnold, Inc. **34.20** Kevin Somerville. **34.21** (a) Kenneth Garrett; (b) Kevin Somerville. **34.22** Kevin Somerville. **34.23** C. Yokochi and J. Rohen, *Photographic Anatomy of the Human Body,* second edition, Igaku-Shoin, Ltd., 1979. **34.24** Left, art, Palay/Beaubois, after Penfield and Rasmussen, *The Cerebral Cortex of Man,* © 1950 Macmillan Library Reference. Renewed 1978 by Theodore Rasmussen. Reprinted by permission of The Gale Group; right, Colin Chumbley/Science Source/Photo Researchers, Inc. **34.25** (a) art, Raychel Ciemma; (b) Marcus Raichle, Washington University School of Medicine. **34.26** Lisa Starr. **34.30** Robert Demarest. **34.31** Left © David Stoecklein/CORBIS; center, © Lauren Greenfield; right, AP/Wide World Photos. **34.32** (a,b) PET scans from E. D. London et al., *Archives of General Psychiatry,* 47:567–574, 1990; Photograph. **34.33** © Kathy Plonka, courtesy The Spokesman-Review

CHAPTER 35

35.1 (a) Eric A. Newman; (b) Merlin D. Tuttle, Bat Conservation International. **35.2** (a) © David Turnley/CORBIS; (b) Robert Demarest; (c–e) Kevin Somerville and Preface, Inc. **35.3** From Hensel and Bowman, *Journal of Physiology,* 23:564–568, 1960. **35.4** (left) Palay/Beaubois after *The Cerebral Cortex of Man,* by Penfield and Rasmussen, Macmillan Library Reference, © 1950 Macmillan Library Reference. Renewed 1978 by Theodore Rasmussen. Reprinted by permission of The Gale Group; (right) © Colin Chumbley/Science Source/Photo Researchers, Inc. **35.5** Raychel Ciemma. **35.6** Gary Head. **35.7** Precision Graphics. **35.8** Micrograph, (a) Omikron/SPL/Photo Researchers, Inc.; art, Robert Demarest. **35.9** (a,c) Kevin Somerville; (b) Lisa Starr. **Page 613** © AFP Photo/Timothy A. Clary/CORBIS. **35.10** Gary Head. **35.11** (a) Art, Robert Demarest; upper right photo image by Mireille Lavigne-Rebillard (INSERM, U.254, Montpellier), from *Promenade around the cochlea* (www.the-cochlea.info), by R. Pujol, S. Blatrix, T. Pujol and V. Reclar (CRIC, University Montpellier 1); lower right photo, Focus On Sports; (b) art, Precision Graphics; (c) Medtronic Xomed; (d) micrograph from Dr. Thomas R. Van De Water, University of Miami Ear Institute; art, Robert Demarest. **35.12** (a,b) Robert E. Preston, courtesy Joseph E. Hawkins, Kresge Hearing Research Institute, University of Michigan Medical School. **35.13** (a–c) Raychel Ciemma; (d) Keith Gilbert/Tom Stack & Associates. **35.14** Raychel Ciemma after M. Gardiner, *The Biology of Vertebrates,* McGraw-Hill, 1972; E. R. Degginger. **35.15** G. A. Mazohkin-Porshnyakov (1958). Reprinted with permission from *Insect Vision,* © 1969 Plenum Press. **35.16** Left Chris Newbert; right, Raychel Ciemma. **35.17** Robert Demarest. **35.18** Chase Swift. **35.19, 35.20** Kevin Somerville. **35.21** (a) www.2.gasou.edu/psychology/courses/muchinsky and www.occipita.cfa.cmu.edu; (b) Lennart Nilsson © Boehringer Ingelheim International Gmbh. **35.22** Left, Palay/Beaubois after S. Kuffler and J. Nicholls, *From Neuron to Brain,* Sinauer, 1977; right, Preface, Inc. **35.23** © Lydia V. Kibiuk. **35.24, 35.25** Art, Kevin Somerville; photo by Gerry Ellis/The Wildlife Collection. **Page 624** Robert Demarest. **35.26** Douglas Faulkner and Sally Faulkner Collection. **35.27** Edward W. Bauer © 1991 TIB

CHAPTER 36

36.1 (a,b) Hugo van Lawick/The Jane Goodall Institute. **36.2** Kevin Somerville. **36.3, 36.4** Art, Lisa Starr. **36.5, 36.6** Art, Robert Demarest. **36.7** (a) Mitchell Layton; (b) Syndication International (1986) Ltd.; (c) Courtesy Dr. William H. Daughaday, Washington University School of Medicine, from A. I. Mendelhoff and D. E. Smith, eds., *American Journal of Medicine,* 1956, 20:133. **36.8** Leonard Morgan. **36.9** (a,b) Raychel Ciemma; (c) Gary Head; (d) © Bettmann/Corbis. **36.10** Precision Graphics. **36.11** Biophoto Associates/SPL/Photo Researchers, Inc. **36.12** Leonard Morgan. **36.13** (a) John S. Dunning/Ardea, London; (b) Evan Cerasoli. **36.14** The Stover Group/D. J. Fort. **36.15** R. C. Brusca and G. J. Brusca, *Invertebrates,* © 1990 Sinauer Associates. Used by permission; Photographs, (c) Frans Lanting/Bruce Coleman; (d) Robert and Linda Mitchell Photography. **Page 642** Robert Demarest. **36.16** Roger K. Bernard

CHAPTER 37

37.1 John Brandenberg/Minden Pictures. **37.2** L. Calver. **37.3** (a) Dr. John D. Cunningham/Visuals Unlimited; (b) Robert Demarest; (c) © CNRI/SPL/Photo Researchers, Inc. **37.4** (a) M. P. L. Fogdon/Bruce Coleman; (b) W. J. Weber/Visuals Unlimited. **37.5** Robert Demarest. **37.6** (a) Lisa Starr; (b) Frank Trapper/Corbis Sygma; (c) © AFP/CORBIS. **37.7** Michael Keller/FPG. **37.8** Linda Pitkin/Planet Earth Pictures. **37.9** Above, D. A. Parry, *Journal of Experimental Biology,* 1959, 36:654; below, © Stephen Dalton/Photo Researchers, Inc. **Page 649** Precision Graphics. **37.10** (a,c,d) D. & V. Hennings; (b, left) Bone Clones®, www.boneclones.com; (b, right) © Dr. Paul A. Zahl/Photo Researchers, Inc. **37.11** Art, Raychel Ciemma; Photograph, C. Yokochi and J. Rohen, *Photographic Anatomy of the Human Body,* second edition, Igaku-Shoin, Ltd., 1979. **37.12** Art, Joel Ito; micrograph, Ed Reschke. **37.13** K. Kasnot. **37.14** (a,b) Prof. P. Motta Dept. of Anatomy/University "La Sapienza" Rome/Science Photo Library/Photo Researchers, Inc. **37.15** Raychel Ciemma. **37.16** (a–c) N.H.P.A./ A.N.T. Photolibrary. **37.17** Robert Demarest. **37.18** Raychel Ciemma. **37.19** (a,c) Robert Demarest; ballet photograph from the Dance Theatre of Harlem, by Frank Capri; micrographs (b, c) © Don Fawcett/Visuals Unlimited, from D. W. Fawcett, *The Cell,* Philadelphia; W. B. Saunders Co., 1966; compilation by Gary Head. **37.20** (a, above), Robert Demarest; (a–g) Nadine Sokol and Gary Head. **37.21** Robert Demarest. **37.22** Lisa Starr. **37.24** Kevin Somerville and Gary Head. **37.25** Gary Head. **37.26** Painting, Sir Charles Bell, 1809, courtesy of Royal College of Surgeons, Edinburgh. **37.27** (a) © Sean Sprague/Stock, Boston; (b) Michael Neveux

CHAPTER 38

38.1 (a) From A. D. Waller, *Physiology: The Servant of Medicine,* Hitchcock Lectures, University of London Press, 1910; (b) Courtesy The New York Academy of Medicine Library; (c) Preface, Inc. **38.2** Gary Head. **38.3** (a,c) Precision Graphics; (b, d) Raychel Ciemma; (e) after M. Labarbera and S. Vogel, *American Scientist,* 1982, 70:54-60. **38.4** Precision Graphics. **38.5** Art, Palay/Beaubois. **38.6** Photograph, © David Scharf/Peter Arnold, Inc. **38.7** Lisa Starr, with art references from Bloodline Image Atlas, University of Nebraska-Omaha/Sherri Wicks, Human Physiology and Anatomy, University of Wisconsin Biology Web Education System, and others; (photo) © EyeWire, Inc. **38.8** Photographs, (a,b) Lester V. Bergman & Associates, Inc.; (c) Gary Head after F. Ayala and J. Kiger, *Modern Genetics,* © 1980 Benjamin-Cummings. **38.9** Nadine Sokol after G. J. Tortora and N. P. Anagnostakos, *Principles of Anatomy and Physiology,* sixth edition. © 1990 by Biological Sciences Textbook, Inc., A& P Textbooks, Inc. and Elia-Sparta, Inc. Reprinted by permission of John Wiley & Sons, Inc. **38.10** Precision Graphics. **38.11** Kevin Somerville. **38.12** (a) Photograph, C. Yokochi and J. Rohen, *Photographic Anatomy of the Human Body,* second edition, Igaku-Shoin, Ltd., 1979; (b,c) Raychel Ciemma. **38.13** Art, Precision Graphics. **38.14** (a) Art, Lisa Starr, micrograph by Don W. Fawcett; (b) Raychel Ciemma. **38.15** Robert Demarest, after A. Spence, *Basic Human Anatomy,* Benjamin-Cummings, 1982. **38.16** Precision Graphics. **38.17** Sheila Terry/SPL/Photo Researchers, Inc. **38.18** Micrograph, © Biophoto Associates/Photo Researchers, Inc.; art, Lisa Starr. **38.19** (a,b) Kevin Somerville; Micrograph, Dr. John D. Cunningham/Visuals Unlimited; right, Lisa Starr, using © 2001 PhotoDisc, Inc. photograph. **38.20** (a) © Ed Reschke; (b) © Biophoto Associates/Photo Researchers, Inc. **38.21** L. Calver. **38.22** Precision Graphics. **38.23** Photograph, Prof. P. Motta/Dept. of Anatomy/University "La Sapienza" Rome/Science Photo Library/Photo Researchers, Inc.; art, Gary Head. **38.24** (a,b) Raychel Ciemma; (c) Lisa Starr. **Page 684** Raychel Ciemma; Robert Demarest. **38.25** Lennart Nilsson from *Behold Man,* © 1974 by Albert Bonniers Förlag and Little, Brown and Company, Boston

CHAPTER 39

39.1 (a) The Granger Collection, New York; (b) Lisa Starr with, PDB ID: 1IGT Harris, L. J., Larson, S. B., Hasel, K. W., McPherson, A. *Biochemistry* 36 pp. 1581, 1997. (c) Lennart Nilsson © Boehringer Ingelheim International GmbH. **39.2** Robert R. Dourmashkin, courtesy of Clinical Research Centre, Harrow, England. **39.3** Lisa Starr. **39.4** (a) NSIBC/SPL/Photo

Researchers, Inc.; (b) Biology Media/Photo Researchers, Inc. **39.5** Raychel Ciemma. **39.6** Art, Lisa Starr; photograph, © David Scharf/Peter Arnold, Inc. **39.7** Art, Lisa Starr; photograph © Ken Cavanagh/Photo Researchers, Inc. **39.8** Precision Graphics. **39.9** Lisa Starr. **39.10** Lisa Starr. **39.11** Preface, Inc. **39.12** Raychel Ciemma. **39.13** Lisa Starr. **Page 696** Lisa Starr. **39.14** Art, Lisa Starr (also see citation for 39.1b). **39.15** Lisa Starr. **39.16** © Dr. A. Liepins/SPL/Photo Researchers, Inc. **39.17** Above, Lowell Georgia/Science Source/Photo Researchers, Inc.; below, Matt Meadows/Peter Arnold, Inc. **39.18** Left, David Scharf/Peter Arnold, Inc.; right, Kent Wood/Photo Researchers, Inc. **39.19** Ted Thai/TimePix. **39.20** © Zeva Olbaum/Peter Arnold, Inc. **39.21** (a) Art, Raychel Ciemma, after Stephen Wolfe, *Molecular Biology of the Cell*, Wadsworth, 1993; (b) micrographs, Z. Salahuddin, National Institutes of Health. **Page 705** Raychel Ciemma

CHAPTER 40
40.1 © Christian Zuber/Bruce Coleman, Ltd.; (inset photo) courtesy of Dr. Joe Losos. **40.2** Gary Head and Precision Graphics. **40.3** Precision Graphics. **40.4** Photographs, (a) Peter Parks/Oxford Scientific Films; (b) Herve Chaumeton/Agence Nature; art, Precision Graphics. **40.5** Micrograph, Ed Reschke; art, Precision Graphics. **40.6** Raychel Ciemma and Precision Graphics. **40.7** Raychel Ciemma. **40.8** Lisa Starr. **40.9** Micrograph, H. R. Duncker/Justus-Liebig University, Giessen, Germany; art, Raychel Ciemma. **40.10** Kevin Somerville. **40.11** Photographs courtesy of Kay Elemetrics Corporation; art, modified from A. Spence and E. Mason, *Human Anatomy and Physiology*, fourth edition, 1992, West Publishing Company. **40.12** (a) Lisa Starr with photograph © 2000 PhotoDisc, Inc.; (b,c) Art, Lisa Starr; x-rays from SIU/Visuals Unlimited. **40.14** From L. G. Mitchell, J. A. Mutchmor, and W. D. Dolphin, *Zoology*, © 1988 by The Benjamin-Cummings Publishing Company. Reprinted by permission. **40.15** (a) © R. Kessel/Visuals Unlimited; (b,c) Lisa Starr. **40.16** Leonard Morgan. **40.17** Above, © CNRI/SPL/Photo Researchers, Inc.; photograph below, Lennart Nilsson from *Behold Man*, © 1974 by Albert Bonniers Förlag and Little, Brown and Company, Boston. **40.18** (a,b) © O. Auerbach/Visuals Unlimited; inset photo courtesy of Dr. Joe Losos. **40.19** Gary Head. **40.20** © 2002 Stuart Westmorland/Stone/Getty Images. **40.21** Courtesy of Ron Romanosky. **Page 722** Kevin Somerville. **40.22** Precision Graphics

CHAPTER 41
41.1 Hulton Getty Collection/Stone/Getty Images. **41.2** Gary Head and Precision Graphics. **41.3** Raychel Ciemma. **41.4** (a,b) Adapted by Lisa Starr from A. Romer and T. Parsons, *The Vertebrate Body*, sixth edition, Saunders Publishing Company, 1986; photograph, D. Robert Franz/Planet Earth Pictures. **41.5** Gunter Ziesler/Bruce Coleman. **41.6** Kevin Somerville. **41.7** Nadine Sokol. **41.8** Art, Robert Demarest; micrograph, Omikron/SPL/Photo Researchers, Inc. **41.9** Above left, Kevin Somerville; (a,b) After A. Vander, et al, *Human Physiology: Mechanism of Body Function*, fifth edition. McGraw-Hill, © 1990. Used by permission of McGraw-Hill; (c) redrawn from *Human Anatomy and Physiology*, fourth edition, by A. Spence and E. Mason, 1992, Brooks/Cole. All rights reserved. **41.10** Art, Lisa Starr after Sherwood and others; (b) micrograph courtesy Mark Nielsen, University of Utah; (e) micrograph © D. W. Fawcett/Photo Researchers, Inc. **41.11** Raychel Ciemma. **Page 733** After A. Vander, et al, *Human Physiology: Mechanisms of Body Function*, fifth edition, McGraw-Hill, 1990. Used by permission. **41.12** Photograph, Ralph Pleasant/FPG; art, Precision Graphics. **41.13** Precision Graphics and Kevin Somerville. **41.14** Art, Gary Head; photographs above, © 2001 PhotoDisc, Inc.; center, © Ralph Pleasant/FPG; below, Elizabeth Hathon/CORBIS. **41.16** Gary Head. **41.17** Dr. Douglas Coleman, The Jackson Laboratory; art, Precision Graphics and Gary Head. **Page 742** Kevin Somerville. **41.18** Raychel Ciemma

CHAPTER 42
42.1 Photographs, background left, David Noble/FPG; below left, Claude Steelman/Tom Stack & Associates; right, Gary Head. **42.2** Gary Head and Precision Graphics. **42.3** Robert Demarest. **42.4** (a, left) photograph © Ralph T. Hutchings; (a–d) art, Robert Demarest. **42.5** Art above, Robert Demarest; below, Precision Graphics. **42.6** Left, Precision Graphics; photograph, Gary Head. **42.7** (a,b) From T. Garrison, *Oceanography: An Invitation to Marine Science*, Brooks/Cole, 1993. All rights reserved; (c) © Thomas D. Mangelsen/Images of Nature. **42.8** © Bob McKeever/Stack & Associates. **42.9** (a) Everett C. Johnson; (b) © S. J. Krasemann/Photo Researchers, Inc.; (c) © David Parker/SPL/Photo Researchers, Inc. **42.10** Left, Robert Demarest; right, Kevin Somerville. **42.11** © Dan Guravich/CORBIS. **Page 756** Robert Demarest. **42.12** © Bettmann/CORBIS

CHAPTER 43
43.1 (a) Hans Pfletschinger; (b), Raychel Ciemma; (c) John H. Gerard; (d,e) © David M. Dennis/Tom Stack & Associates; (f) John Shaw/Tom Stack & Associates. **43.2** (a) © W. Perry Conway/CORBIS; (b) © Ron Watts/CORBIS; (c) © Anthony Bannister/CORBIS; (d) © George D. Lepp/CORBIS. **43.3** (a) Frieder Sauer/Bruce Coleman; (b) © Matjaz Guntner; (c) © George D. Lepp/CORBIS; (d) © Doug Perrine, seapics.com; (e) Fred McKinney/FPG; (e-inset) Carolina Biological Supply Company; (f) Gary Head. **43.4** Palay/ Beaubois and Precision Graphics. **43.5** Art, L. Calver; micrographs, Carolina Biological Supply Company. **43.6** Art, Robert Demarest. **43.7** (a,b) Precision Graphics; (c) Gary Head. **43.8** Dr. Maria Leptin, Institute of Genetics, University of Koln, Germany, computer-enhanced by Lisa Starr. **43.9** (Photographic series) Carolina Biological Supply Company; photograph far right, Peter Parks/Oxford Scientific Films/Animals Animals; Art, Precision Graphics. **43.10** (a) Art, Lisa Starr; (b) Raychel Ciemma after B. Burnside, *Developmental Biology*, 1971:26:416-441. Used by permission of Academic Press. **43.11** (a,b) Art, Lisa Starr; (c) photograph, Prof. Jonathon Slack. **43.12** Photograph left, Peter Parks/Oxford Scientific Films/Animals Animals; (a,b) Raychel Ciemma after S. Gilbert, *Developmental Biology*, fourth edition. **43.13** (a) Precision Graphics; (b–d) F. R. Turner. **43.14** (a) Palay/Beaubois after Robert F. Weaver and Philip W. Hedrick, *Genetics*. © 1989 W. C. Brown Publishers; (b): Art, Precision Graphics after Scott Gilbert, *Developmental Biology*, fourth edition. Sinauer; (c) reprinted from *Mechanisms of Development*, Vol. 49, 1995, John Reinitz and David Sharp, *Mechanism of Formation of Eve Stripes*, pages 133–158, © 1995, with permission from Elsevier Science. **Page 773** © David Seawell/CORBIS. **Page 774** Carolina Biological Supply Company. **43.15** Palay/Beaubois adapted from R. G. Ham and M. J. Veomett, *Mechanisms of Development*, St. Louis, C. V. Mosby Co., 1980

CHAPTER 44
44.1 Left © Minden Pictures; right, © Charles Michael Murray/CORBIS. **Page 777** Lennart Nilsson from *A Child Is Born*, © 1966, 1977 Dell Publishing Company, Inc. **44.2** (a,b) Raychel Ciemma with Lisa Starr. **Page 778** © Laura Dwight/CORBIS. **44.3** (a) Raychel Ciemma; (b) Ed Reschke. **44.4** Raychel Ciemma. **44.5** Raychel Ciemma and Precision Graphics. **44.6** (a,b) Raychel Ciemma. **44.7** (a) Left, Robert Demarest; right, Raychel Ciemma; (b,c) Lennart Nilsson from *A Child Is Born*, © 1966, 1977 Dell Publishing Company, Inc. **44.8** Raychel Ciemma and Precision Graphics. **44.9** Robert Demarest, Kevin Somerville, and Preface, Inc. **44.10, 44.11** Raychel Ciemma. **44.12, 44.13** Raychel Ciemma. **44.14** Art, Raychel Ciemma; photographs, Lennart Nilsson from *A Child Is Born*, © 1966, 1977 Dell Publishing Company, Inc. **44.15** Raychel Ciemma, modified from K. L. Moore, *The Developing Human: Clinically Oriented Embryology*, fourth edition, Philadelphia: Saunders Co., 1988. **44.16** (a) Lennart Nilsson from *A Child Is Born*, © 1966, 1977 Dell Publishing Company, Inc.; (b) James W. Hanson, MD. **44.17** Robert Demarest. **44.18** Raychel Ciemma. **44.19** Raychel Ciemma, adapted from L. B. Arey, *Developmental Anatomy*, Philadelphia, W. B. Saunders Co., 1965; Photograph, Lisa Starr. **44.20** Preface, Inc. **44.21** © CNRI/SPL/Photo Researchers, Inc. **44.22** (a) Dr. John D. Cunningham/Visuals Unlimited; (b) David M. Phillips/Visuals Unlimited; (c) © Kenneth Greer/Visuals Unlimited. **Page 803** Raychel Ciemma; Robert Demarest.

Page 805 UNIT VII Alan and Sandy Carey

CHAPTER 45
45.1 (a) Gary Head; (b) Antoinette Jongen/FPG. **45.2** Precision Graphics. **45.3** (a) E. R. Degginger; (inset) Jeff Foott Productions/Bruce Coleman; (b) © Paul Lally/Stock, Boston. **45.4** Photograph, © Jeff Lepore/Photo Researchers, Inc.; Precision Graphics art. **45.5** (a) Art, Precision Graphics; (b) micrograph, Stanley Flegler/Visuals Unlimited. **45.6** Gary Head. **45.7** Photograph, E. Vetter/Zefa; art, Gary Head. **Table 45.1** Photograph, Eric Crichton/Bruce Coleman. **Table 45.2** Compiled by Marion Hansen, based on data from U. S. Bureau of the Census, *Statistical Abstract of the United States*, 1992 (edition 112). **45.8** (a) © Joe McDonald/CORBIS; (b) © Wayne Bennett/CORBIS; (c) © Douglas P. Wilson/CORBIS; art, Preface, Inc. **45.9** (a) Photograph, Helen Rodd; (b, above, c, above) David Reznick, University of California-Riverside; computer-enhanced by Lisa Starr; (b, below, c, below) Hippocampus Bildarchiv. **45.10** (a,b) John A. Endler. **45.11** Precision Graphics. **45.12** Art, Precision Graphics; photograph, NASA. **45.13** (b) Adapted from *Environmental Science* by G. Tyler Miller, Jr., page 247. © 2003 by BrooksCole, a division of Thomson Learning. **45.14** Art, Precision Graphics and Preface, Inc.; photograph, United Nations. **45.15** Data from Population Reference Bureau after G. T. Miller, Jr., *Living in the Environment*, eighth edition, Brooks/Cole, 1993. All rights reserved. **45.16** Photograph, United Nations; art, Precision Graphics. **45.17, 45.18** After G. T. Miller, Jr., *Living in the Environment*, sixth edition, Brooks/Cole, 1997. All rights reserved

CHAPTER 46
46.1 Left, John Bova/Photo Researchers, Inc.; right, Robert Maier/Animals Animals. **46.2** (a) from L. Clark, *Parasitology Today*, 6(11), Elsevier Trends Journals, 1990, Cambridge, UK; (b) Jack Clark/Comstock, Inc. **46.3** (a) Eugene Kozloff; (b, c) Stevan Arnold, Inc. **46.4** Photograph, Hans Reinhard/Bruce Coleman; Spectogram, from G. Pohl-Apel and R. Sossinka, *Journal for Ornithologie*, 123:211–214 **46.5** (a) Evan Cerasoli; (b) from A. N. Meltzoff and M. K. Moore, "Imitation of Facial and Manual Gestures by Human Neonate," *Science*, 1977, 198:75–78. © 1977 by the AAAS **46.6** Eric Hosking. **46.7** © Nina Leen/TimePix. **Page 831** Left, © 2001 PhotoDisc, Inc.; right, © Ed Reshke. **46.8** Marc Bekoff. **46.9** Above, Edward S. Ross; below, E. Mickleburgh/Ardea, London. **46.10** Art, D. & V. Hennings; photograph, © Stephen Dalton/Photo Researchers, Inc. **46.11** (a) John Alcock, Arizona State University ; (b) Ray Richardson/Animals Animals. **46.12** (a) Michael Francis/The Wildlife Collection; (b) © Layne Kennedy/CORBIS. **46.13** Frank Lane Agency/Bruce Coleman. **46.14** © Paul Nicklen/National Geographic/Getty Images. **46.15** John Alcock, Arizona State University. **46.16** (a) John Dominis, *Life Magazine*, © Time, Inc.; (b) A. E. Zuckerman/Tom Stack & Associates. **46.17** John Alcock, Arizona State University. **46.18** Kenneth Lorenzen. **46.19** Gregory D. Dimijian/Photo Researchers, Inc. **46.20** Yvon Le Maho. **46.21** Lincoln P. Brower. **46.22** F. Schutz. **46.23** Eric Hosking

CHAPTER 47
47.1 Left, Donna Hutchins; right, Edward S. Ross; far right, © B. G. Thomson/Photo Researchers, Inc. **47.2** (a, c) Harlo H. Hadow; (b) Bob and Miriam Francis/Tom Stack & Associates. **47.3** Art, Precision Graphics; photograph, Clara Calhoun/Bruce Coleman. **47.4** Preface, Inc., after G. Gause, 1934. **47.5** (a,b) Stephen G. Tiley. **47.6** Precision Graphics after N. Weland and F. Bazazz, *Ecology*, 56:681–688, © 1975 Ecological Society of America. **47.7** Gary Head, after Rickleffs & Miller, *Ecology*, fourth edition, page 459 (Fig 23.12a) and p. 461 (Fig 23-14); right photograph © W. Perry Conway/CORBIS. **47.8** (a)

Index of Applications

A

ABO blood typing, 184, 184i, 192, 671
Abortion
  bioethics of, 798, 802
  birth control and, 799
  genetic disorders and, 212, 213
Abstinence, in fertility control, 798, 799i, 800
Achondroplasia, 203t, 204, 204i, 205
Acid deposition (acid rain) 31, 31i, 918–919, 919i
Acid stomach, 30, 730
Acidosis, 31
Acne, 647
  anti-acne drugs and, 795
Acromegaly, 634, 634i
Acute otitis media, 625
Acyclovir, 801
ADA deficiency, 701, 701i
Adam (MDMA), 602–603
Adenovirus infection, 357t
Adoptive behavior, human populations, 841
Aerobic exercise, 661
African sleeping sickness, 368
Africanized bees, 132–133, 859
Afterbaby blues, 796–797
Afterbirth, 796, 796i
Agar, 380
Age spots, 97, 97i
Agent Orange, 555
Aging
  aerobic exercise and, 661
  autoimmunity with, 772
  cardiovascular disorders and, 680
  collagen and, 648, 772
  colon malfunctioning in, 735
  cumulative environmental assaults hypothesis, 772
  disorders related to, 264
  Down syndrome and, 208–209, 209i
  eye problems with, 623
  free radicals and, 96–97, 97i, 772
  in human life cycle, 797, 797t
  in progeria, 203t, 206
  programmed life span hypothesis, 772
  skin, 648
  supplements and, 739
  telomeres and, 772
  Werner's syndrome and, 772
Agriculture
  animal-assisted, 921
  aquifer depletion, 926, 926i
  artificial selection and, 252–253
  crop production in, 260
  desalinization and, 926
  desertification and, 925, 925i
  domesticated species, 400, 482, 818
  erosion and, 525
  fungi and, 406–407, 412, 412i
  genetic engineering in, 267–268
  green revolution in, 267–268, 921
  herbicides, 263, 267–268, 562, 562i
  history of, 818–819, 916
  insects and, 439, 439i
  irrigation and, 926, 926i
  major crops worldwide, 400
  marginal lands for, 921, 921i
  mechanized, 921
  pesticides, 267–268, 282–283, 439, 562, 562i
  pine farms, 519i
  plant hormones and, 550–551, 550i, 555, 555i
  red tides and, 374
  salinization and, 492, 492i, 926
  seed banks and, 260

shifting cultivation in, 922–923
  soils and, 525–535, 525i, 896, 896i
  subsistence, 921
  tea harvesting, 400i
  tropical forests and, 922–924, 923i
  vernalization in, 561
  wheat harvesting, 400i
AIDS
  in Africa, 702, 702t
  caseloads, 361t, 702, 702t
  characteristics, 701–703
  HIV and, 356, 686i, 701–703, 703i
  immune response to, 702
  needle sharing and, 602, 663, 702
  opportunistic pathogens and, 372
  social behavior and, 800
  testing for, 800
  transmission of HIV, 702–703
  treatments, 113, 703
  vaccines for, 703
Air pollution
  fungi and, 404, 411
  impact on forests, 411, 919, 919i
  major classes, 918, 918t
  methyl bromide in, 920
  sources, 918–920, 918i–920i
  stomata and, 531, 531i
  weekly variation in, 18
Ajellomyces infection, 412, 412t
Alachlor, 562i
Alar, 562i
Albinism, 186, 187i, 193, 193i, 203t
Alcohol consumption, 65
  birds and, 142–143, 143i
  detoxification in liver, 109
  diseases and, 109
  effects on brain, 603
Alcoholic cirrhosis, 109
Alcoholic hepatitis, 109
Alexander the Great, gene flow and, 471
Alkalosis, 31
Allergic rhinitis, 539
Allergies, 539, 700, 701i
Aloe vera, 401
Alu transposon, in human evolution, 315, 315i
Alzheimer's disease, 209, 264, 599, 772
Ama (Asian divers), 721
Amanita poisoning, 406i
Amniocentesis, 212, 212i
Amoebic dysentery, 361t
Amphetamines, 602
Anabolic steroids, 643, 663
Analgesics, 603
Anaphylactic shock, 700
Androgen insensitivity syndrome, 203t, 630
Anemias, 213, 670, 738t, 739t
Angina pectoris, 681
Angioplasty, 681
Anhidrotic ectodermal dysplasia, 245, 245i, 577
Anorexia nervosa, 724
Anthrax, 687
Anti-acne drugs, 795
Antianxiety drugs, 584
Antibiotic resistance, 11, 110–111, 260, 283
Antibiotics, 11, 14–15, 283, 354, 409, 818
  pregnancy and, 795
Antioxidants, 96
Apnea, 717
Appendicitis, 735
Appetite, hormones controlling, 741
Apple scab, 412, 412i
Aquifer depletion, 926, 926i

Arctic tundra, vulnerability to development, 903
Arenavirus infection, 357t
Argentine fire ants, 859, 859t
Arrhythmias, 681
Arteriosclerosis, 680
Arthritis, 653, 740
Artificial pacemaker, 23, 665
Artificial reef, 486, 486i
Artificial selection
  designer dogs, 270–271, 271i
  foods, 252–253
  yeasts, 409
Aspergilloses, 412
Aspergillus
  commercial uses, 409
  infection, 412, 412t
Aspirin, 750
  fevers and, 754
  in renal failure, 750
Asteroids, human evolution and, 304–305
  future impacts, 305
Asthma, 412, 562, 636, 697, 700
Astigmatism, 622
Atherosclerosis, 680–681, 680i, 750, 772
  bacterial infection and, 681
Atherosclerotic plaques, 252, 252i, 680i
Athlete
  anabolic steroid use, 643, 663
  blood doping by, 670
  cardiac rates, 681
  strength training, 661
Athlete's foot, 412, 412i
Atrazine, 562i
Atrial fibrillation, 681
Atropine, witches and, 401
Autoimmune responses
  aging and, 772
  Graves' disorder, 637, 701
  multiple sclerosis, 499, 586, 701
  rheumatoid arthritis, 653, 701
  toxic goiter, 637
  type 1 diabetes, 639
  women vs. men, 701
AZT, 113, 703

B

Babesiasis, 439
Baby boomers, 820, 821i
Bacillus anthracis infection, 687
Back pain, 651
Bacterial infections (*See* Infections, bacterial agents)
Balantidium coli infection, 371
Bald eagle
  black-market, 478i
  DDT and, 872
Balloon angioplasty, 681
Basal cell carcinoma, 238i
Beer, liver and, 109
Beer belly, 740
Behavior (*See* Human behavior)
Belladonna, humans and, 401
Bengal tiger, black-market, 478i
Bighorn sheep, black-market, 478i
Biodiversity
  as biological wealth, 482–483
  efforts to sustain, 484–486
  hot spots, 895
  losses, 474–485, 887
  sustainable economic development and, 483–485
Bioethics
  abortion, 798, 802
  cloning, 224, 262, 264–265

competition in science, 13, 217, 221, 222
  creating life in test tubes, 268
  deforestation, 924
  fertility control, 798, 802
  gene therapy, 264–265
  genetically engineering food, 261, 261i, 263, 267–268
  human cloning, 264–265
  human impact on biosphere, 917, 931
  human population growth, 806–807, 823
  in vitro fertilization, 802
Biofilms, 80
Biological controls
  nature of, 283
  sea slugs as, 861
Biological magnification, DDT, 872, 872i
Biological therapy, antibiotics vs. bacteriophages, 14–15
Biotech barnyards, 262
Birth (labor), 796, 796i
Birth control, 798–799, 799i
Birth control pill, 799, 799i
Black bread mold, 408, 408i
Black widow bite, 438, 438i
Blood cell disorders
  categories, 670
  leukemic, 194, 195i, 670
  sickled, 49, 49i, 287
Blood clotting
  collagen in, 682i
  diet and, 738t
  in hemophilia, 201, 203t, 205, 205i
Blood doping, 670
Blood fluke infections, 423, 424, 424i
Blood pressure, measuring, 677i, 678–679
Blood transfusions, 670–671
Blue offspring, 203t, 291
Body-builder psychosis, 663
Body-building programs, 148, 663
Body piercing, 663, 663i
Body weight, human
  factors influencing, 724–725, 736, 740, 740i
  pregnancy and, 794
Borrelia infection, 355, 438–439, 438i
Botulism, 113, 142, 354, 661
Bovine spongiform encephalopathy (BSE), 359, 363
Bradycardia, 681, 681i
Breast cancer, 161, 268, 740, 797, 799
Bronchitis, 718
Brown alga, commercial uses, 376, 377
Brown pelican, DDT and, 872
Brown recluse bite, 438, 438i
BSE, 359, 363
"Bubble boys," 264
Bubonic plague, 813, 819
Bulimia, 724–725
Bunyavirus infection, 357t
Burial alternative, human, 486

C

Caffeine, 602, 705
Calicivirus escape from laboratory, Australia, 860
California encephalitis, 357t
Caloric intake, human, 240, 724, 740
Camptodactyly, 188, 203t
Cancer
  basal cell carcinoma, 238i
  bladder, 718, 719i
  bone marrow, 194, 670, 699

More Applications in the Text

For additional big-picture applications, see the list of CONNECTIONS ESSAYS on page xxiii of the Preface